# A Dictionary of
# ENGLISH
# SURNAMES

*by* P. H. REANEY

*Litt. D., Ph.D., F.S.A.*

THIRD EDITION WITH CORRECTIONS
AND ADDITIONS BY

## R. M. WILSON

M.A.

Oxford
OXFORD UNIVERSITY PRESS
1995

*Oxford University Press, Walton Street, Oxford* OX2 6DP

*Oxford New York*
*Athens Auckland Bangkok Bombay*
*Calcutta Cape Town Dar es Salaam Delhi*
*Florence Hong Kong Istanbul Karachi*
*Kuala Lumpur Madras Madrid Melbourne*
*Mexico City Nairobi Paris Singapore*
*Taipei Tokyo Toronto*
*and associated companies in*
*Berlin Ibadan*

*Oxford is a trade mark of Oxford University Press*

*British Library Cataloguing in Publication Data*
*Data available*

*Library of Congress Cataloging in Publication Data*
*Data available*

*ISBN 0–19–863146 4*

*10 9 8 7 6 5 4 3 2 1*

*Printed in Great Britain*
*by Biddles Ltd.*
*Guildford and King's Lynn*

# CONTENTS

# PREFACE TO THE THIRD EDITION

THIS edition of *A Dictionary of English Surnames* contains some 4,000 additional names with their variants, and constitutes a third edition of P. H. Reaney's *A Dictionary of British Surnames*. The change of title reflects a concentration on surnames of specifically English rather than Celtic origin, which has been increasingly apparent in successive editions. As a rule, Scottish, Welsh, and Irish names are only included when forms for them are found in English sources, or when they coincide in form with specifically English surnames. Scottish surnames have been adequately dealt with by G. F. Black, Irish names by E. Maclysaght, and Welsh border names by T. E. Morris, and there seemed little point in reproducing information which could be found in their works.

So far as English surnames are concerned, the coverage of the various counties is inevitably uneven. For some of these counties, mainly the more northern ones, early records are comparatively rare. For others, although the records are more abundant, few of them have as yet been published. This is the case for Cornwall, where there is little in print, apart from the 1297 Ministers' Accounts for the Earldom of Cornwall, and especially for Hampshire where few of the official documents appear to have been published. As a result comparatively few specifically Hampshire names are included. To a lesser degree, that is true also of Cheshire, Herefordshire, Norfolk, and some of the Midland counties.

In general the additional entries follow the same plan as those in the second edition, practically the only difference being that the new entries, when dealing with the origin of surnames derived from local names, give abbreviated forms of the county names, as found in the 'List of Abbreviations'.

Some of the material used in this volume comes from the files of P. H. Reaney preserved in the Library of the University of Sheffield, but most is from my own collections. Similarly the etymologies suggested are usually my own, and from the nature of the surnames included tend to be either obvious or highly speculative, but experience has shown that as many enquiries are received concerning the former type of surname as for the more difficult ones.

R. M. WILSON

# PREFACE TO THE SECOND EDITION

SOME seven hundred names have been added to this second edition, mostly fairly common ones omitted from the original edition from considerations of space; the list of abbreviations has been rewritten; and various necessary corrections have been made. Some of the corrections, and many of the additional names, had already been included by Dr Reaney in his own copy of the book in preparation for a new edition. In addition much of the material for the other names has been taken from Dr Reaney's extensive files, now in the Sheffield University Library, though other sources, not at the time available to him, have also been used. The additional entries follow the pattern of those in the first edition, and it is hoped that their inclusion will make rather more comprehensive a work which has already become the standard book on the subject.

R. M. WILSON

1976

# PREFACE

---

O F previous Dictionaries of Surnames, Lower's *Patronymica Britannica* (1860) is obviously out of date, Barber's *British Family Names* (1902) is a mere collection of guesses unsupported by evidence, whilst Harrison's *Surnames of the United Kingdom* (1912–18) only very occasionally gives any evidence and a large number of his etymologies are clearly based on the modern form. Still the most reliable is Bardsley, whose *Dictionary of English and Welsh Surnames*, published posthumously in 1901, firmly laid the foundations on which future study of surnames must be built. He insisted on the need for the collection of as many early examples of the surname as possible, dated and localized, on which the etymology must be based. These principles he put into practice, so far as he could, but he has suffered the inevitable fate of all pioneers. The last fifty years have seen an enormous increase in the material available in the publications of the Public Record Office, the Pipe Roll Society, county Record Societies, etc., much of it earlier than Bardsley's basic source (the late thirteenth-century Hundred Rolls), with a steadily improving standard of editing. The same period has seen, too, a marked advance in our knowledge of the English language, particularly in the history of its dialects, personal names and place-names.

The earlier literature of surnames has been adequately discussed by Weekley, Ewen and Tengvik. Whilst subscribing to the only sound principles, Weekley, in his published works, seldom gives the evidence on which his etymologies are based, and even then, usually an undated reference to the Patent or the Close Rolls. For many names he clearly had no evidence. He fails to distinguish between Old English, Scandinavian and continental Germanic personal names and is uncritical in his use of Searle. Ewen attempted an impossible task. Generalizations on surnames are valueless if an unimpeachable etymology has not been established. He fails to distinguish between sound and spelling, and postulates impossible forms of Old English names. Worst of all, he rejects sound etymologies which do not fit in with his preconceived theories.

The present work is based on an independent collection of material begun in 1943 to beguile the tedium of the quieter periods of fire-watching. A complete Dictionary of Surnames cannot yet be produced, partly because for many of the large number of surnames surviving material is at present scanty or lacking, partly because of the high cost of such a production. This has meant a strict economy in examples and in exposition and the elimination from the first draft of some 100,000 words and 4,000 names. All surnames included are known to survive. The great majority of those eliminated are local surnames such as Manchester, Wakefield, Essex, etc., which can easily be identified from the gazetteer. When a local surname has been traced to its source, the surname-student's task is finished. The meaning of the place-name is a problem for others and those who wish for further information should consult the *Oxford Dictionary of Place-names* or the county volumes of the English Place-name Society whose latest publication, A. H. Smith's *English Place-name Elements* (2 vols, 1956) is a comprehensive treatment of the subject.

The most valuable modern work on English surnames has been produced in Sweden. Olof von Feilitzen's book on the pre-Conquest personal names in Domesday Book has been invaluable. Tengvik has dealt with Old English bynames, Löfvenberg with Middle English local surnames, Fransson and Thuresson with Middle English occupational names. Ekwall, too, turning aside from English place-names, has already made valuable additions to our knowledge of surnames, particularly those of London. Here I would take the opportunity to express my appreciation of a very generous gesture which I still regard as a private act indicative of national sentiment. Shortly after the war, I mentioned casually to Dr. Gösta Langenfelt, on one of his visits to London, that all my books had been destroyed. Later I received from ten or more Swedish scholars signed copies of their books. With these constantly at hand, work on this book has been greatly facilitated. I am also indebted to the Librarian of the Royal Library, Stockholm, for depositing temporarily in the Library of the University of London certain inaccessible books; to Dr. von Feilitzen, who first suggested the transfer and has kept me informed of new Swedish publications likely to be of use; to E. H. Brandt, for access to his collection of surname material and for many useful discussions; to J. E. B. Gover, for forms from unpublished MSS; and to F. G. Emmison, of the Essex Record Office, for a copy of his unpublished transcript of the 1662 Hearth Tax Returns for Essex. To Mr. Gover and to Professor R. M. Wilson of the University of Sheffield my grateful thanks are due for the time they have spent in reading the proofs. Their comments and criticisms have been invaluable in eliminating errors and inconsistencies. For those that remain the responsibility is mine alone.

P. H. REANEY

Hildenborough
January 1958

# INTRODUCTION

THE purpose of a Dictionary of Surnames is to explain the meaning of names, not to treat of genealogy and family history. The fact that Robert le Turnur lived in Staffordshire in 1199 and that there was a William de Kouintre in London in 1230 does not mean that they were the ancestors of all or any of the modern Turners or Coventrys. To establish this, a fully documented pedigree would be required and very few families can carry back their history so far. Throughout the Middle Ages surnames were constantly changing. William Tyndale was known as Huchyns when living in Gloucestershire. Oliver Cromwell was a Williams and David Livingstone was a McLeay. Even today families change their names. Blackden has become Blacktin, Hogg has been changed to Hodd and Livemore has superseded Livermore—all within living memory.

The modern form of many of our surnames is comparatively recent, often preserving a phonetic spelling found in a seventeenth- or eighteenth-century parish register. When some of the Sussex Bourers migrated to Kent in the seventeenth century they adopted the form Bowra. *Pharaoh* is a reconstructed spelling of *Faro*, originally *Farrer*, found also as *Farrey*, *Farrah* and *Farrow* in the seventeenth century. The Suffolk Deadman is a corruption of Debenham and Tudman of Tuddenham. Each surname has its pedigree which must be traced before the meaning can be discovered, and even then the true origin cannot be decided unless the family pedigree can be carried back far enough to fix definitely the original medieval form. A modern White may owe his name to an ancestor bearing the Anglo-Saxon name of *Hwīta*, or to one nicknamed 'the fair', or to an original home in the bend of a river. The original Howard may have been a ewe-herd or a hayward, or he may have borne either the French name *Huard* or the Old German name *Howard*. The modern forms often conceal rather than reveal information.

The English language lacks terms corresponding to the French *sobriquet* and *nom de famille*. Today, surname means an inherited family name; originally it meant simply an additional name and it is used in this sense in this book.[1] Only very occasionally can early medieval surnames be proved to be hereditary, and any attempt to distinguish them would end in inaccuracy and confusion.

## VARIATION OF SURNAMES

Celts, Anglo-Saxons, and Scandinavians, all originally had a single name for each individual, e.g. Welsh *Llewellyn*, Gaelic *Donald*, Anglo-Saxon *Cuthbert*, Scandinavian *Gamall*. So, too, had the Normans who were ultimately of Danish descent. Already in England before the Norman Conquest we find a number of bynames, and these were increased after the Conquest by those used by Normans. In the twelfth century we have an unsettled and varied type of nomenclature, often by way of description rather than of an actual name, as in the Holme Cartulary, where we find men named by their font-name alone, or by this plus (i) their father's name in full, (ii) their father's christian name, (iii) the name of their estate or of their place of origin, or (iv) a byname, descriptive of office, occupation, or a nickname, e.g.

Odo balistarius, arbalistarius, or de Wrthstede

Osbernus decanus (de Turgetona), presbiter (de Turgertona), de Turgartona, de Tweyt, filius Griffini (de Tweyt)

Guarinus minister noster, Gwarinus dispensarius noster, Warinus dispensator, Warinus de Thoftes

Willelmus filius Hermanni, Willelmus Hermannus, Willelmus de Caletorp, Willelmus de Hobosse

Similar descriptions are found in other twelfth-century documents, the most common being a variation between the name of the father and a place-name or byname. The names of clergy varied with a change of incumbency or office, or as they rose to higher orders.

From twelfth-century Danelaw charters:

Ralph de Cheurolcurt, de Clachesbi

Johannes filius Herberti, de Orrebi

Adam filius Remigii (de Hakethorn), Wodecok

Gaufridus filius Bertranni (de Snelleslund), del Lund, de Lund, filius Bertrammi de Lund

Reginald Lequite, de Courtenay

In the thirteenth and fourteenth centuries similar but less elaborate variations are found, and here we often have real alternative surnames:

Robert Hastif, Robert de Disun 1202–3 Cur (Wa)

Ricardus filius Walteri, de Cliue 1221 AssWo

Milo de Verdun, de Creyton 1222 AssSt

Roger Waudin, Rogerus Anglicus 1243 Fees (Sa)

Adinet de Bidyk 1276 Fine, Adnettus le Taylur 1285 *Ass* (Ess)

Ralph le Verrer, Ralph Vicar 1311 ColchCt

Jordan de Newintone, Jordanus Pistor 1317 AssK

This variation of surname is sometimes implied:

Hugo de Burgo le Mazun 1257 Oseney

Johannes Gyffard dictus le Boef 1297 MinAcctCo
John Bulichromp called le Binder 1300 LLB C
Adam de Sutton, called 'Ballard', sadler 1303 LoCt
William Cros le Bole 1305 LoCt
Ralph de Eyr called Proudfot of Havering 1393 HPD (Ess)[2]

The following names of freemen of York are so entered on the roll:

Thomas le Walche, de Selby, girdeler (1329)
Alan Hare, de Acastre, carnifex (1332)
Rogerus filius Johannis de Burton, de Eton in le Clay, boucher (1343)
Johannes, filius Roberti de Gaunt, de Duffeld, mercer (1356)

With the fifteenth century such names become less common, but variation of surname continued and is found sporadically until the seventeenth century or later:

Robert Curson alias Betele 1410 AD iv (Lo)
Nichol Wigh oþerwise callyd Nicholas Ketringham oþerwise callyd John Segrave otherwise callyd Nicholl' Pecche 1418 LondEng
Henry Lordessone alias Henry de la Heus 1460 AD iv (Nth)
John Morys alias Rede alias Sclattere 1474 Oseney
Roger Harflete otherwise called Roger Cheker son and one of the heirs of Christopher Harflete otherwise called Christopher Atcheker . . . Raymond Harflete also called R. Atcheker 1508 ArchC 40
Richard Bishop alias Hewson of London 1671 EA (NS) iv

This variation may be merely scribal. In Domesday Book Robertus *blundus* is also called *albus*, *flauus*, *blancard*, all meaning 'fair'. The tenants of Woolfin (Devon) were Gregory *Lupus* (1222), Richard *le Low* (1303) and Walter (*le*) *Wolf* (1359 PN D 368). Here the real surname was *Wolf*, translated by the earlier scribes into Latin and French. Allard *Smyth* is identical with Alaerd *le Fevre* (1382 LoPleas).

There is, however, evidence that the surname in the document is not always that used by the man himself:

Robert le Botiler of Hertford. *Seal:* Robertus filius Willelmi (1275 AD iv)
Agnes daughter of Rogerus piscator of Coventre. *Seal:* Agnes filia Petronille (1299 AD v)
Thomas le Diakne of Ikelyntone. *Seal:* Thomas filius Ricardi de Fonte (1300 ib.)
Ralph de Westred. *Seal:* Radulfus filius Willelmi (13th AD iv)
Agnes de Humet. *Seal:* Agnes de Bellomonte (ib.)
Seuual de Walcfare. *Seal:* Sawale filius Petri (13th AD v)
Angerus called Humfrey of Lapworth. *Seal:* Aungerus de Bispwod (1319 ib.)
Katharine daughter of John le Jay, wife of Roger Prodhome. *Seal:* Katerina Franceis (14th ib.)

## TRANSFER OF SURNAMES

In London, in the thirteenth and fourteenth centuries, it was quite usual for the surname of an apprentice to be replaced, either temporarily or definitely, by that of his master.[3] In most cases where sufficient material is available, the new surname displaced the old one altogether, a matter of some importance for genealogists.

Sewald, son of Sewald de Springefeld (1311 LLB B), is identical with Sewal, son of Sewal de Sprengewell, apprentice of Richard de Godesname, paternostrer (1311 LLB D) and with Sewallus de Godesname (1319 SRLo). Robert Podifat (1288 LLB A) was an apprentice of Roger le Fuyster (1312 LLB D), who is also called Roger Podifat (1320 LLB E). Robert, therefore, assumed his master's nickname as his surname. Thomas de Cavendisshe, son of William atte Watre de Ewelle, late apprentice of Walter de Cavendisshe, mercer, was admitted a freeman of the city in 1311–12 (LLB D). His original surname would have been *atte Watre* or *de Ewelle*. From 1319 to 1349 he is regularly called *de Cavendish* and, in the enrolment of his will, Thomas de Cavendych, mercer or draper.

The same custom seems to have existed also at York, though less well evidenced: William Payne serviens John Payne (1323 FrY); Roger Storre, servant of Johan Storre (1379 PTY); Richard Redhode, draper, serviens Willelmi Redhode (1386 FrY); Thomas Gauke, cocus, filius Roberti Nyd servientis Simonis Gauke (1424 FrY). Here, Thomas bore the surname of his father's master which had probably been assumed earlier by his father.

## CLASSIFICATION OF SURNAMES

Surnames may be divided into four groups:
1. Local Surnames
2. Surnames of Relationship
3. Surnames of Occupation or Office
4. Nicknames

Within these groups there is considerable overlapping and a full and accurate classification is impossible. In dealing with names, we are concerned with an intimate possession and with the thoughts and idiosyncrasies of those who bestowed or adopted the names. They were not concerned with rules but with satisfying an immediate need. Nicknames, in particular, were often the result of a spontaneous reaction to a particular occasion.

Local surnames may be occupational. The Panter worked *atte panetrie*. John *atte Gate* may have lived near the town-gate, or he may have been a gate-keeper or porter.

Surnames of office, such as Abbot, Bishop, Cardinal and King, are often nicknames. Ralph Vicar was a glassworker, not a clergyman, and is also called *Verrer*. A single modern name may belong to more than one class. *Low* may be a French nickname from the wolf, a Scandinavian nickname for a small man, a pet-name from *Laurence*, or a local surname, from *hlāw* 'hill'. *Waller* may be a nickname, 'coxcombe, spark', occupational, 'a builder of walls' or 'a salt-maker', or local, 'dweller by a wall' or 'by a stream'. *Mew* may be a patronymic, a nickname from the sea-mew, or occupational, either metonymic for *Mewer*, 'keeper of the hawks', or from a local surname, with the same occupational meaning. It is impossible to fit surnames into a strait-jacket.

# LOCAL SURNAMES

Local surnames, by far the largest group, derive from a place-name, indicating where the man held land, or the place from which he had come, or where he actually lived. Richard de Tonebrige (1086 DB) was so called from his castle of Tonbridge, but he was also called Richard de Clara from the Suffolk Clare which became his chief seat and gave the family its definitive surname. Richard de Hadestoke, a London alderman c1240, had left Hadstock (Essex) and settled in London. Thomas atte Forde lived near a ford.

These local surnames derive (with occasional exceptions) from English, Scottish or French places and were originally preceded by a preposition *de*, *at*, *by*, *in*, etc. A certain number of Old English formations are found before the Conquest: Ælfweard æt Dentune (972), Ælfstan on Lundene (a988), Godcild of Lamburnan (c970), Leofnað in Broðortun (c1050).[4] After the Conquest the usual preposition is *de*, which is used before both English and French place-names. In French names beginning with a vowel, this *de* has often coalesced with the name: Damerell, Danvers, Daunay, Disney, Doyley, etc., and occasionally with English names, as Dash, Daysh, Delafield, Delamere. Many of the French place-names denote the seat of noble families, but many of the modern surnames merely indicate migration from a French place. There was a constant stream of merchants, workmen and others from the English provinces of France.

The earliest local surnames of French origin are chiefly from Normandy, particularly from the departments of Calvados, Eure, Seine-Inférieure and La Manche.[5] Some of the Frenchmen early acquired surnames from English places, e.g. Roger de Berchelai (1086 DB). Philip de Poswyc (c1147) was a son of Richard Basset.[6]

English local surnames may derive from the manor held (Adam de Cokefeld 1121–48 Bury); from the place of residence (Ralph de Nordstrate 1197 P, Goduy ad Westmere c1095 Bury), William Attebroc (1199 P); or from the place from which a man had come: Brihtmarus de Haverhell' (1158 P), who had moved from Suffolk to

London, where his son became alderman and sheriff.[7] Occasionally we have a surname from a sign (*atte Lamb*, *atte Raven*), but these are usually late and less common than has been supposed. Some of the 'signs' really refer to topographical features (*Ball*, *Cock*).

The local surname, even when changes in form or pronunciation have occurred in the modern place-name, is usually straightforward enough. It is more difficult to trace the minor names. A general meaning can usually be assigned to them, but whether, for example, Richard *del Helde* 1246 AssLa lived near a nameless slope, or whether at a place called Heald, is not always easy to discover. The counties surveyed by the English Place-name Society contain fairly complete lists of minor names, and there are similar comprehensive works for Lancs, Kent, and the Isle of Wight. But for other counties, it is not always possible to trace such minor names. As yet there are no historical surveys of the place-names of Cornwall, Hants, Norfolk, Suffolk, Lincs, Herefordshire, Shropshire, Leicestershire, or of the Welsh counties. It is probable that as these counties are surveyed, it will be possible to assign to a particular place more of the local surnames for which, as yet, a general interpretation is all that is possible. A historical survey is necessary since we must know that the place in question was in existence at the time when surnames were coming into use, and also that the medieval forms of the place-name are similar to those for the surname. For example, *Barnacle* is often derived from Barnacle (Wa). But the place-name was Bernhangre in medieval times, and does not appear as Barnacle before 1547, long after the period when a surname from this form could have developed. Similarly, *Brighton* is unlikely to derive from the Sussex town, which is usually Brightelmeston in medieval sources. It first appears as Brighton in the reign of Charles I, but this form of the name did not come into common use before the early nineteenth century. The surname must derive from Breighton (ERY), appearing as Bryghton from 1298 to 1567. Again, *Bristol* must usually come from Burstal (ERY), or Birstal (WRY), since the normal medieval vernacular form for Bristol (Gl) is Bristow, and Bristol does not become regular before the sixteenth century.

## Ash, Nash, Rash; Oakes, Rook

A very large number of English local surnames derive from small places, or denote residence by a wood, in the marsh, by oaks, elms, ash, etc. These occur as *atte wode*, *atte fenne*, etc., and the preposition is usually lost later but is preserved in such names as Attlee, Byfield, Uphill, Underdown, etc. The names are from OE *æt þǣm āce*, *æt þǣm œsce*, etc., which became ME *atten ake*, *oke*, *atten ash*; *atte oke*, *atte ash*. The latter became *Oak*, *Ash*, etc.; the former became *atte noke*, *atte nash*, the preposition was dropped and the surnames became Noke, Nash. Rook, Rash, etc., derive from OE *æt þ̄ēre āce*, *æt þ̄ēre æsce*, which became ME *atter oke, atter ash*; *atte roke, atte rash*. In *Tash, Tesh*, ME *atte ashe, eshe* became *at tash, at tesh*.

## Loss of the Preposition

The absence of the preposition in early forms of local surnames (and of the article before occupational names) has been regarded as a sign that the surname had become hereditary. Such a supposition cannot be upheld. The preposition began to disappear much earlier than has been thought and examples are too numerous to be regarded as scribal errors.

Ekwall remarks that the preposition begins to be dropped shortly after 1300, is mostly preserved through the fourteenth century, but after 1400 is usually absent. His earliest example is 1318.[8] Fransson states that in York, *de* disappears in the early fifteenth century; in Lancashire it sometimes occurs c1450; whilst in the south it is regularly dropped at the end of the fourteenth century.[9]

Tengvik has noted in 1066 in Domesday Book 163 examples of local surnames consisting of a simple place-name without a preposition.[10] There are a few also in 1086: Rogerus Blaneford (Blandford), Roǵerus Povrestoch (Powerstock), Rogerus Povertone (Poorton), all in Dorset, William Tochingeuuiche (Tingewick, Bucks). A further 100 have been noted in twelfth-century documents from both English and French places, 28 in the Curia Regis Rolls (1201–21), mostly English, Alan Cheles 1219 AssL (Keal, Lincs), Richard Sulee 1221 AssWo (Sudley, Glos), and others.

In the London Subsidy Rolls for 1292 and 1319, where about half those assessed have local surnames, the preposition is always retained. In 1332 there are 23 without a preposition. In those for Sussex, local surnames without a preposition (mostly English) increase from 119 in 1296 to 319 in 1327 and 418 in 1332. In the 1327 Somerset Subsidy Roll about one-third of the surnames are local and of these 7 per cent have no preposition, a proportion very similar to that in Suffolk for the same year (6 per cent). In 1332 in Surrey about 20 per cent of the local surnames have no preposition, whilst in Lancashire in 1332 all the 1255 local surnames except 28 retain the preposition. It seems clear, therefore, that there was a definite tendency to drop the preposition from 1066; by the end of the thirteenth century the tendency was marked in Sussex and steadily increasing. In the first quarter of the fourteenth century Surrey shared this tendency, but it had not reached London. The process had begun in Suffolk and Somerset, was almost completely absent in Lancashire and non-existent in Yorkshire. 'The *de* before the surname is in constant use well into the reign of Henry IV.'[11]

## Toponymics

A common form of local surname of which many examples survive consists of an adjective or noun denoting nationality or the country, province, county, town or district from which the individual came: English, Scott, Breton, Fleming, Angwin, Loring, Poidevin, all of which are found in Domesday Book. Later surnames are Irish, Welsh, Wallas, Gall, Norman, Brabazon, Cornish, Cornwallis, Devenish, with Norris, Surridge, Sotheran, Western, Westridge. An early example which has not survived is Wluuardus *le Doverisshe* 'of Dover' (1125 ELPN).

Toponymics formed by the addition of *-er* to some topographical term, e.g. Bridger, Brooker, etc., are particularly common in Sussex at the beginning of the fourteenth century. They are also found in the neighbouring counties of Kent and Surrey, in Essex and Hampshire, but are less common elsewhere. The meaning is 'dweller by the bridge, brook, etc.', or, occasionally, at a particular place, *Rumbridger*, from Rumbridge (Sussex).[12] The names, at times, interchange with names in *atte* and compounds of *-man*. In the Sussex Subsidy Rolls, Hugo *atte Broke* (1296) is identical with Hugo *le Broker* (1327); John *atte Combe* (1327) with John *le Coumber* (1332); John *atte Gore* (1296) with John *Gorman* (1332); William *atte Gate* (1296) and John *Gateman* (1327) both lived in Goring.

Interesting survivals of Scandinavian formations are the local surnames Sotherby, Westoby, from ON *suðr*, *vestr i bý* (the man who lived) 'south or west in the village', and the anglicized Dunnaby, Easterby. Similar English formations survive in Astington, Norrington, Sinton, Uppington, Westington.

# SURNAMES OF RELATIONSHIP

Surnames of this class are often called patronymics, but a more comprehensive term is needed, partly because many modern surnames are formed from women's names, partly because in early sources other relationships are expressed: Alwinus *Childebroder*, Alwin' *pater Cheping'* (1066 Winton), Baldgiua *soror Osuuardi*, Lefuine *frater Toui*, Goduin *Aluuini nepos*, Wluin *Brune stepsune*, Sibbe *Ædesdohter* (c1095 Bury), Willelmus *gener Arnwi* (c1200 DC).

Such names are not uncommon in the twelfth century and are found later: Gilbert *Fathevedsteppeson* (1307 Wak), John *le Personesneve* (1324 FFEss), Richard *Hannebrothir*, Ameria *Ibbotdoghter* (1324 Wak), Amabilla *Hannewyf* (1327 ib.), William *Maisterneue* (1327 SRSf), John *Prestebruther*, Johanna *Raweswyf* (1332 SRCu), Emma *Rogerdaughter*, Robert *Prestcosyn*, Marjoria *Vicar'neys* (1381 PTY), Isolda *Peersdoghter* (1430 FeuDu). The only names of this type to survive are a few compounds of *-magh* 'brother-in-law': Hickmott, Hitchmough, Hudsmith, Watmough.

## Patronymics

### Names in -son

In Old English, patronymics were formed by adding *-ing* to the stem or *-sunu* to the genitive of the personal name: *Dudding* 'son of Dudda', *Ēadrīcessunu* 'son of Ēadrīc'. The latter type was used as a patronymic adjunct: Hering Hussan sunu (603 ASC), a type found also in the eighth and ninth centuries and not uncommon in the names of the festermen of Peterborough (963–92): Godwine Ælfrices suna. This was also a common Scandinavian formation: Þurferð Rolfes sune. In his *Old English Bynames*,[13] Tengvik has collected 146 examples, of which 111 are English and 24 Scandinavian. In twelfth-century London Ekwall has noted a further eleven examples.[14]

## Metronymics

Similar formations, though less common, were based on the mother's name: Eadric Wynflæde sunu (c1015 ASCh), Siwardus Leuerunessone (1066 Winton), Edric' Modheuesune (1137 ELPN).[15]

## Johnson, Williamson, etc. Distribution and Origin

The frequency of names in *-son* in the North has been commonly attributed to Scandinavian influence,[16] but examples are rare or non-existent from the twelfth to the fourteenth centuries in north and south alike. The common form of both patronymics and metronymics in the twelfth and thirteenth centuries is Willelmus *filius Hugonis*, Ricardus *filius Agnetis* (1185 Templars), a form found side by side with compounds of *-sunu* in the eleventh century and, less commonly, of *-son* in the fourteenth. It is hardly conceivable that surnames like *Godricsone* (1066 DB) ceased to be used for a couple of centuries and were then suddenly revived. The formula Rogerus *filius Radulfi* may be merely a description, 'Roger son of Ralph', or it may be a translation of Roger *Fitz Ralph*, a form rare in documents. Radulfus *filius Godrici* may similarly be a translation of *Godricson*. That the formula was merely descriptive is proved by the fact that a man could be named both Willelmus *filius Hermanni* (1134–40) and Willelmus *Hermannus* (1141–9 Holme), whilst there are a few examples of the equation of a simple christian name as a surname with a compound of -*sunu*: Aluuinus *Dode*, Aluuinus *Dodesune* (1066 DB). Names like Willelmus *filius fabri* (1219 AssY), Hugo *filius clerici* (1185 Templars, Gl) are common and descriptive, 'son of the smith or the clerk'; they are found in the fourteenth century as *Smythson* and *Clerkessone* and still survive.

In the Cumberland Subsidy Roll (1332) we find Alan *Malleson*, John *Diksson*, etc., side by side with Adam *son of Alan*, John *son of Robert* (presumably translations of *filius Alani*, *filius Roberti*) and Thomas *Prestson*. It would appear that the form in common use was *Diksson*, *Helewisson*, *Heliotesson*, etc., that the twelfth- and thirteenth-century scribes translated all such names by *filius Roberti*, etc., and that by the fourteenth century there was a growing tendency for the clerk to use the spoken form, particularly with the common pet-names *Dick*, *Hob*, etc. Thirteenth-century examples noted are: Adam *Saresone* 1286 LLB D, William *Marysone* 1298 ELPN (Willelmus *filius Marie* 1292 SRLo), William *Paskessone* 1293 FFC, Thomas *Wummanesone* 1297 Coram (C).

In the Sussex Subsidy Rolls there are no examples of -*son* in 1296, 4 in 1327, 13 in 1332, all metronymics. In other Subsidy Rolls we have in 1327, in Worcestershire 11 (including 4 metronymics); Somerset 8 (5 metronymics); Cambridgeshire 17 (6 metronymics); in 1332, in Surrey 6 (4 metronymics); Lancashire 23 (7 metronymics); Cumberland 22 (7 metronymics); in Yorkshire 2 in 1297, 5 in 1301, 10 in 1327, including 3 metronymics; in FrY (1272–1381) 14, all patronymics—earliest example 1323; in Suffolk (1327) 7 (3 metronymics), (1381) 12 (5 metronymics). In all these sources there are additional examples of *Reveson*, *Smithson*, etc. Occasionally the surname is based on the father's surname. In Cambridgeshire John *Brunnison* and William *Broun* occur in the same parish as do Richard and William *Lawisson* and Henry *Law*. In Cumberland, Hugh *Moserghson* was presumably the son of Thomas *de Mosergh*. *v*. also p. xliv.

The surnames in -*son* form a very small proportion of the whole and are more common in the north. In Lancashire the number assessed is only about one-fifth that in Suffolk. In the northern counties the number of individuals with no surname or described as *filius*——is much greater than in the south where the development of surnames was more advanced. But it is clear that in the fourteenth century, when surnames in -*son* begin to appear again, they were not limited to the north. It is unlikely that in Somerset, Sussex and Surrey these names should be due to Scandinavian influence. Tengvik notes that in his Old English material examples are found at a date when 'we can hardly reckon with any important Scandinavian influence'. The local distribution of the type, too (especially in Devonshire), points to a native origin.[17]

In the north we may have to reckon with Scandinavian influence also, but the frequency of the type may be due, in part at least, to the late development there of hereditary surnames. We find such names as Henry Dicounesson de Clesnesse 1359 Pat

(Nb), 'Henry, son of Dicoun de Clesnesse', Richard Jeffson Nanneson de Radford 1385 NottBR, a type found also in Yorkshire and Lancashire: Robert Tomson Watson, Robert Stevenson Malynson, Thomas Robynson Richardson 1381 PTY, John Robynson Diconson 1408 AD v (La), John Atkynson Jonson 1433 ib. (Y). It is doubtful whether the latter should be interpreted 'John, son of Atkyn, son of John' or 'John, son of Atkyn Jonson'. The occurrence together of John Prestson, Agnes ancilla Johannis Prestson and Geoffrey Jonson Prestson (1379 PTY) points clearly to a surname but that it was hereditary is doubtful. The frequency of the type and the common addition of -*wyf* (sometimes added to the christian name), -*doghter*, -*man*, -*maiden*, -*servant*, give a very strong impression that these were not real surnames in the modern sense but patronymic descriptions in a constant state of flux.

## Font-names as Bynames

Less common in the early twelfth century than names like Symon *filius Ricardi*, but steadily increasing in number, are names of the type Johannes *Gerard*, Henricus *Bertram*, in which a font-name is added to the christian name as a byname. Tengvik has noted ten examples in the eleventh century in which, in seven instances, the byname is Scandinavian, in one, Old English, and in two, French. In Domesday Book there is a great increase in the use of French personal-names (40), as well as Old English (28) and Scandinavian (18), with three Celtic and five Latin.[18]

The general opinion is that such surnames are due to the dropping of *filius*.[19] Tengvik has noted five instances which seem to support this view: Osbern *Hauoc*, Osbern *filius Hauoc*; Rainaldus *Croc*, Rainaldus *filius Croc*, in which two of the personal names (*Hauoc* and *Dudde*) are Old English, two (*Baderon* and *Clamahoc*) Breton, and so introduced by Normans, and one (*Croc*) Scandinavian, though the christian name *Rainaldus* suggests a possible Norman origin.

It is difficult to believe in this 'dropping of *filius*' theory. A name of the type *Johannes filius Willelmi* was never used in everyday life by either Englishmen, Frenchmen or Scandinavians. It is Latin and a documentary form. Where the font-name is French, it might be a translation of Fitzwilliam, though such names are unknown in France and rare in English sources. If the font-name is English, *filius Dudde* might be a translation of *Duddesunu*, but it is unlikely that one-sixth of the Suffolk peasants of c1095 bore such names. There seems no alternative to regarding these forms as scribal descriptions. Walter Dudde was known to be the son of Dudda and he was so described in writing, in the clerk's Latin, *filius Dudde*. But in ordinary conversation, when his full name was needed, he was called Walter Dudde. Three early examples of this type have been noted by von Feilitzen: the Scandinavian Sendi Arfast (c1044), the French William Ingelram (1088) and Ieduue Ialdit (c1100–30), from OE *Ealdgȳð*.[20] Nine other examples occur in

Suffolk (c1095 Bury), all English: Aldwine Ælfuine, Lemmer Brihtmer, Ordric Wihgar, etc.

The origin of surnames of this type cannot, at present, be definitely decided. The majority of such names are not, as Smith states, from personal names of Scandinavian origin. There are numerous examples from English and French personal names and a smaller number from Celtic. Scandinavian influence may be partly responsible, especially in the Danelaw. The type does not become common in England until after the Conquest and we may be concerned with a French custom introduced by the Normans. Similar names are found in northern France in the tenth century and in the south and south-west in the previous century.[21] The fact that similar formations from Old English personal names are common in the south of England in the late thirteenth and the fourteenth centuries and in eleventh-century Suffolk suggests an independent formation in English. The frequency of the type may be due to all three influences, combined with its simplicity for everyday use and the analogy of similar simple attributes in the form of nicknames and occupation names. It is noteworthy that such surnames from very common christian names like William, John, etc., are late formations.

## Post-Conquest Survival of Old English Personal Names

The Norman Conquest revolutionized our personal nomenclature. The Old English name-system was gradually broken up, Old English names became less and less common and were replaced by new names from the Continent, a limited number of which gradually became more and more popular. The general trend is well known, but many of the definite statements on the relative frequency of various names are based on insufficient evidence, often from late sources which can have little or no bearing on the history of surnames. Fashions in names varied among different classes and in different parts of the country. Most of the early documents deal with the upper classes. Names of peasants are less common, rarely occur in large numbers, and have largely been ignored. Intensive work on the abundant material in manuscript and on that already printed will ultimately throw much new light on the history of our names. For the present, we must be content with a study of selected material from varied districts.

In the Holme Cartulary, some 75 per cent of the twelfth-century names are those of witnesses and grantees of charters, monks, clerics, and tenants of the abbey, and these reflect the new nomenclature introduced by the Normans. The personal names are of French or continental Germanic origin, with Norman forms of Scandinavian names brought from Normandy. The remaining 25 per cent, the names of peasants, represent some 30 English and 35 Scandinavian names, some more than once repeated.

In the twelfth century the most popular names were William (10 per cent) and

Robert (7 per cent), followed, with variation of order in different documents and counties, by Richard, Ralph, Roger, Hugo and Walter. John (3 per cent) was much less popular.[22] In a thirteenth-century collection of deeds relating to Aveley (Essex), John shares the top place with William (20 per cent) and these, with Robert, Richard, Geoffrey and Thomas, were the names of 160 out of 250 individuals. There were 41 other French names in use, including the Breton Alan, Hervey and Wygan, shared by 86 persons. In addition, 28 persons shared 16 English names, including an unrecorded *Weorðing*.[23] In the fourteenth century Old English names were fewer and much smaller in proportion, 7 persons sharing 5 names. Some 460 persons shared 35 French names. John (34 per cent) was now much more popular than William (18 per cent). Then came Thomas (9 per cent), Richard and Robert (6 per cent), Henry, Roger and Geoffrey, these 8 names being borne by 375 persons. The remaining 32 names were shared by 85 persons.

Unrecorded Old English personal names are found in Middle English documents. Most of these conform to the traditional *Wulfstan*-type. Some contain elements not noted before (*Geongwine*, *Weorðgifu*), others elements rare in Old English names (*Mǣðwynn*, *Wudubeorht*). Some, especially the original bynames, may have been formed in the Middle English period. Many of these names were those of peasants, among whom the native habits of name-giving survived longer than among the upper classes.[24] A number of personal names which are not recorded in Old English after the eighth or ninth centuries reappear in Middle English. Some of these names are evidenced only by their occurrence as surnames, others by their first record in the eleventh, twelfth or thirteenth century. Five (including *Uhtrīc*, *v.* Outridge) are not found before the fourteenth century and one (*Cwēnrǣd*) only in the fifteenth. Some survive as surnames: *Cwēnhild* (1086) in Quennell; *Ēadwacer* (1066) in Edicker; *Wæcerhild* (c1130) in Wackrell. Others are of importance in confirming the existence of personal names postulated to explain place-names. *Pacchild* noted in Essex in 1166 must be a compound of OE *Pæcc(i)* and the common theme *-hild*. No example of *Pæcc(i)* is known, but it has been assumed as the base of Patching (Essex, Sussex), Patcham, Pashley (Sussex) and other related place-names. Similarly *Wlanchild*, recorded only as the name of a peasant-woman in 1206 in Cambridgeshire, and reappearing as a surname in Suffolk (Warin *Wlankild* 1277 Ely), is a compound of *-hild* with OE *wlanc* 'proud', postulated as a personal name to explain Longslow (Salop).[25]

This vitality of Old English names is confirmed by the number of fourteenth-century surnames formed from Old English personal names no longer then in use. We can only suppose that these personal names had continued in use long after the Conquest and that the surnames had already become hereditary. In the 1327 Subsidy Roll for Somerset, 66 per cent of those assessed were named: John (23 per cent), William (16 per cent), Robert (8 per cent), Richard (7 per cent), Walter (6 per cent) or Thomas (6 per cent). The rest shared 95 different names of which 8 were English and 5 Scandinavian. But there were some 200 surnames formed from English or Scandinavian personal names no longer in use, e.g. Thomas Ailmer, Richard Leverich, John Sefoghel, William

Serich, Robert Outright, Philip Thorbarn, Edith Thurkyld.

The rate at which this change from English to Norman christian names proceeded varied from class to class and among different families of the same class. It was slowest among the peasantry. In 1115, at Winchester, two out of six English fathers gave their children English names. In 1148 all the children named bore French names whether their fathers had English, Scandinavian or French names. At King's Lynn in 1166 the process was less advanced. Of 17 fathers with Scandinavian names and 18 with English names, only half followed the new fashion.

Occasional pedigrees put forward to support or resist a claim that a man was a villein shed some light on the names used by peasants. In one from Lincolnshire which must take us back to 1100 or beyond (1200 Cur), a man with the English name of Elric had a daughter Oise or Osse and two sons Agge and Siuuerd, all probably bearing Scandinavian names. *Oise* may be ODa *Ása* (f), *Agge* is ODa *Aggi*; *Siuuerd* may be ODa *Sigwarth* or OE *Sigweard*. Siward had a son Uhtred (OE) and Agge a son named Elric after his grandfather and a daughter with the French name Beatrice who married a Walter and named her son (living in 1200) after his father. Oise had a son William Belle and two grandsons Roger and Robert Belle. Here the change to French names seems to have been due to women.

A similar Suffolk pedigree (1200 Cur) takes us back a generation farther. Godwin named his three children from three different languages: Turgod (Scandinavian), Goda (English) and Watcelina (French). Goda gave his son the French name of Robert and his daughter Agatha married a man named Thomas. Watcelina named her daughters Einilda (OG *Aginildis*) and Langiva (OE *\*Langgifu*). Turgod named both his sons in English, Godwin and Edric, the latter continuing the English tradition in Alfridus (probably OE *Ælfrēd*, possibly *Æðelfrið*), whose son (alive in 1200) was Osbert (OE *Ōsbeorht*). Thus Godwin's descendants in the male line kept to the native tradition in names for five generations to 1200.

From a number of similar pedigrees going back for two to four generations from 1200 it would appear that among the peasants at the beginning of the thirteenth century Old English personal names were being replaced by names of French origin, but some families were more conservative than others. The variety of personal names used, both English and French, is noteworthy, as well as the general absence of bynames. The vitality of the Old English name-system is revealed by the evidence of the persistence of the repitition of either the first or second theme in names of the same family and by the existence of names otherwise unrecorded.[26] Both English and Scandinavian personal names were still common on the Suffolk manors of the Bishop of Ely in 1277.

The number of unrecorded forms of Old English names that have to be assumed for the surnames dealt with in the following entries emphasizes how little we really know about Old English names, but to the reader it may appear an easy way of providing an origin for a surname, and he may wonder what evidence there is for such assumptions. In most cases there is definite evidence from place-names, or from the existence of the name in early Middle English, to indicate that it was probably current in Old English though not recorded in the surviving records. At the least the various elements or the general form of the name may be comparable with extant names from the period.

## Old English Personal Names Surviving in Modern Surnames

### Monothematic

*Bada* (Bade), *\*Beald* (Bald), *\*Becca* (Beck), *Bēda* (Beade), *Beorn* (Barne), *Bil* (Bill), *Bisceop* (Bishop), *\*Blīða* (Bly), *Boda* (Bode), *Botta* (Bott), *Brūn* (Brown), *Budda* (Budd), *\*Butt(a)* (Butt), *Bynni* (Binns), *Cada* (Cadd, Cade), *Cana* (Cane), *Ceadda* (Chadd), *\*Cēne*, *\*Cyne* (Keen), *Cniht* (Knight), *\*Cocc(a)* (Cock), *\*Codd(a)* (Codd), *Cola* (Cole), *Creoda* (Creed), *Cyng* (King), *\*Cyppe* (Kipps), *\*Dæcca* (Dack), *Dēora* (Dear), *Dodd(a)*, *Dudd(a)*, (Dodd) *\*Ducc* (Ducket, Duxon), *\*Dylla*, *\*Dylli* (Dill), *\*Flint* (Flint), *Fugol* (Fowl), *\*Glæd* (Glade), *\*Glēaw* (Glew), *Goda* (m), *Gode* (f) (Good), *Golda* (m), *Golde* (f) (Gold), *\*Grante*, *\*Grente* (Grant), *\*Grēne* (Green), *\*Hand* (Hand), *Heafoc* (Hawk), *Hēaha* (Hay), *\*Heard* (Hard), *Hunna* (Hunn), *Hwīta* (White), *\*Lemma* (Lemm), *Lēof* (Leaf), *Lēofa* (m), *Lēofe* (f) (Leaves), *\*Leppe* (Lipp), *\*Lutta* (Lutt), *Mann* (Man), *Mawa*, *\*Mēawa* (Maw), *\*Mēaw* (Mew), *\*Milde* (f) (Millsom), *\*Modd* (Mudd), *Odda* (Odd), *Pæga* (Pay), *Pymma* (Pim), *Scot* (Scott), *\*Sida* (Seed), *Snel* (Snell), *\*Sprott* (Sprott), *Swan* (Swan), *Swēt(a)* (m), *Swēte* (f) (Sweet), *Swift* (Swift), *\*Tæppa* (Tapp), *\*Tāt* (Tate), *\*Tetta* (Tett), *Tunna* (Tunn), *\*Þēoda* (m), *\*Þēode* (f) (Theed), *\*Ucca* (Huck), *\*Ugga* (Hug), *Wada* (Wade), *Wine* (Winn)

### Derivatives in -ing

*\*Bealding* (Balding), *Billing* (Billing), *\*Botting* (Botting), *Brūning* (Browning), *\*Budding* (Budding), *\*Cypping* (Kipping), *Dēoring*, *Dȳring* (Dearing), *Dēorling*, *Dȳrling* (Darling), *Dūning* (Downing), *Dunning* (Dunning), *\*Dylling* (Delling, Dilling), *\*Fugeling* (Fowling), *\*Glæding* (Gladden), *Goding* (Gooding), *Golding* (Golding), *Hearding* (Harding), *\*Hræfning* (Ravening), *Hunning* (Hunning), *\*Hwætling* (Whatlin), *Hwīting* (Whiting), *\*Lēofecing* (Lucking), *Lēofing*, *Lȳfing* (Levinge), *\*Lēofring* (Lovering), *Manning* (Manning), *\*Munding* (Munnings), *\*Pening* (Penney), *Snelling* (Snelling), *Swēting* (Sweeting), *\*Tæpping* (Tappin), *\*Tipping* (Tipping), *\*Tylling* (Tilling), *\*Utting* (Utting), *\*Wealding* (Walding), *\*Weorðing* (Worthing), *\*Wihtling*, *\*Hwītling* (Whitling), *\*Wilding* (Wilding), *\*Wulfing* (Woolving)

*Dithematic*

\*_Ācmann_ (Oakman)

_Ælfflǣd_ (f) (Alflatt), _Ælfgār_ (Algar), _Ælfhēah_ (Alphege), _Ælfhere_ (Alvar), \*_Ælfmann_ (Elfman), _Ælfnōð_ (Allnatt), _Ælfrǣd_ (Alfred, Averay), _Ælfrīc_ (Aldrich), _Ælfsige_ (Elsey), _Ælfstān_ (Allston), _Ælfweald_ (Eliot, Ellwood), _Ælfweard_ (Allward), _Ælfwīg_ (Alaway, Allvey), _Ælfwine_ (Alven, Alwin)

_Æscwine_ (Ashwin)

_Æðelbeorht_ (Albright), \*_Æðeldæg_ (f) (Allday), _Æðelflǣd_ (f) (Alflatt), _Æðelfrið_ (m), \*_Æðelfrīð_ (f) (Alfrey), _Æðelgār_ (Algar), _Æðelgēat_, _Æðelgȳð_ (f) (Aylett), _Æðelgifu_ (f) (Ayliff), _Æðelheard_ (Adlard), _Æðelmǣr_ (Aylmer), _Æðelnōð_ (Allnatt), _Æðelrǣd_ (Aldred, Allred), _Æðelrīc_ (Aldrich, Allright, Etheredge), _Æðelstān_ (Allston, Aston, Athelstan), _Æðelðrȳð_ (f) (Audrey), _Æðelweard_ (Allward, Aylward), _Æðelwīg_ (Alaway), _Æðelwine_ (Alven, Alwin, Aylwin)

_Beadurīc_ (Badrick), \*_Bealdgȳð_ (f) (Baldey), \*_Bealdrīc_ (Baldree), \*_Bealdmann_ (Balman), \*_Bealdstān_ (Balston)

_Beorhtgifu_ (f) (Berriff, Brightiff), _Beorhtsige_ (Brixey), _Beorhtmann_ (Brightman), _Beorhtmǣr_ (Brightmore), _Beorhtwīg_ (Brighty), _Beorhtwine_, _Beorhtwynn_ (f) (Brightween)

\*_Bīedlufu_ (f) (Bedloe)

_Blæchere_ (Blacker), _Blæcmann_ (Blackman), _Blæcstān_ (Blackston)

_Brūngār_ (Brunger), _Brūnstān_ (Brunsdon), \*_Brūnsunu_ (Brownson), _Brūnwine_ (Brunwin)

_Burgheard_ (Burchard, Burrard), _Burgrǣd_ (Burrett), _Burgrīc_ (Burridge), _Burgstān_, \*_Bucstān_ (Buxton), _Burgweald_ (Burall), _Burgweard_ (Burward)

\*_Cēnweard_ (Kenward), \*_Cēnwīg_ (Kenway), _Cēolmund_, \*_Cildmann_ (Chillman)

_Cūðbeald_ (Cobbald), _Cūðbeorht_ (Cuthbert), _Cūðrīc_ (Cutteridge), _Cūðwulf_ (Culf)

_Cwēnhild_ (f) (Quenell)

_Cynebeald_ (Kemble), _Cynemǣr_ (Kenmare), _Cynemann_ (Kinman), _Cynerīc_ (Kerrich), _Cyneweard_ (Kenward), _Cynewīg_ (Kenway)

\*_Dægmǣr_ (Daymer), \*_Dægmann_ (Dayman), \*_Denebeald_ (Denbow), _Dēormann_ (Dearman), _Dēorwine_ (Darwin), _Dudemann_ (Dodman), _Dūnstān_ (Dunstan)

_Ēadgār_ (Edgar), _Ēadhūn_ (Eaden), _Ēadmǣr_ (Admer), \*_Ēadmann_ (Edman), _Ēadmund_ (Edmond), _Ēadrǣd_ (Errett), _Ēadrīc_ (Edrich), _Ēadstān_ (Aston, Easton), _Ēadwacer_ (Edicker), _Ēadweard_ (Edward, Ewart), _Ēadwīg_ (Eddy), _Ēadwine_ (Edwin), _Ēadwulf_ (Eddols)

_Ealdgār_ (Algar), \*_Ealdnōð_ (Allnatt), _Ealdrǣd_ (Aldred, Allred), \*_Ealdstān_ (Allston, Elston), _Ealdwīg_ (Alaway, Aldway), _Ealdwine_ (Alden, Alwin)

_Ealhhere_ (Alger, Alker), _Ealhstān_ (Allston, Elston), _Ealhðrȳð_ (f) (Audrey)

_Earnwīg_ (Arneway), _Ēastmǣr_ (Eastmure), _Ēastmund_ (Eastman)

_Ecgbeorht_ (Egbert), _Ecgwulf_ (Edgell)

_Eoforwacer_ (Earwaker), _Eoforwine_ (Erwin)

*Forðrǣd* (Fordred), *Frēobeorn* (Freeborn), *Friðulāf* (Freelove)

*Gārmund* (Garman), *Gārwīg* (Garraway), *Gārwulf* (Gorrell), *\*Geongwine* (Yonwin), *\*Glædmann* (Gladman), *\*Glædwine* (Gladwin)

*Godgifu* (f) (Goodeve), *Godhere* (Gooder), *Godlamb* (Goodlamb), *Godlēof*, *\*Godlēofu* (f) (Goodliffe), *Godmǣr* (Gummer), *Godmann* (Goodman), *Godrīc* (Goodrich), *Godsunu* (Godson), *Godweard* (Godward), *Godwīg* (Goodway), *Godwine* (Godwin, Goodwin)

*\*Goldbeorht* (Goldbard), *\*Goldburg* (f) (Goldburg), *\*Goldheafoc* (Goldhawk), *\*Goldhere* (Golder), *\*Goldmann* (Goldman), *Goldstān* (Goldston), *\*Goldwīg* (Goldway), *Goldwine* (Goldwin)

*\*Gūðbeald* (Gubell), *Gūðlāc* (Goodlake), *\*Gūðmǣr* (Gummer), *Gūðmund* (Godman)

*\*Heardmann* (Hardman), *Heaðuwīg* (Hathaway), *Hereweald* (Harold), *Hereweard* (Hereward), *\*Holdbeorht* (Holbert), *\*Hūngār* (Hunger), *\*Huntmann* (Huntman), *Hūnwine* (Unwin), *Hwǣtmann* (Whatman), *\*Hwītheard* (Whittard), *\*Hwıtmann* (Whiteman), *\*Hygemann* (Human)

*Landbeorht* (Lambrick), *Lēodmǣr* (Lemmer)

*Lēofdæg* (Loveday), *Lēofeca* (Levick, Livick, Leffek), *Lēofgār* (Loveguard), *Lēofgēat* (Levet), *Lēofgōd* (Lovegood), *Lēofmǣr* (Lemmer), *Lēofmann* (Loveman), *Lēofrǣd* (Leverett), *Lēofrīc* (Leverage), *Lēofsige* (Lewsey), *Lēofsunu* (Leveson), *Lēofweald* (Leavold), *Lēofweard* (Livard), *Lēofwīg* (Leavey), *Lēofwine* (Lewin)

*\*Leohtwine* (Litwin), *\*Lȳtelmann* (Lilleyman)

*Mǣrgēat* (Merrett), *Mildburh* (f) (Milborrow), *Norðmann* (Norman)

*Ordgār* (Orgar), *Ordrīc* (Orrick), *Ordwīg* (Ordway)

*Ōsgār* (Hosker), *Ōsmǣr* (Osmer), *Ōsweald* (Oswald), *Ōswine* (Oswin)

*\*Pīcstān* (Pickstone), *\*Rǣdwīg* (Redway), *Rǣdwine* (Readwin), *\*Rimhild* (f) (Rimell)

*Sǣbeorht* (Seabert), *\*Sǣbeorn* (Seaborn), *\*Sǣburh* (Seaber), *\*Sǣfaru* (f) (Seavers), *Sǣfrīð* (Saffery), *\*Sǣgār* (Sagar), *\*Sǣgēat* (Sait), *\*Sǣgōd* (Seagood), *\*Sǣlēofu* (f) (Sealeaf), *\*Sǣlida* (Sallitt), *\*Sǣlwīg* (Salway), *Sǣmǣr* (Seamer), *Sǣmann* (Seaman), *\*Sǣrīc* (Search), *Sǣweald* (Sewell), *Sǣweard* (Seward), *Sǣwulf* (Self)

*\*Sidumann* (Seedman)

*Sigeflǣd* (f) (Siffleet), *Sigegār* (Siggers), *\*Sigemǣg* (Simey), *Sigenōð* (Sinnatt), *Sigerǣd* (Sired), *Sigerīc* (Search), *Sigeweald* (Sewell), *Sigeweard* (Seward)

*\*Smēawine* (Smewing), *\*Snelgār* (Snelgar), *Spearheafoc* (Sparrowhawk)

*\*Stānburg* (f) (Stanberry), *Stānheard* (Stannard), *\*Stānhild* (f) (Stanhill), *Stānmǣr* (Stammer)

*\*Stubheard* (Stubbert), *\*Sunnmann* (Sunman)

*\*Swētlufu* (f) (Sweetlove), *Swētmann* (Sweetman), *\*Swētrīc* (Swatridge)

*\*Trumbeald* (Trumble), *\*Tūnheard* (Tunnard), *\*Tūnhild* (f) (Tunnell)

*Þēodbeorht* (Tebrich)

*Unwine* (Unwin), *Uhtrǣd* (Oughtred), \**Uhtrīc* (Outridge)

\**Wǣcerhild* (f) (Wackrill), *Wǣrmund* (Warman), \**Wealdwine* (Walwin)

*Wīgbeorht* (Wyberd), *Wīgbeorn* (Wyborn), *Wīgburh* (f) (Wyber), *Wīgheard* (Wyard), *Wīgmǣr* (Wymer), *Wīgmund* (Wyman)

*Wihtgār* (Widger), *Wihtheard* (Whittard), *Wihtlāc* (Whitelock), \**Wihtmann* (Wightman), *Wihtrīc* (Whitteridge)

*Wilbeorht* (Wilbert), *Wilrīc* (Wildridge)

*Winebeald* (Winbolt), *Winegār* (Wingar), *Winemann* (Winman)

\**Wudufugol* (Woodfull), \**Wuduheard* (Huddart, Woodard), \**Wudulāc* (Woodlake)

*Wulfbeald* (Wolbold), *Wulffrīð* (Woolfrey), *Wulfgār* (Woolgar), *Wulfgēat* (Woolvett), *Wulfgifu* (f) (Wolvey), *Wulfmǣr* (Woolmer), *Wulfnōð* (Woolner), *Wulfrǣd* (Orred), *Wulfrīc* (Woolrich, Hurry), *Wulfsige* (Woolsey), *Wulfstān* (Woolston), *Wulfweard* (Woollard), *Wulfwīg* (Woolway), *Wulfwine* (Woolven)

*Wynrīc* (Windridge)


## Scandinavian Personal Names


The vitality of the Scandinavian name-system in the Danelaw has been discussed and illustrated by Sir Frank Stenton.[27] In addition to Scandinavian names like Thorald, Swain, Haldan, etc., which might appear in southern texts, there are characteristic northern names like Gamel, Gille, Ketel, and others of characteristic rarity, as Ketelbern, Airic, Ailof, etc. Particularly noteworthy are such diminutive forms as *Hasti* or *Asti*, a colloquial diminutive of ON *Ásketell*, surviving in *Hastie*, *Steinki*, a short form of compounds of *Stein*, *Anke*, a diminutive of names in *Arn-*, the source of *Hanks*.

Though less extensive than in Lincolnshire, Scandinavian influence was not negligible in East Anglia. Some 8 per cent of the peasants of the Bury manors c1095 bore Scandinavian names of which *Lute* and *Challi* are not recorded elsewhere in England.[28] The vitality of these names is shown by the formation of such Anglo-Scandinavian compounds as *Lefchetel*, *Ketelbert* and *Þurwif*, recorded in 962 and reappearing in Yorkshire in 1166 (P), and by the survival in *Kilvert*, *Ketteridge* and *Tureff* of the unrecorded hybrids *Cylferð*, *Cytelrīc* and *Þorgifu* (f). The pet-form *Suarche*, from Anglo-Scandinavian *Swartcol*, has its parallel in the otherwise unknown *Samke*, still found in the rare surname *Sank*. Other noteworthy survivals are *Goodhew* from the previously unknown *Guðhugi*, a parallel to the *Illhugi* found at Thorney and the Suffolk *Tovell* from ON *Tófa-Hildr*, a rare type of compound, 'Hildr the daughter of Tofi'.[29]

## Scandinavian Personal Names Surviving in Modern Surnames

*Aggi* (ODa) (Agg), *Aki* (ODa) (Okey), *Álfgeirr* (Alger), *Algot* (ODa) (Allgood), *Álfgrímr* (Allgrim), *Alli* (ODa) (Alley), *Arnkell* (Arkell), *Ásbiǫrn* (Osborn), *Ásgautr* (Osgood), *Ásketill* (Ashkettle), *Áskell* (Askell), *Áslákr* (Haslock), *Ásvaldr* (Oswald), *Auti* (ODa) (Autie)

*Bjǫrn* (Barne), *Bóndi* (Bond), *Bóthildr* (f) (Bottle), *Brandr* (Brand), *Bróðir* (Brothers) *Dólgfinnr* (Dolphin), *Dreng* (Dring), *Drómundr* (Drummond)

*Elaf* (ODa) (Ayloffe), *Eiléfr*, ODa *Elef* (Ayliffe), *Eiríkr* (Herrick)

*Farmann* (Farman), *Farðegn* (Farthing), *Fastúlfr* (Fastolf), *Fathir* (ODa) (Fathers), *Finnr* (Finn), *Fótr* (Foot)

*Gamall* (Gambell), *Gauki* (Gookey), *Geiri* (Garey), *Gilli* (Gill), *Greifi* (Grave, Greavey), *Grímr* (Grime), *Grímhildr* (f) (Grimmet), *Gunnr* (Gunn), *Gunnildr* (f) (Gunnell), *Gunvor* (f) (Gunner), *\*Guð(h)ugi* (Goodhew), *Guðmundr* (Goodman), *Guðrøðr* (Goodread)

*Hafleikr* (Havelock), *Haghni* (ODa) (Hagan), *Haki* (Hake), *Hákun* (Hacon), *Hálfdan* (Haldane), *Hámundr* (Oman), *Haraldr* (Harold), *Hasteinn* (Hasting), *Hávarðr* (Haward), *Hemmingr* (Hemming), *Hrafn* (Raven), *Hrafnhildr* (f) (Ravenhall), *Hrafnkell* (Rankill), *Hróaldr* (Rowat), *Hrólfr* (Rolf)

*Ingialdr* (Ingall), *Ingiríðr*, ODa *Ingrith* (f) (Ingrey, Ingley), *Ingólfr* (Ingell), *Ingvar* (ODa) (Ingar), *Ívarr* (Ivor)

*Karl(i)* (Carl), *Karman* (Carman), *Kaupmaðr* (Copeman), *Kel* (Kell), *Ketilbiǫrn* (Kettleburn), *Ketill* (Kettle), *Knútr* (Knott), *Kolbein* (Colban), *Kolbrandr* (Colbran), *Koli* (Cole), *Kalman* (Coleman), *Kollr* (Coll), *Kollungr* (Colling), *Kouse* (Couse), *Kupsi* (Copsey)

*Lag(h)man* (ODa) (Lawman), *Langabein* (Langbain), *Lax* (Lax)

*Magnus* (Magnus), *Móðir* (Mothers)

*Oddr* (Odd), *Óleifr* (Olliff), *Ormr* (Orme), *Ottár* (Otter)

*Rannulfr* (Randolph)

*\*Samke* (Sank), *\*Sandi* (Sandey), *Segrim* (ODa) (Seagrim), *Sigarr* (Siggers), *Sigga* (f) (Siggs), *Sighvatr* (Suett), *Sigmundr* (Simmonds), *Sigríðr* (f) (Sired), *Snari* (Snarey), *Sǫrli* (Sarl), *Steinn* (Stein), *Stigandr* (Stigand, Styan), *Stóri* (Storey), *Sumarliðr* (Summerlad), *Svanhildr* (f) (Swannell), *Sveinn* (Swain)

*Þóraldr* (Thorold), *Þorbiǫrn* (Thurban), *Þorfinnr* (Turpin), *\*Þorfrøðr* (Tollfree), *Þorgautr* (Thurgood), *Þorgeirr* (Thurgar), *Þorgils* (Sturge), *Þórhildr* (f) (Turrill), *Þórir* (Thory), *Þorkell* (Thurkell), *Þormundr* (Thurman), *Þorsteinn* (Thurston)

*\*Tófa-Hildr* (f) (Tovell), *Tófi* (Tovee), *Tóki* (Took, Tookey), *Tóli* (Tooley), *Topi* (Toop), *Tubbi* (Tubb), *Tunni* (ODa) (Tunney), *\*Turk* (Turk)

*Úlfr* (Ulph), *Úlfketel*, *Úlfkell* (Uncle)

*Vestmaðr* (Westman), *Vígarr* (Wigger), *Vigot* (ODa) (Wiggett), *Víkingr* (Wicking), *Víðarr* (Wither)

## Anglo-Scandinavian Survivals

*Cytelferð (Kilvert), *Cytelrīc (Ketteridge), Healfdene (Alden), *Sæfugol (Saffell), *Spracaling (Sprackling), *Þorbert, *Þurbert (Turbard), *Þorgifu (f) (Turreff), Þurcytel (Thurkettle), *Toll (Toll), *Tukka (Tuck), Walþēof (Waddilove, Wallett, Walthew)

## Norman Names

Scandinavian names were used by Normans in France where ON Ás- occurs as An- which survives in Anketel, Ankin, Antin, Angood, Angold. Norman diminutives are found in Asketin and Turkentine. Norman Turstin (for Thorstein) survives as Tustin, Tutin, Dusting. Initial T for Th may also represent a Norman pronunciation in England, especially of names not found in Normandy, e.g. Tory for Thory.

## Personal Names in Medieval London

Ekwall's discussion of early London personal names (ELPN) is an outstanding example of what can be achieved by a detailed study of the names of a particular locality, and a perpetual challenge to others to do the same for other areas. In the early twelfth century, Old English personal names were still in living use in London, but gradually grow rarer and after 1200 are found only occasionally, apart from a few names which lived on and are still in use. Particularly common were such names as Ailward and Ailwin, Brichtmar, Godric and Godwin, Leofric and Leofwin, Wulfric, Wulfweard and Wulfwine. The only Old English woman's name at all common was Edith. Rare in Old English were Eadwacer and Smeawine, and the feminine Eastorhild, whilst a few such as Godleofu and Wacerhild, both feminine, are unrecorded in Old English. We find a number of compounds in -ing: Bruning, Hearding, Sweting, and the unrecorded Funding and Sperling. Short forms were rare but we have Golde (f), Milde (f), Hunna, But and Werth.

Some Scandinavian names from late Old English times must have been current in twelfth-century London. Some may have been introduced direct from Normandy. Such names as Turgis are Norman in form. It is noteworthy that few of the Scandinavian names recorded in Domesday Book are found in London sources, but some 30 personal names (e.g. Askill, Esger, Ketel) are probably Scandinavian rather than Norman in origin. Names like *Thurstan*, *Thorold*, when spelled *Tursten*, *Torold*, may be Norman in origin.

There is reason to believe that Old English names survived longer in the provinces than in the capital, where the fashion set by Normans would be followed more quickly. Old English names in London were often those of immigrants from the provinces. The old names were superseded by names introduced by the Normans and many of those with French names in the first two or three decades of the twelfth century must have been Normans by birth. Those with English names at the same period were as a rule of English descent, as, probably, were those with English names later in the century. But it does not follow that a French name necessarily denotes French descent. As early as c1100 it was quite common for English people to give French names to their children whilst there are only a few examples of sons or daughters of parents with French names being given English names. The earliest instances are found among the upper classes, both the clergy and patrician families. Some Englishmen with French names must have been born c1090 or earlier. After 1100 it became a fashion for English families to give French names to their children. Some families were more conservative than others and continued to use the old names. Some gave French names to one or more of their children and English names to another or others. Thus, in a very few generations the Old English christian names were altogether disused in London, apart from a few special names, Alfred, Edmund, Edward and Godwin. Edmund is frequent in London between 1250 and 1350 but Edward occurs only occasionally. Edward I does not seem to have been popular in London and the few Edwards were probably named after the saint, Edward the Confessor. It is unlikely, therefore, that the popularity of Edmund was due mainly to Edmund, son of Henry III. Some, at least, of the London Edmunds came from East Anglia: Edmund de Suffolk 1309, Edmund de Bery 1346 (Bury St Edmunds), and others from places in Norfolk. These Edmunds were, no doubt, named after St Edmund, the martyr-king of East Anglia and founder of the monastery of St Edmundsbury, to whom a London church was dedicated.

The Norman-French names given by apparently English people to their children were generally the names most commonly used by the Normans and the names still most frequent in England: Geoffrey, Gilbert, Henry, Robert, Peter, John, etc., and the women's names Agnes, Alice and Maud. The personal nomenclature of twelfth-century London was well on its way to the modern stage which was, in the main, reached in the thirteenth century.

## The Breton Element

The large Breton contingent which fought at Hastings was rewarded with lands in England. At their head was Earl Alan of Richmond, a cadet of the ducal house, with a fee of the first importance in Lincolnshire, East Anglia and neighbouring counties. In the south-west, Judhael of Totnes had a fief which in the twelfth century owed service of 70 knights. In thirteenth-century Suffolk was a 'Breton soke'. 'There is, in fact, hardly a county in which this Breton element is not found, and in some counties its influence was deep and permanent . . . the Breton colony founded by Earl Alan of Richmond can still be traced, late in the twelfth century, by the personal names which give a highly individual character to records relating to the country round Boston, itself a town of Breton creation, and Louth. In these districts, as also in the North Riding of York, the Breton settlers of the eleventh and twelfth centuries preserved their ancient personal nomenclature with a conservatism resembling that of the Anglo-Scandinavian peasants among whom they lived . . . it was something more than the establishment of a few score knights and sergeants in military tenancies. It must have had the character of a genuine migration, though a migration upon a small scale.'[30]

In twelfth-century Lincolnshire Alan was as common a name as Simon and more popular than Henry and Adam. Other common Breton names were Brian, Conan, Constantine (with its short form Coste), Jarnegon, Justin (with its pet-form Just), Mengi, Samson, and Tengi, all surviving as modern surnames. The christian name of Judhael de Totnes is still found as a surname in Devonshire as Jewell, and elsewhere as Jekyll and Joel. In Essex, Helion Bumpstead, and in Devon, Upton Helions, owe their attributes to Tihel de Herion, their Domesday lord who came from Helléan in Morbihan. His christian name survived in Essex until the thirteenth century and is found as a surname at Barking in 1206 (Roger *Tihell*), whilst his surname, though rare, still lives on in Essex and Suffolk, in Devon and Somerset (*v.* Elion). Wiggens Green in Helion Bumpstead owes its name to the family of John *Wygayn* whose eponymous ancestor may well have been an actual follower of Tihel the Breton. Bretts in Aveley owes its name to John le Bret 'the Breton'. In Aveley is a field, Bumpstead Mead, the last relic of a Bumpsted Hall named from Gilbert de Bumsted ad Turrim who seems to have been accompanied to Aveley by Bumpstead men of Breton descent whose names are found in the district in the thirteenth century (*Wygan, Hervey, Alan, Bryce*).[31] In Essex, too, we find a twelfth-century, *Mingghi*. This Breton influence has left no small mark on our modern surnames.

## The Celtic Element

Although Welsh surnames, as distinct from characteristic Welsh patronymic descriptions, were very late formations, the not inconsiderable number of immigrants from Wales into the border counties found their personal names treated exactly like English names in the formation of surnames. Thus surnames were formed from Welsh personal names and became hereditary in England long before hereditary surnames were known in Wales. e.g. Kemble (1185), Meredith (1191), Morgan (1221), Owen (1221), Cadogan (1273), Maddock (1274), etc.

About 890–3 a body of Norwegians from Ireland entered Yorkshire and were followed by a greater number, probably between 919 and 952. These Norwegians had been settled in Ireland sufficiently long to become partly Celticized and they have left their mark on the modern map of Cumberland and North Yorkshire in a series of place-names containing Irish loan-words and in inversion compounds in which the defining element comes last: Aspatria, Kirkoswald, Kirkbride. They had also adopted Goidelic personal names some of which survive both in place-names and modern surnames. e.g. Coleman, Duncan, Gill, Murdoch, Neal, Patrick, Troyte.[32] Some of these surnames are more common in Scotland where they originated independently.

### The Final -s in Jones, Parsons, Stocks, etc.

Weekley has remarked[33] that 'the majority of monosyllabic, and many dissyllabic, local names are commonly found with -s, originally due to analogy with *Wills, Jones*, etc., where -s is the sign of the genitive. It will be found that this addition of -s in local names generally takes place whenever it does not involve an extra syllable or any exertion in pronunciation, e.g. *Birks* but *Birch*, *Noakes* but *Nash*, *Marks* but *March*, *Meadows* but *Field*, *Sykes* but *Sich*. The only important exception to this phonetic rule is *Bridges*, which is usually derived, not from *bridge*, but from Bruges, once commonly called Bridges in English. This -s is also added to specific place-names, e.g. *Cheales* from Cheal (Linc.),[34] *Tarbox* from Tarbock (Lanc.), *Burls* from some spot in Essex formerly called Berle,[35] *Rhymes* from Ryme (Dors.), etc.' Elsewhere he asks, 'but why always *Summers* or *Somers* with s and *Winter* without?'[36]

Generalizations on surnames are always dangerous. Both *Summer* and *Winters* survive, as does *Fields*. The final -s was formerly found in such names as (Ralph) *Saches* Hy 2 DC, (Richard) *Ryches* 1296 SRSx, (Alice) *la Gegges* 1310 ColchCt, and survives in *Hedges* and *Latches*.[37] In a number of local surnames, plurals are found quite early: *Hales* (1180), *Coates* (1190), *Howes* (1212), *Holmes* (1219). The final -s of surnames from French place-names is retained or dropped quite arbitrarily, the variation, perhaps, being due to the difference between the English and French pronunciation: *Caliss* (Calais), *Gamage* (Gamaches), *Danvers* (Anvers), *Amyas* (Amiens), *Challen* (Chalons), *Sessions* (Soissons).

The final -s in surnames like *Williams, Parsons, Carters*, is a different problem. It cannot be a sign of the plural. For *Parsons, Vicars*, etc., there are two origins:

(i) Alicia *le Parsones* 1327 SRWo, Margery *le Vikers* 1332 SRWa, Ralph *le Prestes* 1327 SRWo, where we have an elliptic genitive, 'the parson's (servant)', etc. cf. Henricus *homo Vicarii* 1297 SRY. Malyna *la Roperes* (1311 ColchCt), described as a servant, was either the servant of the roper or of a man named Roper. Surnames like John *Alysaundresman* 1297 Coram (Bk), Robert *Nicholesman* 1309 AssSt, with others in *-knave*, *-sergeant*, etc., are not uncommon, so that, whilst Gilbert *le Potteres*, Richard *le Cokes* (1327 SRWo) may mean 'son of the potter or of the cook', they might also denote his servant. But Philip *le Redes* (ib.) must be 'servant of a man named *Rede*'. Thus, too, John *Pastons* (1327 SRWo), John *Byltons* (1327 SRC), where the surname is local, 'servant of Paston or of Bylton'.

(ii) William *atte Personnes* 1327 SRSf, again elliptic, '(servant) at the parson's (house)', etc. Similarly Beadles, Stevens, etc. (Margaret *ate Budeles*, Sibilla *ate Stevenes* 1332 SRSo) may also mean 'servant at the beadle's (house)' and 'servant at Steven's'.

No satisfactory explanation has been given of this final *-s* in surnames formed from personal-names. Fransson's examples are late (1310). He regards them as elliptic genitives. As all his examples but one (Roger *le Persones*) are women, he must take all four to mean 'servant of Robert, the parson, etc.'.[38]

Ewen's account is confused.[39] He cites Willelmus *Johannis* (1159–60) and three similar forms of 1229–35 as examples of 'inflected genitives' due to '*filius* and *filia* having fallen into disuse'. But surnames of the type Willelmus *filius* Johannis are common long after 1235. He notes also two undated metronymics, Johanna *Mariote* and Willelmus *Margarete*, adding 'but the English nominative form, as Henry Maynard or John Rogers was also used, and the genitive ending (*es*, *is*, *ys*, or *s*) also begins to be noticeable, and at first most frequently in the names of women, thus Robertus filius Radulfi became Robertus Rolle (Raoul), but Matilda filia Radulfi was written Matilda Rolles . . . There was no precise rule, many surnames of women are without the final sibilant, which is occasionally found added to the second names of men.'[40] He does not explain why *filius Radulfi* becomes *Rolle* when the name is a man's, but *Rolles* when it is that of a woman. He repeats the argument later: 'John, Robin's son, would be called John Robin, but Margaret, Robin's (daughter), would be known as Margaret Robines.'[41] That in the early instances (his earliest example is 1230) the terminal *s* was due to the influence of the French nominative singular ending, seen in names like Jacques, Gilles, Jules, etc.,[42] is most unlikely. Such forms are rare in English and would not be employed for surnames which his own Latin examples prove were in the genitive. Nor can it be accepted that *Driveres* and *Smithes* 'may possibly exhibit the feminine agential suffix'.[43]

Tengvik cites four much earlier examples of surnames like Ulmer *Æltredes*, where an Old English (or Scandinavian) personal name is in the genitive, and two in Latin, Ælfuine *Goduini*, all from Bury (c1095), and regards them as due to the omission of OE *suna* and Latin *filius* respectively.[44] Twenty other examples of this type in *-es* have been noted between 1100 and 1230, all names of men, formed usually from Old English or Scandinavian masculine names: Walter *Ricaldes* c1100 MedEA (Nf), Ranulph *Godes* 1186–1210 Holme, Edricus *Keteles* 1188 BuryS (Sf). Two are Old French: Eudo *Luueles* 12th DC (L), Stephen *Paynes* 1230 Pat, one Welsh, Robert *Howeles* 1210 Cur (Nth), and three from women's names, two English, Æilric *Osuuennes* c1095 Bury (Sf),

Segarus *Aileves* 1188 BuryS (Sf), and one French, Walter *Auices* 1186–1210 Holme (Nf).

The Latin type is more common than Tengvik and Ewen would lead us to think. Over 50 examples have been noted between 1130 and 1240, all except one (Emma *Philippi* 1240 Rams, Nf), names of men, usually from French personal names: Hugo *Oillardi* 1130 P (Sr), Willelmus *Walkelini* 12th DC (Lei), Willelmus *Luce* 1185 Templars (K), Johannes *Jeremie* 1196 P (Y); occasionally from Old English or Scandinavian names: Willelmus *Ailrici* 12th DC (L), Robertus *Edwini* 1229 Pat (So), Alanus *Torberti* 1212 Cur (Ha). Three are formed from names of women: Arnaldus *Mabilie* 1185 Templars (Ess), Robertus *Margerie* 1195 P (Gl). These names can only mean 'John, son of Jeremiah', 'Arnold, son of Mabel', etc., literal translations of the vernacular, just as the clerk translated 'Edward of Salisbury' by Edwardus *Saresberiae* 1100–35 Rams (Hu).

The English forms are early examples of the elliptic genitive, Edricus *Keteles* 'Edric Ketel's (son)', parallel to *Personnes, Prestes* above. That this interpretation is correct is proved by the following. In 1281 we have mention of Robert de Rokesle junior who is twice called Robert Dobes in 1305. His father was Robert de Rokesle senior who must often have been called by his pet-name *Dob*. Hence his son's surname *Dobes* which must mean 'son of Dob', i.e. of Robert.[45]

Toward the end of the thirteenth century, this type of name becomes more common and steadily increases in the fourteenth, but there is a marked difference in its frequency in different counties. In the Worcestershire Subsidy Roll for 1275 there are only 7 examples, 5 being names of women; in that for 1327 we find 138 men and 30 women so named. In Somerset in 1327: 128 men, 73 women; in Warwickshire (1332), 161 men, 20 women; in Suffolk (1327), 27 men, 1 woman; in Surrey (1332), 5 men, 11 women. In other Subsidy Rolls the number is negligible: Sussex (1296), 6; (1327) 2; (1332) 3; Cambridgeshire 4; Lancashire 2; Cumberland 2; Yorkshire (1297) 6; (1301) 3; (1327) 3, all men. In Essex there are 5 men in 1295 ParlR, a number in the Colchester Court Rolls (1311–45) and about 12 (all women) in the 1349 Fingrith Hall Court Rolls. The surnames are usually the common christian names in use, often pet-forms, rarely Old English and almost invariably masculine. Some few are nicknames or occupational names (these sometimes preceded by *le*, occasionally *la*): Ysabella *Barones* 1275 SRWo, Hugh *Rabuckes* 1301 SRY, Claricia *le Parkeres* 1327 SRSo, Juliana *la Kinges* 1285 *Ass* (Ess), Amiscia *la Wrihtes* 1333 ColchCt. Occasionally we have a place-name: John *Dounes* 1327 SRWo.

The interpretation of these surnames is more difficult than one would expect. It is clear that in the twelfth century Segarus *Aileves* meant 'son of Aileve' and that is probably the meaning in the fourteenth century in the names of men. But John *le Cokes* (1327 SRWo) may well have been the cook's servant or assistant and names like William *Hogges* may have lost the article and have a similar meaning. So with women's names. Claricia *le Parkeres* may have been the servant of the parker and Isabella *la Chancelers* 'the servant of (a man named) Chanceler'. But such an interpretation is unsatisfactory for Avice *la Schepherdes* (1311 ColchCt) and Juliana *le Smithes* (1279 RH), for shepherds and blacksmiths were unlikely to have servants. Where the surname is a place-name, 'servant of a man named Bylton', etc., is probable.

In the Colchester Court Rolls (1311 ff.), large numbers of women were regularly fined at court after court for selling ale at too high a price. They were usually described as 'the wife of John Carpenter', etc., but a certain number are mentioned by name, which almost invariably ends in *-es* (Joan *la Warneres*, Alice *Sayheres*). It is a reasonable presumption that they were widows and that this type of women's surname denoted either a widow or a married woman. Matilda *Candeles* (1327 SRSx) was probably the wife of Ralph *Candel*, for in 1332 she is described as 'Matilda relicta Candel'. Margery *la Mazones* was the wife of Walter *le Mazoun* (1311 ColchCt). Agnes *Rickemannes* (1329 Husting) was probably the widow of *Rickeman* le Chaumberleng (1292 SRLo) and if so, we have to reckon with the fact that some of these women's names denote the christian name and not the surname of their husbands. It is not uncommon in these documents to find pairs of names like Nicholas *le Knyt* and Cecilia *Knyctes* (1297 MinAcctCo), who, we may fairly assume, were husband and wife. Amisia *Hugines* (1327 SRWo) was probably the wife of William *Hugyns* and here the surname means 'son of Hugyn'. Thus, a surname like *Stevens* may mean 'son of Stephen', 'servant of Stephen', or 'servant at Stephen's house', or it may be a metronymic derived from a form *Stevenes* 'Stephen's wife'. The only certainty is that *atte Stevenes* means 'servant at Stephen's house'. The *-s* of local surnames may be a plural inflexion (or a sign of French origin), but more often falls into one or other of the above classes. Sometimes, in late additions, it may be a dialectal pronunciation, with excrescent *s*.[46]

## Pet-names

Already in Old English we find pet-names in use: *Tuma* for *Trumwine* in the seventh century and *Ælle* for *Ælfwine* in the tenth, and such forms as *Wine* and *Wulfa* for *Winefrīð* and *Wulfwine*.[47] Names of this type continued to be formed and a number still survive in surnames, some otherwise unrecorded. But most of the pet-names in modern surnames are of post-Conquest formation and some are difficult to identify. Examples are found in the twelfth century; they become more numerous in the middle of the thirteenth and in the fourteenth are common. They are found among all classes and are derived from Old English, Scandinavian and French personal names alike. *Cudd* (1358) and *Cutt* (1279) are from OE *Cūðbeorht*, *Ugga* (1212) from *Uhtrǣd*; *Asti* (1203) is a pet-form of ON *Ásketill*; *Lamb* (1161) is for *Lambert*, *Gibbe* (1179) for *Gilbert*, *Lina* (1181) for *Adelina* or *Emelina*, whilst the Breton *Sanson*, *Samson* has given *Sanne* (1260) and *Samme* (1275). Not all pet-names are so easy to identify. *Hudd* (1177) and

*Hulle* (1227) are undoubtedly for *Hugh*, but *Hudd* is also used for *Richard*. *Pelle* (1274) is a pet-name for *Peter*. *Hann* is undoubtedly for *John* (from *Jehan*), but is just as certainly for *Hanry* (Henry) and is said to have been used also for *Randolph*.

Some of these forms follow normal phonetic laws of assimilation: *Judd* from *Jurd* (Jordan); *Fippe* from *Philip*, *Bette* from *Bertin* and *Bertelmeu* (Bartholomew), *Penne*, a shortening of *Pennel*, from *Pernel*, *Cuss* from *Cust* (Custance, Constance), *Ibb* for *Isabel*. In others, the name begins by anticipating the following medial consonant: *Dande* (1246) for *Andrew*, *Biby* (1240) for *Isabel*. Some pet-names are formed from the second syllable of the full name: *Pot* (1115) from *Philipot*, itself a diminutive of *Philip*; *Coll* (1247) from *Nichol*, *Belle* (1279) from *Isabel*; *Sander* (1248) from *Alexander*. Voiced and voiceless consonants were used indiscriminately: *Dicke*, *Digge*; *Hikke*, *Higge*; *Gepp*, *Gebbe*; *Judde*, *Jutte*. Vowels were unrounded: *Rob*, *Rab*; *Dobb*, *Dabb*; or rounded: *Malle*, *Molle* (Mary); *Magge*, *Mogge* (Margaret), whilst the changes were rung on the consonants: *Robb* (1196), *Hobbe* (1176), *Dobbe* (1202), *Nabbe* (1298), all for *Robert*. In some names we find a combination of more than one of these features: *Libbe* (Elizabeth), *Pogge* (Margaret).

The clue to the explanation of these pet-names is given by Napier and Stevenson when they suggest that OE names such as *Lilla*, *Bubba* and *Nunna* are due to 'regressive assimilation' and have their origin in the speech of children.[48] Scandinavian scholars call them 'Lall-names'. According to this theory, *Lilla* is a short form of some compound of the stem *Bil-*, such names as *Bilheard* or *Bilnoth*. Stenton is disinclined to accept this on the ground that it implies the contemporaneous existence of two sharply contrasted conceptions of nomenclature. 'The state of mind which produced the compound names with their far-fetched significance is hardly compatible with one which allowed infantile attempts at expressing a name to pass into permanent use.' He admits, however, that this theory has 'the great merit of proposing an intelligble connection between these meaningless names and compounds of the normal Germanic type. Its chief weakness is the remoteness of the sound-association between the original compound name and the suggested simple derivative'.[49]

Children are children and parents are parents, whether we are concerned with the eighth or the twentieth century. The process of learning to speak is the same—trial and error by imitation of sounds heard and there are innumerable examples of common words which have been corrupted in form through misdivision, mispronunciation and misunderstanding. When an Anglo-Saxon named *Æðelstan* and his wife *Wulf*gifu deliberately named their son, the future Bishop of Worcester, *Wulfstan*, a combination of the first theme of the mother's name and the second of the father's, they were not concerned with the meaning of the compound—'wolf-stone', any more than were those who named their children *Friþuwulf* 'peace-wolf' or *Wīgfrið* 'war-peace'. Names had become names and their meaning was a matter of no concern. Association—here, a perpetuation of themes common to the two families—is more important than meaning.

Detailed studies of the early speech-habits of children would throw much light on the eccentric forms of many pet-names. A wreath sent recently by the Queen was from

*Lilibet*, the name by which Her Majesty is known in the family circle, a deliberate perpetuation of her early attempts to pronounce her own name. My own daughter still answers to the name of *Titt*, a shortening of *Titter*, her first attempts at *sister*. A newly-wedded wife of my acquaintance regularly addressed her young husband as *sweetheart*, which gradually became *weetheart*, *sweetie*, *weetie* and finally *weet*, a pet-name used for many years. There must be many similar pet-names confined to a particular family and never seen in print. With such developments, it is not difficult to realize that a pet-name may have more than one origin, and that a single name may give rise to a variety of pet-names.

## Diminutives

A few diminutives in *-uc* of OE origin survive (*Haddock*, *Whittock*, *Willock*), but most are derivatives of French names. The most common suffixes are *-ot*, *-et*, *-un*, *-in*, *-el*: Philpot, Ibbott; Hewett, Jowett; Paton, Dickens; Rankin, Higgins; Pannel, Pottell.

Double diminutives are formed from these suffixes:

*-el-in:* Hamlin, Hewlins, Jacklin
*-el-ot:* Giblett, Roblett
*-in-ot:* Adnett, Rabnott
*-et-in* (rare): Turkentine

The variety of surnames resulting from these different forms of pet-names may be seen from the following (varieties of spelling ignored):

*Richard* (pronounced *Rich-ard* and *Rick-ard*):
Rich, Richings, Ritchie; Hitch, Hitchcock, Hitchen, Hitchman, Hitchmough
Ricard, Rick, Ricky; Hick, Hicken, Hicklin, Hickman, Hickmott, Hickox; Higgett, Higgins, Higgs; Dick, Dickels, Dicken, Dickin, Dicketts, Dickie; Digg, Diggen
*Robert:* Rabb, Rabbets, Rabjohn, Rablen, Rabnott; Robb, Robbie, Ropkins, Robins, Robjant, Roblett, Roblin; Dabbs, Dabin, Dabinett; Dobb, Dobbie, Dobbin; Hob, Hobbins, Hobday, Hobgen, Hoblin, Hopkin; Nabb, Nap, Nobbs, Nopp
*Hugh:* Hugo, Hue, Hew, How; Hewell, Hewett, Hewlett, Hewlins; Houchen, Howett, Howlett, Howlin(g); Hudd, Hudden, Huddle, Hudman, Hudsmith; Huelin, Huett, Huot; Huggett, Huggin, Huggon, Huglin; Hukin, Hewkin, Howkins; Hull, Hullett, Hullot; Hutchin
*William:* Will, Wilkin, Wilcock, Willet, Willott; Willmott, Wellemin, Wellerman, Willmin, Willament

Gill(ham), Gilliam, Gillet, Gillman, Guillerman, Gelman
*v.* also Henry, Jack, John, Maud, Paul, Philip.

## *The Suffixes* -cock *and* -kin

These two suffixes are used to form diminutives of the more common names and are very frequently used as personal names, sometimes to distinguish son from father, sometimes as pet-names. John and Jankin, William and Wilkin, are both used as names of the same man. Compounds of *-cock* are less common and later than those of *-kin*, which are found already in the twelfth century: *Potechin* 1166, *Hardekin* 1175, *Lambekyn* 1178, *Wilekin* 1180, *Adekin* 1191; *Hellecoc* 1202, *Alecoc* 1204, *Adecok*, *Wilcok* 1246. Occasionally they are compounded with women's names: *Edekin* 1279, *Malkyn* 1297, *Marekyn* 1390; *Becok*, *Geuecok* 1332.

Such names become more common from the middle of the thirteenth century and are very frequent in the fourteenth, particularly among the lower classes. The earliest examples of *-kin* are names of Flemings: *Derechin* (1158, Essex). *Wilechin* (1166, Newcastle) was the son of a moneyer who may have been a foreigner. This supports the common view that the suffix was brought from the Netherlands but there seems to be no concentration in the east, whilst *-kin* names were common in Cheshire at the end of the thirteenth century.

## *Classical Names*

In addition to the usual sources, Old English, Old French, Old German and Old Norse, of the personal names, a few classical names appear: *Eneas* de Baddeby 1383–4 FFWa; *Aristotile* 1196 P (Hu); *Ciprianus* 1182–1211 BuryS; *Eusebius* Ailbrit 1279 RH (Hu); *Hercules* Loveden 1592 AD v (Berks); *Oratius* presbiter 1193 P (Ess); *Ignatius* filius Athelwaldi 1207 Cur (Nf); *Juvenalis* 1208 Cur; *Lucianus* de Scille 1212 Cur (Db); *Menelaus* 1202 AssNth; *Omerus* 1196 P (Ha); *Uirgilius* 1177–93 CartNat. Many of the saints' names were also of Greek or Latin origin, and probably owe their use in medieval times to this fact. In addition, the popularity of *Alexander* is probably due to the medieval romances dealing with the hero, and the appearance of *Achilles* de la Bech' 1221 AssSa, and *Hector* de Hilleg' 1222 Cur (Sf) to the romances on the Troy legend. Other names which probably owe their use to medieval romance include *Charlemayn* 1230 P (Wo); Rauf *Lancelot* 1506 TestEbor; *Eglamore* Muston 1476

IpmNt, and perhaps *Diggory* Watur 1461 SaAS 2/xi; *Digorie* Maker 1600 AD v (Co, D) to *Sir Degarre*.

Three names are of particular interest. In Old English the name *Beowulf* is known only from the Old English epic of which he is the hero. Since there are no other medieval references to the poem, it is impossible to know whether it or the name of its hero were at all widely known during the Old English period. But the name of *Beowulf* certainly survived until at least the end of the thirteenth century: *Bowulf* 1195 PN D 604; *Bowulf* de Rugeberge 1196 P (D); William *Bewlf* 1264–5 FFSx; William *Bewolf* 1296 SRSx; William *Beowoulf* 1297 MinAcctCo. This would suggest either that a knowledge of the poem and of its hero long survived the Conquest, or that *Beowulf* was a normal Old English name, and not simply an invention by the author of the poem. In the romance of *Havelok the Dane*, written towards the end of the thirteenth century, one of the minor characters is a certain King Birkebayn. The name is usually taken to be derived from ON Birkibeinar, the name given to the followers of King Sverrir who fought his way to the throne of Norway in 1184. But it is found as a surname in England as early as the end of the twelfth century: William *Birkebein* 1199 Pleas (Nf); Rener *Birkebayn* 1232 Pat (L); Isabella *Birkebayn* 1297 SRY; John *Birkebayn* 1379 PTY. These would seem to indicate that it was probably an Old Norse nickname of a not uncommon type with *-beinn* as a second element, and not necessarily connected in any way with the *Birkibeinar* of *Sverrissaga*. The *Geste of Robin Hood* is usually thought to have originated in the North or Midlands, and to be especially connected with Sherwood Forest. Yet the only examples of the use of the name as a surname come from the south: Gilbert *Robynhod* 1296 SRSx; Katherine *Robynhod* 1325 CorLo; Robert *Robynhoud* 1332 SRSx. It would seem probable that these surnames must be connected with the famous outlaw, but no explanation for their presence in the south at this date can be offered.

On the whole medieval feminine personal names were rather more varied than the masculine ones. Most of the latter had feminine equivalents, and whilst today a distinction is usually made between the two, e.g. Denis but Denise, Nicholas but Nichola, this was not the case in medieval England. In the records women's names are normally given a final *-a*, but in the vernacular the pronunciation of the names was usually much the same. Hence such names as Paulina, Eustacia, Andrea, Jurdana, Dionisia, were indistinguishable from the masculine forms, and have probably contributed to the resulting surnames.

Some classical feminine names were in use, though they have rarely given rise to surnames: *Camilla* 1208 Cur (Ess); *Caesaria* 12th Rams (Hu); *Cassandra* de Bosco 1283 SRSf; *Diana* 1256 AssNb; *Felicia* de Winterburn' 1208 P (W); *Olimpias* 1207 Cur (Gl); *Philomena* 1202 FFY; *Prudencia* de Pavely 1210 Cur (Nf). In addition, some classical names were also the names of saints, and probably owe their use in medieval times to this fact: Agatha, Anastasia, Helen, Juliana, Katherine, Margaret, Euphemia, etc.

In the Middle Ages there was a fashion for fanciful feminine names, few of which have survived, or given rise to surnames: *Admiranda* 1231–2 FFK; *Amicabilis* 1232–3

FFWa; *Argentina* 1204 FFO; *Bonajoia* 1319 LLB E; *Clariandra* 1248 AssBerks; *Damisona* a1290 CartNat; *Desiderata* 1385 AD iv; *Diamanda* 1221 Cur (Mx); *Eglentina* 1213 Cur (Sx); *Epicelena* 1208 Cur; *Estrangia* 1202–3 FFK; *Finepopla* 1203 Cur (Sf); *Fousafia* 1218 AssL; *Imagantia* 1219 Cur (Sf); *Ynstauncia* Lyoun 1327 SRY; *Joya* 1195 FFEss; *Jolicia* 1219 Cur (K); *Melodia* 1212 Cur (Sf); *Modesty* 1269 FFY; *Orabilia* 1221 Cur (K); *Plesantia* West 1274 RH (Nf); *Popelina* 1212 Cur (L); *Preciosa* 1203 Cur (Herts); *Primaveira* 1226 FFWa; *Splendora* 1213 Cur (D); *Topacia* 1243 Glast (So).

# SURNAMES OF OCCUPATION AND OFFICE

In early post-Conquest documents, the innumerable surnames of this type—almost invariably in Latin—refer to actual holders of the office, whether of church or of state: Abbot, Prior, Chancellor, Chamberlain, Steward (dapifer), or to ecclesiastical or manorial status: Monk, Dean, Reeve, Sergeant. Among the Normans some offices of state such as steward, constable, marshal, etc., became hereditary and gave rise to hereditary surnames, but the terms were also commonly used of lesser offices, whilst *marshal* was a common term for a farrier and such names frequently denoted the actual occupation. Abbots, priors, monks and nuns were bound by vows of celibacy and thus could not found families. As medieval surnames, these must be nicknames, 'lordly as an abbot', 'meek as a nun', often, too, bestowed on one of most unpriestly habits. Only occasionally do we find in the sources some indication that this is the case, e.g. Geoffrey *le Moyne* was constable of Newcastle in 1219 AssY, and so is unlikely to have been a monk. Similarly, cf. John *le prest* le chaucer c1250 Clerkenwell; William *Priour*, cossun 1283 LLB B; William called *le Clerk*, butcher 1336 Husting; Richard *Priur* lindraper 1300, Roger *le Mounk*, baker 1318 NorwDeeds.

Occupational surnames originally denoted the actual occupation followed by the individual. At what period they became hereditary is a difficult problem. In addition, such names as Pope, Cardinal, Legate, can never have been surnames of office in England, and must have been originally pageant-names. It has often been held that the absence of the article points to a hereditary surname, a supposition which cannot be upheld for early in the twelfth century the article is frequently omitted and the same man is called both Richard *turnur* and *le turnur* (12th DC). It is unlikely that, as Fransson suggests, trade-names were used as nicknames and that a man might be called 'the shoemaker' because he mended his own boots. But it is difficult to account satisfactorily for names like Mower, Ripper (reaper), Sawer (sower), which must have been only seasonal occupations.

A marked feature is the surprising variety and specialized nature of medieval occupations, particularly in the cloth industry where Fransson (p. 30) has noted 165 different surnames, whilst the metal trades provide 108, and provision dealers 107 different names. Many of these were clumsy and have disappeared but other surnames still recall occupations or occupational terms long decayed: Arkwright, Ashburner, Barker (tanner), Billeter (bell-founder), Chaucer (shoe-maker), Cheesewright (cheese-maker), Deathridge (tinder-maker), Harbisher (maker of hauberks), Lister (dyer), Slaymaker (shuttle-maker), Thrower (silk-winder), Whittier (white leather-dresser).

Many of the occupation names were descriptive and could be varied. A worker in metal could be called both *Seintier* or *Bellyeter* from the bells he cast, or simply *Sporoner* from the spurs he made, or 'moneyer' if he made coins. William *le Pinour* 'maker of combs' was also called *le Horner* from the horn he used. Adam *le Marbrer* who paved part of St Paul's and Peter the *Pavier* who paved St Stephen's Chapel, Westminster, both followed the same occupation. William *Founder* cast both bells and cannon.

Surnames of occupation are more common than the modern forms suggest. Many surnames, previously regarded as nicknames difficult to explain, are really occupational. Apart from mere shortening by which *Cofferer* and *Coverer* became *Coffer* and *Cover*, the name of the article made or the commodity dealt in was used by metonymy for the maker or dealer. Modern *Garlick* represents not only medieval *Garlek* but also *Garleker* and *Garlekmonger*. *Cheese* is found as a surname in the twelfth century but, whilst *Cheser* has disappeared, both *Cheeseman* and *Cheesewright* survive. Of *Cheverell*, *Chevereller* and *Cheverelmonger*, only the metonymic *Cheverell* still exists. This frequent use of metonymy gives a satisfactory explanation of such names as *Death*, *Meal*, *Pouch*, etc. So, too, the man in charge of the colts or the palfreys was called not only *Colter* or *Coltman*, *Palfreyer* or *Palfreyman*, but also *Colt* or *Palfrey*. Thus, *Bull* and *Lamb* (sometimes from a personal name) are not always nicknames. They may be metonymic for *bull-herd* and *lamb-herd*.

*Brooker* and *Brook* (*atte Broke*) are undoubtedly local surnames, 'dweller by the brook'. *Bridge*, *Bridger* and *Bridgeman* may similarly be local, but as the keeper of the bridge, especially where tolls had to be collected, also lived near the bridge, the surnames may be occupational also. But names like *Kitchen* (*atte Kechene*), *Kitchener*, *Pantry*, *Buttery*, etc., must be occupational. The man worked in or had charge of the kitchen or the pantry or the buttery, but he certainly did not live in them. Similarly, *Hall*, *Haller*, *Hallman*, probably denote a servant at the hall, where he also may have lived. But the owner—probably the lord of the manor—would have a different surname, one commemorating his possessions or an ancestor.

# NICKNAMES

That many modern surnames were originally nicknames is proved conclusively by the material in the following pages. No full and satisfactory classification can be attempted. Some are unintelligble; the meaning of many is doubtful. Nicknames arise spontaneously from some fortuitous chance. The schoolboy's 'Tiny' is usually a hefty giant in the first eleven, but 'Tubby' is more often an accurate description. In my schooldays, 'Feet' was the nickname of a tall, lanky individual, with heavy boots on large feet which caused havoc in the unorthodox football played during breaks. The chemistry master rejoiced in the name of 'Bublum Squeaks', a corruption of 'Bubble and Squeak'. He was excitable, no disciplinarian, with a voice which rose higher and higher to a shrill squeak as he vainly tried to make himself heard above the uproar in the laboratory. But why a colleague of his was known as 'Joe Plug' no one ever knew. His christian name was Arthur and his surname Watson. Even when the origin of a nickname is known, it is difficult to see why it should stick. A schoolboy, called on to translate a Latin Unseen about Polyphemus, was thenceforth 'Polly' to his friends. Why should one schoolmaster be called 'Wally' and another 'Mike', names impossible to associate with either christian name or surname? 'Kip' had an interesting history. Originally 'Skipper'—Why, nobody knew—it quickly became 'Kipper', later shortened to 'Kip'. It is not surprising, therefore, if we frequently fail to get behind the mentality of the thirteenth and fourteenth centuries and cannot interpret their nicknames.

Nicknames are common in medieval records, but comparatively few have given modern surnames. For many of them only a few examples of the nickname occur, and often enough there is only a single instance. This is not surprising; after all a nickname refers essentially to the characteristics, habits, or appearance of a particular individual, and it is only rarely that any peculiarity will be inherited by his children.

Many medieval nicknames—some cruel and indescribably coarse—have disappeared. Some are simple and obvious, describing physical attributes or peculiarities: Head, Neck, Mouth, Leg, Foot, Shanks, and, with attributes, Broadhead, Redhead, Coxhead, Ramshead, Barefoot, Cruickshanks, Sheepshanks, Goosey, Hawkey, Pauncefote 'arched belly', Vidler 'wolf-face', Chaffin 'bald', Hurren 'shaggy-haired', Garnham 'moustache', Grelley 'pock-marked', etc.

Mental and moral characteristics are often particularized: Good, Moody 'bold', Sharp, Wise, Root 'cheerful'; Daft 'foolish', Grim 'fierce', Musard 'stupid', Sturdy 'reckless', Proud, Prowse 'doughty', Vaisey 'playful', Gulliver 'glutton'; abstract nouns, as Comfort, Greed, Lawty 'loyalty', Sollas, Verity, Wisdom. Here, too, belong such names as Gutsell 'good soul', Thoroughgood, Goodenough, Careless, Pennyfather 'miser', Girling 'lion-heart', Gaine 'trickery', Fairweather, Milsopp.

Names of animals may be nicknames, descriptive of appearance or disposition. *Lamb* may denote meekness, *Bull* strength or a headstrong nature, *Colt* a lively, frisky individual, but they may often be used of a keeper of these animals. Bird names are not always easy to interpret: Raven 'black', Heron, Stork 'long legs', Nightingale and Thrush 'songsters', Kite 'ravenous'. Plant names may refer to a grower or seller, but may be nicknames: Cardon 'thistle', obstinate, stubborn, Pinnell 'tall and upright as a young pine'.

Names derived from dress and equipment are often occupational: Cottle 'cutler', Hood, Capp, Mantell, probably makers of these, but some are nicknames from a partiality for a particular type of dress: Greenhead 'green hood', Hussey 'booted', Gildersleeve 'golden sleeves', Shorthouse or Shorthose 'short boot'.

Many names, originally nicknames, were undoubtedly used as occupation names: Besant 'banker', Blampin 'white-bread', a baker, Collop 'ham and eggs', a cook-house keeper, Drinkwater, sometimes a taverner, Goodale 'beer-seller'.

Particularly interesting are what have been called 'phrase names', a term not entirely satisfactory, as there are two distinct types, the first consisting chiefly, but not entirely, of oath names, the second of 'imperative names', again an unsatisfactory term, as the verb may be merely the verbal stem. Oath names are chiefly French: Debney 'God bless you', Dugard 'God protect you', Pardew, Purday, Purefoy, Pepperday. Of English origin are: Godbehere, Godsave 'for God's sake', and, sometimes, Mothersole. From habitual expressions: Goodday, Goodenday, Goodyear, Drinkale, Bonnally, and the French Bonger 'bon jour'.

'Imperative names' consist of a verb plus a noun or an adverb. A few examples are found in Domesday Book but they are not common until the thirteenth and fourteenth centuries. Most are of French origin but the majority of those surviving are English, with some translations of the French: Crakebone (Brisbane), Cutbush (Tallboys), and a few hybrids: Bindloss, Pritlove, Shakesby 'draw sword'. Many of these nicknames are more or less derogatory occupation names: Bendbow (archer), Copestake (wood-cutter), Waghorn (trumpeter), Wagstaff (beadle), Catchpole (constable), Fettiplace (usher). There are various such names for 'butcher': Knatchbull 'fell bull', Killebolle, Hackbon 'hack bones', Fleshacker, Hoggsflesh. Others denote a fishmonger, Rottenherring, Oldherring, Freshfish, while a wolf-hunter appears as Findlow, Catchlove, Prichlove, Bindloes, all with Old French *louve* 'wolf' as the second element. Crawlboys 'fell wood', Tallboys 'cut wood', Warboys 'guard wood', and Hackwood, are all nicknames for a forester, Whitepayn and Blampin 'white bread', Havercake 'oat bread', and Buntflower 'sieve flour' for a baker. In addition, we find Wendout and Startout for a messenger, Shakelance and Lanceleaf for a soldier, Packstaff for a pedlar, Treadwater and Trenchemer 'cut the sea' for a sailor, Treacle for an apothecary, and Wagpole for a minor official. Others indicate some quality or characteristic: Scattergood (spendthrift), Sherwin (speed), Makepeace, Turnbull (strength or bravery), Bevin (drinker), Crawcour (break-heart), Dolittle, Hakluyt (lazy), Parlabean (good-speaker), Standfast, Standalone, etc.[50]

The main difficulty with nicknames lies in the interpretation of them. There may be more than one possible meaning, e.g. *Quant*, from ME *quoint, queynte*, had various meanings in medieval England, 'strange, curious, ingenious, clever, crafty', and we can rarely tell which sense is intended in any particular case. Similarly *Hare* may mean a fast runner, or a timid person. Sometimes the nickname means the opposite of what it says, so that *Little John* may refer to a giant, and this could often be the case with other nicknames.

Certainly the actual meanings of many nicknames are unknown. It is usually possible to give a literal meaning to the name, but exactly what it meant when attached to a particular individual it is impossible to say. So, for example, with those nicknames which have *-rose* as a second element, Pluckrose, Portrose, Ringrose, Spurnrose, Woodrose. Nicknames involving money may refer to the value of a holding, e.g. Andrew *Tenmark* 1279 RH (C), but it is unlikely that this can be the explanation for Thomas *Quatresoz* 1300 LLB C 'four sous'. Other names seems to refer to age, but it is difficult to know what to make of William *Two yer old* 1311 Ronton; Thomas *Twowynterold* 1327 SREss; Margaret *Tenwynter* 1476 SIA xii; Laurence *Sixweeks* 1570 FrLei, which can hardly refer to the actual ages of the persons concerned. Other difficult names are Robert *Cristendom* 1429 AssLo, Adam *Grenelef* 1327 SRSf 'green leaf', John *Dubbedent* 1160 P 'polish teeth', John *Hurthevene* 1288 CtW 'harm heaven', Thomas *Monelight* 1470 RochW, Geoffrey *Trailwing'* 1200 P (Y). In order to give some indication of the variety of nicknames in medieval records, a good many are included in the entries below, although they may not have given rise to modern surnames.

# HEREDITY OF SURNAMES

The rise of surnames, according to the accepted theory, was due to the Norman Conquest when Old English personal names were rapidly superseded by the new christian names introduced by the Normans. Of these, only a few were really popular and in the twelfth century this scarcity of christian names led to the increasing use of surnames to distinguish the numerous individuals of the same name.[51] This is an oversimplification. Bynames—both English and Scandinavian—are found in England before the Conquest. Some Normans had hereditary surnames before they came to England. Evidence is accumulating that the Old English personal names lived on longer than has been supposed, a fact confirmed by the large number of modern surnames to which they have given rise and which must have been in living use after the Conquest. The new French personal names, too, were more varied than is commonly believed. A few, William, Robert, Richard and John, certainly became much more popular than the rest, but it was not from these that the earliest patronymic surnames were formed. It is often assumed that men 'adopted' their surnames. Some certainly did, but the individual himself had no need for a label to distinguish him from his fellows. The development of the feudal system made it essential that the king should know exactly what service each knight owed. Payments to and by the exchequer required that debtors and creditors should be particularized. The lawyers saw to it that the parties to transfers of land or those concerned in criminal proceedings could be definitely identified. Monasteries drew up surveys and extents with details of tenants of all classes and their services. And later the net was thrown wider in the long lists of those assessed in the subsidy rolls. It was the official who required exact identification of the individual. His early efforts often consisted of long-winded descriptions attached to a personal name. Any description which definitely identified the man was satisfactory—his father's name, the name of his land, or a nickname known to be his. The upper classes—mostly illiterate—were those with whom the officials were chiefly concerned and among them surnames first became numerous and hereditary. It is noteworthy that in London, with its organized government and elaborate records, surnames became fixed early among the patrician classes.

There is evidence that surnames would have developed in England even had there been no Norman Conquest. Towards the end of the Old English period, a limited number of personal names were becoming particularly popular. In the Suffolk Domesday the names of 217 freemen in the Hundred of Colneis are given, only four having bynames. In nine villages there were two or more men of the same name and the clerk was driven to occasional descriptions such as *alter Vluric* 'and a second Wulfric'. In Burgate 4 out of 15 men were named *Godric*, of whom one had a nickname *Godric long*. In Burgh, of 16 persons, two were named *Almer* and three *Godric*. At Micklegate, *Goda*, at Trimley, *Derstan*, each occurs twice in four names. The inevitable result of this state of affairs can be seen from a list of names of Suffolk peasants (c1095).[52] Of 660, more than half (359) had a single name only; 104 were described by their father's name (Ailuuard Goduini filius); 163 had bynames of the various types (Brihtmer Haiuuard, Aluric Godhand, Lemmer Brihtmer, Ulfuine de Laueshel)—a clear indication of the rise of surnames among peasants of English ancestry and name.

The only serious discussion of the heredity of surnames is that of Fransson.[53] His material is late and some of it inconclusive. His general conclusions are sound but require some modification: 'Hereditary surnames existed among the Norman noblemen already in the early 12th century. Among people in general they began to come into use in the following century, and by the end of this they were fairly frequent (especially local surnames and nicknames). This custom increased rapidly in the course of the fourteenth century, and by the end of it practically all people were provided with hereditary surnames.'

His suggestion that one reason for the rise of surnames was that a need was felt to unite the members of a family by means of a common surname is unlikely. It assumes that surnames were adopted and not given and would hardly apply to nicknames. Nor does it explain the varied surnames found in the twelfth century for different members of the same family. Whether local surnames, because of their frequency, had any influence on the fixing of surnames is doubtful. For barons and important land-holders to derive their surnames from their fiefs or manors was natural, but these form only a small proportion of the whole. When surnames like Nash, Wood, etc., became hereditary is a problem for which material is seldom available. In London, local surnames indicated the place from which the man had come, and became hereditary early.

Surnames of various types found in Domesday Book became hereditary at once:

Bruce, Glanville, Montgomery, Percy (from French fiefs), Giffard, Peverell (patronymics), Basset and Gernon (nicknames).[54] Robert de Stafford, a brother of Ralph de Toeni (a surname surviving into the fourteenth century), took his surname from the head of his English barony. The fact that father and son bore the same surname is not always, as assumed by Tengvik and Fransson, a proof that the surname was hereditary. Robertus *Balistarius* held Worstead (Norfolk) in 1086 by serjeanty of performing the duties of *arbalistarius*. His son, Odo *arbalistarius*, inherited the office and the lands (c1140 Holme) and owed his surname either to inheritance or to his office. He is also called Odo *de Wrthesteda* (c1150 Crawford) and his son Richard and his grandson Robert were both called *de Worsted* (1166, 1210 Holme).

Throughout the twelfth and thirteenth centuries surnames of the type Johannes *filius Hugonis* are common, side by side with Johannes *Hugo*, where the son has his father's christian name as his surname. Such names indicate the beginning of a hereditary surname, but proof that it became established is often lacking:

Hugo filius Wisman, Hugo Wisman 1166–7 P (Nf)
Walterus filius Abelot, Walter Abelot 1195–6 P (Sa)
John le fiz michel 1292 SRLo, John Michel 1301 LoCt
Paganus le Cachepol, father of William Payn 1285 *Ass* (Ess)
John Gerveis son of Gervase de Pelsedun 1299 AD vi (K)

Such names as the following are probably already hereditary:

Reginald Ridel son of Hugh Ridel 1156–80 Bury (Nth)
Ralph Belet son of William Belet 1176 P (Sr)
William Brese son of Roger Brese 1210 P (Nf)
Gote Ketel, brother of Peter Ketel; Thomas Ketel son of Peter Ketel c1200, 1218–22 StP (Lo)

Clear evidence of heredity:

A charter of 1153 of Agnes de Sibbeford, wife of Ralph Clement, is witnessed by Hugo Clement and William, son of Ralph Clement, who is later called Willelmus Clemens, with a brother Robertus Clemens (1155 Templars).

Thomas Noel, founder of Ronton Priory, is so called in 1182–5. His father was Robert Noel (ib.), who is called Robertus Noelli filius (c1150 StCh).

Probably hereditary:
William Shepescank, Gilbert his brother, John Sepesank' 1224 Cur (Nf)
John Caritas, Simon Caritas, brothers 1265 FrLeic
William Lefthand, Ralph Lefthand 1268 FrLeic
Peter Wedercok son of Symon Wedircok 1302 Miller (C)

The twelfth century was also a period of vacillation and change in surnames:

Ralph, son of Robert Puintel de Walsham, had two brothers, William de Criketot and Ralph Cangard (12th Holme). He is also called Ralph de Crichetot (1141–9 ib.), with a son Hubert de Criketot (1163–6 ib.).

Philip de Powyk (1147–54 Holme) was a brother of Geoffrey Ridel (1153–68 ib.), a son of Richard Basset, and is called Philip Basset in 1185 (RotDom).

Stigand the priest (1126–7 Holme) had three sons: Thurbern the dean (1126–7), Simond de Ludham (1153–66) and Robert de Ludham or de Ling (1141–9). Simon's son and grandson were Thomas and Stephen de Walton (1175–86).

Griffin de Tweyt (1153–68 Holme) had a son Osbern de Thurgerton (1140–53) who married Cecilia, daughter of Roger de Curcun. Their son was Robert de Thweyt (1153–6) or de Curcun or Robert de Curcun de Thweyt (1186–1210). His son was Robert the Clerk.

In London, surnames of all kinds, patronymics, local, occupational and nicknames, became hereditary among the patrician classes in the twelfth century. They steadily increase in number and are frequent by the end of the thirteenth century.[55] At the same time, there are many later examples which are not hereditary, especially among the lower classes:

Luke le Ayler father of Walter le Mazerer 1278 LLB A, 1306 LLB B

Henry called Cros, son of William le Hornere 1303 LLB C

Amiel de Honesdon, late chandler, or Amiel le Chaundeller had two daughters: Johanna Amyel and Cristina la Chaundeller 1349 Husting

Bartholomew Guidonis (1357 LLB G) or Castiloun (1369 ib.) was father of John Chaungeour (1384 ib.)

Definite information on the development of surnames among the common folk is difficult to find. Their names mostly occur in isolation, with little or no indication of relationship. Fransson has suggested several methods by which heredity can be inferred when relationship is not given. When two men of the same name are distinguished by the addition of *senior* and *junior*, it is a fair assumption that they were father and son.[56] Such examples are fairly common in the subsidy rolls and later.

Further, he notes that in the subsidy rolls it is not uncommon to find several men of the same name assessed in the same village and suggests, very plausibly, that where the surname is a nickname, it has become hereditary. The same might be said of patronymics. Local surnames are not safe instances. There is no proof that the man did not actually live at the place. Similarly, an occupational name may well mean that the man followed that particular occupation. But when a trade-name, different from the surname, follows it, we may safely assume that the surname was hereditary. His caution that a man might have had two trades or that a trade-name might be a nickname seems unnecessary. Examples are found in London in the thirteenth century and elsewhere later, though they are rare where they would be most valuable—in the subsidy rolls: John le Spencer, spicer 1306 LoCt; John Pistor, Taillour 1319 SRLo; John Mariner, hatter 1327 Pinchbeck (Sf); John le Fyssher, pistor, Robert Muleward, carpentarius 1353 Putnam (W).

Where material is available, a further test is to compare different documents of different periods relating to the same village. For Suffolk we have two surveys of the manors of the Bishop of Ely for 1221 and 1277, full of names. Many of these peasants had no surname and most bore English or Scandinavian names. In fifteen parishes, we find the same surname in 1221 and 1277 as in the subsidy of 1327 and these can safely be regarded as hereditary. Only two or three parallels are found, as a rule, in any one parish, but in Glemsford, six surnames occur in both 1221 and 1277, 2 in 1277 and 1327, and one (*Curteis*) in all three years. In Rattlesden, seven surnames occur both in 1221 and 1327, *Haliday* twice in each year, *Barun* twice in 1221, *Hardheved* twice in 1327.

For the Bury manors we have a subsidy of 1283 for Blackburn Hundred, unfortunately damaged, with the loss of many names, and surveys of the Hundreds of Thedwestrey, Thingoe, Blackburn, Cosford and Babergh c1188–90. These surveys are much less detailed than those of the Ely manors and contain many fewer names. In 27 parishes we find some of the surnames of 1327 also in one or both the earlier documents, cumulative evidence that surnames were becoming hereditary throughout the county. In Stanton, N. *Wluric* of 1283 probably owed his surname to *Wuluricus* filius fabri of 1188. *Cat* is found in all three documents, *Hubert* and *Kenne* in 1283 and 1327, *Cauel* and *Brunston* in 1188 and 1327. In Hopton, Honington and Troston six, in Walsham five, and in Culford, Rickinghall and Ixworth Thorpe four surnames occur in both 1283 and 1327.

A noteworthy feature of the southern subsidy rolls is the large number of surnames formed from Old English personal names no longer in use in the county: Worcestershire (1275) 203, Somerset (1327) 208, Suffolk (1327) 441, Surrey (1332) 85; Yorkshire (1297) 17, Lancashire (1332) 1. The complete disappearance of these personal-names proves that the surnames must have become hereditary. There is also evidence of a marked difference between north and south and a hint of a variation in the rate of development in the southern counties themselves.

Much detailed work remains to be done before the full facts can be known. But it appears that surnames among the common people became hereditary later than those of the upper classes. They are found in the thirteenth century and are well established in the south by the middle of the fourteenth. But there is clear proof that many men still had no surname and that many were still not hereditary. In 1381 SRSf, whilst 5 men followed a trade different from that indicated by their surname, there were 20 whose surname denoted their occupation (John Soutere, soutere; Walter Webb, webber). Later examples of the instability of surnames are: William Saukyn *alias* Archer (1442), Philip Daunce *alias* Defford (1473); John Walworth, called Mundis (1502), John Bullok *alias* Byde (1527), Richard Bolle *alias* Bronde (1568 ER 61); Richard Johnson *alias* Jackeson, whose daughters were Margaret Richardson and Elizabeth Richardson 1568 AD v (Ch); Richard son of Geoffrey *Reynald* of Edmascote otherwise called Richard *Ryvelle*, otherwise Richard son of Joan, daughter of William *Ryvell* 1408 Cl.

Another example of a surname from a mother's name is: John *Organ* of Treworian, son of *Organa*, wife of Ives de Treworian 1327 AD v.

## Yorkshire Names

The editor of *Freemen of York*[57] notes that surnames were chiefly from place-names or trade-names. In the earlier years, patronymics were non-existent except as 'Thomas filius Johannis de Wistow' (1295), or, 'Thomas filius Johannis praepositi de Wistow' (1295). Names such as 'Johannes filius Davidis, pulter' (1277) were exceptional. The earliest name in *-son* is 1323. 'It is still later [than the reign of Henry IV] before we find the son invariably taking his father's name; one of the last, if not the last instance to the contrary, occurs in [1431] when we find Robertus de Lynby, fil. Thomae Johnson.' Derogatory nicknames survived late: Henry Scrapetrough, molendinarius 1293; William Whitebrow, plasterer 1333; John Nevergelt, goldsmyth 1431; William Heteblack, baker 1460.

A tenement in Nawton was acquired by a certain Abraham and passed to his son Robert and so to John Abraham grandfather of William Abraham who held it in 1298.[58] This surname goes back, therefore, to 1200 or earlier. In the thirteenth century, William Samson owed his surname to his great grandfather Sansom de Alreton (Kirkstall). There is some evidence of heredity of surnames, too, in York where a number of freemen followed occupations different from those denoted by their surnames: Richard le warner, carnifex 1319; Richard le sauser, pelter, son of John le Sauser 1331; Thomas le hosteler, mariner 1331; Adam Fetheler, mercer 1360. But there is much evidence to the contrary: William Belle, son of Andrew le taillour 1316; William Candler, son of Robert de Stoke 1324; Thomas le parchemyner, son of John le hatter 1334; Johannes filius Willelmi filii Ricardi de Carleton, draper 1339; William Whitehals son of Henry de Marston, webster 1369; John Byller, baxter, son of Henry Holtbyman, milner 1427.

The Yorkshire Subsidy Rolls confirm the impression that surnames were transient and ephemeral. There are only occasional hints of heredity. Most surnames were local, occupational or nicknames. In 1297 (3,160 names) 17 per cent had names of the *filius*-type; in 1301 (8,400 names) 21 per cent; in 1327 (4,500 names) 13 per cent; christian name plus christian name (e.g. Robert Reyner) accounted for 6 per cent in 1297, 3 per cent in 1301 and 1327. Names like Johnson were very rare: 2 in 1297, 5 in 1301, 12 in 1327.

The West Riding Poll-tax of 1379 (19,600 names) provides material quite unlike that found in the south and paralleled only by the East Riding Poll-tax of 1381. The *filius*-

type of name is much less common than in 1327; that in -son much more common. What is noticeable is the frequency of names in -wyf, and -doghter and those of servants in -man, -servant, -woman, -mayden, besides names indicating other relationships in -brother, -cosyn, -syster, -stepson:

Matilda Hanwyfe, Elena Hobsonwyf, Beatrice Clerkwyf, Alice Caresonewyf, Dionisia Raulynwyf, Johanna Jackewyf

Matilda ffoxdoghter, Isabella Shephirddoghter, Johanna Rosedoghter, Johanna Malkyndoghter, Magota Stevendoghter, Johana Robyndoghter

In two instances we have a man's surname: Robertus ffelisdoghter et Cecilia vxor ejus; Richard Wryghtdoghter

John Websterman, Thomas Masonman. Husband and wife were at times servants of the same man: William Mathewman, Magota Mathewwoman; Adam Parsonman, Emma Parsonwoman

Richard Hogeservant, Johanna Vikarservant, Elena Houchounservant

Isabella Vikerwoman, Johanna Prestewoman, Margareta Hallewoman

Matilda Marschalmaydyn, Alice Gibmayden, Elisot' Milessonmayden, Alice Martynmayden

Robert Parsonbrother, Henry Parsoncosyn, Agnes Vikercister, Alice Prestsyster servant, John Robertstepson

In these names, the suffix was often added to the surname and the master, etc., may be named separately:

John Odson, Alice Odsonwyf; William de Bilton, Roger Biltonman; Robert de Wallerthwayte, Margareta Wallerthwaytdoghter; Emma Hurle, Johanna Hurlemayden; Ellota de Helagh, Agnes Helaghmayden; John Whitebred, Adam Wytbredman, John Adamson Whitebredman

Similarly, names in -son were also based on the surname:

John Payg, John Paygson; Richard Parlebene, Robert Parlebeneson; Matilda Millot', Roger Millotson; John Websterson; Adam Souterson. cf. also Roger Taylourson, Agnes Taylourdoghter; William Saunderson, Alice Saunderdoghter; William Milnerson, Agnes Milnerwyf. The wife of Roger *Wright* was Elena *Wrightwyf*; his son, John *Wrightson*.

The sons of William Jonson are named William Willeson Johanson and Benedict Willeson Johnson; that of Robert Hudson was William Robynson Hudson. Wives were similarly named: Margareta Wilkynwyf Raulynson, Agnes Dycounwyfdowson.

It is abundantly clear that in the north surnames became hereditary much later than in the south. There is a fair amount of evidence that a number of occupation names had become hereditary, but many certainly had not. In his *Memoirs of the Wilsons of Bromhead*,[59] Joseph Hunter demolishes the earlier pedigrees by proving errors of heralds and forging of documents. The family descended ultimately from William, father of John de Hunshelf or de Waldershelf (b. c1320) but owed their surname, not to this William, but to William (1369–87), father of John Wilson de Bromhead who is called John son of William son of John de Waldershelf in 1398. Hunter notes that John Dyson de Langeside derived his surname from his mother Dionysia de Langside (1369) and that a Thomas Richardson was the son of Richard de Schagh (1409). 'This (1380)', he concludes, 'was the age at which that class of surnames, which end in -son, began to be assumed', a conclusion not inconsistent with the evidence above.

# WELSH SURNAMES

Hereditary surnames in Wales are a post-sixteenth century development. Many of the modern surnames derived from old Welsh personal names arose in England where they became hereditary in the fourteenth century or earlier, long before such surnames were known in Wales; some, in the eastern counties, derive from Breton immigrants. The normal type of Welsh name was a patronymic: Madog ap Jevan ap Jorwerth, 'Madoc, son of Evan, son of Yorwerth', a type which resulted ultimately in such names as Pumfrey, Benian, Bevan, etc. In 1292, 48 per cent of Welsh names were patronymics of this kind (in some parishes, over 70 per cent); others included nicknames, occupation-names and some local surnames. The great majority of the surnames in the Extent of Chirk (1392–3) were of this patronymic type, with occasional nicknames (Jevan Gough, Ithel Lloit, Grono Vachann), rare occupational-names (Madog Taillour), and a few simple personal-names (Jevan Annwyl, Jevan Gethin), none of which were hereditary.

In later Chirk documents these patronymics are still the normal form. In 1538, all the thirteen men of a jury had names like: *John ap Madog ap Gryffyd ap Res junior*. There is evidence that a change had begun. Edward *ap Richard* and Edward *ap Robert* point to the future preponderance of the Welsh Jones, Williams and Roberts. In 1536 we find one such name already hereditary: John Edwards son of William Edwards. It was only in the reign of Henry VIII that surnames began to be hereditary among the gentry of Wales and the custom spread only slowly among the common people. Even in the nineteenth century, in Merionethshire, it was still not uncommon for a man to take his father's christian name as his surname: e.g. William Roberts son of Robert Williams. The three sons of Evan Thomas and Gwen Jones were known as Howal Thomas, Hugh Evans and Owen Jones, surnames derived (i) from the father's surname, (ii) from his christian name, (iii) from the mother's surname. In the nineteenth century, the frequency of Jones, Williams, etc., brought a need for further distinction and a tendency developed to create double surnames by prefixing the name of a house, parish or the mother's surname, as Cynddylan Jones, Rhondda Williams, etc.[60] In the following generation a hyphen was often introduced, hence Nash-Williams, etc.

## SCOTTISH SURNAMES

The earliest surnames in Scotland, found in the reign of David I (1124–53), were those of Normans: Robert de Brus, Robert de Umfraville, Gervase Ridel, etc., surnames which had already become hereditary in England and were later to be reinforced by such names as Balliol, Cumin, etc. In the towns, the burgesses bore English or continental personal names, with trade-names or occasional nicknames.

In Scotland, early material for the study of surnames is much later than in England. Many names are undocumented before the fifteenth or sixteenth centuries, a period so late that definite etymologies are often impossible. 'The largest and most authentic enumeration now extant of the nobility, barons, landowners and burgesses as well as of the clergy of Scotland, prior to the fourteenth century,' is the Ragman Roll[61] which records the deeds of homage made to Edward I in 1296—an English official document. 'No part of the public records of Scotland prior to that era has been preserved from which any detailed information of the kind might have been derived.'[62] The surnames in the Ragman Roll are, for the most part, of the same type as those found in English sources of the same period—local, patronymic, occupational and nicknames. A number of local surnames derive from places in Scotland. Gaelic surnames occur, but form a distinct minority: e.g. Fergus Mac Dowilt, Macrath ap Molegan, Huwe Kenedy, Dovenal Galbrathe.

In 1382, of 56 tenants of Fermartyne excommunicated by the Bishop of Aberdeen,[63] 23 had names like Robertus filius Abraam. Celtic personal names were rare. The solitary Gaelic surname is Adam Kerde (Caird 'craftsman'). Of the 23 surnames, 4 are patronymics from OE personal-names, Bronnyng, with three examples of Freluf (v. Browning, Freelove); 12 are local; one is Scandinavian, John Grefe (v. Grave); 3 are occupational (Cissor, Barkar, Faber); 2 are nicknames (Gray, Mykyl).

The surnames of 50 members of the Guild of Ayr (c1431)[64] have all the appearance of modern surnames: 16 patronymics (Neil Neilsoun, Patrik McMartyn, Patrik Ahar, Patoun Dugald); 6 local; 12 occupational (Listar, Walkar) and 5 nicknames (Petit, Cambell, Broun, Lang).

The paucity of Gaelic names in these sources is noteworthy. They are from the Lowlands, from Ayr and from Aberdeen, which 'was already predominantly English-speaking in the twelfth century'.[65] The Lowland Scots dialects derive from Northern English, though they have developed their own pronunciations and characteristic vocabularies, and Lowland surnames developed on the same lines as those in England, though they were slower to become hereditary. At the end of the fifteenth century and later we find clear evidence that surnames were not generally fixed. In 1473 the son of Thomas Souter was named David Thomson and in 1481 Alexander Donaldson was the son of Donald Symonson. The frequent patronymics were not permanent. They changed with each succeeding generation and in the Highlands it was not until the eighteenth century that this custom was abandoned.

It was a common practice in Scotland for a laird to take his name from the estate, which itself was often named from its owner. The lands of Hugh de Paduinan (1165–73) were called from him *villa Hugonis* or *Huwitston* 'the estate of *Hewitt*', a pet-form of *Hugh*. His descendants took hence their surname, Fynlawe *de Hustone* (1296 CalSc), now *Houston*. Similarly, the modern *Symington* derives from *Symoundestone* (now Symington, Lanarkshire), the barony once held by *Symon* Locard (c1160). Owing to the frequency of such territorial names, lairds and farmers were often called by the name of their estate or farm and signed their letters and documents by their farm-names. In the seventeenth century an Act was passed forbidding any except noblemen and clerics of high office so to sign themselves but such estate-names long persisted in speech.

In the Highlands, hereditary surnames developed late. The clan system resulted in large numbers of people with the same name, but no specific surname of their own. The desire for protection in unsettled times caused men to attach themselves to a powerful clan and to assume its name. Chiefs of clans and heads of landed families increased the number of their followers by conciliation or coercion, and all took the name of the clan. In the sixteenth and seventeenth centuries the rapid increase in the Clan Mackenzie was due to the inclusion of the old native tenants on lands acquired from time to time by the chiefs. 'Frasers of the boll of meal' were poor Bissets who had changed their name to Fraser for a bribe. Oppressed people from the neighbouring districts sought the protection of Gilbert Cumin who adopted them as clansmen by baptizing them in the stone hen-trough at his castle door. Henceforth they were 'Cumins of the hen-trough' to distinguish them from Cumins of the true blood.[66]

In 1603 an Act was passed ordering the McGregors to renounce their name under pain of death. Some took the names of Johnestoun, Doyle, Menzies or Ramsay. For loyalty to Charles II the Act was repealed in 1661 but revived in 1693.[67] In 1695 'Evan, formerly called M'Grigor' was granted permission to resume his surname of McGregor for life, but only on condition that he gave his children a different surname, for which he chose *Evanson*.[68]

After the Battle of Culloden (1746), Gaelic names began to creep into the Lowlands and were often anglicized to overcome Lowland hostility. English or Lowland surnames were adopted. Names were translated, Johnson for Maciain, Livingstone for MacLevy, Cochrane for Maceachrain. Macdonald became Donald or was translated Donaldson. In the years after 1820, began a steady influx of Irishmen into south-west Scotland, especially into Lanarkshire, Renfrewshire, Ayrshire and Galloway, with further corruption of Gaelic names: Doyle for O'Toole, Swan for McSweeney, Graham for McGrimes and Cuming for McSkimming.[69]

The modern bearer of a clan surname, therefore, is not necessarily a member of the clan by blood or heredity. Nor does a Gaelic or English surname prove descent. A Celtic surname may be borne by one with very little Celtic blood in his veins, whilst a man with an Anglo-Saxon name may be almost a pure Celt.[70]

## Clan-names

The varied origins of Scottish surnames is well illustrated by the names of the clans.

*From Scottish places:* Baird, Brodie, Buchanan, Chisholm, Cunningham, Douglas, Drummond, Erskine, Forbes, Gordon, Innes, Keith, Leslie, Livingstone, Murray, Ogilvie, Ramsay, Ross, Skene, Sutherland, Urquhart

*Gaelic:* Cameron, Campbell, Clan Chattan, Duncan, Farquharson, Ferguson, Gow, Kennedy, Maclennan, Macalpine, Macarthur, Macbean, Maccallum, Maccoll, Macdonald, Macdonnell, Macdougall, Macduff, Macewen, Macfarlane, Macfie, MacGillivray, Macinnes, Macintosh, Macintyre, Mackay, Mackenzie, Mackinnon, Maclaren, Maclean, Maclaine, Macmillan, Macnab, Macnaughton, Macneill, Macpherson, Macquarrie, Macqueen, Macrae, Malcolm, Matheson, Munro, Rob Roy, Shaw

*French:* Anderson, Bruce, Cumin, Davidson, Fletcher, Fraser, Grant, Hay, Henderson, Johnston, Macalister, MacGregor, MacNicol, Menzies, Montgomery, Morrison, Napier, Oliphant, Robertson, Sinclair

*English:* Armstrong, Barclay, Elliot, Graham, Hamilton, Lindsay, Scott, Stewart, Wallace

*Scandinavian:* Gunn, Kerr, Lamont, Macaulay, MacIvarr, MacLachlan, MacLeod

# IRISH SURNAMES

Surnames appear in Ireland in the middle of the tenth century. These were patronymics formed by prefixing *O* or *Ua* to the grandfather's name or *Mac* to the father's, whether a personal or an occupation name. Of these, the *Mac*-names are the later. It has been claimed that these surnames became hereditary by an ordinance of Brian Boru (1002–14) but neither Brian himself nor his sons had hereditary surnames. It was only in the time of his grandsons that O'Brien first came into existence. The development of such surnames was slow and spread over several centuries. Definite evidence is lacking and patronymics have been confused with hereditary family names. Woulfe considers that by the end of the twelfth century surnames were universal in Ireland but admits that they were not at first of a lasting character, and in some instances were laid aside after a generation or two in favour of new surnames taken from less remote ancestors. He admits, too, that some surnames are not older than the thirteenth or fourteenth centuries, whilst a few originated in the sixteenth century.[71]

The Anglo-Norman invasion of Ireland had a similar effect to that of the Norman Conquest in England. New personal names were introduced, and a new type of surname. Some Normans took names from Irish places: John de Athy, Adam de Trim, both now rare surnames in Ireland. Burke, Birmingham and London derive from Anglo-Normans, as do Bassett, Bissett, Savage; Hammond, Hewlett, Sampson.

A list of about 1,500 Dublin surnames of the end of the twelfth century contains very few that could not appear in an English list of the same period. The personal names are mostly French, with a fair sprinkling of English and Scandinavian (Godwin, Ailward, Cristraid, Edwacer; Torsten, Swein, Toki, Turchetel). Of the few Celtic names, Bricius, Samsun and Cradok are found in England. Only Padin, Cullin and Gillamorus are pure Irish. So, too, with the surnames: local names from Colchester, Leominster and Worcester; common occupational names, *lorimer*, *turnur*, etc.; nicknames as Holega, Litalprud, le Crespe, le Gentil, Prudfot, Unred, Philip Unnithing, etc. Walter palmer was the son of David de Tokesburi. There is no hint of hereditary surnames. Two similar lists of some 550 free citizens (1225–50) and of about 200 members of the Dublin Guild Merchant (1256–7) are similar in nature.[72] The names are Anglo-Norman names established in Ireland.

At Limerick and Cork, in 1295, surnames were Irish:[73] O'Kynnedy, Ofechan, Omoriharthy, Maccloni, Maccarthen. Irishmen were beginning to use French christian names: Reginald, Maurice, Thomas, Walter.

One result of the Anglo-Norman settlement was that names acquired two forms, one Irish, one English. Some English settlers adopted Irish names. The Birminghams took the surname of MacFeeter from Peter de Bermingham and the Stauntons that of Mac an Mhíliadha (MacEvilly) from Milo de Staunton. After the murder of William de Burgo, third earl of Ulster, in 1333 and the lessening of English power in Ireland, many Anglo-Norman families in Connaught and Munster adopted the Irish language and assumed Irish surnames (MacWilliam, MacGibbin, etc.) and became so thoroughly hibernicized that in 1366 an Act was passed ordering that 'every Englishman use the English language, and be named by an English name, leaving off entirely the manner of naming by the Irish'.[74] In 1465 an attempt was made to stamp out the use of Irish names among the Irish themselves. Every Irishman living in the counties of Dublin, Meath, Louth, and Kildare was ordered to assume 'an English surname of a town, as Sutton, Chester, Trim, Skreen, Cork, Kinsale; or a colour, as White, Black, Brown; or an art, as Smith or Carpenter; or an office, as Cook, Butler'. The name so selected was to be used also by his issue under penalty for failure so to do.[75]

After the revolution of 1688 the change of Irish into English names increased. This process of anglicization followed very much the same course as in Scotland. Ó Cobhthaigh became Coffey, Cowie or Cowhey, whilst Coffey may represent Ó Cobhthaigh, Ó Cathbhadha, Ó Cathbhuadhaigh or Ó Cathmhogha. *O'* and *Mac* were frequently dropped. Ryan may be for O'Mulryan. Names might be translated (Badger for Ó Bruic; Johnson for MacSeáin); or attracted to a better-known surname (Ó Blathmhaic, anglicized as Blawick and attracted to Blake), or assimilated to a foreign name (Summerville for Ó Somacháin; De Moleyns for Ó Maoláin; Harrington for Ó hArrachtáin); or by substitution (Clifford for Ó Clumháin; Loftus for O Lachtnáin).[76] Such surnames were carried by Irish immigrants to England, Scotland and America where they were often further corrupted in pronunciation and spelling, thus adding endless complications to the difficulties of tracing their origin.


## MANX SURNAMES


Manx surnames reflect the history of the island. Orosius tells us that in the fifth century both Ireland and the Isle of Man were inhabited by *Scoti*—Gaels, of the same name as those from whom Scotland derived its name. In the ninth century Norsemen subdued the island which was mainly ruled by Norwegians from Dublin. In 1266 Norway ceded Man to Scotland who held it about a hundred years, though it was frequently in the hands of the English. The Scandinavian settlers, already partly celticized, intermarried with the native Gaels and added Norse to the Celtic personal-names in common use. Patronymics were formed by prefixing *Mac* to the father's name. The Irish *O'* never took root.

Of modern surnames, Moore estimates that 68 per cent are pure Celtic, 9 per cent pure Scandinavian, 6 per cent Celtic-Scandinavian, 5.4 per cent pure English, 3.3 per cent English-Celtic and 1.3 per cent English-Scandinavian.[77] Early in the sixteenth century the prefix *Mac* was almost universal; a hundred years later it had almost disappeared.[78] In pronunciation, the *Mac* was unstressed and the final consonant tended to coalesce with the first consonant of the following personal-name and became the initial consonant of the surname when the *Mac* was lost. Hence the characteristic Manx surnames beginning with C, K, or Q: Caine (MacCathain), Curphey (MacMurchadha), Kay (MacAedha), Kermode (MacDermot), Kneen (MacCianain), Quine (MacCuinn); Corkhill (MacThorketill), Cowley (MacAmhlaibh, Macaulay, from ON *Óláfr*), Crennell (Macraghnaill, ON *Rögnvaldr*). Some names of this type are from Anglo-Norman personal-names: Clucas (MacLucas), Costain (MacAustin), Kissack (MacIsaac), Quail (MacPhail, Paul), Qualtrough (MacWalter), Quilliam (MacWilliam).


## NOTES

1   cf. *Groups*, 75.
2   For other examples, *v*. MESO 21 (1275–1533).
3   For a full discussion, *v*. Ekwall, *Variation* and *Two Early London Subsidy Rolls* (notes, *passim*).
4   *v*. OEByn 31 ff, 121 ff.
5   ibid., 59 ff.
6   Holme 231.
7   For the light thrown by these surnames on immigration from the provinces into thirteenth- and

fourteenth-century London, *v.* Ekwall, *Two Early London Subsidy Rolls*, 49–71.

8   *Groups*, 92.
9   MESO 25.
10  OEByn 125–30, 137–8.
11  FrY, p. xvi.
12  *v.* Mawer and Stenton, *Place-names of Sussex* 35, MESO 192–208.
13  OEByn 146–66.
14  ELPN 127–8.
15  OEByn 228, 232.
16  Ewen 56, MESO 26, A. H. Smith, *Saga Book* XI (1934), 17.
17  OEByn 147.
18  OEByn 209 ff.
19  ibid., 209 (with references).
20  *Namn och Bygd,* vol. 27 (1939), p. 128.
21  OEByn 210–11.
22  Reaney, *Essex Review,* vol. 61, pp. 135–8.
23  ibid., pp. 142, 202–4, 209–11.
24  von Feilitzen, *Namn och Bygd,* vol. 33, pp. 69–98 (1945).
25  *v.* Reaney, *Survival of OE Personal Names in ME* (Studier i modern språkvetenskap, vol. 18, pp. 84–112), 1953; *Pedigrees of Villeins and Freemen* (NQ, vol. 197, pp. 222–5), 1952; *Three Unrecorded OE Personal Names of a late type* (Modern Language Review, vol. 47, p. 374), 1952.
26  For details, *v.* Reaney, *Pedigrees of Villeins and Freemen*.
27  Stenton, *Danelaw Charters*, pp. cxii ff.
28  *v.* Douglas, *Feudal Documents . . . of Bury St Edmunds*, cxvii–cxx.
29  *v.* also Whitelock, *Scandinavian Personal Names in the Liber Vitae of Thorney Abbey* (Saga Book, vol. 12 (1940), pp. 127–53), and, for Norfolk, West, *St Benet of Holme*, vol. 2, pp. 258–60.
30  Stenton, *English Feudalism*, 24–6.
31  *Essex Review*, vol. 61, p. 140.
32  *v.* Ekwall, *Scandinavians and Celts*: A. H. Smith, *Irish Influence in Yorkshire* (Revue Celtique, vol. 44, pp. 34–58), *Danes and Norwegians in Yorkshire* (Saga Book, vol. 10); Armstrong, Mawer, Stenton and Dickins, *Place-names of Cumberland*, vol. 3, pp. xxii–xxv.
33  *Surnames*, p. 71, n. 1.
34  Keal occurs as *Keles* 12th DC.
35  There was never any such place in Essex. *Berle* is probably Barley (Herts).
36  *Romance of Names*, 90.
37  And, from personal names, in *Hitches, Hodges, Riches*.
38  MESO 27.
39  *History of Surnames*, 120, 246–8.
40  ibid., 120.
41  ibid., 247.
42  ibid., 247.
43  ibid., 252.
44  OEByn 207–9.
45  *Variation*, 44.
46  For a similar genitival formation in -*en*, *v.* Geffen.
47  Stenton, *Personal Names in Place-names*, in Mawer and Stenton, *Introduction to the Survey of English Place-names*, 169–70.
48  *Crawford Charters* 51.
49  Stenton, op. cit., 174.
50  *v.* also Passmore, Perceval, Gotobed, and, for similar obsolete names, Weekley, *Surnames* 270–7.
51  ODCN xxviii, OEByn 8–9.
52  Bury 25–44.
53  MESO 33–41.
54  For other examples, *v.* OEByn 14 ff.
55  ELPN 119, 124, 130, 178.
56  MESO 36ff.
57  p. xvi.
58  1327 SRY, p. 50, n. 2.
59  *Yorkshire Archaeological Journal*, vol. 5 (1879), pp. 63–125.
60  Ewen 208, 387.
61  *Cal. Docs. Scotland*, vol. 2, pp. 193–214.
62  T. Thomson, *Instrumenta Publica* (Bannatyne Club), p. xiv.
63  List printed by Black, *Surnames of Scotland*, p. xxiii.

64  ibid., p. xxiv.
65  F. C. Diack, *Scottish Gaelic Studies*, vol. i, p. 92.
66  *v.* Black, pp. xxxvii–xxxviii.
67  Ewen 421–2.
68  ibid., 423.
69  *v.* Black, pp. xl, xlv.
70  *v.* Black, pp. xlii–xliii.
71  Woulfe, *Sloinnte Gaedheal is Gall*, pp. xviii, xix.
72  Gilbert, *Historic and Municipal Documents of Ireland*, pp. 3–48; 112–23; 136–40.
73  Lists printed by Ewen, p. 129.
74  ibid., 425.
75  ibid., 426.
76  *v.* Woulfe, 36–39.
77  Moore, *Surnames and Place-names of the Isle of Man*, 11.
78  ibid., 9.

# ABBREVIATIONS

| | |
|---|---|
| a | *ante* |
| Abbr | *Placitorum . . . abbreviatio*, London 1811 |
| AC | J. H. Round, *Ancient Charters* (Pipe Roll Soc. 10, 1888) |
| Acc | H. M. Colvin, *Building Accounts of King Henry III*, Oxford 1971 |
| AccM | S. Challenger, *Accounts for Works on the Royal Mills and Castle at Marlborough, 1237–8 and 1238–9* (Wilts Arch. Soc. 12, 1956) |
| AD | A Descriptive Catalogue of Ancient Deeds (in progress) |
| *Add* | BM, Additional MSS |
| *AddCh* | BM, Additional Charters |
| AFr | Anglo-French |
| AN | Anglo-Norman |
| ANF | L. C. Loyd, *The Origins of some Anglo-Norman Families* (Harleian Soc. 103, 1951) |
| ArchC | *Archaeologia Cantiana* (in progress) |
| ASC | *Anglo-Saxon Chronicle* |
| ASCh | A. J. Robertson, *Anglo-Saxon Charters*, Cambridge 1939 |
| *Ass* | Assize Rolls (unpublished) |
| Ass | Assize Rolls: Beds (Beds Hist. Rec. Soc. 1, 3, 1913, 1916); Berks (Selden Soc. 90, 1972–3); Bucks (Bucks Rec. Soc. 6, 1945); Cambs (W. M. Palmer, *The Assizes held at Cambridge A.D. 1260*, Linton 1930; Cambs Antiq. Soc. 55, 1942); Ches (Salt Arch. Soc. (OS) 15, 1894; Chetham Soc. 84, 1925); Devon (Ass. Rolls for 1332, 1359, ed. A. J. Howard, 1970); Durham (Surtees Soc. 127, 1916); Essex (Essex Arch. Soc. 1953); Glos (Selden Soc. 59, 1940); Kent (Kent Rec. Soc. 13, 1933); Lancs (Lancs and Ches Rec. Soc. 47, 49, 1904, 1905); Lincs (Lincs Rec. Soc. 22, 30, 36, 49, 56, 65, 1926–71; Selden Soc. 53, 1934); London (London Rec. Soc. i, 6, 12, 1965–76); Norfolk (Norfolk Rec. Soc. 44, 1976); Northants (Northants Rec. Soc. 5, 11, 1930, 1940); Northumb (Surtees Soc. 88, 1891); Som (Som Rec. Soc. 11, 36, 41, 44, 1897–1929); Staffs (Salt Arch. Soc. (OS) 3–7, 1882–6; Selden Soc. 59, 1940); Warwicks (Dugdale Soc. 16, 1939; Selden Soc. 59, 1940); Wilts (Wilts Arch. Soc. 16, 26, 33, 1961, 1971, 1978); Worcs (Selden Soc. 53, 59, 1934, 1940); Yorks (Yorks Arch. Soc. 44, 100, 1911, 1939; Selden Soc. 56, 1937) |
| AssNu | *The London Assize of Nuisance 1301–1431* (London Rec. Soc. 10, 1973) |
| ASWills | D. Whitelock, *Anglo-Saxon Wills*, Cambridge 1930 |
| ASWrits | F. E. Harmer, *Anglo-Saxon Writs*, Manchester 1952 |
| Bacon | *The Annalls of Ipswiche 1654*, ed. W. H. Richardson, Ipswich 1884 |
| Balliol | *The Oxford Deeds of Balliol College* (Oxford Hist. Soc. 64, 1913) |
| BarkingAS | *Transactions of the Barking Antiquarian Society* (in progress) |
| Barnwell | *Liber Memorandorum Ecclesie de Bernewelle*, ed. J. W. Clark, Cambridge 1907 |
| Bart | N. Moore, *The History of St. Bartholomew's Hospital*, 2 vols, London 1918 |
| Battle | *Custumals of Battle Abbey* (Camden Soc. (NS) 41, 1887) |
| BCS | W. G. de G. Birch, *Cartularium Saxonicum*, 3 vols, London 1885–99 |
| Bec | *Select Documents of the English Lands of the Abbey of Bec* (Camden Soc. 3rd Series 73, 1951) |
| Beds | Bedfordshire |
| Berks | Berkshire |
| *Binham* | BM. Cotton Claudius D xiii |
| BishamPR | *The Register of Bisham, co. Berks*, London 1898 |
| Bk | Buckinghamshire |
| Black | G. F. Black, *The Surnames of Scotland*, New York 1946 |
| BM | *Index to the Charters and Rolls in the . . . British Museum*, 2 vols, London 1902 |
| Boldon | *Boldon Book* (DB vol. iv; Surtees Soc. 25, 1852) |

| | |
|---|---|
| Bosville | T. W. Hall, *A Descriptive Catalogue of . . . the Bosville and the Lindsay Collections*, Sheffield 1930 |
| Building | L. F. Salzman, *Building in England down to 1540*, Oxford 1952 |
| Burton | *Abstract of the Contents of the Burton Chartulary* (Salt Arch. Soc. (OS) 5, 1884) |
| Bury | D. C. Douglas, *Feudal Documents from the Abbey of Bury St. Edmunds* London 1932 |
| BuryS | *The Kalendar of Abbot Samson of Bury St. Edmunds and related documents* (Camden Soc. 3rd Series 84, 1954) |
| BuryW | *Wills and Inventories from . . . Bury St. Edmunds* (Camden Soc. (OS) 49, 1850) |
| Butley | A. G. Dickens, *The Register or Chronicle of Butley Priory, Suffolk, 1510–1535*, Winchester 1951 |
| Buxhall | W. A. Copinger, *History of the Parish of Buxhall*, London 1902 |
| c | *circa* |
| C | Cambridgeshire |
| *Caen* | Cartulary of Holy Trinity Abbey, Caen (Paris, Bib. Nat., MS Lat. 5650) |
| CalSc | *Calendar of Documents relating to Scotland*, 4 vols, London 1881–8 |
| Calv | *The Calverley Charters* (Thoresby Soc. 6, 1904) |
| CantW | *Index of Wills and Administrations . . . in the Probate Registry at Canterbury, 1396–1558 and 1640–1650* (Index Library 50, 1920) |
| CAPr | *Proceedings of the Cambridge Antiquarian Society* (in progress) |
| CarshCt | *Court Rolls of the Manor of Carshalton* (Surrey Rec. Soc. 2, 1916) |
| CartAntiq | *Cartae Antiquae Rolls* (Pipe Roll Soc. (NS) 17, 33, 1939–60) |
| CartNat | *Carte Nativorum. A Peterborough Cartulary of the Fourteenth Century* (Northants Rec. Soc. 20, 1960) |
| CathAngl | *Catholicon Anglicum*, ed. J. H. Herrtage (EETS (OS) 75, 1881) |
| Ch | Cheshire |
| Ch | *Calendar of the Charter Rolls*, 6 vols, London 1903–27 |
| ChambAccCh | *Accounts of the Chamberlains . . . of the County of Chester, 1301–1360* (Lancs and Ches Rec. Soc. 59, 1910) |
| ChertseyCt | *Chertsey Abbey Court Rolls* (Surrey Rec. Soc. 38, 48, 1937, 1954) |
| ChesterW | *Chester Wills* (Lancs and Ches Rec. Soc. 107, 1961) |
| Chetwynd | *The Chetwynd Chartulary* (Salt Arch. Soc. (OS) 12, 1891) |
| Chirk | G. P. Jones, *The Extent of Chirkland, 1391–1393*, London 1933 |
| ChR | *Rotuli Chartarum*, London 1837 |
| ChW | *An Index of the Wills . . . at the Diocesan Registry, Chester, from 1487–1620 inclusive* (Lancs and Ches Rec. Soc. 33, 1875) |
| ChwWo | *The Churchwardens' Accounts of St. Michael's in Bedwardine, Worcester* (Worcester Hist. Soc. 1896) |
| Cl | *Calendar of the Close Rolls* (in progress) |
| Clerkenwell | *Cartulary of St. Mary, Clerkenwell* (Camden Soc. 3rd Series 71, 1949) |
| ClR | *Rotuli litterarum clausarum*, 2 vols, London 1833 |
| Co | Cornwall |
| Cockersand | *The Chartulary of Cockersand Abbey* (Chetham Soc. 38–40, 43, 56, 57, 64, 1898–1909) |
| Colch | *Cartularium Monasterii S. Joh. Bapt. de Colecestria*, 2 vols, Roxburghe Club 1897 |
| ColchCt | *Court Rolls of the Borough of Colchester*, 3 vols, Colchester 1921–41 |
| Combermere | *The Book of the Abbot of Combermere 1289–1529* (Lancs and Ches Rec. Soc. 31, 1896) |
| Copinger | W. A. Copinger, *Materials for the History of Suffolk*, 5 vols, London 1904 |
| *Cor* | Coroners' Rolls (PRO) (unpublished) |
| Cor | Coroners' Rolls: (Selden Soc. 9, 1896); London (R. R. Sharpe, *Calendar of Coroners' Rolls of the City of London A.D. 1300–1378*, London 1913); Notts (Thoroton Soc. 25, 1969) |
| *Coram* | Coram Rege Rolls (PRO) (unpublished) |
| Coram | *Placita coram domino Rege . . . 1297* (British Rec. Soc. 19, 1898) |
| CoramLa | *South Lancashire in the reign of Edward II* (Chetham Soc. 3rd Series 1, 1949) |
| CoWills | *Calendar of Wills . . . relating to the Counties of Cornwall and Devon* (Index Library 56, 59, 1929, 1932) |
| CPR | *Cambridgeshire Parish Registers*, 8 vols, ed. W. P. W. Phillimore, London 1907–27 |
| CR | Pipe Roll, Chancellor's Copy |
| Crispin | J. A. Robinson, *Gilbert Crispin, Abbot of Westminster*, Cambridge 1911 |
| Crowland | F. M. Page, *The Estates of Crowland Abbey*, Cambridge 1934 |
| Ct | Court Rolls (unpublished) |
| CtH | B. Farr, *The Rolls of Highworth Hundred 1275–1287* (Wilts Arch. Soc. 21, 22, 1966, 1967) |
| CtSt | *Alrewas Court Rolls 1259–61, 1266–1269* (Salt Arch. Soc. (NS) 10, 1907; *1272–1273* 3rd Series 1, 1910) |

| | |
|---|---|
| CtW | R. E. Latham and C. F. Meekings, *The Veredictum of Chippenham Hundred, 1281* (Wilts Arch. Soc. 12, 1956); R. B. Pugh, *Court Rolls of the Wiltshire Manors of Adam de Stratton* (Wilts Arch. Soc. 24, 1970) |
| Cu | Cumberland |
| Cullum | J. Cullum, *The History and Antiquities of Hawsted*, London 1784 |
| Cur | *Curia Regis Rolls* (in progress; Pipe Roll Soc. 14, 1891) |
| CurR | *Rotuli Curiae Regis*, 2 vols, London 1835 |
| Cust | T. J. Hunt, *The Medieval Customs of the Manors of Taunton and Bradford on Tone* (Som Rec. Soc. 66, 1962) |
| CWAS | *Transactions of the Cumberland and Westmorland Antiquarian Society* (in progress) |
| D | Devonshire |
| Db | Derbyshire |
| D | Domesday Book |
| DbAS | *Journal of the Derbyshire Archaeological and Natural History Society* (in progress) |
| DbCh | I. H. Jeayes, *Descriptive Catalogue of Derbyshire Charters*, London 1906 |
| DBStP | *The Domesday of St. Paul's of the year 1222* (Camden Soc. (OS) 69, 1858) |
| DC | F. M. Stenton, *Documents illustrative of the Social and Economic History of the Danelaw*, London 1920 |
| DenhamPR | *Denham Parish Registers, 1539–1850* (Suffolk Green Books 8, 1904) |
| DEPN | E. Ekwall, *The Concise Oxford Dictionary of English Place-Names*, Oxford 1960 |
| Dickin | E. P. Dickin, *A History of Brightlingsea*, Brightlingsea 1939 |
| DKR | *The Forty-First Annual Report of the Deputy Keeper of the Public Records*, London 1880 |
| DM | D. C. Douglas, *The Domesday Monachorum of Christ Church, Canterbury*, London 1944 |
| Do | Dorset |
| Du | Durham |
| Dublin | J. T. Gilbert, *Historical and Municipal Documents of Ireland*, A.D. *1172–1320* (Rolls Series 53, 1870) |
| Dugd | W. Dugdale, *Monasticon Anglicanum*, London 1817–30 |
| DWills | *Calendar of Wills and Administrations in Devon and Cornwall* (Index Library 35, 1908) |
| e | early |
| EA | *The East Anglian*, 17 vols, 1858–1910 |
| EAS | *Transactions of the Essex Archaeological Society* (in progress) |
| EChCh | *Facsimiles of Early Cheshire Charters* (Lancs and Ches Rec. Soc. 1957) |
| Edmunds | T. W. Hall, *Descriptive Calendar of the Edmunds Collection*, Sheffield 1924 |
| EET | *Early English Text Society* (in progress) |
| Ek | E. Ekwall, 'Some Early London Bynames and Surnames' (*English Studies* 46 (1965), 113–18) |
| ELPN | E. Ekwall, *Early London Personal Names*, Lund 1947 |
| *Ely* | BM. Cotton Claudius C xi |
| *ElyA* | BM. Cotton Tiberius B ii |
| *ElyCouch* | Liber R (Ely Diocesan Registry) |
| EME | Early Middle English |
| EngFeud | F. M. Stenton, *The First Century of English Feudalism, 1066–1166*, Oxford 1932 |
| EngLife | L. F. Salzman, *English Life in the Middle Ages*, Oxford 1950 |
| EPNE | *English Place-Name Elements* (English Place-Name Society 25, 26, 1956) |
| ER | *The Essex Review* (in progress) |
| *ERO* | Unpublished documents in the Essex Record Office, Chelmsford |
| Ess | Essex |
| Ewen | C. L. Ewen, *A History of Surnames in the British Isles*, London 1931 |
| EwenG | C. L. Ewen, *A Guide to the Origin of British Surnames*, London 1938 |
| Ewing | G. Ewing, *A History of Cowden*, Tunbridge Wells 1926 |
| Exon | Exeter Version of DB |
| Eynsham | *Eynsham Cartulary* (Oxford Hist. Soc. 49, 51, 1907, 1908) |
| FA | *Inquisitions and Assessments relating to Feudal Aids*, 6 vols, London 1899–1921 |
| Fanshawe | H. C. Fanshawe, *The History of the Fanshawe Family*, Newcastle upon Tyne 1927 |
| Fees | *Liber Feodorum*, 3 vols, London 1920–31 |
| FeuDu | *Feodarium Prioratus Dunelmensis* (Surtees Soc. 58, 1872) |
| *FF* | Feet of Fines (unpublished) |
| FF | Feet of Fines: (Pipe Roll Soc. 17, 20, 23, 24, 1894–1900); Bucks (Bucks Rec. Soc. 4, 1942); Cambs (Cambs Antiq. Soc. 27, 1913); Essex (Essex Arch. Soc. 4 vols, 1899–1964); Hunts (Cambs Antiq. Soc. 37, 1913); Kent (Archaeologia Cantiana 11–15, 18, 20, 1877–93; Kent Rec. Soc. 15, 1956); Lancs (Lancs and Ches Rec. Soc. 39, 46, 50, 60, 1899–1910); Lincs (Lincs Rec. Soc. 17, 1920; Pipe Roll Soc. (NS) 29, 1954); Norfolk |

(Pipe Roll Soc. (NS) 27, 32, 1952, 1958); Oxford (Oxford Rec. Soc. 12, 1930); Som (Som Rec. Soc. 6, 12, 17, 22, 1892–1906); Staffs (Salt Arch. Soc. (OS) 3, 4, 11, 12, (NS) 10, 1882–1907); Suffolk (Pipe Roll Soc. (NS) 32, 1958; Cambs Antiq. Soc. 1900); Surrey (Surrey Arch. Soc. 1894); Sussex (Sussex Rec. Soc. 2, 7, 23, 1903–16); Warwicks (Dugdale Soc. 11, 15, 18, 1932–43); Wilts (Wilts Arch. Soc. 29, 1974); Yorks (Yorks Arch. Soc. 2, 5, 7, 8, 42, 52, 62, 67, 82, 121, 127, 1887–1965)

| | |
|---|---|
| Finchale | *The Charters . . . of the Priory of Finchale* (Surtees Soc. 6, 1837) |
| Fine | *Calendar of the Fine Rolls*, 22 vols, London 1911–62 |
| FineR | *Rotuli de Oblatibus et Finibus . . . tempore regis Johannis*, London 1835 |
| *For* | Pleas of the Forest (PRO) (unpublished) |
| For | *Select Pleas of the Forest* (Selden Soc. 13, 1901) |
| Forssner | T. Forssner, *Continental-Germanic Personal Names in England*, Uppsala 1916 |
| ForSt | *Staffordshire Forest Pleas* (Salt Arch. Soc. (OS) 5, 1884) |
| France | J. H. Round, *Calendar of Documents preserved in France*, London 1899 |
| FrC | *The Earliest Canterbury Freemen's Rolls 1298–1363* (Kent Rec. Soc. 18, 1964) |
| FrLei | H. Hartopp, *Register of the Freemen of Leicester, 1196–1770*, Leicester 1927 |
| FrNorw | J. L'Estrange, *Calendar of the Freemen of Norwich from 1317 to 1603*, ed. W. Rye, London 1888 |
| FrY | *Register of the Freemen of the City of York* (Surtees Soc. 96, 102, 1897, 1899) |
| FrYar | *A Calendar of the Freemen of Great Yarmouth* (Norfolk and Norwich Arch. Soc. 1910) |
| FS | *Two estate Surveys of the Fitzalan Earls of Arundel* (Sussex Rec. Soc. 67, 1969) |
| Gage | J. Gage, *The History and Antiquities of Suffolk. Thingoe Hundred*, Bury St. Edmunds 1838 |
| Gardner | T. Gardner, *An Historical Account of Dunwich*, London 1754 |
| Gascon | *The Gascon Calendar of 1322* (Camden Soc. 3rd Series 70, 1949) |
| GeldR | The Geld Roll of 1084 (in DB) |
| Gilb | *Transcripts of Charters relating to the Gilbertine Houses* (Lincs Rec. Soc. 18, 1922) |
| GildC | *The Register of the Guild of the Holy Trinity . . . of Coventry* (Dugdale Soc. 13, 19, 1935–44) |
| GildY | *Register of the Guild of the Corpus Christi in the City of York* (Surtees Soc. 57, 1872) |
| Gl | Gloucestershire |
| Glam | Glamorgan |
| Glapwell | *The Glapwell Charters* (Derbyshire Arch. and Nat. Hist. Soc., 1957–9) |
| Glast | *The Great Chartulary of Glastonbury* (Som Rec. Soc. 59, 63, 64, 1947–56) |
| GlCh | *Earldom of Gloucester Charters*, ed. R. B. Patterson, Oxford 1973 |
| Goring | *A Collection of Charters relating to Goring, Streatley, and the Neighbourhood, 1181–1637* (Oxford Rec. Soc. 13, 14, 1931, 1932) |
| GreenwichPR | *Greenwich Parish Registers, 1615–1637* (Trans. Greenwich and Lewisham Antiq. Soc. 2, 1920) |
| Groups | E. Ekwall, *Studies on the Genitive of Groups in English*, Lund 1943 |
| Guisb | *Cartularium prioratus de Gyseburne* (Surtees Soc. 86, 89, 1889, 1894) |
| Ha | Hampshire |
| Hartland | *The Register of Baptisms, Marriages and Burials of the parish of Hartland, Devon, 1558–1837* (Devon and Cornwall Rec. Soc. 1930–34) |
| He | Herefordshire |
| HeCh | *Charters of the Earldom of Hereford 1095–1201* (Camden Soc. 4th Series 1, 1964) |
| HeMil | *Hereford Militia Assessments of 1663* (Camden Soc. Series 4, 10, 1973) |
| Herts | Hertfordshire |
| Holme | *St. Benet of Holme, 1020–1240* (Norfolk Rec. Soc. 2, 3, 1932) |
| HorringerPR | *Horringer Parish Registers* (Suffolk Green Books 4, 1900) |
| Hoskins | W. G. Hoskins and H. P. Finberg, *Devonshire Studies*, London 1952 |
| HPD | H. F. Westlake, *Hornchurch Priory. A Kalendar of Documents in the possession of the Warden and Fellows of New College, Oxford*, London 1923 |
| HT | Hearth Tax Returns: Dorset (C. A. F. Meekings, *Dorset Hearth Tax Assessments 1662–1664*, Dorchester 1951); Oxford (Oxford Rec. Soc. 21, 1940); Suffolk (Suffolk Green Books 11, 1905); Yorks (Thoresby Soc. 2, 1891) |
| *HTEss* | Essex Hearth Tax Returns, 1662, transcribed by F. G. Emmison (unpublished) |
| Hu | Huntingdonshire |
| Husting | R. R. Sharpe, *Calendar of Wills . . . in the Court of Husting, London, A.D. 1258–A.D. 1688*, 2 vols, London 1889, 1890 |
| Hylle | *The Hylle Cartulary* (Somerset Rec. Soc. 68, 1968) |
| ICC | *Inquisitio Comitatus Cantabrigiensis . . . Subjicitur Inquisitio Eliensis*, ed. N.E.S.A. Hamilton, London 1876 |

| | |
|---|---|
| IckworthPR | *Ickworth Parish Registers* (Suffolk Green Books 3, 1894) |
| InqAug | A. Ballard, *An Eleventh-Century Inquisition of St. Augustine's, Canterbury*, London 1920 |
| InqEl | *Inquisitio Eliensis*. v. ICC |
| InqLa | *Lancashire Inquests, Extents, and Feudal Aids* (Lancs and Ches Rec. Soc. 48, 54, 1903, 1907) |
| *Ipm* | *Inquisitiones post mortem* (unpublished) |
| Ipm | *Calendar of Inquisitiones post mortem* (in progress); Glos (Index Library 30, 40, 47, 1903–14); Lancs (Chetham Soc. 95, 1875); Notts (Thoroton Soc. 6, 12, 1939, 1952); Wilts (Index Library 37, 48, 1908, 1914); Yorks (Yorks Arch. Soc. 12, 23, 31, 37, 59, 1892–1918) |
| IPN | A. Mawer and F. M. Stenton, *Introduction to the Survey of English Place-Names*, Cambridge 1923 |
| Ipsw | *Historical Manuscripts Commission, Report IX, Appendix*, pp. 221–61 |
| Ir | Irish |
| Jackson | K. Jackson, *Language and History in Early Britain*, Edinburgh 1953 |
| JMac | J. B. Johnston, *The Scottish Macs*, Paisley 1922 |
| K | Kent |
| KB | *Select Cases in the Court of the King's Bench* (Selden Soc. 55, 57, 58, 74, 76, 82, 88, 1936–71) |
| KCD | J. M. Kemble, *Codex diplomaticus aevi Saxonici*, 6 vols, London 1839–48 |
| KentW | *Testamenta Cantiana* (Kent Arch. Soc. 1906–7) |
| Kirk | *Kirkstall Abbey Rent Roll 1459* (Thoresby Soc. 2, 1891) |
| KirkEllaPR | *Register of Kirk Ella, co. York.* (Parish Register Soc. 11, 1897) |
| Kirkstall | *The Coucher Book of the Cistercian Abbey of Kirkstall* (Thoresby Soc. 8, 1904) |
| KPN | J. K. Wallenberg, *Kentish Place-Names*, Uppsala 1931 |
| Kris | G. Kristensson, 'Studies on Middle English Local Surnames containing Elements of French Origin' (*English Studies 50* (1969), 465–86) |
| l | late |
| L | Lincolnshire |
| La | Lancashire |
| LaCt | *Some Court Rolls . . . of Thomas, Earl of Lancaster* (Lancs and Ches Rec. Soc. 41, 1901) |
| Lacy | *Two 'Compoti' of . . . Henry de Lacy* (Chetham Soc. (OS) 112, 1884) |
| Landwade | W. M. Palmer, *Landwade and the Cotton Family* (Cambs Antiq. Soc. 38, 1939) |
| LaPleas | *Pleadings and Depositions in the Duchy Court of Lancashire* (Lancs and Ches Rec. Soc. 32, 35, 40, 1896–99) |
| Lat | Latin |
| LaWills | *A List of the Lancashire Wills proved within the Archdeaconry of Richmond* (Lancs and Ches Rec. Soc. 10, 1884) |
| Layer | W. M. Palmer, *John Layer (1586–1640) of Shepreth, Cambridgeshire, a seventeenth-century local historian* (Cambs Antiq. Soc. 53, 1935) |
| LedburyPR | *The Registers of Ledbury, co. Hereford* (Parish Register Soc. 1899) |
| Lei | Leicestershire |
| LeiAS | *Transactions of the Leicestershire Archaeological Society* (in progress) |
| LeiBR | M. Bateson, *Records of the Borough of Leicester*, 3 vols, London 1895–1905 |
| Lewes | *Lewes Chartulary* (Sussex Rec. Soc. 38, 40, 1933, 1935; Norfolk Rec. Soc. 12, 1939; J. H. Bullock and W. M. Palmer, *The Cambridgeshire Portion of the Chartulary of St. Pancras of Lewes*, Cambridge 1938) |
| LewishamPR | *The Register of . . . St. Mary, Lewisham, 1558–1750* (Lewisham Antiq. Soc. 1891) |
| LibEl | *Liber Eliensis* (Camden Soc. 3rd Series 92, 1962) |
| Lichfield | *Catalogue of the Muniments . . . of the Dean and Chapter of Lichfield* (Salt Arch. Soc. (OS) 6, 1886) |
| LitSaxhamPR | *Little Saxham Parish Registers* (Suffolk Green Books 5, 1901) |
| LLB | *Calendar of Letter Books . . . of the City of London*, 11 vols, London 1899–1912 |
| Lo | London |
| LoCh | *The Church in London 1375–92* (London Rec. Soc. 13, 1977) |
| LoCt | *Calendar of Early Mayor's Court Rolls*, ed. A. H. Thomas, Cambridge 1924 |
| LondEng | R. W. Chambers and M. Daunt, *A Book of London English*, Oxford 1931 |
| LoPleas | *Calendar of Plea and Memoranda Rolls preserved among the Archives of the . . . City of London*, ed. A. H. Thomas, 3 vols, Cambridge 1926–32 |
| Loth | J. Loth, *Chrestomathie bretonne*, Paris 1890 |
| LP | *Letters and Papers, Foreign and Domestic, of the reign of Henry VIII*, 23 vols, London 1862–1932 |
| LuffCh | *Luffield Priory Charters* (Northants Rec. Soc. 22, 26, 1968, 1975) |

| | |
|---|---|
| LVD | *Liber Vitæ Ecclesiæ Dunelmensis* (Surtees Soc. 136, 1923) |
| LWills | *Calendars of Lincoln Wills* (Index Library 28, 1902) |
| Malmesbury | *Registrum Malmesburiense* (Rolls Series 72, 1879–80) |
| MCh | *Charters of the Honour of Mowbray*, ed. D. E. Greenway, London 1972 |
| ME | Middle English |
| MED | *Middle English Dictionary*, ed. H. Kurath, S. M. Kuhn, and J. Reidy, Ann Arbor 1954– (in progress) |
| MedEA | D. C. Douglas, *The Social Structure of Mediaeval East Anglia*, Oxford 1927 |
| MedInd | L. F. Salzman, *English Industries in the Middle Ages*, Oxford 1923 |
| MELS | M. T. Löfvenberg, *Studies on Middle English Local Surnames*, Lund 1942 |
| MemR | *Memoranda Roll* (Pipe Roll Soc. (NS) 11, 21, 1933, 1943) |
| MEOT | B. Thuresson, *Middle English Occupational Terms*, Lund 1950 |
| Merton | *Merton Muniments* (Oxford Hist. Soc. 86, 1928) |
| MESO | G. Fransson, *Middle English Surnames of Occupation, 1100–1350*, Lund 1935 |
| METrade | L. F. Salzman, *English Trade in the Middle Ages*, Oxford 1931 |
| Miller | E. Miller, *The Abbey and Bishopric of Ely*, Cambridge 1951 |
| MinAcctCo | *Ministers' Accounts of the Earldom of Cornwall* (Camden Soc. 3rd Series 66, 68, 1942, 1945) |
| Misc | *Calendar of Inquisitions Miscellaneous* (in progress) |
| MLWL | *Revised Medieval Latin Word List from British and Irish Sources*, ed. R. E. Latham, London 1965 |
| Mon | Monmouthshire |
| Moore | A. W. Moore, *Manx Names*, London 1903 |
| Morris | T. E. Morris, 'Welsh Surnames in the Border Counties of Wales' (*Y Cymmrodor* 43 (1932), 93–173) |
| Moulton | H. R. Moulton, *Palæography, Genealogy and Topography*, London 1930 |
| MPleas | *Select Pleas in Manorial and other Seignorial Courts* (Selden Soc. 2, 1889) |
| MScots | Middle Scottish |
| Musters | Muster Rolls: Norfolk (Norfolk Rec. Soc. 6, 7, 1935, 1936); Surrey (Surrey Rec. Soc. 3, 1914–19) |
| MW | Middle Welsh |
| Mx | Middlesex |
| Nb | Northumberland |
| NED | *A New English Dictionary*, Oxford 1888–1933 |
| Newark | *Documents relating to the Manor and Soke of Newark-on-Trent* (Thoroton Soc. 16, 1955) |
| Nf | Norfolk |
| NI | *Nonarum Inquisitiones*, London 1807 |
| NIWo | *Nonarum Inquisitiones 1340 for the County of Worcester* (Worcs Hist. Soc. 1899) |
| NLCh | *Newington Longeville Charters* (Oxford Rec. Soc. 3, 1921) |
| NME | Northern Middle English |
| NoB | *Namn och Bygd* (in progress) |
| NorwDeeds | *A Short Calendar of the Deeds relating to Norwich 1285–1306; A Calendar of Norwich Deeds . . . 1307–1341* (Norfolk and Norwich Arch. Soc. 1903, 1915) |
| NorwDep | *Depositions taken before the Mayor and Aldermen of Norwich, 1549–67* (Norfolk and Norwich Arch. Soc. 1905) |
| NorwLt | *Leet Jurisdiction in the City of Norwich during the XIIIth and XIVth Centuries* (Selden Soc. 5, 1892) |
| NorwW | *Index to Wills proved in the Consistory Court of Norwich* (Index Library 69, 73, 1945, 1950) |
| NottBR | *Records of the Borough of Nottingham, vol. I*, Nottingham 1882 |
| NQ | *Notes and Queries* (in progress) |
| NS | New Series |
| Nt | Nottinghamshire |
| Nth | Northamptonshire |
| NthCh | *Facsimiles of Early Charters from Northamptonshire Collections* (Northants Rec. Soc. 4, 1930) |
| O | Oxfordshire |
| OBret | Old Breton |
| OCo | Old Cornish |
| ODa | Old Danish |
| ODCN | E. G. Withycombe, *The Oxford Dictionary of English Christian Names*, Oxford 1950 |
| OE | Old English |
| OEByn | G. Tengvik, *Old English Bynames*, Uppsala 1938 |

| | |
|---|---|
| OFr | Old French |
| OFris | Old Frisian |
| OG | Old German |
| OIr | Old Irish |
| ON | Old Norse |
| ONFr | Old Norman French |
| Oriel | *Oriel College Records* (Oxford Hist. Soc. 85, 1926) |
| Orig | *Rotulorum originalium . . . abbreviatio*, 2 vols, London 1805, 1810 |
| OS | Original Series |
| Oseney | *Cartulary of Oseney Abbey* (Oxford Hist. Soc. 89–91, 97, 98, 101, 1929–35) |
| OSw | Old Swedish |
| OW | Old Welsh |
| Oxon | *Register of the University of Oxford* (Oxford Hist. Soc. 1, 10–12, 14, 1885–9) |
| p | *post* |
| P | Pipe Rolls: Record Commission, 3 vols, London 1833–44; Pipe Roll Soc. (in progress); *The Great Roll of the Pipe for the twenty-sixth year of Henry the Third*, ed. H. L. Cannon, Yale Hist. Pub. 1918 |
| ParlR | *Rotuli Parliamentorum*, 7 vols, London 1767–1833 |
| ParlWrits | *The Parliamentary Writs*, 2 vols, London 1827, 1834 |
| Paston | *Paston Letters and Papers of the Fifteenth Century*, 2 vols, ed. N. Davis, Oxford 1971, 1976 |
| Pat | *Calendar of Patent Rolls* (in progress) |
| PatR | *Rotuli Litterarum Patentium*, London 1835 |
| PCC | *Index of Wills proved in the Prerogative Court of Canterbury* (Index Library, 12 vols, 1893–1960) |
| Percy | *The Percy Chartulary* (Surtees Soc. 117, 1911) |
| *Petre* | Petre Documents in *ERO*. Calendared by C. J. Kuypers |
| Pinchbeck | *The Pinchbeck Register*, 2 vols, ed. Lord Francis Hervey, Oxford 1925 |
| Pleas | *Pleas before the King or his Justices 1198–1202* (Selden Soc. 67, 68, 1952, 1953); *1198–1212* (Selden Soc. 83, 84, 1966, 1967) |
| PN | *Place-Names of* (e.g. PN Bk, *Place-Names of Buckinghamshire*, &c. English Place-Name Society) |
| PNDB | O. von Feilitzen, *The Preconquest Personal Names of Domesday Book*, Uppsala 1937 |
| PN Do | A. Fägersten, *The Place-Names of Dorset*, Uppsala 1933. When no volume number is given, the reference is to this work, otherwise to the EPNS text |
| PN K | J. K. Wallenberg, *The Place-Names of Kent*, Uppsala 1934 |
| PN La | E. Ekwall, *The Place-Names of Lancashire*, Manchester 1922 |
| PN NbDu | A. Mawer, *The Place-Names of Northumberland and Durham*, Cambridge 1920 |
| PN Wt | H. Kökeritz, *The Place-Names of the Isle of Wight*, Uppsala 1940 |
| PR | Parish Register(s) (of the specified place) |
| Praes | *Praestita Roll 14–18 John* (Pipe Roll Soc. (NS) 37, 1964) |
| PrD | Protestation Returns for Devon (from a transcription by A. J. Howard) |
| PrGR | *The Rolls of Burgesses at the Guilds Merchant of the Borough of Preston* (Lancs and Ches Rec. Soc. 9, 1884) |
| PromptParv | *Promptorium Parvulorum* (Camden Soc. (OS) 25, 54, 89, 1843–65) |
| PrSo | *A Somerset Petition of 1641*, ed. A. J. Howard, 1968 |
| PTW | Clerical Poll-Taxes in the Diocese of Salisbury, 1377–81 (Wilts Arch. Soc. 12, 1956) |
| PTY | Yorkshire Poll Tax Returns: Yorks Arch. Journal 5–7, 9, 20, 1879–1909; Trans. East Riding Antiq. Soc. 30) |
| Putnam | B. H. Putnam, *The Enforcement of the Statutes of Labourers, 1349–1359*, New York 1908 |
| PWi | N. R. Holt, *The Pipe Roll of the Bishopric of Winchester, 1210–11*, Manchester 1964 |
| QW | *Placita de Quo Warranto*, London 1818 |
| R | Rutland |
| Rad | *The Priory of Saint Radegund, Cambridge* (Cambs Antiq. Soc. 31, 1898) |
| RamptonPR | *The Parish Registers of Rampton, Cambridgeshire, A.D. 1599–1812*, n.d. |
| Rams | *Cartularium monasterii de Rameseia* (Rolls Series 79, 3 vols, 1884–94) |
| RamsCt | W. O. Ault, *Court Rolls of the Abbey of Ramsey and of the Honor of Clare*, Yale 1928 |
| RBE | *The Red Book of the Exchequer* (Rolls Series 99, 3 vols, 1896) |
| RBWo | *The Red Book of Worcester* (Worcs Hist. Soc. 1934–50) |
| Redin | M. Redin, *Studies on Uncompounded Personal Names in Old English*, Uppsala 1919 |
| RegAntiquiss | *Registrum Antiquissimum* (Lincs Rec. Soc. 10 vols, 1931–68) |
| *Req* | Court of Requests (PRO) |
| RH | *Rotuli Hundredorum*, 2 vols, London 1812, 1818 |

| | |
|---|---|
| Riev | *Cartularium Abbathiæ de Rievalle* (Surtees Soc. 83, 1889) |
| RochW | *Index of Wills proved in the Rochester Consistory Court* (Kent Rec. Soc. 9, 1924) |
| Ronton | *The Chartulary of Ronton Priory* (Salt Arch. Soc. (OS) 4, 1883) |
| RotDom | *Rotuli de Dominabus* (Pipe Roll Soc. 35, 1913) |
| RothwellPR | *The Registers of the Parish Church of Rothwell* (Yorks Parish Register Soc. 27, 34, 51, 1906–14) |
| RushbrookPR | *Rushbrook Parish Registers* (Suffolk Green Books 6, 1903) |
| Rydeware | *The Rydeware Chartulary* (Salt Arch. Soc. (OS) 16, 1895) |
| Sa | Shropshire |
| SaAS | *Transactions of the Shropshire Archaeological and Natural History Society* (in progress) |
| SaG | *The Merchants' Guild of Shrewsbury. The Two Earliest Rolls* (Trans. Shrops Arch. and Nat. Hist. Soc. 2nd Series 8, 1896) |
| SagaBk | *Saga-Book of the Viking Society* (in progress) |
| SaltAS | *Transactions of the William Salt Archaeological Society* (in progress) |
| Seals | L. C. Loyd and D. M. Stenton, *Sir Christopher Hatton's Book of Seals*, Oxford 1950 |
| Selden Soc. | Publications of the Selden Society |
| Selt | B. Seltén, 'Some Notes on Middle English By-names in Independent Use' (*English Studies* 46 (1965), 165–81) |
| Sf | Suffolk |
| SfPR | *Suffolk Parish Registers. Marriages*, 3 vols, ed. W. P. W. Phillimore, London 1910–16 |
| Shef | T. W. Hall and A. H. Thomas, *Descriptive Catalogue of the Charters . . . forming the Jackson Collection at the Sheffield Public Reference Library*, Sheffield 1914 |
| ShefA | T. W. Hall, *Sheffield, Hallamshire: A Descriptive Catalogue of Sheffield Manorial Records*, 3 vols, Sheffield 1926–34 |
| ShotleyPR | *Shotley Parish Registers, 1571–1850* (Suffolk Green Books 1911, 1912) |
| SIA | *Proceedings of the Suffolk Institute of Archaeology* (in progress) |
| So | Somerset |
| Sol | *Liber Henrici de Soliaco, Abbatis Glaston*, ed. J. H. Jackson, Roxburghe Club 1882 |
| SPD | *Calendar of State Papers Domestic*, 48 vols., London 1856–86 |
| SPleas | *Select Pleas of the Crown* (Selden Soc. 1, 1888; *Select Civil Pleas*, Selden Soc. 3, 1890) |
| Sr | Surrey |
| *SR* | Subsidy Rolls (unpublished) |
| SR | Subsidy Rolls: Beds (Beds Hist. Rec. Soc. 2, 1914; Suffolk Green Books 18, 1925); Cambs (C. H. Evelyn White, *Lay Subsidy for the Year 1327*, n.d.); Cumb (J. P. Steel, *Cumberland Lay Subsidy . . . 6th Edward III*, Kendal 1912); Derbyshire (Journal of the Derbyshire Arch. and Nat. Hist. Soc. 24, 30, 44, 1902–22); Devon (A. J. Howard, *1524 Lay Subsidy Roll County of Devon*, 1967–8); Dorset (Dorest Rec. Soc. 4, 1971); Kent (Arch. Cant. 12); Lancs (Lancs and Ches Rec. Soc. 27, 31, 1893, 1896); Leics (Associated Architectural Societies: Reports and Papers, 19.2, 20.1, 1888, 1889); London (G. Unwin, *Finance and Trade under Edward III*, Manchester 1918; E. Ekwall, *Two Early London Subsidy Rolls*, Lund 1951); Northumb (Archaeologia Aeliana 3rd Series 13, 1916); Shropshire (Trans. Shropshire Arch. and Nat. Hist. Soc. 2nd Series 1, 4, 5, 8, 10, 11, 3rd Series 5–7); Somerset (Som Rec. Soc. 3, 1889); Staffs (Salt Arch. Soc. (OS) 7, 10, 1886, 1889); Suffolk (E. Powell, *The Rising in East Anglia in 1381*, Cambridge 1896; *A Suffolk Hundred in 1283*, Cambridge 1896; Suffolk Green Books 9, 10, 12, 1906–10); Surrey (Surrey Rec. Soc. 18, 33, 1922, 1932); Sussex (Sussex Rec. Soc. 10, 56, 1910, 1956); Warwicks (Dugdale Soc. 6, 1926); Wilts (Wilts Arch. and Nat. Hist. Soc. 10, 1954); Worcs (Worcs Hist. Soc., 4 vols., 1893–1900); Yorks (Thoresby Soc. 2, 1891; Yorks Arch. Soc. 16, 21, 74, 1894, 1897, 1929) |
| St | Staffordshire |
| StaplehurstPR | *The Old Register of Staplehurst, 1538–1558*, Canterbury 1907 |
| StarChSt | *Staffordshire Suits in the Court of Star Chamber* (Salt Arch. Soc. (NS) 10, 1907) |
| StCh | *The Staffordshire Chartulary* (Salt Arch. Soc. (OS) 2, 3, 1881, 1882) |
| StGreg | *Cartulary of the Priory of St. Gregory, Canterbury* (Camden Soc. 3rd Series 88, 1956) |
| *StJohn* | BM. Cotton Nero E vi |
| *Stoke* | BM. Cotton Appendix xxi |
| Stone | *The Stone Chartulary* (Salt Arch. Soc. (OS) 6, 1885) |
| StP | *Early Charters of the Cathedral Church of St. Paul, London* (Camden Soc. 3rd Series 58, 1939) |
| Str | *Accounts and Surveys of the Wiltshire Lands of Adam de Stratton* (Wilts Arch. and Nat. Hist. Soc. 14, 1959) |
| StratfordPR | *Registers of Stratford-on-Avon, co. Warwick. Marriages 1558–1812* (Parish Register Soc. 16, 1898) |

| | |
|---|---|
| StThomas | *A Chartulary of the Priory of St. Thomas, the Martyr, near Stafford* (Salt Arch. Soc. (OS) 8, 1887) |
| StudNP | *Studia Neophilologica* (in progress) |
| Suckling | A. I. Suckling, *The History and Antiquities of the County of Suffolk*, London 1846–8 |
| Surnames | E. Weekley, *Surnames*, London 1936 |
| Sx | Sussex |
| SxAnt | *Sussex Archaeological Collections* (in progress) |
| SxWills | *Sussex Wills* (Sussex Rec. Soc. 41, 43, 45, 1935–41) |
| Templars | B. A. Lees, *Records of the Templars in England in the Twelfth Century*, London 1935 |
| TestEbor | *Testamenta Eboracensia* (Surtees Soc. 4, 30, 45, 53, 1836–68) |
| Trentham | *A Chartulary of the Augustine Priory of Trentham* (Salt Arch. Soc. (OS) 11, 1890) |
| Variation | E. Ekwall, *Variation in Surnames in Medieval London*, Lund 1945 |
| W | Wiltshire |
| Wa | Warwickshire |
| Wak | *Court Rolls of the Manor of Wakefield* (Yorks Arch. Soc. 29, 36, 57, 78, 1901–45) |
| WBCinque | F. Hull, *A Calendar of the White and Black Books of the Cinque Ports, 1432–1955*, London 1966 |
| We | Westmorland |
| Wenlok | *Documents illustrating the rule of Walter de Wenlok, Abbot of Westminster, 1283–1307* (Camden Soc. 4th Series 2, 1965) |
| WhC | *The Coucher Book or Chartulary of Whalley Abbey* (Chetham Soc. 10, 11, 16, 20, 1847–9) |
| Wheat | T. W. Hall, *Descriptive Catalogue of charters . . . of the Wheat Collection at the Public Reference Library*, Sheffield 1920 |
| Whitby | *Cartularium Abbathiæ de Whiteby* (Surtees Soc. 69, 72, 1879, 1881) |
| Winton | 'The Winton Domesday', ed. F. Barlow (in *Winchester Studies I*, ed. M. Biddle, Oxford 1976) |
| WiSur | 'Thirteenth-Century Surveys of Winchester', ed. D. J. Keene (in *Winchester Studies I*, ed. M. Biddle, Oxford 1976) |
| Wo | Worcestershire |
| WoCh | *The Cartulary of Worcester Cathedral Priory* (Pipe Roll Soc. (NS) 38, 1968) |
| WordwellPR | *Wordwell Parish Registers, 1580–1850* (Suffolk Green Books 7, 1903) |
| Works | C. T. Flower, *Public Works in Mediaeval Law* (Selden Soc. 32, 40, 1915–23) |
| WRS | *English Surnames Series I: Yorkshire West Riding*, by G. Redmonds, London 1973 |
| WStowPR | *West Stow Parish Registers 1558–1850* (Suffolk Green Books 7, 1903) |
| W'stowW | G. S. Fry, *Abstracts of Wills relating to Walthamstow* (Walthamstow Antiq. Soc. 1921) |
| Wt | Isle of Wight |
| Y | Yorkshire |
| YAJ | *Yorkshire Archaeological Journal* (in progress) |
| YCh | W. Farrer and C. T. Clay, *Early Yorkshire Charters*, 12 vols., 1914–65 |
| YDeeds | *Yorkshire Deeds* (Yorks Arch. Soc. Record Series 39, 50, 1909, 1915) |
| YWills | *Yorkshire Wills* (Yorks Arch. Soc. 73, 93, 1928, 1937) |
| * | a postulated form |

## Other Works Consulted

| | |
|---|---|
| Bain, R. | *The Clans and Tartans of Scotland*, London 1953 |
| Bardsley, C. W. | *A Dictionary of English and Welsh Surnames*, London 1901 |
| Björkman, E. | *Nordische Personnamen in England*, Halle 1910 |
| | *Zur englischen Namenkunde*, Halle 1912 |
| Boehler, M. | *Die altenglischen Frauennamen*, Berlin 1930 |
| Carnoy, A. | *Origine des noms de familles en Belgique*, Louvain 1953 |
| Cottle, A. B. | *The Penguin Dictionary of Surnames*, Harmondsworth 1967 |
| Dauzat, A. | *Dictionnaire étymologique des noms de familles et prénoms de France*, Paris 1951 |
| | *Les noms de famille de France*, Paris 1945 |
| | *Les noms de personnes*, Paris 1946 |
| Davies, T. R. | *A Book of Welsh Names*, London 1952 |
| Ekwall, E. | *Scandinavians and Celts*, Lund 1918 |
| | *Street-Names of the City of London*, Oxford 1954 |
| Fabricius, A. | *Danske Minder i Normandiet*, Copenhagen 1897 |
| Feilitzen, O. von. | 'Notes on Old English Bynames' (*Namn och Bygd* 27 (1939) 116–30) |

|  | 'Some Unrecorded Old and Middle English Personal Names' (*Namn och Bygd* 33 (1945), 69–98) |
|  | 'Some Continental Germanic Personal Names in England' (in *Early English and Norse Studies*, ed. A. Brown and P. Foote, 46–61, London 1963) |
|  | 'Notes on some Scandinavian Personal Names in English 12th-Century Records' (*Personnamns Studier: Anthroponymica Suecana* 6 (1964), 62–8) |
|  | 'Some Old English Uncompounded Personal Names and Bynames' (*Studia Neophilologica* 40 (1968), 5–16) |
| Forssner, T. | *Beiträge zum Studium der neuenglischen Familiennamen*, Göttingen 1920 |
|  | *De l'influence française sur les noms propres anglais*, Skövde 1920 |
|  | *Deutsche und englische Imperativnamen*, Ostersund 1920 |
|  | *Imitative Alterations in Modern English Personal Nomenclature*, Skövde 1920 |
| Förstemann, E. | *Altdeutsches Namenbuch*, Bonn 1900–16 |
| Förster, M. | 'Keltisches Wortgut im Englischen' (in *Festgabe für E. Liebermann*, Halle 1921) |
| Gautries, J. A. des | *Les Noms de Personnes Scandinaves en Normandie de 911 à 1066*, Lund 1954 |
| Godefroy, F. | *Dictionnaire de l'ancienne langue française*, Paris 1881–1902 |
| Guppy, H. B. | *Homes of Family Names in Great Britain*, London 1890 |
| Harrison, H. | *Surnames of the United Kingdom*, London 1912–18 |
| Lebel, P. | *Les Noms de personnes en France*, Paris 1946 |
| Loth, J. | *Les Noms de saints bretons*, Paris 1910 |
| Lower, M. A. | *Patronymica Britannica*, London 1860 |
| McKinley, R. A. | *Norfolk and Suffolk Surnames in the Middle Ages*, London 1975 |
|  | *The Surnames of Oxfordshire*, London 1977 |
| MacLysaght, E. | *The Surnames of Ireland*, Dublin 1969 |
| Matthews, C. M. | *English Surnames*, London 1966 |
| Michaëlsson, K. | *Études sur les noms de personne français d'après les roles de taille parisiens*, Uppsala 1927–36 |
| Moisy, H. | *Noms de famille normands*, Paris 1875 |
| Nicholson, E. W. B. | *The pedigree of 'Jack' and of various allied names*, London 1892 |
| Reaney, P. H. | 'Notes on Christian Names' (*Notes and Queries* 196 (1951), 199–200) |
|  | 'Onomasticon Essexiense' (*Essex Review* 61 (1952), 133–42, 202–15) |
|  | 'Pedigrees of Villeins and Freemen' (*Notes and Queries* 197 (1952), 222–5) |
|  | 'Three Unrecorded Old English Personal Names of a late Type' (*Modern Language Review* 47 (1952), 374) |
|  | 'The Survival of Old English Personal Names in Middle English' (*Studier i modern språkvetenskap* 18 (1953), 84–112) |
|  | *The Origins of English Surnames*, London 1967 |
| Ritchie, R. L. G. | *The Normans in Scotland*, Edinburgh 1954 |
| Searle, W. G. | *Onomasticon Anglo-Saxonicum*, Cambridge 1897 |
| Seltén, B. | *The Anglo-Saxon Heritage in Middle English Personal Names: East Anglia 1100–1399*, Lund 1972 |
| Smiles, S. | *The Huguenots*, London 1880 |
| Smith, A. H. | 'Some Aspects of Irish Influence on Yorkshire' (*Revue Celtique* 44 (1927), 34–58) |
|  | 'Danes and Norwegians in Yorkshire' (*Saga-Book* 10 (1929) 188–215) |
|  | 'Early Northern Nicknames and Surnames' (*Saga-Book* 11 (1934), 30–60) |
| Smith, E. C. | *American Surnames*, New York 1969 |
| Ström, H. | *Old English Personal Names in Bede's History*, Lund 1939 |
| Sundby, B. | 'Some Middle English Occupational Terms' (*English Studies* 33 (1952), 18–19) |
| Wagner, A. R. | *English Genealogy*, Oxford 1960 |
| Weekley, E. | *Jack and Jill*, London 1948 |
|  | *Romance of Names*, London 1922 |
| Whitelock, D. | 'Scandinavian Personal Names in the Liber Vitæ of Thorney Abbey' (*Saga-Book* 12 (1940), 127–53) |
| Woulfe, P. | *Sloinnte Gaedheal is Gall*, Dublin 1923 |

# A

**Aaron, Aarons:** *Aaron* Iudeus 1189 DC (L); Robert *Aaron* 1185 Eynsham; John *Aaron* 1259 ForNth, 1327 SRSa. The name of the brother of Moses. Rare in medieval England, and usually Jewish.

**Abadam, Badams, Baddams, Badham:** Hoel *ab Adam* 1255 RH (Sa); John *Apadam, Ab Adham* 1308, 1310 ParlWrits (Gl); Mary *Baddam* 1597 Bardsley. 'Son of *Adam*', Welsh *ab, ap*. cf. BOWEN.

**A'Barrow:** Alured *atte Berewe* 1242 Fees (Wo); Adam *a la Barewe* 1275 SRWo; William *Abarowe* 1525 SRSx; Rychard *A Barow* 1545 SRW. 'Dweller by the mound or hill', OE *bearg. v.* also BARROW.

**Abb, Abba, Abbe, Abbs, Labbe:** (i) Ralph *le Abe* c1150–66 YCh; William *le Abbe* 1220 Cur (D); Walter *le Abbe* 1297 MinAcctCo. OFr *abe, abet* 'abbot'. (ii) Ralph *Abbe* 1200 FFO; Walter *Abbe* 1249 AssW; John *Abbe* 1327, 1332 SRSx. Either shortened forms of *Abel, Abraham*, or perhaps further examples of (i).

**Abba, Abbay, Abbey, Abbie, Abby, Labey:** (i) Ralph *le Abbe* 1177 P (Lo); Geoffrey *Labbe* Hy 2 DC (Lei); John *Abby* 1297 MinAcctCo. OFr *abe, abet* 'abbot'. *v.* ABBATT. The Scottish *Abbie* (*Abbe* 1178–80, *del Abbeye* 1370) derives from the office of lay-abbot of a monastery which was hereditary in the leading family of the district (Black). (ii) William *del Abbay* 1283 FrY; Adam *dil Abbeye* 1327 SRSf. From employment at an abbey.

**Abbatt, Abbett, Abbitt, Abbot, Abbots, Abbott, Abbotts, Labbet:** Walter *Abbot* 12th DC (L); Walter *Abat* 1219 AssY; Peter *le Abbot* 1237 HPD (Ess); Ralph *Abbod* 1272 AssSo. OE *abbod*, late OE *abbat, abbot* 'abbot'. Early Latin examples such as Alfwoldus *abbas* 1111–17 Holme (Nf) are names of actual holders of the office of abbot and can hardly have given rise to surnames. Later examples are nicknames.

**Abbay, Abbeys, Abbie, Abby:** John *de Abbatia* 1190 P (Lo); William *del Abbay* 1283 FrY; Adam *dil Abbeye* 1327 SRSf; Roger *Abby* 1364 LoPleas. 'Worker at the abbey', OFr *abbaye*.

**Abbe:** *v.* ABB

**Abbess, Abbis, Abbiss:** Jamys *Abbys*, Richard *Abes* 1524 SRSf. Probably OFr *abe, abet* 'abbot', of which these would be possessive forms, hence 'son of the abbot'. Or, possibly, OFr *abbesse* 'abbess' used ironically. cf. Abbess Hall in Abbess Roding (Ess), *Abbes Hall* 1544.

**Abbay, Abbeys, Abbie:** *v.* ABBAY
**Abbey, Abbie:** *v.* ABBA
**Abbis, Abbiss:** *v.* ABBESS

**Abbs:** *v.* ABB
**Abby:** *v.* ABBAY

**Abdey, Abdie:** Robert *del Abdy* 1379 PTY. 'One employed at an abbey', ME *abbodie*, Lat *abbatia* which meant both 'abbacy' and 'abbey'.

**Abear, A'Bear:** John *atte Bere* 1332 SRSr; John *Abere* 1517 PCC (K). 'At the swine-pasture', probably 'swineherd'. *v.* BEAR.

**Abel, Abell, Abele, Abeles, Able, Abelson, Ableson:** *Abellus* Hy 2 DC (Lei); *Abel* de Etton' 1221 AssWa; William *Abel* 1197 P (Ess); Thomas *Abelle* 1301 SRY; Richard *Abelson* 1623 Bardsley. Hebrew *Abel*, probably 'son', a common 13th-century christian name.

**Abercrombie, Abercromby:** William, John *de Abercromby* 1296, 1305 Black (Fife). From the barony of Abercromby (Fife).

**Abery, Abra, Abrey, Abry:** John, Stephen *Albo(u)rgh* 1327 SREss, 1456 ER 72; John *Aberry* 1662 HTEss. These are probably from Abery House in Ilford (Ess), *v.* PN Ess 98. But the name may also come from Avebury (W), *Abery* 1535.

**Abethell:** *v.* BITHELL

**Abinger:** Gilbert de *Abingeworth'* 1208 Cur (Sr); James *de Abingeworth* 1327 SRSx. From Abinger (Sr), *Abingewurd* 1191.

**Abington:** Fulcho *de Abinton* 1194 P (Nth); William *de Abbinton* 1260 AssC; Thomas *de Abigton*, William *de Abynton* 1296 SRSx. From Abington (C, Nth), or Abington Pigotts (C).

**Ablett, Ablitt, Ablott:** *Abelota* loue 1277 *Ely* (Sf); *Abelot* 1279 RH (C); William *Abelot* 1279 RH (C); William *Ablot* 1335 FrY. *Abel-ot*, a diminutive of *Abel*, sometimes used as a woman's name.

**Ablewhite:** *v.* APPLEWHAITE

**Abney:** Roger *de Abbeneye* Ed1 DbCh. From Abney (Db).

**Aborn:** Robert *Abourne* 1379 LLB G; Jamys *A Bourne* 1467 ArchC xi. For *atte borne. v.* BOURNE.

**Abra, Abrey, Abry:** *v.* ABERY

**Abraham, Abrahams, Abram, Abrams, Abrahamson, Abramson:** *Abraham* de Strattuna 1170–5 DC (L); John *Abraham* 1193 P (Nth); Robert *Abram* 1252 Rams (Hu). Hebrew *Abram* 'high father', changed to *Abraham* 'father of a multitude'. *Abraham*, the name of a priest in DB (1086), was not confined to Jews.

**Abraham, Abram, Adburgham:** John *de Abburgham* 1246 AssLa; Gilbert *de Abram* Hy 4 Bardsley. From Abram (Lancs), *Adburgham* a1199 DEPN.

**A'Brook:** Roger *Attebrok* 1297 MinAcctCo; John *A'Brooke* 1542 SxWills. v. BROOK.

**Absalom, Absolem, Absolom, Absolon, Aspenlon, Aspland, Asplen, Asplin, Aspling, Ashplant:** *Absolon* filius *Apsolon* 1199 FFC; *Aspelon* 1252 Rams (Hu); Johannes *filius Asplom* 1302 SRY; Stephen *Abselon* 1208 Cur (O); John *Asplon'* 1279 RH (Hu); Thomas *Absolon, Aspelon* 1281, 1297 LLB B; William *Aspland, Aspline* 1684, 1690 CPR iii. *Absalom*, from Hebrew *Abshálôm* 'father of peace'. Used as a nickname by Chaucer for a man with a fine head of hair: 'Crul was his heer and as the gold it shoon'.

**Abson:** William *Abson* 1379 PTY, 1595 Shef. Probably 'son of *Ab*', a shortened form of names in *Ab-*, but occasionally, perhaps, from Abson (Gl).

**Aburn:** John *Aburne* 1572 FFHu. v. BOURNE.

**Acaster, Akaster, Akester, Akister:** Ragenild *de Acastr'* 1219 AssY; John *de Acastre* 1344 FFY; William *Akester* 1672 HTY. From Acaster (WRY).

**Ace, Aze:** *Azo* 1086 DB; *Asce* Halvecniht 1213 Cur (Ha); Benedict, John *Ace* 1230 Cl (Ha), 1246–89 Bart (Lo); Geoffrey *Aze* 1296 SRSx. OFr *Ace, Asse*, OG *Azo, Atso*, hypocoristics of compounds in *Adal-*.

**Achard, Ackert, Ashard, Hatchard:** *Acardus* de Lincolnia c1150 DC (L); *Achardus* de Sproxton' 1207 Cur (Lei); William *Achard* 1160 P (Berks); William *Achart* 1177 P (Ha); Willelmus *Achardi* 1190 P (Berks); William *Hachard* 1279 RH (C); Thomas *Acard* 1401 Shef (Y). OFr *Achart, Aquart*, probably from OG *Agihard, Akihart*.

**Acheson, Aicheson, Aitcheson, Aitchison, Atcheson, Atchison:** Scottish and border forms of *Atkinson* found in Cumberland as *Attchison* in 1596 (CWAS viii) and in Scotland as *Atzinson* 1475, *Achesoun* 1497, *Atyesoun* 1540, and *Aicheson* 1590 (Black). The *z* was pronounced *y* and *Atyeson* became *Acheson* as the colloquial 'got yer' became 'gotcha'.

**Ackary, Ackery:** *Acharias, Accarisius* filius Copsi 1155, Hy 1 FeuDu; *Achari* Hy 1 Rams (Hu); Robert *Akari* 1199 CurR (Hu). Hebrew *Zacharias* 'Jehovah has remembered', surviving also as *Zachary*. Roger son of *Zacharias* de Burdun (1217 FeuDu) is also called Rogerus *filius Acharisiae, Akariae, Acris, Akaris* (c1200 ib.).

**Acke, Ackes:** *Aky* prior 1168–75 Holme; Bernard *filius Acke* 1212 Fees (La); Eluiua *Ackes* 1202 AssNth; Margaret *Acke* 1327 SRC. ODa *Aki*, OSw *Ake*, or a shortened form of ON *Áskell*. v. also AKE.

**Ackerman, Acraman, Acreman, Akerman:** William *Acreman* Hy 1 Rams (Hu); Robert *le Akerman* 1233 HPD (Ess). OE *æcermann* 'farmer', a husbandman or ploughman.

**Ackers:** v. ACRES

**Ackert:** v. ACHARD

**Ackery:** v. ACKARY

**Acket, Acketts:** v. HACKETT

**Acklam:** William *de Acclum* 1185 Templars (Y); Robert *de Aclom* 1334 LLB E; Robert *Aklam* 1534 FrY. From Acklam (ERY, NRY).

**Ackland, Acland:** The Devonshire Acklands are said to owe their surname to a grove of oaks near their seat of Acland Barton in Landkey (Devon). Here lived in 1275 William *de Ackelane* (RH). The real meaning of Acland is 'Acca's lane', which is only some four miles from Accott in Swimbridge, 'Acca's cottage' (PN D 342, 351). Robert *de Acland* 1257 Oseney (O) took his surname from land on which oaks grew (OE *ãc, land*).

**Ackling:** v. HACKLING

**Ackroyd, Acroyd, Akeroyd, Akroyd, Aykroyd, Ackred, Akred, Ecroyd:** Hugo *Aikroide* 1612 FrY; Henry *Ackroyd* 1645 ib.; Henry *Akeroyd* 1648 ib.; Christopher *Acred* 1721 ib. 'Dweller by the oak-clearing' (OE *ãc, rod*), a Yorkshire name, preserving the dialectal pronunciation *royd*.

**Ackworth, Acworth:** Thomas *de Akewrth'* 1219 AssY; Adam *de Ackeworth'* 1379 PTY; William *Akworth* 1421 IpmNt. From Ackworth (WRY).

**Acomb, Acum, Akam:** Philip *de Akum* 1210 P (Y); Peter *de Acom* 1325 Wak (Y); John *A Combe* 1525 SRSx. From Acomb (Nb, NRY, WRY), Acombe in Churchstanton (D), or for *atte Combe* 'dweller in the valley', OE *cumb*.

**Acraman, Acreman:** v. ACKERMAN

**Acres, Ackers, Acors, Akers, Akess, Akker:** William *del Acr'* 1214 Cur (Sx); Adam *de Acres* 1346 LLB F. 'Dweller by the plot of arable land' (OE *æcer*). Or from Castle, South or West Acre (Norfolk).

**Acton:** Hugh *de Acton'* 1194 P (Sa); Warin *de Aketon* 1202–3 FFEss; John *de Acton* 1327 SRWo; Richard *Acton* 1421–2 FFWa. From one of the many places of this name.

**Acworth:** v. ACKWORTH

**Adam, Adames, Adams, Addams, Adem, Adhams:** *Adam* Warenarius 1146–53 DC (L); Alianor *Adam* 1281 AssCh; William *Adames* 1327 SRWo. Hebrew *Adam* 'red', found in DB, and common thereafter, with diminutives *Adcock, Adkin, Adnett*.

**Adamson:** John *Adamsone* 1296 Black, 1327 SRDb; Richard *Adamessone* 13th AD i (Nf). 'Son of *Adam*.'

**Adburgham:** v. ABRAHAM

**Adcock, Adcocks, Atcock, Hadcock:** *Adecok* Kay 1246 AssLa; Richard *Hadecoc* 1226 FrLeic; Robert *Adekok* 1275 SRWo; John *Atkoc* 1327 SRSt. *Adecoc*, a pet-form of *Adam*.

**Addams:** v. ADAM

**Addekin:** v. ADKIN

**Addey, Addie, Addy, Addess, Addis, Adds, Haddy:** *Addy* 1290 AssCh; *Addy* de Beuerlay 1297 SRY; John *Haddy* 1279 RH (Beds); Richard *Addy* 1301 SRY; John *Adies* 1327 SRWo; William *Addes* 1379 PTY. A pet-form of *Adam*.

**Addington:** William *de Adinton* 1176 P (Bk); Hugh *de Adinton'* 1202 AssNth; Gilbert *de Adintun'* 1226 Cur (Sr). From Addington (Bucks, Kent, Northants, Surrey).

**Addison:** John *Addisone* 1308 Pat (Y); Robert *Addeson* 1498 GildY. 'Son of *Addy*.'

**Addlestone:** v. ATHELSTAN

**Adds, Addy:** v. ADDEY

**Addyman:** *Adiman* 1204 P (Y); John *Addeman* 1379 PTY. 'Servant of *Addy*.'

**Ade, Ades, Adeson:** Thomas *Ade* 1327 SRSx; William *Adeson'* 1379 PTY. A pet-form of *Adam*.

**Adeane, A'Deane:** John *Adene* 1484 RochW; Thomas *a Dene* 1516 KentW. From *atte dene. v.* DEANE.

**Adeline:** *Adelina* joculatrix 1086 DB (Ha); William *Adeline* 1260 AssC. OG *Adalina, Adelina.* The seal of *Adaleide,* wife of William Peverel (1107–13 NthCh), bears the legend: SIGNVM ADELINE, which was thus used as a pet-form of OG *Adelhaid,* in ME usually *Adeliz, Alesia, Aalis. v.* also ALIS, EDLIN.

**Adem:** *v.* ADAM

**Adey, Adie, Ady, Adye:** Thomas *Ady* 1327 SRWo; William *Adee* 1524 SRSf. A pet-form of *Ade* (Adam).

**Adger:** *v.* EDGAR

**Adhams:** *v.* ADAM

**Adkin, Adkins, Addekin, Atkin, Atkins, Atkyns, Attkins, Hadkins:** *Adekin* filius Turst' 1191 P (Nf); John *Adekyn* 1296 Crowland (C); William *Atkyns* 1327 SRWo; John *Adekynes* 1332 SRWa; William *Atkyn* 1441 ShefA. *Ade-kin,* a pet-form of *Adam.*

**Adlam:** *Adelelmus* 1066 DB (K); Walterus *filius Adelam* 1191 P (Sa); Robertus *Adelelmus* 1130 P (Ess); Roger *Adalem* 1260 AssC. From c940 onwards OE *Æðelhelm* regularly appears as *Æðelm* and occurs in DB as *Ailm* (PNDB). Hence the above forms are probably all from OG *Adalhelm, Adelelm* 'noble protector'.

**Adlard, Allard, Allarde, Allart, Allars, Aylard, Ellard, Ellert, Hallard, Hallett:** *Ailardus, Ælard* 1066 DB (D, Sx); *Adelardus* Hornewitesinne 1125 (c1425) LLB C; *Aeilardus* 1143–7 DC (L); *Halardus* de Weres a1150 ib.; *Elard* de Beisebi 1161 P (L); Walterus *filius Eilardi* 1205 Cur (W); Rogerus *filius Alardi, Adelard', Athelardi* 1212, 1242 Fees (L); Roger *Aillard* 1205 P (Gl); Richard *Alard'* 1209 P (Gl); Nicholas *Adelard* 1275 SRWo; Stephen *Eyllard* 1296 SRSx; Richard, William *Athelard* 1327 SRC, SRSx; John *Adhelard, Allard* 1327, 1332 SRSx; John *Allerd* 1327 SRC. The DB forms are probably from OE *Æðelheard* 'noble-hard', but may be for OG *Adelard* or *Agilard. Adel-* is usually continental in origin, *Ayl-* usually from OE *Æðel-,* but may occasionally be for OG *Adel-* or *Agil-.* Both the native and the continental names are represented but cannot be safely distinguished except that *Ellard* and *Ellert* are probably of English origin. *Alard* the Fleming (1198 P), whose name is certainly of continental origin, also occurs as *Ayllard* (1193 ChR) and *Adthelard* (Ric 1 (1279) RH). *Alard* de Cotekyn of Zeeland (1311 Pat) was also a Fleming. Both personal-names may later have become *Aylett, Aliatt.*

**Adlington:** Walter *de Adelminton'* 1204 P (La); Richard *de Adligton* 1274 RH (L). From Adlington (Ch, La), or Allington (Do, L, W), *Adeling(e)tone* DB.

**Admans:** *v.* EDMAN

**Admer:** Lefstan *filius Ædmeri* c1095 Bury (Sf); *Admerus* le Burgeis 1203 Cur (Sx); Walter *Edmer* 1275 SRWo. OE *Ēadmǣr* 'prosperity-famous'. The surname is also local in origin: Reginald *de Addemere* 1296 SRSx, Nicholas *de Admare* 1344 FrY.

**Adnet, Adnitt:** *Adènet* le Wayder 1293 MESO (Nf); *Adinet* del Forest 1379 PTY; Robert *Adynet, Adinot* 1366 AD ii (Lei), 1428 FA (Sr). OFr *Adenet, Adenot,* diminutives of *Adam. Adam* de Bidyk (1286 ChancW) is also called *Adinet* (1276 Fine).

**Adown, A'Down:** Walter *Adoon* 1424 LLB K; John *Adowne* 1488 *Petre.* For *atte dune* 'dweller by the hill', OE *dūn. v.* also DOWN.

**Adrain, Adrian, Adrien:** *Adrianus* janitor 1186–1210 Holme (Nf); Walter *Adran', Adrian* c1232 Clerkenwell (Lo); Ralph *Adrien* 1277 LLB A. Lat *Hadrianus* 'of the Adriatic', the name of a Roman emperor and several popes, including Nicholas Brakespear, the only English pope (d. 1159).

**Adwick:** Ralph *de Aldewic'* 1219 AssY; Gilbert *de Athwik* 1340, Robert *de Addewyke* 15th Shef. From Adwick le Street, or Adwick upon Dearne (WRY).

**Ady(e):** *v.* ADEY

**Aeneas:** *Eneas* filius Hugonis Hy 2 Seals (Y); *Eneas* de Baddeby 1381 AssWa; David *Enyas* 1379 LoCh. The name of the Trojan hero. Used in Scotland to translate Gaelic *Aonghus,* OIr *Oeng(h)us.*

**Aers:** *v.* AYER

**Affery, Affray:** *v.* ALFREY

**Agar, Agars, Ager:** *v.* ALGAR, EDGAR

**Agass, Aggas, Aggass, Aggis, Aggiss, Aggus, Aguss:** Robertus *filius Agacie* 1279 RH (C); Roger *Agace* 1275 SRWo; Thomas *Agas* 1480 AD iii (Sx); Matthew *Agess,* Thomas *Aggis* 1674 HTSf. OFr *Agace* (f), the vernacular form of the learned *Agatha,* from Greek *ἀγαθός* 'good'.

**Agate, Agates:** John *a gate, atte Gate* 1296, 1327 SRSx. From residence near a gate.

**Agg, Aggs:** Simon *filius Agge* 1195 P (L); William *Agge* 1275 RH (L). ODa *Aggi.*

**Agget, Aggett, Agott:** Gilbert *Agote* 1301 ParlR (Ess). A diminutive of *Agg,* a pet-form of *Agnes,* or *Agace* (Agatha).

**Aglion:** *v.* AGUILLON

**Agnes, Agness:** *Agnes* de Papewurda 1160 P (C); Hugo *Agneis* 1219 AssL; Robert *Agnes* 1230 P (L). Fr *Agnes,* from Greek *ἁγνός* 'pure, chaste', the learned form of the vernacular *Anes, Anneis. v.* ANNAS.

**Agnew:** (i) Mabel *de Aignaus, de Agniws* 1208–9 Pl; Robert *de Ayneaus* 1227 Cur (Sf); Robert *de Aygnel* 1249 AssW. From Agneaux (La Manche). (ii) Thomas *Agnel* 1201–12 RBE (Sf); Susanna *Agniel* 1206 Cur (Bk); Lawrence, John *Agnel* 1254, 1284 IpmW. A nickname from Fr. *agneau, agnelle* 'lamb'.

**Agrove, Atgrove:** John *atte Grove* 1323–4 FFEss; William *atte Groue* 1392, Michael *Agrove* 1443 CtH. For *atte Grove* 'dweller by the grove', OE *grāf. v.* also GROVE.

**Aguila, Aguilar:** Henry *Laguillier* 1188 P (Ha); Godfrey *Aigillier* 1206 P (Sr); Robert *le Aguiller* 1221 AssSa. OFr *aiguillier, agullier* 'a maker of needles'.

**Aguillon, Aglion:** Adelard *aculeus* 1148 Winton (Ha); Geoffrey *Ageilun* c1150 DC; Roger *Aguillon* 1219, William *Aglyon* 1361 AssY; Richard *Aglon* 1642 PrD. Lat *aculeius,* OFr *aiguillon* 'goad', perhaps in the sense 'warrior'. *v.* OEByn 372.

**Aguss:** *v.* AGASS

**Aicheson:** *v.* ACHESON

**Aiers:** *v.* AYER

**Aiken(s), Aikin, Aickin:** *v.* AITKEN

**Aikett:** *v.* AKED

**Ailward:** *v.* AYLWARD

**Ailwyn:** v. AYLWIN

**Aimer(s):** v. AYMER

**Ainger:** v. ANGER

**Ainley, Aynley:** Richard *Aynlay* 1548 FrY. From Ainley House in Ovenden (WRY).

**Ainsley, Ainslee, Ainslie, Aynsley:** William *de Anslee* c1220 Black (Glasgow); Adam *de Aynesleye*, Thomas *de Ainslay* 1357 ib. (Roxburgh), Putnam (Db); Adam *Aynesley* 1652 RothwellPR (Y). From Annesley (Notts), or Ansley (Warwicks).

**Ainsworth, Aynsworth:** John *de Ainesworth* 1281, *de Aynesworth* 1285, John *de Aynesworth* 1401 AssLa. From Ainsworth (La).

**Air(e)s:** v. AYER

**Airey, Airy, Ary:** Robert *de Hayra* 1301 IpmLa (We); John *Ary* 1617 FrY; Christopher *Airy* 1647 ib. Robert *de Ayrawe* 1332 SRCu, assessed in Threlkeld (Cumb), must have owed his name to Aira Beck or Aira Force (Ullswater), *Arey* 1522, with a surname *de Ayraw* 1301 (PN Cu 254). Airy's Bridge in Borrowdale is named from Christopher *Arraye* 1603 and Jane *Araye* 1634 (ib. 352), whilst Airygill Lane in Great Strickland (PN We ii, 150), *John Airey Gill* 1838, commemorates the local family of *Airy* (*Arey*, *Arraye*, *Arra* 1586–1731). From 1508 to 1774 the surname is usually spelt *Airy*, with variants *Ayraie*, *Ayree*, *Arye*, *Aery*, etc., and occasionally *Eyrye*, *Eyree*. The persistent *Ai-* and the final *-ay*, *-ey* dissociate the name from the suggested ON *erg* 'shieling'. It is probably 'dweller by the gravel-bank', cf. ON *eyrará* 'gravel-bank river', or from Aira Force or some similarly named place.

**Airmin, Airmine:** Robert *de Eyrminne* Hy 3 IpmY; William *de Ayrminne* 1327 SRY; Isabel *Ayremyn* 1381 AssL. From Airmyn (WRY).

**Aish:** v. ASH

**Aishford:** v. ASHFORD

**Aislabie, Aislaby:** Henry *de Aslaby* 1379 PTY; William *Aslakeby* or *Aislaby* 1572 Bardsley. From Aislaby (Du, NRY).

**Aitcheson, Aitchison:** v. ACHESON

**Aitken, Aitkens, Aitkin, Aitkins, Aiken, Aikens, Aickin, Aikin:** *Atkyn* de Barr 1340 Black; *Aitkane* of Dunsleson 1482 ib.; Andrew *Atkin* 1469 ib.; William *Ackin* 1476 ib.; Robert *Aykkyne* 1539 ib.; Robert *Aitkins* 1674 HTSf; James *Aitkine* 1685 Black. Scottish forms of *Atkin*.

**Aiton:** v. AYTON

**Aizlewood:** v. HAZELWOOD

**Akam:** v. ACOMB

**Akaster:** v. ACASTER

**Ake:** William *de Ake* 1327 SRWo; William *del Ake* 1364, John *de Ake* 1384 FrY. From Aike in Lockington (ERY), or 'dweller by the oak', OE *āc*.

**Aked, Akett, Akitt, Aikett:** Richard *de Aykeheved* 1280 Riev (Y). 'Dweller by the oak-covered headland' (OE *āc, hēafod*).

**Akehurst, Akhurst:** John *de Ochurst* 1296 SRSx; John *Akeherste* 1525 SRSx. From Akehurst Fm in Hellingly (Sx).

**Akeman:** Heruey *filius Akeman* 1188 P (Gl); Henry *filius Akemon* 1246 AssLa; William *Akemon* 1275 SRWo. OE *\*Ācmann*, recorded only in place-names.

**Akerman:** v. ACKERMAN

**Akeroyd:** v. ACKROYD

**Akers, Akess:** v. ACRES

**Akester:** v. ACASTER

**Akett, Akitt:** v. AKED

**Akhurst:** v. AKEHURST

**Akister:** v. ACASTER

**Akker:** v. ACRES

**Akred, Akroyd:** v. ACKROYD

**Alabastar, Alabaster, Albisser, Arblaster:** Geoffrey *Arbalister* 1198 Cur (Ess); Richard *le Arbelaster* 1198 P (R); Ralph *Alebaster* c1200 HPD (Ess); Peter *le Arblaster* 1278 FFSf. AFr *alblaster*, *arblaster*, OFr *arbalestier*, *arbelestier* 'a soldier armed with a crossbow, a crossbowman' (c1325 MED). v. also BALLASTER. The surname is also due to office. Robertus *Arbalistarius, Balistarius* (1086 DB) and his son Odo *albalistarius* (c1140 Holme) held their land of the king by serjeanty of performing the duties of *arbalistarius*. Others of the same surname held their land by serving at Wallingford Castle with an arbalest, by guarding Exeter gaol, or by providing two arbalests. As London *arblasters* are stated to have had apprentices, the surname may also mean 'maker of cross-bows'.

**Alabone:** v. ALBAN

**Alais:** v. ALIS

**Alan:** v. ALLAIN

**Aland, Alland:** Gerard *Attelaunde* 1281 IpmGl; Robert *de Allandes* 1335 FrY; Thomas *Aland'* 1379 PTY; Thomas *Alonde* 1545 SRW. For *atte lande* 'dweller at the glade or pasture', OFr *laund*. Sometimes, perhaps, 'dweller at the old cultivated land(s)', OE *eald, land*.

**Alaway, Alway, Allaway, Alleway, Alloway, Allawy, Allway, Allways, Elloway, Hallaway, Halloway, Hallway:** (i) *Ailwi, Aluui* 1066 DB; *Willelmus filius Ailwi* 1206 P (Nth); Godfrey *Ailwi* 1188 BuryS (Sf); Peter *Athelwy* 1302 SRf. OE *Æðelwīg* 'noble war'. (ii) *Æluui, Eluui* 1066 DB; *Willelmus filius Alewi* 1185 P (Bk); *Aleway* Pote 1297 SRY; Roger *Alewy* 1200 P (Mx), 1221 AssWo; Henry *Alwi* 1221 *ElyA* (Sf); Richard *Alawy, Halewy* 1260, 1264 Eynsham (O); William *Halawey* 1279 RH (Hu); John *Always* 1301 SRY; Henry *Halloway* 1674 HTSf. The DB forms may stand for OE *Æðelwīg*, 'noble-war', *Ælfwīg* 'elf-war' or *Ealdwīg* 'old war' (rare). v. also ALDWY, ALLVEY.

**Alban, Albon, Albone, Allbon, Allbones, Alabone, Allebone, Alliban, Allibon, Allibone, Aubon:** *Albanus filius Willelmi* 1212 Cur (Y); Walter *Alban* 1250 Fees (Bk); Adam *Albon* 1275 SRWo; Hugh *Alybon* 1297 Coram (Db); William *Albon, Albone* 1376 LLB G; William *Albanes* 1379 PTY; William *Alybone* 1397 FrY. Latin *Albanus* 'of Alba', the name of the first British martyr. St. Albans is recorded as *Seynt Albones* in 1421 (PN Herts 87).

**Albe, Aube:** Nicholas *le Albe* 1230 MemR (Nf); Geoffrey *Albe* 1327 SRLei. Lat *albus*, OFr *albe, aube* 'white'. Used also as a feminine name: *Alba* (f) 1315 AssNf.

**Alberry, Albery, Allberrey, Allberry:** These surnames may be late variants of either AUBRAY, or ALBURY.

**Albert, Alberts, Aubert:** *Albertus* 1066 DB (Beds); Philippus *filius Alberti, Auberti* 1211 Cur (Do); Robert *Alberd, Albert* 1221 AssWa; Isabella *Aubert* 1327 SRSf. OG *Albert* (Fr *Aubert*), from OG *Adalbert*. The cognate OE *Æðelbeorht* 'noble bright' occurs in DB in 1066 as *Ailbertus, Ailbriht* and once as *Albrict*. The *Ailbertus* of 1066 InqEl (C) appears in DB as *Albertus*. Thus *Albert* may sometimes be identical with *Albright*.

**Albin, Albinson, Aubin, Obin:** *Albinus* 1148–53 Bury (Sf), Hy 2 Gilb (L); *Obin* Kinel 1202 AssL; Roger *Albin* 1194 P (Wo); Gilbert *Aubin* 1210 Cur (Nf); John *Obin*, James *Aubyn* 1275 SRWo. OFr *Albin, Aubin*, Lat *Albinus*, a derivative of *albus* 'white'.

**Albisser:** *v.* ALABASTAR

**Albon(e):** *v.* ALBAN

**Alborne, Albourne, Alburne:** John *de Aleburn'* 1177 P (Sx); Ailwin *Alburne* 1187 P (Ess); Nicholas *Alebourne* 1332 SRSx. From Albourne (Sx).

**Albright, Allbright:** *Ailbricd* c1160 DC (L); *Ailbrict* 1202 FFC; Alexander *filius Ailbriht* 1205 Cur (Sr); William *Albrich'* 1275 SRWo; Eusebius *Ailbrit, Ailbric* 1279 RH (Hu); Herriesservant *Albryght* ('servant of Harry Albryght') 1378 Pat (Beds). OE *Æðelbriht*, a metathesized form of *Æðelbeorht*. *v.* ALBERT.

**Albury:** David *de Aldebury* 1278 RH (Sa); Stephen *de Aldebury* 1278 RH (O); William *Albury* 1683 Bardsley. No doubt usually from Albury (Herts, O, Sr), but possibly also a variant of AUBRAY.

**Albutt:** *v.* ALLBUT

**Alcock, Alecock, Allcock, Aucock, Awcock:** *Alekoc, Alecoc* 1204 Cur (Nt), P (Y); *Awcok* de Leuer 1246 AssLa; *Alkok* 1332 SRCu; Alexander *Alecoc* 1275 SRWo; John *Alekok, Auecoc* 1296, 1327 SRSx. A pet-form of some short name in *Al-*.

**Alcoe:** Probably for *Alcock*. cf. the pronunciation *Coeburn* for *Cockburn* and *Coeshot* for *Cockshott*.

**Alcorn, Allcorn, Alchorn, Alchorne, Allchorn, Allchorne, Alchen, Alchin, Allchin:** John *de Alcheshorn*, Adam *de Alchehorn* 1296 SRSx; John *Alchon* 1420 LLB I; John *Alchorn* 1525 SRSx. From a lost *Alchehorne* in Buxted (Sx), last mentioned in 1592. *Allchin* and *Allcorn* survive in West Kent, whilst *Alchin* is found in Godstone (Sr) near the Sussex boundary. *v.* also OLDCORN.

**Alcott, Allcott, Allcoat, Aucott, Aucutt:** Philip *de Allecote* 1255 RH (Sa); Thomas *de Aldecote, de Alecote* 1275 SRWo. 'Dweller at the old cottage' (OE *eald, cot(e)* ). Often, no doubt, confused with *Alcock*. John *Alkot* 1290 AssCh may be a misreading of *Alkoc*.

**Alcrist:** John *Allecrist, Aldecrist* 1221 AssSa; John *Aldecrist* 13th AD iii (He); Walter *Oldecrist* 1258 AD iii (He). Evidently a nickname, 'old Christ', OE *eald, crīst*, the particular significance of which is unknown.

**Aldam, Aldham:** Oslac *de Haldham* c1095 Bury; Adam *de Aldeham* 1206 Cur (Sr); Isaac *Aldham*, William *Aldum* 1674 HTSf. From Aldham (Ess, Sf). *v.* also OLDHAM.

**Alden, Aldin, Aldine, Aldins, Allden, Auden, Olden:** (i) *Aldanus, Alden(e)* 1066 DB; Gamel *filius Alden* 1196 P (We); Walterus *filius Aldan, Aldein*, Haldein 1218–19 AssL; Alexander *Aldeyn* 1279 RH (O); John *Alden* 1524 SRSf. Anglo-Scand *Healfdene*.

*v.* HALDANE. (ii) Ælfwine *Aldine* c1095 Bury (Sf); Osgotus *Aldwinus* 1196 P (Berks); Reginald *Aldyne* 1275 RH (Nf). OE *Ealdwine*. *v.* ALWIN. cf. *Aldenesawe, Aldwynshawe* c1200 DEPN for Audenshaw.

**Alder, Alders, Allder, Nalder, Nolder, Nolda, Older:** Ralph *de Alre* 1221 Cur (Berks); William *atte Nalre, atte Naldhres* 1277 AssSo, 1313 FFEss; John *atte Alre* 1327 SRSo; Richard *atten Alre* 1332 MELS (So); Alexander *Aldres* 1332 SRWa. 'Dweller by the alder(s)' (OE *alor*).

**Alderman:** Adam *le alderman* 12th DC (L); John *Alderman* 1175 P (Sx). OE *ealdormann* 'alderman', also headman, governor of a guild.

**Alderton:** Alexander *de Alreton'* 1191 P (Y); John *de Aldrinton'* 1221 AssWo; Richard *Alderton* 1525 SRSx. From Alderton (Ess, Gl, Nth, Sf, W), Allerton (La, WRY), or Ollerton (Ch, Nt). *v.* also ALLERTON, OLLERTON.

**Aldhous(e):** *v.* ALDIS

**Aldin(s):** *v.* ALDEN

**Aldington:** Richard *de Aldington'* 1204 P (W); William *de Aldinton* 1275 SRWo. From Aldington (K, Wo).

**Aldis, Aldiss, Alldis, Aldhous, Aldhouse, Aldous, Aldus, Audas, Audiss, Audus, Oldis:** Radulfus *filius Alduse* 1168 P (Y); Willelmus *filius Aldus* 1202 AssL; *Aldusa* filia Cristine 1219 AssY; Peter, Robert *Aldus* 1230 P (Nf), 1301 SRY; Peter *Aldous* 1327 SRSf; Robert *Aldhous*, William *Aldowes* 1524 SRSf; Francis *Aldhowse* 1647 Shef (Y); Goody *Aldis* 1674 HTSf. *Aldus* (f), a pet-form of some woman's name in *Ald-*, e.g. *Ealdgýð*. *v.* EDIS. The Scottish *Aldis* is from Auldhous in Eastwood (Renfrewshire): Reginald *de Aldhous* 1265 Black.

**Aldon:** Euota *de Aldone* 1317 AssK; Thomas *de Aledon* 1321 LLB F. From Aldon (Sa).

**Aldred, Alldread, Alldred, Alldritt, Audritt, Eldred, Eldrett, Eldrid, Elldred:** (i) *Ældred, Ældret, Aldret, Eldred, Eldret* 1066 DB; *Aldret* de Windegate c1145–65 Seals (Nb); *Eldredus* 1161–77 Rams (Herts); Roger *Aldredus* 1207 P (D); Golding *Aldred* a1224 Clerkenwell (Mx); William *Aldret* 1275 SRWo. OE *Ealdrǽd* 'old counsel'. Forms in *Alred* may also belong here. (ii) *Eilredus* de Mannest' 1198 Cur (K); William *Eyldred* 1317 AssK; Maud *Aildred* 1327 *SR* (Ess). OE *Æðelrǽd* 'noble counsel', which appears in DB as *Ailred, Alret, Aldret*, and in InqAug (11th) as *Elred*. Forms in both *Alred* and *Aldred* may, therefore, also belong here. *v.* also ALDRITT and ALLRED.

**Aldren:** Thomas *in the Aldren*, Henry *in the Alren* 1327 SRSo. From residence among the alders (OE *alor*).

**Aldrich, Aldrick, Aldridge, Alldridge, Allderidge, Elderidge, Eldridge, Elrick, Oldridge:** These surnames may be local in origin, but usually derive from OE *Ælfrīc* 'elf-ruler' or *Æðelrīc* 'noble ruler'. Both survived the Conquest, by which time the first element had often been reduced to *Al-* or *El-* and consequently cannot be distinguished. A common post-Conquest form of *Æðelrīc* was *Ailric* or *Eilric*. *v.* ALLRIGHT. (i) From OE *Ælfrīc*: *Ælfric, Alfric, Aluric* 1066 DB; Hugo *Ælurici filius* c1095 Bury (Sf); Ricardus *Alurici* 1209 P (W); William *Alfric* 1212 Cur

(Berks); William *Alfrich* 1296 SRSx; John *Alfryg* 1327 SRSo. (ii) From OE *Æðelrīc*: *Adelric, Agelricus, Egelric, Ailric, Aelricus, Alricus, Aluric* 1066 DB; *Ailricus, Ældricus* de Burc 1066 DB (Sf); *Aldrich (Ailrich)* 1220 Cur (Sr); William *Ailric* 1209 P (W); Henry *Aldric*, Geoffrey *Aldrich* 1275 SRWo; Robert *Aylrich, Alrych* 1279 RH (Hu); William *Eldrich* 1336 AD ii (Sr); James *Aldridge* 1674 HTSf. (iii) From OE *Ælfrīc* or *Æðelrīc*: *Ælric, Alric, Alrich, Elric* 1066 DB; *Elricus* filius Leppe Hy 2 DC (L); Roger *Elrich* 1279 Barnwell (C); Robert *Alriche* 1327 SRC; Seman *Olrich* 1327 SRSf; Amicia *Alric* 1346 FFC; Alexander *Elrick* 1510 Black; Peter *Oldryk* 1527 SRSf. (iv) Local. From Aldridge (Staffs): Drogo *de Alrewic* 1201 P (St); from Aldridge Grove (Bucks), *Eldrigge* 1227 PN Bk 151; or from an unidentified place in or near Worcester: Hugo *Elrugge* 1327 SRWo.

**Aldritt, Alldritt, Eldrett, Eldritt, Naldrett, Neldrett:** Robert *atte Nalrette* 1305 FFSx; Robert *atte Aldratte*, Walter *ate Naldratte*, Gilbert *ate Nelrette* 1327 SRSx; John *atte Alrette* 1332 SRSx. OE *alrett, *elrett* 'alder-grove', a derivative of OE *alor* 'alder'. Common in Sussex, usually as *Naldrett(s)* and in Nalderswood Fm (PN Sr 298). v. also ARLET. *Aldritt* and *Eldrett* may also derive from OE *Ealdrǣd*. v. ALDRED.

**Aldwinckle, Aldwinkle:** Henry *de Audewincle* 1230 P (Nth); William *Aldewynkyl* 1386 AD i (Nth); William *Aldewyncle* 1468 IpmNt. From Aldwinkle (Nth).

**Aldwy:** *Aldui* 1066 DB (So); Geoffrey *Aldwi* 1221 ElyA (Sf); William *Aldwy* 1279 RH (O). OE *Ealdwīg* 'old-war'. v. also ALAWAY.

**Ale, Alle:** *Ala* de Bedingh' 1204 Cur (Sx); Nicholas *filius Ale* 1218 P (St); *Alle* Dockett 1642 PrD; John *Ale* 1296 SRSx; Thomas *Alle* 1379 PTY. ME *Ala*, a short form of names in *Al-*.

**A'Lea, A'Lee:** John *Alee* 1521 FrY; John *a Laye, a Lye*, Thomas *a Lee, Aley*, William *at Lee* 1525 SRSx; John *a Lyghe* 1544 Musters (Sr). For *atte Leye* 'dweller at the clearing', OE *lēah*.

**Alecock:** v. ALCOCK

**Alefounder:** Walter *le Alefondere* 1275 Cl; John *Alefondere* 1381 FFEss; Widow *Alefounder* 1674 HTSf. 'Inspector of ale', one appointed by the Court Leet to examine it as it was poured out (Lat *fundere* 'to pour out'). The name still survives in Essex and Norfolk.

**Alexander:** *Alexander* c1150–60 DC (L); Thomas *Alexander* 1283 SRSf. Greek 'Αλέξανδρος 'defender of men'. v. SANDARS.

**Alfill, Alfille:** *Alfilda* 1201 Cur (Sx); John *Alfild* 1309 EAS 23 (Ess); Thomas *Alfild* 1327 SREss. OE *Ælfhild* (f).

**Alfiat, Alflatt, Allflat, Elfleet, Elflitt:** *Elfled* 1222 Cur (Sf); Adam *Ailflet* 1221 ElyA (Sf); Thomas *Alfled* 1279 RH (C). *Æfled, Alfled, Alfleta* (1066 DB) may stand for either OE *Ælfflǣd* 'elf-beauty' or *Æðelflǣd* (f) 'noble beauty'. The latter is also *Ædelflete* in DB and *Ailflet* is certainly from this. *Alf-* or *Elf-* may be from either.

**Alford, Alforth, Allford:** Alan *de Alforde* Hy 2 DC (L); Robert *de Auford* 1202 FFL; Thomas *de Aldeford* 1275 SRWo; Henry *Alforde* 1642 PrD. From Alford (L, So, Aberdeen), or Aldford (Ch).

**Alfred, Alfreds, Allured, Alured:** *Alured* 1066 DB; *Elfredus* pelliparius Hy 2 Gilb (L); *Alfredus* Folkeredus 1204 Cur (Sr); Alexander *filius Alvredi* 1206 Cur (Nf); Walter *Alured* 1260 AssC; Thomas *Alfred* 1294 RamsCt (Beds). OE *Ælfrǣd* 'elf-counsel'. This personal name is not so rare after the Conquest as has been suggested. Michaëlsson has shown that in the *Roman de Rou* the name of Alfred the Great appears as *Alured, Aluered, Auuered, Alvere*, and *Auuere* and that between 1292 and 1313 the French form *Auveré* is found in Paris both as a personal name and a surname. This pronunciation was also used by Normans in England. *Alfred* is probably concealed in many examples of the Latin *Alvredus*, though this certainly includes examples of *Avery*. The father of Normannus *filius Alfredi* (1242 Fees) is also called *Averay* (1236 ib.). cf. also Hugo *filius Alfredi, Aufrey, Alveredi* (1242 ib.). *Alvredus* is usually spelled Aluredus which, curiously enough, survives as *Al(l)ured. v.* ALFREY and AVERAY.

**Alfrey, Alfry, Allfree, Allfrey, Affery, Affray, Elvery:** Ricardus *filius Aufridi* 1212 Fees (L); Richard *Aufrey* 1275 RH (Wo), 1277 Ipm (Nth); John *Aufred* 1279 RH (Hu); Robert *Alfray* 1296 SRSx; Gilbert *Alfrey* 1317 AssK; John *Alfreth* 1327 *SR* (Ess); Adam *Alfrid* 1327 SRSo; William *Alfred, Alfreth, Alfride* 1327 *SR* (Ess). The Lincs *Aufridus* and the Somerset and Essex *Alfrid(e)* suggest that we are not always concerned with *Alfred*, but rather with a name ending in *-frith*, perhaps OE *Æðelfrið* (m), DB *Elfridus*, or *Æðelfrīð* (f), *Egelfride*, both rare after the Conquest; or, possibly, OG *Adalfrith. Alfred* may also belong here, the *-frey* being due to analogical influence from compounds in *-frið*.

**Algar, Alger, Algore, Augar, Augur, Agar, Elgar, Elger:** (i) *Ailgarus* filius Lucie 1151–3 DC (L); *Ailgarus, Agare* (abbot of Faversham) 1193–9, 1200–4 StGreg (K); *Eylgar* de Berwe 1273 RH (Sx); *Elgarus, Ailgarus* King c1232 Clerkenwell (K); Robert *Elger* 1271 AD iv (Nf); Walter *Eylgar, Elgar* 1317 AssK. OE *Æðelgār* 'noble spear' which occurs in DB only as *Algar*. (ii) *Alfgerus, Ælger, Algerus* 1066 DB (L, Nf, Sf); *Ælfger* de Brademere, *Alger* c1095 Bury (Sf); *Algerus* faber 1150 DC (L); Simon, Thomas *Alger* 1221 ElyA (Sf), 1297 MinAcctCo (Y). ON *Álfgeirr*, ODa *Alger*. Some of these forms may be from OE *Ælfgār* or *Ealhhere*, but all are from counties where Scandinavian influence was strong. (iii) *Algar* c1095 Bury (Sf); *Algarus* Dalling 1210 Cur (C); William *Algar* 1221 AssWo; Walter *Elgar* 1234 FFSf; Thomas *Algor* 1260 AssC. OE *Ælfgār* 'elf-spear' occurs once in DB as *Alfgarus* or *Elgar*, otherwise as *Ælgar, Algar*. OE *Ealdgār* 'old spear' is DB *Ealgarus. Ælfgār* is much the more common name and both appear later almost regularly as *Algar*. These surnames may derive from either or from *Æðelgār*.

**Algate:** Edricus *de Alegat'* 1219 Cur (Mx). From Aldgate (London).

**Alger, Aljer, Auger:** *Alcher (Alg(h)erus* Exon) 1066 DB (D); *Alcherus* Venator 1166 P (Ess); *Auchere* filius Henrici 1327 SRC; Willelmus *filius Auger* 1346 SRWo; Ailwin *Alher, Alcher* 1180, c1216 Bart (Lo); Walter *Alger* 1275 SRWo; Henry *Auger* 1279 RH

(Hu); John *Aucher* 1428 FA (W); William *Awger* 1498 FrY. OE *Ealhhere* 'temple-army', which, through Anglo-Norman vocalization of *l*, became *Aucher*, *Auger*. cf. AYER. *Alger* (with a hard *g*) is from ON *Álfgeirr*. *v*. ALGAR.

**Algore:** *v*. ALGAR

**Alin, Aling, Allin:** *Adhelina* 1138, *Aelina* de Rodlos 1160–5 NthCh (L); *Aalina* (f) Hy 2 DC (Nt); *Alina* 1211 Cur (Mx); *Alina* del Hele 1248 MPleas (Nth); Richard *Alyne* 1275 SRWo; Richard *Aelyng* 1296 SRSx. OG *Adalina*, *Adelina*, *Agellina*, *Aillina*. There is also evidence to show that *Aline* was used as a diminutive of OFr *Aalis*. *v*. Michaëlsson ii 2. *v*. also ADELINE.

**Alington:** *v*. ALLINGTON

**Aliman:** *v*. ALLMAN

**Alis, Alise, Allies, Allis, Alliss, Allish, Alais, Hallis:** *Adeliz* de Raimes Hy 2 DC (Nth); *Aliz* Martel ib. (L); *Ahelis, Ahelissa* 1188 BuryS (Sf); Willelmus *filius Alis* 1214 Cur (Beds); *Alicia* filia Engrami 1219 AssY; *Atheleisia, Aelesia, Aeleis* (identical) 1219–20 Cur (Nf); Roger *Alys* 1221 AssWa; William *Aliz* 1297 MinAcctCo. OFr *Aalis, Aliz*, OG *Adalhaidis* 'noble kind, sort', modern *Alice*.

**Alison, Allison, Alleeson, Hallison:** (i) *Alison* c1386 Chaucer; *Alison* Home 1524 Black (Berwick); William *Alisun* c1248 Bec (Bk); John *Allison* 1332 SRCu. OFr *Alison, Alizon*, a pet-name for *Alice*, common both in England and in Scotland from the 13th to the 17th centuries. In Scotland, it became *Elison*: *Elison* Dalrymple 1514; *Alysone* or *Helysoune* Rouche 1535 (Black). *v*. ELLISON. (ii) Patrick *Alissone* 1296 Black (Berwick); John *Aliceson* 1324 Wak (Y); William *Aleissone* 1381 SRSt 'Son of *Aleis* or *Alice*'. Also a late form of *Allanson*. *v*. also DALLINSON.

**Alister, Allister:** Hugo *de Alencestr'* 1221 AssWo; Adam *de Alicestr'* 1275 SRWo. From Alcester (Warwicks). Also for MACALASTER.

**Alker:** (i) William *de Altekar* 1341 LLB F; William *Alker* 1630 Bardsley. From Altcar (Lancs). (ii) William *Alker* 1212 Cur (Nf). This cannot be local in origin and is from OE *Ealhhere*. *v*. ALGER.

**Alkin, Allkins, Aukin:** *Alkin* the Jonge 1296 AssCh; John *Alkyn* 1307 ParlWrits (He). Probably a pet-form of *Allan* or *Alexander*, *Al-* plus *-kin*. cf. ALCOCK.

**Allain, Allan, Allans, Allen, Alleyn, Alleyne, Allin, Allon, Alan, Alen, FitzAlan:** *Alanus* 1066 DB (Sf), c1150 (Lei); *Alain* 1183 DC (L); Geoffrey *Alein* 1234 FFC; Roger *Alain* c1246 Calv (Y); Richard *Aleyns* 1309 AssSt; John *Fitz Alan* 1416 FrY; Richard *Alen* 1544 FFHu; Matthias *Allyn* 1593 ib. OFr *Alain, Alein*, OBret *Alan*, the name of a Welsh and Breton saint, which was popular with the Bretons who came over with the Conqueror, particularly in Lincolnshire, where it ranked eighth in popularity in the 12th century, level with Simon and more numerous than Henry. From 1139 it was common in Scotland where the surnames also derive from Gaelic *Ailéne, Ailín*, from *ail* 'rock'.

**Allaker:** *v*. ELLERKER

**Allam, Allams, Allom, Allum:** Richard *Alum* 1327 SRLei; Robert *Alome* 1379 LLB H; Thomas *Alom* 1524 SRSf. From Alham (So), or for HALLAM.

**Allamand:** *v*. ALLMAND

**Allanby, Allenby, Allonby, Allamby, Allemby:** Adam *de Aleynby* 1332 SRCu; John *Alomby* 1522 FrY. From Allonby (Cumb), *Aleynby* 1285, *Alanby* 1306, *Allonby* 1576.

**Alland:** *v*. ALAND

**Allanson, Allenson, Allinson, Allison:** Henry *Aleyson* 1332 SRWa; Nicholas *Aleynesson* 1383 AssC; John *Alanson* 1395 Whitby (Y); Thomas *Alynson* 1401 AD i (Db); Allison *Allasoune* 1554 Black (Glasgow). 'Son of *Alain* or *Alan*.' *v*. ALLAIN. Occasionally 'Son of *Alwin*': Otho *Alwinessune* 1130 P (Lo). *v*. ALWIN and DALLINSON.

**Allard:** *v*. ADLARD

**Allars, Allart:** *v*. ADLARD

**Allason:** Giles *Alardson* 1421 LLB I. 'Son of *Alard*'. *v*. ADLARD. It may also derive from *Allanson* or *Allatson*.

**Allatson:** *v*. ALLETSON

**Allatt:** *v*. ADLARD, AYLETT

**Allaway, Allawy:** *v*. ALAWAY

**Allberrey, Allberry:** *v*. ALBERRY

**Allbon, Allbones:** *v*. ALBAN

**Allbright:** *v*. ALBRIGHT

**Allbut, Allbutt, Albut, Albutt:** *Alboda* 1114–20 Rams; *Ailboda* c1155 DC (L); John *Albot* 1275 RH (D); Symon *Albod* 1275 SRWo. OG *Albod, Albot, Adelbodo*.

**Allchin, Allchorn, Allchorne, Allcorn:** *v*. ALCORN

**Allcoat:** *v*. ALCOTT

**Allcock:** *v*. ALCOCK

**Allcott:** *v*. ALCOTT

**Allday:** Nicholas *Alday* 1327 *SR* (Ess); John *Aldaye* 1455 ArchC 34; Thomas *Aldy* 1534 ib. 37. From OG *Aildag* or possibly OE *\*Æðeldæg*. cf. *Aildeig* 1066 DB (Nf). Here, too, probably belong: John, William *Aldeth* 1524 SRSf, from OE *Ealdgyð* (f), 'old combat', found in DB in 1066 as *Ældiet, Ældit, Aldgid, Aldid*, and still in use in the 13th century: *Aldeth* Fin 1190 BuryS (Sf); *Alditha* de Pelham 1212 Cur (Herts). For the development to *-ey*, cf. ALFREY.

**Allden:** *v*. ALDEN

**Allder:** *v*. ALDER

**Allderidge:** *v*. ALDRICH

**Alldis:** *v*. ALDIS

**Alldread, Alldred, Alldritt:** *v*. ALDRED

**Alldridge:** *v*. ALDRICH

**Alle:** *v*. ALE

**Allebone:** *v*. ALBAN

**Alleeson:** *v*. ALISON

**Allemby, Allonby:** *v*. ALLANBY

**Allen:** *v*. ALLAIN

**Allenson:** *v*. ALLANSON

**Allerston:** John *de Aluerstan'* 1219 AssY; Adam *de Allerstan* 1349 FrY; Mary *Alertson* 1672 HTY. From Allerston (NRY).

**Allerton:** Richard *de Aluerton'* 1193 P (Y); William *de Allerton'* 1312 FFY; John *Allerton* 1416–7 IpmNt; Jacob, Mary *Alderton* 1817, *Allerton* 1819 LitWelnethamPR (Sf). From Allerton (La, So, WRY), or Alderton (Sf).

**Alletson, Allatson:** Adam *Allotesone* 1301 SRY; Agnes *Aletson* 1516 GildY. 'Son of *Allott*'. Also, perhaps, 'son of *Aylett*'.

**Allett:** *v.* AYLETT

**Alleway:** *v.* ALAWAY

**Alley, Ally:** Johannes *filius Alli* 1205 Cur (Nf); Hugo son of *Ally* 1332 SRCu; Alice, Richard *Ally* 1275 SRWo, 1352 FrY; Walter *Alleye* 1279 RH (O). ODa *Alli*, OSw *Alle*, found as *Alli* in DB (1066) in Bucks, Beds, Northants and Staff.

**Alleyn(e):** *v.* ALLAIN

**Allfield:** *v.* OLDFIELD

**Allflatt:** *v.* ALFLAT

**Allford:** *v.* ALFORD

**Allfree, Allfrey:** *v.* ALFREY

**Allgood, Augood:** *Algod* 1066 DB (Nt); Iordanus *filius Algodi* c1174 Clerkenwell (Lo); *Algotus* de Banneberi 1178 P (O); Philip *Halgot* 1190 Oseney (O); Ralph *Alegod, Halgot, Halegod* 1200 ib.; Thomas *Algod* 1225 Pat; Robert *Algood* 1327 SRSf. ODa, OSw *Algot, Algut*.

**Allgrim:** *Algrim* 1066 DB (Y); *Algrim* de Frisemareis 1195 P (Y); John *Algrym* 1402 YWills. ON *\*Álfgrímr*. The modern form may also be a corruption of *Allgroom*. cf. Richard *Aldegrom* 1198 P (K), 'the old servant', ME *grom*.

**Alliban, Allibon(e):** *v.* ALBAN

**Allies:** *v.* ALIS

**Alliker:** *v.* ELLERKER

**Allimant:** *v.* ALLMAND

**Allin:** *v.* ALIN, ALLAIN

**Allingham:** Cost *de Alingeham* 1191 P (L). From Allingham (K), or perhaps from Alvingham (L).

**Allington, Alington:** William *de Alinton'* 1192 P (Y); Peter *de Alingeton* 1235 Fees (W); William *Alyngton* 1479 Paston. From Allington (D, Ha, K, L, W).

**Allinson:** *v.* ALLANSON

**Alliott:** *v.* AYLETT

**Allis, Allish:** *v.* ALIS

**Allison:** *v.* ALISON, ALLANSON

**Allister:** *v.* ALISTER

**Allitt:** *v.* AYLETT

**Allix:** A Huguenot name. Peter *Allix*, b. Alencon 1641, d. London 1717, was minister of the Church of Charenton near Paris. On the Revocation of the Edict of Nantes, he fled to England, became minister of the Temple of the French hospital in Spitalfields, and was later canon and treasurer of Salisbury Cathedral (Smiles 259). The name is a variant of *Alis*, i.e. *Alice*.

**Allkins:** *v.* ALKIN

**Allman, Allmann, Allmen, Aliman, Alman, Almon, Almond:** (i) *Alemannus* 1101–25, 1125 Holme (Nf); Henricus *filius Aleman* 1219 AssY. As there is no known OE personal name from which this can be derived, it is probably OFr *aleman* 'German' used as a personal-name. (ii) John *Aleman* 1199 MemR (Nb); Walter *le Aleman* 1200 Cur (Y); Roger *Alemon* 1275 SRWo; Robert *Alman* 1327 SRC. OFr *aleman*, modFr *allemand* 'German'. With an excrescent *d* (as in modern French), this would become *Almond*. cf. Bardsley's 'Hanekin *Almond* and John *de Almann*,

valets of the countess of Surrey' in 1358, and 'the almond leap', a German dance (1611 NED). (iii) Thomas *de Alemayne* 1320 LLB E; Inglebright *de Alman* 1332 FrY. 'The man from Germany', commonly called *Almeyne* in the 14th century. This, too, would become *Almond*.

**Allmand, Almand, Alment, Almond, Allamand, Allimant:** Terricus *le Alemaund* 1276 RH (Bk); John *le Alemaund* 1284 LLB A. OFr *alemaund* 'German'.

**Allmark, Almack, Awmack, Hallmark:** Robert *Alfmarck* 1279 RH (Hu); Adam *Halfmark* 1296 Wak (Y); Emma *Halmark* 1324 LaCt; John *Awmack* 1722 YWills. 'Half-a-mark', a nickname from the money of account.

**Allnatt, Allnutt, Allner, Alner, Elnaugh, Elnough:** *Ælnod, Alnoth, Alnot, Alnod, Elnoc, Elnod* 1066 DB; *Alnodus* (*Ailnoð*) ingeniator 1177 P (Mx); *Alnotus* Papillun 1197 FF (Nth); *Ailnothus* Peni 1204 Cur (Sr); William *Aylnoth, Alnoth* 13th Lewes (Nf); Gilbert *Elnod* 1312 FFK; Richard *Eylnoth* 1317 AssK; Thomas *Alnowe* 1427 FFEss. The DB forms may be for OE *Ælfnöð, Æðelnöð, \*Ealdnöð* or *\*Ealhnöð*, but the surname is chiefly, if not solely, derived from OE *Æðelnöð* 'noble daring'. cf. WOOLNER.

**Allom:** *v.* ALLAM

**Allon:** *v.* ALLAIN

**Allonby:** *v.* ALLANBY

**Allott:** *Alote* c1191 BuryS (Sf); Adam *filius Alot* 1203 P (L); William *Allot* 1275 SRWo; Walter *Alote* 1296 SRSx. *Allot* (OFr *Aalot*), a hypocoristic of *Alis*.

**Allpress:** (i) *Ældeprest* 1189, *Aldeprest* 1194 P (Co). 'The old priest', OE *eald, prēost*. (ii) Thomas *Alprest* 1278 RH (C). cf. Fr *Auprêtre* 'son of the priest'.

**Allred, Alred:** *Alret* 1066 DB (K); Godardus *filius Ailred'* 1205 P (Lo); *Alredus* de Wicham 1206 Cur (K); Robert *Alred* 1198 P (K); Walter *Ailred*, Martin *Alred* 1279 RH (Hu). OE *Æðelræd*. *v.* ALDRED. Possibly also from OE *Ealdræd*.

**Allright, Allwright, Oldwright:** William *Ailricht, Ailriht* 1250 Fees (Beds), 1279 RH (C); Roger *Alright* 1457 LLB K; Elizabeth *Aldwright* 1720 Bardsley. OE *Æðelrīc*. *v.* ALDRICH.

**Allsep, Allsepp, Allsop, Allsopp, Allsup, Allsupp, Alsop, Alsopp, Elsip, Elsop:** Gamel *de Haleshoppe* 1175 P (Db); Philip *Alsope* 1279 RH (C). From Alsop en le Dale (Derby), *Elleshope* DB.

**Allston, Alston, Alstone:** (i) *Ælfuine filius Ælstani* c1095 Bury (Sf); Hugo *filius Alstani* 1209 P (Lei); Henry *Alston* 1279 RH (C); William *Alstan* 1283 SRSf. OE *Æðelstän* appears in DB as *Alestan*. *v.* ATHELSTAN. The DB *Alstan* may also be derived from OE *Ælfstän* 'elf stone', *\*Ealdstän* 'old stone' or *Ealhstän* 'temple stone', all of which may have contributed to these surnames: cf. Thomas son of *Aldeston* 1332 SRCu, Richard *Aldston* 1185 Templars (Herts). *v.* also ELSTON. (ii) These surnames may also be local in origin: Roger *de Alleston* 1246 AssLa, from Alston in Ribchester (Lancs); John *de Alnethestan* 1230 P (D), from Alston in Holberton (Devon); William *de Alsiston'* 1221 AssWo, Richard *de Alston'* 1275 SRWo, from Alstone (Glos); Richard *de Aluredeston'* 1194 P (St), from Alstone in Hill

Chorlton (Staffs); and possibly also from Alston (Cumb, Som).

**Allum:** v. ALLAM

**Allured:** v. ALFRED

**Allvey, Alvey, Alvy, Elvey, Elvy, Elphee:** Godric *filius Ælfuuii* c1095 Bury (Sf); *Alfwy* 1212 Fees (Berks); Swein, William *Alvi* 1212 Cur (O); Thomas *Alfy* 1279 RH (C); Simon *Elphey* 1279 RH (D); Adam *Alfwy* 1296 SRSx; John, Agnes *Aluy* 1327 SRSx; Edward *Eluy* 1327 SRSx. OE *Ælfwīg* 'elf war'. v. ALAWAY, ELVEY.

**Allward, Alward, Elward:** *Aluardus, Alfwardus* pistor 1182–6, 1200–11 BuryS (Sf); Willelmus *filius Eluard, Ælfwardi* 1191, 1192 P (Nf); Robert *Elward* 1275 RH (Sf); Gilbert *Allward* 1279 RH (C). The DB *Æluuard, Aluuard, Eluuardus* may represent OE *Ælfweard* 'elf guard' which is clearly represented above or OE *Æðelweard* 'noble guard' which survives as *Aylward* and also appears as *Alward*. In the 12th century the two names were confused. *Ailwardus* grossus and *Ælward* judex occur also as *Alfward* and *Ailward* respectively (c1116 ELPN).

**Allway:** v. ALAWAY

**Allweather:** William *Alweder* 1395 AssL. Robert *Alwether* 1500 NorwW. 'All kinds of weather', OE *eall, weder*. cf. John *Strangweder* 1249 AssW 'strong weather'; Richard *Wetweder* 1392 IpmGl 'wet weather'. cf. FAIRWEATHER.

**Allwood:** v. ELLWOOD

**Allwork:** Robert *de Aldwerc'* 1275 RH (L). From Aldwark (Derby, NRYorks, WRYorks).

**Allwright:** v. ALLRIGHT

**Ally:** v. ALLEY

**Almack:** v. ALLMARK

**Alman, Almon:** v. ALLMAN

**Almand:** v. ALLMAND

**Almond:** *Ælmund, Almund, Æilmundus, Ailmundus* 1066 DB; *Alward filius Elmund* 1086 DB (Sa); *Eilmund Sperie* 1224 Cur (Ess); Thomas *Ailmun* 1279 RH (C); Thomas *Awmond* 1562 FrY; Laurence *Almond* 1564 LaWills. OE *Æðelmund* 'noble protector' or *Ealhmund* 'temple-protector'.

**Almot, Almott:** Robert *Almot* 1298 AssL; Thomas *Almot* 1304–5 IpmY; Alan *Almot* 1312 FFY. OE *Æðelmōd*.

**Alner:** Roger *Alnard* 1317 LLB D; John *Alnard* 1332 SRSx. 'One who tests the measurement of cloth', from a derivative of French *aune* 'ell'. v. also ALLNATT.

**Alnwick, Annick:** John *de Alnewyc'* 1230 P (Nt); Henry *de Aunewyk* 1278–9 PN Berks 324. From Alnwick (Nb), or Antwicks Manor in Letcombe Regis (Berks).

**Alp, Alpe, Alps:** Matilda *Alpe* 1275 RH (Nf); James *Alpes* 1664 FrY. A nickname from the bull-finch, ME *alp(e)* (c1400 NED).

**Alphege, Elfick, Elphick, Elphicke, Elvidge:** *Ælfec* (Ha), *Ælfech* (Sx), *Ælfhag* (Nt), *Alfeg* (Co), *Alfah* (Nf), *Elfeg* (Db), *Elfac* (Sa) 1066 DB; *Alfegus* 1137 ELPN; *Elfegus* de Ernington 1166 P (Gl); *Elphegus* a1216 StGreg (K); *Alfeg'* ater *Legh* 1296 SRSx; John *Elpheg'* 1297 Coram (K); William *Alfegh* 1318 AD vi (K); Robert *Elfegh* 1526 KentW; William *Elphike*

1549 RochW; Margaret *Elvishe* 1609 YWills. OE *Ælfhēah* 'elf-high'. *Alphege* is a Norman form.

**Alpin, Alpine, Elfin:** *Elphin* or *Elpin* c1225 Black; *Alpinus* 1287, *Alpin* mac Donald 1295 ib. GalloLat *Alpinus*, Welsh *Elffin*, the name of two or three Pictish kings.

**Alshioner:** v. ELESENDER

**Alsop(p):** v. ALLSEP

**Alstead:** v. HALSTEAD

**Alston(e):** v. ALLSTON

**Altham:** Adam *de Eluetham* 1246 AssLa; Symon *de Aluetham* 1304 WhC; John *Altham* 1372 LaPleas. From Altham in Whalley (Lancs), *Elvetham* c1150.

**Althorp, Althorpe:** Gimpe *de Aletorp'* 1179 P (L). From Althorpe (L), or Althorp (Nth).

**Alton:** Simon *de Altun* c1141–54 RegAntiquiss; John *de Alton'* 1219 AssL; Peter *de Alton* 1325 IpmNt; Thomas *Alton* 1508 CorNt. From one or other of the many places of this name.

**Alty:** v. AUTIE

**Alured:** v. ALFRED

**Alvar, Elvar, Elver:** *Ælfere* (Nf), *Alfer* (K, Sx), *Elfer* (Sx) 1066 DB; *Alfare* de Neteltuna Hy 2 Gilb (L); Godwinus *(filius) Elfare* 1221 ElyA (Nf); Thomas *Elvare* 1499 ArchC 37. OE *Ælfhere* 'elf-army'. The surnames are rare.

**Alven, Alvin, Elven, Elvin, Elvins:** Hugo *filius Elfwin* 1193 P (He); *Eilwinus, Elfinus* de Benindenne 1214 StGreg (K); John *Alvene* 1279 RH (C); Richard *Elvene* 1296 SRSx; Thomas *Eluyn* 1327 SRWo; Thomas *Aluene, Alwyne* 1351 AssEss, 1357 FFEss. OE *Ælfwine* 'elf-friend' or *Æðelwine* 'noble-friend'. v. AYLWIN, ALWIN. cf. WOOLVEN from *Wulfwine*.

**Alverton:** Jukel *de Alvertun* 1160–9 MCh; Robert *de Alverton* 1290 IpmY; Decima *de Alverton* 1327 SRY. From Alverton (Co, Nt).

**Alveston:** Philip *de Alverstayn* 1276 IpmY. From Alvaston (Ch, Db), or Alveston (Gl, Wa).

**Alv(e)y:** v. ALLVEY

**Alway:** v. ALAWAY

**Alwin, Alwen, Alwyn, Allwyn, Elwin, Elwyn:** (i) *Alduin, Elduinus* 1066 DB; Walter *filius Heldewin* 1191 P (L); *Aldwinus* de Vivario 1207 Cur (Ess); Osegod *Aldwini* 1195 P (Berks); Cristina *Aldwyne* 1275 SRWo; William *Aldwen* 1327 SRC. OE *Ealdwine* 'old friend'. (ii) *Æluuinus* (*Eiluuinus* Exon) 1066 DB (D); Robertus *filius Ailwini, Alwini* 1213–14 Cur (Ess). OE *Æðelwine* 'noble friend', which survives as *Aylwin*, appears in DB as *Aluuine, Eluuinus* and such forms as *Alwin, Elwin* infra may derive from this. (iii) *Æluin, Alfuuinus, Aluuinus, Eluuin* 1066 DB; Goduine *Ælfuini filius* c1095 Bury (Sf); *Elfwinus* de Bekeringe 1165 DC (L). OE *Ælfwine* 'elf friend'. From this, too, may derive *Alwin, Elwin*. There was early confusion between *Ælfwine* and *Æðelwine*, both of which are found as *Alwine*: *Ælfwin Finche* is probably identical with *Ailwinus, Alwinus* Finch 1168, 1173, and *Alfwinus filius Leofstani* with *Ailwin* son of *Leofstan* (ELPN 12). cf. also Robertus *filius Ailwini, Ailfwini* 1214 Cur (Nth). (iv) *Æluuin, Aluuin(e), Eluuin(e)* 1066 DB; *Elwine Ecses* 1101–7 Holme (Nf); *Ælwine* presbiter 1127–34 ib.; *Alwinus Blundus* 1207 Cur (Sx);

John *Allewin* 1219 AssY; Thomas *Alwine* 1260 AssC; Geoffrey *Elwyne* 1274 RH (Nf); William *Helewyn*, Thomas *Alwyn* 1296 SRSx. All these may derive from OE *Ealdwine, Æðelwine*, or, least common, *Ælfwine*. v. also ALVEN.

**Amabell, Amable:** *Amabilia* (f) 1202 AssL; *Amabillia* de Brunham 1208–9 Pleas: Richard *Amable* 1275 SRWo. A feminine name from Lat *amabilis* 'loving'. As a christian name it was later superseded by the shortened form *Mabel*.

**Amand, Aman, Amann, Amans, Ament, Ammann:** Magister *Amandus* 1199 Cur (Lei); *Amanda* filia Johannis 1221 AssWa; *Amanus* de Preston' 1221 AssSa; Richard *Amand* 1279 RH (C); Robert *Amant* 1332 SRSx; Simon *Ament* 1674 HTSf. OFr *Amand, Amant*, Lat *amandus* 'meet to be loved', the name of a 5th-century Bishop of Bordeaux and of four saints. Also used as a woman's name.

**Amar:** v. AYMER

**Amberley, Amberly:** Ralph *de Ambrelee* 1207 Cur (Sx); William *de Amberleg'* 1225 PN W 209; Robert *de Amberlye* 1296 SRSx. From Amberley (Gl, He, Sx).

**Ambler:** (i) William *le Amayler* 1303 LoCt; Simon *le Amelour* 1344 MESO (So); Robert *Ambeler* 1375 LoPleas. OFr *esmailleur* 'enameller', with intrusive *b*. (ii) Thomas *le Amblur* 1276 RH (L); Nicholas *le Aumbleour* 1307 Wak (Y); Alexander *Ambler* 1474 FrY. NED has *amblere* c1386 in the sense 'an ambling horse or mule', a derivative of *amble*. The surname may mean 'keeper of the stable' or be a nickname for one with an ambling gait or a facetious nickname for a fuller. cf. John *Ambeler*, walker 1440 FrY.

**Amblin:** v. EMBLEM

**Ambrose, Ambrus:** *Ambrosius* 1168–75 Holme (Nf); Henry *Ambreis* 1279 RH (O); William *Ambroys* 1332 SRSx. Lat *Ambrosius*, Fr *Ambroise*, from Gk ἀμβρόσιος 'divine, immortal'.

**Amelot, Amlot:** *Emelot* Minne 12th NthCh (Nth); *Ricardus filius Amelot* 1275 RH (D); Robert *Emelot* 1183 P (Lo); Nicholas *Amelot* 1275 RH (W). OFr *Amelot, Emelot*, a hypocoristic of *Ameline*. v. EMBLEM.

**Amery, Amori, Amory, Emary, Emery, Emory, Embery, Emberry, Embrey, Embry, Embury, Emeric, Emerick, Emerig, Imbery, Imbrey, Imbrie, Imery, Imray, Imrie, Hemery, Hembrey, Hembreye, Hembry:** *Amalricus* 1086 DB; *Ymerus* filius Reineri c1160 DC (L); *Haimeri* 1170 P (St); *Haimericus* filius Gidhe c1190–5 DC (L); *Eimericus* uinitor 1191 P (Wa); Robertus *filius Amalrici, Almarici, Aumarici* 1207–14 Cur (Nt); *Amauricus, Amaricus, Ailmarus, Almarus, Aumaricus, Almaricus* de Sancto Amando 1221 Cur; *Hemericus, Eymericus, Heimericus* Buche 1222 Cur (Do); *Emeric* Orchard 1241 AssSo; *Emery* de Roche Chaward 1269 AssSo; *Aymery* de Rupe Cawardi 1278 AssSo; Roger *Hemeri* a1182 Clerkenwell (Ess); Robert *Amalri* 1207 Cur (O); Richard *Aumauri* 1221 Cur (Do); Robert *Emeri* 1223 Pat (Lei); Robert *Heymeri* 1240 Rams (C); Simon *Amarik* 1260 AssY; William *Emeric* 1276 LLB A; Robert *Amary, Amory* 1279 RH (Bk); Walter *Ymery* 1513 Black; John *Imbrie* 1611 ib. OFr *Amauri, Emaurri*, from OG *Amalric* 'work-rule'.

**Ames, Amess, Amis, Amiss, Amys, Amos, Amoss:** Robert *Amis* 1221 ElyA (Sf); Joan *Amice*

1279 RH (O); John *Amize, Amys* 1309–10 LLB D; Gregory *Amys* or *Amisse* 1525 Oxon. Fr *Amis* (m) or *Amice* (f). *Amis* cas-sujet, *Ami* cas-régime, is from Lat *amicus* 'friend', used in late Latin as a name for the lower classes, especially for slaves. There were also derivatives of this, *Amicius* (m) and *Amicia* (f). Both were in use in England: *Amisius* 1200 Cur (D), 1213 ib. (Sr), *Amisius* de Hospitali 1211 Cur (Herts): *Amicia* 1189 DC (L), 1207 Cur (Ess), 1210 ib. (Wa). *Amos(s)* is due to the influence of the Biblical name which was not used in England before the Reformation. v. AMIAS, AMIET. Occasionally the surname is from the cas-sujet of the noun: William *le Amiz* 1275 SRWo.

**Amey, Amy, L'Amie, Lamey:** (i) William *Amy* 1219 AssY; William *Lamy* 1275 RH (Lo); William *le Amy* 1282 LLB A. Fr *ami* 'friend'. (ii) William *Ame* 1248 *Ass* (Ess). Fr *Amé*, from Lat *amatus* 'beloved'. (iii) *Amia* cameraria 1193 P (L); Martinus *filius Amye* 1297 SRY. The Latin form of OFr *Amée*, from *Amata*, the feminine of *Amé*. Occasionally we may have the masculine *Ami*, cas-régime of *Amis*. cf. Rogerus *filius Ami* c1250 Rams (Nf) and v. AMES.

**Amias, Amyas:** Richard *Amias* 1185 Templars (Wa); Willard de *Amiens* 1193 P (Y); Roger *de Amias* 1276 LLB A; John *Amyas* 1296 SRSx. From Amiens. cf. 'merchants of *Amias*' 1326 LLB E. In the 16th century the surname was confused with *Ames*. Thomas *Amyas* of Wethersfield (1521) was, no doubt, of the same family as Robert *Amys, Amyse, Amyce, Ameys* (1462–78 ER 61).

**Amiel:** Alicia *Amyel* 1327 SRSf; Joan *Amyel*, daughter of *Amiel* de Honesdon 1349 Husting. A diminutive of *Ami* or *Amia*. cf. AMIET.

**Amies:** v. AMES

**Amiet, Amiot, Amyot:** William *Amiot* 1195 P (Gl), 1279 RH (O); Stephen *Amyot* 1317 AssK. *Ami* (m) or (f), plus *-ot*. *Amiot* de Wudestoch' 1191 P (Berks) is identical with *Amisius* filius Amisii de Wodestoke 1250 Eynsham (O), a clear case of the pet-form of the masculine *Amis* or *Ami*. v. AMES, AMEY.

**Amis, Amiss:** v. AMES

**Amison, Amson:** John *Amysone* 1358 Putnam (Nth); John *Amysson* 1384 Cl; Matthew *Amson* 1568 Bardsley. 'Son of *Amice, Amis* or *Ami*'. v. AMES, AMEY.

**Amlot:** v. AMELOT

**Ammann:** v. AMAND

**Ammon, Ammonds, Amon, Amond:** *Agemund, Aghemundus* 1066 DB; *Hagemundus* c1150 DC (L); *Agmundus* de Gutario 12th DC (L); *Amund* de Cotes Hy 2 DC (L); Alexander, Ralph *Aghemond* 1327 SRSx; Margeria *Awomond* 1327 SRSo; Henry *Amond* 1380 AssC; James *Amon*, Charles *Amons* 1674 HTSf. ON *Qgmundr*, OSw *Aghmund*.

**Amoore:** John *of Moore* 1467 BuryW; William *A moore* 1479 SIA xii. 'Dweller by the marsh'. v. MOOR.

**Amor, Amour:** Adam *Amour* 1327 SRSf. OFr *amo(u)r* 'love'.

**Amori, Amory:** v. AMERY

**Amos(s):** v. AMES

**Amy:** v. AMEY

**Ancell:** v. ANSELL

**Ancliff(e):** v. ANTCLIFF

**Ancy:** William *de Anesi* c1110 Winton (Ha). From Anisy (Calvados). v. also DANCEY.

**Anderby:** Robert *de Anderbi* c1200 RegAntiquiss; Alan *de Anderby* 1260 FFL; Hugh *de Anderby* 1300, Thomas *Andyrby* 1411 IpmY. From Anderby (L).

**Anders:** *v.* ANDRE

**Anderson, Enderson:** Henry *Androsoun* c1443 Black; John *Andrewson* 1444 ib.; Robert *Androwson, Androson* 1455, 1482 GildY; Thomas *Anderson* 1471 FrY; Thomas *Enderson* 1674 HTSf. 'Son of *Andrew*'.

**Anderton:** William *de Anderton* 1260 AssLa; Robert *of Anderton* 1401 AssLa; George *Anderton* 1642 PrD. From Anderton (Ch, La).

**Andison:** Geoffrey *Andisone* 1376 Black. 'Son of *Andie* or *Andy*', a pet-form of *Andrew*.

**Andre, Andrea, Andree, Andrey, Andress, Andriss, Anders:** Josep *Andree* 1229 Cl; Geoffrey *Andre* 1279 RH (C); John *Andres* 1326 LoPleas. Michaëlsson notes that in the Paris tax-rolls of 1292–1313 the common form of *Andrew* is *Andri* whilst *Andriu* is rare. *Andri* he explains as influenced by the Germanic *Andrik* or the Graeco-Latin *Andricus*. In England and Scotland all early forms of the christian name are in the learned form *Andreas*. In surnames we find both *Andre* and *Andreu* at the beginning of the 13th century. *Andre* is, no doubt, this French *Andri* with a lowering of *i* to *e*. The modern *Andre* often appears as *André*, sometimes a recent introduction from France, though one suspects that the accent is not always original.

**Andrew, Andrewes, Andrews, Andros, Andrus:** *Andreas* 1086 DB, a1242 Black (Moray); *Andreas filius Hugonis* 1147–53 DC (Nt); William *Andreu* 1237 Fees (Bk); Moricius *Andrewys* 1275 SRWo; Richard *Andrew* 1317 AssK; William *Andro* 1399 Black (Aberdeen); John *Andrus* 1510 NorwW; Humfrey *Andros* 1552 FrY; Anthony *Androwes* 1543 FFHu. Greek Ἀνδρέας from ἀνδρεῖος 'manly', was latinized as *Andraeus, Andreus*, whence Fr *Andrieu*, English *Andrew. v.* ANDRE.

**Anflis, Anfliss:** *Amphelisia* vidua 1198 Cur (Lei); *Anfelisa* (f) 1221 AssGl; Roger *Anflis* 1285 IpmY; Hugh *Aunflys* 1327 SRLei. *Amphelisia* is found as a woman's name from the 12th to the 18th century, but the etymology is unknown.

**Anford, Anforth:** Richard *de Aneford'* 1222 AssWa; John *de Aneford* 1278 IpmGl. Probably for HANFORD.

**Angear:** *v.* ANGER

**Angel, Angell:** Warinus *Angelus* 1193 P (K); Ralph *Angel* 1221 ElyA (Sf). A nickname, from OFr *angele*, Lat *angelus* 'messenger, angel'.

**Anger, Angear, Angier, Aunger, Ainger:** Ansgarus, Anser, Angarus, Angerus 1066 DB; *Angerus de Middelton* 1191 P (Sf); *Aunger* the Pheliper 1277 LLB A; Lefuine *Anger* c1095 Bury (Sf); Willelmus *Angeri* 1197 P (Wa); John *Aunger, Anger* 1279–80 AssSo. A continental personal name, either a Norman form of ON *Ásgeirr*, or Fr *Angier* from OG *Ansger*.

**Angers, Aungiers:** Hugo *de Angiers, de Angers* 1207–8 Cur (W). From Angers (Maine-et-Loire).

**Angle, Angles, Nangle:** William *del Angle* 1208 P (Gl); Richard *in the Angle* 1327 SRWo; Peter *Nangle* 1571 Oxon. 'Dweller in the nook or outlying spot', ME, OFr *angle*.

**Angless, Angliss, Anglish:** Nicolaus *Angleis* c1200 StP (Lo). Fr *anglais* 'Englishman'. *v.* ENGLISH.

**Angold:** Alice *Angold* 1326 Wak (Y); Stephyn *Angold* 1568 SRSf. There is no second element *-gold* in OE or ON personal-names. This surname is identical with *Angood*, with a change of *-god* to *-gold* on the analogy of such ME names as *Alwold, Albold*, etc. *v.* OSGOOD.

**Angood:** *v.* OSGOOD

**Angove:** George *Angove* 1591 CoWills. Cornish *an* 'the' and OC *gof* 'smith'.

**Anguish:** Margaret *Anguisshe* 1530, Erl of *Anguyshe* 1563 Bardsley. For ANGUS.

**Angus:** Gael, Irish *Aonghus* 'unique choice'. The surname is also local, from Angus: Serlo *de Anegus* 1229, Andrew *Anguis* 1573 Black.

**Angwin:** (i) William *Angeuin* 1150 Oseney (O); Reginald *Langeuin* 1194 P (K); Godfrey *Aungewin* 1247 AssSt. 'The Angevin', the man from Anjou. (ii) Also a Cornish name, with the article *an* and *gwynn* 'the white'.

**Anke:** *Anke de Ankinton'* 1188 P (L); John *filius Anke* 1277 Ely (Sf); Roger *Anke* 1275 RH (Nf). A shortened form of ON names in *Arn-*.

**Anker, Ankers, Ankier, Anchor, Annacker, Annercaw:** *Anker de Fressenvill'* 1208 Cur (Nth); Alice *Anker* 1395 NottBR. OFr *Anchier*.

**Anketell, Ankettle, Anquetil, Ankill, Antell, Antill:** *Anketillus* c1155 DC (Lei), 1207 Cur (L); Elyas *filius Ankil* 1210 P (So); Geoffrey *Anketil* 1209 P (Nf); John *Antell* 1524 SRSf. A Norman form of ON *Ásketill. v.* ASHKETTLE, ASKELL.

**Anketin:** *Anketin* c1175–99 Black; *Anketin* Madlure 1245 FFY; Roger *Anketin* 1209 P (Nf); Simon *Anketyn* 1249 AssW. ON *Ásketill. v.* ANKETELL. The ending *-in* usually replaced *-il* outside Normandy.

**Ankin:** *v.* ANTIN

**Ankrett, Akritt:** *Angharad* 1207 Cur (He); *Anachorita* 1221 ElyA (Sf); *Angaret* 1246 AssLa; *Ankharet* 1322 AD vi (Ch); *Ancreta* Dibney 1571 ER 62; Gylbart *Ancret* 1580 ChwWo. A Welsh woman's name, *Angharad*, from *an* 'much' and *cariad* 'loved one', recorded in Wales in 877.

**Annable, Anable, Annible:** *Amabilis* c1150–60 DC (L), 1197 FFEss; *Amable* de Creuequer Hy 2 DC (L); *Amabilia*, predicta *Mabilia* 1185 RotDom (Ess); *Amabilla* (*Amable*), *Mabillia* (*Amabilis*) 1200 Cur (Db); *Anabel* de Nostrefeld 1204 AssY; *Annabel* 1260 AssC, (*Anilla*) 1374 Ipm (La); *Anabilia* 1284 FFHu; *Anabilla* 1305 FFEss; *Anabella* c1308 Calv (Y); *Anabel, Amabel* 1312, 1313 AssSt (identical); Walterus *Amabilis* 1166 P (Nf); Robert, John *Anable* 1259 CtSt, 1282 Pat (Gl); John *Amable* 1275 RH (C). Lat *amabilis* 'lovable'. *Amable* was shortened to *Mabel* (*v.* MABLE) and also, apparently by a process of dissimilation, became *Anabel*. Occasionally the surname may be local in origin, from Amble (Northumb), *Anebell* 1256, *Anebelle* 1292: Henry *de Anebelle* 1256 AssNb. *v.* HUNNABLE.

**Annacker:** *v.* ANKER

**Annakin, Annikin:** *v.* ANTIN

**Annand:** *Anundus, Anunt dacus, Anand, Anant* 1066 DB (Sf, Ess, Nf); *Anund* 1101–7 Holme (Nf); Godefridus *filius Anandi* 1182 P (Sf); Roger *Anant* 1275 RH (Nf). ODa, OSw *Anund*.

**Annas, Anness, Annis, Anniss:** *Annes* 1170–76 YCh; *Agnes, Anneis* (identical) Hy 2 DC(L); *Annas* de Preston 1194 P (La); Adam *Anice* 1275 SRWo; John *Anneys* 1296 SRSx. OFr *Anés*, the vernacular form of *Agnes*.

**Annatt, Annett, Annetts, Annott:** Annote Resthanbe 1297 Wak (Y); Thomas *filius Anot* 1357 ShefA; Robert *Anot* 1275 Wak (Y); John *Annot* 1327 SRC. *Ann-ot*, a diminutive of *Ann*, a pet-form of *Annes* (Agnes).

**Anne:** Ralph *de Anne* 1200 P (Ha). From Ann (Hants).

**Annercaw:** *v.* ANKER

**Annesley, Ansley:** Reginald *de Aneslega* 1176 P (Nt); Henry *de Anesley* 1290 KB (Lo); John *Anneslay* 1404 IpmY. From Annesley (Nt), or Ansley (Wa).

**Anness:** *v.* ANNAS

**Annett, Annetts:** *v.* ANNATT

**Annick:** *v.* ALNWICK

**Annis, Anniss:** *v.* ANNAS

**Annison:** (i) William *Anyssone* 1332 SRSt. 'Son of *Annis*'. (ii) Roger *Annotson* 1430 FeuDu; Thomas *Annetson* 1547 FrY. 'Son of *Annot*'. *v.* ANNATT.

**Annott:** *v.* ANNATT

**Anquetil:** *v.* ANKETELL

**Ansell, Anshell, Ansill, Ancell, Hancell, Hansel, Hansell, Hansill:** *Anse(l)lmus* archiepiscopus 1094–5, 1108 StP; *Anselmus, Ansellus* de Ixew(o)rth' 1182–1211 BuryS (Sf), 1208 Cur (Nf); *Ansel, Anselmus* Candau' 1161 P (Ess); *Ancelmus* de Montegni 1166–89 Bec; *Anselmus* de Shelton', *Aunsell'* de Sheldon', *Ansellus* de Seldon' 1221–2 AssWa; *Ancell, Anselm* de Gornay 1269 AssSo; *Aunsel* le Furbur 1296 LLB A, *Anselm* forbisor 1300 ib. B; Petrus *Anselmus* 1192 P (Sx); Roger *Aunsel* 1271 AssSt; William *Ansel* 1279 RH (C); John *Auncel* 1327 SRSx; William *Hansell* 1495 FrY. *Anselm*, a Lombard name, from OG *Ansehelm* 'god-helmet', found in France as *Ansellus*, where, as in England, *Anselm* and *Ansell* were used of the same man.

**Anslyn:** William *Hanselyn* 1379 PTY. OFr *Anselin*, a diminutive of *Ansell*.

**Anson:** John *Anson* 1401 AssLa; Henry *Anson* 1461 FrY. Either 'son of *Hand*', or 'son of Hann'. *v.* HAND, HANN.

**Ansthruther:** Geoffrey *de Einstrother* a1214 Black; William *de Heynstrother* alias *de Aynstrother* 1287 IpmNb; Henry *de Anstrother* 1296 Black. From the lands of Anstruther (Fife).

**Anstee, Anstey, Anstie:** Richard *de Anesti* 1164 P (Ess). From Anstey (Devon, Dorset, Hants, Herts, Wilts) or Ansty (Warwicks).

**Ansteys, Anstice, Anstis, Anstiss:** *Anstasius* de Schirbec 1188 P (L); *Anastasia* 1221 Cur (Herts); *Anstice* 1602 Bardsley; Osegod, William *Anastasie* 1222 FFBk; Ralph *Anstayse* 1327 SRSo. The first surname above certainly derives from the feminine *Anastasia*, though the masculine *Anastasius* also existed. Both are from adjectives from Greek ἀνάστασις 'resurrection'.

**Antcliff, Antcliffe, Antliff, Ancliff, Ancliffe:** Thomas *de Arnecliv'* 1207 Cur (Y); Luke *Antcliff* 1748 Bardsley. From Arnecliff, Ingleby Arncliffe (NRYorks) or Arncliffe (WRYorks).

**Antell, Antill:** *v.* ANKETELL

**Anthoney, Anthonies, Anthony, Antoney, Antony:** *Antonius* Hy 1 Rams (Hu), 1149 NthCh (Nth), 1214 Cur (D); John, Richard *Antoyne* 1275 SRWo; William *Antony* 1306 FFSf. Lat *Antonius*, Fr *Antoine*.

**Antin, Ankin, Annakin, Annikin:** *Hanchetin* de paruo Stepinge Hy 2 DC (L); *Anketin* 1219 AssY; Roger *Anketin* 1209 P (Nf). A Norman form of ON *Áskell*. *v.* ASKELL. *Hanccetin* de Lud Hy 2 DC (L) is also called *Asketillus*. Annakin and Annikin, rare Yorks names, are probably for *Ankin*. cf. *Ankill* and *Antill* from *Anketell*.

**Antliff:** *v.* ANTCLIFF.

**Antrobus:** Richard *Antribussh* 1380–1 PTW. From Antrobus in Higher Whitley (Ch).

**Anwyl, Anwyll:** Jevan *Annwyl* 1391 Chirk. Lewis ap Robert of Park, Llanfrothen, Merioneth, d. 1605, is said to have been the first of his family to adopt *Anwyl* as his surname; his son was Lewis *Anwill* 1612 Reg. Oxon. *v.* Morris 118. Welsh *anwyl* 'dear, beloved'.

**Anyan, Anyon:** *v.* ENNION

**Apehead:** John *Apeheved* 1305 AssW. 'Ape head', ME *ape*, OE *héafod*. cf. James *Horsed* 1641 PrSo 'horse head'; William *Shepesheved* 1276 AssLo 'sheep's head'; Randulf *Hundesheved* 1176 P (D) 'hound's head'.

**Apley, Appley:** Roger *de Appelea* 1190 P (Ha); Nicholas *de Appleye* 1268, *de Apperleg, de Appeleg'* 1269 AssSo. From Apley (L, Sa, So, Wt), Appley in Chicksands Priory (Beds), or Appley Bridge (La).

**Aplin, Applin, Appling:** Thomas *Abelyn* 1275 RH (K); John *Applyn* 1547 FrY. *Ab-el-in*, a double diminutive of *Ab*, a pet-form of *Abel*.

**Appelbe(e):** *v.* APPLEBY

**Apperley, Apperly:** Richard *de Apperlee* 1221 AssGl; Thomas *de Apperleye* c1280 SRWo; Walter *Appurlee* 1372 IpmGl. From Apperley (Glos, Northumb, Som), or Apperley Bridge (WRYorks).

**Appleby, Applebe, Applebee, Applebey, Appelbe, Appelbee:** Vlf *de Appelbi* 1163 P; Hugh *de Apelby* 1204 Pl (Y); Thomas *Appelby* 1366 AssLo. From Appleby (Leics, Lincs, Westmorland).

**Appleford:** John *de Appelford* 1206 Cur; William *de Appelford* 1247 FFEss, 1285 FFO. From Appleford (Berks, Wt).

**Applegarth, Applegath, Applegate:** William *de Apelgart* c1115 Bury (Sf); Robert *Appelgarth* 1279 Ipm (Y); Richard *del Appelgarthe* 1297 MinAcctCo (Y). From Applegarth (NRYorks, ERYorks, Cumb), or from residence near an apple-orchard (ON *apaldr(s)garðr*) in a Scandinavian county.

**Appleton, Napleton:** Hemeri *de Lepeltone* c1182 RBWo; Tomas *de Appleton'* 1196 P (Y); Thomas *ate Napeltone, ate Apiltone* 1317 AssK. From Appleton (Cheshire, Kent, Yorkshire, etc.), or from residence near an orchard (OE *æppeltūn*), or at a homestead where apples are grown.

**Applewhaite, Applewhite, Ablewhite:** Stephen *Appeltheit* 1327 SRSf; Robert *Apylweyte* 1524 SRSf; George *Applewhite* 1674 HTSf; Sarah *Applewhait* 1678 SfPR; Henry *Ablewhite* 1797 ib. From

Applethwaite (Cumb, Westmorland), or a lost place in Suffolk.

**Appley:** v. APLEY

**Appleyard:** Elias *del Apelyerd* 1275 Wak (Y); John *del Apelyard* 1315 ib. From Appleyard (WRYorks) or from residence near an orchard, OE *æppel* 'apple' and *geard* 'enclosure'.

**Applin(g):** v. APLIN

**Apps, Aps, Asp, Epps, Happs, Hesp, Hespe:** John *de Apse* 1214 Cur (Sr); Robert *atte Hepse* 1296 SRSx; Robert *atte Apse* 1327 SRSx. 'Dweller by the aspen', OE *æpse*, a metathesized form of *æspe*.

**April, Averill, Avril:** Thomas *Averel* 1275 SRWo; Robert *Aprill'* 1301 SRY; Richard *Averil* 1322 AssSt; John *Aueril* 1327 SRSx. According to ODCN, *Averil* is to be associated with *Everild*, a christian name not uncommon in the Middle Ages, from OE *Eoforhild* or OG *Eburhilt*. These, however, would become *Everil*. Bardsley has no example of the surname before 1626 and confuses the name with *Avenell*. Harrison, without evidence, explains the surname as 'dweller at the wild-boar hill', which would also become *Everil*. There can be no doubt that here we are concerned with the name of the month, OFr *avrill*, Lat *aprīlis*, which appears in English as *aueril* in 1297 and as *averylle* c1450 (NED). The word was refashioned after the Latin and is found as *apprile* in 1377. There is no evidence for its use as a christian name and the surname must be regarded as a nickname, perhaps with reference to the changeable weather of the month, 'changeable, vacillating', or with reference to spring or youth. Dauzat explains the French *Avril, Abril* thus, with an alternative 'child found in April'. It might denote one born in that month. The modern christian name *Avril* is of recent origin.

**Apsley:** Simon *de Apsele* 1297 MinAcctCo; Stephen *de Apsele* 1327 SRSx. From Apsley Guise, Apsley End in Shillington (Beds), or Apsley Fm in Thakeham (Sx).

**Apthorp, Apthorpe:** Thomas *de Abetrop* 1180 P (Nth), 1197–8 LuffCh; Juetta *de Abethorp* 1201 Cur. From Apethorpe (Nth).

**Apton:** William *de Appelton* alias *de Appetone* 1268 IpmEss; John *de Appeton* 1279 FFEss. From Apton Hall in Canewdon (Ess).

**Araby:** (i) Robert *Arrabi* 1198 Cur; Ralph *Araby* 1221–2 FFWa; Pance *Arabi* 1288, *Raby* 1290 LLB A. 'The Arabian', perhaps a nickname for one with a swarthy complexion. (ii) Robert *de Areby* 1195 P (Lei). From Harby (Lei).

**Aram, Arram, Arrum, Arum:** Henry *Arowme* 1456, Robert *Arwome* 1500 FrY; Robert *Aram* 1649 RothwellPR (Y). From Arram in Leconfield (ERY), or perhaps from Averham (Nt), *Aram* 1280.

**Arber:** v. HARBER

**Arbery, Arberry, Arbury:** Henry *Erburgh* 1332 SRSx. From Arbury (La), Arbury Banks in Ashwell (Herts), or 'dweller by the earthwork', OE *eorð-burh*.

**Arblaster:** v. ALABASTAR

**Arborn, Arbon, Arboune:** Erneber, Ernebernus, Gernebern 1066 DB (Y, L, Db); Arbernus 1190–5 DC (L); Roger *Arborn* 1279 RH (C); William *Arbon* 1329 Rams (Nth). ON *Arnbiqrn*, ODa, OSw *Arnbiorn*.

**Arbuckle, Hornbuckle:** John *Arnbukle, Arbukile* 1499, 1511 Black. From Arbuckle (Lanarkshire).

**Archambault, Archanbault, Archbald, Archbell, Archbold, Archbould, Archbutt, Archibald, Archibold, Archibould:** *Archembold* Wiverun 1130 P; Robert *Archenbold* 1210 P (Gl); William *Erchebaud* 1239 FFSf; Thomas *Herchebaud* 1302 SRY; Agnes *Archebald* 1327 SRSf; Seath *Archbutt* 1616 FrY. OFr *Archamboult*, from OG *Ercanbald* 'precious-bold', found in DB (1086) as *Erchenbaldus, Arcenbaldus* and *Arcebaldus*.

**Archdeacon, Arcedeckne:** Walter *le Ercedekene, le Ercedeiakene, le Archedekene* 1268–71 AssSo; Roger *le Arcedekne* 1297 MinAcctCo. OE *arce-, erçediacon*, OFr *arc(h)ediacne* 'the chief deacon, chief of the attendants on a bishop'.

**Archer, Larcher, L'Archer:** Edward *Archier*, Robert *Larchier* 1166 P (Ha, W); Hugh *le Archer* 1199 FFC. ME *archere*, AFr *archer*, OFr *archier* 'bowman' (c1300 MED).

**Arches:** Peter *de Arches* c1190 DC (L); Alan *de Archis* 1211 Cur (Y); John *Arches* 1327 SRWo. From Arques (Eure, Pas-de-Calais, Seine-Maritime).

**Archibald, Archibold, Archibould:** v. ARCHAMBAULT

**Arckoll:** v. ARKELL

**Arculus:** v. HERCULES

**Ardeley:** v. ARDLEY

**Arden, Ardin, Arding, Hardern:** Turkill *de Eardene* c1080 OEByn; Adam *de Arden* 1268 AssSo; Ranulph *de Ardene* 1318–9 FFEss. From Arden (Ch, Wa, NRY).

**Ardern, Arderne:** William *de Arderne* 1219 AssL; Thomas *de Arderne* 1301 FFY; William *de Arderne* 1372–3 FFWa. From Arden (Ch), *Arderne* 1260.

**Ardin, Arding:** v. ARDEN

**Arding:** v. HARDING

**Ardley, Ardeley:** John *de Ardeleye* 1327 SRSf; John *Ardeleigh* 1417 FFEss. From Ardleigh (Ess), Ardeley (Herts), or Ardley (O).

**Argent, Hargent, Largent:** Geoffrey *Argent* 1180 P (Nth); John *Largeant* 1524 SRSf. OFr *argent* 'silver', probably for one with silvery-white hair.

**Argentine:** (i) *Argentina* 1196 FFO, 1258 Acc; *Argenten* Roost, *Argintyne* Twiggs 1642 PrD. OFr *Argentin* (f), a diminutive of OFr *argent* 'silver' used as a personal name. (ii) Reginald *de Argentein* 1274 PN Ess 392; William *Argenten* 1439–40 Paston. From Argenton (Indre).

**Aries, Aris, Ariss:** v. ARRAS

**Arkell, Arkill, Arkle, Arckoll, Arkcoll:** *Arnketel* 1019 Rams (Nf); *Archetel, Archel, Archil* 1066 DB; *Archil* de Corebr' 1159 P (Nb); Richard *Arkill* 1256 AssNb; Roger *Arketyl* 1279 RH (Hu); William *Harkill, Arkel* 1316, 1331 Wak (Y); John *Arcle* 1455 FrY. ON *Arnkell*, ODa *Arnketil*, OSw *Arkil*.

**Arkless:** v. HERCULES

**Arksey:** Walter *de Archeseia* Hy 2 DC (L); Walter *de Arkesay* 1297 AssNb; Adam *de Arkesay* 1383 FrY. From Arksey (WRY).

**Arkus, Arkush:** v. HARCUS

**Arkwright, Artrick, Hartwright, Hartrick, Hattrick:** Gilbert *de* (sic) *Arkewright* 1246 AssLa;

John *de* (sic) *Arcwryt* Hy 3 DbCh; Thomas *the Arkewrytte* 1286 AssCh; George *Arkewright* 1450 FrY; William *Hartwright* 1494 GildY; George *Arthwright* 1582 PrGR; Thomas *Artwright* 1649 LaWills; Alice *Arthricke* 1670 ib.; Christopher *Arkrick* 1673 ib. OE *arc* 'ark' and *wyrhta*, 'a maker of arks, chests, etc.'.

**Arles:** *v.* ARLISS

**Arlet, Arlett:** Thomas *Attenerlette* 1276 MELS (Sx); Philip *atte Arlette*, Robert *aten Erlette* 1296 SRSx. A metathesized form of OE \**alrett*, \**elrett*, 'alder-grove'. *v.* ALDRITT.

**Arley:** (i) William *de Arle* 1275, Thomas *de Arleye* 1332 SRWo. From Arley (Ch, La, Wo), or Areley Kings (Wo). (ii) Thomas *de Erlyde* 1332 SRSt. Arley (St), *Erlide* 1221.

**Arling:** *v.* HURLIN

**Arlington:** Aelic' *de Erlynton* 1296, John *de Erlington* 1327 SRSx. From Arlington (D, Gl, Sx).

**Arliss, Arles, Harliss:** Henry *Earles* 1295 Barnwell (C). OE *ēare* and *lēas* 'ear-less'.

**Arlott, Arlotte:** Geoffrey *Herlot* 1193 P (Nf); Ralph *le Harlot* 1246 AssLa; John *le Harlet* 1276 RH (C); Muriel *Arlot* 1279 RH (C). OFr *herlot*, *harlot*, *arlot* 'lad, young fellow', found in English as a masculine noun in the 13th century in the sense 'vagabond, beggar, rogue' (c1230 MED). It was used by Chaucer both as a term of derision 'ye false harlot', and also playfully for a good fellow, 'he was a gentil harlot and a kynde'. In the 14th century it was used of an itinerant jester, buffoon or juggler, and also of a male servant, attendant, menial. The modern sense is not found before the 15th century.

**Arlow:** *v.* HARLOW

**Arm, Arme:** *v.* HARM

**Armatage, Armatys:** *v.* HERMITAGE

**Armenters, Darmenters:** David *de Armenteriis* 1169 P (St); Henry *de Armentiers* 1204 FFSt; John *de Armenters* 1298 LLB A, *Darmenters* 1299 LLB C. From Armentières (Nord).

**Armer, Armor, Armour, Larmer, Larmor, Larmour:** Gwydo *le Armerer* 1279 RH (O); Simon *Larmourer* 1334 LLB E; John *Armar* 1519 Black. AFr *armurer*, OFr *armurier* 'armourer, maker of arms'.

**Armes, Arms:** *v.* HARM

**Armett:** *v.* HERMITTE

**Armiger, Arminger:** William *Armiger* 1279 RH (O); Thomas *Arminger* 1674 HTSf. Lat *armiger* 'armour-bearer, squire'.

**Armin, Ermen:** *Ermyn* Donetoun 1327 SRC; Hervicus *Ermin* 1279 RH (O); Thomas *Armyn* 1481 FrY. *Ermin*, a hypocoristic of names in *Ermen-*, *Ermin-*, such as OG *Ermenald*, *Ermingaud*.

**Arminson, Armison, Armson:** Robert *Armison* 1669 FrY; Mary *Armson* 1749 Bardsley. 'Son of *Ermin*.'

**Armistead, Armitstead, Armstead:** Laurence *del Armetsted* 1379 PTY; William *Armistead* 1642 PrGR. 'Dweller by or at the hermitage,' OFr *ermite*, and OE *stede* 'place'.

**Armit(t):** *v.* HERMITTE

**Armitage:** *v.* HERMITAGE

**Armour:** *v.* ARMER

**Armson:** *v.* ARMINSON

**Armstead:** *v.* ARMISTEAD

**Armstrong:** Adam, William *Arm(e)strang* 1250 CalSc (Cu), 1279 AssNb. OE *earm*, *strang* 'strong in the arm', a well-known Border name.

**Armytage:** *v.* HERMITAGE

**Arnald, Arnall, Arnatt, Arnaud, Arnell, Arnhold, Arnold, Arnoll, Arnot, Arnott, Arnould, Arnout, Arnull, Harnett, Harnott:** (i) *Ernold*, Rogerus *filius Ernaldi* 1066 DB; *Arnoldus*, *Hernaldus* de Bolonia 1212 RBE; Puntius *Arnaldi* 1196 P (D); William *Arnold* 1277 FFSf; John *Arnoud* 1279 RH (C). OFr *Arnaut*, *Ernaut*, *Hernaut*, from OG *Arnald*, *Arnold*, *Arnolt* 'eagle-power'. (ii) Richard *de Ærnhale* 1191 P (Y), from Arnold (ERYorks); Roger *de Arnhal'* 1212 Fees (Nt), from Arnold (Notts).

**Arnason:** *v.* ARNISON

**Arndell:** *v.* ARUNDALE

**Arnet, Arnett:** *Erniet*, *Ernet* 1066 DB; *Ærnyeth* Hachchebuters 1185 P (Gl); *Arnet* le Mercer 1279 RH (O); John, Lecia *Arnet* 1279 RH (C). OE *Earngēat* (m), or OE *Earngȳð* (f).

**Arneway:** *Ernui*, *Erneuui*, *Arnui* 1066 DB; *Ernwi*, c1150 DC (Nt); *Arnwi* Hy 2 ib. (L); Peter *Ernewy* 1243 AssSo; Richard *Arnwy* c1272 HPD (Ess); Hugo *Ernawey*, Walter *Ernowey* 1279 RH (Bk). OE *Earnwīg* 'eagle-warrior'.

**Arnhold:** *v.* ARNALD

**Arnison, Arnason, Arnson:** William *Arnaldson* 1460 FrY; Matthew *Arnison* 1680 ib. 'Son of *Arnald*.'

**Arnold, Arnoll:** *v.* ARNALD

**Arnot, Arnott:** Michael *de Arnoth* 1284 Black. From Arnot in Portmoak (Kinross-shire). *v.* also ARNALD.

**Arnould, Arnout:** *v.* ARNALD

**Arnson:** *v.* ARNISON

**Arnull:** *v.* ARNALD

**Arpin:** *v.* HARPIN

**Arram:** *v.* ARAM

**Arrandale:** *v.* ARUNDALE

**Arras, Arres:** John *de Aroz* 1296; Adam *de Airwis* 1328; Adam *de la Arus* 1333; John *Arres* 1525 Black. Probably, as suggested by Black, from Airhouse in Channelkirk (Berwicks), *Aras* 1655, *Arreis* 1630.

**Arras, Aries, Aris, Ariss:** (i) Simon *de Araz* 1202 P (L); Robert *de Arras* 1281 LLB A. From Arras (France). (ii) Hugo *de Erghes* 1347 FrY; John *Aras* 1421 FrY. From Arras (ERYorks), *Herges* 1156. For *Aries*, cf. 'One bede coveringe of *aries* (i.e. *arras*) worke' 1562 Bardsley.

**Arridge:** Ralph *de Arnregg* 1296 SRSx; William *a Rygge* 1525 SRSx. From Eridge (Sx), but the second form may be for *atte Rigge* 'dweller by the Ridge'. RIDGE.

**Arrow:** William *de Arewe* 1204 P (Gl); William *Arwe* 1310 LLB D; Raff *Arrow* 1542 StaplehurstPR (K). From Arrowe (Ch), or Arrow (Wa).

**Arrowsmith, Arsmith, Harrismith, Harrowsmith:** Roger *le Aruesmuth* 1278 AssSt; William *le Arwesmyth* 1324 FFEss; Richard *Arsmith* Eliz Bardsley. OE *arwe* 'arrow' and *smith*. A smith who makes arrows, especially iron arrow-heads (1278 MED). *Harrismith* and *Harrowsmith* are rare but exist side by side with *Arrowsmith* in Yorks and Lancs.

**Arrum, Arum:** v. ARAM

**Arscott:** (i) William *de Ardescote* 1255 RH (Sa). From Arscott (Salop). (ii) John *Aryscote, Adescote, Addyscote, Adescote* or *Addyscoote* 1513 LP (D); John *Arscot* 1523 ib. From Addyscote in South Tawton (Devon), *Arscott* alias *Addiscott* 1656, with a surname *Atherscote* in 1297 (PN D 448). (iii) William *de Hassecote* 1201 Pl (Co); Alnet *Arscot* 1642 PrD. From Arscott in Ashwater or Arscott in Holsworthy (Devon).

**Arsnell:** v. HORSNAIL

**Arson:** v. HARSANT

**Arter:** v. ARTHUR

**Arthington:** Peter, Serlo *de Ardington'* 1175–94 YCh; William *de Arthyngton* 1283 FrY; Robert *Arthington* 1459 Kirk. From Arthington (WRYorks).

**Arthur, Arthurs, Arthars, Arter:** *Erturus* 1130 P (Ha); Henricus *filius Arturi, Artur* 1187 P, 1212 Cur (Cu); *Ærturus* 1192 P (Y); Normannus *filius Arcturi* 1196 P (Y); Geoffrey *Artur* 1135 Oseney (O); Robertus *Arcuri* 1197 P (He); Adam *Arthur* 1246 AssLa. The DB forms *Artor, Azer, Azor*, given by ODCN, are wrongly assigned. They are for ON *Arnþórr*, ODa *Azur* respectively. *Artur(us)* 1086 DB (Ess, Wo) is probably for *Arthur*, the well-known Celtic name, of disputed etymology.

**Artis, Artiss, Artist, Artois, Artus:** William *Artoys* 1327 SRSf; John *Artes* 1524 SRSf; Thomas *Arteis* 1568 SRSf; Robert *Artis* 1674 HTSf; Abraham *Artus* 1724 FrY. Man from Artois.

**Artrick:** v. ARKWRIGHT

**Arundale, Arundel, Arundell, Arondel, Arrandale, Arrundale, Arndell:** (i) Rogerius *Arundel, Harundel* 1086 DB (Do, So); Robert, Roger *Arundel* 1130, 1159 P (Do, So); Osbert *Arundel, Harundel* 1154 Riev (Y). (ii) Roger *de Arundelle* 1148 Eynsham (O); Thomas *Arundel (de Arundel)* 1178 P (Y); John *de Arundel* c1198 Bart (Lo); Roger *Arundel (de Arundel)* 1204 AssY; Nicholas *(de) Arundel*, Roger *de Arundel* 1225, 1268 AssSo; Robert *Arundel*, Robert *de Arundell* 1327, 1332 SRSx; Richard *Arrandell, Arrendale* 1631, 1638 RothwellPR (Y). The most usual source of these surnames must be Arundel (Sussex), *Harundel* 1086 DB, *Arundell'* 1087 France, *Arndle* 1788, locally pronounced *Arndell*. The earliest bearer of the name, Roger *Arundel*, the Domesday tenant-in-chief, has left his name in Sampford Arundel (Som) which he held in 1086. His byname cannot derive from the Sussex place but must be a nickname from OFr *arondel* 'little swallow'. So, too, with Osbert *Arundel*. Though they were, presumably, the respective ancestors of the Somerset and Yorkshire families, the later *de* introduced into their name suggests a Sussex origin. The problem is further complicated by the fact that Arundel itself early lost its initial *H*, whilst the nickname equally early acquired an inorganic initial *H*. The influence of Arundel with its castle was probably too strong for the nickname, the meaning of which was soon forgotten.

**Ary:** v. AIREY

**Asch:** v. ASH

**Ascham:** v. ASKAM

**Ascher:** v. ASSER

**Ascombe:** v. ASHCOMBE

**Ascot, Ascott:** Richard *de Askote* 1375 LLB H. From Ascot (Berks), or Ascott (Bk).

**Ascough:** v. ASKEW

**Ascroft:** v. ASHCROFT

**Ash, Ashe, Asch, Asche, Dash, Daish, Daysh, Nash, Rasch, Rash, Tasch, Tash, Esh, Tesche, Tesh, Aish, Aysh, Naish, Nayshe:** Richard *del Eshe* 1221 AssWo; Ralph *de Asche* 1296 SRSx; Alice *aten Eysse* ib.; William *atte Nasche* ib.; John *ater Aysse* ib.; Henry *Aten Assche* 1301 MELS (Wo); Richard *Dasche, de Ayssh* 1320 LLB E, 1327 LoPleas; Roger *ate Assh* 1327 SRSx; Alan *Tassh* 1327 SRSf; John *atte Naysshe* 1349 LLB F; William *Rash* 1381 SRSf; William *Teshe* 1599 FrY. From a place called Ash or Nash or 'dweller by the ash-tree' (OE *æsc*). v. Introd., p. xiv and NESS. *Dash* retains the French *de*.

**Ashard:** v. ACHARD

**Ashbee, Ashbey:** v. ASHBY

**Ashbery, Ashberry, Ashbury:** William *de Asseberia* 1190 P (Berks); Godfrey *de Asseberge* 1221 AssWo; Thomas *de Asshebury* 1349 FFY. From Ashbury (Berks, D), Ashborough in Bromsgrove (Wo), or Ashberry Hill in Old Byland (NRY).

**Ashbolt, Ashpool, Ashpole:** Robert *Archpoole* 1523 RochW (K); Edward *Archepoll, Archepole* 1591, 1595 ChwWo; Nicholas *Archepoll, Archebold* 1591, 1593 ib. *Ashbolt* is probably a corruption of *Archbolt*, i.e. *Archibald*, through *Archpoll*, which also became *Ashpole*.

**Ashborn, Ashbourn, Ashbourne, Ashburn:** Robert *de Asshebourne* 1311 LLB D; John *de Ascheburn* 1349 FrY; Henry *Assheburn* 1469 FFEss. From Ashbourne (Db), Ashburnham (Sx), *Esseborne* DB, or Ashburton (D), *Æscburnan* 1008.

**Ashbrook, Ashbrooke:** William *de Assebroc* 1190 P (Ess/Herts); Hugh *de Aschbroc* 1218 P (Lei/Wa); John *de Asshebrok* 1353–4 FFSr. From Ashbrook (Gl).

**Ashburn:** v. ASHBORN

**Ashburner:** Robert, William *le Askebrenner* 1278 AssLa, 1308 Wak (Y); Robert *Askbrinner* 1332 SRCu; Thomas *Asborner* 1332 SRSx; Edward *Ascheburner* 1532 FrY. ON *aska* 'ashes' plus a derivative of ON *brenna* or OE *brinnan* 'to burn', 'a burner of ashes' or maker of potash from the ashes of wood, bushes, straw, etc. In Sussex we have the native OE *æsc* plus a derivative of OE *beornan*, which had replaced the Scandinavian word in York by the 16th century.

**Ashbury:** v. ASHBERY

**Ashby, Ashbee, Ashbey:** Robert *de Aschebi* 1200 Pl (Nf); Alexander *de Assheby* 1309–10 FFWa; Richard *Assheby* 1332 SRSx; William *Ashbee* 1633, *Eashbee* 1637, *Eshbee* 1639 LeiAS xxiii. From one or other of the many places of this name, or from Asby (Cumb, Westmorland).

**Ashcomb, Ashcombe, Ascombe:** John *de Asshcomb* 1327 SRSo; Robert *Ascombe* 1376 LLB H; Thomas *Aishecombe* 1641 PrSo. From Ashcombe (D).

**Ashcroft, Ascroft:** Margar' *de Asecroft* 1275 RH (Nf). 'Dweller at a croft with an ash-tree.'

**Ashdown, Ashdowne, Ashdon, Ashdoun:** John *de Essedon'* 1294 KB (O); John *de Asshesdoune* 1327 SRSx; Thomas *Asshedon* 1363 FFEss. From Ashdown (Berks), Ashdon (Ess), or Ashdown Forest (Sx).

**Ashenden:** Gilbert *de Asshendone* 1303 AssNu; Philip *de Asshendone* 1339 CorLo; John *de Asshendene* 1342 AssLo. From Ashendon (Bk).

**Asher:** Thomas *Aisher* 1641 PrSo. Probably 'dweller by the ash-tree', from a derivative of OE *æsc* 'ash-tree'. v. also ASSER.

**Ashfield:** Sparhauoc *de Æssefelde* c1095 Bury (Sf); Robert *de Asshefeld* 1375 FFEss. From Ashfield (Sa, Sf).

**Ashfold, Ashfull:** Richard *de Asshefold* 1305 FFSx. 'Dweller by a fold near an ash', OE *æsc, falod.*

**Ashford, Ashforth, Aishford, Ayshford:** Reginald *de Asford'* 1221 AssSa. From Ashford (Devon, Derby, Salop) or Ayshford (Devon).

**Ashhurst, Ashurst:** Warin *de Asherst* 1192 P (Sx); Robert *de Ashurst* 1305–6 FFSr; Edmund *Asshurst* 1525 IpmNt. From Ashurst (K, Sx), or High Ashurst (Sr).

**Ashken:** v. ASKIN

**Ashkettle:** *Asketillus* de Kedham 1101–25 Holme (Nf); *Aschetillus* Bardel 1158 P (Ess); Robert *Asketill'* 1200 P (Ha): Alexander *Asketell'* 1205 Cur (Nf); Sarah *Ashkettle* 1785 SfPR. ON *Ásketill,* common before the Conquest and found in DB as *Aschil.* It survives as ASKELL. *Ash-* is a late dialectal development. v. also AXTELL.

**Ashley, Ashlee, Ashleigh:** Walter *de Esselega* 1162 P (Gl); Robert *de Ashley* 1281 CtW; Thomas *Assheley* 1401 AssLa. From one or other of the many places of this name.

**Ashlin, Ashling:** v. ASLIN

**Ashlock:** Walter *Aslak* 1426–7 Paston; Thomas *Ayshlocke* 1545, Thomas *Aysshelock* 1576 SRW. Probably from ON *Áslákr.* v. also OSLACK.

**Ashman, Ashment, Asman:** *Assemanus* 1066 DB (Sf); *Asseman, Asman* 13th Rams (Hu); Robert *Asheman* 1275 RH (Sf); Roger *Asman* 1279 RH (C); John *Essheman* 1317 AssK. OE *Æscmann,* perhaps an original byname from OE *æscmann* 'shipman, sailor, pirate'. As the surname is found at Gorleston and in Rochford Hundred (Essex), it may sometimes be an occupation-name.

**Ashmore, Ashmere:** Elias *de Asmere* 1296, *de Ashmer* 1327 SRSx; William *Asschmere* 1349 IpmW. From Ashmore (Do).

**Ashplant:** v. ABSALOM

**Ashpole, Ashpool:** v. ASHBOLT

**Ashton:** John *de Esssheton* 1306 IpmY; John *de Ashtone* 1323, 1326 CorNth; Thomas *Assheton* 1431 FFEss. From one or other of the many places of this name, or from Aston (Glos, Hereford, Salop).

**Ashurst:** v. ASHHURST

**Ashwell:** Ernulf *de Assewell'* 1230 P (Beds); Roger *de Asshewell* 1331 FFEss; Richard *Asshewell* 1391 IpmGl. From Ashwell (Ess, Herts, R).

**Ashwin:** William *Ashwyne* 1332 SRSr. OE *Æscwine* 'ash-, spear-friend', the name of a King of Wessex (674–6). It was not common in OE but clearly survived the Conquest.

**Ashworth, Ashword:** Richard *de Aschewor th* 1285 AssLa. From Ashworth (La).

**Askam, Askem, Askham, Ascham:** Conan *de Ascham* 1201 P (Y); Richard *de Ascum* 1204 AssY;

John *de Askham* 1290 FrY. From Askham (Notts, WRYorks, Westmorland).

**Aske, Ask:** Roger *de Ask* 1208 FFY; Roger *del Ask* 1230 P (O); John *de Ask* 1327 SRY. From Aske (NRY), or 'dweller by the ash-tree', ON *askr.*

**Askell, Astell, Astil, Astill, Eskell, Haskel, Haskell:** *Aschil, Osketel, Anschil* 1066 DB; Robertus *filius Askel* 1180 P (Y); Alanus *filius Askil* 1186–1211 BuryS (Nf); *Astillus* 1202 AssL; Ricardus *filius Eskill'* 1219 AssY; William *Astil* 1227 AssBeds; Thomas *Askil* c1248 Bec (O); Robert *Astel* 1260 AssC; Hugo *Astyl* pro *Asketyl* Ed 1 Battle (Sx). ON *Áskell,* ODa *Eskil,* OSw *Æskil,* a contracted form of *Ásketill (v.* ASHKETTLE), common both before and after the Conquest. The name was also popular in Normandy in the forms *Anschetill* and *Anketill (v.* ANKETELL) and elsewhere in northern France as *Anquetin* and *Asketin (v.* ASKIN). The various forms are often used of the same man in England. In the Whitby cartulary *Aschetillus* de Houkesgard (c1155) is also called *Aschetinus* (c1145), *Astillus* (c1170) and *Anketinus* (12th). In DC (Lei), in the late 12th century, *Anketinus* persona de Prestwalda appears also as *Asketinus* filius Hugonis de Prestewalda and as *Anketillus* de Prestwalda. cf. also *Aschetillus, Ancatillus, Anquetillus* Malore(i) Hy 2 DC (Lei). The name is found in Scotland as MACASGILL and in the Isle of Man as CASTELL. v. also AXCELL.

**Askew, Haskew, Ascough, Haskow, Ayscough, Askey, Askie, Haskey:** William *de Aykescoghe* 1366 SRLa; Robert *Ascowe* 1390 LLB H; Simon *Ascogh* 1488 FrY; John *Ascow,* William *Askew* 1488 GildY; Richard *Askoo* 1533 FrY; Amy *Askie* 1618 Bardsley; William *Ayscough* 1675 FrY; John *Ashkey* 1674 HTSf. From Aiskew (NRYorks).

**Askey, Askie, Haskey:** *Aschi* Wara (Herts), *Asci* (Sa), *Aschi* (Wa) 1066 DB; Amy *Askie* 1618 Bardsley; John *Ashkey* 1674 HTSf. ODa *Aski. v.* also ASKEW.

**Askham:** v. ASKAM

**Askin, Askins, Astin, Astins, Ashken, Haskin, Haskins, Hasking, Haskings, Hastin, Hastins:** *Asketinus* filius Od 1163 DC (L); Robertus *filius Astin* 1219 AssY; *Hastinus* caretarius 1223 Pat (Y); John *Astin* 1230 P (D); Hugh *Astyn* 1297 AssY; John *Asketyn,* William *Hastin* 1317 AssK; John *Haskyn* 1524 SRSf; John *Askin* 1674 HTSf. A Norman form of ON *Ásketill.* v. ASHKETTLE, ASKELL.

**Aslam, Aslen:** v. HASLAM

**Aslet, Aslett, Astlett:** Rogerus *filius Aselot'* Ric 1 Cur (L); Johannes *filius Ascelot* 1221 Cur (O); Hugh *Asselote* 1327 SRSf; William *Asselot* 1327 SRSx. OFr *Ascelot* (m), a double diminutive of OG *Azo. v.* ACE and ASLIN.

**Aslin, Asling, Ashlin, Ashling, Astling:** *Ascelinus* de Wudecroft 1162 P (Nth); *Acelina* de Stanfelde 1195 P (Nf); Walterus *filius Acelini, Ascelini* 1206 Cur (Nth); *Ascelina* 1214 Cur (Mx); Henricus *Ascelinus* 1204 P (W); Nicholaus *Asceline* 1255 Rams (Hu); Richard *Asselyn* 1279 RH (Bk); Henry *Asshelyn* 1374 LLB G. OFr *Acelin, Ascelin* (m), from OG *Azilin,* a double diminutive of *Azo.* cf. ASLET. There was also a feminine *A(s)celine.* The surname survives in Ashlyns and Ashlings in High Ongar which occur as *Astelyns* 1456–9, *Astlynge* 1568 (PN Ess 72).

**Asman:** *v.* ASHMAN

**Asp:** *v.* APPS

**Aspel, Aspell:** Geoffrey *de Asphal* 1275 RH (Sf); Alan *Aspal* 1320, John *de Aspale* 1330 LLB E. From Aspall (Sf), Aspal (La), or 'dweller by the land overgrown with aspens', OE *æspe, healh*, or 'dweller on aspen hill', OE *æspe, hyll*.

**Aspenlon:** *v.* ABSALOM

**Aspinal, Aspinall, Aspineli, Aspinwall, Haspineall:** Henry *de Aspenewell* 1246 AssLa; Miles *Haspinall* 1578 YWills; Edmond *Aspenall* 1599 FrY. From Aspinwall (Lancs).

**Aspland, Asplen, Asplin, Aspling:** *v.* ABSALOM

**Aspole:** For *Aspall* or ASHBOLT.

**Asquith, Askwith:** Ulf' *de Askwith'* 1219 AssY; Adam *de Askequid* 1297 SRY. From Askwith (WRYorks).

**Ass, Asse:** John *Asinus* 1202 Cur; John *Asse* 1248 FFK; Thomas *Arsse* brother of John *Asse* 1379 PTY. A nickname from the ass, Lat *asinus*, OE *assa*.

**Assan:** *v.* HARSANT

**Assard:** *v.* HAZARD

**Asser, Ascher, Asscher, Asher:** Outi *filius Azer*, Ulf *filius Azor* 1066 DB (L, Nth); *Ascherus* 1143–7 DC (L); John *ap Asser* 1218 Bardsley; John *Asser*, Richard *Aseyr* 1249 AssW; John *Asser* 1331 Rams (Hu); Ralph *Asher* 1674 HTSf. ON *Qzurr*, ODa, OSw *Azur*, Welsh *Asser*. But some of the forms appear to represent OE *Æschere* found only once, in *Beowulf*.

**Astall:** Walter *de Astalle, de Estalle* 1279 RH (O). From Asthall (Oxon).

**Astbury:** William *de Astbury* 1287 AssCh. From Astbury (Ches).

**Astell, Astill, Astle, Astles:** Simon *de Astell* c1225 Lichfield (St); Richard *de Asthul* (*Astell*) 1349 FrY. From Astle (Ches) or from residence near the east hill (OE *ēast, hyll*). *v.* also ASKELL.

**Aster:** *v.* ASTOR

**Astin:** *v.* ASKIN

**Astington:** *v.* SINTON

**Astle(s):** *v.* ASTELL

**Astlett:** *v.* ASLET

**Astley:** Gerard *de Astele* 1202 P (Nf); James *de Astlye* 1300 AssSt; Thomas *Asteley* 1377 IpmGl. From Astley (La, Sa, Wa, Wo, WRY).

**Astling:** *v.* ASLIN

**Astman:** *v.* EASTMAN

**Aston:** (i) *Asthone* de Sancto Luca c1140 DC (L); Tomas *filius Adestan* 1187 P (Y); *Astanus* de Hune 1190 P (Y); Lewin *Adstan* 1221 ElyA (Nf); Thomas *Astan* 1276 RH (L). *Adestan* may be from OE *Æðelstān*. *v.* ATHELSTAN and ALLSTON. Robertus *filius Adestani* and Walterus *filius Alstan* 1206 Cur (Sf), both mentioned in the same document relating to the same parish were both probably sons of *Æðelstan*. Or we may occasionally have the less common OE *Ēadstān* 'prosperity-stone', DB *Edstan*. The DB *Æstanus, Estan* may represent either name. (ii) Very often, this surname is local in origin, from one of the numerous places named Aston. cf. Richard *de Aston'* 1206 Cur (Gl). (iii) William *Stone, Astone* 1494, 1500

WBCinque; Thomas *A Stone* 1525 SRSx; Thomas *A Stone*, John *a stone* 1569 Musters (Sr). This is probably for *atte stone* 'dweller by the stone', cf. AMOORE, AVANN, AGATE.

**Astor, Aster:** William *Aster* 1275 RH (L); John *Aster* 1327 SRC; Nicholas *Aster* 1356 FFEss. Probably forms of EASTER. Used as a christian name in the 17th century: *Aster* Foxe 1642 PrD.

**Astringer:** *v.* OSTRINGER

**Astwell:** Matilda *Astwell* 1327 SRSo. From Astwell (Nth).

**Atack:** *v.* ATTACK

**Atberry:** John *Ateburi* 1279 (O); Hugh *ate Bery* 1327 SRC. 'Dweller or servant at the manor-house' (OE *burh*).

**Atbridge:** Walter *Attebrigge* 1290 IpmW; John *atte Brige* 1325–6 FFEss; Richard *atte Brigge* 1327 SRSo. 'Dweller at the bridge', OE *brycg*.

**Atbrook, Atbrooke:** Osbert *Attebroc* 1226–7 FFWa; John *Attebroke* 1291 FFO; Benedict *atte Broke* 1332 SRSx. 'Dweller by the stream', OE *brōc*.

**Atcheson, Atchison:** *v.* ACHESON

**Atcliff, Atcliffe:** John *atte Cliue* 1327 SRSo; William *Attclyff* 1470 Paston; George *Atclyff* 1496 LLB L. 'Dweller by the cliff or slope', OE *clif*.

**Atcock:** *v.* ADCOCK

**Atcot, Atcott:** Ralph *atte Cote* 1327 SRSo; Geoffrey *Attecot'* 1327 SRY. 'Dweller at the cottage', OE *cot*.

**Atfield, Attfield:** Stephen *Attefeld* 1262 FFEss. 'Dweller by the open field' (OE *feld*).

**Atford, Attford:** John *Atteford* 1282, Richard *Ateford* 1297 IpmW; Geoffrey *Ateford* 1453 FFEss. 'Dweller at the ford', OE *ford*.

**Atgrove:** *v.* AGROVE

**Atheis:** Thomas *atte Heye* 1327 SRSo. 'Dweller at the enclosures', containing the plural of OE *(ge)hæg*.

**Athell:** *v.* ATHILL

**Athelstan, Addlestone, Edleston:** *Adelstanus* 1195 P (K); Hugo *filius Athelstan, Adelstan'* 1218, 1219 AssL; Herbertus *filius Edelstani* 1240 Rams (Nf); William *Athelston* c1198 Bart (Lo); Geoffrey *Athelstan* 1219 AssL. OE *Æðelstān* 'noble stone'. *v.* ALLSTON and ASTON.

**Atherden:** William *ater Dene*, Peter *in ther Dene* 1296 SRX; William *Atherdonne*, Ruben *Atherton* 1568 SRSf. 'Dweller in the valley', OE *denu*.

**Atherfold:** cf. John *atte Fold* 1327 SRWo. 'Dweller or worker at the fold', preserving the variant ME *at ther folde*.

**Atherley:** Alfeg' *ater Legh* 1296 SRSx; John *Atherlee* 1419 LLB I. A variant of ATLAY.

**Athersmith:** John *atter Smythe* (1330 PN D 386) lived at the smithy (OE *smiþþe*) and was presumably a blacksmith. The surname might also derive from ME *at ther smethe* 'dweller at the smooth, level place' (OE *smēðe*). cf. William *del Smethe* 1327 SRSf.

**Atherstone:** Adam *de Atherston'* 1275 SRWo. From Atherstone (Warwicks).

**Athersuch, Athersych, Athersytch:** The obvious derivation of the last two names would be from ME *at ther siche* 'dweller by the brook or drain'. As *Athersytch* is found in Sheffield, where *siche* occurs as

*Sykes*, all the names are probably dialectal pronunciations of *Hathersage*, a Derbyshire village some ten miles distant. cf. HATHERSICH.

**Atherton:** Henry *de Athertone* 1332 SRLa; William *de Atherton* 1384 IpmLa; Humphrey *Addertone* alias *Athurton* 1470 Combermere (Ch). From Atherton (Lancs), *Aderton* 1212.

**Athey, Athy:** Lecia *Ateheye* 1279 RH (C). 'Dweller at the enclosure', OE *(ge)hæg*.

**Athill, Atthill, Athell:** Richard *Athill* 1255 RH (Sa); Ralph *atte Helle* 1319 *SR* (Ess). 'Dweller by the hill', OE *hyll*.

**Athoke:** John *atte Hok* 1254 *Ass* (Ess). 'Dweller by the bend', OE *hōc* 'hook'.

**Athol, Atholl:** Robert *atte Hole* 1296 SRSx. 'Dweller in the hollow', OE *holh*. The Scottish *Atholl* derives from Atholl in Perthshire.

**Athorn, Athorne:** William *atte Horne* 1332 SRSx. 'Dweller by the spit of land', OE *horn*. Or 'dweller by the thorn-bush': Emma *Attethorn* 1297 MinAcctCo.

**Athowe, Atthowe:** John *ate How* 1296 SRSx. 'Dweller by the ridge', OE *hōh*. v. HOW.

**Athridge:** v. ETHEREDGE

**Athy:** v. ATHEY

**Atkey:** William *atte Keye* 1370 LoPleas. 'Dweller or worker at the quay'. v. KAY.

**Atkin, Atkins, Atkyns:** v. ADKIN

**Atkinson, Ackenson:** John *Adkynsone* 1381 SRSt; John *Atkinson* 1402 FA (We). 'Son of *Adkin*.'

**Atlay, Atlee, Atley, Attle, Attlee:** Robert *Atte lee* 1275 SRWo; James *Attele* 1276 RH (Berks); William *atte Leye* 1296 SRSx; Thomas *Attlee* 1327 Pinchbeck (Sf). 'Dweller by the wood or clearing', OE *lēah*. cf. ATHERLEY.

**Atlem, Attlem:** William *Athlem* 1347 IpmW; Richard *Athelam* 1380 IpmGl. OE *Æðelhelm*.

**Atlow, Attlow:** John *Attelowe* 1332 SRSt; Robert *Atlow* 1340–1450 GildC; John *a Lowe* 1489 Paston. 'Dweller by the hill', OE *hlāw*.

**Atmeare, Atmer, Attmere:** Cecil *Atemer'* 1279 RH (C); John *Atmer* 1524 SRSf; Edmund *Attmear* 1568 SRSf. 'Dweller by the mere or pool' (OE *mere*) or near the boundary (OE *(ge)mǣre*).

**Atmore:** Jordan *Attemore* 1276 LLB A. 'Dweller near the marsh', OE *mōr*.

**Atread, Attread, Attreed:** Robert *atte Reed* c1295 MELS (Sx). 'Dweller in a clearing.' v. READ.

**Atrill:** v. ATTRILL

**Attack, Attoc, Attock, Atack:** Geoffrey *atte Ock* 1296 SRSx; Thomas *Atack* 1709 Bardsley. 'Dweller by the oak' (OE *āc*).

**Attale:** Robert *Attehal'* 1212 Cur (L); William *atte Hale* 1235 *Ass* (Ess). 'Dweller by a nook or in a remote valley' in the South and Midlands, or by flat, alluvial land near a river in the North (OE *healh*). v. HALE.

**Attaway:** John *ate Wey* 1279 RH (C). 'Dweller by the road' (OE *weg*) or at a place called *Atteweye*, the name in 1306 of Way in Thorverton (Devon), or at Atway (Devon). v. PN D 573, 468.

**Attawell:** v. ATTWELL

**Attenbarrow, Attenborough, Attenborrow, Attenbrough, Attenbrow, Attenburrow:** These surnames can hardly derive from Attenborough

(Notts), earlier *Adinburcha*, which was not a village in the Middle Ages, the name being that of a church only. They are probably all from ME *atten* plus the dative of OE *bearu* 'a grove' or *beorg* 'hill, mound'. The second element was, as often, changed to *borough* through association with *burh*. Hence 'dweller at the grove, hill or mound'.

**Atter:** Edguinus *atre*, Edwin *Atter* 1111, c1130 ELPN. A nickname from OE *ātor*, ME *atter* 'poison, venom; gall, bitterness'.

**Atteridge:** John *atte Rygge* 1333 PN D 567. 'Dweller by the ridge', OE *hrycg*. v. also MELS 173 and ETHEREDGE.

**Atterton:** William *Atterton* 1476–7 FFWa; Widow *Atterton* 1674 HTSf. From Atterton (K, Lei).

**Atterwill:** For ME *at ther wille* 'dweller by the spring or stream'. cf. John *ater Welle* 1296 SRSx. *Will* is a south-western form of *well*. v. ATTWELL.

**Attewell, Attewill, Attiwill:** v. ATTWELL

**Attford:** v. ATFORD

**Attick:** v. ATTWICK

**Attle, Attlee:** v. ATLAY

**Attleborough, Attlebrough, Attleborrow, Attleburrow:** William *Attylburgh* 1340–1450 GildC; John *Attelburgh* 1397 AssWa; William *Attulburgh* 1406–7 FFWa. From Attleborough (Wa).

**Attlem:** v. ATLEM

**Attlow:** v. ATLOW

**Atto, Attoe:** Roger *Atteho* 1236 FFSx. 'Dweller at the ridge', OE *hōh*. v. HOW.

**Attoc, Attock:** v. ATTACK

**Attom:** Adam *de la Homme* 1275 SRWo. 'Dweller by the water-meadow.' v. HAM.

**Attree, Attrie, Attrey:** Thomas *Attere* 1272 PN C 15; Walter *at Reghe* 1287 PN Sr 330; Thomas *Atry* 1320 FFHu; Matthew *atte Ry* 1389 PN Ess 387; Richard *Atre* 1545 SxWills. 'Dweller by the stream (OE *ēa*) or low-lying land (OE *ēg*).' v. REA. Also 'dweller by the enclosure' (OE *(ge)hæg*), Simon *ater Hegh* 1296 SRSx; or 'by the tree' (OE *trēo*), John *A'Tree* 1558 SxWills. v. TREE.

**Attrell:** v. ATTRILL

**Attride, Attryde:** John *at Ride* (1446) and Henry *at Ryde* (1524) took their name from a clearing (OE *\*rīed*, *\*rȳd*), but Thomas *at Ride* (1524) and John *Attryde* (1588) owed theirs to a streamlet (OE *rīþ*, *rīðe*). v. PN Sr 242, 148.

**Attridge:** v. ETHEREDGE

**Attrill, Attrell, Atrill:** Walter *atter Hille* 1330 PN D 477; John *at Ryll* 1524 ib. 562. 'Dweller by the hill', a not uncommon Devon development, surviving in place-names as both Rill and Rull.

**Attru:** Hugo *de la Truwe* 1250 Fees (So); Agnes *atte Trewe* 1333 PN D 595. 'Dweller by the tree', OE *trēow*. v. TRUE. Also from OE *rǣw* 'row, hedgerow', used also of a row of houses, a street or hamlet: Maurice *atte Rewe* 1333 PN D 464.

**Attryde:** v. ATTRIDE

**Attwater, Atwater:** William *Atewatr'* 1198 CurR (Herts); Marye *Atwaters* 1660 ArchC xxx. 'Dweller by the water', OE *wæter*.

**Attwell, Attwill, Attwool, Attwooll, Atwell, Atwill, Atwool, Attawell, Attewell, Attewill, Attiwill:** Gilbert

*Attewell* 1274 RH (Ess); Richard *atte Wille* 1333 PN D 450. 'Dweller by the stream or spring', OE *wiella*. *Atwill* is a Devon and Somerset form, *Attwool* a Dorset one. *v.* WOOLL.

**Attwick, Attick:** Walter *ate Wyk* 1327 SRSx. 'Dweller or worker at the dairy-farm', OE *wīc*. *Attwick* still survives in Sussex.

**Attwood, Atwood:** Thomas *Attewode* 1243 AssSo; Robert *Atwode* 1457 Oseney (O). 'Dweller by the wood', OE *wudu*.

**Atty:** John *atte Tye* 1327 *SR* (Ess); Thomas *Attye* 1568 SRSf. 'Dweller by the enclosure, close or common pasture', OE *tēag*. Also from OE *(ge)hæg* 'enclosure': Geoffrey *Atteheg* 1327 SRDb.

**Atyeo:** Roger *atte Yo* 1333 PN D 263. 'Dweller by the river.' *v.* YEA.

**Aube:** *v.* ALBE

**Aubert:** *v.* ALBERT

**Aubertin, Auberton:** A Huguenot name from a family from Metz (Lorraine). On the Revocation of the Edict of Nantes the original emigrant fled to Switzerland, and his great-grandson came to England c1767. The name is a diminutive of *Aubert*, the popular form of *Albert*.

**Aubin:** *v.* ALBIN

**Aubon:** *v.* ALBAN

**Auborn, Auburn:** Henry *of Auburn* 1226 FFY; William *de Auburne* 1388 IpmNt; Peter *Aubone*, John *Auborne* 1674 HTSf. From Auborn (L), or Auburn (ERY).

**Aubray, Aubrey, Aubry, Aubery, Aubury, Awbery, Obray:** (i) Walterus *filius Alberici*, *Albrici*, *Alberi* 1086 DB (Sf); *Albricus* de Capella 1214 Cur (C); *Aubri* Bunt 1279 RH (C); John *Aubri* ib.; Robert *Aubri* 1308 FFSf; Adam *Albry* 1327 SRSf; Geoffrey *Aubry* 1380 AssC. OFr *Aubri*, *Auberi*, OG *Albericus* 'elf-ruler'. (ii) Osbertus *filius Alberade* 1115 Winton (Ha); *Albreda* 1198 Cur (Herts), 1205 Cur (Wa); Ricardus *filius Albrei* 1199 FrLeic; *Aubreda* 1219 AssY; *Albreda*, *Albree* de Jarpenvill' 1221 Cur (Bk); Johannes *filius Aubre* 1279 RH (O); *Albray*, *Aubray* 1379 PTY; John *Albre* 1243 AssSo; Robert *Aubrey* 1279 RH (Bk); William *Aubray* 1324 Wak (Y); John *Aubery*, *Awberry*, *Aubry* 1460 Bardsley (Nf). OFr *Albree*, *Aubree*, *Auberee*, OG *Alb(e)rada* 'elf-counsel' (f).

**Auckland, Aukland:** Geoffrey *de Aukelaunde* 1269 FFY; Robert *de Aukland* 1327 SRY; William *de Aukland* 1351 FrY. From Auckland (Du).

**Aucock:** *v.* ALCOCK

**Aucott, Aucutt:** *v.* ALCOTT

**Audas, Audiss:** *v.* ALDIS

**Auden:** *v.* ALDEN

**Audlay, Audley, Audeley:** Adam *de Aldeðalega* 1185 P (St); James *de Audeley* 1272 AssSt; Hugh *Daudelegh* 1333 IpmW; John *de Audeleye* 1377 AssWa. From Audley (St).

**Audrey, Awdrey, Awdry:** *Aldreda* 1066 DB (Nf), 1219 AssY; Robert *Audrey* 1279 RH (O). The DB personal-name may be for OE *Ælfðrȳð*, *Æðelðrȳð* or *Ealhðrȳð*, all feminine. Of these the most common was the second, 'noble strength', popular through the reputation of St. Etheldreda, Queen of Northumbria and foundress of the convent at Ely. This became *Audrey* and is found in the 14th century in Essex

and Suffolk as *Etheldreda* (1304 AD i (Ess), 1381 SRSf).

**Audritt:** *v.* ALDRED

**Audus:** *v.* ALDIS

**Augar, Auger, Augur:** *v.* ALGAR, ALGER

**Aughton:** Thomas *de Autun'*, *de Aucton'* 1212 Cur (Y); William *de Aghton* 1354 FrY. From Aughton (La, ERY, WRY).

**Augood:** *v.* ALLGOOD

**Augustin:** *Augustinus* cantor 1153–68 Holme (Nf); Nicholas *Augustin* 1275 SRWo. Lat *Augustinus*, from *augustus* 'venerable'. The christian name is very common, usually in the Latin form, but as a surname is rare, both in ME and today. The medieval surname is common in the vernacular form *Austin*. *v.* AUSTEN.

**Auker:** *v.* ALKER

**Aukin:** *v.* ALKIN

**Aukland:** *v.* AUCKLAND

**Auld, Aulds, Ault, Awde:** John *Alde* 1284 Black (Perth); Johannes *dictus Ald* 1302 ib.; John *Auld* 1488 ib. A Scottish form of OLD, from Northern English *ald*.

**Aumonier:** Adam *le Augmoner* 1297 FFHu; Michael *le Aumouner* 1297 SRY; Adam *Aumener* 1327 SREss. OFr *aumoner*, *aumonier* 'almoner'.

**Aungier(s):** *v.* ANGER, ANGERS

**Austen, Austin, Austing, Austins, Auston:** *Austinus* de Bernardeston 1224 FFSf; *Austinus* de Beston' 1230 P (Nf); *Ostin* atte Putte 1327 SRSo; Henry *Austin*, Edith *Austines* 1275 SRWo; Avice *Augstyn* (*Austynes*) 1327 *SR* (Ess). OFr *Aoustin*, the vernacular form of *Augustine*. *v.* AUGUSTIN.

**Austwick, Austick:** Rose *de Austwic* 1202 FFY; Roger *de Oustewyk* 1341, John *Austewyk* 1425 FrY. From Austwick (WRY).

**Autie, Auty, Awty, Alty:** *Auti*, *Alti*, *Outi* 1066 DB; *Outi* de Lincol' 1166 P (Nf); Walterus *filius Aulti* 1177 P (L); Willelmus *filius Auti* 1200 P (Y); John *Oty* 1251 Rams (Hu); John *Awty* 1524 AD vi (Sf). ODa *Auti*.

**Avann:** John *Avanne* 1527 SxWills. 'Dweller by the fen.' *v.* FANN.

**Avel:** *Auel* de Wappeham 1176 P (Nth); Peter *Auel* 1296 SRSx. A diminutive of OG *Avo*.

**Aveley, Avely:** William *de Auele* 1202 P (So). From Aveley (Ess), or Avely Hall (Sf).

**Aveline, Aveling:** *Auelina* 1175–86 Holme (Nf), 1327 SRC; Henry *Avelin* 1279 RH (C); Reginald *Auelyn* 1296 SRSx. OFr *Aveline*, OG *Avelina* (f), a double diminutive of OG *Avo*.

**Avenall, Avenel, Avenell:** *Avenel* 1086 DB (Sa), 1166 RBE (Ess); *Auenellus* 1190 P (Y), 1196 Eynsham (O); Robert, Paganus *Auenel* 1139 Templars (O), 1195 P (He). A double diminutive of OG *Avo*. Both christian name and surname are common.

**Avent:** *Avenant* de Twipp' 1204 Cur (D); Osbert, Ralph *Auenant* 1156–80 Bury (Sf), 1198 FFNf. OFr *auenant*, pres. part. of *avenir* 'to arrive, happen, befit, become'; 'suitable' (1300 NED), 'handsome, comely' (1340). Used also as a personal-name.

**Avery, Avery, Averies:** Rogerus *filius Alvredi* 1166 RBE (Y); Hugo *filius Auveray* 1275 RH (Nt); Willelmus *filius Averay* 1275 SRWo; *Alvrei* venator 1294 Ch (Y); Nicholas *Auverey* 1273 RH

(Wo); William *Auure* 1275 RH (W); Walter *Averay* 1275 SRWo; Cust *Alvere* 1279 RH (C); Edmund *Avered* 1279 RH (C); Walter *Avery* 1279 RH (O); Richard *Avore* 1428 FA (Sx). A French pronunciation of ALFRED.

**Averill:** v. APRIL

**Averley:** John *de Averle* 1279 RH (Hu). Probably from Aversley Wood in Sawtrey (Hu).

**Aves, Aveson:** Willelmus *filius Aue* 1198 FFNf; Elizabethe *Aves* 1568 SRSf. OG *Avo.*

**Aveston:** William *de Alueston* 1190 P (Wa). From Alveston (Glos, Warwicks).

**Avins:** *Avina* 1221 Cur (Nt); *Avin'* de Eyton' 1255 RH (Sa). *Avina,* wife of Robert de Turuill' (1221 AssGl) was also called *Auicia. Avina* is therefore probably a hypocoristic of *Avis.*

**Avis, Aviss, Avison:** *Auicia* c1175–86 Holme (Nf), 12th DC (Nt), 1199 FFEss; *Auizia* Hy 2 DC (Lei); Ricardus *filius Avice* 1332 SRSt; Walter *Auices* 1186–1210 Holme (Nf); Thomas *Avyce* 1220 Fees (Berks); Thomas *Avis* 1524 SRSf; Ralph *Avyson* 1674 FrY. OFr *Avice,* sometimes derived from OG *Aveza* but Michaëlsson (ii. 79–82) has shown that it probably derives from Lat *Avitia* which, with the masculine *Avitius,* was used in Gaul. cf. *Amis* s.n. AMES.

**Avner:** Roger, Alexander *le Avener* 1230 P (Ha), 1231 Pat (Lo). OFr *avenier, avener* 'oat-merchant'. Used also of the chief officer of the stable who had charge of the provender for the horses (1282 NED).

**Avril:** v. APRIL

**Awbery:** v. AUBRAY

**Awcock:** v. ALCOCK

**Awdas:** v. ALDIS

**Awde:** v. AULD

**Awdrey, Awdry:** v. AUDREY

**Awmack:** v. ALLMARK

**A'Wood, Awood:** William *A Wode* 1485 LLB L; Robert *a Wode* 1525 SRSx. For *atte Wode* 'dweller by the wood', OE *wudu.*

**Awty:** v. AUTIE

**Axbey:** *Aksby,* a metathesized form of *Askby* from some place Ashbury. Castle Ashby (Northants) is *Axeby* in 1235 (PN Nth 142).

**Axcell, Axel, Axell:** Probably metathesized forms of *Askell* through the series *Askell, Aksell, Axell.* cf. AXTELL.

**Axleby, Exelby, Eshelby, Hasselby:** John *de Eskelby* 1327 SRY; Richard *Exilby* 1490 FrY; Thomas *Eshleby* 1672 FrY. From Exelby (NRYorks), a metathesized form of *Eskelby.* A similar metathesis of the DB *Aschilebi* would become *Axleby.* Both surnames might also derive from Asselby (ERYorks) which is found as *Askelby* 1282, *Eskilby* 1199 and *Axilbye* 1551. v. PN ERY 248, PN NRY 226.

**Axon:** Thomas *Acson* 1561, Thomas *Axon* 1635, John *Ackson* 1641 Bardsley (Ch). This might be 'son of *Acke*', OE *Acca.* But forms are late and we may equally well have a metathesis of *Askin,* giving *Aksin, Akson.* v. also AXTEN.

**Axtell:** Elizabeth *Axtell,* John *Axstell,* Richard *Axtill* 1683–90 Bardsley. These forms, though late, are almost certainly from ON *Ásketill,* with loss of the unstressed medial vowel to *Asktill,* and metathesis to *Akstill, Akstell.* v. ASHKETTLE.

**Axten, Axtens, Axton:** Agnes *Axton* 1524 SRSf, Laurence *Axton, Axon* 1561 Bardsley. It is impossible to decide whether *Axton* is from *Axon* (a metathesis of *Askin*), with intrusive t, or whether *Axon* is a simplified form of *Axton.* An original *Axton* would be a metathesized form of *Ashton.* The Northants Ashton is *Axton* in 1253 (PN Nth 229).

**Ayer, Ayers, Ayr, Ayre, Ayres, Ayris, Aiers, Air, Aires, Airs, Aers, Eayrs, Eayres, Eyer, Eyers, Eyre, Eyres, Hair, Haire, Hayer, Heyer, Hoyer:** (i) Ralph *le Eir* 1208 FFEss; Robertus *Heres* 1220 Cur (C); Robert *leyre* 1245 FFC; Richard *le Heyer* 1274 RH (Gl); Henry *Ayer, le Eyer* 1275 RH (L, O); Adam *le Hayre* 1275 Wak (Y); Robert *le Heir* 1281 Eynsham (O); Robert *Air* 1281, John *Ayr* 1296 Black; William *Hoyre* 1302 SRSf; Francis *Eyre* alias *Eare* alias *Aire* 1724 DKR 41 (Co). ME *eir, eyr,* etc., OFr *eir, heir,* CentFr *oir, hoir,* AFr *heyr,* Lat *heres* 'heir' (c1275 NED). Also *leyre* is one source of LAYER. (ii) Occasionally we may have a personal name: Robertus *filius Aier* 1166 RBE, *filius Aer, Aier* 1180, 1196 P (Sa), the latter being identical with Robert *Aier* 1201 P (Sa). cf. Aston Eyre (Salop), held in 1212 by Robert, grandson of *Aer,* a name which Ekwall suggests may be derived from that of the DB tenant *Alcher,* i.e. OE *Ealhhere.* v. ALGER. v. also HAIR. (iii) Reginald *of Ayr* 1287 Black (Ayr); Albinus *de Are* 1315–21 ib. From the royal burgh of Ayr.

**Aykroyd:** v. ACKROYD

**Aylard:** v. ADLARD

**Aylen:** v. AYLING

**Ayler:** Luke *le Ayler, le Ayeler* 'peverer' (i.e. 'pepperer') 1287, 1304 LLB B. OFr *aillier, -iere* 'garlic-seller'.

**Ayles:** Richard *le Eyel* 1275 SRWo; Ralph *Ayl* 1279 RH (C); Richard *Aylis* 1302 FA (Sf). Either OE *Ægel,* found only in place-names, or a nickname from OE *egle* 'loathsome, troublesome'.

**Aylesbury:** Richard *de Æilesberia* 1188 Eynsham; John *de Aylesbury* dictus le Tauerner de Oxonis 1307 Balliol. From Aylesbury (Bk).

**Aylesford:** Edward *de Ailesford* 1202 FFK. From Aylesford (K).

**Aylett, Aylott, Aliatt, Allett, Allitt, Alliott, Ellyatt, Eliot:** Galfridus *filius Ailghieti* a1176 Colch (Ess); *Ailletus* 1180–1207 Rams (Nf); Simon *filius Alet* 1199 P (L); *Ailleth, Ailed* (f) 1198 FF (Nf); *Æliot* Grim 1202 AssL; *Ailhiet* (f) 1202 FF (Nf); Gilebertus *filius Aillith* 1204 P (C); *Ailith, Ailleth* filia Godwini 1207 Cur (Sf); *Alettus* Prepositus 1212 Cur (Nth); Simon *filius Aileda* 1279 RH (C); *Alyott* de Symondston 1311 Lacy (La); Boydin *Ailet* 1212 Fees (Ess); Walter *Aliot, Aylet,* Thomas *Ailot,* John *Ayllyth* 1279 RH (C); Ralph *Alyet* 1286 Pinchbeck (Sf). In DB *Ailiet* and *Aliet* are found for both the common OE *Æðelgȳð* (f) 'noble combat' and OE *\*Æðelgēat* (m) 'noble Geat', which is not recorded before the Conquest but was certainly in use later, and it is impossible to distinguish between them when the sex of the bearer is unknown. The confusion is increased by the variety of forms found for both themes of each name, but it is clear that all the above surnames may derive from either of these personal names. In DB *Æðelgȳð* is *Adelid, Ailiet,*

*Ailith, Ailid, Ailad, Alith,* and *Alid,* all except the first referring to *Æðelgyð* wife of Þurstan. *Æðelgēat* is found as *Ailiet, Ailet, Ælget, Aliet, Elget* and *Eliet,* all of which might stand for *Æðelgyð* where the gender is doubtful. The last two forms make it clear that these OE names have contributed to the frequency of ELIOT in its various spellings. *v.* ADLARD.

**Ayliff, Ayliffe, Ayloff, Ellif:** (i) *Ailef* de Palestun 1175 P (Nb); Willelmus *filius Eilaf* 1191 P (Nth); Robertus *filius Egelof* 1196 P (L); *Egelaf* 12th MedEA (Sf); Ricardus *filius Ailof* 1203 Cur (Nth); Nicholas *Eiluf, Ailof* 1221 AssWa; Julian' *Aylif* 1279 RH (O); Geoffrey *Ayllef,* John *Aylofh* 1327 SRSf. The DB *Eilaf* (*Egilaf, Ailaf* Exon), *Ailof, Elaf* are probably from ODa, OSw *Elaf* (hence *Ayloffe*), but they may also represent ON *Eileifr,* ODa, OSw *Elef* or ON *Eilifr,* ODa, OSw *Elif* with substitution of OE or EScand *-lāf* for *-lef, -lif* (hence *Ayliff, Ellif*). *v.* also ILIFF. (ii) *Eilieua* de Kerletona Hy 2 DC (Lei); Edwardus *filius Eileve* 1206 Cur (Sx); Rogerus *filius Aelive* 1214 Cur (C); Segarus *Aileves* 1188 BuryS (Sf); Robert *Aylgive* 1275 SRWo; Edelina *Ayleve* 1279 RH (Hu). OE *Æðelgifu* (f) 'noble gift', which appears in DB as *Æilleua, Eileua, Aileua* and *Eleua.* For *-iff* from *-gifu,* cf. BRIGHTIFF, here, perhaps, influenced also by *Ailiff.* Ayloffe can only be included here by assuming influence from the Scandinavian name. cf. Richard *Aylyaue* 1332 SRWa.

**Ayling, Aylin, Aylen:** Eadmund *Æðeling* 1006 KCD 1302 (Do); Ædwardus *Aðeling* 1176 P (K); Gilbert *Ædeling* 1177 P (Y); Reginald *Aylyng* 1296 SRSx. OE *æðeling* 'noble, prince of the royal blood', used occasionally as a personal-name: *Ailligg'* (*Eiling*) buttarius 1230 P (Nf). Edgarus *Adeling* 1086 DB (Herts) is also called Eadgar *Cild. v.* CHILD.

**Aylmer, Aylmore, Elmar, Elmer, Elmers:** *Ailmar, Æilmar, Eilmerus, Aimar, Almer, Elmar, Elmer* 1066 DB; Godwinus filius *Elmari* 1115 Winton (Ha); *Hælmerus* Hy 2 DC (L); *Ailmerus* le Bercher 1212 Cur (Herts), quidam *Ailmerus* villanus ib. (Y); Henry

*Ailmer'* 1208 Cur (Berks); Roger *Ailmar* 1221 AssWa; William *Elmer* 1316 FA (Sx). OE *Æðelmǣr* 'noble famous'. *Elmer* is also local in origin. *v.* also AYMER.

**Aylward, Ailward:** Godric *filius Æilwardi* c1095 Bury (Sf); *Egelwardus* 1126–7 Holme (Nf); *Ailwardus* presbiter 1153–68 ib.; Robert *Ailward'* 1201 P (Ha); Robertus *Ailwardi* 1229 Cl (Gl); Nicholas *Eylward* 1243 AssSo. OE *Æðelweard* 'noble protector', DB *Aegelward, Ailuuard. v.* also ALLWARD.

**Aylwin, Aylwen, Ailwyn:** *Adelwinus, Ailwinus, Aluuin(e), Eluuinus* 1066 DB; *Ailwinus* Neht Hy 2 DC (L); *Eilwinus* de la Berne 1211 Cur (Sr); Hubert *egelwin* 1194 Cur (Bk); Walter *Athelwin* 1205 P (Gl); Simon *Aylwyn* 1230 P (Beds); Alice *Eylwyn* 1297 MinAcctCo. OE *Æðelwine* 'noble friend'. *v.* also ALWIN, ALVEN.

**Aymer, Aimer, Aimers, Amar:** *Eymer* Thurberd 1260 AssC; *Aymar* de Valence 1298 Gascon; Philip *Aimer* 1180 P (Ess). In DB *Aimar* is one of the forms for OE *Æðelmǣr. v.* AYLMER. Here we have also a continental personal-name either OG *Agimar* or OG *Hadamar, Adamar.*

**Aynley:** *v.* AINLEY.

**Aynsley:** *v.* AINSLEY.

**Aynsworth:** *v.* AINSWORTH.

**Ayre(s):** *v.* AYER.

**Ayrton:** John *de Ayrton* 1379 PTY; William *Ayreton* 1460 FrY. From Airton in Kirkby Malham (WRY).

**Ayscough:** *v.* ASKEW

**Aysh:** *v.* ASH

**Ayshford:** *v.* ASHFORD

**Ayton, Aytoun, Aiton:** (i) Helias *de eitun* c1166 Black (Dunbar); William *de Eytone* 1296 ib. (Berwick); John *de Aytoun* 1300 ib. From the lands of Ayton in Berwick. (ii) William *de Atune* c1174 YCh; John *de Aiton'* 1219 AssY; John *de Ayton'* 1300 FFY. From Ayton (NRYorks), *Aton* DB. *v.* also EYTON.

**Aze:** *v.* ACE

# B

---

**Babb, Babbs:** Alwinus, Richard *Babbe* 1198 FF (Sf), 1230 P (D); Ralph, Walter *le Babb(e)* 1199 MemR (W), 1327 SRSx. A pet-form of *Barbara.* cf. Margery *Babel,* Nicholas *Babelot* 1279 RH (C) and BABOT. *Le Babbe* is a nickname from *babe* (c1230 MED) 'infant, young child'.

**Babbel, Babbell, Babble:** Isabel *Babel* 1276 KB (Sx); John *Babell'* 1379 PTY; Hugh *Babell* 1642 PrD. *Bab-el,* a diminutive of *Bab,* a pet-form of *Barbara.*

**Babbington, Babington:** Eva *de Babington'* 1201 AssSo; Henry *de Babbyngton* 1379 PTY; Thomas *Babyngton* c1464 Paston. From Babington (So), or Babbington in Kimberley (Nt).

**Babington:** *v.* BABBINGTON

**Babot:** Geoffrey *Babeth* 1279 RH (C). A diminutive of *Bab* (Barbara).

**Babson:** Adam *Babson,* Brabson 1380 AssWa. 'Son of *Bab',* a pet-form of *Barbara.*

**Babthorp, Babthorpe:** Robert *Babthorp'* 1414 KB; William *Babthorp* 1439–40 FFWa. From Babthorpe (ERY).

**Baccas, Bacchus:** *v.* BACKHOUSE

**Bach, Bache, Batch:** Reiner *de Bache* 1212 Cur (L); Ralph *de la Bache* 1252 Rydware (St); William *atte Bache* 1327 SRWo. 'Dweller near a stream', OE *bæce.* *v.* BASH.

**Bachelor, Bachellier, Batchelar, Batcheler, Batchellor, Batchelor, Batchelour, Batchlor, Batchelder, Batcheldor:** Roger *Bachelere* c1165 StCh; Stephen *le bachilier* 1203 FFSf; Walter *le Bachelor* 1248 FFSr; Thomas *Batcheller,* Widow *Bachelder* 1674 HTSf. ME, OFr *bacheler* 'a young knight, a novice in arms' (1297 NED). *v.* also BACKLER.

**Bacher:** Philip *Bacher* 1255 RH (Bk); William *le Bachiere* 1280 MESO (Ha). 'Dweller by a stream.' cf. BACH.

**Bachus:** *v.* BACKHOUSE

**Back, Backes, Bax:** Godwine *Bace* c1055 OEByn (So); Godwin the clerk, called *Bak* 12th ELPN; Richard *Bac* 1182 P (Co); Richard *Backe* 1277 *Ely* (Sf); Henry *le Bak* 1297 Coram (K). Tengvik explains the OE example as from OE *Bacca* or as a nickname from OE *bæc,* in the sense of one with a prominent chin or back or one of a fat, rotund appearance. Ekwall takes the London example as perhaps from OE *bæc* 'back'. We may have a personal-name. OE *Bacca* was in use in Suffolk after the Conquest (*Baccce* (dat.) c1189–1200 BuryS). The nickname is probably, as suggested by Weekley, ME *bakke* 'a bat', either 'blind as a bat' or with reference to their

nocturnal habits; retiring by day to dark recesses, 'they hate the day and love the night'. Also local: Joan *atte Back* 1327 SRSo, 'dweller by the ridge' (OE *bæc).*

**Backer:** *v.* BAKER

**Backhalder:** John *Bakhalder* 1447 CtH; John *Bakeholder* 1525 SRSx. Probably a late form of BACHELOR.

**Backhouse, Baccas, Bacchus, Bachus, Backus:** Walter *de Bakhous* 1306 LLB E; Richard *del Bakhous* 1332 SRLa; Thomas *Bachous* 1334 LLB E; Charles *Baccus* 1544 AD v (Y); Edward *Bacchus* 1725 DKR 41 (Beds). 'One employed at a bakery', from OE *\*bæchūs* 'bakery, bakehouse' (a1300 MED).

**Backler:** John *de* (sic) *Bakalur* 1196 Cur (D); Nicholas *le Bakelere* 1320 Cl (Sa); Edmund *Bacler* 1524 SRSf. Identical with *Bachelor,* with dissimilation of *chl* to *kl.*

**Backman:** Walter *Bakman* 1279 RH (C); John *Bakeman* 1327 *SR* (Ess). OE *(ge)bæc* 'bakemeats', and *mann,* a maker or seller of pastries, pies, etc. cf. Walter *le Bakmonger* 1314 MEOT (Herts).

**Backner:** *v.* BACON

**Backshall, Backshaw, Backshell:** Philip *de Bacselve* 1296 SRXx; John *Bakshelue* 1327 ib.; Henry *Backshyll* 1525 ib.; John *Backshell,* Mary *Backshall* 1591, 1713 Sx Rec. Soc. ix. These are probably forms of Backshells in Billingshurst (Sussex). For the development of the forms with *-shall* and *-shaw,* cf. Gomshall (Surrey), *Gomeselve* 1154, *Gunshal* 1675, and Bashall Eaves in Great Mitton (WRYorks), *Bacschelf* DB, *Basshall* 1562, *Bashawe* 1591.

**Backton, Bacton:** John *de Baketon'* 1212 P (He); Richard *de Baketon* 1212 P (Ess/Herts); William *Bakton* 1444 Paston. From Bacton (He, Nf, Sf).

**Backwell:** Thomas *de Bacwell* 1225 AssSo. From Backwell (So). Sometimes, perhaps, from Bakewell (Db).

**Bacon, Bakon:** William, Richard *Bacun* c1150 StCh, DC (L); Nicholas *Bachun* 1226 Burton (St); Geoffrey *Bacon* 1296 SRSx. OFr, ME *bacon, bacun* 'buttock, ham, side of bacon' is not recorded in England before c1330 (MED), though it may well be older. It refers usually to the cured flesh, occasionally to fresh pork, but is seldom used of the live pig. Hence a nickname must be metonymic and refer to a pork-butcher, as does the Fr *Baconnier* and probably the English *Backner,* though no early forms have been found. The surname is common and early, used of Norman knights, and is probably the accusative of

OG *Bacco*, the nominative of which occurs as *Bacus* c1113 Burton (St).

**Bacton:** v. BACKTON

**Badams, Baddams, Badham:** v. ABADAM

**Badbury:** Herbert *de Badebiri* 1212, *de Baddebir'* 1218 P (Sx). From Badbury (Do, W).

**Badby:** William *de Baddebi* 1202 AssNth; William *Badby* 1388–9 FFSr; Thomas *Badby* 1425 AssLo. From Badby (Nth).

**Badcock, Batcock:** *Batecok* 1288 AssCh; *Badekoc* Korneys 1296 SRSx; Edrich' *Bathecoc* 1221 AssWo; Richard *Batcok* 1285 AssCh; William *Badecok* 1297 MinAcctCo (Do), 1327 SRDb. Both names are usually explained as a compound of *Bat(e)*, a pet-name for Bartholomew, and *cock*, but as both occur frequently side by side, the *d* of *Badecok* may well be original. In 'de catellis *Badde'* (1230 P) we have probably a survival of OE *Bada* (cf. BADE) which may also be the source of the surnames of William *Badde* 1221 AssWo and John *Badde, Bade* 1317 AssK. Though the forms are late, OE *Baduca* probably survives in *Baddock*, whilst an unrecorded OE *\*Badding* occurs as a surname in Robert *Badding* 1197–1221 AD i (Mx) and William *Bading* 1275 SRWo. A formation *Badecoc* is, therefore, not impossible.

**Badcoe:** Probably for *Badcock*. cf. ALCOE.

**Baddeley, Badderley, Baddeley, Badely, Badley:** Robert *de Badelea* 1187 P (Ha); Gilbert *de Badele* 1227 AssLa; John *de Baddyleye* 1327 SRSt. From Baddeley (St), Badley (Sf), Baddiley (Ch), or Baddesley (Ha, Wa).

**Badder:** v. BATHA

**Baddick, Baddock, Badock:** William *Baddoke* 1534 FrY. OE *Baduca*. v. BADCOCK.

**Badding:** Henry *Badding'* 1249 AssW; Walter *Baddyng* 1287–8 NorwLt; John *Baddyng* 1468 *ERO*. OE *\*Badding*. v. also BADCOCK.

**Bade:** *Bado* 1066 DB (Db); Bictricus *Bade* 1066 Winton (Ha). OE *Bada*. cf. BADCOCK.

**Badeley, Badely, Badley:** v. BADDELEY

**Badger, Badgers, Bagger:** (i) Ivo, Richard *le Bagger* 1246 AssLa, 1297 Wak (Y); Adam *Badger* 1324 Wak (Y); Ralph *Baghere* 1348 DbAS 36. The interpretation of *Bagger* is uncertain. It may stand for *Bagger* or *Badger*. The former would be a derivative of ME *bagge* 'bag, small sack', hence *bagger* 'a maker of bags'. *Badger*, not recorded before 1467–8 in MED and of doubtful origin, means 'a hawker, huckster'. Fransson's arguments in favour of a change in pronunciation from *bagger* to *badger*, partly on the grounds that there is no modern surname *Bagger*, cannot be accepted. Though very rare, *Bagger* is still found in Sevenoaks. As often, the metonymic *Bagg(s)* is more common. (ii) William *de Beggeshour'* 1221 AssSa. From Badger (Salop).

**Badham:** v. ABADAM

**Badkin:** v. BODKIN

**Badman:** Simon, John *Bademan* 1279 RH (C), 1375 LoPleas. OE *\*Beadumann*, a late compound of the common theme *Beadu-* and *-mann*.

**Badnall, Badnell:** Robert *de Badenhale* 1242 Fees (St). From Badnall (St).

**Badock:** v. BADDICK

**Badrick, Badrock, Batrick, Battrick, Betteridge, Bettridge, Betteriss:** Robert *Baderich'* 1275 SRWo; William *Betrich* 1279 RH (C); John *Betryche, Betrich* 1296, 1327 SRSx. OE *Beadurīc* 'battle-powerful', a personal name found in Battersea (Surrey) and Bethersden (Kent), but very rare in independent use.

**Badsworth:** John *de Baddesworth* 1334–7 SRY; John *Baddesworth* 1420 IpmY. From Badsworth (WRY).

**Bagby:** Robert *de Bagby* 1327 SRY; John *de Bagby* 1362 AssY. From Bagby (NRY).

**Bagehot:** v. BAGGETT

**Bagenal:** v. BAGNALL

**Bagg, Bagge, Baggs:** William, Nicholas *Bagge* 1166 P (Nf), 1214 Cur (Wa). Although no examples of its independent use have been noted, this may be the cas-sujet of OG *Bago* (*Baco*) surviving in the French *Bague* with its diminutive *Baguelin, Baglin*, which is found also in England: William *Bagelin* 1327 SRSo. v. also BAGGETT. The surname was common in ME and may also be metonymic for *Bagger*, 'a maker of bags', from ME *bagge* 'bag, pack, bundle'. It might also have been used for a beggar. cf. 'Hit is beggares rihte uorte beren bagge on bac & burgeises for to beren purses' (c1230 NED).

**Baggallay, Baggalley, Baggally, Baggarley:** v. BAGLEY

**Bagger:** v. BADGER

**Baggeridge, Baggridge:** William *de Bagerigge* 1201, *de Baggerugge* 1207, FFO; Walter *de Baggerigg* 1274 RH (Do). From Bageridge (Do), Baggridge (So), or a lost *Bagridge* in Woodlands (Do).

**Baggett, Baggott, Bagot, Bagott, Bagehot:** (i) *Bagot* c1125 StCh; Herueius *filius Bagot* 1130–2 Seals (St); Hereveus *Bagod* c1159 StCh; Ingeram *Bagot* Hy 2 DC (L); Hereficus *Bachot* 1195 Cur (Wa); Simon *Baghot* 1198 FFSt; Walter *Bagot* 1201 Cur (Y). A diminutive of OG *Bago*. cf. BAGG. (ii) Robert *Baggard* 1191 P (Sf); Geoffrey, Richard *Bagard* 1275 SRWo, 1279 AssSo. OG *Bago* plus the suffix *-(h)ard*.

**Bagilhole, Baglole:** v. BEAGLEHOLE

**Bagley, Baguley, Baggallay, Baggalley, Baggally, Baggarley, Bagguley:** Peter *de Baggeleg'* 1260 AssCh; Thomas *de Baggeleghe* 1327 SRSo; Walthev *de Baglay* c1345 Calv; John *Baguley* 1527 CorNt. From Bagley (Bucks, Salop, Som, WRYorks), or Baguley (Ches).

**Bagnall, Bagnell, Bagnold, Bagenal:** William *de Bagenholt* 1299 AssSt; John *Bagenelle* 1379 LLB H; Ralph *Bagnall, Bagnold, Bagenall alias Bagnald* 1561 Pat. From Bagnall (Staffs).

**Bagpuis, Bagpuss:** Ralph *de Bagpuize* 1086 DB (Berks); Robert *de Bakepoiz* 1219 P (Lo/Mx); John *Bakeputz* 1327 SRLei. From Bacquepuis (Normandy).

**Bagshaw, Bagshawe:** Robert *de Baggeshagh* 1327 SRDb; William *de Baggeshaugh* 1339 DbCh; Thomas *Bagsha* 1565, *Bagshae* 1572 Petre. From Bagshaw (Db).

**Bagster:** v. BAXTER

**Bagwell:** Geoffrey *Bagwell* 1374–5 NorwLt; John *Bagwell*, Nicholas *Baggwell* 1642 PrD. 'Dweller by badger stream', OE *\*bagga, wiella*. Sometimes, perhaps, for BACKWELL.

**Baiker:** v. BAKER

**Bail, Baile, Bailes, Bails, Bale, Bales, Bayles:** Richard del Baille c1190 Bart (Lo); Eudo del Bayle 1301 SRY; John Bayl 1382 FFSx; Thomas Bale 1524 SRSf; William a Bales 1537 FFHu; Zacarias Bailes 1629 FrY. OFr, ME bail(e) 'the wall of the outer court of a feudal castle', later used of the courts themselves (a1200 MED). The surname is probably identical with Bailward, 'the guardian of the courts or bailey'. The London examples refer to the Old Bailey.

**Baildon, Bayldon:** Hugh de Beyldon 1251 AssY; Henry de Bayldon 1372 FFY; Janie Baildon 1672 HTY. From Baildon (WRY).

**Baileff, Bailiff, Bayliff, Bayliffe, Baylyff:** Richard le Baillif 1242 Fees (He); Gilbert le Balif 1280 AssSo; John Bayllif, Baylly 1296 SRSx. OFr bailif, cas-régime of baillis, originally 'carrier', later 'manager, administrator'. Used of the public administrator of a district, the chief officer of a Hundred (1297 NED) or of an officer of justice under a sheriff, a warrant officer, pursuivant, a catchpoll (1377 NED). This form of the name is much less common than Bailey. v. BAYLIS.

**Bailess:** v. BAYLIS

**Bailey, Bailie, Baillie, Baily, Baly, Baylay, Bayley, Baylie, Bayly:** (i) Roger le baylly 1230 P (Sf); John Baly 1274 Wak (Y); Thomas le Baly 1327 SRSx; Thomas Bailie 1327 SRSf. OFr bailli, a later form of baillis, baillif. v. BAILEFF. The term baillie, now obsolete in England, is still the common form in Scotland, where it was used of the chief magistrate of a barony or part of a county, a sheriff. (ii) John ate Baylie 1317 AssK. A ME variant of bayle (a1200 MED). v. BAIL. Dyonisya en la baillye owned houses and shops in the Old Bailey, London (1319 SRLo). The earliest examples of the Scottish Baillie, William de Bailli 1311–12 (Black), seem to belong here rather than to the noun above. (iii) Ralph de Baylegh 1246 AssLa. From Bailey (Lancs).

**Bailhache:** Baylehache marescallus 1154 Eng-Feud (Sf); John Baillehache 1418 DKR 41. OFr baille hache 'give axe', a name for an executioner.

**Bailward:** v. BAIL

**Bain, Baine, Baines, Bains, Baynes, Bayns:** (i) William Banes 1246 AssLa; Hewerard Ban, Cristiana Bane 1279 RH (C, O); Thomas Baynes 1446 FrY; Alexander Banys 1541 Black; Andrew Baines 1676 ib. OE bān 'bone' in the North and in Scotland became ME bān and later bain; in the Midlands and the South it became ME bōn, modern bone and survives as BONES. A nickname, probably usually plural. (ii) Richard Beine 1279 RH (C); John Bayne 1301 SRY; William Bayn 1323 AssSt. ME beyn, bayn, ON beinn 'straight, direct', also 'ready to serve, hospitable'. Always singular. (iii) Thomas Ban 1324 Black (Perth); David Bane or Bayn 1456–60 ib.; Ewir Bayne alias Quhyte 1623 ib. Gael bàn 'fair, white'. Always singular. (iv) Serlo Baynes 1219 AssY; John de Bayns 1275 RH (L); Roger Bayns 1277 AssLa; Thomas de Bainnes 1333 FrY. ME bayne, Fr bain 'bath'. Probably 'attendant at the public baths'. The forms without a preposition are too early to be regarded as northern forms of bān 'bone'.

**Bainbridge, Bainbrigge:** Matilda de Baynbrigg' 1301 SRY; Robert Baynbryg' 1379 PTY. From Bainbridge (WRYorks).

**Bainton, Baynton, Bayntun:** Osgode on Badingtune 972 BCS (Nth); Turstan de Baynton' 1219 AssY; William de Baynton' 1361 FFWa; William Baynton 1597 SRY. From Bainton (Nth, O, ERY), or Baynton (W).

**Bairnsfather, Barnfather, Banfather:** William Barnefader 1246 AssLa, 1260 AssY; Henry Barnefathir 1392 Shef (Y); Adam Barnisfader 1502 Black. OE bearn 'child' and fæder 'father'. cf. ON barnfaðir 'a child's alleged father', which might well be the direct source of the second form.

**Bairstow, Barstow, Baistow, Bastow:** Ralph de Bayrestowe 1277 Wak (Y); Thomas de Barstowe 1348 DbAS 36. From Bairstow (WRYorks).

**Baisbrown, Bisbrown:** William Basbroun 1332 SRCu; Christopher Baysbrown 1494 FrY; Laurence Besbrowne 1595 LaWills; Hugh Bisbrowne 1667 ib. From Baisbrowne (Westmorland).

**Baish:** v. BASH

**Baiss:** v. BASS

**Baister:** v. BASTER

**Baitson:** v. BATESON

**Baker, Baiker, Backer:** William le Bakere 1177 P (Nf); Robert Bakere 1246 AssLa; Walter le Backere 1280 MESO (Ha). OE bæcere 'baker'. cf. BAXTER.

**Bakon:** v. BACON

**Balaam, Ballaam:** (i) Hamelinus de Baalon 1176 P (D); John Balum 1212 Cur (W); Rosa Balam 1275 SRWo. From Baalon in Meuse. (ii) John Balam 1568 SRSf; Eliza Baalam 1830 LitWelnethamPR (Sf); Sophia Baylham 1834 ib. From Baylham (Suffolk).

**Balch:** v. BELCH

**Balcock, Ballcock:** William Balcok 1263 FFL; Henry Balcok 1332 SRWa; Richard Balcock 1440 ShefA. From Bald, a short form of the common OG Baldwin, or from OE *Beald, plus the diminutive -cock. v. also BAWCOCK.

**Balcomb, Balcombe, Bawcombe:** John de Balecomb 1327, Thomas Balcombe 1525 SRSx; William Bawcom 1591 ArchC 48. From Balcombe (Sx).

**Bald, Bauld:** (i) Simon le Bald' 1178 P (Ess); Hugo Calvus 1198 FF (Herts); John Balde 1221 AssWo; William le Ballede c1248 Bec (W). ME ballede, used first of rotundity or corpulence (1287 NED), later, of baldness (c1386 ib.). cf. Madoc le Balled 14th AD vi (Ch), whose seal was inscribed s. MADOCI CAL[VI], and v. BALLARD. (ii) Balde c1150 DC (L), 1191 P (Lo), 1198 P (Beds); Bald' filius Bald' 1199 P (Herts); Boold 1332 SRLa. A short form of the common OG Baldwin or Baldric or of OE *Beald. Balt 1066 DB (Y) may be OE or continental. cf. BALDING. Survives occasionally as BOLD.

**Balder:** Balder' Martell' 1160–2 Clerkenwell (Do); Thomas Balder 1524 SRSf. A rare surname from the not common OE Bealdhere 'bold army'.

**Balderson, Bolderson, Boldison:** Probably assimilated forms of Balderston and Bolderston rather than from the rare Balder.

**Balderston, Balderstone, Bolderston, Boulderstone:** William de Baldreston 1292 WhC (La). From

Balderston (Lancs) or Balderstone (Lancs).

**Balderton:** Robert *de Baldertona* 1218–9 RegAntiquiss. From Balderton (Ch, Nt).

**Baldey:** *v.* BALDREE

**Baldey, Baldie, Baldy, Boldy:** William *Baldith* 1204 P (Gl); Simon, John *Baldy* 1274 RH (Sf), 1332 SRSx. OE *Bealdgȳð* (f) 'bold combat', a woman's name not recorded before the Conquest but noted once in *Baldith uxor* c1170 Rams (Hu) and, possibly, in a corrupt form in *Baldethiva* 13th AD iv (Wa). In Suffolk, we may have ON *Baldi*. In Scotland, *Baldie* is a pet-form of *Archibald* or *Baldwin* and, as a surname, late: Thomas *Baldy* 1540 Black.

**Baldick, Baldock:** Hugh *de Baldoca* 1185 Templars (Beds); Thomas *Baldac* c1280 SRWo; Robert *Baldec* 1331 IpmW; William *Baldocke* 1460 IpmNt. From Baldock (Herts), a town founded in the 12th century by the Knights Templar, and given the name of Baghdad in its OFr form.

**Balding, Belding, Bolding, Boulding:** Joscius *filius Balding'* Ric 1 Cur (L); Gilebertus *filius Balding* 1212 Cur (L); William, Joan *Bolding* 1255 RH (Sa), 1327 SRSf; Alice, John *Baldyng* 1327 SRSx; Robert *Beldyng* 1332 SRSx. OE *Bealding*, a derivative of *Beald*, not recorded before the Conquest. In 1674 HTSf Anthony *Baldin* and Bartholomew *Baldinge* occur side by side with *Baldwin*, so that *Balding* is sometimes a late development of *Baldwin*. A possible earlier example is Alexander *Baldyne* 1251 Rams (Hu).

**Baldree, Baldrey, Baldry, Baldrick, Baudrey, Baudry, Baldery, Boldero, Bolderoe, Boldra, Boldry:** *Baldric* 1066, 1086 DB; *Baldricus* 1127–54 Holme (Nf); *Baldri* de Grendal Hy 2 DC (L); *Baudricus* de Lawdecote 1208 Cur (Sr); William *Baldri* 1185 Templars (Herts); Aluredus *Baltriht* 1197 P (W); Henry *Belrich* 1203 P (O); Alexander *Baudri* 1205 Cur (Sf); Richard *Balrich* 1238 Oseney (O); Walter *Baldrich* 1275 SRWo; Robert *Baldrik*, Henry *Baudrik* 1327 *SR* (Ess); Francis *Baldry*, *Boldery*, James *Baldery* 1674 HTSf; Mrs *Balderoe*, Stephen *Bolderowe* ib.; Martin *Boldroe*, Widow *Boldery*, *Boldry* ib. OG *Baldric* 'bold rule', common in the French forms *Baldri*, *Baudri*. The cognate OE *Bealdrīc* is unrecorded but such forms as *Belrich*, *Balrich* and *Baldrich* suggest that it did exist and was used after the Conquest.

**Baldwin, Balwin:** *Baldewinus* c1095 Bury (Sf); Randulfus *filius Balduini* Hy 2 DC (L); Stephen *Baldewin* 1200 P (Ha); John *Baudewin* 1260 AssC. OG *Baldwin* 'bold friend', a popular Flemish name, common in England both before and after the Conquest.

**Baldy:** *v.* BALDEY

**Bale, Bales:** *v.* BAIL

**Balfour:** John, William *de Balfure* 1304, 1331–5 Black; Michael *de Balfoure* 1365 ib. From the barony of Balfour (Fife).

**Ball, Balle, Balls, Le Ball:** (i) Godwin *Balle* 1137 ELPN; Norman *Balle* 1183 P (Nth). *Balle* is here disyllabic and may be from *Balle*, an ODa personal-name found in Scandinavian place-names in England and possibly surviving as *Balla* 1250 Fees (Ha). If a nickname, it might be OFr *balle* or OE *bealla* 'ball', or an adjective *ball* might have developed from *ball* 'a bald place', *balle* being the weak form. In later examples, *balle* became monosyllabic and is a form of *bal*. (ii) Robert, Matilda *le Bal* 1296, 1327 SRSx. Either ME *bal*, *ball(e)* 'the rotund one' or an adjective *ball* in the sense 'bald' from *ball* 'a white streak, a bald place'. *v.* NED s.v. *ball* and ELPN 137. (iii) Alfwin *attebal* 1166 P (Nf); Henry *atte Balle* 1327 SRSo. From residence near a *ball*, 'a knoll', 'a rounded hill' (1166 MED). *v.* MELS 5–6, PN W 422.

**Ballaam:** *v.* BALAAM

**Ballance, Ballans:** Roger *Balance*, *Balaunce* 1196 FF (Wa), 1221 AssWa. Metonymic for 'balancer' from AFr *balancer*, OFr *balancier* 'one who weighs with a balance'. cf. Thomas *le Balauncer* 1283 LLB A.

**Ballard:** Peter, Adam *Ballard* 1196 Cur (Nth), 1210 Cur (C). ME *ball* plus *-ard*, 'a bald-headed man'. Where Wyclif has 'Stye up, ballard!', Coverdale translates, 'Come vp here thou balde heade'. cf. Robert *Balheved* 1316 FFEss, Thomas *Balhefd* 1402 FA (Sf).

**Ballaster, Ballester, Ballister, Balster:** William *le Balister* 1293 AssCh. OFr *balestier* 'cross-bowman' (c1450 MED). *v.* ALABASTAR.

**Ballcock:** *v.* BALCOCK

**Baller:** Alan *le Baller* 1243 AssSo; Geoffrey *Ballar* 13th Rams (C). A derivative of ME *ball*, either 'maker of balls' or 'dweller by a rounded hill' *v.* BALL.

**Ballet, Ballett:** (i) Cristina *Balet* 1327 SRSo; John *Ballett* 1641 PrSo. A diminutive of OG *Ballo*. (ii) Robert *Balheved* 1316 FFEss; Richard *Balleheved* 1327 SRSo; Thomas *Balhefd* 1402 FA (Sf). 'Round-headed', OE *beall*, *hēafod*.

**Bailey, Bally:** (i) Richard *Balli* 1176 P (K); Margaret *Bally* 1314 LLB D. ON *Balli*, an original nickname from ON *ballr* 'dangerous'. (ii) Ralph *de Balley* 1327 SRSf. Probably for BAILEY.

**Ballinger:** William *Balinger* 1221 AssSa. AFr *balinger* 'a small fast naval vessel', hence 'a sailor'. Or for BERRINGER.

**Balliol:** Rainald *de Balgiole* 1086 DB (St); John *de Ballio*, *de Baillio* a1187 DC; Ely *de Bailleul* 1235 FFEss. From Bailleul (Somme), or Bailleul-la-Gouffern (Orne). *v.* OEByn 70.

**Ballock, Ballox:** Lewin *Balloc* c1110 Winton (Ha); Simon *Ballok* 1227 Cur (Sx); William *Ballok* 1381 AssL. OE *Balloc*, probably an original nickname from OE *bealluc* 'testicle'. cf. Robert *Blakehalloc* 1243 AssSo 'black testicle'; Roger *Gildynballokes* 1316 Wak (Y) 'golden testicles'.

**Ballon, Balon:** *Baloun*, *Balun* 1276 RH (Lei); John *le Balun* 1275 RH (He); John *Ballon* 1297 MinAcctCo; Thomas *le Balon* 1327 SRWo; Walter *Ballun*, *Ballom* 1296, 1327 SRSx. OFr *Ballon*, cas-régime of OG *Ballo*. The *le* proves we have also a noun, OFr *balon* 'little ball or pack' (Cotgrave), 'package', which may be metonymic for 'a packer' or a nickname for a little man of rotund form.

**Bally:** *v.* BALLEY

**Balman:** *Baldeman* filius Fabri 1188 BuryS (Sf); Nicolaus *filius Baldeman* ib.; Adam *Baleman* 1332 SRSx; Richard *Balman* 1415 LLB I. OE *Bealdmann*.

**Balmer:** Richard *le Baumere* 1305 LoCt. A derivative of ME *balme, bawme,* an aromatic substance prized for its medicinal qualities; a spice-dealer or seller of ointments.

**Balne, Balme:** Robert *de Balne* 1175 P (Y); Alice *de Balne* 1297 MinAcctCo; Thomas *de Balme* 1379 PTY. From Balne (WRY).

**Balsam, Balsham, Balsom, Balson:** Pinna *de Belesham* 1086 InqEl; Margaret *de Balsham* 1260 AssC; William *Balsham* 1317 AssK; Alan *Balsam* 1523 ArchC 41; John *Balson* 1642 PrD. From Balsham (C), or Balstone (D).

**Balshaw:** Robert *de Balsagh* 1246, Robert *Balshagh* 1401 AssLa; Thomas *Balshay* 1512, William *Baldshawe* 1522 FrY. From Balshaw (La).

**Balsom, Balson:** *v.* BALSAM

**Balster:** *v.* BALLASTER

**Balston, Balstone:** Robert *Balston* 1327 SRSf. Probably OE *Bealdstān* 'bold stone'.

**Balwin:** *v.* BALDWIN

**Baly:** *v.* BAILEY

**Bamber:** Malger *de Bænburc* 1190, *de Bamburc'* 1202 P (L); John *Bamburgh* 1428 FFEss; William *Bamber* 1524 SRSf. From Baumber (L). Sometimes, perhaps, from Bamborough (Nb).

**Bamfield, Bampfylde:** Richard *de Bamfeld* 1272 PN Herts 56; Thomas *Bamfeld* 1462, Matthew *Bampfeld* 1492 FFEss; John *Bampfyld* 1642 PrD. From Bamville Fm in Wheathampstead (Herts), Bampfylde Lodge in Poltimore (D), or 'dweller by the bean field', OE *bēan, feld.*

**Bamford, Bamforth, Bampford, Bampforth:** William, Thomas de *Bamford* 1228 Cur (Sf), 1312 WhC; Christopher *Bamfurth* 1539 CorNt. From Bamford (Derby, Lancs).

**Bampton:** Jordan *de Bampton'* 1208 Cur (O); Thomas *de Bampton* 1332 SRCu; John *Bampton* 1642 PrD. From Bampton (Cu, D, O, We).

**Banbery, Banberry, Banbury:** Algot *de Banneberi* 1178 P (O); Henry *de Bannebury* 1310 LLB D; James *Banbury* 1426 Black. From Banbury (O).

**Bancroft, Bangcroft, Bencroft:** Stephen *de bancroft* 1222 DBStP; John *atte Bencrofte* 1296 SRSx; Thomas *Bancrofte* 1481–2 FFWa. From Bancroft in Ardeley (Herts), Bancroft Field in Soham (C), or 'dweller by the bean field', OE *bēan, croft.*

**Banden, Bandon:** Richard *de Bandon'* 1206 Cur (Sr); John *de Bandone* 1353 LLB G; Arthur *Banden* 1642 PrD. From Bandonhill in Beddington (Sr).

**Bane(s):** *v.* BAIN

**Baney, Boney:** John, William *Bani* 1279 RH (Bk); Agnes, Alice *Bonye* ib. (O). Early examples of *bony* (a1515 NED), from ME *bān, bōn* 'bone'. cf. BAIN, BONES.

**Banfather:** *v.* BAIRNSFATHER

**Bangcroft:** *v.* BANCROFT

**Banham:** Geoffrey *de Banham* 1206 Cur (Nf); Thomas *de Banham* 1337 LLB E; Robert *Banham* a1466 Paston. From Banham (Nf).

**Banister:** *v.* BANNISTER

**Banker:** Albrice *le Baunker* 1245 FFC; Thomas *Bankar* 1358 Putnam (Nth). 'Dweller by a bank.'

**Bankes, Banks:** Walter *del Banck'* 1297 SRY; Matthew *Banke* 1327 SRSf. 'Dweller by a slope, bank, hillside' (ME *banke*). The plural form may conceal an original *-hous:* William *Bankhous* 1482 FrY; Robert *Bancus* 1513 GildY.

**Bann, Banne:** Brucstan *Banne* 1066 Winton (Ha); Richard *Banne* 1249 AssW; William *bann* 1327 SRLei. Probably an unrecorded OE *\*Banna, v.* OEByn 150. Sometimes, perhaps, from OFr *bane, banne* 'hamper, pannier', metonymic for a maker of these.

**Bannerman:** Dovinaldus *Banerman* 1368 Black. A Scottish name, 'standard-bearer, ensign' (1450 MED).

**Bannister, Bannester, Banister:** Turstan, Richard *Banastre* 1149–53 EngFeud (L), 1186 Eynsham (O); Richard *Banester* 1459 AD vi (St); John *Banyster* 1554 FrY. OFr *banastre* 'basket'. Metonymic for a basket-maker.

**Banstickle:** Adam *Banstikel* 1275 SRWo. A nickname from ME *banstikel* 'a kind of fish, the three-spined stickleback'.

**Banwell:** John *de Banwell* 1327 SRWo; Nicholas *de Benewell* 1332 SRSx; Walter *de Banewell* 1335 Glast (So). From Banwell (So), or Banwell Fm in Mundham (Sx).

**Banyard, Bunyard:** Ralph *Baignard, Bangiard, Baniardus* 1086 DB (Herts); Robert *Baniard, Bainard* 1207, 1208 Cur (Nf); William *Banyard* 1346 FA (Sf). A variant of *Baynard,* due to the Englishman's difficulty in pronouncing the French *n* mouillé. cf. the English *onion* from the French *oignon.*

**Baram:** *v.* BARHAM

**Barbarel, Barbarell, Barberell:** *Barberella* c1210 Cur (Nt); Geoffrey *Barberell'* 1192 P (K); William *Barberel* 1225 AssSo. *Barbar-el,* a diminutive of *Barbara.* William *Barbet* 1212 is also called William *Barberel* 1219 Fees (Berks).

**Barbary, Barbery:** *Barbery* Marbeck 1581 Bardsley; Richard *Barbary* 1327 SRLei; John *Barbery* 1674 HTSf. The usual vernacular form of *Barbara.*

**Barbe:** Bernardus *Barb* 1086 DB (Ha); Willaim *Barbe* 1229 Pat (K). A pet-form of *Barbara,* or OFr *barbe* 'beard'. *v.* BEARD.

**Barber, Barbour:** Alan, John *le Barbur* 1221 AssWa, 1248 FFEss; Thomas, Richard *le Barber* 1281 LLB A, 1298 LoCt; Seykin, Robert *le Barbier* 1299 LLB C. *Barbour* is from AFr *barbour,* OFr *barbeor* (c1320 NED), *Barber* from OFr *barbier* 'barber'. The barber was formerly a regular practitioner in surgery and dentistry.

**Barberell:** *v.* BARBAREL

**Barbery:** *v.* BARBARY

**Barbet, Barbett, Barbot, Barbotte:** *Barbetta* 1190, *Barbeta* 1191 P (K); *Barbota* (f) 1240 FFEss; William *Barbette* 1195 P (Berks); Henry *Barbot* 1206 Cur (L); Richard *Barbot* 1303 FFY. *Barb-et, Barb-ot,* diminutives of *Barb,* a pet-form of *Barbara.* Occasionally, perhaps, a diminutive of OFr *barbe* 'beard'.

**Barbon, Barebone:** Robert *Barebayn* 1301 SRY; Thomas *Barbon* 1494–5, Wyllyam *Barebone* 1569 LedburyPR (He). The first example makes it clear that this is a nickname, 'bare bone', OE *bær,* ON *beinn/*OE *bān,* presumably for a thin man. Sometimes, perhaps, from Barbourne (We). But the surname is usually Huguenot from a refugee family living at Wandsworth. Praise-God *Barebone* belonged to this family (Smiles 361).

**Barbot, Barbotte:** v. BARBET

**Barby:** *Barby Barby* 1642 PrD; Thomas *Barby* 1641 PrSo; John *Barby*, Nicholas *Barbey* 1642 PrD. A pet-form of *Barb*, a short form of *Barbara*.

**Barchard:** v. BURCHARD

**Barclay, Berkeley, Berkley:** Roger *de Berchelai, de Bercleia* 1086 DB (Gl, So); Henry *de Barcley* 1327 SRDb; Helewys' *de Berkele* 1327 SRSx. From Berkeley (Glos), Berkley (Som), or Barklye in Heathfield (Sussex). William *de Berchelai*, Chamberlain of Scotland in 1165, and the Scottish Barclays probably came from Berkeley (Glos).

**Barcroft, Bearcroft:** William *de Bercroft* 1274 RH (Y). From Barcroft in Bingley, or Barcroft in Haworth (WRY).

**Bard, Barde:** (i) Ralph *Bard* c1155 DC (L); Hugh *Bard'* 1219 P (Y); John *Barde* 1327 SRSo. OFr *barde* 'horse armour'. Metonymic for a maker of this. (ii) Simon *le Bard* 1364 Black. Gaelic *bàrd* 'poet'.

**Bardell, Bardill:** Aschetill *Bardel* 1159 P; William *Bardell* 1327 SRC. Perhaps OFr *bardelle* 'pack-saddle', and metonymic for a maker or user of this. But late forms are probably for BARDOLPH.

**Barden, Bardens, Bardin:** Abraham *de Barden'* 1176 P (L); William *de Bardene* 1327, John *Barden* 1332 SRSx. From Barden (NRY, WRY).

**Bardeney, Bardney:** Richard *de Bardeney* 1306 RegAntiquiss. From Bardney (L).

**Barder:** (i) Hugh *Bardur* 1202–3 FFWa; Nicholas *le Barder* 1328 KentRecs 18; OFr *barde* 'armour', hence a nickname for an armourer. (ii) Herbert *Barbe de Auril* 1187 P (R); William *Barbe de Or* 1230 P (C); Roger *Barbeder*, William *Barbedor* 1279 RH (C). 'Golden beard', OFr *barbe, or*. v. also GOLDBARD.

**Bardill:** v. BARDELL

**Bardin:** v. BARDEN

**Bardney:** v. BARDENEY

**Bardolph:** *Bardolfus* de Fotipoi 12th DC (Nt); Hugo *Bardulfus, Bardol* 1142–53 DC (L); Thomas *Bardolf* 1184 Gilb (L); William *Bardoul* 1418 DKR 41. OG *Bartholf*.

**Bardsey:** John *de Berdeshey* 1334–7 SRY; Hugh *de Berdesey* 1404 IpmLa; Robert *Bardsey* 1469–70 FFSr. From Bardsea (La), or Bardsey (WRY).

**Bardsley, Beardsley:** William *de Berdeslega* 1195 P (Gl); Thomas *de Bardesleg'* 1242 Fees (Sa); Peter *Bardeslay* 1453 FrY. From Bardsley (La).

**Bardwell, Beardwell:** Tedricus *de Berdewelle* 1190 P (Sf). From Bardwell (Suffolk).

**Bare:** John, Roger *Bare* 1274 RH (Sf), 1327 SRSf. OE *bær* 'bare' in one of its early senses, 'unarmed, defenceless, deserted, indigent'.

**Barebone:** v. BARBON

**Barefoot, Barfoot:** Robert *barefot* c1160 EngFeud (Nth); Reginald *Berfot* 1203 P (Cu); John *Barfot* 1317 AssK. OE *bær* and *fōt* 'with bare feet', 'barefooted', used of friars, pilgrims and those doing penance. cf. Simon *Barleg* 1297 MinAcctCo, Emeloth *Baresanke* 1221 ElyA (Nf) 'bare-legged'.

**Bareham:** v. BARHAM

**Barff:** v. BARGH

**Barfield, Barefield:** Nigel *de Bereuile* 1086 DB (Bk); Robert *de Bereuill'* 1204 P (D); Philip *de Bardefeld* 1275 SRWo; Simon *de Berdefeld* 1312 LLB D. The first two examples are perhaps from Berville-la-Champagne (Eure), but there are six Bervilles in Normandy, three in Eure, two in Seine-Maritime, and one in Calvados. Other possible sources are Barfield Copse in Godalming (Sr), or Bardfield (Ess).

**Barford:** Henry *de Bereford* 1204 P (Gl); William *de Berford* 1325, John *Berford* 1419 FrY. From Barford (Beds, Nf, Nth, O, Sr, W, Wa), or Barforth (NRY).

**Bargain, Bargaine, Bargayn, Bargayne:** Thomas *Bargayn* 1297 SRY; Richard *Bargayne* 1365 FrY; Philip *Bargaine*, Walter *Bargin* 1642 PrD. ME *bargaine* 'a business arrangement or agreement'. Probably metonymic for a merchant or trader'.

**Bargate:** Adam *de la Bargate* 1275 SRWo. 'Dweller by the gate that bars entry into the town.' The surname may also refer to the keeper of the Bargate.

**Barge:** Peter *del Barge*, mariner 1359 FrY. 'Bargeman, seaman.' OFr *barge* was originally used of a small sea-going vessel with sails.

**Bargh, Barff, Barugh:** Robert *de Bargh* 1310 FFSf; John *de Bergh* 1365 FrY. The modern northern form of BARROW, ME *bergh*, OE *beorh*. 'Dweller by the hill' as at Barff Hill (ERYorks) and Barugh, pronounced *Barf* (NRYorks).

**Barham, Baram, Bareham, Bearham:** John *de Barham* 1238–9 FFEss; Helewys *de Berham* 1296 SRSx. From Barham (C, D, Hu, K, Sf, Sx).

**Bark, Barke, Barks:** Jordan *le Barc* 1197 P (Nf/Sf); Ralph *le Berk'* 1249 AssW; William *Barke* 1327 SRY. ON *borkr* 'bark'. Metonymic for BARKER.

**Barkas:** v. BARKHOUSE

**Barkaway:** v. BARKWAY

**Barke, Barks:** v. BARK

**Barker, Berker:** (i) Ralph *Berker* 1185 Templars (Y); Aluredus *le berkier* 1193 P (L); John *le Bercher* 1212 Cur (Ha). OFr *berchier, bercher, berkier, berker* 'shepherd'. OFr also had the form *barcher* which may well be represented below. Later, when ME *-er* had become *-ar, barker* 'shepherd' would be indistinguishable in form from *barker* 'tanner'. (ii) Jordan *le Barkere* 1255 Ass (Ess); John *le Barker* 1260 AssC. A derivative of ME *bark* 'to tan', a tanner.

**Barkhouse, Barkas, Barkis:** Thomas *del Barkhous* 1379 PTY. 'A barkhouse' (1463 DbCh, 1483 NED) was a tannery. In 1383 Hugh *de Barkhowse* (*del Barkhous* 1384 DbCh) granted all his goods and chattels in his tannery at Beauchief to Ralph de Dore. The surname is thus occupational, 'a tanner'.

**Barkshire:** v. BERKSHIRE

**Barkston, Barkstone:** Richard *of Barkeston* 1218–9 FFY. From Barkston (L, WRY), or Barkestone (Lei).

**Barkway, Barkaway:** Walter *de Berqueie* 1141–51 Colch (Ess); John *de Berkwey* 1281 LLB B; Richard *Barkaway, Barkway* 1674 HTSf; James *Barkaway* (signed *Barkerway*) 1776 SfPR. From Barkway (Herts).

**Barkwith, Barkworth:** Robert *de Barcword* c1150

Gilb; Roger *de Barkworth* 1339 LoPleas; John *Barkword* 1371 AssL; Richard *Barkwith* 1524 SRSf. From Barkwith (L), *Barcuurde* DB.

**Barlas:** *v.* BURLES

**Barlett:** Willelmus *filius Berlet'* 1219 AssY; Robert *Berlet* 1206 P (Nt); John *Barlet* 1242 Fees (W); Adam *Berilot* 1327 FrY. *Ber-el-ot*, a double diminutive of *Ber-*, from OG *Berard*.

**Barley, Barlee:** (i) Jordan *Barlie* 1221 AssWa; John *Barlich*, Reur' *Barliche* 1279 RH (O, C). OE *bærlīc*, ME *barlich, barli* 'barley', used for BARLEYMAN or by metonymy for a maker or seller of barley-bread or cakes. cf. Josce *Barlibred* 1185 P (Nf) (c1320 NED), Roger *Barliwastel* 1210 FFL. (ii) Leofric *de Berle* c975 OEByn (Herts); Henry *de Berel'* 1219 AssY. From Barley (Herts, Lancs, WRYorks). *v.* also BARLOW.

**Barleycorn:** William, Godfrey *Barlicorn* 1233 Cl(L), 1279 RH (C). *Barleycorn* (1382 NED) was used of both the plant and the grain. The surname may refer to a grower of or dealer in barley. cf. GRANDAGE.

**Barleyman:** Peter *Barlyman* 1332 MEOT (L). A dealer in barley.

**Barling:** Baddewin *de Barling'* 1240–1 ForEss; William *de Berling* 1327 SRSx; John *Barling* 1461 PN K 214. From Barling (Ess), Barlings (L), Barling Green Fm in East Sutton (K), or Birling Fm in Eastdean (Sx), *Barlyng* 1363.

**Barlow:** Thomas *de Barlowe* 1260 AssLa; John *de Berlowe* 1379 PTY; Margery *Barley* or *Barlowe*, William *Barlee* or *Barlowe* 1509 LP (Db, Ess). From Barlow (Derby, Lancs, WRYorks), but there seems also to have been some confusion with BARLEY.

**Barlthrop, Barltrop:** *v.* BARTHORP

**Barman, Bearman, Berman:** Alsi, Gilbert *Berman* 1137 ELPN, 1222 Cur (Sr); Ralph *Bareman* 1275 RH (Beds); Simon *le Berman* 1281 MEOT (L); Geoffrey *le Barman* 1301 SRY. OE *bērmann* 'bearer, porter'. *Berman*, without the article, may also be personal in origin. Walterus *filius Bereman* 1198 P (K) may have been the son of a porter, but his father may have borne the name of *\*Beornmann*, unrecorded in OE, but of a type common in the 11th and 12th centuries. Occasionally we may have the rare OE *Beornmund*. cf. Adam *Beremund* 1204 P (Lo); William *Beremund* 1272 *Ass* (Ha).

**Barmby, Barnby:** Roger *de Barnebia* 1179 P (Y); Robert *de Barnneby* 1282 IpmY; William *de Barneby* 1347 FFY. From Barnaby, Barnby in Lythe (NRY), or Barmby on the Moor, on the Marsh (ERY). The place-names may also have contributed to *Barnaby*. *v.* BARNABÉ.

**Barnabé, Barnaby, Barneby:** (i) Roger *Barnabe* 1327 SRC; Roger *Barnaby* 1331 FFC. The English form of *Barnabas*, not common in the records, but found in the 14th century (ODCN) and surviving until the 19th century as in Dickens's *Barnaby Rudge*. (ii) Richard *de Bernaldeby* 1160 Guisb (Y). From Barnaby in Skelton (NRYorks).

**Barnacle, Barnacal, Barnikel, Burnikell:** Richard *Bernikel* 1344 Cl (K); Richard *Barnakyll* 1514 Oxon; John *Barnacle* 1545 Bardsley. Barnacle (Warwicks), DB *Bernhangre*, did not reach its modern form before 1547. We are, therefore, clearly concerned with a

nickname from ME *bernacle, barnakyll*, a diminutive of ME *bernak*, OFr *bernac* 'a kind of powerful bit or twitch for the mouth of a horse or an ass', used to restrain a restive animal, also used as an instrument of torture. The nickname might have been applied to an expert in taming horses or to a torturer or it might have been given to a man of savage, unrestrained temper who needed such restraint. A further possibility is a nickname from the barnacle goose, ME *barnakyll*, a species of wild goose (cf. WILDGOOSE).

**Barnaclough:** *v.* BARRACLOUGH

**Barnard, Barnet, Barnett, Bernard:** *Bernardus* 1066 DB (Hu), 1101–16 Holme (Nf); Ricardus *filius Bernardi* 1205 Cur (So); Hugo *Bernard'* 1130 P (L); Thomas *Bernhard* 1260 AssC; Robert *Barnard* 1446 FrY. OFr *Bernart*, OG *Bernard* 'bear-brave'.

**Barne:** (i) Siuuard *Barn* 1066 DB (Wa), *Bearn, Barn* 1071, 1072 ASC D, E; Gamell' *Barn* 1166 P (Y); Adam *le Barn* 1212 Cur (Y); William *le Barne* 1232 Pat (L). ON *barn* 'child'. Used in DB as a byname of men of the upper classes, it might also have had the meaning 'a young man of a prominent family'. cf. the English CHILD. (ii) *Beornus* 1066 DB (Sf); *Bern* 1066 DB (Do); Tomas *filius Bern'* 1177 P (St); Þirne *Beorn* c1050 YCh; William, Simon *Bern* 1190 P (Wo), 1202 AssL. In Yorks, Lincs, Staffs and Suffolk we have the Scandinavian personal-name *Biǫrn*, anglicized as *Beorn*. In Dorset and Worcs we may have OE *Beorn*. The source may occasionally be OE *beorn* 'warrior'. (iii) Eilwin *de la Berne* 1211 Cur (Sr); Richard *atte Berne* 1275 SRWo; Peter *del Barne* 1316 Wak (Y). From residence near or employment at a barn (OE *bere-ærn*). *v.* BARNES.

**Barnecut, Barnecutt:** *v.* BARNICOT

**Barnell, Barnhill:** Richard *de Bernhull* 1275, *atte Bernhull* 1332 MELS (Wo). 'Dweller at the hill with a barn on it', OE *bere-ærn, hyll*.

**Barnes, Barns:** Philip *de Bernes* temp. John Seals (Sr); Peter *del Bernes* 1327 SRDb; William *Bernes* 1380 AssC; Joan *Barnes* 1450 Rad (C). From Barnes (Surrey) or residence near or employment at the barns. cf. BARNE.

**Barnet, Barnett, Barnitt:** Brictnod *de la Bernet* c1200 MELS (Sx); William *atte Bernette* 1296 SRSx; Jordan *atte Barnette* 1310 LLB D. From residence near land cleared by burning (OE *bærnet* 'burning') or from Barnet (Herts, Middlesex), or Barnett Fm in Wonersh (Surrey).

**Barnfather:** *v.* BAIRNSFATHER

**Barnfield:** William *de Bernefeld* 1195 P (K); Robert *de Bernefeld* 1296 SRSx; Thomas *Barnefeild, Barnefilde* 1642 PrD. From Barnfield Shaw in Mayfield (Sx), or Barnfield Fm in Luppitt (D).

**Barnham, Barnum:** Walter *de Bernham* c1191 BuryS (Sf); Robert *de Bernham* 1296, Thomas *Barnam* 1525 SRSx. From Barnham (Sf, Sx), or Barnham Broom (Nf).

**Barnhill:** *v.* BARNELL

**Barnhouse:** William *Bernehus* 1147–61 CartAntiq; William *de Bernehus* 1278–9 FFSx; John *Barnehowse* 1524 SRD. From Barn House in Brightling, Barnhouse Fm in Shipley (Sx), or 'dweller at the house by the barn', OE *bere-ærn, hūs*.

**Barnicot, Barnicott, Barnicote, Barnicoat, Barnecut, Barnecutt:** Thomas *de Bernecot'* 1279 RH (O); Walter *de Bernycot'* 1297 MinAcctCo. From Barnacott in Stoke Rivers, in Westleigh (D).

**Barnikel:** v. BARNACLE

**Barningham:** Leomer *de Berningeham* 1121–38 Bury; Walter *de Berningham* 1203 Cur (Sf); Peter *de Berningham* 1219 AssY. From Barningham (Sf, NRY), or Little, Winter, Town Barningham, Barningham Norwood (Nf).

**Barnish:** Ralph *barnage* 1130 P (Do); William *Barnage* 1270 AssSo; Reginald *Barnage* 1311 PN Do ii 112. OFr *barnage*, a contraction of OFr *baronage* 'the qualities or attributes of a baron', hence 'courage, nobleness, &c'. cf. Fr Bernage. v. also BURNAGE.

**Barnsley:** John *de Barnusley* 1340–1450 GildC; Alice *de Berneslegh* 1354 Putnam (Ch); Adam *de Barnesley* 1440 ShefA. From Barnsley (Do, Gl, WRY, Wt).

**Barnum:** v. BARNHAM

**Barnwell:** Clac *on Byrnewillan* 972 BCS 1130; Eustace *de Bernewell'* 1177 P (C/Hu); Thomas *de Bernewell'* 1270 Acc; John *Bernewell, Barnewell* 1475 FFEss. From Barnwell (C, Nth).

**Baron, Barron:** Lefuine *Barun* c1095 Bury (Sf); Geoffrey *le Barun* 1236 *Ass* (Ha); John *Baron* 1296 SRSx. EME *barun*, OFr *barun, -on* 'baron', sometimes, no doubt, denoting title or rank, but more often, especially when applied to peasants, a nickname, proud or haughty as a baron. The term was anciently applied to freemen of the cities of London and York who were homagers of the king and also to the freemen of the Cinque Ports who had the feudal service of bearing the canopy over the head of the sovereign on the day of coronation. Gervase *le Cordewaner* or *camerarius* was also called Gervase *Baronn*, no doubt because he was alderman of Aldgate Ward 1250–6 (ELPN 137). This was an old surname in Angus where it originated from the small baronies attached to the Abbey of Coupar-Angus. The tenant of the barony of Glenisla became Robert *Barrone*, tenant of Glennylay (1508), etc. Elsewhere in Scotland 'barons' were land-owners who had a certain amount of jurisdiction over the population of their lands (Black).

**Barr, Barrs, Le Barr:** (i) Anger *de la Barra* c1216–17 Clerkenwell (Lo); Peter *de Bar* 13th Lewes (Nf); John *ate Barre* 1283 Battle (Sx). OFr, ME *barre* 'barrier, gateway' (c1220 NED). cf. Temple Bar and v. BARRER. In the fens *bar* was used of an obstruction (perhaps a weir) in a stream. The Scottish Barrs derive from Barr in Ayrshire or Barr in Renfrewshire. Atkyn *de Barr* was baillie of Ayr in 1340 (Black). (ii) Edricius *de la Barre* 1170 P (St); William *de Barre* 1199 AssSt. From Great Barr (Staffs). This is from Welsh *bar* 'top, summit' and refers to Barr Beacon. (iii) Richard *de Barra* 1086 DB (So). From Barre-en-Ouche (Eure), or, perhaps, from Barre-de-Semilly (La Manche). (iv) Hugo *Barre* 1155 DC (L); Alexander *Barre* 12th Riev (Y). OFr *barre* 'a piece of any material long in proportion to its thickness or width', a bar or stake, used as a nickname for a tall, thin man, or metonymic for a maker of bars. cf. Robert *Barremakere* 1347 LLB F.

**Barraclough, Barrowclough, Barrowcliff, Barrow-**

cliffe, Barnaclough, Berecloth, Berrecloth, Berrycioth: Peter *del Baricloughe, de Barneclogh* 1315, 1316 Wak (Y); Robert *Bereclough* 1493 GildY; Henry *Barrayclught* 1561 RothwellPR (Y); Thomas *Baraclough* 1588 ib.; Anne *Beraclough* 1606 ib.; Francis *Baroclough, Barrowclough* 1612, 1631 ib.; Elizabeth *Barraclue* 1627 Bardsley; Edward *Barracliff* 1765 ib. From an unidentified place, probably near Wakefield (WRYorks). The pronunciation is *Barracluff*, in London *Barraclow* or *Barraclue*.

**Barras, Barrass:** Richard *Barras* 1672 HTY; Joseph *Barrass* 1713 FrY. Perhaps 'dweller by the outwork of a fortress', OFr *barrace*. But cf. Fr *Barras* 'a seller, dealer'.

**Barrat, Barratt, Barrett, Barritt, Barrott:** Matthew *Baret* c1150–5 DC (L); Robert *Barate* 1165 P (Nt); Jordan *Barat* 1185 Templars (Herts); Seman *Barette* 1207 P (Ha); William *Barrette* (*Barat*) 1327 SR (Ess). This is a difficult name. There seems no evidence for a derivation from OG *Beroald*, OFr *Beraud*, as has been suggested. ON *Bárǒr* is found in Yorks and Lincs in DB as *Bared, Baret*, but there is no proof of its continued use. The commonest form is *Barat* and this must be from OFr *barat*, ME *bar(r)at, bar(r)et(te)*, which accounts for all the forms. The original sense in Romanic seems to have been 'traffic, commerce, dealing' and in ME 'trouble, distress' (c1230); 'deception, fraud' (1292); 'contention, strife' (c1300), from any of which a nickname could arise. Occasionally we may have OFr *barrette* 'a cap, bonnet', as an occupation name, 'a maker of caps'.

**Barrell:** Turstin *Baril* 1166 P (Nf); William *Baril* 1185 P (Wo). OFr *baril* 'a barrel, cask'. Perhaps chiefly for a maker of barrels, a cooper. cf. Stephen *le Bariller* 1224 Pat. It may also have been used as a nickname for a man with a well-rounded belly. cf. 'the ydell and *barrell bealies* of monkes' (1561 NED), *barrel-belly'd* (1697 ib.); or, perhaps, of a man with the capacity of a cask. cf. 'olde barel ful of lies' (1386 ib.), 'to drinke a barelle fulle of gode berkyne' (1436 ib.), *barrel-fever*, a disease caused by immoderate drinking. Also a late form of BARWELL: John and Susan *Barrell, Barwell* 1688, 1691 Bardsley.

**Barrer, Barrere:** Gilbert (*le*) *Barrer* 1221–2 Cur (D, Do), 1229 Cl (Sx); William *Barrer* 1332 SRSx (in the town of Arundel). Equivalent to *atte Barre* 'dweller by a town or castle gate'. v. BARR. Gilbert *le Barrier* 1210 P (Sx) is probably identical with Gilbert *Barre* 1221 Cur (K) and Walter *atte Barre* 1296 SRSx with Walter *le Barrer'* 1327 ib.

**Barrick:** v. BERWICK

**Barrie:** v. BARRY

**Barrington:** Fulk *de Barenton* 1198 FFEss; Geoffrey *de Barrington'* 1219 P (Do/So); Nicholas *de Baryngton* 1344 FFEss; John *Barrington* 1642 PrD. The first example is probably from Barentin (Seine-Maritime), the later ones from Barrington (C, Gl, So).

**Barrow, Barrows, Berrow:** Adam *de Barewe* 1192 P (L); John *de la Berewe* 1242 Fees (Wo); William *del Berwe* 1260 AssC; John *atte Barwe* 1327 SRSo. Either 'dweller by the grove', OE *bearu*, dative *bearwe*, giving modern Barrow, or 'dweller by the hill', OE *beorg*, ME dative *berwe, barwe*, modern Berrow, Barrow.

**Barrowclough, Barrowcliff:** v. BARRACLOUGH

**Barrowman:** v. BORROWMAN

**Barry, Barrie:** (i) Nest *de Barri* 1185 P (Sx); Richard *Barri* 1195 FFSf. Though most examples are without a preposition, the surname must, in the absence of any evidence for a personal-name or any suitable attribute, be local in origin. It was probably brought over from France where it survives as *Barry* and *Dubarry*, from OFr *barri* 'rampart', later applied to the suburb below the rampart, hence 'dweller in the suburb' (Dauzat). The Irish Barry is also chiefly Anglo-Norman, deriving from Philip *de Barry* (1179). It is also for *Ó Báire* 'descendant of *Báire*', short for *Fionnbharr* 'fair-head' or for *Ó Beargha* 'descendant of *Beargha*', 'spear-like'. (ii) William *de Barry* 1360 Black. The Scottish surname derives from Barry in Angus.

**Barryman:** v. BORROWMAN

**Barsham, Basham, Bassham:** Henry *de Barsham* 1198 FFNf; Martin *Barsham, Bassham* 1367 ColchCt. From Barsham (Norfolk, Suffolk).

**Barson:** Hugo *Bertson* 1332 SRCu. 'Son of *Bert*', a short form of *Bertelmeu* 'Bartholomew'. v. BART.

**Barstow:** v. BAIRSTOW

**Bart:** Award *Bart* 1246 AssLa; William *Barte* 1420 LLB I. A short form of *Bartelmew* (Bartholomew).

**Barter, Bartter:** Hugh *le Bartur* 1279 RH (O); Thomas *Bartour* 1360 FFW; Thomas *Bartyr* alias *Bartour* 1561 Pat (Do). OFr *barateor, barateur* 'a fraudulent dealer, cheat, trickster'. Sometimes, perhaps, a derivative of ON *barátta* 'one who fights, a hired bully, quarrelsome person'.

**Bartholomew, Berthelemy:** *Bartholomeus* canonicus 12th DC (Nt); Robert *Bartelmeu* 1273 RH (Hu); Nicholas *Bertelmev* 1296 SRSx; Walter *Berthelmeu* 1334 LLB E. *Bartholomew*, Hebrew 'son of Talmai' ('abounding in furrows'), a common medieval name, with numerous diminutives.

**Barthorp, Barthorpe, Bartrap, Bartrip, Bartrop, Bartropp, Bartrup, Bartrupt, Bartup, Bathrup, Barlthrop, Barltrop:** William *de Baretorp* 1200 P (L); William *de Barkentorp'* 1219 AssY; Walter *Berthrop* 1327 SRWo; John, William *Baltrip* 1341 LLB F, 1351 AssEss; Bartholomew *Balthroppe* 1586 DenhamPR (Sf); Jonathan *Barthrope* 1673 Shef (Y); Hester *Bartrap* 1687 Bardsley; Christopher *Barthrup* 1706 FrY. From Barthorpe Bottoms (ERYorks).

**Bartie:** v. BARTY

**Bartin:** v. BERTIN

**Bartindale:** John *Bartendale* 1424 FrY. From Bartin Dale, a depopulated place in Hunmanby (ERYorks).

**Bartle, Bartill:** *Bartill* Laurenson 1625 Black; *Bartholomew* Chastiloun and Sarah his wife had a son known as John *Bartyll* and a daughter known as Alice Busche 1384 Husting; William *Bartle* 1672 HTY. A shortened form of *Bartilmew*, i.e. *Bartholomew*. Sometimes, perhaps, local: John *of Bartale* 1401 AssLa.

**Bartlet, Bartlett, Bartleet, Barttelot, Bertalot:** Godricus, Walter *Bertelot* c1157 Holme (Nf), 1296 SRSx; Thomas *Bartelot* 1294 FFC, 1327 SRSx; Thomas *Bartlot* 1379 PTY. *Bart-el-ot, Bert-el-ot*, double diminutives of *Bart-, Bert-*, from *Bartelmew, Bertelmew* (Bartholomew).

**Bartley:** Francis *Bartley* 1571 Oxon (So); Richard *Barkeley* or *Bartley* 1592 Oxon (Gl); Andrew *Bartley* 1642 PrD. The forms are late and could be local from Bartley Regis (Ha), Bartley Fm in Wadhurst (Sx), or Bartley Green (Wo). They could also be late forms of BARCLAY, or from BARTHOLOMEW.

**Barton, Barten:** Ælfric *æt Bertune* 1015 OEByn; Paganus *de Barton* 1163 P; John *de Barton'* 1300 FFY; Thomas *Barten, Bartyn* 1586, 1609 Shef. From one or other of the many places of this name.

**Bartram, Bartrum, Barttrum, Bertram, Bertrand, Batram, Batrim, Batterham, Battram, Borthram, Buttrum:** *Bertrannus* 1086 DB; *Bertram* c1150–60 DC (L), identical with *Bertrannus* a1183 ib.; William *Bertram* 1086 DB (Ha); Henry *Bertran* c1155 DC (L); John *Bartram* 1278 LLB A; John *Bartrem* 1332 SRSt; Mariota *Berteram* 1332 SRSx; Nycolas *Bartrum* 1524 SRSf; William *Battram*, George *Bartrom*, *Bateram* 1674 HTSf. OFr *Bertran(t)*, OG *Bertram, Bertran(d)* 'bright raven'.

**Bartrap, Bartrip, Bartrop, Bartropp, Bartrup, Bartrupt:** v. BARTHORP

**Barttelot:** v. BARTLET

**Bartter:** v. BARTER

**Bartup:** v. BARTHORP

**Barty, Bartie:** Robert *Barty* 1552 Black (Dundee); John *Bairty* 1587 ib. (Edinburgh). A Scottish diminutive of *Bartholomew*.

**Barugh:** v. BARGH

**Barways, Barwise:** Robert *de Beriwis* 1246 Misc (Cu); Henry *de Barweis* 1291 Cl (We); Anthony *Barwis* 1561 Pat (Cu). From Barwise (We).

**Barwell:** Elyas *de Barewelle* 12th DC (Lei). From Barwell (Leics). v. also BARRELL.

**Barwick:** v. BERWICK

**Bascot, Bascott:** Adam *de Bascote* 1206–7 FFWa; Adam *de Baskote* 1373 Oriel (O). From Bascote (Wa).

**Base:** v. BASS

**Baseley, Basley, Bazeley, Bazley, Basil, Bassil, Bassill, Bazell, Bazelle:** *Basilia* 1134 40 Holme (Nf), Hy 2 DC (L); Willelmus *filius Basilie* 1219 AssY; *Basill'* Vidua 1296 SRSx; Ralph *Basille* 1251 Rams (Hu); John *Basyly* 1252 Rams (Hu); Walter *Basely* 1275 SRWo; Thomas *Bazill* 1674 HTSf. OFr *Basile, Bazile, Basyle, Basille* (f), from Lat *Basilia*, feminine of *Basilius*, from Greek βασίλειος 'kingly'. The English form was *Basil* or *Bassilly*. The masculine *Basil* is found occasionally: Ricardus *filius Basilii* 1252 Rams (Hu).

**Basford, Bashford, Bashforth:** John *Basheford* 1525 SRSx; Edmond *Bashford* 1695 Bardsley. From Basford (Nt).

**Bash, Baish:** Robert *de Basche* 1199 AssSt. For *Bach*, with a change of *sh* for *ch*.

**Basham:** v. BARSHAM

**Basil:** v. BASELEY

**Basing:** (i) *Besing* c1150–60 DC (L); *Basing* de Blaikemare c1200 DC (L); Robert *filius Basing* 1202 AssL. OE *Basing*. (ii) Cola *de Basinga* 1066 DB (Ha); John *de Basing'* 1200 P (Ha); Henry *de Basyng'* 1297 MinAcctCo. From Basing (Ha).

**Bask:** Henry, Roger *Baske* 1332 SRSt; 1357 AssSt.

ME *baisk, bask,* ON *beiskr* 'bitter, acrid', 'ungrateful or irritating to the senses'.

**Baskerville, Baskerfield, Basketfield, Baskeyfield, Basterfield, Baskwell, Paskerful, Pasterfield, Pesterfield:** Roger *de Bascheruilla* 1127 AC (Gl); James *Baskerfield, Baskervyle* 1530 StarChSt. From Boscherville (Eure).

**Baskett:** (i) William, Henry *Basket* 1191 P (Sr), 1198 CurR (Ess). ME *basket,* here used of a basket-maker. cf. BASKETTER. Or used for one who carried the *baskettes* full of stones to the lime-kiln (Building 151). (ii) Basilia *de Besecot'* 1221 AssWa; Adam *de Baskote* 1373 Oriel (O). From Bascote (Warwicks). (iii) Margeria *atte Bascat* 1319 SRLo; Thomas Kent *atte Basket* beside Billyngesgate 1424 LondEng 184. One who lived or worked at the sign of the Basket. Probably a basketmaker.

**Basketter:** cf. William *Basketwricte* 1229 Pat (L). 'A maker of baskets.' *Basketter* is ME *basket* plus *-er.*

**Baskin:** Adam *Baskyn* 1274 RH (Db). OG *Basso,* with the diminutive ending *-kin. v.* Michaëlsson 86, 88.

**Baskwell:** *v.* BASKERVILLE

**Basley:** *v.* BASELEY

**Bason:** *v.* BATESON

**Bass, Base, Baiss:** (i) Aelizia *Bass'* 1180 P (Wa); Dauid *le Bas* 1205 P (Gl); Geoffrey *Base* 1274 RH (L). OFr *bas, basse,* ME *bass* 1393, *bace* c1440, *base* 1425 'low, of small height'. A man with short legs. cf. BASSETT. (ii) Osbert *Bars* 1207 P (Gl); Richard *le Bars* 1327 SRSx. OE *bærs,* now *bass,* a fish; cf. *Bace,* fysche c1440 PromptParv.

**Basset, Bassett:** Ralph *Basset* 1086 DB (Herts, Beds), 1115 Winton (Ha); Milo *Basseth* 1139 Templars (O); Philip *le Basset* 1260 LLB B. OFr *basset* 'of low stature', a diminutive of *bas* 'low'; 'a dwarf or very low man' (Cotgrave). According to Ordericus Vitalis, Ralph Basset was raised by Henry II from an ignoble stock and from the very dust, 'de ignobili stirpe ac de pulvere'.

**Bassham:** *v.* BARSHAM

**Bassick:** *v.* BASTICK

**Bassil(l):** *v.* BASELEY

**Bassinder:** *v.* BAXENDALE

**Bassingthwaighte, Bassingthwait:** John *Bassynge-whytt* 1568 SRSf. From Bassenthwaite (Cumb).

**Bastable:** Richard *de Bardestapel* 1219 FFEss; Ralph *Barstaple* 1327 SRSo. From Barstable Hall (Essex) or Barnstaple (Devon).

**Bastard:** Robert *Bastard* 1086 DB (D); William *le Bastard* 1201 AssSo. OFr *bastard* (1297 NED). Not always regarded as a stigma. The Conqueror himself is described as 'William the Bastard' in state documents.

**Basten:** *v.* BASTIAN

**Baster, Baister:** Baldwynus, Peter *le bastere* 1230 P (D); 1327 SRSt. OFr *bastier* 'saddler'.

**Basterfield:** *v.* BASKERVILLE

**Bastian, Bastien, Basten, Bastin:** *Bastianus* a1200 Dublin, 1221 AssWo; Colin *Bastin* 1225 Pat; John *Bastian* 1317 AssK. A pet-form of *Sebastian,* from Lat *Sebastianus* 'man of Sebastia', a city in Pontus.

**Bastick, Bassick:** Geoffrey *de Bastwyke* 1335 AssC. From Bastwick (Norfolk).

**Baston, Bastone:** (i) Ernaldus, Richard *Bastun* 1191 P (Sf), 1203 AssNth; Nicholas *Baston* 1279 RH (O). A nickname from OFr *bastun* 'a stick', used as a personal-name in the first element of Bassenthwaite (Cumb). (ii) Turstan *de Baston'* 1191 P (L). From Baston (Lincs).

**Bastow:** *v.* BAIRSTOW

**Baswin:** *Basuin* 1066 DB; *Basewinus* 1203 P (Nth); Richard *Baswyn* 1160 RegAntiquiss; Robert *Basewin* 1202 AssL; Osbert *Basewine* 1208–9 Pleas. OG *Basuin. v.* Forssner 282.

**Batch:** *v.* BACH

**Batchelar, Batchelder, Batchlor:** *v.* BACHELOR

**Batcock:** *v.* BADCOCK

**Batcomb, Batcombe:** John *Batcumbe* 1327 SRSo. From Batcombe (Do, So).

**Bate, Bates:** (i) *Bate* 1275, 1286 Wak (Y); Rogerus *filius Bate* 1327 SRDb; Roger *Bate* 1275 SRWo; Richard *Bates* 1297 MinAcctCo (Y); John *Bat, Bate* 1394, 1396 LLB H. A pet-form of *Bartholomew,* found also as *Batt.* For the variation between *Bat* and *Bate,* cf. *Add* and *Ade* for *Adam,* and *Pat* and *Pate* for *Patrick.* (ii) Thomas *del Bate* 1270 Ipm (Nb); William *of Ye Bate* 1297 SRY. This might be OE *bāt,* Northern ME *bat* 'boat', used for a boatman. More probably we have ON *bati* 'dweller by the fat pasture'. *v.* BATT. It can have no connexion with the common Northern *bait* 'food', etc., which always appears as *bayt, beyt.*

**Bately:** *v.* BATLEY

**Bateman, Baitman, Batman, Battman:** *Bathemanus de Staunford'* 1222 Cur (R); *Bateman le Keu* 1267 Pat; *Batman* d'Appleton 1313 FrY; Alexander *Bateman* 1260 AssC; William *Batemon* 1275 SRWo; John *Baytman* 1553 FrY. *Bateman* 'servant of Bartholomew' is a type of surname formerly common in Yorkshire. cf. ADDYMAN, HARRIMAN, etc. Here it is used early and often as a christian name, perhaps on the analogy of such names as *Blæcmann, Dēormann,* etc. *Pateman* was similarly used in Scotland in the 15th century as an alternative for the christian name *Paton* (Patrick).

**Bater:** Edmund, Robert *le batur* 1199 P (Gl), 1210 P (Ha). OFr *bateor* 'one who beats' has been taken to mean a beater of cloth or fuller, or as a short form of *orbatour,* a beater of metals. It probably means 'a coppersmith or dealer in *baterie,* i.e. beaten copper or brassware' (LoCt). Stephen *le Coperbeter* (1286) was identical with Stephen *le Batur* (1292 LLB A). *v.* BEATER.

**Bateson, Baitson, Batson, Battson, Bason:** John *Batessone* 1327 SRDb; Richard *Bateson* 1327 Wak (Y); John *Battson* 1467 GildY; William *Baitson* 1662 PrGR. 'Son of *Bate* or *Batt'. v.* also BEATSON.

**Batey, Batie, Baty:** William *Baty* 1277 AssSo; Hugo *Baty* 1301 SRY. A pet-form of *Bate* (Bartholomew). cf. BEATY.

**Bath, Bathe:** Walter *de la Bathe* 1275 SRWo; Alexander *Bathe* 1327 SRSf; Robert *A Bathe* 1545 SRW. From Bath (So), or Bathe Barton in North Tawton (D).

**Batha, Bathe, Batho, Bather, Badder:** *Atha ap Atha,* William *ap Atha* 1327 SRSa; *Atha Gogh* 1332 Chirk;

Jevan *ap Atha*, *ap Adda* 1391 ib.; John *Bathowe* 1537 Morris (Haverfordwest); Jevan ap John ap Gryffyd *Batto* 1538 Chirk; Humffrey *Bathowe*, John *Batowe* 1538 SaAS 3/viii; Richard *Bathaw* 1574 Bardsley (Ch); William *Batha*, Adam *Batho* 1610, 1613 ib.; Elizabeth, Hannah *Bather* 1683, 1782 ib. Morris (155) gives *Batha*, *Batho*, *Bather* as Shropshire and Cheshire variants of *Batha*, i.e. *ap Atha* 'son of *Atha*', probably correctly, since the Welsh personal name was common in Shropshire and Chirkland in the 14th century. *Badder* may be from the by-form *ap Adda*, and the variation in the unstressed final syllable can be paralleled in other names.

**Bathley:** Henry *de Bathele* c1220–7 RegAntiquiss; William *of Bathele* 1246 FFY. From Bathley (Nt).

**Bathrup:** v. BARTHORP

**Bathurst:** William *de Batherst* 1296 SRSx; Geoffrey *de Bathurst* 1327 SRSx; Katherine *Batherst* 1392 CtH. From Bathurst in Warbleton (Sussex).

**Batisson:** Remiger *Batisson* 1332 SRCu. 'Son of *Bate* or *Batey*.'

**Batkin:** *Batekyn clericus* 1274 RH (Ess); Thomas *Batekyn* 1325 FFEss; Hugh *Batkyn* 1332 SRSt. A diminutive of *Bate* (Bartholomew).

**Batley, Battley, Bately:** Oto *de Battelay* 1191 P (Y); John *de Bateley* 1274 Wak (Y); Ephraim *Batley* 1672 HTY. From Batley (WRY).

**Batman:** v. BATEMAN

**Baton:** v. BATTEN

**Batram, Batrim:** v. BARTRAM

**Batrick:** v. BADRICK

**Batsford, Battisford:** Richard *de Batesford* 1182–1211 BuryS (Sf); Everard *de Bateford', de Batesford'* 1202 FFSf; John *de Batesford* 1300 Eynsham. From Batsford (Glos), Batsford in Warbleton (Sussex), or Battisford (Suffolk).

**Batson, Battson:** v. BATESON

**Batt, Batte, Batts:** (i) Reginald, Richard *le Bat* 1275 RH (Y), 1296 SRSx. A nickname from the bat, a form first found c1575 and replacing an earlier *bakke*. v. BACK. (ii) William, Herbert *Bat* 1170–87 ELPN, 1182 P (Sa); Matilda *Battes* 1279 RH (C); John *Bate* or *Batt* 1570 Oxon. Without the article, the surname is common and may be a nickname 'the bat' or a pet-name of Bartholomew. Both Gascoigne and Gabriel Harvey addressed their friend Bartholomew Withypoll as *Batt(e)*. We have also to take into account the byname of a Winchester monk: Ælfricus qui *Bata* cognominabatur (c1051 OEByn). This has given rise to various conjectures, none wholly satisfactory. Tengvik suggests a nickname from OE *batt* 'a cudgel', as does Ekwall for the first form above. Tengvik considers the reference is to a person of stout heavy appearance. For the byname *Bata* Ekwall suggests a personal name OE *\*Bata* which he finds as the first element in Batcombe (Dorset, Som), and other place-names, but, in view of the triple occurrence of Batcombe, he suggests also the possibility of a common noun *bata*, corresponding to ON *bati*, OFris *bata* 'profit, gain', in some transferred sense such as 'fat pasture' (v. below), or even 'good husbandman'. With this surname we must also take BATTOCK. This is clearly a diminutive, either OE

*\*Batuc*, from *\*Bata*, or a noun *\*batuc* 'the little good husbandman'. OE *\*Battoc* is the source of Battisborough (Devon). (iii) Walter *atte Batte* 1327 SRSo. This form seems to confirm Ekwall's derivation of Batcombe (Som) from a topographical term. 'Dweller by the fat pasture'.

**Battell:** v. BATTLE

**Batten, Battin, Batting, Baton:** *Batin* Bythemore, *Bathon* Mayster 1327 SRSo; Walter *Batun* 1248 FFEss; Robert *Batin* 1261 AssSo; William *Baton* 1275 SRWo; John *Batten* 1327 SRSt. Diminutives in *-in*, *-un* of *Bat* (Bartholomew).

**Batterham:** v. BARTRAM

**Battersby:** William *de Bathresby* c1170–89 YCh; Roger *de Batersby* 1401 AssLa; John *Badersby* 1428, Edmund *Battersby* 1501 FrY. From Battersby (NRY), or Battersby Fm in Slaidburn (WRY).

**Batterson, Batteson, Battison, Battisson:** Andrew *Batenson* 1561 Bardsley (Du); Abraham *Battison* 1699 FrY; George *Battison alias Pattison*, son of John *Pattison* 1758 FrY. 'Son of *Batten*', later confused with *Patterson*. Also, no doubt, 'Son of *Batty*'.

**Battiscombe:** John *Battiscombe* 1440 AD ii (Do). From Bettiscombe (Dorset).

**Battle, Battell, Battyll:** Hubert *Bataile* c1140 AD i (Ess); William *de la Bataille* 1196 Cur (Nth); John *de Labatil* c1245 Black (Inchaffray); Simon *le Batel* 1327 SRSx. OFr *de la bataile* '(man) of the battle-array, warrior'.

**Battley:** v. BATLEY

**Battman:** v. BATEMAN

**Battock:** Turchil *Batoc* 1066 DB (Wa); Thomas, John *Battok* 1327 SRSf, 1362 Shef (Sf). v. BATT.

**Battram:** v. BARTRAM

**Battrick:** v. BADRICK

**Battrum:** v. BARTRAM

**Battson:** v. BATESON

**Batty, Battye, Battey, Battie:** Johannes *filius Batti* 1332 SRLa; John *Batty*, William *Batti* 1308, 1316 Wak (Y). A pet-form of *Batt* (Bartholomew).

**Battyll:** v. BATTLE

**Baty:** v. BATEY

**Baucutt:** v. BAWCOCK

**Baud:** Simon, Reginald *le Baud* 1219 Cur (Nth), 1239 FFC. OFr *baud* 'gay, sprightly' (c1400 NED).

**Baudechon:** *Baudechon* le Bocher 1274 RH (Lo); *Baudechon* le Chaucer 1311 LLB B; Robert *Baudechum* 1249 AssW; John *Baudechon* 1325 CorLo. OFr *Baudechon*, a hypocoristic of BALDWIN.

**Baudr(e)y:** v. BALDREE

**Baugh:** Madog, Jevan *Bach* 1391 Chirk; Madog Lloit *Bach* 1391–3 ib.; Geoffrey *Bagh'* 1450 SaG; Rychard *Bawgh* 1545 SRW. Probably Welsh *bach* 'little'.

**Baulch:** v. BELCH

**Baverstock, Baveystoch, Beverstock, Bavastock:** Walter *Baberstooke* 1576 SRW. From Baverstock (W).

**Bavridge:** v. BEVERIDGE

**Bawcock, Bawcutt, Baucutt, Bowcock, Bowcott, Bocock, Boocock:** Geoffrey *Balcok* 1276 RH (Y); Alan *Balkok* 1279 RH (Hu); Walter *Boucok* 1297

MinAcctCo; Ibbot *Bolkok* 1379 PTY; Sara *Bawcoke* 1627 Bardsley; William *Bo(o)cocke* 1627, 1641 RothwellPR (Y). *Bald*, a short form of *Baldwin* or *Baldric* (v. BALD, BALDREE) and the diminutive suffix -*cock*.

**Bawcombe:** v. BALCOMB

**Bawden, Boaden, Boden:** *Boden* or *Bawden* Maylle 1591–5 Bardsley; *Bawden* Richards 1642 PrD. Late forms of BALDWIN.

**Bawtree, Bawtry:** Peter *Bautre* 1298 AssL; Nicholas *de Bautre* 1316 FFHu; John *Bawthrie* 1576 SRW. From Bawtry (WRY).

**Bax:** v. BACK

**Baxby:** John *Baxby* 1410–1 IpmY; Thomas *Baxby* 1432 TestEbor. From Baxby in Husthwaite (NRY).

**Baxendale, Baxendall, Bassinder:** From Baxenden (La).

**Baxter, Bagster:** Lieuger *se Bacestere* a1093 OEByn (D); Hanne *Bakestre* 1260 AssCh; William *le Baxtere* 1333 FFSf. OE *bæcestre*, fem. of *bæcere* 'baker'. *Baxter* is found mainly in the Anglian counties and is used chiefly of men. Only two examples have been noted with a woman's christian name. Fransson found only four.

**Bay:** (i) Robert *filius Bay* 1275 RH (Y). OE *Bēaga* (m), *Bēage* (f). (ii) Gilbert *le Bay* 1317 AssK; Agnes *le Bay* 1332 SRWa. OFr *bai* 'reddish-brown', of hair or complexion. (iii) John *ate Bey* 1279 RH (C); Roger *Attebege* 1327 SRY; William *Bay* 1373–5 AssL. 'Dweller at the bend', OE *bēag*.

**Bayard:** Ralph *baird* (*baiart*) 1086 InqEl (Herts); Godfrey *Baiart, Baiard* 1161–2 P (Y); Simon *Bai(h)ard* 1203, 1206 Cur (Herts). OFr *baiart, baiard* 'bay-coloured', used generally of a bay horse, but in particular of the bright-bay-coloured magic steed given by Charlemagne to Renaud and hence as a mock-heroic allusive name for any horse. cf. Chaucer's 'proud Bayard'. Later, 'Bayard' was taken as the type of blindness or blind recklessness. As a surname, this may be used of reddish-brown hair or complexion, but more probably of a proud, haughty or reckless disposition. cf. 'Þay blustered as blynde as bayard' (c1325 NED); 'But as Bayard the blinde stede . . . He goth there no man will him bidde' (1393 ib.). The surname may be occupational in origin, from OFr *bayard, baiart* 'a hand-barrow used for heavy loads' (1642 NED), used by metonymy for OFr, AFr *baïardeur* 'a mason's labourer'. NED doubts the use of this in England. *Bayardours*, however, is found in 1359 (Building 439) and *baiard* 'hand-barrow' in 1278 (ib. 243). In the Vale Royal accounts of 1278 the *bayarders* or *bairdores* are defined as 'men carrying with barrows large stones to be carved into the workshop and out' (Building 353).

**Bayer, Beyer:** John *Beyer* 1261–2 FFWa; William *le Beier* 1327 SRSx; Alice *Bayer* 1351 ColchCt. A derivative of OFr *baies*, ME *bayes*, from the adjective *bai* 'chestnut-coloured, bay', the name of a cloth, probably so-called because of its original colour. It was a coarse woollen stuff, with a long nap, now used chiefly for linings, coverings, curtains, &c., but in warmer countries for articles of clothing, e.g. shirts, petticoats. Formerly, when of lighter and finer texture

used also as a clothing material in Britain. Its manufacture is usually said to have been introduced into this country by immigrants from France and the Netherlands in the 16th century, but the word certainly appears much earlier in English. Sometimes, perhaps, a derivative of OE *bēag* 'bend', hence 'dweller by the bend'.

**Bayfield:** Adam *de Baifeld* 1208 Cur (Nf); Simon *de Bayfeld* 1390–1 NorwLt; Alan *Bayfeld* 1461 Paston. From Bayfield (Nf).

**Bayhouse, Bayus:** Randulf *de Baiwes* 1143–7, *de Baius* c1155–60 DC; Matilda *de Baiocis* 1185 Templars (L); Adam *Bayous* 1277 IpmY; Robert *Bayhuse* 1326 AssY; John *Bayhouse* 1404 IpmY. From Bayeux (Calvados).

**Baylay, Bayley, Baylie:** v. BAILEY

**Bayldon:** v. BAILDON

**Bayles:** v. BAIL

**Bayliff(e):** v. BAILEFF

**Baylis, Bayliss, Bayless, Bailess:** Thomas *Baillis* 1547 FrY; Samuel *Baylles* 1635 ib. OFr *baillis*, nominative of *bailliff*.

**Baynard:** Rotbert' *homo bainardi* 1086 InqEl (Sf); Ralph *baignart, bainard* 1086 DB (Ess), InqEl (Nf); Robert *Bainard* 1182 Guisb (Y); John *Baynard, Beynard* 1317 AssK. OG *Beinhard, -hart*, probably compounded of ON *beinn* 'ready, willing' and OG *hart* 'hart'. v. also BANYARD.

**Bayne(s):** v. BAIN

**Baynham:** v. BEYNON

**Baynton:** v. BAINTON

**Bayus:** v. BAYHOUSE

**Baz(e)ley, Bazell(e):** v. BASELEY

**Beabey, Beaby:** v. BEEBY

**Beach:** v. BEECH

**Beacham, Beachamp:** v. BEAUCHAMP

**Beacock:** Henricus *filius Becok* 1332 SRLa; Stephen *Becoc* 1279 RH (O); John *Beckson* 1366 SRLa. *Be*, a pet-form of *Beton* or *Beatrice*, plus the diminutive suffix -*coc*.

**Beade, Beed:** Alanus *filius Bede* de Swainton' 1230 P (Y); Raymond *Bede* 1260 AssC; Robert *Beda* 1275 RH (W). OE *Bēda*. The name of the Venerable Bede remained in use, though rare, until the 13th century, long enough to become a surname.

**Beadel, Beadell, Beadle, Beadles, Beddall, Bedell, Bedells, Bedle, Beedell, Beedle, Biddell, Biddle, Biddles, Buddell, Buddle, Buddles:** Brun *Bydel* 11th KCD 1353 (So); Brictmarus *Bedel* 1066 DB (Sf); Erneis *bedel*, Luinus, Richard *budel* 1148 Winton (Ha); Ailsi *le Bedell'* 1175 P (Lei); Robert *le Budel* 1327 SRSt; Margaret *ate Budeles* 1332 SRSr; Richard *Bedle* 1541 RochW; Richard *Byddell* 1559 FFHu; John *Biddle* 1655 FrY; William *Beadle*, John *Beddall* 1674 HTSf; Adam *Buddle* 1676 EA (OS) iv. OE *bydel* 'beadle'. OE *y*, in ME dialects, became *i*, *e* or *u*, all of which have survived. Some examples of *bedel* are from OFr *bedel* (Lat *bedellus*), especially in counties such as Hants where OE *y* became *u*. These surnames may also be late forms of BEDWELL. For *Buddle*, v. also BOODLE.

**Beadman:** William *Bedman* 1327 SRSo; William *Bedemon* 1381 SRSt. ME *bedeman* 'a man of prayer', one who prays for the soul of a benefactor.

**Beadnell, Bednall, Bednell:** Thomas *de Bedenhale* 1194 StCh; Thomas *de Bedenhal'* 1230 P (Nb); Adam *de Bedenhale* 1279 AssNb. From Beadnell (Nb), or Bednall (St).

**Beadon, Beedon:** Robert *de Bedon* 1297 MinAcctCo; Nicholas *Beaden* 1642 PrD. From Beedon (Berks).

**Beaglehole, Bagilhole, Baglole:** Henry *Bagelhole* 1560, Thomas *Baggilhole* 1631 HartlandPR (D); Charles *Bagelhole* 1642 PrD; Jane *Bagalhole* 1667 HartlandPR (D). Probably from Bagley Hill in Axminster (D).

**Beal, Beale, Beales, Beals, Beall, Beel, Beels:** (i) *Bele* 1194 Cur (Sx); Alexander *filius Bele* 1203 P (L); *Bella, Bele* Coty 1275 RH (L); Thomas *Bele* 1206 Cur (Ess); John *Bele* 1275 SRWo; Ralph *le Bele* 1279 RH (C); Joan *Beles* 1327 SRSo; William *Beall* 1379 PTY. OFr *bele* 'beautiful', used also as a woman's name. (ii) Simon *de Beel* 1275 SRWo; Thomas *de Behil* 1382 Bardsley (Nb); John *Bele* 1517 ib. (Nb). From Beal (Northumberland), earlier *Behill*, or Beal (WRYorks), *Begale* DB.

**Bealey:** *v.* BEELEY

**Beam:** (i) Osbert *la Beme* Hy 3 HPD; Agnes *Bem* 1319 SRLo. OE *bēam* 'beam (of a loom)', and metonymic for a weaver. (ii) Osbarn *Atterbeame* 1274 RH (Ess); Henry *atte Beme* 1332 MELS (Sr). 'Dweller by the tree or post', or 'by the footbridge', OE *bēam*.

**Beaman, Beamand, Beament, Beamont:** *v.* BEAUMONT

**Beamer:** *v.* BEEMER

**Beames, Beamish, Beamiss, Beams:** William *Baumis, de Beaumis* Hy 2 DC (L); Richard *de Beames, de Belmes* 1191–2 P (Sa); Robert *de Beaumeis* 1208 FFHu. From Beaumais-sur-Dive (Calvados). *v.* ANF.

**Bean, Beane, Been:** (i) Robertus *filius Biene* 1168 P (Cu); Ricardus *filius Bene* 1278 AssLa; Gerard, Ailwardus *Bene* 1166 P (Nf), 1180 P (Lo); Juliana *Bean* 1301 SRY. *Bene* is an original nickname from ME *bēne* 'pleasant, genial, kindly' (a1200 NED) which itself is also used as a nickname. We have also OE *bēan* 'bean', used like *Barley*, of a grower or seller of beans. cf. John *le Bener* 1282 LLB A. Also a nickname. The bean was regarded as typical of things of small value. cf. 'Al nas wurth a bene' c1325 MED. cf. Adam *Benecod* 1221 ElyA (Nf). Or we may have reference to the Twelfth-night custom when the man in whose portion of the cake the bean was found was appointed King of the Company. (ii) The Scottish *Bean* is from Gael *beathán*, a diminutive of *betha, beatha* 'life'.

**Beanland:** John *Beanland* 1672 HTY; Edward *Beanland* 1746 FrY. 'Dweller by the land on which beans are grown', OE *bēan, land.* cf. Beanlands Park (Cu).

**Bear, Beara, Beare, Beer, Beers, Bere, De La Bere:** (i) Ordric *de Bera* 1168 P (D); William *de la Bera* 1168 P (Ha); Nicholas *Attebere* 1247 AssSo; Henry *del Beer* 1327 SRDb. Walter *de la Bere* lived in 1263 at Beare Green in Capel (Surrey) and owed his name to his residence near a swine-pasture (OE *bǣr*). *v.* PN Sr 267. But the real home of this name is in the south-west. In Devon there are 18 places called Beare

or Beara and 17 named Beer, Beera or Beere, from most of which surnames were derived, usually in the form (Robert) *atte Beare* (1330). These are from OE *bearu* 'grove', the normal dative of which (*bearwe*) would become *barrow*. In Devon and the neighbouring counties of Somerset and Dorset, it had a dative *beara*, ME *bere*. (ii) Tedric' *Vrs'* 1130 P (O); Theodoricus *le Bere* 1166 Oseney (O); Ralph *Bere* 1177 P (Nf); Nicholas *le Urs* 1219 AssSt; Robert *le Beer* 1296 SRSx. OE *bera* 'bear' (translated by Lat *ursus*, OFr *urs*).

**Bearaway:** John *Beraway* 1260 AssCh. 'Carry away', OE *beran, onweg.* cf. Gilbert *Beritaway* 1279 RH (O) 'bear it away'; John *Berebac* 1290 IpmW 'carry back'; William *Berecorn* 1327 SRSo 'carry corn'.

**Bearcroft:** *v.* BARCROFT

**Beard:** (i) *Ælfsige mid þam berde* c1100–30 OEByn (D); Hugo *AlaBarbe, Barbatus* 1086 DB (Ha); Baldeuuinus *cum barba* 1086 ICC (C); Alsi *berd* 1086 InqEl (C); Alwine *bierd* 1148 Winton (Ha); Alfwin' *berd* 1155 P (Herts); Robert *a la barbe* 1178 P (Bk); Thomas *Ouelabarbe* 1280 AssSo; William *od la Barbe* 1311 LLB D. OE *beard*, frequently translated by Fr *barbe*, and often in a prepositional form, '(the man) with the beard'. *v.* BARBE. (ii) Adam *de Berd* 1327 SRDb. From Beard (Derby).

**Beardfield:** Ralph *de Berdefelde* 1337 CorLo. From Bardfield (Ess).

**Beardless:** Thomas *Berdles* 1225 FrLei; Robert *Berdeles* 1342 Glapwell (Db). 'Without a beard', OE *beard, lēas.* cf. Richard *Shaveberd* 1286 AssCh 'shave beard'; Matilda *Shereberd* 1306 IpmGl 'shear beard'.

**Beardsley:** *v.* BARDSLEY

**Beardwell:** *v.* BARDWELL

**Bearham:** *v.* BARHAM

**Bearward:** Fulk *le Bereward* 1208 Cur (C); Stephen *Bereward* 1275 SRWo; Edward *Bereward* 1356 LLB G. 'Keeper of the bear', OE *bere, weard.*

**Bearryman:** *v.* BERRIMAN

**Beasley:** *v.* BEESLEY

**Beaston:** *v.* BEESTON

**Beat:** *v.* BEET

**Beatell, Beatle:** *v.* BEETELL

**Beater, Better:** Richard *Batere* 1166 P (Berks); Jordan *le Bettere* 1200 Cur (L); John *le Betere* 1275 RH (W). OE *bēatere* 'beater, fighter, champion'. cf. CHAMPION. It may also be a short form of the common *Coperbeter, Flaxbeter, Goldbeter, Ledbeter, Wodebeter, Wolbeter.*

**Beatey:** *v.* BEATY

**Beaton:** *v.* BEETON

**Beatrice, Bettriss:** Richard *filius Beatricie* 1212 Cur (Y); Geoffrey *Beatriz* 1210 Cur (C); John *Baytrise* 1662 HTEss. OFr *Beatris, Bietriz.*

**Beatson:** John *Batisoun* 1458 Black; William *Beatisoun* 1627 ib.; Thomas *Beatson* 1691 FrY. A Scottish form of BATESON, found also as *Battison.*

**Beaty, Beatey, Beattie, Beatty:** *Baty*, Flessor c1340 Black; David *filius Bety* 1342 ib.; John *Bety* 1558 ib.; Hew *Batie* 1569 ib. A Scottish form of BATEY.

**Beauchamp, Beachamp, Beacham, Beachem, Beecham:** Hugo *de Belcamp* 1086 DB (Herts);

William *de Bellow Campo* 1161 Templars (Lo); Robert *de Beauchamp* 1203 FFEss; John *Bechaumpe* 1376 LLB H; Oliver *Beacham* 1674 HTSf. The DB family came from Beauchamps (La Manche). Others may have come from other French places named Beauchamp.

**Beauford, Beaufort, Bewfort:** Henry *Bewefort* 1340–1450 GildC; Gylbert *Bowfort* 1545 SRW. From one or other of the numerous places in France called Beaufort.

**Beaufoy, Boffee, Boffey, Boffy, Buffey:** Ralph *de Bellafago, de Belfago* 1086 DB (Nf, Sf); William *Belfou, de Belfou* ib. (Herts, W); Nicholas *de Bealfo* 1114–16 Holme (Nf); Thomas *de Beaufow* 1185 RotDom (R); Robert *de Biaufey* 1210 Cur (Db); Emma *de Beaufey, de Beaufo* 1212, 1236 Fees (Nt); Thomas *Buffy* 1276 RH (O); William *Bouffaye* 1544 FFHu; Anne *Boffey* 1793 Bardsley. The DB tenants came from Beaufour (Calvados), *Belfou, Beaufou* 1100, *Bellefai* c1160 OEByn.

**Beaufront:** Alan *Beaufrunt* 1281 IpmY; Adam *Beaufront* 1327 SRY; John *Beaufront* 1382 AssLo. A nickname, 'beautiful forehead', OFr *beau, front*. cf. Henry *Beaubraz* 1228 FFO 'beautiful arms'; John *Beucol* 1327 SRY 'fair neck'; Ivo *Beaudonte* 1327 SRSo 'beautiful teeth'; Richard *Beaupel* 1218 P (D) 'beautiful skin'.

**Beaulah, Beaulieu:** v. BEWLAY

**Beauman:** v. BEAUMONT, BOWMAN

**Beaumont, Beaument, Beumant, Beaman, Beamand, Beament, Beamont, Beauman, Bemand, Belmont, Bemment:** Rogerius *de Belmont, de Bellomonte* 1086 DB (Do, Gl); Ralph *de Belmunt* 1187 P (O); *John Bemund* 1274 RH (Sf); Godfrey *de Beumund* 1275 RH (Nf); William *Beumound, Beumon* 1279 RH (O); John *Bomund* 1300 FFSf; Robert *Beaumond* 1332 SRSx; Laurence *Beamond* 1369 LLB G; Wedow *Beament*, Mrs *Beamonte* 1568 SRSf; Mrs *Bemant*, Peter *Beaman*, Widow *Bomant* 1674 HTSf. From one of the five places in Normandy named Beaumont. The DB family came from Beaumont-le-Roger (Eure).

**Beausire, Bellsyer, Bowser:** Geoffrey *Beusire* 1226 Cur (Ess); John *Belsire* 1274 RH (K); Gregory *Bousyre* 1314–16 SRSt; Alexander *Belsier, Belshyre* 1542 Oseney. OFr *bel, beu* and *sire* 'fair sir', a term of address (cf. GOODSIR, SWEETSER), confused in the 16th century with BELCHER.

**Beautement, Beautyman:** v. BUTEMENT

**Beavan, Beaven, Beavon:** v. BEVAN

**Beaver, Beavers, Beavors, Beever, Beevers, Beevor, Beevors, Bevar, Bever, Bevers, Bevir, Biever:** (i) Ralph *de Belueeir* 1170 P (Y); John *de Beauveir* 1204 AssY; William *Bever, de Beuver* 1207–8 Cur (Lei, Do). From Belvoir (Leics), pronounced *Beever*. (ii) Godwyn *Beure* 1084 (c1300) ELPN; Adam *Bever* 1274 RH (So); Thomas *le Bevere* 1327 SRSx. A nickname from the beaver (OE *beofor*).

**Beaves, Beavis, Beevis, Beves, Bevis, Beviss, Bovis:** (i) Goisbert *de Beluaco* 1086 DB (Herts); Thomas *Beueys* 1317 AssK; Philip *de Beauveys* 1321 QW (La); Robert *de Beueys* 1327 SRC. From Beauvais (Oise). (ii) Odo *Belfiz* 1176 P (Ha); William *Beaufiz, Biaufiz* 1208 Cur (Gl); Hugo *Beaufiz, Beauuiz*

1221 AssWa. OFr *bel, biau, beau* 'fine' and AFr *fiz* 'son'. *Bel* was often used as a term of affection, hence 'dear son'.

**Beavill:** v. BEVILL

**Beavin:** v. BEVIN

**Beazley:** v. BEESLEY

**Bebbington, Bebington:** Adam *de Bebyngton* 13th WhC; Hugh *Bebynton'* 1403 KB (Lo); Peter *Bebynton* 1492 PN Ch iv 24. From Bebington (Ch).

**Bec:** v. BECK

**Beccle:** v. BECKLES

**Bech:** v. BEECH

**Becher:** v. BEECHER

**Beck, Becke, Bec:** (i) Walter *Bec* 1086 DB (Bk); Geoffrey *de Bech* ib. (Do); Robert *de Becco* 1199 AssSt. The DB under-tenants probably came from Bec-Hellouin (Eure). Others may have come from one of the numerous places in France named Bec. (ii) Adam *del Bec* 1207 Cur (L); Henry *Delebec*, Ralph *del Bek* (his son) 1263 Ipm (Ess); Robert *Attebek* 1297 SRY. 'Dweller by the brook', ME *bekke*, ON *bekkr*, common in the North, the Danelaw, and in Scotland. (iii) Æluuin *Becce filius*, Brun *Becce filius* c1095 Bury (Sf); Robertus *filius Beck'* 1297 MinAcctCo (Y). OE *\*Becca*, from *becca* 'pick-axe', or OE *Beocca*. (iv) Osbert *Becche* c1140 ELPN; Terricus *Becce* c1166 ib.; Robert *Becke* 1296 SRSx. Either from the personal-name above or from OE *becca* 'mattock', metonymic for a maker, seller or user of mattocks. (iv) Henry *Bec* 1196 P (L); Bartholomew *Beck* 1297 MinAcctCo (W). OFr *bec* 'beak, bill of a bird'. According to Suetonius, Antonius Primus, as a boy, had a nickname *Beccus*, 'id valet, *gallinacei rostrum*', a nose like a cock's beak. cf. also *Naso adunco*, a beake-nose 1598 Florio.

**Beckard:** John *Bekard* 1242–3, Philip *Bekard* 1330 FFY; William *Bekard* 1402 IpmY. OE *Becca* plus the suffix *-ard*, or a derivative of OFr *bec* 'beak'. It is also probably one of the sources of BECKETT.

**Becker:** Nicholas *Bekar'* 1327 SRSx; Alice *Beckar'* 1379 PTY. A derivative of OE *becca* 'mattock'. A maker or user of mattocks.

**Becket, Beckett, Beckitt:** William *Bechet, Beckett* c1155 DC (L); Robert *Beket* 1176 FF (Berks); Robert *Becket* 1379 PTY. This surname, common in the 12th and 13th centuries as *Beket*, without article or preposition, must be a diminutive of OFr *bec*, 'little beak or mouth' (Moisy). v. BECK (iv). The only evidence noted that might possibly be 'at the beck-head' is: Elezabeth *Becked* 1549 RothwellPR (Y). It may occasionally be local, from Beckett (Berks): John *de Beckcote* 1279 RH (O); or from Beckett (Devon), from a 1333 surname *Bykecote* (PN D 179).

**Beckford:** Nicholas *de Beckeford* Hy 3 IpmGl; Robert *de Becford'* 1245–50 RegAntiquiss. From Beckford (Gl).

**Beckles, Beccle:** Tankard *de Beccles* 1191 P (Nf); Richard *de Bekles* 1278 LLB B; John *Bekyllis* 1487 TestEbor; Lancelot *Beckle* 1642 PrD. From Beccles (Sf).

**Beckley, Beckly:** Ralph *de Beckele* 1211 Cur (Sf); Richard *de Beckele* 1327 SRSx; John *Beklay* 1446 FrY. From Beckley (K, O, Sx).

**Beckwith:** John *Bekwyth'* 1379 PTY; Richard *Bekwyth* 1415 IpmY; Adam *Bekwith* 1423 FrY. From Beckwith (WRYorks).

**Bedale, Bedall:** Leticia *de Bedale* 1348 DbAS 36; John *de Bedale* 1351 FrY; John *Bedale* 1412 IpmGl. From Bedale (NRY).

**Bedd, Bedde:** Roger *Bedde* 1248 AssBerks; Thomas *of the Bedde'* 1312 Pat; Roger *de la Bedde* 1327 Misc (Mx). 'Dweller at the plot of ground where plants are grown', OE *bedd*.

**Beddall:** *v.* BEADEL

**Beddard:** *v.* BEDWARD

**Beddingfield, Bedingfield:** Walkelin *de Bedigfelde* 1198 FFEss; Adam *de Beddingefeld* 1200 Cur; Roger *de Bedyngfeld* 1332 SRLo. From Bedingfield (Sf).

**Beddingham, Bedingham:** Robert *de Bedingham* 1206 Cur (Nf); William *de Bedyngeham* 1296 SRSx; Richard *Bedyngham* 1461 PN C 241. From Beddingham (Sx), or Bedingham (Nf).

**Beddoe, Beddoes, Beddow, Beddowes, Beddows:** *Bedo* ap Richard 1493 SaAs 2/xi; Johanna *Bedowe* 1577 Bardsley; John *Beddoe* 1641 SaAS 3/iv. From *Bedo*, a pet-form of *Meredith*.

**Bedell, Bedle:** *v.* BEADEL

**Bedford, Bedforth, Bedfer:** Osgar *de Bedeford* 1066 DB (Beds); Robert *de Bedeford* c1180 Bury (L); John *de Bedforth* 1379 PTY; William *Bedford* 1465 Paston. Usually from Bedford (Beds), but sometimes from Bedford (La), or Bedforth in Thornhill (WRY).

**Bedingfield:** *v.* BEDDINGFIELD

**Bedingham:** *v.* BEDDINGHAM

**Bedloe:** William *Bedeluue, Biedeluue* 1191–3 P (C). OE \**Biedlufu*, an otherwise unknown woman's name, from OE *beodan* 'to command' and *-lufu* 'love'.

**Bednall, Bednell:** *v.* BEADNALL

**Bedser:** Alice *de Bedesore*, John *Badesore* 1296, 1327 SRSx. From a lost place, possibly near Bexhill (Sussex).

**Bedward, Beddard:** Dafydd *ap Edward* ap Hoell 1498 Chirk; John *Bedhard* 1643 FrY. Welsh *ap* 'son' of *Edward*. cf. BOWEN.

**Bedwell, Bidwell, Biddwell, Bidewell:** Stephen *de Bedewell'* 1229 Cl (Ess). 'Dweller by the spring or stream in a shallow valley' (OE \**bydewelle*), as at Bedwell (Essex, Herts), Bedlar's Green (Essex), Bidwell (Northants, Beds, Devon, Som), or Biddles Fm (Bucks). *v.* PN Nth 222. Bedwellhay in Ely is *Bedelhey* 1576, *Beddlehay* 1615 PN C 127. Later forms have been confused with BEADEL.

**Bedwin, Bedwyn:** Walter *de Bedewynde* 1309 LLB D; Richard *Bedewynd* 1392 LoCh; Henry *Bedwyn* 1452 FFEss. From Bedwyn (W).

**Bee:** Walter *le Be* 1195 Oseney (O); Robert *Be* 1198 CurR (Y); William *le Beo* 1243 AssSo. OE *beo* 'bee', used, no doubt, of a busy, industrious person.

**Beeby, Beebee, Beabey, Beaby:** John *de Beby* 1327 SRLei; Richard *Bebie* 1596 FrY; Robert *Beeby* 1674 HTSf. From Beeby (Lei).

**Beech, Beach, Bech:** John *de la Beche* 1236 Fees (Wo); Idonea *de Beche* 1240 FFEss; Jacob' *atte Beche* 1296 SRSx; William *de la Beche* 1340 FFSt. *Beche* may be for OE *bece* 'stream', *bece* 'beech', or from OE

*bæce* 'stream', and without further evidence it is impossible to distinguish these in ME. In Worcs and Staffs, where *bæce* normally survives as *Bach(e), beche* is probably a variant of this. Robert *de Beche* (1327 SRC) came from either Landbeach or Waterbeach, both earlier *Beche* 'stream, valley'. Elias *ater Beche* (1296 SRSx) probably lived at Beech Fm in Battle. 'Dweller by the stream or the beech-tree'.

**Beecham:** *v.* BEAUCHAMP

**Beecher, Becher:** John *Becher(e)* 1279 RH (C), 1428 FA (Sx). 'Dweller by the beech-tree' (OE *bece*). *v.* BEECHMAN.

**Beechey:** Thomas *de la Bechey* 1279 RH (O) 'Dweller by the beech-enclosure', OE *bece, (ge)hæg*.

**Beeching:** John *Bechyng* 1471 CantW; Elizabeth *Bechinge* 1585, Godley *Beechinge* 1610 StaplehurstPR (K). Either a derivative of OE *bece, bæce* 'stream', hence 'dweller by the stream', or of OE *bece* 'beech-tree', hence 'dweller by the beech-tree'. Perhaps also a derivative of OE *Becca*.

**Beechman:** John *Becheman* 1332 SRSr. Identical with BEECHER.

**Beedell, Beedle:** *v.* BEADEL

**Beeding:** John *Bedyng* 1392 LoCh. From Beeding (Sx).

**Beedon:** *v.* BEADON

**Beeld:** *v.* BELD

**Beeley, Bealey, Bealy, Bely:** Thomas *de Beghley* 1316, Geoffrey *de Beley* 1357 DbCh; Mark *Bealy* 1642 PrD. From Beeleigh (Ess), Beoley (Wo), or Beeley (Db).

**Beel(s):** *v.* BEAL

**Beeman, Beman, Beaman:** R. *Beman* 1283 SRSf; William *le Bemon* 1324 LaCt. OE *beo* 'bee' and *mann*, 'bee-keeper'. The modern forms may also be for BEAUMONT.

**Beemaster:** *v.* BEMISTER

**Beemer, Beamer:** Normannus *Bemere* 1160–5 ELPN. OE *blemere* 'trumpeter'.

**Been:** *v.* BEAN

**Beer(e):** *v.* BEAR

**Beesley, Beasley, Beazleigh, Beazley, Beisley, Bezley:** Thomas *de Besleg* 1246 AssLa. From Beesley (Lancs).

**Beeston, Beaston, Beeson:** (i) William *de Beston(e)* 1153–66 Holme (Nf), 1205 P (Nt). From Beeston (Notts), pronounced *Beeson*, or one of the other Beestons, explained as *Beostun* 'tun where bent-grass grew' (DEPN). (ii) Andrew *de Bieston'* 1203 P (Y); Herbert *de Beston'* 1219 AssY; Richard *de Bestayn* 1297 MinAcctCo (Y). From Beeston (WRYorks), *Bestayn* 1297 MinAcctCo, a place called 'by the stone', OE *bi, be* and ON *steinn*, alternating with OE *stan*. (iii) Ralph *de Bestune* 1279 RH (C). Ralph came from The Beesons in Sutton (Cambs), *Estounesende* 1302, *Beestoun* 1348 '(the place) to the east of the hamlet', in contrast to *Westounesende* de Sutton (PN C 239). (iv) William *Besteton*, Ralph *Biesteton* c1248 Bec (Ha); Ralph *Byeston* 1256 RamsCt (Hu). '(The man who lived) to the east of the hamlet', OE *bi eastan tune*.

**Beet, Beat:** Adam *Bete* 1298 DbCh, 1332 SRLa. *Bete* is a pet-form of *Beton* (Beatrice).

**Beetell, Beetle, Beatell, Beatle:** Gilbert *Betyl'* 1248 AssBerks; John *Betel* 1317 AssK; William *Betill* 1502, John *Betyll* 1544 FFEss. Anglian *bētel*, West Saxon *bītel*, *bỹtel* 'a beetle, an instrument for driving in wedges, ramming down paving-stones, &c.'. Metonymic for a maker or user of this.

**Beetham, Betham:** Ralph *de Bethum* 1279 AssNb; Robert *de Bethum* 1379 PTY; Stephen *Betham* 1541 FFEss. From Beetham (We).

**Beetle:** *v.* BEETELL

**Beeton, Beaton:** *Beton* de Wath 1379 PTY; *Bete* or *Betune* (Betryse) c1440 PromptParv; John *Betoun* 1311 ColchCt; Richard *Beton* 1327 SRDb. *Beton*, a diminutive of *Bete* (Beatrice), still used as a christian name in Cornwall in 1630 (Bardsley).

**Beevens:** *v.* BEVAN

**Beever(s):** *v.* BEAVER

**Beevis:** *v.* BEAVES

**Begg, Begge, Beggs:** (i) Edwin *le bege* 1214 P (D); Henry *Begge* 1327 SRSo; Robert *Begge* 1503 TestEbor. Perhaps OFr *bègue*, a variant of OFr *béguin* 'a member of a 13th-century religious sect'. (ii) Malcolm *beg* c1208–14, Malise *Beg* 1300 Black. Gaelic *beag* 'little, of small stature'.

**Beggar, Begger:** Richard *Beggere* 1210–11 PWi; Adam *le Beggare* 1275 SRWo; Adam *Beggere* 1314 IpmW. OFr *begart*, *begar(d)* 'beggar'. The feminine form also appears: Avelina *Beggestere* 1301 FS.

**Begge:** *v.* BEGG

**Begger:** *v.* BEGGAR

**Beggs:** *v.* BEGG

**Beilby:** *v.* BIELBY

**Beild, Bield:** *Begild* (f) 1202 AssL; *Begilda* (f) 1271–2 FFL; Henry *Beyhild* a1290 CartAntiq; Geoffrey *Beilde* 1332 PN Do i 130. OE *Bēaghild*.

**Beisley:** *v.* BEESLEY

**Belch, Belk, Balch, Baulch, Boalch:** William *Belch* 1185 Templars (O); William *le Belch* 1295 ParlR (Ess); Robert *Balch* 1327 SRSo; Richard *le Balch* 1332 SRSx. ME *balche*, *belche*, *belke*, from OE *bælc* (*bælce*), (i) a belch, *eructatio*, (ii) stomach, pride, arrogance. From this latter sense a surname could arise. cf. PRIDE. The word probably had also in ME the same meaning as OE *balca* 'balk, beam, bank, ridge', and *le Balch*, *le Belch* may have meant 'the beam', used metaphorically for a man of stout, heavy build. *Belk* was also used as a topographical term: Henry *del Belk* 1252 Ipm (Nt), probably 'dweller by the bank or ridge'.

**Belcham, Belchem, Belsham, Belshem, Bellsham:** These Essex surnames preserve the correct *-ham* and the local pronunciation (*Belsham*) of Belchamp Otton, St Paul's and Walter (Essex).

**Belchambers:** *v.* BELLCHAMBER

**Belcher, Belsher, Belshaw, Beuscher, Beushaw, Bewshire, Bewshaw, Bewshea, Beaushaw, Bowsher:** Thomas *Belcher* 1219 AssY; Richard *Belecher* 1274 RH (Gl); Alexander *Belcher* 1453 FFEss; Margaret *Bewcher* 1530 SIA (Sf); William *Bewshawe* 1539 FrY; Henry *Bowschere* 1575 Oxon; Henry *Belsher* 1662 HTEss. OFr *bel(e)*, *beu* and *chiere*, originally 'fair face', later 'fair look', one of a cheerful, pleasant demeanour. The surname was often

confused with BEAUSIRE, and in the York Plays is used as a term of address, often derogatory: Herod addresses a messenger, 'Bewcher! wele ye be', and when Annas orders a boy who has been bound to be brought in, the soldier announces, 'Lo, here is the belschere broght that ye bad bring'. cf. GOACHER, GOODFAR.

**Beld, Beeld, Bield:** Roger *le Belde* 1317–18 FFSr; William *Belde* 1378 LLB F; Henrie *Beld* 1545 SRW. OE *beald* 'brave, courageous'. *v.* also BEILD.

**Beldam, Beldan, Beldham, Beldom:** Godfrey, Richard *Beledame* 1296, '1332 SRSx. AFr *beledame* 'fine lady', a derogatory nickname. ME *beldam* 'grandmother' is not recorded before c1440 and the sense 'aged woman, hag' not until the 16th century.

**Belden, Beldon:** Hugh *de Beldon* 1204 AssY; John *Beledon* 1371 FFEss. From Beldon Hill in Manningham (WRY).

**Belding:** *v.* BALDING

**Belenger:** *v.* BERRINGER

**Belford:** James *de Beleford* p1147 Black; Thomas *de Belfford* 1390 FrY; William *Belford* 1421 IpmY. From Belford (Nb, Roxburgh).

**Belfrage:** John *Belverage* 1685, Thomas *Belfrage* or *Beveridge* 1690 Black. A Scottish form of BEVERIDGE, with intrusive *l* as in *Calmeron* for Cameron and *Chalmers* for Chambers. In Fife, the name also occurs as BERRIDGE: John *Berrage*, *Berrige* 1675, 1711 Black.

**Belgian, Belgion, Belgions:** William *Belegambe* 1185 Templars (So); Nicholas *Belejaumbe* 1221 AssWo; John *Belgeam* 1492 Black. AFr *bele jambe* 'fine leg'.

**Belgrave, Belgrove, Bellgrove:** Reginald *de Belegraue* Hy 2 Seals (Lei); Henry *de Belgrave* 1241 FFO; John *Belgraue* 1365–6 FFWa. From Belgrave (Lei).

**Belham, Belhomme:** *v.* BELLHAM

**Belk:** *v.* BELCH

**Bell:** (i) Ailuuardus *filius Belli* 1086 DB (Sf); Ricardus *filius Bell* 1279 RH (Hu); Osbertus *filius Belle* 1297 SRY. *Bell* may be a pet-form of *Isabel*. *Bella* is probably a latinization of *Bele*, OFr *belle* 'beautiful'. *v.* BEAL. *Bellus* is a Latin form of OFr *Bel* 'beautiful', otherwise unknown as a personal-name. (ii) Seaman *Belle* 1181–7 ELPN; Serlo *Belle* 1190 P (Y). OE *belle* 'bell', probably metonymic for BELLMAN or BELLRINGER. (iii) Hugo *bel* 1148 Winton (Ha); Robertus *bellus* ib.; Robert *le bel* 1186–1200 Holme (Nf). OFr *bel* 'beautiful, fair'. (iv) Roger *del Bel* 1209 P (Nf); Robert *de la Belle* 1222 DBStP; John *atte Belle* 1332 SRLo. The last example denotes one who lives at the sign of the Bell. This type of name is not so common as has been suggested and the other examples are unusually early. They may denote a dweller by the church or town bell or bellhouse or be metonymic for the bellman or bellringer.

**Bellaby:** *v.* BELLERBY

**Bellam:** *v.* BELLHAM

**Bellamy:** Walter *Belami* 1185 Templars (Y); Ralph *Belamy* 1214 Cur (Nf). OFr *bel ami* 'fair friend'.

**Bellanger:** *v.* BERRINGER

**Bellar, Bellars, Bellers:** Hamo *Beler* c1166 DC (L); Hamond *Beler* 1271–2 FFWa; John *Bellars* 1432 FFEss. Fr *bélier* 'ram', a nickname. cf. Kirby Bellars (Lei). Sometimes, perhaps, a derivative of OE *belle* 'bell', and metonymic for a bellringer or a bell-founder.

**Bellas:** *v.* BELLHOUSE

**Bellasis:** Gregory *de Belassis* 13th Lewes (Nf); Robert *de Beleassise* 1305 FrY; Peter *Belassise* 1351 AssL. From Belasis, Bellasis (Du), Bellasis (Nb), Bellasize (ERY), or Belsize (Herts, Nth). There was also a Jewish name which may have contributed to the surname: *Beleasez* Judea 1181 P (O); Jacobus *Belasez* 1209 P (Ess).

**Bellby:** Bussell *de Bellebi*, Leftham *de Belleby* 1202 FFY. From Belby (ERY).

**Bellchamber, Bellchambers, Belchambers:** Thomas *Belchambre* 1369 LLB G. This surname has been regarded as a corruption of *Bellencombre*. The family of William *Belencumbre* (1235 *Ass*) settled in Essex and has long been extinct. Their name survives in Belcumber Hall in Finchingfield and there is no evidence that the name ever took the form of *Belchamber*. The solitary example above (nearly 500 years earlier than the first example in NED) was earlier *atte Belchambre*. The bellhouse was originally a detached structure. The belfry (c1440 NED) was generally attached to the church and later (1549) was used of the room or storey where the bells were hung. This must also have been called the bellchamber. A man could hardly live in this and if he lived by the bellchamber, a more natural name would have been *atte church*. Hence, the surname probably refers to the keeper of the bellchamber or the ringer of the bells, often, no doubt, the same man.

**Bellenger:** *v.* BERRINGER

**Beller:** Richard *le Beller* 1281 MEOT (L); Henry *Beller* 1332 SRCu. A derivative of OE *belle* 'bell', a bell-founder.

**Bellerby, Bellaby:** Elyas *de Belreby* 1251 AssY; Robert *de Bellerby* 1327 SRY; John *Bellerby* 1421 FrY. From Bellerby (NRY).

**Bellers:** *v.* BELLAR

**Bellett, Bellot, Bellott, Bellotte:** *Belet* 1188 BuryS (Sf); William *Belet, Belot* 1086 DB (Do); Herueus *belet* 1130 P (O); Adam *Belot* 1279 RH (Hu). *Belet*, which was very common, may be a nickname from a diminutive of OFr *bel* 'beautiful'. Both *Belet* and *Belot* are diminutives of *Bel*, a pet-form of *Isabel*.

**Bellew, Bellewes:** Gilbert *de Beleawe* c1160 Black; Thomas *de Bellew, de Bella Aqua* 13th PN Ch iv 95; John *de Belewe* 1274 IpmY; John *Belewe* 1367 FFY. From Bellou (Orne). Later examples may be from Belleau (L).

**Bellgrove:** *v.* BELGRAVE

**Bellham, Belham, Bellam, Belhomme:** Reginald *Belhume, Belhome* 1179, 1180 P (C, Sx); William *Belhom* 1279 RH (C). OFr *bel* 'beautiful, fair' and *homme* 'man'.

**Bellhanger:** *v.* BERRINGER

**Bellhouse, Bellas:** Ernald *Belhus* 1167 P (Nf); Richard *de Bellus, de Bellehus* 1206 P (Y), 1230 P (Ess); Walter *atte Belhous* 1266 LLB C; Richard

*Bellus* 1572 RothwellPR (Y); William *Bellas* 1653 FrY. From residence near a detached bell-house or tower, OE *bell-hūs*.

**Bellinger:** *v.* BERRINGER

**Bellingham:** William *de Belingham* 1274 RH (Nf); John *de Belyngham* 1327 LLB E; Harry *Belyngham* 1401 Paston. From Bellingham (K, Nb), or Bellingham Fm in Highworth (W).

**Bellington:** Thomas *Belynton'* 1275 SRWo. From Bellington Fm in Chaddesley Corbett (Wo).

**Bellis, Belliss:** (i) John *ap Elys* 1513 Chirk; John *Bellis* 1747 Bardsley (Ch). 'Son of Ellis.' cf. BOWEN. (ii) For BELLOWS, of which the regular form was *bellies* until the 16th century, whilst *bellis, bellice* are still found in the dialects (NED). cf. Ursula *Bellies* 1610 Bardsley.

**Bellmaine:** Nicholas *Belesmains* 1210 Cur (Herts); John *Belemeins* 1237 Colch (Herts). OFr *belle* and *mains* 'beautiful hands'.

**Bellman, Belman:** William *Belman* 1300 Crowland (C); Katerina *Beleman* 1327 SRC; Gilbert *Belman* 1398 Black. OE *belle* and *mann* 'bellman', in York and Scotland, used of the town-crier. *Beleman* may mean 'servant of *Bele*' (*v.* BEAL) or it might be a hybrid, OFr *bele* 'beautiful, fair' and ME *man*. cf. FAIRMAN, BELLHAM. In the Denham Parish Register (Suffolk) the surname appears as *Beleman, Belleman, Belliman* (1585–1606), *Billeman* (1776), *Billyman* (1784) and in that of Rushbrook as *Billeman* and *Billerman* (1760, 1791). The persistent medial vowel suggests that this is for ME *beli-man*, from ME *beli* 'bellows', 'bellows-blower'. cf. BELLOW.

**Belloc, Bellock:** Peuerel *de Belloc'* 1137 Eynsham; Bidan *de Beauluc* merchant of Burdeux, Bydan *de Beaulok* of Gascony 1305 LLB B. A dialectal form of the common French place-name Beaulieu. There seems also to have been a personal name: Gospatric *filius Beloc* 1163 P.

**Bellot(t):** *v.* BELLETT

**Bellow, Bellows, Beloe, Billows:** William *Beli* 1178 P (Wa); Ralph *Belewe* 1253 Oseney (O); John *Below* 1379 PTY; John *Byllow, Below, Bellow* 1464–79 Oseney (O). ME *beli, belu, below*, from OE (*blǣst*) *bel(i)ʒ* 'bellows', used only as a singular until 1400; here, metonymic for a bellows-blower. cf. William *Belymuð* 1275 RH (Nf), 'bellows-mouth'.

**Bellringer:** Richard *Belringer* 1216–72 MEOT (Sr). 'Bell-ringer' (1543 NED).

**Bellsham:** *v.* BELCHAM

**Bellson, Belson:** (i) Thomas *Beleson* 1317 AssK; John *Belessone, Bellesson'* 1339, 1341 Crowland (C). 'Son of *Bele*' (*v.* BEAL) or of *Bell* (Isabel). (ii) *Belsante, Belsant* 1185, 1190 Oseney (O); William *Belsent, Belesent* 1327, 1332 SRSx. OFr *Belisant, Belisent* (f), OG *Belissendis*.

**Bellsyer:** *v.* BEAUSIRE

**Bellwood, Belwood:** Wibald *de Belwoda* 1138–54 MCh; Henry *de Bellewode* 1341 FFY; William *Belward* 1524 SRSf. From Belwood (L).

**Belmont:** *v.* BEAUMONT

**Beloe:** *v.* BELLOW

**Belsham, Belshem:** *v.* BELCHAM

**Belshaw, Beisher:** *v.* BELCHER

**Belson:** v. BELLSON

**Belt:** Arnold, Robert *Belt*(e) 1203 FFEss, 1327 SRC. OE *belt*, used of a belt-maker, cf. Benedict *le beleter* 1295 FrY.

**Belton:** Turgis *de Beltona* 1179 P (L); Thomas *de Belton'* 1298 AssL; John *Belton* 1398-9 FFWa. From Belton (L, Lei, R, Sf), or Bilton (Wa), *Beltone* c1155.

**Beluncle:** William *Beluncle* 1227 Cur (Sf); William *Belvncle* 1240 PN K 120; Adam *Beluncle* 1338 FFC. 'Uncle, great uncle', OFr *bele, oncle*.

**Belwood:** v. BELLWOOD

**Bely:** v. BEELEY

**Beman:** v. BEEMAN

**Bemand:** v. BEAUMONT

**Bemister, Beemaster:** The local pronunciation of Beaminster (Dorset).

**Bemment:** v. BEAUMONT

**Benbow, Benbough:** William *Bendebowe* 1349 LLB F; John *Benbowe* 1545 FFHu. 'Bend bow', a nickname for an archer.

**Bence:** Ernisius *filius Bence* 1175 P (Y); Ærnulfus *filius Benze* 1178 P (Nb); Osmund *Benz* 1066, 1086 DB (Db); William *Bence* 1279 RH (O). OG *Benzo*. A diminutive *Benzelinus* occurs in DB (O, W) and gave rise to a surname in Suffolk (Seman *Bencelyn* 1327 SRSf), side by side with *Bence* (Roger *Bence* 1327 SRSf), whilst the feminine *Bencelina* is found in Kent (1207 Cur).

**Bench:** Robert *Benche* 1279 RH (C); Thomas *atte Bench* 1327 SRSo. 'Dweller by the terrace, bank, river-bank', from OE *benc* 'bench'. cf. Roger *le Bencher* 1279 RH (O).

**Bencher:** Roger *le Bencher* 1279 RH (O); Adam *le Benchere* 1296 SRSx; Robert *Bencher* 1674 HTSf. 'Dweller by the terrace, river-bank, or bank', from a derivative of OE *benc* 'bank'.

**Bencraft:** v. BANCROFT

**Bendall, Bendell:** v. BENTALL

**Bendbone:** Nicholas *Bendbone* 1296 SRNb. 'Bend bone'. cf. HACKBON. But perhaps an error for *Bendboue* 'bend bow'. v. BENBOW.

**Benden:** William *de Benden* 1269 FFEss; Robert *Bendyn* 1316 Hylle. Perhaps from Benenden (K), *Bennedene* c1100.

**Bender:** Robert *le Bendare* 1327 SRSx. A derivative of OE *bendan* 'to bend' (the bow). Perhaps synonymous with BENBOW.

**Bendish, Bendysh:** Thomas *de Benedish* 1315-16 FFEss; Thomas *Bendissh* 1393 FFEss; John *Bendyssh* 1450 Paston. From Bendish (Herts), or Bendysh Hall in Radwinter (Ess).

**Bendle:** v. BENTALL

**Beneck:** John *Beneneke* 1332 SRDo; Agnes *le Benek*, Richard *le Beneke* 1332 SRSx. A nickname, 'bean neck', OE *bēan, hnecca*.

**Benedict, Bennedik:** Geoffrey *Benedicite* 1221 AssWa; William *Benedicite* 1322 AssSt. Lat *benedicite* 'bless (you)', a nickname, no doubt from a favourite saying.

**Benfield, Benefield, Benfell:** Goduin *de Benefelle* 1066 DB (Herts); Robert *de Benefeld* c1160-9 YCh; Jul' *de Benefelde* 1296 SRSx; John *Benefeld* 1381 AssWa. From Benville Manor in Corscombe (Do),

Benfield Fm in Hangleton, a lost Benfield in Twineham (Sx), or 'dweller by the bean-field, or by the bent-grass field', OE *bēan/beonet, feld*.

**Benger:** v. BERRINGER

**Bengough:** Robert *Benghok* 1332 SRCu; Walter *Bengough* 1663 HeMil. Welsh *pencoch, bengoch* 'red-haired'.

**Benham:** Eudo *de Benham* 1176 P (Nt); John *de Benham* 1296 SRSx; Robert *Benham* 1384 IpmGl. From Benham (Berks), or Benhams in Horsham (Sx).

**Benian:** v. BEYNON

**Beningion:** v. BENNINGTON

**Benjamin, Benjaman:** *Beniamin* 12th MedEA (Nf); Roger *Beniamin* 1166 P (Nf). Hebrew *Benjamin* 'son of the south', interpreted in Genesis 'son of the right hand', much less frequent than *Adam*. The surname is also found in Berkshire, Cambridgeshire and Wiltshire before 1250.

**Benley:** Henry *de Benleia* 1203 P (Lo); Adard *de Beneleg'* 1221 SPleas (Wo); Henry *Benle* 1332 SRSx. From Benley Cross in Chumleigh (D), or 'dweller at the bee frequented wood or clearing', OE *bēo*, pl. *bēon, lēah*.

**Benn, Bennis, Benns:** *Benne* de Ecclesille 1246 AssLa; John son of *Benne* 1306 Wak (Y); Leuuinus *Benne* 1066 DB (Sf); Siuard *Benne* c1190 Gilb (L); Thomas *Ben* 1275 SRWo; Thomas *Bennes* 1524 SRSf. Whether the later Suffolk surname (1327, 1381 SRSf) is identical with the DB *Benne* is doubtful. This Tengvik explains as a nickname 'the plump, lumpish one', or from OE *Bynna* or *Beonna* for the post-Conquest use of which there is no clear evidence. The 13th-century *Benne* is more likely to be a pet-name of some common christian name, of *Bennet* rather than of *Benjamin*.

**Bennedik:** v. BENEDICT

**Bennell, Benwell:** Walter *de Benhala* c1165 StCh; Nicholas *de Benewelle* 1279 RH (C); John *Benhale* 1379 AssEss. From Benwell (Nb), Benhall (Sf), or Banwell Fm in North Mundham (Sx).

**Bennet, Bennett, Bennette, Bennetts, Bennitt, Bennitts:** *Beneit* Kernet 12th DC (Nt); *Beneit, Benedictus* Kepeherm 1193, 1200 Oseney (O); Robertus *filius Benite* 1301 SRY; Robert *Beneyt* Hy 2 Gilb (L); William *Benet* 1208 ChR (Du); Robert *Benyt* 1327 SRC; Thomas *Benetes* 1327 SRSt. OFr *Beneit, Beneoit*, Lat *Benedictus* 'blessed', a common christian name from the 12th century.

**Benneworth, Benneyworth, Benniworth:** Roger *de Benyngworda* c1150 Gilb (L); Margaret *de Benigwrða* 1214 P (Nb); Sibilia *Benigworyd* 1275 RH (L). From Benniworth (L).

**Bennie:** v. BENNY

**Benning, Bennings:** William *Bennyng* 1328 IpmW, 1332 SRSx; Alexander *Benyng* 1463 FrY. OE *\*Benning*.

**Benninger:** v. BERRINGER

**Bennington, Benington:** Almar *de Benintone* 1066 DB (Herts); Alan *de Benington'* 1218 AssL; William *Benyngton* 1406-7 IpmY. From Bennington (Herts, L).

**Bennion:** v. BEYNON

**Bennison:** John *Benettesson* 1396 FrY; Robert *Benyson* 1517 GildY. 'Son of *Bennet*'.

**Benniworth:** v. BENNEWORTH

**Benny, Bennie:** (i) Ralph *Benny* 1301 SRY. A diminutive of *Benn*, a short form of *Bennet*. (ii) Hugh *de Benne* or *Bennef* c1201–7, James *Beny* or *Bynne* 1321 Black. From Bennie in Almond (Perth).

**Bensington:** Thomas *de Bensintone* 1200 FFO; Peter *de Bensinton*' 1208 Cur (O). From Bensington (O).

**Benskin, Binskin:** Thomas *Beynsskyn* 1473 CantW; William *Benskyn* or *Benchekyn* 1508 CantW; Jamys *Bynskyn* 1525 SRSx; Edmund *benchkyn* 1548, *benskyn* 1549 StaplehurstPR (K). A diminutive of *Benn*, a short form of *Bennet*.

**Bensley, Bensly:** John *Benssley* 1524 SRSf; David *Bensley* 1641 PrSo. 'Dweller by the clearing where beans grow', OE *bēan*, *lēah*.

**Benson:** (i) John *Benneson*, Adam *Bensome* 1326 Wak (Y); Alan *Benson* 1332 SRCu; John *Benesson* 1393 FrY. 'Son of *Benn* (Bennet)'. (ii) Peter *de Bensinton* 1208 Cur (O); Henry *de Benson* 1269 Oseney (O). From Benson or Bensington (Oxon).

**Benstead, Bensted, Binstead, Binsted:** John *de Bentestede* 1200 P (K); John *de Benstede* 1311 LLB D; Edward *Benstede* 1402–3 FFWa. From Benstead (Ha), Banstead (Sr), *Benestede* DB, Binstead (Wt), or Binsted (Ha, Sx), all three *Benestede* DB.

**Bent:** Nicholas *Bent* 1256 AssNb; Henry, Adam *del Bent* 1327, 1332 SRSt. 'Dweller by the grassy plain, heath', ME *bent*, from OE *beonet* 'stiff grass'.

**Bentall, Benthall, Bendall, Bendell, Bendle:** Robert *de Benethal*' 1221 AssSa; Philip *de Benedhal* 1255 RH (Sa). From Benthall (Salop).

**Bentham:** William *de Benetham* 1205 P (Y); William *de Beneteham* 1268 IpmGl; John *Bentam* 1340–1450 GildC; John *Bentham, Benton, Bentom* 1681–5 WRS. From Bentham (Gl, WRY).

**Bentley, Bently, Bintley:** William *de Benetlega* 1176 P (Db); William *de Benteley* 1316–17 FFWa; John *Benteley* 1388–9 FFSr. From one or other of the many places of this name.

**Bentliff, Bintliff, Bintcliffe:** Thomas *Bentclyff* 1455 FrY; Joseph *Bentiliffe* 1716 Shef. From Bentcliffe in Saddleworth (WRY).

**Benton:** William *de Benton* 1234 FFSf; Simon *de Beneton*' 1275 SRWo; Richard *Benton* 1450 AssLo. From Little Benton, Longbenton (Nb), or 'dweller at the farm where bent grass or beans grow', OE *beonet*/*bēan*, *tūn*.

**Bentworth, Bintworth:** Martin *de Bintewrda* 1148 Winton (Ha). From Bentworth (Ha).

**Benwell:** v. BENNELL

**Benyon:** v. BEYNON

**Berald, Beraud:** Stephen *filius Beroldi* 1159 P; John *Berald* 1327 SRSo. OG *Berwald*, OFr *Beraut*.

**Berard:** *Berardus* c1125 Bury (Sf), 1143–7 DC (L); Reginaldus *filius Berard*' 1221 AssGl; William, Ralph *Berard* 1208–12 Cur (Y). OFr *Berart*, OG *Berard* 'bear-strong'.

**Berden, Berdon:** Robert *de Berdon* 1296 SRSx; John *de Berdene* 1323 CorLo; Thomas *de Berden* 1362 FFY. From Berden (Ess), Bardown in Ticehurst (Sx), *Berdowne* 1410, or Barden in Skipton (WRY), *Berdene* c1140.

**Berecloth:** v. BARRACLOUGH

**Beresford, Berresford:** William *de Beresford* 1279 RH (C); Adam *de Beresford* 1327 SRSt; John *Berysford* 1447 DbAS 30. From Beresford (St).

**Bergh:** Roger *de le Bergh*' 1221 AssWo. 'Dweller by the hill', OE *beorg*.

**Beriman:** v. BERRIMAN

**Beringer:** v. BERRINGER

**Berk(e)ley:** v. BARCLAY

**Berkenshaw:** v. BIRKENSHAW

**Berker:** v. BARKER

**Berkshire, Barkshire:** William *Berksir* 1249 AssW; William *de Barkescire* 1317 AssK; John *Barksher* 1525 SRSx. 'The man from Berkshire.'

**Berman:** v. BARMAN

**Bermingham:** v. BIRMINGHAM

**Bernard:** v. BARNARD

**Bernays:** John *Bernes* 1356, *de Bernes*, *Biernes* 1364, *Bernys* 1369 LLB G. Perhaps OFr *Bearnais* 'the man from Béarn'.

**Berner:** Tetbaldus *filius Bernerii* 1086 DB (D); *Bernerus* 1211 Cur (Bk); William *Berner* 1150–60 DC (L); Robert *le berner*, *le Bernier* 1190–1 (L); Walter *Berner*, Nicholas *le Berner* 1219 AssY. (i) OFr *Bernier*, OG *Berner* 'bear-army'; (ii) OFr *berner*, *bernier* 'keeper of the hounds'; (iii) A derivative of OE *beornan* 'to burn', a limeburner or charcoal burner. cf. ASHBURNER; (iv) A toponymic, equivalent to *atte berne*. v. BARNE. v. also BRENNER, BOURNER.

**Berners:** Hugo *de Berneres* 1086 DB (Mx, Ess); Goda *de Berners* 1185 Templars (Lo). From Bernières (Calvados).

**Berney, Burney, Burnie:** Ralph *de Bernai* 1086 DB (He, Wo, Sf); Henry *de Berney* 1268 Bardsley (Nf); Walter *Berneye* 1362 LLB G; Roger, Thomas *Burney* 1451 Bardsley (Nf), 1524 SRSf. From Bernay (Eure) or Berney (Norfolk).

**Berrecloth:** v. BARRACLOUGH

**Berresford:** v. BERESFORD

**Berrey:** v. BERRY

**Berrick:** v. BERWICK

**Berridge:** v. BELFRAGE

**Berridge:** Walter *Berich* 1279 RH (Hu); John *Berage* 1427, Richard *Beryge* 1482 FrY; Richard *Berridge* 1601 FFHu. The first example suggests a personal name as the origin, though most of the forms are too late for any certainty to be possible. Perhaps OE *Beornrīc*, or possibly from Berridge Fm in Woodchurch (K). In Scotland usually a variant of BEVERIDGE.

**Berrie:** v. BERRY

**Berriff:** v. BRIGHTIFF

**Berriman, Berryman, Beriman, Bearryman:** Edricus *Buriman* 1148 Winton (Ha); Alexander *Beriman* 1176 P (Bk). 'Servant at the manorhouse', from ME *buri*, the manorial use of OE *burh*.

**Berringer, Beringer, Bellenger, Belenger, Beillinger, Bellanger, Belfhanger, Benninger, Benger:** *Berengerus* 1086 DB (Nf); Robertus *filius Berengarii* c1150 EngFeud (L); *Bengerus* 1203 Cur (Beds); *Belingar*' 1207 ChR (Do); *Berenger faber* 1219 AssY; Hugo

*Berengeri* Ric 1 DC (L); Walter *Beneger* 1208 Cur (Gl); Reginald *Beringer* c1260 Lewes (C); John *Berenger* alias *Beniger* 1271 Ipm (W); John *Belinger* 1275 SRWo. OFr *Berengier*, OG *Beringar* 'bear-spear', the name of one of the paladins in the Charlemagne romances and fairly common in England in the 12th and 13th centuries. The various forms are due to the common Anglo-Norman interchange or loss of liquids in names containing *l, n* and *r*. *Berenger* became *Belenger* or *Benenger*, and in the latter the second *n* was lost, giving *Beneger*, later shortened to *Benger*.

**Berrington:** Alexander *de Beriton'* 1221 AssWo. From Berrington Green in Tenbury (Wo).

**Berriwin:** *v.* BERWIN

**Berrow:** *v.* BARROW

**Berry, Berrey, Berrie, Bury:** Gilbert *de la Beri* 1202 P (Co); Adam *Biry* 1257 Ipm (Y); Roger *Bury* 1260 AssC; Hubert *Bery* 1268 FFSf; Walter *del Bury* 1275 SRWo; William *ate Bery* 1327 SRSx. OE *byrig*, dative of *burh* 'fort', surviving in Berry Pomeroy (Devon), Bury St Edmunds (Suffolk) and Bury (Hunts, Lancs). ME *beri, biri, buri* was used of a manor-house and the surname must often mean 'servant at the manor-house'. Occasionally *Bury* may be 'dweller by an enclosure near the bower' (OE *būr*) or the fort (OE *burh*): Hugo *de Burhey*.

**Berrycloth:** *v.* BARRACLOUGH

**Berryman:** *v.* BERRIMAN

**Bert:** *Berta* 1101–21 Holme (Nf), 1143–7 DC (L); Robertus *vir Berte* 1196 P (Sx); Richard, John *Berte* 1327 SRSf, 1332 SRSx. OFr *Berte* (f), OG *Berhta, Berta*.

**Bertalot:** *v.* BARTLET

**Bertenshaw:** *v.* BIRKENSHAW

**Bertin, Bertim, Bartin:** *Bertinus* 1112 Bury (Sf), 1203 Cur (Sx); *Bertinus* Ruffyn 1322 FFK; *Bartyn* de Rankesley 1441 ShefA (Y); Peter *Bertin* 1204 ChR; John *Bertyn* 1296 SRSx. OFr *Bertin*, OG *Bertin*, a diminutive of names in *Berht-*. Michaëlsson notes its use as a hypocoristic of *Bertrand*.

**Bertram, Bertrand:** *v.* BARTRAM

**Bertwistle:** *v.* BIRTWHISTLE

**Berwick, Berrick, Barwick, Barrick:** Laurence *de Berewyke* 1278 RH (O); Edward *Barwyk* 1463 FrY; John *Barycke* 1547 EA (NS) ii. 'Dweller at an outlying grange', OE *berewīc* 'barley-farm', as at Berwick (Dorset, Northumb, etc.), Berrick (Oxon), Barwick (Norfolk, WRYorks) and Barricks in High Easter (Essex).

**Berwin, Berwyn, Berriwin:** Goscelinus *Beruinus* 1086 DB (D); Matthew *Berewynson* 1428 FrY. OG *Bernwin*.

**Besant, Bessant, Bessent, Beszant, Bezant, Bezzant:** Lefwin *besant* 1147–68 Bart (Lo); Robert *Besant* 1186–92 Clerkenwell (Lo), 1195 P (Lo). ME *besant, besand*, OFr *besan*, pl. *besanz*, Lat *byzantius* (*nummus*), a gold coin first minted at Byzantium (c1200 NED). *Bizantia* is first recorded in MLWL in 1187. cf. *unam bisanciam* c1179 Bart. Lefwin *Besant* was a moneyer (1168 P).

**Bessacre, Bessaker:** John *of Besacre* 1240 FFY. From Bessacar (WRY).

**Bessel, Bessell:** *Besellus* de Hibaldestowe 1177 P (L); William *Besell'* 1205 Cur; Matthew *Beselle* 1327 PN Do i 249. OG *Besel*.

**Bessemer:** Ingulf *besmere* 1148 Winton (Ha); John *le Besmere* 1263 MESO (Sx); William *Besemere* 1279 RH (O). A derivative of OE *besma* 'besom', a maker of besoms.

**Bessick, Bestwick, Beswick:** Thomas *de Bessewik'* 1297 SRY; Thomas *Besewyk* 1379 LoCh; John *Beswyk* 1411 FrY. From Beswick (La, ERY).

**Best, Beste:** William *Best* 1201 AssSo; Wilkin *le Best* 1260 AssCh. ME, OFr *beste* 'beast', used of a brutal, savage man, in earlier examples often connoting stupidity or folly. Also metonyic for BESTER.

**Bestar, Bester:** J. *le Bestere* 1279 RH (Hu); R. *Bestare* 1311 ColchCt. A derivative of ME *beste*, a herdsman.

**Bestman:** John *Besteman* 1327 *SR* (Ess). ME *beste* 'beast' and *man*. Equivalent to BESTER.

**Betchley:** William *de Bechely* 1296, John *Becheley, Belchley* 1525 SRSx. Probably from Beckley (Sx), *Becheleya* 1253.

**Betham:** *v.* BEETHAM

**Bethel, Bethell:** Amicia, Walter *Bethel* 1279 RH (O). The surname is often for ABETHELL, but these forms are too early for such a development. They must be diminutives of *Beth* (Elizabeth). cf. *Betha* de Bureswelles 1176–9 Clerkenwell (Ch).

**Bethson:** 'Son of *Beth*.' *v.* BETHEL.

**Bethune, Betton, Beaton, Beeton:** Baldwinus *de Betton, de Betun* 1195–7 P (Berks); William *de Bettoyne, de Betton* 1282–8 LLB A. From Béthune (Pas-de-Calais).

**Betley, Betteley:** Roger *de Beteleie* 1208 Cur (St); Philip *de Betle* 1296 SRSx; Thomas *de Betteley* 1332 SRSt. From Betley (St), or Betley in Henfield (Sx).

**Betson:** Richard *Bettessone* 1329 AD i (Hu); William *Bettesone* 1332 SRSt. 'Son of *Bett*.'

**Bett, Bette, Betts:** Reginald *filius Bette* 1197 FF (Bk); *Betta* Caperun 1247 FrLeic; Terri *Bette* Hy 2 Bart (Lo); Richard *Bette* 1175 P (Nt); Emma *Bettes* 1279 RH (O). Usually explained as a pet-form of *Beatrice*, which is possible. cf. BEET. There is, however, evidence that *Bette* was at times, at least, masculine: 'Bettinus Beaumond and *Bette* his brother' 1289 LLB A. '*Bette* the bocher' in *Piers Plowman* was also, presumably, a man. This was a pet-name for *Bertram* or *Bertelmew* (Bartholomew). cf. Robert *Bet* son of *Bartholomew Bette* 1312 LLB D, where the surname may be due to a family partiality for the christian name *Bertelmew*. The form *Bettes* is more frequent than usual and is probably often a toponymic with loss of the preposition. cf. John *del Bettis* 1379 NottBR, from OE *bytt* 'butt, cask, flagon, womb', used topographically like OE *byden* 'butt' (cf. BEDWELL) and *trog* 'trough'. Hence, probably, 'dweller by the hollows'.

**Bettany, Betteny, Betney, Bettoney:** William *Betany* 1524 SRSf; Gilbert *Beteny* 1598 ER 62; Susan *Beteny* 1734 RamptonPR (C). The forms are late, but perhaps a plant-name from OE *betonice* 'betony'.

**Betteley:** *v.* BETLEY

**Bettenson:** *v.* BETTINSON

**Better:** v. BEATER

**Betteridge:** v. BADRICK

**Betterton:** Ylbert *de Betreton'* 1185 P (Berks); Ilbert *de Betelintun, de Betertun'* 1211 Cur (Berks). From Betterton (Berks).

**Bettesworth, Bettsworth:** Edward *de Bechesworth* 1296, Thomas *Bettisworth*, William *Bettysworth* 1525 SRSx. From Bechworth (Sx).

**Bettinson, Bettenson, Bettison:** Adam *le fit Betun* 1285 FFEss; Roger *Betonessone* 1316 ib.; Roger *Betissone* 1327 SRSf. 'Son of Beton.' v. BEETON.

**Betton:** v. BETHUNE

**Bettoney:** v. BETTANY

**Bettridge:** v. BADRICK

**Bettriss:** v. BEATRICE

**Bettsworth:** v. BETTESWORTH

**Beumant:** v. BEAUMONT

**Beuscher, Beushaw:** v. BELCHER

**Bevan, Bevans, Beven, Bevens, Beavan, Beaven, Beavon, Beevens:** Edenevet *ap Ieuan* 1287 AssCh; Howel *ap Evan* c1300 Bardsley; Thomas *Bevans* 1680 ib. 'Son of *Evan*.' cf. BOWEN.

**Bevar, Bever:** v. BEAVER

**Beveridge, Bavridge:** Wido, William *Beverage* 1212 Cur (Bk), 1230 P (Sr); Richard *Bevereche* c1240 Rams (Hu); William *Bauerich* 1280 AssSo; William *Beuerege* 1297 MinAcctCo; Robert *Beuerich* 1315 FFHu. ME *beuerage*, OFr *bevrege, buverage* 'drink, liquor for consumption' (1275 MED), also used of a drink or beverage which binds a bargain. Bailey in 1721 has '*To pay Beverage*, to give a treat upon the first wearing of a new Suit of Cloths' and Dr Johnson in 1755 '*Beverage*, a treat at first coming into a prison, called also *garnish*'. These must be survivals of a much older custom. cf. 'Bargeyns and beuerages bigonne to aryse' (Langland 1362). At Whitby in 1199 the purchaser of land paid by custom 4*d*. for seisin and 1*d*. to the burgesses *ad beuerage* (ChR i, 14). At a court of the Abbot of Ramsey in 1275, Thomas de Welles complained that Adam Garsoppe unjustly detained a copper he had previously bought from him for 6*d*., of which he had paid Adam 2*d*. 'et beuerech' and a drink in advance. Later he went and offered to pay the rest but Adam refused to take it and kept the copper 'to his damage and dishonour 2*s*.'. Adam was fined 6*d*. and pledged his overcoat (Selden i, 138). The nickname may well have been bestowed on a man who made a practice of getting free drinks for clinching bargains he had no intention of keeping. This custom of *beverage* was an old one on the continent where it was called *vin du marché* (Du Cange). v. BELFRAGE.

**Beverley:** Albert *de Beuerli* 1145–52 YCh; John *de Beverlay* 1204 AssY; John *Beverlee* 1327 *SR* (Ess); William *Beverlay* 1401 IpmY. From Beverley (ERYorks).

**Beverstock:** v. BAVERSTOCK

**Beves:** v. BEAVES

**Bevill, Beville, Beavill:** Robert (*de*) *Beyville, de Beville* c1225 Rams (Hu). From Beuville (Calvados) or Bouville (Seine-Inférieure). v. BOVILL.

**Bevin, Beavin, Bivins:** Simon *Beivin* 1203 FFEss; William *Bevin* 1236 Fees (Do); Nicholas *Boyvin* 1243

AssSo. OFr *bei vin, boi vin* 'drink wine'. cf. DRINKWATER.

**Bevington:** Nicholas *de Beuington'* 1221 AssWa. From Bevington (Wa).

**Bevir:** v. BEAVER

**Bevis:** v. BEAVES

**Bew:** Robert *le Beu, le Bel* 1242 AssSo; John *le Beu* (*Bew*) 1327 *SR* (Ess). OFr *bel, beu* 'beautiful, fair'.

**Bewes, Bews:** John *de Baiocis* 1212 Fees (L); William *Baiues* 1235 Fees (Bk); Henry *de Beause, de Beuse* 1242 ib. From Bayeux (Calvados).

**Bewfort:** v. BEAUFORD

**Bewick, Bewicke, Bowick, Buick:** John *de Bewic'* 1219 AssY. From Bewick (ERYorks, Northumb), the latter being *Bowich* 1167 P.

**Bewlay, Bewley, Beaulieu, Beaulah:** William *de Beulu* 1273 IpmGl; Philip *de Beauleu* 1329 Hylle; Thomas *Bewley* 1545 SRW. From Beaulieu (Ha), Bewley (Du), or Bewley Castle (We).

**Bewshaw, Bewshea, Bewshire:** v. BELCHER

**Beyer:** v. BAYER

**Beynon, Baynham, Benian, Benians, Bennion, Benyon, Binnion, Binyon, Pinion, Pinnion, Pinyon, Pinyoun:** Cadugan *ap Eynon* 1285 Ch (Radnor); Iorworth *ap Egnon* 1287 AssCh; David *Abeinon* 1313 ParlWrits; John *Baynham* 1455 AD i (Wo); John *Beynon* 1507 Oxon; Daniel *Benion* 1610 FrY; George *Benyon* 1621 Bardsley. Welsh *ap Einion, ap Eynon* 'son of *Einion*'. v. ENNION. *Pinnion*, etc., preserve the *p* of Welsh *ap* 'son', *Beynon*, etc., the *b* of later *ab*. In spite of appearance, *Baynham* is not local in origin.

**Bezley:** v. BEESLEY

**Biart:** v. BYARD

**Bibbey, Bibby:** *Biby* de Knolle 1240 AssLa; Dobbe son of *Bibby* ib.; William *Bibbe* 1196 P (Sa); William *Bibbi* 1228 AssSt; Robert *Bybby* 1284 Wak (Y). A diminutive of *Bibb*, a pet-name of *Isabel*.

**Bible:** *Bibele* 1200 Oseney (O); Godwin, Robert *Bibel* Ric 1 Cur (Sf), 1283 SRSf. *Bib-el*, a diminutive of *Bibb* (Isabel).

**Bick:** Geoffrey *Bike* 1210–1 PWi; William *le Bike* 1221 AssWo; James *le Byke* 1327 SRWo; John *Byk* 1359 IpmNt. Probably metonymic for BICKER, OE *bēocere* 'bee-keeper'. cf. ME *bike* 'nest of wasps, wild bees'.

**Bicker, Bickers, Bikker:** (i) Robert *le Biker* 1176 P (St); Reginald *le Bikere* 1207 P (Sf); Jordan *the Bekere* 1286 AssCh; John *Bickers* 1721 FrY. OE *bēocere*, Anglian *bīocere* (ME *bīker*) 'beekeeper'. (ii) Richard *Bicre* 1185 Templars (L); Roger *de Bikere* 1193 P (L). From Bicker (Lincs).

**Bickerdike, Bickerdyke, Biggadike:** Henry *Bikerdik* 1379 PTY; Barnerd *Biggerdicke* 1584 Musters (Sr); Robert *Bickerdike* 1647 FrY. 'Dweller by the disputed ditch', ME *bicker*, OE *dīc*.

**Bickerstaff, Bickerstaffe, Bickersteth, Biggerstaff:** Alan *de Birkestad, de Bikerstath* 1246 AssLa; Henry *Bekerstaff* 1397 IpmNt; Robert *Byggerstafe, Beckerstaffe* 1539, 1557 ArchC xxxiv. From Bickerstaffe (Lancs), *Bickerstath* 1226.

**Bickerton:** Adam *de Bikerton* 1191 P (Y); William *de Bykerton* 1287 IpmY; William *Bykerton* 1504–5 FFWa. From Bickerton (Ch, He, Nb, WRY).

**Bickford:** Henry *Bickford*, Richard *Bicford* 1642

PrD. From Bickford (St), Bickford Town in Plympton St Mary (D), or Beckford Fm in Membury (D), *Bykeford* 1408.

**Bickley:** William *de Bikeleia* 1187 P (Do/So); Huward *de Bikeleg'* 1227–8 FFK; William *de Bickele* 1310 LLB D. From Bickley (Ch, K, Wo), or Bickleigh (K).

**Bickman:** William *Bykeman* 1301 SRY. 'Bee-keeper', ME *bike* 'nest of wasps, wild bees', OE *mann*.

**Bicknell, Bignell, Bignall, Bignold:** Thomas *de Bikenhulle* 1214 Cur (Wa); John *de Bikenhull*, Walter *Bykenhulle* 1327 SRSo; Nicholas *Bicknell* 1642 PrD. From Bickenhall (Som), or Bickenhill (Warwicks). *v.* also BIGNAL.

**Bidaway:** *v.* BYTHEWAY

**Biddell, Biddle:** *v.* BEADEL

**Bidden, Biddon:** Walter *de Bydun* c1143–7 Black; Trihon *de Bidon* 1177 P (Bk); William *Bidun* 1279 RH (Beds). 'Dweller by the down', OE *bī, dūn*.

**Biddick:** Adinet *de Bidyk* 1276 Fine; William *de Bydik* 1305 AssNu; John *Bidyk* 1332 SRDo. From Biddick (Du), or 'dweller by the ditch', OE *bī, dīc*.

**Biddlecombe:** Galopin *de Bitelescumbe* 1180 P (So); Richard *Biddelcome* 1576 SRW. From Bittiscombe (So).

**Biddolph, Biddulph:** Thomas *de Bidulf* 1199 AssSt; Roger *de Bydoulf* 1285 FA (St); John *de Bydulfe* 1332–3 SRSt. From Biddulph (Staffs).

**Biddon:** *v.* BIDDEN

**Bidgood:** David *Bydgood* 1524 SRD; John, Richard *Bidgood* 1642 PrD. A nickname, 'pray God', OE *biddan, God*.

**Bidlake:** John *Bidlake* or *Bithelake* 1509 LP (D). From Bidlake in Bridestow (D).

**Biddwell:** *v.* BEDWELL

**Bielby, Beilby:** Walter, John *de Beleby* 1202 AssL, 1372 FFY; Sicillia *Beilby* 1464 FrY. From Beelby (Lincs), or Bielby (ERYorks).

**Bield:** *v.* BEILD, BELD

**Bier(s):** *v.* BYARS

**Biever:** *v.* BEAVER

**Bifield:** *v.* BYFIELD

**Bigby:** Thomas *de Bekebi* 1219, John *de Bekeby* 1298 AssL; Beatrix *Bigby* 1379 PTY. From Bigby (L), *Bekebi* 1212.

**Bigg, Bigge, Biggs:** Ægelric *Bigga* c1036 OEByn (K); Walter, Henry *Bigge* 1177, 1195 P (Sf, Gl). ME *bigge* 'large, strong, stout' (c1300 MED). A topographical origin is also possible, though the meaning is obscure: William *de Bigges* 1327 SRC; Laurentia *atte Bigge* 1327 SRSo.

**Biggadike:** *v.* BICKERDIKE

**Biggar:** Baldwin *de Bigre* p1170 Black (Lanark); Henry *de Bygar* 1330 ib. From Biggar (Lanarkshire).

**Bigger, Biggers:** John *le Biggere* 1307 MEOT (Herts); Richard *Bygor* 1321 FFEss. A derivative of OE *bycgan* 'to buy', a buyer, purchaser. Hollinshed's 'He came here as a bier, not as a begger' implies a play on the pronunciation *bigger* and *begger*.

**Biggerstaff:** *v.* BICKERSTAFF

**Biggin, Biggins:** Thomas *del Biggyng* 1391 FrY; William *atte Byggyngge* 1397 PN C 191. ME *bigging*

'dwelling-place, home', used also of an outbuilding as distinct from a house.

**Bigland, Biglen, Biglin:** William *Bigland* 1672 HTY; Robert *Bigland* 1716 FrY. From Biglands (Cu), or Bigland (La).

**Bignal, Bignall, Bignell, Bignold:** William *de Bigenhull* 1279 RH (O); William *Bignolle*, John *Bygnold* 1525 SRSx; Thomas *Bignell* 1727 Bardsley; Robert *Bignall* 1758 FrY. From Bignell in Bicester (Oxon), *Bigenhull* 1220, *Bikenhulle* 1285. Without early forms it is impossible to derive *Bignal* and *Bicknell* with any certainty. *v.* also BICKNELL.

**Bigott, Bygott:** *Bigot* de Loges 1086 DB; Thomas *filius Bigot* a1187 DC (L); Roger cognomentus *Bigot* 1076–9 EngFeud; William *le Bigot* 1214 P (Sf); Richard *Bygot* 1249 AssW; Alice *Bigod* 1332 SRSx; Thomas *Bygood* 1392 LoCh; Thomas *Bygod* 1561 Pat (So). According to Dauzat, a derogatory name applied by the French to the Normans, probably representing the English oath 'by God'.

**Bigrave:** *v.* BYGRAVE

**Bikker:** *v.* BICKER

**Bilborough, Bilbrough:** Serlo *de Billeburg'* 1219 AssY; Richard *de Bilburgh'* 1308 FFY; Matthew *Bilbroughe* 1621 SRY. From Bilborough (Nt), or Bilbrough (WRY).

**Bilby, Bilbie, Billby:** Ralph *de Billebi* 1198 Cur; Robert *de Bilby* 1379 PTY; George *Bylby* 1541 CorNt. From Bilby (Nt).

**Bilcliff, Bilcliffe:** *v.* BILLCLIFF

**Biles, Byles:** Walter, Nicholas *Bile* 1185 Templars (Gl), 1176 P (St); Robert *atte Byle*, Walter *atte Bille* 1275, 1327 SRWo. OE *bile* 'bill, beak of a bird' used both as a nickname (cf. BECK) and as a topographical term 'dweller at the beak-like projection, promontory, hill'. *v.* MELS 13.

**Bilham, Billham, Billam:** Hugh *de Bilham* 1204 AssY; William *de Bilham* 1306 FFY; John *de Bilham* 1379 PTY. From Bilham (WRY).

**Bilicliffe:** *v.* BILLCLIFF

**Bill, Billes, Bills:** Willelmus *filius Bille* 1301 SRY; Griffin, William *Bil* 1188, 1194 P (Wa, Nf); Sewinus *Bille* 1221 AssWa. OE *Bil*, from OE *bil(l)* 'sword', or, possibly, ON *Bildr*, but more commonly metonymic for BILLER.

**Billam:** *v.* BILHAM

**Billaney, Billany:** *v.* BILNEY

**Billby:** *v.* BILBY

**Billcliff, Bilcliff, Billcliffe, Bilicliffe, Bilitcliffe:** Thomas *de Billeclyf* 1308 Wak (Y); John *Bilclif* 1617 FrY; James, Hannah *Bilcliffe, Bincliffe, Bintley* 1788–93 WRS. From Bilcliffe (WRY).

**Biller:** John *Billere* 1275 RH (Sf); Henry *le Billere* 1279 RH (C). A derivative of OE *bil* 'a bill', a maker of bills (halberds or billhooks). cf. Thomas *Bilhouk* 1327 SRSo.

**Billericay:** John *Billerica* 1300 CorLo. From Billericay (Ess), or a lost *Billerica* in Lympne (K).

**Billet, Billett, Billot, Billette:** John *Bilet* 1279 AssNb; Thomas *Bilett'* 1361 AssY; William *Billet* 1642 PrD. *Bill-et*, either a diminutive of OE *Bil*, or of ME *bille* 'piece of wood', and so a nickname for a woodcutter. Sometimes, perhaps, from OE *bile-hwīt* 'pure, innocent': Edwin, John *Bilewit* 1198 FFMx.

**Billham:** *v.* BILHAM

**Billing, Billings, Billins:** (i) Osebertus *Billing* 1188 BuryS (Sf); Geoffrey *Billyng* Hy 3 Gilb (L); Thomas *Billinge* 1282 Oseney (O). Probably OE *Billing*, though rare in OE and not noted in independent use after the Conquest. (ii) Osbertus *de Parua Billing'* 1201 P (Nth). From Billing (Northants). He is probably not identical with the Suffolk sokeman above.

**Billinge:** Mary *de Billinge* Edw 1 Bardsley (La). From Billinge (Lancs).

**Billingford:** John *Byllyngford* 1460, *Bylyngforthe* 1462 Paston; Edmund *Billingford* 1557 NorwDep. From Billingford (Nf).

**Billingham, Billinghame:** John *de Billingham* 1327 SRY; William *de Bilyngham* 1349 FrY; Edward *Billingham* 1672 HTY. From Billingham (Du), or Billingham in Udimore (Sx).

**Billington:** Robert *de Billington* 1203 AssSt; Adam *de Bilyngton* 13th WhC; Thomas *Billyngton* 1483 FFSr. From Billington (Beds, La, St).

**Billiter:** William *le Belyotar'* 1247 Oseney (O); Robert *le Bellegeter* 1283 FrY; Alexander *le Belleyeter* 1377 AD vi (Ch); John *Bellitour* 1534 LP. OE *belle* and *gēotere* 'bell-founder' (1440 NED). The corresponding French term survives as SENTER which Stahlschmidt confuses with *ceinturier* 'girdler'. Salzman, however, notes a 13th-century Worcester family called indifferently *Ceynturer* and *Belleyeter*. 'The demand for bells could hardly have been large enough to enable a craftsman to specialize entirely in that branch; a bell-maker would always have been primarily a founder, and according as the main portion of his trade lay in casting buckles and other fittings for belts, or pots, or bells, he would be known as a girdler, a potter, or a bell-founder.' Most of the known London bell-founders used the title 'potter'. Ekwall notes that Ædmund *Seintier* 1168 (ELPN) is called a moneyer. Most moneyers were goldsmiths, but occasionally other metal-workers had a die in the mint, and a bell-founder may have acted as a moneyer. Several bells were cast for Westminster Abbey by Edward FitzOdo, the famous goldsmith of Henry III. William *Founder* cast both bells and cannon. His trade stamp, bearing his name and a representation of two birds and a conventionalized tree, appears on a number of bells and hints at his real surname—clearly *Woodward*. In two successive entries in 1385 he is called William *the founder* and William *Wodeward* and in 1417 cannon were supplied by William *Wodeward, founder*. At Exeter c1285, Bishop Peter de Quivil assured the proper care of the bells of the cathedral by granting a small property in Paignton to Robert *le Bellyetere* as a retaining fee, Robert and his heirs being bound to make or repair, when necessary, the bells, organ and clock of the cathedral, the chapter paying all expenses, including the food and drink of the workmen, and these obligations were duly fulfilled for at least three generations. In 1454 a Norwich bell-founder was called Richard *Brasier*. *v.* MedInd 145–54.

**Billot:** *v.* BILLET

**Billows:** *v.* BELLOW

**Bilney, Billaney, Billany:** Robert *de Bilneie* 1202–3 FFWa; Cristiana *de Bilneye* 1288–9 NorwLt; Reginald *de Bilneye* 1327 SRSf. From East, West Bilney (Nf).

**Bilsby:** Richard *de Bilesbi* c1155 DC (L); Henry *de Bilesbi* Hy 2 RegAntiquiss; Ralph *de Bilesbi* 1202 AssL. From Bilsby (L).

**Bilsdon:** William *Billesdon* 1415–6 FFWa; Nichol *Billesdon* 1426 Paston; Robert *Billesdon* 1473–4 FFSr. From Billesdon (Lei).

**Biltcliffe:** *v.* BILLCLIFF

**Bilton:** Robert *de Bileton'* 1190 P (Y); John *de Bilton* 1297 SRY; John *Bilton* 1340–1450 GildC. From Bilton (Northumb, Warwicks, ER, NRYorks).

**Binder:** Hugh *le binder* 1219 AssY; Richard *le Byndere* 1278 LLB B. OE *bindere* 'binder', probably of books. cf. William *le Bokbynder'* 1323 MESO (Ha).

**Bindloes, Bindloss, Bindless, Binless:** Alan *Byndlowes* 1301 SRY; John *Byndeloue* 1327 SRSf; Robert *Byndlowys* 1379 PTY; Christopher *Byndelase* 1461 PN Ess 641, *Bindlos* 1582 Oxon. 'Bind wolves', a hybrid from OE *bindan* and OFr *lou*. Probably a wolf-trapper. cf. TRUSLOVE.

**Bindon:** Robert *Bindon* 1384 IpmNt. From Bindon (Do). A member of the family settled in Ireland, and the name is particularly common in co. Clare.

**Binfield:** Reginald *de Benetfeld'* 1230 P (Berks). From Binfield (Berks).

**Binford:** John *Bynford* 1524 SRD, 1642 PrD. From Binneford in Crediton, in Sandford (D).

**Bing, Byng:** Robert *Bing* 1274 RH (D); John *Byng* 1317 AssK; Roger *Bynge* 1384 LLB F. From Byng (Sf), or 'dweller in the hollow', OE *\*bing*, or 'dweller by the rubbish heap or slag heap', ON *bingr*, cf. *Bynge*, theca, cumera, c1444 PromptParv.

**Bingham:** William *de Bingeham* 1175 P (Nt, Db); William *de Bingham* 1257 FFL; William *Byngham* 1433 AssLo. From Bingham (Notts).

**Bingley:** Aliz *de Bingeleia* 1185 Templars; William *de Byngeleye* 1339 CorLo; Richard *Bingley* 1541 CorNt. From Bingley (WRYorks).

**Binham:** Robert *de Binham* 1200 P (Nf). From Binham (Nf).

**Binks:** Robert *de Benkys*, Thomas *del Binkys* 1297, 1301 SRY; Simon *at the benk* 1327 Pinchbeck (Sf). 'Dweller by the banks.' ME *benk*, a northern form of *bench*, now *bink*. cf. BENCH.

**Binless:** *v.* BINDLOES

**Binley:** Ralph *de Bineleg'* 1224–5 FFEss; Isabella *de Bynnelegh* 1330 PN D 325; Richard *Bynleye* 1340–1450 GildC. From Binley (Wa), or Beenleigh in Harberton (D).

**Binne, Binnes, Binns:** Robert *Binns* 1275 RH (L); Missa *Binne* 1279 RH (O); Robert *Byn* 1327 SRSx. OE *binn* 'manger, bin', metonymic for BINNER. Also used topographically: William *de Bynns* 1279 AssSo. 'Dweller by the hollows.'

**Binner:** Walter *le Bynere* 1296 SRSx; Thomas *Binere* 1373 ColchCt. 'Maker of bins' (for corn, meal, bread, etc.). *v.* BINNE.

**Binney, Binnie, Binny:** (i) Ricardus *filius Bini* 1220

Cur (L); Robert, William *Bynny* 1297 SRY, 1379 PTY. From a personal name, probably OE *Bynni*. (ii) Robert *de Binay* 1210 P (C); Ralph *de Bynne* 1317 AssK. 'Dweller on land enclosed by a stream', OE *binnan ēa* 'within the stream', as at Binney Fm in Hoo All Hallows or Binny Cottages in Tonge (Kent). (iii) William *de Binin* 1243 Black; Simon *de Bynninge, de Beny* 1396, 1399 ib.; John *Binnie* 1574 ib. From Binney in Uphall (West Lothian). The older form BINNING also survives.

**Binning:** *v.* BINNEY
**Binnion:** *v.* BEYNON
**Binskin:** *v.* BENSKIN
**Binstead, Binsted:** *v.* BENSTEAD
**Bintcliffe, Bintliff:** *v.* BENTLIFF
**Bintley:** *v.* BENTLEY
**Bintworth:** *v.* BENTWORTH

**Binwood:** Robert *Binnewde* 1199 Pleas (Db/Nt); Robert *Bynwod* 1363 FFEss. 'Dweller within the wood', OE *binnan, wudu*.

**Biram:** *v.* BYROM
**Birbeck:** *v.* BIRKBECK

**Birch, Burch, Byrch:** Walter *de la Birche* c1182 MELS (Wo); Richard *de Birches* 1246 AssLa; Ralph *atte Birche*, Richard *del Birche* 1275 SRWo; Robert *Birch* 1275 RH (Sf); William *de la Burch* 1275 MELS (So); John *Burch* 1309 RamsCt (Sf); William *in le Byrchez* 1332 SRSt. 'Dweller by the birch(es)', OE *birce, byrce*.

**Birchall, Birchell:** Richard *de Byrchulle* 1293 AssSt. From Birchill (Db), or Birchills (St).

**Bircher, Burcher:** Geoffrey *de Byrchover*, Richard *de Birchowe* 1327 SRDb; William *de Birchovere* 1331 Shef; Jeremy *Birtcher* 1663 HeMil. From Birchover (Db), or Bircher (He).

**Birchett, Burchett:** (i) Ralph *atte Birchett* c1280 PN Sr 139; William *atte Burchett'* 1296 SRSx; Adam *Byrchet* ib. 'Dweller by the birch-grove', OE *\*birchett, \*byrcett*, very common in minor names in Sussex and found also in Kent, Surrey and Essex. (ii) Henry *Burrcheued* 1204 P (L); Thomas *de Bircheued* 1327 SRDb (Norton); Robert *Birchehed* 1447 Shef (Y); Catharine *Birchett* of Birchett Hall (Norton) 1622 Fanshawe. 'Dweller by the birch-covered headland', OE *birce* and *hēafod*.

**Birchley:** William *de Byrchleye* 1332 SRWo; John *Bircheleye* 1361, Thomas *Byrchelegh* 1395 FFEss. From Birchleys in Pebmarsh (Ess), or Birchley Fm in Bockleton (Wo).

**Birchwood:** Alan *de Birchwude* 1177 P (Y); Eustace *de Birchwde* 1204 Cur; Saier *atte Birchwode* 1342 PN Ess 223. 'Dweller by the birchwood', OE *birce, wudu*.

**Bircumshaw:** *v.* BIRKENSHAW

**Bird, Birds, Byrd, Byrde, Bride:** Ernald, William *Brid* 1193 P (Y), 1221 *ElyA* (Sf); Ralph, Robert *le Brid(d)* 1235 FFEss, 1243 AssSo; Richard *Bird* 1260 AssC; John *Bride* 1332 SRCu; Richard *Bride* alias *Birde* 1568 SRSf. OE *bridd* 'bird', a nickname. Sometimes, no doubt, metonymic for *birdclever*: Robert *Birdclever* 1427 Calv (Y), William *Burdclever* 1495 FrY, 'bird-catcher'.

**Birdsall, Birdsell:** William *de Briddesale* 1167 Kirkstall; Robert *de Brideshal* 1251 AssY; Thomas

*Birdsall* 1446 FrY. From Birdsall (ERY).

**Birdwood:** Henry *de Bridewode* 1306 AssSt; Richard *Burdwood* 1642 PrD. 'Dweller by the wood frequented by birds', OE *bridda, wudu*.

**Birel:** *v.* BIRREL

**Birkbeck, Birbeck:** George *Birkebek* 1464, Robert *Birtbek* 1471 FrY. From Birkbeck Fells in Orton (We).

**Birkby:** William *de Bretteby* 1219 AssY; William *Birtby* 1462 FrY; Robert *Birtbye* 1540 RothwellPR (Y). From Birkby (Cu, NRY, WRY).

**Birkenshaw, Birkinshaw, Bircumshaw, Birtenshaw, Berkenshaw, Bertenshaw, Burkenshaw, Burkinshaw, Burkinshear, Burkimsher, Burtinshaw, Burtonshaw, Buttanshaw, Buttenshaw, Buttonshaw, Brigenshaw, Briggenshaw, Briginshaw, Brigginshaw, Brockenshaw, Brokenshaw, Brokenshire, Bruckshaw:** William *del Birkenschawe* 1274 Wak (Y); Roger *Birchynshawe* 1408 LLB I; Richard *Brekynshawe, Burtenshaw* 1500, 1637 PN Sx 314; Leonard *Byrkenshay, Byrtynschaw, Byrkynshay* 1542–58 RothwellPR (Y); Thomas *Birkenshire* 1739 FrY. From Birkenshaw (WRYorks).

**Birkett, Birkhead, Brickett:** John *de Birkhaved* 1301 SRY; John *Birkehede* 1442 FrY; Henry *Brikket, Byrkett* 1524 SRSf. 'Dweller by the birch-covered headland', OE *bi(e)rce, hēafod*, surviving as Birkett in late minor names in Cumberland and Lancashire.

**Birkin:** John *de Birekin* 1199 P (Y); John *de Byrkyn* a1228 YCh; John *Birkyne* 1379 PTY. From Birkin (WRY).

**Birks, Burks:** Richard *del Birkes* 1275 Wak (Y). 'Dweller by the birches', from the northern form of OE *birce*.

**Birley, Byerley:** Simon *de Byrley* 1169 Templars (L); Robert *de Birle* a1260 Calv (Y); John *Byrley* 1407 IpmY. From Birley (Db, He), or East, North Bierley (WRY).

**Birmingham, Bermingham, Burmingham:** Peter *de Bremingeham* 1170 P (St); Gilbert *de Birmingeham* 1271–2 FFL; John *de Burmyngham* 1333 KB (Wa); John *Bermyncham* 1340–1450 GildC. From Birmingham (Wa).

**Birn:** *v.* BYRNE
**Biron:** *v.* BYRON

**Birrel, Birrell, Birel:** William *Birril* 1253 IpmGl; John *Byrill'* 1379 PTY; Andrew *Birrell* or *Burrell* 1540 Black. Probably a variant of BURREL.

**Birt:** *v.* BURT
**Birtenshaw:** *v.* BIRKENSHAW

**Birtles:** Robert *Byrtylles* 1537 CorNt; John *Birtles* 1624 PN Ch i 72; Thomas *Birtles* 1672 NorwDep. From Birtle (La), or Birtles (Ch).

**Birtwhistle, Birtwisle, Bertwistle, Burtwistle:** John *de Briddestwysill* 1285 AssLa; Adam *de Briddestwyssle* 1329 Kirkstall (Y); John *Brittwissill* 1397 PrGR; Thomas *Birtwisill* 1460 FrY; Thomas *Burtwisle* 1618 RothwellPR (Y). From a lost *Birtwisle* in Padiham (Lancs), Briestwistle in Thornhill (WRYorks), *Brerethwisel* 1243 PN WRY ii, 211, or a lost *Breretwisel* in Wath-on-Dearne (1253 ib. i, 120).

**Bisacre, Bisiker, Bisseker:** Ralph *de Beseacra* 1182 P (Y); Robert *Bysacle* 1382 FrY. From Bessacar (WRYorks).

**Bisbrown:** v. BAISBROWN

**Biscoe:** William *Birscowe* 1441, Robert *Biscowe* 1463 FrY. Probably from Burscough (La).

**Bisgood:** v. PEASCOD

**Bish, Bysh:** John *de Labisse* 12th MELS (Sr); John *Bische* 1316 FA (Sx); Ralph *ate Byshe* 1327 SRSx. 'Dweller by a thicket', from OE *(ge)bysce*, surviving in Bish Wood (Sussex) and Bysshe Court (Surrey).

**Bishell:** v. BUSHELL

**Bishop, Bishopp, Bisshopp:** *Biscop* 1066 DB (Nth); *Bissop* 1195 P (Nf); *Bissop* atte Combe 1327 SRSo; Algar *se Bisceop* c1100–30 OEByn (D); Lefwinus *Bissop* 1166 P (Nt); Thurstan *le Byssop* 1240 FFEss; Thomas *le Byscop* 1297 MinAcctCo. OE *Bisc(e)op*, or a nickname for one with the appearance or bearing of a bishop, or a pageant-name from the custom of electing a boy-bishop on St Nicholas's Day.

**Bishton:** William *de Bissopeston'* 1199 MemR (Wo); Frarin *de Bissopeston'* 1221 AssGl; Matilda *de Bissopestun'* 1227 AssSt. From Bishton (Gl, Monmouth, Sa, St), Bishopstone (Bk, He, W), or Bishopton (Du, Wa, WRY).

**Bisiker:** v. BISACRE

**Bisler:** v. BUSHELL

**Bisley:** Leofric *de Biselai* 1180 P (Nf/Sf); William *de Bysleg'* 1252–3 FFEss; Thomas *Byseleye* 1372 IpmGl. From Bisley (Gl, Sr).

**Bismire:** William *Bysmere* 1407 LLB I, *Bysmare*, 1412 ib. OE *bīsmer*, *bīsmor*. Originally 'shame, disgrace', it developed the sense of 'a person worthy of scorn; a lewd person, a pander or bawd'.

**Biss:** Stephen, William *Bys* 1327 SRC, SRSf. OFr, ME *bis* 'brownish or dark grey', of complexion or dress. cf. BISSETT.

**Bisseker:** v. BISACRE

**Bissell, Bissill:** v. BUSHELL

**Bissett:** Manasser, Ernulf *Biset* 1155–8 Bury (Sf), 1176 P (Bk). OFr *biset* 'dark', a diminutive of *bis*. cf. BISS.

**Bissey, Bissy:** Thomas *Byssye* 1576 SRW; Andrew *Bissie* 1641 PrSo. A nickname from OE *bysig* 'busy'.

**Bisshopp:** v. BISHOP

**Bithell, Abethell:** Lewlyn *ap Ithel* 1325 ParlWrits (Radnor), 'Son of *Ithel*'. v. ITHELL, BOWEN.

**Bittleston, Bittlestone:** Robert *de Bitlesden'* 1220 Cur (Berks). From Biddlesden (Bucks).

**Bitton:** Baldewin *de Bitton* 1275 RH (K); Robert *de Byttone* 1371 LLB G. From Bitton (Gl).

**Bivins:** v. BEVIN.

**Bixley:** Peter *de Bixle* 1206 Cur (O); Robert *de Bixle* 1327 SRSx. From Bixley (Nf), or Bixley Heath (Sf).

**Blaber:** Walter *Blaber* 1230 P (Y); John *Blabour* 1379 PTY; David *Blabir* 1408 Black. A derivative of ME *blabben* 'to tell secrets', a gossip.

**Blabey, Blaby:** Stephen *de Blaby* 1219 P (Y); Richard *de blaby* 1327 SRLei; John *Blaby* 1508–9 FFWa. From Blaby (Lei).

**Black, Blacke:** Wulfhun *pes Blaca* 901 OEByn (So); Wulfsie *se blaca* 964 ib. (K); Willelmus *Blac*, *Niger* 1086 DB (Herts, D); Godebertus *leblac* 1130 P (Caermarthen); Thomas *Blac* 1198 Cur (Nf);

Edericke *le Blacke* 1275 RH (L). OE *blæc* 'black', dark-complexioned. Wlfricus Niger (c1080 OEByn) is said to have received this nickname because he once went unrecognized among his foes as he had blackened his face with charcoal. The inflected form OE *blaca* became ME *blāke* which was often confused with ME *blāk(e)* from OE *blāc* 'bright, shining; pale, wan', so that the exact meaning of *Blake* is doubtful. *Black-* and *Blake-* frequently interchange in place-names and other surnames. v. BLATCH.

**Blackadder, Blacketer, Blaketter:** Adam *of Blacathathir* 1477 Black; Robert *Blackader* 15th ib.; Charles *Blakater* 1486 ib. From Blackadder (Berwickshire).

**Blackah:** Alice *Blaca* 1475 GildY. From Blacker (WRYorks).

**Blackale, Blackall:** v. BLACKHALL

**Blackaller, Blackler:** John *Blakaller* 1431 AD i (D); John *Blackaller* alias *Blacklawe* 1693 DWills; Agnes *Blacklar* 1715 ib. From Blackler (Devon) where the surname occurs as *Blakalre* in 1333 (PN D 521).

**Blackbird:** Brunstanus *Blachebiert* 1066 Winton (Ha); William *Blacberd* 1206 AssL; Thomas *Blakeberd* 1275 SRWo; William *Blakebird* 1279 AssSo. OE *blæc, beard* 'black beard'. cf. William *Brounebeard* 1379 PTY.

**Blackborn, Blackborne, Blackbourn, Blackbourne, Blackburn, Blackburne:** Henry *de Blakeburn'* 1206 Cur (La). From Blackburn (Lancs).

**Blackboro, Blackborough, Blackborow, Blackbrough, Blagbrough, Blakeborough, Blakebrough:** Robert *de Blakeberg'* 1201 P (Gl); Simon *de Blakeberwe* 1275 RH (Sf). From Blackborough (Devon, Norfolk).

**Blackbrook, Blackbrooke:** Gilbert *de Blakebrok* 1279 PN Sx 505; Henry *Blakebroke* 1392 CtH. 'Dweller by the black stream', OE *blæc, brōc*.

**Blackcliff, Blackcliffe, Blackliffe:** Robert *de Blaclif* 1219 P (Y); John *de Blakeclif* 1289 PN Nt 245. From Blackcliffe Hill in Bradmore (Nt), or 'dweller by the black hill', OE *blæc, clif*.

**Blackden, Blagden, Blagdon:** Roger *de Blakeden'* 1275 SRWo; John *Blakedowne* 1327 ib.; Walter *de Blakedon* 1327 SRSo; Sarah *Blacdon, Blagden* 1688–9 Bardsley. 'Dweller in the dark valley' as at Blackden (Ches) or at Blagdon (Northumb), or by the black hill as at Blagdon (Devon, Som) or at Blagden Fm in Hempstead (Essex).

**Blackell:** Gilbert *ate Blakehulle* 1327 SRSx; Robert *de Blakhill* 1347 FrY; Thomas *Blakell* 1456 ib. 'Dweller by the black hill.'

**Blacker, Blaker:** *Blakere* 1047–64 Holme (Nf); Roger *Blacker* 1246 AssLa; Ralph *le blaker'* 1291 MESO (Ess); William *Blaker* 1296 SRSx; Roger *le Blackere* 1312 ParlWrits. (i) OE *Blæchere* 'black-army'; (ii) A derivative of ME *blāken*, OE *blǣcan* 'to bleach', bleacher, cf. BLATCHER. Fransson explains this as 'one who blacks', from *blæc*, but does not specify the occupation.

**Blackers:** Walter *Blachers* 1189 P (Co). A nickname, 'black arse', OE *blæc, ears*. cf. Ralph *Withars* 1173–6 GlCh 'white arse'; Godwin *Bredhers* 1137 ELPN, 'broad arse'.

**Blacketer:** v. BLACKADDER

**Blackett:** (i) Thomas *Blakeheuede* 1301 SRY; Adam *Blakhed* 1332 SRLa; 'Black head' or 'fair head'. cf. BLACK. (ii) Ralph *Blachod* 1327 SRSf; Robert *Blakhod* 1327 Pinchbeck (Sf). 'Black hood'. (iii) Ralph, Robert *Blachet* 1208 Cur (C), 1274 RH (So); William *Blaket* 1275 RH (Herts), 1332 SRCu. This, the most common form, cannot be from *Blakeheved*, whether as a nickname or a place-name. It must be a diminutive of *Black*, with the French suffix *-et*.

**Blackford:** Robert *de Blakeford'* 1211 Cur (Ha); Roger *de Blakeforde* 1296 SRSx; Richard *atte Blakeforde* 1314 MELS (Wo). 'Dweller by the black ford', as at Blackford (Som).

**Blackhall, Blackale, Blackall:** Robert *de Blakehall* 1221 AssWo; Ralph *de Blackhale* 1332 SRCu. 'Dweller by the dark nook', ME *hale*, with the common confusion with *hall*.

**Blackham:** Benedictus *de Blakeham* 1135–48 Bury; Benedict *de Blakeham* 1212 P (Nf/Sf). From Blakenham (Sf).

**Blackhurst:** Robert *atte Blakhurst, de Blakehurst* 1296 SRSx. 'Dweller by the black wooded hill', OE *hyrst*.

**Blackie:** (i) Henry *Blackeye* 1275 RH (Nf); Roger *Blakheye* 1327 SRSf. A nickname, 'black eye', unless these are from place-names where the preposition has been lost. If so, 'dweller by the black low-lying land or enclosure', OE *ēg* or *(ge)hæg*. (ii) John *Blakye* 1506 Black. A Scottish diminutive of BLACK.

**Blackiston:** v. BLACKSTON

**Blackledge, Blacklidge:** John *del Blakelache* 1332 SRLa; Richard *Blacklach* 1473 DbAS 30; Evan *Blaklidge* 1662 PrGR. 'Dweller by the dark stream', OE *lœcc* 'a stream flowing through boggy land'.

**Blackler:** v. BLACKALLER

**Blackley, Blacklee, Blaikley, Blakeley, Blakely, Blakley:** William *de la Blekelegh* 1301 ParlWrits (St); Robert *atte Blakeley* 1337 AssSt; John *Blaklay* 1543 FrY; Mungo *Blaikley* 1687 Black. 'Dweller by the black wood or clearing', as at Blackley (Lancs), pronounced *Blakeley*.

**Blacklock, Blaiklock, Blakelock:** Peter *Blacloke* 1275 RH (W); Adam *Blakelok* 1332 SRCu; Robert *Blaykelok* 1431 FrY. Though *Blayke-* might mean either 'black' or 'fair' (cf. BLACK), all are probably for 'black lock', OE *blœc, locc*, the man with black hair, as distinct from WHITELOCK.

**Blackman, Blakeman:** *Blacheman* filius Ædwardi 1166 P (Nf); Jordanus *filius Blakeman* 1188 P (Ha); John *Blakeman* 1206 P (Sr); Henry *Blacman* 1279 RH (O). OE *Blæcmann* 'dark man', a personal name fairly common until the 13th century.

**Blackmer, Blackmere, Blakemere:** (i) *Blachemer* 1066 DB (Sa). OE *Blæcmǣr*. (ii) William *de Blakemere* 1275 SRWo; Kateryna *de Blakemere* 1296 PN Herts 16. From Blakemere (He), Blackmore End in Kimpton (Herts), or 'dweller by the dark mere', OE *blœc, mere*.

**Blackmore, Blakemore:** (i) Baldewin *de Blakomor* 1200 P (D); Nicholas *de Blakemore* 1307 AssSt; John *Blakemore* 1547 CorNt; Henry *Blackmore* 1576 SRW.

From Blackmoor (D, Do, Ha), Blackmore (Ess, Herts, W, Wo), or Blakemoor (D). (ii) William *le Blacomer* 1375 NorwLt; John *Blakomor* 1379 PTY; John *Blackamore* 1556 CorNt. 'Black as a Moor, dark-complexioned', OE *blæc*, ME *Mor* 'a Moor'.

**Blackoe, Blacoe, Blacow:** John *Blakow* 1542 PrGR; John *Blacoll* 1562 ib. From Blackhall (Lancs), pronounced *Blackow*.

**Blackstock:** William *de la Blakestok* 1296 SRSx; Adam *Blakstok* 1332 SRCu. 'Dweller by the black stock or stump', OE *stocc*.

**Blackston, Blackstone, Blackiston, Blakeston, Blakiston, Blaxton:** (i) *Blackstan* 1086 DB (Ess); William *Blacston', Blakeston', Blackstan* 1235–42 Fees (Bk). OE *Blæcstān* 'black stone'. (ii) Philip *Atteblakeston'* 1275 SRWo; William *de Blakstan* 1316 FFK. 'Dweller by the black stone' as at Blackstone Edge (Lancs) or Blaxton (WRYorks).

**Blackstrode:** Simon *atte Blakestrode* 1296 MELS (Sx). 'Dweller by the black marshy land', OE *blæc, strōd*.

**Blackthorn, Blackthorne:** John *de Blakethorn* 1276 AssLo; John *Blakethorn* 1379 PTY; William *Blakthorn* 1442 FFEss. 'Dweller by the blackthorn', OE *blæc, þorn*.

**Blacktin:** About two generations ago, two branches of a family of Greystones (Sheffield) adopted different forms of their surname, originally *Blackden*: (i) *Blagden*, (ii) *Blacktin*.

**Blacktoft, Blacktop:** Richard *de Blaktoft* 1324 FrY. From Blacktoft (ERY).

**Blackwall:** For BLACKWELL. v. WALL.

**Blackwell:** Leofric *æt Blacewellan* 1012 OEByn (Wo); Mauricius *de Blacwella* 1175 P (Db); Benedictus *de Blakewelle* 1243 AssDu; Robert *atte Blakewell* 1296 SRSx. From Blackwell (Derby, Durham, Worcs) or from residence near a dark well or stream.

**Blackwin, Blackwyn:** *Blacchewynus* monachus c1125 Bury; *Blakewinus* de Thornham 1198 FFNf; Henry *Blacwin'* 1199 P (Wo); Roger *Blakwyne* 1327 SREss. From an unrecorded OE *\*Blæcwine*.

**Blackwood:** William *de Blacwode* 1327 SRSt; Robert *Blakwode* 1384 Black. 'Dweller by the dark wood', as at Blackwood (ERYorks, WRYorks, Dumfries, Lanarks).

**Bladen, Bladon:** Hugh *de Bladene* 1279 RH (O). From Bladon (O), or Blaydon (Du).

**Blader:** Andrew *le Blader* 1305 LoCt. A derivative of OE *blæd* 'blade', a maker of blades, a bladesmith. cf. Nicholas *Bladsmith* 1357 FrY.

**Blades, Blaydes:** Jacke *Blade* 1297 Wak (Y); Robert *Blades* 1460 FrY; James *Blaydes* 1506 ib. Metonymic for *Blader* or *Bladesmith*. Also topographical in origin, from some unidentified place: Alan, Hugh *de Bladis* 1230 P (Lei), 1258 AssSt; William *de Blades* 1301 SRY.

**Blagbrough:** v. BLACKBORO

**Blagden, Blagdon:** v. BLACKDEN

**Blagge, Blagges:** Thomas *Blagge* 1286 AssCh; Robert *Blage* 1500 FFEss; Thomas *Blages* 1642 PrD. Possibly, as suggested by Harrison, a voiced form of BLACK. cf. Blagdon (D, Do, So), all meaning 'black hill'.

**Blagrave, Blagrove, Blagroves:** Alice *de Blacgrava* 1220 Fees (Berks); William *Blackgroves* 1545 SRW. From Blagrave (Berks), or Blagrove in East Worlington (D).

**Blaikie:** Patrick *Blaikie* 1660 Black. A Scottish diminutive of BLAKE.

**Blaikley:** *v.* BLACKLEY

**Blaiklock:** *v.* BLACKLOCK

**Blain, Blaine, Blayn, Blayne, Blane:** (i) Andrew *Blayn, Bleyn* 1219 AssY; Nicholas *Bleyne* 1275 RH (Sf); John *Blayne* 1507 FFEss. OW *Bledgint*, Middle Welsh *Blethyn. v.* PNDB 204. (ii) Hamo *del Blein* 1219 Cur (K). OE *blegen*, ME *bleyne* 'an inflamatory swelling on the surface of the body', here in some topographical sense. (iii) Patrick *Blane* 1561, John Blain 1674 Black. For MACBLAIN.

**Blaise:** Robert *Blase* 1272 FFY; Robert *le Bleys* 1297 MinAcctCo; William *Blase* 1403 TestEbor; John *Blaze* 1642 PrD. OE *blase, blæse*, ME *blase, blese, bleis* 'a torch, firebrand'. cf. FLAMBARD. *v.* also BLOIS.

**Blake:** Walter *le Blake* 1167 P (D); Adam *Blake* 1219 AssY. *v.* BLACK.

**Blakeley:** *v.* BLACKLEY

**Blakemere:** *v.* BLACKMER

**Blakemore:** *v.* BLACKMORE

**Blakeney, Blakeny:** Thomas *de Blakenia* 1201 Pleas (Gl); Peter *de Blakenheye* 1332 SRDo; John *Blakeneye* 1392 LoCh. From Blakeney (Gl, Nf), or Blackney Fm in Stoke Abbott (Do).

**Blaker:** *v.* BLACKER

**Blakesley, Blaksley, Blaxley:** William *de Blakesle* 1199 AssSt. From Blakesley (Northants).

**Blaketter:** *v.* BLACKADDER

**Blakeway:** Hugo *de Blakewey* 1221 AssSa. 'Dweller by the dark road.'

**Blakey:** Geoffrey *de Blakey* 1388 IpmLa; Richard *Blakey* 1442 TestEbor; Robert *Blakey* 1621 SRY. From Blakey (NRY).

**Blakiston:** *v.* BLACKSTON

**Blakley:** *v.* BLACKLEY

**Blaksley:** *v.* BLAKESLEY

**Blamire, Blamires, Blamore, Blaymire:** William *de Blamyre* 1250 CalSc (Cu). 'Dweller by the dark, swampy place' (ON *blá(r), mýrr*), as at Blamires (WRYorks).

**Blampey, Blampied:** Richard *Blancpie* 1198 P (Nf). 'White foot', OFr *blanc, pied*.

**Blamphin, Plampin:** Nigellus *Blanpein* 1184 Oseney (O); Henry *Blancpain* 1191 P (Nth); Thomas *Plampeyn* 1496 SIA xii; John *Plampen* 1564 EA (OS) i (Sf); Robert *Plampin* 1568 SRSf; Thomas *Blampyn* 1662 DWills. OFr *blanc pain* 'white bread', a nickname for a baker.

**Blampied:** *v.* BLAMPEY

**Blanc, Blanck, Blank, Blanks:** Nigellus *Blanke* 1196 Cur (Lei); John *Blaunk* 1293 LLB C. OFr *blanc* 'white, fair', with reference to hair or complexion.

**Blanch, Blanche, Blaunch, Blaunche:** Alexander *Blanche* 1208 FFL; Matilda *Blaunche* 1270 FFO; Thomas *Blanch* 1312 ColchCt; Matilda *Blansche* 1379 PTY. OFr *blanche* (f) 'fair, white'. Used as a personal name in France where it was fairly common.

**Blanchard, Blanshard:** *Blanchard* de Morba 1180 P (D); Rotbertus *blancard*, Rotbertus quippe *blancard* 1086 InqEl (Sf); Richard *Blanchard'* 1177 P (L); William *Blanchart* 1198 P (L); Thomas *Blansherde* 1552 FrY. OFr *Blancart, Blanchart*, OG *Blankard, Blanchard*. The 1086 example is, however, certainly a nickname, probably identical with Robertus *Blancardus* (1086 DB, Nf), who is probably identical with Robertus *Blundus, Albus, Flavus*. OFr *blanchart* 'whitish', probably with reference to the hair. Only one example of the personal name has been noted.

**Blanchet, Blanchett, Blanket, Blankett, Branchett:** (i) Robert, John *Blanket* 1275 SRWo, 1365 LLB G. OFr *blankete*, ME *blankett* 'white or undyed woollen stuff used for clothing', first recorded in MED c1300, but much older. cf. 'ix ulnis de blanchet' 1182 P. A nickname for a maker or seller of this white cloth. (ii) Jocelin *Blancheved* 1203 Cur (L). A hybrid from OFr *blanc* and OE *hēafod* 'white head'. Rare.

**Blanchflower, Branchflower:** Cecilia *Blaunchflur* 1228 Cl (He); John *Blanchflur'* 1275 SRWo; Jeffrey *Branchflower* 1654 SfPR. OFr *blanche flour* 'white, fair flower', a suitable nickname for a woman. Applied to a man, it was probably derogatory, fair as a woman, of effeminate appearance.

**Blancowe:** *v.* BLENCOWE

**Bland:** John *de Bland'* 1297 SRY; John *Bland* 1447 FrY. From Bland (WRYorks).

**Blandamore:** Christiana *Pleynamur* 1275 RH (Sf); Richard *Playndeamours* 1284 AssLa. OFr *pleyn d'amour* 'full of love'. cf. FULLALOVE.

**Blanden, Blandon, Blondin:** William *Blaundyn* 1327, Symon *Blaundyn* 1332 SRSx; Robert *Blanden* 1672 HTY. Probably connected with OFr *blandir* 'to flatter'.

**Blandford:** John *Blandford* 1642 PrD. From Blandford (Do).

**Blane:** *v.* BLAIN

**Blank(s):** *v.* BLANC

**Blankau, Blanko:** *v.* BLENCOWE

**Blankett:** *v.* BLANCHET

**Blankley:** Simon *de Blankeneia* 1202 AssL. From Blankney (Lincs).

**Blanshard:** *v.* BLANCHARD

**Blatch:** Geoffrey (*le) Blache* 1164–5 P (Nf); Roll' *Blecch'* 1200 P (Nf); Richard *Blatche* Ed 2 FFEss. OE *blæc*. A palatal form of BLACK.

**Blatcher:** Robert *le Blacchere* 1305 MESO (So); Roger *le Blakkere, le Blecchere* 1313 ParlWrits (W). A variant of BLEACHER. cf. BLACKER.

**Blatchford:** John *Blachford, Blatchford* 1642 PrD. From Blatchford in Sourton (D).

**Blatchman:** Stephen *Blacheman* 1210–1 PWi. A variant of BLACKMAN.

**Blatherwick:** Adam *de Blarewic* 1198, Ranulf *de Bladrewyc* 1230 P (Nth). From Blatherwycke (Nth), *Blarewic* DB.

**Blaunch, Blaunche:** *v.* BLANCH

**Blaw, Blow, Blowe:** Randulf *Bla* 1202 AssL; Roger *Blowe* 1271 Rams (Hu); Thomas *Blaue* 1327 SRY. ON *blá*, ME *blaa, bloo, blowe* 'pale, livid'.

**Blaxall, Blaxill, Blacksell:** Adam *de Blakesale* 1308 LLB C; Roger *de Blaxhale* 1324 FFess; John *Blaksell*

1674 HTSf. From Blaxhall (Suffolk).

**Blaxley:** v. BLAKESLEY

**Blaxter:** William de (sic) Blakestere 1199 AssSt; Richard le Blakestare 1275 SRWo; John Blakestre ib. The feminine form of ME blaker 'bleacher', but applied to men. v. BLACKER, BLATCHER.

**Blaxton:** v. BLACKSTON

**Blay:** v. BLOIS

**Blaydes:** v. BLADES

**Blaymire:** v. BLAMIRE

**Blayn, Blayne:** v. BLAIN

**Bleacher:** Robert le Blechere 1327 MESO (Ha). A derivative of OE blǽcan 'to bleach', a bleacher.

**Blear, Blears:** Ælfstanes ðys Blerian 901 BCS 591 (W); Richard Blere 1181 P (Nth); Walter le Bler 1316 IpmGl; William Blere 1450–2 Pleas (K). ME blere 'bleary-eyed'.

**Bleary:** Alice Bleregh, Blereheye 1276 AssLo; John Blary 1327 SRSo; Simon Bleri 1375 AssL. ME bleri 'bleary-eyed', but the London example is probably for ME blere and OE ēage 'eye', with the same meaning.

**Bleasby:** Alexander de Bleseby c1155 Gilb. From Bleasby (L, Nth).

**Bledlow, Bledlowe:** William Bledelauwe 1253 Acc; Thomas de Bledelawe 1361–2 FFSr; Thomas Bledlowe 1379 PTY. From Bledlow (Bk).

**Blench:** Rogert Blenc 1153–68 Holme (Nf); Thomas, Richard Blench 1178 P (Y), 1214 Cur (Ess). OE *blenc 'a trick, stratagem' (a1250 NED).

**Blencowe, Blenko, Blinco, Blincoe, Blincko, Blincow, Blancowe, Blankau, Blanko:** Adam de Blencow 1332 SRCu; Thomas Blincoe 1623 ERO. From Blencow (Cumb).

**Blenkarn, Blenkarne, Blenkhorn, Blenkiron, Blinkhorn:** William Blenkarn 1429 FrY; Robert Blynkarne 1547 FFHu; William Blinkhorne 1664 FrY. From Blencarn (Cumb).

**Blenkin:** John Blenkynson 1553 FrY. The etymology of Blenkinsop is obscure. The first element may be a personal name which these names suggest continued in use.

**Blenkinsop, Blenkinship:** Symon de Blanchainesop, de Blencaneshop, de Blenkensope 13th Riev (Y). From Blenkinsopp (Northumb).

**Blenko:** v. BLENCOWE

**Blennerhasset, Blennerhassett:** John Blenerhasset, Raff Blaundrehasset 1473 Past; Henry Blanerhasset 1495 FrY; Thomas Blenayrhasset 1524 SRSf. From Blennerhassett (Cumb).

**Blessed, Blest, Blissett:** Alicia Iblessed 1297 MinAcctCo; John le Blessed 1327 SRSt; John le Blest 1332 SRSx; Thomas Blesset 1380 SRSt. ME iblescede, past participle of OE blētsian 'to make sacred', in the sense 'happy, fortunate' (c1175 NED). From 1300 onwards the word occurs as blisced, blissed, a form surviving, no doubt in Blissett, which may also derive from a woman's name: Blissot atte Pole 1327 SRSo.

**Bletcher:** Either for BLATCHER or for BLEACHER, with a shortening of the vowel.

**Bletchingdon, Bletchingden, Blissingham:** Elizabeth Blechenden, Blissingham, Blisinggum 1727 ER 52. From Bletchingdon (O).

**Bletchley:** Michael de Blechelai 1181–2 NLCh;

Robert de Blecheleg' 1254 RH (Sa); Ralph de Blecheleghe 1317 AssK. From Bletchley (Bk, Sa), or Bletchingley in Staplehurst (K), Blecchelegh' 1334.

**Blethyn, Blevin, Blevins, Pleaden, Pleavin, Pleven, Plevin, Plevins:** Blepgent 1063 ASC D; Bledienus filius Keneweret' 1173 P (Sa); Madoc son of Bledena 1246 AssLa; Hugh son of Bleuin ib.; Blethin ap Maddoc 1287 AssCh; Robert Blevyn 1275 RH (Nf); Llewellyn ap Bledyn 1313 ParlWrits; William Blethyn 1366 SRLa; Dauid ap Plethyn 1391 Chirk; Hugh Plethen 1524 SRSf; William Plevin 1685 Bardsley (Ch). OW Bledgint, MW Blethyn. Ap Blethyn was assimilated to ap Plethyn.

**Blewett, Blewitt, Blouet, Bluett:** Ralph Bloiet, Blouet, Bloet 1086 DB (Ha, So); Tedbald Blauet 1185 Templars (Herts); Walter Blohet ib. (So); William le Blut ib. (L); Robert Bluet 1196 Cur (W); Geoffrey le Bleuit 1327 SRC. OFr bleuet, blouet 'bluish', a diminutive of bleu 'blue'.

**Blick:** Aluin Blic 1185 Templars (Ess); John le Blyk 1249 IpmY; John le Blyke 1327 SRSo; Richard Blyk 1333 ColchCt. Obviously a nickname, perhaps connected with OE blīcian 'to shine, gleam, glitter'.

**Blickling:** Nicholas de Blykelyng 1348–9 FFWa. From Blickling (Nf).

**Bligh, Blight:** v. BLY

**Blighton:** v. BLYTON

**Blincko, Blinco(e), Blincow:** v. BLENCOWE

**Blind:** Edricus Cecus 1066, 1086 DB (Sf); Agnes Blinde 1221 ElyA (Sf); Ralph le Blinde 1274 RH (Nf). OE blind, Lat caecus 'blind'. cf. BORN.

**Blindell:** Blindel Hy 2 Gilb (L) is identical with Blundel (ib.). cf. Alured Blindel (Blundel) 1221 AssGl. v. BLONDEL.

**Blinkhorn:** v. BLENKARN

**Bliss:** William Blisse 1240 Rams (Hu); Thomas Blysse 1260 AssY. OE blīðs, ME blisse 'gladness, joy'. Also occasionally from Blay (Normandy): Hugo de Blez 1275 SRWo. cf. Stoke Bliss (Worcs).

**Blissett:** v. BLESSED

**Blissing:** Thomas Blissing' 1301 SRY; John Blissyng 1466 FrY; William Blyssyng 1558 Pat (Y). OE blētsung 'blessing, joy'.

**Blissingham:** v. BLETCHINGDON

**Block:** Robert Bloc 1199 Cur (W); Benedict Blok 1327 SRSf. Probably metonymic for blocker: Henry le Blocker 1212 Cur (Y); Deodatus le Blokkere 1275 RH (Nf); one who blocks, especially in shoemaking and bookbinding.

**Blockley, Blockly:** Reginald de Blockeleg' 1221 Cur (Wo); John Blokley 1340–1450 GildC; John Blokle 1364, de Blockeley 1368 LLB G. From Blockley (Wo).

**Blocksidge:** v. BLOXIDGE

**Blofield, Blofeld, Blowfield:** Edward de Blafeld' 1198 FFNf; Geoffrey de Blofeld 1287–8 NorwLt; John Blofeld 1405 FFEss. From Blofield (Nf).

**Blogg, Bloggs:** Ralph Blog 1359 AssD. Probably a voiced form of BLOCK.

**Blois, Bloyce, Bloys, Bloss, Blowes, Bloice, Blaise, Blay:** Tedbalde de Blais 1116, Stephen de Blais 1135 ASC E; Robert de Bloy 1205 Cur (Ess); Robert de Bleys 1219 Cur (Lei); John Blosse 1327 SRSf; John

*Bloyce* or *Blowes* 1497 Bardsley (Nf). From Blay (Calvados), or Blois (Loir-et-Cher).

**Blomefield, Blomfield, Bloomfield, Blumfield:** William *de Blunuill'* 1207 Cur (Sf); Thomas *de Blumuill'* 1230 P (Nf); John *Blumfeilde* 1582 EA (NS) i (Nf). From Blonville-sur-Mer (Calvados).

**Blomer:** *v.* BLOOMER

**Blondel, Blondell, Blundal, Blundell:** *Blundel* 1115 Winton (Ha), c1150 DC (Nt); Walterus *filius Blundelli* 1203 Cur (L); John *Blundel* c1140 StCh; John *Blondel* 1297 MinAcctCo. OFr *blondel*, a diminutive of *blond* 'fair', of hair or complexion. cf. BLUNT. Also used as a personal name.

**Blondin:** *v.* BLANDEN

**Blood:** William, John *Blod* 1256 AssNb, 1328 LLB E. OE *blōd* 'blood', used as a term of address in Chaucer: 'Now beth nought wroth, my blode, my nece'; also 'child, near relative', 'one dear as one's own offspring'. Also metonymic for a blood-letter. Uluric, Walter *Blodletere* c1095 Bury (Sf), 1221 *ElyA* (Nf), OE *blōdlǣtere*; or for *blooder*: Adam *Blodyr* 1441 GildY, from ME *blōden* 'to let blood'. In Ireland, for *ab Lloyd* 'son of *Lloyd*'.

**Bloom:** Anselm, Walter *Blome* 1177, 1198 P (Sx, Lo). Metonymic for BLOOMER.

**Bloomer, Blomer, Blumer:** William *Blomere* 1202 P (Db); Robert *le Blomere* 1279 AssSt. A derivative of OE *blōma* 'an ingot of iron', hence 'maker of blooms, iron-worker'.

**Bloomfield:** *v.* BLOMEFIELD

**Bloor, Bloore, Blore, Blure:** Robert *de Blora* c1165 StCh; Ralph *de Blore* 1208 Cur (St). From Blore (St).

**Bloss:** *v.* BLOIS

**Blossom:** Walter *Blosme* 1195 P (Wa); Peter *Blostme* 1297 MinAcctCo. OE *blōstm(a)*, *blōsma* 'blossom', used in the 15th century of one lovely and full of promise.

**Blossworth:** Laurence *Bloseworth* or de Ware 1379 AssNu. Probably from Bloxworth (Do).

**Blouet:** *v.* BLEWETT

**Blount:** *v.* BLUNT

**Blow, Blowe:** *v.* BLAW

**Blower, Blowers:** William *le Blowerre* 1199 P (Sr), *Blouer* 1219 AssY; Lucia *Blowere* 1317 AssK; Reginald *le Blawere* 1327 *SR* (Ess). OE *blāwere* 'blower' of the horn or bellows. cf. Gilbert *Blouhorn* 1275 RH (L).

**Blowes:** *v.* BLOIS

**Blowfield:** *v.* BLOFIELD

**Bloxam, Bloxham, Bloxsom, Bloxsome:** Robert *de Bloxeham* 1130 P (L); William *de Blokesham*, Alexander *de Bloxam* 1279 RH (O). From Bloxham (Oxon) or Bloxholm (Lincs).

**Bloxidge, Bloxsidge, Blocksidge:** Clement *de Bloxwych* 1332 SRSt. From Bloxwich (Staffs).

**Bloyce, Bloys:** *v.* BLOIS

**Bluett:** *v.* BLEWETT

**Blumer:** *v.* BLOOMER

**Blumfield:** *v.* BLOMEFIELD

**Blumsom, Blumson:** Thomas *Blumsome* 1661 Bardsley. For BLUNSOM or BLUNSEN.

**Blundal, Blundell:** *v.* BLONDEL

**Blunden, Blundun:** William *Blonden* 1524 SRSf.

Probably a diminutive of OFr *blond* 'fair'.

**Blundstone, Blunstone:** Robert *de Blundeston* 1327 SRSf; Nicholas *Bluneston* 1593 FFHu. From Blundeston (Suffolk).

**Blunsden, Blunsdon:** Brian *de Bluntesdon'* 1255 RH (W). From Blunsdon (Wilts).

**Blunsen, Blunson:** For BLUNDSTONE or BLUNSDEN.

**Blunsom, Blunsum:** John *de Bluntisham* 1229 FFHu; John *Bluntsham* 1406 Bardsley; John *Blonsham* 1734 ib. From Bluntisham (Hunts). Also a possible late form of *Blunsen*.

**Blunt, Blount:** Rodbertus *Blon*, *Blondus*, *Blundus* 1086 DB; Robert *se Blund* c1100–30 OEByn (D); Ralph *le Blund* Hy 2 DC (Lei); John *le blunt* c1194 StCh; Hamelin *Blund* 1201 AssSo; Richard *le Blount* 1279 RH (O). OFr *blund*, *blond* (Lat *blondus*) 'blond, fair, yellow-haired', used also of complexion (1481 NED). cf. BLONDEL and Joce *Blonthefed* 1195 P (L) 'fair head'. In DB, Robert *Blundus* is also called *Albus*, *flauus* and *blancard*.

**Blure:** *v.* BLOOR

**Bly, Blyde, Blyth, Blythe, Bligh, Blight:** (i) William *de Bliða* 1177 P (Ess); Gilbert *de Blie* 1200 P (Nt). From Blyth (Northumb, Notts) or Blythe (Warwicks). *Bly* is due to Anglo-Norman loss of *th*. (ii) *Blide* 1101–7 Holme (Nf); Willelmus *filius Blie* 1188 P (La); *Blithe* de Ryseford 1276 RH (Y); Robert *Blithe* 1221 *ElyA* (Nf); John *Blythe* 1296 SRSx. Either a nickname from OE *blīðe* 'gentle, merry' or from an unrecorded personal name, OE *\*Blīða*, derived from this. The adjective is found as *bliht* and *bligh* in the 13th century (NED).

**Blyman:** Dorathea *Blithman* 1505 GildY; Ninian *Blythman* 1530 FrY; Robert *Blythman* 1621 RothwellPR (Y). Either a nickname, 'merry man', OE *blīðe, mann*, or 'the man from Blyth'. *v.* BLY.

**Blyton, Blighton:** Jacobus *de Bliton'* 1202 AssL; Robert *de Blyton* 1280–90 RegAntiquiss; John *Blyton* 1372 FFEss. From Blyton (L).

**Boaden:** *v.* BAWDEN

**Boalch:** *v.* BELCH

**Boaler:** *v.* BOWLER

**Boar, Boor, Bor, Bore:** Æilmar *Bar* c1095 Bury (Sf); Godwin *bar* 1148 Winton (Ha); Walter *Bor* 1255 Rams (Hu); Robert *le Bor* 1287 Ipm (Bk); John *le Boor* 1312 AD iv (D). OE *bār* 'boar'.

**Board, Boord:** Nicholas, William *Borde* 1230 P (Do), 1296 SRSx. Metonymic for BOARDER.

**Boarder, Border, Borders, Bordier:** Thomas *le Border* 1201 AssSo; Robert *le Bordere* 1296 SRSx. Bardsley and Thuresson derive this from OFr *bordier* 'bordar, cottager', a word found only in the medieval Latin form *bordarius* in DB and, as an English word, only in modern historians (1776 NED). The surname may be a derivative of OE *bord* 'board, plank, ? table', ME *\*border* 'maker of boards or tables'. cf. Robert *Bordmakere* 1356 LLB G, William *le Bordhewere* 1327 MESO, Richard *Bordwreghte* 1332 SRSx. Or it may be from ME *bourd(e)our*, AFr *bourd(e)our*, OFr *bordeor* 'a jester, joker, buffoon' (1330 NED), though we should have expected some examples of *bourder*. cf. Bordyoure or pleyare. *Lusor, joculator* c1440 PromptParv (*bordere* 1499).

**Boarer:** v. BOWRA

**Boast, Bost:** Walter *Bost* 1279 RH (O); Walterus *dictus Bost* c1300 Balliol (O); Ralph *Boste* 1327 SRSf; Walter *Boost* 1327 SRSx. ME *bōst* 'vaunt, brag, "tall talk"; vain-glory'.

**Boater:** John *Botere* 1279 RH (Hu), 1317 AssK, *le Botere* 1301 LLB C. A derivative of OE *bāt* 'boat', a boatman (1605 NED).

**Boatman, Bottman:** Thomas *Bootman* 1225 Gardner (Sf); Thomas *Botman* 1378 MEOT (So). OE *bāt* and *mann* 'boatman' (1513 NED).

**Boatswain, Boeson:** Wicing *Batswegen* 1050–71 OEByn (D); Peter *Botsweyn, le Botsweyn* 1327, 1332 *SR* (Ess). Late OE *bātswegen*, from ON *bátsveinn* 'boatman' (1450 NED). Used also as a personal name: *Batsuen* 1055 DB (Sa), Walterus *filius Batsuein* 1190 P (L).

**Boatte:** Alice, William *atte Bote* 1327, 1332 SRSx. Richard Beselin *atte Bote* MELS (Sx) had to 'ferry the Bishop and his carriages and all the men in his service and all avers coming from Busshopestone Manor', thus owing his attribute to the services due from his holding. The surname might also be occupational, 'boatman', 'ferryman'.

**Boatwright, Botwright:** John *Botwright* 1469 SIA xii; John *Botewrighte* 1524 SRSf. OE *bāt* 'boat' and *wyrhta* 'wright', a maker of boats. Common in Suffolk.

**Boayer:** v. BOWYER

**Bobb, Bobbe:** William *Bubbe, Bobbe* 1219 Cur (Do); Reginald *Bobbe* 1294 IpmW; Thomas *Bobbe* 1428 FA (W). A variant of BUBB.

**Bobbett, Bobbitt:** Robert *Bobat* 1327 SRSf; William *Bobbete* 1332 SRSt. *Bobb-et*, a diminutive of BOBB.

**Boby:** v. BOOTHBY

**Bock:** v. BUCK

**Bockett:** v. BURCHARD

**Bocking, Bockings:** Æðelric *æt Boccinge* c997 OEByn; Reginald *de Bokings* 1203 FFEss; William *de Bockyng* 1337 CorLo; John *Bokkyng* 1450 Paston. From Bocking (Ess).

**Bocock:** v. BAWCOCK

**Boddington, Bodington:** Hugh *de Botendune* c1160–7 RegAntiquiss; Robert *de Botenden'* 1202 AssNth; Adam *de Bodington* 1291 AssSt. From Boddington (Gl), *Botingtune* DB, or Boddington (Nth), *Botendone* DB.

**Bode:** *Boda* 1066, 1086 DB (Ha); *Bode* 1066 ib. (W); Hugo *filius Bode* 12th Rams (Nf); Walter *Bode* 1220 Fees (Berks); Robert *Bode* 1221 *ElyA* (Nf); William *le Bode* 1296 SRSx. OE *boda* 'herald, messenger'. Also used as a personal name (Redin).

**Bodecott, Bodicot:** Walter *de Bodicot'* 1279 RH (O). From Bodicote (O).

**Boden:** v. BAWDEN

**Bodenham, Bodnam, Bodham:** Hugh *de Bodeham* 1180 P (He); William *de Bodeham* 1206 Cur (Nf); Margaret *de Bodham* 1212 P (Sx). From Bodenham (He), *Bodeham* DB, or Bodham (Nf).

**Boder:** Andrew *le Bodere* 1296 SRSx; Bartholomew *Bodyr* 1327 SRC. OE *bodere* 'announcer, messenger'. Ralph *le Bodere* 1212 Fees (Ess) is called *le Criur* 1227 ib. cf. CRIER.

**Bodfish:** Joane *Botfishe*, John *Botefyshe*, Alice *Bootefishe* 1523–41 ArchC 41. ME *butte* (c1300 MED), cognate with Sw *butta* 'turbot', Du *bot* 'flounder', flatfish, as sole, fluke, plaice, turbot, etc. cf. But, fysche, *Pecten* PromptParv, Butte fysshe, *plye* Palsgrave. It would appear that *butfish* was used as a generic term and the surname probably denotes a seller of flatfish. cf. *butt-woman* 'a fish-wife' 1620 NED.

**Bodicot:** v. BODECOTT

**Bodin, Boyden:** *Bodin* 1066 DB (St); *Bodin* prior de Parco, prior *Boydin* de Parco Hy 2 DC (L); *Bodinus* or *Boydinus* 1156–80 Bury (Sf); Turstan, Robert *Bodin* 12th DC (L), 1200 P (O); Mainard *Boidin* 1208 Cur (Sx); John *Boydon* 1401 FrY. OFr *\*Bodin*, OG *Baudin*.

**Bodington:** v. BODDINGTON

**Bodkin, Badkin:** Robert *Bodekin* 1274 FFO; Robert *Bodekyn* 1297 MinAcctCo; Peter *Bodekyn* 1331 AssSt. ME *bodkin, bodekin* 'a short pointed weapon or dagger'. Metonymic for a maker or seller of these. *Badkin* may also be a variant of *Batkin*, a diminutive of *Bat*, a pet-form of *Bartholomew*. cf. John *Badekyn* 1312, *Batekyn* 1327 PN K 492.

**Bodley:** Hamelin *de Bodlei* 1196 P (D); Roger *de Bodele* 1269 FFO; Emma *Bodely* 1457–8 FFSr. From Bodley (Sr), or Bodley in Parracombe (D).

**Bodman:** Ralph, Thomas *Bodeman* 13th Guisb, 1316 Wak (Y). OE *boda* 'messenger' or OE *(ge)bod* 'message' and *mann*. 'Servant of the messenger' or equivalent to BODER.

**Bodnam:** v. BODENHAM

**Body, Bodey, Boddy:** Hugh *Body* 1219 AssY; Richard *Body* 1277 FFEss; Roger *Body* 1340 AssSt. A nickname from OE *bodig* 'trunk, frame, bodily presence'. cf. GOODBODY, TRUEBODY.

**Boeson:** v. BOATSWAIN

**Boff, Leboff:** Robert *le Buef* 1169 P (L); Walter *Beof, le Boef'*, *le Bof* 1219 Cur (K). OFr *boef* 'bullock', a nickname for a big, powerful man, a great lubberly fellow.

**Boffee, Boffey:** v. BEAUFOY

**Bogg:** (i) Giles *Bog* 1327 *SR* (Ess); Robert *Bogge* 1504 FrY. Probably early examples of *bog*, adj., 'bold, proud, saucy' (1592 NED). (ii) John *atte Bogge* 1327 SRSo. 'Dweller by the bog' (a1450 MED).

**Boggers, Boggis, Boggs:** Elyas, John *Bogeys* 1260 AssY, 1327 *SR* (Ess); John *Bogays* 1301 SRY, 1327 SRSf; William *Bogace* 1309 Wak (Y); William *Bogges* 1327 SRSf; Richard *Bogas* 1366 FrY. ME *bogeys* 'inclined to bluster or brag, puffed up, bold'; cf. 'bogeysliche as a boy' a1375 MED.

**Boice:** v. BOYES

**Bolam, Bollom:** Gilbert *de Boolum* 1205 P (Nb); Richard *de Boleham* 1279 RH (O); John *Bollom* 1420 IpmY. From Bolam (Du), or Bolham (Nt).

**Bolax, Bolas, Bolasse, Bulax, Bulasse:** Reginald *Bulax* 1202 FFNf; John *Bolax* 1296 SRSx; William *Bolasse* 1379 PTY, *Bulasse* 1418 IpmY. ON *bol-oÞx* 'poleaxe'. Probably metonymic for a butcher. cf. John *Handax* 1327 SRY; Euerard *Brodax* 1197 P (Y) 'broad axe'; William *Brokax'* 1226 Cur (Sx) 'broken axe'.

Bold, Boud, Bould, Boulde, Bowld: (i) Richard *Bolde* 1206 P (La); Henry *Bolde, le Bolde* 1317 AssK, 1327 SRSx; William *Boulde* 1428 FrY; Robert *Bowde* 1563 FrY. ME *bold*, OE *bald* (WS *beald*) 'stouthearted, courageous'. *v.* also BALD. (ii) Heremer *de la Bolde* 1176 P (St); Geoffrey *de Bold* 1199 AssSt; Herbert *de la Boude* 1200 Cur (Sa); Robert *Attebolde* 1332 SRSt. OE *bold* 'dwelling-house'. From Bold (Lancs), The Bold (Salop), or local. 'At the house' is not particularly distinctive. It might refer to residence at a small farm or to employment at the manor-house.

Bolden, Boldon: Roger *de Boldun* 1143–52 FeuDu; Robert *de Boldon* 1242 AssDu. From Boldon (Du).

Bolder, Boulder, Bowder: Albric *Buldur* 1203 AssNth; Bate *Bolder* 1286 Wak (Y); Richard *Buldur* 1379 PTY. A nickname from ME *bulder* 'boulder, cobblestone'.

Boldero, Bolderoe: *v.* BALDREE

Bolderson: *v.* BALDERSON

Bolderston: *v.* BALDERSTON

Bolding: *v.* BALDING

Boldison: *v.* BALDERSON

Boldon: *v.* BOLDEN

Boldra, Boldry: *v.* BALDREE

Boldron, Bowran, Bowron: Robert *de Bolroun* 1332 SRLa. From Boldron (NRYorks).

Boldy: *v.* BALDEY

Bole, Boles: *v.* BULL

Boler: *v.* BOWLER

Bolgar, Bolger: *v.* BOULGER

Boling: *v.* BOLLING

Bolingbroke, Bollingbroke: William *de Bulingbroc* 1170–8 P (L); John *de Bolingbrok* 1275 RH (Y); John *Bullyngbroke* 1476 IpmNt; William *Bolyngbroke* 1503 CorNt. From Bolingbroke (L).

Boll, Bolle: *Bolla* 1066 Winton (Ha); Walter *Bolle* 1185 Templars (K); Walter *Bolle* 1254 Oseney. OE *Bolla*, or ON *Bolli. v.* OEByn 294.

Bolland, Bollands, Boland: John *de Bolland* 1237 HPD; John *de Boughland* 1351 FrY; William *de Bowland* 1370 FrY; John *Bolland* 1482 FFEss. From Bolland (Devon), Bowland (Lancs, WRYorks), or Bowlands (ERYorks).

Bollard: John *Bollard* 1327 SREss; William *Bollard* 1367 IpmNt; John *Bollard* 1689 FrY. Said to be originally Dutch. Perhaps a variant of BALLARD.

Bolle: *v.* BOLL

Bollen: *v.* BULLEN

Bollett: *v.* BULLETT

Bolley: *v.* BULLEY

Bolling, Boling, Bowling: (i) Robert *de Bolling* c1246 Calv (Y); Thomas *Bollyng* 1459 Kirk; John *Bowlinge* 1662, Ann *Bowlin* 1737 Calv (Y). From Bowling (WRY). (ii) William *bolling* 1189 Sol; Robert *Bolling* 1264 Eynsham, 1273 RH (So). A nickname, either from ME *bolling* 'pollard', *v.* POLLARD, or from ME *bolling* 'excessive drinking'.

Boilingbroke: *v.* BOLINGBROKE

Bollington: Alexander *de Bolinton* 1199 FFEss. From Bollington (Ch), or Bolington Hall in Ugley (Ess).

Bollom: *v.* BOLAM

Bolmer: *v.* BULMER

Bolsover, Boulsover: Adam *de Bolesouer* 1202 AssL; Nicholas *de Bolisouere* c1250 Glapwell (Db); John *de Bollesore* 1384 FrY. From Bolsover (Db).

Bolt, Bolte, Boult: Godinc *Bolt* 1066 ICC (C); Walter, Roger *Bolt* 1202 Cur (Sr), c1248 Bec (W). OE *bolt*, 'bolt, bar'. For the first example, Tengvik compares the ON nickname *Boltr*, which might have been applied to a short, heavy person. Later instances are probably metonymic for BOLTER, a maker of bolts. cf. John *Boltsmith* 1346 MESO.

Boltby, Boultbee: Odo *de Boltebia* 1142–5 YCh; Nicholas *de Bolteby* 1256 AssNb; Robert *de Boltebi* 1327 SRY. From Boltby (NRY).

Bolter, Boulter: Roger *le Buleter, le Buletor, le Bolter* 1246, 1248, 1253 Oseney (O); Roger *le Boletere* 1261 Oriel (O); Geoffrey *le Bolter* 1276 RH (Berks). OFr *buleteor* 'a sifter of meal'. *Bolter*, which is the most common form, is probably often a derivative of OE *bolt*, 'a maker of bolts'. cf. BOLT.

Bolton, Boulton: Robert *de Boulton'* a1191 YCh; Thomas *de Bolton* 1262–3 FFWa; Robert *Bolton* 1371 AssL. From one or other of the many places of this name, or from Boulton (Derby).

Boltwood, Boultwood, Boughtwood, Boutwood: Adam *Bolthod* 1265–72 RegAntiquiss; Robert *Bolthoud'* 1332 SRDo; John *Bolthood* 1430 FFEss. OE *bolt* 'a roll of woven fabric', often apparently a fabric suitable for sifting, and OE *hōd* 'hood'. Presumably for the wearer of a hood made from this particular kind of cloth.

Boman: *v.* BOWMAN

Bomfield: *v.* BONFIELD

Bompas, Bompus, Bumpus: William *Bonpas* 1175 P (Gl); Anne *Bompase* 1616 Bardsley; James *Bumpus* 1670 ib. OFr *bon pas* 'good pace'. cf. LIGHTFOOT, GOLIGHTLY.

Bomphrey: *v.* BOUMPHREY

Bonafont: *v.* BONIFANT

Bonally, Bonallo, Bonella, Bonello, Bonnalie, Bonnella: William *Bonaylay* c1570 Black; Robert *Bonalay* 1637 ib.; David *Bonallo* 1818 ib. Black derives these surnames from Bonaly (Midlothian) or a lost Banaley (Fife). They are probably a Scottish equivalent of the English *Drinkale*, from MScots *bonalai, bonnaillie*, from Fr *bon* 'good' and *aller* 'to go, going', 'good speed, farewell!' as in 'to drink one's bonallie' (c1470 NED).

Bonamy: *v.* BONNAMY

Bonar: *v.* BONNAIRE

Bonas: *v.* BOWNAS

Bonaventure: Emma *Boneauenture* c1215 Clerkenwell; John *Bonauenture* 1316 AssNth; John *Bonaunter* 1406–7 Hylle. Fr *bon aventure* 'good fortune', a phrase name. used as a christian name in the 17th century: *Bonaventure Cowle* 1642 PrD.

Bonchristian: Stephen *Bonecristien* 1200 Cur, *Bonecristian* 1209–10 FFSr. 'Good Christian', OFr *bon, crestien*. cf. *Nequam Christianum* 1206 Cur 'hardly Christian'; *Mal Christien* 1206 Cur (L) 'bad Christian'.

Bonchurch: Philip *Boncherche* 1327 SRSx; John *Bonechurche* 1363 IpmGl. From Bonchurch (Wt).

**Bond, Bondi, Bonds, Bondy, Bound, Bounds, Boundy, Bunday, Bundey, Bundy:** *Bonde, Bondi, Bunde, Bundi* 1086 DB; Albertus *filius Bund', Bonde* 1199, 1202 FFNf; Norman *le Bonde* 1180 P (Wa); William *Bonde* 1185 Templars (Wa); Robert *Bunde* 1198 P (Beds); Henry *le Bounde* 1297 MinAcctCo (Herts). *le Bonde* is clearly from OE *bōnda, būnda*, ON *bónde, bóndi* 'husbandman, peasant, churl', later 'unfree tenant, serf'. The simple *Bonde* may be from the same source or from ON *Bóndi*, ODa *Bondi, Bundi*, OSw *Bonde*.

**Bondfield:** *v.* BONFIELD

**Bondgate, Bongate:** Nicholas *de Bondegate* 1303 IpmY. From Bongate (We), or Bondgate in Harewood, in Otley, in Ripon, in Selby (WRY).

**Bondman:** Philip *Bonddeman* 1290–1300 RegAntiquiss; Thomas *Bondman* 1297 SRY. Probably 'servant of *Bond*', rather than 'bondman'.

**Bone, Bonn, Bonne, Bunn:** Edward *le Bon* 1204 Cur (O); Rocelin *le Bun* 1255 RH (W); Walter *le Bone* 1296 SRSx; Thomas *Bonne* 1379 LLB H. OFr *bon* 'good'. For *Bone, v.* also BOON.

**Bonella, Bonello:** *v.* BONALLY

**Boner:** *v.* BONNAIRE

**Bones:** Alicia *Bones* 1327 SRSf. *v.* BAIN.

**Boness:** *v.* BOWNAS

**Boneter, Bonetta:** 'A maker of bonnets.' *v.* BONNET.

**Boney:** *v.* BANEY

**Bonfellow:** *v.* BOUTFLOUR

**Bonfield, Bondfield, Bomfield:** Richard *de Bondauilla* 1131 FeuDu; Robert *de Boneville* 1197 P (Y). From Bonneville (Normandy) where there are three places of the name, two near Rouen.

**Bongate:** *v.* BONDGATE

**Bonger, Bongers, Bonjour:** Osbert *Bonior* 1199 FF (Nth); Roger *Boniur* 1275 RH (Sf); Alice *Bonjour* 1327 SRSx. OFr *bon jour* 'good day!'. cf. GOODDAY.

**Bonham, Bonhomme, Bonome:** Ralph'' ? Boniface; ? Bonome Luscus 1177 P (Lei); Bonhom fullo 1219 AssY; Randulfus *bonus homo* 1148 Winton (Ha); Nigel *Bonhume* 1247 AssBeds; William *Bonum* (*Bonhom*) 1327 SR (Ess). OFr *bon homme* 'good man'. cf. GODMAN. This may also occasionally be local in origin: William *de Bonham* 1225, 1269 AssSo.

**Boniface, Bonifas, Bonniface:** *Bonefacius uinitor* 1193 P (Y); *Bonifacius* 1208 Cur (Ha); Tomas *Boniface* 1190 P (Y); Alis *Boneface* 1200 P (Ha). Contrary to the common opinion, this name derives not from Latin *bonifacius* 'well-doer', but from *bonifatius*, from *bonum* 'good' and *fatum* 'fate'. The change to *Bonifacius* was due to the pronunciation and from this was deduced a false etymology. *Bonifatius* is frequent on Latin inscriptions. *Bonifacius* is found only twice and these late (*Thesaurus*). In Latin the name was given chiefly to ecclesiastics, rarely to men of the lower orders. In ME the form was *Boniface*, but the name was never very popular in England, though it had enough vitality to produce a surname.

**Bonifant, Bonafont, Bullivant:** William *Bonenfant, Bonenfand* 1207–8 Cur (O); Henry *Bonefant* 1279 RH (Bk); John *Bon Effaunt* 1332 SRSx; Roger *Bonyfaunt* 1472 LLB L; Henry *Bolyvaunt* 1524 SRSf; William *Bonyvant* 1540 RochW; Elizabeth *Bullivant* 1707 SfPR. Fr *bon enfant*, identical in meaning with GOODCHILD.

**Bonjour:** *v.* BONGER

**Bonn, Bonne:** Ṽ. BONE

**Bonnaire, Bonnar, Bonner, Bonnor, Bonar, Boner:** John *Boneyre* 13th Rams (Hu); Thomas *Boner* 1281 Black (Aberdeen); Walter *Boneyre* 1297 FFEss; Robert *Boner* 1332 SRSx; Alexander *Bonour* 1413 FrY; William *Bonare, Bonere, Bonour* 1451–3 Black (St Andrews); Walter *Bonnar* 1527 ib. ME *boner(e), bonour*, OFr *bonnaire* 'gentle, courteous', shortened from *debonnaire* (c1300 MED).

**Bonnalie:** *v.* BONALLY

**Bonnamy, Bonamy:** William *bon ami* c1162 DC (L); William *Bonami* 1203 AssL. OFr *bon ami* 'good friend'.

**Bonnan:** Peter *Bonhand* 1327 SR (Ess). OFr *bon* 'good' and OE *hand* 'hand'. 'Good hand.'

**Bonnar:** *v.* BONNAIRE

**Bonnard:** Walter *Bonard* 1327 SRSx. OFr *bon* 'good' with the intensive suffix *-ard*.

**Bonnella:** *v.* BONALLY

**Bonner:** *v.* BONNAIRE

**Bonnet, Bonnett:** Isabella *Bonet* 1201 AssSo; John *Bonet* 1212 Cur (W), 1219 AssL, 1230 P (Sa). ME, OFr *bonet*, used for a maker of bonnets. cf. BONETER.

**Bonnick:** *v.* BONWICK

**Bonniface:** *v.* BONIFACE

**Bonnington:** (i) Roger *de Boninton* 1222–3 FFK; John *de Bonnington* 1353 IpmNt; John *Bonyngton* 1393 CtH. From Bonnington (K). (ii) William *de Bondington* 1258 (Glasgow), Andrew *de Bonynton* 1442 (Linlithgow) Black. From Bonnytoun (West Lothian), or Bonnington, formerly Bondington, (Peebles).

**Bonnor:** *v.* BONNAIRE

**Bonome:** *v.* BONHAM

**Bonsall, Bunsaul:** Osbert *de Bontisal* Hy 3 DbCh. From Bonsall (Db).

**Bonser, Bonsier, Bonsor:** Roger *bonsire* 1246 Bart (Lo); Robert *Bonsir* 1332 SRSx. OFr *bon sire* 'good sir'.

**Bonvalet, Bonvallet, Bonvalot:** William *Bonuaslet* 1086 DB (Bk); Nigel *Bonvalet* 1199 Cur; Wiliam *Bonvadlet* 1232 Pat (L); John *Bounvallet* 1327 SRSo. OFr *bon, vaslet/valet* 'good servant'.

**Bonwick, Bonnick:** Walter *Bonwyk*, William *de Bonwyk* 1296, 1332 SRSx; John *de Bonnewyk'* 1363 AssY. From Bonwick (ERYorks). Walter probably gave his name to Bonwicks Place in Ifield (Sussex).

**Boocock:** *v.* BAWCOCK

**Boodle, Buddle:** Robert, John *atte Bothele* 1327 MELS (So), 1330 PN D 484. From Buddle Oak in Halse (Som), Buddle in Fordingbridge (Hants), Buddle in Niton (Isle of Wight) or Budleigh in Moretonhampstead (Devon), all from OE *\*bōðl*, cognate with OE *botl, bold*. cf. BOLD. *\*bōðl* 'dwelling-house' probably denoted a homestead of some size.

**Bookbinder:** William *ligator librorum* 1273 Oseney; Robert *le Bokbyndare* 1292 Wenlok; Richard *Bokbynder* 1381 LoCh. 'Bookbinder', OE *bōc, bindere*.

**Booker:** (i) William *le Bocer* 1255 RH (Sa), 1296 SRSx; John *Boker* 1275 RH (Nf). OE *bōcere* 'writer of books, scribe'. (ii) Robert *le Bukere* 1229 FFsx; William *le Buker* 1246 AssLa; Elias *le Boukere* 1296 SRSx. A derivative of ME *bouken* 'to steep in lye, to bleach' (1377 NED), a bleacher.

**Bool(s):** *v.* BULL

**Boon, Boone, Bone, Bown, Bowne:** Hunfridus *de Bohum* 1086 DB (Nf); Wnfridus *de Bowhun* 1120–3 EngFeud; William *de Boun* 1119 Colch (Ess); Matildis *de Bohun* Hy 2 DC (L); John *de Bown* 1275 RH (Sx). Reginald *Boon'* 1279 RH (C). From Bohon (La Manche). Bohun's Hall (Essex) is *Boneshall* 1540, *Bowneshall* 1604 (PN Ess 305) and is now pronounced *Boon's* Hall.

**Boor:** *v.* BOAR
**Boord:** *v.* BOARD
**Boorer:** *v.* BOWRA
**Boorman:** *v.* BOWERMAN

**Boosey:** Richard, Roger *Bosy* 1327 *SR* (Ess), 1376 FrY. Late forms, probably for *atte Bosy* 'at the ox- or cow-stall', OE *bōsig*. 'A cowman.'

**Boosie, Bousie, Bowsie:** John *Bousie, Bowsie* 1566, 1580 Black. From Balhousie (Fife) 17th *Bowsie*, and still so pronounced.

**Boot, Boote:** Hugh, William *Bote* 1186 P (Wa), 1279 RH (C); Adam *Boot* 1345 AD i (K). OFr, ME *bote* 'boot', a maker or seller of boots.

**Booth, Boothe:** Gilbert *Bothe, del Both'* 1274, 1297 Wak (Y); John *de la Bouthe* 1287 AssCh; William *atte Bothe* 1297 Coram (Nf). ME *bōþ(e)*, from ODa *bōth* 'cow-house, herdsman's hut'. An occupational name for a cowman or herdsman, identical with BOOTHMAN.

**Boothby, Boby:** Hugo *de Boebi, de Bothebi* 1190, 1205 P (L). From one of the three places named Boothby in Lincs.

**Boothman:** Roger *Bothman* 1279 RH (Hu); Nicholas *the Bouthman* 1287 AssCh. ME *bōþ(e)* and *man*. *v.* BOOTH.

**Boothroyd:** John *del Botherode*, Adam *de Buderude* 1274, 1296 Wak; Richard *Buthroid* 1627 RothwellPR (Y). From Boothroyd (WRYorks).

**Bootham:** Laurence *de Bouthom* 1287–8 IpmY, *de Bothum* 1303 FFY. From Bootham in York.

**Bootman, Bootyman:** *v.* BUTEMENT

**Boozer, Boucher, Bouchier, Bourchier, Bowser, Bowsher:** cf. Boose's Green in Colne Engaine (Ess), Robert, John *de Burser, Bousser* 1285, 1303, *Burghcher* 1349; Bourchier's alias Bouchier's Hall in Aldham (Ess), John *de Bourchier* 1311; Bouchier's Grange in Great Coggeshall (Ess), John *de Bousser* 1326; Bouchiers Chapel in Tollesbury (Ess), John *de Bousser* 1328; Bourchier's Hall in Messing (Ess), John *de Busser* 1309; Boarstye Fm and Bowser's Hall in Rivenhall (Ess), Robert *de Bouser, Boussier, Bourchier* 1327, 1339. Perhaps 'dweller in the place planted with bushes', Fr *bussière*.

**Boram:** *v.* BOREHAM
**Borar:** *v.* BOWRA

**Borden, Bordon:** Alice *de Borden'* 1206 Cur (K); Richard *Bordon* 1296 SRSx; Henry *Borden* 1375 IpmGl. From Borden (K).

**Border(s):** *v.* BOARDER

**Bore:** *v.* BOAR

**Boreham, Boram, Borum, Borham:** Maurice *de Borham* 1192 P (Ess); Adam *de Borham* 1249 AssW; John *Boram* 1429 AssLo; Daniel *Borham*, Henry *Borum* 1674 HTSf. From Boreham (Ess), Boreham Wood in Elstree (Herts), Boreham Street in Wartling (Sx), or Boreham in Warminster (W).

**Borel:** *v.* BURREL

**Borer:** Robert *le Boriere* 1318 LLB B; Hugo *le Borer* 1332 MESO (L). A derivative of OE *borian* 'to bore', one who bores or pierces. cf. Adam Cok *borer* 1366 MESO (Lei), *v.* also BOWRA.

**Borger:** *v.* BURGER
**Borges:** *v.* BURGES

**Borgonon, Burgin:** Robert *Burguignon* 1160 P; John (*le*) *Burguinun* 1173 P (Lo); John *le Burguinn* 1214 P (Mx); John *le Burgenun* 1259 Acc; John *Burgoynoun* 1330 Trentham (St). OFr *Bourguignon, Bourgoin* 'the Burgundian'. *v.* BURGOIN.

**Borkett, Borkwood:** *v.* BURCHARD
**Borlace:** *v.* BURLES
**Borland:** *v.* BURLAND
**Borlas, Borlase:** *v.* BURLES

**Borley:** Almar *de Barlea* 1066 DB (Ess); Roger *de Borle* 1327 SRSx. From Borley (Ess), or Boreley in Ombersley (Wo).

**Borman:** *v.* BOWERMAN

**Born, Borne:** William, Walter *le Borne* 1164, 1185 P (Nf, Do); Simon *Monoculus* 1212 Cur (Berks). OFr *borgne* 'one-eyed, squint-eyed'.

**Borner:** *v.* BOURNER

**Borodale, Borradaile, Borrodell, Borrowdale:** John *Borowdale, Borowdall* 1433, 1483 FrY. From Borrowdale (Cumb, Westmorland).

**Borrage, Borrich:** Henry *Borrich* 1327 SRSo; Hugh *Borage* 1546 FFEss; Walter *Borrage* 1642 PrD. Perhaps OFr *bourgage* 'a freehold property in a town', for the holder of such a property. cf. Borrage Green in Ripon (WRY). It could also be a plant-name from OFr *bourage* 'borage'.

**Borrell, Borrill:** *v.* BURREL
**Borrett, Borritt:** *v.* BURRARD, BURRETT
**Borrow(s):** *v.* BURROUGH(ES)

**Borrowman, Borroman, Burkman, Burman, Barrowman, Barryman:** John *Burgman* de Eboraco 1219 AssY; William *Burman* (*Burghman*) 1221 AssGl; Robert *le Borekman* 1279 RH (Bk); John *Burgman, Burkman* 1281, c1284 NottBR; Geoffrey *Borughman* 1309 ib.; Thomas *Burghman* 1314 FFK; William *Borovman* 1437 Black (Montrose); Alexander *Burowman* 1468 ib. (Edinburgh); Lokky *Barrowman* 1570 ib. (Aberdeen). OE *burhmann* 'townsman, citizen, burgess', in some Yorkshire boroughs 'a burgage tenant'. Both meanings are found in Scotland. Black explains *Barrowman*, also found as *Barraman*, as 'one who helps to carry a handbarrow', but, apart from his first example which is found as *Baroumane* and *Borrowman* in two different MSS, his earliest evidence is in 1570, whilst the first example in NED is c1675. There can be no doubt that this is a late variant of *Burowman*. *v.* also BOWERMAN.

**Borshell:** Walter *de Borselle* 1296 SRSx. From Boarzell in Ticehurst (Sussex).

**Borthram:** v. BARTRAM

**Borton:** v. BURTON

**Borum:** v. BOREHAM

**Bosanquet:** John and David Bosanquet of Lunel in Languedoc came to England as Huguenot refugees in 1685. Their surname is probably the Languedoc *Bouzanguet* 'dwarf'.

**Boseley, Bosley, Bossley:** John *de Boseley* 1336 AssSt. From Bosley (Ch).

**Bosence:** v. BOSSOM

**Boshell:** v. BUSHELL

**Bosher, Boshere, Boshier, Busher:** William *Boschier* 1205 P (Do); Henry *Boscher* 1221 AssWa; Robert *le Buscher* 1276 LLB A. OFr *buischier* 'woodmonger'.

**Boskin:** v. BUSKENS

**Boss:** *Bosse* filius Edrici 1196 P (Sx); Radulfus *filius Bosse* 1210 P (Nf); Hugo, Walter, William *Bosse* 1179 P (C), 1191 P (Y), 1227 AssBeds. A common surname in the 12th and 13th centuries, with more than one origin. The personal name is OG *Boso* 'wicked', Fr *Bos, Boson*. (ii) A nickname from ME *boce, bos*, OFr *boce* 'protuberance, swelling'; 'a hunch or hump on the back', metonymic for a hunch-back. cf. 'crumpled knees and boce on bak' a1300 NED. (iii) In 1333 a vessel called *bos* was used for carrying mortar at Westminster. cf. 'a boket called *le bosse*' (1423 Building 338, 353). This must be *boss* sb. 4, 'a plasterer's tray or hod' (1542 NED), for a maker or a user of these.

**Bossal, Bossall:** William *of Boscehal* 1240 FFY; John *de Bossale* 1290 IpmY; Edmund *de Bossal* 1376 FFY. From Bossal (NRY).

**Bossard, Bosshard:** Henry *Bossard'* 1221 AssSa; Geoffrey *Bussard, Bosard* 1283 SRSf; William *Bosard* 1350 ColchCt. OG *Boshard, Bossard*.

**Bossel:** v. BUSHELL

**Bossey, Bossy:** Simon, Robert *le Bocu* 1196 P (C), 1202 FFK; Robert *le Bossu* 1275 RH (Do). OFr *bocu* 'hunch-backed'.

**Bossiney:** Walter *de Boscinny* 1297 MinAcctCo. From Bossiney in Tintagel (Cornwall).

**Bossley:** v. BOSELEY

**Bossom, Bosson, Bossons, Bosence:** John *Botswayne*, Armand *Bosome*, John *Bottswaine* 1639, 1644 EA (OS) iii, 53, 118 (all in Beccles). A late development of BOATSWAIN. The Sussex *Bossom* is from Bosham (Sussex).

**Bost:** v. BOAST

**Bostel, Bostle:** Martin *de Borstall'* 1198 P (K); Robert *atte Borstalle* 1296 SRSx. OE *borgsteall* 'place of refuge', later 'pathway up a steep hill', common in Kent: Borstal, Bostall Wood, Borstal Hall, Borstalhill Fm; also at Bostal Road in Poynings (Sussex), Boarstall (Bucks) and Boshill (Devon).

**Bostock, Bostick:** William *de Bostok* 1259 AssCh; John *Bustok* 1394 CtH; Robert *Bostocke* 1559 Pat. From Bostock (Ch), or Bostock's Fm in Ewhurst (Sr).

**Boston:** Thomas *de Boston* 1290 FFY; John *de Boston* 1384–5 IpmNt; William *Boston* 1412 FrY. From Boston (L).

**Boswall, Boswell:** William *de Boesavilla* c1130 StCh; Michael *de Bosevill'* 1176 P (Ess); Hugo *de Beseuilla* 1199 FF (Sx); William *Bosevyll* 1362 Shef (Ravenfield); John *Boswell* 1379 PTY. From

Beuzeville-la-Giffard (Seine-Inférieure), *Boesevilla, Bosavilla* 12th ANF.

**Bosworth:** Richard *de Baresworth'* 1206 Cur (Nth); William *de Boreswurth'* 1230 Cur (Lei); Alexander *de Boseworth'* 1298 AssL; Henry *Boseworth* 1327 SRWa. From Bosworth (Leics), *Baresworde* DB.

**Botham:** v. BOTTOM

**Bothamley:** v. BOTTOMLEY

**Bothell, Bottle:** William *Bothel* 1296 SRSx; Robert *atte Bothele* 1327 SRSo; John *of Botill* 1401 AssLa. From Bothel (Cu, Nb), or 'dweller at the hall or manor-house', OE *\*bōþl*.

**Botler:** Sarra *le Bottler* 1332 SRWa; Johan *Bottler* 1351 AssEss. A derivative of ME *botel*, OFr *bouteille* 'bottle', a maker of leather bottles. cf. Stephen, Thomas *Botelmaker* 1346 ColchCt, 1420 LLB I.

**Botley, Botly, Botteley, Bottley:** Walter *de Botele* 1279 RH (O); Robert *de Bottelegh* 1296 SRSx; John *Botlee* 1379 LoCh. From Botley (Berks, Bk, Ha, Wa).

**Botsford, Botsforth:** John *Botesforth* 1465 Paston. From Bottesford (L, Lei).

**Bott:** *Botte* Buny 1222 AssWa; Aldred *Bot, Alrebot* 1066, 1086 DB (K); Walter, William, Richard *Botte* 1189 P (O), 1214 Cur (Wa), 1225 Lewes (Nf); Richard *Bot* 1212 Fees (Ha); Walter *le Botte* 1279 RH (O). Tengvik takes the DB name to be OE *\*Butt*, a nickname, but it should, no doubt, be taken with the other forms. We have clearly a nickname from OFr *bot* 'toad' and are also concerned with a personal name, probably OE *Botta*, found in Botley (Bucks) and in BOTTING.

**Bottams:** v. BOTTOM

**Botteley:** v. BOTLEY

**Bottell:** v. BOTTLE

**Botten:** v. BUTTON

**Botterell, Botterill, Bottrell, Bottrill, Butteris, Butteriss, Buttress, Buttriss:** (i) Hamo, Rannulf *Boterel* c1155 DC (L), 1198 P (Nf); Reginald *Boterell* 1193 P (Y); (ii) Emma *des Boterell'*, *des Boteraus* 1197 P (D), 1211 Cur (So); William *de Botereus* 1277 AssSo; Thomas *Buttris* 1639 YWills. These surnames are difficult and complicated. We seem to have a nickname from OFr *boterel* 'toad' but Peter *Boterel* 1127–47 Bec (W) is also called *Boter* (1107–33 ib.). He was one of the family of *Butery*, tenants of Ogbourne (Wilts), other members of which were Geoffrey *Boter* (1107–33), William *Boterel* (1122–47), James *Butery* and William *Buteri* (c1248). The various forms of the surname must have the same meaning. *Buteri* is for *atte buteri* 'keeper of the buttery', from OFr *boterie*, late Lat *botaria*, from *bota*, a variant of *butta* (OFr *botte*) 'cask, bottle'. cf. BUTTERY. *Boter* is a derivative of *bota*, one in charge of the casks or bottles. The diminutive *Boterel* is curious but we may compare Fr *Pasturel* from *Pasture* 'shepherd or owner of the pasture' and *Peverell*, a diminutive of *peivre* 'pepper'. Boscastle (Cornwall) is *Boterelescastel* in 1302 and was then held by William *de Botereus* whose family presumably took its name from Les Bottereaux in Normandy (DEPN). The place-name means 'Boterel's castle' and is, no doubt, to be associated with William *boterel* 1130 P (Co). One would take this to be a nickname from the toad were he not called *de* or *as Boterell'* in 1178 (P). Aston Botterell (Salop) was held in 1203 by William *Boterell* whose surname is taken as the nickname in DEPN. He is probably identical with William *des Boterels, des Botereals* 1197–8 P (Sa) and of the same family as Albreda *Boterell', de Botereus, de Boterell'* 1221 AssSa, *de Botereaus, de Boterels* 1226, 1242 Fees. Here we seem to have early examples of the loss of the preposition and the substitution of the singular *Boterel* for the plural form of the place-name which appears to mean 'the toads'. The modern surnames may represent all these varieties. *Buttress, Buttriss,* and *Butteriss* are certainly from Les Bottereaux.

**Botting:** John *Bottyng* 1277 AssSo; William *Botting* 1327 SRSx. OE **Botting* 'son of *Botta'. v.* BOTT.

**Bottle:** *v.* BOTHELL

**Bottle, Bottell:** Godwinus *filius Botild* 1188 BuryS (Sf); Johannes *filius Botill'* 1219 AssY; Adam *Botild'* 1221 AssGl; Richard *Botyld* 1296 SRSx; Cuthbert *Bottyll* 1565 Oxon. ON *Bóthildr*, ODa, OSw *Bothild* (f). Also, probably, metonymic for BOTLER.

**Bottlerell:** A curious corruption of BOTTERELL.

**Bottly:** *v.* BOTLEY

**Bottman:** *v.* BOATMAN

**Bottom, Bottome, Bottoms, Bottams, Botham, Bothams:** Dowe *de Bothemes* 1246 AssLa; Richard *del Bothom* 1307 Wak (Y). 'Dweller in the dell or hollow(s)', OE *botm* 'bottom, lowest part of a valley'.

**Bottomley, Bothamley:** Hanne *de Bothemley* 1277 Wak (Y); Peter *Botumley* 1524 SRSf. From Bottomley near Halifax (WRYorks).

**Bottrell, Bottrill:** *v.* BOTTERELL

**Bouch, Buche, Budge:** Ralph *Buche* 1160–70 Templars (Y); Fegga *Buche, Bucca, Bugga* 1165–7 P (L); Hugo *Buche, Bucca* 1199, 13th Guisb (Y); Alexander *Buche, Bugge* 1221 AssWo; Michael *od la Buche* 1225 Pat; Geoffrey *Bouche* 1226 FFBk; John *Bouge,* Walter *Bugge* 1327 SRSf; John *Bougge* 1327 *SR* (Ess). OFr *bouche* 'mouth', a nickname. In ME this also became *bouge* and later *budge,* especially in the sense 'an allowance of victuals granted by a king or nobleman to his household or attendants on a military expedition' (c1440 MED). This sense may be older and may account for some of the alternatives above. *Buche* is common. The form *Bugge,* also common, is ambiguous. It may be for ME *bogue,* OFr *bouge,* also *boulge, buche* (Godefroy) 'a small leather bag or wallet, a skin-bottle' and denote a maker of these. Or it may be for BUGG, where doubtful forms are given.

**Bouchard:** *v.* BURCHARD

**Boucher:** *v.* BUTCHER

**Boucher, Bouchier:** *v.* BOOZER

**Boud:** *v.* BOLD

**Bouffler:** James *Beauflour* 1313 Cl (Beds), 1322 ParlWrits (Lo). OFr *beau flour* 'fair flower'.

**Bough:** *v.* BOW

**Boughtflower:** *v.* BOUTFLOUR

**Boughton:** Geoffrey *de Bocton* 1202 FFY; Walter *Bugheton* 1255 AssSo; Henry *de Boketon'* 1314–6 AssNth; Thomas *Boughton* 1440–1 FFWa. From Boughton (Hu, L, Nf, Nt, Nth), Great Boughton

(Ch), or Boughton Aluph, Malherbe, Monchelsea, under Blean (K).

**Boughtwood:** *v.* BOLTWOOD

**Boulder:** *v.* BOLDER

**Bould(s):** *v.* BOLD

**Boulding:** *v.* BALDING

**Boulderstone:** *v.* BALDERSTON

**Boule:** *v.* BOWL

**Boulger, Boulsher, Bolgar, Bolger, Bulger:** John *Bulgere* 1300 MESO (Wo). OFr *boulgier* 'maker of leather wallets or bags', from OFr *boulge,* ME *bulge* 'leathern bag'.

**Boullen, Boullin:** *v.* BULLEN

**Bouller:** *v.* BOWLER

**Boulsher:** *v.* BOULGER

**Boulsover:** *v.* BOLSOVER

**Boult:** *v.* BOLT

**Boultbee:** *v.* BOLTBY

**Boulter:** *v.* BOLTER

**Boultwood:** *v.* BOLTWOOD

**Boumphrey, Bomphrey, Bumphries:** Roger Bomfrey 1633 Bardsley. Welsh *ab Humphrey* 'son of *Humphrey' v.* PUMFREY.

**Bound(s), Boundy:** *v.* BOND

**Bourchier:** *v.* BOOZER

**Bourde:** Robert *Bourde* 1327 SRSo. OFr *bourde* 'jest', metonymic for *bourder* 'jester'.

**Bourdillon:** A Huguenot name. James *Bourdillon,* descendant of a Huguenot who left France in 1685, was minister of the Artillery Church in Spitalfields. From OFr *borde* 'an isolated country house', or, more rarely, 'man from Bordeaux'. *v.* Dauzat.

**Bourdon:** *v.* BURDEN

**Bourgein, Bourgoin:** *v.* BURGOIN

**Bourke:** *v.* BURK

**Bourne, Born, Borne, Burn, Burne:** Godric *æt Burnan* 1044 OEByn (K); Almarus *de Brunna* 1066 ICC (C); Basilia *de la Burne* 1219 FFEss; William *Atteburn* 1256 AssNb; Richard *Atteburne* 1261 AssSo; Adam *de Burne* c1280 Black (Ayr); Richard *atte Bourne* 1327 SRSx; Robert *del Burn* 1332 SRCu. The first reference above is to Bishopsbourne (K), OE *burna* 'stream', the second to Bourn (Cambs), ON *brunnr* 'stream'. In the North and Scotland *burn* is still the living word for a stream. In the rest of England it was early replaced by *bróc* 'brook' and in the south *bourn* came to be used of a stream which flows only in winter or at long intervals, a meaning still found in the dialects of Kent, Surrey and Wilts. Here, in surnames, the reference is usually to an old stream called *burna,* a name often surviving as that of a farm, etc. Bourne (Surrey) is named from an intermittent stream. *v.* BORN.

**Bourner, Borner, Burner:** Walter *Bournere* 1318 LLB E. 'Dweller by a stream', equivalent to *atte burne. v.* BOURNE.

**Bourton:** *v.* BURTON

**Bousfield, Busfield, Busfeild:** James *Bowsefell* 1567, James *Busfeld,* Christopher *Bowsfeild* 1615 FrY; Alder' *Busfeild* 1672 HTY. From Bousfield in Orton (We), *Bowesfell* 1279.

**Bousie:** *v.* BOOSIE

**Bouskell, Bouskill, Bowskill:** Giles *Bowskille* 1560

Pat; John *Borkenskale* 1583, William *Borrenscale* 1602, John *Burascale* 1649, George *Buskill* 1653 FrY. From Bowscale in Ulpha, Borrowscale in Matterdale, or Borrowscale in Torpenhow (Cu).

**Boutall, Boutell:** *v.* BULTEEL

**Boutcher:** *v.* BUTCHER

**Boutflour, Boughtflower, Bonfellow:** John *Bult(e)flour* 1303 Surnames 258, 1430 FrY; Adam *Bonteflour* 1332 Sundby; Helen *Bonfela* 1438 NorwW (Sf); John *Bownflower* 1505 ArchC 41; John, Thomas *Boutflower* 1511 ib.; William *Buntflowre* ib.; Robert *Bonfelow, Bunfellow* 1521 NorwW (Nf); Ralph *Bultflower* 1568 SRSf. ME *bulte* 'to sift' and *flour* 'flour', 'sift flour', a nickname for a miller. cf. BOLTER. *Buntflowre,* pronounced *Bunfler, Bunfeler,* was reconstructed as *Bonfellow.*

**Boutwood:** *v.* BOLTWOOD

**Bouverie, Bouvery:** Laurence *Buveries* 1199 FFEss. 'Dweller at the place where oxen are reared', OFr *bouverie.*

**Bover, Bovier:** Daniel *le buuier* 1191, *le bouier* 1197 P (Y); John *Bovier* 1327 SRSx. OFr *bovier, buvier, bover,* 'ox-herd'. This would usually appear as *bouer* and be indistinguishable from the forms for BOWER.

**Bovey:** Walter *Boue* 1185 Templars (L); William *Bouy* 1219 AssY; Richard *Bovie* 1642 PrD. ON *Bófi,* ODa *Bovi,* OSw *Bove.*

**Bovier:** *v.* BOVER

**Bovill, Boville, Bovell:** William *de Bu uilla, de Boe uilla, de Bee uilla, de Boeuilla,* Humphrey *de Buivile,* Sahala *de Bou uilla* 1086 DB (Sf); William *de Bouilla, de Buiuilla* c1150 DC (L); William *de Bowile* 1179 Clerkenwell (Ess); John *de Bouilla* 1182 Eynsham (O). Probably from Bouville (Seine-Inférieure). Early forms of two places named Beuville in Calvados make these less certain identifications (OEByn). Four places in Essex preserve this surname: Bovill's Hall (2), Uplands and Marsh, the latter occurring as *Bowelles* (temp. Henry VIII), so that the surname may also have become BOWELL. *v.* BOWLES.

**Bovington:** Walter *de Boventon'* 1200 Cur, *de Bovinton* 1204 AssY. From Boynton (ERY), *Bovintone* DB, Bovington (Do), or Bovington Hall in Bocking (Ess).

**Bovis:** *v.* BEAVES

**Bow, Bowe, Bough:** Henry *atte Bowe, atte Bogh* 1298, 1304 PN D 513; Richard *atte Bowe* 1306 LLB B; Nicholas *atte Boghe* 1327 SRSo. From Bow (Devon, Middlesex) or from minor places of the same name. 'Dweller near a bridge', from OE *boga* 'bow, arch, vault', here 'an arched bridge'.

**Bowbrick:** Wigor *Buuebroc* 1221 ElyA (Sf); John *Abovebrok* 1279 RH (Hu); John *Bowebroke* 1453 SRSr. 'Dweller above the brook', OE *būfan brōce.*

**Bowcher:** *v.* BUTCHER

**Bowcock, Bowcott:** *v.* BAWCOCK

**Bowden, Bowdon:** *Bowden* is frequent but often represents an older *Bowdon.* (i) John *de Boghedon* 1333 PN D 205. There are 17 places called Bowden in Devon and one Bowdin, all 'curved hill'; Bowden Edge (Derby) has the same origin; (ii) Thomas *Bovedon'* 1279 RH (O); OE *būfan dūne* '(dweller) above the hill', as at Bowden (Wilts); (iii) also from

Great Bowden (Leics) or Bowdon (Ches), earlier *Bugge-, Bogedone;* (iv) from Bowden (Roxburghshire): Richard *de Boulden* 1200–40 Black.

**Bowder:** *v.* BOLDER

**Bowdler:** Richard *le Boudler* 1274 RH (Sa); William *Bowdeler* 1493, Andrew *Bowdler* 1644 SaAS 2/xi, x. Perhaps, as suggested by Harrison, a derivative of dialectal *buddle* 'to wash ore', and hence a nickname for a miner.

**Bowell:** Peter *de Boelles* Hy 3 HPD (Ess); Ralph *de Bueles* 1249 FFEss; Walter *de Bowell* 1275 RH (Herts). In 1086 Shellow Bowells (Essex) was held by Lambert *de Buella* who probably came from Bouelles (Seine-Inférieure). *v.* also BOVILL.

**Bowen:** Madocus *ap Oweyn* 1292 QW (Sa); John *Bowen* 1305 FrY; Riseus *Abowen* alias *apowen* 1558 AD v (Gl). Welsh *ab Owein* 'son of Owen'.

**Bower, Bowers, Bour:** (i) Matthew *de Labur'* 1194 Cur (Sr); Mayfflin *Attebur'* 1280 AssSo; Henry *del Boure* 1287 AssCh; Gilbert *atte Boure* 1296 SRSx; Lorence *atte Bure* 1296 Black (Peebles); Peter *ate Boures* 1327 SRC. From minor places called Bower (Som, Sussex, Peebleshire, etc.) or equivalent to CHAMBERS 'chamber-servant', from OE *būr* 'cottage, chamber'. (ii) Teodricus *Bouer* 1187 P (He); Peter *le Bouer* 1296 SRSx; John *Bour* 'bowyer' 1325 Pat; Robert *le Bowyere, le Bower'* 1327, 1332 SRSt. A form of ME *bowyere,* identical with BOWYER.

**Bowering:** *v.* BOWRING

**Bowerman, Boorman, Borman, Burman:** William *Bureman* 1204 P (Y); Robert *Boreman* 1279 RH (O); William *Bourman* 1327 SRSx; Walter *Burman* 1327 SRC. OE *būr* and *mann* 'a servant at the bower'. Identical in meaning with *atte Boure. v.* BOWER and BOWRA. cf. Alice *Bourwyman* 1301 SRY, Marion *Bourswain* 13th AD i (Sx), Alice *Bourgrom* 1327 SRSo. *v.* also BORROWMAN.

**Bowes, Bows:** Gerard *de Bowes* 1269 AssNb; John *de Boughes* 1341 FrY; John *Bowys* 1423–4 FFWa. From Bowes (NRYorks), or 'dweller at the arches or bridge', OE *boga.*

**Bowgen, Budgen, Budgeon:** Robert *Bonjohan* 1297 MinAcctCo; Thomas *Bowgeon* 1454 Fr Norw; John *Bowgyn* 1524 SRSf. Fr *bon Johan* 'good John', perhaps 'the good servant'.

**Bowie, Buie, Buy, Buye:** John *Boye* alias *Bowy* alias *Boee* 1481 Black; Donald *Buy* 1671 ib. Gael *buidhe* 'yellow or fair haired'.

**Bowker:** *v.* BUTCHER

**Bowl, Bowle, Boule:** John *le Boul* 1296 SRSx; James *Boule* 1297 MinAcctCo. OFr *boule* 'round'. cf. BULLETT. Probably also metonymic for *Bowler.*

**Bowland:** Robert *of Bowland* 1339–40 Black; William *de Bowland* 1380 TestEbor; Adam *Boweland* 1415–6 FFSr. From Bowland Forest (La, WRY).

**Bowler, Bouller, Boaler, Boler:** John *le Bouller* 1316 FFSo; Robert *le Bollere* 1332 SRSt. A derivative of OE *bolla* 'bowl', a maker or seller of bowls. Also 'one who continues at the bowl, a tippler' (c1320 NED).

**Bowles:** John *de Boweles* 1292 FFHu; Thomas *Bowles* 1553 ib. *v.* BOWELL.

**Bowling:** *v.* BOLLING

**Bowman, Bowmen, Boman, Beauman:** Adam

*Bogheman* 1223 Cur (We); Thomas *Bouman* 1279 AssNb; Nicholas *the Bowemon, the Bouman* 1286–7 AssCh. OE *boga* 'bow' and *mann*, a bowman, a fighting man armed with a bow (1297 NED).

**Bowmar, Bowmer:** *v.* BULMER

**Bown:** *v.* BOON

**Bownas, Bownass, Bowness, Bonas, Bonass, Boness:** William *Bownus* 1592 FrY; Matthew *Bownas* 1633 ib.; Richard *Bowness* 1758 ib. From Bowness (Cumb).

**Bowra, Bowrah, Boarer, Boorer, Borer, Burra:** Hugo *le Burer* 1218 AssL; Alice *Burrer* 1279 RH (C); William *le Bourere* 1332 SRSr; John *Bourere* 1375 FFSx; John *Bowrer* 1498 AD vi (Sr); William *Bowrar* 1535 SxWills; John *Bowra* 1683 ArchC 53; Thomas *Borer* 1697 DKR 41 (Sx). The name of Thomas *atte Boure*, MP for Horsham in 1320, eventually took the form of *Bourer* or *Borer*, whence the family of William *Borrer*, High Sheriff of Sussex (SxAS viii, 274). The meaning is identical with that of *Bower* and *Bowerman* 'dweller or servant at the bower' (OE *būr*). *Bowra* is a specifically Kentish form. Thomas *Bowra*, surgeon, of Sevenoaks, who was born at East Grinstead, Sussex, abandoned the earlier forms of his name, *Boorer* or *Bowrer*, after migrating to Kent during the Commonwealth. The relatives he left behind in Sussex continued to use various older forms, *Boorer, Boreer, Boorer, Borar, Borra* (ArchC 58, 77).

**Bowran, Bowron:** *v.* BOLDRON

**Bowrick:** John *le Boghewrichgte* 1292 MESO (La); Robert *Bowwright* 1332 SRCu. OE *boga* and *wyrhta* 'bow-maker'. cf. ARTRICK for ARKWRIGHT.

**Bowring, Bowering:** Henry *Bouryng* 1302 DbCh; Walter *Bowryng* 1327 SRSo. OE *\*būring*, a derivative of *būr* 'bower', probably synonymous with BOWRA and BOWERMAN.

**Bowser:** *v.* BEAUSIRE

**Bowser, Bowsher:** *v.* BOOZER

**Bowsie:** *v.* BOOSIE

**Bowskill:** *v.* BOUSKELL

**Bowtell, Bowtle:** *v.* BULTEEL

**Bowton, Bufton:** Robert *Buveton'* 1222 Cur (Beds); Roger *Abovetoun*, John *Aboventoun* c1240 Rams (Hu). '(Dweller) above the village' (OE (*on*) *būfan tūne*).

**Bowyer, Boyer, Boayer:** Ailwardus *le Bogiere* 1183 P (Lo); William *le Boghier, le Bowiere* 1275 RH (Lo); William *Boyer* 1279 RH (Hu); Henry *le Bowyere* 1296 FFSf. ME *bowyere* 'maker of or trader in bows' (c1300 MED), from OE *boga* 'a bow'. *v.* BOWER.

**Box:** (i) Adam *Box* 1276 LLB A, 1317 AssK; John *Box* 1327 SRC. OE *box* 'box-tree', 'box-wood' was used in ME of colour, or of teeth yellow as box, and associated with jaundice. (ii) From Box (Glos), William *Box* 1181 P (Gl); Box Hall (Herts), Alan *de Boxe* 1198 P (Herts); or Box (Wilts), Ebrard *de La Boxe* 1182 P (W). Or from residence near a box-tree: Thomas *atteboxe* 1263 PN Sr 270.

**Boxall, Boxhall, Boxshall:** John *de Bokeselle* 1296, John *Boxole*, Vmfrid *Boxholt* 1525 SRSx. From Bugsell Fm in Salehurst (Sx), *Bokeselle* c1260, or a lost *Boxholte* in Kirdford (Sx).

**Boxer:** Apparently from a lost place in Oxfordshire: Alice *de Boxore* 1279 RH (O), 'box-tree bank'.

**Boxley:** John *de Boxelee* 1325 LLB E; Thomas *Boxly* 1561 PN Do ii 191. From Boxley (K).

**Boxshall:** *v.* BOXALL

**Boyce:** *v.* BOYES

**Boycott, Boykett:** Gerard, William *de Boycote* 1256 FFK, 1278 RH (Bk); John, Adam *de Boycote* 1334–5 SRK. From Boycott (Bucks, Salop).

**Boyd, Boyde:** (i) John *Boyde* 1301 SRY. Gael, Ir *buidhe* 'yellow'. The Manx name is from *Mac Giolla Buidhe* 'the yellow-haired youth's son': Conn *Mac Gillabhuidhe* 1100, *McOboy, McBooy* 1511, *Boid* 1617 Moore. (ii) In Scotland and Ireland, from Bute (Gael *Bod*): Robert *de Boyd* 1205 Black; Walter *de Boht* c1272 ib.

**Boydell:** Hugh *de Boidel* c1200 WhC; William *Boydell'* 1382 AssL; Laurence *Boydell* 1401 AssLa. From an unidentified place of this name, probably in Lancashire.

**Boyden:** *v.* BODIN

**Boyer:** *v.* BOWYER

**Boyes, Boys, Boyse, Boice, Boyce:** (i) Nicholas *del Bois* 1201 P (L); Thomas *Boys* 1296 SRSx; Roger *du Boys* 1327 SRSf; John *Boyse* 1396 FrY. Fr *bois* 'wood'; equivalent to *atte Wode. v.* WOOD. (ii) Godui *Boie filius* c1095 Bury (Sf); Stephanus *filius Boie* 1202 P (Nth); Aluuinus *Boi* 1066 DB (Sr); William *Boie* 1166 P (Nf); Ivo *le Boye* 1232 Pat (L). The personal-name may be OG *Boio* or OE *\*Boia*, derived from the ancestor of *boy* which is not recorded before 1300. This is common in place-names and is probably of native origin. In ME it meant 'young man' or 'servant' (c1300 MED). *Boie* was the name of a 'border' 1221–26 AD iv (So). cf. LADD. It may also have been used as a nickname, 'knave, rogue, wretch' c1300 MED.

**Boyland, Boylan:** (i) Ralph *de Boilund* 1198 FFNf; John *Boilonde* 1349 FFW; Robert *de Booylond* 14th Hylle. From Boyland (Nf), or Boyland in Dunsford (D). (ii) Irish *Boylan* is from *Ó Baoighealláin*, the meaning of which is unknown.

**Boyle, Boyles:** John *Boyle* 1340–1450 GildC; William *Boyle* 1378 LoCh; Robert *Boyll* 1545 SRW. Perhaps from Boyville, Boeville (Seine-Maritime). In the 17th century used as a christian name: *Boyle, Boyell* Hall 1642 PrD. In Scotland from Boyle (Ayr, Wigtown). It is also a common Irish name, the derivation of which is uncertain. *v.* MacLysaght.

**Boyman:** Geoffrey, Robert *Boyman* 1268 AssSo, 1357 Black. 'Bowman.' cf. BOWYER.

**Boynton:** Walter *de Bouyngton* 1210–26 YCh; Ernulf *de Boynton* 1243 FFK; William *Bointon* 1365 IpmW; Thomas *Boynton* 1408 FrY. From Boynton (ERY), *Bovintone* DB, Bovington Court in Swingfield (K), *Bointon* 1207, or Boyton (W), *Boynton* 1366.

**Boys(e):** *v.* BOYES

**Boyten, Boyton:** Richard *de Boitona* 1198 FFSf; Robert *de Boyton* 1268 AssSo; Ralph *de Boytone* 1317 AssK. From Boyton (Co, Ess, Sf), or Boyton Court in East Sutton (K).

**Braban, Brabant, Braben, Brabin, Brabon, Brabyn, Brabban, Brabben, Brabbins, Brabham:** Richard

*Braban* 1260 AssC; Heliseus *de Brabayn* 1275 RH (L); Raban *de Braban* 1281 LLB A; John *le Braban* 1283 LLB A; John *Brabon* 1296 SRSx; Richard *Brabyn* 1549 FrY. Flemish *Brabant* 'a native of Brabant' (Flanders), alternating with the name of the Duchy.

**Brabazon, Brobson:** Thomas *le Brabacum* c1200 Gilb (L); Thomas *Brabezon* 1276 RH (Y); Thomas *le Brabazun* 1294 AssSt; Roger *le Brabanzon* 1301 LLB C; Adam *Brabson* 1381 AssWa; Geoffrey *Brabysson*, *Brabbesson* 1472 NorwW (Sf); Richard *Bropson* 1531 KentW. AFr *Brabançon* 'a native of Brabant'. cf. BRABAN. In the 13th century, *les brabançons* were companies of adventurers of various nationalities who devasted several French provinces.

**Brabban, Brabben, Brabbins, Brabham:** *v.* BRABAN

**Brabiner, Brabner, Brebner:** Peter *Brabaner*, *Brabaynner* 1379 PTY. 'A native of Brabant', an English formation by the addition of *-er* to the name of the Duchy.

**Brabrook:** *v.* BRAYBROOK

**Brace:** *v.* BRASS

**Bracebridge:** John *de Bracebrig'* 1218 P (Lei/Wa); Thomas *Bracebrigg* 1410 IpmY. From Bracebridge (L).

**Bracegirdle:** John *Brachgyrdyll* or *Brecchegirdle* 1544 Oxon; Roger *Brachegirdle* or *Brasgirdell* 1556 ib.; John *Bretchgirdle* 1561 Pat (Wa); Timothy *Brasegirdell* 1620 Bardsley; Roger *Bracegirdle* 1649 ChW. Metonymic for 'a maker of breech-girdles', cf. William *Brigerdler* 1281 LLB B. ME *brēc*, OFr *braie* 'breeches' and OE *gyrdel*.

**Bracer:** *v.* BRASSEUR

**Bracewell, Brazewell:** Gilbert *de Braycewell* 1251 AssY. From Bracewell (WRY).

**Bracey:** *v.* BRASSEY

**Brach, Breach, Brech, Britch:** Peter *de la Breche* 1221 AssSa; Peter *de la Brach* 1248–52 AD iii (Sr); Philip *atte Breche* 1296 SRSx; Rose *atte Brache* 1309 SRBeds. From residence near a piece of newly cultivated land (OE *brǣc*), as at Breach in Maulden and Brache in Luton (Beds).

**Bracher, Bratcher, Breacher, Brecher, Britcher:** Robert, William *le Brechere* 1245 Oseney (O), 1297 MinAcctCo. Identical in meaning with *atte brache*. *v.* BRACH.

**Brachett, Brackett:** Ralph, Richard *Brachet* 1214 Cur (C), 1327 *SR* (Ess); William *Braket* 1524 SRSf. OFr *brachet*, a diminutive of *brach*, from which *braket* was also formed. *v.* BRACK.

**Brack:** Relicta *le Brak* 1296 SRSx; William *Brak* 1327 SRSx; Thomas *Brakke*, *Braxez*, *Brax* 1484, 1496, 1532 FrY. ME *braches* plur. is probably OFr *brachés*, plur. of *brachet*. From this plural was apparently educed an English singular *brache* 'a hound which hunts by scent' (c1340 NED). *braches* occurs as *brackes* in 1490.

**Bracken:** Stephen *del Bracyn* 1219 AssY; William *Braken* 1332 SRSt; Richard *Braken* c1540 PN We 142. From Bracken in Kilnwick (ERY), Bracken Fold in Whinfell (We), or 'dweller in the bracken', ON *brakni*.

**Brackenbury, Brackenborough, Brackenberry, Brackenboro:** Ranulph *de Brachingberge* c1163 Gilb; Jordan *de Brakenberge* 1202 AssL; Thomas *Brakenborgh* 1388 PN Mx 35; Jacobus *Brakenbury* 1478 FrY. From Brackenborough (L), or Brackenbury Fm in Harefield (Mx).

**Brackenridge, Braikenridge, Breckenridge, Brekonridge:** Nicholas *de Bracanrig* 1332 SRCu; John *Brakanryg*, *Breckinrig* 1454, 1634 Black; William *Braikinrig* 1629 ib. 'Dweller by the bracken-covered ridge.' There are five places named Brackenrigg in Cumberland and one in Lanarkshire.

**Brackett:** *v.* BRACHETT

**Brackley:** Richard *de Brackele* 1202 AssNth; Robert *de Brackeley* 1332 SRSt; Richard *Brackley* 1672 HTY. From *Brackley* (Nth), or 'dweller at the clearing in the bracken', OE *bracu*, *lēah*.

**Brackner:** Walter *le Brakener* 1309 RamsCt (Ess). OFr *braconier* 'a keeper of hounds', from OFr *bracon* 'hound' (1490 NED).

**Bracy:** *v.* BRASSEY

**Bradbrook:** William *de Bradebrok* 1327 SRSf. 'Dweller by the broad brook.'

**Bradbourn, Bradbourne, Bradburn, Bradburne, Brabourn:** John *de Bradeburn* 1275 RH (K); Roger *de Bradeburn* 1286 AssSt. From Bradbourne (Db), Brabourne (K), or 'dweller by the broad stream', OE *brād*, *burna*.

**Bradbury, Bradbery, Bradberry:** William, Richard *de Bradbury* 1288 AssCh, 1327 SRDb; Robert *Bradbury* 1401 AssLo. From Bradbury (Ches, Durham).

**Bradcot, Bradcott:** Richard *de Bradecote* 1275, Thomas *Bradecote* 1332 SRWo. 'Dweller by the spacious cottage, or shelter for animals', OE *brād*, *cot*.

**Bradd:** Robert, Stephen *Bradde* 1275 RH (Sf), 1332 SRCu. A shortening of OE *brād*, 'broad'.

**Bradden, Braddon:** (i) William *de Bradden'* 1297 MinAcctCo, *de Braddon* 1330 PN D 128; John *Braddon* 1642 PrD. From Bradden (Nth), or Braddon in Buckland Brewer, in Ashwater (D). (ii) Richard *Bradhand* 1202 AssL. A nickname, 'broad hand', OE *brād*, *hand*.

**Braddle:** *v.* BRADWELL

**Braddock:** Geoffrey *Brodhok* 1275 RH (K); Thomas *del Brodok* 1282 AssSt; Thomas *Broddock* 1341 NI (Ess). 'Dweller by the broad oak', OE *brād*, *āc*.

**Brade:** *v.* BROAD

**Brader:** Robert *le Bredere* 1286 MESO (Nf); John *le Brayder* 1324 ib. (La); William *Brader* 1382 FrY. A derivative of OE *breȝdan* 'to braid, plait'. cf. William *le Lacebreyder'* 1329 MESO (Nf), 'a maker of cords', from ME *lace* 'cord', and Neet Breydare PromptParv.

**Bradfer:** Matthew, William *Braz de fer* 1205 P (Gl), 1230 P (Berks); Matthew *Bradefer* 1212 Rams (Hu). Fr *bras de fer* 'Iron arm'. Now largely absorbed by *Bradford*.

**Bradfield:** Richard *de Bradefeld* 1182–98 BuryS (Sf); William *de Bradefeld* 1256–7 FFEss; Thomas *Bradfeld* 1454 IpmNt. From Bradfield (Berks, Ess, Nf, Sf, WRY).

**Bradford, Bradforth, Braidford:** Alexander, Brun *de Bradeford'* 1206 Cur (D), 1219 AssY; Thomas *de Bradforth* 1358 FrY. From one of the numerous Bradfords.

**Bradgate:** v. BROADGATE

**Bradie, Brady, Broady:** Roger *Bradeie, Bradege, Brodege* 1170, 1184, 1200 Oseney (O); Geoffrey *Bradege* 1188 P (K); Walter *Bradeye* 1243 AssDu; Simon *Brodhegh* 1247 AssBeds; Robert *Brody* 1275 RH (Nt); William *Brodeie* 1279 RH (C); Agatha *de Brodheghe*, Peter *Brodeye* 1327 SRSf; Richard *Brady* 1430 Black (Dundee). Clearly a nickname, 'broad eye', OE *brād, ēage*; also 'dweller by the broad island' (OE *ēg*), or the broad enclosure (OE *(ge)hæg*). *Bradie* is Scottish. cf. BROAD.

**Bradlaugh:** Robert *de Bradlawe* 1275 RH (Db); Godfrey *Bradlaugh* alias Jacob, William Jacobo alias *Brodleye* 1568 SRSf. From Bradley (Derby).

**Bradley, Bradly, Bratley, Broadley, Brodley:** William *de Bradelai* 1170 P (L); William *Brodelegh* 1379 PTY. From Bradley (Lincs, WRYorks, etc.). v. BRADLAUGH.

**Bradman, Braidman:** William *Bradman* 1275 RH (Nf), 'Broad man'.

**Bradmead, Broadmead, Brodmead:** Roger *atte Brodmed* 1327 MELS (So); Richard *Bradmead*, Thomas *Brodmeade* 1642 PrD. 'Dweller at the broad meadow', OE *brād, mǣd*. cf. William *Brodemedowe* 1340–1450 GildC; John *Bradmedowe* 1356 LLB G, with the same meaning.

**Bradmore, Breadmore:** Ælfger *de Brademere* c1095 Bury (Sf); Aluredus *de Brademor* 1185 Templars (Wa). From Bradmore (Notts) or residence near a broad lake (OE *mere*).

**Bradnam, Bradnum:** Suift *de Bradenham* 1177 P (Nf); Francis *Bradnam, Bradnum* 1612, 1628 WStowPR (Sf). From Bradenham (Norfolk).

**Bradney:** Simon *de Bradneghe* 1327 SRSo; William *de Bradeny* 1296 SRSx. From Bradney (Som) or Bradness Wood (Sussex).

**Bradridge, Broadridge:** Hugo *de Braderugg'* 1275 SRWo. 'Dweller by the broad ridge', OE *brād, hrycg*.

**Bradshaw:** William *de Bradesaghe* 1246 AssLa; Simon *de Bradeshaghe* 1303 FFY; Roger *Bradschawe* 1418 IpmNt; Wylliam *Bradsha* 1554 DbAS xxiv; Robert *Bradshey* 1646 RothwellPR (Y). From Bradshaw (Derby, Lancs, WRYorks).

**Bradstock, Brastock:** Simon *de Bradestoke* 1279 RH (O). From Bradenstoke (Oxon).

**Bradstreet:** Roger *de Bradestret* 1327 SRSf. 'Dweller in the broad street', OE *brād, strǣt*.

**Bradway:** Filbert *Bradewei* 1212 P (Nth); Jordan *de Bradewey* 1235–50 Hylle; John *Bradeweye* 1332 SRWo. From Bradway (Db), or 'dweller by the broad road', OE *brād, weg*.

**Bradwell, Braddle:** Walter *de Bradewelle* 1275 SRWo; John *Braddell* 1622 PrGR. 'Dweller by the broad stream' (OE *brād, wella*), or from Bradwell.

**Brady:** v. BRADIE

**Brafferton:** Reginald *de Braferton* 1212 P (Y); Richard *de Brafferton* 1327 SRY. From Brafferton (Du, NRY).

**Brafford:** William *of Braforde* 1246 IpmY; Benjamin *Brafford* 1642 PrD. Probably for BRADFORD.

**Brafield:** v. BRAYFIELD

**Bragg, Bragge:** Walter *Bragge* 1243 AssSo; Henry *Brag* 1260 AssC, 1275 RH (W). ME *brag(ge)* 'brisk, lively, mettlesome' (c1325 MED).

**Braham, Brahame, Braime, Brame, Bramham, Bramman, Brayham, Bream, Breem:** Eustace *de Braham* 1189–99 Colch (Ess); Matthew *de Bramham, de Braham* 1219 AssY; Agnes *de Brame* 1379 PTY; Elizabetha *Bramam* 1628 RothwellPR (Y). From Bramham (WRYorks), Braham (ib.), *Bram* DB, *Braham* 1242 Fees, Braham Fm in Ely (C) or, occasionally, from Brantham (Suffolk), *Braham* 1200 Cur *et freq*. Braham Hall (Essex) is *Braham* 1314, *Brames* 1429, *Bream* 1777, and Bream's Fm (Essex) is *Braham* 1314, *Braeme* 1540, *Breame's farm* 1680 (PN Ess 333, 257). For the development, cf. GRAHAM.

**Braid, Braide, Bread:** (i) Henry *de Brade* 12th Black; Helen *Braid* 1638 ib. From Braid Hills near Edinburgh. In Fife and Perthshire, this is now *Bread*. (ii) Geoffrey *Braid'* 1198 FFNf; William *Breyd* 1275 RH (L). These forms are too early to be regarded as forms of BRADE 'broad'. We must have OE *brægd* 'a sudden jerk', used in 1530 for 'a plait, braid', a sense which must have developed much earlier as in the verb. Metonymic for BRADER.

**Braidford:** v. BRADFORD

**Braidman:** v. BRADMAN

**Braikenridge:** v. BRACKENRIDGE

**Brailer, Brayler:** Roger *le Braeler* 1275 RH (Lo); William *le Braeler* 1281 LLB B; Stephen *le Brayeler* 1311 LLB D. A derivative of OFr *braiel* 'a belt, girdle', for a maker of these.

**Brailey:** v. BRAYLEY

**Brailsford, Brellisford, Brelsford:** Henry *de Brailesford* 12th DbCh; Jeffray *Brelsforth* 1624 Shef. From Brailesford (Derby).

**Brailsham:** Walter *de Braylesham* called 'le Cok' 1341 LLB F. From Brailsham (Sx).

**Braime:** v. BRAHAM

**Brain, Braine, Braines, Brane:** Ketell' *Brain* 1166 P (Nf); Walter *Brayn* 1275 SRWo; John *Braine* 1379 PTY; Thomas *Brayne* 1462, David *Brane* 1477 Black. Perhaps a nickname from ME *brain* 'furious, mad'. In Scotland for MACBRAYNE.

**Brainch:** v. BRANCH

**Brainswood, Brainwood:** Robert *Braynwod* 1327 SREss; John *Braynwode* 1345 PN Ess 235; John *Braynwode* 1488 FFEss. ME *brainwod* 'frenzied, mad'.

**Braithwell:** Adam *de Braythewell'* 1379 PTY; John *Braythewelle* 1401 IpmY. From Braithwell (WRY).

**Braithewaite, Braithwaite, Breathwaite:** Reginald *de Braidewad* 1185 P (Y); Adam *de Braythwayt* 1301 SRY; Peter *Braytwayt* 1364 AssY; Robert *Braithwaite* 1642 PrD. From Braithwaite (Cumb, ER, NR, WRYorks).

**Braizier:** v. BRASIER

**Brake:** Alan *de la Brake* 1176 P (L); John *de Brake* 1275 RH (Nf); Robert *Brake* 1279 RH (Hu). 'Dweller by the copse or thicket', OE *bræc*, ME *brake*.

**Brakefield:** David *de Brakefeld* 1275 RH (Nf). 'Dweller by the bracken-covered open land', ME *brake*, corresponding to the northern *bracken*.

**Brakspear, Braksper:** v. BREAKSPEAR

**Bramah, Bramald, Bramall, Bramalt, Brameld, Bramhall, Brammall, Brammer:** Robert *de Bramhal'* 1221 AssWo; Thomas *Bramall, Brammall* 1543, 1566 ShefA. 'Dweller by the broom-covered nook', OE *brōm*, shortened to *bram*, and *healh*, as at Bramhall (Ches). In Sheffield, where there is a Bramall Lane, *Bramah, Bramall* and *Brammer* are common.

**Bramble, Brambles, Brambell:** Thomas *Brembel* 1296 SRSx; Mabel *Bremeles* 1327 SRSo; Henry *Bremble* 1641 PrSo. cf. John *le Brembestere* 1297 MinAcctCo. The regular absence of any preposition suggests that this is probably a nickname for someone as prickly as a bramble, OE *brǣmel*, rather than for 'dweller among the brambles'.

**Brambley:** v. BRAMLEY

**Brame:** v. BRAHAM

**Brameld, Bramhall:** v. BRAMAH

**Bramford, Bramfit, Bramfitt:** Herbert *de Bramford'* 1205 Cur (L). From Bramford (Sf), but *Bramfit, Bramfitt*, could also be from Bramfield (Herts, Sf).

**Bramham:** v. BRAHAM

**Bramley, Brambley:** Nicholas *de Bramle* 1219 AssY; Goda *de Bremblegh* 1296 SRSx; Richard *Bramley* 1527 CorNt. From Bramley (Db, Ha, Sr, WRY).

**Brammall:** v. BRAMAH

**Bramman:** v. BRAHAM

**Brammer:** v. BRAMAH

**Brampton:** Hermer *de Branton'* 1198 FFNf; Michael *de Brampton* 1275 RH (Y); Thomas *Brampton* 1476 Paston. From Brampton (Cu, Db, He, Hu, L, Nf, Nth, Sf, We, WRY).

**Bramson, Bramston:** v. BRANSTON

**Branaghan, Branigan:** For Irish *Ó Branagáin* 'descendant of *Branagán*', a diminutive of *bran* 'raven'.

**Branch, Branche, Brainch:** Hugh *Branche* 1169–87 P (Sf); William *Branche* 1238 AssSo; Peter *Braunche* 1331 ChertseyCt (Sr). OFr *branche, braunche* 'branch', probably in the sense 'descendant, offspring'.

**Branchett:** v. BLANCHET

**Branchflower:** v. BLANCHFLOWER

**Brand, Brandt, Brant, Braund, Braun, Brauns, Bront:** *Brand* 1193 P (Lo); Jacobus *filius Brand* 1206 AssL; William *Brant* 1086 DB (Nf); Ralph *Brand* 1184 P (Lo); Hamo *Brand, Brant, Braund* 1203–4 P, StP (Lo), 1219 Cur (Beds); Richard *Brawne* 1661 Bardsley. ON *Brandr*, ODa *Brand*, 'fire-brand, sword', found also in Normandy.

**Brandom, Brandon:** Leofric *de Brandune* c975 LibEl (Nf/Sf). From Brandon (Durham, Norfolk, Suffolk, Warwicks) or Brundon (Essex).

**Brane:** v. BRAIN

**Branford, Branfort:** Osbert *de Braneforda* 1175 P (St); Hubert *de Branford* 1200 P (Nf); Henry *de Branford* 1230 P (Mx). From Bramford (Sf), *Branfort* DB, or Brampford Speke (D), *Branford* DB.

**Brangwin, Brangwyn:** *Brangwayna* 1250 FFSf; Adam *Brangwyne* 1283 SRSf; Robert *Brangwayn* 1300 LoCt. Welsh *Branwen, Brangwain* (f), from *bran* 'raven' and *(g)wen* 'fair'. Branwen, daughter of Llyr, is one of the legendary heroines of Wales. In the Tristan

legend Brangwain the Fair was the handmaid and companion of Queen Isolde.

**Branigan:** v. BRANAGHAN

**Brann:** *Bran* 1154–86 Black (Galloway); Philip *Bran* 1275 RH (Sf); William *Bran* 1275 SRWo, 1629 Black. Gael, Ir, Welsh *bran* 'raven'. The name was also used in Brittany, hence, probably, the Suffolk surname.

**Bransby:** Walter *de Brandesby* 1296 Riev; William *Braunceby* 1369 LLB G; William *Brandesby* 1410 GildY. From Bransby (L), or Brandsby (NRY).

**Branston, Branson, Bransom, Bramston, Bramson:** Gilbert *de Branteston'* 1200 P (Sf); Haim *de Branzton'* 1202 FFL; Helte *de Brandeston'* 1210 P (Sf); Richard *de Braunteston'* 1221 AssSa (St); Robert *Braunston*, William *Branston*, Adam *Branson*, Roger *Bramston* 1568 SRSf. From Branston (Leics, Lincs, Staffs), Brandeston (Suffolk), Brandiston (Norfolk), or Braunston (Leics, Northants). v. also BRIMSON.

**Brant:** v. BRAND

**Branthwaite, Branwhite:** Alan *de Braunthwait* 1332 SRCu; Simon *Branthwite* 1523 FrY; Bloss *Branwhite* 1764 Bardsley. From Branthwaite (Cumb).

**Branton, Brantown, Brantom, Braunton:** Ralph *de Branton'* 1162 P (Nth). From Branton (WRYorks) or Braunton (Devon).

**Brashaw:** v. BRAYSHAW

**Brashour:** v. BRASSEUR

**Brasier, Brasher, Braizier, Brazier:** William *Brasier'* 1327 SR (Ess); Thomas *Brasyer* 1381 AssC. A derivative of OE *brasian* 'to make of brass', a worker in brass (a1425 MED). cf. Thomas *le Brasgetere* 1333 MESO (So), 'brass-founder'.

**Brass, Brace:** William *bras* 1127 AC; John *Braz* 1218 P (He); Nicholas *Brace* 1327 SRSo; William *Brasse* 1440 ShefA. OFr *brace, brase* 'arm', for some peculiarity of the arm, or for one or other of its various senses: 'a piece of armour covering the arm', 'part of a horse's harness', or 'a pair of hounds'. Sometimes, perhaps, from Breton *braz* 'big', or local from Brace (Sa).

**Brassett:** Richard *Brassehevede* 1301 SRY. OE *brǣs* 'brass', used as a type of hardness, insensibility (c1330 MED) and *hēafod* 'head'. cf. *brassehead buls* 1613 NED. cf. Roger *Brasenhed* 1434 FrNorw.

**Brasseur, Le Brasseur, Bracer, Brasher:** Azo *le Brascur* 1168 P (K); Richard *le Brazur* 1199 FFEss; Ralph *Bracur'* 1202 Cur (Sr); German *le Brassur* 1296 SRSx. OFr *braceor, brasseur* 'brewer'.

**Brassey, Brassy, Bracey, Bracy:** Hugh *de Braci* 1190 P (Wa); Robert *de Bracy* 1275 FFEss; John *Bracy* 1369 IpmW; George *Brasie* 1642 PrD. From Brécy (Aisne, Ardennes).

**Brassington:** Robert *de Brassyntone* 1348 DbAS 36. From Brassington (Db).

**Brassy:** v. BRASSEY

**Brastock:** v. BRADSTOCK

**Bratcher:** v. BRACHER

**Bratley:** v. BRADLEY

**Bratton:** William *de Bratton'* 1195, *de Braton'* 1219 P (Do). From Bratton (Sa, So, W), or Bratton Clovelly, Fleming (D).

**Braun, Braund:** v. BRAND

**Braunton:** v. BRANTON

**Brawn:** v. BRAND

**Bray, Braye:** (i) Alnod *de Braio* 1084 GeldR (D); Richard *de Brai* 1135 Eynsham (O); Ralph *de Bray* 1225 Cur (D); Daniel *de Bray* 1297 MinAcctCo. From Bray (Berks), High Bray (Devon), or some unidentified place in Cornwall. (ii) Roger *le Bray* 1202 AssNth; Hugh, William *le Brey* 1275 RH (C), c1304 Glast (So); John, Henry *le Bray* 1327 SRC, SRSf. Cornish *bregh* 'fine, brave'. The great home of the Brays is Cornwall; in the eastern counties the name is of Breton origin. (iii) There was also a woman's name *Braya*, which may derive from the Cornish nickname, cf. *Braya*, daughter of William a Istetone of West Angmering 1316 AD i (Sx), and Peter *Braya* of Ist Angmerynge 1324 ib. was probably of the same family. (iv) Godfrey *de Bra* 1400 Black (Aberdeen); Thomas *de Bra* 1438–48 ib. (Dunfermline); Agnes *Bray* 1617 ib. From one of the many places in Scotland called Brae. (v) In Ireland, either a toponymic, *de Bré*, or Ir *Ó Breaghdha*, a native of Bregia, a territory in Meath (MacLysaght).

**Braybrook, Braybrooke, Braybrooks, Brabrook:** Robert *de Braibroc* 1199–1200 FFWa; Henry *de Brabroc* 1221 AssWo; William *de Braibrok* 1315 LLB E; William *Braybrok* 1432–3 FFWa. From Braybrooke (Nth).

**Brayham:** v. BRAHAM

**Brayler:** v. BRAILER

**Brayley, Brailey:** Walter *de Braly* 1275 SRWo; John Brailey, William *Brayley* 1642 PrD. From Brayley Barton in East Buckland (D).

**Brayshaw, Brayshay, Brashaw:** Elizabeth *Brawshawe* 1571 RothwellPR (Y); Henry *Brashay*, *Brayshawe*, *Brayshay* 1602–17 ib. Local pronunciations of the Yorkshire *Bradshaw*.

**Brayton:** Elias *de Braiton'* 1205–14 RegAntiquiss; Martin *de Brayton'* 1308 FFY; John *de Brayton* 1379 PTY. From Brayton (Cu, WRY).

**Brazewell:** v. BRACEWELL

**Brazier:** v. BRASIER

**Breach:** v. BRACH

**Breacher:** v. BRACHER

**Bread:** v. BRAID

**Breadmore:** v. BRADMORE

**Breadon:** v. BREEDON

**Breaklance:** Denis *Brekelaunce* 1334 SRK. 'Break lance', OE *brecan*, OFr *lance*, perhaps a nickname for a soldier. cf. John *Brekpole* 1447 CtH 'break pole'; William *Breksekyll'* 1379 PTY 'break sickle'.

**Breakleg:** Adam *Brekeleg* 1243 AssSo. 'Break leg', OE *brecan*, ON *leggr*. cf. Thomas *Sortleg* 1284 CtW 'short leg'; John *Stifleg* 1363 IpmW 'stiff leg'.

**Breakspear, Brakspear, Braksper:** Alexander *Brekespere* 1199 CurR (L); Geoffrey *Brekespere* 1206 Cur (Sr); Thomas *Brekaspere* 1227 AssBeds. 'Break spear', which, as Bardsley remarks, 'would be

cheerfully accepted as a nickname by the successful candidate in the tournament'. It could also have reference to achievement in actual battle. cf. Stephen *Bruselaunce* 1308 RamsCt (Sf); Martin *Briselaunce* 1312 LLB D; Richard *Brekeswerd* 1195 P (L) 'break sword', and v. SHAKESPEAR.

**Bream, Breem, Brim, Brims, Brimm:** *Breme* 1066 DB (Sf); Hugo *Brem* 1221 AssWa; Symon *Brim* 1279 RH (C); Agatha *Breme* 1327 SRC; Robert *Brymme* 1327 SRSx. OE *brēme*, ME *brem(e)*, *brim(me)* 'vigorous, fierce', earlier 'famous, noble', or OE *Brēme*. v. also BRAHAM.

**Brear, Breare:** Walter, Richard *le Brer* 1255 RH (Sa), 1279 RH (O); William *Brere* 1346 FrY. OE *brǣr* 'prickly thorn-bush', modern *briar*, used as a nickname. cf. 'sharp as brere' (Chaucer). v. BRIAR.

**Brearey, Breary:** Robert *de Brerehaga* 13th Kirkstall; William *Brerehay* 1443 Calv (Y); Richard *Brerey* 1534 FrY. From Breary (WRY).

**Brearley:** v. BRIERLEY

**Brearton, Breerton, Brereton:** Richard *de Brertona* 1176 YCh; Alexander *de Breretone* 1242 AssDu; William *de Brereton* 1356 FFY. From Brereton (Ches, Staffs), Brierton (Durham), or Brearton (WRYorks).

**Breathwaite:** v. BRAITHEWAITE

**Brebner:** v. BRABINER

**Brech:** v. BRACH

**Brecher:** v. BRACHER

**Breckenridge:** v. BRACKENRIDGE

**Brede, Breed, Breede, Breeds:** Brian *de Brede* 1195 FFNf; William *de Bredes*, John *de Brede* 1296 SRSx; Elena *atte Brede* 1317 AssK; Marjery *Brede* 1352 ColchCt. 'Dweller by the plain or flat expanse' as at Brede (Sussex). OE *brǣdu* 'breadth', used topographically in ME of a broad strip of land.

**Breeder:** Equivalent to *atte Brede*. v. BREDE.

**Breedon, Breeden, Breadon, Bredon:** Ysolt *de Bredon'* 1204 P (Db); Richard *de Bredon* 1306 IpmY; Robert *de Bredone* 1345 LLB F. From Bredon (Wo), or Breedon on the Hill (Lei).

**Breem:** v. BRAHAM, BREAM

**Breese, Breeze:** Roger *Brese* 1210 P (Nf); William *Brese* 1275 Wak (Y). Usually explained as *ab Rees* 'son of *Rees*'. This may be the origin of Robert *Breese* 1666 Bardsley (Ch), but the above forms are too early for this development. They are nicknames from OE *brēosa* 'a gadfly'.

**Breffitt:** v. BREVETOR

**Brekonridge:** v. BRACKENRIDGE

**Brellisford, Brelsford:** v. BRAILSFORD

**Bremner, Brimner, Brymner:** Walter *Brabounare* 1418–26 Black (Ayrshire); Agnus *Brebner* 1489 ib. (Elgin); Finlay *Brembner* 1649 ib.; James *Brimner* 1630 ib. cf. BRABINER. Artisans and traders from Brabant settled early at Aberdeen and on the east coast of Scotland. *Bremner* is still at times pronounced *Brembner* in Caithness.

**Brend, Brent, Brind:** (i) Gilbert *Brende* 1273 RH (St); Adam *Brend*, John *Brent* 1327 SRWo; Hugo *le Brent* 1279 RH (O); Geoffrey *le Brende* 1327 SRSf. *Brend*, *brent* and *brind* are ME forms of the past participle of OE *beornan* 'to burn'. 'The burnt', a nickname for a criminal who had been branded. cf.

Henry *Brendcheke* 1279 AssNb, Cutte *Brendhers* 1279 RH (C). cf. BRENNAN. (ii) Symon *del Brend* 1318 FrY. Here *brend* is used topographically of 'burnt land', hence 'dweller by the burnt land' as at Brind (ERYorks), *Brende* 1188 P. *v.* also BRENT.

**Brennan, Brennans, Brennand, Brennen, Burnand:** Reginald *Brennehand'* 12th DC (L); Walter *Brenhand* 1229 Whitby (Y); William *Brennand* 1277 Ipm (Nt); Joan *Byrnand* 1475 GildY; Anna *Burnand* 1512 ib. 'Burn hand', a nickname for the official who carried out the harsh punishment of medieval law. Matilda *Brendhand* 1295 Barnwell (C), whose surname would also become modern *Brennan*, had suffered this punishment. *Brennan* may also be for Irish *Ó Braonáin* 'descendant of *Braonán*', a diminutive of *braon* 'sadness'.

**Brenner:** Jordan, John *le Brenner* 1280 AssSt, 1327 SRSf; John *le Brynner* 1327 Pat. OFr *brenier* 'keeper of the hounds', or a derivative of ON *brenna* 'to burn', 'burner' of lime, bricks or charcoal. cf. BERNER.

**Brenson:** *v.* BRIMSON.

**Brent:** Robert *de Brente* 1269 AssSo. From South Brent (Devon) or East Brent (Som). *v.* also BREND.

**Brereton:** *v.* BREARTON.

**Bret, Brett, Britt, Britts:** Edward *Brit* 1066 DB (D); Tihellus *Brito* 1086 DB (Ess); Walter *Bret* 1164 StCh; Alan *le Bret* 1177 P (C); William *le Brit*, *le Brut* 1230, 1256 Oseney (O); Matthew *le Brut* 1242 Fees (He); Henry *Brit* 1275 SRWo; Ralph *le Brut* 1296 SRSx. The most frequent form of these common surnames is *Bret* which is usually from OFr *Bret*, nominative of *Breton* 'a Breton'. The variation between *e*, *i* and *u* points to OE *Brit*, *Bryt*, *Bret*, which meant 'a Briton' and continued to be applied to the Strathclyde Britons until c1300. In the Welsh border counties it may have been used of the Welsh as *Waleys* was of the Strathclyde Britons. *v.* WALLIS.

**Bretherton:** Robert *de Bretherton'* 1203 SPleas (St); Warin *de Bretherton* 1324 CoramLa; John *of Bretherton* 1401 AssLa. From Bretherton (La).

**Breton, Le Breton, Bretton, Briton, Britton, Brutton:** Geoffrey *(le) Bretun*, *le Briton* 1164–6 P (Ess, Y); Lowis *le Briton* 1166 RBE (Ess); Ralph *Bretun* 1166 Oseney (O); William, John *le Bruton'* 1248 Fees (Ess), 1279 AssSo. *Bretun*, the most common form, is the cas-régime of OFr *Bret* 'a Breton'. For the variation between *Breton*, *Briton* and *Bruton*, cf. BRET.

**Brettel, Brettell, Brettle, Breteile:** *Bretel* 1066 DB (Co); *Bretellus*, *Britellus* 1086 DB (D, Do, So); *Bretellus* de Amber' 1130 P (Ha); Reginald *Bretel* 1169 P (Hu); John *Brutel* 1235 Oseney (O); Richard *Britel* 1243 AssSo. OCornish *Brytthael*, OBret *Brithael*: Godwine *Brytæl* 1035 OEByn (Do).

**Bretton:** Osbert *de Bretton'* 1193 P (Y). From Bretton (Derby, WRYorks).

**Brettoner, Britnor, Britner, Bruttner:** John *Bretener* 1379 PTY. ME *Bretoner*, *Brutiner* 'a Breton', used as a term of reproach. cf. 'A Brutiner, a Braggere, a-bostede him also' (a1376 MED).

**Brevetor, Brevitt, Breffit, Breffitt:** Alexander *Brevetur* 1221 ElyA (C); Adam *le Breuetor'* 1275 SRWo; Walter *le Brefeter* 1285 Ass (Ess); Joan *Breftour* 1327 *SR* (Ess); John *Breuet* 1357 ColchCt. A derivative of OFr, ME *brevet*, a diminutive of OFr, ME *bref*, 'an official or authoritative message in writing', especially papal indulgences, used also by metonymy for a bearer of these (1275 MED).

**Brew:** A purely Manx name from *MacVriw* 'son of the judge', now the Deemster: *McBrow* 1408, *McBrewe* 1417, *Brew* 1616 Moore.

**Brewell, Brewill, Browell, Bruell:** Osbern *de Broilg* 1086 DB (Beds); Letia *de Brouilla* 1194 P (Do); Richard *Bruille*, *Brulle* 1206 Cur (Wo); William *Brule*, *Brulle* 1206 Cur (Gl). cf. Breuil (Calvados), but (La) Breuil is too common as a place-name in France to allow of a safe derivation from any one or other of them.

**Brewer, Brewers, Brouwer, Brower:** (i) Richard *Briwerra* 1192 AC (Ha); William *Bruwere* (*Briwere*) 1148 Winton (Ha); John *Browere* 1201–12 RBE (Sf); Henry *le Brewere* 1278 AssSo. ME *brewere*, a derivative of OE *brēowan* 'to brew', a brewer (a1300 NED). *v.* BREWSTER. (ii) Ralph *de Brueria* 1086 DB (D); Nicholas *de la Bruiere* 1195 P (Gl); Thomas *de la Bruera* 1221 FFSt. The DB under-tenant may have come from Bruyère (Calvados), from OFr *bruière* 'heath', a term used also in England: Temple Bruer (Lincs), *la Bruere* Hy 2. Bruera (Ches) is a translation of OE *hǣð*, *Heeth* 12th, *Bruera* c1190.

**Brewis, Brewse, Browse, Bruce:** (i) William *de Briouze* a1080 France; William *de Braiose* 1086 DB (Sx, Sr, Ha, Berks, W); William *de Breosa* 1154 Templars (Sx); *de Braiuse* 1169 P (Sx), *de Braosa*, *de Breusa*, *de Brewse*, *de Breiuse* 1206 Cur (He, Sr), *de Brews'* 1212 Fees (Sx); Reginald *de Brause*, *de Brawose* 1206 Cur (Sx), *de Brause* 1212 Fees (D), *de Breus'*, *de Breius*, *de Breuis* 1219 Cur (Sx), *de Breaus'* 1226 Fees (He); Maria *de Brewes* 1296 SRSx; Robert *Brous* 1327 SRSx. From the time of Domesday the family of Briouze were lords of the rape of Bramber (Sussex). They came from Briouze (Orne) and their surname survives in Manningford Bruce (Wilts) and Wickhambreux (Kent). Its later forms are inextricably confused with those of BRUCE. (ii) Hugh *del Breuhous* 1302 FrY; Robert *del Brewhus* 1332 SRCu; Richard *del Bruhous* 1401 FrY. 'Worker at the brew-house', a brewer.

**Brewitt, Browett:** Robert *Bruet* 1207 Cur (W); John *Brouet* 1268 Pat (Wa); John *Brewett* 1524 SRSf. OFr *brouet*, *broet* 'soup made of flesh broth', a diminutive of OFr *breu*, earlier *bro*, ME *browet*, *bruet* (1399 NED). Used for a maker or seller of broth.

**Brewster, Broster, Bruster:** Roger *Breuestere* 1221 ElyA (Sf); Emma *le Breustere* 1279 RH (Bk); Geoffrey *Brouster* 1283 SRSf; John *Browster*, Margaret *Brewster* 1381 SRSf; Henry *Bruster* 1383 FrY. ME *brewestere*, *browestere* 'a woman brewer' (1308) NED. cf. BREWER. Three-quarters of the examples are names of men. cf. BAXTER. *Broster* may sometimes be from *broudester* 'a female embroiderer', from ME *broudin* from OFr *brouder* 'to embroider' (1450 NED): Gelisius *Browdester* 1377 FrY; Nicholas *Broudester* 1381 PTY.

**Brian, Briance, Briand, Briant, Brien, Brient, Bryan, Bryand, Bryans, Bryant, O'Brian, O'Brien:** Radulfus *filius Brien* 1086 DB (Ess); *Briendus* de Scal' 1086 ICC

(C); *Brien* 1088 StCh, 1114–19 Bury (Sf); *Brient* 1130 P (W); *Brianus* filius Radulfi, Alani 1207 Cur (Sr, Y); Ralph *Brien* 1160 Bury (Sf); Ralph *Brian'* 1205 P (Y); Bennet *Briant* 1524 SRSf. A Breton name introduced into England by the Normans. In the north, it is OIr *Brian*, brought by Norsemen from Ireland (where the name was common) to Cumberland and across the Pennines into Yorkshire. It is found in ON as *Brján*.

**Briar, Briars, Brier, Briers, Bryar, Bryars, Bryer, Bryers:** John *in le Breres* 1279 RH (Hu). 'Dweller among the brambles', OE *brǣr* 'prickly thornbush', formerly including the bramble.

**Brice, Bryce:** *Bricius* de Kyrkebi 12th DC (L); *Bricius* judex c1189–99 Black (Abernethy); *Brice* 1208 Cur (Y); William, Roger *Brice* 1240 FFEss, 1277 *Ely* (Sf). The name of St *Britius* or *Brice*, successor of St Martin as Bishop of Tours, was popular in England and Scotland in the 12th and 13th centuries. It is probably Celtic in origin.

**Brick, Bricks, Brix:** Hereward *Bric* 1201 P (Nth); John *Brik* 1327 SRC; John *Brix* 1340 PN Do 75; Richard *Brykys* 1456 FFEss. Perhaps OE *brȳce* 'brittle, fragile, worthless'. The later examples may perhaps be variants of BRIGGS.

**Brickett:** *v.* BIRKETT

**Bricklebank:** *v.* BROCKLEBANK

**Bricknell:** *v.* BRIGNALL

**Bricks:** *v.* BRICK

**Brickstock:** *v.* BRIGSTOCKE

**Briddock, Brideoke:** *Bridoc* c1190 Black; Joan *Brydok* 1379 PTY. Gaelic *Brideoc*, a diminutive of *Bride*, Ir, MScots *Brighid*.

**Bride:** *v.* BIRD. In Scotland, for MCBRIDE.

**Bridell:** *v.* BRIDLE

**Briden, Bryden, Brydon, Brydone:** Richard *Bridene* 1221 AssGl; Henry *Brydon* 1332 SRSt; John *Briden* 1424 FrY. OFr *bridon* 'bridle'. Metonymic for a maker of bridles.

**Bridewell:** Walter *de Bridewill'* 1297 MinAcctCo (Co). 'Dweller by a well dedicated to St Bride or near a spring or stream frequented by birds' (OE *bridd*). There can be no reference to Bridewell as a prison. The London Bridewell was a 'lodging' built by Henry VIII near St Bride's Well and later given by Edward VI as a hospital. The modern meaning arose later, when the hospital was converted into a house of correction.

**Bridge, Delbridge, Dellbridge, Dealbridge:** Gilbert *atte Brigge* 1272 PN Sr 143; Nicholas *de la Brugge* 1275 SRWo; William *ater Bregg* 1296 SRSx; Roger *dil Brigge* 1327 SRSf. 'Dweller near or keeper of the bridge', OE *brycg*. *v.* BRIDGEMAN.

**Bridgeland, Bridgland:** Robert *Bregelonde* 1296 SRSx; Henry *de Bregelonde* 1317 AssK. 'Dweller by the land near the bridge.'

**Bridgeman, Bridgman:** John *Brygeman* 1296 SRSx; John *Bregman* 1310 PN Ess 217; John *Bruggemon* 1332 SRWa. 'Dweller by or keeper of the bridge' (1648 NED). William *Breggeman* is identical with William *atte Bregge* (1332–3 *ERO*). *v.* BRIDGER, PONTER.

**Bridgen, Bridgens, Bruggen:** Aylward *Attare-brughend'* 1239 MELS (So); John *At Brugeende* 1279

RH (O); John *Attebriggende* 1280 AssSt; Richard *atte Bruggende* of Kingston 1377 LoPleas; Thomas *Bridgin* 1640 SaltAS xv. 'Dweller at the end of the bridge.' Occasionally, 'man from Bruges': Alexander *Brugeyn* 1260 AssC.

**Bridger, Brugger:** Walter *le Briggere* 1327 SRSo; Walter *Bregger* 1327 SRSx; John *Bruger* 1332 SRSr. A derivative of OE *brycg* 'bridge', with the same meaning as BRIDGEMAN. John *le Bruggere*, also called John *de Ponte* (1294) lived at Bridge End in Ockham (Surrey). *v.* PN Nth xlvi.

**Bridges, Brydges:** William *de Bruges*, *de Brieges* 1205 Cur (O). From Bruges (Belgium).

**Bridgstock:** *v.* BRIGSTOCKE

**Bridle, Bridel, Bridell:** Walter *Bridel* 1196 FrLeic; Richard *Brydel* 1266 FFEss. OE *brīdel* 'bridle', as an occupation name. cf. *bridelsmyth* 1321–4 Pat, 'Bridle-Cutters . . . and all other Makers, Dressers, or Workers in Leather' (1697 NED).

**Bridson, Brydson:** *Giolla Brighde* 1146, *McGilbrid* 1511, *Bridson* 1609 Moore. A Manx name, originally *Mac Giolla Brighde* 'Bridget's servant's son', from St Bridget. Pronounced *Brideson*. cf. Irish KILBRIDE, Gaelic MCBRIDE.

**Brier:** *v.* BRIAR

**Brierley, Brierly, Brearley:** Ivetta *de Brerelay* 1248 AssSt; Roger *de Brerley* 1275 Wak (Y). From Brierley (Staffs) or Brierly (WRYorks).

**Brigden:** Potier *de Brikendon'* 1176 P (Herts); John *de Brikedene* 1296 SRSx. From Brigden Hill in Ashburnham, or Brigdene Fm, Hill in Glynde (Sx). Sometimes, perhaps, from Brickendon (Herts), *Brygyndon* 1346.

**Brigenshaw, Briginshaw:** *v.* BIRKENSHAW

**Brigg, Briggs:** Robert *del Brig* 1275 Wak (Y); Alexander *del Brigg* 1332 SRCu; Robert *atte Brig* 1379 PTY. 'Dweller by the bridge', ME *brig(g)*, ON *bryggia*, the northern and Scottish word for *bridge*.

**Brigham:** Hugh *de Briggeham* 1200 P (Y); Adam *de Brigham* 1307 FFY; Thomas *Brygham* 1409 IpmY. From Brigham (Cu, ERY).

**Brighouse:** Richard *de Briggehuses* 1275 Wak (Y); Walter *Bryghowse* 1548, Robert *Brighouse* 1569 FrY. From Brighouse (WRY).

**Bright:** John *Briht* 1252 Rams (Hu); William *le Brythe* 1278 AssSo; Herveus *Brite* 1279 RH (C); Adam *Bright* 1296 SRSx. OE *beorht* 'bright, beautiful, fair'.

**Brighten:** *v.* BRIGHTON

**Brightiff, Beriff:** *Bricteva*, *Britheue* 1066 DB; Godric *Brihteue filius* c1095 Bury (Sf); *Birghiva* 1208 Cur (Herts); Angerus *filius Brihtiue* 1219 AssL; Adam *Brightyeue*, Thomas *Brytheue* 1326, 1327 FFSf; Edmund *Brightyeve* or *Britiff* 1467 Bardsley (Nf); John *Beriffe* 1496 Dickin (Ess); Richard *Brygthewe*, *Brighteve* 1479 SIA xii, 1508 NorwW (Nf); Thomas *Bereve* 1522 SIA xii; John *Brightif* 1559 Bardsley (Nf). *Berriff* is from OE *Beorhtgifu* 'fair gift'; *Brightiff* is from the metathesized form *Brihtgifu*, a woman's name still in use in the 13th century. *v.* BRIGHTY.

**Brightley, Brightly:** Peter *de Brihtleg'* 1199 P (D). From Brightley (Devon).

**Brightman:** *Brihtmanus* 1066 DB (Sf); Robert

*Brightman* (*Brithman*) 1327 *SR* (Ess). OE *Brihtmann, Beorhtmann* 'fair man'.

**Brightmore, Brimmer, Burkmar:** *Brihtmarus, Britmar* 1066 DB; *Ædmer Brihtmari filius* c1095 Bury (Sf); *Brichmerus* filius Hunne 1193 P (Nf); Lemmer *Brihtmer* c1095 Bury (Sf); John *Brictmer* 1221 AssWo; John *Britmar* 1309 SRBeds; William *Brightmer* 1332 SRSx; Robert *Brykemare* 1568 SRSf. OE *Beorhtmǣr, Brihtmǣr* 'fair-famous', a personal-name common in DB and throughout the 12th century. *Britmer* became *Brimmer* and *Brykemare* became, through metathesis, *Byrkmar, Burkmar*.

**Brighton, Brighten:** Richard *de Brighton* 1328 FrY; Adam *de Bryghton* 1341 Pat (Cu); Nicholas *de Brighton* 1342 Cl (Nt). From Breighton (ERYorks), *Bristun* DB, *Bryghton* 1298–1567, *Breighton* 1636 (PN ERY 239). The surname can have no connexion with the Sussex Brighton, earlier *Brightelmeston*. This occurs as *Brighton* in deeds of the reign of Charles I but did not come into common use until the early nineteenth century (PN Sx 291).

**Brightrich:** *Brictric* Biscop 1202 FFNf; Osbert *Brihtrich* 1189 Sol; Simon *Brightrich* 1317 AssK; Hugh *Bryghtrich* 1327 *SR*Ess. OE *Beorhtric.*

**Brightween, Brightwen:** *Brictuinus, Brithuinus* 1066 DB; *Brithwen* uidua 1066 Winton (Ha); *Brictwen* 1148 ib.; *Brihtwinus* de Ixwrthe 12th MedEA (Sf); *Brichtwenne* 1222 DBStP (Ess); Bartholomew *Bryctwyne* 1296 SRSx; John *Bryghtwyne* 1332 ib. The surname forms are from OE *Beorhtwine* (m) 'fair friend'; the modern forms point rather to OE *Beorhtwynn* (f) 'bright joy'.

**Brightwell:** Robert *de Brictewell'* 1205 Cur (O); Robert *de Brichtewell'* 1221 AssGl; William *Brightwell* 1439 FFEss. From Brightwell (Berks, Sf), or Brightwell Baldwin (O).

**Brighty:** *Brictui* 1066 DB (Do); *Brichwi* 1188 BuryS (Sf); *Berdwi* faber 1192 P (K); Osbertus *filius Britwi* 1221 ElyA (Ess); Alexander *Brictwi* 1210 Cur (C); Agnes *Britwy* 1277 Ely (Sf); Thomas *Brightwy* 1327 SRSx. OE *Beorhtwīg, Brihtwīg* 'bright war'. The surname may also be from BRIGHTIFF, from the 1479 *Brygthewe.*

**Brignall, Brignull, Brignell, Bricknell:** Simon *de Brigenehall* 1313 FrY; John *de Brigenhale* 1327 SRSf; William *Brigkenhall* 1400, Robert *Brignall* 1409 FrY. From Brignall (NRY).

**Brigstocke, Brickstock, Bridgstock:** Walter *Brigestok* 1275 RH (L). From Brigstock (Northants).

**Brill, Brille, Brills:** Walter *de Brehille* 1190 Oseney; Richard *Brylle* 1411 *Petre*; Giles *Bryllys* 1527 FFEss. From Brill (Bk), *Brehilla* 1230.

**Brim:** *v.* BREAM

**Brimblecombe, Brimacombe, Brimicombe, Brimmacombe, Brimmicombe, Brinicombe:** From Brimblecombe (Devon), where the surname was *Brumelcome* in 1281 and *Bremylcomb* in 1330 (PN D 343).

**Brimmer:** *v.* BRIGHTMORE

**Brimner:** *v.* BREMNER

**Brimson, Brinson, Brenston:** John *de Brinzun* 1240 FFEss; Joan *de Bryaunesoun* 1297 ib.; Bartholomew *Brinsun* 1274 RH (Ess). This family came from

Briençun (Normandy) and has left its name in Brimstone Hill in Little Wakering (Essex), *Breaunsons* 1419, *Bremsons, Bramsons* 1549, *Brendston* 1553. New Hall in Purleigh was formerly called from them; *Brymshams* 1527, *Bremstones* 1537, *Brempsons* 1554 (PN Ess 204, 223). The surname may also be contributed to *Bramson, Branson, Branston.*

**Brind:** *v.* BREND

**Brindell, Brindle:** Peter *de Burnhull'* 1206 P (La). From Brindle (La), *Burnhull* 1246.

**Brine, Brines, Brynes:** Matilda *Brine* 1279 RH (O); William *Bryne* 1358 FFY; John *Bryne* 1641 PrSo. Perhaps OE *bryne* 'burning', in one of its various senses. Sometimes, perhaps, from Welsh *bryn* 'hill'.

**Brinicombe:** *v.* BRIMBLECOMBE

**Brinkler:** *v.* BRINKLOW

**Brinkley, Brinkly:** Oliver *de Brincle* 1200 Cur; John *de Brinkele* 1303 LoCt. From Brinkley (C).

**Brinklow, Brinkler:** William *de Brinkelawa* 1190 P (Y). From Brinklow (Warwicks).

**Brinkworth:** Geoffrey *de Brinkewrthe* 1194 Cur (W); William *de Brynkeworth* 1280 AssSo. From Brinkworth (W).

**Brinsley, Brinsly:** Geoffrey *de Brunnesley* 1198, Roger *de Brinseleye* 1293 Fees (Nt); Thomas *Brinsley* 1641 PrSo. From Brinsley (Nt).

**Brinson:** *v.* BRIMSON

**Brinton:** Peter *de Brinton'* 1190 P (Nth); Adam *de Brinton'* 1221 AssSa; Anker *de Brimyngton* 1387 Shef. From Great Brington (Ess), *Brintone* DB, Brimpton (Berks), *Brintone* Db, Brinton (Nf), Brineton (St), or Brimington (Db), *Brineton* 1239.

**Brisbane, Brisbourne:** Ralph *Briseban* 1275 AD i (Mx); John *Brusebon* 1297 MinAcctCo; William *Brisbone* 1298 Black. A hybrid, from OFr *brise* 'break' and OE *bān* 'bone', 'break bone'. *Brisbourne* is due to the common pronunciation of *-bourne* as *-bone*. cf. CRAKEBONE.

**Brisco, Briscoe, Brisker, Briskey:** Robert *de Briscaw* 1332 SRCu. From Briscoe (Cumb, NRYorks, WRYorks).

**Brissenden, Brissendon:** Alice *de Bresinden* 1274 RH (K); William *Bryssendenne* 1348 FFK; Symon *Brisenden* 1525 SRSx. From Brissenden Fm in Ickford (Bk), or Brissenden in Frittenden, in Tenterden (K).

**Bristol, Bristoll:** John *de Brystall* 1392 FrY. From Burstall (ERYorks), *Bristall* 1160–2, or Birstal (WRYorks), *Bristal* 1292. The modern form has been influenced by that of the better known Bristol.

**Bristow, Bristowe, Bristo, Brister:** Lia *de Bristou* 1191 P (Gl); Peter *de Bristo* 1195 P (O). From Bristol (Glos), originally *Brycgstow*, DB *Bristou*. The form in common use until at least the 16th century was *Bristow*. The modern *Bristol* is scribes' Latin. Occasionally the source may be Burstow (Surrey), *Brystowe* from 1486, or Bristow Fm in Frimley (Surrey), *Brister* 1765. *v.* PN Sr 286, 127.

**Britain, Brittain, Brittan, Britten, Brittin, Brittian, Brittney:** John *de Bretagne* 1291 AssSt; Thomas *de Brytannia* 1297 MinAcctCo; John *de Bretayne* 1327 SRC. From Brittany, OFr *Bretagne*. Some of the modern forms are from the ME *Brytane, Brittan* 'Breton'.

**Britch:** v. BRACH
**Britcher:** v. BRACHER
**Britner, Britnor:** v. BRETTONER
**Briton, Britton:** v. BRETON
**Britt:** v. BRET
**Brittlebank:** v. BROCKLEBANK
**Brittney:** v. BRITAIN
**Britwell:** Guy de Brytevilla 1141–72 Hylle. From Britwell (Bk), or Britwell Prior, Salome (O).
**Brix:** v. BRICK
**Brixey, Brixhe:** Alsi filius Brixi 1066 DB (Ha); Godric filius Brichsii c1095 Bury (Sf); Alanus filius Brixi 1209 P (Nf); Stephen, Roger Brixy 1275, 1276 RH (Nf, Beds). OE Beorhtsige, Brihtsige 'bright victory'.
**Broad, Brade, Braid, Braide:** John le Brade 1212 Cur (K); Gilbert le Brode 1235 FFEss. OE brād, ME brod(e) 'broad'. In the north and Scotland, the ā remained in ME, later spelled ai, giving modern Brade, Braid.
**Broadbelt:** Adam Bradbelt 1379 PTY; Richard Broadbelt 1682 FrY. 'Broad belt', one who wore a broad belt, a nickname for a stout man. cf. William Brodgirdel 1275 RH (Nt).
**Broadbent:** Robert Brodebent 1513 FrY; William Brodbent 1528 KentW. 'Dweller by the broad grassy plain.' v. BENT.
**Broadbridge:** Ærnald de Bradebrige 1192 P (Sx); William de Bradebrugge 1296, John de Bradebrugge 1327 SRSx. From Broadbridge in Bosham (Sx).
**Broadfoot:** Turgis Bradfot 1157 P (Cu); Roger Brodfot 1247 AssBeds. OE brād and fōt 'broad foot', a common early nickname.
**Broadgate, Bradgate:** Ralph de Bradgate 1275 RH (Lei); Nicholas atte Brodegate 1344 LLB F. 'Dweller by the broad gate' (ME gate) as at Bradgate (Leics).
**Broadhead:** (i) Thomas Bradheuid 1243 AssDu; John Brodheved 1281 Rams (Hu); Daniel Broadhead 1664 FrY. 'Broad head', OE brād, hēafod. William Brodhod 1327 SRDb may have borne the same nickname, with ON hofuð, Da hoved for OE hēafod; or he may have worn a broad hood (OE hōd). cf. GREENHEAD. (ii) Henry de Bradeheved 1272 AssSt; Alan del Brodeheued 1332 SRLa; Robert de Bradehed 1332 SRSt. 'Dweller by the broad headland.'
**Broadhouse, Broadis:** Richard de Bradhus 1214 Cur (L). 'Dweller at the large house', OE brād 'broad, extensive' and hūs.
**Broadhurst, Brodhurst:** Roger de Brodhurst 1281 AssLa. 'Dweller by the broad wood', OE hyrst.
**Broadis:** v. BROADHOUSE
**Broadley:** v. BRADLEY
**Broadmead:** v. BRADMEAD
**Broadribb, Brodribb:** Peter Brodribbe 1327 SRSo; Richard Brawdrib 1533 RochW. Clearly a nickname, 'broad rib'.
**Broadridge:** v. BRADRIDGE
**Broadway:** Hugh del Brodweye 1276 RH (C). From Broadway (So, Wo), or 'dweller by the broad road', OE brād, weg. v. also BRADWAY.
**Broadwood:** Walter de Brodwode 1274 RH (So). 'Dweller by the broad wood', OE brād and wudu.
**Broady:** v. BRADIE

**Broatch:** v. BROOCH
**Brobson:** v. BRABAZON
**Brobyn:** v. PROBIN
**Brocas:** John Brocas 1337–8 FFSr; Arnold Brokas 1399–1400 IpmY; William Brockas 1642 PrD. From Brocas (Les Landes). Sometimes, perhaps, for Brookhouse.
**Brock, Brocks:** (i) Ralph Broc 1119 Colch (Ess); Joel le Broc 1222 Pat (D); Richard Brock 1275 SRWo. OE brocc 'badger'. From c1400 brock is often used with the epithet stinking and in the 16th century meant 'a stinking or dirty fellow' (ELPN). Or from OFr broque, brocke 'a young stag'. v. BROCKET. (ii) Laurence del Brock 1267 Abbr (So); Nigel de Brocke 1285 FA (Sx); Robert de la Brockes 1286 Ipm (Wo); Alma ate Brock', Imeyna du Brock' 1297 MinAcctCo. OE brōc 'brook, stream'. For the short vowel, cf. the river-name Brock (Lancs). In Kent and Sussex, brook still means 'water-meadow' and in the plural 'low marshy ground'. Hence 'one who lives by a stream or by the water-meadow(s)'. v. MELS and BROOK.
**Brockbank:** v. BROOKSBANK
**Brockelbank:** v. BROCKLEBANK
**Brockenshaw:** v. BIRKENSHAW
**Brocket, Brockett:** (i) Osbert Brochard 1175 P (Ha); John Brocard 1327 SRC. (ii) Henry Broket 1279 RH (O); John Broket 1297 MinAcctCo. OFr brocart, brocquart from OFr broque, brocke, from which was formed ME broket, brocket 'a stag in its second year with its first horns' (c1410 MED).
**Brockhouse, Brockis:** v. BROOKHOUSE
**Brockhurst, Brockherst:** Richard de Brocherst' 1201 Pl; Robert, William de Brokhurst 1296 SRSx; Thomas Brokhurst 1381 AssWa. From Brockhurst (Hants, Sussex, Warwicks).
**Brockhall:** Jul' atte Brochole 1275 MELS (Wo). From Brockhall (Nth), or 'dweller by the badger hole', OE brocchol.
**Brocklebank, Brockelbank, Bricklebank, Brittlebank:** Nicholas Brikilbank 1524 SRSf; William Brokylbanke 1532, Nicholas Brykelbank 1546 FFEss; Ralph Brocklebanke 1645 FrY. 'Dweller by the bank in which there is a badger hole', OE brocchol, ME banke.
**Brocklehurst:** Ralph de Brockolhurst 1246 AssLa. From Brocklehurst (La).
**Brocklesby:** William de Broklesby 1338 Balliol; William de Brokelsby 1351 TestEbor; William de Broclesby 1374 AssL. From Brocklesby (L).
**Brockless, Brockliss:** Broclos, Brocles 1066 DB (L); Roger Broclaus Ric 1 Gilb (L). ON bróklauss 'breechless, without breeches', a nickname used also as a personal name which is the first element of Brocklesby (Lincs). cf. Robert Brekeles 1276 RH (Y), the English equivalent.
**Brockley:** Raulf de Brocleg' 1121–48 Bury; Jordan de Brocleg' 1200 P (Ha); William de Brokele 1311 LLB D. From Brockley (K, Sf, So), or Brookley (K).
**Brockman:** v. BROOKMAN
**Brocksopp, Broxup:** Richard Brokesop 1599 SRDb; John Brocksopp 1704 Shef. 'Dweller in the valley frequented by badgers', OE brocc, hop.

**Brockton, Brocton:** Nesta *de Broketon* 1254–5 FFWa; William *of Brockton* 1297 IpmY. From Brockton (Sa), or Brocton (St).

**Brockway:** Walter *de Brokweye* 1255 RH (W). 'Dweller by the road near the brook.'

**Brockwell:** Walter *de Brocwelle* 1298 IpmGl; John *Brokewell* 1542 SRD. 'Dweller by the badger stream', OE *brocc, wiella.*

**Brockwood, Brookwood:** Dyota *Brokewode* 1319 SRLo; Henry *Brokwode* 1370–1 FFSr. From Brookwood in Woking (Sr), or 'dweller by the wood in which small streams rise', OE *brōc, wudu.*

**Brockworth:** Ernisius *de Brocwurth* 1268 IpmGl; Henry *de Brokworth* 1339 FFW; William *Brockeworth* 1393 IpmGl. From Brockworth (Gl).

**Brocton:** v. BROCKTON

**Brodhurst:** v. BROADHURST

**Brodie:** Michael *de Brothie* 1311 Black. From Brodie (Moray). v. also BRADIE.

**Brodley:** v. BRADLEY

**Brodmead:** v. BRADMEAD

**Brodribb:** v. BROADRIBB

**Brodrick, Broderick:** Bethell *Brothericke* 1672 HTY. Welsh *ap Rotheric* 'son of *Rhydderch*'.

**Brodsworth:** William *de Brodesworth'* 1302 FFY; Robert *Brodesworth* 1340–1450 GildC, *Broddesworth* 1412–3 FFWa. From Brodsworth (WRY).

**Brogden:** Dionisius *Brokden* 1470 FrY; William *a Brokeden* 1525 SRSx; John *Brogden* 1597 FrY; Samuell *Broggdin* 1689 RothwellPR (Y). From Brogden (WRY), or Brook Dean in Fittleworth (Sx).

**Broke:** v. BROOK

**Brokenshaw, Brokenshire:** v. BIRKENSHAW

**Broker:** Constantine *le Brokour* 1276 AssLo. AFr *brocour* 'agent, purveyer'.

**Brokus:** v. BROOKHOUSE

**Bromage:** v. BROMWICH

**Broman:** v. BROOMAN

**Bromby:** v. BRUMBY

**Brome:** v. BROOM

**Bromehead, Bromet:** v. BROOMHEAD

**Bromfield:** v. BROOMFIELD

**Bromford, Brumfitt:** William *de Brumford*, Richard *de Bromford* 1210 FFEss; Juliana *de Bromford* 1321–2 CorLo. 'Dweller at the ford by the broom', OE *brōm, ford.*

**Bromhall:** v. BROOMHALL

**Bromham, Brumham:** Alsi *de Bruneham* 1066 DB (Beds); Robert *de Bromeholme* 1274 FrY; Simon *de Bromhamme* 1296 SRSx; James *Bomholme* 1642 PrD. From Bromham (Beds, W), Broomham in King's Nympton (D), or Bromholm (Nf).

**Bromhead:** v. BROOMHEAD

**Bromige:** v. BROMWICH

**Bromley, Bromly, Brumbley:** Hubert *de Brumle* 1194 StCh; Edward *de Bromleghe* c1200 ArchC vi. From Bromley (Essex, Herts, Kent, Staffs).

**Brommage:** v. BROMWICH

**Brompton, Brumpton:** Geoffrey *de Brunton'* 1205 P (Y); William *de Bromptone* 1312 LLP D. From Brompton (Middlesex, Salop, ERYorks, NRYorks).

**Bromwich, Bromage, Bromige, Brommage:** Adam *de Bromwiz* 1221 AssWa; William *de Bromwic* 1225 AssSt; Thomas *Bromidge* 1581 Bardsley; John *Bromage* 1583 ib. From West Bromwich (Staffs) or Castle and Little Bromwich (Warwicks).

**Bron:** v. BROWN

**Bronsdon, Bronson:** Chanaan *de Bronteston* 1176 P (St). From Branson (Staffs).

**Bront:** v. BRAND

**Brooch, Broatch:** Metonymic for *Brocher.* cf. Ralph *Brocher* 1222 FFSf; John *Brocher*, Roger *le Brochere* 1281 LLB B. ME, OFr *broche* 'a tapering, pointed instrument or thing, a lance, spear, bodkin, etc.', also 'a brooch'. Hence, a maker of *broaches* (lances, spears, etc.) or of brooches. cf. William *Bruchemaker* 1381 PTY and William *ploghbrocher* 1281 MESO (L), probably a maker of ploughshares.

**Brook, Brooke, Brookes, Brooks, Broke, Bruck:** Eustace *delbroc* 1130 P (Nth); Rand' *de Broc* 1157 P (Ha); William *de la Broke* 1208 Cur (Sr); Emma *de Brokes* 1220 Cur (Sf); Peter *Attebroke* 1262 *For* (Ess); William *aboventhebroc* 1276 MELS (Wo); William *atte Brouk* 1296 SRSx; William *in le Broke, Ithebroke* 1317 AssK; Sarra *Bithebrok* 1327 SRSo; William *atte Bruck* 1327 SRC; William *del Brokes* 1332 SRLa; John *Bethebrokes* 1332 SRWo. From Brook (Kent, Rutland), Brooke (Norfolk) or from residence near a stream or by the water-meadow(s). v. BROCK.

**Brookbank(s):** v. BROOKSBANK

**Brooker, Brucker:** William *Brokere*, John *le Broker* 1296 SRSx; John *le Brouker* 1327 ib. 'Dweller by the brook.' William *le Broker* (1332 ib.) was probably a descendant of Anger *atte Broke* (1296 ib.)

**Brookfield:** Thomas *de Brokefeld'* 1199 MemR (O); Adam *del Brokefeld* 1332 SRLa. 'Dweller by the field near the brook', OE *brōc, feld.*

**Brookhouse, Brockhouse, Brockis, Brokus:** Ralph *del Brokhouses* 1297 SRY; Hugh *de Brokehous'* 1379 PTY; William *Brokkus* 1562 Black. 'Dweller at the house by the brook.'

**Brooking, Brookings, Bruckin:** William *Brokyng* 1525 SRSx; Christopher *Brooking* 1642 PrD. From a derivative of OE *brōc* 'stream', hence 'dweller by the stream'.

**Brookland:** Thomas *del Broklondon* 1257 MELS (Sx); Osbert *de Broklonde* 1296 SRSx; Richard *atte Broclonde* 1327 MELS (Sx). From Brookland (K), one or other of the Brookland(s) in Sussex, or 'Dweller by the marshy land', OE *brōc, land.*

**Brookman, Brockman:** Stephen *Brokman* 1296 SRSx; John *Brookman* 1372 ColchCt. Equivalent to BROOKER.

**Brooksbank, Brookbank, Brookbanks, Brockbank:** Thomas *Brokesbank'* 1379 PTY; John *Brockbank* 1700 FrY; Charles *Brookbank* 1758 ib. 'Dweller by the bank of the brook.'

**Brookwood:** v. BROCKWOOD

**Broom, Broome, Brome:** Robert *de Brome* 1193 P (Lei); Alexander *Brom'* 1221 AssWa; Eustace *de la Brome* 1275 RH (K); Richard *atte Brome* 1285 *Ass* (Ess); Richard *del Brom* 1297 SRY. From Broom (Beds, Durham, Worcs), Broome (Norfolk, Salop, Warwicks) or Brome (Suffolk), or from residence near a place where broom (OE *brōm*) grew.

**Brooman, Broman:** *Bruman(nus)* 1066 DB (K, Bk); *Brummanus* 1140–53 Holme (Nf); Gilebertus *filius Brunman* 1211 Cur (Cu); Brihtmar *Bruman, Brunman* 1199, 1200 P (Nf); Ralph *Broneman* 1296 SRSx; John *Bromman* 1327 ib.; Alexander *Brounman* 1327 *SR* (Ess); John *Broman* 1372 ColchCt. OE *Brūnmann*.

**Broomer:** Thomas *de Brommor*, John *Brommor* 1296 SRSx; Robert *Bromere, Bromor* 1327, 1332 ib. All probably came from Bramber Fm, Bremere Rife or Broomer Fm in Birdham (Sussex). *v.* PN Sx 73, 80.

**Broomfield, Bromfield, Brumfield:** Hamo *de Bromfeld* 1275 RH (K); William *atte Bromfeld* 1296 SRSx; John *de Bromfeld* 1327 SRSf. From Broomfield (Essex, Kent, Som), Bromfield (Cumb, Salop) or from residence near broom-covered open-land (OE *brōm, feld*).

**Broomhall, Bromhall:** Matthew *de Bromhale* 1182 P (Ch); Godwin *de Bromhal'* 1182–1200 BuryS (Sf). From Broomhall (Ches, WRYorks), Bromhall (Berks) or residence by a broom-covered nook (OE *healh*).

**Broomhead, Bromehead, Bromhead, Bromet, Brumhead, Brummitt:** Henry *de Bromeheuede* 1290 ShefA; John *Bromehed* 1440 ib.; Mary *Brummett* 1717 Bardsley. From Broomhead in Hallamshire (Sheffield).

**Broster:** *v.* BREWSTER

**Brothers:** *Broder* (St, Sf), *Brodor* (St), *Brodre* (D) 1066 DB; Willelmus *filius Brother* 1202 AssL; John *Brother* 1272 AD i (Mx); Nicholas *le Bruthre* 1279 AssSo; Adam (*le*) *Brother* 1280 ib.; John *Brothers* 1621 Bardsley. ON *Brōðir*, ODa *Brothir* or OE *brōðor* 'brother', used in ME of a kinsman (1382) and a fellow-member of a guild or corporation (1362).

**Brotherton:** Leofnað *in Broðortun* c1050 YCh; Martin *de Brodthreton'* 1230 P (Sf); John *Brotherton* 1379 PTY. From Brotherton (Sf, WRY).

**Brough, Brugh, Bruff:** Daniel *de Burg'* 1219 AssY; William *de Bruggh* 1275 RH (Nf); Arthur *Browghe* 1567 Bardsley. From Brough (Derby, Notts, Westmorland, ERYorks, NRYorks), all ancient camps (OE *burg*), usually Roman, and pronounced *Bruff* or *Broof*.

**Brougham:** Simon, Richard *de Broham* John SP1 (Y), 1244 AssLo; Robert *de Brouham* 1332 SRCu. From Brougham (Westmorland), *Broham* 1176.

**Broughton:** Gerald *de Broctun* 1189 CartAntiq; Richard *de Broghtone* 1247 FFO; William *Broghton* 1332 SRSx. From one or other of the many places of this name.

**Brouncker:** *v.* BRUNGER

**Broune:** *v.* BROWN

**Brounfield:** *v.* BROWNFIELD

**Brounlie:** *v.* BROWNLEA

**Brouwer, Brower:** *v.* BREWER

**Browell:** *v.* BREWELL

**Browett:** *v.* BREWITT

**Brown, Browne, Broun, Broune, Bron:** *Brun, Brunus* 1066 DB; *Brun* 1185 Templars (Wa); Conan *filius Brun* 1209 AssL; Richard *Brun, le Brun* le mercer 1111–38, c1140 ELPN; William *le Brun* 1169 P (Nth); William *Brun* 1182–1205 BuryS (Sf); Hugh *Bron* 1274 RH

(Sa); Agnes *Broun* 1296 SRSx; John *le Browne* 1318 FFC. Occasionally we may have a personal name, OE *Brūn* or, possibly, ON *Brúnn*, but neither was common after the Conquest. Usually we have a nickname, OE *brūn*, or, perhaps occasionally, OFr *brun* 'brown', of hair or complexion.

**Brownbill, Brumell, Brummell, Brumhill:** Adam *Brownbyll* 1401 AssLa; Elizabeth *Browmbell* 1561, Joane *Brownbell* 1565, Nicholas *Brombill* 1608 Bardsley. 'Brown bill', i.e. chopper, axe, OE *brūn, bill*. Used also of a long-bladed slashing weapon. Metonymic for a maker or user of this.

**Brownell, Brownhall, Brownhill:** Alan *de Brunhale* 1275 RH (L); William *de Bronehill* 1329 YDeeds I; Adam *Bronhulle* 1441, Nicholas *Brownell* 1566 ShefA. 'Dweller by the brown hill or corner of land', OE *brūn, hyll/healh*.

**Brownett, Brunet:** Richard *Brunote, Brounnhote* (*Brounote*) 1310 LLB D, 1327 *SR* (Ess); Robert *Brunet*, John *Brownet* 1674 HTSf. *Brun-ot, Brun-et*, diminutives of OFr *brun* 'brown'.

**Browney, Brownie:** Richard *Brunye* c1250, *Browneye* c1290 CartNat; Henry *Brouneye* 1301 FS; Roger *Brownye* 1335 CartNat. 'Brown eye', OE *brūn, ēage*.

**Brownfield, Brounfield:** Ralph *de Brounfeld* 1309 LLB D; Henry *Brounfeld* 1384 ib. H. A variant of BROOMFIELD. The Essex Broomfield is *Brounfeld* 1349 Ipm.

**Brownfleet:** Warin *Brunflet* 1289 FFSf; Roger *Brounflete* 1379 PTY; Thomas *Brounfflete* 1426–7 FFWa. 'Dweller by the brown stream, estuary, or inlet', OE *brūn, flēot*.

**Brownford, Brumford:** Robert *Browneford* 1641 PrSo; George *Brounford*, Thomas *Brownford* 1642 PrD. 'Dweller by the brown ford', OE *brūn, ford*.

**Brownhall, Brownhill:** *v.* BROWNELL

**Brownie:** *v.* BROWNEY

**Browning, Brownings:** *Bruning'* de Cestretona 1086 InqEl (C); Robertus *filius Bruning* 1203 AssSt; *Brunyng* Dypres, *Brounyng*, Otelond 1296, 1327 SRSx; Hugo, Robert *Bruning'* 1198, 1199 P (Nf, Sf); William *Brouning* 1291 FFC; Richard *Brownyng* 1522 FrY. OE *Brūning*, a derivative of *Brūn*, not uncommon after the Conquest, surviving in use until the 14th century.

**Brownjohn:** John *Browneion* 1349 *ERO*. 'Brown John'. cf. *Brun Edrith* (i.e. Edrich) 1255 RH (Sa); Adam *Brounadam* 1329 ColchCt, and *v.* DUNBABIN, HORRABIN.

**Brownlea, Brownlee, Brownlees, Brownless, Brownley, Brownlie, Brounlie, Brunlees:** Robert *de Brownlegh* 1310 AssSt; Cuthbert *Brownles* 1485 FrY; Wedow *Brounlees* 1563 Black; John *Brownelesse* 1636 FrY. From Brownlee (Ayr, Lanark), or 'dweller at the brown clearing', OE *brūn, lēah*.

**Brownlock:** *Brunloc* de Lafham 1190 BuryS (Sf); William *Brunloc* 1221 ElyA (Sf); Peter *Brounlok'* 1332 SRDo. OE *\*Brūnlāc*.

**Brownnutt, Brownutt:** Alexander, Robert *Brunenote* 1279–94 RamsCt (Hu). 'Brown as a nut', OE *brūn, hnutu*. cf. NUTBROWN, and John *Beribroun* 1505–6 FFWa 'brown as a berry'.

**Brownridge, Brownrigg:** William *de Brounrig* 1332 SRCu; William *Brounrige* 1510 FrY; Margaret *Brounrig* 1684 Black. From Brownrigg (Cu).

**Brownsmith:** Thomas *le Brounesmyth* 1296 Wak (Y); William *Brounsmyth* 1327 SRSo. OE *brūn* and *smið* 'brown smith', 'a worker in copper and brass'.

**Brownson, Brunson:** *Brunesune* 1066 DB (K); Alric *Brunesune* 1066 InqEl (C); Alstan *Brune sune* c1095 Bury (Sf); William *Brounsone* 1297 MinAcctCo (Y); Arnulph *Brounessone* 1318 FFEss; John *Brunnisson* 1327 SRC; Hugh *Brouneson* 1327 SRDb. *Brune sune* is 'son of OE *Brūna* or of ON *Brúni*', and has probably not contributed to the surname. The InqEl form is OE *\*Brūnsunu* 'brown son', a personal name not recorded before the Conquest and rare thereafter. *Brounessone* is 'son of *Broun*'. The later forms with a single *s* may be from the personal name but more probably mean 'son of a man named Brown'.

**Brownsword, Brownsort, Brownswood:** William *Brounesword* 1662 PrGR; Joyce *Brownsward* 1673 ER 54. Probably late forms of *Brownsworth* from Brownswolds in Congleton (Ch). cf. Thomas *Brownsworth* 1593 PN Ch ii 297.

**Brownutt:** *v.* BROWNNUTT

**Broxfield:** John *Broxfilde* 1576 SRW. From Broxfield (Nb).

**Broxton:** William *de Broxton'* 1230 P (Nf). From Browston (Suffolk), *Brockestuna* DB.

**Broxup:** *v.* BROCKSOPP

**Bruce:** Robert *de Bruis* 1086 DB (Y), 1185 Templars (L); Robert *de Brus* c1110 Whitby (Y), 1152 Clerkenwell (Mx); John *de Briwes*, *de Brues* 1225, 1277 AssSo; Richard *le Brewys* 1275 SRWo; Robert *le Brus*, Richard *de Brus* 1274–5 RH (Ess). This is usually derived from Brix (La Manche), a derivation accepted without reserve by the *Complete Peerage*. There are certainly some remains of an old castle but no early forms of the place-name have been found. L. C. Loyd cancelled his entry for Brus and discarded the derivation from Brix as the evidence 'hardly seems sufficient' (ANF viii). Tengvik called attention to Le Brus (Calvados), early forms of which (*Bruis* 1177, *Bruix* 1234) fit in with the DB form, though nothing is known of the early history of the place. The son of the Domesday baron, a friend of David I, king of Scotland, was granted by him in 1124 the Lordship of Annandale and his second son Robert became the founder of the Scottish house of Bruce. Later forms of the name have become inextricably confused with BREWIS.

**Bruck:** *v.* BROOK

**Brucker:** *v.* BROOKER

**Bruckin:** *v.* BROOKING

**Bruell:** *v.* BREWELL

**Bruff:** *v.* BROUGH

**Bruggen:** *v.* BRIDGEN

**Brugger:** *v.* BRIDGER

**Brugh:** *v.* BROUGH

**Bruin:** William *Bruin* 1209 Pl (Nf); Patrick *le Bruin* 1269 AssNb; William, Maurice *Bruyn* 1330 IpmW, 1425 FFEss. The name suggests connexion with Dutch *bruin* 'bear', but it may simply show attempts to indicate the pronunciation of Fr *brun* 'brown'.

**Bruiser, Bruser:** Margery le *Bruzre* 1278–9 CtH; William *le Bruser* 1332 FFY; John *le Brusere* 1333 ChertseyCt (Sr). 'The bruiser, breaker', from a derivative of OFr *bruiser* 'to break'.

**Brumbley:** *v.* BROMLEY

**Brumby, Brunby, Bromby:** Walter *de Brunby* 12th Gilb; Geoffrey *Bronby* 1319, Gerard *de Bronby* 1351, John *de Brunby* 1371 FrY. From Brumby in Frodingham (L).

**Brumell:** *v.* BROWNBILL

**Brumfield:** *v.* BROOMFIELD

**Brumfitt:** *v.* BROMFORD, BRUNTHWAITE

**Brumford:** *v.* BROWNFORD

**Brumham:** *v.* BROMHAM

**Brumhill, Brummell:** *v.* BROWNBILL

**Brummer:** *Brumarus*, *Brumerus* 1066 DB (Sf); Ulmerus *Brumari filius* c1095 Bury (Sf); Robert *Brommer* 1510 Butley (Sf). OE *Brūnmǣr* 'brown-fame', of which the above are the only examples known.

**Brummitt:** *v.* BROOMHEAD

**Brumpton:** *v.* BROMPTON

**Brunby:** *v.* BRUMBY

**Brundell, Brundle:** Cecilia *de Brundal'* 1206 P (Nf); John *de Brundall* 1400 FrY; John *Brundale* 1524 SRSf. From Brundall (Nf).

**Brundish:** Robert *Brundische* 1327 SRSx; John *Brundische*, William *Brundisshe* 1524 SRSf. From Brundish (Sf).

**Brundle:** *v.* BRUNDELL

**Brunell:** *v.* BURNEL

**Brunet:** A Huguenot name, from a Protestant family of La Rochelle who fled to England after the Revocation of the Edict of Nantes (Smiles 370).

**Brunet:** *v.* BROWNETT

**Brunger, Brunker, Brouncker:** *Brungar(us)* 1066 DB (Do), 1111–38 ELPN; *Brunger* atte Yate 1327 SRSo; Thomas, William *Brunger* 1275 RH (Nf), SRWo; Robert, Simon *Brungor* 1311 ELPN, 1327 SRSf; William *Brunker* 1572 PN W 127. OE *Brūngār* 'brown spear'. *Brunger* may be from OG *Brunger*.

**Brunham:** Geoffrey (*de*) *Brunham* Hy 2 Gilb (L). From Burnham (Lincs, Norfolk), both with early forms *Brunham*.

**Brunlees:** *v.* BROWNLEA

**Brunning:** William *Brunning* 1260 AssC; John *Brunnyng* 1317 AssK. OE *Brūning*, with shortening of the vowel. *v.* BROWNING.

**Brunsdon, Brunsden:** *Brunstanus* blachebiert 1066 Winton (Ha); *Brunstan(us)* c1095, 1121–48 Bury (Sf); Ralph *Brunston* 13th Rams (Hu); Roger *Brunstan* 1230 P (Nf). OE *Brūnstān* 'brown stone', recorded before the Conquest only in the 11th century as the name of a lay brother of Hyde and of a moneyer.

**Brunthwaite, Brumfitt:** Richard *Brunthwait* 1673 WRS. From Brunthwaite in Silsden (WRY).

**Brunton:** Robert *de Bruntun* c1160–74 YCh; Edmund *de Brunton* 1234 FFSf; Marke *Brunton* 1672 HTY. From Brunton (Nb), or from one or other of the places called Brompton. In Scotland from Brunton (Fife).

**Brunwin:** *Brunwinus, Brunnuinus* 1066 DB (Sf); Laurence, Richard *Brunwyn* 1247 AssBeds, 1276 RH (L); William *Brounwyne*, Geoffrey *Bronwyne* 1327 SRSf. OE *Brūnwine* 'brown friend', a late and rare OE name.

**Bruser:** *v.* BRUISER

**Brush, Brushe:** Alice *Brusch* 1327, John *Brosche*, Robert *Brusshe* 1524 SRSf; Richard *Brush* 1665 FrY. OFr *brosse* 'brushwood', ME *brush* 'brush'. Metonymic for a maker of brushes.

**Bruton:** William *de Briweton* 1271 AssSo. From Bruton (Som). *v.* also BRETON.

**Bryan, Bryand, Bryant:** *v.* BRIAN

**Bryar(s):** *v.* BRIAR

**Bryce:** *v.* BRICE

**Bryceson, Bryson:** John *Briceson* 1332 SRCu; Thomas *Bryson* 1524 SRSf. 'Son of *Brice*'.

**Bryden, Brydon, Brydone:** *v.* BRIDEN

**Brydson:** *v.* BRIDSON

**Bryer(s):** *v.* BRIAR

**Brymner:** *v.* BREMNER

**Brynes:** *v.* BRINE

**Bryson:** *v.* BRYCESON

**Bubb, Bubbe:** Brictmar *Bubba* 1066 DB (Sf); Henry *Bubba* 1173 P (Ha); William *Bobbe, Bubbe* 1219 Cur (Do); Thomas *Bubbe* 1348 IpmW. OE *Bubba*. cf. Melbury Bubb (Do).

**Bubwith:** John *Bubwith* 1437 PN WRY ii 67; Alice *Bubwith* alias Pekeman 1463 TestEbor. From Bubwith (WRY).

**Buchan:** (i) Mariedoc *Bohhan*, Lewelin *Bochan* 1160, 1198 P (Sa). Welsh *bychan* 'small'. (ii) Richard *de Buchan* 1207–8 Black (Aberdeen). From Buchan (Aberdeenshire).

**Buchanan:** Alan *de Buchanan* c1270 Black; Walter *de Buchanan* 1373 ib.; George *Buchanan* 1506–82 ib. From Buchanan (Stirling).

**Buchard:** *v.* BURCHARD

**Buche:** *v.* BOUCH

**Bucher:** *v.* BUTCHER

**Buck, Bucke:** (i) Godwig *se Bucca* c1055 OEByn (So); Herbert *Bucke* 1195 P (Sx); Robert *Buc* 1200 P (Sf); Walter *le Buk* 1243 AssSo. A nickname, OE *bucca* 'he-goat', 'as wild as a buck', or *bucc* 'a male deer', perhaps denoting speed. The surnames may also be metonymic for longer occupational names: Roger *le Bucmanger'* 1221 AssWa, a dealer in bucks or venison; Walter *Bucswayn* 1327 SRSo, perhaps a goat-herd. (ii) Hugo (*del*) *Buc* 1221 *ElyA* (Nf); Peter *atte Buk* 1327 SRSf. 'Dweller by the beech-tree', OE *bōc*.

**Buckell:** *v.* BUCKLE

**Buckenham, Buckenam, Bucknam:** Thomas *de Bukenham* 1245–6 FFEss; John *de Bokenham* 1338 CorLo; Roger *Buckenham* 1642 PrD. From Buckenham (Nf).

**Buckerell:** Geoffrey *bucherellus* 1130 P (Lo); Richard *Bukerelle* 1222 FFO; Richard *Bokerel* 1340 NIWo. cf. Fr *Bouquerel* which Dauzat explains as a double diminutive of *bouc* 'he-goat', in the sense 'lecherous, wanton'.

**Buckett:** Ralph *Bukat* 1275 RH (Herts); Godfrey *Bucket* 1279 RH (C); John *Boket* 1327 SRWo. These forms are too early to be derived from *Burchard*. They may be identical with Fr *Bouquet, Bouquerel*, which Dauzat explains as diminutives of *bouc* 'he-goat' in the sense 'lecherous, wanton'. *Buckerel* is also found in England: Gaufridus *bucherellus* 1130 P (Lo); Andrew *Bukerel* 1174 ib.; William *Bukerel* 1276 LLB A; and these London merchants may well have borne occupational-names. ME *buk* means both 'he-goat' and 'deer' and both *Bukerel* and *Bucket* may be synonymous with *Buckskin* (*v.* BUSKENS), a worker or dealer in buck-skin or leather-goods. cf. CHEVERALL.

**Buckhurst:** Richard *de Buchurst* 1200 P (Bk); John *de Bocherst* 1296 SRSx; Cecilia *de Boukhurst* 1332 SRSx. From Buckhurst (Essex, Sussex).

**Buckingham:** Gerard *de Bukynham* 1183 Eynsham; William *de Bokingham* 1272 IpmGl; John *de Bukyngham* 1384–5 FFSr. From Buckingham (Bk).

**Buckland:** Ælfgyð *of Boclande* c970 BCS (D); Hugh *de Boclanda* 1169 P (Berks); William *de Bocland* 1219 FFO; Thomas *de Boukeland* 1362 IpmW; Richard *Buklond* 1427–8 FFWa. From one or other of the many places of this name which, apart from a single example in Lincs, is found only in the south.

**Buckle, Buckell, Buckels:** John *Bokele* 1296 SRSx; Wymark *Bokel* 1327 SRSf; John *Bukle* 1524 SRSf. ME *bokel* 'buckle', metonymic for *Buckler*.

**Buckler:** Anschetil *buclar'* 1148 Winton (Ha); Peter *le Bucler* 1203 Cur (Y); William *Bokeler* 1317 AssK. OFr *bouclier* 'maker of buckles'. cf. John *Bokelsmyth(e)* 1384 LLB H. Perhaps also metonymic for *bokeler-player* 'fencer' (1339 MED). cf. Nicholas *Bokelereplayere* and Ralph *Bokelerpleyer* 1379, 1381 AssWa.

**Buckley, Bucklee, Buckleigh:** Alan *de Buckeleg'* 1235–6 FFWa; William *de Bockeleye* 1332 SRWa; John *Buckley* 1545 SRW. A common minor placename. cf. Buckleigh in Abbotsham (Devon), Buckley Heath (Sussex), Buckley Green (Warwicks), etc.

**Buckman:** (i) Roger *Bukkeman* 1278 Ewen (Cu). OE *bucca* 'goat' and *mann*, 'goat-keeper'. (ii) Alan *Bokeman* 1279 RH (O); Adam *Bocman* 1294 AD ii (Sf). OE *bōc* 'book' and *mann*, 'a scholar, student' (1583 NED). Perhaps also 'a copier of books, a bookbinder'. cf. William *le Bokmakere* 1293 MESO (Y).

**Buckmaster:** Adam *de Bucemenistre* 1180 P (Lei); John *Buckmuster* 1623 Bardsley; William *Buckmaster* 1629 ib. From Buckminster (Leics).

**Bucknall, Bucknell, Bucknill:** Robert *de Bukenhal'* 1230 P (St); Nicholas *de Buckenhalle* 1347–8 FFWa; William *Bucknolle* 1642 PrD. From Bucknall (L, St), Bucknell (O, Sa), or Bucknowl (Do).

**Bucknam:** *v.* BUCKENHAM

**Bucknell, Bucknill:** *v.* BUCKNALL

**Buckroyd:** Dionisia *Bokerode* 1379 PTY. 'Dweller at the clearing frequented by deer', OE *bucca, rod.*

**Buckston:** *v.* BUXTON

**Buckton:** Walter *de Buketon'* 1206 Cur (Nt); William *de Bucton'* 1303 FFY; Peter *de Bukton* 1340–1450 GildC. From Buckton (He, Nb, WRY).

**Buckwell:** Richard *de Bocwell* 1248 FFK; William *de Bocwell* 1332 SRWo. From Buckwell (K, Sx).

**Budd, Budds:** Brihtmerus *Budde* c1025 OEByn (C);

Leofwinus *Budda* 1135 Oseney (O); Ralph *Budde* 1170 P (Ha). OE *Budda*, from OE *budda* 'beetle', is not recorded in independent use after the Conquest, but may have been used. BUDDING may well be an OE derivative 'son of *Budda*' and BUDDY is probably 'servant of *Budda*'. We have also Simon *Budecok* 1275 RH (Nf) and William *Budekin* 1279 RH (C), both surnames regularly formed diminutives of a personal name, though -*kin* is also added to common nouns. *Beetle* is a frequent school-boy nickname. The root meaning of *budda* is 'to swell', as in *bud*, and the nickname might also denote a fat, corpulent individual. The first example above is glossed *pro densitate sic cognominatus* 'so called because of his thickness'.

**Buddell:** *v.* BEADEL

**Budden, Buddon:** William *Budun* c1200 DC; William *Budden* 1641 PrSo; Philip *Buddens* 1664 PN Do i 278. *Budd-en*, a diminutive of OE *Budda*.

**Budding:** Ralph *Budding* 1225 AssSo; Julian' *Buddyyng* 1296 SRSx. *v.* BUDD.

**Buddle(s):** *v.* BEADEL, BOODLE

**Buddy:** Andrew *Budday* 1279 RH (C). 'Servant of *Budd*', OE *dǣge*.

**Budge:** *v.* BOUCH, BUGG

**Budgen, Budgeon:** *v.* BOWGEN

**Budgett:** *v.* BURCHARD

**Budgey, Buggey, Buggy:** William *Buggi* 1242 Fees (W); Thomas *Buggy* 1275 SRWo; John *Bogi*, 1327 SRSx. The ON byname *Buggi*, with hard *g* cannot be the source of this surname, pronounced with a *j*-sound, all the early examples of which are from the south. We have ME *bugee, bug(g)e, buggye*, a dissyllable, of obscure origin, 'a kind of fur, lamb's skin with the wool dressed outwards' (1382 NED *s.v. budge* sb. 1). cf. 'gounes . . . furryd with bogey' (1465 ib.) The surname probably denotes a maker of *bugee*. BUGGE may also sometimes belong here. cf. BOUCH, BUGG.

**Budley, Budleigh:** Robert *de Budele* 1268 FFO; Richard *de Buddlegh* 1327 SRSo. From Budleigh (D).

**Buffard:** Richard, Reginald *Buffard* 1221 AssSa, 1274 RH (Sa). OFr *bouffard* 'often puffing, much blowing, swelling vp, strouting out; also swelling with anger' (Cotgrave), c1430 NED.

**Buffey:** *v.* BEAUFOY

**Bufton:** *v.* BOWTON

**Bugden:** William *de Buggeden* 1195 P (L); Claricia *de Buggeden* 1219 AssY. From Buckden (Hunts, WRYorks).

**Bugg, Bugge, Buggs:** Walter *Bugge* 1169 P (L); Osbert *le Bugge* 1327 *SR* (Ess). ME *bugge, bugg*, 'hobgoblin, bogy, scarecrow' (1388 NED). *v.* also BOUCH.

**Buggey:** *v.* BUDGEY

**Buggins:** Thomas *Buggynges* 1380 SRSt; William *Buggyns* 1560 Pat (D). A diminutive of ME *bugge* 'hobgoblin, bogey, scarecrow'.

**Bugle:** Bernard *Bugel* 1199 Pleas (Nf); Thomas *le Bugle* 1296 SRSx; John *Bewgle* 1641 PrSo. A nickname from ME *bugle* 'bugle-horn, wild ox'.

**Bugley:** Robert *Bouggeleghe* 1327 SRSo. From Bugley (W, Do). The exclusively Dorset names

*Bugler, Buglar, Buglear*, probably derive from Bugley (Do).

**Buick:** *v.* BEWICK

**Buie:** *v.* BOWIE

**Bulasse, Bulax:** *v.* BOLAX

**Bulbeck, Bulbick:** Hugo *de Bolebec*, Hugo *Bolebec* 1086 DB (Beds); Ralph *de Bolebec* 1197 P (Y). From Bolbec (Seine-Inférieure).

**Bulcke:** *v.* BULK

**Bulcock:** Edwin *Bulecoc* 1221 AssWo; John *Bulekock* 1301 SRY. A compound of *bull* and the diminutive suffix *cock*.

**Bulford, Bullford:** Robert *of Buleford* 1226–7, James *de Bulford* 1334 FFY; William *Bulford* 1641 PrSo. From Bulford (W), or a lost *Bulford* in Strensall (NRY).

**Bulger:** *v.* BOULGER

**Bulgin:** Walter *Bulekin* 1200 P (Sx). A diminutive of *Bull*. cf. BULCOCK, BULLOCK.

**Bulk, Bulke, Bulcke:** William *le Bolc* 1213–4 FFSr; Robert *Bulk* 1324 CoramLa; Henry *Bulke* 1642 PrD. ME *bulk* 'a bulging part of the body', hence 'plump, fat'.

**Bulkeley, Bulkley:** Robert *de Bulkelegh* 1259 AssCh; Richard *de Bulkele* 1339 CorLo; Thomas *Bulkley* 1489 FFEss. From Bulkeley (Essex).

**Bull, Bulle, Bool, Boole, Bools:** Wulfwin *Bule* 1170 P (Ha); Hulle *le Bule* 1201 P (St); William *le Bole* 1214 Cur (Sr); William *Bull* Hy 3 Gilb (L); Ralph *le Bulle* 1288 Ipm (Nth); Robert *le Bool* 1327 SRSx; Robert *Boole, Bull* 1524 SRSf. OE *bula* 'bull'. Occasionally from a sign: Simon *atte Bole* 1377 LLB H.

**Bullan, Bullant, Bullent:** *v.* BULLEN

**Bullar, Buller:** Thomas *le Bulur* 1203 AssSt; John *Bolur, Bolour, le Bulur* 1296, 1327, 1332 SRSx; Philip *Boler* 1348 DbAS 36; Hugh *Bullour* 1354 FrY. OFr *bouleur*, ME *bullere* 'a publisher of false bulls, a deceiver, cheat' (a1300 NED). *v.* BULLARD.

**Bullard:** Henry *Buliard* 1198 FFSf; Fulco *Bulard* 1275 RH (K); Geoffrey *Bolhard* 1275 RH (W); Marke *Buller, Bullard* 1653, 1672 FrY. These forms can have no connection with *bull-herd* as commonly explained. That word developed into *bullard* very late (1825 NED). The surname might occasionally be from *bull-ward* (1614 NED): John *Bulward* 1524 SRSf. The early forms are probably from OFr *boul(e), bole*, ME *bole* 'fraud, deceit' (c1300 MED) with the addition of the suffix -(*h*)*ard* and may be a synonym for BULLAR, a common early surname. NED records the verb *bul* and *bulling* 'fraudulent scheming' in 1532. Or we may have a similar derivative of OFr *boule* 'round, rotund'. cf. BULLETT.

**Bullas, Bullass, Bullus:** John *de la Bulehuse* 1224 Pat (Ha); Henry *de Bolus* 1327 SRDb; William *Bolehouse* 1327 SRSo; Thomas *Bulluse*, John *Bolouse* 1478 ShefA (Y); William *Bullos*, Thomas *Bullus* 1502, 1564 ib.; Robert *Bullas* 1673 ib. 'One employed at the bull-house' (1807 NED).

**Bulled, Bullett:** Robert *Buleheued* 1195 P (Sx); Richard *Bolehead* 1317 AssK. OE *bula* 'bull' and *hēafod* 'head', a nickname for one noted for his bull-headed impetuosity.

**Bullen, Bullent, Bulleyn, Bullin, Bullon, Bollen,**

**Boullin, Boullen, Bullan, Bullant:** Helias *de Bolonia* 1121–48 Bury (Sf); William *Bulein* 1204 P (L); Richard *de Boloygne* 1255 FFEss; Thomas *de Bolenne* 1289 AssCh; John *Boleyn*, Robert *Bolen*, Thomas *Bollyng*, Rychard *Bullyng* 1524 SRSf. From Boulogne, the English pronunciation of which was *Bullen* or *Bullin*.

**Bullett, Bollett:** Robert *le Bulet* 1194 CurR (Sr), *le Bolete* 1290 SRSr; Robert *Bullet* 1268 AssSo. Whilst this may occasionally be a late development of BULLED 'bull-head', as regularly explained, the main source is clearly different. We have probably a diminutive of OFr *boule* 'round' found in the French surnames *Boule, Boulle, Boulot, Boulet, Boullot* and *Boullet* which Dauzat explains as 'un individu gros, arrondi'. The noun *bullet* is from Fr *boulet*, a diminutive of *boule* 'ball' (1557 NED). cf. BOWL, BULLARD and ROUND.

**Bulley, Bully:** (i) Gilbert *de Buili* 1086 DB (Wa); John *de Bulli, de Builli* 1208 Fees (Y, L). Gilbert *de Buili* may have come from Bouillé (La Manche) or perhaps, from Bouillé (Maine-et-Loire, Mayenne). Roger *de Busli* 1086 DB (Wa) may have come from Bully-en-Brai (Seine-Inférieure). *v.* OEByn. (ii) Reginald *Bulega* 1185 P (Wo); Roger *(de) Bullege* 1197 FFK; Wyot *de Bulleye* 1275 SRWo; Maurice *Boleye* 1275 RH (L); Thomas *de Bolleye* 1327 SRWo. The surname is from a place-name; e.g. Bulley (Glos) or Bulleigh Barton in Ipplepen (Devon). As it means 'bull-clearing', it was probably common. There was land called *Buleia* in Worcestershire in 1204 (P). Mainfenin *de Buleheia* 1201 P (Herts) and Nicholas *de Buleheye* 1218 Cur (Bk) took their names from unidentified places meaning 'bull-enclosure'. They may have lived near-by or worked there as bull-herds.

**Bullfinch:** Ralph *Bulvinch* 1218–19 FFK; Robert *Bolefynch* 1332 SRSx. A nickname from the bird.

**Bullfoot, Bullfitt:** Ailric *Bulefot* 1176 P (D). 'Bull foot', OE *bula, fōt*. A common type of medieval nickname, cf. Tydeman *Coufot* 1347 LLB F 'cow foot'; Laurence *Hundefot* 1298 AssL 'hound foot'; Godwin *Oxefot* 1137 ELPN 'ox foot'; Henry *Piggesfot* 1228 Oseney 'pig's foot'; Henry *Rofot'* 1319 SRLo 'roe foot'.

**Bullford:** *v.* BULFORD

**Bullick:** Henry *de Bolewic* 1198 FFNf. From Bolwick (Nf) or Bulwick (Northants), pronounced *Bullick*.

**Bulling:** Richard *Bulling* 1219 P (C, Hu); Robert *Bullyng* 1327, Richard *Bullyng* 1524 SRSf. Probably OE *\*Bulling*, a derivative of OE *\*Bulla*, *\*Bula*. But some of the later forms may belong under BULLEN.

**Bullinger, Pullinger, Pillinger:** Terricus *le Bulenger* 1180 P (Sx); William *Pullenger* 1553–74 ArchC 49. OFr *bolonger, boulengier* 'baker'.

**Bullivant:** *v.* BONIFANT

**Bullman, Bulman:** Ailward *Buleman* 1209 P (Nf); John *Boleman* 1279 RH (C). OE *bula* 'bull' and *mann*, a bull-keeper or bull-herd.

**Bullock, Bullocke:** Walter *Bulluc* c1170 Moulton (Ha); Robert, William *Bulloc* 1195 FF, 1225 AssSo. OE *bulluc* 'a bull calf'. Perhaps also occupational: cf.

Richard *le Bollocherde* 1281 Eynsham (O); Roger *Bullokman* 1332 SRCu.

**Bullus:** *v.* BULLAS

**Bulman:** *v.* BULLMAN

**Bulmer, Bolmer, Bowmar, Bowmer:** Anketin *de Bulemer* 1128 Black; Robert *de Bulemer* 1219 AssY; Hugh *Bulmer* 1375 FFY. From Bulmer (Ess, NRY).

**Bulpin:** John *de Bulepenne* 1274 RH (So). 'Worker at the bull-pen', OE *pen(n)*, 'pen, enclosure'. cf. Robert *at Bulryng* 1381 PTY.

**Bulston, Bulstone:** Edric *Bulstan* 1196 P (St); Roger *Bulstan* 1210–1 PWi; Peter *Bulston* 1275 SRWo. This looks like a late OE personal name, *\*Bulstān*, not otherwise known.

**Bulstrode:** Richard, Ralph *de Bulestrod'* 1205 Cur (Sr), 1221 AssGl; John *Bulstrode* 1407 AssLo. From Bulstrode (Bucks).

**Bulteel, Boutall, Boutell, Boutle, Bouttell, Bouttle, Bowtell, Bowtle:** Ralph *Buletell* 1205 Cur (L); Richard *Bultel* 1280 LLB A; John *Bultell* 1524 SRSf; John *Buttell, Boultell, Bowtyle* 1568 SRSf. OFr *\*buletel*, earlier *buretel* 'a meal-sieve'. In ME *buletel* was used of a kind of cloth specially prepared for sifting. The surname is synonymous with BOULTER. The Bulteels were reinforced by Huguenots. James *Bulteel* of Tournai came to London in 1634 and a family of this name became prominent at Plymouth.

**Bumphries:** *v.* BOUMPHREY

**Bumpus:** *v.* BOMPAS

**Bunbury:** Patrick *de Bunnebury* 1259 AssCh; Elizabeth *de Bunbery* 1279 RH (C). From Bunbury (Ch).

**Bunce:** *v.* BONE

**Bunch:** Geoffrey *Bunch* 1195 P (Nth); William *Bunche* 1327 SRC. Perhaps ME *bunche* 'a protuberance, a hump on the back' (c1325 NED). cf. *bunch-back* 1618 ib.

**Bunclarke:** William *Bonclerk* 1327 SRSf. 'Good clerk', OFr *bon, clerc*.

**Buncombe:** Richard *de Bounecombe* 1327 SRSo. 'Dweller in the reed-valley', OE *bune* 'reed'.

**Bund(e)y:** *v.* BOND

**Bungay, Bunge, Bungey:** Joscius *de Bungeia* 1195 P (Sf); Adam *de Bungeye* 1327 SRC; Thomas *Bungy* 1327 SRWo. From Bungay (Sf).

**Bunker:** Reginald *Bonquer* 1229 Cur (Sr); William, John *Bonquer* 1257–8 FFK, 1298 AssL; Peter *Bonkere* 1381 SRSt. OFr *bon quer* 'good heart'. Bunker's Hill in Charlestown, Mass., first mentioned as *Bunker Hill*, was land assigned to George Bunker of Charlestown (1634) who came originally from Odell near Bedford.

**Bunn:** *v.* BONE

**Bunnell:** (i) John *Bunel* 1221 AssWo; Francis *Bunnell* 1664 PN Do i 206; Benjamin *Bunnell* 1674 HTSf. cf. Fr *Bunel*, the meaning of which is unknown. (ii) Richard *de Bunewell* 1327 SRC. From Bunwell (Nf).

**Bunney, Bunny:** Botte *Buny* 1222 AssWa; Richard *Bunny* 1309 Wak (Y). Perhaps OFr *bugne, beugne* 'a swelling'. cf. 'Bony or grete knobbe . . . *gibbus*' c1440 NED; 'bownche or bunnye, *gibba*' 1552 ib. *v.* BUNCH, BUNYAN.

**Bunsaul:** *v.* BONSALL

**Bunt, Bunte:** Aluric *Bonte* 1176 P (K); Richard *le Bunt* 1249 AssW; Laurence *le Bont* 1332 SRSx. ME *bonte, bunte* 'sieve'. Metonymic for a maker or user of sieves.

**Bunten:** *v.* BUNTING

**Bunter:** Thomas *Bonter* 1576 SRW; Leonard *Bunter* 1641 PrSo. ME *bunter*, a derivative of ME *bunte* 'sieve', hence a maker or user of sieves. cf. Adam *Bunteflour* 1334 SRK 'sieve flour'.

**Bunting, Buntin, Bunten:** Wluric *Bunting* 1188 BuryS (Sf); William *Bunting* 1260 AssC; Henry *Buntyng* 1332 LoPleas; William *Bontyne* 1489 Black. A nickname from the bird, the etymology of the name being unknown, but cf. Scots *buntin* 'short and thick', Welsh *bontin* 'rump', *bontinog* 'large-buttocked'.

**Bunyan, Bunyon:** Henry *Buniun* 1204 Cur (Beds); Simon *Boynon* 1309 SRBeds. As Roger and John *Buignon* were sons of John *Buignon*, the surname, which has been noted only in Beds, was already hereditary in 1227 (AssBeds), when it appears as *Buingon, Buinon, Buignon* and *Bungnon* in the neighbourhood of Ampthill and Bedford. John *Bunyan* was baptized in 1628 as 'sonne of Thomas *Bonnioun*' who was himself baptized as *Bunyon* in 1603. John signed his name *Bunyan* in 1653 and *Bunyon* in 1672 and was twice called *Bunnion* in the general pardon granting him release from Newgate Gaol. Of the numerous spellings noted, only three, and those late, have no *u*: *Bynyon, Binyan, Binnyan*. Bardsley, retracting his earlier etymology *Bonjean* 'Good John', adopted that of Lower, from Welsh *ab Enion, Benyon*, and was followed by Harrison, Ewen and Weekley, although Dr J. Brown in his *Life of John Bunyan* (1885) had solved the problem. Adducing the 1219 form *Buignon*, he equated it with OFr *beignet* 'fritter' and cited from Godefroy 'Et bone char et granz buignons' with the comment: 'The word signifies a little raised pattie with fruit in the middle'. His only error was to regard this as the original instead of a derived meaning. OFr *bugnon* is a diminutive of *bugne* 'a bile, blane' (*v.* BUNNEY) and came to be applied to any round knob or bunch, and later to a raised pattie and also to a bunion, first recorded in NED c1718. The simple *bugne* is probably the origin of *bun*. The existence of this sense in OFr can now be assumed from that of *bugnon*. cf. also: 'bugnets, little round loaves, or lumpes made of fine meale, oyle or butter, and reasons; bunnes, Lenten loaves'. The surname might be a nickname for one disfigured by a knob, lump or hump, or it might be occupational, a pastry-cook, though the latter hardly fits in with the activities of the 1227 Bunyans.

**Bunyard:** *v.* BANYARD

**Burall:** (i) Robertus *filius Burewoldi* 1184 Oseney (O); Gilebertus *filius Buroldi* 1198 ib.; Hugh *Burewald* 1200 ib.; Lambert *Borewald* 1279 RH (O). OE *Burgweald* 'fortress-power', first recorded in the 11th century. (ii) William *de la Burhalle* 1275 RH (L). 'Dweller or servant at the bower-hall', OE *būr* and *heall*.

**Burbage, Burbidge, Burbudge:** Ralph *de Burebeche* 1172 P (Db); William *de Burbache* 1200 Cl (Ha); John *Burbache, Burbage* 1340–1450 GildC; Helene *Burbadge* 1576 SRW. From Burbage (Derby, Leics, Wilts).

**Burbank:** Symon *Burbank* 1524 SRSf; Martin *Burbancke* 1598 FrY. From Burbank House in Dacre (Cu).

**Burch:** *v.* BIRCH

**Burchard, Burchatt, Burghard, Burckitt, Burkart, Burkett, Burkitt, Buchard, Butchard, Buckett, Budgett, Bockett, Borkett, Borkwood, Bouchard, Barchard:** Gaufridus *nepos Bocardi* c1150 DC (L); *Burekardus* de Burewelle Hy 2 DC (L); *Buchardus* 1196 P (Du); Reginaldus *filius Burchardi* 1220 Cur (Sf); *Burchardus, Burcardus* 1222 Cur (Do); Robert *Bocard'* 1207 Cur (Sf); Ralph *Bochard'* 1219 Cur (Nth); Thomas *Burchart* c1248 Bec (Sf); Walter *Buchard* 1255 RH (W); Warin *Burchard*, Robert *Burghard* 1275 RH (Sf); Richard *Burkett* 1524 SRSf; Peter *Barchard* 1702 FrY. The frequency of these surnames and the variety of their forms are due to two names, both ultimately of Germanic origin. OE *Burgheard* 'fortress-hard' is found in DB as *Burchardus, Burkart, Burchart*, and *Bucardus*. The cognate OG *Burghard, Burcard* became OFr *Bouchart* and was brought across the Channel by the Normans.

**Burcher:** *v.* BIRCHER

**Burchett:** *v.* BIRCHETT

**Burckitt:** *v.* BURCHARD

**Burcote, Burcot, Burcott:** Nicholas *de Bridicote* 1207 P (O); Walter *de Birecot'* 1221 AssWo; Geoffrey *de Burecote* 1243 AssSo. From Burcot (O, Wo), or Burcott (Bk, Sa, So).

**Burd, Burds:** William *Burde* 1275 RH (Sx), 1285 *Ass* (Ess). ME *burde* 'young lady, maiden', probably a derogatory nickname.

**Burdall:** Arkell *de Breddal'* 1219, Nicholas *de Breddal* 1251 AssY; John *de Bourdale* 1332 SRCu. From Burdale (ERY), *Breddale* DB.

**Burdas, Burdass, Birdis, Burdus:** Alan *de Burdeus* 1297 SRY; Christopher *Burdus* 1519 GildY; John *Burdas* 1662 FrY. From Bordeaux.

**Burdekin:** John *Burdycane, Burdekin* 1597, 1636 Shef (Y). A diminutive of *burde* 'little young lady', undoubtedly derogatory.

**Burden, Burdon, Burdoun, Bourdon:** (i) Ralph *Burdun* 1128–9 Holme (Nf); Ilger *Burdun* 1166 P (Y). (ii) Arnulf *Burdin* 1115 Winton (Ha); Bruni *Burdin* 1180 P (Bk). With these forms we must include BURDETT. All may be diminutives of OG *Burdo*, introduced from France or, possibly, of Lat *burdo* 'mule'. *Burdonus* and other derivatives of the Celtic *Burdo* are found in France, but in view of the rarity of this name, Michaëlsson prefers to derive the French surname from OFr *bourdon* 'a pilgrim's staff'. The personal name does occur in England, but is very rare: Hamo *filius Burdun* 1166 RBE (Nf). Only two examples of *Burdin* have been noted. *Burdon* is very common in early English sources and in Paris c1300. *Burdet* is equally frequent in England. (iii) Zacharias *de Burdun* 1217 FeuDu. From Great Burdon (Durham) or from Burdon Head (WRYorks) or Burden near Harewood (WRYorks). (iv) Ælfsige *Burðen* 968 BCS 1212 (K); Hugelyn *Bourþeyn*, Hugelinus *cubicularius* 1052–65 Rams (Hu); Nicholas *Burþein, Burdon* 1242 Fees (W). OE *būrðegn* 'bower-servant', 'chamberlain'. This is probably the least common source.

**Burdett, Burditt:** Robert *Burdet, Burdel,* Hugo *Burdet* 1066 DB (Lei); Ralph *Burdet* c1160 DC (Lei); George *Burditt* 1690 FrY. *v.* BURDEN.

**Burdge:** *v.* BURGE.

**Burfield:** Robert *de Beregefeld* 1205 P (Berks); Peter *de Burefeud* 1297 MinAcctCo; John *de Burghefeld* 1341 Goring (O). From Burghfield (Berks), or Burfield in Bosham (Sx).

**Burfitt, Burfoot:** Stephen *Burford* of Cowden (Kent) signed his name *Burfoot* in 1673 (Ewing 83).

**Burford:** Clement *de Bureford* 1186 P (O); John *de Burreford* 1327 SRSx. From Burford (Devon, Oxon, Salop) or Burford Bridge (Surrey).

**Burge, Burdge:** Robert *de la Burge* 1200 P (Ha); Roger *Burge* 1479 AD vi (Ess). William *atte Brugge* lived at Burge End (Herts) in 1302 (PN Herts 21). This metathesis of *r* in OE *brycg* 'bridge' has also been noted in Bucks, Cambridge, Northants, Notts, Oxford, Somerset, Wilts and Warwicks. *v.* PN Nth xxxii.

**Burger, Burgher, Burker, Borger:** Henry *le Burger* 1275 RH (Lo); Robert *Burger* 1327 SRSf. ME *burger* 'inhabitant of a borough', 'a citizen' (1568 NED).

**Burges, Burgess, Burgis, Burgiss, Borges:** Geoffrey *burgeis* 1115 Winton (Ha); Ralph *le Burgeis* 1195 P (Sx); Philip *Burgis* 1199 FrLeic; Philip *Burges, Burgeis* 1220, 1234 Oseney (O); Walter *le Borgeys* 1296 SRSx. OFr *burgeis* 'inhabitant of a borough', strictly one possessing full municipal rights, 'a freeman of a borough' (c1230 MED).

**Burgett, Burgot:** Absalon *de Burgate* 1198 P (K); Ralph *atte Burgate* 1260 AssC; Robert *de Burhtȝete* 1274 RH (Gl); Ingram *atte Burghȝete* 1333 MESO (So). From Burgate (Hants, Suffolk, Surrey) or 'dweller by the castle or city gate'. OE *burggeat* (Burgett), ME *burgate* (Burgot).

**Burgh:** Ailricus *de Burg* 1066 DB (Sf); Ralph *de Burc* 1177 P (Nf). From one of the many places named Burgh.

**Burghard, Burkart:** *v.* BURCHARD

**Burghersh:** *v.* BURWASH

**Burghley:** *v.* BURLEIGH

**Burgin:** *v.* BORGONON, BURGOIN

**Burglin, Burlin, Burling, Burlong:** John *Burgelun* Ric I Bart; William *le Burguillun* 1243 AssSt; William *Burgelun* 1275 RH (Nf); John *Burgoillioun* 1327 SRSf; Widow *Burling* 1674 HTSf. 'The Burgundian', probably a variant of OFr *Bourguignon. v.* BORGONON.

**Burgoin, Burgoine, Burgoyne, Burgon, Burgin, Burgwin, Bourgein, Bourgoin:** Walter *Burgoin* 1086 DB (D); Simon *Burgunie* 1210 Cur (C); Adam *de Burgoigne* 1319 SRLo; Elizabeth *de Burgon* 1379 PTY; John *Burgin* 1638 Bardsley. 'The man from Burgundy', Fr *Bourgogne.* Various forms were used of the same man: John *le Burgoyn* 1301 LLB B, *le Borgiloun* 1310 ib., *de Burgoyne* 1319 SRLo; Ralph *le Burginon* 1314, *Burgillon* 1323 AssSt, *le Burgoynon* 1330 FFSt. Nicholas *Burgoin* was the son of John *Burgoin* and both are also called *Burgwine,* while the seal of Felicia, wife of Nicholas *Burgelun,* bears the legend S. FELICIA BURGUNUNG 1212–23 Bart. *v.* also BURGLIN, BORGONON.

**Burk, Burke, Bourke:** Ailricus *de Burc* 1066 DB (Sf); John *de Burk* 1274 RH (So). The first example refers to Burgh (Suffolk). *Burk* is an Anglo-Norman pronunciation of *Burgh* (which survives as BURGH) and it is doubtful whether the Norman pronunciation took root in England. *Burke* is a very common Irish name which derives from the family of *de Burgh.* William *de Burgo* went to Ireland in 1171 with Henry II and later became Earl of Ulster.

**Burkenshaw:** *v.* BIRKENSHAW

**Burker:** *v.* BURGER

**Burkett, Burkitt:** *v.* BURCHARD

**Burkimsher, Burkinshaw, Burkinshear:** *v.* BIRKENSHAW

**Burkman:** *v.* BORROWMAN

**Burkmar:** *v.* BRIGHTMORE

**Burks:** *v.* BIRKS

**Burlace:** *v.* BURLES

**Burland, Borland:** Robert *de la Burilonde* 1268 MELS (So); William *atte Borland* 1303, *atte Burland* 1346 FA (So); Richard *Burland* 1672 HTY. From Burland (Ch, ERY), or 'dweller on land belonging to the borough', OE *burg-land.*

**Burle, Burles, Burls:** John *Byrle* 1327 SR (Ess). OE *byrele, byrle* 'cup-bearer, butler'.

**Burleigh, Burley, Burghley:** Wihenoc *de Burli* 1086 DB (Nf); John *de Burgeley* 1198 FFHerts; John *of Burlay* 1249 AssW; Henry *Burleigh,* John *Burley* 1642 PrD. From Burley (Ha, R, Sa, WRY), Burleigh in South Huish (D), or Burley Hill (Db).

**Burler:** James *Burler* 1369 FFEss. A derivative of ME *burle* 'to burl', from ME *burle,* OFr *bourle* 'flock of wool'; a burler, one who dresses cloth by removing knots and extraneous particles (c1475 MED). *v.* also BURREL.

**Burles, Burlace, Barlas, Borlace, Borlas, Borlase:** Hugh *le Burdeleis* 1186 P (Sf); Geoffrey *de Burdeleys* 1261 FFC; William *Burdeleys* 1327 SRC; Thomas *Burlas, Burlace, Burlase, Burlaas* or *Borlas* of Burlas Burges 1509 LP (Co). 'The man from Bordeaux', OFr *bordelais.*

**Burley:** *v.* BURLEIGH

**Burlin, Burling, Burlong:** *v.* BURGLIN

**Burlington:** Martin *de Breddelington* 1204 AssY. From Bridlington (ERYorks), pronounced locally *Burlington.*

**Burman:** *v.* BORROWMAN, BOWERMAN

**Burmingham:** *v.* BIRMINGHAM

**Burn:** *v.* BOURNE

**Burnaby, Burnby:** Peter *de Brunnebi* 1204 AssY; John *de Burneby* 1275 RH (Nth); John Burneby 1417 IpmY. From Burnby (ERY).

**Burnard:** *Burnardus* 1211 Cur (Nf); Rogerus *burnardus* 1130 P (Beds); Odo *Burnard* 1192 P (Ess); Richard *Burnhard* 1279 RH (Beds). *Burnhard,* a compound of OFr *brun* and *-hard,* with the same metathesis as in *Burnel* and *Burnet.* A nickname for one of brownish, dark hair or complexion, used occasionally, like *Burnel,* as a personal name.

**Burnel, Burnell, Brunell:** *Burnellus* de Aumeivill 1200 Cur (Y); Robertus *burnellus* 1130 P (O); Roger,

William *Burnel* Hy 2 DC (L), 1197 FF (Sa). ME *burnel*, a metathesized form of OFr *brunel*, a diminutive of OFr *brun* 'brown'. A nickname for one of brownish complexion, used also as a personal name. *Brunell* is probably recent, Fr *Brunel*.

**Burner:** v. BOURNER

**Burnes, Burness, Burns:** (i) Robert, Gervase *Brenhus* 1208 P (Y), 1275 RH (Nf). A nickname, 'burn house'. cf. BRENNAN. (ii) David *Burnis* 1526 Black. The forefathers of Robert Burns migrated from Burnhouse in Taynuilt to Forfarshire where they were called Campbells of Burnhouse, and later *Burness* or *Burns* (MacBain, *Inverness Names*). The stress in *Burness* was on the first syllable and as the name was pronounced in Ayrshire as if written *Burns*, Robert and his brother agreed to drop *Burness* and to assume *Burns* in April 1786 (Black).

**Burnet, Burnett, Burnitt:** Robert *Burnet* 1219 AssY; Richard *Bornet* 1279 RH (Bk); Cristina *Burnete* 1365 LoPleas. OFr *burnete, brunette*, a diminutive of *brun* 'brown', 'dark brown' (c1200 NED), used like *Burnell* of complexion. *Burnete* was also used of a wool-dyed cloth of superior quality, originally of dark-brown colour (1284 NED). cf. 'pro . . . caligis *de burneta*; caligas *de burneto*' 1200 Oseney i, 64, 66 'hose of burnet or coloured cloth' (Ed.). The surname might denote a maker or seller of this.

**Burney, Burnie:** v. BERNEY

**Burnford:** Gilbert *de Burneford'* 1221 AssWo; John *de Burnford*, William *de Burneford* 1275 SRWo. From Burnford in Bromsgrove (Wo).

**Burnham:** Geoffrey, Roger *de Burnham* 1193 P (Nf), 1262 FFO; Thomas *Burnham* 1446 IpmNt. From Burnham (Bucks, Essex, Lincs, Norfolk, Som).

**Burnhill, Burnill:** Thomas *de Burnul* 1212 Fees (La); Peter *de Burnil, de Burnhill* 1281 AssLa; Thomas *Burnehill* 1373 IpmNt. From Brindle (La), *Burnhull* 1206. The name probably usually fell in with BURNEL.

**Burnikell:** v. BARNACLE

**Burns:** v. BURNES

**Burr:** Hugo, Samson *Burre* 1185 Templars (Y), 1206 Cur (C). ME *burre* 'a bur' (c1330 MED), used by Shakespeare of one who sticks like a bur, a person difficult to 'shake off'. This sense may well be older.

**Burra:** v. BOWRA

**Burrage:** v. BURRIDGE

**Burrard:** Robert *Borard'* 1219 AssY; Simon *Borhard, Borart* 1235, 1242 Fees (Lei); Nicholas *Burhard* 1327 SRSf. OE *Burgheard.* v. BURCHARD. This may also have become *Borrett, Burrett*, etc.

**Burrel, Burrell, Burrells, Borel, Borell:** Roger, William *Burel* 1194, 1196 P (W, L); Simon *Borel* 1296 SRSx. ME *borel*, OFr *burel* 'reddish-brown', used of a coarse woollen cloth of this colour (a1325 MED). The surname might refer to dress or complexion, or it may denote a maker of *borel*, a *bureller.* cf. Alfred *le Bureller* 1277 LLB A; John *Burelman* 1311 LLB D. *Borel* had also in ME the meaning 'belonging to the laity' (c1390 MED), 'unlearned, rude' (1513 NED). It was also used as a personal name, perhaps an original nickname in the sense 'dark': Johannes *filius Borelli* 1205 Cur (R); *Burellus* de Rathesnese 1274 RH (Nf).

**Burrett, Borrett, Borritt, Boret:** (i) *Burredus* 1114–18,

*Burret* 1161–77 Rams (Hu); Hugo *filius Buret* 1199 FF (R); Koleman *Burred* 1133–60 Rams (Hu); Nicholas *Bured* 1275 SRWo. OE *Burgrǣd* 'fortress-counsel', found in DB as *Burghered, Burgret, Burred, Burret, Borgered, Borred* and *Borret.* (ii) John *Bureheued* 1219 AssY; William *Burreheud* 1308 Wak (Y); Agnes *Borheued* 1327 SRSf; William, Robert *Borhed, Borrett* 1403, 1577 Shef (Y); Robert *Borhed*, Henry *Boret* 1524 SRSf. OFr *bourre* 'rough hair, flock of wool' and OE *hēafod* 'head', a nickname for one with rough, shaggy hair.

**Burrey, Burry:** (i) Gamel *Burri* 1166 P (Y); Beatrix *Burry* 1279 RH (Beds). ME *burry* 'rough, shaggy'. (ii) Hugh *de Burhey* 1260 CtSt; William *Burry* 1332 SRSt. 'Dweller by the borough enclosure', OE *burg*, (*ge*)*hæg.*

**Burridge, Burrage:** *Burcheric, Burchricus, Burricus, Buric* 1066 DB; *Burrich* de Bradefeld 1203 AssNth (Sf); William *Borrich* 1327 SRSf; Henry *Borrich* 1327 SRSo; John *Burrage* 1568 SRSf; William *Burrydge* 1587 FFHu. OE *Burgrīc* 'fortress-powerful'.

**Burrough, Burrow, Borrow:** John *atte Boroghe*, Thomas *Burewe* 1327 SRSo; Henry *Borowe* 1527 FrY. 'Dweller by the hill' (OE *beorg*), as at Burrow (Devon, Som), or from Burrow (Leics, Lancs), OE *burg* 'fort'.

**Burroughes, Burrows, Burrus, Burris, Burriss, Borrows:** John *de Burhus* 1440 ShefA; Margaret *Burrous* 1564 ib.; William *Burrowes* 1585 ib.; William *Burrosse* 1572 ib.; William *Burroughs* 1742 Bardsley. 'Dweller at the bower-house', or one employed there, OE *būr, hūs.* cf. *Bourhouse* Edw 4 EA (OS) i, a manor of Waltham Abbey.

**Bursacott:** v. BUZZACOTT

**Bursell, Bursill:** Thomas *de Bristhull* 1251 AssY; Alexander *Bursell* 1642 PrD. From Boarshill in Bigbury (D), or Burshill in Brandesburton (ERY), *Bristhil* 12th.

**Burser:** Geoffrey *Burser*, Alan *Bursarius* 1168 P; Roger *Borser* 1253 Acc; Robert *le Burser* 1311–2 FFWa. OFr *borsier* 'treasurer, bursar', or, perhaps, 'a maker of purses'.

**Bursey, Bersey:** (i) Serlo *de Burci* 1084 GeldR (W); Serlo *Borci* 1086 DB (So); Hugo *de Burci* 1185 Templars (So). From Burcy (Calvados), (ii) Rannulf *Bursi* 1195 P (He); Hugh *Bursey* 1275 RH (Nf). These forms are from OE *Beorhtsige*, found as *Birhsie* in Devon (BCS 1248). v. BRIXEY.

**Bursicott:** v. BUZZACOTT

**Burstall:** Peter *de Burstall'* 1206 Cur (Sf); Robert *de Burstall'* 1230 P (Bk); Richard *de Birkestalle* 1316 Wak (Y). From Burstall (Sf, St, ERY), Birstal (WRY), Birstale (Lei), or Boarstall (Bk), *Burcstala* 1161.

**Burstler, Bustler:** William *le Bustlere* 1319 FFC; Robert *Burstlere* 1336 FFEss; Richard *Bustlere* 1355–9 AssBeds. A derivative of OE *byrst* 'bristle', a maker of things from bristles.

**Burston:** Nicholas *de Burgeston* 1199 AssSt; Stephen *de Briddesthorn'* 1230 P (Bk); William *de Burstone* 1275 RH (Nf). From Burston (Bk, Nf, St). Sometimes, perhaps, from an unrecorded personal name, OE *\*Burgstān: Burstanus* 1171–2 MedEA (Nf); Godric *filius Burstan* 12th Rams (Hu).

**Burstow, Burstowe:** Hugh *de Burstowe* 1210 P (Ha);

John *de Burgstowe* 1249 Misc (Sx); Richard *de Burstowe* 1310 LLB D. From Burstow (Sr).

**Burt, Burtt, Birt:** Thomas *Burt, Burd* 1229 Pat (He); Roger *Burt* c1285 StThomas; James *Birt* 1505–6 FFWa. Variant forms of BRIGHT or of BIRD.

**Burton, Bourton, Borton:** Ioluard *in Burhtun* c1150 YCh; Gerard *de Burton* 1178 P (Wa, Lei); William *de Borton'* 1275–6 RegAntiquiss; William *Burton* 1327 SRSx; John *Borton* 1332 SRWo. From one or other of the many places called Burton or Bourton.

**Burtonshaw:** *v.* BIRKENSHAW

**Burtwistle:** *v.* BIRTWHISTLE

**Burward:** *Burewardus* 1206 P (Sf); Ralph *Borgward* 1299 LLB E; Robert *Boreward* 1327 SRSf; William *Burghward*, Austen *Burward* 1524 SRSf. OE *Burgweard* 'fortress-guard'.

**Burwash, Burghersh:** William *de Burwash* 1291 QW (K); Bartholomew *de Burghersh* 1355 PN K 31; Richard *Burwish*, William *Burwash* 1525 SRSx. From Burwash (Sx), *Burgersa* 12th.

**Bury:** *v.* BERRY

**Busby, Bussby:** Eustace *de Buskeby* 13th Guisb; Robert *Busby* 1379 PTY; Andrew *Busby, Busbe, Bushby* or *Bussheby* 1509 LP (St). From Busby (Lei, NRY). In Scotland from the lands of Busby in Carmunnock (Renfrew).

**Buscall, Buskell:** Sigar *Buzecarl* 1111–38 ELPN; Nicholas *Buscecarle* 1205–6 FFEss; John *Buscarl* 1326 *PetreA*; Robert *Buskell* 1680 CWNS 57. OE *butsecarl* 'boatman, mariner'. The *butsecarls* stand in the same relation to the *scip-fyrd* that the housecarls occupy to the *land-fyrd*, i.e. they are the king's standing force as opposed to the national levies.

**Busfeild, Busfield:** *v.* BOUSFIELD

**Bush, Bushe, Busk:** Richard *de la Busce* 1181 P (Y); Henry *del Busk* 1275 RH (Nf); Roger *atte Buske, del Bushe* 1305 SIA iii; Richard *Bussh* 1309 FFSf; Roland *atte Bushe* 1384 LoPleas. 'Dweller by the bush', ME *busk, busche*. Bush is from OE **busc* (*v.* MELS), *Busk* (less common), from ON *buskr*.

**Bushell, Bushill, Bussell, Boshell, Bossel, Bishell, Bissell, Bissill:** Roger *Buissel* 1086 DB (So); Alan *Buscel* c1140 YCh; Richard *Bussell* 1200 P (Beds); Richard *Buschel* 1243 AssSo. ME *buyscel, busshel, bysshell*, OFr *boissell, buissel* 'bushel', probably for one who measured out corn, etc., in bushels, or for a maker of bushel-vessels. cf. Stephen *Busselman* 1327 SRSo, Robert *le Busselar* 1243 AssSo, Peter *Boseler* 1305 MESO (L), OFr *boisselier* 'maker of vessels (baskets) holding a bushel'. This may survive in the very rare BISLER. *Bussell* may also be from OFr *bucel* 'small barrel', for a maker of these.

**Busher:** *v.* BOSHER

**Buskell:** *v.* BUSCALL

**Buskens, Buskin, Boskin:** (i) Nicholas *atte Busken, atte Bosken* 1329–30 PN D 140. From Buskin (Devon) or 'dweller by the bushes', from the dative plural of OE **busc*. (ii) Roger *Buckeskyn* 1281 FFEss; Walter *Buskyn* 1281 Cl; Katharine *Bukeskyn* 1295 Ipm (Nf). 'Buck-skin', skin of a buck, used particularly of 'breeches made of buckskin' (1481–90 NED), and as a surname, for a maker of these or for a worker in buckskin or leather. Richard de Gravele called

*Bokskyn* was an apprentice of Walter Polyt *fuyster* 1311 LLB D. *v.* FEWSTER. *Buckskin* would inevitably come to be pronounced *Buskin*.

**Busky:** Remig' *de Buskeheye* 1296 SRSx. 'Dweller by the bush-enclosure', OE **busc-(ge)hæg*.

**Buss, Busse:** Walter, Richard *Busse* 1195 P (Nf), 1220 Cur (Berks). OFr *busse* 'cask'. cf. BARRELL.

**Bussby:** *v.* BUSBY

**Bussell:** *v.* BUSHELL

**Bussey, Bussy:** Robert *de Buci, de Boci* 1086 DB (Nth); Robert *Buscy* 1208 Cur (Sx); William *Bussy* 1310 EAS xx. The DB under-tenant came from Bouce (Orne). *v.* ANF. Others may have come from Boucey (La Manche) or Bucy-le-Long (Aisne).

**Busson:** John *Buzun* 1197 P (Gl); Thomas *Bussun* 1242 Fees (Db); John *Boson, Bozon* 1536 FFEss. OFr *buzon* 'one connected with the law'.

**Bustard:** Walter *Buistard* 1159, *Bustard* 1162 P; Robert *Boistard* 1231 FFY; Robert *Bustarde* 1343 Whitby. A nickname from the bustard, OFr *bistarde, bustarde*.

**Bustler:** *v.* BURSTLER

**Buszard:** *v.* BUZZARD

**Butcher, Butchers, Bucher, Boucher, Boutcher, Bowcher, Bowker:** Ailwardus *le Bochere* 1184 P (Lo); Richard *le Bucher* 1240 FFEss; William *Bochier*, Alan *le Boucher* 1327 SRSx; Thomas *le Bouker* 1332 SRLa. AFr *bocher, boucher*, OFr *bochier, bouchier* 'butcher'.

**Butement, Beautement, Beautyman, Bootman, Bootyman:** John *de Botemont* 1172 ANF; Hugh *de Buttemund* 1212 Cur (Lei); Nicholas *Botemond* 1327 SRSf; William *Botyman* 1525 SRSx; John *Booteman* 1609 SfPr. From Le Boutimont (Pas-de-Calais), or Boutement (Calvados).

**Butler, Buttler:** Hugo *Buteiller* 1055 France; Alexander *le butiller* 1174–84 Seals (Hu); Baldwin *le Buteilier* 1200 P (K); William *le Boteler* 1260 AssC; Henry *le Butler* 1327 SRWo. AFr *butuiller*, OFr *bouteillier* 'servant in charge of the wine-cellar', usually the head servant (c1250 NED). In some early examples, an officer of high rank nominally connected with the supply and importation of wine (1297 NED). Forms like *Boteller* may occasionally be for BOTLER.

**Butlin, Bucklin:** Robert *Butevilain* 1130 P (Nf); Robertus *Buteuillanus* archidiaconus 1147–53 DC (Nt); Ernis *Buteuilein* 1205 P (D); William *Butveleyn* 1429 Pat (Nth). OFr *boute-vilain* 'hustle the churl'.

**Butner:** Henry, Geoffrey *le Botoner* 1274 RH (Lo), 1282 LLB A. OFr *botonier* 'maker of buttons'.

**Butt, Butts:** (i) *But* (a moneyer) Wm 2; his son was Robertus *filius But'* 1137 ELPN; Godlambus filius *But* 1133–60 Rams (Nf); *But* 1170 P (Ha); Walter, Hubert *But* 1114–30 Rams (Nf), c1150 ELPN; Leuricus *Butte* 1185 Templars (Beds); Robert, William *le But* 1198 P (Sx); 1214 ELPN; Margery *Buttes* 1275 SRWo. OE **Butt* is found in Butsash (Hants) and **Butta* in Butley (Ches, Suffolk), a personal name which, though unrecorded in OE, was in use in the 12th century. It is one source of the surname, particularly of the early examples without the article. cf. Richard *Buttyng* 1327 SRSo 'son of *Butt*'. We have also clearly a nickname from ME *butt* 'thicker end, stump', probably used of a thickset person. (ii) William *de Butte* 1200 Oseney (O); Henry *atte Buttys* 1380 NorwW (Nf). ME *butt*, OFr *but* 'a goal; mark for shooting'. One who lived near the archery butts or, perhaps, an archer. cf. FURLONG.

**Buttanshaw, Buttenshaw:** *v.* BIRKENSHAW

**Butter, Butters, Buttar:** (i) Henry *Butor* 1169 P (Y); Henry *le Butor* Cur (D). ME *botor*, OFr *butor* 'bittern', noted for its 'boom' in the breeding season and called 'bull of the bog', hence, perhaps, the nickname. (ii) William *le Buter* 1243 AssSo; John *le Buttare* 1275 SRWo; William *le Buttere* 14th AD v (Wa). 'Keeper of the buttery.' *v.* BOTTERELL. (iii) Turchetillus, William *Butere* 1130 P (Do), 1198 FF (Nth); Geoffrey *Butter* 1327 SRWo; John *Buttere* 1327 *SR* (Ess). OE *butere* 'butter', metonymic for a maker or seller of butter. cf. William *le Buterar'* 1327 SRSx, John *Butercharl* c1192 HPD (Ess), Thomas *Butterman* 1302 SRY, Henry *Botreman* 1327 Pinchbeck (Sf), Margaret *le Buttermonggere* 1306 LoCt. Forms for (ii) and (iii) cannot always be distinguished.

**Butterfield, Butterfill:** Hugh *de Buteresfeld'* 1199 Pl (Bk); Philip, Adam *de Butterfeld'* 1231 Cur (Bk), 1379 PTY; William *Boterfeld* 1423 LLB K. From Butterfield (WRYorks), or from other minor places of the name.

**Butterick:** *v.* BUTTERWICK

**Butteris(s):** *v.* BOTTERELL

**Butterley, Butterly:** Eduuin *de Buterleio* 1084 OEByn (D); Roger *de Buterle* 1221 AssSa; Stephen *de Butterleye* 1329 WoCh; John *Butterlegh* 1375 FFW. From Butterley (Db, He), or Butterleigh (D).

**Butterwick, Butterick:** Gusa *de Buttirwic* c1155 Gilb; William *de Boterwyk'* 1262 FFL; Hugh *de Buterwyk* 1327 SRY; Thomas *Boterwyk* 1392 LoCh. From Butterwick (Du, L, We, ERY, NRY).

**Butterworth, Butterwood:** Roger *de Butterword* 13th WhC; Alexander *de Boterworth* 1389 IpmLa; Thomas *Butterworth* 1456 FrY. From Butterworth (La).

**Buttery, Buttrey:** William *Buteri* 1177 P (Bk); Reginald *Boteri* 1211 Cur (He); William *de Buteri* 1219 Cur (Sf); John *de la Boterye* 1334 FFSt. OFr *boterie*, originally 'place for storing liquor', but early used of a 'room where provisions were laid up' (1384 MED). 'Keeper of the buttery.' *v.* also BOTTERELL

**Buttler:** *v.* BUTLER

**Button, Botten:** William, Robert *Boton* 1296 SRSx, 1317 AssK; Stephen *Botun* 1327 SRSx. OFr *boton* 'button', metonymic for BUTNER.

**Buttonshaw:** *v.* BIRKENSHAW

**Buttress, Buttriss:** *v.* BOTTERELL

**Buttrey:** *v.* BUTTERY

**Buttrum:** *v.* BARTRAM

**Buxton, Buckston:** (i) Henry *de Bucstanes* 1230 P (Db). From Buxton (Derby). William *Buckeston* 1279 RH (Hu). Probably from Buxton (Norfolk), DB *Bukestuna*. (ii) Ailricus *Bucstan* 1170 P (L); John *Bucstan* 1221 AssWa; Richard *Bocston* (*Bokston*) 1327 *SR* (Ess); John *Bucstone* 1377 LLB H. The distribution and the frequency of this name, regularly in the singular and without sign of a preposition, suggest a personal-name otherwise unknown, perhaps

OE *\*Bucstān*, a combination of OE *Bucca* and the common theme *-stān*. cf. PICKSTONE. Or we may have an OE *\*Burgstān*, with early loss of *r* as in *Burgheard*. cf. BURCHARD.

**Buy(e):** *v.* BOWIE, BY

**Buyers:** *v.* BYARS

**Buzzacott, Bursicott, Bursacott:** (i) From Buzzacott in Combe Martin (D), *Bursecot* 1399, *Bussacott* 1667. (ii) Mlle Anne *Boursequot*, fiancée of Jacques Fontaine, a Huguenot refugee from Bordeaux, landed at Appledore on 11 Dec. 1685. In the BarnstaplePR the marriage is given as that of Mr James Fontaine and Mrs Ann *Bursicott*.

**Buzzard, Buszard:** Robert *Boszart* 1177 P (He); William *Bozard* 1258 AD vi (W); Peter *Busard* 1274 RH (Sf). OFr *busart*, ME *busard*, *bosard* 'buzzard', an inferior kind of hawk, useless for falconry, used also of a worthless, stupid, ignorant person (1377 NED).

**By, Bye, Buy, Buye:** Hugh *de la Bye* 1243 AssSo; Alicia *de By* 1250–70 Black (Berwick); John *ate Bey* 1279 RH (C); William *in the By* 1327 SRSo; John *Bye* 1327 SRC. 'Dweller in the bend', OE *byge*. We are also occasionally concerned with a personal name of obscure origin: Thomas, Henricus *filius Bye* 1279 RH (C).

**Byam:** *v.* BYHAM

**Byard, Byart, Biart:** Thomas *Byerd* 1296 SRSx. 'Dweller by the enclosure', ME *bi yerd*.

**Byars, Byers, Byre, Byres, Bier, Biers, Buyers:** Elias *de la Byare* 1275 RH (D); Willelmus *del Byre* 1301 SRY; John *de Byres* 1309 Black (Newbottle); John *Buyres* 1327 SRSo. 'One employed at the cow-house' (OE *bȳre*), cow-man, or from Byers Green (Durham) or the old barony of Byres (East Lothian).

**Byas, Byass:** Adam *de Byus* 1275 RH (L); John *de Bayhus* ib. (Beds); John *Byas* 1699 FrY. 'Dweller at the house in the bend', OE *byge*, *hūs*.

**Byatt, Byatte, Byott, Bygate:** Nicholas *Byate* 1297 MinAcctCo; Ralph *Bytheyate* 1379 PTY. 'Dweller by the gate', OE *geat*, ME *yat*, *gate*.

**Byerley:** *v.* BIRLEY

**Byfield, Bifield:** Nigel *de Bifeld'* 1202 FFNf; Robert *de Byfeld'* 1314–16 AssNth; Adam *Byfelde* 1367 IpmGl. From Byfield (Northants), or 'dweller by the open country'.

**Byfleet:** Hugh *de Byflete* 1276 LLB A. From Byfleet (Sr).

**Byford:** Geoffrey *de Biford* 1222–3 FFEss; John *Byford* 1371 LoPleas, 1381 FFEss. From Byford (He), or 'dweller by the ford', OE *bī*, *ford*.

**Bygott:** *v.* BIGOTT

**Bygrave, Bygraves, Bigrave, Bygreaves, Bygrove:** Leommær æt *Biggrafan* c1015 ASWills (Herts); William *Bygrave* 1312 LLB D. From Bygrave (Herts) or residence near a grove. *v.* GREAVES, GROVE.

**Byham, Byam:** William *de Biham* 1202 P (Y). Probably from Bytham (Lincs), *Biham* c1100 DEPN.

**Byles:** *v.* BILES

**Bylow:** Roger Cherche *otherwyse callyd Bylaugh* 1452 Paston. From Bylaugh (Nf).

**Byng:** *v.* BING

**Byott:** *v.* BYATT

**Byrch:** *v.* BIRCH

**Byrde:** v. BIRD

**Byre(s):** v. BYARS

**Byrne, Byrnes, Birn:** Ir *Ó Birn* 'descendant of *Biorn*' (ON *Bjǫrn*) or *Ó Broin* 'descendant of *Bran*' (raven).

**Byrom, Byram, Biram:** Roger *de Birum* 1240 FFY; John, Simon *de Byrom* 1342 FFY, 1401 AssLa. From Byram (WRYorks), *Birum* c1170, *Byrun* 1268, or 'dweller at the cowsheds'.

**Byron, Byran, Biron:** William *de Byrun* 1240 FFY; Richard *de Birune* 1242 AssDu; John *de Byron* 1294–5 IpmY; Richard *Byron* 1401 AssLa. A variant of BYROM. The forms give no support to Harrison's derivation from Fr *bu(i)ron*.

**Bysh:** v. BISH

**Bysouth:** Henry *Bisuthe* 1221 AssWo; Maurice *Bi Suthe* 1279 RH (O). One who lived 'to the south'.

**Bythesea:** William *Bythesee* 1336 MESO (So); John *Bethesee, Bitheseo* 1363–4 ib. 'Dweller by the watercourse or drain', OE *\*sēoh*, a Somerset term,

referring to inland places near Bridgwater and to Sea near Ilminster. In Robert *Bythse* 1333 ib. (So), the reference is again to an inland place, OE *sǣ*, 'dweller by the lake or pool'. The surname might also refer to residence near the sea. v. MELS.

**Bytheseashore:** An interesting surname, both because of its survival and for its pronunciation, *Bitherseyshore*, with stress as in Battersea.

**Bytheway, Bythway, Byway, Bidaway:** Richard *Bithewaye* 1243 AssSo. 'Dweller by the road.' Gervase *Bethewy* 1244 Rams (Hu) is also called *de la Rode*.

**Bywater, Bywaters:** Thomas *Bithewater* 1219 AssY; Elyas *Biþewatere* 1279 RH (O); John *Beyewatyr* 1327 SRC. 'Dweller by the water.'

**Bywell:** Richard *de Bywell* 1296 SRNb; Roger *Bithewelle* 1296 SRSx; John *Bywell* 1379 LoCh. From Bywell (Nb), or 'dweller by the stream', OE *bī, wiella*.

**Bywood:** Edward *Bythewode* 1275 RH (Do). 'Dweller by the wood.'

# C

---

**Cabbage, Caboche:** Richard *Caboche* 1280 FFY; John *Cabage* 1304–5 RegAntiquiss; William *Cabbage* 1662 HTEss. AFr *caboche*, ME *caboche, cabage* 'head of a cabbage', also the name of a fish, the English bullhead. Either of these might have given rise to a nickname.

**Cabel, Cabell, Cable, Cabble:** Richard *Cabel* 1212 Cur (W); William *Cabbel* 1297 MinAcctCo; Richard *Cabell* 1576 SRW. Probably from a personal name: *Kabell'* filius Willelmi 1286 Pinchbeck (Sf), perhaps OE *Ceadbeald*. It could also be from AFr *cable* 'cable, rope', metonymic for a ropemaker, or from ME *cabal* 'horse', hence 'horseman.'

**Caboche:** v. CABBAGE

**Cadbury:** Walter *de Cadbury* 1319 SRLo; Thomas *Cabbury*, William *Cadbury* 1524 SRD; James *Cadbery* 1642 PrD. From Cadbury (D, So).

**Cadby, Cadeby:** Martin *de Kaddeby*, Walter *de Cadebi* 1201 AssSo; Gilbert *de Cathebi* 1218 AssL. From Cadeby (L, Lei, WRY).

**Cadd, Cade:** Wigan *filius Cade* c1155 DC (L); Thomas *filius Kade* 1219 AssY; William, Eustace *Cade* c1140 ArchC iv, 1186 P (L); William *le Cade* 1327 SRSx; Richard *Cadde* 1327 SRWo. Some of these examples point to the survival of OE *Cada* but *le Cade* is clearly a byname, found also in (Wulfwine cognomento) *Cada* (a1050 OEByn) which Tengvik derives from a Germanic root meaning 'something lumpy or protruding', hence 'a stout lumpish person'. *Cade* may be identical with *cade* (sb. 2, MED) 'a young animal cast or left by its mother and brought up by hand as a domestic pet', 'a pet lamb': or it may be ME, OFr *cade* 'cask, barrel' (1337 MED), either a nickname for one round as a barrel or metonymic for a maker of casks.

**Caddell:** v. CALDWELL, CAUDELL

**Caddey, Caddie:** v. CADDY

**Caddick, Caddock:** Richard *Caddok* 1260 AssC; William *Caddouc'* 1327 SRSx. OFr *caduc* 'infirm, decrepit, frail', used by Trevisa of 'men that haue the fallyng euyll', epileptics.

**Caddington:** Leofwine *de Cadentune* c1060 OEByn (O); Roger *de Cadintone* 1206–8 Clerkenwell; Nicholas *de Cadington* 1327 SRSx. From Caddington (Beds).

**Caddow:** John, Richard *Caddo* 1327 SRC. ME *cad(d)aw, cad(d)owe* 'jackdaw' (1440 MED).

**Caddy, Caddey, Caddie:** Robert *Cadi* 1185 Templars (Y); Roger *Cadye* 1296 SRSx; Henry *Cadey, Cady* 1327 SRSf; Nicholas *Caddy* 1641 PrSo. A diminutive of OE *Cada*.

**Cade:** v. CADD

**Cadeby:** v. CADBY

**Cadel, Cadle:** v. CAUDELL

**Cadge:** v. CAGE

**Cadman:** Robert *Cademan* 1279 RH (C); Geoffrey *Cademon* 1327 SRDb. Either 'servant of *Cade*' or 'maker of casks'. There is no evidence for the post-Conquest use of OE *Cædmon*.

**Cadney:** Richard *de Cadeney* 1275 RH (L); Robert *de Cadenay* 1374 AssL; George *Cadny* 1674 HTSf. From Cadney (L).

**Cadogan:** *Caducan* 1161 P (Wo); *Cadegan* de Middelton' 1191 P (Sa); Jevan *ap Cadugon* 1287 AssCh; Richard *Cadigan* 1273 RH (Wa). OW *Cadwugaun*.

**Cadwall, Cadwell, Kadwell, Kadwill:** (i) Nicholas *de Cattedale* 1202 AssL. From Cadwell (Lincs). (ii) Robert *de Cadewelle* 1279 RH (O). From Cadwell (Devon, Herts, Oxon). v. also CALDWELL.

**Cadwallader, Cadwalader:** *Cadewadladre* Meraduc 1166 P (Sa); *Cadewadlan* 1176, *Cadawallan* 1180 P (He); Charles *Cadweleder* 1641 SaAS 3/iv; Thomas *Cadwellader* 1664 HTSo. OW *Catguallon*, Co *Caduualant*, OBret *Catuuallon*. v. PNDB 213.

**Caesar:** *Cesar* clericus 1185 Templars (Y); *Cesar* Walpole d. 1613 ODCN; Henry *Sesare* 1334 SRK; John, Harriet *Caesar* 1705, 1748 Bardsley. In ME this is a pageant name, v. CAYZER, and though rare is found occasionally both as a first name and as a surname, cf. also *Caesaria*, sister of the wife of William de la Rode 13th Rams (Hu). The chief family of the name, still with modern descendants, was that of Sir Julius *Caesar* (1558–1636), of Italian extraction. His grandfather, Pietro Marie *Adelmare*, married Paola, daughter of Giovanni Pietro *Caesarini*, and one of his sons, Cesare *Adelmare*, emigrated to England c1550 and became physician to Queen Mary and to Queen Elizabeth. He died in 1569 and his children adopted *Caesar* as their surname (DNB).

**Caff:** v. CHAFF

**Caffery, Caffray, Caffrey:** for MacCAFFRAY.

**Caffin, Caffyn:** v. CHAFFIN

**Caffinch:** v. CHAFFINCH

**Cage, Cadge:** (i) Gervase, Matilda *Cage* 1211 FrLeic, 1279 RH (C); Richard *Cagge* 1275 SRWo; Robert *Cadge* 1524 SRSf. ME, OFr *cage* 'cage', either metonymic for CAGER or equivalent to *atte Cage* below. (ii) Jacobus *dil Cage* 1327 SRSf; John *atte Cage* 1327 SRSo. *Cage* was used of 'a prison for petty malefactors' c1500 (NED), but the metaphorical meaning of confinement was much earlier (1300 ib.). 'Dweller near, or keeper of the Cage.'

**Cager, Caiger:** William, Geoffrey *Cager* 1319, 1327 *SR* (Ess). OFr *cagier* 'a maker or seller of cages' or equivalent to *atte Cage* above.

**Cain, Caine, Kain, Kaine, Kayne, O'Kane, Cane, Kane:** (i) *Keina* mater Berte 1202 AssL; Godfrey *Kein* 1198–1200 BuryS (Sf); Thomas *Kayne* 1260 AssC. *Keina* is a woman's name, perhaps a short form of such Welsh names as *Ceindrych, Ceinlys, Ceinwen*, all feminine, from Welsh *cain* 'beautiful'. cf. St Keyne (Cornwall). The Manx name is a contraction of *Mac Cathain* 'son of *Cathan*', from *cath* 'a battle', 'a warrior': *McKane* 1408, *MacCann* 1430, *MacCane* 1511, *Cain* 1586 Moore. Irish *Ó Catháin* 'descendant of *Cathán*'. (ii) Geoffrey *de Chain* Hy 2 DC (L); Richard *de Kain* 1275 RH (Nf). From Caen (Calvados). cf. CANE, CAM.

**Caines, Cains, Kaines, Keynes:** William *de Cahaignes, de Cahanges* 1086 DB (C, Bk, Nth, Sx); William *de Caynes* 1222 Cur (Nth). From Cahaignes (Eure) or Cahagnes (Calvados).

**Cainham:** John *Cainham* 1275 SRWo. From Cainham (Sa).

**Caird:** Gilfolan *Kerd* 1275, John *Caird* 1613 Black. Gael, Ir *ceard* 'craftsman, smith, tinker'.

**Cake, Cakes:** Alured *Cake* 1210 P (Nf); Gilbert *Kake* 13th NthCh. ME *kake, cake*, a comparatively small flattened sort of bread, originally round or oval, usually baked hard on both sides by being turned in the process (c1225 MED). Metonymic for a cakemaker: John *le Kakier* 1292 SRLo, Symon *Cakyer* 1332 SRSx.

**Cakebread:** Ædwinus *Cacabred* 1109–31 Miller (C); Richard *Cakebred* 1327 SRSf. Metonymic for a maker of *cakebrede*, bread made in flattened cakes, or of the finer and more dainty quality of cake (1377 NED).

**Caker:** Godwin *Kaker* 1210–11 PWi; John *le Kakier* 1292 SRLo; Richard *Caker* 1524 SRD. A derivative of ON *kaka* 'cake', a maker of bread-cakes.

**Calcott:** *v.* CALDECOT.

**Calcraft, Chalcraft, Chalcroft, Choldcroft:** William *Caldcroft* 1441 FrY. 'Dweller at the cold croft', OE *ceald, croft. v.* also CHALCRAFT.

**Caldbeck, Coldbeck, Colbeck:** Alan *de Caudebec* 1214 P (Cu); Thomas *de Caldebek* 1321 FrY; Henry *Caldebek* 1453 FFEss. From Caldbeck (Cu), or 'dweller by the cold stream', OE *ceald*, ON *bekkr*.

**Caldecot, Caldecott, Caldecourt, Caldicot, Caldicott, Callicott, Calcott, Calcut, Calcutt, Callcott, Caulcutt, Caulkett, Cawcutt, Corcut, Corkett, Corkitt, Coldicott, Colicot, Collacot, Collacott, Collecott, Collicutt, Colcott, Colcutt, Collcott, Collcutt, Colkett, Colocott, Chaldecott, Chalcot:** Simon *de Caldecot'* 1195 P (C); Geoffrey *de Caudecot'* 1206 Cur (K); William *de Cheldecot* 1225 PN W 210; Edmund *de Caldicote* 1275 RH (Bk); Richard *de Coldecote* 1275 SRWo; John *Caldekote* 1296 SRSx; John *Calicot* 1524 SRSf. From a common place-name 'cold huts', OE *ceald, cote*, such as Chaldicotes Fm (Wilts), Chalcot(s) (Wilts, Middlesex), and Chollacott

(Devon, Wilts). The northern and midland *cald* survives in the common Caldecote, Caldecott, and in Calcot (Berks), Calcutt (Warwicks), Caulcott (Oxon) and Coldcotes (WRYorks). In Calcutt (Wilts) and Collacotts (Devon) we have 'Cola's hut': Thomas *de Colecote* 1275 RH (D).

**Calder, Caulder:** (i) Adam, Thomas *de Calder* 1246 AssLa, 1332 SRCu. From Calder (Cumb). (ii) Hugh *de Kaledouer* c1178–89 Black; Donald *of Calder* 1419 ib.; Farchard *de Caldor* 1461 ib. From Calder or Cawdor (Caithness).

**Calderon, Cauldron, Cawdron, Coldron:** Stephen *Caldron* 1289 FrY. AFr *caud(e)ron* 'cauldron'. For a maker of cauldrons, OFr *chalderonnier, cauderonnier*: Roger *le calaroner* 1299 FrY.

**Caldwell, Calwell, Cauldwell, Caudwell, Caudell, Caudle, Cawdell, Caddell, Cadel, Cadle, Cadwall, Cardwell, Coldwell, Couldwell, Chadwell, Cholwell:** Adam *de Caldewella* 1195 P (Db); Richard *de Coldewell* 1379 PTY; Richard *Cauldwell* 1381 PTY; John *Cadwewelle* 1524 SRSf. 'Cold spring or stream', OE *ceald, wielle*, surviving as Caldwell (Warwicks, NRYorks), Caldwall (Worcs), Cauldwell (Beds, Derby, Notts), Caulde Green (Glos), Cauldle Ditch and Cawdle Fen (Cambs), *Cadewell* 1417, *Cadle* 1591, Chadwell (Essex, Herts, Leics, Wilts), Chardwell (Essex) and Chardle Ditch (Cambs), *Kadewelle* 13th, *Cadwel* 14th. For forms, cf. CALDECOT. *v.* also CADWALL, CADWELL. In Scotland *Caldwell* and *Coldwell*, formerly pronounced *Carwall*, are from Caldwell (Renfrewshire).

**Cale:** David, Walter *Cale* 1275 SRWo. OFr *cale* 'a woman's head-dress' (1588 NED). *v.* CALL.

**Calendar:** *v.* CALLENDAR

**Caley, Calley, Callie, Cayley, Kaley, Kayley:** (i) William *Cailgi, de Cailgi, de Calgi* 1086 DB (Berks), *de Caillei, de Callei* 1086 ICC; William *de Kailli, de Caly* 1210 Cur (Nf); Adam *de Caly, de Kally* 1212 Cur (Wa). From Cailly (Seine-Inférieure). Walter *de Cayeley* 1332 SRSt probably came from Cayley in Winwick parish (Lancs). (ii) Caley, a common Manx name, is contracted from *Mac Caolaidhe* 'son of *Caoladh*', a personal name from Gael *caol*, Ir *cael* 'slender': *McCaley* 1511, *Cally* 1605, *Caley* 1642 Moore.

**Calf, Calff:** Robert *Calf* 1163 DC (L); Ailwin *Calf* 1176 P (Bk). OE *cealf*, Anglian *calf* 'calf'.

**Call, Caules:** Swanus *Calle* 1275 RH (W); John *Calle* 1279 RH (C). ME *calle*, Fr *cale* 'a close fitting cap worn by women' (1327 NED), metonymic for CALLER. cf. CALE. cf. also Walter *atte Calle* 1307 LLB C, from ME *calle* 'sheepfold' (1483 NED), hence 'shepherd'.

**Callan, Callen:** in Scotland is from *Macallan*. In Ireland, *Callan* is for *Ó Cathaláin* 'descendant of *Cathalán*', a diminutive of *cathgal* 'battle-mighty'. *v.* CALLIN.

**Callander, Callendar, Callender, Calendar:** (i) Bartholomew *le Calendrer* 1311 LLB B; Paganel, Walter *le Kalendre* ib. OFr *calendrier, calendreur* 'one who calenders cloth', i.e. passes it through rollers for smoothing (1495 NED). (ii) Alwyn *de Calyntyr* c1248 Black; George *Kallender* 1631 ib. From Callander (Perthshire).

**Callaway, Calloway, Calway, Kellaway, Kelleway, Kelway:** Philip *de Chailewai* 1165 P (Gl); Thomas *de Kaillewey* 1242 Fees (W); *William Calleweye* 1242 Fees (D). This surname is the source of Kellaways (Wilts) and may derive from Caillouet (Eure). *v.* DEPN, PN W 99.

**Calicott:** *v.* CALDECOT

**Callender:** *v.* CALLANDER

**Caller, Callear, Callier:** Walter *Calyer* 1275 RH (K); Henry *le Callere* 1281 LLB B. A derivative of ME *calle*, from Fr *cale* 'a kind of cap', 'a maker of cauls or coifs for the head'. cf. CALL.

**Calley, Callie:** *v.* CALEY

**Callicott:** *v.* CALDECOT

**Callin:** A Manx name from *MacCathalain* 'son of *Cathalan*', from Gael *cathal* 'valour'. *v.* CALLAN. Occasionally from *Mac Allen: McAleyn* 1511, *Callin* 1623 Moore.

**Callis:** William *de Caleio* 1086 InqEl; Richard *de Caliz* 1190 P (Gl). From Calais.

**Callister, Collister:** A Manx name for *Mac Alister* 'son of Alexander': *Mac Alisandre* 1417, *Mac Alexander* 1429, *Callister* 1606, *Collister* 1799 Moore.

**Callow, Calow:** Brichric *se Calewa* 1070 OEByn (So); Philip *Calewe* 1260 AssC; Simon *Calu*, John *le Calue* 1296 SRSx. OE *calu* (*calewa*) 'bald'. cf. CHAFF.

**Calloway:** *v.* CALLAWAY

**Calman, Calmon:** William *Caleman* 1327 SRC; John *Calman* 1382 AssC. ON *Kalman* from OIr *Colmán*. *v.* PNDB 216.

**Calpin:** a contraction of *MacAlpin*.

**Calthorp, Calthorpe, Calthrop:** William *de Caletorp* 1134–40 Holme; William *de Calthorp* 1259 FFL; John *Colthorp* 1411 FrY; Fraunces *Calthrop*, *Calthrop* 1524 SRSf. From Calthorpe (Nf, O).

**Calton, Kalton:** William *de Caltone* 1275 RH (Db); Ralph *de Calton* 1336 AssSt; Philip *Calughton* 1621 SRY. From Calton (St, WRY), or Calton Lees (Db).

**Calver:** David *de Caluenore, de Caluoure* 1200 P (Db); John *aKarffer* 1560 RothwellPR (Y); Jonathan and Mary *Carver, Calver* 1702, 1705 ShotleyPR (Sf). From Calver (Derby). Indistinguishable in pronunciation from CARVER.

**Calverley, Calverleigh:** Godric *de Calodeleia* 1084 OEByn (D); Iordan *de Caluerlai* 1200 P (Y); Johanna *de Caluerley* 1379 PTY; Walter *Calverley* 1466 TestEbor. From Calverley (WRY), or Calverleigh (D).

**Calvert, Calverd, Calvard:** Warin *le Calfhirde* 1269 FFY; William *Calvehird'* 1297 SRY; John *Calverde* of York 1309 LLB D; James *Calvart* 1596 FrY; William *Calvert* 1620 FrY. OE (Anglian) *calf* and *hierde* 'the calf-herd'. cf. John *Caluekave* 1284 RamsCt (Hu) and William *Caluerknave* 1327 SRSf.

**Calverton:** Henry *Calverton* 1442 IpmNt. From Calverton (Bk, Nt).

**Calway:** *v.* CALLAWAY

**Calwell:** *v.* CALDWELL

**Cam, Camm:** (i) William *de Cada, de Cadam, de Cadomo* 1086 DB (Sf); Ralph *de Caham, de Cadomo* c1162 DC (L); Walter *de Cam* (*Cadam'* CR) 1205 P (Y). From Caen (Calvados): *Cadum* 1040, *Cadomum* 1080. William *Cam* 1205 P (Ha) probably belongs

here. cf. William *de Cadumo* 1148 Winton (Ha); Fabian *de Cam'* 1184 P (Ha). cf. also CANE. Winterborne Came (Dorset) belonged to the Abbey of Caen in 1086. (ii) Hugh, William *de Camme* 1221 AssGl, 1214 Cur (So). From Cam (Glos). (iii) Walter *le Camm* 1260 AssY; Richard *le Kam* 1282 Oseney (O); Hector *Cam* 1541 Black (Skye). Gael *cam* 'crooked, deformed, one-eyed, cross-eyed'.

**Camb:** *v.* CAMBER

**Cambden, Camden:** Ebrard *de Campeden'* 1190 P (C); Walter *de Campeden* 1220 AssGl; John *de Campeden* 1260 AssCh. From Broad, Chipping Campden (Gl).

**Camber, Cammer, Comber, Comer, Le Comber, Lecomber, Camb:** Ralph (*le*) *Cambere* 1201–2 P (L); Reginald *Combere* (*le Camber*) 1220 Cur (Herts); Ralph *le Combere*, William *le Comere* 1286 MESO (Nf); John *Camere, Camber, Comber* 1359–60 ColchCt. A derivative of OE *camb, comb* 'comb', a maker of combs. CAMB is metonymic. Outr. *Cambmaker* is also called *Camber* (1373, 1379 ColchCt). cf. Thomas *Kambesmyth* 1381 PTY, Alice *Comsmyth* 1590 RothwellPR (Y).*v.* also COMBER.

**Camble:** *v.* CAMPBELL

**Cambray, Cambrey, Gambray, Kembery:** Godefridus *de Cambrai* 1086 DB (Lei); Simon *Camberay, Cambrey* 1296 SRSx. From one of three places called Cambrai (Calvados) or from Cambrai (Nord).

**Cambridge:** Richard *de Cambrige* 1182 P (St) and Alan *de Cambrigge* 1227 AssSt must have come from Cambridge (Glos), *Cambrigga* 1200–10. Picot *de Grantebrige* 1086 DB (C) and William *de Cantebregge* 1338 LLB F certainly owed their names to the University town, but it was not until late in the 14th century that the form *Cambrigge* became common: Stephen *de Caumbrigge* 1348 Works (C), John *Caumbrigge* 1376 LLB H.

**Camden, Cambden:** Ebrard *de Campeden'* 1190 P (C); John *de Campeden* 1260 AssCh. From Broad, Chipping Campden (Glos).

**Camel, Camell, Cammell:** (i) Robert *de Camel* 12th Seals (So); Richard *de Cammel* 1319 FFC. From Queen or West Camel (Som). (ii) Walter, Ivo *Camel* 1200 P (D), 1220 Cur (W); John *le Camule* 1332 SRSx. The last form is clearly a nickname, perhaps in the sense 'a great, awkward, hulking fellow' as used by Shakespeare: 'A Dray-man, a Porter, a very Camell.' *v.* CAMPBELL.

**Cameron:** The Highland clan name is Gael *camshrón* 'wry or hook nose'. The Lowland name is from Cameron (Fife): Adam *de Kamerum* 1214–49, Hugh *Cambrun* 1219, John *de Cameron* 1421 Black.

**Camidge, Cammidge:** *v.* GAMAGE

**Cammack, Cammock:** Robert Chamoke alias *Cammock* 1547 FFEss; John *Cammok* 1557 Black. From Cammock in Settle (WRY), or perhaps a nickname from OE *cammoc* 'a thorny shrub'.

**Cammell:** *v.* CAMEL

**Cammer:** *v.* CAMBER

**Cammis, Camis, Cammish, Camous, Camoys, Camus, Keemish:** (i) Adam *le Camhus* 1256 AssNb.; Robert *Cambysshe* 1455 FrY; William *Cammas* 1620 ib.; George *Camisse* 1632 ib. ME *cammus, camois*, OFr *camus* 'having a short, flat nose, pug-nosed' (c1380 MED). (ii) Bartholomew *le Camisur* 1282 LLB B. A derivative of ONFr *camise, kemise*, OFr *chemise*, MedLat *camisia*, an undergarment worn by both men and women, a shirt; used also of a priest's surplice, a herald's robe. Metonymic for a maker of shirts, etc. (iii) Stephen *de Cameis* 1200 P (Nth); Matillis *de Camois* 1205 Cur (Sr). Perhaps from Campeaux (Calvados).

**Camp, Campe:** Alricus *campe* (*cemp*) 1066 ICC (C); Robert *Campe* 1195 P (Wa); Tomas *le Campe* 1200 P (Ha); John *Campe* (*Kempe*) 1205 P (Do). OE *cempa* 'warrior'. *v.* KEMP. *Camp* may be due to the influence of OE *camp* 'battle', *campian* 'to fight'.

**Campaign, Campain, Campen, Campin, Camping:** Gilbert *de Campania, de Champanie* Hy 2 DC (L); Graves *de la Campaine* ib.; Roger *de Campen* ib. (Lei); William *Campaignes* 1180 P (Do). From one of the places named Campagne (Pas-de-Calais (several), Oise), or a Norman form of Champaigne.

**Campbell, Camble:** (i) Colin *Campbell* 1282 Black; Neel *Cambel* 1296 CalSc; Duncan *le Cambell* 1447 Black. Gael *caimbeul* 'wry or crooked mouth'. The surname occurs as *Camille* (1451), *Cammell* (1473), *Camble* (1513). (ii) Thomas *Campell* 1524 SRSf; John *Camell* 1612 FrY. John *Camell* (1667 FrY) or *Cambell* (1697 ib.) was a son of Michael *Camell* and father of Michael *Cambell* and Daniel *Camell* is also called Daniel *Campbell* (1691, 1719 ib.). Their real surname was probably *Cammell*.

**Campion:** *v.* CHAMPION

**Camplejohn:** Robert *Camplyon* 1454 Paston; Thomas *Camplechon* 1589, *Campleion* 1611, Edward *Campleshon* 1630 FrY. The first element may be connected with ME *camplen* 'to fight', and the meaning of the name would then be 'fighting John'. cf. John *Campleman* 1680 YWills.

**Camplin, Campling:** Stephen *Camelyn* 1230 Cl; William *Campelin'* 1275 RH (Nf); James *Camplen* 1664 FrY; Daniel *Camplin*, John *Camplinge* 1674 HTSf. OFr *camelin*, a kind of stuff made (or supposed to be made) of camel's hair (c1400 NED). Either a maker or a wearer of cameline.

**Camps:** Geoffrey *de Campes* 1206 Cur (Ess); Richard *de Caumpes* 1294 LLB B. From Camps (C), or Camps in Nazeing (Ess).

**Camwell:** Richard *de Camuilla, de Canuilla* 1148 Eynsham (O), c1155 Holme (Nf); Thomas *Camuille* 1327 SRSx; Samuel *Camwell* 1713 FrY. From Canville-les-Deux-Eglises (Seine-Inférieure). *v.* ANF.

**Canaan:** *Chanaan* de Bronteston' 1176 P (St); William *Canaan* 1203 AssSt. *v.* CANNAN (ii).

**Cancellor:** *v.* CHANCELLOR

**Candish:** *v.* CAVENDISH

**Candeland, Candland, Candlin:** Hugh *filius Kandelan* 1196 P (Cu); Robert *Candelane* 1332 SRCu; Robert *Candelayn* 1379 PTY; Thomas *Candland* 1575 SaAS 2/i. ME *Candelin*, a variant of ME *Gandelin, Gandelayn*, corruptions of *Gamelin*, a diminutive of ON *Gamall*.

**Candle:** (i) Ailuuin *Candela* c1095 Bury (Sf); Adam *Chandelle* 1196 P (Sr); Samson *Candeille, Candel*

1197, 1207 P (W). OE *candel*, Lat *candela*, or ONFr *candeile*, OFr *chandeile* 'candle', metonymic for a maker or seller of candles. (ii) Ralph *de Candel* 1176 P (So). From Caundel (Dorset).

**Candlemass:** Matilda *Candelmes* 1379 PTY. OE *Candelmæsse* 'the feast of the purification of the Virgin Mary, Feb 2'. A name for one born on that day.

**Candler:** *v.* CHANDLER

**Candlin:** *v.* CANDELAND

**Cane, Kane:** (i) *Cana, Cane, Cano* 1066 DB (Sr, Sx); *Cane* 1160–70 MedEA (Nf); Willelmus *filius Cane* c1213 Fees (Berks); Leofwine *Kana* 11th OEByn; Leouuinus *Chane* 1066 Winton (Ha); Herueus *Cane* 1177 P (Sf); Hugo *Kane* 1210 P (He); William *le Cane* 1332 SRSx. The personal-name is probably OE *Cana*. *v.* also CAIN. The nickname is ME, OFr *cane* 'cane, reed', used, probably, for a man tall and slender as a reed. (ii) Alan *de Cane* 1230 P (Y). From Caen (Calvados). cf. CAIN, CAM. Kirby Cane (Norfolk) was held in 1205 by Walter *de Cadamo* and in 1242 by Maria *de Cham* (DEPN).

**Canfield:** Thomas *de Canefeld* 1310 LLB D; William *de Canefeld* 1321 CorLo. From Great, Little Canfield (Ess). Sometimes, perhaps, from Canville-les-Deux-Églises (Seine-Maritime). *v.* CAMWELL.

**Canham:** Henry *de Cauenham* 1327 SRSf. From Cavenham (Suffolk).

**Cann:** (i) Bartholomew *Canne* 1327 SRSf; Richard *Can* 1327 *SR* (Ess). OE *canne* 'can'. Metonymic for CANNER. (ii) Richard *de Canne* 1276 RH (O). From Cann (Dorset).

**Cannan, Cannon:** (i) The Manx name is MacCannon 1511, *Cannan* 1638, from Ir *Mac Cannanain* 'son of *Cannanan*' (*Ceann-fhionn* 'white head', recorded in 950 (Moore). (ii) *Canan* 1296 CalSc; Fergus *Acannane* 1562 Black; David *Cannane* 1624 ib. Ir *O'Canáin*, descendant of *Canán*, a diminutive of *Cano* 'wolf-cub'.

**Cannel, Cannell:** Simon, John *Canel* 1314 LLB D, 1327 SRC; Richard *Cannell* 1428 FA (Wa). ME, OFr *canele* 'cinnamon'. A seller of cinnamon, a spicer.

**Cannell** (Manx): Mac *Connell* 1511, *Cannell* 1606. From Mac *Conaill* 'Conall's son' (Moore).

**Canner:** John *le Kannere* 1305 MESO (Ha); William *le Cannere* 1327 SRSt. A derivative of OE *canne* 'can', a maker or seller of cans.

**Canning, Cannings:** Lucas *de Canninges* 1200 Cur; Philip *de Caning* 1280 IpmW; Thomas *Canynges* 1450 AssLo; John *Canning* 1642 PrD. From Cannings (W).

**Cannock:** Jordan *de Kanoc* 1199 AssSt; Gilbert *Cannok* 1327 SRSx. From Cannock (St).

**Cannon, Cannons, Canon, Channon:** Aluric *se Canonica* 11th OEByn (D); Reginald *Canun* 1177 P (C); Nicholas *le Chanone* 1332 SRSt; William *Canons* 1401 FrY. The first example is from OE *canonic*, the later ones from ME *canun*, ONFr *canonie*, later *canoine*; ME *chanun*, central OFr *chanoine* 'a clergyman living with others in a clergy house' (c1205 NED). *v.* CANNAN.

**Cant, Caunt, Chant:** Richard *Cante* 1327 SRSf, *Caunt* 1357 FFHu. ONFr *cant*, OFr *chant* 'singing, song', metonymic for CANTER, CHANTER.

**Cantellow, Cantello, Cantelo, Cantlow:** Waterus (sic) *de Cantelupo* c1135 DC (L); Roger *de Cantelo* 1185 Templars (So); William *de Cantelowe* 1320 Eynsham (O); William *Cantlowe* 1448 LLB K. One family came from Canteleu (Seine-Inférieure), another from Canteloup (Calvados). *v.* ANF 24.

**Canter, Cantor, Caunter, Kanter:** Augustinus *cantor, precentor* 1153–68, 1186–1210 Holme (Nf); Walter *le Canter* 1230 Eynsham (O). The early examples are all from Latin *cantor* and refer to precentors in cathedrals or monasteries. The last is from AFr *caunter, cauntour* 'singer, one who leads the singing'. cf. CHANTER.

**Cantes:** *v.* KENTISH

**Cantley, Cantlay:** Wimer *de Cantele* 1198 FFNf; William *Cantly* 1508 Black; Peter *Cantley* 1581 FrY. From Cantley (Nf, WRY).

**Cantock:** Henry *Cantoc* 1280 IpmGl; John *de Cantok*, Robert *Cantok* 1327 SRSo. From the Quantock Hills (So), *Cantok* 1274.

**Cantrell:** Robert *de Canterhulle* 1330 PN D 284. From Cantrell (Devon). *v.* also CHANTRELL.

**Canworthy:** John *Canworthie* 1642 PrD. From Canworthy in Rackenford (D).

**Cape, Cope:** Ailward, Walter *Cape* 1190 P (K), 1221 AssGl; Walter, Maud *Cope* 1275 RH (Lo), SRWo. OE *\*cāpe*, ME *cope* 'a long cloak or cape' (a1225 MED). *Cope* is the normal development, *Cape* the early form retained in the north.

**Capel, Capell, Capelle, Caple, Cappel, Cappell:** Jacob *de Capel* 1193 P (He); Ralph, Philip *Capel* 1214 Cur (Nth), 1285 Ass (Ess); Robert atte Capele 1296 SRSx. From Capel or Capel Le Ferne (Kent), Capel St Andrew or St Mary (Suffolk), Capel (Surrey) or How or King's Caple (Hereford), or from residence near or service at a chapel (ME *capel*, ONFr *capele*). Occasionally also from ME *capel, capul* 'a nag': Rogerus *Caballus* 1230 ArchC 6. *v.* CAPPLEMAN.

**Capelen, Capelin, Capeling:** *v.* CHAPLAIN

**Capener, Capner:** Ædward *Capenore* 1180 P (Sx); John *de Kapenor* 1296 SRSx; Richard *Capenore* 1332 SRSr. From Capenor in Nutfield (Sr).

**Capers:** William, Thomas *Caper* 1200–50 Seals (Sf), 1327 SRWo. A derivative of ME *cape*, 'a maker of capes or copes'. *v.* CAPE.

**Caplan, Caplen, Caplin:** *v.* CHAPLAIN

**Capler:** Baldwin *le capeller* 1216–20 Clerkenwell; Robert *le Capeller* 1270 SaAS 3/vii; John *le Cappeler* 1298 LLB B. A derivative of ME *capele* 'chapel', one who works as a chapel. Sometimes, perhaps, a derivative of ME *capel* 'horse', hence one who looks after horses.

**Capman:** William *Kapman* 13th NthCh; Robert *Capman* 1320 HPD; William *le Capman* 1331 ChertseyCt (Sr). 'A maker of caps', OE *cæppe, mann*.

**Capner:** *v.* CAPENER

**Capon, Cappon:** Simon *Capun* 1227 FFC; Thomas *Capoun* 1382 LLB H. OE *capun* 'a castrated cock', metonymic for a seller of capons.

**Caporn, Capern:** *v.* CAPRON

**Capp, Capps:** William *Cappa* 1111–38 ELPN; Alward *Cappe* 1178 P (K); Roger *Caps* 1327 SRSo.

OE *cæppe* 'cap', metonymic for CAPPER.

**Capper, Kapper:** Nicholas *le Capyare* 1275 SRWo; Symon *le Cappere* 1276 RH (O); William *Capier* 1285 *Ass* (Ess). A derivative of OE *cæppe* 'cap', a maker of caps (1389 NED).

**Cappleman:** Walter *Capelman* 1327 SRSx. Either from ME *capel, capul* 'horse', one who looks after horses (cf. PALFREYMAN), or one who lives near or is employed at a chapel. *v.* CAPEL, CHAPPEL and cf. TEMPLEMAN.

**Cappon:** *v.* CAPON

**Capron, Capuron, Capern, Caporn:** *Caperun* 1130–32 ELPN, 1148 Winton (Ha), 1185 Templars (L); Robert *Caperun* 1130 P (Berks); Roger *Caperun, Chaperon* camerarius Henrici regis 1154–64, 1173–83 Bury (Sf); William *Capron* 13th Gilb (L); John *Capurne* 1503 NorwW (C). ONFr *capron*, OFr *chaperon* 'hood or cap worn by nobles' (c1380 NED). Roger Chaperon was the royal chamberlain whose duties included those of Master of the Robes. cf. 'As hys chamberleyn hym broȝte vorto . . . werye, a peyre hose of say' (1279 NED) and 'Hys (the king's) chaumberlayn hym wrappyd warm' (c1325 ib.). *Caperun* may be a name of office, 'the robe-master', but its chief meaning is, no doubt, 'a maker of hoods'. cf. William *Caperoner* 1327 SRSo. The modern meaning of 'chaperon' is not found before the 18th century.

**Capsey:** Roger *Capsi* 1221 AssSa; Walter *Capsi* 1275 SRWo; William *Caspy* 1394 IpmGl. Probably a byform of ON *Kopsi*.

**Capstack, Capstick:** *v.* COPESTAKE

**Carass:** *v.* CARUS

**Caraway:** *v.* CARRAWAY

**Carbonell, Charbonell:** *Carbunel(lus)* 1086 DB (He); Durandus *Carbonellus* 1130 P (O); Robertus *Charbonellus* c1145 DC (L); William *Carbonel* 1175 P (D). OFr *carbon, charbon* 'charcoal', probably an affectionate diminutive for one with a swarthy complexion or hair black as coal, the essential characteristic of charcoal. The name was sometimes confused with CARDINAL and became CARNALL. cf. Cardinal's Fm (PN Ess 429): *Carbonels* 1381, *Cardynals* 1577, *Carnals* 1777.

**Carboner:** Nicholas *carbonarius* 1221 AssSa; Robert *le carboner* 1247 AssBeds; John *le Carboner* 1277 AssSo. AFr *\*carbonner*, OFr *charbonnier* 'a maker or seller of charcoal'.

**Carbott:** William *Carbott* 1412 FrY; Matthew *Garbutt* son of Robert *Carbutt* 1672 ib. For GARBUTT.

**Card, Carde:** Arnald, Laurence *Carde* 1221 AssSa, 1297 MinAcctCo. OFr *carde* 'teasel-head, wool-card'. Metonymic for CARDER or for *Card-maker*: Richard *Cardemakere*. 1346 FrNorw.

**Carden, Cardon, Carding:** William *Cardon, Cardun* 1086 DB, InqEl (Ess); Richard *Cardun* 1121–48 Bury (Sf). OFr *cardon* 'thistle', used perhaps for one of an obstinate, stubborn character. *Carden* is very common and must sometimes derive from Carden (Ches).

**Carder:** John *le carder* 1332 FrY. 'One who cards wool' (c1450 NED).

**Cardew, Cardy, Carthew:** John *de Carthew* 1332

SRCu; Richard *Cardew* 1376 FrY. From Cardew (Cumb).

**Cardiff:** Richard *de Cardif* 1176 P (Bk); Hugh *de Cardif* 1203 AssNth; John *Cardif* 1275 RH (Ha). From Cardiff (Glam).

**Cardinal, Cardinall, Cardnell:** Ingelrannus *Cardinal'* 1190 P (Y); Geoffrey *Cardinell'*, *Cardinal* 1208 Cur (Y), 1327 SRC. OFr *cardinal* 'cardinal', a pageant-name or a nickname for one like (or unlike) a cardinal or with a partiality for dressing in red. *v.* CARBONELL.

**Carding, Cardon:** *v.* CARDEN

**Cardis, Cardus:** *v.* CARRUTHERS

**Cardwell:** *v.* CALDWELL

**Cardy:** *v.* CARDEW

**Care:** *v.* KEAR

**Careless:** *v.* CARLESS

**Caress, Cariss:** *v.* CARUS

**Carew:** Alice *Careu* c1147–57 MCh; William *de Carreu* 1347–9 FFSr; William *Carewe* otherwise Cooke 1653 EA (NS) ii. From Carew (Pembroke), or for CAREY.

**Carey:** *v.* CARY

**Carker, Charker:** Wulgor *le Carkere* 1166 P (Ha). ME *cark(e)*, AFr *kark(e)*, a northern French form of *carche*, *charche* 'a load, a weight of three or four hundredweights' (a1300 NED), a kark of pepper, ginger, etc. (c1502). Probably 'carrier'. *Karck* is metonymic.

**Carl, Carle, Karl, Karle:** Godric *filius Carle*, *Carli*, Godric *Carlesone* 1066 DB (K); Edmundus *filius Carle* 1205 Cur (Sf); Robert *le Karl* 1202 AssL; William *Carl* 1296 SRSx. The personal name may be ON, ODa *Karli*, ODa *Karl* or OG *Karl*. A more common source is probably ME *carl*, ON *karl* 'man', used in ME with various meanings at different times: man of the common people, a countryman, husbandman; a free peasant; by 1300 it meant 'bondman, villain' and also 'a fellow of low birth or rude manners, a churl'.

**Carless, Careless, Carloss, Carlos:** Richard, Reginald *Carles* 1141 ELPN, 1200 P (Gl); Alan *Karelees*, *Margaret Kareles* 1260 AssC. OE *carlēas* 'free from care', or more likely 'unconcerned, careless'.

**Carleton:** *v.* CARLTON

**Carley, Carly:** Drogo *de Carleg'* 1230 P (D). From Carley in Lifton (D).

**Carlisle, Carlile, Carlill, Carlyle:** Odard *de Carlyle* 1158–64 Black; Thomas *de Karlisle* 1310–11 LLB D; Adam *Carlelle*, *Carlille* 1363, 1370 ib. G. From Carlisle (Cumb).

**Carlos:** *v.* CARLESS

**Carlton, Carleton:** Elsi *de Carleton* 1031 FeuDu; Osmund *de Carleton'* 1163 Cur; Hugh *de Carleton* 1240–1 FFWa; Thomas *de Carleton* 1379 PTY. From Carlton (Beds, C, Du, L, Lei, Nt, Nth, Sf, ERY, NRY, WRY), or Carleton (Cu, La, Nf, WRY).

**Carly:** *v.* CARLEY

**Carlyle:** *v.* CARLISLE

**Carman:** Simon *nepos Kareman* 1196 Cur (Nth); Hamo *filius Karlman* 1201 Cur (K); Robert *Kareman* 1184 P (Lei); Henry *Carman* 1275 RH (Sf); Robert *Carleman*, *Karleman* 1279 RH (C). ON *karmann*, a

variant of *karlmann* (nom. *karmaðr*) 'male, man, an adult male', used as a personal name. *v.* CHARMAN.

**Carn, Carne, Karn, Karne:** Andrew *Karn'* 1275 RH (Nf), James *Carne* 1493 Black; Valentine *Karne* 1642 PrD. From Carn Brae (Co), the River Cairn (Cu), or 'dweller by the heap of stones', Welsh *carn*.

**Carnaby, Carnaby:** Heruey *de Kerneteby* 1219 AssY; Roger *de Carneby* 1370 FrY; John *Carnaby* 1448 TestEbor. From Carnaby (ERY).

**Carne:** *v.* CARN

**Carnall, Carnell, Carnelley, Crenel, Crennell:** William *de la Kernel*, *de la Karnaile* 1244, c1250 Rams (C); Hugo *de la Karnell* 1247 FFHu. ONFr *carnel*, a variant of *kernel*, OFr *crenel* 'battlement, embrasure'. The reference is, no doubt, to arbalesters whose post was on the battlements, *v.* also CARBONELL, CARDINAL.

**Caro, Caroe, Carrow:** Robert *de Carov* 1159 P (Y); Alice *de Carrow* 1275 RH (C); Umfrey *Carow* 1524 SRSf. From Carraw (Nb), or Carrow (Nf, Nt).

**Caron, Carron:** Peter *Carun* 1199 P (Nb); Hugh *de Carun* 1208 P (Lo/Mx); John *Caron* 1642 PrD. From Cairon (Calvados), or the Norman-Picard form of OFr *charron* 'cart', hence metonymic for a carter.

**Carothers:** *v.* CARRUTHERS

**Carp:** William *Carpe* 1275 RH (Nf); Eva *Carpe* 1359 AssD; John *Carpe* 1524 SRSf. A nickname from the carp, OFr, *carpe*.

**Carpenter:** Godwin *carpentar'* 1121–48 Bury (Sf); Ralph *carpenter'* 1175 P (Y); Robert *le carpenter* 1212 Cur (Sf). AFr *carpenter* 'carpenter' (c1325 NED).

**Carr, Ker, Kerr:** Osbert *de Ker* c1200 Riev (Y); Robert *Ker* 1231 Pat (Nb); William *Carre* 1279 RH (O); John *del Car* 1332 SRLa; John *Atteker* 1375 NorwW (Nf). 'Dweller by the marsh or fenny copse', ME *kerr*, ON *kjarr* 'brushwood, wet ground'.

**Carradice:** *v.* CARRUTHERS

**Carran, Carine, Karran:** *Mac Ciarain* 1136, *McCarrane* 1430, *Carran* 1648, *Carine* 1729. Manx names from *Mac Ciarain* 'son of Ciaran', one of the twelve great saints of Ireland, from *ciar* 'mouse-coloured' (Moore).

**Carras:** *v.* CARUS

**Carratt:** *v.* CARRITT

**Carraway, Caraway:** William *Careaway* 1332 SRSr; Stephen *Caraway* c1405 FS; Robert *Carrawey* 1524 SRSf. OFr *carvi*, *caroi*, ME *carewei* 'caraway'. Probably for a seller of spices.

**Carreck:** *v.* CARRICK

**Carrell, Carroll, Caryl, Caryll:** (i) Stephen *Caryl* 1332 AssD; William *Carell'* 1379 PTY; William *Carrell* 1642 PrD. OFr *carrel* 'pillow, bolster'. Metonymic for a maker or seller of these. (ii) Duncan *Carrol* 1663 Black. OIr *Cearbhail*.

**Carrett:** *v.* CARRITT

**Carriage:** *v.* KERRICH

**Carrick, Carreck:** Rolland *de Karryk* 1260 Black; John *Carroc* 1279 RH (O); John *Cayrek* 1462, Thomas *Carrokk* 1536, William *Carrak* 1599 FrY. Usually from the district of Carrick (Ayr), but the Oxford example suggests that there was also another source of the surname.

**Carrier, Carryer:** Robert *de* (sic) *Carier* 1332 SRCu; Roger *le Cariour* 1332 SRLa. ONFr *carier* 'carrier, porter'.

**Carrington:** (i) Thomas *de Karington* 1219 AssLa; John *de Carrington* 1294 AssCh; Richard *Carington* 1523 CorNt. From Carrington (Ches). (ii) Wautier *de Keringtone* 1296 Black; William *Keringtoun* 1506 ib. From Carrington (East Lothian).

**Carris(s):** *v.* CARUS

**Carritt, Carratt, Carrett:** Ailw' *Karet* 1193 P (Nth); John *Carrat* 1642 PrD; Robert *Carritt*, Widow *Carrett* 1672 HTY. OFr *carotte*, ME *carete, carote, carat* 'carrot'. Metonymic for a grower or seller of these.

**Carroll:** *v.* CARRELL

**Carron:** *v.* CARON

**Carrow:** *v.* CARO

**Carruthers, Carrothers, Carothers, Carradice, Carrodus, Cardis, Cardus, Crothers, Crowdace, Cruddace, Cruddas:** John *de Carutherys* c1320 Black; William *of Carruderys* 1460 ib.; William *Corrodas* 1625 RothwellPR (Y); Bertha *Cruddas* 1888 Bardsley. From Carruthers (Dumfriesshire), pronounced *Cridders*.

**Carsbrook:** John *ate Kersbrok'* 1332 MELS (Sr). 'Dweller by the brook where watercress grows', OE *cærse, brōc*.

**Carsey:** *v.* KERSEY

**Carslake, Caslake, Karslake, Kerslake, Keslake:** Ranulph *de Carselak'* 1279 AssSo. 'Dweller by the cress-stream', OE *cærse, lacu*.

**Carslaw, Carsley:** *v.* KEARSLEY

**Carswell, Casewell, Casswell, Caswall, Caswell, Caswill, Crasswell, Craswell, Cressall, Cressell, Cresswell, Creswell, Crisswell, Criswell, Crissell, Kerswell, Kerswill:** Basilia *de Caswella* 1165 P (D); Tomas *de Cressewella* 1190 P (St); Reginald *de Kersewell'* 1212 Cur (O); William *de Kereswell'* 1221 AssWo; Richard *de Carswall*, William *de Karswille* 1275 RH (D); Robert *de Carswell* 1327 SRSo; John and Alice *Cresshills, Creswell* 1816, 1822 ShotleyPR (Sf). 'Dweller by the water-cress-stream', OE *cærse, wiella*, surviving in Carswell (Berks, Devon), Carsewell (Renfrew), Caswell (Dorset, Northants, Som), Crasswall (Hereford), Cresswell (Derby, Staffs), Kerswell (Devon, Worcs) and Kerswill (Devon).

**Cart, Carte:** (i) William *Cart* 1176 P (W); John *Cart* 1524 SRSf. OE *cræt*, ON *kartr* 'cart'. Metonymic for CARTER. (ii) Bartholomew *atte Carte* 1327 SRSx. 'Dweller at the rough ground', OE *ceart*.

**Carter, Cartter, Charter:** (i) Fulco *carettarius* 1177 P (C); Rannulfus *carettator* 1191 P (Hu). (ii) Rannulf *carettier, le caretier* 1192–3 P (Hu); Odo *le careter* 1210 Cur (Nth). (iii) Hugh *Karter* 1225 Lewes (Nf); Robert *le Carter* 1240 FFEss, 1275 SRWo. (iv) Henry *le Charetter* 1222 Cur (So), 1225 AssSo (*le careter* 1225 ib.); William *le Chareter*, John *le Charetter, le Charter*, Walter *le Charettier* 1275 SRWo. NED derives *carter* from ME *cart(e)*, of native or Scandinavian origin, plus -*er* (a1250). The history of the name is more elaborate. Of the above forms, (i) is MedLat (*carettarius* 1213 in MLWL, which does not contain

*carettator*); (ii) is ONFr *caretier*, not in NED, but surviving in the modern French surnames *Carratier, Carretier* and *Cartier* of Norman and Picard origin; (iii) as in NED; (iv) OFr *charetier* 'charioteer' (c1340 NED), but clearly used in English for 'carter'.

**Carteret, de Carteret:** John *Cartrett* 1596 Musters (Sr). A derivative of OFr *cartier* 'quarter' in one or other of its senses. *v.* Dauzat.

**Carters:** Margerie *le Carteres* c1275 St Thomas (St). *v.* CARTER and pp. xxix–xxxii.

**Carthew:** *v.* CARDEW

**Cartland:** Thomas *de Carkelande* 1195 P (D). From Cartland in Alwington (D), *Karkelonde* 1330.

**Cartledge, Cartlidge:** Richard *de Cartelache* 1290 AssCh; Richard *de Cartlege* 1435 DbCh; Thomas *Cartlidge* 1641 Shef. From Cartledge in Holmesfield (Derby).

**Cartmell, Cartmall, Cartmill:** Vckeman *de Kertmel* 1188 P (La); William *de Kertmel* 1260 AssLa; William *Cartmell* 1438 FrY; Peter *Cartmayle* 1504 FFEss. From Cartmell (La).

**Cartmole:** *v.* CATTERMOLE

**Cartrick, Cartridge:** John *Carkerege* 1522, Laurence *Karcharege* 1540 CantW; Thomas *Cartrige, Cartridge* 1583 Musters (Sr). Probably late forms of CARTWRIGHT.

**Cartwright, Kortwright:** John *le Cartwereste* 1275 SRWo; Richard *the Cartwrytte* 1290 AssCh; William *le Cartewryght* 13th Guisb (Y). OE *cræt* or ON *kartr* 'cart' and OE *wyrhta* 'wright', a maker of carts.

**Carus, Carass, Caress, Cariss, Carras, Carris, Carriss:** Robert *del Karhouses* 1332 SRLa; Thomas *de Carrehous'* 1379 PTY; James *Carous* 1555 FrY; William *Carus* 1619 ib.; Robert *Carus* 1709 ib. 'Dweller at the marsh-house.' cf. CARR, and 'road from Balby to *Carhouse*' 1638 Shef.

**Carvel, Carvell:** Roger *de la Keruel* c1204–14 Black; Alexander *Carvayl* 1318 FFSf; Thomas *Carvell* 1524 SRSf. ME *carvel, kervel* 'a small ship'. A nickname for a sailor.

**Carver:** (i) Peter *le caruier* 1203 P (Nt); Gerard *le Carver* 1209 FFEss. OFr *charuier, caruier* 'ploughman'. (ii) Richard *le Kerver(e)* 1275 RH (L), 1277 FFC; William *Keruer* 1327 SRSx. A derivative of OE *ceorfan* 'to cut, carve', one who carves, usually in wood, sometimes in stone; 'wood-carver, sculptor' (c1385 MED). This would later become *Carver*.

**Carvill, Carville:** Walter *de Careuilla* 1195 P (W). From Carville (Calvados, Seine-Inférieure).

**Carwardine:** Richard *de Carwardyn* 1302 ChambAcctCh. From Carden (Ches).

**Cary, Carey, Carye:** Hamo *de Kari* 1205 Pl (So); Richard *de Kary* 1242 Fees (D); Robert *Karye* 1296 SRSx; Thomas *Cary* 1375 AssLo. From Carey Barton (Devon), Castle, Lytes Cary, Cary Fitzpaine, or Babcary (Som).

**Caryl, Caryll:** *v.* CARRELL

**Casbolt:** Stephen *Casebolt* 1327 SRC. ME *casbalde* 'bald-head', a term of reproach: cf. 'Go home, casbalde, with þi clowte' (c1440 *York Plays*).

**Casborne, Casborn, Casbon, Caseborne, Casebourne:** John *de Caseburn* 1275 RH (K). From Casebourne Wood in Hythe (Kent).

**Case:** William, Richard *Case* 1274 RH (Sf), 1327 SRC. Metonymic for 'a maker of boxes, chests, or receptacles', ONFr *casse*, cf. Clais *Case-maker* 1367 MEOT. *v.* also CASS.

**Caselaw, Caseley:** *v.* KEARSLEY

**Casement:** *Mac Casmonde* 1429, *Casymound, Casmyn* 1540, *Casement* 1612. From *Mac Asmundr*, from Celtic *Mac* 'son' and ON *Ásmundr* 'god protector' (Moore).

**Casewell:** *v.* CARSWELL

**Casey:** John *Casey* 1524 SRSf. From Kersey Marsh in South Benfleet (Ess). In Ireland for *O'Casey*, Ir *Ó Cathasaigh*, from *cathasach* 'watchful'.

**Cssh:** Mariota *Chash* 1277 Ely; Roger *Cashe* 1560 RothwellPR (Y); William *Cash* 1642 PrD. OFr *casse* 'box, chest to keep wares in'. Metonymic for a maker of these.

**Cashen, Cashin, Cassin:** Ir, Gael *Caisín*, a diminutive of *cas* 'crooked'. The Manx name is a contraction of *Mac Caisin: McCashen* 1511, *Cashen* 1641, *Cassin* 1687 Moore.

**Casier:** Ascellin, Peter *Casier* 1220 Cur (Nf); William *le Casiere* 1260 IpmW. 'Cheese-maker', ONFr *casier*.

**Caskie, Casky:** For MACASKIE.

**Caskin:** *Caschin* (Db), Elsi *filius Caschin* (Nt) 1066 DB; Henry *Cassekyn* 1332 SRSx. Perhaps an original nickname from ME *cask* 'joyful, lively', with the diminutive suffix *-in*. But it could equally well be a diminutive of *Cass*, a short form of *Cassandra*.

**Caslake:** *v.* CARSLAKE

**Casley, de Casley:** *v.* KEARSLEY

**Cason:** Wolfren *Cawson*, Ro. *Caston, Cason*, Stephen *Corson* 1674 HTSf; Elizabeth *Casen, Cawston* 1788, 1793 LitWelnethamPR (Sf). A local pronunciation of Cawston (Norfolk).

**Cass:** *Casse* Rumpe 1279 RH (C); John, William *Casse* 1170 P (Y), 1200 P (Ess). A pet-form of *Cassandra*, a common 13th-century woman's name: *Cassandra* 1182–1211 BuryS (C), 1208 Cur (Y), 1275 RH (Nf).

**Cassel, Cassell:** Henry *de Cassel* 1168 P (Lo); Gerard *de Cassell* 1218 FFHu; William *Casel* 1327 SRSx. From Cassel (Nord).

**Cassin:** *v.* CASHEN.

**Cassingham:** William *de Casinghamme* 1275 RH (K). From Kensham Green in Benenden (Kent).

**Cassler:** *v.* CASTLEHOW

**Cassley:** *v.* KEARSLEY

**Casson:** William *Casson* 1601 FrY. (i) Ralph *Cattessone* 1115 Winton (Ha), John *Catessone* 1366 FFSf. 'Son of *Catt.*' (ii) Robert *Casseson* 1327 SRC. 'Son of *Cass.*'

**Casswell:** *v.* CARSWELL

**Castell, Caistell:** MacAskel 1311, *MacCaskel, Caskell* 1511, *Caistil* 1669, *Castell* 1750 Moore. For *MacAskell* (Manx). cf. ASKELL. *v.* also CASTLE.

**Castellan, Castellain, Castelein, Castling, Chatelain:** (i) Hugh *le Chastelein* 1235 FFEss; Osbert *Casteleyn* c1240 ArchC viii; Walter *Castelyn* 1255 *Ass* (Ess); Warin *Castellan* 1311 ColchCt. ME, ONFr *castelain*, OFr *chastelain* 'governor or constable of a castle', also 'warden of a prison'. (ii) William *de castellon* 1086 DB (Bk); Hugo *de Castelliun* 1206 Cur (Bk); Robert *de Chastellun* 1220 Cur (Lo). From Castellion (Eure).

**Caster:** *v.* CASTOR

**Casterton:** Richard *de Casterton'* 1298 AssL; Richard *de Casterton* 1306 AssW. From Casterton (R, We).

**Castle, Castles, Castell, Castells:** Richard, Robert *Castel* 1148–54 Bec (Sx), 1201 P (Lei); John *del Castel* 1307 Wak (Y); William *ate Castele* 1317 AssK. ME, ONFr *castel* 'castle'. One who lived near or was employed at a castle. Sometimes from services or rent due to a castle. Henry *de Castell* (1260 AssC) owed rent to Cambridge Castle.

**Castleford:** Nicholas *de Castelford* 1292 IpmY; John *de Castylford* 1340–1450 GildC; Thomas *Castleford* 1375 FFY. From Castleford (WRY).

**Castlehow, Castlehowe, Castello, Castelly, Castley, Castlo, Cassler:** John *Castlehow* 1667 BamptonPR (We); Elizabeth, Mary *Castley* 1683, 1685 ib. There are seven places of the name in Westmorland, usually recorded very late, the earliest being Castle Howe in Kendal (1577) and Castlehow Scar in Crosby Ravensworth (1629). *de Castello*, not infrequent in medieval sources, is probably a latinization of CASTLE, but may have contributed to these names.

**Castleman:** Ralph *Castelman* 1327 SRSo. One employed at a castle. cf. TEMPLEMAN.

**Castling:** *v.* CASTELLAN

**Castlock:** Reginald *Casteloc* 1202 P (Y); Wolnet *Castelok* 1317 AssK; Robert *Castelok* 1388 FrY. 'Cast lock', ON *kasta*, OE *locc*. Perhaps for one who was losing his hair.

**Caston:** (i) Geoffrey *de Caston* 1327 SRSf; John *de Caston* 1350 FFSf. From Caston (Nf). (ii) Amabli *Casteyn* 1327 SRC. 'Dweller by the chestnut tree', AFr *casteyn*.

**Castor, Caster:** Osgod *on Castre* 972 OEByn; Odberd *de Castra* 1134–40 Holme; Adam *de Castre* 1219 AssL; Robert *Caister* 1446 FrY. From Caistor (L), Caister (Nf), or Castor (Nth).

**Caswall, Caswell, Caswill:** *v.* CARSWELL

**Catanach, Catnach, Cattanach, Cattenach:** Thomas *Kethirnathie* 1407 Black; Arthur *Catanache* 1623 ib. Gael *Catanach* 'belonging to (Clan) Chattan' which claims descent from *Gillacatain* 'servant of (St) *Catan*', 'little cat'.

**Catch:** Godfrey *Cacch'* 1225 AssSo; Margery *Cach* 1326 SRC. ME *cache* 'the act of catching', from AFr *cachier* 'to chase', in the sense of 'chase, pursuit', metonymic for *Catcher*.

**Catcher, Ketcher:** (i) Richard *Kaccher* 1200 P (L); Jordan *Cachere* 1221 AssWa. ME *cachere* 'one who chases or drives', 'a huntsman' (c1340 NED). It is probably also used in the same sense as the diminutive *cacherel* which is common both as a name of office and as a surname in Norfolk in 1275 (RH): Alexander *le Cacherel*, Hugh *le Chacherel*, Adam *Kacherel*, Richard Wyche, *cacherel*. The *cacherels* were the bailiffs of the hundred and had an unpleasant reputation for extortion and oppression. (ii) William *Kacchehare* 1204 P (C); Edhiva *Cachehare* 1240 Rams (Hu). 'Catch hare', perhaps 'speedy as a hare', cf. CATCHPOLE and TURNER.

**Catchlove:** Bernard *Cachelu* 1189 Oseney (O); William *Cacheluve* 1208 Cur (Y). ONFr *cachelove*, *cacheleu* 'chase wolf', 'wolf-hunter'. cf. CATCHPOLE.

**Catchpenny:** William *Cachepeni* 1278 AD v (W). 'Chase penny', Afr *cachier*, OE *penig*. Perhaps a nickname for an avaricious man. There were a good many such names, but only *Catchlove* and *Catchpole* have survived. cf. Thomas *Chacchehors* 1304 Shef 'chase horse'; Walter *Cachemayd* 1392 LoPleas 'chase maid'; William *Caccheroo* 1315–16 FFWa 'chase roe'; Robert *Kachevache* 1221 *ElyA* (Sf) 'chase cow'.

**Catchpole, Catchpoll, Catchpool, Catchpoole, Catchpoule:** Aluricus *Chacepol* 1086 DB (Mx); Robert *le Chachepol* Hy 2 AD i (Mx); Hugo *le Cachepol* 1221 AssSa. ME *caccepol*, *cachpol*, OFr (central) *chacepol*, ONFr *cachepol* 'chase fowl', 'a collector of poultry in default of money'; 'a tax-gatherer' (a1050); 'a petty officer of justice, a sheriff's officer or sergeant, especially a warrant officer who arrests for debt' (1377 NED). cf. CATCHLOVE and Geoffrey *Cacemoine* 1198 Cur (Y) 'chase monk', Robert *Kachevache* 1221 *ElyA* (Sf) 'chase cow', Walter *Chacefreins* 1195 P (D), Emma *Cachefrensh* 1327 SRSx 'get hold of the reins', Robert *Chacecapel* 1201 P (D) 'chase nag', Peter *Cachefis* 1279 RH (C), 'catch fish'.

**Cater, Cator, Chater, Chaters, Chaytor:** William *le Chatur*, identical with William *Emptor* 1220 Cur (Beds); Robert *le Achatour* 1229 Rams (C); Amicia *Lakature*, Elias *le Katur* 1271 RamsCt (C); William *le Catour*, *le Chatur*, *le Katour* 1310 Balliol (O); John *Chayter* 1667 FrY. AFr *acatour*, early OFr *acateor*, central OFr *achatour* 'buyer' (c1386 NED), ME *catour*, aphetic form of *acatour*, *acater* 'buyer of provisions for a large household' (c1400 NED). *Cator* is also local, from Cator (Devon): Laua *de Cadatrea* 1167 P (D).

**Caterer:** William *Katerer* 1279 RH (Hu). *Cater* (sb. or vb.) plus *-er*. 'One who caters or purveys provisions for a household' (1469 MED).

**Cates, Kates:** Thurstan *Cati* c1095 Bury (Sf); Osbert *Kate* 1183 Boldon (Du); Geoffrey *Cates* 1332 SRSx. ON *Káti* (nickname) 'the merry one', or, less likely, ODa *Kati*, OSw *Kate* (pers. names).

**Catesby:** Ralph *de Catebi* 1176 P (Y); Richard *de Catesby* 1316 AssNth; William *Catesby* 1446 FFEss. From Catesby (Nth).

**Catford:** Alexander *de Cateford* 1275 RH (K); John *of Catford* 1401 AssLa; Robert *Catford* 1642 PrD. From Catford (K), Catforth (La), or Catfords Fm in Halberton (D).

**Cathcart:** Reginald *de Cathekert* c1200 Black; William *de Kathkerte* 1296 ib.; Adam *Cathcart* 1622 ib. From Cathcart (Renfrew).

**Catherall, Cathrall:** v. CATTERALL

**Catin:** John *Catin* 1222 AssWa; Henry *Catyn* 1353 Hylle. *Cat-in*, a diminutive of *Cat*, a pet-form of *Catherine*.

**Catley, Cattley:** William *de Chateleia* 1148 Winton (Ha); John *de Catteley* 1275 SRWo; William *de Cattele* 1339 CorLo. From Catley (He, L), or Catlees, a tenement in Froyle (Ha).

**Catlin, Catling, Cattlin:** *Katelina* de Walcote 13th Rams (Hu); *Katerina*, *Katelina* de Sauston 1275 RH (Hu); Gervase, Robert *Caterin* Hy 2 Seals (Sr), 1247 AssBeds; William *Catelin*, *Katelin* 1198 FFNf; Robert *Catyln* 1441 ShefA; Richard *Catlyng* 1653 EA (OS) iv (C). OFr *Caterine*, *Cateline*, the French form of *Catharine*, introduced into England in the 12th century when it became popular, usually in the form *Catelin(e)*.

**Catmore, Catmur:** Adam *de Catmera* 1165 P; Henry *de Catmere* 1317 AssK; William *de Cattemere* 1327 SRSf. From Catmore (Berks), *Catmere* DB.

**Catmull:** v. CATTERMOLE

**Catmur:** v. CATMORE

**Catnach:** v. CATANACH

**Caton:** Richard *Caton* 1279 RH (C); Peter *Catoun* 1327 SRSf. *Cat-un*, a diminutive of *Cat*, a pet-form of *Catelin*. cf. KATIN.

**Catt, Katte, Chatt:** *Lufmancat* 1066 DB (Ha); Robert *le Cat* 1167 P (Nf); Geoffrey *Chat* 1190–1200 Seals (Sf); Margaret *Kat* 1202 AssL; Adam *le Chat* 1203 P (W). A common nickname from the cat (OE *catt*, ONFr *cat*, OFr *chat*). *Catt* is probably also a pet-form of *Catelin*, from which were formed the diminutives *Catell*, *Caton*, *Katin*.

**Cattanach:** v. CATANACH

**Cattel, Cattell, Catell, Cattle:** *Cattle* Bagge 1279 RH (C); Hervey, Geoffrey *Catel* 1275, 1279 RH (Nf, Hu); John *Cattle* 1653 FrY; John *Cattell*, *Cattall* 1683, 1707 ib. A diminutive of *Cat*, a short form of *Catelin*. cf. CATON, KATIN and Fr *Catelet*.

**Catten:** v. CATTON

**Catterall, Catterell, Catterill, Catteroll, Cattrall, Cattrell, Cattroll, Catherall, Cathrall:** Robert *de Caterell* 1222 Cur (Ha); John *de Caterhale* 1332 SRLa; Lawrence *Cattrall* 1462 Calv (Y); Richard *Caterall* 1500 FrY. From Catterall (Lancs) and, apparently, also from a place in Hants with a second element *-hill*. William *Katerel* 1203 AssSt suggests also a pet-form of *Caterin*.

**Catterick:** Roger *de Cateric* 1185 Templars (Y); Thomas *Catryk* 1400 FrY; William *Catryk* 1452–3 IpmNt. From Catterick (NRY).

**Cattermole, Cattermoul, Cattermull, Cartmole, Catmull:** William *Cakyrmoll* 1478 SIA xii; John *Cakytmoll*, *Catermoll* 1524 SRSf; Guy *Cackamoule* 1668, Elizabeth *Cackamole* 1743, Sarah *Cattermole* 1749, Susan *Catermoul* 1780 SfPR; Thomas *Cattermoul* 1748 FrYar; Benjamin *Catmull* 1786 SfPR. The late appearance of this name which seems to be found only in Suffolk would suggest a foreign origin for it, probably Dutch or Flemish.

**Cattley:** v. CATLEY

**Catton, Catten:** Ylger *de Catton'* 1181 P (Y); Thomas *de Cattone* 1296 Black; John *of Catton* 1401 AssLa. From Catton (Db, Nb, Nf, ERY, NRY), or Caton (La).

**Caudell, Caudle, Cawdell, Cadel, Caddell, Cadle:** Godfrey, John *Cadel* 1187 P (Gl), 1276 RH (O); Walter, William *Caudel* 1198 FFNf, 1279 RH (C); William *Kaldel* 1277 LLB B. ME *caudel* (*cadel*), ONFr *caudel*, MedLat *caldellum* 'a hot drink', a thin gruel, mixed with wine or ale, sweetened and spiced, given chiefly to sick people, especially to women in childbed; also to their visitors (c1325 MED). Probably a derogatory nickname given to a man who could not hold his drink and so should stick to this invalid beverage, or, conversely, for a toper who scorned anything but a man's drink. *v.* also CALDWELL.

**Caudray:** *v.* CAWDREY
**Caudwell:** *v.* CALDWELL
**Caulcutt:** *v.* CALDECOT
**Cauldron:** *v.* CALDERON
**Cauldwell:** *v.* CALDWELL
**Caules:** *v.* CALL
**Caulkett:** *v.* CALDECOT
**Caunt:** *v.* CANT
**Caunter:** *v.* CANTER
**Cause, Cawse:** Robert *Cause* c1212 RegAntiquiss; Adam *Cauce* 1298 AssL; Thomas *Cawse* 1516 PN Db 221. Perhaps a NFr variant of OFr *chausse* 'stocking', metonymic for a maker or seller of these. Late examples of the name could also be from Cause (Sa), or the Pays de Caux (Seine-Maritime). *v.* also CAW.
**Causey, Cawsey:** (i) Robert *le Cauceis, le Calceis* 1166 RBE, 1166–73 ANF (L); William *le Cauceis* 1212 Cur (Nt); Robert *Causeys, Causay* 1327, 1332 SRSx. A man from the Pays de Caux (Seine-Inférieure). cf. Fr *Cauchois* and *v.* CAW. (ii) Robert *de Calceto* 1202 AssL; Henry *atte Cauce* 1356 Putnam (So); Nycolas *Cawsey* 1524 SRSf. 'Dweller by a causeway', ME *cauce*.
**Causton:** *v.* CAWSTON
**Cautley:** Peter *Cawtley* 1649 FrY. From Cautley in Sedbergh (WRY).
**Cavalier:** Probably Huguenot. John *Cavalier*, the Cevennes leader, was afterwards a brigadier-general in the British army and lieutenant-governor of Jersey, d.1740 (Smiles 234, 372). cf. also Zacheriah *Cavelier* 1739 FrY.
**Cave, Kave:** Nigel *de Caua* 1185 Templars (Y); Hugh *Cave*, William *de Cave* 1212 Cur (L). From Cave (ERYorks). *v.* also CHAFF.
**Cavell:** Roger *Caluel, Chauuel* 1190, 1195 P (K); Enger' *Cauuel'* 1199–1277 Seals (W); Adam *Cavel* 1275 SRWo. A diminutive of OFr *chauf, cauf* 'bald'.
**Cavendish, Candish:** Simon *de Cauendis* 1201 Pl (Sf); Richard *de Cavendish* 1344 FFEss; William *Caundyssh* 1416 FFEss. From Cavendish (Suffolk).
**Cavill:** Tomas *de Kauill', de Cauill'* 1190, 1194 P (Y). From Cavil (ERYorks).
**Caw, Caws, Cawse:** Robert *de Chauz* 1166 RBE (Nt), *de Calz, de Cauz* 1206 Cur (Sx). From the Pays de Caux (Seine-Inférieure). cf. CAUSEY.
**Cawcutt:** *v.* CALDECOT
**Cawdell:** *v.* CALDWELL, CAUDELL
**Cawdrey, Cawderey, Caudray, Cawthra, Cawthran, Cawthrow, Cawthray, Cawtheray:** William *de Caudrey* 1278 RH (O); Robert *Caudray* 1379 PTY; Thomas *Cawdrey* 1597 SRY; Robert *Cawdrey*, Francis *Catherey*, William *Cowtherey, Cawthrey, Gauthrey* 1672 HTY; William, Mary *Cawdry* 1703, William, Anne *Cordery* 1797 BishamPR (Berks). 'Dweller by the hazel copse', OFr *coudraie*. There has been late confusion with CORDEREY. *v.* also COWDREY.

**Cawdron:** *v.* CALDERON
**Cawker:** *v.* CHALKER
**Cawley:** Thomas *Cauly* 1330 IpmNt; William *de Cawlay* 1397 FrY; Rauphe *Cawleye* 1576 SRW. Probably for COWLEY. In Scotland for MACAULEY.
**Cawood:** John *de Cawude* 1219 AssY; Roger *de Cawod* 1303 IpmY; Robert *Cawode* 1430–1 FFSr. From Cawood (WRY).
**Cawse:** *v.* CAUSE
**Cawsey:** *v.* CAUSEY
**Cawston, Causton:** Heroldus *de Caustuna* 1066 DB (Sf); William *de Causton'* c1125 MedEA (Nf). From Cawston (Norfolk).
**Cawtheray:** *v.* CAWDREY
**Cawthorn, Cawthorne, Cawthron:** William *de Calthorn* 1175 P (Y), 1357 Calv; Robert *de Cauthorne* 1379 PTY. From Cawthorn (NRYorks), *Calthorn* 1175, or Cawthorne (WRYorks), *Calthorn* c1125.
**Cawthorpe, Cawthrup:** Roger *de Cautorp'* 1219 AssY. From Cawthorpe (L).
**Cawthra, Cawthran, Cawthray, Cawthrow:** *v.* CAWDREY
**Cawthrup:** *v.* CAWTHORPE
**Caxton:** Geoffrey *de Caxton'* 1202 Pleas (Ess); William *de Caxtone* 1301 LoCt; Richard *Caxton* 1438 IpmNt. From Caxton (C).
**Cayley:** *v.* CALEY
**Cayton:** Robert *de Keyton'* 1219 AssY; Robert *de Cayton* 1289 IpmY. From Cayton (NRY, WRY).
**Cayzer, Kayser, Kayzer, Keyser, Keysor, Keyzor:** Henry *le Caisere* 1172 P (Wa); William *le Keiser* 1195 Oseney (O). ME *caisere*, ultimately from Lat *Caesar* 'emperor'. A pageant name.
**Cazenove:** A Huguenot name. Several members of the family *de Cazenove* de Pradines fled to England on the Revocation of the Edict of Nantes (Smiles 372–3).
**Cecil, Saycell:** *Saissil* 1066 DB (He); *Seisil* 1188 P (Sa); William *Seisil, Seysel* 1205 P (He), 1275 SRWo; Sir Thomas *Cecill* 1591 Bardsley (Nf). OW *Seisill*, said to derive from Lat *Caecilius*.
**Cedervall:** *v.* SIRDIFIELD
**Ceeley, Ceely, Cely, Ceiley:** *v.* SEALEY
**Cendrer:** Ædwin *le cendrer* 1195 P (Db/Nt); Rannulf *le cendrer* 1219 AssY; Siward *le Cendrer* 1222 Pat (Herts). A derivative of OFr *cender* 'a costly fabric of linen or cotton'. A maker or dealer in this.
**Center, Senter:** Agnes *la Ceintere*, John *le Ceinture*, Henry *le Ceynter, le Seynter* 1275 SRWo. girdle' (1595 NED). Metonymic for CENTURY. *v.* also SENTER.
**Century:** John *le Ceinturer* 1275 MESO (Wo); Robert *le Ceinturer* 1298 LoCt. OFr *ceinturier, sainturier* 'maker of waist-belts'. This may also have become SENTRY.
**Cerf:** Adam *le Cerf* 1260, William *le Cerue* 1295 IpmY; Ralph *Cerf* 1336 FFY; William *Cerff* 1416–17 IpmY. OFr *cerf* 'hart, stag', a nickname for a fast runner.
**Chace:** *v.* CHASE
**Chadburn, Chadbourne, Chadborn, Chadbon, Chadbone, Chadband, Chatburn:** John *de Chatteburn* 1379 PTY; William *Chatburn* 1449 FrY; John *Chadbourne* 1660, *Chatband* 1788, *Chadband* 1802 Bardsley. From Chatburn (La).

**Chadd:** *Cedda* de Alrewys 1275 RH (St); *Chad* Maryon 1524 SRSf; Hugo, Henry *Chadde* 1190 P (Wa), 1247 AssBeds; Ralph *Chad* 1219 AssY; Henry *Ced* 1379 PTY; Joan *Chedde*, William *Shedde* 1524 SRSf. OE *Ceadd(a)*, which, though the evidence is slight, seems to have remained long in use.

**Chadderton, Chatterton:** Geoffrey *de Chaderton* 1281 AssLa; William *de Chaderton, de Chaterton* 1324 CoramLa; William *Chatterton* 1641 PrSo; Francis *Chadderton* 1659 FrY. From Chadderton (La).

**Chaddesley, Chadsley:** John *de Chadesley* 1275 SRWo; Adam *de Cheddesleye* voc. *de Cliderowe* 1325 CorLo; Richard *de Chaddesleye* 1340 NIWo. From Chaddesley Corbett (Wo).

**Chadwell:** *v.* CALDWELL

**Chadwick, Chadwyck, Chaddock, Shaddick, Shaddock, Shadwick, Chattock:** Richard *de Chadeleswic'* 1221 AssWa; Andrew *de Chadewyke* 1328 WhC; Pers *Chadick* 1553 WhC; John *Chadwikke* alias *Chaddokke* 1554 CorNt; Thomas *Chadeck* 1704 FrY. From Chadwick (Lancs, Warwicks, Worcs), or Chadwich (Worcs), *Chadeleswik* 1212.

**Chafen:** *v.* CHAFFIN

**Chafer:** John *del Chaufeur* 1301 SRY. OFr *chauffour* 'limekiln', hence 'worker at a limekiln'. cf. William *le Limbrener* 1305 MESO (L), 'lime-burner'. The surname may also survive as CHAFFER.

**Chaff, Chaffe, Chave, Caff, Cave, Kave:** Roger *le Chauf, le Cauf, Calvus* 1214, 1220 Cur (Co); William *Caff* 1214 Cur (L); William *le Cave* 1280 AssSo; Richard *Chafe* 1649 Bardsley. OFr *chauf, cauf*, Lat *calvus* 'bald'.

**Chaffer:** John *Chaffar* 1327 SRC, 1359 FrY. OE *ċēapfaru*, ME *chaffere* (chaffar) 'traffic, trade', also 'merchandise, wares', used by metonymy for a dealer, merchant. cf. *chafferer* 'dealer' (1382 NED). *v.* also CHAFER.

**Chaffin, Chafen, Chauvin, Caffin, Caffyn:** Richard *Chaufin* 1273 RH (Nt); Richard *Caffyn* 1327 SRSx; Thomas *Chafyn* 1505 Oxon. A diminutive of OFr *chauf, cauf* 'bald'.

**Chaffinch, Caffinch:** Walter *Chaffins* 1249 AssW; Peter *Cheffink* 1262 MPleas (W); Simon *Cafynche* 1525 SRSx. A nickname from the chaffinch, ME *chaffinche*.

**Chainey:** *v.* CHEYNEY

**Chalcot:** *v.* CALDECOT

**Chalcraft, Chalcroft, Calcraft:** Thomas *de Chalvecroft* 1272 *Ass* (Ha); John *de Chalfcroft* 1296 SRSx; Robert de *Calvecrofth* 1327 SRSf. From Chalcroft in South Stoneham (Hants) or 'dweller by the calves' croft', OE *cealf. v.* also CALCRAFT.

**Chaldecott:** *v.* CALDECOT

**Chalfont:** Robert *de Chalfhunte* 1284 LLB A; Henry *Chalfhunte* 1334 SRK; Christopher *Chalfehunt* 1662 *HTEss.* From Chalfont (Bk).

**Chalgrave, Chalgrove:** William *de Chalgrave* 1261 FFO. From Chalgrave (Beds), Chalgrove (O).

**Chalice:** *v.* CHALLIS

**Chalk, Chalke, Chaulk:** Walter *de Chelka* 1177 P (W); Ralph *de Chalke* 1268 ArchC 5; William *atte Chalke* 1296 SRSx. From Bower or Broad Chalke (Wilts) or Chalk (Kent), or from residence near a

chalk down. OE *cealc* 'chalk', here 'chalky soil' or an E *\*cealce* 'chalk down'. *v.* MELS.

**Chalker, Kalker, Cawker:** Robert *Calchier* 1195– 1215 StP (Lo); Thomas *le Chalker* 1275 RH (W); Nicholas *le Calkere* 1327 SRSf. A derivative of OE *(ge)cealcian* 'to whiten', 'whitewasher'. But in Wilts and Kent this might also mean 'dweller on the chalk'.

**Challen, Challens, Challin:** Peter *de Chalun, de Chaluns* 1194–5 P (D); Godfrey *Challon* 1275 RH (D); Robert *Chalons* 1428 FA (W). From Chalon-sur-Saône or Châlons-sur-Marne. Or metonymic for CHALLENER.

**Challener, Challender, Challenor, Challinor, Chaloner, Chawner, Channer:** John *le Chaloner* 1213 Cur (Sr); Ralph *le Chaluner* 1224 FFSf; Thomas *Chalander* 1485 RochW; Thomas *Chauner* 1583 AD vi (St). A derivative of ME *chaloun* 'blanket', from its place of manufacture, Châlons-sur-Marne, 'maker of or dealer in *chalons*, blankets or coverlets' (1372 NED). 'Chalons of Guildford' were bought for the king's use at Winchester Fair in 1252 (MedInd).

**Challenge:** John *del Chaleng'* 1327 Kris. 'Dweller by the disputed land', OFr *chalenge*.

**Challenger:** Philip *le Chalengur* 1202 AssL. A derivative of ME *chalangen*, OF *chalonger* 'to challenge', 'an accuser, plaintiff, claimant' (1382 NED).

**Challis, Challiss, Challice, Chalice, Challes:** Henry *de Scalers, de Scalariis* 1086 DB; Henry *de Shallers* 1153–85 Templars (Herts); Geoffrey *de Chaliers* 1203 FFC; Thomas *de Chalers* 1340 AssC; Thomas *Chales* 1524 SRSf; William *Challice* 1642 PrD. From Eschalles (Pas-de-Calais). *v.* OEByn 87.

**Chalmers** is a Scottish form of CHAMBERS: *chalmer* 1375 NED. The *mb* was assimilated to *mm*; the *l* was purely graphic, indicating that the preceding *a* was long, and did not affect the pronunciation: Robert *de la Chaumbre* 1296 Black (Lanarks); Alexander *Chaumir* 1475 ib. (Aberdeen); Robert *Chamer* 1472 ib. (Angus); John *Chalmyr* 1555 ib. (Glasgow). We also find *Chalmer* in Suffolk and Worcester in the 13th century: Roger *le Chalmere* 1255 FFSf; Ralph *le Chalmer* 1275 SRWo. This is probably for *Challoner*, with dissimilation of *ln* to *lm* in *Chalner*.

**Chaloner:** *v.* CHALLENER

**Cham:** Semar *Chamme* 1181 P (D); Werreis *de Cham, de Cam* 1204, 1207 P (Sf). From Caen (Calvados).

**Chamberlain, Chamberlaine, Chamberlayne, Chamberlen, Chamberlin, Champerlen:** Henry *le canberlain* Hy 2 DC (L); Geoffrey *le Chaumberleng* 1194 Cur (W); Robert *canberlenc* 1195 FF; Thomas *Chamberleng' seruiens Regis* 1196 P (C); Martin *le Chamberleyn* 1232 FFC; Thomas *le Chaumberlyn* 1293 AssSt. OFr *chamberlain, -len, -lanc, -lenc* 'officer charged with the management of the private chambers of a sovereign or nobleman' (a1225 NED).

**Chambers:** Nicholas *de Chambres* 1219 Cur (Db); Stephen *de la Chambre* 1240 FFEss. ME *chaumbre*, OFr *chambre* 'room (in a house)', 'reception room in a palace' (a1225 NED). Originally official, identical with *Chamberlain*. To pay in *cameram* was to pay into the exchequer of which the *camerarius* was in charge. The surname also applies to those employed there. cf.

Nicholas *atte Chambre* dictus *Clerk* 1351 AssEss. It was later used of a chamber-attendant, 'chamberman, chambermaid'.

**Chambley:** *v.* CHOLMONDELEY

**Champ, Champe:** Martin *de Champz* 1296 SRSx; William *del Chaumpe* 1341 FrY; John *Champ* 1396–7 FFSr. From Champs in Harting (Sx), or 'dweller by the meadow', OFr *champ*. Sometimes, perhaps, from one or other of the numerous French places of this name, e.g. Champ (Isère, Maine-et-Loire), Champs (Seine-et-Marne).

**Champain:** Peter *de Champaigne* 1195 P (L). From the French province of Champagne. cf. CHAMPNESS.

**Champerlen:** *v.* CHAMBERLAIN

**Champernowne:** Jordan *de Campo Arnulfi* 1172 RBE; Pichard *de Cambernof* 1189–99 France; Geoffrey *de Chaumbernum* 1230 P (D). From Cambernon (La Manche).

**Champflower:** Henry *de Champflur* 1219 P (Do/So); Henry *de Campo Florido* c1265, *Chamflour* 1275 Glast (So). From Champfleury (Aube, Marne).

**Champion, Campion:** Herbert *campion* 1148 Winton (Ha); Geoffrey *Champiun* 1154–69 NthCh (Nth); Roger *le Campion* 1197 P (O); William *le Champiun* 1220 Cur (Sf). ONFr *campiun, campion*, central OFr *champiun, champion* 'a combatant in the campus or arena', 'one who "does battle" for another in wager of battle', 'a champion' (a1225 NED). In the ordeal by battle, in criminal cases, the accuser and the accused took the field themselves, but in disputes about the ownership of land, the actual parties to the suit were represented by 'champions', in theory their free tenants, but in practice, hired men, professional champions, and very well paid. In 1294 the Dean and Chapter of Southwell incurred a prospective liability of about £750 in modern money in hiring a champion to fight a duel to settle a law-suit about the advowson of a church. A *pugil* or champion was a regular member of the household of more than one medieval bishop, Thomas Cantilupe, Bishop of Hereford (1275–82), paying his champion, Thomas de Bruges, a salary of half a mark a year. *Champion* may sometimes be a corruption of CHAMPAIN (cf. also CHAMPNESS). Champion Wood (PN K 287) owes its name to a family variously called *de Chaumpayne* or *Champeyneys* (1278) and *de Campayne* (1332).

**Champness, Champney, Champneys, Champniss:** William *le Champeneys* 1219 Cur (Nf); John *Champenay* 1333 FrY; John *Champness* 1520 ArchC 33. AFr *le champeneis* 'the man from Champagne'.

**Chance:** Robert, Ralph *Chance* 1209 P (Ess), 1310 FFEss. ME *chea(u)nce*, OFr *cheance* in one of its senses, 'fortune, accident, mischance, luck' (c1300 MED). Perhaps used of a gambler. cf. HAZARD.

**Chancellor, Cancellor:** Reinbald *Canceler* 1066 DB (He); Richard *le Chaunceler* 1214 Cur (Berks). ME, AFr *canceler, chanceler*, 'usher of a lawcourt', 'custodian of records', 'secretary'.

**Chancey, Chauncey, Chauncy:** Simon *de Chanci* 1218 AssL; Roger *de Chauncy* 1230 P (Y); John *Chauncy* 1293–4 IpmY; Roger *Chansi* 1361 IpmGl. From one or other of the French places called Chancé (Ille-et-Vilaine, &c.).

**Chandler, Chantler, Candler:** Matthew *le Candeler* 1274 RH (Lo); William *le Chandeler* 1285 *Ass* (Ess). Afr *chandeler*, OFr *chandelier, candelier* 'maker or seller of candles' (1389 MED). Both *Chandler* and *Candler* are used of the same man in 1756 and 1759 in IckworthPR (Sf). cf. William *Candleman* 1268 AD ii (Sf).

**Chandos:** Roger *de Candos* 1086 DB (Sf); Walter *de Chandos* c1148 EngFeud (He). From Candos (Eure). *v.* ANF.

**Chaney:** *v.* CHEYNEY

**Change:** Robert *del Change* 1207 P (Lo); William *de la Chaunge* 1269 Kris; John *del Chaunge* 1327 FrY. 'Dweller at the place where merchants meet for business', OFr *change, chaunge*.

**Changer:** Rolland *le Changur* 1200 P (Ha); Symon *le Changur* 1275 RH (L); John *Chaungeour* 1384 LLB G. 'Money-changer', OFr *chaungeor, chaungeour*.

**Channel, Channell:** Henry *de Chanel* 1219 Cur (Sf); John *Channel* 1270 AssSo; Peter *Chanel* 1327 SRSo. 'Dweller by the estuary, drain, or ditch', OFr *chanel*.

**Channer:** *v.* CHALLENER

**Channon:** *v.* CANNON

**Chant:** *v.* CANT

**Chanter:** Hugh *le Chantur* 1235 Fees (Lei); Walter *le Chauntur* 1285 FFC. ME, AFr *chantour*, OFr *chanteor* 'enchanter, magician' (c1300 MED), 'singer, chorister, precentor' (a1387 ib.). cf. CANTER.

**Chantler:** William *Cantecler* 1192–1218 YCh; Roger *Chauntecler* 1307 LLB C; Robert *Chaunteclere* 1371 FFY. 'Sing loudly', OFr *chaunter, clere*. For similar names, cf. Walter *Chanteben* 1206 AssL 'sing well'; Philip *Chante Merle* 1176 P (Bk) 'sing like a blackbird'; Robert *Chantemerveille* 1203 Cur 'sing marvellously'. Also a variant of CHANDLER.

**Chantrell, Chantrill, Cantrell, Cantrill:** Walterus *Canterellus* (*Chanterel*) 1177 P (Sa); Philip *Canterel* 1203 AssSt; Robert *Chanterel* 1221 AssWa; John *Cantrel* 1297 MinAcctCo. Perhaps OFr *chanterelle* 'a small bell', 'the treble in singing, a treble string or bell' (Cotgrave), probably used of a bellman. Or a diminutive of CANTER, CHANTER.

**Chantrey, Chantry:** John *del Chauntre* 1379 PTY; Thomas *off the Chantery* 1524 SRSf. OFr *chanterie*, originally 'singing or chanting of the mass', then applied successively to the endowment of a priest to sing mass, the priests so endowed, and finally to the chapel where they officiated. The surname might refer to the chantry-priest, but more probably to his servant.

**Chanue:** Ralph *Chanu* 1201 FFK; William *le Chanu* 1243 AssSo; Agnes *le Chanus* 1327 SRSx. OFr *chanué* 'grey-haired'.

**Chaplain, Chaplin, Chapling, Chaperlin, Chaperling, Caplan, Caplen, Caplin, Capelen, Capelin, Capeling, Kaplan, Kaplin:** William *Capelein* 1203 Cur (Ha); Thomas *le Chapelyn* 1241 FFC; Nicholas *le Chapelain* 1260 AssC. ONFr *capelain*, OFr *chapelain* 'priest, clergyman, chantry-priest'.

**Chapman, Chipman:** Hugh *Chapman* 1206 Cur (Y); Alice *Chepman* 1207 P (Db); Thomas *le Chapman* 1266 AssSt; Nicholas *le Chipman* 1320 MESO (So);

Henry *le Chupman* 1327 ib. (Ha). OE *cēapmann, cȳp(e)mann, cēpemann* 'merchant, trader'. *Chapman* is a general form, *Chipman* is West Saxon.

**Chapp:** Reginald *Chape* 1297 MinAcctCo. OFr *chape* 'a churchman's cope'. cf. Thomas *le chapemaker* 1351 FrY.

**Chappel, Chappell, Chappelle, Chapple, Chappill, Chapple, Chapell:** John *Chapel* 1202 P (Nf); Richard *de la Chapele* 1296 SRSx; Eymer *atte Chapele* 1312 LLB D; William *del Chapell* 1380 FrY. From residence near or service at a chapel, ME, OFr *chapele*. cf. CAPEL.

**Chapper, Cheeper, Chipper:** John *Chaper* 1200 Cur (Sr); Geoffrey *Chipere* 1254 ELPN; William *le Chappere* 1327 MESO (Ha); Ralph *Chiper* 1327 SRC. A derivative of OE *cēapian* 'to bargain, trade, buy'. The early examples of *Chiper* are rather from OE *cīepan, cīpan* 'to sell'. In both cases, the meaning is 'barterer, trader'.

**Charbonneau, Charbonnel:** v. CARBONELL

**Charbuckle:** Cecilia *Charbocle* Hy 3 IpmY; John *Charbocle* 1332 PN Cu 216. A nickname from OFr *charbocle* 'carbuncle'.

**Charcrow:** Adam *Charthecrowe*, Adinet *Charricrowe* 1286 AssCh. 'Drive away the crow', OE *cærran, crāwe*. cf. Andrew *Charrehare* 1245 FFL 'turn hare', a nickname for a speedy runner.

**Chard:** John *ate Charde* 1281 AssW; Hugh *de Cherde* 1335 Glast (So); William *Chard* 1641 PrSo. From Chard (So), or 'dweller by the rough land', OE *ceart*.

**Charge:** Nicholas *Charge* 1674 HTSf. OFr *charge* 'impost, levy'. Metonymic for a tax-collector.

**Charity:** Herluin *Caritet* 1148–54 Bec (Sx); Geoffrey *de Caritate, de la Carite* 1185–7 P (Ha); Turstan *Charite* 1195–7 P (Nf); John *de la Charite* 1203 Cur (Ha); Richard *Charite, Karite* Hy 3 Gilb (L); Robert *Karitas* 1236 FrLeic. ONFr *caritedh, caritet*, later *carité*, OFr *charitet, charité*, Lat *cāritātem*, found in ME in various senses: 'alms-giving, hospitality' (a1160 MED), 'man's love for God' (a1225), 'Christian love' (c1200), from which a nickname could easily arise. The local origin is from ONFr *carité*, OFr *charité* 'hospice, refuge'.

**Chark:** William *Charke* 1221 AssGl; Robert *Chark* 1327 SRSo; William *Charke* 1415–16 Hylle. OFr *charche* 'a load'. Probably metonymic for a carrier or a porter. v. CARKER.

**Charker:** v. CARKER

**Charlcot, Charlcott:** William *de Cherlecot'* 1220–1 FFWa; Geoffrey *de Cherlecot* 1221 SPleas (Wa). From Charlecote (Wa).

**Charlemayn, Charlemayne, Charlman:** *Charlemayn* 1230 P (Wo); Gregory *Cherlemayn* 1261–74 Glast (D); Nicholas *Scharlemayn* 1292 IpmGl; John *Charlemayn* 1353 Putnam (W). Lat *Carolus Magnus*, OFr *Charlemagne*. The name probably owes its presence in England to the popularity of the Charlemagne romances.

**Charles:** (i) *Carolus, Karolus* 1208 Cur (Sf, Nf); *Karolus filius* Gerberge 1210 Cur (Nf); *Carolus filius* Willelmi 1212 Cur (Nf); Colina, Nicolas *Charles* 1250 Fees (Sf), 1253 Bart; Robert *son of Charles*, Thomas

and Joan *Charles* 1274 FFSf. OFr *Charles*, from a latinization *Carolus* of OG *Karl* 'man', introduced into England by the Normans, but never common until the Stuart period. (ii) Osbert *Cherle* 1193 P (Wa); Frebesant *Cherl* 1221 *ElyA* (C); John *Charl* 1296 SRSx. OE *ceorl*, originally 'a freeman of the lowest rank'; in ME 'a tenant in pure villeinage, serf, bondman'; also 'countryman, peasant'. *Charl* would later be inevitably assimilated to the personal name.

**Charlesworth:** Jacobus *Charlesworth* 1550 FrY; Thomas *Charlsworth* 1642 PrD. From Charlesworth (Db).

**Charley, Charly:** Walter *de Cherlelai* 1202 P (Bk); Dobyn *de Charlag* 1276 AssLa; Henry *Charley* 1524 SRD. From Charley (Lei), or Charley Fm in Stanstead St Margarets (Herts).

**Charlford:** Roger *de Cherleford* 1275, William *de Charleforde* 1327 SRWo. From Charford in Bromsgrove (Wo), *Cherleford* 1231.

**Charlick, Charlock:** Richard *Carloc* 1279 RH (C); Peter *Charloc* 1317 AssK; Roger *Charleloke* 1497 ArchC 42. OE *cerlic, cyrlic*, ME *carlok, charlok* 'wild mustard'. Metonymic for a grower or seller of this.

**Charlot:** (i) John *Charlot* 1275 SRWo. *Charl-ot*, a diminutive of *Charles*. (ii) Usually Huguenot. Charles *Charlot*, a converted Catholic curé, fled to England and was minister of the Tabernacle in 1699.

**Charlton:** Jordan, Robert *de Cherleton'* 1193 P (Gl), 1230 Cur (Beds); Hugh *de Charleton* 1333 IpmNt; Robert *Charletone* 1372 CorLo. From one or other of the many places of this name.

**Charlwin, Charlwyn:** John *Cherlewyne* 1327 SRSo. OE *Ceorlwine*.

**Charlwood:** Alexander *de Cherleswode* 1296 SRSx. From Charlwood (Sr), or Charlwood in Forest Row (Sx).

**Charlwyn:** v. CHARLWIN

**Charly:** v. CHARLEY

**Charman:** William *le Charman* 1293 AssSt; Adam *le Charman* 1310 LLB B. OFr *char, car*, 'car, cart' and OE *mann*, 'carter, carrier'. *Carman*, which is not uncommon, may belong here as well as to CARMAN, to which occasional early examples of *Charman* belong: Robert *Chareman* (*Careman*) 1183 P (Wa).

**Charnell:** Gilbert *de Charneles* 1170–5 DC (Lei); Hugh *de Charnell* 1246 FFC; Maud *de Cimiterio* 14th Rad (C). ME, OFr *charnel*, 'burial-place, mortuary chapel, cemetery', denoting one in charge of this.

**Charner:** Thomas *le Charner* 1279 RH (C); Richard *le Charner* 1280 AssSo; John *le Charner* 1327 SRC. OFr *charner, charnier* 'burial-place'. Probably metonymic for a grave-digger.

**Charnock:** Richard *de Chernok* 13th WhC; Henry *de Chernok* 1332 SRLa; William *Charnok* 1403 FrY. From Charnock (La).

**Charpenter, Charpentier:** (i) Gilbert *le Charpenter* 1227 FFHu; William *le Charpenter* 1276 AssLa, 1346 LLB F. OFr *charpentier* 'carpenter', v. also CARPENTER. (ii) Also Huguenot, from John *Charpentier* who fled to England and was minister of the Malthouse Church, Canterbury, in 1710 (Smiles 375).

**Charsley, Chartesley, Cheasley:** Geoffrey *de Chardesle* 1279 RH (O). From Chearsley (Bucks).

**Chart:** Richard *de Chert* c1200 ArchC vi; John *de la Chert* 1241 PN Sr 306; Ralph *Atte Chert* 1290 MELS. From residence near a rough common (OE *cert, ceart*), as at Chart (Kent, Surrey) or Churt (Surrey).

**Charter:** v. CARTER

**Charteris, Charters, Chartres, Chatteris, Chatters:** (i) Alcher *de Chartris* 1179 P (Sx); Robert *de Chartres* 1296 (Black); James *Charterhouse* 1556 ib.; John *Charters* 1790 ib. From Chartres (Eure-et-Loire). (ii) Ralph *de Chateriz* 1259 FFC; Alan *de Chartres, de Chartris* 1279 RH (Hu), 1293 Ipm (Hu); John *Charteres* 1417 FFHu; John *Chateryse* 1445 NorwW. From Chatteris (Cambs), earlier *Chateriz, Chatriz* 1086, *Chartriz* 1200 (PN C 247–8). cf. Abbatissa *de Charters* 1279 (Hu).

**Charville, Charwell:** (i) John *de Cherville* 1302 *PetreA*; John *Charvel* 1531 FFEss; John *Charvolle* 1537 *PetreA*. cf. Charville's Fm in West Hanningfield (Ess). (ii) John *Charfowle* 1462 FFEss; John *Charfoule* 1483 *PetreA*; John *Charfowle* 1538 FFEss. A nickname 'turn fowl', OE *cærran, fugol*.

**Chase, Chace:** Robert *Chace* 1327 *SR* (Ess); John *Chase* 1393 FrY. Probably metonymic for *chaser*, from OFr *chaceur, chaceour* 'hunter'. cf. Stephen *le Chacur* 1204 Cur (Nt), Walter *Chacere* 1327 SRSf.

**Chaser:** Stephen *le Chacur* 1204 Cur; Simon *le Chacer* 1275 SRWo; John *Chasour* 1327 *SR*Ess. OFr *chaceur, chaceour* 'hunter'. cf. Philip *Chaceboef* 1218 P (D) 'chase ox'; Robert *Chacecapel* 1201 P (D) 'chase horse'; Peter *Chaceporc* 1253 CartAntiq 'chase pig'; Walter *Chacero* 1261–2 FFWa 'chase roe'.

**Chasey:** Simon *de Caucy* 1205 Cur (L); Geoffrey *de Chausi* 1206 Cur (O). From Chaussy (Seine-et-Oise) which may also have become CAUSEY. cf. Chazey Fm in Mapledurham (Oxon), from Walter *de Chauseia* c1180. Chazey Wood is *Causies Wood* 1658 (PN O 60).

**Chasteney, Chasney, Chastney:** v. CHEYNEY

**Chaster:** v. CHESTER

**Chaston:** Adam *Chasteyn*, Robert *Chasten* 1279 RH (C); John *Chasteyn* 1327 SRSf. ME *chastein*, OFr *chastaigne* 'chestnut-tree'. 'Dweller by the chestnut-tree', originally *atte chastein*.

**Chatburn:** v. CHADBURN

**Chatelain:** v. CASTELLAN

**Chater:** v. CATER

**Chatfield:** William *de Chattefeld* 1296, Richard *Chatfeld* 1525 SRSx. From Chatfields in Bolney, or Chatfield Fm in Cowfold (Sx).

**Chatham:** Hugh *de Chatham* 1206–7 FFEss; Henry *de Chatham* 1218–19 FFK. From Chatham (K), or Chatham Green in Great Waltham (Ess).

**Chatley:** Peter *de Chattele* 1202–3 FFEss. From Chatley in Great Leighs (Ess).

**Chatt:** v. CATT

**Chatteris, Chatters:** v. CHARTERIS

**Chatterley:** Roger *Chaturley* 1388–9 FFWa. From Chatterley (St).

**Chatterton:** v. CHADDERTON

**Chatto:** William *de Chetue* c1198–1214 Black;

Alexander *de Chatthou* c1225 ib.; Eustace *of Chattow* 1358 ib. From the lands of Chatto (Roxburgh).

**Chattock:** v. CHADWICK

**Chaucer, Chauser:** Ralph *le Chaucer* 1220 Cur (Lo); Robert *le Chauser* 1256 AssNb. OFr *chaucier* 'maker of *chausses*', from OFr *chauces* 'clothing for the legs, breeches, pantaloons, hose'. In 1484 these were 'chauces of yron or legge harneys' (NED), but ME *chawce* was a general term for anything worn on the feet, boots, shoes, etc. As Baldwin *le Chaucer* (1307 LLB B) was of Cordwanerstrete, the early chaucer was probably a worker in leather, a maker of leather breeches, boots, etc.

**Chaumont:** Hugo *de Chaumont* 1200 P (Y). From Chaumont (Orne).

**Chauncey, Chauncy:** v. CHANCEY

**Chave:** v. CHAFF

**Chavler:** Richard *le Chaueler* 1221 AssWa; Thomas *Chauler* 1249 AssW. 'Gossip, chatterer', from a derivative of OE *ceafl* 'cheek, jowl', cf. ME *chavlen, chaulen* 'to wag the jaws, chatter'.

**Chawner:** v. CHALLENER

**Chaytor:** v. CATER

**Cheadle, Cheatle, Cheedle:** Hobbe *de Chedel* 1297 Wak (Y); Agnes *de Chedle* 1356 AssSt. From Cheadle (Ch, St).

**Cheal, Cheale, Cheales, Cheel:** (i) Gilbert *de Chele(s)* 1275 RH (L). From Cheal (Lincs). (ii) William, Robert *Chele* 1275 SRWo, 1327 SRSx. OE *cele, ciele* (sb.) 'cold, coldness', ME *chile, chele* 'cold (of the weather), frost'. cf. FROST.

**Chear, Cheer, Cheere, Cheers:** Abraham *filius Chere* 1214 Cur (Ha); Reginald *chere* 1189 Sol; Henry *Chere* 1327 SRSo; John *Chere* 1524 SRSf. 'Precious, dear, worthy', OFr *chier, cher*. Used also as a personal name.

**Cheasley:** v. CHARSLEY

**Cheater, Chetter:** John *Chetour* 1327 SRSf. ME *chetour*, an aphetic form of *eschetour* 'escheator, an officer appointed to look after the king's escheats' (c1330 MED).

**Cheatle:** v. CHEADLE

**Checker, Chequer:** Laurence *de le Eschekere* 1256 *Ass* (Ha); Roger *de la Checker* 1279 RH (C); Gilbert *le Cheker* 1316 Wak (Y); Roger *Cheker*, son of Christopher *Atcheker* 1508 ArchC 40. ME *cheker*, an aphetic form of ME, AFr *escheker*, originally a chess-board, later the table that gave name to the king's exchequer; a table for accounts; the Court of Exchequer. Laurence *de Scaccario* (1279 RH), who has left his name in Chequers (Bucks), was, no doubt, one of the leading officials of the Exchequer. As a surname, it probably meant, as a rule, a clerk in the exchequer.

**Checkley:** Thomas *de Chekele* c1190 StCh; Henry *de Chekele* 1281 AssSt; Robert *Checkley* 1447 IpmNt. From Checkley (Ch, He, St).

**Cheddar, Chedder:** Robert *Cheddre* 1377 IpmGl; Richard *Cheddre* 1404 KB (Lo). From Cheddar (So).

**Cheedle:** v. CHEADLE

**Cheek, Cheeke, Cheke:** Æluric *Chec* c1095 Bury (Sf); Adam, Walter *Cheke* 1202 P (W), 1243 AssSo. OE *cēace, cēce* 'jaw-bone', a nickname for one with a prominent jaw. v. CHICK.

**Cheel:** v. CHEAL

**Cheeld:** v. CHILD

**Cheer, Cheere, Cheers:** v. CHEAR

**Cheesborough, Cheesbrough, Cheeseborough:** Edward *Cheseburgh* 1526 FrY; Robert *Cheesbrough* 1611 RothwellPR (Y). From Cheeseburn (Nb), *Cheseburgh* 1286.

**Cheese, Chiese:** Ailwin *chese* Hy 2 Bart (Lo); Willelmus *cum frumento* 1176 P (Y); John *Chese*, William *Chuse* 1279 RH (Hu, O); Robert *Chuse*, Michael *Chouse* 1332 SRSx. OE *cēse* (Anglian), WSaxon *cȳse* 'cheese', used of a maker or seller of cheese. cf. John de Hugat, *cheser* 1316 FrY, Walter *le cheser* 1366 AD i (He).

**Cheeseman, Cheesman, Cheesman, Cheeseman, Chesman, Chessman, Chiesman, Chisman, Chismon:** Henry *le Cheseman* 1260 AssC; William *le Chesman* 1311 Battle (Sx); Thomas *Chesman*, *le Chusman* 1327 SRSx; Adam *le Chisman*, Alice *Chisman* 1327 SRSo. 'Maker or seller of cheese.' cf. Robert *le Chesemaker* 1275 RH (L), Baldwin *le Chesemangere* 1186 P (K).

**Cheeser:** Ralph *le Chesar* 1332 SRWo; John *le Cheser* 1350 MESO; Walter *le Cheser* 1366 AD i (He). cf. John de Hugat, *chesar* 1316 MESO (Y). 'A maker or seller of cheese', from a derivative of OE *cīese*.

**Cheeseright, Cheesewright, Cheeswright, Cheswright, Chesswright:** Richard *Chesewricte* 1228 Pat (L); Augustin *le Chesewryghte* 1293 MESO (Y); John *Cheswright* 1478 LLB L; Margaret *Chestwright* 1795 SfPR. OE *cȳswyrhte* (f) 'cheese-maker', perhaps also OE *\*cēsewyrhta* (m).

**Cheetham, Cheetam:** Geoffrey *de Chetham* 1246 AssLa, 13th WhC; Thomas *de Cheteham* 1394 IpmLa. From Cheetham (Lancs).

**Cheever, Cheevers, Chevers, Chivers:** William *Cheure, Capra* 1086 DB (D, W); Hamelin *Chieure* 1186 P (L); Nicholas *le Chiuer* 1327 SRSx. AFr *chivere, chevre*, OFr *chievre*, Lat *capra* 'she-goat' (1491 NED), probably denoting agility.

**Cheke:** v. CHEEK

**Chelsham, Chelsam, Chelsom:** John *de Chelesham* 1250 FFK; John *Chelsham* 1359 AssD; John *Chelsam* 1583, James *Chelsom* 1596 Musters (Sr). From Chelsham (Sr).

**Chene, Cheney:** v. CHEYNEY

**Chenery, Chinnery:** Robert *Chenery* 1327, George *Chenery* 1524 SRSf; Peter *Chynnery* 1662 *HTEss*. Perhaps from Chenevray (Haut-Saône).

**Chenevix:** A distinguished Lorraine family, dispersed on the Revocation of the Edict of Nantes, some of them eventually settling in Waterford and Lismore. Philip *Chenevix* fled to England, his grandson becoming Bishop of Killaloe in 1745, whilst his great-grandson, Richard *Chenevix* Trench, was Archbishop of Dublin.

**Chequer:** v. CHECKER

**Cherington, Cheriton:** v. CHERRINGTON

**Cherill:** Adam *de Churhylle* 1275 RH (D); John *Chirel* 1275 RH (K). From Cherhill (W).

**Cherrett, Cherritt:** A Suffolk pronunciation of CHEESERIGHT: Nicholas and Jane *Cheswright, Cherit,*

*Cheritt* 1655, 1656, 1660 DenhamPR (Sf), Nicholas *Cherit* 1674 HTSf.

**Cherrington, Cherington, Cheriton:** William *de Cerinton* 1201–2 FFK; Stephen *de Cheriton* 1260 AssC; Richard *Cheryngton* 1383 AssLo; Peter *Cheryton* 1524 SRD. From Cherrington (Gl, Sa, Wa), or Cheriton (D, K, So).

**Cherry, Cherriman, Cherryman:** Robert *Chyry* 1284 AssLa; Hugh *Chirie*, Richard *Chery* 1524 SRSf. ME *chirie, cherye* 'cherry'. Probably 'grower or seller of cherries'. cf. PERRY, PERRIMAN.

**Chertsey:** John *de Cherteseye* 1367 FFY; John *Chartesey* 1392 LoCh. From Chertsey (Sr).

**Cheselden, Chesilden:** Richard *de Chesilden* 1281 AssW. From Chisledon (W), *Cheselden* 1242.

**Chesham, Chessam, Chessum:** Burchard *de Cestresham* 1200 P (Bk); William *de Chesham* 1297 MinAcctCo (W): William *Chessam* 1525 SRSx; John *Chessum* 1728 Bardsley. From Chesham (Bucks), *Cestreham* DB, or Chestham Park in Henfield (Sussex), *Chesham* 1657.

**Cheshire, Chesher, Cheshir, Chesshire, Chesshyre, Chesser, Chessor:** Richard *de Cestersir'* 1219 AssY; Agnes *Chessyr* 1442 GildY; Nicholas *Chesshyre* 1575 SfPR. From Cheshire.

**Cheshunt:** Hamo *de Cestrehunte* 1212 P (Ess). From Cheshunt (Herts), *Cestrehunte* DB.

**Chesilden:** v. CHESELDEN

**Chesnay, Chesney:** v. CHEYNEY

**Chessel, Chessell, Chessells:** Richard *de Chesthull'* 1221 AssSa; William *Chesul* 1275 SRWo; William *de Chesele* 1322 CorLo. From Chesthill (Sa), or Chesil Bank (Do).

**Chessex:** v. CHISWICK

**Chessman:** v. CHEESEMAN

**Chest:** William *Chest* 1185 Templars (K); Alice *Chest* 1341 FFY. Either OE *ceast* 'strife', for a contentious person, or OE *cest, cyst* 'chest, box', metonymic for a maker of these.

**Chester, Chesters, Chaster:** Richard *de Cestre* 1200 P (L); John, William *de Chester* 1332 SRWa; John *Chestre* 1366–7 FFWa; Barbara *Chesters* 1611 RothwellPR (Y). Usually from Chester (Ches), but occasionally from Little Chester (Derby), Chester le Street (Durham), or Chesters (Northumb).

**Chesterfield:** Pagan *de Cestrefeld* 1172 P (Db/Nt); Robert *Chesterfeld* 1340–1450 GildC; Robert *de Chesterfeld'* 1379 PTY. From Chesterfield (Db).

**Chesterton:** Bruning *de Cestretona* 1086 InqEl (C); Robert *de Chesterton'* 1227 Cur (O); Edward *Chestreton* 1416–17 FFWa. From Chesterton (Cambs, Glos, Hunts, Oxon, Staffs, Warwicks).

**Chestney:** v. CHEYNEY

**Cheswick:** v. CHISWICK

**Chetman:** Thomas *Cheteman* 1443–4 FFSr. 'Dweller in the hut', OE *cēte, cȳte* 'hut, cabin', and *mann*.

**Chetter:** v. CHEATER

**Chettle:** John *Chetel'* 1379 PTY; William *Chetill* 1464–5 IpmNt; Henry *Chettle* 1546 PN Do ii 74; William *Chettle* 1641 PrSo. Anglo-Scandinavian *\*Cytel*. Sometimes from Chettle (Do).

**Chetwin, Chetwind, Chetwyn, Chetwynd:** Richard *de Chetewynde* 1268 AssSt; William *de Chetwynde* 1343–4 FFWa; William *Chetwyn, Chetwynd* 1415 IpmY. From Chetwynd (Sa).

**Chetwode, Chetwood:** Robert *de Chetewod'* 1206 Cur (Wa); Ralph *de Chetwode* 1262 FFK; John *de Chetwode* 1346–7 FFSr. From Chetwode (Bk).

**Chetwyn, Chetwynd:** *v.* CHETWIN

**Cheval:** Roger *Cheval* 1208 ChR; Herbert *le Cheval* 1220 Cur (Beds); William *Cheval* 1241 FFEss. Either a nickname from OFr *cheval* 'horse', or metonymic for CHEVALIER.

**Chevalier:** (i) Robert *le Chevaler* 1205 Cur; Nicholas *Chivaler* 1221 ElyA (Sf); William *Cheualer* 1332 SRSx. AFr *chevaler, chivaler,* OFr *chevalier* 'horseman, mounted soldier'. (ii) Probably usually Huguenot. Antoine-Rodolphe *Chevalier,* born at Monchamps in 1507 and Professor of Hebrew at Cambridge, returned to France, but fled after the Massacre of St Bartholomew and died in Guernsey in 1572. His son, Samuel *Chevalier,* came to England from Geneva. In 1591 he was minister of a French church in London, and later of the Walloon church at Canterbury (Smiles 376–7).

**Chevenger:** Geoffrey *le Chevanchur* 1226 Cur (Sr); John *Chevenger* 1672 HTY. OFr *chevaucheor* 'horseman'.

**Cheverall, Cheverell, Cheverill, Chiverall, Chiverrell, Chivrall:** (i) William *Cheuerel(l)* 1195–6 P (Berks); Tristram *le Cheverer, le Cheverell* 1278, 1289 LLB A, *le Cheverel(l)er* 1291–2 ib. C. ME *chevrelle,* OFr *chevrele, chevrelle* 'kid' (a1400 NED), but in ME always used in the sense of the full *cheverel-leather* 'kid-leather'. The earliest example may be a nickname from the kid, but *cheverel* was certainly sometimes used for *chevereller* 'a maker or seller of kid-leather goods'. cf. Ralph *le Cheverelmongere* 1310 LLB B. (ii) Simon *de Chiverell'* 1200 Cur (W); John *Chiverel* 1275 RH (W). From Great or Little Cheverell (Wilts).

**Chevers:** *v.* CHEEVER

**Cheverton, Chiverton:** Ralph *de Chevereston* 1275 RH (D). From Chiverstone Fm in Kenton (Devon).

**Chevery, Chivery:** William *Chevry* 1327 SRSo. 'Dweller at the goat-fold', OFr *cheverie.*

**Chevins, Chevis:** Margaret *Chyuin* 1295 Barnwell (C); Simon *Cheuyn* 1327 SRC. OFr *chevesne,* Fr *chevin,* a fish, the chub (c1450 NED). 'The cheuyn is a stately fish' (1496). 'Chevins and Millers thumbs are a kind of jolt-headed Gudgeons' (1655).

**Chew, Chue:** (i) Randal *de Chiw* 1201 AssSo. From Chew (Som). (ii) Geoffrey *Chiue* 1203 Cur (C). OE *cīo, cēo* (not found in ME), a bird of the crow family, a name applied to all the smaller chattering species, especially the jackdaw.

**Cheyney, Cheyne, Chainey, Chaney, Cheeney, Chene, Cheney, Chasney, Chasteney, Chastney, Chesnay, Chesney, Chestney:** Radulfus *de Caisned* 1086 DB (Sx); Hugh *de Chaisnei, de Cheisnei* 1140, 1166 Eynsham (O); Robert *de Cheinnei* a1183 DC (L); Hugh *de Chennei* 12th ib.; William *de Chesnei* 1205 Cur (L); Bartholomew *del Chennay* 1212 Fees (Sr); William *de Cheny* 1235 Fees (Sf); Roger *del Chesne* 1236 FFEss; Alexander *de Cheyny* 1242 Fees (Beds);

Alexander *de Cheyne* 1296 SRSx; John *de Chene* 1317 AssK. The DB under-tenant came from Le Quesnay (Seine-Inférieure). *v.* ANF. Others may have come from Quesnay (Calvados, La Manche) or Quesnay-Guesnon (Calvados). All derive ultimately from MedLat *casnetum* (OFr *chesnai*) 'oak-grove' and the surname may also denote an immigrant from France who lived by an oak-grove or came from a place Chenay, Chenoy, or Chesnoy.

**Chibnall:** Alexander *de Chebenhale* 1315 FFC (Sf); Thomas *Chibbenhall* 1559 Pat; Joane *Chibnolle* 1576 LewishamPR (K); From Chippenhall (Sf), *Chebenhale* 12th.

**Chicheley:** Henry *Chichele* p1279 LuffCh; Robert *Chychely* 1332 SRSt; William *Chychely* 1392 CtH. From Chicheley (Bk).

**Chichester:** James *de Cicestrie* 1225 Cur (Sx); Walter *de Chichestre* 1301 CorLo; William *Chichestre* 1382 AssLa. From Chichester (Sussex).

**Chick:** Richard *Chike* 1198 P (Do); Richard *le Chike* 1317 AssK. ME *chike* 'chicken' (c1320 NED), used as a term of endearment. *Chike* occasionally becomes *Cheke,* which would become CHEEK: William *Chike, Cheke* 1278 PN K 269. Cheek's Fm in Bentley (Hants), *Cheakes* c1550 *Req,* owes its name to William *Chike* (1333 *SR*).

**Chicken:** Amindus, Roger *Chiken* 1210, 1212 Cur (Sf, Berks). OE *cicen* 'chicken'. cf. Alexander *Chikehed* 1301 SRY, John *Chykenmouth* 1327 Pinchbeck (Sf).

**Chickerel:** Nicholas *Chikerel* 1230 P (So); John *Chikerel* 1332 SRDo. From Chickerell (Do).

**Chidd, Chide:** Alfgar *Cida* 1086–1100 OEByn; William *Chide* 1166 P (Nf); Norman *Chyde* 1459 IpmNt. OE *Cidda.*

**Chiese:** *v.* CHEESE

**Chignall, Chignell:** Richard *de Chigenhale* 1311 LLB D; Robert *Chicknell* 1662 *HTEss.* From Chignall (Ess).

**Chilcot, Chilcott, Chillcot, Chillcott, Chilcock:** Baldwin *de Chillecota* 1169 P (Gl); John *Chilcott* 1641 PrSo; Robert *Chilcot, Chilcott* 1642 PrD. From Chilcote (Lei, Nth), or Chilcott (So).

**Child, Childe, Childs, Chiles, Chill, Chilles, Cheeld:** (i)Ægelmerus *Cyld,* Æluricus *Cyld,* quod intelligitur *puer* c975 LibEI (Herts); Aluric *Cild, Cilt* 1066 DB (Ess, C); Rodbertus *Puer* 1086 DB (Do); Gode *Cild* c1095 Bury (Sf); luinus *child* 1148 Winton (Ha); Willelmus *Infans* 1159 P (Ess); Roger *le Child* 1204 Cur (Berks). OE *cild* 'child'. In the earliest examples it probably denotes one comparable in status to the drengs of the northern Danelaw, the sergeants of Norman times. Ekwall has shown that Robert *Child* (1202 ELPN) may have been called by the pet-name of *Child* because he was the youngest child or a minor at the time of his parents' death. cf. *puer* and *Infans* supra. In the 13th and 14th centuries *child* appears to have been applied to a young noble awaiting knighthood (MED). It may also mean 'childish, immature' (c1200 MED), 'a page attendant' (1382 ib.). (ii) Peter *de la Child* 1262 ArchC iii; Richard *Attechilde* 1267 FFK. From residence near a spring, OE *celde.*

**Childerhouse, Childers:** Hemericus *de Childerhus* 1230 Cl (Nf); William *atte Childerhous* 1275 RH (Nf); Philip *del Childirhus* 1295 AssCh; John *Chyldirhous, Childurous* 1415, 1450 AD iv, v (Sf); Thomas *Childers* 1675 Shef. From an unrecorded OE \**cildra-hūs* 'children's house, orphanage'. cf. *childermas* and CHILDREN.

**Children:** John *atte Children* 1267 Pat (K); Daniel *Chyltren, de Chiltren* 1298, 1300 LoCt; Peter *ate Children* 1317 AssK; Thomas *Children* 1477 RochW; Robert *Achildren* 1560 ib. Identical in formation and meaning with CHILDERHOUSE, from an unrecorded OE \**cildra-ærn* 'children's house, orphanage'.

**Chiles, Chill:** v. CHILD

**Chilman, Chilman:** (i) *Chilmannus* Lenner, *Chilleman* Dilly 1327 SRC; Nicholas *Childman* 1239 FFC; William *Childeman* 1253 AssSt; William *le Childesman* 1276 AssSo; Walter *Chileman* 1311 ColchCt. *Childesman* is 'the servant or attendant of the young noble'. cf. CHILD. *Childman* may have the same meaning, or it may be from an unrecorded OE personal name \**Cildmann*, one of the late OE personal names compounded with *-mann*. Nicholas *Childman* was the son of *Childman* 1279 RH (C). (ii) William, Henry *Chilemound(e)* 1327 SRSo; John *Chylemonde* ib.; Agnes *Chilmon* 1580 Bardsley. This must be OE *Cēolmund*, 'ship-guardian', common in the 8th and 9th centuries and recorded once later, c1050, in Herts. It must have continued in use after the Conquest, at least in Somerset. For *-man* from *-mund*, cf. OSMAN.

**Chiltern:** Robert *de Cilterne* 1296 MPleas (Mx); John *Chilterne* 1360, Richard *Chylterne* 1397 FFEss. From the Chiltern Hills (Bk, O).

**Chilton:** William *de Chilton* 1195 P (Nb). A common surname from one of the many places named Chilton. Editha *de Childton', de Childhamton'* 1297 MinAcctCo (W) came from Chilhampton (Wilts), a development of the name not hitherto noted.

**Chilver, Chilvers:** Robert *Chilver* 1674 HTSf. From Chilvers Coton (Wa).

**Chimney:** William *de la Chymene* 1275–6 IpmY; John *de la Chemene* 1332 Kris; William *Chimnay* 1453, *Chymney* 1457 FrY. 'Dweller by or worker at the furnace', OFr *cheminee*.

**Chin, Chinn:** Stephen *Chinne* 1243 AssSo; John *Chynne* 1276 RH (Hu). OE *cin* 'chin'. The nickname may be for one with a prominent or long chin or for one with a beard. cf. 'Swor bi his chinne þat he wuste Merlin' (c1205 NED); 'Forked fair þe chine he bare' (a1300 ib.).

**Chine:** Richard *Chyne* 1275 SRWo; Henry *de Chine* 1279 RH (C); Ryner *Attechine* 1298 LLB B. OE *cinu* 'fissure, cleft, chasm'; in Hants 'a deep narrow ravine'. 'Dweller by the ravine.'

**Chinley:** Walter *de Chynleg'* 1286 AssCh. From Chinley (Db).

**Chinneck, Chinnick, Chinnock:** William *de Cynnoc* 1201 P (So); Roger *de Chinnock* 1275 AssW; Adam *Chynnok* 1327 SRSo; Peter *Chinnike* 1642 PrD. From Chinnock (So).

**Chinnery:** v. CHENERY

**Chinnick, Chinnock:** v. CHINNECK

**Chipchase:** William *de Chipchesse* 1379 FrY;

William *Chipchese* 1568 SRSf; John *Chipchase* 1686 RothwellPR. From Chipchase (Nb).

**Chipp, Chipps, Chips:** Isabella *Chippes* 1275 SRWo; John *Chip* 1327 SRSo; William *Chippe* 1606 PN Do i 343. ME *chip, chippe* 'a small piece of wood chipped or cut off', a nickname for a carpenter or a wood-cutter. Occasionally, perhaps, from ME *Chepe* 'Cheapside': Alan *de Chepe* 1311 LLB D; William *Chepe* 1369 Shef.

**Chippendale, Chippindale, Chippindall:** Dyke *de Chipendale* 1246 AssLa; Richard *Chipyndale* 1379 PTY; John *Chippendale* 1434 FrY. From Chippingdale (Lancs).

**Chipper:** v. CHAPPER

**Chipperfield:** Katherine *Cheperfeld* 1516, Thomas *Cheperfeld* 1534 FFEss; Robert *Chipperfeild* 1662 HTEss. From Chipperfield (Herts).

**Chipping:** William *Chipping* 1178 P (Wa); Adam *atte Chepingg* 1327 SR (Ess). 'Dweller near, or trader in the market', OE *cēping, cī(e)ping*.

**Chipps, Chips:** v. CHIPP

**Chisenhale:** v. CHISNALL

**Chisholm, Chisholme, Chisham:** John *de Chesehelme* 1254 Black; John *de Cheseholm* 1313 FeuDu. From the barony of Chisholm in Roberton (Roxburgh).

**Chislett, Chizlett:** (i) Anciet *Chiselet* 1576 SRW. From Chislet (K). (ii) Roger *Chislock* 1271 AssSo. 'Choice, excellent hair', OE *cīs, locc*.

**Chisman, Chismon:** v. CHEESEMAN

**Chisnall, Chisenhale:** Roger *de Chysenhale* 1285 AssLa; Thurstan *de Chisenhale* 1362 LLB G. From Chisnall Hall (La).

**Chiswick, Chissik, Cheswick, Chessex:** Henry *Chesewic* 1170–87 AD iv (Lo); Peter *de Cheswyk* 1275 RH (Ess). From Chiswick (Essex, Middlesex) or Cheswick (Northumb), all 'cheese-farm'.

**Chittenden, Chittendon:** Thomas *de Chetyndone* 1331 LLB E; Roger *Chittinden* 1525 SRSx; James *Chetenden* 1560 StaplehurstPR (K). From Chittenden (K), Chidden (Ha), *Chitteden* 1241, or Cheddington (Bk).

**Chitter:** Lemmer *Citere* 1095 Bury; Walter *Chittere* 1176 P (Y); Thomas *Chitters* 1674 HTSf. 'A player on the cithar', OE *citere*. v. OEByn 374.

**Chitterling:** Alexander *Chiterling'* 1221 AssWo; Simon *Chyterling* 1275 SRWo. ME *cheterling, chiterling* 'the smaller intestines of beasts, especially as an article of food, either fried or boiled'. Metonymic for a maker or seller of this.

**Chittleburg, Chittleburgh:** (i) Matthew *de Chytelesber'* 1275 RH (D). From Chittleburn (Devon), where *-burn* is, as not uncommonly, for earlier *-burgh*. (ii) Matthew *Chettleborowe* 1653 EA (NS) ii (Sf); Thomas *Chittleborough*, Henry *Chetleburgh*, Robert *Chickleborowe* 1674 HTSf. From Kettleburgh (Suffolk), *Chetelbiria, Kettleberga* DB, 'Ketil's hill' or possibly a Scandinavianized form of an OE \**cetel-beorg* 'hill by a narrow valley' (DEPN). The persistence of the *Ch-* in the local pronunciation suggests that the latter is the correct etymology.

**Chittock:** Roger, Henry *Chittok* 1279 RH (Hu), 1327 SRSf. *Chitt-ok*, a diminutive of ME *chitte* 'young of a beast, cub, kitten' (a1382 MED).

**Chitty:** Richard *Chiddy* 1327 SRSf; Thomas *Chittye* 1583 Musters (Sr). From Chitty in Chirlet (K).

**Chive, Chives:** Albert *Chiue* 1185 P (C); Geoffrey *Chiue* 1211 Cur (Hu); William Chive 1227 Cl. ONFr *chive*, OFr *cive* 'the smallest cultivated species of Alluin, the leaves of which were used in soups, stews, &c.' Metonymic for a grower or seller of this.

**Chiverrell:** *v.* CHEVERALL

**Chivers:** *v.* CHEEVER

**Chiverton:** *v.* CHEVERTON

**Chivery:** *v.* CHEVERY

**Chives:** *v.* CHIVE

**Chizlett:** *v.* CHISLETT

**Cholle, Cholles:** Absolon *Chioll'*, *Cholle* 1199 Pleas (C); Ketel *Cholle* 1245 FFL; John *Cholles* 1576 SRW. A nickname from OE *ceole* 'throat, dewlap'.

**Cholmondeley, Cholmeley, Chumley, Chumbley, Chambley:** Hugh *de Chelmundeleg'* 1288 AssCh; Thomas *Cholmeley* 1567 Bardsley; William *Chombley* 1666 ib.; Susanna *Chumbly* 1689 ib.; John *Chumley* 1726 ib. From Cholmondeley (Ches). Roger *Chomley* (1493 GildY) was a son of Richard *Chamley* (1502 ib.) or Sir Richard *Cholmley* (Ed.).

**Cholwich:** William *de Colwiche* 1328, Isobel *Choldeswych* 1411, Walter *Cholleweche* 1528 Hoskins. From a lost *Cholwich*, possibly surviving as Cholwich Town in Plymouth (D).

**Chone:** *v.* CHOWN

**Chopin, Choppen, Choppin, Chopping:** Walter *Chopin* 1219 Cur (D); Henry *Choppin* 1280 AssSo. OFr *chopine*, an old measure (1275 NED). '*Chopine* a chopine; or the Parisien halfe pint; almost as big as our whole one' (Cotgrave). cf. Fr *chopiner* 'to tipple'. A nickname for a tippler.

**Chorley:** Walter *de Cherlelaie* 1201 P (Bk); Elias *de Chorlegh* 1350 Putnam (La); Robert *Chorley* 1642 PrD. From Chorley (Bk, Ch, La, St), or Chorley Wood (Herts).

**Chorlton:** Muriel *de Chorlton* 1327 SRWo; William *de Chorleton* 1380 IpmGl; Thomas *Chorleton* 1419 IpmNt. From Chorlton (Ch, La), or Chapel Chorlton (St).

**Chose:** Thomas *Chose* 1327 SRSo; Hamund *Chose* 1361 Husting; Hamo *Chose* 1365 LLB G. OFr *chois* 'noble, handsome'.

**Chow:** Thomas *Chow* 1297 SRY; Peter *Chow* 1361 AssY; Robert *Chow* 1376 IpmW. A nickname from the chough, OE *cēo*, or OFr *choue*.

**Chown, Chowne, Chowen, Chone:** *Chuna* (f) 1248 AssBerks; *Chun* Pimme, *Chunne* Mervyn 1279 RH (C); Norman *Chone* 1257–8 FFSx; Thomas *Choune* 1327 SREss; John *Chowne* 1524 SRD; Nicholas *Chowne* alias *Chone* 1559 Pat (Lo). Evidently from a ME personal name, *Chun*, *Chunn*, both masculine and feminine, and not otherwise known. Sometimes, perhaps, the Cornish form of *John*.

**Crichton:** *v.* CRICHTON

**Chrimes:** *v.* CRIMES

**Chrippes, Chrisp:** *v.* CRISP

**Chrispin:** *v.* CRISPIN

**Christ, Crist:** John *Crist* 1308 SRBeds; Thomas *Crist* 1327 SRSt; John *Crist'* 1379 PTY. OE *Crīst* 'Christ', a pageant name.

**Christal:** *v.* CHRYSTAL

**Christendom:** Thomas *Kyrystendome* 1379 PTY; Robert *Cristendom* 1429 AssLo; Andrew *Cristendome* 1559 Pat (Beds). OE *crīstendōm* 'Christendom', but its meaning as a surname is unknown.

**Christian:** *Christiana* 1154 Bury (Sf), 12th DC (L); *Cristianus* 1201 Cur (Berks); Thomas *filius Cristian* 1228 FFEss; Robert *Crestien* 1163–9 Miller (C); Philip *Cristian* temp. John HPD (Ess). *Cristian*, Lat *Christianus*, was common in Britanny. In England the masculine name was less frequent than the feminine, which was also common as *Cristina*, the native form. *v.* CHRISTIN.

**Christie, Christey, Christy:** Thomas *Crysty* 1412 FFY; John *Chrysty* 1457 Black (Aberdeen). A Scottish and Northern English pet-form of *Christian* or *Christine*. cf. *Cristine* or *Cristy* de Carvan(t) 1296 Black.

**Christin, Christine:** *Cristina* 1219 AssY; *Cristina* de Burlingeham 1221 AssGl; Peter *Cristyn* 1296 SRSx; Thomas *Crystyn* 1332 SRSt. *Cristin* is the English form (from OE *crīsten* 'christian') of *Cristiana*. *v.* CHRISTIAN.

**Christison:** John *Cristenesone* 1312 FFSf; Robert *Cristianson* 1324 Wak (Y); Henry *Cristeson* 1412 Black (Stirling); Alexander *Cristisone* 1446 ib.; John *Cristerson* 1514 FrY. 'Son of *Christian* or *Christine*'.

**Christman:** William *Cristeman'* 1202 Cur (Ha); Walter *Cristesmon* 1275 SRWo. 'Servant of *Christ*', probably a pet-form of *Christian*, *Christine* or *Christopher*. cf. John *Crist* 1309 SRBeds.

**Christmas, Chrismas:** Ralph *Cristemesse* 1185 RotDom (Ess); Roger *Cristemasse* 1191 P (Sf); Richard *Cristesmesse* 1308 LLB C. A frequent surname for one born at Christmas. cf. Matilda *Candelmes* 1379 PTY.

**Christopher, Christophers:** *Christoforus* 1209 P (Hu); *Cristoferus* Murdac 1221 AssWa; Roger *Cristofore* 1379 PTY; Laurence *Cristofore* 1396 AssWa. Gk Χριστοφόρος, Lat *Christopherus* 'Christ-bearing'. The christian name does not appear to have been common and examples of the surname are late. The earliest noted denotes residence: Thomas *Cristofre* (1319 SRLo), son of William *de Sancto cristoforo* (1292 ib.), also called William *Cristofre* (1317 Husting), who left his son a tenement in St Christopher (St Christopher le Stocks, London).

**Christopherson:** John *Cristovirson* 1514 GildY. 'Son of *Christopher*.'

**Christred:** *Cristeredus*, *Cristiredus*, *Cristredus*, *Cristred* 1189 Sol; Robert *filius Cristraed* 1195 Cur (So); Robert *Cristred* 1207 P (Gl); John *Cristred* 1332 SRSx. A late OE personal name, *Crīstrǣd*, not otherwise known.

**Christy:** *v.* CHRISTIE

**Chrystal, Chrystall, Christal, Crystal, Crystall, Crystol, Kristall:** *Cristall* Knowis 1549 Black; *Christall* Murray 1561 ib.; William *Cristole* 1470 ib.; John *Cristall* 1491 ib. A Scots diminutive of *Christopher*.

**Chubb:** Richard, Gilbert *Chubbe* 1180 P (D), 1202 AssL; William *Chubbe*, *Chuppe* 1230 P (Lo). ME *chubbe*, a fish, 'chub' (c1450 MED), was also used of a 'lazy spiritless fellow; a rustic, simpleton; dolt, fool' (1558), whilst Bailey has '*Chub*, a Jolt-head, a great-headed, full-cheeked Fellow', a description reminiscent of that of the chevin, another name for the chub. *v*. CHEVINS. Thus the nickname may have meant either 'short and thick, dumpy like a chub' or 'of the nature of a chub, dull and clownish'.

**Chuck:** *v*. SUCH

**Chudley, Chudleigh:** William *Chuddeleghe* 1359 AssD; James *Chuddelegh* 1379 Hylle; John *Chudley* 1642 PrD. From Chudleigh (D).

**Chue:** *v*. CHEW

**Chum(b)ley:** *v*. CHOLMONDELEY

**Church:** Thomas *Attechirche* 1275 SRWo; Henry *atte Churche* 1296 SRSx; Henry *of the Chirche* 1368 FrY. Usually 'dweller near the church', but *of the Chirche* suggests an official, verger, sexton, etc., rather than residence near the church. John *Atte-cherch*, rector of Metton, Norfolk, in 1338 (Bardsley), can hardly have owed his surname to his residence. As rector, his attribute would have been *Parson*. *Attecherch* is here probably a hereditary surname.

**Churchard, Churchyard:** John *atte Chircheyerde* 1298 AssSt; Henry *del Churcheyard* 1332 SRWa. This can hardly mean 'dweller by the churchyard'. The natural expression would be 'at the church'. It probably denotes one responsible for the upkeep of the churchyard. Richard *de la Chirchard* (1291 MELS) is identical with Richard *atte Church* (1289 ib.), both surnames being occupational. Similarly, Reginald *le Churchedoor* (1300 Bardsley) was the church door-keeper.

**Churchers, Churches, Churchouse, Churchus:** Iuo *de Cherchous* 1327 SRSf; William *del Chyrchehous* 1332 SRSt. The church-house was formerly a house adjoining the church where church-ales, etc., were held, a parish-room. Again probably occupational, 'care-taker of the parish-room', though he may also have lived there.

**Churchey:** Walter *atte Chircheheye* 1275 SRWo; Peter *atte Churcheheye* 1327 SRSo; Agnes *atte Churcheye* 1327 PN W 155. 'Dweller by the church enclosure, i.e. the churchyard', OE *cyrice*, (*ge*)*hæg*.

**Churchfield:** Ismena *de Chirchefeld'* 1199 Pleas (Nth); Henry *de Chirchefeld'* 1253 ForNth. From Churchfield Copse in Bosham (Sx), Churchfield Fm in Benefield (Nth), or 'dweller by the church field', OE *cyrice*, *feld*.

**Churchgate:** William *Attechurchegate* 1269 CorBeds; Thomas *atte Chirchegate* 1326, John *ate Cherchegate* 1332 MELS (Sr). 'Dweller by the church gate', OE *cyrice*, *geat*. cf. William *atte Churchstighele* 1306 AssW 'dweller by the church stile'.

**Churchill:** John *de Chirchehul* 1221 AssWa; Matilda *de Chirchull'* 1275 SRWo. From Churchill (Devon, Som, Worcs, Warwicks). The name may also denote residence on the church-hill. William *atte Churchull'* 1333 MELS (So) lived near the church-stile and was also called William *Churchestyele* (1327 ib.).

**Churchley:** Ailwin *de chercheslega* 1189 Sol; Wylke *de Chyrchele* 1246 AssLa. 'Dweller by the clearing with a church', OE *cyrice*, *lēah*.

**Churm, Churms:** Ælfric *Cerm* c1055 KCD 937

(So); Geoffrey *Cherm* 1211 Cur (So); William *Chyrme* 1538, Benjamin *Churme* 1658 SaAS 3/viii, 2/iv. OE *cirm*, *cierm* 'noise, uproar', a nickname for a noisy, boisterous person.

**Churchman:** Ælflled *Cerceman* c1095 Bury (Sf); Ralph *Chircheman* 1259 RamsCt (Hu); Ouse *le Chercheman* 1279 RH (C); John *Churcheman* 1307 RamsCt (Hu). OE *cyriceman* 'custodian or keeper of a church'. The meanings 'ecclesiastic, clergyman' (a1400 MED) and 'churchwarden' (1523) are much later.

**Churton:** William *de Chirtona* 1180–90 StP; William *Churton* 1642 PrD. From Chirton (W), or Churton (Ch).

**Churchward:** Oscetel *Circwærd* 949 BCS 882 (Gl); Ælfnoð *Cyrceweard* 11th OEByn; William *le Chirchewart* 1275 SRWo. OE *ciricweard* 'custodian of a church (building)'.

**Churchyard:** *v*. CHURCHARD

**Cinnamond, Sinnamon:** Edmund *de Sancto Amando* 1213 Lewes (Nf); Almaric *de Sancto Amando* 1235 Fees (D); Aimery *de Sancto Amando* 1280 AssSo. From Saint-Amand (Cotentin).

**Ciprian:** Ciprianus *monachus* 1121–35 P (Sf); Ciprianus 1219 AssY; Nigel *Ciprian* 1327 SREss; Henry *Ciprane* 1338 KB (C). Lat *Ciprianus*.

**Circuitt, Cirket, Serkitt, Sirkett, Surkett:** John *de Suthcote* 1297 MinAcctCo (Herts). The rare Bedfordshire *Cirket*, formerly *Surcot*, *Surcoate*, etc., derives from Southcott in Linslade (Bk), *Surcote* 1826, now pronounced *Cirket*, or from a lost Southcott in Stone (*Sircotes* 1511 PN Bk 80, 165). The Berkshire Southcot occurs as *Circuit* c1728 (NQ 196). The surname may also be a nickname from the surcoat: John *Surcote* 1327 SRSf.

**Clack:** *Clac* de Fugelburne c975 LibEl (C); Godwinus *Clec* 1086 DB (W); Godricius *Clacca* 1169 P (Berks); Simon *Clac* 1327 SRSo. OE *Clacc* or ODa *Klak* surviving as *Clak* in Lincolnshire temp. Hy 2 (Gilb), 1193 FF (L). The surname must sometimes be metonymic for *Clacker*: Walter *le Clackere* Ed 1 Malmesbury, Roger *Clackere* c1250 AD iv (W). This may be a nickname 'chatterer', from ME *clacken* (a1250 MED), or for a miller, from the clack or clatter of his mill, or for a bell-ringer, though these senses are not recorded until much later. cf. William *Clacyere* 'belleyetere' (c1425 Building 327), bell-founder.

**Clackett, Claget, Claggitt:** *v*. CLAYGATE

**Claddish, Cladish:** John *Cleydych* 1387 FFEss; John *Claydiche* 1499, Edward *Claydyche* 1494 RochW. 'Dweller by the clay ditch', OE *clǣg*, *dīc*.

**Clague:** *MacLiag* 1014, Gilla *Macliag* 1173, *MacClewage* 1511, *Cloage* 1601, *Claige* 1622, *Clague* 1655 Moore. Ir and Manx *MacLiaigh* 'son of the leech' (*liagh*).

**Clamp:** Roger *Clampe* 1298 IpmY; Robert *Claumpe* 1348 DbAS 36; John *Clamp* 1524 SRSf. ME *clamp* 'clamp, brace'.

**Clandon:** William *de Clandone* 1327 SRSo. From Clandon (Sr).

**Clanfield:** Richard *de Clanefeld* 1279 RH (O). From Clanfield (Hants). *v*. GLANDFIELD.

**Clapham:** Alexander *de Clapehamme* 1204 AssY;

William *de Clapham* 1331 FFY; Richard *Clapham* 1416–17 IpmNt. From Clapham (Beds, Surrey, Sussex, WRYorks).

**Clapp:** Simon *Clapp'*, *Cloppe* 1206 Cur (O); Adam, Godard, William *le Clop* 1219 AssY, 1222 Cur (C), 1227 AssBeds; Laurence *Clappe* 1230 P (O); John *Clap* 1327 *SR* (Ess). There is no evidence for the post-Conquest use of OE *Clappa* which has been suggested as the source of this name. *Le Clop* is a nickname, identical with OE *\*clop* 'lump, hillock, hill' which, Ekwall has shown, is found in the common place-names Clapham, Clapton and Clopton. In some of these, *Clop-* had become *Clap-* early in the 13th century. The surname means 'bulky, heavily-built'.

**Clapper:** John *atte Clapere* 1332 SRSx. 'Dweller by the clapper bridge', a rough or natural bridge across a stream. Matilda *de la Claper'* (1330 PN D 359) lived at Clapper Bridge. Ingram *Clapere* 1267 Pat may be an early example with loss of the preposition, or a nickname from ME *clappe* 'chatter' (c1230 MED), 'chatterer'.

**Clapton:** Turstan *de Cloptuna* 1154 Bury; Alan *de Clapeton* c1185–1210 YCh; Robert *de Cloptun* 1228 FFEss; Roger *de Clapton* 1298 AssL. From Clapton (Berks, Cambs, Glos, Middlesex, Northants, Som), or Clopton (Glos, Suffolk, Warwicks, Worcs), both names having OE *clop* 'rock, hill' as the first element.

**Clarabut:** *v.* CLARINGBOLD

**Clarage, Claridge:** Robert *de Claurugge* 1327 SRSx; John *Claridge* 1665 HTO. From Clearhedge Wood in Waldron (Sussex), *Clavregge* 1288, *Claregge* 1429.

**Clare:** (i) *Clara* 1210 Cur (Hu); Robertus *filius Clarae* 1240, 1255 Rams (Nf, Hu); Goditha, Richard *Clare* 1317 AssK, 1327 SRSo. Fr *Claire*, Lat *Clara*, 'bright, fair', a woman's name, common, probably, owing to the popularity of St Clare of Assisi. (ii) Richard *de Clara* 1086 DB (Sf). From Clare (Suffolk). (iii) Simon *le Clayere* 1279 RH (C); Richard *le Cleyere* 1305 Pinchbeck (Sf). A derivative of OE *clǣg* 'clay'. The clayer was engaged in plastering with mud in wattle and daub work, called *torching* in 1278, *plastering* in 1368 and *claying* in 1486 (Building 189).

**Clarel, Clarell:** *Clarell'* 1219 AssY; Ralph *Clarel* 1187 P (Y); William *Clarel* 1276 IpmY; John *Clarell* 1496–7 FFWa. OFr *Clarel*, perhaps a diminutive of *Claire*.

**Claremont, Clermont:** (i) *Cleremundus* 12th Gilb (L); *Claramunda*, *Cleremunda* (f) 1249 AssW; Hugh *Clermond* 1275 RH (O); Margeria *Clermond* 1297 MinAcctCo. OG *Clarmunt*, *Claremunda*, OFr *Claremonde*. (ii) Bernard *de Claromonte* 1148 Eynsham; Aelicie *de Clermunt* 1185 Templars (Beds); Hugh *de Clermund* 1279 RH (O). From Clermont (Calvados), or Clairmont (Auvergne).

**Clarence:** Richard *Clarence* 1453 LLB K. The name of a dukedom created in 1362 for Lionel, the third son of Edward III, who had married the heiress of Clare in Suffolk. Clarence did not come into use as a christian name until the end of the 19th century.

**Claret, Clarett:** John *Clarrot* 1279 RH (Hu); Magota *Claret* 1379 PTY. *Clar-et*, *Clar-ot*, diminutives of Lat *Clara*, OFr *Claire*.

**Clarges:** Richard (*le*) *Clergis* 1279 RH (Beds); Nicholas, Richard *Clergys* 1309 SRBeds. In ME *clergy* was used as a singular, with a plural *clergis* 'clergymen'. cf. A clerge, *clerus*, *clerimonia* CathAngl. The surname is probably a parallel to *Parsons* and *Vickers*, 'the clergyman's servant'.

**Claringbold, Claringbould, Claringbull, Clarabut:** *Clarumbald* medicus a1116 ELPN; *Clarebaldus* c1150 DC (L); *Clarenbaldus* pincerna 1177 P (Sf); *Clerebaud* del Aune 1200 Cur (Sf); Roger *Clerenbald* 1223 FFSf; Nicholas *Clarebold* 1274 AD i (Ha); William *Cleribaud* 1275 RH (K); John *Claringbold* 1485 KentW; Pascall *Clarebote* 1565 Bardsley; Elizabeth *Clarabutt*, *Claringbull* 1735–6 Bardsley. OFr *Clarembald*, *Clarebald*, *Clarembaut*, from OG *Clarembald*, a hybrid from Lat *clarus* 'famous' and OG *-bald* 'bold'.

**Claris:** *Claricia* 1150–60 DC (L); *Clericia* 1175–86 Holme (Nf); *Clarice* (nom.) 1191 P (Mx); Rogerus *Claritia* Hy 3 Colch (Sf); Walter *Clarice* 1275 SRWo; Richard *Clarisse* 1279 RH (O). *Clarice*, a well-established woman's name. It has been regarded as a derivative of *Clara*, but the formation is difficult. It may be from an abstract noun *claritia*, based on Lat *clarus* 'bright, shining', used as a personal name on the analogy of *Lettice*, from *laetitia* 'joy, gladness'.

**Clark, Clarke, Clerk, Clerke:** Richerius *clericus* 1086 DB (Ha); Willelm *ðe Clerec* c1100 OEByn (So); Reginald *Clerc* 1205 Cur (R); John *le Clerk* 1272 Gilb (L). OE *clerec*, *clerc* (Lat *clericus*), OFr *clerc* 'clerk'. The original sense was 'a man in a religious order, cleric, clergyman'. As all writing and secretarial work in the Middle Ages was done by the clergy, the term came to mean 'scholar, secretary, recorder or penman'. As a surname, it was particularly common for one who had taken only minor orders. *Clerk(e)* is now rare.

**Clarkin, Clerkin:** *Clarekin* le Lumbard 1287, *Clarekin* Felin 1287 LLB A; *Clarkin* de Wolcherchehagh 1290 AssNu. *Clare-kin*, a diminutive of *Clare*, used both as a masculine and a feminine name.

**Clarkson:** Alan *le Clerkissone* 1306 FFSf; William *Clerksone* 1332 SRCu; Ralph *Clarkson* 1491 GildY. 'Son of the clerk.' *v.* CLARSON.

**Clarson:** (i) Johannes *filius Clarae* 1240 Rams (Nf); John *Clareson* 1326 Wak (Y). 'Son of *Clara*'. (ii) William *Clariceson* 1327 SRSo; Thomas *Clerysson* 1467 Shef. 'Son of *Clarice*.' (iii) Thomas *Clerson* 1431 FrY; John *Clarson* 1553 RothwellPR (Y); Margaret *Clarsome* wife of Francis *Clarsom* 1588 ShotleyPR (Sf); Francis *Clarkson* husband of Margaret *Clarkson* 1608 ib. Identical with CLARKSON.

**Clarvis:** Michael *de Clervaus*, *de Clereuals* 1208 FFHu, 1209 P (Hu); John *Clervaus* 1320 FFHu; Margaret *Clarevas* 1446 GildY; Henry *Clerves* 1524 SRSf. From Clairvaux (Aube, Jura).

**Class:** Gilbert *Classe* 1297 MinAcctCo; Robert *Classe* 1576 SRW; Humphrey *Classe* 1642 PrD. A pet-form of *Nicholas*.

**Clater:** John *Clatere* 1327 SRWo. ME *clater* 'noisy talk, gabble', metonymic for *claterer* 'babbler, tattler' (c1390 MED).

**Claughton:** Gregory *de Clacthon* 1228 AssLa; William *de Claughton* 1297 IpmLa; John *de Claghton'* 1342 KB (Mx). From Claughton (Ch, La).

**Clavel, Clavell:** John *Clavel* 1218 P (Sx); Robert *Clauel* 1296, William *Clauel* 1327 SRSx. OFr *clavel* 'keystone of an arch, lintel over a fireplace, especially a beam of wood so used', but the particular sense in which this is used as a surname is not known.

**Claver:** Jordan *le Claver* 1211 Cur (Nf); Simon *le Claver* 1270 Ipm; Robert *le Clauyr* 1327 SRSx. OFr *clavier* 'door-keeper'.

**Clavering:** Cynric *æt Clæfring* c1050 Earle (Herts); Simon *de Claueringg'* 1230 P (Ess); Roger *Claveryng* 1375 AssLa. From Clavering (Ess).

**Clavinger:** Walter *clavigerus* 1195 FFSf; Herbert *Clavigerus* 1210 Cur (Do); William *Clauigerus* 1296 SRSx. The Lat form of OFr *clavier* 'keeper of the keys, mace-bearer'.

**Claxton:** Geoffrey *de Claxton'* 1219 AssL; Roger *de Clakeston* 1287–8, *de Claxton* 1288–9 NorwLt; Hamo *Claxton* 1379 FFEss. From Claxton (Du, Nf, NRY).

**Clay, Claye:** Ralph *de Clai* 1172 P (Sf); Reginald *de la Claie* 1200 P (Ess); William *Cley* 1221 ElyA (C); Richard *atte Cley* 1296 SRSx; Nicholas *del Clay* 1302 SRY. 'Dweller on the clay' (OE *clǣg*). As this would not be distinctive in a county like Essex, the surname may sometimes be occupational, for a worker in a claypit, *v.* CLAYMAN, CLARE.

**Claybrook, Claybrooke, Claybrooks:** Adam *de Cleybrok* 1231–2 FFWa; Richard *de Claybrokes* 1327 SRLei; Ralph *Claybroke* 1340–1450 GildC. From Claybrooke (Lei).

**Clayden, Claydon, Cledon:** William *de Cleydon'* 1275 SRWo; John *de Claydon* 1370 IpmW; John *Claydone* 1372 CorLo. From Claydon (Bk, O, Sf).

**Claygate, Clackett, Claget, Claggitt, Cleggett:** Roger *de la Claigate* 1198 FFK; Robert *de Cleygate* 1215 Wenlok; Gilbert *ate Claygate* 1317 AssK; Joane *Cleget* wife of William *Clegget* 1660 ArchC xxx. From Claygate (Surrey), or from some similarly named minor place, cf. Claygate Farm in Buxted (Sussex), Clacket in Tabsfield (Surrey), Clackett's Place in Ryarsh (Kent), etc.

**Clayman:** John, Thomas *Clayman* 1327 *SR* (Ess), 1365 AD ii (Mx). 'One who prepares clay for use in brickmaking' (NED). PromptParv equates *cleymann* with *dauber*. *v.* CLAY, CLARE.

**Claypole, Claypoole:** Geoffrey *de Cleipol* 1185 Templars (L); Symon *de Claypoll* 1255 Black; John *de Claipole* 1374 AssL. From Claypole (L).

**Clayton:** Jordan *de Claiton* a1191 YCh; Walter *de Clayton* 1332 SRSx; Richard *Clayton* 1452 FFEss. From Clayton (Lancs, Staffs, Sussex, WRYorks).

**Clayworth:** Ralph *de Claworth'* 1205 Cur (Nt); Osanna *de Clawurthe* 1219 AssL; Thomas *Claworth* 1437 IpmNt. From Clayworth (Nt), *Claworth* c1130.

**Clean:** (i) Hugh *Clene* 1195 P (Db/Nt); Richard *Clan'* 1327 SRLei; William *Clene* 1642 PrD. OE *clǣne* 'clean, pure, chaste'. cf. Cristina *Clenemayde* 1297 MinAcctCo 'pure maiden'. (ii) In Scotland for MacLEAN.

**Clear, Cleare, Clere:** (i) Gilbert *filius Cler'* 1279 RH (O); William, Geoffrey *Clere* 1279 RH (Hu), 1296

SRSx. *Clere* was the French popular form of *Clare. v.* CLARE. (ii) Ralph *de Clere* 1319 SRLo. From Clere (Hants).

**Cleasby:** Robert *de Clesebi* 1202 FFY; Hasculph *de Cleseby* 1300 ForSt; Richard *Cleseby* 1415–16 IpmY. From Cleasby (ERY).

**Cleater, Cleator:** Richard *Cletter* 1545 SRW; John *Cleator* 1642 PrD. It could be a derivative of either OE *clēat* 'wedge', or of OE *clāte* 'burdock', but is probably usually from Cleator (Cu): Richard *de Cletre* 1361 FFY.

**Cleatherow:** *v.* CLITHEROE

**Cleave:** *v.* CLIFF

**Cleaveland:** *v.* CLEVELAND

**Cleaver, Cleever:** Richard, John *le Cleuar* 1332 SRSx. A derivative of OE *clēofan* 'to cleave, split', one who split boards with wedges instead of sawing. These were called *clouenbord* 1345, *clofbord* temp. Edward III (Building 243). The surname may also mean 'dweller by the cliff'. *v.* CLIFF.

**Clee:** (i) William *de Clee* 1359 LLB G; John *Clee*, Richard *Cle* 1327 SRSf. From Clee (Lincs, Salop), from OE *clǣg*. The Suffolk name is for *atte cley.* (ii) John *atte Cleo* 1332 SRWo; William *atte Clee* 1349 MELS (Wo). 'Dweller by the river-fork or fork in a road', from OE (Anglian) *clēo* 'claw, cloven hoof'. *v.* MELS.

**Cledon:** *v.* CLAYDEN

**Cleef:** *v.* CLIFF

**Cleever:** *v.* CLEAVER

**Cleft:** *v.* CLIFF

**Clegg, Clegge:** (i) Robert *del Cleg* 1246 AssLa; Henry *de Cleg* 1309 Wak (Y). From Clegg (La). (ii) Matthew *Clege* 1285 AssLa; Richard *Clegge* 1525 SRSx. A nickname from ON *kleggi* 'gadfly, horse-fly'.

**Cleggett:** *v.* CLAYGATE

**Cleland:** (i) Nicholas *Cleland* 1642 PrD. Perhaps from Cleveland in Dawlish (D). (ii) William *Kneland* 1464, Andrew *Cleland* (seal: S. ANDREE KNELAND) 1612 Black. From the lands of Cleland or Kneland in Dalziel (Lanark).

**Clem, Clemm, Clemans, Clemence, Clemens, Clement, Clements, Clemons, Clemas, Climance, Climas, Clemmans, Clemmens:** *Clemens monachus* 1153–68 Holme (Nf); *Clementia* 1162 DC (L); *Clemencia* 1210 Cur (Herts); *Clemens* filius *Clementis* 1212 Cur (Ess); William, Richard *Clement* 1153 Templars (O), 1202 AssL; Robertus *Clemens* 1155 Templars (O); William *Clement*, *Climent* 1275 RH (Nf); Richard *Clemence* 1279 RH (Hu); Robert *Clymant* 1327 SRSx. Fr *Clement*, Lat *Clemens* 'mild, merciful', the name of several popes, was popular in England from the middle of the 12th century and the corresponding woman's name *Clemence*, Lat *Clementia* 'mildness' from about 1200. Both have contributed to the surnames. *Clem* is a pet-form found in 1273 (RH). *Clim* is not much later: John *Climme* 1327 SRC. *v.* also CLEMO, CLEMMEY.

**Clementson, Cleminson, Clemerson, Clemetson, Clempson, Clemson, Climenson, Climpson:** Peter *Clementson* 1379 PTY; John *Clementesson* 1392 FrY; Roger *Clempson* Eliz Bardsley; John *Clympson* 1577 ib. 'Son of *Clement*.'

**Clemmey, Climie:** *Clemie* Croser 1581 Bardsley (Cu); George *Clemy* 1553 Black (Glasgow); William *Climay* 1652 ib. A Scottish pet-name for *Clement*.

**Clemo, Clemow, Clemmow, Climo, Clyma, Clymo:** John *Clemmowe* 1544 Bardsley; Morice *Clymow* 1630 DWills (Co); Ots *Clemow* 1634 ib. A Cornish form of *Clement*.

**Clench, Clinch:** Hugh *Clinche* 1223 Cur (L); Richard, Robert *Clench(e)* 1275 RH (Sf), 1327 SRSf; John *de la Clenche* 1275 RH (W); Isabell *atte Clenche* 1332 SRSx. 'Dweller by the hill' as at Clinch Green (Sussex) or on elevated (dry) land in a fen as at Clenchwarton (Norfolk), from OE *\*clenc* 'lump, mass', found in several minor place-names. *v.* MELS. *v.* also CLINK.

**Clendenin, Clendennen:** *v.* GLENDENING

**Clendon:** Richard *de Clendon'* 1207 SPleas; Alexander *de Clendon* 1219–20 FFEss. From Glendon (Nth), *Clendon* DB.

**Clenfield:** Richard *de Clenefeld* 1275 SRWo; Laurence *de Clenfeld* 1327 SRLei. From Clanfield (O), *Clenefeld* 1195.

**Clennan:** God' *Clenehand* 1066 Winton (Ha); Auice *Clenhond* 1319 SRLo; John *Clenaunt, Clenhond* 1387 LoPleas; Thomas *Clenehonde* 1418 LLB H. 'Clean hand', OE *clǣne, hand.*

**Clent:** Richard *de Clent* 1273 RH (Wo); Petronilla *de Clent* 1275 SRWo. From Clent (Wo).

**Clere:** *v.* CLEAR

**Clerkin:** *v.* CLARKIN

**Clermont:** *v.* CLAREMONT

**Cletch:** Osbert *Clech'* 1221 AssSa; Peter *Clech'* 1221 AssWo. From OE *\*clǣcan* 'to clutch, grasp'. Perhaps a nickname for an avaricious man.

**Cleugh:** *v.* CLOUGH

**Cleve:** *v.* CLIFF

**Cleveland, Cleaveland:** Peter *de Cliuelanda* 1160–80 YCh; John *de Cliueland* 1327 SRY; William *Cleveland* 1530 FFEss; Richard *Cleeveland* 1642 PrD. From Cleveland (NRY), Cleveland in Dawlish (D), or Cleaveland's Fm in Colchester, Cleveland's Fm in Felsted (Ess).

**Clew, Clew(e)s:** *v.* CLOUGH

**Clewer, Cluer:** Richard *de Clifwara* 1156 P (Berks); John *Cluer* 1307 KB (L); Richard *Cluer* 1674 HTSf. From Clewer (Berks, So). *Cluer* could also be a derivative of OFr *clou* 'nail', a maker of nails.

**Clibburn, Cliburn:** Robert *de Cleburne* 1364, Oliver *Clibburn* 1475 FrY. From Cliburn (We). There was also a personal name: *Clibern* Biscop c1150–69 MCh; *Clibernus* 1202 FFY.

**Cliff, Cliffe, Clive, Cleave, Cleaves, Cleef, Cleeve, Cleeves, Cleve, Clift, Cleft:** Gislebertus *de Cliua* 1084 GeldR (W); Alecok *del Clif* 1274 Wak (Y); John *Clif* 1279 RH (O); Adam *del Clef* 1290 AssCh; John *del Clyfes* 1315 Wak (Y); William *Undertheclif* 1327 SRDb; Walter *atte Cliue* 1327 SRSx; John *ate Clif* 1327 SRC; John *de Cleue* 1327 SRSf; Richard *Clyft* 1524 SRSf; Mary *Cleft, Clift* 1755, 1757 DenhamPR (Sf). OE *clif* 'cliff, rock, steep descent' is found in numerous place-names as Cliff(e), Cleeve, Cleve and Clive, any of which may have given rise to a surname. Its most common meaning seems to be 'slope' (not

necessarily a steep one) or 'bank of a river' and many of the surnames are due to residence near such a slope or bank. John *atte Clyve* or *atte Cleve* 1361 ColchCt is called *Clever* in 1365 (ib.). His descendants may have been called *Clive, Cleeve* or *Cleaver* and his name proves that CLEAVER may mean 'dweller by the slope'.

**Clifford, Clifforth:** Fulk, Roger *de Clifford* 1182 P (Wa, Lei), 1269 IpmW; John *Clifford* 1387–8 FFSr. From Clifford (Devon, Glos, Hereford, WRYorks), or 'dweller at the ford by the steep bank'.

**Clifton, Cliffton:** William *de Cliftona* c1145–65 Seals (Nb); Ignatius *de Clifton'* 1249 AssW; James *Clifton* 1375 IpmGl. From Clifton (Beds, Ch, Cu, Db, Gl, La, Nb, Nt, O, We, NRY, WRY), Clifton Reynes (Bk), Clifton Maybank (Do), Clifton Hampden (O), Clifton Campville (St), Clifton on Dunsmore (Wa), or Clifton on Teme (Wo).

**Climance, Climas:** *v.* CLEM

**Climenson:** *v.* CLEMENTSON

**Climie:** *v.* CLEMMEY

**Climo:** *v.* CLEMO

**Climpson:** Ælfled *of Clymestune* c970 OEByn (D); Arthenbaldus *de Climeston'* 1207 P (Co). From Climson (Cornwall). *v.* also CLEMENTSON.

**Clinch:** *v.* CLENCH

**Clindening:** *v.* GLENDENING

**Cline:** *v.* CLYNES

**Clink, Clinker:** Martin *Clink*, Roger *Clynk* 1327 SRSf. In the 14th century, *clinch* and *clench* were used of door-nails secured by clinching or riveting. In 1323 Richard Spark, *clenchar'*, was paid 4½d. per day for clinching and riveting great nails. cf. also *clencher, clenchours* 1363, *clencheres* 1375 (Building 309). The ME verb *clenchen*, from OE *clenc(e)an* is found from 1250 and the corresponding northern form *clink* from 1440. *Clink*, and also CLENCH and CLINCH, are metonymic for 'clincher, riveter'.

**Clint:** Thomas *de Clint* 1206 P (Y); John *de Clynt* 1332 SRSt; Richard *Clynt* 1406 FrY. From Clint in Ripley (WRY).

**Clinton:** William *de Clintona* c1125 StCh; Henry *de Clinton* 1202 AssNth; William *Clynton* 1428 FFEss. From Glinton (Nth), *Clinton* 1060.

**Clipsby:** Richard *de Clipesbi* 1196 Cur (Nf); Maud *de Clipseby* 1256 FFL; William *Clippisby* 1452 Paston. From Clippesby (Nf). Used as a christian name in the 16th century: *Clippesby* Gawdy 1590 SfPR.

**Clipson, Clipston, Clipstone:** Hervey *de Clipston* 1199 AssSt; William *de Clipston'* 1327 SRLei; William *Clipston* 1392 FFEss; Christopher *Clepson* 1641 PrSo. From Clipston (Nt, Nth), or Clipstone (Beds, Nf, Nt).

**Clisby, Clixby, Clizbe:** William *de Clissebi* 1202 AssL. From Clixby (L).

**Clist:** Hereberd *on Clist* a1093 Earle (D); Henry *de Clist* 1230 P (D); Edward *Clist* 1642 PrD. From Clyst (D).

**Clitheroe, Clitherow, Cleatherow, Cluderay:** Thomas *de Cliderhou* 1176 P (Y); Robert *Cletherowe* 1439 FrY; Richard *Cludre, Clydero* 1526, 1541 GildY. From Clitheroe (Lancs).

**Clive:** *v.* CLIFF

**Clivedon, Clivedon:** Reymund *de Clivedon'* 1272 Glast (So); John *de Clyvedon* 1329 FFW; William *Clyvedon* 1382 IpmGl. From Clivedon (Bk).

101

**Clixby, Clizbe:** v. CLISBY

**Cloak, Cloake, Cloke, Cloche:** Robert *Cloche* 1210–11 PWi; Nigel *le Cloc* 1327 SRSx. OFr *cloche*, *cloke* 'cloak'. Metonymic for a maker or seller of these.

**Clodd:** Achì, Adam *Clod* 1166 P (Nf), 1275 RH (Sf). ME *clodde* 'clod of earth', vb., 'to free land from clods by harrowing, rolling, etc.' (c1440 MED). 'Harrower.' cf. William *le Clodder* 1221 AssWo.

**Clodhammer:** Adam *Clodhamer* c1260 *ERO*; John *Cledhamer* 1433 W'stowW. A nickname from the fieldfare, OE *clodhamer*.

**Clogg:** Symond *Clogg*, Joan *Clog* 1524 SRSf. ME *clog(ge)* 'a wooden-soled shoe' (1390 MED), used for a maker of clogs: cf. Matthew *Clogmaker* 1367 ColchCt.

**Cloke:** v. CLOAK

**Close, Closs:** (i) Nicholas *de Clos* 1296 SRSx; Thomas *del Close* 1327 SRY. ME *clos(e)*. OFr *clos* 'enclosure' (a1325 MED), 'farm-yard' (1386). 'Dweller by the enclosed place' or. possibly, 'worker in the farm-yard'. (ii) William *le Clos* 1214 Cur (C); John *Cloos* 1409 LoPleas. ME *clos* (adj.) 'practising secrecy, reserved, reticent' (c1400 NED).

**Clot, Clott:** Nicholas *Clot* 1251 FFY; John *Clot* 1274–5 FFWa; William *Clot* 1301 FS. A nickname from ME *clot*, *clott* 'cold, lump of earth'.

**Clothier:** Robert *le Clother* 1286 MESO (Nf). A derivative of OE *clāþ* 'cloth', maker or seller of cloth. cf. Richard *le Clothmongere* 1296 Oseney (O), Thomas *Clothman* 1416 LLB I.

**Clotworthy:** John *Cloteworthy* 1327 SRSo. From Clatworthy in South Molton (D).

**Cloude:** Wimarc *de la Clude* 1199 P (So); Robert *atte Cloude* 1327 SRSo. 'Dweller by the rock or hill' (OE *clūd* 'mass of rock, hill'), as at Temple Cloud (Som), Cloud Bridge (Warwicks), Clouds Wood (Herts).

**Cloudesley, Cloudsley:** Anthony *Clowdsley* 1573 YDeeds I; Vxor *Cloudesly* 1597 SRY. From Cloudesley Fm in Withybrook (Wa).

**Clough, Cleugh, Cluff, Clow, Clew, Clewes, Clews, Clue, Clues:** Alan *Bouenthecloue* 1261 AssLa; Richard *Clowe* 1275 SRWo; Roger *Clough* 1279 RH (O); John *del Clogh* 1298 Wak (Y); Richard *Clewe* 1327 SRSf; Robert *del Clough* 1327 SRDb; Richard *Cluff* 1428 FA (St); Esabell *Clughe* 1555 RothwellPR (Y). 'Dweller in a ravine or steep-sided valley', OE *\*clōh*. *Cleugh* is a Scottish form. For the development, cf. *enough* and *enow*, *dough* and *(plum)duff*.

**Cloughton:** Abraham *Cloughton* 1672 HTY. From Cloughton (NRY).

**Clout, Cloutt:** Enial *Clut* 1175 P (He); Walter *Clut* 1207 P (Gl). OE *clūt* 'patch'. v. CLOUTER.

**Clouter:** Adam *le Clutere* 1286 MESO (Nf); Robert (*le*) *Clutere* 1301 LLB C; Adam *Clouter* 1307 Wak (Y). A derivative of OE *clūt* 'patch', patcher, cobbler. cf. Robert *le Cloutkeruer* 1327 Pinchbeck (Sf) 'patch-cutter'. Also, possibly, but less likely, from OFr *cloutier* 'nail-smith'.

**Clouting, Cloughting:** William *Cloutyng* 1327 SRSf; William *Clowting* 1524 SRSf; John *Cloughting* 1568 SRSf; John *Cloutinge*, Henry *Clouton*, Thomas

*Clouten* 1674 HTSf. ME *clouting* 'the action of patching, mending, etc.' (1382 NED). Synonymous with CLOUTER.

**Cloutman:** An alternative for CLOUTER.

**Clover:** Robert *le Clovier* 1300 LoCt; Alen *Clover*, John *Clovier*, William *Clovyer* 1524 SRSf. A variant of CLEAVER, from OE *clēofan* 'to split', with change of stress to *cleófan*, *cloven*, whence *Clover*.

**Clow:** v. CLOUGH

**Clower:** William *le Cloer*, *le Cloier* 1201 P (So). A derivative of OFr *clou* 'nail', hence 'nailer'. Stephen *le cloer* 1292 SRLo is identical with Stephen *le Naylere* 1300 LoCt.

**Clowes, Cluse:** Nicholas *de Cluse* 1275 SRWo; Thomas *atte Cluse* 1332 SRSx. 'Dweller at the enclosure' or 'keeper of the mill-dam or sluice', OE *clūse*, ME *cluse*, *clowse* 'enclosure, narrow-passage', 'dam for water, sluice or flood-gate'.

**Clown, Clowne:** Ralph *de Clune* 1230 P (Sa); William *de Clune* 1332 SRSx; Thomas *Cloun* 1405 FFEss. From Clun (Sa), Clowne (Db), or a lost *Clun* near Carburton (Nt).

**Clowser:** Rychard *Closer* 1526 SxWills; John *Clouser* 1549 ib.; Moulde *Clowser* 1546 ib. These three, with Thomas *atte Cluse* above, all lived in Warnham (Sussex), and their surnames are identical in meaning. v. CLOWES.

**Club, Clubb:** William, Gilbert *Clobbe* 1166 P (C), 1202 Cur (Sf); Stephen, Walter *Clubbe* 1204 P (C), 1279 RH (Hu). ME *clubbe*, *clobbe* 'club', metonymic for *clubber* 'maker of clubs': Richard *clobbere* 1222 DBStP; Walter *le Clubbere* 1260 AssC. In 1198–1212 (Bart i, 267), Robert *clobbere* of the text witnesses as Robert *clobbe*. By the Assize of Arms, every adult man had to be provided with at least a knife and a staff or club.

**Clucas:** *MacLucas* 1511, *Clucas* 1643 Moore. From *MacLucais* 'son of *Lucas*'.

**Cluderay:** v. CLITHEROE

**Clue, Cluff:** v. CLOUGH

**Cluer:** v. CLEWER

**Clunch:** Richard *Clunch* 1194 Cur (W); Alice *Clunch* 1277 Ely (Sf); Gilbert *Clunch* 1360 FFEss. A nickname from ME *clonch*, *clunch* 'lump'.

**Clunie, Cluney, Cluny:** (i) Ralph *de Cluneia* 1086 InqEl; William *de Cluini* 1195 P (D). From Cluny (Saône-et-Loire). (ii) William *de Cluny* of Perthshire 1296 Black. Probably from Clunie in Stormont (Perthshire).

**Cluse:** v. CLOWES

**Clutterbuck:** Michael *Cloterbuck* 1560 Pat (Ch); Toby *Clutterbuck* 1608 Oriel; Samuel *Clutterbook* 1662 *HTEss*; Freame *Clutterbuck* 1707 DKR. 'The Clutterbucks . . . originally of Dutch origin, had fled from Holland in the sixteenth century'. H. P. Finberg, *Gloucestershire Studies*, Leicester 1957.

**Clyma, Clymo:** v. CLEMO

**Clynes, Clyne, Cline:** (i) Adam *de Cleynes* c1280 SRWo; John *Clynes* 1327 IpmGl; Benjamin *Cline* 1664 HTSo. From North Claines (Worcs), *Cleynes* 1234, *Clynes* 1293. (ii) William *de Clyn* 1375 Black; Malcolm *de Clyne* 1390 ib. From Clyne (Sutherland).

**Coad(e):** v. CODE
**Coady:** v. CODY
**Coaker:** v. COKER
**Coales:** v. COLE
**Coard:** v. CORDES
**Coate(s):** v. COTE
**Coatman:** v. COTMAN

**Cobb, Cobbe:** Cobba 1201 Pl (Co); Leuric Cobbe 1066 DB; Walter Cobbe 1234–5 FFEss; John Cobbe 1327 SRSo. OE *Cobba 'big, leading man', an original nickname, unrecorded in OE but not uncommon from the 12th century onwards, v. OEByn 305. In the eastern counties it may represent ON Kobbi, while a shortened form of Jacob is a further possibility.

**Cobbald, Cobbold, Cutbill:** Cotebaldus de Wigornia a1200 Dublin; Aluuinus Cubold 1066 DB (Nth); Ricardus Cubaldus 1174 P (He); John Cubald 1219 AssL; Thomas Cutebold', William Cotebold 1292, 1332–57 PN K 492; John Cobald 1309 FFSf. OE Cūðbeald 'famous-bold'.

**Cobbard, Cobber:** Thomas Coberd 1275 RH (Lo); William Cobard 1294 KB (Sx); Robert de Cobbar' 1332 SRSx. ME cobbard 'support for a spit', perhaps a nickname for a cook or a scullion.

**Cobbe:** v. COBB

**Cobbel, Cobbell, Cobble, Coble:** Adam Cobel 1301 CorLo; William Cobell 1482 FrY. ME cobel 'a rowboat', a nickname for a sailor.

**Cobber:** v. COBBARD

**Cobbet, Cobbett:** Roger Cobet 1275 RH (Nf); John Cobat 1327 SRSf; Robert Cobbett 1583 PN Sr 104. Cob-et, Cob-ot, diminutives of Cob, a pet-form of Jacob.

**Cobbing:** Henry Cobbing 1202 AssL; John Cobbyng 1298 KB (L); Richard Cobbyng 1345 FFEss. 'Son of Cobb', a derivative of either OE *Cobba or ON Kobbi. v. COBB.

**Cobble:** v. COBBEL

**Cobbledick, Cobeldick, Coppledick, Cuppledick:** Robert de Cubbeldick' 1242 Fees (L); John de Cupeldike 1276 RH (L); Roger de Cobeldyk 1320 AD iv (L); John Copuldick Hy 6 AD v (L); Ambrose Cobledicke 1642 PrD. From an unidentified place, presumably in Lincs.

**Cobbler:** Emma le Cobelere 1289 NorwLt; Richard le Cobelere 1339 LoPleas; Hugh Cobeler Hy 4 AD iv (Nf). 'Cobbler', ME cobeler.

**Cobden:** John Cobden 1525 SRSx. From Cobden Fm in Sullington (Sx).

**Cobeldick:** v. COBBLEDICK

**Cobham, Cobbam:** Robert, Nicholas de Cobeham 1199–1200 FFK, 1332 SRSx; Thomas Cobham 1400 AssLo. From Cobham (Kent, Surrey, Sussex).

**Coble:** v. COBBEL

**Cobley, Cobleigh:** John de Cobeley 1316 NottBR; Richard de Cobbelaye 1324 IpmNt; John Cobley 1642 PrD. From Cobley in Lapford, in East Worlington (D).

**Cochran, Cochrane, Cochren, Colqueran:** Waldeve de Coueran 1262 Black; William de Coughran 1296 ib.; Robert de Cochrane c1360 ib. From Cochrane (Renfrewshire).

**Cock, Cocke, Cocks, Cox:** (i) Coc de domo Abraham 1192 P (Lo); Koc filius Pertuin 1230 P (L); Cock le Botiller 1281 LLB C; Koc Forester, Kok de mari 1296 SRSx; Aluuinus Coc 1066 DB (C); Osbern Cocc 1175–95 Seals (Db); Aki Coc 1177 P (Nf); Nicholas Cock 1297 MinAcctCo; Petronilla Cockes 1327 SRWo; John Cocks 1332 SRCu; Walter Cocks, Cox 1515 Oxon. The first example is the name of a Jew and is probably a diminutive of Isaac in its Hebrew form (Jacobs). Cock, a common personal name still in use about 1500, may partly be from OE Cocc or Cocca, found in place-names, although not on independent record. But as cock became a common term for a boy, it may also have been used affectionately as a personal name. (ii) William, Godard le Cock 1271 ForSt, 1281 LLB A; Thomas le Cok 1285 Ass (Ess); John le Cockes 1327 SRWo. OE cocc 'cock', a nickname for one who strutted like a cock. This became a common term for a pert boy and was used of scullions, apprentices, servants, etc., and came to be attached to christian names as a pet diminutive (Simcock, Wilcock, etc.). Forms without the article may belong here; cok is ambiguous and may be for Cook. The surname may also mean 'watchman, leader' and, according to Welsh writers, may also be from Welsh, Cornish coch 'red'. (iii) Hugh ate Cocke, ate Coke 1319 SRLo, 1320 LLB E; William dil Cok 1327 SRSf; Thomas atte Cok 1380 FFSf. 'Dweller by the hill', OE cocc 'haycock, heap, hillock'. In London it probably derived from the sign of a house or inn. Sometimes we may have ME cock 'small ship's boat' (1319 MED), name for a boatman. cf. BARGE.

**Cockayne:** William Cokein, Cocaine 1193 P (Wa), 1221 Cur (Bk); Hawisa de Cokaingne 1219 AssY; Geoffrey de Cokaygne 1228 FFEss; John Cokkayn 1332 SRCu. ME cokaygne, OFr coquaigne, the name of an imaginary country, the abode of luxury and idleness. Cocken (Lancs), pronounced Cockin, is said to have been named in jest as the land there was cleared by the monks of Furness Abbey. The surname was probably given to one whose habits and manner of life suggested he had come from the fabulous land of Cockaigne. It has also become COCKIN and COCKING.

**Cockbain:** William Cokben 1545 FrY. 'Cock's bone', OE cocc, ON beinn. The second element is not uncommon in medieval nicknames: Walter Coltbayn 1256 AssNb 'colt's bone'; Richard Schortbayn 1327 SRY 'short bone'; William Longerbayne 1296 SRNb 'longer bones'.

**Cockcroft, Cockroft:** Richard de Cocckecroft 1296 Wak (Y). 'Dweller at the cockcroft.'

**Cockell, Cockill, Cockle:** Simon, Thomas Cockel 1198 P (Nth), 1202 P (K); Richard le Cokel 1279 RH (O); William Cockyl 1327 SRSf; Ralph Kokyl 1327 SRC. Various explanations are possible: OE coccul, coccel 'cockle', a weed particularly common in cornfields. For the taris of Wyclif's 1388 version, that of 1382 has dernel or cokil, and the Rheims and Authorized Versions have cockle. cf. DARNELL. Or ME, OFr cockille 'a shell', also 'cockle', the bivalve. cf. William le Cokeler 1281 MESO (L), which Fransson takes as OFr coquillier 'fabricant de coquilles' (Godefroy) a maker of head-coverings for women. Thuresson explains (William) *Cockeler* 1332 MEOT (L) as a gatherer of cockles. cf. MULLET. The Fr *Coquille* Dauzat takes to be a surname applied to pilgrims to the shrine of St James of Compostella who sewed shells on their clothes as a sign of pilgrimage. cf. *cockle-hat* (1834 NED), a hat with a cockle or scallop-shell stuck into it, worn for the same reason.

**Cocker:** Henry *Cokere* 1198 P (K); Geoffrey, Alexander *Cokkere* 1237 Fees (Bk), 1327 SRSf; Adam le *Kockere* 1327 SRSf. Either a derivative of ME *cocken* 'to fight', a fighter, wrangler, or of ME *coke* 'to put up hay in cocks', haymaker.

**Cockerell, Cockerill, Cockarill, Cockrell, Cockrill:** Stephen *Cokerel* 1166 P (Y); Adam *Cokerell* 1200 P (Sf). *Cockerell* 'a young cock' is not recorded in MED before 1440, though that is not an insuperable objection to its appearance as a surname much earlier. The surname is common and early and is often, no doubt, from OFr *cocherel, cokerel* 'cock-seller', 'poultry-dealer'.

**Cockerham, Cockram, Cockrem, Cockran, Cockren:** John *de Kokerham* 1349 FrY; Thomas *Cokeram* 1450 ArchC vii; Thomas *Cockrom* 1596 RothwellPR (Y); Nathaniel *Cockerum* 1674 HTSf; William *Cockran* 1756 FrY. From Cockerham (Lancs).

**Cockett:** v. COCKIN

**Cockfield, Cofield:** Lucia *de Kokefeld* 1198 FFO; Robert *de Cockfeld* 1236–47 YCh; Nicholas *Cokefeld* 1327 SRSx; Lewis *Cofield*, *Coefield* 1611 ER 61. From Cockfield (Du, Sf), or Cuckfield (Sx), *Cokefeld* 1232.

**Cockhill, Cockle:** Thomas *Cockehyll* 1547 RothwellPR (Y); John *Cockhill* 1642 PrD. From Cockhill in Berrynarbor (D), or from one or other of the eight Cockhills in the West Riding.

**Cockin, Cockett, Cockitt:** Henry, John *Cokin* 1207 Cur (Ess), 1273 RH (Nt); Richard *Cochet* 1170 P (Ess); John *Coket* 1221 AssWa. These two surnames are metonymic for bakers. cf. Ralph *Cocunbred* 1209 FrLeic, Adam, Ralph *Cokinbred* 1265, 1299 ib. *Cockin-bread* is presumably the same as *cocket-bread*, a leavened bread or loaf slightly inferior in quality to the wastell or finest bread (1266 MED). It has been suggested (without evidence) that the bread was stamped with a seal or cocket. This was a seal (AFr *cokette*) belonging to the King's Custom House (1293 NED) and might have been used by metonymy as a surname for a sealer or a customs' house officer. v. also COCKAYNE.

**Cocking, Cockings:** (i) William *Cocking'* 1327 SRSx. From Cocking (Sussex). (ii) William *Coccing* 1266 LeiBR. Probably OE *Coccing* 'son of *Cocc*'. v. also COCKAYNE.

**Cockle:** v. COCKELL

**Cocklin, Cockling:** *Kokelinus* carectarius 1295 Barnwell (C); Ralph *Cokelin* 1279 RH (C); Reginald *Kokelin* 1284 FFHu. A double diminutive of *Cocc*. v. COCK.

**Cockman, Cookman:** William, Reynballus *Cokeman* 1276 AssSo, 1297 MinAcctCo; John *Cookman* 1374 ColchCt. Either 'servant of Cook' or 'the

cook's servant'. *Cockman* is from *cōkman*, with shortening of the vowel before OE *cōc* became ME *couk, cook*.

**Coknage:** Algor *Cochenoc* 1066 DB (Herts); William *Cockenage* 1428 LLB K; William *Cocknedge* 1591 AssLo. From Cocknage (St), or Cockenhatch in Barkway (Herts).

**Cockney:** Edmund *Kokeney* c1290 ERO; Emma *Cokenay* 1379 PTY; John *Cokenay* 1413 TestEbor. ME *cokenei* 'an effeminate youth, a weakling'.

**Cockram, Cockran:** v. COCKERHAM

**Cockrell, Cockrill:** v. COCKERELL

**Cockroft:** v. COCKCROFT

**Cocksfoot, Coxfoot:** Aluuald cognomine *Cockesfot* 1067 Ek. Perhaps 'cock's foot', OE *cocc, fōt*, though this was also a name for columbine. But cf. Alice *Cokschanke* 1265–72 RegAntiquiss 'with legs like a cock'.

**Cockshed, Coxhead:** Roger *Kockesheued* 1227 AssBeds. 'Cock's head', a nickname. cf. William *Cokkesbrayn* 1275 RH (Sx), Henry *Cockeshank* 1323 Wak (Y).

**Cockshot, Cockshott, Cockshoot, Cockshut, Cockshutt:** Symon *de Cokshute* 1296 SRSx; John *Cokschote* 1312 ColchCt; Alice *atte Cocshete* 1327 SRSx; John *Cocke Shoute* 1562 AD vi (Berks). Sometimes pronounced *Coeshot*. OE *\*coccscȳte*, 'a place where nets were stretched to catch woodcock' as at Cockshoot Fm (Worcs), Cockshot (Kent), Cockshut (Lancs), etc.

**Cockson:** v. COXON

**Cockspur:** Richard *Cokespur* 1232–3 FFSr; John *Cockespore* 1307 Wak (Y); Robert *Cokspour* 1379 PTY. A plant-name, perhaps wild clary.

**Cockton, Cocton:** Auic' *de la Coctune* 1296 SRSx. 'Dweller near the enclosure for cocks', OE *cocc, tūn*.

**Cockwell, Cockwold:** v. CUCKOLD

**Cocton:** v. COCKTON

**Codd:** Alanus *filius Chod* 1150–5 DC (L); Osbert *cod* 1148 Winton (Ha); John *Lecod* 1219 AssY. *Chod* may be an example of OE *\*Codda*, unrecorded, but found in place-names, or an original nickname. The surname was usually a nickname but the exact meaning is obscure. It may be OE *cod(d)* 'bag, scrip', used a1250 of the belly or stomach, hence, perhaps, for a man with a belly like a bag. Or we may have ME *codd(e)* 'cod'. cf. John *le Codherte* 1297 MinAcctCo 'cod-heart', Robert *Codbody* 1332 SRSx, either 'with a body lik a cod' or with one like a well-filled bag, and Robert *Codhorn* 1202 P (Y), where the meaning is not clear. cf. also *cod's-head* 'a stupid fellow' 1566 NED. In the 16th century *codder* denoted a worker in leather, a saddler or a peltmonger so that the surname may, perhaps, also be metonymic for a maker of leather bags or a saddler.

**Coddington:** William *de Codington* 1230 P (Lo); Richard *de Codington* 1287 AssCh. From Coddington (Ch, Db, He, Nt).

**Code, Coad, Coade:** Robert *Chode* 1182 P (L); Nicholas *Code* 1275 SRWo. ME *code* 'pitch, cobbler's wax', a name for a cobbler.

**Codell:** William *Codele* 1306 AssW; Reginald *le Codele* 1327 SRSx; John *de Codel* 1327 SRWo;

Gylemyne *Codel* 1375 IpmW. A nickname from the cuttlefish, OE *cudele*, ME *codel*.

**Codlin, Codling, Quadling, Quodling, Girling, Gurling:** (i) John *Kodling* 1208 Cur (Y); Robert *Codling* 1275 RH (L); Emma *Codelingg'* 1297 SRY. ME *codling* 'a young or small cod' (1289 MED), either a seller of these fish, or, perhaps, as the only examples noted are from Yorkshire and Lincolnshire, for a fisherman, or used as an affectionate diminutive. This would survive only as *Codlin* or *Codling*. (ii) Richard *Querdeleon* 1247 AssBeds; Adam *Girdelyon, Girdelion* 1296–7 Wak (Y); William *Gerlyn* 1296 SRSx; William *Querdelion* 1304 LLB C; John *Qwerdeling* 1327 SRSf; Robert *Gerling* 1327 SRC; Thomas *Querdelyng* 1365 LoPleas; John *Kodlyng*, George *Codlyng* 1524 SRSf; Wyllyam *Gyrlinge* 1547 EA (NS) ii (Sf). Fr *cœur-de-lion* 'lion-heart'. cf. the development of *codling* 'a hard kind of apple' (sound to the core), *querdlynge* c1400, *codlyng* 1530, *quodlinge* 1586, *quadlin* 1625 NED. *Girdelion* would become *Girdling* and then *Girling*. *Codling* is common in Yorkshire (North and East Ridings), *Quodling* and *Quadling* are Norfolk and Suffolk names, whilst *Girling*, particularly common in Suffolk, is frequent also in Essex and Norfolk.

**Codman:** William *Codeman* 1327 SRC; Robert *Codman* 1524 SRSf. An occupational name, either a worker in leather, a saddler, a catcher or seller of codfish, or a cobbler. *v.* CODD, CODE.

**Codmer:** John *de Codemere* 1327 SRSx; John *Codmer* 1642 PrD. From Codmore Fm in Pulborough (Sx), or Cudmore Fm in Bampton (D).

**Codner:** *v.* CORDNER

**Cody, Coady:** Geoffrey *Codi* 1210 Cur (Y); Stephen *Cody* 1297 SRY; Roger *Cody* 1364 AssY. Perhaps a diminutive of OE *Coda*. In Ireland for *Mac Óda*.

**Coe, Coo:** Osbert *Ka* 1188 P (L); John *Co* 1221 AssWa; Gilbert *le Co* 1252 Rams (Hu); Beatrice *le Coe* 1274 RH (L); Roger *le Coo* 1327 SRC. ME *co, coo*, the midland form corresponding to northern *ka*, ON *ká* 'jackdaw' (c1325 MED). *v.* also KAY.

**Coey:** *v.* COY

**Coff, Coffe:** Godwin *Coffe* 1218–19 FFK; William *Cof* 1276 AssW; Henry *Coffe* 1370 LuffCh. OE *cōf* 'eager, bold, quick'.

**Coffer:** Thomas *le Coffer* 1298 LoCt; John *Coffere* 1299 LLB B. OFr *cof(f)re* 'box, chest' (c1250 MED), here used for OFr *coffrier* 'maker of coffers' (1402 MED), also 'treasurer' (a1338 MED); John *le Cofrer* 1275 AssSo, John *le Cofferer* 1290 LLB A.

**Coffin:** Richard *Cofin* 1169 P (Gl); John *Coffyn* 1270 AssSo; Richard *Coffyn* 1327 SRSx. OFr *cofin, coffin* 'basket'. Metonymic for a basket-maker.

**Cofield:** *v.* COCKFIELD

**Cogan, Coggan, Coggin, Coggins:** (i) William *de Cogan* 1185 P (Glam); Richard *Cogayn* 1271 AssSo; Peter *Coggane* 1642 PrD. From Cogan (Glamorgan). (ii) In Ireland also for *MacCogan*, Ir *Mac Cogadháin* 'son of the hound of war'.

**Cogg, Cogge, Cogges, Coggs:** (i) Robert *Cog* 12th DC (Nt); John *le Cogge* 1328 IpmW; Robert *Cog* 1375 AssNu. ME *cogge* 'cog of a wheel'. Perhaps a nickname for a wheelwright, or for a miller. (ii) Alice *de Cogges* 1279 RH (O); Peter *de Coges* 1275 SRWo;

Alexander *atte Cogge* 1387 MELS (So). From Cogges (O), or 'dweller by the hill', *v.* MELS 43.

**Cogger:** Arnaldus *coggorius, coggarius* 1191–2 P (L); Osbert (*le*) *Coggere* 1195–7 P (Do). The Latin forms are derivatives of MedLat *coga, cogo* 'boat' (c1200, 13th MLWL), for ME *cogge*, OFr *cogue* 'small ship, cock-boat', used by Chaucer of the ships in which Jason and Hercules sailed. A *cogger* (c1450 NED) may have been a builder of cogs but was more probably a sailor or master of the cog. Roger *le Cogere* and John *le Cogger* were bailiffs of Dunwich in 1218 and 1219 respectively (Gardner). The only examples that do not come from coastal counties are from Cambridgeshire, Herts and Surrey where the Cam, the Lea and the Thames were important waterways, so that Thuresson's alternative suggestion 'maker of cogs for wheels', is an unlikely origin.

**Coggeshall, Coxall:** William *de Choggeshala* 1181 P (Ess); Wlfgarus *de Cokesale* 1232 Colch (Ess). From Coggeshall (Essex).

**Coggrave:** *v.* COPGRAVE

**Coggs:** *v.* COGG

**Coghill:** Ralph *de Coghull* 1286 AssCh; William *Coghhyll* 1427 FrY; John *Coghill* 1576 SRW. Perhaps from Cogill in Aysgarth (NRY).

**Cogman:** Benjamin *Cogman* 1647 Bardsley (Nf). Identical in meaning with COGGER.

**Coit:** *v.* COYTE

**Coke:** *v.* COOK

**Coker, Coaker:** Geoffrey *de Cocre* 1195 P (So); Robert *de Coker* 1262 Hylle; Thomas *de Coker* 1270 AssSo. From Coker (So).

**Colban:** *Colben* 1066 DB (Ch); Richard *Colbain* 1170 P (D); William *Colbeyn* 1235 Fees (Nth). ON *Kolbeinn*, ODa, OSw *Kolben*.

**Colbeck:** *v.* COLDBECK

**Colbeck, Coldbeck:** *v.* CALDBECK

**Colbert:** *Colbert* 1066 DB (D, Ch, L); John *Colbert* 1205 P (D). OG *Colbert*.

**Colborn, Colborne, Colbourn, Colbourne, Colburn, Colburne, Colbon, Colbond:** Geoffrey *de Colebrunn'* 1208 Pl (Y); William *de Colburn* 1386 FrY; John *Colborne* 1642 PrD. From Colburn (NRYorks), or Colesborne (Glos). Occasionally a personal name may be involved. ON *Kolbrún, Kolbiorn*, cf. Robert *filius Colbern* 1185 P (D).

**Colbran, Colbron, Colbrun, Coalbran:** *Colbrand, Colebran* 1066 DB (D, Wa); *Colebrandus* c1200 DC (L); Malger *Colebrond* 1275 RH (Sx); Walter *Colebrand* 1297 MinAcctCo. ON *Kolbrandr*, OSw *Kolbrand*.

**Colby, Colbey, Coleby:** Ralph *de Colebi* 1192–1218 YCh; William *de Colby* 1332 IpmNt; William *Colbe* 1525 SRSx. From Colby (Norfolk, Westmorland), Coleby (Lincs), or Coulby (NRYorks).

**Colcock:** Peter *Colcock* 1379 LLB H. *Col*, a pet-name for *Nicholas*, and *cock*, an affectionate diminutive. *v.* COCK.

**Colcott, Colcutt:** *v.* CALDECOT

**Coldbeck, Colbeck, Colebeck, Coulbeck:** Thomas *de Caldebek* 1321 FrY. From Caldbeck (Cumb).

**Colden:** (i) William *de Colden* 1255 RH (Bk); William *Colden* 1327 SRY. From Colden in Hebden

Bridge (WRY). (ii) David *Colden* 1459 Black. From the lands of Colden near Dalkeith (Midlothian).

**Coldicott:** *v.* CALDECOT

**Coldron:** *v.* CALDERON

**Coldwell:** *v.* CALDWELL

**Cole, Coles, Coales:** *Cola, Cole* 1066 DB; *Cole, Cola* filius Lanterii c1145 EngFeud (K); Robertus *filius Cole* 1206 AssL; Geoffrey, Richard *Cole* 1148 Winton (Ha), 1185 Templars (Wa); George *Coles* 1555 FrY. The personal name may be ON, ODa *Koli*, a short form of names in *Kol-*, but the distribution in DB suggests that it is more often OE *Cola*, an original byname from OE *col* 'coal' in the sense 'coal-black, swarthy'. The surname may also be a nickname with the same meaning: John *le Col* 1321 FFEss.

**Colebrook, Colebrooke:** Alexander *de Colebroc* 1160 P (D); Fucherus *de la Colebrok* 1241 MELS (Sx). From Colebrook (Devon), or 'dweller by the cool brook' (OE *cōl, brōc*). *v.* MELS

**Colegate:** *v.* COLGATE

**Coleham, Colham:** Simon *de Coleham* 1210 FFO; Robert *de Colham* 1340 NIWo. From Colham (Mx).

**Coleman, Colman, Collman, Coulman:** *Coleman* 1066 DB; *Colemannus de Eston'* 1176 P (Bk); Hervicus, Richard *Coleman* 1166 RBE (Y), 1176 P (Sr). The surname is early, frequent and widely distributed. In the north it is usually from OIr *Colmán*, earlier *Columbán*, adopted by Scandinavians as ON *Kalman*, and introduced into Cumberland, Westmorland and Yorkshire by Norwegians from Ireland. In DB the personal name is southern and south-eastern and is probably OG *Col(e)man*. In the Sussex Subsidy Rolls, where both *Coleman* and *Collier* are frequent surnames, both probably mean 'charcoal-burner'.

**Coleridge, Colridge:** Cristian *de Colrig* 1275 RH (D); John *Colregge* 1327 SREss; Humphrey *Coleridge*, William *Colridge* 1642 PrD. From Coleridge in Egg Buckland, or Coleridge House in Stokenham (D).

**Coleson, Colson, Coulson, Coulsen:** (i) Algarus *Colessune* 1138–60 ELPN. 'Son of *Col*', probably ODa *Kol*, or possibly ODa *Koli*. (ii) Ælstan *Cole sune* c1095 Bury (Sf); Bruning *Cola suna* 1100–30 OEByn (D); William *Colesone* 1332 SRSt; John *Colson* 1379 PTY. The first two examples are from either ON *Koli* or OE *Cola*. *v.* COLE. Later examples may have the same origin or may belong below. (iii) John *Collesson'* 1339 Crowland (C), 1379 PTY; John *Colleson* 1386 FFSf; John *Collesson, Colson* 1401, 1408 FrY. 'Son of *Coll*', a pet-form of *Nicholas*.

**Coleyshaw:** *v.* COWLISHAW

**Colfox:** John *Colfox* 1221 AssWa; Richard *Colvox* 1274 RH (Sa). ME *colfox*, from OE *col* 'coal' and *fox*', 'coal-fox', the brant-fox, a variety of fox distinguished by a greater admixture of black in its fur. According to Chaucer, the tail and both ears were tipped with black, unlike the rest of his hairs. A nickname. cf. 'a collfox, ful of sly iniquitee' (c1390 MED).

**Colgate, Colget, Colegate:** Stephen *de Colegate* 1300 LoCt. From Colegates in Shoreham (K), or Colgate in Lower Beeding (Sx).

**Colham:** *v.* COLEHAM

**Colicot, Colkett:** *v.* CALDECOT

**Colin:** *v.* COLLIN

**Coling:** *v.* COLLING

**Colkin:** Hamo *Colekyn* 1242 Fees (K); Roger *Colkyng* 1296 SRSx; Robert *Colkyn* 1327 ib. A diminutive of *Cole* or *Coll*.

**Coll, Colle, Colls, Coull, Coule, Coules, Cowl, Cowle, Cowles:** (i) *Col* 1066 DB (L); *Colle* serviens Henrici 1204 Cur (Y); *Colle* Rikmai 1247 AssBeds; Robert *Cholle* 1166 P (Nf); Osbert, William *Colle* 1196 P (L), 1200 P (Wo); Thomas *Colles* 1327 SRSf; Robert *Coule* 1341 FrY; Thomas *Cowles* 1568 SRSf. The DB *Col* is ON *Kollr*, OSw *Koll*, or ON *Kolr*, ODa, OSw *Kol*. *Colle* may have the same origin or it may be for ON, ODa *Kolli*, but, especially in later examples, it is a pet-form of *Colin*, found from the beginning of the 13th century as a diminutive of *Nicholas*. (ii) Robert *atte Cole* 1275 SRWo; Adam *atte Colle* ib. Probably 'dweller by the hill', from OE *\*coll* 'hill'. *v.* MELS.

**Collacot(t):** *v.* CALDECOT

**Collar, Coller, Colla:** John *Coller'* 1362 Crowland (C); William *Coler* 1377 LLB H. A form of COLLIER (a1375 MED).

**Collard:** *Colard* le Fauconer 1264 *Ipm* (Ess); *Colard* Hariel 1275 RH (Gl); Richard *Colard* 1332 SRSx. A pet-form of *Nicholas*, *Col*, plus the French suffix *-ard*.

**Collcott, Collecott:** *v.* CALDECOT

**Colledd:** John *Colhod* 1327 SRSf. 'Wearer of a black hood', OE *col* 'coal, coal-black' and *hōd* 'hood'.

**Colledge, Collidge:** Brian *de Colewich* 1210 Cur (Nt). From Colwich (Staffs) or Colwick (Notts).

**Collet, Collett, Collete:** *Colet* 1202 AssNth; Richard, Robert *Colet* 1213 Lewes (Nf), 1243 AssSo; Adam *Collette* 1332 SRSt. *Col-et*, a diminutive of *Col* (Nicholas) plus *-et*. There was also a feminine form: *Collette* 1379 PTY. Occasionally the surname is an aphetic form of *acolyte*: Simon *Colyte* 1294 RamsCt (Beds).

**Colley, Collie:** Hugh *Coly* 1212 Cur (Y); Dande *Colly* 1219 AssY; Philip *Coli* 1275 SRWo. OE *\*colig* 'coaly, coal-black'. The original short vowel is retained in the 16th century *colly* 'dirtied with coal and soot'. cf. 'a colie colour' (1565), colley sheep (with black faces and legs) 1793, and *colley* the Somerset dialect name for a blackbird. The surname probably meant 'swarthy', or, perhaps, black-haired.

**Collick:** Reginald *de Colewic* 1202 P (Nt). From *Colwick* (Notts).

**Collicutt:** *v.* CALDECOT

**Collidge:** *v.* COLLEDGE

**Collier, Colliar, Colliard, Colleer, Collyear, Collyer, Colyer:** Ranulf *colier* a1150 DC (L); Bernard *le coliere* 1172 P (So). A derivative of OE *col* 'coal', a maker or seller of charcoal (a1375 MED).

**Collin, Collins, Collen, Collens, Collyns, Colin:** *Colinus* de Andresia 1191 P (Berks); *Colinus* 1196 FrLeic; John *Collin* 1221 Cur (D); William *Colin* 1246 AssLa; Roger *Colynes* 1327 SRSo. *Col-in*, a diminutive of *Col*, a pet-form of *Nicholas*. *Colinus* Harrengod 1207 Cur (Sf) is identical with *Nicholaus* Harengot (1206 ib.). There was also a feminine form: *Colina* Charles 1250 Fees (Sf).

**Colling, Collinge, Collings, Coling, Cowling:**
*Collinc* 1066 DB (Sa, Db); Gerardus *filius Colling*
1185 P (Y); Aluuardus *Colling, Collinc* 1066, 1086
DB (W); Griffin *Collingus* c1114 Burton (St); Roger
*Kolling* c1125 MedEA (Nf); John *Collynges* 1376 AD
iv (Sa). ON *Kollungr*. The distribution does not
support Tengvik's opinion that the name is of native
origin. Scandinavian personal names are found in the
south to 1066. *Colling(s)* may also be a late
development of *Collin(s)*.

**Collingbourn, Collingburn:** Ruald *de Colingeburna*
1179 P (W); Sarah *of Colingburn'* 1249 AssW; John
*Colyngborn* 1373 FFW. From Collingbourne Ducis,
Kingston (W).

**Collingham:** Robert *de Colingeham* 1195 P (Lei);
Richard *de Kollyngeham* 1296 SRSx; John *Colyngham*
1437 IpmNt. From Collingham (Nt, WRY).

**Collingridge:** John *Colyngridge* 1464 Cl (Lo).
Perhaps from Cowan Bridge (La), *Collingbrigke*
c1200.

**Collingwood, Collingworth, Collinwood:** Richard
*de Calangwode, de Chalaungwode* 1323 AssSt, 1327
SRSt; John *atte Calengewode* 1349 DbCh; Ralph
*Colyngwood* 1516 ib.; William *Colynwod* 1512 GildY.
From Collingwood (Staffs) 'the wood of disputed
ownership'.

**Collinson, Collison:** Thomas, John *Colynson* 1349
Whitby (Y), 1382 FFHu; John *Colisson* 1381 SRSf;
Clement *Collyngson* 1524 SRSf; Thomas *Collison*
1622 RothwellPR (Y). 'Son of *Colin*'. *Colisson* is 'son
of *Coll*' or, possibly, of *Cole*.

**Collis, Collishe, Colliss:** Juliana *Colles* 1334
ColchCt; John *Collys* 1442 Eynsham; Harry *Colles*,
Robert *Collys* 1524 SRSf. Variants of *Coles*, or *Colls*.
*v.* COLE, COLL.

**Collishaw:** *v.* COWLISHAW

**Collishe, Colliss:** *v.* COLLIS

**Collister:** *v.* CALLISTER

**Collman:** *v.* COLEMAN

**Collop:** John *Collop* 1279 RH (C); Henry *Colhoppe*
1290 FFEss. ME *colhope, col(l)hop* 'an egg fried on
bacon; fried ham and eggs' (1362 NED). Probably a
name for a cook-house keeper.

**Collyns:** *v.* COLLIN

**Collwell:** *v.* COLWELL

**Colman:** *v.* COLEMAN

**Colmer, Colmor:** William *Colmer* 1189 Sol; John *de
Colmore* 1327 SRWo; Nicholas *Colemar'* 1332 SRSt.
From Colmore (Ha), or Colmer's Fm in King's
Norton (Wo).

**Colocott:** *v.* CALDECOT

**Colqueran:** *v.* COCHRAN

**Colridge:** *v.* COLERIDGE

**Colson:** *v.* COLESON

**Colston, Coulston:** (i) Reginald *filius Colstan* 1190
P (L); *Colstan* 1213 Cur (Do); Roger *Colstayn* 1297
SRY; Adam *Colstan* 1332 SRCu. OE *Colstān*, ON
*Kolstein*. (ii) Roger *de Coleston'* 1208 FFY; Robert *de
Colstone* 1352 LLB G; John *Colston* 1426 FrY. From
Colston Basset, Carcolston (Notts), or Coulston
(Wilts).

**Colswain, Colswayn:** *Colsweinus* 1109 Miller (C);
*Colseinus filius Godwini* 1219 Cur (Herts); Edward

*Colswein* 1189 Sol; John *Colswein* 1293 AssW; Agnes
*Colsweyn* 1361 CarshCt (Sr). ON *Kollsveinn,
Kolsveinn*.

**Colt, Coult:** Godric, Anselm *Colt* 1017 OEByn,
1188 BuryS (Sf); Henry *le Colt* 1227 AssSt. OE *colt*
'colt', either a nickname 'frisky, lively', or metonymic
for COLTER, COLTMAN.

**Coltard, Coltart, Colthard, Colthart, Coulthard,
Coulthart:** Peter *Coltehird'* 1301 SRY; John *Coltart*
1627 Black (Dumfries); John *Coultart* 1679 FrY. OE
*colt* and *hierde* 'keeper of colts'. In Scotland, often
pronounced *Cowtart*.

**Colter:** Robert *le Coltier* 1285 MEOT (O); John *le
Coltere* 1327 SRSf. A derivative of OE *colt*. 'A keeper
of colts.'

**Colthurst:** Thomas *Colthirst* 1574 FrY; Thomas
*Colthurst* 1615 PN Ch ii 84. From Colthurst Mill in
Over Peover Chapelry (Ch), or Colthirst in Great
Mitton (WRY).

**Coltman:** Anote *Coltman* 1332 SRCu; John *le
Coltmon* 1365 DbCh. OE *colt* and *mann*. cf. COLTER.

**Colton, Coulton:** Thomas, Pagan *de Colton'* 1176 P
(St), 1214 P (Y); Roger *Colton'* 1371 AssL. From
Colton (Lancs, Norfolk, Som, Staffs, WRYorks).

**Columbell:** Thomas *Columbel* 1327 SRSo; Stephen
*Columbel* 1332 SRDo; John *Columbell* 1409 PN Db
193. A diminutive of OFr *columbe* 'dove, pigeon'.

**Colvill, Colville:** William *de Colevil(e)* 1086 DB
(Y); Gilbert *de Colauilla* ib. (Sf); William *Coleuille*
1142–53 DC (L). From Colleville (Seine-Inférieure).

**Colvin:** *Coluinus* 1066 DB (D); Godwyne *Colwynes
suna* 1100–30 OEByn (D); Wlfwinus *Colewin* 1210
Cur (Db); Ralph *Coluin* 1296 SRSx. OW *Coluin,
Colwin*.

**Colwell, Colwill, Collwell:** Richard *de Collewele*
1268 AssSo; Robert *de Kolewelle* 1296 SRSx; William
*de Colwell* 1384 FrY. From Colwell (D, Nb), Colwall
(He), or Colwell House in Wivelsfield (Sx).

**Colyer:** *v.* COLLIER

**Combe, Combes, Coom, Coombes, Coombs:**
Richard *de la Cumbe* 1194 FF (Sx); Alan *in la
Cumbe* 1269 AssSo; Robert *atte Cumbe* 1296 SRSx;
Thomas *de Combe* 1317 AssK; John *ate Combe* ib.
From one of the many places named Comb, Combe or
Coombe, or from residence in a small valley (OE
*cumb*).

**Comber, Coomber, Coomer, Cumber, Cumbers:**
William *le Combere* 1260 AssC; Roger *le Coumber*
1276 RH (Berks); John *Comber* 1296 SRSx; Walter
*Cumbar'* 1332 SRSx. 'Dweller in a valley.' *v.* COMBE
and also CAMBER.

**Comer:** *v.* CAMBER

**Comfort, Comport:** William, Richard *Cumfort*
1269 AssSo, 1279 RH (O); Richard *Counfort* 1375
LoPleas. ME *cumfort, comfort*, OFr *cunfort, confort*
'strengthening, encouragement, aid, succour,
support', used of 'one who strengthens or supports, a
source of strength' (1455 NED). *Comport* seems to be
a late development in Kent where both forms are still
found. Bardsley notes Edward *Comport* alias *Comfort*
of Chislehurst.

**Comings, Comins, Comyns:** *v.* CUMING

**Commander:** Roger *le cumandur de templo* 1176–85

Templars (C); William *le Comandur* 1274 RH (So), 1297 MinAcctCo (W). ME *comander, comando(u)r* 'one who commands, ruler, leader' (1300 NED), sometimes 'officer in charge of a commandery' e.g. of the Knights Templars (OFr *comandeor*).

**Commin(s), Commings:** *v.* CUMING

**Common:** Charles *Commons* 1641 PrSo; William *Common* 1642 PrD. OFr *comune* 'common, widely-known'.

**Compain:** Odo *Compyn*, William *Compayn* 1327 *SR* (Ess). OFr *compain*, originally the nominative of *compagnon* 'chum' (1643 NED). A very rare surname. cf. Ralph *Cumpainun* 1221 AssWo.

**Comper:** Elyas *Cumper* 1224 Cur (So); Walter *Compere, le Compeyre* 1332 SRSx. ME, OFr *comper* 'companion, associate, comrade'.

**Comport:** *v.* COMFORT

**Compton, Cumpton:** Gladwin *de Cumtuna* 1167–c1175 YCh; William *de Compton'* 1212 Pl; Nicholas *de Cumpton* 1263 FFL; Richard *Compton* 1376 FFEss. From one or other of the many places of this name.

**Conan, Conant, Conen:** *Conanus* dux Britanniae et comes Richemundie a1155 DC (L); Henricus *filius Conani, Cunani* 1196 P (Nth), 1208 Cur (Y); *Connand, Conian* Gossipe 1479–86 FrY; Gilbert, Thomas *Conan* c1198 Bart (Lo), 1297 MinAcctCo (Y); Robert *Connand*, Adam *Conand* 1379 PTY. OBret *Conan*, the name of Breton chiefs, kings and of a saint; one of the Breton names introduced at the Conquest and common among tenants of the Richmond fee in Lincs and Yorks.

**Cone:** Henry *Cone* 1210–11 PWi; John *Cone* 1297 MinAcctCo; Robert *Cone* 1524 SRD. OFr *coing, coin*, ME *coin, cone* 'wedge, corner'.

**Coner, Condor:** Robert *de Conedour'* 1221 AssSa; William *Conder* 1275 SRWo. From Condover (Salop).

**Condict:** *v.* CONDUIT

**Condie, Condy, Cundey, Cundy:** Roger *de Cundi* c1150 Riev (Y), *de Condeio* Hy 2 DC (L); Aliz *de Condi* 1185 Templars (L); Nicholas *Cundy* 1200 P (L). From Condé (Nord, Oise, Orne, etc.) *v.* also CONDUIT.

**Conduit, Condict, Condy, Cundick, Cundict, Cunditt, Cundy:** Robert *atte Conduyt* 1334 LLB F; William *atte Conduit* 1340 AssC; Walter *atte Condut* 1342 LLB F. ME *conduit, condit, cundit*, OFr *conduit*, originally an artificial channel or pipe for conveying water, later a structure from which water was distributed, a fountain or pump. The surname probably refers to the latter. *v.* also CONDIE.

**Coney, Conie, Cony:** Richard *le Cony* 1296 SRSx; Robert *Cony* 1327 SRC. ME *conig, cony* 'rabbit', a nickname. cf. CONNING. The fact that Thomas *Cony* (1323 FrY) was a pelter suggests that the surname may also have denoted a dealer in rabbit-skins, perhaps also a furrier.

**Congdon:** Peter *Congdon* 1642 PrD. From Congdon (Co).

**Congrave, Congreave, Congreve:** Alan *de Cungrave* 1203 AssSt; Robert *Cungreve* 1381 SRSt; Thomas *Congreve* 1466 FFEss. From Congreve (Staffs).

**Conibeare, Conibeere, Conybeare:** Gawen *Conyber* 1641 PrSo; Josias *Conibeere* 1642 PrD. 'Dweller by

the wood frequented by rabbits', ME *cony*, OE *bearu*.

**Coningham, Conyngham:** *v.* CUNNINGHAM

**Conisby, Conisbee:** Humfrey *Conyngesby* 1487 Paston; Thomas *Coningsby* 1663 HeMil. From Conisby (L).

**Coniston:** Richard *Koniston* 1641 PrSo; John *Coniston* 1672 HTY. From Coniston (ERY), Cold Coniston (WRY), or Church Coniston (La).

**Conner, Connah:** Robert *le Conner* 1297 LLB A; Geoffrey *le Conner* 1327 SRSf. OE *cunnere* 'examiner, inspector', especially an ale-conner.

**Conning:** Geoffrey *Conyng*, Ralph *Konyng* 1296 SRSx. ME, AFr *coning* 'rabbit'. cf. CONEY.

**Conquer, Conquerer:** William *Conquerur* 1275 RH (Nf); Robert *Conqueraunt* 1276 RH (O); Thomas *Conquer* 1421–2 FFSr. OFr *conquereor* 'victor, conqueror'.

**Conquest:** Geoffrey *Conquest* 1248 AssBerks; John *Conquest* 1298 IpmGl; William *Conquest* 1355–9 AssBeds. OFr *conquest* 'conquest'.

**Conroy:** Henry *Cunrey* 1212 P (Ha); Robert *Conreys* 1359 AssD. AFr *cunrei*. OFr *conroi* 'a detachment of troops'. Probably for the leader of such a detachment.

**Consedy:** Richard *Counsedieu* 1319 SRLo; Richard *Consedieu* 1339 CorLo. A phrase name, 'May God begin it', OFr *commencer, dieu*.

**Consell:** *v.* COUNCEL

**Considine:** *v.* CONSTERDINE

**Constable:** Richard *Conestabl'* 1130 P (C); Alice *Cunestabl'* 1200 P (L), *la Konestabl', Constabl* 1202–3 AssL. ME, OFr *cunestable, conestable*, representing late Lat *comes stabuli* 'count or officer of the stable'. 'Chief officer of the household, court' (1240), 'governor of a royal fortress' (1297), 'military officer' (c1300), 'parish constable' (1328 NED).

**Constance, Custance:** (i) *Custancia, Constancia* Ric 1 Gilb (L); *Custans* 1379 PTY; Robert *Custance* 1207 P (C); John *Custaunce* 1279 RH (C). Fr *Constance*, from Lat *constantia* 'constancy', a common woman's name, usually anglicized as *Custance*. *v.* also CUST. Occasionally we may have the masculine form: Hugo *filius Constantii* 1086 DB (Wa), Willelmus filius *Custancii* 1196 Cur (Lei). (ii) William *de Constenciis* c1150 DC (L); Walter *de Constanc'*, Walterus *Custancie* 1173–80, 1181 Bury (Sf); William *de Custanc'* 1206 Cur (O). From Coutances (La Manche). A less common source.

**Constant:** Robert *le Costent'* 1194, 1197 P (L), *le Constent'* 1196 ib. Probably from OFr *constant*, Lat *constans* 'steadfast, resolute' (c1386 NED).

**Constantine, Cossentine:** (i) Willelmus *filius Constantini* 1086 DB (Bk); *Constantin* (*Costetin* CR) filius Godric 1166 P (Nf); Willelmus *Constantinus* c1150 Riev (Y); Geoffrey, Richard *Costentin* 1195 P (W), 1221 AssWo; Henry *Constantin* 1272 FFSf. OFr *Constantin, Costantin*, from Lat *Constantinus*, a derivative of *constans* 'steadfast'. The real pronunciation is represented by COSTINS, COSTAIN, and the pet-form COSTE. *Cossentine* is due to assimilation of *st* to *ss* in *Costentin*. (ii) Geoffrey *de Costentin* 1153 StCh, *de Constantin'* 1156–80 Bury (Sf); Henry *Costentin, de Costentin* 1166, 1180 Oseney

(O). From the Cotentin (La Manche). A less common source. *v.* OEByn, ANF.

**Consterdine, Considine:** From *Constantine* through the forms *Constentin, Constertin, Consterdine* and *Constetin, Consetin, Considine.*

**Conte:** Walter, Robert *Conte* 1296 SRSx, 1297 MinAcctCo. OFr *cunte, conte* 'count', in AFr in the sense 'earl'.

**Convent:** Gabriel *Couent* 1327 SRSx. For *atte couent*, servant at the convent, probably a monastery, but possibly a nunnery (ME, AFr *covent*, OFr *convent*).

**Converse:** Hugo *conuersus* Hy 2 DC (L); Emma *la Converse* 1214 Cur (Ha); Peter *le conuers* 1219 AssY. OFr *convers*, Lat *conversus*, adj. 'converted' (a1300 NED), sb. 'a convert' (1388), used of one converted from secular to religious life in adult age, 'a lay member of a convent' (14..), a sense much older. The Cistercian and Augustinian *conversi* were men living according to a rule less strict than that of the monks or canons, engaged chiefly in manual work, with their own living quarters and their own part of the church. They were numerous among the Cistercians in the 12th and 13th centuries, often outnumbering the monks and were, by rule, illiterate. These lay-brothers were employed on the monastic manors and granges where they were liable to fall into the sin of owning private property. They acquired a reputation for violence and misbehaviour—at Neath in 1269 they locked the abbot in his bedroom and stole his horses—and they were gradually replaced by more manageable paid servants.

**Conway:** (i) John *de Conweye* 1268 Glast (So); David *Coneway* 1340–1450 GildC. From Conway (Caernarvon). (ii) In Ireland an anglicization of various Celtic names, e.g. *Mac Connmhaigh* 'head smashing', *Mac Conmidhe* 'hound of Meath', or *Ó Conbhuidhe* 'yellow hound', *v.* MacLysaght.

**Cony:** *v.* CONEY

**Conybeere:** *v.* CONIBEARE

**Conyer:** John *le Conyare* 1327 SRSx; Henry *Coyner* 1327 SRSf. A derivative of OFr *coignier* 'to stamp money, to mint', a coiner of money, minter. Common from 1202 as *Coner, Cuner. v.* MESO.

**Conyers:** Roger *de Coyners* (Y), *de Coisneres, de Coisnieres* 1196 P (Y); William *de Coniers* 1208 Cur (Nb). From Coignières (Seine-et-Oise) or Cogners (Sarthe).

**Coo:** *v.* COE

**Cooch:** *v.* COUCH

**Cooil:** *v.* COOLE

**Cook, Cooke, Cookes, Coke:** Ælfsige *ðene Coc* c950 ASWills; Galter *Coc* 1086 DB (Ess); Walter *le Kuc* 1260 AssC; Richard *Cok* 1269 AssSt; Henry *Coke* 1279 AssSo; Ralph *le Cook* 1296 SRSx; Joan *Cokes* ib.; Robert *le Couk* 1327 SRSx; Roger *le Kokes* 1332 SRSt. OE *cōc* 'cook', often, no doubt, a seller of cooked-meats, etc. *v.* also KEW.

**Cookham:** Michael *de Cokham* 1255 FFO; John *Cookham* 1356 FFEss. From Cookham (Berks).

**Cookman:** *v.* COCKMAN

**Cooksley:** John *Cookesley* 1641 PrSo; John *Cooksley* 1641 PrD. From Coxleigh in Shirwell (D).

**Cookson, Cuckson, Cuxon, Cuxson:** Hugo *filius Coci* 1208 Cur (Sf); Gilbert *le Fiz Kew, Fiz le Keu* 1279 AssNb; Agnes *Cukeson* 1511 GildY. 'Son of the cook.'

**Coole, Cooil:** *McCoil, McCole* 1511, *Coole* 1666, *Cooile* 1711 Moore. Manx *MacCumhail* 'son of *Cumhall*', from *comhal* 'courageous'.

**Cooling, Cowlin, Cowling:** William *de Culinges* 1203 Cur (K); Matthew *de Couling* 1260 AssC; Avice *Couling* 1327 SRSo; William *Cowlyng* 1520 FrY. From Cooling (K), Cowling (NRY, WRY), or Cowlinge (Sf).

**Coom:** *v.* COMBE

**Coomber:** *v.* COMBER

**Coomb(e)s:** *v.* COMBE

**Coomer:** *v.* COMBER

**Coope, Coupe:** William *le Coupe* 1296 SRSx; Hugh *le Coupe* 1327 SRLei. Lat *cūpa* 'tub, cask'. Metonymic for COOPER. Sometimes, perhaps, from an inn-sign: Thomas hatchere, *atte coupe* beside Wolkeye 1424 LondEng 183.

**Cooper, Coopper, Copper, Couper, Cowper, Cupper:** Robert (*le*) *Cupere* 1176–7 P (Sr); Selide *le Copere, le Cupere* 1181–2 P (Nf); William *le Coupere* 1296 SRSx; Geoffrey *Cowper* 1377 FrY; Walter *Cuppere, Couper* 1378, 1391 LLB H; John *Copper* 1424 FrY. ME *couper* 'maker or repairer of wooden casks, buckets or tubs' (c1400 MED). *v.* also COPPER.

**Coopman:** *v.* COPEMAN

**Coot, Coote, Cootes, Coots:** Reginald, John *Cote* 1201 P (L), 1219 AssY; William *le Coot* 1327 SRC. ME *cote, coote* 'a coot' (c1300 MED), originally the name of various swimming or diving birds, especially the Guillemot, later restricted to the Bald Coot, whose appearance and traditional stupidity ('as bald (or as stupid) as a coot') would readily give rise to a nickname. 'The mad coote, With a balde face to toot' (a1529 Skelton).

**Cope:** *v.* CAPE

**Copeland, Copland, Coopland, Coupland, Cowpland:** Samson *de Copland* 1204 Pl (Nth, R); William *de Coupeland, de Copeland* 1256 AssNb; Thomas *Copeland* 1376 FFEss. From Copeland (Cumb), or Coupland (Northumb).

**Copeman, Coopman, Coupman:** *Copmannus Clokersuo* 1141–6 Holme (Nf); Johannes *filius Copeman* 1256 AssNb; John *Copman* 1205 P (Nf); Eustace *Coupman* 1230 P (Nf). ON *kaupmaðr* 'chapman, merchant', used also as a personal name.

**Copestake, Copestick, Capstack, Capstick:** Geoffrey *Coupstak* 1295 FrY. Henry *Coupestack'* 1301 SRY; John *Copestake* 1474 FrY; A hybrid from OFr *couper* 'to cut' and OE *staca* 'a stake'. 'Cut-stake', a name for a wood-cutter.

**Copgrave, Coggrave:** Richard *de Coppegrave* 1277 IpmY. From Copgrove (WRY).

**Copinger, Coppinger:** Seman *Copinger* 1327 SRSf; William *Copenger* 1383 FFSf; William *Copynger* 1489 FFEss. Perhaps 'dweller at the top of the hill', a derivative of ME *copping* from OE *copp* 'top, summit'.

**Copleston, Coplestone, Coppleston, Copplestone:** Hugh *de Copleaston* 1275 RH (D); Adam *Copelston* 1359 AssD; Arthur *Copleston*, Henry *Coplestone* 1642 PrD. From Copplestone (D).

**Copley:** Adam *de Coppelay* 1297 SRY; Adam *Coplay* 1379 PTY; William *Copley* 1449 FFEss; Susane *Coppla* 1559, Rychard *Copplay* 1560 RothwellPR (Y). From Copley in Halifax (WRY), Copley Plain in Loughton (Ess), or Copley Hill in Babraham (C).

**Copnall:** Walter *de Coppehale* c1147, Helyas *de Copenhale* c1165 StCh. From Coppenhall (Ch, St).

**Copner:** Robert, Richard *le Copener* 1242 Fees (D). OE *copenere* 'paramour, lover'.

**Copp:** (i) Eduinus *coppa* 1148 Winton (Ha); Robert, Geoffrey *Coppe* 1192 P (St), 1212 Cur (Sr). OE *cop, copp* 'top, summit', used of the head: 'Sire Simond de Montfort hath suore by ys cop' c1264 NED. (ii) Roger *de la Coppe* 1221 AssWa; John *atte Coppe* 1332 SRWa. 'Dweller at the top of the hill.'

**Coppard, Coppeard:** William *Copard* 1327 SRSx. OE *cop* 'top, head' plus OFr -*ard*. cf. TESTAR.

**Copped, Coppet, Coppett:** Alestan *Coppede* 1066 Winton (Ha); Richard *le Coppede* 1231–2 FFWa; Hugh *le Coppede* 1276 RH (Lei). ME *copped* 'peaked, pointed, haughty'. v. OEByn 306.

**Coppell:** v. CUPPLES

**Copper:** Juliana *la Copper* 1275 SRWo; Bartholomew, John *le Copper(e)* 1327 SRSf; William *le Copperer* ib. In the 12th century, *copere* is certainly a variant of *cupere* 'cooper', found late as *copper*, but it may sometimes be from OE *coper* 'copper'. The short vowel is clearly evidenced in *Copper* above, 'a worker in copper' used by metonymy for *copperer*. cf. COPPERSMITH, COOPER.

**Coppersmith:** Richard *Copersmid* 1212–23 Bart (Lo); Robert *copersmith* eHy 3 ib.; John *le copersmyth* 1305 LoCt. 'Maker of copper utensils' (1327 MED). cf. Hugo *Coperman* 1202 P (We), Stephen *le Coperbeter* 1286 LLB A.

**Copperthwait, Copperthwaite, Copperwhite, Cowperthwaite:** Mary *Copperwhite* 1672 HTY; Arthur *Copperthwaite* 1675 FrY. From Copperthwaite (NRY).

**Coppet, Coppett:** v. COPPED

**Coppin, Coppins, Coppen:** *Copin* 1188 P (L); *Copin* de Sancto Ædmundo 1188 P (Nf); Nicholas *Copin* 1243 AssSo; William *Copyn* 1275 SRWo. *Copin* is a pet diminutive of *Jacob. Jacob* de Troye (1273 RH) is identical with *Copyn* de Troys 1275 LLB A.

**Copping:** Gilbert *Copping* 1188 BuryS (Sf); Henry *Copping* 1202 AssL. Apparently a derivative of OE *copp* 'top, summit', the exact meaning of which is obscure. For similar formations, cf. Robert *Badding* 1187–1221 ELPN, William *Fatting'* 1297 ib., from the adjectives *bad* and *fat*.

**Coppinger:** v. COPINGER

**Copple:** v. CUPPLES

**Coppledick:** v. COBBLEDICK

**Coppleston, Copplestone:** v. COPLESTON

**Copsey:** *Cofsi, Copsi* 1066 DB (Y); Acharias filius *Copsi* 1155 FeuDu; *Copsi* 1177 P (Nf); Hugo, Robert *Copsi* 1170 P (Wa), 1182 P (St). ON *Kupsi*, OSw *Kofse*.

**Copthorn, Copthorne:** William *Coppethorn* 1359–60 FFWa; William *Copthorn'* 1364 KB (Wa). Probably 'dweller by the pollarded thorn', ME *coppede*, OE *porn*. cf. Copthorne Wood in Rick-

mansworth (Herts), and Copthorne in Burstow (Sr).

**Corb:** Angot *le Corb* 1206–7 Cur (Bk) is identical with Angod' *le Corf* ib. OFr *corb, corp, corf* 'raven'. v. CORBET, CORBIN, CORFE.

**Corbell, Corble:** (i) Richard *Corbeille, Corbell'* 1180, 1197 P (K); John *Curbeyle* 1327 SRC. OFr *corbeille* 'basket' (1706 NED). cf. Robert *Corbiller* 1225 MESO 171, from OFr *corbeillier* 'basket-maker'. (ii) Eudo *Corbel* 1198 P (Y). This might be identical with the above, but might also be from OFr *corbel*, now *corbeau* 'raven'. cf. CORBET, CORBIN.

**Corbet, Corbett, Corbitt:** Rogerius filius *Corbet* 1086 DB (Sa); Roger *Corbet* ib., 1158 P (Sa), 1221 AssWo; Thomas *le Corbet* 1323 Eynsham (O). OFr *corbet* 'raven' (c1384 NED), probably a nickname for one with dark hair of complexion. v. CORB, CORBELL, CORBIN, CORFE.

**Corbey, Corby:** Hugh *de Corebi* 1185 Templars (L); Robert *de Corby* 1255 ForNth; John *Corby* 1448–9 IpmNt. From Corby (Cu, L, Nth).

**Corbin, Corben, Corbyn:** (i) *Corbin(us)* 1086 DB (Wa, K); Roger *Corbin* 1201 AssSo; Walter *Corbyn* 1219 AssY. OFr *corbin* from *corb* 'raven' (a1225 NED). (ii) William *Corbun* 1086 DB (Ess); Hugo *de Corbun* ib. (Nf, Sf). From Corbon (Calvados) or, possibly, Corbon (Orne). The Essex example is probably due to early loss of the preposition, but might, possibly, be an early example of ME *corbun* 'raven' (a1300 NED). cf. CORBET.

**Corcoran, Corkan:** Cathasach *Ua Corcrain* 1045, Donagh *Mac Corcrane* 1576, *Corkan* 1611 Moore. *Mac Corcráin* 'son of *Corcrán*', the red-complexioned.

**Cordell, Cordall, Cordle:** Hugo *Cordel* 12th NthCh (Nth); Margery *Cordel* 1213 Cur (Nf). OFr *cordele*, a diminutive of *corde* 'cord'. The diminutive is not found in ME, but cf. ME *cordilere* (c1430) 'a Franciscan friar of the strict rule', so called from the knotted cord round his waist (OFr *cordelier*).

**Corden, Cordon, Cordwent, Corwin:** Robert *Corduan* 1221 AssWo; Walter *Kordewan* 1296 SRSx; William *Cordiwant* 1327 SRSo. ME *corduan, cordewan,* OFr *cordoan* 'Spanish leather made originally at Cordova', much used for shoes. Metonymic for CORDNER. For forms, cf. *corden* a1400, *corwen* 1483, *cordiwin* 1593 NED.

**Corder, Cordier:** Ralph *Corder* 1207 FineR (Ha); Osbert *le Corder* 1227 FrLeic. OFr *cordier* 'maker of cords'. cf. Augustinus *Cordemaker* 1199 Cur (Sf).

**Corderey, Corderoy, Cordurey, Cordeary, Cordero, Cordery, Cordray, Cordrey:** Rober *de Querderai* c1200 Riev (Y); Hugh *Queor de Rey* 1246 AssLa; Thomas *le Cordrey* 1275 SRWo; Richard *Cordray* 1287 Fees (Wt); William *Corderei* 1327 SRWo; Peter *Kerderey* 1332 SRSr; Robert *Querderey* 1347 WhC (La). *Corderoy* corresponds to Fr *cœur-de-roi* '(with the) heart of a king'; the early forms have AFr *rei* 'king' which survives as -*ray*, -*r(e)y*. *Corder(e)y, Cordray* and *Cordrey* may also derive from OFr *corderie* 'rope-walk': John *de la Corderie* 1292 Bardsley, a worker at the ropery.

**Cordes, Cords, Coard:** Walter, Osmer *Corde* 1182–1211 BuryS (Sf), 1185 P (Co). OFr *corde* 'string' (c1300 MED). Metonymic for CORDER.

**Cordle:** v. CORDELL

**Cordner, Codner:** (i) Randolf *se Cordewan'* 1100–30 OEByn (D); Richard *Cordewaner* 1170 P (St); Walter *Lecordewaner* 1173 P (Gl); Maurice *le corduaner* 1221·AssWo; Bartholomew *le Cordenewaner* 1281 MESO (L). AFr *cordewaner*, OFr *cordoanier* 'cordwainer, shoemaker'. *Corduaner*, no doubt, became *Cord(e)ner*. (ii) Peres, Stephen *le Cordener* 1292 SRLo, 1312 Gardner (Sf). A derivative of OFr *cordon* 'cord, ribbon'. v. CORDER.

**Cordon:** Robert *Cordon* 1327 SRSf. OFr *cordon* 'cord'. Metonymic for *cordoner*. v. CORDNER.

**Cordray, Cordrey:** v. CORDEREY

**Cordwell:** John *de Caldewell* 1327 SRDb. From Cordwell in Holmsfield (Derby).

**Cordwent:** v. CORDEN

**Corey:** v. CORY

**Corfe:** (i) Alard *de Corf* 1195 P (Gl). From Corfe Castle or Corfe Mullen (Dorset), or Corfe (Som). (ii) Angod' *le Corf* 1208 Cur (Bk). OFr *corf* 'raven'. v. CORB.

**Cork, Corke, Corck:** Geoffrey *Cork* 1278 LLB B. ME *cork* 'cork', metonymic for CORKER.

**Corkan:** v. CORCORAN

**Corker:** Adam, Geoffrey *le Corker* 1297 Wak (Y), 1338 MESO (La). A derivative of ME *cork* 'a purple or red dye-stuff', one who sells purple dye; or synonymous with (William) *le Corklittster* 1279 MESO (Y), a dyer of cloth with 'cork'.

**Corkett:** v. CALDECOT

**Corkhill, Corkill:** Donald *MacCorkyll* 1408, Edward *Corkhill* 1532 Moore. A Manx contraction of *Mac Þorketill*, *MacÞorkill*. cf. the Gaelic MCCORQUODALE, McCORKELL, and v. THURKETTLE, THURKELL.

**Corlett:** *Corlett* 1504, *MacCorleot* 1511 Moore. A Manx contraction of *MacÞorliótr*, from ON *Þorliótr* 'Thor-people'.

**Corley:** Roger *de Corleia* 1221 AssWa. From Corley (Wa).

**Cormack, Cormick:** Irish *Ó Cormaic* 'descendant of *Cormac*'. 'Son of the chariot.'

**Cormade:** A variant of Manx KERMODE.

**Cormell:** Gozelin, Anfrid *de Cormelies*, *de Cormeliis*, *de Cormel* 1086 DB (Ha, He); Ralph *de Cormeilles* 1197 P (Gl); Roger *de Cormell'* 1222 Cur (Ess). The DB tenants derive from Cormeilles (Eure) but other families may have come later from Cormeilles-en-Vexin. v. ANF.

**Corn, Corne, Cornes, Corns:** Obbe *Corne* 1203 Pleas (Sa); William *Corns* 1250 RegAntiquiss; John *Corn* 1332 SRSx. Either a nickname from OE *corn* 'crane', or a variant of OE *cweorn* 'hand-mill', metonymic for a maker or user of this.

**Cornah, Cornall, Cornell, Corney:** all common in Lancs, derive from a lost place in Lancs. cf. William *de Cornay* 1332 SRLa; John *Cornall* of Cornall (1672), Adam *Corney* (1666), Richard *Corney* of Greenhall (1571), Richard *Cornah* of Greenall (1737 LaWills).

**Cornberg, Cornborough, Cornbury:** William *de Corneburc* 1204 P (Y); Nicholas *de Cornbury* 1260 AssC; Auery *Cornburght* p1462 Paston. From Cornbrough in Sheriff Hutton (NRY), or Cornbury Park (O).

**Corne:** v. CORN

**Cornelius:** Leuekyn *Cornelys* (a Flemish weaver) 1354 ColchCt; Richard *Cornelius*, Thomas *Cornellis* 1568 SRSf; Lambert *Cornelius* 1662–4 HTDo. Lat *Cornelius*, not found as an English name until the 16th century, when it was brought back by returning sectaries from the Low Countries, where it was a popular name. v. ODCN.

**Cornell:** A weakened form of CORNALL, CORNHILL, CORNWALL or CORNWELL. Cornhill (Nb) is *Cornhale* 12th, *Cornelle* 1539 FeuDu. Henry *de Cornell'* (1229–31 StP) took his name from Cornhill (London). Thomas *Cornell* (1722 HorringerPR) is also called *Cornwall* (1736), *Cornwell* (1740) and *Cornhill* (1766). A further source is Fr *corneille* 'rook, crow', a nickname for a chatterer: Herbert *corneilla* 1148 Winton (Ha); William *Corneille* 1206 Cur.

**Corner, Coroner:** v. CROWNER

**Corner, Cornner:** (i) John *de Cornera* 1204 P (Ess); Dyonisia *Attecornere* 1297 MinAcctCo; Agnes *de la Cornere* in Bredstrate 1299 LoCt. AFr, ME *corner* 'angle, corner' (a1300 NED), 'place where two streets meet' (1382). 'One who lived near the corner', as, no doubt, did Roger *Cornirer* 1218–22 StP (Lo). cf. BRIDGER. Alys or Agnes *Acorner* gave a close called *Cornerwong* to the Abbey of Shelford a1483 (NottBR). (ii) Herueus *Cornur* 1179 P (Sf); William *le Cornur* 1185 RotDom (L); Agnes *le cornier* 1209 P (Nt); Augustine *le corner* 1230 P (Db). AFr *cornier*, OFr *corneor* 'hornblower', from OFr *corn* 'a musical instrument, horn' (c1477 NED).

**Cornes:** v. CORNISH

**Cornes, Corns:** v. CORN

**Cornet:** Durand, Alured *cornet* 1148 Winton (Ha), 1195 P (Co). OFr *cornet* 'a wind instrument made of horn or resembling a horn' (a1400 NED), a diminutive of *corn*. cf. CORNER (ii). 'A player of the cornet.'

**Corney:** Benedict *de Corneye* 1260 AssC; Robert *Cornay* 1301 SRY. From Corney (Herts, Cumb). v. CORNAH.

**Cornford, Cornforth:** Thomas *de Corneford* 1242 AssDu; Michael *de Cornford* 1339 CorLo; William *Cornefurth* 1469 FrY; Richard *Corneforth* 1514 FFEss. From Cornford (Durham).

**Cornhill:** Geruase *de Cornhill'* 1179 P (K); William *Cornhel* 1214 Cur (Y); Henry *de Cornhell'* 1229–31 StP (Lo); Bartholomew *ate Cornhell* 1311 ColchCt. From Cornhill (London) or one of the many places of that name, some of which are no longer remembered. This often becomes CORNELL and has been confused with CORNALL, CORNWALL and CORNWELL.

**Cornish, Cornes:** Badekoc *Korneys* 1296 SRSx; John *Corneys* 1327 SRSf; Henry *Cornysh* 1375 LoPleas. *Cornish* 'a Cornish man' is first recorded in NED in 1547. There must have been a ME *Cornish* formed on the analogy of *English* which was usually Normanized as *Corneys*. Adam *Cornys* (1300 LoCt) is probably identical with Adam *le Cornwalais* 1275 RH. v. CORNWALLIS.

**Cornwall:** William *de Cornewale*, *de Cornewayle* 1305 LoCt, LLB B. From Cornwall.

**Cornwallis:** Henry *le Cornwaleys* 1256 *Ass* (Ha);

Stephen *le Cornewalleys* 1260 AssC; Walter *le Cornwallis* 1280 LLB A. A Normanizing of an unrecorded ME *Cornwalish*. Walter *le Cornewaleys*, sheriff of London in 1277 (LLB A), is also called Walter *de Cornwall* in 1280 and Walter *le Engleys* in 1277. As Ekwall notes, his real name must have been English (*Engleys*) which was changed to *Cornwaleys* or *Cornwall* after his removal to London.

**Cornwell:** Roger *de Cornwelle* 1161 Eynsham. From Cornwell (Oxon). v. CORNELL.

**Corp, Corpe:** Walter, William *le Corp* 1177 P (Y), 1231 Oseney (O); James *Corp* 1297 MinAcctCo (Sf). In Yorks and Suffolk, the source is ON *korpr* 'raven', in Oxfordshire, probably OFr *corp* 'raven'.

**Corran, Corrin:** *M'Corrane* 1422, *M'Corrin, Corrin* 1504 Moore. A Manx contraction of *Mac Oran* from *Mac Odhrain* 'son of *Odhran*' pale-faced, Ir *Odar*. St Odhran was St Patrick's charioteer.

**Corrie, Corry:** (i) Walter *Correye* 1279 RH (C); John *Cory* 1389 FrY. 'Dweller at the shepherd's hut', ME *cori*. Perhaps also for CURREY. (ii) In Scotland from the lands of Corrie (Dumfries).

**Corse:** Lucia *de Cors* 1221 AssGl; Walter *de Cors* 1275 SRWo; Simon *Corsse* 1524 SRD. From Corse (Gl).

**Corsellis, Cusselle:** (i) Wandring *de Curcel*, Wandregesil *de Curcelles* 1201 AssSo. From Courcelles (Aisne). (ii) Also Huguenot, from Nicholas *Corcellis*, son of Zeager *Corcellis* of Ruselier (Flanders), who fled to England from the persecution of the Duke of Alva.

**Corser:** Anketill *le Corser* 1227 AssSt. ME *corser* 'jobber, horse-dealer' (c1383 MED).

**Corson:** v. CURZON.

**Corston:** William *de Corstonn* 1275 RH (Nf). From Corston (Sa, So, W).

**Corte:** v. COURT.

**Corteen:** Ceallach *Mac Curtin* 1376, *Cortin* 1652, *Corteen* 1659 Moore. A Manx contraction of *Mac Cruitin* 'son of *Cruitin*' (hunch-backed), metathesized to *Mac Curtin*.

**Cortes, Cortis:** v. CURTIS.

**Corton:** Amanenus *de Cortone* 1299 LLB C. From Corton (Do, Sf, W).

**Corwin:** v. CORDEN.

**Cory, Corey:** Robert *Cori* 1266 FFEss; Henry *Cory* 1297 MinAcctCo (W), 1327 SRC. This might be ON *Kori*, the first element of Corby (Lincs, Northants), or ON *Kári*. cf. *Cari* 1066 DB (Lei), Walter *Cari* 1200 P (Sf), and Corton (Suffolk), DB *Karetun*, 1266 *Corton*.

**Coryat, Coryot:** Nicholas *Coryot* 1328 IpmW; Walter *Coriot* 1361 IpmGl; John *Coryat* 1545, Anne *Corriett* 1576 SRW. From Coryates (Do).

**Cosens, Cosin, Cosyns:** v. COUSEN.

**Cosgrave, Cosgreave, Cosgrove:** William *de Couesgraue* 1255 ForNth; John *de Couesgraue* 1303 Balliol. From Cosgrove (Nth).

**Cosh, Coyish, Coysh:** Lucas *de la Kosche* 1248 Ass (Ess); Roger *de Coyssh* 1296 SRSx; Robert *Cosh* Ed 1 AD ii (Lei); Philip *atte Cossh* 1327 *SR* (Ess). ME *cosche, cosshe* 'small cottage, hut, hovel' (c1490 NED).

**Cosier, Cozier, Cowser:** Jone *Cosyer* (*Cowser*)

1561, John *Cosare* (*Cowser*) 1567 LedburyPR (He); Robert *Cosier* 1665 HTO. OFr *cousere* 'tailor'.

**Cossar, Cosser:** Ralph *le Kosser'* 1299 MESO (Ess); John *Cosser, Cossier* 1392 LLB H. ME *cosser* 'dealer, broker', 'horse-corser' (1300 MED).

**Cossentine:** v. CONSTANTINE

**Cossey:** Blitha *de Costeseye, de Coteseye* 1230 Cur (Nf); John *Cosseye* 1568 SRSf. From Cossey (Nf), *Costeseia* DB.

**Cossins:** v. COUSEN

**Cosson, Cossons, Cossom:** Robert le Marescal '*Cossun*' 1280, Hugh Pope '*cossun*' 1292 LLB A; John de Kent '*cozon*' 1306 LLB B. Probably a dealer in horses. Robert de Kent 1311 LLB B is styled *mercator equorum* Husting, and Robert le Marescal in 1280 owed 66s.8d. for a horse. v. also COUSEN.

**Costain, Costean, Costen:** *Costane, Costan* 1583 ODCN (Y); *Costaine* or *Constantine* 1586 ib. (Y); Hen *Costen, Costein* 1182, 1197 P (Lei); Alex *Costein* 1219 Cur (Lei). *Costein* is from *Costetin* by dissimilatory loss of *t*. cf. CONSTANTINE. In the Isle of Man, the surname is a contraction of *Mac Austeyn*, from *Mac Augustin* 'son of Augustin': *Costeane* 1507, *Mac Coisten, Mac Costen, Coisten, Costen* 1511, *Costain* 1715 Moore.

**Costard, Coster, Custard, Custer:** Alan *filius Costard* c1160 RegAntiquiss; Alexander *filius Costard* 1203 P (L); Roger *Costardus* 1175–86 Holme (Nf); Richard *Costard* 1249 AssW; Fraunceys *Costard* 1449 Paston. ME *costard* 'a prominently ribbed apple, a kind of large apple'. In the 16th century used of the head. But the word was evidently also known as a personal name.

**Coste:** *Coste de* Widkale 1175 P (L); *Costus* Falconarius 1180 P (Nt); Osbert, Hugo *Coste* 1218 AssL, 1317 AssK. A short form of *Constantine*, common as *Costantin, Costetin*. v. CONSTANTINE, COSTINS.

**Costean, Costen:** v. COSTAIN

**Costins, Costons, Costings:** Herbert *filius Constantini, Costin* 1207 Cur (Nf); *Costinus* 1221 ElyA (Sf); William *Costin* ib. (Nf); Elycia *Costantyn, Costyn* 1311, 1329 ColchCt. A short form of CONSTANTINE. v. also COSTAIN.

**Coston:** Hugh *de Cotston', de Costun'* 1221 AssSa; Stephen *de Coston* 1255 RH (Sa); Richard *de Costone* 1273 RH (Wo). From Coston (Lei, Nf, Sa).

**Cote, Cotes, Coat, Coate, Coates, Coatts, Cottis, Dallicoat, Dallicott, Delicate:** William *de Cotes* 1190 P (L); Walter *de la Cote* 1210 Cur (O); Godfrey *Cote* 1214 Cur (K); Roger *atte Kote* 1296 SRSx. From Coat (Som), Cote (Oxon), Coates (Lincs), Cotes (Leic), or one of the numerous similarly named places, all from OE *cot, cote* 'cottage', also 'shelter', sometimes 'a woodman's hut'. In ME, when the term was common, the surname may denote a dweller at the cottage(s) or, as it was used especially of a sheep-cote, one employed in the care of animals, a shepherd.

**Cotgrave, Cotgreave, Cotgrove:** Robert *de Cotegraue* 1202 FFL; Thomas *de Cotegrave* 1259 AssCh; Richard *Cotgrave* 1458 IpmNt. From Cotgrave (Notts), or Cotgreave in Mapperley (Derby).

**Cotherill:** v. COTTEREL

**Cotman, Coatman:** Ulkillus *cotmannus* 1183 Boldon (Du); William *Cotman, Coteman, Mercator* 1206–8 P (Sx); William *Coteman* 1275 RH (Nf). OE *cot* 'cottage' and *mann* 'a cottager', 'cotset', 'coterell', in Scotland 'a cottar' (*cotmannus* DB, *cotman* 1559 NED), corresponding to MedLat *bordarius*. cf. COTTER, COTTEREL. The equation with *Mercator* points to an alternative origin. OFr, ME *cote* 'outergarment, coat' (c1300 NED), 'seller of coats'. cf. *Capman* 'maker or seller of caps' MESO 116.

**Cotmore:** John *Cotmore* 1642 PrD. From Cotmore in Stokenham (D).

**Coton, Cottam, Cottom, Cotten, Cotton:** Randulf *de Cotton'* 1185 P (Wo); Ralph *de Cottum* 1212 Cur (Y); Stephen *de Coten'* 1297 MinAcctCo (L); John *de Cotome* 1310 LLB B; John *de Cotun* 1325 ib. D; Brian *Cotham, Cotam* 1569, 1596 FrY. OE *æt cotum* (dweller) 'at the cottages', as at Coton (Cambs), Cotton (Ches), Coatham (Durham, NRYorks), Cotham (Notts), Cottam (Notts, ERYorks). The *-um* is preserved only in Durham, Lancs, Notts and Yorks; *Cot(t)on* is found in the midlands, in Cambs, Ches, Derby, Leic, Lincs, Northants, Salop, Staffs, Oxon, Warwicks. cf. COTE.

**Cott, Cotts:** *Cota* atte Stapele 1296 SRSx; Adam *filius Cote* 1307 Wak (Y); Olmenus *Cota* 1066 DB (D); Blakeman *Cot* 1202 FFNf; Thomas *Cote* 1312 ColchCt. OE *Cotta*.

**Cottel, Cottell, Cottle, Cuttell, Cuttill, Cuttles:** Beringarius *Cotel* 1084 GeldR (W); Adam *Cotella* 1167 P (Do); Eilwinus *Kutel, Cutel* 1185 Templars (Ess); Walter *Cotel* 1206 Cur (O). The first form is probably, as suggested by Tengvik, OFr *cotel* 'coat of mail'. The later examples may also derive from OFr *cotel, coutel* 'a short knife or dagger' and are probably metonymic for a cutler.

**Cottenham:** Sturmid *de Cotenham* 1086 InqEl (C); Richard *de Cotenham* 1177–87 NLCh (C); Walter *de Cotenham* 1206 Cur (C). From Cottenham (C).

**Cotter, Cottier:** Robert *le Cotier* 1198 P (Sx); William *le Coter(e)* 1270 HPD (Ess), 1297 MinAcctCo. OFr *cotier* 'cottager' (1386 NED), DB *cotarius* 'villein who held a cot by labour-service'. v. COTMAN, COTTEREL. Both names are found in Isle of Man, pronounced *Cotchier* (*MacCotter, MacCottier* 1504, *Cottier* 1616, *Cotter* 1625 Moore), from *Mac Ottar*, 'son of Ottar' (ON *Óttarr*).

**Cotterel, Cotterell, Cotterill, Cotherill, Cottrell, Cottrill:** William, Gerard *Coterel* 1130, 1170 P (Lo, Berks); Honde *Cotrell* 1288 AssCh. OFr *coterel*, a diminutive of OFr *cotier* 'cottager' (1393 NED), DB *coterellus*. cf. COTTER.

**Cottis:** v. COTE

**Cottle:** v. COTTEL

**Cottom, Cotton:** v. COTON

**Cottrell, Cottrill:** v. COTTEREL

**Cotts:** v. COTT

**Cotwell, Cotwill:** William *Cottewell* 1642 PrD. From Cutwellwalls in Ugborough (D), *Cotewill* 1219.

**Cotwin, Cutwin:** Geoffrey *Cuttwine* 1228 AssSf; Henry *Cotewyne* 1327 SRSf; John *Cotwyn* 1553 NorwDep. OE *Cūðwine*.

**Coubrough:** David *Cowbratht* 1515 Black, Euphame *Cubrughe* 1669 ib. For MACCOUBRIE.

**Couch, Cooch, Cough:** Geruerd *Coh* 1160 P (Sa); John *Coh* 1167 P (He); Meriaduc *Choch* 1170 P (Sa); Thomas *Cuche* 1305 AssSt. Welsh *coch* 'red'.

**Couch, Couche:** Simon, John *Couche* 1279 RH (O), 1386 LoPleas. ME, OFr *couch* 'couch, bed'. Metonymic for COUCHER.

**Coucher, Coucha, Cowcher:** Stephen *Cuchur* Ed 1 Battle (Sx); Nicholas *le Couchur* 1295 MESO (Wo). AFr *coucheour* 'maker of couches, upholsterer'.

**Couchman:** Thomas *Cocheman* 1374 Ct (Ha); William *Cowcheman* 1500 KentW. Identical in meaning with COUCHER.

**Coudray:** v. COWDRAY

**Cough:** v. COUCH

**Coulbeck:** v. COLDBECK

**Couldwell:** v. CALDWELL

**Couling, Coulling:** v. COWLING

**Coulman:** v. COLEMAN

**Coulsdon:** Richard *de Coulesdon'* 1275 SRWo; Richard *de Coulesdon* 1332 SRWo. From Coulsdon (Sr), or Cowsden in Upton Snodsbury (Wo), *Coulesdon* 1198.

**Coulsen, Coulson:** v. COLESON

**Coulston:** v. COLSTON

**Coult:** v. COLT

**Coultar, Coulter, Culter:** Alexander *de Cultre* c1248 Black; John *Coulter* 1686 ib. From Coulter (Lanarkshire, Aberdeenshire).

**Coultard, Coultart:** v. COLTARD

**Coultas, Coultass, Coultous, Coultish, Cowtas:** William *Cowthus* 1562 FrY; John *Coultas, Coultus, Cowtus* 1657, 1671, 1691, 1733 FrY. 'Worker at the colt-house', colt-keeper.

**Coulthard:** v. COLTARD

**Coulton:** v. COLTON

**Councel, Councell, Council, Counsel, Counsell, Consell:** William *Cunseil* 1208 P (Berks), *Consell* 1208 Cur (Bk); John *Counseil* 1310 LLB D. AFr *counseil*, OFr *conseil, cunseil* 'consultation, deliberation' (c1290 NED).

**Count, Le Count, Lecount:** Ralph *le Cunte* 1196 P (Du); Walterus *Comes, le Conte* 1204–5 Cur (Sf); William *Counte* 1225 AssSo. AFr *counte*, OFr *conte, cunte*, Lat *comitem* 'count' (1553 NED).

**Counter:** Matthew *Cunter* 1250 Fees (Ha); John *le Cuntur, le Cunter, le Counter* 1289, 1301 AssSt. AFr *countour*, OFr *conteor* 'one who counts, reckons', 'accountant, treasurer' (1297 NED).

**Countess:** Agnes *Cuntasce* 1279 RH (C); John *Cuntesse* 1279 RH (Beds); John *le Contesse* 1327 SRSf. OFr *contesse* 'countess', when applied to a woman, probably 'proud, haughty as a countess'; applied to a man as a nickname for an effeminate dandy.

**Coupe:** v. COOPE

**Couper:** v. COOPER

**Coupland:** v. COPELAND

**Courage:** (i) Walter, William *Curage* 1254, 1260 FFEss. ME *corage*, OFr *corage, curage*, used as an adjective, 'stout' of body. cf. Corage or Craske, *Crassus, coragiosus* c1440 PromptParv. (ii) John *de*

*Courugge* 1309 SRBeds. From Cowridge End in Luton (Beds) which came to be pronounced *Courage* and is now pronounced *Scourge End*. *v.* also KERRICH.

**Coursey:** *v.* DECOURCY

**Court, Courts, Corte, Curt:** (i) William *de la Curt, de la Cort* 1242 Ipm (Sa); Richard *atte Curt*, William *de la Court* 1296 SRSx. From residence or employment at a large house or manor-house, castle, from OFr *cort, curt*, ME *curt, courte* (1297 NED). cf. COURTMAN. (ii) Reginald *Corte* 1181 P (Sf); Richard *le Curt* 1199 FF (Sr); Richard *le Cort* 1279 RH (O). OFr *curt* 'short, small'.

**Courtald, Courtauld:** A Huguenot name which Dauzat explains as a diminutive of *court* 'short, small'.

**Courtenay, Courteney:** Reginald *de Curtenay, de Courtenay* 1164–9 Bury, c1182 Gilb (L). From Courtenay (Loiret, Isère).

**Courthope:** William *de Curtehope* 1296 SRSx. cf. Courtup Fm, *Curtehope* (1310 PN Sx 231), but, as the well-known Sussex family is commonly found in East Sussex, it derived, perhaps, from an unidentified *Curting(e)hope* in the eastern part of the county.

**Courtice, Courtis:** *v.* CURTIS

**Courtier:** Nicholas *le Curter* 1279 RH (O). A derivative of ME *curt*, identical in meaning with COURT and COURTMAN, rather than the common *courtier* (ME *courteour, courtyer*), which has influenced the spelling.

**Courtman:** Adam, Robert *Curtman* 1275 RH (C), 1296 SRSx; John *Courtman* 1327 ib. 'Dweller near or one employed at a castle or manor-house.' *v.* COURT, COURTIER.

**Courtney:** *v.* COURTENAY

**Cousans:** *v.* COUSEN

**Couse:** Robertus *filius Cous* 1297 MinAcctCo (R); Robert, William *Couse* 1185 P, 1211 Cur (L). ON *Kouse, Kause*, corresponding to ON *Kausi*, a nickname meaning 'tom-cat', the first element of Cowesby (NRYorks).

**Cousen, Cousens, Cousans, Cousin, Cousins, Couzens, Cosens, Cosin, Cosyns, Cossins, Cossons, Cozens, Cozins, Cusins, Cussen, Cussins, Cussons, Cuzen:** Æthelstano *chusin*, id est, cognato suo (i.e. of Wlfstan) c977 (c1200) LibEl (C); Sumerda, Roger *Cusin* 1166, 1169 P (Nf, L); Simon *Cosyn* 1260 AssC; Thomas *Cossin* 1275 RH (Lo); Agnes *Cousseyns* 1327 SRSf. OFr *cusin, cosin*, in ME 'a kinsman or kinswoman', 'cousin' (c1290 NED). *v.* CUSSEN.

**Cove:** John *de Cove* 1219 P (Nf/Sf); Henry *de Cove* 1355 LLB G; John *Cove* 1642 PrD. From Cove (D, Ha), or North, South Cove (Sf). Sometimes, perhaps, from OE *cōf* 'bold, eager': Walter *Cove* 1249 AssW; Robert *Cove* 1282 LLB A.

**Covell:** Robert *Covell* 1476 FrY. A nickname from OE *cufle* 'cloak'.

**Coven:** (i) Roger *Coueyn* 1373–4 FFWa; John *Coueyne* 1373–5 AssL. OFr *covine* 'fraud, deceit'. (ii) Margaret *de Covene* 1286 ForSt. From Coven (St).

**Coventry:** Alan *de Couintre* 1194 P (Nth); Henry *de Coventre* 1262–3 FFEss; John *Coventre* 1366 IpmW. From Coventry (Warwicks).

**Cover:** (i) Robert *le Cuver* 1210 FrLeic; Richard

*Couer* 1219 AssY. A derivative of ME, OFr *cuve* 'cask, vat', or OFr *\*cuvier* 'cooper'. (ii) Walter *le Cuverur* 1200 Cur (Sr); Hamund *le Coverur* 1262 *For* (Ess). OFr *couvreor, covreor* 'one who covers or roofs buildings' (1393 NED). This would inevitably become *Cover*.

**Coverdale:** Reginald *de Coverdall* 1245 FFL; Thomas *de Coverdale* 1297 SRY; John *Couerdale* 1379 PTY. From Coverdale (ER, NR, WRYorks).

**Covert:** Roger *le Couert* 1276 KB (Sx); Robert *le Couert* 1296 SRSx; Jane *Covert* 1662–4 HTDo. OFr *covert* 'reserved, guarded, crafty'.

**Cowan:** *v.* MACOWEN

**Coward, Cowherd:** Thomas *le Cuherde* 1255 MEOT (Ess); John *Kuhirde* 1274 RH (Hu); Adam *le Couherd* 1317 AssK; John *Cowherde* 1327 SRWo; John *Coward* 1540 Whitby (Y). OE *cūhyrde* 'cow-herd'. A rare variant is evidenced in the forename of *Cuward* de Blakepet 1198 FF (Bk). OE *\*cū-weard* 'cow-guard'. *Cowherd* is uncommon.

**Cowburn, Cowban:** Laurence *Cowbron* 1563, William *Cowban* 1585 LaWills; Richard *Cowburne* 1624 OtleyPR (Y); James *Cowbone* 1662, Francis *Cowborne* 1663 LaWills. From Cowburn (La).

**Cowcher:** *v.* COUCHER

**Cowden:** From Cowden (K, Nb, ERY), or Green Cowden in Bakewell (Db). In Scotland from Cowden in Dalkeith (Midlothian).

**Cowdray, Cowdrey, Cowdry, Cowdroy, Cowderoy, Cowdery, Coudray:** Engelram *de Coudrai* c1170 Riev (Y); Richard *de Coudrey* 1220 Cur (Ha); Henry *de la Coudrey* 1279 AssSt. OFr *coudraie* 'hazel-copse'. The earliest bearers of the name came from France, e.g. Coudrai (Seine-Inférieure), Coudray (Eure), etc. As Cowdray (Sussex), which has replaced the earlier English name of *Sengle*, is found as *la Codray* in 1285 (PN Sx 17), the French *coudraie* was also used in England and the surname may derive from the Sussex place or denote residence near a hazel-copse. Later forms show confusion with CORDEREY.

**Cowell, Cowlell:** (i) Henry *de Cuwell* 1196 MemR (Nth); Thomas *de Cuhull'* 1221 AssGl; John *Cowell* 1401 AssLa. From Cowhill (Lancs, Glos), or Cowleigh Park (Worcs). (ii) In Manx for *Mac Cathmaoil* 'son of *Cathmaol*', Cionaidh *Ua Cathmhaoil* 967, Conor *Mac Cawel* 1252, *McCowle, McCowell, Cowle* 1511, *Cowell* 1690 Moore.

**Cowey, Cowie, Cowee:** Robert *de Cowhey* 1275 PetreA; John *de Cowey* 1270 FFC; Felicia *de Coweye* 1279 RH (Hu). From Cowey Green in Great Bromley (Ess). Scottish *Cowie* is from the barony of Cowie (Kincardine).

**Cowherd:** *v.* COWARD

**Cowick:** John *Cowyk* 1440 FFEss. From Cowick (WRY), Cowicks in Sheering (Ess), or Cowick Barton in Exeter St Thomas (D).

**Cowie:** *v.* COWEY

**Cowler, Cowlman:** Richard *le Coulare* 1333 MEOT (So). A derivative of OE *cug(e)le, cūle* 'a garment worn by monks, a cowl', a maker of cowls.

**Cowley:** Osbert *de Couela* 1167 P (O); Juliana *de Kulega* 1199 AssSt; John *de Couele* 1230 P (Mx); William *de Coule* 1314 LLB E; William *de Colley* 1327

SRDb. From Cowley (Bucks, Devon, Oxon, Staffs; Derby, Lancs; Glos, Middlesex), of varied origins.

**Cowley, Kewley:** Flann *MacAulay* 1178, *MacCowley* 1504, *Cowley* 1587, *Kewley* 1611 Moore. A Manx contraction of MACAULAY.

**Cowlin, Cowling:** *v.* COOLING

**Cowling, Cowlin, Couling, Coulling:** Mathew *de Couling* 1260 AssC; William *Cowlyng* 1520 FrY. From Cowling (WRYorks). *v.* also COLLING.

**Cowlishaw, Coleyshaw, Collishaw:** Thomas *Collyshaw* 1641 PrSo; Joseph *Colishaw* 1680, William *Cowlishaw* 1704 DbAS 36. From Cowlishaw (Lancs, Derby).

**Cowlman:** *v.* COWLER

**Cowper:** *v.* COOPER

**Cowperthwaite:** *v.* COPPERTHWAITE

**Cowpland:** *v.* COPELAND

**Cowser:** *v.* COSIER

**Cowstick, Cowstock:** John *de Coustok* 1296, Agnes *Cowstoke*, Thomas *Coustoke* 1525 SRSx. From Cowstocks Wood in Danehill (Sx).

**Cowtas:** *v.* COULTAS

**Cowwell:** *v.* COWELL

**Cox:** *v.* COCK

**Coxall:** *v.* COGGESHALL

**Coxeter:** Adam *le Cocsetere* 1260 AssCh. 'One who sets the cocks in a cock-fight' (1828 NED).

**Coxfoot:** *v.* COCKSFOOT

**Coxhead:** Thomas *Cokkeshed* 1424 LondEng. 'Cock's head', a nickname.

**Coxon, Coxen, Cockson:** Godui *Coccesune* c1095 Bury (Sf); Roger *Cokson* 1332 SRCu; John *Coxson* 1539 FeuDu; William *Coxon* 1631 FrY. 'Son of *Cock*', an original nickname from OE *coc(c)* 'cock'.

**Coxton:** Henry *Cokstan* 1298 AssL; William *Cokston* 1327 PN Ess 460; John *Coxton* 1407 IpmNt. From Little Cockstones in Stebbing (Ess).

**Coy, Coey:** Walter *le Coi* 1203 AssNth; Walter *le Coy* 1296 SRSx; William *Coye* 1301 SRY. Fr *coi*, earlier *quei* 'quiet, still', 'shy, coy' (c1330 NED).

**Coyish:** *v.* COSH

**Coyne:** Allan *Coigne* 13th Ronton (St); John *Coyne* 1242 FFSt; John *Coyn* 1327 SRC. ME *coyn*, *coigne*, AFr *coigne*, Fr *coin* 'a die for stamping money' (1362 NED), 'a piece of money' (c1386). Metonymic for coiner, minter, a common occupation name: William *le Coiner* 1327 SRSo. cf. CONYER.

**Coysh:** *v.* COSH

**Coyte, Coit:** Walter *le Coyt* 1275 RH (O); John *Coyt* 1327 SRSf; William *Coyte* 1681 ER 62. OFr *coit* 'flat stone'. cf. Coyter, or caster of a Coyte c1440 PromptParv. cf. also Alice *Coyteman* 1327 SRSf; John *Coiter* 1327 SRSx. Metonymic for a player of the game.

**Cozens, Cozins:** *v.* COUSEN

**Cozier:** *v.* COSIER

**Crabb, Crabbe, Krabbe:** Walter, Steffanus *Crabbe* 1188 P (Do), 1217 Pat. OE *crabba* 'crab', either for one who walked like a crab (cf. *Crabeleg* 1148 Winton) or, as in German and East Frisian, for a cross-grained, fractious person; or ME *crabbe* 'wild apple' (c1420 NED), of persons 'crabbed, cross-grained, ill-tempered' (1580).

**Crabtree:** John *atte Crabbetrywe* 1301 ParlR (Ess). 'Dweller by the wild apple-tree' (c1425 NED).

**Crace:** *v.* CRASS

**Crackbone:** Simon *Crakebone* 1279 RH (C); William *Crakebon* 1378 FFEss; Thomas *Crackborne* 1635 ER 61. 'Break bone', OE *cracian*, *bān*. A nickname for a quarrelsome person, or for the official who inflicted this punishment of medieval law. cf. 'Quikliche cam a cacchepol and craked a-two here legges' (Langland). Also Richard *Crakepole* 1242 AssDu 'crack pole'; William *Crakepot'* 1299 FFY 'break pot'; Andrew *Crakescheld* 1378 KB (Nf) 'break shield'; Simon *Craketo* 1279 RH (Hu) 'break toe'. *v.* also BRISBANE.

**Crackel:** Alan *de Crachale* 1204 AssY; Thomas *Crakall* 1414 FrY. From Crakehall or Crakehill (NRYorks).

**Cracknall, Cracknell:** Elias *de Crackenhal'* 1220 Cur (Y); Robert *Craknell* 1524 SRSf. Crakehall and Crakehill (NRYorks) are explained by Smith as 'Craca's nook'. This would be OE *Cracanhale*, a form which has survived in the surname although not evidenced in the place-name forms. cf. CRACKEL.

**Cracknot:** Adam *Crakenot* 1296 SRNb. *Crak-en-ot*, a double diminutive of OE *Craca*.

**Craddock, Cradduck, Cradock, Cradick:** *Cradoc* (*Caradoch'*) 1177 P (He), 1185 P (Glam); *Craddoc* Arcuarius 1187 P (Sa); William, Philip *Craddoc* 1205 P (Wo), 1296 SRSx; Robert *Cradock* 1301 SRY. Welsh *Caradawc*, *Cradawc*, *Caradoc*, *Caradog*, an old and famous name, familiar in its Latin form *Caractacus* for *Caratācos* who was taken as prisoner to Rome c. 51 A.D.

**Craft, Crafts:** (i) Aluric *Craft* 1185 Templars (Ess); Basil *Craft* 1283 SRSf. OE *cræft* 'skill, art', especially 'guile, cunning'. (ii) Roger *de Craft* 1213 Cur (Wa), *de Croft* (*Craft*) 1214 ib.; Robert *de Craft* 1222 AssWa. From Croft (Lei), earlier *Craft*. Probably also local. *v.* CROFT.

**Cragg, Craggs:** Henry *Crag* 1204 AssY; Hudde *del Crag* 1260 AssLa; Peter *del Kragg*, John *Cragges* 1301 SRY. 'Dweller by the steep or precipitous rugged rock(s)' (ME *crag*).

**Craig:** John *del Crag* 1143–1214 Black; John *of the Craig* 1335 ib. A Scottish form of CRAGG.

**Craigie:** Brice *de Cragy* 1317 Black; John *de Craigie* 1371 ib. From Craigie (Ayrshire, West Lothian, Angus, Perthshire).

**Craik, Crake, Creyk:** Henry *de Crake* (Dumfries), James *de Crake* (Selkirk) 1296 CalSc; Andrew *Craik* 1453 Black. Scottish Lowland surnames from Crayke (NRYorks).

**Crain, Craine:** *McCroyn* 1408, *McCraine* 1422, *MacCarrane* 1422, *Craine* 1586 Moore. A Manx contraction of *Mac Ciarain* 'son of *Ciaran*'.

**Craise:** For CRASS.

**Crake:** Ralph, Henry *Crake* 1276, 1279 RH (Y, C). ME *crake* 'crow or raven'.

**Crakebone:** Simon, Walter *Crakebon* 1279 RH (C), 1327 *SR* (Ess). 'Crack bone', 'break bone', a nickname for the official who inflicted the cruel punishment of medieval law. cf. 'Quikliche cam a cacchepol and craked a-two here legges' (Langland). cf. BRISBANE.

**Craker:** *v.* CRAWCOUR

**Cramer, Kramer:** A Huguenot name. Jean-Louis *Cramer*, a Protestant refugee from Strasburg, became a captain in the English army, while Jean-Antoine *Cramer* was a professor at Oxford and Dean of Carlisle (Smiles 380). Flemish *kramer* 'merchant, colporteur'.

**Cramp:** Walter *Crampe* 1200 Oseney (O); William *Crampe*, Richard *le Crompe* 1275 SRWo. A variant of CRUMP. cf. OHG *chrampf*, Ger *krampf* 'curved, a hook'.

**Crampon:** William *Cramphorne* 1324 LLB K; Abraham *Cramppone*, Thomas *Crampporne* 1642 PrD. OFr *crampoun* 'a grappling iron', or a nickname, 'curved horn'. *v.* CRAMP.

**Cran:** *v.* CRANE

**Cranbore, Cranbourne:** William *de Cranburne* 1210–11 PWi; Giles *of Craneburne* 1249 AssW. From Cranborne (Do), or Cranbourne (Ha).

**Cranden, Crandon:** William *de Crandene* 1276 RH (O); Walter *de Crandene* 1298 AssL. From Crandon (So), or Crandean in Falmer (Sx).

**Crane, Cran:** Osbert *Crane* 1177 P (C); Jordan *Cran* 1219 Cur (Ess); William *le Crane* 1235 FFEss; Thomas *le Cran* 1243 AssSo. OE *cran* 'crane', no doubt 'long-legged'.

**Craneshanks:** William *Craneschank* 1383 *ERO*; John *Craneshank* 1507 FFEss. 'Crane shanks', OE *cran, scanca*, a nickname for a long-legged person. cf. Nicholas *Cranebayn* 1219 AssY 'crane bone'.

**Cranford:** Roger *de Cranford* a1150–83 MCh; Nicholas *de Cranford* 1259 FFO; Thomas *Cranford* 1466 FFEss. From Cranford (D, Ess, Mx, Nth).

**Crank, Cronk:** Godric *Cranc'* 1121–48 Bury (Sf). ME *cranke* 'lusty, vigorous', 'in high spirits', 'merry' (1398 NED).

**Crankshaw, Cranshaw:** *v.* CRONKSHAW

**Cranley, Cranleigh:** Alan *de Cranle* 1247 FFO; Alan *de Cranlai* 1307 IpmY; John *de Cranle* 1338 FFW. From Cranley (L, Sf), or Cranleigh (Sr).

**Cranmer, Cranmore:** Nigel *de Cranemore* 1235 PN Wt 209; Hugh *de Cranemere* 1275 RH (Herts); Thomas *de Cranmer* 1373–5 AssL; Edmund *Cranmere* 1422, John *Cranmer* 1447 IpmNt. From Cranmere (Sa), Cranmore (So), or Cranmore in Shalfleet (Wt).

**Crannis:** Thomas *Cranewys* 1418 BuryW; Andrew *Cranewys* 1524 SRSf; Charles *Crannis* 1662 *HTEss*. From Cranes in Nevendon (Ess).

**Cransick:** *v.* CRANSWICK

**Cranston:** Robert *Cranston* 1327 SRSo; Andrew *de Cranstoun* a1338, Thomas *de Cranstoun* 1423 Black. From the barony of Cranston (Midlothian).

**Cranswick, Cransick:** Henry *de Crancewic'* 1219 AssY; John *de Cranncewyk'* 1351 AssL; Andrew *Crauncewyk* 1293 KB (Y). From Cranswick (ERY).

**Cranwell:** Richard *de Cranwella* 1176 P (L); William *de Cranewelle* 1298 AssL; John *Cranwell* 1442 IpmNt. From Cranwell (L), or Cranwell in Waddesdon (Bk).

**Crapp:** *v.* CROPP

**Crapper:** *v.* CROPPER

**Craske:** Godith *Crasc* 1197 P (Nf); Ralph *Craske* 1207 Cur (Nf). 'The fat, lusty'. cf. c1440 PromptParv 'craske, or fryke of fatte (K. crask, or lusty), *crassus*'.

**Crass, Crace, Craise, Craze:** Normannus *Crassus* 1086 DB (L); Hervey *le Cras* 1130–2 Seals (St); Rogerus *Crassus*, Roger *le Cras* 1203 Cur (Lei); Robert *Krase* 1277 *Ely* (Sf). OFr *cras* 'fat, big', Lat *crassus*. *v.* also GRACE and GROSS with which this name was early confused: Rogerus *Crossus, Crassus, Grassus* 1202 AssL; Hugo *Grassus, Crassus* 1211–12 Cur (W).

**Crasswell:** *v.* CARSWELL

**Crassweller:** John *Cressweller* 1558 SxWills. 'Man from Cresswell.' *v.* CARSWELL.

**Craster:** Albert *de Craucestre*, Ivo *de Crawecestre* 13th Guisb. From Craster (Nb), *Craucestr'* 1242.

**Crate, Crates:** *v.* CREET

**Crathorne, Crathern, Craythorne:** William *de Crathorne* 13th Guisb; William *de Crathorn* 1345 FFY; Robert *Crauthorn, Craythorn, Crattorn, Crawthorn* or *Crathorn* 1509 LP (L). From Crathorne (NRY).

**Crauford, Craufurd, Crawford, Crawforth, Crawfurd:** John *de Crauford* 1147–60 (Black). From Crawford (Lanark). The surname appears early in England: Nicolaus *de Crauford* 1205 P (So).

**Craven:** Torfin, John *de Crauene* 1166 Cur (Y), 1242 AssDu; John, Agnes *Craven* 1332 SRCu. From the district of Craven (WRYorks).

**Crawcour, Craker, Croaker, Croker, Crocker, Creegor, Cregor:** Helias *de Creuequor* 1158 P (Sf); Robert *(de) Creuequoer* 1195 P (K); Robert *de Crouequoer* 1200 P (K); Rainald, Alexander *de Creuker* 1212 Fees (L); Robert *de Crequer* 1284 FA (C). From Crèvecœur (Calvados, Oise, Nord). The baronial family came from Calvados. Occasional examples with *le* suggest the possibility also of a nickname, Fr *crève-cœur* 'break heart, heart-breaker'. cf. Richard *Brekehert* 1327 SRSo. The surname has probably been partly absorbed by *Craker, Croaker, Croker* and *Crocker*. Hamo de *Creueker* has left his name in Crockers (Sussex) and The Creakers in Great Barford (Beds), *Crewkers* 1539, *Crecors* 17th, *Crakers* 1766 (PN Sx 524, PN BedsHu 52).

**Crawlboys:** Ralph *Crouleboys* 1251–2 FFWa; Peter *Croilleboys* 1290 IpmW; Thomas *Croyleboys* 1344 FFW. 'Overturn the wood', OFr *crouler, bois*, a nickname for a wood-cutter. cf. Fr *Croullebois*.

**Crawley, Crowley:** Pagan *de Craweleia* 1130 P (Bk); Thomas *de Crowele* c1280 SRWo; William *Craweley* 1397 IpmGl. From Crawley (Bucks, Essex, Hants, Oxon, Sussex), or Crawley in Membury (Devon).

**Crawshaw, Crawshay, Croshaw, Crowsher:** John *de Crouschagh* 1308 Wak (Y); Adam *de Craweshaghe* 1332 SRLa; Ralph *de Croshawe* 1379 PTY; Susanna *Crawshay* 1760 Bardsley. From Crawshaw Booth (Lancs).

**Cray, Kray:** Gunnilda *de Craie* 1203 FFK; Simon *de Creye* 1317 AssK; William *Cray* 1372 IpmW. From Foots, North, St Mary, St Paul's Cray (K).

**Craythorne:** *v.* CRATHORNE

Craze: v. CRASS

Crearer: v. CRERAR

Crease, Crees, Creese: Cenric *Cres* c1095 Bury (Sf); Richard *le Cres* 1275 RH (Nf); Hugh *Crees* 1316 Wak (Y). OE *crēas* 'fine, elegant'.

Crebbin: v. CRIBBIN

Crecy, Cressy: Hugo *de Creissi* 1171 P (La); Alexander *de Crecy*, *de Cressi* c1182 Gilb, 1185 Templars (L); Beatrix *Cressy* Hy 3 Gilb (L). From Cressy (Seine-Inférieure). v. ANF.

Creech: (i) John *ate Creche* 1327 SRSx; Robert *Creche* 1327 SRSf. 'Dweller by the creek'; OE *cricc*, ME *crich(e)*, with lengthening of ī in the open syllable became *crēche*. v. PN C 254–6. (ii) Peter *de Cryche* 1327 SRSo. From Creech (Dorset, Som). (iii) Douenaldus *de Creych* 1204–41 Black. From Creich (Fifeshire).

Creed: (i) *Creda* 1198 FFNf; *Crede* 1279 Barnwell (C); Wadin *Crede* 1191 P (Wa); Theynewin *Crede* 1242 AssSo. OE *Creoda* (Redin). (ii) John *de Crede* 1370 LoPleas. From Creed Fm in Bosham (Sussex).

Creedy: John *Credy* 1411–12 FFSr; John *Creedy* 1642 PrD. From North Creedy in Sandford, or Lower Creedy in Upton Hellions (D).

Cre(e)gor: v. CRAWCOUR

Creek, Creeks: (i) Bartholomew *de Crek* 1187 P (Nf); John *de Creke* 1298 PN C 117; John *Creek* 1365 LoPleas. From Creake (Nf). (ii) Godwin *Cræc* 1166 P (Nf); Algar *Chrech'* 1179 P (Nf); Thomas *Crek* 1268 AssSo. OFr *creche*, ME *creke* 'basket'. Metonymic for a maker of baskets.

Creeper: Robert *Crieper'* 1221 AssWo. 'Crawler', OE *crēopere* 'cripple'. cf. William *Crepeheg'* 1238–9 FFEss 'creep by the hedge'.

Crees(e): v. CREASE

Creet, Crate, Crates: John *Cret* 1202 FFNf; Thomas *le Creat* 1275 RH (K); Sisilla *la Crete* 1281 CtW. OE *cræt* 'cart'. Metonymic for a carter.

Creighton: John *de Creghton* 1327 SRDb. From Creighton (Staffs). v. CRICHTON.

Crellin: A metathesized form of *Crennell*, a Manx name from *MacRaghnaill* 'son of *Raghnall*', from ON *Rognvaldr* 'ruler of the gods', the name of several kings of Man: Godfrey *MacMicRagnaill*, king of Dublin, 1075, *MacReynylt* 1511, *Crenilt* 1627, *Crennil* 1646, *Crellin* 1610 Moore.

Crenell, Crennell: v. CARNALL, CRELLIN

Crepin: v. CRISPIN

Crepping: Walter *de Creping* 1202, *de Crepping* 1209 FFEss; Robert *de Creppinges* 1260 AssC. From Crepping Hall in Wakes Colne (Ess).

Crerar, Crerer, Crearer: John *McAchrerar* 1541 Black; William *Crerar* 1554 ib. Gael *criathrar* (miller's) 'sifter' or 'sievewright'.

Cresner, Cressner: Adeliz *de la Kersunere* c1190 BuryS (Sf); John *de la Cressonere* 1331 FFY; Alexander *Cressener* 1479, Thomas *Cresner* 1487 FFEss. From La Cressonière (Calvados), v. ANF, or 'dweller by the cress-bed', OFr *cressonière*.

Crespigny, Champion de: The first immigrant, Claude *Champion* and his sons, after the family settled in England, were still named *Champion* only. Two of Claude's sons became British officers, and adopted the

name of *Crespigny*, but without the *de*. Gabriel *Crespigny* had a commission in the Foot Guards in 1691, and Thomas *Crespigny* was a cornet of dragoons and captain in a regiment of foot in 1710. Far into the 18th century *Crespigny* without the *de* remained the family name, the first baronet's father and mother being so named in the obituaries. v. J. H. Round, *Family Origins* 109–20.

Cressacre: Thomas *de Cresacre* 1303 FFY; James *de Cressaker* 1407 IpmY; Thomas *Crisaker* 1464 TestEbor. 'Dweller by the field where water-cress grows', OE *cresse*, *æcer*.

Cressall, Cressell, Cresswell: v. CARSWELL

Cressner: v. CRESNER

Cressweller, Cresweller: John *Kerswellere* c1405 FS; Austen *Cressweller* 1525 SRSx; John *Cressweller* 1558 SxWills. 'The man from Cresswell' (Db, Nb, St), or 'dweller at the stream where water-cress grows', from a derivative of Cresswell, OE *cresse*, *wiella*.

Cressy: v. CRECY

Crew, Crewe: Thomas *de Crue* 13th WhC; Thomas *Crewe* 1327 SRSa; Henry *Crewe* 1535 FrY. From Crew (Cu), or Crewe (Ch).

Crew(e)s: v. CRUISE

Crewther: v. CROWTHER

Cribb, Cribbes: Hugo, Osbert *Cribbe* 1195, 1200 P (So). OE *crib(b)* originally 'a barred receptacle for fodder in cow-sheds' (used of the manger of Christ c1000), 'a stall or cabin of an ox' (a1340 NED). cf. dial *crib* 'cattle-fold'. The surname is metonymic for a cow-man.

Cribbin, Crebbin, Gribbin, Gribbon: *MacRobyn* 1511, *Crebbin* 1640, *Cribbin* 1666 Moore. Ir *MacRoibin* 'Robin's son'. *Crebbin* is Manx.

Crichlow, Critchlow: Vkke *de Crikelawa* 1176 P (Nb); John *de Cruchelowe* 1342 LaCt; William *Chrichlowe* 1642, *Critchley* 1682 PrGR. From Critchlow (La). Perhaps also one source of CRUTCHLEY.

Crichton, Crighton, Chrichton, Creighton: Turstan *de Crectune* c1128 (Black); Thomas *de Creitton* c1200 ib.; William *de Crichton* c1248 ib.; Alisaundre *de Creightone* 1296 ib.; Margaret *Chrightone* 1685 ib. From Crichton (Midlothian).

Crick, Cricks: Robert *Crike* 1189 Sol; Walter *Cricke* 1276 RH (BK); Thomas *Cricke* 1364 ColchCt; William *Atkrik* 1379 PN ERY 220. From Crick (Nth), or 'dweller at the inlet', ON *kriki*. But the usual lack of any preposition would suggest that there is also another source of the surname.

Cricket, Crickett: Ida *Criket* 1195 Cur (Mx); John *Criket* 1305 AssW; Thomas *Creket* 1470, Robert *Cryket* 1495 LLB L. A nickname from the cricket, OFr *criquet*.

Cricks: v. CRICK

Cridde: Robert *Cryde* 1296 SRNb; Richard *Cridde* 1327 SRSo; Richard *Cryde* 1332 SRSx. OE *Crioda*.

Criddel, Criddell, Criddle: Robert *Cridel* 1327 SRSo; William *Cridill'* 1345 KB (Lo). OE *Cridela*.

Crier, Cryer: Geoffrey, Ralph *le Criur* 1221 Cur (Herts), 1221 AssWo; Robert *le Crieur* 1269 AssNb. ME *criere*, OFr *criere*, nominative of *crieur* 'crier', 'officer of the court of justice who makes public announcements' (1292 NED), 'common or town crier' (1387 NED).

**Crighton:** v. CRICHTON

**Crimes, Chrimes, Crymes, Scrime:** John *Crime* 1275 RH (Nf); Richard *Crymes* alias *Cremes* 1554 Pat (Lo); Ellis *Crymes* 1642 PrD. Probably a variant of GRIME.

**Crine:** Adam *Crin* 1221 AssWa; Robert *Cryne* 1327 SRSo; Thomas *Cryn* 1404 FrY. A nickname from OFr *crin* 'hair'.

**Cripp, Crippe, Cripps:** Gilbert *Crippe* 1249 AssW; Robert *le Crip* 1275 SRWo. ME *crippe* 'pouch'. Metonymic for a maker of these. v. also CRISP.

**Crippen, Crippin:** v. CRISPIN

**Cripps:** v. CRISP

**Crisp, Crispe, Chrisp, Cripps, Crips, Chrippes, Scripps:** Benedictus *Crispus* c1030 OEByn; Henry *le Cresp* c1200 ELPN; Walter *Crips* 1273 RH (Hu); Richard *Crysp* 1275 SRWo; Richard *Crispe, Crips* 1289 AssCh; Joan *le Crypse* 1297 MinAcctCo; Gilbert *le Crispe* 1311 Battle (Sx); John *Chrispe* 1589 SfPR. OE *crisp, cryps*, Lat *crispus* 'curly, curly-haired' or OFr *crespe* 'curled'. *Crisp* may also be a short form of *Crispin*. cf. Odin *Crispi filius* c1095 Bury (Sf), Roger *filius Crispi* c1150–60 DC (L). *Scripps* is for *Cripps*, with inorganic initial *S* as in STURGE.

**Crispin, Chrispin, Crepin, Crippen, Crippin:** Stanmer *Crispini filius* c1095 Bury (Sf); *Crespinus* 1207 Cur (L); Willelmus *filius Crispian* 1273 RH (O); *Creppinus le Seller* 1319 SRLo, *Crispin la Seeler* 1336 LLB E; Milo *Crispinus* 1086 DB; Turstin' *Crispin* 1166 P (Y); Ralph *Crespin* 1169 P (D); Ralph *Crispun* 1208 Cur (Y); Elias *Crepun* 1208 P (W); Roger *le Crespin* 1268 AssSo; Edmund, Walter *Crepyn* 1312, 1317 FFC. *Crispinus*, a Roman cognomen from Lat *crispus* 'curly', was the name of the patron saint of the shoemakers who was martyred at Soissons c285 along with *Crispinianus*, in French, SS. *Crépin* and *Crépinien*. The former survives as *Crepin* or *Crippin* which may also be nicknames from OFr *crespin*, a derivative of *crespe* 'curly'. cf. Ralph de Alegate called *Crepyn*, Ralph *Crepyn* called de Alegate 1306 LLB B. *Crispun* and *Crepun* are hypocoristics of *Crispin*, *Crepin* and cannot be associated with OFr *crespon*, Fr *crépon* 'crape', a material with a crisped or minutely wrinkled surface which is unknown before the 16th century. The surnames may derive from the saint or from a nickname 'curly-haired'. According to Lanfranc (d. 1089), Gilbert Crispin was the first man to receive this nickname and two of his sons adopted it as their surname. His grandson Gilbert Crispin was abbot of Westminster.

**Crisplock:** William *Crisploc* 1228 FFEss; William *Cripslok* 1317 AssNth; Richard *Cripslok* 1396 FFEss. 'Curly-haired', OE *crips, locc*.

**Crissell, Crisswell, Criswell:** v. CARSWELL

**Crist:** v. CHRIST

**Critchlow:** v. CRICHLOW

**Crittall, Crittell, Crittle:** William *Crotall* 1487 Cl (Sx); John *Crotehole* 1525 SRSx. From Crit Hall in Benenden (K), *Crotehole* 1292.

**Croaker:** v. CRAWCOUR

**Croasdale, Croasdell:** v. CROSSDALE

**Crocker, Croker:** John *le Crockare* 1275 SRWo; Simon *le Crockere* 1279 RH (O); Henry *le Crokere* 1288 MESO (Sx). A derivative of OE *croc(c), crocca* 'an earthen pot', hence 'potter' (a1333 MED). The surname might also be identical with CRAWCOUR.

**Crocket, Crockett, Crockatt:** Margeria *Croket* 1332 SRSt; Richard *Croket* 1403 IpmNt; Thomas *Crokket* 1461 PN Ch iv 69. A nickname from AFr *croket*, OFr *crochet* 'a curl or roll of hair'. In Scotland, the Galloway *Crockett* is said to be from *MacRiocaird* 'son of *Richard*'. v. Black.

**Crockford:** Richard *de Crocford* 1214 P (Sr); William *de Crockford* 1332 SRSr; Thomas *Crockford* 1576 SRW. From Crockford Bridge in Chertsey (Surrey).

**Croft, Crofts, Cruft:** Hugo *de Croft* 1162 P (He); Richard *de la Croft* 1230 P (Ha); William *del Croft* 1288 AssCh; Robert *del Croftes* 1332 SRSt; Richard *atte Crofte*, William *Craft, Cruft* 1353 ColchCt; John *Craft, Croft* 1361, 1367 ib. From Croft (Hereford, Lincs, NRYorks) or 'dweller by the croft(s),' OE *croft*. In ColchCt, *Craft* is the almost invariable form. It must often be for *Croft*, but sometimes, probably, for CRAFT.

**Crofter:** For *atte Croft*. cf. Matill *Croftman* 1327 SRSx.

**Crofton:** Reginald *de Crofton* 1190 P (K); Walter *de Crofton'* 1219 AssY; William *Crofton* 1455 FFEss. From Crofton (Cu, Ha, K, W, WRY).

**Croker:** v. CRAWCOUR, CROCKER

**Crokern:** John *Crokern* 1401 Hylle. 'Dweller at the pottery kiln', OE *crocca, ærn*.

**Crole, Croll:** v. CROWL, CURL

**Cromb, Crome, Croom, Croome, Croombes, Croombs, Crumb:** (i) Robert *le Crumbe* 1199 AssSt; Maud *le Crombe*, John *Croume* 1275 SRWo; Simon *Crumbe* 1296 SRSx; Luke *Croom* 1309 FFEss; Geoffrey *Crombe* 1327 SRSx. OE *crumb* 'bent, crooked, stooping' or OE *\*cramb, \*cromb*, ME *crome, cromb* 'a hook, crook', also in the forms *croumbe, cromp*. As we also find Richard *le Crombere* 1327 SRSf, the surname may be either a nickname 'bent, stooping' or occupational 'maker of hooks or crooks'. (ii) Adam *de Crumbe* 1199 MemR (Wo); Simon *de Crombe* 1275 SRWo; Stephen *de Crome* 1275 RH (W); John *de Crome* 1349 FrY. From Croom (ERYorks) or Croome (Worcs).

**Crombie, Cromie, Crumbie, Crummay, Crummey, Crummie, Crummy:** Patrick *of Cromby* 1423 Black; Robert *Crumby* 1450 ib.; David *Crommy* 1516 ib. From Crombie (Aberdeenshire), in which the *b* is not pronounced.

**Crommelin:** A Huguenot name. Louis *Crommelin* from Armancourt near St Quentin, settled in Holland, and was invited by William III in 1698 to superintend the linen industry in Ireland, the family having been linen manufacturers in France for over 400 years (Smiles 296–8, 380).

**Crompton, Crumpton:** Richard *de Crompton* 1246 AssLa; Thomas, Widdow *Crompton* 1592 AssLa, 1672 HTY. From Crompton (Lancs).

**Cromwell:** Ralph *de Cromwella, de Crumwella* 1177, 1195 P (Nt); John *de Crombewelle* 1310 LLB D;

Ralph *Cromwell* 1454 IpmNt. From Cromwell (Notts).

**Cronk:** *v.* CRANK

**Cronkshaw, Cronshaw, Cronshey, Crankshaw, Cranshaw:** William *de Crounkeshawe* 1412 WhC (La). From Cronkshaw (Lancs), *Cronkshay* 1507 PN La.

**Crook, Crooke, Krook:** (i) Rainald *filius Croc,* Rainald *Croc* 1086 DB (Ha); *Crocus* venator Wm 2 (1235) Ch (Ha); Lefwin *Croc* 1066 DB (Sf); Walter *Chroc* ci130 EngFeud (W); Matthew *Croc* 1158 P (Ha); John *le Cruk* 1269 AssSo; Philip *le Crok* 1288 Pat. ON *Krókr,* ODa *Krōk,* which may have been introduced into England from Denmark or Normandy, or the ON nickname *Krókr* 'hook, something crooked', referring to crook-backed or sly and cunning persons. The surname may also derive from the common noun *krókr,* a Scandinavian loan-word in English, in the latter sense. (ii) John, William *del Crok* 1310–33 InqLa, 1332 SRLa. 'Dweller at a nook or bend', ME *crok,* ON *krókr.*

**Crooker:** 'Dweller at a bend.' cf. CROOK.

**Crookes, Crooks:** Robert *de Crokis* 1297 SRY. From Crookes in Sheffield (Yorks).

**Crookfoot:** Arkil *Crocfot* 1190 P (Y); Bartholomew *Crocfot* 1231 Cur (Herts). 'Crooked foot', ON *krókr,* OE *fōt.* cf. John *Bightfoate* 1642 PrD 'bent foot'; John *Crocbayn* 1246 AssLa 'crooked bone'.

**Crookshank(s):** *v.* CRUICKSHANK

**Crookson:** 'Son of Crook.' *v.* CROOK.

**Croom(e), Croomb(e)s:** *v.* CROMB

**Croote:** *v.* CROTE

**Cropp, Crapp:** Ailwin *Crop* 1205 P (Sx); Hervey *Crappes* 1219 AssY; William *Croppe* 1327 SRSf, 1534 FFEss. Usually metonymic for CROPPER, but sometimes, perhaps, local: Isabella *del Crop* 1327 SRSf. 'Dweller on the hill-top', ME *cropp.* *v.* EPNE.

**Cropper, Crapper:** Roger *le Croppere* 1221 AssWo; John *Crapere* 1275 RH (Nf); William *Croper* 1276 RH (Y); Alice *le Crappere* 1315 Wak (Y). A derivative of ME *croppen* 'to crop, pluck', a cropper, reaper. cf. A cropper, *decimator* CathAngl. For the unrounding of *a* in *Crapper,* cf. CRAFT and CROFT.

**Crosby, Crosbie, Crossbee:** (i) Gillemichel *de Crossebi* 1176 P (We); Adam *de Crosseby* 1227 Cur (L); Henry *Crosseby* 1383 AssWa. From Crosby (Cumb, Lancs, Lincs, Westmorland, NRYorks). (ii) Iuo *de Crosseby* 1178–80 Black; Richard *de Crossebi* c1249 ib.; Robert *de Crosby* 1347 ib. From Crosbie or Corsbie (Ayr, Kirkcudbright, Berwick).

**Croshaw:** *v.* CRAWSHAW

**Crosier, Crozier, Croser:** William *le Croyser* 1264 Eynsham (O); William *le Crocer* 1305 MEOT (Sf); Thomas *Croser* 1393 FFEss. OFr *crosier, crocier, crosser* 'crosier', the bearer of a bishop's crook or pastoral staff, or of the cross at a monastery. The name might also denote a seller of crosses or a dweller by a cross. *Croyser* is the common early form.

**Cross, Crosse, du Cros:** Richard *del Crosse* 1285 AssLa; William *atte Cros* 1327 SRSf; Robert *Cros* 1354 ColchCt. 'Dweller by the cross.'

**Crossdale, Croasdale, Croasdell, Croysdill:** John *de Crosdale* 1379 PTY; John *Crosedill* 1688 FrY. From Crossdale (Cumb).

**Crossland, Crosseland, Crosland:** Christiana *de Crosseland* 1308 Wak; George *Crosland,* William *Crosselonde* 1536, 1538 CorNt. From Crosland (WRYorks).

**Crossley, Crosley, Crosleigh:** Peter *de Crosseley* 1298 IpmY; Johamma *de Crosselay* 1379 PTY; Richard *Crossley* 1481 FrY. From Crossley in Mirfield (WRY), or 'dweller at the clearing with a cross', ON *kross,* OE *lēah.*

**Crossman:** Philip *Crosman* 1327 SRSo. 'Dweller by the cross.'

**Crosthwaite, Crosswaite, Crostwight:** Henry *de Crostweyt* 1242 Fees (Nf); John *de Crosthuaite* 1332 SRCu. From Crostwight, Crostwick (Norfolk) or Crosthwaite (Cumb, Westmorland, NRYorks).

**Croston:** Hugh *de Croston'* 1190 P (Bk); Andrew *de Crostone* 1296 Black. From Croston (La). *v.* also CROXTON.

**Crotch:** *v.* CROUCH

**Crote, Croote, Crute:** Nicholas *Crote* 1275 SRWo; John *Crut* 1305 FFEss; Bartholomew *Croote* 1642 PrD. A nickname from ME *crot, crote* 'lump, clod'.

**Crother:** *v.* CROWTHER

**Crothers:** *v.* CARRUTHERS

**Crouch, Crowch, Crotch, Crutch:** Gilbert *Cruche* 1221 Cur (D); William *Attecruche* 1290 *Ass* (Ess); Laurence *atte Crouch* 1327 SRSx; Thomas *Crouch* 1327 *SR* (Ess). 'One who lives near a cross', from OE *crūc,* cf. CROUCHER, CROUCHMAN.

**Croucher, Crutcher:** Dauid *Crucher* 1220 Cur (So); Christina *le Crochere* 1297 MinAcctCo (Beds); John *Crouchere* 1383 AssC. 'Dweller by the cross.' *v.* CROUCH, CROUCHMAN.

**Crouchman:** Richard *Crucheman* 1255 *Ass* (Ess); Ralph *Crocheman* 1260 AssC; John *Croucheman* 1327 SRSx. 'Dweller near the cross.' *v.* CROUCH, CROUCHER.

**Croudace:** *v.* CARRUTHERS

**Crouse:** *v.* CRUISE

**Crow, Crowe:** Ailwin *Crawe* 1180 P (Wa); Nicholas *Crowe* 1187 P (Nf); John *le Crowe* 1332 SRSx. OE *crāwe* 'crow'. In Ireland and the Isle of Man *Crow(e)* is a translation of *Mac Fiachain* 'son of *Fiachan*', 'the crow'.

**Crowdace:** *v.* CARRUTHERS

**Crowde:** Gynuara *Croude* 1327 SRDo; William *Crowede* 1475 FFEss; John *Crowde* 1642 PrD. ME *crouth, croude* 'fiddle'. Metonymic for a fiddler.

**Crowell:** Richard *de Crowell* 1275 RH (L); Deonisia *de Crawel* 1276 RH (Beds); William *Crowell* 1416 IpmY. From Crowell (O), or a lost *Crowell* in Spofforth (WRY).

**Crowfoot:** Godfrey *Crowfote* 1524 SRSf. 'Crow-foot', OE *crāwe, fōt,* though ME *crou-fot* was also a name for the buttercup. But cf. John *Hennefot* 1306 IpmGl 'hen-foot'; John *Cayfot* 1275 SRWo 'jackdaw-foot'; Roger *Pefot* 1202 Pleas (C) 'peacock-foot'.

**Crowhurst, Crowest:** Roger *de Croherst* c1200 ArchC vi (K); William *de Crouherst* 1296 SRSx. From Crowhurst (Surrey, Sussex).

**Crowl, Crowle, Crole:** Walter *de Crul* 1201 P (L); Hugo *de Croul'* 1221 AssWo; Richard *de Crol*, Richard *Croll* 1275–6 RH (L). From Crowle (Lincs, Worcs). *v.* also CURL.

**Crowley:** *v.* CRAWLEY

**Crowmer:** Robert *Crowmer* 1392 LoCh; William *Crowemer* 1406 AssLo. From Cromer (Nf), *Crowemere* 13th.

**Crown, Crowne, Crowns:** (i) Wido *de Credun* 1086 DB (L, Lei); Maurice *de Creun* Hy 2, *de Creona* c1190, *de Croun* c1200 DC (L); Peter *de Croun* 1230 P (Nth); Thomas *Crowne*, William *Croune* 1327 SRWo. From Craon (Mayenne). *v.* OEByn 84. (ii) Richard *Attecroune* 1420 LLB I. 'Dweller at the sign of the crown', OFr *corone*, *corune*.

**Crowner, Corner, Coroner:** Henry *le Coroner* 1255 AssSo; Alice *le Crounor* 1323 AD iv (He); John *Crownere* 1327 SRLei; John *Crouner* 1458 FrY. 'An officer charged with the supervision of the pleas of the Crown'.

**Crowns:** *v.* CROWN

**Crowsher:** *v.* CRAWSHAW

**Crowther, Crowder, Crother, Crewther:** Richard *le Cruder* 1275 RH (K); Hugo *le Crouder* 1278 FrLeic; Kenwrick *le Cruther* 1289 AssCh; Adam *le Crouther* 1296 Wak (Y). A derivative of ME *crouth*, *croude* 'fiddle', a fiddler.

**Croxby:** William *de Croxby* c1200 Gilb; Simon *de Croxeby* 1304 IpmY. From Croxby (L).

**Croxley:** John *de Croxleghe* 1280 IpmY. From Croxley (Herts).

**Croxton, Cruxton, Croston, Croxon:** Godric *de Crocestuna* 1086 ICC (C); Richard *de Croxton* 1277–8 FFEss; John *Croxton* 1403 IpmY. From Croxton (C, Ch, L, Nf, St), Croxton Kerrial, South Croxton (Lei), or Croxton Green in Cholmondeley (Ch).

**Croyden, Croydon:** Stephen *de Croyden'* 1275 RH (W); John *Croidon* 1381 LoCh; Daniel *Croyden* 1641 PrSo. From Croydon (C, Sr).

**Croyle:** *v.* KERRELL

**Croysdill:** *v.* CROSSDALE

**Crozier:** *v.* CROSIER

**Crudd, Crudde:** William *Crude* 1201 Pleas (So); Hervey *Crudde* 1327 SRSf; Thomas *Crudd'* 1379 PTY. ME *crud* 'curds, cheese'. Metonymic for a maker or seller of these.

**Cruel:** William *le Cruel* 1251 AssY; John *Cruel* 1303 FFY; Richard *Cruel* 1315 AssNf. A nickname from OFr *cruel* 'cruel'.

**Cruddace, Cruddas:** *v.* CARRUTHERS

**Cruft:** *v.* CRAFT, CROFT

**Cruickshank, Cruickshanks, Cruikshank, Crook-shank, Crookshanks:** John *Crokeshanks* 1296 CalSc (Haddington); Christin *Crukschank* 1334 Black (Aberdeen). A Scottish name, from ON *krókr* 'hook, something bent', and OE *sceanca* 'shank, leg', 'crooked leg', in early forms always singular.

**Cruise, Cruse, Crewes, Crews, Cruwys, Crouse:** (i) Nicholas *le Criuse* 1213 Cur (Beds); *le Cruse* 1279 RH (Beds); Robert *Creuse* ib. ME *crus(e)*, northern *crous(e)* 'bold, fierce'. (ii) Richard *de Crues* 1214 Cur (D). Perhaps from Cruys-Straëte (Nord).

**Crull, Crulle:** Burewold *Crul* 1066 Winton (Ha);

Roger *Crull* 1219 P (Ha); William *Crul* 1300 CorLo; John *Crull* 1443 CtH. ME *crul* 'curly-haired'.

**Crumb:** *v.* CROMB

**Crumbie:** *v.* CROMBIE

**Crummack, Crummock, Cromack:** John *de Crumbok* 1379 PTY. 'Dweller by the twisted oak', OE *crumb* 'crooked' and *āc* 'oak'.

**Crummay, Crummey, Crummie, Crummy:** *v.* CROMBIE

**Crump:** Peter *Crumpe* 1176 P (Berks); Adam *le Crumpe* 1203 AssSt. OE *crump* 'bent, crooked'.

**Crumplock:** William *Crompeloc* 1275 SRWo. 'Twisted hair', OE *crumb*, *locc*. cf. Hugh *Cromphand* 1302–3 FrC 'twisted hand'.

**Crumpton:** *v.* CROMPTON

**Crundall, Crundle:** Robert *ate Crundle* 1279 RH (O); Thomas *de la Crundle* 1280 AssSo. From Crondall (Hants), Crundale (Kent) or from residence near a chalk-pit or hollow (OE *crundel*).

**Crupper:** Roger *Crupere* 1210–11 PWi. AFr *cruper* 'a cover for the hindquarters of a horse'. Metonymic for a maker of these.

**Cruse:** *v.* CRUISE

**Cruso, Crusoe:** Aquila *Crusoe* 1635 SxAS 86; Francis *Crusoe* 1682 NorwDep. From John *Crusoe*, a refugee from Hownescourt (Flanders), who settled in Norwich.

**Crust:** Herveus *Cruste* 1109 Rams (C); Robert *Crust* 1208 FFL; Nicholas *Crouste* 1275 SRWo. OF *crouste* 'crust of bread', used by metonymy of one hard as crust, obstinate, stubborn.

**Crutch:** *v.* CROUCH

**Crutcher:** *v.* CROUCHER

**Crutchley, Crutchlow:** William *le Crouchele* 1327, John *Crowcheloue* 1525 SRSx. 'Dweller at the clearing or hill with a cross', OE *crūc*, *lēah/hlāw*.

**Crute:** *v.* CROTE

**Cruttenden:** William *Crotynden* 1451 Pat (K). A Kent and Sussex name from a lost place *Cruttenden* in Headcorn (Kent).

**Cruwys:** *v.* CRUISE

**Cruxton:** *v.* CROXTON

**Cryer:** *v.* CRIER

**Crymes:** *v.* CRIMES

**Crystal, Crystol:** *v.* CHRYSTAL

**Cubbell, Cubble:** Henry *Cubbel* c1260–8 LuffCh; John *Cubbel* 1281 MPleas (Bk); Luke *Cubbel* 1296 SRSx. A nickname from ME *cubbel* 'block, stump'.

**Cubbin, Cubbon:** *Gybone* 1429, *M'Cubbon* 1430, *MacGibbon* 1511, *Cubbon* 1605, *Cubbin* 1645 Moore. Manx contractions of MᶜCUBBIN, MᶜGIBBON.

**Cubbison, Cubison:** William *filius Corbucion* 1086 DB (Wo); Peter *Corbezun* Hy I EngFeud; Peter *Corbisoun* 1316 FA (Wa), *Corbyson* 1329 AD vi (St). Evidently from a personal name, and Tengvik suggests an OFr *\*Corbucion*, otherwise unrecorded. *v.* OEByn 178.

**Cubble:** *v.* CUBBELL

**Cubitt:** Henry *Cubyt* 1288 NorwLt; Oliuere *Kubyte* 1454 Paston; Joseph *Cubitt* 1642 PrD. A nickname from OE *cubit* 'elbow'.

**Cuckold, Cockwell, Cockwold:** Uluric *Cucuold* c1095 Bury; William *Cucuel* 1221 AssGl; Henry

Cokewald 1324 CoramLa. OFr *cucuald, cucualt*, ME *cukeweld, cokewold* 'a cuckold'.

**Cuckow:** Warin *Kuku* 1195 P (Y); Gilbert *cuccu* 1195 P (L). ME *cuccou, cuckkow*, OFr *coucou* 'cuckoo'.

**Cuckson:** *v.* COOKSON

**Cudbird:** *v.* CUTHBERT

**Cudd:** *Cudde* Robson 1587 Bardsley (Nb); John *Cudde* 1358 AssSt. A pet-form of *Cuthbert*.

**Cuddamore, Cuddimore:** John *Cuddymore* 1576 SRW; Edward *Cudmore*, John *Cuddamore* 1642 PrD. From Cudmore Fm in Bampton (D).

**Cuddington:** Reginald *de Cudintone* 1161 Eynsham; Fray *de Cudington* 1230 P (O); Elizabeth *Cuddyngton* 1568 SRSf. From Cuddington (Beds, Bk, Ch, Sr).

**Cudlip, Cudlipp:** Roger *Cudlip* 1642 PrD. From Cudlipptown in Petertavy (D), *Cudelipe* 1114–19.

**Cudworth:** William *de Cudewurth'* 1243 AssSo; John *de Cudworth* 1384 IpmLa. From Cudworth (WRY).

**Cuff, Cuffe:** Walter *Cuf* 1210 P (W); Roger *Cuffe* 1275 RH (Nf); Kateryne *Cuffe* 1524 SRSf. Either OE *Cuffa*, or from ME *cuffe* 'mitten', metonymic for a maker or seller of these.

**Culf:** *Cudulf, Codolf, Cuulf, Coolf* 1066 DB; Thomas *Couthulf* 1275 SRWo; William *Cuttwlf* 1299 Ipm (L); Richard *Culfe* 1327 SRWo. OE *Cūðwulf* 'famous wolf'.

**Culham:** *v.* CULLUM

**Cull, Culle:** Richard *cule* 1189 Sol; Lucas *Culle* 1258 IpmW; John *Culle* 1368 FFEss; Stephen *Cull* 1545 SRW. OE *Cula*.

**Cullabine:** Richard *Cullebene* 1275 RH (Nf); Sarah *Cullabine* 1765 PN Gl ii 11. A nickname, 'pick bean', OFr *cuille*, OE *bēan*. cf. John *Cullebole* 1332 SRSt 'pick bull'; William *Culfis* 1230 Pat 'pick fish'.

**Culle:** *v.* CULL

**Cullen, Cullin, Cullon:** Bertram *de Coloigne* 1307 LLB D; John *de Coline* 1340 ib. F; John *de Culayn* 1447 FrY; John *Cullan* 1487 ib.; John *Cullen* 1524 SRSf. From Cologne. The Scottish *Cullen* is from Cullen (Banffshire): Henry *de Culane* 1340 Black. In Ayrshire and Galloway it is probably Irish *MacCullen*.

**Culley, Cully:** Hunfrid *de Cuelai* 1086 DB (Nf); Hugh *de Cuilly* 1313, *de Cully* 1314, Roger *de Kuly* 1318, *de Kuylly* 1322 ParlWrits. From Culey-le-Patry (Calvados).

**Culling:** *Culling* 1086 DB, 1198 P (Nb); Warner *Culling* 1196 P (W); Tresmund *Culling'* 1207 ChR (Do). A personal name not recorded before 1086, perhaps to be identified with *Colling*.

**Cullum, Kullum, Culham:** Vincent *de Culeham* 1212 Cur (Berks); John *Cullum* 1524 SRSf; Francis *Cullam* 1736, John *Culham* 1749 ShotleyPR (Sf). From Culham (Berks, O).

**Cully:** *v.* CULLEY

**Culpepper:** Walter *Colpeper* 1313 AD vi (K); John *Coulpeper* 1332 SRSx. 'Cull (gather) pepper', a name for a spicer.

**Culter:** *v.* COULTAR

**Culvard, Culvert:** Roger *Culuert* 1221 AssWo;

John *Culuerd* 1278, *Culuard* 1331 Oseney. OFr *culvert* 'base, treacherous'.

**Culver:** Geoffrey *Kuluer* 1215–19 RegAntiquiss; Thomas *Colvere* 1334 SRK; William *Culvere* 1423 LLB K. OE *culfre* 'dove', used as a term of endearment.

**Culverhouse:** Richard *Attekulverhuse* 1266 FFEss; Adam *Colverhous* 1309 SRBeds. OE *culfrehūs* 'dovecote', for the man in charge.

**Culvert:** *v.* CULVARD

**Cumber(s):** *v.* COMBER

**Cumberland:** William *de Cumb'land, de Cumberlande* 1191 P (Cu), 1301 SRY; Thomas, John *Comerlond* 1524 SRSf. 'The man from Cumberland'.

**Cuming, Cumings, Cumine, Cummin, Cummins, Cummine, Cumming, Cummings, Comings, Comins, Comyns, Commin, Commins, Commings:** Godwinus *filius Cumine* 1173 P (Nf); Eustachius *filius Cumini* 1219 AssL; Petrus filius *Kymine* 1301 SRY; William *Comyn* 1133 Black; Hugh *Coumini* 1157 France; Walter *Cumin* 1158 P (Wa); John *Comin* 1175–9 DC (L); William *Cumyn* 1230 P (Ha). These forms lend no support to the common derivation from Comines given by the *Scots Peerage*, the *Dictionary of National Biography* and Freeman. This derivation must be based on the form of the name in Ordericus Vitalis, Rodbertus *de Cuminis*, the only form noted with the preposition apart from Balduinus *de Comminis* (1197 France), who may have been of a different family. Robert (d. 1069), one of the companions of the Conqueror and ancestor of the Scottish Comyns, is elsewhere named Rod bearde *eorle* (ASC D *s.a.* 1068) and Robertus *cognomento* (*cognomine*) *Cumin* (Symeon of Durham). Comines (Nord) is near the Belgian border. Le Prevost's suggestion that the family came from Bosc-Benard-Commin (Eure) is probably correct. The place is just south of the Seine, on the edge of the Forêt de Rouvray, near Rouen, where Hugh *Coumini* (1157) and Bernard *Comin* (1175) held land (France), in the heart of the district from which came numerous Domesday barons. The surname is clearly from a personal-name which, surviving in Norfolk, Lincolnshire and Yorkshire in the 12th and 13th centuries, may be of Breton origin. cf. OBret *Cunmin* (895 Loth). There was a 7th-century abbot of Iona named *Cumin*, whilst Woulfe derives the Irish *Comyn, Cummin(g)* from Ó *Coimín* or Ó *Cuimín* 'descendant of *Coimin* or *Cuimín*', diminutives of *cam* 'bent, crooked'. The Scottish Comyns may have owed their name ultimately to a Breton ancestor. cf. MALLET for evidence of Breton influence in Rouen.

**Cumner, Cumnor:** William *de Cumenore* Ric I Cur; Roger *de Comenore* 1266 Oseney. From Cumnor (Berks).

**Cumpton:** *v.* COMPTON

**Cuncliffe, Cunliffe:** Nicholas *de Cumbecliue* 1246 AssLa; Robert *de Cundeclif* 1276 RH (Y); Thomas *Cunclyff* 1411 FrY. From Cuncliffe (La).

**Cundall, Cundell:** Ralph *de Cundale* 1176 P (Y); Richard *de Kundale* 1301 SRY; Thomas *Cundal* 1394 TestEbor. From Cundall (NRY).

**Cundey, Cundick, Cundict, Cunditt:** *v.* CONDIE, CONDUIT

**Cunliffe:** v. CUNCLIFFE

**Cunningham, Cunnighame, Cuningham, Cuninghame, Cunynghame, Coningham, Conyngham:** Richard *de Cunningham* 1210–33 Black. From Cunningham (Ayrshire).

**Cunnington:** Pagan, Richard *de Cuninton'* 1193 P (Lei), 1210 Cur (C). From Conington (Cambs, Hunts).

**Cupper:** v. COOPER

**Cupping:** *Cupping* 1148 Winton (Ha); John *Cupping* 1296 SRSx; John *Cuppyng* called atte Forde 1345 ChertseyCt (Sr). OE *Cypping*.

**Cuppledick:** v. COBBLEDICK

**Cupples, Coppell, Copple:** (i) Walter *Curtpeil* 1200 Cur (Sf); Roger *Curpeil* 1210 Cur (Nf); John *Curpel* (*Curtpail*, *Curpeil*) 1221 Cur (Herts); William *Corpoyl* 1275 SRWo; Robert *Coppayl* 1381 *SR* (Ess). OFr *curt peil* 'short hair'. cf. Coupals Fm from a family of this name (PN Ess 462). (ii) John *de Cophull* 1243 InqLa; Robert *de Cuphull* 1275 SRWo. 'Dweller by the peaked hill' (OE *copp* and *hyll*) as at Coppull (Lancs).

**Curl, Croll:** Burewoldus *Crul* 1066 Winton (Ha); Ralph *Crul* 1191 P (Nf); William *Curle* 1202 AssL; Roger *Crolle* 1221 AssSa. ME *crull(e)*, *curl(e)*, 'curly (hair)'. v. also CROWL.

**Curlew:** Richard *Curlu* 1269 IpmNf; Thomas *Corlew* 1327 SRSx; Richard *Curlewe* 1430 FrY. AFr *curleu*, OFr *corlieu* 'curlew', a nickname from the bird.

**Curley:** (i) Rannulf *de Curleio* c1110 Winton (Ha); Robert *de Curli* 1190 P (O); William de *Curly* 1227–8 FFWa. From Corlay (Côtes-du-Nord, Indre, Saône-et-Loire), or Corlieu, the old name of La-Rue-Saint-Pierre (Oise). (ii) Benedict *le Curly* 1271 ForSt; Thomas *Curly* 1332 SRWa; Oliver *Curley* 1642 PrD. Probably variants of CURLEW.

**Curling:** Robert *Crullyng* 1296 SRSx; John *Crollyng* 1327 SRSx. ME *crulling* 'the curly one.'

**Curphey:** *McCurghey* 1422, *Courghey* 1601, *Curphey* 1643 Moore. A Manx contraction of *Mac Murchadha* 'son of *Murchad*', the sea-warrior.

**Curr:** William *le Curre* 1180 P (Ess); Hugo *le Cur* 1212 Cur (L); Richerus *Canis* 1212 Cur (Herts). ME *curre* 'cur, dog'.

**Currant:** William *Curaunt* c1180 Lichfield (St); John *Corant* 1260 FFHu. Present participle of OFr *courir*, 'running'. For Fr *Courant*, Dauzat assumes an ellipse for *chien courant*, a name for a hunter.

**Currer, Curror:** Nichol *Corour* 1296 CalSc; Thomas *Currour* 1430 FeuDu; William *Currer* 1621 SRY. OFr *corëor*, *courreour* 'messenger'.

**Currey, Currie, Curry:** (i) Dodda *æt Curi* c1075 OEByn (So); Richard *de Cury* 1212 Fees (So). From Curry (Som). (ii) Robert *atte Curie* 1327 SRSx, *atte Corye* 1332 ib.; William *Curry* ib. OFr *curie* 'kitchen'. cf. Curry (PN Ess 210), Petty Cury (PN C 47). (iii) Philip *de Curry* 1179 Black. From Currie (Midlothian).

**Currier, Curryer:** Thurstan *conreor* c1220 Bart (Lo); Richard *le Curur* 1256 AssNb; Henry *le Coureer* 1281 LLB B; Maurice *le couraour* 1293 FrY; William *le Coureour* 1314 LLB D; William *Curreyour*, Robert *Curreour*, John *Curreior* 1375–6 ColchCt; Andrew *Curier* 1400 FrY; Robert *Curryar* 1546 FFHu. Fr *couraieur* (16th cent), OFr *conreeur* 'currier' (1286 MED), a leather-dresser.

**Curror:** Nicol *Corour* 1296 CalSc (Berwicks); Thomas *Corour*, *Currour* 1331, 1430 FeuDu. OFr *coreor* 'runner, courier'.

**Cursham:** John *Cursom*, Robert *Cursumme* 1524 SRSf. For CURZON, with the frequent change of final *n* to *m*.

**Curson:** v. CURZON

**Curt:** v. COURT

**Curthoys, Curthose, Curtois:** Walter *Curtehose* 1210–11 PWi; John *Curthose* 1287–8 NorwLt; Hugh *Curthose* 1392 CtH. A nickname, 'short boots', OFr *curt, hose*. The name probably usually fell in with CURTIS. cf. also Geoffrey *Curtemanche* c1284 Lewes (Nf) 'short sleeves'; William *Curtbac* 1226 Cur (Herts) 'short back'; Michael *Curtcol* 1193 P (L) 'short neck'; Robert *Curtesmains* 1208 Misc (Lei) 'short hands'; John *Courtpe* 1393 FFEss 'short foot'.

**Curtin:** Robert *le Curten* 1275 RH (Nf); Robert *Curtyn* 1311 ColchCt. A diminutive of OFr *curt* 'short'. Also for *MacCurtin*, a metathesized form of *MacCruitin* 'son of *Cruitín*', the hunch-backed.

**Curtis, Curtiss, Curteis, Curtice, Curties, Curthoys, Curtois, Cortes, Cortis, Courtice, Courtis, Kertess:** *Curteis de Capella* 1130 P (Wa); *Curteis* de Cantebr' 1200 Cur (C); Richard *Curteis* 1166 P (Beds); Robert *le Curteis* 1168 P (D); Ralph *le Curtoys* 1230 P (L); John *le Korteys* 1238 Kirkstall (Y); Henry *Courteys* 1297 MinAcctCo; John *Corties* 1327 SRSx; William *Curtes* 1542 ChwWo; John *Curthoise*, Edmund *Curtice*, *Curtis* 1674 HTSf. OFr *corteis*, *curteis*, later *cortois*, *courtois* 'courteous' (a1300 MED), in feudal society denoting a man of good education. Used also as a personal name.

**Curtler, Kirtler:** Geoffrey *le Cultelier* 1186 P (Ess); William *le Curtillier* 1199 P (W); Ralph *Curtiler* 1296 SRSx; William, Roger (*le*) *Corteler* 1327 SRSf. OFr *cortiller*, *courtillier*, *cultilier* 'gardener'; or a derivative of ME *curtil* 'kirtle', a maker of kirtles. cf. Alicia *Curtle* 1231 Cl (D).

**Curton:** William *de Curtune* 1119 Colch (Ess); Gilbert *de Curton'* 1205 Cur (Nf); Oger *de Curton* 1218 FFEss. From Courtonne (Calvados).

**Curwen:** Gilbert *de Culewen* 1262 Black; William *de Colven* 1298 CalSc (Dumfries); Gilbert *de Colwenn* 1332 SRCu; Robert *Curwen* 1379 PTY; William *de Curwen* 1388 FrY. From *Culewen*, now Colvend (Kirkcudbrightshire). The family was early established in Cumberland.

**Curzon, Curson, Cursons, Corson:** (i) Robert *de Curcon* 1086 DB (Nf); Ralph *de Curtesun* (*Curcun*) 1127–34 Holme (Nf); William⋅(*de*) *Cursun* 1198 Cur (Nf). From Notre-Dame-de-Courson (Calvados). v. ANF. (ii) Robert *le Curezun* 12th Gilb (L); Richard *Cursun* c1180 DC (L); William *le Curcun* 1202 FFNf; Katherine *la Curzoun* 1316 FFEss; Thomas *le Curson* 1332 FFSf. There has been some confusion in early forms between *le* and *de*, but the name was also clearly a nickname from OFr *courson*, a diminutive of *curt* 'short'. cf. *courçon*, *courchon* 'a piece of land shorter than the others' (Godefroy). Weekley's alternative suggestion from 'a cursen man', i.e. christian as opposed to heathen, is unlikely as this form is not found before the 16th century.

**Cusack:** William *Cusac* 1214, *de Cusac* 1218 P (Sx). From Cussac, a common French place-name.

**Cush, Cuss, Cusse:** v. KISS

**Cushing, Cushion, Cushen, Cusheon:** Emme *Cusschon, Cusshoun* 1507 NorwW (Nf); Mr *Cushinge* 1674 HTSf. A dialectal form of CUSSEN or COUSEN.

**Cushworth:** v. CUSWORTH

**Cusins:** v. COUSEN

**Cuss:** v. CUST

**Cusselle:** v. CORSELLIS

**Cussen, Cusson, Cussons:** (i) Robert *Custson* 1332 SRCu; William *Custeson*, Richard *Cusson* 1379 PTY. 'Son of *Cust*', i.e. Constance. (ii) Henry *Cuttesone* 1329 ColchCt. 'Son of *Cutt*', i.e. Cuthbert.

**Cust, Cuss:** *Custe* filia Willelmi 1219 AssY; *Cus* nepta Johannis Frost 1279 RH (C); *Cuss* Balla ib. (Hu); *Custa* atte Halle 1296 SRSx; Richard *Cust* 1279 RH (O). Short forms of *Custance* or *Constance*. cf. *Cussata* (*Constancia*) 1230 P (D).

**Custance:** v. CONSTANCE

**Custard, Custer:** v. COSTARD

**Custer:** Sibilla *la Custere* 1254 AssSt. OFr *coustier* (m), *coustiere* (f) 'a maker of feather-beds or cushions'.

**Custerson:** Adam *Custanson* 1379 PTY. 'Son of *Custance*.' cf. *Dickinson* and *Dickerson*, *Stevenson* and *Steverson*.

**Cusworth, Cushworth:** Robert *de Cusworth* 1358 FrY. From Cusworth (WRYorks).

**Cutbili:** v. COBBALD

**Cutbush:** Henry *Cutbussh* 1450 ArchC vii. 'Cut bush.' cf. TALLBOY.

**Cuthbert, Cudbird:** Austinus *filius Cudberti* 1202 P (Y); Laurencius *filius Cutberti* 1207 FFHu; William *Cutbright, Cudbriht* 1260 AssC, 1276 RH (C); John *Cutbert* 1279 RH (Hu); Robert *Cudbert* 1301 SRY; John *Cutberd* 1327 SRC; William *Cuthbert* 1469 FrY. OE *Cūðbeorht* 'famous-bright', a common OE name and popular after the Conquest in the north and the Scottish Lowlands.

**Cuthbertson:** John *Cutberdson* 1410 FrY. 'Son of *Cuthbert*.'

**Cutlack:** v. GOODLAKE

**Cutlard:** William *Cuttelard* 1486 *ERO*. 'Cut lard', OE *\*cyttan*, OFr *larde*..Probably a nickname for a butcher. Such names are not unusual, but few have survived. cf. Robert *Cuttekaple* 1247 AssBeds 'cut horse'; Thomas *Cuttegos* 1247 AssBeds 'cut goose'; Alan *Cuteharing* 1206 AssL 'cut herring'; Symon *Cuttepurs* 1275 Burton 'cutpurse'.

**Cutler, Cuttler:** Ralph *le Cuteiller* 1212-23 Bart (Lo); Peter *le Cutelir* c1216 Clerkenwell (Lo); Dauid *le Cutiller* 1219 AssY. OFr *coutelier, cotelier* 'cutler', 'maker, repairer or seller of knives, etc.' (c1400 MED).

**Cutley:** John *Cutley* 1642 PrD. From Cotleigh (D), *Cuttelegh* 1449.

**Cutliffe:** Joan *Cutloff* 1512 AD vi (Y); Francis *Cutloffe* 1559 SRDb; William *Cutlove* 1674 HTSf. 'Cut loaf', OE *\*cyttan, lāf*. Probably a nickname for a baker.

**Cutt, Cutts:** *Cutte* Brendhers 1279 RH (C); *Cuttie* or *Cuthbert* Armorer 1589 Black; John, Robert *Cut* 1185, 1188 P (He, Db). A pet-form of *Cuthbert*.

**Cuttell, Cuttill, Cuttles:** v. COTTEL

**Cutter:** Robert *Cutre* 1207 Pleas (Ess); Adam *Cutter* 1379 PTY; Henry *Cutter* 1674 HTSf. A derivative of OE *\*cyttan* 'to cut', perhaps a nickname for a tailor or a barber.

**Cutteridge, Cuttridge, Cutress, Cuttriss, Gutteridge, Gutridge:** Ailric *Cuterich* 1176 P (O); Joan *Cudrich* 1279 RH (O); William *Cutrich* 1327 SRSf; Hugh *Coterich* 1327 SRSo; William *Gutheridge* 1597 PN Herts 49; Arthur *Gutterich, Cutterice* 1606, 1611 IckworthPR (Sf); William *Cutteridge*, Widow *Cutteris* 1674 HTSf; Marten *Gutteridge*, Henry *Guttrage* ib. OE *Cūðrīc* 'famous ruler', a rare OE personal name. *Gut*(*ter*)*idge* is due to late confusion with *Goodrich*.

**Cutting:** Herlewin, William *Cutting* 1221 *ElyA* (Nf); Richard *Cutting* 1235 FFEss; William *Cutton, Cutting* 1767, 1775 ShotleyPR (Sf). OE *\*Cutting* 'son of *Cutt*', a pet-form of *Cūðbeorht* or *Cūðbeald*. cf. CUTHBERT, COBBALD.

**Cutwin:** v. COTWIN

**Cuxon, Cuxson:** v. COOKSON

**Cuzen:** v. COUSEN

# D

---

**Dabbs:** John *Dabbe* 1524 SRSf. *Dobb* (Robert), with unrounded vowel.

**Daber:** v. DAUBER

**D'Abernon:** Roger *de Abernon* 1086 DB (Sr); Jordan *Dabernun* 1197 P (Wa). From Abenon (Calvados).

**Dabinett:** Walter *Dobinet* 1301 SRY. *Dob-in-et*, a double diminutive of *Dob*. v. DABBS.

**Dabney:** v. DAUBENEY

**Daborne:** v. DAWBARN

**Dabson:** For DOBSON. cf. DABBS.

**Dace:** Ralph *Dace* 1202 AssL; William *Dase* 1305 AssW; Richard *Dase* 1376–7 FFSr. A nickname from the fish, OFr *dars*.

**Dack, Dax:** *Dacke* 1250 AssSt; Alexander *Dacke* 1275 RH (Nf). Perhaps OE *\*Dæcca*, the first element of Dagenham (Essex), surviving in Hugo *filius Decche* 12th (L). v. IPN 186.

**Dacken:** *Dacken alias dictus* David ap Lewes 1459 SaG. Presumably a pet-form of *David*.

**Dacre, Daker, Dakers:** (i) Rannulf *de Daker* 1212 Cur (Cu); Ralph *de Dacre* 1272 WhC; William *de Dacre* 1360 FFY; Richard *Dakers* 1496 LLB L. From Dacre (Cu, WRY). (ii) Roger *de Acra* 1201 Pleas (Nf); William *de Acre* 1281 LLB A; Adam *de Acres* 1346 LLB F. From Castle, South, West Acre (Nf).

**Dadds:** William *Dad* 1279 RH (C). v. DOD, with vowel unrounded.

**Dade:** v. DEED

**D'Aeth:** v. DEATH

**Daffe:** Lefeke *Daffe* 1279 RH (Beds); Geoffrey *daffe* 1296 SRSx; Roger *Daffe* 1327 SRSo. ME *daff* 'simpleton, fool'.

**Daft:** John, Robert *Daft* c1230 NottBR, 1242 Fees (Nt). OE *gedæft* 'gentle, meek', ME *daffte* 'foolish, stupid'.

**Dafter, Dafters, Daftors:** v. DAUGHTERS

**Dagg:** Robert *Dag* 1275 SRWo; Ralph *Dagg* 1327 SR (Ess). OFr *dague* 'dagger', an ellipse for one who carried a dagger.

**Daggar, Dagger, Daggers, Daggett:** Henry *Daget* 1219 AssY; Peter *Dagard* 1279 RH (C). *Daget* is a diminutive of OFr *dague* 'dagger'. v. DAGG. cf. Richard *Dagun* 1203 P (Y), John *Dagenet* 1185 Templars (Herts), William *Dagenet, Dagunet* 1210 Cur, 1221 AssWa and the French surnames *Dagon*, *Dagot, Daguet, Dagonet, Daguenet* (Dauzat). The ultimate origin of *dagger* is unsettled. It occurs as *daggere* a1375, *daggardus* 15th, *dagard* 1535. If the last form could be regarded as

*original, the word would be* OFr *dague* plus *-ard* (NED). The 1279 form above suggests that this may be the origin.

**Dagleas, Dagless, Daglish:** v. DALGLEISH

**Dagnall, Dagnell:** Peter *de Daggingehal* 1204 AssY; William *de Dagenhale* 1260 AssC. From Dagnall in Edlesborough (Bk).

**Dagworth:** Richard *Daggewurth* 1234 PN Ess 153; John *de Dageworth* 1308–9 FFEss. From Dagworth (Sf).

**Daile:** v. DALE

**Dailey, Dailley:** v. DOYLEY

**Dain, Daine, Daines, Dayne, Daynes, Deyns, Dines, Doyne, Dyne:** Robert *le Dine* 1201 P (Sr); Richard *le Digne* 1222 Cur (Sr); Gilbert *le Dyne, Dynes* 1275, 1284 Wak (Y); Nicholas *Dain* 1275 SRWo; Matilda *Deine* 1279 RH (O); John *le Dyen* 1296 SRSx. William *Dien* 1297 MinAcctCo; William *le Deyne* 1327 SRSx; Richard, Walter *Doyn* 1327 SRSx; John *Deynes* 1327 SRC. (i) ME *digne, deyn(e)*, Fr *digne* (11th), perhaps OFr *\*dein* 'worthy, honourable' (1297 NED); (ii) ME *dain(e)*, OFr *\*deigne*, Burgundian *doigne*, Fr *digne* 'haughty, reserved' (c1500 NED); (iii) OFr *deien, dien*, modFr *doyen* 'dean'. cf. DEAN. The forms are inextricably confused.

**Dainteth, Daintith, Dentith:** Agnes *Deynteth* 1379 NottBR; Thomas *Dentithe* 1591 SfPR. OFr *daintiet, deintiet*, from Lat *dignität-em*, ME *deinteth*, an archaic form of DAINTY.

**Dainton:** From Dainton in Ipplepen (D).

**Daintree, Daintrey, Daintry, Daventry:** Philip *de Dauintrie* 1162 P (K); Gilbert *Dantre* 1369 LLB G; Thomas *Daintree* 1637 Fen DraytonPR (C). From Daventry (Northants), the correct local pronunciation of which is *Daintree*.

**Dainty, Denty:** Osbert *Deintie* 1199 P (Nth); Henry, William *Deinte* 1227 AssBk, c1248 Bec (O). ME *deinte*, OFr *deintié, daintié, dainté*, sb. 'pleasure, tit-bit' (a1225 NED), adj. 'fine, handsome, pleasant' (c1340 NED). v. DAINTETH.

**Dairson:** v. DEARSON

**Daisey:** Roger *Dayseye* 1306–7 FFWa; William *Deyseye* 1332 PN Sr 287; John *Deisy* 1534 CorNt. A nickname from the daisy, OE *dæges-éage*.

**Daish:** v. ASH

**Daker, Dakers:** v. DACRE

**Dakers:** v. ACRES

**Dakin:** v. DAYKIN

**Dalby, Daulby, D'Aulby:** Matthew *de Dalbi* Hy 2 DC (L). From Dalby (Leics, Lincs, NRYorks).

**Dalderby:** John *de Dalderby* 1299, Roger *de Dalderby* 1316 RegAntiquiss. From Dalderby (L).

**Dale, Dales, Daile:** Ralph *de la Dale* 1275 RH (Sf); William *en le Dale* 1318 ShefA (Y); John *atte Dale* 1327 SRSx; Nicholas *Daile* 1481 FrY. 'Dweller in the dale', OE *dæl*.

**Daley:** *v.* DALLY

**Dalgleish, Dalgliesh, Dalglish, Dagleas, Dagless, Daglish:** Symon *de Dalgles* 1407 Black; Adam *Dalgleisch*, Andrew *Dawgles* 1507–10 ib. From Dalgleish (Selkirkshire).

**Dalham, Dallam:** Richard *de Dalham'* 1198 P (St); Robert *Dalam* 1522 PN We i 68. From Dalham (K, Sf).

**Dall, Dalle:** Rannulf *Dal* 1202 Cur; William *Dalle* 1275 SRWo; John *de Dalle* 1362 FFY; John *de Dall* 1398, John *Dall* 1419 FrY. Probably usually for DALE, but other sources may also be involved.

**Dallam:** *v.* DALHAM

**Dallamore:** *v.* DELAMAR

**Dallas:** (i) Roger *del Dalhous* 1301 SRY; William *de Dalhous* 1327 SRY. 'Dweller at the house in the dale.' (ii) Archibald *de Doleys* 1262 Black; John *de Dolas* 1429 ib.; Henry *Dallas* 1513 ib. From the old barony of Dallas (Moray).

**Dallaway:** *v.* DALLOWAY

**Dalle:** *v.* DALL

**Dallicoat:** *v.* COTE

**Dallimore:** *v.* DELAMAR

**Dalling, Dallin, Dallyn:** Turoldus *de Dallenges* 1108 MedEA (Nf). From Dalling (Norfolk).

**Dallinger:** John *Dallinger* 1674 HTSf, *Dalinger* 1722 SfPR. From Dallinghoo (Sf).

**Dallingridge:** Richard *Dalyngrigg* 1419 IpmY; Richard *Dalyngregge* 1455–7 CtH. Probably 'dweller by the ridge in Dalling'. cf. Dalling (Nf), Dallingho (Sf).

**Dallinson, Dallison:** Bernard *de Alencon* 1086 DB (Sf); John *de Alecon* 1189 Whitby (Y); Alexander *Dalencun*, William *Dalizun* 13th Lewes (Nf); Nicholas *Dalasson* 1378 LLB H. From Alençon (Orne). The surname has also contributed to ALLANSON and ALISON.

**Dallman, Dalman:** Thomas, Hugh *Daleman* 1327 SR (Ess), SRC. 'Dweller in the dale' (OE *dæl*).

**Dalloway, Dallaway:** John *Daliwey* 1375 RH (L); John *Daleway* 1305 AssW; Ralph *Daleway* 1327 IpmW. 'Dweller by the road in the dale', OE *dæl*, *weg*.

**Dally, Daley, Daly:** (i) Richard *Daly* 1275 RH (K); Richard *Dally* 1293 MPleas (Hu); Richard *Daly* 1392 CtH; John *Daley*, Albert *Dally* 1642 PrD. Perhaps from Ailly (Eure, Meuse, Somme), or from Dally Fm in Forest Row (Sx). (ii) Ir *Daly* is from *Ó Dálaigh*, from *dáil* 'assembly'.

**Dalton, Daulton, Daughton, Dawton:** William *de Daltone* 1155 FeuDu; William *Dawton* 1518 FrY. From Dalton (Durham, Lancs, Northumb, Westmorland, ER, NR and WRYorks).

**Daltry:** *v.* DAWTREY

**Dalyell, Dalyiel, Dalzell, Dalziel:** Hugh *de Dalyhel* 1288 Black; James *Deell* 1684 ib.; James *Dyell* 1689 ib. From the old barony of Dalziel (Lanarkshire). Also

spelled *Deyell, De Yell*. Pronounced *Diyell* or DL, sometimes *Dal-yell* and now also *Dalzeel, Dalzell*. The old form *Dalȝel* was printed *Dalzel*, hence the incorrect pronunciation with *z* instead of *y*.

**Damary, D'Amery, Dammery, Damry, Amori, Amory:** William *de Dalmari, de Dalmereio, de Almereio* 1086 DB (Do); Richard *de Ameri, Dameri* 1159, 1166 P (Beds, L); Roger *Damery, Dammary, Daimary, de Ammary, de Aumary* 1280 AssSo; Roger *de Amory, Damori* 1274 RH (Bk, O). From Daumeray (Maine-et-Loire). The initial *D* was often regarded as a preposition and the name wrongly divided as *De Aumari*. This preposition was then lost, hence *Amori, Amory*. The name was probably then confused with AMERY.

**Dambell:** William *Damebele* 1303 AssW; John *Dambelle* 1642 PrD. 'Lady *Bele*', OFr *dame*, and *Bele*, a pet-form of *Isabel*. Such surnames are not uncommon in medieval times, but have rarely survived. cf. Walter *Dame Alis* 1327 SRWo 'lady *Alice*'; Richard *Dam Anne* 1327 SRSo 'lady *Anne*'; Thomas *Dameclarice* 1332 SRDo 'lady *Clarice*'; Walter *Damablie* 1327 SRSo 'lady *Mabilia*'; Cristine *Damolde* 1277–8 CtH 'lady *Maud*'. The probable meaning of such names is 'servant of the partular lady'.

**Dame:** Henry *Dame* 1279 RH (O); Agnes *Dame* 1327 SRSf. Fr *dame* 'lady'. Perhaps an ironical attribute.

**Damer:** *v.* DAYMAR

**Damerell, Damiral, Dammarell, Damrel, Damrell:** Robert *de Alba Marula, de Albemarle* 1086 DB (D); Reginald *de Aumarle* 1243 AssSo; Bernullus *de Aumeryl* 13th WhC (La); Thomas *Damarell* 1568 SRSf. From Aumale (Seine-Inférieure), earlier *Alba Margila*. The fuller form is preserved in the title of the Duke of Albemarle. Hinton Admiral (Hants) preserves the surname without the preposition.

**Damet, Damett:** *Dameta* 1130 P (O); *Dametta* 1279 RH (Beds); Alan *Damet* 1280 IpmY; Elias *Damet* 1298 AssL; Simon *Damet* 1327 SRSf. OFr *Damette*, a feminine personal name of unknown origin.

**Dammery:** *v.* DAMARY

**Dampier, Damper:** William *de Damper* 1225 Pat; William *Damper* 1229 ib. (Y). From Dampierre, the name of numerous places in France, two of which are in Normandy.

**Damrell:** *v.* DAMERELL

**Damry:** *v.* DAMARY

**Dams, Damms:** Peter *del Dam* 1221 ElyA (Nf); Thomas *ate Dam* 1327 SRC; Alice *Dam* 1327 SRSf. 'Dweller near a dam', from OE *\*damm* 'dam' (a1400 MED).

**Damsell:** Ralph *Damisel, Dameisele* 1191, 1204 P(Y); Henry *Damisel* 1204 P(Gl); Roger *Damisele* 1214 Cur (Bk). OFr *dameisele, damisele* (f) 'a maiden', originally of noble birth and OFr *dameisel* (m) 'a young squire, page'. Both seem to be represented, the former, probably, in the sense 'effeminate'.

**Damson:** Geoffrey *Dammessune* 1186 P (Nth); Henry *Dameson* 1276 AssSo. OFr *dame*, earlier *damme*, 'the dame's son'.

**Damyon:** *Damianus* 1199 MemR (Nf), 1206 Cur

(Mx); William, John *Damyen* 1294 FFEss, 1327 SRSf. St *Damianus* was martyred in Cilicia in 303 under Diocletian. His name, perhaps to be associated with the goddess Damia, was not common in England.

**Danby:** Rand *de Danbi* 1189 P (L); Robert *de Danebi* 1212 P (Y); John *Danby* 1392 IpmNt. From Danby (NRYorks).

**Dance:** William, Robert *Daunce* 1247 AssBeds, 1301 SRY. ME, OFr *dance* (c1300 NED), metonymic for a dancer or *dawnceledere* (c1440). Robert *de la Daunce* 1305 LoCt was probably a professional dancer, chief of 'a dancing party', a meaning recorded c1385 NED.

**Dancer:** Godwin *Dancere* 1130 P (Herts); Ralph (*le) Dancere* 1240 Rams (Nf), *le Dauncer* 1327 SRSx. A derivative of ME *dancen* 'to dance', 'a dancer, especially a professional dancer in public'.

**Dancey, Dancy, Dansey, Dansie, Dauncey:** William *de Anesi* 1086 Winton (Ha); Milo *de Dantesia* 1177 P (W), *de Andesie, de Dantesie, de Dantesia* 1208 Cur (W); Richard *Danesi* 1210 Cur (K), *de Anesye* 1236 Fees (W), *de Danteseia* 1242 ib., *de Anesy* alias *Daneseye* 1249 Ipm (W); Thomas *de Aunteseye* 1269 AssNb. From Anisy (Calvados), *de Anesi* became *Danesi* and, with an intrusive *t, Dantesi*. This was identical in form with the DB and later forms of Dauntsey (Wilts) where Roger *Dantesie* held ¾ of a fee in 1242 (Fees). The surname was often thought to derive from the Wiltshire place and an additional *de* inserted (*de Dantesie*). The confusion was increased as the family also left its name in Winterbourne Dauntsey in the same county. The modern surname may derive independently from Dauntsey.

**Dand, Dandie, Dandy, Dandison:** *Dande* de Hale, de Leuer 1246 AssLa; *Dandi* ballivus 1275 RH (L); Richard *Dande* 1279 RH (Hu); Adam *Dandy* 1312 FrY; Thomas *Dandisone* 1332 SRLa. *Dand* and *Dandie,* pet-forms of *Andrew,* are generally regarded as Scottish, but the English examples are much earlier than Black's earliest: *Dand* or Andrew Kerr (1499), Andrew alias *Dandie* Cranston (1514).

**Dandelion:** Maurice *Daundelin* a1290 CartNat; William *Dawndelyon* 1363 FrY; William *Daundeleyn* 1425 FFHu. A nickname from the dandelion, OFr *dent-de-lioun* 'lion's tooth', so named from the toothed outline of its leaves.

**Dandely:** Maurice *Daundely* 1251 ForNth; Thomas *Daundely* 1305 AssW; William *Dandele* 1334 SRK. From Grand, Petit Andely (Eure).

**Dandison:** Thomas *Dandisone* 1332 SRLa; William *Dandeson* 1374 AssL. 'Son of *Dand* or of *Dandy*', pet-forms of *Andrew.*

**Dando, Daddow, Daunay, Dauney, Dawnay, Dawney, Delaney, Delany, de Launay, Delauney, De Launey:** William *de Alno* 1086 DB (Sf); John *de Alnai* 1150–60 DC (L); Robert *del Aunei* Hy 2 Gilb (L); Henry *de Launei* 1159–85 Templars (Lo); Helias *de Aunou* 1201 AssSo; William *del Alnei* 1206 P (Nf), *del Aune* 1212 Cur (Ess); Geoffrey *de Alno, de Alneto, Dauno* 1225–54 AssSo; Jordan *del Aunney* 1225 AssSo; Reginald *de Auney* 1242 Fees (D); Mathew *Dauney* 1251 Whitby (Y); Alexander, Richard *Dando* 1274 RH (So), 1296 SRSx. The Somerset family came

from Aunou (Orne) and has left its name in Compton Dando (Som). The surname may also derive from Aunay (Calvados, Eure-et-Loir, Seine-et-Oise, etc.) or Laulne (La Manche), from Lat *alnetum,* Fr *aunaie* 'alder-grove'.

**Dandy:** *v.* DAND

**Dane, Danes:** Henry *Bithedane* Edw 1 Battle (Sx); William *de la Dane* 1275 RH (K); William *atte Dane* 1327 *SR* (Ess). OE *denu* 'valley', found as *dane* in place-names in Essex, Herts, Beds, Kent and Sussex. Roger *ate Dene* (1294) and Walter *ate Dane* (1296) both lived at Dane End (PN Herts 79). *v.* DEAN.

**Danecourt:** *v.* d'EYNCOURT

**Danell, Danels:** *v.* DANIEL

**Danford, Danforth:** Nicholas *de Darneford* 1279 RH (C); Robert *de Derneforde* 1327 SRSf; James *Danford* 1568 SRSf; Robert *Danforth* 1524 SRSf. From Darnford (Suffolk) or Dernford Fm in Sawston (Cambs); or for DURNFORD.

**Dangar, Danger:** Alric *Dangier* c1200 ELPN; Reginald *Danger* 1223–5 ib.; Alexander *Daunger* 1246 AssLa. OFr *dangier, danger* in one of its early senses: 'power, dominion' or 'hesitation, reluctance, coyness'. cf. Gerard *Daungerous* 1275 RH (L).

**Dangerfield:** William *de Angeruill'* 1205 P (Do); Foulke *Dangerfeild* 1659 Bardsley. From one of the places named Angerville in Normandy.

**Dangerous:** Richard *Dangerus* 1201 Pleas (Co); Robert *le Dangerus* 1243–4 IpmY; Gerard *Daungerous* 1275 RH (L). A derivative of OFr *dangier, danger,* in one or other of its early senses 'power, arrogance, reluctance'.

**Daniel, Daniels, Daniell, Daniells, Danniel, Danell, Danels, Dannel, Dennell, Denial:** Eudo *filius Daniel* 1121–48 Bury (Sf); Roger *Daniel* 1086 DB (Sx); Walter *Danyel* 1268 FFSf; Cecilia *Denyel* 1279 RH (C); John *Danyeles* 1319 SRLo; Matthew *Danel* 1327 SRSx. Hebrew *Daniel* 'God has judged'. *Denial* is pronounced *Denyel.*

**Dankin, Dankinson:** Adam *Dankyn* 1327 SRSo; William Jakson *Dankynson* 1401 AssLa; Henry *Dankyn* 1424 LLB K. *Dan-kin,* a diminutive of *Dan,* a pet-form of *Daniel.*

**Danks:** Thomas *Danke* 1501 RochW; John *Danks* 1551 ChwWo; Francis *Dankes* 1674 HTSf. A shortened form of DANKIN.

**Dann:** Geoffrey *atte Danne* 1327 SRSx; Simon *Dann* 1332 SRSx. 'Dweller in the valley.' *v.* DANE.

**Dannatt, Dannett:** Cristiana *Danet* 1275 RH (Nf); John *Danet* 1332 SRWa; Leonard *Danett* 1560 Pat (K). *Dan-et,* a diminutive of *Dan,* a pet-form of DANIEL.

**Dannay, Danny:** Richard *Danney* m13th, William *Danay* a1290 CartNat; Michael *Danneye* 1305 AssW; William *Danney* 1379 ColchCt. OFr *daneis* 'Danish'.

**Dannel, Danniel:** *v.* DANIEL

**Dannett:** *v.* DANNATT

**Danny:** *v.* DANNAY

**Dansel, Dansell:** William *Danzel* 1148 Winton (Ha); Robert *Daunsel* 1260 IpmY; William *Dansel* 1275 SRWo. OFr *danzel* 'young man, young noble'.

**Dansey, Dansie:** *v.* DANCEY

**Danson:** Robert *Dandeson* 1332 SRCu; John

*Dandsone* 1363 FrY; Robert *Danson* 1381 PTY. 'Son of *Dand*' or Andrew. Perhaps also 'son of *Dan*'.

**Danvers:** Ralph *de Anuers, Danuers* 1230 P, MemR (Berks). From Anvers (Antwerp).

**Darben:** Gilbert *Derebarn* 1277 FrY. ME *derebarne* 'dear child', cf. Alicia *Derechild* 1275 SRWo.

**Darblay:** *v.* DOBREE

**Darby, Darbey, Derby:** Roger *de Derby* 1160–82 RegAntiquiss; Edelota *Darby* 1278 RH (O); Simon *Derby* 1377 AssEss. From Derby.

**Darbyshire, Darbishire, Derbyshire:** Geoffrey *de Derbesire* 1203 AssSt; Henry, Richard *de Derbyshire* 1307 AssSt, 1394–5 IpmLa. 'The man from Derbyshire'.

**Darcey, Darcy, D'Arcy:** Norman *de Adreci, de Areci* 1086 DB (L); William *Daresci* 1166 P (L); Roger *Arsi* 1173–82 DC (L); Thomas *Darcy* 1276 Gilb (L). From Arcy (La Manche). The Irish Darcy derives from John *d'Arcy* (14th) but is also an anglicizing of *Ó Dorchaidhe* 'descendant of the dark man'.

**Darch, Dark, Darkes:** Osbern *de Arches, de Arcis,* William *Arcs* 1086 DB; Juelina *de Arches* 1201 Cur; Walter *Darch,* William *Darche* 1642 PrD. From Arques-la-Bataille (Pas-de-Calais), or Argues (Eure, Seine-Maritime. cf. Thorpe Arch (WRY), William *de Arches* c1150. *v.* also DARK.

**Dare:** Walter *Dare* 1243 AssSo; Richard *le Dare* 1327 SRSo; Richard *Dare* 1332 SRDo. OE *Dēor,* or a nickname from OE *dēor* 'wild animal'.

**Darfield:** *v.* DARVAL

**Dark, Darke, Darkes, Durk:** Robert *Derck* 1221 AssWa; Richard *Durk* 1229 Pat (So); Godewynus *Derc* 1230 P (Ess); John *Darke* 1362 LLB G. OE *deorc* 'dark' of complexion.

**Dark, Darkes:** *v.* DARCH

**Darker:** John *le Darkere* 1349 AD i (Wa); John *Darker* 1524 SRSf. This must be an occupational name, 'one who darkens'. cf. BLACKER, WHITER, and 'Every coriar shall well and sufficiently corie and blacke the said Lether tanned' (1532–3 NED), '*Noircisseur,* a blacker . . . darkener, obscurer' 1611 Cotgrave.

**Darkin, Darking:** *Derechin* de Acra 1159 P; *Derkyn* de Wyflingham 1228–32 Gilb; *Derkin* 1279 RH (C); William *Derkyn* c1250 Gilb (L); Henry *Derkyn* 1379 PTY; Richard *Darkyng* 1524 SRSf; An' *Darkin* 1674 HTSf. *Der-kin,* a diminutive of OE *Dēor.*

**Darknell:** *v.* DURTNALL

**Darley:** Warin *de Derleg'* 1200 P (Ess); William *de Derlay* 1379 PTY; John *Darley* 1541 CorNt. From Darley (Derby).

**Darling, Dearling, Dorling, Durling:** Oter *Dirlinges sunu* 1100–30 OEByn (D); *Derling* 1133–60 Rams (Beds), 1177 P (D); *Derling* de Arfdift a1177 Black (Berwick); *Durling* atte Forde 1330 PN D 433; Ælmær *Deorlingc, Dyrling* 1016 ASC E, D; William *Dierling, Derling* 1195–6 P (D); Henry *Durling* 1242 Fees (W); Emma *Derlyng* 1244 Rams (Beds); Ralph *Durlyng* 1327 SRSo; Richard *Dorling, Dorlynges* 1327 SRWo; Adam *Darlyng* 1379 PTY. OE *Dīerling, Dȳrling, Dēorling,* from OE *dēorling* 'darling', 'one dearly loved', both as a personal name and as an attribute.

**Darlington:** Odo *de Derlintone* 12th FeuDu; Nicholas *de Derlington'* 1258–9 RegAntiquiss; John *de Derlingtone* 1318 LLB E. From Darlington (Du).

**Darlow, Dearlove:** William *Derneluue* 1206 Pleas (Y); Hugh *Dernelove* 1279 RH (O); William *Dernelof* 1327 SRY. A nickname, 'secret love', OE *dierne, lufu.*

**Darmenters:** *v.* ARMENTERS

**Darnbrook, Darnbrough:** William *Dernbroke* 1361, Cecilia *Dernebroke* 1379 WRS. From Darnbrook on Malham Moor (WRY).

**Darnell:** Goduine *Dernel* c1095 Bury (Sf); Godwin *Darnel* 1177 P (Sf); Tomas *Darnele* 1193 P (Nf). OFr *darnel* 'darnel' (c1384 MED), a plant formerly believed to produce intoxication (Weekley). Occasionally also local, from Darnall in Sheffield (Yorks): William *de Darnale* 13th Shef.

**Darras:** Ailwin *de Arraz* 1176 P (Ess); William *de Araz* 1235 FFEss; Peter *Darraz* 1322 LLB E. From Arras (France).

**Darree, Darry, Denre, Denry, Derry:** William *Darri, Derri* 1200 Cur (Nth); Nicholas *Darre* 1288 FFSf; Robert *Darre,* John *Dary,* John *Deree* 1327 SRSf. AFr *darree,* OFr *denree* 'penny-worth'. cf. Fr *Danré, Danrée* 'surnom probable de marchand' (Dauzat).

**Darrell:** *v.* DAYRAL

**Darreyn, Darreyns:** Geoffrey *de Arennis* 1202 AssL; Thomas *Dereyns* 1303 IpmY; Hugh *Darreyns* 1342 AssSt. From Airaines (Somme). cf. Darras Hall (Nb), Wydo *de Araynis* 1242.

**Darrington:** John *de Darinton'* 1220 Cur (L); Richard *Daryngton* 1545 FFHu. From Darrington (WRY).

**Dart, Darth:** (i) Walter *Dert* 1221 AssGl; Hugh *Dart* 13th Guisb; John *Dart* 1524 SRD. OFr *dart* 'a pointed missile thrown by hand', perhaps metonymic for a soldier or a hunter. (ii) Ralph *de Derth* 1242 Fees (D); Juhelinus *de Derte* 1275 RH (D). From Dart Raffe in Witheridge (D). ˙

**Darter:** *v.* DAUGHTERS

**Dartnall, Dartnell:** *v.* DURTNALL

**Darval, Darvall, Darvill, Darville, Derville, Dorville, Darfield, Derfield, Darwall, Darwell, Durval:** Robert *de Durevill* 1201 AssSo; Hugh *Durival* 1300 Eynsham; Sibill *Doryual* 1332 SRSx; Thomas *Deryvall* 1577 ER 56; William *Derrivall* 1662 *HTEss.* From Orville (Orne, Pas-de-Calais), Urville (Auche, Calvados, La Manche), or Orival (Charente, Seine-Maritime, Somme). It is impossible to separate out the forms.

**Darwen, Darwent, Derwent:** Syward *de Derewent* 1246 AssLa. From Darwen (Lancs), on the river Derwent.

**Darwin:** *Derwen* 1170 P (Ess); *Derewinus* Purs 1176 P (Bk); John *Derewin* 1219 Fees (Ess); William *Derwyne* c1248 Bec (Bk). OE *Dēorwine* 'dear-friend', recorded in the 10th century, but rare. In 1225 (AssSo) Mabel, daughter of *Derwin',* had as pledges William, Nicholas, Henry and Hugh *Derwin',* probably her brothers, who owed their surname to their father.

**Darwood:** *v.* DURWARD

**Dash:** *v.* ASH

**Dashfield:** For *De Ashfield.* cf. DASH.

**Dashper, Disper:** Henry *Duzepers* 1203 P (Nth); Alb(e)ricus *Duzepers, Duzeper* 1221 Cur (Nth), 1221 AssGl; William *Duzeper* 1279 RH (O); Roger *Dozeper* 1293 Fees (D). OFr *doce, duze pers* 'twelve equals, twelve peers' (*dyssypers* 1503, *duchepers* a1400 NED, of which *Dashper* is a corruption). The reference is to the twelve peers or paladins of Charlemagne, said to be attached to his person as being the bravest of his knights. Later, the term was applied to other illustrious nobles or knights (c1330 MED) and a singular was formed some 200 years before the earliest example in MED (c1380).

**Dashwood:** Zachary *Dashwood* 1693 DWills. For *De Ashwood*.

**Datchet, Datchett:** William *Dachet* 1272 ForNth; William *Dachet* 1326, Richard *Dachet* 1337 CorLo. From Datchet (Bk).

**Daubeney, Daubeny, Daubney, D'Aubney, Dabney, Dobney:** Nigel *de Albengi, de Albingi, de Albinie, de Albinio* 1086 DB (Beds, Berks, Bk); Willelmus *Brito* 1086 DB (Hu), William *de Albinneio* 1115 Winton (Ha), Willelmus *Albineius* Brito 1116–20 France; Nigel *de Albuniaco* 1100–23 Rams (Hu), *de Albeni* 1114–23 ib.; William *de Aubeneio* 1124–30 Rams (Beds), *de Aubini, de Aubeni* 1199 MemR (Ha, Bk); William *Daubenny* 1212 Fees (Berks); Ralf *de Dabeney* 1269 AssSo; Thomas *Dabeney* 1524 SRSf. William, founder of the line of Aubigni, earls of Arundel, and Nigel of that of Cainhoe (Beds) came from Saint-Martin d'Aubigny (La Manche). The family of Aubigny (Brito) of Belvoir came from Saint-Aubin d'Aubigny (Ille-et-Vilaine). *v.* ANF. There is also another Aubigny in Calvados, of identical origin, which may have contributed to the surname.

**Dauber, Dawber, Daber, Dober, Doberer:** Hugo *Daubur* 1219 AssY; Robert *le Daubar* 1221 Cur (Berks); Nicholas *le Doubur* 1260 AssLa; Walter *Dobere, le Daubere* 1319, 1327 *SR* (Ess); Peter, Roger *le Daber* 1332 SRSx; Joseph *Dauber, Douber* 1346 ColchCt. AFr *daubour*, OFr *\*daubier* 'whitewasher, plasterer'. In the Middle Ages walls of 'wattle and daub' were extremely common. Wattling consisted of a row of upright stakes the spaces between which were more or less filled by interweaving small branches, hazel rods, osiers, reeds, etc. On one side, or more usually on both sides of this foundation, earth or clay was daubed and thrust well into the interstices, the surfaces being smoothed and usually treated with plaster or at least a coat of whitewash. Closely allied to daubing was pargetting or rough-casting in which mortar or a coarse form of plaster was used instead of clay or loam. At Corfe in 1285 there is a reference to 'Stephen the Dauber who pargetted the long chamber' and it is not always possible to decide whether the daubers were really daubing or whitewashing (Building 188, 190, 191). cf. PARGETER.

**Daugherty:** *v.* DOUGHARTY

**Daughters, Dauter, Darter, Dafter, Dafters, Daftors, Doctor:** Katheryn *Doctor* 1570 ChwWo is, no doubt, from OE *dohtor* 'daughter'. The surname is ill-documented, the modern forms chiefly colloquial spellings or dialectal pronunciations, and the reference may be to a sole heiress who would ultimately inherit her father's land. Early examples clearly indicate an actual relationship: Joan *Tomdoutter*, Rose *Anotdoghter*, Alice *Wilkynsondoghter* 1379 PTY, but these were not likely to survive though they were used as men's surnames, cf. Richard *Wryghtdoghter*, Robert *ffelisdoghter* 1379 PTY.

**Daughtery, Daughtry:** *v.* DAWTREY

**Daughton:** *v.* DALTON

**Daukes:** *v.* DAWKES

**Daulby, D'Aulby:** *v.* DALBY

**Daulton:** *v.* DALTON

**Daultrey:** *v.* DAWTREY

**Dauncey:** *v.* DANCEY

**Dauney:** *v.* DANDO

**Daunt:** Geoffrey *Daunte* 1229 Cl (Nf); Alan *Daunte* 1290 IpmY; Matilda *Daunt* 1379 PTY. From ME *daunten* 'to subdue, intimidate, tame, soothe'.

**Davage:** *v.* DAVIDGE

**Davall, Davolls, Deavall, Deaville, Devall, De Vile, Devill, Deville, De Ville, de Ville, Divall, Divell, Evill:** (i) Walter *de Davidisvilla* 1107 ANF; Robert *de Aiuilla, de Daiuill'* 1175, 1195 P (Y); Walter *Daiville* 1184 Templars (L), *de Daeuill'* 1190 P (R); Roger *de Divill'* 1198 Cur (Nf); Hugh *Davilla* c1200 Riev (Y); Roger *Deyvill* 1251 AssY; John *de Eyvill* 1260 AssY; Robert *de Hevill', de Heyvill', de Deyvill', de Aivill'* 1235, 1242 Fees (Lei, Nt); Richard *Divill* 1553 WhC (La); Francis *Devall* 1571 FrY. From Déville (Seine-Inférieure). The correct form was *de Daiville*. When the preposition was omitted, *Daiville* was taken to be for *de Aiville*. Hence *Evill*, from *de (H)eyville*. John *le Deyvile*, an alternative name for John *Devile* (1305 SIA) is probably an error for *de Deyvile*. For *Devall, Davall*, cf. Cotes de Val (Leics), *Cotesdeyvill* 1285 FA, held by a family from Déville. All the modern forms may be of topographical origin but some of them are also undoubtedly due to a desire to dissociate the name from *devil* which was certainly used as a nickname. (ii) Aluuinus *Deule* 1066 DB (Beds, Hu); Roger *le Diable* 1230 P (Ess); Laurencius dictus *diabolus* alias Stanford 13th *St John* (Ess); Robert *Dyvel* 1301 SRY; William *Deuel* 1310 ColchCt; John *le Deuyle* 1327 SRSf; John *Deuile* 1327 SRC. OE *dēofol* 'devil', which may be a nickname as a pageant name.

**Davenant:** Thomas *Davenaunt* 1327 PN Ess 442; John *Dauenant* 1379 AssEss. From Davenants in Sible Hedingham (Ess).

**Davenport:** Richard *de Deveneport* 1162–73 P (Ch); Richard *de Taueneport, de Daveneport* 1203 AssSt; Thomas *Davenport* 1642 PrD. From Davenport (Ches).

**Daventry:** *v.* DAINTREE

**Davers:** Robert *de Alvers* 1086 DB (Nth); Ralph *de Auuers* 1205 P (Berks); Geoffrey *Dauuers* 1209 Fees (O); Ralph *de Avers* 1235 Fees (Mx). From Auvers (La Manche) or Auvers-le-Hamon (Sarthe).

**Daves:** *v.* DAVIES

**Davey, Davie, Davy:** *Daui* Capriht 1292 SRLo; Walter *dauy* 1198–1212 Bart (Lo); Richard *Davy* 1275 SRWo. In Scotland, *Davie* is a pet-form of *David*. Here it is rather the French popular form which still survives as *Davy* and was common in England from the 13th century.

**David, Davitt:** *Dauid* clericus 1150–60 DC (L); *Davit* Burre 1278 RH (C); Thomas *Davit* 1275 RH (Nf); Robert *David* 1276 RH (Lei). Hebrew *David* 'darling, friend', a name common in both England and Scotland from the 12th century and in Wales much earlier.

**Davidge, Davage** are for *David's* (son): Richard *Davydge* 1591 SPD.

**Davidson, Davison, Davisson, Davson:** Thomas *Davyson* 1327 SRY; John *Davideson* 1350 AD vi (Wa); William *Daveson* 1500 FrY. 'Son of David.'

**Davies, Davis, Daviss, Daves, Davys:** John *Dauisse* 1327 SRC; Richard *Davys* 1402 FrY. 'Son of *Davy*', i.e. David.

**Davill:** *v.* DAVALL

**Davison:** *v.* DAVIDSON

**Davitt:** *v.* DAVID

**Davolls:** *v.* DAVALL

**Davy:** *v.* DAVEY

**Daw, Dawe, Dawes, Daws:** *Dawe* 1212 Fees (La), 1219 AssY; Ralph *Dawe* 1211 Cur (Wo), 1275 RH (D); Lovekin *Dawes* 1279 RH (O). *Dawe* is a pet-name for *David* which shares this common surname with OE *\*dawe*, ME *dawe* 'jack-daw' (1432 NED).

**Dawbarn, Dawbarne, Dawborn, Daborne:** John *Dawborn* 1569 Musters (Sr); Agnes *Dawborne* 1579 LewishamPR (K); Edmund *Daborne* 1583 Musters (Sr). 'Child of *Daw*, i.e. *David*', OE *bearn*.

**Dawber:** *v.* DAUBER

**Dawkes, Daukes, Daux:** Simon *Dawkes* 1431 FA (Wo). A contracted form of *Dawkins*. cf. JENKS.

**Dawkins:** Willelmus filius *Daukyn* 1332 SRLa; William *Daukyn* 13th WhC (La); Richard *Daukyns* 1354 AssSt; Magister Doctor *Dawkyns* 1534 GildY, identical with John *Dakyn*, LL.D., vicar-general of York. 'Little David', from *Dawe* plus *kin*. *v.* DAW, DAYKIN.

**Dawn:** Roger *Dawen* 1332 SRWa. *v.* DAW, GEFFEN.

**Dawnay, Dawney:** *v.* DANDO

**Dawson:** Thomas *Daweson* 1326 Wak (Y); Richard *Dauewesone* 1332 SRWa. 'Son of *Dawe*', i.e. David, or of *David* himself: Roger *Daudeson* 1372 DbCh.

**Dawton:** *v.* DALTON

**Dawtrey, Dawtry, Daughtery, Daughtrey, Daughtry, Daltry, Daltrey, Daultrey, Dealtry, Doughtery, Dowtry, Hawtrey, Hatry:** William *de Alta ripa* 1166 P (Y); Robert *de Halteripa* c1155 DC (L); Philip *de Hauteriue* 12th DC (L); Philip *de Alteriva* c1200 Riev (Y); Nicholas, Walter *Dautre* 1379 PTY, 1386 LLB A; Robert *Hawtry* 1524 SRSf; George *Daltry, Dealtry* 1671, 1679 FrY. From Hauterive (Orne), Lat *Alta ripa* 'high bank'.

**Day, Daye, Dey, D'Eye, Deyes:** Leofgife *ða Dagean* c1055 OEByn (So); Godiua *Daia* c1095 Bury (Sf); Aluric *Dai* 1196 P (Bk); Ralph *Deie* 1211 FrLeic; Gunild' *daiam* domini episcopi 1221 AssWo; Walter *le Daye* 1269 AssSo; Thomas *le Deye* 1277 Ely (Sf). OE *dǣge* (fem), ME *day(e)*, *dey(e)* 'kneader of bread, bread-baker', later 'dairy-maid', 'female-servant' (cf. the 1221 example). Originally used only of women, it was later used of men (1271 MED, but clearly much

earlier). Women's christian names are rare. The first two early examples are, no doubt, used in the original sense. According to Bardsley, *Day* is also a pet-name for *David*. cf. Roger filius *Daye* 1300 Guisb (Y). *D'Eye* is, no doubt, an affected spelling for *Deye*, rather than for *de Eye*.

**Daybell:** Thomas *Daybell* 1435 Shef. cf. 'Thei daunsyd all the nyȝt, till the son con ryse; The clerke rang the day-bell, as it was his gise' (15.. NED). Probably a nickname for one who turned night into day.

**Dayhouse, Dayus:** Thomas *Dayhouse* 1672 HTY. 'Dye-house', OE *dēag*, *hūs*. But *Dayus* may also be from Welsh *Deiws*, a pet-form of *David*, anglicized as *Dayus*. *v.* Morris 103.

**Daykin, Dakin:** *Daykenus* judaeus 1275 RH (R); *Daykin* de Wich 1290 AssCh; Richard *Deykin* 1344 AD vi (Sa); Thomas *Dakyn* 1551 AD vi (Sr). *Day-kin* 'Little David', cf. DAWKINS.

**Dayley:** *v.* DOYLEY

**Daylove:** Richard *filius Dayluue* 1227 Reg-Antiquiss; Geoffrey *Daylof* 1271–2 FFL; William *Daylof* 1276 RH (Lei). Presumably from an unrecorded OE *\*Dæglufu* (f).

**Dayman:** *Dayman* Buntyng 1221 ElyA (C); Stephen *Deyman* 1224 Cur (Bk); Richard *le Deymon* 1332 SRSt. Bardsley notes that in 1363 *deyes* were coupled with cow-herds, shepherds, swineherds and other keepers of live stock. Hence, *Dayman* probably meant herdsman or, possibly, dairy-man. This was also used as a personal name like *Flotmann* 'sailor' and *Glīwmann* 'ministrel' (DB).

**Daymar, Damer:** *Daimer* de Rodinton' 1203 Cur (Sa). An unrecorded OE *\*Dægmær* 'day-fame'. cf. OE *Dæghēah*, *Dæghelm*.

**Dayne(s):** *v.* DAIN

**Dayral, Dayrell, Darell, Darrell:** Thomas *de Arel* 1166 Cur (Y); Marmaduc *Darel* 1182 P (Y); Ralph *Darel(l)*, *Dairel(l)* 1204–5 Cur (Mx); Henry *de Ayrel*, *Dayrel* 1235 Fees (Bk). From Airelle (Calvados).

**Daysh:** *v.* ASH

**Dayson, Deason:** Henry *Deyesone* 1366 Eynsham (O); John *Deyesone* 1381 SRSt. 'Son of *Day*' (David) or, possibly, of the herdsman.

**Dayus:** *v.* DAYHOUSE

**Deacon, Deakan, Deakin, Deakins:** Richard *le Diakne* 1212 Cur (Sf); Richard *le Deken(e)* 1247 AssBeds, 1256 AssNb; John *Dekne* 1327 SRSx; William *le Dekon* 1332 SRSt. OE *diacon*, *dēacon*, ME *deakne*, OFr, ME *diacne* 'deacon' (*v.* NED).

**Deaconson:** Rainald *filius decani* Hy 2 DC (Nt); John *Deconson* 1374 Black, 1400 IpmNt. 'Son of the deacon', OE *dēacon*.

**Deadman:** *v.* DEBENHAM

**Deakes:** *v.* DEEKES, DITCH

**Deal:** Roger *de le Dele*, John *Dele* 1275 RH (Nf); John, Hubert *de Dele* 1317 AssK, 1327 SRSf. From Deal (Kent) or from residence in a valley (OE *dæl*).

**Dealbridge:** *v.* BRIDGE

**Dealtry:** *v.* DAWTREY

**Deamer:** *v.* DEEMER

**Dean, Deane, Deanes, Deans, Deen, Dene:** (i) Ralph *de Dene* 1086 DB (Sx); William *de la Dena* 1193 P (Sr);

Simon *in la Dene* 1271 AssSo; Robert *ater Dene*, Peter *in ther Dene*, William *atte Dene* 1296 SRSx; Thomas *del Denes* 1297 MinAcctCo; John *Dene* 1366 ColchCt. Very common, from East or West Dean (Sussex), Deane (Hants), etc., from minor places as Dean Fm (Sussex), Deans (Essex), Dene Fm (Surrey), etc., or from residence in or near a valley (OE *denu*). *v.* MELS 49. cf. DANE. (ii) Dauid *Decanus* 1160 P (Nf); Reiner *Dene* 1177 P (Nf); Geoffrey *le Dean* quondam persona (de Whalleye) 1278 WhC (Y); Robert *le Deen* 1279 RH (C); Richard *Dien* 1327 SRWo. ME *deen*, OFr *deien, dien*, modFr *doyen* 'dean' (c1330 MED). *v.* DAIN. Willelmus *filius Dene* 1301 SRY is, no doubt, parallel to Thomas *filius Decani* 1210 Cur (O), 'the dean's son'. *v.* DENSON.

**Deaner:** Henry *Deaner'* 1279 RH (C). 'Dweller in the valley.' *v.* DEAN.

**Dear, Deare, Deares, Deer, Deere, Deerr:** Goduine *Dere filius* c1095 Bury (Sf); Rogerus *filius Dere* 1221 *ElyA* (Nf); Goduui *dere* 1066, 1086 DB (Beds); Matthew *Dere* 1196 FrLeic; Robert *le Dere* 1279 RH (O). The personal-name may be OE *Dēora*, a short form of names in *Dēor-*, or an original nickname from OE *dēore* 'beloved' or OE *dēor* 'brave, bold'. The surname may be from any of these or from OE *dēor* 'wild animal', 'deer', probably 'the swift'. *v.* DEARSON.

**Dearchild:** Hugh *Dierechild'* 1176 P (Nth); Alice *Derechild* 1275 SRWo; Amos *Derchilde* 1642 PrD. Either 'child of *Dēor*', or a nickname, 'dear child', OE *dēore, cild*.

**Dearie, Deary, Derry:** (i) Hugh *le Deray* 1275 RH (Nf); Walter *Dyry* 1321–2 FFSr; Robert *Dyrry* 1540 CorNt. AFr *desrei, derei*, OFr *desroi* 'trouble, noise, injury'. (ii) Henry *de Derheye* 1275 RH (K); John *Derhey* 1275 RH (Sf). 'Dweller by the deer enclosure', OE *dēor, (ġe)hæg*.

**Dearing, Deering, Dering, Doring:** *Derinc* 1066 DB (K); *Diering(us)* 1185 Templars (K); Willelmus *filius Derinch* 1190–4 Seals (Bk); Ælfsige *Dyring* 955 BCS 917 (W); Richard *Dering* c1250 Rams (Nf); John *Dyring'* 1275 SRWo; Henry *During* 1327 SRSo. OE *Dēoring, Dȳring*, patronymics of OE *Dēor, Dȳre*.

**Dearling:** *v.* DARLING

**Dearlove:** *v.* DARLOW

**Dearman, Dorman, Durman:** *Derman* 1066 DB (Wa); *Dermannus* clanewatere c1130 AC (Lo); William, Robert *Derman* 1201 Cur (Y), 1300 LoCt. OE *Dēormann* from OE *Dēor* (*v.* DEAR) and *mann*.

**Dearsley, Dearsly:** Alan *de Deresle* 1279 RH (C). From Derisley Fm in Woodditton (Cambs).

**Dearson, Dairson:** William *Dereson* 1327 SRC. 'Son of *Dēor(a)*.' *v.* DEAR.

**Deary:** *v.* DEARIE

**Deas, Dees:** Stephen *Dees* 1285 WiSur; Robert *Des* 1327 SRY; William *Deyse* 1508 CorNt; Thomas *Dease* 1642 PrD. OFr *dez* 'dice'. Metonymic for a player at dice, or for a seller or maker of them.

**Deason:** *v.* DAYSON

**Death, Deeth, Dearth, D'Eath, D'Eathe, De Ath, De'Ath, D'Aeth:** Robert *Deth* 1196 P (Beds); Roger *Deth* 1221 *ElyA* (C), 1327 SRSf; Robert *Death* Ed 1 Battle (Sx); Gilbert *Deth* 1272 AssSo; Alice *Deth* (*Ded*), William *Deth* (*Det*), Geoffrey *Deth* (*Deet*) 1327

*SR* (Ess); Richard *Deeth* 1346 Pat (Sf). The common explanation that this surname derives from *Ath* (Belgium) is just possible. One example has been noted: Gerardus *de Athia* 1208 Cur (Gl), but the numerous other examples are certainly not topographical in origin. *de Ath, D'Eath*, etc., must be regarded as affected spellings designed to dissociate the name from *death*, OE *dēap*, ME *deeth, deth*, which fits the above forms. The pronunciation is *Deeth*, the normal development of *dēap*. As Weekley has suggested, the name may be derived from the pageants. Death was personified in the Chester plays. cf. *Mort* and the French *Lamort*. Fransson has noted a rare occupational name, John *le Dethewright'*, *le Dedewrithe, le Dedewrighte* in Essex in 1299 and 1327 (MESO), from OE *dȳð* 'fuel, tinder'. This survives as *Deathridge* and *Deth(e)ridge*. *Death*, a medieval and modern surname common in Essex and East Anglia, where *dȳð* would occur in ME as *deeth*, is probably metonymic for a maker of tinder, an etymology confirmed by the survival of *Deether* and *Deetman*, which must be synonymous with *Deathridge*. Bardsley's topographical examples are errors for *Deche*. *v.* DEEKES.

**Deathridge:** *v.* DEATH

**Deavall, Deaville:** *v.* DAVALL

**Deave, Deaves, Deeves:** Rogerus *Surdus* 1196 Cur (Nth); Ralph *le Deue* 1251 Oseney (O); Sarra *le Deaue* (*la Deafe*) 1317 AssK. OE *dēaf* 'deaf'; *deave* is from the inflected forms *deafum, deafe*.

**Deavin:** *v.* DEVIN

**Debbage:** William *Debeche*, Robert *Debedge* 1568 SRSf. From Debach (Suffolk), pronounced *Debbidge*.

**Debdale:** Robert *Debdale* 1504, Anna *Dybdale* 1511 ArchC 41. From Debdale (La, NRY), or 'dweller in the deep valley', OE *dēop, dæl*.

**Debell, Deeble:** William *Debel* 1197 P (Y); Alexander *le Deble* 1221 AssSa; Mary *Deeble* 1699 DWills. Fr *debile*, Lat *dēbil-is* 'weak, feeble' (1536 NED).

**Debenham, Debnam, Deadman, Dedman:** John *de Debenham* 1279 RH (Hu); Robert *Debenham, Debnam* 1674 HTSf; Francis *Debnam, Deadman* ib.; John *Deadman, Debingham* ib.; Stephen *Dedman*, Thomas *Dednum* ib. From Debenham (Suffolk).

**Debney:** Robert *deulebeneie* 1162 P (Nb); Laurence *Deubeneye* 1328 LLB E; Ralph *Dieubeneye* 1341 LLB F; Widow *Debny* 1674 HTSf. A French phrase-name 'God bless (him)'. cf. Olive *Goadebles* 1269 Pat.

**Debonnaire:** Henry *le Deboner'* 1221 Cur (Hu); William *le Deboner* 1247 AssBeds; John *Deboneir'* 1275 MPleas. OFr *debonaire* 'mild, gentle, kind'.

**Debutt:** For TEBBUTT, with voicing of initial *T*. *v.* THEOBALD, DIBBLE.

**de Carteret:** *v.* CARTERET

**Decker:** *v.* DICKER

**Decourcy, de Courcy, de Courcey, Coursey:** Richard *de Curci* 1086 DB (O); Thomas *de Curci* c1150–60 DC (L). From Courcy (Calvados).

**Dede:** *v.* DEED

**Dedman:** *v.* DEBENHAM

**Dedden:** Richard *Desdans* 1197 FFNth; Thomas *Dendanz* 1276 AssW; Drueta *Dedan* 1327 SRWo.

'Dweller within (the town or village)', OFr *desdans*.

**Dedicoat:** *v.* DIDCOCK

**Dee:** Roger *Dee* 1642 PrD. An anglicized form of Welsh *dhu* 'black'. *v.* Morris 120.

**Deeble:** *v.* DEBELL

**Deed, Deedes, Dede, Dade:** Richard *Ded* 1195 P (Bk); Hugo *Dede* 1210 P (D); Thomas *Dade*, Adam *Ded* 1275, 1285 Wak (Y); Roger *Dade* 1275 RH (L). OE *dǣd* 'deed, exploit'.

**Deeker:** *v.* DEEKES, DICKER

**Deekes, Deeks, Deex, Deakes:** (i) *Dike* Marescallus 1212 Cur (Y); *Dyke* de Chypendale 1246 AssLa; Richard, John *Dike* 1195 P (Sf), 1279 RH (Hu); Robert *Dykes* 1327 SRWo; John *Deke* 1332 SRSx; Thomas *Deeke* 1568 SRSf. *Dick* often appears as *Dike* which, with a lowering and lengthening of the *i*, would become *Deke*, *Deek*. *Dickins* is found as *Deekins* in 1728 (Bardsley). (ii) Similarly, OE *dīc* 'ditch, dike' occurs both as *dīke* and *dēke* in ME. cf. *Dekeleye* (1377) for Dickley (PN Ess 344) and Henry and Alice *de Deche* 1279 RH (C). This is printed *Dethe*, but they undoubtedly lived near the Devil's Dyke. *Atte Deke* would become *Deek* and correspond to *Deeker* 'the dweller by the dike'. *v.* DITCH.

**Deem, Deam:** Gilbert, Richard *Deme* 1279 RH (O). OE *dēma* 'judge'. cf. DOME.

**Deemer, Deamer, Demer, Demers:** Leuric *Demere* c1095 Bury (Sf); Alan *Demur* 1250 Trentham (St); Richard *le Demor*, *le Demur* 1301, 1357 ib. OE *dēmere* 'deemer, judge', one who pronounces the verdict or doom. cf. '*Demar*, Iudicator' PromptParv, and *v.* DEEMING, DEMPSTER, DOMAN.

**Deeming, Demings:** Reginald *Demung* 1246 Seals (W); John *Domyng* 1351 AssEss. OE *\*dēmung* 'judging, judgement', from OE *dēma* 'judge'. *Domyng* may be an error for *Demyng*, otherwise from OE *\*dōmung* 'judgement', from OE *dōm* 'doom, judgement'. The surname is identical in meaning with DEEMER and DOMAN.

**Deen:** *v.* DEAN

**Deeping:** Geoffrey *de Depinge* 1189–1203 RegAntiquiss; Warin *de Deping* 1271 FFL; John *Depyng* 1437 IpmNt. From Deeping (L), or Deeping Gate (Nth).

**Deer, Deere, Deerr:** *v.* DEAR

**Deerhurst:** Yuo *de Derherst* 1191 P (Gl); William *de Derhurste* 1275 SRWo; Robert *de Derhirst* 1304 LLB B. From Deerhurst (Gl).

**Deering:** *v.* DEARING

**Dees:** *v.* DEAS

**Deetch:** *v.* DITCH

**Deetcher:** *v.* DICKER

**Deeth, Deether, Deetman:** *v.* DEATH

**Deever:** William *Deever*, John *Dever* 1642 PrD; Mr *Devers* 1672 HTY. Perhaps a variant of DIVER.

**Deeves:** *v.* DEAVE

**Deex:** *v.* DEEKES, DICK, DITCH

**Defender:** Henry *le Deffendur* 1221–2 FFWa. OFr *defendëor* 'defender'.

**de Fraine, de Freyne:** *v.* FRAIN

**de Haviland:** *v.* HAVILAND

**Deighton:** Thomas, Henry *de Dicton* 1204 AssY,

1259 IpmY; Richard *de Dyghton* 1327 FFY; John *Dyghton* 1419 IpmY. From Deighton (ER, NR, WRYorks).

**Delacour:** A Huguenot name from an aristocratic family *De la Cour*. The first refugee, a distinguished officer in the French army, settled near Portarlington, his descendants afterwards removing to the county of Cork (Smiles 383).

**Delafield, de la Feld:** *v.* FIELD

**Delahaye, De La Haye, de la Hey, Delhay:** Ranulf *de Lahaia* 1119 Colch (Ess); John *del Haye* 1275 RH (Nth); Richard *del Heye* 1275 SRWo. Robert *de Haia* (1123), founder of Boxgrove Priory (Sussex), came from Haye-du-Puits (La Manche). *v.* ANF. The surname is commonly English in origin, 'dweller by the enclosure'. *v.* HAY.

**Delahooke:** *v.* HOOK

**Delamar, Delamare, de la Mare, Delamere, Delamore, Dallamore, Dallimore, Dillamore, Dolamore, Dollamore, Dollemore, Dolleymore, Dollimore, Dollymore:** Henry *de Lamara* 1130 P (O); Coleman *de Lamora* 1135–85 Seals (Nth); Robert *de la Mare* 1190 BuryS (Sf); William *de la Mere* 1260 FFEss; Henry *Dalamare* 1385 FrY; Thomas *Dallamour* 1732 FrY; John *Dallamore* 1733 FrY. Early bearers of the name came from one of the numerous French places named La Mare 'pool'. Many later names are of English origin, 'dweller by the mere, lake, marsh or moor', OE *mere*, *mōr*, with common confusion of these words. *v.* MOOR. Dollyman's Fm in Rawreth (Essex), *Dallamers* alias *Dalymers* in 1600, owes its name to John *de la Mare* (1342 PN Ess 193).

**Delamond:** Beatrix *Delamond* 1441 IpmNt. 'Dweller by the hill', OFr *de la mont*.

**Delamotte:** A Huguenot name. Joseph *de la Motte*, born at Tournai, fled to Geneva during the persecution by the Duke of Alva. He returned to Tournai, but was forced to flee to St Malo, thence to Guernsey, and so to Southampton, where the name still survives.

**Delane:** A Huguenot name. Peter *de Laine* obtained denization in 1681. J. H. *Delane*, editor of *The Times* was a descendant (Smiles 383).

**Delaney, Delany, de Launay, Delauney:** *v.* DANDO

**Delaware:** Sarah *de la Ware* 1201 SPleas (K). From Delaware in Brasted (K), or 'dweller by the weir', OE *wer*.

**Delbridge:** *v.* BRIDGE

**Delf, Delph, Delve, Delves:** Richard *de la Delphe* 1295 MELS (Sx); Mabel *de la Delue*, John *atte Delue* 1296 SRSx; John *Delves* 1376 AD vi (St). 'Dweller by the ditch(es), quarry or quarries', OE *(ge)delf* 'digging, excavation'. Sometimes metonymic for 'excavator, quarrier' (OE *delfere*): William *le Deluer* 1230 P (So). Lettice *atte Delue* (1357 ColchCt) is probably to be associated with John *Delvere* (1359 ib.).

**Delford, Dellford:** John *Delforde*, *del Forde* 1324 CoramLa. 'Dweller by the ford', OE *ford*.

**Delhay:** *v.* DELAHAYE

**Delicate:** *v.* COTE

**De Lisle, De L'Isle, de Lyle:** *v.* LISLE

**Dell:** William, Robert *atte Delle* 1296 SRSx, 1309 LLB D. 'Dweller in the dell' (OE *dell*). *v.* also DIL.

**Delland:** Philip *Deneland* 1275 RH (D); Roger *Denlond* 1296 SRSx. 'Dweller by the valley-estate' (OE *denu, land*).

**Dellar, Deller:** Ralph, John *Dellere* 1275 RH (Nf), 1347 AD iii (Sr). 'Dweller in the dell.' cf. DELL.

**Dellaway, Dilloway, Dolloway, Dilliway:** John *Delewey* 1306 AssW; Matthew *Deloway* 1662 HTEss. 'Dweller by the road or path', OFr *de la*, OE *weg*.

**Dellbridge:** *v.* BRIDGE

**Delleman:** Equivalent to DELLAR.

**Dellew:** Henry, John *del Ewe* 1250 Oseney (O), 1274 RH (Sa). 'Dweller by the water' (Fr *eau*).

**Dellford:** *v.* DELFORD

**Delling:** Henry, Jordan *Dellyng* 1296 SRSx, 1327 SRSo. *v.* DILLING.

**Dello, Dellow:** (i) Walter *Delho* 1275 RH (Herts); William *Delhou* 1279 RH (O). (ii) William *Dellowe* 1275 RH (W). 'Dweller by the ridge or hill' (OE *hōh, hlāw*). *v.* HOW, LOW. *Dell*, rare in place-names, is apparently unknown as a first element.

**Delph, Delve(s):** *v.* DELF

**Del Strother:** *v.* STROTHER

**Delver:** William *le Deluer* 1230 P (So); Walter *le Delvere* 1300 LoCt; John *Delvere* 1359 ColchCt. A derivative of OE *(ge)delf* 'excavation, quarry', hence 'digger, quarryman'.

**Deman, Demant:** *v.* DIAMANT

**Demers:** *v.* DEEMER

**Demings:** *v.* DEEMING

**Demmar:** William *le Demmere* 1296 SRSx. A derivative of OE *\*demman* 'to dam, obstruct the course of water'. A maker of dams.

**de Montmorency:** Herueus *de Munmoreci* 1177 P (Sx). From Montmorency (Seine-et-Oise, Aube).

**Dempsey, Dempsy:** James *Dempse* 1526 GildY. Irish *Ó Díomasaigh* 'descendant of *Díomasach*' (proud).

**Dempster:** Haldan *Deemester* 1296 Black (Perth); Walter *Demester* 1313 MEOT (La); Andrew *Dempstar* 1360 Black (Brechin). A feminine form of *Deemer*, like *Baxter*, etc., used of men. A northern and Manx term for 'judge', common in Scotland for the judge of the Parliament, shire or baron-bailie. Until 1747 every laird of a barony had power to hold courts for the trial by his dempster of certain offenders within his barony.

**Denbow:** Eudo, Philip, William *Denebaud* 1214 Cur (So), 1276 RH (D), 1298 Ipm (Do); John *Dembel* 1338 FFSf. OE *\*Denebeald*, an unrecorded compound of *Dene-*.

**Denby:** Simon *de Denebi* 1191 P (Y); Jordan *de Deneby* 1219 AssY; William *de Denby* 1357 IpmNt; John *Denby* 1533 FFEss. From Denby (WRY), or Denaby in Mexborough (WRY).

**Dence, Denns, Dench:** Thomas *Dench* 1327 SRWo. OE *denisc*, ME *denshe, dench*, Scottish *Dence, Dens* 'Danish'.

**Dendale, Dendle:** William *Dendale* 1283–4 IpmY; William *Dendle* 1642 PrD. From Dentdale in Dent (WRY).

**Denford:** Roger *de Deneford* 1242 Fees (Nth); John *Denforde* 1340–1450 GildC; John *Denford* 1642 PrD. From Denford (Berks, Nth). cf. also John *Denfote* 1332 SRWa. 'Dweller at the lower end of the valley', OE *denu, fōt*. This would probably fall in with *Denford*.

**Denham, Denholm, Denholme:** Richard *de Deneham* 1176 P (Bk); John *de Deneholme* 1332 SRLa; John *Denham* 1466–7 FFSr. From Denham (Bucks, Suffolk), Denholme (WRYorks), or 'dweller at the farm in the valley'.

**Denholm, Denholme:** John *de Deneholme* 1332 SRLa. 'Dweller by the *holm* in the valley.' *v.* HOLME.

**Denial:** *v.* DANIEL

**Dening:** *v.* DENNING

**Denington:** *v.* DENNINGTON

**Denis, Denise, Dennis, Denniss, Dennys, Dennes, Denness, Dinnis:** (i) *Dionisius de Chotum* Hy 2 DC (L); *Dionisia* Hy 2 DC (L); *Denis de Sixlea* 1176 P (L); *Deonisia* 1212 Cur (Y); *Denise* 1321 FFEss; *Deonis* 1327 SRSo; *Dionis* ate Brome 1332 SRSr; *Denes* Lister 1379 PTY; Walter *Denys* 1272 AssSt; Walter *Dyonis* 1297 MinAcctCo; Ralph *Denys, Dynis* 1308 EAS xviii; Robert *Deonis'* 1317 AssK. Lat *Dionysius*, Gk *Διονύσιος*, 'of Dionysos', the name of several saints, common in England after the 12th century. The feminine *Dionysia, Denise* was equally popular and both are represented in the surnames. Robert *Denys* was the son of *Dionisius* de Grauntebrigge (1321 Cor). (ii) Radulfus *Dacus* 1176 P (Ha); Robertus *Danus* 1193 P (Nf); Rannulf *le Daneis* 1193 P (Wa); William *le Daneys, le Deneys* 1232, 1241 FFHu. ME *danais*, OFr *daneis* 'Danish', 'the Dane', with the vowel influenced by that of ME *denshe*, OE *denisc*. cf. DENCE.

**Denison, Dennison:** (i) Roger *Deneyson* 13th *Binham* (Nf); Adam *Deynissone* 1381 SRSf; Henry *Dennesson* 1450 Rad. 'Son of *Denis*.' (ii) Walter *Denizen* 1275 RH (Ess). AFr *deinzein*, a burgess who enjoyed the privileges of those living *deinz la cité* 'within the city'.

**Denleigh, Denley, Denly:** Thomas *de Denley* 1279 RH (O). 'Dweller by the valley-clearing' (OE *denu, lēah*).

**Denman:** William *Deneman* 1314 FFEss; Adam *Deneman* 1332 SRSr. 'Dweller in the valley.'

**Denmer, Denmore:** Alan *Denmere* 1296 SRSx. 'Dweller by the valley-lake' (OE *denu, mere*).

**Denn, Denne:** (i) Thomas *de Denne* c1160 ArchC 5; Baldwin *de la Denne* 1275 RH (K); John *atte Denn* 1296 SRSx. A swineherd, one who worked in the woodland or swine-pasture (OE *denn*). (ii) William *Denn* 1296 SRSx. This may be for *atte Denne*, but as *Denis* had a diminutive *Denet*, a pet-name *Den* must have existed side by side with *Din*.

**Dennell:** *v.* DANIEL

**Denner:** 'Dweller in the valley.' cf. DENMAN.

**Dennes:** *v.* DENIS

**Dennet, Dennett, Dennitts:** Alice *Denet* 1279 RH (Beds); Richard *Dynot* 1279 RH (O); John *Denot* 1332 SRSx. Diminutives of *Den*, a pet-form of *Denis*. In *Piers Plowman*, *D note* is a woman's name, from *Denise*.

**Denney, Denny, Dinnie, Dinny:** (i) *Denny* Bocher 1374 AssL; John *Denye* 1275 RH (Sf); William *Dyny* 1298 AssL; John *Denny* 1379 PTY. A pet-form of *Denis*. (ii) Robert *de Denye* 1296 SRSx. From Denny (C). (iii) John *of Deny* 1424 Black. From Denny (Stirling).

**Denning, Dening:** Leonus *Dennyng* 1286 ForC; Henry *Dennyng* 1367 Crowland; John *Denninge* 1642 PrD. 'Son of *Dynna*', or perhaps 'dweller in the valley', from a derivative of OE *denu* 'valley'.

**Dennington, Denington:** Stephen *de Deninton* 1199–1200 FFK; Richard *de Denynton'* 1259 Acc; Johanna *de Denyngton* 1379 PTY. From Dennington (Sf, WRY).

**Dennis, Dennys, Denness, Denniss:** *v.* DENIS

**Dennish:** for DENIS or DEVENISH.

**Dennison:** *v.* DENISON

**Denniston:** *v.* DENSTON

**Denns:** *v.* DENCE, DENN

**Denny:** *v.* DENNEY

**Denre, Denry:** *v.* DARREE

**Denson, Densum:** Henry *le Deneson* 1295 AssSt; Adam *Densone* 1362 FrY. 'Son of the dean.'

**Denston, Denstone, Denniston:** Henry *de Denneston* 1199 AssSt; Hemfrey *de Denarston* 1275 RH (Nf); John *Denston* 1453 FFEss; Robert *Denstone* 1641 PrSo. From Denston (Sf), or Denstone (St).

**Dent:** Waltheef *de Dent'* 1131 FeuDu; William, John *de Dent* c1200 Riev, 1356 FFY; John *Dent* 1403–4 IpmY. From Dent (WRYorks).

**Denterlew, Dentrelew:** Richard *Dentrelewe* 1207 Pleas (Bk). 'Dweller between the streams', OFr *entre l'ewe*.

**Dentith:** *v.* DAINTETH

**Denton:** Ælfweard *æt Dentune* 972 BCS (Nth); William *de Denton* 1271 FFL; Richard *Denton* 1403 IpmNt. From one or other of the many places of this name.

**Denty:** *v.* DAINTY

**Denver, Denvir:** Geoffrey *de Denever'* 1206 Cur (Nf); Walter *de Denevere* 1275 RH (Nf); John *Denver* 1642 PrD. From Denver (Nf).

**Denzil, Denzill:** *Denisel* 1189 Sol; Bartholomew *Densell* 1641 PrSo. *Denis-el*, a diminutive of *Denis*.

**Derby, Derbyshire:** *v.* DARBY, DARBYSHIRE

**Dereham:** *v.* DERHAM

**Derfield:** *v.* DARVAL

**Derham, Dereham:** Anketill *de Derham* 1177 P (Nf); Andrew *de Derham* 1278 Oseney; William *de Dereham* 1393 IpmNt; Thomas *Derham* 1426–7 Paston. From East, West Dereham (Nf).

**Dering:** *v.* DEARING

**Derkin:** *v.* DARKIN

**Dermott:** *Dermot* an Irishman 1243 AssSo. *v.* MACDERMOT.

**Derolf:** Robert *filius Derolf* 1210 Cur (C); Geoffrey *Derholf* 1230 P (Nth); Matilda *Derolf* 1279 RH (C). OE *Dēorwulf*.

**Derrick:** Richard *Deryk*, Ducheman 1525 SRSx; John *Dericke* 1583 Musters (Sr); William *Derrik* 1641 PrSo. A late borrowing from the Low Countries of OG *Theodoric*.

**Derrington:** Walter *de Dyryngtone* 1184 Gilb;

Stephen *de Derington* 13th Guisb. From Derrington (St).

**Derry:** *v.* DARREE, DEARIE

**Derville:** *v.* DARVAL

**Derwent:** John *de Derwente* 1279 AssNb. From the River Derwent (Cu, Db, Du, Nb, NRY, WRY).

**Desbois:** A Huguenot name. Lazarus *Desbois* fled to Amsterdam in 1692, and in 1699 to England. He became a cabinet-maker in Soho (Smiles 385).

**Desborough, Desbrow, Disborough:** John *de Desburgh'* 1216 AssNth; John *de Deseburgh* 1363 FrY; Thomas *Desborrough* 1584 Musters (Sr). From Desborough (Bk, Nth).

**Desert:** Roger *de Deserto* 1199 MemR (Lo); William *Desert'* 1304 IpmY; Lucian *desert* Ed 3 Rydeware (St). 'Dweller in the barren area, wilderness', OFr *desert*.

**Despard, Dispard:** A Huguenot name. Philip *d'Espard* escaped to England at the Massacre of Saint Bartholomew. He was sent to Ireland by Elizabeth, and his grandson William was Colonel of Engineers under William III.

**Despencer, Despenser:** Gilbert *le Despenser* 1198–1212 Bart; William *le Despenser* 1256 AssNb; Robert *le Despencer* 1300 AssSt. OFr *despensier* 'dispenser (of provisions)', a butler or steward. *v.* also SPENCER.

**Detchfield:** *v.* DITCHFIELD

**Detheridge, Dethridge:** *v.* DEATH

**Dethick:** Geoffrey *de Dethek* c1273 Glapwell (Db); John *de Detheke* 1327 SRDb; John *Dethicke* 1506 Pat (Nt). From Dethick (Db).

**de Trafford:** *v.* TRAFFORD

**Deudney, Dewdney, Doudney, Dowdney, Dudeney, Dudney:** *Deodonatus* 1206 Cur (Sr); Dominus *Deudenay* capellanus 1327 SRSx; Richard *Deudone* 1175 P (K); John *Dewdenay* Eliz Musters (Sr); John *Dowdney*, Lancelot *Dudney*, Henry *Dudeny* 1642 PrD. Lat *Deodonatus*, Fr *Dieudonné*.

**Deuters, Dewters:** Martin *Dutere* 1191–1212 Bart; Christopher *Dewtris*, *Dewtrice* 1671, 1682 FrY. Perhaps a derivative of ME *duten*, *douten*, OFr *duter*, *douter* 'to hesitate', used of one who is timid, wavering in opinion, one who dilly-dallies.

**Devall:** *v.* DAVALL

**Devas, Devis:** Robert *Dewias* 1185 Templars (W); John *Deuyas*, *Dewyas* 13th WhC; Hugh *Deuyas* 1359 IpmNt. From Ewyas Harold, Lacy (He), or 'dweller in the sheep district', Welsh *ewig*, *-as*.

**Devenish, Devonish:** Robert *le Deueneis* 1205 P (L); William le *Deveneys* 1243 AssSo; John *le Devenisshe* 1337 LLB F. OE *defenisc* 'the man from Devon'.

**Deverall, Deverell:** Robert *de Deuerel* 12th Seals (So); Peter *de Deverell* c1240 Glast (So); Robert *Deverel* 1362 IpmW. From Brixton, Hill, Kingston, Longbridge, Monkton Deverill (W). These all appear as *Devrel* in DB, and are named from the River Deverill.

**Deveraux, Devereaux, Devereu, Devereux, Deveroux, Deverose, Everix, Everiss, Everest, Everist:** Roger *de Ebrois* 1086 DB (Nf); Walter *de Eureus* 1159 P (He); Stephen *de Euereus* 1199 MemR (Wo); Osmund *de Deuereals* ib. (W); Eustace *de*

*Deueraus* 1204 P (So); Thomas *de Euereus, Deuereus* 1279 AssSo; John *de Ebroicis* 1297 AssSt; John *Deveros* 1385 LLB H; Robert *Everis* 1495 GildY. From Evreux (Eure), from the Celtic tribal name *Eburovices* 'dwellers on the Ebura or Eure River'.

**De Vile, Deville, De Ville, de Ville:** *v.* DAVALL

**Devin, Devine, Devinn, Deavin, Divine:** Nicholas *le Deuin, le Diuin* 1187–8 P (He). Me *devin, divin,* OFr *devin* 'divine', used of persons 'of more than ordinary excellence' (c1374 NED). The Irish *Devin(e)* is for *Ó Daimhin* or *Ó Duibhin* 'descendant of *Daimhin* or of *Duibhin*', diminutives of *damh* 'ox, stag' and *dubh* 'black'.

**Devlin:** Adam *de Divelyn* 1256 AssNb; William *Develyn* 1380 LoPleas. From Dublin. The Irish *Devlin* is *Ó Dobhailein* or *Ó Doibhilin* 'descendant of *Dobhailen*'.

**Devis:** *v.* DEVAS

**Devon, Devons:** Adam *de Devoun* 1275 RH (Nf). From Devon.

**Devonish:** *v.* DEVENISH

**Devonshire:** William *de Devenschyre* 1288 NorwDeeds I; William *de Deveneshire* 1339 LLB F; William *Devenschyr* 1420 TestEbor. 'The man from Devonshire.'

**Dew, Dewe, Dewes, Dews:** (i) William *Deu* 1200 Pleas (W); John *de Eu* 1278 Oseney; John *de Ewe* 1327 SRSf. From Eu (Seine-Maritime). cf. Willingale Doe (Ess), Hugh *de Ou, D'Eu* Hy 2, and Dews Hall in Lambourne (Ess), John *Deu* 1248. (ii) John *atte Dywe* 1327 SRSx. OE *dēaw* 'dew', probably in the sense 'damp ground'. (iii) Ithel *Du* 1327 SRSa. Welsh *dhu* 'dark, swarthy'.

**Dewar:** (i) Gael *Deoir, Deoireach* 'pilgrim, sojourner'. The medieval *deoradh* had custody of the relics of a saint. *v.* MACINDEOR. (ii) Thomas *de Deware* 1296 CalSc (Edinburgh). From Dewar (Midlothian).

**Dewberry:** Robert *Dewbery* 1509 CorNt. From Dewberry Hill in Radcliffe on Trent (Nt).

**Dewdney:** *v.* DEUDNEY

**Dewe, Dewes:** *v.* DEW

**Dewey, Dewy:** (i) Roger *de Duaco* (*Duay*) 1220 Cur (Sf); Henry *Dewy* 1279 RH (Bk); Thomas *Dewy* 1428 FFEss. From Douai (Nord). (ii) David *Dewy* 1297 MinAcctCo; Jeuan *Duy* 1392 Chirk; Mortagh *Dewye* 1642 PrD. Welsh *Dewi,* i.e. *David.*

**Dewhirst, Dewhurst, Dewhurst:** Roger *de le Dewyhurst* c1300 WhC (La). From Dewhurst (Lancs).

**Dewsbury, Duesbury, Jewesbury, Jewsbury, Joesbury:** Tomas *de Dewesberi* 1204 P (Y). From Dewsbury (WRYorks). Jews- is a common colloquial pronunciation of *Dews-.* cf. *Chuesday* for *Tuesday.*

**Dewsnap:** Thomas *de Deuysnape* 1285, Nicholas *de Dewysnap* 1286 AssCh; Thomas *Dewsnop* 1599 SRDb. From Dewsneps in Chinley (Db).

**Dewy:** *v.* DEWEY

**Dexter:** John, Ralph *le Dextere* 1262 FrLeic, 1327 SRSf; Roger, Simon *le Dykestre* 1305 SIA; William *Dexter* 1378 AssWa. A form of *Dyster*, noted in Essex, Leicester, Suffolk and Warwicks (14.. NED).

**Dey, Deyes:** *v.* DAY

**Deykes:** *v.* DITCH

**d'Eyncourt, Danecourt:** Walter *de Aincurt* 1086 DB

(Nth, Db, L, Y); Ralph *Daincurt* c1157 Gilb (L); Oliver *Deyncourt, de Aincurt, de Eyncurt* 1243 AssSo. From Aincurt (Seine-et-Oise).

**Deyns:** *v.* DAIN

**Diamant, Diamond, Deman, Demant, Diment, Dimond, Dimont, Dyment, Dymond:** *Diamant* and *Diamond* may occasionally derive from *Diamanda* 1221 Cur (Mx), 1349 Husting, one of the fanciful names given to women in the Middle Ages. Hence, probably, William *Dyamond* 1332 SRD, who gave his name to Dymond's Bridge in Whitestone (Devon). But these examples will hardly explain Thomas *Dymande* 1332 SRSr; Robert *Dymond'* 1379 PTY; William *Demaunde, Dymaunde* 1391, 1392 CarshCt (Sr), or the numerous and varied forms in the HartlandPR (D): Edmund, Joanna *Demon* 1564, 1566; Thomas *Deymon* 1581; Joanna, Samuel *Dymon* 1582, 1826; John *Deman* 1632; Elizabeth *Daymand* 1685; Charles, Ann *Daymond* 1686, 1688; Grace *Dyamond* 1753; Joannah *Dyman,* Mr Wm *Dayman's* wife 1765; Margaret *Daimant* 1801; Humphrey *Dyment* 1817; Susan *Dayment* 1833. cf. also John *Deyman* alias *Dymond* 1698 DKR (D), while Daymond's Hill in Tiverton (Devon) owes its name to John *Dayman* 1589 SRD. The excrescent *t* or *d* is not common before the 17th century, and all probably derive from *Dayman,* a personal and occupational name 'herdsman'. *v.* DAYMAN.

**Diaper:** *v.* DIPPER

**Dibb, Dibbs:** John *del Dybbe* 1469 FrY. 'Dweller by the hollow', from *dib,* a northern dialect form of *dip,* a small hollow in the ground (1847–78 NED). The surname may also be from *Dibb,* a short form of *Dibble.*

**Dibben, Dibbens, Dibbins:** Thomas *Dybyn* 1332 SRDo; Thomas *Dybben* 1476 SRW; William *Dibben* 1664 HTSo. Equivalent to DIBBS, with the descendant of the OE weak genitive singular. *v.* GEFFEN.

**Dibble, Dible, Diboll, Dybald, Dyball, Dybell, Dyble, Dipple:** William *Dibel* 1275 RH (Lo); Walter *Dipel* ib. (Sf); Ralph *Dibald* 1276 RH (Y); William *Dybald* 1277 AssLa; William *Dybel* 1277 LLB B; William *Dypel* 1327 SRWo; Edmund *Dyboll* 1524 SRSf; Sarah *Dipple, Dibble* 1678, 1680 SaltAS (OS) x. *Dibald* is from *Tibald* (*Theobald*), with a voicing of the initial consonant. *Dibel* is a diminutive of *Dibb,* a pet-name for *Theobald.* For *Dipple,* cf. TIPPELL.

**Dibden:** . . . *de Depedene* 1270 PN K 63. From Dibden in Riverhead (K).

**Dice, Dyce:** Richard *Dyse* 1327 SRSf; John *Dyce, Dys* 1412, 1418 LLB B, I. ME *dyse, dyce* 'dice', both singular and plural; also 'chance, luck', a nickname for a gambler. cf. HAZARD.

**Dick, Dicke, Dicks, Dix:** *Dicke* Smith 1220 Cur (L); *Dik* 1260 AssCh; *Dik* de Hyde 1286 ib.; Richard *Dike* Hy 3 Colch (Ess); William *Dik* 1356 LLB G; William *Dyckes* 1362 AD iii (Nf). A pet-form of *Richard,* found also as *Dike, v.* DEEKES.

**Dickason:** *v.* DICKENSON

**Dickels:** *v.* DIGGLE

**Dicken, Dickens, Dickin, Dickins, Dickings, Dickons, Dykins, Dekin, Dekiss:** *Dicun* Malebiss' 1207 Cur (Y); Richard *Dicun* 1203 AssSt, 1230 P

(Beds); John *Dycon* 1327 SRSt; Maud *Dyconnis*, David *Dyccons* 1327 SRWo; Robert *Dekoun* 1327 SRC. A diminutive of *Dick*, *Dic-un*. Other diminutives were *Dic-el*, *Dic-el-in*, *Dic-et*. *v*. DIGGEN, DEEKES. *Dekoun* would become *Deekon*, *Deekin*. *Dekiss* is from *Deakins*, cf. HODGKINS and HODGKISS.

**Dickenson, Dickinson, Dickerson, Dickeson, Dickison, Dickason:** William *Dykounson* 1366 SRLa; John *Dykonesson* 1388 FrY; Henry *Dicason* 1518 GildY; Gilbert *Dyckenson* 1585 ShefA; Nicholas *Dikersone* 1598 AD vi (Nf). 'Son of *Dicun*.' John and Henry '*Dicounesson* de Clesnesse' were sons of *Richard*, son of Henry de Clesnesse 1359 Pat (Nb).

**Dicker, Dikkers, Decker, Deeker, Ditcher, Deetcher:** (i) John *Dicher* 1210 Cur (Ess); Simon *le Dykere* 1296 SRSx; Hugh *le Dykere* (*Dikkere*) 1327 *SR* (Ess); Agnes *le Dycher* 1354 Putnam (Ch). OE *dīcere* 'one who digs ditches, ditcher', or 'one who lives by a dike'. cf. DICKMAN, DEEKES. (ii) Henry *ater Dykere* 1296 SRSx; John *atte Diker* 1327 ib. *Dicker*, still surviving in Sussex, may also derive from The Dicker (PN Sx 439, MELS 51).

**Dicketts:** John, Robert *Dyket* 1296 SRSx, 1329 ColchCt. A diminutive of *Dick*.

**Dickie, Dickey:** Robert *Dikky* 1504 Black (Glasgow). A Scottish diminutive of *Dick*.

**Dickin(son):** *v*. DICKEN(SON)

**Dickman, Digman:** (i) Richard, John *Dikeman* 1206 Cur (L), 1227 AssSt; Thomas *Dekeman* 1327 SRC. OE *dīc* 'ditch, dike' and *mann*, 'one who lives near, or works on a ditch or dike'. cf. DICKER. *Dekeman* would become *Deekman*. cf. DEEKER. (ii) Robertus *serviens Ricardi*, Thomas *Dikman*, *Richardman* 1379 PTY. 'Servant of *Dick*.'

**Dicksee:** *v*. DIXEY

**Dickson, Dixon, Dixson:** Thom *Dicson* 1307 Black (Castle Douglas); John *Diksson*, *Dikson* 1332 SRCu; Nicholas *Dyxon* 1425 FFEss; Robert *Dixson* 1429 LLB K. 'Dick's son.'

**Didcock, Didcote, Didcott, Dedicoat:** Mistris *Dedycott* 1595 ChwWo. From Didcot (Berks), or Didcote (Gl).

**Didsberry, Didsbury:** Roger *de Diddesbiri* 1260, William *de Dyddesbyry* 1276 AssLa. From Didsbury (La).

**Dieppe:** Saladinus *de Depe* 1224 Pat. From Dieppe.

**Dieulesait:** Thomas *Deuleseit* 1237–8 AccM; Richard *Deusscit* c1250 Rams (C). OFr *Deu le seit* 'God knows it'. There were a number of such surnames in medieval times, none of which seems to have survived: John *Dieugarde* 1377 IpmW 'may God guard (you)'; Stephen *Deusaut* 1219 P (Sr) 'may God preserve (you)'; *Deuleward'* 1219 AssY 'may God guard (you)'.

**Difford:** Hugh *de Difford* 1276 RH (Y). 'Dweller by the deep ford', OE *dēop*, *ford*.

**Digan:** *v*. DIGGEN

**Digby:** Roger *de Digby* c1160–5 RegAntiquiss; Geoffrey *de Dyggeby* 1250 FFL; Simon *Digby* 1497 FFWa. From Digby (Lincs).

**Digg, Digges:** William *Dyg* 1296 SRSx; Thomas *Dygge* 1327 SRSf; John *le Digge* 1327 SRSx; Christina *Dygges* 1335 AD i (Do). ME *digge* 'a duck' (c1450

MED). Also from a voiced form of *Dick*: *Digge* de Torot 1246 AssLa.

**Diggan:** *v*. DIGGEN

**Diggatt:** William *Diggard* 1209 P (Nf); Henry *Digard* 1275 RH (Nf). OFr *digard* (Normandy) 'a spur', hence 'a maker or seller of spurs'.

**Diggen, Diggens, Diggines, Diggins, Digings, Digan, Diggan, Digance, Diggon:** John, Richard *Digun* 1227 AssBk, 1247 AssBeds; Richard *Digon* 1273 RH (Lo); Richard *Digoun*, *Dicoun* 1375–6 LLB H; Fr *Diggins* 1674 HTSf. *Digun*, a common variant of *Dicun*. *v*. DICKEN.

**Diggery, Diggory:** *Degory* Watur 1461 SaAS 2/xi; *Digorie* Baker 1600 AD v. (Co). Apparently the name of the hero of the medieval romance *Sir Degarre*. *v*. ODCN.

**Diggle, Diggles, Dickels:** Agnes *Diggell'* 1219 AssY. *Dikel*, *Digel*, diminutives of *Dick*. cf. *Dikelin* 1275 RH (Nf). This double diminutive survived as a surname until the 17th century when Thomas *Diglin* gave name to Diglin's Drove in Parson Drove (Cambs), called *Diggles* and *Diglings* Drove in 1864 (PN C 279).

**Diggon:** *v*. DIGGEN

**Dightham:** A Leeds and Bradford name, for *Dighton*.

**Dighton:** William *de Dicton'* 1207 P (Y). From Deighton (ER, NR, WRYorks).

**Digman:** *v*. DICKMAN

**Dignam:** Thomas *Dygname* 1540 FFEss. Perhaps a late corruption of Dagenham (Ess).

**Dil, Dill, Diller, Dilcock, Dilling, Dulcken:** Johannes *filius Dulle* 1279 RH (C); Godwin *Dul* 1185 P (Ha); William *Delle* 1195 Cur (Nf); Roger, Godfrey *Dulle* 1202 FF (Nf), 1232 Pat (L); Robert *Dille* 1279 RH (Bk); Alice *le Dul* ib. (C); Warin, Geoffrey *Dylle* 1283 Battle (Sx), 1327 SRSo; William *Del* 1297 MinAcctCo; John *Dyl* 1301 SRY; Thomas dictus *Dyll* c1360 Black (Inverness); Marjorie dicta *Dyll* 1361 ib. A difficult group of names. An OE *\*Dulla*, postulated to explain Dullingham (Cambs), seems to be confirmed by the 13th-century *Dulle* in the same county, possibly surviving in *Dulson*. A mutated form of this, OE *\*Dylli* or *\*Dylla* is found in Dillington (Hunts) and Dilton (Wilts). This would give ME *Delle*, *Dille*, *Dulle*, which may account for some of the surnames *Dell* and *Dill*. An unrecorded compound *\*Dylwine* appears to have existed in Norfolk (Philip, Mariot *Dylewyne* 1275 RH), where we also find *Delle* and *Dulle*. Harrison derives *Dilnott* from *\*Dilnōð*, a possible but unrecorded form. *Dilcock*, *Dolcok* (Robert, Roger *Dolcok* 1327 SRWo) and *Dulcken* (from *Dulekin*) are pet-forms in which the suffixes are usually added to short forms of personal names. It may be, too, that *Dilke* is a contracted form of an OE *\*Dylluc*. ME *dull* 'dull, foolish' is found once in the 13th century, but is not usual before 1350; *dill* is found in the same sense from 1200 to 1440 and the two forms point to an original OE *\*dyl*, *\*dylle* (MED) which would give ME *dell*, *dill*, *dull*, thus accounting for *Dell*, *Dill* and the obsolete *Dull*. The nickname certainly occurred in 1279 as *le Dul* in Cambridgeshire where *Dulle* is also found as a personal-name side by side with *Dill, Dilke* and *Dilcock*. The plant *dill* was commonly cultivated in the Middle Ages for its carminated fruits or seeds and OE *dile, dyle* is the first element of Dilham (Norfolk), Dilicar (Westmorland) and Dilworth (Lancs). This might well be one source of *Dill*, an occupation name for a grower or seller of dill, metonymic for *Diller*, found in 14th-century Somerset where both then and today we find *Dill* and where OE *\*Dylle, \*dylle* would become *Dulle* as in (Robert, Roger) *Dolcok* (1327 SRSo). *Dilnott*, of which no examples have been found, is probably ME *dilnote* (a1400 MED), an old name of the Earth-nut, used in the herbals for cyclamen. *Dilling*, noted in 14th-century Yorkshire by the side of *Dill* and *Dilcock*, and *Delling*, found in Sussex in 1296, may be for *\*Dylling* 'son of *Dylla*' or *\*dylling* 'the dull one'.

**Dilcock:** Geoffrey *Dilcok* 1327 SRC, Adam *Dilkoc* 1379 PTY. *v.* DIL.

**Dilger:** Stannard *Dilker* 1275 RH (Nf). Perhaps 'dill-field'. cf. Dilicar (Westmorland).

**Dilke, Dilkes, Dilks:** Adam *Dylke* 1278 AssSo; Nicholas *Dilkes* 1279 RH (C). *v.* DIL.

**Dillamore:** *v.* DELAMAR

**Diller:** Robert *Dyler* 1301 SRY; Thomas *Diller* 1327 SRSo. *v.* DIL.

**Dilley, Dilly:** William *Dilli* 1279 RH (C); Roger *Dilly* 1359 AssSt; Thomas Dilly 1381 AssC. Perhaps a pet-form of OE *\*Dylla*, unrecorded but found as the first element of Dillington (Hu) and Dilton (W).

**Dilling:** John *Dillyng* 1275 RH (K); Henry *Dyllyng* 1379 PTY. *v.* DIL.

**Dillinger:** John *Delynger* alias Denche 1545 SRW. From Drellingore in Alkham (K), *Dillynger* 1264.

**Dilliway, Dilloway:** *v.* DELLAWAY

**Dillon, Dyllon:** (i) Geoffrey *Dilun* 1203 SP1 (Sa); Richard *Dylon* 1332 SRWa. Perhaps OG *Dillo* with the diminutive *-on*. (ii) Robert, John *de Dilun* 1203 Pl (Sa), 1247 FFO. From Dilwyn (Hereford), *Dilun* 1138.

**Dilly:** *v.* DILLEY

**Dilnot, Dilnott, Dilnutt:** *v.* DIL

**Dilworth:** William *de Dilleworthe* 1332 SRLa; William *Dilworth* 1442 FrY; Widow *Dilworth* 1672 HTY. From Dilworth (La).

**Dimblebee, Dimbleby, Dimbledee:** Eustace *de Denbelby* c1190 Gilb (L). From Dembleby (Lincs). *Dimbledee* is a Welsh corruption.

**Dimbleton:** John *de Dymelton* 1363 FFY. From Dimlington (ERY), *Dimelton* DB.

**Diment, Dimond, Dimont:** *v.* DIAMANT

**Dimm, Dime, Dimes:** Nicholas *Dym* 1289 AssW; Thomas *Dymme* 1359 IpmGl; Arnould *Dymes* 1583 Musters (Sr). OE *dim* 'dark, dusky'.

**Dimmack, Dimmick, Dimmock, Dimock, Dymick, Dymock, Dymoke:** Nicholas *de Dimmoch* 1169 P (Gl). From Dymock (Glos).

**Dimmer:** *v.* DISMORE

**Dimsdale:** Richard *de Dymesdale* 1324 AssSt, 1332 SRSt. From Dimsdale (St).

**Din, Dinn:** *Dyn* Hunte 1296 SRSx. A pet-name of *Dinis* (*Denis*).

**Dinan:** Rolland *de Dinan* c1155 DC (L); Josce *de*

*Dinant* 1158 P (Berks). From Dinan (Côtes-du-Nord).

**Dincock:** Roger *Dyncok* 1303 IpmW. *Din*, a pet-form of *Denis*, plus the diminutive *-cock*.

**Dineley, Dinely, Dinley, Dyneley:** John *de Dynley* 1342 FFY; John *de Dynelai* 1357 Calv (Y); John *de Dynlay* 1358 FFY. From Dyneley (La).

**Dines:** *v.* DAIN

**Dingain:** *v.* GAIN

**Dingle:** Richard *Dingyl* 1246 AssLa; Hugh *de la Dingle* 1275 SRWo; John *ate Dyngle* 1299 MELS (Wo). 'Dweller in the deep dell or hollow', ME *dingle*. cf. Dingle (Lancs).

**Dingley:** Alured *de Dingelai* 1197 FFNth; Nicholas *de Dyngele* 1298 AssL; William *Dyngley* or *Dyneley* 1509 LP (Wo). From Dingley (Nth).

**Dinham:** Oliver *de Dineham* 1275 RH (D); Oliver *de Dinham* 1341 Hylle; James *Dynham*, John *Dinham* 1642 PrD. From Dinham (Monmouth).

**Dinley:** *v.* DINELEY

**Dinn:** *v.* DIN

**Dinnie, Dinny:** *v.* DENNEY

**Dinnis:** *v.* DENIS

**Dinsdale:** Geoffrey *Dynnesdale* 1496 FrY; John *Dinsdale* 1621 SRY. From Low Dinsdale (Du), or Over Dinsdale (NRY).

**Dinsley:** Reginald *de Dinesleg'* 1242 Fees (Ess); John *of Dinesle* 1251 FFY. From Temple Dinsley (Herts).

**Diplock, Duplock:** John *Depelak* 1327 SRSx; Samuel *Deeplocke* 1674 HTSf. Found as *Diplock, Duplake, Deeplake* (16th PN Sx 438). 'Dweller by the deep stream' (OE *dēop, lacu*).

**Dipper, Diaper:** (i) William *de Ipra* 1140–1 Seals (L), *dipre* 1185 Templars (Beds); Geoffrey *de Ipres* 1243 AssSo; Bruning *Dypres* 1296 SRSx. From Ypres (Belgium). (ii) Geoffrey *Dipere* t John Seals (Sr); William *le dipere* 1227 AssBeds; Davy *Dipper* 1520 ChwWo; Rose *Diaper* 1687 Buxhall (Sf). ME *dipper* 'a diving bird, water ousel, kingfisher' (1388 NED). cf. DIVER.

**Dipple:** *v.* DIBBLE

**Diriday:** Richard *Diriday* 1283 AssSt. Presumably a favourite exclamation of the speaker. Such names were not uncommon, cf. Amya *fistifasti* 1297 SRWa; Peter *Hidifidi* 1221 AssWa; Agnes *Houdydoudy* 1326 CorLo; John *Placidacy* 1370 FFEss; Thomas *Tarilari* 1276–7 CtH.

**Disborough:** *v.* DESBOROUGH

**Disher:** Roger *le Disser* 1273 RH (Wa); Richard *Dysser* 1301 SRY; John, Robert *le Disshere* 1304–47 LLB C, E, F. 'Maker or seller of dishes' from OE *disc* 'dish' (1304 NED). In London dishers were makers of wooden measures for wine and ale who had to have each a mark of his own placed on the bottom of each measure, samples of the marks to be submitted to the Chamberlain. They were also called *turnours* (John, Thomas *le Turnour*) 1347 LLB F 160.

**Dishford, Dishforth:** William *de Dyschforde* 1332 SRWo; William *de Disshford* 1340 NIWo; Henry *Dishforth* 1672 HTY. From Dishforth (NRY).

**Dismore, Dimmer:** Robert *Dimars* 1220 Fees (Berks); Roger *Dismars* 1225 FrLeic. OFr *dix mars* 'ten marks'. cf. Alan *de duabus marcis* 1202 AssL,

Robert *Deumars* 1280 LLB A, Bartholomew *Dewmars* 1334 ib., E, from *deux*, 'two marks', which may have been absorbed by *Dimmer*.

**Disney:** William *de Ysini* c1150 DC (L), *de Yseigni* 1177 P (L); Adam *Dyseni* 1202 AssL; Anthony *Dysney* 1552 AD vi (W). From Isigny (Calvados).

**Dispard:** *v.* DESPARD

**Disper:** *v.* DASHPER

**Ditch, Deetch, Deykes, Dyke, Dykes:** Joc. *de la Dike* c1250 MELS (Sx); John *attedich* 1260 AssC; Matilda *in Dich* 1279 RH (C); Roger *de la Diche* ib. (O); John *del Dike* 1297 SRY; William *del Dikes* 1332 SRCu. 'Dweller by the ditch(es) or the dike(s)', OE *dīc. v.* also DICK, DEEKES.

**Ditcham, Ditchum:** Walter *de Decheham* 1296 SRSx; Thomas *Ditcham* 1674 HTSf. From Ditcham in Buriton (Hants), half-a-mile from the Sussex boundary.

**Ditcher:** *v.* DICKER

**Ditchfield, Detchfield:** John *de Dychefeld* 1332 SRLa; Henry *of Dichefeld* 1401 AssLa. From Ditchfield in Widnes (La).

**Ditchford:** William *de Dichforde* 1327 SRWo, *de Dicheford* 1340 NIWo, *de Dychford* 1346 SRWo. From Ditchford (Nth), Lower Ditchford (Wa), or Upper Ditchford (Wo).

**Diter, Dyter:** Helewisa *Ditur* 1327 SRSx. OFr, ME *ditour* 'author', used of a composer, public crier; summoner, indicter (1303 NED).

**Ditton:** Leofwine *æt Dictune* c1060 KCD 929 (K); Henry *de Ditton'* 1212 Fees (La); John *de Ditton'* 1304–5 FFWa; Roger *Ditton* 1395 IpmLa. From Ditton (Bk, K, La), Fen, Wood Ditton (C), or Thames, Long Ditton (Sr).

**Divall, Divell:** *v.* DAVALL

**Diver, Divers, Divver:** Robert *Dyvere* 1252 Rams (Hu); Gunnilda *Divere* 1279 RH (C); Richard *Diverse* 1597 RothwellPR (Y). A derivative of *dive*, 'a diving bird', 'a diver' (c1510 NED).

**Dives:** Boselinus *de Diue* 1086 DB (C); William *de Dyves, de Dyve* 1242 Fees (O). From Dives (Calvados), earlier *Diva*.

**Divine:** *v.* DEVIN

**Dix:** *v.* DICK

**Dixey, Dixcee, Dixcey, Dixie, Dicksee:** Laurence *Dixi* 1279 RH (C); Robert *Dysci* 1301 FFHu; Alice *Dixi*, 1379 PTY. Lat *dixi* 'I have spoken', like the French *Dixi*, a name for a chorister derived from the beginning of a psalm (Dauzat). *Dysci*, by metathesis, for *Dycsi*.

**Dixon:** *v.* DICKSON

**Doag:** *v.* DOIG

**Doak:** *v.* DOIG, DUCK

**Doar:** *v.* DORE

**Dobb, Dobbe, Dobbs:** *Dobbe* filius Iuonis 1202 AssL; *Dobbe* de Deneby 1219 AssY; Reginald, William *Dobbe* 1275 SRWo, 1275 RH(Nf); Robert *Dobes* 1279 RH (O). A pet-form of *Robert*, very common in the 13th century in Yorks, Lancs, Cheshire and Staffs. *Dobbe* de Whitemore (1307 AssSt) is identical with *Robert* de Whitemore (1318 ib.). Richardus *filius Dobbe* was the same man as Richard *Dobbe* (1297 MinAcctCo).

**Dobbie, Dobby, Dobbie, Dobey, Doby:** *Dobbei* 1212 Cur (Y); *Dobi* Spendluf 1457 Black (Peebles); John *Doby* 1275 RH (L); Walter *Dobby* 1327 SRWo; Thomas *Doby* 1471 Black (Peebles). A pet-form of *Dobb*. Common in Scotland.

**Dobbin, Dobbins, Dobbing, Dobbings, Dobbyn, Dobing:** *Dobin* de Hatton 1203 AssSt; *Dobin* Cusin 1221 Cur (D); Hugo, Robert *Dobin* 1207 Cur (He), 1227 AssBk; Thomas *Dobinge* 1539 FeuDu; Samuel *Dobbins* 1674 HTSf. A common diminutive of *Dobb* (Robert).

**Dobbinson, Dobbison, Dobinson:** John *Dobynson* 1379 PTY; Robert *Dobyson* 1507 FrY. 'Son of *Dobbin*.'

**Dobel, Dobell, Doble, Doubell, Double, Doubble:** Richard *dublel* 1115 Winton (Ha); Robert *Dub(b)le* 1191–6 P (Sf); Adam *le Dobel* 1296 SRSx; Richard *Double* 1336 LLB E. OFr *doublel* 'a twin', used as a nickname in OFr. cf. French *Jumeau* and *v.* GEMMELL.

**Dobey, Dobie:** *v.* DOBBIE

**Dobieson:** John *Dobysoun* 1429 Black (Lanarks). 'Son of *Dobie*.'

**Dobinson:** *v.* DOBBINSON

**Dobney:** *v.* DAUBENEY

**Dobree, Dobrée, Darblay:** The ancestor of the family fled to Guernsey at the Massacre of St Bartholomew. His descendant, Peter *Dobrée*, merchant of London, was the father of the Rev. Peter Paul *Dobrée*, Regius Professor of Greek at Cambridge (Smiles 387).

**Dobson, Dopson:** Henry *Dobbesone* 1327 SRWo; Roger *Dobbessone* 1356 Putnam (Ch). 'Son of *Dobbe*.'

**Doby:** *v.* DOBBIE

**Docharty, Docherty, Dockerty:** *v.* DOUGHARTY

**Dock:** Wulfric' *Doc* 1177 P (Nf); Richard *Docke* 1221 Cur (Mx). Perhaps for DUCK, or from OE *\*Docca. v.* DOGGETT.

**Docker:** (i) John *Docker* 1293 FFY; Ralph *Doker* 1379 PTY; Milo *Docker* 1568 SRSf. From Docker (La, We). (ii) Agnes *Dokehare* c1280 *ERO*. A nickname, 'cut the hare's tail', ME *dok, hare*.

**Dockeray, Dockery, Dockray, Dockree, Docwra:** John *de Dokwra* 1332 SRCu. From Dockray (Cumb).

**Dockett:** John *Dockeeved'* 1212 Cur (Y). A nickname 'duck-head' *v.* also DOGGETT, DUCKET.

**Docking, Dockings:** Hamo *de Docking'* 1177 P (Sf); Richer *de Dokkyng'* 1298 AssL; Simon *Dockyng* 1409–10 FFSr. From Docking (Nf).

**Docksey:** *v.* DOXEY

**Doctor:** *v.* DAUGHTERS

**Dod, Dodd, Dodds, Dods:** Brictricus *filius Doddi* 1066 DB (Wo); Balterus *filius Dudde*, Walterus *Dudde* 11th OEByn (Wo); *Dodde* de Lismanoch 1194 P (D); Johannes *filius Dode* 1332 SRLa; Ælfweard *Dudd* c1030 OEByn (Ha); Aluric *Dod* 1066 DB (Do); Aluinus *Dode, Dodeson* ib. (Herts); Ælfric *Dodde* c1095 Bury (Sf); Gamel *Dod* 1175 P (Y); Ailricus *Dodde* 1176 P (Herts). Tengvik regards the OE bynames as nicknames from a Germanic root *\*dudd-* (*\*dodd-*) 'something rounded', denoting a rounded, lumpish man or a stupid person, or from the root of

OE *dydrian* 'to deceive', 'deceiver, rascal', or, possibly, 'the hairless, close-cropped one' from *dod* 'to make bare, lop, cut off'. This is possible and may be supported by the solitary Thomas *le Dode* (1327 SRSx). But we have also clear evidence of derivation from a personal name, OE *Dodd(a)*, *Dudd(a)* which was in use from Lincolnshire to Devonshire and from Essex to Lancashire until the 14th century and is probably the usual source.

**Dodderidge, Dodridge, Duddridge:** Richard *de Doderig'* 1275 RH (D); Richard *Doderugge* 1353 Putnam (D); Richard *Dodrigge* 1524 SRD. From Doddridge in Sandford (D).

**Doddimead:** Geoffrey *Doddyngmed* 1327 SRSo; Edward *Dodymead* 1576 SRW; Robert *Dodimead* 1641 PrSo. 'Dweller by *Dodding*'s meadow', OE *mǣd*.

**Dodding:** *Dodding* 1201 P (L); *Doding'* 1208 Cur (Gl); William *Doding* 1251 *ElyCouch* (C), 1268 AssSo; Andrew *Doddyng* 1296 SRSx. OE *Dodding* 'son of *Dodda*'. *v.* DOD.

**Doddington, Dodington:** Reginald *de Dodinton'* 1198 Fees (Sa); Symon *de Dodingtun* 1275 RH (K); John *Dudyngton* 1340–1450 GildC. From Doddington (C, Ch, K, L, Nb), Dry Doddington (L), Great Doddington (Nth), or Dodington (Gl, Sa, So). In addition, Derrington (St), Detton Hall (Sa), Dunton (Bk), and Denton (Nth), all appear in DB as *Dodintone*, and may have contributed to the surname.

**Doddrell:** a variant of DOTTRELL.

**Dodell, Duddell, Duddle:** Simon *Dodul* 1275 SRWo; John *le Doudel*, William *Dudel* 1296 SRSx; Richard *Dodel* 1310 ColchCt. OE *Duddel*, a diminutive of *Dudda*, or a diminutive of the *dod*, *dud* discussed *s.n.* DOD, perhaps early examples of the 17th-century *doddle*, *doodle* 'foolish fellow'.

**Dodford:** Reginald *de dodford* 1189 Sol; Henry *de Dodford* 1275 FFO; Nicholas *de Doddeford* 1327 SRWo. From Dodford (Nth, Wo).

**Dodge:** *Doge* filius Arnaldi 1196 P (Y); Wigot *filius Doge* 1214 Cur (L); *Dogge* 1246 AssLa; Robert, Nicholas *Dogge* 1206 Cur (Gl), 1279 RH (O). A pet-name for *Roger*, rhymed on *Rodge* and *Hodge*.

**Dodgen, Dodgeon, Dodgin, Dudgeon:** *Dogynus* de Sourhulle 1327 SRSo; Thomas *Doioun* 1327 *SR* (Ess); John *Dugeoun* 1536 Black. *Dogg-in*, *Dogg-un*, diminutives of *Dodge*. Bardsley notes that *Dodgson* and *Hodgson* were pronounced *Dodgin* and *Hodgin* in north Lancashire.

**Dodgson, Dodgshon, Dodgshun, Dodson, Dudson:** Henry *Doggeson* 1332 SRLa; Robert *Dogesson* 1385 FrY; John *Dogeson*, *Dodshon* 1489, 1530 GildY; William *Dodggson* 1642 FrY; William *Dodgshon* 1713 FrY; John *Dodson* 1720 FrY. 'Son of *Dodge*.' There appears to be no post-Conquest trace of the DB *Dodeson*; the modern *Dodson* is due to a simplification of the medial *dgs* of *Dodgson*.

**Dodington:** *v.* DODDINGTON

**Dodman, Dudman:** Goduine *filius Dudeman* 1066 DB (K); Goduine *Dudumanni filius* c1095 Bury (Sf); *Dudeman* 1206 P (Co), 14th AD vi (Sx); Thomas, Walter, John *Dudeman* c1179 Bart (Lo), 1199 FrLeic, 1221 AssSa (Wa); Gilbert *Dodemon* 1275 SRWo; Simon *Dodeman*, *Dudeman* 1296, 1297 Wak (Y);

Andrew *Dodeman*, *Dudeman* 1327, 1332 SRSx. OE *Dudemann*, recorded in the 9th century, and then not until the 11th.

**Dodridge:** *v.* DODDERIDGE

**Dodson:** *v.* DODGSON

**Dodworth, Dodsworth:** Lefode *de Dodesuurda* 1086 InqEl; Adam *de Dodworth* 1375 FFY; Walter *de Dodworth'* 1379 PTY. From Dodworth (WRY).

**Doe:** (i) Robert *le Do(e)* 1188–90 P (Sf); William *Do* 1220 Cur (Ess); Henry *le Daa* 1332 SRLa. OE *dā* 'doe'. (ii) Occasionally, perhaps, local. cf. Willingale Doe (PN Ess 500), from Hugh de *Ou, D'eu* t Hy 2. From Eu (Seine-Inférieure).

**Dogerty:** *v.* DOUGHARTY

**Doggett, Dockett:** *Dogget* 1199 P (Hu); *Doget* 1203 Cur (Herts); William *Doget* 1206 AssL; John *Doget*, *Doket* 1212, 1219 Fees (Y); Roger *Doket* 1225 AssSo. *\*Doket*, a diminutive of OE *\*Docca*, with voicing of the medial consonant to *Doget*, which may also be a diminutive of *dog*. cf. Roger *le Doge* 1296 SRSx, Rogerus *Canis* 1200 Cur (C), and *v.* CURR. *Doggett* may also be a late development of (Syward) *Dogheafd* 1177–95 Seals (Db) 'dog-head'. cf. Matthew *Doggenecce* 1275 RH (L), Symon *Doggeschanke* c1246 Calv (Y), and Lucas *Dogge(s)tayl* 1279–80 PN Wt 10.

**Doggrell:** Nicholas, William *Doggerel* 1249 AssW, 1277 AssSo; Alice *Dogerel* 1321 EwenG (Beds). These names carry the word back over 100 years, but lack of context forbids a definite explanation or etymology though the form suggests comparison with COCKERELL, PICKERELL, and PUTTERILL. The root is probably the word for 'dog' and the nickname may refer to some characteristic of the puppy, later applied derisively to wretched poetry, its irregular rhythm perhaps suggesting the clumsy antics of the puppy.

**Dogood:** Robert *Dogode* 1388 AssL; John *Dogood* 1399 IpmNt; William *Doogood* 1662–4 HTDo. A nickname, 'do good', OE *dōn, gōd*.

**Dogshanks:** Simon *Doggeschanke* c1246 Calv (Y). 'Dog shanks', OE *dogge, sceanca*. cf. Matthew *Dogeneck* 1268–73 RegAntiquiss 'dog neck'; Walter *Doggeskin* 1247 MPleas (W) 'dog skin'; Richard *Doggetall'* 1201 Pleas (Co) 'dog tail'.

**Doherty:** *v.* DOUGHARTY

**Doidge, Doige:** Richard *Doegson*, Alice *Doegewyf*, John *Doegeman* 1379 PTY. A northern form of *Dodge*. cf. William *Doide* 1469 Black (for *Dod*).

**Doig, Doag, Doeg, Doak:** Alexander *Dog* 1491 Black; David *Dogg* or *Doig* 1653–67 ib. *Dog*, for *Gille Dog* 'St Cadoc's servant'.

**Doke:** *v.* DUCK

**Dolamore:** *v.* DELAMAR

**Doland, Dolan:** John *Dolaund* 1260 AssC; Richard *Doland* 1430 FrY; Thomas *Dollen* 1642 PrD. From Dowland (D).

**Dolbey, Dolby:** William *Dolbe* 1327 SRLei. Probably for DALBY.

**Dole, Dool:** William *de la Dole* 1279 RH (C). OE *dāl*, ME *dole* 'portion or share of land', especially in the common field; also used of a boundary-mark and as a unit of area. Perhaps 'dweller by the boundary-mark'.

**Doley:** v. DOYLEY

**Dolittle, Doolittle, Dollittle:** Hugo *Dolitel* 1204 P (R), 1230 Pat (Nb); Walter *Dolitle* 1219 AssY; John *Do Littel* 1275 SRWo; Juliana *Dolutel, Dolute* Ed 1 Malms. A nickname for an idler. For *Dolute*, cf. HAKLUYT. One Molly *Dolitle* (b. c1900) deliberately changed her name to *De Little*.

**Doll:** Lewingus *Dol* 1066 Winton; Robert *Dol* 1198 P (Bk); Letitia *Dolle* 1279 RH (C). OE *dol* 'foolish'.

**Dollamore, Dollemore, Dolleymore, Dollimore:** v. DELAMAR

**Dollar:** Matthew *de Doler* 1316 Black. From Dollar (Clackmannanshire).

**Dolley:** v. DOYLEY

**Dolling, Dowling:** William *Dolling* 1243 AssSo; Peter *Dollyng* 1275 SRWo; Edmund *Dowlinge* 1674 HTSf. OE *dolling* 'the dull one'. v. DOLL. Also from a late assimilation of *rl* to *ll* in DORLING: Samuel *Dorling*, Dolling 1770, 1782 WStowPR (Sf).

**Dollman, Dolman:** William *Dolman* 1260 AssC; Richard *Doleman* 1279 RH (O). 'Dweller by the boundary-mark' (cf. DOLE) or 'dull, foolish man': Richard *Dolle* (1280 AssSo) is regularly called *Dolleman* in another roll. cf. DOLL, DOLLING.

**Dollond:** A Huguenot name. John *Dollond* was the son of Protestant refugees from Normandy who came to England shortly after the Revocation of the Edict of Nantes.

**Dolloway:** v. DELLAWAY

**Dolph:** Thomas *Dolf* 1275 SRWo. A variant of DELF.

**Dolphin, Duffin:** *Dolfin* de Cluttun' 1193 P (Y); Adam *filius Dolfin* 1256 AssNb; Geoffrey, Richard *Dolfin* 1171, 1182 P (Ha, Herts); William *Duffin* 1279 RH (Hu), 1674 HTSf; Robert *Dolphin, Dewfine* 1606, 1609 Bardsley (La). ON *Dólgfinnr*, a common name in 11th-century northern England. *Duffin* may also be from ON *Dufan*, a loan from OIr *Dubán*: William *Duffane* 1379 PTY.

**Dolties, Doltis, Doult:** Walter *Dolte* 1219 FrLei; Robert *Dolt* 1275 RH (L); Peter *Dolte* 1279 RH (C). 'Stupid, dull', from the past participle of ME *dullen, dollen*.

**Doman, Dooman:** John *Doman* 1327 SRSf. OE *dōm* and *mann* 'doom-man', 'judge'. cf. Roger *Dememan* 1246 Seals (W), William *Deman* 1285 AssLa, from OE *dēma* 'judge'. cf. DEEMER.

**Dome:** Henry *Dome* 1275 RH (Lo). OE *dōm* 'judgement', metonymic for DEEM 'judge'.

**Domesday, Dumsday:** Reginald *Domesday* 1297 Coram (Nf); John *Domisday* 1327 SRC; John *Demysday* 1479 NorwW (Nf). 'Servant of the judge', OE *dēma* and *dæge*. *Dome* occurs once in ME for 'judge'. This NED considers perhaps an error of transcription for *deme* but there seems to have been a tendency to substitute *dome* for *deme*. v. DEEMING, DOMAN. cf. Hugh *Domesman* 1155–75 Barnwell (C), William *Domesman* 1260 AssC 'man of judgement, judge'.

**Dominic, Dominick:** Robert *Domenyk* 1405 LLB I; William *Domynyck* 1545 SRW; Edward *Dominye* 1641 PrSo. Lat *dominicus* 'of the Lord', probably a name given originally to a child born on a Sunday. Never a common name in England.

**Domm, Domb:** Thomas *Domme* 1297 MinAcctCo; Robert *le Doumbe* 1309 SRBeds. OE *dumb*, ME *domm(e)* 'mute, speechless'.

**Dommett:** v. DUMMETT

**Domvil(l)e:** v. DUMVILLE

**Don:** v. DUNN

**Donaghie, Donaghy, Donahue:** v. O'DONOGHUE

**Donald:** Haket *Donald* 1328 Black (Kinross). Gael *Domhnall*, OIr *Domnall* (pronounced *Dovnall*), PrCeltic *Dubno-valos* 'world-mighty', found in 13th-century Scotland as *Dofnald, Douenald, Dufenald* and in 1376: Willelmus *filius Donaldi*. v. Black.

**Donaldson:** Henry *Donaldson* 1339–40 Black. 'Son of Donald.'

**Donat, Donnet, Donnett, Donnay:** Geoffrey *Doneth*, Roger *Donet* 1296 SRSx; Nicholas, Bernard *Donat* 1362 LoPleas, 1364 FFSf. Lat *donatus*, Fr *donat* 'given', the name of several saints which has given rise to the French surnames *Donat, Donnat, Donnet, Donnay*. 'Nom mystique ou affectif, donné' (Dauzat). The name has, no doubt, the same meaning as the fuller *Deodatus, Deodonatus*.

**Donbavand:** v. DUNBABIN

**Doncaster:** Laurence *de Doncastrie* 1183–99 YCh; William *de Danecastre* 1259 FFL; John *Doncastre* 1379 PTY. From Doncaster (WRYorks).

**Dongate:** v. DUNGATE

**Donkin:** v. DUNCAN

**Donley:** Alfridus *de Donnleye* 1327 SRWo. From Dunley (Worcs).

**Donn(e):** v. DUNN

**Donnelly:** Irish *Ó Donnghaile* 'descendant of *Donnghál*' (brown-valour).

**Donner:** Gilbert *Donur* 1267 FFL; Simon *le Donnere* 1327 SRSx; Maud *Donner* 1355 FFW. AFr *donour* 'giver, granter'.

**Donnet:** v. DONAT

**Donnington:** Ralph *de Donyngton* 1380 LoCh. From Donnington (Berks, Gl, He, Sa), Donington (L, Sa), Donington on Bain (L), or Donington le Heath, Castle Donington (Lei).

**Donovan:** Irish *Ó Donnabháin* 'descendant of *Donndubhán*' (dark-brown).

**Dooks:** v. DUCK

**Dool:** v. DOLE, DOYLE

**Doolan:** Irish *Ó Dubhláin* 'descendant of *Dubhfhlán*' (black-defiance).

**Dooley, Dowley:** Irish *Ó Dubhlaoich* 'descendant of *Dubhlaoch*' (black-hero).

**Dooman:** v. DOMAN

**Dopson:** v. DOBSON

**Doran, Dorran:** Martha *Dorrane* 1685 Black (Kirkcudbright). Ir *Ó Deoradháin*, a diminutive of *deorádh* 'exile, stranger'.

**Dorant:** v. DURRAND

**Dore, Doar:** (i) Aluuard *Dore* 1066 DB (Ess); Martin *Dore* 1185 Templars (Ess); Walter *Dore* 1268 AssSo. A nickname from OE *dora* 'humble bee'. (ii) John *de Dore* 1255 PN Wt 240; Roger *ate Dore* 1346 AD i (Herts). From Dore (Db), Abbey Dore (He), or 'dweller at the gate', OE *dor*.

**Dorey, Dory:** Fulco *Dory* 1275 RH (L); John *Dorey* 1314 IpmW; Andrew *Doreye* 1525 SRSx. OFr *doré* 'golden', probably of hair. cf. John *Doret* 1230 P (Nf). OFr *doret* 'golden'.

**Doring:** v. DEARING

**Dorington, Dorrington:** William *de Dorinton* 1221 AssSa; Philip *de Dorington* 1296 SRSx; Elyas *de Deoritone* 1327 SRSa. From Dorrington (L, Sa).

**Dorling:** v. DARLING

**Dorking, Dorkings, Dorkins, Dorkis:** Renilla *de Dorking* 1220 Cur (Sr); Robert *de Dorkyng* 1317 AssK; Henry *Dorkin* 1674 HTSf. From Dorking (Sr).

**Dorman, Dormand:** v. DEARMAN

**Dormer:** Geoffrey *Dormour* 1327 SRSf. Fr *dormeur* 'sleeper, sluggard'.

**Dorn, Dorne:** Henry *de Dorne* 1281 Eynsham. From Dorn in Blockley (Wo).

**Dornford:** v. DURNFORD

**Dorracott:** Hugh *Dorracott* 1642 PrD. Probably from Darracott in Georgeham (D), *Dorracott* 1617.

**Dorran:** v. DORAN

**Dorrance, Dorrins:** v. DURRAND

**Dorrett:** v. DURWARD

**Dorrington:** v. DORINGTON

**Dorset, Dorsett, Dossett:** Alexander *de Dorset* 1225 AssSo; Ralph *Dossett* 1602 FrY. From Dorset.

**Dorton:** William *de Dorton* 1190 P (W); Roheisia *de Dorton* 1198 P (Wa); John *de Dorton* 1278 AssSt. From Dorton (Bk).

**Dorville:** v. DARVAL

**Dorward, Dorwood:** v. DURWARD

**Dory:** v. DOREY

**Dosi:** Walter, William *Dosy* c1200 Colch (Sf), 1327 SRSx. ME *dosye*, an obsolete by-form of *dizzy* (OE *dysig*) 'foolish, stupid'.

**Dosser, Dossor:** William *le Dosser* 1259 AssCh; Gilbert *le Dosser* 1275 RH (Ess); John *le Dossere* 1327 SREss. A derivative of ME *dossen* 'to push with the horns, butt, toss, gore'. Perhaps a nickname for a bellicose man.

**Dossett:** v. DORSET

**Dossettor, Dossettour:** Robert *de Dorkecestre*, *de Dorceste* 1297 MinAcctCo; John *Dossetyr* 1524 SRSf; William *Dorcetor* alias *Dossytor* 1568 FFHu. From Dorchester (Oxon).

**Dott:** Philip *Dot* c1231 LuffCh; William *Dot* 1273 IpmGl; Elias *Dote* 1327 SRSf; John *Dotes* 1360 IpmW. On *Dottr* or OE *Dott*. v. PNDB 226. Probably original bynames to be associated with ON *dottr* 'lazy', or OE *dott* 'head of a boil'.

**Dottel, Dottle:** William *Dotel* 1642 PrD. A diminutive of ME *dote* 'fool'.

**Dotton, Dotten:** Philip *de Dotton* 1286 AssSt; William *Dotten* 1642 PrD. From Dotton Fm in Colaton Raleigh (D), Dutton (Ch), *Dottona* 12th, or Dutton (La), *Dotona* 1102.

**Dottrell:** William, Roger *Doterel* 1182–1211 BuryS (Sf), 1292 Burton (St); Ralph *Dotrell* 1301 SRY. ME *dotterel, dottrel* 'a species of plover', also 'a silly person' (1440 MED), cf. 'Dottrelle, fowle, *idem quod* Dotarde', c1440 PromptParv. A derogatory diminutive 'little dotard' (cf. ME *doten* 'to be foolish') from the plover, 'a very foolish bird, easily caught'.

**Doubell, Double:** v. DOBEL

**Doubleday, Douberday:** William, Robert *Dubelday* 1219 AssY, 1223 Pat (Y); William *Dubilday* 14th AD iii (Lei); John *Dubbelday* 1388 AD ii (Nf). Probably 'the servant of *Do(u)bel*' the 'twin'. cf. DOBEL, DAY and Nicholas *Dob(b)eday* 1294, 1305 RamsCt (Hu), 'servant of *Dobbe*' (Robert).

**Doubler:** William *Dubler, Dublier* 1191, 1192 P (L); Adam *le dubblier* 1207 P (Ha); Ralph *Doublere* 1334 SRK. AFr *dobler, dubler*, OFr *doblier, doublier* 'a kind of dish'. Metonymic for a maker of plates or dishes, cf. Adam *le Dublerwrite* 1249 AssW, Hugh *le dublerwrith* 14th YDeeds I.

**Doublet, Doublett:** Alexander *Dublet* 1192 P (Berks); John *Doublet* 1297 MinAcctCo. ME *dobbelet*, OFr *doublet* 'doublet' (1326 NED), a maker or seller of doublets.

**Doubtfire:** Robert *Doughtfire* 1509 LP (L). Either a nickname, 'fear fire', OFr *doute*, OE *fŷr*, or perhaps a late corruption of DOUGHTY.

**Douce, Dowse, Duce:** *Duze* filia Rannulfi fabri 1219 AssY; *Dulcia* Vidua 1275 SRWo; Hugo *Duce, le Duz, Dulcis* 1200 Oseney (O); Geoffrey *Duz* 1200 P (L); Gilbert *le Dus*, Godfrey *le Douz* 1296 SRSx; Walter *Dous, Douce* 1327 SRWo. ME *douce, dowce*, OFr *dolz, dous*, later *doux* 'sweet, pleasant' (c1325 MED). Often used as a woman's name and occasionally of a man: *Douce* de Moster 1274 RH (Ess), *Douce* of Chedle 1307 AssSt.

**Douceamour, Duzamour:** William *Ducamor* 1214 Cur (Sr); William *Duceamur* 1230 Cur (Ess); William *Duzamur* 1296 SRSx. OFr *douce amour* 'sweetheart'. cf. William *Doucedame* 1301 SRY 'sweet lady'; Adam *Dusseberd* 1327 SRWo 'soft beard'.

**Doucet:** v. DOWSETT

**Doudney:** v. DEUDNEY

**Dougal, Dougall, Dougill:** *Dufgal* filius Mocche c1128 Black; *Duuegall* 1208–14 ib.; *Dugall* 1261 ib.; Edward *Dougall* c1552 ib. Gael, OIr *Dubhghall* 'black stranger', a name originally applied by the Irish to Norwegians, now chiefly a Scottish Highland name.

**Dougan:** v. DUGAN

**Dougdale:** v. DUGDALE

**Dough:** William *Dugh* 1327 SRLei; Margery *Dogh* 1334 SRK. 'Dough', OE *dāh*, a nickname for a baker.

**Dougharty, Dougherty, Doherty, Docherty, Dockerty, Dogerty, Daugherty:** Donnall *O'Dochartaigh* 1119, *Daugherdy* 1630, *Dougherty* 1666 Moore. Originally *O' Dochartaigh* 'descendant of *Dochartach*' 'the stern'. Very common in Ireland but now rare in the Isle of Man.

**Doughty, Douty, Dowty, Dufty:** William *Douti* 1247 AssBeds; William *Doughty* 1300 FrLeic; John *Dughti* 1314 FrY; Han. *Doubty* 1674 HTSf. OE *dohtig, dyhtig* 'valiant, strong'.

**Douglas, Douglass:** William *de Duglas* 1175–99 Black (Kelso), *de Duglasse* 1256 AssNb. From Douglas (Lanarkshire), 'the black water' (Gael *dubh, glas*).

**Doull:** *Doull* Macgilleduf 1502 Black; James *Doull* 1623 ib. For *Domhnall*. v. DONALD.

**Doult:** v. DOLTIES

**Doulting:** Richard *Doultyng*, William *Doultynge* 1327 SRSo. From the River Doulting (So).

**Doust:** v. DUST

**Douthwaite, Dowthwaite, Douthet:** John *de Doventhuayt* 1332 SRCu; John *Dowthwayt* 1540 Whitby. From Dowthwaite (Cumb) or Dowthwaite Hall (NRYorks).

**Dove:** *Duua* Hy 2 DC (L), 1175–86 Holme (Nf); Robertus *filius Duue, Doue* 1166 P (Y), 1195 P (Nt); Ralph, William *Duue* 1197 P (Nf), 1289 Barnwell (C); Nicholas *le Duv* 1322 ParlWrits (D). OE *\*Dūfe* (f), from OE *dūfe* 'dove', occasionally from ON *dúfa* 'dove', used also as a by-name for one gentle as a dove. In Suffolk *Dove* is pronounced *Dow*. In Scotland *Dove* is also a variant of *Dow*.

**Dover:** Theoloneus *de Doure* 1086 DB; John *de Dover* 1223–4 FFK; Walter *Douer* 1332 IpmW. From Dover (Kent).

**Doverey:** Roger *Dovery* 1359 AssD. From Doverhay (So).

**Dow, Dowe:** (i) *Dowe* de Bothemes 1246 AssLa; *Dow* 1332 SRCu; William *Dowe* 1194 P (Nth); Laurence *Dow* 1254 AssSo. A pet-form of *David*. cf. DAW. (ii) Ede *Douw* 1366 Black. Gael *dubh* 'black'. Angus son of John *Dubh* Macallister is called Angus John *Dowisoun* (1516 Black). *v.* also DOVE.

**Dowall, Dowell:** Symon *Dowele* 1408 Black. For DOUGAL.

**Dowbiggin, Dowbigging:** John *de Dowbyggyng* 1406 IpmY; John *Douebiggyng* 1460 Paston; Jane *Dousbiging* 1489 TestEbor. From Dowbiggin in Sedbergh (WRY). The name also gives *Dobkin*, cf. *Dowbikin* 1596, *Dabkin* 1633, *Dobikin* 1652 WRS.

**Dowcett:** *v.* DOWSETT.

**Dowd:** *Doude* (1290 AssCh) is a variant of the more common *Daud* 'David': *Daud* Jonson, William *Daud* 1379 PTY. cf. DAW, DOW. *v.* also O'DOWD.

**Dowdall, Dowdell, Dowdle:** John *de Uvedale* alias *de Ovedale* 1304 FFC; Peter *Douedale* 1336 PN C 57; Hugh *de Uuedale* 13th, Hugh *Doudale* 1430 FeuDu; Adam *Dowedall* 1401 FrY. From D'Ovesdale Manor in Litlington (C). Sometimes from Dovedale (Db): Agnes *Doghdale* 1379 PTY.

**Dowding:** Gillian *Doudyng* 1303 AssW; Robert *Doudyng* 1380–1 PTW; Robert *Dowdyng* 1576 SRW. OE *Dudding*.

**Dowdle:** *v.* DOWDALL

**Dowdney:** *v.* DEUDNEY

**Dowdswell, Dowswell:** William *de Dowdeswelle* 1185 Templars (Gl); Anthony *Dowswell* 1576 SRW. From Dowdeswell (Gl).

**Dowell:** *v.* DOYLE

**Dower, Dowers:** Richard, John *le Douar'* 1332 SRSx; William *le Doghere* 1333 SRSr. A derivative of OE *dāh* (gen. *dāges*) 'dough'. 'A maker of dough, baker' 1483 (MED).

**Dowey, Douie:** Walter *de Doai, de Duuai, de Duaco* 1086 DB; Roger *Dowai* Hy 2 DC (L). From Douai (Nord).

**Dowle:** Thurstan, William *le Doul* 1247–8 FFSr, c1280 SRWo; Michael *Doul* 1334–5 SRK. ME *dul* 'stupid', cf. Dowles Farm (PN K 426) from Alan *le Dul* 1254, *le Doul* 1257. *v.* DULLARD.

**Dowler:** John *le Doulare*, Roger *le Douler* 1275 SRWo; William *Dowler* 1431 FA (Wo). A derivative of ME *dowle, doule* 'dowel', 'maker of dowels' (headless pins, pegs or bolts).

**Dowley:** *v.* DOOLEY

**Dowling:** *v.* DOLLING

**Dowman:** Adam *Douman, Doweman* 1332 SRCu, 1379 PTY. 'Servant of *Dow*' or David. *v.* DOW.

**Down, Downe, Downes, Downs:** Thomas *da la Duna* c1170 MELS (Sx); John *atte Doune* 1296 SRSx; William *Bythedoune* 1327 SRSo; John *Dun*, Emylyna *Doun* 1332 SRSx; Reginald *del Downes* 1407 Bardsley (Ch). 'Dweller by the down(s)' (OE *dūn*). We may also be concerned with a personal name: Ricardus *filius Dune* 1220 Cur (L), OE *Dūn*.

**Downen:** John *de la Dunhende* 1276, William *ate Dounende* 1333 MELS (So); John *atte Dounende* 1341 PN W 166. 'Dweller at the end of the down', OE *dūn, ende*.

**Downer:** Ralph *le Douner*, Stephen *le Downar* 1327 SRSx. 'Dweller by the down.' cf. DOWN, DOWNMAN.

**Downey:** (i) Matthew *Dounay* 1327 SRY; Richard *atte Dunye* 1330 MELS (Sr); Justinian *Downey* 1642 PrD. 'Dweller by the hilly island', OE *dūn, īeg*, or 'by the dark island', OE *dunn, īeg*. (ii) In Scotland from the barony of Duny or Downie in Monikie (Angus).

**Downham:** Siverth, Humphrey *de Dunham* c975 LibEl, c1115 Bury (Sf); John *de Downham* 1327 SRWo; Hugh *Downam* 1382 IpmNt. From Downham (Cambs, Essex, Lancs, Northumb, Norfolk, Suffolk).

**Downing:** Richard *Duning* 1197 FF (O); Geoffrey *Dounyng* 1311 ColchCt; Alice *Downyng* 1379 PTY; OE *Dūning* 'son of *Dūn*'. Later confused with DUNNING: John *Dunning* or *Downing* 1432 Bardsley (Nf).

**Downman, Dunman:** Geoffrey, William *Dunman* 1199 P, 1279 RH (O). 'Dweller by the down.' *v.* DOWNER. Also a late corruption of DOWNHAM: Robert *Downnam* or *Doneman* or *Downeman* 1579 Oxon. cf. DEADMAN.

**Downsall:** Richard *Dounesole* 1327 SRSx. 'Dweller down at the muddy place', OE *dūn, sol*.

**Downton:** *v.* DUNTON

**Dowse:** *v.* DOUCE

**Dowsett, Dowcett, Doucet:** John *Dousot* 1315 FFHu; John *Dousete* 1376 LLB H; William *Doucet* 1411 LoPleas. ME *doucet, dowcet, dulcette*, AFr *doucet*, a diminutive of *doux* 'sweet to eye or ear, pleasing, agreeable' (c1425 MED).

**Dowsing:** Robert *Dusing* 1199 P (Nf); Wido *Dusing* 1202 AssL; William *Dowsyng* 1376 AD ii (Nf). This cannot, as suggested by Bardsley, be an -*in* diminutive of *Douce*. His sole evidence is Nicholas *Doussin* (1669) in London. This is clearly the Fr *Doucin, Doussin*, which could not have become *Dusing* in England by 1199. Nicholas may have been a Huguenot. The modern surname is common in Norfolk and Suffolk and was particularly common in Norfolk in the 13th century. It seems to have arisen especially in the eastern counties where Scandinavian influence was strong and is probably an -*ing* derivative of ODa *Dūsi* which forms the first element of Dowsby (Lincs). '*Dūsi*'s son.' It was later confused with DOWSON: John *Dowsyng* 1584 IckworthPR (Sf), *Dowsen* 1589 ib.

**Dowson:** (i) Henry *Doucesone* 1320 AssSt; William *Douceson* 1327 *SR* (Ess); Thomas *Dousone* 1376 AD vi (St). 'Son of *Douce*.' (ii) John *Dowson* 1349 Whitby

(Y); Robert *Doweson* 1379 PTY. 'Son of *Dow*.' *v.*
DAVID, DAW, DOWMAN.

**Dowsell:** *v.* DOWDSWELL

**Dowthwaite:** *v.* DOUTHWAITE

**Dowtry:** *v.* DAWTREY

**Dowty:** *v.* DOUGHTY

**Doxey, Doxsey, Docksey:** Richard *de Dokeseye* 1298 AssSt. From Doxey (Staffs).

**Doyle, Dowell, Doole:** Irish *Ó Dubhghaill* 'descendant of *Dubhghall*' (black stranger, Dane).

**Doyley, D'Oyley, Doley, Dolley, Duley, Duly, Dailey, Dailley, Dayley, Olley, Ollie:** Robert *Oilgi, Olgi, de Oilgi, de Olgi, de Oilleio* 1086 DB; Robert *de Olleyo, de Oili* 1135 Oseney, 1140 Eynsham (O); Henry *de Olli, Doilli* 1156, 1163 ib.; Henry *de Oly, Dolly* 1212 Cur (O); Reginald, Thomas *Duly* ib. (Y), 1297 MinAcctCo (L); John *Dolye* 1272 AssSt; Henry *Dayly* 1279 RH (O); Robert *de Doley* ib.; John *Deyli* 1293 AssSt; Thomas *Duly, Doyllye* 1327, 1332 SRSx; Robert *Oylly, de Oylly* 1378 Oseney (O). There are five Ouillys in Calvados. The DB tenant probably came from Ouilly-le-Basset, or, possibly, from Ouilly-le-Vicomte. The name may also derive from Ouilly-du-Houlley, Ouilly-la-Ribaude or Ouilly-le-Tesson. *v.* OEByn.

**Doyne:** *v.* DAIN

**Drabb, Drabbe:** *Drabbe* 963–92 ASC; Robert *Drabe* 1162 P (L); Walter *Drabbe* 1327 SRSf, 1331 FFEss. OE *Drabba*, recorded only as the name of a Peterborough festerman.

**Drabble:** Matthew, Geoffrey *Drabel* 1273, 1279 RH (Wa, C); Robert *Drabel* 1293 AssSt; Ralph *Drabelle* 1302 SRY. OE *Drabba*, recorded as the name of a Peterborough festerman in 963–92 (ASCh), may survive in the surnames of Robert *Drabe* 1161 P (L) and Walter *Drabbe* 1327 SRSf, and the modern surnames may be from a diminutive of this, *\*Drabbel*. But we may also have early examples of *drab* 'a dirty, untidy woman' and a diminutive of this. *Drab*, not recorded before the 16th century, is of uncertain derivation and has been associated with Gael *drabag*, Irish *drabog* 'slattern' and also with LG *drabbe* 'dirt, mire'.

**Draff, Draffe:** Walter *Draf* 1198 FFNf. A nickname from OE *dræf* 'refuse, chaff'.

**Drage, Dredge, Drudge:** Walter *Drage* 1210–11 PWi; Robert *Drage* 1354 ColchCt; John *Dredge* 1545 SRW. OFr *dragie, dragé* 'a mixture of grains sown together'. For a grower of this crop.

**Dragon:** Walter *Dragun* 1166 P (Y); Walter *Dragon* 1221 FrLeic, *le Dragon* 1275 RH (Sx). Occasionally a nickname, but usually metonymic for *dragoner*: Adam *le Dragoner* 1338 LLB F, John *Dragenir* 1343 AD ii (Wt). OFr *dragonier* 'a standard-bearer'. Standards were often emblazoned with a dragon and were carried not only in battle but in pageants and processions. cf. (at Westminster) 'Hym that beryth the Dragon on Easter Evyn' (c1540 NED). Occasionally also from a sign: William Strode called *atte Dragon*, brewer 1374 AD ii (Mx).

**Drain, Draine:** William *Dren* 1201 Pleas (Co); John *atte Drene* 1327 SRSo. 'Dweller by the drain or ditch', ME *dreine*. cf. Richard *dranar* 1539 StaplehurstPR (K).

**Drake, Drakes, Drakers:** Leuing *Drache* 1066 Winton (Ha); Robert, David *Drake* 1185, 1190 P (Nt, Wa); Wimund *le Drake* 1205 ChR (Do); Geoffrey *le Drak'* 1225 AssSo. OE *draca* 'dragon', like OFr *dragon*, was used in ME of a battle-standard as well as of a serpent or a water-monster. The surname is clearly sometimes a nickname and is also metonymic for *Draker* 'standard-bearer': Godman *the Drakere* 1260 AssC. cf. also the ODa nickname *Draki* 'dragon' which may sometimes occur in the Scandinavian counties. The surname may also be from ME *drake* 'male of the duck' (c1300 MED).

**Drakeley, Drakely:** Robert *Drakelowe* 1397 TestEbor. From Drakelow (Db), Drakelow in Wolverley (Wo), or Dragley (La).

**Draker, Drakers:** Godman *the Drakere* 1260 AssC. 'Standard-bearer', a derivative of OE *draca* 'dragon'. Medieval standards were often emblazoned with a dragon, and were carried not only in battle but also in pageants and processions.

**Drane, Dron:** Roger *Drane* 1276 RH (Y); Adam *le Dron* 1275 SRWo; Walter *le Dran* 1285 *Ass* (Ess). OE *drān* 'drone', a nickname, 'the lazy'.

**Dransfield, Dronsfield:** Reginald *de Dreinesfeld* 1219 AssL; Thomas *de Dronesfeld* 1309 Wak (Y); John *Dronsfeld* 1424 FFEss. From Dronsfield (Db), or Dransfield Hill in Mirfield (WRY).

**Drape:** OFr *drap* 'cloth'. For DRAPER.

**Draper, Drapier, Drapper:** Hugo *drapier*, Walter *draper* 1148 Winton (Ha); Robert *(le) Drapier* 1181–2 P (L). OFr *drapier*, AFr *draper* 'maker or seller of woollen cloth, draper' (c1376 MED).

**Drave:** Ellis *de la Drave* 1249 AssW; John *Draue* 1362 FFEss. 'Dweller by the road along which cattle were driven', OE *drāf*.

**Drawater, Drawwater:** Richard *Drawater* 1279 RH (Hu). 'A drawer of water.'

**Drawer:** Sayhere *Draghere* 1327 *SR* (Ess); John *Drahere* 1327 SRSf; John *le Drawere* 1332 SRLo. A derivative of OE *dragan* 'to draw'. 'One who draws' in various senses. cf. Richard *le Pakkedrawere* 1332 SRLo, a drawer or carrier of packs of wool. He was employed at the woolwharf in Tower Ward; David *Tothedrawer* 1422 LoPleas, Elias *Wyndrawer* 1373 ib., Roger le *wirdragher* 1313 FrY, Alan de Wifestow, *mukdragher* 1340 FrY, a leader of dung, or, perhaps, a scavenger.

**Drawsword:** Richard *Dragheswerd* 1240 AssSf; Maurice *Draweswerd* 1316 Misc; John *Draweswerd* 1327 SRSf. 'Draw sword', OE *dragan, sweord*. Perhaps a nickname for a quarrelsome person. cf. William *Drawelaunce* 1449 AD i (Do) 'draw lance'; William *Dragespere* 1204 Pleas (Y) 'draw spear'.

**Dray, Drey, Dry, Drye:** Roger *Drie* 1219 P (Nf/Sf); Geoffrey *Dreye* 1292 ELPN; William *Drye* 1321 FFEss; Walter *Dreye* 1399 FFC. ME *dreȝ, dregh*, in one or other of its various senses, 'enduring, patient', 'doughty, fierce', 'slow, tedious'. Sometimes, perhaps, from OE *drȳge* 'dry, withered'. cf. Roger *Siccus* c1110 Winton (Ha).

**Draycot, Draycott:** Robert *de Draicote* 1188 P (Db); John *de Draycote* 1275 SRWo; William *Draycote* 1365 LoPleas. From Draycott (Db, O, So,

Wo), Draycott Moor (Berks), Draycott in the Clay, in the Moors (St), Draycot Cerne, Fitz Payne, Foliatt (W), or Draycote (Wo).

**Drayton:** Ledmar *de Draiton* 1086 InqEl; Alice *de Draiton'* 1203 AssNth; William *de Drayton* 1358 SRWo; John *Drayton* 1446 IpmNt. From one or other of the numerous Draytons, or from Dreyton (D).

**Dredge:** *v.* DRAGE

**Dreng, Dring:** *Dreng* de Calualea 1161 P (Nb); Roger, William *Dreng* 1155 FeuDu, 1201 P (L); Roger de Brunham *le dreng* Hy 3 Seals (L); Robert *Dring* 1379 PTY. Occasionally we may have the ON personal name *Drengr*, found in DB, but usually the surname is OE *dreng*, from ON *drengr* 'young man', used of a free tenant, especially in ancient Northumbria, holding by a tenure partly military, partly servile.

**Dresser:** Robert *le Dressour* 1324 LaCt; Adam *Dressur* 1332 MEOT (Y). A derivative of ME *dressen*, OFr *dresser* 'to arrange' in various special and technical senses. Perhaps 'a dresser of textile fabrics', a finisher who gives them a nap, smooth surface. cf. 'Shermen, dressers, carders and spynners' (1520 NED).

**Dreux:** *v.* DRUCE

**Drew, Drewe, Dru:** (i) *Drogo* 1086 DB; Radulfus *filius Drogonis* 1159 P (L); *Driu* filius Radulfi Hy 2 DC (R); Radulfus *filius Dru* 1185 Templars (L); Anschet' *driue* 1130 P (Caermarthen); Ralph *Dreu* 1188 BuryS (Sf); William *Dryw* 1275 SRWo; John *Drew* 1327 SRC. OG *Drogo*, the name of a son of Charlemagne, probably of Frankish origin (OSax (*gi*)*drog* 'ghost, phantom'), became OFr *Dreus*, *Drues* (nom.) and *Dru*, *Driu*, *Dreu* (acc.). *Drew* was also used as a petname of *Andrew* which was confused with the French *Drew*. In 1400 Drew Barentyn appealed to the Council to correct his christian name to *Drew* in the list of freemen of the City where it had been entered as *Andrew* (Riley). (ii) Henry *le Dru* 1255 RH (W); John *le Drew* 1279 RH (O); Henry *le Druie*, William *Dru* 1275 SRWo. OFr *dru* 'sturdy', later 'lover' and in this sense common as a surname.

**Drewell:** William *Drewel* 1176 P (Beds); Baldwin *Druell'* 1196 Cur (Bk); William *Druel* (v.l. *Druett*) 1221 Cur (Beds). A diminutive of OFr *Dru*, *Drewel* interchanging with *Drewet*. *v.* DREW, DREWETT.

**Drewer, Drower:** *Druardus* miles c1150 Gilb (L); Druardus *de Bedeford* 1159 P; William *Druer'* 1208 FFY. OG *\*Droghard*, OFr *Droart*, *Drouart*. *v.* Forssner 61.

**Drewery, Drewry, Druery, Drury:** Alexander *Druri* 1200 Cur (Sf); Robert *Druerie* 1204 P (La). OFr *druirie*, *druerie* 'love, friendship', 'love-token' (c1225 MED), 'sweetheart' (a1393 ib.).

**Drewes:** *v.* DRUCE

**Drewett, Dreweatt, Drewitt, Drouet, Druett, Druitt:** *Druet* 1206 Cur (Sf); Walter *Druet* 1185 Templars (O); William *Drauet* 1198 P (K); Walter *Drywet* 1309 AD v (Sx). A diminutive of OFr *Dru*. *Druettus* Malerbe 1273 RH (Nth) is elsewhere called *Drogo*. In Mabila *Druet* c1248 Bec (Mx), we may have a diminutive of *dru*, 'little sweetheart'. *v.* DREW.

**Drewton:** Simon *de Dreuton* 1305 IpmY. From Drewton in North Cave (ERY).

**Drey:** *v.* DRAY

**Dreyer:** *v.* DRYER

**Dribbel, Dribbell:** Maurice *le Dribel* 1245–72 Colch (Ess). A nickname from ME *drevel*, *dribil* 'slaver, saliva'.

**Driffield, Driffil:** Stephen, William *de Driffeld* 1192–1218 YCh, 1271–7 Str; John *Dryffeld* 1366 AssLo. From Driffield (Glos, ERYorks).

**Dring:** *v.* DRENG

**Dringer:** Richard *Drenger* 1256 AssSo; Richard *le Drynkere* 1296 SRSx; Walter *le Drynkar* 1327 SRWo. OE *drincere* 'drinker'. Used of a tippler a1225 (MED).

**Drinkale, Drinkald, Drinkall, Drinkhall:** *Drinc hala* 1200 *Caen* (Ess); Thomas *Drinkhale* 1281 LLB B; Henry *Drinkale* 1301 SRY; John *Drynkall* 1509 GildY; William *Drynkell* 1559 FrY; Nicholas *Drinkeld* 1677 LaWills. A nickname from the customary courteous reply to a pledge in drinking, *drinc hail* 'drink good health or good luck'. *Hala* probably points to OE *hǣl*, an anglicizing of ON *heill* 'good luck'. As ale was the usual drink, there was no point in a nickname *drink-ale*. *v.* DRINKWATER.

**Drinkpin:** *Drinkpin* balistarius 1205 P (Do); Osbert *Drinchepyn* 1108 (c1425) LLB C; Walter *Drinkepin* 1242 P (Beds/Bk). 'One who drinks a peg', OE *drincan*, *pinn*. Pegs were fixed on the inside of drinking vessels. Perhaps a nickname for a heavy drinker. cf. Henry *Drink al up* 1282 FFEss 'drink all up'; Adam *Drincorawe* 1327 SRY 'drink in turn'.

**Drinkwater:** John *Drinkewater* 1274 RH (Sa); Thomas *Dreinkewater*, *Drynkewater* 1300, 1310 LLB B. 'Ale for an Englysshe man is a naturall drynke.' It was drunk at all times, taking the place not only of tea, coffee, etc., but also of water. A 13th-century writer describing the extreme poverty of the Franciscans when they first settled in London (1224) writes: 'I have seen the brothers drink ale so sour that some would have preferred to drink water' (MedInd 286). The surname was perhaps applied to a man so poor that he could not afford to drink ale even when it was four gallons a penny. It was also used ironically of a tavern-keeper and, perhaps, of a tippler: Margery *Drynkewater*, wife of Philip le Taverner (1324 LLB E), Thomas *Drinkewater*, of Drinkewaterestaverne (1328 Husting). cf. Adam *Drinckmilke* 1674 HTSf.

**Drinkwell:** William *Drinkwell* 1309 SRBeds. A nickname. cf. Henry *Drink al up* 1282 FFEss, Fulredus, William *Drinkepin* 1206 ChR, 1207 Rams 'one who drinks a peg' (pegs were fixed on the inside of large drinking vessels), William *Drinkepani* 1224–46 Bart, 'drink-penny' (cf. BEVERIDGE).

**Driscoll:** Irish *Ó hEidersceoil* 'descendant of *Eidirsceól*' (interpreter).

**Driver, Drivers:** Alice *le Driveres* 1279 RH (C); Gilbert *le Drivere* 1283 SRSf. A derivative of *drive*, 'driver' (14.. NED).

**Dromant:** Archelbus *Dromant* 1642 PrD. AFr *dromund*, OFr *dromont* 'a large, fast sea-going vessel'. Metonymic for a sailor.

**Dron:** *v.* DRANE

**Dronsfield:** *v.* DRANSFIELD

**Droop:** Adam *Drope* 1177 P (Nf); Edric *Drup* 1197 P (Gl); Robert *Drop* 1202 AssL. ME *drup* 'dejected, sad, gloomy', a rare adjective. *v.* MED. The surname may also be local: William *atte Thrope* lived at Drupe Fm in 1330 (*Droope* 1679 PN D 587).

**Drouet:** *v.* DREWETT

**Drought:** William *Droth'* 1209 Pleas (Nf); John *Drought* 1377 AssEss; Thomas *Drout* 1434 FFEss. OE *drugaδ*, ME *drowþe, droghte* 'dryness, lack of moisture, withered', perhaps 'thirsty, addicted to drinking'.

**Drover:** Hugh *Drouere* 1294 MEOT (Herts); Henry *le Drovere* 1327 SRSt. A derivative of OE *dráf* 'drove, herd', a drover.

**Drower:** *v.* DREWER

**Druce, Drewes:** (i) Herman *de Dreuues* 1086 DB (W); Hugo *le Droeis* 1225 Pat (W); Robert *de Dreus* 1230 P (O); William *Drueys, Drois, le Droys, le Droes* 1242 Fees (W). From Dreux (Eure-et-Loir); alternating with the toponymic. (ii) Elias *de Ruwes, de Riueus, Driues, de Ripariis, Droeys* 1235, 1242 Fees. (Bk). From some French place Rieux 'streams'.

**Drudge:** *v.* DRAGE

**Druery:** *v.* DREWERY

**Druett, Druitt:** *v.* DREWETT

**Drummond:** (i) Agge *Dromund* 1221 *ElyA* (C); John *Drommund* 1327 SRC. ON *Drómundr*. (ii) Gilbert *de Drummyn* c1199 Black; Malcolm *de Drummond* or *Drumman* 1270–90 ib. From the barony of Drummond, probably identical with the parish of Drymen (Stirling).

**Drunkard:** Maurice *Druncard* 1275 RH (D). ME *druncard* 'drunkard'. cf. John *Drunken* 1301 SRY 'drunken'.

**Drury:** *v.* DREWERY

**Druval:** Robert *de Druval* 1219 P (Berks); Ralph *Druval* 1241 FFO. From Druval (Calvados).

**Dry:** *v.* DRAY

**Dryden:** Philip *de Dryden* 1296 (Forfar), Henry *de Driden* 1329 Black; John *Driden* 1553 AD vi (Wa). Perhaps from Dryden near Roslin (Midlothian).

**Drye:** *v.* DRAY

**Dryer, Dreyer:** Roger *le Dreyere* 1318 Oseney (O). A derivative of OE *drȳgean* 'to dry', perhaps a drier of cloth. 'The cloth, having been fulled, had to be stretched on tenters to dry' (MedInd 224). The feminine form, used also of men, seems to have been more frequent: Alice *le Dreyster'* 1292 MESO (La), Adam *Dreyster* 1301 SRY.

**Drysdale:** Gawine *Dryfesdale* 1499 Black; Thomas *Dryisdaill* 1619 ib. From Dryfesdale (Annandale), pronounced *Drysdale*.

**Drywood:** Robert *Dryewode* 1299 IpmY; William *Driwode* 1532, John *Dryewood* 1545 FFEss. 'Dweller by the dry wood', OE *drȳge, wudu*.

**Dubbedent, Dubedent:** John *Dubbedent* 1160, *Adubedent, Adubbedent* 1162 P; Alice *Dubedent* 1221 AssWo; Robert *Dubbedent* 1277 AssW. 'Repair, polish teeth', OFr *aduber, dent*. Perhaps a nickname for a dentist.

**Dubber:** Godwin *Ladubur* 1168 P (Lo); Salamon *Leadubur* 1204 P (Lo); William *le Dubbere* 1210 FrLeic; Paganus *le Dubbour* 1226 NED. Bardsley notes the Company of Dubbers of York and suggests that they embellished dresses with gold lace, etc.

Halliwell suggests they were trimmers or binders of books. OFr *adubeour, adubur* 'repairer', renovator of old clothes, perhaps also used of a polisher of arms or harness, ME *dubber* 'fripperer'.

**Dubois, Du Bois:** (i) John *du Boys* 1279 RH (O). 'Dweller in the wood', OFr *du bois*. cf. Theydon Bois (Ess), Hugh *de Bosco* 1240, Bois Hall in Navestock (Ess), John *de Bosco* 1229. (ii) Also Huguenot from a Protestant family of Brittany, many members of which fled to England, settling in Thorney, Canterbury, Norwich, and London. In addition, Francois *Dubois* came to England after the Massacre of St Bartholomew, settled in Shrewsbury, and founded a ribbon manufactory. In the fourth generation this family changed its name to Wood (Smiles 387–8).

**Ducat:** Wolveva, Alice, William *Dukat* 1314 FFSf. OFr *ducat*, It *ducato* 'a gold coin', late Lat *ducatus* 'duchy', so called because, when first coined c1140 in the Duchy of Apulia, they bore the legend: 'sit tibi, Christe, datus, quem tu regis, iste *ducatus*'. Ducat is a 'restored form', spelled *duket* by Chaucer and *ducket* by Shakespeare. The surname has been confused with DUCKET.

**Duce:** *v.* DOUCE

**Duck, Duckes, Doak, Doke, Dooks:** Robert *Ducke* 1260 AssCh; Hugo *Doke* 1279 RH (C); Richard *Duck, Dooke, Doke* 1510 Oxon. ME *duk(ke), duck, doke, dook* 'duck'. *v.* DUCKET.

**Duckels:** John *Dukel*, Richard *le Dukel* 1296 SRSx. A diminutive of *duk* 'duck'. cf. John *Duklyng* 1483 LLB L.

**Duckenfield, Duckinfield, Dukenfield:** Robert *de Dukingfeld* c1250 Calv (Y). From Dukinfield (Ch).

**Ducker, Duker:** Philip *Duker* 1365 FrY. ME *dokare, douker* 'a bather' or 'a diving bird'. cf. DIVER.

**Duckers:** *v.* DUCKHOUSE

**Ducket, Duckett, Duckit, Duckitt:** Rannulf *Duchet* 1130 P (Ha); Herbert *Duket* 1176 P (Herts). The surname is common, usually as *Duket*, and the persistent *-et* dissociates it from DUCAT. *Duchet* 1148 Winton is probably identical with Ralph *Duchet* (ib.) but we seem to have a clear example of a personal name in *Duchet* 1185 Templars (K), *Duket* 1198 P (K). This is probably a diminutive of OE *\*Ducc(a)* found in Duxford (Cambs), and Duckington (Ches), with a diminutive *\*Duccel* in Ducklington (Oxon) and a compound *\*Ducemann* in Duckmanton (Derby). Willelmus *filius Duket* (1301 SRY) and *Doket* Flasby (1379 PTY) are probably diminutives of *Marmaduke*, a purely northern name, from a pet-form, *Duke* Shillito 1634 Bardsley, with a shortened vowel in *Ducke* Duy 1246 AssLa. This is the source of the Yorkshire surname Agnes *Duket* (1301 SRY), John *Doket* (1379 PTY). But in general the surname is either an unrecorded diminutive of *duck* or OFr *ducquet*, a diminutive of *duc* 'leader, guide', one of the names of the owl, so called because thought to serve as guide to certain birds (Littré). *v.* DOCKETT.

**Duckhouse, Duckers:** Richard *Duckhouse* 1553 WhC (La), a name for one in charge of the ducks (1699 NED).

**Duckling:** John *Duklyng* 1483 LLB L; Richard

*Ducklinge* 1568 SRSf. *Duc-el-in*, a double diminutive of OE *Duca*, or of OE *duce* 'duck'.

**Ducksworth:** v. DUXWORTH

**Duckworth:** Henry *de Dukeworth'* 1379 PTY; John *Dukworth* 1439 FrY; Laurence *Dokeworth* 1488 KentW. From Duckworth (La).

**Duddell, Duddle:** v. DODELL

**Dudden:** John *de Duddene* 1298 AssSt. From Dudden (Ch).

**Duddin:** *Dodin* 1153–65 Black; *Dudyn* de Broughtune c1190 ib.; Alexander *Dudyn* 1296 ib. OFr *Dodin*, a diminutive of *Dode* (OG *Dodo*).

**Dudding:** Andrew *Doddyng* 1296 SRSx; Thomas *Duddyng* 1514 FrY. OE *Dudding* 'son of *Dudd(a)*'. cf. DOD.

**Duddridge:** v. DODDERIDGE

**Duddy:** Robert, Reginald *Dudehay* 1332 SRSx. From some lost place in Sussex, 'Dudda's enclosure'.

**Dudeney, Dudney:** v. DEUDNEY

**Dudgeon:** v. DODGEN

**Dudley:** Gladewin *de Duddeleia* 1176 P (St); Herbert *de Dudeleg'* 1225 Cur (St, Wo); William *Dudely* 1379 PTY. From Dudley (Worcs).

**Dudman:** v. DODMAN

**Dudson:** v. DODGSON

**Duesbury:** v. DEWSBURY

**Duff:** Duncan *Duff* c1275 Black. In Scotland 'The family name Duff is merely the adjective *dubh* "black" used epithetically' (Macbain). v. also DOVE. In Ireland, also for *Ó Duibh* and *Mac Giolla Dhuibh* 'son of the black youth', the latter surviving also as *MacIlduff*, *Gilduff*, *Kilduff*.

**Duffes:** v. DUFFUS

**Duffet, Duffett, Duffitt:** Richard *Dovefote* 1301 SRY; William *Dowfhed* 1355 FrY. A nickname, either 'dove-foot' or 'dove-head'.

**Duffey, Duffie, Duffy:** v. MCFEE, O'DUFFY

**Duffield, Duffell, Duffill:** Roger *de Duffeld'* 1190 P (Y); Geoffrey *de Duffeld* 1276 LLB A; William *Duffeld'* 1383 AssL. From Duffield (Derby, ERYorks).

**Duffin:** v. DOLPHIN

**Duffus, Duffes:** Robert *del Dufhus* 1275 RH (Sf). 'Keeper of the dovehouse.' In Scotland, from Duffus (Moray): Arkembaldus *de Duffus* 1222 Black; Agnes *Duffers* 1586 ib.; James *Duffes* 1633 ib.

**Dufton:** William *de Dufton* 1276 IpmY; Thomas *de Dufton* 1332 SRCu; Thomas *de Dufton* 1341 FrY. From Dufton (We).

**Dufty:** v. DOUGHTY

**Dugan, Dougan:** John *Dugan* 1413 Black; Adam *Dougane* 1665 ib. Irish *Ó Dubhagáin* 'descendant of *Dubhagán*', a diminutive of *dubh* 'black'.

**Dugard:** Richard *Deugard* 1322 AD vi (D); William *Dugard* 1327 *SR* (Ess). A phrase-name, Fr *Dieu* (*te*) *garde* 'God guard (you)'. cf. Fr *Dieutegard*.

**Dugdale, Dugdall, Dugdill, Dougdale:** Joan *Dukdale* 1479 GildY; John *Dugdayle* 1566 WRS. From Dug Dale in Warter (ERY), or Dugdales in Great Mitton (WRY).

**Duggan, Duggon:** *Ó Dubhagain* 1372, *Dogan* 1540, *Duggan* 1723 Moore. A Manx name for *O' Dubhagáin*. v. DUGAN.

**Duggel, Duggell, Duggle:** Gilbert *Dugel* 1210–11

PWi; Henry *Dugel* 1218 P (Lei/Wa); Ralph *Duggel* 1327 SRSo. Probably OE *\*Duccel*, found in Ducklington (O).

**Duguid:** John *Dogude* 1379 Black; Robert *Dugoude* 1536 ib.; Francis *Duiguid* 1675 ib. A nickname 'do good'. Black's other examples, *Doget*, *Dowcat*, etc., cannot belong here.

**Duke, Dukes:** Herbert *le Duc* 1185 Templars (Sa); Adam *Duke* 1198 P (Bk); Henry *Dukes* 1214 Cur (Wa); Osbert *le Duk* 1230 P (D). ME *duc*, *duk(e)*, *douk*, *doke*, OFr *duc* 'leader of an army, captain', often, no doubt, a nickname. ME forms cannot always be distinguished from those of DUCK. In Yorkshire, this may be a pet-form of Marmaduke. v. DUCKET, DUXON.

**Dukenfield:** v. DUCKENFIELD

**Duker:** v. DUCKER

**Dulcken:** John *Dolekyn*, *Dulekyn* 1301 ParlR (Ess), 1311–12 ColchCt. ME *Dul-kin*, a diminutive of OE *\*Dylla*. v. DIL.

**Duley, Duly:** John *Deu Le* (*del Le*) 1230 P (La); John *Dulay* 1279 RH (Hu); *John du Lay* 1327 SRSf. 'Dweller by the clearing', Fr *del*, *du* and OE *lēah*. cf. LEA. v. also DOYLEY.

**Dullard:** Richard *Dullard*, *Dollard* 1231 Cl, 1329 ColchCt. ME *dull* plus *-ard* 'a dull or stupid person' (c1440 MED).

**Dulson:** v. DIL

**Dumbell, Dumbill, Dumble:** (i) Thomas *Dumbel* 1273 RH (Wa); William *Dumbulle* 1367 LLB G. A diminutive of *dumb*. cf. 'A Dumel, *stupidus* . . . A Dummel, *mutus*' 1570 NED, *dummil*, a slow jade (Halliwell) and Hants dialectal *dummell* 'slow to comprehend', *dumble* 'stupid'. *Dumbulle* may be for *dunn bull*. cf. DUNCALF. (ii) The Liverpool *Dumbell*, *Dumbill* are probably Cheshire corruptions of *Dumville*: James *Domvile* 1521, James *Dowmbill* 1522, Elizabeth *Dombell* 1568, Hannah *Dumbell* 1660 (Bardsley).

**Dumbelton, Dumbleton, Dumpleton:** Bernard *de Dumbelton'* 1206 Cur (Gl); William *de Dumplynton* a1238 WhC; Geoffrey *de Dumbleton* 1298 IpmGl. From Dumbleton (Gl).

**Dumbrell, Dumbrill:** James *Dunbrell*, Thomas *Dombrell* 1525 SRSx. Probably connected with dialectal *dummel* 'stupid, dumb'.

**Dumjohn:** v. DUNJOHN

**Dummer, Dummere, Dumper:** Henry *de Dumera*, *de Dommere* 1115 Winton (Ha). From Dummer (Hants).

**Dummett, Dommett:** Gerald *de Domnomedardo* c1150 Seals (Nth); Aliz *de Dunmart*, *de Dummart* 1200–1 P (Wa). From Dumart-en-Ponthieu (Somme).

**Dumping:** Odo *Dumping*, *Dumfyng* c1190–1290 RegAntiquiss. A nickname from ME *domping* 'a diving bird'.

**Dumpleton:** v. DUMBELTON

**Dumsday:** v. DOMESDAY

**Dumville, Dunville, Domvile, Domville, Dunfield:** Hugh *de Donvil* 1274 RH (Sa). From Donville (Calvados). v. also DUMBELL.

**Dunbabin, Dunbobbin, Dunbavin, Dunbavand, Dunbebin, Dunbevand, Donbavand:** All these surnames are found in Liverpool and Manchester. Bardsley notes Ralph and Anthony *Dunbabin* of

Warrington (1597–8) where he found *Dunbobbin* and *Dunbobin* also surviving. *Dunbabin* is a variant of *Dunbobbin*, *Dunrobin*, with unrounding of the vowel. 'Dun (dark) Bobbin or Babbin.' cf. *Bob* and *Rab* for *Robert*, *Broun Robin* 1309 Wak (Y), *Robert* called *Brounerobyn* 1311 ColchCt and *v.* HORRABIN and BROWNJOHN. The other surnames are colloquial pronunciations of an uncommon, unintelligible name.

**Duncalf, Duncalfe, Duncuff:** Adam *Duncalf* 1354 Putnam (Ch); John *Doncalfe* 1449 DbCh. 'The dun calf', a nickname. cf. Robert *Dunnebrid* 1183 P (Y), Thomas *Dunfugell* 1291 AssCh 'dun bird'.

**Duncan, Dunkin, Donkin:** *Donecan* 1086 DB (So); *Dunecan'* 1130 P (W), c1135 Black, 1181 P (Y); *Donchadus* c1150 Black; Hugh *Dunkan, Dunken* 1275–6 RH (L); Ralf *Donekan* 1280 AssSo; John *Dunkan* 1327 SRSt, 1367 Black (Berwick); Richard *Donykyn* 1327 SRSo; William *Donekyn* 1332 SRSx; Richard *Donkin* 1674 HTSf. *Dunecan*, from OIr, OGael *Donnchad* 'brown warrior'. This common Scottish name is found in Yorkshire and Cumberland in the 12th century. Here it was probably introduced from Ireland as in Somerset where the name occurs in 1086 and had become a surname in the 13th century. In 1327 (SRSo) there are several men named *le Ireys* and one *Irlonde*.

**Duncanson, Dunkinson:** Nicholas *Donecandonesoune* 1296 CalSc (Dunbar); John *Dunkanson* 1367 ib. (Berwick); John *Dunkysoun* 1603 Black. 'Son of *Duncan*.'

**Dunce, Duns, Dunse:** William *le Duns* 1327 SRSx; William *Douns, Cristina le Douns* 1332 SRSx. *Dunce* is not recorded in NED before 1527. It is derived from the name of John *Duns* Scotus, the scholastic theologian, and in the 16th century was used of a disciple or follower of his, a schoolman, a subtle, sophistical reasoner, and, when his system was attacked and ridiculed by the humanists and reformers, of a 'blockhead incapable of learning or scholarship', 'a dull pedant', 'a dullard'. The surname, found so soon after the death of the 'Subtle Doctor' in 1308, cannot have any derogatory meaning. It must be equivalent to *Duns man* or schoolman, scholar. The Scottish *Duns(e)* is from Duns (Berwick): Hugh *de Duns* c1150 Black. It may be that John *Duns* Scotus came from here.

**Dunch:** Matilda *Dunche* c1200–10 RegAntiquiss; Stephen *Dunche* 1279 RH (C); William *Dunch* 1314 FFSf. ME *dunch* 'a push, knock, bump', perhaps a nickname for a quarrelsome person. cf. Matilda *Dunchedeuel* c1200–10 RegAntiquiss 'hit the devil'.

**Duncklee, Dunckley:** *v.* DUNKLEY

**Duncombe, Duncum, Duncumb, Duncumbe, Dunkum:** James *Duncombe* 1674 HTSf. From Duncombe in Sherford (D).

**Duncuff:** *v.* DUNCALF

**Dundraw, Dundrow:** Adam *de Dundraw* 1332 SRCu. From Dundraw (Cu).

**Dunfield:** *v.* DUMVILLE

**Dunford:** William *de Dunneford'* 1302 FFY; William *Dunford* 1332 SRWa; Richard *Dunford* 1421 IpmY. From Dunford Bridge in Thurlstone, Dunford House in Methley (WRY), or Durnford (So, W).

**Dung, Dunge:** Hugh *Dung* 1184 P (Ess); Peter *Dung'* 1274 RH (Ess). Either a nickname from OE *dung* 'manure, filth', or metonymic for DUNGER. cf. John *Dungebien* 1393 FFEss 'manure well'.

**Dungate, Dongate:** Geoffrey *de Doungate* 1275 SRWo; Richard *ate Dungate* 1296 SRSx. 'Dweller by the gap in the hill', OE *dūn, geat*.

**Dunger:** Roger *Dunger'* 1221 AssWo; William *Donger* 1327 MEOT (L). A derivative of ME *dungen* 'to dung', one who manures the ground.

**Dungworth:** John *Dungeworthe, de Dungeworth* 1440 ShefA. From Dungworth in Bradfield (WRY).

**Dunham:** Richard, William *de Dunham* 1190 P (Nf), 1221 AssWa; William *Dunham* 1332 SRSx. From Dunham (Ches, Norfolk, Notts), or Dunholme (Lincs).

**Duncliffe:** *v.* TUNNICLIFF

**Dunjohn, Dumjohn:** Laurence *de (del) Duniun, de Dungun* 1204, 1208 P (Ha, K); Ralph *Dungun* 1255 RH (Bk). From employment at 'the chief tower of a castle' (OFr *donjon*), probably 'jailer'. cf. Dungeon Fm, pronounced 'Don Johns', from Peter *Dungun* (1272 PN Ess 436).

**Dunkin:** *v.* DUNCAN

**Dunkinson:** *v.* DUNCANSON

**Dunkley, Dunckley, Duncklee:** Roger *de Dunkele* 1332 SRLa; Paul *Dunkeley* 1642 PrD. From Dinckley (La), *Dunkythele* 1246.

**Dunkum:** *v.* DUNCOMBE

**Dunley:** Laurence *de Duneleia* 1221 AssWo. From Dunley in Areley Kings (Wo).

**Dunlop:** William *de Dunlop* 1260, Constantyn *Dunlop* 1496 Black. From the lands of Dunlop in Cunningham (Ayr).

**Dunman:** *v.* DOWNMAN

**Dunmow, Dunmo:** Ralph *de Dunmauue* 1119 Colch (Ess); Richard *de Dunmawe* 1270 Acc; Roger *de Dunmowe* 1339 LLB F; John *Dunmowe* 1374 FFEss. From Dunmow (Ess).

**Dunn, Dunne, Don, Donn, Donne:** William *Dun* 1180 P (Gl); John *le Dunn* 1198 FF (Herts); Jobin *Don* 1271 ForSt; Adam *le Don, le Dun* 1275 SRWo, 1308 AssSt. OE *dunn* 'dull brown', 'dark, swarthy'.

**Dunnaby:** Semon *Donbyby* 1381 PTY. *Dūne i bý* (the man who lived) 'down in the village'. A partly anglicized form (OE *dūne* 'down') of a Scandinavian phrase. cf. WESTOBY.

**Dunnage:** Walter *de Dunewic* 1182–1200 BuryS (Sf); Jasper *Dunwich*, Thomas *Dunage* 1674 HTSf. From Dunwich (Suffolk).

**Dunell, Dunwell:** (i) Adam *de Dunwelle* 1330 YDeeds; Laurence *Donell* 1390, Robert *Donwell* 1489, *Dunel* 1533 FrY. 'Dweller by the dark stream', OE *dunn, wiella*. (ii) *Doneuuald* 1086 DB (Y); Geoffrey *filius Dunualdi* c1160 DC (Lei); Richard *Dunwald* 1275 RH (Sf); Ivo *Donewold* 1327 SRLei. OE *Dunweald*.

**Dunnet, Dunnett:** Walter *Dunheued, de Duneheued* 1201, 1208 P (So); John *de Dunheued* 1246 Seals (W); Jeffery *Dunnett*, Robert *Dunnit* 1674 HTSf. From Downhead (Som) or Donhead (Wilts).

**Dunnicliffe(e):** *v.* TUNNICLIFF

**Dunning:** (i) *Dunning, Dunninc, Donninc* 1066 DB; *Dunning* 1086 (Ch, Gl); Gilbertus *filius Dunning* 1166

P (C); *Dunnigus* (*Dunningus*) 1205 P (Gl); Æluuard *Dunning* 1086 DB (Sa); Roger, Nicholas *Dunning* 1166 P (C), 1201 P (So); Symon *Donnyng* 1332 SRSx. OE *Dunning* 'son of *Dunn*' or 'the dark, swarthy one'. (ii) Gillemichel *de Dunin* c1208 Black; John *Donnung* 1321 ib. From Dunning in Lower Strathearn (Perthshire).

**Dunnington:** Birlecca *de Dunnigton'* 1202 FFY; John *de Duninton'* 1205 P (Y). From Dunnington (ERY).

**Dunrobin:** *v.* DUNBABIN

**Dunsby, Dunsbee:** Warin *de Dunnesby* 1275 RH (L); Thomas *de Dunesby* 1314 AssSt. From Dunsby (L).

**Duns(e):** *v.* DUNCE

**Dunsford:** Robert *de Dunesford'* 1219 AssL; John *de Dunsford* 1348 FFY; Henry *Dunsford* 1524 SRD. From Dunsford (D), or Dunsforth (WRY).

**Dunsley:** John *de Duneslay* 1304 IpmY. From Dunsley in Whitby (NRY), or Dunsley in South Kirkby (WRY).

**Dunstable:** Simon *de Dunstapele* 1221, William *Donestaple* 1253 Acc; William *de Dunstaple* 1285 WiSur. From Dunstable (Beds), *Dunestapele* 1123.

**Dunstall:** *v.* TUNSTALL

**Dunstan, Dunston, Dunstone:** (i) *Dun(e)stan, Donestan* 1086 DB; *Dunstan* de Berstede 1275 RH (K); William *Dunestan* 1212 Cur (Db); Thomas, Margaret *Dunstan* 1275 RH (K), 1327 SRSf. OE *Dūnstān* 'hill-stone'. (ii) Alexander *de Duneston'* 1190 P (L); Hugo *de Dunestun'* 1202 FF (Nf); Reyner *de Dunstan* 1242 Fees (Nb). From Dunstan (Northumb), Dunston (Derby, Lincs, Norfolk), or Dunstone (Devon).

**Dunster:** Hugh *de Dunsterre* 1275 RH (Sf); John *Dunsterre* 1332 SRDo; John *Dunster* 1642 PrD. From Dunster (So).

**Dunsterville:** Adeliz *de Dunstanuilla* 1130 P (Lei); Walter *de Dunstervill'* 1235 Fees (Co). From Dénestanville (Seine-Inférieure).

**Dunt:** Richard *Dount* 1327 SRWo; Dyonisia *Duntes* 1349 Oseney; John *Dunt* 1366 Eynsham. OE *dynt* 'a blow', perhaps a nickname for a quarrelsome person.

**Dunton, Downton:** Richard *de la Duntun'* 1275 SRWo; Ralph *de Dunton* 1296 SRSx. From one of the Duntons or 'dweller at the homestead on the hill' (OE *dūn, tūn*).

**Dunville:** *v.* DUMVILLE

**Dunwell:** *v.* DUNNELL

**Duplock:** *v.* DIPLOCK

**Duppa:** Rys *Doppa* 1397 Morris (Radnor); Hugh *Duppa* 1497 CantW; Thomas *Duppa* 1663 HeMil. Welsh *Duppa*, a personal name of obscure origin and meaning.

**Durable:** John *Durable* 1296 SRSx; John *Durable* 1343 IpmGl. OFr *durable* 'lasting, steadfast'.

**Duran, Durand, Durant:** *v.* DURRAND

**Durden, Durdin:** William, Roger *Durdent* 1148–54 Bec (Sx), 1176 P (St); William *Durdent* 1272 FFSt; Nicholas *Durdon* 1428 FA (W). OFr *dur, dent* 'hard-tooth'.

**Durford:** Walter *Dureford'* 1332 SRDo. From Durford (Sx).

**Durham, Durram:** Osbert *de Dunelm'* 1163 P; William *de Durham* 1236–7 FFEss; Adam *de Duram* 1327 SRSx; Lawrence *Durham* 1400 AssLo. From Durham.

**Durk:** *v.* DARK

**Durling:** *v.* DARLING

**Durman:** *v.* DEARMAN

**Durnford, Dornford:** Roger *de Derneford'* 1190 P (W); William *de Durneford* 1255 RH (W). From Durnford (Wilts). cf. DANFORD.

**Durrad:** *v.* DURWARD

**Durran, Durren:** William *Dureweyin* 1249 AssW; Ralph *Durewyn'* 1297 MinAcctCo; John *Durewyne* 1347 IpmW. OE *Dēorwine. v.* also DURRAND.

**Durrand, Durrant, Durran, Durrans, Durrance, Duran, Durand, Durant, Doran, Dorant, Dorran, Dorrance, Dorrins:** *Durandus* 1066 Winton (Ha); Robertus *filius Durand* 1115 ib.; *Durant* stabularius Hy 2 DC (L); *Doraunt* de Moreby 1312 FrY; Walter *Durand'* 1196 MemR (We); Robert *Duraund* 1221 AssWa; John *Durant* 1222 Cur (Sr); Robert *Duran* 1275 SRWo; William *Doraunt* 1285 FrY; Gabriel *Durance* 1676 Bardsley. OFr *Durant*, perhaps 'obstinate'.

**Dursley:** Thomas *Dursley* 1423 AssLo. From Dursley (Gl).

**Durston:** Giles *Durston* 1641 PrSo. From Durston (So).

**Durtnall, Durtnal, Durtnell, Dutnall, Dartnall, Dartnell, Darknell:** Robert *Darkynhole*, William *Durkynghole* 1435, 1453 Streatfeild *MSS*; William *Darknold* 1505 KentW; William *Derkynhole* 1512 EdenbridgePR; Robert *Darkenhole* 1539 LP (Lo); Robert *Darknoll* 1540 LP (K); John *Dartnoll* 1551, *Durtenall* 1554, Alice *Darknall* 1555, John *Dyrknall* 1556 StaplehurstPR (K); Robert *Darknall*, *Darkenall*, *Darkenoll* 1553 Pat (K). A west Kent and Sussex name, found as *de Durkinghol'* in 1240, from a lost place in Penshurst, now perhaps Doubleton's Farm, *v.* PN K 89–90.

**Durval:** *v.* DARVAL

**Durward, Dorward, Dorwood, Darwood, Durrad, Dorrett:** Reiner *Dureward'* 1208 Cur (Nf); William *Doreward* 1230 P (Ha); Richard *le Doreward* 1255 FFEss; Thomas *Durart* 1620 Black; George *Durrat* 1724 ib.; John *Dorrat* 1753 ib. OE *duru-weard* 'door-keeper, porter'. In Scotland the office of door-ward to the king, *Hostiarius regis*, was hereditary. Alan *Durward* (*Ostiarius, le Usher*), justiciar of Scotland (d. 1268), was the son of Thomas *Ostiarius. Dorward*, still common around Arbroath, is probably from the office of door-ward of the Abbey. Others, both in Scotland and England, derive from a less exalted office.

**Dury:** (i) Richard *Dury* 1344 PN D 198. From Dury in Lydford (D). (ii) Also Huguenot from Paul *Dury*, officer of engineers under William III. From Dury (Aisne, Pas-de-Calais, Somme).

**Duryard:** Richard *Dureyerd* 1289 IpmGl. From Duryard in Exeter St Davids (D).

**Dust, Doust:** Ulf *Dust* c1030 OEByn; Walter *Dust* 1203 P (Y); Robert *le Doust* 1316 FFK. OE *dūst* 'dust', with reference to a dust-coloured complexion or hair, or to a person of little worth, or to a workman (cf. Dusty Miller). cf. Thomas *Dustiberd* 1229 Pat (So), William *Dustifot'* 1221 AssWo, a term used later (a1400) of a wayfarer, especially a travelling pedlar. We may also have an unexplained topographical term: Richard *del Doustes* 1332 SRLa.

**Dustifoot:** William *Dustifot'* 1221 AssWo. 'Dusty foot', OE *dūstig, fōt*, a term used later of a wayfarer, especially of a travelling tinker. cf. John *Liftfot* 1284 CtW 'left foot'; Thomas *Slafot* 1379 PTY 'slow foot'.

**Dusting:** *v.* THURSTAN

**Duston:** Aelisia *de Duston* 1190 P (Nth); William *de Duston'* 1212 P (K); William *de Dustun'* 1219–26 P (Nth). From Duston (Nth).

**Dusty:** Richard *Dusty* 1379 PTY. OE *dūstig* 'dusty'. cf. Thomas *Dustiberd* 1229 Pat (So) 'dusty beard'.

**Dutch:** John *Duch(e)* 1360 ColchCt, 1427 FFEss. 'Dutch', a name given to immigrant Dutch weavers.

**Dutchman:** Henry, John *Duch(e)man* 1354 FrY, 1366 ColchCt. *v.* DUTCH. Also applied in the 15th century to Flemish brick-makers (Building 142).

**Dutnall:** *v.* DURTNALL

**Dutt:** Wulvet *Dutte* c1175 Newark; William *Dut* 1230 P (L); John *Dut* 1269 AssSo; Stephen *Dutte* 1546 FrYar. A nickname from ME *dut* 'joy, delight'.

**Dutton:** Richard *de Duttona* c1150 StCh; Richard *de Dutton* 1246 AssLa; John *Dutton* 1468 IpmNt. From Dutton (Ch, La).

**Duval, Duvall:** (i) William *Duvel* 1224 FFEss. Perhaps a diminutive of OE *\*Dūfe* (f). (ii) Usually Huguenot, from a refugee of that name from Rouen who settled in England (Smiles 392).

**Duxbury:** Siward *de Dokesbir'* 1204 P (La). From Duxbury (La).

**Duxon, Duxson:** (i) Sawi *Duchessune* 1168 P (W). 'Son of Ducc', OE *Ducc*. *v.* DUCKET. (ii) John *Dokeson* 1379 PTY; Thomas *Duckesson* 1586 Bardsley. 'Son of *Duke*' or Marmaduke.

**Duxworth, Duckesworth:** Hervey *de Dukeswurda* 1195–6 P (Berks); John *de Dukesworth* 1307–8 FFEss; Robert *Dokesworth* 1402–3 FFSr. From Duxford (C), *Dukeswrth* c950.

**Duzamour:** *v.* DOUCEAMOUR

**Dwarf, Dwarrow:** Walter *le Dwarew* 1249 AssW; Daniel *Dwarfe* 1642 PrD. OE *dweorg* 'dwarf'.

**Dwelley, Dwelly:** Robert *Dweyle* 1255 RH (W); Richard *Dwelie* 1275 RH (W); Robert *Dwole*, Walter *Dwelie* 1327 SRSo; John *le Dwole*, Richard *le Dwele*, Alice *Duolie* 1332 SRSr. OE *\*dweollīc* 'foolish, erring, heretical', a variant of *dwollīc*, recorded in the same sense, from *dwelian, dweolian, dwolian* 'to err'. cf. *dweola* 'error, heresy', *dwolung* 'dotage'. This surname

explains Dwelly Fm in Lingfield (Surrey) where all the early forms are from surnames. In PN Sr 328 the name has wrongly been taken·as a compound of *lēah*. The name is manorial, 'Dwelly's Fm'.

**Dwight:** Henry *Duyhts* 1327 SRDb; John *Dwight* 1524 SRSf; Josiah *Dwight* 1665 HTO. Usually from *Diot*, a diminutive of *Dye*, a pet-form of *Dionisia*. Sometimes, perhaps, a late corruption of THWAITE.

**Dwyer:** Richard *Dwyer* 1386 PN Do i 62. Ir *Ó Duibhir* 'dun-coloured'.

**Dyas, Dyos:** Thomas *Deyas* 1324 LaCt. Perhaps 'dweller at the dye-house', OE *dēag, hūs*.

**Dyason, Dyerson:** John *Dyosone* 1332 SRSt; William *Dyotson* 1381 PTY; Richard *Dyatson* 1520 FrY. 'Son of *Dyott*.'

**Dybald, Dyball, Dybell, Dyble:** *v.* DIBBLE

**Dyce:** *v.* DICE

**Dye:** *Dye* 1301 SRY, 1316 Wak (Y); Walter *Dye*. 1316 Wak (Y). A pet-form of *Dionisia*.

**Dyer, Dyers:** Henry *le Deghar* 1260 MESO (So); Robert *le Deyare* 1275 SRWo; Alexander *Dyghere* 1296 SRSx; Henry *le Dyer* 1327 SRDb. OE *dēagere* 'dyer'.

**Dyerson:** *v.* DYASON

**Dyett, Dyott:** *Diota* 1296 Wak (Y); *Dyota* 1319 SRLo; *Dyot* 1332 SRCu; William *Dyot* 1348 DbAS 36; Robert *Diotte* 1396 AD v (Y). A diminutive of *Dye*, a pet-form of *Dionisia*.

**Dyke(s):** *v.* DITCH

**Dykins:** *v.* DICKEN

**Dyllon:** *v.* DILLON

**Dyment, Dymond:** *v.* DIAMANT

**Dymick, Dymock, Dymoke:** *v.* DIMMOCK

**Dymott:** Adam *Dyemogh* 1332 SRLa. Probably 'relative of *Dye*', a pet-form of *Dyonisia*. *v.* DYE and cf. HICKMOTT.

**Dyne:** *v.* DAIN

**Dyneley:** *v.* DINELEY

**Dyos:** *v.* DYAS

**Dyson:** Richard *Dysun* 1275 RH (Lo); Ralph *Dyson* 1296 SRSx; John *Dysone, Dyesson* 1327 SRWo, 1387 FrY. 'Son of *Dye*.' John *son of Dionysia de Langside* is also called John *Dyson de Langside* 1369 YAJ v, 76. cf. also Robert *Dyotson* 1379 PTY.

**Dyster:** Henry *le Deystere* 1280 MESO (Wo); Eua *Dygestre* 1296 SRSx; William *le Deyghester* 1325 AssSt; Robert *le Deyster, le Dyster* 1332 SRWa; Thomas *Diestere* 1355 AD iii (O). OE *\*deaȝestre*, *\*deȝ(e)stre* 'dyer' (f), but used also of men. cf. DEXTER, DYER.

**Dyter:** *v.* DITER

# E

**Eacersall:** v. ECCLESHALL

**Ead, Eade, Eades, Eads, Ede, Edes:** *Eda* Hy 2 DC (L), 1194 P (Wo); Roger, William *Ede* 1275 SRWo, RH (Nf); Gilbert *Eadis* 1279 RH (O). *Eda*, always feminine, is rather a pet-form of *Edith* (OE *Ēadgȳð*) than a survival of OE *Ēda*.

**Eaden, Eadon, Eden, Edens, Edon:** (i) Tomas *filius Edon'* 1203 P (O); Edon *le Poleter* 1270 FFSf; Roger *Edun* 1327 SRC; Stephen *Edoun* 1327 SRSf. OE *Ēadhūn*. (ii) Nicholas *de Edune* 1178 P (Nb); William *de Eden* 1256 AssNb. From Castle Eden or Eden Burn (Durham).

**Eadie, Eady, Eadey, Eaddy:** *Eddiva, Ædiva* pulchra 1086 DB; Stephen *Edy* 1278 IpmGl; Edmund *Edye* 1642 PrD. OE *Ēadgifu* (f).

**Eadiman, Eadyman, Eddiman, Eddyman:** Simon *Ediman* 1301 FS; John *Edyman* 1370 LuffCh; Robert *Edyman* 1438 IpmNt. 'Servant of Eadie', i.e. *Ēadgifu*.

**Eadington, Eadmeades:** v. EDDINGTON, EDMEAD

**Eadlesfield, Eaglisfield:** John *de Eglesfeld* 1307 Black; Robert *de Egglesfeld* 1335 FrY; John *de Eglesfeld* 1399 ShefA. From Eaglesfield (Cu).

**Eagar, Eager, Eagers, Egar, Egarr, Eger:** v. EDGAR

**Eagland:** v. EGLIN

**Eagle, Eagles, Eagell:** (i) Ralph *Egle* 1230 P (Y); Robert *le Egle* 1297 Coram. A nickname from the bird. (ii) Gilbertus *de Aquila* 1196 P (Y); Richer *del Egle* 1210 Cur (Nth). From Laigle (Orne).

**Eaglen, Eagling:** v. EGLIN

**Eagleton:** v. EGGLETON

**Eaglisfield:** v. EAGLESFIELD

**Eakin:** v. EDKINS

**Ealand, Eland:** Henry *de Eland'* 1194–1211 Seals (Y); Robert *de Eland* 1296 SRNb; William *Eland* 1439 IpmNt; Richard *Ealand* 1687, *Eland* 1739 FrY. From Ealand (L), Little Eland (Nb), Elland (WRY), or 'dweller at the land by the water', OE *ēaland*.

**Eale, Eales, Eals:** v. EELE

**Ealey:** v. ELY

**Eames, Heams, Hemes:** Andrew *le Em* 1274 RH (Nf); John *Eame* 1280 Bart (Lo); Hugo *Eem* 1281 Rams (Hu); Robert *Heme* 1524 SRSf; William *Eames* 1540 Whitby (Y). OE *ēam*, ME *eme* 'uncle'. Ralph *Heam* (c1200 ELPN) is identical with Radulfus *auunculus*.

**Eamond, Eman, Emmens:** Robert *Eemond* 1446 Paston. OE *Ēanmund*.

**Eamondson, Emmison:** Robert *Emanson* 1594 SPD; John *Eamansson* 1621 SRY. 'Son of *Ēanmund*'.

**Earl, Earle, Harle, Hearl, Hearle, Hurl, Hurle, Hurles, Hurll:** Lefuin *Eorl* c1095 Bury (Sf); Ernald, William *Hurl* 1182 P (Do), 1221 AssWa; Hervicus *Herl* 1210 Cur (C); William *Erl* 1230 P (Ha); Hugh *le Erl* 1255 FFSf; Emma *lē Heorl*, Thomas *le Erl*, Margeria *la Horl* 1275 SRWo. OE *eorl* 'earl'. Leofric *Eorl* and Harold *Eorl* (1038–44 OEByn) bore titles of rank; the surnames above are nicknames or pageant-names.

**Earley, Early, Erleigh, Erley, Erly:** John *de Erlea* 1162 P (So); William *de Erlega* 1190 P (Sx); John *de Erley* 1230 P (Berks). From Earley (Berks), Earnley (Sussex), Arley (Ches, Lancs, Warwicks, Worcs), all 'eagle-wood'. Occasionally a nickname: Geoffrey *le Urrly* 1275 RH (Nf). OE *eorlīc* 'manly'.

**Earnshaw:** Richard *de Erneschaghe* 1316 Wak (Y). 'Dweller by the eagle-wood', OE *earn, sceaga*.

**Earp:** Matthew *Yrp* 1200 Cur (L); Henry *Erpe* 1304 IpmY; John *Irp* 1332 Bacon (Sf). OE *\*ierpe*, a byform of OE *earp* 'swarthy'.

**Earth:** Thurgar *de Erthe* 1202 FFK; Roger *Earth* 1576 SRW. 'Dweller by the ploughed land', OE *erþ*.

**Earwaker, Earwicker, Erricker:** *Euerwacer* de Gepeswic 1130 P (Sf); Edmundus *Erwak'* 1230 P (Do); Richard *Herewaker* 1247 AssBeds; Hamo John *Euerwaker* 1327 SRSf. OE *Eoforwacer* 'boarwatchman', recorded in 1061. *Earwicker* is pronounced *Erricker*.

**Easby:** Thomas *de Eseby* 1230 P (Lei/Wa). From Easby (Cu, NRY).

**Easdale:** v. ESDAILE

**Easey, Easle, Eassie, Easy, Essie, Essy:** (i) Adam *Essy* 1327 SRSo; John *de Esey* 1395 AssL. Perhaps from Eisey in Latton (W), *Esy* 1259. (ii) Hugh *de Essi* a1200, David *Essie* 1495 Black. From the barony of Eassie or Essie (Angus).

**Easingwald, Easingwood, Easinwood:** Ingelot *de Esingwald* 1204 P (Y). From Easingwold (NRY).

**Easman:** v. HEASEMAN

**Eason, Easom, Easun, Easson, Esson:** *Aythe* filius Thome c1360 Black; *Aye* Jonson 1395 ib.; John *Ayson'* 1392 ib.; John *Asson* or *Easson* 1681 ib. 'Son of Adam', Gael *Adamh*. Common in Angus.

**Eassie:** v. EASIE

**East, Eastes, Este:** Ralph *del Est* 1196–1237 Colch (Ess); Walter *Est* c1220 Gilb (L); Osbert *Upest* c1240 Fees (Beds). 'Man from the east' or 'dweller to the east' of the village.

**Eastabrook:** v. EASTBROOK

**Eastacre:** John *Estacre* 1545 SRW. 'Dweller by the east field', OE *ēast, æcer*.

**Eastaff:** cf. Simon *de Essetoft* 1331 FrY. From Eastoft (WRYorks).

**Eastall, Eastell, Estall, Estell:** Walter *de Easthalle, de Estalle* 1279 RH (O); Jonathan *Eastall* 1674 HTSf. 'Dweller at the East Hall.'

**Eastaugh, Easto, Eastoe, Estaugh:** John *de Esthawe,* Geoffrey *de Esthagh* 1327 SRSf. 'Dweller at the east enclosure', OE *ēast, haga.*

**Eastaway:** John *Eastaway* 1642 PrD. 'Dweller by the eastern road', OE *ēast, weg.*

**Eastbrook, Eastabrook, Easterbrook, Esterbrook:** William *Bestebroke* 1296 SRSx; Alan *bi Estebrouk,* Matilda *Estbrok* 1327 SRSx. OE *be ēastan brōce* '(one who lives) to the east of the brook'. Or from residence at a place called Eastbrook. William *de Estbrok* (1254 *Ass*) lived at Eastbrook Fm in Dagenham (Essex).

**Eastburn:** Simon *de Esteburne* c1200 Calv (Y); Alexander *de Esteburne* 1304 IpmY. John *Eastburne* 1672 HTY. From Eastburn (ERY, WRY).

**Eastby:** Henry *de Estby* 1394 FrY. From Eastby (WRY).

**Eastchurch:** Walter *Byestecherche* 1330 PN D 407; John *Bestchirche* 1390 LLB H; John *Estcherche* 1524 SRD. From Eastchurch (K), Eastchurch in Crediton (D), or 'dweller to the east of the church', OE *bī, ēastan, cirice.*

**Eastcott, Eastcourt, Escot, Escott, Estcourt:** Gundwinus *de Estcota* 1190 P (Beds). From Eastcott (Middlesex, Wilts), Eastcotts (Beds) or Eastcourt (Wilts). Robert *atte Estcote* (1327 SRSx) lived at the east cottage, now Eastcourt, near Estcot's Fm in East Grinstead (Sussex).

**Eastell:** *v.* EASTALL

**Easter:** Jordan *del estre* 12th DC (Lei); Robert *del Estre* 1272 FFC; John *de Estre* 1345 FFEss. From Good, High Easter (Ess), or 'dweller by the sheepfold', OE *ēowestere.* Occasionally from a personal name: William *filius Estur* 1212 Cur (Ha).

**Easterbrook:** *v.* EASTBROOK

**Easterby:** Peter *Austebi* 1204 P (Y); Robert *Oustinby* 1297 SRY; Martin *Oustiby* ib.; William *Estyby* ib.; Hugh *Estinby* ib.; Richard *Esterby* 1474 GildY. ON *austr i bý,* influenced by OE *ēast,* (one who lives) 'east in the village'. cf. WESTOBY.

**Easterford:** John *atte Esterford* 1327 MELS (So); Edmund *Easterford* 1662 HTEss. From *Easterford,* now Kelvedon Bridge in Kelvedon (Ess), or 'dweller by the eastern ford', OE *ēasterne, ford.*

**Eastfield:** Henry *Estfeld* 1327 SRSo; William *Estfeld* 1423 AssLo. 'Dweller by the east field', OE *ēast, feld.*

**Eastgate:** Walter *de Estgat'* 1200 P (Nf). From residence near the east gate of some town or castle.

**Eastham:** John *de Estham* 1212 P (Nf/Sf); Ralph *de Estham* 1275, Richard *de Estham* 1358 SRWo. From Eastham (Ch, So, Wo).

**Easthope, Eastop:** John *de Esthop'* 1275 RH (Sa); William *Estoppe* 1327 SRSx. From Easthope (Salop) or 'dweller in the eastern valley' (OE *hop*).

**Eastick:** *v.* EASTWICK

**Eastland:** Simon *de Estlande* 1198 P (K). From residence near one of the many minor places so called.

**Eastley:** Henry *de Estleia* 1219 P (Y). 'Dweller at the east clearing', OE *ēast, lēah.*

**Eastman, Eastment, Astman, Esmond, Esmonde:** *Estmunt* 1066 DB (Sf); Ricardus *filius Estmund* 1195 P (Ess); *Esmond* 1313–14 ODCN; Stephen *Estmund* 1227 AssBk; Geoffrey *Astmund* 1275 SRWo; John *Eastmunde* 1277 AssSo; Alan *Esmund* 1285 FFEss. OE *Ēastmund* 'grace or favour-protection'.

**Eastmure:** Alicia *filia Eastmer* 1212 Cur (Bk); *Estmar* le Carbonier 13th AD ii (Mx); Walter *Estmer* 1221 AssWo; Robert *Eastmer* 1309 FFK; William *Estmor* 1311 Rams (Nf); John *Estmar* 1312 LLB B. OE *Ēastmǣr* 'grace or favour-famous', recorded only as the names of moneyers *temp.* Edward the Confessor to William I.

**Easto, Eastoe:** *v.* EASTAUGH

**Easton:** (i) John *de Eston* 1299 LoCt. From one of the many Eastons. (ii) Philip *atte Estone* 1327 SRSo. 'Dweller at the eastern homestead.' (iii) William *Byestone* 1297 MinAcctCo. OE *be ēastan tūne* (one who lived) 'to the east of the village'. cf. EASTBROOK. (iv) William, John *Anesteton* 1296 SRSx. OE *on ēastan tūne* (dweller) 'on the east of the village'. (v) *Edstan, Estan* 1066 DB (Nf, L); Roger, Robert *Estan* 1275 RH (Nf, Sf). OE *Ēadstān* 'prosperity-stone'.

**Eastop:** *v.* EASTHOPE

**Eastwater:** Reginald *Biestewatere* 1300 IpmW; Richard *Bi estewatere* 1332 SRSx. 'Dweller to the east of the water', OE *bī, ēast, wæter.*

**Eastwell:** Gilbert *de Eastwelle* 1279 RH (O). From Eastwell (K, Lei), or Eastwell in Potterne (W).

**Eastwick, Eastick:** Wluuinus *de Esteuuiche* 1066 DB (Herts); William *de Estwyke* 1296 SRSx. From Eastwick (Herts, WRY), Eastwick in Great Bookham (Sr), or Eastwick Barn in Patcham (Sx).

**Eastwood:** Adam *de Estwde* 1221 ElyA (C). From a place named Eastwood or from residence east of a wood (John *Byestewode* 1339 PN D 98). Also 'dweller to the eastward': Hugh *Enesteward* 1279 RH (O).

**Easy:** *v.* EASEY

**Eatmeat:** Robert *Etemete* 1279 RH (O). A nickname, 'eat meat', OE *etan, mete.* cf. William *Etebred* 1301 LLB C 'eat bread'; Walter *Etecroue* 1361 LLB G 'eat crow'.

**Eaton, Eton:** Ulmar *de Etone* 1066 DB (Herts), *de Ettone* 1086 DB (Beds); William *de Eton* 1374–5 NorwLt. From Eton (Bk), or from one or other of the many minor places called Eaton, either 'farm on a river', OE *ēa-tūn,* or 'farm on an island', OE *ēg, tūn.*

**Eatwell:** Hugo *de Hetewelle* 1187 P (Db). From Etwall (Derby).

**Eaves, Eavis, Eves, Reeves:** John *atte Euese* 1275 SRWo; John *atte Reuese* 1327 ib. (identical); Robert *del Eves* 1332 SRLa. 'Dweller by the border or edge' of a wood or hill (OE *efes*). *v.* also EVE, REEVES.

**Eayres, Eayrs:** *v.* AYER

**Ebbatson:** John *Ebotson* 1485 GildY. 'Son of *Ebbot.' v.* EBBETTS.

**Ebberton:** John de Sharryngworth called *Eberton* 1355 LoPleas; Roger *Eburton* 1414 KB; Herry *Eberton* 1474 Paston. From Edburton (Sx).

**Ebbetts, Ebbitt, Ebbutt:** *Ebbot* (f), *Ebota* 1379 PTY; *Ebete* (f) 1381 SRSf; Simon *Ebet* 1195 P (Nf); Adam *Ebboth* 1327 SRSo; Ralph *Ebottys* 1534 FrNorw; John *Ebitt* 1675 FrY. *Ebbot* and *Ebbet*, as women's names, are diminutives of *Ebb* for *Ibb*, a pet-name for *Isabel*. Occasionally they may be masculine, from *Ebbi*. v. EBBS.

**Ebblewhite:** v. HEBBLETHWAITE

**Ebbs:** Alicia Relicta *Ebbe* 1296 SRSx; Robert, John *Ebbys* 1524 SRSf, 1528 FrNorw. *Ebbe* must be the pet-form of *Isabel* from which *Ebbot* was formed. It is not clear whether *Ebbe* above is the surname or the christian name of Alice's husband. If the latter, it must be masculine and is a pet-form of *Herbert*. cf. *Ebbi* le Estreis 1218 AssL, where *Ebbi* is an interlineation for the cancelled *Herebertus*.

**Eccersley:** v. ECKERSLEY

**Eccles, Ekless:** Adam, Warin *de Eccles* c1170 Black, 1212 Cur (K); Peter *Ekeles* 1378 AssLo. From Eccles (Kent, Lancs, Norfolk, Berwick, Dumfries).

**Ecclescliff, Ecclescliffe:** Walter *de Eclescliue* 1242 AssDu. From Egglescliffe (Du).

**Eccleshall, Eckersall, Eacersall:** William *de Eccleshull* 1246 AssLa; Robert *de Eccleshale* 1251 AssY, *de Ekilsale* 1297 SRY. From Ecclesall (WRYorks), Eccleshall (St), Eccleshill (Lancs, WRYorks).

**Eccleston, Ecclestone:** Geoffrey *de Eckleston'* 1230 P (Nt); Robert *de Eccleston* 1385 IpmLa; Anthony *Eccleston* 1537 FFEss. From Eccleston (Ch, La).

**Eckersley, Eccersley:** Henry *de Ecclesleye* 1301 QW (Y); Jane *Eckersley* 1603 Bardsley (La). From a lost place in Lancashire. v. PN La 101.

**Eckhard, Eckart, Eckert, Eckett:** Simon *filius Echardi* 1219 Cur (L); Adam *Ecard* 1275 RH (Nf). OG *Eckhard, Eckard* 'edge-hard'.

**Eckington:** Stephen *de Ekinton'* 1199 MemR. From Eckington (Db, Sx, Wo).

**Eckton, Ecton:** Peter *de Eketon* 1305 AssW; Thomas *Ecton* 1380 LoPleas; Thomas *Ecton* 1527 FFEss. From Ecton (Nth, St).

**Ecroyd:** v. ACKROYD

**Eday:** v. EDITH

**Edbro, Edbrough:** John *Edborowe* 1479 SxAS 45; Thomas *Edborowe* 1525 SRSx; John *Edbroughe* 1641 PrSo. These could be from OE *Eadburg* (f), or shortened forms of Edburton (Sx). v. also EBBERTON.

**Edden:** Robert *Edden* 1332 SRWa. v. EDDS, GEFFEN.

**Eddiman:** v. EADIMAN

**Eddings:** John *Edyng* 1327 SRC; William *Eddings* 1641 PrSo. 'Son of *Eadda*', OE *Eadda* plus *-ing*.

**Eddington, Edington, Eadington:** Jordan *de Edingeton'* 1201 Pl (Wa); Gilbert *de Edington* 1327 SRSo; Thomas *Edyngton* 1448 IpmNt. From Eddington (Berks), or Edington (Northumb, Som, Wilts).

**Eddis:** v. EDIS

**Eddison:** v. EDESON

**Eddyman:** v. EADIMAN

**Eddolls:** Robertus *filius Edolfi* 1206 Cur (Nb); *Edulfus* 1214 Cur (Ess); *Edolf'* 1219 AssY; Ralph *Edolf* 1276 RH (Berks); Simon *Edulf* 1327 *SR* (Ess). OE *Eadwulf* 'prosperity-wolf', a common OE name.

**Eddowes:** v. EDIS

**Edds:** *Edde* 1279 RH (O); John *Edde* 1524 SRSf. Perhaps a pet-form of *Edwin* or *Edward*, both surviving in use in the 14th century. cf. *Edda* rusticus (968) and *Edwine* qui et *Eda* dictus est (801 Redin). Or for EAD: *Edda* mater Johannis 1379 PTY.

**Eddy, Eddie:** *Edwy* Morwy 1221 AssWo; Willelmus *filius Edwy* 1250 Fees (Sr); John, Robert *Edwy* 1254 ArchC xii, 1275 SRWo; Adam *Eadwy*, *Edwy* 1275 RH (Nf), 1327 SRSf. OE *Eadwīg* 'prosperity-war', a common OE name. The surname is fairly common in ME and has probably become *Eddy*.

**Ede(s):** v. EAD

**Eden(s):** v. EADEN

**Edenborough:** v. EDINBOROUGH

**Edeson, Eddison, Edson:** William *Eddesone* 1314 AssSt; Geoffrey *Edessone* 1328 AD i (Hu); William *Edison* 1394 DbCh. 'Son of *Ead*' or of *Edd*.

**Edgar, Edgars, Edger, Eagar, Eager, Eagers, Eagger, Egar, Egarr, Eger, Eggar, Egger, Eggers, Adger, Agar, Agars, Ager:** *Edgar* 1066 DB (Hu), 1148–67 AD i (Lo); *Ædgar* c1095 Bury (Sf); *Adger* 1182–8 BuryS (Sf); Walter *Eagar* 1250 FFSf; Thomas *Edgar* 1250 Fees (Sr); John *Adger* 1309 LLB D; William *Agar* 1521 FrY; Richard *Egger* 1563 Black; John *Egar* 1664 RothwellPR (Y). OE *Eadgār* 'prosperity-spear'. *Eager* is a Northern Irish, *Egarr* a Dumfries form. cf. *Aeggar* King of Scots a1189 Black.

**Edgbrook, Edgbrooke:** v. EDGEBROOK

**Edgcliff, Edgcliffe:** v. EDGECLIFF

**Edgcombe, Edgcome, Edgcumbe:** v. EDGECOMBE

**Edge:** Henry *del Egge* 1221 AssWo, *sub Egge* 1290 ShefA (Y); John *de Egge* 1260 AssCh, *atte Egge* 1327 SRWo. 'Dweller near or below some prominent edge, ridge, steep hill or hill-side' (OE *ecg*).

**Edgebrook, Edgebrooke, Edgbrook, Edgbrooke:** Robert *Eggebrok* 1327 SRSo. Perhaps 'dweller by Ecca's brook'.

**Edgecliff, Edgecliffe, Edgcliff, Edgcliffe:** Walter *de Eggeclive* 1280 IpmY. From Edge Cliff in Stocksbridge (WRY).

**Edgecombe, Edgecumbe, Edgcombe, Edgcome, Edgcumbe:** William *de Egghacombe* 1275 RH (D); Ser Pers *Egecom* 1494 Paston; William *Edgcombe* 1642 PrD. From Edgcumbe (D).

**Edgefield, Edgfield:** Bartholomew *de Egefeld'* 1198 FFNf; William *de Eggesfeld* 1208 P (Nf/Sf); John *de Eggefeld* 1374–5 NorwLt. From Edgefield (Nf).

**Edgeham, Edgham:** Richard *de Eggeham* 1208 P (Do/So). From a lost *Egham Hill* in Hamfallow (Gl).

**Edgeley, Edgley:** John *de Eggelye* 1296 SRSx; William *de Eddesleye* 1348 AssSt; Henry *Edgly* 1674 HTSf. From Edgeley (Ches), or Edgeley (Salop), *Edesleye* 1327.

**Edgell:** Roger *Eggolf* 1275 SRWo; William, Richard *Eggel(l)'* 1278 FFEss, 1327 *SR* (Ess); Robert *Eggolfe*, William *Eggculf* 1327 SRWo. OE *Ecgwulf* 'sword-wolf', a common OE name. *Eggell* is from OE *\*Ecgel*, a diminutive of *Ecg-*.

**Edgerley, Edgerly:** Albert *de Edgardle* 1189 Sol; Hugh *Eggerley* 1440 PN O 108. From Edgerley (Sa), Edgerley in Hurstpierpont (Sx), Edgerley Fm in Clanfield, or Hedgerley Wood in Chinner (O).

**Edget, Edgett:** Stephen *Egede* c1230 Newark; Robert *Ejote* 1409, John *Egeott* 1490, William *Egiot* 1540 *PetreA*; Robert *Egett* 1674 HTSf. OE *Ecggēat*, but sometimes, perhaps, OE *Ecgheard*.

**Edgeworth, Edgworth:** Peter *de Egewurth* 1221 AssGl; Stephen *de Eggeworth* 1352 AssSt; Robert *Egeworthe* 1524 FFEss. From Edgeworth (Glos, Lancs).

**Edgfield:** *v.* EDGEFIELD

**Edgham:** *v.* EDGEHAM

**Edhouse:** *v.* EDIS

**Edicker:** *Edwacre* 1066 Winton (Ha); *Eduuacher* 1111–38 ELPN; *Edwaker* 1200 FFK; Godwinus *filius Edwacher* 1135 Oseney (O); Adam, Thomas *Edwaker* 1279 RH (O). OE *Ēadwacer* 'prosperity-watchman' or, perhaps, OE *ēadwacer* 'watchman of property' used as a personal name; a late name, recorded only twice before the Conquest.

**Edinborough, Edenborough, Edingborough, Edinbry, Edynbry:** Alexander *de Edynburgh, de Edenburg* 1233–55 Black; Thomas *of Edynburgh* 1396 ib. From Edinburgh.

**Edis, Ediss, Edhouse, Eddis, Eddowes:** *Eddusa* 1196 P (Wo); *Edus* 1197 ib., 1227 AssBk; *Edusa* 1196 P (Y), 1211 Cur (Gl), 1219 AssL; John *Edus* 1277 Ely (Sf); William *Eddows, Edhouse* 1744, 1749 WStowPR (Sf); Thomas *Eddis* 1744 Bardsley. This is one of a difficult series of names ending in *-us* or *-usa*, including *Sigus, Hacus* and *Ingus*, the last taken by Smith as a short form of ON *Ingirīðr* (PN NRY 261). All are feminine, as is *Aldus* (*Aldous*) which is taken in ODCN as masculine from OG *Aldo* 'old'. The only example given is *Aldus* (RH) which is certainly feminine for in RH we find also Hugo *filius Aldus* (not *Aldi*), and Hugo *filius Alduse*, i.e. 'son of *Aldusa* (f)'. cf. *Aldusa Vidua* 'the widow' 1212 Cur (L). Smith notes that one *Edus* is also called *Eadgifu*. *Edusa* soror Renburge 1198 Cur (Sf) is identical with *Ediz*, with alternative readings *Editha, Edit* (1200 Cur). cf. Ricardus *filius Eddiz* 1207 Cur (Nf). *Edus* is therefore a short form of a woman's name in *Ead-*, either OE *Eadgifu* or *Eadgyð* and *Aldus* similarly of a feminine name in *Eald-*, e.g. OE *Ealdgyð*.

**Edith, Eday, Edy, Edye:** Ralph *filius Edihe* 1188 BuryS (Sf); Everard *filius Edithe* 1210 Cur (C); Gerard *Edyth* 1279 RH (C); John *Idyth'* 1327 SRLei. OE *Eadgyð*, a fairly common OE name whose survival was probably due in part to the popularity of St *Eadgyð* (962–84). *Edith* remained in use throughout the Middle Ages, was rare from the 16th to the 18th century, but came back into fashion in the 19th century.

**Edkins, Eakin, Ekins:** *Edekin* Gomey 1279 RH (O); Joan *Edekin* ib.; Elena *Edkynes* 1327 SRSo; Mary *Eakyn* 1598 Bardsley. *Edekin*, a diminutive of *Eda*. *v.* EAD.

**Edland, Edlund:** William *de Edelond* 1279 RH (Beds); William *Edelune* 1351 AssEss. Perhaps from a lost *Ealdland* in Tillingham (Ess), *Edelond* 1337.

**Edleston:** Peter *de Edelauston* 1288 AssCh. From Edlaston (Derby) or Edleston (Ches). *v.* also ATHELSTAN.

**Edlin, Edling:** *Edelina* 1214 Cur (K); *Athelina* or *Edelina* 1221 AssWa; John *Edelin'* 1239 Rams (Nf);

Thomas *Edling* 1660 FrY. OG *Adelina* (Rom). *Edelina* (f) is due to the influence of OE names in *Æðel-* which became *Adel-, Edel-* in ME.

**Edlington:** Andrew *de Edlington'* 1195 P (L). From Edlington (L, WRY).

**Edlund:** *v.* EDLAND

**Edman, Edmans, Edmands, Admans:** *Ediman* Cumin, Canun (sic) 1295–7 P (Nth); William *Edemon, Edeman* 1275 SRWo, 1298 LoCt; William *Edmon* 1332 SRWa; William *Adman* 1644 FrY OE *\*Eadmann* 'prosperity-man'.

**Edmay:** Roger *Edemay* 1293 AssSt; Robert *Edemay* 1327 SRWo. 'Servant of *Eada*', from a short form of OE names in *Ead-*, and OE *mæge*.

**Edmead, Edmed, Edmett, Edmott, Edmeades, Edmeads, Eadmeades, Eadmeads:** Gervase *Eadmede*, Roger *Edemede* 1334 SRK; John *Edmed* 1485 CantW; William *Edmete*, Robert *Edmets* 1577, 1604 StaplehurstPR; John *Edmeads* c1700 ArchC xlix. OE *ēadmēde* 'humble'.

**Edmond, Edmonds, Edmons, Edmund, Edmunds:** *Ædmundus* presbyter, *Edmund(us)* 1066 DB; *Edmund* Wedertihand Hy 2 DC (L); Nicholaus *Edmundus* 1210 Cur (C); John *Edmond*, Sibil *Edmund* 1275 SRWo; William *Admond* 1349 Crowland (C). OE *Eadmund* 'prosperity-protector'.

**Edmondson, Edmundson, Edmenson, Edminson:** Robert *Edmondson* 1379 PTY; John *Edmundson* 1414 FrY; Richard *Edmonson* alias Jonson 1558 Shef (Y). 'Son of *Edmund*.'

**Edney, Edony, Ideny, Idony:** *Idony* 1273, *Ideny* 1644, *Edney* 1754 ODCN; Richard *Edney* 1576 SRW. Lat *Idonea, Idonia* (f), a not uncommon medieval name, the origin of which is unknown, these being the vernacular forms of the name.

**Edon:** *v.* EADEN

**Edrich, Edridge:** *Ædricus, Edric(us)* 1066 DB; *Edrich* Buck 1275 RH (Nf); *Edericke* le Blacke 1275 RH (L); Thomas *Hedricus* 1185 Templars (Gl); Robert *Edrich'* 1200 P (Ha); Robert *Eadric* 1221 Oseney (O); William *Ederyge* 1332 SRSt; Edmund *Edryk* 1381 SRSf. OE *Eadrīc* 'prosperity-powerful'.

**Edrupt:** *Hettrop* Crestien c1166 NthCh (Sa); *Aitrop, Attropus* de Merc 1178 P (Ess); *Eutropius* Alanus 1193 P (Gl); Richard *Edrop* 1332 SRLo. A rare surname from an uncommon personal name of Greek origin, *Εὐτρόπιος, from εὔτροπος* 'probatis moribus', one of proved character. Aythorpe Roding (Essex), *Aydroprothyng* 1351 PN Ess 491, owes its attribute to a 12th-century benefactor of the parish, *Aeitrop* or *Eutropius* son of Hugh, also named *Eitropus, Autropus, Eydrop, Eiltrop* (1206–20 Cur).

**Edsall, Edsell, Edscer, Edser:** William *de Egeshawe* 1332 SRSr; Robert *Eggyshaw* 1372 FFSr; John *Edsawe*, Thomas *Edsall* 1583 Musters (Sr); Robert *Edsurs* 1684 ib. From High Edser in Ewhurst (Surrey). *v.* PN Sr 239.

**Edsforth, Edsworth:** Stephen *de Eddeswurthe* 1275 RH (W), *de Edeworth* 1276 AssLo. From Edworth (Beds), *Edesworth* 1284.

**Edson:** *v.* EDESON

**Edward, Edwardes, Edwards:** *Eaduuardus, Eduuard(us), Æduuardus* 1066 DB; *Edwardus* serviens

1206 Cur (Co); William *Edward'* 1219 Cur (Sf); Cristina *Edwardis* 1279 RH (Hu); John *Edwards* 1498 Chirk. OE *Éadweard* 'prosperity-guard'.

**Edwardson:** William *Edwardson* 1518 KentW. 'Son of *Edward.*'

**Edwin, Edwing, Edwyn:** *Edwin'* Wridel 1066 Winton (Ha); Adam *filius Edwini* 1206 Cur (Nf); William *Edwin'* 1221 AssWa; Stephen *Edwyn* 1252 Rams (Beds). OE *Éadwine* 'prosperity-friend'.

**Edy, Edye:** v. EDITH

**Edynbry:** v. EDINBOROUGH

**Eele, Eeles, Eels, Eale, Eales, Eals:** (i) John *Ele* 1332 SRSr; Edward *Eles* 1554, *Elles* 1560 ChwWo. Perhaps a shortened form of OE names in *Ægel-, Æðel-,* or *Ealh-.* (ii) Laurences *del Eles* 1332 SRCu. From Eel Sike (Cu).

**Eeley:** v. ELY

**Effemy, Effeney, Effeny:** *Eufemia* 1188 BuryS (Sf), 1206 Cur (Gl); *Eufemmia* de Neville 1275 RH (L); Katerina *Eufemme* 1275 RH (Nf); Michael *Effemme* 1327 SRSf. Gk *Εὐφημία* 'auspicious speech', the name of a 4th-century Bithynian martyr who was canonized.

**Efford:** Roger *Efford* 1275 SRWo; Walter *Eford* 1332 SRSx; William *Efford* 1642 PrD. From Efford (Co, Ha), or from one or other of the four places of the name in Devon.

**Egbert:** *Egbert* 1086 DB; Martin *Egebrith* 1279 RH (C). OE *Ecgbeorht* 'sword-bright'. A rare surname.

**Eggar, Egger(s):** v. EDGAR

**Eggerton, Egerton:** David *Eggerton* 1282 AssCh; Joan *de Egerton'* 1327 SRSa; Ralph *Eggerton* 1662 PrGR. From Egerton (Ches, Kent), Egerton in Scammonden, or Edgerton in Huddersfield (WRYorks).

**Eggleston, Egglestone, Egleston:** Roger *de Egleston'* 1196 P (Du); Ralph *de Egliston* 1260 AssLa; John *Egleston, Egglestone* 1530, 1545 CorNt. From Eggleston (Durham), or Egglestone (NRYorks).

**Eggleton, Egleton, Eagleton:** Robert *de Egelton* 1225 FrLei; Simon *de Egilton* 1255 RH (Hu); Roger *de Egilton'* 1297 MinAcctCo. From Eggleton (He), or Egleton (R).

**Eglin, Eglen, Eaglen, Eagland, Eagling:** *Egelina* de Curtenay 1207 Cur (O); Alexander *Egelin* 1185 P (Sf); Peter *Eglyn* 1369 LLB G. *Egelina* (f) is probably a Norman form of OG *\*Agilina.*

**Eglinton, Eglington:** (i) Walter *de Egglinton'* 1204 P (Sa); John *de Egelinton'* 1206 Cur (Sx). From Eggleton (He), *Eglingtone* 1212. (ii) Brice *de Eglunstone* 1205 (Irvine), Rauf *de Eglyntone* 1296 (Ayr) Black. From the lands of Eglinton in Cuningham (Ayr). This is the usual source of the name.

**Egremont:** Richard *de Egremunt* 1200 P (Cu). From Egremont (Cu). Used as a christian name in the 17th century: *Egremont* Thynne 1610 AD vi.

**Eisdale, Eisdell:** v. ESDAILE

**Ekins:** v. EDKINS

**Ekless:** v. ECCLES

**Elam, Ellam, Ellams, Ellum:** John *Ellam* 1231 Pat (La); Henry *de Ellham* 1275 RH (Lo); Henry *de Elham*

1324 FFK; John *Elum* 1501 Pat (Y); Robert *Elam* 1744 FrY. From Elham (K), a lost *Elham* in Crayford (K), or Elam Grange in Bingley (WRY).

**Eland:** v. EALAND

**Elberry, Elbury:** John *Elburye,* Richard *Elberry* 1642 PrD. From Elberry in Churston Ferrers (D).

**Elce:** v. ELL

**Elcock, Elcocks, Elcoux, Ellcock, Ellicock:** *Elecocc* (son of Elias) 1246 AssLa; *Elkoc* Habraham 1297 Wak (Y); Roger *Hellecok* 1275 RH (Gl); Richard *Elcok* 1379 PTY. *Eli-coc,* a diminutive of *Elie* (Elias or Ellis). v. HELLCAT.

**Elcy:** v. ELSEY

**Elder, Elders:** Hugo *le Heldere* 1212 Cur (Herts); Ricardus *ye Elder* 1379 PTY. 'The elder, senior.'

**Elderidge:** v. ALDRICH

**Elderton:** v. ELLERTON

**Eldin, Eldon:** (i) Roger *de Elvedene* 1309 LLB D; Bartylmewe *Elden* 1561, Edward *Elden* 1596, *Eldinge* 1630 FrY. From Elveden (Sf), or Eldon (Du). (ii) Richard *Eldhyne* 1327 SRSf. 'The old servant', OE *eald,* ME *hine.*

**Eldred, Eldrid:** v. ALDRED

**Eldrett:** v. ALDRED, ADRITT

**Eldridge:** v. ALDRICH

**Elen:** v. ELLEN

**Elesender, Elshenar, Elshener, Elshender; Alshioner:** Roger *Elysandre* 1327 SRSo; John *Alshenour* 1596 Black; Katherine *Elshenour* 1597 ib.; Archibald *Alexshunder* or *Alschunder* 1596 ib.; Deacon *Elshender* 1840 ib. Scottish forms of *Alexander.*

**Eley:** v. ELY

**Elfick:** v. ALPHEGE

**Elford:** William *de Elleford* 1195 P (K); Thomas *de Eleford* 1291 FFO; Thomas *Elleford* 1410–11 FFWa. From Elford (Nb, St), or Yelverton in Buckland Monachorum (D), *Elleford* 1291.

**Elgar, Elger:** v. ALGAR

**Elgee, Elgy:** Walter *Elgy* 1269 AssNb; John *Elgy* 1296 SRNb. Pet forms of *Elgar.* v. ALGAR.

**Elgood:** *Helgot* 1086 DB (Bk, D, St); *Helgod* ib. (Sa); Philippus *filius Helgot* 1185 P (St); John *Elgood* 1524 SRSf. OFr *Helgot,* OG *Helgaud,* from OG *Hildegaud* or OFr *Eligaud.*

**Elias, Eliasson:** v. ELLIS

**Elin:** v. ELION, ELLEN

**Elingham:** v. ELLINGHAM

**Elinson:** Richard *Eleyneson* 1327 SRWo; Thomas *Elenessone* 1359 AssCh; John *Elynson* 1379 PTY. 'Son of *Ellen.*' v. also ELLISON.

**Elion, Elin, Helin, Hellen, Hellin, Hellins, Helling, Hellings, Hellon, Hillen, Hilling, Hillings:** Robert *de Helion* 1190 P (Ess); William *de Elyon* 1200 P (D); John *Heline* 1251 Rams (Hu); Robert *de Helun'* 13th Seals (Sf); John *Helon* 1524 SRSf; Edward *Hellynge* 1568 SRSf. Helion Bumpstead (Essex), recorded as *Bumpstede Elyns* (1544), *Heleyns* (1554), *Hellyn* (1570) and locally known as *Helen's* Bumpstead, was held in 1086 (DB) by Tihel *de Herion,* called also Tihel *Brito* 'the Breton', who came from Helléan (Morbihan). The family also had land in Haverhill (Suffolk) where Andrew *de Helin'* held a fee in 1275

(RH) and John *Elioun* half a fee in 1346 (FA). Bumpstead and Haverhill are near the Essex-Suffolk boundary. *Elin, Elion* and *Hillen* are rare surnames still found in Essex and Suffolk. *Helin*, also rare, survives in London. The same family has also left its name in Upton Hellions (Devon) where the attribute occurs as *Hyliun* (1270), *Hylon* (1385) and *Helling* (1557). *Hellings, Hellins* and *Hilling*, all rare, are found today in Somerset. *v.* PN Ess 508–9, PN D 419. Some of these surnames have alternative origins. *v.* ELLEN, HELLING, HILLING.

**Eliot, Eliott, Elliot, Elliott:** (i) *Heliot* de Slohebi 12th DC (L); *Elyot* 1188 BuryS (Sf); William *Elyot* 1257 AssSo; William *Eliot* 1327 SRSx. A diminutive of *Elias*, from OFr *Élie* and *-ot*. The frequency of these surnames is due partly to the absorption of OE *Æðelgēat* (m) and *Æðelgȳð* (f). *v.* AYLETT. (ii) According to Black, the Scottish *Eliot*, with the same four modern spellings, appears regularly as *Elwald* or *Elwold* until the end of the 15th century. This is from OE *Ælfweald* 'elf-ruler'. cf. *Elewald* 1279 CalSc (Cu). From the 16th century, there is a bewildering variety of forms, some of which are difficult to account for. The development, in general, seems to have been: *Elwaud, Elwat, Eluat, Eluott, Elioat, Eliot*, with variants surviving as ELLWAND, ELLWOOD.

**Elis:** *v.* ELLIS

**Eliston:** Nicholas *Elestones* 1444 FrY. From Elston (La, Nt).

**Elivant:** Richard *Elyvant* Hy 3, *Elyvaunt* 1275 IpmGl. ME *elefant* 'leper'.

**Elke, Elkes, Elkies:** William *le Elk* 1279 RH (C). Apparently a nickname from the elk, ME *elk*.

**Elkin, Elkins:** *Elekin* 1279 RH (O); John *Elekyn* 1310 FFEss; Thomas *Elkyns* 1447 Balliol (O). *Eli-kin*, a diminutive of *Elie* (Elias), or of *Ela* (Ellen).

**Ell, Ells, Else, Elce:** *Ala* 1208 Cur (Sx); *Ela* 1208 Cur (Ess), 1214 ib. (Nf); Roger, William *Elle* 1221 ElyA, 1327 SRSo. *Ala* is a variant of *Ela*, both forms being used of the same woman: *Ala* or *Ela* de Eching(e)ham 1207 Cur (Sx); *Ela* de Marci (1207, 1210 Cur) is identical with *Ala* de Saukevill', widow of William de Marcy (1207 ib.). *Ela* is a variant reading of *Elena*, widow of Edwin de Tuggel' 1221 Cur (Sr) whilst *Elena*, wife of the Earl of Salisbury, is also called *Ela* and once *Eda*, the latter being indexed as *Ela* (1212–22 Cur). *Ela* is thus a short form of *Elena* and possibly also for *Eleanor*, the less common *Ala* being due to the influence of *Alienora*.

**Ellaby:** *v.* ELLERBY

**Ellam, Ellams:** *v.* ELAM

**Ellarby:** *v.* ELLERBY

**Ellard:** *v.* ADLARD

**Ellcock:** *v.* ELCOCK

**Elldred:** *v.* ALDRED

**Ellen, Ellens, Elleyne, Ellin, Ellins, Ellings, Elen, Elin, Hellen, Hellens, Hellin:** (i) *Elena* ostiaria 1204 P (So); *Elena* . . . ipsam *Helenam* 1219 AssY; *Helena* de Sutton 1210 Cur (Nth); Roger *Heleyne* 13th StJohn (Ess); Robert *Helene* 1275 SRWo; Walter *Eleyn* 1279 RH (O); Ralph *Elene* 1314 FFEss; Robert *Elyn* 1327 SRSf; William *Helyns* 1332 SRWo. Ellen (f) is the earlier English form of *Helen*, Greek Ἑλένη, feminine

of Ἕλενος 'the bright one'. The popularity of the name is due to St Helena, mother of the Emperor Constantine, said to have been the daughter of a British king. (ii) Nicholas *in the Ellene* 1327 SRSo; Robert *atte Hellene, atte Ellene* 1327, 1332, SRSx. 'Dweller by the alder(s)', OE *ellen*.

**Ellenor, Ellinor:** *Alienor* c1202 NthCh (Nth), 1211 Cur (Beds); Richard *Elyanor* 1327 SRSf; Richard *Elynoreson* 1375 AD vi (Lei); Elizabeth *Ellener* 1674 HTSf. Provençal *Aliénor*, a form of *Helen*, introduced into England by *Eleanor* of Aquitaine (1122–1204), wife of Henry II.

**Elleray:** *v.* HILLARY

**Ellerbeck:** Matthew *de Ellerbeck* 1276 RH (Y); Thomas *de Ellerbek* 1360 FFY; William *of Ellerbek* 1401 AssLa. From Ellerbeck in Osmotherley (NRY), or Ellerbeck in Ingleton (WRY).

**Ellerby, Ellaby, Ellarby:** Nicholas *de Ellerby* 1385 KB; William *de Ellerby* 1410 IpmY; Miles *Ellerby* 1672 HTY. From Ellerby in Swine (ERY), or Ellerby in Lythe (NRY).

**Ellerington:** *v.* ELRINGTON

**Ellerker, Elliker, Allaker, Alliker:** Denis *de Elreker* 1204 AssY. From Ellerker (ERYorks), *Allerker* 1180–95 PN ERY 222.

**Ellerman:** *v.* ELLIMAN

**Ellers:** William *de Elleres* 1247 IpmY; John *Ellers* 1401 FFEss. From High, Low Ellers in Cantley (WRY).

**Ellerson:** Robert *Ellerson* 1672 HTY; George *Ellerson* 1709 FrY. Late forms of ELLISON.

**Ellert:** *v.* ADLARD

**Ellerton, Elderton:** Ralph *de Elreton* 1204 AssY; Roger *de Ethelartone* 1307 Ronton (St). From Ellerton (ERYorks, Salop).

**Ellery:** *v.* HILLARY

**Elles:** *v.* ELL, ELLIS

**Ellesmere, Ellsmore, Elsmore:** David *de Ellysmere* 1377 AssWa. From Ellesmere (Sa).

**Elletson:** John *Elotson* 1332 SRCu. 'Son of *Elot*.' *v.* ELLETT.

**Ellett:** *Ellot* 1332 SRLa, SRCu; *Ellota* 1379 PTY; William *Elot* 1323 AssSt; Thomas *Elote* 1327 SRSf; Rychard *Ellet* 1568 SRSf. A diminutive of *Ellen*.

**Elleyne:** *v.* ELLEN

**Ellice:** *v.* ELLIS

**Ellick:** *v.* ELWICK

**Ellicock:** *v.* ELCOCK

**Elliff:** *v.* AYLIFF

**Elliker:** *v.* ELLERKER

**Elliman, Ellerman, Elman:** Thomas *Elyman* 1377 MESO (Lei); Etheldreda *Elyman* 1381 SRSf. OE *æle* 'oil' and *mann*, 'oil-man', seller of oil. cf. Roger *le Elymaker* 1344 MESO (So).

**Ellin, Ellings, Ellins:** *v.* ELLEN

**Ellington, Elington:** William *de Elinton'* 1206 P1 (Hu); John *de Ellington* 1274 RH (L); John *Elyngton* 1576 SRW. From Ellington (Hunts, Kent, Northumb, NRYorks).

**Ellingham, Elingham:** Geoffrey *de Elingeham* 1191 P (Nf); Gilbert *de Elingham* 1287–8 NorwLt; John *Elyngeham* 1400 AssLa. From Ellingham (Ha, Nb, Nf), or Great, Little Ellingham (Nf).

**Elliot:** v. ELIOT

**Ellis, Elliss, Ellice, Elles, Elis, Elys, Heelis, Helis, Hellis, Elias:** *Helias* scriptor c1150 DC (L); *Elyas* de Westone c1160 ib.; *Heliseus* de Brunne c1175 ib., *Helyas, Elyas* de Brunna Hy 2 ib.; *Willelmus filius Helis* 1212 Cur (So); *Elis* de Adham 1220 Cur (Mx); *Elice* de Cheindue 1224 AssSt; William *Elyas* 1200 P (Y); William *Elis* 1202 P (L); Roger *Elys, Helys* Hy 3 HPD (Ess); Andrew *Elice* 1309 SRBeds. ME *Elis* 'Elias', the Greek form of the Hebrew *Elijah*. John *Elys* was the son of *Elias* de Bampton (1318, 1331 Husting). *Elias* is not common as a surname. v. also ELY. *Ellice* may also be a pet-name for Elizabeth. v. 1319 SRLo 278.

**Ellison:** (i) Rogier *Elyssone* 1296 Black (Berwick); Adam *Elisson* 1379 PTY; John *Ellyson* 1487 FrY. 'Son of *Ellis*.' v. also ALISON. (ii) Adam *filius Elysant* 1190 P (He); *Helisent* de Twining' 1221 AssGl; Henry *Elesant* (*Elesand*) 1327 *SR* (Ess). OFr *Elissent, Elisant* (f), OG *Elisind* (f).

**Elloway:** v. ALAWAY

**Ellsmore:** v. ELLESMERE

**Ellstead, Elstead, Elsted:** Robert *de Ellested'* 1219 P (Sx). From Elstead (Sr), or Elsted (Sx).

**Elliston:** v. ELSTON

**Ellsworth:** v. ELSWORTH

**Ellum:** v. ELAM

**Ellwand:** *Elwand* is found in 1502 and later for *Elwald*, now the Scottish ELIOT.

**Ellwood, Elwood, Allwood:** *Aluuoldus* episcopus 1066 DB (Do); *Aluuolt* (*Alfuuold*) 1066 InqEl (Sf); Arkillus *filius Aluoldi* 1131 FeuDu; *Alfwold* presbiter 1212 Fees (Berks); Robert *Elwald, Elwaud,* 1469, 1579 FrY; William *Elwold* 1524 SRSf; Thomas *Elwod* 1572 FrY. OE *Ælfweald* 'elf-ruler'. In Scotland this has become ELIOT.

**Ellworth, Elworth:** William *de Elleworth* 1342 FFW. From Elworth (Ch, Do).

**Ellyatt:** v. AYLETT

**Elm, Elmes, Elms, Nelmes, Nelms:** John *atte Elme* 1316 Ipm (Wo); Stephen *ate Nelme* 1317 AssK; John *atte Elmes* 1322 MELS (Sr); Thomas *Elm* 1327 SRSf; Semannus *atte Nelmes* 1339 Cl (Ess); William *ate Thelmes* 1356 HPD (Ess). 'Dweller by the elm(s).' v. also HELM.

**Elman:** v. ELLIMAN

**Elmar, Elmer, Elmers:** v. AYLMER

**Elmley:** Walter *de Elmeleye* 1301 CorLo. From Elmley (K), or Elmley Castle, Lovett (Wo). The name may also have fallen in with EMLEY.

**Elmslie, Elmsleigh, Elmsley, Elmsly, Emslie, Emsley:** William *de Elmysley* 1333 (Aberdeen), James *Emslie* 1715 Black. From some, as yet, unidentified place in England.

**Elmswell:** John *Elmyswell* 1379 LoCh. From Elmswell (Sf).

**Elnaugh, Elnough:** v. ALLNATT

**Elphee:** v. ALLVEY

**Elphick:** v. ALPHEGE, ELVEY

**Elphinston, Elphinstone:** John *de Elphinstone, de Elphinston* c1250, a1340 Black; Alexander, William *Elphynston* 1486 ib. From the lands of Elphinstone in Tranent (Midlothian).

**Elray:** v. HILLARY

**Elrick:** v. ALDRICH

**Elrington, Ellerington:** John *Elrynton* 1474 ArchC 41; Edward *Elryngton* 1537 FFEss. From Elrington (Nb).

**Elsam:** v. ELSOM

**Elsden, Elsdon:** Edmund *Elsden* 1674 HTSf. From Elsdon (Nb).

**Else:** v. ELL

**Elsey, Elsie, Elcy:** *Alfsi* 1066 DB (Sf); *Alsi* ib. (Ha); Robertus *filius Elfsi* 1191 P (K); *Elsy* faber 1230 P (L); Reginald, John *Elsi* 1155–77 Templars (Y), 1275 RH (Nf). OE *Ælfsige* 'elf-victory'.

**Elshenar, Elshener, Elshender:** v. ELESENDER

**Elsip:** v. ALLSEP

**Elsley:** John *de Eltesley* 1310 LLB D; William *Elsley* 1674 FrY. From Eltisley (Cambs).

**Elsmore:** v. ELLESMERE

**Elsom, Elsome, Elsam:** Thomas *Elsam* 1413 FrY. Probably from Elsham (L).

**Elson:** Richard *Elson* 1379 PTY; William *Elson* 1524 SRSf. From Elson (Ha, Sa), or perhaps a late form of ELSTON.

**Elsop:** v. ALLSEP

**Elstead, Elsted:** v. ELLSTEAD

**Elstob, Elstub:** Philip *de Ellestob*, William *de Ellestobe* 1235–6, Philip *de Ellestob* 1269 AssDu. From Elstob (Du).

**Elston, Elstone, Elliston:** (i) Galfridus *filius Elstan* 1193 P (Nf); Warinus *filius Elstani* 1230 P (Nb); Henry *Elstan* 1279 RH (O); John *Elston* 1411 FrY. OE *Ealdstān* or *Ealhstān*. cf. ALLSTON. (ii) William *de Ethelistone* 1332 SRLa. From Elston (Notts, Lancs, Wilts).

**Elsworth, Ellsworth:** William *de Ellesworde* 1230 P (C); Alice *de Eliswurth'* 1258 MPleas (Hu). From Elsworth (C).

**Eltham, Eltome:** Adelold *de Elteham* 1066 DB (K); Richard *de Eltham* 1259 Acc; Simon *de Eltham* 1310 LLB D. From Eltham (K).

**Elton:** Thomas *de Eleton* 1230 P (Berks); William *de Elton'* 1327 SRLei; James *Elton* 1434–5 IpmNt. From Elton (Berks, Ches, Derby, Durham, Hereford, Hunts, Lancs, Notts).

**Elvar, Elver:** v. ALVAR

**Elven:** v. ALVEN

**Elverson:** Probably 'Ælfred's son'. cf. William *Elvered* alias *Elluerd* 1696 DKR 41 (W).

**Elvery:** v. ALFREY

**Elvet, Elvett:** John *Eluet* 1396–7 FFWa; William *Elvete* 1401 AssLa. From Elvet Hall (Du), or a nickname from the swan, OE *ylfete*.

**Elvey, Elvy:** *Elviva* 1325 NorwDeeds; Richard *Elvy* 1338 FFY; John *Elphy* 1450 ArchC v; Thomas *Elveve*, John *Elvew* 1488, 1518 CantW. OE *Ælfgifu* (f) 'elf-gift'. v. also ALLVEY.

**Elvidge:** v. ALPHEGE

**Elvington:** John *Elvington* 1459 Kirk. From Elvington (ERY).

**Elvin(s):** v. ALVEN

**Elward:** v. ALLWARD

**Elwes, Elwess:** *Helewis* 1086 DB (Nf); *Heilewisa Extranea* 1160 P (Gl); *Elewisa* uidua 1221 *ElyA* (Sf);

John *Helewis* 1274 RH (Ess); William *Helewys* 1297 MinAcctCo; Thomas *Ellwes* 1625 FrY. OFr *Heluïs, Heloïs*, from OG *Heilwidis, Helewidis* (f) 'hale or sound-wide'.

**Elwick, Ellick:** Ralph *Elwyke* 1512 FrY; John *Ellyk* 1569 ib. From Elwick (Durham, Northumb).

**Elwin, Elwyn:** *v.* ALWIN

**Elwood:** *v.* ELLWOOD

**Elworth:** *v.* ELLWORTH

**Elworthy:** Dimon *de Elleworde* 1201 P (So); Christopher *Elworthie*, David *Elworthy* 1642 PrD. From Elworthy (So), *Elleworthe* 1166.

**Ely, Ealey, Eeley, Eley, Hely, Heley:** (i)*Hely* de Amandeuilla 1150–60 Gilb (L); Philippus *filius Helie* 1213 Cur (So); William *Heli* Hy 2 DC (Nt); John *Elye* 1327 SRC. OFr *Elie* 'Elijah'. *v.* ELLIS. (ii) Huna *de Ely* 1086 ICC (C); Geoffrey *de heli* 1133 Bart (Lo). From Ely (Cambs).

**Elys:** *v.* ELLIS

**Eman:** *v.* EAMOND

**Emans:** *v.* EMENEY

**Emberson:** *v.* EMMERSON

**Emberton:** William *de Embertun'* 1211 Cur (Bk). From Emberton (Bk).

**Embery, Embra, Hembrow, Hembra, Hembry, Hembury:** These Somerset surnames may all derive from Emborough (Som): William *de Enneberg* 1258 AssSo. Or from Hembury (Devon) or Henbury (Dorset): John *Hembury* 1491 PCC (D/Do). *v.* also AMERY.

**Emblem, Emblin, Embling, Emeline, Emlyn, Amblin:** Anschitil *filius Ameline* 1066 DB (Do); Godefridus *filius Emeline* 1115 Winton (Ha); *Emmelina* Hy 2 DC (L); *Emelina* ib. (Db), 1221 Cur (O); *Amelina* 1221 AssSa; John *Emelin* 1208 ChR (Sf); Geoffrey *Amelyn* 1296 SRSx. OFr *Ameline, Emmeline*, a hypocoristic formation of OG names in *Adal-*.

**Embleton:** Roger *de Emilton'* 1230 P (Du); Richard *de Emeldon* 1326 FeuDu; Norman *de Embleton* 1332 SRCu. From Embleton (Cu, Du, Nb).

**Embley:** *v.* EMLEY

**Emblott, Emblot:** Richard *Emelote* 1305 AssW. *Em-el-ot*, a double diminutive of *Em*, a pet-form of *Emeline*.

**Embra, Embr(e)y, Embury:** *v.* AMERY, EMBERY

**Emeline:** *v.* EMBLEM

**Emeney, Emeny, Emmony, Emney, Emmans, Emmence, Emmens, Emmins, Emmons, Emans, Emons, Immink:** *Ysmeine* de Cherchefeld 1199 P (R); *Ismenia, Ysmeina, Hismena* 1206–12 Cur (Sf, Nt); Rogerus *filius Immine* 1219 AssY; *Emoni* Turberd 1275 RH (Y); Hugh *Imayn* 1276 RH (Lei); John *Ymanie* 1279 RH (K); *Imanie* Spring-old 1279 RH (C); *Hemin'* uxor Bercarii ib.; *Imaigne* de Rothynge 1319 SRLo; Roger *Emaygne* 1352 ColchCt; Nicholas *Eman* 1524 SRSf; William *Immings* 1658 Bardsley; Francis *Immynes* 1707 ib.; William *Emmines* 1717 ib.; Ann *Emens* 1783 ib. *Isemeine* (1227 AssBk) is also called *Isemay* but the two names can hardly be identical. *Ismaine, Ismenia*, a common woman's name, found as *Emonie, Emeny* until the end of the 18th century, is a difficult name. In ODCN a Celtic origin is suggested. It may be that we have an OG

compound *Ismagin* 'iron-strength'; both elements are common though *-magin* is not known as a second element and *Ismeine* seems to be unknown in France.

**Emeric(k), Emerig:** *v.* AMERY

**Emerley:** *v.* EMLEY

**Emerson:** *v.* EMMERSON

**Emery:** *v.* AMERY

**Emeryson:** William *Emeryson* 1411 Finchale; George *Emeryson* 1506–7 FFSr. 'Son of *Emery*'. *v.* AMERY.

**Emes:** *v.* EAMES

**Emett:** *v.* EMMATT

**Eminson:** *v.* EMMISON

**Emley, Emly, Embley, Emerley:** John *de Emlay* 1304 FrY. From Emley (WRYorks).

**Emlyn:** *v.* EMBLEM

**Emm, Emms:** *Ema* c1160 DC (Lei); *Emma* 1187 DC (L), 1219 AssY; *Emm* in the Hurn 1327 SRSo; William *Emms* 1274 RH (Sa); John *Emme* 1279 RH (O). *Emm* was the English form of the popular Norman *Emma* (OG *Emma, Imma*). *v.* EMMATT.

**Emmans:** *v.* EMENEY, EMMINGS

**Emmatt, Emmet, Emmett, Emmitt, Emmott, Emmert, Emett, Hemmett:** *Emmote* 1279 RH (Bk); *Emmota* 1327 SRSo, SRY; *Emmete* de Fur' 1279 RH (C); John *Emote* 1327 SRWo; Ranulph *Emmot*, William *Emmoten* 1332 SRWa. *Emmot* was a very common pet-name for *Emma*. The names were interchangeable: *Emma* wife of Adam faber is called also *Emmota* 1353 AD vi (Mx); *Emma* alias *Emmote* Bochere 1413 AD iii (K). *Emmott* may also be from *Emmot* (Lancs): William *de Emot* 1324 LaCt.

**Emmens:** *v.* EAMOND

**Emmerig:** *v.* AMERY

**Emmerson, Emberson, Emerson:** William *Emeryson* 1411 Finchale; Cuthbert *Emerson* 1498 FrY. 'Son of *Emery*.' *v.* AMERY, EMPSON.

**Emmert:** *v.* EMMATT

**Emmings:** Thomas *Emmyng*, John *Hemmyng* 1296 SRSx; John *Emings* 1674 HTSf. 'Son of *Hemming*', and in later examples sometimes for *Emans*, etc. *v.* HEMMING.

**Emmison:** William *Emmotson* 1381 PRY; Robert *Emmysson* 1479 NorwW (Nf); John *Emyson* 1489 FrY. 'Son of *Emmott*.' Also 'Son of *Eman* or *Emeny*'; Robert *Emanson* 1594 SPD, surviving as EMINSON; and for EMMERSON. *v.* EMPSON. Also variant of EAMONDSON.

**Emney:** *v.* EMENEY

**Emonson:** Christopher *Emondson* 1507 FrY. Probably for *Edmondson*.

**Emory:** *v.* AMERY

**Emperor:** Gilbert *Lempereur* 1225 Pat; Richard *le Empereur* 1230 Cur (K). OFr *empereur* 'emperor'. Probably a pageant name.

**Empsall, Emsall:** Richard *Emsall* 1631 RothwellPR (Y). From North, South Elmsall in South Kirkby (WRY).

**Empson:** Richard *Empson* 1498 Chirk, 1510 Butley (Sf). For EMSON. Also for EMMERSON: Richard *Emryson, Emereson, Empson* 1490–5 Bardsley.

**Emsley, Emslie:** *v.* ELMSLIE

**Emson, Hemson:** Richard *Emmeson* 1327 SRSx; Robert *Emson* 1381 PTY. 'Son of *Emma*.' Also for

*Emmerson*, an intermediate stage between *Emereson* and *Empson*.

**End, Ende:** William *atten Hend* 1305 AssW. 'Dweller at the end (of the village)', OE *ende*.

**Enderby:** Robert *de Enderbi* 1170–98 P (L); Thomas *de Enderby* 1298 AssL; Robert *Enderby* 1384 AssWa. From Enderby (Leics, Lincs).

**Enderson:** v. ANDERSON

**Enderwick:** v. INDERWICK

**Endicott, Endecott, Enticott:** Abraham *Endecote*, Nicholas *Endecott* 1642 PrD. From Endicott in Cadbury (D). 'Dweller at the end cottage', or 'beyond the cottages', OE *ende, cot*.

**Enfield, Envill:** William *de Enefeld'* 1190 P (Lo). From Enfield (Middlesex).

**Engeham:** v. GAIN

**Engelbert:** v. INGLEBRIGHT

**Engineer:** William *Lenginnur* 1202 Pleas (Sf); John *le Enguigniur* 1221 AssWo; William *le Enginur* 1251–2 FFWa. 'The engineer', OFr *engineor*.

**England:** Nicholas *de Engelond* 1260 AssC; William *de Engelond* 1295 AssCh; John *Ingelond* (*Engelond*) 1327 *SR* (Ess). There is no authority for the *ing-land* 'meadowland' of Bardsley and Weekley. The reference must be to the name of the country, a surname which appears curiously out-of-place in England. v. ENGLISH.

**Engledew, Engledow:** Henry *Angel Dei* 1275 RH (L). Lat *angelus dei*, OFr *angele, Dieu* 'angel of God'.

**Englefield:** William *de Engelfeld* 1211 Cur (Berks); William *de Englefeld* 1251–2 FFSx; Philip *de Englefeld* 1342–3 FFWa. From Englefield (Berks, Sr). But the name probably usually became INGLEFIELD.

**English, Inglis, Inglish:** Gillebertus *Anglicus* 1171 P (He); Robert *le Engleis* 12th DC (Lei); Adam *le Englis* 1194–1211 Black (Cumb); William *le Engles* 1205 Oseney (O); Gilbert *Engleis* 1208 Cur (Ess); Nicholas *le Engleys* 1269 AssNb; John (*le*) *Englisshe* 1317 AssK; Willelmus dictus *Ingles* a1321 Black (Moffat); John *Inglis* 1402 ib. (Aberdeen); Alexander *Inglyssh* 1478 ib. OE *Englisc* 'English', originally referring to Angles as distinct from Saxons, a meaning not to be considered for the surname. *Inglis* is a Scottish form denoting an Englishman as opposed to the Scottish borderer or the Celtic Scot, whilst the northern *English* probably referred to an Englishman living among Strathclyde Welsh. But the name was not confined to this district. In the Welsh border counties the name would be given to an Englishman in a preponderatingly Welsh community, in the Danelaw to a newcomer who did not share the common Danish origin with its flourishing Danish customs and names. For some generations after the Conquest an official distinction was made between *Angli* and *Franci*, the native, defeated English and the conquering Normans, and this may account for the name in Essex, Kent and Sussex, where it was probably at first derogatory. At the end of the 13th century *l'Englois* is found as a surname in Paris and this, given by Frenchmen in France, may well have been retained when the emigrant returned home. Baudet *le Engleis* (1311 LLB D) was the valet of John Hauekyn, merchant of St Quentin. v. also INGLIS.

**Ennet, Ennett:** *Enota* Rened 1306 IpmGl; William *filius Enota* 1332 SRCu. *En-ot*, the diminutive of a feminine personal name. But the surname of Geoffrey *Ened* 1332 SRCu is probably a nickname from OE *ened* 'duck'.

**Enoch, Enock:** *Enoc* 1148 Winton (Ha); Robert *filius Enoc* 1182 P (Co); Richard *Ennok* 1255 RH (W); John *Enoce* 1327 SRSf; Walter *Ennock* 1329 IpmW. Hebrew *Enoch*.

**Ennion, Enion, Eynon, Inions, Anyan, Anyon, Onians, Onion, Onions, O'Nions, Onyon, Hennion:** *Ennian* filius Gieruerð 1159 P (Sa); *Ennion de Caple* 1205 Cur (He); *Eynon, Eynun* 1221 AssSa; *Eignon* 1287 AssCh; *Anian* bishop of Bangor 1284 Ch; Gruffydd ap Madog *Vnyon* 1392 Chirk; Gode *heynon* 1221 *ElyA* (Sf); William *Anyun* 1279 RH (Bk); Andrew *Heizhnon* 1327 *SR* (Ess); John *Eynon* 1327 SRWo; Richard *Anyon* 1512 AD iii (Ch); Robert *Onnyon* 1568 SRSf; William *Ineon* 1593 Bardsley; Hugh *Inniones* 1622 ib. OW *Enniaun*, an old Welsh name, ultimately from Lat *Anniānus*, also associated with Welsh *einion* 'anvil' for 'stability, fortitude' and, doubtfully, with *uniawn* 'upright, just'.

**Ensoll:** v. INSALL

**Ensom, Ensum:** Robert *de Eygnesham* 1260 Oseney (O). From Eynsham (Oxon).

**Ensor:** Adam *de Ednesovere* 1247 AssSt. From Edensor (Derby).

**Enston, Enstone:** Richard *de Ennestan'* 1279 RH (O). From Enstone (O).

**Enticknap:** John *de Anekecnappe* 13th, Thomas *de Anteknappe* 1332 PN Sr 235; George *Enticknap* 1696 DKR 41. From Enticknaps Copse (Surrey).

**Entwistle, Entwisle, Entwhistle:** Elias, Roger *le Ennetwysel* 1276 AssLa, Misc (La); John *Entwisell* 1332 SRLa. From Entwisle (Lancs).

**Enticott:** v. ENDICOTT

**Envill:** v. ENFIELD

**Envious, Envis, Enviss:** Jordan *le Envaise* 1163–9 MCh; Hamo *le Enueise* 1211 FFNf; Walter *Len Veyse* 1355–9 AssBeds. OFr *envious* 'envious, jealous'.

**Epps:** *Eppa* mulier 13th, *Eppe* c1250 Rams (Hu); Roger *Eppe* 1275 RH (Nf). OE *Eoppa*, or ODa *Øpi*. cf. also Epps Fm in Bentley (Wa) from John *Hebbe* 1327.

**Epsley:** Stephen *de Epsle* 1332 SRSx. From Apsley Fm in Thakeham (Sx), *Epseley* 1316.

**Epworth:** Robert *de Epworth* 1347 FrY; William *de Eppeworthe* 1357, Richard *Ippeworthe* 1412 *ERO*. From Epworth (L), but sometimes, perhaps, for HEPWORTH.

**Erdington:** Thomas *de Erdington* 1194 StCh; Thomas *de Erdinton'* 1218 P (Sa). From Eardington (Sa), or Erdington (Wa).

**Erett:** v. ERRETT

**Eridge:** v. ERRIDGE

**Erkenbald, Erkenbold:** *Erkenbaldus* le Messer 1180 P (L); *Erkenbaldus* Wesdier 1191 P (Y); Robert *Erkenbald* 1204 P (Gl); Walter *Erkenbald* 1212 P (Ha). OG *Erconbald*.

**Erlam:** Ralph *de Erlham* 1275 RH (Nf). From Earlham (Nf).

**Erleigh, Erley:** v. EARLEY

**Erlick:** John *Erlyche* 1514 LP (C). A nickname from OE *ǣrlīce* 'early'.

**Ermen:** v. ARMIN

**Erne:** Martin *Ern* 1312–13 NorwLt. A nickname from OE *earn* 'eagle'.

**Ernest, Harness:** (i) Hamo *filius Erneis* 1115 Winton (Ha); *Herneis* c1127 AC (Gl); *Hernis* de Neuella, *Arnisius* de Neuille 1154–72 Gilb (L); *Ernis* de Besebia 12th DC (L); Geoffrey *Ernis* 1207 P (Ha); Roger *Arneis* 1235 FFEss; Robert *Herneys* 13th WhC (La); Charles *Earnest* 1758 FrY. OG *Arn(e)gis*, OFr *Erneïs, Ernaïs, Hernaïs, Hernays.* (ii) Philip *Harneis* 1285 FFEss; John *Harneys* 1310 FFSf. OFr *harneis*, ME *harnais* 'harness', for a maker of harness or suits of mail. cf. William *le Hernesemaker* 1300 LoCt, William *Duble Harneys*, saddler 1276 LLB A.

**Erpingham:** Thomas *Erpyngham* 1426–7 Paston. From Erpingham (Nf).

**Errel:** v. ERROL

**Errett, Erett:** Gamel *filius Edredi* 1131 FeuDu; *Edredus* Cuperius 1220 Fees (Berks); Maurice *Edred* 1226 Pat; Thomas *Edret* 1276 RH (Berks). OE *Ēadrǣd* 'prosperity-counsel'.

**Erricker:** v. EARWACKER

**Erridge, Eridge:** William *de Eyrugge* 1327, *de Eregg'* 1332 SRSx. From Eridge in Frant (Sx).

**Errington:** John *Erington* 1672 HTY; Edward *Errington* 1674 HTSf. From Errington (Nb).

**Errol, Erroll, Errel:** Roger *Euerolf* 1327 SRSf; William *Eryll* 1545 SRW; Duncan *Errol* 1574, Thomas *Errole* 1592 Black. In England from OE *Eoforwulf*, in Scotland from Errol (Perth).

**Erskine:** Henry *de Erskyn* 1225, John *de Ireskyn* c1280–90, George *Erskine* or *Askine* 1666 Black. From the barony of Erskine (Renfrew).

**Ervin(g):** v. IRVIN

**Erwin, Everwin, Irwin, Irwine, Irwing, Urwin:** *Eueruinus* 1066 Winton (Ha), 1086 DB (Nf); Gilchrist *filius Eruini* 1124–65 Black; Willelmus *filius Irwine* 1185 Templars (Ess); Augustinus *filius Erwin'* 1255 RH (Sa); Eustace *Everwyn* 1310 LLB D; Thomas *Erwyn* 1459 FrY. OE *Eoforwine* 'boar-friend'. There has been confusion with IRVIN.

**Escott:** v. EASTCOTT

**Escreet, Escritt, Eskriett:** William *de Escrik* 1307 FrY. From Escrick (ERYorks).

**Escuyer:** v. SQUIRE

**Esdaile, Esdale, Easdale, Eisdal, Eisdell, Isdale, Isdell:** John *de Esdale* 1413 Black; Margaret *Eskdale* 1472 ib.; James *Esdaill* 1493 ib.; John *Aisdaill* 1599 ib.; John *Isdaill* 1669 ib. From Eskdale (Dumfriesshire).

**Esgar, Esger:** *Esgar* (Berks, So), *Esgarus* (Ess) 1066 DB; Godwin *filius Esgari* c1130 ELPN; John *Esgar*, Adam *Esgor* 1235 FFEss; William *Esgar* 1332 SRWo. ON *Ásgeirr*, ODa *Esger*.

**Eshelby:** v. AXLEBY

**Esherwood:** v. ISHERWOOD

**Eskell:** v. ASKELL

**Eskriett:** v. ESCREET

**Esling:** William *de Eslynge* 1296 SRSx. From Elsing (Nf), with metathesis of *l-s* to *s-l*.

**Esmond(e):** v. EASTMAN

**Essex:** Suein *de Essexa* 1114–16 CartAntiq; William *de Essex* 1246–7 FFWa; John *Essex* 1340–1450 GildC. 'The man from Essex.'

**Essie, Essy:** v. EASEY

**Esson:** Sometimes for EASTON. v. EASON.

**Estall, Estell:** v. EASTALL

**Estaugh:** v. EASTAUGH

**Estcourt:** V. EASTCOTT

**Esterbrook:** v. EASTBROOK

**Estridge:** Peter *Estreis* 1148 Winton (Ha); Henry *le Estreis* 1185 P (Cu). OFr *estreis* 'eastern'. In 13th-century London a name applied to Germans, Easterlings: Hereman *le Estreys* was appointed attorney to receive money on behalf of Bartholomew de Hamburk (1282 LLB A). But the name must also have been used of a newcomer from the east. cf. NORRIS, SURRIDGE, WESTRICH.

**Estrild, Estrill:** *Estrilda* de Gloecestria c1150 ELPN; *Estrilda* de Lopham 1288–9 NLCh; Geoffrey *Estrild* 1279 RH (O); Gilbert *Estrild* 1327 ChertseyCt (Sr); John *Estrild* 1346 FA (Sf). OE *Ēastorhild* (f).

**Etcham:** John *Echeham* 1395 CtH. Probably for Etchilhampton (W).

**Etchells, Neachell:** Richard *de Echeles* 1269 AssSt; William *Attecheles* 1299 ib.; Richard *Atte Echeles* 1332 SRSt. 'Dweller on a piece of land added to an estate' (OE *\*ēcels*), frequently found in Warwickshire as the name of small farms and hamlets, e.g. Echills Wood, Nechells; also in Nechells and Neachill (Staffs), in Hitchells (Yorks), and occasionally in Derbyshire and Cheshire. v. PN Wa 30.

**Etheredge, Etheridge, Ethridge, Etteridge, Etridge, Ettridge, Athridge, Ateridge, Attridge:** *Aethericus* 1066 DB; Jacobus *Atteriche* 1276 RH (Berks); John *Etherych* 1524 SRSf; Henry *Etheridge*, William *Etteredge* 1674 HTSf. *Aethericus* is for OE *Ǣðelrīc*. v. also ALDRICH.

**Eton:** v. EATON

**Etty:** John *Ety* 1327 SRSf; William *Ettys* 1537 FFEss; James *Ettie* 1628 FrY. Perhaps a diminutive of OE *Ēata*, found in Etal (Nb), and Etloe (Gl).

**Eubanks:** v. EWBANK

**Eustace, Eustice:** *Eustachius* 1066 Winton (Ha); *Eustacius camerarius* 1166–75 DC (L); Richard *Eustas(e)* 1275 SRWo, 1279 RH (C); Margery *Eustace* 1297 MinAcctCo. Lat *Eustachius, Eustacius*, of Greek origin, probably from εὔστχος 'fruitful'. Occasionally, perhaps, from the feminine *Eustachia* 1214, 1222 Cur (Sr, Wa). v. also STACEY.

**Evans, Evens, Evins, Heaven, Heavens:** John *Evens, Evans* 1568 SRSf, 1679 FrY; James *Hevens* 1674 HTSf. *Evan* is the Welsh form of *John*, dating from about 1500.

**Evason, Everson, Evison:** William *Evoteson* 1325 Wak (Y); Thomas *Eversome* 1634 Buxhall (Sf); Edward *Everson, Eversham* 1639, 1644 ib. 'Son of *Evot*.' cf. IBBERSON.

**Evatt, Evett, Evetts, Evitt, Evitts:** *Evot* 1314 Wak (Y); *Euota* 1317 AssK; *Evette* 1420 Shef; William, Walter *Euote* 1295 Barnwell (C), 1327 SRY; William *Evett* 1555 Pat; John *Evatts* 1624 Oriel (O). *Evot* or *Evet*, diminutives of Eve.

**Eve:** *Eva* Hy 2 DC (L), 1206, 1211 Cur (Sr, Lei); Gregory *Eve* 1279 RH (C). *Eve* (Lat *Eva*) of Hebrew origin.

**Eveleigh:** Richard, Robert *Eveleigh* 1642 PrD. From a lost *Eveleigh* in Broad Clyst (D).

**Eveling, Evelot, Evelott:** Richard *Evelot* 1305 AssW; John *Evelot* 1327 SRSo; John *Eveling* 1642 PrD. *Ev-el-in, Ev-el-ot*, double diminutives of *Eve*. cf. Walter *filius Evelune* 1218 P (Lei/Wa), from the double diminutive *Ev-el-un*.

**Evening:** Walter *Evenyng* 1275 RH (Sx); Walter *Evening* 1302 FA (Sx). ME *evening* 'an equal, a match', a neighbour in the scriptual sense.

**Everard, Everatt, Evered, Everett, Everid, Everitt:** *Ebrard, Eurardus* 1086 DB; *Eurardus* de Langetona 12th DC (L); *Eborardus* 1222 Cur (C); *Everedus* c1250 Rams (C); Richard, William *Everard* 1204 Cur (Beds), 1225 AssSo; William *Euerrad* 1230 P (Lei); Symon *Eborard* 1275 RH (Nf); Geoffrey *Everad* 1300 Rams (Nf): Mary *Everet* 1725, *Everard* 1728, *Evered* 1732, *Everrad* 1734 IckworthPR. This might occasionally be from OE *Eoforheard* but many of the bearers of this name were undoubtedly from the Continent and the surname is usually from the cognate OG *Eburhard, Everhard* 'boar-hard'.

**Everden, Everdon:** Silvester *de Everdon* 1246, John *Everdon* 1449 FFEss. From Everdon (Nth).

**Everest, Everiss, Everix:** v. DEVERAUX

**Everidge:** Richard *Everich* 1289 NorwLt. OE *Eoforīc*.

**Everingham, Evringham:** Thomas *de Eueringeham* 1191 P (Y); Robert *de Everingham* 1259, Adam *de Everingham* 1348 FFY. From Everingham (ERY).

**Everington, Evrington:** Walter *de Eurinton'* 1190 P (Berks). From Everington (Berks).

**Everley, Everly:** Ailward *de Euerlay* 1200 P (W); William *de Everley* 1247 IpmY; William *de Euerleye* 1346 SRWo. From Everley (W, WRY).

**Evers, Evors:** (i) *Evor* de Dalling' 1221 Cur (Nf); John *Euers* 1465–6 FFWa; John *Evers* 1687 RothwellPR (Y). OE *Eofor*. (ii) William *de Eure* 1296, Adam *atte Eure* 1327, Robert *atte Euere* 1332 SRSx. 'Dweller at the edge or brow of the hill', OE *\*yfer*. (iii) John *Everose* 1386 LLB H. For DEVERAUX.

**Eversden, Eversdene, Eversdon:** Ralph *de Euresdon'* 1279 RH (C). From Eversden (C).

**Evershed, Eversheds:** Thomas *de Everesheved* 1255 PN Sr 276. From Eversheds Fm in Ockley (Sr).

**Eversley, Eversly:** John *de Everslye* 1275 RH (K); Walter *de Evereslegh'* 1278 Glast (So). From Eversley (K), or a lost *Eversley* in Charing (K).

**Everson:** v. EVASON

**Everton:** Alured *de Euerton'* 1212 P (C, Hu); William *de Everton* 1348 IpmW; Robert *Everton* 1474 FFEss. From Everton (Beds, Lancs, Notts).

**Everwin:** v. ERWIN

**Eves:** v. EAVES

**Evesham:** Celestria *de Euesham* 1221 AssWo; John *Evisham* 1663 HeMil. From Evesham (Wo).

**Eveson:** John *Evessone* 13th Rams (C); Hugh *Eveson* 1373 ColchCt. 'Son of *Eve*.'

**Evilchild, Ivelchild:** Thomas *Ivelchild* 1203 FFEss; Richard *Euelchyld* 1327 Pinchbeck (Sf). 'Wicked child', OE *yfel, cild*. cf. Osbert *Evelgest* 1199 P (W) 'evil guest, stranger'; Ralph *Vuelhering'* 1272 ForNth 'evil herring'.

**Evill:** v. DAVALL

**Evins:** v. EVANS

**Evington:** John *de Iuinton'* 1221 AssGl; William *de Euinton'* 1221 AssWa (Lei). From Evington (Gl, Lei).

**Evitt:** v. EVATT

**Evors:** v. EVERS

**Evringham:** v. EVERINGHAM

**Evrington:** v. EVERINGTON

**Ewan, Ewen, Ewens, Ewing, Ewings, Ewin, Ewins, Hewin, Hewins, Uwins, Yewen, Youens, Youings:** *Ewen, Ewein* Brit(t)o 1086 DB (He); *Ewain* 1164 Black (Scone); *Ywein* Ladde 1177 P (Nf); *Eugene* c1178 Black (Moray); *Ywanus surdus* 1181 P (L); *Ewanus* filius Walteri 1192 P (L); Robertus *filius Ywein* 1200 P (Wa); *Yowayne* de Bulling 1246 AssLa; *Ewyn, Iwin* de Salt 1271 AssSt; Dovenaldus *Ewain* a1165 Black; Walter *Ywain* 1202 P (Wa); Robert *Ywein* c1248 Bec (Berks); John *Iuuen* 1297 MinAcctCo; John *Youn* 1346 LLB F; Lucia *Iwynes* 1359 Putnam (Co); Bartholomew *Ewinge* 1555 Black (Glasgow); Richard *Ewyns, Ewens, Ewinges* 1605, 1639, 1681 FrY; George *Yewing* 1664 Black; David *Yewine* 1717 ib. *Ewen* and *Owen* have been regarded as different names and varied etymologies have been advanced for each. The post-Conquest forms suggest we have only one name, ultimately from Greek εὐγενής 'well-born', Gael *Eòghann*, MGael *Eogan*, OIr *Eogán*, OW *Eugein, Ougein*, MW *Ewein, Ywein, Oue(i)n*. In Scotland, Ireland and Wales the name was frequently Latinized as *Eugenius*. It occurs in Scotland as *Eugene* in 1178 and Macewen is *M'Eogan* in 1355. The form *Ewen* is not confined to Scotland, whilst both *Ewein* and *Ywein* appear side by side along the Welsh border. In the eastern counties the name was introduced from Brittany where it occurred as *Even* and has given modern Fr *Ivain*, Breton *Yven, Ivin*. cf. *Ivo* or *Iwein* Titneshoue 1243 Fees (St); already in 1086 we find *Ewen* the Breton in Hereford. *v.* also OWEN.

**Ewart, Yeoward, Yeowart, Youatt:** (i) *Ewart* aurifaber 1084 ELPN; *Eward* filius Gaufridi 1189–1200 BuryS (Sf); Hugo *filius Esward* 1199 ChR; Gillemichel *filius Eward* 1212 Fees (La); Thomas *Eward* 1279 RH (C); John *Ewarde* 1512 RochW. *Ewart*, a French form of *Edward*. The first example is probably identical with *Eadward*, a moneyer of William I. (ii) Adam *Yowehirde* 1297 SRY; John *Ewehird* 1379 PTY; Robert *Yowarde* 1381 PTY; William *Yeoward* 1607 RothwellPR (Y). OE *ēowu* and *hierde*, 'ewe-herd', keeper of the dairy-farm. Ewes were kept for their milk from which cheese was made. (iii) Robert *de Ewrth'* 1242 Fees (Nb). From Ewart (Northumb).

**Ewbank, Eubanks, Ubank:** Waldef *de Yuebanc* c1258 PN Cu 453; William *Hughbank* 1464 FrY; Robert *Hewbank* 1475 GildY; Oswald *Ewbanke* 1592 FrY. 'Dweller by the yew-bank' (OE *īw*), as at Yew Bank in Whitehaven (Cumb).

**Ewe:** Hugh *de Auco* 1148 Winton (Ha); John *Ewe* 1576 SRW. From Eu (Seine-Maritime). There was also a feminine name which may have contributed to the surname: *Ewe* filia Hugonis 1199–1200 FFWa.

**Ewell:** Gilbert *de Ewell* 1238–9 FFSr; Richard *de Ewelle* 1258 IpmW; Hugh *Ewel* 1327 SRSo. From Ewell (Sr), or Temple Ewell, Ewell, Minnis (K).

**Ewen(s):** v. EWAN

**Ewer, Ewers, Lewer, Lower:** Richard *le Ewer* 1185 Templars (Wa); Richard *Lewer* 1219 Fees (Sr); Alexander *Euer* 1309 SRBeds; Robert *Lower* 1513 FrY. OFr *ewer* 'servant who supplied guests at a table with water to wash their hands' (1361 NED).

**Ewin(s):** v. EWAN

**Exall, Excell, Exell:** Richard *de Ecleshal'* 1221 AssWa; Robert *Exall'* 1542 FFHu. From Exhall near Coventry or near Alcester (Warwicks).

**Exelby:** v. AXLEBY

**Exeter, Exter:** Goscelin *de Execestre* 1086 DB (D); Ralph *de Exeter* 1272 FFEss; John *Exetter* 1525 SRSx; John *Exter* 1796 HartlandPR (D). From Exeter (D).

**Exley:** Magot *de Exelay* 1324 LaCt; William *Exley* 1672 HTY. From Exley in Southowram (WRY).

**Exton:** William *de Exton'* 1206 Cur (R), 1242 Fees (D); Robert *de Exton'* 1327 SRLei; John *Exton* 1525 SRSx. From Exton (D, Ha, R, So).

**Eye, D'Eye:** Peitevin *de Eya* 1191 P (Sf); Peter *Ege* ib.; John *de la Eye* 1327 SRWo; Thomas *ate Eye* 1341 NI (Sx). From Eye (Hereford, Northants, Suffolk) or from residence near low-lying land (OE *ēg*). v. RAY, REA, RYE, NYE, YEA.

**Eyer(s), Eyre(s):** v. AYER

**Eynon:** v. ENNION

**Eynulph:** Simon *Einulf* 1195 P (Sx). OG *Einulf*. v. PNDB 246.

**Eyston:** John *de Eyston* 1299 IpmY. From Easton (ERY), Great, Little Easton (Ess), or Eyston Hall in Belchamp Walter (Ess).

**Eyton:** Robert *de Eiton'* 1185 P (Sa); John *de Eyton'* 1280 FFY; John *Eiton* 1384 IpmGl; William *Eyton, Eton* or *Heyton* 1509 LP (Nth). From Eyton (Salop), Eaton (Ches, Leics, Staffs), *Eyton* 1313, 1236, 1293, Eaton Bray (Beds), *Eitona* 1130, Cold, Long Eaton (Derby), *Coldeyton* 1323, *Long Eyton* 1288.

**Ezard:** v. ISSARD

# F

**Fabell, Fable:** John *Fabel* 1329 FFEss. Either a diminutive of a shortened form of *Fabian*, or a nickname from OFr *fable* 'lie'.

**Faber:** Aluricus *filius Fabri* 1066 DB (Sf); Godui *faber* c1095 Bury (Sf); Henry *Fabers* 1279 RH (C). Lat *faber* 'smith', usually occupational, but also used as a personal name: *Faber* de Suterton 1177 P (Y).

**Fabian, Fabien:** *Fabianus* de Cam 1184 P (Ha), de Hulmo 1191 P (Nf); Willelmus *filius Fabiani*, identical with William *Fabian* 1220 Cur (Ess); William *Fabien* 1231–53 Rams (Nf). Lat *Fabianus* 'of Fabius', probably from *faba* 'a bean'.

**Facey:** v. VAISEY

**Facket, Fackett:** Gilbert *Faket* 1193 P (Ess); Jordan *Faket* 1296 SRSx. OFr *fagot*, ME *faget, faket* 'a bundle of firewood'. Metonymic for a seller of firewood. cf. William *Faget* 1317 AssK.

**Fadder:** v. FATHERS

**Fadmore, Fadmoor:** Nisant *de Fademor* 1208 FFK. From Fadmoor in Kirkby Moorside (NRY).

**Faers:** v. FAIR

**Fage:** John *Fage* 1327 SRSo, 1388 FFC. ME *fage* 'the action of coaxing or deceiving' (c1421 MED); the verb 'to coax, flatter' is found c1400. cf. *fagere*, 'flatterer' (1435).

**Fagg, Fagge, Vagg, Vaggs:** Daniel, William *Fag* 1202 FFK, 1269 AssSo; Richard *le Vag* 1286 PN W 144; Ivelo *Vag'* 1317 AssK. cf. Peter *Clanvagg'* baker (1268 ELPN) which Ekwall explains as 'clean loaf', a name for a baker who sold clean bread, from OE *facg*, recorded only in the sense 'a flat fish, plaice', but probably used also of a 'flat loaf'. cf. early Mod *fadge* 'a large flat loaf or bannock'. The nickname may be metonymic for fishmonger or a baker.

**Faggeter, Faggetter:** Simon *le fagotter* 1279 MEOT (Sr). A derivative of ME, OFr *fagot* 'a bundle of sticks', hence a maker or seller of faggots.

**Faiers:** v. FAIR

**Fail, Failes, Fayle:** Richard *Faill* 1548 FrY; Edward *ffaile* 1672 HTY. A nickname from OFr *faille* 'erring, failure'. cf. Turstan *le Faillant* c1216–7 Clerkenwell; Thomas *le Failur* 1279 RH (C).

**Fainer, Feiner:** Roger *le Fener* 1271 MEOT (Sx); Gilbert *le Feyner* 1299 LLB C. OFr *fenier*, 'haymonger'.

**Fair, Faire, Faires, Fairs, Faiers, Faers, Fayer, Fayers, Fayre, Feyer, Fyers, Phair, Phayre:** Edeua *Faira* 1066 DB, *Pulchra* 1086 ib., *Bella* 1086 ICC; Robert *faier* 1191 FF (Sf); Henry *le Vayre* 1297 MinAcctCo; Thomas *le Fayre* 1332 SRSx. OE *fæger*

'fair, beautiful', used occasionally as a personal name: Johannes *filius Fair* 1203 Cur (C).

**Fairall:** v. FAIRHALL

**Fairbain:** John *ffairban* 1332 SRCu. 'Fair bone' (OE *bān*), one with fine limbs.

**Fairbairn, Fairbairns, Fairbarns:** Augustin, Robert *Fayr(e)barn(e)* 1297 SRY, 1379 PTY. 'Beautiful child' (OE *bearn*).

**Fairbank, Fairbanks:** Robert *Fairebank* son of Richard *Farebank* 1583 FrY. 'Dweller by the fair bank(s).'

**Fairbard, Fairbeard:** Thomas *Fairberd* 1331 Wak (Y); Thomas *Fayrebeard* 1590 Oxon. 'Fair beard' (OE *beard*).

**Fairbody:** Gregory *Feyrbody* 1332 SRWa. 'Fair person', OE *fæger, bodig*. OE *fæger* was a common first element in medieval nicknames, but comparatively few of them have survived. cf. Albreda *Fairamay* 1327 KB (R) 'fair friend'; William *Fayrandgode* 1301 SRY 'fair and good'; Adam *Fayrarmful* 1246 AssLa 'fair armful'; Roger *Fayrboie* 1221 AssWa 'fair lad'; Thomas *Fairdam* 1372 IpmW 'fair lady'; William *Fayrefelagh* 1327 SRSx 'fair fellow'; Richard *ffeyrmayden* 1340–1450 GildC 'fair maiden'; Henry *Fairsire* 1309 LLB D 'fair sire'; Cecilia *le Fairewif* 1254 Oseney 'fair wife'.

**Fairbourn, Fairbourne, Fairburn, Fairburne:** Simon *de Fareburne* 1185 Templars (Y); Margery *de Fareburn* 1274 RH (K); John *de Farburn'* 1351 AssL. From Fairbourne (Kent), or Fairburn (WRYorks).

**Fairbrass:** v. FIREBRACE

**Fairbrother, Farebrother, Farbrother, Fayerbrother:** John *Fayerbrother* 1524 SRSf. Perhaps 'brother of Fair'. cf. p. xliv.

**Fairchild:** Lefui *Fæger Cild* c1095 Bury (Sf); Stephen *Veirchill'* 1214 Cur (Ess); Robert *Fairchild* 1250 Fees (Sr). 'Beautiful child' (OE *cild*). cf. CHILD.

**Fairclerk, Fairclark, Fairclarke:** Robert *Feirclerk* 1275 SRWo; Thomas *Fayreclerk* 1327 SRSx. 'Fair cleric', OE *fæger*, OFr *clerc*. Either handsome, or with fair hair.

**Fairclough, Faircloth, Fairtlough, Faircliff, Faircliffe:** Simon *de ffairclogh* 1332 SRLa; William *Fayercliffe*, Robert *Fayreclought* 1568 SRSf; John *Fayercloth* 1629 LitSaxhamPR (Sf); Samuel *Fear Cloth* 1655 Bardsley; Allfabell *Farcloe* 1669 ib. 'Dweller in the fair hollow' (ME *clo(u)gh*).

**Faircock:** William *Veyrecok* 1327 SRSx; John *Fairecok* 1387 Misc (Gl); William *Fayrcok* 1405 AD vi (Nf). A diminutive of OE *fæger* 'fair' used as a personal name. v. FAIR.

**Fairer, Fayrer:** Henry *Fayrher* 1279 RH (C); Gervas' *Vayr Hyer* 1296 SRSx; Robert *Fairer*, William *Farher* 1327 ib.; William *Feyrere* 1375 ColchCt. OE *fæger* 'fair' and *hær* 'hair', 'one with fair hair'.

**Fairest:** v. FAIRHURST

**Fairey, Fairrie, Fairy, Farey:** (i) Lifwine *Feireage* 1050 KCD 1338 (K); Thomas *Feireie* 1231 Pat (O); Ralph *Feyrighe* 1275 RH (W). OE *fæger* 'fair' and *ēage* 'eye'. (ii) Also local, from residence near or employment at some small place such as Fairyhall in Felsted or Fairy Fm in Wethersfield, 'ox or pig enclosure' (PN Ess 422, 467).

**Fairfax:** Nicholas *Faierfax* 1195 P (Y); William *Fairfax, Fayrfex* 1208 Cur (Y), 1219 AssY. OE *fæger* 'fair' and *feax* 'hair'. cf. FAIRER.

**Fairfield, Fearfield:** Ralph *de Feirfeld* 1327 SRDb; Edmund *de ffeyrefeld'* 1331 FFK. 'Dweller by the fair field.'

**Fairfoot, Farfort:** Gundwi *de Faierford* 1203 P (Gl); Beatrix *de Faierford* 1209 P (L). From Fairford (Glos) or Farforth (Lincs). Also a nickname, 'fair foot': Adam *Fairefot* 1328 WhC (La).

**Fairfoul, Fairfoull, Fairfull:** John *Fayrfowel* 1277 AssSo. 'Fair bird' (OE *fugol* 'fowl').

**Fairhall, Fairall:** Thomas *atte Fayrehale* 1332 SRSx. 'Dweller in the fair nook' (OE *healh*).

**Fairham:** v. FAIRHOLME

**Fairhead:** (i) William *Fairhevid* 1279 RH (Hu); John *Feyrhed* 1332 SRWa. 'Fair head' (OE *fæger, hēafod*). (ii) John *Fairhod, Fayrhode* 1299–1311 LLB C, D. A nickname for one noted for his beautiful hood (OE *hōd*). cf. GREENHEAD

**Fairholme, Fairham:** Richard *Fairhome* 1379 PTY. 'Dweller by the fair island' (ON *holmr*).

**Fairhope:** John *atte Fayrehope* 1332 MELS (Sr). 'Dweller at the fair valley', OE *fæger, hop*.

**Fairhurst, Fairest:** Henry *de Fairhurst* 1260 AssLa. 'Dweller by the fair wood or wooded hill' (OE *hyrst*).

**Fairlamb, Fairlem:** Found in Manchester, this surname is probably for FARLAM, but a nickname is possible. cf. WHITELAM.

**Fairlaw:** v. FARLOW

**Fairlock:** William *Fayrlok* 1243 AssSo; Geoffrey *Fairlok* 1306 FFSf; William *Feyrloke* 1327 SRSa. 'Fair hair', OE *fæger, locc*. cf. Richard *Flaxennehed* 1279 RH (C) 'flaxen head'.

**Fairmain:** John *Fayrmayn* 1376 KB. 'Fair hand', OE *fæger*, OFr *mains*.

**Fairman, Fayerman, Fierman, Fireman:** Nicholas *Fairman* 1201 Cur (Sf); Reginald *Feierman, Fareman* 1208 ib. (Nf); Henry *Fayrman* 1297 MinAcctCo. OE *fæger* 'fair' and *mann* 'man'. As *Fareman* clearly occurs for both *Fayerman* and *Farmann*, it cannot be definitely assigned to either. Robert *Farman* or *Fareman*, 1222 Cur (So), was known also as Robertus *Senex* and Robert *le Bel*, so that *Farman* may be a translation of *le Bel* 'the fair'. v. FARMAN.

**Fairrie:** v. FAIREY

**Fairs:** v. FAIR

**Fairson:** Edward *Fairsone* 1332 SRDo; John *Fairesone* 1359 AssD. 'Son of *Fair*', OE *fæger* used as a personal name.

**Fairtlough:** v. FAIRCLOUGH

**Fairweather, Fareweather:** Agnes *Fairweder, Farrweder* c1248 Bec (Nf); Hugh *Fairweder* 1274 RH (L); John *Fayrewether* 1477 AD vi (D); William *Farewedder* 1547 FrY. ME *fair weder* 'weather not wet or stormy', probably 'one with a bright and sunny disposition'. cf. FOUWEATHER, *Manyweathers*, MERRYWEATHER, and John *Coldwedre* 1327 SRC, Alexander *Starkweder*, William *Skoneweder* 1327 SRSf, Alexander *Ilwedyr* 1316 Wak (Y).

**Fairwin, Fairwyn:** *Fairwinus* de Hanlee 1212 Cur (Wa); Robert *Fayrewyne* 1375 IpmW; William *Feyrewyn* 1477 RochW. From an unrecorded OE *\*Fægerwine*.

**Fairwood:** Thomas *Faierwood* 1576 SRW. From Fairwood Fm in Dilton (W), or 'dweller by the fair wood', OE *fæger, wudu*.

**Fairwyn:** v. FAIRWIN

**Fairy:** v. FAIREY

**Faith:** William *Feythe* 1389 FrNorw. Probably for *Faithfull*, now fairly common.

**Faiting, Fayting:** William *Fayting* 1275, John *Fayting'* 1327 SRWo. A nickname from ME *faiting* 'deception, fraud'.

**Faizey:** v. VAISEY

**Fake, Falck, Falco:** v. FAWKE

**Fakenham:** Robert *de Fakeham* 1296 SRSx; John *Fakenham* 1380 LoCh. From Fakenham (Nf), or Fakenham Magna, Little Fakenham (Sf).

**Falcon, Faucon, Facon:** Reginald *Falcun* 1187 P (Nf); Martin *Faukun* 1318 FFSf; Walter *Fakoun, Faucoun* 1346 FA (Sf). ME *faucon, faukun*, OFr *faucon, falcun* 'falcon'. A nickname from the bird or metonymic for *Falconer*.

**Falconer, Falconar, Falkner, Faulconer, Faulkener, Faulkner, Fawckner:** Henry *Falkenar* 1194 Cur (W); Richard *facuner* 12th DC (Lei); Henry *le fauconer'* 1219 AssY. OFr *fau(l)connier* 'one who hunts with falcons or follows hawking as a sport'; also 'keeper and trainer of hawks' (1424 MED). The name may also be from a rent. Colard *le ffauconer* in 1264 paid one falcon's hood and 1*d*. yearly for 26 acres of land at Walthamstow (*Ipm*). The surname may also mean 'worker of a crane'. *Faukonarii* worked at Caernarvon Castle in 1282 at 6*d*. per day in summer and 5*d*. in winter. In 1257 a carpenter was paid for making a *faucon*, a kind of crane or windlass, which the *falconarii* worked (Building 70, 324).

**Falding, Faulding:** Achethe *falding* c1155 DC. ME *falding* 'mantle, cloak', 'a kind of woollen cloth'. Metonymic for a maker or seller of this.

**Falgate:** v. FOLGATE

**Falk:** v. FAWKE

**Falkingham:** John *de Falkingham* 1275 RH (L); Geoffrey *de Folkingham* 1285 RegAntiquiss; John *ffalkingham* 1596–7 SRY. From Falkenham (Sf), or Folkingham (L).

**Falkner:** v. FALCONER

**Falkous, Falkus:** v. FAWKE

**Fall, Falle, Falls, Faull, Fawle:** William *de Fall* 1255 RH (O); Gilbert *de la Falle* 1263 AssLa; Geoffrey *del Falles* 1297 SRY; Richard *Fawle* 1579 FrY. 'Dweller by the fall' (ME *falle*), the slope(s) or waterfall(s).

**Fallas, Fallis:** (i) William *de Faleise, de Falisia* 1086 DB (W). From Falaise (Calvados). *v.* OEByn. (ii) Isabella *atte Faleise* 1327 SRSx; John *de la Faleyse* of Dunwich 1300 FFSf; Augustine *de la Faleys* of Westleton 1316 FA (Sf); William *Faleys* 1337 ColchCt. Fr *falaise* 'cliff' seems to have been adopted into ME and the surname is also local; the above individuals, like Edmund *de la Klyf* of Brampton, 1346 FA (Sf), lived on the cliffs of Sussex or Suffolk.

**Fallover:** cf. Joan *Falladun* 1279 RH (O).

**Fallow, Fallows:** Ralph *de la Falewe* 1272 *Ass* (Ha); Henry *de Falg(h)* 1327, 1332 SRSx; Thomas *del Falghes* 1376 Bardsley (Ch); Rendul *Fallowes* 1563 ib. 'Dweller by the newly cultivated land', OE *fealg*, ME *falwe*.

**Fallowell, Fallwell:** John *Falinthewol* 1301 SRY; Henry *ffalliwolle* 1327 SRC; John *Falleinthewall, Falleinthewelle* 1332 SRSt, 1343 AssSt. 'Fall in the well', an interesting nickname.

**Fallowfield:** William *Faleufeld* 1455 FFEss. From Fallowfield (La, Nb), or 'dweller by the fallow land', OE *fealu, feld*.

**False:** Robertus *falsus* presbiter 1191 P (Nth); John *le fals* 1193 P (Beds); Ralph *le Faus* 1204 Cur. OFr *fals, faus* 'fals'. It may also have contributed to *Falls*. *v.* FALL, and to VAUS.

**Falshaw:** Henry *Falshawe* 1597 FrY. From Falshey in Buckden (WRY).

**Fancourt, Fancott, Fancutt:** Elias *de Fanecurt* 1203 Cur; William *de Fanecurt* 1240 FFEss; Edward *Fannecourt* 1376 FFY. From Fan Court in Chertsey (Sr).

**Fane, Fayne, Faynes, Vaines, Vane, Vanes, Vayne:** William *le Vain* 1242 Fees (W); Robert *Fane* 1279 RH (C); Richard *le Feyn* 1378 FFC. OE *fægen*, ME *fein, fayn, fane* 'glad, well-disposed', proverbially opposed to fools: 'Fayne promys makyth folys fayne' (1471 NED). Also used as a personal name: Ivo *filius Fane* 1208 FFL.

**Fann, Van, Vann, Vanne, Vanns, Vance:** John *del Fan* 1199 MemR (Ess); Richard *ate Fanne* 1297 MinAcctCo (Beds); John *in the Vanne* 1327 *SR* (Ess); Henry *atte Vanne* 1341 NI (Sx). A form of OE *fenn* 'marsh, fen', common in Essex, Herts and Surrey, and noted also in Beds, Cambs, Suffolk, Middlesex, Kent and Sussex. *v.* FENN.

**Fanner, Vanner, Vannah:** (i) William *le Fanner* 1285 *Ass, le Vannere* 1319 *SR* (Ess); John *le Fannere, le Vannere* 1294, 1307 PN Sr. In the south-eastern counties, often a derivative of ME *fann* meaning 'dweller by the marsh'. *Fanner* is East Saxon, *Vanner* the southern form of this. In the Sussex Subsidy Roll, Ralph *le Fanner* of 1327 is called Ralph *atte Fanne* in 1332. John *Vanner* (1373 ColchCt) is probably identical with John *Fanner* (1377 ib.) and may be of the same family as John *atte Vanne* (1310 ib.) who is identical with John *Vanneman* (1336 ib.). It seems clear that *Vanner, atte Vanne* and *Vanneman* are synonymous terms for a marsh-dweller. But examples of (ii) may also be found in these counties. (ii) Walter *Fannere* 1279 RH (Bk); John, Walter *le Vanner(e)* ib. (O); Thomas *le vannere* 1297 MinAcctCo (Do); William *le Fanner* 1332 SRWa. In these counties, *fenn*

does not become *fann*, hence we have a derivative of OE *fannian* 'to fan, winnow', one who fans, a winnower (c1515 NED), or more probably of OE *fann* 'maker of fans or winnowing-baskets'. cf. John *Fanwryghte, Vanwrighte* 1379 AssEss. In the south *Fanner* was pronounced *Vanner*, which might also be from OFr *vannier* 'basket-maker'. (iii) Roger *le Vanur* 1275 SRWo. A clear example of 'basketmaker' from Fr *vanneur*.

**Fanning:** Thomas *Fannyng* 1405 FFEss. Probably a variant of FENNING

**Fanshaw, Fanshawe:** John *ffawnchall, Fanshawe,* 1490, 1566 DbAS 30. From Fanshaw Gate in Holmesfield (Derby).

**Fanson, Fanston:** *v.* FANTSTONE

**Fant, Fantes, Faunt, Font, Vant:** Thomas *Lenfaunt* 1230 P (Sa); William *le Faunt* 1271 ForSt; John *Faunt* 1277 LLB A; Walter *le Font* 1327 SRSx; Alan *Fant* 1327 SRSf. Fr *enfant* 'child'. *Le enfant* became *Lenfant*; this, with dissimilatory loss of the first *n*, became *Lefant*, which was re-divided as *le Fant* and the new article was finally dropped. *Vant* is due to the southern retention of the voiced sound of OE initial *f*. cf. FENN and VENN. Here, it is irregular as French initial *f* was retained. The origin of the shortened *Fant* was forgotten and the word treated as English.

**Fanthorpe:** John *de Fenthorpe* 1202 AssL. From Fanthorpe (L), Fenthorp 1202.

**Fantstone, Fanston, Fanson:** John *Fanteston* 1642 PrD. From Faunstone in Shaugh (D).

**Fanwright:** John *Fanwryghte* 1351 FFEss; John *Fanne wright* 1376 LoCh; John *Fanwryghte* 1379 AssEss. 'A maker of winnowing baskets', OE *fann, wyrhta*.

**Faraday, Fereday, Ferriday:** William *Fairday* 1327 SRSo; John *Fayrday* 1378 ColchCt. 'Servant of Fair', OE *fæger* used as a personal name, and OE *dæge*.

**Faragher, Fargher, Faraker:** *Farker* 1504, *MacFargher* 1511, *Faragher* 1649 Moore. A Manx equivalent to Gaelic FARQUHAR.

**Faraway:** *v.* FARWAY

**Farbrace:** *v.* FIREBRACE

**Farbrother:** *v.* FAIRBROTHER

**Fare, Phare:** Peter *ate Fare* 1341 MELS (Sx). 'Dweller by the track or road' (OE *fær*).

**Farebrother:** *v.* FAIRBROTHER

**Fareweather:** *v.* FAIRWEATHER

**Farewell, Farwell, Varwell:** Bartholomew *de Faierwelle* 1180 P (Y); Stephen *Farwel* 1224 Cur (C); Richard *Farewel* 1275 RH (Sf). The first example may derive from *Farewell* (Staffs) or from a Yorkshire place. *Farewell* appears frequently in Suffolk from 1275 to 1417, always without a preposition, and is, no doubt, a phrase-name, 'Fare well!'.

**Farey:** *v.* FAIREY

**Farfort:** *v.* FAIRFOOT

**Fargher:** *v.* FARAGHER

**Farindon, Farenden, Farringdon:** Elsi *de Ferendone* 1066 DB; John *de Faringdon'* 1278 RH (O); Roger *Faryndon* 1327 SRSo. From Faringdon (Berks, Dorset, Hants, Oxon), or Farringdon (Devon).

**Farington:** *v.* FARRINGTON

**Farlam:** Richard *de ffarlham* 1332 SRCu. From Farlam (Cumb).

**Farley, Farleigh:** William *de Ferlecheia* 1189 Sol; Richard *de Farlegh'* 1222 Acc; John *Farleye* 1332 SRWo. From one or other of the many places of these names.

**Farlow, Varlow, Fairlaw:** Philip *de Farlawe* 1255 RH (Sa); Richard *Farelowe* alias *Farely* 1706 DKR 41 (Y). From Farlow (Salop) or Fairley (Salop).

**Farman:** *Farmannus* 1066 Winton (Ha), 1086 ICC (C), *Fareman* 1168 P (So); Richard *Fareman* 1086 ICC (C); Roger *Fareman, Farman* 1260 AssY. ON *Farmann*, ODa, OSw *Farman. v.* also FAIRMAN.

**Farmar, Farmer, Fermer, Fermor:** William *le Fermer* 1238 FFEss; William *le Farmere* 1279 RH (C); Richard *Fermor* 1293 Fees (D). AFr *fermer*, OFr *fermier*, MedLat *firmārius* 'one who undertakes the collection of taxes, revenues, etc., paying a fixed sum for the proceeds' (c1385 NED), or, perhaps more frequently, 'one who cultivates land for the owner; a bailiff, steward' (1382 NED).

**Farmery:** William *del enfermerie* 12th FeuDu; Robert *de la Fermeria* 1203 P (Lei); William *del Fermere* 1301 SRY; Geoffrey *ate Fermerie* 1327 SRSx; Robert *Firmarie* 1366 LLB H. 'Worker at the (monastic) infirmary' (OFr *enfermerie*).

**Farn, Farnes:** *v.* FERN

**Farnall, Farnell, Farnill, Farnhill, Fearnall, Varnals, Vernall:** Richard *de Farenhull'* 1214 Cur (Berks); William *de Fernhulle* 1263 Ct (Ha); Hugh *de la Fernhull'* 1275 SRWo; John *de Farnhull* 1275 RH (W); William *atte Farnhulle* 1298 PN Sr 286; Alexander *Farnell* 1414 LLB I; Francis *Fearnall* 1679 Bardsley (Ch). 'Dweller by the fern-covered hill' (OE *fearn, hyll*), as at Farnhill (WRYorks), Fernhill (Ches), Farnell Wood (Kent), Farnell Copse (Wilts), or Vernal in Tichborne (Hants).

**Farnborough, Farnbrough:** Lefred *de Ferenberga* 1190 P (Berks); Nicholas *de Farneberge* 1205 P (Wa). From Farnborough (Berks, Ha, K, L, Wa).

**Farncombe, Vearncombe:** Robert *de Ferncumb* 1296 SRSx. From Farncombe (Surrey).

**Farndale, Farndell, Varndell:** John *de Farnedale, de Farndall, Fernedill* 1363, 1397, 1409 FrY. From Farndale (NRYorks). *Varndell* must be from some southern place of the same name.

**Farnden, Farndon:** Chyldeluve *de Farnedon'* 1248 AssBerks; Sarra *de Farndon* 1327 SRSx; Thomas *Pharndon* 1562 Pat (Y). From Farndon (Ch, Nt), or East, West Farndon (Nth).

**Farnell, Farnill:** *v.* FARNALL

**Farnfield:** William *atte Fernefeld* 1327 SRSo. 'Dweller by the fern-covered land.'

**Farnham, Varnham, Varnam:** Richard *de Farenham* 1205 P (Sr); Robert *de Farnham* 1219 AssY; John *de Farnam* 1323 FrY; Stephen *Varneham* 1674 HTSf. From Farnham (Bucks, Dorset, Essex, Suffolk, Surrey, WRYorks).

**Farnhill:** *v.* FARNALL

**Farnley, Fearnley, Fernley:** Hugh *de Fernlee* 1206 Cur (Wo); Hugh *de Fernelay* 1316 Wak (Y); William *ffearneley* 1621 SRY. From Farnley, Farnley Tyas (WRY), Fernilee (Db), or 'dweller by the fern-covered clearing', OE *fearn, lēah*.

**Farnorth, Farnsworth, Farnworth:** Leising' *de Farnewurd'* 1185 P (Ch). From Farnworth (Lancs).

**Farquhar, Farquar:** *Ferchart* a1178 Black; *Farquhar* Macintosh 1382 ib.; Andro *Farchare* 1450 ib. Gael *Fearchar* 'very dear one'.

**Farquharson:** David *Farcharsoun* 1440 Black. 'Son of *Fearchar*.'

**Faro:** *v.* FARRAR

**Farr:** Nicolas *le ferre* Hy 2 Bart (Lo); Simon *Farre* 1381 ArchC iii. OE *fearr* 'bull'.

**Farra, Farrah:** *v.* FARRAR

**Farraker:** *v.* FARAGHER

**Farrand, Farrant, Farrants, Farrance, Farran, Farren, Ferrand, Ferran, Ferrans, Ferens, Ference:** *Ferrandus* clericus 1190 P (Ess); Herebertus filius *Feran* 1198 FF (Herts); *Ferant* Arblastarius 1249 Fees (Ha); Gilbert, Peter *Ferrant* 1188 P (Wa), 1202 AssL; John *Farrant* 1573 RothwellPR (Y); Roger *ffarrand* 1642 PrGR (La); George *Farrance* 1674 HTSf. OFr *Fer(r)ant* (Ferdinand), or OFr *ferrant* 'iron-grey'.

**Farrar, Farrer, Farrah, Farra, Farrey, Farrow, Faro, Pharaoh, Pharro:** Hugo *Farrour* 1379 PTY; Roger *Farrer* 1613 FrY; William *Farrar* 1675 ib.; Magister doctor *Pharor*, identical with Dom. Will. *Farar* 1517–18 GildY; James *Farro* 1525 FrY; William *Farrowe* 1528 GildY; Alys *Farray* 1559 RothwellPR (Y); John *Farry* 1674 HTSf; Tristram *Farrey, Farrer, Farrah* 1632, 1641, 1679 RothwellPR (Y); William *Pharrow*, Fr. *Pharoe* 1674 HTSf; Elizabeth *Pharao* 1702 Bardsley; Giles *Pharaoh* 1760 ib. All these names are variations of FERRER. The unstressed *-er* was slurred in pronunciation and variously spelled *-ey, -ah, -a*, giving *Farrey, Farrah* and *Farra*. This was regarded as an incorrect dialectal pronunciation and the name was re-spelled *Farrow* on the analogy of *barrow*. At Hoxne (Suffolk) in 1835, Dinah *Farrer* signed the marriage register *Farrow* (SfPR). Initial *Ph-* for *F-* is common and *Pharrow, Pharoe* were associated with the biblical *Pharaoh*, which, however, may occasionally be a pageant name or a nickname. cf. Rogero *Pharaone* (abl.) c1158 EngFeud (Nth). *v.* also VARROW.

**Farrell:** Andrew *Farell*, Roger *Farel* 1642 PrD. Probably a variant of FAREWELL, i.e. either from Farewell (St), or a phrase-name 'fare well'. In Ireland from Ir *Ó Fearghail* 'descendant of the man of valour'.

**Farren, Farrin, Varran:** John *Farhyn* 1297 MinAcctCo; William *Varyn* 1332 SRSx; Nathaniel *Farren* 1674 HTSf. The second element is ME *hine* 'servant'. *v.* HINE. The first may be OE *fæger* 'fair', found in the 13th century as *Far(e)-* (cf. FARSON), or OE *fearr* 'bull'. Probably 'fair servant', possibly 'bull-herd'.

**Farringdon:** *v.* FARINDON

**Farrington, Farington:** William *de Farington* 1376 IpmLa; William *de Faryngton* 1402–3 FFWa; Thomas *Farrington* 1591 AssLo. From Farington (La), or Farrington Gurney (So).

**Farrow:** *v.* FARRAR

**Farson:** Richard *ffayrson* 1327 SRC; Adam *ffarson* 1332 SRCu. 'Fair son.'

**Farthing:** (i) *Fardein, Fardan* 1066 DB (Nth, Y); Willelmus *filius Fardain, Fardein* 1163–6 DC (L),

William *Farthain* 12th ib. (identical); Reinfrid *filius Farthein* 1166 P (So); Þurlac *Ferðeng* 972–92 OEByn (Nth); Robert *Fardenc* 1086 DB (Sf); Hugo *Ferðing* 1176 P (Y); William *Farding* 1219 AssY. ON *Farðegn*, ODa *Farthin*. This should give *Far-* but as it became late OE *Færðegn*, it might appear as *Far-*, *Fer-*. *Ferthing* cannot be distinguished from OE *fēorðing* 'a fourth part', 'a farthing'. Both John *Ferthinge* and Margery *Halpeny* were assessed in Pershore in 1327 SRWo. cf. HALFPENNY. In ColchCt, both *Ferthing* 1312–52 and *Farthing* 1334–52 are common. The first example of *farthing* in MED is 1442. (ii) Robert *de la Ferthing* 1279 PN Sx; John *atte fferthynge* 1333 MELS (So). OE *fēorðing* 'a fourth part', used as a measure of land, *ferdin* in Exon, see MED. The meaning was probably 'homestead consisting of a *fēorðing*'. The term survives frequently in minor place-names in Devon, Essex, Surrey and Sussex.

**Farway, Faraway:** Simon *Fareweye* 1340 Hylle; Thomas *Farewey* 1365 FFW. From Farway (D).

**Farwell:** *v.* FAREWELL

**Fassell:** Robert *Fassell* 1258 Acc; Matilda *Fassel* 1281 CtW; William *Fassel* 1327 SRWo. OE *Fassel*.

**Fasset:** *v.* FAWCETT

**Fastiger:** Gilbert *Fastigurt* 1162, *Fastigert* 1163 ELPN. 'Firmly girdled', OE *fæste, gegyrded*.

**Fastolf:** *Fastolf* 1066 DB (L); Alexander, William *Fastolf* 1291 FFSf, 1295 AD iv (Nf); Henry *Forstalff* 1451 Rad (C). ON *Fastúlfr*, ODa, OSw *Fastolf*.

**Fates, Laffeaty, Laffitte, Lafitte:** William *Affaitied* c1162 BM, *Lafeite* 1180 P (Lo); Henry *la Faitie* 1189–99 AD i (Lo); Thomas *Fayte* c1239 Bart (Ess); William *la Feyte* 1262 Fr (Ess); Robert *le Affete* 1276 LLB B; Simon *la Affayte* 1277 FFEss; John *ffaiti* 1319 SRLo; Walter *Laffete* 1327 *SR* (Ess); William *Fayt* 1359 FrY. OFr *afaitié* 'affected, skilful, prudent'. For *Fate*, instead of *Fatey*, cf. PETCH and PETCHEY.

**Fatherless:** William *Faderles* c1170 Riev; Ralph *Faderles* 1219 AssY; Miles *le Vaderlese*, *le Faderlese* 1305 AssW; William *Faderles* 1382 Misc (Y). OE *fæderlēas* 'fatherless', perhaps for a posthumous child.

**Fathers, Fadder:** *Fader* 1066 DB (Nf, Sf) c1095 Bury (Sf); Roger, Alan *Fader* c1095 Bury (Sf), 1167 P (Ha); Robert, Walter *le Fader* 1201 AssSo, 1244 Rams (Hu); John *Father* 1327 SRSo; Juliana *le Faderes* 1332 SRSr. ODa *Fathir*, OSw *Fadhir* is certainly the source of some of the names, but *le Fader* is clearly from OE *fæder* 'father', used in ME of one who exercises protecting care like that of a father, and also of a religious teacher or a confessor.

**Fatt:** Æðestan *Fætta* 1049–58 OEByn (Wo); Richard *Fatt* 1260 AssC; Hugh *le Fatte* 1327 SRSo. OE *fætt* 'fat'.

**Fatting:** Robert *Fattinge* 1259 IpmY; Matilda *Fattynge* 1303 LLB C; John *Fattyng* 1448 FrY. 'Son of *Fætta*', a personal name from OE *fætt* 'fat'.

**Faucett:** *v.* FAWCETT

**Faucon:** *v.* FALCON

**Faugh:** Hugh *Faluo*, *Faue*, *le Fauue* 1222–42 Oseney (O). OE *fealu* 'of a pale brownish or reddish yellow colour', probably of the hair. *Fauue* may be OFr *fauve*. cf. FAVEL.

**Faulconer:** *v.* FALCONER

**Faulder:** *v.* FOLDER

**Faulding:** *v.* FALDING

**Faulds:** *v.* FOLD

**Faulkener, Faulkner:** *v.* FALCONER

**Faulk(e)s:** *v.* FAWKE

**Faull:** *v.* FALL

**Faun, Faunce, Fawn, Fawns, Fownes:** William filius *Faun'* 1230 P (Ha); John *Foun* 1180 P (Nt); Richard *le Foun* 1299 AssSt; Robert *Faun* 1390 LLB H. OFr *faon*, *foun*, ME *faun*, *foun* 'a young animal, cub', 'a young fallow deer' (a1382 MED), used, probably, of a lively, frisky youth. Also used as a personal name. The Scottish *Fawns* is from the lands of Fans or Faunes (Berwicks): Richard *de Fawnes* c1150–90 Black.

**Fauntleroy:** Roger *Le Enfaunt le Roy* 1244 PN Do 215; William *Fauntleroy* 1332 SRDo; John *Fauntleroy* 1408 AD ii (So); Thomas *Fontleroy* 1662 HTEss. 'Son of the king', OFr *enfant, roi*.

**Faure:** William *Faure* 1286 LLB A. A Provencal form of Fr *fevre* 'smith'.

**Faux:** *v.* FAWKE

**Favel, Favell, Favelle, Favill:** Eudo *Faluel* 1160–76 Seals (Y); Rannulf *fauuel'* 1195 P (He); William *Fauel* 1346 AD ii (Wt). OFr *fauvel* 'fallow-coloured, tawny', perhaps with reference to the hair, but also used as a symbol of cunning, duplicity or hypocrisy (c1325 NED).

**Faver, Favers:** Matthew *Favour* 1660 FrY. A nickname from OFr *favor*, *favour* 'help, mercy, beauty'. cf. William *Fauerles* 1373–5 AssL.

**Fawcett, Fawcitt, Fawssett, Faucett, Fausset, Faussett, Fasset, Fossett, Fossitt:** Alan *de Fausyde* 1238 Black; Elias *de ffagheside* 1332 SRLa; John *del ffawside* 1332 SRCu; Richard *Fascet* 1398 FrY; Henry *Fauset*, *Faucet* 1457, 1481 ib.; Robert *Fawsett* 1548 ib.; Robert *Faucit* 1591 RothwellPR (Y); William *Faucett*, John *Fasset*, Thomas *Fossett* 1674 HTSf. From Fawsyde (East Lothian), Fawcett (Westmorland) or Facit (Lancs). Some of the numerous Yorkshire Fawcetts probably derive from Forcett (NRYorks) which is indistinguishable in pronunciation from *Fawcett* and would naturally become *Fossett*: Thomas *de Forset* 1327 SRY.

**Fawckner:** *v.* FALCONER

**Fawcus:** *v.* FAWKE

**Fawden, Fawdon:** William *de Fawdon* 1374 FrY; Nathaniel *Fawdinge*, William *Fauden* 1601, William *Fawden* 1618 LewishamPR (K). From Fawdon (Nb).

**Fawell:** John *Fawvell* 1412 FrY; Richard *Fawell* 1530 ib. For FAVEL.

**Fawke, Fawkes, Fawcus, Faux, Falck, Falco, Falk, Falkous, Falkus, Faulkes, Faulks, Fake, Fakes:** *Falco* le Taverner 1274 RH (Lo); *Falc'* de Fonte 1296 SRSx; Tomas *Falch* 1182 P (Wo); Walter *Falc* 1221 AssWo; Geoffrey *Faukes* Hy 3 Gilb (L); William *Faukus* 1251 Rams (Hu); John *Falk* 1275 SRWo; Richard *Fauke* 1305 SIA iii; John *Fakes*, *Faukys* 1327 *SR* (Ess); Robert *Faukes*, *Faucous* 1327, 1332 SRSx; Roger *Faux* 1443 Oseney (O). OFr *Fauque*, *Fauques* (nom.), *Faucon* (acc.), OG *Falco* 'falcon'. *Faucon* may sometimes be the source of FALCON which is usually from the name of the bird. The personal name is best known from the famous (or infamous) Falkes de Breauté whose name occurs as *Falco* 1219 Fees, *Falk* Hy 3 Ipm, *Falkesius* 1219 Cur, *Falkasius* 1222 ib., *Falkes* 1233 Cl and survives in Vauxhall (Surrey). The modern VAUX may thus also derive from this personal name which has also contributed to FOX: Foxhall Fm in Steeple (Essex) owes its name to one John *ffaukes* (1319 PN Ess 228). The name has also been confused with FOLK: Folkes in Cranham (Essex) is named from Thomas *ffakys, ffaukes* (1463, 1473 ib. 125).

**Fawle:** *v.* FALL

**Fay, Faye, Fey:** (i) Ralph *de Faia* 1194 P (Sx); Richard *de Fay, de Fago* 1242 Fees (He). From a French place named *Fay* (OFr *fay*, Lat *fagus* 'beech'). (ii) Margaret *le Fey* 1332 SRSr. OFr *fae* 'fairy'.

**Fayerbrother:** *v.* FAIRBROTHER

**Fayerman:** *v.* FAIRMAN

**Fayers:** William *del Fehus* 13th Guisb. 'Dweller near the shed for livestock', OE *feoh, hūs*. Also variant of FAIR.

**Fayle:** *McFalle* 1408, *Fail* 1511 Moore. Manx for *Mac Giolla Phoil* 'son of Paul's servant'. cf. QUAYLE.

**Fayle:** *v.* FAIL

**Fayne:** *v.* FANE

**Fayrer:** *v.* FAIRER

**Fayter:** *v.* FETTERS

**Fayting:** *v.* FAITING

**Fazackerley, Fazakerley, Phizackerley, Phizackerly, Phizakarley, Phizaclea:** Henry *de Fasakerlegh* 1276 AssLa; Lawrence *Phezackerley* 1792 Bardsley. From Fazakerley (Lancs).

**Feacey:** *v.* VAISEY

**Feakes:** *v.* FICK

**Feakins:** *v.* FICKEN

**Fear, Feare, Phear:** Walter *Fere* 1279 RH (O); Roger *le Feer* 1327 *SR* (Ess). ME *fere* (*feare, feer*) 'companion, comrade' or ME *fere, feer*, OFr *f(i)er* 'fierce, bold, proud'. In some instances we must have the noun *fear*: cf. Rogerus *Timor* 1200 Cur (Bk).

**Fearby:** Thomas *de Fegerbi* 1200 AssY. From Fearby (NRY).

**Fearenside:** *v.* FEARNSIDE

**Fearfield:** *v.* FAIRFIELD

**Fearman:** *v.* FAIRMAN

**Fearn(e):** *v.* FERN

**Fearnall:** *v.* FARNALL

**Fearnehough, Fearnyhough, Fearnyough:** *v.* FERNEYHOUGH

**Fearnley:** *v.* FARNLEY

**Fearnside, Fearnsides, Fearenside:** Nicholas *del ffernyside* 1324 LaCt; Edward *Fearnesyde* 1587 RothwellPR (Y). 'Dweller on the ferny slope' (OE *sīde* 'side').

**Fearon, Feron:** Walter *le ferrun* c1179 Bart (Lo); Hervey *le Feron* 1277 AD ii (Mx). OFr *ferron, feron* 'ironmonger, smith'.

**Feasey:** *v.* VAISEY

**Feather, Fedder:** Juliana *la Fethere* 1296 SRSx; Adam *ffethir* 1332 SRCu; Antony *Fedder* 1544 RothwellPR (Y). OE *feðer* 'feather', metonymic for a buyer or seller of feathers. cf. Richard *Fethermongere* 1282 LLB A, William *le fetherman* 1275 MESO (Nf), Robert *Le Feþerbycger* 1304 ib. (Ess). But also clearly a nickname for one light as a feather.

**Featherby:** Ralph *de Fecherbi* 1185 Templars (Y). From Fearby (NRYorks).

**Featherman:** William *le fetherman* 1275 MESO (Nf). 'A dealer in feathers', OE *feðer, mann*. cf. John *le Fethermonger* 1276 AssLa.

**Featherston, Featherstone, Fetherston:** William *de Federestan* 1187 P (St). From Featherstone (Staffs, WRYorks).

**Featherstonehaugh, Featherstonhaugh, Fetherstonhaugh:** Elias *de Fetherestanehalg'* 1204 Cur (Nb). From Featherstone (Northumb). The surname preserves the full original form of the place-name.

**Feaver, Feavers, Feaviour, Fever, Fevers, Lefeaver, Lefever, Le Fever, Lefevre, Le Fevre, Le Feuvre:** Roger *le Fevere* 1243 AssSo; Abraham *le Fevre* 1248 FFEss. OFr *fevere, fevre*, Lat *faber* 'smith'.

**Feaveryear, Feavearyear, Feaviour, Feveyear, Fevyer:** John *Feveryer*, Robert *Feveryere* 1524 SRSf; Edmonde *Feueryeare* 1568 SRSf; Zimri *Fevier* 1813 SfPR; John *Fevyear* 1826 ib.; Robert *Feaviour* 1829 ib. ME forms of *February*. *v.* FEVEREL.

**Fedder:** *v.* FEATHER

**Fee:** William *de Feodo* 1220 Cur (Nth); Peter *Fee* 1327 SRSf; Thomas *Fee* 1642 PrD. Lat *feodum*, AFr *fee* 'fief'. Perhaps for the holder of a fief.

**Feehally:** Thomas *Fitz Harry* 1449 FFEss. A Liverpool form of Irish *Feeharry* 'son of Harry'.

**Feek(s):** *v.* FICK

**Feesey:** *v.* VAISEY

**Fegg, Fegge, Feggs:** *Fegge* Buche 1165–7 P (L); *Fegge* Fort 1200 Cur; Richard *filius Fegge* 1202 AssNth; Adam *ffegge* 1287–8 NorwLt; Thomas Fegh 1327 SRSx; William *ffeygs* 1379 PTY. ODa *Feggi*.

**Feiling:** William *de Fieling* 1203 P (Y). From Fitling (ERYorks).

**Feiner:** *v.* FAINER

**Feirn:** *v.* FERN

**Felbridge:** Roger *de Felebrig'* 1232 Fees (Nf). From Felbrigg (Norfolk).

**Feldon:** Richard *de Feldon* 1415 IpmY. From Feldom in Marske (NRY).

**Feldwick, Feldwicke, Fieldwick:** Roger *de Feldwyke* (1296 SRSx) lived at *Feldwycks*, now Old House in West Hoathly (Sussex). *v.* PN Sx 273.

**Felgate:** *v.* FIELDGATE

**Felice, Fillis:** *Felicia* 1194 P (Wo), 1207 Cur (Wa), 1218 FFEss; *Felis* 1381 SRSf; Gilbert *Felice* 1279 RH (C); Margaret *ffelys* 1469 SIA xii. *Felis, Phelis* was the English form of Fr *Felise* (f), from Lat *Felicia*, feminine of *Felix*.

**Felix:** *Felix* monachus 1122 AC (Sx); *Felix* filius Hamonis, Hamo *Felix* 1229 Pat (K); Richard *Felix* 1352 ColchCt. Lat *Felix* 'happy', less common than *Felicia* and also *Felis* in the vernacular, so that the surname may have partly coalesced with *Felice*.

**Fell, Fells:** (i) Roger *del Fel* 1318 AD iv (Cu); Robert *of the Fell* 1421 FrY. 'Dweller on the fell', ON *fell, fjall* 'fell, mountain'. (ii) William *Fel* 1275 RH (Nb), 1279 ib. (O). Black notes that *Fell* was an old surname in Dundee where Finlay Fell, butcher, passed on his trade and his name to his descendants for over a century from 1533. The surname is metonymic for *fell-monger*, a dealer in hides or skins:

Mabel *Felmonger* 1332 SRSr. The Northumbrian example might be for (i) with early loss of the preposition, but that from Oxford certainly belongs here.

**Feller, Fella:** Robert le *Felur* 1275 SRWo. Either a derivative of OE *fell* 'skin, hide', a fell-monger, or of the verb 'to fell', a feller of timber.

**Fellgate:** *v.* FIELDGATE

**Felliper:** Aunger *le Feleper* Hy 3 Rad, *le Pheliper* 1277 LLB A; Godwin *le Feliper* 1406 AssLo. 'A dealer in second-hand clothes or furniture', AFr *feliper*.

**Fellowes, Fellows:** Richard *Felawe* 12th Lichfield (St); Walter *Felagh* 1256 AssNb; Robert *le Felagh* 1327 SRSx. Late OE *feolaga*, ON *félagi* 'partner, co-worker, companion'. *v.* FIELDHOUSE.

**Fellowship:** William *Feliship* 1449 GildY. *Fellowship*, originally 'companionship', came to mean 'membership of a society', perhaps of a guild.

**Felmingham:** Matilda *de Felmingeham* 1194 P (Nf); Peter *de Felmingham* 1212 Cur (Nf). From Felmingham (Nf).

**Felstead, Fellstead:** Ralph *de Felesteda* 1130 P (Ess); Baldwin *de Felested* 1203, John *de Felstede* 1365 FFEss. From Felstead (Ess).

**Felt:** Metonymic for FELTERS.

**Felters:** Roger *Feltere* 1220 FrLeic; Robert, John *le Feltere* 1275 RH (L), 1280 AssSo. A derivative of OE *felt* 'felt'. A maker of felt.

**Feltham, Veltom:** Richard *de Feltham* 1255 FFK; Thomas *Feltam* 1545, Walter *Felthame* 1576 SRW. From Feltham (Mx, So), or 'dweller near the hay-meadow', OE *filipe, hamm*.

**Felton:** Adam *de Felton'* 1208–9 Pl (Nf); Richard *de Felton* 1301–2 FFSr; Ranulph *Feltone* 1413 CorSt. From Felton (Hereford, Northumb, Salop, Som).

**Feltwell:** Geoffrey *de Feltewelle* 13th Lewes (Nf); Roger *Feltewell* 1377 IpmW. From Feltwell (Nf).

**Femister, Fimister, Phemister, Phimester, Phimister:** Alexander *Feemaister* 1458 Black; Christian *Phemister* 1595 ib.; James *Fimister* 1684 ib.; Robert *Phimister* 1737 ib. 'The fee-master', one in charge of the flocks and herds.

**Fenby:** William *de Fenneby* 1274 RH (L); William, Robert *de Fenby* 1381, 1383 AssL. From Fenby (Lincs).

**Fencot, Fencott:** Walter *de Fencotes* 1327 SRY; John *Fencote* 1352 FFSr. From Fencott (O), Fencote (He), or Great, Little Fencote in Kirkby Fleetham (NRY).

**Fendall:** Aibrond *de Fendhale* 1191 P (Nf). From Fenhale (Norfolk).

**Fender:** Robert *le Fendur* 1267 FFHu; Thomas *le Fendour* 1301 SRY. A derivative of *fend*, a shortened form of *defend*, 'defender'.

**Fendick:** Adam *de Fendyk* 1375 FrY. 'Dweller by the fen-dike'.

**Fenemore, Fenimore, Fennemore:** *v.* FINNEMORE

**Fenn, Venn:** Godwin *de la Fenna* 1176 P (D); Thomas *attefenne* 1185 Templars (Wa); Ralph *de Fenne* 1190 P (L); Herveus *del Fen* 1190 BuryS (Sf); John *atte Venne* 1327 SRSo; Walter *en la Fenne* 1340 SRWo. OE *fenn* 'marsh, fen'. From Fen (Lincs), one of the Devon places called Venn, or from residence near a marsh. cf. FANN.

**Fennel, Fennell:** (i) William *Fenigle* 1327 SRSx; Henry *atte Fenegle* 1332 SRSx. 'Grower of fennel', a plant cultivated for its use in sauces. OE *finugl, fenol*. (ii) William, Cristiana *Fenel* 1327 SRC, SRSo. ME *fenel* used metonymically. (iii) A development of *FitzNeal*. Fennells Wood (PN Bk 201), *Fenelgrove* 1391, is named from Robert *FitzNeel* (1283); Fitznells (PN Sr 75), *Fenelles* 1450–3, derives from Robert *le Fitz Nel* (1332).

**Fenner:** *v.* VENNER

**Fenning, Venning, Vening:** David *Fenning* c1248 Bec (Nf); John *ffening* 1290–1 NorwLt; George *Venynge*, Thomas *Venning* 1642 PrD. 'Dweller in the fen', from a derivative of OE *fenn* 'fen'. *Venning, Vening*, are distinctively southern forms.

**Fentiman, Fenttiman:** John *ffentonman* 1379, Alice *Fentyman* 1509 WRS; Robert *Fentiman* 1656 RothwellPR (Y). 'The man from (Church) Fenton'.

**Fenton:** (i) Walter, Adam *de Fenton'* 1199 Pl (Y), 1230 Cur (Db), William Fenton 1382 AssL. From Fenton (Cumb, Lincs, Northumb, Notts, Staffs), Church Fenton (WRYorks), or Venton (Devon). (ii) John *de Fenton* 1261 Black (Forfar); Alexander *de Fentoun* c1330 ib. From the barony of Fenton (East Lothian).

**Fenwick, Finnick, Vinick:** Robert *de ffenwic* c1220 Black (Kelso); Walter *del Feneweke* 1275 RH (L); Thomas *de Fenwyk* 1279 AssNb; Nichol *of Fynnyk* c1431 Black (Ayr). From Fenwick (Northumb, WRYorks, Ayrshire) or 'dweller at the dairy-farm in the fen'. *Vinick* is Southern.

**Fereday:** *v.* FARADAY

**Ference, Ferens:** *v.* FARRAND

**Fergus:** Gillebertus *filius Fergusi* 1180 P (Cu); *Fergus* filius Suein 1188 P (Y); Gilbert *Feregus* 1199 P (Cu); John *Fergus* 1251 AssY. Gael *Fearghus*, OIr *Fergus* 'man-choice'.

**Ferguson, Fergyson:** John *Fergusson* 1466 Black. 'Son of *Fergus*.'

**Fermer, Fermor:** *v.* FARMAR

**Ferminger:** *v.* FIRMINGER

**Fern, Ferne, Ferns, Fearn, Fearne, Feirn, Farn, Farnes, Fairn, Varnes, Varns, Vern, Verne:** John *de la Ferne*, Henry *atte Verne* 1275 SRWo; Joceus *de Ferne* 1296 SRSx. OE *fearn*, used collectively in the singular. 'Dweller among the ferns.' *Verne* is the southern form.

**Ferneyhough, Fernihough, Fernyhough, Fearne-hough, Fearnyhough, Fearnyough:** Adam *de Fernyhough* 1332 SRSt. 'Dweller in the ferny hollow' (OE *holh*).

**Fernley:** *v.* FARNLEY

**Feron:** *v.* FEARON

**Ferraby, Ferrabee, Ferriby:** John *de Ferieby* 1212 Cur (L); William *de Feriby* 1301 LLB B; Wylliam *Feraby* 1524 SRSf. From North Ferriby (ERY), or South Ferriby (L).

**Ferrand, Ferran(s):** *v.* FARRAND

**Ferrer, Ferrar, Farrer:** Henry *le Ferrur* 1196 Cur (Lei); Picot *le Ferur* 1200 P (L, Ha); Hugo, Thomas *Farrour* 1379 PTY. OFr *ferreor, ferour* 'worker in iron, smith' (c1400 MED). *v.* also FARRAR, FAIRER.

**Ferrers:** Henry *de Feireres, de Fereres, de Ferrariis*, 1086 DB; Henry *Ferreres, Ferrieres* ib.; Hugo *de Ferrers* 1222 Cur (He). The Ferrers, earls of Derby, came from Ferrières-Saint-Hilaire (Eure) and the Ferrers of Bere and Newton Ferrers (Devon) from Ferrières (La Manche). *v.* ANF.

**Ferret:** Walter *Feret* 1296 SRSx; John *Furet* 1327 *SR* (Ess). ME *fyrette, ferette,* OFr *fuiret, furet* 'ferret', lit. 'little pilferer' (c1350 MED).

**Ferreter:** Walter *le Furettour* 1318 Cl. A derivative of the verb 'to ferret', 'one who searches for rabbits, etc., with a ferret', also 'one who searches closely, a rummager' (1601 NED).

**Ferriday:** *v.* FARADAY

**Ferrier, Ferrior:** Sibyll *le Feryere* 1279 RH (Hu); John *Feryour* 1390 LoPleas. A derivative of *ferry* 'one who keeps or looks after a ferry' (c1440 NED).

**Ferries, Ferris, Ferriss:** Richard *Ferreys* 1557 Pat (O); John *Ferris* 1642 PrD. Late forms of *Ferrers.*

**Ferriman, Ferryman:** Robert *ferriman* 1192 P (Berks); Richard *le Feriman* 1246 AssLa. ME *feri* and *man,* identical in meaning with FERRIER (1464 MED).

**Ferry, Ferrie, Ferri, Ferrey:** Walter *de Ferie* 1217 FeuDu; John *del Fery* 1379 PTY. From Ferrybridge (WRYorks), DB *Ferie,* or from residence near a ferry, ME *feri* (c1425 NED), or 'ferryman'.

**Fetherston:** *v.* FEATHERSTON

**Fetherstonhaugh:** *v.* FEATHERSTONEHAUGH

**Fett, Fettes:** Emma *Fete* 1227 Cur (O); John *le Fette* 1294 MPleas (Sx); John *ffettys* 1379 PTY; William *Fete* 1473 IpmNt. OFr *fait,* ME *fet* 'suitable, becoming, comely'.

**Fetters, Fayter:** Walter *le Faytour* 1255 Rams (Beds); Adam *le Feytur* 1272 FFSt; Ysabella *le Feter* 1279 RH (C). ME, AFr *faitour* 'doer, maker', in the special sense 'impostor, cheat'.

**Fettiplace:** Thomas *Faiteplace* 1210 Oseney (O); Robert *Fetesplace* 1227 AssBeds; Adam *Feteplace, Fetteplace* 1259–60 Oseney (O); Peter *Fetiplace* 1427 AD vi (O). AFr *fete place* 'make room', probably a name for an usher. A specifically Oxford name, borne by a 14th-century mayor.

**Fever:** *v.* FEAVER

**Feverel:** Roger *Feuerelle* 1153–85 Templars (Herts); Robert *Feuerel* 1286 FFEss; Simon *Féverel* 1334 LLB E. ME *feoverel, feverel* (a1225 MED) is a form of *February,* apparently of English origin. 'February is paynted as an olde man sittynge by the fyre' (1389 Trevisa) suggests a possible origin for the nickname. cf. FEAVERYEAR.

**Feveyear, Fevyer:** *v.* FEAVERYEAR

**Fewster, Foister, Foyster, Fuster:** Durand, Walter *le fuster* c1179 Bart (Lo), 1272 AD ii (Mx); William *Fuyster* 1557 FrY; William *Fuister,* son of Nicholas *Fewster* 1627 ib. AFr *fuster, fuyster,* OFr *fustier, fuyster, fustrier* 'saddle-tree maker' (1309 NED). *v.* also FORSTER, FOSTER.

**Fey:** *v.* FAY

**Fick, Ficke, Feakes, Feaks, Feek, Feeks:** *Ficka* 14th AD vi (St); Semann' *Fike* 1197 P (Nf); Stephen *Fykke,* Walter *Feke* 1332 SRSx, SRSt. *v.* FITCH. *Feek* is from *Fēke,* with lowering and lengthening of the vowel of *Fike.*

**Ficken, Feakins:** Richard *Ficun, Fycun* 1219 AssY. A diminutive of FICK.

**Ficker:** William *Fikere* 1230 P (K), John *ffeker* 1327 SRC. *v.* FITCH, FITCHER.

**Fickett:** Alexander *Fiket* 1206 P (Sr); Richard *Fyket* 1284 FFSf. *v.* FITCH.

**Fiddes:** Edmund *de Fotheis* 1200–7 Black; Walter *de Fothes* 1328–9 ib.; William *Fudes,* Elizabeth *Fiddes* 1524, 1600 ib. From the barony of Fiddes in Foveran (Kincardine).

**Fiddian:** *v.* VIVIAN

**Fiddy, Fido, Fidoe:** John, Walter *Fizdeu, fideu* 1327 Pinchbeck (Sf). Fr *fitz deu* 'son of God'. *v.* FITHIE.

**Fidge:** Richard, William *Fige* 1198 P (Sx), 1230 P (Lo). *v.* FITCH.

**Fidgeon, Fidgen:** *v.* VIVIAN. Also for *FitzJohn.* Fidgeons Croft in High Easter is named from Peter *ffitz John* 1403 PN Ess 638. Other Essex farms named Fitzjohns occur as *Figeons* alias *Fitzjohns* 1630, *Phigons* 1776, *Phidgeons* 1777.

**Fidgett:** *v.* FITCH

**Fidkin, Fitkin:** Philip *Fidekyn* 1324 Wak (Y). A diminutive of *Fidd,* probably a pet-form of *Vivian.*

**Fidler, Fiddler, Vidler:** (i) John *le Fithelare* 1275 SRWo, *le ffithelere* 1285 FF (Ess); Simon *Le Vythelar'* 1327 SRWo; John *Fydeler* 1379 PTY. OE *fiðelere* 'fiddler', 'one who plays on the fiddle', especially for hire. *Vidler* is the southern pronunciation. (ii) Hunfridus *Uis de leuu* 1086 DB (Berks), *Vis de lupo* ib. (Ha); Walchelinus *Visus lupi* 1130 P (O), *Videlu* 1198 FF (Sf); William *Visdelou* 1160 P (Sf). OFr *vis de leuu* 'wolf's face', from OFr *vis* (Lat *visus*) 'face' and *leuu* (Lat *lupus,* 'wolf'). *Videlu* would become *Vidloe* and *Vidler,* then be confused with *fiddler,* now the usual form in place-names.

**Fidling, Fidlin:** Geoffrey *de Fiteling, Fitling* 1193, 1230 P (Y). From Fitling (ERYorks), DB *Fidlinge.*

**Fido(e):** *v.* FIDDY

**Field, Fields, Feild, de la Feld, Delafield:** Robert *de Felde* 1185 Templars (Gl); Hugo *de la Felde* 1188 P (Beds); John *del Feld* 1190 BuryS (Sf); James *atte Felde* 1296 SRSx; William *othe felde* 1327 SRWo; John *in theffelde* 1333 MELS; Baldwin *Felde* 1428 FA (Wo). OE *feld,* but, here, probably, with reference to cultivated land or the open fields.

**Fielden, Feilden, Velden:** William *de la Felden* 1286 Ipm (Wo). The dative plural of OE *feld,* 'one who lives in the fields'.

**Fielder:** Geoffrey *le Felder* 1327 SRSx. 'Dweller by the field', equivalent to *atte felde* (cf. BRIDGER) or 'worker in the field': *Felders* is an alternative reading for *sowers* in *Piers Plowman* (1393 NED).

**Fielders:** *v.* FIELDHOUSE

**Fieldfare:** Ralph *Feldefare* 1284 Wak (Y); John *Feldefer* 1461, Richard *Feldefair* 1483 Black. A nickname from OE *feldware* 'a kind of large thrush'.

**Fieldgate, Felgate, Fellgate, Fellgett, Filgate:** Godwin *de Feldegat* 1203 P (Nf); Elias *de Felgate* 1327 SRWo. From residence near a field-gate, OE *geat* 'gate' (cf. Felgate PN Wo 241), or by a road leading to a field (ON *gata* 'road').

**Fieldhouse, Fielders:** Thomas *de Feldeshous,* Henry *de Felhouse* 1332 SRSt. 'Dweller at the house in the fields' cf. WOODHOUSE. Felhouse may have become *Fellowes.*

**Fieldhurst:** William *de Feldhurst* 1296 SRSx. 'Dweller by the wood in the field', OE *feld, hyrst.*

**Fielding:** Ralph *Felding* 1279 RH (Hu); Roger *Fylding* 1327 SRDb; Richard *Feildeinge* 1630 RothwellPR (Y). OE *\*felding* 'dweller in the open country'.

**Fieldman:** Thomas *Feldman* 1327 SREss. 'Dweller in the open country, or by the field', OE *feld, mann.*

**Fieldsend:** John *attefeldesende* 1270 MELS (Wo). 'Dweller at the end of the field.'

**Fieldwick:** *v.* FELDWICK

**Fiennes, Fynes:** Sibilla *de Fiednes* 1199 MemR (Berks); William *de Fiennes, de Fiesnes* 1212, 1216 ChR; John *de Fienes* 1332 SRSx. From Fiennes (Pas-de-Calais).

**Fierman:** *v.* FAIRMAN

**Fife, Fyfe, Fyffe, Phyfe:** Ele *de Fyfe* 1296 CalSc (Fife); John *de Fyff* 1436 Black. From Fife (Scotland).

**Fifehead, Fifefield, Fifield, Fifett, Fifoot:** Richard *de Fifhida* 1170 P (Berks); William *de Fiffide* 1236 Fees (W); Richard *Fiffede* 1317 AssK. From Fifehead (Dorset), Fivehead, Fitzhead (Som), Fifield (Oxon, Wilts) or Fyfield (Berks, Essex, Glos, Hants, Wilts), all 'five hides'.

**Figg, Figge:** John *le Figge* 1327 SRSx; Simon *Figge* 1381 LLB H. *v.* FITCH.

**Figgess, Figgis:** Thomas *le Fykes, Fykeis* 1281 LLB B, William *Fykays* 1317 AssK. A Norman-French form of OFr *ficheis* 'faithful'.

**Figgett, Figgitt:** *v.* FITCH

**Figgins:** Robert *Fygen,* John *Fygyn* 1545 SxWills. *v.* FITCH.

**Fighter:** Thomas *Fighter* 1350 AssLo. OE *feohtere* 'warrior, foot-soldier'.

**Filbert, Philbert:** Temannus *filius Filberti* 13th Rams (C); Henry *Filberd* 1488 LLB L; Lewis *Fylberte* 1642 PrD. OG *Filuberht,* OFr *Philibert.*

**Filby, Filbee, Filbey, Philby, Philbey:** Nicholas *de Filebi* 1202 FFNf; Adam *de Phyleby* 1287 Wenlok; Simon *de Filby* 1310 LLB D. From Filby (Norfolk).

**Filderay:** William *Fylderay* 1332 SRSx. OFr *fils de rei* 'son of the king'.

**Fildes:** Dike *del Filde* 1281 AssLa. From The Fylde (Lancs).

**Filehewer:** John *le Fylehewere* 1322 CorLo. 'A maker of files', OE *fil, hēawere.*

**Filer:** John *le Filur* 1275 SRWo; John *le Fyler* 1309 SRBeds. OFr *fileur* 'spinner', from *fil* 'thread' or a derivative of ME *filen* 'to file', one who files, smooths, polishes, or of OE *feol, fil* 'file', a maker of files.

**Filgate:** *v.* FIELDGATE

**Filkin, Filkins:** (i) Geoffrey *de Filking* 1185 Templars (O); John *Filekins, Philekinge* 1257–68 Oseney (O). From Filkins (Oxon). (ii) Philip *Filkyn* 1373 AD vi (Ch). A diminutive of *Phil* (Philip). cf. German *Filcoke* 1298 Wak (Y).

**Fill, Fills, Filson, Philson:** John son of *Fylle* 1344 AD vi (Sa); William *Fille, Phille* 1317 AssK; William *Filleson* 1327 SRWo. *Phill, Fill,* a short form of *Philip.*

**Fillary, Fillery:** Richard *fiz le Rey* 1245 FFEss; Henry *Filleray, Fyleray, Fiz le Rey* 13th Percy (Sx); William *Fylderay* 1332 SRSx. 'The king's son', AFr *fiz, fil le (de) rei,* with variation in the forms between *fiz* from *fils* from Lat *filius* and *fil* from *filium.* cf. Fitzleroi Fm and Philderayes (PN Sx 127, 177).

**Filleul, Filliel:** Roger *filiolus* 1130 P (W); William *Filliol* 1175 P (Beds/Bk); Robert *Phillol* 1252 FFEss; William *Filiol* 1363–4 FFSr. Fr *filleul* 'godchild'. Apparently the name is also one of the sources of FELIX, since Felix Hall in Kelvedon (Ess) takes its name from Baldwin *Filoil* Ric I.

**Filley:** John *le ffylye* 1278 PN K 375; John *Fillie* 1327 SRSx; Richard *Filly* 1380 AssWa. ME *filli* 'a young mare, female foal', perhaps in the sense 'skittish'.

**Filliel:** *v.* FILLEUL

**Fillingham:** Alan *de Filingham* 1172 Gilb; Ralph *de Filingham* 13th RegAntiquiss; John *Fillyngham* 1298 AssL. From Fillingham (L).

**Filmer:** John *de Fillemere* 1317 AssK; John *Filmer* 1498 PN K 239. From Filmer Wood in Wichling (K).

**Filson, Philson:** William *Fillesone* 1327 SRWo; Richard *ffillesone* 1332 SRLa; Thomas *Filleson* 1374 AssL. 'Son of *Fill*', a pet-name for *Philip.*

**Finch, Fink, Vinck, Vink:** Godric *Finc* 1049–58 OEByn; Ælfwin *Finche,* Allwin *Finke* 1148–67 ELPN; Gilbert *le Finch* 1205 Cur (Nf); Walter *le Vinch* 1275 SRWo; John *Vynk* 1373 ColchCt. OE *finc* 'finch', as a nickname, perhaps 'simpleton'. *Vinch, Vynk* are southern forms. *Fink* may sometimes be of recent German origin, but is often certainly old. Ælfwin (Aylwin) *Finche, Finke* was of the family which gave name to Finch Lane and St Benet Fink (London), *Finkeslane* in parochia *Sancti Benedicti Finke* 1231 Clerkenwell.

**Fincham:** Robert *de Fincheham* 1202 AssNth; John *de Fincham* 1305 LLB B; Gilbert *Fyncheham* 1394 AssL. From Fincham (Nf).

**Finchcock:** Richard *ffynchkoc* 1332 PN K 309; Richard *Fynchecok* 1334 SRK; William *Fynchecok* 1392 CtH. A diminutive of FINCH, OE *finc, cocc.*

**Finden, Findon:** Roger *de Findon* 1262–3 FFSx; John *de Fyndone* 1317 AssK; William *de Fyndon* 1348–9 FFSr. From Findon (Sx).

**Findlater, Finlator, Finlater:** Geoffrey *de Fynleter* 1342 Black. From Findlater (Banffshire).

**Findlay, Findley, Findlow:** *v.* FINLAY

**Findlow:** William *Findeloue* 1281 FFY; William *Fyndeloue* 1302 RegAntiquiss; Walter *Fyndelove* 1346 Pat (L). 'Find wolf', OE *findan,* ADr *louve.* A nickname for a hunter. *v.* also FINLAY.

**Findon:** *v.* FINDEN

**Fine, Fines:** *Fina* 1293 FFEss, 1327 *SR* (Ess); Richard *Fine* 1196 P (Wa). OFr *fin* 'delicate, tender', or a woman's name *Fina.*

**Finegold:** William *Findegold* 1202 FFL. 'Find gold', OE *findan, gold.*

**Finer, Finar:** Walter *Finor* 1189 FFC; Ralph *le finur* 1218 AssL. OFr *fineur* 'refiner of gold, silver, etc.' (1489 NED).

**Finesilver:** Hugh *Findsilver* 1279 RH (C). cf. William *Findegold* 1202 FFL. Both may be nicknames or both may be occupational, 'refiner of silver, gold', with intrusive *d.*

**Fingard:** Robert *Finegod* 1279 RH (Beds); John *fyngod* 1332 SRCu. The earliest example is *Finde-, ffyn(d)godeshurne* 1232, 1285 PN Ess 159. The nickname may be a parallel to *Gathergood* and the opposite of *Scattergood* but the *d* appears to be intrusive and we may have 'fine goods', a nickname for a seller of choice merchandise.

**Finger:** William *Finger* 1219 AssY; Peter *Fynger* 1310 ColchCt; Richard *Fynger* 1327 SRSf. OE *finger* 'finger'. A nickname from some peculiarity of the finger.

**Fingle:** William *Findegale* 1202 AssL; William *de Fingale* 1327 SRY. From Finghall (NRY).

**Fink:** v. FINCH

**Finkel, Finkle:** (i) John *Fenekele* 1327 SRSf; John *Fenkell* 1485 LLB L; William *Fynkell* 1540 FFEss. Perhaps a nickname from ME *fenkel* 'fennel'. cf. William *Fenkelspire* 14th Whitby. (ii) Robert *de Finkehal'* 1219 AssY. From Finchale (Du).

**Finlater:** v. FINDLATER

**Finlay, Finley, Findlay, Findley, Findlow, Finlow:** *Fionnlaoich*, *Findlaech* c1070 Black; *Fynlayus clericus* 1246 ib.; Andrew *Fyndelai* 1526 ib.; Robert *Finlaw* 1567 ib.; John *Findlo* 1639 ib. Gael *Fionnlagh* 'fair hero'.

**Finlayson, Finlaison, Finlason:** Brice *Fynlawesone* 1296 Black; John *Finlason* 1511 ib. 'Son of *Finlay*.'

**Finmore, Fynmore:** Gilbert *de Finemere* 1206 P (O); Nicholas *de Fynmore* 1249 AssW; Hugh *de Fynemere* 1347 KB (Beds). From Finmere (O).

**Finn, Fynn, Phin, Phinn:** *Fin*, *Phin danus* 1066 DB; *Fin* de Haltun 1143–7 DC (L); Aldeth, Norman *Fin* 1190 BuryS (Sf), 1202 AssL. ON *Finnr*, ODa, OSw *Fin*. Hugo *filius Fin* is identical with Hugo *Fin* 1178–9 P (Y). In Ireland, for *Ó Finn* 'descendant of *Fionn*' (fair).

**Finnemore:** Richard, William *Finamur* 1204 Cur (D), 1237 *FF* (Ha). OFr *fin amour* 'dear love'.

**Finney, Viney:** John *de Fyney* 1274 Wak (Y); Philip *de Fineye* 1318 AD v (St); Thomas *ate Vynhaghe* 1327 MELS (Sx); Thomas *Finney* 1642 PrD. From Finney (La), Fenny Rough in Chaddesley Corbett (Wo); Fenay in Almondbury (WRY), *La Fineia* 12th, Vinals Fm in Cuckfield (Sx), or Viney's Wood in Crundale (K).

**Finnick:** v. FENWICK

**Finning, Finnings, Vining, Vinning:** (i) Robert *Fininge* 1210–11 PWi; Thomas *Finning* 1228 Cl (Sf); Alan *Fynyng* 1332 SRSx. OE *\*Finning*. (ii) Robert *atte Finnyng* 1296, Ralph *de Vynynge* 1327 SRSx; Robert *Vining* 1641 PrSo. From Fyning in Rogate (Sx).

**Finningley:** Harry *Fenyngley* 1459 Paston. From Finningley (Nt).

**Firbank, Furbank:** Henry *Firthbank* 1470 FrY; Richard *ffirbanck* 1657 EA (OS) iv. 27 (Sf). 'Dweller by the wooded slope' (OE *(ge)fyrhþe*).

**Firby, Furby:** William *de Friby* 1219, Robert *de Fritheby* 1251 AssY; Nicholas *Furby* 1296 SRSx. From Firby (ERY, NRY).

**Firebrace, Fairbrass, Farbrace:** John *Fierebrache* 1190 P (O); John *Fierbrace* 1196 P (Ess); Robert *Ferbraz* 1221 Pat (Bk); Walter *Firbras* 1280 LLB A; John *Farbrace* 1533 KentW; Charles *Firebrace* 1723 DKR 41 (Sf). OFr *fer, fier*, ME *feer, fere* 'bold, fierce, proud' and Fr *bras* 'arm'.

**Fireman:** v. FAIRMAN

**Firk(s):** v. FRITH

**Firmage(r):** v. FIRMINGER, FURMAGE

**Firmin, Firman, Furman:** *Firmin'* lifget 1086 ICC (C); Alexander *filius Firmin* 1214–32 AD i (Mx); Richard *Fermin* 1200 Cur (Sf); Stephen *Furmin* 1219 AssY; William *ffyrman* 1522 SIA xii. OFr *Firmin* from Lat *Firminus*, a diminutive of *firmus* 'firm, strong'.

**Firminger, Ferminger, Furmenger, Furminger, Firmager:** Henry *Furmaiger* 1198 P (K); William *le Furmager* 1219 AssY; John *Furmonger* 1490–1508 ArchC 49. OFr *fromagier, formagier* 'maker or seller of cheese'. cf. MESSENGER from *Messager*. v. also FURMAGE.

**Firth, Frith, Frid, Fridd, Fryd, Freeth, Freed, Vreede, Frift, Thrift, Fright, Freak, Freake, Freke, Firk, Firks:** All these surnames derive from various developments of OE *firhþe*, *(ge)fyrhþe*, *ferhþe* 'frith, wood, woodland'. (i) The combination *-rhþ-* was difficult to pronounce. The medial *h* was dropped, *firhþe* became ME *firthe*, *ferthe*: Robert *atte Verthe* 1295 MELS (Sx); Nicholas *atte Ferthe* 1296 SRSx. *Firth* is common in Lancashire and Yorkshire. Vert Wood (Sussex) preserves the southern initial *V*. (ii) By metathesis, ME *firthe* became *frithe*, *frethe*, *frid*, *frede*, and later, dialectal *freeth* and *vreath* (Devon, Glos, Som): Ralph *delfrid* 1176 P (Sr); Wlmar *de Frith* 1195 P (K); John *del Frith* 1201 P (Nf); William *in le Frith* 1276 *For* (Ess); Nicholas *atte Frithe* 1275 SRWo; Edith *Ythefrithe* 1300 MELS (Wo); Denis *Frede* 1327 *SR* (Ess); Thomas *atte Vrythe* 1333 MELS (So); Richard *atte Frethe* 1377 FFSx. *Frith* is frequent in Essex, Herts, Sussex, Kent, Wilts; cf. Frid Fm and Wood (Kent), Freath Fm (Wilts), Frieth (Bucks). *Thrift* is a common late development, not so far noted before the 18th century. (iii) If the *h* was preserved, *-hþ-* became *-ht-*: *firhþe* became *friht*, later *fright*, especially in Kent where the surname is common: Serlo, John *del Friht* 1197, 1203 P (K, Nf); Henry *de fricht* c1248 Bec (Bk); John *atte Frizte* 1327 SRSx. cf. Fright Fm (Sussex). (iv) Or, *-hþ-* became *-kþ-*: *fyrhþe* became *fyrkþe*, ME *ferkthe*, and, by metathesis, *frekthe*, *freek*: Alexander *de la Frike* 1275 SRWo; Robert *atte Ferghe* 1327 SRSx; John *atte Ferkche* 1332 ib. cf. Freek's Fm and Frag Barrow (PN Sx 260, 301) and v. MELS 72–3.

**Fish, Fysh, Fisk, Fiske:** Ernis *Fish* 1202 AssL; Daniel *Fisc* 1208 ChR (Sf); Robert *Fisk'* 1230 P (Nt); Robert *le Fysch* 1297 MinAcctCo. OE *fisc* 'fish', a nickname. *Fisc* occurs as a personal name in DB (Nf) and is probably ON *fiskr* 'fish' used as a byname.

**Fishacre, Fisaker:** Martin *de Fisacre* 1214–18 Praes; Martin *de Fysacre* 13th Hylle; Thomas *Fyshacre* 1359 AssD. From Fishacre in Broadhempston (D).

**Fishbourn, Fishbourne, Fishburn, Fishburne:** Richard *de Fisseburn* 1206 Cur (Lei); Ranulf *de Fissheburne* c1250 FeuDu; William *Fysshebourn* 1332 SRSx. From Fishbourne (Sussex), or Fishburn (Durham).

**Fisher:** (i) Richard *le Fischer* 1263 FFEss. OE *fiscere* 'fisherman', the common source of this surname. (ii) Ralph *de Fisshar'* 1296 SRSx; Martin *atte Fisshar* ib. 'Dweller by an enclosure for catching fish', OE *\*fiscgear*. v. Fisher (PN Sx 74) and MELS 65.

**Fisherman:** William *Fisserman* 1203 AssNth. 'Fisherman', OE *fiscere, mann*.

**Fishleigh, Fishley:** Turstin *de Fislia* 1148 Winton (Ha); John *Fysely* 1332 SRSx; Leonard *Fishly*, Daniel *Fishley* 1642 PrD. From Fishley (Nf), Fishley Barton in Tawstock, Fishleigh in Hatherleigh, in Aveton Giffard (D).

**Fishlock:** Simon *de Fislake* 1204 AssY; William *Fysshlak* 1274 SIA xiii. From Fishlake (WRYorks). John *Fischlake* (1366 ColchCt) is called *Fisshlaker* in 1367, 'dweller by the fish-stream'. cf. BRIDGER.

**Fishman:** Randolph *Fissheman* 1474 MEOT (Sx). 'Fisherman, or seller of fish.'

**Fishwick:** Henry *de Fiswich* 1203 P (La); Robert *de Fisshewyk* 1327 SRY; Roger *Fysshewyk* 1431 FFEss. From Fishwick (Devon, Lancs).

**Fisk(e):** *v.* FISH

**Fiskerton:** Thomas *de Fiskerton'* 1217 LuffCh. From Fiskerton (L, Nt).

**Fitch, Fitchell, Fitchen, Fitchet, Fitchett:** Hugh, Roger, William *Fiche* 1243 AssSo, 1297 SRY, 1327 SRSf; *Fichet* 1201 Cur (K); Hugh, Robert, Walter *Fichet* 1176, 1183 P (Do, Nth), 1194 Cur (He). The common derivation of *Fitch* and *Fitchet* from the polecat is untenable. This word is from OFr *fissell*, plural *fissiaulx*, later *fissau* with forms *fitchewes* 1394, *fycheux* 1418, *fechets* 1535, *fichat* 1653 (NED). This late development of *-et* cannot account for the 12th century *Fichet*. We may compare the French *Fiche, Fichet, Fichot*, which Dauzat derives from OFr *fiche* 'an iron point', from *ficher* 'to fix, plant'. *Fitch* is thus 'an iron-pointed implement', used by metonymy for *Fitcher*, the workman who uses this, and *Fitchen* and *Fitchet* are diminutives. As Hugh *Malet* is said to have abandoned for a time his nickname 'little hammer' in favour of *Fichet* (*v.* MALLET), *fiche* must have been used of a pointed weapon, a spear or lance, and *Fitch* and *Fitchett* of a spearman or a knight famous for his exploits with the lance. By the side of these, French has the Norman *Fiquet*, whence the English *Fickett*, the simple *Fick*, with *Feakes* and *Feek(s)*, the diminutive *Ficken*, the occupative *Ficker* and the apparently obsolete Ædredus *Fikeman* 1180 P (Wo), Nicholas *Fekeman* 1327 SRSf. Richard *fikberd* (1277 Ely) probably wore a pointed beard. With these names must be taken *Figg, Figgett, Figgitt*, and *Figgins*, with the non-surviving diminutive Alfwinus *Figgel* 1166 P (K), Richard *Figel* 1279 RH (Beds), where we have the voicing of *k* to *g* as in *Picot* and *Piggott*. Here, too, must belong *Fidge*, and *Fidgett*, with a similar voicing of *ch* to *dg*. No early examples of *Fidgett* have been found. It cannot be associated with the common *fidget* which occurs too late to have given a surname, apart from the difficulty of derivation. *Fitch*, like *Fick* and *Fichet*, may have been used as a personal name. cf. also *Fechel* de Fercalahn 1225–50 Dublin. This would account for the diminutives. *Fitcher, Ficker* and *Fikeman* must be occupational.

**Fitchelden, Fitcheldon:** *Ficheldene* 1148 Winton (Ha). From Figheldean (W), *Ficheldene* 1115.

**Fitcher:** Roger *Fychere*, Richard *Fichere* 1327 SRSf. *v.* FITCH.

**Fitchew, Fitchie:** *v.* FITHIE

**Fithie, Fithye:** (i) Alice *Feithew* (widow of Henry, lord *Fitzhugh*) 1473 GildY; Edward *Fythewe, Fitzhugh* 1492, 1497 ib.; John *Fyddie* 1609 FrY. A colloquial pronunciation of *FitzHugh* 'son of Hugh'. This probably also became *Fitchew* and *Fitchie*. (ii) In Scotland, from Fithie in Farnell (Angus): Henry *of Fythie, de Fethy* c1328 Black.

**Fitkin:** *v.* FIDKIN

**Fitt, Fitte:** Robert *Fitte* 1201 Cur (Sa); Roger *Fitte* 1301 SRY; James *Fitte* 1484 LLB L. ME *fit* 'fitting, suitable'.

**Fittel, Fittell, Fittle:** Estan *Fitel* 1175–86 Holme; William *Fyttel* 1249 AssW; Richard *Fytele* 1327 SRSf. OE *\*Fitel*, an original byname from the first element of OE *fitel-fōte* 'white footed', in ME a name for the hare.

**Fitter:** Geoffrey, Hugh *le Fittere* 1195 P (Wa), 1231 Cl (Gl). 'A fitter.' Bardsley suggests 'carpenter'. In northern dialect the term meant one 'who vends and loads coals, fitting ships with cargoes' (Halliwell).

**Fitton:** (i) Richard *ffyton* 1188 WhC (La); Richard *Fitun* 1195 P (Wa). In ME *fitten* was a common term for lying, deceit. (ii) Alan *de Fittun* c1213 Fees (C). From Fitton Hall in Leverington (Cambs).

**Fitz, Fitze:** Reginald *le Fiz* 1193 P (Wa); Robertus *Filius*, Robert *Fiz* 1212 Cur (Y); Bartholomew Burghassh *le Fitz* 1346 FA (Sf). AFr *fiz* 'son'. The last example perhaps gives a clue to the meaning. It is apparently used to distinguish the son from the father. Names like *Fitzalan, Fitzwilliam*, etc., have acquired an aristocratic flavour. Early examples are rare. Osbertus *le fiz Fulco* 12th EngFeud (Y), etc., usually appear in documents as Osbertus *filius Fulconis* and at this period were descriptions rather than surnames. This Latin form has often been translated into French by historians and genealogists, and, unfortunately, sometimes by editors of documents who have thus created families of FitzOdo, FitzThomas, etc., at a time when the family had no fixed surname. In colloquial speech, the *fitz* was liable to various forms of assimilation to the following consonant, thus giving *Fithie* and *Fitchew* for *FitzHugh*, *Fidgeon* for *Fitzjohn, Filleray* for *FitzRoy, Fennell* for *FitzNeal, Feehally* for *FitzHarry*. For FITZGERALD, FITZPAYN, etc., *v.* GERALD, PAYN, etc.

**FitzAlan:** *v.* ALLAIN

**FitzAucher:** *v.* ALGER

**FitzHerbert:** William *Fitz Herbert* 1295 Husting; Margery *Fitzharberd* 1421 IpmNt; Antony *Fitzharbard* 1516–17 GildC; Richard *Fitzherbert* 1641 PrSo. 'Son of *Herbert*', AFr *fiz*.

**FitzHugh:** *v.* FITHIE, HUGH

**FitzJames:** Thomas *Fitzjames* 1345 Hylle; James *Fytzjames* alias *Fytjames* alias *Fiejames* alias *Fysejames* 1559 Pat (So). 'Son of *James*', AFr *fiz*.

**FitzJohn:** John *Fis-John* le Irreis 1268 AssSo; John *le Fitz Johan* 1355 LoPleas; John *Fyion* 1453–5 Oseney. 'Son of *John*', AFr *fiz*. *v.* also FIDGEN.

**FitzNeal:** *v.* FENNEL, NEAL

**Fitzroy:** Henry *fis le Rey* 1296 SRSx. cf. FILLARY.

**FitzSimon, FitzSimons, Fitzsimmons, Fitzsimmond:** Walter *le fiz Simon* Hy 2 DC (L); Richard *FitzSymond* 1387 Misc (Ess); John *Fysimond* 1392 LoCh. 'Son of *Simon*', AFr *fiz*.

**Fitzwater:** v. WALTER, WATER

**FitzWilliam, FitzWilliams:** Rauf *le fuiz William* 1299 Whitby; Edmund *Fitzwilliam* 1424 TestEbor; Margaret *Fethwilliam* 1509 GildY. 'Son of *William*', AFr *fiz*.

**Flack:** Robert *del Flac* 1276 RH (Y); Christopher *Flack*, Thomas *Flaack* 1642 PrD. 'Dweller at a place where turfs are cut', ME *flak* 'turf'.

**Flacker:** William *le Flacuer* 1212 Cur (Y); William *Flacker* 1641 PrSo. 'A turf cutter', from a derivative of ME *flak* 'turf'.

**Fladgate:** v. FLOODGATE

**Flagg:** Peter *de la Flagge* c1280 SRWo; Henry *del Flagg* 1327 SRDb. ON *flaga* 'a flagstone', or ON *flag* 'a turf, sod'. Hence 'dweller at a place where turfs were cut or flagstones obtained', as at Flagg (Derby), or Flags (Notts).

**Flain, Flaine:** Robert *Flain* 13th Gilb (L); Hugh *Flayne* 1381 AssL. ON *Fleinn*, an original nickname for a sharp-tongued person, from ON *fleinn* 'dart, arrow, hook'.

**Flaherty:** v. O'FLAHERTY

**Flamank:** v. FLEMING

**Flambard, Flambert:** Robert *Flambard* 1158 CartAntiq; John, Matthew *Flambard* 1218, 1303 FFEss. OFr *flambard* 'a flaming torch'. The earliest and best known of the name was Ranulf Flambard, bishop of Durham 1099–1128. He appears as Ranulfus *capellanus* 1087–99 StP, but his original nickname appears to have been *Passeflambard*, cf. Randulf *Passeflambard* of Dunholme 1128 ASC E, Radulfus *Passeflambard*, Randulfus *Basseflambard* 12th LibEl. This is a phrase-name 'pass on the flaming torch'.

**Flanagan:** Ir *Ó Flannagáin* 'descendant of *Flannagán*' (red).

**Flanders, Flinders:** Euerdai *de Flandria* a1191 DC (L); Thomas *Flaundres* 1327 SRSo. From Flanders.

**Flann:** v. FLAWN

**Flanner:** Richard *le Flauner* 1211 FrLeic; Reginald *Flawner* 1222–32 Seals (Herts); Walter *le flaoner* 1246–89 Bart (Lo); Simon *le Flanner* 1260 AssC. OFr *flaonnier*, *flaunier* 'maker of flawns', a kind of custard or pancake. v. FLAWN.

**Flash, Flasher, Flashman:** Felicia *Flasche* 1279 RH (C); Habram *de Flaskes* 1301 SRY; William *del Flosche* 1314 Wak (Y). 'Dweller by the pool or marshy spot', ME *flasshe*, *flask*.

**Flashby:** William *Flascheby* 1394 TestEbor. From Flasby (WRY).

**Flather:** Simon *Flather* c1265 Calv (Y); John *Flathir* 1324 CoramLa; John *Fladder* 1530 FrY. 'A maker of flathes or flawns'. v. FLANNER.

**Flatman:** Thomas *Flatman* 1568 SRSf. v. FLATT.

**Flatt:** Geoffrey *del Flate* 1327 SRDb; Thomas *del Flat* 1349 FrY. 'Dweller on the level ground.' ON *flatr*.

**Flawn, Flann:** Elena *fflaun* 1327 SRC; John *Flaoun* 1357 LLB G. OFr *flaon*, ME *flaun*. Metonymic for FLANNER.

**Flaxley:** William *Flaxleye* 1327 SRWo. From Flaxley (Gl, WRY).

**Flaxman, Flexman:** William *Flexman* 1279 RH (Hu); Nicholas *Flaxman* 1332 MESO (Nf). v. FLEXER and cf. Cristina *Flexwyf* 1378 LLB H; Crispin *Flaxbeter* 1219 AssY; Richard *le flexmongere* 1294 RamsCt (Hu); John *Flax* 1327 SRY.

**Flay, Flaye, Fleay:** William *Fleie* a1233 FeuDu; Margaret *la Flegh* 1332 SRSx; Thomas *Flea*, Robert *Fley* 1642 PrD. A nickname from OE *fleah* 'flea', or from OE *flēoge* 'a flying insect'.

**Fleck:** (i) Roger *Fleke* 1210–11 PWi; Peter *Fleke* 1327 SRSo; Richard *Fleck* 1642 PrD. ON *flaki*, *fleki* 'hurdle'. Metonymic for a maker of these. (ii) John *de Fleckhe* 1342–3 FFWa. Probably from Flecknoe in Wolfhampcote (Wa).

**Flecker:** (i) Julian *Flekeher* 1327 SRSx; William *Fleiker* 1641 PrSo. A derivative of ON *flaki*, *fleki* 'hurdle', a maker of wattled hurdles. (ii) Simon *le Fleckere* 1279 AssNb. The northern form of FLETCHER.

**Fleckney:** Robert *de Flecenege* 1185 P (Lei); Richard *de Fleckeneye* 1230 P (Lei). From Fleckney (Lei).

**Fleckno, Flecknoe:** Richard *de Flecho* 1221 AssWa. From Flecknoe in Wolfhampcote (Wa).

**Fleeman:** v. FLEMING

**Fleet:** (i) Richard *de Flet* c1158 Gilb (L); Simon *ate Flete* 1297 MinAcctCo; Alicia *del Flete* 1327 SRSf. From Fleet (Lincs) or 'dweller by the estuary or stream', OE *flēot*. (ii) John *le Fleet*, *le Fleot* 1327 SRSo. ME *flete* 'swift'. cf. OE *flēotig*.

**Fleetham:** Thomas *de Fletham* 1204 P (Nb). From Fleetham (Nb).

**Flegg, Flegge:** Ralph *de Fleg* 1198 Pleas (Nf); Reginald *of Fleg* 1238 FFY; John *Fleg* 1325 CorLo. From Flegg (Nf).

**Fleming, Flemming, Flemons, Flemyng, Fleeming, Fleeman, Flamank, Flament, Flement, Le Fleming:** Serlo *le Flemyng* c1150 Gilb (L); William *le Flamanc*, *Flandrensis*, *Flamanc* 1219 AssY; Adam *Flemyng* 1296 SRSx; Thomas *Flemin* 1644 FrY; Richard *Fleaming* 1648 ib.; Isaac *Fleeminge* 1674 ib.; David *Fleemen* 1697 ib. AN *fleming*, OFr *flamanc* 'a Fleming' (c1430 NED). cf. FLANDERS.

**Flerd, Flert:** Richard *Flerd* 1178 P (Ha); William *Flert* 1327 SRLei. OE *(ge)flēard* 'falsehood, deceit', a deceiver.

**Flesh:** William *Fles* 1205 Pleas (Sx); John *Fles* 1308 AssNf; John *Flesshe* 1418 LLB I. OE *flǣsc* 'flesh'. Metonymic for FLESHER.

**Fleshacker:** John *le Flesackere* 1275 AssW. OE *flǣsc*, and a derivative of ME *hakken* 'to cut', a butcher.

**Flesher:** (i) Richard *le Fleshewere* 1268 MESO (Y); Albred *le Flesshewere* 1311 FFC; Richard *Flesseure* 1374 AD ii (Bk); Thomas *Flesshour*, *Flesshuer* 1453, 1455 GildY. OE *flǣsc* 'flesh' and a derivative of *hēawan* 'to cut', 'a butcher'. (ii) Adam, John *Flescher* 1379 PTY; Richard *Flescher* 1410 GildY. A derivative of *flesh*, with the same meaning, 'butcher'. The surname has been confused with *Fletcher*, Fletcher Gate at Nottingham was formerly Flesher Gate (PN Nt 17).

**Fletcher:** Robert *le Flecher* 1203 AssSt; William *Flecher* 1203 Cur (La); Peter *le flechier* 1227 AssBeds. OFr *flechier*, *flecher* 'maker or seller of arrows' (c1330 MED).

**Fleury, Florey, Flory, Flury:** (i) Ranulf *de Flury* 1201 AssSo; Hugh *de Flori*, Giles *Florey* 1286, 1295 FFEss. From Fleury, a common northern French place-name. (ii) *Fluri* 1192 P (Ha); *Floria* 1193 P (Lo); 1221 AssWa; Gilbert *Fluri* Hy 2 DC (L); John *Flory* 1230 P (Nf). A woman's name, ultimately from Lat *flos* 'flower'. There was a Spanish saint Floria, martyred in the 9th century.

**Flew:** Thomas *Flew* 1641 PrSo; Edmund *Flew*, William *Flue* 1641 PrD; John *Flew* 1662–4 HTDo. ME *flue* 'a kind of fishing net'. Metonymic for a fisherman.

**Flewett, Flewitt, Flowitt:** Hubert *flohardus* 1130 P (Lei). A rare name, probably OG *Hlodhard* 'glory-strong', surviving in France as *Floutard*. For the loss of *d*, cf. *Floheld*, *Fluold* (Forssner 90).

**Flexer:** Richard *Le Flexere* 1316 MESO (Ess); Roger *Flaxer* 1329 ib. (Nf). A derivative of OE *fleax*, *flex*, 'dresser or seller of flax'. cf. Robert *Flexhewer* 1367 AD i (Lo).

**Flexman:** v. FLAXMAN

**Flinders:** v. FLANDERS

**Flinn:** v. O'FLINN

**Flint, Flindt:** (i) *Flint* 1066 DB (Sf); John, William *Flint* c1248 Bec (O), 1250 Fees (Bk). OE *\*Flint*, an original nickname from OE *flint* 'rock', or a nickname, 'hard as a rock'. (ii) William *del, de Flynt* 1291–2 AssCh; John *del Flynt* 1345 FrY. From Flint or 'dweller by the rock'.

**Flintard:** Thomas *Flinthard'* 1219 Cur (Sf); William *Flinthard* 1301 CorLo; Henry *Flynthard* 1320 LLB E. A nickname, 'as hard as flint', OE *flint, heard*.

**Flinton:** Robert *de Flinton'* 1204 P (Y); Walter *de Flinton* 1279 IpmY. From Flinton (ERY).

**Flitter:** Robert *le Flittere* 1196 FF (Nf). OE *flītere* 'a disputer', later 'a scold', from OE *flītan* 'to contend, wrangle'.

**Flixton:** Thomas *de Flixton* 1282 IpmY. From Flixton (La, Sf), or Flixton in Folkton (ERY).

**Float, Floate, Flote:** Ralph *Flot* 1155–8 Holme (Nf); Roger *Flote* 1166 P (Y). OE *flota* 'boat, ship', metonymic for FLOATER, FLODMAN.

**Floater:** William *le Flotyere* 1249 MEOT (Sx); Ralph *le Floter* 1281 ib. (L). A derivative of OE *flota* 'ship', a sailor.

**Flock, Flocke:** John *Flocc* c1160 ELPN; Geoffrey *Floc* 1296 SRSx; Walter *Flocke* 1524 SRSf. Either OFr *floc* 'lock of wool', perhaps referring to woolly hair, or OE *flocc* 'herd, company', metonymic for a shepherd. cf. John *Flocker* 1642 PrD, from ME *flokker* 'shepherd'.

**Flodman:** Walter, John *Floteman* 1215 Oseney (O), 1524 SRSf. OE *flotmann* 'sailor'.

**Flood, Floud, Flude:** Wigot *de la Flode* 1198 P (Berks); Roger *flod* c1200 DC (L); Adam *del Flod* 1244 Rams (Beds); Thomas *atte Floud* 1327 SRSx. 'Dweller by the stream (OE *flōd*) or channel, gutter (OE *flōde*)'.

**Floodgate, Fladgate:** Walter *atte Flodgate* 1327 SRSo. 'Dweller by (or keeper of) the flood-gate' (OE *\*flōdgeat*).

**Floor, Floore:** Cecilia *de Flore* 1202 P (Nth); Hugh *Flor'* 1298 AssL; Richard *Flore* 1380 LoCh. From Floore (Nth).

**Florence:** (i) *Florentius* 1130–2 Seals (St); *Florentius de Grotene* 1201 Cur (Sf); *Florentia, Florencia* 1207–8

Cur (Sr, W); Richard *Florenz* 1220 Oseney (O); Gilbert *Florence* 1250 FFSf. *Florence*, both masculine and feminine, Lat *Florentius, Florentia*, from *florens* 'blooming'. (ii) Bartholomew *de Florence* 1273 RH (Y), Skelmynius *de Florentia* 1334 FrY. From Florence (Italy).

**Florey:** v. FLEURY

**Florkin:** William *Flurekin* 1230 P (Lo); John *Florkyn* 1327 SRSx. Diminutives of *Floria, Fluria* (f).

**Flote:** v. FLOAT

**Floud:** v. FLOOD

**Flower, Flowers:** (i) William *Floere*, John *le Floer* 1275 RH (D). ME *flōer*, a derivative of ME *flō*, OE *flā* 'arrow', an arrow-maker. (ii) William *Flur* 1203 P (Y); Edmund *Flour* 1313 FFEss; John *Flower* 1517 FrY. ME *flur, flour*, OFr *flur* 'flower', used already in the 13th century of persons, usually with the epithet 'sweet', and as a woman's name: *Flur', Flour'* 1297 MinAcctCo. cf. Robert *Flouressone* 1325 AssSt. The same forms were used for *flour*, the 'flower' of the meal and the surnames may denote a maker of flour: Robert *le fflourmakere* 1332 SRLo, who may be identical with Robert *Flourman* 1338 LoPleas, though this latter surname may also mean 'servant of Flower'. cf. Walter *Floureman* or *Floureknave* 1309 RamsCt (Ess).

**Flowitt:** v. FLEWETT

**Floyd, Floyde, Floyed, Flude:** Richard, John *Floyd* 1509, 1532 LP; Griffin *Floyde* 1544 Musters (Sr); William *Floyd* alias *Flowde* 1560 Pat (Lo). A form of LLOYD.

**Flude:** v. FLOOD

**Flury:** v. FLEURY

**Flush:** Richard *Flusche*, John *Flusshe* c1405 FS. 'Dweller by the swamp', ME *flosshe* 'swamp'.

**Flute:** Metonymic for *fluter*. v. FLUTTER.

**Flutter:** Arnulf *Flouter* 1224 Pat (Db); William *the Floutere* 1268 FFY. A derivative of ME *floute* 'to play on the flute', flute-player.

**Foad, Foat, Food:** Elfred *fode* 1221 ElyA (Nf); Richard *Food, Vod, Wod, le Vod* 1237 HPD (Ess); Thomas *Vood, le Vod* 1251, 1262 ib.; Hugo *le Vode, le Wode* 1317 AssK. OE *fōda* 'food', later 'that which is fed', 'a child' (a1225 MED). *Vod* shows the southern pronunciation, *Wod(e)* early examples of the dialectal change of initial *V* to *W*. This has probably contributed to the frequency of *Wood*.

**Foaden:** v. FODEN

**Foakes:** v. FOLK

**Foale:** Robert, Reginald *Fole* 1193 P (Wa), 1279 RH (Hu). OE *fola* 'foal', perhaps a nickname, 'frisky', but in the 14th century used of persons in the sense 'ragged: rough, shaggy'.

**Foard:** v. FORD

**Foat:** v. FOAD

**Fobster:** Geoffrey *Fobbestor* 1327 SRSo. A derivative of ME *fobber* 'cheat, trickster'.

**Focke:** v. FOLK

**Foden, Fodden, Foaden:** Thomas *Fodyn* 1296, John *Fodyn* 1327 SRSx. cf. ME *foden* 'to feed, nurse'. Probably a foster parent.

**Fodor:** Roger *le Fodere* 1327 SRSf; Walter *Fodere, Fodier* 1327, 1332 SRSx. A derivative of ME *fōdien* 'to feed', a feeder of cattle.

**Foers:** v. FOWER

**Fogarty, Fogerty, Fogaty, Foggarty:** Ir Ó Fógartaigh 'descendant of Fógartach' (banished).

**Fogden:** Edward Fogden 1525 SRSx. Perhaps from a lost Folkindenne in Sandhurst (K).

**Fogg, Fogge:** Thomas Fog 1381 ArchC iii; John Fogge 1473 Paston; Nicholas Fogg 1524 SRSf. A petform of FULCHER.

**Foister:** v. FEWSTER

**Foker:** v. FULCHER

**Fokes:** v. FOLK

**Folbarbe:** Richard Folebarbe 1204 Pleas (Herts). 'Foolish, absurd beard', OFr fol, barbe. cf. William Folbarun 1263 IpmY 'foolish baron'.

**Fold, Foldes, Folds, Fould, Fouldes, Foulds, Fowlds, Faulds:** Hugh del Foldis 1275 Wak (Y); Adam in le Fold 1327 SRDb; John atte Fold 1327 SRSo; Adam de Falde 1332 SRSt; John del ffald 1332 SRCu. 'Worker at the fold(s) or cattle-pen(s)', OE falod, later fāld, ME fold.

**Folder, Faulder:** Iuo faulder 1332 SRCu. v. FOLD.

**Folgate, Falgate:** Peter de le Falgate, Richard de Faldgate 1275 RH (Nf). 'Dweller by the foldgate', OE falod, geat.

**Folger:** v. FULCHER

**Foliot, Foliott:** v. FOLLIOT

**Foljambe, Fulljames:** William Foleiambe 1172 P (Db); Tomas Folegambe 1206 P (Nth); John Fulgeam 1533 Bardsley; Godfrey Fuljambe 1588 Shef (Y). OFr fol 'foolish, silly' and jambe 'leg', probably, as suggested by Lower, for a useless, maimed leg, the opposite of BELGIAN. In OFr fol was used of something useless or of little value: farine folle 'mill-dust', figue folle 'a good-for-nothing fig'.

**Folk, Folke, Folkes, Folks, ffoulkes, Foulkes, Foulks, Fulk, Fulkes, Fuiks, Fewkes, Foakes, Focke, Fokes, Fookes, Fooks, Foukx, Foux, Fowke, Fowkes, Fuke, Voak, Vokes, Volk, Volke, Volkes, Voiks:** Folco, Fulco 1086 DB; Folche Ribalt c1155 Gilb (L); Folc' filius Folc' 1156–80 Bury (Sf); Alan filius Fuke 1166–95 Seals (Lei); Willelmus filius Fulk 1177 P (L); Ricardus filius Fuc 1196 Cur (He); Folco, Fulko picus 12th DC (L); Fouke de Coudrey 1255 RH (Bk); Folkes 1279 RH (C); Foke Odell 1503 ParlR (Nth); Fooke Edmonds 1611 Bardsley; Peter Fulch' 1198 FF (So); Richard Fulc t John HPD (Ess); Robert Fuke 1209 P (Nf); Richard Foke 1221 AssWo; Robert Fulco 1227 AssSt; Hugo Fouke 1275 SRWo; Robert Folke, Juliana Folkes 1279 RH (C); John Fukes 1296 SRSx; Richard Foulk 1297 MinAcctCo; Alice Fouke, Foukes 1307, 1314 Balliol (O); Walter Fowke 1311 ColchCt; John Fokes 1317 AssK; William Fulke 1333 ERO (Ess); Blanch Fokes, James Fooxe, John Foockes, Raphe Fookes 1564 SRSf; Thomas Foakes 1796 Bardsley. OFr Fulco, Fouques, OG Fulco, Folco 'people'.

**Folkard, Folkerts, Foucard:** Fulcardus sacerdos 1101–25 Holme (Nf); Folcardus presbiter 1121–48 Bury (Sf); Fucard (Fulcardus) filius Johannis 1198 FF (Sf); Rogerus filius Folkart 1219 AssY; Rau Folkard c1100 OEByn (D); John, Richard Folkard 1327 SRSf. OG Fulcard 'people-brave'.

**Folker:** v. FULCHER

**Folkred:** Robert filius Fulcheredi c1110 Winton (Ha); Alward folcred 1189 Sol; William Fulkered 1221 ElyA (Nf); Robert Folkred 1327 SRSf. OG Fulchered.

**Folkson, Foxon, Foxen:** Roger Folkesune Ed 1 NottBR. 'Son of Folk.'

**Foll:** William le fol 1202 AssL; Robert Folle 1202 P (L); Simon le Fu 1273 Ipm. OFr fol, modFr fou 'foolish, silly'.

**Follenfant:** Hugo Folenfant 12th DC (Nt); John Folenfant, Fol Enfant, Foleinfant 1200–12 Cur (Ess). OFr fol 'foolish' and enfant 'child'.

**Follet, Follett, Follit, Follitt:** William Folet 1086 DB (K); Roger Folet 1158 P (K). OFr folet, a diminutive of fol 'foolish'.

**Folley, Folly:** Richard de la Folia 1176 P (W); Thomas de la Folie 1214 Cur (Nf). Folly, common in minor place-names, especially in Warwicks and Wilts, originally had reference to some example of human folly, amusement or pleasure, but occasionally is used of a small plantation, a usage difficult to explain. The above examples are clearly from Fr folie 'foolishness' and are earlier than any previously noted (1228). The first example probably refers to one of the places called Folly in Wilts. v. PN Wa 382–5, PN W 451, PN C 353.

**Follick:** v. FULLICK

**Follifoot, Follifatt:** (i) Alan de Fulifet 1167 P (Y); Martin de Folyfeyt 1279 AssNb; Thomas Folifatt 1362 AssY. From Follifoot in Spofforth, or a lost Follifoot in Wighill (WRY). (ii) Ralph Fullifed 1212 RBE (C); Agnes Folifet 1225 Cur (Berks). 'Well fed', OE full, gefēdde. cf. John Fulfair 1208 Pleas (C) 'very fair'; William Fulredy 1342 IpmGl 'full of good advice'.

**Follin:** John Folin 1206 P (Co). cf. Fr Follin, a hypocoristic of OFr folet 'fool' (Dauzat).

**Folliot, Folliott, Foliot, Foliott:** William Foliot c1150 DC (L); Henry Foliot 1214 P (Berks); Richard Folyot 1353 YAJ xx. A derivative of OFr folier 'to play the fool, to dance about'. cf. Chilton, Draycott Foliat (W), Tamerton Foliot (D), Norton Folgate (Mx), Nortonfolyot 1433. Perhaps one of the sources of FOLLEY: William Foliatt alias Follye 1598 FFHu.

**Follower:** William le Folegere 1205 Cur; Walter le Folewar 1299 IpmGl; William le Folwer 1332 SRWa. OE folgere 'follower, plaintiff'. cf. Gudytha ffoloufast 1379 PTY 'follow quickly'.

**Folsham, Foulsham:** John de Fulsham 1185 Templars (Lo); William de Folesham 1219 P (Nf/Sf); Robert Folsham 1355 FFY; James Fulsam 1420 IpmY. From Foulsham (Nf).

**Fontaine:** v. FOUNTAIN

**Food:** v. FOAD

**Fook(e)s:** v. FOLK

**Foord:** v. FORD

**Foort:** v. FORT

**Foot, Foote:** Seild filia Fot 1212 Cur (Ha); Ernui, Goduin Fot 1066 DB (Ch, K); Robert Fot 1166 P (Y); Gregorious cum pede 1271 FrLeic; Henricus Pes 1290 ShefA. Either the ON nickname Fótr 'foot' or OE fōt 'foot', translated by the Lat pes.

**Footing:** Geoffrey Foting 1275 RH (Nf); Gilbert Fotying 1305 FFEss. ME foting 'dance steps', a nickname for a dancer.

**Footit, Footitt, Foottit:** Roger *Vothot* 1291, *Fothot* or *Vothot* 1294–5 ELPN. ME *fot-hot* 'quickly, suddenly'. cf. Robert *Fotmul'* 1201 Pleas (Co), OE *fōtmǣlum* 'foot by foot', perhaps 'cautiously'.

**Footman:** Thomas *Fotman* 1327 Pinchbeck (Sf), 1332 SRCu. OE *fōt* and *mann*, 'footman', probably 'foot-soldier' (c1325 MED).

**Foray, Forey:** Robert *Foray* 1296, Roger *Foray* 1332 SRSx. A nickname from ME *forrai* 'plunder'.

**Forber:** *v.* FURBER

**Forbes:** Duncan *de Forbeys* c1272 Black. From Forbes (Aberdeenshire).

**Forcer:** Ralph *le forcier, le Forcer* 1210 P, 1221 AssWa; William *le Forcir* 1228 Fees (Sa). A derivative of *force*, from OFr *forcer* 'to clip or shear wool', from *forces* 'clipping-shears'; 'one who forces wool'. cf. 'Sheer-men and Dyers, Forcers of Wools, Casters of Wools and Sorters of Wools' (1533 NED).

**Ford, Forde, fforde, Foard, Foord, Forth:** Bruman' *de la forda* 1066 Winton (Ha); Eadric *æt Fordan* 1100–30 OEByn (So); Reginald *de la Forthe* 1273 RH (Sf); Geoffrey *atte Forde* 1296 SRSx; William *Foorde* 1418 LLB I. 'Dweller by the ford' (OE *ford*) or from Ford (Som, etc.).

**Forder:** William *Forder* 1327 SRSx. 'Dweller by the ford.'

**Fordham:** Henry *de Fordham* 1198 FFEss; Reginald *de Fordham* 1219 P (C); William *de Fordham* 1291–2 FFEss. From Fordham (C, Ess, Nf).

**Fordman:** Robert *Fordman* 1327 *SR*Ess; Robert *Fordman* 1371 FFEss. 'Dweller by the ford', OE *ford, mann*.

**Fordred:** *Fordretus* 1084 Exon (So), Ivo *Fordred* 1275 RH (K). OE *Forðrǣd* 'forth-counsel', common in the 9th century, but rare thereafter.

**Fordwin, Fordwyn:** Richard *filius Fortwin* 1221 AssWa; William *Fordwin* 1208 FFO; Nicholas *Forethewynt* 1296 AssCh. OE *Forðwine*.

**Fordyce:** John *Fordise* 1460 Black; William *Fordyce* 1567 ib. From Fordyce (Banff).

**Foreman, Forman, Fourman:** Robert, Cristina *Foreman* 1296 Black, 1301 SRY; Alan *Forman, Robert Fourman* 1327 SRY. OE *fōr* 'pig' and *mann*, 'swineherd'. The surname may also have absorbed *Fordman*: Robert *Fordman* 1327 *SR* (Ess), 'dweller by the ford'.

**Foreshaw:** *v.* FORSHAW

**Foresight:** *v.* FORSYTH

**Forestal, Forestall:** William *de Forestal* 1255 ForNth; John *de Forestal* 1333 FFW. From The Forstall in Rotherfield (Sx), or 'dweller by the paddock or way in front of the farmhouse', from dialectal *forestall*.

**Forester, Forestier, Forrester, Forrestor:** John *Forester* 1183 P (Sr); Richard *le Forester* 1240 FFEss; Robert *le Forestier* 1322 LLB E. OFr *forestier*, ME *forester* 'officer in charge of a forest' (c1325 MED), also used of one employed in a forest. At Pulham (Norfolk) in 1222 a free tenant held 20 acres *per servicium forestare*. In 1277, at Hitcham (Suffolk), Robertus *forestarius*, a free tenant, paid 4*s*. 8*d*. and suit of court for 17 acres. He had to guard all the woods of his lord, the Bishop of Ely, and had various

forest privileges including one log for his fire at Christmas, all trees and branches in the woods blown down in a storm, and the right to keep his pigs in the wood (*ElyA*). *v.* also FORSTER, FOSTER.

**Forey:** *v.* FORAY

**Forfeitt:** Sexi *Forfot* 1137 ELPN. OE *fōr* and *fōt* 'pig-foot'.

**Forge:** Ralph *del Forge* 1297 Coram (Y). 'Worker at a forge', blacksmith.

**Forley, Fortly:** Stephen *Forle* 1279 RH (Hu). From Fordley (Suffolk).

**Forman:** *v.* FOREMAN

**Formby:** Adam *de Forneby* 1332 SRLa. From Formby (La).

**Former:** Simon *le Formur* 1219 AssY. ME *former* 'maker, fashioner' (c1340 NED), 'Brycke former or maker' (1552 ib.).

**Forn:** Robert *le Forn* 1288 Oseney; William *Forn* 1298 AssL; Roger *le Vorn* 1332 SRWa. A nickname from OE *foran, forne* 'ahead, in front'. But there was also a personal name: William *filius Forn* 1212 Cur (Y); *Forne* filius Siolf 1212 Fees (Cu).

**Fornell:** Robert *Fornel* 1287 IpmY. 'Worker at the small oven or furnace', ME *fornel*.

**Forrest:** Hugh *de Foresta* 1204 Cur (Ha); Adam *ate Forest* 1300 Ipm (K); Anabilla *del fforest* 1354 Kirkstall (Y). 'Dweller or worker in the forest.' *v.* FORESTER.

**Forrester, Forrestor:** *v.* FORESTER

**Forsbrook:** *v.* FOSBROOKE

**Forscott, Foskett:** William *Forscott* 1545 SRW. From Foscott (Bk), or Forscote (So).

**Forsdick, Forsdike, Forsdyke:** *v.* FOSDIKE

**Forshaw, Foreshaw:** Richard *de Forshaghe* 1327 SRSf; James *fforshagh* 1542 PrGR. From Forshaw Heath in Solihull (Wa).

**Forse:** *v.* FOSS

**Forsey:** *v.* FURSEY

**Forst:** *v.* FROST

**Forster:** John *Forster* 1315 AssC; William *le Forster* 1341 AD iv (L); Richard *Forstier* 1381 AD ii (Mx). Various origins are possible: (i) A shortened form of FORESTER (*forster* c1320 NED): Walter *Forster, Forester* 1356 LLB G. (ii) Equivalent to FEWSTER: Warin *le forstere* 1199 Bart (Lo) may have been a forester, in which case we have a very early example of this development, but more probably he followed the same occupation as Durand *le fuster* (1179), 'the saddle-tree maker'; OFr *fustrier* could, by metathesis, become *furster* or *forster*; cf. Gilbert *Le Furstare* or *le Fuster* 1305 MESO (Wo); John *Forstar'* 1202 Cur (Y); Walter *le Forster* 1275 SRWo; Robert, Adam *Forster*, Robert *Foster* 1327 ib. (iii) OFr *forcetier* 'maker of scissors, shearer, cutler': Richard *le Forseter, le Forceter* 1311 ColchCt. cf. AFr *forcettes* 'scissors', a diminutive of *forces* (1474 NED). *v.* FORCER. *Foster* is common in ColchCt, *Forester* rare. cf. William *Forset* 1214 Cur (Bk), 1441 FrY from *forcettes*, 'a maker of scissors'. This does not seem to have survived. It has probably been absorbed by Fawcett.

**Forsyth, Forsaith, Forseith, Foresight, Forsyde:** (i) Osbert *filius Forsyth* c1308 Black (Stirling); *Fersith,*

*Forsyth* 1364 ib.; *Fersithi* mag Uibne 1464 Black; Robert *Fersith* 1420 ib. (Stirling); James *Fersith, Forsithe* 1446 ib. Gaelic *Fearsithe* 'man of peace'. (ii) William *de Fersith* 1365 Black (Edinburgh); Robert, Thomas *de Forsith* 1426, 1487 ib. From an unidentified place of this name.

**Fort, Forte, Foort, Lefort:** Fegge *Fort* 1200 Cur (L); Isabel *le Fort* 1268 AssSo. AFr, OFr *fort* 'strong'.

**Fortescue, Foskew:** Richard *Fort Escu, Fortescu* 1177, 1185 P (D); Oliver *Fortescue* 1212 Cur (Co). OFr *fort escu* 'strong shield'.

**Fortey, Forty, Fortye:** William *de Forteye*, Thomas *de la Fortheye* 1275 SRWo; John *ate Fortheye* 1297 MinAcctCo (Berks), John *de la Forteye* 1318 Cl (Berks). OE *forþ-ēg* 'forth-island', land standing well out from surrounding marsh or fen, or, possibly, OE *forþ-tēag*, from *tēag* 'enclosure', a paddock in front of or near the farm-house. cf. Forty Green (Essex, Worcs), The Forty (Wilts), Forty Hill (Middlesex) and *v.* PN Wo 202, PN Ess 23, MELS 69–70.

**Forth:** *v.* FORD

**Fortin:** *v.* FORTUNE

**Fortly:** *v.* FORLEY

**Fortman:** William *Fortesmains* 1219 AssY. 'Strong hands.'

**Fortnam, Fortnum:** Nicholas *Fortanon* 1279 RH (O). OFr *fort anon* 'strong young ass'.

**Forton:** Gilbert *de Forton* 1204 AssY; Jordan *de Fortun* c1206 Seals (La); John *ffortoun* 1327 SRC. From Forton (Ha, La, Sa, St).

**Fortune, Fortin:** (i) *Fortunus* 1185 Templars (Lo); *Fortune* South 1663 HeMil; Davy *Fortune* 1524 SRSf; Robert *Fortune* 1641 PrSo. Lat *Fortunus*. (ii) Hugh *Fortin* 1176 P (Beds); William le Jeune *Fortin* 1202 AssBeds; Denys *Fortin* 1242 AD ii (Ha). OFr *fortin*, a diminutive of OFr *fort* 'strong'. (iii) John *de Fortun* 1200 (Kelso), John *de Fortone* 1297 Black. From the lands of Fortune (East Lothian).

**Forty:** *v.* FORTEY

**Forward:** *v.* FROWARD

**Forward, Forwood:** Bartholomew *Forward, Forreward* 1279 RH (C), Florence *Forewardes* 1327 SR (Ess). OE *fōr* 'pig, hog' and *weard* 'guard, watchman', 'swineherd'. cf. Thomas *Blleward* 1319 SR (Ess), from *bull* and *v.* COWARD.

**Fosbrooke, Forsbrook:** Osbert *de Fotesbroc* c1190 StCh. From Fosbrook (Staffs).

**Fosdike, Fosdyke, Forsdick, Forsdike, Forsdyke:** Walter *de Fotesdik* 1202 AssL; John *Fosdyke* 1524 SRSf. From Fosdyke (Lincs). cf. FOSBROOKE.

**Foskett:** *v.* FORSCOTT

**Foskew:** *v.* FORTESCUE

**Foss, Forse, Vos, Voss:** John *del Fosse* 1199 MemR (Sx); John *atte, de la Fosse* 1295, 1330 MELS (So). 'Dweller by the ditch' (OE *foss*). In Somerset the surname is recorded from Doulting and Shepton Mallet, on each side of the Fosse Way, along which lie three farms named Fosse in Wilts, four in Warwicks and two in Notts. The initial *V* in *Voss* is southern. cf. Voss (Robert *atte Fosse* 1330 PN D 254).

**Fosser:** Gilbert *le Fosser* c1250 LuffCh; John *le Fossur* 1256 AssNb. 'Ditch digger', OFr *fosseur*.

**Fossett, Fossit, Fozard, Fozzard:** Nigel *Fossard* (Y); Robert *Fossart* 1086 DB (Y); William *Fossard* 1196 P (Y); William *Fossard* 1241 AssSo. OFr *Fossard*.

**Fossey:** Arnold *de la Fossie* 1282 LLB A; John *del Fossey* 1297 SRY; Richard *Fossey* 1279 RH (O). 'Dweller by the low-lying land near the dike', OE *foss, ēg*. The Ampthill *Fossey* is from a lost place *Fotsey* in Aspley Guise, *Foteseige* 969 PN BedsHu 114.

**Foster:** John *Foster* 1373 ColchCt; Edward *Foster* 1381 AssC. This may be ME *foster* 'foster-parent, nurse' (a1225 MED), found once in the compound OE *cild-fōstre*. But it may also be: (i) a development of FORESTER (1386 NED) or FORSTER 'forester'. The seal of Walter *Forestier* (1371 AD v, Lo) bore the legend: SIGILLVM. WALTERI. LE. FOSTER. (ii) from *Forseter* 'shearer', which would inevitably become *Forster* and then *Foster* which is common in the ColchCt. (iii) equivalent to FEWSTER, OFr *fustrier* becoming *furster*, *forster* by metathesis, and then *foster*. *v.* FORSTER.

**Foston:** Richard *de Fozton* 1202 P (Lei); Henry *de ffoston* 1379 PTY. From Foston (Db, L, Lei, NRY), or Foston on the Wolds (ERY).

**Fotherby:** Ranulf *de Foterby* 1250 Gilb; William *ffotherby* 1672 HTY. From Fotherby (L).

**Fothergill:** Henry *Fodyrgyll* 1514 CorNt; William *Futhergill*, John *Fothergill* 1583, 1611 FrY. From Fothergill Well (WRYorks), or from other minor places of the name.

**Foucar:** *v.* FULCHER

**Foucard:** *v.* FOLKARD

**Fouger:** Ralph *de Felgeriis* c1110 Winton (Ha); William *Fouger* 1327 SRWo. From Fougères (Ille-et-Vilaine).

**Foukx:** *v.* FOLK

**Foulcher, Foulger, Foulker:** *v.* FULCHER

**Fould(e)s:** *v.* FOLD

**Foulkes:** *v.* FOLK

**Foulser:** *v.* FULCHER

**Foulsham:** *v.* FOLSHAM

**Foulston, Foulstone, Fowlestone, Fowlston, Fullstone:** Robert *de Fugeleston* 1197 FFK; Robert *de Foweleston* 1251 AssY. From Fulston Manor in Sittingbourne (Kent) or Fulstone (WRYorks).

**Founder:** Richard *Fundator* 1194 Cur (W); Richard *le fundor* c1198 Bart; John *le Fundour* 1300 LoCt; John *le Foundour* 1346 FrY. OFr *fondeur*, *fundeur* 'one who founds or casts metal', a metal-founder.

**Fountain, Fountaine, Fontaine:** Hugo *de Funteines* 1202 P (K); John *de Funtayne* 1270 FFEss; John *de la Funtayne* 1275 AssSo; John *Fowntens* 1426 FA (Sr). 'Dweller near the spring(s)', OFr *fontaine*, ME *fontayne* (a1450 MED). Probably from a French place named Fontaine, Fonteyne or Lafontaine.

**Fouracre, Fouracres, Foweraker:** William *Fourakre* 1327 SRSo; William *Foweracres* 1682 DWills. 'Occupier of a holding of four acres.'

**Fourman:** *v.* FOREMAN

**Fournel:** *v.* FURNELL

**Fourniss:** *v.* FURNACE

**Fouweather, Foweather, Fowweather:** John *Fulweder* 1284 FFHu; William *Foulweder* 1342 Rams (Hu). ME *foul wedir* (c1380 NED) 'wet and stormy', the opposite of FAIRWEATHER.

**Foux:** v. FOLK

**Fow:** Hugh *le Fohe* 1296 SRSx; John *Fow* 1336 IpmNt. OE *fāh, fāg* 'spotted, variegated'.

**Fowell(s):** v. FOWLE

**Fower, Foers:** Ansger *Focarius, Fouuer* 1084 GeldR, 1086 DB (So); Turkillus *le foer* 1190 P (Wa); Hugo (*le) Fuer* 1219 AssY; Ralph *Four* 1219 AssL; Roger *le Fower* 1279 RH (Nth). OFr *fouuer*, Lat *focarius* 'hearth-keeper'. Occasionally in the form: Hugh *de la Four* 1275 RH (Nth).

**Foweraker:** v. FOURACRE

**Fowke(s):** v. FOLK

**Fowlds:** v. FOLD

**Fowle, Fowles, Fowell, Fowells, Fowls, Fuggle, Fuggles, Voules, Vowell, Vowells, Vowels, Vowles:** *Fugel* 1066 Winton (Ha); *Fugel* de Hoilanda 1177 P (Y); 'Wuluard' *Fugel* 1166 P (K); Robert (*le) Fugel* 1186–7 P (So); William *le Foul* 1271 FFC; Agnes *Foweles*, Nicholas *le Fowel* 1275 SRWo; Roger *Fogel* 1296 SRSx; Nicholas *Vogel* 1327 SRSo; William *Vouell* 1578 Oxon; Thomas *Fuggill* 1632 YWills; William *Fugghill* 1685 ib. OE *Fugel*, from OE *fugol* 'fowl, bird', used both as a personal-name and a nickname. The southern form survives in Vuggles Fm (Sussex).

**Fowler, Fugler, Vowler:** Richard *Fugelere* 1218 AssL; Roger *le Fugler* 1227 Pat (Nf); John *þe Fogelere* 1275 RH (W); Ralph *Vouler* 1279 RH (Bk); Edward *le Vowelar* 1327 SRSo; Ralph *le Foweler* 1329 ColchCt. OE *fugelere* 'hunter of wild birds, fowler'.

**Fowl(e)ston(e):** v. FOULSTON

**Fowling:** Alan, Elpher *Fogheling, Fughelyng* 1327–32 SRSx. OE *\*Fugeling* 'son of Fugel'. v. FOWLE.

**Fownes:** v. FAUN

**Fowweather:** v. FOUWEATHER

**Fox:** Toue *fox* Hy 2 DC (L); Hugo *le Fox* 1297 MinAcctCo. A nickname.

**Foxall, Foxhall:** William *de Foxole* 1197 P (K); John *de Foxales* 1276 RH (Y); John *Foxholes* 1406 TestEbor. From Foxhall (Sf), Foxholes (La, ERY), or Foxholt in Swingfield (K), *ffoxole* 1254.

**Foxcot, Foxcott:** Edulf *de Foxcote* 1189 Sol; John *de Foxecotes* 1348 PN Ch iv 38. From Foxcote (Gl, Wa), or Foxcott (Ha).

**Foxhall:** v. FOXALL

**Foxlee, Foxley:** John *de Foxle* 1230 P (Nth); Benet *de Foxley* 1318 Calv (Y); William *Foxleigh* 1372 IpmW; John *Foxleye* 1382 IpmGl. From Foxley (Nf, Nth, W), or Foxleighs in Bray (Berks).

**Foxton:** Simon *de Foxtone* 1159 P (Lei); Robert *de Foxton* 1303 IpmY; John *Foxton* 1382 AssLo. From Foxton (C, Du, Nb, La, Lei, NRY), or Foxdon in Witheridge (D).

**Foy:** John *Foie* 1212 FFSf; Simon *le Foy* 1327 SRSx; John *Foys* 1359 AssD; Magota *ffoy*, William *ffoye* 1379 PTY. OFr *foi* 'faith'. cf. John *Faith* 1380 LoCh.

**Foyle:** (i) Henry *de Foyle* 1249 AssW; John *atte Foyle* 1311 Battle. From Foyle Farm in Oxted (Surrey), or 'dweller by the pit', OFr *fouille* 'excavation'. (ii) Howel *Voyle* 1285 Morris; Ralph *Foil* 1327 SRSo; Samuel *Foyle* 1642 PrD. Welsh *moel* 'bald'.

**Fozard, Fozzard:** v. FOSSARD

**Frail, Freel:** Richard *Fresle* 1086 DB (Nt); Robert *Fresle* 1115 Winton (Ha); Robert *Frelle* 1130 P (Ha). ME *freil, frele, frelle*, OFr *fresle (fraisle)* 'frail, weak' (a1340 NED).

**Frain, Frane, Frayn, Frayne, Frean, Freen, Freyne, de Fraine, de Freyne:** William *de Fraisn'* 1156 P (Sf); Thomas *del Freisne* 1206 Cur (He); Peter *de Frane* 1228 Cl (Lo); Richard *del Frene* 1271 ForSt; Cristina *Freen* 1275 SRWo; William *a la Freyne* 1279 RH (O); John *del Freyn* 1280 AssSo. OFr *fraisne, fresne* 'ash-tree'. 'Dweller by an ash-tree.' cf. FRANEY.

**Frais, Frose:** Geoffrey *Freys, Frois* 1275 RH (Lo); Richard *Froyse Frose* 1403, 1467 FrY. OFr *freis* (m), *fresche* (f) 'full of vigour, active' (c1385 MED), 'blooming, looking healthy or youthful' (c1385 ib.).

**Fram, Framm, Frame:** John *filius Frame* 1250 Fees (D); Geoffrey *Frame* 1196 P (Nf/Sf); Adam *Frame* 1495 Black. OE *fram* 'bold, active, strong'. Also used as a christian name.

**Frampton:** Alesanus *de Frantone* 1066 DB (L); Alan *de Frampton* 1245 FFL; Robert *Frampton* 1427 FFEss. From Frampton (Do, Gl, L), or Frampton Cotterell, Mansel, on Severn (Gl).

**Frances, Francies, Francis, Franses:** Robertus *filius Franceis* 1207 Cur (Sx); *Francais* de Irtona c1230 Whitby (Y); Hugo *Francus, Franceis* 1135, 1166 Oseney (O); Robert *le Franceis* 1169 P (D); Roger *Franceis* 1177 P (Ha); Robertus *Francigena, Franciscus* 1199 P (Wo, Nf); Adam *le Francess* 1201 AssSo. OFr *Franceis*, modFr *François*, the popular form of *Franciscus*, originally a 'Frank', though later used to denote a Frenchman. The surname is usually from the adj. *Franceis* 'Frenchman'.

**Francke:** v. FRANK

**Franckeiss:** v. FRANKIS

**Francklin, Francklyn:** v. FRANKLIN

**Franckton:** v. FRANKTON

**Francom, Francombe:** v. FRANKHAM

**Frane:** v. FRAIN

**Franey, Freeney, Freney:** William *de Freisneto, de Fraisneto* 1170, 1176 P (Cu, R); Ingelrammus *del Freisnei* 1204 P (He); Alicia *de Fresnei* 1205 P (R); Ingelram *del Freidnei, del Frednei* 1205 P (He); William *de Freiney* 1207 Cur (C); Reginald *de Freney* 1244 Fees (R). From residence near an ash-wood, OFr *fraisnaie, fresnay*, or from one of the French villages so called. cf. FRAIN.

**Frank, Franks, Franck, Francke:** *Franco, Francus* 1086 DB (Sa, Y, Nf, Sr); Ricardus *filius Franc, filius Franke* 1172 DC, 1188 P (L); *Franke* de Cnihteton' 1221 AssWo; Ricardus *Franc'* 1201 Cur (Ess); Walter *le Franc* 1221 Cur (C); Richard *le Fraunc* Hy 3 HPD (Ess); Thomas *Frank* 1270 AssSo; Geruas' *le Fraunk*, Adam *le Fronk* 1296 SRSx. OG *Franco* 'a Frank' was not uncommon from the 11th to the 14th centuries. The surname is, perhaps, more commonly descriptive, from ME, OFr *franc* 'free', not a serf or slave (c1325 MED).

**Frankham, Frankom, Frankcombe, Frankum, Francom, Francomb, Francombe:** Thomas *le Franchume* 1234 FFC; Robert *Frankeham* 1243 AssSo; Robert *Frankhom* 1260 AssC. OFr *franc* 'free' and *homme* 'man', equivalent to FREEMAN.

**Frankis, Frankiss, Frankish, Franckeiss:** Roð ðe *Frenccisce* c1100–30 OEByn (So); William *Franckeche* 1240 Fees (K); William *le Frenkisse* 1289 AssCh; Richard *Frankis* 1297 SRY; Henry *Frankissh* c1310 Calv (Y). OE *frencisc* 'French(man)', ME *frankys* (a1325 MED).

**Frankley:** Bernard *de Frankele* 1206 P (Wo), Simon *de Frankele* 1224 AssSt; John *Fraunkeley* 1394 AssL. From Frankley (Wo).

**Franklin, Franklyn, Franklen, Francklin, Francklyn, Frankling:** Ralph *Frankelein* 1195 P (Y); Luke *le Franckeleyn* 1234 FFC; Roger *le Franklyn* 1274 RH (D); John *ffranklyng* 1522 ArchC 33. AFr *fraunclein*, ME *frankeleyn* 'a freeman', 'a land-owner of free but not noble birth' (c1300 MED).

**Frankton, Franckton:** Roger *de Franketon'* 1203 P (Wa); Alexander *de Franketon'* 1221 AssSa; Robert *de Franketon'* 1253 AssSt. From Frankton (Wa), or English, Welsh Frankton (Sa).

**Frankum:** *v.* FRANKHAM

**Franses:** *v.* FRANCES

**Fransham, Frensham, Frenchum:** Agnes *de Franesham'* 1198 FFNf. From Fransham (Norfolk).

**Frary:** *v.* FREDERICK

**Frater:** Richard *del Fraytour* 1301 SRY. 'One in charge of the refectory of a monastery', OFr *fraitur*.

**Fray, Fraye, Frey:** *Fray* de Cudington' 1230 P (O); Fray *Punchard* 1232 FFO; William *Frei* 1275 RH (K); Thomas *Fray* 1332 SRSx; John *Fray* 1435 AssLo. OFr *Fray*, a personal name of which the origin is unknown. *v.* Dauzat.

**Frayne:** *v.* FRAIN

**Freak, Freake:** Ralph *Freke* 1210–11 PWi; Hugh *le Freke* 1248 FFO; Robert *le Freke* 1332 ChertseyCt (Sr); John *Freake, Phreake* 1641 PrSo. OE *freca* 'man, warrior'. *v.* also FIRTH.

**Freake(s):** *v.* FIRTH

**Freaker:** Equivalent to *atte freke*. *v.* FIRTH.

**Frean:** *v.* FRAIN

**Frear, Freear, Freer, Frere, Friar, Frier, Fryer:** Robert (*le*) *Frere* 1196–7 P (Y); Roger *le Frier* 1243 AssSo. OFr *frere* 'friar'.

**Frears:** William *del Freres* 1314 FrY. 'Servant at the friars'.'

**Frearson:** John *le Frereson* 1335 AD vi (St). 'The friar's son.'

**Frecheville, Freshfield:** Cardo *de Frescheuill'* 1204 P (Nf); John *de Fressefeld* 1296 SRSx; Ralph *de Frechefeld* 1327 SRSx. From one of the French places named Frècheville.

**Frederick, Fredericks, Frederic, Frary:** *Frederic* (K), *Fredri* (Sx) 1066 DB; *Frethericus, Fredricus, Fedricus* 1086 DB (Nf); *Frari* filius Willelmi 1198 FF (Nf); *Fredericus, Fraricus, Frarius, Frarinus* de Bisshopesdon' 1221 AssWa, AssGl; Walter *Frethryk* 1275 RH (Sf); John *Frereric* 1279 RH (C); Ralph *Fretherik* 1315 FFSf; John, William *Frary* 1372 FrNorw, 1378 AssEss. OG *Frideric, Frederic* 'peace-rule', not a common medieval name in England, found chiefly in the eastern counties.

**Free:** Walter *le Free* 1255 RH (W); Adam *le Fre* 1282 LLB B. OE *frēo*, ME *fre* 'free'. cf. FRY.

**Freebairn:** William, Maurice *Frebarn(e)* 1248 FFEss, 1318 LLB E. OE *frēo, bearn* 'free child'.

**Freebert:** William *Frebert* 1222 FFEss. OE *Frēobeorht.*

**Freebody:** William *Frebodi* 1275 SRWo; Henry *Frybody* 1296 SRSx. OE *frēo, frīo* 'free' and *bodig* 'body', used in ME of a person, 'a free man'.

**Freeborn, Freeborne, Freeberne, Freeburn:** (i) *Frebern* c1095 Bury (Sf), 1211 Cur (Berks); *Frebern* de Eshcot' 1221 AssWa; Robert, William *Frebern* 1163 P (Nb), 1193 P (Berks). OE *Frēobeorn*, 'free-man'. (ii) Richard *Freborn'* 1256 AssNb; Robert *ffreborn* 1327 SRC. OE *frēo* 'free' and *boren* 'born', 'one born free, inheriting liberty'.

**Freece, Frees, Freese:** (i) Lambin *Frese* 1181 P (K); Alice *Frees* 1275 PN O 267; George *Freece* 1642 PrD. OFr *frise* 'a kind of coarse woolen cloth, with a thick nap on one side only'. Metonymic for a maker or seller of this. (ii) Walo *de Frise* 1204 P (Berks); William *de Frisa* 1230 P (Wo). 'The man from Frisia.'

**Freed:** *v.* FIRTH

**Freeder:** Equivalent to *atte frede*. *v.* FIRTH.

**Freel:** *v.* FRAIL

**Freeland:** Walter *Freland'* 1198 FF (Nf); Ralph *Frieland* 1202 FFC; Richard *de la Fryelonde* 1296 SRSx; Richard *atte Frilonde* 1317 AssK. A holder of 'land held without obligation of rent or service' OE *\*frēoland, \*frīgland.*

**Freelove:** *Frelof* Pollard 1235 Ch (Nf); Nicholas, Joanna *Frelove* 1279 RH (Beds), 1568 SfPR. OE *Friðulāf* 'peace-survivor', recorded twice in the 10th century.

**Freeman, Friman, Fryman:** *Freman* Sceil 1188 P (Ess), William *Freman* 1196 FFNf; Reginald *le Freman* 1221 AssWo; Osbert *Friman* 1240 Fees (Beds). OE *frēomann, frīgmann* 'freeman', 'free-born man', used also as a personal name.

**Freemantle, Fremantle:** Reginald *Freitmantel* 1190 Oseney; Gamel *Freimantell* 1212 Cur (Y); Robert, John *Fremantel* 1360–2 FrC, 1396–7 FFSr. The name is borrowed from France where *Fromentel* 'cold cloak', a common place-name, was also the name of a forest. As a nickname it probably meant 'poorly clad, ragged, cold'. Later examples of the name may come from Freemantle (Hants), *Freitmantell* 1181.

**Freemason:** Nicholas *le Fremason* 1325 CorLo; John *le Fremassoun* 1332 IpmGl. 'Master mason', ME *fremasoun.*

**Freen:** *v.* FRAIN

**Freeney:** *v.* FRANEY

**Freer:** *v.* FREAR

**Frees, Freese:** *v.* FREECE

**Freeth, Freke:** *v.* FIRTH

**French, ffrench:** Simon *le Frensch* 1273 RH (W); John *le Frenche* 1278 Ronton (St). OE *frencisc*, ME *frennsce, frenche* 'French'.

**Frenchum:** *v.* FRANSHAM

**Frend:** *v.* FRIEND

**Freney:** *v.* FRANEY

**Frensham:** *v.* FRANSHAM

**Frere:** *v.* FREAR

**Fresel:** *v.* FRISELL

**Freshet, Freshett:** Hervey *Freschet* 1148 Winton (Ha); Hervey *Freschet* 1212 P (Gl). OFr *freschet* 'fresh, lively, alert'.

**Freshfield:** v. FRECHEVILLE

**Freshfish:** John *ffresfyssh* 1319 SRLo. 'Fresh fish', ME *fresh*, OE *fisc*. A nickname for a fishmonger. cf. Reginald *Fresheryng* 1276 AssLo 'fresh herring'.

**Freshney:** v. FRISKNEY

**Freston:** Agnes *de Freston'* 1221 Cur (Sf); John *Freston* 1483 FFWa. From Freston (Sf).

**Fretgoose:** Alan *Fretegos* 1207 Pleas (Sf). A nickname, 'eat goose', OE *fretan, gōs*. cf. Hagamunt *Freteharing* 1198 P (K) 'eat herring'; Elias *Fretoxe* 1259 IpmW 'eat ox'.

**Frett:** Metonymic for *Fretter*.

**Fretter:** Henry *le Fretter* 1332 SRWa. A derivative of ME *frette*, OFr *frete* 'interlaced-work', used of ornaments (especially for the hair) consisting of jewels or flowers in a network, hence a maker of these. cf. a frette of goold (c1385 NED), frette of perle (1390).

**Fretton:** Wychman *de Freton* 1248 FFEss. From Fritton (Nf, Sf), the former of which appears as *Freton* in 1199, the latter in 1224.

**Fretwell:** Milo *de Freteuill'* 1204 P (O); Stephen *de Fretewell* 1219 FFO; Roger *de Fretewelle* 1240 Eynsham. From Fritwell (O), *Fretewell* 1203.

**Frewen, Frewin, Frewing, Frowen, Frowing, Fruen, Fruin:** *Freowinus, Freuuinus* 1066 DB (Sf, Ess); *Frewinus* c1150 Gilb (L), 1175 P (Ha); Richard, Henry *Frewine* 1221 ElyA (Sf), 1230 P (Ha); Walter *Frewyne* 1313 Glast (Do); John *Froweyn* 1394 LLB H; Edward *Frewinge* 1665 HTO. OE *Frēowine* 'noble-friend'. Occasionally we have forms with *Fra-*, William *Frawin* 1221 AssWa, from OE *Frēawine* 'ruler-friend', which may also appear as *Frewinus* (PNDB 253).

**Frewer:** *Freware* 1180 P (Do); Edmunde *Frewer* alias Meller 1568 SRSf. OE *Frēowaru* (f).

**Frey:** v. FRAY

**Freyne:** v. FRAIN

**Friar:** v. FREAR

**Friby:** Michael *de Frithebi* 1219 P (Y). From Firby in Westow (ERY), *Frithebi* 1170–80, or Firby in Bedale (NRY), *Fritheby* 1184. v. also FIRBY.

**Frick, Fricke:** Osbert *Frike* 1226 Cur (Sr); Ralph *Frike* 1255 RH (W); Peter *Fryk* 1367 FFEss. ME *frik, frike* 'brisk, vigorous'. cf. Thomas *le Friker* 1225 AssSo.

**Frid, Fridd, Fryd:** v. FIRTH

**Friday, Fridaye:** Chetel *Friedai* 1086 DB (Nf); Ralph *Fridai* c1167 AC (Lei); William *Fridey* 1214 Cur (Ess). OE *frīgedæg* 'Friday', perhaps for one born on that day. If the 17th-century phrases *Friday-face* 'a gloomy expression', *Friday-food, Friday-feast* 'a fast-day meal' are old, we might have a nickname for one as solemn and gloomy as a Friday fast-day.

**Friend, Frend:** Robert *Frend'* 1166 P (Nt); Gervase *Lefrend* 1221 Cur (Mx). OE *frēond* 'friend'.

**Friendless:** Henry *Frendles* 1246 AssLa; John *Frendeless* 1525 SRSx. 'Friendless', OE *frēondlēas* cf. William *Friendshippe* 1642 PrD 'friendship'.

**Frier:** v. FREAR

**Frift, Fright:** v. FIRTH

**Frigg, Frigge:** Randulf *Frig', le Frigg'* 1238–9 AccM; Richard *Frig'* 1275 SRWo; Matilda *Fryg* 1332 SRSx. Perhaps connected with ME *friggen* 'to quiver', in the sense 'timid'.

**Friman:** v. FREEMAN

**Fripp, Frippe:** Robert *Fripp* 1205 P (Herts). Perhaps a shortened form of ME *fripperer* 'dealer in old clothes'.

**Frisbee, Frisby:** Ralph *of Friseby* 1226 FFY; Nicholas *de Friseby* 1327 SRLei; William *Frysby* 1434–5 FFWa. From Frisby by Galby, or Frisby on the Wreak (Lei).

**Frisell, Frizell, Frizelle, Frizzle, Froysel, Fryzell, Fresell, Fresel:** William *Fresel* 1200 Cur (Herts); Caterina *Freysel* 1327 SRSf; Robert *Frysell* 1429 FrY. OFr *fresel* 'lace, ribbon'. Metonymic for a maker or seller of these.

**Friskney, Freshney:** Adam *de Freskenay* 1193 P (L). From Friskney (Lincs).

**Friston:** Reginald *de Friston'* 1193 P (L); Alexander *de Friston'* 1202 AssL; Pagan *de Friston'* 1221 Cur (Sr). From Friston (Sf, Sx), or Frieston (L), *Fristun(e)* DB.

**Frith:** v. FIRTH

**Frizell, Frizelle, Frizzle:** v. FRISELL

**Frobisher:** Geoffrey *le Furbisur* c1260 LeiBR; Henry *le Fourbissor* 1306 AD v (Sa); Richard *Forbour, Forbysschour* 1359–60 ColchCt; John *Frobyser* 1517 GildY; Christian *Fribysher* 1582 RothwellPR (Y); William *Frubbisher* 1714 FrY. OFr *forbisseor* 'furbisher'. cf. FURBER.

**Frocke:** Cristine *Frok* 1305 AssW; John *Frocke* 1349 IpmGl. OFr *froc* 'a man's outer garment'. Metonymic for a maker or seller of these. Sometimes, perhaps, a variant of FROGG.

**Frodsham, Frodsom:** Peter *de ffrodesham* c1200 WhC; William *ffrodesam* 1340–1450 GildC; John *de Frodusham* 1377 FFSt. From Frodsham (Ch).

**Frogg, Frogge:** Nicholas *Frog'* 1207 Pleas (Sa); William *le Frogge* 1275 SRWo; William *Frogge* 1332 SRDo; Edmund *Frogs* 1642 PrD. A nickname from OE *frogga* 'frog'.

**Froggatt, Froggett:** Roger *de frogcot'* 1348 DbAS 36; Samual *Froggart* 1776 DbAS 34. From Froggat (Db).

**Frogge:** v. FROGG

**Froggett:** v. FROGGATT

**Frogmore, Frogmoor:** Reginald *de Froggemore* 1238–9 AccM; William *de Froggemere* 1249 PN W 333; William *de Froggemor* 1275 SRWo. From Frogmore (Berks, D, Do, Ha, Herts), or Frogmore Fm in Great Bedwyn (W).

**Frognall:** Henry *de ffrogenhole* 1262 PN K 309. From Frog's Hole in Goudhurst (K).

**Fromant, Froment, Fromont:** *Frumond* 1086 DB (Y); *Fromont* de Macels Hy 2 DC (L); Richard Thomas *Fromund* 1203 P (Do), 1243 AssSt. OFr *Fromont*, OG *Fromund*.

**Frome, Froom, Froome, Vroome:** *Æðelweard æt Frome* c1038 OEByn (He); Peter *de Froma* 1170 P (So). From Frome (Dorset, Hereford, Som).

**Frosh, Frosk, Froske:** Matthew *Frossh* 1312 LLB D; Dionisius *Froysshe* 1338 CorLo; John *Frossh* 1347 AssLo. A nickname from ME *frosh* 'frog'.

**Frost, Forst:** Aluuin, William *Forst* 1066 DB (Ha), 1199 AssSt; Lefstan, Gilbert *Frost* c1095 Bury (Sf), 1195 P (Wa). OE *forst, frost* 'frost', perhaps used in the sense of 'frosty', 'cold as frost', 'without ardour or warmth of feeling' (c1385 NED), or 'with the appearance of being covered with frost', 'with hoary, white hair' (*frosty berd* 14.. NED).

**Froud, Froude, Frowd, Frowde, Frood, Frude:** (i) Richard, Siward *Frode* 1184 P (D), 1187 P (Sx); William *Froud* c1203 StCh; William *le Frode* 1334 SRK; Robert, John *Frowde* 1525 SRSx, 1545 SRW. OE *frōd* 'wise, prudent'. (ii) *Frodo liber homo* 1066 DB (Sf); *Frodo prepositus* c1115 Bury (Sf); *Hugh filius Frodonis* 1121–48 Bury (Sf); *Frode* 1296 Wak (Y). ON *Fróði*, ODa *Frothi*. OE *Frōda* is found only once, and OG *Frodo* is rare. In Normandy *Frodo* may be ON (PNDB 256); *Frodo* 1086 DB was a brother of the abbot of Bury and was *genere Gallus*. Although there is no clear evidence, the possibility that the personal name gave rise to a surname cannot be excluded.

**Frow:** John *filius Frowe* 1202 P (L); William *Frowe* 13th Rams (C); John *Frouwe* (*Frowe*) 1327 SREss; Truda *Frowe* 1420 KB (Mx). According to Weekley, a shortened form of OE *Frēowine*.

**Froward, Forward:** Gille *Fraward* 1230 P (Y); Henry *Frowarde* 1332 SRWa; John *ffraward* 1379 PTY. cf. John *ffrawardson* 1379 PTY. ME *froward* 'wicked, contrary, willful'.

**Frowen, Frowing, Fruen, Fruin:** v. FREWEN

**Froyle:** Henry *de Froille* 1230 P (Ha); John *Froile* 1372 AssNu. From Froyle (Ha).

**Froysel, Fryzell:** v. FRISELL

**Frowen, Frowing, Fruen, Fruin:** v. FREWEN

**Fry, Frye:** William *Frie* 1195 P (Sx); Robert *le Frye* c1248 Bec (W). OE *frīg* 'free'. cf. FREE.

**Fryer:** v. FREAR

**Fryman:** v. FREEMAN

**Fuche, Fudge, Fuge, Fuidge:** Robertus *filius Fuche* a1170 Gilb (L); *Fuche* Bassat c1200 Seals (Nth). A pet-form of *Fulcher*, or *Fucher*. Gilebertus *filius Fulcheri* (c1130 Whitby) is also called Gilebertus *filius Fuche* (c1125 ib.). Robertus *filius Fuch'* is identical with Robertus *filius Fulcheri* 1210 Cur (Db), whilst Henry *Fulcher* is also called Henry *Fouch* 1297 Coram. v. FULCHER.

**Fucher, Fudger:** v. FULCHER

**Fudge, Fuge:** v. FUCHE

**Fugeman:** 'Servant of *Fuge* or *Fouche*.' v. FUCHE.

**Fugere:** v. FULCHER

**Fuggle, Fuggles:** v. FOWLE

**Fugler:** v. FOWLER

**Fuidge:** v. FUCHE

**Fuke:** v. FOLK

**Fulberd, Fulbert:** Aluric *Fulebiert* 1066 Winton (Ha); Cambin *ffoulberd* 1319 SRLo, Cambyn *Fulberd* 1326 CorLo. 'Dirty beard', OE *fūl, beard*. Sometimes, perhaps, for OFr *Fulbert*.

**Fulbrook, Fullbrook:** Walter *de Fulebroc* 1205 Cur (Sr); Edward *de Fulebroc* 1221 AssWa; William *de Fulebroc* 1235 FFO. From Fulbrook (Bk, O, Wa).

**Fulcher, Fulger, Fulker, Fulscher, Fucher, Fudger, Fugere, Futcher, Folger, Folker, Foker, Foulcher, Foulger, Foulser, Foucar, Volker, Fullagar, Fullager:** *Fulcher* 1066 DB, c1095 Bury (Sf); *Seuuale filius Fulgeri* Hy 2 DC (L); *Fulcarius* 12th DC (Lei); Rogerus *filius Foukere* 1201 Cur (O); Roger *Fulchier*

1167 P (Ha); Ralph *Fulcher* 1182 P (Sf); Peter *Fulker'* 1212 Cur (W); Eustace *Folchir* 1212 Fees (Ha); Nicholas *Fuker'* 1234 Fees (D); Warin *Fucher* 1235 Fees (Ess); John *Foucher'* 1242 Fees (W); Robert *Folgar* 1327 SRSf; William *Fouger* 1327 SRWo; Robert *Fowcher* 1524 SRSf. OFr *Foucher, Fouquier*, from OG *Fulchar, Fulcher* 'people-army'. OE *Folchere*, from which these surnames have sometimes been derived, is not recorded after 824. Occasionally we may have ODa *Folkar*.

**Fuldon, Fulton:** Richard *Fulton* 1218–19 FFEss; Roger *de Fuldon'* 1255–8 RegAntiquiss. From Foulden (Nf).

**Fulfair:** v. FULLFAIR

**Fulford:** Robert *de Fulfort* 1219 AssY; Richard *de Fulford* c1280 SRWo; Thomas *Fuleford, Fullefford* 1327, 1332 SRSx. From Fulford (Devon, Som, Staffs, ERYorks).

**Fulgent:** Robert *Fulgens* (*Fulgent*) 1177 P (Ess); Alexander *Fulgent* 1239–40 FFEss. ME *fulgent* 'shining, resplendent'.

**Fulk, Fulkes, Fulks:** v. FOLK

**Fulker, Fullagar, Fullager:** v. FULCHER

**Fullalove, Fullelove, Fullerlove, Fulleylove, Fullilove:** Henry *ffulofloue* 1327 SRC; William *ffuloflof* 1332 SRCu. A nickname, 'full of love', a translation of the French *pleyn d'amour*. cf. BLANDAMORE.

**Fullbrook:** v. FULBROOK

**Fullen:** v. FULLOON

**Fuller, Voller, Vollers:** Roger *Fulur* 1219 AssY; Reginald *fullere* 1221 ElyA (Sf); William *le Fulur* 1221 Ass (Wa); Simon *le Voller* 1316 Oseney (O); John *Follere* (*Vollere*) 1317 AssK. OE *fullere* OFr *fouleor, foleur* 'a fuller of cloth'. The raw cloth had to be *fulled*, i.e. scoured and thickened by beating it in water, a process known as *walking* because originally done by men trampling upon it in a trough. Hence *Walker*, by the side of *Fuller* and *Tucker* from OE *tūcian*, originally 'to torment', later 'to tuck', 'to full'. These three surnames seem to be characteristic of different parts of England. In general, in ME, *Fuller* is southern and eastern; *Walker* belongs to the west and north; *Tucker* is south-western. The French form *fuller* occurs in the whole of England and is often a translation of *walker* or *tucker*. v. MESO 100–1.

**Fullerlove, Fulleylove:** v. FULLALOVE

**Fullfair, Fulfair:** John *Fulfair* 1208 Pleas (C); Roger *Fulfayr* 1279 RH (C). 'Very fair', OE *full, fæger*. cf. William *Fulredy* 1342 IpmGl 'very wise, well-advised'.

**Fullick, Fullicks, Follick:** Adam *de Fullewyk* 1296 SRSx. From Fulwick's Copse in Lurgashall (Sussex), *Fullykland* 1586 PN Sx 113.

**Fulljames:** v. FOLJAMBE

**Fulloon, Fullen:** Ralph *le fullun* 1219 AssL; William *le Fulun* 1223 Pat (Y). OFr *fulun* 'fuller'.

**Fullshawe:** Henry *de ffulshaghe* 1332 SRLa. 'Dweller by the muddy wood', OE *fūl, sceaga*.

**Fullstone:** v. FOULSTON

**Fullwood, Fulwood:** Adam *de Foulewode* 1326 Wak (Y). From Fulwood (Lancs, WRYorks). Also becomes *Fullward* and *Fuller*: Avory *Fullward, Fuller* alias *Fullwood* 1703, 1719 SaltAS (OS) x.

**Fulmer:** Thomas *de Fulmere* 1239–40 FFEss; Henry *de Fulmere* 1325–6 CorLo; Richard *Fulmer* 1392 IpmGl. From Fulmer (Bk), or Fowlmere (C).

**Fulscher:** *v.* FULCHER

**Fulton:** *v.* FULDON

**Funk:** William *Funke* 1314 NorwDeeds; John *le Funke* 1327 SRSo; Nicholas *Founk* 1396 FrY. ME *funke, founck* 'a spark of fire' (MED a1393), cf. 'Funke or lytylle fyyr, *igniculus, foculus*' c1440 PromptParv. As a nickname, perhaps 'fiery, hotheaded'.

**Funnell:** *v.* FURNELL

**Furbank:** *v.* FIRBANK

**Furber, Forber:** Elfwin *Furbor* 1180 Oseney (O); Roger *le Furbur* 1199 P (Lo); Adam *le Forbour* 1333 ColchCt. OFr *forbeor, fourbeor, furbeor* 'furbisher of armour, etc.' (c1415 NED). *v.* also FROBISHER.

**Furby:** *v.* FIRBY

**Furel, Furell, Furrell:** Peter *Furel* 1247 FFO. OFr *forrel, fourrel* 'sheath, case'. Metonymic for a maker or seller of these.

**Furlong, Furlonge, Forlong, Furlonger:** Robert *Furlang* 1242 Fees (D); John *de Forhlangh* 1250 Fees (Sf); Richard *de Furlang* 1260 AssLa; John *Forlong* 1316 Oseney (O); John *atte Forlange* 1327 SRSx. This can hardly mean 'dweller by a furlong'. OE *furhlang*, lit. 'a furrow long', came to mean the length of a field and a division of an unenclosed field. In the 14th century it was used to translate Latin *stadium* in the sense of 'the course for foot-races'. cf. 'þei that rennen in þe ferlong for þe pris' c1380 Wyclif; 'Yif a man renneþ in the stadie or in the forlong for the corone' c1374 Chaucer. Hence *furlonger* (and *atte furlong* as a surname) may have been used of a runner famous for his exploits in the 'forlong'.

**Furmage, Furmedge, Firmage:** Herbert *furmage* 1160 P (L); Luke *Formage* 1280 AssSo; Henry *Firmage* 1524 SRSf. OFr *fourmage, modFr fromage* 'cheese' (14.. NED), metonymic for FIRMINGER.

**Furman:** *v.* FIRMIN

**Furmenger, Furminger:** *v.* FIRMINGER

**Furnace, Furnass, Furness, Furniss, Furnish, Fourniss, Varnish:** Michael *de Furneis* 1171 P (La); Anselm *de Furnes* 1198 P (We); John *Fornace* 1505 FrY. From Furness (Lancs). *v.* FURNEAUX.

**Furneaux:** Anketillus *de Furneis* 1086 ICC (C); Odo *de Fornelt* 1086 DB (So); Simon *de Furneaus, de Furnellis* 1220, 1235 Fees (Herts); John *de Furneals* 1238 Fees (D); Henry *de Furnaus, de Furnell', de*

*Fornell', de Furnellis* 1210–12 Cur (Ess, So). From Fourneaux (Calvados, La Manche) 'furnaces'. Furneux Pelham (Herts), named from Richard *de Furneals* (12th century), occurs as Pelham *Furnelle* (1240), *Forneys* (1291), *Furneus* (1293), *Furnysshe* (1541) and *Funish* (1700 PN Herts 184). There has thus been confusion with *Furnace* and *Furnell*.

**Furnell, Fournel, Funnell, Funnelle, Funel:** Alan, Richard *de Furnell'* 1191 P (O), 1206 Cur (Y); John *Funnell* 1553 SxW. OFr *fournel* 'furnace' or from one of the French places named Fournal or Fournel. *v.* FURNEAUX.

**Furner, Fournier:** William *le Furner*, Simon *Furner* 1208, 1212 Cur (Y); Martin *le Forner* 1283 LLB A. OFr *fornier, furnier* 'baker' (1441 MED).

**Furnival, Furnivall, Furnifall:** Gerard *de Furniualla* 1171–84 DC (L). From Fournival (Oise, Orne).

**Furr:** Reginald, Robert *Furre* 1193, 1202 P (L, Ess). ME *furre* 'a coat or garment made of or trimmed with fur' (c1375 MED), for a wearer, maker or seller of this.

**Fursdon:** (i) William *de Fursdon* 1281 PN D 521; George *Fursdon* 1642 PrD. 'Dweller by the furze-covered hill', OE *fyrs* and *dūn*. (ii) John *de la Fursen* 1318 Pat (D). 'Dweller in the furze' (weak pl.), cf. William *indefurse* 1189 Sol.

**Furse, Furze, Furseman, Furzeman, Furzer:** John *de la Fursa* 1168 P (D); John *atte Furse*, Sibel *atte Ferse* 1296 SRSx. 'Dweller by the furze-covered land' (OE *fyrs* 'furze').

**Furser, Furzer, Furzier:** Daniel *Furshewer* 1560 HartlandPR (D); Daniel *Fursyer* 1642 PrD; Ann *Fourshere* 1692 HartlandPR (D). 'Furze cutter', OE *fyrs*, and a derivative of OE *hēawan* 'to cut'.

**Fursey, Fussey, Fuzzey, Forsey:** John *Forshay* 1431 AD ii (Do); John, Roger *Fursey* 1583 Musters (Sr), 1642 PrD. 'Dweller by the furze-covered enclosure', OE *fyrs* and *(ge)hæg*.

**Fussel, Fussell:** (i) Walter *Fusil* 1260 AssC; John *Fusel* 1376 IpmNt; John *Fussell* 1641 PrD. OFr *fuisil* 'a casting'. Probably for an iron-worker. (ii) Richard *de Furshelle*, David *de Furshille* 1297 MinAcctCo. From Furzehill in Lynton, or Furze Hill in Sidbury (D).

**Fuster:** *v.* FEWSTER

**Futcher:** *v.* FULCHER

**Fyfe, Fyffe:** *v.* FIFE

**Fynes:** *v.* FIENNES

**Fynmore:** *v.* FINMORE

# G

---

**Gabb:** Walter, Thomas *Gabbe* 1275, 1327 SRWo. ME, OFr *gab* 'mockery, deceit', for *gabber* 'deceiver, liar': William, Stephen *le Gabber* 1230 P (D), 1279 RH (O).

**Gabber:** William *le Gabber* 1230 P (D); Gerard *le Gabur* 1275 RH (Sf); Stephen *le Gabber'* 1279 RH (O). OFr *gabeor* 'liar, cheat'.

**Gabbett, Gabbott, Gabbutt:** Garin *gabot* c1198 Bart; Robert *Gabbat* 1576 SRW. *Gabb-et, Gabb-ot,* diminutives of *Gabb,* a pet-form of *Gabriel.*

**Gabby, Gaby:** John *Gaby* 1275, William *Gabby* 1332 SRWo; Robert *Gaby* 1502 FrY. ME *gaby* 'simpleton, fool'.

**Gabler:** Goda *Gabblere* 1193–1212 Clerkenwell (Lo); Reginald *le Gabler* 1230 P (Ha). OFr *gabelier* 'tax-collector' or OFr *gableor* 'usurer'.

**Gabriel:** Gabriel *filius Reginaldi* 1212 Cur (Sx); *Gabriele* Spyg 1296 SRSx; Roger, Nicholas *Gabryel* 1296 SRSx, 1327 SRSf. Hebrew *Gabriel,* never a common name.

**Gache:** Francis *Gach,* Anthony *Gache,* Thomas *Gaich* 1642 PrD. OFr *gache* 'lock'. Metonymic for a locksmith. cf. Andrew *Gachard* 1298 AssL, with the depreciative suffix.

**Gadbury:** Petronilla *de Gadebergh* 1296, William *de Gatebergh* 1327 SRSx. From Cadborough alias Gateborough in Rye (Sx).

**Gadd:** Adam *Gad* 1188 P (He); Lucy *la Gadde* 1277 Misc (W); Robert *Gad* 1327 SRSf. ON *gaddr* 'goad'. Metonymic for GADDER.

**Gadder, Gadders:** John *le Gadder* 1285 AssLa; Henry *Gadere* 1371 ImpNt. ME *gadder* 'a maker of goads'.

**Gadling, Gadeling:** Margaret *Gaddling* 1260 AssC; Roger *Gedelyngs* 1327 SRLei; Thomas *Gadlyng'* 1381 AssWa. OE *gædeling* 'kinsman, companion', but also in a derogatory sense 'a low fellow'.

**Gadsby:** William *de Gadesby* 1361 FFEss. From Gaddesby (Lei).

**Gadsden, Gadsdon:** William *de Gatesden'* 1230 P (Beds); John *de Gadesdene* 1325 CorLo. From Great, Little Gaddesden (Herts), *Gatesdene* DB.

**Gaff, Gaffe:** William *Gaff'* 1275 RH (Nf); George *Gaffe* 1642 PrD. OFr *gaffe* 'an iron hook'. Metonymic for a maker of these. cf. William *Gafer'* 1379 PTY.

**Gage, Gauge:** Alice *Gage* 1310 ColchCt; Robert *Gauge* c1315 Calv (Y). ONFr *gauge* 'a fixed measure' (1432 MED), used by metonymy for a measurer or tester.

**Gager, Gaiger:** Andrebode *Gangeor* (sic) 1066

Winton (Ha); Henry *le Gaugeor* 1305 LoCt; Laurence *Gauger* 1327 *SR* (Ess). AFr *gaugeour,* OFr *gauger* 'gauger', 'exciseman' (1444 MED).

**Gagg, Gagge:** Alviva *filia Gag* 1208 Pleas (K, Nf); Eudo *filius Gagge* 1210 Cur (C); Ketel *Gag* 1214 P (C/ Hu); Richard *Gag* 1275 SRWo; John *Gagge* 1642 PrD. OE *\*Geagga.*

**Gailey, Gaily:** Henry *Gayly* 1376 AssEss; Richard *Gayley* 1663 HeMil. From Gailey (St).

**Gaillard:** *v.* GALLIARD

**Gain, Gaine, Gaines, Gains, Gayne, Dingain, Engeham, Ingham:** William *Ingania, Inganie* 1086 DB (Hu, Nth); Vitalis *Engaine,* Richard *Ingaine* 1130 P (Nth); Ralph *Engaigne* 1158 P (Cu); William *de Engain* 1208 FFHu; Richard *Ingan* 1228 Cl (Gl); John *en Gayne* alias *den Gayne* 1271 Ipm (Sf); John *le Gayne* 1275 Wak (Y); William *Denganye, de Engann'* 1279 RH (C); Richard *Ingayn* 1310 LLB D. OFr *engaigne,* Lat *ingania* 'trickery, ingenuity'. The family has left its name in Colne Engaine and Gaynes Park (Essex), in d'Engaine's Fm (Cambs) and in Aston Ingham (Hereford). The family name is only rarely spelled *de Engaine.* For this, cf. the French surnames *Dejour, Denuit, Damour,* elliptic for *(homme) d'amour,* etc. (Dauzat). *Ingham* has other derivations.

**Gainsboro, Gainsborough, Gainsbury:** William *de Gainesbury* 1166 MCh; Ralph *de Gainesburch* 1177 P (L); John *de Gaynesburgh* 1354 FFY. From Gainsborough (L).

**Gainsford:** John *de Gainesford* 1327 SRSx; John *Gaynesford* 1432 FFEss. From Gainsford Hall in Toppesfield (Ess).

**Gaish:** *v.* WACE

**Gait, Gaite, Gaites, Gaitt:** (i) Stephen, Thomas *Gayt(e)* 1297 SRY, 1331 FrY; John *Gaytt,* Gate 1390, 1416 FrY; Richard *Gaites* 1561 ib. These northern forms are probably from ON *geit* 'goat', but some may belong below. (ii) Reginald *Gayt* 1139 Templars (O); Robert *le Gayt* 1205 Cur (O); William *le Guaite* 1208 FFSt. OFr *gaite* 'watchman', a doublet of WAIT.

**Gaitbane:** William *Gaytebane* 1301 SRY. 'Goat bone', ON *geit.* OE *bān.* cf. Robert *Gaythals* 1244 Bart 'goat neck'.

**Gaiter, Gayter, Gaytor, Gayther, Geater, Geator:** Michael *le Geytere* 1279 RH (Hu). A derivative of ON *geit,* ME *gayte* 'goat', hence 'goat-herd'. The frequency of the names may be due to the absorption of *gaythird* 'goat-herd': John *le Gaythirde* 1301 SRY, Robert *Gayterd* 1466 FrY.

**Gaitskell, Gaitskill, Gaskell, Gaskill:** Benjamin *de*

*Gaytscale* 1332 SRLa; Edward *Gaskell* 1560 LaWills; Richard *Gatskell*, *Gaytscalle* 1595–6 ib.; Richard *Gaitskell* 1632 ib. From Gatesgill (Cumb), earlier *Geytescales*.

**Galbraith, Galbreath:** Gillescop *Galbrath*, *Galbraith* 1208–46 Black; William *Galbrath* 1269 AssNb. OGael *Gall-Bhreathnach* 'stranger-Briton', a name given to Britons settled among Gaels.

**Galby:** William *Galby* 1327 SRWo; Robert *Galby* 1379 PTY; John *Galby* 1430 FrY. From Galby (Lei).

**Gale, Gayle:** (i) *Gualo* legatus 1219 AssY; *Galo* 1230 ClR (Nantes); Gilbert, Alicia *Gale* 1202 AssL, 1275 SRWo; William *le Gal* 1285 *Ass* (Ess); Juliana *le Gale* 1327 SRC. Chiefly, no doubt, from OE *gāl* 'light, pleasant, merry; wanton, licentious'. But also from the central French form of OG *Walo* (v. WALE). (ii) Robert *le Geil* 1186 P (Wo). Identical with modern French *Gail* 'gay, joyous', from the simplex of *gaileard*. v. GALLIARD. (iii) Philip *de la Jaille* 1208 P (Ha); William *ate Gaole* 1317 AssK; Thomas *del Gayle* 1365 FrY; Robert *de Gale* 1402 FrY. The modern surnames represent the Norman pronunciation of ONFr *gaiole*, *gaole*, ME *gay(h)ole*, *gayll(e)*, *gaile*, surviving in the official spelling *gaol*, but replaced in pronunciation by the OFr *jaiole*, ME *jaiole*, *jayle*, which is also represented in the above forms. 'At the gaol', 'jailer'. cf. GALER.

**Galer, Gayler, Gaylor, Jailler:** Robert *le Gaoler* 1255 *Ass* (Ess); Richard *le Gaylor* 1275 SRWo; Richard *le Gayoler* of Newgate 1300 LoCt; William *le Gaoler* of Nottingham 1302 NottBr. ONFr *gayolierre*, *gaiolere*, OFr *jaioleur* 'gaoler, jailer', v. GALE.

**Gales:** Mager' *Galeys* 1279 RH (C); Henry *le Galeys*,02 le Waleis 1299, 1305 LoCt. The central French form of *Wallis*, or of OG *Walo*. v. WALES.

**Galey:** v. GALLEY

**Galilee, Gallally, Galley:** William *de la Galilye* 1337 Cl; Ralph *Galile* 1539 FeuDu; John *Gallale* 1539 Bardsley (Du). OFr *galilee* 'porch or chapel at the entrance to a church', those at Durham and Ely being particularly famous. The reference may be to the keeper or to one who lived near. Bardsley cites John and Robert *Galley* (1551) whom the historian of Newcastle identifies as members of a family of *Galilee*.

**Gall, Gaul, Gaw:** Walter *Galle* c1170 Gilb (L); Adam, Richard *Galle* 1221 AssWa, 1275 SRWo; John *Gal* 1334 Black; William *Gaw* 1397 ib. Celtic *gall* 'foreigner, stranger', found in Irish, Gaelic and Breton. It occurs in Perthshire and Aberdeen where the common pronunciation was *Gaw* and was used of Lowlanders. In England it is found in the Welsh border counties and also in Lincolnshire where it was of Breton origin. In Brittany where the name was common, it was applied to immigrants from France.

**Gallacher, Gallagher, Gallaher:** Ir *O Gallchobhair* 'descendant of *Gallchobhar*' (foreign help).

**Gallant:** *Waland* 1066 DB (Wo); *Galant* fitz Richard 1210 FFEss; Adam *Galland* 1274 RH (Ess); Thomas *Galaunt* 1275 RH (Sf); John *Galant* 1326 FFEss. OFr *Galand*, *Galant*, OG *Waland*. The surname is also from OFr, ME *galant* 'dashing, spirited, bold'.

**Gallard:** v. GALLIARD

**Gallatly:** v. GOLIGHTLY

**Gallear, Galler:** v. WALLER

**Galleon, Gallyon, Galliene, Galling, Galen:** *Galionus* 1130 P (Herts); *Galiena* 1207 Cur (Do); *Galienus* filius Willelmi 1212 Cur (Y); Turstin *Galien* 1190 P (Y); Roger *Galion* 1222 FFEss; John *Galling* 1296 SRSx; William *Galian* 1327 SRWo. Lat *Galienus* (m), *Galiena* (f), not uncommon names in the 13th and 14th centuries.

**Galletley, Galletly:** v. GOLIGHTLY

**Galley, Gallie, Gally, Galey:** Henry *Galye* 1219 AssY; Adam *del Galay* 1304 FrY. Either 'man from the galley' or OFr *galie*, *galee*, ME *galai*, *galy(e)* 'a galley' used by metonymy for a galley-man, rower, v. also GALILEE.

**Galliard, Gaillard, Gallard, Gaylard, Gaylord:** *Gaylardus* 1206 PatR; Robert *Gaylard* 1225 ClR; John *Galard* 1232 FineR; Sabina *Geylard* 1295 ParlR (Ess); Alexander *Galyard* 1426 FrY. OFr *Gaillart*, OG *Gailhard* 'lofty-hard', or OFr *gaillard*, *gaillart*, ME *gaillard*, *galiard* 'lively, brisk; gay, full of high spirits' (c1390 MED).

**Gallier(s):** v. WALLER

**Galliford:** v. GULLIVER

**Galling:** Margaret *Gaddling* 1260 AssC; John *Gallyng* 1296 SRSx. OE *gædeling* 'companion', ME *gadeling* 'fellow, vagabond'.

**Galliot, Galliott:** Wydo *Galiot* 1230 P (D); William *Galyot* 1275 AssSo. OFr *galiot* 'sailor in a galley, galley-slave, pirate', ME *galyot* 'pirate' (c1425 (rare) MED). Noted also in the sea-board counties of Essex, Kent, Sussex and Suffolk (1296–1327).

**Galliver:** v. GULLIVER

**Gallon:** Walter *Galun* 1220 Cur (So); Thomas *Galon* 1275 SRWo. OFr *Galon*, cas-régime of OG *Walo*. v. GALE, WALE. *Galun* may be a diminutive of *Galo* (*Gal-un*), or has been influenced by ME *galun* 'gallon'.

**Galloway, Galway, Gallwey:** Thomas *de Galweia* 1208 P (Wo); Robert *de Galewaye* 1284 IpmY; Michael *Galeway* 1359 AssD; John *Galway* 1405, William *Galloway* 1541 Black. In Scotland from the district of Galloway. This may sometimes be the source of the English names, but there was probably also an English place-name Galway which has not yet been identified.

**Gally:** v. GALLEY

**Gallyer:** v. WALLER

**Galpin, Galpen:** Robert, William *Galopin* 1195 P (Wa), 1201 AssSo; William *Galpyn* 1279 AssSt. OFr *galopin* from *galoper* 'to gallop', 'messenger, page; turnspit, scullion in a monastery' (1567 NED), especially the latter.

**Galsworthy, Golsworthy, Galsery:** Thomas *Gallysworthy* 1524 Hoskins; Thomas *Galsworthie* 1558 HartlandPR; Nectanus, Elizabeth *Galsworthe* 1560, 1571 ib.; William, Joan *Galsery* 1598, 1625 ib. From Galsworthy (Devon).

**Galt, Gault, Gaught, Gaut, Gaute:** Godfrey, William *Galt* 1198 FFNf, 1202 P (Y); William *Galt* 1367 Black (Perth). In England, a nickname from the boar or hog, ME *galte*, *gaute*, *gault*, ON *gǫltr*, *galte*. In Scotland, like *Gall*, it is found in Perth and

Aberdeen (as *Gaut* in 1649) and is regarded by Scots as a variant of *Gaul* or *Gauld*.

**Galton:** John *de Galton* 1296 SRNb; John *Gaulton* 1340 PN Do i 305; John *Galton* 1664 PN Do ii 62. From Galton (Do).

**Galway:** *v.* GALLOWAY

**Gamage, Gammage, Gammidge, Camidge, Cammidge:** Godfrey *de Gamages* 1158 P (He); Philip *de Camiges* 1275 RH (W); Alicia *Gamage* 1279 RH (O); William *Camage* 1583 FrY. From Gamaches (Eure). *v.* ANF.

**Gaman:** *v.* GAME

**Gambell, Gamble, Gambles, Gammell, Gammil:** *Gamel* 1066 DB; *Gamel* Auceps 1158 P (Y); Simon, Adam *Gamel* 1202 AssL, 1260 AssY; Jordan *Gambel* 1297 MinAcctCo; John *Gamyll* 1597 FrY. ON *Gamall*, ODa, OSw *Gamal* 'old'.

**Gambier:** A Huguenot name from a French refugee family at Canterbury. James *Gambier*, b. 1692, was a distinguished barrister, and his grandson was James, Admiral Lord *Gambier* (Smiles 396).

**Gamblin(g):** *v.* GAMLIN

**Gambray:** *v.* CAMBRAY

**Game, Games, Gaman, Gamon, Gamman, Gammans, Gammon, Gammond, Gammons:** Richard *Gamen* 1251 FFEss, 1275 RH (Nf); Roger, John *Game* 1268 AssSo, 1377 LLB H; John, Hugh *le Gamene* 1275 RH (Ha), 1309 SRBeds; John *Goodgame* alias *Game* 1549 FFHu. OE *gamen*, ME *gamen*, *game* 'game', a nickname, no doubt, for one fond of or good at games, cf. GOODGAME, but also one given to the winner of 'the gamen', the prize in a foot-race. cf. 'Men usen ofte þis gamen, þat two men . . . rennen a space for a priis, and he þat comeþ first to his ende shall have *þe gamen* þat is sett, wheþer it be spere or gloves [v.l. gleyves] or oþir þing þat is putt' (c1380 NED). cf. GLEAVE, and *v.* GAMMON.

**Gamlin, Gamlane, Gamlen, Gamblin, Gambling:** Odo *filius Gamelin* 1086 DB (So); *Gamelyn* de Cottyngwith 1347 AD iv (C); James, Ralph *Gamelin* 1262 *For* (Ess), 1279 RH (O). *Gamel-in*, a diminutive of *Gamel*. *v.* GAMBELL. *Gamel* and *Gamelinus* are used of the same man (1218 AssL).

**Gammage:** *v.* GAMAGE

**Gamman(s):** *v.* GAME

**Gammell, Gammil:** *v.* GAMBELL

**Gammidge:** *v.* GAMAGE

**Gammon, Gamon:** Roger, Margery *Gambun* 1209 P (Wa), 1222 Cur (O); Richard, John *Gambon* 1260 AssC, 1327 SRC. A diminutive of *gamb*, the Norman form of *jambe* 'leg'; 'little leg'. cf. Fr *Gambet*, *Gambin*, *Gambon*. *mb* would be assimilated to *mm*, hence *Gammon*. *v.* also GAME.

**Gamson, Gameson:** William *Gambeson* 1297 MinAcctCo. OFr *gambeson* 'a quilted jacket or tunic usually worn under the armour'. Metonymic for a maker of these.

**Gander:** Roger, William *Gandre* 1275 RH (Sf), 1327 LLB E; Reginald *le Gandre* 1327 SRSo. OE *gan(d)ra* 'gander'.

**Gandey, Gandy:** John *Gameday* 1327 SRSf. 'Servant of *Game*.' *v.* DAY. A Suffolk name.

**Ganger:** Ralph *Gangere* 1267 FFL; Richard *le Ganger* 1301 FS. A derivative of OE *gangan* 'to walk'.

In ME also for 'a go-between'. cf. also Isabella *Gange* 1327 SRSo; Richard *Gangishider* 1173 P (Lo) 'goes here'.

**Gannaway:** *v.* JANAWAY

**Gannet, Gannett:** John *Ganet* 1208 P (Glam); Lucas *Ganet* 1275 RH (D); William *Ganet* 1360 FFHu. A nickname from OE *ganot* 'the solan goose'.

**Gannock:** John *Gannok'* 1327 SRLei; John *Gannok* 1392 AssL; Katerine *Gannok* 1437 Paston. ME *gannok* 'shelter, alehouse', hence 'dweller at the shelter', or 'keeper of the alehouse'.

**Ganson:** *v.* GAWENSON

**Gant:** *v.* GAUNT

**Ganter:** *v.* GAUNTER

**Ganton:** John *Galmeton* 13th Cust; Robert *de Galmeton* Jn SPleas (Y); Matilda *de Galmeton'* 1234 YCh. From Ganton (ERY), *Galmeton* DB.

**Gape, Gapes:** William *Gape* 1243 AssSo. OFr *gape* 'weak, enfeebled'.

**Gapp:** Richard *gappe* 1198–1212 Bart; Savatus *del Gap* de Torp 1275 RH (Nf); Nicholas *Gap* ib.; Thomas *atte Gap* 1327 SRSf. ME *gappe*, ON *gap* 'chasm', 'breach in a wall or hedge', in Norfolk and Suffolk 'dweller by a gap in the cliffs'.

**Gapton:** Henry *de Gapeton* 1219 P (Nf/Sf). From Gapton (Sf).

**Garard:** *v.* GERARD

**Garbe, Garber:** Henry *Garbe* 1275 RH (Nf). ONFr *garbe* 'wheatsheaf', first recorded in NED in 1502. The word must be much older. The Latin *garba* is found in 1103 (MLWL) and the term must have been in common use in reference to tithes. *Garber* may mean a binder of wheat into sheaves or a collector of tithes. *Garbe* is metonymic.

**Garbett, Garbutt:** (i) *Gerbodo* 1086 DB; *Gerbodo del Scalt* 1185 P (L); William *Gerbode* 1185 P (Hu); Thomas *Gerbot* 1302 AssSt; John *Garbot* 1397 FrY; James *Garbutt* 1602 FrY. OG *Gerbodo* 'spear-herald'. (ii) *Gerbertus, Gereberct, Girbertus* 1086 DB; *Gerbertus* de Sancto Claro 1199 Cur (Sf); Robert *Gerbert* 1100–13 Rams (Herts), 1174 P (Berks), 1187 P (W); William, Elyas *Girbertus* 1130 P (W). OFr, OG *Gerbert, Girbert* 'spear-bright'. (iii) *Gerboldus* de Curcy, Hugo *filius Gerbold'* 1219 AssY; William *Gerebald* 1221 *ElyA* (C); Adam *Gerbaud* 1300 Rams (Hu). OG *Gerbald, Gerbold* 'spear-bold'. *Gerbodo* and *Gerbald* were early confused: cf. *Gerbod'* . . . et Willelmus *filius Gerboud* (both of Grindleton) 1219 AssY; William *Garbode, Garbolde* 1275 RH (Nf).

**Gard, Guard:** Richard, John *le Gard* 1275 SRWo, 1279 RH (C); William *la Garde* 1309 SRBeds. OFr *garde* 'guard, watchman'.

**Garden, Gardyne:** William *del Gardin* c1183 AC (O); William *Gardin* 1220 Cur (Hu); John *atte Gardyne* 1296 SRSx. ONFr *gardin* 'garden' (a1325 MED). Metonymic for *Gardener*. cf. JARDIN.

**Gardener, Gardenner, Gardiner, Gardinor, Gardner, Gairdner:** Anger *gardiner* a1166 Seals (L); William *le gardinier, le Gardenier* 1199, 1201 P (R); Andrew *Gairner* 1663 Black; Thomas and Rose *Gardiner, Gardner* 1730, 1733 HorringrPR (Sf). ONFr **gardinier* corresponding to OFr, modFr *jardinier* 'gardener' (a1325 MED). Walter *le gardiner* (1292

SRLo) was a fruiterer. *Gairdner* is Scottish. *v.* GARNER.

**Gardham, Gardam:** From Gardham in Cherry Burton (ERY).

**Gare:** Alvred *de la Gare* 1219 Cur (K); John *Gare* 1303 IpmY; William *del Gare* 1343 FrY. Early or northern forms of GORE, OE *gāra*.

**Garey, Geary:** *Gheri* 1066 DB (Sa); *Geri* 1203 AssSt; *Geri* filius Gunni 1150–60 DC (L); Ralph *Gari* 1196 Cur (Nth); Richard, Aluredus *Geri* 1200 P (Sa), 1221 ElyA (Nf); Ralph *Gary* 1249 FFC; Roger *Gairy* 1301 SRY. ON *Geiri*, ODa *Geri* 'spear'. *Gari* is probably a Norman form. *Geri*, with hard *G* (as here) is difficult to distinguish from OFr *Geri* (with *j*-sound) in *Geary, Jeary*.

**Garfin:** *v.* GARVIN

**Garford, Garforth, Garfath, Garfirth, Garfit:** William *de Gereford* 1219 AssY; Henry *de Garforde* 1260 Oseney; Richard *Garford* 1447 FFEss; Rycherd *Garforth* 1558 RothwellPR (Y). From Garford (Berks), or Garforth (WRYorks).

**Gargate, Gargett:** Reginald, Richard *Gargate* 1130, 1166 P (O, Co); John *Garget* 1210 Cur (So). OFr *gargate, garguett* 'throat'.

**Gargrave:** Bartholomew *de Geirgraue* 1196 MemR (Y); Arnald *de Garregraue* 1210 Cur (Y); William *de Gairegraue* 1219 AssY; John *Gargrave* 1446–7 FFSr. From Gargrave (WRY).

**Garland, Garlant:** (i) Robert *Gerland* 1221 AssWa; Adam *Garlaund* 1293 FFEss; William *Garland* 1293 Fees (D). Metonymic for *garlander* 'a maker of garlands', metal chaplets or circlets for the head adorned with gold or silver: cf. William *Garlander* 1319 SRLo. (ii) John, Robert, William *de Garland(e)* 1190 P (Ha), 1225 Pat (Y), 1208 Cur (Ess). 'Dweller by the gore-land', OE *gāra* 'triangular piece of land'.

**Garle:** Cecilia *Garle* 1279 RH (O); John *Garle* 1327 SRC; William *Garl* 1327 SRSx. ME *girle, garle* 'a child of either sex'. *v.* also GIRLE.

**Garlic, Garlick, Garlicke:** Robert *Garlec* 1273 RH (C); Gilbert *garlek* 1277 Ely (Sf); John *Garlyke* 1491 FrY. OE *gārlēac* 'garlick', used of a seller of garlick: cf. Luke *le Garlekmongere* 1309 LLB D; Thomas *Garleker* 1387 FrY.

**Garman, Garmons, Garment:** *Gormundus* 1208 Cur (So); Alanus *Garmundus* 1177 P (Nf); William *Gormund* 1255 RH (W). OE *Gārmund* 'spear-protector'. *v.* GORMAN.

**Garmonsway, Garmondsway:** Ralph *de Garmundeswaie* 12th FeuDu; Robert *de Garmundeseye* 1242 AssDu; William *Garmonsway* 1446 FrY. From Garmondsway (Durham).

**Garn, Garne:** Hugh *de Gerne* 1221 AssGl; Gideon *Gerne* 1642 PrD. From a lost *Garn* in Westbury on Severn (Gl).

**Garnays, Garneys, Garniss, Garnish:** Robert *Garnoise* 1194 ChR (Sf); Gilbert *le Garneys* 1269 IpmNf; Rayment *Garnishe* 1568 SRSf; Thomas *Garnis* 1641 PrSo. Perhaps from a derivative of OFr *gernon* 'moustache'. *v.* GARNON.

**Garne:** *v.* GARN

**Garnell:** John *Gardinal* 1325 FFEss. For CARDINAL.

**Garner, Garnar, Garnier, Gerner:** (i) *Garnerius* de Nugent 1170–83 NthCh (R); Geoffrey *Gerner* 1272 FFEss. OFr *Garnier*. *v.* WARNER. (ii) William *del Gerner* 1332 SRLa. OFr *gerner, gernier* 'storehouse for corn', hence 'keeper of the granary'. (iii) *Garner* (1540 Black) is a late form of *Gardner*: Matthew *Gardner, Garner* 1706–9 Bardsley.

**Garnesy, Garnsey:** Peter *Garnesey* 1524 SRD; Thomas *Garnsey* 1524 SRSf; Richard *Garnesey* 1642 PrD. From Germisay (Haut-Marne), or perhaps sometimes from Guernsey (Channel Isles).

**Garnet, Garnett, Warnett:** *Garnet* 1409 Bardsley; Ricard *Gernet* 1086 DB (Ess); Roger, Benedict *Gernet* 1170, 1195 P (Nt, La); Robert *Warinot* 1275 RH (Hu); John *Garynot* (*Gerynot*) 1327 *SR* (Ess); William *Garnett* 1379 PTY. A diminutive of OFr *Guarin, Warin*: OFr *Guarinot, Warinot*. *Garnet* is also occupational: Joceus, William *le Gerneter* 1296 SRSx, 1327 MEOT (La), from OFr *garnetier, garnetier* 'superintendent of the granary' (1454 MED), or 'maker of garnets', T-shaped hinges, of which the short cross-bar was fastened to the frame of a door, perhaps from ONFr *carne* 'hinge' (*kernettes* 1275, *gornettes* 1325, *garnettes* 1532 Building 297–8; *garnettes* 1459 NED).

**Garneys, Garnish, Garniss:** *v.* GARNAYS

**Garnon, Garnons, Gernon, Garnham, Grennan:** Robert *Gernon, Gernun, Greno, Grenon, Grino, Grinon* 1086 DB; William de Perci cognomento *Asgernuns, ove les gernuns* Wm I Whitby (Y); Willelmus *cum grinonibus* 1104 ELPN 159; Adam *as Gernuns* 1166 P (Ess); William *Asgernuns* 1175 P (Y); William de Perci *Ohtlesgernuns* c1180 Whitby (Y); Alexander *Gernum* (bis) 12th Gilb (L); Richard *gernun* 1221 ElyA (Sf); Abel *Agernun* 1248 FFEss; William Bought called *Gernon* 1310 LLB D; John *Gernon* 1327 SRDb; Henry *Garnon* 1524 SRSf; John *Garnham* ib.; Thomas *Garnome* 1533 SIA xii; Nicholas *Garnom* 1568 SRSf; Thomas *Garnam, Garnham* ib. OFr *grenon, gernon* 'moustache'. Early examples are often prepositional, 'with the moustache(s)', a distinguishing characteristic which readily gave rise to the nickname. All the Normans on the Bayeux Tapestry are represented as clean-shaven. This nickname in the form *Algernon* was ultimately adopted as a christian name. Already in the 12th century *Gernun* is found occasionally as *Gernum*. This later became *Garnom* and *Garnam* and finally *Garnham* as if derived from a place-name. In the 1524 Subsidy Roll for Suffolk four men of this name were assessed in Bacton: Thomas *Gernon*, Nicholas *Gernown*, John and Edmund *Garnon*. In the same parish, in 1568, all four men of this family are called *Garnham*, the only form found in the 1674 Hearth Tax. This is the usual modern form, *Gernon* and *Garnon* being rare. *Grennan* is from the variant OFr *grenon*.

**Garnsey:** *v.* GARNESY

**Garrad, Garred, Garratt, Garrett, Garretts, Garritt, Gerrad, Gerratt, Gerred, Gerrett, Jared, Jarrad, Jarred, Jarratt, Jarrett, Jarritt, Jerratt, Jerreatt, Jerred:** *Geraddus* filius Simonis 1242 Fees (L); *Gerad* 1272 Ipm; *Jaret* Havelok 1539 FeuDu; John *Gerad* (*Gerard'*) 1230 P (So); Peter *Geraud* son of Robert

*Gerold* 1250–1 Fees (Nth); Thomas *Gerad* 1332 SRCu; Robert *Garad* 1540 NorwW (Sf); Thomas *Jarred* 1540 SxWills; John *Garrat*, William *Garrett* 1553–5 RochW; Robert *Jarratt* 1597 SRSr. For GERALD or GERARD. Through loss of *l* or *r* respectively, both became *Gerad*, *Jerad* and developed alike.

**Garrard:** v. GERARD

**Garraway, Garroway:** (i) Ricardus *filius Garwi* 1188 P (Gl); William, Walter *Garwy* 1236 Ipm (Ess), 1327 SRWo; John *Gorewy* 1317 AssK. OE *Gārwīg* 'spear-war'. (ii) Walter *de Garewy* 1228 Cl. From Garway (Hereford).

**Garrick:** Usually Huguenot. Pierre Bouffard, Sieur de la Garrigue, was head of a family at Castres near Bordeaux. A member of the family fled to England in 1685, and adopted the name of the family estate. His grandson and namesake, the actor, David *Garrick*, was born at Hereford in 1716. Fr *carrigue* 'a place covered with oaks' (Smiles 396–7).

**Garrish:** v. GERISH

**Garrison:** (i) Nicholas *de Gerdeston* 1204 AssY. From Garriston (NRYorks). (ii) James *Gerardson*, *Garardson* 1423–9 FrY; Thomas *Garratson*, John *Garretson* 1550 Pat. 'Son of *Gerard*.'

**Garrod, Garrood, Garrould:** v. GERALD

**Garrow:** Thomas *Garrov* 1235–6 AssDu. From Garway (He), *Garou* 1138.

**Garside, Gartside:** Richard *de Garteside* 1285 AssLa; Hugh *Garthesyde* 1498 FrLei; James *Garthside* 1553, Alice *Garside* 1597 WhC. From Garside (La).

**Garson:** Reginald *Garcon* 1177 P (Y); Hugh *le Garchun* 1242 P (Ess). OFr *garcon*, *garchon* 'valet'.

**Garstang:** Adam *de Gahersteng* 1206 P (La); John *of Garstang* 1401 AssLa; Henry *Gairstang* 1443–4 FFWa. From Garstang (La).

**Garston, Garstone, Gaston:** (i) John *de la Garston'* 1210 Cur (Sr); John *at(t)e Gaston*, *Garston* 1327, 1332 SRSx. 'Dweller by the grass-enclosure' (OE *gærstūn*) or, possibly, 'warden of the *garston*', an enclosure near the village where cattle were kept. (ii) Adam *de Gerstan* 1264 Ipm (La). From Garston (Lancs).

**Garth:** John, Mariota *del Garth* 1297 MinAcctCo (Y), 1332 SRCu. 'One in charge of enclosed ground, a garden or paddock', ME *garth*, ON *garðr*.

**Garton:** Robert *de Gertuna* c1163 Gilb; Robert *de Garton* 1249 IpmY; William *Garton* 1346 FFEss. From Garton (ERYorks), or from minor places of this name.

**Gartside:** v. GARSIDE

**Garvin, Garfin:** Godric *filius Gareuinæ* 1066 DB; Ulf *filius Garuini* 12th DC (L); Robert *Gervyne* 1372 FrY. OE *Gārwine*. v. OEByn 182–3.

**Gasche:** v. WACE

**Gascoign, Gascoigne, Gascoin, Gascoine, Gascoinge, Gascoyne, Gasquoine, Gaskain, Gasken, Gaskin, Gaskins, Gasking:** (i) Bernard *Gascon'* 1206 Cur (Nth); William *le Gascun* 1208 P (Y). AFr *Gascon*, ME *Gascoun*, *Gascon* 'a Gascon', 'a native of Gascony' (a1375 NED). This does not seem to have survived, unless it has been absorbed by *Gasken* and *Gaskin(s)*. (ii) Philip *le Gascoyn* 1266 FineR (Sa); Peter *Gascoying* 1274 RH (D); Geoffrey *Gascoyne* 1275 RH (Nf); Joan *Gaschoyn* 1469 GildY; William *Gasquyn*

1473 ib.; John *Gasqwyn* 1486 GildY, John *Gascoyng* 1505 ib., Dom. John *Gascon* 1509 ib., Mag. John *Caswin* 1528 ib.; John *Gascon*, Thomas *Gaskyn* 1524 SRSf. OFr *Gascuinz* adj., 'of Gascony', 'a Gascon'. In the 15th century and later, *Gascon* occurs as *Gaskin*, *Gascogne*, *Gascoigne*, which NED suggests may be due to the influence of *Gascogne* 'Gascony'. Some of the above forms are certainly from the adjective. (iii) Robert *de Gascoin* 1243 AssSo; Nicholas *de Gascoigne* 1340 LLB F; William *Gascoigne* 1389 FFHu. From Gascony (Fr *Gascogne*).

**Gaselee, Gazelee, Gazeley:** Alexander *de Gasele* 1275 RH (Nf); William *de Gayslee* 1326 SRC. From Gazeley (Sf).

**Gash, Gashion:** v. WACE

**Gaskell, Gaskill:** v. GAITSKELL

**Gaslin:** Geoffrey *Gaselyn* 1281 CtW; Walter *Gasselyn* 1327 SRSo; Walter *Gasselyne* 1333 IpmGl. *Gas-el-in*, a double diminutive of *Gass*. v. WACE.

**Gason, Gass:** v. WACE

**Gasper:** v. GAYSPUR

**Gassman, Gastman:** Adam *Gasman*, Richard *Whasman* 1346 FA (Sf). 'Servant of *Gass* or *Wace*', with intrusive *t* in *Gastman*. v. WACE.

**Gasson:** v. WACE

**Gast:** v. GHOST

**Gastall:** v. WASTALL

**Gastman:** v. GASSMAN

**Gaston:** v. GARSTON

**Gatcombe, Gatcum:** Nicholas *de Gaticumbe* 1261, William *de Gatecumbe* 1279 AssSo. From Gatcomb (So), or Gatcombe (Wt).

**Gate, Gates:** Ailricus *de la Gata* 1169 P (D); Ralph *de Gates* 1206 Cur (O); Gilbert *atte Gate* 1260 AssC; Cristina *Gate* 1275 SRWo; Richard *Overthegate* 1327 SRDb; Thomas *atte Gate*, Custancia *del Gates* 1379 PTY. 'Dweller by the gate(s)', OE *gatu*, plural of OE *geat*. v. YATE. In the Scandinavian counties we may also have ON *gata* 'road'. cf. STREET.

**Gatecliff, Gatliff, Getliff:** Thomas *Gaytclyff* 1457 FrY. From Gatley (Ches), earlier *Gateclyve*.

**Gateley, Gately, Gatley, Gatlay:** Ralph *de Gateleia* 1086 DB (Nf); Simon *de Gatle* 1203 FFK; Henry *Gateleye* 1327 SRWo. From Gateley (Norfolk), or Gatley (Ches, Hereford).

**Gateman, Gatman:** William *Gateman* 1296 SRSx; Ralph *Gateman* 1327 SRSf; Roger *Gateman* 1327 SRSo. 'Dweller by the gate', OE *geat*, *mann*, or 'dweller by the road', ON *gata*, OE *mann*.

**Gaten, Gatend:** John *Attegatehend* 1288–9 NorwLt; Stephen *atte Gathende* 1297 SRY; William *Attegateshende* 1308 AssNf. 'Dweller at the end of the road', ON *gata*, OE *ende*.

**Gatenby, Gatensby:** Stephen *de Gaitenbi* c1190–1212 YCh; Thomas *de Gaytenby* 1327 SRY; Peter *de Gaytenby* 1405–6 IpmY. From Gatenby (NRY).

**Gatend:** v. GATEN

**Gatensby:** v. GATENBY

**Gater:** Walter *le Gater* 1279 RH (O); Robert *Gater* 1301 SRY; William *le Gatier* 1332 SRSx. (i) A derivative of OE *gāt* 'goat', hence 'goat-herd', with preservation today of the unrounded northern *ā*. cf. GOATER, GAITER. (ii) In the north, 'dweller by the road', from ON *gata*. (iii) 'Dweller by the gate', *v.* GATE.

**Gaterell:** *v.* GATHERAL

**Gathard:** William *the Gatherde* 1287 AssCh; Reginald *le Gateherd* 1327 SRSt; Henry *le Gatherd* 1331 AssSt. 'Goat-herd', OE *gāt, hierde.*

**Gatheral, Gatherall, Gaterell, Gatterel, Gatrall, Gatrell, Gatrill, Gattrell:** Weekley takes *Gatherall* literally as a nickname, but all are probably voiced forms of CATTERALL, etc.

**Gathercole:** William, Richard *Gaderecold* 1327 SRSf; Thomas *Gadercold* 1524 SRSf; James *Gathercoal, Gathercole* 1597, 1611 Gardner (Sf). There can be no association with charcoal or cabbage as the second element is clearly *cold.* To the medieval physiologist, melancholy and choler were dry and cold. When old, a man's blood was said to wax dry and cold and cold was associated with weakness. Hence, probably, a nickname for an old man, one who 'gathered cold' and weakness.

**Gatherer:** Roger *le gaderer* 1275 RH (L). A derivative of OE *gaderian* 'to gather', a collector of money. ME *gather* was used of collecting dues from gild-brethren, rent, tithes, etc.

**Gathergood:** Thomas *Gadregod* 13th Lower (Sx). 'Gather goods', OE *gaderian, gōd,* a nickname for one with acquisitive and thrifty habits. cf. Abraham *Cathermonie* 1193 Riev 'gather money'; Adam *Gaderpeynye* 1285 AssLa 'gather pennies'.

**Gatland:** William *Gatelond* 1327 SRSx; Ralph *atte Gatelond* 1332 ib. 'Dweller on the land with or by a gate' or at the goat-land.

**Gatliff:** *v.* GATECLIFF

**Gatling, Gattling:** Geoffrey *Gatelin*, Joan *Gatelyn* 1274 RH (W). Probably for GADLING, or for CATLIN.

**Gatman:** *v.* GATEMAN

**Gatorest:** William *Gatorest* 1302 FFY. 'Go to rest', OE *gān, rest,* probably a common expression by the speaker. cf. Serlo *Gotokirke* 1279 RH (C) 'go to church'; Olive *Goadebles* 1269 Pat 'go to the devil'; William *Gawetobedde* 1332 SRSx 'go we to bed'; John *Go inthe Wynd* (a foot soldier) 1334 LLB E; John *Latethewaterga* 1242 AssDu 'let the water go'.

**Gatrell, Gatterel:** *v.* GATHERAL

**Gatsden, Gatsdon:** Aldulf *de Gatesdene* 1212 P (Beds/Bk); William *de Gatesden* 1212 P (D); John *de Gatesdene* 1255–6 FFSx. From Great, Little Gaddesden (Herts), *Gatesdene* DB.

**Gattiker:** William *de Gatacra* 1161 P (Sa). From Gatacre (Salop) or Gateacre (Lancs).

**Gattling:** *v.* GATLING

**Gatton:** Thomas *de Gatton'* 1219 Cur (Sf); Mabilia *de Gatton'* 1219 P (Beds/Bk); Hamo *de Gattuna* 1220 ArchC iv. From Gatton (Sr).

**Gatward:** William, Richard *le Gateward* 1255 Ass (Ess), 1262 MEOT (Herts); Walter *Gatward* 1524 SRSf. OE *geat-weard* 'gate-keeper', or 'goat-ward, goat-herd' (OE *gāt*).

**Gaubert:** Gervase *Gaubert* 1212 P (Gl), *Gauberd* 1225 FFO. OFr *Gaubert,* from OG *Waldberht.*

**Gauche:** John *Gauch* 1260 AssC. OFr *gauche* 'left-handed, awkward'.

**Gauchy:** Ralph *Gauchy* 1248–9 RegAntiquiss. From Gauchy (Aisne).

**Gaudin, Gauden:** *Waldinus* 1066 DB (So), 1086 DB (Y, L); *Gaudinus* 1188 P (Nf); Peter, Roger *Gaudin* 1237 Fees (D), 1275 RH (Sf); Roger *Waudin* 1243 Fees (Sa). AFr *Waldin,* OFr *Gaudin,* a diminutive of OG names in *Wald-. v.* WALDING.

**Gaudy, Gawdy:** Reginald *Gaudi* 1221 AssGl; John *Gaudy* 1297 SRY; Thomas *Gawdy* 1430 FeuDu; Clippesby *Gawdy* 1590 SfPR. Either ME *gaudi* 'trick', cf. William *Gaude* 1332 SRLei, from ME *gaud* 'trick, fraud', or ME *gaudi* 'yellow, yellowish', of hair or complexion.

**Gaught:** *v.* GALT

**Gaukrodger:** *v.* GAWKRODGER

**Gaul:** *v.* GALL

**Gauld:** Jas *Gald* 1550 Black; John *Gauld* 1686 ib. Gael *gallda* 'pertaining to the Lowlands'. *v.* GALL, GALT.

**Gault:** *v.* GALT

**Gaulter, Gaultier:** *v.* WALTER

**Gaumbe:** *v.* JAMBE

**Gaunson:** *v.* GAWENSON

**Gaunt, Gant:** (i) Richard *le Gaunt* 1219 Cur (K); Maurice *le Gant* 1225 AssSo. ME *gaunt, gant* 'slim, slender', or 'haggard-looking, tall, thin and angular' (1440 MED). (ii) Metonymic for GAUNTER 'glove-maker'. (iii) Gilbert *de Gand, de Gant* 1086 DB; Gilbert *de Gaunt* 1219 Cur (L). From Ghent (Flanders).

**Gaunter, Ganter:** Lefwinus *le Wanter* 1172 Oseney (O); Adam *le Gaunter* 1220 Cur (Mx); Adam *le Ganter* 1279 RH (O). OFr *gantier, wantier* 'maker or seller of gloves' (1263 MED).

**Gaussen:** A Huguenot name from a distinguished French Protestant family. David *Gaussen* from Lunel in Languedoc fled to Ireland in 1685, and other members of the family came to England (Smiles 397– 8).

**Gaut, Gaute:** *v.* GALT

**Gautier:** *v.* WALTER

**Gavin, Gaven, Gauvain, Gauvin, Gawen, Gawn, Gawne, Wawn, Wawne:** *Wawanus* filius Ricardi 1208 Cur (Ess); Goseline *filius Gawyne* 1279 RH (C); *Gawynus de Suthorp'* 1332 FFK; *Gaven* Richardson 1631 Bardsley; *Gawne* Gilpin 1653 ib.; William *Walwein* 1169 P (Do); Robert *Wawayn* 1255 RH (W); Matilda *Wawwayne* 1315 Wak (Y); John *Walweyn,* Alice *Waweins* 1327 SRC; Adam *Wawayne* c1372 Black; Emma *Gawyn* 1379 PTY; William *Wawne* 1502 Black; Alexander *Gavin* 1647 ib. *Gavin,* still a common christian name in Scotland, is the Scottish form of the English *Gawayne* and the French *Gauvain,* the name of the son of King Arthur's sister. In the Welsh Arthurian romances it is found as *Gwalchmai,* perhaps 'the Hawk of the Plain' (Jackson), and is latinized by Geoffrey of Monmouth as *Walganus,* a form which has led to the suggestion that there was an alternative name *Gwalchgwyn* 'white falcon'. The name was also common in Brittany which may partly account for its frequency in the eastern counties. In central French the Anglo-Norman *Walwain* became *Gawain* or *Gauvain,* with the common substitution of *G* for *W. Walwain* may still survive in *Walwin. v.* also GAWENSON.

Gaw: v. GALL

Gawdy: v. GAUDY

Gawenson, Gaunson, Ganson: Alexander *Gawensone* 1563 Black (Nairn). 'Son of *Gavin*.'

Gawkrodger, Gawkroger, Gaukrodger, Gaukroger: Rauffe, William *Gawkeroger* 1539 RothwellPR (Y); John *Gawkrycher* 1553 FrY; Danyell *Corkroger* 1685 FrY. 'Awkward or clumsy Roger', a Yorkshire name, from dial. *gawk*.

Gawler, Gowler: Ralph *Gauelere* 1206 P (Do); William *le Gavelere* 1305 FFEss. ME *gaveler, goveler, gouler* 'usurer', from OE *gafol* 'tribute, rent, interest'.

Gawn, Gawne: A Manx name from *Mac-an-Gabhain* 'the smith's son': *MacGawne* 1422, *Gawen* 1517 Moore. cf. Irish *McGowan* and Gaelic *McGavin*. v. also GAVIN.

Gawthorp, Gawthorpe, Gawthrop, Gawthrope: Crystofer *Gawthrop* 1541 FrY; William *Gawthropp* 1672 HTY. From Gawthorpe in Osset, in Lepton (WRY), or Gawthorpe Hall (La).

Gay: (i) Osward *le Gay* 1176 P (Sr); Gilbert *Gay* 1191 P (L). ME *gai(e)*, OFr *gai* 'full of joy, lighthearted' (c1390 MED). (ii) Hilda *de Gay* 1192 P (O); William *de Gaia* 1203 AssSt. From some French place; perhaps Gaye (La Manche).

Gaydon: Peter *Geydun* 1275 RH (Nf); Alexander *Gaydonn*, Christopher *Gaydon* 1642 PrD. From Gaydon (Wa). There was also a personal name: *Gaidun* filius Willelmi 1198 P (Co).

Gaylard: v. GALLIARD

Gayle: v. GALE

Gayler, Gaylor: v. GALER

Gaylord: v. GALLIARD

Gayman: Robert *Gayman* 1674 HTSf. Probably 'servant of *Gay*'. v. GAY.

Gaymar, Gaymer: Rannulf *filius Gaimar'*, *Gaimer* 1193–5 P (L); Rannulf *Gaimer* 1197 ib.; Peter *Gaymer* 1219 FFEss. OFr *Gaimar*, OG *Weimar, Waimer, Guaimar*.

Gayner, Gaynor: Osbert *le Gaynere* 1319, John *Gayner* 1381 IpmGl. OFr *gaaineur* 'farmer'.

Gayspur, Gasper: Walter *Gaispor'* 1203 P (Wa); ffulco *Gayspore* 1319 SRLo; Walter *Gaispore* 1442 AssLo; Edward *Gayspur*, James *Gaspur* 1545 SRW. 'One who wore showy spurs', OFr *gai*, OE *spura*. cf. Matilda *Gaypas* c1248 Bec (Sf) 'one who walked with gay steps'.

Gayton, Geyton: Robert *de Geiton'* 1193 P (L); Henry *de Gayton* 1285 FFO; Robert *de Gayton* 1382 IpmNt; James *Gayton* 1672 HTY. From Gayton (Ch, Nf, Nth, St), or Gayton le Marsh, le Wold (L).

Gaywood: John *de Geiwode* 1206 Cur (Nf); Richard *de Gaywode* 1366 FFSt. From Gaywood (Nf).

Gayter, Gaytor: v. GAITER

Gaze: v. WACE

Gazelee, Gazeley: v. GASELEE

Geake, Jeeks, Jecks, Jex: Walter *Geek* 1275 RH (L); Henry *le Geke, le Gekke* 1279 RH (O); Walter *Jekkes* 1524 SRSf. ME *geke, gecke* 'fool, simpleton' (1515 NED). v. also JACK.

Gear, Geare, Geear, Geer, Geere: Albert, Joscelin *Gere* 1133–60 Rams (Sf), 1221 AssWo. ME *gere* 'sudden fit of passion', 'wild or changeful mood', whence GERISH.

Gearing, Geering: William *Gering* 1202 AssL; Sewal *Geryng* 1296 SRSx; Hugh, Peter *le Geryng* 1327 SR (Ess). ME *gering*, recorded twice (c1300) apparently in the sense 'villain'.

Geary, Gerrey, Gerrie, Gerry, Jaray, Jearey, Jeary, Jary, Jarry: *Gericus* de Gilling 1208 Cur (Y); Willelmus *filius Gerici* 1214 Cur (L); John, Richard *Geri* 1195 P (O), 1221 AssWa; John *le Geri* 1276 RH (Berks); Richard *Jery* 1279 RH (Hu). OFr *Geri*, OG *Geric*; also a nickname from ME *ge(e)ry* 'changeable, giddy', applied by Chaucer to Venus. v. also GAREY.

Geater, Geator: v. GAITER

Geaves: v. JEEVES

Gebb: v. JEBB

Gedde, Geddes: William *Ged* 1230 P (Sa); John *Gedde* 1296 IpmNt; William *Gedes* 1525 SRSx. OE *gedda* 'pike', a nickname from the fish. In Scotland from the lands of Geddes (Nairn).

Geddie: v. GIDDY

Gedge: Thomas *Geg* 1205 Cur (Nf); Walter *le Geg* 1221 ElyA (Hu); William *Gegge* 1263 AssLa; Alice *la Gegges* 1310 ColchCt; John *Gedge* 1568 SRSf. ME *gegge* a contemptuous term applied to both men and women. cf. 'Thus arraied gooþ þe gegges, And alle wiþ bare legges' (1387 NED).

Gedling: Robert *de Gedelyng* 1373 IpmNt; William *Gedelyng* 1541 CorNt. From Gedling (Nt), but sometimes, perhaps, for GADELING.

Gedney, Gedny, Gidney: Richard *of Gedeneye* 1258 FFY; Godfrey *de Geddeney* 1345 FrY; John *Gydeny* 1379 LoCh; John *Gedeney* 1416 FFEss. From Gedney (L).

Gedye: v. GIDDY

Geen: v. JEAN

Geer(e): v. GEAR

Geering: v. GEARING

Geeves: v. JEEVES

Geffen, Giffen, Giffin: Peter, Rose *Geffen* 1332 SRWo, SRWa; Nycolas *Gyffen* 1524 SRSf; Robert *Giffen* 1660 ArchC 30. The Subsidy Rolls for Worcestershire (1327) and Warwickshire (1332) contain at least 40 surnames in -*en* of a type hitherto unnoted. They are usually names of men, occasionally of women, and are found side by side with the more common type *Geffes, Hobbes, Tomes, Watts*. The suffix is added to pet-forms of French personal-names common among the peasants, usually masculine (*Jacken, Kytten, Nicken, Tommen, Watten*), occasionally feminine (*Maggen, Molden* (Maud), *Magoten*). In addition to *Geffen*, this suffix survives in the modern *Dawn, Edden, Hobben, Hudden, Judden* and *Tibbins* and has contributed to *Gibben, Hicken, Hitchen* and *Hullin*. It seems clear that we have a genitive singular, *Geffen, Dawn*, etc., parallel to *Jeffs, Dawes*, etc., but the form is difficult. Our knowledge of the 12th- and 13th-century written language of this area is scanty. Of the spoken language we know nothing. In early ME the genitive singular -*an* of the weak declension would have become -*en*. But at an early date the whole of the singular appears to have been re-formed on the analogy of the nominative plural, giving a singular in *-e* and a plural in *-en*, with a genitive plural *-ene*. The *-e* of the genitive singular was then supplanted by *-es* from the strong declension. In the West Midlands and the south the weak declension survived, but only in the plural. Presumably these pet-names of French origin were adopted at a time when the Old English system of declensions was breaking up and the inflexions were in a state of flux. They would be used most commonly in the nominative and when a possessive was required the common form of genitive in *-es* was normally used. It may be that in the West Midlands, when genitives of these pet-names were first required, the genitive singular in *-en* was still in use in the spoken language and persisted, side by side with that in *-es*, just as strong and weak plurals interchange in the Devonshire place-names *Hayes* and *Hayne*.

**Geldard, Geldart:** William *le Geldehyrde* 1284 Wak (Y); John *Gelderd* 1494 GildY; John *Geldert* 1511 FrY. ME *gelde* 'sterile, barren', and OE *hierde* 'tender of the "geld" cattle'.

**Gelder:** *v.* GILDER

**Geldgris:** Robert *Geldgrise* 1301 SRY. 'Geld pig', ON *gríss.* cf. Geoffrey *Geldecrowe* 1323 Ch (Sf) 'geld crow'; Henry *Geldesowe* 1332 SRWo 'geld sow'.

**Gelding, Gilding:** Geoffrey *Geldyng* 1296 SRSx; Thomas *Gelding* 1327 ib. ON *geldingr* 'a gelded person, eunuch' (a1382 MED), an opprobrious nickname.

**Gell, Gelle, Jell:** *Gelle* Bakur 1275 Wak (Y); *Gelle* Winter 1279 RH (C); John *Jelle* 1296 SRSx; John *Gelle* 1301 SRY. *Gell,* a pet-form of *Jelion* or *Juliana*. *v.* GILLIAN.

**Gellan:** Reginald *Gelon* 1359 Putnam. *Gell-on,* a diminutive of *Gell,* a pet-form of *Julian* or of *Juliana*.

**Gellatly:** *v.* GOLIGHTLY

**Gelley, Gellie:** *v.* GELLY

**Gellibrand:** *v.* GILLEBRAND

**Gelling:** *Mag Gelain* 1222, *Mac Gilleon, Gellen* 1511, *Gelling* 1626. Manx for *Giolla-Guillin* 'Guillin's servant' (Moore).

**Gellion:** *v.* GILLIAN

**Gellman, Gelman:** *v.* GILLMAN

**Gellner, Gellender:** Richard *gelinier* 1214 P (C); Richard *le Geliner* 1242 MESO (Lei); John *le Gelyner* 1286 MESO (Nf). OFr *gelinier* 'poulterer'.

**Gelly, Gelley, Gellie:** *Gelleia* (f) 1221 Cur (Ha); Richard *Gelee* 1209 ForNth; Laurence *Gelley* 1576 SRW. A pet-form of *Gellion,* i.e. *Gillian*.

**Gelston:** Peter *de Geueleston'* 1219, William *de Geueleston'* 1298 AssL. From Gelston (L), *Geueleston* 1202.

**Gem:** *Gemma* 1219 AssY; *Jemma* 1283 FFEss; Alan *Gemma* 1198 P (Nth); Semannus *Gemme* 1205 Cur (Nf); Simon *Gimme* 1232 Pat (L); Walter *Gemma* 1301 AD ii (Y); Agnes *Jemme* 1379 PTY. The above personal-names are those of women as is that of *Jimme* wife of Robert Feysant (1286 AssCh), but the name was also undoubtedly masculine: *Gemme* brother of Gilbert (1251 AssLa); *Gemme* Campion, whose wife Magge is mentioned, and *Gemme* the gardiner, presumably a man (1306 Wak). cf. *Jimme*

Balder 1286 AssCh. This must be the vernacular pronunciation of *James,* used for men and women alike. The surname is indebted to both the masculine and the feminine name.

**Gemmell, Gemmill:** Gabriel *Gymmill,* William *Gemmill* 1599 Black. A late Scottish pronunciation of *Gamel. v.* GAMBELL.

**Gemson, Gimson, Jimpson:** Robert *Gimson, Gemson* 1332 SRCu, 1379 PTY. 'Son of *Jimme* or *Gem.*' *v.* GEM.

**Gender, Genders, Ginder, Ginders:** Peter *Gindur* 1332 SRSr; Beatrice *Gyndours* 1334 SRK; John *Gendor* 1482 KentW. Perhaps equivalent to Fr *Gendre, Gindre*. According to Dauzat this is a name of relationship which became a patronymic when a son-in-law inherited the house of his father-in-law.

**Genn:** *v.* JEAN

**Genner, Genower:** *v.* JENNER

**Gent, Jent:** Robert *le Gent* 1195 P (Ha); John *Gent* 1200 P (Wo). OFr *gent,* ME *gente* 'well-born', 'noble', hence 'noble in conduct, courteous' (c1300 MED).

**Gentle, Gentles, Gentile, Jentle:** Osbert *le Gentil* 1202 P (Ha); John *(le) Gentil* 1242 Fees (W). OFr, ME *gentil* 'high-born, noble' (c1225 MED).

**Gentleman:** Nicholas, William *Gentilman* 1273 RH (Beds), 1301 SRY. 'A man of gentle birth' (a1275 NED).

**Gentrey, Gentry:** Richard *Gentri* 1348 FFEss; John *Gentre* 1524 SRSf; Bartholomew *Gentry* 1674 HTSf. OFr *genterie* 'nobility of birth or character'.

**Genway:** *v.* JANAWAY

**Geoffrey, Geoffroy:** *v.* JEFFRAY

**George, Georgeson:** *Georgius* Grim 12th DC (L); Hugo *filius Georgii* 1222 Cur (Nf); William *George* 1412 LLB I; William *Georgeson* 1471 Black (Coupar). Fr *Georges,* Lat *Georgius,* from Greek Γεωργός 'land-worker, farmer'. The name was introduced into England by the Crusaders but was never common before the Hanoverian succession in the 18th century.

**Gepp, Jepp, Jeppe, Jepps:** *Geppe* Werri 1228 FeuDu; *Geppe filius* Hugonis 1258 Kirkstall (Y); *Jep* le Wyte 1296 SRSx; William *Jep* 1225 Pat; William *Geppe* 1327 SRSf. A pet-form of *Geoffrey*.

**Gepson:** *v.* JEPPESON

**Gerald, Gerold, Garrould, Garrod, Garrood, Jarraud, Jarrod, Jarrold, Jarrott, Jerrold, Fitzgerald:** Robertus *filius Geraldi, Geroldi, Giroldi, Girold* 1086 DB (Do, W); Philippus *filius Geroldi* 1119 Bury (Sf); *Geroudus* filius Radulfi 1166–75 DC (L); William *Geroud* 1200 P (Y); Adam *Jeroldus* 1221 *ElyA* (Sf); Robert *Gerold* 1250 Fees (Nth); John *Jerald* 1275 RH (Sf); Richard *Gerald* 1277 *Ely* (Sf); Thomas *Garold* 1524 SRSf; Elizabeth *Jarrald* 1730 SfPR. OG *Gerald, Girald* 'spear-ruler', a name introduced by the Normans, fairly common, but less so than *Gerard* with which it was early confused: Adam *Gerold, Girardus, Gerard* 1219–46 Fees (La), Nicholas *Garolde* alias *Garard* 1535 NorwW (Nf). Garret *Fitzgarret* 1586 (SPD) is for *Fitzgerald. v.* GARRAD.

**Gerant, Gerrans, Gerring, Jerrans, Jarrand:** (i) Turstinus *de Giron, de Girunde* 1086 DB (K); Walter *de Gyrunde, Gerun, de Gerunde, Gerunde* 1279–92 FFEss. From Gironde (Gascony). (ii) *Gerinus, Girinus*

1086 DB (Wa, Ha); Serlo *filius Gerini* 1208 Cur (Ess); Henry *Girun* 1210 Cur (C); Richard *Gerrun* c1250 Rams (C); John *Geryn* 1300 LoCt; Robert *Jeryn* 1319 SRLo; Richard *Geroun*, *Geround* 1327 SRC. OFr *Gerin*, *Jerin*, OG *Gerin*, a diminutive of a name in *Ger-*, plus the suffix *-in*, and, occasionally, *-un*. One *Gerin* was a pet-name for *Gervasius* (Forssner). *v.* also JEROME.

**Gerard, Gerrard, Garard, Garrard, Jarrard, Jerrard:** *Gerardus, Girardus* 1086 DB; *Gerardus* 1134–40 Holme (Nf), 1149–62 DC (L); *Jerard* filius archidiaconi Hy 2 ib.; John, Hugo *Gerard* Hy 2 DC (L), 1199 P (Nth); William *Gerart* 1281 AssSt; Henry *Jerard* 1284 FFEss; John *Gerrard*, Thomas *Garard*, William *Garrarde* 1412, 1429, 1458 FrY. OFr *Gerart*, *Girart*, OG *Gerard*, *Girard* 'spear-brave'. Willelmus *filius Gerardi* de Kiluington is also called William *Gerard'* (1219 AssY). *v.* also GARRAD, GERALD.

**Gerham:** *v.* JEROME.

**Gerish, Gerrish, Garrish:** Umfrey *le Gerische* 1275 RH (O); William *Girisshe* 1370 Eynsham (O). ME *gerysshe* 'changeful, wild, wayward' (c1430 NED). cf. 'Madnesse that doth infest a man ones in a mone the whiche doth cause one to be geryshe, and wavering witted, not constant, but fantasticall' (1547 NED).

**German, Germann, Germain, Germaine, Germing, Jarman, Jarmain, Jarmains, Jerman, Jermine, Jermyn, Jermynn:** (i) *Germanus* 1086 DB, c1150 DC (L); *Jerman* filius Willelmi 1248 FFEss; John *Jarman* 1227 Pat (Nf); Philippus *Germani* 1236 Fees (Do); William *Jermain* 1279 RH (O); Nicholas *Germeyn* 1292 LLB A; Richard *Germeyn*, *Jermin* 1317 AssK; Margery *Jerman* 1524 SRSf. OFr *Germain*, Lat *Germanus* 'a German'. *v.* JERMEY. (ii) Gilbert *le German* 1318 Pat (Sf). 'The German.'

**Gernon:** *v.* GARNON

**Gerrad, Gerratt, Gerred, Gerrett:** *v.* GARRAD

**Gerrans:** *v.* GERANT

**Gerrard:** *v.* GERARD

**Gerrey, Gerrie, Gerry:** *v.* GEARY

**Gerring:** *v.* GERANT

**Gerrish:** *v.* GERISH

**Gersome:** Robert *Gersum* 1226–7 FFY. A nick-name from OE *gersuma* 'treasure'.

**Gervase, Gervis:** *v.* JARVIS

**Gervers:** Probably a late form of GERVASE.

**Gesling:** Nelle *Geseling* 1298 Wak (Y); Robert *Geslyng* 1379 PTY; William *Geslyng* 1552 FrY. ON *gæslingr* 'gosling'.

**Gething, Gettens, Gettings, Gettins, Gitting, Gittings, Gittins:** Eynon *Gethin* 1332 Chirk; Robert *Gettyns* 1500 FrY; Richard *Gittin* 1603 Bardsley (Ch). Welsh *Gethin*, perhaps from *cethin* 'dusky, swarthy'.

**Getliff:** *v.* GATECLIFF

**Geyton:** *v.* GAYTON

**Ghost, Gast:** Aluiet *Gæst* 1066 ICC (C); John *le Gast* 1275 SRWo; Bartholomew (*le*) *Gost* 1327 SRSf. OE *gǣst* (*gāst*) 'demon'.

**Giant:** Rannulf *le Geaunt* 1219 Cur (Nt); Reginald *Gigan* 1219 P (Nf/Sf); Alice *Gigaunt'* 1335 Glast (So). OE *gigant*, OFr *geant* 'giant'. Either for a very tall person, or ironically for a very small one.

**Gibb, Gibbes, Gibbs:** *Gibbe* de Huckenhale 1179 P (Nt); *Gibbe* 1183 DC (L); Winc' *Gibbe* 1290 AD iv

(Nf); Thomas *Gibbes* 1327 SRSo. *Gibb*, a pet-form of *Gilbert*. cf. *Gibertus* le Vag' 1332 SRSr. *v.* GIPP.

**Gibbard, Gibberd:** Robert *filius Giberti* Hy 2 DC (L); Gislebert *Gibart* c1110 Winton (Ha); Richard *Gybard* 1298 AssL; Alice *Giberd* 1327 SRLei. OG *Gebehard, Gibard*. Sometimes, perhaps, from *Gibb*, a pet-form of *Gilbert*, plus the suffix *-(h)ard*.

**Gibben, Gibbens, Gibbin, Gibbins, Gibbings, Gibbon, Gibbons:** (i) Ralph *Gibiun* (*Gibbewin*) 1176 P (O); Geoffrey *Gibewin*, *Gibwin*, *Gibeuin*, *Gubewin* 1196, 1208, 1213 Cur (So, Nth, Bk). OFr *Giboin*, OG *Gebawin* 'gift-friend', surviving in Marsh Gibbon (PN Bk 54). (ii) *Gebon*, Gilbert *Peke* 1466–7 Oseney (O); Vitalis, Richard *Gibun* c1176 Bart, 1202 P (Sx); Thomas *Gibon* 1317 AssK. *Gibb-un* a diminutive of *Gibb* (Gilbert). (iii) William *Gibben* 1332 SRWa. *v.* GIBB, GEFFEN.

**Gibberich, Gibberish:** Gilbert *Gibriche* 1332 SRCu; William *Gebrych*, *Gebrysh*, *Gyberyssh* 1525 SRSx; William *Gibberishe* 1558 Pat (Sx). OG *Giboric*. The surname has no connection with *gibberish* 'meaning-less speech', which appears to be of imitative origin.

**Gibert:** A Huguenot name. Etienne *Gibert* came to England in 1771. He became minister of the French church of La Patente in London in 1776, and later of the Chapel Royal of St James (Smiles 399).

**Giblett, Giblin, Gibling:** *Gibelinus* serviens 1199 P (Nth); Segerus *Gibelin* 1206 Cur (Co); Dera *Gibelot* 1279 RH (C); John *Gebelot* 1327 SRC; Simon *Gibeloun* 1327 SRSf. *Gib-el-in*, *Gib-el-un*, *Gib-el-ot*, double diminutives of *Gibb*.

**Gibson, Gibbeson:** Henry *Gibsone* 1311 NottBR; Richard *Gibbeson* 1327 SRWo. 'Son of *Gibb*.' May also be a contraction of (Robert) *Gibonson* 1484 FrY.

**Giddings, Giddins, Gidden, Giddens:** Richard *de Geddinges* 1205 P (Hu); Henry *Geddynge* 1337–8 FFWa; John *Giddins* 1642 PrD. From Great, Little, Steeple Gidding (Hu), Gidding (K), or Gedding (Sf).

**Giddy, Geddie, Gedye:** Herbert *Gidi* 1115 Winton (Ha); Walter *le Gidye* 1219 AssY; Roger *Geddy* 1230 P (Y). OE *gydig*, *gidig* 'possessed of an evil spirit', 'mad, insane'. cf. William *Gidyheved* 1347 LLB F.

**Giddycock:** Roger *Gydicok* 1334 SRY. A diminutive of OE *gydig* 'insane, foolish'. cf. William *Gidyheved* 1347 LLB F 'foolish head'.

**Gidney:** *v.* GEDNEY

**Giffard, Gifford, Jefferd, Jefford:** *Gifardus* 1086 DB, Hy 2 DC (L); Johannes *filius Giffard'* 1200 P (C); Walter *Gifard*, *Gifhard*, *Giffart* 1084–6 GeldR, DB, InqEl; William *Giffard* 1158 P (Wa); Reginald *Gifford* 1260 AssC; Philip called *Giffard* 1311 LLB D. OG *Gifard*, or OFr *giffard* 'chubby-cheeked, bloated'.

**Giffen, Giffin:** *v.* GEFFEN

**Gigg, Giggs:** Hugo *Gigge* 1220 Cur (Sf). ME *gigge* 'a flighty, giddy girl' (a1225 NED).

**Gigger, Giggers:** Philip *le Gigur* 1203 Pleas (Sa); Roger *le Gygur* 1274 IpmO; John *Gigger* 1584 PN Do i 161. OFr *gigueor* 'fiddler'. cf. Gigant St in Salisbury (W), *Gigorstrete* 1455.

**Giggle:** *v.* JEKYLL

**Gilbank, Gilbanks:** *v.* GILLBANK

**Gilbert, Gilbard, Gilbart, Gilbeart, Gilburt, Gillbard, Guilbert, FitzGilbert:** *Gislebertus* 1066 DB

(W); *Gilbertus* presbiter c1150 Gilb (L); Willelmus *Gilberti* 1202 Cur (W); Robert *Gylebert* 1235 AD v (Nf); Robert *Gilberd* 1240 Rams (C); William *Gilbert* 1290 Crowland (C); John *Gilbard* 1304 FFSf; Margery *Gilberdes* 1330 AD vi (W); Henry *Gylbart* 1332 SRSt. OG *Gisilbert*, *Gislebert* 'pledge- or hostage-bright', OFr *Gislebert*, *Gil(l)ebert* (*Guilbert*). The frequency of *Gilberd* suggests that this, with *Gil(l)bard*, may at times belong to *Goldbard*, though *Gilbeard* certainly occurs for *Gilbert*.

**Gilbertson:** Nicholas *Gilberdson* 1379 PTY. 'Son of *Gilbert*.'

**Gilbey, Gilby, Guilbey, Gilbie, Gillbee:** Geoffrey *Gilbe* 1327 SRLei; John *Gilby* 1414 TestEbor; George *Gylby* 1561 Pat. From Gilby (L).

**Gilbride:** *Ghilebrid* 1066 DB (Y); *Gilbryde* Macgideride 1147–60 Black. Gael *Gille Brighde* 'St Bride's servant'.

**Gilchrist, Gilcriest, Gillcrist, Gillchreest, Gillchrist, Gilgryst:** *Gillecrist* 1202 P (Cu); Kilschyn *Gilcrist* 1296 Black. Gael *Gillacrist* 'servant of Christ'. *Gillchreest* is Manx.

**Gildea:** Ir *Giolla Dé* 'servant of God'.

**Gildeney:** Roger *Gildeneye* 1292 AssCh. 'Golden eye', OE *gylden*, *eāge*. cf. Roger *Gildynballokes* 1316 Wak (Y) 'golden testicles'; Robert *Gildenfot* 1209 Pleas (Nf) 'golden foot'; John *Gildenher* 1263 IpmY 'golden hair'; William *Gyldynhels* 1376 AssEss 'golden heels'; John *Gyldentoo* 1455 Cl (Lo) 'golden toe'.

**Gilder, Gilders, Gelder, Guilder:** Stephen *le Gelder* 1281 LLB B; John *le Gilder* 1306 ib. C. A derivative of OE *gyldan* 'to gild', a gilder (1550 NED).

**Gilderdale:** William *de Gildusdal'* 1219 AssY; Geoffrey *de Gildresdale* 1332 SRCu. From Gilderdale Forest (Cu), or Gildersdale in Warter (ERY).

**Gilders:** Ralph *del Gildhus* 1176 P (Y), George *Gildus* 1638 Bardsley. Probably 'caretaker of the guildhall'.

**Gildersleeve, Gilderslelve:** Roger *Gyldenesleve* 1275 RH (Nf); John *Gildensleve*, Jeffrey *Gildsleeve*, Robert *Geldyngsleffe*, Richard *Gyldersleve* 1524 SRSf. 'Man with the golden sleeve' (OE *gylden*).

**Gilding:** *v.* GELDING

**Gildon:** Wlgar *se Gildene* c945 OEByn; Henry *le Gulden*, *Gilden* 1325 ParlWrits (Do). Simon *le Geldene* 1327 *SR* (Ess). OE *gylden* 'golden', perhaps 'golden-haired'.

**Gilduff:** *v.* DUFF

**Giles, Gyles, Jiles:** *Gilo*, *Ghilo* 1086 DB; Wido *filius Gisel* Hy 2 DC (L); *Gisle*, *Egidius*, *Gilo*, *Gile* de Gousle Hy 2, 1183–7 DC (L); *Gile* de Costeshal' 1196 P (L); *Gilo* Hose 1210 Cur (W); Ailward, Godfrey *Gile* 1176, 1191 P (Bk, Nth); William *Gyles* 1296 SRSx; William *Gilis* 1317 AssK; Nicholas *Gisel* 1346 FA (Sf). The DB *Gilo* has been identified with OG *Gilo*, equivalent to *Gislebertus*, and this is supported by the forms *Gisel*, *Gisle*. The latinization of this by *Egidius* shows that the scribe associated the name with *Giles*, a difficult name, regularly translated *Egidius*, from Greek αἰγίδιον 'kid'. The name of the 7th-century Provençal hermit St *Ægidius* spread widely and survives as *Gidi*, *Gidy* in southern France, as *Gili*, *Gilli* in the Alpes-Maritimes, elsewhere as *Gile*, *Gille*. The popularity of

this form in England is proved both by the number of churches dedicated to St Giles and by the frequent medieval *Egidius*. In the 12th century it was confused with and absorbed OG *Gilo* and *Gisel*, a short form of *Gilbert*. *v.* JELLIS.

**Gilfillan:** *Gillefalyn* c1214 Black; Ewin *Gilfillane* 1516 ib. Gael *Gille Fhaolàin* 'servant of (St) Fillan'.

**Gilford:** *v.* GUILFORD

**Gilfoyle:** Ir *Giolla Phóil* 'servant of (St) Paul'.

**Gilgryst:** *v.* GILCHRIST

**Gilham:** *v.* GILLAM

**Gilkes, Gilks, Jelke, Jelks:** William *Jalkes* 1291 RegAntiquiss; Adam *Gylke* 1483 AD vi (C); Robert *Gelkes* 1665 HTO. Perhaps a contracted form of GILKIN.

**Gilkin, Gilkins:** *Gilkinus* de Braban 1296 FrY; Richard *Gylekyn* 1317–18 FFSr; Richard *Gylkyns* 1332 SRWo. *Gill-kin*, a diminutive of ON *Gilli*. *v.* GILL.

**Gilks:** *v.* GILKES

**Gilkison:** Robert, John *Gilc(h)ristson* 1332 SRCu, 1499 Black; William *Gilkersone* 1559 ib.; James *Gilkisone* 1672 ib. 'Son of *Gilchrist*.'

**Gill:** (i) *Ghille*, *Ghile*, *Ghil* 1066 DB (Y); Gamel *filius Gille* 1185 Templars (Y); Henricus *filius Gilli*, *Gille* 1200 DC, P (L); Johannes *filius Gille* 1297 SRY; Ralph, Robert *Gille* 1202 AssL, 1332 SRCu, 1379 PTY. OIr *\*Gilla*, ON *Gilli* 'servant'. (ii) Elias *de la Gyle* 1269 Ipm (Y); Thomas *del Gile* 1246 AssLa; Michael *del Gill* 1332 SRCu. 'Dweller in the deep glen or ravine', ME *gille*, ON *gil* (a1400 MED). Both these northern names are pronounced with a hard *g*. (iii) Samuel, Roger *Gille* 1206 Cur (Ess), 1279 RH (O); Thomas *Gylle* 1296 SRSx; John *Jylle* 1312 LLB D. The Irish-Norwegian personal name *Gille* was not used in the south, but both in the north and the south *Gille* is found as a pet-form of *Gillian* and gave rise to surnames *Gill*, *Gell* and *Jell*. *v.* GELL.

**Gill, Gell, MacGill, MacKill:** Harald *Gille* (an Irishman) 1103, *MacGylle* 1430, *McGell* 1504, *McGill*, *Gell* 1511 Moore; Patrick *M'Kille* 1475, James *M'Gill* 1572 Black. Manx *Gill* and *Gell*, Irish *Gill* and Galloway *MacGill*, *MacKill* are all from Ir, Gael *gille* 'servant'.

**Gillam, Gillham, Gilliam, Gillum, Gilham:** *Giliaum* Spyser 1379 PTY; Peter *Gillame* 1276 LLB B; Arnold *Gilleme* 1283 ib. A; William *Giliam* 1379 PTY; Robert *Gylham*, Thomas *Guyllam* 1524 SRSf. English forms of French *Guillaume*. *v.* WILLIAMS.

**Gillard, Jillard, Jellard:** *Gillard*, with a hard *g*, is Fr *Guilard*, corresponding to Norman WILLARD. With a *j*-sound, it is a derivative of *Gille* (Giles), Fr *Gillard*. *v.* GILES.

**Gillbank, Gillbanks, Gilbank, Gilbanks:** Thomas *Gylbank* 1489 FrY; William *Gylbank* 1525 SRSx. From Gill Bank in Bewerley (WRY).

**Gillbee:** *v.* GILBY

**Gillebrand, Gillibrand, Gellibrand:** William *Gilebram'* 1219 AssL; John *Gilibrond* 1278 AssLa; Richard *Gelybrond* 1562 PrGR (La). OG *Giselbrand* 'hostage-sword'. In Suffolk, this became *Gillibrowne* (1576) and *Gillibourne* (1620 LitSaxhamPR).

**Giller, Guiler:** Hugh *le Gilur* 1242 ForEss; Nicholas *le Gilor* 1275 SRWo; William *Gilur* 1302 IpmY. OFr *guileor* 'deceiver, traitor'.

**Gillespie, Gillespey, Gillaspy, Gilhespy:** Ewan *filius Gillaspeck* 1175–99 Black; *Gillescop* de Cletheveys 1220–40 ib. Gael *Gilleasbuig* 'the bishop's servant'.

**Gillet, Gillett, Gillatt, Gillitt, Gilliat, Gilliatt, Gilyatt, Gilyott, Jillett, Jillitt, Jellett:** (i) *Wiotus* (*Guillotus*) 1175 P (Berks); *Guillotus* le Hauberger 1319 SRLo. Fr *Guillot*, a hypocoristic of *Guillaume* (William), a doublet of WILLET. *Gilot* de Portesmue (1292 SRLo) and *Guylote* Belebouche (1319 SRLo) are both called also *William* (1290 LLB G, 1300 LoCt). Surnames would appear as *Gilot*, *Gillot* and cannot be safely assigned here. Pronounced with *G* hard. (ii) Robert *del Gilheved* 1332 SRCu. 'Dweller at the head of the glen' (ON *gil*). This would become *Gillett*, with hard *G*. (iii) *Gilot* de Lackenby 13th Guisb (Y); *Gilot* de Paris 1280 Oseney (O); John *Gilet* 1243 AssDu; Peter *Gyllot* 1260 AssY; William *Gilot* 1296 SRSx. Whilst these forms cannot be distinguished from (i) above, *Gilot* must also correspond to Fr *Gillet*, *Gillot*, diminutives of *Gille* (Giles) and would become *Gillett*, etc., with a *j*-sound. (iv) *Gillota* 1198 FF (Sf), 1221 AssWa; *Gelot* (m) Webester 1379 PTY; *Gillot* (f) 1379 PTY; John *Gillote* 1279 RH (C); Richard *Gilote* 1324 AD vi (K); Peter *Giliot* 1333 Kirkstall (Y); William *Geliot* 1379 PTY; Robert *Jelott* 1512 GildY. *Gillota* is a diminutive of *Gilia*, the feminine of *Giles*: *Egidia* de Clere 1203 Cur (Sf). *Giliot* is a diminutive of *Gillian* or *Juliana* and has given *Gilliat* and *Gilyott* which belong only here and have also become *Gillett*, etc. *Gelot* (m) is a diminutive of *Giles* or of *Julian*. *Gillot* (f) may also be a diminutive of *Gill* (Gillian). (v) *Gillatt*, in particular, and all except *Gilliat* and *Gilyott* may be late forms of *Gillard*.

**Gillham, Gilliam:** *v.* GILLAM

**Gillian, Gillion, Gillions, Jillions, Gellion, Jellings, Jillings:** *Giliana* 1198 FF (Sf); *Jilianus* filius Geroldi 1206 Cur (Y); *Jeliana* Falcard 1274 Ipm (Cu); *Gilianus* de Levekenore 1279 RH (Bk); Adam *filius Jelion'* 1279 RH (C); *Giliana* Damyhan 1315 AD i (Nth); Robert *Gilion* 1279 RH (C); John *Gilian* 1327 SRDb; Matilda *Jeliane* 1379 PTY; John *Jelyon* 1524 SRSf. Colloquial pronunciations of *Julian* or, more commonly *Juliana*. *v.* JULIAN. *Juliana* de Scaccario (1194 Cur) is also called *Giliana*, as is Juliana, wife of Henry Escrop (13th Whitby). The seal of *Juliana*, wife of Robert de Vendore, is inscribed SIGILL' GILIANE (12th DC) and a deed of Osbert Cirotecarius and *Juliana* his wife has a seal with the same form of the wife's name (1230 Oseney).

**Gilliat(t):** *v.* GILLET

**Gillibrand:** *v.* GILLEBRAND

**Gillies, Gillis:** *Gillise* c1128 Black; John *Gyllis* 1521 ib. Gael *Gille Iosa* 'servant of Jesus'.

**Gilligan:** *v.* GILLINGHAM

**Gilling:** Grim *de Gilling'* 1198 P (Y); John *de Gilling* 1306 IpmY; John *Gilling* 1392 LoCh. From Gilling (NRY).

**Gillingham, Gilligan:** Robert *de Gillingham* 1203 Cur (K); Adam *de Gilingham* 1298 LoCt; Nicholas *Gyllyngham* 1392 IpmGl. From Gillingham (Do, K, Nf).

**Gillingwater:** Clement *Gyldenewater* 1275 RH (Nf); Goselin *Gildenewater* 1287 NorwDeeds I; Matthew *Gillingwater* 1644 FrYar. Probably 'dweller by the yellow stream', OE *gylden, wæter*.

**Gilliver:** William *Girofre, Gylofre, Gelofre* 1210 P (Sf), 1275, 1327 SRWo. OFr *girofle, gilofre*, ME *gylofre, gilliver* 'clove', used for sauces, hence, perhaps, a grower of gillyflower or a sauce-maker. It might also refer to one who held land by a rent of a clove of gilly-flower.

**Gillman, Gilman, Guillerman, Gellman, Gelman, Wellemin, Wellerman, Weilman, Welman, Willmin, Willman, Wilman, Willament, Williman, Williment, Willimott, Willmett:** *Wileminus* 1220 Cur (O); Anketinus *filius Gilmyn* 1279 RH (C); *Gilmyn* de Perham 1327 SRSx; John *Wilemyn* 1275 RH (Lo); William *Ghylemyn* 1297 MinAcctCo (Co); John *Wylemyn* 1301 LLB B; William *Gillemyn* 1317 AssK; Ralph *Welemyn* 1327 SRC; Matilda *Gelemyn* ib.; Thomas *Wylman* 1524 FrY; Richard *Williman* 1544 RothwellPR (Y); William *Gylman* son of William *Gylmyn* 1549 FrY. OFr *Guillemin*, Norman *Willemin*, diminutives of *Guillaume, Willelm. v.* WILLIAMS.

**Gillmore:** *v.* GILMOUR

**Gillow:** Thomas *Gyllowe* 1402 IpmNt; Nycholas *Gyllow* 1545, Richard *Gilloe* 1576 SRW. From Gillow (He).

**Gillson, Gilson:** Thomas, John *Gilleson* 1324 Wak (Y), 1332 SRCu; John *Gylesson(e)* 1348 LLB F, 1372 LoPleas. 'Son of *Gill* or of *Giles*.'

**Gillum:** *v.* GILLAM

**Gilly:** *Gilia* Fontisbraldi priorissa Hy 2 DC (Wa); Thomas *filius Gyly* 1332 SRCu; Nicholas *Gyly* 1327 SRSo; William *Gyli* 1332 SRCu. Lat *Gilia*, either a feminine form of GILES, or a pet-form of GILLIAN.

**Gilman:** *v.* GILLMAN

**Gilmichael:** *Gillemichel* de Crossebi 1176 P (We); Ralph *filius Gillemihel* 1202 P (La); *Gillemichil* pretor 1294 Ch (Y); Henry *Gillemighel* 1218–19 FFY. 'Servant of (St) *Michael*', Gaelic *gille*. In the surname the vernacular forms, *Michel, Mighel*, have been replaced by the Latin *Michael*.

**Gilmour, Gilmore, Gilmer, Gillmor, Gillmore:** (i) *Gilmor* 1133–56 Black (Cu); Richard *Gilemor* 1228 FFHu; Gillechad *Gillamor* 1304 Black. Ir, Gael *Gille Moire* 'servant of (the Virgin) Mary'. (ii) William *de Gyllingmor* 13th Guisb (Y). From Gillamoor (NRYorks).

**Gilpin, Gilping:** Ralph *Gilpen* 1387 IpmGl; George *Gilpin* 1563 Pat (Lo); Elizabeth *Gilpinge* 1587 LaWills. From the River Gilpin (We).

**Gilroy, Gildroy:** Ewen *Gilry* 1331 Black. Ir, Gael *Giolla rua* 'servant of the red-haired lad'.

**Gilson:** *v.* GILLSON

**Gilyatt, Gilyott:** *v.* GILLET

**Gimblett, Gimblette, Gimlette:** Robert *Gymlot* 1420 FrY; Robert *Gimlett* 1664 HTSo. *Gim-el-et, Gim-el-ot*, double diminutives of *Gemme, Gimme*, the vernacular pronunciation of *James*, used for men and women alike.

**Gimmer:** Walter *Gymere* 1230 Cur (Sr); John *Gymer* 1327 SRSo. A nickname from ME *gimur* 'a ewe lamb'.

**Gimson:** v. GEMSON

**Ginder, Ginders:** v. GENDER

**Ginger:** Roger *Gingiure* 1221 AssGl; William *Gyngeuer* 1262 FFEss; Roger *Ginger* c1280 St Thomas (St). ME *gingivere, gyngure, gingere* 'ginger', for a dealer in ginger, or, possibly, also for a hot-tempered or a reddish-haired man.

**Ginn, Gynn:** Henry, Roger *Gin* 1191 P (Nf), 1221 FFSt; Walter *Gynn* 1275 SRWo. ME *gin, ginne*, an aphetic form of OFr *engin* 'skill, ingenuity' (c1200 MED), 'snare, trap' (a1250). cf. JENNER.

**Ginner:** v. JENNER

**Ginney, Ginny:** Robert *Ginnay* p1225 YCh; Roger *Gynay* 1293–4 FFEss; John *Gyneye* 1470 Paston. Perhaps ODa *Ginni*. v. PNDB 261.

**Giot:** v. WYATT

**Gipp, Gipps, Gypp, Gypps:** *Gippa* de Neuhus 1150 DC (L); *Gippe* filius Ailsi 1202 FFL; Walter, Edmund *Gippe* 1188 P (L), 1227 FFSf; Agnes *Gyppes* 1352 ColchCt. A pet-name of *Gilbert*. cf. *Gilebertus* filius *Gippe* de Cheyle, *Gilebertus* de Cheiles Hy 3, 1256 Gilb (L). cf. GIBB.

**Gipson, Gypson:** Robert *Gipson* 1524 SRSf. 'Son of *Gipp*.'

**Girdel, Girdle:** Euerard *Gurdel* 1177 P (Nf). OE *gyrdel* 'girdle'. Metonymic for a maker or seller of girdles.

**Girdler, Gurdler, Gurtler:** Luke *le Gerdler* 1277 LLB A; William *Gurdeler* 1279 RH (O); Henry *le Girdlere* 1295 MESO (Nf). A derivative of OE *gyrdel* 'girdle', a girdle-maker (1356 MED).

**Girle:** Geoffrey *Gurle* 1275 RH (Sf); William (*le*) *Gurle* 1296, 1327 SRSx. ME *gurle* 'youth, maiden' (c1300 MED).

**Girling:** v. CODLIN

**Giron:** William *Gyron* 1327 SRWo. OG *Gero*, OFr *Geron, Giron*.

**Girton, Girtin, Gurton, Gretton, Gritton, Gritten:** Godmarus *de Grettona* 1086 InqEl (C); Engenulfus *de Grettona* 1172 P (Nth); Richard *de Grittone* 1279 RH (C). From Girton (Cambs, Notts), Gretton (Glos, Northants).

**Gisborne, Gisbourne, Gisburn, Gisburne:** Nigel *de Giseburn'* 1219 AssY; Thomas *de Giseburn'* 1265 Acc; Hugh *de Giseburn* 1274 IpmY. From Gisburn (WRY).

**Gise:** v. GUISE

**Gisell:** *Gisla* (f) c1110 Winton (Ha); Wido *filius Gisel* Hy 2, William *filius Gisle* 12th DC (L); John *Gysell* 1360–1 AssL; Samuel *Gysle* 1642 PrD. From a shortened form of OG *Gislebert*, or from OG *Gisla* (f).

**Gislam:** Walter *de Gisilham* 1327, Richard *Gislam* 1568 SRSf. From Gisleham (Sf).

**Gissing:** Roger *de Gissing'* 1195 FFNf; John *de Gissyng'* 1327 SRLei; Edmund *Gissyng* 1371 FFEss. From Gissing (Nf).

**Gitting(s), Gittins:** v. GETHING

**Gitton:** William *Gitun* a1290 CartNat. AFr *guidon*, ME *gitoun* 'a military banner, also the banner of a guild'. Metonymic for a bearer of this.

**Givenot:** Godwin *Givenout* c1200 Riev; Alan *Giuenoaut* c1250 RegAntiquiss; Agnes *Gyuenot* 1327 SREss. 'Give nothing,' OE *giefan, nâht*. Presumably a nickname for a stingy person. cf. John *Yivegoud* 1332 SRSx 'give goods'.

**Gladden, Gladding, Glading:** *Gledingus* 1196 Cur; Robert *Glading*, Susan *Gladden*, Stephen *Gladin* 1674 HTSf. OE *\*Glæding*, from OE *glæd* 'glad'. cf. GLADE, GLADWIN.

**Gladders:** Richard *Gladiere* 1296 SRSx; Alice *le Galdier* 1332 SRWa. ME *gladere* 'one who gladdens, cheers'. Or perhaps 'dweller in a glade', from a derivative of OE *glâd* 'glade'.

**Gladdish:** v. GLADWISH

**Glade:** (i) Willelmus *filius Glade* 1202 AssL. OE *\*Glæd*, the simplex of *\*Glæding* and *\*Glædwine*. (ii) Richard *in the Glade* 1275 SRWo. 'Dweller in the glade', the more likely source.

**Gladhill:** v. GLEDHILL

**Glading:** v. GLADDEN

**Gladman:** *Glademanus* 1066 DB (Sf); William *Gladman* 1327 SRY. OE *\*Glædmann*.

**Gladstone, Gledstone, Gledstanes:** Herbert *de Gledstan* 1296 (Lanark), Andrew *de Gladstan* 1364 Black. From Gledstanes (Lanark).

**Gladwin:** *Gladuin(e)* 1066 DB (L, St); *Glad(e)wine* c1113 Burton (St), c1180 ArchC vi, 1207 AssBeds; Henry, Peter *Gladewine* 1148 Winton (Ha), 1210 Cur (C); Robert *Gledewyne* 1317 AssK. OE *\*Glædwine*, unrecorded in OE, but fairly common in the 12th century.

**Gladwish, Gladdish:** Roger *de Gledewysse* 1296 SRSx. From Glydwish in Burwash (Sussex).

**Gladwy, Gladway:** William *Gladewy* c1210–23 RegAntiquiss; John *Gladewy* 1275 SRWo; Richard *Gladwy, Gladwyn* 1373–5 AssL. OE *\*Glædwîg*, apparently confused with GLADWIN.

**Glaisher, Glaysher, Glazyer, Glazier:** Thomas *le glasyer* 1297 MinAcctCo; Robert *le Glasiere* 1327 SR (Ess). A derivative of OE *glæs* 'glass', glass-maker (1296–7 MED).

**Glandfield, Glanfield, Glenfield:** These names may derive from *Clanefeld*: cf. Clanfield, Clanville (Hants), Clanvill (Som), Clanville (Wilts), Glanvill Fm (PN D 629), Glenfield (Leics), all identical in origin; or from *Glanville*. Ann *Glanvill* (1706) is called *Glanfield* in 1712 in the ShotleyPR (Sf). cf. Glandfield's Fm in Felsted: *Glaunvylles* 1360, *Glandfield* 1593, from William *de Glaunvilla* 1200 (PN Ess 423). The surname Glanfield is still found at Witham, not far from Felsted.

**Glanvil, Glanville, de Glanville:** Robert *de Glamuilla, de Glanuill', de Glauilla* 1086 DB (Nf, Sf); William *de Glanvile* 1127–34 Holme (Nf). From Glanville (Calvados). v. OEByn, ANF.

**Glascock, Glascott, Glasscock, Glasscote:** Walter *de Glascote* 1332 SRSt; from Glascote (Warwicks), *Glascote* Hy 2, *Glascocke* 1667 PN Wa 26.

**Glasgow, Glasscoe:** John *de Glasgu* 1258 Black. From Glasgow.

**Glason, Glasson:** Richard *Glazoun* 1314–16 AssNth; John *Glasyn* 1435, Thomas *Glasien* 1501, John *Glasyn* son of Thomas *Glayson* 1545 FrY. Perhaps a derivative of Breton *glaz* 'grey-haired'. Sometimes from Glasson in Cockerton (La), or Glasson in Bowness (Cu).

**Glass, Glasse:** William *Glase* 1641 PrSo; Thomas *Glasse* 1642 PrD; John *Glass* 1674 Black. In England from OE *glæs* 'glass', probably metonymic for a maker of glass. In Scotland from Gaelic *glas* 'blue, green, grey', probably of the eyes.

**Glassbrook, Glazebrook:** Geoffrey *de Glasebroc* 1246 AssLa; William *de Glasebrok* 1350 FFY; Richard *Glasbroke* 1525 SRSx. From Glazebrook (Lancs), or from minor places of this name.

**Glassman:** Richard *le Glasmon* 1327 SRSt. OE *glæs* and *mann*, 'dealer in glass-ware'.

**Glasson:** *v.* GLASON

**Glasswright:** Thomas *le Glasewright* 1286–7 FFWa; Thomas *le Glaswryghte* 1356 LLB G; Robert *Glasewright* 1524 SRSf. 'A worker in glass', OE *glæs*, *(ge)wyrhta*.

**Glaston:** Emma *de Glaston* 1298 AssL. From Glaston (R).

**Glave(s):** *v.* GLEAVE

**Glaysher, Glayzer, Glazier:** *v.* GLAISHER

**Glazanby, Glazenby:** Meriholt *de Glassenebi* 1191 P (Cu). From Glassonby (Cu).

**Glazebrook:** *v.* GLASSBROOK

**Gleadell, Gleadle:** *v.* GLEDHILL

**Gleave, Gleaves, Glave, Glaves:** William *Glaiue* 1202 FFNf; William *Gleiue* 1227 AssBeds; William *Gleue* 1283 SRSf. OFr *glaive, gleive*, ME *gleyve, gleve* 'lance', a name for a spearman or for a winner in a race in which the lance set up as a winning-post was given as a prize. *v.* GAME.

**Gledhill, Gledall, Gleadell, Gleadle, Gladhill:** Adam *de Gledehyll* 1277 Wak (Y); Robert *Gledall, Gledhell, Gleydell* 1571–1603 RothwellPR (Y); Launcelot *Gleadell, Gleadall* 1600–10 ib. From Gledhill (WRYorks).

**Gledstanes, Gledstone:** *v.* GLADSTONE

**Gleed:** *v.* GLIDE

**Gleeman, Glewman:** *Glemanus* de leuetuna 1066 DB (Sf); Robert *filius Gleuman* 1168 P (Nf); William *Gleman* 1201 FFEss; William *le Gleuman* 1306 AD v (Nf); Thomas *Gleman* 1486 ShotleyPR (Sf). OE *glīwman, glēoman* 'minstrel'. Used also as a personal name.

**Glen, Glenn:** (i) Gregory *Glen* 1230 Cur (Sf); Adam, William *de glen* 1327 SRLei; Thomas *Glenne* 1340–1450 GildC. From Glen (Leics). (ii) Colban *del Glen* 1328 Black; John *de Glene* c1377 ib.; William *Glen* 1452 ib. (Paisley). From the lands of Glen in Traquair (Peebles).

**Glenant, Glennant:** Henry *Glenaunt* 1334 IpmGl. 'Gleaner', OFr *glener* 'to glean'.

**Glendening, Glendenning, Glendinning, Clendenin, Clendennen, Clendinning, Clindening:** Adam *de Glendonwyn* a1286 Black; Robert *Glendonying* 1599 ib.; Isabel *Glindinine* 1667 ib. From Glendinning in Westerkirk (Dumfriesshire).

**Glendon:** Richard *de Clendon'* 1219 AssL; John *de Glendon* 1332 SRLa. From Glendon (Nth), *Clenedone* DB.

**Glenfield:** *v.* GLANDFIELD

**Glennant:** *v.* GLENANT

**Glentham:** Richard *de Glentham* 1212 P (Ch), 1219 P (Y). From Glentham (L).

**Glew, Glue:** Radulfus *filius Gleu* 1219 AssY; William *Gleu* 1168 P (Nf); Rannulf *Glowe* 1201 P (So);

Peter *Glew* 1301 SRY. OE *glēaw* 'wise, prudent', occasionally used as a personal-name.

**Glewman:** *v.* GLEEMAN

**Glide, Glyde, Glede, Gleed:** Walter *Glide* 1225 AssSo; Thomas *le Glede* 1277 ib.; Richard *le Glyde* 1296 SRSx. OE *glida*, ME *glede* 'kite' (a1398 MED).

**Glimme, Glymme:** Thomas *Glymme* 1275 SRWo; William *Glym* 1380–1 PTW; William *Glymme* 1403 FFEss. ME *glimme* 'shining'.

**Glin:** *v.* GLYN

**Glind, Glynd:** John *atte Glynde* 1341 MELS (Sx); Geoffrey *le Glynd* 1358 SRWo. 'Dweller by the enclosure', OE *glind* 'hedge, fence'. cf. Henry *Glinder* 1327 SRSx.

**Glinn:** *v.* GLYN

**Glint:** Henry *Glynt* 1292 IpmGl. ME *glint* 'slippery'.

**Glorious:** Geoffrey *Gloriosus* 1186 P (Nf); William *Glorius* 1212 Cur (Bk); Robert *le Glorius* 1228 Cur (Bk). OFr *glorios* 'renowned, magnificent'. Probably often ironical.

**Glory:** (i) John *de Glorie* 1242 Fees (Bk); William *de la Glorie* 1326 Oseney. From Glory Fm in Penn (Bk). (ii) Walter *Glori* a1184 Gilb; Robert *Glorie, le Glorius* 1214 Cur (Bk). For GLORIOUS.

**Gloss, Glosse:** Reiner *Gloz* 1210 Cur (Nf); William *le Glosse* 1279 AssSo; William *Glos* 1442 IpmGl. A nickname from OFr *glose* 'flattery'.

**Gloster:** Durandus *Gloecestra, de Glouuecestre* 1086 DB; William *de Glocestria* 1242 Oseney (O). From Gloucester.

**Glover:** Gilbert, William *le Glouere* 1250 MESO (Nf), 1278 AD iv (Sr). A derivative of OE *glōf* 'glove', maker or seller of gloves (1355 MED).

**Glozier:** Roger *Glosere*, William *Glosur'* 1279 RH (C); Robert *Glosier* 1674 HTSf. A derivative of ME *glosen*, OFr *gloser* 'to make glosses upon, expound, interpret', a commentator (c1400 MED).

**Glue:** *v.* GLEW

**Glutton:** Simon *le Glutun* 1201 P (Nt); Gilbert *le Glutun* 1250 Fees (Nt); Robert *Glotoun* 1309 SRBeds. OFr *gloton* 'glutton'.

**Glymme:** *v.* GLIMME

**Glyn, Glynn, Glin, Glinn:** John *de Glin* 1183 P (Co); Ralph *de Glin* 1297 MinAcctCo; John *Glyn* 1382 IpmGl; John *Glynn* 1552 Chirk. From Glynn (Co).

**Glynd:** *v.* GLIND

**Glynn:** *v.* GLYN

**Gnat, Gnatt:** Walter *le Cnate* 1249 AssW; Roger *le Gnat'* 1318 FFK; William *Gnat* 1364 FFW. A nickname from the insect, OE *gnæt*.

**Goacher, Goucher:** Willelmus dictus *Godechere* 1343 Black. ME *chere*, OFr *chier* 'face', one of good aspect, cheerful appearance.

**Goadby, Godby:** Henry *de Godeby* 1274 Wak (Y). From Goadby, or Goadby Marwood (Lei).

**Goaman:** *v.* GOMMON

**Goate:** (i) William *le gat* 1139 DC (Nt); Suein *Got* 1166 P (Nf); Hugo *Gat* 1219 AssL; John *le Got* 1254 Pat. OE *gāt* 'goat'. *v.* also GAIT. (ii) Alexander *de Gote* 1288 PN Sx 356, Peter *atte Gote* 1327 SRSx, John *de la Gote* 1329 FrY. 'Dweller by the watercourse or sluice'. ME *gote*. *v.* PN C 327, MELS 80. As *gott* is found later for both 'goat' and 'watercourse', these surnames may have contributed to *Gott*.

**Goater:** Thomas, John *le Goter(e)* 1327, 1333 MEOT (O, Ha). A derivative of OE *gāt* 'goat', goatherd. *v.* also GATER.

**Goatley, Goatly, Gotelee:** Robert *de Gotele* 1275 RH (Sx); Goda *de Gotelye* 1332 SRSx; Henry *Gotlee* 1412 SxAS x. From Goatley in Northiam (Sx).

**Goatman:** Ralph *Gateman* 1183 P (Nth); Nicholas *Gooteman* 1455 GildY. OE *gāt* and *mann*, goatherd. cf. Osbert *Gotenecherl* 1183 P (Wa), Reiner *le Gotegrom* 1335 FFSf.

**Gobat:** *v.* GODBOLD

**Gobbett, Gobet:** Robert *Gobet* 1203 FFSf; John *Gobet* 1327 *SR*Ess; Agnes *Gobet'* 1379 PTY; William *Gobett* 1566 NorwDep. Probably a nickname from OFr *gobet* 'lump, morsel', but it may sometimes be for 'go better'. OE *gān, bet*. cf. Walter *Gobiforn* 1212 Pleas (Y) 'go before'; John *Gobisid'* 1379 PTY 'go beside', and GOLIGHTLY.

**Gobel, Gobell, Goble:** Osbert *Gobel* 1297 MinAcctCo; Walter *le Gobel* 1327, Henry *Gobill*, Thomas *Goble* 1525 SRSx. OFr *gobel* 'cup'. Metonymic for a maker or seller of cups.

**Gobert:** *v.* GODBERT

**Gobet:** *v.* GOBBETT

**Gobeley, Gobley:** Robert *Gobelay* 1317 AssK. 'Go by the clearing', OE *gān, bī, lēah*. This is a not uncommon type of medieval nickname. cf. Julian *Gobegrund* 1296 SRSx 'go along the ground', but in OE *grund* also meant an outlying farm or field, and this may be the sense here; Thomas *Gobykerke* 1525 SRSx 'go by the church'; Henry *Gobithesti* 1208–9 Pleas (L/Nf) 'go by the path'; Robert *Gobytheway* 1383 Calv (Y) 'go by the road'.

**Gobey, Goby:** Nigel *Gobey* 1275 RH (Nf); Thomas *Goby* 1327 SRWo; William *Gobie*, Nicholas *Gobye* 1642 PrD. OFr *gobi* 'gudgeon'. According to Dauzat, probably a nickname for a fisherman or a fishmonger.

**Goble:** *v.* GOBEL

**Gobley:** *v.* GOBELEY

**Goblin:** Hugh *Gobelyn* 1274 Wak (Y). OFr *gobelin* 'devil, goblin'.

**Goby:** *v.* GOBEY

**Godbear, Godbeer, Godber, Godbehear, Godbeher, Godbehere, Goodbeer, Goodbehere:** (i) Robert *Godebere* 1296 SRSx. A nickname 'good beer'. cf. GOODALL. (ii) John *Godbehere* 1456 Cl (Lo); Geoffrey *Godbeherinne* 1277 LLB B. Either *God be her(inne)* 'May God be in this house' or *Gōd be her(inne)* 'May it be well in this house'. Such names were common but are usually obsolete. cf. Gilbert *Godbiemidus* 1219 AssL. *v.* GODSAFE.

**Godbert, Gobert:** *Godebert* 1086 DB (Ess); Robertus *filius Goberti* ib.; Ricardus *filius Godebert* 13th Lewes (Nf); Roger *Godberd*, *Godbert* 1200–1 P (Y, Nt); William *Godebert* 1221 AssSa; William *Godebrich* 1262 *For* (Ess); Gilbert *Godebrith* 1327 SRSf; John *Gobard* 1335 AssSt. OG *Godebert* 'god-bright'. *Godebrich* (rare) suggests the possibility of an unrecorded OE \**Godbeorht*.

**Godbold, Godbolt, Godball, Gobat:** *Godeboldus* 1086 DB; Adam *filius Goboldi* Hy 3 Colch (Ess);

William *Godebald'* 1206 Cur (Beds); Henry *Gobaud*, *Gubaut* 1242 Fees (D); John *Gobaut* 1316 FA (Wa); John *Godebold* 1317 FFEss; John *Godball* 1485 RochW. OG *Godebald* 'god-bold'.

**Godby:** *v.* GOADBY

**Godcock, Goodcook:** Henry *Godecoke* 1327 SRC; William *Godecoke* 1327 SRWo. A diminutive of OE *Goda*. *v.* GOOD.

**Goddard, Godard, Godart:** *Godardus* de Clakesbi c1160–6 DC (L); Robert *God(d)ard* 1208 Cur (Ha), Wlfrich *Godard* 1221 *ElyA* (Nf); Symon *Godhard* 1299 LLB C. OFr *Godard*, OG *Gotahard*, *Godhard* 'god-hard'.

**Godel, Godell:** *Godel* 1066 DB (K); Daniel *Godel* 1327 SRSx; Robert *Godell* 1524 SRSf. OG *Godilo*, *Godila*. *v.* PNDB 263.

**Godelin, Godlin, Godling:** *Godelina* 1148 Winton (Ha); Simon *filius Godelini* 1212 Cur (K); William *Godelin* 1249 AssW; Ralph *Godeling* 1327 SRC. OG *Godelin* (m), *Godelina* (f).

**Godfrey, Godfray, Godfree, Godfery, Gotfrey:** *Godefridus* 1086 DB; *Godefridus* filius Baldewini 1138 NthCh (L); Symon *Godefrei* 1221 *ElyA* (Sf); Alan *Godefre* 1252 Rams (Hu); Maud *Godefray* 1277 Ipm (Nt); OFr *Godefroi(s)*, OG *Godefrid* 'god-peace'.

**Godhelp, Godshelp:** Walter *Godeshelp* 1332 SRDo; William *Godeshelp'* 1387 AssL; William *Godhelp* 1596 Musters (Sr). 'By the help of God', or 'may God help (us)', OE *God, help/helpan*, no doubt a favourite expression. cf. Gilbert *Godbiemidus* 1219 AssL 'may God be with us'; Walter *Goduspart* 'may God part us', and *v.* GODSAFE.

**Godin:** *Godun* le Bere 1279 RH (C); Philip *Guodun*, *Guodin* 1280 AssSo; Henry *Godyn* 1386 LLB H. OFr *Godin, Godun*, diminutives of OG names in *God-*.

**Godkin, Goodkin:** *Godekin* de Coufeld 1275 RH (Lo); Walter *Godekin* a1272 Colch (Sf); Henry *Godken* 1605 PN Db 469. Diminutives of OE *Goda*, *Gode*.

**Godley, Godly, Goodley:** Richard *de Goddeley* 1275 RH (D); Robert *de Godely* 1296 SR (Sx). From Goodley (Devon), Godley (Ch), Godley Bridge (Surrey) or Godley's Green (Sussex).

**Godman, Goddman, Goodman, Goudman, Gutman:** (i) *Godman(nus)*, *Godeman(us)* 1066 DB; *Godeman* de Waledena 1176 P (Herts); *Godman* de Offenchurche 1221 AssWa; Astcelinus *Godeman* 1115 Winton (Ha); Nicholas *Godman* 1188 BuryS (Sf). OE *Godmann* or OG *God(e)man*. (ii) Henry *le Godman* 1275 RH (C); Gerard *Gudman* 1352 FrY; John *Goodman* 1365 ColchCt. 'Good man' (cf. Anketin *Bonhomme* 1274 FFC), or 'master of a household'. *v.* BONHAM. (iii) *Guthmund, Gudmund, Godmund* 1066 DB; *Gudmund* del celer 1121–48 Bury (Sf); William *Gudmund* 1190 P (Wo); Hugh *Gothemund* alias *Gutmund* 1265 Ipm (Wo); John *Gotemund* 1274 RH (Ess); Richard *Guthmund, Godmund* 1275 RH (Sf); Adam *Guthmond* 1327 SRSf; Nicholas *Gooteman, Goteman, Godman* 1466, 1467, 1472 GildY. ON *Guðmundr*, ODa *Guthmund* or OE *Gūðmund*.

**Godleman:** Philip *de Godelminges* 1243 Colch (Sf). From Godalming (Sr).

**Godless:** William *Godless* 1525 SRSx. OE *gōdlēas* 'without goods, poor'.

**Godlin, Godling:** v. GODELIN

**Godolphin:** Alexander *de Woldholgan* 1201 Pleas (Co); Dauid *Godolghan* 1359 Putnam (Co); John *Godolghan* of Godolghan 1508 Pat (Lo). From Godolphin (Co), *Wulgholgan* 1194.

**Godrich, Godridge:** v. GOODRICH

**Godsafe, Godsave, Godsalve, Godseff, Godsiff:** Nicholas *Agodeshalf* 1206 P (Sx); Alexander *Agodeshalve* 1260 AssC; Hugo *Ogodishalve, Ongodishalve* 1264, 1271 LeicBR; William *of e Godeshalve* 1311 NottBR; William *a Codeshalf* 1332 SRLo; Henry *Godsalve* 1337 FrY; John *Godeshalf* 1376 ib.; Eustace *Godsafe*, John *Godsaff* 1662 *HTEss*. ME *on* (a) *Godes half* 'in God's name, for God's sake', a nickname, due no doubt, to the frequent use of the phrase. The surname was common and widely distributed. The seal of Matthew son of Robert *a Godeshalf* (1202–16 ELPN) bore the legend 'Sigill. Mathei *ex parte Dei*', so that already at the beginning of the 13th century the surname had become hereditary in London. For similar oath-names, v. GODBEAR. cf. Geoffrey *Godesgrace* 1220 Cur (Sr), Robert *Godesblescinge* 1225 AssSo, Richard *Godesname* paternoster 1298 LoCt, William *Goddesbokes* 1308 FrY, Roger *Godesfast* 1392 LLB H, Henry *Godsake* 1514 NorwW, Olive *Goadebles* 1269 Pat, William *Godthanke* 1275 RH (Nf), William *Godwotte* 1352 PatR, William *Godhelp* 14th AD iv (Sr). For similar names of French origin, v. PARDEW.

**Godsal, Godsall, Goodsell, Gotsell, Gutsell:** Godwin *Gotsaule* 12th ELPN; William *God saule* 1197 FF (Nf); Ralph *Godsoule* 1219 AssY; Amicia *Godsol* 1279 RH (C); Ralph *Godsale* 1379 PTY; Thomas *Godsalle* 1518 FrY. 'Good soul', 'an honest fellow'.

**Godsell, Godsil:** Hugh *de Godeshill* 1225 AssSo, *de Godeshull'* 1230 P (Ha); Thomas *de Godeshelle* 1309 LLB D; Richard *Godeshulle* 1327 SRSx. From Gadshill (Kent), Godshill (Isle of Wight), *Godesill* 1393, Godshill (Hants) or Godsell Fm (Wilts), all originally *Godeshyll* 'god's hill'.

**Godshelp:** v. GODHELP

**Godsman, Goodsman, Goodisman:** *Godesman* 1066 Winton (Ha); *Godesman* de Spalding 1158 DC (L); Astin, Thomas *Godesman* 1214 Cur (L), 1248 FFEss. OG *Godesman* 'the man of the god', or a description, 'the man of God', a name for a clergyman or for one noted for his piety. cf. Willelmus *homo dei* 1195 P (L), Richard *Homedeu* 1216 Oseney (O).

**Godsmark:** Henry *Godesmerk* 1296 SRSx; William *Godesmark* 1327 ib. 'One bearing God's mark, the sign of the plague.' cf. 'He also was full of Godys markys' 1531; 'the markes of the plague, commonly called Goddes markes' 1558; 'Some with God's markes or Tokens doe espie, Those Marks or Tokens shew they must die' 1630 NED.

**Godsname:** Richard *Godesname* 1298 LoCt, 1311 LLB B; John *de Godesname* 1319 SRLo. 'In God's name'. Such nicknames are not uncommon in medieval times, but have rarely survived. cf. Robert *Godesblescinge* 1225 AssSo 'God's blessing'; Geoffrey *Godesgrace* 1220 Cur (Sr) 'by God's grace'; Richard *Godeswowes* 1340–1450 GildC 'by God's grief'; Richard *Godeswones* 1378 AssWa 'by God's dwellings'.

**Godson, Goodson:** (i) *Godsune* 1086 DB (Ess), 1191 P (K); Alwinus *Gode sunu* 1066 DB (So); William *Godsune* 1200 P (Lo); Benedict *Godsone* 1298 FFC; Richard *Goddesone* 1332 SRSt. *Gode sunu* may be 'son of *Goda*' (masc. or fem.) or, possibly, OE *Godsunu* (either OE *god* 'God' or *gōd* 'good' and *sunu* 'son'), the probable source of the rest, though OE *godsunu* 'godson' is also possible. (ii) Thomas filius *Gode* 1302 SRY; Roger *Godessun'* 1205 Cur (Ess); Adam *Godesone, Godessone* 1294, 1303 LLB A; Richard *Godissone* 1402 FFSf. Either 'son of *God*', a short form of such names as *Godwine, Godrīc*, etc., or 'God's son', or 'son of *Gode* (f)'. cf. GODSMAN. Ekwall suggests (ELPN 41) that Adam *Godesone* (*Godessone*) 1294–1305 was a son or descendant of *Godeson* le Megucyer (1235). If so, he was called both 'Godson' and 'son of *Gode*, a pet-form of *Godsunu*. (iii) William *Godesone* 1252 Rams (Hu); Ralph *le Godesone* 1296 SRSx; Sarra *Goudesone*, William *le Goudesone* 1332 SRSx; Robert *Goodsone* 1377 AD vi (He). OE *gōd* 'good' and *sunu* 'son'. *Godson* in NED is always *godson*, never *godeson*. In the York plays, *Gud sonne* 'good son' is a common term of address. cf. William *Goodefader* 1418 Pat (Ess).

**Godston, Godstone:** John *Godstone* 1397 IpmGl. From Godstone (Sr).

**Godward:** Wlmarus filius *Godwardi* 12th DC (L); Alicia *Godward* 1252 Rams (Hu); John *Godward* 1568 SRSf. Late OE *Godweard*, 'God or good' and *weard* 'protector'.

**Godwin, Godwyn, Goodwin, Goodwins, Goodwyn:** Ailmar filius *Goduini* 1086 DB; Ricardus filius *Godwini* 1219 AssY; Walter *Godwin'* 1177 P (Nf); Nicholas *Godwyn* 1239 FFC; Richard *Godwynn* 1296 FrY; William *Goudwyne* 1327 SRSx; Robert *Gudwen* 1327 SRC; William *Godewynes* 1327 SRWo; John *Gudwyn* 1388 FrY. The common OE personal name *Godwine*, from *god* 'god' or *gōd* 'good' and *wine* 'friend, protector, lord', which survived the Conquest. *Gōdwine* became Godwin and *Gōdwine* Goodwin. If *Godwynn* and *Gudwen* are reliable, we have occasionally the feminine OE *\*Godwynn*.

**Godwold:** William *Godwold* 1189 Sol; Robert *Godwald* 1327 SRSf. OE *Godweald*.

**Goff, Goffe:** Bertram, Nicholas, Thomas *Goffe* 1208 FFL, 1327 SRWo, 1332 SRWa. Ir, Gael *gobha*, Welsh, Breton *gof*, Cornish *gov* 'smith'.

**Goffin:** Adam *Goffyn* 1327 SRWo; George *Goffyng* 1524 SRSf. A diminutive of *Goff*, a pet-form of *Godfrey*.

**Gogger:** Walter *le Goggar'* 1277 CtW; Peter *le Gogger* 1305 AssW; Ralph *Goggere* 1332 SRDo. 'Dweller in the swamp', from a derivative of ME *gogge* 'swamp'.

**Goggle:** (i) Robert *le Gogel* 1307 IpmW; John *Gogel* 1311–12 FrC; Walter *Gogul* 1327 SRSo. cf. ME *gogel-eie* 'squinting', probably a nickname for one who squints. (ii) John *atte Gogele* 1327, *atte Goghele* 1332 SRSx. 'Dweller by the swamp', ON *gogli*.

**Goggs:** Ralph *Gog* 1193 P (L); Adam *Gogge* 1275, William *Gogge* 1332 SRWo. OG *Gogo*.

**Golbourn:** v. GOLDBOURN

**Golbrook:** v. GOLDBROOK

**Gold, Goold, Gould:** (i) Hugo *filius Goldæ* 1086 DB (Sf); Ralph *filius Golde* 1193 P (Beds); *Golde* Bassat 1279 RH (C); Walter *Golde* 1165 P (D); Ralph *Golde* 1268 AssSo; John *Gulde* 1297 MinAcctCo. OE *Golda* (m), or *Golde* (f). (ii) Thomas *le Gold* 1327 SRSx; John *le Goulde* 1332 ib. A nickname, 'golden-haired'. (iii) Thomas *withe Gold* 1220 Pat (St). Presumably 'rich'.

**Goldbard, Goldbart:** (i) *Goldebrict'* de Dunewic' 1203 Cur (Sf); Richard *Goldberd* 1230 P (Wa); Odin *Goldeberd, Guldebort'* 1327, 1332 SRSx. OE *Goldbeorht* 'gold-bright'. (ii) William *Gulberde* 1220 FFEss; Richard, Philip *Geld(e)berd* 1296 SRSx. 'Golden beard', OE *gylden, beard*.

**Goldbeater:** Odo *le Goldbeter* 1305 LoCt; John *le Goldbeter* 1327 SRWo; Jamys *Golbeter* 1466 Paston. 'A goldsmith, especially a worker in gold leaf', OE *gold, bēatere*.

**Goldbourn, Golborn, Golbourn, Goulborn, Goulborne:** Richard *de Goldburn* 1332 SRLa. From Golborne (Lancs, Ches).

**Goldbrook, Goldbrooke, Golbrook:** John *atte Goldbrok* 1332 MELS (Sx). 'Dweller by the stream where marsh marigolds grow', OE *golde, brōc*.

**Goldburg, Goldburgh:** Ricardus *filius Goldburg* c1179 Bart (Lo); Richard *Goldburc* 1210 Cur (Lei). OE *Goldburg* (f) 'gold-fortress'.

**Goldcorn:** Edward *filius Goldcorn* 1224–5 FFEss; *Goldcorna* 1271–2, *Goldcorn* (f) 1277–8 ELPN; Johanna *Goldcorn* 1314 LLB C; John *Golcorn* 1491 FrYar. From a ME feminine name, *Goldcorn*.

**Golden, Goolden, Goulden:** Walter *Guldene, le Gelden'* 1212 Cur (Ha); Hilde *Golden'* 1279 RH (C). OE *gylden* 'golden-(haired)'. cf. Richard *Guldenheved* 1222 DBStP.

**Golder, Goulder:** (i) Henricus *filius Goldere* 1197 P (Ess); John, Adam *Golder(e)* 1296 SRSx, 1327 SRSf. OE *Goldhere* 'gold-army', of which only one other example is known, in 1086, at Colchester. (ii) Matthew *de Goldore* 1275 RH (Herts); Richard *de la Goldore* 1295 MELS (Sr). 'Dweller at the bank where marigolds grow' (OE *golde, ōra*), as at Goldor (Oxon).

**Golderon, Goldron:** *Golderon* 1066 DB (Beds); Edmund *Goldrun*, John *Goldren*, Walter *Guldron* 1255 RH (W); Alexander *Golderon* 1298 AssL. From an unrecorded OE feminine name, *Goldrūn*.

**Goldey:** *v.* GOLDIE

**Goldfinch:** William *Goldfing* 1260 AssC; Robert *Goldfynch* 1277 AssSo; Philip *le Goldfynch* 1307 AssSt. A nickname from the bird, OE *goldfinc*.

**Goldfoot:** Robert *Geldenefot, Gildenefot* 1188 P (Nf), 1213 Cur (Nf). OE *gylden* and *fōt* 'golden foot', a curious nickname.

**Goldhawk:** *Goldhac, Goldhauek* Hy 2, 1212 Bart (Lo); *Goldhevek* de Lond(on) 1212 Cur (Sr); Robert *Goldhauek* 1219 FFEss; Swein *Goldhauc* 1235 FFEss. Late OE *Goldh(e)afoc* 'gold hawk'.

**Goldhead, Goldenhead:** Richard *Guldenheved* 1222 DBStP, *Goldhavet* 1226 FFK. 'Golden head', OE *gylden/gold, hēafod*. cf. Henry *Goudenhond* 1332 SRSx 'golden hand'.

**Goldhill:** Henry *Goldhulle* 1386 IpmGl. From Gold Hill in Shoreham, or Goldhill House in Hadlow (K).

**Goldicott, Goldocote:** Michael *de Goldocot'* 1221 AssWo. From Goldicote House in Alderminster (Wo). Sometimes, perhaps, for *Coldicott*. *v.* CALDECOTT.

**Goldie, Goldey:** *Goldiva* (f) 1189 Sol; *Goldiua* 1219 Cur (Nth); *Goldyeua* 1285 AssEss; Alan *Goldhiue* 1221 AssWo; William *Goldgeve*, Richard *Goldyve* 1275 SRWo; Johanna *Goldey* 1524 SRSf. From an unrecorded OE feminine name, *Goldgiefu*.

**Golding, Goolding, Goulding:** *Goldinc* 1086 DB (Ess); *Golding* Aldred a1224 Clerkenwell (Mx); William, Richard *Golding* 1202 AssL, 1210 Cur (C); William *Gulding* 1327 SRSx. Late OE *Golding*.

**Goldingham:** Hugh *de Goldingham* 1204 P (Ess); Alan *de Goldingham* 1285 FFEss. From Goldingham Hall in Bulmer, or Goldingham Fm in Braintree (Ess).

**Goldlock:** William *Goldlok* 1277 AssW; Walter *Guldeloc* 1279 RH (O). 'Golden locks', OE *gold, locc*.

**Goldman:** Adam *Goldeman* 1297 SRY; Maud *Goldman* 1393 LoPleas. OE *Goldmann*, first recorded on an Essex coin c1066 (Grueber).

**Goldney:** Richard *Goldeneghe* 1345 LLB F. 'Golden eye.'

**Goldocote:** *v.* GOLDICOTT

**Goldrich:** *Goldrich* 1066 DB (Ess); *Golricus* de Skegeness 1218 AssL; William *Goldrych* 1344 ChertseyCt (Sr); Robert *Goldwright* 1500 *ERO*. OE *Goldrīc* 'gold-ruler'.

**Goldring:** Gilbert *Goldring* 1227 AssBeds; John *Goldryng* 1305 AssW; Hugh *Goldryngge* 1376 *ERO* (Ess). This looks like an unrecorded personal name, ME *Goldring*, though it could, perhaps, be a nickname for the wearer of a gold ring, OE *gold, hring*.

**Goldron:** *v.* GOLDERON

**Goldsborough, Goldsbrough, Gooldsbury, Gouldesborough, Gouldsbury, Goulsbra:** Walter *de Goldisburc* 1206 P (Y); John *Goldsborough*, Lawrence *Goldesberry* 1674 HTSf; John *Goldsbrow* 1784 SfPR. From Goldsborough (NR, WRYorks).

**Goldsby:** Ralph *de Golkesbi* 1202 AssL. From Goulceby (Lincs).

**Goldsmith, Gouldsmith:** Roger *Goldsmiz* 1250 MESO (Nf); Thomas *Goldsmith* 1255 *Ass* (Ess); John *le Goldesmethe* 1309 LLB D. OE *goldsmið* 'goldsmith'.

**Goldspink:** Robert *Goldspynk* 1524 SRSf. 'Goldfinch.'

**Goldston, Goldstone, Gouldstone, Goulston, Goulstone, Golston, Goldson, Golson, Gulston, Gulson:** (i) *Goldstan, Golstan* 1066 DB (K, Ess); Wulfric filius *Goldston* 1180 P (K); Richard *Golstan* 1185 Templars (Ess); Robert, Walter *Goldstan* 1202 AssBeds, 1214 Cur (Bk); Robert *Goldstone* 1312 LLB D; Thomas *Golston, Golson* 1524 SRSf; John and Mary *Goldson, Goldston* 1795, 1798 HorringerPR (Sf). OE *Goldstān* 'gold stone'. (ii) John *de Goldeston* 1312 FFEss; Ambrose *Golson* 1523 SRK. From Goldstone (Kent), earlier *Goldstaneston*, or Goldstone (Salop).

**Goldway:** Tocwy *filius Goldwy* 1182 RBWo; Thomas *Goldweg* 1297 MinAcctCo. OE *Goldwīg* 'gold-warrior', not otherwise known.

**Goldwell:** Nicholas *Goldewell* 1478 Paston. From Goldwells in Horndon on the Hill (Ess).

**Goldwin, Goldwyn:** *Golduinus* 1066 DB (Sx); *Goldewinus* 1221 *ElyA* (Sf); William *Goldwyn* 1256 AssNb; Richard *Goldwyne* 1317 AssK. OE *Goldwine* 'gold-friend'.

**Golf:** Robert *Golf* 1249 AssW; Walter *le Golf* 1252 FFO. A nickname from OFr *golf* 'belly'.

**Golightly, Gallatly, Galletley, Galletly, Gellatly, Gelletly:** Rannulf *Golicthli* 1196 P (C); Adam *Golictlik* 1201 Cur (Nt); Howe *Golichtly* 1221 AssWa; Henry *Gellatly*, *Galighly* a1291 Black, 1296 CalSc; Gilbert *Galetly* 1592 Black. 'Go lightly', a nickname for a messenger.

**Goll, Golle:** Thomas *Golle* 1297 SRY; John *le Gol* 1327 IpmW; John *Golle* 1416 IpmY. ME *golle* 'an unfledged bird, a silly fellow'. But there was also a feminine name: *Golla filia Ulf* 1219 AssY.

**Golland:** v. JOLIN

**Golledge:** Richard *Golache* 1221 AssWo; Godfrey *Galoch* 1256 FFL; John *Golledge* 1664 HTSf. OFr *galoche*, ME *galoche*, *galoge* 'a kind of footwear'. Metonymic for a maker or seller of these. The surname may have fallen in with COLLEDGE.

**Golley:** v. GULLY

**Gollin(s):** v. JOLIN

**Golsworthy:** v. GALSWORTHY

**Gomer:** v. GUMMER

**Gomersal, Gomersall, Gumersall, Gumerson:** Adam *de Gumersale* 1276 RH (Y); Hugh *de Gomersall* 1382 Calv (Y); Edward *Gumersall* 1617 YRec 74; Thomas *Gummersall* signs as *Gumerson* 1765 WRS. From Gomersall (WRY).

**Gomme, Goom, Gum, Gumb, Gumm:** Simon *Gumme* 1247 AssBeds; Henry *le Gome* 1275 SRWo; John *Gom* 1279 RH (C); Walter *Gomme* 1297 MinAcctCo; Nicholas *Gumes* 1327 SRSf; John (*le*) *Goom* 1340 AssC. OE *guma* 'man'.

**Gommon, Gommen, Goaman:** *Gomanus* sutor, John *filius Goman* 1221 AssWa; Alan *Gomman* 1202–3 FFWa; William *Goman* 1279 RH (O); John *Gommen* 1332 SRWo. Perhaps a late OE *\*Gumman*.

**Gonville:** Hugh *de Gundeuilla*, *de Gundeuille* 1155 NLC (Bk), 1225 AssSo; Adam *Goundevyle* 1332 SRDo. From Gondouville (Seine-et-Marne).

**Gooch, Goodge, Goudge, Gouge, Gudge, Gutch:** Robert, Felicia *Goch* 1203 Pl (Sa), 1306 IpmGl; John *Guch*, William *Gugge* 1327 *SR* (Ess); John *Gooch* 1374 ColchCt; William *Goch* or *Gouche* 1509 LP (Lo). Welsh *coch* 'red'. v. also GOUGH.

**Good, Goode, Goude, Gudd, Gude, Le Good, Legood, Leegood:** *God* 1066 DB (K, Sx, W); *Goda, Gode* (m), *Gode, Goda* (f) ib.; *Goda* de West Wicumbe 1194 Cur (Bk); Odo *se Gode* 960 OEByn; Æilgyuu *Gode* 1050–71 ib.; Edwardius *bonus* 1204 P (O); Gilbert *le Gode* 1212 Cur (Bk); Robert *Gode* 1221 AssGl; William *Goude* 1297 MinAcctCo; Thomas *le Goude* 1327 SRSx; Andrew *Gude* 1537 FrY. As *le Gode* is much the most common form, this is usually a nickname from OE *gōd* 'good', but a personal name cannot be ruled out, either OE *Goda* (m) or *Gode* (f).

**Gooda:** v. GOODER

**Goodall, Goodale, Goodayle, Goodhall:** Toka *Godala* 1181 P (Sf); Roger *Godhal'* 1221 AssSa; William *Godale* 1244 Rams (Beds); John *Godhale* 1297 MinAcctCo; John *Gudale* 1379 PTY; James *Goodall* 1581 RothwellPR (Y); William *Guidaill* 1657 Black (Kirkwall). 'Good ale', a brewer or seller of good ale.

**Gooday, Goodey, Gooddy, Goody:** For GOODDAY, or 'dweller at the good enclosure' (OE *gōd*, (*ge*)*hæg*): John *de Goday* 1327 SRSt; or for GOODIFF.

**Goodbairn, Goodban, Goodband:** William *Godbarn* 1203 Cur (L); Robert *Gudbarn* 1379 PTY. ME *gōd-barn* 'good child', confined to the Scandinavian counties. cf. GOODCHILD.

**Goodbeer, Goodbehere:** v. GODBEAR

**Goodbody, Goodboddy:** Richard *Godbodi* 1221 AssGl; John *Godbody* 1308 AssSt; John *Goodbody* alias *Goodman* 1468 Ewen (Nf). 'Good body' (OE *gōd, bodig*), used as an epithet of courteous address, also equivalent to French *Beaucorps* and to *Goodman*.

**Goodboy:** William *Godeboie* 1206 P (Y); William *Bonus Puer* 1250 Rams (C); Robert *Godeboye* 1334 SRK. 'Good boy, servant', OE *gōd*, ME *boie*.

**Goodcheap, Goodchap:** Walter *Godchep* 1166 P (Nf); William *Gudchep* 1236 Bart (Lo); Hamo *Godchape* 1315 ib. OE *gōd* 'good' and *cēap* 'barter, price'. ME *good cheap* was used of a state of the market good for the purchaser, when prices were low (a1325 MED), hence 'one who gives a good bargain'. The nickname might also have been taken from the vendor's cry.

**Goodchild:** Henry *Goddechild'* 1211 Cur (Ess); Robert *Godchild* 1237 FFEss; Willelmus *Bonus Puer* c1250 Rams (C); Richard *Godechild* 1327 SRC; Gilbert *Godechilde* 1366 LLB G. OE *gōd* 'good' and *cild* 'child'. cf. GOODBAIRN. The first form suggests that the source may occasionally be 'godchild'.

**Goodcock:** v. GODCOCK

**Goodday:** Stephen *Godeday* 1301 SRY; Wylliam *Goodday* 1524 SRSf. For GOODENDAY or, possibly, 'good servant', v. DAY.

**Goodenday:** ME *godne dæie* c1205, *godun dai* c1250, *god dai* a1300 NED, 'Good day', a salutation on meeting or parting: 'Have a good day!' 'God give you good day!' cf. Walter *Godemorwe* (*Godemorwen*) 1327 *SR* (Ess).

**Goodenough, Goodnow, Goodner, Goodanew:** (i) Alan *Godinoch*, *Godinogh'* 1212 Cur (Y), 1297 MinAcctCo; William *Godinogh* 1225 AssSo; William *Godinowe* 1309 NottBR. A nickname, probably for one easily satisfied. cf. Robert *Welynogh* 1260 AssC, William *Welynough* (*Welynou*) 1327 *SR* (Ess). (ii) Roger *Godecnaue* 1220 Oseney (O); Hervicus *Godcnave* 1225 Pat (Nt); Hugo *Gudknave* 1269 AssNb; John *Goddeknawe* 13th Shef (Y). OE *gōd* and *cnafa*. 'boy, servant'. It is not surprising, in view of the apparent contradiction in terms and the unpleasant associations of the modern *knave*, that this name survives only in disguise. For *knave*, NED gives the forms *knaf* c1375, *knaffe* 1481 and *knawe*: *knave-line* became *knauling*. With the loss of *k*, *Godenaf* would easily be confused with Goodenough and *God(e)nawe* with Goodnow and Goodanew, which preserve the now dialectal *enow* for *enough*. The synonymous OE *cnapa* 'boy, servant', also gave rise to a surname: Ernald *le Godeknape* 1308 AD vi (Sf), Martin *Godeknape* 1327 SRC.

**Gooder, Goodere, Gooders, Gooda, Gouda:** *Godere*
1066 DB (Ess, Sf); John *Godere* 1317 AssK; John
*Gooder* 1564 ShefA. OE *Gōdhere* 'good army'.

**Goodered:** *v.* GOODREAD

**Gooderham, Goodram, Goodrum, Guthrum:**
*Guðrum* 12th Whitelock; Gerard *filius Gudram, Guð*
*ram* 1201 P (L); Brictiua *filia Guderam* 1202 AssL;
*Guderam* Gleve 1283 SRSf; Hugh *Godrum* 1260 AssL;
Thomas *Guderam, Godram* 1283, 1327 SRSf; Thomas
*Goodrum* 1577 Musters (Nf); Widow *Guderum*,
*Goddram* 1674 HTSf; Godfrey *Gooderham* 1681
NorwDep; Geoffrey *Goodrum* 1765 FrYar. ON
*Guðormr* 'battle-snake'. *Guðrum, Gudram*, the name
of the first Danish king of East Anglia, is preserved in
the York street-name Goodramgate (PN ERY 289).
There has probably also been some confusion with
ON *Guðrún* (f), OG *Godram* (m), and OE *Gōdrūn* (f),
the last of which is the origin of Gutter Lane
(London), *v.* ELPN 41. But for these as surnames the
evidence is slight.

**Gooderidge:** *v.* GOODRICH

**Goodes, Goodess:** *Godise* 1066 DB (Y); *Godehese*
1198 P (K); Johannes *filius Goduse* 1190 P (Y); *Godusa*
1208, 1220 Cur (Y, L); Roger *Godhus* 1279 RH (C);
Hugh *Godhos* ib. (Bk); William *Godhose* 1296 SRSx;
John *Godhous* 1327 SRSf; William *Goudese* 1327
SRSx. *Godus* (f) must derive from the solitary OE
*Godhyse* (966 BCS 1181), which suggests that *Edus*
and *Aldus* may go back to OE *\*Eadhyse*, *\*Ealdhyse*. *v.*
EDIS, ALDIS.

**Goodesmith, Goodsmith:** John *Godesmaghe* 1351
FrY; Margaret *Goodsmyth* 1525 SRSx. 'Relative of
*Gode*', cf. HUDSMITH.

**Goodeve:** *Godgeua, Godiua, Godeue* 1066 DB;
*Godiua* 1221 AssWa; William *Godyeue* 1327 SRSf;
Edward *Godyf* 1527 RochW; Edmond *Goodeve* 1662
HTEss. OE *Godgifu* (f) 'god or good gift'. *v.* also
GOODIFF.

**Goodfar, Goodfare:** Gilbert *Goudefeyre* 1332
SRSx; Robert *Gudefeir* 1379 PTY. One of 'good
appearance, demeanour', from ME *feyr*, an aphetic
form of *effeir* from OFr *aferir* 'to be proper, meet'.

**Goodfear, Goodfer:** Ketel *Godefere* 1154–76 YCh;
William *Godefere* 1379 PTY. 'Good comrade', OE
*gōd*, (*ge)fēra*.

**Goodfellow:** Richard *Godfelage* 1192 P (Nth);
Roger *Godfelawe* 1274 RH (Ess). ME *god felawe*
'good fellow', a good companion. cf. John *Trewfelagh*
1379 PTY.

**Goodfriend:** William *bonus amicus* c1160 ELPN;
William *Godfrend* 1210 P (W); Robert *Godfrend*
1306–7 FFWa. 'Good friend', OE *gōd, frēond*.

**Goodgame:** Robert *Godgame* 1185 Templars
(Herts); Walter *Godgamen* 1279 RH (Hu). 'Good
sport.' *v.* GAME.

**Goodger, Gudger:** Robert *Gudger* 1616
RothwellPR (Y); Joseph *Goodger, Goodyer* 1738,
1760 WordwellPR (Sf). A late development of
GOODYEAR.

**Goodgroom:** Osbert *Godgrom* 1200 Cur (C); Peter
*le Godegrom* 1306–7 FFSx; John *Godegrome* 1443
CtH. 'Good servant', OE *gōd*, ME *grom*. cf. Robert
*Godemay* 1302 FA (Sf), with the same meaning.

**Goodhall:** *v.* GOODALL

**Goodhand:** Aluric *Godhand* c1095 Bury (Sf). 'Good
hand.' *v.* HAND. Henry *Goudenhond* 1332 SRSx. This is
probably 'golden hand'.

**Goodhart, Goodheart:** John *Godhierte* 1221 Cur
(Herts); John *Goudhert* 1327 SRSo. 'Good heart.'

**Goodhead:** (i) William *Godheved* 1222 Pat (Herts).
'Good head', OE *hēafod*. (ii) William *Godhed* 1327
SRWo. ME *godhede* 'goodness'. cf. Richard *Godnesse*
1327 SRSf.

**Goodheal:** John *Godhele* 1360 IpmGl. A greeting,
'good health', OE *gōd, hǣlu*. cf. Henry *Godehitud*
1396 AssL 'may good come to you'; Walter
*Godemorwe* 1327 SREss 'good morning'; Adam
*Goudtyd* 1332 SRDo 'may good betide you'.

**Goodhew, Goodhue, Goodhugh:** *Got hugo* 1086 DB
(Ess); *Godhuge* c1095 Bury (Sf); Robertus *filius*
*Godhu*, Robert *Godhuge* 1181, 1222 DBStP; Roger
*Godhou* c1240 *ERO* (Ess); Robert *Godhewe* 1311
ColchCt; Richard *Godhowe*, William *Godhewen* 1327
*SR* (Ess); John *Godhiwe, Goudhywe* 1327, 1332 SRSx;
John *Godhewyn* 1333 FFSf; William *Goodhewe* 1383
LLB G. ON *\*Guð(h)ugi*, no doubt the opposite of
*Ill(h)ugi, Illhuge* 11th Saga Bk xii, 137, Radbode *filius*
*Illhuge* 1166 P (Nf). The surname may also mean
'good servant', ME *god* and *hewe*, from OE *hīwan*
(plural) 'members of a household'. cf. GOODHIND.

**Goodhind:** William, Geoffrey *Godhine* a1224
Clerkenwell (Mx), 1230 P (Nth). ME *god* 'good' and
*hine* sg, from OE *hīwan* pl. 'member of a household',
'a lad, boy, stripling'; 'a good lad'. *v.* also GOODHEW.

**Goodhusband:** Agnes, Nicholas *Godhosbonde* 1279
RH (Hu). 'Good husbandman', OE *gōd, hūsbonda*.
cf. Hugh *Godhoweswyf* 1301 FS 'good housewife'.

**Goodier:** *v.* GOODYEAR

**Goodifer:** Sibilla *Godyfer* 1288–9 LuffCh; William
*Godifer* 1382 AssY. 'Good comrade', OE *gōd, gefēra*.
*v.* also GOODFEAR.

**Goodiff, Goodey, Goody:** (i) Nota *Godwyf* 1311
ColchCt; Margaret *Godewyf* 1374 FFHu. ME *good-*
*wife* 'mistress of a house'. cf. GOODMAN. *Goodey* is a
shortened form. cf. *hussey* from *hūs-wīf* and *Goody* or
*Goodwyff* Wilkes (1559 NED). (ii) John *Godyf* 1279
RH (O); Thomas *Godif* 1297 SRY. These forms are
too early to be regarded as a form of *good-wife*. They
are probably from OE *Godgifu* (f). *v.* GOODEVE. This
was confused with OE *Godgȳð* (f) and also with *good-*
*wife*: William *Godith(e), Godyth, Gwodyf* 1317 AssK.
*Godyf* widow of John Clare is also called *Goditha*
Clare de Strode ib.

**Gooding, Goodinge, Goodings, Godding, Goding:**
*Godinc* 1066 DB; *Goding* ib., c1113 Burton (St);
Robert *Goding* 1185 Templars (Ess); Peter *Godding*
1274 RH (Sf). *Gooding* is from OE *Gōding, Godding*
from OE *Gōding*. In 1208 P (Sf), *Goding* is an
alternative for *Godmer*, and so a pet-form of names in
*God-*.

**Goodison:** Nicholas *Godithson* 1332 SRCu; Robert
*Guditson*, John *Godyeson* 1379 PTY. 'Son of *Godith*',
OE *Godgȳð* 'god or good battle', a woman's name
long in use: *Godith* 1206 Cur (Beds), *Godit* 1199 ChR,
*Gudytha* Foloufast 1379 PTY.

**Goodkin:** *v.* GODKIN

**Goodlad, Goodlatte, Goodlet, Goodlett:** Robert *Godelad* 1301 SRY; William *Godlad* 1332 SRCu; William *Goodlad* 1464 GildY; Robert *Guidlett* 1574 Black (Ross); John *Guidlat* 1602 ib. 'Good lad', 'good servant'.

**Goodlake, Goodluck, Gullick, Gulick, Cutlack:** *Gotlac* 1066 DB (Ch); Robertus *filius Guthlach* 1187 P (La); Symon *Goddeloc, Godeloc* 1247 FFC; Hosebert *Gotloc, Guzloc,* John *Gudloc* 1279 RH (O); Isabella *Gullake* 1327 SRSo; Henry *Cutlake* 1538 NorwW. OE *Gūðlāc* 'battle-play', or, possibly, ON *Guðleikr.*

**Goodlamb:** *Godlamb* 1086 ICC (C); *Godlambus* 1114–30 Rams (C); *Godlamb* de Well', de Lenn' 1166 P, 1206 Cur (Nf); Herveus *filius Godlamp* 1221 *ElyA* (C); Roger *Godlamb* 1230 P (Nf); William *Godlomb* 1274 RH (Nf). OE *Gōdlamb,* an original nickname 'good lamb', recorded once before the Conquest. Probably also a nickname.

**Goodland:** (i) Gilebertus *filius Godland'* 1201 Cur (Nf); Hugo *Godlond* 1279 RH (O). OG *Godland.* (ii) Henry *de Godland* 1379 PTY. 'Dweller by the good land.'

**Goodlet:** *v.* GOODLAD

**Goodley:** *v.* GODLEY

**Goodliff, Goodliffe:** *Gothlif* 1086 ICC (C); *Godelief* 1197 P (K); Maud *Goodleef* 1272 Bart (Lo); John *Godelef* 1296 SRSx; William *Godeloues* 1327 SRWo. OE *Godlēof* (m) or OE *\*Godlēofu* (f) 'good or god dear', long used in Kent: *Godlefe* (f) 1508 ArchC xxx.

**Goodlud:** This may be a corruption of *Goodlad,* but is more probably for 'good lord'. cf. the legal 'M'lud': John *Godlord* 1332 AD v (Wa); Ralph *Godlord* 1387 NorwW (Nf).

**Goodman:** *v.* GODMAN

**Goodner:** *v.* GOODENOUGH

**Goodness:** Ranulf *Godnese* 1251 IpmY; Richard *Godnesse* 1327 SRSf. 'Goodness', OE *gōdnes.*

**Goodram:** *v.* GOODERHAM

**Goodread, Goodred, Goodreds, Goodered:** *Gudret, Godred* 1066 DB (L, Y); *Gudred* de Cnappewelle 1102–7 Rams (C); *Godred* 1200 FFL; Ragemerus *filius Gutred* 1204 P (L); Henry *Guthred* C1214 StGreg (K); John *Godred,* Ysabell *Godrad, Godrid* 1279 RH (C); Alexander *Godered'* 1325 FFK; William *Goodrede* 1460 FrY. ON *Guðrøðr.* The surname may also be ME *god rede* 'good counsel'. cf. the Scottish *Meiklereid,* Robert *Smalred* 1176 P (Y), Philip *Lytylred* 1372–6 Gaunt.

**Goodrich, Goodridge, Gooderidge, Gooderick, Goodrick, Goodricke, Godrich, Godridge, Goodwright:** *Godric* 1066 DB; Gaufridus *filius Godrici* 1207 Cur (Bk); Ambrosius *filius Godrige* 1279 RH (C); Ralph *Godric'* 1199 P (Wo); Hugo *Godriche* 1221 *ElyA* (Sf); John *Godrige* 1279 RH (C); John *Godryk* 1313 FFEss; James *Goodrich* 1341 ColchCt; William *Godright* 1363 LLB G; Albreda *Goderik* 1381 PTY; Simon *Goderich* 1388 LLB H; William *Guderyk* 1475 GildY; John *Guddrig* 1477 FrY; Elizabeth *Gutteridge, Goodritch* 1659, 1666 HorringerPR (Sf); John *Goodridg* 1662 *HTEss.* OE *Godrīc* 'good or god ruler'. *v.* also CUTTERIDGE. Occasionally the surname is local in origin: Thomas *de Goderigge* 1275 RH (W).

**Goodrum:** *v.* GOODERHAM

**Goodsell:** *v.* GODSAL

**Goodsir:** Thomas *Goudsyre, Goodsire* 1375 LoPleas, 1384 LLB H. ME *gudsire, goodsir* 'grandfather' (c1425 NED). Probably a term of address. cf. GODSON, SWEETSER.

**Goodsman:** *v.* GODSMAN

**Goodson:** *v.* GODSON

**Goodspeed:** Ralph *Godisped* 1275 RH (Herts); Walter *Godspede* 1275 SRWo. 'God speed (you)', a wish for success for one setting out on an enterprise (1526 NED).

**Goodswen, Goodswin, Goudswen:** Herbert *Godsuain* c1155 DC (L); William *Godswein* 1206 Cur (L); Roger *Gudswen* 14th Lewes (Nf). ME *god swain* (ON *sveinn,* ODa *sven*) 'good servant'. cf. Roger *Godyoman* 1297 SRY.

**Goodway:** *Goduui* 1066 DB (C); *Godwi* 1188 BuryS (Sf); Henry *Godwi* 1212 Cur (Berks); William *Gudeway* 1448 FrY. OE *Godwīg* 'good or god-strife'.

**Goodwin:** *v.* GODWIN

**Goodwood:** Geoffrey *de Godenywode* 1296 SRSx; William *de Godywode* 1332 SRWo. From Goodwood in Boxgrove (Sx).

**Goodwright:** *v.* GOODRICH

**Goodyear, Goodyer, Goodier, Goudier:** Cest', Henry *Godyer* 1279 RH (Hu), 1295 ParlR (Ess); Henry, Annc' *Godyar* 1285 *Ass,* 1301 ParlR (Ess); John *Godhyer* 1296 SRSx; William *Godeyer,* Maud *Godeyiere* 1301 SRY; Henry *Goudhier,* William *Godier* 1327 SRSx; Agnes *Goudyer* 1327 SRSx; John *Goodyer* 1456 FrY; John *Godeyere* 1467 AD vi (Mx); Robert *Gudyere* 1513 GildY; William *Goodeare* 1566 FFHu; Jeremiah *Goodyear* 1682 FrY. ME *goodyeare, goodier, goodere, goodye(e)re* 'good year', an expletive used in questions, 'What the good year?'. Possibly elliptic for 'as I hope to have a good year' (c1555 NED).

**Gook:** Alan *Gok* 1219 AssL; Thomas *Gouk',* Richard *Gouc* 1219 AssY; Thomas *Gauke* 1424 FrY; John *Gook* 1432 ib. ON *gaukr* 'cuckoo', ME *goke, gowke* (a1300 MED).

**Gookey:** Willelmus *filius Goci* 1121–38 Bury (Sf); *Goche* 1166 P (Nf), 1186–1210 Holme (Nf); Roger *Goki* 1191 P (Cu); Ralph *Goky* 1218 AssL. ON *Gauki.*

**Goold:** *v.* GOLD

**Goolden:** *v.* GOLDEN

**Goolding:** *v.* GOLDING

**Gooldsbury:** *v.* GOLDSBOROUGH

**Goom:** *v.* GOMME

**Goose:** Hugo, Richard *Gos* 1176, 1191 P (L, Y), 1210 Cur (C); Hugh *le Gos* 1227 AssBeds; Hamo *le Gous* 1231 FFC; Alice *Gous* 1297 MinAcctCo. OE *gōs* 'goose'. *v.* also GOSS.

**Goosebeard:** William *Goseberd'* 1225 AssSo. 'Goose-beard', OE *gōs, beard.* cf. Roger *Gosefot* 1212 Fees (Berks) 'goose-foot'; Richard *Goseheued'* 1208–9 Pleas 'goose-head'; Godelin *Gosethrote* 1200 Oseney 'goose-throat'; Richard *Gosetunge* 1323 Ch (Sf) 'goose-tongue'; Thomas *Gosonthegrene* 1289 NorwLt 'goose on the green'.

**Gooseman:** Gilbert *Gosman* 1246 AssLa. OE *gōs* and *mann* 'tender of the geese'. cf. GOSSARD.

**Goosey, Goozee:** (i) Robert *Gosege* 1167 P (D); Peter *Goseie* 1199 P (Nth); Henry *Goseye* 1327 SRWo.

'Goose-eye.' cf. John *Gosebody* 1297 MinAcctCo, Roger *Gosefot* 1212 Fees (Berks), Gocelin *Gosethrote* 1200 Oseney (O). (ii) Walter *de Gosey* 1242 Fees (Berks). From Goosey (Berks).

**Gorch, Gordge:** Herveus, Osbert *Gorge* 1185 P, 1221 Cur (Sf). OFr *gorge* 'throat'. *Gorgerer*, from OFr *gorg(i)ere*, was a not uncommon occupational term for a maker of gorgers or gorgets or armour for the throat and possibly also for a maker of chin-cloths or wimples for covering the throat: Alexander *le Gorgerer* 1220 Cur (Mx), called also *le Gorgeur* (1221 ib.), Andrew *le Gorgerer* Hy 3 Colch (Ess). The surname is metonymic for *gorgerer*. cf. Simon *Gorget* 1327 SRC.

**Gordon, Gurden:** The home of the Scottish Gordons was in Berwickshire where there is a place Gordon from which the family probably took its name. The earliest member noted is Richer *de Gordun* lord of the barony of Gordon in the Merse (1150–60). His son was Thomas *de Gordun* (Black). The surname is found early in England: Adam *de Gurdun* 1204 P (Ha) who probably came from a French place named Gourdon (Saône-et-Loire, etc.). The names of Geoffrey *Gurdun* 1220 Cur (K), Adam *Gordon* 1279 RH (C), etc., are probably identical with Fr *Gourdon* which Dauzat takes as a diminutive of OFr *gourd* in the sense 'dull, stupid, boorish'.

**Gore:** Ralph *de la Gare* 1181 P (K); John *de Gore* 1257 ArchC iii; Alan *ate Gore* 1274 RH (Ess). 'Dweller by the triangular piece of land' (OE *gāra*), as at Gore Court (Kent) and Gore (Wilts).

**Goreham, Gorham:** (i) Henry *de Gorham* 1192 P (Ess); Giles *of Gorham* 1246 FFY; Hugh *de Gorham* 1314–16 AssNth. From Gorron (Mayenne). (ii) William nempne *filius gorham* 1086 InqEl (Sf). Perhaps OBret *\*Goron*, cognate with Welsh *gwron* 'valiant'.

**Gorer:** Equivalent to *ate Gore*.

**Goring, Goringe, Gorin:** Geoffrey *de Garinges* 1192 P (Sx); Orvietus *de Goringes* 1255 Cur (Y); Robert *Goring* 1327 SRSx; William *Goringe* 1583 Musters (Sr). From Goring (Oxon, Sussex).

**Gorley, Gorly:** Henry *Gorly* 1327 SRSo. From North, South Gorley (Ha).

**Gorman:** William *Gorman* 1296 SRSx; Adam *Garman* 1327 ib. *v.* GORE, GORER and GARMAN. Also for Irish *MacGorman*, *O'Gorman* 'son or descendant of *Gorman*', a diminutive of *gorm* 'blue'.

**Gorony, Goronwy:** Owen *son of Gronow* 1280 SaAS 2/xi; *Gorgonow* 1327 SRSa; Kenwrig *ap Grono*, Houa *ap Gorgene* 1393 Chirk; Adam *Wronow* 1327 SRSa; William *Gronnowe* 1663 HeMil. OW *Guorgonui. v.* Morris 111.

**Gorrell, Gorrill:** (i) William, Henry *Gorel* 1176 P (K), 1319 SRLo, 1221 Cur (D). ME *gorrell* 'a fat-paunched person', from OFr *gorel* 'pig'. (ii) Margeria *Gorulf* 1296 SRSx; John *Goroulfe* 1327 SRWo. OE *Gārwulf* 'spear-wolf'. (iii) 'Dweller by the muddy spring', as at Gorwell (Essex) or Gorrel (Devon): Walter *de Gorewell* 1274 RH (Ess); or by the muddy hill: Cecily *de Gorhull* 1246 AssLa; or at a muddy nook as at Gorrel Fm (Bucks), earlier *Gorhale*.

**Gorst:** Roger *de la Gorste* 1275 SRWo; Agnes *del Gorstes*, Ralph *in le Gorstes* 1281 MELS (Wo). 'Dweller among the gorse' (OE *gorst*).

**Gorstidge, Gorsuch, Gostage, Gostige, Gossage:** John *de Gosefordsik* 1332 SRLa; Henry *Gorsage* 1579 Bardsley; Roger *Gorsuche* 1602 PrGR; Henry *Gorstich* 1669 Bardsley. From Gosfordsich (Lancs).

**Gorton:** Thomas *de Gorton* 1332 SRLa; John *Gorton* 1642 PrD. From Gorton (La), or Gortonhill in Bishop's Nympton (D), *Gorton* 1524.

**Gosden:** William *de Gosedenn* 1296 SRSx; John *Goseden* 1364 PN Sr 226; Nicholas *Goseden* 1525 SRSx. From a lost *Gosden* in Slaugham (Sx).

**Gosford:** Robert *Goseford* 1346 Hylle. From Gosford (D, O, Wa), or Gosforth (Cu, Nb).

**Goshawk, Goshawke, Gosshawk:** Robert *Goshauek* 1332 SRWa; William *Goshauk* 1400 TestEbor; Henry *Goshauke* 1524 SRSf. 'Goshawk', OE *gōshafoc*.

**Gosling, Gossling, Gostling:** Henry *Goseling* 1260 AssC; Robert *Goseling*, Maud *Gosselyng* 1327 SRC. ME *goslyng* 'gosling'. *Gosling* is, no doubt, often a late development of *Goslin* (*v.* JOCELYN), but these forms are much too early for such a development. They are earlier, too, than the earliest *o*-form in MED (*gos;elyng* c1350).

**Gosnay, Gosney:** John *Gosnay* 1447 FrY. 'Dweller at the island frequented by geese', OE *gōs, īeg*.

**Goss, Gosse, Joass, Joce, Jose, Joss, Josse:** *Gotso* dapifer 1130 P (St); *Gosce, Joce* de Brunna Hy 2, 1195 DC (L); Hugo, Geoffrey *Gosse* 1202 AssL, 1251 Rams (Hu); Richard *Goce* 1205 P (Sf); Thomas *Joce*, William *Josse* 1327 SRSx. OFr *Joce, Josce, Gosse*, OG *Gozzo, Gauz*. Also a pet-name of *Gocelin*: *Gosse* or *Gocelin* de Mealtun c1160 DC (L); *Jose* or *Jocelinus* de Baillol 1164–66 Bury (Sf). *Joss(e)* is also for Breton *Judoc. v.* JOYCE, and also GOTT.

**Gossage:** *v.* GORSTIDGE

**Gossard, Gozzard, Gozzett:** Walter *Goseherd* 1236 Ass (Ha); John *le Goshurde* 1327 SRSx. OE *gōs* and *hierde* 'goose-herd' (*gozzard* 1771 NED).

**Gossell:** Thomas *de Gorsthul* 1272 FFSt; Walter *de Gosehale* 1276 ArchC vi; Gilbert *de Gorsthull* 1341 AssSt. From Gosshill in Sutton at Hone (K), or 'dweller by the hill covered with gorse', OE *gorst, hyll*.

**Gosselin:** *v.* JOCELYN

**Gosset, Gossett, Jossett:** Robert, Isabella *Josset* 1279 RH (C). A diminutive of *Gosse, Josse. v.* GOSS.

**Gosshawk:** *v.* GOSHAWK

**Gossip, Gossop:** Ulfketil *Godsib* 1229 Cl (Nf); William *Godsip* c1248 Bec (Nf); John *le Gossib*, *Gossyp* 1319 AssSt. OE *godsibb* 'godfather, godmother', later 'a familiar acquaintance or friend'.

**Gosson:** William *Gosson'* 1327 SRC. Perhaps an assimilated form of GODSON. cf. 'Godson or gosson, *Filiolus'* c1440 PromptParv. Or for (William) *Gotson* 1314 Wak (Y). 'Son of *Gotte.' v.* GOTT.

**Gostage, Gostige:** *v.* GORSTIDGE

**Gostling:** *v.* GOSLING, JOCELYN

**Gotelee:** *v.* GOATLEY

**Goter, Gotter:** Thomas *de la Gotere* 1275, Andrew *de la gotere* 1327 Kris; Adam *atte Goter* 1339 AssSt. 'Dweller by the water-course', OFr *goutiere*.

**Gotham:** Stephen *de Gotham* 1291–2 FFEss; Henry *de Gotham* c1320 Calv (Y); William *Gotham* 1381 LoCh. From Gotham (Nt).

**Gothard:** Geoffrey *Gothirde* 1229 Cl (O); Reginald *le Gateherde* 1275 SRWo; Geoffrey *le Gotherde* 1332 SRSx. OE *gāt, hyrde* 'goat-herd'.

**Gotobed, Gotbed:** John, Richard *Gotobedde* 1269 Barnwell (C), 1309 SRBeds; William *Gawetobedde* 1332 SRSx; Charles *Godbed* 1760 Bardsley. There can be little doubt that this nickname means what it says. cf. William *Gatorest* Hy 3 Gilb (L), John *Gobiside* 1379 PTY, Nicholas *Gabyfore* 1430 FeuDu, Serle *Gotokirke* 1279 RH (C).

**Gotsell:** *v.* GODSAL.

**Gott, Gotts:** *Gotte* filius Wulfrici 1188 P (Y); Gilbertus *filius Gotte* 1195 P (L); Godui *Gott* c1095 Bury (Sf); Haldan' *Gotte* 1202 AssL; Geoffrey *Gottes* 1346 FA (Nf). *Gocelin*, pronounced *Gotselin*, had petforms *Gosse* (*v.* GOSS, JOYCE) and *Got*, both from *Gots*: cf. *Gocelin* Bouer; Thomas *Gotsone* Bouer 1319 Pat (Hu), i.e. Thomas son of *Got* Bouer, where *Got* is clearly short for *Gocelin*. In Yorkshire and Lincolnshire, this may be of Breton origin. But the 11th-century Godui *Gott* is too early for this development and is probably from OE *gutt*, ME *gutt*, *gotte* 'gut, guts', formerly in polite use, meaning also 'a corpulent or greedy person'.

**Gottacre:** John *Gottacre* 1202 AssL. 'Dweller by the goat field', OE *gāt, æcer*.

**Gotter:** *v.* GOTER

**Goubert:** *v.* JOBERT

**Goucher:** *v.* GOACHER

**Goud-:** *v.* GOOD-

**Goudge, Gouge:** *v.* GOOCH

**Goudie, Goudy:** Scottish variants of *Goldie*, found in Edinburgh from 1598 as *Gowdie, Gaudie, Goddie* (Black).

**Gough, Goff, Goffe:** Griffin, Robert *Gogh* 1287 AssCh, 1327 SRSo; Thomas *Goughe* 1576 SRW. Welsh *coch* 'red'. *v.* also GOOCH.

**Goul(d)-:** *v.* GOLD-

**Goule:** John *Goule* 1327 SRSo; John *le Goul* 1341 ChertseyCt (Sr); John *Goule* 1357 FFW. A nickname from either ON *gulr* 'yellow, pale', or from ME *goule* 'gullet, greed'.

**Goundry:** *v.* GUNDREY

**Goupil:** *v.* GUPILL

**Gourd:** Henry *Gourdemaker* 1327 SRY. A maker of bottles or cups (OFr *gourde*).

**Gourlay, Gourley, Gourlie:** Ingelram *de Gourlay* c1174 Black; Eustace *de Gurleye* 1279 AssNb; John *Gourley* 1327 SRDb. A Scottish name, probably originally from North, South Gorley (Ha).

**Gover:** (i) Gilbert *Gofar'* 1223 Cur (O); Gilbert, John *Gofaire* 1240 Oseney (O), 1260 Husting; Geoffrey *Gouayre* 1242 Ct (Ha); James *Gafaire* 1327 FrY. 'Go fairly', ME *faire* 'beautifully', 'gently', 'quietly'. Either one who walks beautifully or, more probably, one who goes gently, uses gentle means (ELPN). cf. Margerie *Gangefeyr* 1451 Rad (C). (ii) Walter *gouyere* 1296 SRSx; John *Gouere*, William *Gover*, Richard *le Gofiar* 1327 SRSo; Thomas *Gover* 1380 SRSt. This form, which will also account for

*Govier*, is difficult. It cannot be the East Anglian dialectal *goave* 'to store grain', as suggested by Weekley and Harrison, for that word derives from ME *golfe* and retains its *l* until the 16th century. These surnames, too, are southern and western which dissociates them also from Harrison's alternative suggestion, ME *gofe* 'to stare stupidly', which is a Scottish word.

**Gow:** Richard *Gowe* 1230 P (So); George *Gow* 1580 Black. Gael *gobha* 'smith'.

**Gowell:** William *Gowel* 1206 Pleas (W); Richard *Gowel* 1217 GlCh; William *Gawell* 1284 AssLa. 'Go well', OE *gān, wel*. cf. GOVER (ii).

**Gowen, Gowing:** *v.* MACGOWAN

**Gower, Gowar, Gowers, Guwer:** (i) Willelmus *filius Goer* c1160 Gilb (L); *Goerus* Pellipar' 1296 SRSx; William *Goer, Guer* c1150, 1187 Gilb (L); Adam *Goiher, Guher* 1176, 1179 P (Y); Jordan *Guuer* 1202 P (So); Thomas *Guwer, le Goher* 1221 Cur (O); Stephen *Gower* (*le Goher*) 1230 P (Y); Adam *Goier* 1327 SRSf. OFr *Go(h)ier*, OG *Godehar* 'good army'. *Gohier* also denoted an inhabitant of the *Goelle*, the country north of Paris, anciently *Gohiere*, hence *le Goher*. The term seems also to have been applied to a man from Gouy. William *de Goiz* is also called *le Goeir* 1212 Cur (Beds). (ii) Walter *de Guher* 1130 P (Carmarthen). From Gower (Glamorgan).

**Gowing, Gowings:** *Gowin* de Martun Hy 2 RegAntiquiss; *Gowinus* c1250 Rams (Nf); Henry *Gowyng* 1524 SRSf; Richard *Gowing* 1730 FrYar. OFr *Gouin*, a contraction of OG names in *Gudin-*, or perhaps from Breton *gwen* 'fair'. *v.* Dauzat.

**Gowland:** John *Gowland* 1583 FrY. From Gowlands in Moor Monkton (WRY).

**Gowler:** *v.* GAWLER

**Gowthorpe:** William *de Goutorp* 1219 AssY; Sosanna *de Goukethorp* 1298 Wak (Y). From Gowthorpe in Bishop Wilton (ERY), or Gawthorpe in Lepton, Gawthorpe Hall in Crigglestone (WRY).

**Goy, Goys:** Robert *le Goiz* 1201 Cur; Roger *de Gouiz* 1221 Cur (Do); Nicholas *de Goys* 1249 AssW; Walter *Goys* 1327 SRSo. From Gouy (Aisne, Pas-de-Calais, Seine-Maritime).

**Gozzard, Gozzett:** *v.* GOSSARD

**Grace, Gras, Grass:** (i) William *le Gras* 1199 P (Gl), 1202 AssL, 1219 AssY; Roger *le Gras*, Rogerus *Crassus* (identical) 1200 Cur (St); Simon, William *Grace* 1310 ColchCt, 1332 SRSt. OFr *gras* 'fat'. *v.* CRASS, GROSS. Great Graces in Little Baddow owes its name to Nicholas *le Gras* (1275 PN Ess 235). Or from OFr *grace*, ME *grace, gras* 'a pleasing quality', i.e. 'attractive, charming' as opposed to John *Gracemauuais* 1247 AssBeds 'the ungracious'. (ii) William *atte Grase* 1327 SRSo; Robert *atte Gresse* 1381 SRSt. OE *græs* 'grass', also 'pasture', hence 'one who puts out cattle to grass', 'a grazier'.

**Grace, Gracey, Gracie, Grece:** Robertus *filius Grecie* 1188 BuryS (Sf); *Grecia* 1201 P (We), 1221 AssGl, AssWa; *Gracia* 1213 Cur (Sr), 1247 AssBeds; *Gratia* 1221 Cur (Berks); *Gracia* de Saleby 1232–35 Gilb (L); Henry *Grece* 1275 RH (Nth); Gilbert *Gracye* 1296 SRSx; William *Grece* 1297 SRY; Adam *Grace* 1302 SRSf. We are clearly concerned with a single woman's name, perhaps a derivative of OG *grisja* 'grey'. OFr *gris*, 'grey' is found in ME as *grece*, *greyce*. The name was latinized by the scribes as *Gratia* and popularly associated with OFr *grace*. *v.* also GRASSICK.

**Graddige, Gradidge:** From Graddage Fm in Clayhidon (Devon), earlier *Greatediche*, where lived William *atte Graddich* in 1330 (PN D 611).

**Graeme:** *v.* GRAHAM

**Graff:** Robert *Graf* 1279 RH (O); John *Graff* 1414 FrY. cf. Thomas *Graffard* 1200 P (Y). This latter was a Norman term for a public scribe or scrivener, from OFr *grafer* 'to write'. OFr *grafe* 'stylus', 'pencil' is here used for the wielder of the pen.

**Graffard, Grafford:** Thomas *Graffard* 1200 P (Y); William *Graffard* 1290, *Grafford* 1302 IpmY. ONFr *graffard* 'a public scribe or scrivener'.

**Grafham, Graffham:** Gilbert *de Grafham* 1159 P (Hu); Gocelin *de Grafham* 1296, Robert *de Grofham* 1327 SRSx. From Grafham (Hu), or Graffham (Sx).

**Grafton:** William *de Graftona* 1130 P (Lei); John *de Grafton'* c1280 SRWo; John *Grafton* 1443 FFEss. From one or other of the many places of this name.

**Graham, Grahame, Graeme, Grayham, Greim:** The first of the Grahams in Scotland was William *de Graham* (1127 Black). A Norman, he took with him the Norman form of his surname which derives from Grantham (Lincs), found in DB as both *Grantham* and *Graham*. In Scotland it occurs as *Grame* in 1411 and *Graym* in 1447 (Black).

**Grain, Grein:** William *del Greyn* 1297 SRY; John *atte Greine* 1327 SRSx; William *Grayne* 1362 AssY. 'Dweller at the inlet, or at the fork of a river', ON *grein*.

**Grainge:** *v.* GRANGE

**Graley:** *v.* GREALEY

**Grammer:** Robertus *Gramaticus* 1201 Cur (Y). In the margin: 'Robertus clericus et persona est'. Richard *le Gramarie*, *le Gramary*, *le Gramayre*, William *Gramayre* 1219 AssY; Andrew *le Gramare* 1284 FA (Y). OFr *gramaire* 'grammarian, scholar, astrologer'.

**Gramshaw:** *v.* GRANSHAW

**Granby:** Robert *de Graneby* 1356 FFEss; William *Granby* 1455 IpmNt. From Granby (Nt).

**Grand:** *v.* GRANT

**Grandage, Grandorge, Grandidge:** William *Grain de Or* (*Graindorge*) 1166 RBE (Sf); Thomas *Grandage* 1379 PTY; Roger *Grandorge* 1459 Balliol (O); Widow *Grandish* 1674 HTSf. OFr *grein de orge* 'barleycorn'.

**Grandamore:** Robert *Grantamur* 1202 AssL. 'Great love', OFr *grant*, *amour*. cf. John *Graundben* 1295 IpmGl 'great good'; Robert *Grauntfei* 1225 Cur (Hu) 'great faith'; Henry *Grauntfoly* 1379 PTY 'great folly'; Henry *Grauntservis* 1312 FrY 'great service'.

**Grandison:** Otho *de Grandissono* 1280 Glast (So); Peter *de Grandisson* 1335 IpmW; Roger *Graundisson* 1397 FrY. From Granson on the lake of Neuchâtel, *v.* Wagner. cf. also Stretton Grandison (Hereford) from William de Grande Sono 1303.

**Grange, Grainge:** William *de la Graunge* 1275 RH (Ess); Laurence *atte Grange* 1296 SRSx. 'Dweller near, or worker at a grange', AFr *graunge*, OFr *grange* 'granary, barn' (a1325 MED).

**Granger, Grainger:** William *le grangier* c1100 MedEA (Nf); Reginald *le Granger* 1219 FFSf; Walter *le Graunger* 1247 AssBeds; John *Grainger* Hy 6 AD vi (Wo). AFr *graunger*, OFr *grangier* 'one in charge of a grange, a farm-bailiff'.

**Granshaw, Gramshaw, Greenshaw, Grenshaw:** Hugh *de Greneshagh* 1393 FrY; Agnes *Greneshagh* 1400 TestEbor; William *Greneschawe* 1400–1 IpmY. From Garnshaw House in Hebden (WRY), or 'dweller by the green wood', OE *grēne*, *sceaga*.

**Grant, Grand, Le Grand, Legrand:** (i) Hugo *Grandis* 1084 GeldR (W); Gilbert *Grant*, *Grandus* 1183, 1192 Eynsham (O); William *le grant* 1189–1210 Holme (Nf); Thomas *le Graunt* 1219 AssY; Agnes *Grant* 1221 ElyA (Sf). AFr *graund*, *graunt*, OFr *grand*, *grant* 'great'. Ekwall notes (ELPN) that William *grandis* or *le grant* (c1150–60) was a son of Wlfwin *Graunt*, *Grand*, *grant* (1108–30) and inherited his byname, which was evidently given him for distinction from Wulwinus *juvenis* (c1130), so that *grand* here means 'elder, senior'. In most instances it probably means 'tall'. (ii) Petrus *filius Grente* 1166 P (Y); Robert *filius Grante* 1208 P (Gl); *Grante* le Chapman 1274 RH (D); William *Grent* 1204 P (O); Robert *Grente* 1327 SRSx. This must be a survival of the OE *Grante*, *Grente* found in several place-names. *v.* PN ERY 89, DEPN.

**Grantham:** Thomas *de Grantham* 1220 Cur (Herts). From Grantham (Lincs).

**Grantland:** Robert *Grantland*, John *Grantlande* 1642 PrD. From Grantland in Poughill (D).

**Grantley:** Elias *de Grantelay* 1277 IpmY. From Grantley (WRY).

**Grapinell, Grapnell:** *Grapinell'* 1206 P (Nth); Nicholas *Grapinel* 1177 P (Ess); Robert *Grapinel* 1229–30, Henry *Grapinel* 1279–80 FFEss. ME *grapenel* 'grappling iron'. Probably metonymic for a maker or user of these.

**Grasby, Grassby:** Hugh *de Gresby* Hy 2 Gilb; Ælfward *de Grassebi* 1190 P (Y); Adam *de Grossebi* 1219 AssL. From Grasby (L), *Grosebi* DB.

**Grason:** *v.* GRAVESON

**Graspeys, Graspis:** Henry *Graspays* 1318 KB (Lo); Geoffrey *Graspeys* 1334 SRK; Henry *Graspais* 1338 CorLo. A nickname from OFr *graspeis* 'grampus, seal'.

**Grassby:** *v.* GRASBY

**Grassick, Grassie, Grass, Gracey:** John *Grasse* alias Cordonar 1539 Black; Donald *Grasycht* 1548 ib.; John *Graissie* or *Grecie* 1632–3 ib. Gaelic *greusaich*, *griasaich*, originally 'decorator, embroiderer', later 'shoemaker'.

**Grassman:** Walter *Graysman* 1297 MinAcctCo; Mable *Gresman* 1319 *SR* (Ess). OFr *graisse*, *greisse*, *gresse* 'grease' and *man*, 'a seller of grease'.

**Grater:** John *the Gratere* 1223 MESO (Y); Stephen *le Gratier* 1305 Pinchbeck (Sf); Juliana *la Gratour* 1327 MESO (Ha). OFr *grateor*, *gratour*, *\*gratier* 'one who grates', probably a furbisher.

**Gration:** *v.* GRAVESON

**Gratton, Grattan, Gratten:** Robert, William *de Gratton* 1327 SRDb, 1337 IpmNt; John *de Grattone*

1348 DbAS xxxvi. From Gratton (Derby, Devon), probably also one of the sources of Irish *Grattan*.

**Gratrix:** *v.* GREATOREX

**Gravatt, Gravett, Grevatt:** Ralph *de Grauette* 1208 P (Ha); Ascelina *de Lagravate* 1210 Cur (Ha); John *atte Grevette* 1288 PN Sx 17; William *de Greuett'*, Richard *de la Greuett'* 1296 SRSx. OE *\*grǣfet*, *\*grǣfet*, a derivative of OE *grǣf*, *grǣfe* 'grove' and the diminutive suffix *-et*, from residence near a little grove. The term is found chiefly in Sussex and Surrey. *v.* MELS.

**Grave, Graves:** Lefsi *filius Greiue* 1161–77 Rams (Nf); Æthewold *filius Greui* ib.; Adam *filius Graiue* 1221 ElyA (C); Greive de Pincebec 1232 Pat (L); Robert, Walter *Greyue* 1255 NottBR, 1327 SRC; Hubert, Thomas *le Greyve* Hy 3 Colch (Sf), 1275 RH (Nf); William *le Grayue* 1334 FFLa; John *Graue* 1379 PTY; Hugo *Graves* 1540 FrY. This is usually from ME *greyve*, from ON *greifi* 'steward', 'a person in charge of property', in its later forms confused with GRIEVE. In South Yorks in the 16th century, *grave* (*grayves*) interchanges with *grieve* (*grevys*) in the same sense (NED). We have also to deal with a personal name ON *Greifi*, a byname from ON *greifi*, ODa, OSw *grefe* 'count, earl', which may survive in *Greavy*. This appears in DB in 1066 as *Greue* (L) and in 1130 P (L) in Turstin' *filius Greue*, where the *e* is due to the OEscand change of *ei* to *e* and is one source of *Grieve*. The forms *filius Graiue*, *Greiue* may mean 'son of the *greyve*' and survive as GRAVESON.

**Graveley, Gravely:** Richard *de Grauelea* 1200 P (C); Geoffrey *de Grauele* 1296 SRSx; John *Gravele* 1339–40 CorLo. From Graveley (C, Herts).

**Graveling, Gravelling:** John *de Grauelinges* 1205, *de Grauelingges* 1230 P (K). From Gravelines (Nord).

**Gravely:** *v.* GRAVELEY

**Graveney, Greveney:** Alard *de Gravenni* 1207 Cur (Sf); Robert *de Graveney* 1230 Cur (K); John *Graveneye* 1376 AssLo. From Graveney (Kent).

**Gravenor, Gravener:** Robert *le Grant Venur* 1293 AssSt; John *Gravener* 1524 SRSf. OFr *grand* 'great' and *veneor* 'hunter'.

**Graver:** William *le Grevere* 1275 SRWo; Roger *le Graueur* 1279 MEOT (Sr); Peter *le Grauere* 1293 MESO (Y); John *Grafour*, carver 1414 FrY. OE *grafere*, *grǣfere*, OFr *graveur* 'engraver, sculptor'. cf. Robert *le Orgrauer* 1308 Wak (Y) 'gold engraver' and Adam *le Selgraver* 1332 MESO (Lo) 'engraver of seals'. *Gravere* may also mean 'digger', from OE *grafan* 'to dig'. Piers *le Graver* was killed by the collapse of the (coal)pit in which he was working by himself at Silkstone (Db) in 1290 (MedInd).

**Graveson, Grayson, Grason, Grayshan, Grayshon, Gration, Graveston, Grayston, Graystone:** Richard *Grayveson* 1327 Wak (Y); John *Graiveson* 1332 SRCu; John *Graveson* 1379 PTY; John *Grayfson* 1381 PTY; Thomas *Grayson* 1426 FrY; John *Grason* 1542 GildY. 'Son of the *greyve* or steward.' *v.* GRAVE and cf. GRIEVESON. A Lancs and Yorks name, all the modern variants being still found in the Leeds district. *Grayveson*, *Graysoun* and *Grayveston* are all found in the Preston Guild Rolls.

**Gray, Grey, Le Grey:** (i) Baldwin *Grai* 1173 P (Bk); William *Grei* 1198 FFHu; William *le Greie*, Sewyn *le Gray*, Philip *le Grey* 1296 SRSx. OE *grǣg* 'grey', probably 'grey-haired'. (ii) Anschitill *Grai*, *de Grai* 1086 DB (O); Henry *de Gray* 1196 P (Nt). From Graye (Calvados).

**Graybeard:** *Greiberd* 1207 Cur (Ha); Richard *Greyberde* 1279 RH (O); William *Greyberd* 1332 SRWa. 'Grey beard', OE *grǣg*, *beard.* cf. Ralph *Greyeye* 13th CartNat 'grey eyes'; Artur *Grayfot* 1243 AssDu 'grey foot'; Robert *Greyleg* 1327 SRSo 'grey leg'; Gilbert *Greyschanke* 1279 RH (C) 'grey shanks'.

**Graygoose, Graygos:** Richard *Graigos* 1249 AssW; Bartholomew *Gregos* 1305 SIA iii; John *Greygose* 1524 SRSf; Mary *Graygoose* 1662 HTEss; Mary *Graggiss* 1771 DenhamPR (Sf). A nickname from the wild goose, OE *grǣggōs.* cf. Thomas *Graydere* 1373–5 AssL 'grey deer'; William *Grehound* 1327 SREss 'greyhound'.

**Grayham:** *v.* GRAHAM

**Grayland, Graylan:** *Graelanð* de Marisco 1198 FFEss; *Gralandus* 1214 Cur (Sr); *Graelant* Templarius 1224 Pat; Robert *Greylond* 1290–1 FFEss; William *Greyland* 1314–15 AssNth; John *Greyland* 1338 CorLo. OFr *Graelent.*

**Grayling, Greyling:** *Graelencus* de Runcamp 12th DC (L); *Gralang* filius Willelmi 1206 Pleas (Ess); Richard *Grailing* 1205 P (Ess); Robert *Greyling* 1317 AssK; John *Graylyng* 1392 CtH. A variant of GRAYLAND, from OFr *Graelent.*

**Grayshan, Grayshon, Grayson:** *v.* GRAVESON

**Graystoke, Graystock, Greystoke, Greystock, Gristock:** Thomas *de Greystok* 1251 AssY; William *de Graistok* 1332 SRLa; William *Graystoke* 1408 IpmY. From Greystoke (Cu).

**Grayston, Graystone:** William *Graistaine*, *de Graystanes* 1332 SRCu, 1380 FeuDu. 'Dweller by the grey stone(s)', as at Greystones (Sheffield). *v.* also GRAVESON.

**Grazier:** William *le Grasiere*, *le Grazur* 1279 RH (Bk), 13th Guisb (Y); Richard *le Grasiere* 1325 Oseney (O). 'One who grazes or feeds cattle for market', a derivative of OE *grasian* 'to graze'.

**Greader, Greeder:** Wluèrd *Legredere* 1188 BuryS (Sf); Robert *le Gredere* 1243 AssSo; William *Greder* 1360 CarshCt (Sr). 'Crier, town-crier', ME *gredere*.

**Grealey, Greally, Grealy, Greeley, Greely, Graley, Grayley, Gredley, Gridley, Gridly, Greasley, Gresley, le Gresley, Legresley, Grisley, Grealis, Grellis:** Albert *Greslet* 1086 DB (Ch); Robertus *Greslatus* 1127 Seals (La); Robert *Greslet* 1130 P (Y); Robert *Greilli* 1133–60 Rams (Beds); Alexander *Grisle* 1148 Winton (Ha); Albert *Gresley*, *Gresle* 1153–68 Holme (Nf), 1182 P (L); Robert *Gresle*, *de Gredlei*, 1196–7 P (L), *Gredlei* 1200 Cur (L); Albert *Grelli* c1200 WhC (La); Henry *Gredle*, *Greslei* 1212 Cur (K); Robert *Gresle*, *Grelle*, *de Grellay*, *Gredlegh* 1212 Fees (La, Sf, L, O); Robert *Grelay* (*Gresle*) 1230 P (La); Thomas *Greley*, *Graley* (son of Robert) 1235 Fees (La); Thomas *Greyley*, *Grayly*, *Gresley* 1235, 1242 Fees (La, O, Sf); Thomas *Gredley* alias *Grelley* 1271 Ipm (R); Margaret *le Grele* 1295 Barnwell (C); Thomas *Gridly* 1674 HTSf. OFr *greslet* 'marked as by hail', i.e. pitted or pock-marked. *le Gresley* is still found in the Channel Islands and Greslé, Grêlé in France. *Greasley* and *Gresley* are also local in origin, from Greasley (Notts) and Gresley (Derby), pronounced *Greasley*, identical in meaning: William *de Greseleg'* 1198 P (Nt/Db), Nigel *de Greseleg'* 1204 P (Db). As both places also appear as *Griseleia*, William *de Griseleia* 1130 P (Nt/Db), this may sometimes have given modern *Grisley*. In early forms, the nickname has been confused with the place-name. *Grealley, Greeley, Greely*, are also for Ir *Mag Raoghallaigh* 'son of *Raoghallach'*.

**Greasley, Gresley, Gresly:** William, Nigel *de Greseleia* c1130 StCh, 1208 P (La); William *de Gresley* 1373 IpmNt; John *Gresley* 1445–6 FFWa. From Greasley (Notts), or Gresley (Derby). *v.* also GREALEY.

**Greatorex, Greatrex, Greatrix, Gratrix:** William *Gretorex* 1743 BishamPR (Berks). From Greterakes (Db).

**Greathead, Greathed, Greated:** John *Gretheved* 1278 LLB B; Thomas *Gretehed* 1351 Whitby (Y). 'Big head' (OE *grēat, hēafod*).

**Greatshanks:** Robert *Greteschanke* 1296 SRNb. 'Big legs', OE *grēat, sceanca*. cf. John *Shortshank'* 1379 PTY 'short legs'; Sefar *Brokesanke* 1202 Pleas (Nf) 'broken leg'.

**Greatworth:** Richard *Gretword* 1196 P (Nt); Gilbert *de Gretewurthe* 1214 Cur (Nth). From Greatworth (Northants).

**Greaves, Greeves, Greve, Greves:** Geoffrey *de la Greue* 1203 P (Lei); Walter *in the Greve* 1220 Pat (St); Richard *del Greues* 1246 AssLa; Richard *de Greves* 1259 AssCh; Alexander *del Greue* 13th WhC (La); Adam *del Grefes* 1314 Wak (Y); John *del Grayfe* 1363 FrY. From Greaves in Preston (Lancs) or from residence in or near a grove or groves (OE *grǣfe* 'brushwood, thicket, grove'). This appears as *greyve* c1475 and there has been confusion with *Grave* and *Grieve* which occurs as *greeve* (1629) and *greave* (1844).

**Greavison:** *v.* GRIEVESON

**Greavy:** *v.* GRAVE

**Grece:** *v.* GRACE

**Gredley:** *v.* GREALEY

**Greed:** William, Robert *Grede* 1185 P (Y), 1242 Fees (D). According to NED, *greed* is a back-formation from *greedy* and does not occur before 1609. The back-formation appears to have taken place much earlier.

**Greeder:** *v.* GREADER

**Greedy:** Matildis *Gredi* 1209 P (L); Helyas *le Gredie* 1269 LeiBR. OE *grǣdig* 'ravenous, voracious, gluttonous'.

**Greel, Greell:** Thomas *Greelle* 1306–7 FFSx. ME *grille, grelle* 'fierce, cruel'.

**Greeley, Greely:** *v.* GREALEY

**Green, Greene, Grene:** Geoffrey *de Grene* 1188 P (K); Richard *de la Grene* 1200 P (Nf); Geoffrey *Attegrene* 1206 AssL; Walter *ad Grenam* 1210 Cur (L); William *del grene* 1221 ElyA (Nf); Alexander, William *Grene* 1230 P (Y), c1248 Bec (O); John *en le grene*, Robert *Othergrene*, Henry *on the grene* 1274–99 MELS (Wo); John *super le Grene* 1327 SRDb. OE *grēne* 'green', usually from residence near the village green. Occasionally we may have *green* in the sense

'young, immature'. cf. the French *Vert, Levert* which Dauzat refers to 'la *verdeur* de l'homme, sa vigueur, sa jeunesse, sa vivacité'; or a personal name. cf. Godwin *Grenessone* 1115 Winton (Ha), Peter *filius Grenii* 1196 P (Y), Matthew *filius Grene* 1202 Cur (Sr).

**Greenacre:** Richard *de Grenacres* 1332 SRLa, *de Grenacr* 1335 WhC (La). 'Dweller by the green field(s)' (OE *æcer*).

**Greenall:** *v.* GREENHALGH

**Greenberry, Greenburgh, Greenbury:** William, Turstan *de Greneberge* 1205 P (Y), 1207 Cur (Lei); John *de Grenberwe* 1279 RH (O); Ed. *Grenebury* 1544 FrY. From Grandborough (Bucks, Warwicks), earlier *Greneberge*, or 'dweller by the green hill'.

**Greenden:** *v.* GRENDON

**Greenep:** *v.* GREENOP

**Greener:** (i) Robert *de la Grenehore* 1275 RH (Sf). 'Dweller by the green bank' (OE *ōra*). *v.* GREENHALGH. (ii) John *le Greener'* 1332 SRWo. 'Dweller by the green.'

**Greenfield:** Peter *de Grenefeld* 1242 Fees (Sf); John *de Grenefeud* 1296 SRSx. 'Dweller by the green field.'

**Greenford:** John *Greneford, de Greneford* 1259 Acc. From Greenford (Mx), or 'dweller by the green ford', OE *grēne, ford*.

**Greengrass:** Richard, Alice *de Grenegres* 1275 RH (Sf), 1327 SRSf. 'Dweller at the green grassy-spot' (ON *gres*). All examples are from Suffolk, with *-gres*.

**Greenhalgh, Greenhalf, Greenhall, Greenhill:** Richard *de Grenhal* 1230 P (Sa); William *de Grenol* 1246 AssLa; Matill' *de Grenehalgh*, William *de Grenolf, de Grenholl'* 1332 SRLa. From Greenhalgh (Lancs), pronounced *Greener*. Probably also from Greenhaugh (Northumb).

**Greenham, Grinham:** Simon *de Gryndham* 1268 AssSo; Ralph *de Greneham* 1275 RH (Sf). From Greenham' (Som, Berks).

**Greenhead:** (i) Robert *Grenehod* 1221 ElyA (Sf); William *Grenhoud* 1327 SRSx; John *Grenehed*, Alice *Greenhood* 1423, 1449 NorwW. OE *grēne, hōd* 'green hood', from a partiality for this head-dress. (ii) Robert *de le Greneheved* 1290 Black. From Greenhead (Peeblesshire, Northumberland) or from residence near a green hill.

**Greenhill, Grinnell:** William *de Grenehill'* 1200 P (Beds); Symon *de la Grenehell* 1270 ArchC v; Gilbert *ate Grenehelle* 1317 AssK. From one of the small places called Greenhill or from residence near a green hill.

**Greenhough, Greenhow, Greenhowe, Greenough:** Toka *in Grenehoga* 1066 DB (Nf); Geoffrey *de Grenhou* 1219 AssY; Robert *de Grenehowe* 1332 SRCu; John *Greenough* 1682 PrGR. From Greenhoe (Norfolk) or Greenhow (NR, WRYorks).

**Greenhouse:** William *de Grenhous* 1279 RH (C). 'Dweller at the house on the green.'

**Greenidge:** Richard *de Grenehegge* 1289 PN Nt 220. From Greenhedge Fm in Aslockton (Nt).

**Greening:** Roger *Grenyg* 1275 RH (Nf); John *Greeninge* 1641 PrSo. 'Dweller by the green hill', OE *grene, *ing. v.* EPNE.

**Greenist:** Richard, Giles *de Gren(e)hurst* 1246 AssLa, 1251 AssY; Walter *de Greneherst* 1296 SRSx. 'Dweller by the green wooded-hill' (OE *hyrst*).

**Greenland:** John *Greneland* 1400 TestEbor; Alexander *Greneland* 1525 SRSx; John *Greneland* 1576 SRW. From Greenland (Co, Caithness), or 'dweller by the green meadow', OE *grēne, land.*

**Greenleaf, Greenleaves:** Adam *Grenelef* 1327 SRSf; John *Grenelefe* 1441 ShefA; Thomas *Grenelefe* 1577 Musters (Nf). 'Green leaf', OE *grēne, lēaf,* but the exact meaning as a nickname is unknown. cf. William *Leafgrene* 1317 AssK.

**Greenlees, Greenleas:** (i) Adam *Grenelese* 1460 FrY; Thomas *Greenleese* 1584 ArchC 35; Acon *Greenlease* 1663 HeMil. 'Dweller by the green pasture', OE *grēne, lǣs.* (ii) John *Greynleis* 1574 Black. From East, West Greenlees (Lanark).

**Greenley, Greenly, Grindlay, Grindley, Grinley:** William *de Grenlay* 1275 RH (Nt); William *de Grinle* 1279 RH (C). From Little Gringley (Notts), Grindley (Staffs) or residence near a green clearing.

**Greenman, Greenmon:** John *Greneman* 1357 ColchCt. 'Dweller by the green.'

**Greenop, Greenopp, Greenup, Greenep:** Richard *Grenepp* 1523, William *Greneope* 1552 FrY; Thomas *Greenhope* 1616, Agnes *Greenup* 1667 LaWills. 'Dweller at the green valley', OE *grēne, hop.*

**Greenough:** v. GREENHOUGH

**Greenshaw:** v. GRANSHAW

**Greenside, Greensides, Grenside:** Brian *Grenesydes* 1599 FrY. From Greenside in Waverton (Cu).

**Greenslade, Grinslade:** William *Greneslade* 1524 SRD; Thomas *Greeneslade* 1641 PrSo; John *Greenslade, Grinslade* 1642 PrD. From Greenslade in North Tawton (D), or 'dweller in the green valley', OE *grēne, slæd.*

**Greensmith:** Thomas *Grenesmyth* 1523 RochW. 'Coppersmith.'

**Greensted, Grinstead, Grinsted:** William *de Grenested', de Grinstede* 1230 P (Sx), 1327 SRSf. From Greenstead, Greensted (Essex) or Grinstead (Sussex). v. GRIMSTEAD.

**Greenstreet:** Bartholomew *de Grenestrete* 1257 ArchC iii; Peter *atte Grenestrete* 1327 SRSx. 'Dweller by a green road.'

**Greenup:** v. GREENOP

**Greenward:** Adam *Grenewerde* 1276 RH (Y). Apparently 'keeper of the (village) green' (OE *weard*).

**Greenway, Greenaway:** William *de Greneweie* 1214 Cur (K); Robert *Grenewey* 1279 RH (O), 1327 *SR* (Ess); John *atte grenewey* 1327 SRSo. 'Dweller by the green way.'

**Greenwell:** Thomas *de Grenewille* 1279 RH (O). 'Dweller by the green (grassy) spring or stream.'

**Greenwood:** John *del Grenewode* 1275 Wak (Y). 'Dweller by the green wood.'

**Greer:** v. GRIERSON

**Greeson:** v. GRIEVESON

**Greet:** (i) Philippus *filius Grete* 1201 Cur (Nth); *Greta* 1219 AssY, 1260 AssC; Æðelwold ðes *Greta* c900 OEByn; Æðelmæres *Greatan* (gen.) 1017 ib.; Maud, William *Grete* 1243 AssDu, 1296 SRSx; Henry, Gilbert *le Grete* 1279 RH (Bk), 1287 AssCh. OE *grēat* 'big, stout' or, less commonly, from *Greta* (f), probably short for *Margareta.* (ii) Philip *de Grete* 1204 Cur (St); Rannulf *de Grete* 1207 Cur (Gl); Peter

*de Grete* 1255 RH (Sa). From Greet (Glos, Salop, Worcs). Or local. cf. Alexander *de la Grete* 1240 Fees (K), from residence at some gravelly spot, or, perhaps more probably, from working in a gravel-pit.

**Greeter:** John *le Grether* 1327 SRSo; John *le Greter'* 1327 SRWo; Ralph *le Greter'* 1340 NIWo. Probably 'dweller near or worker at the gravel-pit', from a derivative of OE *grēot* 'gravel'.

**Greetham, Gretham:** Roger *de Gretham* e13th RegAntiquiss; Hugh *de Gretham* 1297 MinAcctCo; Alice *de Greteham* 1363 AssY. From Greetham (L, R).

**Greethurst:** Stephen *de Gruthurst* 1221 AssWo. From a lost place *Greethurst* in Yardley (Worcs).

**Greeves:** v. GREAVES

**Greg, Gregg, Greggs:** William *Gregge* 1234 Fees (Do); Henry *Gregge* 1306 FFEss; John *Greggez* 1504 FrY. A short form of *Gregory.*

**Greggor, Gregor:** Robert son of *Gregor* c1240 Black (Coldingham); John son of *Gregor* 1332 SRCu; Richard *Gregour* 1373 ColchCt. The vernacular form of *Gregory.*

**Gregory, Gregori, Grigorey:** Willelmus *filius Gregorii* 1143–7 DC (L); *Gregorius* c1150 Black; *Grigarius* de Bristwic' 1214 Cur (Y); *Grigori* de Bertune 13th AD iv (Wa); Peter *Gregory* 1279 RH (Beds); John *Grigory* 1280 AssSo; John *Gregory* 1296 SRSx. Fr *Grégoire,* Lat *Gregorius,* from Gk *γρηγόριος,* a derivative of *γρηγορέω* 'to be watchful', at times confused with Lat *gregarius* from *grex, gregis* 'herd, flock'. Common in both England and Scotland.

**Gregson, Grigson:** William *Griggesson* 1327 SRC; Richard *Gregson* 1332 SRCu. 'Son of *Greg* or *Grig*', i.e. Gregory.

**Greif:** v. GRIEVE

**Greim:** v. GRAHAM

**Grein:** v. GRAIN

**Grelling:** v. GRILLING

**Grendon, Grindon:** Robert *de Grendune* 1185 Templars (L); Ranulf *de Grendon* 1242 Fees (Nb); Robert *de Grendon'* 1345–6 FFWa. From Grendon (Nth, Wa), Grendon Underwood (Bk), Grendon Bishop, Warren (He), or Grindon (Du, Nb, St).

**Grenford:** v. GREENFORD

**Grennan:** v. GARNON

**Grenshaw:** v. GRANSHAW

**Grenside:** v. GREENSIDE

**Grenville, Grenfell:** Gerard *de Grenvill'* 1161 P (Bk); Eustace *de Greinuill'* 1197 P (Ess); William *de Grenefell* 1363 FrY. From Grainville-la-Teinturière (Seine-Inférieure). v. ANF.

**Grerson:** v. GRIERSON

**Gresham:** William *de Gresham* 1199 Pl (Nf); Thomas *Gresham* 1446 AssLa; Richard *Gressam* 1551 FrY. From Gresham (Norfolk).

**Gresley:** v. GREALEY, GREASLEY

**Gretham:** v. GREETHAM

**Gretton:** v. GIRTON

**Grevatt:** v. GRAVATT

**Greveney:** v. GRAVENEY

**Greves:** v. GREAVES

**Greville:** William *de Greiuill'* 1158 P (Nb). From Gréville (La Manche).

**Grew, Grewe:** Roger *le Grue* 1230 MemR (So); Gerard *la Grue* 1246 Ipm (Y). OFr *grue* 'crane'.

**Grewal:** Thomas *Gruel* 1327 SRC. OFr *gruel* 'fine flour, meal', a name for a miller or a baker.

**Grewcock, Grocock, Grocott, Groocock, Groucock, Groucutt, Growcock, Growcott:** Margeria *Groucok* 1275 SRWo; William *Grucock* 1312 AssSt. ME *grew* 'crane' plus *cock*, a formation similar to that of *Peacock*.

**Grey:** *v.* GRAY

**Greyling:** *v.* GRAYLING

**Greystock, Greystoke:** *v.* GRAYSTOKE

**Gribben, Gribbin, Gribbon:** *v.* CRIBBIN

**Gribble, Gribbell:** Walter *atte Gribbele* 1330 *SR* (D). From residence near a crabtree or blackthorn (ME *gribbele*). *v.* also GRIMBLE.

**Grice, Grise, Griss, Le Grice, Le Grys:** Richard, Eustace *Gris* 1176, 1193 P (Nth, K); Robert, John *le Gris* 1198 FF (Nf), 1202 AssL; Leticia *Grise* 1317 AssK; Thomas *Grys* 1327 SRSx; Walter *Griss* 1337 ColchCt; Richard *Grice* 1413 FrY. ME *grise, grice* from ON *griss* 'a pig' (c1230 MED) or ME *gris, grice* from OFr *gris* 'grey', grey-haired (c1395 MED).

**Gricks:** *v.* GRIGG

**Gridley:** *v.* GREALEY

**Grief:** *v.* GRIEVE

**Grier:** *v.* GRIERSON

**Grierson, Grerson:** Gilbert *Greresoun* 1411 Black; William *Grerson* 1451 ib.; Gilbert *Greir, Greirsone* 1676–7 ib.; *Grierson alias M'Gregor* 1704 ib. 'Son of *Gregor*.'

**Grieve, Grieves, Greif, Grief, Grieff, Greeff:** Farðain *Greva* c1050 YCh; Johan *Greve* 1199 P (W), 1296 Black; William *Greue* 1327 MEOT (L); John *le Greue* 1332 ib. (Nf); John *Grefe* 1470 Black; Laurence, Richard *Greif* 1493 ib., 1529 FrY; John *Grieff* 1605 Black; John *Grieve* 1782 ib. Scottish and Northern English *grieve* is the normal representative of ONorthumb *grēfa*, corresponding to WS *gerēfa* 'reeve' (NED). Originally 'governor of a province', it came to mean 'overseer, manager, head workman on a farm, farm-bailiff'. This is the origin of the Scottish *Grieve* and probably of the first Yorkshire example, though this might be a byname from the personal name discussed under *Grave*. One would not expect an Old Northumbrian form in Wilts, Lincs or Norfolk. The Wilts example is also probably from the personal name. The Norfolk example is certainly and the Lincs one probably the occupational term. Thuresson explains these as due to the influence of *reve* 'reeve' on the ground that the Scandinavian change of *ei* to *e* occurs only in personal names. As the personal name is the occupational term used as a byname, there seems no reason why both should not undergo the same development and these forms from the eastern counties where Scandinavian influence was strong should be derived from ON *greifi* 'steward'. cf. GRAVE.

**Grieveson, Greavison, Greeson, Greson:** Emma *Grefeson* 1379 PTY; Robert *Greveson* 1498 FrY; John *Greason* 1637 Bardsley (La). 'Son of the bailiff.' *v.* GRIEVE.

**Griff, Griffe:** (i) Walter *de Griff* 1219 AssY; John *del Gryf* 1383 IpmNt. From Griff (Wa), Griffe (Db),

Griff Fm in Rievaulx (NRY), or 'dweller by the pit or hollow', ON *gryfja*. (ii) Raffe *Griffe* 1545 SRW. A short form of GRIFFIN.

**Griffin, Griffen:** *Grifin* 1066 DB (Co, Ch, St, Wa); *Griffin filius Gurgan* 1130 P (Pembroke); Osbertus *filius Griffini* 1153–68 Holme (Nf); *Gruffin* son of Oweyn 1285 Ch (Radnor); Robert *Grifin* 1148 Winton (Ha); Godfrey, Robert *Griffin* 1197 P (Wa), 1219 AssSo; John *Griffen* c1230 StThomas (St). *Gruffin*, *Griffin*, a pet-form of MW *Gruffudd*; in the Welsh border counties introduced direct from Wales, in the eastern counties by the Bretons who came over with the Conqueror and were numerous there. cf. *Griffinus* Bret Hy 2 DC (L).

**Griffith, Griffiths:** *Gruffyd ap Madog Vnyon* 1392 Chirk; Jone *Gryffyth* 1524 SRSf. OW *Griph-iud*, MW *Gruffudd*, noted in Wales in 1150. The second element is *iud* 'chief'. cf. *Griffinus seu Griffith Kynaston* 1428 FA (Sa).

**Grigg, Griggs, Gricks, Grix:** *Gricge, Grigge* 1275 RH (Sf); *Grigge* Suel 1327 SRC; William *Grig* 12th DC (L); Warin *Grygge* 1282 FFEss; John *Gregge* 1282 FFEss; Robert *Grigges* 1341 AD iii (Hu); John *Gregge, Grigge* 1343, 1345 LLB F; *Grig* is a short form of *Gregory*. cf. GREG. John *Grygges* was the son and heir of *Gregory* le Bacer 1345 AD i (Herts). The frequency of *Grigg* as compared with *Greg* is due to the fact that *Grig* was also a nickname: Stephen *le Grig'* 1327 SRSx: ME *grigg(e)*, 'a diminutive person, a dwarf' (a1400–50 NED). There appears to be no connexion with 'a merry grig'.

**Grigorey:** *v.* GREGORY

**Grigson:** *v.* GREGSON

**Grill, Grills, Grylls:** John *Grylle* 1327 SRSf; John *Grille* 1346 FA (Sf). ME *grill(e)* 'fierce, cruel', from OE *gryllan* 'to gnash the teeth, rage'.

**Grilling, Grelling:** Goduine *Grelling* c1095 Bury; Roger *Grilling, Grelling* 1201 Cur; Robert *Grellinge* 1286 Rams (Nth). OE *\*Grylling*, an original nickname from OE *\*grylle* 'fierce, cruel'. *v.* OEByn 143.

**Grimble, Grumble, Grumell, Gribbell, Gribble:** *Grimbald* 1066, 1086 DB; *Grimbaldus, Grumbaldus* Pancefot 1272 Forssner; *Grymbald* Fraunceys 1310 AssSt; Robert, William *Grimbald* 1153–63 Templars (O), 1207 Cur (Nth); Radulfus *Grumbaldus* 1185 Templars (O); Warin *Grimboll* 1275 RH (Sf); Martin *Grumbold* 1327 SRC; Richard *Grymbyll* 1524 SRSf. OG *Grimbald* 'helmet-bold'. *Grimbald* became *Gribald* through assimilation of *mb* to *bb*: cf. *Gribaut*, Guido *Gribaud* 1275 RH (Nf). *v.* also GRIBBLE.

**Grimby:** William *de Grenebi* 1178 P (L). From Granby (Notts).

**Grime, Grimes, Grimm, Grimme:** (i) *Grim, Grimus, Grimmus* 1066 DB; *Grim* de Leuertona 1175 P (Nt); Godwin, Bernard *Grim* 1170, 1183 P (Nf, C); Alan *Grime* 1279 RH (C); William, Thomas *Grym* 1309 FFSf, 1332 SRSt; Geoffrey, John *Gryme* 1327 SRY, 1379 PTY. ON *Grimr*, ODa, OSw *Grim*. This should give *Grime(s)*. *Grimmus* may belong below. The common early form *Grim* as a surname cannot be definitely assigned. It is certainly at times for *Grimm*. (ii) Edricus *Grim, Grimma, Salvage* 1066 DB (Sf); Peter *le Grim* 1327 SRSx; John *le Grymme* 1332 SRSt.

*le Grim* is from OE *grim* 'fierce, grim', a nickname clearly borne by Edricus Grim.

**Grimley, Grimbley, Grimbly:** Iuo *de Grimesl'* 1221 AssWo; Geoffrey *Grymely* 1329 SRSf. From Grimley (Wo).

**Grimmet, Grimmett, Grumitt, Grummett, Grummit, Grummitt:** (i) Roger, John *Grymet* 1251 AssY, 1297 SRY; Vyncent *Grumet* 1275 RH (W). A diminutive of *Grim*, a short form of *Grimbald*. v. GRIMBLE. (ii) Thomas *Grimuld*, William *Grimald* 1327 SRSf. OG *Grimbald*, becoming *Grimaud*, *Grimet*. (iii) John *Grymyld*, *Grymyd* 1379 PTY. ON *Grimhildr* (f) 'helmet-battle'.

**Grimmond, Grimond:** John *Griman* 1534 Black (Perth); Anna *Grimen* 1657 ib.; John *Grimmond* 1665 ib. A form of M'CRIMMON with excrescent *d*.

**Grimsdell:** Adam *de Grimisdale* 1332 SRCu. From Grinsdale (Cumb).

**Grimsell:** Robert *de Grimeshull'* 1221 AssWo. From Grimes Hill (Worcs).

**Grimshaw:** Richard *de Grymeschagh*, *de Grymeschawe* 13th WhC, 1284 AssLa; Henry *de Grimeshagh* 1400 IpmLa. From Grimshaw (Lancs).

**Grimson:** Arne *Grimsune* Wm 2 Whitby (Y). 'Son of *Grimr*.' v. GRIME.

**Grimstead, Grimsted:** Walter *de Grimestede* 1201 P (W). From Grimstead (Wilts). Also from Grinstead (Sussex): Thomas *de Grimstede* 1327 SRSx. v. GREENSTED.

**Grimston, Grimstone:** Walter *de Grimeston* 1186–90 MCh; Sissota *de Grimmestone* 1291 IpmY; Thomas *de Grymeston* 1332 AssD. From Grimston (Lei, Nf, Sf, NRY, WRY), Hanging, North Grimston, Grimston in Dunnington (ERY), or Grimstone in Stratton (Do).

**Grimward, Grimwade, Grimwood:** *Grimward* de Cuthmund 1199 FineR; *Grimwardus* 1222 ClR; Robert *Grimward* 1247 FineR (L); Thomas *Grymward* 1327 SRSf; William *Grymwade* 1524 SRSf; Edmund *Grimwood* 1674 HTSf. OG *Grimward* 'helmet-guard'.

**Grindal, Grindall, Grindel, Grindell, Grindle:** (i) Ædricus *Grendel* 1180 P (Wo); Robert *de Grenedala* 1166 P (Y); Walter *de Grendale* 1242 Fees (L); Stephen, Benedict *de Grindale* 1297 AssY, 1332 SRCu. From Greendale (Devon), Grindale (ERYorks) or residence in a green valley. (ii) Richard *de Grenehull'* 1221 AssSa. From Grindle (Salop) or residence near a green hill.

**Grinder:** William *le grindere* 1230 P (So); Stephen *le Grindar* 1274 RH (Sa). OE *grindere* 'grinder of corn, miller'. In medieval England the reference may also have been to a sharpener of iron tools (grindstones and grindelstones are mentioned 1228); or to a grinder of colours for painters (1352 Building 171).

**Grindlay, Grindley, Grinley:** v. GREENLEY.

**Grinham:** v. GREENHAM.

**Grinnell:** v. GREENHILL.

**Grinslade:** v. GREENSLADE.

**Grinstead, Grinsted:** v. GREENSTED.

**Gripp:** (i) Hugo *filius Grip* 1086 DB (Do); Robert, John *Grip* 1195 P (Y), 1275 SRWo. ON *Grip*, ODa *Grip*. (ii) Walter *del le Grip* 1279 AssNb; John *atte Grype* 1332 SRSx. 'Dweller in the deep valley', as at

Crypt Fm (Sussex), OE *gripu* 'kettle, caldron', used topographically, v. MELS.

**Grisdale, Grisedale:** Simon *de Grisdale* 1332 SRLa; William *de Gresdale* 1359 FrY; William *Grisedayle* 1526 FrY. From Grisedale (Cumb).

**Grise, Griss:** v. GRICE.

**Grisethwaite:** v. GRISTHWAITE.

**Grisewold:** v. GRISWOLD.

**Grisley:** v. GREALEY.

**Grissom, Grissin, Grisson:** John *Grisun* 1221 AssWa. OFr *grison* 'grey'.

**Grist:** John *Grystes* 1524 SRSf; Robert *Gryste* 1576 SRW; John *Grist* 1674 HTSf. OE *grist* 'grain to be ground', perhaps a nickname for a miller.

**Gristhwaite, Grisethwaite:** Lambert *de Gristhwait* 1327 SRY; William *Grysthwayte* 1417 IpmY; Richard *Grysethwayte* 1520 FrY. From Gristhwaite in Topcliffe (NRY).

**Gristock:** v. GRAYSTOKE.

**Griswold, Grisewold:** Roger *Grysewold'* 1381 AssWa; Thomas *Gresewold'* 1412 KB (Bk); Thomas *Greswold* 1418–19 FFWa. From Griswolds Fm in Snitterfield (Wa).

**Grithman:** Thomas *Grithman* 1279 AssNb; Paulinus *Grithman* 1297 SRY; John *Gritheman* 1332 SRCu. The name given to one who has taken sanctuary, OE *grið* 'peace', and *mann*.

**Gritton:** v. GIRTON.

**Grix:** v. GRIGG.

**Grocock, Grocott:** v. GREWCOCK.

**Groin:** Richard *Groin* 13th Guisb. 'Ugly nose', OFr *groin*. cf. Ralph *Groyneporck'* 1297 SRY 'pig's snout'.

**Gronow:** Kenwrig *ap Grono* 1391 Chirk. Welsh *Goronwy*, *Gronw*, *Gronow*, OW *Guorgonui*.

**Groocock:** v. GREWCOCK.

**Groom, Groome:** Richard *Grom* c1100 MedEA (Nf); Ernald *le Grom* 1187 P (Gl); Robert *Groum* 1327 SRWo; Roger *le Groom* 1351 AssEss. ME *grom(e)* 'serving-man, manservant', often applied to shepherds. cf. Richard *le Gotegrom* 1335 FFSf, John *Lambegrom* 1279 RH (C), John *Schepgrom* 1327 *SR* (Ess), Richard *Plougrom* 1319 *SR* (Ess).

**Groombridge, Grumbridge:** William *Grumbridge* 1560 StaplehurstPR (K); Abraham *Grumbridg* 1677 LewishamPR (K). From Groombridge (K).

**Groser:** John *Grocer* 1350 ColchCt. OFr *grossier* 'wholesale-dealer'.

**Grosker:** Robert *Groscuer* 1192, Richard *Grosquoer* 1197 P (O). 'Big, brave heart', OFr *gros coeur*.

**Gross, Grose, Groos, Gros, Le Gros:** Willelmus *filius grosse* 1086 InqEl (Sf); Gerardus *filius Grossi* 1176 P (Y); Willelmus *Grossus* 1086 DB (Nf); Adam *le Gros* 1186–1210 Holme (Nf); William *le Groos* 1314 FFEss. Lat *grossus* 'thick', OFr *gros* 'big, fat', used occasionally as a personal-name. v. CRASS, GRACE.

**Grosset:** Richard *Grosset* 1279 RH (O). A diminutive of GROSS.

**Grosvenor:** Robert *le gros Venour* c1200 WhC (La); Warin *le Grovenur* 1259 AssCh. Fr *gros veneur* 'chief huntsman'.

**Groucock, Groucutt:** v. GREWCOCK.

**Grout, Grut, Grute:** Edwin, Geoffrey *Grut* 1066 DB (Ess), 1199 P (L); Walter, Robert *Groute* 1297

MinAcctCo, 1447–8 FFSr. OE *grūt* 'groats, coarse meal', cf. ON *Grautr*, an original nickname from ON *grautr* 'porridge'. *v.* OEByn 376.

**Grote:** (i) William *Grote* 1279 RH (Beds); William *Grote* 1428 LLB K; John *Grot* 1496 Black (Caithness); Phyllypp *Grote*, Jerneseyman 1524 SRSf. Perhaps a nickname from ME *grot* 'a groat', or Dutch *groot* 'great'. (ii) Also Huguenot. Merchants from Antwerp of this name fled to England from the Spanish persecutions, whilst others of the name had earlier settled in England, e.g. the Flemish Ambrose, Peter *de Grote*, whose denization appears in 1510 LP. George *Grote*, the ancient historian, was the grandson of a *Grote* from Bremen (Smiles 322).

**Grove, Groves:** Ralph *de Graua* 1119–27 Colch (Ess); Osbert *de la Graua* 1197 FF (Bk); John *de la Grove* 1275 SRWo; Robert *ate Groue* 1317 AssK; William *Grove* 1327 SRDb. 'Dweller by the grove', OE *grǣf*.

**Grover:** William *Grouar* 1332 SRSx. 'Dweller by the grove.'

**Growcock, Growcott:** *v.* GREWCOCK

**Grubb, Grubbe:** Richard, John *Grubbe* 1176 P (Bk), 1203 P (Sx). The name was fairly common and is probably ME *grubbe* 'grub', used in the 15th century of a short, dwarfish individual.

**Grubber:** Robert *le Grubber* 1308 Wak (Y); William *le Grobber* 1332 SRSt; Henry *Grubur* 1375 AssL. ME *grubbere* 'digger'.

**Grugge, Grugges:** Godwin *Grugge* 1176 P (So); Arnulph *Grugge* 1275 RH (Do); Richard *Gruggs* 1621 SRY. ME *grucche, grugge* 'discontent, resentment'.

**Grumble, Grumell:** *v.* GRIMBLE

**Grumbridge:** *v.* GROOMBRIDGE

**Grumitt, Grummett:** *v.* GRIMMET

**Grundy:** Aicusa *filia Grundi* 1204 Pleas (Y); Robert *Grundy* 1296 Black; William *Grundy* 1642 PrGR; James *Grundy* 1674 HTSf. Probably a metathesized form of OG *Gundric*, OFr *Gondri*. *v.* GUNDREY.

**Grut(e):** *v.* GROUT

**Grylls:** *v.* GRILL

**Guard:** *v.* GARD

**Gubb:** Robert, Richard *Gubbe* 1296 SRSx, 1319 Bart (Lo). A short form of *Gubiun*. *v.* GUBBIN.

**Gubbin, Gubbins:** Hugo *Gubiun* 1130 P (Nth), 1256 AssNb; Richard *Gubun, Gubiun, Gybun, Gubyun* 1219 Cur; Hugo *Gobyon* 1242 Fees (Beds); Thomas *Gubin* 1279 RH (O). OFr *Giboin*, with *u* for *i* through the influence of the neighbouring sounds. *v.* GIBBEN.

**Gubell:** *Gubaldus* c1130 ELPN; *Willelmus Guboldus* 1192 P (L); Alice, Robert *Gubald* 1201 P (L), 1219 AssY. OE *\*Gūðbeald* 'battle-bold'.

**Gude:** *v.* GOOD

**Gudge:** *v.* GOOCH

**Gudgeon, Gudgin:** Peter *Guggun* 1206 Cur (L); Robert *Guiun* 1221 Cur (W); Henry *Gojun* 1332 SRSt; Thomas *Goodgeon* 1613 FrY. ME *gojon, gogen*, OFr *goujon* 'gudgeon', perhaps 'greedy', or 'big-headed', or one who will take any bait, 'credulous, gullible'.

**Gudger:** *v.* GOODGER

**Guest:** Benwoldus *Gest* c1100 OEByn; Richard, Thomas *le Gest* c1248 Bec (Mx), 1275 SRWo. ON *gestr* 'guest, stranger'.

**Guilbey:** *v.* GILBEY

**Guilder:** *v.* GILDER

**Guilding:** *v.* GELDING

**Guiler:** *v.* GILLER

**Guilford, Guildford, Gilford:** Peter *de Guildeford* 1275 RH (Lo); Robert *de Guldeford* 1307 LLB D; Thomas *Gyldeford* 1375 IpmW; Robert *Gylford* 1583 Musters (Sr). From Guildford (Sr).

**Guillerman:** *v.* GILLMAN

**Guiness, Guinness:** For MAGUINESS.

**Guise, Gise, Gyse:** Robert *de Guuiz* 1207 P (Gl); John *de Gyse* 1230 P (Berks); Thomas *Guise* 1663 HeMil. From Guise (Aisne).

**Gull:** Alured, Geoffrey *Gulle* 1200 FFEss, 1218 AssL; Richard *le Gul* 1279 RH (O). ME *gulle* 'gull' or ME *gull*, ON *gulr* 'pale, wan'.

**Gullick, Gulick:** *v.* GOODLAKE

**Gulling:** Roger *Gulling* 1203 P (Ha). ME *\*gulling* 'the pale one'. *v.* GULL.

**Gulliver, Gulliford, Galliford, Galliver:** William *Gulafra* 1086 DB (Sf); Philip *Golafre* 1166 RBE (Sf); William *Golefer* 1459 Cl (Mx); Henry *Gullifer*, *Gulliford* 1654, 1670 Bardsley. OFr *goulafre* 'glutton', a very common name.

**Gully, Gullyes, Golley:** John, Thomas *Gulias* 1202 AssL, 1229 Cl (So); Hugh *Golie* 1206 P (O); Jordan *Gulie* 1225 AssSo; Geoffrey *Golias* 1275 RH (Nf). ME *golias* 'a giant', from the Hebrew *Goliath*. Wyclif uses both *Goliath* and *Golie*; Chaucer and Shakespeare have *Golias*.

**Gulson, Gulston:** *v.* GOLDSTON

**Gumb, Gumm:** *v.* GOMME

**Gumbel:** Durandus *Gumboldus* 1148 Winton (Ha); William *Gumbald'* c1216 Seals (Sf). OFr *Gombaut*, OG *Gundbald* 'battle-bold'.

**Gumbley, Gumley:** Elias *de Gumundel'* 1199 MemR (Ess). From Gumley (Lei).

**Gumersall, Gumerson:** *v.* GOMERSALL

**Gummer, Gomer:** (i) *Gumer* 1091–1102 Rams (Hu); *Guthmarus* clericus c1170 Rams (Nf); Johannes *filius Gudemer'* 1219 AssL; Abraham *filius Gumari* 1219 Cur (C); Simon *Gumer* 1276 RH (Y). OE *\*Gūðmār* 'battle-famous'. (ii) *Godmarus* 1066 DB (Sf), 1255 Rams (Beds), 1279 RH (Hu); Willelmus *filius Gomer* 1191 P (C); Colin *Godmar* 1255 Rams (Hu); Robert *Gomar* 1279 RH (Hu); William *Godmer* 1332 SRWa. OE *Gōdmār* 'good-famous'.

**Gun, Gunn, Gunns:** *Gunne* 1142 NthCh (L); *Gun* Hy 2 DC (L); Warin, William *Gun* 1218 AssL, 1275 RH (Sf); Robert *Gunn* 1297 Wak (Y). ON *Gunnr* 'battle', or, perhaps, *Gunne*, a pet-form of *Gunnhildr*. *v.* GUNNETT.

**Gunby, Gunbie:** Oliver *de Gunby* 1194–1204 YCh; Thomas *of Goneby* 1268 FFY; John *Gunby* 1422 TestEbor; George *Gunbee* 1633 LeiAS 23. From Gunby (ERY), or Gunby St Nicholas, St Peter (L).

**Gundrey, Gundry, Goundry:** *Gundricus* 1100–13 Rams (Herts); William *Gundrey* 1296 SRSx. OG *Gundric* 'battle-ruler', OFr *Gondri*.

**Gundy:** William *Gundi* 1279 RH (Hu); John *Gondy* 1327 SRC. OG *Gaundi*.

**Gunfrey:** Ralph *filius Gunfridi* 1086 DB (Sx); Hugh *filius Gumfrei* 1179 P (Nth); Adam *Goumfrei* 1327

SRDb; John *Gounfrey* 1327 SRLei. OG *Gundfrid*, or Anglo-Scandinavian *\*Gunnfrøðr*.

**Gunhouse:** *v.* GUNNIS

**Gunnell:** *Gunnild* 1066 DB (Sx), c1172–80 DC (L); *Gunnilla* 1214, 1219 Cur (Sr, L); Robert *Gunnilt* Hy 2 DC (L); Stephen, John *Gunnild* 1240 Rams (Nf), 1327 SRC; Simon *Gunel* 1240 FFEss. ON *Gunnhildr* (f) 'battle-battle'. *v.* GUNNETT.

**Gunner:** (i) *Gunneuare*, *Gunnor* 1066 DB; *Gunware* Hy 2 DC (L), 1219 AssY; *Gunnora* 1207–8 Cur (So, Nb); William, Simon *Gunnore* 1275, 1279 RH (Sf, Bk); William *Gunwar* 1279 RH (C). ON *Gunvor*, ODa *Gunwor*, *Gunnor* (f), a common Norman name, usually latinized as *Gunnora*. (ii) Thomas *le gonner* 1285 PN Ess 21. ME *gonner* 'gunner' (1391 MED).

**Gunnett:** *Gunnota* 1279 RH (O); William *Gonnote* 1327 SRSf. *Gunnot*, a diminutive of *Gunne*, a pet-form of *Gunnhildr*. *Gunnilda*, daughter of Fulco, is also called *Gunnota* (1207 Cur).

**Gunning:** *v.* GUNWIN

**Gunnis, Gunhouse:** Herveus *de Gunnesse* 1202 AssL. From Gunness (Lincs).

**Gunson:** Eustace *Gunson* 1279 RH (C). 'Son of *Gunn*.' Henry *Gunneson* (1278 AssLa) is identified by the editor as Henry son of *Gunnild*.

**Gunston, Gunstone:** Bigod *de Goneston* c1184 StCh; Hugh *de Gonestone* 1320 AssSt; Henry *de Gunston'* 1332 SRSt. From Gunstone (St).

**Gunter, Gunther:** *Gunterus* 1094–1100 NthCh (Nth), 1214 Cur (Sr); *Guntier* filius Herberti 1165 P (Ha); William *Gunter* 1205 Cur (Berks), 1221 *ElyA* (Sf). OFr *Gontier*, OG *Gunter* 'battle-army'.

**Gunthorp, Gunthorpe, Guntrip:** Yvo *de Gunethorp* 1207 Cur (Nth); William *Gunthropp* 1623 Bardsley. From Gunthorpe (L, Nf, Nth, Nt, R).

**Gunton:** Bartholomew *de Guneton* 1195 P (Nf, Sf); Matthew *de Gunetun'* 1226 Cur (Nf); Henry *Gonton* 1309 Wak. From Gunton (Norfolk, Suffolk).

**Gunwin, Gunning:** *Gundewinus cortinarius* 1130 P (Sr); Thomas *Gundewin* 1228 Cl; William *Gundewyne* 1296 SRSx; Jo. *Guninge* 1674 HTSf. OG *Gund(e)win* 'battle-friend'. In Corton (Suffolk) in the reign of John, *Gundewyn'* de Nethergate held land which was held in 1275 by Gerald *Gunwine* or *Gundwyne* by heredity (RH).

**Gupill, Goupil:** *Gupill'* 1148 Winton (Ha); Guppyl *de Gelverton* 1276 RH (L); Godfrey *Gupill* 1200 Oseney; Alan *le Gopil* 1275 SRWo; Robert *le Gopyl* 1327 SRSf. A nickname from OFr *golpil*, *goupil* 'fox'. Also used as a personal name.

**Guppy:** Nicholas *Gopheye* 1327 SRSo; Robert *Guppey* 1392 LoPleas. From Guppy (Dorset).

**Gurden:** *v.* GORDON

**Gurdler:** *v.* GIRDLER

**Gurling:** *v.* CODLIN

**Gurnett:** Alexander, Robert *Gurnard* 1275 RH (D), 1327 SRSx. ME *gurnard*, *gurnade*, a fish with a large spiny head and mailed cheeks, with a throat almost as big as the rest of its body, so called because of the grunting noise it makes.

**Gurney:** Hugo *de Gurnai* 1086 DB (Ess, So); Adam *de Gurnay* 1196 P (Nth). From Gournaien-Brai (Seine-Inférieure).

**Gurtler:** *v.* GIRDLER

**Gurton:** *v.* GIRTON

**Guss, Gush:** *Gusa* de Butrewich Hy 2 DC (L); Arnulf filius *Gusse* 1200 Cur; John *Gusse* 1327 SRSo; Richard *Guse* 1363 AssY. OSw *Guse*, *Gusse*.

**Gut, Gutt:** Ædwin *Gut* 1188 P (Do/So); William *le Gut* 1198 P (Lo); Thomas *Gut* 1269 AssSo; John *Gutt'* 1338–9 CorLo. ME *gut* 'belly', a nickname for a greedy person. There was also a personal name: Ralph filius *Gutte* 1196 P (L), probably a short form of OE *Gūðhere*.

**Gutch, Guthrum:** *v.* GOOCH, GOODERHAM

**Gutman:** *v.* GODMAN

**Gutridge, Gutteridge:** *v.* CUTTERIDGE, GOODRICH

**Gutsell:** *v.* GODSAL

**Gutter:** (i) Ralph filius *Guttere* 1197 P (L); William *Guthere* 1203 P (Y); John *Guttere* 1311 ColchCt. OG *Gothari*, or OE *Gūðhere*. (ii) William atte *Gotere* 1316 AssNth; Margeria *Attegutere* 1381 SRSt. 'Dweller at the water channel or drain', OFr *gotier*. *v.* also GOTER.

**Guy, Guye, Guys, Gye, Why, Whye, Wye:** (i) Willelmus filius *Widonis*, *Guidonis* 1086 DB; Turstanus filius *Witdo*. ib.; *Why* de la Haie c1200 MELS (Sx); *Guido* de Hathfeld' 1200 Cur (Mx); Richard *Wi* 1188 P (O); William *Wy* 1297 MinAcctCo; John *Gy* 1317 AssK, 1319 SRLo; Richard *Guy* 1384 LoPleas. OFr *Guy*, OG *Wido*, a common French name, usually latinized as *Wido*, occasionally as *Guido*. *Why*, much rarer than *Guy* today, is the Norman form. (ii) John *le Gy* 1327 *SR* (Ess). OFr *gui* 'guide'.

**Guyan, Guyen:** Adam *de Gianne* 1214 Cur (So); Geoffrey *Gyen* 1273 IpmGl; Nicholas *Gyan* 1327 SRSo; Thomas *Gyen* 1545 SRW. 'The man from Guienne', AFr *Gienne*.

**Guyat, Guyatt:** *v.* WYATT

**Guyer, Gyer:** Henry *le Gyur* 1271 Ch (Do); Thomas *Gaiour* 1327 *SR* (Ess). OFr *guyour*, *guieor* 'guide'.

**Guymer, Gymer:** *Gyomarus* 1101–16 MedEA (Nf); Robertus filius *Guimer* 1204 P (Y); *Guiomarus* filius Warnerii 1210 P (Y); Robert *Guymer* 1277 Ely (Sf). OG *Wigmar* 'battle-famous'. *v.* also WYMER.

**Guyon:** *v.* WYON

**Guyot, Gyatt:** *v.* WYATT

**Gwilliam, Gwillim:** Hova, Jankyn ap *Gwillym* 1391 Chirk; Lewis *Gwillam* 1630 SaAS 2/iv. Welsh forms of *William*.

**Gwinn, Gwyn, Gwynn, Gwynne, Wyn, Wynne:** Thomas filius *Wᴧn* 1255 RH (Sa); *Wyn*, *Win* 1280 SaAS 2/xi (Ellesmere); *Gwyn* 1327 SRSa; *Gwynne* ap Griffud ap Tudur 1474 Chirk; William ap *Guyn*, David ap *Gwyn*, Philip *Wyn* 1327 SRSa; Thomas *Gwynne* 1481–2 FFSr. Welsh *gwyn* 'fair, white'. *v.* also WINN.

**Gye:** Roger *del Gye* 1327, *de la Gye* 1332 Kris. 'Dweller by the salt-water ditch', OFr *guie*. *v.* also GUY.

**Gyer:** *v.* GUYER

**Gyles:** *v.* GILES

**Gymer:** *v.* GUYMER

**Gynn:** *v.* GINN

**Gypp(s):** *v.* GIPP

**Gypson:** *v.* GIPSON

**Gyse:** *v.* GUISE

**Gyves:** *v.* JEEVES

# H

---

**Habberjam:** v. HABERGHAM, HABERSHON

**Habbeshaw, Habeshaw, Habishaw:** William *le Haubergier* 1201 P (Nth); Reginald *le Hauberger*, *le Haubeger* 1275 RH (Lo), *le Haberger* 1281 LLB A. OFr *haubergier* 'maker of hauberks or coats of mail' (1481 NED). cf. HARBISHER.

   **Habbijam:** v. HABERGHAM, HABERSHON

   **Haberer:** v. HARBERER

**Habergham, Habberjam, Habbijam:** Matthew *de Habercham* 1269 InqLa; Lawrence *Haberjam* 1551 FFLa; John *Habergam* 1574 LWills. From Habergham Eaves (Lancs). The modern pronunciation is *Habbergam* (DEPN), but was once clearly *Habberjam*, which has been confused with the later pronunciation of *Habershon*.

**Habershon, Habberjam, Habbijam:** Edmund *Habirgent* 1416 NorwW (Nf); Edward *Haberjon* 1565 ShefA; John *Haberdejohn* 1592 ArchC 48; William *Haberjamb* 1735 FrY. Metonymic for a maker of habergeons, sleeveless coats or jackets of mail or scale armour, worn also as a rough garment for penance. ME, OFr *haubergeon*, a diminutive of *hauberc*. cf. HABBESHAW. *Habershon* and *Habbijam* are both still found in Sheffield, the latter being a direct descendant of the 1565 *Haberjon*. cf. HABERGHAM.

**Habeshaw, Habishaw:** v. HABBESHAW

**Habgood, Hapgood:** John *Hauegod* 1280 KB (So); John *Habbegod* 1343 IpmGl; John *Abgod* 1545 SRW. 'May he have good', OE *habban*, *gōd*. cf. John *Havejoy* 1523 SRK 'may he have joy'; John *Havelove* 1259 Acc 'may he have love'.

**Hack:** v. HAKE

**Hackbon, Hackbone:** Ralph *Hackebon* 1277 IpmGl. 'Hack bone', ME *hakken*, OE *bān*. Perhaps a nickname for a butcher. cf. Ærnyeth *Hachchebutere* 1185 P (Gl) 'hack butter'; Geoffrey *Hakkeches'* 1227 AssBeds 'hack cheese', nicknames for dealers in these.

**Hacker:** Adam *le Hakkere* 1262 MEOT (Herts); John *Hakyere* 1296 SRSx. A derivative of ME *hacken* 'to hack', one who hacks, a cutter, probably a wood-cutter. Or, perhaps, 'a maker of *hacks*', used in ME of agricultural tools such as mattocks and hoes.

**Hackett, Haggett, Haggitt, Acket, Acketts:** *Haket de Ridefort* c1160–66 DC (L); *Haket* filius Clac 1193 FF (L); Walter *Achet* 1086 DB (Bk); Ralph *Hacget* 1131 FeuDu; Rolland' *Haget*, *Haket*, 1158, 1179 P (Y, L); Geoffrey *Haget*, *Hachet* 1191 P (Y). An AN diminutive of ON *Haki*, OSw *Haki*, ODa *Hake*. Occasionally we may have a nickname from a fish: John *Hakede* 1327 SRSf, Roger *Hakat* 1327 SRC, from *hacaed* (*haket*) 'a kind of fish' mentioned in a 14th-century copy of the foundation charter of the Abbey of Ramsey.

**Hackford:** Robert *de hacforð* 1196 FFNf; Walter *de Hakeford* 1262 HPD; John *Hakfurth* 1472 FrY. From Hackford (Nf), or Hackforth (NRY).

**Hacking:** William *de Hakkyng* c1283 WhC (La). From Hacking (Lancs).

**Hackling, Haclin, Ackling:** *Hakelinus* 1192 P (Lo); *Hakelinus* filius Jurneth 1199 MemR (Sf). A double diminutive of ON *Haki, Hak-el-in*. cf. HACKETT.

**Hackman:** William *Hakeman* 1197 P (C); William *Hacman* 1204 Cur (D). 'Servant of *Hake*.'

**Hackney, Hakeney, Hakney:** (i) Benedict *de Hakeneye* 1275 RH (Lo). From Hackney (Middlesex). (ii) Adam called *Hakenay* 1316 Black; John *Hakeney* 1327 *SR* (Ess), 1329 ColchCt; William Mopps *Hakeneyman* 1327 Pinchbeck (Sf). OFr *haquenée*, ME *hakenei* 'an ambling horse or mare, especially for ladies to ride on'. cf. PALFREY and PALFREYMAN.

**Hackshall, Hackshaw:** (i) Walter *Hakesalt* 1212 P (Ha); Roger *Hackesalt* 1297 MinAcctCo; Edward *Hacsalt* 1387 AssL. 'Hack salt', ME *hakken*, OE *sealt*. A nickname for a salt-worker. (ii) John *de Hacunsho* 1264 Bardsley (La); John *de Hacschawe* 1379 PrGR. From Hackinsall (La). (iii) Roger *de Hakkesalt* 1323 PN Herts 28. From Haxters End in Great Berkhamstead (Herts).

**Hacksmall:** William *Hackesmal* e13th Reg-Antiquiss; Thomas *Hacsmal* 1301 SRY; Richard *Hacksmal* 1327 SRLei. 'Hack small', ME *hakken*, OE *smal*. cf. Hugh *Hakepetit* 1202 P (So), with the same meaning; Hugh *Haccemus* 1148 Winton (Ha) 'hack mouse'.

**Hackwood:** John *Hackewude* 1230 P (Sx); Thomas *Hackewode* 1327 SRSx. 'Hack wood', a nickname for a wood-cutter.

**Haclin:** v. HACKLING

**Hacon:** *Hacun, Hacon* 1066 DB; *Hacon* de Crokestun c1160 DC (L); Robert *Hacun* 1221 AssSa; Semann' *Hacon* 1275 RH (Sf). ON *Hákun* 'high race'.

**Hadaway:** v. HATHAWAY

**Hadcock:** v. ADCOCK

**Haddelsey, Haddlesey, Haddesley:** Thomas *Hadilsey* 1402 IpmY; Edward *Haddesly*, Robert *Hadgsly* 1662 HTEss. From Chapel, West Haddlesey (WRY).

**Hadden, Haddin, Haddon:** (i) Ailwin *de Haddun'* 1159 P; Philip *de Haddon* 1267 AssSo; John *de Hadden* 1323 IpmNt; Thomas *Haddun* 1379 PTY. From

Haddon (Derby, Dorset, Northants, Hunts), or Haddon in Shute (Devon). (ii) Ulkill, Bernard *de Hauden* 1165–71, 1296 Black (Roxburgh); William *Haulden* 1328 ib.; Silvester *Hadden* 1514 ib. From the barony of Hadden or Halden (Roxburgh).

**Haddington:** (i) Walter *de Hadyngton* 12th FeuDu; John *de Hadyngton* 1303 RegAntiquiss; Richard *de Hadyngton* 1351 AssL. From Haddington (L). (ii) In Scotland from Haddington (East Lothian).

**Haddlesey:** v. HADDELSEY

**Haddock, Haddacks, Haddick:** Richard, William *Haddoc* 1208 ChR (Sf), 1228 FeuDu; John *Haddock'* 1302 SRY. *Haddock* is the local pronunciation of *Haydock*, but this will not account for these early forms. We may have a nickname from the fish, first recorded in 1307 (MED), not inappropriate in Suffolk for a seller of haddocks. The Durham name is probably from a personal name, a diminutive of OE *Æddi* (Bede), with inorganic *H.* cf. *Addoch* de Eselinton 1187 P (Nb); *Addoc* c1220 FeuDu, William *Addoc* 1243 AssDu.

**Hadenham:** v. HADMAN

**Hadfield:** Wido *de Hadfeld'* 1190 P (Ess); Matthew *de Hadefeld* 1288 AssCh; Richard *Hadefeld* 1401 AssLa. From Hadfield (Derby).

**Hadgkiss:** v. HODGKIN

**Hadham:** Peter *de Hadeham* 1238–9 FFEss; John *de Hadham* 1324–5, Thomas *de Hadham* 1340 CorLo. From Much, Little Hadham (Herts).

**Hadkins:** v. ADKIN

**Hadley, Hadlee:** Matilda *de Hadlega* 1194 P (Sf); Warin *de Hadlai* 1212 P (Y); Richard *de Hadlege* 1311–12 FrC; John *Hadley* 1390 FFEss. From Hadleigh (Essex, Herts, Salop, Suffolk), or Hadley (Worcs).

**Hadlow:** Nicholas *of Hadlow* 1255 FFY; Nicholas *de Hadlo* 1275 RH (K); John *Hadlowe* 1464 FrY. From Hadlow (K), or Hadlow in Skircoat (WRY).

**Hadman:** John *Hadenham, Hadman,* William *Hadnam* 1524 SRSf. From Hadenham (Bucks): Geoffrey *de Hadenham* 1235 Fees (Bk).

**Hadridge:** John *Haderich* 1275 SRWo; John *Hadridge* 1642 PrD. OE *Æðelrīc.*

**Hafter:** Hugh, Robert *le Haftere* Hy 3 AD ii (Mx), 13th Lewes (Nf). A derivative of OE *hæft* 'handle', a maker of handles for tools (1598 NED).

**Hagan, Hagen, Hagon, Hain, Haine, Haines, Hains, Hayne, Haynes, Heyne, Heynes, Hanes:** *Hagene, Hagana* 1066 DB (He, Nf); *Hagena* Jugement 1130 P (Sf); Rogerus *filius Hane* 1198 FFNf; *Hagan(us)* 1199 FFSt, 1240 Rams (Nf); Alicia *filia Hahen* 1202 FFNf; *Heyne* le Forester 1274 Wak (Y); Adam *filius Hayne* 1332 SRLa; Ulricus *Hagana* 1066 DB (Sf); Peter *Hain* 1200 P (Do); Amfridus *Hane* 1209 P (St); Walter *Hein* 1279 RH (C); John *Hagun* ib. (O); Richard *Hagyn* 1304 LoCt; Alice *Heynes* 1327 SRSo; Margery *Haynes* 1352 *ERO*; Robert *Haynesson* alias *Hayn* 1398 FrY. ODa *Haghni,* OSw *Hagne,* or OG *Hagano, Hageno.* v. also HAIN. *Hagan* may also be for *O'Hagan.*

**Hager, Hagger:** Thomas *Hager* 1221 AssSa; Robert *hagger'* 1362 AssY; Henry *Hagger* 1544 FFEss. A derivative of ME *haggen* 'to cut, chop'. Metonymic for a woodcutter.

**Hagerty, Haggerty:** (i) John *Hagerday,* . . . *Haguday* 1674 HTSf; Matthew *Huggoday* 1677 SfPR. ME *haggaday* 'a kind of door latch'. Rings 'with lacches and kacches' (1363), forming the handle by which the latch was raised, were known as 'haggadays', and the word is still in use in the north for a thumb-latch. The entry in King's Hall accounts for 1414: 4s.3d. Pro . . . Hafgooddays . . ., suggests that the popular, and possibly correct, etymology of the name for this type of latch was from the words of greeting spoken when the door was opened (Building 299, 300). (ii) In Ireland, a variant of *O'Hegarty,* from *Ó hÉigceartaigh,* from *éigceartach* 'unjust'.

**Hagg:** Roger *filius Hagge* 1166 P (Nf); Sæwi *Hagg* c1100 OEByn (So); Henry *Hagge (Agge)* 1230 P (So); William *Hag* 1379 PTY. ON *Hagi.* v. OEByn 316.

**Haggard, Haggart, Hagard:** Alice *Haggard* 1275 SRWo. ME, OFr *hagard* 'wild, untamed'.

**Haggas, Haggis, Haggish:** Richard *del Haghous* 1327 SRY; Gilbert *of Haggehouse* 1394 Black; John *Hagas* 1401 FrY; William *Haggus* 1427 Black; Richard *Haggas* 1486 GildY; Christopher *Haghus* 1532 LWills; John, Richard *Haggis* 1547 Black, 1637 SfPR. NED has *hag-house* (1733) '? a place for storing firewood'. *Hag* is a northern and Scottish word from ON *hogg* 'a cutting blow, hewing of trees', common in northern place-names. A *hag-house* was probably a wood-cutter's hut and the surname one for a wood-cutter. *Haggis* is a common place-name in the Scottish Lowlands.

**Hagger:** v. HAGER

**Haggerty:** v. HAGERTY

**Hagley:** Robert *de Haggel'* 1221 AssWo; William *Hagley* 1642 PrD. From Hagley (Wo).

**Haggett, Haggitt:** v. HACKETT

**Hagtharp:** Robert *de Hakethorp* 1251 AssY. From Hackthorpe (Westmorland).

**Haig:** Peter *del Hage* 1162–6 Black, *de la Haga* 1162–90 ANF. The Haigs of Bemersyde derive from the district of La Hague (La Manche). v. ANF.

**Haig, Haigh, Hague:** Jollan *de Hagh'* 1229 Cl (Y); William *del Haghe* 1327 SRY. From Haigh (Lancs, WRYorks) or 'dweller by the enclosure' (OE *haga* or ON *hagi).*

**Haighton:** Hugh *Haighton* 1642 PrGR. From Haighton (La).

**Hail, Haile:** Ralph *Hayl* 1260 AssC; Robert *Heyle* 1327 SRSf; William *Hayl* 1360 IpmGl. ON *heill* 'healthy, sound'.

**Haile, Haill, Hails, Hayle, Hayles:** (i) *Haile(s)* and *Hayle(s)* are Scottish and Northern English forms of HALE(S). cf. *Haile* (Cu): *Hale* c1180 and William *de Hales* 1349 FrY; William *Hayles* 1456 ib. The Scots *Hailes* derives from Hailes in Colinton (Midlothian, East Lothian): William *de Halis* 1189–98 Black; John *Hayles* 1408 ib. (ii) William, Richard *de Hayles* 1273, 1279 RH (Wa, O). From Hailes (Glos), distinct in origin from HALE. v. also HAYLOCK.

**Hailey, Haily, Haley, Haly, Hayley:** William *de Hayleg'* 1251–2 FFWa; Roger *de Hayleye* 1328 IpmGl; Thomas *Haley* 1420–1 IpmNt. From Hailey (Oxon), or 'dweller at the hay clearing'.

**Hailing, Hayling:** (i) Robert, Philip *filius Heilin* 1255 RH (Sa). OW *heilyn* 'cup-bearer'. Used also as a personal name. (ii) Robert *de Hayllyng* 1296 SRSx. From North, South Hayling (Ha).

**Hailsham:** Peter *de Haylesham* 1255–6 FFSx. From Hailsham (Sx).

**Haim, Haimes, Haymes, Hamon:** *Hamo, Haimo, Hamon* 1086 DB; *Hamon* de Hauton c1135 DC (L); *Haim* de Branzton' 1202 FFL; *Heimes* de Henford' 1203 Cur (Nf); Richard, John *Haym* 1279 RH (O), 1327 SRSo; John *Hamon'* 1279 RH (C); Thomas *Hamun* 1296 SRSx. OFr *Haim, Haimon, Aymes,* OG *Haimo* 'home', a popular Norman name. *Hamo* is the OFr cas-sujet, *Hamon* the cas-régime. *v.* HAMMOND.

**Hain, Haine, Hayne, Heyne:** (i) William *de Heghen* 1289 PN D 544; William *atte Heyene* 1327 SRSo; Martin *de la Heghen* 1333 PN D 129. *Hayes* and *Hayne,* respectively strong and weak plurals of *hay,* are common place-names in Devon, particularly in the south-east, and are found elsewhere in the west country. *v.* PN D 129–30. 'Dweller on the farm or holding.' cf. HAY. (ii) William *le Heyne* 1327 SRSt. ME *heyne, haine, hayn* 'a mean wretch, niggard' (c1386 NED).

**Hain(e)s:** *v.* HAGAN

**Hainworth, Hainsworth:** From Hainworth in Keighley (WRY).

**Hainton:** Hugh *de Hainton'* 1219 AssY. From Hainton (L).

**Hair, Haire, Hare, O'Haire, O'Hare, O'Hear:** William *Hare* 1366 Black (Edinburgh); Patrick *Ahayr(e), Ahaire, Ahar, Hayre, Hair, Hare* 1415–38 ib. (Ayr). Ir *Ó hÍr* 'descendant of *Ír*' (sharp, angry). *v.* also AYER, HARE.

**Hairon:** *v.* HERON

**Haizelden:** *v.* HAZELDEN

**Hake, Hakes, Hack:** Turkil *Hako* 1066 DB (Nf); Leuiua *filia Hacke* 1218 AssL; Gilbert *Hake* 1257 FFSf; Robert *Hakke* 1375 Lewes (Nf). ON *Haki,* ODa *Hake* 'hook, crook', an original nickname.

**Hakeney:** *v.* HACKNEY

**Hakluyt:** Walter *Hakelutel* 1255 RH (Sa); John *Hackelut* 1323 AssSt. 'Hack little', a nickname for a lazy wood-cutter.

**Haldanby, Haldenby:** Gundr' *de Haldeneby* 1208 FFY; Richard *de Haldenby* 1298 AssL; Robert *de Haldanby* 1379 PTY. From Haldenby (WRY).

**Haldane, Halden, Haldin, Hallding, Holdane:** *Haldein, Haldanus, Halden* 1066 DB; *Haldeynus* 1109–31 Miller (C); *Halden* filius Eadulf a1124 Black (Glasgow); *Haldane* de Hemingburgh 1208 AssY; Goduinus *Haldein, Halden* 1066, 1086 DB (Nf); Robert *Haldein* 1170 P (Ess); Roger *Haldane* 1255 Black; William *Halden* 1286 AssCh; Simon *Holdeyn* 1358 FFSf. ON *Halfdanr,* ODa *Half(f)dan,* Anglo-Scand *Healfdene* 'half Dane'.

**Halden, Haldon:** John *atte Haldon* 1317 AssK. From High Halden (K). *v.* also HALDANE.

**Haldenby:** *v.* HALDANBY

**Haldon:** *v.* HALDEN

**Hale, Hales:** William *de Hales* 1180 P (Sa); Robert *Attehal'* 1212 Cur (L); Morus *de la Hale* 1214 Cur (K); John *del Hale* 1220 Cur (Herts); Nicholas *en la Hale* 1275 SRWo; John *in the Hale* 1296 SRSx. From residence in a nook, recess or remote valley. OE *halh* (nom.) becomes *Haugh, Hauff, Hallowes.* The dative *(atte) hale* becomes *Hale. v.* also HAILE, HEAL.

**Haler, Hayler, Haylor, Hayllar:** William *le Haliere* 1279 RH (O); John *Haler* 1373 ColchCt; Thomas *Hayler,* Elizabeth *Haler* 1555, 1556 SxWills. Perhaps a derivative of ME *halien* 'to hale, haul, drag', from OFr *haler* 'to pull', a carrier, porter (1479 NED); also, probably equivalent to *atte Hale* 'one who dwelt in a retired spot', etc. *v.* HALE.

**Halesworth, Hailsworth:** John *de Halesworth* 1327 SRSf. From Halesworth (Sf).

**Haley:** *v.* HAILEY

**Halfacre, Halfacree:** Henry *Halfacr'* 1297 MinAcctCo; Alexander *atte Helfakere* 1333 MELS (So). From Halfacre in Northill (Cornwall), or one who lived on a homestead consisting of a half-acre.

**Halfhead, Halfhide, Halfhyde:** Robert *de Halvehid'* 1212 Cur (Herts); Roger *atte Haluehyde* 1332 MELS (Sr); John *de Halfhyde* 1332 SRSt; William *Halfhed* 1536 Eynsham (O). From Half Hyde (Herts), Half Hides (Essex) or from residence on a homestead consisting of half a hide (OE *healf, hīd*).

**Halfnight, Halfnights:** William *Halfchniet, Halfchnigt* 1169–70 P (Gl); Asce *Halvecniht,* Ace *Dimichevaler* 1213 Cur (Ha); Thomas *le Halveknight* 1327 AD ii (Ha). OE *healf* and *cniht* 'a half-knight', probably one who held half a knight's fee, one who held his land by service of paying half the cost of providing a knight or armed horseman in the army for forty days. The surname may also be a derogatory nickname for one who was ony 'half a knight'. Such compounds were not uncommon: Godfrey *le Halveloverd, -laverd* 1207 Cur (Ess), 'half-lord', a lord to whom only half a man's service was due. cf. also Simon *Halvedievel* 1212 Cur (Mx), William *Halfhund* 1203 P (Ha) 'half-hound', Michael and Roger *Halfplow* 1327 SRSf who probably each owned a half-share in the plough.

**Halford:** Thomas *de Haleford* 1200 P (K); Robert *de Halford* 1327 SRLei; Robert *Halford* 1500–1 FFWa. From Halford (Devon, Salop, Warwicks), or the lost Haleford in Kent (PN K 187).

**Halfpenny, Halpenny:** Adam *Halpeni* 1275 SRWo; Richard *Halfpany* 1296 SRSx. Halfpenny Field in Elm (Cambs) was so called because the tenants had to contribute a half-penny for each acre to the repair of the neighbouring Needham Dyke (PN C 268). Some such custom or a similar rent may have given rise to the name.

**Halfrankish:** Richard *Halffrenkise* 1248 AssBerks. 'Half French', OE *healf, frœncisc.* cf. Adam *le Halfnaked* 1324 KB (He) 'half naked'; John *Halfpound* 1335 LLB E 'half pound'; Richard *Halfsnode* 1247–8 FFK 'holder of a portion of woodland'; Robert *Alfwytlof* 1296 SRSx 'half a white loaf'.

**Halfyard:** Stephen *atte Halvezerd* 1327 SRSo. OE *healf* 'half' and *gierd* 'yard', one who lived on a homestead consisting of half a yard-land. In Great Waltham (Essex) the virgate or yard-land of 30 acres was called a *yerd* and divided into halves and quarters. cf. *Camyshalfeyerde* (1497 PN Ess 580).

**Haliburton, Halliburton:** Henry *de Haliburtoune* c1200, Philip *de Haliburton* c1260 Black; Henry *Haliburton'* 1306 FFY. From the lands of Haliburton (Berwick).

**Haliday, Halladay, Halladey, Halleday, Halliday, Hallidie, Holiday, Holyday, Holladay, Holliday:** Reginald *Halidei* 1179–94 Seals (Beds); Suein *Halidai* 1188 P (Nt); Thomas *Holidaie* 1524 SRSf; John *Halladay* 1666 FrY; John *Holliday*, Robert *Holladay* 1674 HTSf. OE *hāligdæg* 'holy-day, a religious festival', a name frequently given to one born on such a day. cf. CHRISTMAS.

**Halifax, Hallifax:** Jordan *de Halifax* 1297 Wak (Y); William *de Halifax* 1382 FrY; William *Halyfax* 1454 TestEbor. From Halifax (WRY).

**Halifield, Hallifield:** Gilbert *de Halifeld* 1178 CartAntiq; John *de Halifeld* 1316–17 FFEss. 'Dweller by the holy field', OE *hālig, feld*, perhaps land belonging to the church.

**Halksworth:** *v.* HAWKESWORTH

**Hall, Halle, Halls:** Warin *de Halla* 1178 P (Ess); Robert *de la Hall'* 1199 MemR (Ha); Alan *atte Halle* 1296 SRSx; Roger *de Hall* 1327 SRDb; Richard *in the Halle* 1332 SRWa. 'Worker at the hall' (OE *heall*). *v.* HALLMAN.

**Hallday, Halladey:** *v.* HALIDAY

**Hallam, Hallum, Hallums:** (i) Adam *de Hallum* 1297 SRY; John *de Hallum* 1328 WhC (La). From Hallam in Sheffield (WRY), *Halhm* DB. (ii) Richard *de Halom* 1327 SRDb; Henry *de Halom* 1392 FrY. From Kirk or West Hallam (Db), *Halun* DB. As both these are from dative plurals, respectively 'at the rocks or slopes' and 'at the nooks', *Hallums* must have a different origin, from Hallams Fm in Wonersh (Sr), *Hullehammes* 1418, where in 1327 lived John *atte Hulhamme*, by the enclosure(s) near the hill (PN Sr 254).

**Hallard:** *v.* ADLARD, HALLWARD

**Hallas:** *v.* HALLOWES

**Hallaway:** *v.* ALAWAY

**Hallawell:** *v.* HALLIWELL

**Hallbarn:** Nicholas *Hallebern* 1195 P (L); Juliana *Hallebarn'* 1219, John *Halbarne* 1361 AssY. ON *Halbiorn*.

**Halfding:** *v.* HALDANE

**Halleday:** *v.* HALIDAY

**Haller:** Robert *Haller* 1332 SRSr. For *atte Hall*. *v.* HALL.

**Hallett:** *v.* ADLARD

**Halley, Hally, Hawley:** Robert *de Hallai* 1166 RBE (Y); John *Hally* 1230 P (Db); Christopher *Halley* 1454–5 CtH. From Hawley (Ha), or 'dweller at the clearing with a hall', OE *heall, lēah*.

**Halleybone, Hallybone:** Stephen *de Haliburn'* 1222 Cur (Ha). From Holybourne (Hants).

**Hallfield:** Robert *Hallefeud* 1296 SRSx. 'Dweller by the field near the hall.'

**Hallgarth:** Henry *de Hallegarth* 1298 AssL. From Hallgarth (Du, ERY), or 'dweller by the hall enclosure', OE *heall, ON garðr*.

**Hallgate:** William *ate Hallegate* 1297 MinAcctCo; Stephen *atte Halleyate* 1351 AssSt; William *Hallegate* 1408 IpmY. 'Dweller by the gate or road to the hall', OE *heall, geat*/ON *gata*.

**Halliburton:** *v.* HALIBURTON

**Halliday:** *v.* HALIDAY

**Hallifax:** *v.* HALIFAX

**Hallifield:** *v.* HALIFIELD

**Halline:** William *le Hallehyne* 1332 SRSt. 'Servant at the hall.' *v.* HINE.

**Halling, Hallings, Hawling:** Carlo *de Halling'* 1195 P (K); Ralph *de Halling'* 1206 Cur (Sr); Robert *de Halling* 1288 FFEss. From Halling (K), or Hawling (Gl), *Hallinge* DB.

**Hallis:** *v.* ALIS

**Hallison:** *v.* ALISON

**Halliwell, Hallawell, Hallewell, Hallowell, Holliwell, Hollowell:** Osbert *de Haliwell'* 1200 P (Sf); Robert *Halwewoll*, Martin *de Halgewelle* 1275 RH (Do, D): Editha *atte Holywelle* 1327 SRSo. From Halliwell (Lancs), Holwell (Dorset, Oxon), Holywell (Hunts, Northumb), or 'dweller by a holy spring'.

**Hallman, Halman:** (i) Æluric *Halleman* c1095 Bury (Sf); Gilbert *le Halleman* 1301 NottBR. 'Servant at the hall.' (ii) William, John *Haleman* 1327 *SR* (Ess), 1379 AssEss. 'Dweller in the *hale*.' *v.* HALL, HALE.

**Hallmark:** *v.* ALLMARK

**Halloway:** *v.* ALAWAY

**Hallowbread, Hollowbread:** William, John *Halibred* 1309 SRBeds, .1317 AssK; — *Hollibred*, Widdow *Hollowbred* 1674 HTSf. 'Holy bread.' cf. Richard *le Halywaterclerc* 1285 *Ass* (Ess).

**Hallowell:** *v.* HALLIWELL

**Hallowes, Hallows, Hallas:** William *in le Halowe* 1276 RH (C); Simon *del Halghe* 1337 FrY; Thomas *Hallowes* 1446 ib. 'Dweller in the *halh* or deep valley.' *v.* HALE.

**Hallsworth, Hallworth:** *v.* HOLDSWORTH

**Hallum(s):** *v.* HALLAM

**Hallward, Hallard:** Ricardus *Halwardus* 1185 Templars (Wa); Gervase *le Halleward* 1250 FFEss. 'Keeper of the hall', OE *heall* and *weard*.

**Hallway:** *v.* ALAWAY

**Hally:** *v.* HALLEY

**Hallybone:** *v.* HALLEYBONE

**Halman:** *v.* HALLMAN

**Halpenny:** *v.* HALFPENNY

**Halsall:** Simon *de Haleshal* 1225 AssLa; Richard *de Hallesale* 1335 FrY; Richard *of Halsall* 1401 AssLa. From Halsall (La).

**Halse:** (i) Alfric *Hals* c1100 OEByn; Robert *Hals* 1182 P (D), 1185 Templars (L). OE *heals* 'neck', a nickname. cf. SHORTHOSE. (ii) Sampson *Attehalse*, *Athalse* 1251 AssSo. 'Dweller at the neck of land' (OE *heals*), as at Halse (Devon, Som).

**Halstead, Halsted, Alstead:** (i) Ralph *de Halsteda* 1181 P (Sf); Henry *de Alsted* 1274 RH (Ess). From Halstead (Essex, Kent, Leics) or Hawstead (Suffolk). (ii) Willyam *del Hallested*, *del Hallstudes* 1340 WhC (La). 'Worker at the hall-buildings', horseman, cowman, etc.

**Halston:** (i) Thomas *Halstein* 1227 ClR (C); John *Halsteyn* 1279 RH (C); William *Halsteyn* 1327 SRC. ON *Halsteinn*. (ii) John *de Halston* 1332 SRWo. From Halston (Sa).

**Halt:** Leowerd *Healte* 1051–71 OEByn (D); Richard *le Halt* 1205 P (Ha); John *le Halte* 1310 EAS xx; Ivo *Halte* 1332 SRDo. OE *healt* 'limping, lame'.

**Halter:** Robert *le Haltrere* 1296 LLB B. A derivative of OE *hælftre* 'halter', maker or seller of halters. In 1301 Henry *le Haltrehere* was owed 62s. by the Commonalty of London for halters for horses (LLB C).

**Halton:** Algar *de Haltona* 1084 GeldR; Walter, Richard *de Halton* 1270 FFL, 1332 SRCu; Henry *Halton* 1407 AssLo. From Halton (Bucks, Ches, Lancs, Lincs, Northumb, Salop, WRYorks).

**Haly:** *v.* HAILEY

**Ham, Hamm, Hamme, Hams:** Robert *de la Hamme* 1275 RH (Sx); Robert *atte Hamme* 1296 SRSx; Henry *Ham*, William *Hamme* 1279 RH (C, O); Hugh *in the Hamme* 1327 SRSo. 'Dweller on the flat, low-lying land by a stream' (OE *hamm*).

**Hambleden, Hambledon, Hambleton, Hambelton:** Godfrey *de Hameleden'* 1185 P (Bk); William *de Hameldon'* 1227 Cur (Berks); Thomas *de Hameldone* 1367 IpmNt. From Hambleton (La, WRY), Upper Hambleton (R), Hambleton Hill, Black Hambleton (NRY); Hambleden (Bk), Hambledon (Ha, Sr), Hambledon Hill (Do), Great Hameldon (La), or Hamilton (Lei), *Hameldon* 1220–35.

**Hamblet:** *v.* HAMLET

**Hambleton:** *v.* HAMBLEDEN

**Hamblett:** *v.* HAMLET

**Hambley, Hambly, Hamley:** Henry *de Hambelegh* 1296 SRSx; John *de Hamelee* 1305 LLB B; Osbert *Hameley* 1361 FFW. From Hamly Bridge in Chiddingly (Sx).

**Hamblin(g):** *v.* HAMLIN

**Hambro, Hamburgh:** Richard *de Hambir'*, Robert *de Haneberg'* 1279, 1281 Eynsham (O). From Handborough (Oxon).

**Hambrook, Hambrooke:** Ralph *Hambrook* 1296 SRSx. From Hambrook (Gl), or Hambrook in Chidham (Sx).

**Hamby, Hanby:** William *de Hambi* c1150–80 YCh; Jollan *de Hanby* c1240, *de Hamby* 1242 Fees (L); Geoffrey *de Hanby* 1355 FFY. From Hanby (L), *Hambi* 1212.

**Hambury:** *v.* HANBURY

**Hamer:** (i) John *of the Hamore* 1401 AssLa. From Hamer (La). (ii) Richard *Hamer* 1296 SRSx. Perhaps metonymic for a maker or seller of hammers, OE *hamor*. Sometimes from a shop sign: An Brewer *atte Hamere* 1426 LondEng 191.

**Hames:** Thomas *del Hames* 1332 SRCu. From Hames Hall in Papcastle (Cumb).

**Hamill:** *v.* HAMMEL

**Hamilton:** Richard *de Hamelton'* 1195 P (Nf/Sf); Gilbert *de Hameldun* 1272 Black (Paisley); William *de Hamil'ton* 1327 SRLei; David *de Hamilton* 1378 Black. From Hamilton (Lei).

**Hamlet, Hamlett, Hamblet, Hamblett:** *Ham(b)lett* Asshelos 1622, 1642 PrGR; *Hamlett* Clarke 1651 ERO; Anne *Hamlett* 1568 SRSx. *Ham-el-et*, a double diminutive of *Hamo*. *v.* HAIM.

**Hamley:** *v.* HAMBLEY

**Hamlin, Hamlen, Hamlyn, Hamblen, Hamblin, Hambling:** Robertus *filius Hamelin* 1130 P (D); *Hamelin* 1166 RBE (Y), 1206 Cur (L); Thomas, Walter *Hamelin* c1230 Barnwell (C), 1243 AssSo; Thomas *Hamblyn*, Bartholomew *Hamling* 1568 SRSf. *Haimelin, Hamelin*, a diminutive of OG *Haimo*.

**Hamm(e):** *v.* HAM

**Hammel, Hamill:** Aldan *Hamal* c1055 OEByn; Gregory *Hamel* c1170 Riev (Y). OE *hamel* 'scarred, mutilated'.

**Hammer:** John *le Hammer'* 1332 SRSx. 'Dweller in the *hamme*.' *v.* HAM, HAMMING. Or metonymic for a maker or user of hammers, perhaps a hammer-smith. cf. Geoffrey *wythe Hameres* 1303 Pat.

**Hammerton:** Fulk *de Hammertona* 1142–c1154 MCh; John *de Hamerton* 1255 IpmY; John *de Hamerton* 1379 PTY. From Hamerton (Hu), Hammerton (WRY), or Green, Kirk Hammerton (WRY).

**Hammett:** William *Hamet* 1297 MinAcctCo. *Hamet*, a diminutive of OG *Hamo*.

**Hamming:** Richard *Hammyg* 1275 SRWo; James *Hammyng* 1327 SRSx; Robert *Hommyng* 1327 SRSo. OE *\*hamming* 'dweller in the *hamme*'. *v.* HAM. The surname appears side-by-side with *atte hamme*.

**Hammond, Hammand, Hammant, Hamman, Hammon:** (i) *Hamandus Cocus* 1140–5 Holme (Nf); *Hamo, Hamandus elemosinarius* c1140 DC (L); Walter, Adam *Hamund* 1242 Fees (He), 1280 AssSo; Richard *Hamond, Hamon* 1327, 1332 SRSx. OFr *Hamond*, OG *Hamon* with excrescent *d*. *v.* HAIM. (ii) *Haimund'* Peccatum 1121–48 Bury (Sf); *Hamo* Pecce 1156–60 ib.; Willelmus *filius Haymundi* 1221 Cur (Sr); Roger *Haymund* 1275 RH (Ha); John *Hemund* ib. (K); William *Heymund, Heymund* 1327 SRSf, 1346 FA (Sf). *Haimund*, from OG *Haimon*. This probably survives also as *Hayman*.

**Hamnett:** (i) *Hamonet* de Waltham 1180–99 P (Berks); *Hamnet* Legh 1517 AD vi (Wa). *Ham-un-et*, a double diminutive of *Hamo*. *v.* HAIM. (ii) Roger *de Hamptonet* 1327 SRSx; William *Hamenet* 1334 SRK; John *Amnettes* 1576 SRW. From Hampnett (Gl), or East Hampnett, Westhampnett (Sx).

**Hamon:** *v.* HAIM

**Hampden:** Alexander *de Hamden, de Hampeden* 1274 FFO, 1275 RH (Bk); Simon *de Hamden* 1334–5 SRK; Edmund *Hampden* 1413 Goring. From Hampden (Bucks).

**Hamper:** John *le Hanaper* 1236–41 StP (Lo); John *Hanapier* c1260 *ERO*; Geoffrey *le Hanaper* 1279 RH (C). OFr *hanapier* 'maker or seller of goblets'.

**Hampshire, Hamshere, Hamshar, Hamsher, Hamshaw, Hampshaw:** (i) Thomas *de Hallumschire* 1296 Wak (Y); Richard *de Halumpschyre* 1348 DbAS 36; John *Hamshaw, Hampshawe* 1506, 1555 FrY; John *Hamschyer* 1507 GildY. From Hallamshire (WRYorks). (ii) Robert *Hamptessire* 1296 SRSx; Thomas *Hamshere* 1523 SRK. From Hampshire.

**Hampson, Hamson:** John *Hammonson* 1332 SRCu; Robert *Hamsone, Hameson* 1354 FrY, 1379 PTY; Henry *Hampson* 1540 FrY. 'Son of *Hamo* or *Hamon(d)*.'

**Hampstead, Hamstead:** Emma *de Hamesteda* 1130, Simon *de Hamstede* 1190 P (Berks). From Hamstead

(St), Hampstead (Mx, Wt), or Hampstead Marshall, Norris (Berks).

**Hampton:** Philip *de Hamtona* 1166 Oseney; Edith *de Hampton'* 1221 AssWo; Richard *Hampton* 1327 SRSx. From one or other of the many places of this name.

**Hanbury, Handbury, Hambury:** William *de Haneberge* 1196 FFSt; Euller' *de Hambir'* 1221 AssWo; Philip *de Hannbir'* 1230 P (Wo); John *de Hambury* 1327 AssSt. From Hanbury (Staffs, Worcs).

**Hanby:** *v.* HAMBY

**Hance:** *v.* HAND, HANN

**Hancell:** *v.* ANSELL

**Hancock, Hancox, Handcock:** *Hanecok* Birun 1276 RH (Y); Thomas *Hancoc* 1274 RH (Sa); John *Hancokes* 1316 Wak (Y); Thomas *Handcocke* 1632 RothwellPR (Y). A diminutive of HANN. *Hannecok* le Nunne (1297 Wak) is also called *Henricus* (1297 SRY).

**Hand, Hands:** *Honde* Cotrell 1288 AssCh; Johannes *cum manu* c1200 Dublin; Richard *Hand* 1279 RH (Beds); Robert *Hond* 1296 SRSx; Walter *Handes* 1332 SRWa. For Handsacre (Staffs) and Handsworth (WRYorks) Ekwall postulates an OE *\*Hand*, originally a nickname from *hand*. cf. FOOT. *Honde* may well be a survival of this. The surname may also refer to some peculiarity of the hand or to skill in its use. cf. GOODHAND.

**Handas:** John *Handax* 1327 SRY. 'Hand-axe', OE *hand, æcs*. Metonymic for a maker or user of this. cf. Geoffrey *Handhamer* 1296 SRSx 'a small hammer'; Ralph *Handsex* Ed I Malmesbury 'a dagger'.

**Handasyde, Handisyde, Handysyde, Hendyside:** Richard *de Hanggandsid*, Adam *Hangalsyde* 1398, William *Hangitsyde* 1494, John *Handisyde* 1665. From Handyside near Berwick (Black).

**Handbury:** *v.* HANBURY

**Handcock:** *v.* HANCOCK

**Handford, Handforth, Hanford, Hanfirth:** Simon *de Hanford'* 1230 P (So); John *de Hondford* 1325 AssSt; Richard *Hanford* 1468–9 FFWa; John *Handford* 1642 PrD. From Hanford (Dorset, Staffs), Handforth (Ches), or Hannaford (Devon).

**Handley, Handlay, Hanley, Hanly:** Gilbert *de Hanlega* 1185 P (Wo); Laurence *de Hanlaye* 1219 AssY; Robert *de Handlegh* 1314 IpmGl; John *Hanley* 1426 IpmNt; Robert *Handley* 1611 Bardsley. From Handley (Ches, Derby, Dorset, Northants), Hanley (Staffs, Worcs), or Handley Farm in Clayhanger (Devon).

**Handscombe:** *v.* HANSCOMB

**Handsley:** *v.* HANSLEY

**Handslip:** *v.* HANSLIP

**Handson:** *v.* HANSON

**Handy:** *v.* HONDY

**Handysyde:** *v.* HANDASYDE

**Hanes:** *v.* HAGAN

**Hanfirth, Hanford:** *v.* HANDFORD

**Hanger:** (i) *Hangere* 1148 Winton (Ha). ME *hangere* 'hangman'. cf. Roger *Hangeman* 1310 AssNf 'hangman'; Adam *Hanggedogge* 1262 ForSt 'hang dog'. (ii) Richard *de Hangre* 1297 MinAcctCo;

Richard *atte Hanger* eHy 4 Cl. 'Dweller by the wood on a steep hillside', OE *hangra*.

**Hanham, Hannam:** Roger *de Hanam* 1327 SRSo; John *Hanham* 1437–8 FFWa; Edmond *Hannam* 1545 SRW. From Hanham (Gl).

**Hanke, Hanks:** *Anke* Hy 2 DC (L); *Hanke* 12th ib.; *Anke* de Ankinton' 1194 P (L); *Hank'* carpentarius 1280 Oseney (O); Roger *Hanke* 1275 RH (Nf); Ralph *Hancks* 1642 PrGR. *Hank* is usually regarded as a Flemish pet-form of *John*. The early examples above are undoubtedly of Scandinavian origin, from ON *Anki*, a diminutive of some name in *Arn-*.

**Hankin, Hankins, Hanking:** (i) *Hankynus* 1285 Oseney (O); *Hanekyn* de London 1300 LLB C; Hugh *Hankyn* 1327 SRC; Alice *Hankynes* 1337 ColchCt. A diminutive of *Hann*. cf. HENKIN. (ii) *Hondekin* the Barbur 1286 AssCh; *Handekyn* Starky 1287 ib. A diminutive of *Hand* (*Hond*). *v.* HAND. (iii) Willelmus *filius Hamekin* 1232 Pat (L); Thomas *Hamekyng* 1327 SRSx. A diminutive of *Hame* (Hamo). *v.* HAIM.

**Hanley, Hanly:** *v.* HANDLEY

**Hanmer, Hanmore:** John *de Hanmore* 1275 SRWo; Ranulph *Hanmer* 1376 IpmLa; John *Hanmer* 1464 Paston. From Hanmer (Flint).

**Hann:** *Hann, Hancock* and *Hankin* are usually regarded as Flemish forms of *John*, but they are probably English formations, sometimes from *Johan* 'John'. *Henecoc* Cotun is probably identical with *John* de Coton (1319 SRLo). *Hanne* was a very common christian name in 13th-century Yorkshire (1274–97 Wak). Camden takes *Hann* as a pet-name of *Randolph*, rhymed on *Rann* which is possible, though *Randolph* was by no means a popular name. *Hann* is certainly in some instances a pet-name for the common *Henry* (cf. *Harry*): cf. *Hanne* or *Henry* de Leverpol 1323 InqLa. Richard *Hannesone* (1379 NottBR) is identical with Ricardus *filius Henrici* (1365 ib.).

**Hannam:** *v.* HANHAM

**Hanney, Hannay:** William *de Hanney* 1234–5 FFEss; Thomas *de Hanneye* 1339 LoPleas; John *Hannay* 1339–40 CorLo. From Hanney (Berks).

**Hannibal:** *v.* HUNNABLE

**Hannington, Hannyngton:** Alfuuin *de Hamingetuna, de Haningatuna, de Hanningetuna* 1086 OEByn; John *de Haninton'* 1176 P (Ha); Cristofyre *Hanyngton* 1461 Paston. From Hannington (Ha, Nth, W).

**Hansard:** Gilbert, Roger *Hansard* c1170 Riev (Y), 1243 AssDu; William *Haunsard* 1230 P (Sr). OFr *hansard* 'cutlass, poniard', for a maker or seller of these. The surname is much too early to denote a Hanseatic merchant.

**Hanscomb, Hanscombe, Handscomb, Handscombe:** Alan *de Hanscombe* 1255 Rams (Beds). From Hanscombe End in Shillington (Beds).

**Hansell, Hansill:** *v.* ANSELL

**Hansley, Handsley:** John *de Hannesley* 1275 RH (Nt). From Annesley (Notts).

**Hanslip, Handslip:** Winemarus *de Hanslepe* 1086 DB (Nth); John *de Hannslap* 1227 AssBk. From Hanslope (Bucks).

**Hansom:** John *Hansom* bricklayer 1659 FrY; Thomas *Hanson* son of John *Hansome* bricklayer 1664 ib. For HANSON.

**Hanson:** (i) John *Handson* 1327 SRY; William *Hondesone* 1332 SRSt. 'Son of *Hand*.' (ii) John *Hanson* 1332 SRCu; Ralph *Hanneson* 1350 Putnam (La). 'Son of *Hann*.'

**Hanton:** Gervase *de Hanton'* 1200 P (Ha). From Southampton (Hants).

**Hanwell:** Thomas *Hanwell* 1393 IpmGl; William *Hanewell* 1419 FFEss; William *Hanewell* 1425–6 FFSr. From Hanwell (Mx).

**Hanwood:** Robert *de Hanewode* 1209 ForSa; John *de Hanewode* 1275 SRWo; Adam *de Hanewod* 1327 SRWo. From Hanwood (Sa).

**Hanworth:** Osbert *de Hanewurð'* 1195 P (L); John *de Haneworth'* 1298 AssL; William *Hanworthe* 1621 SRY. From Hanworth (L, Mx).

**Hapgood:** *v.* HABGOOD

**Happe:** Aldwin' *Happe* 1194 P (Sx); John *le Hap* 1296 SRSx. OE *gehæp* 'fit'.

**Happs:** *v.* APPS

**Hapsford:** John *Happesford* 1402–3 FFWa. From Hapsford (Ch).

**Harbach:** *v.* HERBAGE

**Harber, Harbor, Harbour, Arber:** Geoffrey, John *Herbour* 1279 RH (Bk). OE *herebeorg* 'shelter, lodging', metonymic for HARBERER.

**Harberd, Harbert, Harbird:** *v.* HERBERT

**Harberer, Haberer:** Geoffrey *le Arbrer* 1205 Cur (Mx); Richard *le Erbrer* 1315 Bart (Lo); Augustine *le Herberer* 1319 Husting. A derivative of ME *hereber(e)ʒen* 'to provide lodgings', a lodging-house keeper.

**Harbey:** *v.* HARBY

**Harbidge:** *v.* HERBAGE

**Harbinson:** *v.* HERBERTSON

**Harbisher:** Edric (*le*) *Herberg(e)or* 1184–5 P (Wo); Thomas *le Harbergur* 1198 P (Sx); William *le Herbejour, le Herbyiour* 1298 Ipm (Db), 1343 DbCh. OFr *herbergeor*, also a derivative of OFr *herbege*, 'host, lodging-house keeper'. cf. HABBESHAW.

**Harbison:** *v.* HERBERTSON

**Harblott:** *v.* HERBELOT

**Harbolt, Herbolt:** Gilbert, Ralph *Harebald* 1279 RH (O), 1297 MinAcctCo. OG *Haribald* 'army-bold', OFr *Herbaud*.

**Harbor:** *v.* HARBER

**Harbord:** Wulfwine *Hareberd* c1100 OEByn (So). OE *hār* and *beard*, 'grey-beard'. *v.* also HERBERT.

**Harbott:** *v.* HERBERT

**Harbottle:** Richard *de Herbotell* 1323 FrY; Thomas *Hardebotyll* 1513 RochW; John *Harbottle* 1543 Bacon (Sf). From Harbottle (Northumb).

**Harbour:** *v.* HARBER

**Harbud:** *v.* HERBERT

**Harby, Harbey:** Coleman *de Harebi* 1170 P (L); Henry *de Harebi* 1185 Templars (L); Duneward' *de Herdeby* 1222 AssWa. From Harby (Lei, Nt), or Hareby (L).

**Harcourt:** William *de Harewcurt* 1055 OEByn; Philip *de Harecourt* 1139 Templars (Sx). From Harcourt (Eure).

**Harcus, Harkess, Harkus, Arkas, Arkush:** Adam *of Harcarres* 1216 Black; John *de Harkers* 1329 ib.; William *Harkes* 1444 ib.; Alexander *Harkass* 1547 ib.; Janet *Arcus* 1668 ib. From Harcarse (Berwickshire).

**Hard, Hardes, Hards:** (i) *Hardo* de Acolt 1212 Cur (Sf); Roger *Hard* 1275 RH (L). OE *\*Heard* 'hard', or a nickname 'harsh, severe'. Sometimes for HARDY: Robert *Hard, Hardi* Hy 3 Gilb (L). (ii) Gilbert *del Hard* 1232 Pat (L). 'Dweller on the hard, firm ground' (1576 NED).

**Hardacre, Hardaker, Hardicker:** William *Hardaker* 1379 PTY; William *Hardiker* 1662 PrGR; Widow *Hardacree* 1672 HTY. From Hardacre in Clapham (WRY).

**Hardcastle:** John *Hardcastle* 1621 SRY; William *Hardcastell* 1657 PN WRY v 147. From Hardcastle in Bewerley (WRY).

**Hardell:** William *Hardell'* 1208 P (Lo/Mx); Robert *hardel* 1292 SRLo; William *Hardel* 1350 AssL. OFr *hardel* 'youth, good-for-nothing, rascal'.

**Harden:** (i) Adam *de Hardene* 1214 P (Nb); Richard *de Hareden'* 1275 RH (W); Thomas *de Hardén* 1298 AssL. From Harden (WRY), or Haredene Wood in East Tisbury (W). Sometimes, perhaps, for ARDEN. (ii) There was also a personal name: Adam *filius Harden* 1212 P (Du). (iii) John *de Hardene* 1296, Richard *de Harden* 1312 Black. From the lands of Harden (Roxburgh).

**Harder:** John, William *Harder(e)* 1220 Cur (Do), 1332 SRLa. A derivative of OE *heardian* 'to make hard', used of metals by Chaucer and in ME of hardening dough with heat. A hardener of metals or a baker.

**Hardern:** *v.* ARDEN

**Hardey, Hardie:** *v.* HARDY

**Hardfish:** Clemencia *Hardfysche* 1379 PTY; William *Hardfissh'* 1383 AssL. 'Hard fish', OE *heard, fisc.* Probably dried fish. cf. William *Hardmete* 1275 RH (L) 'dried meat'.

**Hardham, Herringham:** Richard *de Heryngham* 1296 SRSx; Margaret *Hardham,* Richard *Hardam* 1525 SRSx. From Hardham (Sx), *Heringham* 1189. Apparently the place had two entirely different names, cf. *Heryngham al. dicta Herdham* 1399. *v.* PN Sx 128.

**Hardhead:** Ralph *Hardheued* 1202 Pleas (Sf); Robert *Hardheved* 1307 IpmGl; William *Hardhede* 1379 PTY. 'Hard head', OE *heard, hēafod,* though in exactly what sense is uncertain. cf. Hugh *Hardfist* 1287 AssCh 'hard fist'; Walter *Hardepate* 1249 AssW 'hard pate'; John *Hardrybb* Pypere 1352 Putnam (Ess) 'hard rib'; John *Hardsco* 1197 P (Nth) 'hard shoe'.

**Hardicker:** *v.* HARDACRE

**Hardiman, Hardeman:** Walter *Hardiman* 1327 SRSx. 'Bold man.' *v.* HARDY.

**Harding, Arding:** *Hardingus* c1095 Bury (Sf); *Ardingus* 1200 Oseney (O); Roger, Richard *Harding* 1199 P (Nth), 1204 AssY; Hugh *Arding* 1244 Oseney (O). OE *Hearding* 'hard'.

**Hardingham:** William *de Hardingeham* 1161 P. From Hardingham (Nf).

**Hardington:** Jordan *de Herdyton* 1258 AssSo. From Hardington, or Hardington Mandeville (So).

**Hardisty, Hardesty:** John *de Hardolfsty* 1379 FrY; William *Hardosty* 1450 ib.; John *Hardesty* 1659 ib.

From Hardistys in Nesfield or Hardisty Hill in Fewston (WRYorks).

**Hardman:** *Hardman(nus)* c1095 Bury (Sf), 1168–75 Holme (Nf); Robert *Hardman* 1188 BuryS (Sf). OE *Heardmann* 'hard man', a late formation with *-mann*, a type common in the eastern counties.

**Hardstaff:** From Hardstoft in Ault Hucknall (Db).

**Hardware:** Robert, Samuel *Hardware* 1551 ArchC 42, 1682 PrGR. Metonymic for *hardwareman*: Lambert *Hardwareman* 1473 GildY.

**Hardway:** Robert *filius Hardwy* 1208 Cur (Nt); Hugh *Hardwy* 1250 Fees (Beds). OE *Heardwīg*.

**Hardwick, Hardwicke, Hardwich, Hardwidge:** Anketill' *de Herdewic'* 1221 AssWa; Richard *de la Herdewyk* 1243 MELS (So). From one of the numerous Hardwicks (OE *heordewīc* 'sheep-farm'), or 'shepherd'. cf. WICH.

**Hardwin, Hardwyn:** Robert *Hardwin'* 1218 P (Ha); John *Hardewyne* 1327 SRLei; John *Herdewyne* 1332 Selt. OE *Heardwine*.

**Hardy, Hardee, Hardey, Hardie, Le Hardy:** William *Hardi* 1194 P (Y); William *le Hardy* 1206 P (L), 1220 Ass (Berks). ME *hardi* 'bold, courageous'.

**Hare, Hares:** (i) Walter (*le*) *Hare* 1166, 1171 P (Sr); John *le Hare* 1197 P (Nf). OE *hara* 'hare', a nickname for speed or timidity, cf. 'lizons in haile, and hares in the feld' (c1330 MED). *v.* also HAIR. (ii) A nickname from the hair (OE *hǣr*): Godefridus *cum capillo* c1200 Dublin; Henry *Mytehare, Myttehere* 1253–4 ELPN. *mid the here* 'with the hair'. cf. BEARD, HEAD. (iii) Ralph *del Hare* 1309 SRBeds. 'Dweller on the stony-ground', OE *hǣr. v.* DEPN, s.v. *hār*.

**Harenc:** *v.* HERRING

**Hareshead:** William *Hareshefd* 1321 Black. 'With a head like that of a hare', OE *hara, hēafod.* cf. John *Harebroune* 1379 PTY 'brown as a hare'; Peter *le Harlyppede* 1305 AssW 'with a hare-lip'; William *Haretayl* 1237 FFC 'hare tail'.

**Harewood:** *v.* HARWOOD

**Harfoot:** Harold *Haranfot* 1038 OEByn; Robert, Samson *Harefot* 1170 P (L), 1221 AssWo. The byname of Harold, son of Cnut, is, no doubt, an anglicizing of ON *harfótr* 'hare's foot', a nickname for a swift runner. cf. 'Harald, Godwyne sone; He was cleped Harefot, For he was urnare god' (c1330 MED). The surname may also be English in origin.

**Harford:** *v.* HEREFORD

**Hargent:** *v.* ARGENT

**Harger:** *Hardgarus* 1112 Bury (Sf); Willelmus *filius Hargari* 1121–48 ib.; Thomas *Hardgare* 1260 AssC; Baldewynus *Haregar*, Rogert *Hargar'* 1279 RH (C). OG *Hariger, Harger* 'army-spear'.

**Hargood:** Colin *Harrengod'* 1207 Cur (Sf); William *Heryngaud, Herigod, Heringod* 1275 RH (K); John *Harigod* 1327 SRSf. OFr *herigaut, hergaut, hargaut* 'an upper garment or cloak worn by both men and women in the 13th and 14th centuries'. Metonymic for a maker or seller of these.

**Hargreave, Hargreaves, Hargrave, Hargraves, Hargreves, Hargrove, Hargroves:** Geoffrey *de Haregraue* 1188 P (Db); Henry *de Hargreue* 1332 SRLa; William *Hargreave*, Richard *Hargreaves* 1672 HTY. From Hargrave (Ches, Northants, Suffolk), or Hargreave Hall (Ches).

**Harker:** Robert *le Herkere* 1280 AssSo; Ralph *Harka* 1479 GildY. A derivative of ME *herkien* 'to listen', an eavesdropper.

**Harkin, Harkins, Harkiss:** Edwinus, Rogerus *filius Hardekin* 1175 P (Nf); John *Hardekyn, Herkyn* 1327 SR (Ess). A diminutive of HARD.

**Harkus:** *v.* HARCUS

**Harland:** Peter *de Herland'* 1221 AssWa; Adam *Herlond* 1332 SRWa; Thomas *Harland* 1525 SRSx. From Harland Edge (Derby), Harlands Wood (Sussex), or 'dweller by the boundary wood'.

**Harle:** *v.* EARL

**Harley, Harly:** Juhel *de Harelea* 1166 P (Y); Henry *de Hareleye* c1280 SRWo; William *Harley* 1379 IpmGl. From Harley (Salop, WRYorks), or 'dweller at the hares' wood'.

**Harlin, Harling:** *v.* HURLIN

**Harliss:** *v.* ARLISS

**Harlock, Horlock, Horlick:** Borewoldus *Horloc* 1066 Winton (Ha); Edwardus *Horloch, Harloch* 1187–8 P (Lo); Richard *Horloc* 1206 P (Co). 'Grey lock', OE *hār, locc. v.* OEByn, ELPN.

**Harlow, Harlowe, Arlow:** Osbern, Thomas *de Herlaue* 1121–48 Bury, 1205 Pl (Herts); Walter *de Harlow* 1327 SRY; Thomas *Harlowe* 1442 AssLo. From Harlow (Essex), or Harlow Hill (Northumb, WRYorks).

**Harm, Harme, Harmes, Harms, Arm, Arme, Armes, Arms:** Walter *Harm'* c1150–6 RegAntiquiss; Richard *Harm* 1234–5 FFSr; John *Harm*, William *Arm* 1327 SRLei; Thomas *Armes* 1674 HTSf. OE *hearm* 'evil, hurt, injury'. Metonymic for a doer of harm.

**Harman, Harmand, Harmon, Harmond, Hearmon, Herman, Hermon:** *Hermannus* dapifer 1101–25 Holme (Nf); Alexander *filius Hermanni* 1191 P (Sr); Willelmus *Hermannus* 1141–9 Holme (Nf); Robert *Hereman* 1196 P (Nf); William *Heremond* 1296 SRSx; John *Harman* 1327 SRSf. OFr *Herman(t)*, from OG *Hariman, Her(e)man* 'warrior'.

**Harmer, Harmar, Hermer:** (i) *Heremerus* de la Bolde 1176 P (St); Willelmus *filius Hermeri* 1208 Cur (L); Walter *Hermer* 1327 SRSf; Richard *Harmer* 1524 SRSf. OG *Her(e)mar* 'army-famous'. (ii) William *de Hermer'* 1207 Cur (Sx); Simon *de Haremere* 1296 SRSx; William *Harmere* 1428 FA (Sx). *Harmer*, still found in Burwash, Battle, Hastings, and Winchelsea, and over the border in Kent, at Tunbridge Wells, is here local in origin, from Haremere Hall in Etchingham (PN Sx 456).

**Harmond:** *v.* HARMAN

**Harmston, Harmiston, Hermiston:** William *de Hermedeston'* 1202 AssL; Geoffrey *de Hermeston'* 1218 AssL; Richard *de Hermistone* 1297 Calv (Y). From Harmston (L).

**Harmsworth:** Robert *de Hermodesworthe* 1340 CorLo. From Harmondsworth (Middlesex), *Hermodesworth* 1233.

**Harn:** *v.* HERN

**Harness:** *v.* ERNEST

**Harnett, Harnott:** *v.* ARNALD

**Harniman:** *v.* HERNAMAN

**Harold, Haroll, Harald, Harrald, Harrold, Harral, Harrall, Harrel, Harrell, Harrad, Harrod:** (i)

*Haraldus, Harold, Herald, Herold, Horoldus, Haral,*
*Heral* 1066 DB; *Haraldus* Hy 2 DC, 1212 Cur (L);
Ralph *Harold* 1171 P (Y); Radulfus *Haroud'* . . .
predictus Radulfus *Harold'* 1196 FF (Y); Philip
*Harald* 1327 SRSx; Lawrence *Horold* 1333 FFSf. (ii)
*Hereuuoldus* 1066 DB (Sf); Gacobus *Herewald* 1275
RH (Sf); David *Harwald* ib.; Thomas *Harwold* 1331
LLB E; Thomas and Dorathy *Harwold, Harrold* 1664,
1666 Great WelnethamPR (Sf); George and Elizabeth
*Hurrell, Harrold* 1780, 1782 ib. The chief source of
these names is ON *Haraldr,* ODa, OSw *Harald,* a
name very common in DB, but usually referring to
King Harold. *Harrod* is from *Haroud,* whilst *Herald*
and *Herouldus* would give *Herald* and *Herod.* The
latter might also derive from OG *Hairold, Herold.*
There has also been confusion with OE *Hereweald*
'army-power', particularly in Suffolk where there was
also a late confusion with *Hurrell.*

**Harp:** Roger *atte Harp* 1327 SRSo; John *atte*
*Harpe* 1361 LLB G. OE *hearpe* 'harp', in the
compound *\*sealthearpe,* was used of a harp-shaped
contrivance for sifting and cleansing salt and survives
in South Harp in South Petherton (Som). cf. also
Harp House in Eastwood (Essex), near the marshes
where there were salt pans. *v.* MELS 92. The second
reference, from London, is probably due to residence
at the sign of the harp. Ralf *Harpe* (1241 FFBk) is
probably metonymic for 'harper'.

**Harpford:** Jordan *de Harpeford* 1268 AssSo. From
Harpford (D, So).

**Harper, Harpour, Harpur:** Robert *le Harpur* 1186 P
(Ha); Reginald *le Harper* 1275 SRWo. The surname is
common and widely distributed. *Harper,* now the
usual form, is from OE *hearpere* 'harper', *Harpour*
and *Harpur* from AFr *harpour,* OFr *harpeor,* the
common form in early sources.

**Harpham, Harpum:** John *de Harpham* 1203 P (Y);
Hervey *de Harpham* 1346 FFY; Robert *de Harpeham*
1390 IpmNt. From Harpham (ERY).

**Harpin:** Geoffrey *Harpin* eHy 2 DC (L); William
*Harepin* 1185 Templars (Wa), *Harpin* 1231 Guisb (Y),
1243 AssDu. OFr *harpin* 'harper'.

**Harpley:** Geoffrey *de Harpele* 1250 Fees (Nf); John
*Harpeleye* 1332 SRWo. From Harpley (Nf, Wo).

**Harpsfield:** Nicholas *de Harpesfeld* 1357 AssNu;
Thomas *Harpesfeld* 1412–13 FFWa; John *Harpesfeld*
1514 FFEss. From Harpsfield Hall in St Peters
(Herts), or 'dweller by the harp-shaped field', OE
*hearpe, feld.*

**Harpum:** *v.* HARPHAM

**Harraden, Harradine, Harradence, Harridine:**
William *de Harewedon* 1327 SRSx. From Harrow-
den (Beds, Northants).

**Harrall, Harrell:** *v.* HAROLD

**Harre:** *Herre* de Camera 1176 P (Nf); Ailwin *Herre*
1195 P (Do/So); Richard *le Harre* (*Herre*) 1219 Cur
(Beds); Henry *le Harre* 1297 MinAcctCo; Roger
*Harre* 1327 SRSf. A nickname from OE *hēarra* 'chief,
lord', but sometimes probably a pet-form of *Herebert,*
cf. *Here* (*Herebertus*) de Oxhei 1230 P (He). *v.*
HERBERT.

**Harrhy, Harrie, Harries:** *v.* HARRY

**Harriman, Harryman:** Henry *Henriman* 1332
SRCu. 'Servant of *Henry.*'

**Harrington, Harington:** William *de Harinton'* 1202
AssL; Richard *de Harington* 1274 RH (L); John
*Harington'* 1327 SRLei. From Harrington (Cumb,
Lincs, Northants).

**Harringworth:** John *de Haryngworth* 1374 FFEss.
From Harringworth (Nth).

**Harrismith:** *v.* ARROWSMITH

**Harrison, Harrisson, Harison:** Henry *Hennerissone*
1354 Putnam (Ch); Robert *Harriesone* 1355 LoPleas;
John *Herryson* 1372 FFHu; William *Henryson* 1376
FrY; John *Herryson, Harryson* 1445 Shef (Y). 'Son of
*Henry.*'

**Harrod, Harrold:** *v.* HAROLD

**Harrop, Harrup, Harrap:** Richard, Hugh *de Harop*
1185 P (Nb); William *de Harrope* 1242 AssDu;
William *Harrop* 1367 FFY; John *Harop, Herrop* 1476,
1478 Cl (Lo). From Harehope (Northumb), or
Harrop, Harrop Dale, Harrop Hall (WRYorks).

**Harrow:** (i) William *de Harrewe* 1275 RH (Lo);
Hugh *de Harhou* 1332 SRCu. From Harrow on the
Hill (Mx), or Harrow Head in Nether Wasdale (Cu).
(ii) John *Harrow* 1408 Black. From Harrow near Mey
(Caithness).

**Harrower, Harrowar, Harwar:** Geoffrey *Haruer*
1255 MEOT (Ess); Geoffrey *le Harewere* 1275 RH
(Nf); Ralph *le Harwere* 1327 SRSf; John *Harower*
1379 PTY. A derivative of ME *harwen* 'to harrow', a
harrower (c1475 MED).

**Harrowsmith:** *v.* ARROWSMITH

**Harry, Harrie, Harrhy, Harries, Harris,**
**Harriss:** *Herre* de Camera 1176 P (Nf); *Harry* Haket
1270 ArchC 5; *Hanry* Wade, *Hary* de Kent, *Herri* de
Merlowe 1292 SRLo; Nicholas *Herri* 1327 SRWo;
William *Herry* 1337 ColchCt; Richard *Harry* 14th
Shef (Y); William *Harrys* 1406 Eynsham (O);
Lawrence *Harryes* 1468 Fine (Herts). *Harry* was the
regular pronunciation of *Henry* in the Middle Ages
but is rarely found in documents where the name is
usually in the Latin form *Henricus. v.* HENRY.

**Harryman:** *v.* HARRIMAN

**Harsant, Harsent, Hersant, Hassent, Arson,**
**Assan:** *Hersent* 1166 RBE (Nf), 1208 Cur (Gl); Roger
*Harsent* 1276 RH (C); William *Herseynt* 1297
MinAcctCo; John *Hersent* 1327 SRSf; John, William
*Arsent* 1327 SRC; John *Harsand,* Henry *Harshand*
1440, 1493 ShefA (Y); William *Areson* 1520 FrY.
OFr *Hersent, Hersant* (f), OG *Herisint* (f) 'army-
truth'.

**Harsnett:** Robert *de Hassnade, de Halstnade* 1215–
27 StGreg (K); Benedict *de Halssnod* 1334 SRK;
Richard *Halsnoth* 1474 CantW; Roger *Harslett* 1513
ib.; Thomas *Halsnoth* alias *Awstnet* 1523 LP (Lo);
William *Alsnoth* 1524 WBCinque; Walter *Hawsnode*
1565 ib.; Marten *Halsnode, Halsnothe* 1566, 1570 ib.;
William *Hassnod* 1647 ib.; Robert *Halsenorth* 1664
HTSo. From Ayleswade in Frittenden (Kent),
*Halsnod'* 1224. Samuel *Harsnett,* later archbishop of
York, b. 1561, son of William *Halsnoth,* baker, of
Colchester, when appointed Master of Colchester
Grammar School in 1586, signed as *Harsnet,* although
his name is entered six times as Mr *Halsnothe* (ER li,
10).

**Harsom:** *v.* HERSOM

**Harston:** Robert *de Harestan* 1204 P (La); Adam *de Herleston* 1208 FFEss; Robert *de Harestan* 1221 AssWa. From Harston (C, Lei). Sometimes, perhaps, from Harleston (Sf), or Harlestone (Nth).

**Harsum:** *v.* HERSOM

**Hart, Harte, Heart, Hort, Hurt:** Ælfric *Hort* c1060 OEByn (Ha); Roger *Hert* 1166 P (Nf); Reginald *Hurt* 1185 Templars (Beds); Simon *le Hert* 1197 FF (K); Godrich' *le Hurt* 1220 Fees (Berks); Richard *Hort* 1221 AssWa. OE *heorot*, ME *hert* 'hart'. In ME *heorot* became *hurt*, roughly west of a line from Dorking to Birmingham, and this has occasionally survived, though often replaced by the standard *hart* from *hert*.

**Hartell, Hartill, Hartle, Hartles:** Robert *de Herthil* 1176 P (Db). From Harthill (Ches, Derby, WRYorks).

**Hartfield, Heartfield:** Hugh *de Hertfeld* 1204 P (Sx); Relicta *atte Hertefeld* 1327 SRSx. From Hartfield (Sx).

**Hartford:** *v.* HEREFORD

**Hartin, Harting:** William *Hartin* 1275 RH (K); Richard *de Hertyngges* 1297 MinAcctCo. From East, South, West Harting (Sx).

**Hartley:** Robert *de Hertlay* 1191 P (Y); Nicholas *de Hertlegh* 1327 SRSo; Richard *Hertlay* 1379 PTY; William *Hartley* 1621 SRY. From Hartley (Berks, Devon, Dorset, Hants, Kent, Northumb, Westmorland), or Hartleigh in Buckland Filleigh (Devon).

**Hartnell, Hartnoll:** Calen *Hartnell*, George *Hartnoll* 1642 PrD. From Hartnell's Fm in Hemyock, Hartnolls in Tiverton, or Hartnoll in Marwood (D).

**Harton:** Roger *de Hartone* 1279 RH (C); John *Arton* 1414 Shef. From Harton (Du, He, NRY).

**Hartrick, Hartwright:** *v.* ARKWRIGHT

**Hartshorn, Hartson:** John *Hierteshorn* 1165–71 Colch (Ess); William *de Herteshorna* 1196 FFDb; Bertram *de Herteshorne* 1226 AssSt; William *Herteshorn* 1355 LuffCh. From Hartshorn (Nb), Hartshorne (Db), or a plant name from swine cress, ME *harteshorn*, cf. James *Hartestonge* 1559 Pat (Nf) 'hart's tongue fern'.

**Hartwell:** Simon *de Hertwelle* 1185 Templars (Wo); Jordan *de Hertwelle* 1259 Oseney; John *Hertwell* 1327 SRSx; Richard *Hartwell* 1570 PN Berks i 56. From Hartwell (Bk, Nth, St), Hartwell in Hartfield (Sx), Hartwells in Maidenhead (Berks), or Hartwell in Lamerton (D).

**Harty:** Benedict *de Herty* 1241 FFK; Manuesimus *de Herty* 1272 RH (Nt). From the Isle of Harty (K).

**Harvest:** *Herefast* 1148 Winton (Ha); Geoffrey *Harevest* 1277 AssW; Juliana *Heruest* 1327 SREss; Thomas *Haruest'* 1332 SRDo. OE *Herefæst*, or ODa *Arnfast*. Sometimes, perhaps, metonymic for *Harvester*, cf. Cecilie *Harvester* 1576 SRW.

**Harvey, Harvie, Hervey:** *Herveus* 1086 DB; *Heruen de berruarius* 1086 DB (Sf); *Heruei* de Castre 1157–63 DC (L); *Herui* de Burc Hy 2 Gilb (L); Willelmus *filius Hervici* 1242 Fees (Sf); William *Hervi, Herevi* 1190 BuryS, 1196 Cur (Sf); William *Hervy* 1232 FFEss; Richard *Herfu* 1327 SRSx. OFr *Hervé*, OBret *Aeruiu*, *Hærviu* 'battle worthy', a name introduced by Bretons at the Conquest. Occasionally OG *Herewig*, *Herewicus* 'army-war'.

**Harwar:** *v.* HARROWER

**Harward:** *v.* HEREWARD

**Harwarth:** *v.* HAWORTH

**Harwell:** Cecilia *de Harewella* 1194 P (O); Thomas *de Harwell* 1325–6 CorLo; John *Harwell* 1496–7 FFWa. From Harwell (Berks, Nt).

**Harwin:** *Harduuinus, Haruinus* 1066 DB (Nf, Sf); *Hardewinus* 1208 Cur (Herts); Walter *Hardwin* 1207 Cur (Gl); Henry *Harwyn* 1456 NorwW. OFr *Harduin*, OG *Hardwin* 'firm friend'.

**Harwood, Harewood:** Hubert *de Harewda* 1176 YCh; Bernard *de Harewode* 1242 Fees; Alice *Harewode* 1327 SRSo. From Harwood (Devon, Lancs, Northumb, NRYorks), Harewood (Ches, Hereford, WRYorks), or Horwood (Devon), *Harewde* 1219.

**Harworth:** John *de Harewordh'* 1208, *of Hareworth* 1226 FFY; Hugh *of Hareworth* 1283–4 IpmY. From Harworth (Nt).

**Hase:** Roger *filius Hase* 1327 SRDb; Arnulf *Hase* 1220 Cur (Sf); William *Hase* 1327 SRDb. OE *hās* 'a hard, hoarse, raucous voice'. Used also as a personal name.

**Haselar, Haseler:** *v.* HASLER

**Haseldene, Haseldine, Haseltine:** *v.* HAZELDEN

**Hasel:** *v.* HAZEL

**Haselgrove:** *v.* HAZELGROVE

**Haselhurst:** *v.* HAZELHURST

**Haselwood:** *v.* HAZELWOOD

**Haskell:** *v.* ASKELL

**Haskey:** *v.* ASKEW

**Haskey:** *v.* ASKEY

**Hasking, Haskin(s):** *v.* ASKIN

**Haskow:** *v.* ASKEW

**Haslam, Haslem, Haslum, Hasleham, Haslen, Heslam, Aslam, Aslen:** Hugh *de Haslum, de Hesellum* 1246 AssLa; Benjamin *Aslin* 1674 HTSf. From Haslam (Lancs) or 'dweller by the hazels', OE *hæsel*, ON *hesli*.

**Haslegrave:** *v.* HAZELGROVE

**Hasleham:** *v.* HASLAM

**Haslehurst:** *v.* HAZELHURST

**Haslen:** *v.* HASLAM

**Hasler, Haselar, Haseler, Heasler:** Robert *de Heselhour', de Haselore* 1221 AssWa, 1287 AssSt. From Haselour (Staffs) or Haselor (Warwicks, Worcs).

**Haslett, Haslitt:** *v.* HAZLETT

**Haslewood:** *v.* HAZELWOOD

**Haslin, Hasling:** Geoffrey *Alselin* 1086 DB (Ess); William *Haslin* 1148–56 Seals (L); Radulfus *Halselinus* 1176 P (Db); Roger *Haucelin* 1196 P (Y); William *Hauselin, Hanselin, Halselin* 1211–12 Cur (L); John *Halselyn* 1327 SRSx. *Alselin* is OG *Anselin*, with assimilatory change of *n* to *l*. *v.* ANSELL, ANSLYN.

**Haslin, Hasling, Hesling, Heslin, Hessling:** Robert *Hasling* 1275 SRWo; Walter *atte Haselyng* 1327 SRSx; Peter *atte Heselyng* 1332 ib. 'Dweller by the hazel-copse', OE *\*hæsling*.

**Hasluck, Haslock, Hasluck:** *Aslac, Aseloc, Haslec* 1066 DB (L, Nf, Sf); *Haselac molendinarius* 1177 P (L); Ricardus *filius Aslac* 1197 P (L); Radulfus *filius Oselach'* 1189 P (Sf); William *Aselach', Aslac* 1189 P

(Y, Nf); Petronill *Oslok* 1327 SRSf; Richard *Hasloke* 1536 NorwW (Nf). ON *Áslákr*, ODa, OSw *Aslak*.

**Haslop, Haslip, Haslup, Haselup, Hazelip, Heaslip, Heslep, Heslop, Heslop, Hislop, Hyslop:** John *Heslop* 1414 FrY. 'Dweller in the hazel-valley', OE *hæsel*, ON *hesli* and OE *hop*.

**Haspineall:** *v.* ASPINALL

**Hassack, Hasseck, Hassock:** Agnes *Hassoc* 1279 RH (C); Robert *Hassok* 1327 SRSo; Robert *Hasok* 1439 IpmNt. From Haske in Upper Hellions (D), or 'dweller by the coarse grass', OE *hassuc*.

**Hassall, Hassalls:** *v.* HASSELL

**Hassard:** *v.* HAZARD

**Hasse:** John *de Hasse* 1255 RH (Bk); Roger *atte Hasse* 1296 SRSx; Gracia *del Hasse* 1309 SRBeds. From The Hasse in Soham (C), or 'dweller by the coarse grass', OE *\*hasse*.

**Hasseck:** *v.* HASSACK

**Hasselby:** From Asselby (ERYorks). cf. AXLEBY.

**Hassell, Hassall, Hassalls:** John *de Hassell* 1279 RH (O); Henry *de Hasshal* 1299 IpmCh; Robert *Hassell* 1568 SRSf. From Hassall (Ch).

**Hassett:** Robert *Hessete* 1379 FFEss; Jane *Hassett* 1473 Paston; William *Hassett* 1568 SRSf. Perhaps late forms of HAZARD.

**Hassock:** *v.* HASSACK

**Hastain:** *v.* HASTING

**Hasted:** William *Halstead*, Thomas *Hawstead*, Thomas *Hastead* 1674 HTSf. A late development of *Halstead*.

**Haster:** John *Haster* 1440 FFEss; Thomas *Haster* 1576 SRW. A derivative of OFr *haste* 'a spit', a turnspit. cf. William *le Haste* 1296 SRSx, a metonymic form.

**Hastie, Hasty:** (i) Robert *Hastif* 1202 Cur (W); Richard *Hasty* 1221 AssWa; Richard *le Hastie* 1326 LaCt. OFr *hastif*, *hasti*, ME *hasty* (c1340 NED) 'speedy, quick'. (ii) Picot *filius Asti* 1203 P (L); William *Asti* 1212 Cur (Nf); Francis *Asty* 1674 HTSf. *Asti* is a pet-form of Norman *Asketin* from ON *Ásketill*. *v.* ASKELL. *Hasti*, like *Hastin* and *Hasluck*, has an inorganic *H*. *Hasti* de Bolebi (12th DC) is identical with *Asketinus* de Bolebi.

**Hastin:** *v.* ASKIN

**Hasting, Hastain:** Walter *Hastinc* 1190 P (Cu); Aitrop *Hasteng* 1194 StCh; Hamo *Hasting'* 1235 Fees (Bk). *Hastang* (*Hastenc*, *Hasten*) is a Norman personal name, doubtless of Scandinavian origin, from ON *Hásteinn*.

**Hastings** may be a patronymic of *Hasting*, 'son of *Hastang*', but is, no doubt, usually local in origin, from Hastings (Sussex): Robert *de Hastinges* 1086 DB, Hugh *de Hasting* 1130 P (Lei).

**Hastler:** Henry *le hasteler* 1190 P (Ha). AFr *\*hasteler*, from *\*hastele*, a diminutive of *haste*, Fr *hâte* 'a spit'. 'One who roasts meat', 'a turnspit', also 'the officer of the kitchen who superintended the roasting of meat'.

**Haston:** Philip *de Haston'* 1274 RH (Sa). From Haston (Sa).

**Hastvillain:** *Hasteuillanus* homo Roberti 12th Seals (Wa); Robert *Hastevilein* 1209 P (Nb); Gerard *Hastevilain* 1225 FFEss. 'Hurry the serf', OFr *haster*,

*vilein*. cf. William *Hastebeuerage* 1182 P (Nb) 'hurry with the drink'; Walter *Hastmanger* 1248 AssW 'hurry with the food'.

**Haswell:** Tiedewi *de Haswella* 1176 P (Herts); John *de Haselwell* 1246 AssLa; William *de Haselwell* 1342 LoPleas. From Haswell (Du, So), Haswell in North Huish (D), or 'dweller by the spring in the hazels', OE *hæsel, wiella*.

**Hatch:** Gilbert *ad Hacce* 1185 Templars (Ess); Adam *del Hach* 1221 ElyA (Nf); Henry *Hache* 1230 P (Sf); Walter *ate Hacche* 1297 MinAcctCo. From Hatch (Beds, Hants, Som, Wilts), or from residence near a hatch or gate (OE *hæcce*), generally one leading to a forest, sometimes a sluice.

**Hatchard:** *v.* ACHARD

**Hatcher:** John *Hetchere* 1296 SRSx; Andrew *Hatcher* 1560 SxWills. 'Dweller by the gate.' *v.* HATCH.

**Hatchett:** William *Hachet* 1165 P; John *Hachet* 1334 LLB E. OFr *hachet* 'a small axe'. Metonymic for a maker or user of this.

**Hatchman:** William *Hacheman* 1196 P (C/Hu); John *Hachman* 1361 ERO (Ess). 'Dweller by the gate leading to the forest', OE *hæcce, mann*, or perhaps 'keeper of the sluice-gate'.

**Hatcliff, Hatcliffe:** William *de Hadecliue* 1204 P (L); Peter *de Hadcliff* 1312 RegAntiquiss; Nycholas *Hatclyff* 1524 SRSf. From Hatcliffe (L).

**Hatfield, Hatfeild, Hatful, Hatfull:** Tata *æt Hæðfelda* c1050 OEByn; William *de Hatfeld* 1119–27 Colch (Ess); Robert *de Hattefeld* 1343 FFY; Thomas *Hatfeld* 1412 AssLo. From Hatfield (Essex, Hereford, Herts, Notts, Worcs, ER, NRYorks), or Heathfield (Som, Sussex).

**Hathaway, Hathway, Hadaway:** (i) *Hadeuui* 1066 DB (He); *Hathewi* 1175 P (Wo); William *Hatewi*, *Hadewi*, *Hathewy* 1178, 1181 P (He), 1221 AssWa; Nigel *Haðewi* 1208 P (Gl); Thomas *Hatheweye* 1380 SRSt. OE *Heaðuwīg* 'war-warrior' (rare), or OG *Hathuwic*, *Hadewic*. (ii) William *de Hathewy* 1294 FFEss. 'Dweller by the heath-way', OE *hǣð, weg*.

**Hatherley, Hatherly:** Henry *Hatherly* 1642 PrD. Simon *Hatherly* 1698 PN Do ii 77. From Down, Up Hatherley (Gl), Hatherleigh in Bovey Tracey (D), or Hatherly Fm in Hilton (Do).

**Hathersich, Heaversedge:** Matthew *de Hauersegge* 1190 P (Y), *de Hathersich'* 1219 AssL, *de Hathersage* 1254 AssSt. From Hathersage (Derby).

**Hathorn:** *v.* HAYTHORNE

**Hatley:** Arnold *de Hateleia* 1198 FFBeds; Agnes *de Hatleye* 1275 SRWo. From Cockayne Hatley (Beds).

**Hatry:** *v.* DAWTREY

**Hatt, Hatts:** (i) Roger, Randulf *Hat* 1148–67 ELPN, 1168 P (Do). OE *hætt* 'hat', metonymic for *Hatter*. (ii) Thomas *del Hat* 1279 RH (O); Richard *atte Hatte* 1327 SRWo. 'Dweller by the hill' (OE *hætt*), as at Hathitch Fm or Hathouse Fm (Worcs).

**Hatten:** *v.* HATTON

**Hatter:** Reginald, William *le Hattere* 1212 Cur (Berks), 1262 *For* (Ess). A maker or seller of hats (1389 NED).

**Hattersley:** Ralph *de Hattresleia* 1211–25 EChCh; Richard *de Hattersleg'* 1260 AssCh. From Hattersley (Ch).

**Hatton, Hatten:** Joseph *de Hattun* a1193 YCh; Simon *de Hatton* 1259 AssCh; William *Hatton* 1470 FFEss. From Hatton (Ch, Db, L, Mx, Sa, St, Wa), Hatton Hall (Ch), or Cold, High Hatton (Sa).

**Hattrell:** Vilein, John *Haterel* 1210, 1221 Cur (Nth, Nt). *Haterel* is used in the York Plays (c1440) of 'dress, attire'. It is probably a derivative of OE *hæteru* 'clothes' and here probably denotes a tailor. cf. Elias *le Hatermongere* 1251 Oseney (O), a dealer in clothing.

**Hattrick:** Richard *Hatheriche* 13th Cust; Claricia *Hatheric*, Alice *Haterich* c1280, 1332 SRWo. OE *Heaðuric* 'war-ruler'.

**Haugh:** John *de Hawhe* 1409 Black; John *Haugh* 1483 FFEss; Thomas *Haugh* 1524 SRSf. From Haugh (L), or 'dweller by the enclosure', OE *haga*.

**Haughton:** Robert *de Haleghton* 1242 AssSt; Simon *Haughton* 1642 PrD; James *Haughton* 1663 HeMil. From Haughton (Ch, La, Nb, Sa, St), or Haughton le Skerne (Du).

**Hauman:** Alan *Hagheman* 1275 RH (Ha). 'Dweller by the enclosure', OE *haga*.

**Havekin:** *v.* HAWKIN

**Havelock:** *Haveloc* c1210 Fees (Co); William, Robert *Havelok* 1327 *SR* (Ess), 1369 LoPleas. ON *Hafleikr* 'sea-port'.

**Haven:** Nigel *de Hauen'* 1200 P (Ha); Philip *atte Hauene* 1269 AssSo. 'Dweller by the harbour', OE *hæfen* 'haven'.

**Havercake:** Matilda *Havercake* 1275 RH (Nf); Walter *Haverkake* 1279 AssNb; John *Haverkake* 1301 SRY. 'A loaf made from oat flour', ON *hafri*, ME *kake*. A nickname for a baker.

**Havercomb, Havercombe:** William *Havercombe* 1539 PN Do i 5; William *Havercome* 1662–4 HTDo. From a lost *Havercombe* in Rowbarrow Hundred (Do).

**Havercroft:** Liulf *de Hauercroft* 1191 P (Y); Walter *de Havercroft* 1200 Cur. From Havercroft (WRY), or 'dweller by the oat field', ON *hafri*, OE *croft*.

**Haverfield:** Matilda *de Hauerfeld* 1327 SRSf. 'Dweller by the oat field', ON *hafri* 'oats'.

**Havering:** Richard *Haveryng* 1343 FFEss; Roger *de Havering* alias atte Sele 1343 LoPleas; Richard *Haveryng'* 1468 PN Do ii 180. From Havering atte Bower (Ess).

**Havers:** Hugh, Simon *Hauer* 1199 P (Nf), 1230 P (Ess). OE *hæfer* 'he-goat'.

**Haverson, Haversum:** Hugo *de Hauersham* 1190 P (Beds); Nicholas *de Haversam* 1279 RH (Bucks). From Haversham (Bucks).

**Haviland, de Haviland:** Arthur *Haviland* 1662–4 HTDo; John *Havyland* 1664 HTSo. From Haveland in Membury (D).

**Havill, Havell, Hovell, Hovil:** Ralph *de Halsuilla* 1084 GeldR (W), *de Haluile* 1086 DB (W); Ralph *de Hauvill'* Hy 2 Fees (Nf), *de Alta Villa*, *de Havill'* 1198 ib.; Henry *de Hautvill'* 1242 Fees (Bk); Lucas *de Houyle* 1327 SRSf; Alice *Havell* 1464 NorwW. The DB tenant-in-chief *de Hal(s)uilla* came from Hauville (Eure), *Alsvilla* 1050, *Hasvilla* 11th, others from Hauteville la Guichard (La Manche), earlier *Hautteville*. Both became *Hauvill* and *Havill*, the former also developing into *Hovell*.

**Haw, Hawes:** (i) Richard *Haga* (*de Haga*) 1175 P (L); Roger *de la Hagh* 1255 RH (W); Peter *in le Hawe* 1279 RH (C); Margery *del Hawe*, Caterina *Hawe*, John *atte Hawe* 1327 SRSf. 'Dweller by the enclosure', OE *haga* or ON *hagi*. (ii) *Hawe* 1260 AssY, 1277 Wak (Y); *Hawe* de Bolinton 1286 AssCh; William, Thomas *Hawe* 1275 SRWo, 1307 Wak (Y). *Hawe* is clearly a pet-form of some common name, but of what it is difficult to decide. It may be for OE *Hafoc*, from the diminutives *Havekin*, *Havecot*. *v.* HAWKIN, HAWKETT.

**Haward:** *Hauuart* 1066 DB (Y); *Haward* de Wihton 1166 P (Nf); William, Stephen *Haward* 1327 SRC, 1332 SRCu. ON *Hávarðr*, ODa *Hawarth*, OSw *Havardh* 'high or chief warden or guardian', cognate with OG *Howard*.

**Hawarth:** *v.* HAWORTH

**Hawes, Hawyes:** *Hawis* 1208 Cur (Sf); *Hawisia* 1214, 1215 Cur (Sf, Beds); Robert *Hawyse*, William *Hawys* 1279 RH (O); John *Hawys*, Margery *Hawis* 1327 SRSf. OFr *Haueis*, OG *Hadewidis* 'battlewide' (f).

**Hawick:** Adam *de Awic* 12th FeuDu; William *Hawick* 1234 Black; Walter *de Hauwyke* 1279 FeuDu. From Hawick (Roxburgh).

**Hawk, Hawke, Hawkes, Hawks, Hauke:** (i) *Hauok* 1066 Winton (Ha); Willelmus *filius Hawoc* 1240–5 Black; Roger *Hauech'* 1176 P (Do); Robert *Hauk* 1269 AssNb; Walter *le Hauek* 1296 SRSx; Thomas *Haukes* 1460 FrY. Osbertus *filius Hauoc* c1115 OEByn (O) is probably to be identified with Osbern *Hauoc* (ib.). His father bore the OE name of *Hafoc* 'hawk'. In *le Hauek* we have clearly a nickname 'hawk', indicative of a savage or cruel disposition. Sometimes the simple *Hauoc* is used by metonymy for 'hawker' or with reference to the holding of land by providing hawks for the lord. In 1130 (P), Ralph *Hauoc* owed the exchequer two 'Girfals', gerfalcons or hawks. *Hawkes* may also be from *Hawkins*: William *Hawkys* or *Hawkyns* 1539 Oxon. (ii) William *del Halk* 1188 BuryS (Sf); Adam *de Halk* 1260 AssC; William *atte Halk*, Alan *Hauke* 1327 SRSf. 'Dweller in the nook or corner' (ME *halke*).

**Hawken:** *v.* HAWKIN

**Hawker:** Robert *le Hauekere* 1214 Cur (Gl); Mabill' *Haueker* 1221 ElyA (Sf); Robert *le Hauker* 1283 SRSf. OE *hafocere* 'falconer, hawker'.

**Hawkeswell:** *v.* HAWKSWELL

**Hawkesworth, Hawksworth, Halksworth:** Laurence *de Houkeswurda* 1194 P (Nt); Robert *de Hauekeswrth* 1226 FFY; John *Hawkesworth'* 1395 AssL. From Hawksworth (Notts, WRYorks).

**Hawkett, Hawkitts:** Henry *Havekot* 1275 RH (Nf). A diminutive of OE *Hafoc*: *Havek-ot*. cf. HAWKIN.

**Hawkey:** *v.* HAWKSEY

**Hawkhurst, Hawkhirst:** Eustace *de Hauekeherst* 1255–6 FFSx; William *de Haukehirst* 1341 FFY; Robert *Haukehirst* 1405–6 IpmY. From Hawkhurst Court in Kirdford (Sx), or 'dweller by the wood frequented by hawks', OE *hafoc*, *hyrst*.

**Hawkin, Hawken, Hawking, Hawkings, Hawkins, Havekin:** *Haukynus* le cotoler 1313 FrY; *Haukyn*

Skinner 1332 SRCu; Robert *Hauekin* 1248 PN Ess 333; John *Havekyn* 1275 RH (K); Richard, Alexander *Haukyn* 1297 MinAcctCo, 1311 LLB D; Margery *Haukyns* 1327 SRWo; John *Hawkynge* 1570 FrY. *Hawkin* has been regarded as a pet-name for *Henry*, from *Halkin*, but no such form has been adduced. Such a formation from such a common name seems likely, but there must be another source. Harrison cites:

> Wo was *Hawkyn*, wo was Herry!
> Wo was Tomkyn, wo was Terry!

But as *Tomkyn* is certainly not a pet-name for *Terry*, it is a fair assumption that *Hawkyn* is not for *Herry*. Our earliest forms are *Havekyn* which persists until at least 1365 (AD ii). This must be for *Havek-in*, a diminutive of OE *Hafoc*, which was still in use in the middle of the 13th century (*v.* HAWK) and had at least one other diminutive, *Havekot* (*v.* HAWKETT).

**Hawkley:** Philip *de Hauecleia* 1221 AssWo; Hugh *de Hauckelegh* 1346 SRWo. From Hawkley (Hants), Hawkley Farm in Pensax (Worcs), or 'dweller by the hawk-wood'.

**Hawkridge:** Nicholas *de Hauekrig* 1280 IpmW; Thomas *Haukeryg* 1516 FrY; John *Hawkridge* 1642 PrD. From Hawkridge (Berks, So), Hawkridge in Heywood (W), Hawkridge in Coldridge (D), or Hawkridge Fm in Hellingley (Sx).

**Hawksey, Hawkey:** William *Hauekesega* 1191 P (Nth); William *Hauekeseye* Hy 3 Colch (Ess). 'Hawk's eye', a nickname. The Manchester Hawkseys may preserve an old form of HAWKSHAW.

**Hawkshaw:** Adam *de Haukesheye* 1285 AssLa; Thomas *Haukshawe* 1375 AD iv (Wa). From Hawkshaw (Lancs).

**Hawksley:** William *de Hauekesle* 1246 AssLa; Richard *de Hauekeslowe* 1321 WoCh. From Hawkesley Hall in King's Norton (Wo), or 'dweller at the clearing frequented by hawks', OE *hafoc*, *lēah*.

**Hawkswell, Hawkeswell:** Daniel *de Hauekeswell* Ric I Cur; Alexander *de Haukeswelle* 1280, John *Haukeswell* 1408 IpmY. From Hauxwell (NRY), or a lost *Hawkswell* in Sevington (K).

**Hawkwood:** John *de Haukwode* 1343 FFEss; John *de Hawkwod* 1351 AssEss. From Hawkwoods in Sible Hedingham, or Hawkwood's Fm in Gosfield (Ess).

**Hawley:** John *de Hauleye* 1275 SRWo; Walter *Hawelee* 1398 FFEss; George *Hawley* 1672 HTY. From Hawley (Ha, K), or a lost *Hawley* in Sheffield (WRY). *v.* also HALLEY.

**Hawling:** *v.* HALLING

**Haworth, Harwarth:** Robert *de Hawrth* 1200 P (Y). From Haworth (WRYorks).

**Hawson:** Simon, Richard *Haweson* 1331 AssSt, 1424 DbCh. 'Son of *Haw*.'

**Hawtayne:** *v.* HAWTIN

**Hawthorn, Hawthorne:** William *de Hagethorn* 1155 FeuDu; Henry *atte Hauthorn* 1327 SRWo. From Hawthorn (Durham) or from residence near a hawthorn (OE *haguþorn*).

**Hawthornthwaite, Haythornthwaite:** Laurence Hauthornthwayt 1401 AssLa. From Hawthornthwaite (La).

**Hawtin, Hawtayne:** (i) Walter *Halteyn* 1134–40 Holme (Nf); Eborard *Halthein* 1146–75 MedEA (Sf);

Theobald *Hauthein*, *Halþein* 1153–68 Holme (Nf), *Hautein* 1198 FF (L). OE *heall-þegn* 'hall-thane', one who resides in or is occupied in a hall. (ii) Robert *Hauteyn*, *Auteyn* 1242 Fees (L); Walter *Haweteyn* 1277 LLB A. OFr *hautain* 'proud, haughty'.

**Hawton:** Steffan *de Houtune* c1175 Newark; William *Hawton* 1454–5 FFWa; Simon *Hawton* 1518 FrY. From Hawton (Nt).

**Hawtrey:** *v.* DAWTREY

**Hawyes:** *v.* HAWES

**Haxall, Haxell:** John *Haxsall* 1524 SRSf. Perhaps from Hackinsall (La).

**Hay, Haye, Hayes, Hays, Hey, Heyes, Heighes:** (i) Richard *de Hay* 1170 P (L); Robert *de la Haie* 1185 Templars (Herts); Roger *del Heys* 1275 RH (Nf); William *de Hayes* ib. (Nth); Henry *del Haye*, Stephen *in the Haye* 1275 SRWo; Thomas *atte Heye* 1327 SRSo. 'Dweller by the enclosure' (OE (*ge*)*hæg*), as at Hay (Hereford) or Hayes (Devon, Dorset). In ME *hay* also meant 'forest fenced off for hunting'. *v.* HIGHE, O'HEA. (ii) Willelmus *filius Hai* 1168 P (Bk); Ricardus *filius Haye* 1205 Cur (Nf); Robert, Roger *Hai* 1150 Eynsham (O), 1163–6 Seals (Sx). There is recorded in a dubious charter of 704 an OE *Hæha* or *Heaha* which Redin derives from OE *hēah* 'high'. Compounds with *Hēah-*, such as *Hēahbeorht*, *Hēahstān*, etc., were common in OE and two otherwise unrecorded compounds, \**Hēahsige* and \**Hēahhild* (f), suggest that the first theme long remained popular: *Hiechsi* 1066 Winton (Ha), Walter *filius Hehilde* 1201 Cur (L). A short form of these names, *Hēah*(*a*), may well have existed in OE and may survive in the post-Conquest *Hai*. *v.* HAYCOCK. (iii) Robert *le Hay*, *le Heye* 1275 RH (Nf, Bk); Roger *le Hay* ib. (Sx). OE *hēah* 'high, tall'. cf. 'He was strong man and hey' (c1300 NED). *v.* HIGHE.

**Haybittle, Haybitter:** William *Heybetyll* 1525 SRSx; Richard *Heybittle* 1583 Musters (Sr); Ann *Haybeetle* 1746 LewishamPR (K). A nickname, 'hay beetle', OE *hēg*, *bētl*.

**Haycock, Haycocks, Haycox, Heycock:** Robert *Heykok* 1296 SRSx; Ralph *Haycok* 1311 FFEss. *Haicoc*, a diminutive of *Hai*. *v.* HAY, HICKOX.

**Haycraft, Haycroft:** Hugo *de la Heycroft* 1279 RH (O); Richard *atte Haycroft* 1332 SRSx. 'Dweller by the hay-croft', OE *hēg*.

**Hayday:** Geoffrey *Hechdai*, *Hegday*, *Heagday* 1180–5 P (Nth); Alicia *Hedei* 1327 SRWo. OE *hēah-dæg* 'a day of high celebration, a solemn or festal day'. cf. HALIDAY.

**Hayden, Haydon, Heyden, Heydon, Heiden:** Thomas *de Haiden'* 1200 Pl (Ess); William *de Heydon* 1303 IpmGl; Walter *Haydon* 1327 SRSo. From Haydon (Dorset, Som, Wilts), Haydon Bridge (Northumb), or Heydon (Cambs, Norfolk).

**Haydock, Haddock:** Hugo *de Haidoc* 1212 Fees (La); Robert *Haddocke*, *Haydock* 1669, 1671 LaWills. From Haydock (Lancs), pronounced *Haddock*.

**Hayer:** Richard *le Heyer* 1274 RH (Gl). 'Dweller by the enclosure.' *v.* HAY and also AYER.

**Hayes, Hease, Heyes:** Hugh *de la Heise* 1197 Eynsham (O); Robert *de Hese* 1209 P (Nf); Henry *Heyse* 1240 Rams (C); Michael *atte hese* 1296 SRSx.

'Dweller by the brushwood' (OE *hǣs), as at Hayes (K, Mx), Hays (Sx), Heaseland and Heasewood Fm (Sx). *Heise* is from OFr *heis* 'brushwood'.

**Hayesman:** *v.* HEASEMAN

**Hayford, Heafford:** Andrew *de Haiford* 1195 Oseney; Richard *de Heyford* 1262–3 FFSr; John *de Heyford* 1327 SRWo. From Nether, Upper Heyford (Nth), Lower, Upper Heyford (O), or Hayford in Buckfastleigh (D).

**Hayhoe, Hayhow, Heigho, Heyo, Higho:** John *Heyhowe, Hihoo* 1524 SRSf. 'Dweller by the high ridge', OE *hēah, hōh.*

**Hayhurst:** Thomas *de Hayhurst, de Heyhurst* 1246 AssLa, 1327 SRSx; John *Hairst* 1662 PrGR. 'Dweller by the high wood.'

**Hayland, Heyland, Highland, Hyland:** Nicholas *de Haylaund* 1255 AssSo; Wulford atte *Heghelonde* 1275 RH (K); Thomas *de Heyeland* ib. (Sx). 'Dweller by the hay-land or the high-land', OE *hēg* or *hēah* and *land*.

**Hayler, Haylor:** *v.* HALER

**Hayles:** *v.* HAILE

**Haylet, Haylett:** *Hegelot* 1134–40 Holme (Nf); *Heylot* 1279 RH (C); Robert, Emma *Heylot* ib. (Beds, Hu). *Hai-el-ot*, a double diminutive of *Hai*. *v.* HAY.

**Hayley:** *v.* HAILEY

**Hayling:** *v.* HAILING

**Haylock:** *Heiloc* 1188 BuryS (Sf); Nicholas *Heilok* ib.; William *Heylok* 1327 SRSf. Probably OE *Hǣgluc*, a diminutive of *Hǣgel*, unrecorded, but found in Hailsham, Hayling and Hazeleigh. In 1327 SRSf, *Hayl* is found seven times as a surname: John *Hayl*, Robert *Heyle*. This might well be OE *Hǣgel*, surviving as *Haile*.

**Hayman, Heaman, Heyman, Highman, Hyman:** William *Hayman* 1312 AD iv (D); Walter *Heyman* 1332 SRWa. The first element is doubtful. It may be OE *(ge)hǣg*, 'dweller by the enclosure', for *atte Hay* (cf. HAY); or keeper of the enclosure (cf. HAYWARD); or OE *hēg, hīg* 'hay', seller of hay. cf. Henry *Heimongere* 1230 P (L); or even 'tall man', OE *hēah*. *v.* also HAMMOND.

**Haymes:** *v.* HAIM

**Haymonger:** Henry *Heimongere* 1230 P (L); Richard *le heymongere* 1292 SRLo; William *le Heymonger* 1300 LoCt. 'A dealer in hay', OE *hēg, mangere.* cf. William *le Heyberare* 1328 LLB E 'a carrier, perhaps a seller, of hay'.

**Hayne:** *v.* HAIN, HAGAN

**Hayselden:** *v.* HAZELDEN

**Haysom, Haysum:** John *Heysone* 1300 LLB C. 'Son of *Hay*.'

**Hayter, Haytor:** William *Haytere* 1260 AssC; Reginald *le Heytur* 1296 SRSx. For *atte Heyte*. *v.* HEIGHT.

**Haythorne, Hathorn, Heathorn, Hethron:** Philip *Haythorn* 1332 SRSr. From OE *hægþorn*. *v.* HAWTHORN.

**Haythornthwaite:** *v.* HAWTHORNTHWAITE

**Hayton:** William *de Haitun* c1147–54 YCh; Adam *de Haytun'* 1202 P (Nt); Henry *de Heyton'* 1327 SRLei; Robert *Hayton* 1415 IpmY. From Hayton (Cumb, Notts, Salop, ERYorks).

**Hayward, Heyward:** Godric *Heiuuard*, Brihtmer *Haiuuard* c1095 Bury (Sf); Richard *Haiward* 1166 P (Nf); William *le Heiward* c1179 Bart (Lo). The original duties of the hayward seem to have been to protect the fences round the Lammas lands when enclosed for hay (Coulton), hence his name, OE *hegeweard* 'guardian of the fence or hedge'. This *heȝe* was a dead hedge easily erected and removed, forming an enclosure (OE *(ge)hæg*) from which, to judge from the early and regular variation between *heiward* and *haiward*, and from his more general duties of preventing cattle from breaking through into the enclosed fields and growing crops, the hayward seems also to have been called *(ge)hægweard* 'enclosure-protector'. cf. also (from *Piers Plowman*): 'Canstow . . . haue an horne and be haywarde, and liggen oute a nyghtes, And kepe my corn in my croft fro pykers and þeeues?'. In the Parish Register of Horringer (Suffolk) c1670–80 *Hayward* is regularly written *Howard*, and in the Walthamstow Toni court-rolls from 1678 to 1882 the marshbaley is often called the *hayward* or *howard*, so that some of our Howards were probably Haywards. *v.* HOWARD.

**Haywood, Heywood, Heawood, Highwood:** Anselm *de Haiwod* c1199 StThomas (St); Adam *de Heyuuode* 1246 AssLa; Arnold *de Hewode* 1275 RH (W). From Haywood (Hereford, Notts, Salop, Staffs) or Heywood (Lancs, Wilts), or from numerous minor places.

**Hayworth, Heyworth:** Walter *de Heywrthe* 1242 AssDu; Thomas *de Hayworthe* 1327 SRSx; John *Heyworth* 1340–1450 GildC. From Haywards Heath (Sx), *Hayworthe* 1261.

**Hayzeldene:** *v.* HAZELDEN

**Hazard, Hazzard, Hasard, Hassard, Haszard, Assard:** (i) Hugo *Hasard* 1170, 1190 P (Ha), Halsard *(Hasard)* 1178 P (Ha); William *Halsart' (Hausard')* 1177 P (Sr); Gilbert *Hausard* 1196 P (Y). Probably OE *hals* 'neck' plus the pejorative suffix *-ard*, a nickname for one with some peculiarity of the neck. cf. TESTAR. (ii) Geoffrey *Hasard* 1185 Templars (L); Walter *Hassard* 1197 FFK; Sybil *Hasard'* 1201 AssSo. *Hasard* is common and cannot always be from *Halsard*. It may be OE *hasu* 'grey' plus *-ard*, 'grey-haired, or Fr *hase* 'a female hare' plus *-ard*, parallel with *Hare*, or, perhaps more frequently, OFr *hasard*, ME *has(s)rd, haz(z)ard* 'a game at dice' (MED c1300), used of a gambler or one prepared to run risks. cf. John *Hasardour* 1366 ColchCt, ME, AFr *hasardour* 'a dicer, gamester' (1368 MED). This metonymy existed already in French. cf. 'vous n'etes que un *hasart*. Et ledit Robin dist: Je ne suis point *hazart*. Cellui est *hazart* qui joue sa femme aux dez' (Du Cange, *s.v.* 'hazardor').

**Hazel, Hazell, Hasel, Hasell, Haisell, Heazel, Heazell, Hessel, Hessels:** Alured *del Hesel* c1182 MELS (Wo); Gamel *Hesel* 1203 P (L); Hugh *de Hesill* 1204 AssY; William *atte Hasele* 1275 SRWo; John *Hasyl* 1327 *SR* (Ess); Henry *de Hessell* 1341 FrY. 'Dweller by the hazel', OE *hæsel*, ON *hesli*, as at Hessle (ER, WRYorks), Heazille Barton, Heazle Fm (Devon), etc.

**Hazelden, Hazeldene, Hazeldeane, Hazeldine, Hazeldon, Hazeltine, Hazledine, Hazzledine,**

Haiselden, Hasleden, Haseldene, Haseldine, Haseltine, Haszeldine, Hayselden, Haizelden, Hayzelden, Hayzeldene, Heselden, Heseldin, Heseltine, Heselton, Hesleden, Hesselden, Hesseltine, Hestleton: Robert *de Heseldene* 1243 AssDu; Alexander *de Haselindene* 1258 Kirkstall (Y); Reginald *de Haselden'* 1275 RH (W); William *de Heseldenn* 1296 SRSx; Richard *Hessletine* 1632 YWills. From Heselden (Durham), Haselden (Sussex), Haslingden (Lancs), Hazledon Fm (Wilts), Hazelton (Glos), *Heseldene* c1130, or from residence in a hazel-valley. There has been some confusion with *Hazelton*.

Hazelgrove, Hazelgreaves, Haselgrove, Haslegrave, Heaselgrave, Heselgrave, Hesselgrave, Hessel-greaves: Richard *de Hasilgref* 1297 SRY. From Hesselgreave (WRYorks), Hazel Grove (NRYorks), or residence in a hazel-valley (OE *hæsel*, ON *hesli* 'hazel' and ON *gryfja* 'hole, pit').

Hazelhurst, Hazlehurst, Haselhurst, Hasle-hurst: Henry, John *de Haselhurst* 1332 SRLa, SRSr. From Hazelhurst (Lancs, Surrey, Sussex) or 'dweller by a hazel-wood'.

Hazelman, Hazleman: For *atte Hasele. v.* HAZEL.

Hazelip: *v.* HASLOP

Hazelrigg, Hazlerigg: Alan *de Hesilrig* 1279 AssNb. From Hazelrigg (Cumb, Lancs, Northumb).

Hazeltine: *v.* HAZELDEN

Hazeltine, Haselton, Hastleton, Heselton, Hestleton: Robert *de Haselton* 1274 RH (Gl); Geoffrey *de Heseleton* 1275 RH (NF). From Hazelton Bottom (Herts), Hazelton Wood (Essex), both originally *Haselden*, Hazelton (Glos), or Hesselton (NRYorks). *v.* HAZELDEN.

Hazelwood, Hazlewood, Hazzlewood, Haselwood, Haslewood, Heaslewood, Heselwood, Heslewood, Hesselwood, Aizlewood: Ernald *de Heselwude* 1191 P (Y); John *Haselwod* 1332 SRSx. From Hazlewood (Devon, Derby, Suffolk, Surrey, WRYorks), Hesslewood (ERYorks), or residence near a hazel-wood.

Hazlett, Hazlitt, Haslett, Haslitt, Heaslett, Hezlet, Hezlett: (i) John *Haselette* 1332 SRSr; William *atte Hasellette* 1333 MELS (Sr). 'Dweller by a hazel-copse', as at Haslett Copse (Sussex), OE *\*hæslett*, a common term in Essex, Kent and Surrey. (ii) William *Hesilheued* 1421 FrY. From Hazelhead (Lancs, WRYorks), or 'dweller by the hazel-hill', OE *hēafod*.

Hazley, Heasley: Peter *de Hasele* 1295 IpmBerks; John *de Hasleye* 1327 SRSf. From Haseley (O, Wa, Wt), or North Heasley in North Molton (D).

Hazzard: *v.* HAZARD

Hazzledine: *v.* HAZELDEN

Hazzlewood: *v.* HAZELWOOD

Heacock: *v.* HICKOX

Head, Heads, Heed: (i) Hubertus *cum testa* 1130 P (D); Ralph *Heued'* 1166 P (Nf); Thomas *Hede* c1246 Calv (Y). A nickname from OE *hēafod* 'head'. (ii) Thomas *del Heved* 1275 RH (Nt); Roger *Byheved* 1285 AssCh; Walter *Othehede* 1331 DbCh. 'Dweller by the promontory or hill, or near the source of a stream or the head of a valley', OE *hēafod*.

Headlam: John *de Hedlum* 1243 AssDu; John *de Hedelam* 1356 FrY; John *Hedlam* 1461 TestEbor. From Headlam (Du).

Headland: John *de* (*del*) *Hevedlond* 1275 RH (Sf). 'Dweller by the headland', here on the Suffolk coast.

Headley, Hedley: Siward *de Hedeleia* 1148 Winton (Ha); Stephen *de Hedleye* 1327 SRWo; William *Headly* 1672 HTY. From Headley (Ha, Sr, Wo, WRY), Hedley (Du), Hedley on the Hill (Nb), or Headley Hall in Bramham (WRY).

Headman: John *Hauedman* 1317 AssK; William *Heuedman* 1327 SRSo; Henry *Headman* 1642 PrD. 'Head man, leader', OE *hēafod, mann*.

Heafield: *v.* HIGHFIELD

Heafford: *v.* HAYFORD

Heal, Heale, Heales, Hele: William *de Lahela* 1130 P (Ha); William *in Thehele* 1234 MELS (So); Roger *de Hele* 1242 Fees (D); Hugh *Attehele* 1306 FFSo. Hele and Heale are very common place-names in Devon, less frequent in Somerset, and occasionally noted in Hants, Wilts, Worcs and Surrey. The surname is from residence in a nook or secluded place, from ME *hēle* from OE *heale*, dative singular of *healh*, corresponding to the Anglian dative *hale*, frequent as HALE.

Heald, Held, Hield, Hields: Adam *de Helde* 1207 P (K); Richard *del Helde* 1246 AssLa; Hamon *Attehelde* c1260 ArchC 34; Eustace *ater Hylde*, Matthew *atte Hulde* 1296 SRSx; Robert *Heild* 1605 FrY; William *Hield* 1706 ib. From residence near a slope, OE *helde, hi(e)lde, hylde*.

Healey, Healy, Heeley: John *de Hely* 1284 AssLa; Ellota *de Helagh* 1379 PTY; William *Heley* 1439 IpmNt. From Healey (La, Nb, NRY, WRY), Heeley (WRY), or Healaugh (NRY, WRY).

Heaman: *v.* HAYMAN

Heams: *v.* EAMES

Heane, Heanes: Robert *Hene* 1275 RH (L); Robert *Heanes* 1655, *Hene* 1660, *Heaines* 1669 LeiAS 23. OE *hēan* 'poor, wretched'.

Heap, Heape, Heaps, Heep: (i) John *Hepe*, Richard *of the Heppe* 1401 AssLa. From Heap Bridge (La). (ii) Adam *de Hepe* 1296, Henry *del Hepe* 1301 Black. From the lands of Hepe (Roxburgh).

Heaphey, Heapy: Robert *de Hepay* 1332 SRLa; Oliver *Hepy* 1527 CantW. From Heapey (La).

Heaps: *v.* HEAP

Heapy: *v.* HEAPHEY

Heard: *v.* HERD

Hearder: *v.* HERDER

Heardman: *v.* HERDMAN

Hearle: *v.* EARL

Hearne: *v.* HERN

Hearnshaw, Hernshaw: John *Hernchagh* 1379 PTY. From Earnshaw in Bradfield (WRY).

Hearsom, Hearson, Hearsum: *v.* HERSOM

Heart: *v.* HART

Heartfield: *v.* HARTFIELD

Hease: *v.* HAYES

Heaselgrave: *v.* HAZELGROVE

Heaseman, Heasman, Hayesman, Easman: Simon *le Heysman* 1275 RH (Nf). 'Dweller in the brushwood'. *v.* HAYES. *Hayesman* and *Easeman* are found side by side in Kent.

Heasler: *v.* HASLER

Heaslett: *v.* HAZLETT

**Heaslewood:** v. HAZELWOOD

**Heasley:** v. HAZLEY

**Heaslip:** v. HASLOP

**Heath:** John *de la Heth* 1248 FFEss; Laurence *atte Hethe* 1296 SRSx; Peter *del Heth* 1297 SRY; Alan *Othehethe* 1332 SRSt. 'Dweller on the heath', OE *hǣð*.

**Heathcock:** Walter *Hathecok* 1274 RH (Ess). A nickname from the heathcock, the black grouse (1590 NED).

**Heathcote, Heathcoat, Heathcott:** Godfrey *de Hetcota* 1166 P (Db); Ralph *de Hethcote* 1227 AssSt. From Heathcote (Derby, Warwicks). In Yorkshire, pronounced *Hethkett*.

**Heathen:** Henry *le Hethene* 1249 AssW; Robert *le Hethene* 1268 IpmGl; Walter called *Hethene* 14th AD vi (He). A nickname from OE *hǣþen* 'heathen'.

**Heather:** John, Henry *le Hether* 1327 SRWo, 1332 SRSr. For *atte hethe*. v. HEATH.

**Heatherley:** v. HETHERLEY

**Heatherington:** v. HETHERINGTON

**Heathfield:** Livesa *de Hethfeld* 1276 RH (O); Adam, Isabel *atte Hethfeld* 1327 SRSo, 1332 SRSx. From Heathfield (Sussex) or dweller on the heathland (OE *hǣð, feld*).

**Heathman:** v. HEATHER

**Heathorn, Heathron:** v. HAYTHORNE

**Heaton:** Vincent *de Heton'* 1219 AssY; John *de Heton* 1296 SRNb; Thomas *de Heton* 1374 IpmLa; John *Heton* 1460 IpmNt. From Heaton (Nb, WRY), Heaton Norris, under Horwich, with Oxcliffe, Great, Little Heaton (La), Capheaton, Kirkheaton (Nb), Cleckheaton, Hanging Heaton, Kirkheaton (WRY).

**Heaven, Heavens:** Richard *Hevyn* 1279 RH (C); Richard *Heuene* 1327 SRWo; Richard *Overhevene* 1334 SRK. For HAVEN. v. also EVANS.

**Heaven(s):** v. EVANS

**Heaver:** Geoffrey *Heuere* 1297 MinAcctCo. A derivative of ME *heve* 'to lift, raise', a porter, carrier. (1586 NED). v. HEVER.

**Heaversedge:** v. HATHERSICH

**Heaward, Heaword:** Robert *Heaward* 1585 Shef; George *Heaword* 1681 ib. v. HOWARD.

**Heawood:** v. HAYWOOD

**Heazell:** v. HAZEL

**Hebb:** *Hebbe* Capell 1284 Wak (Y); Herberdus, Juliana *Hebbe* 1279 RH (Hu), 1296 SRSx. *Hebb*, a short form of *Hebert*, i.e. *Herbert*. v. HEBBARD.

**Hebbard, Hebard, Hebbert:** Henry, Reginald *Hebert* 1279 RH (O); Adam *Hebert* ib. (Beds). *Hebert*, for *Herbert*, with loss of the first *r*.

**Hebblethwaite, Hebblewaite, Hebblewhite, Hepplewhite, Ebblewhite:** Agnes *de Hebletwayt* 1379 PTY; John *Ebyllthwayte* 1455 GildY; John *Epulweyte* 1481 SIA xii; Johanna *Hebbylewhayt* 1526 GildY; Alexander *Hebylthwat* 1528 FrY; Christabel *Hebeltwhitt* 1624 LaWills. From Heblethwaite near Sedbergh (WRYorks).

**Hebborn, Hebburn:** v. HEPBURN

**Hebden, Hepdon:** Elias *de Heppedon'* 1208 Cur (Y); William *de Hebbeden* 1312 FFY; Edmond *Hebden* 1672 HTY. From Hebden, or Hebden Bridge (WRY).

**Hebditch:** John *de Hebbedych'* 1332 SRDo; Richard *Hebditch* 1641 PrSo; Robert *Hebditch* 1664 HTSo. 'Dweller by the ditch where wild roses or brambles grow', OE *hēope, dīc*.

**Heberden:** v. HIBBERDINE

**Hebson:** Thomas *Hebson* 1326 Wak (Y). 'Son of *Hebb*.'

**Heck:** John *del Hek* 1219 AssY, *de Hecke* 1242 Fees (Y). From Heck (WRYorks) or 'dweller by the gate', from *heck*, a northern form of *Hatch*.

**Heckingbottom:** v. HIGGINBOTHAM

**Heckler:** William *le Hekeler* 1297 Wak (Y). A derivative of ME *hekel* 'to heckle', a heckler or dresser of hemp or flax (1440 MED).

**Hector:** *Hector* 1190 P (Sf); *Hector* de Hilleg' 1222 Cur (Sf); Richard *Ector* 1306 AssW; Peter *Ectour* 1524 SRD; John *Hector* alias Jaques 1568 SRSf. The name of the Trojan hero, from the Greek via French. Rare in England, more common in Scotland where it was used for Gaelic *Eachdonn*. v. Black.

**Hedden, Heddon:** Ailwin *de Heddon* 1162 P; Thomas *de Heden* 1249 IpmY; William *de Hedone* 1303 FFY. From Headon (Nt), Headon Hill (Wt), Heddon on the Wall, Black Heddon (Nb), Hedon (ERY), Heddon in Filleigh, in Sutcombe (D).

**Hedge, Hedges:** Ailmar *Hegge* 1227 AssBeds; Robert *atte Hegge* 1256 AssSo; Richard *de Hegges* 1296 SRSx; Robert *del Hegges* 1301 SRY; Roger *in the Hegg* 1327 *SR* (Ess). 'Dweller by the hedge(s) or enclosure(s)', OE *hecg*.

**Hedgecock, Hedgcock, Hedgecoe:** John *Hechecok* 1327 SRC. Probably for HITCHCOCK. cf. ALCOE.

**Hedger:** Henry *le Hegger* 1327 SRSx; John *Heggere* 1332 SRSr. A parallel to HEDGMAN and HEDGE.

**Hedgman:** Richard *le Hegman* c1240 *ERO*; John *Heg(ge)man* 1305 HPD, 1327 *SR* (Ess); Ralph *Hedg(e)man* 1524 SRSf. 'Dweller by the enclosure.'

**Hedley:** v. HEADLEY

**Heed:** v. HEAD

**Heeley:** v. HEALEY

**Heep:** v. HEAP

**Heginbotham, Heginbottom:** v. HIGGINBOTHAM

**Heiden:** v. HAYDEN

**Heiffer, Heffer:** Hugh *le Hayfour* 1327 SRSo; Alexander *Heifers* 1642 PrD. OE *hēafore, hēafre* 'heifer'. Metonymic for a keeper of the heifers.

**Heighes:** v. HAY

**Heigho:** v. HAYHOE

**Height, Hight, Hite:** Henry *de la Heyt* 1275 RH (Db); Robert *atte Heyte* 1279 RH (O). 'Dweller at the height' or top of the hill, OE *hīehþu*, ME *heyt* 'height, top, summit'.

**Heighton:** v. HIGHTON

**Heighway:** v. HIGHWAY

**Held:** v. HEALD

**Helder:** Hugh, Cristiana *le Heldere* 1212 Cur (Herts), 1279 RH (C). Either (i) a derivative of OE *healdan* 'to hold', a tenant, occupier. cf. HOLDER; or (ii) a toponymic from OE *hylde* 'slope', equivalent to *atte held*, 'dweller on the slope'. cf. HEALD; or, possibly, for 'the elder' with inorganic *H*.

**Hele:** v. HEAL

**Heley:** v. ELY

**Helin:** v. ELION

**Helis:** v. ELLIS

**Hell:** Johannes, Fulco *filius Hel'* 1279 RH (Hu); Wulfwin, Roger *Helle* 1188 P (Ess), 1221 *ElyA* (Nf). This may occasionally be for HILL, preserving the south-eastern form: William *atte Helle* 1296 SRSx, but in surnames, as in place-names, this usually takes the standard form. The surname is clearly chiefly from a personal name *Helle*, a pet-form of *Ellis* (v. ELCOCK) or of some name in *Hil(d)-*, or perhaps of *Helen*. v. HELLCAT.

**Hellass:** v. HILLHOUSE

**Hellcat, Hillcoat:** Ralph, Walter *Hellecoc* 1202 AssL, 1219 AssY; Roger, John *Hellecok(e)* 1276 RH (Gl), 1327 Pinchbeck (Sf). v. ELCOCK, HELL.

**Hellen(s):** v. ELION, ELLEN

**Heller:** Alexander *Huller*, *Heller* 1327, 1332 SRSx. A derivative of ME *hell*, *hull*, OE *hyll* 'dweller on the hill'.

**Helliar, Hellier, Hellyer, Helyer, Hilliar, Hillier, Hillyar, Hillyer, Hilyer:** Robert *le Heliere* 1275 RH (K); Gilbert *le Helyere* 1280 MESO (Ha); Robert *le Hillier* 1347 Cl (Beds). A derivative of OE *helian* 'to cover, roof', a slater or tiler. cf. Walter *Helier* vel *Tiler* c1450 NED.

**Hellin(s):** v. ELION, ELLEN

**Helling, Hellings:** Ralph *de Helling'* 1191 P (L). From Healing (Lincs). v. also ELION.

**Hellis:** v. ELLIS: Probably also for HELLASS.

**Helliwell, Hellewell, Hellawell, Hellowell:** Thomas *de Heliwelle* 1285, Richard *de Heliwall* 1297 Wak (Y). From Helliwell in Worsborough (WRY).

**Hellman, Helman:** William *Helman* 1274 RH (Ess). 'Dweller on the hill.' cf. HELLER.

**Hellon:** v. ELION

**Hellyer:** v. HELLIAR

**Helm, Helme, Helms:** Nicholas *de Helm* 1180 P (Sr); William *Helmis* 1279 RH (Bk); Hugh *del Helm* 1296 Wak (Y); Richard *ate Helme* 1327 SRWo; William *del Helmes* 1366 SRLa. OE *helm* 'a covering', later, in dialects, 'roofed shelter for cattle', used, presumably for a herdsman.

**Helmsley:** v. HEMSLEY

**Help, Helps:** *Helpe* Arbalistarius 1181 P (L); Walter *Help* 1230 Pat (Nb); Gilbert *Helpe* 1327 SRSf. ON *Hjalpi*. Perhaps also a nickname from OE *help* 'aid, assistance': Alnoit *le Help* 14th AD iv (K). cf. Simon *Helpusgod* 1296 SRSx 'may God help us'.

**Helston, Helstone:** Adam *de Helleston'* 1177 P (Co); Richard *de Hellestone* 1297 MinAcctCo. From Helston in Kerrier, or Helstone in Trigg (Co).

**Hely:** v. ELY

**Helyer:** v. HELLIAR

**Hembrey, Hembry, Hemery:** v. AMERY

**Hembrow, Hembury:** v. EMBERY

**Hemes:** v. EAMES

**Hemmett:** v. EMMATT

**Hemming, Hemmings:** *Hemmingus* 1066 ICC; *Hemming* de Welega 1166 P (Ess); Walter, William *Hemming* 1170 P (So), 1221 AssWo. ON *Hem(m)ingr* ODa *Hem(m)ing*.

**Hemington, Hemmington:** John *de Hemyngton* 1279 RH (Hu); Reginald *de Hemyngton* 1327 SRSx. From Hemington (Lei, Nth, So).

**Hemmingway, Hemingway:** John *de Hemyngway*, William *Hemyngway* 1379 PTY; Richard *de Hemmyngway* 1309 Wak; Robert *Hemmyngwaye* 1561 Pat (Y). 'Dweller by Hemming's path', from an unidentified minor place, probably in Yorks.

**Hempenstall:** v. HEPTONSTALL

**Hemphrey, Henfrey:** *Heinfrid'* de Colecestr' 1197 P (Ess); *Hemfridus* 1214 Cur (Nth); Margareta *Henfrey* 1327 SRSf; Wylliam *Hemffrey* 1524 SRSf. OG *Hainfrid*, *Heinfrid*, OFr *Hainfroi*.

**Hempson, Hemson:** Roger *Hemson*, *Hempson* 1524 SRSf. For EMSON, or *Hemingson*: Suein *Hemmingessune* 1166 P (Nt).

**Hempstead, Hemstead, Hemsted:** Hamo *de Hemstede* 1193 P (Sf); John *de Hemsted* 1275 RH (Nf); Simon *de Hemstede* 1296 SRSx. From Hempstead (Ess, Gl, Nt), Hemel Hempstead (Herts), or Hempstead in Gillingham, Hemsted in Benenden, in Lyminge (K).

**Hems:** Walter *de Hemma* 1182 P (Sa); Richard *de la Hemme* 1275 RH (W); John *en le Hemme* 1300 AssSt. From The Hem (Salop), Hem (Montgomery), or 'dweller by the border', OE *hemm* 'hem'.

**Hemsley, Hemesley, Hemsworthy:** William *de Helmeslac* c1160–83 YCh; Walter *de Helmeslay* 1260 IpmY; William *de Helmesley* 1373 FFY; William *Hemsley* 1477 IpmNt. From Helmsley (NRYorks), *Elmeslac* DB, *Helmesley* 12th, *Hemesley* 1548.

**Hemson:** v. HEMPSON

**Hemsworth, Himsworth:** Adam *de Himeswurth'* 1219 AssY; Henry *Hemsworthe* 1621 SRY; John *Hemsworth* 1672 HTY. From Hemsworth (WRY).

**Henchcliffe:** v. HINCHCLIFFE

**Hencher:** v. HENSHALL

**Henday, Hendey:** v. HENDY

**Hende:** Henry *Hende* 1327 SRLei; Thomas *Hende* 1343 FFEss; Gresilda *Hend* 1509 LP (Ess). OE *(ge)hende* 'courteous, handsome'. cf. Richard *Hendchyld* 1357 Black 'handsome child'.

**Henden, Hendon:** Gilbert *de Hendon'* 1206 Cur (Mx); Richard *de Hendene* 1327 SRSx; John *Hendon* 1397 IpmGl. From Hendon (Du, Mx). *Henden* may sometimes be from a feminine name: *Hendina* (f) 1219 P (C/Hu).

**Henderson:** William *Henrison* 1374 Black; David *Hennerson* 1504 ib.; James *Hendirsoune* 1553 ib. 'Son of *Henry*.' cf. HENDRIE.

**Hendisson:** William *Hendissone* 1327 SRSf. 'Son of *Hendy*.'

**Hendman:** v. HENDYMAN

**Hendra:** John *del Hendre* 1370 MinAcctCo. From one of the Cornish places named Hendra, *hen dref* 'old homestead'.

**Hendrick, Hendricks:** *Hendricus* 1188 BuryS (Sf); *Hendricus* Prid 1212 Cur (Y); Alice daughter of Henry de Sandford (seal: S. ALIS. FILIA. HENDRICI) 13th AD v (Ess); John *Hendrich* 1279 RH (C). OFr *Henri* with intrusive *d*.

**Hendrie, Hendrey, Hendry:** *Hendrie* Ralstoun 1519 Black; *Hendrye* Stanford 1593 EA (OS) ii (Sf); John *Hendre* 1359 Putnam (Co); Hendyre *Hendry* 1562 Black. OFr *Henri*, with intrusive *d*. v. HENRY.

**Hendy, Henday, Hendey:** Henricus *filius Hendy* 1279 RH (O); William *Hendi* 1198 FF (Sr); Robert *le*

*Hendy* 1275 RH (Nf). ME *hendy* 'courteous, kind, gentle', used also as a personal-name. *Hendyman* was also a common surname and a personal-name: *Hendeman* del Ho 1259 AD iii (K); *Hendeman* Oldlegh 1260 AssC.

**Hendyman, Hendman:** *Hendeman* holdeye 1221 *ElyA* (C); *Hendeman* Oldlegh 1260 AssC; Robert *Hendeman* 1256 AssNb; Richard *Hendiman* 1317 AssK; William *Hendman* 1338 FFY. 'Courteous, handsome man', ME *hende/hendy, mann*. Used also as a personal name. Sometimes, perhaps, 'servant of *Hendy*'.

**Hendyside:** *v.* HANDASYDE

**Heney, Henney:** Richard *de Heneye* 1277 *Ely* (Sf); Edmund *de Heney* 1327 SRC; John *Henney* 1380–1 PTW. From Great, Little Henny (Ess).

**Henfoot:** John *Hennefot* 1306 IpmGl. 'Hen foot', OE *henn, fōt*. cf. Geoffrey *Henneheued* 1195 P (Cu) 'hen head'.

**Henfrey:** *v.* HEMPHREY

**Hengham, Hingham:** Ralph *de Hengham* 1275 SRWo; Ralph *de Hengham* 1303 IpmY. From Hingham (Nf), *Hengham* 1158.

**Henkin, Hinkins:** (i) *Hanekin, Henekin* le Rede 1299 LoCt; William *Henekyn* 1327 *SR* (Ess); John *Hinkins* 1787 Bardsley. A diminutive of *Hen* (Henry). cf. HANN, HANKIN. (ii) Thomas *Hendekyn* 1337 LLB F. A diminutive of *Hendy*.

**Henley, Henly:** Gilbert *de Henlega* 1181 P (Wo); Robert *de Henlay* 1270 IpmY; William *Heneley* 1392 IpmNt. From Henley (Oxon, Salop, Som, Suffolk, Surrey, Warwicks).

**Henlow:** Hugh *de Henlawe* 1201 Cur. From Henlow (Beds).

**Henman:** William *Henman* 1327 SRSo; Richard *Heneman* 1327 SRSf. Probably 'man in charge of the hens'.

**Henn:** (i) *Henna* Curi 1192 P (Lo); Johannes *filius Hen* 1275 RH (Nt); Euerard *Henne* 1202 P (L). *Henn*, like *Hann*, was a pet-name for *Henry*. cf. HENKIN. *Henna* is the corresponding feminine. Forssner cites *Henrica* (without date). (ii) Colemannus, Thomas *le Hen* 1275 RH (Sf). A nickname from the hen.

**Henner, Hennah:** William *de Henner* 1279 RH (C); Henry *de Henouere* Ed I DbCh; Henry *Henner* 1327 SRWo. From Hennor (Do), or Heanor (Db).

**Hennessey, Hennessy, Hennesy, Henesy:** Irish *Ó hAonghusa* 'descendant of *Aonghus*' (one-choice).

**Henney:** *v.* HENEY

**Henniker:** John *Hennacre* 1474 CantW; Robert *Hennaker* 1531 KentW; Charles *Henicar* 1660 ArchC 30. 'Dweller by the hen field', OE *henn, æcer*.

**Hennion:** *v.* ENNION

**Henry, Henrey, Henery, Heneries, Fitzhenry:** *Henricus* 1066, 1086 DB; Thomas *Henery* 1275 RH (K); Richard *Henry* 1293 Fees (D); Adam *Henris* 1323 AssSt; John *Fitz Henrie* 1346 Ipm. OFr *Henri*, OG *Haimric, Henric* 'home-rule', after the Conquest one of the most popular Norman names. The English form was *Herry* or *Harry*, with pet-forms *Henn*, *Hann*, and diminutives *Henkin, Hankin, Henriot*, and a pet-form *Hal* which may survive in *Hawkins. v.* also HARRY, HERIOT, HENDRIE.

**Henryson:** Richard *Henrison* 1343 IpmNt; Nicholas *Henryson* 1381 PTY; William *Henrison* 1438–9 TestEbor. cf. Joan *Henrydoughter* 1379 PTY. 'Son of *Henry*'.

**Hensall, Hensell, Henzell:** (i) Alan *de Hethensal'* 1219 AssY. From Hensall (WRY). (ii) *Hensell* and *Henzell* are also Huguenot, from a Protestant refugee who settled at Newcastle upon Tyne after the Massacre of St Bartholomew. With two other refugees, Tysack and Tittory, he started a glassworks there (Smiles 401–2).

**Henshall, Henshaw, Hensher, Hencher:** (i) Peter *de Henschal* 1332 SRCu. From Henshaw (Northumb). (ii) Richard *de Henneshagh* 1365 Bardsley (Ch); Sysley *Henshall* (*Henshawe*) 1570 ib. From Henshaw in Prestbury (Ches).

**Hensley:** John *de Henselay* 1297 SRY; Henry *Hensleigh* 1662–4 HTDo. From Hensley (D).

**Hensman, Hinckesman:** William *Henxman* 1413 Cl; Thomas *Hengysman* 1460 RochW; John *Hengestman* 1473 SIA xii; Thomas *Hinchman* son of Thomas *Henchman* 1674 Bardsley. OE *hengest* 'stallion' and *mann*, a groom, later a squire or page of honour, and also 'sumpter-man, carrier'.

**Henson:** (i) Hugh Balle *Heyneson* 1304 Cl. 'Son of *Hayne*' *v.* HAGAN. (ii) William *Hendisson* 1327 SRSf; William *Henson* 1381 PTY. 'Son of *Hendy*.'

**Henton:** Adam *de Henton'* 1200 P (Ha); Laurence *de Henton* 1258 AssSo; Thomas *Henton* 1545 SRW. From Henton (O, So).

**Hentwin, Hentwind:** Ralph *Hentewinde* 1292 KB. 'Seize the wind', OE *hentan, wind*. cf. Walter *Henteloue* 1327 SRWo 'seize the wolf'.

**Henty:** Robert, John *de Hentye* 1327, 1332 SRSx. From Antye Fm in Wivelsfield (Sussex).

**Henzell:** *v.* HENSALL

**Hepburn, Hebburn, Hebborn:** Adam *de Hepburne* 1271 Black; Thomas *de Heburn'* 1279 AssNb; John *Hebburn* 1407 FrY. From Hepburn (Northumb), or Hebburn (Durham).

**Hepdon:** *v.* HEBDEN

**Heppastall:** *v.* HEPTONSTALL

**Hepple, Heppel, Heppell, Hepples:** Henry *de Heppal* 1354 Putnam (Ch). From Hepple (Nb).

**Hepplewhite:** *v.* HEBBLETHWAITE

**Heptonstall, Heptinstall, Heppenstall, Heppinstall, Hempenstall:** John *de Heptonstall* 1296 Wak (Y); William *Heptinstall* 1558 RothwellPR (Y); Matthew *Hepinstall* 1668 ib. From Heptonstall (WRYorks).

**Hepworth, Hipworth:** Eadward *de Heppeworde* 1121–48 Bury; Stephen *de Heppeworth'* 1330 FFEss; Richard *Hyppeworthe* 1387 Petre; John *Hepworth* 1414 IpmY. From Hepworth (Suffolk, WRYorks).

**Herald, Heraud, Herold, Herod, Herrald, Herrold, Herrod:** *Herouldus, Hairaudus* 1066 DB; William *Herode* 1279 AssNb; John *Heraud* 1296 SRSx; Peter *Herodes* 1297 SRY; Seman *Erode, Harrold, Herodes* 1297 SIA x. These surnames may derive from ON *Haraldr*, OG *Hairold, Herold* or from OE *Hereweald. v.* HAROLD. They may also be from ME *heraud, herault, herode, haraude* 'herald' (c1385 MED), from OFr *heraut, herault*. Weekley explains *Herod* as a stage braggart. Dauzat similarly derives the rare

French *Herode* from the Biblical name which Michaëlsson considers a suitable nickname from the name of the King of the Jews who slaughtered the Innocents and is often mentioned in the *chansons de geste*. There seems to be no clear evidence for this in England. In Wace's *Roman de Rou*, Harold occurs as *Heraut*.

**Herbage, Harbach, Harbage, Harbidge:** Probably for *atte herberge* from OFr *herberge* 'hostel, shelter'. Equivalent to HARBISHER.

**Herbelot, Harblott:** *Herbelott* Someter 1296 SRSx; Nicholas *herblot'* 1327, Richard *Herbelot* 1393, John *Herblet* 1447 CarshCt (Sr). *Herb-el-ot*, a double diminutive of *Herb*, a short form of *Herbert*.

**Herbert, Herbit, Harbard, Harberd, Harbert, Harbird, Harbord, Harbot, Harbott, Harbud, Harbut, Harbutt, Fitzherbert:** *Herbertus, Hereberd* 1086 DB; *Herebertus* capellanus 1148–56 Bury (Sf); William *Herebert* 1206 P (Do); Richard *Herbert, Herebert, Herberd* 1221 AssWo; Johannes *Herberti* 1230 Cl (Nf); Reginald *le Fitz Herbert* 1347 Ipm; Christopher *Harbart* 1550 FrY; Philip *Herbert* son of John *Harbart* 1609 FrY; William *Harbatt* son of Richard *Harbart* 1626 FrY; William *Harbort* 1626 Bardsley. OFr *Herbert* from OG *Hariberct, Her(e)bert* 'army-bright'. Introduced by the Normans, but much less common than *Hugh*. Diminutives were formed but have not survived: *Herbelott* Someter 1296 SRSx, Richard *Herbelot* 1327 ib., Alexander *Herbelyn* 1376 ColchCt, Robert *Herbekyn* 1296 SRSx.

**Herbertson, Harbertson, Harburtson, Herbinson, Herbison, Harbinson, Harbison:** Archibald *Herbertson* 1525 Black (Glasgow); William *Herbesone* 1555 ib. (Berwick); Richard *Harbertson* 1605 ib. (Glasgow). *Herbisone* is found in Glasgow in 1551 and the development is shown in *Harberson, Harbeson, Harbisone* 1706 (Black). An intrusive *n* in this last gave *Harbinson*.

**Hercock, Hircock:** John, Henry *Hercok* 1327 SRSf. *Her*, a short form of *Herbert* or *Herman* and *cock*. cf. *Here* (*Herbertus* CR) de Oxhei 1230 P (Herts).

**Hercules, Herkless, Arkless, Arculus:** Hercules Dykis 1567 Black (Kelso); *Hercules* Loveden 1592 AD v (Berks); *Hercules* Gibberd 1642 PrD; John *Harklyes* 1569 Oriel. A Greek name which continued in use in Cornwall until the 19th century. In Shetland it was used to render ON *Hákon*, whilst *Arkless* may sometimes be from Erchless in Kiltarlity (Inverness).

**Hercy:** *v.* HERSEY

**Herd, Heard, Hird, Hord, Hurd:** William *Lehird* Ric 1 Cur(L); Thomas *Hord'* 1221 AssSa; Reginald *le Herd* 1243 AssSo; Richard *le Hurde* 1296 SRSx. OE *hierde* 'herd', 'herdsman'. For the dialectal forms, *v.* MEOT 254–7.

**Herder, Hearder, Horder:** Nicholas *le Herder* 1327 SRSo; John *Hurder* 1333 MEOT (So). A rare name. A derivative of ME *herden, hurden* 'to take care of, to tend', from OE *hierde* 'herd'. 'A herdsman.'

**Herdman, Heardman, Hurdman:** Osbert *Hirdman* 1166 P (Nf); Simon *le Hyrdeman* 1181 P (K); Henry *Erdman* 1260 AssC; John *le Herdman*, Walter *Herdman* 1296 SRSx. OE *hierdeman* 'herdman', one who tends sheep, cattle, etc.

**Herdsman:** Robert *Herdesman* 1367 ColchCt (Ess). 'Servant of the herd', or more probably 'of one named Herd'.

**Herdson:** Thomas *Hirdson* 1332 SRCu; Gilbert *Herdson* 1415 PrGR. 'Son of the herd.'

**Here:** Simon *le Here* 1275 RH (Nf); Alan *Here* 1276 RH (Y). OE *hēore, hȳre*, ME *here* 'gentle, mild, pleasant'.

**Hereford, Herford, Harford, Hartford:** Ilbert *de Hertford* 1086 DB, *de Hereforda* 1086 ICC (Herts); Walter *de Hereford* 1158 P (Wa); Richard *de Herfordia* 1185 Templars (Herts). From Hertford (in which the *t* was early lost as in the modern pronunciation *Harford*), Hereford, or Harford (Devon, Glos).

**Hereward, Harward:** *Hereuuard, Heruart* 1066 DB; *Herewardus* de Barneby 1219 AssY; Walter, Alexander *Hereward* 1156–80 Bury (Sf), 1221 AssGl; Reynold *Harward* 1523 SRK. OE *\*Hereweard* 'army-guard'.

**Hering:** *v.* HERRING

**Heriot, Herriot, Herriott, Heryet, Herratt, Herrett, Herrits, Herritts:** German *de Heriet* (*de Herierð*) 1176 P (Ha); Richard *de Heriett, de Herierd* 1197–8 FF (Bk, Beds). From Herriard (Hants). The Scots *Heriot* is from the lands of Heriot (Midlothian): Laurence *de Herryhot* 1221–38 (Black).

**Heritage, Herrtage:** John *Heritag'* 1279 RH (O); John *Erytage* 1279 RH (Hu); William *Heritage* c1280 Whitby (Y). ME *heritage* from OFr *eritage, heritage* 'property, especially land, which devolves by right of inheritance'; used in 1390 of heirs collectively. Here, perhaps, an heir.

**Herman, Hermon:** *v.* HARMAN

**Hermer:** *v.* HARMER

**Hermiston:** *v.* HARMSTON

**Hermitage, Armatage, Armatys, Armitage, Armytage:** Richard *de Ermitage* 1259 AssCh; Hugh *del Hermytage* 1296 PN Wa; John *de Armitage* 1423 Shef (Y). ME (*h*)*ermitage*, OFr *hermitage* 'hermitage' (c1300 MED). From Armitage (Staffs), Hermitage Fm in Little Packington (Warwicks), where there were early hermitages, or 'dweller by or in a hermitage'.

**Hermitte, Armett, Armit, Armitt:** William *lermite* 1196 P (Y); William *le Heremit* 1208 Cur (Y); Andrew *Ermite* 1255 RamsCt (Hu); Thomas *Harmyt* 1526 RochW; John *Harmet* 1549 GildY. ME *hermite, ermite*, OFr (*h*)*ermite* 'hermit' (a1225 MED).

**Hern, Herne, Hearn, Hearne, Hurn, Hurne, Harn:** Gunnora *de la Hurn'* 1212 Cur (Ha); Walter *Atehurne* 1267 AssSo; Henry *en le Hurne* 1279 RH (Bk); Ralph *in þe Hurne* 1279 RH (O); William *del Herne* 1327 SRSf; John *ate Hirne* 1327 SRSx; Robert *Herne*, John *Harne* 1524 SRSf. From residence in a nook or corner of land or in a bend (OE *hyrne*) as at Herne (Kent) or Hirn (Hants). For the variation in the vowel, cf. HERST. *Hern* is also a ME form of HERON.

**Hernaman, Herniman, Harniman, Horniman:** Hugo *Herneman* 1428 FA (W); John *Hernaman*, William *Hurnaman*, Richard *Hernaman*, 1547, 1668 DWills. Identical in meaning with *atte Herne*.

**Hernshaw:** *v.* HEARNSHAW

**Herod, Heroid:** v. HERALD

**Heron, Herron, Herroun, Hairon:** (i) Ralph *Hairun* 12th DC (L); William *Herun* 1212 Fees (Do); Roger *Heirun* 1221 AssWa; John *Heroun* 1298 LLB B. OFr *hairon*, AFr *heron*, ME *heiroun*, *heyron* (a1302 NED), *herne* (13..), *heron* (c1386) 'heron', a nickname for a thin man with long legs. v. HERN. (ii) Drogo *de Hairum, de Harum* 1131, 12th Riev (Y), Drogo *Hairun* 1196 P (Y); William *de Harum, de Herun* c1150, 1175 ib. (Y). From Harome (NRYorks), DB *Harun, Harem, Harum*. This has clearly been confusion with the more common nickname. Jordan *Herum, Hayrun* 1232, 1234 Riev and William *Herun, Heyrun* 1256 AssNb were probably of the same family.

**Herratt, Herrett:** v. HERIOT

**Herrick:** *Eiric, Eric, Erich* 1066 DB; Siwate *filius Airic* Hy 2 DC (L); John *Eirich* 1211 FrLeic; Nicholas *Heyryke* 1524 ib.; John *Eryke* 1568 ib.; Edward *Hericke* 1620 ib. ON *Eirikr*, ODa, OSw *Erik*.

**Herries:** Henry *Heriz* 1214 P (Co); Nicholas *le Herice* 1296 CartNat (Nth); John *Herrys* 1433 FFC; George *Herrys* 1536 FFEss. Fr *hérissé* 'rough, prickly, shaggy'. The name probably often fell in with HARRIS.

**Herring, Hering, Harenc:** Ralph *Hareng* 1166 Eynsham (O); Peter *Harang* 1210 Cur (Y); Nigel *Haring* 1275 RH (C); Roger *Hering* 1279 RH (O). Early examples are from OFr *hareng* 'herring'; later we have ME *hering* from OE *hæring, hēring*. Metonymic for a dealer in herrings. cf. Theobald *le Heryngmongere* 1212 Cur (Berks), Symon *Haryngbredere* 1274 RH (Lo), Walter *le Heringman* 1327 SRSf. *Hering* is common in Suffolk in 1327 SR.

**Herringcarter:** Robert *Heryngkartere* 1332 SRLo. 'A carrier of herrings', OE *hǣring*, and a derivative of ME *carte* 'cart'. cf. Isabel *Heryngquen* 1304 AssW 'herring-woman'.

**Herringham:** v. HARDHAM

**Herrington:** Walter *de Herindenn* c1200 ArchC vi; Thomas *de Heringtone* 1291 FeuDu; William *Heryngton* 1415 IpmY. From Herrington (Du), or Heronden in Tenterden (K).

**Herriott:** Walter *Henriot* 1327 SRSo; Henry *Henryot* 1327 SRSt. A diminutive of *Henry*. v. also HERIOT.

**Herrits:** v. HERIOT

**Herrod:** v. HERALD

**Herroun:** v. HERON

**Herrtage:** v. HERITAGE

**Hersant:** v. HARSANT

**Hersey, Hersee, Hershey, Hercy:** Hugh *de Hersy* 1201 MemR; Richard *Herci* 1275 RH (Nf); Richard *le Hercy* 1351-2 FFWa. From Hercé, or Hercy (Mayenne).

**Hersom, Herson, Hearsom, Hearson, Hearsum, Harsom, Harsum:** William *Herisson* 1212 P (Gl); William *Herlicun, le Herlicun* 1221 Cur (Nth); Robert *Herison* 1292 IpmD. A nickname from OFr *hérisson* 'hedgehog'.

**Herst, Hirst, Hurst, Horst:** Thomas *de Herst* 1066 DB (Sx); Helias *de Hirst* 1177 Templars (Y); Walter *del Hurst* 1196 P (Bk); Robert *de la Hurste* 1214 Cur (Sr); Nicholas *Horst* 1220 AssSt; William *del Hirst* 1275 Wak (Y); Henry *Attehurst* 1277 AssSo; William

*ater*, Robert *ate*, Peter *de la Herst* 1296 SRSx; Joan *upe*, Geoffrey *uppe the Hurst* 1332 SRSx. OE *hyrst* 'wood, wooded hill'. The surname may derive from Hirst (Northumb), Hirst Courtney (Yorks), Temple Hirst (Yorks), Hurst (Berks, Kent, Warwicks) or Hurstpierpoint (Sussex), earlier *Herst*, or, more frequently, from residence near a wood or wooded hill. The variant forms are due to different developments in ME of OE *y* which, in general, became *e* in the south-east, especially Essex and Kent, *i* in the north and the east midlands and *u* in the west and central midlands and the southern counties.

**Herve:** A Jersey name. v. HARVEY.

**Hervey:** v. HARVEY

**Heryet:** v. HERIOT

**Heselden, Heseldin, Heseltine:** v. HAZELDEN

**Heselgrave:** v. HAZELGROVE

**Heselton:** v. HAZELTON

**Heselwood, Heslewood:** v. HAZELWOOD

**Hesketh, Heskett, Heskitt:** Robert *de Heskeythe* 1332 SRLa; John *de Hesketh* 1338 WhC; Thomas *Hesketh* 1390 AssLa. From Hesketh (Lancs, NRYorks), or Hesket (Cumb).

**Heslam:** v. HASLAM

**Heslin, Hesling:** v. HASLIN

**Heslington:** Geoffrey *de Heselington'* 1190 P (Y); William *de Heselington'* 1219 AssY; William *de Haselingden* 1332 SRLa. From Heslington (ERY), or Haslingden (La).

**Heslop:** v. HASLOP

**Hespe:** v. APPS

**Hessel:** v. HAZEL

**Hesselden:** v. HAZELDEN

**Hesselgrave, Hesselgreaves:** v. HAZELGROVE

**Hesselwood:** v. HAZELWOOD

**Hessey:** Robert *de Hessey* 1208 Cur (Y), *de Hessay* 1219 AssY. From Hessay (WRY).

**Hestleton:** v. HAZELTON

**Heston:** William *de Hestone* 1321 CorLo. From Heston (Mx).

**Hetherington, Heatherington:** Richard *de Hetherington* 1298 AssL; Edmund *de Hetherynton* 1316 AssNth; Mr *Etherington* 1672 HTY. From Hetherington (Northumb), or 'dweller at the enclosure on the heath'.

**Hetherley, Heatherley:** Philip *de Hetherlegh* 1246 IpmGl. From Down, Up Hatherley (Gl), *Hetherlegh* 1221.

**Hethersett:** Hamo *de Hidersete* 1287-8 NorwLt; Richard *of Hethersett* 1316 AssNf. From Hethersett (Nf).

**Heugh:** v. HOW

**Hever:** Walter *de Hevre* 1203 Cur (K). From Hever (Kent). v. HEAVER.

**Hew:** Dyonisia *le Hewe* 1279 RH (Bk); Thomas *le Heue* 1283 SRSf. ME *hewe* 'a domestic, a servant', a singular formed from ME *hewen*, OE *hīwan* 'members of a household, domestics'. v. also HUGH.

**Hewar:** v. HEWER

**Heward, Hewart:** v. HOWARD

**Hewartson, Hewertson, Huartson, Huertson:** 'Son of *Huard.*' v. HOWARD.

**Hewat:** v. HEWET

**Hewell:** Thomas *Huhel* 1260 AssC; Sampson *Hugheles* 1296 SRSx. A diminutive of *Hugh.* cf. HEWLETT, HEWLINS.

**Hewer, Hewar:** Alwinus *Heuere* 1066 Winton (Ha); Hugh le *Hewer* 1255 *Ass* (Ess); Benedict le *Huwere* 1279 RH (C). ME *hewere* (*huwere*), a derivative of OE *hēawan* 'to hew', 'a hewer' (a1382 MED), a cutter of wood or stone, probably the latter.

**Hewertson:** *v.* HEWARTSON

**Hewes:** *v.* HUGH

**Hewet, Hewett, Hewat, Hewit, Hewitt, Howat, Howatt, Howett, Howitt, Huet, Huett, Huitt, Huot:** (i) Roger *Huet, Huiet* 1182, 1185 P (D); William *Heuet* 1221 AssSa; Roger *Hughet* 1280 AssSo; John *Howet* 1316 *ERO*; Roger *Howat* 1327 *SR* (Ess); Ricot *Huot,* Thomas *Huwet* 1327 SRSx; William *Howit* 1444 FrY; Alisoune *Hewat* 1662 Black. Diminutives of *Hugh.* cf. HUGGETT. (ii) Occasionally local in origin, from residence in a clearing. At Hewitts in Chelsfield and at Hewitts in Willesborough (Kent) lived families named *de la Hewatte* (1270), *de la hewett* (1301), *atte Hewete* (1338 PN K 16, 421). OE *hīewett* 'cutting', here used of a place where trees had been cut down.

**Hewetson, Hewitson, Hewison, Huetson, Huison, Huitson:** Henry *Hwetsone* 1363 FrY; Richard *Huetson* 1379 PTY; Thomas *Hewetson* 1489 LLB L; John *Hewyttson* 1544 FrY. 'Son of *Hewet.*'

**Hewgill:** *v.* HUGILL

**Hewick:** Nicholas *de Hewic* 1219 AssY; Agnes *de Hewyk* 1327 SRY; Anthony *Hewicke* 1672 HTY. From Bridge, Copt Hewick (WRY).

**Hewin(s):** *v.* EWAN

**Hewish:** *v.* HUISH

**Hewison, Hewitson:** *v.* HEWETSON

**Hewit, Hewitt:** *v.* HEWET

**Hewkin:** *v.* HUKIN

**Hewlett, Hewlitt, Howlett, Hulatt, Huleatt, Hulett, Hullett, Hullot:** Agnes *Hughelot* c1248 Bec (Nf); Thomas *Huwelot* c1250 Rams (Hu); Richard *Hulot* 1275 RH (Sf); Walter *Howlot, Hughlot* 1310, 1311 LLB D; John *Huelot* 1327 SRWo; John *Hoghlot* 1357 Calv (Y); Robert *Hulat* 1381 SRSf. A double diminutive of *Hugh, Hugh-el-ot.*

**Hewlins, Hewlings, Howlin, Howling, Howlings, Huelin, Huglin, Hulance, Hulin, Hullin, Hullins:** *Hugelyn* Bourþeyn, *Hugelinus* cubicularius 1052–65 Rams (Hu); *Hvchelinus* parcherius 1066 Winton (Ha); Galfridus *filius Hugelini* 1169 P (Do); *Huelinus de Fednes* 1221 Cur (Bk); Robert *Huelin* 1202 P (Wa); Hugo *hugelini* 1222 DBStP; William *Hugelin* 1230 P (Ha); William *Huhelyn* 1267 AssSo; Richard *Hulin* 1275 RH (Sf); John *Huwelyn* 1327 SRWo; Thomas *Howelyn* 1327 SRSf; John *Hullyng* 1428 FA (W); Robert *Howlinge* 1549 NorwW (Ess). OFr *Hugelin, Huelin, Hulin,* from OG *Huglin,* a diminutive of *Hugo.* There was also a feminine form: Tomas *filius Hugolinae* 1185 P (C), *Hugolina* 1219 AssY, 1309 FFEss.

**Heworth:** Roger *de Heworthe* 1284, Hamon *de Heworth* 1303 IpmY. From Heworth (Du, NRY).

**Hews:** *v.* HUGH

**Hewson:** *v.* HUGHSON

**Hewster:** Richard *le Heustere* 1327 SRSa; Richard *le Heuster'* 1332 SRSt. A derivative of OE *hīwian* 'to colour', with the feminine ending *-estere,* a dyer.

**Hext:** Walter *Hexte,* Nicholas *Exte* 1327 SRSo. ME *hext,* from OE *hēhst* 'highest', 'tallest'.

**Hey-:** *v.* HAY-

**Heyden, Heydon:** *v.* HAYDEN

**Heyer:** *v.* AYER

**Heyne(s):** *v.* HAGAN, HAIN

**Heyo:** *v.* HAYHOE

**Heyworth:** *v.* HAYWORTH

**Hezlett:** *v.* HAZLETT

**Hiatt:** *v.* HIGHET

**Hibbard, Hibberd, Hibbert:** *v.* ILBERT

**Hibberdine, Heberden:** Ralph *de Iburgedenn* 1279 *Ass* (Sx); William *de Hyburdenn, de Hyburgdenn, de Hybourghedene* 1296, 1327 SRSx; Matthew, Henry *Hiberden* 1507 SxAS xl; Richard, John *Hyberden, Hyberdyn* 1525 SRSx; Roger *Heberden* 1525 FFEss; Nicholas *Hibberdine* 1622 GreenwichPR. From a lost Heberden in Madehurst (Sussex), last recorded in 1795.

**Hibberson:** *v.* IBBERSON

**Hibbin, Hibbins:** James *Hybbyns* 1555 Pat; Thomas *Hebins* 1560 Bardsley. *Hibb-in,* a diminutive of *Hibb,* from *Hebb.*

**Hibbitt, Hibbott:** *v.* IBBOTT

**Hibble, Hible:** *v.* IBELL

**Hibbs:** *v.* IBBS

**Hichens:** *v.* HITCHEN

**Hichisson:** John *Hichsone* 1332 SRLa; John *Hichenson* 1535 KentW. 'Son of *Hitch* or *Hitchen.*'

**Hick, Hickes, Hicks, Hix:** *Hikke de Sauteby* 1276 RH (Y); *Hyk* serviens 1286 Wak (Y); Richard *Hick'* 1302 SRY; William *Hickys* 1332 SRWa; Henry *Hix* 1484 FrY. *Hick,* a pet-form of *Ricard.* *v.* RICHARD, HITCH.

**Hicken:** Isabella *Hicken* 1332 SRWa. *v.* HICK, GEFFEN.

**Hickey:** John *Hicci* 1279 RH (C); Richard *Hykeys* 1394 LLB H; Derby *Hickey* 1642 PrD. A diminutive of *Hick,* a pet-form of *Ricard, v.* RICHARD.

**Hickin, Hicking:** *Hekyn* de Wath 1379 PTY; David *Hicun* 1275 RH (Sf); Alicia *Hykyn* 1379 PTY. *Hicun, Hicin,* diminutives of *Hick.*

**Hickinbotham:** *v.* HIGGINBOTHAM

**Hicklin, Hickling:** (i) Iuo *de Hickelinge* 1163–6 Holme (Nf); William *de Hikelyng* 1327 SRSt. From Hickling (Norfolk, Notts). (ii) Robert *Hikeling* 1212 Fees (La); John *Hykelyng* 1327 SRSx; William *Hykelyn* 1332 SRSt. *Hik-el-in,* a double diminutive of *Hick.* The *Hikeling*-forms may be early examples of loss of *de. v.* above.

**Hickman, Higman:** *Hykeman* 1279 RH (O); *Hykemon* Smert 14th AD iii (Wo); Richard *Hykemon,* Juliana *Hykemones* 1275 SRWo; Walter *Hikeman* 1279 RH (O). 'Servant of *Hick',* used also as a personal-name.

**Hickmott:** Thomas *Hikemot* 1460 RochW. 'Hick's brother-in-law.' *v.* HITCHMOUGH.

**Hickox, Heacock:** *Hikoc* 1279 RH (Hu); *Hekoc de Par* 1366 SRLa; William *Hekoc* 1219 AssY; William *Hycoxe* 1555 Pat. *Hekoc* is probably from *Hēa-coc,* a diminutive of OE *Hēah*. This would become *Heacock*, and, with shortening of the vowel, *Hickock*. cf. HAY, HAYCOCK.

**Hickson, Hickeson, Hixson:** Stephen *Hykson* 1381 PTY; John *Hixon* 1450 Rad (C). 'Son of *Hick*.

**Hidden:** Richard *de Hiddon'* 1200 P (D); Thomas *Hidden* 1576 SRW. From Hidden (Berks), or Clayhidon (D), *Hidone* DB.

**Hiddleston:** *v.* HUDDLESTON

**Hide, Hides, Hyde, Hydes:** (i) Robert *de la Hyda* 1188 P (Do); Avice *atte Hyde* 1296 SRSx; John *of then hyde* 1299 MELS (Wo). OE *hīd* 'a hide of land', probably 'holder of a hide'. (ii) Gilbert *filius Hide* 1185 Templars (Ess); *Hida* de Knapwell 1255 RamsCt (Hu). The personal-name appears to be feminine, possibly *Ida*. *v.* IDA. cf. Gilbert *Hidecock* 1286 FFSf.

**Hider, Hyder:** Robert *le Hider* 1309 MESO (Sx); Hugh *Hyder* 1389 AD vi (Do). 'Holder of a hide.' cf. Roger *Hydeman* 1252 Rams (Hu). *v.* HIDE.

**Hie:** *v.* HIGHE

**Hieatt, Hiett:** *v.* HIGHET

**Hield, Hields:** *v.* HEALD

**Higdon:** Richard son of *Hykedon* 1313 Chamb-AcctCh; John *Hikedun* 1221 AssWo, 1273 RH (Wo); Thomas *Hykedon, Hekedon,* 1327 SRC. *v.* RIGDEN.

**Higford:** Walter *de Hugeford', de Hukeford'* 1221 AssSa; James *Hugeford* c1700 ArchC 49. From Higford (Sa), *Hugeford* 1206.

**Higgett, Higgitt:** *Hyket* 1279 RH (O); John *Hykot, Hicoth* 1317 AssK. A diminutive of *Hick*, with late voicing of *k*.

**Higginbotham, Higginbottam, Higginbottom, Higen-botham, Higenbottam, Higinbothom, Higgenbottom, Heckingbottom, Heginbotham, Heginbottom, Hickin-botham:** Alexander *de Akinbothun* 1246 AssLa; Nicholas *Hichinbothome* 1579 Bardsley (Ch); Mary *Higginbotham* alias *Hickabotham* 1695 DKR 41. From Oakenbottom in Bolton-le-Moors (Lancs), probably originally *\*ǣcen-botme* 'oaken-valley', becoming *Eakenbottom, Ickenbottom.* This was then associated with *hickin* or *higgin,* a Lancashire and Cheshire dialect-word for 'mountain-ash'.

**Higgins, Higgens, Higgon, Higgons:** *Hygyn* de Bowland 1379 PTY; John *Hygyn* 1377 AssEss, *Hygyns* 1452 AD i (Wo). A diminutive of *Higg,* a voiced form of *Hick.* *v.* HICKIN, HIGGS.

**Higginson:** Alan *Hygginson* or *Hickynsone* 1552 Oxon. 'Son of *Hickin* or *Higgin.*'

**Higgs:** *Hegge* Hog 1286 AssCh; Richard *Higge* 1275 RH (Nf). A voiced form of *Hick.*

**Higham, Hioms, Hyam, Hyams:** Osward *de Hecham* 1176 P (Ess); Hugo *de Hegham* 1198 P (K). From one of the Highams.

**Highe, Hie:** (i) Gilbert *le High* 1327 SRSx. OE *hēah* 'high, tall'. (ii) Robert *atte Heghe* 1327 SRSx; Richard *atte High* 1332 ib.; both assessed in Heighton (Sussex). 'Dweller on the height or hill', OE *hēah* 'high' used substantively. *v.* MELS and cf. HEIGHT.

**Highet, Hiatt, Hiett, Hyatt, Hyett, Hieatt, Highatt:** James *Hyet* 1514 LP (Lo); John *Hyett* 1539 ChwWo; William *Hiatt* 1599 FrLei; John *Hyatt* 1641 PrSo. The forms are late, but perhaps 'dweller by the high gate', OE *hēah, geat.*

**Highfield, Heafield:** Robert *de Heghefeld* 1275 RH (Ha); John *de Hefeld'*, William *de Hyefeld* 1332 SRSr. 'Dweller by the high field or open-country.'

**Highland:** *v.* HAYLAND

**Highman:** *v.* HAYMAN

**Hignham:** Godfrey *de Hynam* 1222 Acc; Richard *Hynam* 1382 IpmGl. From Highnam (Gl).

**Highway, Heighway:** Richard *de Heghweye* 1324, Thomas *Heighweye* 1361 IpmW. From Highway (W).

**Higho:** *v.* HAYHOE

**Highstead, Highsted, Histead, Hysted:** Helewis' *de Heghested* 1296 SRSx; Thomas *Highstede,* Lawrance *Histed* 1473, 1494 ArchC 41. From Highstead in Sittingbourne (Kent).

**Hight:** *v.* HEIGHT

**Highton, Heighton:** John *atte Heghetun* 1296 SRSx. From Heighton (Sussex).

**Highwood:** *v.* HAYWOOD

**Higman:** *v.* HICKMAN

**Higson:** Roger *Higson* 1487 LLB L. 'Son of *Higg.*'

**Hilary:** *v.* HILLARY

**Hilbert:** *v.* ILBERT

**Hild:** *Hilda* de Tidintun Hy 2 DC (L); *Hilda* de Gay 1192 P (O); Hereward, Arnald *Hilde* 1168, 1200 P (Do, Ha). *Hilda* (f), a short form of a name in *Hild-,* e.g. OG *Hildigard* (f), or OE *Hildgȳð* (f).

**Hildebrand, Hilderbrand:** *Hildebrand* lorimarius 1086 DB (Nf); Brand *filius Hildebrand* 12th DC (Nt); Adam *filius Heudebrand* 1210 Cur (Nf); Geoffrey *Hildebrant* 1195 P (Nf); Gregory *Hildebrand* 1275 RH (L); John *Heldebrond* 1296 SRSx. OG *Hildebrand, Eldebrand* 'battle-sword'; occasionally, in the Danelaw, also ON *Hildibrandr.*

**Hilder:** William *le Hilder'* 1332 SRSx. 'Dweller on the slope', from OE *hylde.* cf. HEALD.

**Hildith:** Hugh *filius Hildithe* 1208 CurBeds; *Hildith* (f) 1212 Cur (Berks); *Hilditha* (f) 1247 AssBeds; John *Hildyth* 1360 AD vi. OE *Hildgȳð* (f).

**Hildyard, Hilliard, Hillyard:** *Hildiard'* de Trule 1206 Cur (Sr); *Hyldeiard* (f) 1228 FFEss; Robert *Hildyard, Hiliard* 1276 RH (Y); Peter *Hildeyerd'* 1297 SRY. OG *Hildigard, Hildiardis* (f) 'war-stronghold', not a common name.

**Hilger:** (i) *Ilgerus* 1086 DB, 1221 AssWa; *Ilger'* de Wilberfoss' 1219 AssY; Walter *Ylger* 1191 P (Sf); John *Ilger* 1239 FFSf; Emma *Hilger* 1279 RH (O). OG *Hildeger, Hilger* 'battle-spear', not uncommon, invariably as *Ilger.* (ii) Willelmus *filius Hildegaric* 1250 Rams (Hu); William, Julian *Hildegar* 1267–85 Rams (Nf), 1279 RH (Hu); Geoffrey *Hildegor* 1221 ElyA (Sf). ODa *Hildiger.* The name is limited to East Anglia.

**Hill, Hille, Hills:** Gilbert *del Hil* 1191 P (Nf); William *attehil* 1260 AssC; Simon *Hille* 1273 RH (Wo); William *o the hil* 1313 AddCh (C); Matilda *Hilles* 1327 SRSo. 'Dweller on the hill', OE *hyll.* The surname is partly from a personal-name *Hille,* a pet-form of some such name as *Hilger* or *Hillary*: Rogerus *filius Hille* 1221 Cur (D).

**Hillam, Hillum:** Robert *in Hillum* c1050 OEByn (Y); John *de Hillum* 1294 FFY; William *de Hillum* 1302 IpmY. From Hillam (WRY).

**Hillary, Hillery, Hilary, Elleray, Ellery, Elray:** (i) *Hilarius* Brunus 1177 P (Wo); Richard *Ilarie* 1227 AssSt; Willelmus *Hillar'* (*Ylarius*) 1230 P (Wo); Roger, William *Hillari* 1275 RH (L), 1283 AssSt; William *Hillary, Illore* 1308–9 ib. Fr *Hilaire, Hilari,* Lat *hilaris* 'cheerful', the name of several saints, in particular St Hilarius of Poitiers (d. 368). The name was popular in France and not uncommon in England. (ii) *Eularia, Eilaria, Yllaria* Trussebut 1200, 1204 Cur; *Eularia* de Hulle 1200 Cur (Beds); *Eularia, Elaria* 1212 Cur (So, L); *Illaria, Ilaria, Hillaria* 1219 AssY. This woman's name is not, as has been suggested, a feminine of *Hilarius,* but *Eulalia,* a latinizing of Greek εὔλαλος 'sweetly-speaking'. It was the name of the patron saint of Barcelona, not uncommon in Spain and France (*Eulalie*), and survives as a French surname in the popular form *Aulaire.* There was a masculine form, probably absorbed by *Hillary: Eularie* (nom.) de Hegge 1212 Cur (Y); *Eularius* de Syre 1214 Cur (Sr); *Illarius, Hillarius* de Bechamton' 1208 Cur (So); *Eylarius* de Bachamton 1214 ib. (Ha).

**Hillas:** *v.* HILLHOUSE

**Hillcoat:** *v.* HELLCAT

**Hillen:** *v.* ELION

**Hiller:** Peter *Hiller* 1432 FrY. 'Dweller on the hill.' cf. HELLER.

**Hillers:** *v.* HILLHOUSE

**Hillery:** *v.* HILLARY

**Hillhouse, Hilhouse, Hillas, Hillers, Hillis, Hellass:** Nicholas *del Hellus* 1235 RH (Bk); Grace *Hellhouse* 1598 Bardsley (Essex). 'Dweller at the house on the hill.' cf. BULLAS.

**Hilliar:** *v.* HELLIAR

**Hilliard:** *v.* HILDYARD

**Hilling, Hillings:** Robert, Richard *Hilling* 1206, 1212 Cur (Nth, Lei). Probably OE *\*hylling* 'hill-dweller'. *v.* also ELION.

**Hillis:** *v.* HILLHOUSE

**Hillman:** John *Hildman* 1327 SRY. 'Dweller by the slope' (*v.* HEALD) or 'servant of *Hild'.* Also 'hill man'. cf. HELLMAN.

**Hillock:** Ragher *Hilloc* 1205 Cur (Nf); John *Hillok* 1327 SRSo; John *Hillok* 1517 Black (Glasgow). 'Dweller by the small hill', from a diminutive of OE *hyll* 'hill'.

**Hillson:** (i) Alexander *Hildson* 1332 SRCu. 'Son of *Hild'.* (ii) Walter *Hillessone* 1381 SRSf. 'Son of *Hill.'*

**Hillum:** *v.* HILLAM

**Hillyar:** *v.* HELLIAR

**Hillyard:** *v.* HILDYARD

**Hilsden, Hilsdon:** Robert *de Hildisdone* 1279 RH (Bk). From Hillesden (Bk).

**Hilt:** William *filius Helte* 1173 P (K); *Helte* de Boisdele 1176 P (L); Ralph *filius Hilto* 1219 AssY; Richard *Hilte* 1330 FFSf. OG *Helto.*

**Hilton:** Roger *de Hiltun* 1132 Riev; Robert *de Hilton* 1343 FFY; John *Hilton* 1393–4 FFSr. The usual sources of the name are, no doubt, Hilton (Derby, Hunts, Staffs, NRYorks), or Hilltown (Devon), but occasionally a personal name may be involved, cf. Ralph *filius Hilton* 1219 AssY.

**Himsworth:** *v.* HEMSWORTH

**Hince:** *v.* HINTS

**Hinchcliffe, Hinchliff, Hinchliffe, Henchcliff, Henchcliffe, Hinchsliff:** John *de Hengeclif* 1324 Wak (Y); Agnes *de Hingeclif* 1327 ib.; John *Hyncheclyffe* 1441 ShefA; William *Hynseclif* 1485 FrY; Henry *Henseclyf* 1552 ib.; John *Hinchliffe* 1633 ShefA. From Hinchcliff (WRYorks).

**Hinckesman:** *v.* HENSMAN

**Hinckley, Hinkley:** Anfred *de Hinkelai* 1176 P (Lei); Roger *de Hinckelee* 1208 Cur (W); John *de Hynkeley* 1332 SRSt. From Hinckley (Leics).

**Hincks:** *v.* HINKS

**Hind, Hinde, Hindes, Hinds, Hynd, Hynds:** Cristiana, Henry *Hynde* 1285 *Ass* (Ess), 1332 SRSt. OE *hind* 'female of the deer', perhaps 'timid as a hind'. Late examples may be from *Hine,* with excrescent *d* (1520 NED).

**Hindell, Hindle:** William *de Hindal'* 1205 Pleas (Sx); John *de Hyndale* 1255–6 FFSx; John *de Hyndedale* 1332 SRSx. 'Dweller in the valley frequented by hinds', OE *hind, dæl.*

**Hindley, Hindeley, Hyndley:** Simon *de Hindelay* 1219 AssY; Robert *de Hindeley* 1243 AssDu; John *Hyndeley* 1340–1450 GildC. From Hindley (Lancs), or Hiendley (WRYorks).

**Hindon:** Giles *de Hynedon* 1287, Sewyn *de Hynedone* 1331, John *de Hyndon* 1353 IpmW. From Hindon (W).

**Hindson:** John *Hyndson* 1493 LLB L. 'Son of the hind.'

**Hine, Hines, Hyne, Hynes:** William, Robert *le Hyne* 1240 Eynsham (O), 1254 AssSo, 1256 AssNb. ME *hīne* 'hind, servant'. *v.* MED.

**Hingeston, Hingston:** John *de Hyngeston* 1343 AssNu; William *Hingston* 1642 PrD; Justinian *Hingston* 1662–4 HTDo. From Hinxton (C), or Hingston Down (Co, D).

**Hingham:** *v.* HENGHAM

**Hingle:** *v.* INGALL, INGELL

**Hinkins:** *v.* HENKIN

**Hinks, Hincks:** William *Hynke* 1327, *Hynks* 1381 SRSf; John *Hinckes* 1576 SRW. OE *Hynca.*

**Hinsull:** *v.* INSALL

**Hint, Hints, Hince:** Philip *Hyntys* alias *Hynce* 1553 FFSt; John *Hints* 1663 HeMil. From Hints (Sa, St).

**Hinton:** Robert *de Hintona* 1086 ICC; Thomas *de Hyneton'* 1230 Cur (Do); Thomas *Hynton* 1385 FFEss. From one or other of the many places of this name.

**Hipkin, Hipkins, Hipkiss:** cf. *Hyppe* 1275 RH (Nf); Lefsius *Hippe* 1275 RH (Sf); John *Hipecok'* 1243 AssSo. *Hyppe* is an unvoiced form of *Hibb,* from *Heb, Hipkin,* like *Hipecok,* a diminutive.

**Hipper:** William *le Hyppere* 1278 AssSo; John *Hippere* 1359 AssD; John *Hipper* 1443 FrY. A derivative of OE *\*hyppan* 'to hop', a nickname for a dancer. cf. Henry *Hyphup* 1251 FFY 'hop up'.

**Hipperson:** *v.* IBBERSON

**Hipwell, Hipswell:** Robert *de Hispeswell'* 1200 P (Y); Robert *de Heppeswell'* 1219 AssY. From Hipswell (NRY).

**Hipworth:** *v.* HEPWORTH

**Hircock:** *v.* HERCOCK

**Hird:** v. HERD

**Hirst:** v. HERST

**Hiscocks:** v. HITCHCOCK

**Hislop:** v. HASLOP

**Histead:** v. HIGHSTEAD

**Histon:** Robert *de Histona* 1086 InqEl; Ralph *de Huston'* 1202 Pleas (C). From Histon (C), *Huston* 1188.

**Hitch, Hitches, Hytch:** *Hiche* de Sadinton 1198 FrLeic; *Hicche* uenator 1202 P (Hu); Walter *Hicch* 1279 RH (Hu); Stephen *Hiche* 13th Lewes (Nf); John *Hyches* 1319 AD iv (Hu). *Hich*, a pet-form of *Richard*, was common in the 13th century, particularly in Cheshire.

**Hitcham:** Henry *de Hucham* 1227 Acc; Robert *de Hicham* 1257–9 RegAntiquiss. From Hitcham (Bk), *Hucheham* DB.

**Hitchcock, Hitchcox, Hitchcott, Hiscock, Hiscocks, Hiscox, Hiscott, Hiscutt, Hiskett:** *Hichecoc* 1260 AssCh; *Hichecok* 1275 RH (Y); *Higecok* Trente 1310 LLB D; Richard *Hichecokes* 1327 SRWo; John *Higecok* 1327 SRC; William *Hygecok, Hichecok* 1329, 1360 AD i, vi (Do); Nicholas *Hekkox* son of Abraham *Heskcok* 1636 Bardsley. A diminutive of *Hich*. v. HITCH.

**Hitchcoe:** For HITCHCOCK. v. ALCOE.

**Hitchen, Hitchens, Hitcheon, Hitchin, Hitchins, Hitching, Hitchings, Hichens:** (i) *Hichoun* 1286 AssCh; John *Hichun* 1279 RH (O). *Hich-un*, a diminutive of *Hich*. cf. Robert *Hichel* 1327 SRWo. v. HITCH. (ii) John *Hychen* 1332 SRWa. v. HITCH, GEFFEN. (iii) Geoffrey *de Hicchen* 1321 LLB E. From Hitchin (Herts).

**Hitchman:** Richard, William *Hicheman* 1238 AssSo, 1279 RH (O). 'Servant of *Hich*.'

**Hitchmough:** Robert *Hichmughe* or *Hytchmoughe* 1584 Oxon (La). 'Brother-in-law of *Hich*.' cf. HICKMOTT, WATMOUGH.

**Hite:** v. HEIGHT

**Hix:** v. HICKS

**Hixson:** v. HICKSON

**Hoad, Hoath:** John *del Hoth* 1275 RH (Nf); Simon *atte Hothe* 1296 SRSx, 1317 AssK. From Hoath (Kent), or 'dweller on the heath', from OE *$*h\bar{a}p$, ME *hōth*, a by-form of *hǣp* 'heath', a Kent and Sussex form.

**Hoadley:** William *de Hodlegh* 1296 SRSx. From Hoathly (Sussex).

**Hoar, Hoare, Hore:** (i) William *Hore* 1188 BuryS (Sf); Robert, William *le Hore* 1203 AssSt, 1221 AssWo. OE *hār* 'hoar, grey-haired'. (ii) Gilbert *de Hore* 1200 P (Sx); Richard *de la Hore* 1230 P (D). From Ore (Sussex) or 'dweller by the bank', OE *ōra*. v. NOAR.

**Hoather, Hother:** 'Dweller on the *hoath* or heath.' v. HOAD and cf. HEATHER.

**Hob, Hobb, Hobbes, Hobbis, Hobbiss, Hobbs:** *Hobbe* Litel 1176 P (Sa); *Hobb(e)* 1198 Cur (Nth), P (Y), 1205 Cur (Wo); Osbert, Ralph *Hobbe* 1204 P (R), 1230 P (Wa). Agnes *Hobbis* 1279 RH (Hu); Isabella *Hobbes* 1327 SRWo, *Hobb*, a pet-form of *Robert*, rhymed on *Rob*.

**Hobart:** v. HUBERT

**Hobbins:** (i) Henry, Alice *Hobben* 1327 SRWo. 1332 SRWa. v. HOB, GEFFEN. (ii) Richard *Hobbyn*

1408 PN Ess 53. A diminutive of *Hobb*. *Hobbins* may have become *Hobbiss*.

**Hobby, Hobey, Hoby:** (i) Richard *Hobi* 1175 P (Do/So); William *Hobey*, Ralph *Hobay* 1296 SRSx; John *Hobbeye* 1568 SRSf. A nickname from ME *hobi* 'a small species of hawk, or a small horse'. (ii) Gilbert *de Hobi* Hy 2 DC (Lei); Philip *of Hoby* 1315 AssNf. From Hoby (L).

**Hobcraft, Hobcroft:** Richard *Hobcroft* 1436 Oseney. From Hopcroft Spinney in Orlingbury (Nth).

**Hobday, Hobdey:** William *Hobday* 1469 KentW. 'Servant of *Hobb*', '*Hobb* the servant', v. DAY.

**Hobey:** v. HOBBY

**Hobgen:** John *Hobjohn* 1526 SxWills. A combination of *Hobb* and *John*, perhaps 'rustic, boorish John'. cf. *Hobhouchin*, a name for the owl (1682 NED).

**Hobhouse:** John *Hobhouse* 1524 SRD, 1642 PrD. From Hobhouse in Drewsteignton (D).

**Hobkinson:** v. HOPKINSON

**Hobkirk:** v. HOPKIRK

**Hoblet, Hoblett:** Constance *Hobelot* 1279 RH (C); Alice *Hobelotte* 1311 ColchCt; Agnes *Hobelot* 1376 AD vi (Sf). *Hob-el-ot*, a double diminutive of *Hob*, a pet-form of *Robert*.

**Hoblin, Hobling, Hoblyn:** *Hobelyn* Flemyng 1373 ColchCt; William *Hobelyn* 1374 ib. *Hob-el-in*, a double diminutive of *Hob*.

**Hobourn:** v. HOLBORN

**Hobson, Hopson:** John *Hobbessone* 1327 SR (Ess); John *Hobsone* 1327 Wak (Y). 'Son of *Hob*.'

**Hoby:** v. HOBBY

**Hochkins:** v. HODGKIN

**Hock, Hocks:** Ranulf *Hocke* 1279 RH (C); John *Hock* 1327 SRY; Richard *Hockes* 1642 PrD. OE *Hocca*, but sometimes for HOOK (ii) with a shortened vowel, cf. Simon *del Hock* 1279 RH (O).

**Hockaday, Hockerday:** John *Hockedeʒ* Hy 2 Suckling (Sf); Simon *Hochede* 1206 P (Bk); Henry *Hokedey* 1296 SRSx. ME *Hocedei, Hokedey*, the second Tuesday after Easter Sunday, in former times an important term-day, on which rents were paid, etc., Hock-day and Michaelmas dividing the rural year into its summer and winter halves. From the 14th century it was also a popular festival. A name for one born at this time. cf. PENTECOST, CHRISTMAS, MIDWINTER.

**Hockenhall, Hockenhull, Hocknell, Hucknall, Hucknell:** Gibbe *de Huckenhale* 1179 P (Nt); Hamo *de Hukenelle* 1275 RH (Sf); William *Hockenale* 1345, *Huckenale* 1378 IpmGl. From Hockenhull (Ch), Ault Hucknall (Db), or Hucknall Torkard (Nt).

**Hockeridge:** v. HOCKRIDGE

**Hockett, Hoggett:** Stephen *Hoket* 1205 P (Co); Berenger *Hoget* (*Hoket*) 1219 Cur (Nth); Geoffrey *Hoket* 1275 SRWo; Adam *Hoket* 1375 ColchCt. *Hocc-et*, a diminutive of OE *Hocca*.

**Hockin, Hockins, Hocking, Hockings:** Robert *Hokyn* 1297 MinAcctCo; John *Hokyn* 1327 SRSf; John *Hockin*, Christopher *Hockins*, Abel *Hockinge* 1642 PrD. *Hocc-en*, a diminutive of OE *Hocca*.

**Hockley, Hockly:** Michael *de Hockele* 1203 FFEss; Nicholas *de Hockelaye* 1332 SRWo; John *Hokeleye* 1340–1450 GildC. From Hockley (Ess, Wa).

**Hockliffe:** Thomas *de Hocklefe* 1310 LLB D. From Hockliffe (Beds).

**Hockly:** *v.* HOCKLEY

**Hocknell:** *v.* HOCKENHALL

**Hockold:** Adam *de Hockewald* 1275 RH (Nf). From Hockwold (Norfolk).

**Hockridge, Hockeridge:** Stephen *de Hokeregge* 1275 RH (K); Maurice *de Haukerigge* 1333 PN D 365. From Hawkridge (Kent), from OE *hōc*, or from Hawkeridge (Devon, Wilts) or Hawkridge (Devon), from 'hawk'. Hawkridge in Morebath is *Hockerige* in 1596.

**Hocks:** *v.* HOCK

**Hodd, Hoddes, Hodds, Hodes, Hood, Hoods:** Osbernus *Hod* c1100–30 OEByn (D); Walter *Hod* c1200 ArchC 6; Gilbert *Hodde* 1225 AssSo; Robert *Hood* (*Hod*) 1230 P (Y); Philip *Hodde, Hudde* 1305 LoCt; Willeame *Hood* 1320 Merton (O); John *Hoddes* 1524 SRSf. OE *hōd*, ME *hod, hud, hood, hodde* 'hood'. A maker of hoods. Hamo *Hode* is also called *Hodere* (1317 AssK). Occasionally *Hood* is local, from Hood in Rattery (Devon): Jurdan *de Hode* 1242 Fees (D).

**Hodder:** John *le Hoder* 1220 Cur (Ess); William *Hudere* 1279 RH (C). A derivative of OE *hōd* 'hood', a maker of hoods. cf. HODD, HOTTER. John *Hoder* (1361 ColchCt) is also called *Hodmaker* and *Hodman*.

**Hoddinot:** *v.* HODNETT

**Hodge, Hodges:** *Hogge* 1208 FFL, 1212 Cur (Cu); William *Hogge* 1297 MinAcctCo; Alicia *Hogges* 1327 SRSo; William *Hodges* 1524 SRSf. A pet-form of *Roger*. Chaucer's cook 'highte *Hogge* of Ware' is invoked 'Now tell on, *Roger* . . .'

**Hodgeman, Hodgman:** John *Hogemon* 1275 SRWo; John *Hogeman* 1327 SRSo; Walter *Hoggeman* 1332 SRCu. 'Servant of *Hodge*', a pet-form of *Roger*.

**Hodgeon, Hodgens, Hodgins:** *v.* HODGSON

**Hodgett, Hodgetts:** William *Hogettes* 1408 AssCh (St). A diminutive of *Hodge*.

**Hodgkin, Hodgkins, Hodgkiess, Hodgkiss, Hadgkiss, Hodgskin, Hodgskins, Hochkins, Hotchkin, Hotchkis, Hotchkiss:** Robert *Hochekyn* 1327 SRSt; John *Hogekyn* 1453 AD ii (Pembroke); Richard *Hoggekynes* 1445 AD vi (Nf); William *Hochekys* 1470 AD iv (Sa); Robert *Hodgekin* 1524 SRSf; Charles *Hodgskines* t. Eliz Bardsley; John *Hotchkis* 1690 DKR 41 (Sa). *Hogge-kin*, a diminutive of *Hodge*.

**Hodgkinson, Hodgkison:** John *Hogkynson* 1459 PrGR; Ewan *Hodg(e)kynson* 1582 ib. 'Son of *Hodgkin*.'

**Hodgson, Hodgshon, Hodgens, Hodgin, Hodgins, Hodgeon, Hodson:** Henry, William *Hoggeson(e)* 1325 LaCt, 1381 PTY; John *Hodgeson* 1525 FrY; Thomas *Hodshon* 1528 GildY; Richard *Hodson* 1582 PrGR; James *Hodgshon* 1591 LaWills; William *Hodgson* 1602 PrGR; Elizeus *Hodgeon* 1682 ib. 'Son of *Hodge*.' In PrGR, *Hogeson* is common before 1582 and *Hodson* and *Hodgeon* are clearly the same name.

**Hodkin:** *Huddekyn* 1324 LaCt. 'Little Hugh.' *v.* HUDD.

**Hodkinson, Hodkison, Hodskinson:** Ellen *Hodkinson*, Richard *Hodskinson* 1626, 1661 LaWills. 'Son of *Hodkin* or of *Hodgkin*.'

**Hodnett, Hoddinot:** Paulinus *de Hodenet* 1204 P (Sa); Odo *de Hodynet* 1293 AssSt; Henry *Hoddynott* 1641 PrSo; Humphrey *Hodnett* 1663 HeMil. From Hodnet (Sa). *v.* also HUDNOTT.

**Hodsell, Hodsoll:** Maurice *de Hodsell'* 1221 AssWo; William *Hudsoll* 1512 PN K 37. From Hodsoll (St), or Hodsoll Street in Ash (K).

**Hodson:** William *Hodson*, Robert *Odeson* 1379 PTY; John *Hodessone* 1385 LLB H. 'Son of *Odo*.' *v.* also HODGSON.

**Hoe(s):** *v.* HOW

**Hogarth:** *v.* HOGGARD

**Hogben, Hogbin:** Thomas *Huckebone* 1479 ArchC 39; Robert *Hucbone* 1500 ib.; Peter *Hugbone* 1549 ib.; John *Hogben* 1588 ib. 35. From *huck-bone* 'the hip-bone or haunch-bone', huckle-bone, found in Craven dialect (Yorks) as *hug-baan*. The origin of *huck* is obscure, but it may be from a Teutonic root *\*huk-* 'to be bent'. cf. ON *huka* 'to crouch, sit bent, sit on the haunches' and *huck-backed* 'hump-backed, crump-shouldered' (1631 NED). This Kentish surname is, no doubt, synonymous with the Scottish *Cruickshank*.

**Hogg, Hogge:** Ailmer, William *Hog* 1079 Rams (Hu), 1174–80 StP (Lo); Alice, William *le Hog* 1279 RH (O), 1296 SRSx; Adam *le Hogge* 1332 SRLa. OE *hogg* 'pig'.

**Hoggar, Hogger:** John *le Hoggere* 1327 SR (Ess); William *Hogir* 1327 SRSx. A derivative of OE *hogg*; 'hog-herd'.

**Hoggard, Hoggart, Hoggarth, Hogarth, Hoggett:** William *Hoggehird* 1279 AssNb; Richard *le Hoghird* 1327 SRY; John *Hoggard* 1461 FrY; John *Hoggerd* 1509 ib.; Henry *Hogget, Hogged* 1627 Bardsley. OE *hogg* and *hierde*, 'swine-herd'. *v.* HOGG.

**Hoggett:** *v.* HOCKETT, HOGGARD

**Hoggsflesh, Hogsflesh:** Robert *Hoggesflech* 1332 SRSx; Sarah *Hoggesflessh* c1405 FS; William *Hoggisflesh* 1525 SRSx. 'Hog's flesh', OE *hogg, flǣsc*, perhaps a nickname for a pork-butcher. cf. Hugh *Hoglambe* 1218 AssL 'a young sheep before the first shearing'; Jordan *Oggesfot* 1220 Cur (Herts) 'hog foot'; Edward *Hogsed* 1642 PrD 'hog's head'; Widow *Hoggtrough* 1662 HTEss 'hog trough'.

**Hoghton:** *v.* HOUGHTON

**Hoile, Hoiles:** *v.* HOYLE

**Holbeach, Holbech, Holbeche:** Ailfric *de Holebeche* 12th RegAntiquiss; Thomas *de Holebech'* 1298 AssL; Thomas *de Holebeche* 1340 LLB F. From Holbeach (L).

**Holbem, Holben:** William, Walter *de Hol(e)bem(e)* 1242 Fees (D), 1296 SRSx. From Holbeam (Devon), Holbeanwood in Ticehurst, or Holban's Fm in Heathfield (Sussex).

**Holbert, Holbird, Hulbert, Hulburd:** *Holbertus* venator 1168 ArchC 5; William, John *Holdebert* 1205 P (Wa), 1219 AssY; John *Hulberd* 1524 SRSf. These surnames, rare in early sources, probably derive from an unrecorded OE *\*Holdbeorht* 'gracious-bright'. Old English names in *Hold-* are rare and late, only *Holdburh* (f), *Holdfrið* (f), and *Holdwine* being recorded in independent use.

**Holborn, Holbourn, Holbourne, Holburn, Hobourn:** Bald' *de Holeborn'* 1193 P (Lo); John *de*

*Holeburne* 1296–7 FFSr; Stephen *de Holbourne* 1364 FFEss. From Holborn (Mx).

**Holbrook, Holbrooke, Houlbrook, Houlbrooke:** Richard *de Holebroc* 1189 Sol; William *atte Holebrok* 1296 SRSx; John *Holebrok* 1327 *SR* (Ess). From Holbrook (Derby, Dorset, Suffolk), or 'dweller by the brook in the hollow'.

**Holcombe, Holcomb, Holcom:** Brihtmer *æt Holacumbe* c1100–30 OEByn; Adam *de Holecumb* 1256 AssSo; John, Thomas *Holcombe* 1525 SRSx. From Holcombe (Devon, Dorset, Glos, Lancs, Oxon, Som).

**Holcot, Holcott:** Peter *de Holecot'* 1202 AssNth; Robert *de Holecote* 1273–4 FFEss; Henry *de Holecote* 1314–16 AssNth. From Holcot (Beds, Nth).

**Holcroft, Holdcroft, Houlcroft, Houldcroft:** Robert *de Holecroft* 1246 AssLa; John *Holecroft* 1327 *SR* (Ess); Richard *Holdcroft* 1569 Bardsley. From Holcroft (Lancs).

**Hold, Hould:** Ysopa *Hold* 905 ASC A; Aluredus *Holde* 1198 P (K); Elyas *Lehold* 1205 ChR (W); Caterina *le Holde* 1327 SRSf. ON *hǫldr* 'a nobleman in rank beneath a *jarl*'.

**Holdam:** *v.* HOLDHAM

**Holdane:** *v.* HALDANE

**Holden, Holdin, Houlden, Houldin, Howlden:** Robert *de Holden* 1285 AssLa; William *Holeden* 1327 SRSx; Richard *atte Holdene* 1331 FFK. From Holden (Lancs, WRYorks) or from residence in a hollow valley (OE *holh, denu*).

**Holder, Houlder:** Geoffrey *le Holder* 1262 MEOT (Herts); Robert *le Holdere* 1274 RH (Gl). A derivative of OE *haldan* 'to hold', an occupier, possessor, probably a tenant or occupier of land (a1400 MED). *v.* also HELDER.

**Holderer:** Very rare. Probably for HOLDER.

**Holderness, Holdernesse:** Thomas *de Holdernes* c1220–30 RegAntiquiss; Roger *de Holdernesse* 1379 PTY; John *Holderness* 1414–15 FFSr. From the district of Holderness (ERY).

**Holdfast:** John *Haltefast* 1334 SRK. 'Hold firmly', OE *healdan, fæste*. cf. Thomas *Haldebytheheved* 1301 SRY 'hold by the head'; John *Holdshrewe* 1379 PTY 'hold the rascal'.

**Holdford, Holdforth:** *v.* HOLFORD

**Holdham, Holdam, Holdom, Holdum:** Robert *Holdom* 1477 FrY. Probably for OLDHAM.

**Holdman:** Thomas *Holdeman* 1312 ColchCt. Either 'servant of the noble' (*v.* HOLD) or for OLDMAN. *Aldman* and *Holdman* interchange in 1240 *Ass* (Sf).

**Holdroyd:** *v.* OLDROYD, HOLROYD

**Holdship:** William *Holshep* 1327 SRWo. OE *holdscipe* 'loyalty'.

**Holdsworth, Holesworth, Holsworth, Houldsworth, Houdsworth, Hallworth, Hallsworth:** John *de Haldeworth* 1273 RH (Y); John *de Halworth*, Richard *de Haldeworthe* 1379 PTY; Joshua *Houldsworth* 1593 Oxon (Y); Robert *Haulsworth* or *Holdsworth* 1595 ib.; Thomas *Haldsworth*, Jenetta *Halsworth* 1616, 1621 RothwellPR (Y); Thomas *Holdsworth* 1639 ib. This appears to be a specifically West Riding surname from either Holdsworth in Eckington or Holdsworth in Ovenden (WRYorks), both originally *Haldeworth* with a late intrusive *s*.

**Holdwin, Houdwin, Holwin:** Henry *Houdwin* 1219 AssL; Thomas *Holdwine* 1275 SRWo; Walter *Holwyne* 1327 SRSf. OE *Holdwine*.

**Hole, Holes, Houle:** William *de la Hole* 1200 P (D); Alice *Attehole* 1279 AssNb; Hugh *del Hole* 1296 Wak (Y); John *Houll* 1433 FrY. 'Dweller in the hollow', OE *hol, holh*, dat. *hole*.

**Holeman:** 'Dweller in the hollow.' cf. Robert *le Holare* 1275 SRWo.

**Holer, Houler:** William *le Holare* 1275 SRWo; Peter le Taverner called *Holer* 1311 LLB D; Thomas *Hollere* 1454 Paston. 'Dweller in the hollow', from a derivative of ME *hole* 'hollow'.

**Holeyman:** *v.* HOLYMAN

**Holford, Holdford, Holdforth, Houlford:** Jordan *de Holeford* 1179 P (So); John *de Holeford* 1296 SRSx. From Holford (Som) or Holford Fm in Chailey (Sussex).

**Holgate:** John *de Holegate* 1200 P (Y); Gommer *de Holgate* 1343 FFY; John *Holgate* 1525 SRSx. From Holgate (WRYorks), or 'dweller by the hollow road'.

**Holgood:** Philip *filius Holegod* 1207 P (La); Philip *Holegod* 1218, *Holegod'* 1219 P (St). OG *Holegot*.

**Holgreaves, Houlgrave:** Roger *de Holingref* 1297 SRY. From Hollingreave (WRYorks).

**Holiday, Holladay:** *v.* HALIDAY

**Holkham:** William *de Holcham* 1177, Thomas *de Holcham* 1212 P (Nf). From Holkham (Nf).

**Holland, Hollands:** Begmundus *de Holande* c975 LibEl (Ess); William *de Holaund* 1246 AssLa. From Holland (Essex, Lancs, Lincs).

**Hollaway:** *v.* HOLLOWAY

**Holledge, Hollidge:** John *Holynhege* 1379 PTY; Samuel *Hollidge* 1763 Bardsley. 'Dweller by the holly hedge', OE *holegn, hecg*.

**Holleman:** *v.* HOLYMAN

**Hollen(s), Holles, Hollis:** *v.* HOLLIES

**Holler:** *v.* HOLER

**Hollest:** *v.* HOLLIST

**Holley, Holly:** (i) Leouuine, Alfwinus *Holege* c1095 Bury (Sf), 1166 P (Nf); John *Holeghe* 12th ELPN; John *Holey, Holley* 1332 SRCu. OE *hol, ēage* 'hollow-eye'. (ii) Adam William *de Holleye* 1327 SRSf. 'Dweller by the clearing in the hollow', OE *hol* and *lēah*. *v.* also HOLLIES.

**Holleyman:** *v.* HOLYMAN

**Holliday:** *v.* HALIDAY

**Hollidge:** *v.* HOLLEDGE

**Hollier, Hollyer, Hullyer:** Robert *le Holyere* 1309 SRBeds; Adam *Holiere* 1327 *SR* (Ess). OFr *holier, huler*, a variant of *horier, hurier*, ME *holer, polyer, hullor* 'whoremonger, debauchee'.

**Hollies, Holles, Hollis, Holliss, Hollen, Hollens, Holling, Hollings, Hollins, Hollay, Holley, Holly:** Adam *atte Holies, atte Holye* 1275, 1327 SRWo; Robert *del Holins* 1297 Wak (Y); Nicholas *del Holyn* 1301 SRY; John *in the Holis* 1327 SRSf; Richard *del Holyes* 1332 SRSt; John *Holyn* 1428 FA (Wo); John *Holyns* 1431 FA (Wo); William *Holling* 1649 FrY. Dweller by the 'holly, holm-oak', OE *holegn, holen*, ME *holi(e), holin. v.* also HOLLEY.

**Holliman:** *v.* HOLYMAN

**Hollingrake, Hollindrake, Hollinrake, Hollenrake:** Adam *del Holirakes* 1275 Wak (Y); John *of*

*Holynrakes* 1315 ib.; Anna *Hollenrecke* 1644 RothwellPR (Y). From some small spot in the West Riding.

**Holling(s):** *v.* HOLLIES

**Hollingshead, Hollingshed, Hollinshead:** Richard *de Holineside* c1220 FeuDu; John *del Holynshede* 1408 Bardsley (Ch). From Hollingside or Holmside (Durham).

**Hollingsworth, Hollingworth, Hollinsworth, Hollinworth:** Thomas *de Holinewurthe* 1211–25 StP; Thomas *de Holingworth* 1286 AssCh; Laurence *Holyngworth* 1451 TestEbor. From Hollingworth (Ch, La).

**Hollins:** *v.* HOLLIES

**Hollinshead:** *v.* HOLLINGSHEAD

**Holliss:** *v.* HOLLIES

**Hollist, Hollest:** Alice *de Holhurst* 1296 SRSx. From Hollist House in Easebourne (Sussex).

**Hollister, Holister, Ollister:** Robert *Ollister* 1642 PrD. A derivative of OFr *holier* 'adulterer, lecher', hence 'a female brothel-keeper'.

**Holliwell:** *v.* HALLIWELL

**Hollman:** *v.* HOLMAN

**Holness:** Stephen *Holnest* 1474 ArchC 42; Leonard *Holnest, Holneherst* 1527, Agnes *Holnesse* 1548 CantW. cf. Holnest (Do), but since this appears to be a Kentish or Sussex name, it is perhaps rather from a lost *Holmherst* in Smarden (K).

**Hollock, Hollocks, Hollox, Holock:** Walter *de Holoc* 1208 ChR (Sf); Walter *Hollock* 1327 SRSo; John *ate Holock'* 1332 SRSr; John *del Hollokes* 1549 SxWills. 'Dweller by a little hollow', OE *holoc*, a diminutive of *hol*.

**Holloman:** *v.* HOLYMAN

**Hollow, Hollows:** Peter *in le Holwe* 1279 RH (C); Thomas *de Hollowe* 1327 SRWo. 'Dweller in the hollow', OE *holh*.

**Holloway, Hollawa, Hollway, Hollwey, Holoway:** Richard *de Holeweia* 1130 P (D); John *de la Holeweye* 1275 AssSo; John *Holewey* 1279 RH (O); John *del Hollewaye* 1308 Wak (Y); Hugh *atte Holewey* 1310 LLB D; Edward *Hollaway* 1674 HTSf. 'Dweller by a sunken-road', OE *holh, weg*.

**Hollowbread:** *v.* HALLOWBREAD

**Hollowell:** *v.* HALLIWELL

**Hollox:** *v.* HOLLOCK

**Hollway, Hollwey:** *v.* HOLLOWAY

**Hollyer:** *v.* HOLLIER

**Hollyhock:** *v.* HOLYOAK

**Hollyman:** *v.* HOLYMAN

**Hollyoak:** *v.* HOLYOAK

**Holm, Holme, Holmes, Holms:** Roger *de Holm* a1186 Seals (Lei); Urkell' *de Holmes* 1219 AssY; John *atte Holme* 1296 SRSx; William *del Holmes* 1327 SRDb, 1332 SRCu. From residence near a piece of flat land in a fen or by a piece of land partly surrounded by streams (ON *holmr*). ME *holm* also derives from OE *hole(g)n*, ME *holin, holm* 'holly, holm-oak' which survives in Holne (Devon) and Holme (Dorset, WRYorks). *v.* also HULM, HOME, HUME.

**Holman, Hollman:** John *Holman, Holeman* 1327 SR (Ess). OE *holh* and *mann* 'dweller by a hollow'.

**Holmer, Homer:** (i) Andrew, Robert *de Hol(e)mere* 1227 AssBk, 1279 RH (O), 1327 SRSf. From Holmer

(Bucks, Hereford) or 'dweller by a pool in the hollow' (OE *holh, mere*) as at Homer (Devon). (ii) John *Holmar'* 1296 SRSx; Richard *le Holmare* 1332 SRSr; William *le Holmere* 1340 MESO (Ha); Thomas *Holmer*, Edward *Homer* 1603 SRWo. 'Dweller by a holly-bush' (OE *holen*, ME *holm*), or on flat land near water (ME *holm*, ON *holmr*). *v.* HOLLIES, HOLM, HOMER.

**Holmwood:** *v.* HOMEWOOD

**Holock:** *v.* HOLLOCK

**Holoway:** *v.* HOLLOWAY

**Holroyd, Holroyde, Holdroyd, Howroyd:** Thomas, Andrew *Holerode* 1296 SRSx; Gilbert *de Holrode* 1327 SRSf; George *Holroyd* 1709 FrY. 'Dweller at the clearing in the hollow.'

**Holt, Hoult:** Hugo *de Holte* 1185 Templars (K); Simon *del Holt* 1230 P (Wa); Walter *in the Holte* 1260 MELS (So); Hugh *atte Holte* 1268 PN Sr 175. From Holt (Dorset, Hants, Leics, Norfolk, Staffs, Warwicks) or from residence in or near a wood (OE *holt*).

**Holtby, Houltby:** William *de Holtebi* 1208 Pl (Y); William, John *de Holteby* 1303 FFY, 1396 AssL. From Holtby (NRYorks).

**Holter:** Richard *le Holtar'* 1327 SRSx. 'Dweller by the wood.'

**Holtham, Holtom, Holttum, Holtum:** Matilda, Alan *de Holtham* 1200 P (L), 1256 AssNb. 'Dweller at the enclosure in the wood', OE *holt, hamm*. Richard *de Holtham* 1327 *SR* (Ha) came from Holtham in East Tisted (Hants). *v.* also HOTHAM.

**Holtie:** William *de Holetye* 1296 SRSx. From Holtye Fm (Sussex).

**Holton, Holten, Houlten:** Jordan *de Holton'* 1211 Cur (Do); John *de Holton* 1310, William *Holton* 1352 ColchCt. From Holton (Do, Sf, So).

**Holway:** *v.* HOLLOWAY

**Holwell, Holwill:** Leofric *æt Holewelle* 963 OEByn (Herts); Roger *de Holewell'* 1193 P (Nth); Robert *de Holewell* 1323–4 CorLo; Robert *Holewell* 1364 IpmW. From Holwell (Do, Herts, Lei, O), or one of the numerous minor places of the name in Devon.

**Holwin:** *v.* HOLDWIN

**Holyday:** *v.* HALIDAY

**Holyfather:** Philip *Olyfader* 1296, Richard *Holifader* 1327 SRSx. 'Holy father', OE *hālig, fæder*. Probably ironical. cf. Robert *Holydam* 1296 SRSx 'holy lady'; Richard *Holigost* 1354 Putnam (Sa) 'holy spirit'; Adam *Halygode* 1327 SRSa 'holy God'; Roger *Holymoder* Hy 3 Colch (Herts) 'holy mother'.

**Holyman, Holeyman, Holleman, Holleyman, Holliman, Holloman, Hollyman:** Roger *Haliman* 1212 Fees (Berks); William *Holyman* 1276 RH (L); Richard *Hollyman* Eliz Bardsley. OE *halig, mann* 'holy man', a nickname, probably pejorative. cf. Richard *Holifader* 1327 SRSo, Roger *Holymoder* Hy 3 Colch (Herts).

**Holyoak, Holyoake, Hollyoak, Hollyoake, Hollyhock:** Godric *de Haliac* Hy 2 Seals (Sf); Gerard *de Haliach* 1188 P (Lei); Peter *de la Holyok* 1300 MELS (Wo). From Holy Oakes (Leics) or from residence near a holy-oak or gospel-oak, an oak marking a parish-boundary where a stoppage was made for the reading of the Gospel for the day when beating the bounds during the Rogation Days.

**Homan, Homans:** Richard *Homan* 1306 AssW; John *Homan* 1374 FFEss; Christopher *Homan* 1642 PrD. Perhaps 'tall man', ON *há*, OE *mann*, or for HUMAN.

**Home, Homes:** Robert *del Houme* 1222 Cur (Nf); Alan *de Home* 1275 SRWo; Henry, William *Home* 1279 RH (O); Walter *of the home*, John *atte Home* 1327 SRWo; Andrue *Holmes*, John *Home*, Richard *Homes* 1524 SRSf. 'Dweller by a *holm* or by a holm-oak.' *v.* HOLM, HUME. The Scottish *Home* and *Hume* derive from the barony of Home in Berwickshire: William *of Home* 1268, Alexander *de Hume* 1408 (Black).

**Homedy:** *Homdei* 1188 P (Y); William *homo Dei* 1196 P (L); Roger *Homedeu* 1216 Oseney; Maud *Omedy* 1327 SREss. 'Man of God', OFr *homme, dieu*. cf. Roger *Homo Diaboli* 1227 Cur (Ess) 'man of the devil'.

**Homer, Homere:** (i) Identical with HOLMER. (ii) John *le Heaumer* 1220 Cur (Lo); Richard *le Heumer* 1284 LLB A; Mannekin *le Haumer* 1298 ib. B. OFr *heaumier, heumier* 'maker of helmets'.

**Homeward:** Peter *de Homewerth* 1296 SRSx; John *atte Homewerthe*, Robert *Homworth* 1332 ib. From residence at a homestead by the river-meadows (OE *hamm, weorð*).

**Homewood:** William *atte Homewode* 1279 PN Sr 331; Simon *atte Homwode* 1296 SRSx. From residence by a 'wood near the manor-house', OE \**hām-wudu*, surviving as Home Wood, Holmwood, and probably also as HOMEWARD.

**Homfray:** *v.* HUMPHREY

**Hond:** *v.* HAND

**Hondy, Handy:** *Hondy* Williams 1332 SRWa; Thomas *Hondy* 1275 SRWo; William *Hondy* 1391–2 FFWa; Peter *Handy* 1642 PrD. ME *hondi* 'skilled with the hands'.

**Hone, Hones:** Walkelin *de Hone* 1200 P (Ha); John, Thomas *Hone* 1276, 1279 RH (Berks, O); Godfrey *de la hone* 1296 SRSx; James *atte Hone* 1342 MELS (So). 'Dweller by some prominent stone or rock', OE *hān*, often a boundary-stone.

**Honebone:** *v.* HONEYBOURNE

**Honer:** Robert *le honer* 1230 P (Lo). A derivative of OE *hān* 'stone' in the sense 'whetstone' (c1325 NED), a sharpener of tools, a grinder or honer (1826 NED).

**Honex:** *v.* HUNWICK

**Honey:** Geoffrey *Hony* 1275 SRWo; Richard *Honey* 1279 RH (C); Robert *le Hony* 1296 SRSx. OE *hunig* 'honey', used as a term of endearment, 'sweetheart, darling' (a1375 MED).

**Honeybourne, Honeyburne, Honeyban, Honeybun, Honebone, Hunnybun:** Robert *de Hunuburn* 1221 AssGl; William *de Honybourn'* 1327 SRWo; Thomas *Honeybunn* 1802 PN Mx 55. From Cow Honeybourne (Glos) or Church Honeybourne (Worcs), pronounced *Hunnybun*.

**Honeychurch, Honychurch:** Margery *de Hunichurche* 1242 Fees (D); William *Honychurch* 1359 AssD; John *Honychurch* 1642 PrD. From Honeychurch in Sampford Courtenay (D).

**Honeycomb, Honeycombe, Unicum, Unicume:** Robert *de Honicomb'* 1327 SRSo. 'Dweller in the "honey", pleasant, fertile valley.'

**Honeyman, Honneyman:** (i) Robert, Everard *Huniman* 1199 FrLeic, 1235 Fees (Nth); Osbert *Honiman* 1279 RH (O); John *Honyman* 1296 SRSx. OE *hunig* and *mann*, 'seller of honey'. cf. Richard *Honymanger* 1382 AD i (W). (ii) *Hunman* 1066 DB (Bk); *Hunemannus* 12th Rams (C); Gillebertus *filius Honemann'* 1199 FF (W); John *Huneman* 1260 AssC; Thomas *Honeman* 1275 RH (Sf). OE *Hūnmann*.

**Honeywell, Honeywill, Honiwell, Honywill, Honnywill:** William *Honeywill* 1524 SRD; Joshua *Honiwell* 1642 PrD. From one or other of the various minor places of this name in Devon.

**Honiatt, Honiett, Onyett:** William *Honiet*, Reginald *Honyet* c1280 SRWo; Peter *Onyett*, *Honyot* 1279 RH (O), 1327 SRWo; John *Oniet*, Roger *Onyott* 1327 SRSo. OE \**Hungēat* 'young bear-Geat'.

**Honick:** *v.* HUNWICK

**Honneyball, Honneybell:** *v.* HUNNABLE

**Honnor, Honner, Honor, Honour:** Roger *Honnere*, John *Honner* 1332 SRSx; Mawde *Honor* 1547 CantW; Mr *Honner* 1674 HTSf. From Honor End Fm in Hampden (Bk).

**Honnywill:** *v.* HONEYWELL

**Honor, Honour:** *v.* HONNOR

**Honychurch:** *v.* HONEYCHURCH

**Honywill:** *v.* HONEYWELL

**Hood:** *v.* HODD

**Hoodlass, Hoodless, Huddless, Hudlass, Hudless:** William *Hodeles* 1292 FrY; Robert *Hudelesse* 1545 ib.; Thomas *Hoodles* 1568 SRSf; William *Hudlass* 1713 FrY. 'Hoodless', one who did not wear a hood. cf. HOOD.

**Hoof, Hooff:** *v.* HOW

**Hook, Hooke, Hookes, Huk, Huke, Huck, Hucke, Hucks:** (i) Halwun *Hoce* 1050–71 OEByn (D); Hervicus, Richard *Hoc* 1218 AssL, 1230 P (Berks); John *Hook* 1327 *SR* (Ess); Adam *Huke* 1524 SRSf. OE \**Hōca*, or OE *Hōc*, from *hōc* 'hook', probably denoting a man with a hooked or bent figure. Or metonymic for *Hooker*. (ii) Robert *de Hoke* 1192 P (K); Gilbert *de Huc* 1219 AssY; Geoffrey *de la Hoke* 1242 Fees (D); Gervase *ad Hokys* 1244 Rams (Beds); Richard *del Hokes* 1277 Wak (Y); John *atte Houke* 1332 SRSr. From Hook (Hants, Surrey, Lancs, Wilts, WRYorks), Hooke (Dorset), or from residence near a bend or hill-spur (OE *hōc*). This is the common source.

**Hooker:** Osmundus cognomento *Hocere* c975 LibEl; William, Osbert *Hoker(e)* 1199 P (Nf), 1219 AssL; John *le Hoker*, *le Houker* 1327, 1332 SRSx; Richard *Hooker* 1558 SxWills. OE, ME *hōcere*, a maker of hooks or an agricultural labourer who uses hooks. Also equivalent to *atte hoke* 'dweller by a hill-spur or bend'.

**Hookins, Hookings:** *v.* HUKIN

**Hooman:** *v.* HUMAN

**Hooper:** Adam, Philip *le Hoper(e)* 1228 Cl (W), 1297 MinAcctCo; William *le Houper* 1327 SRSo; Richard *Hoper*, couper 1367 FrYk; Ralph *Hooper* 1444 AD vi (D). A derivative of OE *hōp* 'hoop', a maker or fitter of hoops, cooper.

**Hoose:** *v.* HOUSE

**Hooson, Hoosun:** *v.* HUGHSON

**Hope, Hopes:** Robert *de Hope* 1255 RH (Sa); John *atte hop* 1296 SRSx; Robert *del Hope* 1302 SRY. From Hope (Derby, Salop, NRYorks, etc.), or from residence in or near raised land in a fen, or a small enclosed valley (OE *hop*).

**Hopewell, Hopwell:** Aluric *de Hopewella* 1066 DB (Sf); Roger *de Hoppewelle* 1373 IpmNt. From Hopwell (Db).

**Hopkin, Hopkins, Hopkyns:** *Hobekinus* 1224 Cur (St), 1231 Oseney (O); *Hopkyn* 1324 LaCt; William *Hobkyn,* Richard *Hobkyns* 1327 SRWo; William *Hopkyn,* John *Hopkynes* 1327 SRSt. *Hobbe-kin,* a diminutive of *Hobb* (Robert). *v.* HOB.

**Hopkinson, Hobkinson:** Richard *Hobbekynessone* 1354 Putnam (Ch), John *Hopkynson* 1469 FrY. 'Son of *Hobkin.*' *v.* HOPKIN.

**Hopkirk, Hobkirk:** James *Hopkirk, Hobkirk* 1574, 1679 Black. From Hopekirk (Roxburghshire), pronounced *Hobkirk.*

**Hoppe:** Walter *Hoppe* 1249 AssW; John *le Hoppe* 1273 RH (Wo); Alice *Hoppe* 1310 ColchCt. Metonymic for HOPPER. cf. Hugh *Hoppeoverhumbr'* 1220 Cur (Sx) 'hop over Humber'; Robert *Hopperobyn* 1306 AssW 'hop Robin'.

**Hopper:** Waldr', Edric *le Hoppere* 1203 P (W), 1204 Cur (Wo). A derivative of OE *hoppian* 'to hop, leap, dance', a dancer.

**Hopping:** William *Hopping* 1327 SRSo. Probably OE *\*hoping* 'dweller in a *hop*'. *v.* HOPE.

**Hopshort:** *Hoppeschort* 1169 P (Ha); Alured *filius Hoppesort* 1189 Sol; Gilbert *Hoppeshort* 1214 Cur (W); William *Hoppeshort* 1234 FFY; Ralph *Hopshort* 1374 AssL. 'Hop short', OE *hoppian, sceort,* perhaps a nickname for a lame man. cf. John *Hopop* 1435 FFEss 'hop up'.

**Hopson:** *v.* HOBSON

**Hopton:** Thomas *de Hopeton'* 1196 P (Sf); Robert *de Hopton* 1250 IpmY; William *Hopton* 1478 FFEss. From Hopton (Derby, Hereford, Salop, Suffolk, WRYorks).

**Hopwell:** *v.* HOPEWELL

**Hopwood:** William *de Hopwod, de Hopwode* 13th WhC, 1298 AssLa; William *Hopwod'* 1379 PTY; John *Hopwood* 1642 PrD. From Hopwood (Lancs).

**Horam, Horum:** Walter *Horum* 1275 RH (K); Nicholas *de Horham* 1296 SRSx. From Horham (Sf).

**Horberry, Horbury:** Jordan *de Hordbir'* 1204 AssY. From Horbury (WRYorks). This form confirms Ekwall's alternative etymology, *hordburg* 'treasure-burg'.

**Hord:** *v.* HERD

**Horden:** (i) John *Hordwyne* 1260 AssC. OE *Hordwine.* (ii) John *de Horredene* 1296 (Lanark), John *of Hordene* 1408 Black; Adam *Horden* 1642 PrD. From Horden (Du).

**Horder:** Ælfwine *ðe Hordere* 1001 OEByn (W); Ordric *hordere, þe hordor* 11th (1333) ASCh 220 (Sf); Simon *le Horder* 1225 AssSo. OE *hordere* 'keeper of the hoard or treasure', 'treasurer', 'keeper of provisions'. Ordric was the cellarer of the abbey of Bury St Edmund's. *v.* also HERDER.

**Hordern:** Robert *de Horderne* 1327 SRDb; Richard *Hordron* 1441 ShefA. From Horderne (Ch),

Horderns in Chapel en le Frith (Db), Hordron in Langsett (WRY), or 'dweller by the storehouse', OE *hord-ærn.*

**Hore:** *v.* HOAR

**Horey:** William *Horege* 1199 P (C); Alan *Horegh* 1297 SRY. 'Gray eye', OE *hār, ēage.*

**Horler:** Nicholas *le Hurlere* 1312 LLB D; Richard *Horler* 1327 SRSo; Edward *Horlor* 1641 PrSo. ME *hurlere* 'one who contends or strives, one who creates a disturbance'.

**Horley:** Godfrey *de Horelee* 1212 Cur (Wa); John *de Horle* 1296 SRSx; William *de Horle* 1305–6 FFSx. From Horley (O, Sr), or Horley Green Fm in Mayfield (Sx).

**Horlick, Horlock:** *v.* HARLOCK

**Horn, Horne:** (i) Lifwine *Horn* 1066–86 OEByn; Aluuinus *Horne* 1066 DB (Mx); Wulmer, Herui, Robert *Horn* 1166 P (Nf), 1185 Templars (So), 1197 P (Sf). Metonymic for HORNBLOWER or HORNER. (ii) Roger *de Horne* 1208 Cur (Sr); William *de la Horn* 1261 PN Sx 379; Thomas *atte Horne* 1327 SRSo. From Horne (Rutland, Som, Surrey) or from residence near a spur or tongue of land or a bend, OE *horn(a).*

**Hornabrook, Hornbrook:** *v.* HORNIBROOK

**Hornagold, Hornigold:** John *Hornyngold* 1524 SRSf; Widow *Hornigold, Hornegold* 1674 HTSf. From Horninghold (Leic).

**Hornblow, Hornblower, Horniblow, Orneblow:** John, Geoffrey *le Hornblauere, -blawere* 1255, 1285 Ass (Ess); Adam *Horneblawer* 1301 SRY. OE *hornblāwere* 'horn-blower'. In the Middle Ages workmen were called to work by the ringing of bells or by a horn. In 1320, at Caernarvon, Walter de la Grene was paid 1*d.* per week 'for blowing the horn' (Building 62). cf. Thomas *Blauhorn* 1303 Pat, Alice *Blawhorn* 1379 PTY.

**Hornbuckle:** *v.* ARBUCKLE

**Hornby:** William *de Horneby* 1205 Pl (Y), 1296 SRNb; John *Hornby* 1376 IpmGl. From Hornby (Lancs, Westmorland, NRYorks).

**Horncastle:** Simon *de Horncastel* 1262 PN Bk 34; Robert *de Horncastell* 1360 IpmNt; Nicholas *Horncastell* 1406–7 FFWa. From Horncastle (L).

**Horner, Hornor:** William *le Hornare* 1275 SRWo; Matilda, Clement *le Hornere* 1279 RH (O), 1302 LLB C. A derivative of OE *horn,* a maker of horn spoons, combs, etc. Later, also a maker of musical horns, and also one who blows a horn. Adam *le Horner* 1297 SRY is also called *le Harpour.*

**Hornibrook, Hornibrooke, Hornabrook, Hornbrook:** John *Hornabrooke* 1642 PrD; Cicilia *Hornbrooke* 1679, Peter *Hornabrooke* 1798 DWills. From Hornbrook in Kelly (D).

**Horniman:** *v.* HERNAMAN

**Horrabin, Horabin, Horobin:** Henrie *Horerobyn* 1596 DWills; William *Horabin* 1783 ib. 'Grey Rabin or Robin', OE *hār* and a pet-name of *Robert.* cf. *Jolyrobin* 1332 SRCu; Agnes *Greyadam* 1297 MinAcctCo.

**Horrey:** *v.* HURRY

**Horridge:** (i) Richard *de Hawerugge* 1297 MinAcctCo. From Hawridge (Bucks). (ii) Thomas *de*

*Horewich* 1327 SRDb. From Horwich (Lancs). (iii) Richard *de Horugg* 1275 RH (D). From one of the five places in Devon named Horridge.

**Horrocks, Horrox, Horrex, Horrix:** John *Horroc* 1279 RH (Bk); An. *Horex* 1674 HTSf. ME *horrok* 'part of a ship', a nickname for a shipwright or a sailor.

**Horrod:** Richard *Horhod* 1293 AssSt. 'Grey hood', OE *hār, hōd*.

**Horsbrugh, Horsburgh:** Symon *de Horsbroc* 1214–49 Black; William *de Horsbrok* 1329 ib. (Peebles); Alexander *Horsbruik* 1550 ib.; *Horsborrough* 1686 ib. From Horsburgh in Innerleithen (Peebles).

**Horscraft, Horscroft, Horsecroft:** John *de Horscrofte* 1274 RH (Ess); John *atte Horscrofte* 1333 MELS (So). 'Dweller by, or worker at an enclosure for horses.'

**Horse:** Richard *Hors* 1148–79 P (Gl); Gilbert *Horse* 1201 Pleas (Co); Henry *le Hors* 1276 AssLo; Wylliam *Horse* 1545 SRW. Either a nickname from the horse, OE *hors*, or for a worker with horses.

**Horsegood:** *v.* OSGOOD

**Horseman, Horsman:** Hugh *le Horsman* 1226–7 FFWa; John *le Horsman* 1327 SRWo; William *Horsman* 1415 IpmY. 'Mounted warrior, rider, groom, horse-dealer', OE *hors, mann.* cf. Adam *Horsdriuere* 1199 P (We) 'horse-driver'; Thomas *Horsknave* 1401 AssLa 'stable-boy'; Leo *le Horsmongere* 1279 RH (C) 'horse-dealer'.

**Horsey:** William *de Horseia* 1182 P (Ha); William *de Horseye* 1268 AssSo; John *atte Horsee* 1332 SRSr; Gabriel *Horsey* 1642 PrD. From Horsey (Nf), Horsey Pignes (So), or Horse Eye (Sx).

**Horsfall:** William *de Horsfolde* 1327 SRWo; John *Horsfall'* 1379 PTY; Abrahem *Horsfall, Horsfield* 1790–1 WRS. From Horsfall in Todmorden (WRY), or 'worker at the horse enclosure', OE *hors, falod.*

**Horsford, Horsforth:** Nigel *de Horsford'* 1209 Cur (Y); John *de Horsseford* 1287–8 NorwLt; Thomas *Horsford* 1662–4 HTDo. From Horsford (Nf), Horsford in East Worlington (D), or Horsforth (WRY).

**Horsham:** Godfrey *de Horsam* 1255–6 FFSx; Geoffrey *de Horsham* 1326 CorLo; William *Horsham* 1351 AssEss. From Horsham (Sx), Horsham St Faith (Nf), or Horsham in Martley (Wo).

**Horsler:** *v.* OSTLER

**Horsley:** Fulcho *de Horselega* 1170 P (Wo); Roger *de Horsleye* 1279 AssNb; John *de Horsley* 1362 AssY. From Horsley (Db, Gl, St), Long Horsley (Nb), East, West Horsley (Sr), or Horseley Hills Fm in Wolverley (Wo).

**Horsman:** *v.* HORSEMAN

**Horsnail, Horsnaill, Horsnall, Horsnell, Horsenail, Horsenell, Arsnell:** Henry, Richard *Horsnail* 1221 AssSa, c1248 Bec (Berks). *Horse-nail*, a horseshoe-nail (1432 MED), frequently mentioned in medieval accounts. Either a maker of these or a nickname for a shoer of horses.

**Horst:** *v.* HERST

**Horstead, Horsted:** William *de Horsted'* 1206 Cur (Mx); William *de Horsted* 1287 NorwDeeds I; Alan *de Horstede* 1317 AssK. From Horstead (Nf), Horsted (K), or Horsted Keynes, Little Horsted (Sx).

**Hort:** *v.* HART

**Horton:** Leofwine Godwines sunu *æt Hortune* c1018 OEByn; Alan *de Hortun* c1160–80 YCh; Richard *de Horton'* 1255 ForNth; William *Hortone* 1334–5 SRK. From one or other of the many places of this name.

**Horum:** *v.* HORAM

**Horwell, Horwill:** Hugh *de Horewell'* 1195 P (Hu); Nicholas *de Horewelle* 1314 LLB D; John *Horwell'* 1390 KB (Ireland). From a lost *Horwell* in Severn Stoke (Wo), or 'dweller by the muddy stream', OE *horh, wiella.*

**Horwich:** Robert *de Horwich* 1332 SRLa. From Horwich (Lancs).

**Horwick:** Walter *de Horewyk* 1327 SRSx. 'Dweller at a muddy dairy-farm', OE *horh, wīc.*

**Horwood:** Osbert *de Horwude* 1214 Cur (Ha); William *del Horewode* 1332 SRSt. 'Dweller near a muddy wood', as at Horwood (Bucks, Devon).

**Hosbons:** *v.* HUSBAND, OSBORN

**Hosburn:** *v.* OSBORN

**Hose, Huse:** (i) Henry *Hose* 1154 MCh; Roger *le Hose* 1275 SRWo; Walter *Hose* 1327 SRSf. OE *hosa* 'hose'. Metonymic for a maker or seller of hose. (ii) Hugh *de la Hose* 1189 P (Nf); Thomas *de la Hose* 1254 FFHu; Peter *de la Huse* 1327, William *atte Hose* 1332 SRSx. 'Dweller in the brambles or thorns', OE *hos.*

**Hos(e)good:** *v.* OSGOOD

**Hosey:** *v.* HUSSEY

**Hosier:** William *Husier* 1180 Oseney (O); John *le Huser* 1182–7 Clerkenwell (Lo); Nicholas, Yuo *le hosier* 1197 P (Lo, Y); Alexander *le Hoser* 1200 Cur (Sx). A derivative of OE *hosa* 'hose', dealer or maker of stockings and socks (1381 MED). Early *husier* points to an OFr *\*hosier.* cf. OFr *heuse* (*huese, house, hose*) 'boot', hence 'shoe-maker'.

**Hosken, Hoskin, Hosking, Hoskings, Hoskins, Hoskyns, Hoskison, Hoskisson:** Osekin 1274 RH (Lo); Robert *Osekin* ib.; Peter *Osekyn* 1306 FFEss; Thomas *Hoskyns* 1463 AD vi (Berks); William *Hoskyn* 1472 KentW. A diminutive of *Os-*, a short form of such names as *Ōsgōd, Ōsbeorn, Ōsmǣr,* etc.

**Hosker:** *Osgarus* de Bedeford 1066 DB (Beds); Robert *Hoseger* 1199 FF (R); William *Hosker* 1375 ColchCt. OE *Ōsgār* 'god-spear'.

**Hosler:** *v.* OSLAR

**Hosmer:** *v.* OSMER

**Hosell:** William *del Hostel, del Ostel* 1222 Cur (Do); John *del Hostell* 1332 SRCu. OFr *ostel, hostel* 'lodging, inn', hence inn-keeper. cf. OSTLER. William was a brother of the Abbot of Ford and may have been responsible for the abbey's entertainment of guests.

**Host, Hoste, Ost:** Elias, John *le host* 1254 AssSo, 1279 RH (C); Richard *le Ost, Lost* 13th Guisb (Y). OFr *oste, hoste* 'host, guest'; an inn-keeper (c1290 NED).

**Hostage:** William *Hostage* 1309 AssSt; William *Ostage* 1311–12 FFEss. OFr *(h)ostage* 'lodging'. Metonymic for an innkeeper.

**Hoster:** Stephen *Hodestre* 1379 AssEss. A feminine form of HODDER. A rare surname. cf. BAXTER.

**Hostler, Hosteller:** *v.* OSTLER

**Hotchen, Hotchin:** v. HUTCHIN

**Hotchkin, Hotchkiss:** v. HODGKIN

**Hotfoot:** John *Hotfoot* 1327 *SR*Ess. 'Hot foot', OE *hāt*, *fōt*. cf. Simon *Hothede* 1204 P (Bk) 'hot head'.

**Hotham, Holtham:** Robert *de Hotham* 1202 AssL; Walter *de Hothum* 1327 SRY; John *Hotham* 1381 LoPleas. Usually from Hotham (ERYorks), but occasionally perhaps from Holtham or Hougham (Lincs). v. also HOLTHAM.

**Hother:** v. HOATHER

**Hotson:** v. HUDSON

**Hott:** Robert, Geoffrey *Hot* 1195 P (Y), 1275 SRWo; William *Hotte* 1327 SRSf. OFr *hotte* 'basket' (a1400 MED). 'Basket-maker.' v. HOTTER and cf. BANNISTER.

**Hotter:** William, John *le Hottere* 1275 SRWo, 1331 LLB E. A derivative of OFr *hotte* 'basket', a basket-maker. cf. HOTT. Hotters (*hottarii* 1284, *hottatores* 1295) were also carriers of wicker panniers (*hotti* 1208, *hottes de wekere* 1441) full of sand for making mortar. v. Building 353–4.

**Hottot:** Hugo *de Hotot* 1086 DB (C); William *hotot*, *de hotot* 1148 Winton (Ha). From Hottot-les-Bagues or Hottot-en-Auge (Calvados). v. OEByn.

**Houchen, Houchin:** v. HUTCHIN

**Houdsworth:** v. HOLDSWORTH

**Houdwin:** v. HOLDWIN

**Houf, Hough:** v. HOW

**Hougham, Huffam:** William *de Huham* 1207 P (K). From Hougham (Kent), formerly spelled *Huffam* (Lower).

**Houghton, Hoghton, de Hoghton, Howton:** William *de Hoctona, de Hohton'* 1115 Winton (Ha), 1208 Cur (Sx); Symon *de Howtone* 1279 RH (C). From one of the Houghtons.

**Houlbrook(e):** v. HOLBROOK

**Houlcroft, Houldcroft:** v. HOLCROFT

**Hould:** v. HOLD

**Houlden, Houldin:** v. HOLDEN

**Houlder:** v. HOLDER

**Houldershaw:** v. OLDERSHAW

**Houldsworth:** v. HOLDSWORTH

**Houle:** v. HOLE

**Houlford:** v. HOLFORD

**Houlgrave:** v. HOLGREAVES

**Hoult:** v. HOLT

**Houltby:** v. HOLTBY

**Houlten:** v. HOLTON

**Hound:** Bonde *Hund* 1166 P (Nf); Alice *le Hound* 1327 SRSf. OE *hund* 'hound, dog'.

**Houndfoot, Houndsfoot:** Wuluiet *Hundesfot* 1162 P; Walter *Hondesfot* 1279–80 CtH; Stephen *Hundfote* 1317 Misc (Y). A nickname, 'hound foot', OE *hund*, *fōt*. cf. Alfwi *Hundesege* 1180 P (Wo) 'hound's eye'; Randulf *Hundesheved* 1176 P (D) 'hound's head'; Ailmar *Hundesnase* 1210 P (Ess) 'hound's nose'.

**House, Hows, Howse, Hoose:** Simon *Hus* 1226 Eynsham (O); Geoffrey *de la House*, Richard *Hous* 1279 RH (Hu, O); Walter *del Hus* 1289 AssCh; William *atte House* 1331 FFK; Nicholas *Howse*, Robert *Howes* 1524 SRSf. One employed at 'the house' (OE *hūs*), probably a religious house, convent (a1160 MED). *House* and *Howse* may sometimes be for *Howes*.

**Household:** William, Richard *Housold* 1279 RH (Hu), 1327 SRC. Metonymic for *Householder* 'one who occupies a house as his own dwelling' (c1387–95 MED).

**Houseley, Housley:** John *de Houseley* 1440, William *Houseley* 1441 ShefA; Virtue *Housley* 1662–4 HTDo. From Housley Hall in Ecclesfield (WRY).

**Houseman, Housman:** John *Houseman* 1365 ColchCt. v. HOUSE.

**Houser:** v. HOUSE, HOUSEMAN

**Houston, Houstoun:** Finlay *of Huwitston*, alias Finlawe *de Hustone* 1296 Black; John *de Howistone* 1406 ib. (Paisley); Patrick *de Huyston* 1415 ib. (Glasgow); Alexander *Howstoun* 1460 ib. From the barony of Houston (Lanark).

**Hovell, Hovil:** v. HAVILL

**Hovell:** Adam *Houel* 1193 P (Nf); Robert *Hovel* 1248 FFEss; John *Houel* 1327 SRSf. Probably usually for HOWELL, but sometimes, perhaps, 'dweller in the hut or cottage', ME *hovel*. v. also HAVILL.

**How, Howe, Howes, Hows, Hoe, Hoes, Hoo, Heugh, Hough, Hoof, Hooff, Houf, Houfe, Huff, Hughf, Hughff:** William *de Ho* 1121–48 Bury (Ess); Eustace *de Hou* 1190 P (Ess); Alured *de la Ho* 1199 P (Wo); Benedict *de Ho, de Hoes'*, *de Howe*, Marjoria *de Howes* (his wife) 1211–12 Cur (Lei); Walter *de Howes*, *de Hou, de Hoes* 1212 Cur (Nt); Herebertus *Alahoge* 1240 Rams (Nf); Gilbert *ate How* 1296 SRSx; William *ate Howes* 1327 SRC; Thomas *de Hoo* 1332 SRSx; Henry *del Hoe* 1332 SRSt; William *de Huff* 1379 PTY; William *Hough* 1564 Bardsley; Thomas *Hoofe* 1626 FrY; Arthur *Houfe* 1743 ib. OE *hōh* 'heel', 'projecting ridge of land', was common in place-names in the sense of 'a spur of a hill', 'steep ridge', or 'a slight rise'. It is especially common in Bedfordshire and Northants, fairly so in Suffolk, Essex, Herts and Bucks, and very common in Northumberland and Durham. The nominative singular gives *Hough*, in Scotland and Northern England *Heugh*. The dative singular *hō* becomes *hoo* or *hoe*, a later form *hōge* giving *howe*. From the nominative plural *hōs* we have *hose* and from the later *hōgas, howes*. The surname may derive from Hoe (Norfolk), Hoo (Kent), Hooe (Devon, Sussex), Hose (Leics), Heugh (Durham, Northumb), Hough (Ches, Derby), often pronounced *Hoof, Huff*, or from residence near a *hōh*. *Howe* may also be from ON *haugr* 'mound, hill' and without other evidence cannot be distinguished from *howe* 'ridge'. It is certainly the origin of Howe (Norfolk) and of Howe Hill in Kirkburn (ERYorks), near which there is a tumulus beside which lived Robert *atte Hou* in 1333 (PN ERY 166). *Howes* often refers to groups of tumuli. *How* is also a form of *Hugh*, and *Howes* of *Hughes* and *House*.

**Howard, Howerd, Heward, Hewart, Huard, Huart:** (i) *Huardus, Houart* 1086 DB (He, Lei, Sf, W); Willelmus *filius Huward* 1170 P (Nb); *Huwardus, Huardus* de Bikeleg' 1214–19 Cur (D); *Huart* de Noerel 1216 PatR (Ess); *Heward* de Horewell 1279 RH (C); Geoffrey *Hohard* 1194 P (Nf); Geoffrey *Huard* 1209 P (Nf); John *Huward* 1279 RH (O); Robert, John *Heward* 1313 Wak (Y), 1337 ColchCt. OFr *Huard* from OG *Hugihard* 'heart-brave'. As

*Hugh* appears in ME as both *How* and *Hew*, this is the origin of *Heward* and a source of *Howard* but there has been confusion with OG *Howard* below, the cognate *Haward* of Norse origin, and also with *Hayward*. (ii) *Houardus* 1066 DB (Ess); *Howard* 1101–7 Holme (Nf); Willelmus *filius Howard* 1188 BuryS (Sf); *Owardus, Houwardus, Howardus* (identical) 1221–2 Cur (Sf); Geoffrey *Houward* 1210 Cur (C); Robert, William *Howard* 1221 *ElyA* (Nf), 1312 ColchCt. OG *Howard, Howart* 'high or chief warden'. Confused with *Haward*, which is sometimes found for the Norfolk *Howard*. cf. 'my Lord *Haward*' (1503), 'Henry *Hawarde*, Esq., son of Sir Thomas *Hawarde*, Viscount Bindon' who succeeded as Viscount *Howard* of Bindon (1565–6 Bardsley). (iii) John *Howeherde* 1348 DbAS 36. 'Ewe-herd' from OE *ēowu* and *hierde*, a rare but contributory source. *v.* EWART.

**Howarth, Howorth, Howourth:** Randal *Howarth* 1532 Bardsley. The surname is common in Manchester and Leeds. In the former it is from Howarth in Rochdale (Lancs), in Leeds it is, no doubt, often a corruption of *Haworth*.

**Howat, Howatt:** *v.* HEWET

**Howatson:** Jok *Howatson* 1569 Black. 'Son of Howett.' *v.* HEWET.

**Howchin:** *v.* HUTCHIN

**Howden, Howdon:** Robert *de Houedona* 1173 P (Y); Thorstan *of Houeden* 1268 FFY; John *de Houeden* 1406 IpmY. From Howden (Nb, ERY), or Howden Clough (WRY).

**Howel, Howell, Howels, Howells:** (i) *Huwal* West Wala cyning (King of the West Welsh) 926 ASC D; Morganus *filius Hoel* 1166 P (Sa); *Hoelus* de Charlion 1184 P (Glam); *Howell*' filius Ade (Walensis) 1221 AssSa; Geffrei *Hoel* c1100–30 OEByn (D); Bald', William *Hoel* 1183 P (W), 1221 *ElyA* (Nf); Robert *Howeles* 1210 Cur (Nth); John *Houeles* 1280 AssSo; John *Howel* 1313 FFC; William *Huwel* 1331 AssSt. OW *Houel*, OBret *Houuel*, *Huwel*, *Huwal*, *Howael*; in the Welsh border counties from Wales, in the eastern counties, where Bretons were numerous, direct from Brittany. *v.* also POWELL. (ii) Walter *de Huwella* 1165 P (L); Alfredus *de Howella* 1177 P (L). From Howell (Lincs).

**Howers:** Walter, Henry *le Howere* 1275 RH (K), 1327 SRDb. 'Dweller near a ridge or hill.' OE *hōh* or ON *haugr* plus *-er*. Equivalent to atte Howe. *v.* HOW.

**Howetson:** William *Howetson* 1379 PTY. 'Son of *Howett*.'

**Howett:** *v.* HEWET

**Howgrave:** Cecilia *de Hougraue* 1219 AssY; Robert *Howgrave* 1327 SRY. From Howgrave (NRY).

**Howick:** Richard *de Howic* 1210 Cur (Nb); Cecil' *de Houwyke* 1296, William *de Howike* 1327 SRSx. From Howick (Nb), or Howick Fm in Lodsworth, Howick Fm in Rudgwick (Sx).

**Howieson, Howison:** John *Howison* 1450 Black; Alan *Howesson* 1474 NorwW (Sf). 'Son of *Howe*' (Hugh), or, perhaps, an assimilated form of *Howetson*.

**Howitt:** *v.* HEWET

**Howkins:** *v.* HUKIN

**Howlett:** *v.* HEWLETT

**Howlin(g):** *v.* HEWLINS

**Howman:** *v.* HUMAN

**Howroyd:** *v.* HOLROYD

**Howse:** *v.* HOUSE, HOW

**Howson:** *v.* HUGHSON

**Howton:** *v.* HOUGHTON

**Hoy, Hoyes, Hoye, Hoys:** John *le Hoy* 1255 RH (W); Richard *Hoye* 1278 AssSo; Robert *le Hoy* 1327 SRC. Middle Dutch *hoey* 'cargo ship'. Metonymic for a sailor.

**Hoyer:** *v.* AYER

**Hoyland:** Ralph *de Hoyland* 1204 AssY; Robert *de Hoyland* 1278 AssLa; John *Hoiland, Holland* 1564 ShefA. From one or other of the four places of this name in West Yorks. There is late confusion with HOLLAND.

**Hoyle, Hoyles, Hoile, Hoiles:** Henry *att Hoyle* 1379 PTY. A south Yorkshire dialectal pronunciation of HOLE.

**Hoys:** *v.* HOY

**Hoyt, Hoyte:** John *le Hoit* 1327 SREss; Stephen *Hoyte* 1642 PrD. ME *hoit* 'a long stick', a nickname for a tall, thin man.

**Huard:** *v.* HOWARD

**Huban, Huband:** John *Huebarne* 1301 SRY. 'Hugh's bairn.' cf. GOODBAIRN.

**Hubbard, Hubbert:** *v.* HUBERT

**Hubberstey:** John *Hubersty* 1452 FrY. From Hubbersty, a lost place in Cockerham (Lancs).

**Hubbold, Hubball, Hubble:** *Hubald* de Bereford' 1205 P (He); Hugo *Hubald, Hubolt* 1086 DB, ICC (Beds); Bernard *Hubold* 1148 Winton (Ha); Henry *Hubaut, Hubald, Hubalt* 1199, 1205 P (Wa); William *Hubball* 1640 SaltAS (OS) xv. OG *Hugibald, Hubald* 'mind-bold', OFr *Hubaut*.

**Hubert, Hubbard, Hubbart, Hubbert, Hobart:** Eudo *filius Huberti* 1086 DB (Ha); *Hubert* de Bissoppesgate 1292 SRLo; Roger *Hubert* 1199 FF (Nth); Thomas *Huberd (Hubert)* 1230 P (Do); William *Hoberd* 1291 FFSf; Roger *Hubard* 1327 SRSo; John *Hobart* 1346 FA (Sf). OG *Hugibert, Hubert* 'mind-bright'.

**Hubling:** William, Thomas *Hubelin(e)* 1327 SRSf, 1348 DbAS 36. *Hub-el-in*, a double diminutive of *Hubb*, a short form of *Hubald* or *Hubert*. cf. Amyas *Hubbe* 1327 SRSf.

**Hubling:** William *Hubelyne* 1327 SRSf; Thomas *hubelyn* 1348 DbAS 36. *Hub-el-in*, a double diminutive of *Hub*, a pet-form of HUBBOLD, or of HUBERT. cf. Amyas, Thomas *Hubbe* 1327 SRSf.

**Huby:** Nicholas *de Hoby* 1251 AssY; William *Huby* 1381 LoPleas, 1430 TestEbor. From Huby (NRY, WRY).

**Hucheson:** *v.* HUTCHINSON

**Huck, Hucks:** *Hucche* purs c1150 DC (L); Gamel *filius Hucca* 1185 Templars (Y); *Hucke* 1221 *ElyA* (Nf); William *Hukke* 1279 RH (Hu); William *Huckes* 1568 SRSf. OE *Ucca, a pet-form of OE *Ūhtrǣd*. cf. Huctredus *filius Ucke* 1212 Fees (La). cf. HUG, HUGMAN and *v.* HOOK.

**Huckel, Huckell, Huckle:** Reginald, Jocelin *Huckel* 1209 P (C), 1225 AssSo. A diminutive of *Huck*.

**Hucker:** John *le Hukker'* 1307 MESO (So); John *le Hucker'* 1333 ib. A derivative of ME *hucke* 'to bargain', a petty dealer.

**Huckerby:** Robert *de Huccherbi* 13th Guisb; Ralph *de Huckerby* 1252, Ranulph *de Huckerbi* 1333 Riev. From Uckerby (NRY).

**Huckin:** *v.* HUKIN

**Huckle:** *v.* HUCKEL

**Huckman:** *Huckeman* 1181 P (Y); *Hukeman* de Moricebi 1194 P (Cu); *Hucmon* riding 1259 AssLa; William *Hukeman* 1279 RH (Hu). Probably usually a ME personal name, *Huckman*, but it may sometimes mean 'pedlar, petty trader', from ME *hukken* 'to bargain', and *man*. cf. Thomas *Hukechese* 1296 SRSx 'cheese seller'; Aluric *Huchpain* 1177 P (C) 'bread seller'; John *Hucketrout* 1301 SRY 'a seller of trout'.

**Hucknall, Hucknell:** *v.* HOCKENHALL

**Hudd:** *Hude* de Rafningeham 1177 P (Sf); *Hudde* de Bosco 1210 P (Nf); *Hudde* de Pesefurlang 1246 AssLa; William, Reginald *Hudde* 1230 P (Sa, Y). *Hudde* was very common and is a pet-form of *Hugh*: *Hugo* filius Johannis . . . Johannes pater ipsius *Hudde* 1212 Cur (Y); *Hugo* Sturdy 1219 AssY, *Hudde* Sturdi 1230 P (Y). It appears to have been used also for *Richard*: Bardsley's *Ricardus* de Knapton and Cristiana *Hud*-wyf (1379 PTY) is not absolutely conclusive, but his '*Ricardus* dictus *Hudde* de Walkden' (1346 Cl) leaves no doubt. Also found as *Hutte* (1246 AssLa).

**Huddart:** *v.* WOODARD

**Hudden:** William *Hodden* 1332 SRWa. From *Hudde*, or *Hodde* (1211 FFL). *v.* HUDD, GEFFEN.

**Huddle:** *Hudd-el*, a diminutive of *Hudd*. cf. the double diminutive in: William, Richard *Hudelin* 1208 P (L), 1247 AssBeds.

**Huddleston, Huddlestone, Huddelston, Hudleston, Hudlestone, Hiddleston:** Richard *de Hudelesdun* 1200 Cur (Y). From Huddleston (WRYorks).

**Hudgell:** *v.* HUDSWELL.

**Hudlin:** William *Hudelin* 1208 P (L); Richard *Hudelin* 1247 AssBeds; Jordan *Hudelin* 1279 RH (Beds). *Hud-el-in*, a double diminutive of *Hudd*, a pet-form of *Hugh*.

**Hudman:** Roger *Hudman* c1248 Bec (Do). 'Servant of *Hudd*.'

**Hudsmith:** Thomas *Huddemogh* 1332 SRLa; William *Hudmagh* 1379 PTY; Thomas *Huddesmawth* 1464 FeuDu; Cuthbert *Hodgemaght*, *Hodgemaughthe* 1545–6 NorwW (Nf); Ralph *Hudsmyth* 1582 PrGR. 'Hudd's brother-in-law.' cf. Richard *Gepmouthe* and *v.* WATMOUGH.

**Hudson, Hutson:** John *Hudsone*, *Hutson* 1323 Wak (Y), 1568 SRSf. 'Son of *Hudd*.'

**Hudswell, Hudgell:** Geoffrey *de Huddeswell'* 1206 P (Y). From Hudswell (NRYorks).

**Hue:** *v.* HUGH

**Huelin:** *v.* HEWLINS

**Huett:** *v.* HEWET

**Huff:** *v.* HOW

**Huffam:** *v.* HOUGHAM

**Huffington:** Alan *de Huffintone* c1270 ArchC 34; Margery *Uffington* 1553 ib. 37. From Ufton Court in Tunstall or Uffington Fm in Goodneston (Kent).

**Hufton, Huffton:** William *de Hufton'* 1271 Glapwell (Db); Edward *Huffton* 1760 PN Db 502. Probably from Houghton (Db).

**Hug:** Galfridus *filius Hugge* 1301 SRY; Robert, William *Hugge* 1180 P (Y), 1279 RH (O). OE \**Ugga*, a pet-form of *Ūhtrǣd*. cf. *Uctredus* pater Henrici et *Ugge* avus suus 1212 Cur (Y). cf. HUCK, OUGHTRED.

**Huggard:** Richard *Huggard* 1535 LWills. From *Hug-ard*, a variant of *Huard*.

**Huggens:** *v.* HUGGIN, HUGGON

**Hugessen:** A Huguenot name. James *Hugessen* was a refugee from Dunkirk who settled in Dover, later moving to Sandwich (Smiles 320, 403).

**Huggett, Huget:** (i) A variant of *Hewet*, from *Hug-et*, a diminutive of *Hugh*. Huggett's Lane in Willingdon (Sussex) owes its name to Helwis' *Hugot* 1296 SRSx, *Huwet* 1327 ib., and Huggett's Fm in Heathfield (Sussex) to William *Hughet* 1296 SRSx. It is *Huggetts* in 1614 (PN Sx 425, 467). (ii) Robert *de Hugat'* 1219 AssY; Isabel *Hugate*, *Hugitte* 1440 ShefA. From Huggate (ERYorks).

**Huggill:** *v.* HUGILL

**Huggin, Huggins:** Reginaldus *filius Hugin* Hy 1 Fees (So); *Hugyn* 1246 AssLa; Robert, John *Hugyn* 1327 SRSx, 1337 FFSt; William *Hugyns*, Amisia *Hugines* 1327 SRWo. *Hug-in*, a diminutive of *Hugh*. Also for *Huggen*: Robert *Huggen* 1332 SRWa. *v.* GEFFEN.

**Huggon, Huggons, Hugon:** *Hugun* filius Ricardi c1250 Rams (Hu); Richard *Hugons* 1327 SRSo; Hugh *Hugune* 1379 PTY; William *Hugon* 1432 FrY. There seems to be no evidence for the use in England of OFr *Hugon*, cas-régime of *Hugues*. *Hugun* is a diminutive of *Hug*, i.e. *Hugh*. For the series *Hugun*, *Hugin*, *Huget*, cf. that of *Burdun*, *Burdin*, *Burdet*, s.n. BURDEN.

**Huggonson:** 'Son of *Hugun*.'

**Hugh, Hughes, Hugo, Ugo, Hue, Huws, Hew, Hewes, Hews, How, Howe, Howes, Hows, FitzHugh, Fitzhugues:** *Hugo* 1066 DB (Hu, Sf); Willelmus *filius Hugonis* 1084 GeldR (W); Reginaldus *le fiz Hugonis* 1195 P (Lei); *Howe* Golichtly, *Howe* caretarius 1221 AssWa; Edde *filius Hugh* 1279 RH (C); *Hwe* purte 1292 SRLo; *Hue* de Clopham ib.; Rogerus *Hugo* 1185 Templars (Db); Ralph, Robert, William *Howe* 1221 AssWa; Richard *Hue* 1275 SRWo; Richard *Huwe* 1279 RH (C); Constance *Huwes* 1279 RH (O); William *Hewe* 1291 LLB A; John *Hugh* 1296 SRSx; John *Hugh* (*Huwe*) 1327 *SR* (Ess); Thomas *Hughes* 1327 SRSo; John *Fitz Huwe* 1344 Ipm; Thomas *Howe*. *Huwe* 1353–4 ColchCt; Richard *Houwes* 1359 Putnam (So); Elysabeth *Hewes* 1524 SRSf. OFr *Hue*, from OG *Hugo* 'heart, mind', a very popular name after the Conquest. It is usually found in the Latin form *Hugo*, the vernacular forms being *Hewe* and *Howe*, all of which survive as surnames. The popularity of Hugh is clearly shown by its 14 variants in modern surnames and by numerous derivatives in *-in*, *-on*, *-et*, *-ot*, *-kin*, *-son*, *-man*, and by double diminutives in *-el-in*, *-el-et* and the French double diminutive *Huchon*. Of these derivatives there are some 90 different modern forms and the list is probably not complete. In Scotland and Ireland *Hugh* is frequent for an original Gaelic *Aodh*, in Argyllshire for *Èghann* and in north and north-west

Scotland for *Uisdeann*. In Ireland, Hughes and MacHugh are regarded as equivalent to MacKay. In Wales it has become *Pugh*. *Hew*, *How* and *Howes* have alternative origins. *v.* FITHIE.

**Hughf, Hughff:** *v.* HOW

**Hughill:** *v.* HUGILL

**Hughman:** *v.* HUMAN

**Hughson, Huson, Hewson, Hooson, Hoosun, Howson:** Wlfuric *Hugo sune* 1066 InqEl; Richard *Hughson* 1310 LLB D; William *Huggesone* 1327 SRWo; Henry *Howsone* 1332 SRCu; Michael *Hwesone* 1378 ColchCt; Thomas *Hughesson* 1389 FrY; William *Hewson* 1437 FrY; Henry *Hooson* 1635 RothwellPR (Y); Thomas *Hueson* 1653 ib. 'Son of *Hugh*.'

**Hugill, Huggill, Hughill, Hewgill, Howgill:** Matilda *de Hogyll* 1379 PTY (Sedbergh). From Howgill in Sedbergh (WRYorks) or Hugill (Westmorland).

**Huglin:** *v.* HEWLINS

**Hugman:** William *Uggeman*, *Huggeman* 1301 SRY. 'Servant of *Ugge*.' *v.* HUG.

**Hugo:** *v.* HUGH

**Hugon:** *v.* HUGGON

**Huish, Huyshe, Hewish:** Robert *de Hiwis* 1194 P (D); William *de Hewis* 1207 Cur (Do); John *de Hewish* 1278 AssSo. From one of the places named Huish in Devon, Dorset, Somerset or Wilts.

**Huison, Huitson:** *v.* HEWETSON

**Huitt:** *v.* HEWET

**Huk, Huke:** *v.* HOOK

**Hukin, Hukins, Hewkin, Hookins, Howkins, Huckin:** John *Hukyns* 1332 SRWa; John *Hukyn* 1337 BarkingAS ii (Ess); Thomas *Howkyn* 1378 Balliol (O). A diminutive of *How* or *Hew*, i.e., *Hugh* plus *-kin*.

**Hulance:** *v.* HEWLINS

**Hulatt, Huleatt, Hulett:** *v.* HEWLETT

**Hulbert, Hulburd:** *v.* HOLBERT

**Hulin:** *v.* HEWLINS

**Hull, Hulles, Hulls:** (i) Peter *de Hull* 1199 AssSt; John *ate Hulle* 1297 MinAcctCo (Mx); Robert *in le hull* 1327 SRDb. From Hull (Ches). *atte hulle* is a southern and west-midland form of *hill*, 'dweller on a hill', but this has, no doubt, usually assumed the standard form. *v.* HILL. (ii) Henry, Simon *Hulle* 1309 Wak (Y), 1312 ColchCt; John *Hull* 1332 SRCu. The surname is most commonly from a personal name *Hulle*: *Hulle* le Bule 1201 P (St), *Hulle* son of Stephen 1227 AssBeds, *Hulle* de Alperam 1259 AssCh. A petform of *Hugh* or of its common diminutives *Hulin*, *Hulot*. In 1380 John *Hulle*sone Rudde held a tenement formerly *Hugh* Rudde's (AD v). John was the son of *Hulle* or *Hugh* Rudde.

**Hullah, Huller:** Ralph *le Hullere* 1327 SRSx; Richard *le Huller* 1332 SRSr. 'Dweller by a hill' from OE *hyll*, equivalent to *atte hulle*. This would survive as there was no common standard word. The southeastern *Heller* also survives. cf. BRIDGER.

**Hulland, Hulands, Hulance:** John *Hulland* 1489 ArchC 39; Seyches *Huland* 1545 SRW; Richard *Hewland* 1558 LWills. From Hulland (Db).

**Hullcock:** *Hulcok* de Clay 1282, *Hulcoc* de Fernileg' 1286 AssCh; Peter *Hulecoke* c1250 ERO; Hugh *Hullecok* 1327 SRC; Robert *Hulkok* 1524 SRSf. A diminutive of *Hulle*, a pet-form of *Hugh*.

**Hullett:** *v.* HEWLETT

**Hullin, Hullins:** *v.* HEWLINS. Also for *Hullen*: Richard, Thomas *Hullen* 1332 SRWa. *v.* HULL, GEFFEN.

**Hullot:** *v.* HEWLETT

**Hullyer:** *v.* HOLLIER

**Hulm, Hulme:** Turstinus *de Hulmo* 1169 P (Hu); Geoffrey *de Hulm* 1202 P (La); John *de Hulm* 1260 AssCh. ODa *hulm*, corresponding to ON *holmr*. *v.* HOLM. From Hulme (Staffs) or one of the places of the same name in Cheshire or Lancashire. *v.* HUME.

**Hulman:** Roger *Hul(le)man* 1327 SRSo, 1359 Putnam (So). 'Servant of *Hull*' or 'dweller by a hill', equivalent to HULLAH.

**Hulson:** Thomas *Hull(e)son* 1308, 1333 FrY. 'Son of *Hull*.'

**Hulton:** Jelvorth *de Hulton'* 1206 Cur (La); Adam *de Hulton* 1246 AssLa; William *Hulton* 1418–19 FFWa. From Hulton (Lancs, Staffs).

**Hulver:** Isabel *Huluyr* 1473 (Sf), Henry *Hulver* 1479 (Nf) NorwW. 'Dweller by the holly-tree', ON *hulfr*.

**Hum, Humm:** Gilbert *Om* 1177 P (Ha); Ernald *Hum* 1208 Cur (L); John *Humme* 1229 Cl (Sf); Geoffrey *le Home* 1296 SRSx. OFr *homme* 'man'. cf. MAN.

**Human, Hughman, Howman, Uman:** Matill' *filia Hiweman* c1248 Bec (W); *Hugeman* de Assinton 13th *Stoke* (Sf); Nicholaus *filius Howemanni*, *filius Huemanni* 1252 Rams (Hu); Willelmus *filius Howman* 1276 RH (Hu); William *Hiweman* c1248 Bec (W); Humfrey *Hueman* 1277 *Ely* (Sf); John *Human* 1279 RH (C); William *Huweman* (*Howeman*) 1327 *SR* (Ess). The forms would suggest 'servant of *Hugh*' and the surname may sometimes have this meaning, but such a combination as a personal name is rare or unique. In late OE times names in *-mann* were popular and new combinations were formed. This is a similar late formation, *\*Hygemann*, the first element *hyge* 'mind', cognate with the OG *hugi* of *Hugh*, found in OE *Hygebeald* and *Hygebeorht*. Compounds of *Hyge-* were in use after the Conquest: *Hygemund*: Gamel filius *Higmund* 1185 Templars (Y), and *Hygeræd*: Willelmus filius *Hugered* 1198 P (Sx). *Howman* as a surname might also mean 'dweller by a hill-spur or hill', equivalent to HOWERS. *v.* HOW.

**Humber:** John *le Humbre* 1305 IpmW; John *Humbre* 1392 LoCh; Edward *Humber* 1662–4 HTDo. From Humber (He), Humber in Bishopsteignton (D), or the River Humber.

**Humberston, Humberstone, Humblestone, Humerstone, Hummerston, Hummerstone:** Ralph *de Humberstein* 1180 P (L). From Humberstone (Leics, Lincs).

**Humble:** Adam *Homel* 1242 AssDu; Thomas *Humble* 1633 Black. 'Humble, obedient', OFr *humble*.

**Humby:** Walter *de Hunby* Hy 2 Gilb; Thomas *Humbye* 1545 SRW; Cristofer *Humby* 1662–4 HTDo. From Humby (L), or Hanby in Welton (L), *Humbi* DB.

**Hume:** Walter *de Hulmo* (*Humo*) 1221 Cur (Sf); Ralph *de la Hume* 1275 RH (Nf); Walter *Hume* 1275 RH (Sf). ODa *hulm* with loss of *l* before the labial. *v.* HOLM, HOME.

**Humfress, Humfrey:** v. HUMPHREY

**Humm:** v. HUM

**Hum(m)erstone:** v. HUMBERSTON

**Humpherson, Humpherston:** John *Humphreson* 1663 Bardsley (Ch). 'Son of *Humphrey*.'

**Humphery:** v. HUMPHREY

**Humphrey, Humphreys, Humphries, Humphris, Humphriss, Humphry, Humphryes, Humphrys, Humphery, Homfray, Humfress, Humfrey:** *Hunfridus*, *Humfridus* 1086 DB; *Humfridus* 1186–88 BuryS (Sf); William *Humfrey* c1240 Fees (Beds); William *Humfray*, *Umfrey* 1293 AssSt; Roger *Houmfrey* 1311 LLB B. OG *Hunfrid*, *Humfrid*.

**Huncock, Huncot, Huncott:** John *de Hunecot'* 1206 Cur (Y); John *Huntecook* 1342 LLB F. From Huncoat (La), or Huncote (Lei).

**Hundred:** Robert *Hundred* c1102–11 ELPN; Robert *del Hundred* 1275 SRWo; Richard *atte Hundred* 1327 SREss. 'Dweller at the meeting-place of the hundred', OE *hundred*. Sometimes metonymic for *Hundreder*: Walter *le Hundreder* 1280 AssSo 'chief officer of a hundred'. cf. Richard *Hundredman* 1200 P (D).

**Hungate:** Vlf *de Hundegate* 1191 P (L); William *Hungate* 1402 IpmY. From Hungate in Market Weighton (ERY), or Hungate in Methley, in Sawley (WRY).

**Hunger:** *Hungar* Cofrich 1166 P (Nf); *Hunger* 1167 ib.; Ricardus *filius Hundger* 1185 Templars (L); Turstinus *Hunger* 1176 P (Nth); William *Hungere* 1198 P (K). OE *\*Hungār*, unrecorded, but both themes are common in OE.

**Hungerford:** Robert *de Hungerford* 1148 Winton (Ha); Edward *de Hungrefford* 1200 Cur; Thomas *Hungerford* 1354 FFW. From Hungerford (Berks).

**Hunn:** (i) Robertus *filius Hunne* c1155 DC (L); Brictmerus *filius Hunne* 1193 P (Nf); Robert, Elwin *Hunne* 1166 P (C), 12th FeuDu. OE *Hun(n)a*. (ii) Seman', Martin, Robert *le Hunne* 1277 Ely (Sf), 1279 RH (C), 1288 RamsCt (Hu). As the nickname is East Anglian, it may be ON *húnn* 'young bear'.

**Hunnable, Hunneyball, Hunneybell, Hunnibal, Hunnibell, Hannibal, Honeyball, Honneybell, Honniball:** *Anabul* 1379 PTY; Roger, George *Anabull(e)* 1499, 1539 DbCh; Thomas *Hannyball* 1513 Oxon; John *Anyable* 1568 SRSf; Simon *Honeyball* 1792 SfPR; Elizabeth *Hunnibal(d)* 1821, 1824 ShotleyPR (Sf). *Anabul* is clearly for *Anabel*, from *Amabel*. v. ANNABLE. These surnames are late corruptions of this. There is no evidence for the use of *Hannibal* as a christian name in England before 1619 in Cornwall: *Hannibal* Gammon (ODCN). As a surname, we find Matthew *Hanibal* (1255 Cl), Peter *Haniballus* (Bardsley), both described as 'civis Romanus' and obviously Italians. A Roman surname from the Carthaginian Hannibal is somewhat surprising.

**Hunnex:** v. HUNWICK

**Hunnings:** *Huning* 1086 DB; *Hunningus* de Cliva 1176 P (Sa); Alexander, William *Hunning* 1269 AD iv, 1275 RH (Sf). OE *Hun(n)ing*, 'son of *Hun(n)a*'.

**Hunnybun:** v. HONEYBOURNE

**Hunsley:** John *Hunsley* 1672 HTY. From High, Low Hunsley (ERY).

**Hunt:** Humphrey *le Hunte* 1203 FFSx; Ralph *Hunte* 1219 AssY. OE *hunta* 'hunter, huntsman'.

**Hunter:** Simon *Huntere* 1220 Cur (Beds); Juliana *la Hunter* 1312 ColchCt. A derivative of OE *huntian* 'to hunt', 'huntsman' (c1250 MED).

**Hunting:** *Hunting* c1095 Bury (Sf); *Hunting* filius Hanuse 1193 P (Y); Geoffrey *Hunthing* 1209 P (Lei); Roger *Hunting* 1275 RH (K). OE *Hunting* a derivative of OE *Hunta* 'hunter', recorded only once before the Conquest (c1060).

**Huntingdon, Huntington, Huntinton:** Eustace *de Huntedune*, *de Huntendone* 1086 OEByn; Humphrey *de Huntendun'* 1202 SP1 (Beds); William *de Huntinton'* c1280 SRWo; Robert *Huntyngdon* 1375 AssLa. From Huntingdon (Hunts), or Huntington (Ches, Hereford, Salop, Staffs, NRYorks).

**Huntley, Huntly:** Thomas *de Huntelega* 1176 P (Gl); Thomas *de Hunteley* 1280 AssSo; John *Hunteleye* 1374–5 FFSr. From Huntley (Glos), or 'dweller by the hunter's wood'.

**Huntlow:** Ysabella *Hontelove* 1283–4 CtH. 'Hunt wolf', OE *huntian*, OFr *louve*, a wolf-hunter.

**Huntman, Huntsman:** John *Hunteman* 1219 Fees (Sf); John *Hunteman*, *Huntesman* 1347, 1348 FA (Sf). A compound of OE *hunta* 'huntsman' or ME *hunte* 'the act of hunting' and *mann*, either in the sense 'hunter' or 'servant of the hunter'. Occasionally we may have an unrecorded OE personal name *\*Huntmann*, a compound of *Hunta* and *-mann*, an addition to these late compounds in *-mann*. It is found in Suffolk in 1199 (SIA iv).

**Hunton:** Roger *of Hunton* 1263 FFY; William *de Hunton* 1304 IpmY. From Hunton (Ha, K, NRY).

**Hunwick, Hunwicke, Hunwicks, Hunnex, Honex, Honick:** Walter *de Honewyke* 1296 SRSx; Richard *de Hunewyke* 1307 FFEss. 'Dweller at the honey-farm', or, bee-keeper. OE *hunig*, *wīc*.

**Hunwin:** v. UNWIN

**Hunworth:** Thomas *Honeworth*, *Hunworth* 1465 Paston. From Hunworth (Nf).

**Huot:** v. HEWET

**Hupple:** Juliana *Huppehulle* 1279 RH (O); Adam *Uppehull* 1327 SRSo. Dweller 'up the hill'.

**Hurd:** v. HERD

**Hurdler:** William *Hurdeler'* 1249 AssW; John *le Herdler* 1327 SRC; Henry *Hurdler* 1662–4 HTDo. 'A maker of hurdles', from a derivative of OE *hyrdel*.

**Hurdman:** v. HERDMAN

**Hurer:** Alan *le Hurer* 1244 AssLo. 'A maker of caps', from a derivative of OFr *hure* 'cap'.

**Hurl, Hurle, Hurles, Hurll:** v. EARL

**Hurland:** Philip *de Hurlonde* 1296 PN Sx 20; Richard *Hureland* 1321–2 FFSr; Thomas *Hurlond* 1327 SRWo. From a lost *Hurlands* in Fernhurst (Sx).

**Hurlbatt, Hurlbut, Hurlbutt:** John, Robert *Hurlebat* 1327 SR (Ess), 1333 SR (Ha). A nickname from a game. 'Pleying at þe two hande swerd, at swerd & bokelere, & at two pyked staf, at þe hurlebatte' (c1450 MED). It took its name from 'short battes of a cubit long and a halfe, with pykes of yron, and were tied to a line, that when they were throwne, one might plucke them again' (1565–73).

**Hurley, Hurly:** John *de Hurleia* 1210–11 P (W); Richard *de Hurleye* 1327 SRWo; John *Hurleye* 1340–1450 GildC. From Hurley (Berks, Som, Warwicks).

**Hurlin, Hurling, Hurlen, Harlin, Harling, Arling, Urlin, Urling, Urlwin, Harwin, Herwin:** Balduinus *filius Herluini* 1066 DB; *Herluinus, Urleuuine* 1086 DB; *Herlewine* de Sumerfeld' 1191 P (K); *Harlewyn* 1279 RH (C); John *Erlewin* 1225 Pat (K); Roger *Herlewyn* 1230 P (Do); Simon *Arlewine* 1242 Fees (Beds); Peter *Harlewyne* 1279 RH (C); Richard *Urlewyn'* ib. (O); William *Harlyn* 1327 SRSx; Alice *Hurlewyne* 1332 SRSr; Agnes *Herlyng* 1379 PTY; Thomas *Harlewyn*, Robert *Harlyng*, John *Harwyn* 1524 SRSf; Roger *Urlen* 1581 Oxon; Marmadux *Urlin* (siue, ut Atavi scripsere, *Earlewyn*) 1654 CAPr 34; Mr *Urling* 1698 ib. OFr *Herluïn, Arluïn,* OG, *Erlewin, Herlewin* 'earl-friend'. *Url-* is an Anglo-Norman spelling.

**Hurlston, Hurlestone:** (i) Henry *Hurleston* 1474 Bardsley; Philip *Hurlestone* 1642 PrD. From Hurleston (Ches). (ii) John *Huddilston, Hudelston, Hodelston, Hudleston* or *Hurleston* 1509 LP. *Hurleston* is here an unusual development of HUDDLESTON, found side by side with *Hurlston* in Bradford (WRYorks).

**Hurman:** Walter *Hurneman* 1297 MinAcctCo. 'Dweller in the nook.' *v.* HERN.

**Hurne:** *v.* HERN

**Hurran, Hurren:** William *Hurant* 1086 DB (Sf); William *Hurand* 1245 *Ass* (Ess); Edmund *Hurryng* 1524 SRSf. OFr *hurand, hurant* from OFr *hure* 'hair' and the suffix *-and, -ant,* hence 'the shaggy-haired one'.

**Hurrell:** Roger *Hurel* 1154 Riev (Y), 1170 P (Nb). An *-el* diminutive of OFr *hure* 'hair' in the same sense as *Hurran.*

**Hurry, Hurrey, Hurrie, Horrey, Horry, Orrey, Orry, Ourry, Oury, Urry, Urie, Ury:** *Vlfric, Vlfricus, Vlricus* 1066 DB; *Urrius* de la haie c1148 EngFeud (He); Reginaldus *filius vrri* 1182–98 Bart (Lo); *Urric'* Arbelestier 1194 CurR (Ess); *Wlricus* Balistarius 1233 Fees (Ess); *Hurri* c1250 Rams (Hu); Walter, Herueus *Urri* 1208 Cur (Sr), 1209 P (Nf); Gilbert *Uri* 1214 Cur (L); Alan *Hurry* 1219 FFEss; Geoffrey *Orry* 1235 Fees (Sa); Walter *Horry* 1290 Fees (Wt); John *Ourry* 1297 MinAcctCo; John *Ulricus* 14th AD vi (C). A Norman pronunciation of OE *Wulfrīc. v.* WOOLRICH.

**Hurst:** *v.* HERST

**Hurt:** *v.* HART

**Hurtheaven:** John *Hurthevene* 1288 CtW; William *Hurteheuene* 1361 IpmW. 'Harm heaven', OFr *hurter,* OE *heofon.* cf. Richard *Hurtemotoun* 1305 AssW 'harm the sheep'; Robert *Hurteskeu* 1334 FFY 'harm the sky'; Walter *Hurtevent* 1183 P (Lei/Wa) 'harm the wind'.

**Husband, Husbands, Hosbons:** Ernald, Robert *Husebond'* 1176 P (Y), 1231 Pat (Nb); Robert *le Hosebonde* 1279 RH (C). Late OE *hūsbonda* 'householder'; farmer, husbandman.

**Huse:** *v.* HOSE

**Husey:** *v.* HUSSEY

**Husher:** *v.* USHER

**Huskinson, Huskison, Huskisson:** 'Son of *Oskin.' v.* HOSKEN.

**Huson:** *v.* HUGHSON

**Hussey, Hussy, Husey, Hosey:** (i) Walter *Hosed, Hosatus* 1086 DB, *Hosethe* c1100 OEByn (D); Henry *Hoese, Huse* 1153, 1185 Templars (O); Geoffrey *Hoset* (*Hose*) 1168 P (W); William *Hose, Huse* 1221 AssGl; Richard *le Hose* 1243 AssSo; Geoffrey *Husey* 1275 RH (W); John *Hussee* ib. (Nt); Richard *Hussi* 1275 SRWo; Henry *Hosey* 1296 SRSx. OFr *hosed* (*housé*), Lat *hosatus* 'booted'. (ii) Richard, Rose *Husewif* 1192 P (Sx), 1279 RH (C); Margeria *Hosewyf* 1327 SRC; Roger *Huswyffe* 1435 AD i (Herts). ME *hus(e)wyf* 'mistress of a family; wife of a householder; a domestic economist', applied to a woman, a compliment; to men, derogatory. This became *hussie* in the 16th century. The surname is not to be associated with the modern meaning of *hussy.* The word was also used as a woman's name: *Husewyua* (*Husewyf*), widow of Adam Brist (1317 AssK).

**Husting, Hustings:** Robert *Husting* 1275 RH (Sf); William *Hustyng* 1524 SRSf. Metonymic for the officer of a law-court, ON *hús-þing.*

**Hustler:** *v.* OSTLER

**Hustwayte, Hustwitt:** William *de Husthayt* 1219 AssY. From Husthwaite (NRYorks).

**Hutchin, Hutchins, Hutchings, Hutchence, Hutchens, Hutcheon, Hutchons, Hotchen, Hotchin, Houchen, Houchin, Howchin:** *Huchun* Aleyn 1277 Wak (Y); *Hucohun* le Cu 1292 SRLo; Gilbert *Huchun* 1296 SRSx; John *Huchon* 1321 FFSf; John *Howechoun* 1327 SRSf; John *Huchin,* Richard *Huchins* 1327 SRWo; Edith *Huchenes* 1332 SRSt; John *Huchouns* 1337 ColchCt; Cecily *Howchyngs* 1523 NorwW (Nf); John *Howchyn* 1524 SRSf. OFr *Huchon,* a double diminutive of *Hue* (Hugh), common in Picardy, corresponding to OFr *Hueçon, Huesson,* modFr *Husson.* It was used as a christian name in Scotland (*Hucheon* Fraser 1422) where it also became a surname (John *Hutching* 1525, Helen *Huchown* 1547) and was adopted in Gaelic as *Huisdean* or *Uisdean* (Black). *v.* MCCUTCHEON

**Hutchinson, Hutcherson, Hucheson, Hutchingson, Hutchison:** Isota *Huchonson* 1379 PTY; John *Hucheson* 1440 ShefA (Y); John *Huchynson* 1475 GildY; Tomasine *Hutcherson* 1674 EA (OS) i (Sf); Abell *Hutchison* 1679 RothwellPR (Y). 'Son of *Huchun.' v.* HUTCHIN.

**Huth:** Robert *atte Huthe* 1327 SRSx; Adam *Huth* 1352 FFY. 'Dweller by the landing-place', OE *hȳð.* cf. Robert *le Huthereve* 1321 CorLo; Ralph *le Huthward* 1332 SRSx, both 'warden of the landing-place'.

**Huthwaite:** Adam *de Hothwayt* 1260 AssLa; Thomas *de Hothuait* 1332 SRCu. From Howthwaite (Cumb), Huthwaite (NRYorks), or Hucknall under Huthwaite (Notts).

**Hutler:** Wlward *Hutlawe* 1279 RH (C). *v.* OUTLAW.

**Hutley:** *v.* UTLEY

**Hutson:** *v.* HUDSON

**Hutt:** *Hutt* son of Thomas 1246 AssLa; Osbert *Hut* 1203 P (Bk); Simon *Hutte* 1279 RH (Bk); John *Hutte* 1327 SRLei. Probably for *Hudd,* a pet-form of Hugh.

**Huttley:** *v.* UTLEY

**Hutton:** Ernewi *de Hottona* 1175 P (Y); Nicholas *de*

*Hutune* 1246 'AssLa; John *Hutton*, *Huttun* 1403 IpmY, 1502 FFEss. From one or other of the many places of this name.

**Huxham:** Robert *de Hokesham* 1230 P (D). From Huxham (D).

**Huxley:** Robert *de Huxeleg'* 1260 AssCh; Thomas *de Huxeley* 1332–3 SRSt; William *Huxley* 1530 FFEss. From Huxley (Ches).

**Huxster, Huxter:** Lecia *la hucstere* 1277 *Ely* (Sf); Amable *la Hukkester* 1310 ColchCt; William *le Hukester* 1327 SRSf; Richard *Huxster* 1662 HTEss. ME *huckestere*, feminine of *huckere* (v. HUCKER), 'pedlar, hawker', used chiefly of women.

**Huxtable:** John *de Hokestaple* 1330 PN D 35; James *Huxtable*, Richard *Hucstable* 1642 PrD. From Huxtable in East Buckland (D).

**Hyam(s):** v. HIGHAM

**Hyatt, Hyett:** v. HIGHET

**Hyde(s):** v. HIDE

**Hyder:** v. HIDER

**Hyland:** v. HAYLAND

**Hyman:** v. HAYMAN

**Hyndley:** v. HINDLEY

**Hynes:** v. HINE

**Hyslop:** v. HASLOP

**Hysted:** v. HIGHSTEAD

# I

**Iago:** v. JAGO

**I'Anson:** John *Janson, I'anson* 1569, 1673 FrY; Nehemiah *Janson, I'anson* 1615, 1673 ib. For JANSON.

**Ibbelot, Iblot:** Thomas *Ybelote* 1296 SRSx. *Ibb-el-ot*, a double diminutive of *Ibb*, a pet-form of *Isabel*, or of *Ilbert*. v. ILBERT.

**Ibberson, Ibbeson, Ibbison, Ibbotson, Ibbetson, Ibbitson, Ibotson, Ibeson, Ibison, Hibberson, Hipperson:** Henry *Ibbotson* 1379 PTY; John *Ibotessone* 1392 NottBR; Anthony *Ibbison* 1596 FrY; Roger *Ipperson, Ibbatson* 1611–14 Bardsley; George *Ibberson* 1695 Shef. 'Son of *Ibbot*.'

**Ibbott, Ibbett, Ibbitt, Hibbit, Hibbitt, Hibbott:** *Ybot, Ibbota* 1286, 1314 Wak (Y); *Hibbot* 1379 PTY; Roger *Ybott* 1415 LLB I; John *Hybbot* 1550 Pat. *Ibb-ot*, a diminutive of *Ibb*. *Hibbot* is a diminutive of both *Ibb* (Isabel) and of a man's name: *Hibot'* de Buskeby 1379 PTY, probably from *Hilbert* or *Ilbert*. v. ILBERT.

**Ibbs, Hibbs:** *Ibbe* 1324 LaCt; Adam *Ibbe* 1334 ColchCt; John *Hibson* 1442 Bardsley. *Ibb*, a pet-form of *Isabel* or of *Ilbert*.

**Ibell, Hibble, Hible:** cf. John *Ibelsone* 1381 SRSt. *Ib-el*, a diminutive of *Ibb*.

**Ibson:** William, John *Ibbeson* 1324 LaCt, 1332 SRCu. 'Son of *Ibb*.'

**Icemonger, Isemonger:** Ailred *Ismangere* 1165–72 P (K); Hervey *le Ismongere* 1248 FFEss. OE *īsern, īsen, īren* 'iron' and *mangere*, 'ironmonger'. v. IREMONGER.

**Ickeringill, Ikringill:** *Icornegill'* 1379 PTY; Benjamin *Ickeringill* 1670 FrY; William *Ickeringale, Hickeringill* 1679, 1700 FrY. From a lost place in Skipton (WRYorks), *Ecorngill* 1329, *Ickering-Gill* 1822 (PN WRY v, 71).

**Ickford:** John *Yckeford* 1383 LuffCh. From Ickford (Bk).

**Ickhills, Ickles:** Robert *Ikhils* 1416 IpmY. From Ickles in Rotherham (WRY).

**Ida, Ide:** *Ida* 1175 AD i (Mx), 1207 Cur (L); William *Ide, Yde* 1279 RH (C), 1296 SRSx. *Ida*, an OG name from *id-* 'labour', popular among the Normans.

**Iddenden, Iddenten, Iddenton, Idenden:** Godleue *Ydendene* 1442 ArchC xi. From Iddenden Fm in Hawkhurst (Kent).

**Iddins:** v. IDDON

**Iddison:** Thomas *Idonson*, Richard *Ideson* 1379 PTY. 'Son of *Idon*.' v. IDDON.

**Iddon, Iddins:** *Idonea* Hy 2 DC (L), 1205 Cur (Lei); Willelmus *filius Ydeneye* 1297 SRY; Nicholaus *filius Ydon'* 1297 MinAcctCo; *Idone* 1327 SRSo; William *Idony* 1274 RH (Lo); Robert *Ydany* 1300 Ipm (K); Robert *Ydeyn* 1327 SRC. Probably ON *Idunn, Iðuna* (f), from *iðja* 'to do, perform' and *unna* 'to love'. It was latinized as *Idonea* 'suitable'. The vernacular form was *Idony*.

**Idel:** v. IDLE

**Idelson:** 'Son of *Idell*.' v. IDLE.

**Iden:** Osbert *de Idenn* 1296 SRSx. From Iden Green in Benenden or Iden Manor in Staplehurst (Kent).

**Ideny, Idony:** v. EDNEY

**Idle, Idel, Idell:** (i) Wrennus *filius Ydel* 1193 P (He); *Idellus* 1202–7 Black (Glasgow); William, John *Idel(e)* 1199 FF (Nth), 1332 SRSr; William *Ydil* 1343 Black. Welsh *Ithell*, MW *Ithael* 'lord-bountiful'. In England, the surname may also be OE *īdel* 'lazy'. cf. ITHELL. (ii) Ailsi *de Idla* c1190 Calv (Y); John *del Idle* 1317 AssK; John *dil Idle* 1327 SR (Ess). From Idle (WRYorks) or from residence in an island (AFr *idle*, OFr *isle*).

**Ifield:** Scorland *de Yfeld* 1198 P (K); William *de Ifeld* 1259 FFK; John *de Ifeld* 1327 SRSx. From Ifield (K, Sx).

**Ifould:** Benedict *de Ifold* 1296 SRSx. From Ifold Ho (Sussex).

**Iggleden, Iggledon, Igglesden, Iggulden:** Richard *Igolynden*, John *Igulden* 1475, 1536 ArchC 31. From Ingleden (Kent).

**Ikin:** Richard *Idekyne* 1324 FA (W). A diminutive of *Ida*.

**Ikringill:** v. ICKERINGILL

**Ilbert, Hilbert, Hibbard, Hibberd, Hibbert:** *Ylebertus* c1150 Gilb (L); *Hildebertus* 1150–60 DC (L); Willelmus *filius Hiberti* 1163 ib.; Estmundus *filius Hilbert* 1205 Cur (Nf); *Ilbert* de Betelintun' 1212 Cur (Berks); Walter *Ilberd* (*Hilbert*) 1230 P (Y); Martin *Yllebert* 1243 AssSo; Margaret *Hilbert* 1283 SRSf; Roger *Hileberd* 1327 SRSo; John *Heebarde* 1568 SRSf; James *Hibbert* 1591 Bardsley. OG *Hildeberht* 'battle-glorious', OFr *Ilbert*.

**Ilderton:** Walter *de Ildirton* 1397 PrGR. From Ilderton (Nb).

**Iles, Isles:** John *del Ile* 1275 RH (Sf); Ralph *Iles* 1560 FrY. 'Dweller in the isle' (OFr *isle*, Fr *île*). v. ILLES.

**Ilford:** Robert *de Ileford* 1232–3 FFEss; Laurence *Illford* 1642 PrD. From Ilford (Ess).

**Iliff, Iliffe, Ilieve, Ilive:** *Æillovus*, identical with *Illivus* 1212 Cur (Cu, Nb); *Ilyf* de Wyn 13th FeuDu;

Francis *Ilive* 1622 Bardsley; John *Iliff* 1640 ib. A form of ONE *ilifr*. v. AYLIFF.

**Illbode:** William *Ilbode* c1187–1216 Clerkenwell; William *Ilbode* 1230 P (Lo). 'Evil messenger', ME *ille*, OE *boda*. cf. Richard *Illechilde* 1297 MinAcctCo 'wicked child'; Roger *Illefoster* (a Jew) 1192 P (Lo) 'wicked fosterchild'; William *Ilwicht* 1219 AssY 'wicked person'.

**Illes:** (i) Ralph *Ille* 1202 P (Nf), 1275 RH (Sf). ME *ille* 'bad, evil'. (ii) Baldwin, Andrew *del Ille* 1255 RH (O), 1332 SRCu. 'Dweller in the island.' v. ILES.

**Illey:** Adam *de Illega* 1188 P (Sf); Thomas *Iley* 1604 FrY; Elizabeth *Iley* 1747 RamptonPR (C). From Illey (Wo), or Brent, Monks Eleigh (Sf).

**Illing:** Richard *Illing* 1191 P (Lei/Wa); Richard *Illing* 1206–7 FFWa; Walter *Illing* 1276 RH (Lei). OE \**Illing*.

**Illingworth, Illingsworth:** Alice *de Illingworth* 1314 Wak (Y); John *de Illyngworth'* 1379 PTY; Richard *Illyngworth* 1454 IpmNt. From Illingworth (WRYorks).

**Illsley, Ilsley:** Isabella *de Ildesle* 1297 MinAcctCo. From East, West Ilsley (Berks).

**Image:** Richard *Image* 1563 Bardsley. Metonymic for *Imager*: Alexander *le Ymagour* 1305 LLB B. OFr *ymageour*, *ymagier* 'maker of images, carver, sculptor'.

**Imber, Immer:** Robert *de Immer* 1327 SRSx; John *Imber* 1545 SRW. From Imber (W), *Immemer* 1198.

**Imbrey, Imbrie:** v. AMERY

**Imm, Imms, Ims:** Thomas *Imme* 1243 AssSo; Geoffrey *Ymme* 1332 SRSr. OG *Emma*, *Imma*.

**Immink:** v. EMENEY

**Imper:** Adam *Impere* 1299 KB (Y); Thomas *le Impere* 1305 AssW. A derivative of OE *impa* 'a graft, shoot'. Probably 'dweller near the young trees'.

**Impey, Impy:** John *atte Imphage* 1327 SRSx; John *de Impey* 1327 SRSf. 'Dweller near a hedge or enclosure made of saplings, or an enclosure for young trees' (OE \**imphaga*, \**imphæg*), as at Emply (Surrey), Empty (Northants) or Imphy Hall (Essex).

**Imray, Imrie:** v. AMERY

**Ince:** v. INNES

**Inchbald:** *Hengebaldus* (Sa), *Ingebald* (D) 1086 DB; *Ingeboldus* (*Engebald'*, *Ingelbold'*) filius Karlonis 1221 AssGl; William *Ingebald* 1379 PTY; Emma *Inchebald* ib. OG *Engelbald*, *Ingilbald* 'Angle-bold'.

**Ind:** Nicholas *Attehende* 1332 SRSt; Robert *de Ynde* 1369 LLB G. 'Dweller at the end (of the village).'

**Indecome, Indicombe, Indecombe:** John *de Byendecoumbe* 1314, *de Hyndecomb* 1330, Thomas *de Yundecomb* 1333 PN D 35. From Indicombe in West Buckland (D).

**Inderwick, Enderwick:** Roland *de Inuerwic'* c1190 Black; Robert *Inderwick* or *Innerwick* 1652 ib. From Innerwick (East Lothian).

**Ineson:** Thomas *Idonson*, John *Ineson* 1379 PTY. 'Son of *Iddon*.'

**Ing, Inge, Ings, Indge:** (i) *Inga* c1160 DC (Lei); *Inga* filia Thore 1202 AssL; *Inga* of the Hull 1332 SRWa; John *Ing'* 1212 Cur (Gl); William *Ingge* 1283 AssSt; William *Inges* c1436 Paston. ON *Inga*, ODa *Inga* (f), OSw *Inga* (f), *Inge* (m), a pet-form of such names as

*Ingiriðr*. (ii) Reginald *de Inga* 1162 P (Ess); Ralph *de Ging* 1245 HPD; John *de Inges* 1306 IpmY; William *Ynges* 1393 AssL. From Ing (Ess), *Gynges* 1230, or 'dweller by the hill', OE \**ing*.

**Ingall, Inggall, Ingold, Ingle, Ingles, Hingle:** *Ingold* 1066 DB (Y), c1095 Bury (Sf); *Ingoldus* 1114–30 Rams (Hu); *Ingald'* Ledbater 1221 AssWa; *Ingol* Textor c1250 Rams (Hu); Edmund, Peter *Ingold* 1274 RH (Sf), 1312 LLB D; Geoffrey *Ingal* 1279 RH (Hu); Roger *Hyngoll* 1524 SRSf. Anglo-Scand *Ingald*, *Ingold*, OSw *Ingæld*, ON *Ingialdr* 'Ing's tribute'.

**Ingamells:** Walter *de Ingoldemeles* 1219 AssL. From Ingoldmells (L).

**Ingate:** Matilda *de Endegate* 1327 SRSf; Robert *Ingate* 1568 SRSf. From Ingate (Suffolk).

**Ingell, Ingels, Ingle, Ingles, Hingle:** (i) Emma, William *Ingel* 1279 RH (Hu), 1327 SRC; William *Ingelle* 1381 SRSt. ODa *Ingeld*, ON *Ingialdr*. cf. INGALL. (ii) *Ingulf*, *Ingolf* 1066 DB; Adam *Ingulf* 1229 Cl (Sf); Richard, Thomas *Ingolf* 1327 SRSx, 1391 FrY. ON *Ingólfr*, ODa, OSw *Ingulf* 'Ing's wolf'. The surname may also be from OG *Engenulf*, *Ingenulf*: *Engenulfus*, *Ingenulf* 1086 DB, *Ingenolf* 1212 Cur (Nth), *Engenolfus* 1219 Cur (Beds).

**Inger, Inker, Ingerson:** *Ingeuuar*, *Imgarus* 1066 DB; Roger *Inger* 1255 RH (W); Peter *Ingarson* 1546 FrY. ODa, OSw *Ingvar*, ON *Yngvarr*.

**Ingersoll, Inkersole:** Roger *de Hynkersul* 1321 Shef. From Inkersall (Derby).

**Ingham:** Alwinus *de Ingham* 1049–52 OEByn (O); Roger *de Ingham* 1162–8 Holme (Nf). From Ingham (Lincs, Norfolk, Suffolk). v. also GAIN.

**Ingillson:** v. INGLESON

**Ingle, Ingles:** v. INGALL, INGELL

**Inglebright, Engelbert:** *Engelbricus* 1066 DB (Mx); *Engelbricht* de Stanlega 1176 P (So); *Ingelberd* c1190 DC (L); Robert *Ingelberd'* 1230 P (Y); Robert *Ingelbert* 1275 RH (L); Arnald *Ingelbright* 1373 LLB G. OFr *Engilbert*, *Englebert*, OG *Engelbert*, *Ingelbert* 'Angle-glorious'.

**Ingleby, Ingilby:** Goslanus *de Engelby* 1157 Gilb(L); Thomas *de Ingelby* 1280 FFY; John *Ingelby* 1370 IpmGl. From Ingleby (Derby, Lincs, NRYorks).

**Ingleden, Ingledon:** Richard *Ingleden* 1462 CtH. From Incledon in Braunton (D), or Ingleden in Tenterden (K).

**Inglefield:** William *de Englefeld* 1185 P (St); Philip *de Ingelfeld* 1355 LLB G; Stephen *Ingylfeld* 1407 LLB I. From Englefield (Berks), or Englefield (Surrey), *Ingelfeld* 1282.

**Inglesent:** Willelmus filius *Ingelsent'* 1219 AssY; William *Inglesant* 1379 PTY; William *Inglissent* 1447 FrY. OG *Ingilsind*(*is*) (f).

**Ingleson, Ingillson:** Thomas *Ingleson* 1458 FrY. 'Son of *Ingle*.' v. INGALL, INGELL.

**Ingleton:** John *de Ingelton'* 1297 SRY; Robert *de Ingleton* 1379 PTY; Robert *Ingilton* 1457 FFEss. From Ingleton (Du, WRY).

**Inglett:** (i) *Engelardus* de Strattone 1166 RBE (Sa); Robertus filius *Ingelard'* 1221 AssGl; Isabella, John *Ingelard* 1279 RH (Beds). OG *Engel*(*h*)*ard*, *Ingilard* 'Angle-hard'. (ii) Willelmus filius *Ingelot* 1200 P (L); William, Reginald *Ingelot* 1279 RH (O), 1327 SRC.

*Ingel-ot* a diminutive of *Ingel-*, a short form of such names as *Ingelbald*, *Ingelbert*, etc.

**Inglewood:** John *de Ingelwode* 1312 LLB D; John *Inglewode* 1401 KB. From Inglewood (Berks), or Inglewood Forest (Cu).

**Ingley:** Laurencius *filius Ingelieth* 1221 *ElyA* (Nf); Peter *Inglyth* 1395 EA (OS) i (C). For INGREY, through an interchange of liquids in the combination *n—r*; *Ingrith* becoming *Inglith*.

**Inglis, Inglish:** *v.* ENGLISH. Occasionally we may have OFr *Engelais* (f), OG *Engilheid*: *Engelise* 1190 P (W), *Engeleisia* 1200 P (W); *Ingeleis* 1202 AssL, 1205 Cur (Herts).

**Ingman:** *Ingemundus* 1142 NthCh (L); Alan *filius Yngemundi* 1208 Cur (L); Richard *Ingemunt* 1219 AssY; Mark *Hingman* 1642 PrD. ON *Ingimund*.

**Ingold:** *v.* INGALL

**Ingoldby, Ingoldsby:** Nigel *de Ingoldebi* 1208–9 Pl (L); Ralph *de Ingaldesby* 1275 RH (L); William *de Ingoldesby* 1371 AssL; Ralph *Ingoldesby* 1439–40 FFWa. From Ingoldsby (Lincs).

**Ingpen:** *v.* INKPEN

**Ingram, Ingrams, Ingrem:** *Ingelram* de Sayl c1140 Holme (Nf); *Ingelrannus* tanurus 1148–66 NthCh (Nth); *Engelrannus* clericus c1150–60 DC (Nt); Radulphus filius *Engelram* 1158 P (Y); *Ingeram* de Helesham Hy 2 DC (L); *Ingerannus* capellanus 12th DC (Nt); John *Engelram* 1132 Riev (Y); John *Ingelram* (*Ingeram*) c1138 Whitby (Y); John *Ingeram* c1150 Whitby (Y); Robert *Engram* c1220 Guisb (Y); Richard *Ingram* c1250 Calv (Y). OG *Engel-*, *Ingelramnus*, *-rammus* 'Angle-raven', OFr *Enguerran*, *Engerran*.

**Ings:** *v.* ING

**Ingworth:** Robert *de Ingeworth* 1219 P (Nf/Sf); John *of Ingworth* 1316 AssNf. From Ingworth (Nf).

**Inions:** *v.* ENNION

**Inker:** *v.* INGER

**Inkersole:** *v.* INGERSOLL

**Inkpen, Inkpin, Ingpen:** Richard *de Ingepenne* 1255 RH (Bk). From Inkpen (Berks).

**Inland:** Peter *Inlond* c1405 FS. From Inland in Gateby (Nf), or Inlands in Westbourne (Sx).

**Inman:** William *Inman* 1379 PTY. OE *inn* 'abode, lodging' and *mann*. 'Lodging-house keeper.'

**Innes, Ince:** (i) Richard *de Ins* 1324 Coram La; John *de Ines*, *de Ince* 1401 AssLa; Henry *de Ines* 1415 IpmLa; Edmund *Ins*, *Ince* 1474 Cl (Sf). From Ince (Ches, Lancs), or Innes (Cornwall). (ii) Walter *de Ineys* 1226 Black; William *de Inays* 1296 ib.; Robert *de Innes* 1389 ib. From the barony of Innes in Urquhart (Moray).

**Insall, Inseal, Insole, Insoll, Insull, Hinsull:** Richard *de Inneshal'* 1327 SRWo; John *de Insale* 1341 NI (Wo); Philip *Insoll* 1603 SRWo. From a place *Insoll* in Elmley Lovett (Worcs), no longer on the map, first

recorded as *Inerdeshell* in 1275 and last as *Insoll* in 1642 (PN Wo 241).

**Inskip, Inskipp, Inskeep:** Margery *de Inskip* 1246 AssLa; John *Inskip* 1401 AssLa. From Inskip (Lancs).

**Inston, Instone:** Sibilla *de Inardeston'* 1275 SRWo. From Innerstone (Worcs).

**Inward, Inwards:** Roger *de Ynewrðe* 1202 P (Nth). From Inworth (Essex), *Inward* 1467. *Inward* and *Inwood* have been confused. Sarah *Inward*, daughter of Richard *Inwood*, died in 1685 (Weekley).

**Inwood, Innwood:** Thomas *de Inwode* 1327 SRSo. 'Dweller by the 'in-wood', the home-wood, as opposed to the 'out-wood'.

**Ipsen:** William *de Ipstone* 1220 AssSt. From Ipstones (Staffs) or Ipsden (Oxon).

**Ipsley:** Gregory *Ipsley* 1642 PrD. From Ipsley (Wa).

**Ipswell:** Robert *Ipyswell* 1465 Paston. Probably from Epwell (O). *Ippewelle* c1260.

**Irby, Ireby:** Hugh, Ailsi *de Yrebi* 1193 P (L), 1195 P (Cu); William *de Irby* 1280 IpmY; Richard *Yrby* 1341 IpmGl. From Irby (Ches, Lincs, NRYorks), or Ireby (Cumb, Lancs).

**Ireland:** Ralph *de Ibernia*, *de Irlande* 1200, 1210 Oseney; Robert *de Irlonde* 1327 SRSx; John *Ireland* 1458 FFEss. 'The man from Ireland.'

**Iremonger, Ironmonger:** Richard *Yernmonger* Hy 3 Gilb (L); John *le Irmongere* 1255 RH (O); John *Irinmongere* 1279 RH (Bk); Elyas *le Ironmongere* 1294 AssSt; Roger *le Ernmongere* 1327 SRSt; John *le yremongere* 1327 SRC; John *le Yernemanger* 1332 MESO (Y). OE *īren* 'iron' and *mangere*, 'ironmonger'. Identical with ICEMONGER. *Ern-*, *Yern-*, in Yorks, Lincs and Staffs is from the ON *earn*, *jarn* 'iron'.

**Ireton, Irton:** Richard, Henry *de Irton'* 1218 AssL, 1272 AssSt; William *de Yrton'* 1351 AssL. From Ireton (Derby), or Irton (Cumb, NRYorks).

**Irish:** Richard *Ireis* 1169 P (Sa); William *le hyreis* 1227 FFSf; Robert *le Irish* 1356 AssSt. OFr *Ireis* 'Irish', later replaced by the English form.

**Ironfoot:** Roger *Yrenfot* 1251 Rams (Hu). A nickname, 'iron foot', OE *īren*, *fōt*. cf. Henry *Irenherde* 1379 PTY 'hard as iron'; Langa *yreneman* 1327 SRSf 'iron man'; Alviva *Yrento* 1209 FFNf 'iron toe'.

**Ironmonger:** *v.* IREMONGER

**Ironside:** Hugh *Irninside*, Thomas *Irnenside* 1297 Coram (L); John *Irenside* 1333 FrY. OE *īren* and *sīd* 'iron-side, warrior'. *Irnenside* preserves the adjectival inflection. cf. *mid irenen neilen* c1175 NED. The first and most famous bearer of this nickname was Edmund Ironside, so called for his doughtiness; '*Irensid wæs geclypod* for his snell-scipe' 1057 ASC D. cf. '*Edmundes sones wiþ þe Irenside*' c1350 Brut.

**Irvin, Irvine, Irving, Ervin, Erving, Urvine:** Robert *de Hirewyn* 1226 Black; Simon *de Irwyn* 1296 AssCh; Adam *Irvine* 1455 Black. From Irvine (Ayrshire) or Irving (Dumfries). There has been confusion with ERWIN.

**Irwin:** *v.* ERWIN

**Isaac, Isaacs, Isacke, Isaacson:** *Isac* 1086 DB; Willelmus *filius Ysac* 1206 Cur (Ess); Henry *Isaac* 1275 SRWo; Walter *Isak* 1327 SRSo; John *Isakson*

1379 PTY. Hebrew *Isaac*, from a root meaning 'laugh'. The name was not confined to Jews and the medieval surname was certainly not Jewish.

**Isabel, Isbell, Isbill:** *Isabel* 1141–9 Holme (Nf), c1160 DC (Lei); William *Isabelle* 1202–16 StP (Lo); William *Isabel* 1275 SRWo; John *Isbel* 1379 PTY. *Isabel*, a form of *Elizabeth*, which seems to have developed in Provence. A very popular name, with pet-forms *Ibb, Libbe, Nibb, Tibb, Bibby, Ellice*, and various diminutives.

**Isard, Iseard, Isitt:** *v.* ISSARD

**Isdale, Isdell:** *v.* ESDAILE

**Isembard:** *Ysembardus* de Hatel' 1206 Cur (Beds); *Isaberd* of Wykeford' 1221 AssW; Eustace *Isembard'* 1219 P (W); Joan *Ysembard* 1243 AD ii (Ha); John *Issanberd* 1305 AssW. OG *Isanbard, Isanberht*.

**Isham, Isom, Issom:** Henry *de Isham* 1206 Cur (Nth); John *Isom* 1713 FrLeic. From Isham (Northants).

**Isherwood, Esherwood, Usherwood:** William *de Yserwude* 1246 AssLa; Adam *de Esherwode* 1332 SRLa; Robert *Issherewood*, John *Ussherwood* 1524 SRSf. From an unidentified place, probably in Lancs.

**Isles:** *v.* ILES

**Islip:** Ealhstan *æt Isslepe* 972 OEByn; Richard *de Ystlape* 1184 P (Ha); Adam *de Yslep'* 1255 ForNth; William *Islep* 1381 LoCh. From Islip (Nth).

**Ismay:** *Ysemay* de Mult' 1275 RH (L); Ralph *Isemay, Hyssmaye* 13th Shef (Y); William *Ysmay* 1327 SRDb. This somewhat rare woman's name might possibly be an unrecorded OE *\*Ismæg*, but *Is-* is unknown in OE names and *-mæg* is found as a second theme in only three late names. cf. also *Idemay* la Frye 1327 SRSo. The equation of the earliest example, *Isemay* (1227 AssBk), with *Isemeine* (cf. EMENEY) suggests a continental origin. *Is-* occurs in a number of OG names. Förstemann derives *Macharias* from a root *\*mag*, related to OHG *magan* 'might'. Forssner suggests that the OFr *Maissent* may contain the same first element, from OG *\*Magisind, \*Megisind*. Hence, *Ismay* may, perhaps, be from an OG *\*Ismagi, \*Ismegi*.

**Isom:** *v.* ISHAM

**Ison:** William *Ideson* 1583 RothwellPR (Y). 'Son of Ida.'

**Issard, Issett, Issit, Issitt, Issott, Issolt, Isard, Iseard, Isitt, Iszard, Iszatt, Izard, Izat, Izatt, Izant, Izod, Izzard, Izzètt, Ezard:** *Iseldis* 1086 DB (Do); *Ysoude* 1186–1210 Holme (Nf); *Hysode* 12th DC (L); *Ysout* 1200 Oseney (O); *Isolda* 1200 Cur (Lei), 1219 Cur (Sf), AssY; *Ysolt* 1201 Cur (Y); *Isouda, Iseuda* 1214 Cur (Sx, Lei); *Isota* Holebrook 1327 SRSo; *Isata* 1459 FrY; *Izota* 1603 SRWo; Thomas *Isolde* 1275 RH (Sf); Robert *Isaud* 1316 Wak (Y); Thomas *Isoude* 1326 FFEss; Margaret *Isod*, John *Isot* 1379 PTY; John *Esod* 1496 FrNorw; Joane *Isard*, Rachel *Izatt* Henry *Izod* 1626, 1666, 1668 Bardsley; Robert *Issitt* 1763 FrLeic. OFr *Iseut, Isalt, Isaut, Ysole*, OG *\*Ishild* 'ice-battle'.

**Ithell:** *Itel* Karlo 1160 P (Sa); *Ithel* de Landinab 1215 Cur (He); Ralph *Ithel* 1336 AssSt. Welsh *Ithell*, MW *Ithael*. cf. IDLE, BITHELL.

**Ivatt, Ivatts:** *Iuetta* 1166 P (C), 1175 DC (L), 1221 AssWa; Richard *filius Iuotte* 1270 RamsCt (C); John *Ivette* 1262 FFSf; William *Ivet* 1271 ForSt; John *Iuot* 1327 SRC. A diminutive of OFr *Iva*, feminine of *Ivo*. cf. IVE.

**Ive, Ives:** Herbertus *filius Iuonis* 1086 DB; *Iue* de Verdun 1101–7 Holme (Nf); *Iuo, Hiue* de Gausla 1155–60 DC (L); Roger *Yuo* 1175 P (L); Thomas, Henry *Ive* 1274 RH (So), 1280 FFEss; John *Ives* 1327 SRSx. OFr *Ive, Yve(s), Ivon*, particularly common among the Normans and Bretons.

**Ivelchild:** *v.* EVILCHILD

**Ivers:** *v.* IVOR

**Iveson, Ivison:** Adam *Iveson* 1383 Shef (Y). 'Son of *Ive*.'

**Ivey, Ivy:** Geoffrey *de Iuoi, de Iuei* 1161–2 P (O). From Ivoy (Cher).

**Ivor, Ivers, Iverson:** *Juuar* 1066 DB (Sa); *Iware* diaconus 1140–53 Holme (Nf); *Iuor paruus* 1161 P (Wo); Robert *Yuor* 1296 SRSx. ON *Ívarr*, ODa *Iwar*, OSw *Ivar* 'yew-army'.

**Ivory, Ivery:** (i) Roger *de Iuri, de Iueri* 1086 DB (Bk); G. *de Iverio*, Matillis comitissa *de Ibreio* 1215 Cur (O). From Ivry-la-Bataille (Eure), earlier *Ibreium*. (ii) *Ivory* Malet 1270 Bardsley; Thomas son of *Ivorie* 1332 SRCu; Alice *Yuory* 1296 SRSx; William *Ivory* 1364 LLB G. A diminutive of *Ivor*.

**Izard, Izatt, Izant, Izod, Izzard, Izzett:** *v.* ISSARD

# J

Jack, Jacka, Jacke, Jackes, Jacks, Jagg, Jaggs, Jakes, Jeeks, Jecks, Jex: Petrus, Andreas *filius Jake* 1195–7 P (Co); Normannus *filius Jacce* 1218 AssL; *Jake* Heriel 1275 RH (Gl); Robertus, Johannes *filius Jake* 1279 RH (C), c1315 Calv (Y); *Jakes* Flinthard 1292 SRLo; *Jak* del Thorp 1332 SRCu; *Jagge* the jogelour 1377 Piers Plowman; William *Jagge* 1251 Rams (Hu); William *Jake* 1260 AssC; Geoffrey, Richard *Jakes* 1269 AssSo, 1296 SRSx; Agnes, William *Jakkes* 1279 RH (Hu), 1366 Eynsham (O); William, Robert *Jacke* 1302 AssSt, 1332 SRWa; John *Jak* 1327 SRC; Adam *Jeke* 1332 SRWa; John *Jekes* 1346 ColchCt; Walter *Jekkes* 1524 SRSf; Thomas *Jagges* 1568 ib. The ODCN, following E. W. B. Nicholson (1892), derives *Jack*, the commonest pet-name of *John*, from *Jankin*, a diminutive of *Jan, Jehan* (John). According to this theory, *Jankin* became *Jackin* and was then shortened to *Jack*, a process completed by the beginning of the 14th century. Both *Jack* and *Jake* are certainly found for *John* towards the end of the 13th century: *John* or *Jacke* le Warner 1275 RH (C); *John* or *Jakke* de Bondec 1279 RH (Bk); *Jake* or *John* de Couentre 1292 SRLo, 1300 LoCt. *Jankin* was a 14th-century diminutive of *John*. v. JENKIN. But it is difficult to believe that the 1195 *Jake* and the 1218 *Jacce* can derive from *Jankin*. In France, *Jacques* (James) was so common a name that it became the normal term for a *peasant*, just as, in England, *Jack* became a synonym for *man* or *boy*. It would be strange if so popular a French name did not appear in England. Though rare, undoubted examples of *Jacques* are found: *Jacobus* or *Jakes* Amadur 1275 RH (L); *Jakes* or *James* Flinthard 1292 SRLo, 1300 LoCt. The diminutives *Jackett, Jacklin, Jakins*, correspond to the French *Jaquet, Jacquelin, Jacquin*, from *Jacques*. v. also JAMES. *Jakes, Jeeks, Jex*, show the same vowel development as *James, Jeames, Jem*.

**Jackalin:** v. JACKLIN

**Jackaman:** *Jakemyn* le Hatter 1314 LLB D; *Jakemina* 1321 FFEss. A diminutive of *Jacqueme*, a Picard form of *James*, from *\*Jacomus*. Also feminine, and has probably become JACKMAN. *Jakemin* de Sessolu 1302 LoCt is also called *James*.

**Jackett, Jacketts, Jacot:** William *Jaket* 1296 SRSx. A diminutive of *Jack*. The feminine *Jaketta* (1300 LoCt) was a pet-name for *Jakemina*.

**Jacklin, Jackling, Jackalin:** *Jakelinus* 1219 Fees (Y); *Jakelinus* de Boeule 1327 SRSx; Elias *Jakelyn* 1296 SRSx; Edmund *Jakelin* 1327 SRSf. A double

diminutive of *Jack*. *Jacolin* hugelin (1292 SRLo) is also called both *James* and *Jack* (1291 LLB A).

**Jackman, Jakeman:** Robert, William *Jakeman* 1296 SRSx, 1327 SRSf; Robert *Jacman* 1379 PTY. 'Servant of *Jack*.' v. also JACKAMAN.

**Jackson, Jacson, Jagson, Jaxon:** Adam *Jackessone* 1327 SRSf; Adam *Jakson* 1353 AssSt; John *Jakeson* 1438 AD i (L). 'Son of *Jack*.'

**Jacob, Jacobs, Jacubs:** *Jacob* c1250 Rams (Hu); Agnes *Jacobes* 1244 Rams (Beds); Walter, Alan *Jacob* c1250 Rams (Hu), 1324 FFK; Emma *Iacop* 1332 SRSr. The medieval surname was not Jewish. *Jacob* is found before the Conquest as the name of an ecclesiastic. After the Conquest, it is impossible to decide how common the name was as the Latin *Jacobus* was used for both *Jacob* and *James*.

**Jacobson:** William *Jacobson* 1332 SRCu; Peter *Jacobbesson* 1375 ColchCt. 'Son of *Jacob*.'

**Jacoby, Jacobi:** Robertus *filius Jacoby* 1297 SRY; Thomas, Pieres *Jacoby* 1275 RH (Nf), 1338 LoPleas. A preservation of the Latin genitive *Jacobi* '(Son) of *Jacobus*', Jacob or James.

**Jacot:** v. JACKETT

**Jacox:** v. JEACOCK

**Jacques, Jacquet, Jacquot, Jaques, Jaquest, Jacquet, Jaquin, Jaquiss:** Either late introductions from France or a refashioning of *Jakes, Jackett*, etc., after the French. Camden notes such a tendency in the 16th century 'which some Frenchified English, to their disgrace, have too much affected'. *Jaquest* and *Jaquiss* are certainly English forms.

**Jade:** Simon *Jade* 1258 IpmY. A nickname from ME *jade* 'a cart-horse, a hack'.

**Jaffray, Jaffrey:** v. JEFFRAY

**Jagg, Jaggs:** v. JACK

**Jaggar, Jaggars, Jagger, Jaggers:** Thomas *Jager*, John *Jagher* 1379 PTY; Katerina *Jeggar* 1480 GildY. A West Riding name, from *jagger* 'carrier, carter; pedlar, hawker' (1514 NED). It may also be from *Jaggard*, with loss of d.

**Jaggard:** Aldred, William *Jagard* 1194 Cur (W), 1279 RH (C); John *Jakard* 1296 SRSx. *Jack-ard*, a hypocoristic of *Jack*.

**Jago, Jagoe, Jeggo, Iago:** *Jago* filius *Ytel* 1185 P (He); William *Jeago* 1221 AssSa; Thomas *Jagoo* 1524 SRSf. Welsh *Iago*, Cornish *Jago* 'James'.

**Jagson:** v. JACKSON

**Jailler:** v. GALER

**Jaine:** v. JAN

**Jakeman:** v. JACKMAN

**Jakes:** v. JACK.

**Jakins, Jakens, Jeakins, Jeakings:** *Jakin* de Lagefare 1202 FFEss; Robert *Jacun* Hy 3 HPD (Ess); Claritia *Jagun* 1275 RH (Nf); John, William *Jakyn* 1296 SRSx, 1314 FFK; Robert *Jeakins* 1806 Bardsley. *Jak-un, Jak-in*, diminutives of *Jak* (James). v. JACK.

**Jambe, Gaumbe:** Ranulf *Jambe* 1221 AssWa; Adam *Jaumbes* 1256 Gilb (L); Thomas *Gaumbe* 1298 AssL. OFr *jambé* 'having good legs'.

**James:** *Jacobus* 1160 DC (L); *Jam'* de Sancto Hylario 1173–6 NthCh (Nth); *Jamos* (*Jacobus*) de Vabadun 1221 Cur (Sf); *James* or *Jacobus* de Audithleg' 1255 RH (Sa); Walter, Emma *James* 1187 P (Gl); Cristiana, Thomas *Jemes* 1279 RH (C), 1332 SRWa. *Jacob* was latinized as *Jacōbus*, but in late Latin became *Jacōbus* and *Jacōmus*. From the former came French *Jacques*, English *Jacob* and Welsh *Iago*, from the latter, Spanish *Jayme*, OFr *James, Gemmes*, AFr *Jam* and English *James*. In early documents the name is usually *Jacobus*, but *James* is occasionally found in the 12th and 13th centuries, sometimes alternating with *Jack* or its diminutives *Jackamin, Jackett* and *Jacklin*. The vernacular form was *Gemme* or *Jemme* (both masculine and feminine), and occasionally *Jimme*. v. GEM.

**Jameson, Jamieson, Jamison:** William *Jamesson* 1379 PTY; John *Jameson* 1440 FrY; Richard *Jamieson* 1642 PrGR. 'Son of *James*.'

**Jan, Jans, Janse, Jannis, Jane, Janes, Jaine, Jayne, Jaynes, Jeynes:** Simon *Ianes* 1297 MinAcctCo; William *Jan* 1327 SRSo; John *Janne* 1327 SRC; Thomas *Jannes*, John *Janys* 1327 SRWo; Robert *Jans, Jance* 1539–40 Bardsley; John *Jane* 1548 Oxon. *Jan*, from *Johan*. v. JOHN.

**Janaway, Janaways, Jannaway, Janeway, Janway, January, Januarys, Jennerway, Jennery, Gannaway, Genway:** Gilbert *Genewy* 1218–19 FFK; Peter *de Geneva* 1249, *de Geneweye* 1251 Fees (He); William *Janoway* 1562 Pat; John *Jenewaye* 1576 SRW. From Genoa, Lat *Genua*.

**Jancock:** John *Janecok'* 1332 SRDo; Norman *Jancock* 1353 AD vi (Sr); Henry *Janecok* 1396–7 FFSr. A diminutive of *Jan*, i.e. *John*.

**Janet, Jennett:** Willelmus *filius Jonet* 1297 SRY; *Jonot* 1308 Wak (Y); John *Ionet* 1297 MinAcctCo; Robert, Simon *Janot* ib., 1327 SRSx. *Janot, Jonet, Jonot*, diminutives of *Jan, Jon*, from *Johan*. *Jennett* is from *Jehan*. v. JOHN.

**Janeway, Jannaway:** v. JANAWAY

**Janin, Jannings:** v. JENNINGS

**Janks:** v. JENKIN

**Janman, Jenman:** Richard *Janeman* 1327 SRSx; Wyote *Janemon* 1327 SRWo. 'Servant of *Jan*.' v. JOHN.

**Janney:** John *Janny* 1324 LaCt; William *Janny* 1332 SRWa; John *Janny* 1332 SRWo. A diminutive of *Jan*, i.e. *John*.

**Jannis:** v. JAN

**Jansen:** A Dutch name. On the execution of his father by the Duke of Alva, Theodore *Janse*, youngest son of the Baron de Hees, fled to France, and thence to England. His descendants became London merchants and bankers, one of them becoming Lord Mayor of London in 1755 (Smiles 403).

**Janson:** Henry *Jannesonne* 1327 SRSt. 'Son of *Jan*.'

**Janston:** Thomas *de Janeston* 1248 AssSt. From Johnson Hall (St), *Johannestun* 1227.

**January, Janaway:** v. JANAWAY

**Jaques, Jaquest, Jaquet, Jaquiss:** v. JACQUES

**Jaram:** v. JEROME

**Jaray:** v. GEARY

**Jardin, Jardine:** Winfredus *de Jardine* a1153 Black (Kelso); Humphrey *del Gardin* 1194–1211 ib.; Matilda *atte Jardin*, Mabilia *Jardyn* 1296 SRSx. 'Dweller near, or worker at a garden', central OFr *jardin*.

**Jared:** v. GARRAD

**Jarmain, Jarman:** v. GERMAN

**Jarmay, Jarmey:** v. JERMEY

**Jarrad, Jarred, Jarratt, Jarrett, Jarritt:** v. GARRAD

**Jarrand:** v. GERANT

**Jarrard:** v. GERARD

**Jarraud, Jarrod, Jarrold, Jarrott:** v. GERALD

**Jarry:** v. GEARY

**Jarville:** William *de Iarpunuill'* 1173 KRec xviii; John, Laurence (*de*) *Jarpe(n)ville, Charpeneuill* 1258, 1285, 1296 PN Ess 128; Alice *Jarpenvil* 1301 SRY; William *Jerpeville* 1327 *SR* (Ess); John *Gardeuill'* 1376 AssEss; Gerald *de Charpeuile* 14th *Petre* (Ess); John *Gerdevyle, Jardevelde* 1423 ib., 1473 *ERO* (Ess); Richard *Jard(e)feld* 1507, 1527 FFEss. From Gerponville (Seine-Inférieure), cf. Gerpins in Rainham (PN Ess 128).

**Jarvis, Jervis, Jervois, Jervoise, Gervase, Gervis:** (i) *Geruasius* Painel 1158–66 DC (Lei); *Garvasius* Godihalt 1275 SRWo; John *Geruas'* 1202 P (Sa); Thomas *Geruais* 1230 P (Nf); William *Gerveys* 1270 AssSo; Richard *Jerfeys* 1327 SRSx; Thomas *Jerveys* 1360 ColchCt. OG *Gervas*, OFr *Gervais(e)*. (ii) Robert *de Gerewall'* 1275 SRWo; John *Gerveux* 1360 FrY; William *de Gervaux* 1370 ib.; William *Gerveys* 1395 ib.; John, Richard *Gervas* 1435 ib.; Thomas *Jarvis* 1713 ib. Though usually from the personal name, these names, in Yorkshire in particular, are at times from Jervaulx (Yorks), of which the pronunciation (now obsolescent) is *Jarvis*, earlier *Gervaus* 1200, *Jervax* 1400, *Gerveis* 1530, *Gervis* 1577 (PN NRY 250).

**Jary:** v. GEARY

**Jasper:** *Jasper* Pen 1522 LP; Edward *Jasper* 1561 Pat (Ess); Henry *Jesper* 1576 SRW; Edmund *Jasper* 1662–4 HTDo. *Jasper* is the usual English form of *Caspar* or *Gaspar*, the traditional name of one of the three kings. The name has always been rare in England.

**Jaxon:** v. JACKSON

**Jay, Jaye, Jayes, Jays, Jaze, Jeayes, Jeays, Jeyes:** Peter *le iai* 1195 P (C); Gilbert *Jai* (*Gai*) 1202 P (L); Walter *le Jay* 1225 AssSo. OFr *jay, gai*, Fr *geai* 'jay', chatterer.

**Jaycock(s):** v. JEACOCK

**Jayne(s):** v. JAN

**Jeacock, Jeacocke, Jacox, Jaycock, Jaycocks:** William *Jacok* 1327 SRSf; John *Jecok* 1375 ColchCt; John *Jecokes* 1381 AssWa; Elizabeth *Jeacock* 1712 Bardsley. *Jak-coc*, a pet-form of *Jack*.

**Jeafferson, Jeaffreson:** v. JEFFERSON

**Jeakins, Jeakings:** v. JAKINS

**Jealous:** Ralph *Jelus* 1230 Pat (K); Adam *Gelus* 1249 AssW; John *Jelous* 1327 SREss. OFr *gelos* 'jealous, amorous, ardent'.

**Jean, Jeanes, Jeans, Jeens, Jenn, Jenne, Jenness, Jennis, Jenns, Geen, Genn:** (i) William *Gene* 1275 RH (Sf); Thomas *Gennes* 1297 MinAcctCo; Alice *Genne* 1327 SRSf; Thomas *Geene* 1378 FrY; Mary *Jeenes* 1663 Bardsley. *Jen*, from *Jehan*. v. JOHN. (ii) Henry, Francis *de Gene* 1255 RH (St), 1309 LLB D. From Genoa.

**Jearey, Jeary:** v. GEARY

**Jearum:** v. JEROME

**Jeavons:** v. JEVON

**Jeay(e)s:** v. JAY

**Jebb, Gebb:** William *Gebbe* 1327 SRSf; Lucas *Jebbe* 1508 FrY. A voiced form of *Gepp* (Geoffrey).

**Jebordy:** Francis *Jebordy* 1525 SRSx. OFr *jeu parti*, ME *juparti, jeberdi* 'risk, danger; harm, misfortune'.

**Jebson, Jibson:** Robert *Gebbesson, Gebisson* 1442, 1453 Shef. 'Son of *Gebb*.'

**Jeckell(s):** v. JEKYLL

**Jecks, Jeeks:** v. GEAKE, JACK

**Jeeps, Jeapes:** Francis *Geapes* 1662 HTEss; William *Jeapes* 1813 WStowPR (Ess). Late forms of *Gepp*, a pet-form of *Geoffrey*.

**Jeeves, Jeves, Geaves, Geeves:** *Geua* 1120–3 EngFeud, Hy 2 Gilb (L), 1221 AssWa; Willelmus *filius Geue* 1208 Cur (Nb); *Geva* Mullyng 1313 AD iv (Lo); Richard *Geves* 1279 RH (O); Thomas *Geue*, Adam *Geues* 1327 SRWo; Thomas *Jeve* 1327 SRSo. Bardsley takes *Jeeves* as identical with *Geff*, a pet-name of Geoffrey, but this accounts neither for the long vowel nor for the v. It is a pet-name for *Genevieve*, a favourite name in France, where St Geneviève is the patron saint of Paris. Its use in England in the 12th–14th centuries seems to have escaped notice. A 12th-century Lincolnshire deed of a woman named *Geua* has attached to it her seal, with the legend: SIGILL' GENEVEVE F' ARNALD (DC). *Geva*, wife of Segar, is also called *Eva* (1199 CurR).

**Jeff, Jeffe, Jeffes, Jeffs:** *Geffe* 1260 AssCh; Ralph *Jeffe* 1275 RH (D); Richard *Geffes*, William *Geffe* 1332 SRWa. *Geff*, a pet-form of *Geoffrey*. cf. Alicia *Gefraywif*, Alicia *Gefdoghter* 1379 PTY. v. also GEFFEN.

**Jeffcock, Jeffcoat, Jeffcoate, Jeffcote, Jeffcott, Jefcoat, Jefcoate, Jefcott, Jephcott:** (i) Geoffrey *Geffecoke* 1327 SRWo; William *Gefcok* 1332 SRSt; Emota *Jeffecockes* 1380 SRSt; Agnes *Jeffcott* 1616 Bardsley. A diminutive of *Jeff*. cf. Thomas *Gepcok* 1360 Cl (K). (ii) Thomas *Geuecok* 1332 SRWa; John *Jevecok* 1454 RochW. A diminutive of *Geva* (Geneviève). v. JEEVES. *Jevecok* became *Jevcok*, and was then assimilated to *Jeffcock*. As often, the suffix -*cock* has become -*cott* or -*coat*.

**Jefferd:** v. GIFFARD

**Jeffers:** v. JEFFRAY

**Jefferson, Jeafferson, Jeaffreson:** Robert *Geffreysone* 1344 AssSt; Alice *Geffrason* 1488 GildY; John *Jeffrason* 1528 FrY. 'Son of *Geoffrey*.'

**Jeffkin, Jephkin:** Margery *Gefkyn* 1354 LuffCh. *Jeff-kin*, a diminutive of *Jeff*, a pet-form of *Geoffrey*.

**Jefford:** v. GIFFARD

**Jeffray, Jeffrey, Jeffreys, Jeffree, Jeffress, Jeffries, Jeffry, Jeffryes, Jeffares, Jefferey, Jefferies, Jefferis, Jefferiss, Jeffery, Jefferyes, Jefferys, Jeffers, Jeoffroy, Jaffray, Jaffrey, Geffers, Geoffrey, Geoffroy:** *Goisfridus, Gaufridus, Gosfridus* 1086 DB; *Galfridus* c1150 DC (L); Simon *filius Gosfrei* 1210 Cur (Nf); Walter *Geffrei* 1203 Cur (Nf), *Gefray* 1243 AssSo; Agnes *Geffreys* 1283 SRSf; Robert *Geoffray* 1293 AssSt; William *Geffrey* 1296 SRSx; John *Gaffry* 1327 SRSo; Hugo *Jafres* 1327 SRSt; Joan *Jeffrey* 1327 SRC; Henry *Geffreys* 1332 SRSt; John *Geffre* 1333 AD iv (Nth); Symon *Geffris* 1340 SRWo; Roger *Jeffray* 1379 PTY; William *Jaffrey* 1450 SIA xii; John *Jafery* 1499 NorwW; Philip *Jeffereyes* 1566 ChwWo; Robert *Jeffers* 1689 FrY; Robert *Jefferys* 1723 ib. ME *Geffrey*, from OFr *Geuffroi, Jeufroi* or OFr *Jefroi*, representing two or, possibly, three OG names usually latinized in early documents as *Galfridus* or *Gaufridus* (most commonly) and *Goisfridus* or *Gosfridus*. v. Forssner 101–2, 125–6.

**Jeggo:** v. JAGO

**Jeggons:** v. JUGGINS

**Jekyll, Jeckell, Jeckells, Jickells, Jickles, Jiggle, Jockel, Giggle, Jockelson:** *Judichel* uenator 1066 DB (C); *Judicaelis* 1066 ICC (C); *Iuichel', Iuchel* presbiter 1086 DB (Sf); *Gykell, Jukel* de Jertheburc c1170, 1182 Gilb (L); Johannes *filius Jokell', Jukell'* 1218 AssL; *Jukel* Brito 1207 Cur (Nf); *Gigelle* Belle 1276 RH (Y); John *Iekel* (*Iukel*) 1174 P (Ha); John *Gikel* 1201 P (L); Thomas *Jokel* 1296 SRSx; Robert *Jekel* 1312 FFEss. OBret, OC *Iudicael*, which became *Iedecael*, modern *Gicquel, Iezequel* (Loth) and survives in French as *Jézéquel*. The name is found at Bodmin (c1000) as *Gyðiccæl* (PNDB). In England the personal name was particularly common in Yorkshire and Lincolnshire and in districts where the Breton contingent settled after the Conquest. v. also JOEL, GOSS.

**Jelke, Jelks:** v. GILKES

**Jell:** v. GELL

**Jellard:** v. GILLARD

**Jellett:** v. GILLET

**Jelley, Jelly:** Thomas *Jely* 1472 Cl (Ess); Robert *Jely* 1524 SRSf; John *Iellye*, Richard *Gellye* 1583 Musters (Sr); John *Jelley* 1641 PrSo. Late forms of *Giles*, cf. Gillygate (York), *Saintgeligate* 1356, *Giligate* 1373, taking its name from the church of St Giles which has long disappeared.

**Jellico, Jellicoe, Jellicorse:** John Dawson alias John *Jelicoke* 1553 Pat (Db); James *Jelicoe* 1648 ChW; John *Jellicoe* 1713, *Jellicorse* 1730; Thomas *Jellicoe* 1745 SaAS 2/xi. For *Jelli-cock*, a diminutive of JELLY. cf. *Alcoe* for *Alcock*.

**Jellings:** v. GILLIAN

**Jellis, Jelliss:** Robert *Geliss*, John *Gelis* 1527 Black; Andrew *Jeles* 1681 ib. For GILES.

**Jelly:** v. JELLEY

**Jemmett, Jemnett:** Thomas *Jemot* 1524 SRSf; Alice *Jemett* 1544 StaplehurstPR (K); John *Jemmett* 1665 HTO. *Jem-et, Jem-ot*, diminutives of *Jem*, a shortened form of *James*.

**Jempson, Jemson:** v. GEMSON

**Jencken:** v. JENKIN

**Jenckes:** v. JENKS

**Jeneson:** William *Geneson* 1346 AssSt. 'Son of *John*.' v. JEAN.

**Jenkin, Jenkins, Jenking, Jenkings, Jenkyn, Jenkyns, Jencken, Junkin:** *Janekyn* de sancto Iohanne 1260 Oseney (O); *Jonkin* the Turnur 1288 AssCh; *Jenkin* le persones 1327 SRSt; Richard *Janekyn* 1296 SRSx; William *Jonkyn* 1297 Ipm (Do); William *Jankins* 1327 SRWo; Richard *Jenkins* 1327 SRSo; William *Junkin*, George *Jonkyng* 1469, 1530 FrY. A diminutive of *Jan, Jon, Jen* (John). v. JEAN, JOHN. cf. *John* nicknamed *Janekin* de Bocking Ed 3 PN Ess 461; Henry *Janekynesmon* Hastang (1319 SaltAS x), 'servant of John Hastang junior'. His father was alive and his servants were Henry *Jonespaneter* Hastang and Richard *Jonesprest* Hastang, so that *Janekyn* clearly means 'Young John'.

**Jenkinson, Jenkerson, Jenkison, Junkinson, Junkison:** Robert *Jonkinson* 1379 PTY; William *Jenkynson* 1484 FrY. 'Son of *Jenkin*.'

**Jenks, Jenckes, Janks:** Walter *Jenks* 1542 Oseney (O). A contraction of *Jenkin, Jankin*.

**Jenman:** v. JANMAN

**Jenn, Jenne, Jenns:** v. JEAN. The Cornish *Genn, Gynn* may be for *Jennifer*.

**Jennaway, Jennerway, Jennery:** v. JANAWAY

**Jennens:** v. JENNINGS

**Jenner, Jenoure, Genner, Genower, Ginner:** Richard *lengignur, lenginnur* 1191–7 P (Y); William *Enginur* 1202 Cur (Sf); Robert *le enguigniur* 1221 AssWo; Robert *le Ginnur* 1229–30 Clerkenwell (Lo); Ralph *Gynnour* 1301 SRY; William *le Genour* 1324 RamsCt; Robert *Jenour* 1327 SRSf. OFr *engineor, enginior* 'engineer, maker of military machines' (c1380 MED). In the 12th century *ingeniator* was used of men who combined the duties of master-mason and architect. Ailnoth *ingeniator*, a military architect, was surveyor of the king's buildings at Westminster and the Tower in 1157 and was in charge of building operations at Windsor 1166–73. He repaired Westminster Abbey after a fire and superintended the destruction of the castles of Framlingham and Walton (MedInd 108).

**Jenness:** v. JEAN

**Jennett:** v. JANET

**Jennifer, Junifer, Juniper:** Mabilia *Jeneuer* 1296 SRSx; Henry *Juneuyr* 1332 ib.; Joshua *Junefer* 1623 SfPR; Captain *Jenifer* 1667 Pepys; John *Juniper* 1753 Bardsley. Welsh *Gwenhwyvar* (f), from *gwen* 'fair, white' and (*g*)*wyf* 'smooth, yielding', found in Shropshire as *Gwenhevare* in 1431 and in Cornwall as *Jenefer* (1554) and *Junipher* (1691).

**Jennings, Jennens, Jennins, Jenings, Jenyns, Janin, Jannings, Jouning:** *Janyn* le Breton 1332 SRLa; *Jenyn* de Fraunce 1379 PTY; Roger *Jonyng* 1296 SRSx; John *Ianin* 1297 MinAcctCo; Walter *Jannen*, Richard *Janyns* 1327 SRWo; John *Janyng* 1327 SRSx; Thomas *Jenyn* 1428 FA (Sx); Cristofer *Jenyng* 1532 FrY. Diminutives in -*in* of *Jan, Jon, Jen*. v. JOHN. cf. *John* or *Janyn* Nichol(e) 1413–17 LLB I.

**Jennis:** v. JEAN

**Jennison:** Robert *Genyson* 1491 GildY; William *Jenyson* 1518 FrY. 'Son of *Jenin*.' v. JENNINGS.

**Jenns:** v. JEAN

**Jent:** v. GENT

**Jentle:** v. GENTLE

**Jeoffroy:** v. JEFFRAY

**Jephcott:** v. JEFFCOCK

**Jephkin:** v. JEFFKIN

**Jephson:** Hugh *Geffesone* 1327 SRWo. 'Son of *Geff*', i.e. Geoffrey.

**Jepp, Jeppe, Jepps:** v. GEPP

**Jeppeson, Jepson, Gepson:** John *Gepsone* 1326 Wak (Y); John *Jepson* 1379 PTY. 'Son of *Gepp*.'

**Jerams:** v. JEROME

**Jeray:** v. GEARY

**Jerdan, Jerden, Jerdein, Jerdon:** Jon *Jerdein* 1563 Black; Andrew *Jerdon* 1659 ib. Scottish forms of JARDIN.

**Jeremy, Jermyn:** *Iheremias* de Tornhill' 1189 P (Y); *Geremias, Jeremias* Banastre c1200 Rad (C); *Gereminus* de Eclefeld', *Jeremias* de Ecclesfeld' 1219 AssY; Johannes *Jheremie* 1193 P (Y); Thomas *Jeremye* 1225 Pat; William *Germin* 1253 HPD (Ess). *Jeremias* is a latinization of the Greek form of the Hebrew *Jeremiah* 'May Jehovah exalt', in the vernacular *Jeremy*. cf. 'þe propheci þat said was þoru *Jeremi*' (Cursor Mundi). v. JERMEY. *Gereminus* is a hypocoristic.

**Jermain, Jerman, Jermin, Jermine, Jermyn, Jermynn:** v. GERMAN

**Jermey, Jermy, Jarmay, Jarmey, Jarmy:** Johanna *Germye, Jermye* 1303, 1346 FA (Sf); Thomas *Jarmy* 1652 Bardsley (Nf). One would naturally assume that *Jermey* was identical with *Jeremy* and Bardsley cites a late example: *Jeremye* or *Jermy* Gooch from Norfolk (1617, 1652). He also notes that Blomefield, the Norfolk historian, states definitely that the true origin of *Jermy*, a Norfolk surname, is *Jermin* and cites Sir John *Germyn* or *Jermy* (1300). The truth is that *Germin* was used for both *Jeremy* and *Germain* and here the vernacular *Jermy* prevailed. v. GERMAN, JEREMY.

**Jernegan, Jerningham:** *Gernagus* filius Hugonis 1166 P (Y); *Gernagan* de Tanefeld, Hugo *filius Jernagan* 1204 AssY; Hubert *Gernagan* 1166 RBE (Sf); Aluredus *Gernegan* 1170 P (Y); Hubert *Jernegan, Jarnegan* 1218, 1222 FFSf; John *Gernyngham, Jerningham* 1472, 1503 Copinger (Sf); Edward *Jermingham* 1510 ib.; Henry *Jernegam* 1572 Bardsley (Nf). The surname, now rare, is found particularly in Suffolk where Little Stonham, also known as *Stanham Gernagan* (1244 FFSf), was long held by the family of Hubert *Jarnegan* (1222 FFSf). In 1086 (DB) land was held in Stonham by Earl Alan and Iuichel the priest (v. JEKYLL) and there can be little doubt that *Gernagan* is a Celtic name brought over by the Bretons at the Conquest, probably identical with OBret *Iarnuuocon* (*Iarnogon* 1062 Loth) 'iron-famous'.

**Jerome, Jerrom, Jerrome, Jerromes, Jerams, Jerram, Jerrams, Jerrim, Jearum, Jaram, Gerham:** *Geram* Hy 2 DC (L); *Geram* de Curzun 1206 FineR (Bk); *Jeronimus, Geronimus, Gerarmus* de Curzun 1206–11 Cur (Berks, St); *Jerom'* de Ponte Burgi 1219

AssY; *Jeronimus* de Normaneby 1230 P (L); William *Geran'* 1194 CurR (Sa); Roger *Geram* 1333 AD ii (Lei); Peter *Jerrome* 1604 FrY; Joseph *Jerram* 1729 Bardsley. It is clear that two distinct personal names were early confused. OG *Ger(r)am, Gerrannus* 'spear-raven' would certainly be taken as the name of *Geram* de Curzun were it not for the forms *Jeronimus, Geronimus* which point clearly to *Jerome*, Greek 'Ιερώνυμος 'sacred name', Italian *Geronimo*, French *Jérôme*. The name of the translator of the Bible does not seem to have been very popular in England. *Gerrannus* may also have become GERRANS or JERRANS. *v.* GERANT.

**Jerrans:** *v.* GERANT

**Jerrard:** *v.* GERARD

**Jerratt, Jerreatt, Jerred:** *v.* GARRAD

**Jerrold:** *v.* GERALD

**Jerrom(e), Jerrim:** *v.* JEROME

**Jervis, Jervois, Jervoise:** *v.* JARVIS

**Jessey:** Ralph *Jesse* 1275 RH (D); William *Jesse* 1576 SRW; Philip *Jesse* 1642 PrD. Perhaps metonymic for a maker of jesses for hawks. cf. Robert *le Jesemaker* 1275 RH (L).

**Jesson:** *v.* JUDSON

**Jessop, Jessopp, Jessope, Jessup, Jessep:** William *Josep* 1296 SRSx; William *Josop*, John *Jesop* 1379 PTY; Joan *Josopp, Jesopp* 1524 SRSf. A pronunciation of *Joseph*.

**Jester:** John *Gestour* 1377 ColchCt. A derivative of OFr *geste, jeste* 'exploits'; ME *gester* 'a mimic, buffoon, merry-andrew' (c1362); 'a professional reciter of romances' (c1380).

**Jestice:** *v.* JUSTICE

**Jetson:** *v.* JUDSON

**Jeudwin, Jeudwine:** *Jeldewin* 1157 P (Sx); *Joldewinus* filius Sauarici 1158 P (Sx); *Jodewinus* 1214 Cur (Sx); Martin *Gaudewyne* 1296 SRSx; Gilbert *Goldewyne, Geldewine* 1317 AssK; John *Jeldewyne, Jewdewyne*, Agnes *Joidewyne* 1327 SRSx; John *Joldewyne, Judewyne* 1332 ib.; John *Judwyn* 1534 SxWills. AFr *Jeudewin*, a Norman form of OE *Goldwine. v.* GOLDWIN.

**Jeune, June, Lejeune:** Richard *le Jeune* 12th Lichfield (St); Matilda *Jun* 1279 RH (C); John *le June* 1301 SRY. Fr *jeune* 'young'. cf. JEVON.

**Jeves:** *v.* JEEVES

**Jevon, Jevons, Jeavons:** (i) Alexander *le iouene* Hy 2 DC (L); John *le Jofne* 1200 Cur (Herts); Bartholomew *le Joevene, le Juvene* 1254–69 Rams (Beds); Robert *le Jeofne, le Jevene* 1242–55 Fees (Nth); Thomas *le Geven'* 1279 RH (O). OFr *jovene*, Lat *juvenis* 'young'. cf. JEUNE. (ii) *Jeuan* Vachann 1391 Chirk; Ririt *ap Jeuan* ap Eigon 1393 ib.; *Jevan* or *Evan* Thomas 1600 Bardsley (Glam); Thomas *Ieuane* 1436 Oseney (O); John *Jevanne* 1459 ib. Welsh *Ieuan*, later *Evan*, 'John'.

**Jew:** *Gewe* de Sumerkotes 1275 RH (L); Richard *Jue* 1314 FFEss. From *Jull* (Julian or Juliana). Also a nickname: Thomas *le Jeu* 1275 RH (Nt), 'the Jew'.

**Jewell, Jewels:** *v.* JOEL. The surname may also be metonymic for jeweller, goldsmith: Robert *le*, Alicia *la Jueler* 1319 SRLo.

**Jewers:** *v.* JOWERS

**Jew(e)sbury:** *v.* DEWSBURY

**Jewett, Jewitt:** *v.* JOWETT

**Jewhurst:** *v.* DEWHIRST

**Jewison, Jewisson:** Henry *Jewetson* 1533 GildY. 'Son of *Jewet*.' *v.* JOWETT.

**Jewkes:** *v.* JUKES

**Jewry, Jury:** Richard *ate Jewerye, atte Giwerye* 1327, 1332 SRSx; William *Jury* 1495 ArchC 42. 'Dweller in the Jewry, the Jews' quarter, the Ghetto.' AFr *juerie*, OFr *juierie, jurie*.

**Jewson, Juson:** Richard *Juwesone, Jullesome, Jullesone* 1333, 1340, 1341 ColchCt. 'Son of *Jull*.'

**Jewster:** *v.* JUSTER

**Jex:** *v.* GEAKE, JACK

**Jeyes:** *v.* JAY

**Jeynes:** *v.* JAN

**Jibson:** *v.* JEBSON

**Jickells, Jickles:** *v.* JEKYLL

**Jickling:** *Jukelinus, Jukellus* de Smetheton' 1200 Cur (Y); Edward *Jeglin* 1662 *HTEss*. A diminutive of *Jukel. v.* JEKYLL.

**Jiggen(s), Jiggins:** *v.* JUGGINS

**Jiggle:** *v.* JEKYLL

**Jiles:** *v.* GILES

**Jillard:** *v.* GILLARD

**Jillett, Jillitt:** *v.* GILLET

**Jillings, Jillions:** *v.* GILLIAN

**Jimpson:** *v.* GEMSON

**Jinkin, Jinkins, Jinkinson, Jinks:** Johanna *Jinckson* 1578 Bardsley; Mary *Jynckes* 1615 Moulton (Sf). Late developments from JENKIN.

**Joan, Joanes:** *v.* JOHN

**Joass:** *v.* GOSS

**Job, Jobe, Jobes, Jope, Jopp, Jubb, Jupe, Jupp:** *Jop* serviens Osulf 1185 P (Co); *Joppe* filius Hardekin 1199 P (L); *Jubbe* de Donewiz 1275 RH (Sf); John *Molendinar'* 1296 SRSx; William *Job* 1202 FFNf; Elyas *Jubbe* 1230 P (Nf); Henry *Joppe* 1255 RH (W); Walter *Jobbe* 1275 SRWo; Eudo *le Jope* 1290 SRSr; William *Jopes* 1296 SRSx; Richard *Joup, Joop* 1327, 1332 SRSx. Often, no doubt, as suggested by Bardsley and Weekley, from the Hebrew *Job* 'hated, persecuted', a frequent character in medieval plays. The genitive of *Job* in Orm is *Jopess*. For the variation between *Jop* and *Job*, cf. *Gipp* and *Gibb, Gepp* and *Jebb*. There is no evidence for OG *Joppo* in England as suggested in ODCN. We have also clearly a nickname. Jupeshill Fm in Dedham (Essex) derives from Matthew *le Jop* (Ed 3 PN Ess 387), OFr *jobe* 'a fool'. cf. *joppe* 'fool', *joppery* 'folly', from Lat *joppus* (PromptParv). We may also be concerned with ME *jubbe, jobb(e)* (c1386 NED), a large vessel for liquor, holding 4 gallons. Hence a nickname for one who could carry that quantity or a trade-name for a maker of *jubbes*. OFr, ME *jube, jupe* 'a long woollen garment for men' (c1290 NED) may also, as an occupation-name, have contributed to the frequency of these surnames.

**Jobar, Jobber, Jubber:** Alan *Iober* 1317 AssK; Robert *Jobour, Jober* 1356, 1369 LLB G; Thomas *Jobber* 1524 SRSf. Probably occupational, 'maker of woollen garments' (ME *jube*) or of large vessels (ME *jobbe*). *v.* JOB. Perhaps also to be associated with East

Anglian dialect *to job* 'to peck with the beak, or with the mattock'. *Nutjobber* was formerly a name for the nuthatch, which has become a surname.

**Jobbins, Jobin:** *Jobin* Don 1271 ForSt; Ern' *Jobin* 1173 P (He). A diminutive of *Job*..

**Jobbinson, Jobinson:** Isabella Jonwyff *Jobinson* 1381 PTY. 'Son of *Jobin*', a diminutive of *Job*.

**Jobborn, Jobern, Joberns:** Probably a substitution of *-bern* for *-bert* in *Jobert*. *Ketelbern* and *Ketelbert* frequently interchange.

**Jobert, Joubert, Jubert, Goubert:** John, Robert *Joyberd* 1256 *Ass* (Ha), 1327 *SR* (Ess). *Goisbert* de Inge (1141) gave his name to *Ginges Joberti* (1230), now Buttsbury (Essex). OG *Gautbert* became OFr *Jaubert, Joubert, Jobert*, and also *Goisbert, Joibert*, through association with OFr *joie*.

**Jobey, Joby:** Gilbert *Jobi* 1210–11 PWi; William *Joby* 1280 AssSo; John *Joby* 1524 SRSf. A pet-form of *Job*.

**Jobin:** *v.* JOBBINS

**Jobinson:** *v.* JOBBINSON

**Joblin, Jobling, Joplin, Jopling:** Abraham *Joblin* 1652 FrY; Henry *Jobling* 1738 Bardsley; Hannah *Joplin* 1742 ib.; Elizabeth *Jobling* 1763 ib. A double diminutive of *Job, Job-el-in*, used in OFr of a beggar and in English of 'a sot, a fool'.

**Jobson, Jopson:** Joppe son of *Joppeson* 1332 SRCu; Ralph *Jopson* 1382 Whitby (Y); Richard *Jobson* 1491 FrY. 'Son of *Job*.'

**Joby:** *v.* JOBEY

**Joce:** *v.* GOSS

**Jocelyn, Joscelyn, Joscelyne, Josselyn, Joselin, Joslen, Joslin, Josling, Josolyne, Joseland, Josland, Goslin, Gosselin, Goseling, Gosling, Gossling, Gostling, Gosland:** *Gozelinus, Gos(c)elinus* 1086 DB; *Joscelinus, Gotcelinus, Goscelinus* de Stalham 1149–66 Holme (Nf); *Goslanus, Goslein, Goslinus, Joslanus* de Engelby 1154–84 Gilb, DC (L); *Joslinus* Hy 2 DC (L); *Joslenus* But 1175 P (L); Willelmus *filius Jocealini, Jocelini, Gosceami, Josealmi* 1208–12 Cur (L); Robert *Goselin* 1185 Templars (L); Walter *Joslein, Joslani, Goslein* 1195–8 P (Y); Ralph *Jocelin'* 1198 Cur (Nf); Edrich *Gocelin' (Goscelin')* 1204 P (L); Ralph *Joscelin* 1208 Cur (Nth); William *Joclenne* 1243 AssSo; Thomas *Gosselyn* 1327 SRC; John *Gostelen, Gosteleyn* 1462 NorwW (Nf); John *Gostlyng* 1526 ib. (Sf). OFr *Goscelin, Gosselin, Joscelin*, OG *Gautselin, Goz(e)lin*, diminutives of compounds with *Gos-* or *Goz-*, or of OFr *Josse* (OBret *Judoc*). *v.* also GOSS, GOTT, GOSLING.

**Jockel, Jockelson:** *v.* JEKYLL

**Joel, Joels, Joell, Jewell, Jewels, Joule, Joules, Jowle, Juell, Joelson, Jolson:** *Joel, Jool, Jol (Johol)* de Lincolnia c1051 (1334) Rams; *Judhel* de Totenais 1086 DB (D); *Johel* de Helsam c1160 DC (L); *Juhel* de Harelea 1166 P (Y); *Joel* de Creton 1195 FF (C); *Juel(us)* de Vautort 1214 Cur (Co); Richard *Juel* 1247 AssBeds; Alan *Joel* 1256 AssNb; William *Jool* 1332 SRSx; William *Juwel* 1358 AD vi (Beds); Roger *Juylle* 1383 Cl (Co); John *Jewell* 1462 FrY; John *Joulle* 1579 Bardsley; Augustine *Jowles* 1650 ib.; Joane *Jule* 1704 ib. OBret *Iudhael*, from *Iud-* 'lord, chief' and *haël* 'generous'. cf. JEKYLL. The personal name was

common in Devon and Cornwall and the Breton districts of Yorks and the Eastern counties. *Jewell* is now common in North Devon.

**Joesbury:** *v.* DEWSBURY

**Joester:** *v.* JUSTER

**John, Johnes, Johns, Johnys, Joan, Joanes, Jone, Jones, Joynes, Fitzjohn:** *Johannes* c1140 DC (L); Alanus *filius Jene* 1275 RH (L); Willelmus *filius Gene* 1276 RH (Y); Walterus *filius Jone* 1279 RH (Hu); *Jon* 1292 SRLo; Richard *le Fitz Joan* 1327 SRSo; *Johan* 1379 PTY; Petrus *Johannis* 1230 Cl (Sf); Thomas *John* 1279 RH (Bk); Matilda *Jones* ib. (Hu); Arnold *Johan* 1280 LLB A; Robert *Jhoun* 1295 Barnwell (C); Thomas *fiz Jon* 1296 SRSx; Richard *Jon*, Robert *Jone* 1327 SRSx; Roger *Jonis* 1327 SRSf; Robert *Johns* 1327 SRSo; Robert *Joyne, June* 1524 SRSf. Hebrew *Johanan* 'Jehovah has favoured', usually latinized as *Johannes* in early documents, OFr *Johan, Jehan, Jean*. By the beginning of the 14th century *John* rivalled *William* in popularity and has always been a favourite name. The feminine *Joan*, latinized as *Johanna* was also common and had the same forms and pronunciation as *John*. The surname may derive from either. *Johan* became *Jan* and *Jon*, and *Jehan* became *Jen*, with diminutives *Janin, Jonin, Jenin; Janet, Jonet, Jenet; Jankin, Jonkin, Jenkin*. A pet-form was *Han*, with diminutives *Hancock, Hankin*. *Jack* is sometimes for *John*. The Welsh form was *Ieuan, Evan* (*v.* JEVON, EVANS). The form *Ioan* was adopted for the Welsh Authorized Version of the Bible, hence the frequency of the Welsh patronymic *Jones*. For *Joynes*, *v.* JOHNSON. *v.* also JACK, JAN, JEAN, JENKIN, JENNINGS, JINKIN, HANN, HANCOCK, HANKIN, HENKIN.

**Johncock, Johncocks, Johncook:** John *Johncock* 1525 ArchC 30. A diminutive of *John*.

**Johnson, Jonson, Joinson, Joynson:** John *Jonessone* 1287 AD i (Sr); Wautier *Jonessone* 1296 CalSc (Berwick); John *Jonesone* 1321 AD i (Sr); William *Johnson*, Robert *Johanson, Jonson* 1379 PTY; Robert *Joynson* 1582 Bardsley (Ch); Jone *Geynson* 1595 ib.; Thomas *Jeynson* 1667 ib. 'Son of *John*.' *Joynson* is a Cheshire form, from *Jeynson*, from *Jeyn*. *v.* JAN.

**Johnstan, Johnstone:** (i) Caterina *Jonstons* 1327 SRSf. For *Johnson*, with intrusive *t*. (ii) Alan *de Johannestun* 1227 AssSt; Peter *de Jonestone* 1299 ib. From Johnson Hall (Staffs). (iii) Gilbert *de Jonistoune* 1195–1215 Black. From Johnstone in Annandale (Dumfries), named from *John*, father of Gilbert.

**Joice:** *v.* JOYCE

**Joiner, Joyner:** John *Joinur* 1195–1215 StP (Lo); William *le juinnur, Joinier, Joyner* 1204, 1218–33 ELPN; John *le Jeynuur* 1296 FFEss. AFr *joignour*, OFr *joigneor* 'joiner' (1322 MED).

**Joinson:** *v.* JOHNSON

**Joint:** Louelyn *Joynte* 1282 KB (Lo); William *Joynte* 1539 FeuDu; Henrie *Joynte* 1576 SRW. OFr *joint* 'united, joined'.

**Jolin, Jolland, Jollands, Jowling, Gollan, Golland, Gollin, Gollins:** *Jollanus; Jolinus; Joelinus* Hy 2 DC (L); *Goillanus, Gollanus* de Aute(r)barg(e) ib.; *Jollein* 12th ib.; *Iolanus, Joelinus* de Nouilla ib.; *Gollinus (Joelinus)* de Pomereia 1214 Cur (D); Adam *filius Jollein, Joulen* 1219 AssL; *Juelinus* (*Joelus*) de Wic 1220 Cur (O); *Joelus, Joelinus, Jollinus, Jollanus, Joylin* de Sowe 1221 AssWa, 13th AD ii (Wa); Alexander *Jolleen* 1196 Cur (L); William *Goelin* 1212 Cur (O); Richard *Joelan'* 1214 Cur (Beds); Roger *Jollein* 1219 AssL; Geoffrey *Joelin, Jolin* 1221 AssWa; William *Joylin* 1279 RH (C); William *Gollayn* 1408 FrY; Thomas *Gollan, Golland* 1412, 1481 ib. *Joel-in*, a diminutive of *Joel*. For the suffix and development, cf. JOCELYN.

**Jolivet:** William *Jolivet* 1270 Acc; Adam *Joliuet* 1279 AssSo; Thomas *Jolyvet* 1375 IpmNt. *Jolif-et*, a diminutive of JOLLIFF.

**Jolland(s):** v. JOLIN

**Joll(es):** v. JULL

**Jolliff, Jolliffe, Jolley, Jollie, Jolly, Jollye, Joly:** John *le Goly* 1275 RH (W); John *Jolif* 1279 RH (Hu); Walter *Jolyf* 1281 LLB B; John *Joly* 1415 FrY. ME, OFr, *jolif, joli* 'gay, lively'.

**Jollyboy:** Roger *Joliboye* 1308 PN Ess 326; Robert *Goleboye* 1317 AssK. 'Gay, vigorous child', OFr *jolif*, ME *boye*. cf. Ingelram *Jolifaunt* 1349 FFY with the same meaning; John *Jolifion* 1377 FFC 'gay John'; John *Jolyman* 1379 PTY 'gay man'; *Jolyrobin* 1332 SRCu 'gay Robin'.

**Jolyon:** William *Jolyan* 1327 SRSf. Probably for *Jolly-jan*, i.e. *John*.

**Jolson:** v. JOEL

**Jone(s):** v. JOHN

**Jonson:** v. JOHNSON

**Jope, Jopp:** v. JOB

**Joplin, Jopling:** v. JOBLIN

**Jopson:** v. JOBSON

**Jordan, Jordain, Jorden, Jordens, Jordin, Jordon, Jourdain, Jourdan, Jourdon:** *Jordanus* presbiter 1121–48 Bury (Sf); *Jurdanus* de Brakenberge 12th DC (L); Robert *Jurdan* a1182 Seals (Y); John *Jorden* 1202 FFC; Walter *Jourdan* 1327 SRSx; William *Jurdain* 1332 SRSt. The name of the River Jordan, used as a christian name by returning crusaders who brought back with them Jordan water for the baptism of their children. The name became very common. *v.* JURD, JUDD.

**Jorey:** v. JORY

**Jort:** Anschetill *de Iorz* c1110 Winton (Ha); Reginald *Jort* 1328 IpmNt. From Jort (Calvados).

**Jory, Jorey:** Ralph *Jory* 1221 Cur (Lei); William *Jory* 1275 RH (W); John *Jory* 1325 FFK. A diminutive of *Jore*, the northern French form of *George*.

**Joscelyn, Jos(e)land, Joselin:** v. JOCELYN

**Jose:** v. GOSS

**Joseph, Josephs:** *Josephus* 1086 DB; *Joseph* 1141–9 Holme (Nf), 1187 DC (L); Umfridus *filius Josep* 1205 Cur (Herts); Henry, William *Joseph* 1191 P (Ha), 1205 Cur (Sf). Hebrew *Joseph* 'May Jehovah add'. *v.* JESSOP.

**Josephson:** John *Josepsone* 1332 SRCu. 'Son of *Joseph*.'

**Joskin:** Walter *Josekyn* 1285 AssEss; Richard *Josekyn* 1332 SRSr; William *Iosekyn* 1360 CarshCt (Sr). *Jos-kin*, a diminutive of *Joss*. *v.* JOYCE.

**Josland, Joslen, Joslin, Josling, Josolyne:** v. JOCELYN

**Joss(e):** v. GOSS, JOYCE

**Josselyn:** v. JOYCELYN

**Jossett:** v. GOSSET

**Jotcham:** v. JUDSON

**Joubert:** v. JOBERT

**Jouce:** Robert *Jouce* 1253–4 ForNth; Richard *Jouce* 1327 SRLei; Mr *Juce* 1662–4 HTDo. OFr *jus*, ME *juce, jouce* 'a liquid extract obtained by boiling herbs'. Perhaps a nickname for an apothecary.

**Joule(s):** v. JOEL

**Jouning:** v. JENNINGS

**Jourdain, Jourdan, Jourdon:** v. JORDAN

**Jowers, Juer, Jewers:** Adam *Jour* 1293 AssW; William *Jour* 1327 SRC; Robert *Jower*, William *Jowyr* 1524 SRSf. OFr *jour* 'day'. Perhaps metonymic for a journeyman 'one who has served his apprenticeship in a craft and works for a master'.

**Jowett, Jowitt, Jewett, Jewitt, Juett:** *Juetta* Ric 1, 1201, 1214 Cur (L, Lei, Nf); *Juheta* Vidua 1200 Cur (L); *Joetta* 1219 AssY; *Juwete* 1227 AssSt; *Juhota* 1329 AD iv (Wa); William *Juet* 1279 RH (Hu); Robert *Iuwet* 1297 MinAcctCo; William *Jouet* 1299 AssSt; Ludowycus *Gouet* 13th WhC (La); Richard *Jouot* 1300 AssSt; Roger *Guet* 1317 AssK; Robert *Jowet* 1379 PTY; Lawrence *Jewet* 1458 FFHu; Thomas *Jewitt* 1488 GildY. *Jou-et, Jou-ot*, diminutives of *Juwe, Jowe*, from *Jull*. *Juetta*, which was common, was a pet-name of *Juliana*, of which *Jull* was a short form, as *Jell* and *Jill* were of the common pronunciations *Jelian* and *Jillian*. *Juetta* de Locton' 1212 Cur (L), is twice called *Juliana*. *Jull* was also a short form of *Julian*. *v.* JULL, GILLIAN.

**Jowle:** v. JOEL

**Jowling:** v. JOLIN

**Jowsey:** v. JOYCE

**Joy, Joye, Joyes:** Manser *filius Joie* 1186 P (L); *Joia* 1195 FFEss, 1221 AssWa; Robert *Ioie* c1155 DC (L); Lefwin *Joie* 1166 P (Nf); Ricardo *Gaudio* (abl.) c1176 Bart (K). Either from *Joia*, a fairly frequent woman's name, or from the common noun *joy*. *Joie* was also masculine. cf. *Joie*, husband (*vir*) of Joan de Lalleford 1219 Cur (O).

**Joyce, Joice, Joisce, Joss, Josse, Joicey, Joysey, Jowsey:** (i) *Josce, Iocius* c1140–50 DC (L); *Gocius* de Huptuna Hy 2 NthCh (Nth); Isaac *filius Joscei* 1208 Cur (Mx); Johannes *filius Jocey* 1275 RH (Sf); *Josce* or *Joice* 1313, 1320 LLB D, E; Robert *Joysy* 1533 RochW; Swithin *Joyce* 1568 FFHu; William *Joyse* 1574 ib. Breton *Iodoc*, the name of a saint, son of Judicael, who had a hermitage at the modern St Josse-sur-Mer (ODCN). Earlier examples of the surname (*Joce*) cannot be distinguished from those of GOSS. Two farms in Essex, Joyce's Fm; derive respectively from Robert and John *Joce* (1353, 1398 PN Ess 303, 309). (ii) *Jocea* 1199 CurR; *Juicia* 1211, 1212 Cur (Do, Nt); *Jocosa* de Huntynfeld 1346 FA (Sf). A feminine form of *Joceus, Jodoc*. (iii) Occasionally the surname may derive from Jort (Calvados). cf. Burton Joyce (Notts), from Geoffrey *de Jorz* 1234 (*Jorce* 1327, *Joce* 1433, *Joys* 1535 PN Nt 157).

**Joyful:** Richard *Joyful* 1248 *AssHa*; Robert *Joyfull* 1486 FrY. ME *joiful* 'gay'. cf. Peter *Joycors* 1226 Cur (Nf/Sf) 'joyful person'.

**Joyner:** *v.* JOINER

**Joynes:** *v.* JOHN

**Joynson:** *v.* JOHNSON

**Joysey:** *v.* JOYCE

**Jubb:** *v.* JOB

**Jubber:** *v.* JOBAR

**Jubert:** *v.* JOBERT

**Juckes:** *v.* JUKES

**Judd, Jutte:** *Judde* Rampe 1246 AssLa; *Judde* Clubbe 1260 AssCh; *Judde* de Halifax 1309 Wak (Y); Hugo, Henry, John *Judde* 1204 P (He), 1279 RH (C, O); Alan, John *Jutte* 1260 AssC, 1279 RH (C); Thomas *Joudde* 1297 MinAcctCo; Richard *Joddes* 1327 SRWo. *Judd, Jutt,* pet-names for *Jordan,* from *Jurd,* with the common interchange of voiced and voiceless final consonants.

**Jude:** Hugo *filius Jude* 1193 P (L); Herueus, John *Judas* 1191 P (C), 1327 SRWo; William, Reyne *Jude* 1211 Cur (D), 1279 RH (O); Simon *Jude, Judde* 1327, 1332 SRSx. *Judas* is the hellenized form of Hebrew *Judah,* the name of a son of Jacob, of which *Jude* is an abbreviation. There is little evidence for its use in medieval England, but it was occasionally used, either from the name of the Apostle Jude, sometimes called Judas, or through the popularity of the story of Judas Maccabaeus. *Jude* appears to be also a variant of *Judd.* cf. Alicia *Judedoghter* 1379 PTY.

**Juden:** Adam *Judden* 1327 SRWo; Nicholas *Jurdan,* Walter, William *Judden* 1332 SRWa. Either an assimilated form of *Jurdan* (Jordan), or *Judd-en. v.* GEFFEN.

**Judge, Judges:** Adam *le Jugge* 1309 MEOT (Wo); Thomas *Judges* 1524 SRSf. OFr *juge* 'judge'.

**Judgement, Judgment:** Hagena *Jugement* 1130 P (Nf); William *Iugement* 1230–3 WoCh; Richard *Jugement* 1332 SRDo. 'Judgement', ME *jugement.* Perhaps a nickname for a judge.

**Judkins:** *v.* JUGGINS

**Judson, Jutson, Jutsum, Jutsums, Justum, Justham, Jotcham, Jetson, Jesson:** John *Judson* 1324 Wak (Y); Henry *Juddessone* 1370 AD vi (Ch); John *Jutsam* 1581 DWills; Prudence *Jutsham* 1611 ib.; Agnes *Jutson* 1635 ib.; John *Jutsum* 1725 ib.; Barbara *Jetsome* 1735 ib.; John *Jesson* ib. 'Son of *Judd* or *Jutt.*'

**Juell:** *v.* JOEL

**Juer:** *v.* JOWERS

**Juett:** *v.* JOWETT

**Juggins, Jeggons, Jiggen, Jiggens, Jiggins:** *Jecun* de Sutton 1190 P (Sa); *Jukin* le Walshe 1323 AssSt; William *Jokin* 1275 RH (Sf); John *Jokyn* 1296 SRSx, 1327 SRSf; John *Joukyn* 1303 AD v (W); Thomas *Jukyn* 1327 SRSx; Robert *Gygoun* 1377 AssEss; John *Jowkyn* 1379 PTY; John *Jeggon* 1590 Bardsley (Ess); Thomas *Juggins* 1650 ib.; William *Jigins* 1662 *HTEss*. Bardsley derives *Juggins* from *Judkin,* a diminutive of *Jordan,* his earliest example being Thomas *Judkins* (1648). This is a formation one would have expected to be common and early, but in the light of the above forms *Judkin* is probably from *Jukin,* with intrusive *d. Jek-un, Juk-in, Jok-in* are diminutives of *Jok* or *Juk,* a short form of Breton *Judicael,* with its variants *Juk-, Jok-, Jek-, Gik-. v.* JEKYLL, JUKES. cf. Thomas *Juket,* John *Joket* 1296, 1332 SRSx. *Jowkyn* suggests also a derivation direct from *Jull.* cf. Robert *Julkynesone* 1332 SRSx. cf. JOWETT.

**Juggler:** William *le Gugelour* 1250 AssSt; Lucy *Jugeler* 1260 AssC. OFr *jo(u)glere, jougelour* 'juggler'. cf. TREDGETT.

**Jugson:** *v.* JUXON

**Jukes, Juckes, Jewkes, Jugg:** Adam *Jock* 1279 RH (O); Robert *Jokke* 1327 SRSo; John *Iucke, Iukkes* 1360 Eynsham (O); John *Jokes* 1381 SRSt. A short form of *Jukel, Jokel,* from Breton *Judicael. v.* JEKYLL.

**Julian, Julians, Julien, Jullens, Jullion, Jullings:** *Julian* de Horbelinghe Hy 2 Gilb (L); *Juliana* de Haketoren 1185–7 DC (L); Hugo *filius Juliani* 1212 Cur (Y); Gunnilda *filia Juliane* 1211 Cur (K); Walter *Julien* 1200 P (L); John *Juliane* 1275 SRWo; Henry *Julian* 1327 SRSf; John *Julion* 1327 SRC; John *Julyoun* 1346 FA (Sf). In ME, *Julian* was both masculine and feminine and both are represented in the surnames. Lat *Julianus,* from *Julius,* the name of a Roman *gens,* and *Juliana,* its feminine, were both names of saints and both names were popular, the latter particularly as *Gillian.*

**Jull, Joll, Jolles:** Tomas *filius Golle* 1196 P (Y); *Golle* 1203 P (Ess), 1219 AssY (f); *Jowe* 1227 AssBk, 1327 SRDb, SRSt; Richard *Juwe* 1288 Rams (C); John *Jowe* 1297 MinAcctCo; Robert *Julle* 1317 AssK, 1327 SRSx; William *Jolle* 1327 SRSx; William *Goll* 1421 FrY. A short form of *Julian* or *Juliana.* cf. GILL from GILLIAN.

**Jullings:** *v.* JULIAN

**Jumble:** Maria *iumelle* 1186 DC (Lei); John *Jombel* 1327 SRSa. Fr *jumelles* 'twin'.

**June:** *v.* JEUNE

**Junifer, Juniper:** *v.* JENNIFER

**Junkin:** *v.* JENKIN

**Junkinson, Junkison:** *v.* JENKINSON

**Jupe, Jupp:** *v.* JOB

**Jurd:** *Jurdy* 1275 RH (Sf); Adam *Jorde* 1210 P (Nf); John, Robert *Jurd(e)* 1209 P (Hu), 1275 SRWo. A pet-form of *Jordan.*

**Jury:** *v.* JEWRY

**Juson:** *v.* JEWSON

**Just:** Gilebertus *filius Juste* 1203 P (L); *Justus* de Thurlestun(e), Martin *Justuse* of Thurleston Hy 3 AD ii, iii (Sf); Martin, Roger *Justus* of Thurlestone (brothers) 1292 AD ii (Sf). A rare name. The only evidence noted for a meaning 'the just' is Thomas *le Guste* 1327 SRSo. *Justus,* Lat *justus* 'just', the name of a 4th-century bishop of Lyons, has given rise to the French surnames *Just, Juste* and *Jux* and the derivatives *Juteau, Jutot* (Dauzat). It was rare in England but certainly gave rise to a hereditary surname near Ipswich in the 13th century.

**Juster, Jewster, Joester:** John *le justur* 1230 P (Sx); Robert *le Gustur* 1255 Ipm (So); John *Justere* 1279 RH (O); Walter *Justour* 1327 SRSx. AFr *justour,* OFr *justeor, justeur* 'jouster' (c1400 MED). To give young knights practice in deeds of arms and their elders excitement, tournaments or jousts were held. They sprang up in France in the 12th century and were popular with the Anglo-Norman knights. 'These early tournaments were very rough affairs, in every sense, quite unlike the chivalrous contests of later days; the rival parties fought in groups, and it was considered not only fair but commendable to hold off until you saw some of your adversaries getting tired and then to join in the attack on them; the object was not to break a lance in the most approved style, but frankly to disable as many opponents as possible for the sake of obtaining their horses, arms, and ransoms.' William the Marshal, later Earl of Pembroke and regent of England during the minority of Henry III, the most brilliant jouster of his day, 'made a very good income out of his tournaments' (EngLife 202).

**Justham:** v. JUDSON

**Justice, Jestice:** William *la Justis* 12th AD iii (Sf); Thomas *Justic'* 1202 AssL; Robert *la justice* 1220 Bart (Lo); Peter, Roger *le Justice* 1255 RH (Sa). OFr *justise, justice* had various senses, 'uprightness, equity, vindication of right, court of justice, judge', etc., of which the first to be adopted in England was 'the exercise of authority in vindication of right by assigning reward or punishment'. As early as c1172 it was used of judicial officers or judges (*Justices et baruns* NED) and that is often its sense as a surname. cf. William *Justyse* then constable 1253 Lewes (Nf). Sometimes it may be a shortening of OFr *justiceor* (c1330 NED): John, Tiphina *le Justiser* 1322 AD vi (D). The variation between *le* and *la*, though not to be taken too seriously, may point to the use of the abstract *justice*. Weekley (*Surnames* 221n.) points out that in one of the Chester plays the speakers included 'Veritas, Misericordia, Justitia, and Pax' and that here we have a plausible origin of the names *Verity, Mercy* or *Marcy, Justice* and *Peace*.

**Justin, Justins:** *Justinus* Clericus 1175–80 Holme (Nf); *Justina* 1221 AssWo; *Justinus* filius Wakerild 1229 Pat (Sx); Ralph *Justyn* 1327 SRSf; John *Justyn* 1327 SRSx. *Justinus*, a derivative of Lat *justus* 'just', the name of two Byzantine emperors and of Justin Martyr (d. 163 A.D.), has given rise to the French surnames *Justin* and *Jutin* (Dauzat). The association of the name with *clericus* and the existence of *Justina*, which must be associated with the 4th-century martyr St *Justina*, patron saint of Padua, previously not noted in England before the 18th century (ODCN), suggest that *Justinus* was used occasionally in this country. But *Justinus* is also certainly a latinization of ON *Iósteinn*, which is found in DB in 1066 as *Justen, Justan* and *Justinus* (PNDB) and this would certainly be at home among the Scandinavian names of Norfolk and Suffolk.

**Justum, Jutson, Jutsum:** v. JUDSON

**Jutte:** v. JUDD

**Juvenal:** *Juvenal* 1203 Cur (Nth); *Iuvenal* 1204 Cur (Herts); William *Juvenal* 1222 Cur (Nth); Thomas *Juvenal* 1310 LLB D. Lat *Iuvenalis*.

**Juxon, Jugson:** Richard *Juxon* 1573 SxAS 40; John *Juxon* 1674 HTSf. 'Son of *Jukes*'. v. JUKES.

# K

**Kadwell, Kadwill:** *v.* CADWALL

**Kaighan, Kaighin:** A Manx name, contracted from *Mac Eachain* 'son of *Eachan*' 'horse-lord': *MacCaighen* 1422, *Kaighin* 1611 Moore. cf. MACEACHAN, MCGACHAN, common in Galloway.

**Kain, Kaine:** *v.* CAIN

**Kale:** Philip *de Kale* Hy 2 Gilb; William *de Kale* 1208 Cur (L). From East, West Keal (L), *Cale* DB.

**Kaley:** *v.* CALEY

**Kalker:** *v.* CHALKER

**Kalton:** *v.* CALTON

**Kane:** *v.* CAIN, CANE

**Kaneen:** A purely Manx name, contracted from *Mac Cianain* 'son of *Cianan*': *Kynyne* 1422, *Keneen* 1666, *Kaneen* 1740 Moore.

**Kanter:** *v.* CANTER

**Kaplan, Kaplin:** *v.* CHAPLAIN

**Kapper:** *v.* CAPPER

**Karck, Kark:** *v.* CARKER

**Karl(e):** *v.* CARL

**Karn, Karne:** *v.* CARN

**Karslake:** *v.* CARSLAKE

**Kate:** Elizabeth *Kate* 1662–4 HTDo. Probably a pet-form of *Catharine*, but perhaps ON *Káti*, ODa *Kati*, OSw *Kate*.

**Kater:** *v.* CATER

**Katerin:** *Katerina filia Johannis* 1208 Cur (L); *Caterina* 1214 Cur (Hu); Robert *Katerine* 1286 ForC; Thomas *Katherin* 1308 AssNf; William *Keteryn* 1398 TestEbor. Lat *Katerina* from the Greek. A popular name in the Middle Ages due to the legend of the virgin martyr St *Katherine* of Alexandria, d. 307. The *h* was introduced into the modern form in the 16th century.

**Kates:** *v.* CATES

**Katin:** Richard, Walter *Catin* 1177, 1200 P (Wa). A diminutive of *Cat*, a short form of *Catharine, Cat-in.*

**Katte:** *v.* CATT

**Kave:** *v.* CAVE

**Kay, Kaye, Kayes, Keay, Keays, Keeys, Key, Keye, Keyes, Keys, Keyse:** (i) Britius *filius Kay* 1199 P (Nth). OW *Cai*, MW *Kei*, from Lat *Caius*. Not a common name but probably used by the Bretons as well as in Wales and the Welsh border counties. It is the origin of the Cornish *Key*. (ii) Cecilia *de Kay* 1199 P (Gl); John *del Cay* 1207 P (Lo); William *atte Keye* 1372 LoPleas. OFr *kay, cay*, ME *kay(e), key(e), keay* 'quay' (1306 MED), from residence near or employment at a wharf or quay. (iii) Richard *Ka*, Adam *Kay* 1219 AssY; Adecok *Kay* 1246 AssLa; Alicia *Ka, Kay* 1297 MinAcctCo (Y); Donald *Ka* 1399 Black; Thomas *Kay* 1552 ib.; Thomas *Keay* 1758 FrY. ME *ka, kae, kay*, from ON *ká* 'jackdaw'. In the north and Scotland OE, Scand *ā* remained in ME, often spelled *ai, ay*. NED has *Cais* 'jackdaws' c1450. In FrY between 1355 and 1461 the surname occurs as *Caa* (1461), *Kaa* (1433), *Cay* and *Kay* (thrice each). It is difficult to believe, however, that *kā* had become *kay* by 1219. Whilst most of the Yorkshire, Lancashire and Lowland *Kay(e)s* are nicknames from the bird, there is probably some other source. In Lancashire and Cheshire we find a ME *kei*, Da dialect *kei* 'left' (hand or foot) about 1300 and this survived in these dialects until the 19th century. As Cotgrave uses the word in 1611 (*Gauchier* left-handed, key-fisted) it was probably more widely distributed and we have a nickname 'left-handed, clumsy'. The surname may also be metonymic for *key-er* 'a maker of keys' (cf. KEAR) or from office, a key-bearer, key-keeper. (iv) Geoffrey *Cai* 1197 P (Nf); William, Roger *Keys* 1275 SRWo, 1444 AD i (Berks); William *le Kay* 1296 SRSx; Benedict *Cay* 1297 FFSf; Stephen, Edmund *Kay* 1327 SRSf; Hugh *Kegh*, William *le Keye* ib.; Geoffrey *Kay* 1331 AssSt. In these counties *kā* became ME *co(o)*, now COE. A Scandinavian nickname from *kei* 'left-handed' would be at home in Norfolk, Suffolk, Derbyshire and Staffordshire. In Sussex *le Kay* is probably for *de Kay* 'at the quay'. 'Key-bearer' is possible in all, and *Cai* in Worcestershire and possibly in Norfolk and Suffolk. The Manx *Kay* is for *Mackay*.

**Kayfoot, Cafoot:** Thomas *Kafot* 1185 Templars (L); John *Cayfoot* 1275 SRWo; Matilda *Cafot'* 1319 SRLo. 'Jackdaw foot', ME *cā*, OE *fōt*.

**Kayley:** *v.* CALEY

**Kayne:** *v.* CAIN

**Kayser:** *v.* CAYZER

**Keable:** *v.* KEEBLE

**Keach, Keattch:** *v.* KEECH

**Keal, Keale, Keall:** William *de Kele* Hy 2 Gilb; Alan *de Keles* 1218 AssL; Gilbert *de Cheiles* 1256 Gilb. From East, West Keal (L). *v.* also KALE.

**Kear, Keer, Care:** Adam *filius Cheigher* 1178 P (Nb); Robert *le Keyere* 1275 RH (K); William *le Keer* 1303 MESO (Lei); Richard *le Kayer* 1287 LLB A; Richard *Kere* 1322 FFEss. OE *\*cǣgere*, from OE *cǣg* 'key', a maker of keys. The first example probably means 'Son of the key-smith'.

**Kearsey:** *v.* KERSEY

**Kearsley, Keasley, Kersley, Carslaw, Carsley, Caselaw, Caseley, Casley, de Casley, Cassley:** Simon *de Caresle* 1206 Cur (Wo). From Keresley (Warwicks), pronounced *Carsley*, or Kearsley (Lancs, Northumb).

**Kearton:** From Kearton in Grinton (NRY).

**Keat, Keate, Keates, Keats, Keet, Keit, Keyte, Kett, Ketts, Kite, Kyte:** (i) Ailnoð, Richard *Kete* 1166 P (Nf), 1221 *ElyA* (Sf); Robert *Chet* 1183 Boldon (Du); Richard *Kyte* 1243 AssSo; William *Ket* 1275 RH (Nf); Peter *le Kyte* 1327 SRSx; John *Kette* 1327 SRC. OE *cȳta*, ME *kete, kyte* 'kite', a rapacious person. (ii) Ralph *atte Kete* 1292 PN K 493. This survives in Kite Fm which Löfvenberg has shown (StudNP 17) derives from OE *cȳte* (f), *\*cyte* (m) 'hut', some kind of shed or outhouse for cattle or sheep. The surname probably denotes a herdsman.

**Keatley:** *v.* KEIGHLEY

**Keaton:** *v.* KEETON

**Keattch:** *v.* KEECH

**Keay(s):** *v.* KAY

**Kebbell, Kebell, Keble:** *v.* KEEBLE

**Keck, Keek:** Walter *Kek* 1316 AssNf; John *Kek* 1352 ColchCt; John *Keek* 1370 IpmGl. ON *Keikr, Kekkja.*

**Keddel, Keddle:** Gilbert *de Keddel* 1258 IpmY. From Kiddal Hall in Barwick in Elmet (WRY).

**Kedge, Ketch:** Alured *Keg'* 1177 P (Nf); Alexander *Kech* 1221 *ElyA* (Nf); William *Kigge* Hy 3 Gilb (L); Richard *le Keg'* 1332 SRSx. East Anglian dialectal *kedge* 'brisk, lively'. cf. '*Kygge*, or ioly' c1440 PromptParv. Perhaps a palatalized form of Da, Norw *kiæk*, Sw *käck* 'bold, brisk'.

**Kee, Kees:** John *Kese* 1379 PTY; John *Kee* 1481, Robert *Key* 1482 FrY. ME *ke* 'jackdaw'.

**Keeble, Keable, Kebbell, Kebell, Keble, Kibbel, Kibble, Kibel:** Æluric *Chebbel* c1095 Bury (Sf); William, Salomon *Kebbel* 1214 Cur (K), 1263 FFEss; Thomas *Kibel* 1275 RH (L); Henry *Kybbel* 1327 SRC; John *Kebyll*, Richard *Kebull* 1524 SRSf. OE *\*cybbel*, perhaps the source of *kibble* 'cudgell' (1397 MED). A maker or seller of cudgels, or, perhaps, originally, one stout and heavy as a cudgel.

**Keech, Keetch, Keach, Keattch, Keitch:** Reginald, Hugo *Keche* 1206, 1219 Cur (C, Sr). ME *keech* 'a lump of congealed fat; the fat of a slaughtered animal rolled up into a lump', used in the 16th century for a butcher: 'Did not goodwife Keech the Butchers wife come in then?' (Henry IV); 'I wonder, That such a Keech can with his very bulke Take vp the Rayes o' th' beneficiall Sun, And keepe it from the Earth' (Henry VIII), where the reference is to Cardinal Wolsey, a butcher's son.

**Keed:** Henry, Simon *Kede* 1275 RH (Sf), 1327 SRC; William *Keed* 1524 SRSf. ME *kide, kede* 'kid'.

**Keek:** *v.* KECK

**Keel, Keele, Kell:** Richard *Kele* 1246 FFY; John *de Keel* 1332 SRSt; Robert *Keell* 1481 IpmNt. From Keele (St), or East, West Keal (L).

**Keelby:** *v.* KELBY

**Keele:** *v.* KEEL

**Keeling:** Ælfuine *Celing* c1095 Bury; Henry *Keling* c1170–80 RegAntiquiss; Robert *Kelyng'* 1277 *Ely* (Sf); John *Kelyng* 1372 ColchCt. A nickname from ME *keling* 'young codfish'. Sometimes from Keeling (Nf).

**Keell:** *v.* KEEL

**Keemish:** *v.* CAMMIS

**Keen, Keene, Keenes, Keens:** Ricardus *filius Kene* 1188 P (D); Alexander *filius Kene* 1202 AssL; Adam, Hugo *Kene* 1207 Cur (Sf), 1221 AssWo; Richard *le Kene* 1297 MinAcctCo (O). *Kene* is a short form of OE names in *Cēn-* or *Cyne-*, such as survive in *Kenward, Kenway, Kerrich*, etc. *le Kene* is OE *cēne*, ME *kene* 'wise, brave, proud'.

**Keenleyside, Keenlyside, Kindleyside, Keenliside:** From Keenleyside (Nb).

**Keep:** Thomas *ate Kepe* 1327 SRSx; Robert *de Kepe* 1332 SRCu. One employed at a keep or castle (a1586 NED). Perhaps 'jailer'.

**Keepax:** *v.* KIPPAX

**Keeper:** William *le Kepere* 1279 RH (Hu); Simon *Kepere* 1327 SRSx. Identical in meaning with KEEP.

**Keer:** *v.* KEAR

**Kees:** *v.* KEE

**Keet:** *v.* KEAT

**Keetch:** *v.* KEECH

**Keetley:** *v.* KEIGHLEY

**Keeton, Keaton:** Robert *de Keton* 1362 IpmNt; John *de Keton* 1379 PTY; Henry *Keton* 1423 AssLo. From Ketton (Du, R), or Keaton in Ermington (D).

**Keevil, Keevill, Kevill, Kivill:** Hugh *de Chiuilli* c1110 Winton (Ha); William *de Keuill'* 1203 P (Do); Elias *de Kivili* 1205 Cur (K). From Keevil (W).

**Keys:** *v.* KAY

**Kegan, Keggin, Keggins:** A Manx contraction of *Mac Taidhgin* 'son of little *Tadhg*', the poet (Moore).

**Kehoe:** *v.* KEYHO

**Keig, Kegg:** Muircheartach *Mac Taidhg* 1159, *Mac Keg* 1511, *Kegg* 1630, *Keige* 1653 Moore. 'Son of *Tadg*', the poet, philosopher. Irish *Teague, MacTague.*

**Keighley, Keighly, Keightley, Keitley, Keatley, Keetley, Kightly, Kitlee, Kitley:** Henry *de Kythelay* 14th Calv (Y). From Keighley (WRYorks).

**Keit:** *v.* KEAT

**Keitch:** *v.* KEECH

**Keitley:** *v.* KEIGHLEY

**Kekwick, Kekewich:** Walter *de Kekingwich* 1209 P (R); Richard *de Kekwike* 1332 SRLa; John *Kekewiche* 1526 FFEss. From Kekwick (Ch).

**Kelby, Keelby:** Peter *de Kelesby* 1219 AssL; John *de Keleby* 1305 RegAntiquiss; Walter *de Keleby* 1373–5 AssL. From Keelby (L).

**Kelcey:** *v.* KELSEY

**Keld:** William *Attekelde* 1296 IpmY; Edmund *Attekeld* 1327 SRY; George *Keld* 1524 SRSf. 'Dweller by the spring', ON *kelda.*

**Kelham, Kellam:** Peter *de Kellum* 1204 Pl (Nt); William *Kelom* 1327 SRLei; John *Kellom* 1447 IpmNt. From Kelham (Notts).

**Kelk:** Robert *Kelk* 1296 IpmY; Roger *de Kelk'* 1411 KB (Y); Stephen *Kelke* 1470 Paston. From Great, Little Kelk (ERY).

**Kell:** *v.* KEEL

**Kell, Kells:** Reginaldus *filius Chelle* 1219 AssL; *Chel* filius Mabillæ c1250 Rams (C); Ansfredus, Isabella *Kelle* 1176 P (Ha), 1311 RamsCt (Hu). ON *\*Kel*, OSw *Kæl*, shortened from *Ketill*. *v.* KETTLE.

**Kellaway:** v. CALLAWAY

**Keller:** (i) Michael *Keller* 1296 IpmY; Robert *le Kellere* 1327 SRLei; John *Keller* 1379 PTY. Metonymic for *kellerer* 'a maker of caps, cauls, &c'. (ii) Roger *le queller* 1203 Pleas (Sa); Geoffrey *le Quellere* 1249 AssW. OE *cwellere* 'killer, executioner'.

**Kellett, Kellet, Kellitt:** Adam, William *de Kellet* 1194 P (Cu), 1279 FFY; Mariota *Kelitt* 1327 SRC. From Kellet (Lancs), or Kelleth (Westmorland).

**Kellock:** (i) Robert *Chelloc* 1166 P (Nf); Alexander *Kellok* 1275 RH (Nf); William *Kellok* 1395 AssL. Either ON *Ketillaug*, or, more probably, ON *Kiallakr* from OIr *Cellach*. (ii) Robert *de Kellok* 1343, Robert *Killocke* 1662 Black. From Keiloch in Braemar (Aberdeen), or from the lands of Killoch (Ayr).

**Kellog, Kellogg:** Geoffrey *Kyllehog* 1277 FFEss; Walter *Kelehoog* 1369 PN Ess 456; John Ryche alias *Kelhoge* 1541 FFEss. 'Kill hog', a name for a butcher, cf. Gilbert *Killebole* 1327 SRLei; Richard *Kilfole* 1442 AssLa.

**Kellow, Kellough:** William *de Kellawe* 1256 AssNb; Thomas *de Kellawe* 1339 FFY. From Kelloe (Du).

**Kelly, Kelley, Kellie:** In Ireland for *O'Kelly*, Ir *Ó Ceallaigh* 'descendant of *Ceallach*' (war). In Galloway and the Isle of Man, where it is as common as in Ireland, it is for *MacKelly: McKelly* 1429, *Kellye* 1601 Moore. It may also be local in origin, in Scotland from Kelly near Arbroath or Kellie in Fife: John *de Kelly* 1373 Black; in England, from Kelly (Devon), from Cornish *celli* 'wood, grove': Warin *de Kelly* 1194 P (D).

**Kelman:** William *kelman* 1257 MEOT (Y); John *Keleman* 1328 MESO (L). Probably 'a maker of kells or calls', from ME *kelle*. cf. Symon *le keller* 1322 FrY, and CALLER. It might also be from ME *kele* 'ship', a worker on keels, flat-bottomed vessels, a bargeman.

**Kelner:** v. KILNER

**Kelsall, Kelshaw, Kilshall, Kilshaw:** Richard *de Kelsale* 1260 AssCh; John *de Kelshulle* 1340 AssLo; John *Kelesall* 1379 PTY. From Kelsall (Ches), Kelshall (Herts), or Kelsale (Suffolk).

**Kelsey, Kelcey, Kellsey:** Richard *de Kelleseia* 1179–84 RegAntiquiss; William *de Kelesey* 1284 FFY; Robert *Kelsey* 1394 AssL. From North, South Kelsey (L).

**Kelway:** v. CALLAWAY

**Kemball:** v. KEMBLE

**Kember:** Roger *le Kembar'* 1327 SRSx; Peter *le Kembere* 1333 MESO (Ha). A derivative of OE *cemban* 'to comb', a comber of wool or flax. cf. KEMPSTER.

**Kembery, Kembry:** v. CAMBRAY

**Kemble, Kemball, Kimball, Kimbell, Kimble, Kimmel, Kemple:** (i) Turbet *filius Chembel* 1130 P (W); Richard *Cembel* 1185 P (Hu); Roger *Cumbel, Chimbel, Kymbel, Kimbel* 1191–3 P (Wa); John, Richard *Kenebelle* 1327 SRSf; Simon, William *Kymbel* ib. OW *Cynbel*, from *cyn* 'chief' and *bel* 'war'. cf. *Cymbeline (Cunobelinus)*. (ii) Robert *Kinebald* 1215–21 AD iv (So); Ralph *Kenebold* 13th AD ii (Wt); William *Kembald* 1302 SRSf; William *Kenebold*, Nicholas *Kembold* 1327 SRSf; Samuel *Kembull* 1621 Buxhall (Sf); Susan *Kemball* 1638 ib. OE *Cynebeald*

'family-, kin-bold', a name which does not seem to be recorded after the 9th century but which must have continued in use. (iii) Hugo *de Kenebell'* 1196 Cur (Bk); Roger *de Kinebelle* 1255 RH (Bk); Thomas *de Kymble* 1327 SRC. From Kimble (Bucks).

**Kemm, Kemme, Kimm:** *Kima* 1221 Cur (L); *Kemma* Scriuener 1311 MESO (Ess); John *Kymme* 1276 RH (L); Thomas *Kemme* 1332 SRWa. OE *Cymme* (f), perhaps a pet-form of OE *Cyneburh* (f).

**Kemmery:** v. CAMBRAY

**Kemp, Kempe:** Eadulf *Cempa* 902 OEByn (W); Edmund *Kempe* c1100 MedEA (Nf); Ralph *le Kemp* 1296 SRSx. OE *cempa* 'warrior', ME *kempe*, also 'athlete, wrestler'. v. CAMP.

**Kempsey, Kimsey:** John *de Kemesie* 1206 P (Sx); John *de Kemeseye* 1207 Rams (Hu); Boidin *de Kames'* 1221 AssWa. From Kempsey (Wo).

**Kempster:** Agnes *Kembestere* 1252 Rams (Hu); Dionisia *le Kemstere* 1317 Oseney (O); Agnes *Kempster* 1327 SRC. The feminine form of *Kembere* (a1400 NED). v. KEMBER.

**Kempston:** Peter *de Kemeston'* 1190 P (Beds); Simon *de Kempston* 1276 RH (Bk); Symond *Kempston* 1426 Paston. From Kempston (Beds, Nf).

**Kempton:** Robert *de Chenipetona* c1132 StCh; Richard *de Kemtone* 1232 Oseney; Richard *de Kemeton* 1248 FFEss. From Kempton (Sa), or Kempton Park (Mx).

**Ken, Kenn:** (i) Robert *le Chien (Chen)* 1183 P (Co); Willelmus *Chen, Canis* 1212, 1219 Fees (Ess); Henry *le Kenne* 1337 SRSx. AFr *ken, chen*, OFr *chien* (Lat *canis*) 'dog'. (ii) Sewulf, John *de Ken* 1170, 1176 P (D, So). From Kenn (Devon, Som).

**Kendal, Kendall, Kendell, Kendle, Kindell:** John *de Kendale* 1332 SRLa. From Kendal (Westmorland).

**Kendrick:** v. KENRICK, KERRICH

**Kenington, Kennington:** Michael *de Kenington'* 1222 Cur (Nf); Richard *de Kenyngton* 1291 FFEss; Adam *de Kenyngton* 1369 FFW. From Kennington (Berks, K, Sr).

**Kenion, Kennion:** v. KENYON

**Kenley, Kenleigh:** William *de Kenleg'* 1255 RH (Sa); Thomas *de Kenleye* 1327 SRSa; John *de Kenlay* 1410 IpmY. From Kenley (Sa).

**Kenmare, Kenmore, Kenmir:** Richard *Kenemer* 1277 Ely (Sf); John *Kenemer*, Thomas *Kenemere* 1296 SRSx. OE *Cynemǣr*, a personal name found in several early place-names, but recorded only in the reign of Edward the Confessor as the name of a moneyer. It occurs also as the name of a 12th-century cowman: terram que fuit *Chinemeri* bouarii, a1183 EngFeud.

**Kennard:** v. KENWARD

**Kennaway:** v. KENWAY

**Kennedy:** Gilbert *mac Kenedi* c1180 Black; Huwe *Kenedi* 1296 CalSc. Ir *Ó Cinnéide* 'ugly head'.

**Kennerley, Kinnerley:** Hugh *de Kenardel'* 1243 AssSo. From Kennerleigh (Devon).

**Kennet, Kennett:** Nicholas *de Kenette* 1175 P (K); Thomas *de Kenete* 1270 IpmW; John *Kennet* 1525 SRSx. From Kennett (C), or East, West Kennett (W).

**Kenning:** Reginald *Kenyng* 1315 AssNf; Nicholas *Kenyng* 1348 IpmW. Probably an unrecorded OE *Cēning*.

**Kennington:** *v.* KENINGTON

**Kennish:** *v.* KINNISH

**Kenrick, Kendrick, Kenwrick, Kenwright:** Hugo *filius Chenwrec* 1160 P (Sa); Ennian *filius Kenewrec* 1161 ib.; *Kenwrec* Walensis 1195 P (Sa); *Kenrig* de Gretewurth' 1219 Cur (Nth); *Kenwrig* ap Madog Duy 1391 Chirk; *Kendrick* Eyton 1602 Bardsley (Ch); *Kenrick* Evans 1613 ib.; Nicholas *Kenwrec* 1327 SRSo; Richard *Kendrick* 1593 Bardsley (Ch). Welsh *Cynwrig*, from *cyn* 'chief' and (*g*)*wr* 'man, hero' plus the suffix of quality. *v.* also KERRICH.

**Kenston, Kenstone:** *Kenestan* de Westweniz 1214 Cur (Nf); Ailmar *Kenston* 1199–1204 Clerkenwell; William *Kenstone* 1292 SRLo; Richard *Kenston* 1327 SRSf. OE *Cynestān*.

**Kent:** Nicholas *de Kent* 1185 Templars (Wa); William *Kent* 1296 SRSx; John *a Kent* 1524 SRSf, 1525 SRSx. 'The man from Kent.'

**Kentell, Kentwell:** Richard *de Kentewelle* c1165 Bury; Gilbert *de Kentewelle* 1207–8 FFK, *de Kentewell'* 1210 Cur (Sf). From Kentwell (Sf).

**Kenting:** William *Kenting* 1198–1212 Bart; William *Kentyng* 1296 SRSx; John *Kentyngg* 1378–80 FFSr. 'The man from Kent.'

**Kentish, Kintish, Cantes:** William *Centeis* c1165 Bury (Sf); Richard (*le*) *Kenteis* 1176 P (Lei); Ralph *Canteis, le Kenteis* 1208 Cur (Sf); Richard *Kentissh* 1332 SRSx. OFr *Kanteis, Kenteis* 'Kentish (man)', influenced by OE *Centisc*.

**Kenton:** Yvo *de Keneton'* 1208 Cur (Sf); Walter *de Kenton* 1275 RH (D); John *de Kenton* 1379 PTY. From Kenton (D, Mx, Nb, Sf).

**Kentwell:** *v.* KENTELL

**Kenward, Kennard:** *Keneward* lingedraper 1198–1212 Bart (Lo); Walterus *filius Kenewardi* 1214 Cur (Wa); *Kynnard* Delabere 1590 AD v (Gl); Agnes, Walter *Kyneward* c1250 Rams (Beds); Nicholas *Kenward* 1274 Wak (Y); Simon *cuneward* 1297 MinAcctCo (Mx). OE *Cēnweard* 'bold guardian' or *Cyneweard* 'royal guardian'.

**Kenway, Kennaway:** *Kenewi* 12th MedEA (Nf), 1279 RH (C); Robertus *filius Kenewi* 1198 Cur (K); William *Chienewe* 1130 P (Ha); Ralph *Kenewi* 1221 *ElyA* (Sf). OE *Cēnwīg* 'bold war' or *Cynewīg* 'royal war'.

**Kenwood:** Rauði *de Kenewode* 1275 RH (D); Adam *de Kenewode* 1297 MinAcctCo (Mx); John *Kenewode* 1359 AssD. From Kenwood in Kenton (D).

**Kenwrick, Kenwright:** *v.* KENRICK, KERRICH

**Kenyon, Kenion, Kennion:** Robert *de Kenien* 1212 Fees (La); Jordan *Kenyan* 1260 AssLa; Nicholas *Kynion* 1288 AssCh. From Kenyon (Lancs).

**Keown:** *v.* MACOWEN

**Ker, Kerr:** *v.* CARR

**Kerb(e)y:** *v.* KIRKBY

**Kerk:** *v.* KIRK

**Kerkham:** *v.* KIRKHAM

**Kerl(e)y:** *v.* KIRKLEY

**Kerman:** *v.* KIRKMAN

**Kermode, Cormode:** *Mac Kermott* 1430, *Mac Cormot* 1511, *Kermod* 1586 Moore. A Manx contraction of *MacDermot*.

**Kerrell, Kirrell, Kirriell, Kriel, Croyle:** Robert *de Cruel* 1086 DB (Sx); Simon *de Crieil* 1170 StGreg (K);

Bertram *de Criel, de Crioille* 1221 StGreg, 1252 *ERO*; William *de Kiriel* 1287 FFHu; John *Kyriel* 1332 SRSr; John *Kyrrell*, Edward *Kerrell* 1583 Musters (Sr). From Criel-sur-Mer (Seine-Inférieure).

**Kerrich, Kerridge, Kerrage, Carriage, Kerrick, Kirrage, Kenrick, Kendrick:** *Kenricus* (2), *Chenricus, Chericus* (identical) 1066 DB (Sf); *Chenric* ib. (Nth); *Cenric, Chenrich* 1095 Bury (Sf); John *Kendrich* 1279 RH (C); John *Kerrych, Kenrich* 1297, 1299 Ipm (Sf); Thomas *Kenrick* 1299 Gardner (Sf); John *Kenrich, Kerrich* 1327 SRSf, *Kerrick* 1330 Gardner (Sf); Robert *Kerysche* 1524 SRSf; Susannah *Carriage* daughter of Thomas *Kerridge* 1632 Bardsley; Samuel *Kerridge* 1671 SfPR. OE *Cynerīc* 'family-ruler', which in Suffolk would become *Kenrich*. There is little evidence for OE *Cēnrīc*. The surname might also have become *Kenwright* and have been refashioned as *Courage*. *Kerridge* might also be local, from Kerridge (Cheshire, Devon).

**Kersaw:** *v.* KERSHAW

**Kersey, Keresey, Kearsey, Kiersey, Carsey:** Ralph *de Karesey* 1279 RH (C); Adam *de Kersey* 1325 FFEss. From Kersey (Suffolk).

**Kershaw, Kersaw:** Adam *de Kyrkeschawe* 1307 Wak (Y); Geoffrey *del Kyrkeshagh* 1390 Bardsley (La). From Kirkshaw in Rochdale (Lancs) or from residence near the 'church-wood'.

**Kerslake:** *v.* CARSLAKE

**Kersley:** *v.* KEARSLEY

**Kerswell, Kerswill:** *v.* CARSWELL

**Kertess:** *v.* CURTIS

**Kerton:** *v.* KIRTON

**Kerwood:** *v.* KIRKWOOD

**Kesby:** *v.* KISBY

**Keslake:** *v.* CARSLAKE

**Kessell:** *v.* KESTEL

**Kestel, Kestell, Kessel, Kessell:** John *de Kestel* 1297 MinAcctCo. From Kestell (Cornwall).

**Keston:** Geoffrey *de Keston* 1279 RH (Hu); William *Keston* 1327 SRC; Richard *Keston* 1642 PrD. From Keston (K), or Keyston (Hu), *Keston* 1255.

**Ketchell, Kitchell:** William, John *Kechel* 1221 AssWo, 1279 RH (Hu). ME *kechel, kichel* (*v.* NED) 'a cake given as almes in the name, or for the sake, of God'. cf. Chaucer's 'A Goddes *kechel*' where some MSS read *kichel*, so called because godfathers and godmothers used commonly to give them to their godchildren when they asked a blessing. The surname probably denotes a maker or seller of these small cakes.

**Ketchen:** *v.* KITCHEN

**Ketcher:** *v.* CATCHER

**Ketelbey, Ketelby, Kettleby:** John *de Ketelbi* c1200–10 RegAntiquiss; John *Ketylby* 1513 FFEss. From Kettleby, Kettleby Thorpe (L), or Ab, Eye Kettleby (Lei).

**Ketley:** Edward *Kettlye* 1642 PrD. From Ketley (Sa).

**Kett, Ketts:** *v.* KEAT

**Kettell:** *v.* KETTLE

**Ketteridge, Kettridge, Kitteridge, Kittredge:** Adam, Roger *Keterych(e)* 1317 FFEss, 1379 ColchCt; Thomas *Kederych* 1524 SRSf. Probably an Anglo-

Scandinavian hybrid *Cytelrīc, a compound of ON *Ketill* and the common OE second theme -*rīc*. cf. KETTLEBURN.

**Kettle, Kettles, Kettless, Kettel, Kettell, Kittel, Kittle:** Grym *Kytel* 972 BCS 1130 (Nth); *Chetel* (Ha), *Chitel* (Co), *Ketel* (Gl, La, Nf, Sf, Y), *Kitel* (Nf, So) 1066 DB; *Ketel* filius Eutret 1212 Cur (Cu); Roger *Chetel* 1180 P (Nth); Edricus *Keteles* c1188 BuryS (Sf); Hulf *Ketel* 12th Seals (Nf); William *Kitel* 1243 AssSo. ON *Ketill* '(sacrificial) cauldron', anglicized as *Cytel*, from which comes *Kittel*.

**Kettleburn:** *Ketelbern* (Wo), *Chetelbern* (L, Nt, Nf) 1066 DB; *Ketelbern* Hy 2 DC (L), 1211 Cur (Y); Roger, Walter *Ketelbern* 1192 P (Sx), 1221 AssWa; William *Ketelbern'*, *Cutelbern* c1248 Bec (Bk); Robert *Ketelbarn* 1324 FrY. ON *Ketilbiǫrn* '(sacrificial) cauldron-bear', late OE *Cytelbearn*, *Ketelbern*.

**Kettleby:** v. KETELBY

**Kettless:** v. KETTLE

**Kettlewell, Kettelwell:** Richard *de Ketelwel* 1212 P (Y); Thomas *de Ketilwelle* 1327 FrY; William *Kettlewell* 1672 HTY. From Kettlewell (WRY).

**Kettridge:** v. KETTERIDGE

**Kevil:** v. KEEVIL

**Kew, Le Keux, Lequeux:** Roger *le Cu* 1196 FFNf; William *Kue* 1203 P (Lei); William *le Keu* 1231 FFC; Hugh *le Kew* 1246 AssLa. OFr *queu, keu, kieu, cu* 'cook', probably one who sold cooked meat, keeper of an eating-house. v. also KEYHO.

**Kewen, Kewin:** A Manx contraction of *Mac Eoin* 'John's son': *McJon* 1417, *M'Kewne* 1504, *Kewyne* 1540 Moore.

**Kewish:** A similar Manx contraction of *Mac Uais* 'the noble's son': *Kewish* 1618 Moore.

**Kewley, Cully:** Hunfridus *de Cuelai* 1086 DB (Nf); Hugh *de Cuilly, de Cully* 1313–14 Bardsley. From Cully-le-Patry (Calvados). The Manx *Kewley* is for COWLEY.

**Kexby:** William *de Kexby* 1374 FFY; Joan *Kexby* 1467 IpmNt. From Kexby (WRY).

**Key, Keye, Keyes, Keys, Keyse:** v. KAY

**Keyho, Keyhoe, Kehoe, Kew:** Robert *de Cahou, de Cayho, de Kaiho* 1195–6 P (Bk); William *de Kehou, de Caihou* 1205 Cur (C), P (K); William *de Kayu* 1230 P (Ess), Wilkin *de Keu* ib. (C). All these men came from Caieu, a lost town in the vicinity of Boulogne-sur-Mer (Pas-de-Calais), recorded as *Cahu, Kaeu, Kaio, Kayhou, Keu* (Fees). Kew (Surrey), recorded only from the 14th century, has not been noted as a surname.

**Keynes:** v. CAINES

**Keyser, Keysor, Keyzor:** v. CAYZER

**Keyte:** v. KEAT

**Kibbe:** William *Kibbe* 1185 P (Wo); John *Kybe* 1327 SRSx. OE *cybbe* 'clumsy, thick-set'.

**Kibbel, Kibble:** v. KEEBLE

**Kichin:** v. KITCHEN

**Kicker:** Sibald *Kiker* 1166 P (Nf); Ralph *le Kicur* 1212–23 Bart; Nicholas *Kyker* 1230 P (So). A derivative of ME *kiken* 'to watch, spy', a watcher.

**Kid, Kidd, Kidde, Kidds, Kyd, Kydd:** Unfrei *cide* Hy 2 DC (L); William, Ralph *Kide* 1181, 1198 P (Sf, Nth). ME *kid(e)* 'kid'. Also, occasionally, a voiced form of

*Kitt*, a pet-name for Christopher: Robert *Kyd* de Dunde 1357 Black. Or metonymic for KIDDER.

**Kiddell, Kiddle:** Simon *Kidel* 1219 Fees (K); William *Kyddall* 1554 FrY. AFr *kidel, kydel*, OFr *quidel, cuidel* 'kiddle', 'a Wicker engine whereby fish is caught' (Cotgrave); a dam, weir or barrier in a river with an opening fitted with nets, etc., for catching fish. One in charge of a fishing-weir. cf. Katharine *Kydelman* 1327 SR (Ess).

**Kidder:** Ailric (*le*) *Chidere* 1190–1 P (Wa); Roger *Kidere* 1233 FFLa; Thomas *le Kidere* 1301 SRY; Richard *le Kedere* 1310 LLB D. Fransson explains this as 'kiddier, hawker, badger', of obscure origin, and compares MDu *keder* 'one who announces or proclaims', first recorded in NED in 1552. The first example is early for a Dutch loan-word and we may have a derivative of ME *kidde* (of unknown origin) 'a faggot or bundle of twigs, brushwood, etc.' (c1350 MED), a woodman, a cutter or seller of faggots. cf. *Kidberers* 1477 MED.

**Kidgell, Kiggel, Kigelman:** Godwin *Kiggel*, Siger *Kigel* 1221 ElyA (Sf, Nf); William *Kegghel* 1327 SRSf; Margery *Kedgell*, John *Kegel* 1524 SRSf. OE *cycgel* 'cudgel'; a maker or seller of cudgels.

**Kidman:** Richard *Kideman* 1221 ElyA (Nf); Alan *Kydeman* 1275 RH (Nf). Man in charge of the kids.

**Kidnet, Kidnett:** Roger *Kidenot* 1180 P (Sa); Reginald *Kidenot* 1206 Cur (Sr); Thomas *Kydnot* 1394 AssL. *Kid-en-ot*, a double diminutive of *Kid*, a pet-form of *Christopher*.

**Kidson:** John, Richard *Kydson* 1522 FrY, 1531 GildY. 'Son of *Kid*' (Christopher). v. KID, KITSON.

**Kiersey:** v. KERSEY

**Kigelman, Kiggel:** v. KIDGELL

**Kightl(e)y:** v. KEIGHLEY

**Kilbride:** John *de Kilbrid* 1202–7 Black. From Kilbride (Lanarks). v. also BRIDSON, McBRIDE.

**Kilburn, Kilborn, Kilbourn:** Richard *de Killeburne* 1284 IpmY; Thomas *de Kilburn* 1305 FrY; John *Kylberne* 1576 SRY. From Kilbourne (Db), or Kilburn (Mx, NRY).

**Kilby, Kilbey, Killby, Kilbuy:** William *de Kilebi* 1202 AssNth; Thomas *de Kylby* 1327 SRLei; Richard *Kylby* 1396 AssWa. From Kilby (Leics).

**Kilduff:** v. DUFF

**Kilham, Killham:** Stephen, Everard *de Killum* c1160–9 YCh, 1219 AssY; William *de Kilham* 1337 FFEss. From Kilham (Northumb, ERYorks), or Kilham in Kirk Neuton (Cumb).

**Kill, Kille:** *Cille* 1066 DB (Y); Godwine *filius Chille* 1187 P (St); Robert *Kille* 1185 Templars (L); Robert *Kyl* 1327 SRSf. Probably ODa *Killi* or *Kille*.

**Killer:** Siward *le Killere* Hy 3 Colch (Ess); Ranulph *le Kyllere* 1327 SREss. 'The killer', ME *killere*. Probably a nickname for a butcher. cf. Gilbert *Killebole* 1327 SRLei 'kill bull'.

**Killick:** v. KILNWICK

**Killigrew:** John *de Kelligreu* 1176 P (Co). From Killigrew (Cornwall).

**Killin, Killing:** Gilbert *de Kellinges* 1177 P (Sf); Adam *de Killing'* 1221 Cur (Y); Robert *Kyllyng* 1466 FFEss. From Kelling (Nf), *Killinge* c970, or Nunkeeling (ERY), *Killing* 1200.

**Killip:** *M'Killip* 1430, *Killop* 1540 Moore. A Manx contraction of *MacPhilip* 'son of *Philip*'.

**Kilmartin:** Ir *Mac Giolla Mhártain* 'son of (St) Martin's servant'.

**Kilner, Kelner:** William *le Kylnere* 1292 MESO (La); Robert *Kilner* 1305 ib. (L). A derivative of OE *cylene* 'kiln', one in charge of a kiln, lime-burner.

**Kilnwick, Killwick, Killick:** Daniel *de Killingwic* 1219 AssY; John *Kyllyk* 1433 AssLo; Robert *Killicke* 1596 Musters (Sr). From Kilnwick, or Kilnwick Percy (ERY).

**Kilpatrick, Killpartrick:** (i) Stevene *de Kilpatric* 1296 Black; Nigel *Kilpatrick* 1302 ib.; Thomas *de Kylpatrik* 1468 ib. From one or other of the Scottish places of this name, e.g. Kilpatrick in Closeburn (Dumfries), East, West Kilpatrick (Dumbarton). (ii) In Ireland for *Mac Giolla Phádraig*, an older form of *Fitzpatrick* (MacLysaght).

**Kilpin:** Richard *de Kilpin* c1190–1207 YCh; John *Kylpine* 1376 FFY; Helen *Kilpyn* 1392 LoCh. From Kilpin (ERY).

**Kilroy:** Ir *Mac Giolla Rua* 'son of *Gilroy*'.

**Kilshall, Kilshaw:** *v.* KELSALL

**Kilton:** William *de Kilton'* 1206 P (Y); Adam *de Kilton'* 1219 AssY. From Kilton (Nt, So, NRY).

**Kilvert:** *Kilvert* filius Ligulfi c1015 PNDB (Y); *Chiluert* 1066 DB (Y, L); Geoffrey *Culverd'* 1207 Cur (L); Robert *Culuert* 1211 AssWo; John *Culuert, Kiluerd* 1283–4 Balliol (O). Probably an ON *Ketilfrøðr*, ODa *Ketilfrith*, anglicized as *Cytelferð* which became *Cylferð*. *v.* PNDB 215.

**Kilwardby:** William *Kilwardby* 1407 IpmY. From Kilwardby (Lei).

**Kimball, Kimbell, Kimble, Kimmel:** *v.* KEMBLE

**Kimberley, Kimberely, Kimberlee:** William *de Chineburlai* 1161 P; Robert *de Kynmerley* 1300 ForNt; Thomas *de Kymberle* 1338 LLB F. From Kimberley (Norfolk, Notts, Warwicks).

**Kimber:** Nicholas *Kember* 1545 SRW; John *Kimber* 1642 PrD, 1662–4 HTDo. From East, West Kimber in Northlew (D).

**Kimbrough, Kinniburgh:** *Kynborough* (f) 1592 Moulton (Sf); *Kimberrow* Herbert 1609 EA (OS) iv (Sf); *Kinbarrow* Wray 1633 ER 62. OE *Cyneburg* (f).

**Kimm:** *v.* KEMM

**Kimpton:** Walter *de Kimton* 1327 SREss. From Kimpton (Ha, Herts).

**Kimsey:** *v.* KEMPSEY

**Kin, Kinn, Kins:** Saxe *filius Kin* e Hy 2 DC (L); *Kyn* Pestell 1260 AssC; *Kinna* vidua 1271 Rams (C); Lefwin *Kinne* 1180 P (Nth); Adam *Kyne* 1268 FFO; William *Kyn* 1329 FFEss. A shortened form of OE names in *Cyne-*.

**Kincaid, Kincade, Kinkead:** Robert, David *de Kyncade* 1450, 1467 Black; Thomas *Kyncayd* 1545 ib. From the lands of Kincaid in Campsie (Stirling).

**Kincey:** *v.* KINSEY

**Kindell:** *v.* KENDAL

**Kinder:** Philota *de Kender* 1274 RH (Db); Hugh *Kynder* 1419 LLB I; Margaret *Kyndur* 1492 PN Ch i 155. From Kinder (Db).

**Kindersley:** *v.* KINNERSLEY

**Kindleyside:** *v.* KEENLEYSIDE

**King, Kinge, Kings:** (i) *King'* 1201 Cur (C), 1219 ib. (Sr); Mariota filia *King* ... William *King* 1259 RamsCt (Hu). OE *Cyng*, an original nickname from OE *cyning, cyng* 'king'. (ii) Ælwine *se Cyng* 1050–71 OEByn (D); Wlfric *Ching* c1130 ELPN; Geoffrey *King* 1177 P (C); Wuluricus, Gaufridus *le King* 1182–1200 BuryS (Sf); Juliana *la Kinges* 1275 SRWo, 1285 *Ass* (Ess). OE *cyning, cyng* 'king', a nickname from the possession of kingly qualities or appearance. Also a pageant name, one who had acted as king in a play or pageant, or had been King of Misrule, or 'king' of a tournament.

**Kingdom, Kingdon:** Nicholas *de Kingdon* 1276 RH (D); John *Kingdone*, Thomas *Kingdome* 1642 PrD. From Higher Kingdon in Alverdiscott (D).

**Kingett, Kinggett:** Robert, John *Kynget* 1296 SRSx, 1317 FFEss; John *Kyngot* 1327 SRWo. *King-et, King-ot*, diminutives of *King*, the personal name. cf. the double diminutive: William *Kyngelot* 1328 ArchC 33.

**Kinghorn, Kinghorne, Kinghan, Kingan:** Adam *de Kyngorn, de Kynghorn* 1204–11, 1357 Black; James *Kinghorne* 1597 ib. (Dunfermline); Alexander *Kingarn* 1679 ib. (Kirkcudbright); John *Kinging* 1684 ib.; Andrew *Kingan* 1689 ib. (Kirkcudbright). From the barony of Kinghorn (Fife).

**Kingsbury, Kingsberry, Kinsbury:** John *de Kingesberi* 1211 Cur (Herts); William *de Kynnesbir'* 1221 AssWa; John *Kingsbury* 1662–4 HTDo. From Kingsbury (Mx, Wa), or Kingsbury Episcopi, Regis (So).

**Kingsford:** Edwin *de Kingesford* 1185 Templars (Ess); John *de Kyngesford', de Kyngesford* 1221 AssWa, 1327 SRWo. From Kingsford (Devon, Essex, Warwicks, Worcs).

**Kingsley:** William *de Kingesle* 1246 AssLa; Richard *de Kyngesleye* 1327 SRSt; John *Kyngesley* 1421–2 FFWa. From Kingsley (Ches, Hants, Staffs).

**Kingsman:** Godwin *Kingesman* 1166 P (Nf); William *Kingesman* 1184–1215 AD iv (Lo). 'The king's man', a surname common in Norfolk and Suffolk where it probably meant one who had commended his services to the king and not to some baron. cf. Walter *Kingesbonde* 1205 P (Nb), Godwin *Kingesreive* 1208 FFL.

**Kingsmill:** Peter *de Kyngesmulne* 1249 AssW; Hugh *de la Kingesmille* 1275 RH (Ha); John *Kyngesmyll* 1502–3 FFSr. From King's Mill in Marnhull (Do), King's Mill in Barton (O).

**Kingson:** Godricus *Chingessone* 1066 Winton (Ha); Ælfmær *Cynges sune* 1100–30 OEByn (D); Henry *Kingessone* 1224–46 Bart (Lo). 'Son of *Cyng*', from OE *Cyng*, or of *King*. *v.* KING.

**Kingston, Kingstone:** Alan *de Kingistona* 1175 P (Y); Nicholas *de Kyngeston* 1247 FFO; Thomas *Kyngeston* 1398 FFEss. From one or other of the many places of this name.

**Kingswell:** Richard *Kingeswell* or *Kinswell* 1604 Oxon (Ha); John *Kingswill*, William *Kingwell* 1642 PrD. From Kingswell in Whitestone (D).

**Kingswood:** Wlmer *de Kingeswde* 1185 Templars (Ess); William *de Kingeswode* 1275 SRWo; John *de Kingeswode* 1330 PN Sx 199. From Kingswood (Gl,

Sr, Wa), Kingswood in Findon (Sx), or King's Wood in Himbleton (Wo).

**Kinkead:** v. KINCAID

**Kinlay, Kinley:** A Manx contraction of *Mac-Cinfaolaidh* 'son of *Cinfaoladh*' 'wolf-head': *Kinley* 1604 Moore. The name may also be English: Thorold *de Kynely* 1220 FFEss. Probably from Kenley (Surrey).

**Kinman, Kynman:** *Keneman* c1250 Rams (Nf); Richard *Kineman* 13th *Binham* (Nf); Robert *Kynemon*, Richard *Kenemon* 1327 SRWo. OE *Cynemann* 'royal-man', recorded only once, c770 in Worcester. More commonly, OE *cȳna* and *mann* 'cows' man', herdsman. cf. Ralph *Kenegrom* 1235 Bart.

**Kinn:** v. KIN

**Kinnaird:** Richard *de Kinnard* 1204–14 Black; Rauf *de Kynnard* 1296 ib. (Fife); Thomas *de Kynnarde* 1431 ib.; Erche *Kinȝerd* 1567 ib. (Kelso). From the barony of Kinnaird (Perth).

**Kinnerley:** v. KENNERLEY

**Kinnersley, Kinnersly, Kindersley, Kynnersley:** Hugh *de Kinardesle(g)* 1208 Cur (He), 1221 AssWo. From Kinnersley (Hereford, Salop, Surrey, Worcs).

**Kinnett:** *Kinett* 1251 *WAM*; Michael *Kinnet* 1275 RH (L); William *Kynot* 1317 AssK; William *Kinnett* 1662–4 HTDo. *Kin-et, Kin-ot*, diminutives of *Cyne*, a short form of OE names in *Cyne-*.

**Kinniburgh:** v. KIMBROUGH

**Kinnish, Kennish:** A Manx contraction of *Mac Aenghuis* 'son of *Aenghus*': *McInesh* 1511, *Kynnish* 1626, *Kenish* 1649 Moore. cf. MAGUINESS.

**Kins:** v. KIN

**Kinsbury:** v. KINGSBURY

**Kinsey, Kincey, Kynsey:** William *Kynsei* 1306 IpmGl; George *Kynsey, Kyngsey* 1525 SRSx; Robert *Kensaye* 1558 Pat (Sx); Margery *Kynsee* 1584 StratfordPR; Joseph *Kinsey* 1648 FrYar. OE *Cynesige* 'royal-victory'.

**Kinsley, Kynsley, Kinslea:** Ralph *de Kineslea* 1191 Pl (Ess); John *de Kyneslay* 1244–5 IpmY; Robert *Kynslay* 1542 RothwellPR (Y). From Kinsley (WRYorks).

**Kinsman:** William *Kinesman* 1198 FFNf; John *Cunesmon* 1275 SRWo. ME *cunnes, kinnes*, genitive of *kin* and *man*, 'kinsman', a relative by blood (or, loosely, by marriage).

**Kintish:** v. KENTISH

**Kinton:** William *de Kinton'* 1191 P (Nth); William *de Kynton'* 1242 Fees (Wo); Geoffrey *de Kyntone* 1317 AssK. From Kinton (He), Kineton (Wa), or Kyneton in Thornbury (Gl).

**Kippax, Keepax:** Alan *de Kipais* 1190 P (Y); Richard *de Kippax* 1347 AssSt; John *Kypas* 1441 Calv (Y). From Kippax (WRY).

**Kipping, Kippen, Kippin:** Alwinus *filius Cheping* 1086 DB (Berks); Mafrei *filius Kipping* 1170 P (Nth); *Kipping* de Burehamton' 1243 AssSo; John, Richard *Kipping* 1195 P (Wa), 1206 Cur (Ess). OE *Cypping*, an original nickname from the Germanic root *kupp-* 'to swell, be swollen', used of a man of fat, rotund appearance (Tengvik).

**Kipps:** Ulwardus *Cheppe* 1066 Winton (Ha); Wolberne, Robert *Kippe* 1202 AssL, 1279 RH (C); Richard *Keppe* 1327 SRSx. OE *Cyppe*, the simplex of *Cypping*, with the same meaning. v. KIPPING.

**Kirk, Kirke, Kerk, Kyrke:** Reginald *Attekireke* 1209 FFL; Richard *Attekirck* 1301 SRY; Adam *Ofthekirke* 1308 Pat (Sf); John *be ye kyrk* 1438 DbCh. 'Dweller by the church', ON *kirkja*.

**Kirkbride, Kirkbright:** Richard *de Kirkebryd* 1274 Ipm (Cu). From Kirkbride (Cumb).

**Kirkby, Kirkebye, Kirby, Kerbey, Kerby:** Godebold *de Kirkebi* 1121–48 Bury (Sf); Ketellus *de Kerkebi* 1191 P (Y); Richard *Kyrby* 1524 SRSf. From one of the numerous places named Kirby or Kirkby.

**Kirkfield:** Henry *de Kyrkefeld* 1247–8 ForNth. 'Dweller by the church field', ON *kirkja*, OE *feld*.

**Kirkham, Kirkam, Kerkham:** Simon *de Kirkeham* 1219 AssY. From Kirkham (ER, WRYorks).

**Kirkhouse, Kirkus:** John *Kirkhous, Kirkeasse* 1446 FrY, 1618 YWills. 'One employed at the churchhouse.' v. KIRK.

**Kirkland, Kirtland, Kirtlan:** Michael *de Kerkeland'* 1196 P (Cu); John *de Kyrkeland* c1280 Black (Berwick); Samuel *Kirtland* 1802 Oseney (O). From Kirkland (Cumb, Lancs; Ayr, Dumfries, Lanarks).

**Kirkley, Kirkly, Kirtley, Kerley, Kerly:** William *de Kirkelee* 1200 Cur (Sf). From Kirkley (Northumb, Suffolk).

**Kirkman, Kirman, Kerman:** Robert *Kirkeman* 1230 P (Y); Roger *le Kirkeman* 1259 Calv (Y). ON *kirkja* and ME *man*, 'custodian of a church'. cf. CHURCHMAN.

**Kirkstead, Kirksted:** William *de Kirkestede* 1219–22 RegAntiquiss; Ralph *Kyrkestede* 1379 LoCh; Henry *Kyrkestede* 1392–3 Hylle. From Kirkstead (L).

**Kirkus:** v. KIRKHOUSE

**Kirkwood, Kerwood:** John *Kirkwood* 1476, Alexander *Kirkwod* 1526 Black. From Kirkwood (Ayr, Dumfries, Lanark).

**Kirrage:** v. KERRICH

**Kirrell, Kirriell:** v. KERRELL

**Kirtland:** v. KIRKLAND

**Kirtler:** v. CURTLER

**Kirtley:** v. KIRKLEY

**Kirtling:** Hugh *de Kertling* 1218 FFC; Sire John *Kyrtelyng* 1460 Paston. From Kirtling (C).

**Kirton, Kerton, Kurton:** Lambert *de Kirketon'* 1219 AssL; William *Kirton* 1508 FrY. From Kirton (Lincs, Notts, Suffolk).

**Kisby, Kesby:** Ralph *de Kisebi* 1205 Cur (L). From Keisby (Lincs).

**Kiss, Cuss, Cusse, Cush:** Amice, John *Kisse* 1327 SRLei; William, Thomas *Kysse* 1329 FFSf, 1430 FrYar; John, Edmond *Cusse* 1545, 1576 SRW; Laurence *Kyshe*, Sarah *Kish* 1573, 1765 Bardsley. Metonymic for KISSER.

**Kissack, Kissock:** Gilbert *McIssak* 1418, *Kissak* 1599 Moore. A Manx contraction of *MacIsaac* 'Isaac's son'.

**Kisser, Kissa:** William *Kisere* 1224–46 Bart; Richard *le Kissere* 1288 LLB A; Hugh *le Kysser, le Cussere, le Kisehere* 1292 SRLo, 1294 LLB A, 1307 Husting; Margaret *Cusser* 1298 IpmY; Benjamin,

Edward *Kishere* 1738, 1750 Bardsley. 'A maker of (leather) armour for the thighs', from OFr *cuisse* 'thigh', cf. 'cuisses, armour for the thighs' (Cotgrave). The kissers were cordwainers.

**Kitchell:** *v.* KETCHELL

**Kitcheman:** *v.* KITCHINGMAN

**Kitchen, Kichin, Kitchin, Kitching, Ketchen, Ketchin:** Henry *atte Kychene* 1311 ParlWrits (Sx); Nicholas *atte Kechene* 1327 SRSo; Robert *del Kychin* 1359 FrY; Thomas *Kytchyng* 1513 GildY. 'Worker in a kitchen' (OE *cycene*).

**Kitchener, Kitchiner:** (i) William *le Cuchener* 1332 MEOT (Sr); Thomas *Kytchener, Kitchynner* 1472 FrY, 1494 GildY. A derivative of OE *cycene* 'kitchen', one employed in a kitchen, especially in a monastery (c1440 MED). (ii) Robert *de Kechenor'* 1207 Cur (Sx). From Kitchenour (Sussex).

**Kitching:** *v.* KITCHEN

**Kitchingman, Kitcheman, Kitchman:** John *Kychynman* 1379 PTY; Thomas *Kechynman* 1475 GildY; Jenet *Kycheman* 1553 RothwellPR (Y); William *Kitchingman* 1583 FrY. *v.* KITCHEN, KITCHENER.

**Kite:** *v.* KEAT

**Kitebone:** Roger *Kitebein* 1210 P (Nf); Reginald *Kytebon* 1296, John *Kytebon* 1332 SRSx. 'With bones like a kite', OE *cȳte*, OE *bān*/ON beinn.

**Kitling:** Hugh *Kytlyng* Hy 3 Gilb (L); William *Kytling* e 14th CartNat; William *Kitlyng* 1360 FFSf. *Kit-el-in*, a double diminutive of *Kit*, a pet-form of *Christopher*.

**Kitson:** William *Kittesson* 1340 Crowland (C); Thomas *Kytson* 1357 ib. 'Son of *Kit*.'

**Kitt, Kitts:** *Kytte* the Soper 1286 AssCh; Roger son of *Kytt* 1297 Wak (Y); Wigar, Amfridus *Kitte* 1173–9 Clerkenwell (Mx), 1190 Oseney (O). *Kytte* is a pet-form of *Christopher* and of *Katharine*. The surname was common and may also be metonymic for KITTER.

**Kittel:** *v.* KETTLE

**Kitter:** Walter *le Kittere* 1177 P (Y); Holbe *Kittere* 1221 AssWo. A derivative of ME *kitte* (1362 MED) 'a wooden vessel made of hooped staves', a maker of kits (tubs, milking-pails, etc.). cf. Richard *le Kittewritt'* 1275 MESO (Y), 'kitwright'.

**Kitteridge:** *v.* KETTERIDGE

**Kittermaster:** Ida *de Kidministre* 1230 P (Wo); Symond *Kytermyster* 1524 SRSf; Thomas *Kiddermaster* 1637 EA (NS) ii (Sf); William *Kittermaster* 1663 Bardsley. From Kidderminster (Worcs).

**Kittle:** *v.* KETTLE

**Kitto, Kittoe, Kittow:** Henry *Kitto*, Tristram *Kittoe* 1642 PrD. A Cornish diminutive of *Griffith*.

**Kittredge:** *v.* KETTERIDGE

**Kivill:** *v.* KEEVIL

**Knaggs:** Henry *Knag* 1185 Templars (Y); Richard *Knag* 1442, John *Knagges* 1598 FrY. ODa, OSw *Knag*.

**Knape:** John *le Knape* 1332 MEOT (Nf). OE *cnapa* 'youth, servant'.

**Knapman:** *v.* KNAPP, KNAPPER

**Knapp:** William *atte Kneppe* 1294 PN Sr 89; Henry *de Cnappe* 1301 PN D 81; John *Knappe* 1279 RH (Bk). 'Dweller at the top of the hill or on a hillock' (OE *cnæpp*) as at Knapp (Devon) or Knapp Fm (Sussex).

**Knappen:** Henry *Knappyng* 1327 SRSf. OE *\*cnæpping* 'dweller on the hill'. *v.* KNAPP.

**Knapper:** William *Knapper* 1360 FFSx; Andrew *Knapere* 14th Rad (Sf). Identical in meaning with KNAPP, KNAPMAN, KNAPPEN.

**Knappett:** Thomas *Knapet* 1378 LLB H; John *Nappett* 1473 FrY. Perhaps a diminutive of KNAPE.

**Knapton:** Adam *de Knapeton'* 1207 Cur (Y); William *Knapton* 1330 ColchCt; Widow *Knapton* 1672 HTY. From Knapton (Nf, ERY, WRY).

**Knapweed:** John *Knapwedd* 1319 SRLo; John *Knapwed* 1332 SRLo; Thomas *Knopwed* 1524 SRSf. A nickname from the common weed of this name.

**Knatchbull:** John *Knechchebole*, *Knetchebole* 1375 ColchCt; Thomas *Knechboll*, John *Knachbull* 1481, 1504 KentW. ME *knetch*, *knatch* 'to knock on the head, fell' and *bull*, 'Fell bull', a nickname for a butcher.

**Knave, Nave:** Alwin *Cnave* 1210–11 PWi; Henry *le Knaue* 1271–2 FFWa; Adam *le Cnave* 1327 SRSo; Richard *Knave* 1357 FFEss. OE *cnafa* 'boy, servant'.

**Kneal, Kneale:** McNelle 1408, MacNeyll 1430, Kneal 1598 Moore. A Manx contraction of MacNiall 'Niall's son'. *v.* NEAL.

**Kneebone:** John *Knebone* 1469 AD v (Mx); Lucreatia *Knebone* 1623 GreenwichPR (K); Gilbert *Kneebone* 1642 PrD. 'Knee-bone', OE *cnēow, bān*.

**Kneeder:** Robert *le Knedere* 1280 MESO (Ha); John *Kneder* 1296 SRSx. A derivative of OE *cnedan* 'to knead'. A nickname for a baker.

**Kneen:** A purely Manx name, probably identical with KANEEN. It seems to have been confused with NIVEN: Jenkin *M'Nyne* or *Mac Nevyne* 1429, 1430 Moore.

**Knell, Knill:** Alvredus *de Knelle* 1220 Cur (Sx); Gilbert *de Knille* 1279 RH (C); William *atte Knelle* 1296 SRSx. 'Dweller by the knoll', OE *\*cnyll(e)*, as at Kneela (Devon), Knell House (Sussex), Knill (Hereford), Nill Well (Cambs).

**Kneller:** William *Kneller* 1327 SRSx. Identical in meaning with KNELL.

**Knevet, Knevit:** *v.* KNIVETT

**Knife:** Alwyne *Knyf* 1139 Templars (O); John *Knif'* 1277 AssSo. OE *cnīf* 'knife', metonymic for *knifesmith*, cutler. *v.* NAYSMITH.

**Knifton:** *v.* KNIVETON

**Knight, Knights:** (i) *Chenicte* 1066 DB (Y); Johannes *filius Cnith* 1182–1211 BuryS (Sf); Gamell' *filius Cniht* 1219 AssL. OE *Cniht*. (ii) Godefridus *Niht*, Oschetel *Cniht* 1166 P (Nf); Walter *le Knit* 1200 Oseney (O); William *Knicht'* 1221 AssWo; John *Knyght* 1275 RH (Sf); Beatrix *Knictes* 1279 RH (Hu); Alicia *Knyghtes* 1327 SRSo; Henry *le Nyte* 1327 SRSx. OE *Cniht* or *cniht* 'servant', 'knight, feudal tenant bound to serve as a mounted soldier', 'a common soldier'.

**Knightley, Knightly:** Jordan *de Knitteleg'* 1207 Pl (St); Robert *de Knyghteleye* 1351 Ronton; Richard *Knyghtley* 1411 FFEss. From Knightley (Staffs).

**Knighton:** Alexander *de Cnichteton'* 1181 P (Wo); Ralph *de Knichton* 1222 AssSt; Hugh *de Knyghton'* 1327 SRLei; Thomas *Knyghton* 1503 FFEss. From one or the other of the many places of this name.

**Knightson:** Robert *Knyhteson* 1271 IpmGl; Francis *Knythsoun* 1464 Black. 'Son of *Cniht*'. *v.* KNIGHT (i).

**Knightwin:** *Cnichtwin'* 1190 P (Bk); *Chnitwinus* 1203 P (Do); *Knythwyn* 1260 Rams (Nf); Richard *Knihtwine* 1189 Sol; Henry *Knythewyn* 1249 AssW; Richard *Knithwine* 1279 RH (O). OE *Cnihtwine*.

**Knill:** *v.* KNELL

**Kniveton, Knifton:** Matthew *de Knyveton* 1275 AssSt. From Kniveton (Derby), pronounced *Nifton*.

**Knivett, Knyvett, Knevet, Knevit, de Knevett, Nevet, Nevett, Nevitt:** Leuricus, Nicholas *Cnivet* 1087–97 Crispin (Mx), 1185 Templars (Herts); William, Osbert *Knivet* Hy 2 DC (L), 1199 AssSt; Ernald *Knikt* 1275 RH (Nf); Thomas *Knifet* ib. (L); Walter *le Knift* 1279 ib. (O); John *Knyft, Knyvet* 1311, 1337 ColchCt. A Norman pronunciation of *Knight*, owing to the French difficulty with the *h* of *cniht*. Mathew *de Knyvet* 1273 RH (Nt) must be identical with Matthew *de Knyveton* above and William *de Knyvet* 1327 SRDb must similarly be for *Knyvet'*, i.e. *Kniveton*. Alexander *de Knyft* 1279 RH (O) is an error for *le Knyft* (cf. Walter *le Knift* above). There is no evidence for a place *Knevet* and the modern *de Knevett* is either a perpetuation of one of these errors or due to a late prefixing of an unetymological *de*.

**Knock, Knocker:** Thomas *atte Knocke* 1296 SRSx; Nicholas *Knok* 1279 RH (Beds). 'Dweller by the hill' (OE *\*cnocc*) as at Knock Hatch (Sussex), Knock Fm (Kent). *v.* MELS.

**Knoll, Knollys, Knowles, Nowles:** Robert *de la Cnolle* 1185 P (D); Theobald *de Chnolle* 1242 Fees (K); Thomas *Knolle* 1279 RH (C); William *atte Knolle* 1296 SRSx; Adam *del Knol* 1318 AssSt; Christopher *Knolles* 1407 FrY. 'Dweller at the top of a hill' (OE *cnoll*), as at Knole (Kent, Sussex), Knowle (Devon, Dorset, Som, Warwicks).

**Knoop, Knope, Knopp:** Osgod *Cnoppe* 1066 OEByn (Ess); William, Walter *Knop* 1327 SRC, SRSf. OE *\*cnoppa*, ME *knop(p)*, *knope* 'a small rounded protuberance, knob', used also of the knope of the knee and the elbow-joint.

**Knorr:** Richard *de Knarre, de Knorre* 1279 RH (C). From Knarr Fm or Lake (Cambs).

**Knott:** (i) *Cnut, Cnud, Canut* 1066 DB; Randulfus *filius Cnut* 1191 P (D); *Knot* pater Alani, Alanus filius *Knod* 1202 AssL; Radulfus *filius Knut* 1203 P (Y); Walter, Robert *Cnot* 1165 P (Sf), 1185 Templars (Herts); William *Cnotte* 1206 Cur (Beds); William, John *Knotte* 1221 AssWo, 1260 ELPN; Stephen *le Knotte* 1296 SRSx; Hugo *Knout'* 1301 SRY. The personal name is ON *Knútr*, ODa, OSw *Knut*, an original nickname from ON *knútr*, 'knot',

occasionally, perhaps, the rare OE *Cnotta*. *Knut* (Canute) was still in use in the 13th century. The surname is usually a nickname from OE *cnotta* 'knot', used of a thickset person. (ii) Emma *del Knot* 1332 SRCu. 'Dweller on the hill' (ME *knot*) as at Knott End (Lancs).

**Knotter:** John *Knotter* 1524 SRSf. Equivalent to KNOTT (ii).

**Knotting:** John *Knottyng* 1379 PTY. A derivative of OE *cnotta* 'hill', hence 'dweller on the hill'. cf. Thomas *Knottyngman* 1379 PTY.

**Knottson:** 'Son of *Knut*.' *v.* KNOTT (i).

**Knowler, Knowlder, Knowlman:** John, Walter *le Knollere* 1296 SRSx; Wilmore *Knowleman* 1616 Bardsley. Found side by side with *atte Knolle*, *v.* KNOLL.

**Knowles:** *v.* KNOLL

**Knowling:** William *Knollyng* 1327 SRSx. OE *\*cnolling* 'dweller at the top of the hill'. *v.* KNOLL, KNOWLER.

**Knox:** John *de Cnoc* 1260 Black; Hugo *Cnox* c1272 ib., Alan *de Knockis* 1328 ib. From Knock (Renfrew).

**Kortwright:** *v.* CARTWRIGHT

**Krabbe:** *v.* CRABB

**Kramer:** *v.* CRAMER

**Kray:** *v.* CRAY

**Kriel:** *v.* KERRELL

**Krips:** *v.* CRISP

**Kristall:** *v.* CHRYSTAL

**Kroll:** *v.* CROWL

**Krook:** *v.* CROOK

**Kullum:** *v.* CULLUM

**Kurton:** *v.* KIRTON

**Kyd(d):** *v.* KID

**Kyffin:** Hoel ap Madog *Kyffin* 1391, Howel *Kyffin* 1392, Geoffrey *Kiffin* 1475 Chirk; Thomas *Kyffin* 1477 *Petre A*; John *Keffin* 1637 SaAS 2/iv. Cyffin is a hamlet in Mongomery, and the name, Gyffin, of a parish in Carnarvon. cf. also Welsh *cyffin* 'limit, confine'. According to VisSalop 1623, Madoc ap Madoc of Cyndleth, living in 1530, assumed the name of *Kyffin* (Morris 134).

**Kynaston:** Walter *de Kynwardeston'* c1295 Glast (So); Ralph *de Kinastan* t Ed 3 Rydware; William *Kynastone* 1418 LLB I. From Kynaston (Hereford, Salop).

**Kynman:** *v.* KINMAN

**Kynnersley:** *v.* KINNERSLEY

**Kynsey, Kynsley:** *v.* KINSEY, KINSLEY

**Kyrke:** *v.* KIRK

**Kyte:** *v.* KEAT

# L

**Labbe:** Robert *Labbe* 1208 P (Ess/Herts). ON *Labbi*, OSw *Labbe*. v. also ABB.

**Labbet:** v. ABBATT

**Labern:** v. LAYBORN

**Labey:** v. ABBA

**Lace:** OFr *laz, las*, ME *lace* 'cord'; metonymic for *Lacer*, a maker of cords or strings: Richard *le Lacir* 1278 LLB B; William *le Lacer* 1292 SRLo, 1298 Wak (Y).

**Laceby:** Henry *de Lessebi* 1202 AssL; Walter *de Leissebi* 1204 P (L). From Laceby (L), *Leysebi* c1115.

**Lacey, Lacy, Lassey, de Lacey, de Lacy:** Ilbert *de Laci*, Roger *Laci* 1086 DB; Henry *de Lasci* 1185 Templars (L). From Lassy (Calvados).

**Lachlan, Laughlan, Laughland:** *Lohlan* 1158–64 Black; Eugene *filius Loghlan* 1296 ib.; Reginald son of *Lauchlan* 1327 ib.; Adam *Lachlane* 1417 ib.; Robert *Laughland* 1642 ib. Gael *Lachlann*, earlier *Lochlann* 'lake or fjord-land', i.e. Scandinavia, Norway, the personal name denoting 'one from Lachlann'.

**Lackenby:** Gilbert *de Lackenby* 13th Guisb. From Lackenby (NRY).

**Lacklison:** William *Lauchlanesone* 1497 Black. For MACLACHLAN.

**Ladbrook, Ladbrooke:** John *de Ledbroc, de Lodbroc* 1221 AssWa. From Ladbrooke (Wa).

**Ladd, Ladds:** Godric *Ladda* c1100 OEByn (So); Richard *Ladde* c1175 EngFeud (Nth); Walter *le Ladd* 1242 Fees (K). The common English *lad*, of obscure origin, originally 'servant or man of low birth'. cf. 'to make lordes of laddes' (*Piers Plowman*); 'A ladde . . . Borne he was of pouere lynage' (Robert of Brunne).

**Lade:** Richard, Petronilla *de la Lade* 1214 Cur (C), 1275 SRWo. 'Dweller by the road, path or watercourse', OE *(ge)lād*.

**Ladler, Ladler:** Walter *le Ladelere* 1278 Misc (Wa); John *le Ladeler* 1327 SRY; Nicholas *Ladelere* 1377 AD iii (Hu). 'A maker of ladles', OE *hlædel* and *-er*. cf. Nicholas *Ladel* 1187–9 Ek, William *Ladyl* 1337 CorLo.

**Ladye:** William *le Lady* 1340 AssC. A nickname.

**Ladyman:** Rannulf (*le*) *Leuediman* 1202 FFL, 1214 Cur (Nth); Geoffrey *le Leuediman* 13th Guisb (Y); Roger *Ladyman* 1296 MEOT (Herts). 'The lady's servant', OE *hlæfdige* and *mann*.

**Laffeaty, Laffitte, Lafitte:** v. FATES

**Lafflin, Laffling:** An English pronunciation and spelling of *Laughlan*.

**Lafford:** Ralph *de Lafford* 1202 AssL; Adam *de Lafforde* c1280 Hylle. Probably for *de la Ford* 'dweller by the ford'. v. FORD.

**Lainer, Leiner:** Hugo *le Layner* 1279 MESO (Y); Goceus (Joyce) *le Leyner, le Laner, le Wollemongere* 1292 SRLo, 1312 LLB D, 1301 Husting. OFr *lainier, lanier* 'woolmonger'.

**Laing, Layng:** Thomas *Laing* 1357 Black (Dumfries). A Scottish variant of LANG.

**Laird:** Roger *Lawird, lauird* 1257 Black (Berwick); Thomas *Lairde* 1552 ib. (Glasgow). A Scottish form of LORD, 'landlord, land-owner'.

**Laising, Leising:** John *filius Laising* 1212 Cur (Y); *Leisingus* 1219 Cur (La); John *Laysinge* 1288, William *Leysing* 1304 IpmY; John *Leysyng* 1331 FFY. ON *Leysingr*.

**Laister:** v. LEICESTER

**Lake, Lakes:** Richard *de la Lake* 1200 P (Sa); Robert *Attelake* 1242 P (Sr). 'Dweller by the stream', OE *lacu*.

**Lakeman:** John *Lakeman* 1320 HPD (Ess). 'Dweller by the stream.'

**Laker:** (i) William *le Lakere* 1325 MESO (Ha); Adam *Lakyare* 1391 ib. (Sx); Robert *Laker* 1545 SxWills. From OE *lacu* 'stream'. v. LAKE, LAKEMAN. (ii) Osbert *Laycar* 1274 RH (Ess); John *le Leykere* 1309 SRBeds, 1327 FFK. ME *leyker*, ON *leikari* 'player, actor'. cf. Richard *Leyk* 1292 SRLo, from ME *leyk* 'play, sport'. cf. GAME.

**Lamb, Lambe, Lamm:** *Lamb* dispensator 1161 Dugd v. 421; *Lambe* de Harewude 1290 ShefA; Ædward, Wulmar *Lamb* 1195 P (K), 1230 P (Nf); William, Roger *le Lamb* 1279 RH (C), 1296 SRSx. A nickname from the animal or a shortened form of *Lambert*. Occasionally from a sign: William *atte Lamme* 1320 LLB E.

**Lambard, Lambart, Lambarth, Lambert, Lambirth, Lamburd, Lampard, Lampart, Lamperd, Lampert, Lammert, Limbert:** (i) Gozelinus *filius Lamberti* 1086 DB (L, Y); *Lambertus* 1142 NthCh (L), 1219 AssY, 1221 AssWo; Ricardus *filius Lambricht* 1148 Winton (Ha); *Lambricthus, Lambrihtus* 1196–7 P (Bk); Constantinus *filius Lambrichti* 1220 Cur (Sf); *Lambard* 1296 Black; Richard *lambert* 1148 Winton (Ha); Simon *Lamberti* 1212 Cur (Berks); Peter *Lambert, Lamberd'* 1220 Cur (Nf); John *Lambard* 1250 Fees (Sx); William *Lambryt* 1279 RH (Hu); Thomas *Lambrygt, Lambrith* 1280 AssSo; Thomas *Lamparde* 1544 RochW; Mr *Limbert* 1587 EA (NS) ii (Nf). OFr *Lambert*, OG *Lambert, Lanbert* 'land-bright', a popular name from the 12th century, probably introduced from Flanders where St Lambert of Maestricht was highly venerated. Late OE

*Landbeorht* was rare but was used after the Conquest and has contributed to the surname. *v.* LAMBRICK. (ii) William *Lambhyrde* 1255 *Ass* (Ess); Robert *le Lambhurde* 1288 *Ct* (Ha); Hugh (*le*) *Lambehird* 1309 Wak (Y); William *Lambeherde* 1332 SRSx. OE *lamb* and *hierde* 'lamb-herd'.

**Lambden, Lambdin, Lamden, Lamdin:** (i) William *Lambeden'* 1279 RH (C); John *de Lambedenne* 1317 AssK. From Lambden (K). (ii) Roland *de Lambeden* c1261 Black. From the lands of Lambden (Berwick).

**Lamberton, Lamerton:** William *de Lamberton* c1136 Black; John *de Lambertoune* c1200 ib.; William *de Lambirtoun* 1300 ib.; Robert *Lamberton* 1672 HTY. From the barony of Lamberton (Berwick). Occasionally perhaps from Lamerton (Devon), *Lambertone* 1232, and cf. John *de Lamerton'* 1206 P (D).

**Lambie, Lamby, L'Ami, Lammie, Lampey:** Robert, Henry *Lambi* 1203 P (Nth), 1281 Black (Dundee); Mariora *Lammeis* dothyr 1527 ib. (Strathdee); George *Lammie* 1628 ib. ON *Lambi.*

**Lambirth:** *v.* LAMBARD

**Lambkin:** *Lambekyn* Flandrensis 1178 P (Nb); *Lamekin* filius Beatricis 1188 P (Bk); Nicholas *Lambekyn* 1301 SRY; John *Lamkyn* 1379 ColchCt. *Lamb-kin*, a diminutive of *Lamb* (Lambert).

**Lambley, Lamley:** Robert *de Lambeley* a1248 Black; Robert *de Lambelay* 1367, Richard *Lamlay* 1412 FrY. From Lambley (Nb, Nt).

**Lamborn, Lamborne, Lambourn, Lambourne, Lamburn, Lamburne:** Ralph *de Lamburne* 1198 FFEss; Alice *de Lamborne* 1278 CtW; John *Lambourn* 1451 AssLa. From Lambourn (Berks), or Lambourne (Essex).

**Lambrick:** Richard *Lambrich* 1327 SRSf. OE *Landbeorht. v.* LAMBARD.

**Lambrook, Lambrooke:** Robert *de Lambroc* 1212 Cur (So); Roger *de Lambroc* 1221 AssGl; Nicholas *de Lambrok* 1268 AssSo. From Lambrook (So).

**Lambshead:** Suetman, Agnes *Lambesheved* c1130 ELPN, 1279 RH (Hu). A nickname. cf. John *Lomesfot* 1327 SRSo.

**Lambson:** *v.* LAMPSON
**Lamburd:** *v.* LAMBARD
**Lamburn(e):** *v.* LAMBORN
**Lamden, Lamdin:** *v.* LAMBDEN
**Lamerton:** *v.* LAMBERTON
**Lamey, L'Ami, L'Amie:** *v.* AMEY, LAMBIE
**Lamley:** *v.* LAMBLEY
**Lamm:** *v.* LAMB

**Lammas, Lammers:** Brictius (*de*) *Lammasse* 1190–1 P (Nf); William *de Lammers* 1248 FFEss. From Lammas (Norfolk) or Lamarsh (Essex), both DB *Lamers.*

**Lammert:** *v.* LAMBARD
**Lammey, Lammie:** *v.* LAMBIE

**Lammiman, Lamyman:** Alexander *Lamaman* 1463 FrY; Richard *Lambeyman* 1521 GildY; John *Lamyman* 1525 FrY. 'Servant of *Lambie*' or of *Lambin. v.* LAMPEN.

**Lammin(g):** *v.* LAMPEN

**Lammond, Lamond, Lamont:** *Laumannus* filius Malcolmi 1230–46 Black; Molmure *filius Lagman*

1290 ib.; *Lawemund* McGreghere 1292 ib.; John *Lawmond* 1466 ib.; John *Lamond* 1674 HTSf. MIr *Lagmand*, from ON *Lǫgmaðr* 'lawman'. *v.* MACLAMON, LAWMAN.

**Lamotte:** A Huguenot name. Francis *La Motte*, a refugee from Ypres, settled in Colchester as a manufacturer (Smiles 406).

**Lampard, Lampart, Lamperd, Lampert:** *v.* LAMBARD

**Lampen, Lampin, Lammin, Lamming:** *Lambinus* Frese 1181 P (K); *Lambinus* 1197 P (K), 1221 Cur (C); Robert *Lambin* 1292 SRLo; John *Lambyn* 1302 FFSf, 1305 LLB B; Roger *Laming* 1683 Bardsley. *Lamb-in*, a diminutive of *Lamb* (Lambert). *Lambert* de Langham (1283 SRSf) is also called *Lambinus.*

**Lampet, Lampitt, Lamputt:** John *de Lampet*, Agnes *Lampit* 1327 SRSf; William *atte Lamputte* 1332 SRSr. 'Dweller near, or worker at the clay-pit', OE *lām* 'loam', *pytt* 'hollow'.

**Lampey:** *v.* LAMBIE

**Lamplough, Lamplugh:** Robert *de Lamplo* 1181 P (Cu); Cristiana *de Lamploch'* 1213 Cur (Cu); Thomas *Lamplugh* 1432, Elizabeth *Lamplogh* 1437 TestEbor. From Lamplugh (Cu).

**Lamprey:** Richard *Lampreða* 1201 Pleas (Co); Lucy *Lampreye* 1243 AssSo; Nicholas *Lampray* 1524 SRD. A nickname from the lamprey, OE *lamprede.*

**Lampson, Lambson, Lamson:** Adam *Lambeson* 1332 SRCu; Thomas *Lamson* 1464 Cl; William *Lampson* Eliz Bardsley. 'Son of *Lamb* (Lambert).'

**Lamyman:** *v.* LAMMIMAN

**Lancashire, Lankshear, Lankshire:** Richard *de Lancastreschire* 1387 AssL; Robert *Lancashire* 1604 ChW; Robert *Lankshire*, John *Lankshear* 1693, 1805 Bardsley. 'The man from Lancashire.'

**Lancaster, Langcaster, Lankester, Loncaster, Longcaster:** William *de Lonecastre* 1175 StCh; John *de Lancastre* 1327 SRC; Thomas *de Langcastre* 1327 *SR* (Ess); Alice *Longcaster* 1494 GildY; Ales *Lankester* 1565 SfPR. From Lancaster.

**Lance:** *Lance* Bliaut 1185 NthCh (Nth); *Lance* prepositus 1219 AssL; Alice, John *Lance* 1196 FFNf, 1237 Oseney (O). OG *Lanzo*, a hypocoristic of names in *Land-.*

**Lanceleave, Lancelew:** Geoffrey *Lanceleue* 1189 Sol; John *Lanceleue* 1219 AssL; Richard *Launceleve* 1330 IpmNt. 'With lifted lance', OFr *lance levé.* cf. David *Lancea acuta* 1109–22 MCh 'with a sharp lance'.

**Lanceley, Lanslyn:** *Launcelinus* de Thorp 13th CartNat; William *Lancelin* 1163 P; Henry *Launcelyn* 1271 FFL; William *Launcelyn* 1341 ChertseyCt (Sr). OG *Lancelin.*

**Lancelot:** Rauff *Lancelott* 1506 TestEbor. *Lanc-el-ot*, a French double diminutive of OG *Lanzo.*

**Lanchester:** Heruis *de Langecestre* c1150 FeuDu; Walter *de Lancestre* 1344 FFY. From Lanchester (Du).

**Land, Lawn:** Thomas *de la Lande* 1205 P (Nth); James *de la Launde* 1262 AssSt; Widdow *Lawne* 1674 HTSf. 'Dweller by the glade', as at Launde (Leics), ME *launde*, OFr *land.*

**Landel, Landels, Landell, Landells, Landale, Landles:** Robert *de Landeles* 1192–1205 YCh; John

*Laundel* 1326 CorLo; Robert *Laundelles* 1383 AssL. 'Dweller in the glade', from a diminutive of OFr *launde* 'a glade, forest pasture'.

**Lander, Landers:** v. LAVENDER

**Landimore, Landymore:** John *Landimer* 1568 SRSf. From Landermere (Ess), *Landimer* 1211.

**Landies:** v. LANDEL

**Landmott:** Alan *de Landemote* 1303 IpmY; Richard *Landemote* 1379 PTY; Richard *Landmote* 1415 IpmY. From Landmoth in Leake (NRY).

**Landray, Landrey, Landry, Laundrey:** (i) *Landri, Landric(us)* 1086 DB; Willelmus *filius Landrei* 1219 Cur (L); Gerard *Landri* 1198 P (D). OFr *Landri*, OG *Landric(us)* 'land-ruler'. (ii) Ricardus *de la Lavendaria* 1219 Cur (So); Robert *de la Lauendrie* 1278 AssSo. 'Worker in a wash-house', ME *lavendrie*.

**Landseer:** Gilbert *de la Landsare* 1243 AssSo; Thomas *de la Landschare* 1274 RH (So). 'Dweller by the boundary or landmark', OE *landscearu*.

**Landymore:** v. LANDIMORE

**Lane, Lanes, Loan, Lone, Lones:** Ralph *de la Lane* 1176 P (K); Osbertus *in Lane* 1212 Cur (Sr); Adam *Ithelane* 1227 AssBeds; Walter *atte Lane* 1260 AssC; Nicholas *atte Lone* 1275 SRWo; Roger *de la Lone* 1279 AssSt; Bartholomew *Lane* 1292 FFSf; Thomas *in le lone* 1327 SRDb. 'Dweller in the lane', OE *lanu*. *Lone* shows the characteristic *o* of the West Midland dialects.

**Lang, Lange, Long, Lung:** Ætheric *ðes Langa* 972 OEByn (Nth); Leofwine *Lange* 1070 ASC E; Berard *Long* 1121–48 Bury (Sf); Godfrey *Lunge* 1179 P (Gl); Nicholas *le Long* 1290–2 LLB C; William *Lange* 1296 Black; Adam *ye Langge* 1297 SRY. OE *lang, long* 'long, tall'. *Lang* is Scottish and Northern English.

**Langbain:** *Langebeyn* 1101–7 Holme (Nf); John *Langebayne* 1306 Riev (Y). ON *Langabein* (byname), 'long bone', perhaps 'long leg'.

**Langbant:** William *Langbarne* 1327 SRY. 'Tall child', OE *lang, bearn*. cf. GOODBAIRN.

**Langbridge:** William *de la Langebrigge* 1199 Cur (Sr). 'Dweller by the long bridge', OE *lang, brycg*.

**Langbrook, Longbrook:** Robert *de Langebrok* 1330 PN D 216; John *Langgebrok'* 1332 SRDo. From Longbrook in Milton Abbot (D).

**Langdale:** Henry *de Langdale* 1332 SRCu; Patrick *de Langedale* 1362 AssY; John *Langdale* 1402 IpmY. From Langdale (Westmorland).

**Langdo(w)n:** v. LONGDEN

**Langer:** Henry *de Langhar* 1337, Robert *de Langar* 1347 FrY. From Langar (Nt). v. also LONGER.

**Langfield, Longfield:** Robert *de Langefeld* 1207–8 FFK; William *Langefeld* 1367, *Longefeld* 1393 IpmGl; William *Langfeld* 1470 FFEss. From Longfield (K), Longfield House in Hornchurch (Ess), or 'dweller by the long field', OE *lang, feld*.

**Langford:** v. LONGFORD

**Langham:** Walter, William *de Langham* 1201 Pl (Do), 1327 SRLei; Thomas *Langham* 1392 CtH. From Langham (Dorset, Norfolk, Rutland, Suffolk), Langham Row (Lincs), or Longham (Norfolk), *Langham* 1254.

**Langhurst, Longhurst:** Richard *de Langherst'* 1221 Cur (Sr); Robert *de Longehurst* 1332 SRSx. 'Dweller

by the long wooded hill' (OE *hyrst*), as at Longhirst (Northumb).

**Langland(s):** v. LONGLAND

**Langley, Longley, Longly:** Richard *de Langelega* 1191 P (Sa); Simon *de Longelay* 1297 SRY. From one of the many Langleys or 'dweller by the long wood or clearing' (OE *lang, lēah*).

**Langman:** v. LONGMAN

**Langner, Longner:** Thomas *de Langenehalr'* 1221 AssSa; Thomas *de Longenouere* 1275 RH (Db); Alice *de Langenor'* 1279 RH (O). From Longner (Sa), or Longner (Sa, St).

**Langridge, Langrick, Longridge, Longrigg:** Thomas *de Langgerugge* 1175 P (So); Dionisia *de Langerig* 1253 AssSt; Robert *de Longrigge* 1276 ib.; Thomas *de Langerigg* 1332 SRCu. 'Dweller by the long ridge' (OE *lang, hrycg*, ON *hryggr*), or from Langridge (Som), Langrigg (Cumb), or Longridge (Lancs).

**Langrish, Langrishe:** Robert *de Langerisce* 1199 Pl (Ha); Ralph *de Langris* 1222 Acc; Hugh *Langrisch'* 1332 SRDo. Usually no doubt from Langrish (Hants), but occasionally perhaps it may be a nickname, cf. Robert *le Langerus* 1200 Cur, a derivative of OFr *langeuer* 'lassitude, inertia'.

**Langshaw:** v. LONGSHAW

**Langstaff:** v. LONGSTAFF

**Langston, Langstone:** Berengar *de Langestan* 1230 P (D). From Langstone (D).

**Langthorp, Langthorpe, Longthorp:** Stephen *de Langetorp* 1212 Cur (Y). From Langthorpe (NRYorks).

**Langthwaite:** Hugh *de Langetweit* 1199–1200 FFEss; William *de Langethwait* 1277–8 IpmY; William *de Langthwayt'* 1379 PTY. From Langthwaite (La, NRY, WRY).

**Langton, Longton:** Osbert *de Langeton'* 1191 P (L); William *de Longetone* 1332 SRLa. From Longton (Lancs, Staffs) or one of the Langtons.

**Langtree, Langtry:** Siward *de Langetre* 1206 Cur (La). From Langtree (Devon, Lancs).

**Langworth:** Thomas *Langworth* 1463 IpmNt. From East, West Langworth (L).

**Lanier, Lanyer:** Terricus *Lanier* c1195 Clerkenwell; Thomas *Laniar* 1327 SRY; William *Lanyers* 1146 FrY. 'A dealer in wool', OFr *lanier*.

**Lank:** Avice *Lanke* 1275 RH (Nf). OE *hlanc* 'long, narrow', i.e. tall and thin.

**Lankester:** v. LANCASTER

**Lankshear, Lankshire:** v. LANCASHIRE

**Lansdown, Lansdowne:** John *Landsdowne* 1641 PrSo. From Lansdown (So).

**Lanslyn:** v. LANCELEY

**Lanyer:** v. LANIER

**Lapage, Lappage:** John *Lawpage* 1379 PTY; Robert *Lappadge*, Thomas *Lappage* 1568 SRSf. 'Page, servant of *Law* (Laurence).' ME, OFr *page*.

**Lappin, Lapping:** Robert, Beatrix *Lapyn* 1320 FFK, 1481 FrY. OFr *lapin* 'rabbit'.

**Lapworth:** Nicholas *de Lapworth* 1327 SRWo. From Lapworth (Wa).

**Larcher, L'Archer:** v. ARCHER

**Larcomb, Larcombe, Larkom, Larcom, Larcum, Larkham:** Richard *Larcome* 1576 SRW; John

*Larkam* 1641 PrSo; Henry *Larcombe*, Robert *Larkham* 1662–4 HTDo. From Larcombe in Diptford, in Blackawton (D).

**Larder:** Bernard *Larderer* 1130 P (W); Peter *de Larder'* 1173 P (Ha); Thomas *del Larder* 1304 Cl. David *Larderer* is identical with David *le Lardener* (c1170 Riev). *v.* LARDNER. *Larderer* is a derivative of OFr *lardier*, originally 'a tub to keep bacon in', later, 'a room in which to keep bacon and meat'; hence, 'officer in charge of the larder'.

**Lardner:** Dauid *Lardener*, *le Lardener*, *Lardiner* 1161–81 P (Y); Thomas *le Lardiner* 1193 P (Sr). AFr *lardiner* 'officer in charge of a larder', also 'the officer who superintended the pannage of hogs in the forest'.

**Large, Lardge:** Geoffrey (*le*) *Large* 1204–5 P (Nth). OFr *large* 'generous'.

**Lark, Larke, Laverack, Laveric, Laverick:** Juliana *laueroc* 1243 AssDu; Ralph *Larke* 1275 RH (Nf); William *le Lauerk* 1332 SRSx. A nickname from the lark (ME *larke*, *lavero*(*c*)*k*, OE *lāwerce*).

**Larkin, Larking:** *v.* LORKIN

**Larmer, Larmo(u)r:** *v.* ARMER

**Larner, Lerner:** Larner's Wood in Little Saxham (Suffolk) owes its name to the family of Edmund *de Lauueney* (1327 SRSf). Other members were John *Lawney* (1381 SRSf) and James *Larner* (1562 Saxham PR). The wood is *Lawners* Wood in 1638 (ib.).

**Larrett:** Richard *Laryot* 1524 SRSf; Thomas *Larret* 1674 HTSf. A diminutive of *Larry*, a pet-form of *Laurence*.

**Larwood:** Geoffrey *de Larwode* 1299 NorwDeeds; William *Larwoode* 1524 SRSf; Thomas *Larwood* 1577 Musters (Nf). From some minor place, as yet unidentified but probably in Norfolk.

**Lascelles:** Peter *de Laceles* c1150 Riev (Y); Pigot *de Lasceles* 1185 Templars (Y). From Lacelle (Orne).

**Lasenby:** *v.* LAZENBY

**Lasham, Lassam:** Thomas *de Lasham* 1260 AssC; Oliver *Lassam*, John *Lassambe* 1583 Musters (Sr). From Lasham (Ha).

**Lashford:** *v.* LATCHFORD

**Lashmar, Lashmore:** *v.* LATCHMORE

**Laskey, Lasky:** Henry *Laskey* 1642 PrD. From Lesquite (Co).

**Lass, Less:** William, Thomas *Lesse* 1276 RH (Lei), 1327 *SR* (Ess); Ralph *le Lasse* 1332 SRSx; Herry *Lasse* 1524 SRSf. OE *lǣssa* 'less', ME *lesse*, *lasse* 'smaller', perhaps 'the younger'.

**Lassam:** *v.* LASHAM

**Lasseter:** *v.* LEICESTER

**Lassey:** *v.* LACEY

**Last:** Thomas, Richard *Last* 1385 FFSf. ME *last*, *lest* 'a wooden mould of the foot for a shoemaker', used of a maker of such. cf. Thomas *le Lastur* 1275 AssSo, Hugo *Lastemaker* 1395 NottBR.

**Latch, Latches, Letch, Leche, Leach, Leech:** Richard *del Lech* 1177 P (Y); Cristiana *de Lech* 1210 Cur (Gl); Ralph *de la Leche* 1214 Cur (Sr); Peter *de la Lache*, Henry *del Lache* 1297, 1308 Wak (Y). From Lach Dennis or Lache (Ches), Eastleach or Northleach (Glos), or from residence near a stream or some wet place (OE *\*læcc*, *\*lecc* 'stream'). *v.* also LEACH.

**Latchford, Lashford, Letchford:** Philip *de Lecheford* 1279 RH (O). From Latchford (Ches, Oxon).

**Latchmore, Lashmar, Lashmore, Lechmere:** William *Lechemere*, John *Lachemer* 1296, 1327 SRSx. From Lashmars Hall (Sussex).

**Lateways:** Alan *Latewis* e 13th RegAntiquiss; John *Lateways* 1324 CoramLa; Robert *Lateways* 1379 PTY. This looks like a nickname, 'lately wise', OE *læt*, *wīs*.

**Latham, Lathem, Lathom, Laytham:** Robert, Henry *de Latham* 1204 AssY, 1327 SRSo; Robert *Lathom*, *Laytham* 1494 FFEss, 1563 Pat (L). From Latham (WRYorks), Lathom (Lancs), Laytham (ERYorks), or 'dweller at the barns'.

**Lathe, Leath, Leathes:** Gilbert, Richard *del Lathes* 1296 FrY, 1332 SRLa; Adam *del Laythes* 1332 SRCu; John *del Leth* 1379 PTY. 'Worker at the barn(s)', ON *hlaða*. cf. Henry *Latheman* 1278 AssLa. *Leath* is a Lancashire and Cumberland dialect form.

**Latimer, Lattimer, Lattimore, Latner:** Hugo *Latinarius*, *Interpres*, *Hugolinus Interpres* 1086 DB (Ha); Robertus *Latinus*, *Latinarius* 1086 DB (K), *Latimir* 1087 DM (K); Ralph *Latimarus* 1086 DB (Ess); Gocelinus *le Latimer* 1102–7 Rams (C); William (*le*) *Latimier*, *le Latimer* 1163–85 P (Y); Richard *le Latener* 1332 MEOT (Ess); Richard *Latoner*, *Latomer* notary 1485 GildY. Lat *latimarus*, *latinarius*, *latinus*, *interpres*, OFr *latinier*, *latim*(*m*)*ier* 'interpreter', lit. a speaker of Latin (a1225 MED). 'Latonere, or he that usythe Latyn speche' (PromptParv). cf. Hugo *Latyn* 1327 SRSf. *v.* also LATNER.

**Latner:** Alan, Richard *le Latoner* 1306 MESO (Nf), 1311 LLB D; William *le Latouner* 1327 SRSf. ME *latoun* 'brass' from OFr *laton*, 'worker in or maker of latten' (1339 MED). *v.* also LATIMER.

**La Touche:** A Huguenot name. David *de la Touche* fled to Amsterdam and served in the Irish campaigns and at the Boyne. Digues *de la Touche* established a silk, cambric, and poplin factory at Dublin (Smiles 407–8).

**Latta, Lattey, Latto:** James *Lattay*, *Lata* 1677, 1709 Black. Scottish forms of LAWTEY.

**Latter:** Thomas *le Latier* 1199 ChR (Do); Robert *le Latthere* 1318 LLB E; Robert *Latter* 1327 Wak (Y). A derivative of OE *lætt* 'lath', a lath-maker.

**Lauder, Lawther:** William *de Laudre* 1184 P (We); Robert *de Lauedre* Alex 3 Bain (Berwick); Robert *de Lawdre* 1398 Black. From Lauder (Berwick).

**Laufer:** *v.* LAVER

**Laugharne:** Maurice *Lagharn* 1381 Morris. The English form of Llacharn, an old town in Carmarthenshire.

**Laughlan(d):** *v.* LACHLAN

**Laughton:** William *de Lactone* 1185 Templars (L); Walter *de Laughton'* 1327 SRLei; Thomas *Laughton* 1541 CorNt. From Laughton (Leics, Lincs, Sussex, WRYorks).

**Launder:** *v.* LAVENDER

**Laundrey:** *v.* LANDRAY

**Laurence, Laurance, Laurens, Lawrance, Lawrence, Lorence, Lorenz, Lowrance:** *Laurentius* cellerarius 1141–9 Holme (Nf); *Laurencius* molendinarius 1219 AssY; John *Lorence* 1268 FFSf; Benedict *Laurenz*

1292 FFHu; John *Laurens*, William *Lorens* 1296 SRSx; William *Lourance*, Lucus *Lowrance* 1374, 1481 FrY. Latin *Laurentius*, a common name from the 12th century, with pet-names and diminutives, *Lawrie*, *Laurie*, *Lowrie*, *Larry*, *Lorry*, *Law*, *Low*, *Larkin*, *Lorkin*. There was also a feminine form: *Laurencia* 1201 FFEss, 1296 SRSx.

**Laurette, Lorett, Lorette:** *Lauretta* Picot 1185 RotDom; *Loretta* 1219 AssY; *Loreta* de Motecombe 1332 SRSx. *Laur-et*, a diminutive of *Laura*, a short form of *Laurencia* (f).

**Lavell, Lavelle, Laval:** (i) John *de Laval* 1121–48 Bury (Sf); Gilbert *de Laual* 1200 P (Nb). From Laval, a common French place-name. (ii) Also Huguenot, from Etienne-Abel *Laval*, minister of the French church in Castle Street, London, c1730 (Smiles 409).

**Lavender, Launder, Lander, Landers:** Ysabelle *la Lauendere* 1253 Oseney (O); Ralf *la Lavendere* 1268 AssSo; Thomas *Launder* 1331 FrY; Elyzabeth *Lander* 1524 SRSf. OFr *lavandier* (masc.), *lavandiere* (fem.) 'one who washes', 'washerman, washerwoman' (c1325 MED).

**Laver, Lavers, Laufer:** Eustace *de Lagefara* 1190 P (Ess); Reginald *de Laufare* 1276 LLB A; John *Laver* 1327 SRC. From High, Little, Magdalen Laver (Ess), or 'dweller by the bulrushes or the wild iris', OE *læfer*. Used also as a christian name in the 17th century: *Lavers* Tanbline 1642 PrD.

**Laverack, Laverick:** *v.* LARK

**Laverty:** Thomas *de Lauerketye* 1296, William *Lauertye* 1332 SRSx. From a lost place in East Grinstead (Sx).

**Lavin, Lavine, La Vine:** *Lavin* filia Jordani 1201 Cur; *Lavina* (f) 1203 Cur, 1337 AD vi (D); Richard *Lauyn* 1340 ColchCt. Lat *Lavinia*, of unknown origin.

**Lavington:** Bartholomew *de Lauinton'* 1190 P (W); Osbert *de Lavinton'* 1206 Cur (L); William *de Lavyngtone* 1293 AssW. From Lavington, Bishop's, Market Lavington (W), or Woolavington (Sx).

**Law, Lawes, Laws:** John *de la Law'* 1208 Cur (Wo); William *de Lawe* 1229 Cl (Ess); Hugh *del Lawe* 1309 Wak (Y); William *Law* 1279 RH (C); Nicholas *Lawes* 1539 FeuDu. OE *hlāw* 'hill, burial mound', which became *low* in the south but *law* in the north. The surname may also be a pet-name of Laurence: *Lawe* del Park 1275 SRWo, *Lawe Robynson* 1379 PTY. cf. LOW.

**Lawday:** *v.* LAWTEY

**Lawford:** Thomas *de Lalleford'* 1221 AssWa; Henry *Lawford* 1663 HeMil. From Lawford (Ess), *Laleforda* DB, or Church, Long Lawford (Wa).

**Lawless:** Thomas *Lagheles* 1360 FrY; Richard *Lawles* 1533 KentW. ME *laweles*, *laghles* 'uncontrolled by the law, unbridled, licentious'; for *lawless-man* 'an outlaw'.

**Lawley:** Roger *de Laueleg'* 1221 AssSa; Thomas *Laweley* 1419 KB (Ha). From Lawley (Sa).

**Lawman:** *Lagman* 1066 DB (Y); *Laghemannus* ib. (Ess); *Lageman* 1219 AssY; Alwold, Brictric *Lageman* 1066 DB (L); Hugo *Lageman* c1210 Fees (Ha); Geoffrey *Lauman* 1214 Cur (Herts); William *Laweman* 1279 AD i (Nf). ODa, OSw *Lag(h)man*, ON *Lǫgmaðr*. Adam son of *Lagheman* took his father's

name as his surname, Adam *Laweman* (1246 AssLa). Alwold and Brictric *Lageman* were 'lawmen' of Lincoln (Anglo-Scand *lagman* 'one whose duty it was to declare the law').

**Lawn:** *v.* LAND

**Lawrance, Lawrence:** *v.* LAURENCE

**Lawrenson:** Walter *Laurenceson* 1480 Black. 'Son of *Laurence*.'

**Lawrey, Lawrie, Lawry, Lorie, Lorrie, Lory:** Simon *filius Lari* 1197 FFL; William *Larie* 1279 RH (Bk); Hugh *Laurie* 1784 Bardsley. A diminutive of *Laur* (Laurence). *v.* also LOWREY.

**Lawson:** Richard *Lawisson* 1327 SRC; Henry *Laweson* 1379 PTY. 'Son of *Law*' (Laurence). *v.* LAW.

**Lawtey, Lawty, Leuty, Lewtey, Lewty, Loalday, Luty, Lawday:** Huctredus *Leute* 1212 Fees (La); Alan *Leaute* 1256 Rams (C); Thomas *Lawtye* 1613 YWills. OFr *leaute* 'loyalty'.

**Lawther:** *v.* LAUDER

**Lawton:** Adam *de Lauton'* 1205 P (La); Philip *de Lauton* 1281 AssCh; Robert *Lawton* 1642 PrD. From Lawton (Ches, Hereford).

**Lax:** *Lax* de Ludham 1141–9 Holme (Nf); Thomas *Lax* 1351 FrY. ON *Lax*, a nickname from the salmon, otherwise unrecorded in England.

**Laxton:** Robert *de Laxtun* c1240–50 Reg-Antiquiss; Richard *de Laxton* 1376 IpmNt; Thomas *Laxton* 1442 FFEss. From Laxton (Nt, Nth, ERY).

**Layard:** A Huguenot name from an ancient Albigensian family. The original name was Raymond, Layarde the name of their estate near Montpellier. Pierre Raymond *de Layarde*, b. 1666, left France on the Revocation of the Edict of Nantes, and became a major in the army of William III. His descendant, Austin Layard, was the excavator of Nineveh (Smiles 409).

**Lay(e):** *v.* LEA

**Layborn, Laybourn, Labern, Leeburn, Leyborne, Leyburn, Lyburn:** Robert *de Leburn'* 1192 P (K); Richard *de Laibrunn* 1204 AssY; Roger *de Leiburn'* 1214 Cur (K); Henry *Laburn* 1488 FrY. From Leybourne (Kent) or Leyburn (NRYorks).

**Laycock, Leacock:** Ralph *de Laycok*, Roger *de Lacok* 1247, 1250 AssSt. From Lacock (Wilts) or Laycock (WRYorks).

**Layen:** *v.* LAYNE

**Layer, Leir:** (i) Hugh *de Leir* 1275 RH (L); William *de Leyre* 1291 LLB A. From Leire (Leics) or one of the three Layers in Essex. (ii) Alice *la Eyr* (Leyr) 1327 SR (Ess); Gilbert *Leyr* ib.; William *le Eyr* 1327 SRSf. 'The heir.' *v.* AYER. *Leyer* was also a name for the layers or setters who placed in position the stones worked by the (free)-masons: *cubitores* 1252, *positores* 1365, *leggeres* 1282, *leyers* 1412 (Building 31).

**Layfield, Leighfield, Leyfield, Lyfield:** Hugo *Layfeld* 1442 FrY; William *Leyfeld* 1484 LLB L. 'Dweller by the lea-field', the pasture, grass-land, as at Leyfield (Notts).

**Layland:** *v.* LEYLAND

**Layman, Leyman:** William *Leyman* 1327 SRSx; John *Layman* 1524 SRSf. 'Dweller by the wood or clearing.' *v.* LEA.

**Layne, Layen:** William *de la Leyne* 1275 MELS (Sx); Henry *atte Layne* 1327 SRSx. 'Dweller by the

open tract of arable land at the foot of the Downs.' *v.*
MELS.

**Laytham:** *v.* LATHAM

**Layton, Leighton, Leyton, Lighten, Lighton:**
Richard *de Lecton'* 1201 P (Sa); Roger *de Leyton'*
1276 RH (Hu); William *de Leghton* 1287 AssCh.
From Layton (Lancs, NRYorks), Leighton (Beds,
Ches, Hunts, Lancs, Salop) or Leyton (Essex),
of varied origins. John *atte Layhton* 1390 MELS (Wo)
may have lived at a homestead where leeks were
grown or may have been a worker in the kitchen-
garden (OE *lēac-tūn*).

**Lazar:** Thomas *le Lazur* 1280 AssSo; William *le
Lazer* 1297 Wak (Y). 'The leper' (ME *lazare* 1340
NED).

**Lazenby, Lasenby:** Robert *de Leisingeby* 1204 Pl
(Y); John *de Laysingby* 1361 AssY; William
*Laysyngby* 1421 IpmNt. From Lazenby (NRYorks),
or Lazonby (Cumb).

**Lea, Lee, Legh, Leigh, Ley, Leys, Lay, Laye, Lye,
Lyes:** Ailric *de la Leie* 1148–66 NthCh (Nth); Liffild *de
Lega* 1176 P (Ess); Turgod *de la Lea* 1193 P (Wa);
Philip *de Lye* 1198 Fees (W); Henry *del Lea* 1203 P
(La); William *de la Le* 1207 Cur (Berks); Pagan *a la
Legh* 1208 Cur (Y); John *de Leye* 1275 SRWo; Simon
*atte Lee*, Richard *atte Legh* 1296 SRSx; John *del Lee*
1384 FrY; Hugh *atte Leygh* 1392 MELS (Sr). From
one of the many places named Lea, Lee, Leigh,
Leighs, or Lye, or 'dweller by the wood or clearing'.
OE *lēah* (nominative) became ME *legh, leigh*; the
dative *lēa* became *lee*, and the later dative *lēage* gave
ME *leye, lye*.

**Leach, Leech, Leetch, Leche:** Robert *Leche* c1250
Rams (Hu); John *Lache* Hy 3 Colch (C); Edmund *le
Leche* 1279 RH (O). OE *lǣce* 'leech', 'physician'.

**Leachman, Leechman:** Adam *Lacheman* 1210 Cur
(Y), 1212 Cur (Ha); William *Lachman* 1327 SRSf. OE
*lǣce* 'leech' and *mann*, 'servant of the physician'.

**Leacock:** *v.* LAYCOCK

**Leadbeater, Leadbeatter, Leadbetter, Leadbitter,
Ledbetter, Lidbetter:** Ingald' *Ledbater'* 1221 AssWa;
Walter *Ledbeter, le Ledbetere* 1256 AssNb; Henry
*Leadbetter* 1645 YWills; Mr *Leadbutter* 1674 HTSf.
OE *lēad* and *bēatere* 'a worker in lead'.

**Leaden:** *v.* LEADON

**Leader, Leeder, Lader, Ledder:** Ralph *ledere* 1243
AssDu; Henry *le leeder* 1328 Pinchbeck (Sf); Jo.
*Leader* 1674 HTSf. OE *lǣdere* 'leader', in the sense
'driver of a vehicle, carter' (1423 MED). In the north,
'leading coal, manure, etc.' is still used even though
the driver of the loaded vehicle may not be leading a
horse. In the Middle Ages, the particular commodity
carried was often named: Richard *Bredleder* 14th
Whitby, Robert *Cornlader* 1372 ColchCt, Laurence *le
Maltlader* 1294 MEOT (Herts), William *Waterladar'*
1177 P (Wa). The reference may also be to pleasure:
Richard *Pleyledere* 1327 SRSo. cf. PLAYER, and
*dawnceleder* s.n. DANCE. Fransson derives the
surname from OE *lēad* 'lead' in the sense 'plumber'.
This certainly survives in *Ledder* which is rare. cf.
'wages of a *ledder* soldering the gutters over the great
gate' (1344 Building 266).

**Leadham, Ledham:** Robert *Ledeham* 1452, Robert
Lethum otherwyse callyd Robert *Ledham* 1454
Paston. From Leadenham (L).

**Leadon, Leaden:** Simon *de Ledene* 1221 AssWo;
John *Ledene* 1379 IpmGl. From Leadon (He), or
Highleadon, Upleadon (Gl).

**Leadsom:** *v.* LEDSAM

**Leadwell, Ledwell:** John *de Ledwell* 1279 RH (O).
From Ledwell (O).

**Leaf, Leafe, Leefe, Leif, Lief, Life:** Godwin *'Lief*
1198 P (Nf); Alice *le Lef*, Loue *þe Lef* 1279 RH (C,
Hu); John *(le) Leef* 1318 Crowland (C); Henry *Lyf*
1327 SRSf; Henry *Lief (Leef)* 1327 *SR* (Ess); Lucia *le
Lyf* 1327 SRSo; William *Leof* 1332 SRWa. This is
usually from OE *lēof* 'dear, beloved' but occasionally
it may derive from *\*Lēof*, a short form of *Lēofrīc,
Lēofwine*, etc. This is not evidenced with certainty in
OE but its existence is suggested by *Lif* sacerdos 12th
DC (L).

**Leak, Leake, Leek, Leeke, Leeks, Leck:** (i) Walter,
Ralph *de Lek* 1202 AssL, 1219 AssY; Henry *de Leek*
1290 AssCh. From Leak (NRYorks), Leake (Lincs,
Notts), Leek (Staffs) or Leck (Lancs). (ii) John *Leke*
Hy 3 Gilb (L); Ralph *Leecke* 1279 RH (Beds).
Metonymic for LEAKER.

**Leaker:** Adam *lekere* 1279 MESO (Y); William *le
Leker* 1293 ib. A derivative of OE *lēac* 'leek', a seller
of leeks. cf. Hugh *le Lekman* 1319 MESO (Nf).

**Leal, Leale, Lealman:** Andrew *Leal* 1479 Black;
William *Leleman* 1297 SRY, 1363 FrY. ME *lele*, OFr
*leial* 'loyal, faithful'.

**Leaman, Leamon:** *v.* LOVEMAN

**Lean, Leane:** Hugo *Macer* 1208 Cur (O); Walter
*Lene* 1276 RH (Y); John *Leane* 1317 AssK. OE *hlǣne*
'lean, thin' (Lat *macer*). Also for MACLEAN.

**Leaper, Leeper:** Robert *le Lepere* 1185 Templars
(Wa); Henry *Leper* c1200 DC (Nt); William *le Leapere*
1295 AD vi (K). Either OE *hlēapere* 'leaper, dancer,
runner, courier' or a derivative of OE *lēap* 'basket',
'basket maker'. cf. John *le Lepmaker* 1338 MESO
(Nf). *v.* also LEPPER.

**Leaphard:** *v.* LEOPARD

**Leapingwell:** *v.* LEFFINGWELL

**Leapman:** 'Basket-maker'. cf. LEAPER.

**Lear, Leir:** John *Lear*, Ambrose *Leere* 1642 PrD;
George *Leere* 1662–4 HTDo. A nickname from OE
*hlēor* 'cheek, face'.

**Learmond, Learmont, Learmonth, Learmount,
Leirmonth:** William *de Leirmontht* 1408 Black. From
Learmonth (Berwicks).

**Learmouth:** William *Lermouth* 1438 FrY. From
Learmouth (Northumb).

**Leary:** *v.* O'LEARY

**Leason:** *v.* LEESON

**Leat, Leate, Leates, Leatt, Leet, Leete:** William
*Bytherlete* 1279 AssSo; John *atte Lete* 1330 PN D 172.
'Dweller by the conduit or watercourse' (OE *gelǣt*), as
at Leat (Devon) & The Leete (Essex).

**Leath(es):** *v.* LATHE

**Leather, Leathers, Leatherman:** Robert *Lether*
1524 SRSf. Probably a dealer in leather or one who
used leather in his work. cf. John *Lethercarver* 1404
AD iv (Nth). 'Leather-dyer' was an occupation-term
in London in 1373 (MEOT).

**Leatherhead:** Richard *Leddred*, John *de Leddred* 1274 RH (So); John *Ledered* 1381 AssL. From Leatherhead (Sr).

**Leathley:** William *de Lelaia* Ric I Calv (Y); William *de Letheleia* 13th Kirkstall; Adam *Lethelay* 1332 SRCu. From Leathley (WRY), *Leeleai* 1166.

**Leaver:** *v.* LEVER

**Leaves, Leeves:** *Leue* 1066 DB (Hu); *Leva vidua* c1188 BuryS (Sf); Thomas *Leve* 1229 Pat (Do). OE *Lēofa* (m) or *Lēofe* (f). Sometimes, apparently, topographical: Robert *Intheleves* 1322 LLB E.

**Leavett, Leavitt:** *v.* LEVETT

**Leavey, Levey, Levi, Levy, Lewey, Lewy, Lovie:** *Leuui* 1066 DB; Ricardus *filius Lefwi* 1171–2 MedEA (Sf); Agnes *filia Lewi* 1221 AssSa; Thomas *Leui* 1228 Eynsham (O); Robert *Levi* 1275 SRWo; Geoffrey *Leuwy* 1301 ParlR (Ess). OE *Lēofwīg* 'beloved warrior'. Modern *Levi* is usually the Hebrew *Levi* 'pledged' or 'attached'.

**Leavis:** *v.* LEVIS

**Leavold, Levell, Lovold:** *Leuot* 1066 DB (Ess); *Lewold'*, *Liwwoldus*, *Luwold'* 1066 Winton (Ha); Simon *filius Lefwaldi, Lefwoldi, Leualdi* c1165–80 Bury (Sf); William *Leuald* c1250 Rams (Nf); John *Levold* 1275 RH (K); John *Levell* 1568 SRSf; James *Leavold* 1674 HTSf. Late OE *Lēofweald* 'beloved power or ruler'.

**Lebbell:** Robert *Lebel, le Bel* 1235 Fees (So). 'The beautiful.' *v.* BELL.

**Lebby:** *v.* LIBBY

**Lechmere:** *v.* LATCHMORE

**Leck:** *v.* LEAK

**Lecomber:** *v.* CAMBER

**Lecount:** *v.* COUNT

**Ledbright:** Thomas *Ledbricht* 1219 AssL; Thomas *Leodbricht* 1221 Cur (Gl); Robert *Ledberd* 1275 RH (K). OE *Lēodbeorht*.

**Ledbury:** Nicholas *de Ledebur'* 1275 SRWo; John *de Ledeburi* 1297 MinAcctCo; John *de Ledbury* 1328 ArchC 33. From Ledbury (He).

**Ledder:** *v.* LEADER

**Ledger, Ledger:** *Leodegar'* 1192 P (Ha), 1209 FFEss; *Leodegarus* de Dive 1212 Cur (Nth); Adam, William *Leger* 1279 RH (C), 1305 FFEss; Richard *Leggere* 1377 LLB H. OFr *Legier*, OG *Leodegar* 'people-spear', common throughout France and Normandy through the memory of St Leger, a 7th-century bishop.

**Ledham:** *v.* LEADHAM

**Ledingham:** Hugh *de Ledenham* Hy 2 Gilb (L). From Leadenham (Lincs).

**Ledmore:** Peter *Ledmore* 1642 PrD. OE *Lēodmǣr*.

**Ledsam, Ledsham, Ledsome, Ledson, Leadsom:** Nigel *de Ledesham* 1219 AssY; Cuthbert *Ledsome* 1606 Bardsley; Sarah *Ledson* 1809 ib. From Ledsham (Ches, WRYorks).

**Ledster:** *v.* LISTER

**Ledwell:** *v.* LEADWELL

**Ledwich, Ledwidge:** Roger *de Ledewich* 1221 AssSa. From Ledwyche (Salop).

**Lee:** *v.* LEA

**Leeburn:** *v.* LAYBORN

**Leece, Leese:** *Lecia* 1172–80 DC (L); Willelmus

*filius Lecie* 1219 AssY; Godfrey, William *Lece* 1296 SRSx, 1394 AD iv (Sf). OFr *Lece* (f). *v.* LETTICE.

**Leech:** *v.* LEACH

**Leechman:** *v.* LEACHMAN

**Leed, Leede:** Richard *de la Lede* 1296 SRSx; Robert *du Leed* 1297 MinAcctCo; Jordan *Lede* 1327 SRSf. 'Dweller by the loud brook', OE *hlyde*.

**Leeder:** *v.* LEADER

**Leeds:** Paulinus *de Ledes* 1175–6 MCh; Peter *de Ledes* 1198 FFK; Hugh *de Leedes* 1285 Riev; John *Ledys* 1441 ShefA. From Leeds (K, WRY).

**Leefe:** *v.* LEAF

**Leegood:** *v.* GOOD

**Leek(s):** *v.* LEAK

**Leeman:** *v.* LOVEMAN

**Leeper:** *v.* LEAPER

**Lees:** Roger *de Leges, de Lees* 1212 Cur (K). The plural of LEA.

**Leese:** Alan *de la Laese* 1276 ArchC vi; Robert *de la Lese* 1290 FFHu; Robert *atte Leese* 1315 MELS (Sr). 'Dweller by the pasture' (OE *lǣs*). *v.* also LEECE.

**Leeson, Leason:** John, Roger *Leceson* 1332 SRCu, SRSx. 'Son of *Lece*.' *v.* LEECE.

**Leest:** Peter, Thomas *de la Leste* 1275 RH (K), SRWo. 'Dweller by the path or track' (OE *lǣst* '(foot)step, track'.

**Leet(e):** *v.* LEAT

**Leetch:** *v.* LEACH

**Leeves:** *v.* LEAVES

**Le Fanu:** A Huguenot name. Etienne *Le Fanu* of Caen fled to England c1675, eventually settling in Ireland (Smiles 410).

**Lefeaver, Lefever, Lefevre:** *v.* FEAVER

**Leffan:** Robert *Lifthand* 1204 P (Db); Ralph *Lefthand* 1258 FrLei; John *Leftehand* 1390 IpmNt. 'Left hand', ME *lift, left*, OE *hand*, perhaps 'left-handed'. cf. John *Liftfot* 1284 CtW 'left foot'.

**Leffek:** *v.* LEVICK

**Leffingwell, Lepingwell, Leppingwell, Leaping-well:** Avina *de Leffeleswell* 1258 FFEss; William *de Lefheldewelle* 1291 EAS xxi; Richard *Leffende-welle* 1357 FFEss; William *Leffingwell* 1465 EAS xxi; John *lepyngwell* 1539 ib.; William *Leapingwell* 1709 DKR 41 (Ess). *Lepingwell* derives from Leppingwells in Little Maplestead, Essex, which is called *Leffingwelles* in 1561 and owed its name to the possessions here of the family of Robert *de Leffeldewelle* (1302) who is called *Leffingwell* in an Elizabethan transcript of the Court Rolls. The family, *Leffingwell* in the 15th century and *Leppingwell* in the 16th, took its name from a lost place *Liffildeuuella* (DB) which may survive in a corrupt form in Levit's Corner in Pebmarsh into which their possessions extended. *v.* PN Ess 446–7.

**Leffred:** Martin *Leffreud, Lefred, Lyfred* 1255 FFK. OE *Lēofrǣd*.

**Le Fleming:** *v.* FLEMING

**Lefort:** *v.* FORT

**Lefridge:** *v.* LEVERAGE

**Leftwich, Leftwidge:** Richard *Leftwiche* 1505 AD vi (Ch); Richard *Leftwyche* 1530 LP (Ch). From Leftwich (Ch).

**Legard:** Godefridus *filius Leggard* 1204 P (Sx); Osbertus *filius Leggardi* 1205 Cur (Nf); *Leggard*

Joseph 1327 SRC; John, Adam *Legard* 1275 SRWo, 1296 SRSx; Hugh, Thomas *Leggard* 1275 Wak (Y), RH (Nf); William *Lyggard* 1379 PTY. OFr *Legard*, OG *Leudgard, Liudgard* 'people-protection'. There was also a feminine: *Ligarda* 12th DC (L), *Legarda* 1230 P (He), from OG *Leutgarda*.

**Legat, Legate, Legatt, Leggat, Leggate, Leggatt, Leggett, Leggit, Leggott:** Hugolinus *Legatus* 1084 GeldR (So); Peter *Legat* 1199 P (Co); Ralph *le Legat* 1279 AssNb; John, Robert *Legg(e)* 1327 SRC, SRSx. ME *legg*, ON *leggr* 'leg'. ON *Leggr* was used as a personal-name but no examples have been noted in England. The nickname is confirmed by Rannulf *Jambe* 1221 AssWa.

**Leger:** *v.* LEDGER

**Legg, Legge:** Alueredus, Ædwardus *Leg* 1176, 1185 P (Gl, Nb); John, Robert *Legg(e)* 1327 SRC, SRSx. ME *legg*, ON *leggr* 'leg'. ON *Leggr* was used as a personal-name but no examples have been noted in England. The nickname is confirmed by Rannulf *Jambe* 1221 AssWa.

**Legh:** *v.* LEA

**Legood:** *v.* GOOD

**Legrand:** *v.* GRANT

**Legresley:** *v.* GREALEY

**Le Grice, Le Grys:** *v.* GRICE

**Legwood:** Thomas *Legwood*, Benjamin *Leggett*, Edmund *Legitt*, Henry *Legate* 1674 HTSf. A corruption of LEGAT.

**Lehern:** *v.* HERON

**Leicester, Leycester, Lester, Lestor, Lessiter, Lisseter, Laister, Lasseter:** Hugo *de Legrecestra* 1130 P (Lei); Nicholas *de Leycester* 1287 AssCh; Richard *de Laycestre* 1305 FrY; Henry *Lycester* 1381 PTY; William *Leycetter* 1480 GildY; Henry *Lasisture* 1503 ib.; Richard *Lasseter* 1550 SxWills; Nicholas *Lessetur* 1603 SfPR. From Leicester.

**Leif:** *v.* LEAF

**Leifchild, Liefchild:** *Lefchild* de Ranam 1201 SPleas (Co); *Lefchild* filia Johannis 1221 *ElyA*; William *Lefchild* 1197 P (Nf/Sf); William *Levechilde* 1318 LLB E; Thomas *Leefchyld* 1524 SRSf. OE *Lēofcild*, but sometimes, perhaps, a nickname, 'dear child', cf. Cecilia *Leuebarne* 1379 PTY with the same meaning.

**Leigh:** *v.* LEA

**Leighfield:** *v.* LAYFIELD

**Leighton:** *v.* LAYTON

**Leiner:** *v.* LAINER

**Leiper:** A Scottish form of LEAPER (1608 Black).

**Leir:** *v.* LAYER

**Leir:** *v.* LEAR

**Leirmonth:** *v.* LEARMOND

**Leishman, Lishman:** William *Leischman* or *Leschman* 1466 Black; John *Lecheman* or *Lescheman* 1560 ib.; Roger *Lisheman* 1605 FrY. Scottish forms of LEACHMAN.

**Leising:** *v.* LAISING

**Leiston, Leyston:** William *de Leiston'* 1219 P (Nf/Sf); John *de Leyston* 1332 SRWa. From Leiston (Sf), or Layston (Herts).

**Leitch:** Patrick *Leich(e)*, *Lech(e)* 1440–82 Black. Scots for LEACH.

**Lejeune:** *v.* JEUNE

**Le Keux:** *v.* KEW

**Leleu:** *v.* LOW

**Lely:** *v.* LILEY

**Leman, Lemon:** *v.* LOVEMAN

**Lemarchand, Le Marchant:** *v.* MARCHANT

**Le May:** *v.* MAY

**Lemm:** Eustac' *filius Lemme* 1275 RH (Nf); Geoffrey, Thomas *Lem* 1218 AssL, 1296 SRSx. *Lemme*, a hypocoristic of OE *Lēofmǣr* or *Lēodmǣr*. cf. LEMMER.

**Lemmer:** *Lefmer, Leodmar, Ledmær, Leomar, Lemer* 1066 DB; Robertus *filius Lemmer, Lefmer* 1203–4 P (Y); Robert, William *Lemmer* 1221 *ElyA* (C), 1332 SRSx; John *Ledmer* 1316 FFK. OE *Lēodmǣr* 'people-famous' or *Lēofmǣr* 'dear-famous'.

**Lemmon:** *v.* LOVEMAN

**Lempriere:** John *Lemprere* 1420 DKR 41 (Jersey). 'The emperor', a nickname or pageant-name. In England, of Huguenot origin.

**Lemster:** Roger *Lemsterr* 1381 LoCh. From Leominster (He).

**Lenard:** *v.* LEONARD

**Lench:** Roger *de Lench'* 1208 Fees (Wo); Thomas *de Lench* 1327 SRWo; Robert *de Lench* 1402–3 FFWa. From Abbots, Atch, Church, Rous, Sheriffs Lench (Wo), or 'dweller on the hill', OE *\*hlenc*.

**Lenfestey:** *v.* VAISEY

**Leng:** Symon *Lenge* 1275 RH (Nf). The comparative of OE *lang* 'tall'.

**Lenham:** Ralph *de Lenham* 1206 Cur (Y); Thomas *de Lenham* 1317 AssK; Robert *de Lenham* 1348 FFEss. From Lenham (K).

**Lenn:** *v.* LYNN

**Lennard:** *v.* LEONARD

**Lennox, Lenox:** John *of Levenax* 1400 Black; John *de Lenox* 1428 ib. (Glasgow); George *Lennox* 1542 ib. (Glenluce). From the district of Lennox (Dunbarton).

**Lenthal, Lenthall, Lentell, Lentle:** William *Lentale* 1369 LLB G. From Leinthall (Hereford).

**Lenton:** Clemenc' *de Lenton* 1279 RH (Hu); Henry *de Lenton* 1333 FrY. From Lenton (L, Nt), or Lenton in King's Nympton (D).

**Leo, Leon:** *Leo camerarius* 1121–48 Bury (Nf); *Leo, Leon* de Romeslega 1271 AssSt; Hugo *Leo* 1180 P (Lo); William, John *Leon* 12th Seals (So), 1279 RH (O). OFr *Leon*, Greek λέων, Latin *leo* 'lion', the name of several popes who were canonized. *v.* LYON.

**Leonard, Lenard, Lennard:** *Leonardus* de Berhedon' 1219 Cur (R); Radulphus *filius Lennardi* 13th Rams (Hu); Stephanus *Leonardus* 1221 AssWo; William, Agnes *Leonard* 1279 RH (Hu), 1296 SRSx. OG *Leonhard* 'lion-bold', a name not so common in England as one would have expected in view of the number of churches dedicated to St Leonard, the patron saint of captives.

**Leopard, Leaphard, Leppard, Leppert, Lippard:** William *Lepard* 1296 SRSx; John *Lyppard* 1327 SR (Ess). A nickname from the leopard.

**Le Patourel:** Ralph *Pasturel* 1166 RBE (L). A Norman diminutive of *pastour* 'shepherd', surviving in the Channel Islands. v. PASTOR.

**Le Pelley:** v. PELLY

**Lepick:** v. PIKE

**Lepingwell:** v. LEFFINGWELL

**le Poidevan, Le Poidevin:** v. POIDEVIN

**Leppard, Leppert:** v. LEOPARD

**Lepper:** Geoffrey *Lepere* 1221 Cur (Ess); Richard *le Lepor* 1298 Wak; William *Lepper* 1674 HTSf. ME *lepre* 'a leper'. Hospitals for lepers were not uncommon in medieval England and the use of *leper* as a byname is attested by the survival of LAZAR. That this surname is now very rare is not surprising. It may well often be concealed in *Leopard*. The frequent early form *leper(e)* is often for *Leaper*.

**Leppingwell:** v. LEFFINGWELL

**Lepton:** Adam *de Lepton* 1379 PTY; Barba *Lepton* 1672 HTY. From Lepton (WRY). cf. Alan *Leptoneman* 1379 PTY.

**Lequeux:** v. KEW

**Lerner:** v. LARNER

**Lesley, Leslie, Lesslie:** (i) Robert *de Leslie* 1272 Black; Symon *de Lescelye* or *Lesellyn* 1278 ib. From Leslie (Fife). (ii) *Lecelina* de Clinton' 1207 Cur (O); *Lecelina* 1220 Cur (Herts); Ralph *Letselina* 1101–16 MedEA (Nf); Adam *Lestelin* 1327 SRSf; Thomas *Lestelin* (*Lessely*) 1346 FA (Sf); John *Leslyn* 1524 SRSf; Robert *Lesle* 1674 HTSf. *Lec-el-in*, a double diminutive of *Lece*. v. LEECE.

**Less:** v. LASS

**Lessells:** Alan *de Lascels* a1173 Black; David *Lessellis* 1560 ib. A Scottish form of LASCELLES.

**Lessingham:** John *Lesyngham* 1452–3 IpmNt. From Lessingham (Nf).

**Lessiter:** v. LEICESTER

**Lester:** (i) William *de Lestra* 1084 GeldR (So). From Lestre (La Manche). (ii) Andrew *le lestere* 1327 Pinchbeck (Sf). v. LISTER. v. also LEICESTER.

**Lestrange, L'Estrange:** v. STRANGE

**Lesuard:** For *le su-hierde* 'the sow herd'. v. SEWARD.

**Le Sueur:** a Jersey name. 'Shoemaker.' v. SEWER.

**Letch:** v. LATCH

**Letcher:** (i) Henry *Lacher* 1224 Cur (Y); Walter *le Lecher* 1272 MESO (Ha). Fransson explains *Lecher* as a derivative of ME *leche* 'to cure', physician, a possible explanation of the form but unlikely in view of the frequency of *Leche* 'physician', apart from the fact that it would become modern *Leacher* which does not seem to exist. The surname is a derivative of *læcc* 'stream', with the same meaning as LATCH, LETCH. (ii) James *Lecchur* 1269 AssNb; Nicholas, Thomas *Lechour* 1327 SRC, SRSf. ME *lech(o)ur*, OFr *leceor* 'lecher'. The surname is rare, probably usually disguised as *Leger*.

**Letchford:** v. LATCHFORD

**Letmore:** John *Letmogh* 1356 Pat (Nt); Richard *Letmore* 1662, John *Letmore* 1682 PrGR. 'Kinsman of *Lett*', OE *māga*, a pet-form of *Lettice*.

**Letson, Lettsome:** John *Letesson* 1327 SRC; Elizabeth *Letsome* 1775 Bardsley. 'Son of *Lett*.'

**Lett, Letts:** Gerardus *filius Lete* 1208 FF (C); Ralph *Lette* 1296 SRSx. *Lett*, a pet-form of *Lettice*.

**Lettice:** *Leticia* Hy 2 DC (L), 1206 Cur (Ess), 1221 AssSa; John *Letice* 1247 AssBeds; Warin *Letiz*, Margaret *Letice* 1275 RH (Sf). Latin *laetitia* 'joy'.

**Letton:** Symon *de Letton* 1275 RH (Nf); Richard *Letton* 1524 SRSf. From Letton (He, Nf).

**Lettsome:** v. LETSON

**Leuty:** v. LAWTEY

**Levelance:** Roger *Leuelance* 1221 AssWa; Simon *Levelance* 1275 SRWo; Thomas *Levelaunce* 1361 AssL. 'Raise lance', OFr *lever, lance*, perhaps a nickname for a soldier. cf. LANCELEAVE.

**Levell:** v. LEAVOLD

**Leven, Levens:** Ketel *de Levene* 1196 P (We); William *de Leven* 1260 AssY; John *Levans* 1672 HTY. From Leven (ERY), or Levens (We). In Scotland from Leven (Fife, Renfrew). v. also LEWIN.

**Levene:** v. LEWIN

**Levenger:** v. LOVEGUARD

**Lever, Levers, Leaver:** (i) Ralph *le Levere* 1276 RH (Lei); William *le Levere* 1294 AssSt. OFr *levre* 'hare', a nickname either from speed or timidity. Sometimes, no doubt, a shortening of *leverer* 'a hunter of the hare', 'harrier': Roger *Leuerier* c1230 Barnwell (C), Hugo *Leverer* 1242 Fees (Nb). cf. MAULEVERER. (ii) Dande *de Leuer* 1246 AssLa; Ralph *atte Levere* 1387 AD vi (Sx). From Great or Little Lever (Lancs) or from residence near reedy ground (OE *lǣfer* 'rush, reed, iris').

**Leverage, Leverich, Leverick, Leveridge, Lefridge, Livery, Liverock, Loveredge, Loveridge, Loverock, Luffery:** *Lefric, Leuric* 1066 DB; *Leofricus* 1066 InqEl; *Lowric* presbiter 1066 Winton (Ha); Willemus *filius Lefrich'* 1196 Cur (Lei); Thomas *filius Leueric'* 1219 AssY; William *Leuric* 1086 DB (O); Walter *Lufrich* 1206 P (W); Robert *Leffrich* 1240 DBStP (Ess); Robert *Leverik'* 1276 RH (L); John *Levericke* 1279 RH (C); Henry *Leverige* ib.; William *Loverich* 1279 RH (O); Richard *Liuerich* 1309 SRBeds; Richard *Loverik* 1350 LLB F. OE *Lēofrīc* 'beloved ruler'.

**Leverett, Leveritt:** (i) *Leuret* 1066 Winton (Ha), DB (K); Danielis *filius Lefredi* 1202 FFK; Thomas *filius Lofridi* 1229 Pat (K); Emma, Robert *Luuered* 1221 *ElyA* (Sf), 1227 AssBeds; William *Leverad* 1255 RH (W); William *Lefred* 1274 RH (Ess); William *Loverede* 1279 RH (C); John *Leverede* 1327 *SR* (Ess). OE *Lēofrǣd* 'dear-counsel'. (ii) Peter *Leverot* 1224 FFEss; Agnes, William *Leverit* 1279 RH (O). AFr *leveret* 'young hare'.

**Levermore:** v. LIVERMORE

**Leverrier:** Roger *Leuerier* c1230 Barnwell (C); Hugh *Leverer* 1242 Fees (Nb). OFr *levrier* 'a hunter of the hare'.

**Leversage, Leversuch:** v. LIVERSAGE

**Leverton:** Grim *de Leuertona* 1175 P (Nt); Thomas *Leverton* 1400 FFEss. From Leverton (Berks, L), or North, South Leverton (Nt).

**Leveson, Lowson:** *Leofsuna, Lefsune, Leuesuna* 1066 DB; *Leuesune* 12th Gilb (L); Hugo *Lovesone* 1230 Pat (Nb); Richard *Levesone* 1255 RH (O); Adam *Leveson'* 1275 SRWo; Walter *Lufesone* 1279 RH (O); Richard *Livesone* 1279 RH (C). OE *Lēofsunu* 'beloved son'.

Levet, Levett, Levitt, Leavett, Leavitt, Livett, Livitt: (i) *Leuiet, Leueget* 1066 DB; *Lefget* c1095 Bury (Sf); *Levetus* 1146–53 Rams (Hu); *Leuiet* 1166 P (Nf); *Liviet* 1205 Cur (C); Manewine *Leuiet* 1188 BuryS (Sf); Gilbert *Liuet* 1200 P (Gl); William *Liuet* (*Luuet*) 1205 P (L); Adam *Leuet* 1221 AssWa; John *Lefet* 1285 *Ass* (Ess). OE *Lēofgēat* 'beloved Geat'. The personal name may sometimes be the woman's name OE *Lēofgyð* 'beloved battle': *Leued* 1066 DB, *Liuitha, Liuete*, 1221 AssWa, *Leuith'* 1221 AssSa, *Leffeda* Rugfot 1230 Pat (Sf), *Livith, Livid* 14th Rams (Hu). *Levet*(*t*) may also be a nickname from OFr *leuet* 'wolf-cub', a diminutive of OFr *leu* 'wolf'. cf. LOVETT. (ii) William *de Lieuet* 1196 P (Y). From one of the places named Livet in Normandy.

Levey: *v.* LEAVEY

Levibond: *v.* LOVEBAND

Levick, Leffek: (i) Hardekin *filius Leueke* 1175 P (Nf); *Leweke* de Rouington 1246 AssLa; *Lefeke* Daffe 1279 RH (Beds); William *Leuke* 1204 P (O); Amicia *leueke* 1277 Ely (Sf). OE *Lēofeca*, a diminutive of *Lēofa*. This may also have become *Luke. v.* also LIVICK. (ii) Osbert *le Eveske* 1189 Dugd vi; Henry *Leveske* 1200 P (Hu); John *Levick* 1674 HTSf. OFr *eveske*, ModFr *évêque* 'bishop'. In *Levick* the article has coalesced with the noun. This was then re-divided as *le Vick* and survives as VECK, VICK.

Levin, Levine: *v.* LEWIN

Levinge, Levings, Lewing, Liveing, Living, Livings, Loving, Lowing, Lowings: *Leuing, Liuing, Louincus* 1066 DB; *Leouing', Leowing', Leuing', Lewing', Luuing'* 1066 Winton (Ha); *Levingus* 1198 P (L); Richard *Livinge* 1259 FFEss; Henry *Lywing* 1272–80 Bart (Lo); Robert *Levinge* 1275 SRWo; William *Leuing'* 1297 AssY; William *Lovinge* 1674 HTSf. OE *Lēofing, Lȳfing*.

Levingston(e): *v.* LIVINGSTON

Levis, Leavis: *Lefhese* 1066 ICC (C); Steingrim *Leuis* 1197 P (Nf/Sf); John *Lefhese* 1327 *SR*Ess; Richard *Levis* 1674 HTSf. OE *Lēofhyse*.

Levitt: *v.* LEVET

Levy: *v.* LEAVEY

Lew: (i) Adam *de Lewe* 1191 P (Gl); Alice *ate Lewe* 1279 RH (Hu). 'Dweller by the hill', OE *hlǣw*. (ii) Robert *le Leu* 1207 ChR (Do). AFr *leu* 'wolf'. *v.* LOW.

Lewcock: *v.* LOWCOCK

Lewer(s): *v.* EWER

Lewes: John *de Leuwes* 1296 SRSx. From Lewes (Sussex).

Lew(e)y: *v.* LEAVEY

Lewin, Lewins, Lewens, Leven, Levene, Levien, Levin, Levine, Livens, Lowen, Lowin: Wlfricus *filius Leofwini* 1010 OEByn; *Leuuin, Leuinus, Lifwinus, Liuuinus* 1066 DB; *Lowinus* Rex 1185 Templars (Sa); *Leowinus* pater Normanni 1206 Cur (Sr); Iwanus *filius Lofwini* 1212 Fees (La); Hugo *filius Lewyn* 1230 P (Y); Hugo *filius Levine* 1232 Pat (L); John *Lewyn* 1230 P (Nth); Henry *Lowyn* 1275 SRWo; William *Lowen* 1275 RH (Sf); Robert *Lefwyne*, Roger *Livene* 1279 RH (O, C); Robert, Andrew *Leuene* 1327 *SR* (Ess); John *Liuene* ib. OE *Lēofwine* 'beloved friend'. In the Isle of Man, *Lewin* has the same origin as *Gelling*, but the *Giolla* has transferred only *l* to *eoin* instead of *Gill*:

*McGilleon, MacGillewne* 1511, *Lewin* 1627, *Lewne* 1628, *Lewen* 1698 Moore.

Lewing: *v.* LEVINGE

Lewis, Louis, Lowis: *Lowis* le Briton 1166 RBE (Ess); *Lodowicus* clericus 1205 Cur (R); Walterus *filius Lowis* 1209 P (Wa); Robert, Geoffrey *Lowis* 1202 P (La, L); William *Lewys* 1267 FFSf, 1309 LLB D. *Lewis* is an Anglo-French form of Old Frankish *Hlúdwig* 'loud battle', latinized as *Ludovicus* and *Chlodovisus*, gallicized as *Clovis* or *Clouis*, Fr *Louis*. In Wales it was used for *Llewelyn*: Llewelyn ap Madoc alias *Lewis* Rede 1413 Bardsley.

Lewknor, Luckner: Eadgyfu *æt Leofecan oran* c992 OEByn; Henry *de Leucnore* 1211 FFO; Thomas *Lewconore* 1379 LoCh; Richard *Lewknor* 1525 SRSx. From Lewknor (O).

Lewsey, Lewzey: Azor *filius Lefsi* 1066 DB (Nt); *Lefsius* Hippe 1275 RH (Sf); William *Lefsi* 1221 *ElyA* (C); Helena *Lewsay* 1486 GildY. OE *Lēofsige* 'dear victory'.

Lewt(e)y: *v.* LAWTEY

Ley: *v.* LEA

Leyborne, Leyburn: *v.* LAYBORN

Leycester: *v.* LEICESTER

Leyfield: *v.* LAYFIELD

Leyland, Layland: Aldulf *de Leilande* 1203 FFK; Robert *de Layland'* 1219 AssY; William *de Leylond* 1339 LoPleas; Richard *Leyland* 1418–19 FFWa. From Leyland (Lancs), or 'dweller by the untilled land'. Occasionally perhaps from Ealand (Lincs), cf. Walter *de Leilande* 1205 P (L), *de Eiland* 1205 Cur (L).

Leyman: *v.* LAYMAN

Leyston: *v.* LEISTON

Leyton: *v.* LAYTON

Libbet, Libbett: John *Libet* 1319 AD v (Sr); John *Lybet* 1332 SRSr; Simon *Lybbett* 1642 PrD. *Libb-et*, a diminutive of *Libb*, a pet-form of *Isabel*.

Libby, Lebby: John son of *Libbe* 1298 Wak (Y); William *Lybbe* 1506 Oxon. Pet-names for *Isabel. v.* IBBS and cf. the diminutive: John *Libet* 1319 AD v (Sr).

Licence: *v.* LYSONS

Lickbeard: Hugh *Likkeberd* 1230 P (C); John *Lickberd* 1674 HTSf. 'Lick beard', ME *likken*, OE *beard.* cf. Leofric *Liccedich* 1114–19 Rams (Hu) 'lick dish'; William *Lykkedoust* 13th AD i (Ess) 'lick dust'; Geoffrey *Lickefinger* 1206 Cur (Nf) 'lick finger'; Eylwin *Likkepeny* 1231 Cur (Sf) 'lick penny'.

Lickorish, Licrece, Liquorish: *Licorice* the Jew 1252 AD iii (Ha); Ralph *Lycorys* 1348 AssSt; John *Licorishe* 1637 Bardsley. ME *likerous* 'wanton, lecherous'.

Lidbury, Lydbury: Adam *de Ludebir'* 1221 AssSa; John *de Lydebur'* 1327 SRSa. From Lydbury North (Sa).

Liddall, Liddel, Liddell, Lidell, Liddle, Lydall: Richard *de Lidel* 1202–34 Black (Largs); Geoffrey *Liddel* 1266 ib. (Roxburgh); Christiana *de Lydell* 1370 FrY. From Liddel (Cumb, Roxburgh).

Liddiard, Lyddiard, Lydiard, Lydiart: Ralph *de Lidiard'* 1212 Cur (K); Robert *de Lidʒerd* 1276 Glast (So); Thomas *Lidyerd* 1367 FFW. From Bishop's Lydeard, Lydeard St Lawrence (So), or Lydiard Millicent, Tregose (W).

**Liddington, Lidington:** Robert *de Lidinton'* 1204 P (O); Henry *de Lidi'gtun* 1230 P (L); William *de Lydyngton* 1300 FrY. From Liddington (R, W).

**Lidgate, Liggat, Liggatt, Liddiatt, Lidgett, Lyddiatt, Lydiate:** Ralph *de Lidgate* 1230 P (Sx); William *de Lydyathe* 1242 Fees (La); Philip *atte Lidgate* 1274 Wak (Y); Richard *de la Lydeyate* 1280 AssSt; John *atte Lygate* 1332 SRSx. 'Dweller by the swing-gate' (OE *hlidgeat*), as at Lidgate (Suffolk) or Lydiate (Lancs).

**Lidney:** Walter *de Lydeneye* 1377 IpmGl. From Lydney (Gl).

**Lidster:** *v.* LISTER

**Lief, Life:** *v.* LEAF

**Liefchild:** *v.* LEIFCHILD

**Liell:** *v.* LYAL

**Lier, Liers:** Adam *le Liur* 1222 AssWa; Alan *le Lyere* 1298 AssL; Richard *le Lyer* 1327 SRLei. AFr *liur* 'bookbinder'.

**Light, Lyte:** (i) Thomas *de Leht* 1275 RH (K); Richard *atte Lighte* 1317 MELS (Sx). OE *lēoht* 'light', probably 'light place', hence 'dweller in a clearing or glade'. (ii) John *le Lyt* 1266 AssSo. OE *lȳt* 'little'. (iii) Thomas *le Leht* 1275 RH (K); Thomas *Lyght* 1377 ColchCt. OE *lēoht* 'light'; active, bright, gay.

**Lightband, Lightbound, Lightbown:** William *Litebonda* 1195 P (Cu); James *Lightbowne* 1682 PrGR. The first element may be OE *lēoht* 'light', but *Lite-* in compounds is often for *Little*, OE *lȳt*. Hence 'gay, active' or 'little husbandman, peasant'. *v.* BOND.

**Lightbody:** (i) Katerina *litlebod'* 1327 SRC. 'Little body', 'little person' (OE *lȳtel*). cf. BODDY. (ii) William *Leichtbody* 1552 Black, John *Lychtbodie*, *Lightbody* 1574, 1678 ib. 'Active, gay individual' (OE *lēoht*).

**Lightburn:** (i) Geoffrey *Lithbarn* (*Lytbarn*) 1327 *SR* (Ess). 'Active or little child' (cf. LIGHTBAND, GOODBAIRN), or from OE *līðe* 'gentle, mild'. (ii) Rober *Lightborn* 1593 Bardsley (La). From Lightburne in Ulverston (Lancs).

**Lighten, Lighton:** *v.* LAYTON

**Lightfellow:** Robert *Lyghtfelow* 1393 LoPleas. 'Active man', OE *lēoht*, ON *félagi*.

**Lightfoot:** William *Lihfot* eHy 2 Seals (O); Hugh *Lihtfott* 1206 Cur (L); John *Lyghtfot* 1296 AssCh. OE *lēoht* 'light' and *fōt* 'foot', one with a light, springy step, a speedy runner, messenger. cf. *Lyghtefote Nuncius* in the Towneley Play of *Cæsar Augustus* and John *Litefot*, cacher' 1274 RH (Nf).

**Lightlad:** Elias *Lithelad* 1327 SRDb. Perhaps 'mild, gentle lad' (OE *līðe*), or, 'active or little lad'. cf. LIGHTBAND, LADD.

**Lightoller, Lightollers, Lightouler, Lightowler, Lightowlers:** Michael *de Lightholres* 13th WhC (La). From Lightollers (Lancs).

**Lightwood:** John *de Lightwood* 13th Shef. From Lightwood in Norton (Derbyshire).

**Lilbourn, Lilbourne, Lilburn, Lilburne:** Alexander *de Lilleburna* 1170 P (Nb); John *de Lilleburn* 1327 SRY; Stephen *Lylborn* 1434–5 FFSr. From Lilbourne (Northants), or Lilburn (Northumb).

**Lile:** *v.* LISLE

**Liley, Lilley, Lillie, Lilly, Lely:** Thomas son of *Lylie* 1296 Wak (Y); Alan *Lilie* 1247 AssBeds; Hugh

*Lily* 1275 RH (L); William *Lely* 1275 RH (Nf). *Lylie* is probably a pet-form of *Elizabeth*. The diminutive *Lilian* occurs in Geoffrey *Lilion* 1279 RH (Beds) and the modern surnames *Lelliott, Lilliard* point to other early derivatives. Occasionally the surname derives from Lilley (Herts) or Lilly (Berks): Nicholas *de Lilleye* 1342 LoPleas.

**Lilleker, Lillico:** *v.* LITHGOE

**Lilleyman, Lilliman, Lillyman, Lithman, Litman, Littman, Lyteman, Lutman, Luttman, Lulman:** *Lutemannus* 1066 Winton (Ha); *Liteman de Clunton'* 1176 P (D); Rogerus *filius Liteman* 1204 Cur (Sf); *Lillemannus* Pastor 1212 Cur (Y); Reginaldus *filius Luteman* 1221 AssGl; Robert, Geoffrey *Litleman* 1202 AssL, 1221 *ElyA* (Sf); Roger *Lelman* 1212 Cur (Y); Robert *Lulleman* 1235 Fees (C); Adam *Lilleman* 1260 AssY; Richard *Liteman* 1275, 1279 RH (Nth, Beds); Malina *Lyleman* 1297 Wak (Y); Gilbert, William *Lȳteman* 1305 SIA iii, 1327 SRSx. OE *\*Lȳtelmann* 'little man', an original nickname. For the development, cf. Lilly Wood and Plantation (PN Ess 396, 420), both *Littlehei*.

**Lillingston, Lillystone:** Geoffrey *de Lillingestan'* 1200 P (O); Thomas *Lyllingstone* 1350 LLB F; Anthony *Lilliston*, John *Lilliston* 1674 HTSf. From Lillingstone Dayrell, Lovell (Bk).

**Lillywhite, Lilywhite:** John *Lyliewhyt* 1376 ColchCt; Stephen *Lylywythe* 1398 Balliol (O). 'White as a lily', originally used of a woman. For a man, a nickname for one with a complexion fair as a woman's.

**Liman, Limon:** *v.* LOVEMAN

**Limbert:** *v.* LAMBARD, LOMBARD

**Limbrey, Limbery, Limbury:** Gerard *de Linberga* 1130 P (W); Roger *de Limberge* 1212 Cur (L); John *Limbrey* 1662–4 HTDo. From Limbury (Beds).

**Limer, Lymer:** Thomas *Limer, Lymer* 1219 AssY; John *le Limer* 1279 RH (Hu). A derivative of OE *līm* 'lime', a whitewasher.

**Limmer, Lumner:** Ralph *le Liminur*, Robert *le Luminur* 1230 Oseney (O); Reginald *le Eluminur* 1257 ib.; Reginald *le Ylluminur* 1265 FF (O); John *le Leominur*, Richard *Lemner* 1275 SRWo; John *le Lumynour, le Alumynour* 1282, 1291 AD iv (Sr); Edmund *le lumner* 1327 SRSt; Thomas *Lymnour* 1493 LLB L. OFr *enlumineor, illumineor* 'an illuminator of manuscripts'.

**Limmington, Lymmington, Lymington:** Thomas *Lymington* 1662–4 HTDo. From Limington (So).

**Limpett:** Beatrice *atte Lympette* 1332 MELS (Sx). 'Dweller at the lime-pit', OE *līm, pytt*.

**Limsey:** Richard *de Limesia* 1148 Winton (Ha); Nicholas *de Lymesy* 1285 WiSur. From Limésy (Seine-Maritime).

**Linacre, Linaker, Lineker, Liniker, Linnecar, Linnecor, Linnekar, Linegar, Linnegar:** Godwin *de Linacra* 1086 DB (C); Alan *de Linacre* 1227 AssSt. From Linacre (Lancs), Linacre Court *alias* Lenniker (Kent), or 'dweller by the flax-field'.

**Lincey:** *v.* LINDSAY

**Linch, Lince, Linck, Link, Lynch:** Geoffrey *Linch* 1228 FFSf; Gilbert *de la Lynche* 1275 SRWo; Robert *de Bynithelinche* 1278 AssSo; Robert *Lincke* 1279 RH

(C); Simon *atte Lynke* 1296 SRSx; Robert *atte Linch* 1327 ib. 'Dweller by the hill' (OE *hlinc*). *Lynch* may be Ir *Ó Loingseacháin* 'descendant of *Loingseach*' (sailor).

**Lincoln, Linkin:** Aluredus (*de*) *Lincolia* 1086 DB. From Lincoln.

**Lind, Lynde, Lynds:** Robert *de la Linde* 1185 P (Sr); Richard *atte Lynde* 1275 SRWo. 'Dweller by the lime-tree' (OE *lind*).

**Lindall, Lindale, Lindell:** John *Lyndale* 1419 IpmY; Alexander *Lyndell* 1621 SRY; James *Lindall* 1662 LaWills. From Lindal (La).

**Linden, Lindon:** Gilbert *de Lindona* 1086 InqEl (C); Simon *de Lindon'* 1202 AssL; Richard *de Lyndone* 1327 SRLei; William *Lyndon* 1405 FFEss. From Lindon (L, Wo), Lyndon (R), or Linden End in Haddenham (C).

**Linder:** Richard *le Lyndere* 1327 SRSx. v. LIND and p. xv.

**Lindhurst, Lyndhurst:** Herbert *de Lindherst* 1169 P (Ha); Robert *de Lyndhurst* 1268 AssSo; William *Lyndherst* 1392 CtH. From Lyndhurst (Hants), or 'dweller by the lime-wood'.

**Lindley, Lingley, Linley:** Siward *de Lindele* 1204 AssY; Thomas *de Linleia* 1204 P (Do); Matthew *de Linlee, de Lindleg'* 1205–6 Cur (Herts); Martin *de Lingleg'* 1219 Fees (Do); Richard *de Linlegh'* 1221 AssSa, *de Lingl'* 1251 Fees (Sa); Jordan *de Lynley* 1275 Wak (Y). From Lindley (Leics, WRYorks), Linley (Salop, Wilts), or from residence near a lime-wood (OE *lind, lēah*) or a clearing where flax was grown (*līn-lēah*).

**Lindsay, Lindsey, Lincey:** Thomas *de Lindesie* 1207 FFL. From Lindsey (Lincs).

**Lindwood, Linwood, Lynwood:** Richard *de Lindwude* 1196–1203 RegAntiquiss; John *de Lyndewode* 1334 AssNu; Thomas *Lynwod* 1437 FrY. From Linwood (L, Ha).

**Line, Lines, Lyne, Lynes:** *Lina* 1181 P (O), 1195 Cur (So), 1202 FFL; William, Thomas *Lyne* 1296 SRSx, 1332 SRWa; Reginald *Lynes* 1340 AssC. *Lina*, a pet-form of such women's names as *Adelina, Emelina, Lecelina,* etc.

**Linegar:** v. LINACRE

**Lineham, Lynam, Lynham:** Walter *de Linham* 1205 P (D). From Lyneham (Devon, Oxon, Wilts). Also Ir *Ó Laigheanáin* 'descendant of *Laidhghnean*' (snow-birth).

**Linford, Linforth, Linfoot:** Geoffrey *de Lineford'* 1202 FFNf. From Linford (Bucks) or Lynford (Norfolk).

**Ling, Linge, Lings:** Bernard *del Ling* 1207 Cur (Sf); William *atte Lyng* 1327 FrNorw; John *Lyng* 1433 AssLo. From Lyng (Nf, So), or 'dweller in the ling', ON *lyng*.

**Lingard, Linyard, Linger:** Jordan *de Lingarth* 1246 AssLa; William *Lyngard* 1401 AssLa; John *Lynger* 1524 SRSf. From Lingart (La), or Lingards Wood in Marsden (WRY).

**Linge:** v. LING

**Lingen, Lingham, Linghorn:** Ralph *de Lingein* 1183 P (Sa); John *Lyngam* 1332 SRWo; John *Lyngeyn, Lingen, Lingoyn* or *Lyngham* 1509 LP (He). From Lingen (He), *Lingein* 1178.

**Linger:** v. LINGARD

**Lingham, Linghorn:** v. LINGEN

**Lingley:** v. LINDLEY

**Lings:** v. LING

**Link:** v. LINCH

**Linkin:** v. LINCOLN

**Linklater:** Criste *Ælingeklæt* 1424 Black; Andro *Lynclater, Lincletter, Linclet* 1504 ib.; Thomas *Linkletter* 1634 ib. A distinctively Orkney name from Linklater (South Ronaldsay, North Sandwick), or Linklet (North Ronaldsay).

**Linley:** v. LINDLEY

**Linn:** v. LYNN

**Linnecar, Linneker, Linnegar:** v. LINACRE

**Linnett, Linnit, Linnitt:** *Linota* 1279 RH (O); Robert, Thomas *Linet* 1275 RH (Sx), 1317 AssK; Walter *Lynot* 1389 LLB H. *Lin-et, Lin-ot*, diminutives of *Lina* (v. LINE), or nicknames from the linnet (ME *linet*, OFr *linot*).

**Linney, Linny:** *Linniue* 1185 Templars (Herts); Eustace *Linyeve*, Geoffrey *Lyneue*, William *Linyiue* 1279 RH (C); Richard *Lyneue* 1327 SRSf. OE *\*Lindgifu, \*Lindgeofu* (f) 'shield-gift'.

**Linstead, Linsted:** Thomas *de Linsted'* e 12th RegAntiquiss; Walter *de Linstede* 1255 FFK; John *de Lynstede* 1374 FFEss; Richard *Lynsted* 1450 Paston. From Linsted (K), Linstead Magna, Parva (Sf), or Linstead in Pleshey (Ess).

**Lint:** Metonymic for *Linter*.

**Linter:** Robert *le lynetier* 1263 MESO (Sx); Walter *le Lintere* 1275 SRWo. A derivative of ME *lynet, lynt* 'flax', a dresser of flax.

**Linton, Lynton:** Juliana *de Linton'* 1208 FFY; Walter *de Lynton* 1360 IpmGl; John *Lynton* 1488 FFEss. From Linton (C, Db, He, K, Nb, WRY), Linton Grange, West Linton (ERY), Linton upon Ouse (NRY), or Lynton (D). In Scotland from Linton (East Lothian, Peebles, Roxburgh).

**Linwood:** v. LINDWOOD

**Linyard:** v. LINGARD

**Lipp:** Ælfsinus *Lippe* 958 OEByn; Edwardus *Lipe* 1066 DB (Do); Roger *Lippe* 1148 Winton (Ha), 1210 Cur (Nth). The OE example is certainly a nickname from OE *lippa* 'lip'. Post-Conquest examples are common, though the modern name is rare, and we may also have a pet-name *Leppe* or *Lippe*, from an OE name in *Lēof-*. cf. *Leppe* c1190 DC (L), 1198 P (Y); *Lepsy* 1275 RH (Nf), Robert *Lipsi* 1196 P (Co), for *Lēofsige*; *Lepstan(us)* 1066 Winton (Ha), for *Lēofstān*. v. PNDB §P 90, pp. 91, 315.

**Lippard:** v. LEOPARD

**Lippett, Lippitt:** William *Lypet* 1296 SRSx; Roger *Lypat* 1327 ib. A diminutive of *Lipp*.

**Lippiatt, Lippiett, Lipyeat:** Roger *de la Lypiat* 1242 P (So); Thomas *atte Lippȝete* 1333 MELS (So). From Lypiate (Som) or Lypiatt (Glos); or 'dweller by a *leap-gate*, a low gate in a fence which can be leaped by deer, while keeping sheep from straying' (OE *hlīepgeat*).

**Lippincott:** Thomazine *Lippingcott* 1637 PN D 613; John *Lippingcot*, Thomas *Lippincot* 1642 PrD. From Lippingcotts in Culmstock (D).

**Lipsett:** v. LUPSETT

**Lipson:** *Lepstan', Lippestanus* bittecat 1066 Winton (Ha); Juliana *Lepsone* 1283 SRSf. 'Son of *Lēofstān*.' *v.* LIPP. Also from Lipson (Devon).

**Lipton:** From Lipton in East Allington (D).

**Liquorish:** *v.* LICKORISH

**Lishman:** *v.* LEISHMAN

**Lisle, Lile, Lyle, de Lisle, De L'Isle, de Lyle:** Hunfridus *de Insula* 1086 DB (W); Peter *de Isla* 1166 RBE (Y); Ralph *de Lile* 12th DC (L); Robert *del Ile* 1311 FrY; Henry *Lyle* 1319 Crowland (C). AFr *del isle* '(dweller) in the isle'. *Isle* is a common French place-name and the surname may sometimes derive from Lille (Nord), but it may also be of English origin. Robert de Insula, Bishop of Durham in 1274, was the son of poor crofters at Lindisfarne and took his name from Holy Isle.

**Lison:** *v.* LYSONS

**Lisseter:** *v.* LEICESTER

**Lister, Lyster, Litster, Lidster, Ledster, Lester:** Ralph *Litster* 1286 Wak (Y); John *le Letstere* 1305 FFSf; Richard *le Lyster* 1327 SRDb; John *le Listere*, Peter *le lestere* 1327 SRSf. A derivative of ME *lit(t)e* 'to dye', a dyer. An Anglian surname.

**Liston:** (i) Geoffrey *de Liston* 1235–6 FFEss; John *de Liston* 1307–9 FFSr; Thomas *Liston* 1642 PrD. From Liston (Ess). (ii) Roger *de Liston* 1163–85, Thomas *de Lystoune* 1334 Black. From the barony of Liston.

**Litcook, Litson, Litt:** *Litecock* de Salford 1246 AssLa; Simon *Lytecok* Ed 3 Battle (Sx); Roger *Lutesone* 1321 AssSt; John *Litecok* 1327 SR (Ess). *Litecok* is a pet-form of *Litt*, a short form of such names as *Litman*, *Litwin*.

**Lithgoe, Lithgow, Lythgoe, Lillico, Lilleker:** Symon *de Lynlithcu* 1225 Black; John *de Lithcu* 1312 CalSc; James *Lithgo* 1552 Black; Robert *Lynlytgow* 1599 ib. From Linlithgow.

**Lithman, Litman:** *v.* LILLEYMAN

**Litson:** *v.* LITCOOK

**Litster:** *v.* LISTER

**Litt:** *v.* LITCOOK

**Little, Littell, Lytle, Lyttle:** Eadric *Litle* 972 OEByn (Nth); Lefstan *Litle* c1095 Bury (Sf); Thomas *le Lytle* 1296 SRSx. OE *lȳtel* 'little'.

**Littlebond:** Waldev *Littlebond* 1231 Pat (Nb). 'Little husbandman', OE *lȳtel*, ON *bóndi*. *Little* is a common first element in nicknames. In addition to the examples given in Reaney, cf. Stephen *Litilbaker* 1497 SRSr 'little baker'; John *Lytylbrother* 1485 CantW 'little brother'; *Littel Dick* 1640 LeiAS 23 'little Dick'; Agnes *Lyttelfyngre* 1401 AssLa 'little finger'; Hugh *Litilfot* 1209 P (Db) 'little foot'; William *Lytelgrom* 1212–23 Bart 'little servant'; Edwin *Litelhand* 1203 AssSt 'little hand'; Robert vocatus *Litelrobyn* 1324 CorLo 'little Robin'; *Lytle Symme* 1351 AssEss 'little Sim'.

**Littleboy:** Robert *litelboie* 1206 FrLeic; John *Lytelboye* 1384 Add 27671 (Ha). 'Little boy.'

**Littlechild:** Ralph *Litechild* 1209 P (Lei); Richard *Litilchilde* 1524 SRSf. 'Little child.'

**Littledale:** William *de Liteldale* 1332 SRCu; John *Litildale* 1447 FrY. From Little Dale in Pickering (NRY).

**Littlefair:** Agnes *Lutfair* 1381 SRSt. 'Little companion' (ME *fere, feir*).

**Littlehailes, Littlehales:** Kiena *de Litlehale* 1180 P (L). 'Dweller at the small nook of land', OE *lȳtel, healh*.

**Littlejohn, Littlejohns, Litteljohn:** Litel *Jon* 1350 ColchCt; John *Lytelion* 1372 ib. 'Little John', often, no doubt, for a giant.

**Littlepage:** Ralph *Litelpage* 1314 LLB D. *v.* LITTLE, PAGE.

**Littleproud:** Goderun *Litteprot* 1066 Winton (Ha); Matilda *Lytillprowd* 1379 PTY. A hybrid, OE *lȳtel* 'little' and OFr *prut, prud* 'worth, value'. The wife of Henry *Lytilprud* (1301 NottBR) was Hawisia '*Crist a pes*' whose constant cry 'Christ have peace!' suggests her husband deserved his nickname.

**Littleton, Lyttleton:** Driu *de Litletun'* 1210 P (W); Thomas *de Luttelton* 1358 SRWo; John *Lytelton* 1416 IpmNt. From one or other of the many places of this name.

**Littlewood:** Geoffrey *de Litelwode* 1275 Wak (Y); Robert *atte Lytlewode* 1327 SRWo. 'Dweller by the little wood.'

**Littleworth:** Hugo *de Litlewurd* 1180 P (Wa). From Littleworth (Warwicks).

**Littley:** (i) Walter *litel eie* 1162 DC (L). A nickname, 'little eye'. (ii) Juliana *Littlehey* 1327 SR (Ess). 'Dweller at the little enclosure' (OE *lȳtel, (ge)hæg*), as at Littley Green or Littley Wood (Essex).

**Littman:** *v.* LILLEYMAN

**Litton, Litten, Lytton:** Gamel *de Litton* 1175 YCh; Henry *de Lideton'* 1212 Fees (Do); Thomas *de Lytton* 1327 SRDb; Robert *Lytton* 1403 FFEss. From Litton (Derby, Dorset, Som, WRYorks).

**Litwin:** *Lihtuuinus* 1066 DB (Sf); Walterus nepos *Lituini, Lictwine* 1172, c1175 DC (L); *Licthwinus* 1235 CartAntiq (Ess); Guthmarus filius *Lithwini* c1250 Rams (Nf); John *Lutewin* 1194 Cur (Beds); Herveus *Lightwyne* 1285 *Ass* (Ess); Hugh *Lithwyne, Lightwyne* 1312, 1346 ColchCt; John *Lytzwyne* (*Lytwyne*) 1327 SR (Ess); William *Litwyn* 1441 FrY. An unrecorded OE personal name *Lēohtwine* 'bright friend'.

**Livard, Luard, Lyward:** *Leuuard, Liuuard* 1066 DB (K, Nt); Madducus filius *Luardi* 1170 P (L); *Lefwardus* filius Gode 1197 P (K); *Lewardus* filius Ainilde 1207 Cur (Sr); Willelmus *filius Leuardi* 1214 Cur (K); Nicholas *Lefward* 1212 Fees (Wo); Peter *Loward* 1332 SRSr. OE *Lēofweard* 'beloved guardian'. We should have expected *Leward* as the normal development. Perhaps the final *d* was lost (cf. *Lefwarus* c1250 Rams) and the surname absorbed by *Lewer*. After the Revocation of the Edict of Nantes in 1678, *Luard* was re-introduced by a Huguenot from Caen, Robert Abraham Luard, who became the ancestor of the Luards of Essex and Lincs (Lower). This is a Germanic name from the roots *Hlod-* and *-ward* 'glory-protector', French *Louard*.

**Liveing, Living, Livings:** *v.* LEVINGE

**Livelong:** Richard *Lyvelonge* 1236 AssHa; Thomas *Livelong* 1662–4 HTDo. 'May you live long', ME *liven, long*. cf. Roger *Libbesofte* 1374–5 FFSr 'live softly'.

**Livemore:** To the question 'Why do you call yourself *Livemore*? Your name ought to be

*Livermore*', a schoolboy of about 12 in the 1930's replied: 'Yes, sir. My father's name is *Livermore*, but when my mother registered my birth, the registrar left out the *r*, and my name's *Livemore*.'

**Livens:** *v.* LEWIN

**Livermere, Livermore, Levermore:** Raulf *de Lieuremere* 1087–98 Bury (Sf). From Livermere (Cambs, Suffolk).

**Liverock:** *v.* LEVERAGE

**Liverpool:** John *Lyverpole* 1379 LoCh; John *of Lyverpull* 1401 AssLa; Roger *Lyuerpole* 1417–18 FFWa. From Liverpool (La).

**Liversage, Liversedge, Liversidge, Leversage, Leversuch:** Robert *de Luvereseg', de Liverseg'* 1212 Cur (Y). From Liversedge (WRYorks).

**Liverton:** Laurence *de Lyverton* 1376 FrY; Thomas *Lyverton* 1415 IpmY. From Liverton (NRY).

**Livesay, Livesey, Livsey:** Henry, Adam *de Liuesay* 13th WhC, 1332 SRLa; Christopher *Levesey* 1455 IpmNt. Usually no doubt from Livesey (Lancs), but occasionally perhaps from OE *Lēofsige* 'dear-victory'.

**Livett, Livitt:** *v.* LEVET

**Livick, Livock:** Walter *Lyvoke* 1369 LLB G. A variant of *Levick* or *Leffek*. As names in *Lēof-* became *Lief-*, *Live-*, *Lēofeca* would become ME *Liveke*.

**Livingston, Livingstone, Levingston, Levingstone:** Archibald *de Levingestoune* 1296 Black. From Livingston (West Lothian).

**Llewellin, Llewellyn, Llewelyn, Llewhellin:** *Lewelinus* Bochan 1198 P (Sa); *Thlewelin* ap Euer 1287 AssCh; Tudor ap *Llywelyn* 1391 Chirk; Roger *Lewelin* 1255 AssSa; Peter *Lewlyn* 1275 SRWo; Peter *Lewelyne* 1301 SRY. Welsh *Llywelin*, often explained as 'lion-like', but probably from *llyw* 'leader'.

**Lloyd, Lloyds, Loyd:** Richard *Loyt* 1327 SRWo; Ithell *Lloit* 1391 Chirk; Richard *Lloyd* 1524 SRSf. Welsh *llwyd* 'grey'.

**Loach:** Robert *Loche* 1206 P (So); William *Loch* 1327 SRSo; William *le Loche* 1524 FFEss. A nickname from the fish, ME *loche*.

**Loader, Loder:** Emma *la Lodere* 1279 RH (O); Simon *le Loder* 1332 SRWa. Either a derivative of ME *lode* 'to load', a carrier, or equivalent to *atte Lode*. *v.* LOADES, LODDER.

**Loades, Loads:** Wulfward *de la Lada* 1194 P (So); John *ate Lode* 1327 SRSx. 'Dweller by the path, road, or watercourse' (OE *(ge)lād*), as at Long Load (Som) or Load Fm in Kingsley (Hants).

**Loadman, Loadsman:** John *Lademan* 1301 SRY; Petronilla *Lodman* 1341 ColchCt. OE *lād* 'way, course; load, burden' and *mann*. Either 'carrier' or 'dweller by the road or stream'. cf. LOADER.

**Loalday:** *v.* LAWTEY

**Loan:** *v.* LANE

**Loaring:** *v.* LORING

**Lob, Lobb:** Godric, Richard *Lobbe* Hy 1 ELPN, 1236 FFBk. OE *lobbe* 'spider', a nickname. Also from Lobb (Devon, Oxon): Philip *de Lobbe* 1242 Fees (D).

**Lobley:** Adam *Lobley* 1493 FrY; Thomas *Lobley* 1672 HTY. From Lobley Gate in Baildon (WRY).

**Lock, Locke, Locks:** (i) Leuric *Loc*, Leovric *Locc* 1130 P (Wa), 1130 Cur (Ha); Eustace *Loc* 1235

CartAntiq (Ess). OE *loc(c)* 'lock (of hair)', probably a nickname for one with fine curls. (ii) William *de Lok* 1230 P (Berks); Robert *Atteloke* 1300 LoCt. OE *loc(a)* 'enclosure'. ME *loke* was also used of a barrier in a river which could be opened or closed at pleasure (c1300 NED). Along the Lea it was used of a bridge, *unum pontem alias Lok* (1277). *v.* PN Ess 585. The surname may thus have reference to residence near an enclosure or bridge or may be used of the keeper of the lock or bridge.

**Locker:** (i) Henry *le Lokkere* 1292 MESO (La); William *Locker* 1301 SRY; Richard *le Lokker* 1327 SRDb. A derivative of OE *loc(a)*, equivalent to *atte Lock* (*v.* LOCK) and *Lockman*: Scanard *Loceman* 1279 RH (C), Geoffrey *Lockeman* 1334 LLB E. 'Dweller by the enclosure' or 'keeper of the bridge'. cf. Sewal, William *(le) Locward* 1327 *SR* (Ess), from ME *loc* and OE *weard* 'ward, guardian'. (ii) Peter *le Loker* 1221 AssWo; John *Lokar'* 1279 RH (O); Adam *Lokere* 1296 SRSx. Some of these forms, which are common, may belong above. *Lokere* may also be a derivative of OE *loc* 'lock', equivalent to *Locksmith* or *Lockyer*; or of OE *lōcian* 'to look, see', equivalent to *Looker* 'shepherd'.

**Lockett, Lockitt:** *v.* LUCKETT

**Lockeyear:** *v.* LOCKYER

**Lockhart:** (i) Symon *Locard* 1153–65 Black; Warin *Lockard* 1190 BuryS (Sf); Jordan *Lokard* 1203 P (Nf). Probably an OG *Lochard* 'stronghold-hard'. (ii) Uruay *le Lockhert* 1203 AssNth (C). OE *loc* 'enclosure, fold' and *hierde*; herdsman in charge of the sheep or cattlefold. cf. Roger *le Medherd* 1317 MEOT (O), herd in charge of grazing cattle.

**Lockier:** *v.* LOCKYER

**Locking:** Ranulph *Locking'* 1275 RH (L); Peter *de Lokyng'* 1279 RH (O). From Locking (So), or East, West Lockinge (Berks).

**Lockless:** John *Locles* 1327 *SR* (Ess); Austen *Lockeles* 1568 SRSf. OE *loc* 'lock of hair' and *lēas* 'free from', a nickname for one with straight hair, as opposed to *Bald* and *Curl*.

**Lockley:** Peter *de Lockelee* 1199–1200 FFWa; Osbert *de Lockeleg* 1226 AssSt. From Lockleys in Welwyn (Herts).

**Locksley, Loxley:** John *de Lokkesleye* 1275 SRWo; John *de Loxelegh* 1315–16 FFSr; Thomas *de Lokeslay* 1379 PTY. From High Loxley in Dunsfold (Sr), or Loxley in Bradfield, in Plompton (WRY).

**Locksmith:** Walter, Roger *le Loksmyth'* 1255 *Ass* (Ess), 1293 Pinchbeck (Sf); John *Loc Smyth* 1279 AssNb; Robert *Locsmyth* 1279 RH (Hu). OE *loc* 'lock' and *smiδ*, 'locksmith'. cf. William *Locwricht* 13th AD iv (Lo).

**Lockton:** Iuette *de Loctun'* 1212 Pleas (L); Maleta *de Loketon'* 1219 AssY; Thomas *de Lokton* 1359 FFY. From Lockton (NRY).

**Lockwood:** Henry *de Locwode* 1294 AssSt; Adam *de Lokwode* 1352 FFY; Richard *Lokwode* 1455 LLB K. From Lockwood (WRYorks), or 'dweller by the enclosed wood'.

**Lockyer, Lockyers, Lockeyear, Lockier, Lokier:** John, Henry *le Lokier* 1221 AssWa, 1275 RH (Lo); Simon *le Lokyere* 1296 SRSx, 1316 FFK. A derivative of OE *loc* 'lock', 'locksmith'. *v.* also LOCKER.

**Locock:** *v.* LOWCOCK

**Lodbrook, Ludbrook:** John *de Lodebroc* 1206–7 FFWa; William *de Ludebrok* 1256 IpmGl; Adam *de Lodbrok'* 1346–7 FFWa. From Ludbrook (D).

**Lodder:** Richard (*le*) *Lodere* 1234, 1247 Oseney (O). OE *loddere* 'beggar'. *Lodere* cannot be distinguished from forms for LOADER.

**Loder:** *v.* LOADER

**Lodge:** Adam *atte Logge* 1327 SRSf; John *del Loge* 1379 PTY. ME *logge* 'small cottage, place to rest in'. cf. 'Logge, or lytylle howse' PromptParv. 'Dweller at the cottage', but probably often in a more technical sense, with reference to the masons' *logge* (1332, 1351) 'the building on which the life of the temporary community of masons centred'. *v.* Building 39–40. *atte Logge* may often denote the warden of the masons' lodge.

**Lodwick, Ludwick:** William *de Ludewic'* 1221 Cur (Herts); Gilbert *Lodewyk* 1279 RH (O); William *Lodewyk* 1364 FFEss. From Ludwick Hall in Bishops Hatfield (Herts).

**Loft, Lofts:** Matthew *ad le Loft* 1279 RH (Hu); Hugh *ate Lofte* 1317 AssK. ME *lofte* 'upper-chamber, attic'. Hardly 'dweller on the upper-floor', possibly 'servant of the upper-chamber'. *Lofte* may have been used in the same sense as *Lofthouse*.

**Lofthouse, Loftus:** Peter *de Lofthusum* 1166 P (Y); Ralph *de Lofthus* 1219 AssY; John *del Lofthouse* 1365 FrY; John *Lofthus* 1505 GildY. This common Yorkshire name may mean 'dweller at the house with an upper floor' (ON *lopthús*) but is probably usually from Lofthouse (WRYorks (3)), Loftsome (ERYorks) or Loftus (NRYorks).

**Logan:** Robert *Logan* 1204 Black; Walter, Thurbrand *de Logan* 1230, 1272 ib. From Logan (Ayr), or from minor places of this name.

**Loker:** *v.* LOOKER

**Lolle:** Richard *Lolle* 1281 AssCh; William *Lolle* 1314 AssNf; John *Lolle* 1327 SRLei. OE *Lulla*, or a nickname from ME *lollen* 'to droop, dangle'.

**Loller:** Elfred *le Lollere* 1133–60 Rams (Hu), Alfward, Robert *Lollere* 13th ib., 1279 RH (C). A very rare and interesting name. Perhaps from ME *lolle* 'to droop, dangle; to lean idly, to recline in a relaxed attitude', hence 'an indolent fellow'. It is, however, possible that *Lollere* is an early ME *lollere* 'mumbler', applied to pious persons and substituted in the 14th century for *Lollard* as a description of the Lollards. As Alfred's wife Edgiva became a nun at Ramsey, she was a pious woman and it is reasonable to suppose that her husband was also pious (ELPN).

**Lomas, Lomax, Loomas, Lummis, Lummus:** Richard *Lumas, Lumax*, Geoffrey *Lomax* 1602, 1622, 1642 PrGR; John *Lummis* 1674 HTSf. From a lost place Lomax, earlier *Lumhalghs*, the name of a district south of the Roch in Bury (Lancs).

**Lomb, Loombe, Loom, Loomes, Lumb, Lum:** (i) Houward, Simon *Lomb* 1198 FFNf, 1252 Rams (Hu); Ralph, Sibilla *le Lomb* 1327 SRSx, SRSf. *v.* LAMB and cf. Robert *lumbekin* 1277 Ely (Sf), either 'little lamb' or *Lambkin*, the origin of the surname of Tony

**Lumpkin.** (ii) John *del Lumme* 1327 Wak (Y). 'Dweller by the pool', as at Lumb (Lancs), OE *\*lum(m)*.

**Lombard, Lumbard, Lumbert, Limbert:** (i) *Lumbardus* 1203 Cur (Y); William *Lumbart* 12th DC (L); Martin *Lumbard'* 1208 Cur (Nt). OFr *Lambert* (*v.* LAMBARD), with the same vowel-development as in *Lomb.* (ii) Peter *le Lumbard'* 1193 P (Hu); Brankeleon *le Limbard* 1293 AssSt; Pelle *le Lombard* 1319 SRLo. 'A Lombard', a native of Lombardy, originally *Langobardus* 'long beard'. In the Middle Ages these Italian immigrants were moneylenders and bankers and the surname may sometimes mean 'banker'.

**Lomer:** Richard *le lomere* 1258, Henry *Lomere* 1327 PN K 105; John *Lomer* 1327 SRSf. 'A digger of loam', from a derivative of OE *lām* 'loam'. cf. Lomer Fm in Meopham (K).

**Loncaster:** *v.* LANCASTER

**Londesborough, Lownsbrough:** Osward *de Launeburc'* 1219 AssY; Thomas *de Lownesburgh* 1354 FrY. From Londesborough (ERYorks).

**London, Lundon, Lonnon, Lunnon:** Ælfstan *on Lundene* a988 OEByn (K); Leofsi *de Lundonia* 1066 DB; Bernardus *Lundonie* 1086 ib. From London. cf. William *Londoneman* 1309 SRBeds.

**Londsdale:** *v.* LONSDALE

**Lone:** *v.* LANE

**Loney:** *v.* LOONEY

**Long:** *v.* LANG

**Longbone, Longbones:** An anglicizing of LANGBAIN. cf. Walter *le Longebak* 1332 SRLa.

**Longbotham, Longbottom:** Richard *Longbotehom* 1379 PTY; Agnes *Longbothom* 1539 RothwellPR (Y); Thomas *Longbottom* 1557 Bardsley. 'Dweller in the long bottom or valley.'

**Longbrook:** *v.* LANGBROOK

**Longcaster:** *v.* LANCASTER

**Longden, Longdin, Longdon, Langdon, Langdown:** Ælfward *æt Langadune* c1050 OEByn (Wo); Chetelbern *de Longedun* Hy 2 Seals (Wa); Maurice *de Langedun* 1201 FFK; Reginald *de Langedon'* 1221 AssSa (St); Alan *de Longedon'* ib. (Sa). From Langdon (Devon, Dorset, Essex, Kent), Longden (Salop) or Longdon (Salop, Staffs, Worcs).

**Longenow, Longenough:** Richard *Langyno* 1332 SRDo; Richard *Langynow* 1360 Misc (Do); John *Longinow* 1382 AssWa. 'Long enough', OE *lang, genōh*. cf. Walter *le Longebak* 1332 SRLa 'long back'; Isaac *Langabeard* 1642 PrD 'long beard'; Godric *Langhand* c1095 Bury 'long hand'; Reginald *Lungeiaumbe* 1212–23 Bart 'long leg'; Thomas *Langnase* 13th BlackBk (K) 'long nose'; William *Lengeteyll'* 1299 IpmY 'long tail'; William *Longeto* 1291–2 CartNat 'long toe'.

**Longer:** Richard *le Langer* 1257 IpmW; Matilda *Longer* 1316 AssNf; William *le Longer* 1327 SRLei. 'The longer, the taller', OE *langra*. cf. William *Longerbayne* 1296 SRNb 'longer bone'.

**Longfellow:** Jacobus *Langfelley* 1475 FrY; Maria *Langfellow* 1639 RothwellPR (Y). 'Tall fellow.'

**Longfield:** John, Roger *de Longefeld* 1317 AssK, 1332 SRLa. 'Dweller by the long field.'

**Longfield:** *v.* LANGFIELD

**Longford, Langford:** Osm' *de Langeford'* 1130 P (W); Henry *de Longeford'* 1242 Fees (La). From

Langford (Norfolk) or Longford (Derby, Glos, Hereford, Middlesex, Salop, Wilts).

**Longhead:** Gilbert *Langeheved* 1247 AssBerks. 'Long head', OE *lang, hēafod.* cf. John *Lutelheed* 1301 FS 'little head', or perhaps 'little heed'; Richard *Noteheued* 1276 AssLa 'nuthead'.

**Longhurst:** v. LANGHURST

**Longland, Langland, Langlands:** Thomas *de Longelond* 1296 SRSx; Ralph *atte Longelonde* 1332 SRSr; John *Langlandes* 1458 FrY. 'Dweller by the long piece of land.'

**Longley:** v. LANGLEY

**Longliff:** *Langlif* 1188 BuryS (Sf); *Langliuus* 1207 Cur (C); *Langlive* 1294 AD ii (Sf); Cecilia *Langlyue* 1296 SRSx; Walter *Longelif* 1301 FS; Richard *Langeliffe* 1336 NorwDep. This looks like an unrecorded personal name, OE *\*Langlīf*, but it may sometimes be a nickname 'long life', OE *lang, līf.* cf. LIVELONG.

**Longman, Langman:** Alice *Longemon* 1275 SRWo; Agnes *Langeman* 1279 RH (Bk); William *Longeman* 1292 SRLo. 'Tall man.'

**Longmay:** Martin *Longemey* 1296, Ralph *Langemei* 1327 SRSx. 'Tall servant', OE *lang.*

**Longmead:** William *Langmede* 1514 PN Do i 29; James *Langmeade* 1642 PrD. From Langmead in Sampford Courtenay (D), or 'dweller by the long meadow', OE *lang, mǣd.*

**Longmer, Longmore:** (i) Wydo *de Longemer* 1275 SRWo. From Longmorehill Fm in Astley (Wo). (ii) Elice *de la Longmore,* Robert *de Langemore* 1296 Black. From Longmore, or Langmuir (Ayr).

**Longner:** v. LANGNER

**Longridge, Longrigg:** v. LANGRIDGE

**Longshank, Longshanks:** Richard *Longschaunk* 1307, *Longschankes* 1315 Wak (Y); John *Langssonke* 1334 SRK. 'Long legs', OE *lang, sceanca.* cf. John *Craneshank* 1507 FFEss 'crane legs'.

**Longshaw, Langshaw:** John *de Langchawe* 1297 SRY. From Longshaw (Derby).

**Longspey, Longspy:** William *Lungespee* 1166 P (Y); John *Longespee* 1219 P (C/Hu); Richard *Langspey, Longspey* 1298 AssL; John *Longspy* 1375 NorwLt. 'Long sword', OE *lang,* OFr *espee.* cf. Richard *Langknyf* 1332 SRDo 'long knife'; Elfald *Langstirap* 1183 Boldon 'long stirrup'.

**Longstaff, Longstaffe, Langstaff:** Richard *Langstaf* 1210 P (We); Hugo *Longstaf* 1210 FrLei. 'Long staff', probably, as suggested by Bardsley, a nickname for a bailiff, catchpoll, or other officer of the law.

**Longstreet:** Walter *de Langestret'* 1249 AssW. 'Dweller by the long road', OE *lang, strǣt.*

**Longthorn, Longthorne:** John *de Langethorn* 1371 FFY; Robert *Longthorne* 1524 SRSf; Peter *Longthorne* 1672 HTY. From Langthorne (NRY), or 'dweller by the tall thorn-tree', OE *lang, þorn.*

**Longthorp:** v. LANGTHORP

**Longton:** v. LANGTON

**Longueville, Longville, Longfield, Longwell, Longwill:** Henry *de Longauilla* 1185 Templars (L); Henry *de Longavill'* 1229 Cur (Ess); Thomas *de Longevill* 1336 FFEss; John *Longuevyle, Longwyle, Longvyle* or *Longefyld* 1509 LP (Bk). From

Longueville-la-Gifart, now Longueville-sur-Scie (Seine-Inférieure).

**Longworth:** Hugh *de Lungewurth, de Langewurth* 1246, *de Lungewrthe* 1276 AssLa. From Longworth (Berks, La).

**Lonnon:** v. LONDON

**Lonsdale, Londsdale:** Ralph *de Lounisdale* 1260 AssY; John *de Lonesdale* 1301 FFY; Thomas *Lounsdale* 1419 IpmY. From Lonsdale (Lancs, Westmorland).

**Look:** v. LUKE

**Looker, Loker, Luker:** (i) John *le Lokar,* Robert *Louker* 1327 SRSo; William *Lookar* 1582 Oxon (Ha). A derivative of OE *lōcian* 'to look'; one who looks after something, a shepherd or farm-bailiff (NED). v. also LOCKER. (ii) Simon *de Lucre* 1256 AssNb. From Lucker (Northumb).

**Loomas:** v. LOMAS

**Loombe, Loom(es):** v. LOMB

**Looney, Loney:** Gillacrist *O'Luinigh* 1090, *M'Lawney* 1504, *Lownye* 1540, *Looney* 1644, *Loney* 1681 Moore. A Manx contraction of *O'Luinigh* 'descendant of *Luinigh*', from *luinneach* 'armed'. Survives in Ireland as *O'Looney, Looney.*

**Loos, Loose:** Stephen, William *de Lose* 1200 P (Sf), 1204 FFK. From Loose (Kent) or Lose (Suffolk).

**Looseley, Loosley, Loosly:** John *de Loslee* 1297 MinAcctCo (Bk). From Loseley (Sr), Looseleigh in Tiverton, in Tamerton Foliot (D), or Loosley Row in Risborough (Bk).

**Lopham:** Estrilda *de Lopham* 1288–9 NorwLt; Hugh *de Loppeham* 1303 IpmY; Denys *Lopham* 1392 LoCh. From Lopham (Nf).

**Loraine:** v. LORRAIN

**Lord:** William *le Lauerd* 1198 P (Sf); Gilbert *Louerd* c1202 NthCh (Nth); John *le Lord* 1252 Rams (Hu). OE *hlāford* 'lord, master', often, no doubt, a nickname for one who aped the lord, or for a lord's servant.

**Lording:** Geoffrey *Lauerding* 1198 P (Sf); Richard *Louerding* 13th *Binham* (Nf); Thomas *Lording* 1327 SRSf. OE *\*hlāfording* 'son of the lord', or, perhaps 'follower, servant of the lord'.

**Lorence, Lorenz:** v. LAURENCE

**Lorett, Lorette:** v. LAURETTE

**Lorie:** v. LAWREY

**Lorimer, Lorrimar, Loriner:** Goldwin' *lorimar* c1130 AC (Lo); Hernisus *le lorinner* 1166 P (Y); Aimar *lorimer* a1174 Clerkenwell (Lo); Reginald *le Lorimier* 1185 P (Gl); David *le Loriner* 1301 MESO (Wo). OFr *loremier, lorenier* 'lorimer, spurrier'.

**Loring, Loaring:** Geoffrey *(le) Lohareng'* 1158–59 P (St); Dauit *le Loreng* 1197 FFNf; Thomas *Loring* 1280 AssSo. OFr *le Lohereng* 'the man from Lorraine'.

**Lorkin, Lorking, Larkin, Larkins, Larking:** Adam *Lartkyn* 1296 SRSx; Thomas *Lorekyn* ib.; Thomas *Lorkyn* 1327 SRC; Robert *Larkyn* 1524 SRSf. A diminutive of *Lar-, Lor-,* pet-forms of *Laurence.*

**Lorrain, Lorraine, Lorain, Loraine:** Eustache *de Lorreyne* 1333 Black. From Lorraine.

**Lorrie:** v. LAWREY

**Lorriman:** William *Loryman* 1540 FrY; William *Lurryman* 1662 PrGR. 'Servant of *Lorrie*' (Laurence).

**Lorrison:** Robert *Lowryson* 1471 Black; Thomas *Lorisoun* 1551 ib. 'Son of *Lowrie* or *Lorrie*' (Laurence).

**Lory:** *v.* LAWREY

**Lothian, Louthean, Lowthian:** Ranulph *de Louthyan* 1256 AssNb; John *de Loudonia* 1327 Black (Berwick); Alice *Louthian* 1364 AssY. From Lothian (Scotland).

**Loten, Loton:** Simon *Lotun* 1284 FFEss; John *Lotun* 1310 ColchCt; John *Lotoun* 1316 FFEss. Probably a diminutive of the ME feminine personal name *Lota*.

**Lott, Lotte, Lots:** (i) Alwin *Loth* 1162 P (K); William *Lot* 1275 RH (Sf); John *Lotte* 1275 RH (Nf); Andrew, John *Lote* 1279 RH (C), 1317 FFEss. There is no evidence for the early use of the Biblical *Lot* in England, but the first form may be this in its French form *Loth* which was common in Brittany. There was a woman's name *Lota* 1279 RH (C), probably a petform for such names as *Allot* (Alis), *Amelot* or *Emelot* (Amelina), *Ellot* (Ellen), *Gillota* (Gilia), etc., which is probably the usual source. (ii) Richard, William *at(t)e Lote* 1296 SRSx, 1332 SRSr. OE *hlot* 'lot, portion', here used of 'a share of land'. 'Holder of an allotted share of land.'

**Loud, Lowde:** (i) *Lude* 1185 Templars (Y); Richard *Lud* 1221 Cur (D); Reginald *Lude* 1215 Cur (Sr); William *Loud* 1242 Fees (D). *Lude* must be the OE *\*Hlūda* postulated by Ekwall to explain Loudham (Suffolk) and Lowdham (Notts). It is from OE *hlūd* 'loud'. The surname might also be a nickname for a noisy individual. (ii) Robert *de la Lude* 1225 AssSo; Henry *atte Lude* 1275 SRWo; John *ate Lude* 1327 SRC. 'Dweller by the roaring stream', from OE *\*hlȳde*, as at Lyde (Hereford, Som), or from OE *\*hlūde*. *v.* MELS. (iii) Richard *de Luda, de Louthe* 1319 SRLo, 1325 *Cor*. From Louth (Lincs), named from the River Lud.

**Louden, Loudon, Loudoun, Lowden, Lowdon:** James *de Loudun* c1189 Black; Adam *Loudin* c1280 Balliol; John *de Louden* 1332 SRCu. From Loudoun in the district of Cunningham (Ayr), or for LOTHIAN.

**Loudham:** John *Loudeham* 1451 IpmNt. From Loudham (Sf), or Lowdham (Nt).

**Loughton:** Terricus *de Lugetona* 1172 P (Ess); William *de Loughton* 1344 FFY. From Loughton (Bk, Ess, Sa).

**Louis:** *v.* LEWIS

**Loukes:** *v.* LUKE

**Lound, Lount:** *v.* LUND

**Loury:** *v.* LOWREY

**Loutett, Louttett, Loutit, Louttitt:** Richard *de Luvetot* 1161 YCh; Nigel *de Luuetot* 1205 FFHu; John *de Luvetot* alias *de Lovetoft* 1294 IpmSf; Malcolm *Lowtoit* 1459 Black. From Louvetot (Seine-Maritime).

**Louth, Lowth:** *v.* LOUD (iii)

**Lovat, Lovatt:** Thomas *filius Lovota* (*Loveta*) 1277 AssW; William *Lovatt* 1537 CorNt. *Luv-ot*, a diminutive of OE *Lufu* (f), *Lufa* (m).

**Love:** (i) *Lufe* c1095 Bury (Sf); Radulfus *filius Luue* 1176 P (Nt); Galfridus *filius Love* 1208 Cur (Nf); *Love Meel* (f) 1315 Bart (Lo); Gilbert *Luue* 1177 P (C);

Peter *Love* 1255 FFEss. OE *Lufu* (f), from OE *lufu* 'love', a popular and widely distributed woman's name, or OE *Lufa* (m). (ii) Alan, Robert *le Love* 1279 RH (C). AFr *louve*, feminine of *loup* 'wolf', occasionally alternating with the diminutive: Martin *Lovel, Love* 1346, 1348 FA (Sf). cf. LOW.

**Loveband, Lovibond, Levibond:** Matilda *Lovibond* 13th Rams (Hu); William *Lovebounde* 1312 Glast (So); Anthony *Loveband* 1642 PrD. 'A bond of love', OE *lufu, (ge)band*. Apparently a nickname.

**Loveday, Lowday:** *Leuedai* (*Liuedai*) 1066 DB (So); *Leuedæi* c1095 Bury (Sf); *Luveday* Vidua 1205 Cur (Nth); *Lovedaia* 1302 FFC; Æluric *Leuedey* c1095 Bury (Sf); Robert *Luvedey* 12th AD iii (K); Walter *Loveday* 1256 FFEss; Adam *Lowday* 1524 SRSf. OE *Lēofdæg* (f) 'dear day', a common medieval christian name in use from before the Conquest to the 20th century. It still survives in Cornwall and Devon, sometimes as *Lowdy*. cf. *Lowdye* Trelogan 1601 Bardsley. As a surname, it may also be from ME *loveday*, a day appointed for a meeting between enemies, litigants, etc., with a view to the amicable settlement of disputes, being given to children born on such a day. cf. *Loveday* wife of Robert Christemasse 1381 SRSf.

**Lovechild:** *Luuechild* aker 1199 PN Wa 321; *Luuechild* uidua 1221 ElyA; Robert *Luffechild* 1190 P (Berks); William *Lovechild* c1248 Bec (Wa); Henry *Louechild* 1327 SRSx. Apparently a late OE *\*Lufucild* (f).

**Lovecot, Lovecott:** Richard *Lovecot* 1275 SRWo; John *ate Lovecot* 1300 AD v (Sr); John *Louecote* 1378 AssEss. From Lovecott Fm in Debden (Ess).

**Lovegood:** Richard *Livegod* 1204 Cur (Wa); Alexander *Louegod* 1327 SRSf; William *Levegood* 1398 LLB H. The variation in the vowel points to a first element *Lēof-*, hence from OE *Lēofgod* 'beloved god or good', a rare OE personal name recorded only once as the name of a moneyer in the reign of Ethelred II. Its occurrence as a surname proves it must have remained in use after the Conquest.

**Loveguard, Lugard, Lugger, Levenger:** *Lefgar, Leueger* c1095 Bury (Sf); *Leuegarus* mercator 1199 P (Ess); Edward *Leuegar* 1199 P (D); Richard *Leugar* 1272 AD i (Lo); John *Leugor*, William *Leuegor* 1327 SRSf; Richard *Loueger* 1327 SRSo. OE *Lēofgār* 'beloved-spear'. In *Loveguard* and *Lugard* the *d* is excrescent, in *Levenger* the *n* is intrusive as in *Messenger*.

**Lovejoy:** Philipp *Loveioy* 1596 Musters (Sr); Emmanuel *Loveioye* 1642 PrD, *Loveioy* 1665 HTO. 'Love joy, joyful', OE *lufu*, OFr *ioie*. cf. Adam *Luvelavedy* 1297 Wak (Y) 'love lady'; John *Lovelyf* 1326 CorLo 'love life'.

**Lovekin, Lufkin, Lucken, Luckens, Luckin, Luckins, Lukin, Lukins, Lukyn:** *Luuekin* 1221 AssSa; *Lovekinus* 1255 RH (O); *Lovekyn* 1279 RH (C); Osbert *Lovekin* 1275 SRWo; John *Loukin*, Robert *Lukyn* 1297 MinAcctCo; Robert *Lufkyn* 1524 SRSf; Maistres *Lovekyn, Luffkyn* 1542–3 Bardsley. *Love* plus the diminutive suffix *-kin*, 'little Love', either a pet-name or an affectionate attribute. Occasionally, perhaps also 'little she-wolf'. *v.* LOVE, LOW.

**Lovel, Lovell, Lowell:** Ricardus *lupellus* c1118 AC (Sx); William *Luuel* 1130 P (O); Willelmus *Luvel, Lupellus* 1206 Cur (O); William *Luvel, Luel* 1212 Fees (Sf); Philip *Lowel* 1255 RH (O); Richard *Luvel* alias *Lowel, Lovel* 1263 Ipm (So). AFr *lovel* 'wolf-cub', a diminutive of AFr *love* 'wolf'. Like the alternative diminutive *lovet*, it was common and widely distributed. Richard Luvel above was 'of Kari Lowel barony', and was descended from William, Earl of Yvery, who was called *Lupellus* to distinguish him from his father Robert who had acquired the nickname *Lupus* because of his violent temper. Here the diminutive is clearly used in the sense 'the younger'. Used also as a personal name: *Luuellus* 1145–7 Colch (Ess); *Lovel* le Clark 1274 RH (Ess); cf. LOVE, LOW, LOVETT.

**Lovelace, Loveless, Lowles, Lowless:** Edith *Luvelece* 1243 AssSo; William *Luuelaz* c1250 StGreg (K); John, Albricus *Loveles* 1251 FrLei, 1275 RH (Sf); Richard *Lovelas*, John *Lovelace* 1344, 1367 PN K 407; Robert *Lufelesse* 1444 Calv (Y). OE *lufu* 'love' and *-lēas* 'free from, without', 'loveless'. The common form is *Loveles*. Occasionally we may have 'love lass' from ME *las(se)*. cf. Alan *Luveswain* 1166 P (Y). 'Love lace', a dandy, is less likely. OFr *laz*, ME *las* meant 'cord'. The sense *lace* is not recorded in England before 1550 (NED).

**Lovelady:** Adam *Luvelavedy* 1297 Wak (Y); Adam *Loveladi* 1314 ib. An obvious nickname for a philanderer.

**Loveless:** *v.* LOVELACE

**Lovelock, Loveluck:** I. *Luveloc* 1283 SRSf; William *Louelok* 1327 *SR* (Ess). A nickname for a dandy, a wearer of pendant locks of hair falling over the ears and cut in a variety of fashions. They were common in the 16th century and apparently much earlier. Walter (*le*) *Loveloker'* (1279 RH (O)) must have been a maker of these.

**Lovely:** John *Lovely* 1319 AssSt; Henry *Loveliche* 1406 LLB I; William *Loveliche*, Matthew *Louledge*, John *Lowledge* 1524 SRSf. ME *luuelich* 'lovely'.

**Loveman, Lowman, Luffman, Leaman, Leamon, Leeman, Lemman, Leman, Lemon, Liman, Limon:** *Leman* 1066 DB (Ha); *Lufmancat* 1066 Winton (Ha); *Lemannus* de Fordham 1175–86 Holme (Nf); Reiner *Leman* 1185 Templars (Ess); William *Luveman* 1211 FrLeic; Aumfridus *Leofman* 1221 AssWo; Wfricus *Lefman* c1250 Rams (Hu); William *Lemmon* 1275 SRWo; Elyas *Loveman* 1276 RH (Y); Henry *Lemman* 1327 SRSx; Thomas *Leaman* 1568 SRSf. OE *Lēofmann* 'beloved man', a rare personal name of which only three pre-Conquest examples are known. The surname has been reinforced by the common ME *leofman, leman, lemman* 'lover, sweetheart': Robert *le Leman* 1285 Pinchbeck (Sf). *Luve-, Loveman* may mean 'servant of Love'.

**Loven:** (i) *Luuunus* 1066 DB; *Leuunus* 1086 DB; Robertus *filius Louun* 1160 P (Sa); Ralph *Lovyn* 1345 ColchCt. OE *Lēofhun* 'beloved young bear', a rare name, unrecorded in OE. (ii) Jocelin *de Luvan* 1187 Gilb (L); Godfrey *de Luvein* 1195 P (K). From Louvain. The more common source.

**Lover:** William *le Lovere* 1275 RH (Nf); Alice *Louer* 1327 SRSf. ME *lovere* 'lover, sweetheart'.

**Loveren:** Godfrey *filius Leurun* 1194 P (Nf); *Leueruna* 1208 Cur (Wa); Edith *Loverun* 1275 SRWo; Robert *Loueroun* 1327 SRSf; John *Loverin* 1642 PrD. OE *Lēofrūn* (f).

**Loveridge:** *v.* LEVERAGE

**Lovering:** William *Luuering* 1203 P (Sr), Thomas, Martin *Lovering* 1275 RH (D), 1327 SRSo. OE *Lēofhering*, *Lēofring* 'son of *Lēofhere*' 'beloved-army'.

**Loverock:** *v.* LEVERAGE

**Loversall:** Alexander *de Luureshale* 1230 P (Db); William *de Lovershal* 1251 AssY; William *de Loversall* 1338 Shef. From Loversall in Doncaster (WRY).

**Lovett, Loveitt, Lovitt:** William *Louet, Loueth, Luueth* 1086 DB (Berks, Nth); Geoffrey *Luvet* c1125 StCh; Robert *Lovit* 1279 RH (Bk). OFr *louet* 'wolf-cub', a diminutive of *lou* 'wolf', occasionally alternating with a diminutive in *-el*: Robert *Luvet, Luvell'* 1206 Cur (R). cf. LOVEL.

**Lovey, Loveys:** Cecilia *Lovye* 1283–4 CtH; Richard *Loveye* 1349 LLB F; John *Loveye* 1382 FFEss. OE *Lēofwīg*.

**Lovibond:** *v.* LOVEBAND

**Lovick:** This may be (i) from *Loveke* from OE *Lēofeca*. *v.* LEVICK. (ii) from *Lufwic* becoming *Lowik*. *v.* LOWICK.

**Lovie:** *v.* LEAVEY

**Loving:** *v.* LEVINGE

**Lovney:** Mabilia *de Luveny, de Loveny* 1242 Fees (L). From Louvigny (Calvados).

**Lovold:** *v.* LEAVOLD

**Low, Lowe, Lowes:** (i) John *le Lu* 1207 P (Gl); Robertus *Lupus*, Robert *le Lu* 1221 AssWa; Walter *le Lou* 1242 Fees (D). OFr *lou* 'wolf', occasionally interchanging with the diminutive: Gregory *le Lu, Lupus, Lupet* 1221 Cur (D), Martin *Lovel* or *Love* 1346, 1348 FA (Sf). (ii) Turgot *Lag* 1066 DB (Y, L); William *le Low* 1284 AssLa; Martin *le Low* 1275 RH (Lo). ON *lágr* 'low, short'. (iii) Robert *de la Lowe* 1275 SRWo; Roger *del Lowe* 1288 AssCh; Stephen *Atteloue* 1301 SRY. OE *hlāw* 'hill', 'burial mound'. cf. LAW. (iv) From a pet-name of *Laurence*. cf. Simon *Loustepsone* 1297 SRY.

**Lowcock, Locock, Luckcock, Luckcuck, Luckock, Lucock, Lewcock:** *Luuecok, Leucok* Schayf(e) 1246 AssLa; *Lovekoc* de Wlvedale 1275 Wak (Y); *Lokoc* de Heppewrth 1286 ib.; Peter *Luuecok'* 1221 AssWa; Geoffrey *Luvecoc, Lucoc* 1259, c1308 Calv (Y); Henry *Lovecok* 1274 RH (Ess); William *Loukok* 1327 SRDb; Robert *Lukok* 1338 FrY; William *Locoke* 1531 FrY; John *Lewcocke* 1674 HTSf. A diminutive of OE *Lufa*. cf. LOVE.

**Lowday:** *v.* LOVEDAY

**Lowde:** *v.* LOUD

**Lowell:** *v.* LOVEL

**Lowen, Lowin:** *v.* LEWIN

**Lower:** *v.* EWER

**Lowery:** *v.* LOWREY

**Lowick:** Balde *de Lufwic* 1200 Cur (Nth). From Lowick (Lancs, Northants, Northumb).

**Lowing:** *v.* LEVINGE

**Lowis:** *v.* LEWIS

**Lowless:** *v.* LOVELACE

**Lowman:** Richard *de Lumene* 1242 Fees (D). From Uplowman (Devon). *v.* also LOVEMAN.

**Lownsbrough:** *v.* LONDESBOROUGH

**Lowrance:** *v.* LAURENCE

**Lowrey, Lowrie, Lowries, Lowry, Lowery, Loury:** *Lowry* Smith 1467 Black; Robert *Lowri* 1332 SRCu; Gilbert *Lowrie* 1497 Black. A pet-form of *Laurence*. A Border form. cf. LAWREY. *Lowery* may also be from Lowery (Devon), whence the surname *de Loveworþe* (1256 PN D 245).

**Lowrieson, Lowrison:** Robert *Lowryson* 1471 Black. 'Son of *Lowrey*.'

**Lowson:** (i) Geoffrey *le Leuesone* 1309 SRBeds. 'Beloved son', OE *lēof* and *sunu*. cf. LEAF. (ii) Richard *Lowson* 1381 PTY; William *Loweson* 1456 FrY. 'Son of *Low*', a pet-name of *Laurence*. *v.* also LEVESON.

**Lowth:** *v.* LOUD

**Lowther:** William *de Laudre* 1184 P (We); Henry *de Louthere* 1230 Cur (Y); Hugh *de Louthre* 1342 FFY; William *Lowther* 1672 HTY. From Lowther (Westmorland), *Lauder* c1180.

**Lowthian:** *v.* LOTHIAN

**Lowthorpe, Lowthrop:** Walter *de Luitorp* 1161 P (Y). From Lowthorpe (ERYorks).

**Loxley:** *v.* LOCKSLEY

**Loyd:** *v.* LLOYD

**Luard:** *v.* LIVARD

**Lubbock:** Robert *de Lubyck* 1276 RH (L). From Lubeck.

**Lucas:** *Lucas* c1150 DC (Nt); Eueard *Lucas* 1153–85 Templars (Herts). The learned form of *Luke*.

**Luce, Lucia:** *Lucia* c1150 DC (L), 1205 Cur (Sf); Richard, Asselyna *Luce* c1230 Barnwell (C), 1332 SRSx. *Lucia*, feminine of Lat *Lucius*, from *lux* 'light'. St Lucia, martyred at Syracuse under Diocletian, was a popular medieval saint whose name was common in England as *Luce*.

**Lucian:** *Luciana* 1205 Cur (Sx); *Lucianus* de Seille 1212 Cur (Db); Philip *Lucian* c1265 Glast (W). Lat *Lucianus* (m), *Luciana* (f).

**Luck(es):** *v.* LUKE

**Luckett, Locket, Lockett, Lockitt:** Eudo, Walter *Loket* 1275 RH (Nf), 1389 FrY; Matilda *Luket* 1418 Bardsley (Y). A diminutive of *Luke*.

**Luckhurst, Lukehurst:** This Kentish surname must derive from Luckhurst in Mayfield (Sussex), itself a corruption of the first part of *Lukkars Croche* 1553, *Luckers Crouch* 1823, where probably lived Henry *Luggere* in 1296 (PN Sx 382).

**Luckin:** *v.* LOVEKIN

**Lucking:** William *Luueking* 1188 P (Lei); Nicholas *Loveking* 1275 SRWo. OE *\*Lēofecing* 'son of *Lēofeca*'. cf. LEVICK. *Lucking* may also be a late development of *Lovekin*.

**Luckless:** Vincent *de Louueclyve* 1281, William *de Loueclyve* 1333 PN D 534. From Luckless Cottage in Hockworthy (D).

**Luckman:** *v.* LUKEMAN

**Luckner:** *v.* LEWKNOR

**Luckock, Lucock:** *v.* LOWCOCK

**Lucy, Lucey:** (i) William, Cecilia *Lucy* 1297 MinAcctCo, 1327 SRSo *v.* LUCE. (ii) Richard *de Luci* 1135–54 Bury (Sf). From Lucé (Orne).

**Ludbrook:** *v.* LODBROOK

**Ludd:** Yvo *Lude* 1210–11 PWi; Henry *Lud* 1270 IpmW; William *Lude* 1327 SRSo. OE *Luda*.

**Ludford:** Elena *de Ludeforda* 1187 Gilb; Jordan *de Ludeford* 1242 Fees (He); William *de Luddeford* 1335 Glast (So). From Ludford (He, L, Sa), Lydford (D), or East, West Lydford (So).

**Ludham:** Walter *de Ludham* 1221 AssWa; Ralph *de Ludham* 1287–8 NorwLt; William *Ludham* 1387 FFEss. From Ludham (Nf).

**Ludkin, Ludkins:** Adam *Lotekyn, Ludekyn* 1311 LLB D; Hugh *Lutkyn* 1332 SRSt. A diminutive of the *Lote* discussed under LOTT or of *Lutt*.

**Ludley:** Hugh *de Lodeleye* 1275, Richard *de Lodeleye* 1327 SRWo; Richard *Ludelegh* 1392 CtH. From Lutley (Wo), *Ludeleya* 1169.

**Ludlow:** William *de Ludelawa* 1182 P (Sa); John *de Lodelawe* 1327 SRWo; George *Ludlow* 1545 SRW. From Ludlow (Salop).

**Ludwick:** *v.* LODWICK

**Luff:** William, Simon *Lof* 1177, 1185 P (C, Nth); William, Geoffrey *Luffe* 1188 P (Co), 1309 SRBeds. OE *lēof* 'beloved' or *\*Lēof*, with change of stress to *leōf, lōf*. cf. LEAF. Or OE *\*Luffa*, a geminated form of *Lufa*. cf. LOVE.

**Luffingham:** William *de Lufham* 1199 FF (R). From Luffenham (Rutland).

**Luffman:** *v.* LOVEMAN

**Lufkin:** *v.* LOVEKIN

**Lugard, Lugger:** *v.* LOVEGUARD

**Lugg, Lugge:** William *Lug* 1219 AssY; Thomas *Lugge* 1275 RH (K); William *Lugg'* 1332 PN Do i 147. OE *\*Lugga*.

**Luke, Luck, Luckes, Look, Looks, Loukes:** (i) *Luke* 1277 Ipm (Nt); William, Thomas *Lucke* 1279 RH (C), 1332 SRSt; Simon *Luk* 1286 Ipm (Sf); Godfrey, John *Lukke* 1327 SRSf, 1381 ArchC iii; John *Louk* 1327 SRSf; Robert *Lukes* 1376 LoPleas; John *Look* 1379 ColchCt. *Luke* or *Luck* was the popular form of *Lucas*. Greek *Λουκᾶς*, a man from Lucania. *v.* also LEVICK. (ii) Lucas *de Luk', de Lukes* 1274 RH (Lo); Richard *Luike* 1279 RH (C). From Luick (Liège).

**Lukehurst:** *v.* LUCKHURST

**Lukeman, Luckman:** 'Servant of *Luke*.'

**Luker:** *v.* LOOKER

**Lukin(s):** *v.* LOVEKIN

**Lull, Lulle:** *Lulle* Cokeman 1260 IpmY; Richard *Lulles* 1256 AssNb; Christopher *Lullis* 1641 PrSo. OE *Lulla*.

**Lulling:** William *Lulling'* 1219 P (L); John *Lullinge* 1319 NorwDeeds II; William *Lullyng* 1327 SRSx. OE *Lulling*.

**Lulman:** *v.* LILLEYMAN

**Lum, Lumb:** *v.* LOMB

**Lumbard:** *v.* LOMBARD

**Lumby:** Robert *de Lumby* 1219 AssY; Nicholas *de Lumby* 1280 FFY; Simon *Lumbie* 1459 Kirk. From Lumby (WRY).

**Lumley, Lumbley:** William *de Lumeley* 1235–6 AssDu; Marmaduke *de Lomley* 1348 FFY; William *Lumley* 1672 HTY. From Lumley (Durham).

**Lummis, Lummus:** *v.* LOMAS

**Lumner:** *v.* LIMMER

**Lumsden, Lumsdaine, Lumsdon:** William *de Lumisden* 1166–82 Black; Adam *de Lumesdene* 1212 P (Du); John *de Lummysden* c1335 Black. From Lumsden in Coldingham (Berwick).

**Lund, Lunt, Lound, Lount:** Ralph *de la Lunde* 1183 P (Y); Geoffrey *de Lund* 1200–30 Seals (Nf); John *del Lound* 1327 SRSf; John *Lunt* 1524 SRSf. From Lund (Lancs, Yorks), Lunt (Lancs), Lound (Lincs, Notts, Suffolk), or 'dweller by the grove' (ON *lundr*).

**Lundon:** *v.* LONDON

**Lundy:** (i) *Lundi* 12th DC (L); Adam *Londy* 1279 RH (O); Stephen *Lundey* 1641 PrSo. ON, ODa *Lundi*, OSw *Lunde*. (ii) Walter *de Lundy* 1305 (Perth), John *of Lundy* 1499 (Fife) Black. From Lundie (Angus, Fife).

**Lung:** *v.* LANG

**Lunniss:** Richard *Londoneys* 1279 RH (C); Robert *Lundenissh* 1327 *SR* (Ess). 'The Londoner.'

**Lunnon:** *v.* LONDON

**Lupino:** A theatrical family of Italian origin, established in England since 1642. The Victoria and Albert Museum possesses a handbill of that date, beginning: 'Signor Luppino will present Bel and the Dragon newly arrived from Italy . . .' The surname is an Italian diminutive, corresponding to the French *Lovell* and *Lovett*, ultimately from Latin *lupus*, 'wolfcub'.

**Lupsett, Lipsett:** Ralph *de Lupesheved* 1297 Wak (Y); William *de Lupesheved* 1297 SRY; Roger *de Lupsete* 1331 Wak (Y). From Lupsett in Wakefield (WRY).

**Lupton:** Robert *de Lupton'* 1297 SRY; Roger *Lupton* 1506 FFEss; Andrew, Peter *Lupton* 1642 PrD. From Lupton House (Devon), and probably also from an unidentified Lupton in Yorks.

**Lurey, Lurie, Delhuary:** John *del Ewry* 1379 PTY. ME *ewerye*, a room where ewers of water, table-linen, etc., were kept. 'Sergeant or groom of the Ewery.' cf. EWER.

**Lurk:** Godebald *filius Lurc* 1166 P (Nf); Adam *filius Lorch* 1215 Cur (He); Godebald *Lurc* 1166 P (Nf); John *Lurk* 1275 RH (L); Walter *Lurk* 1369 FFEss. ON *Lurkr*, a byname from ON *lurkr* 'a cudgel', hence a nickname for a strong, heavy person. *v.* PNDB 322.

**Luscombe:** Aunger *de Luscumb'* 1230 P (D); William *de Loscomb'* 1332 SRDo; Richard *Luscomb* 1524 SRD. From Loscombe (Do) 'valley with a pigsty'.

**Lusher:** *v.* USHER

**Lute:** Ailric *Lute filius* c1095 Bury; William *Lute* 1216–22 Clerkenwell; John *Lute* 1524 SRSf. ODa *Luti*, OSw *Lute*, a byname from ON *lútr* 'stooping'.

**Luter, Lutter:** (i) Alvredus *le Lutur* 1221 Cur (K); John *le Leuter, le Leutour, Luter* 1304–10 LLB B, C, D; Idonea *Lutier* 1358 AD ii (Mx). OFr *leuteor*, or a derivative of ME *lute* 'to play on the lute', a luteplayer. (ii) William *Lelutre* 1130 P (Ess); Geoffrey, Ralph *Lutre* 1204 StP (Lo), c1208 FrLeic; Ralph, William *le Lutre* 1207 Cur (Sx), 1235 AD i (Mx). OFr *loutre* 'otter', metonymic for *Lutterer* 'otter-hunter'.

**Luther:** Ralph *Luther* 1529 GildY; John *Luthur* 1674 HTSf. 'Lute-player', Fr *luthier*. There is no evidence for the personal-name *Luther* in medieval England.

**Lutman:** *v.* LILLEYMAN

**Luton:** Osbert *de Luton'* 1206 Cur (Co); Roger *Luton'* 1253 Acc. From Luton (Beds, D, K).

**Lutt:** Simon, Seman *Lutte* 1279 RH (Hu), 1327 SRSf. OE \**Lutt(a)*, from the root *lut-* in OE *lȳtel*. cf. *Lutting* c1095 Bury (Sf) 'son of \**Lutt(a)*', Ralph, John *Lutting* 1185 P (Nf), 1357 FrY.

**Lutterer:** Ralph *le Lutrer, le Luterer* 1220 Cur, 1232 Fees (Bk). A derivative of OFr *loutre* 'otter', an otterhunter. cf. LUTER.

**Lutherworth:** Hugh *de Lutherwrd'* 1177–90 MCh; Elias *de Lutterwrthe* 1209 ForR; John *Lutterworth'* 1379 AssWa. From Lutterworth (Lei).

**Luttrell:** Geoffrey *Lutrel* 1194 P (Nt); Alexander *Lutterel* 1268 AssSo. A diminutive of OFr *loutre* 'otter'. Ralph *Lutrel* is also called *le Lutrer* 1219 Cur (Bk). cf. LUTTERER.

**Luty:** *v.* LAWTEY

**Luxford:** Bartholomew *de Luggesford* 1279 PN Sx 373; Isabella *de Lockesford* 1327, Thomas *Luxford* 1525 SRSx. From Luxford in Crowborough (Sx).

**Luxton:** Henry *de Luxton* 1251 AssSt; Bartholomew *Luxton* 1642 PrD. From Luxton in Upottery, or East Luxton in Winkleigh (D).

**Lyal, Lyall, Lyel, Lyell, Liell:** Johannes *filius Lyelli* 1329 Black; *Lyell* Robson 1541 Bardsley; John *Lyell* 1408 FrY; John *Liel* 1411–34 Black. A hypocoristic of *Lyon* (*v.* LYON and cf. John *Lyot* 1327 SRDb), or of its diminutive *Lionel*.

**Lyard:** *Liardus* chascurus 1210 P (Lei); John *Liard* 1307 KB (L); John *Lyarde* 1547 EA (NS) ii. OFr *liard* 'grey'.

**Lycett:** Richard *Licett* 1640 SaltAS (OS) xv; Widow *Licett* 1674 HTSf. Perhaps for Ir *Lysaght*. *v.* MacLysaght.

**Lyburn:** *v.* LAYBORN

**Lydall:** *v.* LIDDALL

**Lydbury:** *v.* LIDBURY

**Lyddiard, Lydiard, Lydiart:** *v.* LIDDIARD

**Lyde:** *v.* LYTH

**Lye(s):** *v.* LEA

**Lyfield:** *v.* LAYFIELD

**Lyle:** *v.* LISLE

**Lymburner:** Geoffrey *Limbarner* John FFEss; Alexander *le Lymbrennere* 1240 FFEss; Robert *Lymbrinner* 1327 SRY. OE *līm* plus a derivative of OE *beornan* or ON *brenna* 'to burn', a lime-burner, a maker of lime by burning lime-stone.

**Lymer:** *v.* LIMER

**Lymington, Lymmington:** *v.* LIMMINGTON

**Lynam:** *v.* LINEHAM

**Lynch:** *v.* LINCH

**Lynde:** *v.* LIND

**Lyndhurst:** *v.* LINDHURST

**Lyne:** *v.* LINE

**Lyner:** Gilbert *le Lyner* 1279 RH (Hu); John *Lynar* 1327 SRSx. OFr *linier* 'maker or seller of linen cloth'.

**Lynham:** *v.* LINEHAM

**Lynn, Lynne, Linn, Lenn:** Ædricus *de Lenna* 1177 P (Nf). From King's Lynn (Norfolk).

**Lynton:** *v.* LINTON

**Lynwood:** *v.* LINDWOOD

**Lyolf:** Roger *son of Lyelf* 1246 AssLa; John *Lyolfe* 1258 IpmY; Alan *Liolf* 1333 FFY. OG *Leutwulf*, OFr *Lyulf*.

**Lyon, Lyons:** (i) *Lyon* son of *Lyon* 1293 AssSt; Ricardus *Leo* 1176 P (O); Thomas *Lioun* 1287 AssCh; William *le Leoun* 1290 Fees (Wt); William *Lyon* 1327 SRSf; Johanna *le Lyon* 1332 SRSx. Either from *Lyon*, the popular pronunciation of *Leo* or *Leon*, or a nickname from the lion. (ii) Azor *de Lions* 1159 P (Nf); Geoffrey *de Liuns* 1170 P (Nf); Henry *de Lyons* 1296 SRSx. As early forms of the surname, which is not uncommon, invariably end in -s, this must be from Lyons-la-Forêt (Eure) and not from the better-known Lyons, earlier *Lugudunum*, Fr *Lyon*.

**Lysons, Licence, Lison:** Godfrey *de Lisun* 1195 P (So); William *Lycens* 1524 SRSf. From Lison (Calvados).

**Lyster:** *v.* LISTER

**Lyte:** *v.* LIGHT

**Lyteman:** *v.* LILLEYMAN

**Lyth, Lythe, Lyde:** (i) Richard *de la Lid'* 1212 Cur (Bk); Robert *de Litha* 1214 Cur (Y); Adam *atte Lythe* 1275 SRWo. Abraham *atte Lyde* 1296 SRSx. 'Dweller by the slope' (OE *hlĭþ, hlid*, ON *hlið*), as at Lythe (NRYorks), Lyth (Westmorland) or Lydd (Kent). (ii) Gonnilda *le Lyth* 1279 RH (Bk). OE *līðe* 'mild, gentle'.

**Lythgoe:** *v.* LITHGOE

**Lytle, Lyttle:** *v.* LITTLE

**Lyttleton:** *v.* LITTLETON

**Lytton:** *v.* LITTON

**Lyveden:** William *de Lyveden* 1330 PN Nth 205. From Lyveden in Aldwinkle (Nth).

**Lyward:** *v.* LIVARD

# M

**Mabane:** *Maban* 1066 DB (Y); Anora *Maban* 1218 P (St). Perhaps Welsh *Mabon*, or OIr *\*Maban*, a diminutive of *mab* 'son'.

**Mabb, Mabbs:** *Mabbe* 1293 AssCh, 13th WhC (La); Ralph, John *Mabbe* 1278 LLB B, 1300 Oseney (O); John *Mabbys* 1309 SRBeds. A pet-name for *Mabel*.

**Mabberley, Maberley, Maberly, Mapperley:** Richard *de Maborlay* 1353 FrY; Thomas *de Mapurleye* 1381 PN Nt 139. From Mapperley (Db), or Mapperley in Basford (Nt).

**Mabbett, Mabbitt, Mabbot, Mabbott, Mabbutt:** *Mabota* Ryder 1379 PTY; Richard *Mabot* 1509 Oxon; William *Mabbett* 1646 Bardsley. *Mabb-ot*, a diminutive of *Mabb*.

**Maben, Mabin, Mabon:** *Mabon* de Oteford 1187 P (K); Richard *Mabun* 1279 RH (C). Welsh *Mabon*.

**Maber:** Nathaniel *Maber* 1662–4 HTDo. From Mabor in Buckland Monachorum (D).

**Maberley, Maberly:** *v.* MABBERLEY

**Mabin, Mabon:** *v.* MABEN

**Mable, Mabley, Mably:** Rogerus *filius Mabilie* 1130 P (Nth); *Mabilia* c1150 DC (L), 1221 AssGl; Arnaldus *Mabilie* 1185 Templars (Ess); John *Mabile* 1274 RH (Ess); John *Mably* 1279 RH (C). *Mabel* or *Mabley* was the popular form of *Amabel*. *v.* ANNABLE.

**Mabson:** William *Mabbeson* 1332 SRLa; William *Mabbesson* 1381 SRSf. 'Son of *Mabb.*' *Mabbesson* may be for *Mabbetson*. cf. William *Mabotson*, *Mabetson* 1379 PTY. 'Son of *Mabbett.*'

**McAdam:** Dolfinus *mach Adam* 1160–2 Black; John *M'Cadame* 1609 ib. 'Son of *Adam.*'

**Macaddie, Macadie:** John *McChaddy* 1596 Black. Anglicized forms of Gaelic *MacAda* or *MacÀdaidh*, the Gaelic dialectal form of Lowland *Adie*. *v.* ADEY.

**Macalaster, Macallaster, Macalister, McAllister, Maccalister:** Ranald *Makalestyr* 1455 Black. Gaelic *MacAlasdair* 'Son of *Alexander*'. *v.* CALLISTER.

**McAlery:** *v.* MCCHLERY

**Macall:** *v.* MCCALL

**Macallan:** Dungall *M'Alayne* 1376 Black. 'Son of *Allain.*'

**Macally, McAlley:** *v.* MACAULAY

**Macalpin, McAlpine:** John *MacAlpyne* c1260 Black. 'Son of *Alpin.*'

**MacAndrew:** Donald *Makandro* 1502 Black. 'Son of *Andrew.*'

**Macara, Maccarra:** John *M'Ara* 1614 Black. 'Son of the charioteer' (OGael *ara*).

**MacArthur, Maccairter, Maccartair, McCarter:** Gyllemechall *M'Carthair* 1569 Black; Dougall *M'Arthour* 1580 ib.; Patrick *McKairtour* 1630 ib. Gaelic *MacArtair* 'Son of *Arthur.*'

**Macartney, Macartnay, Maccartney:** Gilbert *McCartnay* 1529 Black (Galloway); Thomas *McCartnay, Makartnay* 1562 ib.; Helen *Macartney* 1588 ib. (Dumfries). Ir *Mac Cartaine*, a variant of *Mac Artáin* 'son of *Art*'.

**Macasgill, Macaskill, Maccaskell, Maccaskill, Mackaskill:** Gilbert *Mac Askil* 1311 Pat (Du). Gaelic *MacAsgaill* 'Son of *Askell*'.

**Macaskie, McCaskie:** Gilbert *Makasky, Makaskel* 1316, 1318–19 Cl (IOM). Gaelic *MacAscaidh* 'Son of *Ascaidh*', a pet-form of *Askell*.

**McAteer:** *v.* MACINTYRE

**Macaulay, Macauley, McAuliffe, Macalley, Maccally:** Iwar *McAulay* 1326 Black; John *Makalley* 1602 ib. In Dumbartonshire, from Gaelic, Ir *MacAmhalghaidh* 'Son of *Amalghaidh*', an old Irish personal-name; in the Hebrides, from Gael *MacAmhlaibh* or *MacAmhlaidh*, OIr *Amlaib*, ON *Oláfr* 'relic of the gods'.

**Macauslan, Macausland, Macauslane, McCausland:** Malcolm *Macabsolon* 1308 Black; Alexander *Macausland* 1421 ib. 'Son of *Absalom.*'

**McBeath, Macbeth:** An old Gaelic personal-name, 'son of life'.

**MacBrayne:** Eugenius *Makbrehin* 1525 Black. 'Son of the judge', Ir *brehon*, Gael *breithean*.

**McBride, McBryde:** Cristinus *McBryd* 1329 Black. 'Son of (the servant of St) Bride.'

**MacByrne:** *v.* BYRNE

**Maccabe, M'Cabe:** Hugh *MacCabe* 1368 Black (Ireland). Irish *Mac Cába* 'Son of *Cába*' (cape, hood).

**Maccaffie:** *v.* MCFEE

**MacCaffray, MacCaffery, Caffray, MacGaffrey:** Duncan *Macgoffri* 1319 Cl. Irish *Mac Gafraidh* 'Son of *Godfrey*'.

**McCall, Maccaull, Macall, Mackail, Mackall:** Robert *M'Kawele* 1370–80 Black; Finlay *Makcaill* 1506 JMac; John *M'Call* 1583 Black. Gaelic *MacCathail* 'Son of *Cathal*', Ir, Gael *Cathal*, OW *Catgual* 'war-wielder'.

**MacCallum:** Gilbert *MacCalme* 1631 Black (Ayr); Iain *M'Callum* 1647 ib.; Archibald *M'Callome* 1661 ib. Gaelic *MacCaluim* 'son of the servant of *Calum*'.

**M'Calman, Maccalmon, Maccalmont:** Gill. *M'Colemane* c1180 JMac; Gilbert *M'Calmont* 1581 Black; Alexander *McAlman* 1682 ib. Irish *Mac Calmáin, Mac Colmáin*. 'Son of *Colmán*.' *v.* COLEMAN.

**Maccartair, McCarter:** v. MACARTHUR

**MacCarthy,      MacCartie:** Douenald    Roth'
*Mackarthi* 1285 Pat. Irish *MacCárthaigh* 'Son of
*Cárthach*', earlier *Caratācos*. cf. CRADDOCK.

**Maccartney:** v. MACARTNEY

**MacCashin:** v. CASHEN

**McCaskie:** v. MACASKIE

**McCausland:** v. MACAUSLAN

**McChlery, McCleary, McCleery, McAlery:**
Malcom *M'Cleriche, M'Clery* 1461, 1475 Black;
Duncan *McInclerycht* 1537 ib.; Finlay *M'Aclerich*
1638 ib. Gaelic *Mac an chlerich, M'a'chleirich*
'Son of the clerk or cleric'.

**McClarron:** v. McLAREN

**McClean:** v. MACLEAN

**McClement(s):** v. MACLAMON

**McCloud:** v. MACLEOD

**McCloy:** Donald *Clowie* or *Makcloye* 1537–40
Black. Gaelic *MacLughaidh* 'Son of *Lewie*', a pet-
form of *Lewis*.

**McClure, Maclure, MacCloor, Macleur:** (i) John
*McLur* 1526 Black; Thomas *Maklure* 1532 ib.
(Carrick); Gilleane *MacIloure* 1618 ib.; John *M'Clour*
1664 ib. (Appin). Gaelic *M'Ill'uidhir* 'son of the
servant of *Odhar*'. (ii) In Ireland sometimes an
anglicization of *Mac Giolla Uidhir*, from *odhar* 'dun-
coloured'.

**McClymont:** v. MACLAMON

**Maccolman:** v. M'CALMAN

**McComb, McCombe:** Gilchrist *Makcome* 1526
Black. Gaelic *MacThóm* 'Son of Tom'.

**M'Combie, Maccombie:** Ferquhair *McCombquhy*
1556, John *M'Comy* 1571, John *Makthomy*
Makgilleweye 1586 Black. Gaelic *MacComaidh*, from
*MacThomaidh* 'Son of Tommy'.

**McConachie, McConaghey, Macconkey:** Gilbert
*Maccoignache* 1296 CalSc; Angus *M'Conchie* 1493
Black. Gaelic *MacDhonnchaidh* 'Son of Duncan'.

**MacCone:** v. MACOWEN

**Macconnal, McConnell:** The Irish name is
sometimes from *MacConnaill* 'Son of *Conall*', but is
usually, as in Scotland, a variant of *Macdonald*.

**MacCoole:** v. MACDOUGALL

**McCorkell, Maccorkill, Maccorkle:** *M'Thurkill,
M'Kurkull* 1661,    1663    Black.    v.    CORK-
HILL.

**McCormac, Maccormack, McCormick:** Gillecrist
*mac Cormaic* 1132 Black. Irish *Mac Cormaic* 'Son of
*Cormac*'. v. CORMACK.

**McCorquodale, Maccorkindale:** Ewen *Mactor-
quedil* or Ewgyne *M'Corqueheddell* 1430 Black.
Gael *MacCorcadail* 'Son of *Thorketill*'. v. CORKHILL.

**Maccoubrie, McCoubrey:** John *Makcopery* 1513,
Henry *McCowthry* 1539 Black. Gael *Mac Cuithbreith*
'Son of *Cuthbert*'.

**McCrae, Maccraith:** v. MACRAE

**M'Crimmon:** Johne *Mcchrummen* 1533 (Inverness), Hector *M'Crimmon* 1595 Black (Skye). Gael
*Mac Cruimein* 'Son of *Rumun*', ON *Hrómundr*
'famed protector'.

**Maccrindell, Maccrindle, MacRanald:** John
*M'Rynald* or *Makrynnild* 1483, Donald *McRanald*
Vaan 1506, Malcome *McRyndill*, James *Mc Crynnell*

1526 Black. Gael *Macraonuill* 'Son of *Raonull*', i.e.
*Ranald*.

**McCrossan:** Irish *Mac an Chrosáin* 'Son of the
rhymer'.

**McCubbin,    McCubbine,    McCubbing:** Martin
*M'Cubyn* 1376, John *Makcubyng* 1567 Black. 'Son of
*Cubbin*', i.e. *Gibbon*.

**MacCure:** v. MACIVER

**MacCurtin:** v. CURTIN

**McCutcheon, Maccutchen, MacHutchin:** John Roy
*Makhuchone* 1495, James *McCutchone* 1686 Black.
'Son of *Hutcheon*.' v. HUTCHIN.

**MacDermot, McDermott:** Nemeas *Mactarmayt*
1427 Black; Alexander *M'Dermite* 1687 JMac. 'Son of
*Diarmaid*', OIr *Diarmait* 'freeman'.

**Macdonald,      MacDonnell,      McConnell:** Robert
*Dovenald* 1257 Dublin; Dovenilt *Macdencuilt*,
Conoye *Mac Deuenilt* 1264 ib. Gael *Mac Dhómhnuill*,
pronounced *Mak oonil*, 'Son of *Donald*'.

**McDonaugh, McDonogh:** 'Son of *Duncan*.'    v.
MCCONACHIE, DUNCAN.

**Macdougal, McDougall, MacDool, Macdouall,
McDowall,      McDowell,      MacCoole:** Duncan
*Makdougal* c1230 JMac; Omertach *Macdowyyll*,
Molawelyne    *Macduulle*    1264    Dublin;    Fergus
*MacDowilt* 1296 CalSc; Robert *M'Kowele* 1370–80
Black; Nigel *MakCowl* 1497 ib. Gael *MacDhùghaill*
'Son of *Dhubgall*'. v. DOUGAL.

**Macduff:** Roger *Macedugh* 1264 Dublin; Malisius
*mc Duf* 1284 Black. Gaelic *Mac Dhuibh* 'Son of *Dubh*'
(black).

**MacDuffie:** v. McFEE

**MacEachan:** v. KAIGHAN

**Macellar, Maceller:** v. MCKELLAR

**McElroy, MacIlroy:** Michael *M'Gilrey, M'Ylroye*
1376, 1500 Black. Irish *Mac Giolla Rua* 'Son of the
red-haired lad'. cf. GILROY, MILROY.

**MacEwan, McEwen, Macewing:** Gilpatrik *mac
Ewen* 1219 Black. Gael *MacEoghainn* 'Son of *Ewen*'.

**Macey:** v. MASSEY

**McFadden, McFadyean, MacFadyen, McFadzean:**
*Padyne Regane* 1264 Dublin; Malcolm *Macpadene*
1304, John *McFadyeane* 1457, Donald *M'Fadzeane*
1473 Black. Gael *MacPhaidin* or *MacPhaidein* 'Son
of *Paidean*' or Little Pat.

**Macfail, Macfall, Mackfall, MacPhail, MacVail:**
Gillemore *M'Phale* 1414, Donald *M'Pawle* 1490,
Gylleis *Makfaill* 1492 Black. Gael *Mac Phàil* 'Son of
*Paul*'. cf. QUAYLE.

**Macfait, M'Fate, M'Feate:** Finlay *M'Fead* 1485,
Gilcrist *McPaid* 1539, Robert *M'Faitt* 1596 Black.
Gael *Mac Pháid* 'Son of *Pate*'.

**MacFarlan, Macfarlane, Macfarland, Macparlan,
Macpartland:** Malcolm *Mcpharlane* c1385 Black.
Gael *MacPharlain* 'Son of *Parlan*', OIr *Partholon*
(Bartholomew).

**M'Farquhar:** Malmur *Mac Hercar* c1200 JMac.
'Son of *Farquhar*.'

**McFee, Macfie, MacPhee, Macphie, Maccaffie,
Machaffie, McDuffie:** Thomas *Macdoffy* 1296 CalSc;
Archibald *McKofee* 1506, Morphe *mcphe* 1531, Ewin
*McAphie* 1681 Black. Gael *MacDhubhshith* 'Son of
*Dubhshithe*' (black man of peace).

**McGachan:** v. KAIGHAN

**MacGaffrey:** v. MacCAFFRAY

**McGee, McGhee, McGhie, Magee, Maggee:** Gilmighel *Mac Ethe* 1296 CalSc; *McGethe* 1297 Pat; *Macge* or *Mageth* 1339, Robert *Macgye, M'Gy* 1444–9 Black. Irish *Mag Aoidh* 'Son of *Aodh*'. cf. MCKAY.

**MacGeorge:** v. MacINDEOR

**McGibbon, Mackibbon, McKibbin:** Thomas *Makgibbon* 1507 JMac. 'Son of *Gibbon*.'

**McGill:** Maurice *Macgeil* 1231, James *M'Gill* 1550 Black. Gael *Mac an ghoill* 'Son of the Lowlander or stranger'. cf. GALL.

**MacGillivray, MacGilvray:** Archibald *McIluray* 1542, Farquhar *MacGillivray* 1622 Black. Gael *Mac Gillebhrath* 'Son of the servant of judgement'.

**MacGilp:** v. MCKILLOP

**McGinley:** v. KINLAY

**MacGorman:** v. GORMAN

**MacGowan, McGowing, McGown, MacGoun, MacGown:** Gilcallum *McGoun* 1503, Gilbert *Makgowin* 1526 Black. Gael, Ir *Mac a'ghobhainn* 'Son of the smith'.

**McGrath, McGraw:** v. MACRAE

**MacGregor, McGreigor, McGrigor:** Duncan *M'Greg(h)ere* 1292 JMac. 'Son of *Gregor*.'

**McGuinness:** v. MAGUINESS

**McHaffie:** v. MCFEE

**Machen:** v. MASON

**MacHendrie, MacHenry, Mackendrick:** John *M'Henri* 1370–80, *McHenrik* 1590 Black. Gael *MacEanruig* 'Son of *Henry*'.

**Machent, Machon:** v. MASON

**MacHutchin:** v. MCCUTCHEON

**MacIlduff:** v. DUFF

**MacIlroy:** v. MCELROY

**Macilvride, Macilvreed:** Colanus *McGilbride* 1363, Nigel *M'Ylwyrd, Makkilbreid* 1476–8 Black. Gael *Mac Giolla Brighde* 'Son of the servant of St Bride'.

**Macindeor, Mackindewar, Mackinder, M'Jarrow, M'Jerrow, MacGeorge:** Colin *Macindoyr* 13th, Gillaspy *McIndewir* 1541, *McInder* 1613, John *Macjore* 1691, John *M'George* 1726 Black. Gael *Macindeoir* 'Son of the stranger', pronounced *Makinjor*, hence *M'Jarrow, MacGeorge*.

**MacInnes, Mackinnes, Mackiness:** Donald *McKynes* 1514, Allester *M'Callen M'Aneiss* 1574 Black. Gael *MacAonghuis* 'Son of *Angus*'.

**McIntosh, Mackintosh:** Farchard *Mctoschy* 1382, Alexander *Mackintoche* 1390 Black. Gael *Mac an toisich* 'Son of the chieftain'.

**Macintyre, McEntire, McAteer, McTear:** Nicholas *Mac in tsair* 1268, Paul *M'tyr* 1372, Gildow *Makintare* 1506 JMac. Gael *Mac an tsaoir* 'Son of the carpenter'.

**MacIver, Macivor, Maccure, Mackeever:** Malcolm *Mcluyr* 1292 Black. Gael *MacIomhair* 'Son of *Ivar*' (ON *Ivarr* 'bow-army').

**M'Jarrow:** v. MACINDEOR

**Mack:** *Maccus* de Leum 1176 P (Nb); Hugo *Mac* 1188 P (Wo); John *Mack* 1327 SRSo. OC *Maccos*, OIr *Maccus*, an Irish adoption of ON *Magnus*.

**Mackail, Mackall:** v. MCCALL

**Mackay, McKay, McKee, MacKey, Mackey, Mackie, McKie:** Cucail *Mac Aedha* 1098 Moore;

Gilchrist *M'Ay* 1326, Odo *Macidh* 1433, Gilnew *McCay* 1506 Black. Gael *Mac Aoidh* 'Son of *Aodh*'.

**McKellar, MacKeller, Macellar, Maceller:** Patrick *McKellar* 1436 Black; Archibald *Makelar* 1488 ib. (Argyll); Duncan *McCallar, Makcallar* 1500 ib. (Dumbarton). Gaelic *Mac Ealair* 'son of *Ealair*', the Gaelic form of Lat *Hilarius*.

**McKelly:** v. KELLY

**McKenna, Mackinney:** William *M'Kinnay* 1544 Black. Gael *MacCionaodha* 'Son of *Cionaodh*'.

**Mackenzie:** Makbeth *Makkyneth* 1264 Black; Nevin *M'Kenze* 1473 JMac. 'Son of *Coinneach*' (comely).

**McKibben:** v. MCGIBBON

**McKillop, McGilp:** Finlaius *Macpilibh* 1433, William *Makillop* 1526, Donald *M'Gillip* 1532 Black. Gael *Mac Fhilib* 'Son of Philip'.

**McKisack, MacKissack:** v. KISSACK

**Mackown:** v. MACOWEN

**MacLachlan, McLauchlane, Maclaughlan, MacLoughlin:** v. LACHLAN

**Maclaine:** v. MACLEAN

**Maclamon, Maclamont, Macclemment, McClements, Macclymond, McClymont:** Gael *MacLaomuinn* 'Son of *Lamont*'. v. LAMMOND.

**McLaren, Maclauren, Maclaurin, McClarron:** Donald *Maklaurene* 1586, Laran *McLaran* 1592 Black. Gael *Mac Labhruinn* 'Son of *Labhran*'. v. LAURENCE.

**MacLean, Macclean, Maclaine, Maclane:** Lachlan *M'Gilleon* 1436, Alexander *McKlane* 1684 Black. Gael *Mac Gille Eoin* 'Son of the servant of St John'.

**MacLeod, McCloud:** Gillandres *MacLeod* 1227 Black. Gael *MacLeòid* 'Son of *Leòd*' (ON *Liótr* 'ugly').

**Macleur, Maclure:** v. MCCLURE

**McManus:** v. MAGNUS

**McMichael:** v. MICHAEL

**Macmillan, MacMullan:** John *Macmullan, Makmylan* 1454–87 Black. Gael *MacMhaolain* 'Son of the tonsured one'.

**MacNab:** Mathew *M'Nab* 1376 Black. Gael *Mac an Aba* 'Son of the abbot'.

**MacNachtan, MacNaghten:** v. MACNAUGHTON

**McNamara, MacNamara:** *Mac Conmera* 1311 Moore; *Mac Namara* 1511 ib.; *McNameer, McNamear* 1610, 1793 ib. An anglicized form of Gaelic *Mac con-mara* 'son of the hound of the sea'.

**MacNaughton, MacNaghten, MacNachtan:** Gillecrist *Mac Nachtan* 1247 Black; Donald *Macnachtane* 1431 ib. (Dunkeld); Maureis *McNauchtane* 1510 ib. Gaelic *Mac Neachdainn* 'son of *Neachdain*', the Pictish *Nechtan* 'pure'.

**McNeal, McNeil:** v. NEAL

**MacNevin, MacNiven:** v. NIVEN

**McNichol:** v. NICHOLAS

**McOmish, Maccomish:** Donald *McHomas* 1688, Archibald *M'Comash* 1696 Black. Gael *Mac Thomais* 'Son of Thomas'.

**MacOwen, MacCown, MacCone, Mackeown, Cowan, Keown:** 'Son of *Ewen*'. v. MACEWAN, OWEN.

**McParlan(d), McPartland:** v. MACFARLAN

**McPhail:** v. MACFAIL

**McPhee:** v. MCFEE

**McPheeters:** Thomas Moir *McGillifedder* 1607, Duncan *M'Fater* 1694 Black. Gael *Mac Gille Pheadair* 'Son of (St) Peter's servant'.

**Macpherson:** Alexander *Makfersan* c1447, Bean *Makimpersone* 1490 Black. Gael *Mac an Phearsain* 'Son of the parson'.

**McQueen, MacQueen, Macqueen:** Hector *Mac-Souhyn* 1271 Black; Luke *Macquyn* 1403 JMac; Finlay *M'Quene*, Gillereoch *M'Queane* 1541 ib. Gaelic *Mac Shuibhne* 'son of *Suibhne*', from *suibhne* 'pleasant'.

**M'Quisten, McQuistin:** Kenneth *M'Hustan* 1542, John *Macquiestoun* 1662, Donald *M'Houstone* 1664 JMac. 'Son of *Húisdean*.' v. MCCUTCHEON.

**Macrae, McRaith, Macraw, Macray, McRea, McCrae, McCraith, McCraw, McCrea, McCreath, McCreith, Mackereth:** *Maccret* mac Iodene a1200 Dublin; *Macrath* ap Molegan 1296 CalSc; Alexander *Macrad* c1225, Patric *M'Re, M'Rey* 1376, Adam *M'Creich* 1438 Black; Richard *Makereth* 1526 FrY; John *Makcra* 1621 Black. Gael *Macrath* 'son of grace', an old personal-name.

**MacRanald:** v. MACCRINDELL

**M'Sorley:** v. SUMMERLAD

**McTaggart, MacTaggert:** Ferchar *Machentagar, Mackinsagart* 1215 Black; Otes *Mactagart* 1511 Moore. Gael, Irish *Mac an tsagairt* 'Son of the priest'.

**MacTague, MacTeague, MacTigue:** 'Son of the poet.' v. KEIG.

**Mactavish:** Doncan *M'Thamais* 1355, Duncan *McTawisch* 1480, Thomas *McTaevis* 1515 Black. Gael *Mac Támhais* 'Son of *Tammas*', Lowland Scots for *Thomas*.

**McTear:** v. MACINTYRE

**McTurk:** John *Makturk* 1538 Black. Gael *Mac Tuirc* 'Son of *Torc*' (boar).

**Mace:** (i) *Mace* 1236 Oseney; Richard *Mace* 1229 Cl (Beds); James *Mace* 13th AD ii (Nth); William *Mace* 1372 ColchCt. Perhaps a shortened form of *Thomas*, or of *Matthew*, cf. MASSEY. (ii) Perotus *de Mace* 1319 LLB E. From Macé (Orne).

**Macer:** Robert *le Macr'* 1203 Cur; Walter *le Macere* 1275 SRWo; Henry *le Macer* 1332 SRSt. OFr *maissier, massier* 'mace-bearer'.

**Mackarel, Mackerell:** v. MACKRELL

**Mackley:** William *de Mackel'* 1210 Cur (Db); Thomas *de Mackeley* 1327 SRDb. From Mackley (Db).

**Mackrell, Mackarel, Mackerell, Mackrill, Mackriell:** William *Makerel* 1177 P (L); Thomas *Makerel* 1249 AssW; Michael *Makerel* 1343 IpmW; Mary *Mackerill* 1662–4 HTDo. There are three possibilities: (i) OFr *maquerel* 'bawd', (ii) OFr *makerel* 'sea-fish', (iii) OFr *makerel, maqereaux* 'red scorches or spots on legs of such as used to sit near the fire' (Cotgrave). All three may have contributed to the surname.

**Mackworth:** William *de Macworth'* 1204 Cur (Nt); Henry *de Macworth* 1298 AssSt; John *Macworth* 1410 FFEss. From Mackworth (Db).

**Macwilliam:** Edward *Makwillyam* 1461 Fine (Ess). 'Son of *William*.'

**Macy:** v. MASSEY

**Madan, Maddans, Maddams:** v. MATTAM

**Maddams:** John *Madame* 1327 SRY. A nickname.

**Madden:** Tathige *O Madan* 1264 Dublin. Ir *Ó Madáin* 'descendant of *Madadhán*', a diminutive of *madadh* 'dog'.

**Madder, Madders, Mader:** Thomas *Mader* 1221 Cur (Nf); Jacobus *le Madur* 1275 RH (L). OE *mædere* 'madder', used for *madderer* 'dyer with or seller of madder'. cf. Walter *le Maderere* 1317 FFEss, Thomas *(le) Maderman* 1293 · LLB A, 1300 LoCt, Robert *madermanger* 1230 P (Nth).

**Maddison, Madison:** Thomas *Madyson* 1425 FrY; William *Maddison* 1430 FeuDu; Edmund, Lancelot *Madyson* 1532 LP (Nth), 1537 FFEss. Usually a by-form of MATHIESON 'son of Matthew', but occasionally perhaps from *Maddy*, a pet-form of MAUD.

**Maddock, Maddocks, Maddox, Maddick, Madicks, Madocks, Mattack, Mattacks, Mattock, Mattocks, Mattick, Mattuck:** *Madoch* 1066 DB (Gl); Oenus *filius Madoc* 1160 P (Sa); *Maddock* le Waleys 1283 AssSt; William *Madoc* 1274 RH (Sa); Robert *Mattok* 1290 AssCh; Robert *Madduk'*, Stephen *Madek'* 1297 MinAcctCo (W, Co). OW *Matōc*, Welsh *Madawc, Madog* 'goodly'.

**Madel(l):** v. MALE

**Madeley:** Reimund *de Madeleia* 1212 P (Ch); Hugh *de Maddelee* 1318 Husting; Thomas *de Maddeleye* 1332 SRWo; Watkin *Madley* 1663 HeMil. From Madeley (Sa, St), Madely (Gl), or Madley (He).

**Mader:** v. MADDER

**Madge, Maggs:** *Magge* 1246 AssLa, c1248 Bec (Wa); Ailmundus, John *Magge* 1200 P (L), 1279 RH (Hu); John *Magges* 1327 SRSf. *Magge* may be *Magg*, a short form of the common *Magot, Magota* 1208 Cur (Nf), a hypocoristic of *Margaret*, or for *Madge* (*Margery*).

**Madgett:** William *Maggard* 1327 SRSf; Samuel *Maggett* 1647 RothwellPR (Y). *Magge* (Margery), plus the suffix-(*h*)*ard*.

**Madle:** v. MALE

**Maffey:** v. MORFEY

**Mafflin, Maflin:** (i) Hamo *filius Meinfelinus* 1174 LuffCh; Hamo *Meifelinus* 1185 Templars (O); Ralph *Meyfelin* 1242 Fees (Berks); Roger *Mayflyn* 1327 IpmGl. OFr *Meifelin*. (ii) Richard *Martefelun* 1230 P (L); Richard *Matefelun* 1230 P (Nf); Robert *Matefelun* 1232 Cur (W). 'Kill the felon', OFr *mate, felun*.

**Maffre:** *Matfridus, Mafrei* filius Kipping 1185–6 P (Nth); Henry *Matefrei* 1208 Cur (Sf); Francis *Maffry* 1674 HTSf. OFr *Mafreiz*, OG *Mathefrid* 'power-peace'.

**Magee:** v. MCGEE

**Mager:** v. MAUGER

**Maggot:** *Magota* 1208 Cur (Nf); *Maggot* de Worth 1286 AssCh; *Magot* Lomb 1296 SRSx; Robert *Maggote* 1279 RH (C); Henry *Magot* 1327 SRSo; John *Magote* 1367 IpmGl. *Magg-ot*, a diminutive of *Magge*, a pet-form of *Margaret*. cf. Thomas *Magotson* 1379 PTY; John *Maggekin* 1396 PN Wa 64.

**Maggs:** v. MADGE

**Magill:** v. McGILL

**Magnus, Magnusson, Manus, McManus:** *Magnus* de Weitecroft Hy 2 DC (L); Hugo *Magnus* c1114 Burton (St); John *M'Manis* 1506 Black. ON *Magnus*, Lat *magnus* 'great'. The first of this name was Magnus I, King of Norway and Denmark (d. 1047), named after Charlemagne (Carolus *Magnus*), under the impression that *Magnus* was a personal-name. It became the name of many Scandinavian kings and was very popular in Shetland where *Magnusson* also survives as *Manson*. In Scotland it has given *McManus*.

**Magowan:** v. MacGOWAN

**Magrath, Magraw:** v. MACRAE

**Magson:** William *Maggessone* 1327 SRSf; Hugh *Maggesone* 1332 SRSt. 'Son of *Magg* (Margaret).'

**Maguiness, Maguinness, Magennies, Maginniss:** Irish *Mag Aonghuis* 'Son of *Angus*'. cf. Scottish MACINNES.

**Mahood:** v. MAUD

**Mahood:** v. MAWHOOD

**Maiden, Maydon:** Robert *Maiden, Maide* 1197–8 P (Nf); Adam *le Maiden* 1279 RH (C). 'The maiden', a derogatory nickname.

**Maidens:** Geoffrey *Maidens* 1225 AssSo; Thomas *del Maydenes* 1345 AssSt. cf. PARSONS.

**Maides:** Abram *Maide* 1185 P (Co); John *Mayde* 1327 SRWo. 'The maid.' cf. MAIDEN.

**Maidford:** Nicholas *de Maydeford* 1235–6, John *Maydeford* 1429–30 FFWa. From Maidford (Nth).

**Maidman, Maidment, Maitment:** Richard *Maydenemon* 1275 SRWo; Robert *Maideneman* 1327 SRSx; William *Maideman* 1332 SRSr. 'Servant of the maidens.'

**Maidwell:** Henry *de Maydewelle* 1262 FFO; Simon *de Maydewell* 1275 RH (Nth); William *de Maydewelle* 1279 RH (O). From Maidwell (Nth).

**Mail, Maile, Mailes, Mayle, Mayles:** (i) William *Mail* 1221 AssSa; Thomas *Mayle* 1296 SRSx; John *Maile* 1454 Paston. Probably metonymic for an enameller, OFr *esmailleur*, or for a maker of mail armour. (ii) *Maiel* de Hereford' 1165 HeCh; Adam *filius Maelis* 1242 Fees (He); Wildric *Mael* 1205 Cur; Symon *Mael* 1275 SRWo; Robert *Mael* 1304 IpmGl. Ir *mael* 'bald'. Used also as a personal name.

**Mailer, Mailler, Maylor, Meyler:** (i) *Meilerus* frater Laissioc 1160 P (Sa); Robertus *filius Meilir* 1255 RH (Sa); William *le Maillier* 1203 P (Wa), *le Mailur* 1262 HPD (Ess); Ralph *Meillur* 1227 AssBeds; Walter *Meyler* 1255 RH (Sa); Philip *le Mayllur* 1268 AssSo. The surname is usually occupative, from ME *amel*, with loss of *a* in *ameillur* (OFr *esmailleur*), 'enameller'. cf. AMBLER. We have also a Welsh personal-name, *Meilyr*, OW *\*Maglorīx* (Jackson 625). There was a Breton saint *Maglorius*. (ii) In Scotland, from Mailer in Forteviot (Pertshire): Johan *de Malere* 1296 Black (Perth).

**Mailey:** William *Maillie* 1310 AssNf. Perhaps from Mailly (Marne, Somme, &c.).

**Main(e), Mains:** v. MAYNE

**Mainer:** v. MAYNER

**Mainerd:** v. MAYNARD

**Mainland:** Manus *Mainland* 1550 Black (Kirkwall). An Orkney and Shetland name, from Mainland, the principal island of the Shetlands.

**Mainprize, Mainprice, Mainprise, Mempriss, Mimpress, Mimpriss:** Robert *Maineprice* 1613 FrY; John *Memprice* 1661 YWills; . . . *Mimprest* 1674 HTSf. 'a surety, pledge', OFr *main* 'hand', *prise* 'taken'.

**Mainston:** Widow *Mainston* 1662–4 HTDo; Charles *Mainston* 1663 HeMil. From Mainstone (Ha, He, Sa), Mainstone in Heanton Punchardon, in Egg Buckland (D), Mainstone Fm in Buxworth (Db), or Mainstone Gate in Chobham (Sr).

**Mainwaring, Manwaring, Mannering:** Ralph *de Maisnilwarin* 1185 P (L); Thomas *de Meinwaring* 1260 AssCh; Randoll *Manwaryng* Hy 8 StarChSt; Arthur *Mainering* 1714 Shef. From a place named Mesnilwarin 'the manor of Warin'.

**Mair:** Symon *le Mare* 1296 Black (Perth); John *Mair* 1453 ib. Gael *maor* 'officer'. In Scotland 'Mair was the designation of an officer who executed summonses and other legal writs' and was used not only of the king's herald or sergeant but also of such officers as a head forester. v. also MAYER.

**Maisey, Maizey, Maysey, Meysey, Meazey:** Geoffrey *de Maisi* 1130 P (Sr). From Maizy (Aisne) or Maisy (Calvados).

**Maitland:** Richard *Maltalant* 1170 P (Nb); William *Mautalent* 1208 Cur (Nb), 1221 Black; Gilbert *de Maltalent* c1215 Black; Richard *de Mauteland* a1315 ib.; Robert *Matilland, Maitland*, 1417, 1424 ib. In England, where the surname was not uncommon, this appears to be a nickname for one with discourteous, unpleasant manners (OFr *maltalent*), but in Scotland there are sufficient examples with *de* to show that it derives from a place-name. Dauzat derives the French surname from Mautalant (Pontorson) 'peu gracieux'.

**Major:** v. MAUGER

**Majoribanks:** v. MARJORIBANKS

**Makefair, Makefare:** Ralph *Machefare* 1176 P (Beds); Robert *Makefare* 1221 *ElyA* (Sf); Richard *Makefare* 1327 SRSf; Robert *Makefayre* 1344 ChertseyCt (Sr). 'A travelling companion', OE *mæcca* 'companion', *faru* 'journey'.

**Makejoy:** Richard *Makeioie* 1221 *ElyA* (Sf); Bartholomew *Makeioye* 1277 *Ely* (Sf); Richard *Makejoye* 1301 SRY. 'Make joy', OE *macian*, OFr *ioie*. cf. William *Makeblith*' 1208 Cur (Y) 'make joy'; Julian *Makeblise* 1279 RH (O) 'make happiness'; Richedon *Makedance* 1301 SRY 'make dance'; Thomas *Makehayt* 1250 FFL 'make haste'; John *Makelayke* 1379 PTY 'make sport'.

**Makeless:** John *Makeles* 1242 Fees (La); Alice *Makeles* 1327 SRC; John *Makeles* 1425 *ERO*. ME *makeles* 'without an equal'.

**Makepeace, Makepiece:** Gregory *Makepais* 1219 FrLeic; Thomas *Makepays* 1340 FFSt. 'Make peace', peace-maker.

**Makin, Makins, Making, Makings, Meaken, Meakin, Meakins, Meakings, Meekings, Meekins, Mekking:** (i) *Maikin* de Eylesburi 1212–23 Bart; *Maikin* Sutor 1311 HPD (Ess); Peter *Maykyn* 1319 FFEss; John *Maykyn, Makyn* 1362, 1368 LLB G;

John *Mekyn* 1486 KentW; Thomas *Meekin* 1622 PrGR; William *Makin*, Widow *Makinge* 1674 HTSf. *May-kin*, a diminutive of *May*, a hypocoristic of *Mayhew* (Matthew). (ii) Robert *Maidekyn* 1327 PN K 554; Jeva *Maydekyng* 1327 SRC. *Maide-kin*, 'little maid', a nickname.

**Malcolm:** Norman *filius Malcolumbe* 1066 DB (L, Y); *Malculum* 1192 P (Sa); *Maukolum* 1207 Cur (Ha); Aleyn *fitz Maucolum* 1296 Black. Gaelic *Mael Coluimb* 'devotee of (St) *Columba*'. As a surname its use is comparatively modern.

**Malcolmson, Malcomson:** Symon *Malcomesson* 1296 Black. 'Son of *Malcolm*.'

**Mald:** v. MAUD

**Malden, Maldon, Maulden, Mauldin, Mauldon:** Robert *de Maldon* 1236–7 FFK; John *de Maldone* 1337 CorLo; Thomas *Maldon* 1376 FFEss. From Maldon (Ess), or Malden (Sr).

**Male, Males, Madle:** Robert *le Masle* 1187 P (Berks); Osbert *le Madle* 1202 FFEss; Stephen *Male* 1230 P (Ess). OFr *masle* 'male, masculine'. William *le Masle* (1280), *le Madle* (1303), or *le Male* (1305) has left his name in Marles Fm in Epping Upland (PN Ess 24). For forms, cf. GREALEY.

**Malet:** v. MALLET

**Malham, Maleham:** Roger *Malham* 1210–11 PWi; William *de Maleham* 1230 P (Sx); John *de Malham* 1296 SRSx. From Malham (WRY), Malham Fm in Wisborough Green (Sx), or 'dweller at the stony place', ON *mjol*, OE *hām*.

**Malhead:** William *Malheved* 1248 FFO; Baldric *Malleheved* 1248 AssBerks. 'With a head like a hammer', OFr *mal*, OE *hēafod*.

**Malherbe:** Adam *de Malerba* 1130 P (Co); John *Malherbe* Hy 2 DC (L); William *Malherbe* 1249 AssW; Geoffrey *Malherbe* 1359 AssD. From the name of a French domain or hamlet (Dauzat).

**Malin, Malins, Mallin:** *Malina* 1212 Cur (Nt); *Malyn* del Wllehous 1277 Wak (Y); John, Richard *Malin* 1297 SRY, 1327 SRDb; John *Malynes* 1358 Putnam (Wa). *Mal-in*, a diminutive of *Malle* (1297 Wak), a pet-name for *Mary*.

**Malindine:** v. MALLANDAIN

**Maliphant:** Geoffrey *Malenfant* 1205 P (Sf); William *Malefant* 1448 Pat. OFr *mal enfant* 'naughty child'.

**Malise, Meliss, Melles, Mellis, Melliss, Mellish:** *Malis* c1190 Black; Gillemycell *Malys* 1481 ib.; John *Mellish* 1771 ib.; David *Meliss* 1773 ib. Gael *Maol Iosa*, EIr *Mælisu* 'tonsured servant of Jesus'.

**Malkin, Maulkin:** Ricardus *filius Malkyn* 1297 SRY; John *Malekin* 1284 RamsCt (Hu); William *Malkyn* 1297 MinAcctCo. A diminutive of *Mall* (Mary).

**Malkinson:** Walter *Malkinesone* 1327 SRSx. 'Son of *Malkin*.'

**Mall, Malle:** Adam *son of Malle* 1297 Wak (Y); Alured *Malle* 1221 AssWa; William *Malle* 1279 RH (C); John *Malle* 1378 IpmGl. *Mall*, a pet-form of *Mary*.

**Mallandain, Mallandaine, Mallandin, Mallendine, Mallindine, Malidine:** Gerard de *Manegeden* 1208 FFEss. From Manewden (Essex), formerly pronounced *Mallendine*.

**Mallard:** For *Maylard*, or, perhaps, OFr *malard* 'wild drake'.

**Malleson:** William *Malleson* 1332 SRCu. 'Son of *Mall* (Mary).'

**Mallet, Mallett, Mallette, Malet:** Robert *Malet* 1086 DB (Sf); William, Gilbert *Malet* 1166 DC (L), 1185 Templars (So); William *Malait* 1230 P (K); Irveyus *Maleit* ib. (Berks). *Maleit* is OFr *maleit*, past participle of *maleïr*, 'cursed'. This is certainly the origin of some Norman surnames, but the three DB tenants-in-chief, frequently mentioned and invariably as *Malet*, may have brought with them a Norman patronymic, a diminutive of *Malo*, the popular form of the name of St *Maclovius*, a 6th-century Welsh monk who worked in Brittany, which survives in Saint-Malo and in the church of Saint-Maclou in Rouen. *Malo, Malet* and *Maclou* survive in France, the latter, particularly in Seine-Inférieure, Eure and Calvados. Tradition claims that the surname of William Malet, founder of Eye Priory, arose 'ob bellicam fortitudinem eo quod in prœliis hostes ut malleo contunderet'. This is supported by the form *Gulielmo agnomine Mal(l)eto* in Orderic and William of Poiters. When William Malet was banished in 1109, his son Hugh took the name of *Fichet* which was retained by his son Hugh, although his eldest son reverted to *Malet* (A. Malet, *An English Branch of the Malet Family* (1885), 7, 23–4, 102–3). This undoubtedly implies that Hugh regarded both surnames as nicknames with very similar meanings, both suited to a knight expert in the use of lance (cf. FITCH) and mace, the *mailz de fer* of the *Chanson de Roland*. OFr *maillet, mallet*, a diminutive of *mail, mal* (Lat *malleus*) 'hammer'. cf. OFr *mailleor* 'hammerer, smith', and v. MARTEL. In England the chief source of the surname is *Mal-et*, a diminutive of *Mall* (Mary): *Maleta* 1219 AssY. *Mallet* may also be a late development of *Mallard*.

**Malley, Mally:** Peter *de Malley* 1210 Cur (Beds); Peter *de Mallay* 1218 P (Beds/Bk); Robert *Mally* 1327 SRC. Probably from Mailly (Marne, Somme, &c.). v. also MAILEY.

**Mallin:** v. MALIN

**Malling:** Lifing *æt Meallingan* c1060 OEByn; William *de Mallinges* 1209 P (Sf); Juliana *Malyngges* 1297 MinAcctCo; Thomas *Mallyng* 1392 LoCh. From East, West Malling (K), or South Malling (Sx).

**Mallinson, Mallison:** John *Malynessone* 1317 Pat (Y); Thomas *Malisone* 1445 Black (Aberdeen). 'Son of *Malin*.' *Mallison* may also derive from *Mallet* (Mary): William *Malitesoun* 1469 Black (Aberdeen).

**Mallory:** Geoffrey *Maloret* 1086 DB (Do); Richard *Mallorei* c1155 DC (Nt); William *Maleuerei* 1170 P (Y); Henry *Mallore* 1195 P (Db); Robert *Mallory* 1255 Fees (Nth). OFr *maloret* (*maleuré, maloré*) 'the unfortunate', 'the unlucky'.

**Mallot:** *Malota* 1221 AssWa; Isabella *Malot* 1327 SRSf; Mariota *Mallot* 1332 SRCu. *Mal-ot*, a diminutive of *Mall* (Mary).

**Mally:** v. MALLEY

**Malmain, Malmains:** Vitalis *Malesmains* 1195 P (Mx); John *Malemeyns* 1276 FFEss; Richard *Malemeyn* 1340 CorLo. 'Bad hands', OFr *mal, mains*, a nickname, either for a wicked man, or from some deformity of the hands.

**Malone:** Ir *Ó Maoileoin* 'descendant of *Maoleoin*' (servant of St John).

**Malosel, Maloisel:** Hugh *Maloisel* 1214 Cur (Nf); Henry *Maloysell* 1264 AssSo; Richard *Maloysel* 1327 SRSo. A nickname, 'bad bird', OFr *mal, oisel*.

**Malpas, Malpass, Melpuss, Morpuss:** Henry *Malpas, Maupas* 1203 Cur (Y), 1219 AssY; Walter *de Malpas* 1275 SRWo. From Malpas (Ches, Cornwall) or some French Malpas or Maupas.

**Malser, Malzer:** Fulcher *Mala Opera* 1086 DB (R); William *Malesoveres* c1144 Riev; John *Maleshoures* 1235 Fess (Nth). Lat *mala opera*, OFr *malesouvres* 'bad workman'. cf. Thorpe Malsor (Nth), Fucher *Malesoures* 12th, Milton Malzor (Nth), William *Malesoures* 1202. cf. also Roger *Mal Esquier* 1203 Cur 'bad squire'.

**Malster:** Robert *le Maltester'* 1279 MESO (Y); John *le Malstere* 1327 SRSf. A feminine form of *Malter*, as a surname, used only of men.

**Malt:** v. MAUD

**Maltas:** v. MALTHOUSE

**Maltby, Maultby:** Robert *de Maltebi* 1169 P (Nf); Andrew *de Malteby* 1219 AssY; Robert *Maltby, Mawteby* 1378 AssLo, 1439–40 Past. From Maltby (Lincs, NR, WRYorks), or Mautby (Norfolk).

**Malter:** (i) Roger, John *le Malter* 1319 SR (Ess); 1336 ColchCt. A derivative of OE *mealt* 'malt', a maltster. cf. Hugh *le Maltmakere* 1255 RH (Bk), Peter *le Maltmetere* 1271 AD i (Mx), William *Maltmelnere* 1214 FFK, John *le Maltemeller* 1327 Pinchbeck (Sf). (ii) John *Maletere* 1198 FFK; Ralph *de Maleterr'* 1211 Cur (Nb). From some French place *Maleterre* 'poor ground'.

**Malthouse, Malthus, Maltus, Maltas:** Fulk *atte Malthuse* 1297 MELS (Sx); William *Malthouse* 1497 FrY. 'Worker at the malt-house.'

**Maltman:** Richard *Maltmon* 1332 SRWa; John *Malteman* 1471 Eynsham (O). cf. MALTER.

**Maltmeter:** Peter *le Maltmetere* 1271 AD i (Mx). 'Malt tester', OE *mealt*, and a derivative of OE *metan* 'to measure'.

**Malton:** Henry *de Maltun* c1138–43 YCh; John *de Malton'* 1279 FFY; John *Malton* 1423 FFEss. From Malton (NRY).

**Maltravers:** v. MATRAVERS

**Malvenue:** Henry *Mauvenu* 1200 Cur; Robert *Malvenu* 1205 Cur (Sf); Ralph *Mauvenu* 1207 Cur (Nf). 'Evil comer', OFr *mal, venu*. cf. Hubert *Maleuuenant* 1194 Cur (W) with the same meaning.

**Malvern:** Athelard *de Maluerna* 1221 AssSa; William *de Malvernia* 1275 SRWo; John *Maluerne* 1360 IpmGl. From Great, Little Malvern (Wo).

**Malyan, Malyon:** John *Marion, Malyon* 1351–4 PN Ess 483. *Marion*, a diminutive of *Mary*, with change of *r—n* to *l—n*.

**Malzer:** v. MALSER

**Mammatt, Mammett:** Gilbert *Maminot* 1086 DB (K); Emma *Maminet* 1130 P (K); Wauchelin *Maminot* 1157 Templars (L). *Mamin-ot, Mamin-et*, diminutives of *Mamin*, a pet-form of *Maximus*. cf. Fr *Mame* from *Maxine*, and v. OEByn 222.

**Mammen:** v. MAYMON

**Mammett:** v. MAMMATT

**Mammon:** v. MAYMON

**Man, Mann:** *Man* 1066 DB (Y, Sf); Ricardus, Hugo *filius Man* 1188 P (L), Hy 3 Colch (Ess); Algar, William *Man* 1141 ELPN, 1185 Templars (Y); Alan *le Man* 1288 FFSf; Walter *le Mon* 1327 SRDb. The OE personal name *Mann* was still in use, though not common, in the 12th century and was sometimes, no doubt, adopted as a surname. The usual source is OE *mann* 'man' though the exact sense is not always clear. A common meaning is, no doubt, 'servant' as in the compounds *Harriman, Ladyman, Monkman*, etc. Sometimes it may correspond to such a phrase as *homo Bainardi* 'the man of Bainard', one who owed him feudal service. At times it may have reference to a lower rank in the social scale, *nativus* 'bondman'. In 1279 Ralph de Ginges granted to Simon de Duntona (both of Essex) Peter *Man* son of Robert *Man* his *nativus* with all his family and all his chattels for 40*s*. sterling (Bart i. 503).

**Manby:** Antonius *de Manebi* 1181 P (L); Hugh *de Mageneby* 1226 FFY; Thomas *de Mauneby* 1336 FFY; Thomas *Manneby* 1403 IpmY. From Manby (Lincs), *Magnebi* 1212, or Maunby (NRYorks), *Magnebi* 1166.

**Mancell:** v. MANSEL

**Mancer:** v. MANSER

**Manche:** Geoffrey *Maunche* 1253–4, John *Maunche* 1371 FFEss. OFr *manche* 'sleeve'. Metonymic for a sleeve-maker.

**Manchester:** Robert *de Manecestre* 1198 RegAntiquiss; Henry *de Manecestre* 1291 KB (Nb); William *Manchestre* 1392 IpmGl. From Manchester (La). Sometimes from Mancetter (Wa), *Manecestre* 1236.

**Manchip:** v. MANSHIP

**Manclark, Moakler, Mockler, Mokler:** Roger *Malclerc* 1194 Cur (W); Walter *Mauclerc* 1207 Cur (L); Walter dictus *Manclerc* 1279 RH (O); John *Maucler* 1317 AD iv (Wa); William *Manclark* 1524 SRSf. OFr *mal, mau* and *clerc* 'bad cleric'. Walter *Mauclerc* 1275 RH (L) was Bishop of Carlisle. Pierre de Bretagne, similarly nicknamed, was unfrocked (Dauzat). *Manclark* is from *malclerc* with dissimilation of *l—l—r* to *n—l—r; Moakler, Mockler* are from *Maucler(c)*.

**Mander, Manders, Maunder, Maunders:** Thomas *Mander* 1524 SRD; John *Mander*, Giles *Maunder* 1642 PrD; Mary *Maunders* 1662–4 HTDo. ME *mander* 'beggar', or 'basket-maker', from a derivative of OFr *maund* 'basket'.

**Mandeville, Manvell, Manville, Manwell:** Goisfridus *de Magna uilla, de Manneuille* 1086 DB (Ess, K); Ernulf *de Mandeuill'* 1158 P (W); William *de Manevell'* 1210 Cur (Bk); William *de Manewell* 1296 SRSx. The Mandevilles, earls of Essex, came from Manneville (Seine-Inférieure); the Mandevilles of Earl's Stoke and Devon, from Magneville (La Manche); the undertenants of Montfort and the counts of Meulan, from Manneville-sur-Risle (Eure). v. ANF. Others may have come from Manneville (Calvados).

**Manfield:** Roger *de Manefeld* 1209 P (Ess); Roger *de Manfeld* 1303 FFY; Robert *Manfeld* 1413–14 IpmY. From Manfield (NRY).

**Mangan, Mangin, Mangon, Mannion, Mangen, Margan:** (i) Alexander *Maniaunt* (Manducans), *le Mangant* 1199 CurR; Alexander *Mangaunt* 1279 RH (C); Roger *Mangaunt* 1327 SRC. 'Glutton', from OFr *manger* 'to eat'. (ii) Peter Loptz 1313 Pat, surety of the Spanish traders, identical with Peter *Manioun* 1319 SRLo; Ferrand *Manioun* 1329 FFK; Peter Lopice, *mangoun* 1331 Pat; Ferrand *Mangeoun* 1344 LLB F. Spanish *mangon* 'a small trader'. (iii) Also Huguenot, several families of the name from Metz having settled in Ireland. (Smiles 413). (iv) In Ireland for *Ó Mongáin*, from *mongach* 'hairy'.

**Manger, Monger:** William *Manger'* 1255 RH (W); Richard *le Manger* 1275 SRWo; Robert *Monger* 1316 Wak (Y). OE *mangere* 'monger, dealer, trader'.

**Mangnall, Mangold, Manknell:** Geoffrey *Mangwinel* 1204 Cur (Pembroke); Stephen *Manguinel* 1212 Cur (Wa); Thomas *Mangenel* 1327 SRSx; John *Mangel, Mangulle* 1363, 1390 LLB G, H. 'A worker of the mangonel', OFr *mangonelle* 'a war engine for throwing stones'.

**Manhire:** *v.* MENNEAR

**Manistre, Manisty:** (i) William *Mannol(f)stygh* 1315–16 Wak (Y); Richard *Manaste* 1422 FrY. Manesty in Borrowdale (Cumberland) is *Manistie, Maynister* in 1564 (PN Cu 353). Whether the Yorkshire surnames derive thence is not certain but *Manesty* is a normal development of *Mannolfstygh* and Cumberland surnames are not uncommon in Yorkshire. (ii) William *Manitre* 1674 HTSf; John *Manister, Manester* ib.; Edmund *Manistre* ib.; Robert *Mannestey* ib. From Manningtree, on the Essex side of the Stour. Usually *Manitre*, it occurs as *Manystre* in 1291 and 1343 (PN Ess 343) and this appears to have been the local pronunciation which became *Manister, Manisty*.

**Mankin:** *Manekyn* le Heumer 1318 Pat (Lo); Stephen, William *Manekyn* 1242 Fees (K), 1327 SR (Ess). *Man-kin*, a diminutive of *Man*, and, possibly, also of the noun, whence a nickname 'the little man'.

**Manknell:** *v.* MANGNALL

**Manley, Manly, Manleigh:** William *de Manelegh'* 1202 Fees (D); Alexander, James *Manly* 1363 AssY, 1642 PrD. From Manley (Ches, Devon), or 'dweller by the common wood'.

**Mannell:** *v.* MEYNELL

**Mannering:** *v.* MAINWARING

**Manners:** Reginald *de Meiniers* c1180 ANF (Sx); Walter *de Maners* c1230 Barnwell (C). From Mesnières (Seine-Inférieure). *v.* ANF.

**Mannin:** John *le Manant* 1169 P (Gl); John *le maneant* 1200 Pleas (So); Roger *le Maniant', le Manant* 1221 Cur (Sf); William *Mannin* 1642 PrD. Probably 'workman', from OFr *manier* 'to work, handle'.

**Manning:** *Mannicus* 1066 DB (Ess); Algarus *Manningestepsune* c1130 ELPN; Seman *filius Manning* 1181 P (Ess); Ainulf, Richard *Manning* 1190 P (K), 1221 ElyA (Sf). OE *Manning*.

**Manningham:** Robert *de Mayningham* 1285–6 IpmY; Roger *de Manyngham* 1353 FFY; John

*Manyngham* 1427 IpmNt. From Manningham (WRY).

**Mannington:** Dusa *of Maninton'* 1249 AssW; John *Maningeton* 1642 PrD. From Mannington (Do, Nf).

**Mannion:** *v.* MANGAN

**Mansel, Mansell, Mancell, Maunsell:** *Mansell* de Patleshull 1203 AssSt; Thomas *filius Manselli* 1256 AssNb; Turstinus *Mansel* 1148 Winton (Ha); Robert *le Mansel* 1171 P (Ha); Andrew *Maunsel* c1180 Black. From a personal-name or from OFr *Mancel*, an inhabitant of Maine or its capital Le Mans. In France *Mansel* was also a feudal tenant, occupier of a *manse*, land sufficient to support a family.

**Manser, Mancer:** (i) *Manasserus* de Danmartin 1166 RBE (Sf); *Manserus* filius Joia 1186 P (L); *Maneserus* Judeus 1191 P (L); *Manaserus* de Hasting', de Hamwold' 1207–8 Cur (Sx, K); *Manasserus, Manser* Aguliun 1211–12 Cur (Ha); Ranulfus *filius Manser* 1221 ElyA (Sf); *Mancerus* le Parmenter 1296 SRSx; Walter *Manser* 1250 Fees (Sf); Alan *Mauncer* 1296 SRSx; John *Maunser* 1327 SR (Ess). There can be little doubt that this must be the Hebrew *Manasseh* 'one who causes to forget', used undoubtedly of Jews (*Manaser* Judeus 1219 AssY), but also of others. The DB tenant-in-chief *Manasses* was presumably a Norman, whilst *frater Manasserus* of the Hospital of St John of Jerusalem (1219 Cur, Beds) cannot have been a Jew. Mansers Shaw in Battle owes its name to the family of *Manasseh* de Herst, William Fitz *Manser* of Herst and *Manserus* de Scotegny (PN Sx 500). (ii) Nicholas *le Mauncer* 1297 LLB B; Richard *le mancher* 1292 SRLo. A derivative of OFr *manche* 'handle, haft', a maker of hafts for knives. Richard is also called *le Haftere* (1301 Husting).

**Mansfield:** Hugh *de Manesfeld* 1209 P (Nt); Richard *de Maunsfeld* 1321–2 IpmNt; Henry *Maunesfeld* 1371 LoPleas. From Mansfield (Nt).

**Manship, Manchip:** Ailsi *de Menscipe* (*Maneschipe*) 1167 P (Y); Philip *Mansipe* 1189 P (Nf); John *Manshipe* 1247 AssBeds; Robert *Mansipe* 1279 RH (O); Alexander *de Manshipe* 1319 SRLo; John *Manschupe* 1327 SRSo. From Minskip (WRYorks), Manships Shaw (Surrey) or Manchips Field in Bishop's Stortford (Herts), all from OE *gemǣnscipe* 'community, fellowship', used of land held in common, and also in its abstract sense.

**Manson:** (i) John *Mannisson* 1305 FFSf. 'Son of *Mann*.' (ii) Angus *Mangson* 1446 Black (Kirkwall); David *Manson* 1504 ib. For *Magnusson*. Common in Shetland.

**Manston:** (i) Walter *Manstan* 1315 AssNf. OE \**Mannstān*. (ii) Walter *de Manneston'* 1198 Cur (K); Richard *de Maneston* 1290 IpmW; Alverey *Manston* 1419 IpmY. From Manston (Do, K, WRY).

**Mantel, Mantell, Mantle:** Turstinus *Mantel* 1086 DB (Bk); Robert *Mantell'* 1176 P (Berks). OFr *mantel* 'cloak, mantle', a nickname or trade-name.

**Manthorpe, Mantrip:** William *de Manthorp* 1275 RH (Sf); Henry *de Manthorp'* 1298 AssL; John *de Manthorp* 1327 SRSf. From Manthorpe (L).

**Manton:** Odinell *de Maneton'* 1202 AssNth; Thomas *de Maneton* 1227–8 FFEss; William *Mantoun* 1327 SRC. From Manton (Lincs, Notts, Rutland, Suffolk, Wilts).

**Manus:** v. MAGNUS

**Manvell, Manville:** v. MANDEVILLE

**Manwaring:** v. MAINWARING

**Manwell:** v. MANDEVILLE

**Manwin:** *Manewine* Leuiet 1188 BuryS (Sf); *Manuinus* de Mundone Hy 3 Colch (Ess); Marjeria *Manwyne* 1327 SRSf. OE *Mannwine.

**Manwood:** Alice *de la Manewod* 1332 SRSx. 'Dweller by the common wood', OE (ge)*mǣne* and *wudu.*

**Manyweathers:** cf. FAIRWEATHER

**Maples:** Robert *atte Mapele* 1285 *Ass* (Ess); John *Mapel* 1327 SRC; John *del Mapples* 1348 Shef. 'Dweller by the maple(s)', OE *mapul.*

**Maplethorpe, Mapplethorp:** Gilbert *de Maupertorp'* 1219 AssY. From Mablethorpe (Lincs).

**Mapner, Mapnor:** Walter *de Mappenor'* c1233 WoCh. From a lost *Mapnors* in Knightwick (Wo).

**Mapp, Mappes:** Godric *Map* 950–1000 OEByn (Co); Walter *Map* 1165–81 HeCh; Nicholas *Map* 1212 Cur (Wo); John *Mapys* 1406 FFEss. Probably OE *Mappa,* but sometimes, perhaps, OW *map* 'son'. v. OEByn 378.

**Mapperley:** v. MABBERLEY

**Mappin:** Peter *Maupyn* 1302, *Maupin* 1303 LLB. From the name of a French domain.

**Marber, Marbrow, Marbler:** Reginald *le Marbrer* 1230 P (Ess); Godfrey *le Marbeler* alias *le Marberer* 1265 Ipm (Sr); Walter *le Marbeler, le Marbrer, le Marberer* 1281, 1288 LLB B, 1292 SRLo. OFr *marbrier* 'quarrier, hewer of marble', carver or worker in marble. At Westminster in 1385 Thomas Canon of Corfe, marbrer, was paid £30 6s. 8d. for making stone images in the likeness of kings, to stand in the Great Hall. Marble was also used for paving in churches, etc. In 1312 Adam le Marbrer undertook to pave part of St Paul's with squares of marble (Building 32, 147).

**Marble:** Robert *Marbull,* Hugh *Marble* 1479, 1531 AD i (Mx). Metonymic for *Marbler.*

**Marbrook:** Adam *atte Merbrock* 1332 SRWo. From Marbrook (Wo).

**March:** Henry, William *de la Marche* 1295 Barnwell (C), 1307 Oseney (O); Hugo *atte Marche* 1349 LLB G. 'Dweller by the boundary', ME, OFr *marche.*

**Marchant, Marchent, Marchand, Marquand, Merchant, Le Marchand, Le Marquand, Le Marchant, Lemarchand:** Roger *Marcand (Marchand')* 1202 P (Berks); Roger *Marchaunt,* Herueus *Merchant* 1219 AssY; Ranulph *le Marchand* 1240 FFEss; Reginald *le Marchant* 1247 FFC; Thomas *le Markaund* 1274 RH (So). OFr *marchand, marchĕant* 'merchant, trader'.

**Marchbank(s):** v. MARJORIBANKS

**Marchman, Marchment:** William *Marchman* 1583 Musters (Sr). 'Dweller by the boundary', OFr *marche.*

**Marcy:** Ralph (*de) Marci* 1086 DB (Ess). From Marcy (La Manche). v. also MASSEY

**Marden, Mardon:** Walter *de Mahurdin* 1204 P (He); John *de Merdene* 1278 IpmW. Thomas *de Merdon* 1332 SRSx. From Marden (He, K, W), or East, North, Up Marden (Sx).

**Mares:** v. MARRIS

**Margan:** v. MANGAN

**Margary, Margery:** *Margeria* 1219 AssY; Galfridus *filius Margerye* 1221 ElyA (Sf); Robert, John *Margerie* 1195 P (Gl), 1275 RH (Sf); Agnes *Mariory* 1327 SRSf. *Margerie* was a French popular form of *Marguérite* (Margaret).

**Margerison, Margereson, Margerrison:** Robert *Mariorison* 1379 PTY. 'Son of *Margery.'*

**Margerson, Margesson, Margison:** Richard *Margison* 1683 EA (OS) iv (Sf). 'Son of *Margery.'* Also for *Margetson.*

**Margetson, Margitson:** Richard *Margretson* 1381 PTY; Thomas *Margetson* 1425 DbCh. 'Son of *Margaret.'*

**Margetts:** John *Margaret'* 1275 RH (Sf); John *Marget* 1524 SRSf. From *Margaret* 'pearl', a common medieval woman's name, found as *Merget* in 1460 and *Margat* in 1534 (ODCN).

**Marie, Marry:** Godfrey *filius Marie* 1189 Sol; *Maria* 1219 AssY, 1297 FFSf; John *Marie* 1279 RH (Bk); William *Marye* 1367–8 FFWa; John *Marrie* 1642 PrD. Lat *Maria,* Fr *Marie,* ultimately from Hebrew.

**Mariman, Marriman:** Henry *Mariman* 1296 SRNb; Adam *Mariman* 1332 SRCu; Robert *Marimon* 1332 SRWa. 'Servant of *Mary'.*

**Mariner, Marriner, Marner:** Hugo *le marinier* 1197 P (O); Ace *Meriner* 1211–23 Clerkenwell (Lo); Ivo *le Mariner* 1228 Cl (Bk); Peter *le Marner* 1327 SRSx. AFr *mariner,* OFr *marinier, marnier, merinier* 'sailor, seaman'.

**Marion, Marians, Maryan, Maryon, Marrian, Marrion:** *Marion* Lambert 1379 PTY; John *Mariun* 1279 RH (C); Richard *Marioun* 1350 LLB F. A diminutive of *Mary.*

**Maris:** v. MARRIS

**Marjoribanks, Majoribanks, Marchbank, Marchbanks:** John *Marjoribankis* 1550 Black; James *Marchbank* 1664 ib. The barony of Ratho was granted by Robert Bruce as a marriage portion to his daughter Marjorie on her marriage in 1316 to Walter, High Steward of Scotland, ancestor of the royal house of Stewart (Burke). The lands, known as Ratho-Marjoribankis, came into the possession of a family named Johnson who assumed the surname of Marjoribanks (Black).

**Mark, Marke, Marks:** (i) Rogerus *filius Markes* 1207 Cur (Ha); *Marc* le draper 1292 SRLo; Robertus *Marcus* 1148 Winton (Ha); Philip *Marc* 1209 P (Nt); Robertus *Markes* 1288 RamsCt (C). *Mark,* Lat *Marcus.* Never a very common name. (ii) Adelolfus *de Merc* 1086 DB (Ess); Geoffrey *de Merc* 1130 P (Ess). From Marck (Pas-de-Calais). (iii) Simon *del Merc* 1208 Cur (Ess); Matilda *de la Merke* 1227 FFEss; Aylward *atte Merke* 1296 SRSx. 'Dweller by the boundary', OE *mearc.*

**Markby:** Alan *de Markebie* 13th Glapwell (Db); John *de Markeby* 1339 CorLo; Henry *Markeby* 1381 AssLo. From Markby (L).

**Marker:** (i) Symon *filius Markere* 1168 P (D); *Markerus* de Torneberga 1176 P (Bk); Walter, Eborard *Marker* 1297 SRY, 1309 SRBeds. Probably an OG *Marchari* 'boundary-army'. *Mearc-* seems not to occur as an OE theme. (ii) William, Agnes *le Marker(e)* 1185 P (Ess), 1260 Oseney (O); Reginald *le Merkere* 1275 RH (Nf). Thuresson's suggestion that this is a derivative of OE *mearcian* 'to mark', a marker of game or a stamper or brander, is not satisfying. The name may at times mean 'dweller by the boundary' (cf. MARK) but examples are more numerous than usual for such a toponymic. It may be a survival of OE *mearcere* 'notary, writer'.

**Market, Markett:** Edith *atte Markete* 1327 Kris; Henry *Markett* 1411 FrY. 'Dweller by the market-place', ME *market*. cf. Walter *Marketman* 1545 SRW; John *Marketstede* 1414 FFEss, with the same meaning.

**Markham:** Jordan, Roger *de Marcham* 1204 Pl (Nt); Richard, Emma *de Markham* 1259 FFEss, 1371 AssL; John *Markham* 1408 IpmNt. From Markham (Notts).

**Markin, Marking:** Nicholas *Merkyn* 1357 ColchCt; William *Marekyn* 1390, *Markyn* 1435, John *Marykyn* 1458 FrY. *Mary-kin*, a diminutive of *Mary*, but sometimes, perhaps, a diminutive of *Mark*.

**Markland:** John *Markland* 1401 AssLa. 'The holder of land of the annual value of one mark', or 'dweller at the boundary land', OE *mearc*, *land*, cf. John *Markman* 1318 FrY.

**Markley, Markleigh:** Geoffrey *de Merkele* 1210 Cur (He); Robert *de Markelegh* 1296 SRSx; John *Markeley* 1394 IpmGl. From Markly in Heathfield (Sx).

**Markson:** William *Markeson* 1445 FrY. 'Son of *Mark*.'

**Markwick, Marquick:** William *de Merquik* 1332 SRSr; Alan *Markwick* 1525 SRSx; Richard *Markwyke* 1553 CtSx. From Markwich Fm in Hascombe (Sr), or Markwicks in Wadhurst (Sx).

**Marler, Marlor:** Hugo *le Merlere* 1275 SRWo; Thomas *le Marlere* 1277 Ely (Sf); William *marlur* 1297 MinAcctCo (Y). A derivative of ME, OFr *marle* 'marl', a hewer or quarrier of marl. cf. William *le Marlehewere* 1327 SRSt.

**Marley, Marlee:** William *de Merlai* c1145–65 Seals; Thomas *de Mardele* 1208 Pl (Y); John *de Marley* 1285 Riev; William *de Marleye* 1306 FFEss. From Marley (Devon, Kent, WRYorks), Marley Farm in Brede (Sussex), or Mearley (Lancs), *Merlay* 1241.

**Marlin, Marling:** William *Marlyn* 1297 MinAcctCo; John *Marling* 1361 ColchCt; Christopher *Marlin* 1642 PrD. Probably from OFr *Merlin*, Welsh *Myrddhin*. Used as a christian name in the 17th century: *Marling* Gooding 1642 PrD.

**Marlow, Marlowe:** Edric *Merlaue* 1066 DB (Bk); Wido *de Merlaue* 1225 Cur (Bk); Richard *de Merlawe* 1325–6 CorLo; Joan *Marlowe* 1534 CantW. From Marlow (Bucks), but there was evidently also some confusion with MARLEY, cf. Richard *Merleye*, and Alice *Marlowe* his wife, 1488, 1494 ArchC xxxix.

**Marmaduke:** *Marmaduc* de Ar' Hy 2 Gilb (L); *Marmaduc Darell*, *Marmaduc* de Tweng 1219 AssY; Thomas *Marmeduc* 1276 IpmY; John *Marmeduk* 1286 IpmNb; Walter *Marmeduke* 1441 FrY. Perhaps OIr *Maelmaedoc* 'servant of *Maedoc*', *v.* ODCN. In the Middle Ages a distinctively Yorkshire name.

**Marmion:** Robart *Marmion* 1103–15 OEByn; Philip *Marmiun* 1182 P (W); John *Marmyon* 1297 MinAcctCo; William *Marmyon* 1350 FFW. OFr *marmion* 'monkey, brat'.

**Marner:** *v.* MARINER

**Marney, Marnie, de Marney:** Robert *de Mareigni* 1168 P (Ess); William *de Marenni* 1207 Cur (Ess), *de Marny* 1284 FFEss. From Marigni (La Manche).

**Marnham:** Nicholas *de Marnham* e 13th Glapwell (Db); William *de Marnham* 1332 SRSt; Walter *Marnham* 1452–3 IpmNt. From High, Low Marnham (Nt).

**Marple, Marples:** John *de Marpell* 1327 SRDb; Francis *Marples* 1670 PN Db 303. From Marple (Ch).

**Marquand:** *v.* MARCHANT

**Marquick:** *v.* MARKWICK

**Marquis, Marqueez:** Ralph *le Marchis* 1218 AssL; Howel *le Marchis* 1221 AssWo; Elspet *Marquis* 1613 Black (Inverness). A nickname from OFr *marchis* 'marquis'. Found also as a feminine name: *Marchisa* 1202 FFY.

**Marr, Marre, Marrs:** Roger, James *de Mar* 1182 P (Y), 1296 Black (Aberdeen); Ralph *atte Mar* 1297 SRY; William *del Marre* 1302 SRY. From Marr (WRYorks), High or Low Marr in Wheldrake or The Marrs in Swine (ERYorks), Mar (Aberdeenshire), or from residence near a pool or marsh (ON *marr*).

**Marrable:** Roger *Mirabell'* 1230 P (D); Richard *Marabile* 1244 Rams (Beds); Robert *Merable* 1327 PN Wt 253. *Mirabel*, a woman's name latinized as *Mirabilis* 1210 Cur (Wo) 'marvellous, wonderful'.

**Marriage:** Alice *Marriage* 1616 Bardsley (Wa); Samuel *Marredge* 1626 ib. (Mx); Stephen *Marridge* 1709 ib. (Lo). In Essex and Suffolk, probably from a lost place in Finchingfield or Aythorpe Roding, OE *(ge)mǣre* and *hæcc* 'boundary-gate', found as a surname: William *Marhach'*, *Mar(r)ach* 1377, 1379 AssEss. Marriage Fm 'boundary-ridge' (PN K 386) and Marridge 'meadow-ridge' (PN D 285) both gave rise to surnames in the 13th century.

**Marrian, Marrion:** *v.* MARION

**Marriman:** *v.* MARIMAN

**Marrin:** *Marinus* clericus 1192 P (Lo); *Marina* 1230 P (Nt), 1302 FA (Bk); Walter *Maryne* 1327 SRSo. There were several saints named *Marinus* and one *Marina* (Lat *marinus* 'of the sea'). Neither name was common in England.

**Marriott, Marritt, Marryatt:** *Mariota* Hoppesort 1195 FFSf; *Mariota* 1200 P (Lei), 1219 AssY; William, Hervicus *Mariot* 1185 Templars (Wa), 1210 Cur (C). *Mari-ot*, a very common diminutive of *Mary*. *Marryatt* is probably from OE *Mǣrgēat*: Richard *Meryet* 1297 MinAcctCo; William *Mariet* 1327 SRDb. cf. MERRETT.

**Marris, Maris, Mares:** Clarenbodo *de Maresc*, Richard *de Maris* 1086 DB (Bk, K); William *de Mares* 1191 P (K); Aldwin *des Mareis* 1199 P (Gl); Henry *Mareys* 1252 Rams (Hu); Walter *del Mareys*, Richard *de Marise* 1275 SRWo. From (Le) Marais (Calvados) or from residence near some marsh (OFr *marais*, Norm-Picard *marese*).

**Marrow, Marrows:** John *Marwe* c1208 FrLeic; Geoffrey *le Marewe* 1276 RH (C). ME *marwe* 'companion, mate, lover'.

**Marry:** v. MARIE

**Marryatt:** v. MARRIOTT

**Marsden:** Alan *de Marchesden* 1246 AssLa; Johanna *de Merssden* 1379 PTY; Peter *Marsden* 1459 Kirk. From Marsden (Lancs, WRYorks).

**Marsh, Mersh:** Godard *le la Merse* 1194 P (So); Henry *del Merse* 1212 Cur (Y); William *atte Mersche* 1296 SRSx. 'Dweller near a marsh' (OE *mersc*).

**Marshall, Marschall, Marskell, Mascall, Maskall, Maskell, Maskill:** Goisfridus *Marescal* 1084 GeldR (W); Hugo *Maskercal* 1087–97 Crispin (Mx); William *Marescald* 1100–30 OEByn; Roger *Mascherell* 1130 P (L); Rainald *le mareschall* c1140 DC (L); Robert *Maskerel* 1166 RBE (Sf); Henry *le Marscal* 1238 AsSo; John *Marschal* 1296 SRSx. OFr *mareschal, marescal, marescald, marechault,* 'one who tends horses, especially one who treats their diseases; a shoeing smith, a farrier' (1258 NED). Early examples may refer to a high officer of state (cf. the Earl Marshal). Later, the surname is equated with *Smith* and *Faber.* Mascallsbury in White Roding (from Robert *le Marescal* 1235) is *Maskerelesburi, Maskelesbury* 1351, *Mascallesbury* 1476 (PN Ess 495).

**Marsham, Marshom:** Leofstan *æt Merseham* c1060 OEByn; Benjamin *de Merseham* 1236 FFK; John *de Marsham* 1336 CorLo; John *Marsham* 1551 NorwLt. From Marsham (Nf), or Mersham (K).

**Marshfield:** John *Marsfeld* 1327 SRSo; Widow *Marshfeild* 1662–4 HTDo. From Marshfield (Ch, Gl, Monmouth).

**Marshman, Mashman:** Edmund *le Mersman* 1233 HPD (Ess); John *Mershman* 1301 ArchC 9. 'Dweller in the marsh.' John *Mersman* and Richard *ate Mersshe* lived in the same parish (1317 AssK). cf. William *le Mersher* 1327 SRSx.

**Marshom:** v. MARSHAM

**Marskell:** v. MARSHALL

**Marsom, Marson:** Robert *Marson* 1596 Musters (Sr). From Marston, a common place-name, often spelled and pronounced *Marson.* Later, the surname acquired an intrusive *t,* restoring it to its earlier form. *Marson* is now less common than *Marston,* while *Marsom* is rare after about 1720. At Olney (Bk), Thomas and Elizabeth *Marson* (1796–8) appear as *Marston* in 1801 and 1804, whilst their daughter Sophia (d. 1806) is called *Marson.* At Church Leigh (Staffs), William, son of James *Marston* (b. 1685), was baptized as *Marston* in 1721, but is called *Marson* on his tombstone (1787), whilst his son was baptized and married as *Marston* but buried as *Marson* which has since been the family name. v. MARSTON.

**Marsters:** v. MASTER

**Marsterson:** v. MASTERSON

**Marston:** Hugh *de Merstona* 1170 Eynsham; William *de Merstun'* 1231 Cur (Herts); Sylvester *de Marston* 1349–51 FrC; John *Marsten* alias Mason 1565 NorwDep. From one or other of the many places of this name, or from Merston (Kent, Sussex, Wight). v. MARSOM.

**Mart, Marte:** Robert *filius Mart'* 1279 RH (Hu); Aluuard *Mert* 1066 DB (D); William *Marte* 1243–4 FFEss; John *Lemart, Lemert* 1279 RH (C); William *Mart* 1327 SRSx. OE *\*Meort, v.* OEByn 323. Perhaps also a nickname from OE *mearð* 'marten'.

**Martel, Martell:** *Martellus* 1163–6, c1187 DC (L); *Martel* Saladin 1222 Cur (Nf); Goisfridus *Martel* 1086 DB (Ess); William *Martel* 1148 Eynsham (O). The personal name is a hypocoristic of *Martin.* The surname may derive from this (cf. Martin *Martel* Hy 2 DC (L) ) or be a nickname from OFr *martel* 'hammer' (1474 NED), here, no doubt, the *martel de fer,* the iron hammer or mace of medieval warfare. The first bearer of the name was Charles son of Pépin d'Héristal 'qui martela les Sarrasins' at the battle of Poitiers in 732 and received the name of Charles *Martel* (Moisy). Later, the name was probably occupational, 'smith'.

**Marter, Martyr:** Robert *Lamartre* 1130 P (O); Walter *martre* 1148 Winton (Ha); William *le Marter* 1275 SRWo. A nickname from the weasel, OFr *martre,* ME *martre, marter.*

**Martin, Martins, Marten, Martens, Martyn, Martyns:** *Martinus* 1066 Winton (Ha), c1166 NthCh (Nth); Walter, Helewis *Martin* 1166 RBE (C), Ric 1 Cur. MedLat *Martinus,* a diminutive of *Martius,* from *Mars* the god of war. A very popular christian name and an early surname. The Manx *Martin* is a contraction of KILMARTIN.

**Martindale, Martindill:** John *Martyndale* 1476, Nicholas *Martyldall* 1530 FrY; Roberte *Martindale,* Thomas *Markindale* 1672 HTY. From Martindale (We).

**Martineau:** (i) Martin *de Martinaus* 1276 RH (L); Ralph *de Martinell* 1276 RH (Lei). From Martineau (France). (ii) Also Huguenot. Gaston *Martineau* of Bergerac in Perigort was naturalized in 1668, and later moved from London to Norwich.

**Martinson:** John *Martynson* 1463 Black. 'Son of *Martin.'*

**Martley:** Robert *de Martleye* 1275 SRWo; John *de Martelay* 1283 IpmY; William *de Martleye* 1327 SRWo. From Martley (Wo).

**Martock:** Roger *Martok* 1249 AssW; Thomas *Martocke* 1662–4 HTDo. From Martock (So).

**Marton, Martin, Merton:** Adam *de Mertuna* 1189 Sol; Thomas *de Marton* 1212 P (Y); Ellis *de Martin,* William *de Mertton'* 1249 AssW; John *Merten* alias Noble, William *Martene* alias Perham 1545, 1576 SRW. From one or other of the many places called Marton, Merton, or Martin.

**Martyr:** v. MARTER

**Marval, Marvel, Marvell:** (i) Richard *Merveyle* 1275 RH (C); Roger *Marvell* 1524 SRSf. OFr *merveille* 'the marvel'. cf. William *le Merveillus* 1186 P (L) 'the wonderful'. (ii) Ranulph *de Mereville* 1306 FFEss. From Merville (Nord, Somme). (iii) Marvell (Isle of Wight) derives from a family of Merryfield: *de Meriefeld* 1255, *Marfeildes* 1558, *Marvell* 1608 (PN Wt 102).

**Marvin, Marven, Mervin, Mervyn, Mirfin, Murfin:** Ælfwine *Merefinnes sune* c1060 KCD (Nth); *Merefin* (Nth), *Meruin* (Sr), *Mervinus* (C) 1066 DB. Peter *Merevin* 1160–5 ELPN, *Merefin* 1169 P (Lo); Robert *Merwyn* 1298 AssL; Gilbert *Mervyn* 1327 SRSf; John *Murfyn* 1410 IpmY; William *Marvin* 1524 SRSf. There are four possible sources for these names: (i) OE *Merewine.* (ii) OE *Merefinn* from ON *Mora-Finnr.* (iii) Welsh *Mervin* from Welsh *Merlin.* (iv) OE

*Mǎrwynn* (f): *Merewen* (f) 1202 AssL; *Merewina* de Acle 1221 Cur (O).

**Marwell:** James *Marwell* 1662–4 HTDo. From Marwell (Ha).

**Marwood:** (i) William *Marwod, de Marrewod* 1312, 1365 FrY; William *Marwood* 1559 Pat (D). From Marwood (Devon, Durham). (ii) The name may sometimes represent OFr *\*Malregard* 'evil look, evil eye', *v.* OEByn 322, and cf. Richard *Malregard* c1170 Riev; Walter *Malreward* 1200 Cur; Thomas *Maureward* 1208 Cur (Do); Robert *Mareward* 1242 Fees (W). Goadby Marwood (Leics) owes the second part of its name to William *Maureward* who held part of the manor in 1316.

**Maryan, Maryon:** *v.* MARION

**Maryson:** Willelmus *filius Marie* 1292 ELPN is identical with William *Marysone* 1298 LoCt. 'Son of *Mary*.'

**Mascall:** *v.* MARSHALL

**Maser:** Adam, William, Walter *le Mazerer* 1275 RH (Nth), 1307 LLB C. *Maser* is metonymic for *maserer*, a maker of masers (ME *maser*, OFr *masere* 'a maple-wood bowl'). *v.* MASLEN.

**Mash:** Robert *Masshe* 1524 SRSf. Probably for MARSH.

**Masham:** Gospatric *de Massam* 1138–53 MCh; Laurence *de Massham* 1280 FFY; John *Masham* 1401 IpmY. From Masham (NRY).

**Masheder, Masheter, Mashiter, Massheder, Mesheder, Messiter:** Richard *Maschrother* 1498 FrY; Robert *Masherudder* 1517 GildY; Peter *Mashrether* 1584 Bardsley (Ess); Janet *Masheder* Agnes *Masheter* 1637 LaWills; Ellen *Masheder* or *Mashred* 1663 ib. Although the evidence is not quite conclusive, this seems to be a nickname for one who steeped malt. cf. 'Maschel, or rothyr, or mascherel. *Remulus, palmula, mixtorium*' (PromptParv). 'A rudder (so called from its resemblance in shape), or instrument to stir the meash-fatte with' was one of the instruments of the brewhouse. Peter Mashrether came from Chigwell, not far from Mashiter's Hill in Romford, so called from a family of Mashiter.

**Mashman:** *v.* MARSHMAN

**Maskall, Maskell, Maskill:** *v.* MARSHAL

**Maskery, Maskrey:** Henry *le Macegre* 1260 AssY; Syremon *le Macekre* 1275 SRWo; Henry *Maskery* 1327 SRDb. OFr *macegref*, an altered form of *macecrier* 'butcher'. cf. MASSACRIER.

**Maslen, Maslin:** (i) Ricardus *filius Mascelin* 1187 P (Sa); Alexander *filius Macelini, Mazelini* 1203 Cur (Bk); Richer *Mazelin* 1168–75 Holme (Nf); William *Masselyn* 1327 SRC. OFr *Masselin*, OG *Mazelin*, a diminutive of *Mazo*. It might also be a double diminutive of *Maci* or *Masse* 'Matthew'. There was also a feminine *Mazelina* (1212 Cur (Nf), 1221 AssWa), probably a diminutive of *Matilda*. (ii) William *le Mazelin* 1282 AD ii (Mx). OFr, ME *maselin*, 'mazer, bowl of maple-wood', used by metonymy for a maker of these. William *le Mazeliner* 1283 LLB A (perhaps the same man as the above) is also called *le Mazerer* (1281), *le Mazener* (1282), *le Maceliner* (1292 ib.). cf. MASER.

**Mason, Massen, Masson, Machen, Machent, Machin, Machon:** John *Macun* c1130 AC (Lo); Ace *le mazun* 1193 P (He); Richard *machun* 12th DC (L); Roger *le Mason'* 1200 Oseney (O); Godfrey *le Mascun* 1203 FFEss; Osbert *le Masson* 1279 RH (O); Adam *le Machon* 1279 AssNb; Richard *Machen* 1284 AssSt; Thomas *Machyn* 1439 AD iv (Wa). ONFr *machun*, OCentrFr *maçon, masson* 'mason'. *v.* also MAYSON.

**Massacrier:** William *the Massacrer* 1235 FFY; Roger *le Macecrer* 1243 AssSo. OFr *macecrier* 'butcher'. A very rare surname. cf. MASKERY.

**Massday:** Richard *Messedei* 1185 Templars (K); Peter *Messeday* 1276 AssLo; Matilda *Masseday* 1332 SRSx. 'Massday', OE *mæssedæg*, for one born on that day. Sometimes, perhaps, 'servant of *Masse*', a pet-form of *Matthew*.

**Masse:** Osbertus *filius Masse* 1177 P (Sf); *Masse* faber 1219 AssY; Osbert *Masse* 1194 P (Nf). A pet-form of *Matthew*.

**Massen:** *v.* MASON

**Masset:** *Masota* de Middeton 1327 SRSx; *Mazota* Vyncent 1332 ib.; Richard *Masothe* 1279 RH (Bk); Richard *Masote* 1332 SRSx. A feminine diminutive of *Masse*, i.e. *Matthew*.

**Massey, Massie, Massy, Macey, Macy:** (i) *Mathiu, Maci* de Mauritania 1086 DB (Berks, Gl); *Masci* filius Mathei 1198 P (Nf); Alan *Macy* 1275 RH (Sf); William *Massy* 1330 NottBR. A pet-form of *Matthew*. (ii) Hamo *de Masci* 1086 DB (Ch), 1179 P (Db); Hugo *Maci* 1086 DB (Ha), *Mascy* 1221 Cur (Hu); John *de Maci* 1221 Cur (Mx). Hamo de Masci came from Macey (La Manche). Others may have come from Massy (Seine-Inférieure), Macé-sur-Orne (Orne) or La Ferté Macé (Orne). (iii) Ralph *Marci, de Marcei* 1086 DB (Ess); William *de Marsei* 1180 P (Nt). From Marcy (La Manche). Ralph's name survives in Stondon Massey, developing thus: *de Marcy* (1238), *Masse* (1542), *Massie* (1642 PN Ess 81).

**Massingbeard, Massingberd:** Thomas *Massyngberd* 1472 Cl (Sx); John *Messingberd* 1560 Pat; Oswell *Massingberd* 1596 Musters (Sr). Perhaps 'brazen beard', with the first element from ON *messing* 'brass'.

**Massinger:** *v.* MESSENGER

**Massingham, Messingham:** Ralph *de Massingham* 1208 P (Nf/Sf); Adam *de Messingham* 1264–5 RegAntiquiss; Robert *Massyngham* 1374–5 NorwLt. From Great, Little Massingham (Nf), or Messingham (L).

**Masson:** *v.* MASON

**Master, Masters, Marsters:** (i) Robert *le Meistre* 1202 Cur (Berks); John *Maister* 1225 Lewes (C); Hubert *Mastres* 1279 RH (Hu); John *Mastere* 1379 PTY. ME *maister*, OFr *maistre* 'master'. The exact signification is not clear. It may refer to a schoolmaster (cf. 'Maystyr, *Magister, didascolus, petagogus*' PromptParv), to the master of a farm or house, or to the master of an apprentice. (ii) William *atte Maystres* 1327 SRSt. 'One who lived at the master's (house)', his servant, or, possibly, his apprentice. In Scotland the eldest sons of barons and the uncles of lords were called 'Masters'.

**Masterman:** John *Maystreman* 1327 SRC. 'Servant of the Master.'

**Masterson, Marsterson:** Robert called *Maistersone* 1300 Black (Galloway); Hugh *le fiz Mestre* c1320 Calv (Y); John *Maisterson* 1323 FrY. 'Son of the Master.'

**Masterton:** William *de Mastertone* 1296 Black (Fife); Symon *de Maysterton* 1357 ib.; Thomas *Masterton* 1476 ib. From the lands of Masterton (Dunfermline).

**Matchell:** *Machel* 1066 DB (Y); John *Machel* 1327 SREss; Henry *Machell* 1510 GildY. The DB form is perhaps OG *Maghelm, Machelm*, but it is doubtful whether the later examples have any connexion with this.

**Mather, Mathur, Mathers, Matthers:** Alan *le Mathere* 1249 AssW; Beatrix *Matther* 1396 NottBR; Margaret *Mather* 1524 SRSf; Thomas *Mathar* 1621 SRY. OE *mæðere* 'mower, reaper', not, as Guppy thought, confined to Lancs, Northumb, and Derby.

**Mathes, Mathez, Mathie, Mathis, Maths, Matthes, Matthey, Matthys:** *Mathe* 1195 P (Nt); William son of *Mathie* 1332 SRCu; Hugh *Mathi* 1221 Cur (Herts); William *Mathy* 1275 RH (W); Henry *Matthe* 1279 RH (C); John *Mathis* 1674 HTSf. *Math* or *Mathi*, pet-forms of *Matthew*.

**Matheson:** v. MATHIESON

**Mathew:** v. MATTHEW

**Mathieson, Mathison, Matheson, Matthison:** John *Mathyson* 1392 Black (Angus). 'Son of *Mathi*.' v. MATHES.

**Mathis:** v. MATHES

**Matland:** v. MAITLAND

**Maton, Matten, Mattin, Matton:** William *Matun* 1275 RH (Nf); Matheon *Mateon* 1327 SRC; Robert *Maton*, William *Mathon*, Adam *Mathin* 1379 PTY. Diminutives of *Matthew*, from *Math* or *Mat*.

**Matraves, Matraves, Mattravers, Maltravers:** Hugo *Maltrauers*, *Malus Transitus* 1084 GeldR (So); Walter *Maltrauers* 1130 P (Do); Walter *de Matrauers* 1194 Cur (W). OFr *mal travers* 'bad (difficult) passage'. One would assume that this was a place-name, like *Malpas* but no such French place has been noted, although the 1194 form suggests there was one. If a nickname, it probably denotes a man difficult to pass, one who can look after himself, a stout soldier.

**Matt, Mattes, Matts:** Adam *Matte* 1275 RH (W); Simon *Mat* 1309 FFEss. In view of the existence of *Matkin* and *Maton*, there must have been a diminutive *Mat* from *Matthew*. The surname may also be metonymic, for MATTER.

**Mattack:** v. MADDOCK

**Mattam, Mathams, Matthams, Maddams, Maddans, Madan:** Henry *de Matham* 1195 Cur (Nf); William *de Matton'* 1212 Cur (Y); Geoffrey *de Matham* 1230 P (Ess). From Mathon (He), Mattins Fm in Radwinter (Ess), or Martinfield Green in Saffron Walden (Ess), John *de Matham* 1248.

**Mattar:** v. MATTER

**Matten:** v. MATON

**Matter, Mattar:** Arnald *Matere* 1214 Cur (L); Richard *le Mattere* 1310 LLB B. A derivative of OE *matte* 'mat'. cf. John *Mattemaker* 1381 AssWa.

**Matterface:** Richard *de Martinwast* 1166 RBE; Sarra *de Martuast'* 1252 FFEss; Geoffrey *Matterface*,

Gregory *Mattervers*, Thomas *Matterverse* 1642 PrD. From Martinvast (La Manche). v. ANF.

**Matters:** Alexander *Matras* 1379 PTY; John *Matres* 1438 FrY. *Matters* is, no doubt, for *Matres* 'mattrass', metonymic for 'mattrass-maker'. cf. Alice *Matresmaker* 1381 PTY.

**Matthams:** v. MATTAM

**Matthers:** v. MATHER

**Matthes:** v. MATHES

**Matthew, Matthews, Mathew, Mathewes, Mathews:** *Mathiu, Matheus* 1086 DB; *Matheus* Baret c1150–5 DC (L); Alan *Mathew* 1260 AssC; John *Mathows* 1395 Whitby (Y). Hebrew *Mattathiah* 'gift of Jehovah', latinized as *Matthaeus* and *Mathaeus*, which in France became respectively *Mathieu* and *Mahieu* (v. MAYHEW). The name was introduced into England by the Normans and became very popular, with a variety of pet-forms and diminutives. v. also MACEY, MASSE, MASSET, MASSEY, MATHES, MATON.

**Matthewman:** Hugh *Mathewman* 1379 PTY; John *Mathyman* 1504 ShefA. 'Servant of *Matthew*.'

**Matthewson, Mathewson:** John *Matheuson* 1416 FrY; George *Mathowsone* 1539 FeuDu. 'Son of *Matthew*.'

**Matthey, Mathie:** v. MATHES

**Mattick:** v. MADDOCK

**Mattimoe:** William *Matheumogh* 1327 SRDb; Ann *Mathemore* 1619 GreenwichPR. 'Relative of *Matthew*.' cf. WATMOUGH.

**Mattin:** v. MATON

**Mattingley, Mattingly:** Stephen *de Madingel', de Matingel'* 1206 Cur (Sr, Ha); Peter *de Mattingeley* 1249 IpmHa. From Mattingley in Heckfield (Hants).

**Mattingson, Mattinson, Mattison, Matterson:** Robert *Mattison* 1635 FrY; Thomas *Matterson* 1672 FrY. Probably from *Matonson* or *Mattinson* with the same development as *Pattinson*. v. MATON.

**Mattleson:** v. MAUD

**Mattock:** v. MADDOCK

**Matton:** v. MATON

**Mattravers:** v. MATRAVERS

**Maturin:** A Huguenot name. Gabriel *Maturin*, a refugee pastor, escaped from France after 26 years imprisonment in the Bastille. He settled in Ireland, his son Peter becoming Dean of Killaloe, and his grandson Dean of St Patrick's, Dublin (Smiles 414).

**Mauchan:** v. MAUGHAM

**Maud, Maude, Mahood, Mawhood, Mald, Malt, Mault, Mold, Mould, Moulds, Moult, Mowat:** *Mathild, Mathildis* 1086 DB; *Mahald* regina 1119 Colch (Ess); *Matilda* c1140 Holme (Nf); *Matildis* c1150 Gilb (L); *Mactilda* 1154–69 NthCh (Nth); *Matillis* temp. Stephen DC (L); *Mactildis* Hy 2 DC (L); *Mahald* 1172–80 DC (Lei); *Mahalt* Hy 2 Gilb (L), 1197 P (Wo); *Mahaut* vidua 1190 BuryS (Sf); Comitissa *Malt* 1275 RH (Sf); *Moude* de Seint edmund 1292 SRLo; *Mauld* 1303 ODCN; *Maud* 1314 Cl; Robert 'Maldesman of the Ker' 1327 Pat (Y); *Molde, Moolde* 1450 ODCN; *Mawte* 1502 BuryW (Sf); Gilbert, William, Robert *Mald'* 1190 P (Ess, R, W); Smale *Mautild* 1199 P (Nf); Hugh *Mold* 1275 RH (Wo); Geoffrey *Maude* 1279 RH (Hu); John *Malt* 1279

RH (C); Gilbert *Maughtild* 1327 *SR* (Ess); William *Matild'* 1327 SRC; Geoffrey, Agnes *Molt* 1327 SR (C, Sf); Edmund *Mohaut* 1332 SRLo. OG *Mahthildis*, introduced into England at the Conquest. William I's queen, Matilda, is called *Mold* by Robert of Gloucester. *Matilda* is a learned form, rarely found in medieval surnames and now completely lost unless it is the source of the very rare *Mattleson*. A pet-form *Till*, however, survives in *Till(e)*, *Tilley*, *Tillie*, *Tilly*, in the diminutives *Tillet(t)*, *Tillott*, and in *Tillotson*, *Til(l)son*. The common vernacular forms *Mahald*, *Mahalt*, have given *Mald*, *Malt*, *Mault*, *Maud*, *Mold*, *Mould(s)* and *Moult*. In *Mahald* and *Mahalt* the *l* was vocalized, giving *Mahaud*, *Mahaut*, now *Mahood*, *Mawhood*, *Mowatt*. cf. also MAUDSON, MAWSOM, MOLSON.

**Maudling:** Simon *Maudeleyn'* 1279 RH (O); John *Maudeleyne* 1368 FrY. Hebrew *Magdalen* 'woman of Magdala', a somewhat uncommon christian name in the 13th century.

**Maudsley, Maudslay, Mawdsley, Mawdesley:** Adam *de Moudesley* 1257 Bardsley (La); William *de Maudesley* 1401 AssLa; Robert *Mawdesley* 1476 FrY. From Mawdesley (Lancs).

**Maudson, Maulson:** Ralph *Maldesone* 1327 SRSx; John *Mauldsone* 1376 Black (Fife); John *Maltson* 1438 FrY. 'Son of *Mald*' or *Matilda*. *v.* MAUD. cf. MAWSOM.

**Mauduit, Mawditt, Mudditt:** Gunfridus *Maledoctus* 1084 GeldR (W); William *Maldoit*, *Malduith*, *Malduit* 1086 DB (Ha); Otuell *Malduit* 1169 P (Ess); Richard *Maudut* (*Maudit*) 1229 Cur (Bk). OFr *mal-duit*, Lat *male doctus* 'badly educated'.

**Maufe:** *v.* MAW

**Mauger, Mager, Mayger, Major:** Hugo *filius Malgeri* 1086 DB (Ess); Drogo *filius Matelgerii* ib. (D); *Malger* filius Gilleberti 1150–60 DC (Nt); *Maugerus* episcopus 1212 Fees (Wo); John *Mauger*, *Malger* 1250 Fees (So); 1272 AssSo; Thomas *Mauger'* 1260 Oseney (O). OFr *Maugier*, OG *Madalgar*, *Malger* 'council-spear'. Tolleshunt Major (Essex) owes its attribute to the DB *Malger* (Tolshunt *Major* 1480 Pat).

**Maugham, Maughan, Mauchan:** John *de Machen* 1263 Black (Lanark); Adam *Mauchan* 1300 ib.; Richard *Maghan*, John *Mawgham* 1476, 1537 FrY. From earlier Machan, now Dalserf (Lanark).

**Maul, Maule, Maull:** Robertus *filius Malle* 1297 SRY; Gerard *Malle* 1297 MinAcctCo (Bk). A pet-name for *Mary*.

**Maulden, Mauldin, Mauldon:** *v.* MALDEN

**Mauleverer:** William *Malleu(e)rier* 1159, 1166 P (Y); Ralph *Mauleverer* 1204 AssY. OFr *mal leverier*, Lat *malus leporarius* 'poor harrier'.

**Maulkin:** *v.* MALKIN

**Maulson:** *v.* MAUDSON

**Mault:** *v.* MAUD

**Maultby:** *v.* MALTBY

**Maund:** Thomas *de Magene* 1195 P (He); Arnald *de Maundes* 1230 P (O); Alexander *Maunde* 1327 Pinchbeck (Sf). From Maúnd Bryan, Rose Maund (He), *Magene* DB.

**Maunder, Maunders:** *v.* MANDER

**Maurice, Morrice, Morris, Morriss, Fitzmaurice:** *Mauricius* de Edligtona, de Creona c1176, c1190 DC (L); Fulco *filius Mauricii*, *Moriz* 1185 Templars (L); Ricardus *filius Morys* 1297 SRY; Josce *Mauricii* 1191 P (Lo); Richard *Maurice* 1252 Rams (Hu); John *Morice* 1275 RH (Bk); Simon *Morys* 1296 SRSx; Robert *Morisse* 1308 StThomas (St); Nicholas *le fiz Mouriz* 1314 Cl. Lat *Mauritius* 'Moorish, dark, swarthy', from *Maurus* 'a Moor'. *Maurice* is the learned form, *Morice* the common popular one. *Morris* may also be a nickname; Robert *le Moreys* 1274 RH (So), 'the swarthy'.

**Mauvoisin, Mavesin:** *Malveisun*, *Mauveisin* 1211 Cur (Wa); *Mauuaisin* clericus 1230 P (Y); Nicholas *Malvesin* c1159 StCh; Walter *Mauvesin* 1276 AssSo; Henry *Mauesyn* 1392 CtH. 'Bad neighbour', OFr *mauvais voisine*. cf. Berwick Maviston (Sa), Henry *Malveisin* 1166. Also used as a christian name. In addition, *Malveisin* was the name of a castle built by William II against Bamburgh when besieging there Robert de Mowbray, earl of Northumberland: 'on his spæce Malueisin het, þat is on Englisc Yfel nehhebur', 1095 ASC E.

**Maw, Mawe, Mowe, Maufe, Muff:** (i) William *Mawe* 1275 RH (Nf), 1322 FrY; William *Mogge*, *Mugh* Ed 1 NottBR; Robert *Mouth*, *le Mogh* 1336–7 ib.; William *Magh'* 1381 PTY. OE *mãge* 'female relative', ME *maugh*, used vaguely of a relative by marriage; in the north of a brother-in-law. cf. 'Mow, husbondys syster, or wyfys systyr, or syster in lawe' (PromptParv). This will give *Maufe* and *Muff*. (ii) *Mauua* 1066 DB (Sf); Godefridus, Galfridus *filius Mawe* 1199 P (L), 1297 MinAcctCo (Y); Johannes *filius Mawe* 1256 AssNb. Ekwall takes the DB name as OE *\*Mawa* and compares OG *Mawo*. cf. Agnes *Mawedoughter* 1381 SRSt and the diminutive *Moco(c)k* 1297, 1307 Wak (Y). We may also have a nickname from the sea-mew, OE *mãw*: Henry *le Mou* 1327 SRWo. cf. ME *mawe*, *semawe* (c. 1450 NED), 'Mowe, byrd, or semewe' (PromptParv), and *v.* MEW. (iii) Sibilla *de la Mawe* 1275 RH (St); William *de Mawe* ib.; William *atte Mowe*, John *Mowe* 1327 SRSf. 'Dweller by the meadow', OE *\*mãwe*.

**Mawd(e)sley:** *v.* MAUDSLEY

**Mawditt:** *v.* MAUDUIT

**Mawer, Mower:** John *le Mawere*, *le Mowere* 1225 AssSo; Robert *le Mouer* 1263 ArchC iv (K); John *le Mawer* 1297 SRY; Roger *le Mower* 1305 Pinchbeck (Sf); Alan *Mawer* 1332 SRCu. A derivative of OE *mãwan* 'to mow', 'a mower' (c1440 NED). OE *aw* became ME *au* north of the Humber and remained; to the south it became *ou*. *Mower* may also be occasionally a dialectal form of MOORE: Laurance *a Mower* 'at the moor', Ezabell *Mower* 1562 RothwellPR (Y).

**Mawhood:** *v.* MAUD

**Mawhood, Mahood, Maud:** Simon *de Mohaut* 1160 WRS; Richard *de Mohaud* 1208 Cur (Y); Roger senescallus *de Monte Alto* 1212 (L), identical with Senescallus *de Mohaut* 1235 (Db), identical with Roger *de Mohaut* (Sf) 1242 Fees; Idonia *Mohaud* 1342 FFY; Christopher *Monteald* alias *Mawde* 1555 WRS. From Montaut (Dordogne). *v.* also MAUD.

**Mawsom, Mawson:** John *Malteson* 1332 SRSr; William *Mawson* 1382 Calv (Y). 'Son of *Maud*' or 'son of *Maw*'.

**Maxey, Maxcey, Maxcy:** Suein *de Makesia* 1185 P (Nth); Nicholas *de Makeseye* 1297 MinAcctCo; Robert *Makeseye* 1346 LLB F; John *Maxy* 1523 FFEss. From Maxey (Nth).

**Maxton, Maxtone:** (i) Peter *de Makestan* 1200 P (Hu); John *Makeston* 1279 RH (C). From Maxton (K). (ii) Adam *de Makeston'* 1214 P (Nb); Adam *de Makeston* c1250, *de Maxton* 1261, Alexander *de Maxston* 1285 Black. From the barony of Maxton (Roxburgh).

**Maxwell:** Herbert *de Makeswell'* 1190 P (Y); John *de Maccuswell* 1221 Pat (Scotland); Eymer *de Mackisuuell* 1262 Black (Peebles). From Maxwell, a salmon pool on the Tweed near Kelso Bridge.

**May, Le May, Maye, Mayes, Mays, Mayze, Mey, Meye, Meys, Lemay:** Johannes *filius Maie* 1274 RH (L); Elena *filia May* 1301 SRY; *May* de Hindley 1379 PTY; William *Mai* 1167 P (Nf); William *le Mai* 1177 P (Nf); Thomas *le Mey*, Goscelin *Mey* 1221 ElyA (Sf); John *Meys* 1276 RH (Gl); Stephen *Mayes* 1332 SRWa. ME *may* 'young lad or girl' or *May*, a hypocoristic of *Matthew*, from *Maheu, Mayhew*.

**Maybank, Maybanks:** William *Malbeeng* 1084 GeldR (Do); Richard *Malbanc* 1177 P (Ess); William *Maubanc* 1268 AssSo; William *Maybank* 1382 FFSr. A nickname, 'bad bench', but the exact meaning here is unknown, cf. Clifton Maybank (Do), William *Malbeeng* 1084, Nantwich (Ch), sometimes Wich Malbank from William *Melbedeng* DB.

**Maycock, Meacock, Mycock:** *Maicoc* le Crouder 1284 AssLa; William *Maycock, Moycock* 1323 AssSt; Thomas *Macok*, John *Moycok* 1327 SRDb; John *Mecocke* 1585 Oxon. *Mai-coc*, a diminutive of *May*, from *Mayhew* (Matthew).

**Maydon:** v. MAIDEN

**Mayer, Mayers, Mayor, Meier, Meyer, Meyers:** William *le Maier* 1243 AssSo; Henry *Meyer*, Bartholomew *le Meyre* 1275 RH (Nf); David *le Meir* 1276 RH (L); Alan *Mair* 1279 AssNb. Occasionally we may have OFr *maire* 'mayor', but the term was limited in England to mayors of boroughs, much less numerous than the corresponding, but less dignified, 'mayors' of France and Scotland (v. MAIR). It may have been sometimes a nickname for one who aped the mayor. Usually the surname is synonymous with MYER, OFr *mire*, ME *mire, meir, meyre* 'physician'. The modern surname is often Jewish and German, from Ger *meier* 'steward, bailiff; farmer'.

**Mayfield:** Philip *de Maleville* 1210–12 PN K 28; William *Mayfeild* 1642 PrD. From Mayfield (St, Sx).

**Mayger:** v. MAUGER

**Mayhew, Mayo, Mehew:** *Maheo* de Charun 12th DC (Lei); Geoffrey *Maheu* c1240 Fees (Bk); John *Mahyw*, William *Mahu* 1296 SRSx; William *Mayhew* 1351 ColchCt; John *Mayho* 1428 LLB K; John *Mayhow*, Wyllian *Mayo* 1524 SRSf. OFr *Mahieu*, a common Norman form of *Matthew*.

**Mayland, Maylan, Maylon:** Thomas *de la Mailande* 1190 P (Ess); Richard *Maylond* 1316–17 FFEss. From Mayland (Ess).

**Maylard, Maylett, Mallard:** *Meillardus, Maillard* 1221, 1226 Fees (Gl); *Mallard(us)* 1148 Winton (Ha); 1229 Cl (Gl); Gilbert *Maillard* 1185 Templars (Lo); William *Maillard'* 1209 Cur (Nt); John, Andrew *Mallard* 1219 Cur (Nt), 1296 SRSx. OFr *Maillart*, OG *Madalhard* 'council-strong'.

**Mayle, Mayles:** v. MAIL

**Maylon:** v. MAYLAND

**Maylor:** v. MAILER

**Maymon, Mammon, Mammen:** Robert *Maimunde* 1178–1214 Black; Eudo *Maymunde* 1277 *Ely* (Sf); John *Maymond* 1327 SRSf; Robert *Maymon* 1602, Peter *Mamon* 1639 RothwellPR (Y). OFr *Maismon, Maimon*.

**Maynard, Mainerd:** *Meinardus* uigil 1086 DB (Nf); Thomas *filius Meinard* 1202 AssL; Richard, Robert *Mainard* 1195, 1198 P (Sf, Ha). OFr *Mainard, Meinard*, OG *Maganhard, Meginard* 'strength-strong'.

**Mayne, Maynes, Main, Maine, Mains:** (i) *Main* 1135 Oseney (O), 1168 P (W); *Mein* 1198 P (O); Robert *Main* 1204 AssY; John *Mayn* 1255 RH (Bk). OG *Maino, Meino* 'strength'. (ii) Robert *le meyne* 1285 Kirkstall (Y); Richard *le Mayne* 1327 SRSx. OFr *magne, maine* 'great'. (iii) Walter *Asmeins* 1223 Pat (Gl); William *Asmayns* 1230 P (L). A nickname from OFr *mayns*, the man 'with the hands'. (iv) Adam *de Meine* 1205 P (So); Robert *de maine* 1213 Cur (So). From the French province of Maine. (v) Johel *de Meduana, de Mayne* 1212, 1237 Fees (D). From Mayenne. (vi) In Scotland, ON *Magnus* became (*Mac*)*Manus* and *Main* (*M'Manis* 1506, *M'Maines* 1673 Black). *Main* is common at Nairn and in Aberdeenshire.

**Maynell:** v. MEYNELL

**Mayner, Maynor, Mainer:** *Mainerus* filius Roberti 1207 Cur (L); Walkevinus *filius Mainer* 1210 Cur (C); Hugo *Mainer* 1279 RH (C); William *Mayner* 1328 FrY. OG *Maginhari* 'army-might'.

**Mayo:** v. MAYHEW

**Mayor:** v. MAYER

**Maysey:** v. MAISEY

**Mayson:** (i) William *Mayson, Mayeson* 1327 SRY, 1332 SRWa; John *Mayesson* 1369 AD iv (Nf). 'Son of *May*.' (ii) Adam de Morton, *mayson* 1371 PTY; John *Mayson* 1379 PTY; Alexander *Mason, Mayson* blaksmyth 1520, 1542 FrY. A northern spelling of MASON.

**Mayze:** v. MAY

**Meacham, Meachem, Meachim, Meachin, Meecham:** Anne *Machin, Macham* 1615, 1620 LitSaxhamPR (Sf); Robert *Meacham* 1676 ib. Late developments of MACHIN. v. MASON.

**Meacock:** v. MAYCOCK

**Mead, Meade, Meads:** (i) John *Atemede* 1248 FFEss; Richard *inthemede* 1332 MELS (Sr); John *del Mede* 1379 PTY; John *a Mede* 1454 KentW. 'Dweller in or by the mead', OE *mǣd*. (ii) Richard *Mede* 1190 P (Wa). Either an early example of loss of the preposition, or metonymic for MEADER. cf. Matillis *Medewif* 1327 SRY.

**Meaden:** Hugo *ate Medende* 1332 SRSr. 'Dweller by the meadow-end.'

**Meader, Medur:** Alexander *le Meder, le Medier, Medarius* 1180, 1200 Oseney (O), 1188 P (Bk); Thomas *Meder* 1332 SRSr. A derivative of OE *meodu* 'mead', Lat *medarius* 'a maker or seller of mead'. cf. John *Medemaker'* 1332 MESO (Nf). Or 'dweller by the mead'. cf. MEAD. Such toponymics are common in 1332 SRSr.

**Meadfield:** John *of Meadfield* 1311 AssNf. From Meadfield in Haslemere (Sr).

**Meadland, Medland:** Walter *de Medeland* 1279 RH (C); John *Meadland, Medland* 1642 PrD. 'Dweller by the meadow land', OE *mǣd, land*.

**Meadow, Meadows, Medewe:** Henry *I'the, de la Medewe* 1280 MELS (Wo); Henry *del Medue* 1283 Ipm (Db); Richard *atte Medewe* 1327 SRSx. 'Dweller by the meadow', OE *mǣd, mǣdwe*.

**Meadowcroft:** Peter *de Medwecroft* 1246 AssLa. From Meadowcroft in Middleton (Lancs).

**Meadway, Medway:** Alexander *at þe Medewaye* 1302 MELS 129 (Sx); John *atte Medeweie* 1327 ib. 'Dweller by the river Medway.'

**Meager, Meagers, Meagher, Megar:** Geoffrey *le Megre* 1179 P (Ess); Robertus *Macer*, Robert *le Megre* 1210 Cur (Wa), 1221 AssWa. ME, OFr *megre*, Lat *macer* 'thin, lean'.

**Meaken, Meakin(g)s:** v. MAKIN

**Meale, Meals:** Geoffrey, Simon *Mele* 1279 RH (C), 1300 LLB B. OE *melo* 'meal', metonymic for a maker or seller of meal. *Melemakers* in York were millers. cf. Adam *le Melemakere* 1274 Wak (Y), Roger *le Meleman* 1275 SRWo, Gilbert *le Melemongere* 1296 SRSx.

**Mean, Means, Meen:** (i) Hugh *de Menes* c1110 Winton (Ha); Elias *de Mene* 1296 SRSx. From East, West Meon (Ha). (ii) William *le Men* 1279 RH (Hu); William *le Mean* 1334 SRK; John *le Meen* 1340 NIWo. OE *mǣne* 'false, wicked'. There was also a personal name: Roger *filius Mene* 1207 Cur (Nf); William *Menne* 1351–2 FFSr.

**Meaney:** John *atte Meneheye* 1315 MELS (Wo); John *Meneye* 1327 SRWo. From Menithwood in Lindridge (Wo), *Menehey* 1240.

**Meanley, Menley:** Richard *de Menlee* 1279 RH (O); William *de Menley* 1299 FFY. From Meanley in Newton (WRY).

**Means:** v. MEAN

**Mear, Meares, Mears, Meers:** Robert *atte Mere* 1269 AssSo; Adam *del Mere* 1307 Wak (Y). 'Dweller by the pool (OE *mere*) or the boundary (OE *(ge)mǣre*).'

**Measham, Measom, Messam, Messum:** William *de Meysham* 1248 AssSt. From Measham (Leicester). The surname occurs as *de Meysam* 1305, *Measham* 1421, *Measam* 1633, *Messam* 1661 DbAS 36, Thomas *Mesham*, Maute *Messume* 1524 SRSf.

**Meathrell:** v. METHERALL

**Meazey:** v. MAISEY

**Meazon:** v. MESSENT

**Medcalf(e):** v. METCALF

**Medcraft:** William *de Medecroft* 1313 Eynsham (O). 'Dweller at the meadow-croft.'

**Medewe:** v. MEADOW

**Medhurst:** Ailnot *de Medherst* 1275 RH (K); Rose *Medherst* 1525 SRSx. From Medhurst Row in Edenbridge (K), or Madehurst (Sx), *Medhurst* 1255.

**Medland:** v. MEADLAND

**Medley, Medlay:** Thomas *Medlay* 1419 IpmY; Benedict *Medley* 1496–7 FFWa; Christopher *Medelay* 1504 CorNt; Robert *Medley* 1672 HTY. From Medley (Oxon), or perhaps Madely (Glos), *Methlegh* 1234, but the examples are late and a nickname may also be involved.

**Medlicott, Medlycott:** Lewelin *de Modlincot* 1255 RH (Sa). From Medlicott (Salop).

**Medur:** v. MEADER

**Medway:** v. MEADWAY

**Meech:** Symon *le Meche* 1275 RH (Nf); Thomas *Meche* 1489 FFEss; George *Meech* 1662–4 HTDo. OE *mecca* 'companion, friend'.

**Meecham:** v. MEACHAM

**Meek, Meeke, Meeks:** Richard *Mek* 1229 Pat (So); Robert *le Meke* 1300 FrY. ON *mjúkr*, ME *meke* 'humble, meek'.

**Meekin(g)s:** v. MAKIN

**Meen:** v. MEAN

**Meers:** v. MEAR

**Meese:** Richard *del Mes* 1276 AssSt. From Meece (Staffs).

**Meetham, Metham, Mettham, Mettam:** Thomas *de Metham* 1276 RH (Y). From Metham (ERYorks).

**Meffan, Meffen:** v. METHUEN

**Meffatt, Meffet:** v. MOFFAT

**Megar:** v. MEAGER

**Meggat, Meggett, Meggitt:** (i) George *Meggott* 1677 Bardsley. *Megot*, a variant of *Maggot*, a diminutive of *Magg, Megg* (Margaret): *Megota de Rypon* 1309 Wak (Y). (ii) Randulf *de Meggate* c1190 Black. From Megget in Yarrow (Selkirk).

**Megginson:** For MEGGISON with intrusive *n*.

**Meggison:** Robert *Meggotson* 1379 PTY. 'Son of *Meggot* (Margaret).'

**Meggs:** *Megge* (f) 1254 IpmY; John *filius Megge* 1279 RH (O); John *Megge* 1275 RH (Berks); Robert *Megges* 1357 FFW. cf. Alice *Megmayden* 1263 AssY. *Megg*, a pet-form of *Margaret*.

**Megson:** Johannes *filius Megge* 1279 RH (O); Adam *Meggesone* 1332 SRCu. 'Son of *Megg* (Margaret).'

**Mehew:** v. MAYHEW

**Meier:** v. MAYER

**Meikle, Mickle:** William *Mykyl* 1382 Black; Bessie *Mekill* 1609 ib.; William *Meikill* 1616 ib. MScots *mekill, meikill* 'big'.

**Meiklejohn, Micklejohn, Mucklejohn:** *Mekle John* Burne 1495 Black; William *Meiklejohne* 1638 ib. 'Big John.'

**Melady:** v. MELODY

**Melborn, Melborne, Melbourn, Melbourne:** John *de Meleborna* 1086 DB (So); Wido *de Meleburna* 1159 P (Nb); William *de Meleburne* 1257–8 FFSr; Richard *Melburne* 1431 IpmNt. From Melbourn (C), or Melbourne (Db, ERY). Sometimes for MILBOURN.

**Melden, Meldon:** Robert *de Meledon'* 1242 Fees (D); William *de Meldun* 1259, John *de Meldon* 1366 Black. From Meldon (D, Nb).

**Melding:** Remigius de Meldingge 1327 SRSf. From Milden (Sf), *Meldingg* 1254.

**Meldon:** v. MELDEN

**Meldred:** v. MILDRED

**Melford:** Hubert *de Meleforda* c1095 Bury; Godwin *de Meleford'* 1176 P (D); Richard *de Meleford* 1275 SRWo. From Long Melford (Sf), but often for MILFORD.

**Melhuish, Mellhuish, Mellish:** John *de Melewis* 1242 Fees (D); Elinora *de Melhywys* 1274 RH (D); John *Mellishe* 1583 Musters (Sr); George *Melhuish* 1642 PrD. From Melhuish Barton (Devon).

**Meliss:** v. MALISE

**Mellanby:** Alan *de Melmerby* 1316 FrY; Walter *Mellerby* 1412 ib. From Melmerby (Cumb, NRYorks).

**Mellard:** v. MILLWARD

**Meller, Mellers:** Roger *le Meller* 1319 Crowland (C), 1327 SRSf. 'Miller.'

**Mellichap:** v. MILLICHAMP

**Mellin:** v. MELVILL

**Melling, Mellings:** Henry *de Mellinges* 1194 P (La). From Melling (Lancs). v. also MELVILL.

**Mellis, Melliss:** Joan *de Melles* 1327 SRSf; John *Melys*, Thomas *Mellice* 1524 SRSf. From Mellis (Suffolk). v. also MALISE.

**Mellish:** v. MELHUISH

**Mellon:** Robert *de Melun* 1194 P (Wa). From Meulan (Seine-et-Oise).

**Mellor, Mellors:** Richard *de Meluer* 1246 AssLa. From Mellor (Lancs).

**Mellowship:** v. MILLICHAMP

**Mellsop:** v. MILSOPP

**Melody, Melady:** *Melodia* 1212 Cur (Berks); Richard *Melodie* 1279 RH (O). From a woman's name, *Melodia*.

**Melpuss:** v. MALPAS

**Melsom(e), Melson:** v. MILLSOM

**Melton:** Siwate *de Meltona* 1175 P (Y); Walter *de Melton'* 1230–9 RegAntiquiss; Alan *de Melton* 1327 SRSx. From Melton (Sf, ERY), Melton Ross (L), Melton Mowbray (Lei), Great, Little Melton, Melton Constable (Nf), or High, West Melton (WRY). Sometimes confused with MILTON.

**Melvill, Melville, Melvin:** Geoffrey *de Maleuin*, 1161–3, 1165–71; Hugh *de Malleville* c1202 Black. The surname appears as *Meiluill* 1520, *Melwin* 1550 and is commonly pronounced *Melvin*, besides *Mellin, Mellon, Melling*. James Melville the reformer spelled his name indifferently *Melville* and *Melvin* (Black). The surname usually derives from Melville (Midlothian), itself named from Geoffrey de Mallaville who came from Emalleville (Seine-Inférieure).

**Membery, Membry, Membury:** Richard *de Membri, de Membyr'* 1201 AssSo, 1242 Fees (D); John *de Membury* 1327 SRSo. These Somerset surnames derive from Membury (Devon). v. also MOWBRAY.

**Memory:** v. MOWBRAY

**Mempriss:** v. MAINPRIZE

**Mendham:** William *de Mendeham* 1195 P (Sf); William *a Mendham* 1447–8, John *Mendham* 1454 Paston. From Mendham (Sf).

**Menel, Mennell:** v. MEYNELL

**Menley:** v. MEANLEY

**Mennear, Menneer, Manhire:** Walter *Maenhir'* 1293 Fees (D); Ellis *Menheire* 1642 PrD; William *Menheare* 1663 HeMil. 'Dweller by the high stone', Breton *men, hir*.

**Menter:** Geoffrey *le Mentur* 1166 P (Nf); Brese *le Mentur* 1232 Cur (L); Geoffrey *le Menter* 1251 AssY. OFr *mentur* 'liar'.

**Menzies:** A Scottish form of MANNERS, spelled *de Meyneris* or *Meinzeis* 1306–29, *Meygnes* 1425, *Menyhes* 1428, *Mengues* 1487, *Mengzes* 1572. Pronounced *Meeng-us* or *Mingies* (Black). For the *z*, cf. DALYELL.

**Mepham, Meopham, Meppam:** Ælfgar *on Meapaham* 964 OEByn (K); Maurice *de Mepham* 1207 Cur (K); Rycharde *Meppam* 1583 Musters (Sr). From Meopham (K).

**Mercer, Mercier, Merchiers:** Gamel *mercer* 1168 P (L); John *le mercier* 1196 P (Gl); Hamo *le Merchier* 1204 Cur (O); Richer *le Mercher, le Mercer* 1298 LLB B. OFr *mercier, merchier* 'merchant', one who deals in textile fabrics, especially silks, velvets and other costly materials (NED).

**Merchant:** v. MARCHANT

**Meredith:** *Mereduht* ap Grifin 1248 Ipm (Mon); William *Meredich* 1191 P (St). OW *Morgetiud*, MW *Maredudd*, Welsh *Meredydd*.

**Merewether:** v. MERRYWEATHER

**Merifield:** v. MERRIFIELD

**Merioth, Meriott, Meritt, Meryett:** v. MERRETT

**Meriton:** Robert *de Meryton* 1274 FFO. From Kirk Merrington (Du), or Merrington (Sa).

**Merivale:** Geoffrey *de Miriuall'* 1206 Pl (Y); William *de Muryvale* 1332 SRSt; Richard *Meryvale* 1418 LLB I. From Merevale (Warwicks), *Mirival* c1190, or 'dweller in the pleasant valley'.

**Merlin:** Ralph *filius Merlin* 1202 FFL; John *Merlini* c1210 FeuDu; Henry *Merling* 1327 SRC; John *Merlyn* 1347 KB (Nb). OFr *Merlin*, from Welsh *Myrddhin*.

**Merrall, Merrell, Merrill:** v. MURIEL

**Merrett, Merriott, Merrit, Merritt, Merioth, Meriott, Meritt, Meryett:** (i) Alric *filius Meriet*, Ælric *Meriete sune* 1066 DB; Ægelricus *filius Mergeati* c1086 Crispin; *Meriet* c1113 Burton (St), 1200 Oseney (O); Ralph, Symon *Meriet* 1202 AssL, 1255 RH (Bk); John *Meryet* 1316 FA (W); John *Meryatt* 1375 ColchCt. OE *Mǣrgēat* 'famous Geat'. v. MARRIOTT. (ii) Hardinus *de Meriet* 1084 GeldR (So); Henry *de Merieth* 1185 Templars (So). From Merriott (Som).

**Merriam:** Laurence *de Meryham* 1296 SRSx; Robert *Meriham* 1396 PN K 222. From Merriams in Leeds (K).

**Merrick:** (i) *Meurich* filius Rogeri 1187 P (C); *Meuricus* 1207 Cur (He); Jeuan Eigon *ap Meuric* 1391 Chirk; Edward *Merrycke* 1545 SRW; George *Mericke* 1641 PrSo. *Meuric*, the Welsh form of *Maurice*. (ii) In Scotland from Merrick (Kircudbright).

**Merriday, Merredy, Merridew:** John *Meridewe* 1379 PTY. An English and Irish form of MEREDITH. Bardsley gives two examples of men called both Meredith and Meriday in 1596–8 and 1680–2.

**Merriden, Merridan:** Roger *de Meriden'* 1219 Cur (Herts); William *de Meryden* 1258 AssSo; Margaret *Miriden* 1302–3 CartNat. From Meriden (Wa).

**Merrifield, Merifield:** Walter *de Merifild'* 1200 P (D); James *de Meryfeld* 1341 FFEss; John *Merifilde* 1576 SRW. From one or other of the seven places of this name in Devon, or 'dweller by the pleasant field', OE *myrge, feld.*

**Merrielees, Merrilees:** *v.* MERRYLEES

**Merrikin:** William *Marekyn,* John *Marykyn* 1390, 1458 FrY. A diminutive of *Mary.*

**Merriman, Merriment:** Adam *Murymon* 1332 SRSt; John *Meryman* 1359 LLB G; Adam *Myryman* 1379 PTY. 'The merry man.' *v.* MERRY.

**Merrimouth, Murrimouth:** Richard *Murimuth'* 1318 FFK; John *Mirymuth* 1381 LoCh; John *Merymowth* 1418 BuryW. 'Merry mouth', OE *myrge, mūð,* perhaps a nickname for a good singer.

**Merritt, Merriott:** *v.* MERRETT

**Merriweather:** *v.* MERRYWEATHER

**Merrlees:** *v.* MERRYLEES

**Merrow:** Thomas *de Merewe* 1275 RH (Sx); William *de Merewe* 1296 SRSx; John *de Merwe* 1346 LLB F. From Merrow (Sr).

**Merry:** Gilbert *le mirie* 12th DC (Lei); John *Myrie* 1219 AssY; Robert *Merye* Edw 1 Battle (Sx). OE *myrige,* ME *mirie, merie, murie* 'merry'.

**Merrylees, Merryless, Merrilees, Mireylees, Mirrielees, Merrlees, Murless:** (i) Henry *Mariles* 1175 P (St); Henry *Murlis* 1642 PrD. 'Dweller by the pleasant pasture', OE *myrge, lǣse.* (ii) Thomas *Mereleys* 1529, Richard *Mereleis* 1545 Black. From Merrilees (West Lothian).

**Merryweather, Merrywether, Merriweather, Merewether:** Henry *Meriweder* 1214 Cur (Beds); Robert *Muriweder* 1223 Pat (O); Roger *Mirywoder* 1260 AssC; John *Mereweder* 1345 ColchCt. OE *myrige* 'merry' and *weder* 'weather', a common term for fair weather, used as a nickname for a gay or blithe fellow. cf. FAIRWEATHER.

**Merton:** *v.* MARTON

**Mervin, Mervyn:** *v.* MARVIN

**Meryett:** *v.* MERRETT

**Mesheder:** *v.* MASHEDER

**Message:** Metonymic for MESSENGER.

**Messam:** *v.* MEASHAM

**Messenger, Massinger:** Lucas *le mesagier* 1193 P (Nth); Richard, Hugh *le Messager* 1211 Cur (Mx), 1221 AssSa; William *le Messinger* 1293 Guisb (Y); Robert *le Massager* 1317 AssK; Richard *Messenger* 1377 LoPleas; Walter *Massynger* 1428 FA (W). OFr *messagier, messager.* 'A messenger' (a1225 NED).

**Messent, Meazon, Mezen:** *Meisent, Moysent* Hy 2 DC (L); *Maysant* 1197 FF (Y); Helie *Moysant* 12th DC (L); William *Maysaunt* 1297 SRY. OFr *Maissent* (f), OG *Mathasuent(a)* or *\*Magisind, Megisend* (Forssner). Measant's Charity alias Messants Fm in Debden (Essex) owes its name to John *Meysaunt* or *Mesande* (15th PN Ess 524).

**Messer:** Roger *Messer* 1172–80 DC (L); Erkenbald *le Messer* 1180 P (L); William *le Messier* 1187 P (Nt). OFr *messier, messer* 'harvester', or more probably 'hayward'.

**Messin:** Randulf *Meschin* a1153 Black; William *le Meschin* 1219 P (Do/So); William *le Messim* 1221 AssGl. OFr *meschin* 'rogue, rascal'.

**Messingham:** *v.* MASSINGHAM

**Messiter:** *v.* MASHEDER

**Messum:** *v.* MEASHAM

**Metcalf, Metcalfe, Medcalf, Medcalfe, Mitcalfe:** Adam *Medecalf* 1301 SRY; John *Metcalf, Medcalfe* 1423, 1463 FrY; William *Meatecalf* 1560 FrLeic; Mark *Meadcalfe* 1568 Oxon; John *Madcalfe* 1674 HTSf. It is quite clear that Bardsley's derivation of this much discussed Yorkshire name, from *Medcroft,* is untenable. The first form supports Weekley's *mead-calf.* Harrison's 'mad calf' need not be seriously considered. Forms are late and the unfortunate gap between 1301 and 1423 makes any explanation speculative. OE had *mete-corn* 'corn for food' and *mete-cū* 'a cow that is to furnish food'. A similar formation *\*mete-cealf* 'a calf to be fattened up for eating' might have existed and given rise to a nickname 'fat as a prize-calf' (cf. *Duncalf* as a nickname). Such a calf would be turned out in the lush grass of the meadow, the meaning of *mete-cealf* (never a common term) forgotten, and the first element associated with *mead,* hence the variation between *Metcalf* and *Medcalf.*

**Metham:** *v.* MEETHAM

**Metherall, Metherell, Methrell, Meathrell:** John *Methrell* 1588 HartlandPR (D); Ambrose *Metherell* 1642 PrD. From Metherall (Co), or Metherall in Bratton Clovelly, in Chagford (D).

**Methuen, Methwen, Meffan, Meffen:** Robert *de Methven* 1233–55 Black; Roger *de Methfenn* 1296 CalSc; Roger *Meffen* a1320 Black. From Methven (Perthshire).

**Mettam, Mettham:** *v.* MEETHAM

**Meux, Mewes, Mewies, Mewis, Mews, Mewse:** John *de Mehus* 1196 FF (Y); Hugo *de Mues* 1201 P (Lei); Thomas *de Meuse* 1282 FFEss; Andrew *Mewes* 1371 FFHu; Roger *Meux* 1403 PrY; Lodewycus *Mewys* 1428 FA (W). From Meaux (ERYorks), pronounced *mews,* to rhyme with *sluice. v.* PN ERY 43–4, where the different forms are illustrated.

**Meverell:** Richard *Meverel* 1130–2 Seals (St); Stephen *Meverel* 1219 P (St). Francis *Meverel* 1561 Shef. A diminutive of OFr *mievre* 'malicious', *v.* Groups 34.

**Mew:** (i) Algarus *filius Meawes* 1016 OEByn; Robert *Meu* 1275 SRWo; William *Mew* 1284 RamsCt. (Hu); William *le Mew* 1296 SRSx; John *Meau, Meaw* 1312 LLB B. OE *mǣw, mēaw* 'a gull', also 'a sea-mew'. *Mēaw* was used as a personal name in OE (Redin). *v.* also MAW. (ii) Geoffrey *Muwe* 1275 RH (Nf). OFr *mue,* a cage for hawks, especially when mewing or moulting (13.. NED), used by metonymy for *Mewer.* This was probably more common than appears. When OE *ǣw* and OFr *u* had fallen together, this would be indistinguishable from *Mew* above. (iii) William *de la Muie, de la Mue* 1199 P (D). 'One in charge of the mew.' *v.* MEWER.

**Mewer, Meur, Meurs, Muer, Muers, Mure:** Robert, Alan *le Muer* 1195 P (Ess), 12th DC (L); Adam *Muier*

1199 P (L). A derivative of OFr *mue* 'a cage for hawks', used of one in charge of the mews where hawks were kept whilst moulting. cf. William *Meweman* 1279 RH (O). The surname may have been confused with *Muir*.

**Mewes, Mewies, Mewis, Mews, Mewse:** v. MEUX

**Mewett, Mewitt:** Richard *Mewot* 1279 RH (Hu). A diminutive of *Mew*.

**Mewha:** Richard *de Mewy* 1242 Fees (D). From Meavy (Devon).

**Mey:** v. MAY

**Meyer, Meyers:** v. MAYER

**Meyler:** v. MAILER

**Meynell, Maynall, Mennell, Menel:** (i) Robert *de Meinil, de Maisnil, del Maisnil*, 1166–77 P (Y), *del Meisnil* 1195 P (Ha), *del Menill* 1199 P (Y), *de Mesnill'* 1219 AssY; Hugh *de Meynyl* 1260 AssLa; Nicholas (*de*) *Menil* 1276 RH (Y); Henry *Meynel* 1321 FFEss. Meynell Langley (Derbyshire) was held by Robert *de Maisnell* temp. Henry I. The family came from one of the French places named Mesnil, from OFr *mesnil*, Lat *\*mansionile*, diminutive of *mansio* 'abode, habitation', used of the country domain where a noble lived with his family and retainers. As a surname it may also be used of a member of the household, retainer. (ii) *Mehenilda* vidua c1250 Rams (Nf); *Maghenyld* 1275 RH (Nf); Alan *Mahenyld* 1275 RH (Nf); Ralph *Manel* 1279 RH (Hu); Henry *Manild*, Hugh *Maneld* 1327 SRSf; John *Maynild* 1327 *SR* (Ess); Thomas *Maynell* 1426 PCC; William *Menneld* 1546 FrY. OG *Maginhild* (f) 'strength (in) battle'. This has also become MANNELL.

**Meysey:** v. MAISEY

**Mezen:** v. MESSENT

**Miall, Miell, Myall, Myhill:** *Mihil* Cristen 1549 Bardsley; John *Mihell* 1524 SRSf; Priscilla *Miell* 1779 Bardsley. v. MICHAEL.

**Miatt, Miot, Myott:** Probably diminutives of *My*, from *Myhel*, i.e. Michael.

**Mich:** *Miche* de Sancto Albano 1275 RH (Lo); Robert, Adam *Miche* 1279 RH (Hu), 1327 *SR* (Ess). A pet-form of *Michel*, i.e. Michael. cf. Johannes *filius Michecok* 1322 SRLa. Also synonymous with MUTCH.

**Michael, Michaels, McMichael:** *Michaelis* de Areci c1160 DC (L); *Michael* 12th DC (L); John *Makmychell* 1507 Black. Hebrew 'Who is like the Lord?' *Michael* is the learned form. The popular pronunciation was from the French *Michel* which survives as MITCHEL. The OFr popular form was also used in England. v. MIALL. Butler, in *Hudibras*, rhymes *St Michael's* with *trials*. In ME *Michaelmas* is found both as *Mielmasse* and *Mighelmasse*, and the *Cursor Mundi* has a form seynt *Myghell*, surviving as MIGGLES. In Orkney *St Michael's* is pronounced *St Mitchell's*. St Michael's, Cornhill, was *Saint Mihills* in 1626 (Bardsley). Miles's Lane (London) was *Seint Micheleslane* 1303, *Saynte Mighelles Lane* 1548 (Ekwall). *Miles* Coverdale, the translator of the Bible, when in Germany, called himself *Michael* Anglus. Thus *Miles* may derive from *Michael* through the form *Miel*.

**Michaelson:** John *Michaelson* 1646 Black. 'Son of *Michael*.'

**Michell:** v. MITCHEL

**Michelgrove:** Henry *Michelgrove* 1377–8 FFSr. From Michel Grove in Clapham (Sx).

**Michelson:** v. MITCHELSON

**Michie:** *Michy* Nycholson 1446 Black; John *Mychy* 1473 ib. A Scottish diminutive of *Mich*, from *Michael*.

**Michieson:** Tawis *Michison* 1415 Black. An anglicizing of *MacMichie*.

**Mickle:** v. MEIKLE

**Mickleboy:** William *Mikeleboye* 1241 FFEss; Robert *Mikelboy* 1327 SRSf. 'Big servant', OE *micel*/ON *mikill*, ME *boye*. cf. John *Mikelfot* 1249 AssW 'big foot'.

**Micklefield:** William *de Miclefeld'* 1206 Cur (Sf); Geoffrey *de Mikelfeld'* 1219 AssY. From Micklefield (WRY), or Mickfield (Sf), *Mikelefeld* 1242.

**Micklejohn:** v. MEIKLEJOHN

**Micklem, Miklem:** John *de Mychelham, Mykelham* 1296, 1327 SRSx; Fraunces *Mycklem* 1561 BishamPR (Berks). From Mickleham (Surrey).

**Micklethwaite, Micklewaite:** Walter *de Mickelthwayt* 1277, Robert *de Mikelthwayt* 1303–4, *de Mekilthwayt'* 1304–5 IpmY. From Micklethwaite (WRY).

**Midday:** Edmund *Middey* 1288 NorwLt; William *Mydday* 1490 RochW; John *Mydday* 1524 SRSf. 'Mid day', OE *mid, dæg*, perhaps for one born at this time. cf. MIDNIGHT.

**Middle:** (i) Richard *le Midel* 1279 RH (O). A nickname, for one who was neither *Bigg* nor *Small*. (ii) Simon *atte Middele* 1327 SRSo. OE *middel* 'middle' used topographically. Probably 'dweller in the middle of the village'.

**Middlebrook:** Thomas *de Middelbroke* 1334 SRK. 'Dweller by the middle stream', OE *middel, brōc*.

**Middlecoat, Middlecott:** William *de Middelcot* 1200 P (D); Reginald *de Meddelkote* 1297 MinAcctCo; Gilbert *de Middelcote* 1330 PN D 137. From one or other of the various places of the name in Devon.

**Middleham:** Drogo *de Midelham* 1210 Fees (Du); Richard *de Midelham* 1361 FFY; Thomas *Midlum* 1643 RothwellPR (Y). From Middleham (NRY), or Bishop Middleham (Du).

**Middlemarsh:** Thomas *de Meddelmersse* 1297 MinAcctCo. From Middlemarsh (He).

**Middlemas, Middlemass, Middlemiss, Middlemist, Midlemas:** The Middlemasses were vassals and tenants in and around Kelso and took their name from 'lie Middlemestlands' in Kelso. The surname occurs as *de Meldimast* 1406, *Myddilmast* 1439, *Midlemes* 1612, *Midlemist* 1670 and *Midlemiss*, 1685 Black.

**Middlemoor, Middlemoore, Middlemore:** Thomas *Medulmoor* 1340–1450 GildC; Edmund *Middelmore* 1377, Roger *Middelmore* 1466–7 FFWa. From Middle Moor in Ramsey (Hu), Middle Moor in Whittlesford, Middlemoor in Sutton (C), Middle Moor in Renwick (Cu), or Middle Moor in Litton (WRY).

**Middleton, Myddleton:** Robert *de Mideltone* 1166 Eynsham; William *de Midelton* 1327 SRSx; John

*Middilton* 1409–10 FFWa. From one or other of the many places of this name, or occasionally from one of the Milton's, many of which represent an OE *middel-tūn* 'middle farm'.

**Middlewood:** Roger *de Middelwude* 1297 SRY; Adam *de Midelwode* 1365 FFY. From Middlewood (He), Middlewood in Bradfield (WRY), or Middlewood in Dawlish (D).

**Midgley, Midgeley, Midglow, Midgelow:** Adam *de Miggele* 1275 Wak (Y); William *de Miggelay* 1305 FFY; Richard *Midgeley* 1459 Kirk. From Midgley (WRY), or Migley (Du).

**Midnight:** Wlfric *Midnith* 1199–1208 Clerkenwell (Lo); William *Midnicht*, Osbert *Midnihte* 1221 AssWa; John *Midnyght* 1327 *SR* (Ess). The contrasting *Midday* is found from the 14th to the 16th century: Adam *Midday* 1335 FFNf, Robert *Myddaie* 1568 SRSf. Birth at noon or midnight might well have been considered sufficiently noteworthy to warrant commemoration by name.

**Midsummer:** John *Midsumer* 1224 Cur (So). Probably for one born on midsummer day. cf. Henry *Midthefest* 1334 SRK 'in the middle of the feast'.

**Midwinter:** *Midwinter* 1188 P (Berks); John *Midwinter* c1248 Bec (W); Geoffrey *Midewynter* 1275 SRWo. OE *midwinter* 'midwinter, Christmas'. A name for one born at Christmas. cf. CHRISTMAS, NOEL, YOULE, and John *Midsumer* 1224 Cur (Sf). The surname may also be local: John *de Midwinter* (1238), from Midwinter (PN D 435).

**Miell:** *v.* MIALL

**Mier(s):** *v.* MYER

**Miggles, Mighell, Mighill:** *Mighell* Axendall 1598 Bardsley; Adam, John *Miggel* 1327 SRSf; Richard *Mighell* 1327 *SR* (Ess). *v.* MICHAEL.

**Mignot, Mignott, Minet, Minett, Minette, Minnett, Minnitt, Mynett, Mynott:** Peter *Mignot, Minnot* 1191, 1201 P (K); Robert *Mignot, Minot* 1232, 1237 Oriel (O); Nicholas *Mynyot* 1379 PTY; William *Minet* 1524 SRSf. OFr *mignot* 'dainty, pleasing'. Also a diminutive of *Minn: Minnota* (f) 1274 FFEss. *v.* MINN. *Minet* is a Huguenot name but was in use in England centuries earlier.

**Miklem:** *v.* MICKLEM

**Milborrow, Millborough:** Walter *Milburegh, Milburew* 1275, 1279 RH (O). OE *Mildburh* (f) 'mild-fortress', daughter of a Mercian king, an abbess and saint much venerated in the Middle Ages. The christian name *Milborough* was still in use in Shropshire in the 18th century.

**Milbourn, Milbourne, Milburn, Millbourn, Millbourne, Millburn:** Hugh *de Meleburn'* 1201 Pl (Do); Walter *de Milleburne* 1251 IpmY; William *Milbourne* 1465 FFEss. From Milborne (Dorset, Som), Milbourne (Northumb, Wilts), Milburn (Westmorland), or Melbourne (Derby), *Mileburne* DB.

**Milby, Mildby:** Herbert *de Mildebi* 1202, William of *Myldeby* 1245–6 FFY; Roger *de Mildeby* 1276 IpmY. From Milby (NRY), *Mildebi* DB.

**Mild:** (i) *Milda libera femina* 1066 DB (Sf); Reginald *filius Milde* Hy 2 DC (L); Thomas *Milde* 1274 IpmY; Alice *Milde* 1332 SRLa; Thomas *Mylde*

1437 FFEss. ON *Mildi*, or OE *\*Milde* (f). *v.* PNDB 328. (ii) William *le Mild* 1327 SRDb; Walter *le Milde* 1334 ChertseyCt (Sr). OE *mild* 'mild, gentle'.

**Milden, Mildon:** *Mildo, Mildon'* 1219 AssY; John *Mildon* 1642 PrD. OE *Mildhūn*. There was also a feminine name: *Mildoina* (f) 1246 FFY. Sometimes from Mildon in Oakford (D).

**Mildenhall, Mindenhall:** Andrew *de Mildenhale* 1170–83 Bury (Sf). From Mildenhall (Suffolk, Wilts).

**Mildmay:** Walter *Myldemay* 1483 AD i (Gl). 'Gentle maiden.' cf. John *Mildeman* 1381 SRSf and the opposite, John *Wyldmy* 1287 LLB A 'wild lad'.

**Mildred, Meldred:** William *filius Meldreð* 1175 P (La); Richard *Mildryth* 1327 SRSf; John *Myldrede* 1379 LoCh. OE *Mildōryð*. In the Middle Ages the name probably owed its popularity to the fame of St *Mildred*, an abbess and the daughter of a Mercian king.

**Miles, Myles:** *Milo* 1086 DB; Johannes *filius Mile* 1150–60 DC (L); *Milo* Noyrenuyt 1230 P (Berks); Nicholaus *filius Miles* 1297 SRY; Nicholaus *Miles* 1177 P (Sx); Henry, Richard *Mile* c1225 Fees (Beds), 1230 P (Berks); Agnes *Milys* 1297 MinAcctCo; Walter *Miles* 1327 SRSx. OFr *Milon, Miles*, OG *Milo*. The christian name is usually latinized as *Milo*. The early vernacular form in English appears to have been *Mile*, as in modern French *Mile, Mille*. *Miles* has been noted only twice. As a surname, *Miles* is ambiguous, being sometimes Lat *miles* 'soldier'. cf. Ralph *Miles* 1324 Cor, identical with Radulpho *Milite* 1319 SRLo. Another Ralph *Miles*, a fishmonger, of Bridge Ward (1292 SRLo), founded a chantry for his late lord *Milo*, no doubt *Miles* de Oystergate, fishmonger (1291 Husting). Ralph had adopted his master's christian name as his surname. *Miles* may also be from *Miel*, *v.* MICHAEL.

**Mileson:** Elisota *Milesson* 1379 PTY. 'Son of *Miles*.'

**Milford, Millford:** Hugh *de Mileford'* 1208 FFY; William *de Melleford* 1347 AssLo; John *de Milford'* 1379 PTY. From Milford (Derby, Hants, Wilts, WRYorks), or Long Melford (Suffolk).

**Milk:** Ailmar *Melc* 1066 DB (Ess); Henry, Geoffrey *Melc* 1279 RH (C); William *Mylk* 1367 FrY. OE *meolc* 'milk', perhaps as a nickname for one whose drink was milk, effeminate, spiritless. cf. MILSOPP. Or for one with milk-white hair. cf. Hugh *Milkeheued* 12th DC (L). But more probably metonymic for a seller of milk. cf. Robert *le Milker* 1221 AssWa, William *le Milkster* 1246 AssLa, Geoffrey *le melkmakiere*, John *le melkberere* ('carrier') 1285 *Ass* (Ess).

**Milkwhite:** Adam *Milkwhyet* 1307 IpmGl 'Milk-white', OE *meolc, hwīt*, perhaps for one with white hair or complexion, cf. Hugh *milkheued* 12th DC (L); 'milkhead'. But in Hervey *Milkegos* 1288 NorwLt 'milk goose', and Richard *Mylkeshep* 1332 SRWa 'milk sheep', we probably have the verb OE *meolcan*.

**Mill:** (i) Richard *de la Melle* 1200 Cur (Sx); Walter *Attemille* 1242 P (Sx); John *atte Mulle* 1327 SRSo. An assimilated form of OE *mylen* 'mill'. *v.* MILN. (ii) *Mylla* 1246 AssLa; *Milla* c1250 Rams (C); *Mille Cruche* 1275 RH (Nf); Richard *Mille* 1279 RH (C), 1316 Wak (Y), *Mylle* 1296 SRSx. This is usually regarded as a variant of *Miles*, but the evidence points rather to a short form of a woman's name, perhaps *Millicent*.

**Millachip:** v. MILLICHAMP

**Millage, Milledge:** v. MILLIDGE

**Millais:** Rannulf *de Millai* Hy 2 DC (L), *de Millei* 1196 P (L), *de Millay* 1210 Cur (L); Adam *de Millers* 1296 SRSx. From Milliez (Nord).

**Millard:** v. MILLWARD

**Millbourn(e), Millburn:** v. MILBOURN

**Millen:** (i) Ralph *Milun* 1198 FFSf; Adam *Milun* 1200 BuryS (Sf). OFr *Milon*. (ii) Godwyn' *atte Mullane* 1296 (Sx), Gilbert *atte Mullane* 1327 (So) MELS. 'Dweller in the lane leading to the mill', OE *mylen, lanu*.

**Miller, Millar:** Ralph *Muller* 1296 SRSx; Reginald *Miller* 1327 SRSx; John *Millare* 1467 Black. An assimilated form of MILNER. *Millar* is Scottish.

**Millerchip, Millership:** v. MILLICHAMP

**Millett, Milot:** *Milot* c1248 Bec (W); Hugo *Milot* 1206 Cur (Nf); Roger *Millot* 1275 RH (Nt); Richard *Mileth* 1279 RH (O); Richard *Myles* or *Mylot* 1453 RochW. A diminutive of *Mile*. v. MILES.

**Millford:** v. MILFORD

**Millhouse:** Henry *del Melnehous* 1327 SRDb; Joan *del Milnhous* 1331 Wak (Y). 'At the mill-house', a miller.

**Millican, Milliken, Millikin, Milligan, Mulligan, Mullikin:** Molior *Omolegane* 1264 Dublin; Thomas *Ameligane, Amuligane* 1477, 1485 Black; Cuthbert *Amullekyne* 1578 ib.; John *Myllikin* 1593 ib.; John *Mulligane* 1630 ib.; Thomas *Mulliken* 1672 ib. Gael *Maolagan*, OIr *Maelecán*, a double diminutive of *mael* 'bald', 'the little bald or shaven one', monk, disciple. Irish *Ó Maoileacháin, Ó Maoileegáin*, 'descendant of *Maolagán*'.

**Millicent:** *Milisendis* lotrix 1179 P (O); *Milesent, Melisentia* 1208 Cur (St); William *Melisent* 1221 AssSa; Robert *Milisant* 13th Guisb (Y); John *Milicent* 1279 RH (Hu); William *Mylecent, Milsent* 1296, 1332 SRSx. OG *Amalasuintha* 'work-strong' became *Melisenda*, the name of a daughter of Charlemagne, and, as OFr *Milesindis, Milesenda, Miles(s)ent*, was brought to England in the 12th century.

**Millichamp, Millichap, Millicheap, Millachip, Millichip, Millinchip, Millerchip, Millership, Mellichap, Mellowship:** Roger *de Millinghope* 1199 P (Sa); Roger *de Miligehop* 1255 RH (Sa). From Millichope (Salop).

**Millidge, Milledge, Millage:** Roger *de Melewich* 1198 P (St), *de Mulewic* 1219 AssSt. From Milwich (Staffs).

**Millington:** Peter *de Milington'* 1206 FFY; James *de Milyngton* 1371 FFY; Roger, Thomas *Myllyngton* 1525 SRSx. From Millington (Ches, ERYorks), or 'dweller at the farm with a mill'.

**Millmaker:** Adam *le Melemakere* 1274 Wak (Y); William *Melmaker* 1453 Black. 'Maker of meal', miller. v. MEALE.

**Millman, Milman:** Robert *Melneman* 1373 ColchCt; William *Melman* 1447 CarshCt (Sr); John *Myllman* 1642 PrD. 'Worker at the mill', cf. John *milne knave* 1348 DbAS xxxvi, and v. MILLWARD.

**Millner:** v. MILNER

**Millom:** Thomas *de Millome* 1385 KB. From Millom (Cu).

**Mills, Milles, Millis:** John *Myls* 1336 LLB F. *Mills* is much more common than *Mill* and may be from the plural, 'dweller by the mills' or a patronymic, from *Miles* or *Mill*.

**Millsom, Millson, Millsum, Milsom, Melsom, Melsome, Melson:** (i) Richard *Milleson* 1309 AssSt; Thomas *Milsson* 1379 PTY; Robert *Melsam* 1524 SRSf; Thomas *Melson* 1674 HTSf. 'Son of *Miles*' or of *Milla* (f). (ii) Simon *Mildesone* 1327 SRSo. 'Son of *Milde*', OE *Milde* (f): *Milda* 1066 DB (Sf); Osbertus *filius Milde* Hy 2 DC (L).

**Millward, Millwood, Milward, Milliard, Mellard, Mullard:** John *le Milleward* 1279 RH (Hu); Cecilia *le Mulvard* 1286 ELPN; Richard *Meleward* 1296 SRSx; Walter *le Milneward* 1300 LoCt; Robert *le Moleward* 1327 SRDb; Richard *Millard* 1696 Bardsley. OE *myle(n)weard* 'keeper of a mill, miller'. Fransson notes that this is very common in the southern and western counties, elsewhere the common form is *Milner*. Equivalent to *atte Mille*: John *atte Mylle* otherwise called *Mylleward* 1426 AD v (Sx). cf. also the approbrious nickname: William *Offelle* alias *Milleward* 1429 AD i (Do).

**Miln, Milne, Milnes, Milns, Milln, Milln, Mylne:** Richard *atte Mulne* 1275 SRWo; Nicholas *atte Melne* 1296 SRSx; Robert *del Miln* 1332 SRCu. 'Dweller or worker at the mill' (OE *mylen*). v. MILLWARD.

**Milner, Millner:** John *le Mulnare* 1275 SRWo; Robert *le Milner* 1297 SRY. This may be an OE *\*mylnere* or a derivative of ME *mylne* 'mill', hence 'miller'. Examples of ME *melnere, mulnere* are much less frequent than one would expect. The surname is most common in the north and eastern counties, where Scandinavian influence was strong and may thus often derive from ON *mylnari* 'miller'. v. MILLWARD.

**Milnethorpe, Milnthorpe:** Robert *de Milnethorp* 1379 PTY. From Milnthorpe (We, WRY), Milnethorpe (Nt), or Millthorpe (L), *Milnetorp* 1202.

**Milot:** v. MILLETT

**Milroy:** For Manx *Mac Gilroy* or Scots *M'Ilroy*. v. McELROY.

**Milsom, Milson:** v. MILLSOM

**Milsopp, Melsopp:** Ranul *milcsoppe* 1221 ElyA (Sf); Symon *Melksop* 1237-45 Colch (Lo); Roger *Milcscoppe* Hy 3 Stone (St); John *Milesop* 1279 RH (O). ME *milksop* 'a piece of bread soaked in milk' (c1386 NED), used of 'an effeminate, spiritless man or youth'.

**Milton:** Boia *on Mylatune* 972 OEByn; Alan *de Milton* 1235 FFO; Robert *Milton* 1360 IpmGl. From one or other of the many places of this name, sometimes from Middleton, and occasionally from Melton (Leics, Lincs, Norfolk, ER, WRYorks).

**Milverton:** Robert *de Mulverton', de Miluerton'* 1210 Cur (Wa). From Milverton (So, Wa).

**Milward:** v. MILLWARD

**Mimms, Mims:** Andrew *de Mimmes* 1220 Cur (Mx); John *de Mimmes* 1325-6 CorLo; Nicholas *Mymmys* 1369-71 FFSr. From North, South Mimms (Herts).

**Mimpress, Mimpriss:** v. MAINPRIZE

**Mims:** v. MIMMS

**Minchin:** Aluredus, Peter *Minchun* 1190, 1205 P (Wo, D); Robert *Mincin* 1381 SRSf. OE *mynecen* 'nun', a nickname.

**Mindenhall:** v. MILDENHALL

**Miner, Miners, Minor, Minors:** Jordan *le mineur* 1195 P (Co); Adam *le Miner* 1212 Cur (L); Henry *le Minur* 1224 Pat (Db). OFr *mineor, mineur* 'miner'.

**Minet(t):** v. MIGNOT

**Mingay, Mingey:** Johannes *filius Menghi* c1154–5 DC (L); Robertus *filius Mingghi* 1178–81 Clerkenwell (Ess); Andrew *Mengy* 1262 *For* (Ess); Richard *Mingy* 1276 FFEss. *Menguy*, a Breton name, 'stone dog', from Breton *men* 'stone' and *ki* 'dog'.

**Minn, Minns, Mynn:** *Minna* 1202 AssBeds; Gostelynus *Mynne* 1327 SRSf; John *Mynnys* 1524 SRSf. *Minna*, a pet-form of some woman's name, perhaps from *Ameline, Emeline*, or from *Ismenia* (*Imyne*). For a diminutive *Minnota*, v. MIGNOT.

**Minnett, Minnitt:** v. MIGNOT

**Minor(s):** v. MINER

**Minshall, Minshull, Minshaw:** John *Munshawe* 1529 FFEss; Ranulph *Mynshall* 1559 Pat (Ch); Thomas *Mynshawe* 1591, *Mynshewe* 1592 ChwWo. From Church Minshull, Minshull Vernon (Ch).

**Minstrell:** Hugo Harper de Wolpet, tregettour, alias dictus Hugo *Mynstrall* de Bildeston, jogelour 1483 EwenG 90. cf. TREDGETT.

**Minter, Mintor:** William *le Myntere* 1221 AssWa; John *Muneter* 1296 SRSx; Hamo *le Meneter* 1296 FFEss. OE *myntere* 'moneyer'.

**Mintern, Minterne:** William *de Minterne* 1189 Sol; Richard *Mintorne* 1641 PrSo; Elizabeth *Mynterne* 1662–4 HTDo. From Minterne (Do).

**Minton:** Walter *de Muneton*' 1209 ForSa; Richard *de Mineton* 1221 AssSa; John *Minton* 1663 HeMil. From Minton (Sa).

**Minty:** Thomas *Mynty* 1392 IpmGl; Richarde *Myntye* 1545 SRW; Nicholas *Myntye* 1662–4 HTDo. From Minety (W).

**Miot:** v. MIATT

**Mireylees:** v. MERRYLEES

**Mirfield:** Wluric *de Mirfeld* 1212 FFY; Adam *de Mirefelde* 1285–6 IpmY; William *de Mirfeld* 1371 FFEss. From Mirfield (WRY).

**Mirfin:** v. MARVIN

**Mirrielees:** v. MERRYLEES

**Mirth:** Alexander *Mirthe* 1279 RH (C); Peter *Merthe* 1275 SRWo; William *Mirthe* 1327 SRY. OE *myrgð* 'joy, pleasure'.

**Miskin:** William *le Meschin* 1161 P (Wa); Alan de Perci *le Meschin* c1180 Whitby (Y). OFr *meschin* 'young man', used to distinguish son from father.

**Mitcalfe:** v. METCALF

**Mitcham, Mitchem, Mitchum:** Wulfward *de Michham* 1190 P (Sr). From Mitcham (Surrey).

**Mitchel, Mitchell, Michell:** (i) *Michel* de Whepstede 1327 Pinchbeck (Sf); *Michele* Michel 1327 SRSo; *Mitchell* M'Brair 1490 Black (Galloway); Gilbert, Robert *Michel* 1205, 1219 Cur (Nb, So). Examples of *Michel* as a christian name are rare; the clerks regularly used *Michael*. In 1198 *Michael* de Middelton' held a quarter of a carucate in Middleton (Wilts) by service of keeping the king's wolf-hounds (Fees). About 1219 William *Michel* received 3½d. per day for keeping two wolf-hounds and in 1236 Richard *Michel* held a cotset in Middleton for a similar service. The relationship between the three men is not stated but it appears a reasonable assumption that they were of the same family and that *Michael* de Middelton had two sons who derived their surname *Michel* from their father's christian name. The clerk gave the christian name its usual learned form, the surname in the form in use. v. MICHAEL. (ii) Roger *Michil* 1202 Cur (W); Ralph *le Muchele*, Geoffrey *Michel* 1230 Pat; Gilbert, Robert *Michel*, Adam *le Mechele* (*Muchele*) 1280 AssSo; John *Mochel* 1327 SRSo; Thomas *Mechel* Ric 2 PN Ess 320. OE *mycel*, ME *michel, mechel, muchel* 'big'. This survives as *much* and, as a surname, MUTCH. The survival of *Mitchell* alone as a surname is due to the influence of the personal name *Michel*, a form which cannot be distinguished from the adjective when there is no article.

**Mitchelman:** John *Michelesman* 1298 AssL; Thomas *Michelman* 1369 Misc (Cu). 'Servant of *Mitchell*.'

**Mitchelmore:** Philip *Mitchellmore*, Thomas *Mitchelmore* 1642 PrD. 'Dweller at the big moor', OE *mycel, mōr*.

**Mitchelson, Mitchellson, Michelson:** Roger *Michelson* 1383 DbCh. 'Son of *Michel*' (Michael).

**Mitchenall, Mitchener, Mitchiner, Mitchimer:** Henry *de Michenhale* 1347 PN Sr 349; Richard *Mychenall* 1524 SRSx; Thomas *Mychinall*, John *Mychinor* 1583 Musters (Sr); Pawle *Mychenor*, Robert *Michenour* 1596 ib. Probably from Michen Hall in Godalming (Surrey).

**Mitcheson, Mitchison:** Adam *Michesone* 1348 DbAS 36; William *Mitchesonne* 1568 SRSf. 'Son of *Mich* or *Michie*.'

**Mitchinson:** Ann *Mitchinson* 1749 Bardsley. 'Son of *Michel*', with change of *l—n* to *n—n*.

**Mitford:** John *de Midford*' 1196 P (Du); Robert *de Mitford* 1256 AssNb; Hugh *Mitford* 1412 IpmY. From Mitford (Northumb).

**Mitherell:** v. METHERALL

**Mitton, Mitten:** (i) Alan, Jordan *de Mitton*' 1219 AssY, 1246 AssLa. From Mitton (Lancs, Worcs, WRYorks). v. MUTTON. (ii) Simon *a Myddethune* 1296 SRSx; William *a Middeton* 1327 ib.; Richard *Middeton* 1332 ib. 'Dweller in the middle of the village.'

**Mixen:** Henry *OtheMixene* 1332 SRSt. From Mixon (St), or 'dweller by the dung-heap', OE *mixen*.

**Moakler:** v. MANCLARK

**Moase:** v. MOSE

**Moat, Moatt, Mote:** William *de la Mote* 1305 FFEss; Stephen *ate Mote* 1317 AssK. 'Dweller at the castle', ME, OFr *mote*.

**Mobert:** John *Modbert* 1202–3 FFWa; Adam *Mobert* 1214–18 Praes; William *Modbert* 1274 RH (Sa). OE *Mōdbeorht*.

**Mobbs:** Wylliam *Mobbys* 1524 SRSf. A variant of MABBS.

**Moberley, Moberly, Mobberley:** Patrick *de Moberleia* 1190–1200 EChCh; Ralph *de Modberleg*

1260 AssCh; John *Moburlee* 1369 LLB G. From
Mobberley (Ches).

**Mock, Moke:** *Mokke* 1296 SRSx; Ralph *Moke*
1243 AssSo; Robert *Mok'* 1331 FFY; Richard *Mokke*
1401 FFEss. OE *Mocca*.

**Mockett:** William *Moket* 1275 SRWo; Adam
*Mochet* 1327 SRLei; Henry *Mockett* 1662–4 HTDo.
*Mocc-et*, a diminutive of OE *Mocca*. *v*. MOCK.

**Mockford:** Ralph *Mokeford* 1327, Henry
*Mokeford* 1332 SRSx. From a lost *Mockford* in
Henfield (Sx).

**Mocking:** John *Mockynge* 1334 SRK; Thomas
*Mokkyng* 1392 LoCh; John *Mockyng* 1400 FFEss.
'Son of *Mocca*'. *v*. MOCK.

**Mockler:** *v*. MANCLARK

**Mockridge:** *v*. MOGGRIDGE

**Modred:** Robert *filius Maudredi* 1214 Cur (Wa);
Robert *Modred* 1180, *Modret* 1201 P (Co). OE
*Mōdrǣd*.

**Moffat, Moffatt, Moffett, Moffitt, Muffatt, Muffett,
Meffatt, Meffet:** Nicholas *de Mufet* a1232 Black;
Thomas *Moffet* 1296 ib. From Moffat (Dumfries).

**Mogg:** William, Henry *Mogg(e)* 1195 P (Lei, Gl).
*Mogg*, a pet-name for *Margaret*. cf. *Mogota* de Ripon;
*Mogota* Maure 1313, 1325 Wak (Y); for the more
common *Magot*.

**Moggridge, Mogridge, Mockridge, Muggeridge,
Muggridge, Mugridge:** John *Moggrydge* 1544 AD v
(W), John *Moggeridge* 1671 DWills. From Mogridge
(Devon).

**Moir:** Robert *More* 1317 Black (Aberdeen). Gael
*mòr* 'big'.

**Moise:** *v*. MOYSE

**Moke:** *v*. MOCK

**Mokler:** *v*. MANCLARK

**Mold:** *v*. MAUD

**Moldcroft:** Thomas *Moldekroft* 1291 IpmY.
Probably from Molescroft (ERY).

**Mole, Moles, Moule, Moules:** (i) Geoffrey *filius
Mol'* Hy 2 DC (L); Robert *filius Mole* 1191 P (L);
William *Mole* 1279 RH (C); Martin *Mol* 1327 SRSf;
William *Moale* 1642 PrD. OE *Moll* (m). (ii) Ernald *le
Mol* 1210 FFL; Richard *le Mol* 1248 Fees (Ess);
Robert *le Mol* 1327 SRWo. ON *moli* 'a crumb, small
particle', perhaps a nickname for a small man.

**Molesey, Molesley:** Robert *de Molseleye* 1325
CorLo; John *de Mollesley* 1332 SRSt; Thomas
*Mollesseley* 1340–1450 GildC. From East, West
Molesey (Sr).

**Molesworth:** Godfrey *de Molesworth'* 1255 ForHu;
Richard *de Molesworth* 1279 RH (Hu); Simon *de
Molesworth'* 1316 AssNth. From Molesworth (Hu).

**Molin, Molins:** Jon *de*, Adam *del Molyn* 1274 RH
(Ess), 1289 AssSt; William *de Molyns* 1297
MinAcctCo (Co). From a French place *Moline(s)*, or
'dweller by the mill(s)'. *v*. also MULLIN.

**Molineux, Molyneux:** Richard *de Mulinas, de
Molinaus* 1212 Fees (La); Adam *de Mulyneux, de
Mulinas* 1235 Fees (Y); Roger *de Molineus* 1259
AssLa. According to Dauzat, the Fr *Molyneux* is an
alteration of *molineur* 'miller', *v*. MULLINER. This may
be the case with Adam *le Molineus* 1242 Fees (La),
Richard *Mollineux* 1374 IpmLa, but clearly a French
place-name Moulineaux is also involved.

**Moll:** Walterus *filius Molle* 1203 P (Db); *Molle* litel
1277 *Ely* (Sf); Alan, Adam *Molle* Hy 3 Gilb (L), c1250
Rams (Nf). A pet-name for *Mary*.

**Molland:** Simon *de Molland'* 1242 Fees (D);
Thomas *atte Molond'* 1332 FFK; George *Molland*
1642 PrD. From Molland (D), or a lost *Molland* in
Cliffe (K).

**Mollet, Mollett:** *Molot* 1290 AssCh; Roger *Molot*
1275 RH (Nf); Ralph *Molet* 1327 SRSf. *Mol-ot, Molet*, diminutives of *Moll* (Mary).

**Mollinson:** From an unrecorded *Mollin*, a
diminutive of *Moll*. cf. MALLINSON.

**Mollison, Mollyson:** Thomas *Mollysone* 1589
Black (Aberdeen). 'Son of *Molly*.'

**Molson, Moulson:** (i) Robert *Mollesone* 1323
AssSt, 1324 Wak (Y). 'Son of *Moll* (Mary).' (ii) John
*Moldesone* 1327 SRWo. 'Son of *Mold*.' *v*. MAUD.

**Molton:** *v*. MOULTON

**Momerie:** *v*. MOWBRAY

**Mompesson, Mumberson, Mumbeson:** Philip *de
Munpincun* Hy 2 DC (L); Edward *Mounpesoun* 1327
SRC; Edward *Momperson* 1694 Bardsley. From
Montpinçon (Calvados, La Manche).

**Monamy:** William *Monamy* 1305 LoCt; William
*Munamy* 1319 SRLo, *Monamy* 1337 LoPleas. 'My
friend', Fr *mon ami*, perhaps a habitual expression.

**Moncaster:** *v*. MULCASTER

**Monday, Mondy:** Symon *Moneday* 1279 RH (Hu);
William *Monday* 1317 AssK; John *Monenday* 1332
SRSx. OE *mōnandæg* 'Monday'. Perhaps a name
given to one born on that day, or to a holder of
*Mondayland*, land held on condition of working for
the lord on Mondays.

**Mondon:** *v*. MUNDEN

**Moneypenny, Monypenny:** Richard *Monipenie*
1200–11 Black; William, Richard *Manypeny* 1229 Pat
(So), 1256 RamsCt (Hu). 'Many pennies', a nickname
for a rich man, or ironically, for a poor one.

**Monger:** *v*. MANGER

**Monier:** John *monier* c1198 Bart; Gilbert *le
Moneur* 1207 FFSf; Matilda *la Muner* 1283 SRSf;
Hamo *le Moneour* 1302 LLB C; William *le Mouner*
1316 AssNth. OFr *monier, monnier, monnoyer*
'moneyer', cf. Andrew *Monemaker* 1381 PTY. But it
could also be OFr *molnier, mousnier, moonier, monier,
mouner* 'miller', since it is rarely possible to distinguish
the two sets of forms.

**Monk, Monck, Monnick, Munck, Munk:** Aylric
*Munec* c1045 ASWills; Regenolde *ðam Munece* a1103
OEByn; William *Munc* 1222 FFEss; Walter *le Monec*
1243 AssSo; Richard *le Monk* 1327 SRWo. OE
*munuc, munec* 'monk', originally occupational, later a
nickname.

**Monkby:** Cecilia *de Monkeby* 1354 FFY. From a
lost *Monkby* in Hudswell (NRY).

**Monkhouse:** Roger *del Munkhous* 1379 PTY; Alice
*Munckus* 1602 Bardsley; Thomas *Munkas* 1660 ib.
'Worker at the monks' house.'

**Monkman, Munkman:** Robert *Monekisman* 1179 P
(L); Edward *le Munekesman* 1199 Cur (Ha); John
*Monkeman* 1276 RH (Y); John *Munkman* 1332 SRCu.
'Servant of the monks.'

**Monks, Munks:** (i) Agnes, Mariota *le Monekes*
1274 RH (Ess), 1279 AssSo. 'The monk's servant.' (ii)

Henry *de Monkes* 1332 SRLa; William *del Munkes* 1332 SRCu. 'Servant at the monks', one employed at a monastery.'

**Monkton, Munkton:** Edward *de Munketon'* 1202 FFY; Robert *de Moneketon* 1346 IpmW; Thomas *de Munkton'* 1395 AssL. From Monkton (Devon, Dorset, Durham, Kent, Som, Wilts, WRYorks).

**Monmouth:** John *de Monemue* 1218 P (He); John *Monmouth* 1362 IpmGl; Walter *Monemothe* 1387 Misc (Sa). From Monmouth (Wales).

**Monnery, Munnery:** John *Monery* 1525 SRSx. 'Worker at the monastery', OFr *moinerie*.

**Monroe:** *v.* MUNRO

**Monsell:** Adam *de Moncell'* 1219 AssY; Roger *de Moncell'* 1219 P (Sx). From Monceau, Moncel, common French place-names.

**Monsey:** *v.* MOUNCEY

**Montacute, Montagu, Montague:** Drogo *de Montagud, de Monte Acuto* 1084 GeldR (So); William *de Montacute* 1255 AssSo; William *de Montagu* 1312 LLB B. The DB family came from Montaigu-le-bois or from Montaigu (La Manche) and have left their name in its Latin form in Montacute (Som) and in its French form in Shepton Montague and, disguised, in Sutton Montis in the same county. Later, Montacute itself, no doubt, gave rise to a surname.

**Montford, Montfort, Mountford, Mountfort, Mumford, Mundford, Munford:** Hugo *de Montford, de Montfort* 1086 DB, *de Munford* 1086 InqEl; *de Mundfort* 1087 InqAug; Ralph *de Munford* 1159 P (K), *de Muntford* 1200 P (Gl); Simon *de Mumford, de Munford* 1242 Fees (K). The DB undertenant came from Montfort-sur-Risle (Eure). Others later may have come from some other French Montfort. Many of the early forms are indistinguishable from those of Mundford (Norfolk) from which *Mumford* and *Mun(d)ford* may also derive: Richard *de Mundeford* 1247 FFC.

**Montgomerie, Montgomery, Montgomrey:** Hugo *de Montgomeri, de Montgumeri* 1086 DB (St); William *de Mungumeri* c1159 StCh. From Sainte-Foy-de-Montgomery and Saint-Germain-de-Montgomery (Calvados).

**Monyash:** William *de Monyasshe* 1332 SRSt. From Monyash (Db).

**Moodey, Moody, Mudie:** Alwine *Modi* c1100-30 OEByn (D); Godric *Modi* c1150 DC (Nt); William *Mudy* 1365 Black. OE *mōdig* 'bold, impetuous, brave'.

**Moon, Moone, Munn:** (i) William *de Moion, de Moiun, de Moine, de Mouin* 1086 DB (Do); Reginald *de Moyn, de Moyun, de Moun, de Mohun* 1239-53 AssSo. From Moyon (La Manche). (ii) Robert *Mone* 1260 AssY; Thomas *le Mun* 1275 RH (Nf); Geoffrey *le Moun'* 1279 RH (Beds). AFr *moun, mun* 'monk'. cf. MOYNE.

**Moonlight:** Stephen ate Nelm vocatus *Monelight* 1317 AssK; Robert *Monelyght* 1442, Thomas *Monelight* 1470 RochW. 'Moonlight', OE *mōna, lēoht*, possibly for someone given to roaming about at night.

**Moor, Moore, Moores, Moors, More, Mores:** (i) Johannes filius *More* 1185 Templars (L); *Morus* de la Hale 1214 Cur (K); *More* Kalendrer 1332 SRLo; Hugo *Maurus* 1186 P (C); William, Osbert *Mor* 1198 FFEss, 1227 FFSf; Thomas, Hugh *le Mor* 1201, 1205 Cur (K, Beds). The personal-name *More* is from OFr *Maur* (Latin *Maurus*), in the vernacular *More*, either 'a Moor' or 'swarthy as a Moor'. There was a 6th-century saint of this name. *le Mor* is a nickname (OFr *more* 'Moor', swarthy). (ii) William *de More* 1086 DB (Sf); Lefric *de la Mora* 1169 P (Nth); Matthew *del More* 1275 Wak; Thomas *atte More*, Simon *atte Moure* 1296 SRSx; John *Bythemore* 1327 SRSo; John *in le Mor* 1332 SRSt. From Moore (Ches), More (Salop) or from residence in or near a moor (OE *mōr* 'moor, marsh, fen').

**Moorby:** *v.* MOREBY

**Moorcock:** *Morecok* Chepman 1327 SRSo; Walter *Morcoc* c1160 Colch (Sf), 1202 AssL; John *Morekoc, Morecok* 1283 Battle (Sx), 1327 *SR* (Ess). Either a diminutive of *More* (*Morcoc*), *v.* MOOR; or a nickname from the moorcock.

**Moorcroft, Morecraft:** Richard *de Morcroft* 1366 SRLa. 'Dweller at the croft on the moor.'

**Moorey:** (i) Robert *de la Morhage* 1207 P (Db). 'Dweller at the enclosure on the moor', as at Moorhay (Devon). (ii) William *de Morewraa* 1332 SRCu. 'Dweller at the nook on the moor' (ON *vrá*).

**Moorfield:** Roger *Moorefeild* 1642 PrD. 'Dweller at the field by the moor', OE *mōr, feld*.

**Moorfoot, Morfett:** Robert *Morfot* 1373-5 AssL; George *Morfotte* 1525 SRSx. 'Dweller at the foot of the moor', OE *mōr, fōt*.

**Moorhen:** *v.* MOREHEN

**Moorhouse, Morehouse, Morres:** William *de Morhuse* 1180 P (Y); Matilda *del Morhouse* 1301 SRY; Geoffrey *atte Morhouse* 1327 SRSo; William *Morhous* 1440 ShefA; George *Morras* or *Morehouse* 1680 LaWills. From Moorhouse (WRYorks) or from residence at a house on the moor or marsh. OE *mōr* and *hūs*. This may have become MORRIS.

**Mooring, Moreing:** Richard *Moring* 1275 SRWo, 1327 SRSx; Henry *Morynge* 1327 SRWo. Perhaps OE *\*mōring* 'dweller on the moor'.

**Moorman, Moreman, Morman:** Ralph *le Morman* 1287 SRSf; William *le Moreman* 1327 SRSf. OE *\*mōrmann* 'marsh-dweller'.

**Moorshead:** John *Murside* 1260 AssLa. 'Dweller at the edge of the marsh.'

**Moorwood, Morwood:** Ralph *de Morewode* 1275 RH (L). 'Dweller by the moor-wood.'

**Mopp, Moppe, Moppes:** Henry *Moppe* 1243 AssDu; William *Moppe* 1327 SRSo; John *Mopp* 1373 ColchCt. A variant of *Mobb, Mabb*, pet-names for *Mabel*.

**Moppett:** Probably a diminutive of *Mopp*, a variant of *Mobb, Mabb*, pet-names for *Mabel*. cf. Ivetta, William *Mopp(e)* 1243 AssDu, 1279 RH (O), the diminutive *Moppel*, Fulco *Moppel* 1251 Rams (Hu), and the double diminutive *Moppelin*, Ralph *Moppelin* 1195 P (L).

**Moram, Moreham, Moran, Morron, Morum:** (i) Hugh *de Monte Virun* 1130 P (Ess); Richard *de Muvirun* 1197 FFEss; Richard *de Mouirun* 1275 SRWo; Lewis *Moran* 1378 ColchCt. From some, as

yet, unidentified French place. cf. Moreham Hall in
Frating (Ess), John *de Mouviron* 1331. (ii) In Scotland
from Morham (East Lothian).

**Morand, Morant, Murrant:** *Morandus* de Kerkebi
1176 P (Y); *Morant* filius Ernis 1198 FF (Nf); *Moraunt*
Lambard 1317 AssK; Hugo *le Demurant* 1182 P (K);
William *Morand'* 12th DC (L); Richard *Morant* 1200
P (D); William *Morand, Morant* 1210–11 Cur (Sx);
John, Thomas *le Moraunt* 1297 Coram (Sf), 1320
FFSf; Margery *Demoraunt* 1379 PTY. OFr *Moran,
Morant*, probably OG *Modrannus*, or a nickname,
aphetic for OFr *demorant* (*demeurant*), present
participle of *demeurer* 'to reside', residing, staying,
probably 'a sojourner', a stranger who stays.

**Morbey, Morby:** v. MOREBY

**Morce, Morss:** John *Morice, Morce* 1382 AssC.
For MORRIS.

**Mordaunt:** William *mordaunt* 1148 Winton (Ha);
Eustace *le Mordant* 1176 P (Bk). OFr *mordant*,
present participle of *mordre* 'to bite', 'biting,
sarcastic'.

**Morden, Mordan, Mordin, Mordon, Murden:**
Richard *de Morduna* 1086 InqEl (C); Thomas *de
Mordone* 1235–6 AssDu; William *de Mordon* 1341
FFEss; Richard *Morden* 1483–4 FFSr. From Morden
(Do, Sr), Guilden, Steeple Morden (C), or Moredon
(W).

**Mordew, Mordey:** William *Mordew* 1551
NorwDep. 'Death of God', Fr *mort Dieu*, an oath
name.

**Mordin, Mordon:** v. MORDEN

**Moreby, Morbey, Morby, Moorby:** William *de
Moreby* 1280 IpmY; Robert *de Morby* 1367 FFY;
Edmund *of Moreby* 1401 AssLa. From Moreby
(ERY), or Moorby (L).

**Moreham:** v. MORAM

**Morehen, Morehan, Moorhen:** Philip *de la Morend*
1275, William *atte Moreende* 1327 SRWo. 'Dweller at
the end of the moor or fen', OE *mōr, ende*.

**Morel, Morell, Morill, Morrell, Morrill:** *Morel*
1086 InqEl (Nf); *Moræl* of Bæbbaburh 1093 ASC E;
*Morellus* 1191 P (Sf); Peter, Milo *Morel* 1164, 1196 P
(O, Bk). A diminutive of either *More* or OFr *more*
'brown, swarthy as a Moor'. v. MOOR.

**Moreland, Morland:** Edith *de la Morland* 1257
MELS (So); Henry *atte Morlonde* 1296 SRSx;
William *de Moreland* 1327 SRY. 'Dweller on the
moor-land.'

**Moreman:** v. MOORMAN

**Moresby, Morrisby:** Hugh *de Moricheby* 1265
AssL; Hugh *de Moriceby* 1332 SRCu; Christopher *de
Moresby* 1422 IpmY; John *Morisby* 1562 Pat (Db).
From Moresby (Cu), *Moricebi* 1195.

**Moreton:** v. MORTON

**Morey, Mory:** Hugh *Mori* 1195 P (Nf/Sf); Roger
*Mory* 1296 SRSx; John *Morey* 1662–4 HTDo. OFr
*Mory*, a byform of *Maury*, a pet-form of OFr *Amauri*.
v. AMERY.

**Morfett:** v. MOORFOOT

**Morfey, Morffew, Morphey, Morphy, Morphew,
Maffey:** Wido *Malfeth* 1130 P (Nth); William *Malfet*
1163–6 Seals (Sx); Simon (*le*) *Malfe* 1176, 1184 P
(Nth, C); Lucas *Maufe, Malve* 1196, 1205 Cur (Nth);

Simon *Malfei* 1198 FF (Nf); John *Malefay* 1307 Wak
(Y); Thomas *Mauphe, Maufe, Maufie* 1221 AssWa;
Laurence *Maufei*, Andrew *Maufee* 1327, 1332 SRSx;
Elias *Maifai*, William *Mayfai* 1327 SRSx; Richard
*Maffay* 1498 SxWills; John *Morfewe* 1564 SfPR;
Jedion *Morphewe* 1629 ib.; John *Morfee* 1679 ib. OFr
*malfé, malfeü, malfeü*, 'representing a barbarous Latin
*male-fatus* and *male-fatutus*' (Weekley), 'ill-omened',
a term of abuse applied to the Saracens and the devil.
cf. *Hobgoblyng, goblin* mauffé (Palsgrave); *malfé,
maufé, maufeit, maffé, malfait, malfee* 'devil, demon'
(Godefroy). The normal development would be
*Mauffey*. With the loss of *r* in pronunciation, *Mau-*
(*Maw-*) and *Mor-* were indistinguishable. For the *ph*,
cf. FARRAR and PHARAOH.

**Morgan, Morgans:** *Morganus* 1159, 1166 P (Gl,
Sa); *Morgund* 1204–11 Black; Isabella, John *Morgan*
1214 Cur (Berks), 1279 AssNb; John *Morgane* 1419
Black. OBret, OW, Cornish *Morcant*, Welsh *Morgan*,
Pictish *Morgunn*, a very old Celtic name.

**Morhall, Morrhall, Morral, Morrall:** Gilbert *de
Morhalle* 1332 SRLa. 'Dweller at the hall on the
moor.'

**Morin, Morren, Morrin:** Rogerus *filius Maurini,
Morini* 1086 ICC (C); *Morin'* del Pin 1130 P (Lei);
Robert, Ralph *Morin* c1140 DC (L), 1183–4 EngFeud
(Nth). OFr *Morin*, a diminutive of *More*. v. MOOR.

**Morison:** v. MORRISON

**Morland:** v. MORELAND

**Morley, Morely, Moorley:** Milo *de Morleia* 1196 P
(Bk); Ralph *de Morleg* 1230 P (Db); Thomas *Morleigh*
1377 IpmW. From Morley (Derby, Durham,
Norfolk, WRYorks), or Moreleigh (Devon).

**Morman:** v. MOORMAN

**Morphew, Morphey:** v. MORFEY

**Morpuss:** v. MALPAS

**Morrall:** v. MORHALL

**Morrell, Morrill:** v. MOREL

**Morren:** Roger *filius Morewen* 1177 P (L); *Morwin*
1183 P (He); Robert *Morwen* 1205 Cur; John
*Morwyne* 1275 SRWo; Robert *Morren* 1576 SRW.
OE *Mōrwine*. v. PNDB 330. v. also MORIN.

**Morres:** v. MOORHOUSE

**Morrice:** v. MAURICE

**Morrick:** Nicholas, Hugo *de Morewic* 1190 P (He),
1219 AssY. 'Dweller at the dairy-farm in the fen or on
the moor', OE *mōr* and *wīc*.

**Morrin:** v. MORIN

**Morrisby:** v. MORESBY

**Morris(s):** v. MAURICE

**Morrish:** John *Morysch* 1416 LLB I; Humfrey
*Mores, Morrishe* 1569, 1602 Oriel (O). For MORRIS. cf.
NORRISH, PARISH.

**Morrison, Morison, Moryson:** Robert *Morisson*
1379 PTY; Andrew *Morison* 1463 Black. 'Son of
*Maurice*.'

**Morron:** v. MORAM

**Morse:** Thomas *Morse* 1434 FFEss; William
*Morsse* 1524 SRD; Richard *Morse* 1642 PrD. Late
forms of either *Moores*, v. MOOR, or of *Morris*, v.
MAURICE.

**Mort, Morte:** Simon *Mort* 1279 RH (Beds); John
*Morte* 1327 SRC; David *Mort* 1381 LoCh. OFr *mort*
'death', perhaps a pageant name.

**Morten, Mortain, Mortyn:** Macus *de Mauritania, de Moretaine* 1086 DB; Gilbert *de Moretaign'* 1187 P (Sr); Eustace *de Mortaine* 1219 Cur (Db); John *Morteyn* 1330 Rams (Hu). From Mortagne (La Manche). cf. Marston Mortaine (Beds). Sometimes confused with MORTON.

**Mortiboy, Mortiboys:** William *Mordeboice* 1644 BuryW. 'Bite wood', OFr *mordre, bois*, a nickname.

**Mortimer, Mortimore, Mutimer:** Ralph *(de) Mortemer, de Mortuo Mari* 1086 DB; Peter *Mortemer* 1296 SRSx; Elizabeth *Mottimer* 1712 SfPR. From Mortemer (Seine-Inférieure).

**Mortlake, Mortlock:** Walter *Mortelake* 1279 RH (C); Walter *Martelake* 1327 SRC. From Mortlake (Sr).

**Morton, Moreton:** Robert *de Mortone* 1130 P (W); William *de Moreton, de Morton* FFO, 1307 IpmGl; Thomas *Morton* 1432–3 FFSr. From one or other of the many places of this name.

**Mortyn:** *v.* MORTEN

**Morum:** *v.* MORAM

**Morwood:** *v.* MOORWOOD

**Moseley, Mosely, Mosley, Mosleigh, Mossley, Mozley, Mozeley:** Suen *de Moseleia* 1195 P (Wo); William *de Moseleg* 1271 AssSt; Alan *de Moselegh* 1332 SRLa; John *Mosley* 1414 FFEss. From Moseley (Staffs, Worcs), Moseley in West Dean (Glos), or Mowsley (Leics).

**Mory:** *v.* MOREY

**Moss, Mosse:** David *del Mos* 1286 AssCh; Stephen *atte Mos* 1327 SRSt; Robert *del Mosse* 1327 SRDb. 'Dweller by the moss or morass', OE *mos*. *Mosse* was also a common form of Jewish *Moses*: *Mossus* cum naso 1183 P (Nf), Master *Mosse* the Jew of London 1260 AssY and this has probably contributed to the surname. Ailmerus *filius Mosse* or Almer *Mosse* 1153–68, 1186–1210 Holme (Nf) was probably English. cf. Richard, Henry *Mosse* 1250 *Ass* (C), 1275 RH (L).

**Moston:** Robert *de Moston* 1324 CoramLa; Hugh *de Moston* 1389 IpmLa; Alan *of Moston* 1401 AssLa. From Moston (Ch), or Moston (La, Sa).

**Mostyn:** William *Mostyn* 1568 Morris. From Mostyn in Whitford (Wales).

**Mote:** *v.* MOAT

**Motherless:** Hugh *Moderles* 1198–9 RegAntiquiss; Walter *le Moderles* 1275 SRWo; Adam *Moderless* 1327 SRSf. 'Without a mother', OE *mōdorlēas*. cf. ffrancisca *Motherinlawe* 1638 WRS.

**Mothers:** *Moder* c1095 Bury (Sf); Vlfgiet *Moder* 1162 P (Nth); Alicia *le Moder'* 1279 RH (C). ON *Móðir*, ODa, OSw *Modher*. Also a nickname.

**Mothersole, Mothersill:** John *Mothersole* 1674 HTSf. These surnames may be a dialectal pronunciation of Moddershall (Staffs): William *de Modreshalle* 1305 StThomas (St). But we are probably also concerned with a nickname: Richard *Modisoule*; Hugh *Modysowel* 1275, 1308 Wak (Y), 'brave, proud soul', OE *mōdig, sāwol*. In Ralph *Modersoule* 1313 Cl (La), we have an oath-name, 'by my mother's soul'.

**Motherson:** William *Modersone* c1232 Oriel; Lambin *Modersone* 1298 LoCt; Peter *Moderson* 1381 PTY. 'Son of *Mother*', ON *Móðir*.

**Motley, Mottley:** Alys *Motlye* 1525 SRSx; Richard *Motley* c1600 ArchC 49. No doubt usually from

Motley Hill in Rainham (K), but sometimes, perhaps, a nickname from the colour of the garments.

**Mott, Motte:** *Mott* c1248 Bec (Sx); *Motte* 1279 RH (Bk); William *Mot* 1221 Cur (Ess); Robert *Motte* 1298 AssL; Laurence *Motte* 1371 Misc (C). *Motte*, a pet-form of *Matilda*.

**Mottley:** *v.* MOTLEY

**Motton:** *v.* MUTTON

**Mottram, Motteram:** John *de Mottrum* 1287 AssCh; Richard *Motteram* 1541 CorNt; John *Motrum* 1678 NorwDep. From Mottram (Ches).

**Mouat, Mouatt, Mowat, Mowatt:** Robert *de Montealto* 1124–53 Black; Robert *(de) Muhaut* 13th Riev (Y); Alexander *de Mohaut, Mouhat* 1198–1218 Black; Simon *de Munhalt* 1197 P (Y); Richard *de Mohaud* 1208 Cur (Y); Thomas *de Moaud* 1219 AssY; John *Mowet* 1410 Black. From a French place Montaut (several). The surname may also have become *Mahood, Mawhood* (*v.* MAUD) and, occasionally may mean 'meadow-ward', or hayward (cf. MAW): Roger *le Mowerd* 1280 AssSo.

**Moubray:** *v.* MOWBRAY

**Moulder:** John *Molder* 1487 W'stowWills; Edward *Moulder* 1665 HTO. 'A maker of measures', from a derivative of OFr *moule*. cf. Gilbert *le Moldemaker* 1335 FrY.

**Mould(s):** *v.* MAUD

**Moule, Moules:** *v.* MOLE

**Moule(s):** *v.* MULE

**Moulson:** *v.* MOLSON

**Moult:** *v.* MAUD

**Moulton, Molton, Multon:** Ælfgar *de Muletune* c975 LibEl (Sf); Thomas *de Moleton, de Multon* 1166 P (L), 1327 SRLei; John *Multon* 1482 FFEss. From Moulton (Ches, Lincs, Norfolk, Northants, Staffs, Suffolk, NRYorks), or Molton (Devon).

**Mouncey, Mounsey, Mounsie, Monsey, Muncey, Munsey, Munsie, Munchay, Mungay:** William *de Moncels, Moncellis, Muntcellis, Muncellis* 1086 DB; Z—' *de Muncehaus*, Edonea *de Munchaus* 1185 Templars (L); William *Munci* 1198 FF (Gl); Walter *de Mouncy, de Munsy* 1300 LoCt. William de Moncels of DB came from Monceaux (Calvados). Edonea *de Munchaus* is also called *de Herste* and had a son, Waleran, called both *de Herste* and *de Muncaus*. They came from Monchaux (Seine-Inférieure) and their name survives in Herstmonceux (Sussex), pronounced *Hurstmounsies*. *Mungay* may also be for *Mountjoy*.

**Mount, Mounter:** Richard *del Mount* 1301 SRY; John *le Mountere* 1305 MESO (Ha); Richard *le Monter* 1327 SRSo. 'Dweller by the mount, hill' (OE *munt*).

**Mountain:** Richard *Mounteyne* c1240 Glast (So); Joseph *Mountaine* 1672 HTY. 'Dweller by the mountain', OFr *montagne*.

**Mountford, Mountfort:** *v.* MONTFORD

**Mountjoy, Mungay:** Gilbert *de Montgoye* 1219 AssY; Elias *Munjoye* 1243 AssSo; Robert *de Mountgay, de Mungay* 13th WhC (La); John *Mountjoye* 1307 AssSt. From Montjoie (La Manche).

**Mountney, Mounteney:** Robert *de Munteigni* 1177 P (Ess); Michael *de Munteny* 1208 Cur (Nf). From Montigni (Calvados, La Manche, etc.).

**Mousebeard:** William *Museberd* 1198 P (St); Robert *Musberd* 1293 AssSt. 'Mouse-beard', OE *mūs, beard*, perhaps referring to the colour. cf. Walter *Mousetonge* 1302 AssW 'mouse-tongue'.

**Mouth:** (i) Robert *Muth* 1183 P (Ess). OE *mūð* 'mouth'. cf. BOUCH. (ii) William *atte Muthe* 1315 PN Sr 294. 'Dweller at the junction of the streams' (OE *(ge)mȳðe*), as at Meath Green (Surrey).

**Mowat(t):** *v.* MAUD, MOUAT

**Mowbray, Mowbury, Moubray, Momerie, Mulberry, Mulbery, Mulbry, Mumbray, Mummery, Memory, Membry:** Rodbeard *a Mundbræg* 1087 ASC E; Roger *de Mulbrai, de Mubrai* c1130 Whitby (Y), *de Molbrai* c1150 ib.; Paganus *de Moubrai* 1150 Eynsham (O); Roger *de Munbrai, de Moubrai* 1185 Templars (L, Y); William *de Mumbray* 1242 Fees (K); Richard *Mulberye* 1381 SRSf; John *Mowbray, Memory, Membry* 1714, 1725, 1748 FrLeic; Samuel *Mowbery* 1745 ib. From Montbrai (La Manche).

**Mowe:** *v.* MAW

**Mower:** *v.* MAWER

**Mowling:** *v.* MULLINGS

**Mowll, Mowle, Mowles:** John *Moule* or *Moulde* 1584 Bardsley. For MULE or MAUD.

**Mowse:** Æluric *Mus* 1066 ICC (C); Geoffey *le Mus, le Mous* 1296, 1332 SRSx. OE *mūs* 'mouse'.

**Mowsley:** Reginald *de Muselee* 1214 P (Lei); William *de Muslegh* 1222 AssSt; Thomas *de Musle* 1314 LLB E. From Mowsley (Lei).

**Mowson:** Margarita *Mowson* 1549 RothwellPR (Y). Probably for *Mollson* 'Son of *Mary*'.

**Moxham:** Adam *de Mokeshum* 1255 RH (W); John *de Moxham* 1348 IpmW; Nicholas *Moxham* 1642 PrD. From a lost *Moxhams* in Atworth (W).

**Moxon, Moxom, Moxson, Moxsom:** Siuuard *Mocesun* 1087–98 Bury; John *Mokesson* 1379 PTY; John *Moxson* 1499 Calv; Thomas *Moxon* 1588 RothwellPR (Y). The first example is from OE *Mocca*, found twice in early documents, but later examples are more probably for 'son of *Mog*', i.e. Margaret, with -*gs*- assimilated to -*ks*-.

**Moyce, Moyes:** *v.* MOYSE

**Moyle:** William, Richard *Moil* 1275 SRWo, 1359 Putnam (Co); William *le Moil* 1327 *SR* (Co); David, Richard *Moill* 1393 Chirk, 1504 Black. Cornish *moel*, Ir, Gael *maol* 'bald'.

**Moyne:** Robert *le Muine* 1141–51 Colch (Ess); Ralph *Moin* 1168 P (Sx); Henry *le Moygne* 1255 *Ass* (Ess). OFr *moine, muigne, moigne* 'monk', originally of office, later a nickname. Geoffrey *le Moine* was constable of the castle of Newcastle in 1219 (AssY).

**Moyse, Moyses, Moyce, Moyes, Moise:** Gaufridus *filius Moyses* 1210 Cur (Nf); *Moys'* de Bilham 1230 P (Y); Elyas *Moyses* 1196 P (Y); William *Moyse* 1274 RH (Ess). Fr *Moise*, Hebrew *Moses*.

**Moz(e)ley:** *v.* MOSELEY

**Much:** *v.* MUTCH

**Muckle, Mutchell:** Richard *Mukel* 1255 RH (Sa); Agnes *la Muchele* 1279 RH (O). OE *mycel* 'big'. cf. MITCHELL.

**Mucklejohn:** *v.* MEIKLEJOHN

**Mucklow:** *v.* MUTLOW

**Mudd:** *Modde* 1307, 1315 Wak (Y), 14th Shef (Y); Hugo *Mud* 1205 P (Sf); Cristofor *Mod* 1327 SRSf;

Widow *Mud* alias *Mutt* 1625 EA (NS) ii (Sf). Perhaps OE *\*Modd*, a short form of names in *Mōd*-, though these were rare. cf. *Modingus, Modinc* 1066 DB (Beds, Ess); Hugh *Moding'* 1275 SRWo; William *Modbert* 1274 RH (Sa).

**Muddeman, Muddiman:** Richard *Modimon* 1275 SRWo. 'Brave, courageous man.' cf. MOODEY.

**Mudditt:** *v.* MAUDUIT

**Mudford:** Terric *de Mudiford* 1177 P (So); William *Mudford* 1642 PrD. From Mudford (So).

**Mudie:** *v.* MOODEY

**Muers:** *v.* MEWER

**Muffatt, Muffett:** *v.* MOFFAT

**Mugg, Mugge, Mudge:** Geoffrey *Mugge* 1212 P (Do/So); Henry *Mug* 1275 RH (Sf); William *Mugge* 1332 SRWo; John *Mugg*, Robert *Mudge* 1642 PrD. OE *\*Mugga*.

**Muggeridge, Muggridge:** *v.* MOGGRIDGE

**Muggle, Muggles:** Edric *Muggel* 1188 BuryS (Sf); William *Muggel* 1208 Cur (W); Richard *Mugel* 1313 AssNf; Philip *Mugles* 1642 PrD. *Mugg-el*, a diminutive of OE *Mucca, \*Mugga*.

**Muir, Mure:** Thomas *de la More* 1291 Black; Elizabeth *Mure* 1347 ib.; Aleusa *en le muyre* 1348 DbAS 36; John *Muyr* 1470 Black. 'Dweller by the moor.' Chiefly Scottish. Perhaps also for MEWER.

**Muirhead:** William *de Murehede* 1401 Black; David *Muirheyd* 1527 ib. 'Dweller at the head of the moor.' Common in Scotland.

**Mulberry, Mulbry:** *v.* MOWBRAY

**Mulcaster, Muncaster, Moncaster:** Walter *de Mulecastr'* 1219 AssY; John *Muncaster* 1675 HTSf. From Muncaster (Cumb).

**Mule, Moule, Moules:** David *le Mul* 1199 P (Wo); Yuo *Mul* 1206 P (L); Baldewin *Mule* 1225 Pat; William *le Moul* 1327 SRSt. OE *mūl*, ME *moul* 'mule'. This would have become *mowl* but was ousted by OFr *mule* in the 13th century.

**Mulford:** Richard *de Muleford* c1219 Fees (W); John *de Muleford* 1292–3 FFEss; Edmund *de Mulford* 1305 IpmW. From Milford in Laverstock (W), *Muleford* 1236.

**Mulgrave, Mulgreave:** Adam *Mulgreue* 1362 AssY; Thomas *Molgryff* 1403 TestEbor. From Mulgrave (NRY).

**Mullard:** *v.* MILLWARD

**Mullet, Mullett:** John *Mulet* 1275 RH (Sf), 1311 ColchCt. Perhaps a diminutive of OFr *mule* 'mule'. Or, possibly, a nickname for a seller of mussels (OFr *mulet*). Michaëlsson notes that on the Normandy coast he is called 'Moulettes' from his cry. His real name is unknown to most of his customers.

**Mulligan:** *v.* MILLICAN

**Mullin, Mullen, Mullins, Mullens, Mullings, Molins:** Ralph *de Molins* 1159 P; Adam *del Molyn* 1289 AssSt; John *de Molyns* 1341 LoPleas; Edward *Mullens* 1545 SRW; William *Mullyns* 1642 PrD. 'Dweller at the mills', Fr *moulins*, or from one or other of the numerous French places of the name. *v.* also MOLIN.

**Mulliner, Mullinar, Mullinder, Mullinger:** Sancheus *Moliner* 1275 RH (L); Walter *le Moliner* 1283 SRSf; William *Molindiner* 1327 SRSo; William *Mullinder*

1612 Shef (Y); Francis *Mullender, Mullenger*, Sam *Mulliner* 1674 HTSf. OFr *molinier* 'miller'.

**Mullings, Mowling:** William *Molling'* 1292 SRLo; Geva *Mullyng* 1313 AD iv (Lo). OE *\*mulling* 'darling' (c1450 NED).

**Multon:** *v.* MOULTON

**Mumberson, Mumbeson:** *v.* MOMPESSON

**Mumbray:** *v.* MOWBRAY

**Mumby, Munbey, Munby:** Alan *de Munby* 1162 P; Beatrice *de Mumby* 1241–5 RegAntiquiss, *de Mumby* 1245 FFL; John *Munby* 1340 CorLo. From Mumby (L).

**Mumford:** *v.* MONTFORD

**Mummery:** *v.* MOWBRAY

**Muncaster:** *v.* MULCASTER

**Muncey, Munchay:** *v.* MOUNCEY

**Mund, Munde:** Nicholas *Mund* 1275 RH (L); John *Munde* 1327 SREss; Joan *Munde* 1372 ColchCt. OE *\*Munda.*

**Munday, Mundie, Mundy:** Richard, Thomas *Mundi* 1239 Rams (Nf), 1291 AssCh; John, Walter *Mundy* 1327 SRSf, SRY. ON *Mundi.*

**Munden, Mundin, Mundon, Mondon:** Ranulf *de Mundona* 1119 Colch (Ess); Richard *de Munden* 1249 AssW; Robert *Munden* 1447 CtH. From Mundon (Ess), or Great, Little Munden (Herts).

**Mundford, Munford:** Adam *de Mundeford* c1213 PN C 294; Richard *de Mundeford* 1247 FFC. From Mundford (Nf). *v.* also MONTFORD.

**Mundin, Mundon:** *v.* MUNDEN

**Munford:** *v.* MUNDFORD

**Mungay:** *v.* MOUNCEY, MOUNTJOY

**Mungeam, Mungam:** Robert *de Monte Begonis* 1086 DB (Ess); Adam *de Mundegund* 1143–7 DC (L); Cecilia *Mundegoun* 1298 AssL; William *Mundgome* 1334 SRK. From some, as yet, unidentified French place.

**Munkton:** *v.* MONKTON

**Munn:** *v.* MOON

**Munn, Munns:** Ernis *filius Munni* 1166, Godman *filius Munne* 1167 P (Nf); Reginald *Munne* 1271 FFL; Richard *Munne* 1327 SRC; Thomas *Munne* 1663 HeMil. ON *Munni, Munnr*, a byname from ON *munnr* 'mouth'.

**Munnery:** *v.* MONNERY

**Munnings:** *Mundingus* 1066 DB, c1095 Bury (Sf); Maurice *Munning* 1221 AssWa; Gilbert *Monding*, William *Munding*, Robert *Munnyng* 1327 SRSf. OE *\*Munding* 'son of *Munda*'. OE *\*Munda* 'protector' is unrecorded but is the first element of Mundford (Norfolk) and Mundham (Norfolk, Sussex).

**Munns:** *v.* MUNN

**Munro, Munroe, Munrow, Monro:** Gael *Rothach* 'man from Ro'. 'According to a tradition which may be substantially correct the ancestors of the Munros came from Ireland, from the foot of the river Roe in Derry, whence the name *Bun-rotha*, giving *Mun-rotha* by eclipsis of *b* after the preposition *in*' (Watson).

**Munsey, Munsie:** *v.* MOUNCEY

**Murch:** Geoffrey *Morch* 1327 SRSf; Hugh *Murch* 1651 DWills. 'Dwarf.' cf. '*Murche*, lytyll man. *Nanus*' PromptParv.

**Murchie, Murchison, MacMurchie:** Kathel *Mac murchy* 1259 Black; John *Murchosone* 1473 ib. Gael *MacMhurchaidh* 'son of *Murchadh*' 'sea-warrior'.

**Murcott, Murcutt:** Simon *de Morcote* 1177 P (R); Alan *de Morcote* 1239–40 FFWa; Hugh *de Morcote* 1247 AssSt. From Murcot (O), Murcott (W), or Morcote in Minsterworth (Gl).

**Murden:** *v.* MORDEN

**Murdoch, Murdock:** *Murdac, Murdoc, Meurdoch* 1066 DB (Y); *Mariedoc* Bohhan 1160 P (Sa); Geoffrey, Roger *Murdac* 1130 P (Y), 1182 StCh; Ralph, Sibilla *Murðac* 1197, 1199 P (Db, Gl); Nicholas *Murdoc* Hy 3 AD i (Bk). An anglicizing of OIr *Muireadhach*, OW *Mordoc*, MGael *Muireadhaigh*, Gael *Murdoch* 'mariner'. Introduced into Yorkshire before the Conquest by Norwegians from Ireland.

**Murfin:** *v.* MARVIN

**Murgatroyd:** John *Mergetrode* 1379 PTY; Bryan *Murgetroyde* 1647 RothwellPR (Y). From a lost Yorkshire place, 'Margaret's clearing'.

**Muriel, Murrell, Murrells, Murrill, Merrall, Merralls, Merrell, Merrells, Merril, Merrill, Merrills, Mirralls:** (i) Ougrim *filius Miriel'* 1188 P (L); *Mirielis, Muriella* de Stokes 1203 Cur (Bk); Johannes *filius Miriald'* 1208 Cur (Y); Godric *Miriild, Mirild', Mirield* 1184–8 P (L); Robert, Richard *Muriel* 1195 P (L), 1221 *ElyA* (Sf); Simon *Mirield* 1323 AD v (Nf); John *Myrel* 1327 SRSf; Richard *Meryel*, Walter *Merel* 1381 SRSf; William *Meryellys*, Thomas *Merelles* 1524 SRSf. *Muriel*, much the least common form as a surname, is of Celtic origin, found in Welsh as *Meriel, Meryl*, and in Irish as *Muirgheal*, earlier *Muirgel* 'sea-bright'. It was brought from Brittany by the Normans, hence its popularity in Lincolnshire, East Anglia and Essex, and also in Yorkshire where it may also have been introduced earlier by Scandinavians from Ireland. The name occurs in ON as *Mýrgjol*, daughter of a King of Dublin. In Staffordshire and the Welsh border counties, it came, no doubt, direct from Wales. (ii) Adam *de Merihel* 1275 RH (Sf); Nicholas *de Meriel* 1276 RH (Y); Walter *de Merihil* 1283 SRSf; Ralph *Muryhull* 1332 SRSt. From residence near some pleasant hill.

**Murless:** *v.* MERRYLEES

**Murphy:** Ir *Ó Murchadha* 'descendant of *Murchadh*' 'sea-warrior'.

**Murrant:** *v.* MORAND

**Murray, Murrey, Murrie, Murry:** William *de Moravia* 1203 Black; Alan *de Morref* 1317 ib.; Andrew *Moray* 1327 ib. From the province of Moray (Scotland). Some of the modern names may be from ME *murie*. *v.* MERRY.

**Murrell, Murrill:** *v.* MURIEL

**Murrimouth:** *v.* MERRIMOUTH

**Murthwaite:** Richard *Murthwaite* 1672 PN We ii 36. From Murthwaite in Longsleddale (We).

**Murton:** William *de Murton'* 1221 AssWa (Lei); William *Murtone* 1375 ColchCt. From Murton (Durham, Northumb, Westmorland, NRYorks).

**Muscat, Muscott:** Celestria *de Musecot* 1206 Cur (Nth); Richard *de Musecot'* 1275 RH (Nth). From Muscott in Norton (Nth), or Muscoates (NRY).

**Muschamp:** Thomas *de Muscamp* 1190 P (Y); William *de Muscamp* 1219 AssY; Stephen *de*

*Muscamp* 1291 Black. From North, South Muskham (Nt), *Muscampe* 1155.

**Musgrave, Musgrove:** Alan *de Musegrave* 1228 Cur (Nb); Thomas *de Musgraue* 1362 AssY; Robert *Musgrave* 1413 IpmY; Philip *Musgrave, Musgrove* 1642 PrD. Usually no doubt from Musgrave (Westmorland), but occasionally perhaps from Mussegros (Normandy), cf. Charlton Musgrove (Som) from Richard *de Mucegros* t John.

**Muscott:** *v.* MUSCAT

**Mushet, Muskett:** Robert, Osketell' *Muschet* Hy 2 DC (Nt), 1177 P (Sf); William *Musket* c1210 Fees (Nf); William *Mouchet* 1327 SRC. OFr *mouchet, mouschet, mousquet* 'a musket; the tassell of a sparhawke' (Cotgrave), 'a lytell hauke' (Palsgrave).

**Mushroom:** John *Mussheron* 1327 SRSo; John *Mussheroun* 1340–1450 GildC. A nickname from Fr *mousseron* 'a kind of mushroom'.

**Mussard, Mussared:** Hascoit *Musard, Musart* 1086 DB (Berks, Bk); Alfricus *Musard* 1134–40 Holme (Nf). OFr *musard* 'absent-minded, stupid' (c1300 NED).

**Mussel, Mussell:** Mathias *Muscel* 1230 P (L); John *Mussel* 1354 IpmW; John *Mussel* 1545 SRW. OE *muscelle* 'mussel, shell-fish', probably metonymic for a gatherer of these.

**Mussey:** Randulf *Musege* 1180 P (Wo); William *Musegh'* 1230 Cur (He); Peter *Musege* 1234–5 Clerkenwell. 'Mouse eye', OE *mūs, ēage*.

**Musson:** Robert *Mussun'* Hy 2 DC (L); Adam *Muisson* 1207 Cur (Gl); Richard *Mussen* 1662–4 HTDo. AFr *muisson* 'sparrow'.

**Mustard, Mustart:** Adam, William *Mustard'* 1191 P (Y), 1206 P (He). OFr *mostarde* 'mustard', perhaps one with a sharp, biting tongue, but usually, no doubt, a dealer in mustard, by metonymy for *mustarder*. cf. John *le Mustarder* 1327 SRC, Adam *le Mustardman* 1327 SRSf.

**Mustell, Mustill, Mustol, Muzzel, Muzzle:** Robert *Mustail* 1175–90 Seals (Y); Roger *Musteile* 1177 P (L); Hugh *Mustell'* 1208 Cur (Gl); William *Mustol'* 1208 Cur (Nf). OFr *musteile, mustoile* 'weasel'.

**Musters, Mustre:** Robert *de Mosteriis, de Mosters* 1086 DB (L, Nt); Robert *de Mustres* 1185 Templars (Y); Lisiard *des Mustiers* 1197 P (L). From Moutiers-Hubert (Calvados).

**Musto, Mustoe, Mustoo, Mustow:** Ralph *atte Mustowe* 1327 SRSx. 'Dweller by the moot-stow', OE *(ge)mōt-stōw*, a common name for the meeting-place of a hundred.

**Muston:** John *de Museton'* 1219 AssY; Robert *de Muston'* 1373–5 AssL; Thomas *Muston* 1437 IpmNt. From Muston (Lei, ERY).

**Mutch, Much:** William *Moch* 1275 RH (Nf); Richard *Muche* 1374 Ipm (Ess). ME *moche, muche* 'big'.

**Mutchell:** *v.* MUCKLE

**Mutchman:** *Mucheman* Wetebede 1235 Ch (C); Thomas *Mucheman* 13th BlackBk (K); Robert *Muchelman* 1279 RH (O). 'Big man', OE *mycel, mann*.

**Muter, Mutter, Mutters:** Grinchetell' *Mutere*, Hugo *le Motere* 1130, 1175 P (L); William *le Mouter* 1327 SRWo. OE *mōtere* 'public speaker'.

**Mutimer:** *v.* MORTIMER

**Mutlow, Mucklow:** Henry *de Motelowe* 1359 AssSt; William *Mucklow* 1515 PN Wo 178; Richard *Mutlow* 1663 HeMil. From Mutlow (Ch), or Mutlows in Welland (Wo).

**Mutton, Motton:** Stephen *Muton'* 1195 Oseney (O); Robert *Mouton'*, *Mutun* 1219 Cur, 1242 Fees (Lei); William *Moton* 1327 SRSx. OFr *mouton* 'sheep', a nickname, or metonymic for OFr *moutonier*, a keeper of sheep, shepherd: Hugh *Motoner* 1275 RH (Lo).

**Mutton, Mitton, Mytton:** William *de Mitton, de Mutton* 1286–7 AssSt. From Mitton (Lancs, Worcs, WRYorks) or Myton (Warwicks, Salop, NRYorks).

**Muttycombe:** A Somerset name, from Mothecombe (Devon), pronounced *Muddicombe*, where the surname was *de Modecumbe* 1238, *de Muthecumbe* 1244 (PN D 277).

**My, Mye:** John *le My* 1332 SRSt; William *My* 1393 FFWa. A shortened form of OFr *ami* 'friend'.

**Myall:** *v.* MIALL

**Mycock:** *v.* MAYCOCK

**Myddleton:** *v.* MIDDLETON

**Myer, Myers, Mier, Miers:** (i) Herewardus *le Mire, Medicus* 1212 Cur (Berks); Thomas *le Myre* 1256 AssNb. OFr *mire* 'physician'. (ii) Hugo, Adam *del Mire* 1274 Wak (Y), 1332 AssLa. 'Dweller by the marsh', ON *mýrr*.

**Myerscough:** William *de Mirscho* 1246, Walter *de Myreskou* 1277, William *de Mireschow* 1285 AssLa. From Myerscough (La).

**Myhill:** *v.* MIALL

**Mylechreest, Mylchreest, Mylecrist:** *MacGilchreest* 1511, *McYilchrist* 1713, *Mylechreest* 1717. Manx *Mac Giolla Chreest* 'son of Christ's servant' (Moore).

**Myles:** *v.* MILES

**Mylles:** *v.* MILLS

**Mylroy:** Bæthan *McIlroy* 1408, *McYleroij* 1612, *Mylroi* 1741, *Mylroi* 1759 Moore. Manx *Mac Giolla-ruaidh* 'son of the red-haired youth'.

**Mynett, Mynott:** *v.* MIGNOT

**Mynn:** *v.* MINN

**Myott:** *v.* MIATT

**Mytton:** *v.* MUTTON

# N

**Nabb, Nabbs, Napp:** *Nabbe* Brodeye 1298 Wak (Y); *Nabbe* son of Broun 1308 ib.; John *Nap* 1279 RH (C); Gregory *Nabbys* 1524 SRSf. *Nabb* and *Napp* are probably parallels to *Nobb* and *Nopp*, pet-names for *Robert*, rhymed on *Rab* and *Rob*. cf. Robert *Nabelot* 1524 SRSf and *v.* NOBBS, ROBLETT.

**Nacke:** Alured *Nacche* 1189 Sol; Adam *Nacke* 1297 Wak (Y); Thomas *Nak'* 1356 FFY. ON *Hnaki*.

**Nadder:** Robert *Nadder* 1219 AssY; Robert *de la Nedre* 1280–1 FFSr. The first example may be a nickname, 'adder', OE *nǣdre*, the second 'dweller at the place frequented by adders'.

**Nadler:** *v.* NEEDLER

**Naesmith:** *v.* NAYSMITH

**Nafferton:** William *de Nafferton'* 1219 AssY; Thomas *de Naffreton'* 1327 SRY; William *Naferton* 1382 FFEss. From Nafferton (Nb, ERY).

**Nail:** William, Geoffrey *Nayl* 1255 RH (Bk), 1327 *SR* (Ess). Metonymic for NAYLAR.

**Nailer:** *v.* NAYLAR

**Naim:** John *le Neim* 1275 SRWo; Hamo *le Neym* 1296 SRSx; John *le Naym* 1327 SRWo. OFr *nain* 'dwarf'.

**Naisbet, Naisbitt:** *v.* NESBIT

**Naisby:** *v.* NASBY

**Naish:** *v.* ASH

**Naismith:** *v.* NAYSMITH

**Nalder:** *v.* ALDER

**Nalderfan:** Thomas *atte Nalderfan* 1318–19 FFEss. 'Dweller at the alder-fen', OE *alor, fæn*.

**Naldrett:** *v.* ALDRITT

**Nangle:** *v.* ANGLE

**Nanson:** Robert *Nanson* 1379 PTY; William *Nanson* 1441 TestEbor; Thomas *Nanson* 1524 SRSf. 'Son of *Nan*', a pet-form of *Anne*.

**Napier, Naper, Napper:** Peter *Napier* 1148 Winton (Ha); Ralph *(le) Naper, le Napier* 1167–71 P (Ess); Reginald *le Nappere* 1225 AssSo. OFr *napier, nappier* from *nappe* 'table-cloth'. 'Naperer, one who has charge of the napery or table-linen' (1880 NED).

**Napleton:** *v.* APPLETON

**Napp:** *v.* NABB

**Narborough, Narbrough, Narburgh:** Hamo *de Nerburg'* 1242 Fees (Nf); Richard *de Nareburch* 1284 RamsCt (Hu); William *Narbow* 1454 Paston. From Narborough (Nf).

**Nares:** Simon *le Neir* 1221 AssGl; William *le Nayr* 1297 SRY. AFr *neir* 'black'.

**Narracott:** William *de Northecote* 1330 PN D 31. From Narracott (Devon).

**Narramore, Narrowmore:** Reginald *Bynorthemore* 1318 PN D 481. 'Dweller north of the moor', as at Narramore (Devon).

**Narrasty:** William *Narousty* 1361 AssY. 'Dweller by the narrow path', OE *nearu, stīg*.

**Narraway, Narroway:** Nicholas *Bynortheweye* 1333 PN D 453. 'Dweller north of the road', as at Narraway (Devon).

**Nasby, Naseby, Naisby:** Robert *de Naseby* 1321 AssSt; John *Naseby* 1456 FFEss. From Naseby (Nth).

**Nash:** *v.* ASH

**Nasmyth:** Hugh *Nasmith* 1277 IpmY. 'Navesmith', OE *nafa, smiþ*. *v.* also NAYSMITH.

**Nass:** *v.* NESS

**Nassard:** Henry *Nasard'* 1319 SRLo; John *Nasard* 1339–40, *Nazard* 1340 CorLo. 'Speaking with a nasal accent', OE *nosu*, and *-ard*.

**Nateby:** John *de Nateby* 1332 SRLa, 1350 FFC. From Nateby (La, We).

**Natley:** Vincent *de Nateleye* 1275 RH (Ha); Richard *de Natele* 1296 SRSx. From Nately Scures, Up Nately (Ha).

**Natrass, Natress, Nattrass, Nattriss:** Cristina *Nattrys* 1474 GildY; Edmund *Nateres* 1522 LP (D); Edmund *Natres* 1560 Pat (Du). From Nattrass in Alston (Cu).

**Natural, Naturel:** Elyas *Naturel* 1208 P (O); Robert *Naturell* 1301 FS; William *Naturel* 1347 ChertseyCt (Sr). OFr *natural*, possibly in the sense of a naturalized citizen.

**Naughton:** Hugh *de Nawelton, de Nawenton* 1327 SRSf. From Naughton (Sf), *Nawelton* c1150.

**Naunton:** Hugh *de Naunton* 1326 FFEss; Wyllyam *Naunton* 1461, Wyll *Naunton* 1465 Paston. From Naunton (Gl, Wo), or Naunton Beauchamp (Wo).

**Nave:** John *le Cnaue* 1221 AssWa; Simon *Knave* 1296 SRSx. OE *cnafa* 'child, youth, servant', often used of servants: William *Margeriknave* 1307 Wak (Y). *Nave* may also be metonymic for *Naver*: Eadric *Nauere* 13th AD iv (Lo), Richard *le Navere* 1275 RH (Nf), a maker of naves of wheels, from OE *nafa, nafu*. cf. Adam *Nawrith'* 1301 SRY 'nave-wright'.

**Nay:** *v.* NYE

**Naybour:** *v.* NEIGHBOUR

**Nayland, Naylon:** Simon *de Neylond* 1285 FFEss; Geoffrey *atte Nelonde* 1296 SRSx; Thomas *de Neylonde* 1379 LoCh. From Nayland (Sf), or 'dweller at the island', OE *æt þǣm ēgland*.

**Naylar, Nayler, Naylor, Nailer:** Stephen *le Nailere*

1231 Pat (Lo); James *le nayler* 1273 FrY. A derivative of OE *nægel* 'nail', a maker of nails (c1440 NED).

**Nayshe:** *v.* ASH

**Naysmith, Naismith, Naesmith, Nasmyth:** Roger *Knifsmith* 1246–89 Bart (Lo); Adam *Knyfsmith* 1285 AssLa; Saman *le Knyfsmyth* 1310 LLB D; William *Knysmyt* 1326 AssSt; Robert *Knysmithe* 1594 Bardsley; John *Naysmith* n.d. ib. OE *cnīf* 'knife' and *smiδ* 'smith', a cutler.

**Neachell:** *v.* ETCHELLS

**Neal, Neale, Neall, Neel, Neele, Neeld, Neels, Neil, Neild, Neill, Niall, Niel, Nield, Niell, Niles, Nigel, Nihell, Nihill, McNeal, McNeil, McNiel, O'Neal, O'Neill:** Willelmus *filius Nigelli* 1086 DB (Bk); *Neel* 1170–82 YCh (Y); *Niel* de Wellebek 1260 AssY; Achyne *mac Nele* 1289 Black; Willelmus *filius Nele* 1304 SRY; *Neil* Carrick c1314 Black (Galloway); *Nile* Hog 1557 Black (Stirling); Willelmus *Nigelli* 1195 FF (W); Robert *Nel, Neel* 1208–10 Cur (Berks); Nicholaus *Nigelle* 1252 Rams (Hu); Henry *Nel* 1260 AssC; Robert *Neel* 1294 AssSt; Roger *Niel* 1319 FFSf; John *Nihell* 1565 Bardsley; Matthew *Nihill* 1796 ib. OIr *Niáll*, Gael *Niáll* 'champion', latinized by Adamnan as *Nellis*. The name was carried to Iceland by the Scandinavians as *Njáll*, taken to Norway, then to France and brought to England by the Normans. It was also introduced direct into north-west England and Yorkshire by Norwegians from Ireland. It was usually latinized as *Nigellus* through an incorrect association with *niger* 'black'. *Neil* and *Nigel* are now mainly Scottish. In Yorkshire it is found as *Nell*. Robertus *filius Nigelli* and Robertus *Nel* (1221 AssWa) are identical.

**Neam, Neame:** Richard *le Naim* 1170–8 P (L); John *Nepos, le Neim* 1214 Cur (Sr); John *le Neim* c1280 SRWo; Henry *Neem*, John le *Naym* 1327 SRDb, SRWo; John *Naym* 1431 FA (Wo). OE *ēam* 'uncle' with the initial *N*- due to misdivision of syllable. *v.* also EAMES. But Fr *nain* 'dwarf' is probably also a source of the name.

**Neap, Neep:** Peter, John (*le*) *Nep* 1279 RH (Bk, Hu); Richard *Nepe* 1524 SRSf. OE *nǣp* 'turnip'.

**Neat, Neate:** Probably OE *nēat* 'ox, cow', metonymic for 'cowherd'. cf. William *Nethirde* 1301 SRY, Hubert *le Netdriver* 1295 SIA xiii (Sf). Or a nickname: cf. Ascer *Neteheved* 1200 Cur (L), 'ox-head'.

**Neatby:** John *de Nateby* 1350 FFC. From Nateby (Lancs, Westmorland).

**Neave, Neaves, Neeve, Neeves, Neve:** Robert *le Neve* 1242 Fees (K); Andrew *Neve* c1250 Rams (Beds). OE *nefa*, ME *neve* 'nephew'. Also a nickname for a prodigal or parasite: 'neuerthryfte, or wastour' PromptParv.

**Neck:** Henry *Nekke* 1279 RH (C); Richard *Necke* 1327 SRC. A nickname for one with some peculiarity of the neck. cf. Symon *Nekeles*, probably identical with Symon *Chortneke* 'short neck' 1275 RH (Nf), Geoffrey *Neckebon* 1316 FFC.

**Need, Needes, Needs, Nead, Neads:** Richard *Ned* 1228 Cur (Mx); William *le Need* 1296 SRSx; Ralph *le Ned* 1327 SRSx. OE *nēd, nīed* 'need', perhaps 'needy'.

**Needham, Nedham:** John, Robert *de Nedham* 1275 RH (Db), 1305 LoCt; John *Needhom* 1371 FFEss. From Needham (Derby, Norfolk, Suffolk).

**Needle:** Metonymic for NEEDLER.

**Needler, Neelder, Nelder, Nadler:** Peter *le Nedler* 1221 AssSa; Richard *Nedlere, Neldere* 1235 Oseney (O); Robert *le Nadlere* 1309 LLB D, 1327 SR (Ess). OE *\*nǣdlere* 'needler' from *nǣdl* 'needle'. A maker of needles (1362 NED). *Neelder* and *Nelder* are due to metathesis of *dl*. In ME *Nadler* was the regular form in Essex and common in London.

**Neel(e), Neeld, Neels:** *v.* NEAL

**Neep:** *v.* NEAP

**Neeve(s):** *v.* NEAVE

**Neighbour, Naybour:** Ralph *Nechebur* 1222 DBStP (Herts); William *le Neybere* 1309 SRBeds; Bartholomew *Neighebour* 1327 *SR* (Ess). OE *nēahgebūr* 'neighbour'.

**Neil(d), Neill:** *v.* NEAL

**Neilson, Nielson, Nilson:** John *Neylsone* 1510 Black; John *Nilsoune* 1654 ib. 'Son of *Neil*.'

**Nelder:** *v.* NEEDLER

**Neldrett:** *v.* ALDRITT

**Nell:** *Nelle* de Wynter, de Soureby 1274 Wak (Y); Henry *Nelle* 1297 MinAcctCo (R). OIr *Nel, Niáll. v.* NEAL.

**Nelm(e)s:** *v.* ELM

**Nelson:** John, Robert *Nelleson* 1324 Wak (Y), 1332 SRCu. 'Son of *Nell*.' *v.* NELL.

**Nephew:** Thomas *le Neveu* 1268 AssSo; John *Neveu* 1274 RH (Nf). OFr *neveu* 'nephew'.

**Nepicar:** Richard *de Nepakere* 1334 SRK; John *Nepeker* 1548, *Nippeker* 1549, Mathye *Napecker* 1593 StaplehurstPR (K). From Nepicar House in Wrotham (K).

**Nesbit, Nesbitt, Naisbet, Naisbit, Naisbitt, Nisbet, Nisbit:** Robert *de Nesbit* 1160–1200 Black; William *de Nesebite* c1250 FeuDu. From Nesbit (Berwicks, Northumb) or Nesbitt (Durham, Northumb).

**Nesfield, Nessfield:** William *de Nesfeld* 1345 FFY; Richard *de Nessfeld* 1381 PTY; Aliua *Nesfeld* 1395 Whitby. From Nesfield (WRYorks).

**Ness, Nass, Noss:** Robert *de Nesse* 1177 P (Y); William *de Nes* 1275 RH (Sf); Agnes *ate Nass* 1279 RH (O). 'Dweller on the headland', OE *nǣss*, as at Nass (Glos) or Ness (Ches, NRYorks). John *atte Nasse* lived at Noss Point in Brixham in 1330 (PN D 508). Occasionally these modern forms may be for ASH, NAYSHE: Walter *ate Nesse* or *ate Neysshe* 1317 AssK. Nicholas *de la Nesse* in 1279 lived at Ashes Fm, Icklesham (PN Sx 512).

**Nessling, Neslen:** Nicholas *Nestlyng* 1524 SRSf; Peter *Nestlyn* ib.; Widow *Neslin* 1674 HTSf. 'Nestling'.

**Nest:** *Nest* de Barri 1185 P (Sx); *Nesta* (f) 1221 AssWo, 1222 Cur (O); *Nesta* de Broketon 1254–5 FFWa; Adam *Nest* 1185 P (Do); Richard *Neste* 1327 SREss; William *Neste* 1379 PTY. *Nest* (f), a Welsh diminutive of *Agnes*.

**Neston:** William *A Neston* 1447 AD i (Sx); Robert *A Neston* 1525 SRSx. From Neston (Ch), or for EASTON.

**Nethercoat, Nethercoate, Nethercot, Nethercott:** Ern' *de Nethercot'* 1208 Cur (O); Richard *de*

*Nethercote* 1244 PN D 34; Simon *de Nethercote* 1263 IpmGl. From Nethercot (Nth), Nethercote (O), or one or other of the many minor places of this name in Devon.

**Nethersall, Nethersole:** Edmund *Nethersole* 1410 FFEss; John *Nethersole* 1498 FFEss; Thomas *Nethersole* 1508 ArchC 40. From Nethersole Fm in Womenswold (K).

**Netherton:** Petronilla *de la Netherton'* 1275 SRWo; Walter *atte Nuthereton* 1330 MELS (Sr). 'Dweller at the lower farm', as at Netherton (Northumb, Worcs).

**Netherwood:** Thomas *de Netherwode* 1274 RH (Ess). 'Dweller by the lower wood.'

**Nettard, Netter:** (i) William *Nethirde* 1301 SRY; Thomas *le Nethurd* 1353 NottBR; Peter *Nethyrde* 1374–5 NorwLt. 'Cattle herd', OE *nēat, hierde.* (ii) John *de Netesherde* 1327 SRSf. From Neatishead (Nf).

**Nettelfield:** Roger *de Netelfed'* 1221 AssWo. 'Dweller by the land overgrown with nettles.'

**Netter:** John *le Nettere* 1298 LLB B; Christina *Netter* 1367 ColchCt. A derivative of OE *net(t)* 'net', a net-maker. cf. John *le Netmaker* 1336 ColchCt.

**Nettlefold:** Hugh *Netelfold* 1390–1 FFSr. From a lost *Nettlefold* in Dorking (Sr), or 'dweller by the enclosure overgrown with nettles', OE *netele, falod.*

**Nettleton:** John *de Neteltone* c1220–30 RegAntiquiss; Covecok *de Nettelton* 1309 Wak; George *Nettleton* 1621 SRY. From Nettelton (Lincs, Wilts), or 'dweller by the enclosure where nettles grow'.

**Nettleworth:** John *de Nettleworth* 1370 IpmNt. From Nettleworth (Nt).

**Neve:** *v.* NEAVE

**Neverathome:** Simon *Neveratham* 1226 ELPN, 1244 AssLo; Adam *Neverathom* 1276 AssLo. A phrase name, 'never at home'.

**Nevett, Nevitt:** *v.* KNIVETT

**Nevill, Neville, Nevile, Newill:** Ralph *de Neuilla*, Richard *de Nouuilla* 1086 DB; Gilbert *Neuille, de Nouila* 1142–60 DC (L); John *de Newill'* 1235 Fees (W). The Nevilles of Raby came from Néville (Seine-Inférieure). *v.* ANF. Others may have come from Neuville (Calvados) or other French places of the same name.

**Nevin:** *v.* NIVEN

**Nevinson, Nevison:** William *Nevinson* 1493 TestEbor; James *Nevison* 1758 FrY. 'Son of *Nevin*'. *v.* NIVEN.

**New:** (i) William *le Neuwe* 1221 AssWa; Walter *le New* 1234 FFC; John *le Nywe* 1333 AD v (L). OE *nīwe*, ME *newe* 'new', 'the newcomer', the more usual source. (ii) John *atte Newe* 1327 SRSo; Petronilla *ate Nywe* 1332 SRSr. From residence near a yew tree, OE *ēow, īw. v.* p. xiv.

**Newall:** *v.* NEWHALL

**Newbald, Newbold, Newbolt, Newbould, Newboult, Newball:** Robert *de Newebolt* 1175 P (Wa); John *de Neubald'* 1219 AssY; William *de Newbold'* 1299–1300 FFWa; Roger *Neubolt* 1350 LLB F; Thomas *Newball* 1653 FrY. From Newbald (ERYorks), or Newbold (Ches, Derby, Lancs, Leics, Northants, Notts, Warwicks, Worcs). Variants *Newbatt, Newbart, Newbert* are said to be peculiar to Notts (Bardsley).

**Newberry, Newbery, Newbury:** Godwin *de Neweberia* 1190 P (Berks); Henry *de Neubury* 1279 RH (Bk); Giles *Newberry*, William *Newbery* 1642 PrD. From Newbury (Berks), or Winfrith Newburgh (Do).

**Newbiggin, Newbigging, Newbegin, Newbigin:** Walter *de Neubigging'* 1219 AssY; Adam *del Neubigging* 1275 Wak (Y); William *de Neubiggyng* 1332 SRCu. From Newbigging (Cu, Du, Nb, We, NRY, Lanark), Newbegin (NRY), or Biggin in Church Fenton (WRY) *Neuebiggynge* 13th.

**Newbon, Newbond, Newbound:** Hugo *le Neubonde* 1271 Rams (C); William *Newebonde* 1275 SRWo. 'The new *bond*', *v.* BOND.

**Newborn, Newborne:** Roger *atte Neweburn* 1296 SRSx; Richard *Newborne* 1641 PrSo. From Newbourn (Sf), or Newburn (Nb).

**Newbottle:** Adam *de Neubotle* c1177–9 Black; William *de Newebotl* 1253 PN Nth 114. From Newbottle (Du, Nth), or Newbattle (Midlothian).

**Newbury:** *v.* NEWBERRY

**Newby:** Jacobus *de Newebi* 1197 P (We); John *de Newby* 1275 IpmY; John *de Neuby* 1340–1450 GildC; Richard *Newby* 1672 HTY. From Newby (Cu, We, NRY, WRY), or Newby Wiske (NRY).

**Newcastle:** Geoffrey *de Newcastle* 1246–7 FFSr; Agnes *of Newcastle* 1315 AssNf; Adam *de Newcastle* 1340 CorLo. From Newcastle (Nb, Sa), or Newcastle under Lyme (St).

**Newcomb, Newcombe, Newcombes, Newcome:** Alan *le Neucument* 1175 P (L); William *Neucum* 1183 Boldon (Du); Walter *le Neucumen* 1185 Templars (L); Richard *Newecume* 1195 P (L). OE *nīwe* and *cumen, cuma* 'newly-arrived stranger'.

**Newdigate:** Richard *de Neudegat'* 1219 P (Sr); John *de Nywedegate* 1332 SRSr; Francis *Newdigate* 1559 Pat (Mx). From Newdigate (Sr).

**Newell:** Thomas *de Newell* 1201 FFEss. For NEVILL. Ralph *Nuuel* 1209 P (Y). For NOEL.

**Newey, Neway:** Thomas *de Newehawe* 1327 SRSf. 'Dweller at the new enclosure.'

**Newhall, Newall:** Robert *de Niwehal'* 1195 P (Y). 'Dweller at the new hall.'

**Newham:** Roger *de Niweham* 1109–29 MCh; Emma *de Neweham* 1227 FFO; Thomas *de Neweham* 1350 FFY. From *Newham* (Nb, NRY).

**Newhouse, Newis, Newiss:** Ralph *de Niwehus* 1176 P (y); William *ate Neuhous* 1327 SRC; John *Newis* 1672 FrY. 'Dweller at the new house.'

**Newhusband:** John *Nywehosebande, Yongehose-bond* 1286–7 CtH; Thomas *le Newehosebonde* 1327 SRSf. 'New householder', OE *nīwe, hūsbonda*, or equivalent to YOUNGHUSBAND.

**Newill:** *v.* NEVILL

**Newington:** Walter *de Niwenton* 1163 P; Jordan *de Newintone* 1317 AssK; Roger *de Newynton* 1327 SRWo. From Newington (K, O, Sr), Newington Bagpath (Gl), North, South Newington (O), or Long, North Newnton (W).

**Newland, Newlands:** Samson *de la Niwelande* 1188 P (K); William *atte Niwelond* 1327 SRSo. From a place called Newland(s) or 'dweller at the newly cleared or newly acquired land'.

**Newling, Newlyn:** Peter *filius Neuelon* c1170 ELPN, 1192 P (Lo); Andrew *Neuelun* 1207–8 P (Lo); Adam *Newlyn* 1327 SRC; Thomas *Neulyn* 1327 SRSf. OG *Neveling, Nivelung*, OFr *Nevelon, v.* PNDB 331. Sometimes, perhaps, from East Newlyn (Co).

**Newman, Nieman, Niman, Nyman:** Stangrim *Noueman* 1166 P (Nf); Godwin *Nieweman* 1169 P (O); Ailwin *le Newman* 1195 P (Ess); William *þe Niweman* 1227 Eynsham (O); Robert *le Nyman* 1296 SRSx. 'The new man, newcomer', OE *nēowe, nīwe, nīge* and *mann.*

**Newmarch:** Adam *de Neumarche* 1242 Fees (Nt). From Neufmarché (Seine-Inférieure). *v.* ANF.

**Newnham, Newnam, Newnum:** Ralph *de Neunam* 1255 RH (Sa); Robert *de Newenham* 1296 SRSx; George *Newnam* 1642 PrD. From Newnham (Beds, C, Gl, Ha, Herts, K, Nth, Wa, Wo), Newnham Murred (O), or Kings Newnham, Newnham Paddox (Wa).

**Newport:** Ailwin *de Niweport* 1177 P (Ess); Nicholas *Neuport* 1359 AssD; Francis *Nuporte* 1641 PrSo. From Newport (D, Ess, He, Monmouth, Sa, Wt), or Newport Pagnell (Bk).

**Newsam, Newsham, Newsholme, Newsom, Newsome, Newsum:** Robert *de Neusum* 1195 P (Y); Robert *de Neusom* 1275 Wak (Y); Ralph *Newsame* 1598 RothwellPR (Y). From Newsham, Newsam or Newsholme, common north-country names, all 'at the new houses'.

**Newton:** Alward *de Niwetuna* 1066 DB (Nf); Robert *de Neweton'* 1190 P (Y); Stephen *de Neuton'* c1280 SRWo; William *Neuton* 1370 IpmNt. From one or other of the many places of this name.

**Ney:** *v.* NYE

**Niall:** *v.* NEAL

**Nicholas, Nicholass, Nicolas, Nicklas, Nicklass, Nickless, Niccols, Nichol, Nicholds, Nicholes, Nicholl, Nicholls, Nichols, Nickal, Nickalls, Nickel, Nickell, Nickells, Nickels, Nickle, Nickol, Nickolds, Nickolls, Nickols, Nicol, Nicole, Nicoll, Nicolle, Nicolls, Nickolay, Nicolai, Nicolay, McNichol, McNicol, McNickle:** Nicolaus 1086 DB; *Nicholaus* presbiter 1147–66 Gilb (L); *Nicolus* Hy 2 DC (L); *Maucolum fiz Nicol* 1296 Black; Waleram *Nicholai* 1198 Cur (Sf); John *Nichole* c1270 *ERO*; Dovenald *Macanecol* 1294 JMac; William *Nicholas* 1311 AD i (Beds); Robert *Nicholes* 1322 AssSt; Andrew *Nicoles* 1327 SRSo; Gylbryd *Nycholay* 1446 Black; James *Nickle* 1650 Black; William *Nickless* 1783 SfPR. Lat *Nicolaus*, Greek Νικόλαος 'victory-people' was a very popular medieval name as is proved by its numerous forms as a surname, its diminutives and pet-forms. The vernacular form was *Nicol*. Fairly common, too, was the feminine *Nic(h)olaa* 1207–8 Cur (Hu, Gl), corresponding to the French *Nicole*, which may account for surnames with a final *-e*. *Nickolay* is a survival of the Latin genitive: Patrick *Nicholai* (1436 Black) is the same man as Patrik *Nicholsone* (1446 ib.). *v.* COLLIN.

**Nicholetts:** John *Nicholetes* 1327 SRWo; Henry *Nicolet* 1347 AD vi (K). A diminutive of *Nic(h)ol*, i.e. *Nicholas.*

**Nicholin:** *Nicholina* Baker 1524 SRD. *Nichol-in*, a diminutive of *Nichola* (f).

**Nicholl(s):** *v.* NICHOLAS

**Nicholson, Nickelson, Nicolson:** Michael *Nycholson* 1443 Black; Peter *Nicholasson* 1459 Oseney (O); Thomas *Nicolson* 1496 LLB L; Mallie *Niclasson* 1663 Black. 'Son of *Nichol(as)*.'

**Nickall(s), Nickell(s):** *v.* NICHOLAS

**Nickaman, Nickman:** John *Nikeman* 1327 SRSf; Thomas *Nikeman* 1335 FFSf; John *Nykeman* 1375 ColchCt. 'Servant of *Nick*', a short form of *Nicholas.*

**Nicker:** Jordan *Niker* 1214 P (C); William *le Niker* c1240 Colch (Ess); Cecilia *Niker* 1327 SRC; Henry *Nyker* 1553 NorwDep. OE *nicor* 'a water spirit'.

**Nickerson, Nickinson, Nickisson:** For NICHOLSON. cf. HERBERTSON.

**Nickes, Nicks, Nix:** John son of *Nyk* 1316 Wak (Y); Henry *Nix* 1279 RH (O); Nicholas *Nike, Nicks* 1355, 1385 DbAS 36. A pet-form of *Nicholas*. cf. John *Nicholls* alias *Nicks* 1697 DKR 41 (O).

**Nickle:** *v.* NICHOLAS

**Nicklen, Nicklin:** Nicholas, Robert *Nykelin* 1387 AssSt. A double diminutive of *Nick* (Nicholas).

**Nickman:** *v.* NICKAMAN

**Nickol, Nickolay, Nickolds:** *v.* NICHOLAS

**Nickson, Nixon, Nixson:** Robert *Nikkesune* 1309 NottBR; John *Nickeson* 1332 SRWa; John *Nyxon* 1450 Rad (C). 'Son of *Nick*.'

**Nicoll(s), Nicolay:** *v.* NICHOLAS

**Nidd:** Uctred *de Nid* 1219 AssY; Walter *de Nidde* 1297 MinAcctCo; Robert *Nyd* 1424 FrY. From Nidd (WRY).

**Nie:** *v.* NYE

**Nield, Niell:** *v.* NEAL

**Nielson:** *v.* NEILSON

**Nieman:** *v.* NEWMAN

**Nifton:** The local pronunciation of Kniveton (Derby).

**Nigel:** *v.* NEAL

**Nightingale, Nightingall, Nightingirl:** Walter *Nichtegale* 1176 P (Gl); Richard *Nihtingale* 1227 AssBeds; Alan *Nightegale* 1260 AssC; Henry *Nitingale* 1281 LLB B. OE *nihtegale* 'night-singer', 'nightingale', a common nickname for a sweet singer.

**Nighton:** *v.* KNIGHTON

**Nihell, Nihill:** *v.* NEAL

**Niker:** William *le Niker* c1240 Colch (Ess); Cecilia *Niker* 1327 SRC. OE *nicor* 'water-monster, water-sprite'.

**Niles:** *v.* NEAL

**Nilson:** *v.* NEILSON

**Niman:** *v.* NEWMAN

**Nind:** Adam *Attenende* 1260 MELS (So); Thomas *atte Nende, atte Ynde* 1327, 1332 SRWo. 'Dweller at the end' (of the village). OE *ende. v.* p. xiv.

**Nineham, Ninham, Ninnim:** Rare surnames, from *atten innome, atte ninnome*, one who dwelt by a piece of enclosed ground (OE *\*innām*, ME *innom*). The term is common in field-names, survives as Inholms (PN Sx 29), Ninehams (PN Sr 312), and occurs as *Nynnom, Inums, Innims* and (PN Ess 583) as *Nynnams. v.* p. xiv.

**Nisbet, Nisbitt:** *v.* NESBIT.

**Niven, Neven, Nevin, Nevins, MacNevin, MacNiven:** *Nevinus* c1230 Black; *Neuyn* filius Ade 1332 SRCu; Thomas *filius Neuini* 1295 JMac; Thomas *Maknevin* 1528 ib.; Thomas *Nevin* 1538 Black; John *M'Nivaine* 1638 ib.; John *Nivine* 1675 ib. Ir, Gael *Naomhin* 'little saint'.

**Nix:** *v.* NICKES

**Nixon, Nixson:** *v.* NICKSON

**Noad, Noades, Nodes:** Adam *Node* 1297 MinAcctCo; Thomas *Noades* 1692 *ERO*. Richard *ate Node* lived at the Node in Codicote (Herts) c1282–5 and Adam *atten Ode* at the Nodes in Totland (Isle of Wight) in 1311 (PN Herts 110). *Ode* is from OE *ād* 'pile, heap, funeral pile'; its exact sense in these place-names is doubtful. *v. p.* xiv.

**Noah:** Probably for NOAR.

**Noake, Noakes, Noaks:** *v.* OAK

**Noar:** Walter *Nore* 1275 RH (L); John *at(t)e Nore* 1314 LLB E, 1332 SRSt. From residence near a shore, bank, or, very commonly, a steep slope, OE *ōra. cf.* Nore Fm (PN Sr 227), the home of John *Attenore* in 1263. In Sussex, Surrey, Devon and Middlesex it has become *Nower. cf.* ORE. Nore Hill (PN Sx 70), the home of Richard *atte Noure* in 1353, is from OE *ōfer* in the same sense. *cf.* OWER. *v. p.* xiv.

**Nobbs, Nobes, Nopps, Nops:** *Nobbe* Caipe 1202 AssL; Wido *Nobbe* c1248 Bec (O); Philip *Noppe* 1279 RH (Hu); William *Nobys* 1327 SRSf. *Nobb* is a rhymed pet-form of *Robert*, from *Rob. Noppe* is a variant with the common interchange of voiced and unvoiced consonants.

**Noble, Nobles:** Peter *Noble* 1185 Templars (Wa); Robert *le Noble* 1206 Cur (Hu). Fr *noble* 'well-known, noble'.

**Noblet, Noblett:** (i) Ordric, Walter *Noblet* 1187 P (Berks), 1206 P (Hu). A diminutive of *Noble.* (ii) Hugh, John *Nobelot* 1327 SRC, SRSf. *Nob-el-ot*, a double diminutive of *Nobb. v.* NOBBS.

**Nock:** *v.* OAK

**Nodder, Nother, Nothers:** William *Nadder* 1219 AssY; William *Nedder* 1260 AssY; Hugh, John *le Nodder* 1314, 1316 Wak (Y); George, Thomas *Nother* 1601 FrY. The well-known Yorkshire *Nodder* is diffidently explained by Bardsley as 'a nickname for one of sleepy or apathetic habits: one who nodded'. It is probably a nickname, with a typical Yorkshire sting, from OE *nædre*, ME *nadder, nedder* 'adder'. The initial *n* was lost in ME through misdivision of *a naddre* as *an adder*, a combination which would not occur in the surname. *Nedder* is still a northern dialect form. *Nodder*, from *Nadder*, is an inverted spelling and pronunciation, due to the unrounding of *o* to *a*, found in Yorkshire, certainly in the 15th century, and possibly earlier: cf. John *Stapper, Stopper* 1464, 1483 FrY (*v.* STOFFER). The 1314 *Nodder* is too common to be an error for *Nedder*. There are early examples of variation between *Mall* and *Moll*, *Nabb* and *Nobb*, and in the Wakefield Court Rolls of *Magota* and *Mogota. Nother*, also a Yorkshire name, is due to the common dialectal interchange of *d* and *th* in *fadder* and *father*, etc.

**Nodes:** *v.* NOAD

**Nodger:** Herbert *Notgor* 1202 P (Y); Nicholas *Noddegar* 1309, *Nodger* 1313–16 Wak (Y); John *Nogger* 1327 SRSo. OG *Notgar.*

**Noe:** *v.* NOY

**Noel, Nowell, Nowill:** *Noel* 1130 P (Lo), 1200 Cur (Nth); Geoffrey *Noel* Hy 2 DC (R); William *Nowel* 1248 FFHu. OFr *noël* 'Christmas', a name given, like the English *Christmas* and *Midwinter*, to one born at that festival.

**Nogg, Nogge:** *Nogga* (f) 1162 DC (L); William *Nog* 1221 AssWo; Geoffrey *Nogge* 1295–6 IpmY; William *Nogge* 1327 SRC. A shortened form of OG *Norigaud, Norgaud, Northgaud. v.* Forssner 193.

**Nogood:** Robert *Nagod* Hy1 ELPN. 'No good', OE *na, god. cf.* Godwin *Nalad* c1110 Winton (Ha) 'not harmful'.

**Noice, Noise:** *v.* NOY

**Noke, Nokes:** *v.* OAK

**Nolda, Nolder:** *v.* ALDER

**Nopps, Nops:** *v.* NOBBS

**Norbrook:** *v.* NORTHBROOK

**Norburn, Norbron:** Thomas *de Nortburne* 1275 RH (K). 'Dweller by the north brook.'

**Norbury:** Thomas *de Northbir'* 1221 AssSa; John *Norbury* 1401 KB (Lo); Henry *Norbury* 1454–5 FFWa. From Norbury (Ch, Db, Sa, Sr, St).

**Norby:** *v.* NORTHBY

**Norchard:** *v.* ORCHARD

**Norcliffe, Norclyffe:** *v.* NORTHCLIFFE

**Norcombe:** *v.* NORTHCOMBE

**Norcott:** *v.* NORTHCOTE

**Norcroft:** John *Norcroft* 1672 HTY. From Norcroft in Cawthorne, in Hawksworth (WRY), or 'dweller by the north field', OE *norþ, croft.*

**Nordaby:** Walter *Northinbi*, Robert *Northiby* 1297, 1301 SRY; John *Northobe* 1535 GildY; Nicholas *Northaby* 1617 FrY. 'Dweller to the north of the village', ON *norðr i bý. cf.* WESTOBY.

**Nordan, Norden, Nordon, Nording:** Roger *de Norden'* 1198 P (K); John *de Northdene* 1317 AssK; John *Nording* 1662–4 HTDo. From Norton Green in Stockbury (K), *Northdene* 1258, or Norden in West Alvington (D).

**Norem:** *v.* NORTHAM

**Norfolk:** William *de Norfolc* 1154–76 YCh; Robert *de Norfolk'* 1228 Cur (C); John *Northfolk, Norfolk* 1377 AssEss, 1380 AssLo. 'The man from Norfolk.'

**Norgate, Norgett:** Siward *de Nordgat'* 1198 P (K); Gilbert *de Northgate* 1239 Rams (Nf); Trikke *del Nortgate* 1277 Wak (Y); Elvina *de Norgate* 13th Rams (Nf). From residence near the north gate of a town or castle.

**Norgrove:** Thomas *de Norgrave* 1311 Rams (Hu). One who lived near the north grove.

**Norham:** Henry *de Norham* c1218–22 (Brechin), John *de Norham* 1269 (St Andrews), Thomas *de Norham, de Noram* 1329 Black. From Norham (Nb).

**Noridge:** *v.* NORRIDGE

**Norkett, Norkutt:** *v.* NORTHCOTE

**Norland:** John *de Northland* 1296 Wak (Y); John *de Northland* 1305 AssW; Richard *Northland* 1387 PN Wt 130. From Norland (WRY), Norlands in Freshwater (Wt), or 'dweller at the north land', OE *norþ, land.*

**Norley:** Cristiana *de Norleg'* 1204 P (Sa); Honde *de Northle* 1288 AssCh; Thurstan *de Norleigh* 1328 WhC. From Norley (Ch), Norley Fm in Wonersh (Sr), Norley in Calne (W), Northleigh in Goodleigh (D), or 'dweller at the north clearing', OE *norþ, lēah.*

**Norman, Normand:** (i) *Norman, Normannus* 1066 DB; *Norðman* 1066–70 Bury (Sf); *Norman* c1113 Burton (Staffs); *Normannus* 1230 P (Ha); Reginaldus *filius Normandi* 1220 Cur (Ess); Hugo, William *Norman* 1171 P (W), 1185 Templars (Herts); Robert *Norþman* 1279 RH (O); William *Northeman* 1301 SRY. OE *Norðmann* 'dweller in the North, Scandinavian, especially a Norwegian', recorded as a personal name from the second half of the 10th century and fairly common in 1066. (ii) John *Normand* c1216 Calv (Y); John *le Norman*, Nicholas *le Normand* 1221 AssWa; Alexander *le Normaunt* 1273 RH (L). OFr *Normand, Normant* 'a Norman'.

**Normandy:** Robert *Normandie* 1369 FFEss; Thomas *Normandy* 1421 PN Ess 383. 'The man from Normandy'.

**Normanville:** Emma *de Normanuill'* 1195 P (Sx); Isabel *Normafeld* 1535 RochW; John *Normavell* 1571 YWills. From Normanville (Seine-Inférieure).

**Norracott:** Christopher *Norracott* 1642 PrD. From one or other of the six Northcotts in Devon.

**Norridge, Noridge:** This may be identical with NORTHRIDGE. cf. Norridge in Upton Scudamore (Wilts), *Northrigge, Norrigge* 1203 (PN W 156). Or it may be a phonetic rendering of NORWICH.

**Norrie, Norry:** v. NORTHEY

**Norrington:** John *Noryngton* 1523 SRK. From Norringtonend Fm in Redbourn, the home of Alice *de Northington* in 1296 (PN Herts 79), or Northingtown Fm in Grimley (PN Wo 128), where lived Robert *de Norinton'* in 1275 (SRWo), or from Norrington in Alvediston (Wilts), *Northintone* 1227 PN W 199, all originally *norð in tūne* 'to the north of the village'. cf. NORTHINGTON and v. SINTON. This may also be a dialectal pronunciation of *Northampton*: cf. Henry *de Norhantona* 1165–71 Colch (Ess); Peter *Norampton* 1352 ColchCt and Northington (Hants), *Northametone* 903, *Norhameton* 1167.

**Norris, Norriss, Norreys:** (i) Robert *norreis* 1148 Winton (Ha); William *le Norreis* 1163–1200 Seals (Gl); Robert *le Norais* c1170 Riev (Y). AFr *noreis, norreis* 'northerner'. A very common name, particularly in the midlands and the south. Used also as a personal name: *Norreis* 1180 P (Y), *Noreis* de Blida c1200 DC (Nt). (ii) Robert *le Noris* 1297 SRY; Alice *la Norisse*, Agnes *le Norice* 1310, 1337 ColchCt; John *Norice* 1317 AssK. OFr *norrice* 'nurse'. Less common than (i). (iii) Adam *de Northus, de Norhuse* 1206 Cur (Ess). 'Dweller at the north house.'

**Norrish:** Rosa *Noryssh* 1459 GildY; Robert *Nores, Norish, Norysche* 1509 Oriel (O); Raff *Norysche*, John *Noresse* 1555, 1558 RothwellPR (Y). A late development of NORRIS. cf. PARISH.

**Norrison:** Philip *le Noriscun*, John *le Norisoun* 13th AD ii (Wa); John *Norrison* 1672 FrY. OFr *norriscun* 'nursling'.

**Norsworthy:** v. NOSWORTHY

**North:** Aylmar *del North* 1230 P (Sf); John *de North* 1257 FFC; William *North* 1296 SRSx; Agnes

*Bynorth* 1301 ParlR (Ess). 'Man from the north', or 'dweller to the north'.

**Northall:** William *de la Northalle* 1280 MELS (So). 'Dweller at the north hall.'

**Northam, Norem:** William *de Northam* 1214 Cur (Wa). From Northam (Hants, Devon).

**Northampton:** Henry *de Norhantona* 1165–71 Colch (Ess); Peter *Norampton* 1352 ColchCt; John *Northampton* 1524 SRD. From Northampton (Nth).

**Northard:** Probably 'dweller to the northward'. cf. John *Anorthward* 1275 RH (Berks), which may also have become *Northwood.*

**Northbrook, Norbrook:** Hugo *de Nordebroc* 1190 P (K); Gilbert *de Norbroc* 1205 P (Wa); Philip *de Northebrok* 1327 SRSx. From a place called Northbrook or 'dweller north of the brook'.

**Northby, Norby:** William *de Northby* 1327 SRY. From Norby in Thirsk (NRY).

**Northcliffe, Nortcliffe, Norcliffe, Norclyffe:** Henry *de Northclyf, del Northeclif* 1307, 1309 Wak (Y); John *de Norclif* 1327 SRC. 'Dweller by the north cliff.'

**Northcombe, Norcombe:** Stephen *de Northcumb'* 1230 P (D). From Northcombe in Bovey Tracey or in Bratton Clovelly (Devon).

**Northcote, Northcott, Norcott, Norkett, Norkutt, Notcutt:** Nicholas *de Northicote* 1199 AssSt; William *de Nordcote* 1205 P (Gl); John *atte* Northcote 1296 SRSx. From one of the places named Northcote or Northcott in Devon, from Norcott in Northchurch (Herts), or from residence at the cottage to the north. cf. NARRACOTT.

**Northeast:** Jonathan *Northeast* 1642 PrD; Robert *Northeast* 1662–4 HTDo. 'Dweller to the north-east', OE *norþ, ēast.* But cf. Northease Fm in Rodmell (Sx) 'the north brushwood land', OE *\*hǣs.*

**Northend:** Peter *de Northende* 1279 RH (Beds); Thomas *del Northend* 1307 Wak (Y). 'The man from the north end (of the village).'

**Northern, Northen:** William *le Northerne* 1252 Rams (Hu); Richard *le Northryn, le Northren* 1317 Wak (Y); William *Northene* 1327 SRSf. OE *norþern* 'the man from the north'.

**Northey, Norey, Norrie, Norry:** Thomas *de Northie* 1200 P (K); John *de Norhie* 1205 Cur (Sx). From various minor places, compounds of *north* and varied suffixes: (*ge*)*hæg* 'enclosure': Northey Wood in Ugley, *Norhey* (1345 PN Ess 554), Northay in Hawkchurch, John *de Northeheye* (1280 PN D 656), Northey Wood in Anstey (PN Herts 171); *ēa* 'river': Northey in Whittlesey, near the Cat's Water, (*Northee* 1280 PN C 261); *ēg* 'island': Northey in Bexhill (PN Sx 493), Northey in Bury (Hunts), *Northeya* (c1350 PN BedsHu 208); *hōh* 'ridge': Northey Wood in Shudy Camps (Cambs), *Norro* 13th, *Northou* 1260 (PN C 105), Northey in Turvey (Beds), *Northho* 1242 (PN BedsHu 49).

**Northfield:** Hugh *de Northfeld* 1275 SRWo; Thomas *de Northfeld* 1323 AssW; Thomas *de Nortfelde* 1327 SRWo. From Northfield (Wo), or 'dweller at the north field', OE *norþ, feld.*

**Northing:** Adam *Northynne*, Simon *de Northinne* 1327, Simon *Northynne* 1332 SRSx. From North End in Hamsey (Sx).

**Northington:** William *de Northentone* 1327 SRSf. *v.* NORRINGTON and SINTON.

**Northoe, Northow:** John *Northehawe* 1392 LoCh. 'Dweller by the north enclosure', OE *norþ, haga*.

**Northop:**. John, Thomas *Northupp(e)* 1296 SRSx. Probably 'Dweller north up (in the village).'

**Northover:** Richard *Northover* 1189 Sol; Agnes *Northover* 1662–4 HTDo. From Northover (So), or 'dweller on the north bank', OE *norþ, ōfer*.

**Northow:** *v.* NORTHOE

**Northridge:** William *de Northerugg'* 1332 SRSx. 'Dweller on the north ridge.'

**Northrop, Northrope, Northrupp:** Reginald *de Northorp'* 1219 AssY. From Northorpe (ERYorks).

**Northway, Norway:** Richard *de Northweye* 1275 SRWo; Geoffrey *Bynorthweye* 1280 MELS (So). 'Dweller to the north of the road.' William *Bynorthewey* lived at Northway in Widdecombe in the Moor in 1330. At Whitestone (Devon) in 1344, Roger *Bynorthewaye* lived at Norway and Ralph *Bysouthweye* at Southway. They owed their names to the fact that they lived respectively to the north and south of the main road which separates the farms (PN D 529, 457). cf. NARRAWAY.

**Northwood, Norwood:** Painot *de Norwude* 1176 P (D); Alexander *de Nordwuda* 1190 P (K); Geoffrey *Northwud'* 1205 Cur (Nf). From one of the places named Northwood or Norwood 'north wood', or from residence to the north of a wood. William and David *Bynorthewode* (1330, 1333) lived north of woods now represented by Northwood Fm in Okehampton and Morchard Bishop respectively (PN D 205, 409).

**Norton:** Osuuardus *de Nordtone* 1066 DB (K); Leofwin *de Norton'* 1177 P (L). From one of the numerous Nortons. It may also derive from some place called *norð in tūne* (place) 'to the north of the village': Walter *de Northinton'* 1275 SRWo came from Norton Fm in Suckley (Worcs) where the corresponding place to the south of the village is Sindon's Mill. *v.* SINTON. This type was very common in Sussex: William *Anortheton*, Margery *atte Northetone* and John *atte Northon* (both in Combes), Walter *de Northeton* 1296 SRSx. *v.* also NORRINGTON and NORTHINGTON.

**Norwell:** Henry *de Northewelle* 1296 SRSx. 'Dweller by the north spring or stream.'

**Norwich:** Goscelinus *de Norwic* 1086 DB (Nf); Richard *de Northwyco* 1296 SRSx. From Norwich (Norfolk) or 'dweller at the north dairy-farm'.

**Norwold:** John *de Norwold* 1379 LoCh. From Northwold (Nf).

**Norwood:** *v.* NORTHARD, NORTHWOOD

**Nose:** Adam *cum Naso* 1275 MPleas (Hu); John *Nose* 1332 SRCu. A nickname from a peculiarity of the nose! cf. Thomas *Nosbende* 1525 SRSx 'bent nose'.

**Noss:** *v.* NESS

**Nosworthy, Noseworthy, Norsworthy:** John *Nosworthy*, Robert *Noseworthy* 1642 PrD. From Norseworthy in Walkhampton (D).

**Notcutt:** *v.* NORTHCOTE

**Nothard, Nutter:** Nicholas *le noutehird* 1296 FrY; Henry *le Nauthird* 1308 Wak (Y); Margaret *Nutter*

1562 RothwellPR (Y). ON *naut* 'beast, ox' and OE *hierde* 'herd', cowherd.

**Nother(s):** *v.* NODDER

**Notman:** *v.* NUTMAN

**Nott:** Roger *Not* 1100–30 OEByn; Algar *le Notte* 1183 P (So); Henry *le Not* 1210 Cur (St). OE *hnott* 'bald-headed, close-cropped'.

**Nottage, Nottidge:** Thomas *Nuthech* 1220 Fees (Berks); Alan *Nuthach* 1224 Cur (Ess). A nickname from the nuthatch.

**Notting:** Siward *Noting* 1219 AssL; Adam *Nottyng* 1327 SRSf. OE *\*hnotting* 'the bald-headed one'.

**Nottingham:** Thomas *de Notingeham* 1169 P (Db/Nt); William *de Notingeham* 1257–8 FFL; Hugh *de Notingham* 1327 SRLei; Robert *Notyngham* 1427 AssLo. From Nottingham (Nt).

**Notton:** Gilbert *de Noton'* 1207 Cur (Y); William *de Notton* 1351 FFY. From Notton (WRY).

**Nought:** William *Noght* 1327 SRSa; Richard *Nouht* 1334 SRK; Matilda *Naght* 1363 AssY. A nickname from OE *nāht* 'nothing'.

**Noven:** Robert *atte Novene* 1276 AssW; Walter *atte Nouene* 1327 SRWo; William *atte Noven* 1423 LLB K. 'Dweller at the furnace', ME *atten oven* becoming *atte noven*.

**Nourse:** *v.* NURSE

**Nowell, Nowill:** *v.* NEWELL, NOEL

**Nowers:** Roger *de Nuiers* 1199 Fees (Nf); Alexander *de Nowers* 1230 P (Wa). From Noyers (Eure). *v.* ANF, NOAR, OVER, OWER.

**Nowles:** *v.* KNOLL

**Noy, Noyce, Noyes, Noe, Noice, Noise:** *Noe* c1125 StCh; Thomas *filius Noe* 1185 Templars (Wa); William, Simon *Noysse* 1327 SRSf. Hebrew *Noah* 'long lived'. Not a common medieval name. No doubt also from the medieval drama.

**Nugent:** Philip *de Nugent* 1203 Pl (Sa); Philip *Nogent* 13th Ronton; Richard *de Nugent* 1293 AssSt; John *Nogent* 1340 CorLo. From one or other of the many French places called Nogent.

**Nunhouse, Nunniss:** William *Nunhouse* 1379 PTY. 'Worker at the nuns' house', servant of the nuns. cf. MONKHOUSE.

**Nunley:** Helias *de Nuneleg'* 1221 AssWa. From Nunley Fm in Wroxhall (Wa).

**Nunman:** Nicholas *le Nunneman* 1212, Robert *Nunman* 1242 Fees (L); William *le Nounneman* 1333 Riev. 'Servant of the nuns', OE *nunne, mann*. cf. Richard *le Nunnefrere* 1260 IpmY 'brother of the nun'.

**Nunn, Nunns:** (i) Eluiua *nonna* 1155–66 Holme (Nf); Alice *Nunne* 1243 AssDu, *la Nonne* 1275 RH (Nth); Robert, Hannecok *le Nunne* 1252 Rams (Hu), 1297 Wak (Y). Where the christian name is that of a woman, this is a surname of office. Applied to a man (as most commonly), it is a nickname for one meek and demure as a nun. (ii) Robert *del Nunnes*, *othe Nonnes* 1297 SRY, 1309 LLB D; John *atte Nunnes* 1325 Cl. 'Servant at the nuns' ', identical in meaning with NUNHOUSE and *Nunman*: Walter *Noneman* 1217 Pat (Y). Roulf, predecessor of Simon de Kime, gave 60 acres in Lincs to the nunnery of Elstow (Beds) which Nicholas *le Nunneman* held by service of 15s. per annum (1212 Fees).

**Nunney, Nunny:** Robert, Beatrice de Nonhey 1327 SRSf. From residence near an enclosure belonging to the nuns.

**Nunniss:** v. NUNHOUSE

**Nunweek, Nunwick:** Geoffrey de Nunnewich 1193 P (Y). From Nunwick (Northumb).

**Nurrish:** v. NURSE

**Nursaw, Nurser, Nursey:** v. NUSSEY

**Nurse, Nurrish, Nourse:** Joan Nurys, Magota le Nuris 1379 PTY. OFr nurice 'nurse'. v. NORRIS, NORRISH.

**Nurser, Nussey:** v. NURSAW

**Nussey, Nussie, Nursey, Nursaw, Nurser:** Robert de Nussey, John de Nussay 1379 PTY; Richard Nussey 1605 FrY; William Nursay, Thomas Nursaw 1707, 1747 ib. From Nussey Green in Burnsall (WRYorks).

**Nutbeam, Nutbeem, Nutbeen:** John ate Notebem 1279 RH (O); William atte Nhutbyme 1332 SRSx. 'Dweller near a prominent nut-tree or hazel' (OE hnutbēam). John atte Notebeme 1327 SR (Ha) came from Nutbane in Weyhill (Hants).

**Nutbourn, Nutbourne:** Ida de Nuthburn 1296, John Notbourne 1327 SRSx; Robert Notebourne 1377 FFSr. From Nutbourne (Sx).

**Nutbrown:** William Nuttebrun 1185 Templars (Y); Willelmus filius Nutebrun 1203 P (L); Richard Nutebrune 1296 SRSx. 'Nut-brown', a nickname from the complexion, used also as a personal name.

**Nutcomb, Nutcombe:** Thomas Nottecomb 1382 Hylle; Richard Nutcomb 1524 SRD; Nicholas Nuttecombe 1642 PrD. From one or other of the four Nutcombes in Devon.

**Nutkin, Nutkins:** Notekyn (f) 1305 SIA iii; Adam Notekyn 1274 RH (Ess); Thomas Notekyn 1327 SRC; Miles Nutkyn 1544 CorNt. Not-kin, a diminutive of NOTT.

**Nutley:** Adam de Nuteleg 1203–4 FFEss; John del Nutle 1251 AssY; John de Nutlye 1332 SRSx. From Nutley (Ha, Sx).

**Nutman, Nuttman, Notman:** John Noteman, William Nuteman 1275 RH (Nf, L). 'Servant of Nott' or 'dealer in nuts'.

**Nutt:** William Nutte 1181 P (Nth); Richard le Nute 1274 RH (K); John Nutte, Nutt 1379 PTY, 1523 CorNt. OE hnutu 'nut', a man with a round head or a brown complexion.

**Nuttall, Nuthall:** Richard de Nutehal' 1201 Pl (Nt); Stephen de Notehale 1269 RegAntiquiss; John Notehal 1354 ColchCt; Peter Nutill 1375 FFY. From Nuthall (Notts), or Nuttall (Lancs).

**Nutter:** Robert le Notere 1221 AssWa; Adam le Notyere 1293 AssSt; Dyonisia Notur 1302 Oseney (O). OE nōtere 'scribe, writer' or a derivative of OFr note 'note', 'secretary'. v. also NOTHARD.

**Nye, Nie, Nay, Ney:** Robert Nay 1207 FFEss; Robert Atteneye 1269 AssSo; William Atteneye 1276 AssSo; Gilbert de la Nye 1315 FFEss. From residence near some low-lying land. ME atten ye, eye became atte nye, neye. cf. Nye in Winscombe (Som). Neigh Bridge in Somerford Keynes (Wilts) is from OE ēa 'by the stream'. v. REA, RAY, RYE.

**Nyland, Nylund:** From Nyland (Do, So).

**Nyman:** v. NEWMAN

# O

Oade, Oades, Oaten, Oates, Oats, Ott, Otten,
Otton: *Ode, Odo, Otho* 1066 DB; Radulfus *filius Ode*
c1160 DC (L); Willelmus *filius Ote (Otte)* 1177 P (Ess);
William *Ode* 1213 Cur (Sf); Cristiana *Odes* 1275 RH
(Nth); Andrew *Otes* 1275 RH (Nf). OG *Odo, Otto*
'riches', OFr *Odes, Otes* (cas-sujet), *Odon, Otton* (cas-
régime). Both survive, though the latter is seldom
recorded. Belchamp Otton (Essex) owes its attribute
(*Otes* 1254, *Oton* 1255) to *Otto* (temp. Hy 2),
descendant of *Otto* or *Odo* (DB). *Oade* may also be
local in origin. *v.* NOAD.

Oak, Oake, Oaks, Oke, Noak, Noake, Noakes,
Noaks, Nock, Noke, Nokes: Henry *atte Noke* Hy 3
PN Ess 140; Adam *at þe Ock* 1273 RH (Sa); Thomas
*del Oke* 1275 RH (Bk); Thomas *atten Oke* 1296 SRSx;
Henry *atte Nok* 1326 PN Ess 145; John *atte Noke* 1327
SRSx; Robert *atte Nokes* 1332 SRWo; Richard *en le
Okes* 1383 StThomas (St). From residence by an oak
or a group of oaks (OE *āc*), with variation between
*atten oke(s)* and *atte noke(s)*. *v.* also ROKE.

Oakden, Ogden: Elias *de Akeden, de Aggeden* 1246
AssLa; Richard *de Okeden* 1332 SRLa; William *Ogdin*
1612 FrLeic. From Ogden (Lancs).

Oakenroyd, Ockenroyd: Hugh *del Okenrode* 1323
LaCt. From Oaken Royd in Norland (WRY), or
'dweller at the clearing in the oaks', OE *āc, rod.*

Oaker: Henry *Oker* 1275 RH (Sf); Walter *Oaker*
1279 RH (C). 'Dweller by the oak.' cf. OAK.

Oakey: *v.* OKEY

Oakham, Ockham: Robert *de Ocham* 1327 SRLei;
Nicholas *de Ocham* 1327 SRSx; Thomas *Ockham*
1340–1450 GildC. From Oakham (R), Ockham (Sr),
or Ockham House in Ewhurst (Sx).

Oakhurst: Simon *de Okhurste* 1283–4 FFSx; John
*Okhurst* 1395 PN Sx 106. From Oakhurst (Herts), or
Oakhurst in Kirdford (Sx).

Oakley, Oakeley, Okeley, Okely: Hervey, Philip *de
Ocle* 1199 AssSt, 1246 IpmGl; Richard *de Okeley* 1327
SRLei; John *Okelee* 1377 AssEss; Robert *Okeleye*
1545 SRW. From one or other of the many places of
this name, or from Oakle Street (Glos), Oakleigh
(Kent), or Ockley (Surrey).

Oakman: Alexander *filius Okeman* 1219 AssY;
Robert *Okman* 1296 SRSx. OE *Ācmann* 'oakman'.

Oastler: *v.* OSTLER

Oaten, Oat(e)s: *v.* OADE

Oatland, Oteland: Robert *atte Otlond* 1290,
William *atte Otlonde* 1294 PN Sr 99; Adam *aten
Otlond* 1296 MELS (Sx). From Oatlands Park in

Weybridge (Sr), or 'dweller by the land where oats are
grown', OE *āt, land.*

Oatway: *v.* OTTOWAY

Obin: *v.* ALBIN

Obray: *v.* AUBRAY

O'Brian, O'Brien, O'Bryan, O'Bryen: Irish *Ó
Briain* 'descendant of *Brian*'.

O'Byrne: *v.* BYRNE

O'Clery: Ir *Ó Cléirigh* 'descendant of the clerk'. cf.
MCCHLERY.

Ockenden, Ockendon, Okenden: Ailric *de
Wochendone* 1170 P (Ess); William *de Wokenden'* 1204
Cur; William *de Wokindone* 1206 FFEss. From
Ockendon (Ess), *Wokindon* 1230.

Ockenroyd: *v.* OAKENROYD

Ockham: *v.* OAKHAM

Ockilshaw: John *Okylschagh* 1401 AssLa. From
Ockilshaw in Wigan (La).

O'Connell: Ir *Ó Conaill* 'descendant of *Conall*',
Celtic *Kunovalos* 'high-powerful'.

O'Connor: Ir *Ó Conchobhair* 'descendant of
*Conchobhar*' 'high-wish'.

Octon: Odelina *de Hoketon'* 1219 AssY. From
Octon (ERY).

Odam, Odams, Odom, Odhams: Walter, Robert
*Odam* 1313 Wak (Y), 1327 SRSf; John *Odames* 1327
SRSt. OE *āðum*, ME *odam* 'son-in-law'.

Odart: *Odard* 1086 DB; Eilsi *filius Odardi* 1188 P
(Y); Peter, Walter *Odard'* 1206 ClR. 1332 SRCu. OG
*Authard, Othard, Odard* 'riches-hard'.

Odber, Odbert: *Otbertus* (Y), *Outbert* (L) 1066,
*Odbertus* (Sr) 1086 DB; *Odbertus* de Castra 1134–40
Holme (Nf); Margery *Odbert* 1341 FS; Richard *Odber*
1641 PrSo. OG *Odbert, Otbert.*

Odd, Odde: Asketinus *filius Od* 1163 DC (L); Odd
*atte Flete* 1327 SRSx; Gilbert, Richard *Odde* 1225
AssSo, 1284 Wak (Y). OE *Odda* or ON *Oddr.*

Oddie, Oddy, Oddy: John *Ody* c1280 SRWo, 1390
CorW; Robert *Odye* 1576 SRW; William *Oddy* 1657
FrY. A diminutive of ODD.

Oddson, Odson: *Odesune* 1086 InqEl; Alfuuin
*Odesone* 1086 ICC; Adam *Odesone* 1342 FFY; John
*Odson* 1379 PTY. cf. Alice *Odsonwyf* 1379 PTY. 'Son
of Odd', OE *Odda* or ON *Oddr.*

Odell, O'Dell: Simon *de Wahella* 1195 P (Beds);
Simon *de Wahull'* 1212 Pl (K); Robert *de Wodhull'*
1314–16 AssNth; John *Odyll, Odell* 1545, 1576 SRW.
From Odel (Beds), *Wahelle* 1162, *Woodhull* 1276,
*Odyll* 1494.

Oden, Othen: *Oudon* 1066 DB (L); *Odin Goldeberd*

1327, *Othynus* Guldebort 1332 SRSx; Henry *Otheyn* 1275 SRWo; John *Otheyn* 1327 *SR*Ess; Thomas *Oden* 1332 SRSx. Anglo-Scandinavian *Oðin*, ON *Auðun*, ODa *Øthin*, OSw *Odhin*. *v.* PNDB 342.

**Odgear, Odger, Odgers, Oger, Ogier:** *Ogerus* Brito 1086 DB (Lei); *Ogerus* filius *Ogeri* 1196 Cur (D); *Odgarus* de Lind' 1214 Cur (Sr); Nicholas *Oger* 1296 SRSx; Peter *Ogger* 1306 LoCt. OFr *Ogier*, OG *Odger, Og(g)er* 'wealth-spear'.

**Odhams:** *v.* ODAM

**Odierne:** Robert *filius Odiern'* 1197 P (D); *Odierna* 1213 Cur (Berks); *Hodierna* 1286 Glast (So); Henry *Odierne* 1273 IpmGl; John *Odyerne* 1326 CorLo; William *Odierne* 1484 Sibton. OG *\*Audigerna* (f), OFr *Odierne, Hodierne* (f). *v.* Forssner 197.

**Odiham:** Alwin *de Hodiam* c1110 Winton (Ha); Nicholas *Odyham* 1380–1 PTW; Richard *Odyham* 1392 LoCh. From Odiham (Ha).

**Odinell, Odneld:** *Odinellus* de Vnfranvilla 1187 P (Nb); Geoffrey *Odinel* 1276 RH (Y); Alexander *Odneld* 1327 SRY. *Od-in-el*, a double diminutive of Odo. *v.* OADE.

**Odlin, Odling:** *Odolina* 1086 DB; *Odelina* 1198 FFSf, 1207 Cur (Do); *Hodlyn* 1315 Wak (Y); Robert *Odelyn* 1250 Fees (Nf); Henry *Odelin* 1276 RH (Y). OG *Odelin* (m) or *Odelina* (f), double diminutives of *Odo*. *v.* OADE.

**Odom:** *v.* ODAM

**O'Donnell:** Ir *Ó Domhnaill* 'descendant of *Dhomnall*'. *v.* DONALD, MACDONALD.

**O'Donoghue, O'Donohue, Donohue, Donaghy, Donahue:** Ir *Ó Donnchadha* 'descendant of *Donnchadh*'. *v.* DUNCAN.

**O'Donovan:** Ir *Ó Donnabháin* 'descendant of *Donndubhán*' (dark brown).

**O'Dowd:** Ir *Ó Dubhda* 'descendant of *Dubhda*' (black).

**Odson:** *v.* ODDSON

**O'Duffy, Duffy:** Ir *Ó Dubhthaigh* 'descendant of *Dubhthach*'.

**Ofchurch, Offchurch:** Adam *Ofthechuche* 1293 KB (Sf); Elizabeth *Offechurch* 1340–1450 GildC; William *Offechurch* 1376–7 FFWa. From Offchurch (Wa).

**Offer, Orfeur:** (i) William *le Orfere* 1265 Pat; William atte more, *orfrer* 1292 SRLo; John *Ourefre, Ourefyre* 1327 SRSf. *Orfrer* is found as an occupational name in London and probably meant 'a maker of orphrey or gold-embroidery', OFr *orfreis*, MedLat *aufrisium*, a modification of Lat *auriphrygium*, cf. OFFICER. (ii) William *le Orfeuere* 1235 FFHu; Thomas *le Orfevre* a1284 NottsBR; Robert *Lorfeure* 1319 SRLo. OFr *orfevre* 'goldsmith'.

**Officer:** Alicia *aurifrigeria* 13th SRLo; Adam *le Orfreyser* 1302 FrC. Probably a corruption of OFr *orfroisier* 'a maker of orphrey or gold-embroidery'. *v.* OFFER.

**Official:** Thomas *Officialis* 1198 P (Cu); Ralph *Officialis* 1230 P (Nth); John *le Official* 1327 SRDb; John *Official* 1473 NorwW (Nf). Lat *officialis* 'an official of some kind'.

**Offield, Ofield:** *v.* OLDFIELD

**Offord:** Robert *de Offewurth'* 1221 AssWa; Robert *de Offorde* 1327 SRSf; Thomas *de Offord* 1374 FFEss. From Offord (Hunts, Warwicks).

**Offley, Ofley:** Ranulf *de Offele* 1207 Cur (Herts); Richard *de Offeleye* 1275 SRWo; Richard *de Offiley* 1332 SRSt. From Great, Little Offley (Herts), or Bishops, High Offley (St).

**O'Flaherty:** Ir *Ó Flaithbheartaigh* 'descendant of *Flaithbheartach*' (bright ruler).

**O'Flinn, O'Flynn:** Ir *O Floinn* 'descendant of *Flánn*' (red).

**Ogden:** *v.* OAKDEN

**Oger:** *v.* ODGEAR

**Ogg, Ogge:** (i) *Ogge* filius Adam 1199 Pleas (St); Ralph *filius Ogg* 13th FeuDu; William *Og* 1369 LLB G; John *Ogh* 1642 PrD. OE *Oga, Ogga*. (ii) Donald *Oge* 1457 (Aberdeen), Finla *MacJames Uig* 1613 (Abernethy) Black. Gaelic *og* 'young'.

**Ogilvie, Ogilvy, Ogilwy:** Alexander *de Oggoluin* a1232 Black; Patrick *de Oggiluill* c1267 ib.; Walter *de Ogilby* 1425 ib.; Thomas *of Oglevy* 1466 ib. From the barony of Ogilvie in Glamis (Angus).

**Ogle, Ogles:** Robert *de Ogle* 1181 P (Nb); Agnes, Gilbert *Dogel* 1221 Cur (Nb); Robert *Ogill'* 1379 PTY; Richard *Ogle* 1519 FFEss. From Ogle in Whalton (Northumb).

**Oglethorpe:** Hemer *de Oclestorp'* 1219 AssY; William *Ogylsthorpe* 1407 IpmY; Maria *Oglethorpe* 1621 SRY. From Oglethorpe (WRYorks).

**Ogley:** Robert *de Oggele* 1275 RH (Lo); Richard *de Oggele* 1326 CorLo. From Ogley Hay (St). But these could, perhaps, be forms of OGLE.

**O'Gorman:** *v.* GORMAN

**O'Grady:** Ir *Ó Grádaigh* 'descendant of *Gráda*' (noble).

**O'Hagan:** James *Ohagan* 1280 Pat. Ir *Ó hÁgáin* 'descendant of *Ógán*' (young).

**O'Haire, O'Hare, O'Hear:** *v.* HAIR

**O'Hea, Hay, Hayes:** Ir *Ó hAodha* 'descendant of *Aodh*' (fire).

**Oildebof:** Richard *Oildeboef* 1219 P (Sx); Hugh *Oyldebuf* 1286–7 FFEss; William *Oyldeboef* 1355–9 AssBeds. 'Ox eye', Fr *oeuil de boeuf*. cf. William *Oildefer* 1210 P (Lei) 'iron eye'; Alan *Oil de Larrun* 1183 P (Lo) 'eye of a thief'; Adam *Oildeleure* 1205 Pleas (La) 'eye of a hare'.

**Oillard, Oyler:** Hugh *filius Oillardi* 1210 Cur (Nth); Hugh *Oillardus* c1110 Winton (Ha); Hugh *Oillardi* 1130 P (Sr). OG *Odilard, Oilard*.

**O'Kane:** *v.* CAIN

**Oke:** *v.* OAK

**Okeley, Okely:** *v.* OAKLEY

**O'Kelly:** *v.* KELLY

**Okenden:** *v.* OCKENDEN

**Okeover, Okover:** Hugh *de Acoure* 1198 P (St); John *de Okovere* 1332 SRSt; Benet *de Okovere* 1362–3 IpmNt. From Okeover (St).

**Okey, Oakey:** (i) *Aki* (Sf), *Achi* (L, Sf, Wa) 1066 DB; *Oky, Hoky* Ed 3 Rydware (St); Henry *Aky, Oky* 1221 AssGl; William, Elyas *Oky* c1250 Rams (Hu), 1327 SRSf. ODa *Aki*, OSw *Ake*. (ii) Cecilia *de Okeye* 1275 SRWo; Robert *de Okhei* 1327 SRWo. 'Dweller near an oak-copse' (OE *āc* and (*ge*)*hæg*).

**Olarenshaw:** *v.* OLLEREARNSHAW

**Old, Olds, Ould, Oulds:** Boia *þe Ealde* c980 OEByn; Willelmus *Vetus* 1183 P (Lo); Gerardus

*Senex* 1193 P (Berks); Henry *le Olde*, Roger *Old* 1327 SRSf. OE *eald* 'old' (Lat *vetus, senex*), not necessarily implying old age. Wulfstan *Ealda* was so called to distinguish him from Wulfstan *Geonga* 'the young' c1060 OEByn (K).

**Oldacre, Oldaker:** Helyas *de Aldeacris* 1231 Guisb (Y). One who lived by the old ploughed field (OE *æcer*). cf. OLDFIELD.

**Oldale, Oldall:** Robert *del Oldehale* 1275 MELS (Wo); John *atte Oldehalle* 1356 FFEss; William *Oldhalle* 1431 Paston. From Oldhall Fm in Rock (Wo), one or other of the many Old Halls in Essex, or 'dweller near the old hall', OE *eald, heall*.

**Oldbury:** Jacobus *de Oldebur'* 1275 SRWo; John *Oldebury* 1334 IpmW; William *Oldebury* 1402 IpmGl. From Oldbury (K, Sa, Wa, Wo), Oldbury on the Hill, upon Severn (Gl), or Oldbury Castle in Cherhill (W).

**Oldcastle:** Thomas *Oldecastell* 1389 IpmGl; John *Oldcastell* 1407 FFSr. From Oldcastle (Ch).

**Oldcorn:** Godwin *Oldcorn* 1214 P (R); Matilda *Eldcorn* 1219 Cur (Herts); John *Oldcorn* 1327 SRSf; John *Holdcorn* 1520 GildY; John *Aldcorne* 1533, *Awdcorn* 1540 FrY. This looks like an unrecorded personal name, OE *\*Ealdcorn, \*Aldcorn*, but it could also be a nickname, though the meaning would be obscure.

**Olden:** *v.* ALDEN

**Older:** *v.* ALDER

**Oldershaw, Houldershaw:** Hugh *de Alreschawe* 1307 AssSt. 'Dweller by the alder-wood', ME *aldershawe*, OE *alor, sceaga*.

**Oldfield, Offield, Ofield, Allfield:** Agnes *de Aldefeld* 1221 ElyA (Sf); Robert *de le Aldefeld* 1279 RH (C); Adam *del Oldefeld'* 1297 SRY; Salkyn *Eldefeld* 1327 SR (Ess). 'Dweller by the old field', OE *eald, feld*, as at Allfield (Salop), Alfell Fm in Elmstead (Essex), Ofields (Warwicks) and Oldfield Grove in Langley (Essex). In Cambridgeshire, this has become The Offal (Comberton), Offal End (Haslingfield) and Offals Wood (Steeple Morden).

**Oldgate:** Richard *de Oldegate* 1275 SRWo. From Oldyates Fm in Abberley (Wo).

**Oldham, Oldam:** Achard *de Aldeham* 1218–19 FFK; Richard *de Oldham* 1384 IpmLa; Robert *Oldum* 1470 Cl (Lo); Ralph *Oldham, Oldam* 1508, 1514 CorNt; John *Owldam* 1599 SRDb. From Oldham (Lancs), or 'dweller by the long-cultivated river flat'.

**Oldhead:** Robert *Oldheued* 1219 AssL; Roger *Oldheved* c1270, William *Eldheved* c1300 *ERO*. 'Old head', OE *eald/ald, hēafod*. cf. John *Olde Jon* 1380 Misc (C) 'old John'; William *Aldinoch* 1203 P (L) 'old enough'; Mary *Oldmaid* 1720 WStowPR (Ess) 'old maid'; John *le Oldesmyth* 1409 IpmGl 'old smith'.

**Oldherring:** Hacun *Aldharing* 1202 P (Nf); Thomas *Oldhering* 1327 SRSf. 'Old herring', OE *eald, hæring*, a nickname for a fishmonger.

**Oldis:** *v.* ALDIS

**Oldknow:** William *Aldinoch* 1203 P (L). OE *eald, genōh* 'old enough'. cf. GOODENOUGH.

**Oldland, Olland:** Wluard *Oldelond* 1221 ElyA (Nf); Hugh *de la Aldelond', del Heldelonde* 1260–70 MELS (So); William *de Oldlond* 1276 RH (Gl); Henry *de Aldelonde* 1296 SRSx; John *atte Oldelond* 1327 SRSo.

'Dweller at the old estate', as at Oldland (Glos) or Oldland(s) (Sussex).

**Oldman, Olman:** William *Aldeman* 1196–1237 Colch (Ess); Robert *Oldman* 1275 RH (Nf); Marjery, Joan *Oldmanes* 1310 ColchCt. OE *eald* 'old' and *mann*. This has, no doubt, also become ALLMAN.

**Oldmixen:** John *Oldemexen* 1327 SRSo. From Oldmixton (So).

**Oldridge:** *v.* ALDRICH

**Oldroyd:** Adam *de Olderode* 1316 Wak (Y); Robert *Ouldroyde* 1666 RothwellPR. 'Dweller at the old clearing', OE *eald, rod*, Yorks dialectal *royd*.

**Oldwell:** Simon *de Oldewelle* 1237–8 FFEss. 'Dweller by the old stream or spring', OE *eald, wiella*.

**Oldwright:** *v.* ALLRIGHT

**O'Leary, Leary:** Ir *Ó Laoghaire* 'descendant of *Laoghaire*' (calf-keeper).

**Olerenshaw:** *v.* OLLEREARNSHAW

**Oliff(e):** *v.* OLLIFF

**Oliphant, Olivant:** David *Olifard* 1141 Black; Robert *Oliphard* c1148 EngFeud (He); John, Hugh *Olifard* 1179 P (C), 1208 (Wo); William *Olyfat* 1296 Black; William *Holifarth* 1300 ib.; William *Olivant* 1572 LaWills. A difficult name. Derivation from ME *olifant* 'elephant' is impossible. The persistent medial *f* (except for the solitary *terram Olivard* 1186–8 BuryS) dissociates the name from *Oliver*. A compound of ON *Óleifr* and the French suffix *-ard* might be suggested were it not for the fact that the persistent medial *i* of *Oliphant* does not appear in *Olliff* before the 14th century. From OFr *olif* 'olive-branch' (Godefroy) a pejorative *Olifard* may have been formed, a derisive nickname for one who preferred an olive-branch to more martial weapons. The development was *Olyphard* 1249, *Olifaunt* 1317, *Oliphand, Olyfant* 1326 (Black), due to interchange of *r* and *n* in the combination *l*—*r*. *Oliphant* became the common form by association with *olifant* 'elephant'. Sometimes we may have a late surname from a sign: Isabell *del Olyfaunt* 1318 LeicBR.

**Olive, Ollive:** *Olyve* 1202 AD ii (Nth); *Oliva* 1207–8 Cur (O, Wo); Richard *Olive* 1279 RH (O); John *Olyve* 1310 FFEss. Latin *oliva* 'olive'. There were two saints named *Oliva*, one the patroness of olive trees.

**Oliver, Olivier, Olliver:** *Oliverus* 1086 DB; *Oliuerus de Vendouer* 1149–62 DC (L); Jordan *Oliver, Oliuier* 1201 AssSo, 1204 P (Co); Willelmus, Jordanus *Oliueri* 1206 P (Co), 1223 Pat (D); Robert *Olyveyr* 1260 AssC; Ralph *Olifer* 1327 SRWo. The French *Olivier*, recorded in 1011, was at first southern and later became common through the influence of the *Chanson de Roland*. Dauzat takes it as symbolic, from *olivier* 'an olive branch', the emblem of peace. This can hardly be the origin of the name of the peer of Charlemagne which is probably Teutonic. The suggested derivations from ON *Óláfr* and OG *Alfihar* are both difficult phonetically.

**Olland:** *v.* OLDLAND

**Ollerearnshaw, Ollerenshaw, Olarenshaw, Olerenshaw, Olorenshaw:** Adam *del Helerinshagh*, John *del Olrynshagh* 1327 SRDb. 'Dweller by the alder-wood.'

**Ollerton:** William *de Olreton'* 1255 RH (Sa). From Ollerton (Ch, Nt).

**Olley, Ollie:** v. DOYLEY

**Ollier:** v. OYLER

**Olliff, Oliff:** *Olefus (Olafus)* 1221 Cur (Sa); Robert *Olef* 1275 SRWo; John *Olof* 1296 SRSx; Adam *Olif* 1379 PTY. ON *Óleifr*, ODa *Óláfr* 'ancestral-relic', the name of a Norwegian king and saint.

**Ollister:** v. HOLLISTER

**Olman:** v. OLDMAN

**Olney:** Roger *de Olnei* 1086 DB (Bk); Henry *de Olnea* 1186 P (Bk); John *de Olney* 1325 AssSt; John *Olneye* 1385–6 FFWa. From Olney (Bk, Nth).

**O'Looney:** v. LOONEY

**Olton:** Alsneta *de Oltun'* 1275 RH (Sf); John *de Olton* 1327 SRSf. John *de Olton* 1362 FFY. From Olton (Wa). Sometimes for OULTON.

**Oman, Omond:** Leuric *Omundi filius* c1095 Bury (Sf); John *Omond* 1327 SRSf; Edduard *Homondsone* 1546 Black (Orkney); Edward *Omand* or *Omandson* 1576 ib. ON *Hámundr* 'high protector' or *Ámundi* 'great-grandfather-protector'. An Orkney and Shetland name.

**Ombersley:** William *Ombresleye* 1340 NIWo. From Ombersley (Wo).

**O'Neal, O'Neill:** v. NEAL

**Onefoot:** Richard *Onefote* 1283 CtW. 'One foot', OE *ān, fōt*. cf. Stephen *le Oneyede* 1293 AssSt 'one-eyed'.

**Onians, Onion, Onions, O'Nions, Onyon:** Thomas *Onoiun* 1279 RH (C); Robert *Oygnoun* 1295 Husting; Robert *Onnyon* 1568 SRSf; Thomas *Oynion, Onion* 1686–9 ShotleyPR (Sf). OFr *oignon* 'onion', for an onion-seller. More commonly for ENNION.

**Onley, Only:** Hamon *de Onyleye* 1285 AssSt; John *de Onleye* 1332 SRWa; John *Onley* 1488–9 FFSr; William *Only* 1620 LewishamPR (K). From Onley (Nth), or Onneley (St).

**Onraet:** Walter *Unred* a1200 Dublin; Alredus *Vnred'* 1230 P (Nf); John *Onrett* 1275 RH (K). OE *unrǣd* 'evil counsel, folly'. cf. Adam *Huncouthe* 1379 PTY.

**Onslow:** Roger *de Ondeslauwe* 1255 RH (Sa); William *de Ondeslowe* 1327 SRSa. From Onslow (Sa).

**Onyett:** v. HONIATT

**Openshaw, Oppenshaw:** John *Opensha* alias *Openshawe* 1559 Pat. From Openshaw (La).

**Oram:** v. ORME

**Orange, Orringe:** *Orengia, Horenga* 1201, 1204 Cur (O); *Orenga, Orenge* 1226 FFEss, 1247 AssBeds, 1296 SRSx; Sibel, John *Orenge* 1296 SRSx; Alexander *Orrynge* 1327 SRWo. From a woman's name *Orenge*, of doubtful etymology. It is found too early to be associated with the orange. William *de Orenge* (1086 DB) probably came from Orange (Mayenne), but as no later examples have been noted, this topographical source is probably not to be considered.

**Orbater:** Walter *le Orbat'r* 1275 RH (Lo); Alexander *le or Batour* 1311 AD ii (Mx), *le Orbatour* 1313 LLB B. OFr *orbatour* 'goldbeater'.

**Orbell:** *Orabilis* (f) 1221 Cur (K); *Orabla, Orable, Arable* de Meyhamme 1243 Fees (K); *Orabilia* 1273 FFEss; *Orabella* 1275 AssSt; John *Orable* 1279 RH (C); Adam *Orbel* 1327 SRSf. This woman's name can hardly be the Scottish *Arabella*, found in the 17th

century as *Arbell.* Lyford (1655) derives *Orabilis* from Lat *orabilis* 'easy to be entreated', a derivation clearly implied by the 13th-century scribes, and probably correct, being a formation parallel to *Amabilis*. v. MABLE.

**Orchard, Norchard:** Alexander *de Orchyard* 1225 AssSo; Emeric *del Orchyard* 1244 AssSo; William *Orchard* 1279 RH (O); Richard *ate Orchard, Atten Orchard* 1316 Ipm (Wo), *atte Norchard* 1332 SRWo. From residence near or employment at an orchard, OE *ortgeard, orceard*.

**Ord, Orde:** Henry *de Orde* 1209 P (Nb); Adam *of Horde* 1296 Black; Andrew *de Ord* t Robt I ib.; Alexander *Ord* 1596 ib. The first of these is from East Ord (Northumb), the others are from Ord, later Kirkurd (Peebles), or from the lands of Ord (Banff). But a personal name may also be involved, cf. Henry *filius Ord'* 1206 Pl (K).

**Ording:** *Ordingus* 1121–48 Bury; Gilbert *filius Ordingi* 1204 P (L); Jordan *Ordyn* 1285–6 FFEss; Roger *Ording*, Matilda *Ordinge* 1327 SRSf. OE *Ording*.

**Ordish:** John *Oredysshe* 1547 PN Db 193; Robert *Oredishe* 1599 SRDb. From Highoredish in Ashover (Db).

**Ordmer:** *Ordmar* de la Wada 1190 P (Nf); *Ordmerus* 1214 Cur (Sr); Geoffrey *Ordmeri* 1277 Ely (Sf); John *Ordmer* 1331 FFK. OE *Ordmǣr*.

**Ordway:** *Orduui* 1066 DB (Hu); Ricardus *filius Ordwi* 1185 P (Gl); Radulfus *filius Ordewi* c1250 Rams (Hu); Thomas *Ordwy*, Ralph *Ordwey* 1276 RH (Beds). Late OE *Ordwīg* 'spear-warrior'.

**Ore:** William, Simon *de Ore* 1207, 1208 Cur (Sx, K); Walter *Ore* 1210 Cur (C). From Oare (Berks, Kent, Wilts) or Ore (Sussex). No doubt also from residence near a shore or slope (OE *ōra*), though the local names seem usually to have become NOAR or NOWERS.

**O'Regan, Regan:** Padyne *Regane* 1264 Dublin. Ir *Ó Riagáin* 'descendant of *Riagán*' (little king).

**O'Reilly:** William *Orailly* 1451 Pat. Ir *Ó Raghailligh* 'descendant of *Raghallach*' (valiant).

**Orfeur:** v. OFFER

**Orford:** Richard *de Oreford* 1191 P (Sf); Richard *Oreford* 1247–8 FFEss; Matilda *Oreford* 1296 SRSx. From Orford (L, La, Sf).

**Organ, Organe:** Robert, Thomas *Organ* 1210 Cur (K); 1332 SRSt. This is sometimes metonymic for ORGANER. John *Organ* of Trewarian (Co) was the son of *Organa*, wife of Ives de Trewarian (1325 AD v). cf. *Organus* Pipard 1236 Fees (O).

**Organer:** Walter *le Organer* 1332 SRSr; John *Organer*, apothecary 1374 ColchCt. A derivative of ME *organ*, in early use, of a variety of musical (especially wind) instruments. 'A musician, a player or maker of the organ.' cf. John *le Orgeniste* 1241 FFY, Adam *le Organystre* 1327 Pinchbeck (Sf). OFr *organiste, organistre* 'organist'.

**Orgar, Orger:** *Orgar, Orger* c1095 Bury (Sf); Walter *filius Ordgeri* c1150 DC (L); Roger *Orgar'* 1198 FFNf; Geoffrey *Ordgar* 1226 ClR (C). OE *Ordgār* 'point-spear'.

**Orgel, Orgill:** Edmund *Liorgil'* 1198–1211 StP; Walter *Orgouyl* 1297 MinAcctCo; Gerard *Orgul* 1305 LoCt. OFr *orgueil* 'pride'. cf. PRIDE.

**Orgrave:** Thomas *Orgrave* 1381 LoCh; Christopher *Orgrave* 1672 HTY. From Orgrave (La), or Orgreave (St, WRY).

**Oriel:** *Oriholt, Oriolda* de Endrebi 1206–7 Cur (Y); *Oreute* Cur (Nt); Robertus *filius Oriold* 1230 P (He); Roger *Oriel* 1276 LLB A. OG *Aurildis, Orieldis* (f) 'fire-strife'.

**O'Riordan:** *v.* REARDON

**Orlebar:** Alur' *de Orlinberga* 1130 P (Nth); Thomas *Orlibar* 1732 SfPR. From Orlingbury (Northants).

**Orledge, Orlich:** John *Orlog* 1356 ColchCt. OFr *orloge* 'clock', metonymic for *orlogier* 'clock-maker'. cf. John, Adam *Orloger* 1295 Barnwell (C), 1311 ColchCt.

**Ormandy:** Ellen *Ormandie* 1296 LaWills; Agnes *Ormundie* 1552 Bardsley; John *Ormondy* 1675 LaWills. From Osmotherley in Ulverston (La).

**Orme, Ormes, Oram, Orum, Orrom:** *Orm* 1066 DB (Y); *Orm* de Hedoc 1169 P (La); *Orum* 1175 P (Db); John *Orm* 1275 SRWo; Augustin *Orumme* 1327 SRSf. ON *Ormr*, ODa, OSw *Orm* 'serpent'. *Oram* and *Orum* are due to the strongly trilled *r*.

**Ormerod, Ormrod:** John *Ormerod* 1593 Bardsley (Ch). From Ormerod (La).

**Ormesby, Ormsby:** Ulf *de Ormesbi* 1066 DB (L); Robert *de Ormesby* c1150 Gilb; Peter *de Ormesby* 1310 FFY; Arthur *Ormesby* 1457 FFEss. From Ormesby (Nf, NRY), or North, South Ormesby (L).

**Ormondroyd:** William *de Hamundrode* 1354 WRS; William *Hawmunrode* 1379 PTY; Francis *Armeroyd* 1623 ShefA. From Hamundrod in Hipperholme (WRY).

**Ormrod:** *v.* ORMEROD

**Ormsby:** *v.* ORMESBY

**Orneblow:** *v.* HORNBLOW

**Orpen, Orpin:** Elyas *Orpyn* 1298 Seals (Nth); Elizabeth *Orpen* 1665 HTO. cf. *Auripigmentum* . . . a coulour lyke golde, in englysshe *Orpine* (1548 NED). Fr *orpin* is given by Littré as a generic term for *Sedum* (yellow stonecrop), the name of a succulent herbaceous plant, *Sedum Telephinum*, a native of Britain, found in cottage gardens and esteemed as a vulnerary. From its tenacity of life it was called *Livelong* (a1387 NED). 'Orpyn, herbe, *Crassula major, et media dicitur* Howsleek, *et minima dicitur* stoncrop', c1440 PromptParv. One of the medicinal plants that gave rise to surnames.

**Orpet, Orpett:** William *Orpede* 1230 P (Wa); Walter *le Orpede* 1255 RH (Bk). OE *orped* 'stout, strenuous, valiant, bold'. cf. William *Orpedeman* a1223 Clerkenwell (Lo).

**Orr:** (i) Roger, William *Orre* 1202 AssL, 1277 Wak (Y); Robert *Orre* 1334–5 SRK. ON *Orri*, originally a byname 'black-cock'. (ii) Agnes, Peter *de Ore* 1210–11 PWi, 1264–5 IpmW; William *de Ore* 1334–5 SRK. From Oare (Berks, Kent, Som, Wilts), Ore (Sussex), or 'dweller by the bank or ridge'. (iii) Donald *Oure* 1512 Black; John *Or* 1578 ib. Gaelic *odhar* 'dun, of sallow complexion'.

**Orred, Orrett:** *Huredus* c1250 Rams (Hu); William *Orede* 1275 RH (Sf); Nicholas *Ored* 1279 RH (O). OE *Wulfrǣd* 'wolf-counsel'. cf. HURRY.

**Orrell:** Richard *de Orhille* 1206 P (La), *de Horul* 1212 Fees (La); Godith *de Orul* 1246 AssLa. From Orell in Wigan, or Orell in Sefton (La). There was also a feminine name: *Orella* (f) 1192 P (K).

**Orrey:** *v.* HURRY

**Orrey, Orway:** Ralph *de Orweie* 1202 P (D); John *de Orewey* 1341 Hylle. From Orway Fm in Kentisbeare (D).

**Orrick, Orridge:** Reinaldus *filius Ordrici* 1114–30 Rams (Hu); Walter, William *Ordrich* 1193, 1196 P (Wa, Bk); Thomas *Orridge* 1778 HorringerPR (Sf). OE *Ordrīc* 'spear-powerful', a late OE name, common in Domesday Book.

**Orringe:** *v.* ORANGE

**Orriss, Orys:** *Oratius* presbiter 1193 P (Ess); Richard *Oras* 1312 FFEss; William *Orice* 1668 SfPR; Henry *Orris* 1674 HTSf. Latin *Horatius*, the name of a Roman *gens*, best known from Horatius Cocles, the defender of the bridge, and the poet Quintus Horatius Flaccus. The name has not previously been noted in England before the Renaissance when it apparently was introduced from Italy as *Horatio*.

**Orrom:** *v.* ORME

**Orry:** *v.* HURRY

**Orsborn:** *v.* OSBORN

**Orton, Overton:** (i) Æðelweard *of Ortun* c1051 OEByn; Henry *de Orton* 1229 FFO; Hugh *Orton* 1357 IpmNt. From Orton (Cumb), or from Nether, Over Worton (Oxon), *Orton* 1191. (ii) Hugh *de Overtuna* c1150 StCh; Bartholomew *de Ouerton'* 1203 AssNth; Adam *atte Overton* 1275 MELS (Wo); Geoffrey *de Overton* 1324 Oriel (O). From one or other of the many places called Overton, or from Orton (Hunts, Leics, Northants, Warwicks, Westmorland), all with OE *ōfer, ofer, ufera* as the first element. In the absence of medieval forms the two names cannot be distinguished.

**Orum:** *v.* ORME

**Orwell:** (i) Turbert *de Orduuelle* 1066 InqEl (C); William *de Orewell'* 1201 Pleas (C); Alan *de Orewell'* 1212 P (Ess/Herts). From Orwell (C), or Orwell Haven (Sf). (ii) Richard *de Orewell* 1231, John *de Vrwell* 1342 Black. From the lands of Orwell (Kinross).

**Orwin:** *v.* ERWIN

**Osbaldeston, Osbaldiston, Osbaldstone, Osboldstone, Osbiston:** Adam *de Osbaldeston* 1292 WhC (La); Elezabeth *Osbosstone* 1552 RothwellPR (Y). From Osbaldeston (Lancs).

**Osband:** *v.* OSBORN

**Osbert:** John *filius Osberti* 1189 Sol; *Hosebert* le Cur 1222 Acc; *Osbert* Wakerild 1319 FFK; Richard *Osbert* 1246 AssLa. OE *Ōsbeorht*.

**Osberton, Osbiton:** Roger *de Osberton* 1242 Fees (Nt). From Osberton Hall in Worksop (Nt).

**Osbiton:** Roger *de Osberton'* 1242 Fees (Nt). From Osberton (Notts).

**Osborn, Osborne, Osbourn, Osbourne, Osburn, Osbon, Osband, Orsborn, Usborne, Hosburn, Hosbons:** *Osbern* 1066 DB; Osbernus *filius Willelmi*

1221 Cur (Bk); Henry *Osbern* 1260 AssC; Henry *Oseburn* 1297 MinAcctCo; John *Osebarn* 1296 SRSx; Walter *Osborn* 1310 FFC; Andrew *Housborn*, Walter *Hosebarn* 1327 SRSo; Elena *Usburne* 1381 PTY; John *Usbarne* 1467 KentW; William *Osbon* 1792 Bardsley. Late OE *Ōsbern*, from ON *Asbiǫrn*, ODa *Asbiorn* 'god-bear'. The name is found in England before the Conquest and may be of direct Scandinavian origin. It was also common in Normandy whence it was often brought over by Normans after the Conquest.

**Oscroft:** Thomas *de Oxcroft* 1327 SRDb. 'Dweller by the ox-croft.'

**Osen, Ozin:** Robert *Oisun* 1148 Winton (Ha); Martin *Oisun* 1198 FFD; Hugh *Oisun* 1212 P (Ha). OFr *oison* 'gosling'.

**Osgerby:** Sylvester *de Osgoteby* 1219 AssY. From Osgodby (L, ERY, NRY), or Osgoodby (NRY).

**Osgood, Hosegood, Hosgood, Horsegood, Angood:** *Asgot*, *Ansgot*, *Angot*, *Osgot* 1066 DB; *Osgot* c1095 Bury (Sf); *Angothus* clericus c1166 DC (Lei); Henricus *filius Osegod* 1199 FF (Nth); Galfridus *filius Angod'*, *Angot* 1208 Fees, 1212 Cur (Bk), identical with Galfridus *Angod* 1235 Fees (Bk); Matilda *Angot* 1198 FF (Nf); William *Osgot'*, *Osegod* 1202 FF (Nf), 1213 Cur (Sf); Petrus *Angoti* 1209 P (Nf); Robert, John *Hosegod* 1275 RH (W), 13th AD iii (Beds). Late OE *Ōsgod*, *Ōsgot*, from ON *Ásgautr*, ODa, OSw *Asgut*, *Asgot*. This was common in Normandy in the forms *Ansgot*, *Angot*, hence *Angood*, ANGOLD. *Hosegood* is the Anglo-Scandinavian *Osgod* with retention of the inorganic *H* so frequently prefixed by Norman scribes.

**O'Shea, O'Shee, Shea, Shee:** Ir Ó *Séaghdha* 'descendant of *Séaghdha*' (stately, majestic).

**Oslack, Ashlock:** Matilda *Oslac* 1275 SRWo; Geoffrey *Oseloc* 1327 SRLei; Walter *Aslak* 1426 Paston. ON *Áslákr*.

**Oslar, Osler, Hosler:** William *Osselur*, *Oiselur* 1170, 1171 P (Y); Alberic *le Oselur* 1208 FFC; Thorald *le Osiller* 1250 Fees (L); Robert *le Oyseler* 1255 AssSo; Godfrey *le Hoselur* 1279 RH (C); William *le Vsseler* 1332 SRSr. OFr *oiseleor*, *oiseleur* 'bird-catcher, fowler'. Forms in -*er* may be from OFr *oiselier* 'seller of game, poulterer'.

**Osman, Osmant, Osment, Osmint:** *v.* OSMOND

**Osmaston:** John *Osmerson* or *Osmundstone* 1571 CantWills. From Osmondiston (Nf), or Osmaston (Db).

**Osmer, Hosmer, Usmar:** *Osmar*, *Osmer* 1066 DB (Lei, D); *Osmer* ater Dune 1296 SRSx; William *Osmere* 1230 P (D); Chaneys *Usmer* 1589 ArchC 48. OE *Ōsmǣr* 'god-fame'.

**Osmond, Osmon, Osmund, Osman, Osmant, Osment, Osmint:** *Osmund* 1066 DB; *Osemundus* clericus 1222 Cur (Gl); Roger *Hosemund* 1199 FF (Nth); Robert *Osemund* 1221 ElyA (Nf); Richard *Osemond* 1297 MinAcctCo (O); William *Osman* 1367 ColchCt; Edward *Osmand* 1660 Bardsley. OE *Ōsmund* 'god-protector', or Norman *Osmund* from ON *Ásmundr*, ODa, OSw *Asmund*.

**Ossell:** Robert *oisel* c1198 Bart; Robert *Oysel* 1249 AssW; Simon *Oysel* 1296 SRSx; John *Ossel* 1327 SRWo. OFr *oisel* 'bird', probably metonymic for a bird-catcher.

**Ost:** *v.* HOST

**Ostle:** For *atte ostle*, metonymic for *Ostler*.

**Ostler, Oastler, Hosteller, Hostler, Horsler, Hustler:** William *Hostiler* 1190 Eynsham (O); Robert, Edid *le Osteler* 1204 AssSt, 1260 Oseney (O); Robert *le Hostler* 1275 RH (Nf); Henry *Husteler* 1301 SRY; Thomas *Ostler* 1562 FrY; Mr *Husler* 1674 HTSf; William *Oastler* 1738 FrY. OFr *ostelier*, *hostelier*, ME *(h)ostiler*, *(h)osteler* 'one who receives, lodges or entertains guests, especially in a monastery' (c1290 NED), 'keeper of a hostelry or inn' (c1365 ib.). This also came to mean 'a stableman', modern *ostler* (c1386 NED), a meaning less probable here.

**Ostridge:** John *Ossreche*, Matilda *Ostrich* 1327 SRSf; Thomas *Ostriche* 1421 LLB I. OFr *ostrice*, *hostrige* 'a hawk', used by metonymy for *ostricer* 'a keeper of goshawks, hawker, falconer'. cf. William *(le) ostricer* c1160 DC (L).

**Ostringer, Astringer:** Alfgar *ostriciarius* 1135–48 Bury (Sf); William *le ostricier* c1160 DC (L); John *le Ostricer* 1243 AssSo; Robert *le Ostryzer* 1275 FFEss. A derivative of OFr *ostrice* 'hawk', hence a keeper of goshawks, a falconer.

**O'Sullivan, Sullivan:** Ir Ó *Súileabháin* 'descendant of *Súileabhán*' (black-eyed).

**Oswald, Osswald, Oswell:** *Osuuald*, *Osuuoldus* 1066 DB (So, Sr); Robertus *filius Oswaldi* 1240 Rams (Nf); Simon *Aswald* 1279 RH (O); John *Oswald* 1327 SRSx. OE *Ōsweald* 'god-ruler', or, possibly, also ON *Ásvaldr*. The christian name occurs as *Oswell* (1540) and *Oswall* (1620 Bardsley).

**Osward:** *Osward* 1188 BuryS (Sf); *Oswardus* Suete Melt 1206 Cur (Ess); Robert *Oseward* 1257–8 FFEss; Alecok *Oseward* 1275 SRWo; William *Oseward* 1317 AssK. OE *Ōsweard*.

**Oswick:** John *de Oxewic'* 1206 Cur (Nf). From Oxwick (Norfolk).

**Oswin:** *Oswinus* c1250 Rams (C, Nf); Roger *Oswin* 1221 Cur (Ess). OE *Ōswine* 'god-friend'.

**Oteland:** *v.* OATLAND

**Othen:** *Oinus dacus* 1066 DB (Ess); Henry *Othyn* 1296 SRSx. Anglo-Scand *Ōðin*. *v.* ODEN.

**Otley, Ottley:** Tor *de Ottalay* 1148–56 YCh; Thomas *de Otteleg'* 1225 Cur (Sf); Paulinus *de Otteleye* 1301 FFY; Robert *Otteley* 1425 AssLo. From Otley (Sf, WRY).

**Ott, Otten:** *v.* OADE

**Ottaway:** *v.* OTTOWAY

**Otter:** *Otre* 1066 DB (D, Y); Walterus *filius Otheri*, *Other* 1086 DB; *Other* Pilemus 1187 P (Nf); Roger *Oter* 1185 P (C); John *Otir* 1230 P (Ha); Symon *Other* 1279 RH (C); Mariota *Otre* 1297 SRY. ON *Óttar* 'terrible army', anglicized in OE as *Ohthere*; or, more commonly, ME *oter* 'otter', metonymic for 'otter-hunter': Adam *Loterhunt*, *le Oterhunter* 1246 AssLa.

**Otterburn:** Rannulf *de Oterburn'* 1219 AssY; Helias *de Hoterburne* 1274–5 IpmY; Alan *de Ottyrburne* 1426 Black. From Otterburn (Nb, WRY, Roxburgh), or Otterbourne (Ha).

**Otterway:** *v.* OTTOWAY

**Ottery, Ottrey:** Auicia *de Otery* 1204 P (D). From Ottery (D), Mohun's, Venn Ottery, Ottery St Mary, Upottery (D).

**Ottewell, Ottewill, Ottiwell, Otterwell:** Willelmus *filius Otuelis* c1150 DC (L); *Otuell'* Malduit 1169 P (Ess); William *Ottiwell* 1564 Bardsley. OFr *Otuel*, interchanging with *Otoïs*. v. OTTOWAY. cf. *Otuelis*, *Otewicus* de Clipeston' 1207 Cur (Nf). The christian name was still in use in the 17th century.

**Ottley:** v. OTLEY

**Otton:** v. OADE

**Ottoway, Ottaway, Otterway, Ottway, Otway:** *Otewi* 1202 AssL; *Otewicus* le Poher' 1221 Cur (Sx); *Otewi* de Esthall' 1228 Eynsham (O); Roger *Otway* 1260 AssLa; Hugh *Otewy* 1319 SRLo. OFr *Otoïs*, OG *Otwich*.

**Ottrey:** v. OTTERY

**Oughtibridge, Outerbridge:** John *Ughtybrygg* 1440 ShefA. From Oughtibridge (WRY).

**Oughtred, Outred, Ughtred:** *Vctred* 1066 DB (D); *Uthret* ib. (Sf); Ysabella *filia Uhtredi* 1207 Cur (Ess); *Utred* de Norflet' 1214 Cur (K); Roger *Uctred, Hutred* 1260 AssY; Roger *Outred'* 1297 MinAcctCo (O). OE *Ūhtrǣd* from *ūht(e)* 'twilight, dusk, dawn' and *rǣd* 'advice, counsel, wisdom'.

**Oughtright:** v. OUTRIDGE

**Ould:** v. OLD

**Oulton:** Alan *de Oldeton* 1279 AssSt; Nicholas *atte Noulton* 1327 MELS (Wo); Henry *de Oulton* 1328 FFY. From Oulton (Ch, Cu, Nf, Sf, St, WRY). v. also OLTON.

**Ourry:** v. HURRY

**Oustiby:** Hugh *Oustiby* 1230 P (Y); William *Oustiby* 1305 IpmW; John *Owsterby* 1622, Thomas *Owstaby* 1665 FrY. 'Dweller east in the village', ON *austr i bý*.

**Ouston, Owston:** Adam *de Ouston'* 1207 P (Y); Geoffrey *de Oveston'* 1275 RH (Nth); John *Owston* 1636 FrY. From Ouston (Du, Nb), or Owston (L, Lei, WRY).

**Outerbridge:** v. OUGHTIBRIDGE

**Outhwaite:** From Outhwaite (La).

**Outlaw:** Alan *le Vtlage* 1230 FFSf; John *le Utlawe* 1316 FFC; Robert *Outlawe* 1327 SRC, SRSf. ME *outlawe* from ON *útlagi* 'outlaw'. Occasionally used as a personal name: *Hutlage* de Wrmedale Ed 2 AD i (Gl).

**Outram, Owtram, Outtrim, Outran, Outrim:** William *Owtrem* 1493 CorNt; Richard *Owtrem* 1524 SRSf; William *Owtram, Utteram* 1525 SRSx. OG *Othram*.

**Outred:** v. OUGHTRED

**Outridge, Outteridge, Oughtright, Utteridge, Uttridge, Utridge:** John *Outrich* 1333 ERO (Ess); William *Utteredge, Utteridge* 1663 Buxhall (Sf), 1674 HTSf. OE *\*Ūhtrīc* 'dawn-powerful', unrecorded in OE.

**Outsax:** Roger *Houtsex* 1275 SRWo. 'Out with the sword', OE *ūt, seax*. cf. John *Oute with the Swerd* 1402 Black, with the same meaning.

**Outtrim:** v. OUTRAM

**Ovenden, Ovendon:** John *de Ovenden* 1277 Wak (Y). From Ovenden (WRY).

**Ovens:** John *Attenouene* 1276 AssSo; John *atte Oven* 1299 MELS (Wo); William *ate Ouene* 1317 AssK. 'One who lived near, or worked at an oven or furnace' (OE *ofen*), probably an old iron-furnace or a charcoal-furnace.

**Over:** Thomas *atte Overe*, William *del Overe* 1275 SRWo; Walter *atte Novere* 1327 ib. From Over (Cambs, Ches, Glos), or from residence near a bank or steep slope (OE *ōfer*). cf. OWER, NOAR, NOWERS.

**Overal, Overall, Overell, Overill:** William *Oueral* 1217 Pat; William *del Overhall* 1316 Wak (Y); Thomas *Overell* 1632 ERO; Thomas *Overill* 1674 HTSf. 'Dweller at the upper hall.' *Overell* and *Overill* might mean 'over the hill', but may be due to late weakenings of *Overhall*.

**Overbeck:** Thomas *Overthebek* c1270 Whitby (Y); Hugh *Overbecke* 1301 SRY. 'Dweller beyond the stream', ME *beck*, ON *bekkr* 'brook, beck'.

**Overbrook:** Richard *Ouerbroc* 1239 StP; Edith *Ouerbrok* 1328 ChertseyCt (Sr); John *Ouerbrok* 1332 SRSr. 'Dweller beyond the brook', OE *ofer, brōc*.

**Overbury:** William *de Overbury* 1275, John *Overbury* 1327 SRWo; William *Overbury* 1545 SRW. From Overbury (Wo).

**Overend:** William *de Overende* 1279 RH (Bk). 'Dweller at the upper end (of the village).'

**Overton:** v. ORTON

**Overs:** Walter *Overse* 1221 Cur (He); Griffin *des Oueres* 1250 Fees (Sa); Robert *de Oweres* 1255 RH (Sa). From Overs (Sa).

**Overwater:** Richard *Overwater* 1275 SRWo; William *Overyewater* 1301 SRY. 'Dweller across the water', OE *ofer, water*. cf. Vlfkil *Ouerwold'* 1176 P (Y) 'dweller across the wold'.

**Overy:** (i) Robert *Overhe* 1279 RH (C); Richard *Oueree* 1314 Oseney (O). 'Dweller beyond the stream', OE *ēa*. (ii) Alexander *Overye, Overey* 1328 ArchC 33. 'Dweller beyond the low-lying land' (OE *ēg*). v. NYE, REA.

**Owen, Owens:** *Uwen* Wenta cyning 926 ASC D; *Ouen* 1066 DB (He); *Oenus* filius Madoc 1160 P (Sa); Robertus *filius Owen* 1221 Oseney (O); Ralph *Oein, Owein* 1221 AssWa; John *Owain* 1242 Fees (Sa); Henry *Oweyn, Ewayn* 1286–9 Balliol (O); Robert *Owyng* 1430 Oseney (O). OW *Oue(i)n*. v. EWAN.

**Ower, Owers:** Alexander *del Owre* 1240 PN K 72; Richard *de Owre* 1310 Balliol (O); John *Owers* 1524 SRSf. From Ower (Hants) or Owermoigne (Dorset), The Nower (Kent), or from residence near a bank or steep slope (or banks or slopes). OE *ōfer*. cf. OVER, NOWERS. Occasionally this is from OE *ōra* in the same sense. cf. Ores Fm (PN K 508), The Owers (PN Sx 83) and v. NOAR.

**Owles:** Gilbert *Vle* 1176 P (Nt); Stevyn *Owle*, Thomas *Owles* 1524 SRSf. OE *ūle* 'owl'.

**Owlett:** William *Owlete* 1524 SRSf. 'Little owl.'

**Owram:** Adam *de Ouerum* 1274, *de Ourum* 1284, Dobbe *de Overom* 1297 Wak (Y). From Northowram, Southowram (WRY).

**Owston:** v. OUSTON

**Owtram:** v. OUTRAM

**Oxborough, Oxborrow, Oxbrow, Oxburgh, Oxbury:** William *de Oxeburg'* 1275 RH (Nf). From Oxborough (Norfolk).

**Oxendon:** Ralph *de Oxendon'* 1327 SRLei. From Great, Little Oxendon (Nth).

**Oxenford:** v. OXFORD

**Oxenham, Oxnam:** Richard *Oxenham* 1642 PrD. From Oxenham in South Tawton (D).

**Oxer:** Adam *Oxehird* 1327 Wak (Y); William *le Oxherd* 1332 SRLa. 'Ox-herd.' cf. OXNARD.

**Oxfoot:** Godwin *Oxefot* 1199 (1319) Ch (Sf), 1137 ELPN; William *Oxefot* 1231 FFY. 'Ox foot', OE *ox, fōt*. cf. Robert *Oxenose* 1221 AssSa 'ox nose'.

**Oxford, Oxenford:** Ulric *de Oxenford* 1086 DB (K); Walter *de Oxenforde, de Oxford* 1319 SRLo, 1341 LoPleas. From Oxford.

**Oxlade:** Michael *de Ocslade* 1279 RH (O). 'Dweller in the oak-valley', OE *āc, slæd*.

**Oxley, Oxlee:** Robert *de Oxeleia* 1227 FFSt; John *de Oxleye* 1310 LLB D; Thomas *Oxley* 1505, 1541 CorNt. From Oxley (Staffs) or Ox Lee in Hepworth (WRYorks).

**Oxman:** John *Oxeman* 1201 Cur (K); Richard *le Oxeman* 1289 AssCh. OE *oxa* 'ox' and *mann* (NED c1830), 'oxherd'.

**Oxnam:** v. OXENHAM

**Oxnard:** William *Oxenhird'* 1301 SRY; John *le Oxenhurde* 1327 MEOT (Ha); Josias *Oxenerd, Oxnerd, Oxnard* 1645, 1646, 1671 FrY. OE *oxnahyrde* 'oxherd'.

**Oxney:** John *Oxeneye* 1405 KB (Ess). From Oxney (K).

**Oxspring:** Richard *de Ospryng* c1265 Glast (So); William *Oxspring* 1461 TestEbor. From Oxspring (WRY), or Ospringe (K).

**Oxton:** Hugh *of Oxston* 1236–7 FFY; John *de Oxton* 1382 IpmNt; Thomas *Oxton'* 1398 KB (L). From Oxton (Ch, Nt, WRY).

**Oxwick:** Patric *of Oxwick* 1315 AssNf; William *Oxwyk* 1381 LoCh. From Oxwick (Nf).

**Oyler:** v. OILLARD

**Oyler, Ollier:** Richard *Oylere, le Oyller'* 1248 *Ass* (Ha), 1281 MESO (L); Reginald *le Oyler* 1286 ib. (Lei). OFr *olier, huilier* 'maker or seller of oil'.

**Ozanne:** *Osanna* 1160 P (He), 1208 Cur (Do); Reginaldus *filius Osanna* 1180 P (Y); Walter *Ossenna* 13th AD i (Herts); John *Osan* 1279 RH (Bk); Richard *Osan, Ossan* 1296–7 Wak (Y). *Osanna*, a woman's name, from a Hebrew liturgical word used among the Jews, meaning 'save now', or 'save pray'. Latinized as *Hosanna*. Dauzat derives the name from the medieval feast of Palm Sunday, called *Osanne* or *Ozanne* from the hosanna sung on that day.

**Ozin:** v. OSEN

# P

Pace, Paice, Pays, Payze, Peace: John *Pais* 1219 FrLeic; Roger, Ralph *Pays* 1275 RH (Nf); William, John *Pace* 1242 Fees (D), 1269 Ipm (D); John *Pax* 1275 RH (D); Peter *Pece* 1302 SRY. ME *pais, pes(e)*, OFr *pais*, Lat *pax* 'peace, concord, amity'. As ME *pasches* appears also as *paisch, peice, peace*, and Easter eggs are still called *Pace* eggs, later examples may be a variant of PASH.

Pacey, Pacy: Robert *de Peissi* 1158 P (Nt), *de Pacy* 1214 Cur (Wa). From Pacy-sur-Eure (Eure).

Pack, Packe, Paik, Pakes: *Payke* 1260 AssC; Johannes *filius Pake* 1279 RH (C); John *Pac* 1190 BuryS (Sf); Roger *Pake* 1195 P (Lei); Richard *Packe* 1221 *ElyA* (Sf); Richard *Pake, Packe* 1311, 1340 ColchCt; Joan *Pakes* 1312 ib. OFr *Pasques, Paque* 'Easter', *v.* PASH. cf. the diminutives in *Pakerel* 1251 Rams (Hu), Robert *Pakerel* 1201 AssSo, *Paket* 1190 P (Wa), Walter *Paket* 1176 P (O).

Packard: William *Pachart* 1208 P (L); Geoffrey, Henry *Pac(k)ard* 1327 SRSf. *Pack*, plus the suffix *-hard*.

Packer: Henry, Robert *Packer(e)* 1209 FrLeic, 1221 *ElyA* (Sf); John *le Pakkere* 1254 ArchC xii. A derivative of ME *packe* 'to pack', probably a wool-packer.

Packington, Pakington: Robert *de Pakinton'* 1195 P (Db); David *de Pakington* 1244 AssSt; Richard, Robert *de Pakynton* 1332 SRWa. From Packington (Leics, Staffs, Warwicks).

Packman: *v.* PAKEMAN

Packstaff: William *Packestaff* 1340–1450 GildC; William *Packstaff* 1576 SRW; Richard *Packstaffe, Pickstaffe* 1665 HTO. ME *pakestaff* 'a pedlar's staff', a nickname for a pedlar.

Pacock: *v.* PEACOCK

Padbury: Hamo *de Padeberia* 1190 P (Bk). From Padbury (Bk).

Padd, Padde: Osbert *pade* 1189 Sol; Roger *le Pade* 1299 AssW; Walter *Pade* 1327 SRSx. A nickname from OE *padde, pade* 'frog, toad'.

Paddington: William *de Padintone* 1185 Templars (Lo); John *Padyngton* 1410 FFEss. From Paddington (Mx, Sr).

Paddock: Thomas, Walter *Paddoc* 1279 RH (C), 13th AD ii (Berks). OE *\*padduc* 'frog'.

Padfield: Fulcher *de Padefeld* 1298 AssSt; Hugh *de Padfeld* 1327 SRDb; Edward *Padfild* 1642 PrD. From Padfield (Db).

Padgett: *v.* PAGET

Padgham: Hugh, Roger *de Pageham* 1206 Cur

(Ha), 1255–6 FFSx; Edwarde *Pagham* 1576 HTY. From Pagham (Sussex).

Padley: Nicholas *de Paddeleye* 1275 RH (Db); Henry *de Padelay* 1379 PTY; William *Paddeley, Padley* 1504, 1508 CorNt. From Padley (Derby), or Padley Common (Devon).

Padmore: William *Pademor* 1279 RH (Hu); John *Padmor* 1642 PrD. From Padmore in Whippingham (Wt).

Paenson: *v.* PENSON

Paerson: *v.* PEARSON

Paffard, Pafford, Parford: William *Pafard* c1100–30 OEByn (D); Alexander *Paford* 1641 PrSo. OFr *pafard* 'shield', either a nickname for a warrior, or metonymic for a maker of shields. Sometimes, probably, from Pafford in Moretonhampstead, or Parford in Drewsteignton (D).

Pagan: *v.* PAIN

Page, Paige: Ralph *Page* 1230 P (D); William *le Page* 1240 FFEss. OFr *page* 'page'.

Paget, Pagett, Padgett: William, John *Paget* 1327 SRSx, 1359 ColchCt. A diminutive of PAGE.

Paice: *v.* PACE

Paige: *v.* PAGE

Paik: *v.* PACK

Paikin, Paykin: Walter *Paykin* c1300 *ERO*; Richard *Paykyn* 1351 AssEss; John *Paykinge* 1568 SRSf. *Pay-kin*, a diminutive of *Pay*, OE *Pǣga*. *v.* PAY.

Pail, Paile, Payle: Thomas *Payl* 1298 AssL; Ralph *Peyl* 1327 SRLei. OE *pægel* 'a pail'. Metonymic for a maker of pails.

Pain, Paine, Paines, Pane, Panes, Payan, Payen, Payn, Payne, Paynes, Pagan, Pagon, FitzPayn: Edmund *filius Pagen* 1086 DB (So); Reginaldus *filius Pain* 1185 Templars (L); *Payn* de Weston 1268 AssSo; Radulfus *Paganus* identical with Radulf *Pagenel* 1086 DB (So); John *Pane* 1190 P (Wo); Rotrotus *Pagani* 1195 P (Lei); Robert *Pain* 1200 P (Ha); William *Paen* 1220 Cur (So); Ralph *Payn* 1221 *ElyA* (C); Stephen *Paynes* 1230 Pat; John *Pagan* 1275 SRWo; Thomas *Payen* 1296 SRSx; Roberd *le filz Payeng', filz Payn* 1305 SRLa. OFr *Paien*, from Lat *paganus*, originally 'villager, rustic', later 'heathen'. Lebel explains this as a name given to children whose baptism had been postponed. Dauzat prefers to regard it as a derogatory term applied to adults whose religious zeal was not what it should be. Radulfus *filius Pagani* is also called Radulfus *Paganus* and *Pagenel* already in 1086, that is, as son of Paganus he had adopted his father's christian name as his surname, *Pagenel* 'little Payn' distinguishing him from his father. In the 12th and 13th centuries *Payn* was a very common christian name and was, no doubt, given without any thought of its meaning. The surname is probably always patronymic.

**Painter, Paynter:** Richard *the Paintur* 1240 FFY; John *Peynter* 1317 AssK; Thomas *Peyntour* alias *Steynor* 1430 Oseney (O). AFr *peintour*, OFr *peintour*, *peintor* 'painter'.

**Pairpoint:** *v.* PIERPOINT

**Paish:** *v.* PASH

**Paisley:** William *de Passele* c1199, William *Passeley* 1389, John *Paisley* 1616 Black. From Paisley (Renfrew).

**Paiton:** *v.* PAYTON

**Pakeman, Packman:** Henry *Pacheman* c1160 ELPN; Simon *Pakeman* (*Pacchem'*) 1202 P (Nt), 1221 Cur (Lei); William *Pakeman* 1278 AssSo; John *Packeman, Pakeman* 1352, 1360 ColchCt. Perhaps 'servant of *Pake*' (*v.* PACK), but usually ME *pake, packe* 'pack, bundle', plus *man*, 'hawker, pedlar'.

**Pakenham:** William *de Pekenham* 1196–1200 BuryS (Sf); Edmund *de Pakenham* 1327 SRSf; Nicholas *Pakenham* 1492–3 FFWa. From Pakenham (Sf).

**Pakes:** *v.* PACK

**Pakington:** *v.* PACKINGTON

**Palairet:** A Huguenot name, Fr *Pailleret*, a diminutive of *Pailler*, denoting the owner of a *grenier a paille* 'a straw-rick, farm-yard'.

**Palcock:** Henry *Palecoc* c1260–70 LuffCh; Adam *Palecoke* 1275 SRWo; John *Palcok'* 1379 PTY. *Pallcock*, a diminutive of *Pall*. *v.* PALL.

**Paler, Payler:** Ralph *le Paeler, le Payller* 1193 P (We), c1297 MESO (Wo). A derivative of ME *payle* or OFr *paelle* 'pail, pan'. A maker or seller of pails.

**Pales:** Metonymic for PALER.

**Paley, Pally:** (i) Geoffrey *Pallig'* 1249 AssW; Thomas *Pally* 1310 AssNf; John *Paly* 1371 AssL; Robert *Paley* 1560 Pat (Ha). ODa *Palli*, OSw *Palle*. (ii) Adam *de Palay* 1246 AssLa. From Paley Green in Giggleswick (WRY).

**Palfrey, Palfery, Parffrey, Parfrey:** Hunfridus *Palefrei* 1148 Winton (Ha); Ralph *Palefray* 1183 Boldon (Du); Philip *Pawfry* 1605 LitSaxhamPR (Sf); Ro. *Parfry*, An. *Paufry*, Ro. *Pawfery*, Jea. *Payfrey* 1674 HTSf; Francis *Parfree* bapt. 1742, Frances *Palfrey* widow bur. 1797 RushbrookPR (Sf). OFr *palefrei* 'saddle-horse, palfrey', by metonymy for PALFREYMAN. cf. Adam *le Palefreur* 1255 *Ass* (Ess).

**Palfreyman, Palfreeman, Palfreman, Palframan, Palphreyman, Parfrement:** John *le Palfreyman* 1279 RH (C); John *Palfreman* 1333 MEOT (So); John *Palframan* 1379 PTY. 'Man in charge of the palfreys.' For *Parfrement*, cf. PALFREY.

**Palgrave:** Osketel *de Palegrave* 1199 P (Sf); Thomas *de Palegrave* 1339 Crowland; Stephen *Palgrave* 1434 FFEss. From Palgrave (Sf), or Great, Little Palgrave (Nf).

**Palin, Paling, Palling, Payling:** Gernagod *de Paling'* 1156–8 Seals (Sx); Edmund *of Palling* 1318 AssNth; John *Palyng* 1441 AssLo; Richard *Palin*

1524 SRD. From Palling (Nf), or Poling (Sx), *Palinge* Hy 2.

**Pall, Palle:** (i) Gerard *Palle* 1227 Pat (Nf); Alice *Pal* 1275 SRWo; John *Pal* 1332 SRWa. OE \**Palla*. (ii) Roger *de la Palle* 1274, Walter *de la Palle* 1325 MELS (So). 'Dweller by the ledge or terrace', OE \**peall*.

**Pallant:** John *ate Palente* 1285 *Ass* (Sx); John *de Palenta* 1296 SRSx. From The Pallant (OE *palant* 'palace, enclosure'), the south-eastern quarter of the city of Chichester, a peculiar of the Archbishop of Canterbury, with special 'palatine rights'.

**Palling:** *v.* PALIN

**Pallis:** (i) Henry *Paillehus* 1166 P (Y). A hybrid, Fr *paille* 'straw' and OE *hūs* 'house', a worker at the straw-rick. (ii) Augustin *de la Paleysse* 1327 SRSf. Rather OFr *palis, paleis* 'palisade' than AFr *paleis* 'palace'. Hence, 'dweller by the fence or palisade'.

**Palliser, Pallister, Palser:** Roger *Paleser* 1315 Wak (Y); Richard *Palicer* 1381 SRSt; William *Pallester* 1598 FrY. A derivative of OFr *palis, paleis* 'palisade', a maker of palings or fences. cf. William *Palycemaker* 1379 PTY.

**Pally:** *v.* PALEY

**Palmar, Palmer, Paumier:** Sagar *Palmer* 1176 P (D); Wiger *le Palmer* 1191 P (L); Richard *le Paumere* 1198 FF (Mx). OFr *palmer, paumer* 'palmer, pilgrim to the Holy Land', so called from the palm-branch he carried.

**Palphreyman:** *v.* PALFREYMAN

**Palser:** *v.* PALLISER

**Pamber:** *v.* PANBOROUGH

**Pammenter:** *v.* PARMENTER

**Pamplin:** Simon *filius Pampelyn* 13th CartNat; John *Pampiloun* 1310 FFSf; Thomas *Pamphiloun* 1377 AssEss. Perhaps a diminutive of OFr *Pamphile*, the name of a Greek martyr of the 4th century. Harrison's suggestion that it is from Fr *papillon* 'butterfly' is much less likely.

**Pan, Pans:** Lefwin, Aldwin *Panne* 1176 P (Nth), c1250 Rams (Nf). Probably metonymic for PANNER.

**Panborough, Pamber:** Walter *de paneberga* 1189 Sol. From Pamber (Ha).

**Panchard, Paunchard:** William *Panchard* 1194 P (L); Robert *Paunchard* 1339 CorLo; Nicholas *Panchard* 1662–4 HTDo. ONFr *panche* 'stomach', and the suffix *-ard*, a nickname for a man with a large belly. cf. Ralph *Paunch* 1315 AssNf.

**Panckridge, Pankridge:** (i) Pancrace Grout 1532 Bardsley; Richard *Pancras* 1296 SRSx; Robert *Panckridge* 1698 Bardsley (La). From *Pancras*, a saint to whom eight churches are dedicated in the south-eastern counties, three of them in Sussex. The London church of St Pancras was *Pancradge* church in 1630. (ii) William, John *de Panegregg'* (1296 SRSx) were connected with Pannelridge Wood in Ashburnham (Sussex), *Panegregg'* 1296, *Panyngrigge* 1366, while Penkridge (Staffs), *Pancriz* DB, *Pencrich* 1156, may be another source of the name.

**Pancras:** *Pamcras* Berry 1642 PrD; Richard *Pancras* 1296, John *Pancrace* 1525 SRSx. Lat *Pancratius*, the name of a legendary 1st century martyr.

**Pane, Panes:** v. PAIN

**Pankhurst:** v. PENTECOST

**Pannaman:** Thomas *le Paynerman* 1301 SRY; Adam *Panyarman* 1354 Putnam (L). 'A hawker who carries fish or other provisions in a pannier.' v. PANNER (ii), and cf. Ranulf *Panermaker* 1305 *ERO*, Ralph *le Paniermakier* 1310 MESO (Ess), 'a maker of panniers'. The surname may also be a south-eastern form of PENNYMAN.

**Pannell:** v. PAYNELL. Also metonymic for *Panneler:* Simon *Paneller'* 1240 Eynsham (O); Martin *le Panneler* Hy 3 AD ii (Mx). ME *panel* 'a pannel of wainscot, of a saddle, etc.', hence a maker of panels.

**Panner, Panniers:** (i) Peter, John *le Panner* 1262 *For* (Ess), 1268 AssSo. A derivative of OE *panne* 'a pan', 'one who casts pans', synonymous with (Cristin *le) Pannegetter* 1250 MESO. (ii) Æluric *Paner* c1095 Bury (Sf); John *Panier* 1209 P (W); Nicholas *Paner* 1216 ChR; Edith *Panier* 1279 RH (C). OFr *paniere*, AFr *paner*, ME *panier* 'a basket' (c1300 NED). This might be metonymic for 'a basket-maker', or a hawker. v. PANNAMAN.

**Pannett:** *Painotus* de Norwude 1176 P (D); *Painot* 1200 P (Y); Richard *Painot* 1200 P (Y); Thomas *Paynet* 1332 SRSx. OFr *Payen* plus the diminutive suffixes *-ot*, *-et*.v. PAIN, PAYNELL.

**Pannifer:** v. PENNYFATHER

**Panson:** 'Son of *Pain*', v. PAIN, PANNETT, PENSON.

**Pant, Pante:** William *Pante* 1221 Cur (Do); Robert *Pante* 1327 SRLei. OG *Panto*.

**Pantall:** *Pantul* 1086 DB (St); *Pandulfus* 1219 Cur (Nf); William *Pantulf* 1086 DB (Sa); William *Pantulf, Pantul* Hy 2 DC (Lei); Yvo *Pantolf* 1219 P (Sa). OG *Pandulf, Pandolf*. v. OEByn 223.

**Pante:** v. PANT

**Panter, Panther:** Reginald *le Paneter* 1200 Cur (K), *le Panetier* 1206 P (K); Warin *le Paneter* 1230 Cl (Ha). AFr *paneter*, OFr *panetier*. 'A household officer who supplied the bread and had charge of the pantry' (1297 NED). The panter of a monastery also distributed loaves to the poor.

**Panton:** Ralph *de Panton'* 1190 P (L); Thomas *de Panton* 1255 FFK; Michael *de Panton* 1344 FFEss. From Panton (L), or Great, Little Ponton (L), *Pantone* DB.

**Pantrey, Pantry:** John *de la Paneterie* 1274 RH (Lo); Robert *atte Panetrye* 1332 SRSx. 'The officer of the pantry.' v. PANTER.

**Pap, Papp, Pappe, Papps:** Nicholas *Pappe* 1279 RH (C); Walter *Pap* 1301 FS; Roger *Pappe* 1357 Pat (C). OE *\*Papa*, not recorded in independent use, but the first element of Papworth (C).

**Pape:** Stubhard, Blachemannus *Pape* c1095 Bury (Sf), 1178 P (Sr); William, Adam *la Pape* 1207 Cur (Bk), 1268 AssSo. Some of the earlier forms, particularly those with an English christian name may be from OE *pāpa* 'pope' which became ME *pope* and survives as POPE. Those with the article and French christian names are from OFr *pape* 'pope'. A nickname for one of an austere, ascetic appearance, or a pageant-name.

**Papigay, Pebjoy, Pobgee, Pobjoy, Popejoy, Popjoy:** Robert *Papejay* 1321 AssSt; Robert *Papyngeye* 1371 Cl (Nf); Robert *Popungey* 1397 NorwW (Nf); Roger *Popyngeay* 1410 AD i (Mx); William *Popjoy* 1759 Bardsley; John *Pobjoy* 1784 ib. ME *papejai, popingay*, OFr *papegai* 'parrot', either a nickname from the bird or a title for the winner in the sports. cf. '*Papegay*, a parrot or popingay; also a woodden parrot (set up on the top of a steeple, high tree, or pole) whereat there is, in many parts of France, a generall shooting once euerie yeare; and an exemption, for all that yeare, from *La Taille*, obtained by him that strikes downe the right wing thereof (who is therefore tearmed *Le Chevalier*); and by him that strikes downe the left wing (who is tearmed, *Le Baron*); and by him that strikes downe the whole popingay (who for that dexteritie or good hap hath also the title of *Roy du Papegay*, all the yeare following)' (Cotgrave). Stow notes a similar prize of a peacock for running at the quintain.

**Papillon:** Turoldus *Papilio* c1095 Bury (Sf); Ridellus *Papillun* 1162 P (Nth); Thorald *le Papillon* 1250 Fees (Do). OFr *papillon*, Lat *papilio* 'butterfly', 'un surnom d'homme inconstant, imprudent' (Dauzat).

**Papworth:** Elena *de Papwrth'* 1208 Cur (C); Walter *de Pappeworth* 1311 LLB D. From Papworth Everard, St Agnes (C).

**Paradise:** William *Paradys* 1334 SRK; Robert *Paradise* 1545, Thomas *Paradice* 1576 SRW. 'Dweller by the park or garden', OFr *paradis*.

**Parage, Paraige, Parrage:** William *Parage* 1176 P (Ess); Reginald *Parage* 1271–2 FFEss; John *Parage* 1303 AssW. OFr *parage* 'of high birth'.

**Paragreen:** v. PILGRIM

**Paramor, Paramore, Parramore:** John *Paramour* 1296 SRSx, 1372 ColchCt. ME, OFr *par amour* 'with love', 'lover, sweetheart'.

**Parchment:** Walter *Perchamunt* 1200 P (L). Metonymic for OFr *parcheminier* 'maker or seller of parchment': Gille *Parcheminer* 1180 P (Sa), Richard *parchment maker, Parchemener* 1413, 1420 Oseney (O).

**Parden:** v. PARDON

**Pardew, Pardey, Pardy, Pardoe, Perdue:** Richard *Parde* 1228 FFSf; Robert *Pardey* 1296 SRSx; Henry *Pardeu* 1332 SRWa; Walter *Perdu* Ed 3 Rydware (St). An oath-name, *par Dieu*, perhaps shortened from *de par Dieu* (Lat *de parte Dei*) 'in God's name'. Common in ME as *pardee*: 'I have a wyf, pardee, as wel as thow' (Chaucer). Such names were common. cf. PURDAY, PUREFOY, GODSAFE, and Adam *Parmafey* 1327 SRSf, Rychard *Parmafoye* 1568 SRSf, Richard *Parmoncorps* 1332 SRSx, Alan *Par la Roy* 1302 SRY.

**Pardner, Partner, Partener:** Walter *le Pardoner* 1322 ParlWrits (C). 'The pardoner', a licensed seller of indulgences.

**Pardon, Parden:** Alfred, Thomas *Pardon* 1327 SRWo, 1347 Cl (L). Metonymic for PARDNER.

**Parent, Parrent, Parrant:** Gerald *parent* Hy 2 DC (L); Geoffrey *Parent* 1185 P (Nth); William *(le) Parent* 1316 AssSt, 1327 SRSt. Weekley and Dauzat take this to be a surname of relationship. This is vague. We may well have OFr *parant, parent*, adj. used figuratively of a fine appearance, 'powerful' (Godefroy).

**Parfait, Parfett, Parfit, Parfitt, Perfitt, Parfect, Perfect:** Vnfridus *parfait* 1115 Winton (Ha); Richard *Parfeit, Parfet* 1196 Cur (So); William *Perfyt* 1383 LLB H; Richard *Parfytte* 1568 SRSf. ME *perfit, parfit* from OFr *parfeit, parfit* from Lat *perfectus* 'perfect'. *Parfit(t)* and *Parfett* are the most common forms today. *Perfect* is fairly frequent but is late and due to the influence of the reconstructed, learned spelling of the adjective.

**Parfay:** Richard *Parfoy* 1296 SRSx; Richard *Parfey* 1332 SRWa; John *Perfay* 1379 PTY. An oath name, OFr *par (ma) fei* 'by my faith'. cf. Adam *Parmafey* 1327 SRSf with the same meaning; Richard *Parmoncorps* 1332 SRSx 'by my body'.

**Parffrey, Parfrey:** *v.* PALFREY

**Parford:** *v.* PAFFARD

**Parfrement:** *v.* PALFREYMAN

**Pargeter, Pargetter, Pargiter:** Willelmus *Dealbator* 1207 Cur (Nf); James *Pergetor* 1533 Bardsley (Nf); Edmund *Pargitur* 1617 ib. Pargetting at Marlborough is mentioned in 1237. It was allied to daubing, plaster being used instead of clay or loam. The surface of the parget might be finished smooth, with a coat of whitewash, or as rough-cast with sand or small stones. cf. 'Stephen the Dauber who pargetted the long chamber' at Corfe in 1285 (Building 191).

**Parham, Perham, Parram, Perram:** Turmod *de Perham* 1066 DB (Sf); William *de Perreham* 1187 P (Sx); William *de Parham* 1264 RegAntiquiss; John *Parham* 1366 IpmW. From Parham (Suffolk, Sussex), or 'dweller by the homestead where pear-trees grow'.

**Paris, Parris, Parriss:** (i) Geruase *de Paris* 1158 P (Lo); Willelmus *Parisiensis* 1185 Templars (L); William *de Paris* 1238 AssSo. From Paris, the usual derivation. William *de Paris* Hy 2 DC is identical with Willelmo *Paride* c1150 ib. Here *Paris* (abl. *Paride*) appears to be identical with *Parisiensis* 'the Parisian'. (ii) Magister *Parisius* 1157–63 DC (L); Reinger *filius Paridis* 1203 Cur (Sx); *Parisius* Miles 1311 Battle (Sx); Ralph *Paris* c1220 NthCh (Nth); Richard *Parys* 1293 Fees (D); Adam *Parisson* 1332 SRCu. This is from a personal name which, to judge from the number of early examples without a preposition, was not uncommon. It is the French popular form of (St) *Patrick*, Lat *Patricius* 'patrician', the origin of French *Paris, Patrice, Patris*. The variant *Paris, Paridis* must be due to the influence of the Trojan *Paris*, unknown as a personal-name in either England or France.

**Parish, Parrish:** John *Parisshe, Pareys* 1462–4 LLB L. Identical with PARIS. Paris Hall in North Weald Bassett (Essex), named from Richard *de Paris* (1276), is *Parrishe* in 1593 (PN Ess 87). This development of final *sh* from *s* is found already in 1319 in *Wrabenash* for Wrabness (PN Ess 358). Consequently, in William *de Parysch* 1379 PTY we probably have the same development. Derivation from *parish* is very unlikely. cf. MORRISH, NORRISH.

**Park, Parke, Parkes, Parks, Duparc, du Parcq, Duparcq:** Henry *del Parck* 1272 AssSt; Iselota *atte Park* 1285 AD vi (K); Agnes *del Parkes*, Henry *del Parks* 1304 MELS (Wo). ME *parc, parke* (13th NED), from OFr *parc* 'lists', 'enclosed space', 'park',

'enclosure'. The surname may denote residence in or near a park or enclosure but is, no doubt, often synonymous with PARKER. cf. PANTREY. Modern *Park(e)* is sometimes a contraction of PARRACK. *Duparcq* is a Guernsey name.

**Parker, Parkers:** Anschetil *Parcher* 1086 DB (So); Geoffrey *parchier* c1145–65 Seals (Nb); Claricia *le Parkeres* 1327 SRSo. AFr *parker*, OFr *parquier, parchier, parker* 'parker', 'one in charge of a park', 'park-keeper' (1321 NED).

**Parkhouse:** John *del Parkhouse* 1379 PTY. 'Dweller at the house in the park.'

**Parkhurst:** William *de Parkhurst* (1327 SRSx) lived at Parkhurst Fm in Tillington (Sussex). There are two other Parkhursts in Sussex and others elsewhere.

**Parkin, Parkins, Parkyn, Perken, Perkin, Perkins, Purkins:** Edmund *Perkyn* 1327 SRSf; Robert *Parkyn* 1327 SRSt; Walter *Perkyns* 1327 SRWo; Maud *Parkynes* 1332 SRWa. 'Little Peter', from *Per* (*Peres*) and -*kin*. *v.* PIERCE.

**Parkinson, Parkerson:** John *Parkynson* 1379 PTY; William *Perkynsone* 1382 AssWa; Emmot *Parkyngson* 1540 Whitby (Y). 'Son of *Perkin*.'

**Parkman:** Richard *Parcman* 1307 MEOT (Herts); Richard *Parkeman* 1429 LLB K. ME *park* and *man*, identical with PARKER.

**Parlabean, Parlby:** Richard *Parlebien* 1200 P (L); Roger *Parleben* 1231 FFC. Fr *parle bien* 'speak well'. cf. John *Parlefrens* 1283 Battle (Sx).

**Parlane:** *v.* MACFARLAN

**Parle:** Edward *Parole* c1130 ELPN; Richard *Parle* 1327 SRSo; John *Parl* 1382 Hylle. OFr *parole* 'word', perhaps for *ma parole* 'on my word'.

**Parlett:** Herlewin *Perlet* 1180 P (Nf); Walter *Parlet* 1218 FFSf. A double diminutive of *Per*, contracted from *Per-el-et*. cf. PARKIN.

**Parlour:** (i) Robert *Parler* 1086 DB (W); Richard *le Parlour* 1219 Rams (Hu). OFr *parlier*, AFr *\*parlour* 'avocat, discoureur' (Godefroy), 'lawyer', perhaps 'chatterbox'. Also probably 'the servant who attended the parlour', originally the conversation and interview room in a monastery. cf. Richard *atte Parlur* 1296 SRSx. (ii) Henry, Simon *le Perler* 1291 LLB A, 1297 FFEss. A derivative of ME, OFr *perle* 'pearl', a seller of pearls, or perhaps 'maker of glass pearls', 'paternostrer' (Ekwall). *v.* PEARL.

**Parmenter, Parminter, Pammenter, Parmeter, Parmiter:** Robert *Parmenter* 1177 P (Lei); Vmfridus *parmentier* 1198 P (K); William *le Parmenter* 1204 Cur (O); Robert *le Parminter* 1221 AssWo; Thomas *le permeter* c1250 Rams (C); William *Parmater* 1381 SRSf; Thomas *Parmiter* 1674 HTSf. OFr *parmentier, parmetier* 'tailor'.

**Parnacott, Parncutt:** William *Parncott*, Xr *Parncutt* 1655 HTO. From Parnacott in Pyworthy (D).

**Parnall, Parnell, Parnwell, Purnell:** *Petronilla* Ric 1 DC (L); *Purnelle, Petronille, Peronelle* Kepeherme, -*harm* 1249, 1250, 1253 Oseney (O); *Pernel, Parnell* la Brune 1268, 1280 AssSo; Johanna *Peronele* 1250 Fees (C); Agnes *Peronell* 1274 Wak (Y); Roger *Pernel* 1295 Barnwell (C); Edith *Purnele* 1297 MinAcctCo (W);

Robert *Parnel* 1332 SRSt. *Petronilla*, a feminine diminutive of the Latin *Petronius*, was the name of a saint much invoked against fevers and regarded as a daughter of St Peter. The name was accordingly considered to be a derivative of *Peter* and became one of the most popular of girls' names, the vernacular *Parnell* being still used as a christian name as late as the 18th century in Cornwall. *Parnwell*, in spite of its local appearance, is a corruption, in use in 1670 as a christian name, *Parnwell* Graystocke (LaWills) and in 1801 as a surname, William *Parnewell* (Bardsley).

**Parnham, Parnum:** Thomas *Parnum* 1630, *Parnam* 1633, *Parman* 1647, *Parnham* 1650 LeiAS 23. From Parnham in Beaminster (Do).

**Parr, Par:** (i) Robert, Margeria *Perre* 1275 RH (Y), 1327 SRSf. OFr *Perre, Pierre* 'Peter'. (ii) Henry *de Par* 1284 AssLa; Richard *del Par* 1356 LLB G; Thomas *Par* 1402 FA (We). From Parr (Lancs), or 'dweller at the enclosure'.

**Parrack, Parrick, Parrock:** Artur *de Parrok* 1212 Fees (Ha); John *atte Parrok* 1296 SRSx; Walter *Parrok* 1299 LoCt. OE *pearroc* 'paddock, enclosure', 'Parrocke, a lytell parke, *parquet*' (Palsgrave). Sometimes this has given a modern PARK. Park Fm in Great Horkesley, *Parrokyslond* Ric 2, owes its name to the family of Elias *del Parrok* (1276 PN Ess 393).

  **Parrage:** *v.* PARAGE
  **Parram:** *v.* PARHAM
  **Parramore:** *v.* PARAMOR
  **Parrant:** *v.* PARENT
  **Parratt, Parrett:** *v.* PARROT
  **Parrell, Parren:** *v.* PERRIN
  **Parrick, Parrock:** *v.* PARRACK
  **Parris, Parrish:** *v.* PARIS, PARISH

**Parrot, Parrott, Parratt, Parrett, Parritt, Perot, Perott, Perratt, Perret, Perrett, Perrot, Perrott, Porrett, Porritt:** *Perot* 1246 AssLa; William *Peret* 1066 DB (Sf); Ralph *Perot* 1235 Fees (Ess), 1279 RH (Beds); William *Poret* 1301 SRY; John *Parrat* 1344 LoPleas; William *Parotte* 1470 LLB L; John *Porrett, Perott, Parott, Parrett* 1520 Oxon. Diminutives of *Perre* (Peter) plus the suffixes *-et, -ot. Parratt* and *Perratt* are also from *Pirard* or an undocumented *Perard. v.* PIRRET. Also a nickname from the parrot, itself nicknamed from *Perot*: William *le Perot* 1277 ParlWrits (Wo).

**Parry:** John *ap Harry* 1407 AD iii (He); Morres *Parry*, John *Apharry* 1527, 1528 LP; Richard *Upharry* 1545 SRW; Margery *A parry*, Harrye *Aperry* 1556, 1557 LedburyPR (He); Thomas *Parrye* 1557 Pat (Gl). Welsh *ap Harry* 'son of Harry'.

  **Parsey:** *v.* PEARCEY, PERCY
  **Parshall:** *v.* PEARSALL

**Parsloe, Parslow, Parsley, Paslow, Pasley, Pashley, Pashler:** Radulfus *Passaqua* 1086 DB (Bk); Ralph *Passelewa* 1104–6 Bury (Sf); Hamo *passelowe* 1194 Cur (Bk); Edmund *Passeleye* 1314 AssK; Simon *Passelegh* (*Passelewe*) 1327 *SR* (Ess); Nycholas *Passhloo* 1524 SRSf; Edward *Parsley* 1637 EA (NS) ii (Sf). OFr *Passelewe* 'cross the water'. *v.* PASSMORE. cf. Alan *Pasewater* 1439 NorwW (Nf), John *Paswater* 1568 SRSf. The Domesday tenant's name survives in Drayton Parslow (Bucks). cf. also Parsloes in

Dagenham (Ess): *Passelewesmede* 1390, *Pashlewes* 1456, *Parslo(w)es* 1609, 1634 (PN Ess 92).

**Parson:** William *Persun* 1197 P (Nf); Richard *la Persone* 1221 AssWo; Robert *Parson* 1296 SRSx. OFr *persone*, ME *persone, persoun* 'priest, parson'.

**Parsons:** (i) Roger *le Persones* 1323 AssSt; Alicia *le Parsones* 1327 SRWo; Isabella *Parsones* 1327 SRSo. This is an elliptic genitive, 'the parson's servant'. cf. Eudo *homo persone* 1210 Cur (C), Henry *le Personesman* 1327 SRDb. (ii) Gilbert, Stephen *ad Parsons* 1297 MinAcctCo; Ralph *del Persones* 1323 AssSt; William *atte Personnes* 1327 SRSf; Gilbert *atte Parsones* 1332 SRSx. This again is elliptic, one who lived (or worked) 'at the parson's (house)', at the parsonage.

**Parsonson, Parsison, Parsizon:** Stephanus *filius persone* 12th DC (L); Thomas *le Fiz la Persone* 1250 FFL; John *le Personesone* 1312 AssSt. 'Son of the parson.

  **Partener:** *v.* PARDNER

**Partington:** Henry *de Partinton* 1260 AssCh; Hugh *de Partyngton* 1401 AssLa. From Partington (Ches).

  **Partlan:** *v.* MACFARLAN

**Partleton:** Stephen *de Pertlington* 1327 SRSx. From Pattleton's Fm in Westfield (Sussex), *v.* PN Sx 506.

  **Partner:** *v.* PARDNER

**Parton:** (i) Adam *of Perton'* 1249 AssW; Robert *de Perton* 1249 RegAntiquiss; John *Parton'* 1377 AssWa. From Parton (St), or Parton in Churchdown (Gl). (ii) Matheu *de Partone* 1296 Black (Dumfries). From Parton (Kircudbright).

**Partrick, Partridge, Patriche:** Ailward *Pertriz* 1176 P (D); Geoffrey *Pedriz* 1197 FF (Ess); John *Perdrich* 1244 FFSt; Philip *Partrich* 1260 AssCh; Henry *Pertrik* 1274 RH (Nf); Thomas *le Partrich* 1327 SRSo; Sibil *Partryge* 1332 SRSt; John *Pattriche* 1579 Gt Welnetham PR (Sf); John *Pattridge* 1622 ib. ME *pertriche*, OFr *perdrix* 'partridge'. The surname is very common and whilst often a nickname from the bird, it probably also means 'a hunter or catcher of partridges'. cf. Thomas *Pertricour* 1279 RH (Hu), Nicholas *le Pertricour* 1324 Wak (Y), from AFr *\*perdrichour, \*pertricour*. In the 1524 Suffolk Subsidy Roll the name appears as *Partriche* (frequent), *Parterych, Patrick, Pattrik* and *Patryk*, and in the Hearth Tax for 1674 as *Partridge, Pattridge, Pattrige, Pateridge, Pattarage, Pattrage* and *Putteridge. Patriche*, which is rare, is due to the loss of the first *r*, and a similar loss in *Partrick* gave rise to *Patrick*, so that some of our *Patricks* owe their name to the bird and not to the saint.

**Pascall, Paskell, Pasquill:** *Paskell* Langdon 1606 Bardsley (Co); John *Pascal* 1221 AssWa; William *Pascale* 1275 SRWo; William *Paschale* 1327 SRSf. Fr *Pascal*, the name of a 9th-century pope and saint, from Latin *paschalis* 'pertaining to Easter'. cf. OBret *Paschael* (Loth).

**Pascoe:** *Pascow* Meneux 1424 LLB I; Simon *Pascoe* 1372 ColchCt; John *Pascowe* 1443 AD iv (Co). A form of *Pascall*, long surviving in Cornwall. *Paskell* Langdon (1606) is called *Pascowe* in 1571 (Bardsley).

  **Pasfield:** *v.* PASSFIELD

**Pash, Pashe, Paish, Pask, Paske, Pasque:** Hugo *filius Pasch'* 1279 RH (C); Thomas *Paske* 1253 Oseney (O); Felic' *Pasch'* 1279 RH (C); William *Pasques, Paskes* 1311 LLB D, 1319 SRLo; Walter *Passh'* 1327 SRWo. ME *pasche*(s), *paske*(s), OFr *pasche, pasque*(s) 'Easter', from Hebrew *pesakh* 'a passing over', used as a personal name for one born at Easter. cf. OBret *Pasc* (Loth). cf. William *Paskessone* 1293 FFC 'son of *Paske*'.

**Pashen:** Anthony *Passyon*, John *Passhyn*, Nycolas *Passham* 1545 SRW; Elizabeth *Pashen* 1662–4 HTDo. Perhaps from Passenham (Nth), *Passeham* DB.

**Pashler:** *v.* PARSLOE

**Pashley:** Robert *de Passelegh* 1303 FA (Sx). From Pashley in Ticehurst (Sussex). *v.* also PARSLOE.

**Paske:** *v.* PASH

**Paskell:** *v.* PASCALL

**Paskerful:** *v.* BASKERVILLE

**Paskin, Paskins:** *Pasken* de Stafford, Ralph *Pasken, Paschen* 13th St Thomas, Ronton (St). OW *Paskent*, MW *Pascen*, from Lat *Pascentius* (Förster).

**Pasley, Paslow:** *v.* PARSLOE

**Pasmore:** *v.* PASSMORE

**Pasque:** *v.* PASH

**Pasquill:** *v.* PASCALL

**Pass, Passe:** *Passe* de Albeneye 1230 P (Nt); John *Passe* 1275 RH (Nf), SRWo. Probably a pet-form of *Pascall*. cf. *Passe* 'Easter' (1533 NED).

**Passage:** Ralph *de Passagio*, Agnes *del Passage* 1275 RH (Sf). From residence in a passage or narrow lane.

**Passant, Passavant:** *Passeuant* 1155 FeuDu; William *Passavant* 1198 Cur (L); Andrew *Passeavant* 1212 Cur (Ha); Robert *Passaunt* 1314 Pat (Y). OFr *passe avant* 'go on in front', perhaps a herald or messenger.

**Passboys:** Richard *Passeboys* 1297 SRY. 'Dweller past the wood', OFr *passer, bois*. cf. Adam *Passebusck'* 1297 SRY 'dweller past the bush'; Alfw' *Passe Culuert* 1166 P (Sx) 'dweller past the culvert'.

**Passenger:** Robert *le Passager* Hy 3 AD iii (Sr); Rebecca *Passenger* 1771 Bardsley. OFr *passager* 'wayfarer'.

**Passfield, Pasfield:** Andrew, William *de Passefeld'* 1214 Cur (Ess), 1345 LLB F. From *Passefelda* (1062), now Paslow Hall in High Ongar (Essex). *v.* PN Ess 73.

**Passifull:** *v.* PERCEVAL

**Passmore, Pasmore:** William, Richard *Passemer* 1199 P (Nth), 1242 Fees (D); Walter *Passemore* 1266 AsSt. OFr *passe mer* 'cross the sea', seafarer, sailor. In England the second element was associated with *mere*, interchanging, as often, with *more*. A type of surname common both in France and England. *v.* PASSANT, PARSLOE, and cf. Richard *Passeboys* 'wood', Adam *Passebusck* 'bush' 1297 SRY, Alfwin *Passeculuert* 1166 P (Sx) and the French *Passedouet* (canal), *Passefons* (fountain), *Passepont* (bridge), *Passerieu* (stream). These Dauzat plausibly suggests mean 'one who has to cross the bridge, stream or water, or pass the spring or fountain to reach his home'. cf. PERCEVAL.

**Paster:** This is, no doubt, often for PASTOR, but must also be a derivative of OFr *paste* 'paste or

dough', a baker, synonymous with (John) *Pastemakere* 1340 AssC. cf. Tewaldus *Paste* 1159–85 Templars (Lo); Gilbert *Paste* 1210 Cur (C).

**Pasterfield, Pasterful:** *v.* BASKERVILLE

**Paston:** Geoffrey *de Paston'* 1202 AssNth; John *de Pastone* 1321 CorLo; William *Paston* 1414 FFEss. From Paston (Norfolk, Northants, Northumb).

**Pastor:** Gerardus *pastor* 1140–53 Holme (Nf); Godrich *le Pastur* 1227 Pat (Nf); Dionisius *Pastour* 1317 AssK. AFr *pastour*, OFr *pastor, pastur* 'herdsman, shepherd'. In early sources *Pastor* is often Lat *pastor* 'shepherd'. cf. Willelmo *pastore* 12th DC (L).

**Pasturel, Le Paturel, Pastrell:** Ranulf *Pasturellus* c1110 Winton (Ha); Alexander *Pasturel* 1193–5 WoCh; John *Pastrel* c1265 Glast (So); Arnold *Pasturel* 1281 LLB A. OFr *pastorel* 'shepherd, simpleton'. *Le Paturel*, a Norman diminutive of AFr *pastour* 'shepherd', still survives in the Channel Isles.

**Patch:** Seman *Pac, Pach* 1177–8 P (Sx); Nicholas *Pacche* 1248 Oseney (O); Henry *Patche* c1248 Bec (Do); William *Payche* 1327 SRSf. *Pache* is a ME form of *pasches* 'Easter'. *v.* PACK, PASH.

**Patchett:** *Pachet* 1221 AssWa; Gerard, William *Pachet* 1183, 1195 P (Berks, Lei); Richard *Pascet* c1198 Bart (Lo). A diminutive of *Patch*.

**Patchin:** John *Pachun, Pachon* 1279 RH (O); Robert *Pachyns* 1296 SRSx. Diminutives of *Patch*.

**Patching:** Elena, William *de Pacching*(g) 1327 SRSx, *SR* (Ess). From Patching (Sussex) or Patching Hall (Essex).

**Patchcott:** William *Patchcott* 1642 PrD. From Patchacott in Beaworthy (D).

**Patchell, Petchell:** William *filius Pacchild* 1166 P (Ess); Thomas *Pechel* 1278–9 FFWa; John *Pechel* 1326 FFK. OE *\*Pæcchild* (f).

**Pate, Pates:** Aluric, Osbert *Pate* c1100–30 OEByn (D), 1196 P (Bk). ME *pate* 'head, skull', a nickname (cf. HEAD), or a pet-name for *Patrick*. *v.* PATEMAN.

**Pateman, Patman, Pattman:** *Patein* or *Pateman* Broin 1407 CalSc; Jordan *Pateman* 1219 AssY; Peter *Patemon* 1275 SRWo; John *Patman* 1524 SRSf. A pet-form of *Patrick*.

**Patent:** Thomas *Paton, Paten*, John *Patten*, Edward *Patent* 1524 SRSf. Late forms, with excrescent *t*, of PATON, PATTEN, PATTON.

**Paternoster:** *Paternoster* de Mumbi Hy 2 DC (L); Roger *Pater noster* 1185 Templars (L); Robert *Paternoster* 1197 FF (Bk), 1221 AssWa. The first words of the Latin Lord's prayer are recorded as a noun, ME *paternoster* (c1250 NED) in the sense '(bead in a) rosary'. Here we have the word used both as a personal-name and, by metonymy, for 'a maker of paternosters, rosaries, chaplets, beads strung together for pattering aves' (Bardsley), a natural shortening of the fuller *paternosterer*: William *le Paternostrer* 1280–95 LLB A, B, Jordan *paternosterer* 1303 LoCt. There is a Paternoster Row near St Paul's Cathedral and another near Carlisle Cathedral and there was formerly one near Chertsey Abbey, all centres of this once important industry. In the 14th century the term used was *paternostermaker*: Nicholas *Paternostermaker* 1388 LoCt. Sometimes the name is due to the service by which land was held. In the reign of Edward I John *Paternoster* held a virgate of land in East Hendred (Berks) now called Paternoster Bank by service of saying a paternoster every day for the King's soul (PN Sr 380). In the same reign Alice *Paternoster* held land in Pusey in the same county by service of saying five paternosters a day for the souls of the king's ancestors and Richard *Paternoster* on succeeding to an estate in the same parish, instead of paying a sum of money as a relief, said the Lord's Prayer thrice before the Barons of the Exchequer, as John his brother had previously done (Lower).

**Paterson:** *v.* PATTERSON

**Pateshall, Patsall:** Simon *de Pateshylla, de Pateshull* 1190 P (Nth); John *de Pateshull* 1298 AssL; Richard *Patsale* 1506 FFEss. From Pattishall (Northants).

**Patey:** Walter *Pati* 1275 SRWo; John *Patey* 1381 PTY. A diminutive of *Pate* (Patrick).

**Patman:** *v.* PATEMAN

**Patmore, Pattemore:** Walter *de Patemere* 1208 Cur (Herts); Richard *de Patmere* 1332–3 FFSr. From Patmore (Herts).

**Patney:** Ralph *Patenay* 1357 Hylle. From Patney (W).

**Paton, Patton:** *Paton* 1332 SRCu; *Patone* de Hangaldsyde, *Patrick* Hangangside 1467, 1469 Black; *Pattoun* Millar 1492 ib.; Hugo *Patun* 1230 Pat (Nb); James *Patoun* 1279 RH (Beds); John *Paton* 1413 FFC; Thomas *Pattoun* 1538 Black; Thomas *Patton* 1547 ib. *Pat-un*, a diminutive of *Pat(e)*, i.e. *Patrick*.

**Patriche:** *v.* PARTRICK

**Patrick, Pattrick:** *Patricius* 1175–96 YCh (Y); *Patricius* clericus 1214 Cur (Sf); William *Patric* 1130 P (K); William *Patrik* c1180 WhC (La); John *Pateric* 1229 Cl (Ireland). OIr *Patraicc*, Lat *Patricius* 'patrician'. In England the name was chiefly northern. In Scotland *Patrick* was common in the west and became confused with *Peter*. The surname was common in Ayrshire. In Ireland it only became a popular christian name after 1600, due probably to the Scots settlers in Ulster.

**Patry:** *v.* PETRIE

**Patsall:** *v.* PATESHALL

**Patt:** (i) Gerard *de la Patte* 1219 MELS. Probably 'dweller by the marsh', OE *pat(t)e*. *v.* MELS. (ii) Hugo *Pat* 1206 P (He); William *Patte* 1297 MinAcctCo. Either identical with the above, with early loss of the preposition, or a pet-name of *Patrick*.

**Patten, Pattin, Pattern:** Anger, Robert *Patin* 1182 P (So), 1214 Cur (Ha); Rannulf *Aspatins* 1225 Pat. ME *paten* 'patten, clog' (1390 NED), often metonymic for *patoner* or *patynmaker*: Laurence *Patener* 1381 PTY, John *Patynmaker* 1379 ColchCt, 'a patten-maker'. *Aspatins* means 'the man with the pattens'. cf. 'Pateyne, fote vp berynge (pateyne of tymbyre, or yron, to walke with)' PromptParv. Way notes that pattens were used anciently by ecclesiastics, probably to protect the feet from the chill of the bare pavement of the church. *v.* also PATON.

**Pattenden:** Isolda *de Patindenne* 1317 AssK; William *Pattenden* 1594 PN Sx 516. From Pattenden in Goudhurst, or Great, Little Pattenden in Marden (K).

**Patterson, Paterson:** William *Patrison* 1446 Black (Aberdeen); Donald *Patryson* 1490 ib.; Adam *Patersoun* 1499 ib. 'Son of *Patrick*.' This has probably also become *Pattison*.

**Pattin:** *v.* PATTEN

**Pattinson, Patteson, Pattison, Pattisson, Pattyson, Patison:** Nichole *Patonesone* 1305 Black; John *Patonson* 1332 SRCu; Alexander *Patynson* 1475 Black; Henry *Patyson* 1524 SRSf. 'Son of *Paton* (Patrick).'

**Pattman:** *v.* PATEMAN

**Patton:** John *de Patinton', de Patton'* 1221 AssSa, 1297 SRY. From Patton (Salop, Westmorland). *v.* also PATON.

**Patvine:** *v.* POIDEVIN

**Paul, Paule, Paull, Pawle, Pole, Poles, Poll, Polle, Pool, Poole, Powell, Powle, Powles:** *Powel* c1260 ODCN; Thomas *filius Pole* 1275 RH (Db); Stephen *le fiz Pauwel* 1276 AssLa; *Paulus* Godchep 1292 SRLo; Adam *filius Pole* 1323 AD v (St); Haldanus *Paulus* 1182 P (Sf); William *Pol* 1188 BuryS (Sf); William *Polle* 1193 P (L); Richard *Poul* 1224–46 Bart (Lo); John *Pol* 1275 SRWo; Nicholas *Pole* 1276 RH (Db); William *Poll* 1279 RH (O); John *Powel, Paul* 1292 SRLo, 1307 LLB C; John *Poul* (*Pouel*) 1293 Rams (Hu); John *Pool* 1324 FFEss; William *Powl* 1327 SRC; Adam *Poull* 1346 FA (Sf). Lat *Paulus* 'small'. As a christian name, examples are not common but it can hardly be regarded as 'a very rare name in the Middle Ages' (ODCN) in view of its numerous forms and derivatives. It was already a surname in the 12th century. Weekley regards the vogue of the diminutive *Paulinus* as largely due to Corneille but it is found in Domesday Book in 1066, in 1200–5 Cur (where it is indexed as *Paul*), and as a surname in the 13th century. The everyday form was *Pole* or *Poul* (*Piers Plowman*, Wyclif). The attribute of Belchamp and Wickham St Paul's (Essex) appears as *Pol*(*l*) 1285, *Poel* 1327, *Poulis* 1343, *Powel* 1358, *Poles, Poole* 1607. Pole Hill in Chingford is a relic of the manor of Chingford *Pauls* and Paul's Fm in Little Bardfield is *Pooles* in 1593. *v.* PN Ess *s.nn.* cf. also *Pollekin* 1206 Cur (Sr), John *Polekin, Polekyn* 1279 RH (O), 1297 MinAcctCo, Jordan *Polekoc* 1250 Rams (Bk), Isabella *Polecok* 1274 RH (Berks). *v.* POWELL.

**Pauley, Paulley, Pauly, Pawley:** Geoffrey *Pauly* 1275 RH (C); Marjorie *Pawley* 1515 KentW. A pet-form of *Paul*. cf. Wykkam *Pawley* (Wickham St Paul's) 1545 PN Ess 467.

**Paulin, Pauling, Paulling, Pawlyn, Pawling, Pollins:** *Paulinus* 1066 DB (So); *Paulinus* de Kirtlingtun Hy 2 DC (Nt); Johannes *filius Paulini* 1301 SRY; Basilia *Paulin* 1220 Cur (Mx); John *Paulyn* 1291 FFSf; William *Pawelyn* 1397 PrGR; William *Paulinge* 1607 Oxon (Wo). A diminutive of *Paul*. The feminine *Paulina* is occasionally found: *Paulina* 1169 P (K).

**Paulley, Pauly:** *v.* PAULEY

**Pauncefoot, Pauncefote, Pauncefort, Ponsford:** Bernard *Paunceuolt* 1086 DB (Ha); Hunfridus *pancheuot* 1148 Winton (Ha); Richard *Pancefot* c1180 DC (L), 1197 FF (Wo), *Panceuot* 1195 P (Gl); Richard *Pancefot* (*Paceford'*) 1220 Cur (He); Walter

*Pauncefot* 1280 AssSo; Tracy *Pauncefoot, Pauncefort*
1702 DKR 41. Tengvik explains this name as 'paunch-
face' from ME *panche* 'stomach' from OFr *pance*,
ONFr *panche* (1375 NED) and OFr *volt* 'face', not an
attractive etymology. His description 'The nickname
may indicate corpulence and fleshiness' is, however, to
the point, for the second element is, as Harrison has
noted, OFr *volt* 'vaulted, arched', hence 'the man with
the arched and rounded belly'. By the common
Anglo-Norman interchange of liquids *Paunceuolt*
would become *Paunceuort*, hence *Pauncefort* and
there is some evidence of this in the 1220 *Paceford* and
the form *Poncefortt* given by Lower which would give
*Ponsford*. The surname may also derive from
Ponsford (Devon), *Pantesfort* DB, *Pontesford* 1249
(PN D 561), the first form becoming *Pauncefort*, the
second *Ponsford*.

**Paunchard:** *v.* PANCHARD

**Paveley, Pavly, Paveley, Pawley:** Reginald *de
Paueilli* 1190 P (W), Walter *de Pavily, de Pawely*
1235, 1250 Fees (W). From Pavilly (Seine-Inférieure).

**Paver, Pavier, Pavior, Paviour, Pavyer:** Walter
*Pavier* 1212 Cur (Y); John *Pauur* 1252–3 Clerkenwell
(Lo); William *le Pavour* 1281 LLB A, 1299 LoCt;
Henry *le Paveor* 1327 SRWo. OFr *paveur* 'paviour',
'one who lays pavements' (1477 NED). The status and
skill of the paviours varied from the masters of a craft
to mere labourers who could only dig and ram. The
more skilled were men of standing; in 1398 two
paviours supped with the Fellows of New College,
Oxford. The London Paviours were ultimately
organized under the Wardens of the craft and their
ordinances of 1479 are remarkably modern. They
inveighed bitterly against 'fforeyns and laborers not
havying the verry kounyng of pavying' . . . who 'take
upon theyme the saide works of pavying'. Each
paviour was restricted to two labourers and
apprentices were to be few. But, the bad worker was to
'take up his werks agayn and make hit new and pay vjs
viijd'. *v.* further, G. T. Salusbury, *Street Life in
Medieval England* 13–39. Ornamental tiles and marble
were used for paving the floors of cathedrals, palaces,
etc. In 1308 Hugh le Peyntour and Peter the Pavier
were employed 'making and painting the pavement' at
St Stephen's Chapel, Westminster. *v.* Building 146–7,
MARBER, PAVEY.

**Pavett, Pavitt:** John *Pavet* 1591 Bardsley. *Pav-et*, a
diminutive of *Pavey*.

**Pavey, Pavie, Pavis, Pavy:** Ricardus *filius Pauee*
1156–85 Seals (Wa); Gillebertus *filius Pauie* 1172 P
(So); *Pavia* 1206 Cur (Ess); *Paveya* 1219 AssY;
Robert *Paui* 1219 AssY; Richard *Paveye* 1251 Fees
(Wa). *Pavia*, a woman's name, may be OFr *pavie*
'peach' or *Pavie* 'a woman from Pavia' (Italy). The
surname became confused with PAVER: Thomas
*Pavey* alias *Paviour* 1719 DKR 41.

**Pavley:** *v.* PAVELEY

**Paw, Pawe, Pea, Pee, Poe, Powe:** Tedricus *Paue
filius* c1095 Bury (Sf); *Pavo Cocus* 1203 AssSt; Robert
son of *Pawe* 1277 Wak (Y); Walter, Robert *Pa* Ric 1
Gilb (L), 1260 AssY; Robert *Paue* 1202 AssL;
William, Morice *Powe* 1207 P (Y), 1327 SRSf; John,
Richard *Paw(e)* 1312 AD ii (C), 1327 SRSx; John,

William *le Poo* 1324 FFEss, 1332 SRSt; William,
Thomas *Poo* 1327 SRSf, *SR* (Ess). *Paw* and *Powe* are
from OE *pāwa, Pea* and *Pee* from OE *pēa, Poe* from
ON *pá*, ME *pō* 'peacock', used both as a personal-
name and as a nickname.

**Pawle:** *v.* PAUL

**Pawlet, Pawlett:** (i) William *de Poulet* 1296 Glast
(So); James *Pawlett* 1642 PrD. From Pawlett (So). (ii)
Adam *Pauelot* 1317 Wak (Y). *Paw-el-ot*, a double
diminutive of *Paw*, a nickname from *pāwa* 'peacock'.

**Pawley:** *v.* PAULEY, PAVELEY

**Pawling:** *v.* PAULIN

**Pawson:** Simon, Hugh *Paweson* 1379 PTY, 1471
GildY. 'Son of *Pawe* (Peacock).'

**Paxford:** Robert *de Paxford* 1210 Cur (Wo).
From Paxford in Blockley (Wo).

**Paxman:** William *Paxman, Pakysman* 1439, 1443
LLB K. 'Servant of *Pack* or *Pake*.'

**Paxton:** Azelin *de Paxtun* 1180 P (Lei); Richard *de
Paxton* 1251 FFY; John *Paxton* 1489–90 FFWa.
From Paxton (Hunts).

**Pay, Pey:** *Paie* filius Wlstani, *Paie* Blancheard
1142 NthCh (L); Elias, Richard *Pay(e)* 1275 RH (D),
1296 SRSx; Roger *Peye* 1293 AssSt. ME *pē*
developed to *pai* which may explain some late forms
of the surname as in *Paycock, Paybody*, but the above
personal-name is much too early for this
development. Peakirk (Northants) takes its name
from St *Pega*, sister of St Guthlac, who is reputed to
have established a cell there. One, at least of the above
personal-names, is that of a man, and is probably a
survival of OE *Pæga*, recorded only as the name of a
monk and of an abbot.

**Paybody:** *v.* PEABODY

**Paykin:** *v.* PAIKIN

**Payle:** *v.* PAIL

**Payling:** *v.* PALIN

**Payler:** *v.* PALER

**Payman, Peyman, Peaman:** John *Peman* 1327
SRC; William *Peyeman* 1351 AssEss. 'Servant of *Pea*
or *Pay*.' *v.* PAW, PAY, PEACOCK.

**Payn, Payne, Paynes:** *v.* PAIN

**Paynell, Panal, Panell, Pannell, Pennell:** Radulfus
*Pagenel, Paganellus* 1086 DB; Jordanus *Pænellus*
1130 P (Y); William *Painel* 1160 P (Y); Adam *Painell,
Panell'* 1172 Gilb (L); William *Paynel* 1268 AssSo.
Lat *Paganellus*, OFr *Payenel*, diminutives of *Paganus*
and *Payen. v.* PAIN.

**Paynter:** *v.* PAINTER

**Pays, Payze:** *v.* PACE

**Payton, Peyton, Paiton:** Matthew *de Peytun* 1240
FFEss; John *de Payton* 1292 FFEss; Edmund *Peyton*
1389 LoPleas; Ann *Payten* 1603 IckworthPR (Sf).
From Peyton (Essex, Suffolk).

**Pea:** *v.* PAW

**Peabody, Peberdy, Pebody, Paybody:** Thomas
*Paybodie* 1615 Oxon; Francis *Peboddy* 1635
Bardsley. Perhaps 'servant of Pay', OE *Pæga, bodig*.
*Peberdy* may also be local: Cyneweard *æt Pebbeworðy*
c1012 OEByn (Gl). From Pebworth (Gl), *Pebbewurðy*
c1012.

**Peace:** *v.* PACE

**Peach:** *v.* PETCH

**Peacher:** v. PETCHER

**Peachey:** v. PETCH

**Peachurch:** Henry *de Peichirche* 1290 CartNat. From Peakirk (Nth), *Peychirche* 12th.

**Peacock, Peacocke, Peecock, Pacock, Pocock:** *Pecoc* 1086 DB (Ess); *Pecoc de Briminton* 1285 AssCh; Roger *Paucoc* 1194 P (Co); Roger *Paucoc, Pecoc* 1194–5 P (Co); Richard *Pocok* 1225 AssSo; Simon *Pacock* 1297 SRY; Robert *Pecok* (*Paycok*) 1327 *SR* (Ess). OE *pēacocc* 'peacock', ME *pecok, pacok, pocok*, a nickname used also as a personal-name. v. PAW, PAY, PAPIGAY.

**Peacop:** v. PICKUP

**Pead, Peade:** Walter *Pede* 1249 AssW; Thomas *Pede* 1327 SRSf; John *Peede* 1595 AssLo. OE *Pēoda*.

**Peagram, Peagrim:** v. PILGRIM

**Peak, Peake, Peek:** (i) Richard *del Pec* 1192 Eynsham (O); William *de Peke* 1296 SRSx; William *atte Peke* 1321 LLB D. 'Dweller by the peak or hill', OE *pēac*. (ii) Uluric *Pec* c1095 Bury (Sf); Richard, Henry *le Pek* 1297 MinAcctCo, 1327 SRSx; William *Peke* 1327 SRC. Probably OE *pēac* used as a nickname for a stout, thick-set man. The surname is usually topographical in origin and *le* may be an error for *de*.

**Peaker, Peakman:** Walter *le peker* 1212–23 Bart (Lo). 'Dweller by the peak or hill.'

**Peakman:** Alice Bubwith alias *Pekeman* 1463 TestEbor. 'Dweller by the peak or hill', OE *pēac, mann*.

**Peaman:** v. PAYMAN

**Pear, Peear, Peer, Pere:** Osbert *Pere* 1230 P (Berks); Richard *le Pere* 1279 RH (Hu). ME *pere*, OFr *per, peer* 'a peer, paragon', also 'a match, companion'.

**Pearce:** v. PIERCE

**Pearcey, Pearcy, Piercey, Piercy:** Ralph *Percehaie* 1086 DB (Berks); Walter *Percehaie, Percehaye* 1168 P (Y), 1280 Guisb (Y); Ralph *Parcehaye* 1247 AssBeds; Roger *Perseye* 1331 AD vi (Beds); Thomas *Pearcye* 1652 RothwellPR (Y). OFr *percehaie* 'pierce hedge', from OFr *percer* 'to pierce and *haie* 'hedge', which may have been used either of a hedge protecting a forest or enclosure, or of a military work. Hence either 'a poacher' or 'a warrior renowned for forcing his way through fortifications'. This may also have become PERCY, PARSEY.

**Pearch:** Geoffrey *del Perche* 1199 MemR (W); Adam *de Perche* 1221 Cur (Mx). ME *perche*, a measure of land, perhaps for one who cultivated this area. Or it may be used in its original sense 'rod', a measurer of land.

**Peareth:** Gamel *de Penred* 1190 P (Y). From Penrith (Cumb), still locally pronounced *Peereth*.

**Pearl:** Henry, Reginald *Perle* 1259 Oseney (O), 1316 WhC (La). ME, OFr *perle* 'pearl', metonymic for a seller of pearls. cf. PARLOUR.

**Pearlman, Perlman:** cf. PEARL, PARLOUR.

**Pearman:** Robert *Pyrman* 1296 SRSx; Gilbert *Perman* 1376 LLB H. A grower or seller of pears, OE *peru, pyrige*.

**Pearsall, Parshall, Persall:** William *Persale* 1310 ColchCt; Thomas *Persall* 1560 Pat (Ch). From Pearshall (St).

**Pears(e):** v. PIERCE

**Pearson, Pearsons, Paerson, Peirson, Pierson:** John *Pierisson* 1332 SRWa; Robert *Peresson* 1395 Whitby (Y); William *Pierson* 1412 WhC (La); John *Pereson* 1440 ShefA (Y). 'Son of *Piers* (Peter).' v. PIERCE.

**Peart, Pert, Perts:** Richard *Perte* 1227 AssBk; Henry *Perte* 1324 CoramLa; John *Apert* 1344 FFEss; John *Peert* 1453 NorwW (Nf); Henry *Peart* 1665 HTO. A shortened form of OFr *apert* 'ready, skilful'.

**Peartree:** Richard de *Peretre* 1230 P (Nth); Emma *atte Peretre* 1279 RH (Hu); Robert *del Pertre* 1327 SRSf. 'Dweller by the pear-tree.'

**Peascod, Pescod, Pescodd, Pescud, Peasegood, Peasgood, Pescott, Peskett, Bisgood:** Richard *Pisecod* 1221 AssWa; John, Walter *Pesecod* 1279 AssNb, 1332 SRCu; William *Piscod, Puscod* 1327, 1332 SRSx; John *Pesegod* (*Pesecod*) 1317 AssK; John *Pasegude* 1441 GildY. OE *peose, pise* 'pea' and *codd* 'bag', a peascod, peapod; probably for a seller of peas. cf. Richard *pesemongere* c1198 Bart.

**Pease:** Thomas *Pese* 1194 Cur (Bk); Roger *Pise* 1206 Cur (Nf); Margeria *Pyse* 1327 SRSf. OE *peose, pise*, ME *pese* 'pea', metonymic for a seller of peas.

**Peasey:** v. PIZEY

**Peat, Peate, Peet:** Ralph *Peet* 1210–11 PWi; Richard *Peet* 1327 SRWo; William *Peate* 1642 PrD. OE *Pēot*, or a pet-form of *Peter*.

**Peatrie:** v. PETRIE

**Peavot:** Robert *Pefot* 1202 Pleas (C); Alice *Payfot* 1327 SRC. 'Peacock foot', OE *pēa, fōt*.

**Peberdy:** v. PEPPERDAY

**Peberdy, Pebody:** v. PEABODY

**Pebjoy:** v. PAPIGAY

**Pechin:** William *Peccin'* (*Peccat'*), Robert *Peccin* (*Peche*) 1176–7 P (Ha). A diminutive of PETCH.

**Peck:** (i) Richard *Pecke* 1187 P (Ha); Hervicus *Pecke* 1283 FFSf. This, like the topographical term below, might be for PEAK. Or it may be ME *pekke* 'a peck', metonymic for a maker of pecks or vessels used as a peck measure. cf. PECKAR. (ii) John *de Peck* 1275 RH (Lo); Henry *del Peck* ib. (L). 'Dweller by the peak', with vowel shortened.

**Peckar, Pecker:** Henry *Peckere* 1221 ElyA (C); Simon *le Peckere* 1279 RH (Hu). Either 'dweller by the hill' or 'maker of peck measures'. v. PECK. Or we may have a derivative of ME *pekken* 'to peck, snap up', used of one of acquisitive habits.

**Peckett, Peckitt:** William *Pecche, Pekat* 1088, c1100 Rams (C); William *Pechet* 1194 P (Ha); Robert *Pecket* 1198 FF (Nth). A diminutive of PETCH. William *Pekat* was of a Cambridgeshire family usually called *Pecche*.

**Peckham:** Salomon *de Pecham* 1317 AssK; Robert *Pekham* 1351 ColchCt; William *Peckham* 1662–4 HTDo. From Peckham (Sr), or East, West Peckham (K).

**Peckover, Pecover, Pickaver:** Robert *Picauer* c1255 RegAntiquiss; William *Pikhauer* 1321 WRS; William *Pickover* 1566 Oseney. A nickname, 'pick oats', ME *pikken*, ON *hafri*. cf. Simon *Pickebarli* 1200 Cur (Sf) 'pick barley'; John *Pyckeble* 1327 SRSf 'pick corn'.

**Pedcock:** Adam *pie de coc* 1202–3 FFWa; John *Pedekok* 1334 SRK. 'Cock's foot', Fr *pied de coq*.

**Peddar, Pedder:** William *le Pedder* 1165 P (Y); William *Peddere* 1243 AssDu. A derivative of ME *pedde* 'a pannier', hence 'one who carries goods for sale', 'a pedlar' (a1225 NED).

**Pedlar, Pedler, Pidler:** (i) William *Le Pedelare* 1307 MESO (Wo); Ralph *le Pedeler* 1332 SRWa; Martin *Pedeler* 1376 LoPleas. 'Pedlar, hawker' (1377 NED). (ii) Walter *Pedeleure* 12th HPD (Ess); William *Pe de levre* 1242 Fees (So); Cristina *Pedeleuere* 1327 *SR* (Ess). Fr *pied de lievre* 'hare-foot', speedy, nimble. cf. Alan *Pedeken* 1187 Gilb (L) 'dog-foot', Robert *Piedurs* 1199 P (O) 'bear-foot', John *Pedebef* 1381 SRSf. 'ox-foot'.

**Pedley:** Alice *Pedele* 1263 IpmW; Thomas *Pedele* 1385 FFHu. From Pedley Barton in East Worlington (D).

**Pedlow, Pillow:** Simon *Piedeleu*, William *Piedleu* 1367 LLB G. AFr *pie de leu*, Fr *Piédeloup* 'wolf-foot'. This may also have become PELLEW, PELLOW.

**Pedmer, Pedmore:** Alice *Pedmer* 1275 SRWo. From Pedmore (Wo).

**Pedrick:** *v.* PETHERICK

**Pedwell:** Robert *de Pedewelle* 1189 Sol; Ralph *de Pedewell* 1243 AssSo; Agnes *Pedwil* 1571 HartlandPR (D). From Pedwell (So).

**Pee:** *v.* PAW

**Peech:** *v.* PETCH

**Peecock:** *v.* PEACOCK

**Peek:** *v.* PEAK

**Peel, Peele, Peile, Piele:** (i) Walter, Henry *Pele* 1202 AssL, 1238 FFY; Robert *Peel* 1382 IpmNt. OFr *pel* 'stake', a nickname for a tall, thin person. (ii) Robert *de Pele* 1199 MemR (So); John *de Pele*, William *de la Pele* 1301, 1332 Kris. From Piel or Peel Island (Lancs), or 'dweller by the palisade', OFr *pel*, *piel*, cf. John *le Peler* c1280 SRWo.

**Peerless:** Richard *Pereles* 1377 Misc (Sf); John *Pereles* 1472 TestEbor. 'Without equal, peer', OFr *pair*, OE *-lēas*.

**Peers:** *v.* PIERCE

**Peet:** *v.* PEAT

**Peever, Peevers, Peffer:** Roger *Peiure* 1198 FFEss; William *Paiuer* 1219 AssY; William *Peyforer, le Peyfrer, Peyfore* 1293, 1301, 1324 FFEss; Vincent *le Peuerer, le Peuerier* Hy 3 AD ii (Mx); Roger *Peuere* 1297 MinAcctCo. OFr *peyvre* 'pepper', *peyvrier, pevrier* 'pepperer', both used at times of the same man.

**Pegden:** Stephen *de Pehedenn*, Richard *de Peghedenn* 1296 SRSx. From Pegden Fm in Lindfield (Sx).

**Pegg, Peggs:** Æluric *Pegga* 1165 P (Wa); Turbert, Stephen *Peg* 1192, 1195 P (Ha, Y); Ralph, Simon *Pegge* 1243 AssSo, 1301 SRY. ME *pegge* 'peg', metonymic for a maker or seller of pegs. We may also have a pet-name for *Margaret*. cf. Martin *Peggi* 1279 RH (O); John *Peggy* 1338 Oseney (O).

**Pegram, Pegrum:** *v.* PILGRIM

**Peirce, Peire(e):** *v.* PIERCE

**Peirson:** *v.* PEARSON

**Peiser:** *v.* POYSER

**Pelerin:** *v.* PILGRIM

**Pelham:** Ralph, Peter *de Pelham* 1170 P (Herts), 1260 AssC; William *Pelham* 1350 IpmGl. From Pelham (Herts).

**Pelican:** Thomas *Pellican* 1316 FFK; John *Pellycan* 1404 LLB I; Robert *Pellican* 1416 FFC. Either a nickname from the pelican, or from a shop sign.

**Pelissier:** A Huguenot name. Abel *Pelessier*, a refugee Huguenot officer, settled at Portarlington in Ireland.

**Pell, Pells, Pelly:** *Pelle* 1274, 1316 Wak (Y); Walterus *filius Pelle* 1279 RH (Hu); *Pelle* chaundeler 1332 SRLo; Reginald *Pel* 1222 Cur (K); John, William *Pell(e)* 1260 AssC, 1279 RH (O); Robert, William *Pelleson* 1331 DbCh, 1412 GildY. *Pell* is a pet-form of *Peter*, with a diminutive *Pelly*. *Pelle* de Honeden (1296 Ipm (Sf) is indexed as *Peter*, and *Pelly* Wyth (1278 Oseney (O) ) is identical with *Peter* son of John (Ed.). cf. *Pell* Heigham (J. Kirby, *Suffolk Traveller*, 1764). *Pell* may also be OFr *pel* 'skin', for a dealer in skins. cf. FELL, PELLY, PILL.

**Pellatt, Pellett:** William *Pelet* 1297 Wak (Y), 1327 SRSx. Perhaps a diminutive of *Pell* (Peter), or, like Fr *Pelette*, of *pel* 'skin'.

**Peller:** William *le Peller'* 1263 MESO (Sx); John *le Pallere* 1327 SRSt. A derivative of OE *pæll, pell* 'costly cloak, purple cloth', for a maker or seller of these. cf. Rober Dun, *peller* 1332 SRLo. In Sussex, this might also mean 'dweller by the stream'. *v.* PILL.

**Pellew, Pellow:** Anketil *Pel de lu, Pealdelu, Piaudelu* 1195–7 P (L), 1202 AssL. OFr. *pel* (ModFr *peau*) *de lou* 'wolf-skin'.

**Pelling:** William *Pelling'* 1222 Acc; Hammyng *de Pellyng'* 1296 SRSx; Adam *Pelling* 1327 SRSf. From Peelings in Westham (Sx), *Pellinges* DB. Sometimes, perhaps, from Peatling Magna, Parva (Lei), Robert *de Pedlinge* 1193 P (Lei/Wa).

**Pellson, Pelson:** Robert *Pellesone* 1327 SRLei; John *Pelson'* 1384 AssL; William *Pelleson* 1412 GildY. 'Son of *Pell*', a pet-form of *Peter*.

**Pelly, Le Pelley:** Arnald *Pele* 1210 Cur (C); Gerard *le Pele* 1221 AssGl; Guillerm *le Peley* 1419 DKR 41. OFr *pele*, ModFr *pelé* 'bald'. *v.* also PELL.

**Pelter, Pilter:** William *le Peleter* 1219 AssY; Adam *le Peletur* 1296 SRSx; Henry *le Pelter* 1301 SRY; William *Pilter* 1332 SRSx. OFr *peletier* 'fellmonger, furrier' (1389 NED); also 'a dresser of fells'. Now a rare name.

**Pemberton:** Alan, Hugh *de Pemberton* 1212 Fees (La), 1387 IpmLa; John *Pemberton* 1443 FFEss. From Pemberton (Lancs).

**Pembery, Pembury, Pembrey, Pembro:** William *de Pember'* 1220 Cur (Ha). From Pembury (K).

**Pembridge:** Ralph *de Penebrigg'* 1212 P (He); Henry *de Penbrugge* 1272 AssSt; William *de Penebrugge* 1295 PN Sx 190; William *Pembrigge* 1449 AssLo. From Pembridge (He), or Pen Bridge in Shipley (Sx).

**Pembro, Pembury:** *v.* PEMBERY

**Pendant:** Richard *del Pendant* 1274, Margery *del Pendaunt* 1275 Wak (Y); Thomas *Pendaunt* 1379 PTY. 'Dweller on the slope', OFr *pendant*.

**Pendcrow:** Henry *Pendecrowe* 1295 AssSt. A nickname, 'hang the crow', OFr *pendre*, OE *crãwa*. cf. Reginald *Pendeleu* 1301 SRY 'hang the wolf'.

**Pender:** *v.* PINDAR.

**Penderel, Penderell, Pendrell, Pendrill:** Richard *Pendorayl* 1253–63 Seals (Lei). A derivative of Fr *pendre* 'to hang' and *oreille* 'ear', 'hang-ear'.

**Pendergast:** *v.* PRENDERGAST

**Pendlebury:** Adam *de Penhulbur* 1205 WhC; Robert *de Penelbyry* 1246 AssLa; Roger *de Penylburi* 1302 StThomas. From Pendlebury (La).

**Pendleton:** Alexander *de Penelton'* 1181 P (Y). From Pendleton (La).

**Pendock:** Walter *de Pendoch'* 1196 P (Wo); Walter *Pendok* 1242 Fees (W); John *Pendok* 1346 SRWo. From Pendock (Wo).

**Pendred:** Gamel *de Penred* 1189 P (Y); Margaret *Penred* 1487 NorwW; John *Pendred* 1568 SRSf. From Penrith (Cu), *Penred* 1167.

**Pendry:** *v.* PENRY

**Pendue:** Henry *Pendu* c1190–1200 LuffCh. 'The hanged man', Fr *pendu*.

**Peneycad:** *v.* PENNYCAD

**Penfare:** *v.* PENNYFATHER

**Penfold:** *v.* PINFOLD

**Penford:** Robert *de Pendeford* 1220, *de Penneford* 1222, Juliana *de Pendeford* 1340 AssSt. From Pendeford (St).

**Pengelley, Pengelly, Pengilly:** John *(de) Pengelly* 1297 MinAcctCo. From Pengelly (Cornwall).

**Penhale:** Ralph *de Pennal* c1210 Fees (Co). From Penheale in Egloskerry (Cornwall).

**Penicud:** *v.* PENNYCAD

**Penkethman, Penkeyman:** T. *Penkethmonne* 1525 GildY; Thomas *Pingithmoan* 1533 YWills. On the analogy of the numerous 14th-century Yorkshire surnames like *Matthewman*, this is rather 'servant of a man named *Penketh*' than 'the man from Penketh'.

**Penley, Penly:** William *Penle* 1279 RH (C); William *de Penlegh* 1340 FFW. From Penleigh in Dilton (W).

**Penn:** (i) Warin *de Penne* 1176 P (St); Walter *de la Penne* 1196 P (Bk); John *ate Penne* 1297 MinAcctCo. From Penn (Bucks, Staffs) or from residence near a fold or hill, OE *penn* 'pen, fold' or OE *penn* 'hill'. The more usual source. (ii) Adam son of *Penne* 1277 Wak (Y); *Penne* Ronge 1327 SRC; John *Penn(e)* 1327 SRSx, *SR* (Ess); Katherine *Pennes* 1345 ColchCt. *Penne* is probably a pet-form of *Pernel*. *v.* PARNALL, PENNALL.

**Pennall, Pennell, Pennells:** (i) William *Pennel* 1377 ColchCt; Petherick *Pernell, Pennell* 1580, 1583 Bardsley. An assimilated form of PARNALL. (ii) Aluredus *de Penhull'* 1221 AssWo. From Penn Hall in Pensax (Worcs), earlier *Penehull*, or from Penhill (Devon, NRYorks).

**Pennefather:** *v.* PENNYFATHER

**Penner:** John *le Penner* 1327 SRSo. A derivative of ME *pennen* 'to impound', pinder. Or, 'worker at the fold', or 'dweller on the hill'. *v.* PENN.

**Penney, Pennie, Penny, Penning, Pennings:** Gilebertus *filius Pening* 1206 P (L), probably identical with Gilebertus *Penning* 1204 P (L); Ralph *Penig* 1191 P (Hu); Ailnoth *Peni* 1204 Cur (Sr); William, Ralph *Pening* 1206 AssL, 1219 AssY; Alan *Pani* 1219 AssL; William *Peny* 1221 AssSa; Geoffrey *Pennyng'* 1305 Clerkenwell (Lo). This is usually a nickname from the

coin, OE *pening, penig*, ME *peni*. *Pening* is clearly used as a personal name in Lincs in 1206 and may be a survival of a similar, unrecorded use in OE, thus lending support to the derivation of Penistone (WRYorks), earlier *Peningeston*, from this personal name. The surnames may also, at times, be patronymics.

**Pennycard:** *v.* PENNYCAD

**Penniall, Pennyall:** Agnes *Paniale* 1327 SRSx; Ralph *Penyale* 1436 IpmNt; John *Penyall* 1524 SRSf; Leonard *Peniale* 1577 Musters (Nf). 'Penny ale', OE *penig, ealu*, a nickname for a brewer or seller of ale.

**Penniless:** John *Peniles* 1332 SRCu. 'Pennyless', OE *penig, lēas*. cf. Henry *Penistrang* 1297 MinAcctCo 'penny strong', perhaps 'wealthy'; William *Takepeni* 1317 AssK 'take penny'.

**Penniman:** *v.* PENNYMAN

**Penning, Penny:** *v.* PENNEY

**Pennington:** Benedict *de Penington'* 1185 P (Cu); Nicholas *de Penyngton* 1279 AssSt; William *de Penynton* 1340–1450 GildC; John *Penyngton* 1379 PTY. From Pennington (Ha, La).

**Pennyall:** *v.* PENNIALL

**Pennycad, Pennycard, Penneycard, Peneycud, Penicud:** William *Pennycod* 1503–4 FFSr; George *Penycod* 1525 SRSx; Oliver *Penicod* 1634 SxAS 86. The forms are late, and look like OE *penig* 'penny', and *codd* 'bag', perhaps 'money-bag', cf. Aluric *Penipurs* 1066 Winton (Ha) 'penny purse'. But note Fr *Pénicaud* 'runner, boaster'.

**Pennycock, Pennycook, Pennycuick:** David *de Penikok* 1250 Black, *de Penycuke* 1373 ib. From Penicuik (Midlothian). The first two surnames are found early in the southern half of England and must be independent formations: Hugo . . . *Penigcoc* 1202 FFL; Thomas *Penycok'* 1379 PTY; John *Penycok(e)* 1452 AD v (W), 1483 AD i (Sr), 1485 AD iv (Ess). '*Cock* (son) or *cook* of Penny.'

**Pennyfather, Pennefather, Pennyfeather, Penfare, Pannifer:** Godwin *penifeder* 1066 Winton (Ha); Robert *Panyfader* 1296 SRSx; Geoffrey *Penistone* 14th AD iii (Wa); Thomas *Pennyfather* 1563 FFHu; Mary *Panifee*, Salmon *Panivie* 1749, 1781 SfPR. OE *penig* and *fæder* 'penny-father', a miser (1549 NED). cf. 'Pinse-maille, a pinch-pennie, scrape-good, niggard, penie-father' (Cotgrave). Pennyfeathers in High Roding was named from Richard *Panyfader* (1285 PN Ess 493). *Pannifer* preserves the southeastern ME *pani*. For *Pannifer* and *Penfare*, cf. *granfer* for *grandfather*.

**Pennyman, Penniman, Pannaman:** Simon *Penyman* 1268 AssSo; William *Peniman* 1279 RH (C); Ralph *Panyman* 1296 SRSx. 'Servant of *Penny*.' *v.* also PANNER.

**Penrose:** Philip *de Penros* 1195 P (Co). From Penrose (Cornwall, Devon).

**Penry, Pendry:** Cadogann *Ab-Henry* 1294 Bardsley; Howel *ap Henri* 1316 ParlWrits; Joane *Pendrie* 1605 Bardsley (He); Joseph *Penry* 1748 ib. Welsh *ab, ap* 'son' and *Henry*, 'son of *Henry*'.

**Penson, Pensom:** (i) Richard *Paynesone* 1305 SIA iii. 'Son of *Pain*.' (ii) Thomas *Pennesone* 1337 ColchCt. 'Son of *Penn*.' (iii) Edmund *de Penson* 1297 MinAcctCo. From Penson (Devon).

**Pentecost, Pentercost, Pentycross, Perrycost, Pankhurst:** *Pentecoste* de Wendleswurda 1187 P (Sr); Gaufridus *filius Pentecostes* 1201 Cur (Mx); William, John *Pentecost* 1200, 1212 Cur (Sr, Sx); John *Pantecost* 1371 Cl (Mx); Rychard *Pencost* 1560 SxWills. OFr *Pentecost*, a name given to one born on that festival. The seal of Cristina *filia Pentecuste* (1250 Oseney) bore the legend: S. CRISTINE PEDTECOST. She had adopted as her surname the christian name of her father who was probably *Pentecost* de Oxonia (1230 ib.). Pankhurst (Surrey), named from John *Pentecost* (1332 PN Sr 116), is *Pentecost* alias *Panchurst* in 1605.

**Pentlow, Pentelow, Pentolow:** Humfrey *de Pentelawe* 1248 FFEss; John *de Pentelowe* 1287 PN Herts 191; James *de Pentlowe* 1329 FFW. From Pentlow (Ess).

**Penton:** Reginald *de Penton'* 1196 P (Do); Matilda *de Penton* 1297 MinAcctCo; Thomas *Pentan* 1327 SRWo. From Penton Grafton, Mewsey (Ha).

**Pentrich, Pentridge:** Simon *Pentrich* 1306 AssW; William *de Penrich* 1332 SRLa. From Pentrich (Db), or Pentridge (Do), *Pentric* DB.

**Penwarden:** Elisha *Penwarden* 1688 FrYar. From Penwortham (La).

**Pepall, Peaple, Peopall, Peoples:** Wyun son of *Pepel* 1246 AssLa; Hawis' *Pepell* 1301 SRY; John *Pepil* 1324 Wak (Y). Apparently a hypocoristic of *Pepin*.

**Pepdie:** *v.* PEPPERDAY

**Pepin, Peppin, Pepys, Pippin, Pipon:** Ralph, Henry *Pipin* 1086 DB (Lei), 1195 P (Y); John *Pepin* c1160 DC (L), 1202 FF (Nf); William, Walter *Pipun* 1176 P, 1212 Cur (Bk); Reginald *Peppin* 1205 Cur (Lei); William *Pippin* 1279 RH (Beds); John *Pepes*, Richard *Pepis* 1279 RH (C); William *Peps* 1377 AssEss; Widow *Peaps* 1671 HTSf. OFr *Pepin*, OG *Pipin*, *Pepin*, from the root *bib-* 'to tremble', popular in France in memory of the founders of the Carolingian monarchy, Pépin d'Héristal and Pépin le Bref, with the occasional variant *Pipun*. *Pepis* is a nominative form. Later examples of the surname may be from OFr *pepin*, *pipin* 'seed of a fleshy fruit', used for a gardener.

**Peplow, Peploe, Pepler:** Richard *de Peppelowe* 1327 SRSa. From Peplow (Sa).

**Pepp, Pepps:** Elfled *Peps* 1209 FFNf; Anabel *Pep* 1251–2 FFY; William *Peps* 1379 FFEss. OE *Pyppa*.

**Peppar, Pepper:** Robert, Alice *Peper* 1197 P (Nf), 1241 FFEss. OE *pipor* 'pepper', metonymic for a dealer in pepper, a pepperer or spicer. John *Pepper* alias *Peyure* (1298 LoCt) alternated between the English and the French form of the name. *v.* PEEVER.

**Pepperall:** *v.* PEVERALL

**Peppercorn, Peppercorne:** Adam *Pepercorn* 1198–1212 Bart (Lo); Roger *Pipercorn* 1202 AssL; Richard *Pepercorne*, spysar 1379 PTY. OE *piporcorn* 'peppercorn', most probably a name given to a seller of peppercorns, a pepperer or spicer. But, in view of such phrases as 'peppercorn-sized' and 'peppercorn-hair', we may also have a nickname for a little man or one with darkish complexion or hair. Sometimes we may have reference to the holding of land at a peppercorn rent.

**Pepperday, Pepdie, Peberdy, Pipperday:** Stephen *Papedi* 1166 Black; Henry *Papedi* 1180 P (Y); Thomas *Popedy* 1260 AssY; John *Pabdy* 1381 PTY; Nichola *Pepdie* 1403 Black; John *Papeday* 1504 ib. An oath-name, OFr *pape-Dieu* '(by the) Pope-God'.

**Pepperell, Pepperwell:** *v.* PEVERALL

**Pepperwhite:** William *Piperwhit* 1225 AssSo; Roger *le Piperwhite* 1327 SRSx; Margaret *Peperwhyte* 1363 AssY. Apparently a nickname, 'as white as pepper', OE *pipor, hwīt*.

**Peppin:** *v.* PEPIN

**Pepprell:** *v.* PEVERALL

**Pepps:** *v.* PEPP

**Pepys:** *v.* PEPIN

**Perceval, Percival, Percifull, Purcifer, Passifull:** *Perceval* 1224 Pat; *Percival* Soudan 1414 AD vi (W); *Persefall* 1666 ODCN; William *Percevall'* 1229 Cl (Sa); Roger *Perceval* 1286 Ipm (So); John *Perseval* 1297 MinAcctCo; John *Percival* 1372 LoPleas; OFr *perce-val*, from OFr *percer* 'to pierce' and *val* 'valley', 'pierce the valley'. Such names were not uncommon in France. cf. *perce-bois* 'wood-borer', *perce-roche* 'rock-piercer', *perce-forêt* 'forest-piercer' which Harrison interprets as 'keen hunter'. It might also mean 'a poacher'. cf. PEARCEY, and William *Percevent* 1221 Cur (Mx), 'pierce-wind', from OFr *vent* 'wind', no doubt renowned for his speed. The 1224 Perceval was one of the servants of Faukes de Breauté, notorious for breaking into places where he should not. For *Purcifer*, a Yorkshire name, cf. the Yorks *Brammer* from *Bramhall*. cf. also PASSMORE. Occasionally, the surname may be topographical in origin. Richard *de Percevill* (1203 AssSt) may have come from one of the two places in Calvados named Perceval.

**Percil, Persil:** Ranulf *Percesuil* 1166 RBE (Hu); Roger *Percesoll'* 1219 Cur (Bk); John *Percesoil* 1355–9 AssBeds. 'Pierce the threshold', OFr *percer, soel*/*suel*.

**Percy, Persey, Parsey, Pursey:** William *de Perci* 1086 DB (Y), c1133 Whitby; Walter *Perci*, Margaret *de Perci* 1185 Templars (L); Henry *Percy* 1332 SRSx; Rebecca *Parsy* 1702 SfPR. William de Perci, the Domesday tenant-in-chief and under-tenant of Hugh, earl of Chester, and ancestor of the second line of Percy, came from Percy-en-Auge (Calvados). *v.* ANF. Percy (La Manche), his traditional place of origin, and two other Percys in Calvados may also have contributed to the surname. cf. PEARCEY.

**Perdue:** *v.* PARDEW

**Pere:** *v.* PEAR

**Perebourne:** Walter *Perbrun* 1229 FFSf; Roger *Perebrun* 1275 RH (Sf). OE *pere* 'pear' and *brūn* 'brown', 'pear-brown', no doubt, a nickname for one of swarthy complexion. cf. NUTBROWN.

**Peregrine:** *v.* PILGRIM

**Peres:** *v.* PIERCE

**Perfect, Perfitt:** *v.* PARFAIT

**Perham, Perram:** *v.* PARHAM

**Perken, Perkin(s):** *v.* PARKIN

**Perott:** *v.* PARROT

**Perowne:** *v.* PERRIN

**Perraton:** *v.* PERRITON

**Perratt, Perrett, Perrott:** v. PARROT
**Perrell, Perren:** v. PERRIN
**Perrers, Perress:** William *de Perieres* 1196 P (Lei); Hugh *de Perers* 1274 RH (Sa); John *Perers* 1327 *SR*Ess. From Perriers near Rouen.

**Perrie:** v. PERRY

**Perrier, Perryer, Peryer, Purrier, Puryer:** (i) Robert *Perier* 1194 P (Lo); Henry *le Perer* 1217 Bart; Richard *Le Perur* 1288 MESO (Sx). OFr *perrier*, *perrieur* 'quarrier'. (ii) English *FFEss*; John *Pirun* 1166 P (Gl); William *le Piryere* ib.; John *Peryere* 1327 *SR* (Ess). 'Dweller by the pear-tree.' cf. PERRY.

**Perriman, Perriment, Perryman, Periman, Peryman:** Richard *Piriman* 1274 RH (Ess); William *Peryman* 1279 RH (C); Adam *Puryman* 1327 SRSo. 'Dweller by a pear-tree', OE *pyrige*. v. PERRY.

**Perrin, Perring, Perrins, Parrell, Perrell, Parren, Perren, Perron, Perowne:** *Perrinus* uadletus 1207 P (Nth); *Peryna* (f) 1280 FFEss; John *Pirun* 1166 P (Gl); Geoffrey *Perrun* 1185 Templars (L); Henry *Piron* 1194 Cur (O); William, John *Perel* 1222–3 Cur (Nf, Sr); John *Pirun* 1255 RH (W); Alexander *Peryn* 1268 AssSo; John *Perin* 1279 RH (C); Thomas *Paron* 1297 MinAcctCo; William *Peroun* 1327 SRSf; William *Peryng* 1332 SRSx. Diminutives of *Perre* (Peter) plus the suffixes -*in*, -*el*, -*un*. *Pirun* is an east French form from *Pier(r)un.* v. PIRRET. *Perowne* is a Huguenot name.

**Perriton, Perraton, Purton:** Eduuardus *de Periton* 1086 DB (Herts); Gervase *de Piriton'* 1200 P (So); Ascelin *de Periton, de Perton* 1207–8 P (O); Robert *de Puryton* 1269 AssSo; Walter *Piriton'* 1376 AssEss. From Pirton (Herts, Worcs), Pyrton (Oxon), Purton (Som, Wilts), or Purton End in Saffron Walden (Essex), all 'pear-tree farm'.

**Perron:** v. PERRIN

**Perrott:** v. PARROT

**Perry, Perrie, Pirie, Pirrie, Purry, Pury, de Pury:** Henry *de Peri, de Piri* 1176 P (St), 1199 AssSt; Richard *Pirie* 1198 P (K); William *de la Purie* 1243 AssSo; Gilbert *atte Pyrie* 1263 FFSr; Richard *del Piry* 1381 AssSt; Richard *atte Perye* 1392 MELS (Sx). 'Dweller by the pear-tree', OE *pirige, pyrige.* James *atte Pyrie* and John *Peryere* (1327 *SR*) both lived in Dedham (Essex). v. PERRIER, PERRIMAN.

**Perrycoste:** v. PENTECOST

**Perryer:** v. PERRIER

**Perryman:** v. PERRIMAN

**Persall:** v. PEARSALL

**Perse, Persse:** v. PIERCE

**Persey:** v. PERCY

**Persil:** v. PERCIL

**Pert, Perts:** v. PEART

**Peryer:** v. PERRIER

**Peryman:** v. PERRIMAN

**Pescod, Pescott, Peskett:** v. PEASCOD

**Pesson:** Ralph *Peissun* 1196, *Pessun* 1198 P (L); William *Pessun* 1282 IpmY. OFr *poisson* 'fish'. Metonymic for PESSONER.

**Pessoner:** Eilmer *le Pessuner* 1208 Cur (Ess); Albin *le Peschoner* 1219 Cur (Herts); Matthew *le Pessoner* 1296 SRSx. OFr *poissonnier* 'fishmonger'.

**Pestell, Pistol:** Robert, Nicholas *Pestel* 1221 AssSa, 1246 Assla; Symon *Pystel* 1296 SRSx. ME

*pestel* from OFr *pestel* 'an instrument for pounding things in a mortar', 'pestle'. cf. *Pestel*, of stampynge, Pila, pistillus, pistellus' PromptParv. Probably for a user of this instrument, a compounder of drugs, a spicer. For *Pystel*, cf. *pistil*, so called from its resemblance in shape to a pestle.

**Pester, Pistor:** Robertus *pistor* 1115 Winton (Ha); William, Symon *le Pestur* 1239 MESO (L), 1259 FFC; Richard *le Pester, le Pestour* 1279 RH (Beds); John *Pistor*, peleter 1281 LLB B; John *le Pistour* 1307 AssSt. AFr *pestour, pistour*, OFr *pestor, pesteur* 'baker'. In early sources, *Pistor* is common but is often the Latin *pistor* 'baker', clearly so when inflected as: Toli *Pistore* (abl.) c1140 StCh, Roberto *Pistori* (dat.) 1196 FF (D). Certain examples of AN *pistour* are rare. The legend on the seal of Turbernus *Pistor* (1200 Oseney) was SIGIL'L' TVRBERTI LE PESTVR. The modern *Pester* is not common; *Pistor* is rare. The feminine has also been noted: Alicia *le Pesteresse* 1270 RamsCt (C), as well as Robert *de la Pesterye* 1280 AssSo, 'of the bakery', cf. PANTER and PANTREY.

**Pesterfield:** v. BASKERVILLE

**Petch, Petche, Petchey, Pechey, Peach, Peache, Peachey, Peech:** Willelmus *Peccat'* 1086 DB (Ess, Nf); Haimund *Peccatum*, Hamo *Pecce* 1121–60 Bury (Sf); Rotbert *Pecceð* 1123 ASC E; God' *Pecce* 1165 P (Wa); Ralph *Pecche* 1168–75 Holme (Nf); William *Pesche* 1178 P (Y); William, Gilbert *Pechie* 1190, 1200 P (Ha, C); Geoffrey *Pech* 1191 P (L); Richard *Pechee* 1275 RH (Nf); Herbert *Pechy* 1275 RH (Berks). OFr *peche, pechie*, Lat *peccatum* 'sin', a curious nickname for Robert *Pecceð* (more commonly *Peche*), Bishop of Coventry in 1123, v. also PECHIN, PECKETT.

**Petchell:** v. PATCHELL

**Petcher, Peacher:** Adam *le Pechur* 1210 Cur (Do); Nicholas *le Peschur* 1221 AssWo; John *le Pechere* 1242 Oseney (O). OFr *pescheor, pecheour, pecher* 'fisherman'.

**Petchey:** v. PETCH

**Peter, Peters, Petre:** *Petrus* 1086 DB, Hy 2 Gilb (L); Ralph *Peter* 1195 P (Herts); Luke *Petre* 1282 LLB A; Elias *Petri* 1318 LLB E; William *Petres* 1327 SRSo. Lat *Petrus*, from Greek πέτρος 'rock'. *Peter* is the learned form, usually latinized in early documents. The popular form *Peres* (Fr *Piers*) was very common both as a christian name and a surname. v. PIERCE. The frequency of *Peters* as a surname is due to its late adoption by the Welsh.

**Peterborough:** John *de Peterburgh* 1360 FFY; Thomas *Petirburgh'* 1383 KB (Ha); Robert *Petirburgh* 1402–3 FFSr. From Peterborough (Nth).

**Peterfield, Petersfield:** Richard *de Petresfeld* 1320 PN Wt 178; John *Petrusfeld* 1450 AssLa; John *Peterfeild* 1642 PrD. From Petersfield (Ha), or 'dweller by (St) Peter's field'.

**Peterken, Peterkin:** *Petrekyn* le chapler 1319 SRLo; Andrew *Peterkin* 1488, John *Peterkyne* 1537 Black (Aberdeen). *Peter-kin*, a diminutive of *Peter*.

**Petersfield:** v. PETERFIELD

**Peterson:** John *Peterson* 1375 LoPleas. 'Son of *Peter*.'

**Pethard:** William, Reginald *Petard* 1296 SRSx, 1436 CtH. A derogatory nickname from OFr *peter* 'to break wind' and the pejorative suffix *-hard*. *v.* PETTER (i).

**Pether, Pethers, Pither, Pithers:** Thomas *Pither* 1287 QW (Gl); John *Peter* or *Pether* 1526 Oxon. A west-country and Cornish form of *Peter*.

**Petherick, Pethrick, Pedrick, Petterick:** *Pethroke* 1547 Bardsley (Co); *Pethericke* 1579 ib. *Petrock*, a personal name common in Cornwall, commemorating the Cornish St Petrock.

**Petherton:** John *de Pederton'* 1340 Glast (So). From North, South Petherton (So).

**Petifer:** *v.* PETTIFER

**Petijohn, Pettijohn, Pettyjohn:** John *Petijohan* 1327 SRSx; John *Petyjohn* 1386 *ERO*; John *Petyjon* 1443 CtH. 'Little *John*', OFr *petit*. A not uncommon type of name: John *Petytenyn* 1327 SRSo 'little *Ennion*'; *Petinicol* 1279 RH (Hu) 'little *Nichol*'; Roger *Petytwyll* 1303 Misc (Y) 'little *Will*'; William *Petitwillam* 1299 AssW 'little *William*'.

**Petipas, Pettipas:** Baldwin *filius Petitpas* 1208 Pleas (Ess); Adam *Petipas* 1191 P (Y); John *Petipas* 1276 KB (Nth); Roger *Petipas* 1317 AssK. A nickname for one who took small steps when walking, OFr *petit, pas*. Sometimes, perhaps, from Petitpas (Loire-Inférieure, &c.).

**Petley, Pettley:** Ralph *de Petle* 1296 SRSx. From Petley Wood in Battle (Sx).

**Peto, Peyto:** Robert *de Peytowe* 1222 Cur (W); Johannes *Pictavensis*, John *de Peyto* 1238, 1247 Fees (Bk). 'The man from Poitou', AFr *Peitow*. cf. POIDEVIN.

**Petre:** *v.* PETER

**Petrie, Petry, Peatrie, Patry:** Charles *Patre* 1513 Black; Andrew *Petre* 1530 ib.; Hendrie *Petrie*, Henry *Patrie* 1612–19 ib.; George *Paitrie* 1620 ib. Scottish diminutives of *Peter* or *Patrick*. In Arran and Kintyre *Patrick* became Gaelic *Pàdair* and *Pàtair* and was confused with *Peter*. *Patrick* Young, Dean of Dunkeld, is *Patry* Yhong (1452) whilst *Patrick* Roy Macgregor (1667) is usually called *Petrie* (Black).

**Pett, Petts:** *v.* PITT

**Pettegree, Petegree, Petticrew, Petticrow, Pettigree, Pettigrew:** Walter *Peticruw* 1227 *Ass* (Ess); Richard, Roger *Peticru* 1283, 1298 *Ass* (Ess, Sf); Thomas *Petykreu, Peticru* 1296 CalSc, 1297 MinAcctCo; John *Petegrew* 1346 FA (Sf); Robert *Pedgrewe* 1568 SRSf. The common belief that this name derives from a place in Cornwall is clearly untenable. There is no place of the name in that county, early forms have no preposition, and come from the eastern counties. Nor can the name be identical with *pedigree*, Fr *pied de grue* 'crane-foot', as early forms regularly have *t*, whilst the *g* does not appear before the 14th century. It is apparently *petit cru* 'little growth' (OFr *cru* 'growth, increase'), a nickname for a small man of stunted growth.

**Petten, Petton:** John *Petten* 1642 PrD. From Petton (D, Sa).

**Pettengell, Pettingale, Pettingill:** *v.* PORTUGAL

**Petter:** (i) John *le Petour* 1299 LLB B; John *Pethour* 1519 NorwDep. ME *-our* distinguishes this from the more common *Petter* 'dweller in a hollow'. Rolland *le Pettour* held land in Hemmingstone

(Suffolk) by serjeantry of appearing before the king every year at Christmas to do a jump, a whistle, and a fart (*unum saltum, unum siffletum et unum bumbulum*) c1250 Fees. In 1330 the holder of the serjeantry is referred to as Roland *le Fartere* Ipm, and cf. John *le Fartere* 1327 SRLei. *Pettour* is from OFr *peter* 'to break wind', from which come the French names *Pétard, Peton, Petot*, and probably *Pétain*. Dauzat notes that at the end of the 19th century one family of *Pétard* had licence to change its name to *Pérard*. *v.* also PETHARD. (ii) For *atte pette*, *v.* PITT, PUTTER.

**Pettifer, Pettifor, Pettyfar, Pettyfer, Petifer, Pettafor, Pettefar, Pettefer, Pettipher, Pettiver, Pettiford, Pottiphar, Puddepha, Puddifer:** Herbertus *Pedesferri* 11th OEByn; John *Pedefer* 1190 P (Beds); John *Piedefer* 1198 P (Beds); William *Pedifer* 1221 AssWa; William *Petifer* 1327 SRSx; Richard *Pitifer* ib.; William *Pidefyr* 1332 SRSx; William *Putifer* 1382 RamsCt (Hu); William *Petefer* 1392 AD ii (Hu); Thomas *Petyver* 1524 SRSf. *Pedesferri* is Lat *pedes ferri* 'feet of iron', the rest are from OFr *pedefer*, i.e. *pied de fer* 'iron foot', explained by Larchey as 'surnom de bon marcheur'. The name was common and sometimes used as a nickname alone: *Piedefer* 1186 P (Wo), *Pie de Fer* 1185 P (Nf), and may also have referred to an old soldier who had lost a foot. cf. the English translation, Roger and Geoffrey *Yrenfot* 1251 Rams (Hu) and the parallel John *Stelfot* 1301 LLB C. Bardsley notes the development to *Pottiphar* (1777), a natural development of *Putifer* (1382).

**Pettijohn:** *v.* PETIJOHN

**Pettinger:** *v.* POTTINGER

**Pettipas:** *v.* PETIPAS

**Pettipher, Pettiver:** *v.* PETTIFER

**Pettit, Pettitt, Pettet, Petit, Petyt, Pittet:** Aluric *Petit* 1086 DB (Ha); John *le Petit* 1228 FFEss. OFr *petit* 'little'.

**Pettle:** Richard *son of Pettel* 1246 AssLa; Margery *Petul* 1275 SRWo; Richard *Pettel* 1332 SRSx; William *Pettle* 1642 PrD. OE *Pyttel*.

**Pettley:** *v.* PETLEY

**Pettman:** *v.* PITMAN

**Petton:** *v.* PETTEN

**Petty, Pettie:** William *Petie* 1198 FFNt; Walter *le Petiit* 1249 AssW; Thomas *Petyt* 1327 SRLei; Richard *Pettye* 1553 FFSt. Fr *petit* 'small', a not infrequent element in medieval names, cf. Walter *Petitclark* 1304 IpmGl; John *Petijohan* 1327 SRSx; William *Petitpas* 1199 Pl (K); Richard *Petitprudum'* 1230 Cur (L); John *le Petit Smyth* 1351 AssEss.

**Pettyjohn:** *v.* PETIJOHN

**Peutherer:** Lambert *le Peutrer* 1311 AD i (Mx); Geoffrey *le Peautrer* 1319 SRLo; Thomas *Powterer* 1472 GildY. AFr *peautrer*, OFr *peautrier* 'pewterer', a maker of pewter vessels (1348 NED). *Pewter, Pewtress* and *Powter* are metonymic.

**Peverall, Peverell, Peverill, Peperel, Pepperall, Pepperell, Pepperwell, Pepprell:** Geruasius *filius Peurelli* 1130 P (Lo); *Peverel* 1161–77 Rams (Hu), 1205 Cur (Ess); Rannulfus *Peurellus, Piperellus* 1086 DB (K, Ess); Richard *Peuerel* 1186–1210 Ho!me (Nf); William *Peuerell'* 1221 AssWa; William *Peperel* 1224 Clerkenwell (Mx). OFr *Peurel*, Lat *Piperellus*, a diminutive of OFr *pevre*, Lat *piper* 'pepper', used as a personal name. Tengvik suggests the reference may have been to one of small, rounded shape, 'peppercorn-sized', or to a man with a darkish complexion or hair. But it may well have been applied to a small man with a fiery, peppery temper.

**Pew:** William *Piwe* 1202 P (L); William *le Pew* 1327 SRSf; John *Pewe* 1379 PTY. OFr *pi, pis, piu* 'pious'. *v.* also PUGH.

**Pewter, Pewtress:** *v.* PEUTHERER

**Pey:** *v.* PAY

**Peyman:** *v.* PAYMAN

**Peyser:** *v.* POYSER

**Peyto:** *v.* PETO

**Peyton:** *v.* PAYTON

**Pezey:** *v.* PIZEY

**Phair:** *v.* FAIR

**Phaisey:** *v.* VAISEY

**Phalp:** *v.* PHILIP

**Pharaoh, Pharo, Pharro:** *v.* FARRAR

**Phare:** *v.* FARE

**Phasey:** *v.* VAISEY

**Phayre:** *v.* FAIR

**Phear:** *v.* FEAR

**Pheasant:** John *Faisant* 1166 P (Sx); Thomas *Fesaunt* 1221 AssWa; Richard *le Feisant, le Faisant* 1229 Cl (Jersey); Roger *Fesant* 1241 Oseney (O). A nickname from the pheasant, ME *fesaunt*.

**Pheasey, Pheazey:** *v.* VAISEY

**Phelips, Phelops, Phelps:** *v.* PHILIP

**Phethean:** *v.* VIVIAN

**Pheysey:** *v.* VAISEY

**Phibbs:** *v.* PHIPPS

**Philbert:** *v.* FILBERT

**Philb(e)y:** *v.* FILBY

**Philbrick, Philbrock, Philbrook:** Perhaps from Felbrigg (Nf).

**Philcott:** A corruption of *Philcock*. *v.* PHILCOX.

**Philcox, Phillcox:** Betrice *Philekoc* 1283 SRSf; German *Philecok, Filcokes* 1307, 1314 Wak (Y). 'Son of *Phil*', a pet-form of *Philip*, plus *cock*. *v.* FILL.

**Philip, Philips, Philipse, Philipp, Philipps, Philliphs, Phillipps, Phillips, Phillipse, Philp, Philps, Phillp, Phillps, Phelips, Phelops, Phelp, Phelps, Phalp:** *Filippus* de Crochesbi 1142–53 DC (L); *Philipus* Rabode c1150 Gilb (L); *Philippa* de Faia 1195 P (Sx); *Phelippus* 1290 Oseney (O); *Phelip* de Kocfeld 1292 SRLo; *Philp* Gledstanes 1541 Bardsley; Henry *Philip, Phelipe* 1275 RH (Nf); Maud *Philippes* 1279 RH (Hu); John *Felipe*, Gilbert *Phelip* 1296 SRSx; William *Phelipes* 1327 SRSt; Cecilia *Philipp* 1379 PTY; John *Phelpes* 1570 Bardsley; Peter *Philpe* 1698 DKR 41. Greek Φίλιππος 'horse-lover', usually latinized as *Philippus*. The vernacular form seems to have been *Phelip* which is not confined to the south-west as has been suggested. cf. Fr *Phélip, Félip*. It was also used as a woman's name, latinized *Philippa*. *Philip, Phelp* are contracted forms. For *Phalp*, cf. Fr *Phalip*. The popularity of the name is revealed by its pet-forms and derivatives. *v.* FILL, FILKIN, PHIPPS, PHILCOX, PHILLINS, PHILLOTT, PHILPOT, with its hypocoristics POTT, POTTELL, and POTKIN and their varied forms. Other derivatives surviving, but not well-evidenced are: *Philben, Philbin, Philpin, Phippard, Phippen.*

**Philippot:** *v.* PHILPOT

**Philipson, Philippson, Phillipson:** Gilbert *Phelipson* 1396 AssSt; John *Philipson* 1524 SRSf. 'Son of *Philip*.'

**Phillins:** A diminutive of *Phill* (Philip), plus -*in*.

**Phillips:** *v.* PHILIP

**Phillipson:** *v.* PHILIPSON

**Phillott:** *Philota* de Kender 1273 RH (Db). A diminutive in -*ot* of *Phil* (Philip), here feminine.

**Philpot:** *v.* PHILPOT

**Philp(s):** *v.* PHILIP

**Philpot, Philpots, Philpott, Philpotts, Phillpot, Phillpots, Phillpott, Phillpotts, Philippot:** *Phelipot* Herneys 1377 AD ii (Beds); *Philipot* (f) 1379 PTY; John *Philipot* 1327 SRSx; Stephen *Philippot* 1336 AD v (K); John *Philippot, Phelipot, Philpot* 1367 LoPleas, 1378 LLB F, H. *Philip-ot*, a diminutive of *Philip*. From this were formed POTT, POTTELL and POTKIN.

**Philson:** *v.* FILL

**Philson:** *v.* FILLSON

**Phimester, Phimister:** *v.* FEMISTER

**Phin, Phinn:** *v.* FINN

**Phippen, Phippin:** Nicholas *Phippen* 1332 SRWo; George *Phippen* 1607 Oxon; John *Phippen* 1662–4 HTDo. *Phip-en*, a diminutive of *Phipp*, a pet-form of *Philip*.

**Phipps:** William *Fippe, Fyppe* 1227 FFBk, 1250 Fees (Sf); Henry, Geoffrey *Phippe* 1332 SRWa, 1337 ERO; John *Phippes* 1364 LoPleas. *Phip*, a pet-form of *Philip*.

**Phipson:** John *Phippesson* 1373 LLB G; Andrew *Phipson* 1374 Crowland; Richard *Phipson* 1375 AssL. 'Son of *Phip*', a pet-form of *Philip*.

**Phizackerley, Phizaclea, Phizakarley:** *v.* FAZACKERLEY

**Phyfe:** *v.* FIFE

**Physick, Visick:** Richard *Physik* Ed 1 Malmesbury. Metonymic for *Physician*: John *leFisicien* 1269 AssSt.

**Phythian:** *v.* VIVIAN

**Pick:** *v.* PIKE

**Pickard:** (i) Hugh *le Pycard* 1276 AssSt; John *Pikart* 1279 RH (Hu); Michael *le Pykard* 1289 LLB A; John *Picard* 1292 FFSf. 'The Picard', a man from Picardy. (ii) Paganus *filius Pichardi* 1160 P (Ha); *Picardus* filius Pagani 1208 Cur (Ha); Paganus, William *Pichard* 1169 P (Ha), 1198 FF (Sa); John *Pikhard* (*Pikard*) 1230 P (Ha). A French personal-name, compounded of *Pic* (cf. PICKETT) and -*hard*. On the analogy of *Richard*, this would become both *Pickard* and *Pitchard*, the latter, apparently, no longer surviving, but one origin of PITCHER.

**Pickaver:** *v.* PECKOVER

**Pickbourne, Pickburn, Pigburn:** Ralph *de Pekeburn'* 1202 FFY; Richard *de Pykeburne* 1305–6 IpmY; Richard *Pigburne* 1440 ShefA. From Pickburn (WRY).

**Pickenham:** John *de Pykenham* 1334 LoPleas; William *Pykynham* 1479 Paston. From Pickenham (Nf).

**Picker, Pickers:** Richard *le Pickere* 1188 P (Y); Stephen *Pikere* 1199 P (K); Richard *le Pikere* Hy 3 AD iii (Ess). 'Maker or seller of pikes, pick-axes or picks', or 'fishmonger, seller of pike', or 'dweller on the hill'. *v.* PIKE.

**Pickerell, Pickerill, Pickrell, Pickrill:** Yuo *Pikerell* 1199 P (Nf); Richard *Pikerel* 1240 FFEss. ME *pykerel* 'a young pike'.

**Pickering, Pickerin:** Reginald *de Pichering* 1165 P; Henry *de Pikeringes* 1246 FFO; John, William *Pykeryng* 1327 SRSo, IpmW. From Pickering (NRYorks).

**Pickernell:** *v.* SPICKERNALL

**Pickersgill:** Robert *Pickersgill* 1672 HTY; Elizabeth *Picersgill* 1679 Bardsley; Robert *Pickersgale* 1690 FrY. From Pickersgill Lane in Killinghall (WRYorks). ME *pyker* 'thief, robber' and *gil* (PN WRY v, 101).

**Pickett, Picot, Pikett, Pykett, Piggot, Piggott, Pigot, Pigott:** Picot de Grantebrige 1086 DB (C); *Pichot* de Lacele c1155 DC (L); *Pigotus* de Hutun Hy 2 DC (L); Robertus *filius Picot* 1166 RBE (Y); Roger *Picot* 1086 DB (Ch); Robert *Picot* 1140–5 Holme (Nf); William *Piket* 1177 P (Bk); Waubert *Pyket* 1277 LLB A; Peter *Picot, Pycot, Pygot* 1285 Ipm (C). We may occasionally have a nickname from OFr *picot* 'point, pointed object', but the surnames are probably usually from a personal name OFr *Pic*, plus -*ot* and, occasionally, -*et*. cf. PICKARD, of which *Pickett* may also be a late development.

**Pickford:** Alcock *de Pykeford* 1288 AssCh; Thomas *Pikeford* 1332 SRSx; Thomas *Pickford* alias Pickfatt 1649 DKR. From Pickforde in Ticehurst (Sussex).

**Pickin, Picking:** William *Pykyng* 1311 ColchCt; Thomas *Pykyn* 1353 ib. Probably OE *pīcing* 'dweller on the hill'. *v.* PIKE.

**Pickis:** Edmund *Pycoyse* 1524 SRSf. ME *pikois, pikeis* 'mattock'.

**Pickles, Pickless, Pickle, Pighills:** Richard *de Pighkeleys* 1379 PTY; Henry, Arthur *Pickles* 1571 LWills, 1672 HTY. 'Dweller at the small enclosure', ME *pightel, pighel*.

**Pickman:** Ralph *Pykeman* 1259 AD ii (Mx); Robert *Pyckeman* 1300 LLB C. *v.* PICKER, PIKE.

**Picknett:** Robert *Pikenot* 1175 P (Ha); Richard *Pikenet, Pikenot* 1206–8 Cur (Sf). *Pik-en-et, Pik-en-ot*, double diminutives of *Pic*. *v.* PICKETT.

**Picknell:** Thomas *de Pikenhale* 1279 RH (C); John *Pyknoll* 1525 SRSx. From Picknill Green in Bexhill (Sx).

**Pickpease, Pickpeace:** Adam *Pickpese* 1283 SRSf; Walter *Pikkepuse* 1306 AssW; Alexander *Pickepese* 1327 SRSf. 'Pick peas', ME *pikken*, OE *peose*. Perhaps a nickname for a grower or seller of peas. cf. Baldwin *Pikechese* 1209 FFEss 'Pick cheese'; John *Pykehuskes* 1316 Wak (Y) 'pick husks'; Roger *Pikerihe* 1279 RH (O) 'pick rye'.

**Pickrell, Pickrill:** *v.* PICKERELL

**Pickstock, Pickstoke:** Robert *de Pickstoke* 1255 RH (Sa); Robert *de Pykstoke* 1327 SRSa; William *de Pykestoke* 1332 SRSt. From Pickstock (Sa).

**Pickstone, Pixton:** Richard *Pixton'* 1209 P (Nf); Walter *Picxston* 1275 RH (K); Richard *Pikstan* 1296 SRSx. The surname has been noted in some six counties, always without qualification, and a topographical origin appears impossible. We probably

have an unrecorded OE *Pīcstān. Piichel* is found in LVD and in Pickelescott (Sa) and *Pīca* in Picton (Ch, NRY) and Pickworth (L, R).

**Pickthorn, Pickthorne:** William *de Pikelestorn* 1221 AssSa. From Pickthorne (Sa).

**Pickton, Picton:** Geoffrey *de Piketon* 1191–4 YCh; William, John *de Piketon* 1251 AssY, 1327 SRDb. From Picton in Kirkleavington (NRYorks).

**Pickup, Peacop:** Edmund *Pickeupp* 1632 RothwellPR (Y). From Pickup or Pickup Bank (Lancs).

**Pickwell:** Robert *de Pidekeswell'* 1210 P (D); Hugh *de Picwell* 1265–72 RegAntiquiss; John *de Pikewell* 1327 SRSx. From Pickwell (Lei), Pickwell in Georgeham (D), or Pickwell Fm in Cuckfield (Sx).

**Pickwick:** William *de Pikewike* 1275 RH (W); Stephen *de Pykewik* 1297 MinAcctCo. From Pickwick in Corsham (W).

**Pickwoad, Pickwort, Pickworth:** Robert *de Pickewrd* 1212 Fees (L); Hugh *de Pikewode* 1275 RH (L); Richard *de Pykewrth'* 1278 AssSo. From Pickworth (L, R).

**Picot:** *v.* PICKETT

**Pidcock:** Adam *Pydekock* 1301 SRY; Robert *Pydecock* 1306 AssW. A diminutive of OE *Pydda*.

**Piddell, Piddle:** Robert *de Pydele* 1280 PN Do i 113; Laurence *Piddle* 1309 AssNf; Mary *Piddle* 1662–4 HTDo. From Piddle Hinton, Bryants, Piddletrenthide, Turners Puddle, Afpuddle, Tolpuddle (Do), or North, Wyre Piddle (Wo).

**Piddington, Pidington:** John *de Pidintun'* c1181 Goring (O); William *de Pidington'* 1219 Cur (Nth); Emote *Pidington'* 1359 LuffCh. From Piddington (Nth, O).

**Piddle:** *v.* PIDDEL

**Pidington:** *v.* PIDDINGTON

**Pidgen, Pidgeon, Pidgon, Pigeon:** William, Alan *Pigun* 1200 Cur (Nf), 1202 AssL; Ralf *Pyjun* 1268 AssSo. Fr *pigeon*, OFr *pipjon* 'pigeon', perhaps 'one easy to pluck'. Although definite identifications are impossible in Subsidy Rolls, those for Sussex suggest that *Pigeon* may also be *Petit Johan* 'Little John': Relicta *Pygon* 1296, Relicta *Pijohan*, John *Petijohan*, John *Pyion*, John *Petiion*, Thomas *Pyion* 1327.

**Pidler:** *v.* PEDLAR

**Pidsley:** Walter *de Pideneslegh* 1274 RH (D); Isabel *Pyddeslegh* 1524 SRD; John *Pidsleigh* 1642 PrD. From Pidsley in Sandford (D).

**Pie, Pye:** (i) Ralph, Eustace *Pie* 1177 P (Y); 1210 Cur (C); William *le Pye* 1296 SRSx. This is certainly at times a nickname from the magpie, ME, OFr *pye, pie*. The surname is common, usually without the article, and may also be metonymic for a maker or seller of pies (ME *pie*). cf. Peter *Piebakere* 1320 LLB E, Adam *le Piemakere* 1332 SRLo, John *Pyman* 1524 SRSf. (ii) Stephen *atte Pye* 1347 LLB F. 'Dweller at the sign of the *Pie*.' cf. *'atte Pye* on the hope' 1340 LoPleas.

**Piedargent:** Belin *Pe de Argent*, Walter *Pedeargent* 1276 AssLo. 'Silver foot', Fr *pied, argent*. cf. John *Pedebef* 1381 SRSf 'ox-foot'; William *Pedechen* 1242 Fees (L) 'dog-foot'; Robert *Piedurs* 1199 P (O) 'bear-foot'; *Pieferret* 1148 Winton (Ha) 'iron-shod foot'.

**Piegrome:** v. PILGRIM

**Piele:** v. PEEL

**Pierce, Pieris, Piers, Pierse, Pearce, Pears, Pearse, Peers, Peirce, Peirs, Peirse, Peres, Perris, Perriss, Perse, Persse:** *Peris* le ceynturer, *Peres* le cordener 1292 SRLo; Gilbert *Perse* 1198 P (Lo); Geoffrey *Peres, Pieres* 1237 HPD (Ess); Richard *Peris* 1275 SRWo; Adam *Pieris* 1332 SRWa; William *Peers* 1444 AD i (Gl). OFr *Piers*, nom. of *Pierre*, 'Peter', a common name, usually latinized as *Petrus*, with diminutives PARKIN, PARROT, PERRIN, PIRRET.

**Piercey, Piercy:** v. PEARCEY

**Pierpoint, Pierrepoint, Pierpont, Pearpoint, Pairpoint:** Reinaldus *de Perapund* 1086 DB (Nf); Richard *de Pierrepunt* 1178 P (Do); Roger *Perpunt* 1240 NottBR (Nt); Symon *de Perpoynt* 1316 FA (Nb). Godfrey and Robert *de Petroponte*, undertenants of William de Warenne in Suffolk and Sussex in 1086, came from Pierrepont (Seine-Inférieure). v. ANF. Others may have come from Pierrepont (Calvados), Saint-Nicholas-de-Pierrepont (La Manche), or Saint-Sauveur-de-Pierrepont (ib.). v. OEByn.

**Piers, Pierse:** v. PIERCE

**Pierson:** v. PEARSON

**Pieshank, Pieshanks:** William *Pyshank* 1373 FrY; John *Pyshanke* 1379 PTY. 'magpie legs', OFr *pye*, OE *sceanca*. cf. William *Pyfote* 1524 SRSf 'magpie foot'.

**Pigache:** Richard *Pigace* 1176 P (Gl); Ralph *Pigaz* 1219 AssY. Identical with French *Pigasse*, Norman-Picard *Pigache*, a variant of *Picasse*: a nickname from OFr *pic* 'pick-axe, hatchet', probably used of a workman.

**Pigburn:** v. PICKBOURNE

**Pigeon:** v. PIDGEN

**Pigg:** Aluricus *Piga* 1066 DB (D); John *Pig* 1186 P (W); Robert *Pigge* 1277 Wak (Y). OE *picga, *picg*, ME *pigge* (a1225) 'pig'. cf. HOGG.

**Pig(g)ott:** v. PICKETT

**Piggrem, Pigram, Pigrome:** v. PILGRIM

**Pighills:** v. PICKLES

**Pigsflesh:** Roger *Piggesflesh* 1276 AssLo; Reyner *Piggesfles* 1319 SRLo. 'Pig's flesh', OE *picg, flǣsc*. Probably a nickname for a pork-butcher. cf. Henry *Piggesfot* 1228 Oseney 'pig's foot'.

**Pike, Pyke, Pick, Lepick:** (i) Aluric, Alwinus *Pic* 1066 DB (D, So); Fulco *picus* 12th DC (L); Hugo *Pik* 1177 P (O); Robert *le Pic* 1191 P (W); Henry *Picke* 1221 AssWo; William *le Pyc* 1296 SRSx; Nicholas *Pyke* 1344 FFC. These surnames have various origins. The DB examples are from OE *pīc* 'point, pick-axe', and may have the same sense as the corresponding Scandinavian nickname *Pik*, 'a tall, thin person'. Or it may denote a man armed with a *pīc*, a pikeman. Later examples may have the same meaning or may be nicknames from OFr *pic*, Lat *picus* 'wood-pecker' or from ME *pike* 'pike, fish'. Alexander *le pik* (1292 SRLo) was a fishmonger and owner of a ship. William, Robert and Stephen *Pikeman* (ib.) were also fishmongers. Here, *Pike* and *Pikeman* are both from *pike*, the fish, and mean 'sellers of pike'. (ii) Thomas *del Pic* 1220 FFEss; Ralph *del Pik'* 1292 QW (Herts). OE *pīc* 'point' in the sense of 'hill'. From

residence near a hill as at Pick Hill (Ess, K), the former the home of Reginald *de Pike* (t Hy 3 PN Ess 29).

**Pikesley:** Hugo *de Pikesle* 1242 Fees (He). From Pixley (Herefords).

**Pikett:** v. PICKETT

**Pilat, Pillatt, Pilot:** Gilbert *Pilat* 12th DC (L); William *Pylate* 1277 AssSo; John *Pilot* 1392 LoCh. Dauzat derives Fr *Pilat* from Pontius *Pilate* which seems an unlikely origin for this surname, though it could, perhaps, be a pageant name. More probably it represents *Pil-ot*, a diminutive of OE *Pīla*.

**Pilbeam, Pilbean, Pilbin:** Adam *de Peltebhem* 1296 SRSx; Thomas *Peltebem* 1327 ib. This Sussex and West Kent name must derive from a lost place in Sussex. Pilbeams in Chiddingstone (Kent) is named from Her' *Peltebem* (1347 PN K 80).

**Pilch:** Ralph *Pilche* 1251 Rams (Hu); Richard *Pilche* 1275 SRWo. OE *pylece*, ME *pilche* 'a fur garment'. Metonymic for *Pilcher*.

**Pilcher:** Mabilia *Pullchare* 1214 FFSx; Hugh *Pilchere* 1271 FFC; Henry, Nicholas *le Pilchere* 1275 SRWo, 1317 AssK. A derivative of OE *pylece* 'a pilch', a maker or seller of pilches (an outer garment made of skin dressed with hair).

**Pilcock:** John *Pilecoc* 1199 P (Gl); John *Pilecok* 1274 IpmW; Giliana *Pylcock'* 1301 SRY. *Pil-cock*, a diminutive of OE *Pīla*.

**Pile, Pyle:** Henry *de la Pil* 1221 Pat (So); Robert *Pile* 1243 AssSo; Robert *Attepile* 1274 RH (So). 'Dweller by the stake or post', OE *pīl*.

**Pilfold:** William *Pilefold* 1327 SRSx. From Pilfolds in Horsham (Sussex).

**Pilgrim, Pilgram, Pagram, Peagram, Peagrim, Peggram, Pegram, Pegrum, Piggrem, Pigram, Piegrome, Pigrome, Pelerin, Peregrine, Paragreen:** Hugo *peregrinus*, Hago (sic) *le pelerin* 1189–98, 12th MedEA (Nf); Hugo *Pilegrim* 1185 Templars (Wa); William *Pegerim, Pegrum* 1200 Cur (Do); Eustace *Pelrim* 1221 ElyA (C); Robert *Peregrine* 1243 AssSo; William *Pylegrim* 1251 Rams (Hu); Robert *Pelrin, Pelgrim* 1260 AssC; William *Pegrin* 1275 RH (C); Symon *Pegrym* 1327 SRSf. ME *pelegrim, pilegrim*, OFr *pelegrin*, Lat *peregrīn-um* 'one that comes from foreign parts', used of one who had made the pilgrimage to Rome or the Holy Land (*pilegrim* c1200 NED). Occasionally used as a personal name: *Pelerin* 1206 P (Sx), sometimes given to or adopted by a pilgrim; William *Pelerin* de Albana is also called *Pelerine* de Albana, clerk (1272 Lewes (Nf) ).

**Pilk, Pilke:** William *Pylk* 1303 AssW; Peter *Pylke* 1332 SRSt; John *Pilk* 1361 IpmGl. OE *Pileca*, unrecorded, but found in Pilkington (La).

**Pilkington:** Alexander *de Pilkington, de Pilkinton* 1205 WhC, 1285 AssCh; John *Pylkyngton* 1470 Past. From Pilkington (Lancs).

**Pill:** Walter, John *Pille* 1197 P (O), 1275 SRWo. Probably OFr *pile* 'little ball, pill'. cf. BALL.

**Pill, Pell:** Robert *de la Pulle* 1221 AssWo; Hugh *de la Pille* 1225 AssSo; John *atte Pelle* 1332 SRSx. 'Dweller by the creek or stream', OE *pyll*.

**Pillar, Piller, Pillers:** (i) Roger, Dike *le Pilur* 1246 AssLa. OFr *pilleur* 'plunderer'. cf. *'Pylowre*, or he that pelyth other men, as cachpolls or odyre lyk, *pilator,* *depredator'* PromptParv. (ii) John *le Piler* 1327 SRSo; Thomas *Piler'* 1332 SRSr. 'Dweller by the stake or stream.' *v.* PILE, PILL. (iii) Thomas *Attepiler* 1231 Cl (O); Walter *de Piler'* 1279 RH (O). 'Dweller by the pillar', OFr *piler*.

**Pillay, Pilley:** Hugh *Pilly* 1240 FFY; Gilbert *de Pilleghe* 1327 SRSo; John *Pilleye* 1392 NottBR. From Pilley (Ha, WRY).

**Pillet, Pillett:** Thomas *Pilet* 12th DC (L); Nicholas *Pylet* 1275 SRWo; Richard *Pilet'* 1302 SRY. Probably *Pil-et*, a diminutive of OE *Pīla*. But Fr *Pillet* is given by Dauzat as a diminutive of OFr *pile* 'hod', and as a nickname for a workman.

**Pilley:** *v.* PILLAY

**Pillifant:** Judith *Pellefant* 1612, Alse *Pellefant* 1619, Andrew *Pilliphant* 1633 HartlandPR (D). Probably from East, West Pilliven in Witheridge (D), but perhaps sometimes for BULLIVANT.

**Pilling, Pillings:** Adam *Pilling* 1283 SRSf; Emma *Pylyng* 1296 SRSx; Thomas *Pillyng* 1344 AD i (Wa). OE *\*pīling* 'dweller by the stake' or OE *\*pylling* 'dweller by the stream'.

**Pillinger:** Elizabeth *Pillinger* 1663 HeMil. A variant of BULLINGER.

**Pillman:** William *Pyleman* 1286 Wak (Y). 'Dweller by the stake or the stream.' *v.* PILE, PILL.

**Pillock:** *Pilluc* 1066 Winton (Ha); Stephen *Pilloc* 1208 Pleas (Sf); Roger *Pillok* 1298 IpmY; John *Pillokes* 1327 SRLei. A derivative in *-uc* of OE *\*Pilla*, a hypocoristic of names in *Pil-*.

**Pillow:** *v.* PEDLOW

**Pilot:** *v.* PILAT

**Pilpot:** For PHILPOT. Pilpot Wood in Calne (Wilts) is *Filpott* Coppice in 1650 (PN W 260).

**Pilton:** Robert *of Pilton'* 1269 ForR; Thomas *de Piltone* 1338 IpmW; John *Pilton* 1439 IpmNt. From Pilton (D, R, So).

**Pim, Pimm, Pym, Pymm:** *Pimme* filius Sirith 1204 Cur (St); *Pimme* forester 1246 AssLa; William, Henry *Pimme* 1250 Fees (Nt), 1279 RH (C); Agnes, Edmund *Pymm(e)* 1307 Wak (Y), 1316 AD ii (Nf). Bardsley explains this as a pet-form of *Euphemia*. In the two instances where certainty is possible it is clearly a man's name and is presumably the OE *Pymma* recorded once in LVD. cf. Gilbert *Pimming'* 1221 AssWa; William *Pymyng* 1295 FFEss. OE *\*Pymming* 'son of *Pymma*'.

**Pimblett, Pimblott, Pimlett, Pimlott:** Mergret *Pymlot* 1561 Bardsley. *Pim-el-et, Pim-el-ot*, double diminutives of *Pim*.

**Pimbley, Pimley:** Stephen *de Pimbelee* 1221 AssSa. From Pimley (Sa).

**Pimperley:** Nicholas *de Pinperleg'* 1208 Cur (W); Euphemia *de Pymperlegh* 1305 AssW; Anselm *Pymperle* 1377 FFW. From Pimperleaze Road in Mere (W).

**Pinch, Pinck:** *v.* PINK

**Pinchbeck, Pinchback:** Walter *de Pincebec* 1202 AssL; John *de Pinchebeke* 1327 SRLei; William *Pynchebek* 1447 AssLo. From Pinchbeck (Lincs).

**Pinchen, Pinchin, Pinching, Pinchon, Pinshon, Pinson, Pinsent:** Hugo *filius Pinchonis, filius Pincun* 1121–33 Seals (O), Hy 2 DC (L); *Pincun de*

Blacheshola 1166 P (Berks); Ralph *Pincun* 1166 RBE (L); Richard *Pincon* Hy 2 DC (Lei); William *Pynson* 1296 SRSx; Robert *Pynchoun* 1310 ColchCt; John *Pynsent* 1524 SRSf. OFr *pinçon, pinson* Norman-Picard *pinchon* 'finch', used as a symbol of gaiety, 'gai comme un pinson' (Dauzat), both as a nickname and a personal-name. Ekwall suggests a nickname from OFr *pinçon, pinchon* 'pincers, forceps'.

**Pinckney:** *v.* PINKNEY

**Pinckstone:** *v.* PINKSTONE

**Pindar, Pinder, Pindor, Pender:** Richard *le pynder* 1219 AssY; William *le Pendere* 1231 Pat (Ess); Richard *le Pundere* 1296 SRSx. A derivative of OE *(ge)pyndan* 'to impound, shut up'; pinder, an officer of a manor who impounded stray beasts. cf. Thomas *le Pendere* tunc prepositus de Atleford 1275 RH (K). *v.* also PONDER, POUNDER, POYNDER.

**Pine, Pyne:** (i) Bonde, Robert *Pine* 1101–7 Holme (Nf), 1181 P (Sa); Robert, William *Pin* 1208 Cur (Nth), 1221 AssSa; Henry *le Pyn* 1332 SRSx. OE *pīn* or OFr *pin* 'pine', a nickname for a tall, upright man. cf. PINNELL. (ii) Morin *del Pin* 1130 P (Lei); Alexander *de Pinu* 1169 P (D); Thomas *de Pyne* 1277 AssSo; John *atte Pyne* 1327 SRC. Combepyne and Upton Pyne (Devon) were held in the 13th century by a family of *de Pyn* who may have come from Le Pin (Calvados) or some other French place named from a prominent pine. The surname may also have arisen independently in England, 'dweller by the pine'.

**Pinel:** *v.* PINNELL

**Pinfold, Penfold:** Thomas *ate Pundfolde* 1296 SRSx; John *Pennefold* 1332 SRSx. 'One in charge of the pinfold or pound.' cf. PINDAR. The surname has probably also absorbed (Richard, Henry) *Pynfoul* 1322 LLB E, 1327 SRSf, 'pin, pen fowl'. cf. CATCHPOLE.

**Pingstone:** *v.* PINKSTONE

**Pinion:** *v.* BEYNON

**Pink, Pinck, Pinks, Pinch:** Sewine *Pinca* 1100–30 OEByn (D); Adam *Pinc* 1176 P (Y); Hugo *Pinch* 1190 P (L); Robert, Henry *Pinke* 1200 P (Ha), c1248 Bec (W); Lucia *le Pynch*, Dionisia *le Pinch* 1317 AssK. OE *pinca* 'chaffinch', with a variant *pinc*. cf. FINCH.

**Pinkhurst:** Adam *Pynkhurst* 1370–1 FFSr. From Pinkhurst Fm in Abinger (Sr).

**Pinkney, Pinckney:** Ansculfus *de Pinchengi* 1086 DB (Bk); Robert *Pinkenie* 1184 P (Lo). From Picquigny (Somme).

**Pinkstone, Pinckstone, Pingstone:** Hugo *de Penkeston'* 1207 Cur (Nt). From Pinxton (Derby).

**Pinn:** (i) John *Pinne, Pynne* 1211 Cur (O), 1327 SRSf. OE *pinn* 'pin', for a pinmaker. *v.* PINNER. (ii) Geoffrey *de Pinne* 1230 P (Do); Richard *atte Pynne* 1327 SRSo. 'Dweller on the hill' (OE *penn*), as at Pinn or Pin Hill Fm (Devon). *v.* PENN.

**Pinnell, Pinel:** Ralph *Pinel* 1086 DB (Ess); Robert *Pinel* 1185 Templars (Wa). OFr *pinel* 'little pine-tree', a diminutive of OFr *pin* 'pine-tree'. A nickname (not uncommon), applied either affectionately to a tall, thin man, or derisively to a small weedy man.

**Pinner:** (i) Andrew, Richard *de Pinner* 1275 RH (Lo). From Pinner (Middlesex). (ii) Adam *le Pinare* 1244 MESO (Wo); Walter *le Pinnere* 1281 LLB B;

Andrew *le Pynner* 1332 SRWa. A derivative of OE *pinn* 'a peg, pin', a pinmaker. Besides pins, the pinner also made wire articles, especially the small needles inserted in cards used in cloth dressing. (iii) Eborard *Penier* 1275 RH (L); Thomas *le Peniur* ib. (Nf); Maud *Penir* 1279 RH (C); John *le Peynur* ib. (Bk); Robert *le Penyr* 1327 *SR* (Ess). OFr *peigneor, peignier, pignour* 'a maker of combs'. William *le Pinour* (1292 SRLo) was also called William *le Horner* (1295 LLB A) and was, no doubt, a maker of horn combs. cf. also Juliana *le Pineresse, le Peyneresse* 1281 MESO (Wo), a comber of wool or flax (OFr *peigneresse*).

**Pinnick:** *v.* PINNOCK

**Pinning:** Alexander *Pinning'* 1212 Cur (Herts); William *Pynning* 1275 SRWo; John *Pynnyng* 1379 PTY. OE *Pinning.

**Pinnington, Pinington:** Richard *de Pynyngton* 1332 SRLa. From Pennington in Leigh (Lancs).

**Pinnion:** *v.* BEYNON

**Pinnix:** *v.* PINNOCK

**Pinnock, Pinnick, Pinnix:** Nicholas *Pinnoch* 1199 P (W); Walter *Pinnok* 1255 RH (W). ME *pinnock* 'hedge-sparrow'.

**Pinsent, Pinshon, Pinson:** *v.* PINCHEN

**Pintel:** Robert *Pintel* 1177 P (Sf); Richard *Pintell'* 1219 AssY; Matilda *Pyntel* 1251 Rams (Hu). A nickname from OE *pintel* 'penis'. cf. Alan *Coltepyntel* 1276 RH (Y); John *Swetpyntel* 1275 RH (Nf); William *Whytpintel* 1232 Cur (Nf), and note Robert *Grosuit* 1208 FFY 'large penis'.

**Pinwill:** Simon *Pynewelle* 1327 SRSx. From Pingwell Haw in Wilmington (Sussex).

**Pinyon, Pinyoun:** *v.* BEYNON

**Pipe, Pipes:** (i) Swan, William *Pipe* 1221 AssSa, 1227 FFBk; William *le Pype* 1274 AssSo. OE *pīpe* 'pipe', metonymic for a piper, or, occasionally, the man's name *Pipe* recorded in Domesday Book, perhaps from the root of Fr *Pepin*, OG *Pipin*. There was also a woman's name *Pypa* (of Wytlesford 1260 AssC). (ii) Henry *de Pipa* 1152 StCh; John *del Pipe* 1267 Cl. OE *pīpe* was also used of a water-pipe, conduit or aqueduct, and of the channel of a small stream. The name may thus arise from residence near such a *pipe* or from Pipe (Hereford) where there is a brook, or from Pipehill (Staffs), earlier *Pipa*, where there are springs from which water has for centuries been piped to Lichfield.

**Piper, Pyper:** Jordan *Piper* 1185 Templars (So); William *le pipere* 1202 AssL. OE *pīpere* 'a player on the pipe, a piper'.

**Pipon, Pippin:** *v.* PEPIN

**Pipp:** *Pipa* de Witlesford 1260 AssC; Alan *Pippe* 1204 P (L); Hervey *Pippe* 1279 RH (C); John *le Pippe* 1327 SRWo. *Pip*, a short form of PEPIN.

**Pippard:** William *Pipard* 1142–64 Templars (Bk); Geoffrey *Pipard* c1210 NthCh; Richard *Pippard* 1327 SRWo. *Pip-ard*, a short form of OFr *Pepin, Pipin*, plus the suffix -(*h*)*ard*.

**Pipperday:** *v.* PEPPERDAY

**Pipperell:** Roger *Piperel* 1359 AssD; Richard *Piperell* 1407 Hylle. A variant of PEVERALL.

**Pippet, Pippett:** *Pipot* de Turton 1246 AssLa; Roger *Pipet* 1199 AssSt; William *Pipat* 1327 SRC;

Stephen *Pippett* 1641 PrSo. *Pip-et, Pip-ot*, diminutives of *Pip*, a short form of OFr *Pepin, Pipin*.

**Pirard:** *v.* PIRRET

**Pirie:** *v.* PERRY

**Pirkiss:** *v.* PURCHAS

**Pirret, Pirrett, Pirard:** *Pirotus* 1154–61 Colch (Sf); *Pirot* 1204 Cur (Beds); Ralph *Pirot* 1176 P (Beds); Henry *Pirard* 1208 Cur (O); Richard *Pyrot* 1296 SRSx. From east French *Pier(r)ot, Pier(r)ard*, hypocoristics of *Pierre* (Peter). Now rare names, mostly absorbed by PARROT, etc.

**Pirrie:** *v.* PERRY

**Piser:** *v.* POYSER

**Pistol:** *v.* PESTELL

**Pistor:** *v.* PESTER

**Pitcher, Pitchers:** William *le Picher* 1243 AssSo; William *Pycher* 1289 FFSf. A derivative of OE (*ge*)*pīcian* 'to pitch', 'one who covers or caulks with pitch' (1611 NED). *Pitcher* is also a late form of PICKARD. Col. D. G. Pitcher notes an ancestor John *Pychard*, named in 1551 both *Pichard* and *Pichar*. In his will he calls his wife *Pytchard* and his uncle William both *Pitchard* and *Pytcher*. cf. also William *Picard* or *Pitcher* (1648 Harrison).

**Pitchford, Pitchforth:** Hugh *de Pichford* 1176 P (Sa); William *de Pycheford* 1260 AssY; John *de Pichford* 1332 SRWa. From Pitchford (Sa).

**Pitcock:** *Pitecoc* filius Simonis 1221 AssWa; William *Pitecok* 1301 FS; Henry *Pitecok* 1332 SRWa. A diminutive of OE *Pytta*.

**Pitford:** Hugh *de Pittesford* 1203 Pleas; Philip *de Pitesford* 1210 Cur (Nth); Roger *Pitford* 1642 PrD. From Pitsford (Nth), or Great Pitford in Winkleigh (D).

**Pither(s):** *v.* PETHER

**Pitman, Pittman, Pettman, Putman:** Unban *Piteman* 1203 AssNth (Nf); Henry *Putman* 1296 SRSx; Walter *Petman* 1317 AssK. OE *pytt* and *mann*, 'dweller by the pit or in the hollow'.

**Pitney:** John *Pitney* 1641 PrSo; William *Pitney* 1662–4 HTDo. From Pitney (So).

**Pitt, Pitts, Pett, Petts, Putt:** Geruase *de la Puette* 1182 P (Sx); Maurice *de Pette* 1198 Cur (K); Thomas *de la Pitte, de la Pute* 1225 AssSo; Roger *de Pettes* 1276 ArchC vi; William *Bitheputte* 1277 AssSo; Thomas *ithe Putte* 1278 MELS (Wo); Gilbert *atte Pitte* 1294 PN Sr 151; John *ater Puttes* 1296 SRSx; Juliana *atte Pette* 1327 *SR* (Ess); Richard *Pyts* 1395 Whitby. From Pett (Kent), Pitt (Hants), or 'dweller by the pit(s) or hollow(s)', OE *pytt*.

**Pitter:** For *atte pitte*. cf. PITT, PUTTER.

**Pittet:** *v.* PETTIT

**Pittis:** Walter *le Pitus* 1204 P (Lo); Thomas *Pyttose* 1559 Pat (W). OFr *pitous* 'pitiful'.

**Pitwine:** Wigor *Pitewine* 1221 ElyA (Sf); Nicholas *Pitewyne* 1327 SREss; Geoffrey *Pytewyn* 1448 NorwW (Nf). OE *Peohtwine, Pihtwine*.

**Pixton:** *v.* PICKSTONE

**Pizer:** *v.* POYSER

**Pizey, Pizzey, Pizzie, Peasey, Pezey, Pusey:** Adam *de Pesy* 1220 Fees (Berks); Joan *de Pusye, de Pysy* 1277–8 AssSo; John *de Puseye, de Peseie* 1284, 1300 Oseney (O). From Pusey (Berks), *Pesei* DB, *Pusie* Wm 1 DEPN.

**Place, Plaice:** (i) William *de la Place* 1276 RH (L); John *atte Place* 1313 FFSf; Emma *del Place* 1332 SRCu. 'Dweller in the market-place', ME *place*. (ii) William *de Plaiz* 1190 P (Y); Roger *de Plaice* 1200 Cur (La); Richard *de la Pleyse, de la Plesse* 1277 AssSo. OFr *pleix, plais*, Lat *plexum*, 'an enclosure or coppice surrounded by a fence of living wood with interlacing branches'. The surname may be from some French place so named or may denote a forester. Also survives as PLEACE. (iii) Beatrice, Adam *Playce* 1297 SRY, 1332 SRCu; William *Plaice* 1346 Whitby (Y). These might be for *Plaiz*, with loss of the preposition. But they may also be OFr *plaise* 'plaice', so called from its flatness; either a nickname or a seller of plaice or fish in general.

**Plain, Playne, Plane:** William *de Planes* 1200 P (Ess); Roger *Playne* 1293 AssSt; Robert *Playn* 1383 IpmGl. From Plasnes (Eure).

**Plaisted, Playsted, Plaster, Plested:** Osbert *de la Plested'* 1221 AssWo; Alexander *atte Pleystude* 1327 SRSo. From residence near a place for play, OE *plegstede*. Chapel Plaster in Box (Wilts) was the home of John *atte Pleistede* in 1333 (PN W 83).

**Plaistow, Plaistowe, Plaister, Plastow, Plaster, Plester:** Robert *de Plegestoue* 1168 P (D); William *de la Pleystowe* 1275 RH (W); Cristian' *atte Pleystouwe* 1296 SRSx; William *Pleystowe* 1382 PN Herts 98. From residence at Plaistow (Devon, Essex, Surrey, Sussex), Playstow (Herts), Plaistows Fm, *Plasters* 1669, *The Plaisters* 1679 (PN Herts 98), Pleystowe Fm, *Plestore* 1596 (PN Sr 266), Plestowes Fm, *Plestowe* Ric 1, *Plesters* 1821 (PN Wa 249), Plaster Down (Devon), Plastow Green, Plasterhill Fm, or Plestor (Hants), or from residence near the village sports ground. OE *plegstōw* is more common than *plegstede*, and several of these modern places are in or near a large open space in the middle of the village. Plaistow in West Ham is pronounced *Plarstow*.

**Plampin:** *v.* BLAMPHIN

**Plank:** Maud *de la Plank* 1288 Ipm (W). 'Dweller by the plank' or narrow foot-bridge, ME *planke*.

**Plant, Plante:** William *Plante* 1262 *For* (Ess), 1279 RH (C); William *Plauntes* 1275 RH (Nf). Metonymic for a gardener or planter of various plants. cf. Henry *le Plaunter* 1281 Rams (Hu); Ralph *Plantebene* 1199 P (Nf) 'beans' and PLANTEROSE.

**Planter:** Henry *le Plaunter* 1281 Rams (Hu); Richard *le Plonter* 1335 ChertseyCt (Sr). A derivative of OFr *plant* 'a plant', a nickname for a gardener.

**Planterose:** Robert, Alice *Planterose* 1221 AssWa, 1272 RamsCt (C). 'Rose-grower.'

**Plascott:** *v.* PLASKETT

**Plash:** William *Plash* 1273 IpmGl; Hugh *de Plassh'* 1327 SRWo; John *Plassh* 1378 FFEss. From Plash (So), Plaish (Sa), or 'dweller by the muddy pool', OE *plæsc*.

**Plaskett, Plasked, Plascott, Plaskitt:** John *Plaskett* 1570 FrY. 'Dweller by the swampy meadow', OFr *plasquet*.

**Plass:** From *plais. v.* PLACE (ii).

**Plaster, Plaister:** Symon *le Emplastrer*, Robert *le Plastrer* 1276 LLB B; Hugo *le Playstrer* 1293 MESO (Y); Emma *le Plastrer* 1323 Bart (Lo); William

*Plaister* 1646 FrY. OFr *plastrier* 'plasterer'. Richard *le Plasterer* of York and his mates (*sociis*) made the partition walls of the Queen's chamber of *plastre de Parys* at Scarborough in 1284 (Building 156). These names also represent a common pronunciation of PLAISTED and PLAISTOW.

**Plastow:** *v.* PLAISTOW

**Plater, Platter:** (i) William *le plater, le Platier*, armurer 1292 SRLo, 1311 LLB B; John *le Plater* 1300 LLB C; John *le Plattour* 1319 SRLo. A derivative of ME *plate*, a maker of plate-armour or of plates for armour. cf. John *Platesmyth* 1379 PTY. (ii) Herueus *Plaitere* Hy 3 Colch (Ess); Walter *Playtur* 1279 RH (Hu); Philip *le Pleytour* 1327 SRSf. A derivative of OFr *plait* 'plea' or OFr *plaitier* 'to plead', a pleader, advocate.

**Platfoot:** Matilda *Platfot* 1260 AssC. 'Flat-foot.' OFr *plat*.

**Platner:** John *le Platener* 1290 LLB A. A derivative of OFr *platon* 'metal plate', plate-maker.

**Platsmith:** John *Platesmyth* 1379 PTY. 'A maker of plate armour', ME *plate*, OE *smiþ*.

**Platt, Platts:** John *de la Platte* 1242 P (Wo); Geoffrey *de Platte* 1285 AssLa; Henry *Atteplatte* 1327 SRSt; Robert *Plattes* 1590 FrY. 'Dweller by the small patch of land' (OE *plætt*) or by a foot-bridge, dialectal *plat* 'foot-bridge' (1652 NED), from OFr *plat* 'flat surface', as at Platt Bridge (Lancs).

**Platten:** William, Walter *Platon* 1198 FFSf, 1327 SRWo. Metonymic for PLATNER.

**Platter:** William *Platter* 1332 SRSx. Probably 'dweller by the footbridge'. *v.* PLATT, PLATER.

**Play:** John *Play* 1274 IpmY; John *Play* 1355 IpmW; George *Pley* 1662–4 HTDo. OE *Plega*.

**Playden, Playdon:** Thomas *de Pleydenn* 1296 SRSx. From Playden (Sx).

**Player:** John, William *le Pleyer* 1296 MEOT (Herts), 1332 SRSr. OE *plegere* 'player', probably an athlete who distinguished himself on the *Plaisted* or *Plaistow*.

**Playfair, Playfer:** William *Playfayre* 1290 Black, 1570 FFHu. Probably 'play fairly'. cf. GOVER.

**Playfoot:** Richard *Pleyfote, Playfote* 1310–12 ColchCt. Probably 'splay-foot'. Loss or prefixing of initial *s* was common in the Essex dialect. The examples are early to be regarded as forms of *Playford*.

**Playford:** Fulcher' *de Pleiforda* 1130 P (Sf). From Playford (Suffolk).

**Playne:** *v.* PLAIN

**Playsted:** *v.* PLAISTED

**Pleace, Please, Pleass:** For PLACE (ii).

**Pleaden:** *v.* BLETHYN

**Pleader:** Walter *le Plaidur* 1199 P (Nth); Richard *le playdur* 1262 LuffCh; Ralph *le Pledour* 1309 SRBeds; Isabella *Pledour* 1388 LoPleas. OFr *plaideor* 'advocate, lawyer'.

**Pleasance, Pleasaunce, Pleasants, Pleasant:** (i) Reginald *de Pleisauns, de Plesence* 1275 RH (L); John *de Plesaunce, de Plesancia* of Lumbardy 1319 SRLo, 1339 LoPleas. From Piacenza (Italy), earlier *Placentia*. (ii) *Plesantia* West 1275 RH (Nf); Alicia *filia Plesance* 1279 RH (O); *Pleasant* Tarlton (f) 1681

Bardsley. *Plaisance* (OFr *plaisance* 'pleasure'), *Plaisant* (OFr *plaisant* 'pleasing'), women's names long in use.

**Pleavin:** *v.* BLETHYN

**Pledger:** Richard *Plegger* 1309 Glast (So); John *Pleger* 1525 CantW; Elias *Pledger* 1662 *HTEss*. ME *pleggere* 'one who stands bail in a law-suit'. cf. the metonymic forms: William *Plegge* 1435–6 FFSr; Robert *Pledge* 1569 Musters (Sr).

**Plenty:** Simon *Plente* 1230 P (Y); William *Plentee* 1243 AssSo. OFr *plente* 'abundance'.

**Plessis:** Gilbert *de Plesset* 1204 FFEss; Augustine *de Plecy* 1276, *de Pleyci* 1279 AssSo; John *Plecy* 1389 PN Do ii 176. From Pleshey (Ess), Plessey (Nb), or 'dweller at the enclosure made by interwoven branches', OFr *plaisseis, plaissiet*.

**Plested:** *v.* PLAISTED

**Plester:** *v.* PLAISTOW

**Pleven, Plevin:** *v.* BLETHYN

**Plewman:** *v.* PLOWMAN

**Plimsoll:** Peter *Plymsoll*, James *Plymsould* 1642 PrD. A Huguenot name. Several refugees of this name came from Brittany to southern England after the Revocation of the Edict of Nantes, one of them to Bristol (Smiles 421).

**Plock:** Thomas *de la Plocke* 1275 MELS (Wo); Stephen *atte Plokke* 1311 IpmGl; William *atte Plocks* 1332 SRWo. 'Dweller at a small piece of ground'. cf. dialectal *plock*.

**Plomer:** *v.* PLUMER

**Plomley:** *v.* PLUMBLEY

**Plott:** John *atte Plotte* c1280 Bart (Herts); Henry *ate Plotte* 1317 AssK. 'Dweller on the small plot of ground', OE *plot*.

**Plow, Plowes, Plows, Plough:** Alan *de Ploghe* 1306 Riev; John *Plough* 1524 SRSf; John *Plough* 1559 Pat (Gl). Probably for PLOWMAN, but perhaps also for 'dweller at the ploughland'.

**Plowden:** William *de Ploden'* 1203 Pl (Sa); Roger *de Ploeden'* 1254 RH (Sa); Thomas *de Plowedene* 1327 SRSa. From Plowden (Salop).

**Plowman, Plewman:** Robert *Pleueman, Plouman* 1223 Cur (We); Philip *Ploman* 1255 *Ass* (Ess); John (*le) Plouman* 1275 RH (L); John *le Ploghman* 1275 RH (R); John *Plowman* 1345 AD i (L); John *Plewman* 1560 FrY. OE *plōh* 'plough' and *mann*, 'ploughman' (1271 NED).

**Plowright:** William *le Plowritte* 1279 RH (C); Robert *le Plogwryth* 1285 Wak (Y); Baldwyn *le Ploghwright* 1297 FFSf. OE *plōh* 'plough' and *wyrhta* 'wright', 'a maker of ploughs' (1285 NED).

**Pluckett:** Hugh *Pluket* 1147–8 CartAntiq; Alexander *Pluket* 1218 P (Sx); Rychard *Pluket* 1524 SRSf. A diminutive of Norman-Picard *Pluque*, OFr *Pluche*. Probably connected with OFr *peluchier* 'to pick, clean', but the meaning of the surname is uncertain. *v.* Dauzat.

**Plucknett, Plumkett, Plunket, Plunkett:** Hugh *de Plugeneio* 1163 Oseney; Hugh *de Plucheneit, de Plugenet* 1200 Oseney; Alan *Plugenet, de Pluckenet* 1270 AssSo; Robert *Plukenet* 1280 IpmW. From Plouquenet (Ille-et-Vilaine).

**Pluckrose:** William, Robert *Pluckerose* 1275 RH (Sf), 1296 SRSx. 'Pluck rose.' cf. John, William

*Pullerose* 1296 SRSx, 1301 SRY. Lower mentions a friend of his who held land in Ashdown Forest (Sussex) of the Duchy of Lancaster by one red rose. 'On the front of a farm belonging to him is a large rose tree, to which the reeve of the manor periodically comes, and either *plucking* or *pulling* a flower, sticks it into his button-hole, and walks off.'

**Plum, Plumb, Plumbe:** Geoffrey, Simon *Plumbe* 1208 ChR (Sf), 1251 Rams (Hu); John *Ploumbe* 1327 SRSf; Ralph *Plomme* 1327 SRDb; Nicholas *Plumme* 1469 LLB L. Common in early sources, without any sign of either article or preposition, almost invariably in the form *Plumbe*, this must be OFr *plomb*, Lat *plumbum* 'lead', used by metonymy for *plumber*. *v.* PLUMMER.

**Plumbley, Plumbly, Plumley, Plomley:** Alexander *de Plumleg'* 1235 Fees (Db); William *de Plumley* 1275 SRWo; George *Plumleigh* 1642 PrD. From Plumley (Ch), Plumley in Bovey Tracy (D), or Plumbley in Eckington (Db).

**Plume:** Alan *Plome* c1000 MedEA (Nf); Hugh *Plume* 1221 Cur (Do); Walter *Ploume* 1275 SRWo. OFr *plume*, Lat *pluma* 'feather', used of a dealer in feathers. *v.* PLUMER.

**Plumer, Plomer:** Walterus, Rogerus *plumarius* 1176 P (We), 1230 P (Nt); Roger *Plumer* 1185 Templars (Y); Turbertus *plumator* 1230 P (Sf); Simon *le Plumer* 1246 AssLa; Ralph *le Plomer* 1280 Oseney (O). AFr, OFr *plumier*, Lat *plumarius* 'a dealer in plumes or feathers' (1282 NED). MLWL records *plumator* 'flock-puller' in 1281 and *plumarium* 'feather-bed' in 1384, but does not include *plumarius*. John de Cestrehunte, *fethermongere* 1280 LLB A, is called *plumer* in 1281 ib. *Plumer* in early forms may sometimes mean 'plumber'. *v.* PLUMMER.

**Plumkett:** *v.* PLUCKNETT

**Plummer, Plimmer:** (i) Godric *Plumberre* 1102–7 Rams (Hu); Gillebertus *Plumbarius* 1183 P (C); Osbertus *Plumbarius, le Plumer* 1221–2 Cur (Mx); Ernald, Osbert *le plummer* 1225 FrLeic, 1227 Bart (Mx); Robertus *plumberius* 1230 P (Wa); Richard *le plumber* 1230 P (D); Peter *Plumber* 1256 AssNb; Robert *le Plummer* 1260 AssC; Simon *le Plommer* 1280 AssSo. The earliest examples seem to be a direct formation from OFr *plomb* 'lead', later assimilated to OFr *plummier, plommier* 'plumber' (1385–6 NED). *Plumbarius* is first recorded in MLWL in 1290 and *plummarius* in 1380. It has no example of *plumberius*. Early forms *plumer*, often undoubtedly for PLUMER, may sometimes belong here. (ii) This may sometimes be local in origin. Plummers in Kimpton (Herts) was the home of Thomas *de Plummere* in 1272, 'the pool by the plum-tree' (PN Herts 16). Plummer Wood in Thaxted (Essex) is *Plumetonemore* in 1348 (PN Ess 499).

**Plumptree:** *v.* PLUMTREE

**Plumstead, Plumsted:** John *de Plumstede* 1242 Fees (Nf); William *de Plumstede* 1306 FFEss; William *Plumstede* 1450 Paston. From Plumstead (K, Nf), or Great, Little Plumstead (Nf).

**Plumton, Plumpton:** William *de Plumton'* 1174 LuffCh; Nigel *de Plumpton* 1247 FFO; John *de Plumpton* 1379 PTY; Robert *Plumpton* 1432 IpmNt.

From Plumpton (Nth, Sx, WRY), Plumpton Wall (Cu); Fieldplumpton, Woodplumpton (La), Plumpton End (Nth), or a lost *Plumpton* in Kingsbury (Wa).

**Plumtree, Plumptree:** Richard *de Plumptre* 1230 P (Nt); John *Plumtre* 1392, Henry *Plomptre* 1480 IpmNt. From Plumtree (Nt).

**Plunger:** Eudo *de Plungard'* 1185 P (Wa); Ralph *Plungard* 1202 AssNth. From Plungar (Lei), *Plungard* 1186.

**Plunket(t):** *v.* PLUCKNETT

**Plush:** Robert *Plusch* 1327 SRSo; John *Plushe* 1641 PrSo. From Plush (Do).

**Pluthero:** *v.* PROTHERO

**Poad, Poat, Podd, Pods:** Richard *Pode* 1230 P (D); John *le Pod* 1275 RH (K). ME *pode* 'toad'.

**Pobgee, Pobjoy:** *v.* PAPIGAY

**Pocket, Pockett, Pocketts:** Adam, Geoffrey *Poket* 1210 Cur (C), 1258 FFEss. ME *poket*, a diminutive of AFr *poque*, 'a small pouch'. cf. POKE, POUCH.

**Pockley:** Richard *de Pokel'*, William le carbon *de Pokele* 1219 AssY. From Pockley (ERY).

**Pocklington:** William *de Pokelington* 1208 AssY; Robert *de Pokelynton* 1335 FrY. From Pocklington (ERY).

**Pockridge:** Thomas *Pocrege* 1576 SRW. From Pockeredge Fm in Corsham (W).

**Pocock:** *v.* PEACOCK

**Podd, Pods:** *v.* POAD

**Poddington:** Emma *de Podington'* 1242 Fees (Beds). From Podington (Beds).

**Podmoor, Podmore:** Reginald *de Podymore* 1279, William *de Podmore* 1295 AssSt; William *Podymor* 1332 SRWo; Thomas *Podmor* 1417–18 FFWa. From Podimore (So), or Podmore (St).

**Poe:** *v.* PAW

**Poel:** *v.* PAUL, POWELL

**Poggs:** (i) William, Andrew *Pogge* 1286, 1309 Wak (Y). *Pogg* is a pet-name for Margaret, rhymed on *Mogg*. (ii) Imbertus *Pugeis* 1235 Fees (Ha); Robert *le Pugeys* 1286 Ipm (Sf); Alicia *Pogeys* 1327 SRC; Gregory *Pogeis* 1300 LLB B. Imbert *le Pugeys*, called also *de Salvadya* 'of Savoy' and *de Bampton* (Oxon), came to England with Queen Eleanor of Provence, who married Henry III on 20 January 1236 (Fees). He came from Le Puy-en-Velay (Haute-Loire) and has left his name in Stoke Poges (Bucks), *Stoke Pogys* (1514), *Stokbogies* (1526), and in Broughton Poggs (Oxon). The surname means 'man from Puy' or 'dweller by the hillock'. Le Puy is a common place-name in Anjou, Poitou, etc., and the surname may also have been introduced into England from one of these.

**Pogmore:** Cecilia *Pogmore* 1379 PTY; John *Pogmour* 1479 Shef. From Pog Moor in Barnsley (WRY).

**Pogson, Poxon:** Richard *Pogson* 1440 ShefA (Y). 'Son of *Pog* (Margaret).'

**Poidevin, Le Poidevin, le Poideven, Podevin, Patvine, Potvin, Potwin, Portwin, Portwine, Putwain, Puddifin, Puddifant, Puttifent:** *Peiteuin* de Eya 1186 P (Sf); *Peyteuinus* le Esquier 1247 AssBeds; Rogerus *Peteuinus*, *Pictauensis* 1086 DB (Ess); Rogger

*Peiteuin* 1094 ASC E; Ralph *Patefine* Hy 1 Whitby (Y); Odo *Petefin* ib.; Adam *Petevin* 1198 FFK; David *le Poitevin* 1199 AssSt; Rainer *le Payteuin* 1219 AssY; Richard *Patevin* 1242 Fees (Nb); William *Peytefin* 1247 AssBeds; Osemund *Petewin* 1279 RH (C); Preciosa *Potewyne* ib.; Reginald *Peytewin* 1296 SRSx. OFr *Poitevin*, AFr *Peitevyn*, a man from Poitou.

**Pointel:** Tedric *Pointel*, *Puintel* 1086 DB (Ess); Stephen *Puintel* 1166 P (Nf). OFr *pointel* 'point', 'a sharp pointed instrument', a diminutive of OFr *point* 'point', a nickname that might have been applied to a tall, thin man. The complete absence of any sign of the preposition *de* prevents any association with Pointel (Orne).

**Pointer, Poynter:** Benedict, Richard *le Puintur* 1206 P (Berks), 1212 Cur (K); Richard *le Pointur* 1213 FFC; Bartholomew *le Pointer* 1314 Oriel (O). A derivative of ME *poynte* 'a tagged lace or cord, of twisted yarn, silk, or leather' (1390 NED), from OFr *pointe* 'a sharp or pointed extremity'. 'A pointer', a maker of points for fastening hose and doublet together. We may also have a building-term. In roofing, the layers of tiles overlapped and the lowest layers, and sometimes all the layers, were pointed or rendered with mortar. This is called 'pointing' in 1265 and the slaters doing this work may well have been called *pointers* (Building 233–4).

**Pointon, Poynton:** Alexander *de Pochintun* c1155 Gilb; Jordan *de Poyngtun'* c1200–10 RegAntiquiss; Alice *de Poynton* 1344 LoPleas; Robert *Pointon* 1419 FFEss. Usually from Pointon (Lincs), *Pochinton* DB, but occasionally perhaps from Poynton (Ches, Salop).

**Points, Poyntz:** Walterus *filius Ponz* 1086 DB (Berks); Ricardus *filius Puinz* 1185 Templars (So); *Puntius*, *Pontius* Arnaldi 1196 P (D); Reginald *Puinz*, *Poinz* 1176–7 P (Sr); Rannulf *Poinces* 1207 Cur (Sx). OFr *Ponz*, Latin *Pontius*, 'man from Pontus in Asia Minor', the name of a saint (*Pontius*) of Asia Minor whose cult was widely spread in the Middle Ages, surviving also in Fr *Pons*, *Ponce* and the Burgundian *Point*. The surname is also a nickname for a fop, the man with the points (*v.* POINTER): Geoffrey *Aspoinz* 1193 P (L), and may be local in origin: Hugh *de Poinz* 1221 Cur (So), Reginald *de Ponz*, *de Pontibus* c1222 InqLa, from Ponts (La Manche, Seine-Inférieure), Lat *pontes* 'bridges'.

**Poke:** Hugh, Robert *Poke* 1200 P (K); 1279 RH (Hu). ME *poke* 'bag', metonymic for *Poker*, 'maker of bags or small sacks': Ithel (*le*) *Poker* 1314, 1323 AssSt.

**Polbrook, Pollbrook:** Walter *Pollebroke* 1443–4 FFSr. From Polebrook (Nth).

**Polcat:** *v.* POLECAT

**Polden:** Robert *Poleden*, William *Polden* 1662–4 HTDo. From Polden Hill (So).

**Polder, Poulder, Pulder:** Roger *Puldre* 1175 Riev; Robert *Polre* 1198 FFK. From Great, Little Poulders Fm in Woodnesborough (K), or 'dweller by the marshy land', OE *\*polra*.

**Polecat, Polcat:** Roger *Polcat* Hy 3 PN Ess 285; Ranulph *Pulkat*, *Polkat* 1327 Pinchbeck (Sf). A nickname from the polecat, OFr *pole*, OE *catta*, either because of its ferocity or its offensive smell.

**Polecott, Polecutt:** v. POLLIKETT

**Pole(s):** v. PAUL, POOL

**Polglase:** John de Polglas 1297 MinAcctCo. From Polglaze (Cornwall).

**Polkin:** v. POLLKIN

**Poll:** v. PAUL, POOL

**Pollard:** Pollardus Ostiarius 1201 Cur (Sf); Pollardus Forestarius 1207 Cur (Gl); Stephanus filius Pollard 1275 RH (K); William, Peter, Richard Pollard 1181, 1192, 1195 P (Sr, Herts, L). The personal-name Pollhard is a derivative of Paul which seems to have been pronounced Poll by the end of the 12th century. cf. Pol 1188 and Polle 1193 s.n. PAUL. The surname is also from a nickname pollard from ME poll 'to clip', poll 'the head', one with a close-cropped head or a big head. cf. BALLARD, TESTAR.

**Pollbrook:** v. POLBROOK

**Pollikett, Polecott, Pollicott, Polecutt:** Adam Policot 1279 RH (C); Thomas Polket 1327 SRC. From Pollicott (Bk).

**Polling:** Nicholas Pollyng' 1249 AssW; Roger Pollyng 1327 SREss; John Polyng 1524 SRSf. Probably from an unrecorded OE *Polling, but sometimes, perhaps, from Poling (Sx).

**Pollington:** Gregory de Polint' 1219 AssY; William de Polinton' 1219 P (Y). From Pollington (WRY).

**Pollins:** v. PAULIN

**Pollit, Pollitt:** Henricus filius Ypoliti 1171 P (Y); Yppolitus de Pridias 1207 P (Ess); Walter, Geoffrey Polite 1222 Cur (Nth), 1285 Ass (Ess); Christina Polytes 1311 ColchCt. Hippolytus, Gk Ἱππόλυτος 'letting horses lose', the name of a Roman saint martyred in 235, to whom the church of Ippollitts (Herts) is dedicated (Polytes 1412).

**Pollkin, Polkin:** Pollekin 1206 Cur (So); John Polekin 1279 RH (O); Elizabeth Polkyn 1530 NorwW (Nf). A diminutive of Poll, a pet name for Mary.

**Pollock, Pollack, Pollok:** Peter, John de Pollok c1172–8, 1304 Black; John Pullok 1453 ib. From Upper Pollock (Renfrew).

**Polman:** v. POOLMAN

**Polson, Poulson, Poulsom, Poulsum:** Adam Poles(s)on 1323–5 AssSt. 'Son of Paul.'

**Polstead, Polsted:** Hugh de polsted 1189 Sol; Michael de Polstede 1212 P (Sr); Hugh de Polstede 1242–3 FFK. From Polstead (Sf).

**Polton, Poulton:** William de Polton' 1200 P (St); Walter de Pulton 1259 IpmGl; John Polton 1327 SRSx; William Powlton 1576 SRW. From Poulton (Ch, Gl), Poulton (K), Poulton cum Spital, cum Seacombe (Ch); Poulton with Fearnhead, le Fylde, le Sands (La), or Poulton in Mildenhall (W).

**Polworthy:** Robert Poleworthi 1537 Hylle. From Pulworthy in Molland (D), Poleworthi 1300, or Pulworthy in Hatherleigh (D), Poleworthy 1303.

**Pomeroy, Pomery, Pomroy, Pummery:** Ralph de Pomerai 1086 DB (D, So); Ralph Pomeria ib. (So); Samson de la Pumeray 1200 Cur (O); Henry de la Pomereie 1225 AssSo; Robert Pomeroy 1327 SRSo. The DB tenant-in-chief, whose name survives in Berry Pomeroy and Stockleigh Pomeroy (Devon), came from La Pommeraye (Calvados). v. ANF. Others may have come from La Pommeraie (Seine-Inférieure) or

Saint-Sauveur-la-Pommeraie (La Manche). The surname may also denote residence near an apple-orchard (OFr pommeraie). cf. APPLETON.

**Pomfret, Pomfrett, Pomphrett, Pumfrett, Pontefract:** Willelmus de Pontefracto 1191 P (Y); William Puntfreit 1200 Cur (Ess), de Puntfreit 1203 Cur (K); Matthew de Pomfrait 1275 RH (Lo); Robert Pumfret 1275 RH (Nf). From Pontefract (WRYorks), Lat (de) ponte fracto '(at the) broken bridge'. Puntfreit is the Norman-French form, found also in Pomfret Mead (PN Ess 642). v. PUMFREY.

**Pomfrey, Pomphrey:** v. PUMFREY

**Pond:** William del Pond 1190 BuryS (Sf); John de la Ponde 1203 FFEss; Henry Attepond 1260 AssC; John Ponde 1262 For (Ess). 'Dweller by the pond', ME pond.

**Ponder:** William le Pondere 1279 RH (C); John le Ponder' 1306 RamsCt (Hu); Rychard Ponder, Thomas Punder 1524 SRSf. 'Dweller by the pond', cf. Richard Pondeman 1327 SR (Ess), or 'keeper of the pound'. v. POUNDER.

**Ponsford:** v. PAUNCEFOOT

**Ponsonby:** Alexander de Ponsonby 1332 SRCu; Dionisia Pounsounby 1401 LLB I; William Ponsonbye 1600 AssLo. From Ponsonby (Cumb).

**Pont, Ponte, Punt:** Amice de Ponte 1268 AssSo; Peter del Pount 1279 FFEss; William Punt 1316 FFEss; Walter Pont 1327 SRSf. 'Dweller by the bridge', ME pont, punt, OFr pont, Lat pons 'bridge'.

**Pontefract:** v. POMFRET

**Ponter, Punter:** Walter Punter 1214 Cur (Nth); Stephen le Punter 1243 AssSo; John, Richard Ponter 1255 RH (Sa). 'Dweller by, or keeper of the bridge.' v. PONT, BRIDGER.

**Pontiff:** Robert Puntif 1260, James Puntyfe 1292–3 IpmY; James Pountif 1343 FFY. A pageant name, 'pontiff', OFr pontife. cf. John Pontifex 1353 Goring (O) 'the pope or one of the high priests in a play'.

**Pontis, Puntis:** Robert de Punteise 1190 P (L); Adam Pontis 1275 RH (Nf); Henry Ponteys 1279 RH (Hu); James Punteys 1296 SRSx; John de Pounteysse, de Pountoyse 1300 LLB B, 1302 LoCt. From Pontoise (Seine-et-Oise).

**Ponton:** Nicholas de Ponton 1221–2 FFK; Peter de Ponton 1297 Coram (K); Thomas Ponton 1539 Black. From Great, Little Ponton (L).

**Pook, Pouck:** William Puch 1166 P (Nf); Richard le Pouke 1296 SRSx; John le Puk 1332 SRSx. OE pūca 'elf, sprite, goblin'.

**Pool, Poole, Pole, Poles, Poll, Polle:** Mauritius de la Pole 1176 P (D); Roger de Pole 1191 P (W); Thomas del Pol 1260 AssC; John Pool 1324 FFEss; Ralph atte Polle 1327 SRSx; William atte Poule 1327 SRSo; Thomas del Poule 1332 SRCu. From residence near a pool or tidal stream (OE pōl). All the modern surnames may also derive from PAUL.

**Pooler:** Equivalent to atte Pole (v. POOL) or POOLMAN.

**Pooley:** Walter de Polhey 1248 FFEss; William de Poleye 1273 RH (Bk); Elias de Polye, de Polee 1275 RH (Nf); William de Polleye 1275 SRWo; William Polleye 1346 ColchCt; John Poley 1379 AssEss. From residence near low-lying land or an enclosure by a pool (OE *ēg*; (*ge*)*hæg*). This sometimes becomes *Polley*. Hunt's Hall in Pebmarsh (Essex) is *Polheia* (1086), *Polleheye* (1238), *Pollyhall* (1446), *Pooley* t Eliz (PN Ess 450). Polly Shaw (PN K 40) has the same origin.

**Poolman, Polman:** Hugh *Poleman* 1260 ArchC iii; John *Polman* 1327 *SR* (Ess). 'Dweller by the pool.'

**Poor, Poore, Power, Powers:** (i) Drogo *poher* 1127 AC (Gl); Walter *le Poher* 1162 DC (L); Hugo *le Puhier* 1166 P (Sa); Hugo *Puher* 1170 P (Wo); John *le Poer* 1199 FF (Nth); Roger *le Puiher* 1204 Cur (Db); Philip *le Poyer, le Power* 1257 StThomas (St), 1292 AssSt; Richard *le Poier, le Pouer*, William *Pouwer* 1297 MinAcctCo (O); Stephen *le Poer, le Power, Power* 1296, 1327, 1332 SRSx. OFr *Pohier* 'a Picard'. Fairly common and almost invariably preceded by *le*. Without the article (Gilbert *Poer* 1274 FFSf), the source might, theoretically, be OFr *poer* 'power', but examples are so rare that all probably stand for *Pohier*. (ii) Walter *le Poure* 1163 Eynsham (O); Hugo *Pauper, le Pouere* 1191, 1206 Oseney (O); Roger *Pauper, le Poer, le Povr'* 1211–12 Cur (Nf); Roger *Pauper* (*le poure*) 1230 P (O). OFr *povre, poure* 'poor'. This, too, is common, but forms such as *poure, power* cannot safely be assigned to *Poher* or *povre* without other evidence. Geoffrey and William *Pouere*, customary tenants of the Bishop of Ely at Wetheringsett (Suffolk) in 1221 (*ElyA*) were not likely to be Picards. In London, Henry *Puer* or *Poer* (1300 LoCt) was probably of this nationality, but whether Geoffrey *le Power* (1299 ib.) and John *le Poer* (1300 ib.) came from Picardy or were nicknamed 'the poor' cannot be determined.

**Pope:** Agnes, Hugo *le Pope* c1230 Barnwell (C), 1247 AssBeds; Henry *Pope* 1296 SRSx. ME *pope* from OE *pāpa*. *v*. PAPE.

**Popejoy:** *v*. PAPIGAY

**Popham:** Agnes *de Poppham* c1210 Fees (Ha); Hugh *de Popham* 1312 Glast (So); Henry *Popham* 1376 IpmW. From Popham (Ha).

**Poplett, Puplett:** Thomas *Pupelot* 1214 P (C); Roger *Popelote* 1295 Barnwell (C); William *Popellot* 1381 CarshCt (Sr). A double diminutive of OFr *poupee* 'doll, baby'.

**Popley:** Matilda *atte Popele* 1275 SRWo; Richard *de Popelay* 1285–6 IpmY; William *de Popeley* 1390 IpmLa; Robert *Popley* 1641 PrSo. From Popeley House in Birstal (WRY), or 'dweller at the stony clearing', OE *popel, lēah*.

**Popp, Poppe:** *Poppe* Hy 2 PN Sx 25; Stephen *Poppe* 1202 AssL; Robert *Poppe* 1279 RH (Hu); William *Poppys* 1524 SRSf. OE *\*Poppa*, unrecorded in independent use, but found in place-names.

**Popple, Pople:** (i) Nicholas *de Poppehale* 1296, Crystobell *Pople* 1525 SRSx; Roberte *Popple* 1576 SRW. From a lost *Pophall* in Linchmere (Sx). (ii) Robert *Popehull* 1327 SRSo; John *de Pophull* 1332 SRWa; John *Pople* 1641 PrSo. From Pophills in Salford Priors (Wa).

**Poppleston, Popplestone:** Alwin *Poplestan* 1066 Winton (Ha); Jocelin *Pobelstone* 1210–11 PWi; George *Poplestone* 1642 PrD. A nickname from OE *papolstān* 'pebble', perhaps used of a person hard to deal with. *v*. OEByn 369.

**Poppleton:** Ralph *de Popelitun* c1147–c1154 MCh; Adam *de Popelton* 1303 IpmY; William *Popilton* 1392 LoCh. From Nether, Upper Poppleton (WRY).

**Popplewell:** John *de Popelwelle* 1316 Wak (Y); Roger *de Popelwell'* 1379 PTY; Margaret *Popilwell* 1454 TestEbor. From Popplewell in Cleckheaton (WRY).

**Poppy:** Jach', Geoffrey *Popy* 1275 RH (Nf), 1327 SRSf. A nickname from the flower.

**Porcas:** *v*. PURCHAS

**Porcher:** Edric *Porcher* 1185 Templars (Sa); Reginald *le Porker* 1221 AssWa; Thomas *le Porcher* 1275 SRWo. OFr *porcher, porkier* 'swineherd'.

**Pork, Porke:** *Porcus* 1148 Winton (Ha); Robert *le Porc* 1198 AD ii (Ess); Ralph *le Porc* 1221 Cur (So); Thomas *Porke* 1642 PrD. OFr *porc* 'pig', either a nickname, or metonymic for a swineherd.

**Porkiss:** *v*. PURCHAS

**Porrett, Porritt:** *v*. PARROT

**Port, Porte:** (i) Hugo *de Portu, de Port* 1084 GeldR (D), 1086 DB (K); Henry, Hawis *de Port* 1115 Winton (Ha), 1185 Templars (Berks). From Port-en-Bessin (Calvados). (ii) Walterus (*de*) *Extra Portam, de Horslaporte* 1199, 1200 Cur (L); Adam *de la Port* 1243 AssSo; Thomas *in the porte* 1275 SRWo; Alan *ad Portam* 1275 RH (L); Simon *Atteporte* ib. (Sx); Henry *Porte, de Porta* 1297 MinAcctCo. 'Dweller near or outside the town or castle-gate' (OE *port* 'door, gate', OFr *porte* 'entrance, door'); occupational, 'doorkeeper, gate-keeper', equivalent to PORTER; 'dweller in the market-town or port' (OE *port* 'harbour, town').

**Portal:** A Huguenot name, southern French *Portal* 'dweller by the town-gate'.

**Porter, Porters:** (i) Willelmus *Portarius* (*Janitor*) 1183 P (Berks); William *le portier* 1190 P (Berks); Adam *le porter* 1202 AssL; Alice *le Porters* 1330 ColchCt. AFr *porter*, OFr *portier* 'door-keeper, gate-keeper'. Milo *Portarius* (1086 DB, Ha) performed porter-service at the jail or castle of Winchester. (ii) Nicholas, David *le portur* 1263 MEOT (Sr), 1279 AssSo; Andrew *Portour* 1356 ColchCt. OFr *porteour* 'porter', carrier of burdens. *Porter* 'door-keeper' also appears as *portour* in the 13th century (NED).

**Porteous, Porteus, Portas, Portass, Porteas, Portus:** *Portehors* 1178 CartAntiq (Ess); Turgis *Portehors* 1227 Cur (Mx); Reginald *Porthors* 1301 CorNth; William *Portus* 1550 Black. AFr *porte-hors* 'breviary, portable prayer-book'. Metonymic for a writer of these.

**Portjoy:** Walter *Porteioie* 1193 P (Beds); John *Porteioye* 1276 AssLo; Geoffrey *Portejoye* 1332 SRSx. 'Carry joy', OFr *porter, joie*, probably a nickname for a happy man. cf. Richard *Portefer* 1221 Acc 'carry iron'; Thomas *Portefleur* 1202 Pleas (Nf) 'carry a flower'; Robert *Porterose* 1243 AssSo 'carry a rose'.

**Portland:** Richard *de Portland* 1281 AssW; William *Portlond* 1360, Walter *Portlond* 1397 IpmGl. From Portland (Do).

**Portman, Portmann:** John, Henry *Portman* 1225 AssSo, 1297 Ipm (Sf); Thomas *le Portman* 1275 SRWo. OE *portmann* 'townsman', 'portman', one of the body of citizens chosen to administer the affairs of a borough.

**Portno, Portnoy, Portner:** Robert *Portenuit* 1214 Cur (Lei). 'Carry night', OFr *porter, nuit*.

**Portrait, Portrey:** cf. John *le portreor* 1292 SRLo; Nicholas *le Portreour* 1312 LLB D. OFr *portraiour* 'painter (of pictures or portraits)'. The modern surnames appear to be metonymic, from OFr *pourtraict* 'portrait'.

**Portugal, Pettengell, Pettengill, Pettingale, Pettingall, Pettingell, Pettingill, Puttergill:** Walterus *filius Portingalliae* 1201 Cur (Sf); William *Potagal* 1214 Cur (Lei); Symon *Petingal* 1278 Oseney (O); William *Pettingall* 1568 SRSf; John *Portingale* 1569 SPD; Helena *Pettingall* 1607 RothwellPR (Y). ME *Portingale* 'Portugal' (c1386 NED), 'a Portuguese' (1497).

**Portus:** *v.* PORTEUS

**Portway:** Richard *de la Portweye* 1279 RH (Hu); Christopher *Portewey* 1524 SRSf. 'Dweller by the road to the town or harbour', OE *port, weg*.

**Portwin(e):** *v.* POIDEVIN

**Poser:** *v.* POYSER

**Posford:** John *Posford, Potsforde* 1568 SRSf. From Potsford Barn in Letheringham (Suffolk).

**Poskitt:** *v.* POSTGATE

**Posnett:** *v.* POSTLETHWAITE

**Possell:** *v.* POSTLE

**Posselwhite:** *v.* POSTLETHWAITE

**Post, Poste:** William *Post'* 1148 Winton (Ha); Robert *le Post* 1228 FFEss; William *Post* 1379 PTY. A nickname from OE, OFr *post* 'post, pillar'.

**Postan, Postans, Postance, Poston, Postons, Postin, Postings:** Mabill', John *de la Posterne* 1203 P (Ess), 1242 Fees (W); John *Postans alias* Little John 1575 ERO; William *Poston* 1613 FrY. 'Dweller by, or keeper of the postern-gate', OFr *posterle, posterne*.

**Postgate, Posgate, Postkitt, Poskett, Poskitt:** Thomas *Postgate* 1349 Whitby (Y); Richard *Poskett* 1514 GildY; William *Posgate* 1648 YWills; Thomas *Poskitt* 1661 ib. From Postgate (NRYorks).

**Postle, Postles, Posthill, Postill, Possell:** *Postellus* 1176 P (Sr); *Apostollus* 1203 Cur (Mx); William, Richard *Postel* 1170 P (Nb), 1202 AssL; William *La Postle* 1300 LoCt; Ralph *Postle* 1300 LLB B; John *le Pusel* 1332 SRSx; Alice *Postill* 1500 NorwW (Nf). OE *apostol, postol*, OFr *apostle* 'apostle', both as a nickname or pageant-name and as a personal-name. cf. Fr *Apostol, Lapostolle*, which Dauzat explains as 'pape', an ironical nickname for a serious man. In OFr *apostole* was the original word for Pope.

**Postlethwaite, Posselwhite, Posnett, Posnette:** Thomas *Postilltwayte* 1467 GildY; Margaret *Postlewhaite* 1588 RothwellPR (Y); John *Postelwhat* 1584 LaWills; Gerard *Postlet* 1586 ib. From Postlethwaite in Millom (Cumb), now the name of a field in Mirehouse, first mentioned as a manor in 1278 (PN Cu). Bardsley notes that in Furness the name was colloquially pronounced *Poslett*, whence *Posnett*.

**Poston:** *v.* POSTAN

**Pothecary, Potticary:** William *Apotecarius, Ypotecarius* 1283–5 Oseney (O); Richard *Ipotecar'* 1297 SRY; Christopher *Potticary* 1591 Oxon (W). ME *apotecarie*, OFr *apotecaire*, MedLat *apothēcārius* 'storekeeper' (c1386 NED). Originally

---

one who kept a store for spices, drugs and preserves, later one who prepared and sold drugs for medical purposes.

**Potisman:** *v.* POTMAN

**Potkin, Potkins:** William *Potechin* 1166 P (Nf); Roger, Reyner *Potekin* 1191 P (Nf), 1221 *ElyA* (Sf). 'Little Philip', from *Pot*, an apheitc form of *Philipot*, plus *-kin. v.* POTT.

**Potman, Potisman:** Nigellus *filius Poteman* 1185 Templars (K); *Poteman de Rokesakere* 1258 ArchC iii; Stephen, William *Poteman* 1296 SRSx, 1327 *SR* (Ess). Either 'servant of *Pott*' (*v.* POTKIN, POTT), or synonymous with POTTER and *potmaker*. cf. Richard *le Potmaker* 1297 Wak. The earliest example may mean 'son of the potman or potter'. The second example of the personal name is probably an original nickname. A personal name compounded of a French pet-form and English *-man* is unusual. cf. Richard *Potemay* 1332 SRSx, 'maid or servant of *Pot'. v.* MAY.

**Pott, Pot, Potts:** (i) Godwin, Richard *Pot* 1115 Winton (Ha), c1180 ArchC vi; Petronilla *Potes* 1311 ColchCt; Roger *Potte* 1352 ib.; William *Pottes* 1540 Whitby (Y). An apheitc form of *Philipot* 'little Philip' or metonymic for POTTER, 'a maker of pots'. (ii) Richard *de la Potte, Attepotte* 1221 Cur, 1228 Pat (Sx); Gilbert *atte Potte* 1332 SRSr. Margaret *atte Potte* (1296 SRSx) lived at either Pothill Fm or Potcommon in West Grinstead (Sussex). This is OE *pott* 'pot' used topographically of 'a hole or pit'. cf. Pott Hall (NRYorks) and *v.* MELS 156.

**Pottage:** John *Potage* 1296 SRSx. Metonymic for POTTINGER.

**Pottell, Pottle:** Durant, Richard *Potel* 1243 AssSo; 1279 RH (Bk). *Pot-el*, a diminutive of *Pot. v.* POTT. cf. the double diminutive *Potelinus* 1192 P (Lo), Mosse *Potelin* 1198 P (He).

**Potter:** Seuard *le potter* 1172 Gilb (L); Geoffrey *Poter* 1196 Cur (Lei); John *le Potier* 1197 P (Ess); Lambert *le Pottur* 1214 Cur (Ess). Usually late OE *pottere* 'potter', occasionally OFr *potier*, AFr *\*poteor*. A maker of earthenware vessels or of metal pots. The potter was sometimes also a bell-founder.

**Potterell, Potterill:** *v.* PUTTERILL

**Potticary:** *v.* POTHECARY

**Pottin, Potting:** *Potinus* Stele 1229 CR (Ha); William *Potyn* 1226–7 FFK; Alice *Pottyng* 1296 SRSx; John *Potinne* 1642 PrD. *Pot-in*, a diminutive of *Pot*, an apheitc form of *Philipot*, a diminutive of *Philip*.

**Pottinger:** Walter *le Potagier* 1300 LoCt; Walter *le Potager* 1321 ParlWrits (O); John *Potyngar, Petynger* 1356, 1373 ColchCt. OFr *potagier* 'a maker or seller of pottage', a thick soup or broth (1572 NED). For the intrusive *n*, cf. MESSENGER, PASSENGER. *v.* POTTAGE.

**Pottiphar:** *v.* PETTIFER

**Pottle:** *v.* POTTELL

**Pottock:** *v.* PUTTOCK

**Potton:** Hugh *de Potton* 1227 Black (Glasgow); Henry *de Poton* 1312 FFEss; John *Potton* 1381 LoCh. From Potton (Beds).

**Potvin, Potwin:** *v.* POIDEVIN

**Pouch:** Simon, Thomas *Poche* 1327 SRWo. ME *pouche* 'pouch', by metonymy for POUCHER.

Occasionally this may denote a member of the Society of *Pulci* or *Pouche* of Florence, frequent traders with London: Bernard *de la Pouche* 1319 SRLo, *de Pouches* of Florence 1330 Cl, *Pouche* of Lumbardia 1343 Cl.

**Poucher:** Philip, Robert *Poucher* 1275 RH (Sf), 1317 MESO (Lei). A derivative of ME, ONFr *pouche* 'pouch', a pouchmaker (1401 NED). cf. Richard *Pouchemaker* 1349 FrY.

**Pouck:** *v.* POOK

**Pougher:** A derivative of OE *pohha* 'a bag', synonymous with *poghwebbe*. cf. John *le Poghwebbesone* 1306 AssSt, 'son of Poughwebbe', the weaver of bags.

**Poulder:** *v.* POLDER

**Poulson, Poulsom, Poulsum:** *v.* POLSON

**Poulter:** Aunger *le Poltur* 1222 Cur (Sr); Osbert *le puleter* 1230 P (Sa); Gilbert *Poleter* 1234 Oseney (O); Richard *Poulter* 1666 FrY. OFr *pouletier, poletier* 'poultry-dealer' (a1400 NED).

**Poulton:** *v.* POLTON

**Poultney:** *v.* PULTENEY

**Pound, Pounds, Pund:** (i) Ralph *de Punda* 1242 Fees (Ha); Nicholas *Attepounde* 1270 Eynsham (O). OE *pund* 'enclosure'. 'Dweller by the pound' or 'man in charge of the pound'. cf. PINDAR. (ii) Henry, William *Pund* 1206 P (K), 1221 Cur (Nth); Stephen *Pound* 1279 RH (C). Forms are too early and too frequent for this to be for *atte pounde*. On the analogy of other names it should be metonymic for POUNDER which would then mean 'maker or seller of *pounds*'. This *pound* must be OE *pund* 'a weight' and *pounder* 'a maker of weights'.

**Pounder:** Gregory *le pundere* 1176 P (We); Hugo, William *Punder(e)* 1183 Boldon, 1212 Cur (Y). In the west midlands and the south ME *pundere* is for PINDAR and this form is used in the Boldon Book. With the exception of one from Sussex, all the examples of *punder* in Thuresson are from counties where *pinder* would appear as ME *pinder* or *pender* and are from ME *pounde* 'to impound', 'impounder', hence 'pinder'. All may be toponymics, 'dweller by the pound'. *Pounder* may also be 'maker of weights' (*v.* POUND) or occasionally, metonymic for *pundermaker*: Peter, William *le Pundermaker(e)* 1286, 1303 MESO (Nf), a maker of punders or auncels (a kind of balance).

**Poundley:** Rannulf *de Pondeia* 1221 AssWa. From Poundley End in Rowington (Wa).

**Poussin:** *v.* PUSSIN

**Pountney, Poutney:** *v.* PULTENEY

**Powdrill:** *v.* PUTTERILL

**Powe:** *v.* PAW

**Powell, Poel:** (i) Philip *ap Howel* 1285 Ch (Radnor); Hugh *Apowell* 1524 KentW; Richard *ap Hoell* 1544 AD v (Flint). An aphetic form of Welsh *ap Howell* 'son of Howel'. *v.* HOWEL. (ii) The seal of John *Paul* 1296 AD v (Sr) bears the legend S. JOH'IS POWEL. *v.* PAUL. (iii) Ralph *ate Powel* 1288 RamsCt (Hu). 'Dweller by the pool.' *v.* POOL. cf. Jordan *de Powella* 1184 P (Wa), John *de Pawel* 1339 Oseney (O). These unidentified place-names probably have a different etymology.

**Powenall:** *v.* POWNALL

**Power:** *v.* POOR

**Powick, Powicke:** William *de Powic* 1202 FFY; Goda *de Powic'* 1221 AssWo. From Powick (Worcs).

**Powis, Powys:** Ernald *de Powis* c1148 EngFeud (He). From Powis, an ancient district in North Wales.

**Powle, Powles:** *v.* PAUL

**Powlet, Powlett:** Robert *de Powelet* 1214 Cur (Sr); John *Poulet* 1378 IpmGl; John *Poulet* 1428 FA (W). From Pawlett (So).

**Pownall, Pownoll, Powenall, Powner:** Robert *de Pounale* 1286 AssCh; Ezekiel *Pownall* 1641 PrSo; Anne *Poughnell* 1663 HeMil. From Pownall (Ch).

**Powter:** *v.* PEUTHERER

**Poxon:** *v.* POGSON

**Poyle:** *v.* PULLEY

**Poynder:** Isabella *le Puynder* 1279 RH (Bk). Identical with PINDAR. *uy* denotes the long front rounded *u*-sound from OE *y*. The surname may also be for POYNER, with intrusive *d*.

**Poyner, Poynor, Punyer:** Geoffrey *le Poinnur* (*Poignur*) 1220 Cur (Ess); William *le Poinur, le Pungneur* (*Puinur*) 1230 P (He, Y); R. *Poyner* 1283 SRSf; Richard *Punyer* 1327 Pinchbeck (Sf). OFr *poigneor* 'fighter'. cf. CHAMPION.

**Poynton:** *v.* POINTON

**Poyntz:** *v.* POINTS

**Poyser, Poyzer, Peiser, Peizer, Piser, Pyser, Pyzer, Poser:** Simon *le Peser* 1198 P (K); Elyas *Poyser* 1219 AssY; Josceus *le Pesur* 1224 Cur (K); William, John *Poser* 1275 RH (K), 1296 SRSx. AFr *peiser, poiser*, OFr *peseor* 'weigher', official in charge of the weighing machine. cf. I. *de la Payserie* 1317 Oseney (O).

**Prall, Pralle:** Geoffrey *de Praulle* 1204 Cur (D); Geoffrey *de Praule* 1261 AssSo; John *Pralle* 1392 CtH. From Prawle (D).

**Prater:** *v.* PREATER

**Pratt, Prett, Pritt:** Wlfric *Prat*, Withmer' *Pret* 1179 Seals (Sf); Ædmund *Pret* 1192 P (Berks); Dereman *le Prat* 1198 Cur (K); Richard *Pritte* 1295 ParlR (Ess); William *le Pritte* 1332 SRSr. Harrison, Tengvik and PN Bk 46 derive this from OE *prætt* 'a trick'. The earliest example is Lefwinus *Prat* c1080 OEByn which the document explains '(id est) *Astutus*, quod ab inimicis saepe captus caute evaserit'. This is explained in NED as from an unrecorded adjective OE *prætt* 'cunning, astute', the existence of which is supported by the forms *le Prat, le Pritte*.

**Pray:** Willelmus *de Prato* 1176 P (Do); Nicholas *de Pre* 1269 AssSo; Henry *de la Preye* 1279 RH (O). From residence near a meadow, OFr *pray*, Fr *pré*, from Lat *prātum* 'meadow'.

**Prayer:** Robert *de Praiers* 1161 P (Sx); John *de Prayers* 1306 FFEss; Thomas *Praeres* 1316 FA (Sx). From Presles (Calvados), *Praelliae* 1198, *Praeriae* 1269 (ANF).

**Preacher:** William *le precher* c1208 FrLeic; Hugo *le Prechur* 1247 AssBeds; Hugh *Precheour* 1297 SRY. OFr *precheor* 'preacher', probably a derogatory nickname.

**Preater, Pretor, Prater:** Willelmus *pretor* c1150 DC (L). Latin *praetor* used in the sense of 'reeve': Robertus *Pretor, Praepositus* 1208 Cur (Gl).

**Precious, Pretious:** *Preciosa* 1203 Cur (Herts); *Preciosa* Potewyne 1279 RH (C); Willelmus *Precios'* 1301 SRY; John *Precious* 1327 SRSf. Lat *Preciosa* 'of great value', used as a woman's name.

**Preddle:** v. PRIDDLE

**Predgen:** v. PRIDGEON

**Preece, Prees:** (i) Philip *de Pres* 1251 AssSt; Richard *de Prees* 1289 AssCh. From Prees (Salop) or Preese (Lancs). (ii) Griffin *ap Res, Apres* 1309, 1323 ParlWrits; John *Aprees* 1414 LLB I. Welsh *ap Rhys* 'son of Rhys'. (iii) Richard, Leticia *Pres* 1243 AssSo, 1303 FA (Sf). These forms are too early to be regarded as short for *ap Res* and unlikely to be associated with the Shropshire and Lancashire place-names. They may be ME 'Prees, or thronge. *Pressura*'. PromptParv.

**Preen, Preene:** (i) Richard *le Pren* 1275 SRWo; John *Pren* 1297 MinAcctCo. OE *prēn* 'pin', perhaps for a tall, thin man with a small head. (ii) Henry *de Prone* 1221 AssSa. From Church, Holt Preen (Sa), *Preone* 1245.

**Preist:** v. PRIEST

**Prendergast, Prendergrass, Prendergrast, Prenderguest, Pendergast:** (i) Robert *de Prendergat'* 1225 Cur (L); Adam *de Prendrogest* 1354 FFY. From Prendergast near Haverfordwest (Pembroke), but these examples are perhaps rather to be connected with the following. (ii) Waldoev *de Prendergest* c1170 Black (Kelso); Adam *de Prendergest* c1240 ib.; Henry *de Prendregast, de Prendergest* 1296, 1325 ib. From Prenderguest Farm near Ayton (Berwick), which takes its name from the Welsh Prendergast.

**Prentice, Prentis, Prentiss:** John *Prentiz* 1295 Barnwell (C); Adam *Prentys* 1297 MinAcctCo. An aphetic form of ME, OFr *aprentis* 'apprentice', a learner of a craft. Roger and John *Prentiz* were assessed in the subsidy of 1319 (SRLo) and were presumably no longer apprentices. Either the surname had already become hereditary or it was a nickname as Ekwall suggests. cf. John Kyng called *Prentiz*, mercer 1350 LLB F, where *Kyng* appears to be the real surname, *mercer* the occupation, and *Prentiz* a nickname.

**Prentout:** Fulco *Prentut* 1155 FeuDu; William *Prentut* 1255 FFK; Reginald *Prentout* 1327 SRSo. 'Take all', OFr *prendre, tout*, a nickname for an avaricious man. cf. John *Pernezgarde* 1270 AssSo 'take care'.

**Presbury, Pressbury:** John *Presbro* 1327 PN Do i 363. From Prestbury (Ch, Gl).

**Prescot, Prescott, Preskett, Priscott:** Gilbert *de Prestecota* 1175 P (D); Richard *de Prestecot* 1192 WhC (La). From residence or employment at the priests' house (OE *cot(e)* 'cottage', in place-names sometimes 'manor'), as at Prescot (Lancs, Oxon) or Prescott (Glos). In Devonshire there are three Prescotts, four Priestacotts, two Prestacotts, and one Pristacott.

**Preskett:** v. PRESCOT
**Presland:** v. PRIESTLAND
**Presley:** v. PRIESTLEY
**Presman:** v. PRIESTMAN
**Press:** v. PRIEST

**Presser:** v. PRESTER
**Pressey:** William *Presthey* 1377 AssEss. 'Dweller by the priests' enclosure' (OE *(ge)hæg*), as at Priesthaywood Fm in Wappenham (Northants).

**Pressick, Prissick:** Robert *de Prestewyke* 1296 SRSx; William *Presike* 1538 Riev (Y). From Prestwick (Bucks, Northumb, Surrey) or from employment at the priests' dairy-farm (OE *wīc*). Prissick Fm in Marton (NRYorks) is 'priest-stream'. In Scotland, from Prestwick (Ayrshire): Bartram *de Prestwyc* c1272 Black (Paisley).

**Pressland:** v. PRIESTLAND
**Pressley:** v. PRIESTLEY
**Pressman:** v. PRIESTMAN
**Pressney:** v. PRESTNEY

**Presson:** Robert *filius presbiteri* 1219 AssY; William *le Prestson* 1284 AssLa; John *Prestessone* 1356 FFEss; John *Presson* 1583 Musters (Sr); Thomas *Presson* 1662 HTEss. 'Son of the priest', OE *prēost, sunu*. cf. John *Prestebruther* 1332 SRCu 'brother of the priest'; Robert *Prestcosyn* 1381 PTY 'cousin of the priest'; Johanna *Prestdoghter* 1379 PTY 'daughter of the priest'; Henry *Prestesneve* 1327 SRSf 'nephew of the priest'.

**Presswood:** v. PRIESTWOOD
**Prest:** v. PRIEST
**Prestage, Prestedge:** v. PRESTWICH
**Prester, Presser:** Henry *le Prestre, le Prester* 1228, 1248 AssSt; Edmund *le Prestre* c1235 HPD (Ess). OFr *prestre* 'priest'. v. PRIEST.

**Prestidge, Prestige:** v. PRESTWICH
**Prestney:** Richard *de Prestonheye, Prestney* (1291 For, 1346 FA) came from Prestney's Fm in Great Horkesley (Essex).

**Prestoe, Pristo:** Eva *de Presthall* 1278 AssLa. From a lost place *Prestall* in Deane (Lancs), the name of which survives in Presto Lane.

**Preston:** Peter *de Prestun* 1185 Templars (Y); Laurence *Preston* 1327 SRSx. From one of the numerous places named Preston. v. PRIEST.

**Prestwich, Prestage, Prestedge, Prestidge, Prestige:** Thomas *de Prestwyc* 1230 P (La). From Prestwich (Lancs). *Prestage*, etc. show the local pronunciation. cf. Swanage, *Suuanwic* DB.

**Prestwood, Preswood:** v. PRIESTWOOD
**Pretheroe:** v. PROTHERO
**Pretious:** v. PRECIOUS
**Pretlove:** v. PRITLOVE
**Pretor:** v. PREATER
**Prett:** v. PRATT

**Prettjohn, Prettejohn, Prettejohns, Prettyjohn, Prettyjohns:** *Prestreiohan* 1219 AssL; *Prestre Johan* 1301 SRY; John *Prestrejohan* 1346 Pat (So). That England was fully abreast of the news of the world is shown by the occurrence of the name Prestreiohan borne by a Lincolnshire attorney. Tales of a great priest-king who ruled in central Asia were current in the twelfth century in Europe. He is first mentioned by a western chronicler in the middle of the century when Otto of Friesing tells how Johannes Presbyter won a great victory over the Persians and Medes. Between 1165 and 1177 a forged letter purporting to come from him was circulated in Europe. Some rumour of this must have reached Lincolnshire and caused at least one English child to be christened by the name of this supposititious eastern king (Sir Denison Ross in *Travel and Travellers of the Middle Ages*, ed. A. P. Newton, 174–94). No early example of *Prettyjohn* has been noted. The occurrence of *Prestrejohan* as a surname in 1346 suggests that this is the real origin. The name can never have been common. The meaning of *Prestre* would be known and the name assimilated to the later French form, becoming *Pretrejohn, Pretterjohn*, a form no longer suggesting *Prestrejohn* whose story was probably by then unfamiliar to the possessors of the name and they made it intelligible as *Prettyjohn*.

**Pretty, Pritty:** Robert *Prytty* 1327 SRSf; William *Pritty* 1428 FA (Sx); Thomas, Agnes *Praty* 1479 LLB L, 1524 SRSf. OE *prættig* 'crafty, cunning'.

**Prettyman, Pretyman:** Thomas *Pratyman, Pretyman* 1524, 1568 SRSf. Perhaps 'cunning man' or 'servant of a man named *Pretty*'. But cf. PRETTJOHN.

**Prevett:** *v.* PRIVETT

**Prevost, Le Prevost:** John *le Prevost* 1418 DKR 41. *Prevost* is a Huguenot name, *Le Prevost* a Guernsey one. Owing to its frequency, Dauzat regards the French *Prévost* as a nickname rather than from the office of provost.

**Prew, Prow, Prue:** Robert *Prue* 1270 AssSo; Robert *Prowe* 1276 RH (C); Ellis *Prew* 1280 AssSo; John *le Proo* 1332 SRSx. ME *prew, prue*, OFr *prou, preu* 'valiant, doughty'.

**Prewell:** William *atte Prewelle* 1336 MELS (Sr). 'Dweller by the spring in the field', OEr *pre*, OE *wiella*.

**Prewett, Prewitt, Pruett:** Matthew *Pruet* 1202 P (So); Richard *Prouet, Pruet, Prowet* 1278, 1280 LLB A. A diminutive of PREW.

**Prewse:** *v.* PROWSE

**Price, Prise, Pryce, Pryse:** (i) Jorwerth *ap Reys* 1393 LoPleas; John *Aprice* 1492 AD iii (Pembroke). Welsh *ap Rhys* 'son of *Rhys*'. (ii) Robert *Price* 1297 MinAcctCo; Richard *Prys* 1320 FFEss. ME, OFr *pris* 'price', metonymic for a fixer of prices. cf. *Prysare*, or settar at price, ynn a merket, or oþer placys; *Prysin'*, or settyn' a pryce (PromptParv).

**Prichard(s):** *v.* PRITCHARD

**Pricher, Prickman:** *v.* PRYKE

**Prickard, Prickart:** Alexander *Prikehurt* 1208 P (Lo/Mx); John *Prikehert* 1219 P (Sf); Roger *Prikehert* 1232 Cur (Sf). 'Pierce hart', OE *prician, heorot*, a nickname for a hunter. cf. John *Prikehering* 1279 RH (Hu); William *Prikeavant* 1279 RH (Beds) 'spur in front'; Nicholas *Prikhors* 1327 SRY 'spur the horse'.

**Prickett:** William, Laurence *Priket* 1296 SRSx, 1325 AssSt. A nickname from the prickett, a buck in his second year (ME *priket*).

**Prickmore:** Nicholas *Prekemere* 1328, William *Prykemer* 1331 IpmW. From Prickmoor Wood in Bromham (W).

**Priddle, Preddle, Pridell:** John *Pridel* c1320 *PetreA*; John *Priddle* 1641 PrSo. Perhaps Welsh *ap Ridel* 'son of *Ridel*'.

**Priddy:** William *Pridy* 1327 SRWo; John *Pridee* 1642 PrD; Thomas *Predy* 1665 HTO. From Priddy (So).

**Pride, Pryde:** John *le Pride* 1208 P (D); Richard, Robert *Pride* 1221 AssSa, AssWo. No doubt often a

nickname or pageant-name from ME *pride* 'pride', but also clearly an adjective. The surname appears particularly in the Welsh border counties and may be from Welsh *prid* 'precious, dear'.

**Prideaux:** Nicholas *de Pridias* 1182 P (Co). From Prideaux (Cornwall).

**Pridell:** *v.* PRIDDLE

**Pridgeon, Predgen:** William *Prygion* 1392 AssL; George *Pridgeon* 1695 DKR (L). Perhaps Fr *preuxjean* 'wise, brave *John*'.

**Pridham, Prodham, Prudhomme:** Gilbert *Prodhome, Prudume* 1176, 1197 P (Sr, W); Roger *Prodomme, Prodhomme* 1284, 1326 FFEss. OFr *prudhomme* 'upright, honest man; expert', used in the 13th century to distinguish the 'wiser' or 'lesser folk' (*prudhommes*) from the 'greater folk' or general mass. Pridhamsleigh (Devon) owes its name to John *Prodhomme* (1281 PN D 520).

**Prier:** *v.* PRIOR

**Priest, Preist, Prest, Prestt, Press, Prust:** Ælfsige *Preost* 963 OEByn (Herts); Asci *Preost* 1066 DB (Nf); Baldwin, Rodbert *Prest* 1176 P (L), 1188 BuryS (Sf); Robert *le Prest* 1243 AssSo; William *Prost* 1279 RH (O); Henry *Prust* 1279 RH (O), 1327 MEOT (Ha); Robert *le Preest* 1296 SRSx; Hugh *le Prist* 1327 ib.; John *le Preost* 1327 SRSo; Walter *Preyst* 1332 SRSx; Simon *le Prust* 1353 AssSt. OE *prēost* 'priest', in early examples denoting office but later usually a nickname for a man of 'priestly' appearance or behaviour or, no doubt, often for one of a most unpriestly character. *v.* PRESTER. As with *Parsons* and *Vickers*, we also find forms with an elliptic genitive and with *-son*: (i) Thomas *le Prestis* 1326 AD iv (Wa); Peter *Prestes* 1332 AD ii (Do); Cristina *Prestes* 1327 SRSo; 'servant of the priest'. (ii) Sarra *atte Prestes* 1327 SRSx; Richard *del Prestes* 1332 SRCu; 'servant at the priest's'. (iii) Thomas *fiz al prestre* 1230 Pat (Lo); William *le Prestesson* 1293 FFEss; Walter *Preesteson* 1316 Wak (Y); Adam *Prystesonne* 1327 SRSf; Thomas *Prestson* 1332 SRCu; 'son of the priest'. These do not seem to have survived, though both Bardsley and Harrison give *Presson*. The difficult combination of final consonants in *Prestes, Prests* has probably been assimilated to *Press*. In *Prestson*, a similar change would give *Presson*; an intrusive *t* would give *Preston* (cf. HOUSTON) and this common name (and the less common *Priston*) probably often means 'priest's son'. *Prestson* may have been simplified direct to *Preston*. *Prust*, which is rare and has been noted only in Somerset and London (where all varieties arrive), preserves the South-Western and West Midland ME rounded vowel, written *u, o, eo*.

**Priestland, Presland, Pressland:** Roger *de Preslond* 1305 FFEss. From residence near the priests' land as at Priestland Copse in Bampton (Devon), Priestlands in Horley (Surrey), which belonged to Reigate Priory, or Pressland in Hatherleigh (Devon).

**Priestley, Priestly, Presley, Presslee, Presslie, Pressley, Pressly, Prisley:** Samson *de Presteleia* 1198 P (Beds); Richard *de Presteley* 1297 Wak (Y); John *de Presle* 1311 LLB D; Richard *de Prestele* 1327 SRSx. 'Dweller by the priests' wood or clearing.' cf. Priestlie (Beds, Herts, WRYorks), Prestley Wood (Wilts), Presley's Plantation in Norton (Notts), *Prisley* c1840.

**Priestman, Presman, Pressman:** Robert *Prestman* 1275 RH (Y); William *Prestesman* 1283 SRSf; Henry *le Prestesmon, le Prestemon* 1332 SRSt, 1333 AssSt; William *Priestman* 1393 NottBR. 'Servant of the priest' (OE *prēostesmann*) or 'of the priests' (*prēostamann*).

**Priestwood, Preswood, Presswood:** William *de Prestewude* 1176 P (St). From Prestwood (Bucks, Sussex) or 'dweller by the priests' wood'.

**Prigg:** Henry *Prig* 1186 P (Wo); Simon, Adam *Prigge* 1210 Cur (C), 1274 Wak (Y). Probably a voiced form of ME *prikke*. v. PRYKE.

**Prime:** William, Ralph *Prime* 1275 RH (L), 1296 SRSx; Adam *Prymme* 1286 ForSt. OFr *prim(e)* 'fine, delicate'.

**Primmer:** Peter *le Primur* 1279 RH (C); Robert *le Premyr*, Simon *Premir* 1327 SRSx; Robert *Prymer* 1471 Paston. Fr *premier* 'first'.

**Primrose:** (i) Roger *Primerose* 1219 RegAntiquiss; William *Prymerole* 1324 CoramLa; John *Primerose* 1379 PTY. A nickname from the flower, OFr *primerole*. (ii) John *Prymros* 1387, Archibald *Prymrose* 1569 Black. From the lands of Primrose (Dunfermline). (iii) Also Huguenot. Gilbert *Primrose*, of Scots origin, settled in France in 1601 as minister of the Protestant church at Mirambeau, later of Bordeaux. He was banished in 1623 and came to London as minister of the French church in Threadneedle Street. Later Bishop of Ely (Smiles 421–2).

**Prin, Prinn, Prynne:** William *Prin* 1275 RH (Berks); Matilda *Pryn* 1275 SRWo; Nicholas *Prinne* 1327 SRSx. OFr *prin* (Lat *primus*) 'first, superior; small, slender'.

**Prince, Prins:** Robert *Prince* 1177 P (Cu); Robert *le Prins* 1327 SRSx. Fr *prince* 'prince', a nickname.

**Pring, Pringe, Prink:** Æðelgeard *Preng* a958 OEByn; Simon *Pring* 1203 AssSt; Walter *Prink'* 1327 SRWo; John *Pryng* 1524 SRD. OE *Preng*. v. OEByn 330.

**Prior, Prier, Pryer, Pryor:** Roger *Priur* 1205 Cur (Sf); Roger *le Priur* 1237 FFC; Nicholas *le Prior* 1268 AssSo; Robert *Pryer* 1274 RH (Ess). OE *prior* or OFr *priur, priour* 'prior', originally a surname of office, later a nickname. For Editha *le Priores* 1327 SRSo, William *atte Priours* 1327 *SR* (Ess), and Thomas *Priorman* 1332 SRCu, cf. PRIEST.

**Priscott:** v. PRESCOT

**Prisley:** v. PRIESTLEY

**Prison:** Godwin *Prison* c1110 Winton (Ha); Alexander *Prison* 1219 P (Berks); Reginald *Prisun* 1248 AssBerks. OFr *prisun* 'prison'. Metonymic for 'prisoner', or, perhaps, for 'keeper of the prison'.

**Prissick:** v. PRESSICK

**Pristo:** v. PRESTOE

**Pritchard, Prichard, Prichards:** William *Prichard* or *Ap-Richard* 1521 Oxon. Welsh *ap Richard* 'son of *Richard*'.

**Pritchet, Pritchett, Pritchatt:** John *Prichet* 1309 FFSf, 1319 SRLo. Probably for PRICKETT, with palatal *ch* for *k*.

**Pritlove, Pretlove:** Richard *Prykkeloue* 1296 SRSx; Alexander *Frikeloue* 1297 MinAcctCo. ME *prikke*

and OFr *love* 'wolf', 'prick wolf', a nickname, no doubt, for a hunter and killer of wolves. cf. John *Prikehors* 1379 PTY 'prick horse', for a hard rider, John *Prikehert* 1230 P (Sf), a hunter of harts.

**Pritt:** v. PRATT

**Pritty:** v. PRETTY

**Pritwell:** Adam *Prytewell* 1381 LoCh. From Prittlewell (Ess), *Pritteuuella* DB.

**Privett, Privitt, Prevett:** Reimund *de Prevet* 1210 P (Ha); Thomas *Prevet* 1545 SRW; John *Privett* 1662–4 HTDo. From Privett (Ha).

**Probert, Probets, Propert:** Philip *ab Robert* Hy 3 AD iii (He); John *ap Ed ap Dafydd ap Robert* 1538 Chirk; Thomas *Uprobarte* 1540 Bardsley; Joseph *Probert* 1792 ib. Welsh *ap Robert* 'son of *Robert*'.

**Probin, Probyn, Brobyn:** John *Probyn* 1550 SxWills; Edward *Up Robyn* 1565 ShefA. Welsh *ab, ap Robin* 'Son of *Robin*'.

**Prockter, Procter, Proctor:** Johanna *la Proketour* 1301 SRY; John *Proketour* 1326 FeuDu. ME *prok(e)tour*, a contraction of Lat *procurator* 'manager, agent', commonly used of an attorney in a spiritual court (c1380 NED). cf. Johannes *le Procurator* 1279 AssNb.

**Prodgers, Proger:** John *ap Roger* 1538 Chirk; Charles *Proger* 1607 Bardsley. Welsh *ap Roger* 'Son of *Roger*'.

**Prodham:** v. PRIDHAM

**Profitt, Profit:** v. PROPHET

**Propert:** v. PROBERT

**Prophet, Proffitt, Profit:** William *le Profete* 1220 Cur (Bk); Gunnora *Prophete* 1327 *SR* (Ess). OFr *prophete* 'prophet', a nickname.

**Prosser, Prossor:** William *ap Rosser* alias *Approssor* 1553 Pat (Sa); Henry Prosser 1663 HeMil. Welsh *ap Rosser* 'son of *Rosser*'.

**Prothero, Protheroe, Protherough, Pretheroe, Pluthero, Prydderch, Prytherch, Prytherick:** William *Prythergh* or *Protherugh* or *Protherough* 1581 Oxon; Rowland *Prythero* alias *Prothero* alias *Prytherch* 1725 DKR 41 (Brecon). Welsh *ap Riderch* 'son of *Riderch*' 'the reddish brown'.

**Proud, Proude:** Toui *Pruda, Prude* 1033 OEByn; Orgar *le Prude* 1125 (c1425) LLB C; Richard *Prude* 1185 Templars (Sa); William *le Proude* 1275 SRWo. Late OE *prūt, prūd* 'proud, arrogant'.

**Proudfoot, Proudfoot:** Gilbert *Proudfoot, Prudfot, Prutfoot* 1114–30 Rams, c1130 ELPN; Geoffrey *Prudfot* 12th MedEA (Nf); John *Prutfot* 1203 P (Ha). OE *prūd* (*prūt*) 'proud' and *fōt* 'foot', 'one who walks with a haughty step'. Gilbert Proudfoot was sheriff of London c1140 and 'It is interesting to find that the first known bearer of the surname was a sheriff, thus a person who might be justified in walking with a proud step' (ELPN).

**Proudlove, Proudler:** Thomas *Prudelove* 1289 AssCh. A nickname. cf. TRUELOVE, TRUSLOVE.

**Proudman:** William *Proudman* 1327 SRSf. 'Proud man' or 'servant of *Proud*'. cf. Thomas *Prudswain* 1223 Cur (Sr), Olive *Prodemay* 1327 SRSf.

**Proudmay:** Oliua *Prodemay* 1327 SRSf; John *Prodemay* 1359 IpmGl; William *Prudemay* 1395 AssL. Probably 'servant of *Proud*', rather than 'proud servant', OE *mǣge*. cf. Thomas *Proudfoster* 1379 PTY 'foster-father of *Proud*'; Thomas *Prudswain* 1223 Cur (Sf) 'servant of *Proud*'; Hugh *Proud of Noght* 1348 Misc (Y) 'proud of nothing'.

**Prouse:** *v.* PROWSE

**Proust:** Alan *Prouste* 1275 RH (L); Robert, Simon *le Preust* 1297 MinAcctCo, 1314 AD iv (Nth). Contractions of ME *prevost, provost* 'provost'. cf. Fr *Proust, Leproust*.

**Prout:** William *Prute* 1207 P (D); Thomas *le Prute* 1274 RH (Gl); Robert *Proute* 1280 AssSo. ME *proute* 'proud'.

**Provan, Provand, Proven:** Richard *de Prebenda* c1190 Black; Stephen *Provand* 1489 ib.; Robert *Provane* 1549 ib. From the lands of Provan, formerly a possession of the prebendary of Barlanark, one of the canons of Glasgow Cathedral (Black).

**Provender:** Philip *de la Provendre* 1249 AssW; Hugh *atte Provendre* 1345 LLB F; Jeffrey *Provinder* 1576 SRW. A derivative of OFr *provende* 'provisions', 'the man in charge of the provisions for a household'.

**Province, Provins:** William *de Provinc* 12th DC (L); Ralph *de Prouinz* 1202 P (Sx). 'Man from Provence.'

**Provost, Prevost, Provis:** Eanstan *Prafost* a925 OEByn (D); Uui *Provast* 11th ib. (D); Goscelin *Provost* 1200 Cur (O); Geoffrey *le Provost* 1206 Cur (L); William *le Pruvost* 1219 Cur (Ha); Richard *Provest* 1445 FrY; John *Provess*, William *Provise* 1535, 1545 RochW. OE *prafost*, AFr *provost* 'provost'.

**Prow:** *v.* PREW

**Prowse, Prouse, Prewse, Pruce:** Richard *le Pruz* 1207 Cur (Herts); Adam *Pruce* 1225 AssSo; William *le Prouz* 1275 RH (D); Roger *le Prus* 1275 SRWo; William *Prous* 1279 RH (O). ME, OFr *prous, prouz* 'valiant, doughty'.

**Prudence:** Robertus *filius Prudence* 1206 P (Sr); *Prudencia* de Pavely 1210 Cur (Nf); Hugh, Adam *Prudence* 1203 FFC, 1206 P (Sr). *Prudence*, Lat *prudentia* 'prudence', a woman's name.

**Prudhomme:** *v.* PRIDHAM

**Prue:** *v.* PREW

**Prust:** *v.* PRIEST

**Pryce, Pryse:** *v.* PRICE

**Prydderch:** *v.* PROTHERO

**Pryde:** *v.* PRIDE

**Pryer, Pryor:** *v.* PRIOR

**Pryke:** William *Prike, Prikke* 1205–6 P (Wa); Geoffrey, Robert *Pricke* 1221 Ely (Nf), 1341 ColchCt; Alice, John *Prich(e)* 1295 Barnwell (C), 1327 SRC; Simon *Prike* 1340 FFSf. ME *prike, prikke* 'a point, prick', with occasional palatalized forms *priche*, also the name of a pointed weapon. The surname is probably metonymic for a maker or user of these. cf. Hamo *Pricchere* 1175 P (Do); William *Priker* 1256 AssNb; Hugh *le Prichere* 1327 *SR* (Ess); surviving as *Pricher*, and *Prickman* with the same meaning.

**Prynne:** *v.* PRIN

**Prytherch, Prytherick:** *v.* PROTHERO

**Puckle, Puckell:** (i) Richard *Puchel* 1200 P (Beds); Nicholas *Pokel* 1275 RH (K); John *le Pochel* 1297 MinAcctCo. OE *pūcel* 'little goblin, elf, sprite'. (ii)

Robert *de Pukehole* 1296 SRSx. 'Dweller by the elf-hollow', OE *pūca, holh*.

**Puckney:** William *atte Pukenegh*, Richard *atte Poukeneye* 1332 MELS (Sx). 'Dweller at goblin-island', OE *pūcena* g. pl., *ēg*.

**Puckridge:** William *Pukerich'* 1220 Cur (Herts); Richard *Pokeridge* 1545, *Pocrege* 1576 SRW. From Puckeridge (Herts), or Pocheridge Fm in Corsham (W).

**Puddifant, Puddifin:** *v.* POIDEVIN

**Puddifer:** *v.* PETTIFER

**Puddifoot, Pudifoot, Puddefoot, Puddephat, Puddephatt, Puttifoot:** Roger, Ralph *Pudifat* 1188 P (C), 1223 Cur (Herts); Herbert *Pudifot* 1212 Cur (Y); Richard *Pudifed'* 1213 Seals (O); Geoffrey *Putifat* 1221 ElyA (Sf); Robert *Podifat* 1288 LLB A, 1332 LoPleas. It is clear that the second element is not *foot* but *fat*, probably OE *fæt* 'vessel, vat'. The first is probably the dialectal *puddy, poddy* 'round and stout in the belly' from the Germanic root \**pud(d)* 'to swell, bulge' found in *pudding*, OE *puduc* 'a wen' and the dialectal *pod* 'a large protuberant belly'. cf. LG *puddig* 'thick, stumpy'. An early English example may be (Rotberd) *Puddig* c1100–30 OEByn (D), though this may be merely a bad form for *Pudding*. The meaning of the compound is 'round and stout-vessel, cask, barrel', a nickname for a man with a prominent paunch, a parallel to PAUNCEFOOT and reminiscent of the 'barrel-bellied' monks mentioned under BARREL.

**Pudding:** Ailword *Pudding* c1100–30 OEByn (D); Alured, Hugo *Pudding* 1176 P (Nth), 1219 AssY; John *Poddyng* 1297 MinAcctCo. Probably an *-ing* derivative of the root \**pud(d)* discussed under PUDDIFOOT, a nickname for a round, stout man. The surname was common and may also be a nickname for a butcher, either from the sausages or puddings he sold (cf. the Yorkshire 'black-pudding') or from the puddings or offal from his slaughter-house, a perpetual nuisance to his neighbours, which gave name to Pudding Lane in London.

**Pude:** John *Pudde* 1186 P (W); Gilbert *Pud* 1213 Cur (Co). A nickname 'round and stout'. cf. PUDDIFOOT.

**Pudney:** Roger *Podeney* 1360, *Podenho* 1361, *Podenhay* 1365 ColchCt. From Pudney Fm in Rayne (Ess).

**Pudsey:** Roger *of Pudekeshay* 1218–19 FFY; Simon *de Podesay* 1304 IpmY; Henry *de Pudesay* 1409 IpmY. From Pudsey (WRY).

**Pugh, Pughe, Pew:** Richard *ap Hughe* 1563 AD v (Montgomery); John *Apew, Pew* 1642 Bardsley. Welsh *ap Hugh* 'Son of *Hugh*'.

**Pugsley:** Thomas *Puggesley* 1525 SRSx; Davy *Pugsly*, William *Pugsley* 1642 PrD. From Pugsley in Warkleigh (D).

**Pulder:** *v.* POLDER

**Pulford:** Herbert *de Pulford* c1200 RegAntiquiss; Richard *Pulford* 1524 SRD; William *Pulford* 1524 SRSf. From Pulford (Ch).

**Pulham, Pullum:** William *de Pulham* 1214 Cur (Nf); Simon *de Pulleham* 1301 FS; William *Pulham* 1361 IpmGl. From Pulham (Do, Nf), or Pulham in Twitchen (D).

**Pullan, Pullen, Pullein, Pulleine, Pulleyn, Pullin:** Richard *Pulein* 1166 P (Nf), 1195 FF (Beds); Geoffrey *Poleyn* 1266 AssSo; Thomas *Pullan* 1509 GildY; John *Pullen* 1601 FrY. OFr *poulain* 'colt'.

**Pullar, Puller:** (i) Ralph *pullehare* 1189 Sol; William *Pullehare*, John *Pullehar* 1327 SRLei; Thomas *Pullehare* 1367 IpmW. A nickname, 'pull hare', cf. Alexander *Pullegandre* 1249 AssW; Roger *Pullegos* c1170 ELPN. (ii) John *de Pulhore* 1339 Black (Du); Henry *Pulour* 1379 ib. (Perth); Robert *Pullour* 1474 ib. 'Dweller by the bank of the pool or creek', OE *pull* and *ōra*.

**Pulley, Poyle:** (i) William *le Pulleis* 1191, *de Puilleio* 1192 P (Nf); Walter *de la Poille* 1221 Cur (He); John *de Apuelle* 1275 RH (C); Walter *de la Poyle* 1279 RH (O). 'The man from Apulia'. cf. Poyle House in Seale (Sr), from the family of *de la Poille, Puyle, Puilly*, which held the manor from 1299, and Pòyle in Stanwell (Mx) from *de l'Apulie*. (ii) Warin *de Pulileg'* 1221 AssSa; Joyse *Pulley* 1545 SRW; Walter *Pulley* 1642 PrD. From Pulley (Sa).

**Pullum:** v. PULHAM

**Pullinger, Pillinger:** v. BULLINGER

**Pulman, Pullman:** John *Pulman* 1525 SRSx; John *Pullman*, Roger *Pulman* 1642 PrD. 'Dweller by the pool', OE *pull* and *mann*.

**Pulteney, Poultney, Pountney, Poutney:** John *de Pulteneye* 1334 LLB E. From Poultney (Leics), DB *Pontenei*.

**Pumfrett:** v. POMFRET

**Pumfrey, Pumphery, Pumphrey, Pumphreys, Pomfrey, Pomphrey:** Edward *ap Humfrey* 1575 AD v (Sa); Anable *Pumfrey* 1633 Bardsley. Welsh *ap Humphrey* 'Son of Humphrey'. The surname may also be a doublet of POMFRET: John *Pounfrey* 1297 MinAcctCo (Co). Here, *Pounfrey* is for *Pountfreyt*. v. also BOUMPHREY.

**Pummery:** v. POMEROY

**Punch:** Godfrey, Philip *Punch(e)* 1181 P, 1275 RH (Sf); Seman *Ponche* 1327 SRSf. ONFr *Ponche*, OFr *Ponce*. v. POINTS.

**Punchard, Puncher:** Robert *Puncard* 1230 P (O); Oliver, Ralph *Punchard(e)* 1243 AssSo, 1255 Rams (Hu); Geoffrey *Puncard* alias *Punzard* 1265 Ipm (Berks); Fray *Punsard* 1279 RH (O). ONFr *Ponchard*, OFr *Ponsard*, a diminutive of *Ponce*. v. POINTS.

**Punchardon:** Alan *Punchardun* Hy 2 Gilb; Adam *Punchardon'* 1219 AssY; Ivo *Punjardon* 1260 IpmY. *Ponchard-on*, a diminutive of OFr *Ponchard*.

**Pund:** v. POUND

**Punge:** Hugh *Punge* 1203 AssNth; Alexander *Punge* 1319 SRLo; Nicholas *Poungge* 1339 CorLo. OE *pung* 'purse, pouch'. Metonymic for a maker of these.

**Punnett:** Anna *filia Puinant* 1199 FFEss; *Puinant de Chelese* 1212 Cur (Herts); Richard *Poingiant, Pugnant, Puignant, Puinant, Puniant, Punat* 1086 DB; Roger *Puignant, Poinant* 1194 Cur, 1196 P (Bk); John *Poygnaunt* 1203 AssSo; Thomas *Poynaunt* 1296 SRSx. Tengvik and Bardsley explain this as from OFr *puignant, poignant*, 'stinging, biting' (c1386 NED). Such a nickname and personal-name are possible. cf. MORDAUNT. The DB forms in particular, however,

suggest a derivation from the present participle of OFr *poignier* 'to strike with the fist', *poignant* 'the striker, the fighter'. cf. *poniard*, a derivative of *poing* 'fist'. The surname survives in Poynatts Fm (Bucks), *Pinets* 1766 (PN Bk 179), and in Poynetts (Essex), *Poynes* 1412, *Poynattes* 1554 (PN Ess 143).

**Punshon:** Johannes *filius Puncon, Punzun* 1177, 1180 P (Cu); William *Puncun, Punzun* 1210 Cur (Nth). A diminutive of OFr *Ponce*. v. POINTS.

**Punt:** v. PONT

**Punter:** v. PONTER

**Puntis:** v. PONTIS

**Punyer:** v. POYNER

**Puplett:** v. POPLETT

**Pur, Purr, Purre:** Edric *Pur* 1066 DB (C); Walter *Purr* 1219 P (Sx); Adam *le Pur* 1249 AssW; Thomas *Purre* 1360 FFEss. A nickname from the bittern, OE *pūr*. cf. Roger *Purfoghel* 1296 SRSx, OE *fugol* 'bird'.

**Purcell:** Gaufridus *porcellus* 1130 P (Sr); Ralph, William *Purcel* 1159 P (St), 1230 P (Lei). OFr *pourcel* 'little pig'.

**Purchas, Purchase, Purches, Purchese, Purkess, Purkis, Purkiss, Pirkis, Pirkiss, Porkiss, Porcas:** William *Purchaz* 1190 P (Ess); Geoffrey *Purcaz* 1206 Cur (Ess); Roger *Purchas* 1239 Eynsham (O); William *Purkas* 1327 SRSf. OFr *purchas* 'pursuit, pillage', used as a name for messengers and couriers. Bardsley notes 'Purchase the Pursuivant' temp. Henry VI.

**Purcifer:** v. PERCEVAL

**Purday, Purdey, Purdie, Purdy, Purdye, Purdu, Purdue:** Gilbert *Purdeu* 1227 AssBeds; John *Purde* 1279 RH (C); John *Purdew, Purde* 1296, 1332 SRSx; John *Purdy* 1436 NorwW (Nf); Robert *Purdu* 1479 ib. An oath name, Fr *pour Dieu*. cf. Fr *Pourdieu* and v. PARDEW.

**Purden, Purdon, Purdom:** William *Purdome* 1312 ColchCt; Adam *Purdone* 1327 SRSo; Robert *Purdome* 1374 AssL; Samuel *Purden* 1613 FrY. A metathesized form of OFr *prudhomme* 'honest, upright man'. v. PRIDHAM.

**Purefoy:** William, Henry *Parfei* 1195 FF, 1203 AssSt; Robert *Parfoy* 1296 SRSx; William *Purfey* 13th AD ii (Wa); Ralph *Perfai, Parfay* 1327, 1332 SRSx; William *Purefay* 1412 AD ii (Lei). AFr *par fei* 'by (my) faith'. *Parfoy* is central French *par foi*. The present spelling is a popular etymology, 'pure faith', which arose when *per, par* and *pur* fell together in pronunciation. cf. PURNELL and PARNALL. v. PARDEW.

**Purkins:** v. PARKIN

**Purkiss:** v. PURCHAS

**Purley:** John *de Purle* 1221 AssWa; Hugh *de Purley* 1327 SRLei; William *de Purle* 1351 AssEss. From Purleigh (Ess), or Purley (Berks, Sr).

**Purnell:** v. PARNALL

**Purr, Purre:** v. PUR

**Purrier:** v. PERRIER

**Purry:** v. PERRY

**Pursall:** Gilbert *de Pureshull'* 1221 Cur (Sr); John *Pursel, Pursull* 1428 Fees (Wo); John *Purshull* 1431 FA (Wo). From Purshull in Elmbridge (Wo).

**Purse:** Hucche *purs* c1150 DC (L); Derewin *Purs* 1176 P (Bk). OE *purs* 'purse', by metonymy for PURSER.

**Purseglove, Pursglove:** Thomas *Purceglove* 1511 CorNt; John *Pursglove* 1559 Pat (Y); Robert *Pursglove* 1603 SRDb. Perhaps a corruption of Purslow (Sa).

**Purser:** Alexander *Purcir* 1279 RH (Bk); Roger *le Porser* 1299 LLB C; Adam *le Purser* 1332 SRLa. A derivative of OE *purs* 'purse', purse-maker. Robert *le Pursere* (1319 SRLo) was identical with Robert Neel, *burser* (1338 LoPleas) and was an associate of William de Borham, *pouchmaker* (1344 ib.).

**Pursey:** *v.* PERCY

**Pursley:** Hamo *de Puresleg'* 1206 Cur (Herts). From Pursley Fm in Shenley (Herts).

**Purt, Purte:** Wulward *le Purte* 1198 FFEss; Gilbert *Purte* 1276 ELPN; Geoffrey *Purte* 1315–16 FFSr. cf. dialectal *purt* 'sullen'.

**Purton:** *v.* PERRITON

**Purvis, Purves, Purvess:** William *Purveys, Porveys, Purvys, Purvais* 1214–49, 1296 Black; Gilbert, Eva *Purveys* 1400 LLB I, 1450 NorwW (Nf); Thomas *Purvas* 1427 Black, *Purvis* 1524 SRSf; John *Parvysse* 1445 RochW. Neither Harrison nor Weekley give any evidence for their derivation from ME *parvis* 'porch' and Black's forms prove this etymology untenable. The original vowel was clearly *u* and the name goes back to AFr *purveier* 'to provide'. In 1451 John Graunger of the butlery of the Prioress of St Radegund, Cambridge, was paid 3s. 'in regardo *pro officio Purvis domine*' (Rad 174), the editor explaining *Purvis* as 'a servant who acted as purveyor, *provisor*'. cf. OFr *porveor* 'provider' of supplies, especially in a hospital.

**Purvey:** John, Cons' *Purvey* 1279 RH (C). Either for *purveys*, with silent *s*, or metonymic for *purveyor*. cf. Alexander *Purveyance* 1623 Black.

**Pury:** *v.* PERRY

**Puryer:** *v.* PERRIER

**Puscat:** Robert *Pusekat* 1256 AssNb. 'Pussycat', a nickname. One example of the surname occurs in the London Telephone Directory for 1949.

**Pusey:** Henry *de Puset* 1219 P (Y). From Le Puiset (Eure-et-Loire). *v.* also PIZEY.

**Pussin, Poussin:** Richard *Pucin* 1202 FFK; William *Pucin* 1233–4, *Poucyn* 1305 FFEss. cf. Fr *Poussin*, a nickname for a small man.

**Putley:** William *de Putleye* 1230 P (Nth); Richard *de Putlygh* 1296 SRSx. From Putley (He), or Petley Wood in Battle (Sx).

**Putman:** *v.* PITMAN

**Putnam, Puttnam, Puttnum:** Ralph *de Puteham* 1205 Cur (Bk). From Puttenham (Herts, Surrey).

**Putt:** *v.* PITT

**Putter:** John *le Putter* 1327 SRSx. 'Dweller by the pit or hollow.' cf. PITT.

**Puttergill:** *v.* PORTUGAL

**Putterill, Puttrell, Potterell, Potterill, Pottrill, Powdrell, Powdrill:** Roger *Putrel* 1166 RBE (Ess); Henry *Pultrel* 1180 P (Lei); Robert *Puterel, Putrell* 1199, 1200 Cur (Lei); Richard *Poterell'* 1235 Fees (Lei); William *Poutrel* 1316 AD vi (St); William *Powdrell* 1379 PTY; George *Powderhill* 1586 Oxon (Berks); Martin *Powdrill* 1592 ib. OFr *poutrel, potrel, pultrel, putrel, peutrel, poudrel*, MedLat *pultrellus*, 'a colt', probably denoting one of a lively, frisky disposition.

**Puttifent:** *v.* POIDEVIN

**Puttifoot:** *v.* PUDDIFOOT

**Puttock, Puttack, Puttick, Puttuck, Pottock:** Ælfricus *Puttuc* 1034 OEByn; Aluied *Pottoch* 1066 DB (So); Edricus *puttuc* 1148 Winton (Ha); John *Puttok* 1176 P (Herts); Thomas *Puttek'* 1297 MinAcctCo (Co); John *Pottock* 1332 SRSx. OE *\*puttoc* 'kite' (c1400 NED), 'metaphorically applied to a greedy, ravenous fellow' (Halliwell).

**Putwain:** *v.* POIDEVIN

**Pyatt, Pyett, Pyott:** Simon *Pyot* 1297 MinAcctCo; John *Pyet* 1308 LLB C; William *Pyatt* 1327 SRSo. A diminutive of OFr *pye* 'magpie'. *v.* PIE. *Pyatt* may also be a late development of *Pyard*, a pejorative from OFr *pye* and -(*h*)*ard*: John *Pyarde*, Nicholas *Piarde* 1327, 1332 *SR* (Ha), surviving in Pyott's Hill in Basing (Hants).

**Pye:** *v.* PIE

**Pyke:** *v.* PIKE

**Pyle:** *v.* PILE

**Pymm:** *v.* PIM

**Pyne:** *v.* PINE

**Pyper:** *v.* PIPER

**Pyrah:** Joshua *Pyrah* 1703, John *Pyrah* 1789, Benjamin *Pyrrah* 1811 WRS. A West Riding name, perhaps a variant of PERRY.

**Pyser, Pyzer:** *v.* POYSER

# Q

**Quad, Quadd:** Robert *Quadd* 1148 Winton (Ha). A nickname from OE *cwēad* 'dung, dirt'.

**Quadling:** *v.* CODLIN

**Quaif, Quaife:** William *Coyfe, Coif* 1260 AssC, 13th Gilb (L). OFr *coif* 'coif, a close-fitting cap', metonymic for *coifier* 'maker of coifs': William *Coifar'* 1180 P (Bk); Bidan *le Coyfier* 1228 Cl (Ess); Geoffrey *Quayfere* 1301 SRY.

**Quaile:** *v.* QUAYLE

**Quain, Quane:** *MacQuaine* 1429, *MacQuayne* 1540, *Quaine* 1629, *Quane* 1680 Moore. Manx forms of *MacShane* 'son of *John*'.

**Qualter, Qualters, Qualtrough:** Manx names for *MacWalter*: Thomas *MacWalter* 1308, *MacQualtrough* 1429, *Qualtrough* 1430, *MacWhaltragh* 1511 Moore.

**Quant:** William *le Qwointe, le Coynte* 1254–67 Rams (Beds); Richard *le Queynte* 1256 Seals (Ha). OFr *cointe* (*quoint, cuinte*), ME *cointe, queynte, quante* 'wise, skilled, clever', also 'cunning, crafty' (a1225 NED).

**Quantic, Quantock:** Hugh *de Cantoc'* 1220 Cur (So); Walter *de Cantok* c1262 Hylle; Scipio *Quanticke*, Edward *Quantock* 1642 PrD. From the Quantock Hills (So), *Cantok* 1274.

**Quantrell, Quantrill, Quarntrill, Quintrell:** Ailric *Cointerell', Cuinterel* 1176, 1180 P (Co, K); William *Cuinterell' (Cuonterel)* 1214 Cur (So); William *Queinterell'* 1219 AssY; Adam *le Coynterel* 1281 QW (L); Robert *Quyntrel* 1332 SRSx. OFr *cointerel* 'a beau, a fop'.

**Quarles, Warfles:** Martin *de Warfles* 1198 FFNf; John *Quarles* 1561 Pat (Lo); John *Quarles* 1665 HTO. From Quarles (Nf), *Warfles* 1175.

**Quarmby:** Thomas *de Querneby* 1219 AssY. From Quarmby (WRY).

**Quarrell, Quarrelle:** Osbert, Yvo *Quarel* 1175–6 P (So, Hu). OFr *quarel, quarrel* 'a short, heavy, square-headed arrow or bolt for cross-bow or arbalest' (a1225 NED). Metonymic for an arbalester.

**Quarrie, Quarry:** (i) Henry *de la Quarrere* 1279 RH (O); William *atte Quarere* 1332 SRSx. OFr *quarrere*, ME *quarer(e)* 'quarry' (13.. NED). (ii) Thomas *atte Querre* 1428 FA (Sr). ME *quarey, querry* 'quarry' or *quarrer* 'quarry' (c1420 NED). Metonymic for QUARRIER. (iii) Alice Relicta *le Quarye* 1296 SRSx; Thomas *Quarry* 1524 SRSf. OFr *quarré*, ME *quarre* 'square; squarely built, stout' (1297 NED). 'Quarry, thykk mann, or womann, *Corpulentus, grossus*' PromptParv. (iv) The very common Manx *Quarry* is

for *MacGuaire* 'son of *Guaire*': *MacQuarres* 1504 *MacWharres* 1511, *Quarry* 1684 Moore.

**Quarrier:** Henry *le Quarreur* 1275 SRWo; Henry *le Quarreour* 1324 Wak (Y); Walter *Quarer* 1333 MESO (Ha). OFr *quarreour, quarrier* 'quarryman' (c1375 NED).

**Quarrington, Quarrinton:** Alexander *de Querington'* 1219 AssL. From Quarrington (L).

**Quartermain, Quartermaine, Quarterman, Quatermain, Quatermaine:** Herbert *Quatremains*, Robertus *Quatuormanus* 1187 Merton (O); Herbert *Quatermayns* 1230 P (O). AFr *quatremayns, quatremans* 'four hands', i.e. mail-fisted.

**Quatermass, Quatermass:** Ralph *de Quatermars* c1184 Clerkenwell; Colin *de Quatremares* 1219 AssY; Ranulf *Quatremare* 1242 AssDu. From Quatremares (Normandy).

**Quayle, Quail, Quaile:** (i) *MacFayle* 1511, *MacQuayle, Quayle* 1540 Moore. A Manx form of MACFAIL. (ii) Simon *Quayle* 1327 SRC. OFr *quaille* 'quail', noted for its supposed amorous disposition and timidity.

**Queech:** Reginald *Queycche* 1296, John *Quecche* 1327 SRSx; Hugh *Quecche* 1402–3 FFSr. 'Dweller by the thicket', ME *queche*.

**Queen:** Hunfrid *filius Quene* 1176 P (Hu); *Quena* 1276 FFEss; Matilda *le Quen* 1279 RH (O); Richard *Quene* 1301 FS; Agnes *Quene* 1332 SRSx. OE *cwēn* 'woman', used also as a personal name. Occasionally, perhaps, a nickname from OE *cwene* 'queen'.

**Quenell, Quennell, Quinnell:** *Cvenild* monialis 1086 DB (Gl); *Quenilda* uxor Gimpi e Hy 2 DC (L); *Quenill'* 1221 AssWa; *Quenilla* (*Quenild*) 1221 Cur (Nf); William *Quenell'* 1201 AssSo; Thomas *Quenild* 1275 RH (Nf); Richard *Quynel* 1286 Wak (Y). OE *Cwēnhild* (f) 'woman-war', first recorded in the 11th century.

**Queeniff, Queniff:** *Queniva* filia Gaufridi 1208 Cur (L); *Queniva* 1210 Cur (L); John, Maud *Quenyeve* 1279 RH (Hu). OE *Cwēngifu*.

**Queldrick, Queldrake:** Richard *Queldryk* 1411 AssLo. From Wheldrake (ERY), *Queldric* 1190.

**Queniff:** *v.* QUEENIFF

**Quernbeater:** Stephen *Quernebetere* 1271 AD ii (Mx); John *de Quernbetere* 1301 CorLo. 'A maker of millstones', OE *cweorn, bēatere*. cf. Robert *le Quernhacker* 1313 NorwLt.

**Quest:** Minquinus *Quest* 1323 KB (L); Cristiana *Quest* 1351 AssL; William *Quest* 1379 PTY. OFr *queste* 'tax'. Metonymic for a tax-collecter.

**Quick, Quicke:** (i) Robert *Quic* 1279 RH (C); William *Quik* 1282 FFEss. OE *cwic* 'nimble, lively'. (ii) Gedmær *on Cuike*, Ricard *a Cuik* 1100–30 OEByn (D). From Cowick Barton (Devon) or Quickbury (Essex), DB *Cuica*. (iii) Afwardus *de Quike* 1179 P (Y); John *de Quyke* 1278 AssLa; Gilbert *de la Quyk* 1297 SRY; Adam *del Quyk* 1313 Wak (Y). From Quick Mere in Saddleworth (WRYorks), a lost Quick in Prescott (Lancs), or from residence near a poplar or aspen. cf. OE *cwicbēam* 'quickbeam, poplar', *cwictrēow* 'aspen'.

**Quickly:** William *Quiklich* 1260 AssC. 'Nimbly, in lively fashion.'

**Quill:** *McCuill* 1511, *Quill* 1624 Moore. The Manx form of *Mac Cuill* 'son of *Coll*'.

**Quiller:** Roger *le cuillur* 1209 P (Nf); Walter *le cuillour* 1327 SREss. OFr *cuiller* 'spoon, ladle'. Metonymic for a maker of these.

**Quilliam,** a Manx name from Irish *MacWilliam* 'William's son'.

**Quilter:** Richard *le cuilter* c1179 Bart (Lo); Ralph *le Cuiltier* 1186 P (Wa); William *le Quilter* 12th DC (L). A derivative of OFr *cuilte, coilte* 'a quilt', 'a maker of quilts or mattresses' (1563 NED).

**Quin, Quinn:** Geoffrey *Quine, Quinne* 1275 RH (Nf); Robert *Quin* 1394 AssL; Thomas *Quyne* 1443 LLB K. OFr *quin* 'monkey'. *v.* also QUINE.

**Quincey, Quinsee, Quinsey, De Quincey:** Saer *de Quincy* 1153–63 Templars (O); Henry *Quenci* Hy 2 DC (L). Saer de *Quincy*, ancestor of the earls of Winchester, came from Cuinchy (Pas-de-Calais). *v.* ANF. The name may have been reinforced by immigrants from Quincy-sous-Sénart (Seine-et-Oise) or Quincy-Voisins (Seine-et-Marne).

**Quine, Quinn:** *Mac Cuinn* 1027, Luke *Mac Quyn* 1403, *Quine* 1504. From *Mac Coinn* or *Mac Cuinn* 'son of *Conn*', from Ir *conn* 'counsel' (Moore). A Manx name.

**Quinnell:** *v.* QUENELL

**Quinney:** *Quinnye* 1429 Moore. Manx for *MacConnaidh* 'son of Connaidh' 'the crafty'.

**Quinton:** (i) Gladewin, Walter *de Quenton'* 1176 P (Bk), 1221 AssWo. From Quinton (Glos, Northants, Worcs). (ii) Hugo *de Sancto Quintino* 1086 DB (Ha); Robert *de Quentyn* 1268 AssSo. From Saint-Quentin (La Manche) or, possibly, Saint-Quentin-en-Tourmont (Somme). (iii) *Quintinus* 1086 DB; *Quintinus* Taleboth 1200 Cur (Bk); Geoffrey, William *Quintin* 1205 P (O), 1222 Cur (W); William *Quentyn* 1262 FFEss. OFr *Quentin*, Lat *Quintinus* 'fifth', popular in France from the cult of St Quentin of Amiens, and brought to England by the Normans. This was the common source of the medieval surname but seems to have been completely absorbed by *Quinton*. Occasionally the surname may have reference to tilting at the quintaine: John *Quynteyn* 1378 LoPleas.

**Quirk, Quirke:** Ceinnedigh *O'Cuirc* 1043, *McQuyrke, Quyrke* 1511 Moore. Manx for *Mac Cuirc* 'son of *Corc*' (heart). In Ireland, for *O'Cuirc*.

**Quixley:** William *de Quyxelay* 1299, Roger *de Quyxley* 1369 FFY. From Whixley (WRY).

**Quodling:** *v.* CODLIN

# R

**Rabb:** *Rab'* atte Wyk' 1332 SRSr; Walter *Rab* 1199 AssSt; Richard *Rabbe* 1296 SRSx. For *Robb* (*Robert*).

**Rabbatts, Rabbets, Rabbetts, Rabbits, Rabbitt, Rabbitts, Rabett:** (i) *Radboda, Rabbode* 1086 DB (Nf, Sf); *Radbode* filius Ilhuwe 1166 P (Nf); Philippus *filius Rabat* 1203 Cur (Y); *Rabot* de Bovington 1300 Guisb (Y); Philip *Rabot* Hy 2 Gilb (L); Sygwat *Radbode* 1275 RH (Nf). OG *Radbodo, Rabbodo* 'counsel-messenger'. (ii) Andrew *Robat* 1279 RH (Hu); Alexander *Robet* 1317 AssK; Thomas *Robot* 1327 SRC; Reginald *Rabett* 1524 SRSf. *Rob-et, Rob-ot*, hypocoristics of *Robert*. Rabbit's Fm in Dallington (Sussex) owes its name to John *Robet* (1288) and Richard *Rabett* (1482 PN Sx 474).

**Rabey, Raby:** (i) Ysaac *filius Raby* 1196 P (Ess); Abraham *filius Rabi Josei* Jud' 1199 ChR; Thomas *Rabi* 1275 RH (Lo). OFr *rabbin* 'rabbi'. Usually Jewish. (ii) William *de Raby* 1260 AssCh; John *de Raby* 1332 SRCu; John *Rabey* 1524 SRSf. From Raby (Ch, Du).

**Rabjohn(s):** *v.* ROBJANT

**Rablan, Rablen, Rablin:** *v.* ROBLIN

**Rabson:** Gilbert Jakson *Rabbeson* 1401 AssLa. 'Son of *Rab*', a variant of *Rob*, a short form of *Robert*.

**Raby:** *v.* RABEY

**Race:** Roger *Race* 1193 P (Nf); Robert *Race* 1238 FFO; John *le Ras* 1279 RH (Hu); Robert *Raas* 1358 IpmNt. Perhaps a nickname from OFr *ras* 'clean-shaven'. But there was also a personal name: Anketill *filius Race* 1207 Pleas (Sr); Robert *filius Race* 1261 FFO.

**Rackford:** Edward *Rackford* 1642 PrD. From Rackenford (D).

**Rackham:** John *Rakham* 1524 SRSf. From Rackham (Sx).

**Rackliff(e):** *v.* RADCLIFF

**Rackstraw, Raickstraw:** John *Rakestraw* 1544 GildY; Thomas *Raykestray* 1605 FrY; William *Rakestraw* 1665 HTO. 'Rake straw', OE *racian, strēaw*, a nickname for a scavenger.

**Radbourne, Radburn, Radbone, Redbourn, Redburn:** William *de Redburn'* 1202 FFL; Bartholomew *de Redburn'* 1204 P (Herts); William *de Radburn', de Redburn'* 1219 AssY. From Radbourn (Warwicks), Radbourne (Derby), Redbourn (Herts) or Redbourne (Lincs).

**Radcliff, Radcliffe, Radclyffe, Ratcliff, Ratcliffe, Ratliff, Ratliffe, Rattcliff, Rackcliff, Rackliff, Redcliffe, Redcliff, Reddecliff, Reddicliffe:** Walter *de Radecliua* 1182 P (D); Robert *de Radeclyf* 1313 Wak

(Y); Robert *Racclyff* 1496 Ipm (Sf); Henry *Rattcliffe* 1636 RothwellPR (Y); Elizabeth *Ratliff* 1792 SfPR. From Radclive (Bucks), Radcliffe (Lancs, Notts), Ratcliffe (Leics, Notts), Ratclyffe, Ratcliffes (Devon), or Redliff Hill (Warwicks), all meaning 'red cliff'.

**Radcot, Radcott:** John *de Radcote* 1285 FFO. From Radcot (O).

**Raddie:** *v.* READY

**Raddish:** *v.* REDDISH

**Raddle:** *v.* RADWELL

**Raddon:** Rauening *de Raddon'* 1194 P (Beds/Bk); Baldewin *de Raddon* 1221 Cur (Co); Odm' *de Raddenn* 1296 SRSx. From Raddon in Marystow, West Raddon in Stoke, or Raddon Court in Thorverton (D).

**Radford, Radforth, Radfirth, Ratford, Redford, Reddiford, Retford:** John *de Radeford* 1209 P (Nt); Walter *de Redford'* 1230 P (Berks); Hugo *de Retford* 1275 RH (Nt); Geoffrey *de Ratforde* 1296 SRSx; Nicholas *atte Rydeforde* 1296 SRSx, *atte Redeford* 1305 MELS (Sx). From Radford (Devon, Notts, Oxon, Warwicks, Worcs), Ratford Fm (Sussex), Redford (Sussex) or Retford (Notts), all either 'red ford' or 'reed ford'. OE *rēad* 'red' would give both Radford and Redford, OE *hrēod* 'reed' would give Redford. Also 'dweller near a reedy ford'.

**Radish:** *v.* REDDISH

**Radleigh, Radley:** Osbert *de Radelega* 1177 P (Sr); Philip *de Radleg'* 1260 FFK; Adam *de Radelee* 1342–3 FFSr; Richard *Raddeley* 1471 Past. From Radley (Berks, Devon), or 'dweller at the red clearing'.

**Radmall:** Walkelyn *de Radmelde* 1296 SRSx. From Rodmell (Sx), *Radmelde* 1202.

**Radnor, Radner:** John *de Radenoura* 1193 P (He); William *de Radenore* 1255–6 FFSx; John *Radenore* 1398 FFEss. From Radnor (Radnor).

**Radway, Reddaway, Rodaway, Rodway:** Henry *de Radeweie* 1205 P (So); Stephen *Rodweye* or *Radwaye* 1581 Oxon; Richard *Reddaway* 1648 DWills. From Rodway (Som) 'road way', or Radway (Warwicks), Radway, Reddaway, Roadway (Devon), all 'red way' (OE *rēad*). Thomas *de Radeweye* (1242 Fees) came from Roadway, Geoffrey *de Radeweye* (ib.), from Reddaway.

**Radwell, Raddle, Reddall, Reddell:** Robert *de Radewelle* 1185 Templars (Beds); Robert de *Redewelle* 1274 RH (So). Usually from Radwell (Beds, Herts), occasionally from Redwell Wood (Herts), or 'dweller by the reed-stream'. Also 'dweller on the red hill': Richard *atte Redehulle* 1327 SRSo, or 'by the cleared woodland': Brun *de la Redeweld* 1296 SRSx. Certain forms suggest the possibility of a survival of OE *Rǣdweald* or *Rǣdwulf*, but the evidence is inconclusive: Ralph *Redwald*, William *Redolf* 1276 RH (O); Richard *Redwal* 1297 MinAcctCo.

**Rae:** Robert *Raa* c1231 Black. A Scottish form of ROE.

**Raeburn, Rayburn, Reburn, Reyburn, Ryburn:** William *of Raeburn* 1331, Andrew *de Raburn* 1430, Thomas *Reburne* or *Ryburne* 1463 Black. From the lands of Ryburn in Dunlop (Ayr).

**Rafe, Raff:** v. RALF

**Raffell, Raffle, Raffles:** (i) Thomas *Raffale* 1375 AssL; John *Raffel*, *Raphel* 1642 PrD. *Raff-el*, a diminutive of *Raff*, i.e. *Ralph*. (ii) John *de Resfholes* c1215–45, John *de Rafhols* 1361, William *Raffel*, *Raphael* 1684 Black. From Raffles in Mouswald (Dumfries).

**Ragdale:** Stephen *de Raggedal'* 1202 Pleas. From Ragdale (Lei).

**Ragg:** For WRAGG. A correspondent informs me that his pedigree proves that the surname was *Wragg*, but at the end of the 18th century, one son, by mistake of the clerk (presumably at his baptism), had his name entered as *Ragg*, which thereafter became his surname and that of his children.

**Ragg, Ragge:** William *le Ragge* 1198 Pleas (Nf); Simon *Rage* 1327 SRDb; Thomas *Ragge* 1558 Pat (Y). ME *ragge* 'rough stone', or perhaps OE *\*ragge* 'moss, lichen'. Sometimes, perhaps, ODa *Wraghi*. v. also WRAGG.

**Raggatt, Raggett, Ragot:** (i) Hamelin *Ragot* 1177 P (C); William *Ragat* 1279 RH (C); John *Raggett* 1661 FrY. *Rag-ot*, a diminutive of OG *Rago*. cf. Fr *Ragot*, *Raguet*. (ii) Richard *le Raggede* 1197 FFNt; Robert *le Ragget* 1266 IpmY; John *le Ragede* 1332 SRWa. 'The ragged', ME *raggede*.

**Ragge:** v. RAGG

**Raggett:** v. RAGGATT

**Raglan, Ragland:** William *Ragilond* 1260 AssC; Nicholas *Ragalan* 1406 IpmGl; John *Raglond* 1545 SRW. From Ragland Coppice in Corsley (W), or 'dweller by the stony land', ME *ragge*, OE *land*.

**Ragley:** Richard *Rageleye* 1275 SRWo. From Ragley Hall in Arrow (Wa).

**Ragnell:** *Ragenild* Springsus 1198 Pleas (Nf); *Ragenhild* (f) 1208 Pleas (W); *Ragenilda de Bec* 1248 MPleas (Mx); William *Ragenel* 1192 P (W); William *Raghnil* 1327 SRSf; William *Ragonell* 1402 AssLo. OG *Raginhild* (f).

**Ragot:** v. RAGGATT

**Raickstraw:** v. RACKSTRAW

**Raikes:** v. RAKE

**Raiman:** v. RAYMAN

**Raimes, Reames:** Roger *de Rames*, *de Ramis*, *de Raimis* 1086 DB (Mx, Ess, Sf); William *de Reimes*, *de Remes* 1199 MemR (Nf). From Rames (Seine-Inférieure). v. ANF.

**Rain, Raine, Rayne:** (i) *Regina* 1203 Cur (K); *Regin'* Prat 1275 RH (C); *Reina*, *Reyna* vidua 1214 Cur (Nth), 1244 Rams (Beds); Alan *Reyne* 1260 AssC; Hugo *Ragen'* 1275 RH (He); Alice *Reine* 1279 RH (C); William *le Reine* 1332 SRSx; John *Rayn*, Richard *Rayneson* 1379 PTY. Fr *Reine*, Lat *Regina* 'queen', a personal-name found in France; also a nickname, here derogatory. We may also have a short form of names in *Regen-*, such as *Reyner*, *Reynold*, etc. *le Reine* is probably Fr *raine* 'frog'. (ii) Robert *de Ran'* c1180 Black. From Rayne (Aberdeenshire).

**Rainbird, Raynbird, Rambart, Rambert, Ramart:** *Rainbertus* Flandrensis 1086 DB (Gl); Willelmus *filius Reinbert* 1206 Cur (Mx); William *Reinberd*, *Reinbert*, *Renberd* 1208, 1212 Cur (Herts); Simon *Ramberd*, Stephen *Rambard* 1327 SRSx. OFr *Rainbert*, *Reinbert*, *Raimbert*, from OG *Raginbert*, *Reginbert* 'mighty guardian'.

**Rainbow, Raybould, Rambaut, Ramble, Rammell, Rimbault, Renbold:** *Raimbaldus* 1066 Winton (Ha); *Rainbaldus* auribfaber 1086 DB (Nf); Elward *filius Reinbaldi* ib. (Gl); *Reimbaldus* 1148–66 NthCh (Nth); *Rembaldus* 1168–75 Holme (Nf); Reginaldus *filius Renbald*, *Rainbaud* 1212 Cur (Wa); Johannes *filius Rambaldi* 1275 RH (Berks); William *Reimbaud* 1214 Cur (Mx); John *Rambald* 1332 SRSx; John *Reynbald* 1505 NorwW. OFr *Rainbaut*, *Raimbaut*, *Raimbault*, *Rambaut*, from OG *Raginbald*, *Rainbald*, *Reinbald* 'might-bold'.

**Raincock:** Robert, William son of *Raincok* 1332 SRCu; William, John *Rayncok* ib. *Rain-cok*, a diminutive of a short form of names in *Rain-*, e.g. *Rainbald*, *Rainbert*, etc.

**Rainer:** v. RAYNER

**Raines, Rains, Raynes, Rayns, Reynes:** Alveva *de Reines* 1203–4 FFEss; Richard *de Rayns* 1297 SRY; Nicholas *de Reynes* 1301 FFY. From Rayne (Ess), Raines (DB). (ii) Hugh *de Rennes* 13th Lewes (Nf). From Rennes (France).

**Rainey, Rainie, Rainnie, Rainy, Raney, Rannie, Rennie, Renny:** *Rayny* Voket 1409 Black; *Rany* Ra 1446 ib.; Henry *Raney* 1275 RH (Db); Thomas *Renie* 1279 RH (Beds); Symon *Renny* 1362 Black; Margaret *Rany* 1379 PTY; John *Rayny* 1436 Black; Walter *Rannie* 1453 ib.; John *Rany* 1510 ib.; Agnes *Reanie* 1636 ib. *Rayny*, with shortened vowel, *Rannie*, *Rennie*, pet-forms of *Reynold*. These names are Scottish. Similar forms seem to have arisen in northern England. In Ireland the name became *Mac Raighne*, *Ó Raighne*, with modern forms *Rainey*, *Raney*, *Reaney*, *Reanney*, *Reanny*, *Reany*, *Reinny*, *Rennie*, *Reyney*, *Ryney*. Theoretically, all these names could derive from Regny (Loire): Roger, William *de Reigni* 1157, 1195 P (D, Cu); William *de Reingny* alias *de Reyney* alias *de Reny* 1276 Ipm (Cu). The name was common in the 12th and 13th centuries and survives in Ashreigny (Devon) and Newton Reigny (Cumb), but there seems to be no connexion between this and the modern names.

**Rainford, Rainforth, Ranford:** William *de Reynford* 1246, John *or Raynford* 1401 AssLa; Thomas *Rainforth* alias *Rentforth* 1724 FrY. From Rainford (La).

**Rainger:** v. RINGER

**Raingold:** *Rengot* Barat Hy2 DC (L); *Reingotus de Weinflet* 1193 P (L); Andrew *filius Reingod*, *Reingodi* 1221 Cur (L). OG *Reingot*.

**Rainham:** Richard *de Reynham* 1318 KB (Lo). From Rainham (K).

**Rains:** v. RAINES

**Raisbeck, Reasbeck:** William *Raysebeck* 1301 SRY. From Raisbeck (WRY).

**Raistrick:** v. RASTRICK

**Raithby:** Robert *de Reidebi* 1218, Simon *de Rethebi* 1219 AssL. From Raithby (L).

**Rake, Raikes:** John *de Rak'* 1242 Fees (D); Henry *de la Rake* 1275 MELS (Sx); Walter *Rake* 1279 RH (Hu); Beatrice *ate Rake* 1332 SRSr; Thomas *del Rakes* 1332 SRCu; Nicholas *Raikes* son of Thomas *Rakes* 1585 FrY. From residence near a pass or narrow valley (OE *hraca* 'throat') as at The Rake (Sussex), Raikes Fm (Surrey) or Raikes (WRYorks). v. MELS 160–1.

**Raleigh, Raley, Ralley, Rally, Rawley, Rayleigh:** Hugh *de Ralega, de Raalega* 1164, 1169 P(D); John *de Rallye* 1296 SRSx; Thomas *Ralegh* 1386–7 FFWa; Edmund *Rawley, Rayley, Raweleygh, Raleygh, Ralegh* or *Rawleigh* of Exeter 1509 LP. From Raleigh (Devon), Rayleigh House in Morthoe (Devon), or Rayleigh (Essex).

**Ralf, Ralfe, Ralfs, Ralph, Ralphs, Rales, Ralls, Rafe, Raff, Rau, Raves, Raw, Rawe, Rawes, Rawle, Rawles, Rawll:** *Radulf, Radolf* 1066 DB; *Raulf, Raulfus* clericus c1095 Bury (Sf); *Radulfus* de Henlinton c1140 DC (L); Robertus *le fiz Raol* Hy 2 DC (L); *Raul* Cordel 12th NthCh (Nth); *Radufus* de Braibof c1200 DC (L); Edricus filius *Raw* Hy 3 Colch (Ess); *Rauf* c1350 Brut; *Raaph* 1387 Trevisa; Johannes *Radulphus* 1186–8 BuryS (Sf); Richard *Rau* 1212 Cur (Nf); Adam *Rauf* 1275 RH (Nf); Amic' *Raffe* 1279 RH (C); Simon *Raulf* 1296 SRSx; Denis *Rauf, Rolf* 1308 EAS xviii; John *Ralf* 1327 SRSx; Thomas *Raules,* John *Raweles* 1327 SRSo; Henry *Rawe* 1332 SRLa; Richard *Raaf,* Edward *Raphe* 1524 SRSf; Miles *Rawes* 1639 RothwellPR (Y). This may occasionally be ON *Ráðúlfr* but is usually from OG *Radulf,* the source both of Fr *Raoul* and the Norman *Radulf, Raulf.* The name had reached England before the Conquest and may have come direct from Scandinavia but was usually introduced from Normandy. It is very common indeed in the 12th century but almost invariably latinized as *Radulfus.* In spite of, or perhaps because of this popularity, it is rare as a surname before the 13th century when it is very common in a variety of forms. There has been some confusion with *Rolf.*

**Ralling(s), Rallis:** v. RAWLIN

**Rallison:** v. RAWLINGSON

**Ralls, Ralph(s), Ralphson:** v. RALF

**Ralston, Ralstone, Raulston:** Nicholas *de Ralstoun* 1272, John *Raleston* or *Raliston* 1488 Black. From the lands or barony of Ralston near Paisley (Ren& frew).

**Ram, Ramm:** (i) Hendricus *Ram* 1188 Bury (Sf); Geoffrey *Ram* 1212 Cur (Nth), 1274 RH (Ess). A nickname from the ram, OE *ram(m).* (ii) William, Giles *atte Ramme* 1307 LLB C, 1339 LoPleas. From the sign of the ram.

**Ramage, Ramadge:** Robert, Peter *Ramage* c1240 Fees (L), 1321 LLB E, 1304 Black (Perth). ME, OFr *ramage* 'wild', used of a hawk 'living in the branches', MedLat *\*ramaticus,* from *ramus* 'branch'.

**Ramart, Rambart, Rambert:** v. RAINBIRD

**Rambaut, Rammell:** v. RAINBOW

**Ramett:** *Rametta* (f) 1195 P (Wo); *Rametta* filia Roberti 1228 CR (Nf); *Rametta* (f) 1256 AssNb; Walter *Ramet'* 1213 Cur (Do). OFr *Ramette* (f). But a local origin is also possible: Stephen *ate Rammette* 1317 AssK.

**Ramsay, Ramsey, Ramshay:** Æðelstanus *de Rameseia* c1036 OEByn (Ess). From Ramsey (Essex, Hunts). The Scottish Ramsays derive from Simund *de Ramesie* (a1175 Black) who went to Scotland from Ramsey (Hunts).

**Ramsbotham, Ramsbottom:** Roger *de Romesbothum* 1324 LaCt. From Ramsbottom (Lancs).

**Ramsden:** Roger *de Rammesden* 1195 FFL; John *de Rammesdenne* 1334–5 SRK; Peter *Ramsden* 1672 HTY. From Ramsden (Essex, Oxon).

**Ramsey:** v. RAMSAY

**Ramshaw:** v. RAVENSHAW

**Ramshay:** v. RAMSAY

**Ramshead:** William *Romesheued* 1327 SRDb. 'Ram's head', a nickname.

**Ramshire:** v. RAVENSHAW

**Ranacre:** v. RUNACRES

**Ranald:** v. RONALD

**Ranby:** Norman *de Randebi* 1193 P (L); Gilbert *de Randeby* 1202 AssL. From Ranby (L, Nt).

**Rand, Rands, Rance:** (i) *Rande* de Borham 1299 LoCt; Adam *Rand* 1275 SRWo; Thomas *Rande* 1327 SRSf; Richard *Randes* 1379 PTY; Rebecca *Rants, Rance* 1735–6 Bardsley. *Rand,* a pet-form of *Randolph.* (ii) Herlewinus *de Rande* 1176 P (Nth); Robert *de Randes* 1317 AssK. From Rand (Lincs), Rand Grange (NRYorks) or Raunds (Northants).

**Randall, Randell, Randle, Randles, Randoll:** *Randal* 1204 AssY; *Randle* de Stok 1260 AssCh; Thomas *Randel* 1250 FFSf; Richard *Randall* 1547 FFHu; *Rand-el,* a diminutive of *Rand* (Randolph).

**Randerson, Randlesome:** Thomas *Randalson* 1471 Black; William *Randerson* 1734 FrY. 'Son of *Randall.'*

**Randolph, FitzRandolph:** *Randulfus* c1095 Bury (Sf); Nicolaus *filius Randulphi* 1175–86 Holme (Nf); William, Robert *Randulf* 1260 AssC, 1275 RH (L); Ralph *Fetzrandolff* 1498 GildY. *Randolph* is ON *Rannulfr* 'shield-wolf', brought to England by the Normans as *Randulf.* It was confused with the equally common OG *Rannulf* 'raven-wolf', introduced the same time: *Randolphus* de Brachenberch c1155 Gilb (L), *Ranulfus* de Brachinberge 1160–6 ib.

**Randy:** Malie *Randie* 1573 Black (Perth). A pet-form of *Rand* (Randolph).

**Raney:** v. RAINEY

**Ranford:** v. RAINFORD

**Ranigar:** v. RUNACRES

**Rank:** John *Ronk* 1327 SRSf; William *Ronkes* 1332 SRCu. OE *ranc,* ME *rank* 'strong, proud'.

**Rankill:** Rogerus *filius Ranchil* 1130 P (Y); Rogerus *filius Rauenkil* 1170 P (La); Tomas *filius Ramkell'* 1191 P (Y); Robert *Ravenkil* c1138 Whitby (Y); William *Ramkil, Ranckil* 1204 AssY. ON *Hrafnkell,* OSw *Ramkel* 'raven-cauldron'.

**Rankin, Rankine, Ranking, Ranken:** Rankin de Fowlartoun 1429 Black; Reginald *Ranekyn* 1296 SRSx; Ralph *Rankin* 1301 SRY; Robert, John *Randekyn* 1327 SRSx, 1381 SRSf. *Rand-kin*, a diminutive of names in *Rand-*, e.g. *Randulf*, or *Rankin*, for *Rannulf*.

**Rannall, Rannell:** Hugh *filius Rannulfi* 1086 DB (Sx); *Randulfus* de Baiwes 1143–7, identical with *Rannulfus* de Baiocis c1150 DC (L); Richard *filius Ranulfi* 1210 Cur (Ess); Richard *Ranel* 1275 RH (Bk); Adam *Ranel* 1327 SRWo; Richard *Ranel* 1340 NIWo. OG *Rannulf*, sometimes confused with OG *Randulf*.

**Rannard:** v. REYNARD

**Ranner:** v. RAYNER

**Rannie:** An assimilated form of RANDY, or for RAINEY.

**Ranscomb, Ranscombe, Ranscome:** Thomas *de Rennescumbe*, *de Ramescumbe*, *de Ravenescumbe* 1203–6 Cur (K). From Ranscombe in Cuxton (Kent). Also from Ranscombe (Hants), Ranscombe Fm in South Malling (Sussex), one of the five Ranscombes in Devon, or Rainscombe House in Wilcot (Wilts), *Ranscombe* 1581, all 'raven's or ram's valley'.

**Ransdale:** William *Ranesdale* 1477 IpmNt. From Ramsdale (NRY), or Ramsdale Fm in Arnold (Nt).

**Ransford:** John *de Rammesford* 1327 SRSx probably lived at Ramsfold Fm in Lurgashall (Sussex).

**Ranshaw:** v. RAVENSHAW

**Ransley:** v. RAWNSLEY

**Ranson, Ransom, Ransome:** William *Randesson'* 1347 *SR* (C); John *Randson* 1395 Whitby (Y); Elizabeth *Ransom* 1518 NorwW (Sf); Adam *Ranson* 1524 SRSf. 'Son of *Rand*' (Randolph).

**Rant:** For RAND. A back-formation from *Rants*.

**Ranwell:** William *de Ranuill'* 1200 P (W); Adam *de Raineville* c1216 Calv (Y). From Ranville (Vosges).

**Raper:** v. ROPER

**Rapson:** John *Rapson* 1642 PrD; Mary *Rapson* 1662–4 HTDo. 'Son of *Rab*', a variant of *Rob*, a short form of *Robert*.

**Rasch, Rash:** v. ASH

**Rasen:** v. RASON

**Rashleigh:** John *Atterashlegh* 1292 Ipm (D). From Rashleigh Barton (Devon).

**Raskell:** Richard *de Raskel* 1333 FFY; John *Rascall'* 1383 AssL. From Raskelf (NRY).

**Rason, Rasen:** Robert *de Rasene* 1202 SPleas (L); John *de Rasene* 1280 FFY; Robert *de Rasin* 1290 RegAntiquiss. From Market, Middle, West Rasen (L).

**Rasor:** Baldewinus *rasor* 1130 P (O); Alan, Thomas *Rasur* 1159, 1196 P (O). OFr *rasor, rasur* 'razor', metonymic for a maker of razors. cf. Walter *le Rasorer* 1285 Wak (Y).

**Raspberry:** Marioth' *de Radespree* 1242 Fees (D). From Ratsbury in Lynton (D).

**Rastall:** Roger, Walter *Rastel(l)* 1185 P (So), 1206 Cur (Ha). OFr *rastel* 'rake, mattock', either a maker or seller of rakes, or an agricultural labourer.

**Rastrick, Raistrick, Rustrick:** Roger *de Rastric* 1212 P (Y); John *de Rastrik* 1274 Wak; Katerina *Rastrik'* 1379 PTY. From Rastrick (WRYorks).

**Rat, Ratt:** Jordan *Rat* Hy 2 Gilb; Ralph *le Rat* 1210

Cur (Nth); John *le Rat* 1334 SRK. A nickname from the rat, OE *ræt*.

**Ratchford:** v. ROCHFORD

**Ratcliff(e):** v. RADCLIFF

**Ratford:** v. RADFORD

**Rathbone, Radbone, Rathborne, Rathbourn:** Richard, John *Rathebon* 1275 SRWo, 1347 AD vi (Ch); Robert, Richard *Rathebun(e)* 1297, 1327 SRY; Robert *Radbone* 1547 Bardsley. For this difficult name Harrison suggests an Irish or, preferably, a Welsh origin, Ir *Rathbane* 'white fort' or Welsh *Rhathbon* 'stumpy clearing or plain'. Bardsley, whose earliest example is 1547, doubtfully suggests a derivation from Ruabon. Early examples are rare and without any preposition. No satisfactory suggestion can be offered.

**Ratliff(e):** v. RADCLIFF

**Ratnage:** Richer *de Radenache* 1197 FF (Bk). From Radnage (Bucks) or Radnidge (Devon), *Rothenesse* 1292.

**Ratner, Ratter:** Margeria *le Ratonner* 1327 SRSf. A derivative of OFr *raton* 'rat', a rat-catcher.

**Raton, Ratten, Ratton:** Richard *Raton* 1290 AssCh; Godfrey *le Raton* 1293–4, Ranulf *Ratun* 1304 IpmY. A nickname from OFr *raton* 'rat', or perhaps metonymic for a rat-catcher. v. also RAT.

**Ratt:** v. RAT

**Rattenbury, Rattenberry, Rottenbury:** Mycheal *Rottenbury* 1619, Margrette *Rattenbury* 1638 HartlandPR (D); Nicholas *Rattenbury* 1642 PrD. From Rattenbury (Co).

**Ratter:** v. RATNER

**Ratton:** v. RATON

**Rattray:** Thomas *de Rettre* 1253 Black; Adam *de Retref* 1294 ib.; William *Rettre* 1436 ib.; Silvester *Rettray* 1526 ib. From the barony of Rattray (Perth).

**Rau:** v. RALF

**Raulin:** v. RAWLIN

**Raulston:** v. RALSTON

**Ravel, Ravell:** John *de Ravel* 1279 RH (Hu); Nicholas *de Ravele* 1287–8 NorwLt; Nicholas *Rauell* 1327 SRLei. Probably from the common French place-name Ravel, but sometimes, perhaps, from Great, Little Raveley (Hu).

**Raven, Ravens, Revan, Revans, Revens:** Leduuinus *filius Reuene* 1086 DB (L); *Rauen* de Engelbi 1185 P (Y); Godric, William *Raven* 1133–60 Rams (Beds), 1188 P (Nb); Adam *Reven* 1279 RH (O); Elena *Ravenes* 1312 ColchCt; Walter, Alice *le Reven* 1327 SRWo, 1332 SRWa; William *Revance* 1520 NorwW (Nf); Alan *Rivance* 1534 ib. ON *Hrafn* or OE *\*Hræfn* 'raven', or a nickname from the bird, ON *hrafn*, OE *hræfn*. Occasionally from a sign-name: William *atte Raven* 1344 LLB F.

**Ravenhall, Ravenhill, Revnell:** (i) Willelmus *filius Rauenilde* 1297 SRY; William *Ravenild* 1276 RH (Y); Matilda *Rafenild* 1279 RH (C); John *Ravenell* 1700 Bardsley. ON *Hrafnhildr* (f). (ii) Nicholas *de Rauenhull'* 1230 P (He). From Ravenhill (NRYorks) or some other 'raven-hill'.

**Ravening:** *Rauening* de Raddon' 1194 P (Beds); *Ravening* 1212 Cur (Bk); Simon *Ravening* 1248 FFEss; William *Rauenyng* 1297 MinAcctCo. OE *\*Hræfning*, an *-ing* derivative of OE *\*Hræfn*.

**Ravenshaw, Ravenshear, Ramshaw, Ramshire, Ranshaw, Renshaw, Renshall, Renshell:** Stephen *de Ravenshagh* 1332 SRLa; Ralph *Raynshae, Renshae* 1548, 1561 Bardsley (Ch); Robert *Ravenshaw, Ramshaw* 1606, 1617 ib.; Randle *Renshaw* 1613 ib.; John *Renshall* 1679 ib. 'Dweller by the raven-wood', as at Ravenshaw (Warwicks) or Renishaw (Derby).

**Ravensthorpe, Raventhorpe:** Thomas *de Ravenesthorpe* 1271–2, William *de Raventhorpe* 1294 IpmY; William *de Raventhorp* 1348 FFY. From Raventhorpe (L), or Ravensthorpe (Nth, NRY).

**Raw, Rawe:** Edward *Bitherawe* 1279 AssSo; John *de Rawe* 1297 SRY, William *ate Rawe* 1332 SRSr; William *del Rawe* 1429 DbCh. 'Dweller by the hedgerow or in the row of houses or street', OE *rāw* 'row'. *v.* REW, ROW and also RALF.

**Rawbone, Rabone, Rawbins, Raybon:** William *Rabayn* 1301 SRY, 1332 SRLa; Thomas *Rawboon* 1492 FrLeic; Widow *Rawbones* 1674 HTSf. *Rabayn* is ON *rá-bein* 'roe-bone', a nickname for one with legs as speedy as those of a roe. cf. Henry *Coltebeyn* ('colt'), William *Gaytebayn* ('goat'), Alan *Kabayn* ('jackdaw') 1301 SRY. *Raybon* is from OE *\*rā-bān* 'roe-bone', showing the northern spelling of *rā*; *Rawbone* is a corruption of the midland and southern *Roebone*.

**Rawcliff, Rawcliffe, Rawlcliffe, Rockcliffe, Rockliff, Rockliffe, Rowcliffe:** Elsi *de Routecliua* 1170 P (Y); John *de Rouclef* 1332 SRCu. From Rawcliffe (NRYorks), Roccliffe (WRYorks), Rockcliff (Cumb), or Rowcliffe (Devon).

**Rawdon:** Walthef *de Raudon'* 1202 FFY; John *de Rawdon* 1379 PTY; John *Rawdon* 1459 Kirk. From Rawdon (WRYorks).

**Rawes:** *v.* RALF

**Rawkins:** Joane *Rawkyns* t Eliz Bardsley. *Rawkin* 'little Ralph'. *v.* RALF.

**Rawland:** *v.* ROWLAND

**Rawlcliffe:** *v.* RAWCLIFF

**Rawle(s):** *v.* RALF

**Rawley:** *v.* RALEIGH

**Rawlin, Rawlins, Rawling, Rawlings, Rawlence, Rawlyns, Raulin, Raulins, Ralling, Rallings, Rallis:** *Raulyn* le Forester 1277 Wak (Y); William *Raulyn* 1290 Eynsham (O); John *Rawlynes* 1343 AD iv (Wa); Walter *Rawling* 1520 NorwW (Nf); Catherine *Ralling* 1642 Black (Dumfries). OFr *Raul-in*, a diminutive of *Ralf*, both sometimes used of the same man: *Ralph* de Knyghton . . . the above-mentioned *Raulyn* 1377 LoPleas. *v.* RAWLINGSON.

**Rawlingson, Rawlinson, Rawlison, Rallison:** Richard *Rawlinson* 1538 Riev (Y); Robert *Rallinson* 1617 ShefA. 'Son of *Raulin*.' *v.* RAWLIN. Bardsley notes that in Furness and Cumberland, where Rawlinsons are numerous, *Rowland* is pronounced *Rawland* and *Rolland* and these Rawlinsons derive from *Rawlandson*, which became *Rawlinson*. *v.* ROLLINGSON.

**Rawll:** *v.* RALF

**Rawnsley, Ransley:** Margaret *de Rauenslawe* 1379 PTY; Nicholas *Raunsley, Rawnsley* 1564, 1565 ShefA; Joseph *Ransley* 1663 FrY. From an unidentified place, apparently in the West Riding.

**Raworth:** *v.* ROWARTH

**Rawson, Rawsen:** William *Raufson*, John *Rauson*, Richard *Raweson* 1379 PTY. 'Son of *Rauf* or *Rau*.' *v.* RALF.

**Rawsthorn, Rawsthorne, Rawstorn, Rawstorne, Rawstron, Rosten, Rostern, Rosterne, Rosthorn, Rosson, Roston, Rostron:** Richard *de Routhesthorn* 1246 AssLa; James *Rawstorne* 1599 FrY; Jane *Rosthern* 1613 Bardsley. From Rostherne (Ches).

**Ray, Raye, Rey:** William *Lerei* 1195 FFNf; Robert *Raie* 1206 P (C); Henry *le Ray* 1296 SRSx. The most common source of these names is OFr *rei* 'king', occasionally used as a personal name: Thomas *filius Rey* 1279 RH (C). It may be a nickname denoting pride, one with the bearing of a king, or refer to the victor in some competition or sport such as shooting at the popinjay (cf. PAPIGAY). It might also be OE *rǣge* 'female of the roe' or, in the north, the dialectal pronunciation *ray* found in the 16th century for OE, ON *rā* 'roe'. *v.* also REA, RYE, ROY.

**Raybon:** *v.* RAWBONE

**Raybould:** *v.* RAINBOW

**Rayburn:** *v.* RAEBURN

**Rayden, Raydon:** Geoffrey *de Raydon* 1276 RH (L). From Raydon, or Reydon (Sf).

**Rayer:** *Raher'* 1137 Bart; Ricardus *filius Raher* 1206 Cur (Ess); Ralph *Raher* 1275 RH (Berks). OG *Radheri, Rathar, Rather* 'counsel-army', the name of the founder of St Bartholomew's Hospital.

**Rayleigh:** *v.* RALEIGH

**Rayman, Raiman, Reaman, Reeman, Ryman:** Robert *Ryman* 1327 SRSx; Simon *Reman* 1359 ColchCt; John *Rayman, Reyman* 1377 AssEss. 'Dweller by the low-lying land or the stream.' *v.* RAY, REA, RYE.

**Raymond, Raymont, Rayment, Raiment:** *Raimundus* 1086 DB (Ess); *Reimundus* 1121–48 Bury (Sf), 1214 Cur (Wa); Giraldus *Reimundus* 1086 DB (Ess); William *Reimunt* 1207 P (Ha); Ernald *Reimund'* 1208 P (K). OFr *Raimund, Raimond* from OG *Raginmund* 'counsel or might-protection'.

**Raynbird:** *v.* RAINBIRD

**Rayner, Raynor, Rainer, Ranner, Reiner, Reiners, Reyna, Reyner, Renner:** *Rainerus* 1086 DB; *Reynerus* cancellarius 1101–25 Holme (Nf); Ricardus *filius Rainer* 1148 Winton (Ha); *Renerus* c1250 Rams (Nf); Alexander *Reygner* 1229 Cl; William *Rayner* Hy 3 Gilb (L); William *Reyner* 1286 AssSt; Robert *Ranare* 1651 RothwellPR (Y). OFr *Rainer, Reiner, Renier*, from OG *Raginhari* 'counsel, might-army'.

**Raynes, Rayns:** *v.* RAINES

**Rayson:** John *Rayson* 1294 FFSf; William *Rayson* 1327 SRSo; John *Reyson* 1332 SRSt. 'Son of *Ray*', OFr *rei* 'king' used as a personal name.

**Rea, Ree, Ray, Raye, Rey, Rye:** Ralph *de la Reye* 1279 RH (O); William *atte Ree* 1285 *Ass* (C); William *bithe Ree* 1293 MELS (Wo); John *atte Reye* 1327 *SR* (Ess); William *atte Rea* 1327 SRSx. All these are local surnames from residence near a stream or low-lying land near a stream. OE *ēa* 'water', 'stream' had two datives, *īe* and *ēa*. OE *æt þǣre ēa* 'by the stream' became ME *at ther ee, atte ree*, hence *Rea* and *Ree*. OE *æt þǣre īe* became ME *at ther ie, ye, eye, atte rie, rye, rey*, hence *Ray(e), Rey, Rye*. OE *æt þǣre, ēge, īge*, from OE *īeg, īg, ēg* 'island, piece of firm land in a fen', became ME *at ther eye, ye, atte reye, rye* which cannot be distinguished from the similar forms from OE *ēa* without further evidence. Michael *atte Ree* (1327 SRSx) is called *atte Rye* in 1332 (ib.). Thomas *atte Ree* in 1332 probably lived at the same place as William *ate Rye* in 1359 (PN Sr 233), whilst both Walter *at Reghe* in 1287 and John *ate Ree* in 1332 lived at Ray Lodge (PN Sr 330). All these lived near a stream (*ēa*), as did Stephen *atte Ree* (1323 *For*) at Rye Mill in Feering (Essex). *v.* also RYE, NYE, YEA.

**Reace:** *v.* RHYS

**Reach:** Walter *de Reche*, John *Reche* 1279 RH (C); Robert *Reache* 1641 PrSo. From Reach (Beds). In Scotland, a variant of *Rioch*.

**Reacher:** *v.* RICHER

**Read, Reade, Reed, Red, Redd, Reid:** (i) Leofwine *se Reade* 1016–20 OEByn (K); Aluric *þane Reda* c1100–30 ib. (D); William *Red* 1176 P (Gl), *le Red* 1332 SRSx; Hugo *le Rede* 1220 Cur (La); Hamo *le Reed* 1296 SRSx; Thomas *Read* 1327 SRSx. OE *rēad* 'red', of complexion or hair. *Reid* is Scottish. (ii) William *de Reved* 13th WhC; Ralph *de Rede* 1203 Cur (Herts); John *de Rede* 1327 SRSf. From Read (Lancs), earlier *Reved*, Rede (Suffolk) or Reed (Herts). (iii) Alwin *de Larede* c1160 MELS (Sx); Roger *de la Rede* 1208 P (D); Julian *atte Rede* 1296 SRSx; John *ater Rede* ib., identical with John *Rede* 1327 ib.; Thomas *atte Red* 1332 SRSx. From residence in a clearing, OE *rīed, *rȳd v.* RIDE.

**Readdie, Readdy, Readey:** *v.* READY

**Reader, Readers, Reder, Reeder, Reeders:** Adam, Symon *le Redere* 1279 RH (C), 1283 SRSf. A derivative of ME *redyn* 'to thatch with reed', 'a thatcher' (c1440 NED). Particularly common in Norfok (MESO 179). cf. THAXTER.

**Readett, Readitt, Redit, Reditt, Riddett:** Reymbro *ater Redette* 1296 SRSx. 'Dweller by the reed-bed', OE *hrēodet*.

**Readhead:** *v.* REDHEAD

**Reading, Readings, Redding, Reding, Reddin, Ridding, Riding, Rydings, Ruddin:** Grifin *del Ruding* 1246 AssLa; Richard *del Ryding* 1277 Wak (Y); Sara *de Redyngg* 1311 ColchCt; William *atte Rydyng* 1339 AssSt. 'Dweller in the clearing', OE *rydding*.

**Readman, Reedman, Redman:** Robert *Redeman* 1274 RH (Ess); John *Redman* 1275 RH (Ha); Norchinus *le Redman* 1332 SRLa; Richard *le Redemon* 1332 SRSt; Alexander *Reademan* 1474 GildY. 'Reed-man', either a cutter of reeds or a thatcher. cf. READER. Possibly also a nickname, 'red man'. cf. READ.

**Readwin, Redwin:** *Redewinus del Broc* 1185 Templars (Sa); Robert, Richard, John *Redwyne* 1254 ArchC 12, 1327 *SR* (Ess), SRSx; Simon *Redwin* 1275 RH (K). OE *Rǣdwine* 'counsel-friend', a rare OE name.

**Ready, Readey, Readdie, Readdy, Reddie, Reddy, Raddie:** Robert *le Redye* 1260 AssY; John *Rady* 1327 *SR* (Ess). ME *rēdi(ȝ), readi, redi* 'in a state of preparation, prompt, quick'.

**Readyhough:** *v.* RIDEALGH

**Readyman, Reddiman:** Roger *Redyman* 1327 SRSo; Adam *Rediman* 1334 SRK; Diota *redymon* 1348 DbAS 36. ME *redi* 'prepared, prompt, quick', and OE *mann*.

**Reakes, Reaks:** *v.* REEK

**Reaman:** *v.* RAYMAN

**Reames:** *v.* RAIMES

**Reaney:** Henry *de Ravenhowe* 13th PN WRY i, 340; Thomas *de Ranaw* 1379 PTY; John *Ranoe* 1433 Wheat; Robert *Reanye* 1577 Edmunds; George *Rayney* or *Rayner* 1725 ib. This surname, though not common, is a Sheffield name, always with this spelling. The correct pronunciation is *Rainey*, now being replaced by the spelling pronunciation *Reeney*. The above forms are all from around Penistone and Sheffield. The name derives from Ranah Stones in the township of Thurlstone in Penistone parish (WRYorks), called by T. W. Hall 'Raynah or Ranah, a farm-house in Thurlstone': *Reynoe* 1645 Bosville, *Renold stones* 1741 Edmunds. ON *hrafnhaugr* 'raven-hill'. *Rayner* above is a dialectal pronunciation.

**Reaney, Reanney, Reanny, Reany, Raney, Rainey, Reney, Reinny, Reyney, Rennie, Ryney:** Irish spellings of *Mac Raighne*. *v.* RAINEY.

**Reap, Reape, Reep, Reepe:** Roger, William *Repe* 1297 MinAcctCo (D), 1545 SRW; John *Reepe* 1642 PrD. Since the first named was paid *pro equis et aueriis non scriptis*, he was evidently concerned with horses, and was probably a carrier, *v.* RIPPER. A metonymic name from REAPER 'harvester' is much less likely.

**Reaper:** John, Mariota *Reper* 1327 SRSf. A derivative of OE *repan* 'to reap', reaper.

**Reardon, Riordan, O'Riordan:** Ir *Ó Ríoghbhardáin* 'descendant of *Ríoghbhardán*' (royal-bard).

**Reasbeck:** *v.* RAISBECK

**Reaves:** *v.* REEVES

**Reburn:** *v.* RAEBURN

**Reckless:** John *Recheles* 1273 IpmGl; Henry *Reklesse* 1477 IpmNt. A nickname from OE *recelēas* 'reckless, negligent'.

**Recks:** *v.* REX

**Record(s):** *v.* RICKWARD

**Red, Redd:** *v.* READ

**Redb(o)urn:** *v.* RADBOURNE

**Redcliffe, Redclift:** *v.* RADCLIFF

**Redbeard, Redbert:** Osbert *Redberd* c1200 Black; Simon *Redberd* 1246 AssLa; Richard *Redberd* 1327 SRSo; Hugh *Redeberd* 1379 PTY. 'Red beard', OE *rēad, beard.* cf. Richard *Redhond* 1354–5 FFSr 'red hand'; Thomas *Redhose* 1282 CtW 'red hose'; William *Redeknapp* 1464–5 FFSr 'red top'.

**Reddall, Reddell:** (i) Ralph *Redwald*, William *Redolf* 1276 RH (O); Richard *Redwal* 1297 MinAcctCo. OE *Rǣdweald*, or OE *Rǣdwulf.* (ii) Brun *de la Redeweld* 1296 SRSx. 'Dweller at the cleared woodland', OE *(ge)ryde, weald.* *v.* also RADWELL.

**Reddan, Redden, Reddin:** Richard *atte Redene* 1327 *SR*Ess; John *atte Reden* 1327 SRSx. 'Dweller at the clearing', OE *ryden.*

**Reddaway:** *v.* RADWAY

**Reddecliff, Reddicliffe:** *v.* RADCLIFF

**Reddell:** *v.* REDDALL

**Redden:** *v.* REDDAN

**Reddick:** John *de la Redewyke* 1271 AssSo. From Redwick (Gl).

**Reddie:** v. READY

**Reddiford:** v. RADFORD

**Reddihough, Reddyhoff:** v. RIDEALGH

**Reddiman:** v. READYMAN

**Reddin:** v. REDDAN

**Redding:** v. READING

**Reddish, Redish, Raddish, Radish:** John *de Reddich* 1202 FFL. From Redditch (Lancs, Worcs).

**Redfearn, Redfern, Redfarn:** Henry *de Redefern* 13th WhC; William *del Redferne* 1325 LaCt; Richard *Redfern* 1532 CorNt. From Redfern (La).

**Redford:** v. RADFORD

**Redgewell:** v. RIDGEWELL

**Redgrave, Redgraves, Redgrove:** Ebrard *de Redegraue* 1179 P (Sf); John *de Redgrave* 1312 LLB D; John *Redgrave* 1425 FFEss. From Redgrave (Sf).

**Redhead, Readhead:** Adam *Redhed* 1256 AssNb; John *Redheved* 1279 RH (C). 'Red head' (OE *rēad, hēafod*).

**Redihalgh:** v. RIDEALGH

**Redhood:** William *Redhod* Hy 3 IpmY; Richard *Redhod* 1314–16 AssNth; Henry *Redhod* 1396 CorGl. 'The wearer of a red hood', OE *rēad, hōd*. The surname has not survived, probably because it has fallen in with REDHEAD.

**Redit(t):** v. READETT

**Redley:** Thomas *de Redleia* 1242 AssDu; Roger *de Reddeleg* 1274 RH (Ess); Richard *Redley* 1312 IpmW. From Ridley (K), *Redlege* DB, or Ridley Hall in Terling (Ess), *Redleigh* 1385.

**Redman:** v. READMAN, REDMAYNE

**Redmayne:** Norman *de Redeman* 1188 P (La); William *Redmaine* 1674 FrY. From Redmain (Cumb).

**Redmond:** Ir *Réamonn* or *Mac Réamoinn* '(Son of) *Remund*' (Raymond).

**Redmore, Redmoor:** William *de Redmore* 1251 ElyCouch; John *de Redmar* 1276 IpmY; Adam *de la Redemere* 1292 MELS (So). From Redmere (C), Redmoor Bank in Elm (C), or Redmere in Owthorne (ERY).

**Rednall, Rednell:** Stephen *de Redenhale* 1327 SRSf; John *Redenhale* 1376 AssEss; Thomas *Rydnale, Redenhale, Rydnall* or *Redenall* 1509 LP (Sf). From Rednal (Wo).

**Redrose:** William *Rederose* 1301 SRY. Apparently a nickname 'red rose', but the reason for it is unknown. cf. Joan *Dubblerose* 1360 ColchCt 'double rose'; Henry *Woderose* 1332 SRSx 'wild rose'.

**Redshaw:** Nicholas *de Redschaghe* 1297 MinAcctCo (Y); Richard *de Redeshagh* 1379 PTY; John *Redeshawe* 1465 TestEbor. From Redshaw Gill in Blubberhouses (WRY).

**Redvers:** Richard *de Reueris* 1084 GeldR (Do), *de Reduers* 1086 DB (Do); Baldwin *de Reduers* 1135 ASC; Margery *de Redvers* 1280 IpmY. From Reviers (Calvados). v. OEByn 109.

**Redway:** Walterus *filius Radewi, Redwi* 1165, 1169 P (L); Willelmus *filius Redwi* 1185 P (Wo); Gilbert, Andrew *Redwy* 1221 AssWo, 1279 RH (O). Late OE *Rǣdwīg* 'counsel-warrior'.

**Redwin:** v. READWIN

**Redwood:** John *de Redewod'* 1242 Fees (Nb); Nicholas *Redewood* 1527 FFEss; John *Redwood* 1642 PrD. From Redeswood (Nb).

**Ree:** v. REA

**Reece:** v. RHYS

**Reecks:** v. REEK

**Reed:** v. READ

**Reeder(s):** v. READER

**Reedland, Ridland, Rudland, Rudlen:** Laurence *atte Rydelonde* 1296 SRSx; Richard *atte Redelande* 1363 PN Sx 114; Andrew *Rudland* 1524 SRSf; From Redland (Gl), Redlands Fm in North Chapel (Sx), or 'dweller at the cleared land', OE *(ge)rydde, land*.

**Reedless:** Thomas *Redeles* 1524 SRSf. 'The ill-advised', OE *rǣd, lēas*.

**Reedman:** v. READMAN

**Reek, Reeks, Reecks, Reakes, Reaks, Reekes:** Nicholas *atte Reke* 1333 MELS (So). 'Dweller by the heap or stack', OE *hrēac*.

**Reeman:** v. RAYMAN

**Reep(e):** v. REAP

**Rees(e):** v. RHYS

**Reeson:** Roger *Reuesone* 1296 Crowland (C); Adam *le Reuessone* 1327 SRSf. 'The reeve's son.'

**Reeve:** Walter, James *le Reve* 1220 FrLeic, 1281 LLB A. OE *(ge)rēfa* 'reeve'.

**Reeves, Reaves:** (i) Richard *del Reves* 1332 SRLa. 'Servant at the reeve's (house).' (ii) John *atte Reuese* 1327 SRWo. 'Dweller at the border' of a wood or hill (OE *efes* 'edge'; ME *atter evese*). v. EAVES.

**Regan:** v. O'REGAN

**Reid:** v. READ

**Reigate, Reygate:** Agnes *de Reigat* 1209 Cur (Bk); Stephen *de Reygate* 1275 RH (W); Reginald *Reygate* 1296 SRSx. From Reigate (Sr).

**Reiner:** v. RAYNER

**Reinny:** v. RAINEY

**Relf, Relfe, Relph, Realff:** *Richolf* de Gameltun' 1210 Cur (Y); *Riculfus* 1212 ib.; *Ricolfus filius Ailwini* 1219 Cur (Sx); Robert *Reolf* 1296 SRSx; John *Relf* 1327 ib. Probably OFr *Riulf*, from OG *Ricwulf*.

**Remfrey, Remfry, Renfree:** *Rainfridus* 1086 DB (L); *Reinfridus* 12th DC (L); Gilebertus *filius Reimfridi, filius Reynfrey* 1213 Cur (Cu), 1230 P (La); Roger *Reymfrey* 1221 Cur (So). OG *Raganfrid, Rainfrid* 'might-peace'.

**Remington, Remmington, Rimington, Rimmington, Riminton:** Goda *de Rimington'* 1219 AssY; Henry *de Rymington* 1297 SRY; Matilda *Rymyngton, de Remyngton* 1379 PTY; Robert *Remington, Remton* 1649–50 WRS. From Rimington (WRY), *Remington* 1303.

**Renaud, Renaut:** v. REYNOLD

**Renbold:** v. RAINBOW

**Rendall, Rendell, Rendel, Rendle:** (i) Thomas *Rendell*, Richard *Rendle* 1642 PrD. A pet-form of *Randolph* or of *Reynold*. (ii) Rechinald *de Rayndel* 1325–6, James *Rendall* 1516 Black. From Rendall (Orkney).

**Reney:** v. RAINEY

**Renfree:** v. REMFREY

**Rengger:** v. RINGER

**Renham:** John *de Renham* 1214 Cur (Ess); John *de Renham* 1327 SRSx. From Rainham (Ess), *Renham* 1192, or Rainham (K), *Renham* 1130.

**Renhard, Rennard:** *v.* REYNARD

**Rennell:** *v.* REYNOLD

**Renner:** Aluuinus *Rennere* c1134 Black (Govan); Richard *Renner* 1319 *SR* (Ess); John *le Renner* 1340 MEOT (La). A derivative of OE *rennan* 'to run', 'runner', probably a messenger or courier. *v.* also RAYNER.

**Rennick:** *v.* RENWICK

**Rennie, Renney, Renny:** *v.* RAINEY

**Rennison, Renison, Rennilson, Renilson, Renson:** John *Renisson* 1327 SRC; John *Rendelson* 1694 Black. 'Son of *Reynold.*'

**Renshall, Renshell, Renshaw:** *v.* RAVENSHAW

**Renter:** Ralph *le Renter* 1343 Clerkenwell; Bartholomew *le Renter* 1365 LLB F; Robert *Renter* 1396 PN Ess 309. Probably for 'rent-gatherer'. cf. Reginald *Rentegaderarr* 1367 FrNorw.

**Renton:** Robert *de Rentun* c1225, John *de Raynton* 1323, Isobel *Renton* 1548 Black; Robert *Renton* 1671 PN WRY v 125. From Renton in Coldingham (Berwick), or Rainton (Du, NRY).

**Renwick, Rennick:** Rannulf *de Rauenwic* 1191 P (Cu). From Renwick (Cu), *Rauenwich* 1178.

**Renyard:** *v.* REYNARD

**Reppes, Repps:** John *Reppes* 1408 FFEss; Robert *Reppys* 1451 Paston. From Repps (Nf).

**Reppington, Rippington:** John *Rependene* 1398 *ERO*; William *Repyngton* 1434–5 FFWa. From Repton (Db), *Repandon* 1197.

**Repps:** *v.* REPPES

**Reresby:** Ysoria *de Reresby* 1206 AssL; Ralph *de Reresby* 1269 FFY; Ralph *Reresby* 1458–9 IpmNt. From Reasby (L), *Reresbi* DB.

**Ressler:** *v.* RESTLER

**Rest:** Reste c1135 DC (L); Walterus *filius Reste* 1176 P (L); John *Rest* 1447 LLB K. A short form of OG *Restold. v.* RESTALL.

**Restall, Restell:** *Restaldus, Restolt* 1086 DB; *Restold'* 1130 P (O); Gillebertus *filius Restaldi* 1206 Cur (L); Hugh, Peter *Restwald* 1279 RH (O). OG *Restold, Restald,* a Frankish name, perhaps of Celtic origin.

**Restler, Ressler:** Robert, William *le Wrastler* 1317 MEOT (O), 1332 SRWa; Henry *Wrasteler* 1440 ShefA. OE *wrǣstlere* 'wrestler'.

**Reston:** (i) Richard *de Riston'* 1219 AssL; Walter *de Reston* 1327 SRSf. From North, South Reston (L), *Ristone* DB. (ii) Roger *de Reston* 1166, John *de Restoun* 1417 Black. From Reston in Coldingham (Berwick).

**Retchford:** *v.* ROCHFORD

**Retford:** *v.* RADFORD

**Retter:** Alice *le Retour* 1279 RH (O); William *Retour* 1332 SRSx. OFr *retier* 'net-maker'.

**Revan, Revans, Revens:** *v.* RAVEN

**Revel, Revell, Revill, Reville, Reavell:** *Reuel* de Teteneia Hy 2 DC (L); Paganus *revellus* 1130 P (Herts); Robert, Hugh *Reuel* 1177, 1196 P (Ess, Nth); Richard *Revel(l), Rivel(l)* 1210 Cur (So). OFr *revel* 'pride, rebellion, sport', from *reveler* 'to rebel'. The font-name *Revel* was common in OFr and ME and may actually be from Lat *rebellus.* The surname is also metonymic for *Reveler* which survives.

**Revere:** (i) Paul *Reyvere* 1255 RH (Bk); William *le Reuere* 1316 FFK; John *Reuere* 1375 ColchCt. ME *revere* 'reiver, robber'. (ii) John *ater Reure* 1296 SRSx; Robert *atte Revere* 1327 SRSx; Matilda *atte Riuer* ib.; Adam *atte Eure* ib. 'Dweller on the slope or brow of the hill' (OE *\*yfer* 'edge').

**Revis, Rivis:** Adam *de Rievalle, de Ryvaus* 1333 Riev (Y); Richard *Ryvax* 1414 GildY; Thomas *Revis, Rivice* 1588, 1618 FrY. From Rievaulx (Yorks), pronounced *Rivers.*

**Revnell:** *v.* RAVENHALL

**Rew, Rue:** Robert *ate Rewe* 1297 MinAcctCo; William *in therew* 1327 SRSo. 'Dweller in the row of houses or in the street' (ME *rew,* OE *rǣw*), as at Rewe (Devon). cf. RAW.

**Rewald, Rewell:** *Ruald* 1148 Winton (Ha); *Rualdus* de Wodecot' 1206 Cur (Ha); Walter *Rewalt* 1308 AssNf. OG *Hrod(w)ald, Roald.*

**Rex, Recks:** John *Rex* 1279 RH (C); Richard *Rex* 1327 SRWo; William *Reckes* 1576 SRW. 'Dweller by the rushes', ME *rexe, rixe.*

**Rey:** *v.* RAY, REA

**Reyburn:** *v.* RAEBURN

**Reygate:** *v.* REIGATE

**Reyna:** For RAYNER or RAINEY

**Reynard, Renhard, Rennard, Renyard:** Rogerus *filius Rainardi, Rainart* 1086 DB (Nf); Elias *Reynardi* 1205 Holme (Nf); Henry *Renard* 1325 Orig (Ha). OFr *Reinart, Renart,* OG *Raginhard, Rainard* 'counsel-brave'.

**Reyner:** *v.* RAYNER

**Reynes:** *v.* RAINES

**Reynold, Reynolds, Reynalds, Reynell, Renaud, Renaut, Renals, Rennell, Rennels, Rennoll:** Willelmus *filius Rainaldi* 1086 DB; *Reinaldus* camerarius 1121–48 Bury (Sf); *Rennaldus* filius Berwaldi Hy 2 DC (L); *Renalt* de Saint Leger ib.; William *Reynaud* 1272 HPD (Ess); John *Reynald* 1275 SRWo; Henry *Reynel* ib.; Adam *Renaud* 1297 MinAcctCo; William *Reynold* 1299 FFEss; John *Rennold* 1327 SRC; John *Rennels* 1788 Bardsley. OFr *Reinald, Reynaud,* OG *Raginald* 'counsel-might', latinized as *Reginaldus.* Some of the numerous instances of *Rainald* in England may be from ON *Ragnaldr,* but most were introduced from France and Normandy where both the OG and ON forms contributed to its popularity.

**Reynoldson:** Richard *Raynoldson* 1379 PTY. 'Son of *Reynold.*'

**Rhodes, Rhoades, Road, Roads:** Hugh *de Rodes* 1219 AssY; Alexander *de la rode* 1277 Ely (Nf); John *atte Rode* 1294 RamsCt (Beds); Robert *del Rodes* 1332 SRLa; Robert *Roades, Rhoades* 1660, 1676 FrY; Thomas *Rhodes* 1678 ib. 'Dweller by the clearing(s)', OE *rod(u),* as at Roade (Northants), Rhode (Devon), etc.

**Rhys, Reace, Reece, Rees, Reese, Rice:** *Hris* 1052 ASC C; *Rees* 1066 DB (Ch); *Resus* filius Griffini 1178 P (D); Griffinus *filius Res, Ris* 1198–9 P (Sa, Gl); *Rice* or *Rise* Powell 1570 Oxon; William *Res* 1203 Cur (L); John *Rees* 1288 FFSf; Walter *Rys* 1327 SRWo; John *Ryce* 1524 SRSf. OW *Rīs,* Welsh *Rhys* 'ardour'.

**Ribald, Ribell, Ribble:** Ralph *filius Ribaldi* 1159 P (Nf); Folco *Ribald* 1165 P (L); William *Ribald* 1198 FFNf; Richard *Ribald* 1230 Glast (So); John *Ribald* 1373–5 AssL. OG *Ribald, Ripald. v.* Forssner 213.

**Ribber:** Thomas *le Rybbere* 1311 FFSf. ME *ribbere* 'flax-scraper'.

**Ribble, Ribell:** *v.* RIBALD

**Ribson:** Robert *Rybson* 1327 SRY. Probably from Ribston (WRY).

**Ricard, Riccard:** *v.* RICHARD

**Rich, Riche, Riches, Ritch:** (i) Godwinus *Diues, le Riche* 1177, 1185 P (L); Mosse *le Riche* 1195 P (Gl); William *Riche* 1296 SRSx. OFr, ME *riche* 'rich'. (ii) Ricardus *de la Riche* 1200 P (Ha); Alexander *atte Riche* 1296 SRSx. 'Dweller by the stream' (OE *ric), as at Glynde Reach (Sussex). *v.* MELS 166. (iii) *Riche* Algar 1296 SRSx; Richard *Ryches* 1296 SRSx. A diminutive *Rich* from *Richard* to which some examples without the article belong, e.g. Adam *Rych* 1281 FFSf. cf. RICK.

**Richard, Richardes, Richards, Ritchard, Ricard, Ricarde, Ricards, Riccard, Rickard, Rickardes, Rickards, Rickeard, Rickerd, Rickert, Ricket, Rickets, Rickett, Ricketts:** *Ricard* 1066 DB; *Richardus* Basset 1127–34 Holme (Nf); *Ricard* de Linlee c1166 NthCh (Sa); Juliana, Thomas *Richard* 1276 RH (O), 1296 SRSx; Adam *Ricard* 1327 SRSo; Thomas *Richardes* 1327 SRWo; William *Ricardes* 1327 SRSo; Hester *Rickett* 1606 Bardsley; David *Rickart* 1651 Black. AN *Ricard*, CentralFr *Richard*, OG *Ric(h)ard* 'powerful brave', one of the most popular names introduced by the Normans. Usually latinized as *Ricardus*, the common form was *Ricard* whence the pet-form *Rick*, etc. *Richard* has given *Richie, Ritch*, etc. For other pet-forms, *v.* HICK, HITCH, DICK, with their derivatives. cf. also the double diminutive *Richelot* 1188 P (L), 1219 AssY, Robert *Richelot, Rikelot* 1191–2 P (L); Richard *Rykelot* 1315 FFEss. *Ricket*, often a late development of *Rickard*, may also be from the diminutive *Ric-ot: Ricot* Huot 1327 SRSx. *Richard* occasionally interchanges with *Richer: Ricardus, Richerus* filius Bondi, filius Stannardi 1134–45 Holme (Nf). *v.* also HUDD, RICKWARD.

**Richardson, Richarson:** Murdac *Richardesson* 1359 Black; William *Richardson* 1381 PTY; Agnes *Richarson* 1470 GildY; George *Richison* 1683 RothwellPR (Y). 'Son of *Richard*.'

**Richell, Rickel, Rickell, Rickells:** *Rikilda* 1197 FF (Sf); Willelmus *filius Richild, Rikilde* 1212 Cur (Herts); Thomas *Richil* 1275 RH (Nb); Geoffrey, Hugh *Rikild* 1279 RH (Hu, O); Robert *Rykeld* 1279 RH (O); Walter *Rychyld* 1297 MinAcctCo; Thomas *Rikel* 1327 SRSx. OG *Richild, Richeldis* (f), Fr *Richeut.* cf. RICHOLD.

**Richer, Reacher, Ricker, Rickers:** *Richerius* clericus, *Ricerus* 1086 DB (Sf); *Richer* de Brunho 1219 AssY; William, Geoffrey *Richer* 1236 Oseney (O), 1260 AssC; Isabell' *Ryker* 1332 SRSx. OG *Richer(e)*, *Riker*, OFr *Rich(i)er*, *Ricier*.

**Richey, Richie(s):** *v.* RITCHIE

**Richings, Richins, Richens:** Thomas *filius Ricun* 1274 RH (Hu); Richard *Rikon* 1327 SRSo. *Ric-un*, a diminutive of a short form of *Richard*.

**Richlord:** William *Richeloverd* 1221 Cur (Nth). 'Rich, powerful lord', OFr *riche*, OE *hlāford.* cf. William *filius Richefemme* 1148 Winton (Ha) 'noble woman'; Nicholas *Richeangod* 1317 AssK 'rich and good'.

**Richman:** *Richeman* Molendinarius 1205 Cur (Beds); *Richeman* Brutte 1280 AssSo; Richard *Richemannus* 1212–23 Bart; Ralph *Richeman* 1275 RH (Lo); Thomas *Richeman* 1382 IpmGl. Probably an unrecorded OE *Rīcmann, but William *le Richeman* 1199–1200 FFSf would suggest that it is sometimes a nickname, 'rich, powerful man'.

**Richmay:** Domina *Richemaya* 1240 Colch (Ess), *Richemeya* de Crek' 1240–1 FFEss; William *Rechemay* 1240 Colch (Ess). OE *Rīcmǣge* (f).

**Richmond, Richmont:** Roger *de Richemund* 1199 CartAntiq; Adam *de Richemond* 1296 SRNb; William *Richemound* 1326 CorLo. From Richmond (Surrey, NRYorks), but early examples may be from one of the many French places of this name.

**Richold:** *Ricolda* mater Baldeuini 1151–4 Bury (Sf); Johannes nepos *Richold* 1177 P (L); Osbertus *filius Rikoldi* 1208 P (Y); *Richolda* uxor Pagani 1213 Cur (Sr); Walterus *Ricaldes* c1100 MedEA (Nf); Ralph *Richold* 1297 MinAcctCo. *Richolda* is from OG *Richoldis* (f), a form of *Richild*, which survives as *Richell* and *Rickell(s)*. Some of the forms above are from OG *Ricoald, Ricold, Richold* (m), which has certainly contributed to *Richold*, and probably also to *Rickell(s)*.

**Rick, Ricks, Rix:** (i) *Rike* 1260 AssCh; Ralph *Rixe* 1279 RH (C); Ema *Ricke*, David *Rickes* 1327 SRSf; Richard *Rycke*, Milcentia *Ryckes* 1330, 1351 ColchCt. *Rick*, a pet-form of *Ricard. v.* RICHARD. (ii) Osbert, John *de la Rixe* 1274 RH (So); Roger *ate Rixe* 1302 PN D 545; William *atte Ryxe* 1333 MELS (So). 'Dweller by the rushes', from West Saxon *rixe, rexe*, a metathesized form of OE *risc, rysc* 'rush', which survives as *rix, rex* in the dialects of Dorset, Somerset and Devon. cf. Rix, Rixdale (Devon). The plural has become RIXON.

**Rickaby, Rickerby:** Alan *de Ricardby* 1332 SRCu. From Rickerby (Cu).

**Rickard(s), Rickeard:** *v.* RICHARD

**Rickcord:** *v.* RICKWARD

**Rickel, Rickell(s):** *v.* RICHELL, RICHOLD

**Ricker(s):** *v.* RICHER

**Rickerd, Rickert, Rickett(s):** *v.* RICHARD

**Rickman:** (i) John *filius Rikeman* 1279 RH (Hu); Richard *Richeman, Rikeman* c1216–20 Clerkenwell; Robert *Rykeman* 1298 AssL; William *Rykeman* 1327 SRSx, cf. Agnes *Rikemannes*, widow of *Rickeman* le Chaumberleng 1329 Husting. A variant of RICHMAN. (ii) Richard *Rikemund* 1275 RH (Sf). OE *Rīcmund.

**Rickmay:** Colle *Rikmai* 1247 AssBeds; John *Rykmayde*, Thomas *Rikmay* 1524 SRSf. A variant of RICHMAY.

**Rickson:** *v.* RIXON

**Rickward, Rickword, Rickwood, Rickcord, Record, Records:** *Richewardus* de Westberi 1190 P (Bk); *Rikeward* Kupere 1296 SRSx; Robert *Rikeward* 12th DC (L); Walter *Rykeward* 1275 RH (Nf); Roger *Recard* 1393 NorwW (Sf); John *Recorde* 1409

RochW; William *Rekerd* 1455 NorwW (Nf); John *Record*, *Rikecord*, *Rikeworth* 1595, 1599, 1601 Bardsley. OFr *Ricoart*, OG *Ricward* 'powerful-guardian'. *Recard* might also be from *Ricard* (Richard): *Rec'* pentecoste, *Rec'* bene 1292 SRLo. *Rickwood* may also, occasionally, be from OG *Ricoald* 'powerful-might': *Rikewaud'* 1215 Cur (Sx). cf. RICHOLD.

**Ridal, Rideal, Riddall, Riddell:** Alan *de Ridale* c1160–75 YCh; Nicholas *de Ridal* 1222 FFY; William *Rydale*, *Riddall* 1338 FFEss, 1672 HTY. From Rydal (Westmorland), or Ryedale (WRYorks). *v.* also RIDDELL.

**Riddell, Riddel, Riddall, Riddle, Riddles:** *Ridel* Papillun 1163 P; *Ridel* de Kisebi e Hy 2 DC (L); Goisfridus, Geva *Ridel* 1086 DB, 1138–47 P (Ch); William *Ridell'* 1205 Pl (Nth); Reginald *Ridel* 1327 SRLei. A nickname from OFr *ridel* 'a small hill', though its exact significance as a personal name is obscure, *v.* OEByn 380.

**Riddett:** *v.* READETT

**Ridding:** *v.* READING

**Riddiough:** *v.* RIDEALGH

**Riddlesden:** Thomas *Ridilsden* 1379 PTY; John *Ruddilsdene* 1556 WRS. From Riddlesden in Morton (WRY).

**Ride, Ryde, Rude:** Roger *de la Rude* 1176 P (Sr); Robert *de la Ryde* 1294 LLB C; Richard *ate Rude* 1297 MinAcctCo. 'Dweller in the clearing', OE *\*rīed*, *\*rȳd*. Survives also as READ.

**Ridealgh, Ridehalgh, Redihalgh, Reddihough, Reddyhoff, Readyhough, Riddiough:** William *de Redihalgh* 1324 LaCt; Robert *del Riddyough* 1397 PrGR; Edward *Riddihough*, *Riddihalgh* 1682 ib. From Higher Ridihalgh (Lancs).

**Rideout, Ridout, Ridoutt:** Elyas *Rydhut* 1274 RH (So); Ellis, John *Ridut* 1276, 1278 AssSo; William *Rydhowt* 1379 PTY. Probably 'ride out', a nickname for a rider. cf. Adam *Rideway* 1219 AssY.

**Rider, Ryder:** John, Thomas *le Rider* 1204 AssY; Richard *Ridere* 1212 Fees (Cu). Late OE *rīdere* 'rider', probably 'knight, mounted warrior' (c1085 NED). Also 'dweller by the clearing' *v.* RIDE). Henry *at Ryde* is identical with Henry *Ryder* (1599 PN Sr 149).

**Ridge:** Geoffrey *de la Rigge* 1166 P (Ha); Edith *atte Rigge* 1327 SRSo. 'Dweller on the ridge', OE *hrycg*. *v.* also RUDGE.

**Ridgeman, Ridgman:** Richard *Rigeman* 1197 P (Nf/ Sf); John *Regman* 1332 SRSx. 'Dweller at the ridge', OE *hrycg*, *mann*.

**Ridgeway, Ridgway:** Gilbert *de Ruggeweie* a1183 EngFeud (Gl); Walter *de Regewey* 1262 ArchC iii; William *de Rygeway* 1310 DbCh. 'Dweller by the ridgeway' (OE *hrycgweg*) or from a place named Ridgeway.

**Ridgewell, Ridgwell, Redgewell, Redgwell:** Thomas *de Redeswell* 1281 FFEss. From Ridgewell (Essex).

**Riding:** *v.* READING

**Ridland:** *v.* REEDLAND

**Ridler:** Geoffrey *le ridelere* 1230 P (So); Andrew *le Rydelere* 1294 RamsCt (Beds). A derivative of ME *rid(e)len* 'to sift', from OE *hriddel* 'a sieve', a sifter of corn or a sifter of sand and lime in making mortar.

**Ridley:** Elyas *de Redleg'* 1227 FFK; Alexander *de Rydeleye* 1279 AssNb; Henry *Rydleye* 1300 IpmGl. From Ridley (Ches, Essex, Kent, Northumb).

**Ridout:** *v.* RIDEOUT

**Rigby, Rigbey:** Gilbert *de Rigebi* 1208 P (L); Henry *de Ryggeby* 1285 AssLa; Thomas *Rygby* 1453 FFEss. 'Dweller at the farm on the ridge', ON *hryggr* and *bý*.

**Rigden:** Richard *Rikedoun* 1317 AssK; Reginald *Rykedoun* 1327 SRSx; William *Rigden*, 1557 ArchC 34. This name must be taken with HIGDON, earlier *Hikedun*. Both names look like place-names but neither is ever found with a preposition and no place-names of similar form are known. *Ricardun* is a diminutive of *Ricard* (Richard). The -*ar*-, being unaccented, was weakened to -*e*- and then lost, giving *Ricdun*. This, with partial assimilation, would become *Rigden*. *Hikedun* is a variant based on the rhymed *Hick*.

**Rigg, Rigge, Riggs:** William (*de*) *Rigge* 1197–8 P (Sa); John *del Rigg* 1332 SRCu. 'Dweller by the ridge', ON *hryggr*.

**Rigglesford, Riggulsford:** *v.* WRIGGLESWORTH

**Righton:** *v.* WRIGHTON

**Rigley:** *v.* WRIGLEY

**Rigmaiden:** John *de Ryggemayden* 1285 AssLa. From Rigmaiden (Cumb).

**Rigsby:** John *de Ryggesby* 1271 FFL; Thomas *de Rigesby* 1398, Richard *de Riggesby* 1385 AssL. From Rigsby (L).

**Rigton:** Ranulf *de Rigetun* c1175 MCh; John *de Rigton* 1302 IpmY; Robert *de Rigton* 1363 FFY. From Rigton (WRY).

**Riley, Ryley:** John *de Ryeley* 1284 Wak; Henry *de Ryley* 1327 SRDb; John *Ryley* 1488 CorNt. From Riley (Devon), High Riley (Lancs), or 'dweller at the rye clearing'.

**Rimell:** Elias *filius Rimilde* 1201 Cur (So); *Rimilda* Vidua 1201 ib. (Ha); Rogerus *filius Rimilde* 1203 ib. (Nf); Robert *Rimel* 1327 SRWo; Agnes *Rymyld* 1332 SRWa. OE *Rimhild* 'border-war', a rare and late woman's name.

**Rimer, Rimmer, Rymer:** Warin *Rymer* 1229 Pat (Y); Richard *le Rimour* 1277 WhC (La). A derivative of ME *rimen* 'to rime' or AFr *rimour*, *rymour* 'a rimer, poet'.

**Rimington, Rimmington, Riminton:** *v.* REMINGTON

**Ring, Ringe, Rings:** Eilwinus, Robert *Ring* 1207 ChR (Nf), 1279 RH (C). Probably metonymic for *Ringer* (ii) 'bell-ringer', or, perhaps, for a maker of rings, jeweller.

**Ringbell:** Henry *Ringebell* 1275 RH (Sf); John *Rynggebelle* 1327 SREss; Robert *Ryngbell* 1491 LLB L. 'Ring bell', OE *ringan*, *belle*, metonymic for a bell-ringer.

**Ringell, Ringle:** (i) *Ringold* filius Alhich 1188 P (L); William *Ringeld* 1327 SRSf. OG *Ringold*. (ii) *Ringulfus* (Nf), *Ringul* (Sf) 1066 DB; Godric *filius Ringolf* c1250 Rams (C); John *Ryngulf* 1275 SRWo; Agatha *Ryngolf* 1316 AssNf; Robert *Ringolfe* 1329 NorwDeeds II. OE *Hringwulf*.

**Ringer:** (i) Two examples of OE *Hringhere* have been noted: *Ringer* 12th Searle 562; *Ringer'* filius Alani 1219 AssL. There is no evidence for the formation of a surname from this, but it might have contributed to *Ringer* which has other origins. (ii) Hugo (*le*) *Ringere* 1207 Cur (Sf); William *le* *Ringere* 1327 SRSf. A derivative of OE *hringan* 'to ring', 'a bell-ringer' (c1425 NED). cf. Henry *Ringebell* 1275 RH (Sf). (iii) Richard *le Wringar* 1327 SRSx. A derivative of OE *wringan* 'to wring, squeeze', perhaps a wringer or presser of cheese. cf. Richard *le Chesewryngere* 1281 MESO (L).

**Ringer, Rengger, Rainger:** Radulfus *filius Raingeri* 1147–53 DC (Nt); John *Rynger* 1319 *SR* (Ess). OFr *Rainger*, from OG *Reginger*. Richard *Renger*, alderman, sheriff and mayor of London, was the son of *Rainger* (Ricardus *filius Reinger* 1221 StP) and is usually called Richard *Reinger* or *Renger* (1225 Pat), having adopted or acquired his father's name as his surname. This at once became hereditary, for his sons, Richard and John, are both called *Renger* in 1235 (FFEss). The family bought an estate in Essex which is called *Renggers* in 1318, the 'inheritance of John *Renger*' in 1277, and is now represented by Ringer's Farm in Terling (PN Ess 297).

**Ringland:** Agnes *of Ringland* 1316 AssNf; William *Ringelond* 1327 SRC. From Ringland (Nf).

**Ringle:** v. RINGELL

**Ringrose:** John *Ringerose* 1259 NorwDeeds; John *Ringros* 1332 SRCu; Thomas *Ryngotherose* 1332 SRLa; William *Ringrose* 1535 LWills. A nickname 'ring rose', the exact significance of which is obscure, cf. PLUCKROSE. The forms lend no support to Harrison's derivation of the second element from OE *ráew* 'row'.

**Ringshall, Ringshaw:** Godric *de Ringeshale* 1066 DB (Sf); Brian *de Ringeshal* 1275 RH (Sf); Andrew *de Ryngeshagh* 1327 SRSf. From Ringshall (Sf).

**Ringstead, Ringsted:** John *de Ringested'* 1210 Cur (Nf); John *de Ringstede* 1284 RamsCt (Hu); Alexander *de Ringstede* 1299 AD v (Nf). From Ringstead (Do, Nf, Nth).

**Ringston:** Richard *Ringstan* 1167 P (Ha); William *Ringstan* 1210 P (W). OE *Hringstān*.

**Rington:** v. WRINGTON

**Ringwood:** Thomas *Ryngwood* 1642 PrD. From Ringwood (Ha).

**Riordan:** v. REARDON

**Ripley:** Bernard, Roger *de Rippeley* c1175–83 YCh, 1242 Fees (Nb); Richard *Rypplay* 1381 PTY. From Ripley (Derby, Hants, Surrey, WRYorks).

**Ripon, Rippon:** Robert *de Ripun* c1154–7 MCh; Henry *of Ripun* 1268 FFY; Peter *Rypon* 1402–3 FFWa; John *Ripon* 1437–8 FFSr. From Ripon (WRY), or Rippon in Hevingham (Nf).

**Ripp, Rippe:** Eborard *de la Ripe* 1221–30 MELS (Sr); Henry *atte Ryp* 1296 SRSx; William *Ryppes* 1524 SRSf. 'Dweller by the strip of woodland', OE *rip*. v. MELS 166–7.

**Ripper:** Adam *le Ripier(e)* 1279 RH (O); Gilbert *le Riper* ib. (Beds). A derivative of ME (*h)rip* 'basket', maker or seller of baskets. *Rypier* was used in 14th century London of those who brought fish from the sea for sale in the city.

**Rippington:** v. REPPINGTON

**Rippon:** v. RIPON

**Risbridger, Rusbridger:** John *Rysebrigger* 1497 SRSr. 'Dweller by the brushwood-bridge or causeway' (OE *\*hrīs-brycg*), as at Rice Bridge (Sussex), Risebridge (Essex), Rising Bridge (Northants), Ridgebridge (Surrey), etc.

**Risby:** Wlfric *de Rysebi* 1112 Bury; Roger *de Rizebi* 1167 P (Y); Ralph *de Risebi* 1202 AssL. From Risby (L, Sf, ERY).

**Risden, Risdon:** Simon *de Risedene* 1198 P (K); William *Risdon* 1642 PrD. From Riseden in Goudhurst, Risden in Hawkhurst (K), or Risdon in Jacobstow, in Bratton Clovelly (D).

**Rise:** William *de Ris* 1210 Cur (Nt); Robert *del Rys* 1332 SRLa. From Rise (ERYorks) or 'dweller in the brushwood', OE *hrīs*.

**Risely, Risley:** Robert *de Ryslegh* 1284 AssLa; William *de Riseleye* 1327 SRDb; John *de Ryslee* 1375–6 FFWa; John *Riseley* 1417 IpmY. From Risley (Db, La), or Riseley (Beds, Berks).

**Risher:** v. RUSHER

**Rishworth:** v. RUSHWORTH

**Rising:** Hugh *de Rising* c1227 Fees (Nf); Nicholas *Rysing* 1327 SRSx; John *Rysing* 1465 Paston. From Castle, Wood Rising (Nf).

**Risley:** v. RISELEY

**Riston:** William *de Ryston* 1252 FFO; John *de Riston* 1312–13 NorwLt. From Long Riston (ERY), or East, Sco Ruston (Nf), *Ristuna* DB.

**Ritch:** v. RICH

**Ritchard:** v. RICHARD

**Ritchie, Richey, Richie, Richies:** Duncan *Richie* 1505 Black. A Scottish diminutive of *Rich* (Richard).

**Rivel, Rivell:** William *Riuel* 1198 FFC; Payne *Rivel* 1232 FFEss; Thomas *Ryuel* 1339 FFEss. A variant of REVEL. cf. Curry Rivel (So), from Richard *Revel* Ric I.

**Rivers:** Gozelinus *de Lariuera* 1084 GeldR (W), *de Riuere, de Riuaria* 1086 DB (So); Gozelinus *Riueire, Riuere* ib.; Walter *de la Rivere* c1150 Riev (Y); John *de Rypariis* 1286 FFEss. From La Rivière (Calvados, Pas-de-Calais) or other places of the same name. Tengvik suggests also a possible derivation from OFr *rivere*, 'dweller by the river'. How far later examples are from this or from a French place-name can only be decided by detailed investigation in each case. It is noteworthy that the preposition is always *de la*. For Matill' *atte Riuer*, John *Riuer* 1327 SRSx, a different etymology is essential. v. REVERE.

**Rivett:** John *Ryuet* 1327 SRSf. ME *ryvet, revette* 'a rivet', metonymic for *riveter*: Richard *le reveter* 1313 FrY, John *Reveter* 1381 PTY.

**Rivis:** v. REVIS

**Rivlin, Rivling:** Robert *Riueling* 1230 P (Nth); Ivo *Riueling* 1242 AssDu; Elyas *Ryuelynges* 1327 SRLei. ME *riveling* 'rogue, rascal'.

**Rix:** v. RICK

**Rixon, Rixom, Rixson, Rickson:** (i) Richard *Rikson* 1457 LLB K; William *Rixon* 1621 Oriel (O). 'Son of *Rick*.' (ii) Cristiana *atte Ryxen* 1395 FFSo. 'Dweller among the rushes.' v. RICK.

**Roach, Roch, Roche:** (i) John *de Roches* 1086 DB (Beds); Lucas *des Roches* 1249 Fees (Ha). From Les Roches (Seine-Inférieure). An occasional source. (ii) John *Roche* 1195 P (So); Ralph *de la Roche*

1195 P (Co), 1220 Cur (Sx); Roger *atte Roche* 1351 ColchCt. 'Dweller by the rock', ME, OFr *roche*. Robert *atte Roche* lived at Roach Fm in 1330 (PN D 578).

**Road, Roads:** v. RHODES

**Roadnight, Road-Night, Rodnight:** Robert *Rodcnicht*, John *le Rodknicht* 1221 AssWa; John *Rodknyth* 1332 SRWa. OE *rādcniht* 'mounted servant or retainer'.

**Roaf, Roalfe:** v. ROLF

**Roake:** v. ROKE

**Roan, Rone:** (i) William *de Rotomago* 1148 Winton (Ha); William *Ronne* 1327 SRC; John *Rone* 1545 SRW. From Rouen (Normandy). cf. þe cite of Roen, Roan, Rone 1418–20 LondEng 70. (ii) In Scotland from Roan (Ayr, Berwick, Roxburgh). (iii) There was also a feminine name: *Roana* (f) 1212 AssGl.

**Roback:** v. ROEBUCK

**Robart(s):** v. ROBERT

**Robathan:** v. ROWBOTHAM

**Robb, Robbs, Robe:** *Robe* coccus 1196 P (W); *Robbe* Cuth 1199 P (L); Richard *Robbe, Robe* 1177–8 P (Sx); Richard *Robbe* 1212 Fees (So); Simon *Robes* 1319 ELPN; Adam *Robbes* 1327 SRSf. A pet-form of *Robert*.

**Robberds:** v. ROBERT

**Robbie, Robey:** Beatrix *Robbie*, Issobell *Robie* or *Robye* 1597 Black (Aberdeen). A Scottish diminutive of ROBB.

**Robbings, Robbins:** v. ROBINS

**Robchild:** John *Robechild* 1255 RH (O). 'Son of *Robb*', a pet-form of *Robert*.

**Robe:** v. ROBB

**Robearts, Robers:** v. ROBERT

**Robelard:** v. ROBILARD

**Robens:** v. ROBINS

**Roberson:** v. ROBERTSON

**Robert, Roberts, Robart, Robarts, Robberds, Robearts, Robers:** *Rodbertus, Rotbert, Robert* 1066 DB; Willelmus *filius Roberti* 1086 DB (K); *Rotbertus* 1134–40 Holme (Nf); *Rodbert* prest 1188 BuryS (Sf); John *Roberd* 1279 RH (Bk); William *Robert* 1292 FFEss; William *Robbard* 1296 SRSx; Richard *Roberdes* 1327 SRWo; Thomas *Robart* 1332 SRSt; Roger *Robardes* 1341 AD iv (Sa). OFr *Rodbert, Robert*, OG *Rod(b)ert* 'fame-bright'. It was introduced by Normans during the reign of Edward the Confessor and became very popular. Pet-forms and diminutives were formed early, *Robin* being particularly common. v. also DOBB, HOB, NOBBS, HOPKIN, ROPKINS, ROBJANT. The vowel has sometimes been unrounded to *a*. v. DABBS, RABB, RABBATTS.

**Robertshaw, Robinshaw, Robshaw:** Richard *Robartshawe* 1441 ShefA; Edward *Robertschaw* 1502 TestEbor; Jennet *Robertshay* 1565 RothwellPR (Y). From a lost *Robertshaw* in Heptonstall (WRY).

**Robertson, Roberson:** William *Robertsone* 1327 SRDb; John *Roberdeson* 1354 AssSt; Andrew *Robersoun* 1450 Black (Arbroath); John *Robertsoune* 1464 ib. (Leith). 'Son of *Robert*.'

**Robeson:** George *Robesoun* 1633 Black. Perhaps 'Son of *Robe*' (Robert), but also a reduced form of *Robertson* or *Robinson*.

**Robet:** Thomas *Robet* 1242 AssDu; Robert *Robat* 1317 FFEss; Juliana *Robates* 1327 SRWo. *Rob-et*, *Rob-ot*, diminutives of *Robb*, a pet-form of *Robert*.

**Robey, Roby:** v. ROBBIE. May also be local from Robey (Derby) or Roby (Lancs).

**Robilard, Robillard, Robelard:** *Robelard* 1275 RH (Sf); John *Robilard* 1295 Barnwell (C); Joan *Robylard* p1316 LuffCh; Thomas *Robelard* 1327 SRWo. *Rob-el-ard*, a double diminutive of *Robb*, a pet-form of *Robert*.

**Robins, Robyns, Robbins, Robbings, Robbens, Robens:** *Robinus* probator 1198 P (Mx); *Robinus* 1206 Cur (C); Richard *Robin* c1248 Bec (Mx); Walter *Robyn* 1279 RH (O); Margaret *Robines* 1279 RH (C). A diminutive of *Rob* (Robert); *Robin* and *Robert* are used interchangeably: *Robin* or *Robertus* 1206 Cur (L); *Robinus, Robertus* de Leic' 1212 Cur (Nth); *Robert* called de Tylemount . . . the above *Robins* (sic) 1300 LoCt.

**Robinshaw:** v. ROBERTSHAW

**Robinson, Robison:** John, Richard *Robynson* 1324 Wak (Y), 1332 SRLa; Thomas *Robyson* 1379 PTY; John *Robynson* 1426 Black; Donald *Robison, Robertson* 1446 ib.; James *Robison* (signs *Robinson*) 1808 SfPR. Usually 'Son of *Robin*', but also for *Robertson*, through *Roberson*.

**Robjant, Robjohn, Robjohns, Rabjohn, Rabjohns:** *Robezun* de Weleford 1166 P (R); William *Robegance* 1279 RH (C). *Rob-eç-un*, a double diminutive of *Rob* (Robert). For the formation, cf. HUTCHIN, and the French *Robeçon, Robichon, Robuchon*. The *ch* is the Picard form. Robjohns in Finchingfield (Essex) is named from Robert and Simon *Robechun, Robichun, Robchon* (1285, 1299, 1306) and is *Robjents* in 1777 (PN Ess 427).

**Roblett:** Richard *Robeloth* 1221 *ElyA* (Sf); Nicholas *Robylot* 1327 SRWo; Thomas *Roblet* 1524 SRSf. A double diminutive of *Rob* (Robert), *Rob-el-ot*.

**Roblin, Rablan, Rablen, Rablin:** Simon *Robelyn* 1276 RH (Beds). *Rob-el-in*, a similar diminutive of *Rob* or *Rab* (Robert).

**Robotham, Robottom:** v. ROWBOTHAM

**Robsart:** Lewis *Robessart* 1426, *Robessart* 1430 FFEss; Percyvall *Robsart* 1479 Paston. Perhaps from Robirsart (Nord).

**Robshaw:** v. ROBERTSHAW

**Robson:** Willelmus *filius Rob* 1327 SRDb; Richard *Robson* 1379 PTY. 'Son of *Rob*' (Robert).

**Roby:** v. ROBEY

**Roch(e):** v. ROACH

**Rochelle, Rockall, Rockell, Rotchell:** William *de Rokella* (*Rochella*) 1175 P (Beds); Philip *de la Rokele*, *Rochelle* 1203, 1207 FFEss; John *Rockel* 1275 RH (W). From La Rochelle or some French place by a rock (*roche*, Norman *roque*).

**Rochester, Rogister:** Turoldus (*de*) *Rouecestra* 1086 DB (Ess); Robert *Rouchestre* 1377 AssEss. From Rochester (Kent).

**Rochford, Ratchford, Retchford:** Waleram *de Rocheforde* 1198 FF (Nt). From Rochford (Essex, Worcs).

**Rock(s):** *v.* ROKE

**Rockall, Rockell:** *v.* ROCHELLE

**Rockcliffe:** *v.* RAWCLIFF

**Rocker, Rokker, Rooker, Rucker:** William *le Rockere* 1228 Cl (Ireland); Richard *le Roker, le Rockare* 1279 RH (O); John, Lucas *Rockere* 1296 SRSx. Bardsley cites Juliana *Rokster* 1388, a derivative of ME *rok, rocke* 'distaff', a spinner. *Rocker* may be a masculine derivative of this, 'a maker of distaffs'. Or 'dweller by the rock'. *v.* ROKE.

**Rocket, Rockett:** Simon *Roket* 1274 RH (Ess); Richard *Roket'* 1332 SRDo; William *Rokkyt* 1534 CorNt. Probably 'dweller by the small rock', from a diminutive of Norman-French *roque* 'rock'.

**Rockley:** Richard *de Rokele* 1189 Sol; John *de Rocle* 1275 RH (W); Robert *Rockley* 1642 PrD. From Rockley (W), Rookley (Wt), Ruckley (Sa), Rockley in High Bray (D), or 'dweller at the wood frequented by rooks', OE *hrōc, lēah.*

**Rockliffe:** *v.* RAWCLIFF

**Rodaway:** *v.* RADWAY

**Roddam, Rodham:** Yvo *de Rodham* 1204 AssY; Matilda *Rodeham* 1430 TestEbor; William *Roddam* 1673 FrY. From Roddam (Nb).

**Rodding, Roding:** Gilbert *Roding* 1305 RegAntiquiss. From Roding (Ess).

**Roddis:** *v.* ROODHOUSE

**Rodgate:** Cristine *of Rodgate* 1312 AssNf; Adam *atte Rodgate* 1313, Richard *ate Rodgate* 1332 MELS (Sr). From Rodgate in Chiddingfold (Sr).

**Rodge:** John *Rogge* 1275 SRWo; Walter *Rogge* 1286 IpmGl; John *Rogg'* 1332 SRWo. A pet-form of *Roger.*

**Rodger(s):** *v.* ROGER

**Rodgie:** John *Roger, Rogy, Rogie* c1525 Black; Helen *Rogie* or *Rodgie* 1597 ib. A Central Scottish pronunciation of *Roger.*

**Rodgman, Roggeman:** Richard *Rogerman* 1332 SRCu. 'Servant of *Roger*.' cf. William *Robertman* 1332 SRCu, William *Thomasman* 1379 PTY.

**Rodham:** *v.* RODDAM

**Rodhouse:** *v.* ROODHOUSE

**Rodican:** *v.* RUDKIN

**Roding:** *v.* RODDING

**Rodland:** John *Rodlond* 1299–1300 NorwLt; Nicholas *Rodland* 1316–17 FFEss. 'Dweller at the cleared land', OE *\*rod, land.*

**Rodley:** Ralph *de Rodleye* 1290, Thomas *Rodley* 1385 IpmGl. From Rodley (Gl, WRY).

**Rodney:** Richard, William *de Rodeneye* 1304 Glast, 1331 IpmW; Walter *Rodney* 1520 FFEss. From an unidentified place on the borders of Somerset and Wilts, cf. *Radaneheye* 1282 Glast.

**Rodnight:** *v.* ROADNIGHT

**Rodway:** *v.* RADWAY

**Rodwell:** Nicholas *de Rodwell'* 1200 P (R); Warin *de Rodewell* 1208–9 FFWa; Roger *de Rodewelle* 1285 Balliol. From Rothwell (L, Nth, WRY), all *Rodewelle* DB.

**Roe, Roo:** Ului *Ra* 1095 Bury (Sf); William *le Roe* 1170 P (Wa); Reginald *le Ro* 1188 BuryS (Sf);

Benedict *le Roo* 1270 FFEss; Isolda *Ro, Roo* 1314–16 Wak (Y). OE *rā*, ME *rō* 'roe'. These names are midland and southern forms. In the north, *rā* remained unrounded and survives as RAE and RAY.

**Roebotham:** *v.* ROWBOTHAM

**Roebuck, Roback:** Adam *Rabuck* 1246 AssLa; Matilda *Robuc* 1297 SRY. OE *rā* and *bucc* 'roe-buck'.

**Rofe, Roff:** *v.* ROLF

**Roffey:** Flavyen *de Rughehege* 1239 FFSf; Amfr' *de la Rogheye* 1275 RH (K). 'Dweller by some rough enclosure' as at Rolphy Green or Roffy (PN Ess 271, 477).

**Roger, Rogers, Rogger, Rodger, Rodgers:** *Rogerus* marescalcus 1066 DB (Ess); *Rogerius, Rogerus* 1086 DB; Richard *Roger* 1263 ArchC iv; William *Rogger* 1296 SRSx; Henry *Rogeres* 1327 SRWo. OFr *Roger, Rogier*, OG *Ro(d)ger* 'fame-spear'. The name was introduced from Normandy where OG *Rodger* was reinforced by the cognate ON *Hróðgeirr*. In England it was very popular, with pet-forms *Hodge* and *Dodge* and their derivatives, but it was not so productive of direct diminutives as Richard and Robert, only *Roget* and *Rogerun* having been noted: Robert *Rogeroun* 1327 SRSf. *Rogers* is the common form, with *Rodgers* in Scotland.

**Rogerson:** Donald *Rogerson* 1364 Black; William Rogerson 1379 PTY. 'Son of *Roger*.'

**Roget, Rogett:** Robert, Walter *Roget* 1279 RH (O), 1296 SRSx. *Rog-et*, a hypocoristic of *Roger.*

**Roggeman:** *v.* RODGMAN

**Roginson:** John *Rogesoune* 1570 Black. 'Son of *Roger*.' The modern form is eccentric, based on the variation between *Robinson* and *Robison*, etc.

**Rogister:** *v.* ROCHESTER

**Rohan:** Alan *de Roham* 1190 P (Sf). From Rohan (Morbihan).

**Rohard:** *v.* ROWARTH

**Roke, Roake, Rock, Rocke, Rocks, Rook:** Robert *Dellroc* c1182 MELS (Wo); Peter *de la Roke* 1243 AssSo; Richard *del Ak, atte Rok'* 1275, 1327 SRWo; Ralph *ater Oke, atte Oke* 1296, 1327 SRSx; Geoffrey *atte Ock, atter Ok* 1296, 1332 ib.; William *atte Rock, atter Ok* 1296, 1327 ib.; Nicholas *de Aka* called *othe Roke* 1313 LLB E; Robert *of ye Rook* 1318 ib.; William *Attroc* 1327 SRWo; Henry *del Rook* 1332 SRSt; Richard *Bytherok'* 1333 MELS (So). ME *atter oke, atte roke* 'at the oak'. In ME the vowel sometimes became short, *v.* OAK, hence *Rock*, but where there is no confirmatory evidence, this may be from ME *rok* 'rock'. William *de Rok* 1242 Fees (Nb) came from Rock (Northumb) 'rock', but Rock (Devon, Worcs), Roke (Oxon) and Rook (Devon) all derive from *āc* 'oak'. *v.* OAK.

**Rokeby:** Robert *de Rokebi* 1195 P (Y); Thomas *de Rokeby* 1247 FFO; Thomas *de Rokeby* 1347 FFY. From Rokeby (NRY), or Rookby (We).

**Rokewood, Rookwood:** John *Rookwode* 1417 FFEss; William *Rokewode* 1452 Paston; Nicholas *Rokewood* 1535 FFEss. 'Dweller by the wood frequented by rooks', OE *hrōc, wudu.*

**Rokker:** *v.* ROCKER

**Rolance, Roland:** *v.* ROWLAND

**Roleston:** *v.* ROLLESTON

**Rolf, Rolfe, Rolph, Roalfe, Roles, Rollo, Roll, Rolle, Rolles, Rolls, Roff, Roffe, Roaf, Rofe, Roof, Roofe, Rouf, Rouff, Rove, Row, Rowe, Rowes, Rowles, Ruff, Ruff, Rule:** *Rolf* 1066 DB (Nt, Nf); *Roulf* ib. (Lei); Turstinus *filius Rolf, Rou, Roffi* 1086 DB; Robertus *filius Rolui, Roulf* ib.; *Roolf* 1142 NthCh (L) (a peasant); *Rolf* de Ormesby 1147–66 Gilb (L); *Rothof filius Ketelli* c1155–60 DC (L); *Rodulfus filius Ketelli* c1160–66 ib.; Martin *Rof* 1242 Fees (D); Robert *Rolf* Ed 1 Battle (Sx); William *Roulf* 13th Gilb (L); William *Rowe* 1275 RH (Sf); Robert *Rolle* 1279 ib. (Beds); Matilda *Rolles* ib. (Hu); Roger *Rolves* ib. (O); John *Roolf* 1296 SRSx; Robert *Roulfes* 1327 SRWo; Juliana *Roules* 1327 SRSo; John *Rowes* 1333 ColchCt; Thomas *Rowf* 1524 SRSf; William *Rowle* 1662 FrY; Richard *Rowles* 1675 FrY. ON *Hrólfr*, ODa, OSw *Rolf*. Found as the name of a peasant in Danish Lincs, it must sometimes be Anglo-Scandinavian, but the name was also common in Normandy where it became OFr *Roul, Rou*, often latinized as *Rollo* and it is to this that the frequency and variety of the surnames are due. *Roulf* may be a contracted form of ON *\*Hróðwulf*, the ultimate source of *Hrólfr*. There has been confusion with RALF. Some of these names have alternative origins.

**Rolin, Roling:** *v.* ROLLIN
**Rolinson, Rolison:** *v.* ROLLINGSON
**Roll, Roll(es):** *v.* ROLF
**Rolland:** *v.* ROWLAND
**Rollason, Rollerson:** *v.* ROLLINGSON

**Roller:** John *le Rollere* 1274 Wak (Y); Philip *le Roulour* 1327 SRSo, *le Rouller* 1337 FFSo. A derivative of ME *rolle*, OFR *rolle, roolle, roulle* 'a roll, piece of parchment'. Probably a maker or seller of rolls of parchment.

**Rolleston, Rollston, Roleston, Rolston, Rolstone, Roulston, Roulstone, Rowleston, Rowlstone, Rowstone:** Simon *de Roluestona* 1170 P (Lei); Robert *de Rolleston'* 1181 P (Nt). From Rolleston (Notts), *Roldeston* DB 'Hróald's farm', Rolleston (Leics, Staffs, Wilts) or Rowlston (ERYorks), all earlier *Rolvestun* 'Rolf's farm'.

**Rollet, Rollett, Rollit, Rollitt:** *Roelet* 1208 Cur (Ha); William *Raulot* 1279 RH (C); John *Raulot* 1296 SRSx. *Roelet* is a hypocoristic of *Rowland*. cf. ROWLATT, ROWLING. *Raulot* is a diminutive of *Raul* (Ralph). cf. RAWLIN. The normal development would be to *Rawlett*, the spelling *Rollet* being due to the influence of the common *Rolland. Rawlett* survived until 1670 (Bardsley).

**Rollin, Rolling, Rollings, Rollins, Rolin, Roling:** (i) John *Rolins* 1327 SRSf; John *Rolyns* 1327 SRSo. A diminutive of *Roll* (Rolf) or *Roland* (Rowland). *v.* ROLF, ROWLAND. cf. Richard *Rollevilain* 12th Riev (Y), 'Roll's villein'. The paucity of early examples suggests that *Rollin(g)* is often a late spelling for *Rawlin(g)*. cf. ROLLET. It may also be identical with ROWLING. (ii) Roger *de Rolling* 1275 RH (K). From Rowling Court in Goodnestone or Rowling Street in Bilsington (Kent). *v.* PN K 532.

**Rollingson, Rollinson, Rollason, Rollerson, Rollison, Rollisson, Rolinson, Rolison:** William *Rollandson* 1590 LaWills; John *Rollingson* 1596 ib. 'Son of

*Roland.*' Common in Furness, side by side with ROWLINSON, where it has also become RAWLINGSON.

**Rollitt:** *v.* ROLLET
**Rollo:** *v.* ROLF
**Rollons:** *v.* ROWLAND
**Rollston:** *v.* ROLLESTON
**Rolt, Roult:** *Rold* 1066 DB. From ON *Hróaldr*. *v.* ROWAT.

**Rolwright, Roulwright:** Xr *Roulewright* son of Richard *Roulewrite* 1635 GreenwichPR (K); John *Roleright* 1726 LewishamPR (Ess). From Great, Little Rollright (O).

**Romain, Romaine, Roman, Romans, Romayn, Romayne:** *Romanus* le corduaner 1221 Cur (Y); Adam *Romanus, Romayn* 1207–8 Cur (Sr); Reginald *le Romayn* 1275 RH (L); John *Roman* 1367 ColchCt; George *Romans* 1636 FrY. OFr *Romeyn* or *romeyn* 'a Roman'. *Romanus* was the name of two martyrs and a 7th-century bishop of Rouen. Only one English example of this personal-name has been noted, but it may sometimes be the source of the surname.

**Romanes, Romanis, Romans:** Philip *de Roumanoch* c1250 Black; Alexander *Romannois* 1508–20 ib.; John *Romanoss* 1567 ib.; James *Rolmainhous, Romanis* 1642 ib. From Romanno in Newlands (Peebles-shire).

**Rombulow:** *v.* RUMBELLOW

**Rome, Room, Roome:** William *Rome* 1296 SRSx; John *de Rome* 1379 PTY; John *Roome* 1654 FrY. Either an immigrant from Rome or, perhaps, a pilgrim back from Rome. cf. ROMER. *Room* is the old pronunciation of *Rome* which Shakespeare rhymes with *doom* and *groom*, a pronunciation still used by Donne and Pope.

**Romer, Roomer, Rummer:** (i) Cristiana *la Romere* 1274 RH (Sf); John *Romere* 1296 SRSx. ME *romere* 'one who had made the pilgrimage to Rome'. For *Roomer, v.* ROME. *Rummer* shows a shortening of the vowel. (ii) William *de Romara* 1190 P (L). William de Roumare, earl of Lincoln, made a yearly payment to Rouen Cathedral from his rent of Roumare (Seine-Inférieure) near Rouen.

**Romilly:** Alexander *de Romeilli*, Alexander *Rumeilli* 1190, 1193 P (O); Alice *de Romeilli, de Rumelly* 1197, 1230 P (Cu, Y). From Remilly (La Manche). *v.* ANF. Richard *de Romilly* (1198) came from Romilly (Eure), ib. Alan *de Romely* 1262 AssSt may have come from Romiley (Ches).

**Romney, Rumney:** Robert *de Romenel, de Rumenae* 1086 DB (K); Ralph *de Romeney* 1279 RH (O). From Romney (Kent).

**Romsley:** Leo *de Romeslega* 1271 AssSt. From Romsley (Sa).

**Ronald, Ronalds, Ranald:** John *Rannald* 1463 Black; John *Ronald* 1655 ib. A Scottish equivalent of *Reynold*, but derived from ON *Rǫgnvaldr* which became OIr *Ragnall*, Gael *Raonull*.

**Ronaldson:** Doul *Ranaldsone*, Dowill *M'Renyll* (son of *Ronald Alanson*) 1511 Black. 'Son of *Ronald.*'

**Rone:** *v.* ROAN
**Ronson:** *v.* ROWNSON
**Roo:** *v.* ROE
**Roobottom:** *v.* ROWBOTHAM
**Roodhouse, Rodhouse, Roydhouse, Roddis:** Henry

*del Rodehouse* 1379 PTY; John *Roidosse, Roiddus* 1539–40 RothwellPR (Y); Joan *Roydhouse* 1549 ib.; John *Rodhouse* 1552 ib. 'Dweller at the house in the clearing.' v. RHODES, ROYDS.

**Roof, Roofe:** v. ROLF

**Rook, Rooke, Rookes, Rooks, Ruck:** William *Roc* 1185 Templars (O); William *le Roke* 1243 AssSo; William *Ruk* 1296 SRSx; Richard *le Rouke* 1327 SRSo; Adam *Rucke* 1327 SRSf. OE *hrōc* 'rook', a nickname. v. also ROKE.

**Rooker:** v. ROCKER

**Rookledge:** v. ROUTLEDGE

**Rookwood:** v. ROKEWOOD

**Room(e), Room(e)s:** v. ROME

**Roomer:** v. ROMER

**Roope:** v. ROPE

**Roos(e):** v. ROSS

**Root, Roote, Rootes, Roots:** Æðelstan *Rota* 955 OEByn (W); Walter, Ralph *Rote* 1185 Templars (L), 1216–36 Miller (C); John *Rotes* 1643 ArchC 48; William *Root* 1644 FrY. OE *rōt* 'glad, cheerful'. Occasionally also ME *rōte* 'psaltery', metonymic for RUTTER.

**Rootkin:** v. RUDKIN

**Rope, Ropes, Roope, Rupp:** Richard *Rop* 1327 SRC. Metonymic for *Roper*.

**Roper, Rooper, Raper, Rapier:** Roger *Raper* 1219 AssY; Richard *le Ropere* 1220 Cur (Herts); Agnes *Raper* 1430 FeuDu; Alice *Rooper* 1450 Rad (C). A derivative of OE *rāp* 'rope', 'a roper, rope-maker'. *Roper* is the normal southern development. *Raper* persisted in the north.

**Ropkins:** Adam, Joan *Robekyn* 1279 RH (O), 1332 SRSx; Richard *Ropkin* 1524 SRSf. *Rob-kin*, a diminutive of *Rob* (Robert).

**Rosamond:** v. ROSEMAN

**Rose, Royce, Royse:** (i) *Rothais* 1086 DB (Herts); *Rohesia* 1219 AssY; *Roes'* de Killum ib.; Thomas *filius Rose* 1279 RH (C); *Rosa* Weuere 1327 SRSx; Robert, Peter *Rose* 1302 LoCt, 1327 SRWo; Richard *Roys* 1327 SRSf; Richard *Royse, Rose* 1604, 1609 Shef. OG *Hrodohaidis, Rothaid* 'fame-kind' (f), common among the Normans as *Rohese, Roese,* later *Royse,* becoming ME *Rose,* as if derived from *rosa* 'rose'. (ii) Robert *de la Rose* 1242 Fees (O); Adam *atte Rose* 1305 LoCt. From the sign of the rose.

**Roseland:** William *Roseland* 1466 PN K 238. From Roselands Fm in Ulcombe (K).

**Roseman, Rosamond, Rosoman:** *Rosamunda* 1206 Cur (Sf); John *Rosemound* 1356 LLB F; Nicholas *Rosamon* 1359 Putnam (Co). OG *Rosemunda* (f).

**Roskell:** Swein *filius Roskil* 1176 P (Y); Humphrey *Roscale* 1642 PrD. ON *Hrosskell.* v. PNDB 294.

**Roskin:** v. RUSKIN

**Rosling, Rosslyn, Rusling:** Robertus *filius Rozelin, Rotselini* 1086 DB; *Rocelin* de Riggesbi c1150 Gilb (L); Thomas *Roscelin* 1221 AssGl; John *Russelyn* 1316 Wak (Y); Amycia *Roslyn* 1327 SRC. OFr *Roscelin, Rocelin,* OG *Ruozelin, Roscelinus,* a double diminutive of OG *Rozzo.*

**Rosoman:** v. ROSEMAN

**Ross:** (i) *Rozo* 1086 DB (W); *Rosce* de Pileham 1196 P (L); Johannes *filius Rosce* 1197 P (K); Robert *Roce, Rosce* 1195, 1199 P (Nth); Richard *Rosse* 1327

SRC. OG *Rozzo,* a hypocoristic of compounds in *Hrod-.* Willelmus *filius Roce* 1207 Cur (Ess) is also called Willelmus *filius Rocelini* and William *Roce.* (ii) Anschetillus, Serlo *de Ros* 1086 DB (K, Beds); Bernard *de Rosse* 1177 P (Y); Philip *de Roos* 1246 AssLa. The families of Ros of Bedfordshire and of Kent came from Rots (Calvados), that of Roos of Helmsley from Roos (ERYorks), others from Roose (Lancs) or Ross (Hereford, Northumb).

**Rossall:** Robert *de Roshal* 1221 AssSa. From Ross Hall (Salop) or Rossall (Lancs).

**Rossel, Rossell:** Robertus *filius Rocel* 1214 Cur (L); Reginald, William *Rosel* 1182, 1195 P (Y, Do); Patriz *Rozel* 1185 Templars (Db). OFr *Roc-el,* a diminutive of *Rosce.* v. ROSS.

**Rosseter:** v. ROSSITER

**Rossington:** Jeremias *de Rosinton'* 1210 Cur (Y); Peter *of Rosington* 1246 FFY. From Rossington (WRY).

**Rossiter, Rosseter:** Richard *Rocetor* 1563 Pat; Katharine *Rosseter* daughter of Thomas *Rossestir* 1636 GreenwichPR (K); William *Rossiter* 1662–4 HTDo. Late forms of Wroxeter (Sa), or of Rochester (K).

**Rosslyn:** v. ROSLING

**Rosson, Rosten, Rostern, Rosthorn, Roston, Rostron:** v. RAWSTHORN

**Roston:** Adam *of Roston* 1240–1 FFY; Henry *de Roston* 1327 SRY; William *de Roston* 1365 FFY. From Roston (Db). v. also RAWSTHORN.

**Rotchell:** v. ROCHELLE

**Roth:** Adam, Henry *atte Rothe* 1346 ColchCt, 1353 LLB G. 'Dweller in the clearing', OE *\*rop,* as at Roe (Herts) or Rothend (Essex).

**Rotheram, Rotherham, Rudderham, Rudrum:** Richard *de Roderham* 1256 AssNb; Henry *de Rotherham* 1356 AssSt. From Rotherham (WRYorks).

**Rotherforth:** v. RUTHERFORD

**Rothley:** Geoffrey *de Rothele* 1327 SRLei; Walter *de Rothelay* 1355 FFY; Edward *Rotheley* 1459 Kirk. From Rothley (Lei, Nb).

**Rothman:** John *Rotheman* 1327 *SR* (Ess); Richard *Rothemon* 1342 AssSt. 'Dweller in the clearing.' v. ROTH.

**Rothwell, Rowell:** William *de Rowell'* 1200 P (Nth); William *(de) Rowell'* 1212 Fees (L); Robert *de Rothewelle* 1297 MinAcctCo. From Rothwell (Lincs, Northants, WRYorks).

**Rottenbury:** v. RATTENBURY

**Roubottom:** v. ROWBOTHAM

**Rouf(f):** v. ROLF

**Rough, Ruff:** (i) John *le Rug* 1279 RH (O); Geoffrey *Rugh* 1297 SRY. OE *rūh* 'rough'. (ii) John *ate Rogh* 1332 SRSx. 'Dweller by the rough, uncultivated ground', OE *rūh* used topographically.

**Rougham, Roughan:** Lefstan *de Rucham* 1121–48 Bury; Robert *de Ruhham* 1208 P (Sx); John of Rougham 1312 AssNf. From Rougham (Nf, Sf).

**Roughead, Rowed, Ruffhead:** Robert *Ruhaued* 1170–87 ELPN; Alexander *Ruhheued* 1218 FFSf; Robert *Roheued* c1230 P (Hu); John *Roughheved* 1304 AssSt; John *Rouhed* 1384 LoPleas; William *Roughed*

1524 SRSf. OE *rūh* 'rough' and *hēafod* 'head', 'one with rough, shaggy hair'.

**Roughley:** John *atte Roughle* 1351 Putnam (Sr). 'Dweller by the rough clearing', OE *rūh, lēah*.

**Roukin:** Agnes *Roulekin* 1275 RH (L). *Roul-kin*, a diminutive of *Roul* (Rolf).

**Roulson, Rouson, Rowson:** Thomas *Roulfisson'* 1327 SRC. 'Son of *Rolf*.'

**Roulston(e):** *v.* ROLLESTON

**Roult:** *v.* ROLT

**Roulwright:** *v.* ROLWRIGHT

**Rounce:** *v.* ROUND

**Rouncewell:** *v.* ROUNSEFELL

**Round, Rounds, Rounce:** Ralph *Rund'* 1202 FFEss; Alecok *Ronde* 1246 AssLa; Alan *Lerond* 1377 AssEss. OFr *roond* 'rotund, plump'.

**Roundell:** William *Rundel, Roundel* 1150 Oseney (O), 1301 SRY. A diminutive of ROUND.

**Roundtree:** *v.* ROWNTREE

**Rounsefell, Rounsevall, Rounsivelle, Rouncewell, Rounswell:** Geoffrey *de Runceual* 1254 *Ass* (Ess); Richard *Runcyual, Runsifal* 1296, 1327 SRSx. From Roncesvalles in the Pyrenees where there was a Priory of St Mary of which the Hospital of Our Lady of Rouncevale at Charing Cross was a cell.

**Rountree:** *v.* ROWNTREE

**Rous, Rouse, Rowse, Ruse, Russ:** Wilekin *Rous* 1225 AssLa; John *Russe* 1218 Fees (W); Symon *le Rus* 1253 FFHu; Margareta *le Ruse* 1285 FA (St). ME, AFr *rous(e)* 'red'.

**Rouson:** *v.* ROULSON

**Roussell:** *v.* RUSSEL

**Rout:** Clarice *atte Route* 1296 SRSx; Henry *Route* 1317 AssK. Probably 'dweller by the rough ground', OE *\*rūt. v.* MELS 171.

**Routh:** John *Routh* 1327 SRSf; Thomas *de Routhe* 1354 FFY; John *Routhe* 1406 IpmY; Thomas *Rooth* 1672 HTY. From Routh (ERY).

**Routledge, Rutledge, Rudledge, Rookledge, Rucklidge:** Symon *Routlage* 1494 Black; David *Routlesche* 1512 ib.; William *of Retleche*, William *Routleth* Hy 8 PN Cu 25; Jorge *Rutliche* 1524 SRSf; Martin *de Rotheluche* 1537 Black; John *Rutledge, Routledge* 1606, 1639 FrY; Abraham *Roottlidg, Rookledge, Rockleidg* 1657, 1670, 1679 RothwellPR (Y). Harrison's derivation from Routledge (Cumb) is unlikely as the only place of the name in that county is Routledge Burn which is not recorded before the 16th century and probably derives from the surname. This is a border name, English rather than Scottish, from some place which still remains unidentified.

**Rove:** *v.* ROLF

**Rover, Ruffer:** John *le Rofere* 1279 MEOT (Sr); John *Rouer(e)* 1297 MinAcctCo, 1375 ColchCt. A derivative of OE *hrōf* 'roof', a constructor or repairer of roofs.

**Row, Rowe:** (i) Geoffrey *le Ruwe* 1195 P (Lei); Richard, John *le Rowe* 1260 AssCh, 1327 SRC; Walter *le Rowe, le Rouwe* 1262 *For* (Ess), 1320 FFEss; William *Rowe* 1275 RH (Sf). OE *rūh* 'rough'. cf. ROUGH and *v.* also ROLF. (ii) Richard *atte Rowe* 1306 AssSt; John *de Rowe* 1317 AssK; Robert *del Rowe*

1327 SRSf. 'Dweller in the row of houses or in the street', OE *rāw. v.* RAW, REW.

**Rowarth, Roworth, Raworth, Rohard:** *Rohard(us)* 1086 DB (So); Hugo *filius Rohard* 1193 P (He); Benedictus *filius Ruardi* 1219 AssY; Peter *Ruard'* 1220 ClR (L); Alice, Joan *Roward* 1279 RH (O), 1327 *SR* (Ess). OFr *Rohart, Roart*, OG *Hrodhard, Rodhard* 'glory-brave'. This has probably also become ROWAT. *Rohard* is rare.

**Rowat, Rowatt, Rowett:** *Rold* 1066 DB (L); Alanus *filius Ruhald'* 1175 P (Y); *Roaldus* filius Alani 1213 Cur (Nth); Robert *Rohald', Roald, Ruald* 1214 Cur (Wa); John *Makrowat* 1513 Black (Wigtown); John *Rowett* or *Rowatt* 1585 ib. (Glasgow). ON *Hróaldr*, ODa, OSw *Roald. v.* ROLT. From *Roald* we probably also have ROWELL, and from *Roaud*, the vocalized form undocumented, *Rowat*. The name may also derive from *Rohart. v.* ROWARTH.

**Rowberry, Rowbrey, Rowbury, Rubery, Ruberry, Rubra:** Roger *de Rubury* 1327 SRWo. From Roborough (Devon), Rowberrow, Ruborough Hill (Som) or Rowborough (Isle of Wight), all 'rough hill'.

**Rowbotham, Rowbottam, Rowbottom, Roubottom, Robathan, Robotham, Robottom, Roebotham, Roobottom:** Dorythye *Robotom* 1546 Bardsley; Oliver *Robotham* 1592 ib.; Thomas *Rowbotham* 1613 ib. 'Dweller in the rough valley', probably in Lancashire.

**Rowcliffe:** *v.* RAWCLIFF

**Rowden, Rowdon:** Richard *de Roudon* c1250 Calv (Y); Nicholas *de Rouden'* 1297 MinAcctCo; Philip *de Rowedon* 1330 PN K 122. From Rowden (He), one or other of the various places called Rowden in Devon, or Rawdon (WRY), *Roudun* DB.

**Rowed:** *v.* ROUGHEAD

**Rowell:** For ROTHWELL, or 'dweller by the rough hill' (OE *rūh* and *hyll*), as at Rowell (Devon) where the surname is *Rowehill* in 1333 (PN D 491).

**Rower, Royer:** Roger *le Roier* 1176 P (Herts); Gilbert *le Roer* 1185 Templars (Sa); John *le Rohier* 1195 P (Nf); Ralph *le Rowere* 1240 Oseney (O); John *Royer* 1279 RH (Hu); John *le Rouier* 1327 SRSx. OFr *roier, rouwier, roer, rouer* 'wheel-wright'. Dyonisia *la Rowere* was owed money for *wheels* by the Commonalty of London in 1301 (LLB C).

**Rowet, Rowett:** (i) William *Roet* 1219 Cur (Nf). *Ro-et*, a diminutive of *Ro(w)*, i.e. *Rolf*. (ii) Richard *atte Rouette* 1296 SRSx. 'Dweller by the rough ground', OE *\*rūwet*.

**Rowland, Rowlands, Rawland, Rolland, Rollons, Roland, Rolance:** (i) *Rolland* 1086 DB; *Rolandus, Ruelandus* Decanus 1133–60, 1188 Rams (Hu); *Rollandus* de Dinan c1155 DC (L); *Roudlandus* de la Genouerai Hy 2 DC (Lei); *Roulandus* de Withcala 12th DC (L); *Roelandus* 1212 Cur (Berks); *Rotelandus* Dauvers 1214 Cur (Bk); *Rodlandus* 1240 Rams (Nf); Nicholas *Ruel'* (*Rueland'*, *Roelent*) 1219 Cur (Nf, D); Simon *Rolland'* 1218 AssL; William *Roulland'* 1221 AssGl; Geoffrey *Rodland* 1244 Rams (Beds); Nicholas *Roland* 1303 FFEss. OFr *Rollant, Rolant, Rolent, Roulent*, OG *Hrodland, Rodland* 'famous land'. A popular name, largely due to *Roland*, the most famous of the peers of Charlemagne. In the *Chanson de*

*Roland* it appears as *Rollanz* and *Rollant*, and in the *Cursor Mundi* as *Rauland*. It was introduced by the Normans and has numerous derivatives, some of which have been confused with those of *Ralf*. (ii) Ralph *de Rowlond* 1296 SRSx; Alan *de Roland* 1327 SRDb. From Rowland (Derby) or Rowland Wood in Slinfold (Sussex).

**Rowlandson:** 'Son of *Rowland*.' cf. William *Roulandman* 1332 SRCu. *v.* ROLLINGSON, ROW-LINGSON, ROWNSON.

**Rowlatt, Rowlett:** William *Roulot* 1327 SRSf. A hypocoristic of *Rowland*. cf. ROLLET.

**Rowlerson:** *v.* ROWLINGSON

**Rowles:** *v.* ROLF

**Rowleston:** *v.* ROLLESTON

**Rowley:** William, Adam *de Ruelay* 1219 AssY; Geoffrey *de Roweleye* c1280 SRWo; John *Rowley* 1348–9 IpmNt. From Rowley (Devon, Durham, Staffs, ER, WRYorks), or Rowley Hill (Essex).

**Rowling, Rowlins:** *Roelin, Rohelin* 1195 P (Cu); Adam, Geoffrey *Roulin* 1327 SRSf. *Roel-in, Roul-in,* hypocoristics of *Rowland*. *v.* ROLLIN.

**Rowlingson, Rowlinson, Rowlison, Rowlerson:** John *Rowlinson* 1608 LaWills. 'Son of *Rowland*.' *v.* ROLLINGSON, ROWNSON.

**Rowlstone:** *v.* ROLLESTON

**Rowney:** Agnes *de Rowenheye* 1275 SRWo; Walter *ate Roueneye* 1279 RH (O); Walter *de Rounhey* 1309 SRBeds; Beatrice *Roweney* 1327 *SR* (Ess). From Rowney Green (Worcs), Rowney Wood (Essex), or from residence near a rough enclosure (OE *rūh*, *(ge)hæg)*.

**Rownson, Ronson:** Richard *Rowlandson* or *Rownessonn* 1607 LaWills; John *Rowanson* or *Rownson* 1639 ib.; John *Rowlandson* or *Rownson* 1715 ib. A north Lancashire form of ROWLANDSON.

**Rowntree, Rountree, Roundtree:** Robert *Rountre* 1301 SRY; John *Roundtree* 1659 Bardsley. 'Dweller by the rowan-tree.'

**Rowse:** *v.* ROUS

**Rowson:** *v.* ROULSON

**Rowstone:** *v.* ROLLESTON

**Rowstron:** *v.* RAWSTHORN

**Rowton:** Ivo *de Roweton'* 1255 RH (Sa); William *atte Ruetune* 1275 RH (K); Robert *atte Rowetoune, atte Rugheton* 1327, 1332 SRSx. From Rowton (Salop, ERYorks) or 'dweller by the rough enclosure', OE *rūh, tūn*.

**Roxby, Roxbee:** Rannulf *de Rokesbi* 1188 P (L). From Roxby (L), Roxby in Thorntondale, in Hinderwell, or Roxby House in Pickhill (NRY).

**Roy:** Adam *le Roy* 1268 FFSf; Simon *Roy* 1279 RH (C). OFr *roi* 'king', used also as a personal-name: *Roi de Scallebi* 1188 P (L). *v.* RAY.

**Royce:** *v.* ROSE

**Roycraft, Roycroft:** *v.* RYCROFT

**Royden, Roydon:** Robert *de Roydon* 1301 FFEss; Piers *Roidon* c1480 Paston. From Roydon (Ess, Nf), or Roydon Drift (Sf).

**Roydhouse:** *v.* ROODHOUSE

**Royds:** Adam *de Roides* 1379 PTY. A Yorkshire dialect pronunciation of RHODES.

**Royer:** *v.* ROWER

**Royle:** Bernard *de Royl* 1230 Cl; William *de Roille* 1290 AssCh; Nicholas *Royle* 1551 FFSt. From Royle (Lancs).

**Royston:** John *de Roystone* 1310 LLB D; Stephen *de Royston* 1327 CorLo. From Royston (Herts, WRY).

**Ruber(r)y, Rubra:** *v.* ROWBERRY

**Rubin, Ruben, Rubens:** Ralph *Ruben* 1275 RH (K); Roger *Rubens* 1279 RH (Hu); Robert *Rubins* 1279 RH (C). cf. Fr *Rubin*, according to Dauzat, a name for a jeweller.

**Rubottom:** *v.* ROWBOTHAM

**Ruck:** *v.* ROOK

**Rucker:** *v.* ROCKER

**Rucklidge:** *v.* ROUTLEDGE

**Rudd:** Gerard, William *Rudde* 1189 P (Y), 1199 AssSt. OE *\*rud-*, the root of OE *rudig* 'ruddy', *rudduc* 'red-breast'.

**Rudderham:** *v.* ROTHERAM

**Ruddick, Ruddock, Rudduck:** Azor *Ruddoch* 1176 P (Beds); Matilda *Ruddoc* 1275 SRWo. OE *ruddoc* 'robin-redbreast'.

**Ruddiforth:** *v.* RUTHERFORD

**Ruddin:** *v.* READING

**Ruddle:** (i) Elyas *Rudehal'* 1201 P (Gl); Elias *de Rudel* c1220 Fees (Gl). From Ruddle in Newnham (Gl). (ii) Richard *Rudel* 1238–9 AccM; Reyner *Rudel* 1296 SRSx. *Rudd-el*, a diminutive of ME *rudde* 'red'.

**Rude:** *v.* RIDE

**Rudeforth:** *v.* RUTHERFORD

**Rudge:** (i) William *de Rugge* 1196 P (St). From Rudge (Glos, Salop). (ii) Walter *de la Rugge* 1243 AssSo; Alan *atte Rugg* 1296 SRSx; Geoffrey *ate regge* ib.; Roger *othe Rugge* 1327 SRWo. From residence near a ridge (OE *hrycg*). In ME this became *rugge, regge, rigge* in different dialects. *Regge* does not seem to have survived. For *rigge v.* RIDGE. *Rudge* is still used for *ridge* in Worcs where it is found in Rodge Hill. (iii) Roger *Rugge* 1195 P (D); Osbert *le Rugg'* 1275 RH (K); Gilbert *Rougge* 1377 AssEss. AFr *rug(g)e*, Fr *rouge* 'red', i.e. 'red-haired'. cf. ROUS.

**Rudgeley:** *v.* RUGELEY

**Rudgwick, Rudgwick:** William *de Ruggewyke* 1296 SRSx. From Rudgwick (Sx).

**Rudham:** Adam *de Rudham* 1275 RH (Nf); Leonard *of Rudham* 1315 AssNf. From East, West Rudham (Nf).

**Rudkin, Rudkins, Rutkin, Rootkin, Rodican:** Ralph *Rudkyn* Hy 3 Gilb (L); William *Ruddekyn* 1301 SRY; John *Rodekyn* 1346 FA (Sf); Thomas *Rutkin* 1524 SRSf. *Rudd-kin*, a diminutive of ME *rudde* 'red', or of ON *Rudda* (f).

**Rudland, Rudlen:** *v.* REEDLAND

**Rudledge:** *v.* ROUTLEDGE

**Rudston:** Mauger *de Rudestan* 1205 Cur (Y); Robert *de Rudeston* 1341 FFW; John *de Rudston* 1381 PTY. From Rudston (ERY).

**Rudyard:** Richard *Rodyarde* 1473 Glapwell (Db). From Rudyard (St).

**Rue:** *v.* REW

**Ruff:** *v.* ROLF, ROUGH, ROW

**Ruffer:** *v.* ROVER

**Ruffet, Ruffett:** Roger *Rucfot* 1205 P (Y); John *Rofot* 1277 AssSo; Henry *Rugfot* 1297 MinAcctCo;

Adam, Arnold *Roufot* 1327 *SR* (Ess), 1360 ColchCt. OE *rūh* 'rough, hairy' and *fōt* 'foot', the exact meaning being obscure.

**Ruffhead:** *v.* ROUGHEAD

**Ruffin:** Philip *Ruffin* 1203 Cur (K); Simon *Ruffyn* 1296 SRSx; Nicholas *Ruffyn* 1337 CorLo. Lat *Rufinus*, the name of more than one saint.

**Rufford:** William *de Rufford'* 1219 AssY; Geoffrey *of Rufford* 1226 FFY; Alan *de Rufforth* 1287 IpmY. From Rufford (La, Nt), or Rufforth (WRY).

**Rugeley, Rudgeley, Rudgley:** Giles *de Rugeleye* 1301 AssSt. From Rugeley (Staffs).

**Rule:** (i) Alan *de Rule* 1212 Cur (C); Philip *de Rule* 1249 IpmY; William *Rule* 1384 LLB H. From Rule (St). (ii) Adam *de Roule* 1296, Thomas *Roule* 1429 Black. From the lands of Rule in Hobkirk (Roxburgh).

**Rulf:** *v.* ROLF

**Rumball, Rumbell, Rumbold, Rumbolt, Rumbol, Rumboll, Rumble, Rumbles:** *Rumbaldus* 1086 DB (Gl); Robert, Roger *Rumbald* 1191 P (Ess), 1195 P (Cu); William *Rumbol'* 1222 Cur (Ha); William *Rumbolt* 1327 SRC. OG *Rumbald*.

**Rumbellow, Rumbelow, Rombulow:** Robert *Romylow, Rombilow* 1524 SRSf; William *Rumbilow, Rumbelow* ib. Bardsley, Harrison and Weekley agree in taking this name as a nickname for a sailor from the meaningless combination of syllables sung as a refrain whilst rowing, an explanation unlikely in itself, as there is no evidence for the use of *rumbelow* as a generic name for a sailor, whilst the nickname seems quite unsuited to the first known bearer of the name, Stephen *Romylowe*, constable of Nottingham Castle in 1347. Of ten examples of his surname only one contains a *b* (*Rombylou* 1351 Pat) whilst two, *de Romylo* (1346 Pat) and *de Romylou* (1363 NottBR), point clearly to a place-name ending in *-low*. This must be identical in origin with The Rumbelow in Aston (Warwicks), *Rumbelowe* 1461, the home of Richard *de Thrimelowe* in 1334. A second example of the same name is Tremelau Hundred (Warwicks) and a third is Rumbelow in Wednesfield (Staffs), *le Thromelowe* 1339, *Romylowe* 1420 (PN Wa 31). All three are from OE *æt prēom hlāwum* (dweller) 'by the three mounds or barrows'. The Suffolk surname probably originated in Suffolk.

**Rumford, Rumfitt:** Warin *de Rumford* c1233 HPD (Ess). From Romford (Essex).

**Rummer:** *v.* ROMER

**Rumney:** *v.* ROMNEY

**Rump:** Thurstan, Robert *Rumpe* c1095 Bury (Sf), c1170 Rams (C). ME *rumpe* 'buttocks'. cf. dial. *rump* 'an ugly rawboned animal', used contemptuously of a person (EDD).

**Rumsey:** Walter *de Rumesie* 1205 Cur (So); John *de Romesy* 1331 IpmW; George *Rumsy* 1641 PrSo. From Romsey (Ha).

**Runacres, Runacus, Runagle, Runeckles, Runecles, Runicles, Runnackles, Runnacles, Runnagall, Runneckles, Runnicles, Ranacre, Ranigar:** Alan *de Ruynacres* 1246 AssLa; Richard *de Reinacre* 1261 ib.; Richard *de Rynacres* 1332 SRLa; William *Renacles* 1500 FrLeic; Alice *Renakers* 1568 SRSf; Alice

*Runnaker* 1575 SfPR; Widow *Runnicre* 1674 HTSf; Elizabeth *Runicles* 1798 SfPR; Mark *Runnacles* 1815 ib. 'Dweller by the rye acre(s)' (OE *rygen*, *æcer*), as at Renacres (Lancs).

**Runaway:** Thomas *Rennaway* 1276 RH (Db); Richard *Rynaway* 1332 SRCu; John *Renneaway* 1401 AssLa. 'Runaway', ON *renna*, OE *onweg*. cf. Robert *Renandgo* 1309 AssNf 'run and walk'.

**Runcie:** Roger, Laurence *Rouncy* 1230 Eynsham (O), 1276 RH (O). ME *runcy, rouncy* 'rouncy, nag', for RUNCIEMAN.

**Runcieman, Runciman, Runchman:** Adam *Runciman* 1697 Bardsley. 'Man in charge of the rouncies.'

**Rundall, Rundell, Rundle:** (i) Richard, Herbert *Rundel* Hy 2 DC (Lei), 1197 P (Nf); *rondel*, a diminutive of *rond*. *v.* ROUND. (ii) Thomas *de Rundal* 1275 RH (K). From Rundale in Shoreham (Kent).

**Runnackles, Runnagall:** *v.* RUNACRES

**Rupp:** *v.* ROPE

**Rusbridger:** *v.* RISBRIDGER

**Ruse:** *v.* ROUS

**Rush:** John *ate Russh'* 1332 SRSr. 'Dweller among the rushes', OE *rysc* 'rush', used collectively.

**Rushbrook, Rushbrooke:** Wudardus *de Rosshebrok* 1148–56 Bury; William *de Rucchebrok* 1344 FFEss. From Rushbrooke (Sf).

**Rusher, Risher:** John *le Russere*, Geruase *le Rischere* 1296 SRSx. A derivative of OE *rysc* 'rush', a cutter or seller of rushes. cf. Alan *le Rusmangor* 1210 Oseney (O). Also 'dweller among the rushes'. cf. RUSH.

**Rushmer, Rushmere, Rushmore:** William *de Russemere* 1206 Cur (K); John *de Russemere* 1275 SRWo. From Rushmere (Sf), or Rushmere in Charing (K).

**Rushton:** Robert *de Riston'* 1203 AssNth; John *de Russheton'* 1340 Crowland; John *Russheton* 1433–4 FFWa. From Rushton (Ches, Northants, Staffs), the last two are *Risetone* DB.

**Rushworth, Rishworth:** Henry *de Rissheworthe* 1276 IpmY. From Rishworth (WRY).

**Ruskin, Roskin:** *Rosekin* 1220 FFEss; John *Rosekyn* 1389 FFC; John *Ruskyn* 1498 RochW. A diminutive of *Rosce*. *v.* ROSS.

**Rusling:** *v.* ROSLING

**Russ(e):** *v.* ROUS

**Russel, Russell, Russill, Rousel, Rousell, Roussel, Roussell:** *Russel* c1095 Bury (Sf); *Russellus* de Suberton' 1212 Cur (So); Ralph, Robert *Russel* 1115 Winton (Ha), 1169 P (Do); John *Roussel* 1297 MinAcctCo. OFr *rous-el*, a diminutive of *rous* 'red', used also as a personal-name.

**Russinol:** Alan *Russinol'* 1208 FFL; Walter *Russinol*, a Lombard 1278 LLB A. OFr *rosignol* 'nightingale'.

**Russon:** Peter *Russessone* 1308 AD i (Hu); Hugh *Rosesone* 1342 AD vi (St); John *Rossissone* 1365 AD ii (Sf). 'Son of *Ross* or of *Rose*.'

**Rust:** Cenwold *Rust* 1016–20 OEByn (K); Robert, Richard *Rust* 1148 Winton (Ha), 1275 RH (Nf). OE *rūst* 'rust', used of reddish hair or complexion.

**Ruston, Rustom:** Richard *de Ruston'* 1194 P (L); Randle *de Ruston* 1260 AssCh; John *Ruston* 1379

PTY. From Ruston (Devon, Norfolk, NRYorks), or Ruston Parva (ERYorks).

**Rustrick:** v. RASTRICK

**Ruth:** William *Ruth* 1180 P (L), 1275 RH (Nf). ME *reuthe* 'pity'.

**Ruthall:** John *de Rohall'*, Richard *de Rothal'* 1221 AssSa. From Ruthall (Sa).

**Rutherford, Rutherfoord, Rutterford, Rotherforth, Ruddiforth, Rudeforth:** William *de Rwthirford* a1200 Black; Nicholas *de Rothirford* 1296 ib.; Nicholas *Rotherforth* 1645 FrY; John *Rudderforth* 1758 ib. From Rutherford (Roxburghshire, NRYorks).

**Rutkin:** v. RUDKIN

**Rutland:** Hugh *de Roteland'* 1214 Cur (Wa); Simon *of Roteland* 1245–6 FFY; Robert *Roteland* 1395 IpmNt. From Rutland.

**Rutledge:** v. ROUTLEDGE

**Rutley:** John *Rutley* 1642 PrD. From Great Rutleigh in Northlew (D).

**Rutt:** Godric *Rute* 1166 P (Nf); William *le Rutte* 1278 RH (O); James *Rutt* 1524 SRSf. Metonymic for RUTTER 'player on the rote'.

**Rutter:** Simon *le rotur* Hy 2 DC (Lei); Reginald *Ruter* 1210 Cur (Cu); Thomas *le Roter* 1251 Fees (Wa); Thomas *le rotour*, Richard *le routour* 1297 MinAcctCo; John *Rutter* 1524 SRSf. OFr *roteor*, *roteeur*, *routeeur* 'a player on the rote', a musical instrument, a kind of fiddle (a1300 NED). The forms in -*er* may be from OFr *rotier*, *routier* 'robber, highwayman, ruffian' (1297 NED).

**Rutterford:** v. RUTHERFORD

**Ryall, Ryalls:** Richard *de Ruyhale* 1327 SRWo; John *Ryall* 1642 PrD; William *Ryall* 1662–4 HTDo. From Ryall in Bradworthy (D), Royal Fm in Peper Harow (Sr), or Ryall in Ripple (Wo).

**Ryburn:** v. RAEBURN

**Rycroft, Rycraft, Roycraft, Roycroft:** Richard *de Riecroft* c1230 Barnwell (C); Richard *de Ruycroft* 1325 AssSt. 'Dweller by the ryecroft', OE *ryge*, *croft*.

**Ryde:** v. RIDE

**Ryder:** v. RIDER

**Rydings:** v. READING

**Rye:** (i) Hubert *de Ria* 1169 P (Nf). From Ryes (Calvados). If this has survived, it is rare. (ii) Matillis *de la Rye* 1237 CartAntiq (Ha); William *de Rye* 1240 FFEss; Geoffrey *ate Rye* 1297 MinAcctCo. Usually from ME *at ther ye*, becoming *at the rye* 'at the island or low-lying land', as in Rye (Sussex), The Rye (Bucks), Rye House (Herts), etc. v. also REA.

**Ryecart:** Fulco *de Ruycote* 1259 Oseney (O). From Rycote (Oxon).

**Ryland, Rylands, Rylance:** Stephen *de Riland* 1232–45 RegAntiquiss; John *de Rylaundes* 1281 AssLa; Thomas *de Rilond* c1296 AssCh. 'Dweller by the land where rye is grown.'

**Ryle, Ryles:** Henry *de Ryel* 1269 AssNb; John *Ryle* 1441 ShefA; Thomas *Ryles* 1455 LLB K. From Ryal (Nb, WRY), or Ryle in Whittingham (Nb).

**Ryley:** v. RILEY

**Ryman:** v. RAYMAN

**Ryton:** John *de Ruiton'* 1221 AssSa; Lecia *de Ruton'* 1221 AssWa. From Ryton (Du, Sa, NRY), or Ryton on Dunsmore (Wa).

# S

**Saayman:** *v.* SEAMAN

**Saban, Sabben, Saben, Sabin, Sabine:** *Sabina* 1186–1210 Holme (Nf), 1220 Cur (K, Sr); Rogerus *filius Sabini* 1252 Rams (Hu); Richard *Sabin* 1221 AssWa; John *Sabine* 1279 RH (C). There were three saints named *Sabinus* and one *Sabina*. Lat *Sabinus* 'a Sabine'. In England, the woman's name was much the more common.

**Saber:** *v.* SEABER

**Sablin, Sabline:** Willelmus *filius Sabeline* 1182–3 StP (Lo); *Sabelina* 1197 P (Lo), 1211 Cur (Mx); Roger *Sabelyn* 1297 MinAcctCo (O). *Sab-el-in(a)*, a double diminutive of *Sabin(a)*.

**Sach, Sacher, Satch:** Henry *le Sachiere* 1280 MESO (Ha); John *le Sachere* 1294 RamsCt (Hu). OFr *sachier* 'maker of sacks'. cf. SACKER. *Sach, Satch* are metonymic.

**Sacheverell:** John *de Saltcheverel* 1199 AssSt; Nicholas *Saucheverel* 1247–8 FFEss; John *Saucheverell* 1456 TestEbor; Mrs *Sacheverill* 1662–4 HTDo. From Sault-Chevreuil (La Manche).

**Sack:** Symon *Sac'* c1250 Rams (C); William *Sak* 1327 *SR* (Ess). OE *sacc* 'sack'. Metonymic for SACKER.

**Sacker, Sackur, Secker, Saker:** Geoffrey *Sakker* t John ELPN; Hugh *le Saker* 1225 AssSo; Eva *le Seckere* 1277 Wak (Y). A derivative of OE *sacc* or ON *sekkr* 'sack'. 'A maker of sacks or coarse cloth, sackcloth.' cf. Ralph *Sakeman* 1209 P (Hu), Henry *le Sacwebbe* 1279 AssSo.

**Sackville, Sackwild:** Richard *de Sachanuilla, de Sacheuilla* 1086 DB (Herts, Ess); Simon *Sakeuilla* 1154–89 Colch (Ess); Alexander *de Saccauill', de Saccheuilla, de Sakeuill'* 1162–98 StP (Lo); William *de Saucheuilla* 1176 P (Bk). Round's derivation of the Sackvilles, later dukes of Dorset, from Sauqueville (Seine-Inférieure) is accepted by Loyd. Their identification of the Essex DB under-tenant, of a different family, as coming from Secqueville-en-Bessin (Calvados) depends solely on the fact that the place is 11 kilometres from Ryes, the place of origin of Eudo Dapifer under whom he held. The early forms of Secqueville are *Sicca Villa* (1077), *Secheville* (1155), *Secqueville* (1217), which do not fit in with those of Sackville and probably survive as SETCHFIELD. Richard probably came, as suggested by Dupont, from Sacquenville (Eure), recorded as *Sachenville* (1195), *Sakenvilla* (c1210), *Sackevilla* (1220). *v.* OEByn, ANF.

**Sadd:** Roger *Sad* 1086 DB (Nf); Henry *Sadde*

1229–41 StP (Ess). ME *sad(de)* 'serious, discreet, firm'.

**Saddick:** *v.* SADDOCK

**Saddington:** Hiche *de Sadinton* 1198 FrLei; Roger *de Sadington* 1260 AssSt. From Saddington (Lei).

**Saddler, Sadleir, Sadler, Sadlier:** Simon *le Sadelere* 1288 MESO (Sx); Peter *le Sadelare* 1296 Wak (Y). A derivative of OE *sadol* 'saddle', a maker or seller of saddles (1389 NED).

**Saddock, Saddick, Sadick:** Ralph *filius Saddoc'* 1201 Pleas (Hu); *Saddok* de Wautham 1248 AssBerks; Richard *Saddoc* 1201 Pleas (Sa); John *Saddok* 1327 SRSf. Evidently from an unrecorded OE *Sadduc*. For the first part of the name, cf. Ranulf *Sadewi* 1202 AssL; Adam *Sadwyn'* 1250 Selt.

**Saer:** *v.* SAYER

**Saffell, Saffill, Saffle, Safhill:** *Safugel* 1066 Winton (Ha); Willelmus *filius Safoul* 1182–1200 BuryS (L); Rannulfus *filius Safugel, Safuwel, Sefugel* 1196–7 P, 1200 Cur (Sf); *Sefull* the chaplain 1268 AD ii (Sf); Halwær *Sæfugulasuna* c1150 YCh; John *Seful* 1275 SRWo; Robert *Sefoul* 1279 RH (O); John *Safowel* 1327 SR (Ess); John *Sefughel* 1332 SRSx; John *Safoul* 1376 AssEss; John *Saffull* 1434 FFEss. AScand *Sæfugul*, from ON *sæfogl* 'sea-bird', especially the cormorant, not recorded as a personal name in Scandinavia but common in England after the Conquest.

**Saffer, Saffir:** Robert *le Saffere*, John *Saffare* 1275 SRWo. Probably OFr *saffre* 'glutton'. *v.* also SAVARY.

**Saffery, Saffrey:** *Safrei* 1198 FF (Sf); *Sefridus* 1214 Cur (Sx); *Saffredus* 1214 Cur (Bk); Robert *Safrey* c1230 Barnwell (C); William, Roger *Sefrey* 1275 RH (Nf); Bryan *Saffrey* 1327 SRC. OE *Sæfriδ* 'sea-peace', a personal name recorded in the 9th and 10th centuries, and then not until the 12th. *v.* also SAVARY.

**Saffill, Saffle, Safhill:** *v.* SAFFELL

**Safford:** John *de Safford* 1296 SRSx. From Seaford (Sussex).

**Sagar, Sager, Saiger, Saggers, Seagars, Seager, Seagers, Seegar, Seeger, Segar, Seger, Segger:** *Sagar* 1066 DB (D); *Segarus* ib. (Ess); Galfridus *filius Segar* 1222 Pat (D); Walter *Sagar* 1195 P (Do); Ralph *Segar* 1207 Cur (Beds); John *Seger* 1275 RH (Nf). OE *Sægār* 'sea-spear', unrecorded in OE, but fairly common from 1066 onwards. *Seager* would naturally become *Sigger* and be confused with SIGGERS.

**Sage:** Robert *le Sage* 1185 P (Sa); Ralph *Sage* 1190 BuryS (Sf); Petronilla *la Sage* 1206 FFSt. OFr *sage* 'wise'.

**Saggers, Saiger:** *v.* SAGAR

**Sailant, Saillant:** Robert *Saillant* 1208 P (Ha); Robert *le saillant* 1214 P (D). OFr *saillant* 'dancing', a name for a dancer. cf. Robert *Sayleben* 1319 SRLo 'dance well'; Richard *Sayllefest* 1275–6 CtH 'dance quickly'.

**Sailer, Saylor, Seiler, Seiller, Seyler:** Herbert *le Sayllur* 1191–1210 YCh; Hugh *le Saylliur* 1275 RH (Sf); Robert *le Salyour* 1327 SRSf; John *Sayller* 1327 SR (Ess). OFr *sailleor, salleor, sailleur, saillur* 'dancer'.

**Sailes, Sails:** Hugh *de Sailes* 1219 AssY; Agnes *del Sayles*, William *Saylles* 1379 PTY. From Sales (Lancs), *Saylys* c1535, or from residence near a pool or pools (ON *seyla*). *v.* PN La 217.

**Sainer:** *v.* SENIOR

**Saines, Sains:** *Sæuuine, Seuuinus, Sauuin, Sauinus* 1066 DB; *Sewinus* 1207 Cur (Ess); Petrus *Sewinus* 1242 Fees (W); William *Sewine* c1250 Rams (C); Thomas *Sewyne* 1327 SR (Ess). OE *Sǣwine* 'sea-friend'. One would expect this to become *Sewin* or *Sawin* but such forms would inevitably be contracted in everyday speech. cf. Saintbury (Gl), *Seinesberia* 1186, probably from *Sǣwines-burg* (DEPN).

**Sainsberry, Sainsbury:** Reginald *de Seinesberia* 1190 P (Gl). From Saintbury (Glos).

**Saint, Sant, Sants, Saunt:** Roger *le Sent* c1250 Riev (Y); Hugh *Sant* 1270 RamsCt (C). ME *seint, saint*, AFr *seint*, OFr *sant* 'saint', a nickname.

**St Barbe:** *v.* SIMBARB

**Sainter:** *v.* SENTER

**St John, Sinjin:** Thomas *de sancto Johanne* c1110 Winton (Ha); Edward *de Sein' Johan* 1327 SRSx; Laurence *Seintjohan* 1355–9 AssBeds. From Saint-Jean-le-Thomas (La Manche).

**St Nicholas, Sennicles:** John *de St Nicholas* otherwise John *Sennycoles* 1455 AD i (K); John *Seyntnycolas* 1462, William *Sincklas* 1544 ArchC 37. From St Nicholas at Wade (K).

**St Quintin:** Herbert *de Sancto Quintino* c1110 Winton (Ha); John *de Sancto Quintino* 1219 P (Y); John *Seintquyntyn* 1374 PN Do ii 96. From St Quentin-des-Isles (Eure).

**Sait:** *Saiet* 1066 DB (Beds, O), 1115 Winton (Ha); *Saegeat* 1077 PNDB (Wo); Thomas *Seiete* 1115 Winton (Ha); Robert *Seyet* c1240 Fees (Bk). OE *Sǣgēat* 'sea-Geat'.

**Saive, Sayve:** *Seiua* (f) 1189 Sol; Ralph *filius Sageue* Hy 3 Colch (Ess); William *Sayf* 1304 IpmY; John *Sayye* 1468 NorwW (Nf). OE *Sǣgiefu* (f).

**Saker:** *v.* SACKER

**Salaman, Salamon, Salamons, Salomon, Salomons, Salman, Salmen, Salmon, Salmond, Salmons, Samman, Sammon, Sammonds, Sammons:** *Salomon* 1066 DB (Y); Gislebertus *filius Salamonis* 1086 DB; *Salamon clericus* 1121–48 Bury (Sf); *Salomon* (a chaplain) 1159 Black (Roxburgh); Hugo *filius Salman* 1219 AssY; *Salman* Salman 1382 LoPleas; *Salamon* Salmon 1394 ib.; Roger *Salmon* 1210 Cur (Beds); Robert *Salemon'* 1212 Fees (L); Lambert *Saumon* 1242 Fees (Ha); Geoffrey *Saleman* c1248 Bec (Sf); Adam *Sauman* 1275 RH (L); William *Sammoun* 1279 RH (Hu); Willelmus dictus *Salamon* 1287 Husting; William *Saleman* 1296 SRSx; Richard *Salman, Salamon* 1301 LoCt; John *Sawmond* 1494 Black; Thomas *Samond* 1524 SRSf; James *Salmond* 1546 Black. OFr *Salomon, Salemon*, a Hebrew name, from *shalom* 'peace'. *Salomon* was the common medieval form, used in the Vulgate, by Tyndale and Cranmer and in the Rheims version (1582). *Solomon* is the form used in the Geneva Bible and the Authorized Version. The *dictus Salamon* of 1287 shows that it was occasionally a nickname, probably 'the wise', but the name was not uncommon from the 12th to the 14th centuries and was not confined to Jews. It is found as the name of a cleric, a chaplain and a canon. Occasionally the modern *Salamon(s), Salomon(s)* may be Jewish. *Solomon(s)* certainly is. Tengvik and Wallenberg consider the surname may also be OE, ME *\*saleman* 'salesman', evidenced in NED from 1642. This appears unlikely. There is no example of *le saleman*. The form shows a natural weakening of the vowel on the way from *Salaman* to *Salman*, which is certainly found as the personal name. There seems no reason to doubt that Thomas *Salamon, Salomon* (1317), who probably gave name to Salmans in Penshurst (Kent), was a descendant of a *Salamon* (PN K 93). John *Saleman* (1319 SRLo) was a son of *Salomon* Burghard. The *d* of *Salmond* is late and excrescent.

**Salby, Saleby:** William *de Salebi* Hy 2 Gilb; Roger *de Salby* 1327 SRLei. From Saleby (L).

**Salcock, Salcott:** William *de Salcoke* c1224–33 YCh; William *of Salecoc* 1240 FFY; John *Salcot* alias Capon bishop of Bangor 1534 LP. From Salcott (Ess).

**Salcombe:** Walter *de Saltecumbe* 1176 P (D); Roger *Salcombe* 1361 IpmGl. From Salcombe, or Salcombe Regis (D).

**Salcott:** *v.* SALCOCK

**Sale, Sales:** Robert *de la Sale* 1243 AssSo; Nicholas *ate Sale* 1296 SRSx; Edmund *del Sale* 1327 SRSf; John *Sales* 1524 SRSf. Dweller *atte sale*, from OE *sealh, salh* 'sallow', used of certain species of *Salix* of a low-growing or shrubby habit, as distinct from osiers or willows. The surnames may also derive from OE *sæl* 'hall', servant at the hall. Hugh *de la Sale* (1277 So) and Philip *de la Sale* (Wo) are also called *de Aula* (MELS).

**Salesbury:** *v.* SALISBURY

**Saleby:** *v.* SALBY

**Salford:** Hugh *de Saleford'* 1215–16 P (Bk/Beds); Hugh *de Salford* 1326 CorLo; John *de Salford* 1351–2 FFSr. From Salford (Beds, La, O), Abbots Salford, Salford Priors (Wa), or Salfords in Horley (Sr).

**Salingar, Salinger, Sallagar, Seliger, Selinger, Sellinger:** Reginaldus *de Sancto Leodegario* 1176 P (Wo); John *de Sentliger* 1327 SRSx; Thomas *Sillinger* 1327 SRWo; Richard *de Seyntliger* 1332 SRSx; John *Selyngier* 1423–66 Bart (Lo). From Saint Léger-aux-Bois (Seine-Inférieure) or Saint-Léger (La Manche). *v.* ANF.

**Salisbury, Salisberry, Salesbury, Salusbury, Salsbury:** William *de Salesberie* 1115 Winton (Ha). From Salisbury (Wilts). In Lancashire, from Salesbury (Lancs): Bernard *de Salesbiry* 1246 AssLa.

Salkeld, Salkield, Salkild, Salkilld, Sawkill: Hamo de Salkil 1210 P (Cu); Thomas de Salkild 1279 AssNb; John Salkeld 1642 PrD. From Salkeld (Cu).

Salkin, Salkind, Sawkins: Salkyn Eldefeld 1327 SR (Ess); Robert, William Salekyn 1255 RamsCt (Hu), 1307 AD iv (K); John Saukyn 1359 Putnam (Co). This might be from Saul but is probably from the more common Salaman: Sale-kin 'little Salaman'.

Sall, Salles: (i) Warin de Salle 1202 FFNf. From Sall (Norfolk). (ii) Robert del Sal(l) 1297 MinAcctCo; John atte Salle 1327 SRWo; John del Sal 1332 SRCu. 'Servant at the hall' (OE sæl, or in Cumb, ON salr). (iii) Johannes filius Salle 1308 RamsCt (Nf). A petname, perhaps from Salaman.

Sallagar: v. SALINGAR
Sallaway: v. SALWAY
Sallis(s): v. SALLOWS

Sallitt: Salide 1066 DB (Ha); Ailwardus filius Salide 1195 P (Berks); Selide le Cupere 1195 P (Nf); John, Matill' Selide 1206 P (Gl), 1221 ElyA (Sf); Robert Seled' 1252 Rams (Hu); William Saled 1297 MinAcctCo (Mx). OE *Sǣlida, an original nickname, not recorded before the Conquest, from OE sǣlida 'seafarer, pirate'.

Salloway: v. SALWAY

Sallows, Sallis, Salliss: Nicholas de Sallowe 1254 RH (Sa); Robert ate Salwe 1297 MinAcctCo; William Salowes 1524 SRSf. 'Dweller by the willows' (OE sealh), though the examples are chiefly singular.

Salman, Salmen, Salmon, Salmond, Salmons: v. SALAMAN

Salsbury: v. SALISBURY

Salt, Sault: Nicholas de Salt 1199 AssSt; William de Saut 1203 Pl (St); Hugh de Salt 1332–3 SRSt. From Salt (Staffs), Saute 1236.

Saltby: Hugh de Salteby 1298 AssL. From Saltby (Lei).

Salter, Saulter, Sauter, Sautter, Sawter: (i) Robert, Philip le Salter 1243 AssSo, 1262 For (Ess); Thomas le Selter 1296 SRSx. OE sealtere 'maker or seller of salt'. cf. John Saltman 1327 SRSf, Laurence le Saltmetere 1300 LoCt. William le Saltere 1279 AssNb is also called le Salterer, i.e. Psalterer, 'player on the psaltery', a stringed instrument like a harp. Thus Salter is at times identical with sautreor, a derivative of OFr saltere 'psaltery'. (ii) John le Sautreor 1276 LLB B; William le Sautreour, minstrel to the Lady Margaret, Queen of England 1304 LLB C. A derivative of OFr sautere 'psaltery'.

Saltern, Salterne: Henry de Salterne 1333 SR (D); John, Stevyn Saltern 1524 SRSf. 'Worker at the salthouse', OE sealt-ærn, as at Saltren's Cottages in Monkleigh (Devon).

Salthouse, Salters: Adam de Salthus 1274 RH (Nf); John Saltos 1525 SRSx; Janet Saltehowse 1562 LaWills; William Saltus 1662 ib. From Salthouse (Norfolk), from minor places where there were saltworks, as at Salthouse in Lytham and Salthouse in Furness (Lancs), or 'worker at the salt-house'. For Salters, cf. DUCKHOUSE and DUCKERS.

Saltley: Walter de Saltleye 1275 SRWo. From Saltley (Wa).

Saltman: Henry, John Saltman 1315 NorwDeeds, 1327 SRSf. 'Maker of salt.'

Saltmarsh, Saltmarshe: Robert de Saltemers 1130 P (Gl); Geoffrey de Saltmarais 1219 AssY; Robert de Saltemerhs 1332 SRLei; Thomas Saltmarsch 1419 IpmY. From Saltmarsh (Glos, Hereford), Saltmarshe (ERYorks), or 'dweller by the brackish marsh'.

Salton, Saltoun: Matilda de Saleton 1219 AssY; William de Saulton 1357 Black (Linlithgow); John Saulton 1536 ib. From Salton (NRYorks, East Lothian).

Saltonstall, Sattenstall, Sattersall: William de Saltonstal 1275 Wak (Y); Richard de Saltonstall' 1379 PTY; Richard Saltonstall 1610 Oxon (Ess); Anne Salthingston 1642 Bardsley (Herts). From Saltonstall in Warley (WRYorks).

Salusbury: v. SALISBURY
Salvidge: v. SAVAGE
Salvin: v. SAUVAIN

Salway, Salwey, Sallaway, Salloway, Selway: Salewi de Bliþesdune 1185 Templars (So); Walter, Adam Salewy 1275 SRWo; Robert Saleweye 1327 SRSo. This is clearly from a personal-name, apparently an unrecorded OE *Sǣlwīg 'prosperity-war'.

Sam, Samme, Sammes, Samms, Sams: Samme Parvus 1275 RH (L); Richard Sammys 1458 NorwW (Nf). A pet-form of SAMPSON. Samson Fullon' 1265 Calv (Y) is also called Samme (c1260 ib.). There was also a pet-form Sampe: Huelin Sampe 1276 RH (L).

Sambell: v. SEMPLE
Samber: v. SEMPER

Samborne, Sambourne: Philip de Sambourne 1297 MinAcctCo; Peter de Samborne 1327 SRSo. From Sambourne (Wa).

Sambridge: Richard de Samebrugg' 1279 RH (O); Matilda atte Samesbrugg 1296 SRSx. The Sussex place is now Salmons Bridge in Tillington. Probably 'dweller by a bridge used in common' (OE *sam-brycg). v. MELS 177.

Sambrook, Sambrooke: Thomas de Sambrok 1258 FFSt. From Sambrook (Salop).

Sameday, Samedy: Reginald Samedy 13th BlackBk (K). Fr samedi 'Saturday', for a child born on that day. cf. SATURDAY.

Samman, Sammon, Sammonds, Sammons: v. SALAMAN, SEAMAN

Sampe: Huelin Sampe 1276 RH (L); Hugelin Sampe 1299 LoCt. A pet-form of Sampson.

Samper: v. SEMPER
Sampford: v. SANDFORD
Sample: v. SEMPLE

Sampson, Samson, Samsin, Sansam, Sansom, Sansome, Sanson, Sansum: (i) Sanson 1086 DB; Samson, Sansone, Sampson Takel(l) Hy 2 DC (L); Samson Cornuwala c1170 Riev (Y); Hugo Samson 1130 P (Nt); Philip Sampson 1192 WhC (La); Hemericus Samsun 1221 Cur (So); William Sansum 1260 AssCh; Marget Sansom 1524 SRSf. OFr Sanson, Samson, the name of a Welsh bishop (fl. 550) who crossed over to Brittany and founded the abbey of Dol where he was buried and venerated as a saint. Whether his name is the Biblical Samson or one of Celtic origin is uncertain. The name was popular in Yorkshire and the eastern counties where it was introduced by the Bretons after the Conquest, and also in the Welsh border counties where it no doubt came from Wales. (ii) Occasionally the surname may be local in origin: Ralph *de Sancto Samsone* 1087 InqAug (K), Albert *de Samsona* 1086 InqEl (C). Probably from Saint-Samson (Seine-Inférieure), or from Saint-Samson (Calvados), Saint-Samson-de-Bonfosse (La Manche) or Saint-Samson-de-la-Roque (Eure).

**Samter:** *v.* SANTER

**Samuel, Samuels, Samwell:** *Samuel* 1198 Cur (K); *Samuelis* Gille 1206 Cur (Ess); Adolfus *Samuel* c1160 DC (L); William *Samwel* 1279 RH (O). This Hebrew name, 'name of God', is not common in the Middle Ages, but the surnames are not necessarily of Jewish origin.

**Samways:** Robert *Samwis* 14th AD i (Nth). OE *sāmwīs* 'dull, foolish'.

**Sanby, Saunby, Saundby:** William *de Sandebi* 1190 P (Y); Hugh *de Sandebi* 1205 P (Nt); William *of Sandeby* 1234 FFY. From Saundby (Nt).

**Sanctuary, Santry, Sentry:** John *Seyntwary(e)* 1517 Landwade (C), 1524 SRSf. ME *seintuarie*, OFr *saintuarie* 'a shrine'. As suggested by Lower, this is probably for one who has taken sanctuary in a church or monastery. For *Sentry*, cf. 'He hath no way now to slyppe out of my hands, but to take sentrie in the hospital of Warwick' (1590 Nashe).

**Sand, Sandes, Sands, Sandys:** William *de Sandes* 1205 Cur (Sr); Walter *de la Sonde* 1248 FFSr; Andrew *atte Sonde* 1296 SRSx; Reginald *del Sond* 1298 Ipm (Sf); Thomas *Attensandes* 1301 SRY; Gilbert *del Sandes* 1332 SRCu. 'Dweller on sandy soil or by the sands', OE *sand* 'sand'.

**Sandal, Sandall:** (i) Peter *de Sandal* 1188 WhC (La). From Sandal Magna, Kirk or Long Sandall (WRYorks). (ii) Rogerus *filius Sandolf* 1208 P (Gl); *Saundulfus* 1221 AssWa; Thomas *Sandolf* 1327 SRSf. ON *Sandúlfr*.

**Sandars, Sander, Sanders, Saunder, Saunders:** *Sandre* c1248 Bec (O); *Sander* 1255 RH (Sa); Henry *Sandres* 1275 SRWo; William *Sandre, Saundre* 1316–17 AssK; Richard *Saunder* 1332 SRSt. *Sander*, a pet-form of *Alexander*.

**Sanday:** *v.* SANDEY

**Sandbach, Sandbatch:** Richard *de Sandebech* 1227, Roger *de Sandbach* 1254 FFSt. From Sandbach (Ch).

**Sandbeck:** Robert *de Sandebecke* 1303 IpmY. From Sandbeck (WRY).

**Sandcraft:** Robert *de Sandecroft* 1200 FFSf. From residence near a sandy croft.

**Sandeford:** *v.* SANDIFER

**Sandell, Sandells, Sandels, Sandhill:** (i) William *de Sandhull* 1276 AssSo; Walter *de Sandhell* 1327 SRSx; William *atte Sandhille* 1332 SRSx. 'Dweller by a sand-hill.' OE *sand, hyll*. (ii) Alexander *de la Sandhelde* 1275 AD i (Sx); William *ater Sandhyld* 1296 SRSx. 'Dweller by a sandy slope.' OE *sand, hylde*.

**Sandelson:** For SANDERSON.

**Sandeman:** John *Saundirman* 1379 PTY; David *Sandeman* 1628 Black (Perth); Richard *Sandeman, Sandiman* 1645, 1674 FrY. 'Servant of *Saunder*' (Alexander). cf. Robert *Alexsanderman* 1379 PTY.

**Sandercok:** John, William *Sandircok* 1425 AD i (K). A diminutive of *Sander*.

**Sanderford:** Thomas *Sanderford* 1662–4 HTDo. Perhaps from Sandford Orcas (Do).

**Sanders:** *v.* SANDARS

**Sanderson, Saunderson:** Adam *Saunderson* 1349 LoPleas; Robert *Saundreson* 1464 FeuDu. 'Son of *Sander*' (Alexander).

**Sandes:** *v.* SAND

**Sandever:** *v.* SANDIFER

**Sandey, Sandy, Sanday:** (i) Hugo *de Sandeia* 1202 AssBeds. From Sandy (Beds). (ii) Adam *Sandi* 1332 SRCu. ON *Sandi*. cf. SANK.

**Sandford, Sanford, Sampford:** Jordan *de Sandforda, de Samford'* 1175, 1190 P (W); Bartholomew *de Sandford* c1280 SRWo; John *Sandford* 1473 IpmNt; James *Sanford* 1642 PrD. From Sandford (Berks, Devon, Dorset, Oxon, Salop, Westmorland), or Sampford (Devon, Essex, Som).

**Sandham:** Robert *Sandham* c1405 FS; Thomas *Sandam* 1525 SRSx. From Sandown (Wt), *Sandham* 1287–90.

**Sandhill:** *v.* SANDELL

**Sandhurst:** Hugh *de Sandhurst* 1212 P (K); William *de Sandhurst* 1265 IpmGl; Ivo *Sandhurst* 1373 FFEss. From Sandhurst (Berks, Gl, K).

**Sandifer, Sandiford, Sandeford, Sandever, Sandyfirth:** Richard *de Sandiforth* 1286 Wak (Y); John *de Sandeforthe* 1379 PTY. From an unidentified Sandiford, apparently in Yorkshire.

**Sandle:** For SANDAL or SANDELL.

**Sandler:** *v.* SANTLER

**Sandon:** Richard *de Sandun'* 1222 DBStP; Gilbert *de Sandon* 1234–5 FFSr; John *de Sandon* 1284–5 FFEss. From Sandon (Berks, Ess, Herts, St), or Sandown (Sr).

**Sandro:** Adam *de Sandewra* 1332 SRCu. From Sandraw (Cumb).

**Sandwell, Sanwell:** John *de Sandwell* 1250 Fees (Bk). From Sandwell (St), Sandwell in Harberton (D), or 'dweller by the sandy stream or spring', OE *sand, wiella*.

**Sandwich:** Wibert *de Sanwic'* 1221 Acc; Laurence *de Sandwyco* 1297 MinAcctCo; Nicholas *de Sandewyche* 1342 LLB F. From Sandwich (K).

**Sandy:** *v.* SANDEY

**Sandyfirth:** *v.* SANDIFER

**Sandys:** *v.* SAND

**Sangar, Sanger, Songer:** John *le Songere* 1296 MEOT (Herts); Richard *le Sangere, le Songer* 1327 SR (Ess). OE *sangere, songere* 'a church-singer, chorister'.

**Sangster:** Sibilla *Sangistere* 1327 MEOT (Ha); Adam *le Songster* 1327 ib. (La); James *Sankstar* 1452 Black (Aberdeen). OE *sangestre*, feminine of *sangere*, though not always used of a woman. In Scotland 'a chorister'.

**Sanguine, Sangwin, Sangwine:** John *Sanguin* 1194 P (Gl); William *Sangwyn* 1270 AssSo. ME, OFr *sanguin* 'of a sanguine complexion'.

**Sank:** William *samke* 1221 ElyA (Sf). A rare surname. *Samke* is an interesting example, hitherto unnoted, of the vitality of the Anglo-Scandinavian name-system in the eastern counties. It is of a type noted by Stenton in the Danelaw: ON *Steinki*, a short form of a name compounded with *Stein-*, *Anke* (ON *Anki*), from ON *Arnkell*, etc. *Samke* may be from ON \**Sandúlfr* (cf. SANDAL), or, possibly, from *Sanni*, a pet-form of ON *Sandi*, both recorded in 12th-century Lincolnshire charters (IPN 185). *Sandi* is the first element of Saundby (Notts).

**Sankey, Sanky:** Gerard *de Sanki* t Hy I Fees; Roger *de Sonkey* 1246 AssLa; William *Sanky* 1292 AssCh. From Sankey (Lancs).

**Sankin:** John *Sampkin* 1351 *ERO* (Ess); William *Samkyn* 1358 LLB G. 'Little *Samme*', a diminutive of Samson, plus *-kin*.

**Sanne, Sans:** A diminutive of *Sanson*: Eynon son of *Sanne* 1260 AssCh. cf. William *Sampson* ib.

**Sansam, Sansom(e), Sanson, Sansum:** v. SAMPSON

**Sansaver, Sinaver:** Hugh *Sanzaveir* 1219 P (Sx), 'Without goods'. OFr *sans*, *aveir*. OFr *sans* is a common first element in nicknames, though its exact sense is not always clear. cf. Adam *Sauns Buche* 1296 SRSx 'without a mouth'; John *Saunbrays* 1332 SRSx 'without breeches'; Thomas *Saunfayle* 1332 SRSt 'without fail'; Alice *Sans Mauntel* 1236 Barnwell (C) 'without a mantle'.

**Sant:** v. SAINT

**Santer, Saunter, Samter:** John *Sancterre* 1260 AssC; John *Sanzterre*, *Santerre*, *Säuntere* 1275 RH (L). OFr *sanz terre* 'without land'. cf. *Lackland*, the nickname of King John. v. also SENTER.

**Santler, Sandler, Sendler:** (i) Robert *de Sancto Elerio* 1219 AssL; Roger *de Seinteler'* 1235 Fees (Nth); Peter *de Saint Elena*, Hilary, *Leyre* 1268–9 AssSo. From Saint-Hilaire-du-Harcouët (La Manche). v. ANF. (ii) Geoffrey *de Sancto Laudo* 1148 Winton (Ha); Hubert *de Sancto Lot* c1158 EngFeud; Gilbert *de Sanlo* c1150 DC (L); Adam *de Sainlow* 1275 RH (L); Agnes *Senclowe*, *Seyntlowe* 1422, 1444 Rad (C). From Saint-Lô (La Manche, Somme) or Saint-Laud (Maine-et-Loire).

**Santon:** William *de Santon'* 1202 FFY; Peter *de Saunton* 1282, Thomas *de Santon* 1364 IpmY. From Santon (Cu, L, Nf).

**Santry:** v. SANCTUARY

**Sanwell:** v. SANDWELL

**Sape:** v. SOPP

**Sapey:** Bernard *de Sapie* 1184 P (Wo); Roger *de Sapie* 1221 AssWo; Robert *Sapi* 1327 SRSx. From Upper Sapey (He), or Lower Sapey (Wo).

**Sapp, Sappe:** Robert *del Sap* 1199 P (C); William *Sap* 1327 SRC; Gerard *Sape* 1406 IpmY. 'Dweller by the spruce tree', OE *sæppe*. v. SOPP.

**Saper, Sapier:** v. SOPER

**Sapsed, Sapsford, Sapstead, Sapsworth:** Richard *de Sabricheword'* 1230 P (Lo). *Sapsworth* or *Sapseth* was the old pronunciation of Sawbridgeworth (Herts) which is *Sapsforde* in 1568. *Sapseth* became *Sapsed*, and, with an intrusive *t*, *Sapstead*.

**Sara, Sare, Sarra, Sarre:** *Sarra* c1160 DC (L), 1219 AssY; Benedictus *filius Sarre* 1169 P (Nf); Alan *Sare* 1296 SRSx; Adam *Sarre* 1317 FFEss. Hebrew *Sara(h)* 'princess'. *Sare* may also be for SAYER.

**Sarch:** v. SEARCH

**Sare:** v. SARA, SAYER

**Sarel(l):** v. SARL

**Sarfoot:** William *Sarfot* 1297 SRY. 'Sore foot', OE *sār*, *fōt*. cf. Hugh *Spikfot* 1219 AssY; John *Stelfot* 1301 LLB C 'steel foot'.

**Sargaison:** v. SARGEANTSON

**Sargant, Sargeant, Sargeaunt, Sargent, Sargint, Sarjant, Sarjeant, Sarjent, Seargeant, Searjeant, Sergant, Sergean, Sergeant, Sergeaunt, Sergent, Serjeant, Serjent:** Edric *le sergant* Hy 2 DC (L); Thomas *Seriant* 1185 Templars (O); Robert *le Serjaunt* 1221 Cur (Lei); Thomas *le Sergeate* 1266 AssSt; John *Sargeant* 1396 FFSf; Thomas *Sarjeant* 1689 FrY. OFr *sergent*, *serjant*, probably in general 'servant' (c1200 NED). Often latinized as *serviens*. Other possible meanings are: 'an officer charged with enforcing the judgements of a tribunal, arresting offenders, or summoning persons to appear before a court' (a1300 NED) or 'a tenant by military service under the rank of a knight' (c1290 NED).

**Sarge:** v. SEARCH

**Sargeantson, Sargentson, Sargeson, Sarginson, Sargison, Sargisson, Sargaison, Sarjantson, Sergenson, Serginson, Sergison, Surgison, Serjeantson:** William *Sergantson* 1379 PTY; Thomas *Sargenson*, *Sardganson* 1477, 1488 GildY; Thomas *Sarganson* 1538 SfPR; Margareta *Sergison* 1593 RothwellPR (Y); James *Sargeson* 1740 FrY. 'Son of the serjeant.'

**Sark, Sarkes:** v. SEARCH

**Sarl, Sarll, Sarel, Sarell:** *Sarlo Iuuenis* 1091–3 LVH; *Sarle Tinctor* 1274 RH (Hu); Matilda, Thomas *Sarle* 1275 RH (C), 1327 SRSf, SRSx; Robert *Saryll*, clericus 1412 FrY; Thomas *Serle*, capellanus, filius Roberti *Sarle*, clerici 1438 ib.; Edward *Sarel* 1788 Bardsley. Anglo-Scandinavian or Norman *Sarli*, *Sarle*, from ON *Sǫrli*. This is the ON cognate of OG *Sarilo* which survives as SEARL, and is more common. In later examples, *Sarle* may be a development of *Serle*.

**Sarratt, Sarrett:** *Saretus* the serjeant 1130 P; Robert *Sarrot* 1297 MinAcctCo; Robert *Sarote* 1327 SRSa. Usually *Sarr-et*, *Sarr-ot*, diminutives of *Sarre*, i.e. *Sara*, but evidently there was also a masculine name *Saret*.

**Sarre:** v. SARA

**Sarson, Sarsons:** (i) Adam *Sareson* 1285 LLB A; John, Peter *Sarson* 1332 SRCu. 'Son of *Sara*' (cf. Gilebertus *filius Sarre* 1327 SRC), or 'son of *Saer*' (cf. Richard *Saressone* 1327 SRSf). v. SAYER. (ii) Oliver *Sarazin* 12th DC (Lei); Philip *le Saracin* 1201 AssSo; Stephen called *Saracen* 1281 AssCh. OFr *Sarrazin* 'a Saracen', used as a nickname for one of swarthy complexion.

**Sarter, Sartor:** Henry *de Sartis* 1185 Templars (Wa); Gilbert *de Assartis* 1239 FFEss; William *de Sartere* 1310 ColchCt. 'Dweller at the clearing in the woodland', OFr *assart*, *essart*.

**Sartin, Sartain, Sarton, Sattin, Sertin:** William *Certayn* 1394 LLB H; Richard *Sartin* 1693 DKR 41 (W). OFr *certeyn* 'self-assured, determined'.

**Sarvent:** v. SERVANT

**Sarver:** Turstan *le seruier* 1197 P (Nf); Thomas *Server* 1516 LLB F. OFr *serveur* 'servant'.

**Sarvis:** v. SERVICE

**Satch:** v. SACH

**Satchel, Satchell:** Nicholas, Thomas *Sachel* 1243 AssSo, 1327 SRSo. OFr *sachel* 'a little bag', for a maker of these. cf. SACH.

**Satchwell:** v. SETCHFIELD

**Sattenstall, Sattersall:** v. SALTONSTALL

**Satterday:** v. SATURDAY

**Satterley, Satterly:** Edmund *de Saterleye* 1242 Fees (Sf); Andrew *Satterley*, William *Satterly*, *Satterleigh* 1642 PrD. From Satterleigh (D).

**Satterthwaite, Satterfitt, Setterfield:** William *Setterthwaite* 1614 FrY; Charles *Satterthwait* 1625 ib.; Thomas *Satturwaite* 1653 ib. From Satterthwaite in Hawkstead (Lancs).

**Sattin:** v. SARTIN

**Saturday, Satterday:** Alan *Saterdai* 1194 P (L); William *Seterday* 1365 FFY; Hamo *Saturday* 1375 ColchCt. OE *Sæterndæg* 'Saturday', presumably for one born on that day.

**Saturley:** v. SATTERLEY

**Saul, Saull, Sawle:** Ricardus *filius Sawl* 1198 AC (Lo); Ralph *Saule* 1255 Rams (Hu); John *Sawle* 1296 SRSx. Hebrew *Saul* 'asked for'. Not a common medieval name.

**Sault:** v. SALT

**Saulter:** v. SALTER

**Saunby, Saundby:** v. SANBY

**Saunder(s):** v. SANDARS

**Saunderson:** v. SANDERSON

**Sauner, Sawner:** Adam *le saunier* 1209 P (C); Hugh *le Sawner* 1279 RH (Hu); Lambert *Sawner* 1517 ArchC 34. OFr *saunier* 'salter'.

**Saunt:** v. SAINT

**Saunter:** v. SANTER

**Sausby:** Probably a phonetic spelling of *Saursbye* for SORBIE.

**Sauser:** Adam *le Sauser* 1210 Cur (Cu); Robert *le Sauser* 1285 FFHu; Roger *le Sauser* 1340 FFW; Robert *Sawcer* 1525 SRSx. OFr *saucier, saussier* 'a maker of sauces'.

**Saut(t)er:** v. SALTER

**Sauvage:** v. SAVAGE

**Sauvain, Sauvan, Sauven, Savin, Salvin, Selwin, Selwyn, Selwyne, Sylvaine:** (i) *Siluein* de Torp 1170 P (Wa); Ricardus *filius Seluein* 1195 FF (Nth); Robertus *Siluein, Seluenus* 1127–34, c1140 Holme (Nf); Henry *Siluain* Hy 2 Gilb (L); Osbert *Seluein* 1162 P (Y); Robert *Seluan* 1185 Templars (L). *Silvanus*, a Latin cognomen, from *silva* 'wood', the name of the god of the forests and also of a saint. (ii) William *Salvayn, Salven* c1170 Riev (Y); Geoffrey *Selvayn, Salvan'* 1242 Fees (L); Hugo *Salveyn, Selveyn* ib.; Robert *Selveyn, Selweyn* 1244, 1271 AssSt. This may, at times, be from *Silvanus* but is more often from OFr *salvagin* 'wild, savage'. cf. Thorpe Salvin (WRYorks) from Henry *Selvein* (temp. Ric. 1). (iii) *Seleuuinus* 1066 DB (W); Paganus *filius Selewin'* 1203 P (Beds); Roger *Selewyne* Hy 3 AD ii (Mx); Juliana *Selewyne* 1332 SRSx. \**Selewine* 'hall-friend', unrecorded in OE, is found once in DB and again in the 13th century. Always rare, it is one origin of *Selwin*.

**Savagar, Savager, Saveker, Savigar, Savigear:** Humfridus *de Sancto Vigore* 1168 P (W); Thomas *de Saint Vigor* 1268 AssSo. From Saint-Vigor (La Manche, etc.).

**Savage, Savaage, Savege, Savidge, Sauvage, Salvage, Salvidge:** Edric *Saluage* 1066 DB (He); Edricus cognomento *Silvaticus* 1067 OEByn; John *Saluage* 1166 P (Nf); William *le Saluage* 1194 Cur (Sa); Robert *le Sauuage* 1198 FF (Sr); Ralph *le Savage* 1268 FFSf. OFr *salvage, sauvage*, Lat *silvāticus*, in popular Latin *salvāticus*; 'savage, wild' (a1300 NED).

**Savary, Savery, Savory, Savoury, Severy:** Joldewinus *filius Sauarici* 1158 P (Sx); Jocelinus *filius Sephari, Sefar, Safar* 1195 Oseney (O), filius *Saphar, Safari* 1200 ib., *filius Seauary* 1210 ib.; Willelmus *filius Saveric, filz Saviere* 1206–7 Cur (Do); Philip, William *Savery* 1276 RH (Lei), 1327 SRC; Robert *Sauerai, Seffray* 1327, 1332 SRSx. OFr *Savari*, OG *Sabaricus, Savaricus*. This may also have become SAFFER; its forms are confused with those of SAFFERY.

**Saveall, Savell:** v. SAVIL

**Savege, Savidge:** v. SAVAGE

**Savigny:** Ralph *de Sauenie, de Sauigneo, de Sauigni* 1086 DB; Jordan *de Saueigni* 1196 P (Nt). The DB under-tenant came from Savenay (Calvados). v. ANF. Others may have come from Savigni (La Manche) or Savigni-le-Vieux (La Manche).

**Savil, Savill, Savile, Saville, Saveall, Savell, Saywell, Seville:** John *de Sayvill* 1246 FFY; Stephen *de Savile* 1277 FFY; Robert *Sayuill* 1379 PTY; John *Sayvell* 1431 Calv; Rosemunda *Savell* 1549 RothwellPR (Y); Thomas *Savil* 1672 HTY. Perhaps from Sauville (Ardennes, Vosges), or Sainville (Eure-et-Loir). Sometimes maybe for SAFFELL.

**Savoner:** Richard *le Savoner* 1225 AssSo; Nicholas *le Savoner* 1255 RH (Sa); Agnes *le Savoner* 1279 RH (Bk). A derivative of OFr *savon* 'soap', a maker or seller of soap.

**Savory, Savoury:** v. SAVARY

**Saward, Saword:** v. SEWARD

**Sawbridge:** Ysaac *de Salebrigg'* 1221 AssWa. From Sawbridge (Warwicks). v. also SEABERT.

**Sawdon, Sawden:** Ralph *de Saldene* 1257 AssY; Robert *Sawdan* 1502 GildY; John *Sawdon* 1525 SRSx. From Sawdon (NRYorks), *Saldene* 1289.

**Sawell:** v. SEWALL

**Sawers:** Jordan *le Sawer* 1257 MEOT (Y); Baldwin *le Sawere* 1270 AD iii (K); Cetil *Sower* 1332 MEOT (L); William *Sawer*, Robert *Sawyer* 1524 SRSf. This surname is much less frequent, both in early documents and today, than *Sawyer* which is therefore unlikely to be, as suggested by NED, an altered form of *sawer* 'one who saws'. The earliest example of *sawer* in NED is from a 1379 Yorkshire surname which is late and throws no light on the meaning. The early examples above are probably from OE *sāwere* 'a sower (of seeds)'. In the 16th century *sawer* may well be a slurred pronunciation of *sawyer*.

**Sawkill:** v. SALKELD

**Sawkins:** v. SALKIN

**Sawle:** v. SAUL

**Sawner:** v. SAUNER

**Sawter:** v. SALTER

**Sawyer, Sawyers:** Nicholas *le Sagyere* c1248 Bec (Berks); Humfrey *le Sayhare, le Sawyere*, Robert *le*

*Sawyere, le Saweare* 1270 AssSo; Richard *le Sawier'* 1278 LLB B; Philip *le Sagher* 1324 Wak (Y); John *le Saghiere* 1327 SRSx. A derivative of ME *saghe, sawe* 'to saw', 'a sawer of timber, especially in a saw-pit' (1350 NED). ME *saghier* also became *sayier* and survives as SAYER, a form which might also derive from OFr *seieor* 'sawyer'. *v.* also SAWERS, SEWER.

**Sax, Saxe:** *Saxe* de Hotton' 1190 P (Y); *Saxe* de la Hal 1283 SRSf; Godwin *Sax* 1178 P (Ha); Reginald *Saxe* 1212 P (L); John *le Sax* 1327 Pinchbeck (Sf). ON, ODa *Saxi*, OSw *Saxe*.

**Saxby:** (i) Nicholas, Simon *de Saxebi* 1200 P (Lei), 1202 AssL. From Saxby (Leics, Lincs). (ii) Jordan *Sacheespee* 1183 P (Y); Robert *Sakespe* 1206 Cur (L); Thomas *Sakespey* 1296 SRSx; John *Saxepe* 1327 ib. 'Draw sword', a name for a trainer in swordsmanship. cf. Fr *Sacquépée* and John *Draweswerd* 1327 SRSf.

**Saxelby, Saxelbye:** Richard *de Saxelebi* 1202 AssL. From Saxilby (L, Lei).

**Saxon, Saxton, Sexton:** Jordan *de Saxton*, Simon *de Sexton* 1208 Cur (Y, C). From Saxton (WRYorks) or Saxton Hall and Saxon Street (Cambs). *v.* also SEXTEN.

**Say, Saye:** Jordan *de Sai* 1161 Eynsham (O). From Sai (Orne).

**Sayburn:** *v.* SEABORN

**Sayce, Seys:** Em' *Seis* 1255 RH (Sa); Jeven *Sais* 1392 Chirk. Welsh *sais* 'Saxon, Englishman'.

**Saycell:** *v.* CECIL

**Sayer, Sayers, Sayre, Saer, Sare, Seyers, Sear, Seares, Sears, Seear, Seers:** (i) *Saher* de Arcelis 1147–53 DC (L); Stephanus *filius Seir* 1148–52 Bury, *filius Saheri* (*Saieri*) c1160 DC (L); *Seherus, Seiherus* de Quenci c1155 RegAntiquiss (L); *Seyr'* de Quinci 1205 ib.; Robertus *filius Seer* 12th DC (Nt); *Saerus* Thomasseruaunt Clench 1318 Pat (Ess); Richard *Sayer* 1230 P (D); Nicholas *Sare* 1275 SRWo; Godwin *Seer* 1279 RH (C); Stephen *Saer* 1281 LLB A; Thomas *Sare*, John *Sayer* 1292 FFEss; Robert *Saier*, Roger *Sayher* 1322, 1324 ib.; Roger *Seyer* 1302 LLB B. *Saer, Sayer* is a personal-name common in early records in a variety of forms. It is obviously Norman, though of Old German origin, perhaps OHG *Sigiheri* (Ekwall). *Sear*(*s*), *Seers*, etc. seem to derive only from the personal name. *v.* SARA, SEARSON. (ii) Richard *le Saer* 1204 AssY; Humfrey *le Sayhare, le Sawyere* 1270 AssSo; Robert *le Sayer* 1284 LLB A; Geppe *le Sahar* 1285 Wak (Y); Alice *Sayeres* 1310 EAS xx; Robert *Saer, Sawer, Sare, Saare* 1465–87 Combermere (Ch); John *Sare* or *Sayer* 1605 ChwWo. This common occupational name, with its varied early and modern forms, has five different origins. It is clearly a form of SAWYER (*le Sayhare* or *le Sawyere* 1270), or of SEWER (iii), *le Syur* or *le Sayur* 'sawyer' (1286), and may occasionally be a derivative of OE *secgan* 'to say', 'a professional reciter' (c1330 NED, with forms *segger, seiere, saier*). It may also be an aphetic form of ME *assayer* from AFr *assaior, assaiour* 'one who assays or tests', 'an assayer of metals', or 'a fore-taster of food, etc.' (1370 NED). Or, especially in later examples, it may be from OFr *saier*, from *saie* 'silk, serge', 'a maker or seller of say'.

**Sayles:** *v.* SAILES

**Saylor:** *v.* SAILER

**Sayner, Saynor:** *v.* SENIOR

**Sayre:** *v.* SAYER

**Saysell:** *v.* CECIL

**Sayve:** *v.* SAIVE

**Sayward:** *v.* SEWARD

**Saywell:** John *Seywell* 1600 FrLeic, Henry *Seewell* 1754 ib. For SEWALL or SEWELL. *v.* also SAVIL.

**Scafe, Scaife, Skaife:** Geoffrey *Skaif'* 1219 AssY, 1240 AssLa; Robert *Scafe* 1418 FrY. ON *skeifr*, NEng dial *scafe* 'crooked, awry; awkward, wild'.

**Scalby:** John *de Scalby* 1316 RegAntiquiss; William *Scalby* 1407 IpmY. From Scalby (ERY).

**Scale, Scales, Schoales, Scholes, Scoles:** Richard *del Scoles* 1275 Wak (Y); Adam *de Scoles* 1285 AssLa; Thomas *del Scales* 1332 SRCu; William *del Scale* ib.; John *del Scholes* 1379 PTY. 'Dweller by the hut(s) or shed(s)' (ON *skáli*, ME *scale, scole*), as at Scales (Cumb, Lancs), Scholes (WRYorks) or Scole (Norfolk).

**Scales:** Hugh *de Scalariis* 1195 P (C); Robert *de Scales* (K), *de Schales* (Nf), *de Eschales* (Sf) 1242 Fees; Robert *de Scales* 1281 FFEss; Margaret *de Skales*, Isabelle *de Schales* 1327 SRSf. Probably from Escalles (Pas-de-Calais). *v.* OEByn 87.

**Scamell, Scammell:** William *Scamel* 1185 P (W); Symon *del Scameles* 1302 FrY; William *Scammel* 1375 IpmW. OE *scamol* 'a bench (on which meat was exposed for sale)', hence a worker in a slaughterhouse, or in a fish or meat market.

**Scarborough, Scarbrough, Scarbrow, Scarbro:** Hakun *de Scardeburg* 1176 P (Y); Alberic *de Scartheburg'* 1230 MemR (Y); Robert *de Scartheburgh* 1348 FFY; William *Scarbrugh* 1418–19 FFSr. From Scarborough (NRY).

**Scarcliff, Scarcliffe:** William *de Scardecliue* 1212 P (Ch); Roger *Skardeclyve* 1366 TestEbor. From Scarcliff (Db).

**Scarf, Scarfe, Scarff, Scarffe:** John *Scarf* 1260 AssY; Henry *Scharf* 1275 RH (L). ON *skarfr* 'cormorant'.

**Scargill:** Warin *de Scakergill'* 1177 P (Y); William *de Skaregile* 1285 IpmY; John *Scargill* 1459 Kirk. From Scargill (NRYorks).

**Scarlet, Scarlett:** William *Scarlet* 1185 Templars (O); Geoffrey *Escarlata* 1195 P (W); Ralph *le Escarlet* 1220 Cur (So). OFr *escarlate* 'scarlet'. Probably a dealer in 'scarlet', the name of a cloth already in 1182 (P): 'x ulnis de escarleto'.

**Scatchard:** John *Skacher* 1327 SRSf; Thomas *Scochard* 1336 AD v (K); John *Scacharde*, William *Skachade* 1381 SRSt. Probably a derivative of ONFr *escache*, OFr *eschace*, modFr *échasse* 'a stilt'. *Skacher* may be 'maker of stilts'; *scachard* is a pejorative, perhaps, as suggested by Harrison, a nickname for a long-legged bird such as the heron, later applied to a man. Dauzat explains the French *Eschasseriaux* as 'man with a wooden leg'.

**Scathelock:** Adam *Scatheloc* 1315 AD i (Wa); Matilda *Schathelok* 1359 AssD; John *Scathelok* 1402 FrY. Apparently 'burst the bars', OE *sceððan* but. But the second element may be OE *locc* 'hair', while the earliest example of the name, Ranulf *Scathelac* 1196

FFSr, would suggest OE *lāc* 'play, sport', perhaps in some such meaning as 'spoil-sport'.

**Scattergood:** Henry *Skatergot* 1219 AssY; Walter *Skatergod* 1247 AssBeds; Robert *Scatergod* 1327 SRY. 'Scatter goods', a nickname for a spendthrift or, possibly, for a philanthropist.

**Schoales:** *v.* SCALE

**Schoemaker:** *v.* SHOEMAKE

**Schofield, Schoefield, Scholefield, Scholfield, Scolfield, Scoffield, Scofield:** John *de Scholefeld* 1343 WhC (La); Richard *de Skolefeld* 1363 FrY; Anne *Scofeld* 1561 RothwellPR (Y); John *Schofeld* 1592 ib. 'Dweller by a field with a hut' (ON *skáli*, OE *feld*), a common Lancashire and Yorkshire name.

**Scholar, Scholard, Scholer, Schollar, Scholler, Scollard, Scoular, Scouler, Scouller:** Adam *del Scoler* 1332 SRLa; Henry *Scoular* 1525 Black. Probably 'dweller by the shieling with a hut', ON *skáli, erg*.

**Schol(e)field:** *v.* SCHOFIELD

**Scholes:** *v.* SCALE

**Scholey, Schooley, Scollay:** John *de Scolay* 1379 PTY. Probably 'dweller by the low-lying land with a hut', ON *skáli*, OE *ēg*.

**School, Scowle:** *Scule* 1066 DB; Robert *filius Scule* 1196 FFNf; Robert *Scule* c1165 Bury; Richard *Scoule* 1297 SRY; Gilbert *Scul* 1327 SRSx. ON *Skúli*, ODa *Skuli*, OSw *Skule*.

**Schoolcraft, Scowcroft:** Richard *de Schalecroft* 1246 AssLa; Richard *Scowcroft* 1689 Bardsley. 'Dweller by the croft with a hut', ON *skáli*, OE *croft*.

**Schoolmaster:** Geoffrey *Scolmayster* 1226–33 CartNat; Alice *le Scoulmaystr'* 1332 SRSx; Richard *Scolemaystre* 1392 TestEbor. 'The schoolmaster', Lat *scola*, OFr *maistre*.

**Schorah:** *v.* SCORAH

**Schrieve:** *v.* SHERIFF

**Science:** *Sciencia filia Gerardi* 1260 AssC; *Sciencia ate Watere* 1332 SRSr; Roger *Science* 1642 PrD. Spanish and Provençal *Sancha, Sanchia* (f). Introduced into England in 1243 by the marriage of Richard of Cornwall to *Sanchia*, daughter of the Count of Provence. It appears in ME as *Cynthia, Scientia, Science*, and later as *Sens, Sence, Sense, Saints, Sanche*, cf. *Sanche* widow of John Strelley 1502 AD iii (Nf), James Bynde and *Sanctia* or *Sence* his wife 1620 ODCN.

**Scillitoe:** *v.* SHILLITO

**Scoffield, Scofield:** *v.* SCHOFIELD

**Scoles:** *v.* SCALE

**Scollan:** *v.* SCOTLAND

**Scollard:** *v.* SCHOLAR

**Scollas:** *Scolacia* (f) 1261–2 FFWa; *Scolacia* Eustas 1327 SRSx; Eva *Scolace* 1249 AssW; Mabel *Scholace* 1296, Robert *Scolace* 1332 SRSx. *Scolace* appears to be the vernacular form of Lat *Scholastica*, the name of a saint who was the sister of St Benedict and the first nun of the order. It is found as a christian name in England from the late 12th century until the Reformation: *Scolastica* (f) 1195 P (Y), 1207 Cur (C), 1221 Cur (W), 1316 FA (Wa).

**Scollay:** *v.* SCHOLEY

**Scorah, Scorer, Scorrer, Schorah:** William *le Scorur* 1297 SRY; Alice *la Scoriere* 1327 *SR* (Ess). A

derivative of OFr *escorre, escourre* 'to run out', a scout, spy, or of OFr *escurer* 'to scour', a scourer, cleanser.

**Scorby, Scoreby:** Thomas *de Scoreby* 1381 PTY. From Scoreby (ERY). cf. Adam *Scorebyman* 1381 PTY.

**Score:** Henry *Scor* 1297 MinAcctCo; John *atte Score* 1330 PN D 34. From Score in Ilfracombe, Scur Fm in Braunton (D), or 'dweller at the steep place', OE *\*scoren*.

**Scoreby:** *v.* SCORBY

**Scorton:** Hugh *de Scorton* Hy 3 IpmY; Henry *de Scortone* 1296 Black. From Scorton (La, NRY).

**Scotbrook, Scotchbrook:** Hugh *de Scotbroc* 1212 P (Berks); Henry *de Scotesbroke* 1275 RH (W). From Shottesbrooke (Berks), *Scotebroc* 1190.

**Scotford:** John *Scotteforde* 1355–9 AssBeds. From Scotforth (La).

**Scothern, Scothorne, Scothron, Scottorn:** Hugh *de Scotþorn* 1279 RH (O). From Scothern (Lincs).

**Scotland, Scollan:** (i) *Scotlandus* 1081 Bury (K); *Scollandus* 1086 DB (Sx); *Scotlande* 1101–7 Holme (Nf); Gaufridus *filius Scollandi* 1130 P (Ha); William *Escoland* 1155 FeuDu; Jordan *Escotland* 1196 P (L); Thomas *Escollant* 1198 FFK; William *Scotland* 1332 SRSx. *Scotland, Scolland* is a personal name found in Normandy but not in Germany. It is formed from the name of the *Scots*, plus *-land*. (ii) Galfridus *de Scotland'* 1193 P (Ess); Geoffrey *de Scolaund* 1268 AssSo. From Scotland. Much rarer than the personal name. In Scotland, from Scotland Well in Portmoak (Kinross): Richard *de Scocia* 1178–80 (Black).

**Scotney:** Hugo *de Scotini* 1143–7 DC (L); Walter *Escoteni, Escoteigni* 1195–6 P (L); William *de Scoteny* 1219 AssY. From Etocquigny (Seine-Inférieure). Scotney Castle (Sussex) was built c1180 by Walter *de Scotiniis*. *v.* ANF.

**Scotson:** Gilbert *Scotessun'* 1131 FeuDu; Alexander *Scotteson* 1379 PTY; John *Scotchson* 1689 FrY. 'Son of *Scot*.' *v.* SCOTT.

**Scott, Scotts, Scutt, Scutts:** (i) Uchtred *filius Scot* c1124 Black (Selkirk); Roger, William *Scot* c1150–60 DC (L), 1183 Boldon (Du); Gillebertus *Scottus* 1177 P (C); Hunfridus, Robertus *le Scot* 1194 P (W), 1197 P (Wa); Aluredus *Scotticus* 1198 P (K); Adam *le Scot* c1221 Black (Dryburgh); John *le Escot* 1260 AssC. (ii) Godwin, Lefstan *Scut* 1185 P (Nf), 1190–1200 Seals (Sf); John *le Scut* 1248 *Ass* (Ha); William *le Skut* 1327 SRSx. *Scott* is one of the twelve commonest surnames in Scotland, a border name, from OE *Scott*, originally 'an Irishman', later 'a Gael from Scotland'. In the English border counties where the surname is also common the name means 'a man from Scotland', not necessarily a Gael. The surname may also derive from a personal-name, OE *Scott*. According to Guppy, *Scott* is well-established in the eastern counties whilst the Scutts of Dorset are as numerous as the Scotts of Devonshire or Yorkshire. The 12th-century Scots of eastern England may well have been retainers of David I, King of Scotland, who succeeded to the extensive lands of Waltheof, Earl of Huntingdon, on his marriage to Waltheof's daughter. Many of these must have remained in England and their surnames became hereditary for in the 1327 Subsidy Roll for Suffolk there are 35 men called *Skot*. Some may have been retainers of Scottish nobles who accepted Edward I's overlordship and fought against Bruce. The surname, too, may in part have absorbed *Scutt*, also found in the east. This is probably OFr *escoute* 'a spy' which would become ME *scut* and be confused with *scot*.

**Scotter:** Roger *de Scotre* 1202 FFL. From Scotter (Lincs). Also for SCOTTOW.

**Scotter, Scutter:** Alan, John *Scutard* 1279 RH (C, O); Hugh, Walter *Scotard* 1327 SRC, *SR* (Ess). A pejorative of OFr *escoute* 'spy' or of ME *Scot* 'Scott'.

**Scotton, Scotten:** Andrew *de Scotton'* c1200–10 RegAntiquiss; Ralph, John *de Scotton* 1250 IpmY, 1347 FFY. From Scotton (Lincs, NR, WRYorks).

**Scottorn:** *v.* SCOTHERN

**Scottow, Scotto:** Hervei *de Scothoue* 1177 P (Nf). From Scottow (Norfolk). *v.* also SCOTTER.

**Scoular, Scouler, Scouller:** *v.* SCHOLAR

**Scovell, Scovil:** Ralph *de Scouilla* 1194 P (Bk). From Escoville (Calvados).

**Scowcroft:** *v.* SCHOOLCRAFT

**Scowle:** *Scule, Escule* 1066 DB; *Scul* c1155 DC (L); Robert *Scule* c1165 Bury (Sf); William *Scowle* 1273 RH (L); Richard *Scoule* 1297 SRY. ON *Skúli*, ODa *Skule*, probably from *skýla* 'to protect'.

**Scowle:** *v.* SCHOOL

**Scrafield, Scrayfield:** William *de Scraifeld* 1374 AssL; Roger *Skrafeld* 1379 LoCh. From Scrafield (L).

**Scrafton:** Robert *de Skrafton'* 1219 AssY. From East, West Scrafton in Coverham (NRY).

**Scragg, Scragge, Scraggs:** Thomas *Scrag* 1185 Templars (Y); Osebert *Scragg* 1218 P (W); John *Scrag'* 1366 Eynsham; John *Scragge* 1507–8 FFSr. ON *Skragg*. cf. Norwegian dialect *skragg* 'a shrivelled, wretched person'.

**Scrayfield:** *v.* SCRAFIELD

**Screach, Screech:** Robert *Screech* 1279 RH (Bk); John *Screyk* 1335 FFEss; John *Screche* 1525 SRSx. An anglicization of ON *skrækr* 'a shriek, scream'.

**Scrime:** *v.* CRIMES

**Scrimgeour, Scrimgeoure, Scrimger, Scrimiger, Scriminger, Scrimygeour, Scrymgeour, Scrimshaw, Skrimshaw, Skrimshire:** Richer *le schirmissur* 1154–89 DC (L); William *Lescermissur* 1180 P (Sf); Symon *Leskirmisur* 1221 Colch (Ess); Gilbert *le Skermisur* 1222 AssWa; Alexander *Skrymchur* 1387 Black; Thomas *Skrymshire* 1465 ib.; Thomas *Skrymsher*, *Skrymshawe* 1520, 1533 FFSt; James *Skrympshire* 1723 DKR 41 (St). OFr *escremisseor*, *-our*, *eskermisor*, *scremisseur* 'fencing-master'. *Scrimshaw* and *Skrimshire* are more common in England, *Scrimgeour* in Scotland, where Black also gives the forms *Skrimagour* (1411), *Scrymgeoure* (1456), *Scrimigeor* (1503) and *Scrymger* (1541). In spite of the re-enactment in 1285 of the Assize of Arms of 1181, fencing was regarded as unlawful. The keeping of fencing-schools was forbidden in the City of London 'as fools who delight in mischief do learn to fence with buckler, and thereby are encouraged in their follies'. Fencing-masters were legally denominated as rogues and vagabonds and classed with stage-players, bear-wards, gipsies and other undesirable characters. *v.* NQ 198, 231–4. cf. SKIRMER and CHAMPION.

**Scrippe, Scrippes, Scripps:** Robert *Scrippe* 1185 Templars (Lo); Nicholas *Scrippe* 1260 AssCh; Michael *Scrippe* 1286 ForC. ME *scrippe* 'a small bag, wallet, or satchel, especially one carried by a pilgrim, shepherd, or beggar'. Hence a nickname for these. For *Scripps, v.* also CRISP.

**Scripps:** *v.* CRISP

**Scripture:** Ricardus *scriptor* 1158 P (So); John *Scripture* 1686 Bardsley. A refashioning of Lat *scriptor* 'writer, clerk'.

**Scriven, Scrivens, Scrivins, Scrivings:** Richard *le Scrivein* 1208–13 Fees (O); Gervase *le Escriuein* 1278 AssSo; Richard *Scrivin* son of Norman *Scrivin* 1294 AD vi (K). OFr *escrivain, escrivein* 'writer', one who writes and copies books, manuscripts, etc.; also a clerk (a1300 NED).

**Scrivener, Scrivenor:** Simon *Scriuiner* 1218–22 StP (Lo); Kemma *Scriuener* 1311 ColchCt. A derivative of *scrivain* with the same meaning as SCRIVEN.

**Scrogg, Scroggs:** (i) Roger *le Scrog'* 1304 IpmY. 'Dweller by the brushwood', ME *scrogge*. (ii) Adam *of Skrogges* 1296 (Haddington), David *de Scrogis* (Aberdeen) Black. From the lands of Scrogges (Peebles).

**Scrope, Scroop, Scroope:** Richard *Scrupe* 1066 DB; Robert *Escrupe* 1165 P (Gl); Gilbert *Escrop* 1210 Cur (L); Reyner *Scropp* 1317 NorwDeeds II; Geoffrey *le Scrop* 1328 LLB E. ON *Skropi, Scroppa. v.* OEByn 224. But the forms would suggest that an unidentified place-name is also involved: Richard *de Scropes* 1190 P (Gl); Robert *de Scrop* 1196 P (Y); Henry *de Scrupes* 1218 P (Gl).

**Scrouther:** *v.* SCUDDER

**Scruby:** Nicholas *de Scrouteby* 1220 Cur (Nf); Nicholas *de Scroby* 1256 AssLa. From Scratby (Nf), *Scroutebei* DB, or Scrooby (Nt).

**Scrutton, Scruton:** Thomas *de Scruton* 1364 FFY; Daniell *Scrutton* 1568 SRSf; Edward *Scruton* 1672 HTY. From Scruton (NRYorks).

**Scrymgeour:** *v.* SCRIMGEOUR

**Scudamore:** *v.* SKIDMORE

**Scudder, Skudder, Scrouther, Shrouder:** John *le Scoudere* 1289 NorwLt; Thomas *Skudder* 1492, John *Scudder* 1521 RochW; Thomas *Scudder* 1660 ArchC 30. cf. Robert Ose of Norwich, *scouder* 1315 NorwDeeds II. A derivative of OE *scrūd* 'garments, clothes'. Probably a dealer in second-hand clothing.

**Scurrell:** *v.* SQUIRREL

**Scutt, Scutts:** Godwin, Lefstan *Scut* 1183 P (Nf), 1190–1200 Seals (Sf); William *le Scutt, le Skut* 1222 Acc, 1327 SRSx; William *Skutt* 1545 SRW. A nickname from *scut*, originally used of the tail of the hare, particularly noticeable when the animal was fleeing, and later of the hare itself, cf. 'Scut, hare', c1440 PromptParv.

**Scutter:** *v.* SCOTTER

**Sea, See:** Bertram *del See* 1312 ColchCt; Philip *atte See* 1327 SRSx; John *otheSee* 1382 AssL. 'Dweller by the lake or pool', OE *sǣ*, or 'dweller by the watercourse or drain', OE *sēoh. v.* MELS 179.

**Seaber, Seaberg, Seaburgh, Seabury, Sebry, Saber, Sibary, Sibery, Siberry, Sibery, Sibree:** (i) *Seburga filia* Oseberti 1222 Cur (O); William *Sibry* c1227 Fees (Y); Alan *Sibri* 1276 RH (Y); John *Seber* 1279 RH (C); John *Seburuh* 1279 RH (Hu); John *Sabourgh* 1327 *SR* (Ess); William *Seburgh*, Margaret *Seborw* 1327 SRSf. OE *Sǣburh* (f) 'sea-fortress', of which only one earlier example has been noted, in LVD (11th cent.). The early *Sibry* forms may have been influenced by OE *Sigeburh* 'victory-fortress', an early OE woman's name, found in place-names, but not known to have survived the Conquest. (ii) Warin *de Seberg'* 1230 P (Do). From Seaborough (Dorset). Also, perhaps, from Seaborough Hall (Essex), which is found as *Seueberghe* 1293, *Sebergh* 1334, *Sebarhall* 1544, *Sibbery Hall* 1805 (PN Ess 164).

**Seabert, Seabridge, Seabright, Sebert, Sebright, Seebright, Sibert, Sawbridge:** *Sebertus* 1199 MemR (Gl); Ricardus *filius Sebrihi* 1200 P (Lei); Gaufridus *filius Sabricti* 1210 Cur (C); William *Sebright* Hy 2 PN Ess 235; Richard *Sebrihi* 1279 RH (O); Peter *Sabright* 1290 Cl (Ess); Robert *Sebriche*, Michael *Sebryth* 1327 SRSf. OE *Sǣbeorht* 'sea-bright', the name of a 7th-century king of Essex, found occasionally until the 10th century, and then not until 1199. The surname has probably been reinforced by OE *Sigebeorht* 'victory-bright', the name of an early king of Essex and much more common in OE than *Sǣbeorht*. This, too, certainly survived the Conquest but is rare: Wlfwardus *filius Sibrith* 1189–1200 BuryS (Sf), *Sibryth* 1275 RH (Sf), John *Sybrith* 1327 SRSf. William *Sebright* has left his name in Great Seabrights in Great Baddow (Essex), *Sowbridge* 1777 (PN Ess 234). *Sǣbeorht* is the first element in Sawbridgeworth (Herts).

**Seaborn, Seaborne, Seabourn, Seabourne, Sebborn, Sibborn, Sibbons, Siborne, Siburn, Sayburn:** *Sabernus Monachus*, Philippus *filius Seberni* 1114–30 Rams (Hu); Margareta *filia Seberni* 1207 Cur (Mx); Nel, John *Sebern* 1190 BuryS (Sf), 1275 SRWo; Sayer *Sabarn* 1327 *SR* (Ess); John *Sabern* 1377 AssEss. OE *Sǣbeorn* 'sea-warrior', of which the earliest known examples are those above.

**Seabrook, Seabrooke, Seabroke:** Henry *Sebroke* 1438 FFEss. From Seabrook (Bk).

**Seach:** *v.* SYKES

**Seacome, Seacombe, Seckombe:** William *de Sekom* 1285 AssLa. From Seacombe (Ch).

**Seacroft:** Hugh *de Secroft* 1219 P (Y); James *de Secroft* 1334–7 SRY; Richard *Secroft* 1459 Kirk. From Seacroft (WRY).

**Seader:** Peter, Ralph *le sedere* 1221 ElyA (C), 1263 MEOT (Sr). OE *sǣdere* 'sower'.

**Seaford, Seaforth, Seafourth:** Hugh *de Seford'* 1208 Pl (Herts); William *Seford* 1327 SRSx; Humphrey *Seaford* 1642 PrD. From Seaford (Sussex).

**Seagars, Seager(s):** *v.* SAGAR

**Seagood, Segot:** Adam *Segud* 1317 AssK; Alan *Segode* 1327 SRY; Katerine *Seegode* 1450 RochW; John *Saygude* 1473 GildY. An unrecorded OE *Sǣgōd* 'sea-good'.

**Seagram:** Ernald *Segrom* 1307 MEOT (Wo); John *le Segrom*, Richard *Segrom* 1327 SRSx. OE *sǣ* 'sea' and ME *grom* 'servant', probably 'seaman, sailor'. cf. SEAGRIM.

**Seagrave, Seagrief, Seagrove, Segrave:** Thomas *de Segraua* 1180 P (Lei). From Seagrave (Leics).

**Seagrim, Seagrin:** *Sagrim* 1066 DB (Nth, St); *Segrim* de Haltun c1155 DC (L); *Sagrim* carnifex 1190 P (Hu); Seeman *Seegrim* c1248 Bec (Sf); John *Sagrim* 1327 *SR* (Ess); John *Sigrym* 1327 SRSo. ON *Sægrímr*, ODa *Segrim* 'sea-protector'. cf. SEAGRAM.

**Seagrove:** *v.* SEAGRAVE

**Seal, Seale, Seales, Seals, Seel, Seels, Zeal, Zeale:** (i) Ralph *de la Sele* 1168 P (D); Robert *ate sele* 1332 SRSr. From employment at the hall (OE *sele*) or, perhaps, 'dweller by the sallow copse' (OE *\*siele*, *\*sele*). Seal (Kent), Seale (Surrey) and Sele (Sussex) are probably from *sele*. (ii) Ralph *de Sseill* 1185 Templars (Wa). From Seal (Leics). (iii) Hugh *le Sele* c1113 Burton (St); Roger *Sele* 1198 P (Nf). Probably metonymic for SEALER. cf. Everard *de Sigillo* (*del Sael*) 1219 Cur (Sf). But *le Sele* suggests a nickname from the seal (OE *seolh*, ME *sele*).

**Sealeaf:** *Saloua* 1066 ICC (C); Robertus *filius Seluue* 1190 BuryS (Sf); *Salove* 1203 Cur (Mx); Agnes, John *Seloue* 1308 EAS xviii, 1327 SRSf; John *Salove* 1327 *SR* (Ess). von Feilitzen derives the 1066 personal name from an unrecorded OE woman's name *\*Sǣlufu* 'sea-love'. This would account for all the above forms but would give a modern *Sealove*. In DB *Lēof-* usually appears as *Leue-*, but also as *Liue-* and occasionally as *Luue-*. An OE *\*Sǣlēofu* (f) 'sea-love' would give modern *Sealeaf* and this could be the source of all the above. cf. LEAF, GOODLIFF.

**Sealer:** Robert, William *Seeler* 1221 AssGl, 1230 Cl (Db); Walter *le Seler* 1327 SRSf; Richard *le Sealer* 1328 LoPleas. A derivative of ME *sel, seel, seal* from OFr *seel* 'a seal'; 'a maker of seals'. Hugh *le Seler* of York in 1333 made a new seal for the bishopric of Durham (MedInd 132). Numerous examples of *Seler* are found but it is usually impossible to decide whether they belong to *Sealer* or to *Sellar*. *v.* also SEAL. As Cotgrave has *sele* for *saddle* and Palsgrave '*Seale*, horse harnesse', *Sealer* must sometimes mean 'saddler'.

**Sealey, Seally, Sealy, Seeley, Seelly, Seely, Selley, Selly, Silley, Silly, Ceeley, Ceiley, Cely, Zealey, Zealley, Zelley:** Richard *Seli* c1200 Gilb (L); Roger (*le*) *Seli* 1205 P (He); Roger *Cely* 1255 RH (Sa); Richard *Sely*, John *Celi* 1275 SRWo; Thomas *Sali* 1279 RH (Hu); Thomas *Sali, Sely* 1281, 1286 LLB B; Thomas *Zely* 1327 SRWo. OE *sǣlig* 'happy, blessed', whence modern *silly*, often misunderstood in the phrase 'Silly Suffolk'. This was also used as a woman's name: *Sela* 1219 AssL, *Sely* filia Nicholai 1221 AssWo, *Sely* Percy 1327 SRSo.

**Seaman, Seammen, Seamons, Seeman, Seman, Semens, Semmens, Seyman, Saayman:** *Seman* 1066 DB (Sr), c1155 DC (L), 1211 Cur (So); *Seman* le Erl 1327 SRSx; Rufus, Richard *Seman* 1182–1211 BuryS (C), 1235 FFEss; Geoffrey *Semman* 1292 SRLo; Robert *Sayman* 1379 PTY. OE *Sǣmann* 'sea-man'.

**Seamark, Semark:** Henry *Semarke* 1524 SRSf; John *Semarke* 1578 StaplehurstPR (K); John *Seamarke* 1625 GreenwichPR (K). Probably from a ME *Semark*, formed on the analogy of the other names in *Sea-*.

**Seamer, Seamour:** (i) *Sæmar, Samar, Semær, Semar, Semer* 1066 DB; *Semarus* Cham 1179 P (D); *Semer* de Risenberg' 1212 Cur (Bk); Geoffrey *Semare* 1251 Rams (Hu); Henry *Semer* 1275 RH (Do); John *Samar* 1279 RH (C); George *Seama* 1558 RattlesdenPR (Sf). OE *Sæmær* 'sea-famous'. Not quite so common as *Sæmann*. (ii) Ioscelinus, Petronilla *de Semere* 1176 P (Y), 1329 ColchCt. From Seamer (NRYorks) or Semer (Norfolk, Suffolk). (iii) John (*le*) *Semere* 1327 SRSo, 1340 MESO (Ha). OE *sēamere* 'tailor'. cf. SIMESTER.

**Seanor:** v. SENIOR

**Searby:** Geoffrey *de Seuredebi, de Seurebi* 1197 P (L); Robert *de Seuerbi* 1202, Robert *de Seuerbi* 1219 AssL. From Searby (L), *Seurebi* DB.

**Search, Sarch, Sarge, Searight, Seawright, Serrick, Serck, Sark, Sarkes, Surrage, Surridge:** (i) *Saric, Seric* 1066 DB; William *Serych* 1296 SRSx; Reginald *Serich, Serche* 1297 Coram; John *Sarich, Serich* 1296, 1299 FFEss; Richard *Sarich* 1327 *SR* (Ess); Emma, John *Serch(e)* 1332 SRSx, SRSt; Alan *Serech* 1359 FFSt. OE *Sārīc* 'sea-ruler', frequent in post-Conquest sources (PNDB). This would regularly give ME *Saric, Seric*, which would normally appear as *Sarridge, Serridge*. These do not seem to survive but have probably been absorbed by *Surridge* which, with *Serrick*, is still found in Sussex, and is, no doubt, the direct descendant of the 1296 *Serych*. All the southern Surridges cannot have brought back to their native south a name acquired by migration north. For *Serrick* and *Sea(w)right*, cf. ALDRICH and ALLRIGHT. With loss of the unstressed vowel, *Saric, Seric* would become ME *Sarc, Serc*, giving modern *Sarch, Sarge, Search, Serck* and *Sark*. (ii) *Siric, Syric, Seric* 1066 DB; *Siricus* de Lenna 1210 Cur (Nf); R. *Siricus* 1114– 77 Rams (Hu); Hugo *Serich* 1206 P (Ha); Hugo *Sirich(e)* 1230 P (Berks), 1327 SRSf; Aubert, Roger *Syrik* 1275 RH (L). OE *Sigerīc* 'victory-powerful'. Occasionally also ON *Sigrikr*, ODa *Sigrik*. The vowel of *Sirich* was early lowered to *e* in *Serich* which would then develop like *Serich* from *Sārīc*.

**Seares:** v. SAYER

**Seargeant, Searjeant:** v. SARGANT

**Searl, Searle, Searles, Searls, Serle, Serrell:** *Serlo* 1066 DB (Ess); *Serlo* le Flemyng c1150 Gilb (L); *Serle* Gotokirke 1279 RH (C); *Serill'* Pynder 1379 PTY; Adam, Hugo *Serle* 1226 FFBk, 1250 Fees (Do); Hugh *Sereleson* 1320 ParlWrits (Y); William *Serell'* 1379 PTY. OG *Sarilo, Serilo*, (Romance) *Serlo*, probably related to OE *searu* 'armour', cognate with ON *Sǫrli*. v. SARL. The name was frequent in Normandy and common in England after the Conquest.

**Searson:** Ralph 'Johnesseruaunt *Searessone* de Sutton' 1317 Pat (L). Ralph was the servant of John, son of *Seare* (i.e. *Saer*) de Sutton. v. SAYER.

**Seath:** William *Ateseth'* 1275 SRWo. 'Dweller by the pit or pool', OE *sēap*.

**Seaton, Seton:** (i) Adam *de Seton* c1194–1214 Black; Alexander, Serlo *de Seton* 1225, c1250 ib. From Seaton (Haddington). (ii) Ralph *de Seton'* 1207 P (Nb); Robert *de la Setene* 1296 MELS (Sx); John *Seeton* 1557 CorNt. From Seaton (Cumb, Devon, Durham, Northumb, Rutland, ER, NRYorks), or 'dweller by the plantation or the cultivated land', OE *seten*, v. MELS 183. Occasionally a personal name may also be involved, cf. Madoch *filius Setun'* de Sutton 1187 P (Sa).

**Seavers, Sever, Severs:** *Seuare* 1185 Templars (So); *Sephare* 1188 BuryS (Sf); Radulfus *filius Sefare* 1221 *ElyA* (Sf); *Seuar'* Boykin 1277 *Ely* (Sf); William, Hugh *Seuare* 1185 P (Co), 1285 FFEss; Walter *Sefare* 1230 P (Beds); Thomas *Safare* (*Savare*) 1327 *SR* (Ess); William *Sefare*, Ralph *Seffare* 1327 SRSf. An unrecorded OE woman's name, *\*Sǣfaru* 'sea-passage', of which the above are the earliest known examples.

**Seaward:** v. SEWARD

**Seawright:** v. SEARCH

**Sebbage:** Adam *de Seuebech*, William *Seuebeche* 1327 SRSx. From Seabeach in Boxgrove (Sussex), *Sebeche* in 1521, now pronounced *Sibbidge*.

**Sebborn:** v. SEABORN

**Sebert:** v. SEABERT

**Sebley:** v. SIBLEY

**Sebright:** v. SEABERT

**Sebry:** v. SEABER

**Secker:** v. SACKER

**Seckerson, Secretan:** v. SEXTEN

**Seccombe, Secomb, Secombe:** Leonard *Seccomb* 1642 PrD. From Seccombe in Germansweek, in Blackawton (D). v. also SEACOMBE.

**Seckington:** Henry *de Sekindon'* 1221 AssWa; Ralph *de Sekendon* 1262 AssSt. From Seckington (Wa), or Seckington in Hartland (D).

**Seckombe:** v. SEACOMBE

**Secomb, Secombe:** v. SECCOMBE

**Secular, Seckler:** Nicholas *le Seculer* 1197 P (He), *le Secler, le Seccoler* 1250 Fees (He); Walter *le Seculer* 1255 RH (Sa); Robert *Seclermunck'* 1301 SRY. OFr *seculier* 'secular, lay, temporal'. MedLat *secularis*, by which the name is translated (1206 ChR), meant 'secular, of this world', and, as a noun, 'layman, member of the secular clergy'; cf. *secularitas* 'worldliness, worldly life'. *Seclermunck* was probably a nickname for a monk of worldly life and the surname in general 'worldly'.

**Sedany:** *Sedehanna* (f) 1218–19 FFK; *Sedania* (f), *Sedaina* de Selua 1221 AssWo; William *Cedany* 1275 SRWo. Fr *Sedaine*, a feminine form of Lat *Sidonius*. v. Dauzat.

**Sedge:** Richard *de la Seg'* 1230 P (D); Thomas *atte Segh* 1308 Ipm (Wo). 'Dweller by the sedgy or reedy spot' (OE *secg* 'sedge, reed').

**Sedgbrook, Sedgebrook:** Roger *de Segbrok* 1327 SRLei. From Sedgebrook (L).

**Sedgefield, Sedfield:** Peter *de Seggesfeld* 12th FeuDu; William *de Segefeld* 1243 AssDu; John *de Seggefeld'* 1361 AssY. From Sedgefield (Durham).

**Sedgeford, Sedgford:** Edmund *Seggeford* 1340–1450 GildC. From Sedgeford (Nf).

**Sedger:** Osbert, Gilbert *Seggere* 1200 Cur (Sf), 1275 RH (O); Richard *le Segger'* 1292 MESO (La). A derivative of OE *secg* 'sedge'. cf. Richard *le Seggemaker* 1306 AssSt and *v.* SEDGMAN.

**Sedgman:** cf., at King's Hall, Cambridge 'For wages of Brown, *seggeman*, for thatching walls, 7*d.* Also in reward to 6 boys carrying *segge*, 6*d.*' (1428) and in 1439 'for the dinners of 2 *seggethakkers* for 4 days, 16*d.*' (Building 226).

**Sedgwick, Sedgewick, Sedgewicke, Sidgewick, Sidgwick:** John *de Segheswyk*, Thomas *de Sigeswik* 1379 PTY. From Sedgwick (Westmorland) or Sedgewick Castle in Nuthurst (Sussex).

**Sedley:** John *de Sedeleghe* 1296 SRSx; James *Sedle* 1447 CtH; John *Sedley* 1501 FFEss. From Sidley Green in Bexhill (Sx).

**Sedman:** *v.* SEEDMAN

**Sedwick:** For SEDGWICK.

**See:** *v.* SEA

**Seear:** *v.* SAYER

**Seebright:** *v.* SEABERT

**Seed, Seeds:** Johannes *filius Sede* 1210 Cur (Nf); Gilbert, William *Sede* 1275 RH (L), 1327 SRWo. Metonymic for SEEDMAN, but also from an OE \**Sida*, a short form of late names recorded after the Conquest, from OE *sidu* 'custom, manner; morality, purity': OE *Siduwine*: *Seduuinus* 1066 DB (D), Ricardus *filius Sidewini* 1188 P (K); \**Sidumæg*: *Sedemai* 1185 Templars (K), Robert *Sedemai* 1180 P (Sf); \**Sidumægden* (f): *Sedemaiden* c1095 Bury (Sf); \**Sidumōd* (f): Ælfget *Sedemode filius* c1095 ib.; and \**Sidulufu* (f), \**Siduwulf*, found in place-names (NoB 33). *Sede-* may be from \**Seodu-*. *v.* PNDB § 13, 359.

**Seedman, Sedman:** Robert *Sedeman* 1219 Cur (Nf), c1248 Bec (Wa). 'Dealer in seeds.' This might also be from OE *Sidumann*: *Sydeman* 931 BCS 674; *Sideman* presbiter 974 (1334) Rams. cf. SEED.

**Seegar, Seeger:** *v.* SAGAR

**Seel:** *v.* SEAL

**Seel(e)y:** *v.* SEALEY

**Seeman:** *v.* SEAMAN

**Seener:** *v.* SENIOR

**Seers:** *v.* SAYER

**Seeviour:** *v.* SEVIER

**Sefton, Sephton:** Henry *de Sefton* 1285 AssLa. From Sefton (La).

**Segar, Seger:** *v.* SAGAR

**Segger:** *v.* SAGAR, SEDGER

**Seggins:** *v.* SEGUIN

**Segot:** *v.* SEAGOOD

**Segrave:** *v.* SEAGRAVE

**Seguin:** *Segwinus* 1208 Cur (Y); *Segwin'* forestarius 1209 P (Sf); William *Seguin* 1177 P (O); Richard, Alan *Segin* 1276 RH (L), 1279 RH (C). Identical with French *Seguin* which Dauzat derives from OG (Visigothic) *Sigwin* 'victory-friend'. *Segwinus* 1210 Cur (L) was a citizen of Cologne.

**Seignior:** *v.* SENIOR

**Seiler, Seiller:** *v.* SAILER

**Seivwright:** *v.* SIEVWRIGHT

**Selborne, Selbourne, Selburn, Selburne:** Robert *de Seleburn* 1272 FFEss; William *Selebourne* 1327 SRSx; John *Selbourne* 1362–4 FrC. From Selborne (Hants).

**Selby, Selbey, Selbie:** William *de Selebia* 1175 P (Y); Hugh *de Seleby* 1219 AssY; Robert *Selby* 1395 IpmGl. From Selby (WRYorks).

**Selden, Seldon, Seldom:** Roger *de Seldon'* 1199 Pl (Wa); William *de Selkedon* 1296 SRSx; Robert *Selden, Selkeden* 1525 SRSx; Richard *Seldon*, William *Selden* 1642 PrD. From Seldon in Hatherleigh (Devon), or Selden Farm in Patching (Sussex). The latter is the source of the surname of the famous lawyer, John Selden (1584–1654), whose memory is associated with West Tarring (Sussex).

**Self, Selfe:** *Saiulfus, Sahulfus, Saulf, Saolf, Saul, Seulf* 1066 DB; Robert *filius Seulfi* 1185 Templars (Berks); William *Sewolf* 1296 SRSx; John, William *Self* 1327 SRSf. OE *Sæwulf* 'sea-wolf'.

**Selgrave:** Robert *de Segrave* 1253 FFK. From Selgrave in Faversham (K).

**Selifant:** *v.* SILLIFANT

**Seliger, Selinger:** *v.* SALINGAR

**Sell, Selle, Sells, Zell, Zelle:** Humfrey *ater Selle* 1296 SRSx. OE *(ge)sell* 'shelter for animals' or 'herdsman's hut'. *v.* MELS 181. Probably for a herdsman.

**Sellar, Sellars, Seller, Sellers, Sellors, Sellier, Cellier, Zeller:** (i) Alriz *Sellere* 1086 InqEl (C); Alwinus *leseller* 1115 Winton (Ha). Either OFr *selier*, *seller* 'sadler' (1311 NED) or ME *seller* from OE *sellan* 'to give, hand over', 'a seller, dealer' (c1200 NED). Tengvik and Fransson regard the latter as an improbable meaning for the surname. But VENDER still survives. cf. Henricus *Venditor* 1274 RH (L). Sanson *Sellarius* of Yorkshire in 1175 (P) paid 5 marks into the exchequer for selling shields to the king's enemies and Walter *Sellarius* of Warwickshire was fined half-a-mark in 1183 (P) for false description of his wares (*pro falso clamore*). Bodo or Bondo *Sellator* (1175–7 P, Sf) was a sadler (*sellator* 'sadler' 1224 MLWL), but this may be a translation of either the French *seller* or the English *sadler*. Philip *le Celler* (1319 SRLo) is also called *le Sadeler* (1320 LLB E). The frequent form *Seler*, sometimes for *Sealer* and sometimes for *cellar*, may also mean '*sadler*'. Sellars in Barnston (Essex) owes its name to Thomas *Seler* (1396 PN Ess 470). cf. *Sele*, horsys harneys (Cotgrave). (ii) Ordingus *cellerarius* 1121–48 Bury (Sf); William *Sellerarius* 1185 Templars (Wa); Robert *le Celerer* 1297 SRY; William *Sellerer* 1419 LLB I. A form of *cellarer*, identical in meaning with *atte celer* (*seler*). (iii) Gudmund *del celer* 1121–48 Bury (Sf); Richard *ate Celer*, taverner 1308 LLB B; William *atte Seler, Selere* 1365 LLB G, F. AFr *celer*, OFr *celier* 'a cellar' (a1225 NED), 'a store-house or store-room for provisions; a granary, buttery or pantry'. ME, AFr *celerer* (a1300 NED) was the officer of a monastery, etc., who had charge of the cellar and provisions. Later, it was used of a taverner.

**Sellerman:** John *Celereman*, Patric *Celererman* 1332 SRCu. Respectively 'worker in the "cellar" ' and 'servant of the cellarer'.

**Sellex, Sellick:** John *Sellych* 1327 SRWo; Robert *Sellake* 1379, *Selake* 1380 Hylle; John *Sellack*,

*Sellicke* 1642 PrD. From Sellack (He), or Sellake in Halberton, Sellick in Clawton (D).

**Selley, Selly:** Richard *de Selleia* 1203 AssSt. From Selly Oak (Worcs). *v.* also SEALEY

**Sellier:** *v.* SELLAR

**Selling:** John *Sellingge* 1325 CorLo; Emma *atte Sulingg* 1327 MELS (Sx). From Selling (K).

**Sellinger:** *v.* SALINGAR

**Sellis:** John *Selawe* 1443 CtH; Robert *Sevelos* 1558, *Selowes* 1566, John *Sellowes* 1596 Buxhall (Sf). 'Dweller by the willows', OE *sealh*. *v.* also SALLOWS.

**Sellman, Selman, Sillman, Silman:** Ailricius *Seliman* 1169 P (Nth); Thomas *Selman* 1275 RH (W); John *Seliman* 1279 RH (C); Robert *Salyman* 1327 *SR* (Ess); Claricia *Selimon*, Henry *Selmon* 1327 SRWo. OE *sǣlig* and *mann* 'happy man'. Occasionally used as a personal name: *Selmon*' de Copmanford 1279 RH (Hu). Sometimes 'Servant of *Sely*'. John *Sely* and Walter *Selyman* lived in the same parish (1327 SRSo). cf. Joan *Saliwymman* 1276 LLB B and SEALEY.

**Sellwood:** *v.* SELWOOD

**Selton:** Nicholas *de Selton* 1208–9 Pleas (Nf); Richard *de Selton* 1275 SRWo; Robert *de Selton* 1277 IpmY. Either for SHELTON, or for SILTON.

**Selvester:** *v.* SILVESTER

**Selway:** *v.* SALWAY

**Selwin:** *v.* SAUVAIN

**Selwood, Sellwood:** John *Selewode* 1189 Sol; Richard *de Selwode* 1339 LoPleas; Richard *Sellwood* 1662–4 HTDo. From Selwood (So).

**Seman:** *v.* SEAMAN

**Sember:** *v.* SEMPER

**Semens:** *v.* SEAMAN

**Semkin:** *v.* SIMKIN

**Semmence, Semmens:** Usually for SIMMONDS, occasionally for *Seamons*. *v.* SEAMAN.

**Semper, Sember, Samber, Samper, Simper, Symper:** Richard *de Sancto Petro*, *de Sempere*, 1256 AssNb; Richard *Saunper* 1256 FFNb; Simon *Saumper* 1345 AssSt; John *de Seyntper* 1353 AssSt; Ralph *de Seynpere* 1371 DbCh; Urian *Seintpier* 1419 IpmY; John *Semper* 1466 DbCh; John *Simper* 1674 HTSf. Geoffrey de Clinton, chamberlain of Henry I, came from Saint-Pierre-de-Semilly (La Manche). The Count of Meulan and his under-tenants, the Mandevilles, were associated with the abbey of Saint-Pierre-des-Preaux.

**Sempkins:** *v.* SIMKIN

**Semple, Sempill, Sambell, Sample, Simble, Simpole:** Robertus *de Sancto Paulo* 1159 P (Beds); Symon *Sempol* 1271 RamsCt (C); Simon *Senpol* 1274 ib.; Agnes *Seynpol* 1289 ib.; John *Sampol* 1351 FrY; Thomas *Seintpoule, Seinpoull* 1403, 1418 IpmY; Cicily *Sampule* 1413 GildY; Nicholas *Sempool* 1421 NorwW (Sf); Anna *Sampall*, *Sampoll* 1515, 1524 GildY. From one of the French places named Saint-Paul or from Saint-Pol (Pas-de-Calais, Nord). The Scottish *Semple* has a different origin. Black's forms show clearly that it is identical with SIMPLE. From 1315 to 1574 the vowel is invariably *i* or *y*: *Sympil, Simpil, Simple*. The *e* first appears in 1691 (*Sempell*).

**Senchel(l):** *v.* SENESCHAL

**Sendall, Sendell:** John *Sendal* 1303 FA (Sf); John *Sendale* 1374 LoPleas. Among the materials bought for Edward I in 1300 were sindon, or *sendal*, a silk at 16*s.* the yard, samite, also silk at £4 10*s.*, silken cloth of gold at 26*s. 8d.*, cloth of gold of Tartary at 36*s.*, and of Turkey at no less than £7 the yard (METrade 423). The surname probably denotes a merchant who sold sendal or one who made garments of sendal.

**Sende:** William *ate Sende* 1332 (Sr), William *atte Sende* 1333 (So) MELS; Robert *Sende* 1362 IpmGl. 'Dweller at the sandy place', OE *\*sende*.

**Sendler:** *v.* SANTLER

**Seneschal, Seneschall, Senchell, Seneschall, Senecal, Sensicall, Sensicle, Senskell:** Alan *le Senescall* 1194 AssSt; Ralph *le Seneschall*' 1222 Cur (Sx); William *le Seneschel* 1243 AssSo; Thomas *Senycle* 1395 Cl (Lo); Sarah *Senskell* 1693 Bardsley. OFr *seneschal* 'seneschal', an official in the household of a sovereign or great noble, to whom the administration of justice and the entire control of domestic arrangements were entrusted. In a wider use, a steward, 'major-domo' (1393 NED). *Seneschal* and *Senecal* were Norman forms (Moisy). cf. MARSHALL and MASKALL.

**Senett:** *v.* SINNATT

**Sengel:** *v.* SINGLE

**Senhouse:** Robert *de Sevenhowes*, John *de Senehowes* 1346 FrY. From Hallsenna in Gosforth (Cumb), earlier *Sevenhoues* 'seven hills', *Hall Senhouse* 1668 (PN Cu 395).

**Senior, Seniour, Senier, Seanor, Seener, Seignior, Senyard, Sainer, Sayner, Saynor, Seyner, Sinyard, Synyer:** Walter *Seignure* 1164 P (Nf); Hugh *Seinure* 1212 Fees (Nf); Thomas *le Senyur* 1271 AssSt; Robert *le Seynur* 1275 RH (Sf); Henry *Senior* 1279 RH (O); Hugelyn *le Seygnur* 1280 AssSo; Thomas *Senyer* 1332 SRCu; William *Synyer* 1379 PTY. OFr *seignour* 'lord', Lat *senior* 'older'. Perhaps a name of rank, 'lord of the manor', or a nickname for one who aped the lord, or 'the senior, the elder'.

**Senneck, Sinnocks:** John *Seuenok* 1521 RochW; George *Sennok* 1532 ib. *Sennocke* is the old pronunciation of Sevenoaks (Kent), still used as the name of an hotel and an engineering firm.

**Sennett, Sennitt:** *v.* SINNATT

**Sennicles:** *v.* ST NICHOLAS

**Sensicall, Sensicle, Senskell:** *v.* SENESCHAL

**Senter, Center, Sainter, Santer:** Edmund *Sein(e)tier* 1160–8 ELPN; Benedict *le Seintier* 1197 P (Lo), *le Sentier* 1206 P (Y), *campanarius* c1216–22 Clerkenwell (Lo), *le Seincter* c1230–40 ib.; Simon *le Sainter* 1219 AssY; Roger *Santer*, *le Sainte* 1333 Riev (Y). OFr *saintier* 'bell-founder', the French equivalent of the English BILLITER. *v.* also CENTURY.

**Sentry:** *v.* SANCTUARY

**Sephton:** *v.* SEFTON

**Seppings, Sippings:** Nycolas *Sevenpennys*, Wylliam *Sevynpenys* 1524 SRSf; Hamond *Sepens* ib.; Francis *Sipins*, George *Sepings* 1674 HTSf. A nickname, 'seven-pence'. cf. HALFPENNY.

**Serck:** *v.* SEARCH

**Sergant, Sergean, Sergeant, Sergeaunt, Sergent, Serjeant, Serjent:** *v.* SARGANT

**Sergenson, Serginson, Sergison, Serjeantson:** v. SARGEANTSON

**Serkitt:** v. CIRCUITT

**Serlby:** Oliver *de Serleby* 1345 FFY. From Serlby (Nt).

**Serle:** v. SEARL

**Sermin, Sermon, Surman:** Richard *le Sermoner*, Ralph *Sermoner* 1212 Cur (Herts, O); Hugh *le Sarmener* 1269 AssSo; Herbert *le Sarmuner* (*Sermunur*) 1220 Cur (Sf). ME *sermoner, sarmoner*, OFr *sermounier* 'preacher, speaker'. The modern surnames are metonymic.

**Serrell:** v. SEARL

**Serrick:** v. SEARCH

**Sersmith:** v. SHEARSMITH

**Sertin:** v. SARTIN

**Servant, Servante, Servent, Sarvent:** Adam *le Serviant* 1242 P (So). OFr *servant, serviant* 'serviteur' (Godefroy). Probably equivalent to *serviens* 'sergeant'. cf. William *Seruantman* 1379 PTY 'the sergeant's man'.

**Service, Servis, Servais, Servaes, Sarvis:** Walter *Cerveise* 1177 P (O), 1206 Cur (O); William *Ceruaise* 1230 P (Berks); Robert *Cereveyse, Sereveyse* 1279 RH (O). OFr *cervoise* 'ale', for a seller of ale, a taverner.

**Sessions:** Esueillardus *de Seissuns, de Soissuns* 1181, 1190 P (C); Riulfus *de Sessuns* Hy 2 (1212) Fees (Berks). From Soissons (Aisne).

**Setch:** Probably for SACH.

**Setchell:** For SATCHEL.

**Setchfield, Satchwell:** Ralph *de Secchevill'* 1176 P (D); Geoffrey *de Saccheuill'* 1198 P (Nt). From one of the two Secquevilles in Normandy.' v. SACKVILLE. Heanton Satchville (Devon) was held by John *de Sicca Villa* in 1242 (Fees) and Thorpe Satchville (Leics) by Ralph *de Secheville* in 1212 (RBE).

**Seton:** v. SEATON

**Setter, Setters:** (i) Stephen *le Setere* 1262 MESO (Ha); Symon, Philip, Henry *le Setere* 1280 ib. (ii) Roger *le Settere* 1278 LLB A; Robert *Setter* 1292 SRLo; Isabel *Settere* 1354 ColchCt. *Setter* was common in London as a surname and as an occupational term. Riley took it to mean 'an arrow-smith'. Weekley noted that in 1314 John Heyroun, *settere*, and William *le Settere* were called in as experts to value an embroidered cope and explained the term as 'a maker of *sayete*, a kind of silk'. Fransson adopts this etymology, 'silk-weaver', from OFr *saietier*. Ekwall points out that whilst this meaning may suit the Hampshire forms *setere*, it hardly fits the London examples, *setter(e)*, invariably with a double *t*. He further notes that Alexander *le Settere* in 1307 received £10 in part payment of £40 for an embroidered choir cope bought of him and undertook well and fittingly to complete it. He was an embroiderer who made copes. *Setter* is from ME *setten* 'to set', used of placing ornaments, etc., on a surface of metal or on garments. v. SRLo 357. *Setter* was also an occupational term in York. This Thuresson similarly derives from OE *settan* 'to set', one who sets or lays stone or brick in building, and compares *rogh setter* 'a rough-stone mason' (1435 NED). This was a common term in building: 'the layers, *setters*, or wallers, who

placed in position the stones worked by the (free) masons'. They are called *cubitores* at Westminster in 1252, at Corfe in 1280 and at Caernarvon in 1282; *positores* at Windsor in 1365 and Westminster in 1385; *leggeres* and *setteres* at Newgate in 1282; *leggers* at St Paul's in 1382 and *leyers* at Cambridge in 1412. The contemporary term at York was *setters* (Building 31).

**Setterfield:** v. SATTERTHWAITE

**Setterington:** John *de Setrinton'*, *de Seteryngton* 1200 FFY, 1272 FrY; Adam *de Setryngton* 1367 FFY. From Settrington (ERYorks).

**Settle:** Daud *de Setel* 1276 FrY; Richard *Settle* 1621 SRY; Eliza *Settle* 1672 HTY. From Settle (WRY).

**Sevenoaks:** Michael *de Sevenoke* 1258 AssSo; Peter *Seuenok* 1332 SRSx. 'Dweller by the seven oaks', as at Sevenoaks (Kent).

**Sevenstar:** William *Sevensterre* 1355 LoPleas; William *Seuesterrys* 1379 LoCh; John *Sevesterre* 1384 LLB H. 'Seven stars', OE *seofon, steorra*, probably from a shop or inn-sign. cf. Robert *Sevenhode* 1276 AssLo 'seven hoods'; William *Seyvinyere* 1529 FrLei 'seven years'; William *Seuenswaprude* 1305 RegAntiquiss 'as proud as seven'.

**Severn:** John *Seuarne* 1327 SRWo; William *Seuarne* 1362 IpmGl; William *Severne*. From the River Severn.

**Sever(s):** v. SEAVERS

**Severwright:** v. SIEVWRIGHT

**Severy:** v. SAVARY

**Sevier, Sevior, Seviour, Sevyer, Seeviour, Siveyer, Sivier, Sivyer:** Edith *Siuiere* 1274 RH (Ess); Ralph *le Siviere* 1279 RH (C); Walter *le Seuyare* 1327 MESO (Ha). A derivative of OE *sife* 'sieve', a sieve-maker (c1440 NED).

**Seville:** v. SAVIL

**Sewall, Sawell:** *Sauualdus, Sauuold, Seualdus, Seuuold* 1066 DB; *Seuuale* filius Fulgeri Hy 2 DC (Lei); *Sewaldus* de Cornhull' 1214 Cur (Ess); William *Sewald* 1220 Fees (Berks); Roger *Sewale* 1275 SRWo; Nicholas *Sewal* 1279 Calv (Y); Thomas *Sewold* 1317 AssK; John *Sawalle*, Cristina *Sewalle* 1327 SR (Ess); John *Sawold* 1377 AssEss. OE *Sæweald* 'sea-power'. Now rare, largely absorbed by SEWELL. cf. SAYWELL.

**Seward, Sewards, Sewart, Seaward, Saward, Saword, Sayward, Suart:** (i) *Sauuard, Seuuard, Seuuart* 1066 DB; *Sewardus* de Hohton' 1199 Cur (Hu); *Sewarde* 1275 RH (Nf); Richard *Seward* 1275 RH (Sa); John *Sewar* 1275 RH (W); William *Saywart* 1385 Combermere (Ch). OE *Sæweard* 'sea-lord'. (ii) *Siuuard* 1066 DB, *Seuuardus* Exon; *Siuuart* 1066 DB (Nt); *Siuuard* c1095 Bury (Sf), c1150 DC (L); *Sigwardus* peregrinus c1155 DC (L); Richard *Siward, Suard* 1235 Fees (O); Richard *Siward* 1241 FFEss, *Syward* 1260 AssC. OE *Sigeweard* 'victory-lord' and ODa *Sigwarth* are frequent in DB but the forms cannot be distinguished with certainty. In the Danelaw, the latter is more likely. *Siward*, at times, alternates with *Seward*, which may also be from *Sæweard*, from which all the modern surnames may derive. *Seward, Sewart* and *Suart* may also be from *Sigeweard* or *Sigwarth*. (iii) Cecilia *Sueherd* 1379 PTY. OE *sū* 'sow' and *hierde* 'herd', 'sow-herd'. cf.

SWINERD. This may occasionally be the source of *Seward, Sewart* or *Suart*.

**Sewat, Sewatt, Suett:** *Siuuat* (L), *Siuuate* (O) 1066 DB; Gilbert *filius Siwat'* 1180–1218 Rams (Hu); Richard *filius Sewate* 1191 P (Ha); Geoffrey *Siwath'* 1221 *ElyA* (Sf); Geoffrey *Sewhat* 1300 LoCt; John *Suet* 1327 SRDb; Richard *Sewet* 1568 SRSf. ON *Sighvatr*, OSw *Sighwat*.

**Sewell, Sewill:** (i) *Siuuoldus* 1066 DB (So); *Siwaldus* c1250 Rams (Hu); Henricus *filius Sewal'* 1221 AssWa; Richard *Suuel* 1212 Cur (L); John *Sewale* 1277 FFC; Richard *Suel* 1297 MinAcctCo. OE *Sigeweald* 'victory-ruler' or ON *Sigvaldr*, both in ME appearing as *Sewal*, a form indistinguishable from those of *Sǣweald*. (ii) Ebrardus *de Seuewell'* 1193 P (Bk); Walter *de Siwell'* 1196 Cur (Nth); Simon *de Siwelle* 1200 Eynsham (O); Walter *de Seuewell'* 1222 Cur (D). From Sewell (Beds), Seawell (Northants), Sywell (Northants), Showell (Oxon), or Sowell (Devon), all 'seven springs'. Seawell occurs as *Seywell* in 1681 and 1930. Thus both Sewall and Sewell may become SAYWELL.

**Sewer:** (i) William *Suuer* 1185 Templars (Ess); Roger *le Suhur* 1220 Cur (Lo); Geoffrey *le Suur* 1240 FFEss; Peter *le Suour* 1283 SRSf. This rare name has no less than four origins, with a variety of similar forms frequently very difficult to assign definitely. The most common is OFr *suor, suour, suur, seur* 'shoemaker'. *v.* also SOUTAR. (ii) Nicholas *le Seur* 1284 LLB A; Robert *le Seure* 1301 SRY; John *le Seuwour, le Seour* 1311 LLB D. An aphetic form of AFr *asseour*, OFr *asseoir* 'to cause to sit, to seat', 'an attendant at a meal who superintended the arrangement of the table, the seating of the guests, and the tasting and serving of the dishes' (a13.. NED, with forms *sewere, sawere, sewre*). (iii) Gilbert *le Suir* 1222 DBStP; Roger *le Syur* 1225 AssSo; Humfrey *le Syur, le Sawyere, le Sayhare* 1270 AssSo; John *le Syur, le Sayur* 1286 MESO (Nf); Philip *le Ciour, le Syour* 1316 Wak (Y). OFr *seieor, syeeur, scieur, saieur* 'sawyer'. Humfrey le Syur's name appears in both French and English forms which have given modern *Sewer*, *Sawyer*, and *Sayer*. (iv) Richard *le Sewer* 1279 MESO (Y); Margery *Sewer* 1327 SRC. A derivative of OE *sēowian* 'to sew', 'a sewer, tailor' (1399 NED, with forms *sewer, sawer, sower*). The feminine form survives as SOUSTER.

**Sewter:** *v.* SOUTAR

**Sewy:** *Seuui* 1124–30 Rams (Beds); William *filius Sewy* 1189 Sol; William *Seauwy* 1220 Fees (Berks); William *Sewy* 1250 Fees (Bk); Walter *Sewy* 1275 SRWo. OE *Sǣwīg*.

**Sexon:** *v.* SEXTEN

**Sexten, Sexton, Sexstone, Sexon, Seckerson, Secretan:** Tomas *Sekerstein* 1203 P (Y); Gilbert *le Segerstein* 1285 *Ass* (Ess); William *le Sekersteyn* 1296 SRSx; Thomas *Segrestan* 1299 LLB B; Henry *le Secrestein* 1327 SRSf; John *Sekesteyn* 1310 EAS xx; William *Sexstain* 1327 SRSx; Peter *Sexten* 1327 SRWo; Thomas *Sexton, Sexten* 1524 SRSf. AFr *segerstaine*, OFr *segrestein, secrestein*, MedLat *sacristānus*, a doublet of *sacristan* (1303 NED), 'sexton', originally 'the officer in a church in charge of

the sacred vessels and vestments', not, as now, the grave-digger.

**Sexy:** *Sexi* Forfot 1137 ELPN; Warin *filius Sexhiue* 1185 Templars (Ess); John *Sexi* 1210–11 PWi; Robert *Sexy* 1230 P (Ha); John *Sexy* 1508 FFEss. ON *Saxi, Sexi*. But OE *\*Seaxgifu* (f) has probably also contributed to the surname.

**Seyler:** *v.* SAILER

**Seyman:** *v.* SEAMAN

**Seymour, Seymoure, Seymore, Seymer:** Gaufridus *de Sancto Mauro* 1159 P (Hu); Henry *de Seimor* 1203 AssSt; Henry *de Seintmor* 1251 FFHu; Henry *de Seynmor, de Seymmor, de Semmor* 1272 AssSt; Robert *Seymor* 1344 LoPleas. From Saint-Maur-des-Fossés (Seine). Also from Seamer (Yorks): John *Saymor* 1416 FrY.

**Seyner:** *v.* SENIOR

**Seys:** *v.* SAYCE

**Shackcloth:** Probably a corruption of SHACKLOCK.

**Shackel, Shackell, Shackle, Shackles, Skakle:** Robertus *filius Scakel* Hy 2 DC (L), Robert *Scakel* ib.; Herbert *Scakel* ib.; William *Shakelle* 1379 PTY. *Schakel* is an anglicizing of Anglo-Scand *Skakel*, from ON *Skǫkull* (byname), OSw *Skakli*, the latter occurring in Scagglethorpe (ER, WRYorks).

**Shacklady:** *v.* SHAKELADY

**Shackleton:** Hugh *Schacheliton* 1246 AssLa; Hugh *de Shakeldene* 1302 SRY. From Scackleton (NRYorks), a Scandinavianized form of OE *Scacoldenu*. The surname preserves the original English *Sh-*.

**Shackley:** *v.* SHAKERLEY

**Shacklock, Shatlock, Shadlock:** Roger, Richard *Schakeloc* 1187 P (Gl), 1246 AssLa; Robert *Shakeloc* 1227 AssSt; Adam *Schakelokes* 1316 Wak (Y); John *Shadlock* 1524 FrY. 'Shake lock', a nickname for one with a habit (not unknown today) of shaking back his long hair (note the 1316 plural *locks*), or metonymic for a gaoler, either as a shaker of locks or from ME *schaklock* 'fetters'.

**Shackman:** Henry *Shakeman* 1221 FFEss. cf. SHAKELADY.

**Shackshaft:** *v.* SHAKESHAFT

**Shad:** Durand *Scad*, John *Shad* c1140 ELPN, 1326 FFEss; Nicholas, Ida *Schad* 1255 RH (W), 1275 SRWo. A nickname from the fish, OE *sceadd* 'a shad', the importance of which is shown from the existence of a 'shad season' in OE times.

**Shadbold, Shadbolt:** Thomas *Shotebolte* 1559 Pat (Herts). Forms of the name are few and late, but Harrison's 'shot bolt' seems unlikely. Perhaps 'dweller at the building on the boundary', OE *scēad, bōōl*.

**Shaddick, Shaddock, Shadwick:** *v.* CHADWICK

**Shade, Schade:** Lucas *Shadue* 1203 Cur (C); Hugo *Scade* (*Schade*) 1221 Cur (Lei); Andrew *Shadewe* 1314 LLB D; Ralph *Shade* 1296 SRSx. OE *sceadu*, ME *shade, shadwe* 'shadow', perhaps a nickname for a very thin man. Also 'dweller by the boundary' (OE *scēad*): Richard *de la Schade* 1230 P (D).

**Shadford, Shadforth:** Robert *de Shaddesford* 1332 SRSx; Richard *Shadforth* 1549 FrY. From Shadforth (Du).

**Shadlock:** v. SHACKLOCK

**Shadwell:** William, Roald *de Schadewelle* c1176–90 YCh, 1261 FFO; Thomas *de Shadewelle* 1349 IpmW; John *Shadwell* 1576 SRW. From Shadwell (Middlesex, Norfolk, WRYorks).

**Shadworth:** William *de Schaddeworth'* 1291 KB (Nt); William *de Shadworth'* 1373–5 AssL; John *Shadworth* 1410 FFEss. From Shadworth Manor Fm in Swaffham Prior (C).

**Shafe:** v. SHAW

**Shafto, Shaftoe:** John *de Schafthou* 1275 RH (Nb). From Shaftoe (Nb).

**Shafton:** Adam *de Schafton* 1261 IpmY; Adam *Shafton* 1407 FrY. From Shafton (WRY).

**Shail, Shayle, Shales:** Richard *Schayl* 1253–4 FFSr; John *Scheyl* 1327 SRSo; William *Sheyl* 1376 IpmGl. Metonymic for SHAYLER.

**Shailer:** v. SHAYLER

**Shairp:** v. SHARP

**Shakelady, Shacklady:** Richard *Shakelauedy* 1332 SRLa; Richard *Shaklady* 1384 FFLa; Rowland *Shakelady* 1529 ib. 'Known in Lancashire as a corruption of the ancient local surname of Shackerley' (Lower). Clearly an unwarranted local derivation. 'Shake lady.' Forssner suggests 'house-tyrant'.

**Shakelance:** Henry *Shakelaunce* 1275 RH (Nth); Robert *Schakelaunce* 1327 SRSf. 'Shake lance', a spearman. cf. SHAKESPEAR.

**Shakerley, Shackley:** Henry *de Shakresleghe, de Shakerleghe* 1332 SRLa. From Shackerley or Shakerley (both Lancs).

**Shakesby:** Richard, John *Shakespey* 1292 EwenG (Y), 1296 SRSx; Walter *Shakespye* 1342 Ewen (Sr). For SAXBY, *Sakespe* 'draw sword', with early substitution of the English *Shake-* for the French *Sac-*.

**Shakeshaft, Shakesheff, Shackshaft:** William *Shakeshaft* 1332 SRLa, 1542 PrGR; Henry *Shakshaft* 1674 HTSf. 'Shake shaft', OE *sceaft* 'shaft, spear'.

**Shakespear, Shakespeare, Shakspeare:** W. *Sakespere* 1248 Ewen (Gl); W *Shakesper'* 1318 ib. (St); Simon *Shakespere, Schakespere* 1324 AssSt, 1327 SRSt; Robert *Schaksper* 1379 PTY. 'Shake spear', a spearman, a common type of surname. cf. SHAKELANCE, SHAKESHAFT, etc. above, and John *Shakstaf* 1423 LLB K, Geoffrey *Schakeheved* 13th Rams (Hu), John *Shakeleg* 1333 ColchCt, William *Schaktre* 1301 SRY, Richard *Schakerake* 1379 *Coram* (Ha). 'Never a name in English nomenclature so simple or so certain in its origin. It is exactly what it looks—Shakespear' (Bardsley). To which Weekley adds that 'no European philologist of any reputation would dissent from this opinion' (*Words Ancient and Modern* 204). And there the name could be left but for Ewen's elaborate and pretentious ten-page disquisition in support of his preconceived theory that such names cannot be nicknames (*History of British Surnames* 312–322). An ignorance of phonetics and OE nomenclature is bolstered up by a confused collection of examples, some irrelevant, some proving conclusively the very etymology he wishes to disprove, the whole aptly summarized by Weekley as 'etymological moonshine'. Unrepentant, Ewen returned to the fray with

an elaborate pedigree of the name, *partly conjectural* (his own italics) in his *Guide to the Origin of British Surnames* (91–97), beginning with OE *Sceaftloc* and ending with *Shakeberry*. Apart from the fact that *Sceaft* is unknown as an element in OE personal-names, it would become *Shaft* and not *Shake* or *Skathe*. Eliminating from the pedigree all compounds of these, we are left with *Shaftspere* (1501) and *Scapespere* (1523), obviously 16th-century corruptions or errors of transcription, and *Shafsbury* (1688), clearly originating from the Dorset Shaftesbury. This leaves high and dry, in solitary seclusion, the conjectural OE *Sceaftloc*, presumed ancestor of an illegitimate progeny.

**Shales:** v. SHAIL

**Shallcrass, Shallcross:** v. SHAWCROSS

**Shane:** John Fitz Desmond alias *Shane* Fitz John Desmond 1540 LP; George *Shane* 1642 PrD; Giles *Shane* 1662–4 HTDo. The Irish form of *John*.

**Shank, Shanks:** Lefuine *Scanches* c1095 Bury (Sf); Rotbern *Sceanca* 1100–30 OEByn (D); Walter, Robert *Schanke* 1177, 1190 P (D, Herts); Stephen *Schankes* 1275 RH (Nf); John *Shonke* 1349 LoPleas. OE *sceanca* 'shank, leg', a nickname.

**Shann:** Thomas *Shan* 1551 FrY; Henry *Shann* 1672 HTY. From High, Low Shann in Keighley (WRY).

**Shaper:** Julia *Schaper* 1332 SRSx. A derivative of ME *schapen* 'to shape', a tailor. v. SHAPSTER.

**Shapler:** Gervase, Ralph *le Chapeler* 1214 Cur (Sr), 1230 MemR (K). OFr *chapelier* 'maker or seller of hats' (1601 NED). The name is also found in the AFr form *capelier*: Baldwin, Benedict *le capel(l)er* 1216–20 Clerkenwell (Lo), 1291 Cur (Mx).

**Shapley, Shapleigh:** Thomas *Shaplee* 1333–4 FFSr; Edward *Shaply* 1642 PrD. From Shapley in Chagford, in North Bovey (D).

**Shapster:** Matilda *Shapistre* 1275 RH (Sf); Sarra *le Shapester* 1327 SRWo. The feminine form of SHAPER. ME *schapen* is a new formation from the substantive *schap*. The OE verb was *scieppan*, whence SHIPSTER.

**Shard, Sheard, Sheards:** William *atte Sharde* 1275 SRWo; Adam *atte Sherd* 1332 SRWa; Hugo *del Sherd* 1354 Putnam (Ch). 'Dweller by the cleft or gap', OE *sceard*.

**Shardlow, Shardeloe, Shardalow:** Robert *de Scherdelawe* 1230 Guisb; John *de Shardelowe* 1327 SRDb; Henry *Shardelowe* 1417 FFEss. From Shardlow (Db).

**Shargold:** v. SHERGOLD

**Sharland:** v. SHIRLAND

**Sharman, Shearman, Sheerman, Sherman, Shurman:** Roger *sereman* 1207 FrLeic; William *le Shereman* 1281 LLB B; John *Sherman* 1327 SRSf; Philip *Shareman* (*Sharman*) 1327 *SR* (Ess). OE *scēarra* 'shears' and '*mann*', a shearer of woollen cloth.

**Sharp, Sharpe, Sharps, Shairp:** Healðegn *Scearpa* 1026 OEByn (K); Ailmer *Scharp* 1184 P (Herts); Richard *Serp* 1210 Cur (C); Aylmer *Sarp* 1228 FFEss; Alan *Sharp* 1296 SRSx. OE *scearp* 'sharp, quick, smart'. *Shairp* is Scottish.

**Sharparrow:** Robert *Sharparu* 1364 FFY; John *Scherparowe* 1448 NorwW (Nf); William *Sharparrow*

1568 Musters (Sr). 'Sharp arrow', OE *scearp, arwe*, a nickname for a good bowman, or perhaps for a maker of arrows.

**Sharples, Sharpless:** John *de Scharples* 1246 AssLa; John, Adam *de Sharples* 1332 SRLa; Robert *Scharpiles* 1418 IpmY; Richard *Sharplys* 1543 CorNt. From Sharples (Lancs).

**Sharrah, Sharrow:** Richard *de Scharhow* 1299 FrY; Robert *Sharrowe* 1642 PrD. From Sharrow (WRY).

**Sharrock(s):** v. SHORROCK

**Shatlock:** v. SHACKLOCK

**Shave(s):** v. SHAW

**Shaw, Shawe, Shafe, Shave, Shaves, Shay, Shayes, Shea:** Simon *de Schage* 1191 P (Berks); Richard *de la Schawe* 1275 SRWo; John *ate Shaw* 1295 ParlR (Ess); Henry *del Schawe* 1307 Wak (Y); William *Bithe Shaghe* 1333 MELS (So); Richard *Shay*, Hugh *Shey* 1564 ShefA (Y). 'Dweller by the wood', OE *sceaga*, as at Shaw (Berks, Lancs, Wilts). *Shay* is a Yorkshire dialect form, found also in Devon where *shave* and *shafe* are also found. Shaugh (Devon), pronounced *Shay*, is *Shaue* 1281, *Sheagh* 1535, *Shea* 1584, *Shaff* 1584, *Shaye* t Hy 8 (PN D 258). Shave Cross (Dorset) and Shave Fm (Som) both derive from *sceaga*. *Shea* is also for O'SHEA.

**Shawcross, Shalcross, Shallcrass, Shallcross, Shellcross:** Benedict *de Shakelcros* 1327 SRDb; Matilda *de Shalcros* 1348 DbAS 36. From Shackelcross (Derby).

**Shayle:** v. SHAIL

**Shayler, Shaylor, Shailer:** Thomas *Shaylard* 1275 SRWo; Henry *Shailard* 1292 SRLo; William *Schaylard* 1311 ColchCt. cf. ME *shailer* 'shambler'.

**Shead:** v. SHED

**Shean, Sheane:** v. SHEEN

**Shear:** v. SHEER

**Shearburn:** v. SHERBORN

**Sheard:** v. SHARD

**Shearer, Sheara, Sherar, Sherer, Sharer, Shirer:** Robert *le Sherer* 1231 FFC; William *Le Scherer', Le Schirere* 1305 MESO (Ha). A derivative of OE *sceran* 'to cut', 'one who removes the nap of cloth by shearing', identical in meaning with SHEARMAN, SHARMAN.

**Shearman:** v. SHARMAN

**Shearsmith, Sersmith:** Walter *Scheresmythe* 1325 ParlWrits (Gl); Geoffrey *Sheresmyth* 1391 FrY. 'A maker of shears, scissors', OE *scēarra* and *smið*.

**Shearwood:** v. SHERWOOD

**Sheat, Sheate:** v. SHUTE

**Sheather:** Henry, Richard *le Schether(e)* 1302 FrY, 1304 LoCt; Robert, Alexander *le Shether* 1334 LLB E, FrLeic. 'Maker of sheaths' (OE *scǣp, scēap*).

**Shed, Shedd, Shead:** John *Schede* 1301 SRY; John *de Schedde* 1327 SRSf; William *Shed* 1662–4 HTDo. 'Dweller at the hovel', OE *\*scydd*.

**Sheen, Shean, Sheane:** (i) Robert *Schene* 1226 Cur (So); Robert *Sheane* 1641 PrSo; John *Sheene* 1642 PrD. OE *scīene* 'fair, handsome'. (ii) John *de Schene* 1297 MinAcctCo. From Sheen (Sr, St).

**Sheep:** Simon *Shep* 1221 AssWa; Thomas *Sheep* 1321 FFSf. OE *scēap* 'sheep'. A nickname, but also, probably, for a shepherd or a dealer in sheep: John

**Shepgrom** 1327 *SR* (Ess), Alexander *Shepmongere* 1227 AssBeds.

**Sheephead, Sheepshead:** Richard *Sepeheued* 1208–9 Pleas (Hu); William *Shepesheved* 1276 AssLo; Walter *Shepesheued'* 1332 SRDo. 'Sheep's head', OE *scēap, hēafod*, a nickname. cf. Richard *Schepeshe* 1287–8 NorwLt 'sheep's eye'.

**Sheepshanks:** William *Shepescanke* 1224 Cur (Nf), 1379 PTY. 'With legs like a sheep' (OE *sceanca* 'leg'). Yorkshiremen were partial to this type of name: Symon *Doggeschanke* c1246 Calv (Y), William *Pyshank* 1373 FrY 'magpie-legs', John *Philipschank* 1379 PTY 'sparrow-legs'.

**Sheepwash, Shipwash:** Robert *Sepwas, de Sepwass'* 1279 RH (O); Philip *atte Shepewassh* 1332 SRSx. 'Dweller near a place for washing sheep' (OE *scēapwæsce*) as at Sheepwash (Devon, Northumb) or Sheepwash Fm in Nuthurst (Sussex).

**Sheer, Sheere, Sheeres, Sheers, Sher, Sherr, Shere, Shear, Shears:** Walter *Leschir* 1193 P (Berks); Reginald *le Scher* 1327 SRSt. OE *scīr, \*scǣre* 'fair, bright'. *Shear(s)*, if the spelling is to be relied on, may be metonymic for SHEARSMITH.

**Sheering:** v. SHERRIN

**Sheffield, Sheffle:** William *de Shefeld* 1227 AssSt; Thomas *de Sheffeld* 1328 FFY; Ralph *Sheffeld* 1456 Goring (O); George *Sheafeild* 1621 SRY. From Sheffield (Berks, Sx, WRY).

**Shefford:** John *Schefford* 1291 TestEbor. From Shefford (Beds), or East, West Shefford (Berks).

**Sheldon:** Roger *de Scheldona* 1189–90 P (St); Cecily *de Sheldon* 1248 FFK; John *Sheldone* 1367 IpmGl. From Sheldon (Devon, Derby, Warwicks).

**Sheldrake, Sheldrick, Shildrake, Shildrick:** Roger *Scheldrac* 1195 P (Ess); Adam *le Sceldrake* 1275 RH (Sf). 'Sheldrake', a bird of the duck tribe, remarkable for its bright and variegated colouring (c1325 NED).

**Shelf, Shelfe:** John *de Schelf* 1204, Thomas *Scelf* 1230 P (Y); William *atte Shelue* 1327 SRSa. From Shelf (WRY), or 'dweller by the hill', OE *scylf*.

**Shelford:** John *de Shelford* 1220 SPleas (Herts); Walter *de Shelford* 1304–5 FFEss; Henry *Shelford* 1412–15 FFSr. From Shelford (C, Nt), or Shelford House in Burton Hastings (Wa).

**Shell:** John *Shelle* 1340 NIWo. From Shell (Wo).

**Shellabear:** v. SHILLABEER

**Shellcross:** v. SHAWCROSS

**Shelley, Shelly:** Matilda *de Selleg*, Richard *de Selueleg'* 1201 Pl (Sf); Henry *de Schellay* 1297 SRY; Roger *de Shelley* 1379 PTY; John *Shelley* 1473–4 FFSr. From Shelley (Essex, Suffolk, WRYorks).

**Shellito:** v. SHILLITO

**Shelton:** Robert, John *de Schelton'* 1191 P (St), 1260 AssY; John *Shelton'* 1385 AssWa. From Shelton (Beds, Norfolk, Notts, Salop, Staffs).

**Shemmans, Shemming:** v. SHIMMINGS

**Shenley:** Burchard *de Senelai* 1066 DB (Bk); Geoffrey *de Shenlee* 1204 Cur (Herts). From Shenley (Bk, Herts).

**Shenston, Shenstone:** Richard *de Scheneston'* 1275 SRWo; Geoffrey *de Shenstone* 1348 AssSt; William *Shenston* 1426–7 FFWa. From Shenstone (St).

**Shepard, Shepeard, Sheperd, Shephard, Shepheard, Shepherd, Sheppard, Shepperd, Sheppherd,**

**Shippard:** William *Sepherd* 1279 RH (O); Henry *Sephurde* 1296 SRSx; Emma *le Schepherde* 1297 MinAcctCo; Walter *le Shepperde* 1307 FFSf; Alice *the Schiphird* 1307 Wak (Y); Avice *la Schepherdes* 1311 ColchCt; John *Shipherde* 1317 AssK; Walter *le Sheparde* 1327 SRWo; Adam *le Shepherd* 1328 Crowland (C); Robert *le Shephard* 1330 FFSr. OE *scēaphyrde* 'shepherd'. This may also occasionally have absorbed *shipward*; William *le Shipward* 1357 Putnam (He), Hugo *Shipward* 1471 FrY. Probably OE *scēap-weard* 'sheep-ward'. Thomas *Shypward* (1432) and John *Shipward* (1467), both domiciled in Bristol, were probably ship-masters (OE *scipweard*). *v.* MEOT.

**Shepherdson, Sheperdson, Sheppardson, Shepperdson, Shepperson, Shipperdson:** John *Schephirdson* 1332 SRCu; Ralph *Schiperdson* 1509 GildY. 'Son of the shepherd.'

**Sheppick:** Henry *de Schepewic* 1275 RH (Nt). 'Dweller or worker at a sheep-farm' (OE *scēapwīc*). cf. Shapwick (Dorset, Som).

**Sher:** *v.* SHEER

**Sherar:** *v.* SHEARER

**Sherard:** *v.* SHERRARD

**Sheraton:** Robert *Scurveton* 1407 FrY. From Sheraton (Du), *Scurvertune* 1190.

**Sherborn, Sherborne, Sherburn, Shearburn, Shirbon:** Pagan *de scirburna* 1189 Sol; Ralph *of Shyreburne* 1251–2 FFY; John *Shirborne* 1368 IpmW; William *Sherborne* 1576 SRW. From Sherborne (Do, Gl, Wa), Sherborne St John, Monk Sherborne (Ha); Sherburn (Du, ERY), Sherburn in Elmet (WRY), or Shirburn (O).

**Shercliff, Sheircliffe, Shirtcliffe, Shirtliff, Shetliff, Shortliffe:** John *de Shirclyf* 1379 PTY; Thomas *Shirtliffe* 1510 ShefA; Richard *Shirtcliffe* 1564 ib.; Robert *Sheirclyff* 1593 ib. From Shirecliff in Sheffield (Yorks).

**Shere:** *v.* SHEER

**Sherer:** *v.* SHEARER

**Shergold, Shargold:** William *Shergot* 1377 AssEss; John *Shergall* 1545, Robert *Shergolld*, Thomas *Shargall* 1576 SRW; Henry *Shergoll* 1662–4 HTDo. Weekley's 'shear gold', a nickname for a worker at the mint, does not seem likely judging from the available forms. On the whole it looks like an unrecorded personal name, *Scīrgōd*.

**Sheriff, Sheriffs, Sherriff, Sherriffs, Shireff, Shirreff, Shirreffs, Shirrefs, Shiriff, Shreeve, Shreeves, Shreve, Shrieve, Shrieves, Shrive, Shrives, Schrieve:** Æthelwine *Sciregerefa* 1016–20 OEByn (K); Hugo *le Sirreve* 1212 Cur (Lei); Alan *Sciriue* 1219 AssL; Walter *Sherrev'* 1220 Cur (K); Thomas *Shyrreue* 1230 P (Ha); John *Schiref* 1273 RH (Nb); Thomas *Shreeve* 1457 Bacon (Sf); Robert *Shryve* 1568 SRSf. OE *scīrgerēfa* 'sheriff'. *v.* SHIRRA.

**Sherland:** *v.* SHIRLAND

**Sherlock, Shurlock:** Ælfwerd *Scirloc* c1002–19 OEByn; Ralph *Shirloc* 1159 P (Lo). OE *scīr* 'bright, shining' and *loc(c)* 'lock (of hair)'. 'Fair-haired.'

**Sherman:** *v.* SHARMAN

**Sherr:** *v.* SHEER

**Sherrard, Sherreard, Sherrad, Sherratt, Sherred, Sherrett, Sherrott, Sherard:** William *Shirard* 1298

AssSt; Richard *Schirard* 1323 AD v (St); William *Sherard* 1337 AssSt; William *Sherratt* 1578 AD vi (Ch). OE *scīr* 'bright' plus the French intensive suffix -(*h*)*ard*.

**Sherriff:** *v.* SHERIFF

**Sherrin, Sherring, Sheering:** Geoffrey *de Sheringg* 1327 FFEss; Philip *Sherren* 1641 PrSo; John *Sherring* 1662–4 HTDo. From Sheering (Ess).

**Sherringham:** Walter *de Scheringham* 1327 SRY; William *Sheryngham* 1389 AssNu. From Sheringham (Nf).

**Sherrington:** Richard *de Schirinton'* 1190 P (Bk); Margaret *de Sherynton* 1296 SRSx; Richard *de Shiryngtone* 1337 LLB F. From Sherington (Bk).

**Sherston:** William *Sherston'* 1379 AssWa. From Sherston (W).

**Sherwen, Sherwin:** Gilbert *scerewind', scorewint* Hy 2 DC (L); William *Scherewind* 1187 P (Cu); John *Shirwyn* 1479 NorwW (Nf); John *Sherwyn* 1524 SRSf. 'Cut wind', from OE *sceran* 'to cut' and *wind*, used of a swift runner.

**Sherwood, Shearwood:** William *de Shirewude* 1219 AssY; Alan *de Shirewod* 1327 SRSx; John *Shirwod* 1405 IpmY; Elizabeth *Sherwood* 1539 CorNt. From Sherwood (Notts), or 'dweller by the bright wood'.

**Shetliff:** *v.* SHERCLIFF

**Shettle:** *v.* SHUTTLE

**Shield, Shields:** (i) Robert *Scild* 1206 P (Y); William *Sheld* 1267 AssSt. OE *scild, sceld* 'shield', probably for a maker of shields. cf. *scyldwyrhta* c1114 MLWL, Geoffrey *le Seldmakere* 1285 MESO (Ess), Adam *Scheldman* 1327 SRSf. (ii) Roger *ate Schelde* 1332 SRSr. 'Dweller by the shelter' (OE *scild, sceld* 'protection') or 'by the shallow place' (OE *scieldu*).

**Shilbottle, Shillbottle:** John *Shilbotill* 1392 TestEbor; Henry *Shilbotell* 1417 IpmY; Christopher *Shilbotell* 1500 TestEbor. From Shilbottle (Nb).

**Shilleto:** *v.* SHILLITO

**Shilling:** John *Eschelling* 1176 P (So); Geoffrey *Scelling* 1188 P (So); John *Eskeling'* 1221 AssGl; William, John *Schilling* 1275 RH (Nf), 1309 AD iv (Wa). Apparently a nickname from OE *scilling* 'shilling'. There is no evidence for a personal-name as suggested by Bardsley and Harrison.

**Shillingford:** John *Shilingford* 1392 LoCh. From Shillingford (D, O).

**Shillito, Shillitoe, Shilleto, Shellito, Scillitoe, Silito, Sillitow:** William *Shillito* 1374 FFY; William *Shelito* 1398 YDeeds I; John *Shelyto* 1536 FrY; Thomas *Shillito* 1684 RothwellPR (Y). Evidently a Yorkshire name, but there is no likely origin, and the absence of early forms makes it difficult to suggest one.

**Shilton:** Henry *of Shilton* 1231 FFY; William *Shilton* 1502 GildY; Jeffrey *Shilton* 1576 SRW. From Shilton (Berks, O, Wa), or Earl Shilton (Lei).

**Shimmin:** Dermot *MacSimon* 1366, *MacShemine* 1430, *Shimin* 1614 Moore. Manx *McSimeen* 'little Simon's son'.

**Shimmings, Shimmin, Shimmins, Shemming, Shemmans:** Roger *Scymming*, John *Schemmeng* 1279

RH (C); Richard, John *Shymmyng* 1297 MinAcctCo, 1332 SRWa; Robert *Shimmon*, Amy *Shimming*, Ann *Shimman* 1775, 1787, 1793 SfPR. Perhaps OE *\*sciming* 'the fair one', from OE *scīma* 'bright'.

**Shimpling:** Andrew *of Shimpling* 1313 AssNf; Isabella *Shymplyng'* 1364 KB (Nf). From Shimpling (Nf).

**Shine:** William *Shine* 1221 AssWo; Sabine *Schyne* 1279 RH (C); William *Shine* 1327 SRWo. OE *scīene* 'beautiful, attractive'.

**Shingler:** Roger *le Shinglere* 1335 MEOT (Ess); William *Shyngelere* 1381 SRSt. A derivative of ME *schinglen* 'to cover with shingles', a roofer.

**Shingles:** John, Henry *S(c)hingel* 1327 SRSx, SRSf. ME *shingle* 'a wooden tile', metonymic for SHINGLER.

**Shingleton:** Richard *Shyngleton* 1545 SRW; George *Shyngleton* 1571 StratfordPR (Wa); Nycholas *Shingleton* 1583 Musters (Sr). Probably late forms of SINGLETON.

**Shinn, Shynn:** Herveus *Schin* 1165 P (Sf); Hugo *Scin* 1190 P (He). Probably OE *scinn*, ME *shin* 'skin', the native equivalent of ON *skinn*. v. MESO 119–21. Metonymic for a skinner. v. SHINNER.

**Shinner:** John *le Scinner* 1279 RH (C); Geoffrey *le Schinnere* 1296 SRSx; John *Schynnere*, Adam *Le Schinner* 1305 MESO (Ha, So). A derivative of OE *scinn* 'skin', a skinner. v. SHINN, SKINNER.

**Shipdam, Shipdame:** Ralph *de Shipdam* 1308 AssNf; Watkyn *Schipdam* 1449, *Shipdam* 1460 Paston. From Shipdam (Nf).

**Shipden:** William *de Schypeden* 1274 Wak (Y); Emma *de Schepden* 1379 PTY. From Shibden in Southowram (WRY).

**Shipley, Shiplee:** Hugh *de Sciplay* 1219 AssY; Robert *de Sheplay* 1375 FFY; Robert *Shipleye* 1402–3 FFWa. From Shipley (Derby, Durham, Northumb, Salop, Sussex, WRYorks), or 'dweller by the sheep pasture'.

**Shipman:** (i) *Schipemannus* 1130 P (R); *Scipmanus* c1250 Rams (Nf); Hubert *Scipman* 1221 ElyA (Hu); Simon *le Sipman*, *Schipman* 1267, 1290 HPD (Ess). OE *scipmann* 'seaman, sailor'. William *Scipman* (1243 AssSo) was drowned from a boat in the water of the Parret. (ii) Richard *le Schepman* 1296 MEOT (Herts); Adam *le Schepman* 1316 ib. (Ess). OE *scēap* 'sheep' and *mann*, 'shepherd'. In certain dialects, OE *scēap* became ME *ship* and *shipman* may sometimes mean 'shepherd'. v. MEOT 63, 257.

**Shipp:** Roger *de, del Schipp* 1288 AssCh; Walter *del Schippe* 1297 SRY. Richard Stonham *atte Shippe* withouten Crepulgate (1423 LondEng 149) lived at the sign of the ship but early examples of such sign-names are rare and the above examples are probably identical in meaning with SHIPMAN 'sailor'.

**Shippard:** v. SHEPARD

**Shippen:** William *Shepene* 1317 AssK; Adam *atte Shupene* 1327 SRWo; Henry *ate Shypene* 1341 ib. OE *scypen* 'cattle-shed', either from residence at a place named from such, as Shapens (PN Ess 521), Shippen (PN D 495), Shippen (WRYorks) or Shippon (Berks), or from employment there, 'cattle-man, cowman'.

**Shipperbottom:** v. SHUFFLEBOTHAM

**Shipperdson:** v. SHEPHERDSON

**Shippey:** Thomas *Shepey* 1524 SRSf; George *Shippie* 1641 PrSo; Thomas *Shippey* 1662 HTEss. 'Dweller at the isle where sheep are kept', OE *scēap*, *īeg*, or 'dweller by the sheep enclosure', OE *scēap*, *(ge)hæg*.

**Shiprod, Shiprode:** Walter *de la Seperode* 1296, John *ate Shiprode* 1327, Robert *atte Sheperod* 1332 SRSx. From Shiprods Fm in Henfield (Sx), or 'dweller at the clearing where sheep are kept', OE *scēap*, *rodu*.

**Shipster:** Alice *La Seppestre* 1296 SRSx; Matilda *le Scipstere* 1320 Oseney (O); John *Shepster* 1379 ColchCt. A derivative of OE *scieppan*, *sceppan* 'to create', ME *shippen* 'to shape', 'a female cutter-out of material, a dressmaker' (1377 NED). Now a rare surname.

**Shipton:** Osbert *de Scipton* 1200 Cur; Baudewyn *de Shipeton* 1304 IpmY; John *Shipton* 1447 AssLo. From Shipton (Sa, ERY, NRY), Shipton Lee (Bk), Shipton Gorge (Do), Shipton Moyne, Oliffe and Sollars (Gl), Shipton Bellinger (Ha), or Shipton on Cherwell, under Wychwood (O).

**Shipward:** William *le Shipward* 1357 Putnam (He); Nicholas *Shipward* 1375 IpmGl; Hugh *Shipward* 1471 FrY. Probably 'keeper of the sheep, shepherd'. OE *scēap*, *weard*. No doubt this was usually absorbed by SHEPARD. But Thomas *Shypward* 1432, and John *Shipward* 1467 MEOT, both domiciled in Bristol, were probably ship-masters, OE *scipweard*.

**Shipwash:** v. SHEEPWASH

**Shipwright:** Warin *Sipwriet* t John HPD (Ess); Alice *le Schipwrith* 1279 AssNb; John *le Shipwrighte* 1309 FFEss. OE *scipwyrhta* 'a man employed in the construction of ships'.

**Shirbon:** v. SHERBORN

**Shirbrooke, Shirebrooke:** Roger *de Shirbrok* 1368–9 FFSr. From Shirebrook (Db).

**Shire:** Walter *de Schyre* 1296 SRSx; Roger *del Shires* 1327 SRY; Gregory *atte Shire* 1335 LLB E. perhaps 'dweller at the meeting-place of the shire', OE *scīr*.

**Shirer:** v. SHEARER

**Shirland, Sherland, Sharland:** Robert *de Scirland'* 1199 P (K); Alexander *de Syrlund* c1250 Glapwell (Db); Simon *Schyrland* 1379 LoCh. From Shirland (Db), Sheerland in Pluckley (K), or Sharland in Morchard Bishop (D).

**Shirley:** William *de Schirle* 1219 AssY; Ralph *de Shirleye* 1318–19 FFWa; William *Shirley* 1442–3 FFSr. From Shirley (Derby, Hants, Surrey, Warwicks, WRYorks).

**Shirman:** Ufegeat *Scireman* 1000–14 OEByn (K); William *Schiremon* 1275 SRWo. OE *scīrmann*, *scīremann* 'sheriff', also 'bailiff, steward'.

**Shirmark:** Alice *Shirmarke* 1327 SRSx; Giles *atte Shiremarkes* 1379 MELS (Sx). 'Dweller at the boundary of the shire', OE *scīr*, *mearc*.

**Shirra:** The Scottish pronunciation of SHERIFF: Elspeth *Shirra* 1676 Black (Edinburgh).

**Shirt:** Godwin *de la Sirte* 1179 P (Sr); John *atte Shurte* 1296 SRSx. 'Dweller by the detached piece of land', OE *scyrte* 'skirt, piece cut off'.

**Shirtcliffe, Shirtliff:** v. SHERCLIFF

**Shitler:** v. SHUTLAR

**Shobrooke:** John *Shobroke* 1641 PrSo; John *Shobrocke*, Robert *Shobrooke* 1642 PrD. From Shobrooke (D).

**Shocker:** John *le Schokere* 1296 SRSx. An occupational-name from ME *schokken* 'to heap up, arrange sheaves in a shock'.

**Shodd:** Estan *Scodhe* 1066 Winton (Ha); Roger *Schod* 1287–8 NorwLt; Walter *Shodde* 1359 AssD. A nickname from OE *scōda* 'pod, husk'.

**Shoe:** Emma *of the Schoe* 1218–19 FFEss; John *Shoe* 1641 PrSo; John *Shoe* 1642 PrD. 'Dweller at the shoe-shaped piece of land', OE *scōh*.

**Shoebotham, Shoebottom:** *v.* SHUFFLEBOTHAM

**Shoemake, Shoemaker, Schoemaker:** Hugh *S(c)homaker* 1365 ColchCt. 'Shoemaker', from OE *scōh.*

**Shoesmith, Shoosmith:** William *Le Shosmith, Sosmyth* 1288 MESO, 1296 SRSx. OE *scōh, smið* 'shoeing-smith', maker of horseshoes.

**Shoot:** *v.* SHUTE

**Shooter, Shotter, Shuter:** Robert *scotere* 1148 Winton (Ha); Stephen *le Shotiere* 1255 *Ass* (Ess); Henry *Schuetere* 1275 SRWo; William *Schoter* 1316 NottBR; Henry *le Shotter* 1320 AssSt; John *le Schotiere* 1327 *SR* (Ess); John *Schewter* 1379 PTY; Richard *Shoiter, Shooter* 1554 Shef, 1579 ShefA; John *Shuter* 1784 Bardsley. A derivative of OE *scotian* 'to shoot', 'shooter, archer'. *S(c)hoter* may also derive from ME *schōten*, from OE *sceōtan*, through a shifting of the vowel-stress from *scēotan* 'to shoot'. This also gave *sheter*, now *Sheeter*: Ralph *Scheter* 1297 SRY, Robert *Schetere* 1327 SRSf.

**Shoppe:** William *en le Shope* 1271 AssSt; Thomas *atte Shoppe* 1341 LLB F; John *Shoppe* 1380–1 PTW. 'Dweller or worker at the building concerned with the manufacture or sale of goods', OE *sceoppa.*

**Shopper:** Thomas *atte Shoppe* 1341 LLB F; John *Shopper* 1353 Putnam (W). 'Shop-keeper.'

**Shore, Shores:** William *del Shore* 1332 SRLa. 'Dweller by the shore', ME *schore.*

**Shorey:** John *Shorie* 1327 SRSx; William *Shorye* 1341 ChertseyCt (Sr); Thomas *Shory* 1535 FFEss. 'Dweller on the island near the shore', ME *schore*, OE *īeg.*

**Shorrock, Shorrocks, Sharrock, Sharrocks, Shurrock:** Richard *de Shorrok* 1332 SRLa; George *Sharrocke* 1682 PrGR. From Shorrock Green (Lancs).

**Short, Shortt:** Ordric *Scort* 1176 P (Do); Richard *le Sorte* 1269 AssSo; William *Short* 1327 SRSx. OE *sceort* 'short'.

**Shortbayn:** Richard *Schortbayn* 1327 SRY. 'Short bone', OE *sceort*, ON *beinn*. cf. Thomas *Sortleg* 1284 CtW 'short leg'; John *Shortnekke* 1381 SRSf 'short neck'.

**Shorter, Shorters:** Roger *le Shorter* 1306 AssW; Robert *Shorter* 1481 Cl (Sr); Nycholas *Shorter* 1583 Musters (Sr). 'The smaller', ME *shortere*. cf. William *Schortfrend* 1270 ForSr 'small friend'; Robert *Shorthoume* 1401 AssLa 'small man'; Reginald *Scortecnicht* 1219 P (C/Hu) 'short boy'.

**Shorthose, Shorthouse, Shortose, Shorters, Shortis, Shortus:** (i) William *Shorthose* 1260 AssLa; Henry *Schorthose* 1275 Wak (Y); Robert *Shortus* 1585 Oxon; Martha *Shorthouse* 1731 SfPR. A translation of *Curthose* 'short boot'. Trevisa (1387) translates Higden's 'filio Roberti *courtehose*' by 'Robert *Scorthoses* sone' (Groups 62). cf. Adam *Sortkyrtell* 1256 AssNb 'short-gown', Robert *langhose* 1277 Ely (Sf) 'long boot'. (ii) Robert, John *Schorthals* 1198 P (K), 1219 AssL. 'Short neck', OE *h(e)als*. cf. Rogerus *longhals* c1200 Dublin. *v.* WHITEHOUSE.

**Shortliffe:** *v.* SHERCLIFF

**Shortreed:** Martin, Simon *Shortrad(e)* 1274 RH (Ess), 1327 SRY. 'Short counsel', OE *sceort, rǣd*, 'short' in the sense 'limited'. cf. 'My wit is short' (Chaucer) and Robert *Smalred* 1176 P (Y).

**Shortrigg:** Thomas *de Shortrig* 1330, Henry *de Shortrugg'* 1333 PN D 401, 83; James *Shortrige* 1642 PrD. From Shortridge in Crediton, in Yarnscombe (D).

**Shortwood:** Bartholomew *de Sortewud* 1226–7, Thomas *de Schortewode* 1271 FFK; Alan *Shortwode* 1392 CtH. From Shortwood in Throwley (K), or 'dweller by the small wood', OE *sceort, wudu.*

**Shot, Shott:** Roger *Schot* 1279 RH (Hu); Robert *de la Shote* 1341 AssSt; Walter *Shot* 1343 FFW. 'Dweller by the piece of projecting land', ON *skot*, or OE *\*sceota.*

**Shotter:** *v.* SHOOTER

**Shotton:** Robert *de Schottun* c1180 Black (Kelso); Huwe *de Shottone* 1296 ib. (Roxburgh); Ralph *del Schoton* 1327 SRDb. From Shotton (Durham, Northumb), or 'dweller by the enclosure on the slope'.

**Shoulders:** Rau *Sculdur* 1100–30 OEByn (D); Robert *Schuldre* 1275 RH (Nf); Simon *Shulder* 1327 SRSx. OE *sculdor* 'shoulder', a nickname for one with broad shoulders or some peculiarity of the shoulders.

**Shouldham:** *v.* SHULDHAM

**Shouler:** *v.* SHOVELLER

**Shove:** Aluricus *Scoua* 1066 DB (Herts); Leuuinus *Scufe* c1067 OEByn; *Scoua* 1086 DB (Herts); Wulnod *Scoue* 1185 Templars (K). Probably OE *\*scufa*, a derivative of *scūfan* 'to thrust, push'. cf. SHOWERS.

**Shovel, Showl:** Turoldus *scuuel* 1148 Winton (Ha). Metonymic for SHOVELLER.

**Shovelbottom:** *v.* SHUFFLEBOTHAM

**Shoveller, Shouler, Showler:** William *le Schovelere* 1301 MEOT (O); Walter *le Shouelere* 1314 ib. (Nth); Nicholas *Schoueler, Schoveler* 1366 ColchCt. A derivative of ME *schovelyn* 'to shovel'. One who works with or makes shovels.

**Showell:** Benjamin *de Showell* 1275 RH (Sx); Thomas *Showell* 1576 SRW. From Showell (O).

**Showers:** Richard *le Shouere* 1297 MinAcctCo (W). A derivative of OE *scūfan* 'to thrust, push', a nickname, no doubt, for one of violent habits. cf. SHOVE, and SHOWLER for SHOVELLER.

**Showl:** *v.* SHOVEL

**Showler:** *v.* SHOVELLER

**Shrapnel:** Richard *Sharpanel* 1218–22 Cockersand (La); Richard *Shrapnell* 1706 DKR 41 (W). A metathesized form of *Charbonnel*. *v.* CARBONNEL.

**Shreeve(s):** *v.* SHERIFF

**Shrewsbury, Shrosbree:** John *Shrouesbury* 1280 SRSt; William *de Schrouesbir'* 1332 SRLo. From Shrewsbury.

**Shrieve(s), Shrive(s):** v. SHERIFF

**Shrigley:** Pimecok *de Scriggel'* 1285, Silkot *de Shriggel'* 1286 AssCh; Robert *Shrygeley* 1545 PN Ch i 107. From Pott Shrigley (Ch).

**Shropshire:** Edith *de Shropshir'* 1226 Cur; Alice *de Shropshire* 1327 SRWo; Richard *Shropschire* 1414 IpmNt. From Shropshire.

**Shrouder:** v. SCUDDER

**Shrubb:** Robert *Shrub* 1288 FFSf; John *in the Scrobbes* 1327 *SR* (Ess). 'Dweller by the shrub(s)', OE *scrybb*.

**Shrubland:** Adam *de Shrubeland* 1271–2 FFEss; John *Shriblound* 1346 KB (Sf). From Shrubland (Sf).

**Shubotham:** v. SHUFFLEBOTHAM

**Shuckborough, Shuckburgh:** Henry *de Sukeberge* 1214 P (Wa); William *Shukkebourgh* 1340–1450 GildC; John *Shukburgh* 1460 FFEss. From Shuckburgh (Wa).

**Shucksmith:** v. SUCKSMITH

**Shufflebotham, Shufflebotam, Shovelbottom, Shipperbottom, Shoebotham, Shoebottom, Shubotham:** Richard *de Schyppewallebothem* 1285 AssLa; James *Shepobotham* 1579 Bardsley; John *Shippobotham* 1582 ib.; George *Shupplebotham* 1621 ib.; Charles *Shifabothom* 1626 ib.; James *Shipplebotham* 1642 ib.; Richard *Shufflebotham* 1674 ib. From Shipperbottom (Lancs).

**Shuldham, Shouldham:** Simon *de Shuldham* 1177 P (Nf); John *of Shouldham* 1312 AssNf; Thomas *Sholdham* 1454 Paston. From Shouldham (Nf).

**Shurlock:** v. SHERLOCK

**Shurrock:** v. SHORROCK

**Shute, Shutes, Shoot, Sheat, Sheate:** Simon *ater Schute* 1296 SRSx; Alan *atte Shute* 1332 SRWa; Adam *atte Shute* 1340 FFSt. From Shute, Shewte in Bovey Tracy (D), Sheetland in Ambersham (Sx), Shate, Sheat (Wt), or 'dweller at the nook of land', OE *scīete*.

**Shuter:** v. SHOOTER

**Shutlar, Shutler, Shuttler, Shitler, Shittler:** William *le Scuteler* 1177 P (Nf); John *Scutelere* 1296 SRSx; Hugo *le Scotiler* 1327 SRSt. Thuresson explains this as a derivative of OE *scutel* 'dish, platter', 'a scullion' or a dealer in crockery. This would account for the above forms and *Shutler*, but the existence of *Shitler* and the possibility that *Shettle* may be metonymic for *Shutler*, implies an OE *y*. *Shitler*, often *Shutler*, and the implied *Shetler* are derivatives of OE *scytel*, ME *shittle, schetylle, schutylle*, an instrument used in weaving, a shuttle, and the surnames denote a maker of shuttles or, more probably, 'a weaver'. cf.

> 'My godsire's name, I tell you,
> Was *In-and-In Shittle*, and a weaver he was,
> And it did fit his craft; for so his shittle
> Went in and in still, this way and then that way.'
> (Ben Jonson, *Tale of a Tub*, iv. 2.)

**Shutt, Shut, Shutte, Shutts:** Wulfsige *Scytta* c1050 OEByn (Herts); Liuricus *Shitte* 1165 P (Sf); Michael *Shut* 1296 SRSx; John *Schut, Shit* 1327 SRSf. OE *scytta* 'shooter, archer'.

**Shutter:** Simon *le Shutter* 1332 SRWa. OE *scytere* 'archer'.

**Shuttle, Shettle:** William *le Schutel* 1296 SRSx; Simon *Shitel* 1313, 1314 Pat (Y). ME *shytell, shuttle*

'inconstant, variable, fickle, flighty' (c1440 NED). Also metonymic for SHITLER, SHUTLAR.

**Shuttleworth, Shuttlewood:** Henry *de Schutlesworth* 1246 AssLa; John *de Shutelisworth* 1338 WhC; James *Shittleworth*, John *Shettleworth*, Grace *Shettlewood* 1662 *HTEss*. From Shuttleworth (Lancs), Shuttlewood in Bolsover (Derby), or Shuttleworth, now Littleworth, in Rossington (WRYorks).

**Sibary, Siberg, Siberry:** v. SEABER

**Sibbe, Sibbes, Sibbs:** *Sibba* Ædesdohter c1095 Bury; John *filius Sibbe* 1297, *Sibbe* Hamond 1306 Wak (Y); William *Sibbe* 1327 SRC; Robert *Sibbe* 1332 SRSx; Robert *Sybbes* 1568 SRSf. ON *Sibba* (f), but usually from *Sibb*, a pet-form of *Sibyl*.

**Sibbett, Sibbit, Sibbitt:** *Sebode* c1095 Bury (Sf); William *Sibode* 1206 P (So); Walter *Sybod* 1221 AssWa; William *Sibbitt* 1581 FrY; Roger *Sybbett* 1594 ib. OG *Sigibodo* 'victory-messenger'.

**Sibble, Sibbles:** Walter *filius Sibile* 1189 Sol; *Sibilla* 1219 AssY, Cur (Nth); Walter *Sibilla* 1170 P (Berks). Fr *Sibille* (f). v. SIBLEY

**Sibbons, Sibborn:** v. SEABORN

**Sibcy, Sibsey, Sibsy:** John *de Sibeteia* 1185 Templars (L); Ralph *de Sibicy* 1209 P (L); John *Sybsey* 1393 AssL. From Sibsey (L).

**Sibert:** v. SEABER

**Sibley, Sibly, Sebley:** Geoffrey *Sibilie* 1275 RH (Sf); William *Sibeli, Sibli* 1279 RH (Hu); Richard *Sebely* 1327 SRSf. *Sibyl*, Greek Σίβυλλα, Lat *Sibilla*, the name of the priestess who uttered the ancient oracles, a common woman's name after the Conquest, in the vernacular *Sibley*.

**Sibling:** Thomas *Sibilling* e 13th (1443) AD i (Mx); John *Sybeling* 1242 FFK; Thomas *Sybylling* 1287 AD i (Mx). OE *sibling* 'relative, kinsman'.

**Siborne:** v. SEABORN

**Sibree:** v. SEABER

**Sibsey, Sibsy:** v. SIBCY

**Sibson:** John *Sibson, Sibbeson* 1314, 1316 Wak (Y); William *Sibbison* 1327 SRC. 'Son of *Sibb*', a pet-form of *Sibyl*. cf. John son of *Sibbe, Sibbe* Hamond 1297, 1306 Wak.

**Sibthorp, Sibthorpe, Sipthorp:** William *de Sibetorp'* 1207 Cur (Nt). From Sibthorpe (Nt).

**Sibton:** Roger *de Sibbeton'* 1212 P (C/Hu); William *de Sibiton* 1296 NorwDeeds I. From Sibton (Sf).

**Siburn:** v. SEABORN

**Sicely:** v. SISLEY

**Sich:** v. SYKES

**Sicklefoot:** Gilbert *Sykelfot* 1289–90 FFSx; John *Sikilfot* 1327 SRSx. 'Splay-footed', OE *sicol, fōt*. cf. John *Sikelta* 1200 AssL 'with splayed toes'.

**Sickling:** Walter *Sicling* 1208 P (Y); Thomas *Sicling* c1253 Black; John *Siklyng'* 1387 AssL. OE **Siceling*.

**Siddall, Siddalls, Siddell, Siddle, Sidle, Syddall:** Thomas *Sydall* 1379 PTY; William *Siddall* 1672 HTY. From Siddall (La), Siddall in Southowran (WRY), or Siddle in East Harlsey (NRY).

**Sid(d)enham:** v. SYDENHAM

**Side, Sydes:** Herbert *de Side* 1221 AssGl; Henry *Beside* 1290 *ERO*; Michael *Aside* 1327 SRSx; Adam *del Syde* 1332 SRSt. 'Dweller by the slope', OE *sīde*, as at Syde (Glos).

**Sidebeard:** Peter *Sydeberd* 1327 SRSo. 'Wide beard', OE *sīd, beard*. cf. Hubert *Sidekertell'* 1202 Pleas (Sf) 'wide tunic'; Robert *Sidskirte* 1177 P (Y) 'wide skirt'.

**Sidebotham, Sidebottom:** Hugh *de Sidbothume* 1289 AssCh. 'Dweller in the wide valley', OE *sīd*. Sometimes pronounced *Siddybottarm*.

**Sideman:** v. SEEDMAN

**Sidey:** Laurence *de Sydhaghe*, Alan *de Sidhey* 1327 SRSf; John *Sidey* 1524 SRSf. 'Dweller by the wide enclosure.'

**Sidg(e)wick:** v. SEDGWICK

**Sidle:** v. SIDDALL

**Sidley:** John *de Sydelegh* 1327 SRSx. From Sidley Green in Bexhill (Sx).

**Sidlock:** *Sideloc* 1066 Winton (Ha); Roger *Sidlok* 1332 SRSx; Thomas *Sydelok'* 1332 SRDo. OE *\*Sidlāc*.

**Sidnall, Sidnell:** Roger *de Sidenhal'* 1219 AssY; Henry *de Sydenhale* 1339–40 FFWa; Robert *Sydnall* 1545 SRW. From Sidenhales Fm in Tanworth (Wa).

**Sidney, Sydney:** John *ate Sydenye* 1332 SRSr; William *Sydny* 1428 FA (Sx). 'Dweller by the wide well-watered land' (OE *sīd, īeg*), as at Sidney Fm in Alfold (Surrey). The name is usually derived from St Denis (Normandy) but proof is lacking. The only evidence noted is: Roger *de Sancto Dionisio* 1212 Fees (Nf), John *Seyndenys* 14th ArchC 29 and this is certainly not conclusive. Harrison's William *Sidney* (1325) cannot be from St Denis. His name belongs above.

**Sier:** v. SIRE

**Sievwright, Sivewright, Seivwright, Severwright:** Simon *le siuewricht'* 1219 AssY; John *le Syvewryct'* 1301 SRY. OE *sife* and *wyrhta* 'maker of sieves'.

**Siffleet, Siffeet, Siflet:** John *Sifled, Syflede* 14th ADiv (Wa); William *Syflete*, *Syfflett* 1490–1508, c1566 ArchC 49. OE *Sigefflǣd* (f) 'victory-beauty', which must have remained in use after the Conquest.

**Siggers, Siger:** *Sigar* 1066 DB (W); *Sigarus* ib. (C); *Sigar* c1095 Bury (Sf); *Sigherus* 1180–1207 Rams (Nf); *Sigerus* c1250 ib. (Beds); Edricus *Sigarus* 1066 DB (Sf); Eluiua *Sigar* 1221 ElyA (Nf); William *Siger* 1275 RH (Nf). This may be OE *Sigegār* 'victory-spear' or ON *Sigarr*, ODa, OSw *Sighar*, or OG *Sigger*. The form *Sigar* goes better with OE *Sigegār* than with ODa *Sighar*, which should have become *Sigher, Sier* in English (ELPN).

**Siggs:** *Sigga* 1162 DC (L); *Sigge* de Anemere 1275 RH (Nf); Aunketillus *Sigge* 1214 Cur (Do); John *Sygges* 1524 SRSf. *Sigga* (f), a short form of ON *Sigríðr*, ODa *Sigrith*: *Sigreda* 1066 DB (Y).

**Sigsworth:** Anthony *Sigswith* 1587, George *Sigsworth* 1645 SRY. From Sigsworth Grange in Fountains Earth (WRY).

**Sikes:** v. SYKES

**Silabon, Silburn:** v. SILLIBOURNE

**Silby:** William *de Sileby* 1205 Cur (Sf); Robert *de Sileby* 1311–12 FFWa; Roger *de Syleby* 1332–3 FFSr. From Sileby (Lei).

**Silcock, Silcocks, Silcox:** *Silcokkus* de Altricheham 1283 Cl (Ch); *Silkoc* of Middlewich 14th AD iv (Ch); William *Selecok* 1327 SRSo; John *Silcok* 1379 PTY. *Sil-coc*, a diminutive of *Sil*, a short form of *Silvein* (v. SAUVAIN) or *Silvester*.

**Silito:** v. SHILLITO

**Silk:** Ædwardus, John *Selk(e)* 1170 P (So); William, John *Silke* 1350 LLB F, 1353 Putnam (W). OE *seolc* 'silk', metonymic for a worker or dealer in silk. cf. Alice *la Selkwimman* 1334 ColchCt.

**Silkenside:** Adam *Silkenside* 1221 AssSa. 'Silken side', OE *seolcen, sīd*. cf. Warin *Silketop* 1276 AssLo 'silk hair'.

**Silkin:** John *Selekyn* 1327 SR (Ess). *Sil-kin*, a diminutive of *Sil*. cf. SILCOCK.

**Silkston, Silkstone:** Andrew *de Silkiston'* 1195 P (Y); Adam *de Silkeston* 1337 FFY; John *de Silkeston* 1374 Calv (Y). From Silkstone (WRY).

**Sill, Sills:** Robert *Sille* 1397 PrGR. A short form of *Silvein* or *Silvester*. v. SILCOCK.

**Sillet, Sillett:** (i) *Siled* 1219 AssL; Richard *Silat* 1327 SRC; Thomas *Sylet* 1390–1 NorwLt; John *Sillett* 1539 ArchC 40. *Sill-et, Sill-ot*, diminutives of *Sill*, a short form of *Silvester* or of *Silvein*. But other sources are probably also involved. (ii) William *Seliheved* 1315, Richard *Seliheved* 1324 Wak (Y). 'Fortunate head, person', OE *sælig, hēafod*.

**Silley:** v. SEALEY

**Sillibourne, Silabon, Silburn:** Adelina *Salibern* 1279 RH (Hu); John *Selybarn'* 1297 SRY; Robert *Seliberne* 1332 SRSt; Richard *Selbarn* 1407 FrY; John *Sillibarn* 1528 ib.; James *Silburne* 1689 ib. OE *sælig, bearn* 'happy child'. cf. SELLMAN.

**Sillick, Sillock:** *Silac'* 1086 ICC; *Silac* Passur 1221 ElyA (C); Robert *Silac* 1221 AssWo; Cecilia *Silok'* 1395 AssL. Probably from an unrecorded OE *\*Sigelāc*.

**Sillifant, Selifant:** John *Silliphant* 1616, Thomas *Sillafant* 1776 DWills. 'Happy, fortunate child', OE *sælig*, OFr *enfant*. cf. William *Selisaule* 1268 IpmY 'fortunate soul, person'.

**Sillitow:** v. SHILLITO

**Sillman:** v. SELLMAN

**Silly:** v. SEALEY

**Silton:** Adam *de Siltun* c1170–84 MCh; William *de Silton* 1299–1300, 1405 IpmY. From Silton (Do), or Nether, Over Silton (NRY).

**Silver:** (i) Lucas, John *Siluer* 1205–13 Seals (L), 1301 SRY. OE *silfor* 'silver', metonymic for a silversmith. cf. Robert *Silverhewer* 1212 Cur (Y), William *Sylvereour* 1417 FrY. (ii) Thomas *atte Selure* 1327 SRWo; Thomas *of the silvere* 1332 ib. 'Dweller by the silvery stream', OE *\*seolfre*, *\*sylfre*, as at Silver Beck (Cumb) or Silver (Devon). v. MELS.

**Silverlock:** Adam *Seluerloc* 1221 AssWo; John *Silverloc* 1268 AssSo. 'Silver lock', silvery-haired.

**Silverside:** Elena *Siluerside* 1379 PTY; John *de Silversyd* 1397 PrGR; Isaac *Silverside* 1665 HTO. From Silver Side in Farlam (Cu). But a nickname, 'silver side' may also be involved. cf. Richard *Silvereghe* 1414–15 IpmY 'silver eye'; Adam *Siluermouth* 1379 PTY 'silver mouth'; John *Silvertop* 1478 FrY 'silver hair'.

**Silvester, Selvester, Sylvester, Siveter, Siviter:** *Siluester* capellanus Hy 2 DC (Lei); *Seluester* filius Walteri 1204 Cur (Y); Thomas *Silvestr'* 1212 Fees (Ha); William *Silvester* 1250 Fees (La); William *Sevester* 1455 LLB K. Lat *Silvester* 'dweller in the forest', the name of three popes, which seems to have been first used in England by clerics. *Sevester* gives the clue to the origin of *Siveter* and *Siviter*. In modern French we have the popular forms *Sevestre* and *Sivestre* by the side of *Silvestre*. The loss of the second *s* would give *Seveter, Siveter*.

**Sim, Simm, Simms, Sims, Symms, Syms, Sime, Simes, Syme, Symes:** *Sym* Clerk 1446 Black; Ralph *Simme* 1317 AssK; John *Symme* 1345 ColchCt; Robert *Symmes* 1379 PTY. *Sim(m)*, a pet-form of *Simmond*, or *Sime* of *Simon*.

**Simbarb, St Barbe:** Ralph *de Sancta Barba* 1196 Cur (So); Richard *de Seynte Barbe* 1304 Glast (So); John *Sembarb* 1469 *PetreA*; Giles *Seinbarbe* 1555 Pat (So). From Ste Barbe-en-Poitou (Haut-Vienne).

**Simble:** *v.* SEMPLE

**Simco, Simcoe:** Probably for SIMCOCK. cf. ALCOE.

**Simcock, Simcocks, Simcox, Symcox, Sincock:** Robert *Symcot* (sic) 1275 RH (C); Simon *Simecok* 1327 SRSo; Thomas *Symcokes* 1395 AssSt. 'Little *Sim*' (Simon), *Sim* plus *cock*.

**Sime(s):** *v.* SIM

**Simeon, Simeons, Simion:** Ricardus *filius Simeonis* 1219 Cur (Bk); *Simeon* 1220 Cur (Berks); Robert *Simeon* 1254 FFHu; William *Symeon* 1278 FFSf. The Hebrew *Shimeon* 'hearkening' appears in the English versions of the Old Testament as *Simeon*, occasionally *Shimeon*, but in the New Testament as *Simon* (except in one instance). The forms were kept distinct, as separate names, in the Middle Ages, *Simon* being much more common.

**Simer:** Walter *le Simer* 1243 AssSo. A derivative of either OE *sīma* 'cord, rope', a roper, or of OE *sīman* 'to load', a loader. Very rare.

**Simester, Simister:** Peter *le Semester* 1275 RH (L); Alicia *Semester* 1376 MESO (Y); Julia *Semster* 1380 SRSt; Margaret *Sembster* 1381 PTY. OE *sēamestre*, feminine of *sēamere* 'sewer, tailor', sempstress; used also of a man. cf. SEAMER.

**Simey:** Robert *Sygemay* 1275 Wak (Y). OE *\*Sigemǣg* 'victory-kinsman', of which this is the only known example. An addition to the late compounds of *-mǣg*, noted by von Feilitzen: *Sǣmǣg* (1185), *Sidmǣg* (1185), *Wulfmǣg* (1063). *v.* NoB 33, 87. A very rare surname.

**Simison:** *v.* SIMONSON

**Simkin, Simkins, Simkiss, Simpkin, Simpkins, Simpkiss, Semken, Sempkins, Sinkin, Sinkins, Sinkings:** *Symekyn* Sadeler 1378 LoPleas; Anand *Simekin* 1199 SIA iv; William *Symkyn* 1524 SRSf. *Sim-kin* 'Little *Sim*' or Simon. Used as a diminutive: *Symond* or *Symkyn* Skynner 1466 AD v (Wa).

**Simmance, Simmans, Simmens, Simmins:** *v.* SIMMONDS

**Simmer, Simmers, Symmers, Symers:** George *Symmer* 1529, John *Symer* 1641. Scottish forms of SUMMER. Black notes *Sumer* and *Somyr* from c1200–1478, after which the vowel varies between *u* and *i*.

OFr *som(m)ier*, used both of a pack-horse and a sumpter.

**Simmie:** Alexander *Simmey* 1613 Black (Aberdeen). A diminutive of *Sim*.

**Simmonds, Simmons, Simmance, Simmans, Simmen, Simmens, Simmins, Semens, Semmence, Semmens, Simond, Simonds, Symmons, Symonds, Fitzsimmons:** *Simund* danus 1066 DB (Wa); *Simundus* 1134–40 Holme (Nf); *Symundus* de Ludham 1150–60 Holme (Nf), *Simon* de Ludham 1155–89 ib.; *Simund, Simon* clericus de Abi Hy 2 DC (L); *Simond* Moysaunt 1426 LondEng; Robert *Simund* 1222 Cur (W); John *Symont* 1260 AssC; William *Simon, Simond* 1291 LLB A; Margery *Simondes* 1308 EAS xviii; Hugo *le Fiz Simond, filius Simonis* 1325 ParlWrits (Herts). *Simon* 1066 DB (Wo), also *Simundus* (ib.), is elsewhere called '*Simund* quidam, genere Danus' (PNDB). His name was thus ON *Sigmundr*, ODa *Sigmund* 'victory-protector' which was already confused with *Simon*. Some, at least, of the Norfolk and Lincolnshire personal names were Scandinavian and contributed to the surname. Others may have been introduced from Normandy. Here, in addition, we have to reckon with the Hebrew *Simon* and OG *Sigmund*, cognate with ON *Sigmundr*. After the Conquest, in England, *Simund* was a common AFr form of *Simon*. Wycliffe uses *Symound* for Simon Peter; Simon de Montfort is called *Simond* by Robert of Gloucester. Adam *Cimond* (1292 SRLo) is identical with Adam *Simon* (1292 LLB A) and Adam *son of Simon* (1290 Pat). It is thus impossible to distinguish between the various origins. *v.* SIMON.

**Simmonite, Simonett, Simonite, Simnett:** William *Symonet* 1327 SRSt; Asselin *Simonett'* 1346 LLB F. *Simon-et*, a diminutive of *Simon*. *Simon* Corder is also called *Simonet* 1377 LoPleas. Another diminutive, *Simon-el*, has not survived: Philip *Simenel* 1200 P (Ha), Henry *Symnell* 1373 Bart (Lo). cf. also Lambert *Simnel*.

**Simner, Simnor:** For SUMNER.

**Simon, Simons, Simans, Symon, Symons, Fitzsimon, Fitzsimons:** *Simon* 1134–40 Holme (Nf); *Symon* de Cheurolcurt c1150 DC (Nt); William *le fiz Simon* Hy 2 DC (L); William, John *Simon* 1291 LLB A, 1296 SRSx. The Hebrew *Shimeon* may have been influenced by Greek Σίμων, from σιμός 'snub-nosed'. *Simon* and *Simond* were very popular and had a variety of diminutives. *v.* SIMEON, SIMMONDS.

**Simond(s):** *v.* SIMMONDS

**Simonson, Simison, Simyson, Symondson:** Philip *Simondson* 1430 FrY; Marques *Symondesson* 1473 ParlR. 'Son of *Simond* or *Simon*.'

**Simper:** *v.* SEMPER

**Simpkins, Simpkiss:** *v.* SIMKIN

**Simpkinson, Sinkinson:** William *Symkynson* 1381 PTY. 'Son of *Simkin*.'

**Simple:** Reginaldus *Simplex* c1160 Gilb (L); Robert *le Sinple* 1202 AssL; Hugh *le Simple* 1243 AssSo. OFr *simple* 'free from duplicity, dissimulation or guile; honest, open, straightforward'.

**Simpole:** *v.* SEMPLE

**Simpson, Sympson, Simson:** Richard *Symmeson* 1353 AssSt; Adam *Symson* 1395 Whitby (Y); John

Simpson 1397 Calv (Y); Henry *Symmesson* 1450 Rad (C); John *Symson* or *Sympson* 1487 LLB L. 'Son of *Simm*' (Simon). Three places named Simpson in Devon, earlier *Siwineston*, gave rise to surnames in the 13th century.

**Sinaver:** v. SANSAVER

**Sinclair, Sinclaire, Sinclar, St Clair:** Hubertus *de Sancto Claro* 1086 DB (So); Richard *de Sencler* ib. (Nf); Richard *de Seincler* 1194 Cur (Nf); Ralph *de Seintcler* 1197 P (So); Emma *de Sancler* 1198 FF (K); John *de Sentcler* 1327 SRSx. The tenants of Eudo Dapifer in Essex came from Saint-Clair-sur-Elle (La Manche). v. ANF. The Sinclairs of Caithness appeared in Scotland c1124–53, coming, according to Black, from Saint-Clair-l'Evêque (Calvados). The frequency of the surname in Caithness and the Orkneys is due to the adoption by tenants of the name of their overlord. The name appears as *Sincklair* in 1634.

**Sincock:** v. SIMCOCK

**Singer, Singers:** Lucas le *Syngere* 1296 SRSx; William le *Syngur* 1297 SRY. A derivative of OE *singan* 'to sing', a singer.

**Single, Sengel:** Alan *de la Sengle* 1296 SRSx; John *ate Sengle* 1327 ib.; Widow *Singull* 1674 HTSf. 'Dweller by the burnt clearing', OE *\*sengel*.

**Singler:** Arnald *Singlere* 1195–1215 StP; Richard *le Sengler* 1317 FFEss. ME *sengler*, *syngler*, OFr *sengler*, *seingler* 'singular', perhaps 'living alone, solitary'.

**Singleton:** Adam *de Singelton'* 1220 Cur (La); Geoffrey *de Sengelton* 1296 SRSx; Adam *de Syngleton* 1379 PTY. From Singleton (Lancs, Sussex).

**Sinjin:** v. ST JOHN

**Sinker, Sinka:** William le *Sinoker* 1260 AssC; Thomas le *Synekere* 1314 AssNf; William le *Syneker* 1327 SRSo. 'One who engraves figures or designs on dies'. cf. Thomas *Synkman* 1379 PTY.

**Sinkin(s):** v. SIMKIN

**Sinkinson:** v. SIMPKINSON

**Sinnamon:** v. CINNAMOND

**Sinnatt, Sinett, Sinnott, Senett, Sennett, Sennitt, Synnot, Synnott, Synott:** *Synodus* c1095 Bury (Sf); Robertus *filius Sinothi* 1200 P (St); Dionisia *filia Sinod* 1207 Cur (Herts); Stephen *Sinot*, *Sinut* 1275 RH (Sf); Warin *Sinat* 1276 RH (C). OE *Sigenōð* 'victory-bold', DB *Sinod*.

**Sinnington:** William *de Senyngton* 1365 FrY. from Sinnington (NRY).

**Sinnocks:** v. SENNECK

**Sinthwaite:** William *de Synyngtheayt* 1307 IpmY; Richard *Sinnethwaite* 1672 HTY. From Syningthwaite in Bilton (WRY), or a lost *Siningthwayte* in Mallerstang (We).

**Sinton:** Aldich *de Suthinton'* 1275 SRWo; Henry *de Sothinton'* ib.; Adam *a Suthetun*, Robert *atte Suthethun*, William *de Suthetun* 1296 SRSx; John *de Sothynton, de Sutingthun* 1296 SRSx; Thomas *Southetoun* 1327 SRSx; John *de Sotinton* 1327 SRSx; William *de Sotiton* 1332 SRSx; Agnes *de Suthyngton*, Richard *de Sotyngton* ib. Like WESTINGTON, this surname was common in Sussex where it has been noted in some 35 parishes between 1296 and 1332. In Worcestershire it derives from Sodington, Leigh Sinton or from Sindon's Mill in Suckley, (the place) 'south in the *tūn*' (*Suthiton, Sothyntone* 1275) as opposed to that 'north in the *tūn*', now Norton Fm. cf. also Siddington (Glos) and Southington in Selborne (Hants). Similar place-names were formed also from *east*, *north* and *west* and survive in Eastington (Devon, Dorset, Glos, Worcs) and Norrington (Herts, Wilts, Worcs). As surnames, *Astington, Norrington, Uppington* and *Westington* survive. *Southeton* and *Sotiton* would become indistinguishable from SUTTON. For a similar northern formation of Scandinavian origin, v. SOTHEBY and WESTOBY. In Scotland the surname derives from the barony of Synton (now included in Ashkirk, Selkirkshire); Andrew *de Synton* (1165–1214 Black).

**Sinyard:** v. SENIOR

**Sippings:** v. SEPPINGS

**Sipthorp:** v. SIBTHORP

**Sirdifield, Sirdyfield, Cedervall, Surrell:** Richard *de Surdeual* 1086 DB (Y); Robert *de Surdeuall'* 1197 P (We); John *Sowrdewall* 1488 GildY; Richard *Surwald* 1516 ib. From Sourdeval (Calvados, La Manche).

**Sire, Sier, Sirr, Surr, Syer, Syers:** Geoffrey *Sire* 1177 P (Nf); Matheus le *Sire* 1201 Cur (Nt); Ralph le *Seyr* 1296 SRSx; Walter *Surr* 1327 SRC. OFr *sire*, ME *sire, sier* 'master', a nickname; also used of an elderly man. *Surr* may be from OFr *sor, sieur*, oblique case of *sire*.

**Sired, Sirett, Sirette, Sirrett, Syrad, Syratt, Syred, Syrett:** *Sired, Siret* 1066 DB; *Sigreda* 1066 DB (Y); *Sired(us)* 1095 Bury, 1213 Cur (Sf); Willelmus *filius Sigerith* 1197 P (Y); *Sireda* de Kirkby 1332 SRCu; Wulfwine *Sired* 1077 OEByn (So); Robert *Sired* 1198 FFK; Roger *Syred* 1306 FFSf. OE *Sigerǣd* 'victory-counsel' (m), or, at times, ON *Sigríðr* (f).

**Sirkett:** v. CIRCUITT

**Sisley, Sicely:** Cecilia *cellararia* Hy 2 DC (Lei); Henricus *filius Cecilie* 1210 Cur (Db); Willelmus *Cecilie* 13th Kirkstall (Y); Henry *Cecili* 1279 RH (C); Richard *Cecely* 1296 SRSx; Roger *Cysely* 1332 SRWa. OFr *Cecile*, Lat *Caecilia*, feminine of *Caecilius*, from *caecus* 'blind'. The martyred St Cecilia was the patron saint of musicians and her name, introduced into England by the Normans, soon became very popular as *Cecely* or *Sisley*, with pet-forms *Siss, Sissel, Sisselot* and *Sissot*.

**Sismore, Sissmore, Sizmore, Sizmur:** Thomas *Sysmore* 1432 Petre (Ess); John *Sysemore* 1591 ChWo. OFr *sis mars* 'six marks', cf. DISMORE.

**Siss:** Thomas *filius Sisse* 1297 SRY; *Cisse* de Wassedene 1309 Wak (Y); William *Sys* 1327 SRSo. A pet-form of *Cecily*. v. SISLEY.

**Sisson, Sissons, Sissens:** Thomas *filius Sisse* 1297 SRY; Robert *Cisson*, John *Sisson* 1379 PTY; Edward *Sissons* 1641 FrY. 'Son of *Ciss*' (Cecily).

**Sisterson:** *Cissota* 1298, 1309 Wak (Y); Agnes *Sissotson* 1470 GildY. 'Son of *Cissot*', a diminutive of *Ciss* (Cecily).

**Sitch:** v. SYKES

**Siveter, Siviter:** v. SILVESTER

**Sivewright:** v. SIEVWRIGHT

**Siveyer, Sivier:** v. SEVIER

**Sixandtwenty:** Simon *Sixanttwenti* 1253 FrLei; William *Sixandtwenti* 1271 LeiBR. 'Six and twenty', probably the holder of that amount of land. cf. William *fifandtwenty* 1230 Cor (Wa) 'five and twenty'.

**Sixpenny:** John *de Sexpenne* 1340, William *Sexpenne* 1379 PN Do 19, i 371. From Sixpenny Handley Hundred, or Sixpenny Fm in Fontmell Magna (Do).

**Sixtyman:** Roger *Sixtiman* 1206 FFSf. 'Sixty man'. cf. Thomas *Six* 1531 FFEss 'six'; William *Sextynth* 1327 SRSo 'sixteenth'.

**Sizer:** William *Sisar* 1379 PTY; Robert *Syser* 1524 SRSf. ME *sysour* 'assizer', a member of the assize, sworn recognitor.

**Sizmore, Sizmur:** *v.* SISMORE

**Skaife:** *v.* SCAFE

**Skakle:** *v.* SHACKEL

**Skate:** Gotte *Scate* 1202 P (Y); Roger *Skate* 1301 SRY; William *Scate* 1395 AssL. A nickname from the fish, ME *scate, scaite*.

**Skeale, Skeeles, Skeels:** Richard *Skele* 1272 HPD; John *Skele* 1327 SRSf; Robert *Skele* 1376 AssEss. ME *skele, skeyll* 'a wooden bucket'. Metonymic for a maker or seller of these.

**Skeat, Skeats, Skett, Skeet:** *Scet, Schett* 1066 DB (Nf); Ricardus *filius Schet* 1166 P (Nf); Nicholas, Walter *Sket* 1201 P (Sa), 1275 RH (Nf); Robert *Skeet* 1327 SRSf. ON *skjótr* 'swift, fleet', used also as a personal-name.

**Skeffington, Skiffington, Skevington, Skivington:** Simon *de Scheftinton* 1193 P (Lei); Henry *de Skefyngton* 1327 SRLei; John *Skevyngton* 1520 FFSt. From Skevington (Leics).

**Skegg:** Thomas *Skegge* 1379 PTY. ON *skegg* 'beard'.

**Skelding, Skeldon:** Adam *Skelding'* 1332 SRCu; John *Skeldyng* 1463 FrY. From Skelding (WRY).

**Skelhorn, Skellern, Skellorn, Skillern, Skillen, Skillan, Skilland, Skellon:** John *Scalehorne* 1467, Peter *Skelhorne* 1574 PN Ch i 184. From Skeleron in Rimington (WRY), *Skelhorn* 1616.

**Skelton:** Hamo, John *de Skelton'* c1160–93 YCh, 1286 FFY; John *Skelton* 1410 FFEss. From Skelton (Cumb, ER, NR, WRYorks).

**Skepper:** Simon *Sceppere* 1221 ElyA (C); Walter *le Skeppere* 1281 MESO (L). A derivative of ON *skeppa* 'basket', a maker of baskets. *v.* SKIPP.

**Sker, Skerr:** Stephen *Sker* 1206 FFY, *Scer* 1212 Pleas (Y); Ralph *Sker* 1219 AssY. ON *skærr* 'clear, bright'.

**Skermer:** *v.* SKIRMER

**Skerratt, Skerrett, Skerritt, Skirrett:** (i) Eudo *de Skirwiht* 1285 IpmCu. From Skirwith (Cumb), formerly pronounced *Skerritt*. (ii) Alice *Skyrewhit* 1327 SR (Ess); Thomas, Roger *Skyrewyt* 1332 SRSr; John *Skyrwhyt* 1377 ColchCt. ME *skirwhit(e)*, apparently an alteration by popular etymology of OFr *eschervis*, a variant of OFr *carvi* 'carraway', a species of water parsnip formerly much cultivated in Europe for its esculent tubers which were used for sauce or physic, cf. *Skyrwyt, herbe or rote*, c1440 PromptParv. Metonymic for a grower of white parsnips.

**Skett:** *v.* SKEAT

**Skevens:** Walter *Skyuein, Schyuein* 1277 Ely (Sf); Richard *Skeuyn* 1301 SRY. ME *skeuayne, skeuyn*, ONFr *eskevein*, OFr *eschèvin* 'steward of a guild' (1389 NED).

**Skidmore, Skitmore, Scudamore:** Hugh *de Scudimore* 1167 P (He); Peter *de Skidemor* c1170 Glast (So); Geoffrey *Escudemor', Eskidemor'* 1242 Fees (W); Peter *de Skydemore* 1282 FFC; Richard *Skidmore* 1576 SRW. From an unidentified place, probably somewhere in the west or south-west.

**Skiffington:** *v.* SKEFFINGTON

**Skilful, Skillfull:** Geoffrey *Scilful* 1279 RH (C); William *Skylful* 1327 SRC; Richard *Skylful* 1428 FFEss. ME *skilful* 'expert, clever'.

**Skill:** William *Skille* 1279 RH (O); William *Skyl* 1302 AssSt. ON *skil*, ME *skil* 'reason, discernment'. cf. Geoffrey *Scilful* 1279 RH (C), William *Skylful* 1327 SRC and *v.* SKILLMAN.

**Skillan, Skillen, Skillern:** *v.* SKELHORN

**Skilleter, Skilliter:** John *Skellattour* 1327 SR (Ess); John *Skelete* 1332 SRCu. A derivative of ME *skelet* 'skillet' (a1403 NED), a maker of skillets.

**Skillfull:** *v.* SKILFUL

**Skillicorn, Skillicorne:** Adam *de Skillingcorne* c1350 EwenG (La); William *Skillecorne* 1458 Cl (Ha); John *Skilicorn* 1459 PrGR; Sir Philip *Skillicorne* 1521 Moore. A name of doubtful origin, said to be peculiar to the Isle of Man.

**Skillman:** John *Skyleman* 1275 RH (Nf); Matthew *Skylman* 1327 SRC. An anglicization of ON *skilamaðr* 'trustworthy man'.

**Skin:** Roger, Simon *Skin* 1221 AssWo, 1279 AssNb; Cecilia *Skynn* 1301 SRY. ON *skinn* 'skin'. Metonymic for SKINNER.

**Skinner, Skynner, Skyner:** Robert *le Skynnere* 1263 MESO (Sx); Ralph *le Skinnere* 1269 AD i (Herts). A derivative of ON *skinn* 'skin', a skinner. cf. SHINNER.

**Skipp:** Osbert *Sceppe* 1210 Cur (C); John *Skip* 1282 LLB A; William, Walter *Skyp* 1296 SRSx, 1310 ColchCt; John *Skep*, William *Skyp* 1327 SRC. ON *skeppa*, ME *scepp(e)* 'basket', metonymic for SKEPPER and, probably, for SKIPPER.

**Skippen, Skippon, Skippins, Skippence:** John *Skippon* 1314 AssNf; John *Skipen* 1415 IpmY; Thomas *Skipping* 1449 Paston. 'Dweller by the cattleshed', a Scandinavianized form of OE *scypen*.

**Skipper:** Geoffrey *le Skippere* 1285 *Ass* (Ess); Robert *le Skypper* 1286 AssSt. A derivative of ME *skip* 'to jump', a leaper, jumper, or one who moves lightly and rapidly. Later examples may be for SKEPPER (cf. SKIPP) or for ME *skypper* 'master of a ship' (1390 NED).

**Skipsey:** Robert *Skipse* 1415 IpmY; Robert *Skipsy* 1524 SRSf. From Skipsea (ERY).

**Skipton:** William *de Skiptun* 1185 Templars (Y); Robert *de Skipton'* 1258 ForHu; Henry *de Skipton* 1335 FrY. From Skipton (WRY), or Skipton on Swale (NRY).

**Skipwith:** William *de Skipwyth* 1251 AssY; William *de Skipwyth* 1374 AssL; Katerine *Skipwith* 1502 TestEbor. From Skipwith (ERY).

**Skirmer, Skermer, Skurmer:** Richard *le Skiremar'*, Thomas *le Scuremer*, Hugh *Skermere* 1279 RH (O); William *le Skurmere* 1296 SRSx; Robert *le Skirmer* 1332 SRLa. A derivative of ME *skurmen, skirmen*, from OFr *eskirmir, eskermir* 'to fence', 'a fencing-master'. cf. SCRIMGEOUR.

**Skirrett, Skivington:** *v.* SKERRATT, SKEFFINGTON

**Skitmore:** *v.* SKIDMORE

**Skitt, Skitte:** Geoffrey *Skitte* 1229 Cur (Nf); Reginald *Skyte* 1296 SRSx; John *Skitt* 1327 SRSf. ON *skyti* 'archer'.

**Skreven:** Henry *Screuyn* 1297 MinAcctCo (Y); William *de Skrevyn* 1309 FrY. From Scriven (WRYorks).

**Skrimshaw, Skrimshire:** *v.* SCRIMGEOUR

**Skudder:** *v.* SCUDDER

**Skurmer:** *v.* SKIRMER

**Skynner:** *v.* SKINNER

**Slabbard, Slabbert:** John *Slabbard* 1287–8 NorwLt; John *Slabart'* 1319 SRLo. MDutch *slabbaert* 'glutton'.

**Slack, Slacke:** (i) Gerebod *le Slac* 1195 P (L); Thomas *Slak* 1359 LLB G. OE *slēac, slæc* 'lazy, careless, slow'. (ii) Thomas, Nicholas *del Slac, slakk* 1275 Wak (Y), 1331 WhC (La). 'Dweller in the shallow valley', ON *slakki*.

**Slade:** Sabern *de la Slade* 1255 FFEss; Reginald *atte Slade* 1306 AD v (Mx); Walter *in the Slade* 1327 SRSf. 'Dweller in the valley', OE *slæd*.

**Slape:** Adam *de la Slape* 1276 AssSo; Elena *atte Sclape* 1327 SRSo. 'Dweller near a miry place or marsh' (OE *\*slæp*).

**Slapton:** Agnes *Slapton'* 1317 LuffCh. From Slapton (Bk, D, Nth).

**Slate:** Roger *Slat* 1221 AssGl. Metonymic for *Slater*.

**Slater, Slator, Sclater, Slatter:** Thomas *le Sclatere* 1255 MESO (Wo); Saundr' *le Sclattur* 1278 Oseney (O); Roger *Sclatiere* 1279 RH (O); Walter *Sclatter* 1279 RH (Bk); Thomas *Slater* 1297 SRY. A derivative of ME *sclate, slate*, OFr *esclate* 'slate' or ME *sclat, slatt*, OFr *esclat* 'slat', slater.

**Slatford:** Adam *de Slagtreford'* 1210 P (W); William *Slattford* 1665 HTO. From Slaughterford (W).

**Slaughter, Slagter:** (i) Robert *de Scloctres* 1191 P (Gl); Robert *de Sloutre* 1251 Eynsham (O); Mariota *de la Sloghtere* 1296 SRSx; John *Sloutere* 1327 SRWo. OE *\*slōhtre* 'slough, muddy place', the source of Upper and Lower Slaughter (Glos) and minor places in Sussex, as Slaughterford Fm, Slaughterbridge Fm and Slaughterwicks Barn. *v.* MELS 190. (ii) Thomas *le Slaghterere* 1296 FFEss; Henry *le Sclaufterer*, Stephen *le Slawterer* 1327 SRSf; Roger *Slaghtere* 1360 FFSf; Walter *Slautere* 1381 SRSf. A derivative of ME *slahter* 'slaughter' (a1300 NED), 'a killer of animals, a butcher' (1648 NED).

**Slaven, Slavin:** Roger *Sclauin* 1177 P (Sf); Goditha *Sclauine* 1221 AssSa. ME *sclaueyn, slaveyn, slavyne*, OFr *esclavine* 'a pilgrim's mantle'.

**Slay, Slee, Sleigh, Sleith, Sleath, Sligh, Sly:** Walter *Sleh* 1219 FFEss; Thomas *Sleh, Slei, Slegh* 1219 AssL; Robert *Sley* 1221 AssWa; John *le Slege* 1273 RH (O);

Ralph *Sly* 1273 RH (Hu); John *Slygh* 1327 SRSx; John *le Slegh* 1333 FrY; John *Slee* 1446 FrY. ME *sleȝ* from ON *slægr* 'clever, cunning' became *slee* in the north and *sligh, sly* in the south and midlands. 'Skilful, clever, expert' (c1200 NED). For *Sleath, Sleith*, cf. the pronunciation of Keighley. In 1247 AssBeds we have both William *le Slaer* and John *Slay* and in 1311 ColchCt William *Slayare*. The latter is a derivative of ME *slaye* 'an instrument used in weaving', 'a maker of slays', synonymous with SLAYMAKER, and *Slay* must here be used in the same sense. The common forms today are *Slee* and *Sly*.

**Slayforth:** Thomas *Sleyforth* 1465 Paston. Probably from Sleaford (L).

**Slaymaker:** John *Slaymaker* 1379 PTY; Henry *Slaymaker* or *Slymaker* 1594 Oxon. 'A maker of slays', ME *sleȝe, sleye, slay(e)* 'an instrument used in weaving to beat up the weft', 'a weaver's reed or shuttle'. Always a rare name. Formerly more common, but now apparently extinct, was *Slaywright*: Robert *le Sleywrihte* 1334 ColchCt. *v.* MESO 167 and cf. SLEEMAN.

**Slayter, Slaytor, Sleator:** Roger *Sleghtere* 1304 FFEss; William *le Sleghtere* (*Sleyter*) 1327 *SR* (Ess); William *le Sleter* 1327 SRSx. A derivative of OE *slyht* 'slaughter', slaughterer, butcher.

**Sleap:** *v.* SLEEP

**Sleath:** *v.* SLAY

**Sleator:** *v.* SLAYTER

**Sledd:** John *Slede*, Stephen *de Slede* 1275 SRWo; John *Sledde* 1296 SRSx. 'Dweller in the valley', OE *slæd*.

**Sledge:** William *Slech* 1327, John *Slech* 1332 SRSx; Richard *Sledge* 1641 PrSo. OE *slecg* 'sledge-hammer'. Metonymic for a maker or user of this.

**Sledmere:** Ralph *de Sleddemer* 1219 AssY; Agnes *Sledmer* 1432 TestEbor. From Sledmere (ERY).

**Slee, Slay:** Stephen *del Sle* 1274 RH (Ess). 'Dweller by the grass-grown slope', OE *\*slēa*, as at Slay Down (Wilts). *v.* also SLAY.

**Sleeman, Slemmond, Slemming, Slemmings, Slimming, Slimmon, Sliman, Slyman:** Thomas *Sleman* 1277 Wak (Y); Auicia *Scleyman* 1327 SRC. May be 'cunning, sagacious man', 'dweller by the grassy slope', or 'maker of slays'. *v.* SLEE, SLAY, SLAYMAKER.

**Sleep, Sleap:** Coc *de Slepe* 1255 RH (Sa); Cristina *atte Slepe* 1332 Glast (So); Robert *Slepe* 1478 FFEss. From Sleap (Sa), or 'dweller at the slippery place', OE *\*slæp*.

**Sleeper:** Hugh, John *le Slepere* 1212 Cur (K), 1232 FFEss. ME *sleper* 'sleeper', used of an indolent or inactive person (a1225 NED).

**Sleigh, Sleith:** *v.* SLAY

**Sleight, Slight:** (i) William *Sleght* 1297 IpmY; Robert *Sleight* 1394 AssL; Walter *Slyght* 1525 SRSx. ON *sloegð* 'cunning, artfulness'. (ii) Edith *atte Sleyte* 1327 SRY. 'Dweller at the level field', ON *slétta*.

**Sleightholm, Sleightholme:** Amfrid *de Sletholm* 1208 P (Y); Walter *of Sletholm* 1234, Roger *de Sleyghtholme* 1360 FFY. From Sleightholme (NRY).

**Slemmonds, Slemming(s), Sliman, Slimming, Slimmon:** *v.* SLEEMAN

**Slight:** *v.* SLEIGHT

**Slim, Slimm, Slymm:** John *atte Sclyme* 1333 MELS (So). 'Dweller at the muddy place', OE *slīm*.

**Slin, Slinn, Slynn:** Gamel *de Slin* 1185 P (La); Henry *de Sline* 1246 AssLa; Robert *de Slind* 1332 SRSx. From Slyne (La), or 'dweller on the slope', OE *slind*, *slinu*.

**Sling:** Reyner, Robert *Sleng*(e) 1221 *ElyA* (Nf), 1298 Wak (Y); William *Slinge* 1279 RH (Hu); Robert *Slyngg* 1327 *SR* (Ess). ON *Slengr*; cf. Norw *sleng* 'idler'. This would become *Sling*, early examples of which may, however, be metonymic for SLINGER.

**Slinger:** William *Slinger, Slinge* c1248 Bec (Wa); John *Slingere* 1297 MinAcctCo; Adam *le Sclynggere* 1327 *SR* (Ess). A derivative of ME *slingen* 'to sling', slinger, a soldier armed with a sling (1382 NED). As *le slengg', slyngges* is found in 1323 and 1348 for the stout ropes hitched round large blocks of stone and over a hook as a means of lifting them during building operations (Building 322), *slinger* probably also denoted workers of these slings.

**Slingsby:** Roger *de Slingesby* 1219 Cur (Y); Higdon *de Slyngesby* 1379 PTY; Henry *Slingsby* 1672 HTY. From Slingsby (NRYorks).

**Slipper:** William *Slipere* c1248 Bec (Mx); Laurence *Slyper* 1332 SRSx. A derivative of ME *slipe* 'to polish, sharpen', probably a sword-sharpener. cf. William *Suerdsliper* 1313 Wak (Y).

**Slocombe, Slocum:** Andrew *Slocombe* 1641 PrSo; Robert *Slocombe*, John *Slocum* 1642 PrD. From Slocombeslade in Brendon (D), *Slocumb* 1330.

**Sloley:** *v.* SLOWLY

**Sloman, Slowman, Sluman:** William *Sloman* 1327 SRSx; Henry *Sloghman* 1392 AD i (Ess). 'Dweller by the slough.' *v.* SLOUGH.

**Sloper:** Agatha, Geoffrey *le Slopere* 1279 RH (Hu), 1286 MESO (Nf). A derivative of ME *slop*(e) 'slop, outer garment, tunic, etc.', a maker or seller of these. cf. Emma *le Sclopmongere* 1317 Oseney (O).

**Slopp:** Walter *Slop* 1332 SRSx; Simon *Sloppe* 1372 AD ii (W). ME *slop, slope* 'outer garment, tunic'. Metonymic for a maker or seller of these.

**Slot, Slott:** Walter *de la Slot* 1275 RH (Nf). An early example of ME *sclott* (c1475 NED), glossed *limus*. 'Dweller by the muddy place.'

**Slough, Slow, Slowe:** Ulric *a teslo* c1095 Bury (Sf); Hemming *de Slo* 1196 P (Bk); Unfridus *de la Slowe* 1225 Pat (Bk); Hugh *Aterslo* 1228 FFEss; Richard *atte Slouh* 1275 SRWo; John *dil Slow* 1327 SRSf; Walter *atte Slogh* 1337 LLB E. 'Dweller by the slough or miry place' (OE *slōh*). Occasionally, *Slow* may be a nickname from OE *slǣw* 'slow, sluggish, dull': Richard *Slou* 1296 SRSx.

**Slowfoot:** Thomas *Slafot* 1379 PTY. 'Slow-footed', OE *slāw, fōt*.

**Slowly, Sloley:** Juliana *de Sloleg'* 1221 AssWa; Peter *de Sloleye* 1275 RH (Nf). From Slowley (Warwicks) or Sloley (Norfolk).

**Slowman, Sluman:** *v.* SLOMAN

**Sly:** *v.* SLAY

**Slyman:** *v.* SLEEMAN

**Slymm:** *v.* SLIM

**Slynn:** *v.* SLIN

**Smail, Smailes, Smails:** Richard *de Smegle* 1296 SRSx. These surnames, with, probably, *Smiles*, still found in Sussex and West Kent, derive from Broxmead, Brooksmarle (Sussex) or a lost vill of *Smeghel* in the east part of the county. From OE *smēagel* 'burrow', the compound meaning 'badger-hole'. *v.* PN Sx 262. *Smail* is also a later northern and Scottish form of *Small*: Henry *Smailes* 1640 FrY (Christopher *Smales* 1627 ib.), Henry *Smaill* 1657 Black.

**Smaldon:** *v.* SMALLDON

**Smale, Smales, Small, Smalles, Smalls, Smeal, Smeall:** William *Smale* 1221 *ElyA* (C), 1275 SRWo; Alexander *le Smele* 1221 AssSa; William *le Smale* 1294 RamsCt (Hu); Thomas *Smail* 1360 Black. OE *smæl* 'small, slender, thin'. *Smeal* is from ME *smel* (c1275 NED). *v.* SMAIL.

**Smalham:** *v.* SMALLHAM

**Smallacombe, Smallcombe:** Henry *de Smalecomb* 1330 PN D 299; William *Smalcombe* 1366 IpmGl; John *Smalcombe* 1642 PrD. From Smallecomb, Smallecombe, Smallcombe, Smallicombe (D), or 'dweller in the narrow valley', OE *smæl, cumb*.

**Smalland:** Robert *Smaland* 1208 P (Ess/Herts); Richard *Smallond* 1253–4 FFEss. From Smallands Hall Fm in Hatfield Peverel (Ess).

**Smallbent:** William *Smalbyhind'* 1379 PTY; Thomas *Smalbend* 1440 ShefA; Robert *Smallebynde, Smalbehind* alias *Smalbent* 1552 ShefA. Evidently a nickname, 'small behind', OE *smæl, behindan*.

**Smallbone, Smallbones:** Wilk *Smalbon* 1198 FrLeic; John *Smalbayne* 1301 SRY. OE *smæl* and *bān* (in Yorks, ON *bein*) 'small bone', a nickname, probably 'short legs'.

**Smallbrook:** Thomas *of Smalebrok'* 1249 AssW; William *de Smalbroke* 1327 SRWo; Robert *Smalbrook* 1419–20 FFWa. From Smallbrook in Staverton, or Smallbrook Fm in Warminster (W).

**Smallcombe:** *v.* SMALLACOMBE

**Smalldon, Smaldon:** Josias *de Smaldene* 1275 RH (K); Margeria *de Smaldane* 1314 FFK. From Smalldane in Ospringe (K).

**Smalley, Smally:** William *de Smalleghes* 1325 Wak; Richard *de Smalley* 1388–9 FFWa; Alexander *Smalley, Smalley* 1532, 1537 CorNt. From Smalley (Derby).

**Smallfield, Smawfield:** Thomas *de Smalfeld* 1290 ShefA; John *de Smalefeld* 1296 SRSx. 'Dweller by a small field' (OE *smæl, feld*).

**Smallham, Smalham:** (i) Alice *de Smalham* 1327 PN Sx 189. From Smallham in West Grinstead (Sx). (ii) William *de Smalham* c1153–77, Robert *de Smalhame* 1248 Black. From Smailholm (Roxburgh).

**Smallknight:** Arnald *Smalknit* 1229 Cur (Sf); Richard *Smalcnicht* 14th AD vi (Sf). Perhaps 'small servant', OE *smæl, cniht*. cf. Elizabeth *Smalpage* 1579 RothwellPR (Y) 'small page'; William *Smalwriter* 1275 RH (K) 'small scribe'.

**Smallman, Smallsman, Smalman:** Hubert *Smaleman* 1209 FFSf; Richard *Smaleman* 1275 RH (Sf). In Durham the *smaleman* (OE *smælmann*) is coupled with thanes and drengs in an order which suggests he is inferior to both (*de Tainis et Dreinis et*

*Smalemannis* 1130 P). Two Essex examples *Smallemanneslond* 1367, *two Smallmannes lond* (1405 PN Ess 584) seem to confirm that we have here the name of a particular type of tenant.

**Smallpage:** Elizabeth *Smalpage* 1579 RothwellPR (Y). cf. LITTLEPAGE.

**Smallpeace, Smallpeice, Smallpiece:** Geoffrey *Smalpece* 1414 PN Ess 545. Probably 'Dweller by or holder of a small enclosure'. ME *pece*, dial. *piece* 'piece of land, enclosure, field'.

**Smallproud:** Simon *Smalprut* 1221 Cur (Y); Adam *Smalprud* 1243 AssSo; John *Smalprot'* 1332 SRDo. A nickname for someone with little to be proud of, OE *smæl, prūd.* cf. Robert *Smalred* 1176 P (Y) 'of small advice'.

**Smallridge, Smaridge:** Henry *de Smalrug'* 1221 AssWo; William *de Smalerugge* 1330 PN D 96. From Smallridge in Langtree and in Axminster (Devon), *Smarige* 1086, *Smalrigge* 1200 PN D 634, or from residence near a narrow ridge (OE *smæl, hrycg*).

**Smallshaw:** (i) Dobbe *de Smaleschawe*, Roger *Smalchaghe* 1298, 1316 Wak (Y). 'Dweller by a small wood' (OE *sceaga*). (ii) William *Smalsho* 1378 PN Ess 481, 1420 LLB I. Perhaps a nickname 'Small shoe' (OE *scēo*). Or from a lost place *Smelesho* 1297 Coram (Beds).

**Smallstreet:** Adam *Smalstrete* 1412 AssL. 'Dweller in the narrow street', OE *smæl, strǣt.*

**Smallthwaite:** *v.* SMITHWHITE

**Smallwell:** John *Smalwell* 1427 FFEss; Widow *Smallwell* 1662–4 HTDo. 'Dweller by the narrow stream', OE *smæl, wiella.*

**Smallwood:** William *Smalwud'* 1220 Cur (Ess); John *de Smalwode* 1332 LLB E; Richard *Smalwode* 1401 AssLa. From Smallwood (Ches), or 'dweller by the small wood'.

**Smaridge:** *v.* SMALLRIDGE

**Smart:** Lifwinus *Smart* c1180 ArchC viii; William *Smert* 1275 SRWo. OE *smeart* 'quick, active, prompt'.

**Smawfield:** *v.* SMALLFIELD

**Smead, Smeath:** *v.* SMEED

**Smeal(l):** *v.* SMALE

**Smearman:** Philip *le Smereman* 1308 AssNf; William *Smereman* 1327 SRSx. 'A maker or seller of grease or butter', OE *smeoru, mann.* cf. Henry *le Smeremongere* 1327 SRSf.

**Smeathers:** 'Dweller on the level ground.' *v.* SMEED.

**Smeathman:** Robert *Smytheman* 1379 PTY; Christopher *Smethman, Smoothman* 1600, 1628 FrY. 'Smooth, suave man.' *v.* SMEED.

**Smeaton, Smeeton, Smeton, Smieton, Smitton, Smiton:** (i) Henry *de Smithetone, de Smeithtone* 1296, 1359 Black; Alexander *Smethytone* 1506 ib.; Erasmus *Smittoune* 1613. From the lands of Smytheton, now Smeaton (Midlothian). (ii) Johel *de Smetheton'*, Jukel *de Smitheton'* 1201, 1204 Pl (Y); Robert *de Smytheton* 1340 FFY; John *Smeton* 1379 PTY. From Smeaton (NR, WRYorks), Smeeton Westerby (Leics), or Smeetham Hall (Essex), all OE *smiþatūn* 'the smiths' enclosure'.

**Smedmore:** Peter *Smethemor* 1332 IpmW; Ann *Smedmore* 1662–4 HTDo. From Smedmore House in Kimmeridge (Do).

**Smeed, Smeeth, Smead, Smeed, Smee:** (i) Richard *Smethe* 1202 P (Co); William *le Smeth*, Philip *le Smeþe* 1279 RH (O, Hu). OE *smēþe* 'smooth, polished, suave'. (ii) Laurence *de Smethe* 1275 RH (K.); William *del Smethe* 1327 SRSf. From Smeeth (Kent), or 'dweller on the smooth, level place', OE *smēþe.*

**Smelt:** Alwine *smelt* 1040–2 OEByn; Richard *Smelte* 1230 Cur; John *le Smelt* 1334 SRK; Robert *Smelt* 1415 IpmY. OE *smelt* 'sardine, smelt', clearly a nickname, probably also 'a seller of smelts', and possibly a patronymic, though the personal name is not recorded after DB, cf. *Smelt* 1035 Redin (Do), *Esmellt, Esmeld* 1066 PNDB (K, Sx). If the spellings are reliable, John *Smolt* 1318 FFEss, 1405 IpmY, would lend support to a possible nickname from OE *smelt, smolt, smylte* 'mild, peaceful' (PNDB 367).

**Smethley:** Hugh *de Smethelay* 1307 IpmY. Probably from Smithley in Darfield (WRY).

**Smeton, Smieton:** *v.* SMEATON

**Smewing:** *Smeuuin* 1066 DB (So); *Smeuuinus* 1086 ib. (O); *Smewinus* c1130 Ewen (Do); Martin, John *Smewyne* 1256 Ass (Ha), 1327 *SR* (Ess); Matilda *Smywyne* 1332 SRSx. OE *\*Smēawine* 'sagacious friend'.

**Smidman:** Henry *Smythman* 1379 PTY. 'The smith's assistant.'

**Smiles:** William *Smyles* 1301 SRY. Perhaps 'dweller by the burrows', OE *smÿgels.*

**Smisson:** *v.* SMITHSON

**Smith, Smithe, Smyth, Smythe, Smye:** (i) Ecceard *Smið* c975 OEByn (Du); Ælfword *þe Smith* c1100 ib. (So); William *le Smyth* 1275 AssSo; Julian' *le Smithes* 1279 RH (O); John *Smye* 1524 SRSf. OE *smið* 'smith, blacksmith, farrier'. Early examples are common in the Latin form FABER. (ii) William *atte Smithe*, Thomas *de la Smythe* 1313 AD i (Sx); Robert *atte Smyth* 1332 SRSx. 'Worker at the smithy', OE *smiþþe.*

**Smitham:** Robert *de Smetham* 1275 RH (D); Thomas *Smitham* 1642 PrD. From Smytham in Little Torrington (D).

**Smither, Smithers:** John *Smythiere* 1379 AssWa. 'Smith, hammerman.' cf. 'The Jorneymen . . . of all oþer Craftes . . . except hakmen and smythers wurche in hur own houses and nott in hur masters housz' (1435 NED).

**Smitherman, Smythyman:** Robert *Smythyman* 1309 Cl (Y). 'Smithy man', worker at the smithy.

**Smithies, Smithyes, Smythyes:** John *del Smythy* 1332 SRLa, 1385 ShefA; John *Smythes, Smythies* 1568, 1586 FrY; Thomas *Smiddies* 1629 ib. 'Worker at the smithy', ME *smythy.*

**Smithson, Smythson, Smisson:** Reginald *le Smythessone* 1296 SRSx; Henry *le Smithson* 1324 Wak (Y); Peter *Smitson* 1327 SRY. 'Son of the smith.'

**Smithwhite, Smorthwaite, Smorfit, Smurthwaite, Smurfit, Smallthwaite:** Henry *de Smetwayt* 1285 AssLa; Thomas *de Smythuait* 1327 SRY; Matilda *Smarthwate* 1489 GildY; Robert *Smerwhaitt* 1518 ib.; Thomas *Smirthwaite* 1577 FrY; Thomas *Smorthwait* 1627 ib.; Matthew *Smurwhaite* 1657 ib.; Edward *Smurfett* son of Matthew *Smurfett* 1713 ib. In Cumberland there is a Smaithwaite in Keswick and another in Lamplugh, *Smathwaitis* c1280,

*Smerthwayte* 1530, *Smethwayte* 1552 (PN Cu 314, 407), whilst in Guiseley (Y) we have *Smerthwayt* 1323 Calv, 'small clearing' (ON *smár*). In Cumberland, too, we find two Smallthwaites, neither recorded before 1611, which may be identical in origin, with late substitution of English *small* for Scandinavian *smár*. The first two examples above may contain OE *smēðe* 'smooth'.

**Smithwick:** Adam *de Smithewyk* 1327 SRSx; Roger *de Smythewyk* 1332 SRSx. From Smethwick (Ches), or from a lost Smithwick in Southover (Sussex), last recorded c1608 (PN Sx 323).

**Smithyman:** *v.* SMITHERMAN

**Smiton, Smitton:** *v.* SMEATON

**Smock:** Randulf *Smoc* 1170–90 Seals (Mon); Robert *Smok* 1314 AssNf. Metonymic for SMOCKER.

**Smocker, Smooker:** Thomas *le Smokere* 1279 AssSo; Robert *le Smoker*, William *Smoker* 1327 SRSo. A derivative of OE *smoc* 'smock, a woman's under-garment'. For a maker or seller of these.

**Smollett:** John *Smalheued* 1332 SRWa; William *Smalhed*, *Smallyd* 1524 SRSf, 1547 EA (NS) ii (Sf). 'Small head.'

**Smooth, Smoothe:** John *le Smoth* 1296 SRSx; Peter *Smothe* 1325 FFK; Nicholas *Smothe* 1379 LoCh. OE *smōð* 'calm, polished, smooth'. cf. Robert *Smetheberd* 1230 P (Y) 'smooth beard'. *v.* SMEED.

**Smorfit, Smorthwaite, Smurfit, Smurthwaite:** *v.* SMITHWHITE

**Smye, Smythe:** *v.* SMITH

**Snail:** Hamelin *Sneyl* 1221 ElyA (Sf); Geoffrey *Snayl* 1277 Ely (C). OE *snegel, snægel* 'snail', of persons, 'slow, indolent'.

**Snailham, Snailum:** Henry *de Sneylham* 1296 SRSx. From Snailham (Sx).

**Snaith:** Ithelard *de Snayth* 1250–1 IpmY; William *de Snayth* 1338 FFY; Henry *Snayth* 1381 LoCh. From Snaith (WRY).

**Snake:** William *Snak* 1327 SRSx. OE *snaca* 'snake'. A nickname.

**Snape, Snepp:** Agnes *del Snappe* 1242 Rams (Hu); John *atte Sneppe* 1327 SRSx; Robert *de Snape* 1355 FFY; Roger *Snape* 1525 SRSx. From Snape (Suffolk, NRYorks), or 'dweller by the pasture', OE *snæp*, ON *snap*.

**Snarey, Snary:** *Snarri* de Wurðintona 1169 P (Wa); Richard *Snarri* 1200 P (Ha); Andrew *Snary* 1224 FFEss. ON *Snari*, from ON *snarr* 'swift'.

**Snashall, Snazel, Snazell, Snazelle, Snazle:** William *Snawshill* 1416, *Snawsell* 1436 FrY; William *Snawsell* 1455 GildY. Perhaps from Snaizeholme in Hawes (NRY), or Snowshill (Gl), *Snawesille* DB.

**Snawdon:** *v.* SNOWDEN

**Snead, Sneed, Snoad, Snode, Snee:** Ailnoth *de Snode* 1214 Cur (K); Robert *del Sned*, Agnes *Sned* 1275 SRWo; Robert *atte Snede* 1327 SRWo; John *Snode* 1327 SRSf. 'Dweller by a clearing or piece of woodland', OE *snǣd*, hence *Snead*, OE *snād*, hence *Snoad*, as at Snead Fm in Rock (Worcs), Snoad's Hole in Linton, Snoad Farm in Otterden and Snoadhill in Betterden (Kent).

**Sneezam, Sneezum:** Richard *de Snetesham* 1161 P (Nf). A local pronunciation of Snettisham (Norfolk).

**Snel, Snell, Snelle:** Johannes *filius Snel* 1196 P (Y); Alexander *filius Snell'* 1219 AssY; William *Snel* 1185 Templars (K); Edwinus *Snell'* 1195 FF (Nf); Thurstan *le Snel* 1223 Cur (Sr). OE *Snell*, from OE *snel(l)* 'smart, active, bold' or the adjective as a nickname. In Yorks and Norfolk we may have ON *Snjallr*.

**Snelgar:** *Snelgarus* 1191, 1196 P (Sx, Do); *Snelgarus* de Broke 1205 Cur (K); Thomas *Snelgar* 1243 AssSo; Richard *Snelgorr* 1450 ArchC vii. An unrecorded OE *\*Snelgār* 'bold-spear'. cf. SNEL.

**Snelland, Snellen, Snellin:** Thomas *de Snellund'* 1202 AssL. From Snelland (L).

**Sneller:** William, Richard *Snellard* 1205 P, 1275 RH (D). A hybrid compound of *Snel* and Fr *-ard*.

**Snelling:** *Snellinc* 1066 DB (C); *Snelling* 1204 P (Ess); Andrew, Brithmarus *Snelling* 1222 Pat (Herts), c1250 Rams (Nf). OE *Snelling*, a derivative of *Snell*.

**Snellman, Snelman:** John *Snelman* 14th AD iv (Ess). 'Bold, active man', or 'servant of *Snel*'. *v.* SNEL.

**Snepp:** *v.* SNAPE

**Snetterton:** Thomas *de Snetterton* 1311 AssNf. From Snetterton (Nf).

**Snider, Sniders, Snyder, Snyders, Sniderman:** John *Snyther* 1332 SRSx; John *Sniders* 1674 HTSf. A derivative of OE *snīðan* 'to cut', cutter, tailor.

**Snigg:** Aluric *Snig* 1169 P (Ha); Richard *Snyg* 1276 AssW; Richard *Snygg'* 1327 SRWo. Probably from a ME personal name, *Snigge*.

**Snipe:** (i) Matthew *Snype* 1293 Pinchbeck (Sf); Thomas *Snipe* 1672 HTO. cf. Nicholas *Snypewife* 1309 SRBeds. OE *Snīp*, or ON *Snipr*. (ii) William *Snype* 1534, John *Snype* 1575, Phillane *Snype* 1657 Black. From Snipe (Nb).

**Snoad, Snode:** *v.* SNEAD

**Snodden, Snoden:** John *Snodding* 1279 RH (C). OE *Snodding*.

**Snodgrass:** Patrick *Snodgrass* 1578, John *Snodgers* 1621, Andrew *Snodgrasse* 1679 Black. From the lands of Snodgrasse in Irvine (Ayr).

**Snook, Snooks:** Eduuardus *Snoch* 1066 DB (K); Stenesnoc ib. (Ha); John, Thomas *Snok* 1222 DBStP, 1356 ColchSt; Robert *Snouk* 1327 SRSo. Lower, followed by Bardsley and Weekley, derives this surname from the old pronunciation of Sevenoaks and states that Sussex deeds relating to a family of Snooks give all the modes of spelling from *Sevenoakes* down to *S'nokes*. The pronunciation was *Sennocke*, with stress on the first syllable, which is still used occasionally. We should not expect this to become *Snook* and it certainly cannot account for the early forms above. These must go back to an OE *\*snōc* 'a projecting point of land', used as a nickname for a long-nosed individual, or from an OE *\*snōc* 'snake' which appears to have been used as a personal-name in Snorscomb (Northants), *Snoces cumb* 944 DEPN.

**Snoring:** William *Snoryng'* 1351 AssL; Thomas *Snoryng* 1379 LoCh. From Great, Little Snoring (Nf).

**Snow:** Richard *Snow* 1221 AssW; Robert *Snou* 1239 FFSf; Gilbert *Snawe* 1327 *SR* (Ess); William *le Snow* 1327 SRSx. OE *snāw* 'snow', a nickname for one with snow-white hair. Haylwardus *Snew* (c950 OEByn) is said to have been so called *propter albedinem*.

**Snowball:** Robert *Snawbal* 1301 SRY; Roger *Snowbald* 1327 SRSt; Robert *Snaubal* 1332 SRLa. OE *snāw*, ME *snow* and ME *ball*, 'a white streak, a bald place', a nickname for one with a snow-white patch of hair, or, possibly, with a whitish bald spot amid jet-black hair. *Snowbald* may be evidence of the latter meaning, from ME *ballede*, or the *d* may be excrescent. cf. BALD, BALL.

**Snowden, Snowdon, Snawdon:** Henry *de Snewedon* 1277 FFEss; Matthew *de Snoudon'* 1278 AssSo. From Snow End (Herts), *Snowdon* 1362, Snowdon (Devon), or Snowden (WRYorks).

**Snowhite:** John *Snouwhight* 1377, *Snowwhyt* 1379 ColchCt; Matilda *Snouwhyte* 1381 SRSf. 'Snow-white', OE *snāw, hwīt*, a nickname for one with white hair.

**Soal(l):** v. SOLE

**Soame, Soames, Somes:** Warin *de Saham* 1086 InqEl (C); Henry *Somes*, William *Soumes*, Widow *Soames* 1674 HTSf. From Soham (Cambs, Suffolk).

**Soane(s):** v. SON

**Soar, Soares, Soars:** John, Roger *le Sor* 1176 P (Do), 1229 Cl (Co); John *le Soor* 1327 SRSf. OFr *sor* 'reddish-brown'. cf. SORREL.

**Soards:** v. SWORD

**Sock:** Godwin *Socche* 1066 Winton (Ha); Robert *Soc* 1305 IpmW; Thomas *le Sock'* 1355–9 AssBeds. OE *Socca*, or a nickname from OE *socc* 'slipper, light shoe'.

**Sockman:** v. SOKEMAN

**Soden, Soldan:** v. SOWDEN

**Soft, Softe:** Ailwin *Softe* 1195 P (Ess); John *Soft* 1327, 1524 SRSf. OE *sōfte* 'mild, gentle'. cf. God *Softebred* 1066 Winton (Ha), probably 'soft beard', rather than 'soft bread'.

**Softely, Softly, Softley, Softlaw:** Roger *de Softlau* c1235–58, Aimer *de Softlawe* 1292, John *de Softelaw* 1368 Black. From Softley (Du, Nb, Roxburgh).

**Sokeman, Sockman:** Walter *Sochemannus* 1169 P (C/Hu); Reginald *Sokeman* 1247 Fees (Nth); John *Sokman* 1377 AssEss. 'A tenant holding land in socage', OE *sōcn, mann*.

**Sole, Soles, Soal, Soall:** (i) William *de la Sole* 1207 Cur (Sx); Thomas *atte Sole* 1294 PN Sr 101; Hamo *de Soles* 1242 Fees (K). 'Dweller by the miry place', OE *sol* 'mud, wallowing place for animals', as at Soles (Kent). (ii) Osbert, Walter *Sole* 1203 Cur (Nf), 1207 Rams (C); Godfrey, Osbert *le Sol* 1274 RH (Ess), 1275 SRWo. OFr *sol* 'sole, lonely'.

**Sollas, Sollis:** Walter, Robert *Solace* 1269 AssNb, 1372 LLB H. ME *solas*, OFr *solaz* 'comfort'.

**Soller, Sollars:** Henry *de Solariis* 1176 P (He); Cecilia *le Soliere* 1279 RH (C); Walter *atte Solere* 1327 SRSo; John *Solere* 1417 LLB I. OE *solor* 'the upper part of a house', probably for a servant whose duties lie there.

**Sollinger:** Probably usually for SALINGAR, but sometimes from Solinger in Kimble (Bk).

**Soltan:** v. SOWDEN

**Somerfield:** v. SUMMERFIELD

**Somerford:** v. SUMMERFORD

**Somer(s):** v. SUMMER

**Somerscales:** v. SUMMERSCALE

**Somerset, Somersett, Sommersett, Summersett:** Walter *de Sumerset* 1206 Pl (L); John *de Somersete* 1331 IpmW; Edmund *Somerset* 1545 SRW. 'The man from Somerset.' But John, Robert *de Somersete* 1319 SRLo, 1352 AssLo, may owe their names to the London parish of St Mary Somerset.

**Somersham:** Isabel *Somersam* 1379 LoCh. From Somersham (Bk).

**Somerton:** v. SUMMERTON

**Somervell, Somerville, Somervaille, Sommerville, Summerville:** Adam *de Somervila* 1153 Templars (O); William *de Summeruill'* 1158 P (Y). Probably from Graveron-Sémerville (Nord).

**Somes:** v. SOAME

**Sommerfeld, Sommerfield:** v. SUMMERFIELD

**Sommer(s):** v. SUMMER

**Sommerlat:** v. SUMMERLAD

**Somner:** v. SUMNER

**Son, Sonn, Sone, Sones, Soan, Soane, Soanes, Soans:** John *Sune* 1203 P (Wa); James *le Sone* 1275 AD iii (Mx); Roger *le Son* 1327 SRSf; William *Sones* 1327 SRWo; Thomas *Sonne* 1327 SRDb; Robert *Soones, Soane*, John *Soanes* 1674 HTSf. OE *sunu*, ME *sone* 'son', perhaps 'the younger', junior.

**Songer:** v. SANGAR

**Songhurst, Songest:** William *de Sunghurst* 1332 SRSr. From Song Hurst in Ewhurst (Surrey) and from Songhurst Fm in Wisborough Green (PN Sx 135).

**Sonning, Sunning:** Gilbert *Sunning'* 13th Rams (Hu); John *Sonyng'* 1327 SRWo; Robert *Suninge* 1641 PrSo. From Sonning (Berks).

**Soord:** v. SWORD

**Sooth, Soothe:** Haldan *Soth* Hy 3 Gilb (L); Thomas *le Sooth'* 13th FeuDu; John *Sothe* 1379 PTY. OE *sōð* 'truth, justice'. cf. William *Sothman* 1226 Cur (Nf) 'true man'.

**Soothill, Soutell, Suttill, Suttle:** Michael *de Sothil, de Suthill'* 1207 P, 1208 Cur (Y); John *Sootell* 1472 GildY; Katerina *Sottell* 1497 ib.; Robert *Suttell* 1526 ib.; William *Suttle* 1679 FrY. From Soothill (WRYorks). v. also SUTTLE.

**Soper, Soaper, Sopper, Saper, Sapier:** Edgar *le soppier* 1138–60 ELPN; William (*le*) *Sopere* 1195–6 P (Gl); Roger *Sapere* 1243 AssDu; Alexander *le Soppere* c1260 *ERO* (Ess); Emma *la Sapere* 1301 SRY. A derivative of OE *sāpe* 'soap', a maker or seller of soap.

**Sopp, Soppe, Sope, Sape, Sappe:** (i) Walter *Sappe* 1202 FFSf; Gerard *Sape* 1406 IpmY. OE *sāpe* 'soap'. Metonymic for a maker or seller of soap. v. also SAPPE. (ii) William *Soppe* 1210–11 PWi; John *Soppe* 1275 SRWo; John *Sopp'* 1438 FFEss. OE *Soppa*. But some of the forms may belong under (i).

**Sopwith:** Henry *de Sopeworth'* 1206 Cur (W). From Sopwith (W).

**Sorbie, Sorby, Sorsbie, Sorsby, Sowerby, Sowersby, Surbey:** Odierna *de Sourebi* 1195 P (Cu); Richard *Surby* 1381 LoPleas; William *Sourby* 1381 PTY; Isabel *Soreby* 1485 GildY; Thomas *Sowerbye* 1597 Oxon (C); Christopher *Saursbye* 1615 Shef; Robert *Sorsbie* 1626 ib. Sowerby 'farm or village in marshy ground' is a common place-name in Cumberland, Westmorland, Lancashire and Yorkshire. Sowerby near Inskip (Lancs) is *Sorbi* in DB. The unetymological medial *s* in *Sorsby* is found in 1202 in *Kirkesoresbi* for Castle Sowerby (Cumb). The most common form today is *Sowerby*. In Sheffield, *Sorby* is usual.

**Sorley:** *v.* SUMMERLAD

**Sorrel, Sorrell, Sorrill:** William *Sorell'* 1130 P (Sf); Thomas, William *Sorel* 1175 P (Nf), 1185 Templars (Herts). OFr *sorel*, a diminutive of *sor* 'reddish-brown'.

**Sorrie:** Ralph, Fulco *Sori* 1275, 1276 RH (Nf, L); Robert *le Sorei* 1279 RH (O). OE *sārig* 'sorry, sad'.

**Sorrowless:** Symon *Sorweles* 1226 FFY; Olive *Sorweles* 1321 CorLo; John *Sorowles* 1379 PTY. 'Free from sorrow; careless, unconcerned', OE *sorgh, lēas*.

**Sorsbie, Sorsby:** *v.* SORBIE

**Sotham, Sothcott:** *v.* SOUTHAM, SOUTHCOTT

**Sotheby, Suddaby, Sutherby:** Stephen *de Sottebi* 1194 P (Y); John *Suthiby*, Robert *Suthinby* 1297 SRY; William *Sothybe* 1479 GildY; John *Suddebe* 1516 ib.; John *Sotheby* 1674 HTSf. ON *suðr i bý* (the man who lived) 'south in the village'. *v.* WESTOBY.

**Sotheran, Sothern, Sotheron, Southan, Southon, Southerin, Southern, Southorn, Sudran, Sudron, Sutherin, Suthern, Sutherns, Suthren:** Geoffrey *le Sutherne* 1243 AssSt; Robert *le Sotherun, le Sotheren* 1297 SRY, MinAcctCo (Y); John *le Southeren* 1307 Wak (Y); Henry *le Suthreen* 1325 AssSt; Hugh, Henry *le Sotheron* 1327 SRDb; Agnes *le Southeron* 1327 SRWo; Richard *Sotheran* 1352 FrY; Thomas *Sothryn* 1387 AD iv (Y); Robert *Sudurrom* 1496 GildY; William *Sothoryng* 1524 SRSf; Andrew *Suddren* 1755 FrY. *Southern* and *Sothern* are from OE *sūðerne* 'southern', the man from the south. For this NED has one metathesized form *southren* c1386 and *southron*, originally Scottish and northern, from c1470. It regards as an alteration of *southren* with the ending probably modified on the analogy of *Briton, Saxon*. The above forms, much earlier, and mostly from Scandinavian counties, suggest that some may derive from ON *suðrænn* 'southern'. The intrusive *e* of *Sotheran*, etc., found in most forms is probably due to the strongly trilled northern *r*. The sole example of this form in NED is from the northern *Cursor Mundi* (a1300), *sotherin englis*.

**Sott:** Hugh *Sot* 1201 SPleas (Co); Adam *Sote* 1298 AssL; John *le Sot'* 1332 SRDo. ON *Sóti*, or a nickname from OFr *sot* 'fool'.

**Souch:** *v.* SUCH

**Soulby:** John *de Soulby* 1332 SRCu. From Soulby (Cu, We).

**Sound, Sund:** (i) Simon *Sund* 1195 P (Nth). OE *sund* 'sound, healthy, prosperous'. (ii) John *de Sunde* 1296 SRSx; Reginald *atte Sound* 1327 SRSf; John *atte Sunde* 1333 MELS (So). 'Dweller by the water', OE *sund. v.* MELS 205–6.

**Sour, Soures:** Gilbert *le Sour* 1279 RH (C); Roger *le Soure* 1305 AssW; Richard *le Sour* 1310 AssNf. OE *sūr* 'sour, tart'.

**Sourbut, Sourbutts:** *v.* SOWERBUTTS

**Sourmilk:** Ralph *Surmylk* 1290 NorwLt; Thomas *Sourmylk* 1307 Wak (Y); John *Surmelk* 1327 SRSx. 'Sour milk', OE *sūr, meoloc*. cf. William *Sourale* 1301

SRY 'sour ale'; John *Sourappill* 1376 FFY 'sour apple'; Roger *Sourdogh* 1327 SRSo 'sour dough'; William *Surlaf* 1148 Winton (Ha) 'sour loaf'.

**Souster:** Cristiana *Seustere* 1279 RH (Hu); Alice *Sewstere* 1301 SRY; Emma *le Sowester* 1307 Cl; Juliana *le Suster* 1309 SRBeds; Margery *le Sewester* 1326 Wak (Y). ME *sewester, sowester* from OE *sēowian* 'to sew', 'a woman who sews, a sempstress', the feminine of SEWER.

**Soutar, Souter, Souttar, Soutter, Sowter, Sueter, Suter, Sutor, Sewter, Sutters:** Lewinus *sutor* 1066 Winton (Ha); Nicholas *le Soutere* 1263 MESO (Sx); John *le Sutere* 1273 RH (C); William *le Soutare*, Richard *Suter*, Roger *Soutere* 1327 SRSx; Nota *la Soutres* 1312 ColchCt; John *Sowter* 1379 PTY. *Sutor*, which is now rare but was very common in the 12th century, is Lat *sutor* 'shoemaker', often a translation of OE *sūtere*, the source of the remaining forms.

**South:** William *de la Sothe* 1273 RH (D); Isabella *South* 1297 MinAcctCo; William *del South* 1379 PTY. 'Man from the south', a synonym for *Southern* and *Sotheran*. cf. BYSOUTH.

**Southall:** Nicholas *de Suthalle* 1273 RH (Nf). 'Dweller at the south hall.'

**Southam, Sotham:** Thomas *de Suham* 1199 Pl (Nth); Ralph *de Sutham* 1237 WhC; William *de Southham* 1305 IpmGl; John *Southam* 1371 FFEss. From Southam (Glos, Warwicks), or 'dweller to the south of the village'.

**Southan, Southern, Southern:** *v.* SOTHERAN

**Southbrook, Southbrooke:** (i) Walter *de Suthbroc* 1229 Cur (K); John *de Suthbroc* 1275 IpmGl; Stephen *Southbrok* 1327 SRSo. From Southbrook in Kenton (D), or 'dweller by the south brook', OE *sūð, brōc*. (ii) Walter *Bisudebrok* 1238 PN D 501. 'Dweller to the south of the brook', OE *bī, sūð, brōc*.

**Southcombe, Southcomb:** Richard *de Suthcombe* 1317 AssK; Nicholas *Bysouthecombe* 1333 PN D 142. From Southcombe in Holsworthy, in Milton Abbot, in Widdecombe (D), or 'dweller to the south of the valley', OE *bī, sūð, cumb*.

**Southcott, Sothcott:** Geoffrey *de Suthcot* 1229 Cur (K); Stephen *de Southcote* 1305 IpmW; John *Suthcote* 1398 FFEss. From Southcot (Berks), or 'dweller in the southern cottage'.

**Southerland:** *v.* SUTHERLAND

**Southey:** Peter *de Suthag'* 1219 AssY; John *Ofthesuthhey* 1315 FFEss; Sara *Bysoutheya* 1330 PN D 613; Christopher *Southey* 1642 PrD. From Southey in Culmstock (Devon), Southey Green (Essex), Southey Wood in Ufford (Northants), or Southey in Ecclesfield (WRYorks).

**Southgate, Suggate, Suggett, Suggitt:** Osbert *de Sudgate* 1197 P (Ess); Alice *de Southgate* 1327 SRSf; Richard *Suggett* 1516 GildY. 'Dweller near the south gate', especially of some ancient town.

**Southin, Southing:** Reginald *de Suthyne* 1327, William *Southinne* 1326 SRSx; William *Southen* 1652 PN Herts 35. From Southings in Westfield (Sx).

**Southland:** Christiana *ate Southlande* 1317 AssK. 'Dweller by the southland.'

**Southley:** Imedia *de Suthlegh'* 1246–7 FFWa; John *de Suthleye* 1299 IpmGl; Orm *de Southley* 1332

SRWo. From Southleigh (D), South Leigh (O), or a lost *Southley* in Ripple (Wo).

**Southman:** John *Southman* 1332 SRSx. 'The man from the south', OE *sūð, mann*.

**Southon, Southorn:** *v.* SOTHERAN

**Southouse:** Gilbert *atte Suthhuse* 1296 SRSx; William *atte Southouse* 1327 ib. 'Dweller at the southhouse.'

**Southover:** Thomas *Southovere* 1324 IpmW; Thomas *Southoure* 1331 FFW, *Southouere* 1344 KB (W). From Southover (Sx).

**Southward** is a corruption of SOUTHWOOD or SOUTHWORTH. Southward Downs in Aldbourne (Wilts) were probably by *Southwode* (1509 PN W 293).

**Southwell:** William *de Suthwelle* 1287 FFY; Henry *de Suthwell* 1360 IpmNt; Richard *Sowthwell* 1451 Paston. From Southwell (Nt).

**Southwick:** Ogga *æt Suthwycan* 972 OEByn (Nth); William *de Sudwic* 1202 AssNth; Thomas *de Suthewyk* 1332 SRSx; William *Southwyke* 1363 IpmGl. From Southwick (Durham, Glos, Hants, Northants, Sussex, Wilts).

**Southwood, Sowood:** Elfere *de Sudwude* 1202 P (Nf); William *de Suthwud* 1225 FFEss; Margaret *Suwode* 1296 SRSx. 'Dweller near the south wood.' Or 'to the south of the wood': Roger *Bisothewode* 1283 Battle (Sx).

**Southworth:** Gilbert *Southworthe, de Sotheworth* 1281 AssLa. From Southworth (Lancs). Confused with SOUTHWOOD. Southwood Fm in Walton-on-Thames is *Suthwode* 1235, *Sutheworthe* 1337 (PN Sr 98).

**Souttar, Soutter:** *v.* SOUTAR

**Sow, Sowe:** John *de Sowe* 1203 AssSt; William *Sou* 1327 SRY. From the River Sow (St), or from Sowe (Wa).

**Sowden, Soden, Soltan, Soldan, Sultan:** Alan *Soldenc, Soldench'* 1166, 1175 P (Y); *Soutan, Soldangus* Ric 1 Cur (K); Henry *Soldan* 12th FeuDu; Roger *le Soudan* 1208 Cur (Y); Robert *le Sowden'* 1279 RH (Bk). OFr *soudan* 'sultan', occasionally, perhaps, a nickname, but often a pageant-name from the Soldan of the Saracens. cf. 'He that playeth the sowdayn is percase a sowter. Yet if one should ... calle him by his owne name ... one of his tormentors might hap to breake his (one's) head' (Sir Thomas More).

**Sowerbutts, Sourbut, Sourbutts:** Richard *de Sourbuttes* 1401 AssLa; Mary *Sourbuts* 1480 TestEbor; Richard *Sowerbutts* 1631 RothwellPR (Y). From Sowerbutts in Garstang (La).

**Sowerby, Sowersby:** *v.* SORBIE

**Sowman:** John *Southman* 1332 SRSx. 'Man from the south' or 'dweller to the south'.

**Sowood:** *v.* SOUTHWOOD

**Sowter:** *v.* SOUTAR

**Spack:** Walter *Spac* 1209 FFNf; William *Spake* 1249 AssW; Malina *Spak* 1303 IpmGl. ON *Spakr*, ODa, OSw *Spak*.

**Spackman:** *v.* SPEAKMAN

**Spafford, Spafforth:** Richard *Spafford* 1401 AssLa; Joice *Spaford* 1622 GreenwichPR (K). From Spofford (WRY). *v.* also SPOFFORD.

**Spaight:** *v.* SPEIGHT

**Spain:** Richard *de Espaigne* 1177 P (Ess); Thomas *Spane* 1302 SRY; Thomas *de Spaigne* 1318 LLB E; John *Spayne* 1327 SRC. Alueredus *de Hispania* (1086 DB) came from Épaignes (Eure). Herueus *de Ispania* (ib.) was probably a Breton and may have come from Espinay (Ille-et-Vilaine). *v.* ANF. Later, the name probably derives from Spain, especially in the case of London merchants.

**Spalding, Spaulding:** Gilbert *de Spaldingis* 1175 P (L). From Spalding (Lincs).

**Spaldington:** Gerard *de Spaldingtone* Hy 2 Gilb; Robert *de Spaldinton'* 1219 P (Y). From Spaldington (ERY).

**Spalton, Spolton:** Henry *Spalding*, carpenter 1633 FrY; Mathew *Spaldinge*, carpenter, son of Henry *Spaldinge* 1662 ib.; Henry *Spawlden*, son of Henry *Spalden*, carpenter 1672 ib.; Marcus *Spaldinge*, carpenter, son of Mathew *Spalton*, carpenter 1689 ib.; Mathew *Spalding*, son of Mathew *Spalding*, carpenter 1702 ib.; Henry *Spaldon*, son of Henry *Spaldon*, taylor 1713 ib. Most, if not all of the above freemen of York were members of the same family of *Spalding*, of which *Spalton* is one of several dialectal pronunciations.

**Spanton, Spaunton:** Dolfin *de Spanton* 1219 AssY; John *Spanton* 1674 FrYar. From Spaunton (NRY).

**Spare:** Walter, John *Spare* 1186 P (So), 1268 AssSo. 'Frugal', from OE *spær* 'sparing'.

**Sparegood:** Walter *Sparegod* 1271 RH (C). 'Frugal with goods', OE *spær, gōd*. cf. Adam *Sparebutter* 1321 Wak (Y) 'frugal with butter'; John *Sparwatre* 1327 SRY 'frugal with water'.

**Sparham:** Alexander *de Sparham* 1175–86 Holme; John *de Sparham* 1354 FFSt; John *of Sparham* 1448 Paston. From Sparham (Nf).

**Sparhawk:** *v.* SPARROWHAWK

**Spark, Sparke, Sparkes, Sparks:** William *Sperc* 1202 AssL; Ralph *Sparke* 1221 ElyA (Sf); John *Sparkes* 1301 SRY. ON *sparkr, spræk* 'lively, sprightly'.

**Sparkwell:** John *Sparkwell, Sparkwill* 1368 Hylle; Laurence *Sparkwell* 1642 PrD. From Sparkwell (D).

**Sparling, Sperling, Spurling:** Jordan son of *Sparling* 1148–67 ELPN; Alexander *filius Esperling', Sperleng* 1173, 1193 P (Lo); Alexander, Simon *Sperling* a1187 ELPN, 1219 AssY; John *Spurlang* 1221 AssWa; Ralph *Esperlang'* 1221 Cur (Herts); Ælstanus *Sperling* c1250 Rams (Nf). Probably, as suggested by Ekwall, OE *\*Spyrling*, from *\*Spyrtling*, a metathesis of *\*Spryting*, a diminutive of *\*Spryt(t)el*, the first element of Spridlington (Lincs), *Spredelintone, Sperlinctone* DB. *v.* ELPN.

**Sparr, Sparre:** William *de la Sparre* 1263 MELS (Sx); John *atte Sparre* 1296 SRSx. 'Dweller by the enclosure', OE *spearr*.

**Sparrick:** Eudo *filius Spireuuic* (L), *filius Spiruit, Spirvin, Spiruin* (Sf) 1086 DB; *Spirewic Brito* 1170 P (L); Thomas *Spirewit* 1189 Sol; Richard *Spiric* 1198 P (Nth); Michael *Spirewit* 1249 AssW; John *Spyrewygge* 1297 MinAcctCo. OG *\*Spirwic. v.* OEByn 198.

**Sparrow, Sparrowe:** Richard, *Sperewe* 1160–5 ELPN; Ralph *Sparewe* 1182 P (Nf); Ibbota *Sparow* 1325 Wak (Y). OE *spearwa* 'sparrow', lit. 'flutterer'.

**Sparrowhawk, Sparhawk:** *Sparhauoc, Sperhauoc, Sperafoc* 1066 DB (Nt, Sf); *Sparhauec* vtlagus 1172 P (Nth); *Sparhauk* Outlaw 1327 SRSf; Geoffrey *Sparheuec*, Robert *Sperhauec* 1221 AssWa; Thomas *Sparhauk*, William *Sparhawk* 1327 SRSf. OE *Spearh(e)afoc* 'sparrow-hawk', found as a personal name before the Conquest. As a byname, the surname must be both patronymic and a nickname.

**Sparshatt, Sparshott:** Peter *de Sparshete* 1327 SRSx. From Sparshot (Sussex).

**Spashett, Spatchett:** Hugo *Pachat*, Robert, Leticia *Spachet* 1327 SRSf. For PATCHETT, with inorganic initial *S*. cf. STURGE.

**Spaul, Spaull:** William *Spalla* c1100–30 OEByn (D); Anger *Spalle* 1219 Cur (Nf). OE *\*Spalda. v.* OEByn 335.

**Spaulding:** *v.* SPALDING

**Spaunton:** *v.* SPANTON

**Spawforth:** *v.* SPOFFORD

**Speak, Speake, Speaks, Speck, Speek, Speeks, Speke:** William *Spec, Spech* 1086 DB (Beds); Walter *Espec* 1130 P (Y); Ailfwin *Speke* 1180 P (Bk); William *le Spec* 1195 P (Do). OFr *espech(e), espek* 'woodpecker'.

**Speakman, Spackman:** Nigel *Spakeman* 1195 P (Ess); William *Spakeman, Spekeman* 1297 MinAcctCo (O). ME *spekeman* 'advocate, spokesman' (1340 NED).

**Spear, Speare, Speares, Spears, Speer, Speers:** Walter *Speare* 1185 P (So); Henry *Spere* 1246 AssLa. OE *spere* 'spear', for *Spearman*.

**Spearman:** John *Spereman, Spermon* 1327 *SR* (Ess), SRSt. 'Spearman.'

**Spearpoint:** Robert *Sperpoynt* 1541 CantW; Richard *Sperpoynt* 1569 StratfordPR (Wa). 'Spear point', OE *spere*, OFr *point*, a nickname.

**Speck:** *v.* SPEAK

**Speed:** Godfrey *Sped* 1185 P (Sf); John *Spede* 1277 *Ely* (Sf), 1296 SRSx. OE *spēd* 'speed, success, wealth'.

**Speers:** *v.* SPEAR

**Speight, Speaight, Spaight, Spieght:** William *Speyt, Speght* 1297 MinAcctCo (Y), 1332 SRCu; John *Speht, Speght* 1313, 1315 Wak (Y). OE *\*speoht, \*speht*, ME *speight* 'wood-pecker' (c1450 NED).

**Speirs:** *v.* SPIER

**Spellar, Speller:** Robert *Speller'* 1202 AssL; Gerard *le Speller* 1301 ParlR (Ess). Probably a derivative of OE *spellian* 'to speak, discourse', a speaker, preacher or a professional story-teller.

**Spellen, Spelling:** Edmund *Spelyng* 1327 *SR* (Ess). Perhaps a derivative of OE *spellian* with the formative suffix *-ing*, in the same sense as SPELLAR.

**Spellman, Spelman:** John *Speleman* 1273 RH (Nt), 1327 *SR* (Ess). OE *spell* 'discourse, homily, story' and *mann*, equivalent to SPELLAR. *v.* also SPILLMAN.

**Spence, Spens:** Simon *del Spens, de la Despense* 1300 Guisb (Y); Amice *atte Spense* 1327 *SR* (Ess); William *atte Spence* 1327 SRWo. ME *spense, spence*, from OFr *despense* 'larder', one who worked at or was in charge of the buttery. cf. SPENCER. Alan *de la Spense* sometarius Domini Roberti de Shirlaunde (1317 AssK) is also called Alanus *sometarius*, Alanus *seruiens* and Alanus *homo Domini Roberti*. He was the servant or pack-horse driver who carried the provisions to the buttery.

**Spencer, Spenser:** Robert *le Despenser'* 1204 Cur (Sa); William *le Spencer* 1275 RH (Ha); Ralph *le Spenser* 1275 RH (Sf); Richard *le Espenser* 1279 RH (Beds). AFr *espenser*, OFr *despensier* 'dispenser (of provisions)', a butler or steward (a1300 NED).

**Spender:** Henry *le Despendur* 1214 Cur (Wa); Peter *de Spendure* 1227 AssSt; Agnes *Spendure* 1301 SRY; Adam *le Spendur* 1327 SRSt. An aphetic form of ME, OFr *despendour* 'a steward' (1340 NED).

**Spendlove, Spendlow, Spenlow, Spindelow:** Ralph *Spendeluue* 1219 AssL; Robert *Spendelove* 1256 AssNb; Henry *Spendelowe* 1309 SRBeds; Walter *Spenlowe* 1433 GildY; Amy *Spenderlow* 1794 SfPR. 'Spend love', one generous with his love, from OE *\*spendan* 'to spend, employ lavishly, squander' and *lufu* 'love'.

**Spenn, Spenner:** Thomas *del Spen* 1297 Wak (Y); Henry *ate Spene* 1297 MinAcctCo; William *del Spen* 1332 SRLa. From High Spen (Du), Spen Valley (WRY), or a lost *Spen* in Lancs.

**Spens:** *v.* SPENCE

**Spenser:** *v.* SPENCER

**Sperling:** *v.* SPARLING

**Sperrin, Sperring:** *v.* SPURREN

**Spice:** William, Clement *Spice* 1326 FFSf, 1399 FFC. ME *spice*, OFr *espice* 'spice', metonymic for SPICER.

**Spicer:** William *le Espicier* 1184 P (K); Bertram *le Specier* 1200 P (Ha); Robert *le Spicer* 1201 AssSo; Hugo *le Especer* 1214 Cur (C); John *Spicer ElyA* (C). OFr *espicier, especier* 'a dealer in spices; an apothecary or druggist' (1297 NED).

**Spich, Spiche, Spitch:** Aldwin *Spich'* 1148 Winton (Ha); Simon *Spich* 1230 P (Nf/Sf); Geoffrey *Spiche* 1298 IpmY. OE *spic* 'fat bacon', perhaps a nickname for a pork-butcher. cf. William *Spichfat* 1200 P (Nt) 'bacon fat'.

**Spickernall, Spickernell, Spicknell, Pickernell:** Richard, Walter *Spigurnel(l)* 1192 P (Lo), 1205 Cur (Y); John *Spigernel* 1259 Crowland (C); Nicholas *Spikernel* 1275 RH (Nf); Edmund *le Espycurnel'*, Geoffrey *Espigurnel* 1285 *Ass* (Ess); Adam *Sprigunnel* 1297 Wak (Y). ME *spigurnel* 'a sealer of writs'.

**Spieght:** *v.* SPEIGHT

**Spier, Spiers, Speir, Speirs, Spyer, Spyers:** William *le Spiour* 1302 ChambAcctCh; Robert *Spyer* 1379 PTY. A derivative of ME *espyen*, OFr *espier* 'to spy', a spy, watchman. cf. 'The wayte (var. *spiere*) that stood upon the toure of Jezrael' (Wyclif).

**Spike, Spikes:** Geoffrey *Spik* 1275 RH (Sf); Robert *Spikes* 1316 AssNf; John *Spyke* 1322 NorwDeeds II. ME *spike* 'spike', perhaps a nickname for a tall, thin man. cf. Adam *Spikefis* 1276 RH (Berks) 'spike fish', probably a pike.

**Spill:** *Spilo* filius Brette 1197 FFBeds; Robert *Spil* 1189 Sol; Thomas *Spyle* 1327 SRSf. OG *Spilo*, or ON *Spilli*.

**Spillbread:** William *Spillebrede* 13th Guisb, *Spillebred* 1297 MinAcctCo. 'Spoil, waste bread', OE *spillan, brēad*. cf. William *Spillecorn* 1214 Cur (Wa) 'spoil corn'; Adam *Spilgold* 1332 SRCu 'spoil gold';

John *Spillhaver* 1301 SRY 'spoil oats'; Richard *Spiltimbir* 1331 Wak (Y) 'spoil timber'.

**Spillman, Spilman:** Goduine *filius Spilemanni* c1095 Bury (Sf); *Spileman* c1095 Bury (Sf), c1160–65 NthCh (Nth), 1166 P (Nf); Berdic *Joculator*, Adelina *Joculatrix* 1086 DB (Gl, Ha); Edwine *Spileman* 1103–15 OEByn (D); William *Spileman* 1167 P (Ha); Swayn *Spileman* 1204 AssY. The Latin *Joculator* is probably a translation of OE *spilemann* 'jester, juggler'. We are also clearly concerned with a personal name, one of the nicknames used as personal names which were not uncommon in the 12th century. There has been some confusion with SPELLMAN. cf. William *Spileman, Speleman* 1221 Cur (Ha).

**Spilsbury:** William *de Spellesbury* 1180 Eynsham; Philip *de Spillesbury* 1327 SRWo; Thomas *Spellsbury* 1599 ChwWo. From Spelsbury (Oxon), *Spillesbury* 1343.

**Spindelow:** v. SPENDLOVE

**Spindler:** William *le Spinlere* 1236 FFSx; William *le spindlere* 1297 MinAcctCo. A derivative of OE *spinel* 'spindle', a maker of spindles.

**Spine, Spines:** Simon *Spinis* 1296 IpmY; Henry *Spine* 1379 PTY; Michael *Spyns* 1402 IpmY. A nickname from OFr *espine* 'thorn'.

**Spink, Spinke, Spinks:** Roger *Spinc* 1133–60 Rams (Beds); Thomas *Spink* 1256 AssNb. ME *spink* 'a finch', especially a chaffinch.

**Spinner:** John *le Spinner* 1270 MESO (Wo); Thomas *Spinnere* 1356 AD i (D). A derivative of OE *spinnan* 'to spin', 'a spinner of wool, yarn, etc.' (1393 NED).

**Spinney, Spinny:** Robert *de la Spinei* 1198 AC; John *atte Spyneye* 1327 SRLei; John *Spynneye* 1377 FFEss. 'Dweller by the copse or spinney', OFr *espinei*.

**Spir, Spire:** Thomas, William *Spir* 1229 Pat (Ess), 1279 RH (C); Thomas *le Spyr* 1296 SRSx. OE *spīr* 'spike, stalk, tapering stem', probably a nickname for a tall, thin man.

**Spirett, Spirit:** Osbert, Robert, Geoffrey *Spirhard* 1206 Cur (Nf), 1219 AssY, 1221 *ElyA* (Sf). OE *spīr* plus the French suffix *-hard*, perhaps in an intensive sense, 'the tall one'. cf. POLLARD.

**Spiring:** William, Lyme *Spiring* 1229 Pat (So), 1250 Fees (Gl); Henry *Spiryng* 1275 SRWo. OE *spīring*, a derivative of *spīr*, 'the tall one'.

**Spitch:** v. SPICH

**Spitteler, Spittler:** Synonymous with *atte Spitele* and (Adam) *Spitelman* 1176 P (Y). v. SPITTLE.

**Spittle, Spittall, Spittel, Spital, Spittles:** Geoffrey *del Hospital* 1210 Cur (So); Adam *del Spitell* 1307 Wak (Y); Walter *atte Spitele* 1332 SRSx. 'Dweller at or attendant in a hospital', ME *spitel*, OFr *(h)ospital*.

**Spittlehouse:** Robert *de Spytelhous* 1440 ShefA. 'Dweller or worker at the hospital-house.'

**Spivey, Spivy:** William *filius Spiui* c1100 RegAntiquiss; Richard *son of Spivey* 1271 ForSt; William *Spivi* 13th RegAntiquiss; Jordan *Spivi* 1316 Wak (Y); William *Spyvay* 1350 Misc (Y). From a ME personal name, *Spivey, v.* PN WRY iii 253.

**Splatt, Splott:** Walter *Splat* 1209 Pleas (Nf); Walter *de la Splotte* 1262 MPleas (W); William *atte Splotte* 1327 SRSo; John *Splat* 1524 SRD. 'Dweller at the plot of land', OE *splott*.

**Spofford, Spofforth, Spawforth:** Hynge *de Spoufford* 1212 Cur (Y). From Spofforth (WRYorks).

**Spolton:** v. SPALTON

**Spondon:** William *Spondon* 1481–2 IpmNt. From Spondon (Db).

**Spong, Sponge:** Robert *Spong* 1275 RH (Nf); William *Spong* 1327 SRC. 'Dweller by the narrow strip of land', ME *spong*.

**Spooner:** Roger *Lesponere* 1179 P (Y); Thomas, Robert *le Sponere* 1221 AssL, 1254 FFEss. OE *spōn* 'chip, splinter' is recorded in 1316 in the sense 'roofing-shingle' and c1340 as 'spoon'. The surname may therefore mean 'maker of spoons', but, as it is mainly a northern name and *spone* 'shingle' is a northern term (Building 229), the meaning is probably in general 'maker of shingles'.

**Spoor, Spore:** v. SPURR

**Spott:** (i) Geoffrey *Spot* 1268 FFSf; William *Spot* 1313 AssNu; William *Spotte* 1332 SRSr. ME *spot* 'a spot, blemish', perhaps 'the spotty one'. (ii) Thomas *atte Spotte* 1317 AssK. 'Dweller at the small plot of ground', ME *spotte*. (iii) Moyses *de Spot* 1296 (Berwick), Adam *of Spot* 1386, Ninian *Spot* 1437 Black. From the lands of Spott (East Lothian).

**Sprackling, Spratling, Sprankling:** *Sprachelingus* 1130 P (Lo); *Sprakelingus* 1200 P (K); Robertus *filius Sprakeling* 1204 P (W); Vnspac *Spracheling* 1166 P (Nf); John, Gervase *Sprakeling* 1170–87 ELPN, 1204 P (W). OE *Spracaling*, an anglicizing of ON *Sprakaleggr* 'man with the creaking legs'.

**Spragg, Spragge, Sprague, Sprake:** Richard *Sprak* 1327 SRSf; Reginald *Sprag* 1303 FA (Sf); Alice *Sprakes* 1359 Putnam (So). *Sprag* is a voiced form of *sprak*, a metathesis of *spark*. cf. Wilts dial. *sprack* 'lively' and v. SPARK.

**Sprankling, Spratling:** v. SPRACKLING

**Spray, Sprey:** (i) Geoffrey *Espray, Lesprai* 1205 Cur (L); John *Spraye* 1319 SRLo. ME *spray* 'slender shoot or twig'. (ii) William *de Spray, Espray* 1180–1 P (D). From Sprytown (Devon), DB *Sprei*.

**Spriddell, Spriddle:** William *Spridel* 1213 Cur (D); Henry *Spridel* 1275 RH (D); Walter *Spridel* 1359 AssD; Thomas *Spriddle* 1642 PrD. OE *Sprytel*, unrecorded in independent use, but found in Spridlington (L).

**Sprigg, Spriggs:** William *Sprig* 1199 P (Nf); Osbert *Sprigge* 1206 P (D). ME *sprigge* 'twig'. cf. Lonsdale dialect *sprig* 'a small, slender person'.

**Spring:** William *Esprinc* Hy 2 DC (L); Henry *le Springe* 13th FeuDu; William *Spring* 1280 Riev (Y). Probably a nickname for an active, nimble individual, from ME *spring* 'spring', the season when young shoots spring or rise from the ground. There appears to be no evidence for a topographical origin.

**Springall, Springell, Springhall, Springle, Springate:** *Springald* 1195 P (Gl); *Springaldus, Springoud* 1197 P, 1220 Cur (Ess); Robert *Springout* 1245–6 Seals (Nth); Emma *Springhalt* 1272 AssSt; Adam *Springald* 1275 Wak (Y); Walter *Springaud* 1279 RH (O); John *Springot* 1297 MinAcctCo; William *Springold* 1310 ColchCt; Deynes *Springat* 1327 SRC; William *Springate* 1814 SfPR. OFr *espringalle*, AFr *springalde* 'an engine of the nature of a bow or catapult used in medieval warfare for throwing heavy missiles' (a13.. NED). Early examples are probably metonymic for a soldier in charge of such engines but we are also concerned with ME *springal(d)*, of doubtful origin, perhaps a derivative of *spring*, 'a young man, youth, stripling' (c1440 NED). There has, no doubt, been confusion with SPRINGETT.

**Springer:** William, Simon *Springer* 1185 Templars (K), 1296 SRSx. A derivative of OE *springan* 'to jump', a jumper, perhaps a nickname, 'nimble, lively'.

**Springett:** William *Pringet* 1193 P (Wa); William *Springet* 1262 *For* (Ess), 1296 SRSx. A diminutive of SPRING, 'a lively young man'. cf. SPRINGALL.

**Springham:** Ralph *Springham* 1242 Fees (D); Robert *Springham* 1305 AssW; Walter *Sprynghom* 1327 SRSx. From Springham Fm in Hellingly (Sx).

**Sproson, Sproxton:** v. SPROXTON

**Sprot, Sprott:** *Sprot* 1066 DB (Ess), c1130 ELPN; *Sprot* de Spaldintona 1176 P (Y); Willelmus *filius Sprot* ib.; Alduinus, Ralph, William *Sprot* c1140 ELPN, 1176 P (Ess), 1199 P (Beds). OE *\*Sprot(t)* 'sprout, shoot', considered by von Feilitzen to be possibly Scandinavian, but more probably native, from the same root as SPARLING.

**Sprotley:** John *de Sprottele* 1251 FFY. From Sproatley (ERY).

**Sprowston:** Ralph *de Sprouston* 1299–1300 NorwLt; Reginald *de Sprowston* 1308 AssNf. From Sprowston (Nf).

**Sproxton, Sproston, Sproson:** Robert *de Sproxtun* c1160–9 MCh; Richard *de Sproxton* 1214 P (L); William *Sproxton* 1417 IpmY. From Sproxton (L, Lei, NRY).

**Sprunt:** William *Sprunt* 1237 HPD; Matilda *Sprunt* 1314–16 AssNth; William *Sprunt* 1361 ColchCt. ME *sprunt* 'brisk, active'.

**Spurling:** v. SPARLING

**Spurnrose:** Robert *Spurnrose* 1296 SRNb. 'Spurn the rose', OE *spornan*, Lat *rosa*. cf. Alan *Spernecurteis* 1196 P (L) 'spurn courtesy'; Robert *Spornegold* 1327 SRC 'spurn gold'; Alice *Spurneprud* 1296 SRSx 'spurn pride'.

**Spurr, Spoor, Spore:** Peter *Spore* (*Spure*) 1230 P (Sx); Osmund, Alice *Spore* 1274 RH (Ess), 1303 FA (Sf). OE *spura, spora* 'spur', metonymic for (John) *Spureman* 1222 Cur (Nf), 'spurrier'.

**Spurrell, Spurrill:** Thomas *Spurel* 1314 NorwDeeds II; James *Spurwill* 1642 PrD; James *Spurrell* 1672 HTY. From Spirewell in Wembury (D), or a variant of Spurwell, as in Spurrells Cross in Ugborough (D). But other sources are probably also involved.

**Spurren, Spurring, Spearon, Sperrin, Sperring, Sperryn:** Hugo *Esporun, Spurun* 1141–2 StP, 1185 Templars (Ess); William *Sporun* 1212 Cur (Y); William *Espurun* 1227 AssBeds; Thomas *Sperun* 1274 RH (Ess). OFr *esporon, esperun* 'spur', metonymic for SPURRIER. Richard *le sporiere* (1281 LLB B) is also called *Sporon*. cf. Henry *Speroner, le Esporoner* 1296, 1301 LLB C, B.

**Spurrier:** Walter *Spurier* t. John HPD (Ess); Benedict *le Sporier* 1298 LoCt; Roger *Spurreour* 1360 FrLeic. A derivative of OE *spura* 'spur', 'spur-maker'.

**Spyer:** v. SPIER

**Squarey, Squeery:** John *Squerry* 1334 SRK; Richard *Squery* 1403–4 FFSr; Thomas *Squery* 1462 Cl (K). OFr *esquerré* 'recalcitrant, stubborn'.

**Squiller:** Robert *Lescuieler, le Esquieler, le Squieler* 1219–21 Cur (Nth); John *le Squiller* 1334 LLB E. AFr *sculier*, OFr *esculier, escuillier, esquelier* 'maker or seller of dishes' (often of gold or silver); also 'servant in charge of the scullery' (NED).

**Squire, Squires, Squier, Squiers:** Alword *se Scuir* 1100–30 OEByn (D); William *Scuer* c1180 Bury (Sf); Richard *Lesquier* 1197 P (Gl); Simon *Esquier* 1206 Cur (R); Roger *le Esquier* 1228 FFEss; Roger *Squier* 1293 Fees (D). OFr *escuyer, escuier*, ME *squyer* 'shield-bearer, esquire'. In the early examples, used of a young man of good birth attendant on a knight (c1290 NED). The sense of personal attendant or servant is not recorded before 1380.

**Squirrel, Squirrell, Scurrell, Squirl:** Ralph *Squrel* 1221 AssWa; William *Scurell* 1230 P (Nf); Geoffrey *le Esqurel* 1274 RH (Sf); Peter *Squirel* 1301 SRY. AFr *esquirel*, OFr *esquireul, escureul* 'squirrel' (a1366 NED). Used contemptuously of men in the 16th century, but as a surname, probably a nickname denoting agility or thrift. For *Squirl*, cf. the northern dialect form *swirl*.

**Stable, Stables:** (i) Walter *de la Stable* 1275 SRWo; Robert *del Estable* 1270 AssSo; Robert *atte Stable* 1327 SRSx. OFr *estable* 'a stable', equivalent to STABLER, but occasionally, perhaps, denoting one who lives near a stable. (ii) Roger *Estable, Stable* 1199 ChR, 1200 P (So); William *Stable* 1250 Fees (So). OFr *stable, estable* 'steadfast in purpose or resolution; sound in counsel or judgement, trustworthy' (a1275 NED).

**Stableford, Stableforth:** v. STAPLEFORD

**Stablegate:** Robert *de Stablegate* 1271 FFK. From a lost *Staplegate* in Nackington (K).

**Stabler, Stabeler:** Laurence *le Stabler* 1196 FFEss; Alan *le Establer* 1257 MEOT (Y). OFr *establier* 'stable-keeper' (14.. NED) also 'stable-man, ostler'.

**Stace:** *Stacius* 1147–65 Colch (Ess); *Stacius de Hant'* 1187 P (Ha); John, Roger *Stace* 1275, 1279 RH (Lo, Hu). *Stacius* is a latinization of *Stace*, the vernacular form of *Eustace*.

**Stacey, Stacy:** *Stacy* Hernowe 1327 SRSo; Robert *Staci* 1270 Eynsham (O); William *Stacy* 1275 RH (D); Lucia *Eustasy* 1327 SRSo. A diminutive of *Stace*. *Staci* de Gynes (1292 SRLo) is called *Eustace* (1299 LLB C).

**Stack:** Robert *Stac* 1199 P (Bk); Simon *Stakke* 1244 Fees (Ha). ON *stakkr* 'hay-stack', either metonymic for *Stacker*, a builder of stacks, or a nickname for one hefty as a hay-stack.

**Stackard:** Robert *le Stacker* 1264 AD i (Herts); William *le Stackere* 1327 SRSf. A maker of hay-stacks. The modern form is probably very late, with excrescent *d*.

**Stackpoole:** Walter *de Stakepol* a1200 Dublin. From Stackpoole (Pembrokeshire).

**Staddon, Stadden:** John *Staddin* 1642 PrD. From one or other of the six minor places of this name in Devon.

**Staff:** William *Staf* 1177 P (Sf); Thomas *le Staf* 1297 FFEss. OE *stæf*, used by Chaucer as a type of thinness or leanness:

Ful longe were his legges and ful lene,
Ylyk a staf, ther was no calf ysene.

**Stafford, Staffurth:** Robert *de Stadford* 1086 DB; Robert *de Stafford'* 1177 P (St); Adam *de Stafford* ,c1280 SRWo; John *Stafford* 1423 AssLo. From Stafford (Staffs), East, West Stafford (Dorset), Stowford (Devon), *Staveford* 1086, or 'dweller by the ford marked with staves'.

**Staffordshire:** William *de Staffordeshire* 1290 AssCh; Richard *Stafforshyre* 1446–7, Thomas *Staffordshire* 1473–4 FFWa. From Staffordshire.

**Stagg:** Robert *Stagge* 1198 P (Sx); Richard *Stagge* 1243 AssDu. OE *stagga* 'stag'.

**Stain:** *v.* STEIN

**Stainburn, Stainburne:** Henry *de Stainburne* 1263 IpmY. From Stainburn (Cu, WRY). A personal name may also be involved: William *filius Stainburn* 1203 AssNth.

**Stainer, Steiner, Steynor:** Henry *Stainer* 1319 *SR* (Ess); John *le Stainer* 1327 *SR* (Ess); William *le steignour* 1353 FrY; Thomas *Peyntour* alias *Steynor* 1430 Oseney (O). A derivative of ME *steyne* 'to stain' (1388 NED). 'Paid to Thomas *Staynour* of Windsor for painting the fenestral like a glass window, 4*d.*' (1427 Building 174).

**Staines:** Richard *de Stanes* 1275 RH (K). From Staines (Middlesex).

**Stainesby:** *v.* STAINSBY

**Stainford, Stainforth:** Hugh *de Steynford* 1251 AssY; Nigel *de Staynford* 1306 IpmY; William *Steynforth* 1395 Shef. From Stainforth (WRY).

**Stainsby, Stainesby, Stanesby:** Walter *de Steinesbi* 12th Riev; Walter *de Staynesby* 1256 AssNb; Robert *de Stanesby* 1328 PN Db 476. From Stainsby (Db, L), or Stainsby in Stainton (NRY).

**Stainthorp, Stainthorpe:** Gilbert *de Steyndrope* 1346 LLB F. From Staindrop (Durham).

**Stainton:** Hugh *de Staintone* 1185 Templars (L); Margaret *de Staynton* 1277 IpmY; John *de Stayntone* 1324 LLB E. From Stainton (Cu, Du, L, La, We, WRY, NRY), Stainton by Langworth, Market Stainton, Stainton le Vale (L), or Stainton Dale (NRY).

**Stair, Stairs:** William *de Stegre* 1195 P (K); Robert *atte Steghere* 1296 SRSx. OE *stæger* 'stair', used in Kent and Sussex of an ascent, rising ground. cf. Stairbridge Fm in Bolney, Steer's Common in Kirdford (Sussex), the Stair (Kent).

**Staker:** Stephen *Staker* 1242 Fees (Lei). 'One who drives in stakes', or 'dweller by the stake'.

**Stakes:** (i) William *de Lestake* 1214 Cur (Bk); Alice *ate Stake* 1332 SRSr. 'Dweller by the stake' (OE *staca*), probably a boundary-post. (ii) Geoffrey, Hugh *Stake* 1199 P (Nf), 1203 Cur (C). OE *staca* 'a stake'. cf. STAFF.

**Stalbridge:** Hugh *Stalbrugge* 1370 FFW. From Stalbridge (Do).

**Stalham:** Clement *de Stalham* 1213 ChR; Antigonia *de Stalham* 1219 P (Nf/Sf); Richard de Stalham 1288 NorwDeeds I. From Stalham (Nf).

**Stalker:** Walter *le Stalkere* 1202 P (Nt); Nicholas *Stalkere* 1252 Rams (Hu). A derivative of ME *stalke*, OE *\*stealcian* 'to walk stealthily, to pursue game by stealthy approach'.

**Stallan, Stallen, Stallon, Stallion:** Alfwin *Stalun* 1202 FF (Nf); Alexander *Stalon* 1275 SRWo; John *Staloun* 1327 SRSf. OFr *estalon* 'a stallion' (a1388 NED); applied to a person, 'a begetter' (c1305); 'a man of lascivious life'. cf. 'þe monke þat wol be stalun gode . . . He schal hab wiþute danger. xii. wiues euche ȝere' (c1305 NED).

**Stallard, Stollard:** Sybyll *Stalward, Stallard* 1572, 1573 LedburyPR; Thomas *Stollard* 1645 Bardsley. A variant of STALLWOOD.

**Stallibrass, Stallebrass, Stallabrass, Stallybrass, Stollybrass:** William *Stalipres* 1167 P (Ha); Samson *Stalipras* 1247 AssBeds; Matilda *Stalipras* 1309 SRBeds; John *Stallowbrass* 1662 HTEss. Harrison's interpretation 'steel arm' can hardly be correct, since the early examples have a second element *-pres*, and *-brass* does not appear before the 17th century. At first sight it looks like a place-name, but the invariable absence of a preposition would suggest a personal name or a nickname.

**Stallman, Stalman:** Adam *de Stalmyn* 1292 QW (La). From Stalmine (Lancs).

**Stallwood:** Reiner *Stalewurd* 1227 Pat (Nf); John *le Stalwrthe* 1219 RH (O); William *Staleworthe* 1327 *SR* (Ess). OE *stælwierðe* 'sturdy, robust; brave, courageous'.

**Stallworthy, Stolworthy:** Thomas *Stalworthi* 1285 Pinchbeck (Sf); John *Stalworthy* 1327 SRSf. An altered form of *stalworth*, after *worthy* (c1250 NED).

**Stamer:** Æðelwearde *Stameran* 1015 OEByn; John *Stamer* 1296 SRSx; Robert *le Stamere* 1327 SRSx. OE *\*stamera* 'stammerer', from OE *stamerian* 'to stammer'.

**Stammer, Stammers, Stanmore:** *Stanmar* 1066 DB (Sf); *Stanmarus* c1250 Rams (Nf); *Stanmer* 1095 Bury (Sf); Widwinus *Stammere* 13th Bart (Lo); Robert *Stammers* 1674 HTSf. A late OE *Stānmǣr* 'stone-fame', chiefly East Anglian. Also local, from Stanmore (Middlesex) or Stanmer (Sussex): Robert *de Stanmere* 1206 Cur (Mx); Reginald *de Stanmere* 1296 SRSx.

**Stamp, Stamps:** Salamon, John *de Stampes* 1191 P (Lo); Thomas *Stampe* 1424 FFEss. From Étampes (Seine-et-Oise), earlier *Estampes*.

**Stampa, Stamper:** John *Stamper* 1279 RH (C); John *Stamper* 1524 SRSf; John *Stamper* 1653 FrY; Fra. *Stamper* 1671 RothwellPR (Y). A derivative of ME *stampen* 'to stamp'. Perhaps a stamper of coins, a worker at a mint.

**Stanberry, Stanborough, Stanbury, Stanbra:** Alan *Stanborw* 1279 RH (C). This form represents an unrecorded OE woman's name *\*Stānburh* 'stone-fortress': *Stanburc* 1086 DB (Ess), *Stanburch* 12th LVD. *Stanbury* is common in Devon where it is probably from Stanborough. The Lancashire *Stanbury* is from Stanbury (WRYorks).

**Stanbourn, Stanbourne, Stanburn, Stanburne:** Philip *de Stanburn'* 1274–5 FFEss; Ralph *de Stanbourn'* 1327 SRWo; Richard *Stanbourn* 1384

IpmGl. Perhaps from Stan Brook in Thaxted (Ess), or 'dweller by the stony stream', OE *stān, burn*.

**Stanbow:** Hervey *Stanbowe* 1275 RH (Nf); Hervey *atte Stanbowe* 1275 SRWo. From Stonebow (L), or Stonebow in Stock-and-Bradley (Wo).

**Stanbridge:** Gilbert *de Stanbrugge* 1276 RH (Beds); Simon *atte Stanbrugg'* 1332 SRSx. From Stanbridge (Beds, Ha), or Stanbridge Grange in Slaugham (Sx). *v.* also STONEBRIDGE.

**Stanbrook, Stanbrooke:** Richard *de Stanbroc* 1275, John *de Stanbrok'* 1327, Nicholas *de Stanbrock* 1332 SRWo. From Stanbrook in Powick (Wo).

**Stanburn, Stanburne:** *v.* STANBOURN

**Stancliff, Stancliffe:** Elyas *de Stanclif* 1202 P (Y); John *Stancliffe* 1672 HTY. From Stancliffe in Kirkheaton (WRY).

**Standage:** *v.* STANDISH

**Standall:** William *Standelf* 1279 RH (C); Robert *de la Standelve* 1275 (Wo), Jordan *atte Standelue* 1333 (So) MELS. From Standhill (O), *Stangedelf* 1002, or 'dweller by the stone quarry', OE *stāngedelf*.

**Standalone, Standerline:** Richard *Standalan* 1332 SRCu. 'Stand alone', a nickname for a resolute, self-reliant man.

**Standard:** Stephen *atte Staundard* 1327 SRSx; Robert *Standard* 1562 CtH. From Standard Hill in Ninfield (Sx).

**Standen, Standon:** Ralph *de Standon'* 1200 Pleas (Beds); Thomas *atte Standenn* 1296 SRSx; Simon *Standen* 1443 CtH. From Standen (Berks, La, W, Wt), Standen in Biddenden, in Benenden, Upper, Lower Standen in Hawkinge (K), Standeal in Ditchling, Standen in East Grinstead (Sx); Standon (Herts, St), or 'dweller at the stony valley or hill', OE *stān, denu/dūn*. *v.* also STANDING.

**Standeven:** Thomas *Standeven* 1544 GildY, 1559 Pat; John *Standeven* 1574 Shef. Probably 'stand firmly', OE *standan, efne*. cf. John *Stondelongws* 1285 AssCh 'stand with us'; Henry *Standinnough* 1702 LewishamPR (K) 'stand enough'; Adam *Standonhisfot* 1260 AssCh 'stand on his foot'.

**Standfast:** Thomas *Stanfast, Stampfast* 1279 RH (O); Robert *Stantfast* 1296 SRSx. A nickname for one of steadfast, resolute character. cf. Gudytha *ffoloufast* 1379 PTY.

**Standidge:** *v.* STANDISH

**Standing:** Robert *Standing* 1553 WhC (La). Probably from Standen in Lancashire where *Standen* and *Standing* are both found as surnames.

**Standish, Standage, Standidge:** Ralph *de Stanedis* 1206 Cur (La); 'Herriesservant *Standich* de Clyfford' 1377 Pat (Wa). From Standish (Lancs, Glos).

**Standon:** *v.* STANDEN

**Standwell:** Richard *Standwele* 1309 Wak (Y). cf. 'Beo stalewurðe & stond well' a1225 NED, and STANDFAST.

**Stanesby:** *v.* STAINSBY

**Staney, Stanney, Stoney:** Thangustella *de Staneye* 1289 AssCh; Amira *de Staneya* 1328 WhC. From Stanney (Ch).

**Stanfield:** Ascelina *de Stanfelde* 1195 FFNf; Hugh *de Stanefeld'* 1219 AssY. From Stanfield (Nf, St), or 'dweller by the stony field', OE *stān, feld*.

**Stanford, Stanforth:** William *de Stanford* 1190–1 P (St); Thomas *de Stanford* 1252 IpmY; William *Staunforde* 1327 SRLei; John *Stanford* 1332 SRWo. From Stanford (Beds, K, Nf), Stanford Dingley, in the Vale (Berks), Stanford le Hope, Rivers (Ess), Stanford Bishop, Regis (He), Stanford on Avon (Nth), Stanford upon Soar (Nt), or Stanford in Teme (Wo).

**Stanger:** Whilst not common, the name seems to be more frequent in Kent and Sussex than in Yorks and Lancs. The northern surname is from Stanger in Embleton or Stangrah in Whitbeck (Cumb): Robert *de Stangre* 1332 SRCu, or, possibly, from Stanghow (NRYorks), all from ON *stǫng* 'pole'. In the south, from residence near a stony gore: Jordan *de Stangar* 1327 SRSo; Henry *Stangor* 1327 *SR* (Ess). There was a place (*to*) *stangare* in the bounds of Upminster in 1062 (PN Ess 131), OE *stān* 'stone', *gāra* 'triangular piece of land'.

**Stangrave:** Johanna *de Stangrave* 1250–1, John *de Stangrave* 1278–9 FFSr. 'Dweller by the stony grove', OE *stān, grāf*.

**Stanham, Stannum:** Robert *de Stanham* 1177 P (Sf). From Stonham (Suffolk).

**Stanhill, Stonhill, Stonnill:** (i) Robertus *filius Stanilde* 1095 Bury (Sf); *Stonilda* 1185 Templars (K); Adam *filius Stanilde* 1198 FFSf; Robert, Thomas *Stonild* 1305 FFSf. An unrecorded OE woman's name *Stānhild* 'stone-war'. (ii) Henry *atte Stonhill* 1305 PN Sr 154. 'Dweller by a stony hill', OE *stān, hyll*.

**Stanhope:** Swayn *de Stanhop* 1242 AssDu; William *Stannop* 1327 SRSf; Richard *Stanhope* 1427 IpmNt. From Stanhope (Durham), or 'dweller at the stony valley'.

**Stanhurst:** Hugo *de Stanhurst* 1230 P (K). From residence near a stony, wooded hill (OE *stān, hyrst*).

**Stanier:** *v.* STONEHEWER

**Staniford, Staniforth, Stanyforth:** Matild' *de Staniforþe* 1297 SRY. 'Dweller by a stony ford', probably near Sheffield (Yorks) where the surname is common.

**Staniland:** William *atte Stanylond* 1272 AssSt. 'Dweller by the stony land.'

**Stanistrete:** *v.* STONESTREET

**Stanlake:** Thomas *de Stanlac'* 1221 Cur (Sf). 'Dweller by the stony stream.' OE *stān, lacu*.

**Stanley, Stanly, Stanleigh:** Robert *de Stanleya* 1130 P (St); Alan *de Stanlai* 1208 P (Y); Hugh *de Stanleye* 1323 IpmNt; Richard *Stanley* 1398 IpmGl. From Stanley (Derby, Durham, Glos, Staffs, Wilts, WRYorks).

**Stanlow:** Ralph *de Stanlowe* 1327 SRLei; William *de Stanlow* 1442 IpmNt. From Stanlow (Ch).

**Stanmore:** *v.* STAMMER

**Stannah:** *v.* STANNER

**Stannard, Stonard, Stonhard, Stonner, Stannett:** *Stanhard, Stanhart, Stanardus, Stanart* 1066 DB; *Stannard* 1095 Bury (Sf); *Stanardus* ie Couhirde 1327 SRC; Ralph *Stanhard* 1221 *ElyA* (Sf); Richard *Stanhert* ib.; Henry, Walter *Stonhard* 1222 DBStP (Herts), c1250 Colch (Herts); Laurence *Stannard* 1327 SRC. OE *Stānheard* 'stone-hard', a late name, common after the Conquest, especially in the east.

**Stannas, Stanners, Stannis, Stannus:** Robert *de Stanehouse* 1275 RH (Nb). 'Dweller at the stonehouse.'

**Stanner, Stannah:** Nicholas *de Stanhoe* 1146–8 MedEA (Nf). From Stanhoe (Norfolk).

**Stannett:** *v.* STANNARD

**Stanney:** *v.* STANEY

**Stannum:** *v.* STANHAM

**Stansfield, Stansfeld:** Richard *Stanesfeld* 1275 RH (Y); William *de Stanesfeld* 1330 FFEss; Edward *Stanesfeld* 1523 CorNt. From Stansfield (Suffolk, WRYorks).

**Stanstead, Stansted:** Simon *de Stanstede* 1178 CartAntiq; Matthew *de Stansted* 1231–2 FFK; Peter *de Stanstede* 1257–8 FFEss. From Stanstead (Sf, Sx), Stanstead Abbots, St Margaret (Herts), Stansted (K), or Stansted Mountfitchet (Ess).

**Stanton, Staunton:** Godwine *æt Stantune* c1055 OEByn; Walter *de Stanton'* 1199 P1 (Nt); John *de Staunton* 1290 FFO; John *Stanton* 1395–6 FFSr; Robert *Staunton* 1445–6 IpmNt. From one or other of the many places called Stanton or Staunton, all 'enclosure on stony ground'.

**Stanway:** Hawise *de Stanweie* 1206 P (Wa); Johanna *de Stanweye* 1327 SRSx; Richard *Stanewey* 1380 LoCh. From Stanway (Ess, Gl, He, Sa).

**Stanyer:** *v.* STONEHEWER

**Stanyforth:** *v.* STANIFORD

**Stap:** William *a la Stappe* 1275 SRWo; Robert *atte Stappe*, Robert *Stappe* 1332 SRSx; OE *stæpe*, a sideform of *stepe* 'step'. Probably 'Dweller by the stepping-stones or footbridge.' *v.* MELS 198.

**Stapeley, Stapely, Stapley:** Adam *de Stapelea* 1190 P (Ha); John *de Stapeleg'* 1221 Cur (Beds); John *Staplee* 1401 IpmY. From Stapeley (Ch), or Stapely (Ha).

**Staple, Staples:** Walter *de Stapel'* 1275 SRWo; Osmund *atte Staple* 1279 PN Sr 253; Richard *de Staples* 1321 FFEss; John *Stapel* 1325 ib. 'Dweller by a post or posts', OE *stapol* 'post, pillar', as at Staple (Kent) or Staple Fitzpaine (Som).

**Stapleford, Stableford, Stableforth:** Alfgar, Richard *de Stapleford* 1177 P (Ess), 1277 IpmW; John *de Stapulford* 1369 IpmNt; William *Stapulford* 1437–8 IpmNt. From Stapleford (Cambs, Ches, Essex, Herts, Leics, Lincs, Notts, Wilts).

**Stapler:** Richard *Stapeler* 1327 SRSx; Richard *Stapelere* c1405 FS. John *Stapler* 1662 HTEss. 'Dweller by the post', from a derivative of OE *stapol* 'post'.

**Stapleton, Stapylton:** Randulf *de Stapeltuna* 1166 P (Y); John, Robert *de Stapilton'* 1327 SRLei; William *Stapulton* 1460 IpmNt. From Stapleton (Cumb, Glos, Hereford, Leics, Salop, Som, NR, WRYorks).

**Stapley:** *v.* STAPELEY

**Starbuck:** Robert *Starbok'* 1379 PTY. From Starbeck in Harrogate (WRY).

**Starey, Starie:** Aluric *Stari* 1066 DB (Sf); John *Starie* 1275 SRWo. ON *Stari*, a nickname from ON *stari* 'starling'.

**Stark, Starks:** Rannulf, William *Stark* 1222 Cur (Sf), 1314 FFEss. OE *stearc* 'firm, unyielding; harsh, severe'.

**Starkbayn:** Alicia *filia Starkbayn*, Robert *Starkbayn* 1379 PTY. 'Strong bones', OE *stearc*, ON *beinn*. Used also as a christian name. *v.* also Thomas *Starkbaynson*, Margaret *Starkbayndoghter* 1379 PTY. cf. Alexander *Starkweder* 1327 SRSf 'cruel weather'.

**Starkey, Starkie:** Richard *Starky* 1260 AssCh; Lawrence *Starky* 1446 AD v (Ch). A diminutive of STARK. cf. 'Starky, stiff, dry. Westmorland' (Halliwell).

**Starkman:** William *Starckeman, Starcman* 1279 RH (C). 'Strong, stern man.' cf. STARK.

**Starling, Sterling:** *Starlingus, Starlinc* 1066 DB (Sf); Willelmus *filius Sterling* 1133–60 Rams (Hu); Jordan, Wimund *Starling* 1166 P (C), 1203 Cur (Sf); Richard *Sterling* 1230 P (Herts). A nickname from the bird, OE *stærling* 'starling', used also occasionally as a personal name.

**Starman:** Margeria *Starremon* 1327 SRWo. 'Servant of *Star*.'

**Starr:** Leuenot *Sterre* 1066 DB (Db); Simon *Sterre* 1130 P (Nf); William *Sturre, Sterre* 1221 AssGl; John *Sterre, Starre* 1305–6 LLB B. OE *steorra*, ME *sterre* 'star', used, like the ON *Stjarna*, as a nickname, but also, occasionally, as a personal name: *Sterre* 1066 DB (Ha). Occasionally a sign-name: Richard *ate Sterre* 1322 LLB E.

**Start, Starte, Stert, Sturt:** Roger *de Lesturte*, Tomas *de Sterte* 1168, 1197 P (D); Walter *de la Sterte* 1225 AssSo; Richard *de Stirt* 1279 RH (Hu); John *Sterte* 1317 AssK; John *ate Sturte* 1332 SRSr. 'From residence near a promontory, tongue of land or a hillspur' (OE *steort* 'tail'), as at Start (Devon).

**Startifant, Sturdevant, Sturtevant, Sturtivant:** John, Richard, Thomas *Stirtava(u)nt* 1404 Pat (Y), 1413, 1447 FrY; Alan *Stertevaunt* 1445 GildY; John *Sturdyvaunt* 1570 Bardsley. 'Start forward', ME *sterten* and AFr *avaunt*, a nickname for a messenger or pursuivant. cf. forms of *Startup* and John *Startout* 1381 ArchC 4, Robert *Styrtover* 1320 Shef (Y).

**Startout:** William *Startowt* 1389 AD v (Sf); Thomas *Stertowte* c1500 PN Herts 54; William *Stertout* 1525 SRSx. 'Start out', OE *styrtan, ūt*, perhaps a nickname for a messenger. cf. John *Stertewey* 1208–9 FFWa 'start away'; Robert *Stertinthehegge* 1275 SRWo 'start in the hedge'.

**Startup:** Stephen *Sturthup* 1190 P (W); Geoffrey *Sturchup* 1219 Cur (R); William *Stercup* 1275 RH (Nf); Nicholas *Stirtupp* 1359 LLB G; Geoffrey *Startup* 1541 NorwW (Sf). 'Start up', probably with a meaning similar to that of *Startifant*, from OE *styrtan*, *\*steortan* 'to leap, jump' and *upp*. *Startup* was the original form of *upstart* and also the name of a kind of boot worn by rustics. These may both have contributed to the surname, but neither is recorded before the 16th century.

**Statham, Stathum:** Richard *de Stathum* 1413 DbCh; John, Henry *Stathum* 1450 IpmNt, 1488 Cl (La). From Statham (Ches).

**Statheman:** Philip *Statheman* 1314 AssNf. 'Dweller at the landing-place', OE *stæþ, mann*.

**Staunton:** *v.* STANTON

**Staveley, Stavely:** Adam *de Staueleia* 1183–99 YCh; Robert *de Staueleye* 1242 AssDu; Nicholas *de*

*Staveleye* 1275 SRWo. From Staveley (Db, La, We, WRY).

**Staxton:** Henry *of Staxton* 1240 FFY; Richard *de Staxtone* 1280, *de Staxton* 1290, IpmY. From Staxton (ERY).

**Stay:** Henry *le Stay* 1255 RH (W); Adam, Benedict *Stay* 1275 SRWo, 1332 SRCu. ME *staye* 'support, prop', used in the 16th century of persons.

**Staziker:** *v.* STIRZAKER

**Stead, Stede, Steed, Steede, Steeds:** (i) Vchtred *Stede* 1180 P (D); Henry *le Stede* 1281 Eynsham (O). OE *stēda* 'stud-horse, stallion'. As a surname, probably 'a man of mettle, of high spirit'. (ii) Richard *de Stede* 1276 AssLa; Roberd *del Stede* 1336 Calv (Y). OE *stede* of varied meanings; in ME used of 'a property or estate in land, a farm' (1338 NED). 'A farmer or farm-worker.'

**Steadfast:** William *Stedefast* 1296 NorwLt; Thomas *Stedfast* 1505 FFEss. OE *stedefæst* 'steady, firm'.

**Steadman, Stedman, Steedman, Stedmond:** Roger *Stedeman* 1275 RH (Hu); John *Stedemon* 1275 SRWo; Henry *le Stedeman* 1285 *Ass* (Ess); Robert *le Stedman, le Stedeman* 1323 AssSt. The first element of this compound may be ME *stede* in either sense found for STEAD, (i) a man responsible for the care of the war-horses or a mounted warrior, (ii) a farmer or farm-worker.

**Steady, Steddy:** William *Stedy* 1327 SRSx. ME *stedy(e), steadie* 'immovable, steadfast, firm' (1530 NED).

**Stear(s):** *v.* STEER

**Stearman, Sterman:** William *Stereman* 1202 AssL; Simon *Sterman* 1296 SRSx. OE *stēormann* 'pilot, master of a ship', or OE *stēor* 'steer' and *mann*, 'bullock-herd'.

**Stearn(s):** *v.* STERN

**Steavenson:** *v.* STEPHENSON

**Stebbens, Stebbins, Stebbing, Stebbings:** Edith *de Stebbing* 1207 FFEss; Thomas *Stebing* 1279 RH (C); Nicholas *de Stebbinges* 1283–4 FFEss. From Stebbing (Ess).

**Steckles:** *v.* STILE

**Stediford, Stedeford:** Richard *Stedefot* 1298 NorwDeeds I; Richard *Stedefote* 1301 SRY. 'Horse foot', OE *stēda, fōt*. cf. the numerous other animal and bird names with *-foot*.

**Stedman:** *v.* STEADMAN

**Steede:** *v.* STEAD

**Steedman:** *v.* STEADMAN

**Steel, Steele, Steels:** Walter, Robert *Stel* 1206 Oseney (O), 1278 AssLa; Jordan *le Stel* 1324 FFEss; Robert *Steel* 1327 SRSx. OE *stȳle, stēle* 'steel', probably for one hard, reliable as steel. cf. the common phrase 'true as steel' (a1300 NED).

**Steeling:** *v.* STELLING

**Steen, Steene:** Robert *le Steen* 1275 SRWo. Usually, and certainly in Scotland, a shortened form of *Stephen*. cf. Steene Feild alias Steven Field 1610 PN Db 419.

**Steeper:** Alice Relicta *le Stepere*, Robert *le Stupere* 1327 SRSx. A derivative of ME *stepe* (OE *stīepan*) 'to steep', one who steeps flax, cloth, etc., probably a bleacher.

**Steeping:** Robert *de Steping* 1308 AssNf. From Steeping (L).

**Steer, Steere, Steers, Stear, Stears:** Geoffrey *Ster* 1209 P (Wo); Robert *le Steer* 1296 SRSx. OE *stēor* 'a steer'.

**Steff:** Robert *Steff* 1524 SRSf. ME *stef* 'firm, unyielding, strong'.

**Steffan, Steffen, Steffens:** *v.* STEPHEN

**Steggall(s), Steggals, Steggel, Steggles:** *v.* STILE

**Stein, Steyn, Steyne, Stain, Staine, Staines, Stains, Stayne, Staynes:** *Stain, Stein, Sten* 1066 DB (Y, Ch, Sa); *Steyne* de Holton 1246 AssLa; Galfridus *filius Steyne* 1275 SRWo; Alfricus *Stein* 1155–66 Holme (Nf); Richard *Steyn* 1275 SRWo. ON *Steinn*, ODa *Sten*. The modern *Stein* is often of late German origin (OHG *stein* 'stone'). *v.* STAINES.

**Stekel:** *v.* STILE

**Stelling, Steeling:** William *Stelling* 1274 RH (So); Geoffrey *Stelling* 1275 RH (Nf); Robert *Stelyng* 1347 FFY. From Stelling (Nb, K).

**Stembridge:** Robert *Attestenebrugg* 1273 IpmGl; Anthony *Stembridge* 1664 HTSo. 'Dweller by the stone bridge', OE *stānen, brycg*.

**Stennet, Stennett:** Nicholas *Stennett* 1560 Pat (C); Mr *Stennett* 1674 HTSf. *Sten-et*, a diminutive of *Sten*, a pet-form of *Stephen*.

**Stepeney, Stepney:** Richard *Stuppeney* 1540 KentW. From Stepney (Mx).

**Stephen, Stephens, Steffan, Steffen, Steffens, Steven, Stevens, Stiven:** *Stefanus* 1066 DB; *Stephanus* capellanus 1134–40 Holme (Nf); *Steffan(us)* 1142–53 DC (L), 1200 Cur (O); Robert *Stephen* 1260 AssCh; Agnes *Stiven* 1279 RH (Bk); Roger *Stefne* 1283 Oseney (O); Alice *Stevenes, Stephenes* 1279 RH (Hu), 1346 ColchCt; John *Steuene* 1296 SRSx; Sibilla *ate Stevenes* 1332 SRSr. Greek Στεφανᾶς from στέφανος 'crown', a name found already in OE, but only as a monk's name. It became common after the Conquest.

**Stephenson, Stephinson, Steavenson, Steevenson, Stevenson, Stevinson, Steverson:** Adam *Steveneson* 1327 *SR* (Ess); John *Stephenson* 1395 Whitby (Y); Katherin *Steverson* 1656 Bardsley. 'Son of *Stephen*.'

**Steptoe, Stepto, Steptowe, Steptow:** William *Steptoe* 1665 HTO; John *Steptoe* 1674 HTSf. 'Tiptoe', OE *steppan, tā*, probably a nickname for one who treads lightly. cf. John *Stepesoft* 1260 Oseney 'step softly'.

**Sterling:** *v.* STARLING

**Sterman:** *v.* STEARMAN

**Stern, Sterne, Sterns, Stearn, Stearne, Stearns:** Henry, William *Sterne* 1279 RH (C), 1289 FFC. OE *styrne* 'severe, strict; uncompromising, austere'.

**Sterndale, Sterndall:** Robert *Sterndale* 1603 SRDb. From Sterndale (Db).

**Stert:** *v.* START

**Steuart:** *v.* STEWARD

**Steven, Stevens:** *v.* STEPHEN

**Stevenson, Stevinson:** *v.* STEPHENSON

**Stew, Stewer:** *Stew* is a rare surname, metonymic for (William *le*) *Steweman* 1327 SRSf, which itself is probably identical in meaning with (William) *Steur* (i.e. *Stewer*) 1279 RH (Bk). The first element is probably (certainly in the country) ME *stewe* 'fish-pond' (c1386 NED), from OFr *estui*. cf. 'Cuidam valetto custodienti *le Stewe* manerii Episcopi' (139. NED). Or, possibly, from OFr *estuve* 'a heated room used for hot air or vapour baths' (1390 NED). Hence, either 'keeper of the fish-ponds' or 'attendant in charge of public baths'.

**Steward, Stewart, Steuart, Stuart:** Rogere *se Stiwerd* 1100–30 OEByn (D); Alwinus *Stiward* 1148 Winton (Ha); Reginald *le Stiward* 1205 ChR (Do); Martin *Steuhard* 1275 RH (Nf); Phelippe *Styward* 1296 Black (Roxburgh); Henry *Steward* 1327 MEOT (L). OE *stīweard, stigweard* 'steward'. The common derivation 'keeper of the pig-sties' has no authority. *Stiȝ* is of uncertain meaning; the compound probably denotes a 'keeper of the house' (NED). The word had various meanings: 'an official who controls the domestic affairs of a household' (c1000 NED); 'an officer of the royal household' (a955). After the Conquest it was used as the English equivalent of OFr *seneschal*, of 'the steward of a manor' (1303) and of 'the manager of an estate' (a1386). The (Lord High) Steward of Scotland was the first officer of the Scottish King in early times; he had control of the royal household, great administrative powers, and the privilege of leading the army into battle. The office, described as *senescallatus Scotiae* in a charter of 1158, fell in to the crown upon the accession of Robert the Steward as Robert II, whence the name of the royal house of Stuart. According to Black, the spelling *Stuart* was a French spelling adopted by Mary, Queen of Scots, but he himself records the form in 1429. The final *t* is Scottish, *Stewart* occurring c1370–88 (Black) and in 1432 (NED), *Steuart* in 1504 (Black). Both in Scotland and in England the surname derives from the lesser offices. In Scotland the term was used of a magistrate originally appointed by the king to administer crown lands forming a stewartry (1432), but there, as in England, every bishop, earl and manor had a steward, and the surname is no proof of royal descent as James VI (who retained the old spelling) emphasized when he said that all Stewarts were not 'sib' to the King.

**Stewardson, Stewartson:** Osbertus *filius Stiwardi* 1148 Winton (Ha); William *le fiz le Stywarde* of the counte of Berewyk 1296 Black; Richard *Stewardson* 1710 Bardsley. 'Son of the steward.'

**Steyne:** *v.* STEIN

**Steynor:** *v.* STAINER

**Stezeker:** *v.* STIRZAKER

**Stibbard, Stibbards, Stifferd:** Basil' *de Stiberde* 1202 FFNf; James *de Stiberde* 1309 AssNf. From Stibbard (Nf).

**Stick:** Gilbert *Stikke* 1190 P (Y); Wido *Stik'* 1204 P (Nf). OE *sticca* 'a rod or staff of wood', 'a slender branch', probably a nickname for a tall, slender man.

**Stickbuck:** Stephen *Stykebuc* 1230 P (Wo); John *Stykebuk'* 1303 IpmY; John *Stykkebukke* 1344 AssSt. 'Stab the deer', OE *stician, bucca*, probably a nickname for a hunter. cf. Godric *Stichehert* c1095 Bury 'stab the hart'; William *Stykewynd* 1327 SRSa 'stab the wind'.

**Stickels, Stickells, Stickles:** Richard, Stephen *Stikel* 1194 Cur (Sf), 1327 SRC. OE *sticol* 'steep,

rough, difficult', used later (1615) for 'rough, bristly, stickle-haired'. *v.* also STILE.

**Sticker:** (i) Richard *Stiker* 1275 RH (He); John *le Stikkere* 1327 SRSo; John *Stykkere* 1428 FA (Sx). A derivative of OE *stician* 'to stick, kill', a nickname for a butcher. (ii) Aluuin *Stichehare* 1066 DB (Mx); William *Stichehare* 1166 P (Sx); Hugh *Stikehare* 1200 P (Sf). 'Kill the hare', OE *stician, hara*.

**Stickland:** Temperantia *Stickland* 1620 PN Do i 272; Henry *Stickland* 1662–4 HTDo. 'Dweller by the steep land', OE *sticol, land*.

**Stickley:** Walter *de Sticlea* 1195 P (O); Richard *de Stikelauwe* 1230 P (Nb). From Stickley (Nb).

**Stickney:** Alan *de Stikenay* 1202 AssL; Richard *de Stykeneye* 1336 CorLo; William *Stykenay* 1380 AssWa. From Stickney (Lincs).

**Stiddolph, Stidolph:** John *Stithulf*, Richard *Stidolf* 1313 *Ass* (K); Richard *Stydolff* 1525 SRSx; Bryan *Stiddole* 1564 ArchC xxi. OE *Stīðwulf* 'hard-wolf'.

**Stiff, Stiffe:** Robert *Stife* 1275 RH (W); William *Styf* 1332 SRWa; Gye *Stiffe* 1576 SRW. OE *stīf* 'rigid, firm'.

**Stifferd:** *v.* STIBBARD

**Stiffkey:** *v.* STUCKEY

**Stigand, Stigant, Stiggants, Stiggins:** *Stigand, Stigan* 1066 DB; *Stigandus* presbiter 1126–7 Holme (Nf); Radulfus *filius Stigant'* 1222 AssWa; William *Stigaunt* 1296 SRSx; Matthew *Stigan* 1419 LLB I; Anne *Stigans* 1706 Bardsley; Mary *Stiggins* 1747 ib. ON *Stigandr*, also used in Normandy. *v.* also STYAN.

**Stile, Stiles, Style, Styles, Steggall, Steggalls, Steggals, Steggel, Steggell, Steggle, Steggles, Steckel, Steckles, Stekel, Stickells, Stickels, Stickles, Stiggles, Stikel, Stygal, Stygall:** Reginald *atte Stighel* 1227 AssBeds; William *de Stile* 1229 Pat (So); Richard *del Stigele* ib.; Osmund *Atthe Stihele* 1234 MELS (So); Robert *de la Stiele* 1275 SRWo; Elyas *atte Stigle* Ed 1 MELS (Sr); William *ate Stegele* 1296 SRSx; Roger *Attestichell* 1301 SRY; John *del Steghel* 1327 Wak (Y); Isabella *Stiles* 1337 ColchCt; Edmond *Stegyll*, Margery *Stekyll*, William *Stykyll*, William *Steykell*, Richard *Stegold* 1524 SRSf; John *Steckles* 1674 HTSf. 'Dweller by the stile or steep ascent', OE *stigol*. The normal development is to *Stile*, but in surnames, *stigele* became *stigle* and the continuant *g* became a stopped *g*, hence *Stiggle*. As often, *gl* interchanged with *cl*, hence *Stickle*. ME *stegele* is from OE *\*steogol* and similarly developed to *Steggle* and *Steckle*.

**Stileman, Stillman:** Robert *Stegelman* 1270 HPD (Ess); John *Stighelman* 1327 SRSx; John *Stileman* 1327 *SR* (Ess). Identical in meaning with STILE, STILL. cf. STYLER.

**Still, Stille:** (i) Aluuinus *Stilla* 1066 DB (Ha); Lefwinus *Stille* 1166 P (Sx); Richard *le Stille* 1275 SRWo. OE *stille* 'still, quiet'. (ii) John *atte Stille* 1327 SRWo, 1332 SRWa; Reginald *atte Stylle* 1333 MELS (So). This may be for STILE, with a shortened vowel, but there seems to have been an OE *stiell, still* 'place for catching fish' or 'trap for wild animals', found in Stildon (Worcs). Hence, a name for a fisherman or trapper. *v.* MELS 200.

**Stiller:** Identical in meaning with STILL (ii).

**Stillingfleet:** Peter *de Steuelingflet* 1190 P (Y); Stephen *de Stivelingflet* 1204 AssY; George *Stillingfleete* 1662–4 HTDo. From Stillingfleet (ERY).

**Stillman:** *v.* STILEMAN, STILLER

**Stillwell, Stilwell:** William *Stilewell* 1324–5 FFSr; John *Stillwell* 1583 Musters (Sr). 'Dweller by the quiet stream', OE *stille, wiella*.

**Stilton:** Nicholas *de Stilton'* 1214 Cur (W); Richard *de Stilton'* 1255 ForHu; John *de Stiltone* 1336–7 CorLo. From Stilton (Hu).

**Stilwell:** *v.* STILLWELL

**Stimpson, Stimson, Stinson:** Edward *Stynson* 1539 Bardsley (Nb); Daniel *Stimpson* 1674 HTSf. 'Son of *Stephen*.'

**Stinton:** William *of Stinton* 1218–19 FFY; Adam *de Stinton* 1284 FFEss; Ralph *of Stinton* 1316 AssNf. From Market Stainton (L).

**Stirk:** William *Sturc, Stirc* 1202 AssL; Hugh *Stirc* 1208 P (Sf); Richard *Styrke* 1275 Wak (Y); Henry *Sterck* 1279 RH (C). OE *styrc* 'bullock'.

**Stirling:** Gilbert *Stirling* 1269 AssNb; John *Stirling* 1327 SRSo. The first example may be from Stirling (Scotland). Otherwise for STARLING.

**Stirrup, Stirrop, Sturrup:** Rannulf *de Stirap* 1200 P (Nt); John *Stirapp* 1268 AssSo; Simon *Stirop* 1327 SRSx. From Styrrup (Notts). The Sussex and Somerset examples, without the preposition, suggest that this was also, at times, a name for a stirrup-maker.

**Stirzaker, Sturzaker, Staziker, Stezaker:** William *de Steresacre* 1332 SRLa; Thomas *Styrsaker* 1620 Oxon (Leic); Robert *Sturzaker* 1664 Bardsley (La). From Stirzacre (Lancs).

**Stitch, Stych, Styche, Stitcher:** Philip *Sticher* 1235 Fees (W); William *Steche* 1296 SRSx; John *Stiche* 1327 SRSf. *Steche, Stiche* are probably for *atte steche, atte stiche*, from OE *stycce* 'piece of land' found in Cambridgeshire and Essex field-names, and in Essex dialect meaning 'a ploughing land'. Hence, probably, a man who owned or cultivated a 'stitch' of land.

**Stittle, Stuttle:** Wluuin *Stettel* 1095 Bury (Sf); Ralph *Stuttel* 1179 P (Nth); John *Stettel* 1319 *SR* (Ess). Probably, as suggested by Tengvik, an OE *\*styttel*, a diminutive formation from the Germanic stem *\*stu(t)-* 'to knock, strike against, push'. cf. dial *stot* 'a stupid, clumsy fellow', a likely nickname.

**Stiven:** *v.* STEPHEN

**Stiver:** Richard *Stivarius* 1189 Sol; William *Stiuur* 1242 AssDu; Nicholas *le Estivur* 1268 AssSo. Lat *stivarius* 'ploughman'.

**Stoakes, Stoaks:** *v.* STOKES

**Stoate:** *v.* STOTT

**Stobart, Stobbart:** *v.* STUBBERT

**Stobbe, Stobbs:** *v.* STUBBE

**Stock, Stocks, Stok:** John *de la Stokke* 1225 AssSo; Roesia *atte Stocke* 1275 SRWo; William *atte Stokkes* 1310 LLB D (Herts); Rose *atte Stocke(s)* 1316, 1325 FFEss. 'Dweller by the stump(s)' or, in the singular, perhaps also 'by a foot-bridge', OE *stocc* 'stock, trunk or stump of a tree'.

**Stockbridge, Stogbridge:** Thomas *de Stocbrugg'* 1279 RH (O); Richard *de Stogbrige*, Sibilla *de Stokbrig'* 1379 PTY. From Stockbridge (Ha),

Stockbridge in Bentley, in Morton, a lost *Stockbridge* in Spofforth (WRY), Stock Bridge in Owston (WRY), or Stock Bridge in Brantingham (ERY).

**Stockdale, Stockdill, Stogdale:** Alan *de Stokdale* 1332 SRCu; William *de Stokdale* 1379 PTY. From Stockdale (Yorks, Cumb).

**Stocken:** Robert *Stokyn* 1327 SRSf. 'Dweller by the stumps', from the weak plural of *stocc*. cf. STOCK and Stocken in Inwardleigh (Devon). *v.* PN D 151.

**Stocker:** Richard, Elena *le Stocker* 1275 SRWo, 1276 RH (Bk). Identical in meaning with *atte stocke*. *v.* STOCK.

**Stockett:** John *atte Stoket* 1311 FFSr; John *atte Stokette* 1320–1 FFEss; John *de la Stoket* 1322 MELS (Sr). 'Dweller at the clearing with the tree-stumps', OE *\*stoccet*.

**Stockfish:** Roger *Stokfisshe* 1342 LLB F. Synonymous with Luke *le Stockfysmongere* 1293 AD ii (Mx). 'Seller of stockfish.'

**Stockford:** Robert *de Stokeford* 1246 AssLa. 'Dweller by a ford marked by a stump' (OE *stocc, ford*).

**Stockill:** Tomas *de Stochill'* 1212 Cur (Y); Baldewin *de Stokeld* 1222 ib. From Stockeld in Spofforth (Yorks).

**Stockings:** Edmund *del Stocking* 1279 RH (Bk). 'Dweller on ground cleared of stubbs' (OE *\*stoccing*).

**Stockle, Stockleigh, Stockley, Stokeley, Stokely, Stokle:** Pagan *de Stockleye* 1279 RH (O); Robert *de Stokele* 1296 SRSx. From Stockleigh or Stokeley, common place-names.

**Stockman:** Emma *Stokeman* 1279 RH (O); Roger *Stokman* 1327 SRC. Identical with *atte stocke*. *v.* STOCK, STOCKER.

**Stockport:** Robert *de Stokeport* 1204 P (La); Robert *de Stokeport* 1246 AssLa. From Stockport (Ch).

**Stockton:** Herbert *de Stocton'* 1204 P (Sa); Richard *de Stoketon'* 1301 FFY; John *Stokton* 1465–6 FFWa. From Stockton (Ch, He, Nf, Sa, W, Wa, WRY), Stockton Heath (Ch), Stockton on Tees (Du), Stockton on Teme (Wo), Stockton on the Forest (NRY).

**Stockwell:** John *de Stokewell* 1195–6 FFSr; John *de Stokwel* 1297 SRY; John *Stokwell* 1434 FFEss. From Stockwell (Sr, So), or 'dweller by the footbridge over a stream', OE *stocc, wiella*.

**Stockwood:** Robert *Stokwode* 1327 SREss. From Stockwood (Do, So), or Stoke Woods in Stoke Canon (D).

**Stocky:** William *de Stochey* 1276 RH (L). From residence near an enclosure made with trunks or stumps (OE *stocc, (ge)hæg*), as at Stockey in Meeth (Devon).

**Stodart, Stoddard, Stoddart, Stodhart, Studart, Studdard, Studdeard, Studdert:** Vlfus *Stodhyrda* 1195 P (Cu); Geoffrey *Stodhurd'* 1219 Cur (Nth); John *the Stodhirde* 1286 AssCh; Richard *le Stodehard* 1332 MEOT (Y); Thomas *Stoderd* 1481 FrY; John *Stodard* 1482 ib. OE *stōd* 'stud (of horses)' and *hierde* 'herd', 'servant in charge of a stud', 'horse-keeper' (1458 NED).

**Stodden, Stoddon:** Hugh *de Stoddon'* 1200 P (D); Symon *de Stoddenn* 1296, Jacobus *de Stoddenne* 1332

SRSx. From Studding's Fm in Herstmonceux (Sx), and probably also from an unidentified place in Devon.

**Stoffer, Stopher, Stopper:** John *Stopper* 1471 GildY; Gyelles *Stofer* 1568 SRSf. A pet-form of *Christopher*.

**Stogbridge:** v. STOCKBRIDGE

**Stogdale:** v. STOCKDALE

**Stogdon:** Richard *de Stockdon'* 1274 RH (So); Robert *de Stokedon* 1281 PN D 415. From Stockadon (Devon) or residence by a 'hill with tree-stumps on it' (*stocc, dūn*).

**Stokeley, Stokely, Stokle:** v. STOCKLE

**Stoker:** Hugh *le Stoker* 1227 AssBeds. 'One who lived at Stoke', in this instance, Stoke Goldington (Bucks).

**Stokes, Stooke, Stookes, Stoakes, Stoaks:** Ricerus *de Stochas* 1084 GeldR (So); Cnut *de Stoch'* 1166 P (Db); William *Stoc* 1185 Templars (Wa); Ailwin *de Stokes* 1195 P (Nth). From Stoke (Som, Derby, Warwicks, Northants) or one of the other Stokes. Old forms vary between singular and plural.

**Stokesby:** Thomas *of Stokesby* 1316 AssNf. From Stokesby (Nf).

**Stokesley:** John *de Stokesley* 1347 FFY; John *de Stokesley* 1416 IpmY. From Stokesley (NRY).

**Stokoe:** Adam *de Stochowe* 1332 SRCu. From Stockhow (Cumb).

**Stolerman:** Robert *Stalwrthman* 1297 SRY. 'Sturdy, courageous man.' cf. STALLWOOD and STOLLERY.

**Stollard:** v. STALLARD

**Stoller:** Richard *le Stolere* 1343 LoPleas. A derivative of ME *stole* 'stole'. A maker or seller of stoles.

**Stollery:** Mrs, Michael *Stollery* 1674 HTSf, 1763 SfPR. Probably a dialectal pronunciation of STALLWORTHY, found in Suffolk in the 13th century.

**Stollybrass:** v. STALLIBRASS

**Stolworthy:** v. STALLWORTHY

**Stonard:** v. STANNARD

**Stone, Stones:** Robert *Ston* 1212 Cur (O); Roger *del ston* 1277 Ely (Sf); Robert *atte Stone* 1296 SRSx; Elias *atte Stonis* 1327 SRSf; John *in le Stones* 1332 SRSt; William *del stones* 1348 DbAS 36. 'Dweller by the stone(s) or rock(s)' or among the rocks. Walter *de Stanes* c1130 StCh came from Stone (Staffs) and Richard *de Stone* 1275 SRWo from Stone (Worcs). v. also ASTON.

**Stoneage:** Richard *de Stonhach* 1327 *SR* (Ess). From residence near a stone gate (OE *stān, hæcc*).

**Stonebreaker:** Henry *Stonebreker* 1380 SRSt. 'Stone-breaker', OE *stān*, and a derivative of OE *brecan* 'to break'. cf. Alice *Stondelvare* 1278–9 CtH 'dweller by or worker at a stone-quarry'; Henry *Stonthacker* 1363 DbCh 'stone thatcher', i.e. slater.

**Stonebridge:** Walter *de Stanbrugg* 1296 SRSx. 'Dweller by a stone bridge.'

**Stoneham:** v. STONHAM

**Stonehewer, Stonier, Stanier, Stanyer:** Walter *Stanhewer* 13th Kirkstall (Y); Thomas *Stonhewar* 1279 RH (O); John *Stonehewer* or *Stonier* 1605 Bardsley (Ch); Nathaniell *Stanyar* 1689 ib. OE *stān*

and a derivative of *hēawan* 'to hew'. 'Stone-cutter, quarrier of stone.'

**Stonehouse:** Alexander *de Stonhuse* 1257 ArchC iii; Ranulf *del Stonhouse* 1332 SRSt; Henry *atte Stonhouse* 1359 MELS (Sr). From Stonehouse (Devon, Glos) or from residence at a stone house.

**Stoneley, Stonly, Stonley:** Nicholas *Stonle* 1275 RH (L); Richard *de Stonley* 1330–1 FFWa; William *Stonley* 1340–1450 GildC. From Stoneleigh (Wa), or Stonely (Hu).

**Stoneman:** Richard *Staneman* 1327 *SR* (Ess). 'Dweller by the stone', or 'worker in stone', stone-mason. cf. STONEHEWER.

**Stonestreet, Stanistreet, Stanistrete:** Salomon *de Stonstret'* 1275 RH (K); Stephen *de Stonstred* 1276 RH (Berks); Elias *de Stonstrete* 1279 FFEss; Richard *atte Stanstrete* 1293 PN Sx 8. OE *stānstrǣt* 'a paved road', usually Roman. The Essex reference is to the Roman road to Colchester and that from Sussex to Stane Street.

**Stoney:** Amica *de Staneya* 1328 WhC (La). From residence near some gravelly low-lying land (OE *stān, ēg*).

**Stoney:** v. STANEY

**Stonham, Stoneham:** John *de Stanham* 1205 Pl (Sf); Roger *de Stonham* 1333 FFEss; John *Stoneham* 1525 SRSx. From Stonham (Suffolk), or Stoneham (Hants).

**Stonhard, Stonner:** v. STANNARD

**Stonhill, Stonnill:** v. STANHILL

**Stonier:** v. STONEHEWER

**Stonor, Stoner, Stoners:** John *de Stonore* 1323 CorNth; John *Stonere* 1376 *Ass* (Ess); Thomas *Stonor* 1472–3 FFSr. From Stonor (Oxon).

**Stoodley:** v. STUDLEY

**Stooke(s):** v. STOKES

**Stopes, Stoop, Stopps, Stops:** Roger *de Stoppes* 1276 AssLo; Robert *Stope* 1408 IpmY. Perhaps 'dweller by the post or boundary mark', ON *stólpi*, cf. Stoop Bridge in Ilford, PN Ess 99, or 'dweller by the pit or hollow', OE *stoppa* 'pail, bucket', in a topographical sense.

**Stopford, Stopforth:** Robert *de Stokeport* 1204 P (La); Thomas *Stoppforth* 1379 PTY; Henry *Stopford* 1669 FrY. The local pronunciation of Stockport (Ches).

**Stopham:** Ralph *de Stopham* 1249 AssW; Ralph *de Stopeham* 1254–5 FFSx; John *de Stopham* 1334 FFY. From Stopham (Sx).

**Stopher:** v. STOFFER

**Stoppard, Stopard:** For STOPFORD.

**Stopper:** John *Stopper* 1471 GildY. May be for either STOFFER or STOPFORD.

**Stopps, Stops:** v. STOPES

**Storah:** v. STORER

**Stordy:** v. STURDEE

**Storer, Storah, Storrar, Storror, Storrow:** William *le Estorur* 1309 Guisb (Y); Thomas *le Storer* 1332 SRCu; Adam *le Storour* 1357 AssSt; John *Storrer* 1501 GildY; William *Sturror* 1534 Black; Bessie *Storrar* 1732 ib. A derivative of ME *storen* 'to store', from OFr *estorer* 'to build, establish, furnish, stock', or of the corresponding noun ME *stor*, OFr *estor* 'store'. 'Storer', perhaps 'keeper or overseer of the provisions for a household' (1540 NED), or, as suggested by Bardsley, 'wool-storer, warehouseman'. In Scotland, the tenant of a sheepfarm was called a 'storemaster' and the *storour* had charge of the flocks and herds. For *Storah, Storrow*, cf. FARRAR and FARROW.

**Storeton, Storton:** William *de Storgheton* 1327 SRSx. From Storrington (Sx), *Storgetune* DB. *v.* also STOURTON.

**Storey, Storie, Story, Storrie, Storry:** *Stori, Estori* 1066 DB; *Storicus* de Wycham 1132–60 Miller (C); Reginaldus *filius Story* 1219 AssY; Alexander *Story* 1248 FFEss; William *Stori* 1281 Black (Dundee). ON *Stóri*.

**Stork:** Osbert *Storc* 1198 P (K); Hugo *Stork* 1229 Pat (L); Robert *le Stork'* 1280 AssSo. OE *storc* 'stork', a nickname for a man with long legs.

**Storm, Stormes, Storms:** William, John *Storm* 1206 Cur (Nf), 1297 MinAcctCo (Y). OE *storm* 'storm'. cf. TEMPEST.

**Storr:** Geoffrey, John *Stor* 1200 P (Nf), 1290 LLB A. ON *stórr* 'big'.

**Storrar, Storror, Storrow:** *v.* STORER

**Storrie, Storry:** *v.* STOREY

**Storrs:** Matthew *de Stordes*, Robert *de Storthes*, Henry *del Storres* 1275, 1284, 1316 Wak (Y); Anthony *Storres* 1564 ShefA. 'Dweller by the brushwood or young plantation' (ON *storð*), as at Storrs (La) and High Storrs (Sheffield).

**Story:** *v.* STOREY

**Stote:** *v.* STOTT

**Stothard, Stothart, Stothert, Stuttard:** William *Stothard* 1279 RH (C); John *le Stothirde* 1297 SRY; Thomas *le Stothurd* 1306 AssSt; John *Stotard* 1317 AssK; Thomas *le Stotherd* 1327 SRSa. OE *stott* 'an inferior kind of horse', in ME also 'steer, bullock', and *hierde*, 'keeper of horses or bullocks' or 'oxherd'.

**Stott, Stote, Stoat:** Gamel *Stot* 1166 P (Y); John *Stotte* 1296 SRSx; Elena *la Stott* 1312 ColchCt. ME *stott* 'bullock'. cf. Agnes *Stotwylde* 1436 NorwW, perhaps 'wild as a young bullock'.

**Stoughton:** Richard *de Stoghton* 1327 SRLei; John *Stoughton* 1375 IpmGl; Gilbert *Stoughton* 1498–9 FFSr. From Stoughton (Lei, Sr, Sx).

**Stourton, Storton, Sturton:** Geoffrey *de Sturton* 1197 P (Wa); William *de Storton* 1264 Eynsham; Richard *de Stourton* 1332 SRDo. From Stourton (St, W, Wa).

**Stout, Stoute, Stoutt, Stutt:** (i) Osbert *Stute, Stutte* 1190–1 P (Y); William *Stutte, Stute, le Estut, Stut* 1219 AssL, 1221 AssWo; William *Estoute* 1327 SRSx; Adam *Stout* 1373 ColchCt. OE *stūt* 'gnat' or ME *stout*, from OFr *estolt, estout* 'stout, bold'. (ii) Henry *atte Stoute* 1330 PN D 652. Henry lived at Stout Fm in Yarcombe (Devon). 'Dweller by the rounded hill', OE *\*stūt*.

**Stoven, Stovin:** Simon *de Stouene* 1194 Cur (Sf); John *Stoven* 1576 SRW. From Stoven (Sf).

**Stow, Stowe:** Wlnoþus *de Stoue* c975 LibEl (C); Osbert *de Stowa* 1190 P (C); William *de la Stowe* 1315 MELS (Wo). From one of the many Stows. In Worcestershire, OE *stōw* 'place, holy place' was still

in use in ME, and here the surname is probably 'dweller by the monastery or church'. *v.* MELS 203.

**Stowar:** *v.* STOWER

**Stowell:** Henry *de Stowell'* 1242 Fees (Berks); Geoffrey *de Stawell* 1270 Glast (So); Francis *Stowell* 1641 PrSo. From Stowell (Gl, So, W).

**Stower, Stowers, Stowar:** (i) William *filius Stur* p1162 P (Ha); William *Stur* p1162 Seals (Ha); Anger *Sturre* 1206 Pleas (Berks); Adam *Sture* 1642 PrD; Henry *Stower* 1662–4 HTDo. OSw *Sture*. (ii) William *atte Sture* 1332 FFEss. From the River Stour (Ess).

**Stoyle, Stoyles:** William *Stoile* 1194 P (Db); Eliot *Stoyle* 1210 Cur (K). ME *stoyle*, OFr *estoile* 'star'.

**Strachan, Straghan, Strahan, Straughan:** Waldevus *de Stratheihan* c1200 Black; David *Straughin* 1512 ib. From Strachan (Kincardineshire).

**Strachey, Strachie, Stracey, Stracy:** Thomas *Strachy* 1508 FFEss; William *Strachye* 1561 Pat (C); Thomas *Stracie* 1641 PrSo. The forms are late, and are probably corruptions of some other surname. Perhaps for TRACEY, with an inorganic initial *s*. cf. STURGE.

**Stradbrook, Stradbrooke:** Henry *de Stradebrok* 1318 FFEss. From Stradbroke (Sf).

**Stradling:** William *Stradlinge* 1642 PrD. From Strättligen near Thun. *v.* Wagner 7, 211.

**Stradwick:** *v.* STRUDWICK

**Strafford:** Athelstan *de Straford* 1210 Cur (Bk). A variant of STRATFORD.

**Straight:** Godwin *Streit* 1203 FFL; Richard *Streht* Hy 3 AD i (Wo). ME *streʒt* 'not crooked, upright', probably 'erect'.

**Strainge:** *v.* STRANGE

**Straker:** Robert *le Straker* 1246 AssLa; Robert *Stracour* 1332 SRCu; William *Strakour* 1327 SRY. A derivative of ME *strake*, from OE *strācian* 'to stroke', related to OE *strīcan*, hence a variant of *Striker*. *v.* STRIKE.

**Stranders:** Ralph *Stronder*, John *atte, del Stronde* 1327 SRSf, *SR* (Ess). 'Dweller by the strand or shore', OE *strand*.

**Strang:** *v.* STRONG

**Strange, Strainge, L'Estrange, Lestrange:** John *Lestrange, Leestrange*, 1192 P (Nf), 1195 FFNf; Ralph *le Estrange* 1199 CurR (Sf); Hugh *le Strange* 1221 AssSa; Fulco *Strange* 1221 ElyA (C). ME *strange*, OFr *estrange* 'foreign', stranger, newcomer.

**Strangeway, Strangeways, Strangways, Strangwick:** James *Strangwishe, Strangwych, Strangwesh*, c1450, 1467, 1489 GildY; Anna *Strangwish* 1501 ib.; Richard *Strangeways* 1513 ib.; Martin *Strangways* 1527 ib. From Strangeways (Lancs).

**Strangman, Strongman:** *Strangman* c1095 Bury (Sf); Robert, John *Strangman* 1327 SRSf, *SR* (Ess). OE *\*Strangmann* 'strong, bold man'.

**Stransham, Stransom:** Thomas *Stransham* 1584 ArchC 75. From Strensham (Wo).

**Stratfield, Stratful, Stratfull:** Thomas *de Stratfeld* 1204 P (Ha). From Stratfield Mortimer (Berks), or Stratfield Saye, Turgis (Ha).

**Stratford:** Robert *de Stratford* 1086 DB (Sf). From one of the Stratfords.

**Stratton:** Richard *de Stratton* 1199 Pl (Do); Peter *de Stratton'* 1228 Cur (Beds); John *Stratton* 1366

IpmGl. From one or other of the many places of this name. Mainly southern.

**Straughan:** *v.* STRACHAN

**Strawbridge, Strowbridge, Strobridge:** John *Strowbridge* 1559 Pat (D); John *Strobridge* 1662–4 HTDo. From Strawbridge in Hatherleigh (D).

**Streake, Streek:** Godwin, Walter *Streke* 1176 P (Sr), 1275 SRWo. OE *stræc, strec* 'strong, violent'.

**Stream, Streamer:** Henry *ate Streme* 1279 RH (O). 'Dweller by the stream.'

**Streat:** *v.* STREET

**Streater:** *v.* STREETER

**Streek:** *v.* STREAKE

**Street, Streete, Streets, Streat:** Modbert *de Strete* c1100–30 OEByn (D); Roger *Stret* 1197 P (Nf); William *de la Stret'* 1228 Cl (D); Richard *del Strete* 1275 AssSo; Adam *of the Strete* 1284 AssLa; Reginald *atte Strete* 1309 LLB D. OE *strāt* 'street, Roman road'. From Street (Hereford, Kent, Som), all of which are on or near Roman roads, or local, from residence in the village street as opposed, say, to the village green, or in a hamlet, called in some parts of the country a street.

**Streeten, Streetin:** Roger *de la Stratend* 1261–72 MELS (Sr), *de la Stretende* 1262 ib. (Sx); John *Strethende* 1296 SRSx; Robert *ate Stretende* 1297 MinAcctCo. 'Dweller at the end of the street.' cf. STREET.

**Streeter, Streater:** John *Streter* 1332 SRSx. Identical in meaning with *atte Strete*. *v.* STREET.

**Streit:** Roger *Streit* 1367 ColchCt. ME *streit*, AFr *estreit* 'narrow, strict'.

**Strelley, Strelly:** Ralph *Strelley* 1325 FFEss; Sampson *Strelley* 1454 IpmNt; Nicholas Strelley 1599 SRDb. From Strelley (Nt).

**Strensall:** John *Strensall* 1426 TestEbor. From Strensall (NRY).

**Stretch:** Richard *Estrech', le Strech'* 1176 P (Wo), 1221 AssWo; Adam *Strecche* 1210 P (Ha). OE *stræc, strec* 'strong, violent'.

**Stretford:** Thomas *Stretford* 1419 IpmY. From Stretford (He, La).

**Stretton, Stretten:** Hervey *de Strettona* c1165 StCh; Hugh, Robert *de Stretton'* 1230 Cur (Nt), 1327 SRLei; John *Stretton* 1421 IpmY. From one or other of the many places of this name. Usually northern or midland.

**Stribling:** *v.* STRIPLING

**Strick:** Wulnoð *Strices sune* 972 OEByn (Nth); Alestan *Strick* 1066 DB (Ess); Richard *Stric* 1170 P (Y); William *Stric* 1274 RH (Ess). OE *Stric*, ON *Strikr*. *v.* OEByn 337.

**Strickland:** Vchtred *de Stirclanda* 1193 P (We); William *de Strikeland* 1278 IpmWe; Walter *Strykland* 1442 AssLo. From Strickland (Westmorland).

**Stride:** Adam *atte Stryd* 1296 MELS (Sx); Isaac *Stride* 1642 PrD; Sarah *Stride* 1662–4 HTDo. OE *stride* 'stride, pace', probably used of a place where one can stride a brook. cf. The Strid, Bolton Abbey (WRY).

**Strike, Striker:** Nicholas *Stryke* 1296 SRSx; Reginald *le Strikere* 1297 Coram (L); Thomas *Strikere* 1360 FFSf. One of the clauses of Magna

Carta provided that there should be one measure of corn, namely the London quarter. Local custom, however, was stronger than national law and there was a constant struggle between law and custom over the bushel and quarter. By statute the quarter should contain 8 bushels, each of 8 gallons, 'rased' or 'striked', that is to say, filled and levelled with the brim of the measure by passing a flat stick, or 'strike', over it. By custom, however, the quarter consisted of 8 'heaped' bushels, which were equivalent to 9 'striked'; and when this was prohibited the corn-dealers got round this prohibition by a measure of 9 striked bushels, which they called a 'fat' or 'vat' (METrade 43–47). The official responsible for the accuracy of the measure was clearly called a 'striker' and was sometimes named from the instrument he used.

**String:** William, Henry *Streng* 1177 P (Sa), 1275 RH (Sf). Metonymic for STRINGER.

**Stringer:** Walter *Stringere* 1194 Cur (W); Roger *le Strenger* 1293 MESO (Y). Derivative of OE *streng* 'string, cord', maker of strings for bows, a stringer (1420 NED), a common Yorkshire name.

**Stringfellow:** William *Strengfellow* 1286 AssCh; John *le Strengfelagh* 1308 Wak (Y); John *Stryngfelowe* 1489 Calv (Y). ME *streng* and OE *fēolaga*, ON *félagi* 'partner, fellow'. cf. the rare OE *strenge* 'severe'; *strenghefully* 'with might or power' 13.. NED. But *streng* may be for *strang* 'strong', with *e* on the analogy of the comparative *strengra* and OE *strengu*, ME *strenge* 'strength'.

**Stripe, Stripp, Strype:** Robert *Strips* 1276 RH (C); Henry *Strype* 1279 RH (O); John *Strippis* 1383 AssL. 'Dweller by the strip of land', OE *strīp*.

**Stripling, Strippling, Stribling, Stribbling:** Symon *Stripling* 1259 RamsCt (Hu); William *Strupelynge* 1296 SRSx. ME *stryplynge* 'a youth' (1398 NED).

**Strobridge:** *v.* STRAWBRIDGE

**Strode, Strood, Stroud:** Aluina *de Strodes* 1206 FFK; Thomas *de la Strode*, William *Strodde* 1230 P (D); William *atte Strode* 1275 SRWo; Edytha *atte Stroude* 1327 SRSo. 'Dweller by the marshy ground' (OE *strōd*) as at Strood (Kent) or Stroud (Glos).

**Strong, Stronge, Strang:** Richard *Stronge* 1185 Templars (Wa); Ralph *le Esstrang* 1228 FFSf; William *le Stronge* 1276 AssSo; Adam *Strang* 1379 PTY. OE *strang*, ME *strong* 'strong'. *Strang* was preserved in the north.

**Strongbow:** Hugh *Strangboge* 1182 P (Ha); Ranulph *Strongbowe* 1275 RH (Ess); Simon *Strongbow* 1395 AssL. 'Strong bow', OE *strang, boga*. A nickname for a good archer.

**Strongitharm:** John *Strongharme* 1379 FrLeic; Roger *Strongeitharme* 1581 Bardsley (Ch); Richard *Stronge in Arme* 1597 ib. Self-explanatory.

**Strongman:** *v.* STRANGMAN

**Strother, Strothers, Struther, Del Strother:** Henry *del Strothre* 1363 Misc (Nb); John *Strother* 1459 TestEbor; Ninian *Strodre* 1539 FeuDu. From Strother (Du), or 'dweller at the place overgrown with brushwood', OE *strōðer*.

**Stroud:** *v.* STRODE

**Stroulger, Strowger, Strudger:** William *Strowgeor* 1504 NorwW (Nf); John *Strowger* 1568 SRSf; Robert

*Stroger, Strowger* 1654 EA (OS) iv (Nf). Perhaps 'astrologer', from a derivative of OFr *astrologe*.

**Strout, Strouts:** v. STRUTT

**Strowbridge:** v. STRAWBRIDGE

**Strowger, Strudger:** v. STROULGER

**Strudwick, Strudwicke, Stradwick:** Richard *Strudwick* 1524 SRSf; Richard *Strudewyke* 1546 FFEss; John *Strudwicke* 1583 Musters (Sr). From Strudgwick in Kirdford (Sx).

**Strumanger, Struminger:** Thomas *le Strumonger* 1285 Oseney (O); Richard *straumongere* 1294 MEOT (Herts). OE *strēaw* and *mangere*, 'dealer in straw'.

**Struther:** v. STROTHER

**Strutt, Strout, Strouts:** Ailric *Strut* 1188 BuryS (Sf); Walter *Strut* 1242 Fees (Nf); William *Strout* 1327 SRC; John *Strout*, Zachary *Strutt* 1642 PrD. ON *Strútr*, ODa *Strut*.

**Strype:** v. STRIPE

**Stuart:** v. STEWARD

**Stubbe, Stubbs, Stobbe, Stobbs:** (i) Ælfeah *Stybb* c1000 OEByn; Richard *Stubbe* 1185 Templars (Y); William *Stob* 1332 SRSx. OE *stybb* 'stub', a nickname for one of short, stumpy stature. (ii) Geoffrey *de Stubbes* 1199 P (Nf); Robert *del Stobbes* 1288 AssCh; Thomas *de la Stubbe* 1296 SRSx; Richard *atte Stubbe* 1327 ib. 'Dweller by the tree-stump(s)', OE *stubb*.

**Stubber:** Jordan *Stubbere* 1181–91 StP (Lo); William *le Stubber* 1327 SRSx; William *Stubber* 1393 FFEss. A derivative of ME *stubben* 'to dig up by the roots', OE *stubb* 'tree-stump'. Hence 'one who grubs up roots or removes stumps from the ground'.

**Stubbert, Stobbart, Stobart:** *Stubart* 1066 DB (Sf); *Stubhard* Pape c1095 Bury (Sf); Simon *filius Stubbard* 1203 Cur (Nf); Symon *Stubard* 1275 RH (Nf); Richard *Stobard* 1336 ColchCt. OE *Stubheard*, a late formation.

**Stubbing, Stubbings, Stubbins:** (i) William, Astin *Stubbing* 1191 P (Nf), 1219 AssY; Richard *Stubin* 1279 RH (O). OE *stybbing* 'the stumpy one'. v. STUBBE. (ii) Richard *del Stubbyng* 1297 MinAcctCo. 'Dweller by the cleared land', ME *stubbing*.

**Stubbus:** Hugh *de Stubhus* 1228 Cl (Y); Henry *of Stubhuse* 1251 FFY; William *de Stubhus* 1263 IpmY. 'Dweller at the house by the tree-stumps', OE *stubb, hūs*. But the surname was probably usually absorbed by *Stubbs*, v. STUBBE.

**Stubley, Stubbeley, Stubbley:** William *de Stublegh* 1239–40 FFEss; William *de Stubleye* 1313 FFEss; John *Stubliegh* 1385 FFEss. From an unidentified minor place, presumably in Essex.

**Stuckey, Stukey, Stiffkey:** Iwyn *de Stiuekeye* 1108 MedEA (Nf); Geoffrey *de Stivekeye* 1212 Cur (Nf); John *Stukie*, Nicholas *Stuckey* 1642 PrD. From Stiffkey (Nf), pronounced *Stukey*.

**Studd, Studds:** Alnod *Stud* 1066 DB; William *Studd* 1296 SRSx; Peter *Stud* 1334–5 SRK; John *Studde* 1540 FFEss. Probably a nickname, and Tengvik (OEByn 367) suggests OE *stūt* 'gnat', with the *d* due to AN influence. A personal name is also possible, cf. OE *Stūt*, ON *Stútr*, while Walter *de la Stud* 1231 Cur (Sr) may have taken his name from OE *styde*, a by-form of *stede* 'site of a building, a farm'.

**Studdard, Studdeard, Studdert:** v. STODART

**Studder:** Bartholomew *Studdere* 1296 SRSx. A derivative of OE *stōd* 'stud', a worker at the stud.

**Studdy:** Hamo *de Stodhag* 1210 Cur (Y); Roger *Stody* 1275 SRWo. From Studdah in Fingal (Yorks), Stodday (Lancs), Stody (Norfolk), pronounced *Studdy*, or residence near, or employment at an enclosure for horses (OE *stōd*, *(ge)hæg*).

**Studham:** Nicholas *de Stodeham* 1135 Oseney; William *de Stodham* 1204 P (Beds); Adam *de Stodham* 1243 FFO. From Studham (Beds).

**Studholm, Studholme:** Richard *de Stodeholme* 1332 SRCu; Robert *Studholme* 1504 PN Cu 145. From Studholme in Kirkhampton (Cu). Probably often absorbed by STUDHAM.

**Studley, Studeley, Stoodley:** Gamel *de Stodlay* c1144–56 MCh; William *de Stoddleye* 1282 IpmW; Henry *Studley* 1373 IpmGl; James *Stoodly* 1662–4 HTDo. From Studley (O, W, Wa, WRY), Stoodleigh (D), or Stoodley in Holne (D).

**Studman:** Walter, Adam *Stodman* 1297 MinAcctCo, 1332 SRCu. OE *stōd, mann* 'one employed at the stud'.

**Stukey:** v. STUCKEY

**Stukeley, Stukely:** Nicholas *de Stiuecle* 1198 Pl (Hu); Geoffrey *de Stiuecleg'* 1224 FFO; Thomas *Stukele* 1296 SRSx; George *Stukeley* 1642 PrD. Usually from Stukeley (Hunts), *Stivecle* DB, but occasionally from Stewkley (Bucks), *Stiuecelea* 1183.

**Stumbles:** Gervase *atte Stumble* 1296 SRSx; Juliana *Stumbel* 1311 ColchCt. 'Dweller by the tree-stump' (OE *\*stumbel*). v. MELS 205 and cf. Richard *Stombelere* 1327 *SR* (Ess). v. BRIDGER and p. xv.

**Stump:** Martin, William *Stumpe* 1332 SRSx; Henry *Stompe, Stoompe* 1373 ColchCt. 'Dweller by the stump' (ME *atte stumpe*).

**Stunt:** Simon *Stunt* 1230 Pat. OE *stunt* 'foolish'.

**Sturdee, Sturdy, Stordy:** Richard *Estordet, Estordit* 1084 GeldR (W); Hugo *Sturdy* 1219 AssY, *Sturdi* 1230 P (Y); Geoffrey *Sturdy, Sturdi* 1220 Oseney, Oriel (O). OFr *estordet, estordit*, identical with *estourdi, estordi, esturdi* 'reckless, violent'; in ME 'impetuously brave, fierce in combat' (1297 NED).

**Sturdevant:** v. STARTIFANT

**Sturge, Sturges, Sturgess, Sturgis, Turgoose:** Hugo *filius Turgisi* 1086 DB (Sa); *Turgis* Hy 2 DC (L); *Thurgis* filius Owani 1221 AssWa; Henry *Turgis* 1210 P (W); Adam *Thurgis* 1279 RH (Beds); John *Sturgys* 1353 FFC; Joan *Sturge* 1379 ColchCt; Richard *Sturges* 1481 LLB L; John *Turgos* 1524 SRSf. ON *Þorgils*, ODa, OSw *Thorgisl*. 'Thor's hostage.' cf. SPASHETT and Scaldhurst, earlier *Caldhous* (PN Ess 180).

**Sturgeon:** William *Sturjon* 1281 Ipm (Cu); Richard *Sturioun (Storioun)* 1327 *SR* (Ess); William *Sturgeon* 1380 LLB H. A nickname from the fish, OFr *esturgeon*.

**Sturman:** (i) Hugolinus, Stefanus *Stirman* 1066 DB (Berks, Ha, Wa); Turchil *Stirman* Regis E(dwardi) 1066 DB (Wa); Matildis uxor *Sturmanni* 1179 P (Sf). ODa *styreman* 'steersman', probably also 'captain, master of a ship'. Edricus *Stirman* Regis Edwardi (1066 DB), called also **Rector navis**, was the commander of the sea and land forces of the Bishop of

Worcester in the service of Edward the Confessor (PNDB). The term was also used as a personal name and has, no doubt, been confused with STARMAN. (ii) William, Robert *Sturmyn* 1327 SRC, SRSf; John, Thomas *Stormyn* 1327 SRSf. This is probably *Sturmin*, a diminutive in *-in* of *Sturmi. v.* STURMEY.

**Sturmer:** William *de Sturmere* 1286 FFEss; William *Stormer* 1327 SRC. From Sturmer (Ess).

**Sturmey:** *Sturmidus* de Cotenham 1086 InqEl (C); *Sturmi* Hy 2 DC (L); *Sturmy* de Straton' 1201 Cur (So); Richard *Estormid, Estormit* 1084 GeldR (W); Richard *Sturmid, Sturmi* 1086 DB (Sr, Ha); Ralph *Turmit* 1086 DB (Nf); John, Richard *Esturmit* 1130 P (W, Sf); Henry *Esturmi* 1158 P (W); Hugo *Sturmi* 1192 P (Berks); Roger *Esturmy, le Esturmi, le Sturmy* 1236, 1242 Fees (Sf); Richard *Stormy* 1296 SRSx. We have clearly two different origins here, (i) a personal-name, OFr *Estourmi*, OG *Sturmi*, and (ii) OFr *estormi*, 'étourdi, troublé, accablé' (Godefroy). cf. STURDEE.

**Sturrey, Sturry:** Richard *Sturry* 1392 LoCh. From Sturry (K).

**Sturrock:** Robert *Stourok* 1275 RH (K); Walter *de Stonrok* 1283 LLB A; John *Stonrok* 1327 SRSx; Roger *ate Staurokk'* 1332 SRSr. 'Dweller by the high rock' (OE *stānrocc*), as at Stonyrock in Effingham (Surrey), the home of Henry *de la Stanrok* in 1241 (PN Sr 103) or at Starrock Green in Chipstead (Rocius *de Stonrocke* 1265 ib. 291).

**Sturrup:** *v.* STIRRUP.

**Sturt:** *v.* START

**Sturtevant, Sturtivant:** *v.* STARTIFANT

**Sturton:** *v.* STOURTON

**Sturzaker:** *v.* STIRZAKER

**Stutfield:** Rotbert *de Stutteuile* 1106 ASC E; Ralph *Stutville* 1260 AssC. From Estouteville or Etoutteville (Seine-Inférieure).

**Stutt:** *v.* STOUT

**Stuttard:** *v.* STOTHARD

**Stutter, Stutters:** Thomas *Stotere* 1327 SRSf. A derivative of ME *stut(te), stoten* 'to stutter'. 'Stutterer.'

**Stuttle:** *v.* STITTLE

**Styan, Styance, Styants:** Robertus *filius Stiand* 1230 P (Wa); Nicholas *Stiant* 1275 RH (W); John *Stion* 1712 FrY. An anglicizing of ON *Stigandr. v.* STIGAND.

**Styche:** *v.* STITCH

**Styer:** William *Styers* 1674 HTSf. 'One employed at the stye', a pig-herd (OE *stig(u)* 'stye') or 'dweller by the path' (OE *stīg* 'path'). cf. Roger *ate Stye* 1297 MinAcctCo.

**Stygall:** *v.* STILE

**Styler:** William *le stygheler* 1332 SRSr; Roger *Stylere* 1428 FA (Sx). Equivalent to STILEMAN.

**Styles:** *v.* STILE

**Suart:** *v.* SEWARD

**Such, Sutch, Souch, Zouch, Chuck, Chucks:** Alan *de Lachuche, la Zuche, de la Zuche* 1172, 1177, 1198 P (Nth, Sx); Roger *la Zuche, de la Suche, de la Soche* 1212 Fees (D), 1275 RH (D); Alan *La Sutche* 1243 AssSo; Roger *Suche* 1275 RH (W); Walter *le Chuck* 1296 SRSx; William *la Zouche, de la Souche* 1316 FA (Wa, Sx); Roger *Souche* 1316 FA (Lei). OFr *souche*

'tree-stump', identical in meaning with the English STUMP. The name may have been brought from some small French place named La Souche. cf. Ashby de la Zouch (Leics), held by Roger de la Zuche in 1200 (FF). *Chuck* is from the Norman-Picard form *chouque*.

**Sucker:** Thomas *le Sukkere* 1277 Ely (Sf); William *Souker* 1301 SRY. A derivative of OE *sūcan* 'to suck', a sucker, probably, as suggested by Fransson, 'a blood-letter'.

**Suckling:** John *Sokeling* 1195 P (Berks); Richard *Sukeling* 1253 Oseney (O); Mabilia *Sucling* 1283 SRSf. A double diminutive from OE *sūcan* 'to suck'. 'Suckling', a nickname.

**Sucksmith, Shucksmith:** 'A maker of plough-shares', OFr *soc*, OE *smiþ*.

**Suckley:** John *de Suckeleg'* 1221 AssWo. From Suckley (Wo).

**Suckspitch:** William *Sukespic* 1170–80, John *Sokespiche* 1420, *Suckbitch* 17th, *Sucpitch* 1768 Hoskins 105–9. If of French origin 'suck spice', OE *sūcan*, OFr *espice*, if of English origin 'suck fat bacon', OE *sūcan, spic*.

**Sudbury, Sudberry, Sudbery:** Hugh *de Suthberia* 1159 P; John *de Sudbury* 1359 IpmW; Henry *Sudbury* 1367 AssLo. From Sudbury (Derby, Middlesex, Suffolk).

**Suddaby:** *v.* SOTHEBY

**Suddell, Sudell:** Robert *de Sudale, de Suthdale* 1201–2 FFNf. 'Dweller in the south dale' (OE *sūð, dæl*).

**Sudden:** Richard *Soudain* 1219 AssY; Nigel *Soudein* 1231 Cur (Nf); Geoffrey *Soudan* 1327 SRY. OFr *soudain* 'sudden, quick'.

**Suddery:** *v.* SUTHERY

**Sudgrove:** William *Sudgrove* 1383 IpmGl. From Sudgrove in Miserden (Gl).

**Sudley:** William *de Sudlega* c1150–66 GlCh; Bartholomew *de Sudleye* 1274 RH (Gl); Eleanor *de Sudleye* 1361 IpmGl. From Sudeley Manor, or Soudley in East Dean (Gl).

**Sudran, Sudron:** *v.* SOTHERAN

**Suett:** *Siuuat(e)* 1066 DB (L, O); Gilbertus *filius Siwat'* 1180–1218 Rams (Hu); *Sygwat* Radbode 1275 RH (Nf); Gilbert *Siwate* 13th Rams (Hu); Geoffrey *Sewhat* (*Suat*) 1300 LoCt; John, Simon *Suet* 1327 SRDb; Richard *Sewet* 1568 SRSf. ON *Sighvatr*, OSw *Sighwat. Siuuate* reflects ON *\*Sighvati*.

**Suett:** *v.* SEWAT

**Suffield, Suffell, Suffill:** William *de Sudfeld'* 1191 P (Nf); Nicholas *de Sudfeld* 1208 P (Y); Roger *de Suffeld'* 1225 Cur (Nf). From Suffield (Norfolk, NRYorks).

**Suffolk:** Alexander, Edmund *de Suffolk* 1254–5 FFEss, 1301 CorLo; John *de Suthfolke* 1317 Glast (So). 'The man from Suffolk.'

**Sugden, Sugdon:** Adam *de Suggedone* 1327 SRSa; William *de Sugden* 1362 FFY; Richard *Sugden* 1672 HTY. From Sugden (WRYorks).

**Sugg:** Wadnod, Richard *Sugge* 1185 P (Ha), 1196–1237 Colch (Sf). OE *sucga*, ME *sugge* 'a bird', perhaps for *hegessugge* 'hedge-sparrow'.

**Suggate, Suggett, Suggitt:** *v.* SOUTHGATE

**Sulley, Sully:** John *de Sulleia* 1130 P (Gl); Ralph *de Sulleg'* 1221 AssWo; Walter *de Sullye* 1269 AssSo;

John *Sully* 1642 PrD. From Sully (Calvados), but late forms may be from Sudely (Gl).

**Sullivan**: v. O'SULLIVAN

**Sultan**: v. SOWDEN

**Sumeray, Sumray, Sumrie**: Roger *de Sumeri* 1086 DB (Ess); Adam *de Sumeri* 1160 P (C); Walter *Somery* 1296 SRSx. From Sommery (Seine-Inférieure).

**Sumerfield**: v. SUMMERFIELD

**Summarsell**: v. SUMMERSALL

**Summer, Summers, Sommer, Sommers, Somer, Somers**: Adam, Geoffrey *Sumer* 1203, 1205 P (Ess, O); William *Somer* 1275 SRWo, 1295 Barnwell (C); Isabella *Somerys* 1327 SRSo. A personal-name *Sumor* is the first element of Somersall (Db), Somersham (Sf), and, possibly, of Somersham (Hu), but it is unrecorded and there is no evidence for the use of such a personal name in DB or later. Christmas, Noel, Pentecost, etc., are specific times suitable for the commemoration of a birth, but that can hardly be said of a season like summer. These earlier explanations must, therefore, be rejected. The surname is probably identical with the Scottish SIMMER. In Scots, *somer* is 'a sumpter', applied to men and horses alike. Dauzat explains the French *Sommier* 'bete de sommer' as 'a muleteer'. The English name is probably also from OFr *somier* 'sumpter'. Alice *at Someres* 1327 SRSx must be a parallel to *Vickers* and *Parsons*, 'servant at the sumpter's'.

**Summerbee, Summerby, Summersby**: Thomas *de Sumardebi* Hy 2 DC (R); John *de Somerdesby* 1297 MinAcctCo; Walter *de Somerby* 1351 AssL. From Somerby (L, Lei).

**Summerfield, Summerfield, Sommerfield, Somerfield, Sommerfeld**: Herlewine *de Sumerfeld'* 1191 P (K); Geoffrey *of Sumerfeud* 1249 AssW; William *Sommerfeild* 1577 Musters (Nf). From Summerfield (W).

**Summerford, Somerford**: Martin *de Sumerford* 1222 Cur (So); William *of Sumerford* 1240–1 FFY; William *Somerforde* 1372 CorLo. From Somerford (Ch), Somerford Keynes (Gl), or Great, Little Somerford (W).

**Summerhill, Summerill**: William *Somerild* 1277 AssW; Nicholas *Somerhild* 1297 MinAcctCo; Richard *Somerhild* 1364 FFY. ON *Sumarhild* (f).

**Summerlad, Sommerlat**: *Sumerlet, Summerled(e)* 1066 DB (D, Sf, Hu, L); Adam *filius Sumerlad'* 1198 P (Y); Richard, Robert *Sumerlede* 1184 Gage, 1188 BuryS (Sf). ON *Sumarliðr, Sumarliði* 'summer warrior'. The personal-name was common in Scotland where it became Gaelic *Somhairle* and was corrupted by the chroniclers to *Sorli Marlady*, etc., whence the Scottish surnames *Sorley* and *M'Sorley*. v. Black.

**Summerland**: Adam *filius Sumerland* 1200 P (Y). An early corruption of SUMMERLAD.

**Summersall, Summersell, Summersall**: Jacobus *de Sumereshal'* 1200 Cur (Herts); William *de Somersale* 1316 AssSt. From Somersall (Derby), or from *Somersale*, the old name of Hyde Hall in Sandon (PN Herts 165).

**Summersby**: v. SUMMERBEE

**Summerscale, Summerscales, Summersgill, Summerskill, Somerscales**: John *de Somerscale(s)* 1379

PTY; John *Summersgill* 1803 Bardsley. 'Herdsman at the summer-huts' (ON *sumar, skáli*).

**Summersett**: v. SOMERSET

**Summerson**: John *Simmerson* 1667 RamptonPR (C); Ab. *Sumerson* 1674 HTSf. 'Son of *Sumer*', OE *Sumor*. cf. John *Somerswayn* 1327 SRSo 'servant of *Sumer*'.

**Summerton, Somerton**: Wulfwyn *æt Sumortune* 901 OEByn (So); Basilia *de Sumorton'* 1177 P (L); Beatrice *de Sumerton'* 1230 P (Nf); John *Somerton* 1479 Paston. From Somerton (L, O, So, Sf), or East, West Somerton (Nf).

**Summerville**: v. SOMERVELL

**Summerwell, Summerwill**: Jacobus *de Somerwille* 1275 RH (D); John *Somerwell* 1377 IpmGl; John *Sommerwill* 1642 PrD. From Summerwell in Hartland (D).

**Sumner, Sumpner, Somner, Simner, Simnor**: Robert *le Sumonur* 1199 Cur (Lei); Matthew *le Sumener* 1230 Pat (K); William *le Sumnir* 1279 AssSo; John *Somnour* 1327 SRC; William *Sumpnour* 1420 MEOT (La). AFr *somenour, sumenour*, OFr *somoneor, semoneor* 'summoner', a petty officer who cites and warns persons to appear in court (c1325 NED). For *Simner*, cf. SIMNER and SUMMER.

**Sumpter, Sunter**: Roger *le Summeter* 1206 Cur (Bk); William *le Sumeter* 1221 AssGl, AssWo; William *Sompter*, Alan *Sumpter* 1301 SRY. OFr *sometier, sommetier* 'driver of a pack-horse' (c1320 NED).

**Sumray, Sumrie**: v. SUMERAY

**Sund**: v. SOUND

**Sunday**: John *Soneday* c1270 ERO; John *Sunday* 1356 FFEss; Robert *Sonday* 1524 SRD. A nickname for one born on a Sunday, OE *sunnandæg*.

**Sunderland**: Ralph *de Sunderland'* 1230 P (Ess); Robert *Sonderlaunde* 1297 MinAcctCo; Thomas *de Sundirland'* 1379 PTY. From Sunderland (Cu, Du, La), North Sunderland (Nb), High Sunderland in Northowram (WRY), or 'dweller on land set apart for some special purpose', OE *sundorland*.

**Sunman**: *Suneman* 1066 DB (Sf, Y); Estmundus *filius Suneman* 1202 FFNf; Hugo *Suneman* 1200 P (Nf); William, Roger *Soneman* 1275, 1279 RH (Sf, C). Forssner and von Feilitzen take this to be OG *Suneman*. Late compounds of *-mann* are common in the eastern counties and *Sunnmann*, *Sunngifu* (f), and *Sunnwine*, all found in the same area, are probably native compounds of OE *Sunna* found in Sonning (Berks). cf. Robert *Sungyve* 1279 RH (C), *Sunuuinus* 1066 DB (Sf), Simon *Sonewyne* 1326 SRSf.

**Sunning**: v. SONNING

**Sunter**: v. SUMPTER

**Sunwin**: *Sunwinus* 1066 DB (Sf); Robert *Sonewyne* 1327 SRSf. OE *Sunnewine*.

**Super**: *Superius* de Baiocis 1228–32 Gilb (L); Roger *Superius* 1221 AssWo; Robert *Souper* 1301 SRY; Roger *le Supere* 1309 AssNf. Lat *Superius*.

**Surgenor, Surgeoner**: William *Rydale*, *surgener* 1422 FrY; John *Sudgener* 1580 LaWills; Robert *Suggener* or *Sojourner* 1676 LaWills. A late, extended form of AFr *surgien* 'surgeon'. The surname-forms are late and dialectal, showing loss of *r*. The 1676

*Sojourner* is probably a dialectal pronunciation, but such a surname is possible. cf. Walter *Soiournaunt* 1377 *AddCh* (Ess).

**Surgeon:** Thomas *le Surigien* 1255 *Ass* (Ess); Robert *le Surgien* 1279 RH (C). AFr *surgien, sur(r)igien* 'surgeon' (13.. NED).

**Surgerman:** 'Servant of the surgeon', from ·OFr *surgier*, a rare by-form of *surgien* 'surgeon'. cf. Thomas Warde, *surgeour* 1487 FrY.

**Surgison:** *v.* SARGEANTSON

**Surkett:** *v.* CIRCUITT

**Surr:** *v.* SIRE

**Surrell:** *v.* SIRDIFIELD

**Surrey, Surry:** John *de Surreye* 1274 RH (O). 'The man from Surrey', cf. Henry, Roger *le Surreis* 1208 Pl (Y), 1231 Cur (He).

**Surridge, Surrage:** (i) Robert *Surrais* 1143–7 DC (L); Geoffrey *le Surreys* 1219 AssY. OFr *surreis* 'southerner'. *v.* also SEARCH. (ii) Ambrose *Surradge*, 1580 DWills; Robert *Surridge* 1609 ib.; Edward *Surredge* 1625 ib. From Surridge in Morebath (Devon).

**Surtees:** Randulf *de Super Teise* 1174 P (Nb); Ricardus *super Teisam* 1195 P (Nb); Ralph *de Supertay* 1230 P (Nb); Ralph *sur teyse* 1243 AssDu; Nicholas *de Surteys* 1315 Riev (Y). The forms vary between 'dweller by the Tees' and 'dweller at a place called Surtees'. cf. William *Overswale* 1301 SRY.

**Susan:** Susanna 1194 CurR (Sf); *Susanna Agnell'* 1206 Cur (Berks); William *Susann'* 1279 RH (O); Eustace *Susanne* 1327 SRSf. Hebrew *Shushannah* 'lily'. Not a common medieval name.

**Sussams:** Philip *Susson* 1327 SRSf; John *Sussoun* ib.; Samuel *Sussum* 1674 HTSf. *Susan*, with a short vowel and the common change of final *n* to *m*.

**Sussands:** John *Sussaunt* 1524 SRSf. *Susan*, with shortened vowel and an excrescent *t* or *d*.

**Sussex:** William *de Sudsex* c1210 NthCh (Nth); Robert *de Sussex* 1296 SRSx; John *Sussex* 1583 Musters (Sr). From Sussex.

**Sutch:** *v.* SUCH

**Sutcliff, Sutcliffe:** Hugo *de Suthclif* 1274 Wak (Y); Elyzabeth *Sutlyff* 1566 RothwellPR (Y). 'Dweller by the south cliff', a common Yorkshire name.

**Suter:** *v.* SOUTAR

**Sutherby:** *v.* SOTHEBY

**Sutherin, Suthern(s), Suthren:** *v.* SOTHERAN

**Sutherland, Sutherlan, Southerland:** David *de Sothirlandae* 1332 Black; Richard *Sutherlond* 1342 LoPleas; Alexander *Sutherland* 1441 Black. 'The man from Sutherland.'

**Suthery, Sutthery, Suddery:** John *de Sothereye* 1327 SRSf; John *de Sotherei* 1327 SRWo; John *Sotherey* 1435 AssLo. From Southrey (L), or Southery (Nf).

**Sutor:** *v.* SOUTAR

**Suttle, Suttill:** Adam *le Sutel* 1275 RH (Lo); Reginald *Sutel* 1279 RH (O). AFr *sotil* 'subtle, clever, cunning'. *v.* also SOOTHILL.

**Sutton:** Ketel *de Sudtone, Suttune* 1086 DB (L); Alnod *Suttuna* 1086 InqEl (C). From Sutton (Cambs, Lincs, Kent, Suff, Som). *v.* also SINTON.

**Swabey, Swaby:** Richard *de Suabi* 12th Gild; Philip *de Swaby* 1219 P (L); Walter *de Swaby* 1263 FFL. From Swaby (L).

**Swaffer, Swoffer:** Robert *Swoffer, Swoffa* 1518 CantW, KentW; John *Swaffer* 1523 CantW; Robert *Swafford* 1549 ib. From Swatfield Bridge in Willesborough (Kent), *Swatford* 1254.

**Swaffham, Swaffam:** Richard *de Suafham* 1178 Black; William *de Swafham* 1299 RegAntiquiss; Simon *de Swafham* 1327 SRWo. From Swaffham (Nf), or Swaffham Bulbeck, Prior (C).

**Swaffield, Swafield:** Julian *de Swafeld'* 1201 Pleas (Nf); Robert *de Swaffeld* 1406 IpmY; Thamasin *Swaffield* 1662–4 HTDo. From Swafield (Nf).

**Swain, Swaine, Swayn, Swayne:** (i) Suein, Suen, Suuain, Suan, Suuan 1066 DB; Eduuardus *filius Suani* 1066 DB (Mx, Ess); *Swein* de Hecham 1175–86 Holme (Nf); Robertus *filius Swain* 1219 AssY; Osgot *Sveyn* 1045 ASWills (C); Robert *Suein* 1166 P (Y); Walter *Swayn* 1295 FFSf. ON *Sveinn*, ODa, OSw *Sven*, often anglicized as *Swan*. cf. *Suanus* carpentarius, called also *Swein'* carpentarius 1221 AssWa, and *v.* SWAN. (ii) Walter *le Swein* 1221 AssWo; James *le Swayn* 1300 LLB C. ON *sveinn* 'boy, servant', used also of a swineherd, a peasant.

**Swainson:** Thomas *Swaynson* 1332 SRCu. 'Son of *Sveinn* or of the swain.'

**Swale:** (i) Richard *Swale* 1212 P (Ha); Juliana *Swale* 1242 AssDu; Adam *Swal* 1275 SRWo. ON *Svala* (f). (ii) William *Over Swale* 1306–7 IpmY; William *de Swale* 1359 FFY; John *Swayles* 1635 FrY. From the River Swale (K, R, WRY).

**Swallow:** (i) William *Sualewe, Swalewe* 1205 Cur (Y), 1275 SRWo. OE *swealwe* 'a swallow'. (ii) Thomas *de Sualewe* 1175 P (L). From Swallow (Lincs).

**Swan, Swann:** (i) *Suannus* faber 1177 P (Cu); *Suan'* filius Arkill' 1219 AssY; Hugo *Suan* 1176 P (Sf); John *Swann'* 1221 AssSa; Gilbert *Swan* 1260 AssC; Thomas *le Swan*, John *le Swon* 1327 SRSf; Walter *le Swon*, Stephen *le Swan* 1296 SRSx. With the article, this may be either OE *swān* 'herdsman, swineherd, peasant', which became ME *swon*, or a nickname from OE *swan, swon*, 'swan'. Without the article, these cannot be distinguished from OE *Swan*, an anglicizing of ON *Sveinn*. cf. SWAIN. (ii) Godfrey *atte Swan* 1344 LoPleas; Thomas *atte Swan* 1364 LLB G. From the sign of the Swan.

**Swanbourne, Swanburne:** John *Swanburn'* 1394 KB (Mx). From Swanbourne (Bk).

**Swanby:** John *Swanby* 1401 IpmY. From Swainby (NRY), *Suanebi* DB.

**Swancott, Swancott, Swancoat, Swancutt:** Henry *de Swanecote* 1275 SRWo. From Swancote Fm in Chaddesley Corbett (Wo).

**Swanland:** Alan *de Swanlond'* 1311 FFY; Peter *de Swanland* 1400 IpmY; John *Swanland* 1402 AssLo. From Swanland (ERY).

**Swannack:** *v.* SWANNICK

**Swannell, Swonnell:** *Swanild(a)* 1201 AssSo, 1221 AssSa; *Swanhild* de Herteford 1227 AssBk; John *Swonild* 1247 AssBeds, 1277 *Ely* (C); Ralph *Swanyld* 1332 SRSt. From the ON woman's name *Svanhildr*.

**Swannick, Swannack:** John, William *Swaneke* 1279 RH (C). Probably 'swan-neck'. cf. Edgyue

Suanneshals 1066–86 OEByn. The surname is, no doubt, usually from Swanwick (Derby, Hants).

**Swanton:** Ralph *de Suaneton* 1204 P (K); Walter *de Suanton* 1283 SRSf; John *Swanton* 1373 AssLo. From Swanton (Kent, Norfolk).

**Swart:** Suenus, Mannius *Suart* 1066 DB (Sf); Walter *Swart* 1219 AssY; John *le Swart* 1285 LLB A. ON *svartr*, OE *sweart* 'swarthy'.

**Swatman:** *v.* SWEATMAN

**Swatridge, Swattridge:** Adam, William *Swetrich* 1309 SRBeds. OE *Swētrīc* 'sweet-ruler'.

**Swears:** (i) Ralph, Hugh *Swere* 1208 FFEss, 1225 AssSo. OE *swēora* 'neck', a nickname. (ii) Robert *atte Swer'* 1327 SRSx. 'Dweller by the neck of land', as at The Swares or Swear Fm (Sussex).

**Sweatman, Sweetman, Swetman, Swatman:** *Suetman* 1066 DB (Nf); *Suetman* Kempe 1169 P (Bk); *Swetmannus* ate Optone 1327 SRSx; Eche *Suetman* 1134–40 Holme (Nf); Robert *Sweteman* 1222 FFSf; Thomas *Swatman* 1524 SRSf. OE *Swētmann* 'sweet man'.

**Sweet, Sweett:** *Suet, Suot* 1066 DB (D), Winton (Ha); Æluric *Suete filius* c1095 Bury (Sf); *Swet* carbonel 1292 SRLo; Wimund *Svote* 1179–94 Seals (Beds); Adam *Swet* 1211 Cur (Wo); Richard *Swote* 1313 LLB B; William *le Swete* 1327 SRSx. OE *Swēt, Swēta* (m), *Swēte* (f) 'sweet', or a nickname, OE *swēte, swōt* 'sweet'.

**Sweetapple:** Henry *Suetapill* 1259 CtSt; Ralph *Swetappul* 1309 SRBeds. Apparently a nickname.

**Sweetblood:** Thomas *Sueteblod* 1197 P (L); William *le Sweteblod* 1203 AssL; John *Sweteblod* 1327 SRSf. 'Sweet blood', OE *swēte, blōd*. cf. Robert *Swetalday* 1360 FFY 'sweet all day'; John *Swete bi ye bone* 1225 AssSo 'sweet on the bone'; Richard *Swetchicke* 1315 HPD 'sweet chick'; William *Sweteghe* Hy 3 Rad 'sweet eye'; Ranulph *Swetefrond* 1299 AD ii (Ch) 'sweet friend'; Joseph *Sweetlas* 1714 DKR (C) 'sweet lass'; Peter *Swetemilk* 1258 IpmY 'sweet milk'; Nicholas *Swetemouth* 1327 SRY 'sweet mouth'.

**Sweeting:** *Sueting* Cadica 1135 Oseney (O); *Sweting* de Hunebir' 1225 AssSo; Ralph *Suetinge* 1185 Templars (Herts); John *Sweting'* 1250 Fees (W). OE *Swēting*.

**Sweetlove:** *Swetelove* 1279 RH (C); William *Sueteluue* 1197 P (Y); Margeria *Swetelove* 1279 RH (C). OE *Swētlufu* (f) 'sweet-love'.

**Swetman:** *v.* SWEATMAN

**Sweetnam, Swetenham, Swettenham:** Richard *de Swetenham* 1278 AssSt. From Swettenham (Ches).

**Sweetser, Sweetsur, Sweetzer, Switsur, Switzer:** Robert *Swetesire* 1355 Cl (Beds); Richard *Swetesyr* 1467 RochW; Denys *Swetesyre* 1498 RochW; Richard *Sweytsyer, Sweetser* 1553 ib., 1584 Bardsley. 'Sweet sire.' *v.* SIRE.

**Swell:** Ingelram *de Suell'* 1200 P (Gl); Roger *de Swell'* 1206 Cur (Gl); Godfrey *de Swelle* 1221 AssGl. From Swell (So), or Upper, Lower Swell (Gl).

**Swepstone, Swepson:** Ralph *de Suepiston* 1199 FrLeic. From Swepstone (Leics).

**Swetenham:** *v.* SWEETNAM

**Swetman:** *v.* SWEATMAN

**Swiers:** *v.* SWIRE

**Swift:** *Suift* 1166 P (Nf); Nicholaus *filius Swift* 1222 Cur (Sf); William *Swift* 1167 P (Ha), 1219 AssY. OE *swift* 'swift, fleet', used also a personal-name.

**Swillington:** William *Swillington* 1672 HTY. From Swillington (WRY).

**Swinbrook, Swinbrooke:** Robert *de Swynebroke* 1252 FFO; William *de Swynebroc* 1252–3 FFSr. From Swinbrook (O).

**Swinburn, Swinburne, Swinbourn, Swinbourne:** John *de Swynburn* 1256 AssNb; Robert *de Swynebourne* 1325 IpmGl; Robert *Swynbourn* 1382 FFEss; Widow *Swingbourne* 1662 HTEss. From Swinburn (Northumb), or 'dweller by the pig brook'.

**Swindell, Swindells:** James *Swindell* 1621 SRY; Humphrey *Swindells* 1647 PN Ch i 147. From Swindale House in Skelton (NRY).

**Swinden, Swindin, Swindon:** Richard *de Swinden'* 1212 Cur (Y); John *de Swindon* 1271 IpmGl; John *Swyndon* 1380–1 PTW. From Swindon (St, W), or Swinden (Gl, WRY).

**Swinerd, Swinnard:** Adam *Swynherde* 1327 SRSx; Walter *le Swynhurde* 1327 SRWo; Robert *le Swynerde* 1332 MEOT (Nf). Late OE *swīnhyrde* 'swineherd'.

**Swinford:** Geoffrey *de Suinford* 1190 P (Lei); William *de Suyneford* 1250 FFL; Roger *de Swyneford* 1332 SRWa. From Swinford (Berks, Leics), Kingswinford (Staffs), or Old Swinford (Worcs).

**Swinn:** (i) Robert *Swin* 1207 P (Db); Thomas *le Swyn* 1327 SRC; John *Swyne* 1407 IpmY. OE *swīn* 'swine', either a nickname, or metonymic for a swineherd. (ii) Robert *de Swyne* 1297 SRY. From Swine (ERY).

**Swinnerton, Swynnerton:** Ralph *de Suinerton* 1185 P (Sf); Robert *de Swinnerton'* 1221 AssSa; Roger *de Swynnerton* 1295 AssSt, 1322 LLB E. From Swynnerton (Staffs).

**Swinstead:** (i) Gocelin *de Swynested* 1276 RH (L). From Swinstead (L). (ii) Walter *de Swinesheved* 1207 Cur (Hu); Robert *de Swynesheved* 1269 AssSt. John *Swynesheved* 1288 NorwLt. From Swineshead (Beds, L). This would, no doubt, have been absorbed by *Swinstead.*

**Swinton:** William *de Swintona* 1162–c1176 YCh; Nicholas *de Swynton* 1256 AssNb; Thomas *de Swynton* 1379 PTY. From Swinton (La, NRY, WRY).

**Swinyard:** John *de Swynyard* 1332 SRLa. 'Worker at the swine-enclosure', OE *swīn, geard.*

**Swire, Swires, Swiers, Swyer:** Geoffrey *le Swyer* 1275 RH (Nt); John *Swyer* 1297 SRY. A northern form of SQUIRE.

**Swithegood:** William *Swithegod* 1221 AssWa; Agnes *Swythgod'* 1379 PTY. 'Very good', OE *swīþe, gōd.* cf. Alice *Swythered* 1332 ChertseyCt (Sr) 'of very good advice'.

**Switsur, Switzer:** *v.* SWEETSER

**Swoffer:** *v.* SWAFFER

**Swonnell:** *v.* SWANNELL

**Sword, Swords, Soards, Soord:** Robert *Suerd* 1185 P (Berks); Peter *Swerd* 1297 Wak (Y). OE *sweord* 'sword', used of a sword-maker. *v.* SWORDER.

**Sworder:** Walter *le Sorder* 1324 AD vi (Sx);
Richard *Swordere* 1354 PN Ess 483; John *Swerdere* ib.
A derivative of OE *sweord* 'sword', a maker of swords.

**Swyer:** *v.* SWIRE

**Sy-:** *v.* SI-

**Syddall:** *v.* SIDDALL

**Sydenham, Syddenham, Sidenham, Siddenham:**
Ascelin *de Sidenham* 1202 AssNth; Walter, John
*de Sydenham* 1284 FFO, 1327 SRSo; Richard
*Sydenham* 1384 AssL. From Sydenham (Devon,
Oxon, Som).

**Sykes, Sikes, Sich, Sitch, Seach:** Ralph *de Sich*
1166 P (Nf); Roger *del Sik* 1212 Cur (Nf); Robert
*Sitche* 1255 RH (Sa); Richard *del Siche* 1298 AssSt;
Richard *del Sikes* 1309 Wak (Y); John *atte Siche* 1327
SRWo; Richard *in le Syche* 1332 SRSt. From
residence near a small stream or streams, especially
one flowing through flat or marshy ground (OE *sīc*);
also 'a gully, dip or hollow'. This became ME *siche*
in the south and midlands, *sik(e)* in the north. cf. *ditch*
and *dike*. *Sykes* (which is plural) is common in
Yorkshire where it may also be from ON *sik*.

# T

**Tabah:** v. TABER

**Tabard, Tabbitt:** Thomas *Tabart*, Peter *Tabard* 1275 RH (L, K). ME *tabard*, OFr *tabart, tabard* 'a sleeveless coat', formerly worn by ploughmen and noblemen and not restricted, as now, to heralds. Also from a sign-name: John *atte Tabard* 1371 LoPleas.

**Tabb:** Aluuidus *Tabe* 1066 DB (D); Hugh *Tabbe* 1214 Cur (Bk); Henry *Tab*, Peter *Tabb* 1642 PrD. OE *\*Tæbba*. v. OEByn 338.

**Tabberer, Tabborah:** v. TABERER

**Tabberner:** v. TABERNER

**Taber, Tabor, Tabah:** Suein *Tabor* 1185 P (W); Adam *Tabur* 1204 P (Sa). OFr *tabur, tabour* 'drum, tabor', by metonymy for *Taberer.*

**Taberer, Tabberer, Tabborah, Tabrar:** Peter *Taburer* 1222 Cur (Nf); Peter *le Taburur* 1279 AssNb; Bernard *le Taborer* 1280 Ass (Ha). A derivative of ME *tab(e)re* 'to play on the tabor' or from OFr *tabur, tabour* 'tabour'; 'one who plays the tabor, a drummer' (c1400 NED).

**Tabern:** Metonymic for TABERNER.

**Taberner, Tabberner, Tabiner, Tabner:** Benedict *Taberner* 1274 RH (D); Robert *le Taburner* 1301 SRY. A derivative of ME *tabourne* or OFr *tabourner* 'to drum'. 'Drummer.' cf. TABERER.

**Tabler:** Peter *Tabeler* 1181 Templars (Sx); Thomas *le Tabler* 1248, Robert *le Tabler* 1275 AssW. OFr *tablier* 'joiner'.

**Tabletter:** Peter *le Tableter* 1281 LLB B; Richard *Tableter* 1327 SRC; Geoffrey *le Tableter* 1343 LoPleas. OFr *tabeletier* 'a maker of chess or draught boards'.

**Tabois, Ta'bois:** v. TALLBOY

**Tabor:** v. TABER

**Tabrar:** v. TABERER

**Tacey, Tacy:** Jerome *Tacie* 1641 PrSo; Alice *Tacye* 1662–4 HTDo. A pet-form of Lat *Eustacia* (f).

**Tackel, Tackell, Tackle:** v. TAKEL

**Tackley:** William *de Takel* 1200 Pleas (Sf); John *de Takele* 1325–6 CorLo; Alexander *Tacle* 1437 FFEss. From Tackley (O).

**Tacy:** v. TACEY

**Tadcaster:** Nathaniel *Tadcaster* 1663, *Tadcastle* 1665 WRS. From Tadcaster (WRY).

**Tadde:** Thomas *Tade* 1327 SRSo. OE *tadde* 'toad'.

**Tafner:** v. TAVERNER

**Taft:** John *de Taft* 1340 AssC. An unrounded form of TOFT.

**Tagart, Taggard, Taggart, Taggert:** v. MCTAGGART

**Tagg, Tagge:** Suinus *Tagge* 1195 P (Berks); Hugh *Tagge* 1297 Wak (Y); William *Tagge* 1533 DbCh. A nickname from OE *\*tacca* 'a young sheep', or a personal name derived from this.

**Tague:** v. KEIG

**Tail, Tayl:** (i) Edwin *taile* 1189 Sol; William *Tayl* 1249 AssW; John *Taill* 1560 FrY. A nickname from OE *tægl* 'tail'. (ii) Walter *del Teil* 1204 P (Nt); Robert *de la Taill* 1225 AssSo. 'Dweller by the mill-race', OE *tægl* in the sense of that part of the mill-race below the wheel.

**Tailyour:** v. TAYLOR

**Tainter:** Thomas *le Teinturer* 1196 P (Db/Nt); Peter *le Teyntur* 1268 AssSo; Nicholas *le Teyntour* 1331 FFW. OFr *teinturier* 'dyer'.

**Taisez-vous:** Roger *Taiseznus* 1177 P (Nf). 'Be silent', Fr *taisez-vous*, a nickname from a favourite phrase. cf. Thomas *Jeo Vousdy* 1299 AssW 'I tell you'.

**Tait, Teyte:** Ralph *Tait* 1185 Templars (Y); Robert *Teyt* 1279 RH (O); Thomas *Tayte* 1301 SRY. ON *teitr* 'cheerful, gay'.

**Takel, Takle, Tackel, Tackell, Tackle:** John *filius Takell* 1219 AssY; Samson *Takell* c1170–86 MCh; Thomas *Takel* 1245 FFL; John *Tackle* 1662–4 HTDo. Probably an unrecorded OE *\*Tæcela.*

**Talbert:** John *Talbard* 1327 SRSf. A rare name, identical with the French *Talbert*. v. TALBOT.

**Talbot, Talbott, Talbut, Talbutt, Taulbut:** *Talebotus prior* 1121–38 Bury (Sf); *Talebod de Neweham* 1146–8 Seals (Ess); *Talebot* 1185 Templars (K), 1283 FFSf; Geoffrey, Richard *Talebot* 1086 DB (Ess, Beds); Gilbert, Andrew *Talebot* 1190, 1196 P (He, Do); William *Taillebot* 1202 AssL; Richard *Thalebot* 1225 AssSo; William *Taleboth* 1229 FFEss; Gilbert *Talbot* 1332 SRSx; John *Talbut* 1332 SRCu. Harrison explains this name as 'bandit', from OFr *talebot* which Moisy takes as a nickname for robbers who blacked their faces to avoid recognition, *talebot* meaning 'lampblack, pot-black' in the dialect of Normandy. Tengvik associates the name with OFr *talbot* 'a wooden billet hung round the neck of animals to prevent them from straying'. Miss Withycombe, comparing *Telfer*, suggests a compound of *tailler* 'to cut' and *botte* 'faggot'. The 1202 form *Taillebot* lends some support to this, but it is a solitary form and early forms for *Telfer* and *Tallboy* usually have Taille-. We are undoubtedly concerned with a personal-name, not uncommon in the 12th and 13th centuries, the first theme of which is found also in the French *Talabert*, *Talbert* (OG *\*Talaberht*; cf. TALBERT), *Talamon* (OG

*Talamund). v. also TALLON. This Dauzat derives from an obscure root *dala*. Hence, OG *Dalabod*, *Talabod*, OFr *Talebot*.

**Talboys:** v. TALLBOY

**Talen:** v. TALLANT

**Talfourd:** v. TELFER

**Talks, Tawlks:** Gregory *de Talk* 13th Guisb; Robert *Tauk* 1332 SRSx; Ellis *Talke* 1642 PrD. From Talke (St).

**Tallamach, Tallemach:** v. TALMADGE

**Tallant, Tallent, Tallents, Talen:** *Tallant* Faidge, *Talland* Gin 1642 PrD; Odo *de Talent* 1155–60 Templars (Do); Anthony *Tallen* 1642 PrD. From Talland (Co).

**Tallantire, Tallentire:** Alexander *de Tarentir* 1212 P (Cu); Geoffrey *Talentir* 1225 Pat (Cu); Margaret *de Kalantir* 1348 FFY. From Tallantire (Cu).

**Tallboy, Tallboys, Talboys, Tabois, Ta'bois:** Ralph *Tailgebosc, Taillebosc, Tallbosc* 1086 DB (Herts); Yvo *Taileboys* 1133–60 Rams (Hu); Reginald *Tailebois, Tallebos, Tallebys, Taullebois* 1203–21 Cur (Herts); Robert *Tayleboys* alias *Taleboys* 1257 Ipm (Nb). OFr *taillebosc*, from *tailler* 'to cut' and *bosc* 'wood', 'cut wood', a name for a wood-cutter. cf. CUTBUSH. Two forms, if correct, suggest a place-name: Roger *de Tallebois* 1161 P (Herts), Robert *de Talbois* 1279 AssNb. *Taille* and *Taillis* 'copse' are French place-names and Harrison notes Taillebois (Orne).

**Tallent, Tallents:** v. TALLANT

**Tallet, Tallett:** Andrew *Taillard* 1155 CartAntiq; Robert *Taillard* 1208 Cur (Sx); Richard *Taillard* 1340 CorLo. A nickname for the wearer of a sword. v. Dauzat.

**Tallon:** Henry *Talon, Talun* c1160 DC (L); Hugh, Geoffrey *Talun* 1180 P (Y), 1187 Oseney (O); Johanna *Taloun* 1327 SRSf. Probably, as suggested by Dauzat for the French *Talon*, the cas-régime of OG *Talo*, rather than OFr *talon* 'heel', 'claw', cf. TALBOT.

**Talmadge, Talmage, Tallamach, Tallemach, Tammage, Tollemache:** Hugo *Talamasche* 1130 P (O); Robert *Talemasche* 1150 Eynsham (O); William *Talemach* 1297 MinAcctCo; William *Talmege* 1512 FFHu; Elizabeth *Talmage* 1524 SRSf. A nickname from OFr *talemache* 'knapsack'.

**Talwith, Talworth:** Adam *de Talewurde* 1161 P; Richard *de Taleworth* 1323 FFEss; John *Talworth'* 1351 KB (Bk). From Tolworth (Sr), *Talewurtha* 1150–67.

**Tame, Tames, Thame:** Aluered *de Tame* 1086 DB; Claricia *de Thame* 1279 RH (O); Thomas *Tamese* 1371 IpmW. From Tame (Bk, NRY).

**Tamlin, Tamlyn, Tambling, Tamplin:** Peter *Tamelyn* 1327 SRSf. A double diminutive of *Tam* (Tom). cf. TOMLIN.

**Tampson, Tamson:** William *Tamson* 1395 EA (OS) iv (C); Walter *Tampson* 1641 PrSo; John *Tampson* 1642 PrD. 'Son of *Tam*', a variant of *Tom*, a pet-form of *Thomas*.

**Tamworth:** Ralph *de Tamewurthe* 1189–99 Colch (Ess); William *de Tamwurth* 1262 FrLei; Hugh *Tampworth* 1380 LoCh. From Tamworth (St).

**Tancred:** *Tancredus* filius Bernardi, *Tancredus de Witton* 1252 Rams (Hu); John *Tankred* 1389 Crowland (C). OG *Tancrad* 'thought-counsel'.

**Tandy:** *Tandy* de Jay 1221 AssSa; Richard *Tandy* 1275 SRWo; William *Tandy* 1307 AssSt; William *Tandys* 1340 NIWo. Harrison's suggestion that this is an unvoiced form of *Dandy*, a pet-form of *Andrew*, may well be correct, cf. Thomas *Tancocke* 1636 HartlandPR (D) for *Dancock*.

**Taney, Tawney:** Hasculfus *de Tania* 1195 P (Ess); John *Tawny* 1674 HTSf. From Saint-Aubin-du-Thenney or Saint-Jean-du-Thenney (Eure). v. ANF.

**Tanfield:** Matilda *de Tanefeld'* 1203–4 FFY; Elias *de Tanfelde* 1301 SRY; John *Tanfeld* 1463 TestEbor. From Tanfield (Du), or East, West Tanfield (NRY).

**Tanguy, Tangye:** v. TINGAY

**Tankard:** *Tankardus* Flandrensis 1175 P (Nf); Radulfus *filius Tancard'* 1185 Templars (Y); William *Tankard* 1190 P (O); John *Tancart* 1202 AssL. OG *Tancard* 'thought-hard'. The surname may also be metonymic for a maker of tankards: John *le Tanckardmaker* 1298 LoCt.

**Tankerley, Tankersley:** cf. Roger *Tankerlayman* 1387 FrY. From Tankersley (WRY).

**Tanner:** Henry *Taneur, le Tanur* 1166–7 P (Nf); Lemmer *le Tannur* c1175 Whitby (Y); William *le Tanner* 1256 AssNb. Usually OFr *taneor, tanour*; also OE *tannere* 'tanner'.

**Tansley, Tanslye:** Hemming *de Taneslea* 1175 P (Db). From Tansley (Db).

**Tantifer:** John *Tantefer* 1272 AssSo; Walter *Tauntefer* 1303 IpmW. 'Iron tooth', Fr *dent de fer*.

**Tanton, Taunton:** Gilbert *de Tanton'* 1188 P (So); Gilbert *de Tanton* 1222 FFEss; John *Taunton* 1381 LoCh. From Taunton (So), Tanton (NRY), or Taunton Fm in Coulsdon (Sr).

**Tapeley, Tapley:** John *de Tappelega* 1185 P (D); John *Tapelegh'* 1249 AssLo; John *Tapley* 1525 SRSx. From Tapeley (D).

**Taper:** Robert, John *Taper* 1327 SR (Ess), 1332 SRSx. Metonymic for a maker of tapers: William *(le) Tapermaker* 1340–1 Oriel (O).

**Tapner:** Gregory *le Tapiner* 1272 MESO (Ha); Henry *le Tapenyr* 1327 SRSx. OFr *taponner* 'maker of chalons and burel'. v. BURREL, CHALLENER.

**Tapp:** Tappe: John, Roger *Tappe* 1194 P (Do), 1247 AssBeds, 1327 SRWo. OE *Tæppa*, unrecorded, but the first element of Taplow (Bucks), Tapners (Kent), Tappington (Kent) and Tapton (Derby). v. TAPPIN.

**Tapper:** Ulfuine *Teperesune* c1095 Bury (Sf); John, Robert *le Tapper* 1279 RH (C), 1332 SRLa. OE *tæppere* 'tapper (of casks), beer-seller, tavern-keeper'.

**Tappin, Tapping:** Ralph, Geoffrey *Tapping* 1220, 1235 Fees (Berks, Nth); Robert *Tappyng, Tepping* 1221 AssWa. OE *Tæpping* 'son of *Tæppa*'. v. TAPP.

**Tapster:** (i) Elisota *Tapester* 1379 NottBR; Alicia *Tapstere* 1384 AssWa. OE *tæppestere*, feminine of *tæppere*, a woman who sells ale, hostess. cf. TAPPER. (ii) Ralph, Thomas *le Tapicer* 1282 AD ii (Mx), 1306 LoCt. AFr *tapicer* 'maker of figured cloth or tapestry' (c1386 NED). As *tapiner* became *tapner*, so *tapicer* probably became *tapser* and, with intrusive *t*, *tapster*.

**Tapton:** Henry *de Tapton* 1342 PN WRY i 201; John *Tapton* 1451, William *Tapton* 1467 IpmNt. From Tapton (Db), or Tapton Hall in Sheffield (WRY).

**Tarbard, Tarbath, Tarbat, Tarbert, Tarbet, Tarbitt, Tarbutt:** v. TURBARD

**Tarbin, Tarbun:** v. THURBAN

**Tarboc, Tarbox, Tarbuck, Tarbet, Tarbitt, Tarbutt:** Henry de Torbok 1246 AssLa; Ellen Torbok 1324 CoramLa; Thomas Tarbox 1663 HeMil. From Tarbock (La).

**Tardew, Tardif:** William, Geoffrey Tardif 1115 Winton (Ha), 1260 AssC; Hugh Tardy 13th WhC (La); Henry Tardi Ed 1 NottBR. OFr tardif, Fr tardieu 'slow, sluggish', the origin of tardy.

**Tarleton, Tarlton:** Maien de Torleton' 1204 P (Gl); Gilbert de Tarleton 1332 SRLa; Magota de Tarlton 1379 PTY. From Tarleton (Lancs, Glos).

**Tarling, Tarlen, Terling:** Gilbert de Terlinges 1185 Templars (Ess); John de Terling 1303 FFEss; Matthew Tarling 1662 HTEss. From Terling (Ess).

**Tarn:** Hugo del Tern 1332 SRCu. 'Dweller by the tarn', ME terne 'small lake, pool'.

**Tarrant:** Reginald de Tarenta 1190 DC (L); John de Tarent 1212 Cur (Mx); Ralph Taraunt 1296 SRSx. From Tarrant (Dorset).

**Tarrier:** Geoffrey le terrier, le terrer 1193–4 P (Ess, Sx). ME terrere, taryer, Fr (chien) terrier 'a hunting dog' (c1440 NED).

**Tarring:** Reynold Terryng 1305 AssW; Thomas Tarringe 1642 PrD. From Tarring Neville, or West Tarring (Sx).

**Tarrington:** John Tarington 1662–4 HTDo. From Tarrington (He).

**Tarry:** v. TERREY

**Tart, Tarte:** Alwin filius Tarte 1066 DB; Teart Iuo le messagier 1221 AssWa; Walter Tart 1302 IpmY; Margaret Tarte 1481 TestEbor; Thomas Tart 1663 HeMil. OE teart 'sharp, rough'. Used also as a personal name.

**Tarves, Tarvis:** Nicholas Talewaz 1199 FFEss; Robert Talvace 1296 Wak (Y); Thomas Taleuas 1327 SRSf; William Turphas, Turface 1416, 1424 Petre (Ess); Robert Talface 1520 NorwCt; Margaret Talves 1524 SRSf. A maker or a seller of, or one armed with, a talevas, described by Cotgrave as 'a large, massive, and olde fashioned targuet, having in the bottom of it a pike, whereby, when need was, it was stuck in the ground'.

**Tasch, Tash:** v. ASH

**Tasker:** John le Tasker 1279 AssNb; Hugh Tasker, Benedict le Taskur 1279 RH (Bk). A derivative of ME taske 'task, assessment', in ME texts 'one who threshes corn with a flail as task-work or piece-work' (1375 NED).

**Tassel, Tassell:** (i) Wimund de Taissel 1086 DB (Beds). Probably from Tessel (Calvados). v. OEByn 115. (ii) William Tassell' 1206 Cur (Sx); Bartholomew Tassel 1288–9 FFSr. A nickname from the hawk, OFr tiercel, ME tassel.

**Tatam, Tatem, Tatum, Tatham:** William de Tateham 1208 Pl (Y); William, Robert de Tatham 1230 P (La), 1379 PTY; John Tattum, Tatam 1665 HTO, 1693 DKR. From Tatham (Lancs).

**Tate:** Uluric Tates c1095 Bury (Sf); Nicholas, Richard Tate 1279 RH (C), 1345 ColchCt. Tates is from OE *Tāt, the strong form of Tāta, found in

several place-names, with a diminutive in -el: Hugh, John Tatel 1195 P (Sf), 1373 ColchCt.

**Tatlock:** Hugh Tatelok 1332 SRLa. From Tatlock (La).

**Tatman:** Hugh, John Tateman 1195 P (Y), 1327 SRY. Either 'servant of Tate' or an unrecorded OE *Tātmann.

**Tatnall, Tatnell, Tattenhall:** William de Tatenhill c1180 Black; Robert de Tatenhull 1220 AssSt; Roger Tatnall 1524 SRSf. From Tattenhall (Ch), or Tatenhill (St).

**Tattersall, Tattershall, Tattershaw, Tattersill, Tettersell, Tetsall:** Hugo de Tateshal' 1191 P (L); Joan de Tatersale 1301 SRY. From Tattershall (Lincs).

**Tatton:** Avicia de Tatton' 1212 Fees (Do). From Tatton (Ch, Do).

**Taulbut:** v. TALBOT

**Taunton:** v. TANTON

**Taverner, Tavernor, Tavener, Tavenner, Tavenor, Taviner, Tavner, Tafner, Tavender:** William le Tauerner, Tauernier 1175, 1177 P (Y); William le Taverner 1268 Ipm (W); John Tavender 1674 HTSf. AFr taverner, OFr tavernier 'tavern-keeper'.

**Tavy:** John Tavy 1338 CorLo. From Marytavy, Petertavy (D).

**Tawlks:** v. TALKS

**Tawney:** v. TANEY

**Tawyer:** Ralph le Tawyere 1275 RH (W); Martin le Tauier 1300 LoCt; Henry Tawar 1381 PTY. A derivative of OE tāwian 'to taw', a tawyer, one who prepares white leather.

**Taycell:** v. TEASEL

**Tayl:** v. TAIL

**Taylor, Tayler, Tailyour:** Walter Taylur c1180 ArchC vi; William le Taillur 1182 P (So); John le talliur 1202 AssL. AFr taillour, OFr tailleor, tailleur 'tailor' (1296 NED).

**Taylorson, Taylerson:** Hugh le Taylleressone 1280 AssSo; Thomas le Taillourson 1324 Wak (Y). 'The tailor's son.'

**Tazelaar, Tesler, Tessler:** Geoffrey Taseler (Taslere) 1230 P (Ess); Matilda la Taselere 1301 ParlR (Ess); Richard le Tesler 1332 SRSr. A derivative of OE tāsel 'teasel', one who teasels cloth (14.. NED). The teaseler drew up from the body of the cloth all the loose fibres with teasels, the dried heads of the 'fuller's thistle' (MedInd 225).

**Teague:** v. KEIG

**Teal, Teale, Teall:** Ralph, Robert Tele 1201 P (Nt), 1275 SRWo; German le Tele 1327 SR (Ess). A nickname from the water-fowl, ME tele.

**Tealby:** Roger de Tauelesby Hy 2 Gilb. From Tealby (L), Tavelesbi DB.

**Tear, Teare:** M'Teare 1504, Teare 1599 Moore. A Manx contraction of MACINTYRE.

**Tearall:** v. TIRRELL

**Teasdale, Teesdale:** Walter de Tesedale 1235–6 AssDu; Mariota de Tesdale 1332 SRCu; Thomas Tesdall 1525 SRSx. From Teesdale.

**Teasel, Taycell:** Metonymic for TAZELAAR.

**Teaser:** William le Tesere 1275 AssSo. 'A woolcomber', from a derivative of OE tāsan 'to comb'.

**Tebay, Tebbey, Tebby, Tibbey, Tibby:** Herbert *de Tibei* 1192 P (We); Hugh *de Tybbay* 1230 P (Cu); Richard *Tebay* 1376 LuffCh. From Tebay (We), *Tibbay* c1200. *v.* TIBBEY.

**Tebb, Tebbs:** *Tebbe* de Wifardebi 1177 P (Y); *Tebbe* filius Toke 1208 FFL; Adam, John *Tebbe* 1316 FA (Lei), 1327 SRC; John *Tebbes* 1524 SRSf. A petform of *Tebbold*. *v.* THEOBALD.

**Tebbell, Tebble:** Either *Tebb-el*, a diminutive of Tebb (cf. TIBBLE), or a weakening of *Tebald*. *v.* THEOBALD.

**Tebbet, Tebbett, Tebbit, Tebboth, Tebbutt:** *v.* THEOBALD.

**Tebrich, Thebridge:** Robert *Tedbrith*, *Thedbryt* 1275 RH (Sf), 1296 SRSx. OE *þēodbeorht* 'people-bright', rare in OE, but the forms suggest a native origin.

**Tedder, Teddar, Tetther:** Hew *Tedder* 1583 Musters (Sr); Griffine *Tedder* 1632 GreenwichPR (K). The forms are late, but could perhaps represent an OE *þēodhere*. cf. Herbert *Tedmers* 1285 FFO for OE *þēodmǣr*.

**Tee:** Robert *atte Ee* 1351 AssSt. 'Dweller by the stream', OE *ēa*.

**Teece:** *v.* TICE.

**Teffan:** John *Tefan* 1341, Adam *Tefonte* 1376 IpmW. From Teffont Ewyas, Magna (W).

**Tegart:** Stephen *Teghurde* 1396 *Ct* (Ha). 'Shepherd', from OE *\*tegga* 'young sheep'. cf. CALVERT.

**Tegg, Tigg:** Hugo *Tig* 1211 Cur (Sr); Henry *Teg* 1278 AssLa; Thomas *Tegge*, *Tigge* 1327 SRSo. 'Young sheep', probably metonymic for TEGART.

**Telfer, Telford, Taillefer, Talfourd, Tilford, Tilfourd, Tolver, Tulliver:** *Taillefer* de Scaccario 1214 Cur (Mx); *Taillefer* de Fraxineto 1221 AssWo; Dunning *Tailifer* a1103 OEByn (D); Humphrey *Tallifer* 1195 P (He); Richard *Taillefer* 1225 AssSo; Stephen *Tayleford* 1381 *SR* (Ess); John *Talyver*, *Tollfeyre* 1524 SRSf; William *Tylford* 1615 PN Ess 194; William *Tollever* 1674 HTSf; George *Telford* 1685 FrY; George *Tealfer* 1696 Bardsley (Cu). OFr *taille fer* 'cut iron', 'iron-cleaver', used as a personal-name, an original nickname for a man who could cleave clean through the iron armour of his foe.

**Teller:** Normannus *telarius* 1193, Roger *le Telier* 1198 P (L); Thomas *le Teller* 1243 AssSo; Robert *Teller* 1323 Wak (Y). OFr *telier* 'a maker or seller of linen cloth, a weaver'.

**Tellow:** Nicholas *at(te) Telgh* 1297 PN Sx 401, 1327 SRSx; Henry *Telowe* 1292 IpmGl. 'Dweller by the young oak', as at Sweetwillow Shaw (Sussex), from OE *\*telg, telga, telge* 'branch, twig', *v.* MELS 209.

**Tellwright:** Simon *le Tywelwrighte* 1280 MESO (Ha). OE *tigelwyrhta* 'tile-wright', tile-maker, brickmaker. cf. TILER, TILLMAN.

**Tempany:** Henry *Tynnepanne* 1254–67 Rams (Hu); Richard, William *Timpon* 1279 RH (C); Nicholas *Tympanye* 1402 FA (W). OFr *tympan* (Lat *tympanum*) 'a drum, timbrel or tambourine', later *tympany*. 'Praise him with tympany and tabret' (1535 NED). Metonymic for a drummer.

**Temperley, Timperley:** Walter *de Timperleie* 1211–25 StP; John *Tymperley* 1332 SRWa; Thomas

**Temperle** 1340–1450 GildC. From Timperleigh (Ch).

**Tempest:** Roger *Tempeste* 1168 P (Y), 1209 P (Y); Richard *Tempest* 1222 Cur (Y). ME *tempest*, OFr *tempeste* 'a violent storm' (c1250 NED), 'agitation, perturbation' (c1315).

**Templar, Templer:** Johannes *Miles templi* 1150–5 DC (L); Gilbert *Templer* 12th DC (L); Robert *le Templer* 1220 Cur (O). ME *templer*, (c1290 NED). AFr *templer*, OFr *templier* 'a member of a military and religious organization consisting of knights, chaplains and men-at-arms' founded c1118 chiefly for the protection of the Holy Sepulchre and of pilgrims to the Holy Land; suppressed in 1312. The Knights Templar were so called because of their occupation of a building on or near the site of the Temple of Solomon in Jerusalem. In later examples, probably identical in meaning with TEMPLEMAN.

**Temple:** Hugo *of þe temple* a1131 (*s.a.* 1128) ASC E; Paganus *de Templo* 1141 Oseney (O); Matilda *du Temple* 1273 RH (O); John *del Temple* 1360 AD vi (Lei); William *Temple* 1380 LoPleas. OE *templ*, ME, OFr *temple* 'temple'. Hugo Paganus, founder of the Templars was called Hugo of the Temple. *v.* TEMPLAR. Not all those called Temple or Templar can claim an association with Jerusalem. Between '1728 and 1755 no less than 104 foundlings baptized at the Temple [in London] were surnamed Temple or Templar, which will explain the frequency with which the name appears in the *London Directory*' (Ewen 270).

**Templecombe:** Henry *Templecombe* 1309 IpmW. From Combe Temple (D).

**Templeman:** William, Reginald *Templeman* 1240 FFY, 1260 FFC; Ambrose *le Templeman* 1279 RH (C). 'Servant of the Templars', 'one who lived on one of their manors'. The surname was not uncommon in Cambridgeshire where the Templars had manors at Isleham and Duxford, still called The Temple and Temple Fm respectively. The real name of Ricardus *serviens Templariorum*, employed in some capacity at the West Hanningfield manor of the Templars in 1277, was, no doubt, Richard *Templeman*, a variation of that of John *de Templo*, connected with the same neighbourhood in 1248 (PN Ess 252).

**Tenacre, Tennaker:** John *de Tenacre* 1275 SRWo; Philip *de Teneacre* 1316 IpmGl; John *Tenacre* 1317 AssK. From Tenacre in Lynsted (K), or 'a holder of ten acres', OE *tīen*, *æcer*. cf. Walter *Tenmark* 1272 FFC 'holder of a fee worth ten marks'.

**Tench:** Alan, William *Tenche* 1193 P (L), 1221 AssWa. OFr *tenche* 'a tench', a fat and sleek fish.

**Tennyson, Tenneson, Tennison, Tenison:** John *Tennysone*, *Tenison* 1361 CorY, 1632 RamptonPR; Edward *Tennyson* 1691 FrYar. Usually explained as a variant of DENISON. Although no firm proof of this is forthcoming, there is clear evidence for the interchange of initial *t* and *d*, both early and late. *Tunstall* and *Tunnicliff* survive by the side of *Dunstall* and *Dunnicliff*. Tavistock is found as *Davistock* 1220 Cur, Peter *Tyson* or *Dysun* occurs in a Canterbury will of 1528, while Robert *Dredegold* must have been a *Treadgold* or *Threadgold*. *Dandridge* is from Tandridge (Sr), and cf. also John *Dendelyoun* 1313

SRWa and Richard *Taunteloun* 1327 *SR* (Ess), both for 'dandelion', *Tanefeld* and *Danefeld* 1086 for Tanfield (NRY), *Dyddenhamm* 956 and *Tideham* 1086 for Tidenham (Gl), and *Dreamanuuyrða* 824 for Trimworth (K). In view also of such pairs as *Dandy* and *Tandy*, *Dancock* and *Tancock* (Andrew), *Dannett* and *Tannett* (Daniel), the equivalence of *Tennyson* and *Denison* may be accepted.

**Tengue:** *v.* TINGAY

**Tennant, Tennent, Tennents:** Richard *Tenand* 1332 SRCu; Robert *Tenaunt* 1332 SRLa. OFr *tenant* 'tenant', a holder or possessor of a tenement.

**Tenner:** Robert *le Tennore* 1255 RH (Sa); Richard *le Tenur* 1275 SRWo; Henry *Tenere* 1377 AssEss. OFr *teneor*, *teneur*, *tenor* 'tenant'.

**Terling:** *v.* TARLING

**Termday:** Margery *Termeday* 1279 RH (O); William *Termeday* 1420 LLB I. ME *terme day* 'a day appointed for doing something, especially for the payment of money'. Presumably for one born on that day.

**Ternent:** William *le Turneaunt* 1223 Cur (Y); Richard *Turnaunt* 1486 *ERO*. Present participle of OFr *turner* 'to turn', a variant of TURNER. cf. Fr *Tournant* (Dauzat).

**Terrall, Terrell, Terrill:** *v.* TIRRELL

**Terrey, Terry, Terris, Tarry, Todrick, Torrey, Torrie, Torry:** *Theodricus*, *Tedric(us)*, *Teodericus* 1086 DB; Willelmus *filius Teorri* 1114–30 Rams (Lo); *Theodericus* 1124–30 Rams (Beds); *Terri* Vsuarius 1166 P (Nf); *Terricus* de Mudiford 1177 P (D); Ricardus *filius Thedrici* 1185 Templars (Wa); *Theoricus* de Werdesford' 1213 Cur (Do); Jordanus *frater Tedric'* 1219 AssY; *Tyrri* de Athelartone 13th Ronton (St); Ralph *Teri* 1199 FrLeic; Warinus *þedrich* 1221 *ElyA* (Sf); John *Terry* 1221 AssWa; Thomas *Therry* 1243 AssSo; Hugo *Tyry*, *Tyeri*, *Terry* c1250 Rams (Beds); Alice *Tedrich* 1276 RH (Berks); John *Thudrich* 1279 RH (O); Henry, William *Torry* 1276, 1279 RH (Do, Bk); William *Tarri* 1279 RH (O); Richard *Tarry* (*Thary*) 1327 *SR* (Ess); John *Therrich* 1327 SRSf. OFr *Thierri*, *Tierri*, *Terri*, OG *Theudoric* 'people-rule', a common name on the continent. OE *þēodrīc* is well-evidenced in OE and may be the source of *Thedrich*, *Thudrich*, *Therrich*. Many of the early ME bearers of the name were Flemish or German merchants, frequently described as *Teutonicus*, *Alemannus* or *Flandrensis*. The name was at times confused with *Theodore*. Both *Terricus* de Treiagol and *Theoricus* de Wycheford 1221 Cur (Co, Wa) are alternatively called *Theodorus*.

**Terse:** Henry *Ters* 1221 AssWa; John *Ters* 1275 SRWo; Richard *Ters* 1391 LuffCh. OFr *tiers* 'the third (son)'. cf. Fr *Thiers*.

**Tesh:** *v.* ASH

**Tesler, Tessler:** *v.* TAZELAAR

**Tesseyman, Tessimond:** Richard *Tacyman* 1340 Bardsley; Brian *Tesymon* 1537 GildY; Robert *Tesymond* 1558 FrY. ME *Tesmond*, but sometimes 'servant of *Tacey*'.

**Testar, Tester:** Roger, Hugh *Testard* 1135 Oseney (O), 1172 DC (L); Gilbert *Testar* Hy 2 DC (L); William *Testart* 1230 P (Sr); Thomas *Tester* 1551 SxWills. OFr *testard*, a pejorative of *teste* 'head', 'big head'. Fairly common.

**Teste:** Robert *Teste* 1211 Cur (Nf); Matilda *Teste* 1327 SRSo; Robert *Test* 1451 AssLo. A nickname from the head, OFr *teste*.

**Testwood:** Peter *Testewod* 1278 IpmW; Gilbert *Teswode* 1343 IpmW; John *Testewode* 1352 FFW. From Testwood (Ha).

**Tetherley:** *v.* TITHERLEY

**Tetlow:** Richard *de Tetlawe* 1389 IpmLa; Edmund *Tetlowe* 1554 PN Ch i 306. From Tetlow (La).

**Tetsall:** *v.* TATTERSALL

**Tett:** Robertus, Ricardus *filius Tette* 1166 P (Nf); 1180 P (Y); John *Tette* 1275 SRWo. OE *\*Tetta*, found in several place-names, the masculine form of *Tette*.

**Tettersell:** *v.* TATTERSALL

**Tetther:** *v.* TEDDER

**Teversham:** Ulfric *de Teuersham* 1086 ICC; William *de Teversham* 1227 FFC. From Teversham (C).

**Tew:** (i) Hugo *de Tiw* 1130 P (O). From Tew (Oxon). (ii) Hugh *le Tyw*, *le Tuy* 1286–8 AssCh. Welsh *tew* 'fat, plump'.

**Tewer, Tewers:** Nicholas *le Teware* 1275 SRWo; John *Tewer* 1327 SRY. A derivative of ME *tewe* 'to taw'. *v.* TAWYER.

**Tey:** Walter *de Teye* 1294 FFEss. From Great, Little or Marks Tey (Essex); John *del Teie* 1221 ElyA (Sf). 'Dweller by the enclosure or common pasture' (OE *tēag*).

**Teyte:** *v.* TAIT

**Thacker, Thakore:** Richard *the Thaker* 1316 Wak (Y); Roger *le Thacchere*, *le Thakkere* 1339 Crowland (C). A derivative of OE *þacian* 'to thatch' or ON *þak* 'thatch'; 'thatcher' (1420 NED). cf. THATCHER and THEAKER.

**Thackeray, Thackara, Thackra, Thackrah, Thackray, Thackwray:** John *de Thakwra* 1379 PTY; John *Thackerowe* 1548 RothwellPR (Y); Anne *Thackarawe* 1566 ib.; James *Thackerey* 1602 FrY; Thomas *Thackwrey* 1613 ib. 'Dweller by the nook where reeds for thatching grew', ON *þak*, *(v)rá*, as at Thackray Wood in Blindcrake (Cumb).

**Thackway:** Adam *de Thakthuait*, John *de Thaktwait* 1332 SRCu. 'Dweller by the land where thatching-reeds grew', as at Thackthwaite (Cumb, NRYorks).

**Thain, Thaine, Thayne:** Robertus *filius Thein*, Robert *Tein* 1166 P (Nf); Geoffrey *le þein* 1199 P (Nth); Adam *þein* 1221 ElyA (Sf); William *le Theyn* 1243 AssSo. OE *þegen*, *þegn* 'thane', a tenant by military service.

**Thame:** *v.* TAME

**Tharp:** *v.* THORP

**Thatcher:** (i) Reginald *le Thechare* 1273 RH (O); Robert *Thechere* 1327 SRSx. (ii) John *le Thacchere* 1275 SRWo; John *Thatcher* 1327 SRDb. A derivative of OE *þæccan* 'to cover', later 'to thatch', 'a thatcher' (c1440 NED). The modern form is from (ii), which is apparently Anglian. The Saxon *Thetcher* has been absorbed by this. cf. THACKER, THAXTER and THEAKER.

**Thaxter:** Elena *la Thakestere* 1295 MESO (Nf). A feminine form of THACKER (c1440 NED). cf.

BAXTER. Apparently peculiar to Norfolk where it is still found.

**Theadom:** v. THEEDAM

**Theaker:** Walterus *filius Thecker* 1199 MESO (St); William *le Theker(e)* 1273 RH (Nf), 1297 SRY; Anabilla *thekker'* 1327 MESO (L). A derivative of ON *þekja* 'to cover'; 'one who roofs buildings' (14.. NED). The craftsmen of the York mystery plays included the *Tille-thekers* 'men who covered roofs with tiles'. cf. also Thomas *Ledtheker* 1305 MESO (Y), who used lead. cf. THACKER and THATCHER.

**Theakston, Theakstone, Thexton:** William *de Thexton* 1376 FFY; Thomas *Theackstone* 1652 RothwellPR (Y). From Theakston (NRY).

**Theal, Theall:** Ralph *atte Thele* 1296, Robert *atte Thele* 1327 SRSx. 'Dweller by the footbridge', OE *þel*.

**Thebridge:** v. TEBRICH

**Thede, Theed:** Goduuine *Thede filius* c1095 Bury (Sf); Willelmus *filius Thede* 1166 P (Nf); *Thieda* de Westweniz 1198 FF (Nf); *Theda* mater Herlewini 1207 Cur (Nf); Alan *þede* 1277 Ely (Sf); Nicholas *Thede* 1279 RH (C). OE *þēoda*, a short form of names in *þēod-*. Alan, son of *Thede*, held land called *Theduluesmere* in Burstall (Suffolk) temp. Henry III (AD iii). cf. Geoffrey *Thedolf* 1276 RH (Bk), John *Thedwyne* 1317 AssK, and v. TEBRICH. Also occasionally from OE *þēode* (f). cf. *Theberga* vidua c1250 Rams (Hu), Richard *Thedware* 1252 ib., Gerbaga *Thedlef* 1327 SRSf.

**Theedam, Theedom, Theadam:** Richard, William *Thedam* 1308 Cl (Nf), 1319 SRLo; Richard, Gerard *Thedom* 1317 AssK, 1327 *SR* (Ess). ME *þeodam, þeedom* 'thriving, prosperity'.

**Theedolph, Thedall:** Geoffrey *Thedolf* 1276 RH (Bk); Richard *Thedulf* 1327 *SR*Ess; Thomas *Teddoll* alias *Theddol* 1559 Pat (Ha); Edwarde *Thedall* 1576 SRW. OE *þēodwulf*, unrecorded but both elements are common.

**Theobald, Theobalds, Tebbet, Tebbett, Tebbit, Tebbitt, Tebboth, Tebbut, Tebbutt, Tibald, Tiball, Tibballs, Tibbatts, Tidbald, Tidball, Tidboald, Tudball:** *Teobald, Tedbaldus, Tetbald, Tebaldus* 1086 DB; *Theobaldus* 1161–77 Rams (Hu); *Thebaldus, Tebbaldus* de Valeines 1206 Cur (Sf, C); *Tibaldus* Walteri 1212 RBE; *Tybaud* de Valeines 1212 RBE; Tomas *Teobald'* 1199 P (Gl); Hugo *Tebaud* 1202 P (L); William *Theobald* 1250 FFSf; Simon *Tebalde* 1255 Rams (Hu); Stephen *Thebaud* 1275 RH (Bk); Rustikill *Thedbald* 1277 LLB B; Thomas *Tedbald* 1279 RH (C); Margery *Tebbolt* ib. (C); John *Tebold, Tebbel* 1296 SRSx; Philip *Tubald, Tubbald* 1332 SRSx; John *Tedbot* 1338 FFSf; Geoffrey *Tebbaut* 14th AD ii (Nth); William *Tebott* 1405 FrY; Herbert *Tebbell, William Tebball, John Tebbet* 1674 HTSf; John *Tibbatts* 1802 Bardsley. OFr *Theobald, Teobaud, Thiebaut, Tibaut*, OG *Theudobald, Thiotbald,* 'people-bold', a common continental name. The modern surname *Theobald* is a learned form. The vernacular pronunciation in the 16th and 17th centuries was *Tibbald*, even when the spelling was *Theobald*. *Tebbott* is from an AN spelling in which the *th* was pronounced *t*. v. also TEBB, TEBBELL, TIBB, TIBBETS, TIBBLE.

**Thew, Thewes, Thow:** Gilbert, William *Thewe* 1190 P (Y), 1196 P (Nt); Johanna *Thow* 1348 DbAS 36. OE *þēow* 'a slave, bondsman, thrall'.

**Thewles, Thewless, Thewlis, Thouless, Thowless:** John, William *Theules* 1327 Wak (Y); Thomas *Thewelesse* 1379 PTY; William *Thowlas, Thowless* 1464, 1493 Black. OE *þēawlēas* 'ill-mannered', from OE *þēaw* 'custom, usage; manner of behaving or acting; a personal quality (mental or moral)' and *lēas*, 'destitute of morals or virtue; vicious, immoral' (a1327 NED). *Thowless* is a Scottish form though the phonology is unexplained; 'wanton, dissolute' (1375 NED).

**Thexton:** v. THEAKSTON

**Thick, Thicks:** John *le Thike* 1243 AssSo. OE *þicce*, ME *thikke* 'thick-set, stout'.

**Thickett, Thickitt:** Edith *atte Thikkette* 1338 FFW. From Thicket Copse in Mildenhall (W), or 'dweller by the thicket', OE *þiccett*.

**Thickman:** Henry *Thikeman* 14th AD i (Herts); William *Thikeman* 14th AD ii (Ha). A nickname for a thick-set man, OE *þicce, mann*.

**Thicknesse:** Richard *de Thyknes* 1295 AssSt. From Thickness (Staffs).

**Thimblebee, Thimbleby, Thimbelbee:** Odo *de Thimelbi* 1170–98 P (L); Richard *de Thymelby* 1274 IpmY; Elianora *Thimulby* 1483 GildY. From Thimbleby (L, NRY).

**Thimbler:** Robert *le Thumeler* 1332 SRWa; Robert *le Themeler* 1340–1450 GildC; William *le Themelere* 1344–5 FFWa. A derivative of OE *þymel* 'a thimble', a maker of thimbles.

**Thin, Thyne, Thynne:** Thomas *Thin* 1218 AssL; Gilbert *Thynne* 1269 AssNb. OE *þynne* 'thin, lean, slender'.

**Thirkell:** v. THURKELL

**Thirkettle:** v. THURKETTLE

**Thirkhill, Thirkill, Thirtle:** v. THURKELL

**Thirsk:** v. TRASK

**Thistleden:** Henry *de Thisteldene* 1292 IpmW; Gilbert *de Thistelden* 1327 SRSf; John *de Thisteldene* 1329 IpmW. 'Dweller in the valley where thistles grow', OE *þistel, denu*.

**Thistlethwaite:** Miles *Thissilthwate* 1581 FrY. The Penrith Thistlethwaites derive from a lost *Thistelthueyt* in the Forest of Inglewood (1285 Ipm). The name may also have arisen elsewhere.

**Thistleton, Thiselton:** Symon, Adam *de Thistelton* 1250 FFL, 1286 IpmY; William *Thistylton* 1480 IpmNt. From Thistleton (Lancs), or 'dweller at the farm where thistles abound'.

**Thistlewood:** Stephen *de Thysteleswerthe* 1327 SRSx; George *Thistlewood* 1643 FrY; John *Thustlewood* 1678 SRSf. From Thistleworth Fm in West Grinstead (Sussex) or some similar enclosure overgrown with thistles.

**Thoburn:** v. THURBAN

**Thom, Thoms:** John *Thomme* 1311 ColchCt; John *Thoms* 1327 SRSo. A pet-form of *Thomas*. In Scotland, *Thoms* is an anglicizing of *MacThomas*.

**Thomas, Tomas:** Thomas 1086 DB; Walter *Thomas* 1275 RH (W); Hugo *Tomas* 1317 AssK. An Aramaic name meaning 'twin'. Before the Conquest, *Thomas* is found only as the name of a priest. After the Conquest it became one of the most popular christian names.

**Thomason, Thomasson, Thomerson:** Adam *Thomassone* 1327 SRDb. 'Son of *Thomas*.'

**Thomazin:** *Thomaysin* 1301 SRY; Walter *Thomasyn* 1327 *SR* (Ess). *Thomas-in*, a diminutive of *Thomas*. Both *Thomasinus* and *Thomasina* are found in 1346 (FA). The modern surname is rare, having been almost completely absorbed by *Thomason, Thomerson, Thomson* and *Tomson*.

**Thomerson:** *v.* THOMASON and THOMAZIN

**Thomley:** *v.* TOMLEY

**Thomline:** *v.* TOMLIN

**Thomlinson:** *v.* TOMLINSON

**Thommen:** *v.* TUMMAN

**Thompsett, Thomsett, Tompsett, Tomsett:** Peter *Thomasset* 1792 Bardsley; Charlotte *Thomsett* 1801 ib. *Thomas-et*, a diminutive of *Thomas*. The formation must be old, but examples are late.

**Thompson, Thomson, Tompson, Tomsen, Tomson:** John *Thomson* 1318 Black (Carrick); John *Thompson* 1349 Whitby (Y); John *Thomesson* 'Scot' 1375 LoPleas; Ralph *Thommyssone* 1381 SRSf; John *Tommesson* 1382 AssC; William *Tomsone* 1395 EA (NS) ii (C); Alexander *McThomas* alias *Thomsoune* 1590 Black; John *Tompson, Tomson* 1591 ShefA. 'Son of *Tom*', i.e. Thomas. *Thomson* is the Scottish form, that with the intrusive *p* being English.

**Thompstone:** William *de Tomestona* 1175 P (Nf); Geoffrey *de Thomestune* 13th Lewes (Nf). From Thompson (Norfolk).

**Thonger:** Henry *Thonger* 1428 FA (W). 'A maker of thongs' (OE *þwang*).

**Thor, Thore:** Frane *filius Tor* 1066 DB (L); *Thore* prepositus 1219 AssY; Peter *Thore* 1191 P (L); Thomas *Thorr* 1352–3 IpmNt; Peter *Thur* 1313 AssNf. Anglo-Scandinavian *Þōr, Þūr*, probably a shortened form of ON compounds in *Þór-, Þúr-*. There was also a feminine name: *Thura* (f) 1198 FFNf; *Thora* filia Gamel 1210 Cur (Y). Sometimes, perhaps, from a place-name: Robert *de Thore* 1222 Cur (D).

**Thorbrand:** Adam *Thorbrand* 1303–4, William *Thorbrond* 1422 IpmY; Thomas *Thurbrande* 1456 Black. ON *Þorbrandr*.

**Thorburn:** *v.* THURBAN

**Thoresby, Thorsby:** Mabel *de Thoresby* 1245 FFL; Geoffrey *de Thoresby* 1334 ForNt; Richard *Thoresby* 1444 TestEbor. From Thoresby (Nt, NRY), or North, South Thoresby (L).

**Thorkell:** *v.* THURKELL

**Thorley:** Adam *de Thorlee* 1213 Cur (Ess); John *de Thorleye* 1310–11 FFEss; John *Thorley* 1440–1 FFSr. From Thorley (Herts, Wt).

**Thorman:** *v.* THURMAN

**Thorn, Thorne, Thornes, Thorns:** William *Thorn* 1206 Cur (Sx); Magge *de Thornes* 1275 Wak (Y); William *del þorn* 1277 Ely (Sf); Richard *atte Thorn* 1296 SRSx. 'Dweller by the thorn-bush(es)' (OE *þorn*), or from Thorne (Som, WRYorks).

**Thornberry, Thornbery, Thornber, Thornborough, Thornborrow, Thornburgh, Thornburrow, Thornbury:** Markerus *de Torneberga* 1176 P (Bk); Robert *de Thorneberg'* 1208 Cur (Y); Hugh *de Thornburgh* 1327 SRY. From Thornborough (Bucks), Thornbrough (Northumb, NR, WRYorks), or Thornbury (Devon, Glos, Hereford).

**Thornby:** Philip *de Thornby* 1336–7 CorLo; Alan *Thorneby* 1382 IpmNt. From Thornby (Nth).

**Thorndike, Thorndyke, Thorndick:** Simon *Thornedike*, John *Thorneditch* 1674 HTSf. 'Dweller by the thorny ditch.'

**Thorner:** William *de Thorenour'* 1219 AssY; Thomas *de Thorner* 1393 Calv (Y); Richard *Thorner* 1576 SRW. From Thorner (WRY), or 'dweller by the thornbush', from a derivative of OE *þorn*.

**Thornett, Thornitt:** Thomas *de Thornheved* 1332 SRCu. 'Dweller by the thorn-covered headland.'

**Thorney:** Geoffrey *de Thornay* 1275 RH (L); Simon *de Thorneye* 1327 SRSf; Widow *Thorny* 1662–4 HTDo. From Thorney (C, Mx, Nt, Sf, So), or West Thorney (Sx).

**Thornham:** Michael *de Thorneham* 1168 ArchC v; William de *þornham* 1185 Templars (Ess); Blakewin *de Thornham* 1198 FFNf. From Thornham (K, La, Nf), or Thornham Magna, Parva (Sf).

**Thornhill:** John *de Tornhull'* 1212 P (Y); Walter *de Thornhulle* 1327 SRSo; John *Thornhill'* 1379 PTY. From Thornhill (Derby, Dorset, Wilts, WRYorks).

**Thornholme:** Adam *de Thornholm* 1276 RegAntiquiss; Hugh *de Thornholme* 1304–5 IpmY; John *de Thornholme* 1361 AssY. From Thornholm (ERY). The name probably usually fell in with THORNHAM.

**Thorning:** William *Thirnyng* 1397 Hylle; Jogn *Thornyng* 1359 AssD; Henry *Thorninge* 1642 PrD. 'Dweller by the thornbush', from a derivative of OE *þorn*.

**Thornley:** Nelle *de Thornleye* 1274 Wak (Y). From Thornley (Du, La).

**Thornton:** Beatrice *de Thornton'* 1202 FFY; Ralph *de Thorntone* 1312 LLB D; Henry *Thornton* 1362–4 FrC. From one or other of the many places of this name.

**Thorogood, Thoroughgood, Thorougood, Thorowgood, Thorrowgood:** Stephen *Thorghugod* 1301 ELPN, John *Thourgod* 1330 ib.; Walter *Thorougod* 1306 LLB B; William *Thorowgood* 1674 HTSf. Ekwall explains this as 'thorough-good', but perhaps contracted from ME *þurhūt gōd* (ELPN). This, of course, is a possible explanation, but it is curious that clear examples are so late and so few, whilst those of *Thurgood* are both earlier and more numerous. *Thorghugod* is not a complete proof of this etymology. It may stand for *Thurgod*. cf. Matilda *Þorustan* 1279 RH (Beds), Walter *Thorouston* 1327 SRC, i.e. *Thurstan*, Richard *Thurubern* (C) for *Thorbern*, and William *Throwketyll* 1524 SRSf for *Thurkettle*, where we are certainly concerned only with personal names. Note, too, that three of these are earlier than the earliest example of *Thorghugod*. Many of our Thorogoods are probably Thurgoods though some Thurgoods may be Thorogoods. *v.* THURGOOD.

**Thorold, Thorrold, Thourault, Turral, Turrall, Torode:** *Turold* 1066 DB (L); *Turoldus* ib. (Nf, W); Azor *filius Turaldi* 1066 DB (So); *Toroldus, Thoroldus, Turoldus* capellanus 1121–61 Holme (Nf); *Toroudus,*

*Toroldus* presbiter 1143–7 DC (L); Robertus *filius Thoradi* 1185 Templars (Y); William *Turolde* 1190 P (Gl); Simon *Turold'* (*Turoll'*) 1220 Cur (Ess); Henry *Turaud* 1258 Oseney (O); Robert *Thorald* 1261 ib.; Edmund *Thorold* 1279 RH (C); Symon *Turrad* 1279 ib. (Hu); Walter *Turald* 1296 SRSx. ON *Þóraldr*, *Þorváldr*, ODa, OSw *Thorald* 'Thor-ruler', found also in Normandy. The initial *T* is due to the Anglo-Norman pronunciation of *Th-*. *Torald* became *Toraud*, *Toroud*, giving the modern *Torode*.

**Thorp, Thorpe, Tharp:** William *de Torp* 1158 P (Nb); Robert *de Thorp* 1287 AssCh; William *in le Thorp* 1327 SRDb; Jak *del Thorp* 1332 SRCu. From one of the numerous places named Thorp(e) or from residence in a hamlet or outlying dairy-farm (OE *þorp*). *v.* also THROP.

**Thorsby:** *v.* THORESBY

**Thory, Tory:** *Thori, Tori, Thuri, Thure, Turi* 1066 DB; Hugo *filius Thory* 1218 AssL; John *Thori* c1140 ELPN; Reginald *Thory* 1221 Cur (Nth). ON *Þórir*, ODa *Thorir*, *Thori*. A Danish personal name, found chiefly in the eastern counties and not found in Normandy. *Tory* is due to Anglo-Norman pronunciation in England.

**Thoules, Thowless:** *v.* THEWLES

**Thow:** *v.* THEW

**Thoyts:** *v.* THWAITE

**Thrale, Thrall:** John *Thral* 1309 SRBeds; Richard *le Threl* 1332 SRSx. OE *þræl*, from ON *þræll* 'a villein, serf, bondman'.

**Thrasher, Thresher:** Richard *þrescere* 1221 *ElyA* (Sf); Geoffrey *le Thressher* 1319 FFEss. A derivative of OE *þerscan, þrescan, þryscan* 'to thresh'. 'A thresher.'

**Threader, Thredder:** Thomas *Thredere* 1365 LoPleas; William *Treder* 1379 PTY. A derivative of OE *þrǣd* 'thread'. 'One employed to keep the shuttles threaded in weaving.'

**Threadgold, Threadgould, Threadkell, Threadgill, Thridgould, Treadgold, Tredgold:** *Tredegold* 1166 P (Sr); Reginald, Agnes *Tredegold'* 1199 MemR (W), 1202 AssL; Edmund *Thredgall* 1674 HTSf; Daniel *Thredkill*, Widow *Thredkell* ib.; John *Thridgale* 1681 SfPR. 'Thread gold', a name for an embroiderer.

**Threapland:** John *Threpland* 1443 TestEbor; John *Threapland* 1459 Kirk; Joseph *Threapland* 1707 WRS. From Threapland (Cu).

**Thredder:** *v.* THREADER

**Threlfall, Trelfall:** William *de Threliffal* 1246 AssLa. From Threlfall (Lancs).

**Thresh:** *v.* THRUSH

**Thresher:** *v.* THRASHER

**Thresk:** *v.* TRASK

**Thrift:** William *Thrift* 1315 Wak (Y). A name, presumably, for one possessed of this virtue. *v.* also FIRTH.

**Thring:** Robert *de Thring* 1275 RH (K). From Tring (Herts).

**Thripp:** A form of *Throp* or *Thorp* which does not seem to have survived as an independent place-name but is found in the unstressed position in Eastrip (Som), Westrip (Glos) and Williamstrip (Glos).

**Throckmorton, Throgmorton:** Adam *de Throkemerton'* 1221 AssWo; Robert *de Throkemarton* 1327

AssSt; John *Throkmorton* 1442 IpmNt; Thomas *Throgmorton* 1464 AD iii (Wa). From Throckmorton (Wo).

**Throp, Throup, Thrupp:** Ralph *de Trop, de Thorp, de Throp* 1263 Ipm (Nth); Adam *de la Thropp'* 1275 RH (W); Edward *Thrupp* 1618 ArchC 49. From Throop (Hants), Throope (Wilts), Thrup (Oxon), or Thrupp (Berks, Glos, Northants). A metathesized form of THORP.

**Throssell, Thrussell, Thrustle:** Roger *Throsle* 1282 AssCh; Richard *Throstel* 1297 SRY. OE *þros(t)le* 'throstle'. A nickname from the bird.

**Throup:** *v.* THROP

**Thrower, Trower:** Simon *le Throwere* 1293 AD i (Nf); Alice *la þraweres* 1301 ParlR (Ess); Thomas *le Throwere* (*Trowere*) 1327 *SR* (Ess). A derivative of OE *þrāwan* 'to throw', probably 'thread-thrower', one who converts raw silk into silk thread. It might also mean 'turner' or 'potter', though these trades gave rise to many other common surnames.

**Thrupp:** *v.* THROP

**Thrush, Thresh:** William *Thresse, Thrusse* c1204 Clerkenwell, 1230 P (Lo); Clement *Thresshe* 1524 SRSf. OE *þrysce* 'a thrush'.

**Thrussell, Thrustle:** *v.* THROSSELL

**Thruston:** Andrew *Thurston*, John *Thryston* 1524 SRSf. A metathesized form of THURSTAN.

**Thum:** Geoffrey *Thumb* 1232 Pat (L). Perhaps a nickname, 'a Tom Thumb', or from some peculiarity, or perhaps loss, of a thumb. cf. William *Thumbeles* 1260 AssC 'thumb-less'.

**Thunder:** Andrew *Thunre* (*Tundur*) 1221 Cur (Ha); John *le Tundur* 1289 NorwDeeds I; William *Thunder* 1332 SRSx. OFr *tondeur* 'shearer'. But perhaps also a nickname from OE *þunor* 'thunder', cf. William *Tonitruus* 1160 P (Nf/Sf).

**Thurban, Thurbon, Thurburn, Thorburn, Thoubboron, Thoburn, Turbin, Tarbun, Tarbin:** *Thurbernus, Turbern, Torbern* 1066 DB; Thomas *filius Thurberni, Turberni* 1153–86 Holme (Nf); *Thurbarnus* filius Ailiue 1197 FF (Sf); Richard *Turbern* 1198 FF (Nf); William *Thurbern* 1221 AssWa; Richard *Thurubern* 1277 Ely (Nf); Richard *þorebarn*, William *Thorebern* 1279 RH (O); William *Thourubern* 1295 Barnwell (C); Walter *Thurbarn* 1327 *SR* (Ess); — *Thurbin*, Thomas *Thurbon* 1674 HTSf. ON *þorbiorn*, ODa, OSw *Thorbiorn* 'Thor-bear', anglicised as *þurbeorn* 'Thor-warrior'. *Turbin* preserves an Anglo-Norman pronunciation of *Th* and may represent a diminutive, *Turb-in*. cf. Robert son of *Turbyn* 1227 AssLa, Unwinus *filius Turbini* Hy 3 Colch (Ess).

**Thurgar, Thurger, Thurgur:** *Turgar* 1066 DB (Wo, He, Nth); *Thurgarus* 1066 ICC (C); *Turger* filius Ade 1230 P (L); Paganus *Thurgar* 1279 RH (Beds); Hugo *Thurger*, Matilda *Thorgor* 1327 SRSf. ON *þorgeirr*, ODa, OSw *Thorger* 'Thor-spear', found also in Normandy.

**Thurgate:** *v.* THURGOOD

**Thurgell:** *v.* THURKELL

**Thurgood, Thorgood, Thurgate:** Hunfridus *filius Turgoti* 1183 P (St); Magnus *filius Thurgot* 1219 AssL; *Thurgod* de Fynnynghersh 1327 SRSx; Adam *Turgod* 1207 ChR (Do); William *þurgod* 1275 SRWo;

William *Thorgot* 1297 SRY; Thomas *Torgod* (*Torgot*) 1298 LLB B; Adam *Thorgod* 1327 SRSo. ON *Þorgautr*, ODa, OSw *Thorgot* 'ThorGeat'. *v.* also THOROGOOD.

**Thurkell, Thurtell, Thurtle, Thorkell, Thirkell, Thirkill, Thirkhill, Thirtle, Turkel, Turkil, Turtill, Turtle, Tuttle, Thurgell, Turgell, Turgill, Toghill:** *Turkil* 1066 DB (Sf); *Thorkil* c1095 Bury (Sf); *Thorchill, Thurchill* 1066 InqEl (Herts); *Thirkillus* 12th Gilb (L); *Turkil* Palmer 1177 P (Nt); Robert *Turkil* 1190 P (Wo); John *thurkil* 1224–46 Bart (Lo); William *Thorekil* 1279 RH (O); Adam *Turkild* 1283 SRSf; John *Thurkeld*, Alan *Thurkild* 1327 SRSf; John *Togel* 1327 *SR* (Ess); Richard *Turtill*, John *Thyrthyll* 1524 SRSf; Richard *Thyrkle* 1544 NorwW (Nf); John *Tuttell* 1568 SRSf; John *Tirtle*, Henry *Turtle* 1674 HTSf; William *Tuttle* 1674 HTSf. ON *Þorkell*, ODa *Thorkil*, a contracted form of ON *Þorketill*, very common in England and often used in Normandy. *v.* THURKETTLE.

**Thurkettle, Thurkittle, Thirkettle:** *Turketel* 1066 DB (Nf, Sf); *Thurketel* c1095 Bury (Sf); Johannes *filius Turketilli* 1214 Cur (Nf); Robert *Turketil* 1182 P (O); Richard *Turchetel* 1198 FF (Nf); John *Thurketell, Therketell* 1524 SRSf; William *Throwketyll* ib. Anglo-Scand *Þurcytel*, from ON *Þorketill* 'Thor's (sacrificial) cauldron'. Less common than the shortened *Þorkell v.* THURKELL.

**Thurkleby:** Roger *de Turkilleby* 1247 LuffCh. From Thirkleby (NRY).

**Thurland:** Thomas *Thurland* 1472 IpmNt; John *Thurlond* 1524 SRSf. From Thurland (La).

**Thurlow, Thurloe:** John *de Thrillowe* 1278 RH (C), 1327 SRSf; Antony *Thurlowe* 1524 SRSf. From Thurlow (Suffolk), *Thrillauue* c1095.

**Thurlston, Thurlstone:** John *Thoralston* 1672 HTY. From Thurlstone (WRY).

**Thurman, Thurmand, Thorman:** *Turmund* 1066 DB (Do, So); Adam *Thuremund* 1248 *Ass* (Ha); Henry, Walter *Þurmond* 1279 RH (O). ODa, OSw *Þormund* 'Thor's protection'. This is a rare personal name. More common is ON *Þormóþr*, ODa *Thormoth*, OSw *Thormodh* 'Thor's wrath': *Thurmot, Turmod* 1066 DB; *Þurmod* 1221 ElyA (Sf); Hubert *Thurmod* 1212 RBE (Sf); Alan, Hugh *Thormod* 1275 RH (Nf). The second theme *-mod* has probably been assimilated to the more common element *-mund*.

**Thurrell:** *v.* TURRILL.

**Thurstan, Thurstans, Thurston, Thursting, Tustain, Tustian, Tustin, Tusting, Tutin, Tuton, Tuting, Dusting:** *Turstan* (Y), *Turstinus* (St), *Tursten* (Y), *Tostin* (He), *Turtin* (Sa) 1066 DB; *Turstin* (*Thursten*) 1086 ICC (C); *Thurstanus, Turstan* c1095 Bury (Sf); *Turstin'* 1177 P (Nf); *Tutan* Hy 2 DC (L); *Turstanus, Tostenus* Bodin 1180–1200 DC (L); *Turston* 1188 BuryS (Sf); *Tosten'* Basset 1190–4 Seals (Bk); *Thurstan le Brok* 1296 SRSx; *Tustanus cocus, Tostin coc* 14th AD iv (Nth); *Wlmer Þurstan* 1221 ElyA (Sf); John *Turstein* 1250 Fees (So); William *Thurstan*, *Thursteyn* 1278 AssSo; Matilda *Þorustan* 1279 RH (Beds); William *Dusteyn* 1282 Ipm (Nf); William *Thurston* 1297 MinAcctCo (W); William *Thursten* 1354 Oriel (O); John *Tuteing* 1641 FrY; John *Tueton*

1653 RothwellPR (Y); John *Tutin* 1692 FrY. ON *Þorsteinn*, ODa *Thorsten* 'Thor's stone', anglicized as OE *Þurstān*. Common in Normandy as *Turstinus, Turstenus*, where it became Fr *Toustin, Toustain*. This, in England, became *Tustin*(*g*), *Tutin*(*g*). *Thurston* may also be local in origin: Osward *de Turstun'* 1121–48 Bury (Sf); Herueus *de Thurston* 1221 ElyA (Sf). From Thurston (Suffolk).

**Thury:** *Thuri, Thure, Turi* 1066 DB; Alice *filia Thuri* 1205 Cur (O); Adam *Thurry* 1247 FFO; Isabel *Thouri* 1327 SRSx; Walter *Thoury* 1327 SRWo. ON *Þórir*, ODa *Thorir*, OSw *Thore, Thure*.

**Thurtell, Thurtle:** *v.* THURKELL.

**Thwaite, Thwaites, Thwaits, Thwaytes, Twaite, Twaites, Twaits, Tweats, Thoytes:** Ralph *del Thweit* 1206 P (Nf); Ralph *de Thweit* (*Twaeit*) 1221 Cur (Nf); Alan *del Twayt* 1301 SRY; Robert *del Twaytes* 1379 PTY; William *Twaytes* 1492 BuryW (Sf); Matthew *Thwayts* son of James *Twayts* 1618 Oseney (O). From Thwaite (Norfolk, Suffolk), or from residence near a forest clearing, a piece of land fenced off and enclosed, or low meadows. ON *þveit* 'a meadow, piece of land'.

**Thyne, Thynne:** *v.* THIN.

**Tibb, Tibbs:** *Tibbe, Tybbe* 1286, 1290 AssCh; William, Richard *Tibbe* 1327 SRWo, SRY. In the 13th century *Tibbe* was used as a pet-name for both men and women, from *Isabel* or *Tibald* (Theobald). The latter interchanged with *Tebbe*. Tibb's Fm (PN Sx 517) owes its name to Nicholas *Tebbe* (1327 SRSx).

**Tibbets, Tibbett, Tibbetts, Tibbits, Tibbitt, Tibbitts, Tibbatts, Tibbott, Tibbotts:** John *Tybote* 1327 SRWo; Stephen *Tybet* 1332 SRSt; James *Tibbett* 1674 HTSf. *Tib-et, Tib-ot* may be diminutives of *Tibb*, from *Tibaud* (*v.* THEOBALD) or of *Isabel*. cf. *Tibota* Foliot 1279 RH (O). *Tibbott*, like *Tibbatts*, may derive directly from *Tibold* or *Tibaud*.

**Tibbey:** Robert *Tibi* 1279 RH (C); Agnes *Tybi* 1327 SRWo. A diminutive of *Tibb*, a pet-form of *Isabel*, or of Theobald. *v.* TIBB.

**Tibbins:** Thomas *Tibben* 1332 SRWa. *v.* TIBB, GEFFEN.

**Tibble, Tibbles:** Ulketellus *Tibel* 1203 Cur (Nf); William *Tybel* 1309 SRBeds. A diminutive of *Tibb*. It may also be a late weakening of *Tibald* (Theobald).

**Tice, Teece:** *Tiecia* 1148 Winton (Ha); *Tiecia* Ramona 1203 Cur (Sf); Adam *Tice* 1206 P (Ha); John *Tece* 1279 RH (C); John *Tyse* 1486 TestEbor. OG *Tietsa, Teucia, Tezia* (f), hypocoristics of OG names in *Theud-*.

**Tichborne, Tichbourne, Tichbon, Tichband:** Walter *de Ticheburne* 1235 *FF* (Ha). From Tichborne (Hants).

**Tickel, Tickell, Tickle:** William *de Tikehill* 1175 P (Y); Roger *de Tikell* 1327 SRDb. From Tickhill (WRYorks).

**Ticknall, Ticknell:** Geoffrey *de Tykenhale* 1327 SRDb; John *de Tyknale* 1374 DbAS 36. From Ticknall (Db).

**Tidball, Tidball, Tidboald:** *v.* THEOBALD

**Tidbury:** Thomas *Tidbury* 1514 PN Do i 30; William *Tidbury* 1662–4 HTDo. From Tidbury Ring (Ha).

**Tidd:** Gerbod *de Tid* 1190 P (L); Walter *Tyde* 1327 SRSf; John *Tydde* 1392 NottBR. From Tydd St Mary

(L), or Tydd St Giles (C). Sometimes, perhaps, from OE *Tida*.

**Tiddeman, Tidiman, Tidman, Tydeman, Titman, Tittman:** William *tedingman, teðingman* 1193, 1197 P (W); John *Tytheman* 1327 SRSf; John *Tytman* 1524 SRSf. OE *tēoðingmann* 'the chief man of a tithing (originally ten householders), headborough'. For the development, cf. William *Tythynglomb* 1333 LLB E, Agnes *Tidilamb* 1364 LoPleas.

**Tiddswell, Tideswell, Tidswell, Tidsall:** Warin *de Tideswell* 1206 P (Nth); Henry *de Tiddeswelle* 1336 AssSt. From Tideswell (Derby). Locally pronounced *Tidza*.

**Tidey, Tidy, Tiddy:** Margaret *Tydy* 1327 SRC; Robert *Tydy* 1371 Misc (C); Robert *Tydy* 1394 Glapwell (Db). ME *tidif*, the name of a small bird, mentioned by Chaucer as inconstant, and by Drayton as a beautiful singer.

**Tidmarsh:** Adam *de Thedmers* 1297 MinAcctCo; Stephen *de Tydemersh'* 1315 FFK. From Tidmarsh (Berks).

**Tidy:** *v.* TIDEY

**Tier:** *v.* TYER

**Tiffany, Tiffen, Tiffin:** *Tephania* Hy 2 Gilb (L); *Theophania, Teffania, Theffanie* (identical) 1206 Cur (K); *Tiphina* le Justiser 1322 AD vi (D); *Tiffania* 1323 AD iii (Nf); *Tiffan, Teffan* 1379 PTY; Gilbert *Tyffayne* 1288 RamsCt (Nf); Cristina *Typhayn* 1327 SRSo; William *Tyffen* 1524 SRSf; Nicholas *Tiffin* 1674 HTSf. Late Latin *Theophania*, from Greek θεοφάνια 'the manifestation of God', another name for the Epiphany. In French this became *Tiphaine* and occasionally *Tiphine* and was given to girls born on Epiphany Day.

**Tigar:** *Tigerus* 1086 DB (Sf); John *Tygre* 1319 FFEss; Nicholas *Tygre* 1327 SRWo. OFr *Tigier*, OG *Thiodger* 'people-spear'.

**Tigg:** *v.* TEGG

**Tighe, Tigue:** *v.* KEIG

**Tike, Tyke:** Walter *Tike* 1141–9 Holme (Nf); John *Tyke* 1248 AssSo; Thomas *Tike* 1327 SRC. Either OE *Tica*, or a nickname from ON *tik* 'bitch', in the sense 'a low-born, lazy fellow'.

**Tilbrook, Tilbrooke, Tillbrook:** William *de Tilebroc* 1200 Pleas (Hu); Alexander *de Tilebrock* 1245 FFL; Henry *de Tilbroke* 1340 CorLo. From Tilbrook (Hu).

**Tilcock, Tillcock:** William *Tilkoke* 1556 Oxon. *Till* (Matilda) plus *cock*.

**Tilden, Tylden:** Richard *Tylden* 1524 SRSf; John *Tilden* 1576 SRW. From Tilden in Headcorn, Great, Little Tilden in Marden, or Tilden in Benenden (K).

**Tildesley, Tildsley, Tillsley, Tilsley, Tilzey, Tydsley, Tyldesley:** Hugo *de Tildesle* 1212 Fees (La). From Tyldesley (Lancs), with varied simplifications of the difficult consonantal combination *ldsl*.

**Tiler, Tyler, Tylor:** (i) Roger *le Tuiler* 1185 Templars (Lo); Alard *le Tuilur* 1198–1212 Bart (Lo); Richard *le Tiwelur* 1205 P (L); Nicol *tyulur* 1230 P (L); Hamo *le Tewler* 1276 LLB B; John *le Tuler* 1281 ib. OFr *tieuleor; tieulier, tiewelier, tuiweliere, tiuelier, tuilier*. 'Tiler, tile-maker.' This French form, which is common, is earlier than the corresponding English one but does not seem to have survived and is not

given in NED. (ii) Simon *le Tyeler'* 1286 MESO (Nf); Stephen *Le Tygeler'* 1288 ib. (Sx); Richard *le Tyghelere*, Simon *le Teylur'* 1296 SRSx; William *le Tielere* (*le Tiller'*) 1327 SR (Ess). A derivative of OE *tigule, tigele* 'a tile' or of the ME verb *tile* formed from this. 'Tiler.' (iii) Robert *le Tiler* 1222 Cur (Ess); Geoffrey *le Tylere* 1279 RH (Hu); Alan *le Tyliere* 1327 SRSf. NED has *tile* 'tile' once c1300 and *tyle* for both the noun and the verb from c1375 onwards, but it is doubtful whether the form had developed as early as 1222. Some of these forms with a single *l* may belong to TILLER.

**Tiley, Tily, Tyley:** *v.* TILLEY

**Tilford:** (i) John *Tilford* 1715 FrY. From Tilford (Sr). (ii) Henry *Tylfot* 1296 SRSx; William *Tilfot* 1443 CtH. A nickname from some peculiarity of the foot.

**Till, Tille, Tills:** *Tilla* 1246 AssLa; *Tille* 1325 Wak (Y), 1332 SRCu; William *Tyl* 1327 SRWo; John, Richard *Tille* 1327 SRSx. A pet-form of *Matilda*.

**Tillbrook:** *v.* TILBROOK

**Tiller, Tillier, Tillyer, Tilyer:** John *le Tillere* 1299 LLB C; William *le Tiller* 1327 SRWo. ME *tiliere, tilyer*, a derivative of OE *tilian* 'to till' and taking the place of OE *tilia* (*v.* TILLEY), 'one who tills the soil, husbandman, farmer or farm labourer' (c1250 NED). Some of the forms under TILER (iii) may belong here.

**Tillett, Tillott:** *Tillot* Hobwyfe 1379 PTY; Cecilia, George *Tillote* 1279 RH (O), 1303 FA (Sf); John *Tillet* 1674 HTSf. *Till-ot*, a diminutive of *Till* (Matilda).

**Tilley, Tillie, Tilly, Tiley, Tily, Tylee, Tyley:** (i) Ralph *de Tilio* 1086 DB (D); Oto *de Tilli* 1185 Templars (Y); Wulwordus *Tillie* (*de Tilie*) 1230 P (K). From Tilly-sur-Seulles (Calvados) or Tilly (Eure). (ii) John, Matilda *Tilly* 1274 RH (So), 1277 *Ely* (Sf); Simon *Tyly* 1296 SRSx; William *le tylie* 1332 SRSr. OE *tilia*, ME *tilie*, from OE *tilian* 'to till', 'husbandman'. cf. TILLER. This may also be a diminutive of *Till* (Matilda).

**Tilling:** John *Tylling*, William *Tulling* 1279 RH (O); William *Tyllyng* 1280 AssSo; Richard *Tulling* 1327 SRSx. *Tilling* may be an unrecorded OE *\*Tilling* 'son of *Tilli*'; but the alternative *Tulling* suggests OE *\*Tylling* 'son of *\*Tylli*', an unrecorded form of OE *Tulla*. The *Tulling* who was a witness at Exeter c1100 (Searle) bore a Devonshire form of the name.

**Tillingham:** Terry *de Tillingham* 1208 P (Ess); Benedict *de Tyllyngham* 1296 SRSx. From Tillingham (Ess), or Tillingham Fm in Peasmarsh (Sx).

**Tillinghast:** Ralph *de Tellingherst* 1230 P (Sx). From Tillinghurst (Sussex).

**Tillman:** Alexander, Alan *Tileman* 1204 P (Gl), 1260 AssC; Adam *Tilleman* 1301 SRY. Either OE *\*tilman* 'husbandman, farmer', identical in meaning with TILLER, or early examples of *tile-man* 'maker of tiles', *tyleman* 1479–81 NED, *tyll mane* 1609 ib. *v.* also TILER.

**Tillotson:** William *Tillotson* 1379 PTY. 'Son of *Tillot*.' *v.* TILLETT.

**Tillott:** *v.* TILLETT

**Tillsley:** *v.* TILDESLEY

**Tilly:** *v.* TILLEY

**Tillyer:** *v.* TILLER

**Tilney:** William *de Tilnea* 1170 P (Nf); Walter *de tilneye* 1277 *Ely* (Sf); Geoffrey *Tylneye* 1414 FFEss. From Tilney (Nf).

**Tilsley:** *v.* TILDESLEY

**Tilson:** George *Tylson* 1609 FrY. 'Son of *Till*' (Matilda).

**Tilt:** Osbert *de la Tilthe* 1202 P (Sr); Emma *Bytilde* 1275 SRWo; John *ate Tilthe* 1297 MELS (Sx). 'Dweller by the cultivated land', OE *tilðe*.

**Tilton:** Robert *de Tilton'* 1163 P; John *de Tilton'* 1218 P (Lei/Wa); Richard *de Tilton'* 1327 SRLei. From Tilton (Lei).

**Tily:** *v.* TILLEY

**Tilyer:** *v.* TILLER

**Tilzey:** *v.* TILDESLEY

**Timbell, Timble:** Ysaac *de Timbel* 1219 P (Y); Robert *of Timbel* 1259 IpmY; John *Tymble* 1400 PN WRY iv 130. From Timble (WRY).

**Timberlake, Timblick:** John *de Tymberlake* 1281 AssSt. From a lost Timberlake in Bayton, Worcestershire (PN Wo 40).

**Timmerman:** Margeria *Tymberman* 1327 SRSf. 'A dealer in timber' (1429 NED).

**Timmins, Timmings, Timmon, Timmons, Timings:** Gilbert *Timin* 1279 RH (C); Richard *Tymyng* 1332 SRSx; Richard *Tymmyng* 1477 IpmNt. *Tim-en, Tim-on*, diminutives of OG *Thiemmo*, or of an unrecorded OE *\*Tima*.

**Timms, Tims, Timme, Timm, Tymms, Tymm, Tym:** William *filius Tymme* 1285 AssCh; Alicia *Timme* 1327 SRWo; William *Tymmes* 1332 SRWa. OE *\*Tima*, OG *Thiemmo*. Hardly, as suggested by Bardsley and Harrison, from *Timothy*, since that name does not appear in England until after the Reformation.

**Timperley:** *v.* TEMPERLEY

**Timperon:** Adam *de Tymperon* 1332 SRCu. From Tymparon Hall in Dacre (Cumb).

**Tindal, Tindale, Tindall, Tindell, Tindill, Tindle, Tyndale, Tyndall:** Adam *de Tindal'* 1165 P (Nb); Roger *Tindale* 1332 SRCu; John *Tyndall* 1395 Whitby (Y). From Tynedale, the valley of the Tyne, or from Tindale (Cumb).

**Tinegate:** Thomas *Tynegate* 1332 SRCu.

**Tingay, Tingey, Tengue, Tanguy, Tangye:** Rogerus *filius Teingi* 1130 P (C); *Tinghi* 1182 Gilb (L); *Tengi* de angulo 1219 AssL; Ralph *Tenghy* 12th Lichfield (St); Richard *Tengi* 1202 AssL; Alexander *Tingy*, William *Tengy* 1260 AssC. *Tanguy, Tanneguy*, a common name in Brittany, from St Tanguy, one of the entourage of St Pol of Leon, found also in Cornwall. cf. Breton *Tanchi* 859–65, *Tangi* 1086 (Loth). The first element is Bret *tân* 'fire', the second may be *ci* 'dog'. In the 12th and 13th centuries the name is found in England in districts where Bretons are known to have settled, usually as *Tengi*, occasionally as *Tingi*.

**Tingle, Tingler:** Alan *Tingel* 1209 P (C); John *Tyngil* 1275 RH (Nf). ME *tingle, tyngyl* 'a very small kind of nail' (1288 NED), usually *tingle-nail*, used by metonymy for *Tingler*, a maker of these. cf. William *Tingenail* 1205 Cur (Nf) and HORSNAIL.

**Tining:** Thomas *de la Tunyng* 1278 MELS (So); John *atte Tynyng* 1327 SRSo. 'Dweller by the fence', OE *tyning*.

**Tinker:** Laurence *tinekere* 1244–46 Bart (Lo); Robert *le Tinker* 1243 AssSo. 'One who mends pots, kettles, etc.', 'a tinker'. The surname is not confined to the south as Bardsley thought. Examples have been noted between 1275 and 1292 in Yorks, Lancs, Lincs, Norfolk and Cheshire.

**Tinkler:** Roger *le Tinkelere* 1268 MESO (Y); Thomas *le Tinclere* 1279 AssNb. A formation based on *tinker*. 'A worker in metal, tinker' (c1175 NED). Characteristically northern, noted only in Yorks, Lancs, Cumberland and Northumberland.

**Tinner:** William *Tinier* 1327 SRSx. A derivative of OE *tin* 'tin', a worker in tin.

**Tinsley, Tinslay:** Adam *de Tindesle* 1207 P (Sr); Richard *de Tyntesle* 1327 PN Sx 283. From Tinsley (WRY), or Tinsley Green in Worth (Sx).

**Tinton:** Richard *de Tynton* 1298 AssL; John *Tynton* 1327 SRSf. From Tinton (K).

**Tinworth:** Robert *de Tymeworthe* 1327, William *Tynworth* 1524 SRSf. From Timworth (Sf).

**Tiplady:** Johanna *Tippelevedy* 1301 SRY; John *Typlady*, Henry *Tipelady*, 1490, 1494 GildY. cf. TOPLADY, TOPLASS, both probably names for a libertine. cf. *Othello* i, 1.

**Tipler:** Robert *le Tipelere* 1250 MESO (Nf); William *Tipeler* 1275 RH (L). ME *tipeler* 'a seller of ale, tapster' (1396 NED).

**Tipp:** *Tipp'* filius Harding 1204 Cur (L); William *Tip* 1279 RH (C). A variant of *Tibb*, a pet-form of *Theobald.* *v.* TIPPING.

**Tippell, Tipple, Tipples:** Albreda *Tepple* 1185 Templars (Wa); (tenementum) *Tipel* 1275 RH (Nf); John *Typull* 1524 SRSf; Edmund *Tipall*, Robert *Tiple*, Robert *Tipell* 1674 HTSf. *Tipp-el*, a diminutive of *Tipp* (or *Tebb, Tepp*), or a contracted form of *Theobald.*

**Tippenny:** *v.* TWOPENNY

**Tipper:** William *Tipere* 1176 P (Hu); William *(le) Tipper* 1214 Cur (Wa). A derivative of ME *typpe* 'to furnish with a tip', 'a maker or fitter on of metal tips', probably of arrow-heads.

**Tippett, Tippetts:** Wimer *Tippet'* c1250 Rams (Hu); Hugh *Typet*, *Tepet* 1297 Wak (Y). *Tipp-et*, a diminutive of TIPP.

**Tipping:** Robert *Tipping'* 1301 SRY; Thomas *Typpyng* 1381 PTY. *Tipping* would appear to be OE *\*Tipping* 'son of *\*Tippa*', an unrecorded name found in place-names, which may be an occasional source of TIPP.

**Tippins:** Seman *Typoun* 1327 SRSf; Robert *Typpyn* 1524 SRSf. *Typoun* is *Tip-un*, a diminutive of *Tipp* (Theobald). *Typpyn* may be the same name, or *Tip-in*, also a diminutive of *Tipp*, or a weakening of *Tipping*.

**Tipple(s):** *v.* TIPPELL

**Tipson:** Richard *Tibson* 1326 Wak (Y); William *Tybessone* 1327 SRDb. 'Son of *Tibb*.'

**Tiptaft, Tiptoft, Tiptod:** Robert *de Typetot* 1275 FFEss; Elyas *Tybetot* 1327 SRSf; John *Typtofte* 1524 SRSf; Nathaniel *Tiptot*, William *Tiptod* 1674 HTSf. From some Norman place Tibetot.

**Tireboys:** Robert *Tyreboys* 1275 SRWo; Geoffrey *Tyreboys* 1288 IpmGl. 'Drag wood', OFr *tirer, bois*, a nickname for a seller of wood. cf. Maurice *Tireauant*

1185 P (W) 'drag forward'; Simon *Tirhare* 1327 SRY 'drag hare'; William *Tirevache* 1172 P (Ha) 'drag cow'.

**Tirrell, Terrall, Terrell, Terrill, Tearall, Tyrell, Tyrrell:** Walter *Tirel* 1086 DB (Ess); Rocelinus, Simon *Tirel* 1127 AC (Gl), 1194 Cur (W); Hugo, Eudo *Tirell* 1153 Seals (Ess), 1195 P (Wo); Henry *Terel* 1275 SRWo; William *Terryll* 1568 SRSf. The common association with Anglo-Scandinavian *Turold* is impossible. Probably, as suggested by Tengvik, a derivative of OFr *tirer* 'to draw' in the same sense as Fr *Tirand* which Dauzat explains as 'one who pulls on the reins', hence 'obstinate, stubborn'.

**Tisser, Tissier:** Nicholas *Tisser* 1327 SRC; John *Tyssur'* 1327 MESO (L). OFr *tisseur* 'weaver'. *Tissier* is a Guernsey name.

**Tissington:** Robert *de Tiscunton* 1223 AssSt. From Tissington (Db).

**Titch(e)ner:** *v.* TWITCHEN

**Titchmarsh:** Ralph *de Tichemers* 1178 P (Nth); Ralph *de Thychemers'* 1255 ForNth; John *Tichemers* 1327 SRC. From Titchmarsh (Nth).

**Tite:** Richard *tite* 1332 SRDo; Walter *Tyte* 1641 PrSo; Titus *Tite* 1674 HTSf. The French form of *Titus*.

**Tithe:** Richard *atte Titthe* 1279, Peter *atte Tethe* 1296 MELS (Sx). 'Dweller by the land subject to tithe', OE *tēoða*.

**Titherley, Tytherleigh, Tetherley:** John *Tyderle* 1381 LLB H; Humphrey *Tetherley* 1642 PrD; Nathaniel *Titherley* 1662–4 HTDo. From Tytherley (Ha), or Tytherleigh in Chardstock (D).

**Titler, Tytler:** Richard *le Titteler*, Symon *le Titteler*, Symon *le Tuteler* 1275 RH (Sf). A derivative of ME *title*, 'a tell-tale, gossip'. *Tuteler* may be from ME *tutel* 'whisper'. Or we may have ME *titlere* 'hound'.

**Titley, Tittley:** William *de Titteley* 1281, Thomas *de Tytteleye* 1300 AssSt; Peter *Tittley* 1662–4 HTDo. From Titley (He).

**Titman:** *v.* TIDDEMAN

**Titmarsh:** William *Tytemers* 1279 RH (C); Peter *de Tytemersshe* 1339 CorLo. From Tidmarsh (Berks).

**Titmus, Titmuss, Titmas:** Edward *Titmouse* 1662 HTEss. A nickname from the bird, perhaps indicating worthlessness or insignificance.

**Titt:** Stephen *Tit* 1270 RamsCt (C); John *Titte* 1379 PTY; Thomas *Titt* 1576 SRW. OE *Titta*, or a nickname from the bird, ME *tit*.

**Tittersall:** Mrs *Titersall* 1662–4 HTDo. From Titleshall (Nf).

**Titterton:** John *Teterton* 1525 SRSx. From Titterton (Ch), or Titterton in Coldridge (D).

**Tittley:** *v.* TITLEY

**Tiverton:** John *Teverton* 1641 PrSo. From Tiverton (D).

**Tobey, Toby, Tobias:** *Tobias* prior c1142–50 PN Ess 261 (first prior of Thoby Priory); *Tobias* Ric 1 Gilb (L); Simon, William *Toby* 1271 RamsCt (C), 1275 RH (L). Hebrew *Tobiah* 'Jehovah is good'. *Tobias* is the learned, Greek form, *Toby* the vernacular.

**Tod, Todd:** Hugo, Arding *Tod* 1168–75 Holme (Nf), 1225 Oseney (O); Richard *Todd* 1231 Pat (Nb);

Richard *le Todde* 1275 SRWo. ME *tod(de)* 'fox', chiefly, but not solely, northern.

**Toddenham:** Robert *Todenham* 1406–7 IpmY; Thomas *Todenham* 1453 Paston. From Todenham (Gl).

**Toddington:** Roger *de Todington* 1327 SRSf. From Toddington (Beds, Gl).

**Todhunter:** Thomas *Todhunter* 1332 SRCu. 'Foxhunter'. Still a Cumberland surname.

**Todman:** Juliana *Todman* 1275 SRWo. This form cannot derive from Toddenham (Glos) as suggested by Bardsley and Harrison. It is a compound of *tod* 'fox' and *man*, probably a southern equivalent of TODHUNTER.

**Todrick:** *v.* TERREY

**Toe, Toes:** For *at* (*h*)*oe*. *v.* HOW, ATTO.

**Tofield:** *v.* TOVELL

**Toft, Tofts:** Elyas *del Toft* 1197 P (Nth); Robert *de Toft* 1279 RH (Beds); Gilbert *de Toftes* 1327 SRSf. 'Dweller at the croft(s) or homestead(s)' as at Toft (Cambs, Norfolk, Warwicks). cf. TAFT.

**Togge:** Hugh *Togge* 1212 Cur (Y); Thomas *Toge* 1219–20 FFEss; Alan *Togge* 1296 PN Herts 157. OE *Tocga*.

**Toghill:** *v.* THURKELL

**Tointon:** *v.* TOYNTON

**Toke:** *v.* TOOK

**Tole, Toole:** (i) William *filius Tole* 1130 P (Ess); Hugh, Peter *Tole* 1297 MinAcctCo, 1327 *SR* (Ess). OE *Tola*, ON *Tóli*. (ii) In Ireland for Ó *Tuathail*, from *tuathal* 'people-mighty'.

**Tolefree:** *v.* TOLLFREE

**Toley:** *v.* TOOLEY

**Tolhurst:** George *Tolherst* 1525 SRSx. From Tollhurst in Frittenden (K).

**Toll, Tolles, Towle, Towll:** (i) *Tolle* le grangier 1218 AssL; Nicholas, Richard *Tolle* 1275 RH (W), 1296 SRSx. OE \**Toll*, found in Tollesbury and Tolleshunt (Essex), or Anglo-Scand \**Toll*, a pet-form of ON *Þorleifr*, found in Thurleston (Leics, Warwicks) and Tollerton (Notts), or of ON *Þorleikr*, found in Thurloxton (Som). (ii) Robert *atte Tolle*, Roger *Tolle* 1327 SRWo. Probably an early example of the Kent to Hampshire dialectal *toll* 'a clump of trees' (1644 NED).

**Tolladay, Tollady, Tolleday, Tolliday:** Thomas *Towlewardie* 1574 SfPR; Thomas, Mary *Towlardy* 1664, 1668 ib.; Isaacke *Towlworthy* 1672 ib.; Thomas *Tilladay* 1674 HTSf. Suffolk pronunciations of *stalworthy*, with loss of initial *S*. cf. STALLWORTHY, STOLERMAN, STOLLERY, STALLWOOD.

**Tollard:** Henry *de Tollard* 1262 IpmGl. From Tollard Farnham (Do), or Tollard Royal (W).

**Tollemache:** *v.* TALMADGE

**Toller, Toler, Towler:** (i) Robert *Toller'* 1199 P (L); William, John *le Tollere* 1251 AssY, 1255 *Ass* (Ess); William, Richard *Towler* 1624 HorringerPR (Sf), 1639 RothwellPR (Y). OE *tollere* 'tax-gatherer'. (ii) Robert *de Tolra* 1179 P (Do). From Toller (Dorset).

**Tollerton:** Thomas *de Tollerton* 1304 IpmY. From Tollerton (Nt, NRY).

**Tollfree, Tolfree, Tolefree, Turfery, Turfrey, Tuffery, Tuffrey:** *Turfredus* 1049 PNDB 392 (Rouen);

*Toruert, Toruerd* 1066 DB; Godric *Turuerde filius* c1095 Bury (Sf); *Toruard* Cappe 1203 FFL; John *Torfray* 1279 RH (O); Robert, Henry *Tolfry*, William *Tholfry* 1296 SRSx; John *Tolfrut* 1297 MinAcctCo; Mary *Tufferey* 1765 *ERO*. ON *\*Þorfrøðr*, OSw *Thorfridh*.

**Tollworthy:** A variant of STALLWORTHY. cf. TOLLADAY.

**Tolman, Tolmon, Toleman:** Hereward *Tholeman* 1219 AssL; William *Tolman* 1327 SRSo. OE *toll* and *mann*, 'tollman, a collector of tolls'.

**Tolson:** *v.* TOWNSON

**Tolver:** *v.* TELFER

**Tom, Tomes, Toms:** William *Tom* 1245 HPD (Ess); Edmund *Tommys* 1524 SRSf. A pet-form of *Thomas*.

**Tomalin:** *v.* TOMLIN

**Toman:** *v.* TUMMAN

**Tomas:** *v.* THOMAS

**Tombleson:** Wylliam *Thomlesson* 1524 SRSf. From TOMLINSON, with intrusive *b*.

**Tomblin, Tombling:** *v.* TOMLIN

**Tombs, Toombes, Toombs:** Robert *Toume*, Walter *Tomes* 1327 SRWo; Thomas *Tombes* 1701 Bardsley. Identical with TOMS, with late intrusive *b*.

**Tomcock:** Simon *Thomecok* 1327 SREss; Robert *Thomekok* 14th WhC. A diminutive of *Tom*, a pet-form of *Thomas*.

**Tomkin, Tomkins, Tomkies, Tomkys, Tompkin, Tompkins:** William *Thomekyn* 1323 Eynesham (O); William *Tumkyns* 1327 SRSt; Geoffrey *Tomkynes* 1332 SRWa; Richard *Tompkyn* 1566 Bardsley. 'Little Tom.'

**Tomkinson, Tompkinson:** John *Tomkynson* 1393 SaltAS (OS) ix. 'Son of *Tomkin*.'

**Tomley, Thomley:** John *Tompele* 1279 RH (C). From Thomley (Oxon).

**Tomlin, Tomline, Tomlins, Tomalin, Tomblin, Tombling, Tomblings, Thomline:** *Thomelinus* Herys 1380 Black; Robert *Thomelyn* 1327 SRC; John *Thomeling* 1327 SRSx; John *Tomelyn* 1327 SRSo. A double diminutive of *Tom* or *Thom* (Thomas), *Tom-el-in*.

**Tomlinson, Thomlinson:** Henry *Thomlinson*, Richard *Tomlynson* 1379 PTY. 'Son of *Tomlin*.' *v.* TOWNSON.

**Tommis:** Identical with TOM.

**Tompkin, Tompkins:** *v.* TOMKIN

**Tompkinson:** *v.* TOMKINSON

**Tompsett, Tomsett:** *v.* THOMPSETT

**Tompson, Tomsen, Tomson:** *v.* THOMPSON

**Toms:** *v.* TOM

**Tonbridge, Tunbridge:** Richard *de Tonebrige* 1086 DB (K); John *atte Tunbregg* 1345 MELS (Sr). Usually, certainly in Kent where the surname still survives, from Tonbridge (Kent); or from residence near a bridge by the *tūn* or village. Salom' *de Tunbrygge* (1274 RH) lived at Two Bridge, whilst Robert *de Tonebrugge* (1323 *For*) lived at the field still known as Tunbridge in Stapleford Abbots. *v.* PN Ess 279, 604.

**Tone:** *v.* TOWN

**Toner:** Irish *Ó Tomhrair* 'descendant of *Tomhrar*'. In England the surname is early: Hugh *Tunere* 1242

Fees (W); Andrew *le toner* 1327 Pinchbeck (Sf). 'Dweller by the farm or village', equivalent to *atte tune*. *v.* TOWN.

**Toney:** Richard *Tony* 1275 SRWo; William *Toni* 1279 RH (Beds). A pet-form of *Antony*.

**Tong, Tonge, Tongs, Tongue:** (i) Wluricus *Tunge* 1188 BuryS (Sf); Nicholas, Richard *Tonge* 1279 RH (Bk), 1297 MinAcctCo (O). The frequency of early forms without a preposition suggests a nickname from OE *tunge* 'tongue', a chatter-box, scold. (ii) Richard *de Tanga, de Tong* c1200–10 Calv (Y); Elias *de Tong* 13th WhC (La). From Tong (Salop, WRYorks) or Tonge (Kent, Lancs, Leics), varying between OE *\*twang, tang* 'tongs' and *tunge* 'tongue of land'. (iii) William *in la Tunge* 1248 FFSx; James *a Tonge* 1545 ArchC 42. 'Dweller in the tongue of land' (OE *tunge*).

**Tonnell, Tunnell:** Ralph, William *Tonild* 1279 RH (C); Henry *Thonild* 1327 SRC. OE *\*Tūnhild* (f), unrecorded, but both themes are common.

**Tooby:** *v.* TOBEY

**Toogood, Towgood, Tugwood:** Richard, Robert *Togod* 1200 P (L), 1221 *ElyA* (C), 1264 Oriel (O); Roger *Togod, Togot* 1207 Cur (Nth); Agnes, Stephen *Tougod* 1297 MinAcctCo, 1327 SRSx; Stephen *Togoud* 1332 SRSx; Margerie *Toogood* 1473 Rad (C); Mary *Tugwood* 1761 BishamPR (Berks): Hilton *Toogood, Twogood, Tugwood* 1763, 1765 ib. This name cannot be associated with *Thurgood* and *Thorogood* as suggested by Bardsley and Harrison. These names regularly retain the *r*, of which there is no trace in *Toogood*. As Weekley suggests, the surname is really an adjectival nickname 'too good'. cf. Scots 'unco guid', Hamon *Toproud* 1287 AssCh 'too proud', Adam *Overprud* 1222 AssLa, and TRAPNELL.

**Took, Tooke, Toke, Tuke:** *Toc* 1066 DB (Y); Rogerus *filius Toke* 1214 Cur (Nth); Wrange *Tocha* 1166 P (Nf); Henry *Thoche* 12th DC (L); Robert, William *Toke* 1200 P (Sf), 1211 Cur (Nth); Robert *Touk* 1325 AssSt; Wylliam *Took* 1524 SRSf. Anglo-Scand *Tōka, Tōke* (ON *Tóki*). *v.* TOOKEY.

**Tookey, Tuckey:** *Tochi, Toke, Toche, Toca, Tuka* 1066 DB; Ormus, Rogerus *filius Toki* 1183 Boldon (Du); 1206 Cur (R); William, John *Toki* 1200 P (Sf), c1248 Bec (Bk); Thomas *Tookye* 1599 Oxon (Lei); Joane *Tuckey* 1624 Bardsley. ON *Tóki*, ODa *Toki, Tuki*, OSw *Toke, Tuke*.

**Toole:** *v.* TOLE

**Tooley, Toley:** *Toli, Thole, Tholi* 1066 DB; Reginaldus *filius Toli, Tuly* 1214 Cur (L); Richard *Toly* c1150 StThomas (St); Peter *Toli* 1155 P (Nt); David *Toolye* 1631 Bardsley. ON *Tóli*.

**Toomb(e)s:** *v.* TOMBS

**Toon(e):** *v.* TOWN

**Toop, Toope:** *Topi, Tope* 1066 DB (Herts, L); Ulf *Tope sune* ib. (L); Aldene *Tope* ib. (L); Robert *Topi* 1200 P (Nf); Alan, John *Tupe* 1202 P (Y), 1307 Wak (Y). ODa *Topi*.

**Tooth:** Hugo *cum dentibus* c1102–11 ELPN; Robert *Tothe* 1219 AssY; Thomas *Toth* 1275 RH (Nth). OE *tōð* 'tooth', a nickname for one with a prominent tooth or teeth.

**Toothill, Tootal, Tootell, Tootill, Tootle, Tothill, Tottle, Tuthill, Tutill:** Gilbert *de Totehille* 1185

Templars (Wa). 'Dweller by the look-out hill' (OE *tōt-hyll), as at Toot Hill (Essex), Tothill (Lincs, Middlesex), Tootle Height (Lancs), or Tuttle Hill (Warwicks). v. THURKELL.

**Toovey:** v. TOVEE

**Topham:** Hugh *Topeham* 1332 SRSr; ffrancis *Topham* 1672 HTY. From Topham in Sykehouse (WRY).

**Toplady:** John *Taplady* 1400 AD vi (Db). cf. TIPLADY.

**Toplass, Toplis, Topliss:** Robert *Topples* 1596 Shef. cf. TIPLADY.

**Topliffe:** Herueus *de Toppecliue* 1219 AssY; John *Topplyffe* 1491 GildY. From Topcliff (NRYorks).

**Topp, Topps:** Herueius *filius Toppe* 1200 P (L); Robert *Toppe* 1196 P (Gl); Roger *Top* 1208 Cur (Nf). ON *Toppr*.

**Topper:** Walter *le Toppare* 1275 SRWo; Robert *le Toppere* 1327 AD i (W). A derivative of ME *toppe* 'a tuft or handful of hair, wool, fibre, etc., especially the portion of flax or tow put on the distaff. A topper was probably the one who put the 'toppe' on the distaff.

**Toppin, Topping:** John *Toppyng* 1246 AssLa; Thomas *Topyn* 1327 SRSx; Adam *Toppyn* 1382 IpmGl; Robert *Toppyng* 1422 TestEbor. OE *Topping, or a diminutive of OE *Topp, ON *Toppr*.

**Topsfield:** Elias *de Topesfeld* 1185 Templars (Ess); William *de Topesfeld* 1203 FFEss. From Toppesfield (Ess).

**Topsham:** Richard *de Toppesham* 1262 PN D 306. From Topsham (D).

**Torbett, Torbitt:** v. TURBARD

**Torch:** Alueue *Torce* c1095 Bury (Sf); Robert *Torche* Ed 1 Battle (Sx). OFr *torce* 'torch', perhaps for a torch-bearer.

**Toril:** Robert, Ralph *Torel* c1150 DC (Lei), 1185 Templars (Herts). A diminutive of OFr *tor* 'bull'.

**Torney:** (i) Gilbert *de Toreigni* 1195 P (Y); Richard *de Tornei* 1207, *de Torenni* 1208 Cur (Sx); John *de Torny* 1274 RH (So). From Tournay (Calvados, Orne). Sometimes from Thorney (So). (ii) Æilric *Torenega* 1180 P (Wa). 'Torn eye', OE *toren, ēage*. cf. William *Tornemantel* 1279 RH (O) 'torn mantle'.

**Torode:** v. THOROLD

**Torpel, Torpell:** Roger *de Torpel* c1202 NthCh (Nth); Mabel *Torpel* 1269 FFK. From Torpel in Ufford (Nth).

**Torr, Torra, Torre:** (i) Robert *de Torra* 1182 P (Co); Martin *de la Torre* 1242 Fees (D); Walter *atte Torre* 1296 SRSx. 'Dweller by the rocky peak or hill' (OE *torr*). (ii) John, Gilbert *le Tor* 1240 Oriel (O), 1276 LLB A. OFr *tor* 'bull'.

**Torrell:** Ralph *de la Thorail* 1256, Thomas *del Torail* 1275 Kris. 'Dweller at the small tower', OFr *torail, torel*.

**Torr(e)y, Torrie:** v. TERREY

**Torrington:** William *de Torinton'* 1218 P (D); Thomas *Torrington* 1672 HTY. From Black, Great, Little Torrington (D), or East, West Torrington (L).

**Tort:** John *le Tort* 1175 P (Y); Ralph *le Tort* 1268 AssSo; William *le Tort* 1327 SRSx. OFr *tort* 'twisted'. cf. Robert *Tortesmains* 1169 P (Y) 'with twisted hands'; Deudon *cum pedibus tortis* 1191 P (Lo) 'with twisted feet'.

**Tortise, Tortiss:** Hubert *Turtuse* 1200 Cur (Sf); Gilbert *Tortouse* 1327 *SR* (Ess). ME *tortuse* 'tortoise', a type of slowness.

**Tory:** v. THORY

**Tosh:** v. MCINTOSH

**Toshach, Toshack:** Simon *Tuschech* 1219–37 Black; Finlay *Tosscheach* 1516 ib. EGael *toisech*, Gaelic *toiseach* 'chief, leader'.

**Tosty:** Hugh *Tosti* 1193 P (Ha); Roger *Tosti* 1219 AssY; William *Tosty* 1288 IpmY. OE *Tostig*.

**Totham, Tottem:** Roger *de Totham* 1285 FFEss; John *Totham* 1365 ColchCt. From Great, Little Totham (Ess).

**Tothill:** v. TOOTHILL

**Totman, Tottman:** Robert *Toteman* 1202 P (Ess). OE *tōt-mann* 'look-out man', 'watchman'.

**Totnam:** v. TOTTENHAM

**Tott:** Iweinus *Tottes* 1219 AssL; Gilbert *Tot* 1275 RH (Sf). ME *totte* 'simpleton, fool'.

**Totten:** John *Totens* 1278–9 CtH; John *Totyn* 1331 IpmW. From Tottens in Coombe Bissett (W).

**Tottenham, Totnam:** Anfred *de Totenham* 1196 P (K); John *de Totenham* 1340 LLB F; Thomas *Totnam* 1618 *ERO*. From Tottenham, or Tottenham Court (Mx).

**Totteridge:** John *de Tateregge* 1276 AssLo; Henry *Toteriche* 1285–6 FFEss. From Totteridge (Herts).

**Tottle:** Custanc' *Totel* 1279 RH (C); Robert *Totele* 1283 SRSf; Geoffrey *Totel* 1298 AssL. OE *Tottel*. v. also TOOTHILL.

**Totty:** Edwin *Toti* 1200 P (Ha); Robert *Totty* 1327 SRY; Thomas *Totty* 1415–16 FFWa. A pet-form of OE *Tota, Totta*.

**Tough, Tow, Towe, Towes:** Robert, Gilbert *Towe* 1275 RH (L, Ha); Alicia *la Towe* 1275 SRWo; Nicholas *le Toghe* 1275 RH (K); John *Towes* 1685 FrY. OE *tōh*, ME *togh, tow(e)* 'vigorous, steadfast, stubborn'.

**Toulet:** v. TULLETT

**Toulson:** v. TOWNSON

**Toulouse:** Robert *de Tolowse* 1305 LuffCh; Richard *Tolous* 1316 AssNth; William *Toulouse* 1355–9 AssBeds. From Toulouse (Artois), or Toulouse in the south of France.

**Tours, Towers:** Picoth *de Turs* c1150 DC (L); Robert *de Tours* 1297 MinAcctCo. From Tours (Indre-et-Loire).

**Tout:** Richard, William *Tut* 1219 AssY, 1296 SRSx; Simon, Richard *Tute* 1275 RH (Nf); Cissota *Toute* 1307 Wak (Y). OE *tūte*, ME *toute* 'buttocks, rump'. Also used topographically, of a smooth, rounded hillock: John *de la Toute* 1316 AssK.

**Tovee, Tovey, Toovey, Tuvey:** *Toui* 1066 DB; *Toue* fox de Salebi Hy 2 DC (L); *Toui* Hering' 1177 P (Nf); Richard, William *Toui* 1197 P (Nf), 1200 P (Ha). ON *Tófi*, ODa *Tovi*, a diminutive of *Þioðvaldr* 'nation ruler', a name brought to this country by Tovi the Proud, a follower of Cnut.

**Tovell, Tofield, Tuffield, Tuffill:** *Touillda* 1066 DB (Ess); *Touilt* 1086 DB (Sf); Willelmus *filius Touild'* 1175 P (Sf); *Touild'* 1201 Cur (Nf); Stephen *Tovild* 1473 NorwW (Sf); John *Tovell* 1491 ib.; Alice *Tovyld* 1506 ib.; Robert *Toveld*, William *Tuffeld* 1524 SRSf. A rare surname from a rare Scandinavian personal name of an unusual type, ON *Tófa-Hildr* (f), 'Hildr the daughter of Tófi', found chiefly in Suffolk, occasionally in Essex and Norfolk.

**Toward:** Margery *Toward* 1275 RH (Nf). ME *toward* 'compliant, docile'.

**Tow(e), Towes:** v. TOUGH

**Tower:** (i) Elyas *de Toure* 1202 P (So); William *de la Tur* 1260 AssC; Theobald *atte Tur* 1296 SRSx. 'Dweller by the tower', OE *tūr* or OFr *tour* 'tower'. (ii) Gilbert *le Tower* 1255 RH (W); Thomas *le Touere* 1275 RH (Ha); William *le Towyere, le Tawyere* 1280 MESO (Ha). A derivative of OE *tāwian* 'to taw'. v. TAWYER and MESO 122.

**Towers:** v. TOURS

**Towersey, Towerzey:** Widow *Towersie* 1662 *HTEss*; Richard *Towersey* 1665 HTO. From Towersey (Bk).

**Towgood:** v. TOOGOOD

**Towl(e):** v. TOLL

**Towler:** v. TOLLER

**Towlson:** v. TOWNSON

**Town, Towne, Townes, Towns, Tone, Toon, Toone, Tune:** Wistric *Oftun* c1095 Bury (Sf); Peter *de la Tune* a1219 Seals (Sr); Arnold *Inthetune* 1243 AssSo; John *de la Tone* 1275 RH (Sf); Peter *Abopetoune*, Richard *above the toune* 1275 SRWo; Nicholas *Tone* 1279 RH (C); Mabel *atte Tune* 1296 SRSx; John *Douninthetoune* 1327 SRWo; Alan *de Toune*, William *dil Toun* 1327 SRSf; Henry *Town* 1428 FA (D); John *Townes* 1524 SRSf. OE *tūn* originally denoted a 'fence' or 'enclosure' but at an early date came to mean 'enclosure round a house', 'homestead', 'village', 'town'. In these surnames, originally preceded by a preposition, the reference is to 'one who lived in, at, above or down in the *tūn*' which must here mean 'the village' as in *Townend* and *Townsend*.

**Townend, Townen, Townsend, Townshend:** (i) Richard *atte Tounende* 1297 MinAcctCo; Adam *Attetounhend* 1297 SRY. (ii) Wulfric *at te tunesende* 1192 P (Sf); Godmannus *atte Tounsende* 1284 RamsCt (Hu); William *atte Townesend* 1327 SRWo. 'Dweller (i) at the town-end, (ii) at the town's end', a genitival compound always the more common, *town* having the meaning 'village' as in TOWN.

**Towner:** Hugh *le Tolnur* 1221 AssWo. OE *tolnere* 'tax-gatherer'. cf. TOLLER.

**Towning:** Nicholas *Tounyng* 1327 SRWo. OE *tūning* 'dweller in the village'.

**Townley, Townly, Towneley:** Richard *de Tūnleie* 1214 FFSf; Robert *de Tounlegh* 1287 IpmLa; Laurence of *Tounlay* alias Laurence of Lehg 1401 AssLa; Richard *Towneley* de Towneley 1622 PrGR. From Towneley (Lancs), or 'dweller by the farm clearing'.

**Towns(h)end:** v. TOWNEND

**Townson, Tolson, Toulson, Towlson:** Christopher *Towson, Tolson* 1553, 1593 Ipsw (Sf); Edmund *Tollenson, Townson* 1571 LaWills; Robert *Tolneson, Townson* 1580 ib.; Richard *Towlson, Tounson* 1587 ib.; Jenet *Tomlinson, Towenson* 1588 ib.; Catherine *Toinson, Towenson* 1591 ib. Colloquial pronun-

ciations of *Tomlinson*, particularly common in North Lancashire.

**Toy, Toye:** Aluric *Toi* 1066 Winton; Aldwin *Toie* 1184 P (D); Wydo, Alan *Toye* 1274 RH (Sf). Presumably a nickname, the meaning and origin of which remain obscure, v. Tengvik 372. It was evidently used also as a personal name, cf. Godfrey *filius Toye* 1200 ChR (L).

**Toynton, Tointon:** Dauid *de Totingtona* c1158 Gilb; Hugh *de Totinton'* 1205 P (L); Thomas *de Toynton'* 1279 RH (O). From High, Low Toynton, or Toynton All Saints, St Peter (L).

**Tozer:** John *le Tosere*, William *le Thosere* 1249 MESO (Sx), 1280 AssSo; John *Tosere* 1333 PN D 519. A derivative of OE *tāsian* 'to toze, tease', a side-form of OE *tāsan* 'to tease'. cf. William *le Tesere* 1275 AssSo. 'A comber or carder of wool.'

**Tracer:** John *Tracere* 1327 SREss. 'One who makes tracings or drawings for the masons employed in stone-carving'. v. Building 21.

**Tracey, Tracy:** Henry *Traci* 1139 Templars (O); Henry *de Traci* 1148 Winton (Ha); Oliver *de Trazi* 1166 P (D). The Tracies were undertenants in the department of La Manche (ANF). The surname derives from Tracy-Bocage or Tracy-sur-Mer (Calvados).

**Trad(e)well:** v. TREADWELL

**Trafford, Trafferd, de Trafford:** Richard *de Trafford* 1246 AssLa; Richard *de Trafford* 1290 AssCh; John *Trafford* 1481 IpmNt. From Trafford (La), or Bridge, Mickle, Wimbolds Trafford (Ch).

**Traharn, Traherne:** v. TREHARNE

**Traies:** v. TRAY

**Trailwing:** Geoffrey *Trailewing'* 1200 P (Y); John *Trailweng* 1346 TestEbor; William *Trailweng* 1364 FFY. 'Trail wing'. Perhaps a nickname from the lapwing, popularly supposed to pretend to be injured in an attempt to lure people away from its nest.

**Train:** (i) Warin *Traine* 1181 P (Nb); Robert *Trayne* 1243 AssDu. ME *trayne*, OFr *traine* 'guile, trickery', 'a trap or snare for catching wild animals', metonymic for a trapper, hunter. (ii) Richard *Trane* 1301 SRY; Thomas *Tran* 1455 Black; William *Trayne* 1557 ib. A northern and Scottish form of ON *trani* 'crane'. (iii) In Devonshire, from Train in Wembury or Traine in Modbury, the homes respectively of Thomas *atte Trewen* (1311) and John Tirry *atte Trewen* (1370), 'at the trees' (PN D 261, 281). v. TREE.

**Trainell, Traynell:** Robert *Trainell'* 1205 P (Sa); Robert *Traynel* 1327 SRC; Henry *Treynell* 1392 LoCh. Perhaps a derivative of OFr *traine* 'trap', and so metonymic for a trapper of animals.

**Trainer, Trainor, Treanor, Traynor:** Robert *Treiner, Trainer* 1243 AssDu, 1379 PTY. A derivative of ME *trayne* 'to lay a train or snare, to set a trap', a trapper.

**Tranter, Trenter:** Terri *Trauetarius* 1148 Winton (Ha); Philip *Trenter'* 1221 Cur (Ess); Hugo *le Trauenter* 1292 MESO (Ess); Mark *le Traveter* 1306 AD iii (Co); Simon *le Traunter* 1332 SRWa. MedLat *travetarius* 'carrier, hawker'. v. *tranter* (NED).

**Trapnell:** William *Tropisnel* 1183 P (Nf); Walter, William *Tropinel(l)* 1195 P (W), 1201 AssSo; Robert

*Tropnell* 1497 RochW; Anne *Trapnell* 1662 SfPR. OFr *trop isnel* 'too swift'.

**Trapp, Trappe:** John, Henry *Trappe* 1230 P (Wo), c1250 Rams (Hu). OE *træppe, trappe* 'trap, gin, snare', no doubt for a trapper of animals.

**Trapper:** William *le Trappere* 1338 Misc (Berks). A derivative of OE *træppe* 'trap', a trapper of animals.

**Trask, Thresk, Thirsk:** Reginald *de Tresch* 1185 P (Y); Duva *de Tresk'* 1219 AssY; John *Traske* 1662-4 HTDo. From Thirsk (NRY), *Tresc* DB.

**Travell, Travil:** (i) William *le traueillie* 1196 P (Nb); Malger *le Travaile, le Travaillie, le Travallie* 1205-7 Cur (L); (ii) Ralph, Robert *Trauel* 1185 Templars (L), 1203 P (Bk); Thomas *Trauail* 1202 AssL; William *Trauayl* 1296 SRSx. The past part. of OFr *travaillier, traveillier* 'to afflict, vex, trouble', corresponding to ME *traveled, travailed* 'wearied in body and mind, troubled, harassed' (c1420 NED), alternating with ME, OFr *travail* 'exertion, trouble, hardship, suffering' (c1250 NED).

**Travers, Traves, Travis, Traviss:** Robert *Trauers, Traues, trauerse* Hy 2 DC (L); Walter *Travers* 1172 Gilb (L); Margareta *Travas* 1433 GildY; Ann *Travis* or *Travers* 1578 Bardsley; Cordell *Traverse* son of Phillip *Travesse* 1640 ib. ME *travers*, later *travas, traves, travis*, corresponds to two OFr nouns, *travers* (m) and *traverse* (f), 'the act of passing through a gate or crossing a river, bridge, etc.', used of a toll paid on passing the boundary of a town or lordship (*toll-traverse* 1567 NED). In 1285 (QW) the Bishop of Norwich claimed that he and his predecessors were accustomed to take a certain *Travers* at South Elmham (Suffolk) from foreign merchants crossing a certain bridge with merchandise towards Beccles, Yarmouth and Bungay for the upkeep of the bridge. The same custom existed elsewhere in Suffolk and Norfolk, and continued at Hull as late as 1852. The surname, which was common and widespread, probably refers to one who possessed this custom or to one who collected the toll.

**Tray, Trayes, Traies, Treays:** Lefstan *Trege* c1095 Bury; Elfred *Treye* 1221 ElyA (Sf); William *Tray* 1392 IpmGl. A nickname from OE *trega* 'grief, misfortune'.

**Traynell:** *v.* TRAINELL

**Traynor:** *v.* TRAINER

**Treacher:** Walter *le Trichur* 1243 AssSo; Robert *Trechour* 1301 SRY. OFr *trecheor, tricheor* 'deceiver, cheat'.

**Treacle:** William *Triacle* c1211 Guisb; Richard *Triaccle* 1284 IpmW; John *Tryakle* 1296 SRSx. OFr *triacle* 'an antidote to poison', a nickname for an apothecary.

**Treadaway:** *v.* TRETHEWEY

**Treadgold:** *v.* THREADGOLD

**Treadwater:** William *Tredewater* 1279 RH (C). 'Tread water', OE *tredan, wæter*. cf. Walter *Tredefeu* 1304 IpmGl 'tread fire'; Symon *Tredhard'* 1379 PTY 'tread hard'; Eðred *trede wude* c1050 BCS (Y) 'tread wood'.

**Treadwell, Tredwell, Tretwell, Tradewell, Tradwell:** Seeman, Thomas *Treddewel* c1248 Bec (Sf). 'Tread well.' cf. Symon *Tredeven* 1275 SRWo, John *Tredebas* 1327 SRC.

**Treanor:** *v.* TRAINER

**Treasure:** Ansgod *Tresor*, Clement *treisor* 1148 Winton (Ha); Adgar *Treissor'* 1199 FFEss; Laurence *Tresor, Thresor* 1207, 1230 P (Ha); Nicholas *le Tresor* 1243 AssSo. OFr *tresor*, ME *tresor, treysour, thresur* 'wealth, riches' (1154 NED). Metonymic for 'treasurer'. cf. Henricus *Thesaurius* 1066 Winton (Ha), Nicholas *de la Tresorie* 1325 FFEss.

**Treavett:** *v.* TRIVETT

**Treays:** *v.* TRAY

**Tredgett, Tredjett, Trudgett:** William *de* (*sic*) *Treget* 1176 P (He); Robert, Roger *Treget* 1188 P (He), 1327 SRSf; John *Tredgett*, Richard *Tridgitt* 1674 HTSf; — *Trudgett* 1674 HTSf. OFr *tresgiet, treget*, ME *treget, trigit* 'jugglery, trickery, deceit' (a1300 NED), metonymic for a juggler. cf. Robert *le Tregettur* 1203 AssSt, Symon *le Tregetor* 1279 RH (C), ME *tregetour, trigettur*, 'a juggler, mountebank' (a1300 NED).

**Tredgold:** *v.* THREADGOLD

**Tredinnick, Treddinick:** Ralph *de Tredenek'* 1297 MinAcctCo. From Tredinnick (Cornwall).

**Tree, Treece, Trees, Treen, Trewin:** Henry *en le Tres* 1327 SRDb; John *del Trees* 1379 PTY. 'Dweller by the tree(s).' *Treen*, the weak plural form, survives in Trewyn (PN D 148). cf. TRAIN (iii).

**Treeby:** John *Treeby* 1642 PrD. From Treby in Yealmpton (D).

**Treeton:** Alice *Treeton* 1599 SRDb. From Treeton (WRY).

**Trefusis:** James *Trefusis* 1359 Putnam (Co). From Trefusis (Cornwall).

**Tregellas, Tregelles, Tregillus:** Nicholas *de Tregellest* 1297 MinAcctCo. From Tregelles (Cornwall).

**Tregoose, Tregoz:** William *de Tresgos* 1199 Pleas (Ess), 1218 P (Ess); William *Tregose* 1382-4 Hylle. From the name of a castle in the arondisment of St-Lô.

**Treharne, Trehearne, Trehern, Treherne, Traharn, Traherne:** Ann *Treyerne, Treherne* 1573 LedburyPR (He); Thomas *Treherne* 1663 HeMil. From Trehane (Co).

**Trelawney, Trelawny:** Peter *Trelany alias Trelauny alias Trevlauney alias Trelawny alias Treylany* 1499 Pat (Co); Francis *Trelawney*, Henry *Trelawny* 1642 PrD. From Trelawney (Co).

**Trelfall:** *v.* THRELFALL

**Treloar:** Gregory *de Trelouargh'* 1297 Coram. From Treloar (Cornwall).

**Tremayne, Tremain, Tremaine, Treman:** Nicholas *Tremeayne, Tremayn* 1562 Pat (Co); Andrew *Tremayne* 1576 SRW; John *Tremaine* 1642 PrD. From Tremaine (Cornwall).

**Tremelling:** William *Tremillin* 1255 RH (St). From Tremellen (Cornwall).

**Tremenheere:** Thomas *Tremenhir* 1359 Putnam (Co). From Tremenheere (Cornwall).

**Tremlett, Trimlett:** Walter *de Tribus Minetis* 1130 P (D); Joslen *Treismunettes* 1194 P (D); Richard *de Treminettes* 1204 Pl (St); Richard *Treminet* 1275 RH (D); John *Trymlett* 1525 SRSx; Matthew, Thomas *Tremlett* 1642 PrD. From Les Trois Minettes (Calvados).

**Trench:** Richard *Trench* 1275 RH (Nf); Henry *de Trench* 1327 SRWo. From Trench Lane in Huddington (Wo), or 'dweller by the path or track cut through a wood', cf. OFr *trenchir* 'to cut'.

**Trencham:** William *Trenchaunt* 1221 AssWa; Simon *Trenchant* 1311 LLB B; William *Trenchant* 1335 IpmW; Nicholas *Trencham* 1674 HTSf. OFr *trenchant* 'cutting'.

**Trenchard:** Ralph *Trencart, Trenchard* 1086 DB (So); Robert *Trenchart* 1166 P (Ha). A derivative in *-ard* of OFr *trenchier* 'to cut'. Bardsley suggests 'swordsman'. It might be occupational: cf. Walter *Trenchebof* 1214 Cur (Mx) 'cut ox', butcher.

**Trenchmer:** Alan *Trenchemer* 1175 P (Ha); Robert *Trenchemer* 1296 SRSx; Thomas *Trenchemere* 1327 SRSf. 'Cut the sea', OFr *trenchir, mer*, a nickname for a sailor. cf. Robert *Trencheuent* 1199 Pleas (Hu) 'cut the wind', a nickname for a fast runner; Robert *Trenchevot* 1220 Cur 'cut the foot'.

**Trendall, Trendell, Trindall:** Ernald *Trendel* 1177 P (Sf); John *atte Trandle* 1327 SRSx. 'Dweller by the circle', OE *trendel*, used of woods, wells, hills, earthworks, etc. of circular shape.

**Trenham, Trenholm, Trenholme:** William *Trenham* 1674 HTSf. From Trenholme (NRY).

**Trenow:** Clement *Trenowe* 1297 MinAcctCo. From Treknow in Tintagel (Cornwall).

**Trenowath, Trenoweth, Trenouth:** Ranulf *de Trenewyth* c1210 Fees (Co). From Trenoweth (Cornwall).

**Trent:** John *de Trente* 12th Seals (So); William *Trent* 1300 LoCt; Peter *Trente* 1351 FFEss. From Trent (Do), or from the River Trent.

**Trenter:** v. TRANTER

**Trentmars:** *Trentemars* 1148 Winton (Ha); John *Trentemars* 1276 AssLo; William *Trentemars* 1332 SRSx. 'Thirty marks', OFr *trente, mars*, probably for one owning property worth that amount. cf. Adam *Trenttedeux* 1332 SRDo 'thirty two'.

**Treseder, Tresidder, Tressider:** William *de Tresoder* 1297 MinAcctCo. From Tresidder (Cornwall).

**Tresham:** Thomas *Tresham* 1383 IpmGl; William *Tresham* 1439 NLCh; Thomas Tresham 1460 Paston. From Tresham in Hawkesbury (Gl).

**Tresilian:** Robert *Tresilian* 1379 AD iii (Co). From Tresillian (Cornwall).

**Trethewey, Trethewy, Treadaway:** Andrew *de Tredewi* c1210 Fees (Co); Henry *de Trethewy* 1297 MinAcctCo. From Trethewey (Cornwall).

**Trett:** Baldwin *le Tret* 1194 P (Do). ME *tret* 'neat, graceful, handsome', used of the face, nose and neck.

**Tretwell:** v. TREADWELL

**Trevallion, Trevelion, Trevelyan, Trevilian, Trevillian, Trevillion, Trevillyan:** John *Trevilian* 1503 Pat; Thomas *Trevillyan* 1641 PrSo; Peter *Trevillian* 1642 PrD. From Trevelyan in St Veep (Co).

**Treveal:** William *de Treuael* 1297 MinAcctCo. From Treveal (Cornwall).

**Trevellick:** John *de Treuellek* 1297 MinAcctCo. From Trevillick (Cornwall).

**Trever, Trevor, Trevers, Treversh:** Richard *Trevor*, Lewelin *Trefaur* 1538 Chirk. From Trevor (Denbigh, Anglesey).

**Trevilian, Trevillian, Trevillion, Trevillyan:** v. TREVALLION

**Trevis:** Zacharie *Trevis* 1630, *Trauis* 1634 LeiAS 23; Thomas *Trevis* 1681 FrY. Probably a late form of TRAVERS.

**Trevithick:** Geoffrey *Treuethec* 1359 Putnam (Co); Nicholas *Trevittick* 1642 PrD. John *Trevedicke* 1674 HTSf. From Trevethick (Co).

**Trevitt:** v. TRIVETT

**Trevor:** v. TREVER

**Trevvett:** v. TRIVETT

**Trew:** v. TRUE

**Trewick, Treweek, Treweeks:** Marjer' *de Trewyk* 1275 RH (Nb). From Trewick (Nb).

**Trewin:** v. TREE

**Trick, Trix:** *Trikke* del Nortgate 1277 Wak (Y); William *Tryk'* 1332 SRDo; Roger *Trix* 1642 PrD. OFr *trique* 'trick, deception'. Used also as a personal name.

**Tricker:** Gilbert *Trykere* 1260 FFEss; Adam *le Trikur* 1275 Wak (Y). A derivative of ME *trik*, OFr *trique*, Norman-Picard form of *triche* 'trick'. 'Cheat, deceiver.'

**Trickett:** Ralph *Trichet* 1130 P (Mx); Robert *Triket* 1198 Cur (Bk). A Norman-Picard form of Fr *Trichet, Trichot* which Dauzat takes as hypocoristics of *Trichard*, Norman-Picard *Tricard* 'cheat, deceiver'.

**Trickey:** A surname from Trickey (Devon) existed in 1238 (PN D 620).

**Trier, Tryer:** William *le Treyere* 1225 Cur (Sf); Robert *le Treyour* 1299 LLB C; John *le Treyere* 1327 SRSf. OFr *traieor, treieour* 'tapster, wine merchant'. v. MEOT 92.

**Triffit, Triffitt, Trifitt:** v. TRIVETT

**Trigg, Trigge, Triggs:** *Trig* 1185 Templars (Y); William *Trig* 1202 AssL, 1240 Rams (Nf); William *Trigges* 1279 RH (C); Ralph *Trigge* 1332 SRLa. ON *tryggr* 'true, faithful, trustworthy', also used as a personal-name.

**Trilby:** Alan *of Trilleby* 1254 IpmY. Probably from Thurlby (L).

**Trillo:** Philip *de Trillowe* 1279 RH (C). From Thurlow (Suffolk).

**Trim, Trimm:** *Trim* 1196 Cur; Henry *Trym* 1341 FFY; Nicholas *Trym* 1641 PrSo. OE *Trymma*.

**Trimlett:** v. TREMLETT

**Trimmer:** William *le Trymmare* 1327 MESO (Ha). A derivative of *trim* vb. v. NED. 'One who trims', the exact sense being uncertain as the verb has not been found in ME.

**Trimnell:** Thomas *Trymenel* 1259–60 FFWa; William *Trymnel* 1388 LLB H; Robert *Trymnel* 1576 SRW. *Trim-en-el*, a double diminutive of OE *Trymma*.

**Trindall:** v. TRENDALL

**Trinder:** Hugo *le Trinder* 1275 RH (Nf); John *le Trendare* 1278 AssSo. A derivative of OE *trendan* 'to turn round, roll'. As ME *trend* means 'to twist, plait' and *tryndelle* is a spindle, the surname probably denotes a braider.

**Tringham:** Gerard *de Tringham* 1205 P (L); William *Tringham* 1663 HeMil. From Tringham (Nf).

**Tripe:** Herbert *Tripe* 1185 Templars (K); Robert *Tripes* 1302 SRY. ME *tripe, tripis,* OFr *tripe* 'tripe', metonymic for a tripe-seller. cf. Fr *Tripier.*

**Tripp:** Gilbert, Hugh *Trip* 1275 RH (W), 1312 LLB D; Robert *Tryppe* 1296 SRSx. Either a variant of TRIPE (OFr *trippe*) or metonymic for TRIPPER.

**Tripper:** William *le Trippere* 1293 AssSt. ME *trippere* 'dancer' (c1380 NED).

**Trippett:** Ralph, Peter *Tripet* 1204 Cur (Beds), 1256 AssNb; William *Tripat* 1327 *SR* (Ess). ME *tripet, trepett,* OFr *tripot* 'an evil scheme, malicious trick'. *Trypet* alternates with *treget* c1330 (NED). cf. TREDGETT.

**Triprose:** Roger *Tripperose* AssL. Presumably a nickname, 'trip rose', but the particular meaning here is unknown.

**Trist, Trister:** Peter *ate Treste* 1279 RH (Bk); Richard *Trist* 1332 SRCu; William *Trystour* 1394 LLB H. ME *triste, tryster, tristur,* OFr *triste, tristre* 'an appointed station in hunting', denoting, probably, the man in charge of the hounds and the preparations for the hunt.

**Tristram, Trustram:** *Tristram* Cementarius 1204 Cur; *Tristram* Merewyne 1296 SRSx; Henry, Leonard, Richard *Tristram* 1207 Pl (Ess), 1296 SRSx, 1334 SRK; Richard *Trustram* 1577 Musters (Nf). Celtic *Drystan,* from *drest, drust* 'tumult, din'. It appears as a christian name in England from the end of the 12th century, usually in the form *Tristram.*

**Trivett, Trivitt, Treavett, Trevitt, Trevvett, Triffit, Triffitt, Trifit:** Hugh *Treuet* 1206 P (So); William *Trevet, Trivet* 1306 AssW; Thomas *Tryvet* 1391 FFC. Presumably a nickname from ME *trivet, trevet* 'tripod'.

**Trix:** *v.* TRICK

**Troath:** *v.* TROTH

**Trobridge:** *v.* TROWBRIDGE

**Trollop, Trollope:** John, Andrew *Trollop* 1427 AD iii (Du), 1461 Past; Androwe *Trowlopp,* Thomas *Trollopp* 1525 SRSx; John *Trollope* 1577 Musters (Nf). From Troughburn (Northumb), formerly *Trolhop* 'troll-valley.'

**Tromans:** *v.* TRUEMAN

**Tron, Trone:** Walter *Trone* 1255 FFK; John *Trone* 1327 SRSx; John *Troune* 1379 PTY. ME *trone* 'a weighing machine'. Metonymic for one in charge of this.

**Tronson:** *v.* TRUNCHION

**Troop, Troops, Troup, Troupe, Trupp:** John *de Trope* 1275 SRWo. A variant of THROP.

**Troth, Troath:** Roger *Troth* 1327 SRSf. ME *troupe, trothe* 'faithfulness, loyalty'.

**Trotman:** Richard *Troteman* 1224 Cur (Ess); Robert *Trotteman* 1281 Ipm (Sx). Synonymous with TROTTER.

**Trott:** Walter, Robert *Trot* 1206 P (Sr), Cur (Bk); William *le Trot* 1327 SRSx. A verbal substantive from 'to trot' in the sense of TROTTER. Or, perhaps, AFr *trote* 'old woman, hag', though usually *trate* in ME.

**Trotter:** Robert *trotar* 1148 Winton (Ha); Adam *le Troter* 1219 AssY. OFr *trotier* 'a trotter, messenger'.

**Troubridge:** *v.* TROWBRIDGE

**Trough:** William *of the Trogh* 1401 AssLa. From the Trough of Bowland (La), or 'dweller in the hollow', OE *trog,* in a topographical sense.

**Troughton:** Nicholas *Troughton* 1607 FrY; William *Troughton* 1642 PrD. From Troughton (La).

**Trouncer:** Gilbert *le trunchier* 1192 P (Lo); Henry *le Truncer* 1315 Wak (Y). A derivative of OFr *tronche* 'club, cudgel'. A maker or seller of cudgels.

**Trounson:** *v.* TRUNCHION

**Troup(e):** *v.* TROOP

**Trout, Troutt, Trowt, Trute:** William *Trute, Troute* 1202 AssL, 1327 SRSf. OE *trūht* 'trout'.

**Troutbeck:** William *Troutebek* 1421-2 FFWa. From Troutbeck (Cu).

**Trow:** *v.* TRUE

**Trowbridge, Troubridge, Trobridge, Trubridge:** Walter *de Trobrigge* 1184 P (Gl); William *de Trewebrugg'* 1275 SRWo. From Trowbridge (Wilts).

**Trowell, Trowill, Trowles:** Richard *de Trowell'* 1204 P (Nt); Edmund *Trowlles* 1524 SRSf. From Trowell (Notts).

**Trower:** *v.* THROWER

**Trowman:** *v.* TRUEMAN

**Trowt:** *v.* TROUT

**Troy:** Elyas *de Troie* 1200 P (Bk); Copyn *de Troys* 1276 LLB A. From Troyes (Aube).

**Troyt, Troyte:** Robertus *filius Troite* 1159 P (Cu); Robert *Troite* 1166 ib.; Richard *Troyte* 1230 ib. OIr *\*Troit.* cf. Ir *troid* 'quarrel' (Smith).

**Trubridge:** *v.* TROWBRIDGE

**Truce:** Juliana *in Trewes* 1279 RH (Beds). 'Dweller among the trees.' cf. TRUE (ii).

**Trudgett:** *v.* TREDGETT

**True, Trew, Trow:** (i) Rannulfus *Truue* (*Triue*) 1180 P (Wa); Ralph *Truwe* 1185 Templars (K); Roger *Trowe* 1200 Cur (W); William *Trewe* 1301 SRY; Henry *le Trewe* 1327 AD i (W). OE *trēowe,* ME *trew(e), trow(e)* 'faithful, loyal, trustworthy'. (ii) Hugo *de la Truwe, de la Trowe* 1250 Fees (So); Laurence *atte Trowe* 1332 SRSx. OE *trēow,* ME *treow, trew, trow* 'tree', from residence near some prominent tree or from one of the places in Devon named Tree, Trew, True or Trow. (iii) Roger *de Trow* 1207 Cur (W); Jacobus *de Trewe, de Trowe* 1219, 1222 Pat (W). From Trow Fm and Down (Wilts), from OE *trog* 'trough'. This becomes *Trow. v.* PN W 200.

**Trueblood:** Richard *Trewebloude* 1450-1 FFWa. Probably 'loyal person', OE *trēow,* and *blōd* 'blood'. *True-* is a common first element in nicknames. cf. John *Truchilde* 1559 Pat (Berks) 'true child'; John *Trewfelagh* 1379 PTY 'true fellow'; John *Trewpage* 1379 PTY 'trusty servant'; John *Trewpeny* 1472 SIA xii 'true penny', probably 'trusty person'.

**Truebody:** Roger *Trewebodi* 1277 AssSo; William *Treubody* 1301 ArchC ix. 'True, faithful man.'

**Truecock:** Bartholomew *Truecoke* 1332 NorwDeeds II; Geoffrey *Trucok* 1361 AssY. A diminutive of TRUE.

**Truefitt, Truffhitt:** Henry *Trewfoote* 1524 SRSf; Robert *Trewfett* 1669 Kirk EllaPR (Y). William *Trewfitt* 1703 FrY. A nickname, 'true foot', OE *trēowe, fōt.*

**Truelock:** Walter *Treuloc* 1202 P (Co). ME *treulac* 'fidelity'.

**Truelove, Trolove:** Roger *Trewelove* 1275 RH (Sa); Robert *Treuweloue* 1296 SRSx; John *Truloue* 1384

AssWa. OE *trēowe* and *lufu* 'faithful love' in the sense 'faithful lover, sweetheart, beloved' (c1385 NED).

**Trueman, Truman, Trewman, Troman, Tromans, Trowman:** Richard *Treweman* c1215 StGreg (K); William *Trueman* 1279 RH (O); Robert *Trowman* 1327 SRSf. 'Faithful, trusty man' (1297 NED).

**Truett:** *v.* TYRWHITT

**Truffhitt:** *v.* TRUEFITT

**Trugg, Trugge:** William *Trug* 1275–6 FFSr; Robert *Trugge* 1325–6 CorLo; Thomas *Trugg* 1392 LoCh. cf. early modern English *trug* 'prostitute'.

**Trull, Trulle:** Reginald *Trull* 1222 Acc; Robert *le Trulle* 1296 SRSx; Richard *Trulle* 1379 PTY. ME *trull* 'slutton, drab'.

**Trumble, Trumbull:** Alan *Tumbald, Thrumball* 1313, 1316 Wak (Y); Peter, Reginald *Thrumbald* 1315 ib., 1327 SRSf; Alice *Thrumbald, Trumbald* 1316–17 Wak (Y); Elizabeth *Trumble* 1568 SRSf; Ralph *Trumball* 1696 DKR 41 (Berks). OE *\*Trumbeald* 'strong-bold'. cf. OE *Trumbeorht, Trumwine*, and Robert *Trumwine* 1236 Fees (St). *v.* TURNBULL.

**Trump:** Patrick, Nicholas *Trumpe* 1275 Ipm (Cu), 1279 RH (C). Metonymic for *Trumper*.

**Trumper:** Adam *Trumpur* 1253 AD iv (Ess); William *Trompour* 1320 LLB E; John *le Trumpour* 1327 SRY. OFr *trompeor, trompour, trumpeur* 'trumpeter'.

**Trumwin, Trumwyn:** Robert *Trumwine* 1236 Fees (St); John *Trumwyn* 1332 SRSt; Thomas *Tromwin* 1341 AD i (Berks). OE *Trumwine*.

**Trunchion, Tronson, Trounson:** William *Trunchun* 1209 FrLeic; Thomas *Trunson* 1327 SRC. OFr *truncun, tronchon* 'truncheon, club', metonymic for a maker or seller of cudgels. Or, perhaps, for an official who carried a truncheon.

**Trupp:** *v.* TROOP

**Truscott:** Michael *de Trescote* 1272 AssSt. From Trescott (Staffs).

**Trushell:** *v.* TRUSSEL

**Trusher:** William *Trussehare* 1206 Cur (Wo). 'Carry off the hare', a nickname, probably for a poacher. cf. TRUSLOVE.

**Truslove, Trusler, Trussler:** Henry *de* (sic) *Trusseluue* 1221 AssWa; Nicholas, Adam *Trusselove* 1296 SRSx, 1302 SRY; Thomas *Truslowe* 1524 *SR* (W). OFr *trousser, trusser* 'to truss, bind, carry off' and AFr *love* 'wolf'; 'bind wolf', a wolf-hunter. cf. BINDLOES and Walkelin *Trussevilain* 1171 Riev (Y).

**Truss:** Robert, John *Trusse* 1202 FFSf, 1275 SRWo. OFr *trousse* 'bundle, package'. *v.* TRUSSMAN.

**Trussbutt:** William *Trussebut* 1154–74 YCh; Ylaria *Trussebut* 1208 P (Nth); Geoffrey *Trussebutt'* 1363 AssY. 'Load the packhorse', OFr *trusser*, a nickname for a porter or a carrier. cf. William *Trushernays* 1296 SRNb 'bind on the armour'; Walkelin *Trusseuilain* 1175–6 YCh 'bind the villein'.

**Trussel, Trussell, Trushell:** Robert *Trusel* 1195 P (Lei); Godfrey *Trussel* 1204 P (Y); Richard *Trussell* 1221 AssWa; William *Troussel* 1285 FA (St). OFr *troussel* 'packet', in ME also 'the puncheon or mould used in the stamping of coins'. A maker or user of 'trussels'.

**Trussman:** John *Trusseman* 1327 SRSx. Probably a baggage-man, porter. *v.* TRUSS.

**Trustram:** *v.* TRISTRAM

**Trute:** *v.* TROUT

**Try, Trye:** Thomas *Trie* 1274 RH (Sa); Juliana *Trye* 1301 SRY. ME *trie, triȝe* 'excellent, good'.

**Tryer:** *v.* TRIER

**Tubb, Tubbs, Tubby:** *Tubi, Tube* 1066 DB (Bk, Wa); Reginaldus *filius Tobbe* 1166 P (Y); Laurencius *filius Tubb* 1230 ib.; Alan *Tubbi* 1206 Cur (Nf); John *Tub* 1212–23 Bart (Mx); Roger, William *Tubbe* 1243 AssSo, 1296 AssNb. ON, ODa *Tubbi*, OSw *Tubbe*.

**Tubman:** Robert *Tubman* 1430 FrY. A derivative of ME *tubbe*, a maker of tubs, cooper.

**Tuck:** *Tukke* faber 1101–7 Holme (Nf); Radulfus *filius Tokke* 1175 P (Y); Symon *filius Thocche* a1187 DC (L); Johannes *filius Tuch* c1250 Rams (C); Besi *Tuk* 1051 KCD 795 (L); Henry *Thoche* 12th DC (L); Robert *Tucke* 1202 FF (Nf). Tengvik derives (Besi) *Tuk* from ODa *\*Tuk*, a strong form of *Tuki*, which is possible, but the frequent occurrence of the personal name in the 12th and 13th centuries suggests that we have an Anglo-Scand. *\*Tukka*, a pet-form of ON *Porketill*.

**Tucker:** Baldwin *Tuckere* 1236 Battle (Sx); Wolward *le Tukare* 1243 AssSo; Thomas *le Touchere* 1293 Pinchbeck (Sf); Hugo *le Tukker'* 1297 MinAcctCo (Co); Richard *le Touker* 1327 SRSo. A derivative of OE *tūcian* 'to torment', later 'to tuck, to full', 'a tucker, fuller'. *v.* FULLER. Occasionally a nickname for courage (Fr *tout cœur*): Geoffrey *Tutquor, Totquer* 1217 Pat (K), Hy 3 Colch (Ess).

**Tuckerman, Tuckermann:** Richard *Tuckerman* 1647 DWills. 'The tuckerman', a tucker, fuller.

**Tucket, Tuckett:** Alan *Tuchet* 1207 Cur; Nicholas *Tochet* 1298 AssL; James *Tucket* 1642 PrD. Probably a diminutive of Anglo-Scandinavian *\*Tukka*. *v.* TUCK.

**Tuckey:** *v.* TOOKEY

**Tuckwood:** John *de Tuxeford* 1398 Shef. From Tuxford (Nt).

**Tudball:** *v.* THEOBALD

**Tuddenham:** John *de Tudeham* 1191 P (Sf). From Tuddenham (Norfolk, Suffolk).

**Tudman:** John *Tudnham*, Thomas *Tudnam*, William *Tudman* 1524 SRSf. For TUDDENHAM. cf. DEBENHAM.

**Tudor:** *Tudor* 1221 AssSa; David *ap Tudir* 1287 AssCh; *Tudur* ap Llywelyn 1391 Chirk; Christian *Tudor* 1327–9 FrC; John *Tewdre* 1334–5 SRK. The Welsh form of *Theodore*.

**Tuer:** *v.* TEWER

**Tuffery, Tuffrey:** *v.* TOLLFREE

**Tuffield, Tuffill:** *v.* TOVELL

**Tufton:** John *Tufton* 1500 SxAS 45. From Tufton (Ha).

**Tugford:** Richard *de Tuggeford* 1275 SRWo. From Tugford (Sa).

**Tugwood:** *v.* TOOGOOD

**Tuke:** Walter *filius Tuke* 1197 P (L); Henry *Tuke* 1246 ForNth; Richard *Tuke* 1487 Paston. ON *Tóki*.

**Tulet:** *v.* TULLETT

**Tull, Tulle:** Roger *Tulle* 1219 AssY; Peter *Tulle* 1298 AssL; Hugh *Tulle* 1454 IpmNt. OE *Tulla*.

**Tullet, Tullett, Tulett, Toulet:** Walter *Tulet* 1219 P (Nb); Robert *Tuylet* 1295 Barnwell (C); Robert *Tuliet* 1361 FFEss. ME *tuillet*, a diminutive of OFr *tieule* 'tile, plaque'. In medieval armour one of two or more plates of steel covering the front of the thighs. Hence, probably, metonymic for a maker of armour.

**Tulliman, Tullimond:** Thomas *Tutlemund'* 1219 AssY; John *Totlemound* 1301 CorLo; John *Toulemounde* 1337 IpmGl. From a favourite expression, 'all the world', Fr *tout le monde*. cf. William *Altheworld* 1303 AssW.

**Tulliver:** *v.* TELFER

**Tumber:** John *Tumbur* 1276 MEOT (O); Henry *le Tombere* 1327 ib. (Ha). OE *tumbere* or OFr *tombeor*, *tumbeur* 'tumbler, dancer'.

**Tumley:** Ralph *Tumele* 1210 Cur (L). Probably 'dweller at the farm in the clearing', OE *tūn*, *lēah*.

**Tumman, Tummon, Thommen, Toman:** (i) Ralph, Adam *Tuneman* 1279 RH (Beds), 1296 SRSx; Pagan *Toneman* 1327 *SR* (Ess); Nicholas *Tounman* 1365 LoPleas; Thomas *Tonman* 1379 PTY. OE *tūnmann* 'villager'. (ii) Nicholas *Thomasman* 1301 SRY; William *Thomeman* 1379 PTY. cf. Johannes *Tomman* Cisson 1379 PTY, i.e. 'John, servant of Tom Cisson'. Rarer than (i) and, no doubt, the origin of the Yorkshire *Tummon*. cf. ADDYMAN, MATTHEWMAN.

**Tunbridge:** *v.* TONBRIDGE

**Tundew:** Alard *Tundu* 1203–4 FFK; Philip *Tundu* 1289 IpmY; William *Tundowe* 1367 FFY. OFr *tondu* 'shaven, shorn'.

**Tune:** *v.* TOWN

**Tunn:** Hugo *filius Tunne* 1204 P (Y); Robert *Tun* 1218 AssL; Robert, Reginald *Tunne* 1279 RH (O), 1311 RamsCt (Hu). *Tunne* is OE *Tunna* (Redin), a pet-form of such names as OE *Tūnrǣd*, *Tūnwulf* and *Tūnrīc*. cf. Robertus *filius Tunrici* 1182–1200 BuryS (Sf). The surname may also be metonymic for a maker of tuns. cf. Robert *le Tunnewrytte* 1279 AssNb and *v.* TUNNAH.

**Tunnah:** (i) William *le Tunnere* 1280 MESO (Ha). A derivative of OE *tunne* 'a tun', a maker of tuns. (ii) Hugh, William *le Tundur* 1275 RH (Nf), 1296 SRSx. AFr *tundour*, OFr *tondeur* 'shearman'. For the development, cf. *Lunnon* from *London* and *Farrah* for *Farrer*.

**Tunnard:** Augustin *Tunherd* 1279 RH (C); Robert *le Tunherd* 1327 SRC. OE *\*tūn-hierde* 'guardian of the village or town animals'. Johannes *filius Tunherd* 1327 SRC may mean 'son of the town-herd' but we may have an unrecorded OE *\*Tūnheard*. cf. *Tonhardus* 1066–87 Bury (Sf).

**Tunnell:** *v.* TONNELL

**Tunney:** *Tunne* 1066 DB (L); Gillibertus *filius Tunny* 1219 AssY; Simon *Tunnie* 1327 AssSt; Adam *Tunnyson* 1332 SRCu. ODa *Tunni*, OSw *Tunne*.

**Tunnicliff, Tunnicliffe, Dunnicliff, Dunnicliffe:** Henry *de Tunwaleclif* 1246 AssLa. From Tonacliffe (Lancs).

**Tunnock:** Alan *filius Tunnoc* 12th FeuDu; *Tunnok* de Dunum 1246 AssLa; Thomas *Tunnok* 1327 SRC; John *Tunnok'* 1387 AssL. OE *Tunnoc*. It was also used as a feminine name: *Tunnok'* filia Richold 1219 AssY; *Tunnoke* vidua 1240 FFY.

**Tunsley:** Robert *de Tundesle* 1214 Cur (Sr). From Townslow (Surrey).

**Tunstall, Tunstill, Dunstall:** Reginald *de Tunstal* 1185 P (Y). From Tunstall (ER, NRYorks, Suffolk, etc.).

**Tunstead:** Augustine *de Tunsted'* 1198 FFNf. From Tunstead (Db, La, Nf).

**Tuphead:** William *Tupeheved* 1246 AssLa; Robert *Tupeheved* 1299 Misc (Nt). 'Ram's head', ME *tuppe*, OE *hēafod*.

**Tupp, Tuppe:** Martin *Tuppe* 1230 Cur (Y); William *Tuppe* 1329 IpmNt; Thomas *Tup* 1379 PTY. ME *tuppe* 'ram'.

**Tuppenny, Tuppeny:** *v.* TWOPENNY

**Tupper:** Robert *Tupper* 1314 Wak (Y). At York in 1365 men were employed in beating and ramming (*tupant'*) 'the earth and mud, strengthened with straw, with rammers (*tuppis*) and great hammers' (Building 85). As the rams were called *tups*, these workmen may well have been named *tuppers*. The surname may also be a late form of *tup-herd* (ME *tup* 'ram'): Robert *Tophird* 1327 SRY, William *Tuphird* 1379 PTY.

**Turbard, Turbet, Turbett, Turbott, Turbutt, Torbett, Torbitt, Tarbard, Tarbert, Tarbath, Tarbat, Tarbet, Tarbitt, Tarbutt:** *Turbert*, *Torbertus* 1066 DB; *Thurbert* 1066 InqEl (Sf); *Turbertus* c1160–70 NthCh (Beds); Alanus, Gaufridus *Torberti* 1212 Cur (Berks); John *Turbut* 1221 Cur (Herts); William *Turbert* 1248 Fees (Ess); John *Turberd* 1274 RH (Ess); Thomas *Torebat* 1279 RH (C); Eudo *Turbot* 1327 *SR* (Ess). There is no OE or Scand personal name which fits these forms. We are concerned with a hybrid *Þorbert*, *Þurbert* in which the first theme is Scand *Þor-*, *Þur-*, and the second OG *-bert*. The name was probably formed on the Continent and is identical with the Norman *Turbert* found in *Turbertivilla*, probably Thouberville (Eure). Hence the Norman initial *T* and the frequent loss of the second *r*. Only one example of the name is found in England before the Conquest but it is common in DB and continued in use until at least the 13th century. It is sometimes confused with *Þorbiǫrn*. *v.* THURBAN.

**Turbefield, Turberfield, Turberville:** Ralph *de Tuberilli* 1115 Winton (Ha); Robert *de Turbertualla* 1121 AC (He); Hugo *de Turbervilla* 1123 AC (He); William *de Truble villa*, c1125–30 EngFeud, *de Turbertiuilla* 1130 P (Do); Maud *de Turbervill*, *de Trubleuile* 1269, 1279 AssSo. From Thouberville (Eure). *v.* PNDB 391, n. 8.

**Turbett, Turbott:** *v.* TURBARD

**Turbin:** *v.* THURBAN

**Turfery, Turfrey:** *v.* TOLLFREE

**Turfley:** Bonekoc *de Turflegh* 1296 SRSx. From a lost *Turfleigh* in Willingdon (Sx).

**Turgel, Turgill:** *v.* THURKELL

**Turgoose:** *v.* STURGE

**Turk:** *Turch*, *Turcus* 1066 DB (C); *Turche* c1150 DC (L); *Turkus* fugitivus 1172 P (Sx); Ricardus *filius Torke* 1188 P (Y); Ricardus *filius Turk'* 1205 ChR (K); Eadwin *Turcus* (*le Turch*, *Tercus*) c1140 ELPN; William *Turc*, *le Turc* 1193, 1196 P (Gl); Robert *Turk* 1296 SRSx. The DB *Turch* is explained by von Feilitzen as ON *Þorkell*, with AN loss of *-el*. It seems clear that it was also used as a pet-form of this Scandinavian name. Most of the surnames appear to be nicknames from OFr *turc* 'Turk', a word which NED suggests was introduced into England during the third crusade (1187–92). It is found as a nickname in London half a century earlier.

**Turkel, Turkil:** *v.* THURKELL

**Turkentine, Turketine:** Robertus *filius Turketin* c1150 DC (L); Nathaniel *Turquentine* 1654 ShotleyPR (Sf); Mr *Turkenton* 1674 HTSf; Hannah *Turkeytine* 1817 RushbrookPR (Sf). A double diminutive of *Turk*, from ON *þorkell*, *Turk-et-in*. cf. Nicholas *Turkot* 1317 AssK.

**Turnagen:** John *Turneagayn* 1525 SRSx. 'Turn again', OE *turnian*, *ongegn*. cf. William *Turnabute* Acton 1272 AssSt 'turn about'; Thomas *Turnecote* 1524 SRSx 'turn coat'; *Turneteil* 1148 Winton (Ha) 'turn tail'.

**Turnbull, Turnbill:** Willelmus dictus *Turnebule* 1314 Black; William *Turbolle* 1327 SRSf; Walter *Tornebole* c1354 Black; Richard *Turnebull'* 1379 PTY; David *Trumbull* or *Turnbull* 1495 Bardsley. There can be no doubt that this much-discussed surname is a nickname 'turn bull', indicative of strength or bravery. The name appears to be northern, particularly Scottish, but early examples are not common. Black's derivation from *Trumbald* cannot be correct. The early forms of *Trumble* are quite distinct from those of *Turnbull* and there is no proof that any of the 15th-century Scottish Trumbles were Turnbulls. The Fife families of Trimbill, Trombill and Trumble may well have owed their name, as Black suggests, to the same place from which Robert de Tremblee (1296) came. OE *Trumbeald* developed naturally to *Trumball*, also spelled *Trumbull*. It is much more likely that this should be corrupted to *Turnbull* than that *Turnbull* should become an unintelligible *Trumbull*, *Trumble*. The nickname origin of the surname is proved by Ewen himself (despite his antipathy to nicknames) in his reference to a Yorkshire horse named *Turnebull* (1358) and is confirmed by the modern French *Tournebœuf* which Dauzat explains as a name for a drover.

**Turnell:** Wyther cognomento *Turnel* 1134–40 Holme (Nf); William *Turnel* 1202 FFNf; William *Turnell* 1642 PrD. Probably 'dweller at the small tower', OFr *tornele*, but sometimes, perhaps, a nickname from this.

**Turner, Turnor:** (i) Warner *le Turnur* 1180 P (Lo); Ralph *le tornur*, *tornator*, *le turner* 1191–2 P (Lei). OFr *tornour*, *tourneour* 'turner, one who turns or fashions objects of wood, metal, bone, etc., on a lathe' (c1400 NED). This is, no doubt, the common source of this occupational surname. Its frequency is due to the variety of objects that could be turned and to the use of the word in other senses. Lat *tornator* meant 'turnspit' (1308) as well as 'turner' (1327 MLWL). cf. 'Turnowre, *Tornator*' PromptParv, and 'Turnare, or he that turnythe a spete or other lyke, *versor*' ib., *tornerers* 'translaters' 1387 NED and *v.* DISHER. Nor can we exclude OFr *tornoieor*, *tournoieur* 'one who takes part in a tourney or tournament' (Lat *torneator* 'jouster' 1220 MLWL; ME *tourn(e)our* 1303 NED).

cf. JUSTER. (ii) Bernard, Robert *Turnehare* 1224 Cur (St), 1301 SRY. 'Turn hare', one so speedy that he could outstrip and turn the hare. As a surname, the second element would be unstressed and became *Turner*. cf. CATCHER.

**Turney, Tournay:** Goisfridus *Tornai* 1086 DB (L); Thomas *de Turnay* 1192 P (Lo). From Tournai, Tournay or Tourny, all in Normandy. The DB baron came from Tournai (Calvados).

**Turnham:** Robert *de Turnham* 1194 P (Mx); Stephen *de Turneham* 1219 P (K); Simon *de Tournham* 1322 CorLo. From Turnham Green in Chiswick (Mx), or Turnham Hall in Cliffe (ERY).

**Turnpenney, Turnpenny:** Ralf *Turnepeny* 1227 AssBk; John *Tornepeny* 1269 AssSt. A nickname from the phrase 'to turn a penny' (1546 NED) in the sense 'a person who is intent on a profit' (1824 ib.).

**Turp:** Robert *de Turp* 1177, 1230 P (Cu); Adam *Tourpp* 1332 SRCu. A metathesized form of TROOP, for THORP.

**Turpie:** John *Turpy* 1607 Black (Fife). A Scottish diminutive of *Turpin*.

**Turpin:** *Torfin*, *Turfin* 1066 DB (Y); *Turfinus filius Torfini* 1130 P (Y); *Turpin* Hy 2 DC (L); *Turpinus* 1180 P (Ha); *Torphinus* 1196 P (Y), c1227 Fees (Nb); *Turfin* 1202 P (Nb); Gaufridus *filius Thorphini* 1204 Cur (Y); *Thorpinus filius Simonis* 1230 P (D); William, Richard *Turpin* 1187, 1196 P (Ha, Y); Thomas *Thurpin* 1230 P (Y); Simon *Turpyn*, *Tropyn* 1317 AssK. The above forms make it clear that *Turpin* derives from ON *þorfinnr*, from *þórr*, the god, and the ethnic name *Finnr*. The French *Turpin*, *Tourpin* is derived by Dauzat and Michaëlsson from *Turpinus*, a derivative of Lat *turpis* 'disgraceful, base', a name adopted by the early Christians as a token of humility, which came into vogue again, its meaning forgotten, through the influence of the *Chanson de Roland*, where the 8th-century archbishop of Rheims appears as *Turpin*. Forcellini has no example of *Turpinus*. *þorfinnr* is found in Normandy where the earliest examples of *Turpin* occur and the modern surname is common. The English and the French surnames must have the same origin.

**Turpitt:** Stacia *del Torfpet*, William *Turpet* 1327 SRSf. 'Dweller by or worker at a turf-pit' (OE *turf* and *pytt*).

**Turral, Turrall:** *v.* THOROLD

**Turreff, Turriff:** (i) *Turgiua* feneratrix 1169 P (Gl); *Thurrieua* de Cestrehunte 1197 FF (Herts); William *Thuryff* 1404 DbCh. An unrecorded Anglo-Scandinavian hybrid *þórgifu* (f), from ON *þórr* 'Thor' and OE *g(i)efu*, *gifu* 'gift'. (ii) David *Turreff* 1682 Black (Aberdeen). From Turreff (Aberdeenshire).

**Turrill:** *Durilda* 1066 DB (Sf); Claricia *þourild*, Walter *þurild* 1279 RH (O); John *Torild* 1308 Wak (Y). ON *þórhildr*, ODa *Thorild*, a rare woman's name.

**Turtill:** *v.* THURKELL

**Turtle:** Robert *Turtell'* 1176 P (Nth); Walter *Turtel* 1214 Cur (K); Henry *Turtle* 1327 *SR* (Ess). A nickname from the turtle-dove, ME *turtel*, *turtle*, OE *turtle* (f), *turtla* (m). Or this may sometimes be from Fr *tourtel*, a diminutive from Lat *tortus* 'crooked'.

Most commonly no doubt, the modern surname is due to a late assimilation of *kl* to *tl* in *Turkle*. *v.* THURKELL.

**Turton:** Everard *de Turton'* 1208 P (Nf); Nicholas *de Turton* 1278 AssLa; William *Turton* 1478 TestEbor. From Turton (La). Sometimes, perhaps, from Thurton (Nf).

**Turvey:** Stephen *de Torfeia* 1191 P (Beds); John *de Turveie* of co. Hunt 1324 CoramLa; John *Turvey* 1524 SRSf. From Turvey (Beds).

**Turvill, Turville, Turvell:** Geoffrey *de Tureuilla* 1130 P (Bk); Maurice *de Turuill'* 1206 Pl (Ha); Robert *de Turevill'* 1280 FFY; John *Turuyll* 1332 SRWa. From Turville-la-campagne (Eure). The name can have no connexion with Turville (Bucks) since the original *-feld* in this name was not replaced by *-ville* until the 19th century.

**Tustain, Tustian, Tustin, Tusting:** *v.* THURSTAN
**Tutchener, Tutchings:** *v.* TWITCHEN
**Tuthill, Tutill:** *v.* TOOTHILL
**Tutin, Tuting, Tuton:** *v.* THURSTAN
**Tuttle:** *v.* THURKELL, TOOTHILL, TURTLE

**Tuttlebee, Tuttleby:** Roger *de Turkelby*, *de Thirkelby* 1241 Eynsham, 1257 Riev (Y). From Thirkelby (ER, NRYorks). For *Tuttle-*, cf. THURKELL.

**Tuvey:** *v.* TOVEE
**Twaite, Twaits:** *v.* THWAITE

**Tweedale, Tweddell, Tweddle, Tweedle:** Robert *de Twedhall'* 1279 AssNb; John *de Tweddale* 1376 Black. 'Man from the valley of the Tweed.'

**Tweedy:** (i) Richard *Twedy* 1515 FFEss; Roger *Twedie* 1568 SRSf; John *Tweedy* 1665 PN Cu 97. From Tweedyhill in Kingwater (Cu). (ii) Finlay *de Twydyn* 1296, Walter *de Twydi* 1303 Black. From the lands of Tweedie in Stonehouse (Lanark).

**Twell, Twells:** Hugh *Twell* 1566 ShefA; Godfrie *Twelles* 1606 RushbrookPR (Sf). An apheric form of ATTWELL.

**Twentiman, Twentyman:** Thomas *Twentyman* 1662 HTEss. 'One in command of twenty men', OE *twēntig, mann*. At Vale Royal in 1278, Carnarvon in 1282, and Harlech in 1286, the labourers were organized on a semi-military basis under *vintenarii*, though judging from their numbers the gangs must have been considerably more than the 20 implied by the title. *v.* Building 54. cf. Robert *Tuelfmen* 1327 SRY 'twelve men'; William *Twentipayr* 1315 Wak (Y) 'twenty pair'.

**Twiceaday, Twisaday:** Thomas *Twyssaday*, *Twyseaday* 1485 Pat (K), 1487 Cl (Lo); Richard *Twyssoday* 1496 GildY; Thomas *Twysaday*, *Twysdey* or *Twysadey* 1509 LP (Lo); Harry *Twisedaie* 1548 Bardsley (La); Edward *Twiceaday* 1661 LaWills. This is almost certainly a popular perversion of *Tuesday*, for someone born on that day, cf. twysday c1275, Twesdaie 1587 NED, and John *Tewyssnothe*, Denyce *Twesnott*, Robert *Twysnoth* 1493, 1508, 1541 CantW, Thomas *Twisnodde* 1558 Pat, all from Tuesnoad (Kent).

**Twigg, Twigge:** John *Twyg* 1296 AssCh. Northern OE *twigge* 'a slender shoot'.

**Twin, Twinn:** Nicholas *Twin* 1279 RH (C). OE *(ge)twinn* 'twin'. cf. GEMMELL.

**Twine:** Edmund *Twyne* 1422 FFHu. OE *twīn* 'thread, string'. Metonymic for TWINER or (Agnes) *Twynmaker* 1367 ColchCt.

**Twineham, Twinham:** *v.* TWYNAM

**Twiner:** Adam *Tweyner* 1327 SRC; John *Twyner* 1429 LLB K. A derivative of ME *twīnen* 'to twine', one who twines or twists thread.

**Twining:** John *de Twynyng* 1338 FFW. From Twyning (Gl).

**Twisaday:** *v.* TWICEADAY

**Twiselton, Twistleton:** Thomas *de Twisilton'* 1208 FFY; Hugh *de Twiselton* 1260 AssLa; John *Twisilton* 1490 TestEbor. From Twistleton (WRY), or Twiston (La), *Tuiseltun* 1102.

**Twisnod:** Thomas *Tusnoth* 1477, Robert *Twysnoth* 1541 CantW; Laurence *Twisnodde* 1558 Pat (K). From Tuesnoad in Bethersden (K).

**Twiss, Twisse:** Gilbert *de Twisse* 13th WhC; Roger *del Twysse* 1276 AssLa; Jordan *del Twys* 1416 IpmLa. From Twiss (La).

**Twist:** Peter *Twist* 1672 HTO; Samuel *Twist* 1674 HTSf. From Twist in Chardstock, in Petertavy (D), or Twist Wood in Brede (Sx).

**Twistleton:** *v.* TWISELTON

**Twissell:** *v.* TWIZEL

**Twitchen, Twitching, Twitchings, Tutchings, Tutchener, Titchener, Tichner:** Thomas *de la Twichene* 1275 SRWo; Gilbert *ate Thuychene* 1297 MinAcctCo; Richard *Twichener* 1432 LLB K. 'Dweller at the place where two roads meet', OE *twicen(e)*, particularly common in Devon as Twitchen (once as Tuchenor), with forms: *Nitheretochene* 1330, *Tuchyn* 1577, *Towchyn* 1650, *Twitching* 1679, *Titchin* 1809 (PN D 90, 353, 42, 57).

**Twite:** (i) Lambert *del Tuuit* c1190 Gilb (L); Hugh *de Twyt* 1219 AssY; Ymanye *de Thuyt* 1274 RH (Nf). OE *þwīt* or ON *þvít*, a variant of *þveit*, surviving in Twit (Lincs). *v.* THWAITE. (ii) Richard *Twyt* 1275 SRWo; William, Robert *Twyt* 1296 SRSx; John *Twyt* 1315 Wak (Y). A species of linnet called the Mountain Linnet or Twite-finch (1562 NED), found in the hilly and moorland districts of the north. In winter they migrate south and haunt the coast and 'enliven the marshes with their twittering song' (NED). In the 18th century they were sold as singing birds.

**Twizel, Twissell:** Richard *Twisle* 1196 P (Du); Richard *de Twysel* 1272 AssSt. From residence near the fork of a river or land in such a fork (OE *twisla*), or from Twizel Castle or Twizell (Northumb). Magota *atte Twisele* in 1379 lived at Twissell's Mill (PN Sx 467). cf. also Tweazle Wood (ib. 470).

**Twopenny, Tippenny, Tuppenny, Tuppeny:** Thomas *Twopenes* 1260 AssY; Thomas *Twapenis* 1297 SRY; John *Twapens* 1381 PTY. 'Two pence', OE *twā, penig*. cf. Alice *Fouerpenys* 1285 Pinchbeck (Sf) 'fourpence'; John *Fivepeni* 1279 RH (O); Nicholas *Terdepeny* 1327 SRSf 'third penny'.

**Twycross:** William *de Twicros* 1221 AssWa; John *de Twycros* 1354 AssSt. From Twycross (Lei).

**Twyer:** Peter *de la Twyer* 1280 FFY; William *Twyer* 1393 TestEbor; Robert *Twyer* 1435 Shef. 'Worker at the furnace', OFr *tuyere, tuhiere, tuiere, touyere* 'a blast-pipe for a furnace'.

**Twyford:** Juliana *de Twiford* 1221 AssWa; John *de Twyford* 1316 AssNth; Ralph *Twyford* 1381 LoCh. From Twyford (Berks, Bk, Db, Ha, He, L, Lei, Mx, Nf).

**Twynam, Twinham, Twineham:** Adam *de Twynam* 1251 Balliol; John *Twynem* 1327 SRSx; Walter *de Twynham* 1368 FFEss. From Twineham (Sx), or Twinham, now Christchurch (Ha).

**Twysden:** John *de Twysdenne* 1334 SRK; John, Richard *Twysden* 1447 CtH, 1525 SRSx. From Twysden in Goudhurst (Kent).

**Tyas, Tyes:** Everard *le Tieis* 1170–90 Seals (Herts); Fulbrich *Tyes* 1236–7 Clerkenwell (Lo). OFr *tieis* 'German'.

**Tydeman:** *v.* TIDDEMAN

**Tydsley:** *v.* TILDESLEY

**Tye:** Hugo *de la Tye* 1230 P (D); Richard *ater Tye* 1296 SRSx. 'Dweller by a common or enclosure' (OE *tēag*). cf. TEY.

**Tyer, Tyers, Tyars, Tier, Tyre:** Osbert *filius Thiardi* c1110 Winton (Ha); Henry *filius Tihardi* 1178 P (Nth); Osbert *Tiart* c1110 Winton (Ha); Thomas *Tyard* 1275 RH (Nth); Susan *Tyers* 1662 HTEss. OG *Theudhard*, OFr *Thiart*. Sometimes, perhaps, 'dweller by the enclosure', from a derivative of OE *tēag*.

**Tyerman, Tyreman:** John *tyreman* 1332 SRSr; Francis *Tyreman* 1620 FrY; John *Tyreman* 1631 RothwellPR (Y). 'Servant of *Tyer*'. *v.* TYER.

**Tyke:** *v.* TIKE

**Tylden:** *v.* TILDEN

**Tyldesley:** *v.* TILDESLEY

**Tylee, Tyley:** *v.* TILLEY

**Tyler, Tylor:** *v.* TILER

**Tym, Tymm, Tymms:** *v.* TIMMS

**Tyndale, Tyndall:** *v.* TINDAL

**Tyre:** *v.* TYER

**Tyrell, Tyrrell:** *v.* TIRRELL

**Tyreman:** *v.* TYERMAN

**Tyrie:** *v.* TERREY

**Tyrwhitt, Truett:** John *de Tyrwyt* 1256 AssNb; William *Tyrwhit* 1383 IpmGl; William *Truwhet* 1515 GildY. From Trewhitt (Northumb), *Tirwit* 1150–62.

**Tysall:** Lambert *de Tyrssale* Hy 3 Calv (Y). From Tyersall in Pudsey (WRY).

**Tyson:** Gilbert *Tison* 1086 DB (Nt); Adam *Tisun* 1130 P (Y). OFr *tison* 'firebrand'.

**Tysoe:** Ralph *de Tisho* 1204 P (Wa).

**Tytherleigh:** *v.* TITHERLEY

**Tytler:** *v.* TITLER

# U

---

**Ubank:** *v.* EWBANK

**Udal, Udall, Udale, Udell:** *v.* YEWDALL

**Udlin:** *Udelina* 1148 Winton (Ha); John *filius Udelin* 1212 Cur (Nf); Simon *Udeline* 1279 RH (Beds). *Ud-el-in*, a Romance derivative of OG *Uda*.

**Uffmore:** Henry *Uffemor* 1327 SRWo. From Uffmoor Fm in Hasbury (Wo).

**Ufford:** Robert *de Vffewurda* 1179 P (Nth); Turstan *de Vfford* 1199 P (Ha); Thomas *Ufford* 1391 FFEss. From Ufford (Northants, Suffolk).

**Uffworth:** Robert *de Vffewurde* 1179 P (Nth); Robert *de Uffewurda* 1202 AssNth. From Ufford (Nth), *Uffewurda* 1178.

**Ufton:** William *de Ufton'* Hy 3 Glapwell (Db). From Ufton Nervet (Berks), or Uftonfields in South Wingfield (Db).

**Uggle, Ugley:** (i) Walter *Uggel* 1256 IpmW; John *Uggel* 1306 AssW. *Ugg-el*, a diminutive of OE *Ugga. (ii) William *de Uggle* 1276 AssLo. From Ughill (WRY), or Ugley (Ess).

**Ughtred:** *v.* OUGHTRED

**Ugo:** *v.* HUGH

**Ullett, Ulliott:** *v.* WOOLVETT

**Ulley:** William *de Ulley* 1306 IpmY; William *Ulley* 1672 HTY. From Ulley (WRY). There was also a feminine name: *Ulia* (f) 1202 FFY.

**Ullman, Ullmann, Ulman, Ulmann:** John *le Ulemon* 1275 SRWo; Richard *Olmon* 1297 SRY. ME *oli, oyle, uile*, OFr *oile, uille* 'oil' and ME *man*. 'A maker or seller of oil.' Now often for German *Ullmann*.

**Ulimer:** *v.* WOOLMER

**Ullock, Hullock:** John *Ullayk* 1332 SRCu. From Ullock (Cu).

**Ullrich:** *v.* WOOLRICH

**Ullyatt, Ullyett, Ullyott:** *v.* WOOLVETT

**Ulph:** *Vlf, Ulfus, Olf(us)* 1066 DB; *Ulf* c1095 Bury (Sf); Alwinus *Wlf* c1125 Bury (Sf); Ædwin', Robert *Vlf* 1166 P (Nf), 1279 RH (Hu). ON *Úlfr*, ODa, OSw *Ulf* 'wolf'.

**Ulrich, Ulrik:** *v.* WOOLRICH

**Ulting, Oulting:** Thomas *de Ultinge* 1309 FFEss. From Ulting (Ess).

**Ulyate, Ulyatt, Ulyett:** *v.* WOOLVETT

**Uman:** *v.* HUMAN

**Umfreville:** Gilbert *de Vnfranuilla* 1166 P (Nb). From Umfreville (La Manche).

**Umpelby, Umpleby, Umplebye:** William *de Anlauby* 1289 Pat. A Leeds and Bradford name, deriving from Anlaby (Hull), DB *Umlouebi*.

**Uncle, Uncles, Ungles, Ungless:** (i) William *Uncle*

Hy 2 Gilb (L); Eustace *le Uncle* c1200 ArchC vi; William *le Huncle* 1243 AssSo. ME *uncle*, OFr *oncle* 'uncle'. (ii) *Ulfketel, Ulketel, Vlchetel* 1066 DB; *Vlchel, Vlchil* ib.; *Unchel de Mersca* 1170–85 YCh; *Ulketel* 1186–8 BuryS (Sf); *Vlfketellus* presbiter 1189–98 ib.; *Ulfskillus* 1208 Cur (Wa); Willelmus *filius Ulkilli* 1242 Fees (Nb); Robert *Hulfketell* 1199 SIA iv; Robert *Huneketel* 1214 FFK; William *Unketel* 1274 RH (So); Symon, Thomas *Ulfketel* 1275 RH (Sf); John *Unkyl* 1445 NorwW (Sf); Henry *Ulketyll* 1460 ib. Thomas *Unketyll* 1500 ib. (Nf); Myghell *Unketyll*, John, William *Unkyll* 1524 SRSf; Christofer *Ungle* 1568 SRSf. The DB *Hunchil* (Y) is explained by von Feilitzen as ON *Húnketil* or *Hundkell*, or if the *H* is inorganic, as identical with Anglo-Scandinavian *Unketel*. In *le Huncle* the *H* is undoubtedly inorganic as it is in *Hulfketell*. The Suffolk examples make it clear we are concerned with only one name, ON *Úlfketel, Úlfkell* 'wolf-cauldron', which is common in that county where we also have the mutated *Ylfketill* in Ilketeshall. *Ulfketell* and *Ulfkell* became *Ulketel, Ulkell*, and then by dissimilation of *l–l* to *n–l*, *Unketell* and *Unkell*.

**Uncleby, Unkelby:** Robert *de Unkelby* 1327 SRY. From Uncleby (ERY).

**Uncouth:** Philip *le Oncothe* 1277–8 CtH; Adam *Huncouthe* 1379 PTY. OE *uncūð* 'unknown, unkind, rough'. cf. Robert *le Uncuthemon* 1278 AssLa.

**Underborough:** Robert *Underburgh* 1311 AssNf. 'Dweller below the town', OE *under, burg*.

**Undercliff, Undercliffe:** Roger *Underclif* 1210 FFL; Adam *Underclive* 1254 FFK; Thomas *Undertheclyf* 1334 FFY. 'Dweller below the cliff or slope', OE *under, clif*.

**Underdown:** Wulfwinus *Vnderdune* 1185 Templars (K); Theobald *de Vnderdoune* 1316 FFK. 'Dweller at the foot of the hill' (OE *dūn*) or at a place called Underdown.

**Underedge:** William *Vnderegge* 1194 P (He); William *Undereghghe* 1348 DbAS 36. 'Dweller at the foot of the escarpment', OE *under, ecg*.

**Underhay, Underhayes:** Edward *Underhay* 1572 Bardsley; George *Underhaye* 1642 PrD. From Underhays in Ideford (D), or 'dweller below the enclosure', OE *under*, (ge)*hæg*.

**Underhill, Undrell:** Henry *Underhulle* 1275 SRWo; Geoffrey *Undrehille* 13th Rams (Hu); Robert *de Underhull* 1268 AssSo. 'One who lived at the foot of a hill' or at a place called Underhill, as in Devon.

**Undertree:** Stephen *Undertre* 1301 SRY; Thomas

*Underyetre* 1332 SRSt. 'Dweller at the foot of the tree', OE *under*, *trēow*.

**Underwater:** Walter *under Water* 1219 Fees (La). 'Dweller south of the stream.' cf. Undermillbeck (We), '(place) below, i.e. south of Millbeck'.

**Underwood:** William *de Underwode* 1188 BuryS (Sf); William *Underwude* de Clokton' 1219 AssY; William *Under the Wode* 1332 SRSt. Either one who lived below a wood on a hillside, or within a wood (lit. 'below the trees of the wood'), or at a place so named, as Underwood (Derby, Notts).

**Undrell:** *v.* UNDERHILL

**Ungle(s):** *v.* UNCLE

**Unicume:** *v.* HONEYCOMB

**Unkelby:** *v.* UNCLEBY

**Unready:** John *Unnyredy* 1334 SRK. 'Ill advised', from a derivative of OE *unrǣd* 'bad advice'. cf. Ædgar *unniðing* 1170 P (Nb) 'not a rascal'; Leuric *Unsiker* 1188 BuryS (Sf) 'unsure'; John *Unwyse* dictus Kypard 1317 AssK 'unwise'.

**Unstead, Unsted, Ounstep:** John *Unsteade* 1558 Pat (Lo); John *Ownsted* 1583 Musters (Sr); William *Ownsted* 1592 LewishamPR (Ess). cf. Hounstead Bury in Sanderstead (Sr), from the family of *Ovenstede* 1332, later *Ownstead*.

**Unsworth, Hunsworth, Ounsworth:** John *Vnssworthe* 1572 Musters (Sr). From Hunsworth (WRY), or Unsworth (La).

**Unthank:** John *de Unthanc* 1242 Fees (Nb); Alan *de Unthanke* 1332 SRCu. From one of the places named Unthank in Cumberland, Northumberland or the North Riding of Yorkshire. OE *unþances* 'without leave', dweller at a squatter's farm.

**Unwin, Hunwin:** *Hunuuinus*, *Onouuinus* 1066 DB (C, Do); *Vnwine*, *Hunwine* de Batha 1166–7 P (Nf); Hugo *filius Unwini* 1221 AssWo; Rannulf *Vnwine* 1195 P (Nf); William *Unwin* 1221 AssSa; Gilbert *Unwine* 1228 Cl (K); Walter *þonwyne* 1275 SRWo; Reginald *Hunwyn* 1275 RH (C); Edmund *Hunwine* 1283 SRSf; John *Hunwyne*, *Onewyne* 1295, 1301 ParlR (Ess); Thomas *Unwyn* alias *Onyon* 1559 PN Ess 428. OE *Hūnwine* 'young bear-friend', with loss of the initial *H* would tend to be confused with OE *unwine* 'unfriend, enemy' which is certainly a common source of the surname. *þonwyne* is for *þe onwyne*. There may have been some contribution from OE *Unwine*. The story of *Unwen* 'son born beyond hope', with those of Hengest and Horsa, was current till so long after the Norman Conquest that it was possible for these heroes to be classed with Waltheof (Chambers, *Widsith* 254). There has also been some confusion with ONIANS.

**Upcher:** *v.* UPSHER

**Upcliff:** Eylard *Uppeclive* 1254 FFK. 'Dweller up on the slope', OE *uppan*, *clif*. cf. Adam *Opathom* 1327 SRWo 'up at the homestead'; Roger *Upbithebrake* 1275 SRWo 'up by the untilled ground'.

**Upcot, Upcott:** Iohel *Uppa cote* 11th OEByn (D); Adam *de Vppecote* 1199 P (D); Thomas *de Upcote* 1221 AssGl; John *Uppecote* 1359 AssD. From Upcote Fm in Withington (Gl), or from one or other of the numerous Upcot(t)s in Devon.

**Upham:** William *de Vpham* 1199 P (W); Alexander *de Upham* 1210–11 PWi; Thomas *Ophome* 1524 SRD.

From Upham (Ha), Upham Fm in Farringdon (D), or Upham in Aldbourne (W).

**Uphill:** (i) Roger *de Vphull'* 1199 P (Co). From Uphill (Devon). (ii) Henry *Uppenhull* 1255 RH (W); John *Uphulle* 1268 AssSo; William *Vppehelle* 1320 FFK. OE *uppan hylle* '(Dweller) up on the hill'.

**Upholder:** Godwin *le Upheldere* 1294 ELPN; Gilbert *le Upholdere* 1308 AssNf; Robert *le Vpheldere* 1327 SRSf. 'A dealer in second-hand clothes and other articles', OE *ūp*, and a derivative of OE *healdan* 'to hold'. The feminine form also appears: Roysia *la Uphaldestere* 1226–7 FFK.

**Uphurst:** Richard *Oppethehurst* 1327, Geoffrey *uppe the Hurst* 1332 SRSx. 'Dweller up in the wood', OE *uppan*, *hyrst*.

**Upjohn:** Roger *ap-John* 1638 Bardsley (Ch); Richard *Upjohn* 1778 DWills. Welsh *ap John* 'Son of John'.

**Upperton:** Ralph *de Upperton* 1296 SRSx. From Upperton in Eastbourne, in Harting, in Tillington (Sx).

**Uppiby:** Margaret *Uppiby* 1297 SRY; Richard *Upiby* 1297, Walter *Upiby* 1302 IpmY. 'Dweller up in the village', ON *uppr í bý*.

**Uppington:** (i) Adam *de Uppinton'* 1275 SRWo. From Uppington (Salop). (ii) Richard, Geoffrey *Uppinton'* 1275, 1327 SRWo; Nicholas *Up-in-the-ton* of Waterfal 1318 AssSt. OE *upp in tūne* '(Dweller) up in the village'.

**Uprichard:** Robert *Upprichard* 1637 Bardsley. Welsh *ap Richard* 'son of Richard'. cf. PRITCHARD.

**Upridge:** Jordan *Uprigge* 1206 Pleas (Ha). 'Dweller up on the ridge', OE *uppan*, *hrycg*.

**Upright:** William *Uprict* 1210 Cur (Beds); Symon *Upriht* 1279 RH (C); Walter *Upright* 1307 FFEss. OE *ūpriht* 'erect'.

**Upsall, Upshall:** Ralph *de Upshale* 1185 P (Y); Hugh *de Upsal'* 1219 AssY; William *Uppesale* 1349 FFY. From Upsal (NRY).

**Upsher, Upcher:** John *Upchar* 1420 AD i (Ess); George *Upsher* 1674 HTSf. From Upshire (Essex).

**Upson, Upsom:** Probably from UPSTON.

**Upston, Upstone:** Roger *de Ubbeston'* 1279 RH (Sf); John *Upston* 1524 SRSf. From Ubbeston (Suffolk).

**Upton:** Ethestan *on Optune* 972 OEByn (Nth); Richard *de Vpton'* 1162 P (Nth); Swetmann *atte Uppeton* 1332 SRSx. From one of the numerous Uptons.

**Upwell:** Robert *Vpewel* 1332 SRDo; John *Upwell* 1642 PrD. From Upwell (C, Nf).

**Upwick:** Robert *de Upwic'* 1208 Cur (Herts); Henry *de Upwyk'* 1278 FFEss. From Upwick Hall in Albury (Herts).

**Upwood:** Walter *de Upwod'* 1258 MPleas (Hu); Adam *de Upwode* 1279 RH (Hu); John *Upwode* 1327 SREss. From Upwood (Hu).

**Urban, Urben:** *Vrbanus* de Herlingdon' 1197 P (Bk); William *Urban* 1275 RH (Herts). *Urban*, Latin *urbanus* 'of the city', the name of a 3rd-century saint and seven popes, used occasionally as a christian name in the Middle Ages.

**Uren, Urion, U'ren, Urian, Urin:** *Urian* de St Pierre 1272 AssSt; John *filius Urian'* 1279 RH (Hu); William

*Urine* 1301 SRY; Robert *Vryen* 1459 SaG; Gregory *Vryn* 1642 PrD. British *Orbogenos*, OW *Urbgen*, MWelsh *Urien*.

**Urich, Uridge:** Thomas *de Iwrugg'* 1327, Thomas *Euregg* 1332, Richard *Ewreg* 1525 SRSx. From Eridge in Frant (Sx).

**Urin, Urion:** v. UREN

**Urmston:** William *de Vrmestone* 1332 SRLa. From Urmston (La, Ch).

**Urlin, Urling, Urlwin:** v. HURLIN

**Urquhart:** Adam *Urquhart* 1358 Black; Alexander *of Hurcharde* 1381 ib.; Charles *Urquhart* 1669–1734 ib. From the barony of Urquhart (Inverness).

**Urry, Urie:** v. HURRY

**Ursell:** *Ursellus* de Busco 1200 Cur (Y); William *Vrsel* 1163 P (Beds). *Ursel*, a diminutive of Latin *ursus* 'bear'.

**Urswick:** Robert *of Urswyk* 1401 AssLa; Thomas *Urswyk* 1449 FFEss. From Urswick (La).

**Urvine:** v. IRVIN

**Urwin:** v. ERWIN

**Ury:** v. HURRY

**Usborne:** v. OSBORN

**Usher, Ussher, Husher, Lusher:** Richard *Ussier* Hy 3 Colch (Ess); William *le Usser*, *Lussier* 1243 AssSo; Geoffrey *le Uscher* 1300 LLB C; Richard *Lusscher* 1319 SRBeds; Adam *Husser* 1332 SRCu; John *Huscher* 1506 GildY. OFr (*le*)*ussier*, *huissier*, AFr *usser*, ME *usher* 'usher, door-keeper' (NED c1380).

**Usherwood:** v. ISHERWOOD

**Usk:** Thomas *de Uske* 1337 LLB F; Walter *Usk* 1340 NIWo; John *Uske* 1423 AssLo. From Usk (Monmouth).

**Uskell:** Rueland *Huscerl* 1188 P (O); Roger *Huscarl'* 1211 FFO; Rolland *Huscarl* 1305–6 FFSr; John *Huscall* 1593 FFHu. OE *hūscarl* 'a member of the king's bodyguard'.

**Usmar:** v. OSMER

**Ussell, Uzzell:** Matthew *Oisel* 1168 P (Nb); Geoffrey *Oysel* 1262 For (Ess); Margeria *Ussell*, John *Ossell* 1327 SRWo. OFr *oisel* 'bird'. cf. OSLAR.

**Ussher:** v. USHER

**Utley, Uttley, Hutley, Huttly:** Roger *de Huttelege* 1242 AssDu; John *Vtlay* 1379 PTY; Richard *Hutley* 1662 HTEss. From Utley in Keighley (WRY).

**Utteridge, Utridge:** v. OUTRIDGE

**Utting, Uttin:** Willelmus *filius Uttingi* 1183 Boldon (Du); *Utting'* 1219 AssY; Coket *Uttingus* 1183 Boldon (Du); Tomas *Vtting* 1191 P (Y); Anthony *Uttinge*, Goody *Utten* 1674 HTSf. OE *Utting*, a derivative of *Utta*.

**Uwins:** v. EWAN

**Uzzell:** v. USSELL

# V

**Vache, Vatch:** Reginald *Vacca* 1210 Cur (L); Alexander *la uache* 1214–19 Black; Richard *Vache* 1275 SRWo. Either a nickname from OFr *vache* 'cow', or metonymic for a cowherd.

**Vacher, Vacha, Vatcher:** Simon *le Vacher* 1219 AssY; Hugh *le Vachir* 1227 AssSt. OFr *vachier* 'cowherd'.

**Vagg(s):** v. FAGG

**Vail(e):** v. VALE

**Vaillant:** v. VALIANT

**Vaines, Veigne, Vein:** Richard *de Vein (Veym)* 1221 AssGl. From Vains (La Manche). v. ANF.

**Vairo:** v. VARROW

**Vaisey, Vaizey, Vasey, Veasey, Veasy, Veazey, Vesey, Vezey, Veysey, Voisey, Voizey, Voysey, Voyzey, Facey, Faizey, Fasey, Feacey, Feasey, Feazey, Feesey, Foizey, Phaisey, Phasey, Phazey, Pheasey, Pheazey, Pheysey, Lenfestey:** Robertus *Invesiatus, Lascivus* 1086 DB (Ess); Robert *Lenveiset* 1131 Riev (Y); Thomas *le Envaiset* c1150 ib.; Jordan *Veiset, le Envaise* 12th ib.; William *le Enveise (Lenvesie)* 1220 Cur (Nf); Adam *le Veyse* 1270 AssSo; Robert *le Enueysi* 1277 ib.; Alice *Vesy* 1296 SRSx; Peter *le voyse* 1327 SRC; Thomas *Fecy* 1327 SRSo; Beatrice *a Vesy* 1332 SRSx; William *Veysy* 1357 Crowland (C); Nicholas *Vaysi* 1386 AD iv; Robert *Feysy* 1395 NottBR; John *le Vesie* 1420 DKR 41; John *Vasey* 1456 NorwW (Sf); John *Veysey*, or *Vesey*, or *Voysye*, or *Pheysy* 1512 Oxon; Maud *Fyseye* 1541 W'stowWills (Ess); Christian *Facy* 1671 DWills; Philip *Veasy* 1674 HTSf. In spite of Weekley's correct explanation of *Vaisey*, Tengvik has repeated Bardsley's confusion of the name with *Vessey*. True, there has been a late and occasional confusion between these surnames, as in the 1524 Subsidy Roll for Suffolk where we find *Vessy* side by side with *Vasy, Veysy* and *Vaysy*, but all the above surnames undoubtedly derive from AFr *enveisé*, OFr *envoisié* 'playful', latinized in DB as *invesiatus* (cf. MedLat *invasus* 'possessed by a demon') and paralleled by *lascivus* 'wanton'. As with FANT, the first syllable was lost, the meaning of the remaining *Veyse* was forgotten and the initial *V* frequently regarded as a dialectal pronunciation of *F*, hence *Feasey*, etc., and, with the spelling change of *Ph* for *F*, *Phaisey*, etc. The surname is common in all its forms except *Lenfestey* which is rare and preserves the fuller form with an intrusive medial *t*.

**Valance:** v. VALLANCE

**Vale, Vail, Vaile:** Wido *de la Val* 1190 P (Nth);

Robert *del Val* 1221 AssWa; Walter *ate Vale* 1327 SRSx; John *Vale* 1382 AssC; Nicholas *Vayle* 1623 RothwellPR (Y). 'Dweller in the valley', ME *val(e)*, OFr *val*. v. also VEAL, VEIL.

**Valentin, Valentine, Valintine, Vallentin, Vallentine, Vallintine:** *Valentinus* 1198 Cur (W), 1276 RH (C); *Valentyn* le Warner 1327 SRC; Matilda *Valentyn* 1251 Rams (Hu); William *Valentin* 1260 AssC. Lat *Valentinus*, a derivative of *valens* 'strong, healthy', the name of a 3rd-century Roman saint and martyr. Found in England from the end of the 12th century.

**Valet(t):** v. VALLET

**Valiant, Vaillant:** Hugo *le Vaillant* 1185 P (Y); Gilbert *le Valiand* 1207 Cur (Do); Gilbert *Valiant* 1212 Cur (So). OFr *vaillant* 'courageous, sturdy'.

**Valin, Valins:** v. VALLIN

**Vallance, Vallans, Valance:** Reiner *de Valenc'* 1158 P (Lo); Almerus (Eymerus) *de Valencia, de Valence, de Valenz* 1303–46 FA (Sf). From Valence (Drôme).

**Vallentin, Vallentine:** v. VALENTIN

**Valler, Vallier, Vallor:** Nicholas *le Valer* 1263 MESO (Sx); John *le Valier* 1296 SRSx; Thomas *Valour* 1379 PTY. 'Dweller in the vale, from a derivative of OFr *val*.

**Vallet, Valet, Valett:** Walter *Vadlet* 1198 P (Sx); Robert *Valet* 1221 Cur (Gl); Richard *le Vallet* 1243 AssSo. OFr *valet, vallet* 'man-servant'.

**Valley:** Richard atte *Valeye* 1346 LLB F. 'Dweller in the valley.'

**Vallin, Vallins, Vallings, Valin, Valins:** Petrus *de Valoinges* 1086 DB (C); Robert *de Valeynes* 1155–68 Holme (Nf); Robert *de Valuynes* 1238–42 StGreg (K); John *de Ualin'* a1237 ib. From Valognes (La Manche).

**Vallintine:** v. VALENTIN

**Vallis:** Sibil *de Valeyse* 1275 RH (Sf); James *Valloyes, Valleys* 1601–2 Bardsley. From the old province of Valois in the Ile-de-France.

**Van, Vance:** v. FANN

**Van Acker:** John *Vanacker*, a refugee from Lille, became a merchant in London. His grandson, Nicholas, was created a baronet in 1700 (Smiles 429). 'Dweller in the field.'

**Vandervelde, Vandervell:** Henry *Vandelveld* (denization) 1550 Pat. Dutch *van der veld* 'of the field'.

**Vane(s):** v. FANE

**Vann(e), Vanns:** v. FANN

**Vannah, Vanner:** v. FANNER

**Vanston, Vanstone:** John *Fanson* 1629, Janne *Fanstone* 1705, James *Vanstone* 1820 HartlandPR (D). From Fauntstone in Shaugh (D).

**Vant:** v. FANT

**Vantage:** William, Adam *Vauntage* 1319 SRLo, 1327 SRSo. ME *vantage* 'advantage, profit, gain'.

**Vanter:** v. VAUNTER

**Varah:** v. VARROW

**Varden, Vardon:** v. VERDEN

**Varder, Verdier:** Walter *le Verder* 1279 RH (O); Richard *le Verdour* 1327 SRSa. AFr *verder, verdour*, OFr *verdier, verdeur* 'verderer', a judicial officer of a royal forest (1502 NED).

**Vare:** v. FARE

**Varey:** v. VARROW

**Varley, Verley, Virley:** This may be a southern form of *Farleigh*: Thomas *de Varley* 1316 FA (Sf). Both Varley and Varleys (Devon), earlier *Fernlegh*, gave rise to surnames in the 14th century. The surname cannot derive, as has been suggested, from Virley (Essex) which is not so called before the 16th century. It was originally *Salcota*, later *Salcote Verly*, from its Domesday lord Robert *de Verli*, who came from Verly (Aisne), whence the surnames of Hugo *de Verli* 1219 AssL, and Thomas *Virley* 1275 RH (Sf). v. PN Ess 323, OEByn. Hugh *de Verleio* (1166 RBE) is said to have come from Vesly (La Manche). v. ANF.

**Varlow:** v. FARLOW

**Varnals:** v. FARNALL

**Varndell:** v. FARNDALE

**Varnes:** v. FERN

**Varney:** v. VERNAY

**Varnham, Varnam:** v. FARNHAM

**Varnish:** v. FURNACE

**Varnon:** v. VERNON

**Varns:** v. FERN

**Varran:** v. FARREN

**Varrow, Varah, Varey, Vary, Vairo:** John *Vary* 1663 FrY; Humfrey and Frederick, sons of Charles and Mary *Varow* bapt. 1754 HorringerPR (Sf); Humfrey and Frederick *Varoh* infants bur. 1755 ib.; Charles *Varo* 1760 ib.; Mary *Varo* 1843 ib.; Mary *Varer* bur. 1795 ib.; Mary *Varer* mar. 1819 ib.; Lucy *Varer* m. 1837 ib. This is clearly a late development of *Farrow*, due to an unetymological substitution of initial *V* for *F*. v. FARRAR.

**Varty:** v. VERITY

**Varwell:** v. FAREWELL

**Vary:** v. VARROW

**Vasey:** v. VAISEY

**Vass, Vasse:** *Vasse* le Poynur 1275 RH (C). OFr *vasse*, Lat *vassus* 'servant, vassal'. Like *Vassal*, occasionally used as a personal name.

**Vassall:** *Vassallus* de Aunfoilliis 1221 Cur (R); Hugo *Vassall'* 1202 P (Gl); Henry *Vassal* 1221 AssWo. ME, OFr *vassal* 'vassal, servant, dependant'. Occasionally used as a personal-name.

**Vassar:** Thomas *Vasour* 1327 SR (Ess); John *Vasour* 1332 MEOT (Nf). OFr *vasseor, vasseur* 'vassal'.

**Vassie:** Walter *de Vasci* 1203 P (D). From Vassy (Calvados).

**Vatch:** v. VACHE

**Vatcher:** v. VACHER

**Vaughan, Vaugham:** Grifit *Vehan* 1222–64 Seals (Brecon); Rys *Vychan* 1248 Harrison; William *Vachan*

1275 RH (Sa); Gronou *Vahan* 1285 Ch (Radnor); Jeuan *Vachann, Vaghann* 1391 Chirk; Mr *Vaugham* 1674 HTSf. Welsh *fychan*, mutation of *bychan* 'small, little'.

**Vaunter, Vanter:** John *le Vaunteur* 1304 IpmGl; William *Vanter* 1662–4 HTDo. OFr *vaunteur* 'boaster, braggart'.

**Vaus, Vause, Vaux:** (i) Robert *de Vals, de Valibus, de Vaux* (Ess), *de Wals* (Nf) 1086 DB; Robert *de Wals, de Vallibus* 1134–40, 1188 Holme (Nf); Ralph *de Vaus* 1185 Templars (Y); Richard *de Vause* 12th DC (Lei). From Vaux, a common French place-name, plural of *val*, Lat *vallis* 'valley'. The occasional initial *W* survives in Wasse's Fm in Terling, from Nicholas *Vaux* (1522 PN Ess 298) and so may have contributed to *Wass*. We also find an unvoicing of *V* to *F* in Fowe's Fm in Belchamp Otton and Fox Hall in Shopland, both from a *de Vaux* (PN Ess 410, 201) and this may survive as *Fow*. It is an interesting speculation whether this is the clue to the unexplained surname of Daniel Defoe (b. 1661). It is said that he did not add the *De* to his family name of *Foe* until he had reached manhood. In 1524 SRSf we find Edmund *Foo*, James *a Foo* and Robert *a Fowe*. The *a* may be a substitution of an English preposition for the French *de*, or the *d* may have been absorbed in pronunciation by the *t* of Robert. These names, like Fowe's Fm, may derive from *de Vaux* and *Defoe* may have the same origin. Daniel may have been following some family tradition in insisting on the *De*. (ii) A number of surnames have *le* instead of *de*: Agnes *le Vaus* 1275 SRWo, Nicol' *le Vaus* 1296 SRSx; Alexander, Robert, John *le Vaus* 1327 SRSf. There is only one doubtful example of late OE *fals* 'false'. The frequency of the word in 12th-century English and later is due to OFr *fals, faus*. This French word would retain its initial *f* in the south where the above surnames are found but the ME form of OE *fals* would become *vals*. Examples are rare but NED gives a few, including some from the verb (from OFr *falser*): a1225: *ualse* 'mendacious', *ualse* (vb.), *valsinde* (pres. part.); 1340: *uales* moneye, *ualse* lettres, *ualsehedes* 'falsehoods', *ualsnesse* 'falseness', *ualsere* 'falsifier'. Thus *le vaus* probably owes its *v* to OE *fals* and its *au* to OFr *faus*, though the French *f* might have been voiced occasionally in the south. cf. FANT, FIDLER (ii), and the forms of VIVIAN.

**Vavasour, Vavasseur:** William *Vavassur* 1166 RBE (Y); Gilbert *le Vauassur* 1167 P (Nf); John *le Vavaseur* 1288 LLB A. OFr *vavas(s)our, vavasseur*, Lat *vassus vassorum* 'vassal of vassals' (NED 13..). A feudal tenant ranking immediately below a baron. The surname is common in the 12th and 13th centuries. The status of the vavassor varied with time and place. In England at the time of Domesday the vavassors were men of very moderate estate. By the 12th century the whole of military society was divided into two great classes, barons and vavassors. v. EngFeud 16–23.

**Vawser:** Nicholas *Vausour* 1379 PTY. A syncopated form of VAVASOUR.

**Vayne:** v. FANE

**Veaill:** v. VEIL

**Veal, Veale, Veall, Veel:** (i) Richard *le Vele* 1270 AssSo; William *Vel* 1276 FFSf; Thomas *le Veel* 1296

SRSx. OFr *veel* 'calf'. (ii) Reginald *Leviel* 1173 P (Lo); Geoffrey *Viele* 1206 Cur (Sx); William *le Viel* 1218 AssL. AFr *viel*, OFr *vieil* 'old', or OFr *viel* 'calf'. Without the article, *Viel* is indistinguishable from VIAL. There has also been confusion with VALE: Martin *Veal, Vale* 1774, 1780 WStowPR (Sf).

**Vear, Veare:** *v.* VERE

**Vearncombe:** *v.* FARNCOMBE

**Veasey:** *v.* VAISEY

**Veck, Vick:** Henry *le Eueske* 1218 AssL; Richard *Veke* c1248 Bec (W); Robert *Vesk* 1275 SRWo; Robert *le Veck* 1279 RH (C); Richard *le Veske* 1296 SRSx. OFr *le eveske* 'the bishop' became *leveske* which was wrongly taken for *le vesk*. This became *Vesk* and later *Veck, Vick*. The full form survives as LEVICK. cf. BISHOP.

**Veel:** *v.* VEAL

**Veevers, Veivers:** *v.* VIVERS

**Veigne:** *v.* VAINES

**Veil, Veail, Vail:** Roger *de Ueille* 1127–34 Holme (Nf); Richard *la Veille* 1175–86 ib.; Richard *la Veyle* 1242 FFEss; Geoffrey *de Veel* 1256 AssSo; Robert *le Veyle* 1260 AssY; Thomas *le Veyl* 1276 AssSo; Hubert *la Veylle* 1277 FFEss. This must be OFr *de la veille* 'of the watch', 'watchman', with common loss of *de* as in early forms of BATTLE and WARE. There has been confusion with VALE and VEAL.

**Vein:** *v.* VAINES

**Velden:** *v.* FIELDEN

**Veley, Vella, Velley:** Roger *de Velay* 1230 P (D); Thomas *Velle* 1577 HartlandPR (D); John *Velly* 1642 PrD. From Velly (D).

**Vellacott:** John *Vellacote* 1642 PrD. From Cellacott (D).

**Velley:** *v.* VELEY

**Veltom:** *v.* FELTHAM

**Venables:** William *de Venables* c1200 WhC (La). From Venables (Eure).

**Vender:** Peter *le Vendier* 1206 P (K); William *le Vendur* Ed 1 NottBR. AFr *vendour*, OFr *vendeor*, *\*vendier* 'seller, dealer'.

**Venes, Veness, Venis, Venise, Venus:** Robert *de Venuiz* 1130 P (Ha); William *de Venuz, de Venoiz* 1197 P, 1205 ChR (Ha); Robert *de Veniz* 1203 Cur (Ha); William *de Venus* 1230 P (Wa). From Venoix (Calvados). *v.* ANF.

**Vening:** *v.* FENNING

**Venison:** John *Venesoun* 1314 FFK, *Venison* 1334 SRK; John *Vennison* 1654 Black. OFr *venesoun* 'venison', perhaps a nickname for a deer-hunter.

**Venn:** *v.* FENN

**Vennall, Vennel, Vennell:** Reginald *de la Venele* 1240 FFEss; Michael *in la Venele* 1279 RH (Beds). 'Dweller in the alley', Fr *venelle* 'small street, alley'.

**Venner, Venour, Fenner:** Walter *le Veneur* 1195 P (C); David, William *le Venur* 1219 AssY, Cur (Sx); Adam *le venour* 1297 MinAcctCo (Herts). OFr *veneor, veneur* 'hunter, huntsman'. A common surname in early records. In view of the existence of *Fanner* and *Vanner*, one would have expected *Fenner* and *Venner* also to mean 'marsh-dweller', from OE *fenn*. But the only evidence noted is that of Lower who states that Fenn Place in Worth, Sussex, had owners called from

it *Atte Fenne*, who in the time of Henry VI changed their name to *Fenner*, whilst a Kentish branch wrote themselves *Fenour*. As *Fenner* is common and still found in Kent and Sussex, where *Bridger, Brooker*, etc., are frequent, this seems probable.

**Vennicker:** 'Dweller at the dairy-farm in the marsh.' cf. FENWICK.

**Venning:** *v.* FENNING

**Venters, Ventress, Ventriss:** Francis *Ventreys* 1600 FFHu; Edward *Ventres* 1642 SfPR; Mrs *Ventris* 1674 HTSf. 'The venturous', from ME *aventurous*.

**Venton, Ventom:** From Venton (Devon), *Venton* 1301 PN D 255.

**Ventre, Ventur, Venture:** William *Aventur* 1279 RH (Hu); William *le Ventre* 1327 SRSx. ME *aventure* 'chance, hazard', in the same sense as VENTERS and VENTURA.

**Ventura:** John *le Ventrer* 1273 RH (Nt). 'The adventurer, venturesome.'

**Venus:** *v.* VENES

**Verden, Verdin, Verdon, Verduin, Varden, Vardon:** Bertrannus *de Verduno* 1086 DB (Bk); Iue *de Verdun* 1101–7 Holme (Nf); William *de Verduin* 1196 Cur (Lei). From Verdun (La Manche). *v.* ANF. Possibly also from Verdun (Eure).

**Verdier:** *v.* VARDER

**Vere, Veer, Veare:** Alberic *de Ver* 1086 DB (Ess); Roger, Walter *de Ver* 1121–35 Bury, 1208 FFY; Gilbert *de Veer* 1303 LLB C; Robert *Veer* 1403 IpmY. From Ver (La Manche, Calvados).

**Verey:** *v.* VERREY

**Verger:** John *de Verger, del Vergier* 1230 Cl; Matilda *le Virger* 1327 SRSo. A derivative of OFr *verge* 'a rod half-acre', for the owner of that amount of land.

**Vergin(e), Vergo:** *v.* VIRGIN

**Verity, Varty:** Adam, Richard (*le*) *Verite* 1275 SRWo, 1296 SRSx; Thomas *Verty* 1379 PTY. Fr *vérité* 'truth'.

**Verley:** *v.* VARLEY

**Vern(e):** *v.* FERN

**Vernall:** *v.* FARNALL

**Vernay, Verney, Verny, Varney:** Robert *de Vernai* 1221 AssGl. From a French place *Vernay*, probably chiefly Saint-Paul-de-Vernay (Calvados).

**Vernon, Vernum, Varnon:** Richard *de Vernon* 1086 DB (Ch); William *de Vernun* 1130 P (Wt). From Vernon (Eure). *v.* ANF.

**Verrey, Verry, Verey, Very:** Edward *le Verreis* 1202 P (L); Edward *Verey* 1641 PrSo; George *Verry* 1674 HTSf. OFr *verai* 'true'.

**Verrier, Verriour:** Roger *Verer* c1100 MedEA (Nf); Fulko *le Verrier* 1185 P (W); Walter *le verrour* 1313 FrY. AFr *verrer*, OFr *verrier, verrieur* 'worker in glass, glazier' (1415 NED).

**Vertue:** *v.* VIRTUE

**Very:** *v.* VERREY

**Vesey:** *v.* VAISEY

**Vessell:** Metonymic for VESSELO.

**Vesselo:** William, John *le Vesseler* 1296 SRSx, 1327 SRSf. A derivative of ME *vessele*, AFr, OFr *vessel(e)* 'vessell'. A maker or seller of household vessels.

**Vessey:** Robert *de Veci* 1086 DB (Nth); William *de Vescy* 1166 RBE (Ess). From Vessey (La Manche). *v.* OEByn.

**Veutrey, Vewtrey:** Suein *Valtrarius* 1170 P (Ess/ Herts); William *le Vealtre* 1182 P (Sf); Reginald *le Feutrer* 1275 SRWo; Roger *Veuuter* 1451 Paston. OFr *feutrier* 'a maker of felt', or OFr *veautrier* 'keeper of the greyhounds'.

**Veysey, Vezey:** *v.* VAISEY

**Vezin:** *v.* VOICIN

**Vial, Vials, Viall, Vialls, Viel, Vidal, Vidall, Vital, Vitall:** *Vitalis, Vitel, Fitel* 1066 DB; *Vithele* abbode 1076 ASC E; *Fibele* abbode 1077 ASC D; *Uiðel æt Culumtune* 11th PNDB 406 (D); *Vitalis de Colintone* 1086 DB (D); *Viel Luuet* c1150 DC (L); Adam *filius Viell* Hy 2 Whitby (Y); Radulfus *Vitalis* 1086 DB (D); Estanus *Fitele* 1175–86 Holme (Nf); Richard *Viel* 1194 P (D); John *Vitele* 1207 P (Sx); Henry *Vyel* 1275 SRWo; Thomas *Vytele* 1296 SRSx; Thomas *Vyall* 1524 SRSf. *Vitalis,* OFr *Vitel, Viel,* the name of some ten saints (Latin *vitalis* 'pertaining to life, vital'), became common in England after the Conquest both in its learned form *Vitalis* and in the northern French form *Viel*. The absence of early forms of *Vidal* suggests that this was a later immigrant from Languedoc. Black gives no example of the name from Scotland. *Viel* also survives as VEAL. *Vital* may also be attributive from ME *vital* in the sense 'full of vitality': Adam, Matilda *le Vytele* 1327 SRSx.

**Vian:** Ralph *de Viana* 1184 P (K); Hugh *de Vyen(na)* 1286–8 AssSt. Probably from Vienne (Calvados), *Viana* 1198.

**Viant:** William *Viand'* 1161 P (O); Emma *Viaunde* 1324 Bart. OFr *viande* 'meat, food', no doubt for a seller of foodstuffs.

**Vicar, Vicker:** Richard *Vicar* 1251 AssSo; Thomas *le Vicayr* 1279 RH (O); Stephen *le Vyker* 1313 FFEss. AFr *vikere, vicare, vicaire* 'vicar', one acting as parish priest in place of the parson or rector. (c1325 NED).

**Vicarage, Vickerage, Vickridge, Vickress:** Hugh *Vicaries* 1332 LLB E; John *Vicarish* 1547 Bardsley; Alice *Vicaridge* 1665 ib.; Anne *Vickris* 1706 DKR 41 (So). 'Servant of the *Vicary*', with the dialectal change of final *s* to *sh* as in *Parish* for *Paris*.

**Vicarey, Vicari, Vicary, Viccari, Viccary, Vickary, Vickery:** Henry *le Vicarie* 1249 MEOT (Sx); William *Vikery* 1319 SRLo. An adaptation of Lat *vicārius* 'vicar' (1303 NED).

**Vicars, Viccars, Vickars, Vickers:** (i) William *del Vikers* 1327 SRSt; William *de Vykeres* 1332 SRSt; Henry *Attevickers* 1327 SRDb; Joan *Atvicars* 1400 YWills. '(Servant) at the vicar's.' (ii) Margery *le Vikers* 1332 SRWa; Anne *Vickars* 1592 RothwellPR (Y). 'The vicar's servant.' Occasionally, perhaps, also for 'the vicar's son': Gilebertus *filius vicarii* c1248 Bec (O).

**Vick:** *v.* VECK

**Vickerman:** William *Vikarman* 1379 PTY. 'The vicar's servant.' cf. Richard *le Wycarisman* 1275 RH (C) and Isabella *Vikerwoman* 1379 PTY.

**Vickerson, Victorson:** Gilbert *filius vicarii* c1248 Bec (O); John *Vicarson* 1381 PTY; Alexander *Vicarson* 1529 Black. 'Son of the vicar', AFr *vicare. v.* VICAR.

**Vickress, Vickridge:** *v.* VICARAGE

**Vidal(l):** *v.* VIAL

**Videan, Vidgen, Vidgeon:** *v.* VIVIAN

**Vidler:** *v.* FIDLER

**Viel:** *v.* VEAL, VIAL

**Vieler, Viola:** Robert *Vielur* 1210 P (Ha); Adam *le Vielur* 1212 Fees (La); Brun *le Vilur* 1242 Fees (Nb); Richard *le Vyolour* 1311 LLB B. AFr *violour*, OFr *violeur* 'player on the viol, fiddler' (1551 NED).

**Vigar, Viggor, Vigor, Vigour:** Richard *Vigur* 1224 Pat; Peter *Viger* 1284 LLB A. OFr *vigor, vigur* 'vigour, liveliness', by metonymy for one displaying this quality.

**Vigars, Vigers, Viggars, Viggers, Vigors, Vigours, Vigrass, Vigurs, Vigus:** Walter *le Vigrus* 1221 AssWo; Henry *Vigrus* 1256 AssSo; Hugh *Vigeros* 1275 SRWo; William *Vigerus* 1279 RH (O); William *Vigerous* 1305 LLB B; Lewis *Vigures* 1598 Oxon (D); Samuel *Vigars* 1746 Bardsley. AFr *vigrus*, OFr *vigoro(u)s* 'hardy, lusty, strong'.

**Vigeon:** *v.* VIVIAN

**Vigne, Vignes:** *v.* VINE

**Vigrass, Vigus:** *v.* VIGARS

**Villain, Villin, Vilain:** Ernald *Vilein* 1167 P (Nf); Robert *Vilain* 1188 Eynsham (O); Roger *le Vilain* 1196 Cur (Nth). AFr *villein, vilein* 'serf, bondman, servile tenant', an occupier or cultivator entirely subject to a lord or attached to a manor (1303 NED).

**Villar, Villars, Villers, Villiers, Villis:** William *de Vilers* c1130 StCh; William *de Viliers* 1185 Templars (Y); Nicholas *de Vylirs* 1327 SRSx. From Viller(s) or Villier(s), both common French place-names. Roger *de Vilers* (1166 RBE), of Dorset, came from Villiers-le-Sec (Calvados). *v.* ANF.

**Villy:** Vnfridus *de Villi* 1190 P (Y); Robert *de Vili, de Villi* 1208–9 Cur (Db). From Villy-Bocage (Calvados). *v.* ANF.

**Vimpany, Vimpenny:** Thomas *Vimpenny* 1664 HTSo. A corruption of WINPENNY.

**Vinal, Vinall:** John *Vynale* 1378 ColchCt; John *Vynall* 1525 SRSx. From Vynall's Fm in Pebmarsh (Ess).

**Vince:** Antony *Vince* 1674 HTSf. A short form of *Vincent*.

**Vincent, Vincett, Vinsen, Vinson, Vinsun:** *Vincencius* 1206 Cur (Nf); *Vincentius* filius Wuluiet 1222 AssWa; *Vincent'* de Wyke 1296 SRSx; William, Agatha *Vincent* 1230 Oseney (O), 1296 SRSx; Abram *Vincent*, Anthony *Vinson* 1674 HTSf; Thomas *Vincen* 1776 Bardsley. Lat *Vincentius*, from *vincens* 'conquering', the name of a 3rd-century martyr. Introduced into England about 1200. *Vinson,* etc., are due to loss of the final *t*, *Vincett* to assimilation of *nt* to *tt*.

**Vinck:** *v.* FINCH

**Vine, Vines, Le Vine, Vigne, Vignes:** Robert *de Vigne* 1236 Fees (So); Henry *de la Vine* 1283 LLB A; Roger *atte Vine* 1297 LLB B; Richard *atte Vygne*, brewer 1311 LLB D; Richard *Vygn* 1327 SRSo. OFr *vigne* 'vine'. The surname probably means 'worker at a vineyard'. cf. VINE. Vine-growing was of some importance in England in the Middle Ages. There are still places named Vineyards in Essex and Cambridgeshire. *v.* WINYARD. Or the meaning may be 'one living by some prominent vine'. In towns it may denote a wine-seller 'at the sign of the vine'.

**Vinen:** Walter *le vyngnon* 1333 MEOT (So). OFr *vignon, vingnon, viegnon* 'vine-dresser, vine-grower'.

**Viner, Vyner:** Robert *le Vinnur, le Vinior, le Vinur* 1207, 1211 Cur (Hu); Symon *le Vignur* 1221 *ElyA* (Sf); William *le Vinyour* 1279 RH (Hu); Walter *le Vynour* 1309 LLB B; John *Vyner* 1407 FFSx. OFr *vignour, vigneur, vigneour*, AFr *viner* 'vine-grower, vine-dresser' (1390 NED).

**Viney:** *v.* FINNEY

**Vinick:** *v.* FENWICK

**Vinicombe, Vinnicombe, Vinycomb:** Francis *Venycombe* 1664 HTSo. From Vinnicombe in Crediton (D), or Venniscombe (Wt).

**Vining, Vinning:** *V.* FINNING

**Vink:** *v.* FINCH

**Vinnicombe:** *v.* VINICOMBE

**Vinsen, Vinson, Vinsun:** *v.* VINCENT

**Vint:** William *Fynt* 1374 ColchCt; William *Vynt* 1524 SRSf; Christopher *Vint* 1689 FrY. From Vint in Oath (Bk).

**Vinter, Vintor:** Saulfus *uineter* 1170 Oseney (O); Baldwin *le Vineter* 1221 Cur (Sr); John *le vynter* 1327 SRSf. AFr *viniter, vineter, vinter* 'vintner, wine-merchant' (1297 NED).

**Vintiner:** Robert *le uintner* c1179 Bart; Richard *le Vyntener* 1327 MESO (La). An alteration of *vineter* (c1430 NED), 'wine-merchant'.

**Vintsous:** John *Vintsoutz* 1325, *Vyntsouth* 1340 CorLo. 'Twenty halfpennies', OFr *vint, sous*. cf. William *Vintesisdeners* 1251 FrLei 'twenty-six deniers'.

**Vinycomb:** *v.* VINICOMBE

**Viola:** *v.* VIELER

**Vipan, Vipond, Vipont, Vippond:** Robert *de Viezponte, de Vezpunt* 1159, 1178 P (D); Roger *de Vipont* 13th WhC (La); Thomas *Vipond* 1772 Bardsley. From Vieuxpont (Calvados).

**Virgil:** *Uirgilius* clericus 1177–93 CartNat; *Virgilius* Chapman 1296 SRSx; Reginald *Virgil'* 1201 P (Ha); Hamon *Virgil* 1285 IpmW; Isabell' *Virgile* 1332 SRSx. Lat *Virgilius*.

**Virgin, Vergin, Vergine, Virgo, Virgoe, Vergo:** Simon *Virg'* 1275 RH (K); Isabella *Virgo* 1428 FA (Wa); William *Virgyn* 1581 Bardsley; John *Vergine* 1610 ib. Probably, as Bardsley suggests, a name given to one who had played the part of the Virgin Mary in some miracle play.

**Virley:** *v.* VARLEY

**Virtue, Vertue:** Simon *Vertu* 1510 Oxon; George *Vertue* 1674 HTSf. OFr *vertu* 'virtue', probably a pageant name.

**Vise, Vize, Vyse:** Robert *atte Vise* 1327 SRSx; John *de la Vise* 1330 PN D 128. 'Dweller by the boundary' (OFr *devise*), as at Viza in Ashwater (Devon), Vyse Wood in Morthoe (Devon), or from Devizes (Wilts), formerly *The Devise, Vises* and *The Vyse* (PN W 243).

**Visick:** *v.* PHYSICK

**Vital(l):** *v.* VIAL

**Vivers, Veevers, Veivers:** Thomas *Vevere* 1525 SRSx; Richard *Vyvers* 1597, *Vevers* 1621 SRY; Richard *Vivers* 1672 HTY. OFr *vivres* 'victuals', metonymic for a dealer in foodstuffs.

**Vivian, Vivians, Vivien, Vyvyan, Videan, Vidgen, Vidgeon, Vigeon, Fiddian, Fidgen, Fidgeon, Phethean, Phythian:** Johannes *filius Viuian* 1175 P (K); *Vivianus* de Cattele c1200 Gilb (L); *Fithian* alias *Vivian* 1514 ODCN; Henry *Vivien* 1235 Stone (St); William *Phythien* c1250 Rams (C); William *Phivien* 1271 Rams (Herts); John *Vivyan* 1275 RH (Ha); John *Fiuian* 1279 RH (O); Roger *Fidian*, Robert *Fithion* 1279 RH (C); John *Vyvyan* 1296 SRSx; Agatha *Fydion* 1310 FFSf; Roger *Fythien* 1327 SRSf; John *Fifian* 1327 SRSo; David *Fyuyen* 1332 SRSx; William *Fetheon'* 1379 AssC; Joan *Fithyan* 1406 LLB I; Laurence *ffedyan* 1493 SIA xii; James *Vydean, Vydyan* 1513, 1548 ArchC 33, 34; John *Fiffeon, Phiffion, Pfiffian, Phithien, Fitheon* 1604–38 ER 55 (Ess); Thomas *Phydian, Fython, Pythian, Pitheon* 1712–34 ib. Fr *Vivian, Vivien* (m), Lat *Vivianus*, a derivative of *vivus* 'living', the name of a 5th-century martyr not uncommon in England from the 12th century. Its pronunciation appears to have caused difficulty and it is found in a bewildering variety of forms, not all of which have survived. In the south, the *v* was regarded as the normal southern pronunciation of *f* and was replaced by it. As the child says *fum* for *thumb*, and *fevver* for *feather*, and the dialect-speaker *favver* for *father*, *Fivian* became *Fithian*, and this, with the common interchange of intervocalic *th* and *d*, gave *Fidian*. The initial *Ph* is merely scribal. As *Goodier* becomes *Goodger* and *Indian* is often colloquially *Injun*, so *Fidian* became *Fidgeon* and *Vidian, Vidgen*. The normal *Vivian* is much more common than appears from the above forms.

**Vizard:** *v.* WISHART

**Vize:** *v.* VISE

**Vizer, Vizor:** John *le Visur* 1273 RH (Wo). OFr *viseur* 'observer, overlooker'.

**Voak:** *v.* FOLK

**Voice, Voyce:** Thomas *Voyce* 1379 PTY; John *Voyce* 1674 HTSf. From Voise (Eure-et-Loir).

**Voicin, Vezin:** William *Veisin* 1192 P (Gl); William *le Veysin* 1279 RH (O). AFr *veisin*, OFr *voicin* 'neighbour'.

**Voisey:** *v.* VAISEY

**Vokes:** *v.* FOLK

**Volant, Volante:** Henry *le Volaunt* 1221 AssGl; Robert *le volant* 1221 *ElyA* (Sf). The present participle of OFr *voler* 'to fly', in the sense 'nimble, agile'.

**Volk(e), Volkes:** *v.* FOLK

**Volker:** *v.* FULCHER

**Voller(s):** *v.* FULLER

**Voss:** *v.* FOSS

**Voules, Vowells, Vowles:** *v.* FOWLE

**Vowler:** *v.* FOWLER

**Voyce:** *v.* VOICE

**Voyle:** Dauid *Voil* 1332 Chirk; George *Voyell* 1586 Oxon (Pembroke); William *Voile* 1609 ib. (He). A mutated form of Welsh *moel* 'bald'. cf. *pen-foel* 'bald-headed'.

**Voysey:** _v._ VAISEY
**Vreede:** _v._ FIRTH
**Vroome:** _v._ FROME

**Vyner:** _v._ VINER
**Vyse:** _v._ VISE
**Vyvyan:** _v._ VIVIAN

# W

Wace, Waison, Wase, Wash, Wason, Wass, Wasson, Wayson, Gaish, Gasche, Gash, Gashion, Gason, Gass, Gasson, Gaze: Teðion *filius Wasso* c1000 OEByn (Co); *Wazo, Gazo* 1086 DB; Robertus *filius Wazonis* 1087 DM (K); Herebertus *filius Watscon'* 1148 Winton (Ha); *Wace* armiger (seal: *Wazo*) c1150–60 DC (L); *Waszo cocus . . . ipsi Wasconi* 1170 P (Gl); Reginaldus *filius Wasce* 1177 P (W); *Wazo, Wace* de Norton' 1195–6 P (Y, Nth); *Gace* de Gisorz 1225 ClR; William *Wase* 1194 Cur (Ess); Wido *Wasun* 1195 P (So); William *Was* 1210 Cur (C); William *Wace, Waze, Wasce* 1220 Cur, 1235 Fees (O); John *Gace, Wace* 1224–5 Pat (W); John *Gace, Gasce* 1230 P, Cl (Ha, K); Godesman *Gase* 1232 Pat (L); Richard *Wason* 1273 Ipm (So); Simon *Gason* 1498 ArchC 34; James *Gasse* 1568 SRSf; Leonard *Waice* or *Wace* 1568 Oxon; Sarah *Gasson* 1773 Bardsley. The first form is from OCymr *Wasso*, OCornish *Wasō*, which may sometimes survive, but the surnames are usually from OFr *Gace*, OG *Waz(z)o*, Frisian *Watso*, hypocoristics of OG names in *Wad-* or *Warin-*. *Wace*, etc., are Northern, *Gace*, etc., Central French. The forms with *-on* are OFr accusatives. cf. WASTLING.

Wacey, Wasey, Wassey: *Wascius* Cokus 1187 P (Gl); Robertus *filius Wacey* 1275 RH (L); *Gaceus* de Broughton 1322 FrY. *Wascius* Cokus is identical with *Waszo cocus* above, a latinization of *Wazo*. This might possibly survive as *Wacey*, which might also derive from OG *Waswig*, cf. *Wasuuic* 1086 DB (Gl).

**Wacher**: *v.* WATCHER
**Wachman**: *v.* WATCHMAN
**Wackley**: *v.* WAKELEY
**Wackrill, Wakerell**: *Wakerilda* c1130 ELPN, 1211 Cur (Mx); *Wekerild* 1185 Templars (K); Justinus *filius Wakerild* 1229 Pat (Sx); William, Osbert, Agnes *Wakerild* 1188 BuryS (Sf), 1319 FFK, 1327 *SR* (Ess); William *Whaykrylle* 1374 Putnam (Mx). OE *Wacerhild* (f) 'watchful war' (cf. OG *Wagarhilt*) or, in view of the Kent form, OE *Wæcerhild*.

**Wadcock**: William *Wadecok* 1327 *SR*Ess; Walter *Wadekok* 1349 IpmW; Thomas *Wadcock'* 1379 PTY. A diminutive of OE *Wada*. *v.* WADE.

**Waddacor**: *v.* WADDICAR
**Waddams**: *v.* WADHAM
**Waddecar**: *v.* WADDICAR
**Waddell, Waddle**: (i) Ralph *Wadel'* 1249 AssW; Thomas *le Wadel* 1357 IpmW. OE *Wædel*. (ii) Maurice de *Wadehala* 1181 P (K). From Wadden Hall in Waltham (K), *Wadehale* 1179.

**Waddelow**: *v.* WADDILOVE

**Waddicar, Waddicker, Waddicore, Waddecar, Waddacor**: Edmund *de Wadeker* 1275 RH (Nf). Probably 'dweller by the field in which woad is grown', OE *wād, æcer*. The name could also have arisen from Waddicar, Woodacre (La) or Weddiker (Cu).

**Waddie**: *v.* WALTHEW
**Waddilove, Waddilow, Waddelow, Wadlow**: William, John *Wadylove* 1260 AssY, 1321 AD iv (L); Ralph *Wadiluue* 13th FeuDu; John, William *Wadeinlove* 1279 RH (Hu); Thomas *Wadyloue*, Robert *Wadyloef*, Adam *Wadinlof* 1379 PTY; Thomas *Wadloffe* 1564 Bardsley; Susannah *Wadlow* 1684 ib. '*Waddilove*', says Weekley, 'is a phrase-name which seems very out of place in the thirteenth century. . . . In fact *wade in love* is so unlike anything medieval that I am inclined to guess that the first element may belong to ME *weden*, to rage, and that the name may mean rather "furious wolf". . . . cf. Walter *Wodelof* (*Pat. R*), from the related ME *wode*, mad.' The names are rather from Anglo-Scand *Wealdþēof* which is commonly *Waldeve* or *Waldive* in ME. With a metathesis of *ld* to *dl*, these give *Wadleve*, *Wadlive*. Other forms given under *Walthew*, would similarly give *Wadleuf* or *Wadeluf* (Waltheuf), *Wadlief* or *Wadilef* (Waldief) and, by association with the common names in *-lēof* or *-lufu*, would be written *Wadilove*, etc. *Wadeinlove* should, no doubt, be read *Wadeuiloue*, an eccentric form combining *Wadeleue* and *Wadiloue*. Weekley's *Wodelof* probably also belongs here, showing the otherwise unrecorded development of *Waldeof* to *Woldeof*, with metathesis to *Wodelof*.

**Waddingham, Wadingham**: Nicholas *de Wadingham* c1160 Gilb; Alan *de Wadingeham* 1218 AssL; Nicholas *de Wadingham* 1264–5 RegAntiquiss. From Waddingham (L).

**Waddington, Wadington**: Ogis *de Wadintona* 1169 P (L); Walter *de Wadington* 1276 AssLa; Richard *Wadyngton* 1519 FFEss. From Waddington (Lincs, Surrey, WRYorks).

**Waddle**: *v.* WADDEL
**Waddon**: William *de Waddun'* 1194 P (Do); Alexander *de Waddon'* 1207 Cur (Do); Simon *de Waddon'* 1306 KB (Sr). From Waddon (Do, Sr).

**Wadds**: *Wadde* forestar 1332 SRCu. A short form of some common name, probably *Waltheof*. cf. WALTHEW.

**Waddy**: *v.* WALTHEW
**Wade, Waide**: (i) *Wada, Wade, Wado* 1066 DB;

*Wade* de Langad' 1176 P (Sa); *Wade* le fol 1297 MinAcctCo; Godwin, Gilbert *Wade* 1166 P (Ess), 1190 BuryS (Sf). OE *Wada*, from *wadan* 'to go', or OG *Wado*. The persistence of the personal-name may be due, in part, to the tale of Wade, originally a sea-giant, dreaded and honoured by the coast tribes of the North Sea and the Baltic. 'In England the memory of Wade lived longer than that of any of the old heroes of song, Weland only excepted' (*v*. Chambers, *Widsith* 95–100). (ii) Ordmar *de la Wade* 1189 P (Nf); Richard *del Wade* 1327 SRSf. 'Dweller by the ford', OE *(ge)wæd*.

**Wademan, Wadman, Wodeman:** (i) Symon *le Wademan* 1296 MESO (Nf); Richard *de Norham, waddeman* 1375 FrY; William *Wadman* 1417 MESO (Y). OE *wād* 'woad' and *mann*, identical in meaning with WADER. (ii) John *Wodeman* 1296 SRSx. This spelling, common in ME, is indistinguishable in form from *wōdeman* 'woodman'. That ME *wōdman* from OE *\*wādmann* existed is proved by the modern *Wadman* and *Wader* (ii) b.

**Wader, Waider, Weider:** (i) William *le Waisdier* 1185 P (Y); Erkenbald *Wesdier, le Waisder, le Weisdier, le Waisdier* 1191–8 P (Y); Alan *le Wader, le Weider* p1197, 1227–37 Clerkenwell (Lo); Reiner *le Waider* 1202 AssL. OFr *wesdier, quesdier, waisdier* 'dyer with or seller of woad', a blue dye-stuff obtained from the plant, in great demand in the Middle Ages. (ii) (*a*) Thorald de Cawston *le Wader* 1290 MESO (Nf), *le Weyder* 1293 ib.; Robert, Walter *le Wadere* 1296 SRSx; John *le Wadder* 1332 SRLa, SRCu. A derivative of OE *wād* 'woad', 'woad-merchant' (1415 NED). This English form is more common today, but rarer in ME than the French *Waider*. (*b*) Simon *le Wodier* 1206 Cur (K); Andr' *le Wodere* 1275 MESO (Sx). As with *Wodeman*, this form is indistinguishable from ME *wodere* 'woodman'. ME *wōdere* 'woad-merchant' certainly existed; Robert *le Woder* de Merthone is identical with Robert de Merthone, *wayder* (1276 LLB B). OE *\*wādere* became ME *wōdere* and should give a modern *Woader* or *Woder*. The identity of spelling may have led to confusion of the two names and the absorption of the less common *Woder* by the more frequent and more readily understood *Wooder*.

**Wadeson, Waidson:** Tobias *Waidson* 1614 Bardsley; Thomas *Wadeson* 1697 LaWills. 'Son of *Wade*.' cf. William *Wadecok* 1327 *SR* (Ess).

**Wadey, Wadie:** *v*. WALTHEW

**Wadham, Waddams:** John *Wadham* 1397 Hylle; William *Wadham* 1508 PN Do ii 50. From Wadham in Knowstone (D).

**Wadingham:** *v*. WADDINGHAM

**Wadland:** Michael *de Wadelond* 1275 RH (D). From Wadland Barton in Ashbury (D), or Wadland in Liskeard (Co).

**Wadleigh, Wadley, Wadly:** William *de Wadleg'* 1208 P (Ess/Herts); Nicholas *de Wadelegh'* 1242 Fees (D); Roger *de Wadeleye* 1327 SRSa. From Wadley (Berks).

**Wadlow:** John *de Wadelaw'* 1212 Cur (Beds). From a lost place *Wadlow* in Toddington (Beds).

**Wadman:** *v*. WADEMAN

**Wadsley:** Osbert *de Waddeslega* 1195 P (Y); Robert *de Waddeseleye* 1310 FFY; John *Waddysle* 1394–5 FFSr. From Wadsley (WRY).

**Wadster:** William *le Wadster* 1327 SRY; Richard *Wadstare* 1375 AssL; John *le Wadster* 1384–5 IpmNt. 'A grower of or a dealer in woad', OE *wād*, plus the feminine ending.

**Wadsworth, Wordsworth:** Adam *de Waddeswrth* 1275 Wak. From Wadsworth (WRYorks). Bardsley notes that at Silkstone the surname appears as *Waddysworth* (1556), *Wardsworth* (1656), *Wadsworth* (1666) and *Wordsworth* (1668 seq.).

**Wadworth:** Elias *de Waddewurth'* 1202, Hugh *de Waddeworth'* 1313 FFY; Thomas *de Wadworth* 1379 PTY. From Wadworth (WRY).

**Wady:** *v*. WALTHEW

**Wafer:** (i) Simon *le Wafre, le Wafrer* 1212 Fees (He), *le Wafrer* 1221 AssSa, *le Wafre* 1222 AssWa (He); Ralph *le Waverer* c1227 Fees (Y), *le Wafrur* 1250 ib. (Y). AFr *wafre* 'wafer', alternating with AFr *wafrer, \*wafrour* 'waferer, maker or seller of wafers or thin cakes' (1362 NED). The waferers seem to have been chiefly concerned with the provision of the eucharistic bread but also sold sweet, spiced cakes, the 'wafers piping hot' which Absolon the clerk gave to Alison. (ii) Ricardus *filius Waifier* 1180 P (W); Richard *Waifer* 1242 Fees (W); Ranus *Wayfer* 1267 AssSo. OG *Waifar, Waifer*, OFr *Gaifier*, a less common source than (i).

**Wagg, Wagge:** (i) Walter *Wagge* 1230 P (Ha); Robert *Wagg'* 1327 SRSx; Gilbert *Wagge* 1395 AssL. A nickname from OE *wagian* 'to shake, waddle'. (ii) Henry *atte Wagge* 1327 SRSo. 'Dweller by the marsh or bog', ME *wagge*.

**Waghorn, Waghorne:** Roger *Waggehorn* 1332 PN K 185; Peter *Waghorne* 1357 Black (Dumbarton). 'Wag horn', a name for a hornblower or trumpeter.

**Wagland:** William *Waglande* 1576 SRW. From Wagland in Diptford (D).

**Wagstaff, Wagstaffe:** William *Waggestaf* 1219 Cur (Lei); Robert *Waggestaff* 1279 RH (O); William *Waggestave* 1301 SRY. 'Wag staff', perhaps a name for a beadle. cf. WAPLE and Walter *Waggespere* 1227 AssLa, Richard *Wagetail* 1187 P (St), and *v*. SHAKESPEAR.

**Waide:** *v*. WADE
**Waider:** *v*. WADER
**Waidson:** *v*. WADESON
**Waigh:** *v*. WAY
**Weight(s):** *v*. WAIT
**Wailes:** *v*. WALES

**Wain, Waine, Waines, Wayne:** John, Richard *Wayn* 1319 FFEss, 1386 LoPleas. Metonymic for WAINER, WAINMAN or WAINEWRIGHT, or 'dweller at the sign of the wain': John *Attewayne* 1327 SRDb.

**Wainer:** Henry *le Wayner* 1381 LoPleas. A derivative of OE *wægn, wægen* 'wain, wagon'; 'driver of a wain, wagoner' (a1500 NED).

**Wainewright, Wainright, Wainwright, Wainwrigt:** Ailmar *Wanwrecthe* 1237 HPD (Ess); Adam *the Waynwrith* 1285 Wak (Y); Alan *le Waynwright* 1332 SRLa. OE *wægnwyrhta* 'wainwright, wagon-builder'.

**Wainfleet:** Wido *de Waineflet* c1180 Bury; William *de Wayneflet* 1273–4 RegAntiquiss; John *Waynflet* 1461 Paston. From Wainfleet (L).

**Wainman, Wenman, Whenman:** John *Waynman* 1297 SRY; John *Weneman* 1327 *SR* (Ess). OE *wægn, wǣn* 'wagon' and *mann*; 'wagoner' (1392 NED).

**Waison:** *v.* WACE

**Waistcoat:** *v.* WESTCOTT

**Wait, Waite, Waites, Waits, Wates, Wayt, Wayte, Waytes, Waight, Waighte, Weight, Weait, Whait, Whaite, Whaites, Whaits, Whate:** (i) Ailward *Waite* 1170–87 ELPN; Roger *le Wayte* 1221 *ElyA* (Sf); Hugh *le Weyt* 1251 AssSt; Roger *le Wate* 1296 SRSx; Adam *le Whaite* 1349 LLB G; Richard *Waight, Weight* 1595, 1610 Bardsley. ONFr *waite*, OFr *guaite, gaite* 'watchman', either in a fortified place or a town. The town waits combined the functions of watchmen and musicians (EngLife). For *guaite, v.* GAIT. *Wait* may also be a late form of WHEAT. cf. *Whaytley* for *Whatley* 1539 PN Wa 18. *v.* also WHITE. (ii) Roger *la Waite* 1197 FF (Wa); Ralph *laweite* 12th DC (Nt); John *la Wayte* 1243 AssSo. For *de la waite* 'of the watch', with frequent loss of *de* as in early forms of BATTLE, VEIL, and WARR. Robert *la Waite* 1207 P (Ess) is also called *de la Waite* (1206 P), whilst Ralph *la Waite* 1202 Cur (Ha) is identical with Radulfus *Vigil* 'the watchman'.

**Waithman:** Richard *Weythman, Waitman* 1223 Cur (We); Robert *Waythman* 1437 FrY; Jean, Thomas *Wayman* or *Waithman* 1612–13 LaWills. ON *veiðimaðr, veiðimann* 'hunter'. This may also survive as *Wayman*.

**Wake:** Hugo *Vigil* 1133 StCh; Hugo *Wac* 1153 ib.; Hugo *Wach* 1176 P (L); William *le Wacce* 1185 Templars (Y); Henry *le Wake* 1298 LLB C. Clearly a nickname, translated by Lat *vigil* 'watchful, alert'. The most common early form is *Wac*, found chiefly in Staffs, Lincs, Leics and Yorks, where a Scandinavian origin is possible, probably ON *vakr* 'watchful'.

**Wakefield:** Malger *de Wakefeld* 1219 AssY; Thomas *de Wakefeld* 1342 FFEss; Richard *Wakefeld* 1415–16 IpmY. Usually from Wakefield (WRY), but sometimes from Wakefield (Nth).

**Wakeham, Wakem, Whackum:** Henry *de Wakeham* 1296 SRSx; William *Wakeham* alias *Wakam* 1715 DKR 41 (Co). From Wakeham in Aveton Gifford (Devon) or Wakeham in Terwick (Sussex).

**Wakehurst:** Alice *de Wekehurst* 1221 Cur (Sx). From Wakehurst Place in Ardingly (Sussex).

**Wakelam:** *v.* WAKELIN

**Wakeley, Wakely, Wakley, Wackley:** Roger *Wakeley* 1332 SRSt. From Wakeley (Herts).

**Wakelin, Wakeling, Wakling, Wakelam, Walklin, Walkling, Walklyn:** *Walcelin* 1086 DB; Ricardus *filius Walkelini* 1119 Bury (Sf); *Wauchelinus* Maminot 1157 Templars (Lo); *Walclinus* de Normentun Hy 2 DC (L); *Wakelinus* de Roking' 1221 Cur (Wa); John *Wakelin* 1221 Cur (Lo); Nicholas *Walklin* 1225 AssSo; John *Wakelam* 1544 FFSt. From *Walchelin*, a Norman-French double diminutive of OG *Walho* or *Walico*.

**Wakem:** *v.* WAKEHAM

**Wakeman:** William *Wakeman* 1200 Oseney (O); Richard *le Wakeman* 1225–50 Dublin. ME *wake* 'watch, vigil' and *man*, 'watchman' (c1200 NED). *Wakeman* was the title of the chief magistrate of the borough of Ripon until 1604 when it was changed to mayor. He was the head of the body of wakemen whose duty was to blow a cow's horn every night at nine o'clock. If between then and sunrise any burglary took place, it was made good at the public charge.

**Wakenshaw:** *v.* WALKINGSHAW

**Waker:** William *Waker* 1230 P (Berks); Richard *le Waker* 1327 SRSx. OE *wacor*, ME *waker* 'watchful, vigilant'.

**Wakerell:** *v.* WACKRILL

**Wakering:** William *Wakeryng* 1400 AssLo. From Wakering (Ess).

**Wakerley, Wakerly:** Richard *de Wakerle* 1214 Cur (Nth); Henry *de Wakerle* 1297 MinAcctCo. From Wakerley (Nth).

**Wakewell:** Walter *Wakewel* 1225 AssSo; Nicholas *Wakewel* 1306 AssW. 'Watch well', OE *wacian, wel*. cf. William *Wakelevedy* 1302 SRY 'watch the lady'; John *Wakewo* 1279 RH (Hu) 'rouse up sorrow'.

**Wakley:** *v.* WAKELEY

**Walbanck, Walbancke, Walbank, Wallbank, Wallbanks:** Adam *de Wallebonk* 1332 SRLa; Thomas *Walbanke* 1489 GildY. *Wall-* is the West Midland form of *well*, hence 'dweller on the bank(s) of the stream'.

**Walbey:** *v.* WALBY

**Walbourn, Walbourne, Walburn:** *Walebrun* c1155 (DC) (L); Robert *Walebrun* 1296 SRSx; John *Walebron* 1327 SRSx. OFr *Walebron* (Langlois).

**Walbrook, Walbrooke:** Albin *de Walebroc* 1197 Clerkenwell (Lo). From Walbrook (London).

**Walburn:** *v.* WALBOURN

**Walby, Walbey, Wauldby, Waudby:** (i) Tomas *de Waldebi* 1190 P (Y); Peter de *Walebi* 1193 P (L). From Wauldby (ERYorks), DB *Walbi* 'village on the wold'. (ii) William *de Walby* 1332 SRCu. From Walby (Cumb), 'village by the Roman Wall'.

**Walch:** *v.* WALSH

**Walcock:** Peter *filius Walcok'* 1305 IpmY; Alienora *Walkoc* 1297 SRY; Hervey *Walcok* 1327 *SR* Ess; John *Walkok* 1447 FFEss. A diminutive of OG *Walo*.

**Walcot, Walcott:** Stephen *de Walecote* 1199 FFO; Warin *de Walcote* 1259 Ronton; John *Walcot'* 1379 CorO. From Walcot (Lincs, Northants, Oxon, Salop, Wilts, Warwicks), Walcote (Leics), or Walcott (Norfolk, Worcs).

**Wald, Walde, Waud, Weald, Weild, Weld, Wolde, Would, Woulds:** William *de Welde* 1121–48 Bury (Sf); Gilbert *del Wald'* 1206 Cur (Bk); Hugh *in þe Wold* 1279 RH (O); John *Weld* 1279 RH (C); Thomas *de la Waude* 1279 RH (Bk); John *at þe Welde* 1316 MELS (Sx); Richard *of ye Wolde* 1327 Pinchbeck (Sf); John *atte Wolde* 1327 *SR* (Ess); John *Woulde* 1568 SRSf. 'Dweller in the forest or by the woodland.' West Saxon *weald* survives as Weald (Essex, Kent, Hants, Oxon, Sussex), Wield (Hants), and in surnames also as Weld and Weild. Anglian *wald*, later *wold*, survives in Old (Northants) and The Wolds, and in surnames also as Wald(e), Waud, Wold(e) and Would(s).

**Waldegrave, Walgrave, Walgrove:** Robert *de Waldegrave* 1202 AssNth; Robert *de Waldegrave*

1314–16 AssNth; Thomas *Waldegrave* 1375, *Walgrave* 1379 IpmNt. From Walgrave (Nth), *Waldgrave* DB.

**Walden, Waldon, Wallden:** Godeman *de Waldena* 1176 P (Herts); Simon *de Waldene* 1304 IpmY; Thomas *Walden'* 1377 AssEss. From Walden (NRY), Saffron Walden (Ess), or King's, St Paul's Walden (Herts).

**Walder:** Adam *Waldere* 1226 FFBk; Robert *Walder* 1296 SRSx; John *Wolder* 1558 SxWills. A derivative of OE *wald*. 'Forest-dweller.'

**Waldern, Waldron:** Henry *Walderne* 1332 SRSx. From Waldron (Sx), *Walderne* 1197.

**Waldie:** *v.* WALTHEW

**Walding, Welding:** Willelmus *filius Walding* 1171 P (Y); *Walding* c1248 Bec (Bk); Roger, Henry *Walding* 1193 P (Wo), 1210 P (Gl); Henry *Wolding* 1275 SRWo. OE *Wealding*, a derivative of *Weald-* in *Wealdhelm*, etc. There has been some confusion with OG *Waldin. Wadin* or *Waldin* Crede (1191, 1195 P) is also called *Waldingus* 1195 P (Wa). The surname may also be descriptive, from OE *wealding*, a derivative of *weald*, 'forest-dweller'. *v.* GAUDIN.

**Waldman:** William *Waldeman* 1329 Rams (C); Thomas *Waldman* 1349 MESO (Y). OE *wald* (Anglian) and *mann*, 'dweller in the forest'.

**Waldo:** *v.* WALTHEW

**Waldram, Waldren, Waldron, Waldrum, Waleran, Walrond:** Iohannes *filius Waleranni, Galerami* 1086 DB (Ess); *Waleran* t Stephen DC (L); *Walran* Hy 2 ib. (Lei); *Walerannus, Galerannus, Walerandus* de Crikelade 1182, 1191, 1200 Oseney (O); John *Waleram, Walerand* 1196–8, 1218–20 Clerkenwell (Lo); Richard *Walram* 1262 FFEss; Robert *Waldrond* 1275 SRWo; Matilda *Walrond* 1275 RH (W); Roysa *Walraunt* 1297 MinAcctCo; William *Galeram* 1332 SRSx. OG *Walahram, Waleran*, OFr *Galeran(t)*. *v.* also WALLRAVEN.

**Waldwyn:** *v.* WALWIN

**Waldy:** *v.* WALTHEW

**Wale:** (i) *Walo* 1086 DB (Sx); *Walo, Wale* 1219 AssY; Robert, Lewin *Wale*, 1169 P (Y), 1221 *ElyA* (Sf); Hugo *le Wale* 1220 Cur (Wo); Richard *le Wale* 1250 Fees (Ha). *le Wale* is from ME *wale*, a general laudatory expression, 'excellent, noble, goodly', from ME *wale* 'act of choosing, choice', used as in 'men of wale' (a1300 NED), 'men of merit'. The simple *Wale* may be from this or from OG *Walo*, OFr *Gal, Galon*, which has also given *Gale* and *Gallon*. The nominative survives as *Gales* and *Wales*. (ii) Richard *de Wale* 1196 P (Nth); Philip *atte Wale* 1296 SRSx. 'Dweller by the ridge' (OE *walu* 'ridge, bank').

**Wales, Wailes:** (i) Ricardus *filius Wales* (*Walis*) 1175 P (Bk), *filius Wales* 1188 P (Bk), *filius Wale* 1208 Fees (Nth); *Wales* (*Waleis*) de Possbroc 1203 Cur (Sf). The cas-sujet of OFr *Galon*, OG *Walo*, probably from *walh* 'foreign', which has been associated with *waleis* (*v.* WALLIS) and is one source of GALES. (ii) William *le Wales* 1275 RH (L). Identical with WALLIS. (iii) Amicia, Cecilia *de Wales* 1327 SRSf, 1379 PTY; John *Wailes* 1587 LaWills; James *Wales* 1612 ib. Amicia may have come from Wales and her surname may be a variant of *Wallis*. But she may have come from Yorkshire, as certainly did Cecilia, who lived at Wales, near Sheffield.

**Walesby, Walsby:** John *de Walesby* 1202 AssL; William *de Walesby* 1275 RH (Nt). From Walesby (L, Nt).

**Waley, Walley:** John *Waley* 1377 AssEss. From Walley Hall in Fairford (Ess). *v.* WHALLEY.

**Walford:** William *de Waleford'* 1221 AssWa; Henry *de Walford* 1279 IpmGl; Gilbert *Walford* 1327 SRSo. From Walford (Dorset, Hereford, Salop), or Walford Hall (Warwicks).

**Walgrave, Walgrove:** *v.* WALDEGRAVE

**Walker:** Richard *le Walkere* c1248 Bec (Wa); Robert *le Walker* 1260 AssY. OE *wealcere* 'fuller'. *v.* FULLER.

**Walkingham:** Thomas *de Walkingham* 1204 P (Y). From Walkingham (WRY).

**Walkingshaw, Walkinshaw, Wakenshaw:** Robert *Walkyngschaw* 1551 Black; Constantine *Walkinschaw* 1562 ib.; — *Wakinshaw* 1658 ib. From the lands of Walkinshaw (Renfrewshire), *Walkeinschaw* c1235 ib.

**Walkington, Walkinton:** Robert *de Walkynton* 1327 SRWo. From Walkington (ERY).

**Walklate, Walklett:** William, John *Walkelate* 1353 LLB G, 1365 LoPleas; George *Walklot* 1511 Black (Edinburgh). Black unconsciously gives the clue to this surname when he refers to a *domus Wauklet* in a confirmation of the foundation charter of Holy Trinity, Edinburgh. This must be 'the house of Wauklet', i.e. *Walk-el-et*, a double diminutive of OG *Walho, Walico*. cf. *Walkelin s.n.* WAKELIN. *Walkelot* seems also to have existed.

**Walkley:** Robert *de Walkelay* 1219 AssY; John *Walkeley* 1379 PTY; Giles *Walkly* 1662–4 HTDo. From Walkley (WRY).

**Walklin, Walkling, Walklyn:** *v.* WAKELIN

**Walkman:** A variant of WAKEMAN 'watchman'. cf. 'Angels ben called walkmen and wardeyns for they warne men of perylles that may fall' (1398 NED).

**Wall, Walle, Walls:** Robert *de la Walle* 1195 P (Ess); Robert *del Wal* 1213 Cur (St); Alexander *super le Wal* 1279 RH (Hu); Walter *Opethewalle* 1312 ColchCt; John *Bithwalle* 1321 LLB E; Robert *atte Walle* 1327 SRWo; Agnes *Walle* 1327 SRSf. From residence near some wall (OE *weall*). In Essex the reference is probably to the sea-walls of Rochford Hundred, in Oxford, London and Colchester to the town wall, in Northumberland to the Roman Wall. Wall (Staffs) is on the site of a Roman station. In the West Midlands, *walle* is a dialectal form of *welle* and here the reference may be to a stream. cf. WALLER (ii).

**Wallace, Wallas:** *v.* WALLIS

**Wallaker:** *v.* WALLENGER

**Wallbank(s):** *v.* WALBANCK

**Wallden:** *v.* WALDEN

**Wallenger, Wallinger, Wallaker, Walliker:** *Warengerus* 1086 DB (Nf, Sf); *Garengerus* ib. (Ess, Sf); *Waringarus* Frost 1202 FFNf; *Walingerus filius* Hugonis 1221 AssSa; Roger *Wareng'* 1086 DB (Sf); Thomas *Warenger* 1482 LLB B; Robert *Walenger* 1524 SRSf; John *Walycar* 1575 *ERO* (Ess). OG *Warengar, Guarenger*.

**Waller, Wallers:** (i) John *le Walur* Hy 2 DC (Nt); William *Waliere* 1185 Templars (K); Adam *le Walere*

1280 MESO (Ha); Nicholas *le Walour* 1327 SRSf. OFr *galure, galier, gallier, guallier* 'a coxcombe, spark' or 'a man of pleasant temper'. This survives as *Gallear, Galler, Gallier, Galliers, Gallyer*, from the Central French form; *Waller* is from the NE French form. (ii) Henry *le Waller'* 1279 RH (C); John *le Wallere* 1312 LLB C. This name is common in this form and is usually a derivative of ME *walle* 'to furnish with walls'; 'a builder of walls' (c1440 NED). ME *walle* is also a dialectal form of *welle*, found in place-names from Lancashire and from Worcestershire to Cambridge and Suffolk. Here it may also mean 'dweller by a stream': William *le Waller* 1327 SRSf. v. WELLER: or, in Cheshire, Worcs and Staffs, 'salt-weller', 'salt-boiler' (1600 NED), John *le Waller'* 1221 AssWo.

**Wallerthwaite:** Robert *Wallerthwayt*, cf. Margaret *Wallerthwaytdoghter* 1379 PTY. From Wallerthwaite in Markington (WRY).

**Walles:** v. WALLIS

**Wallet, Wallett:** *Wallauiet, Walhauiet, Welleuiet* 1221 AssGl; *Wallet* 1249–86 Black; *Walettus* de Bygestrode 1296 SRSx; Thomas *Waleth* 1296 SRSx; William *Walot* 1327 SRDb; Alexander *Wallet* 1679 Black. Clearly a personal-name, perhaps an *-et* diminutive of Anglo-Scandinavian *Walþēof*. cf. WALTHEW.

**Walley:** v. WALEY, WHALLEY

**Walliker:** v. WALLENGER

**Wallingford:** Wigot *de Walingeford* 1066 DB (Berks); Hugh *de Wallingeword* 1218 FFO; Richard *Wallyngford* 1306 AssW. From Wallingford (Berks).

**Wallis, Wallice, Walles, Wallace, Wallas:** Osbert *Waleis* 1156–85 Seals (Wa); Robert *Waliscus, Waleis* 1166, 1169 P (Nf); Eudo *le Waleis* c1200 DC (L); Richard *le Waleis, le Walles* 1225, 1244 Oseney (O); Maddock *le Waleys* 1283 AssSt; Alice *Walas* 1379 PTY; William *Wallis* 1463 LLB L. AFr *Waleis, Walais* 'Welshman, Celt'. In the border counties of Warwicks, Worcs and Staffs, we clearly have reference to Welshmen. In Norfolk and Lincs, the Celts were probably Bretons, many of whom settled in the Eastern counties after the Conquest. *Wallace* is a Scottish form. Here it means a Briton of the kingdom of Strathclyde where we find Richard *Walensis* or *Waleis* between 1165 and 1173. *Walace* is found in 1432 and *Wallas* in 1497 (Black).

**Wallman, Walman:** Ailwin, John *Waleman* 1166 P (Hu), 1279 RH (C). These forms mean 'servant of *Wale*'. v. WALE. The modern form may also be an assimilated form of *Waldeman* (v. WALDMAN), or equivalent to *atte Wall*: Ralph *Walleman* 1332 SRCu. v. WALL.

**Wallop, Wollop:** Matthew *de Wallop* 1169 P (Ha); Robert *de Wallop* 1293 AssW; Philip *Wallop* 1371 IpmW. From Nether, Over Wallop (Ha).

**Wallraven:** *Walrauen, Welrauen* 1066 DB (L); *Walrafnus* de Muirteus 1275 RH (C); John *Walraven* 1275 RH (Sf); Agnes *Walreuen* 1275 SRWo. A rare name, from *Wælhræfn*, an anglicizing of OG *Walahram*. This is not Scandinavian as sometimes suggested but was probably introduced into England

by Franks and other Germans who came over with the Scandinavian settlers (PNDB 409). After the Conquest it was reintroduced in its later OG and Romance form *Waleram* which survives in WALDRAM.

**Wallwin:** v. WALWIN

**Walman:** v. WALLMAN

**Walmesley, Walmisley, Walmsley, Wamsley, Warmsley, Waumsley:** Roger *de Walmeresleghe* 1332 SRLa; John *Wamesley* 1549 FrLeic; Henry *Walmesley* 1582 PrGR. From Walmersley (Lancs).

**Walpole, Wolpole, Waple, Waples:** Godfrey *de Walpoll'* 1198 Pl (Nf); Henry *de Walepol* 1230 Cur (K); William *Walpol* 1356 FFEss; Andrewe *Walpole*, Thomas *Wapyle* 1524 SRSf; Hillary *Wapolle* 1557 Bardsley. From Walpole (Norfolk, Suffolk).

**Walrond:** v. WALDRAM

**Walsall:** Richard *de Waleshal'* 1219 P (Wo). From Walsall (St).

**Walsby:** v. WALESBY

**Walsgrave:** Richard *de Wallesgrave* 1327 SRY. From Walsgrave on Sowe (Wa).

**Walsh, Walshe, Walch:** Rose *la Walesche* 1277 AssSo; John *le Walsche, le Walche* 1327 SRSf; John *Walshe, Walche* 1360, 1376 Black (Roxburgh). ME *walsche*, OE *wælisc* 'foreigner'. cf. WELSH.

**Walsha:** v. WALSHAW

**Walsham:** Robert *de Wallesham* 1141–9 Holme (Nf); Roger *de Walsham* 1283 SRSf; John *Walsham* 1448 Paston. From North, South Walsham (Nf), or Walsham le Willows (Sf).

**Walshaw, Walsha:** Adam *de Walschagh* 1333 WhC; Adam *Walschawe* 1403 IpmY; John *Walshawe* 1461 LLB L. From Walshaw (La), or Walshaw in Wadsworth (WRY).

**Walshman, Walsman:** Walter *Walseman* 1279 RH (C); John *Walsheman* 1303 LoCt; Walter *Walcheman* 1327 SRY; William *Walisshman* 1367 AD iii (K). 'Welshman.' cf. WALSH. John *Walshman*, tailor 1324 Cor (Lo) is probably identical with John *le Walshe*, tailor 1340 LLB F.

**Walsingham:** Stephen *de Walsingham* 1206 Cur (Nf); Reginald *de Walsingham* 1301 CorLo; Edmund *Walsyngham* 1418 FFEss. From Great, Little Walsingham (Nf).

**Walter, Walters, Fitzwalter, Fitzwater, Gaulter:** *Walterus* episcopus 1066 DB; Robertus *filius Walterii, Galterii* 1086 DB; *Galterus* le Lingedraper 1210 Oseney (O); Petrus *Walterus* 1182 Bury (Sf); Petrus *Walteri* 1191 P (Sf); Geoffrey *Walter* 1296 SRSx; William *Walters* 1327 SRSt; Johan *fitz Waulter* 1350 AssEss. OG *Walter* 'mighty army'. The name was introduced into England in the reign of the Confessor and, after the Conquest, soon became one of the most popular christian names, usually as *Walter* or *Wauter*, the latter representing the actual pronunciation. v. WATER.

**Waltham, Whatham:** John *de Waltham* 1119–27 Colch (Ess); Geoffrey *de Waltham* 1190 P (L). From one of the numerous Walthams.

**Walthew, Waltho, Waldie, Waldo, Waldy, Watthews, Watthey, Waddie, Waddy, Wadey, Wadie, Wady, Wealthy, Wildee, Wildey, Wildy, Wilthew:** *Waltef, Walteu, Walteif, Waldeuus* 1066 DB; *Wallef* filius Arnabol c1153–65 Black (Kelso); Willelmus *filius Walthef* Hy 2 DC (L); *Waldeuius* filius Wesescop 1176 P (Nb); Willelmus *filius Waldeu'* 1192 P (Y); Henricus *filius Waldief* 1200 P (Ha); *Waldef* filius *Waldef* 1212 Cur (Cu); *Walyf* de Derby Ed 3 Rydware (St); Robert *Wlthef* Hy 2 DC (L); Adam *Walthef* 1219 AssY; Hugo *Waldef* 1275 SRWo; Alexander *Waltheuf*, *Walthew* 1297 Wak (Y); Hugh *Waddy* 1316 Wak (Y); Adam *Waldi* 1400 Black (Aberdeen); Thomas *Watho* 1539 ib. (Kelso); James *Waddie* 1543 ib. (Edinburgh); John *Waddo* 1567 ib. (Kelso); Robert *Wadie* 1689 ib. Anglo-Scand *Wælpēof*, ON *Valpiófr*, familiar as *Waltheof*. *Waddy*, etc., are due to the common development of *Wald-* to *Waud-* and *Wad-*. Similarly, *Watthey* is from *Wautheu*. Most of our forms are from the north, where *Wæl-* became *Wal-*. In the south, this would become *Wel-*, and later, *Wil-*, hence *Wealthy*, *Wilthew*, *Wildey*, etc. *v.* also WADDILOVE.

**Walton, Walten, Wauton:** Geoffrey, Osmund *de Waltona* 1189 Sol; Richard *de Walton* 1253 ForNth; Alan *Walton* 1352 IpmW. From one or other of the many places of this name.

**Walwin, Walwyn, Wallwin, Waldwyn:** *Walduinus* 1066 DB (Sf); *Wealdwine* Wm 2 Searle; *Welwin* 1205 Cur (Ess); Mable *Welwyne* 1327 *SR* (Ess). The DB *Walduinus* is derived by von Feilitzen from OG *Waldwin*, a possible explanation, but the existence of an OE *Wealdwine* 'power-friend' is proved by the later *We(a)l-* forms. The surname may also derive from *Walweyn*. *v.* GAVIN.

**Walworth:** John *de Walworth* 1340 CorLo; John *de Walleworth* 1375 FFY; Richard *Walleworth* 1400 AssLo. From Walworth (Du).

**Wambe, Wombe:** Ralph *Wambe* 1214 P (C); Alice *Wombe* 1315 AssNf; John *Wombe* 1375 FFEss. A nickname from OE *wamb* 'belly, stomach'. cf. Hugh *Wambestrong'* 1243 AssSo 'strong in the belly'.

**Wamsley:** *v.* WALMESLEY

**Wand:** Ralph *Wande* 1306 AssW; William *Wonde* 1327 SRSo; John *Wand* 1576 SRW. Perhaps a nickname from OE *wand* 'mole'.

**Wangford:** John *de Wangeford* 1275 RH (Nf); John *Wangford*, Nicholas *de Wangford* 1327 SRC. From Wangford (Sf).

**Wanless, Wanlass, Wanliss, Wanlace, Wandless, Wandloss:** Simon *Wanles* 1451 Black (Montrose); Robert *Wanless* 1538 ib. (Linlithgow); Robert *Wandles* 1638 ib. ME *wanles* 'hopeless, luckless'.

**Wann:** Warner, Robert *Wan* 1297 SRY, 1327 SRSf. OE *wann* 'wan, pale'.

**Wannaker:** Nicholas *de Walacre* 1276 ForEss. From Great, Little Wanacre, fields in Little Tey (Ess).

**Wannock:** (i) William *Wannoc* c1140 ELPN. A derivative of OE *wænn* 'wen, wart'. (ii) Geoffrey *de Wannok*, Martin *de la Wennok* 1296 SRSx; John *Wonnok'* 1332 SRDo. From Wannock in Jevington (Sx).

**Wannop:** Robert *Wanhope* 1486 GildY; Christopher *Wannopp'* 1556 FFHu; Hugh *Wanop* 1600 FrY. 'Without hope', ME *wanhope*. Or perhaps 'dweller in the dark valley', OE *wann*, *hop*.

**Wansey:** Hugo *de Wanceio* 1086 DB (Sf); Roger *de Wanci* 1146–8 Seals (Ess). From Wanchy (Seine-Inférieure). *v.* ANF.

**Wansford:** Thomas *Wandesford'* 1421 KB (Lo). From Wansford (Nth, ERY).

**Want, Whant:** (i) Edwin *Wante* 1207 Cur (Nf); Benedict *le Want* 1327 SRSx. ME *want* 'mole'. (ii) James *atte Wante* 1332 SRSx. 'Dweller by the cross-roads', ME *want*, a dialect form of WENT.

**Wanter:** Lefwin *le Wanter* 1172 Oseney; Ralph *le Waunter* 1221 Cur (W); Hugh *le Wanetere* 1262 IpmGl. OFr *wantier*, *gantier* 'a maker or seller of gloves'.

**Wantley:** Philip *de Wantele* 1327 SRSx; John *Wantelee* 1391 KB (Sx). From Wantley Fm in Henfield (Sx).

**Wantling:** *Wentiliana* 1208 Cur (Sa); *Wantliana* 1221 AssSa; *Wenthelen* 1411 Fine (Glam); Albreda *Wentelien'* 1221 AssWa; Thomas *Wantlyn* 1674 HTSf. Welsh *Gwenllian* (f) 'fair-flaxen', the name of the daughter of a Prince of Powys (c1180) and the subject of a song 'Gwenllian's Repose' composed in 1236.

**Wanton:** William *Wantoun* 1298 AssL; Stephen *Wantoun* 1312 AssNf. ME *wanton* 'gay, merry'.

**Wapham:** Auel *de Wappeham* 1176 P (Nth). From Wappenham (Nth), *Wapeham* DB.

**Waple:** Everard, William *Wagepol(e)* 1169 P (W), 1206 Cur (Sf); Thomas *Waghepol'* 1219 AssY; Geoffrey *Waupol*, *Wagpoll* 1271 LeicBR. 'Wag pole', synonymous with WAGSTAFF. *Wagpoll* suggests an alternative 'wag head'. cf. William *Wauberd* 1295 Barnwell (C), John *Wagheberd* 1297 MinAcctCo. 'wag beard' *v.* also WALPOLE.

**Waraker:** *v.* WARWICKER

**Warboy, Warboys, Worboys:** (i) William *de Wardeboys* 1261 FFEss. From Warboys (Hunts). (ii) Richard, Silvester, William *Wardebois* 1207 P (St), 1227 AssBeds, 1219 AssY; John *Gardeboys* 1280 LLB A. The last form, and the wide distribution of forms without a preposition, show we have also a nickname 'guard wood' for a forester (OFr *garder*, *warder* and *bois*).

**Warbrick:** Robert *de Warthebrek* 1332 SRLa. From Warbreck (La).

**Warburton, Warbleton:** Mabilia *de Warberton'* 1212 P (Ess, Herts); William *de Warbilton* 1278 RH (C); Geoffrey *de Werberton'* 1325 WhC; John *Warberton* or *Warbulton* 1413–14 FFSr. From Warburton (Ches), or Warbleton (Sussex), *Warburton* 1211, *Warbelton* 1273.

**Warcup, Warkup:** Alan *de Wardecop* 1197 P (We); William *de Wardecop'* 1279 AssNb; Richard *Warecope* 1340–1450 GildC. From Warcop (We).

**Ward, Warde, Wards:** (i) William, John *Warde* 1194 P (Y), 1257 MEOT (Sx); Thomas, Simon *le Warde* 1279 RH (Beds, O); Alice *le Wardes* 1352 ERO. OE *weard* 'watchman, guard'. (ii) William *de la Warda* 1176 P (Lei), *de la Warde* 1203 P (Db); Robert *del Warde* 1285 AssSt; Peter *ate Warde* 1332 MELS (Sr). This may go back to an OE *wearde*, *wearda* 'beacon', but is more probably OE *weard* 'watching, guarding', with the same meaning as (i) above, '(man) of the watch', 'watchman'. We may, however, sometimes be concerned with ME *werd, ward* 'marsh', found in Essex, Kent and Herts, hence 'marshman'. cf. WARDMAN and *v.* MELS 220.

**Wardale, Wardall, Wardell:** *v.* WARDILL

**Warden:** (i) Wluric *Uuerdenus* 1066 DB (Herts); Walter *Wardein* 1279 RH (O); John *le Wardeyn* 1289 Barnwell (C). AFr *wardein*, OFr *gardein* 'one who guards, warden', perhaps a gate-keeper, porter or sentinel (a1225 NED). (ii) Symon *de Waredon'* 1232 Clerkenwell (K); Elyas *de Wardon'* 1297 MinAcctCo (R). From Warden (Beds, Durham, Kent, Northumb, Northants).

**Warder:** Adam *Wardour'* c1272 HPD; John *de Werdour* 1341 IpmW. From Wardour (W).

**Wardill, Wardle, Wardale, Wardall, Wardel, Wardell:** (i) William *de Werdale* 1216, *de Werdall* c1217 FeuDu; William *Wardell* 1539 ib. From Weardale (Durham), which may also have developed to *Wardill*. (ii) Thomas *de Wardhill* 1218 AssLa; Alexander *de Wardhull* c1238 WhC (La), *de Werdul* 13th ib.; Richard *de Wardle* 1275 RH (L). From Wardle (Ches, Lancs), both earlier *\*weardhyll* 'watch hill'.

**Wardlaw, Wardlow, Warlow:** Henry *de Wardlowe* 1275 RH (Db); John *de Wardlowe* 1348 DbAS 36; Alexander *Wardlaw* or *Warlaw* 1467 Black. From Wardlow (Db).

**Wardley:** William *de Wardlegh* 1296 SRSx. From Wardley (R), or Wardley in Iping (Sx).

**Wardlow:** *v.* WARDLAW

**Wardman, Wordman:** John *Werdeman, Wirdman* 1351 AssEss, 1374 Pat (Ess); Christopher *Wardeman* 1491 GildY. The Essex name is 'marshman' (ME *werde*). The surname may also mean 'watchman'. cf. WARD.

**Wardrobe, Wardrop, Wardrope, Wardrupp, Whatrup:** Robert *de Warderob* c1210 Black; Joscelin *de la Warderob(e)* 1219, 1220 Cur (Bk); Thomas *de Garderoba* 1286 Wak (Y); John *atte Warderobe* 1327 SRSx; Thomas *Warderope* 1334 LLB E; William *Waddrope* 1608 Black. ONFr *warderobe*, OFr *garderobe*, from *warder, garder* 'to watch' and *robe*. An official of the wardrobe, identical in meaning with WARDROPER.

**Wardroper, Wardropper, Waredraper:** William *Warderober* 1275 RH (K); William *Wardroper* 1301 SRY; Thomas *Garderober, (le) Wardrober* 1351 AssEss; Thomas *Wardroppare* 1465 Black. ONFr *warderobier*, OFr *garderobier* 'an official of the household in charge of the robes, wearing-apparel, etc.' (a1400 NED).

**Wardson:** John *Wardson* 1379 PTY; John *de Hoghton alias* John *Wardson* 1422 IpmLa. 'Son of the watchman, guard', OE *weard, sunu. v.* WARD.

**Ware, Wares:** (i) Peter *le Ware* 1218 ChR (Gl), 1297 MinAcctCo (D). OE *wær*, ME *war(e)* 'wary, astute, prudent'. This nickname certainly existed but examples are rare. (ii) Anschil *de Waras* 1066 DB (Herts), *de Wara, de Wares* 1086 ib.; Aschi *Wara* 1066 DB (Herts); John *de Ware* 1276 LLB B. From Ware (Herts). (iii) William *de la War'* 1194 Cur (Sr); William *ater Ware* 1296 SRSx; Richard *atte Ware* 1327 *SR* (Ess); Nicholas *Betheware* 1369 LLB G. From

residence near or employment at a weir or dam (OE *wær*), one who looked after a fishing-weir.

**Wareham, Warham, Wearham:** Alexander *de Warham* 1207 P (Do); William *de Warham* 1332 SRDo; George *Warum*, Edward *Warham* 1641 PrSo. From Wareham (Do), or Warham (He, Nf).

**Wareing, Waring, Warin, Warring, Wearing, Wharin, FitzWarin:** Robertus *filius Warini* (*Warin*) 1086 InqEl (C); Gislebertus *filius Garini* 1086 DB (Ess); *Warinus* filius Toruerdi 1142 NthCh (L); Willelmus *filius Warin, Gwarini* Hy 2 DC (L); Robert, Gilbert *Warin'* 1198 Cur (Y), c1209 Fees (Du); John *Waryng* 1275 SRWo; Roger *Warenge* 1478 LLB L; John *Warren alias Waryng* 1512 Oxon; William *Wairin* 1665 RothwellPR (Y); Thomas *Warring* 1674 HTSf. AFr *Warin*, OFr *Guarin, Guérin*, from OG *Warin*, a very common Norman personal name. This has been confused with WARREN.

**Warfield:** Simon *de Warefeld'* 1218 P (Berks); John *de Warfelde* 1319 SRLo; Robert *Warfeild* 1641 PrSo. From Warfield (Berks).

**Warfles:** *v.* QUARLES

**Warham:** *v.* WAREHAM

**Warkup:** *v.* WARCUP

**Warleigh, Warley:** Alfuuin *de Werlaio* 1066 InqEl; William *Warlege* 1201 Cur; Walter *Warlagh'* 1332 SRDo. From Warleigh (D, So); Warley Salop, Wigorn (Wo), or Great, Little Warley (Ess).

**Warlock, Werlock, Worlock:** Nicholas *Warloc* 1279 RH (C); Simon *(le) Warlok* 1290–1 Crowland (C); John *Werlok* 1327 SRC. OE *wǣrloga* 'traitor, enemy, devil', not noted as a surname, came to be applied especially to the Devil, sorcerers (with particular reference to the power of assuming inhuman shapes) and monsters. The normal development would be to *warlow*. In Scotland we find *warloche* and *warlok*, used of wizards. As these forms do not appear before the 16th century, NED discounts association with ON *varðlokkur* 'incantations'. *Warloc* has now been noted in the 13th century on Cambridgeshire manors of Crowland Abbey, a district subject to Scandinavian influence.

**Warlow:** *v.* WARDLAW

**Warman, Warmen:** (i) Robertus *filius Waremanni* 1208 P (So); Henry, Richard *Wareman* 1214 Cur (Nth), 1263 ArchC iii; William *Warman* 1338 LLB F. OE *waru* 'articles of merchandise, goods (for sale)' and *mann. Wareman* probably meant a chapman. (ii) *Warmundus* 1199 MemR (Sr); John *Waremund* 1275 RH (Berks); Isabella *Wermund* 1279 RH (C). OE *Wǣrmund* 'faith-protector'.

**Warme:** Matilda, Henry *Warme* 1279 RH (O). OE *wearm* 'warm', probably in the sense 'zealous, keen'.

**Warmington:** Þurfeorð *æt Wermingtune* 972 OEByn (Nth); William *de Warmintona* 1185 Templars; Richard *Warmyngton* 1376 AD i (Sr). From Warmington (Northants, Warwicks).

**Warmoll:** *v.* WORMALD

**Warmsley:** *v.* WALMESLEY

**Warne, Warnes, Warn, Wearn, Wearne:** Jervase, John *de Werne* 1273 RH (So); John, William *Warne* 1524–5, 1545 SRW. From Wearne (Som), or Warne in Marytavy (Devon).

**Warnefield, Warnfield:** Geoffrey *de Warnefeld* 1219 P (Y). From Warmfield (WRY), *Warnefelde* 1119.

**Warneford, Warnford:** Wulfric *æt Wernæforda* c1050 OEByn (Ha); Ralph *de Warneford* 1296 SRSx; John *Warneforde* 1545 SRW. From Warnford (Hants).

**Warner:** (i) *Warnerus* de Lusoriis 1140 Eyns (O); *Garnerus, Gwarnerius, Warnerus* de Campania c1160 DC (L); *Warner* (*Garner*) de Waxtunesham 1221 Cur (Nf); *Wariner* le Botiler 1275 RH (Bk); Robert *Warnier* 1196 P (Do); Geoffrey *Warner* 1203 Cur (Sr). ONFr *Warnier*, OFr *Guarnier, Garnier*, from OG *Warinhar*(*i*), a common Norman personal name. (ii) Peter *le Warner* 1214 Cur (Y). A contracted form of OFr *warrennier* 'warrener'. *v.* WARRENER, GARNER.

**Warnett:** *v.* GARNET

**Warr, Warre:** Herebertus *la Guerre* 1179 P (Do); John *la Werre, la Guerre* 1187, 1195 P (Gl); Tomas *la Werre, la Guerre, de la Warre* 1196, 1199 P (Gl); Peter *le Werre, de Warre* 1199, 1203 P (Gl); George *Warre* 1468 LLB L. OFr *werre, guerre* 'war', originally *de la werre, de la guerre* 'of the war', a warrior. cf. Fr *Guerre, Laguerre*. Geoffrey *le Werreur* is also called *le Werre* (1221 AssWa). For the common loss of the preposition, cf. BATTLE, VEIL.

**Warrack:** *v.* WARRICK

**Warren, Warran, Warrand, Warrant:** William *de Warene, de Warenne, de Garenna* 1086 DB; Hamelinus *de Warrena* 1187 Gilb (L); William *de Warenne* 1285 FFEss. From La Varenne (Seine-Inférieure). There is no reason to suppose that this also means 'at the warren' as stated by Bardsley. There are no examples with *atte*. The surname has been confused with *Wareing*.

**Warrender:** Thomas *Warendier* Hy 3 Colch (Sf); John *Warrender* 1327 *SR* (Ess); William *le Warender* 1332 SRLa. A form of WARRENER with intrusive *d*, found already in OFr *garendier* (Godefroy).

**Warrener, Warriner, Warner:** William *le Wariner* 1198 FF (K); Adam *le Warner* 1218 FFSt; John *le Warner, le Warener, le Wariner, le Warenner* 1280 AssSo; William *le Warrener* 1317 AD iii (W). NE OFr *warrennier*, central OFr *garennier* 'warrener', an officer employed to watch over game in a park or preserve (1297 NED). *v.* also WARNER, which has an alternative origin.

**Warrick, Warrack:** Walter, Roger *Warrock* 1271 ForSt, 1285 *Ass* (Ess). In building, the lashings for scaffolds were tightened by driving in wedges called *warrocks*. cf., at Westminster in 1324, 'for 6 pieces of timber for *warrokis* for binding the scaffolds' and 'for half a hundred of talwode of *ash* for *warrokes* for making scaffolds' (Building 320). The surname is probably metonymic for a maker of warrocks or a builder of scaffolds. It has been assimilated to *Warwick* with which it has been confused.

**Warricker:** *v.* WARWICKER

**Warring:** *v.* WAREING

**Warrington:** Richard *de Warynton'* 1316–17 FFWa; Agnes *de Werynton* 1354 Putnam (Ch). From Warrington (Bucks, Lancs).

**Warrior, Warrier, Wharrier:** Herbert, Jordan *le Werreur* 1202 AssL, 1230 P (He); Thomas *le Werreor*

1324 LoPleas. NE OFr *werreieor, werrieur* 'warrior'. cf. WARR.

**Warriss, Warry:** *v.* WERRY

**Warsap, Warsop:** William *de Wyrshop* 1379, *de Wyrsop* 1397 NottBR. From Warsop (Nt).

**Warter:** Robert *de Warter* 1219 AssY. From Warter (ERYorks).

**Warters, Worters:** Roger *Warters* 1637 EA (NS) ii (Sf). For *Waters*.

**Warton:** Ruffus *de Wertona* c1175 Seals (Nf); Adam, John *de Warton* 1246 AssLa, 1332 SRWa; Peter *Warton* 1527 FFEss. From Warton (Lancs, Northumb, Warwicks). *v.* also WHARTON with which there may have been some confusion.

**Warwick, Warrick, Wharrick:** Turchil *de Waruuic* 1086 DB (Wa); Richard *de Warewic* a1196 Black (Glasgow). From Warwick. Turchil owed his surname to his office as sheriff of Warwickshire. Less commonly, but particularly in Dumfries and Kirkcudbright, from Warwick (Cumb).

**Warwicker, Warricker, Waraker, Woracker, Woraker, Woricker:** *Warwicker* is an alteration of *Warricker* through the influence of *Warwick*. The surname denotes a maker of 'warrocks' or a builder of scaffolds. *v.* WARRICK.

**Wasey:** *v.* WACEY

**Wash:** *v.* WACE

**Washbourn, Washbourne, Washburn, Washburne:** William *de Wasseburn'* 1204 Pl (Wo); John *Washburn'* c1280 SRWo; William *de Wasshebourn* 1333 FFY. From Washbourne (Devon, Glos), or Washburn (WRYorks).

**Washbrook, Washbrooke:** John *de Wassebroc'* 1202 Pl (Sf); Margaret *Wasshebrook* 1374 ColchCt; Caleb *Wasbrook* 1628 Musters (Sr). From Washbrook (Suffolk).

**Washer:** Henry *le Wassere* 1293 MESO (Y); Alan *le Wascere* 1295 ib. (Nf). A derivative of OE *wæscan* 'to wash', 'washer'. cf. LAVENDER.

**Washford:** Richard *de Wasshforde* 1337 CorLo. From Washford (So), or Washford Pyne (D).

**Washington:** Ralph *de Wassingeton* 1327 SRSx; Robert *de Wasshington* 1395 IpmLa; John *Wasshyngton* 1401 AssLa. From Washington (Durham, Sussex).

**Washtell:** *v.* WASTALL

**Waskett:** Elyas *Wasket* 1274 RH (Ess); John *Wasket* 1351 AssEss. Probably a diminutive of *Wask*, the Norman form of OFr *Gasc* 'Gascon'. cf. Fr *Gasquet*. The same development appears in Adam *Waskin* 1276 RH (Db), for *Gaskin*.

**Waslin:** *v.* WASTLING

**Wasnidge:** *v.* WASTNAGE

**Wason, Wass:** *v.* WACE

**Wasp, Waspe:** William *Waps* 1202–16 StP (Lo); Richard *Wasp* 1203 P (Sr); Roger *le Waps* 1296 SRSx. A nickname from OE *wæps* 'wasp'.

**Waspail, Waspel, Waspell:** Roger *Waspeil* 1130 P (W); Henry *Waspail* 1228 Cur (W); Peter *Waschepaille* 1327 SRSf; John *Waspal* 1364 FFW. The northern French form of OFr *gaspail* 'spendthrift, wastrel'. cf. Hartley Westpall (Ha).

**Wassall, Wassell:** *v.* WASTALL

**Wasselin:** v. WASTLING

**Wassey:** v. WACEY

**Wasson:** v. WACE

**Wastall, Wastell, Washtell, Waistell, Wassall, Wassell, Gastall:** (i) Ralph *Wastel* 1182–1211 BuryS (Sf), 1279 RH (O); Thomas *Wassall* 1674 HTSf. OFr *wastel*, north-eastern form of *g(u)astel*, modFr *gâteau*, 'a cake or bread made of the finest flour' (c1300 NED). A common medieval name, metonymic for (John) *Wasteler'* 1327 MESO (L) or (Robert) *Wastelmonger* 1317 MEOT (O), maker or seller of *wastels*. *Gastall* is rare. (ii) Nicholas *de Wasthull'* 1221 AssWo. From West Hills in Alvechurch (Worcs).

**Wastenay:** v. WASTNAGE

**Wastling, Waslin, Wasselin:** *Wacelinus* clericus Hy 2 DC (L); *Wascelinus* de Brunham c1200 DC (L); William *Wacelin, Waselin* c1150, 1157 Gilb (L); Thomas *Wazelin* 1162 P (Y); Samson *Wascelin'* 1176 P (Ha); Adam *Wastelyn* 1327 SRSf; Peter *Waslinge* 1581 FrY. OG *Wazelin, Wa(s)celin*, OFr *Gacelin, Gaselin*, a double diminutive of OG *Wazo*.

**Wastnage, Wastnidge, Wasnidge, Wastenay, Westnage, Westnedge, Westnidge, Westney:** Robert *de Wastenays* c1165 StCh; William *Wastineis* 1177 P (St); William *de Westenay* 1191 P (L); Robert *de Westeneis* 1203 P (L); William *Wastneys* 1249 AD i (Lei); Ralph *le Wasteneys* 1281 AssCh. From le Gâtinais, formerly Gastinois, a district south of Paris and east of Orleans.

**Watch:** Hugh *Wach* 1176 P (L); William *le Wacce* 1185 Templars (L); John *Wacche* 1371 IpmW. OE *wæcce* 'watch, vigil'. Metonymic for a watchman.

**Watcher, Wacher:** Johan *Wachere* 1237 ELPN; Elys *le Wacher* 1279 RH (C). A derivative of OE *wæcce* 'watch, vigil'. A watchman.

**Watchet, Watchett:** John *Wachet* 1327 SRSo. From Watchet (So).

**Watchman:** Robert *Wachman* 1513 Black; James *Watchman* 1564 ib. 'Watchman'.

**Watcock:** John *Watecok* 1333 FFEss. A diminutive of *Wat*, a pet-form of *Walter*.

**Watcyns:** v. WATKIN

**Wate(s):** v. WAIT

**Water, Waters:** (i) *Waterus* de Cantelupo c1135 DC (L); *Walterus, Waterus* filius Herberti Hy 2 DC (L); *Water* Dey 1479 BuryW (Sf); John *Watter* 1214 Cur (Wa); Richard *Wauter*, Roger *Water* 1275 SRWo; Hugh *Water* 1279 AssNb; Alice *Waters* 1327 SRSf; John *Wauters* 1348 AD i (Wa); Thomas *Watter* 1509 GildY. *Water* was the normal medieval pronunciation of *Walter*. Theobald *Walter* is also called Theobaldus *filius Walteri*, Theobaldus *Walteri*, Tebaut *Water* 1212–36 Fees (La). (ii) William *de la Watere* 1245 FFC; Thomas *del Water* 1246 AssLa; William *atte Watere* 1296 SRSx. 'Dweller by the water or stream.' cf. ATTWATER, BYWATER.

**Waterer, Watrer:** Richard *Waterer* 1443 MEOT (Sx). A derivative of OE *wæterian* 'to water, irrigate, lead (cattle) to water'.

**Waterfall:** Herbert *de Waterfale* 1168 P (D); Robert *de Waterfalle* 1242 AssSt; Johanna *Waterfall'* 1379 PTY. From Waterfall (St), Waterfall in Guisborough (NRY), a lost *Waterfall* in Pontefract (WRY), or 'dweller by the waterfall', OE *wætergefeall*.

**Waterfield:** Hugo *of Walteruile* 1137 ASC E; Guefridus *de Walteriuill* c1142 Seals (He); Ralph *de Walteruill'* 1190 P (L); Ascelina *de Watervill'* 1205 Cur (L); Roger *Wateruile* 1327 SRC; John *Waterfield* 1601 Oxon. From Vatierville (Seine-Inférieure).

**Watergate:** William *de Watergate* 1323 KB (Sx); Thomas *atte Watergate* 1341 MELS (Sx); John *Watergate* c1405 FS. 'Dweller by the water-gate', OE *wæter, geat*.

**Waterhouse:** Adam *del Waterhous* 1308 Wak (Y). 'Dweller at a house by the water' or, possibly, at a moated house.

**Waterkeyn:** John *Waterkyn* 1301 LLB C. 'Little Walter.' v. WATKIN.

**Waterman:** (i) Adam *Walterman* 1260 AssY; Geoffrey *Walterman* 1275 RH (Sx). 'Servant of Walter', which would inevitably become *Waterman*. v. WALTER, WATER. (ii) Wilke *Waterman* 1196 FrLeic; William *le Waterman* 1249 Oseney (O). 'Waterman', i.e. waterbearer, water-carrier. cf. William *Waterladar'* 1197 P (Wa) 'water-carrier', one who carts water for sale; Richard *Waterberere* 1381 LoPleas, one who carries water from a spring, etc. for domestic use. 'Boatman' is also a possible sense.

**Watership:** Geoffrey *atte Watershipe* 1332 SRSx; John *atte Weterschipe* 1334 SRK. 'Dweller by the expanse of water, or by the conduit', OE *wæterscipe*.

**Waterson, Watterson:** John *Wauterson* 1379 PTY; William *Watterson* 1507 GildY; Thomas *Waterson* 1674 HTSf. 'Son of *Walter*.' v. WATER.

**Waterton:** Richard *de Waterton* 1341 KB (Mx); Hugh *Waterton* 1400 IpmGl; Robert *Waterton* 1415–16 IpmY. From Waterton (L).

**Watford:** Eustace *de Watford'* 1196 P (Nth); Walter *de Watford* 1287 FFEss; John *de Watford* 1361–2 FFSr. From Watford (Herts, Nth).

**Wath, Wathe, Wathes, Waythe:** Ioseph *de Waða* 1177 (P) (L); Walter *del Wathe* 1301 SRY. From Wath (NRYorks, WRYorks), Wade (Suffolk), Waithe (L), DB *Wada*, or from residence by a ford, ON *vað*.

**Watkin, Watking, Watkins, Watkeys, Watkis, Watkiss, Watcyns, Whatkins:** Adam *filius Walterkini* 1200 Oseney (O); *Watkin* 1252 Bardsley; *Watkynge* Llooyde 1623 ib. (Gl); John *Watkyns* 1327 SRSf; John *Wattkyn* 1552 ShefA; Thomas *Watkys* 1662 PrGR. *Wat-kin*, a diminutive of *Walter*.

**Watkinson:** John *Watkynson* 1485 GildY. 'Son of *Watkin*.'

**Watler, Whattler:** Thomas *Watelere* 1377 AssEss. 'Wattler.' v. DAUBER.

**Watley:** v. WHATELEY

**Watling:** v. WHATLIN

**Watlington:** Peter *de Watlingeton'* 1200 FFNf; William *de Watlyngethon* 1296, *de Watlyngton* 1332 SRSx. From Whatlington (Sx), or Watlington (Nf, O).

**Watman:** v. WHATMAN

**Watmough, Watmore, Watmuff, Whatmaugh, Whatmough, Whatmoor, Whatmore, Whatmuff:** Robert *Waltersmaghe* 1305 Black; Robert *Watmaghe* 1379 PTY; Hugh *Watmoughe* 1581 Oxon (Y); Hugo *Watmouth* 1599 Bardsley (Y); William *Watmer* 1718 DKR 41 (Wo). The first element is *Wat*, a pet-form of *Walter*; the second is ME *maugh, mough* which may be from ON *mágr* 'brother-, father- or son-in-law', OE *mǣga* 'relative, son', OE *mǣge* 'female relative', or OE *magu* 'child, son, servant'. The usual explanation of the compound is 'Wat's brother-in-law', but there seems no special reason for selecting this particular relative, except that *maugh* or *mauf* has this meaning in dialect. The variation between *Mago Healfdenes* and *sunu Healfdenes* (BT) suggests that the meaning may well be 'Wat's son', but this is unsuitable for William *Barnmaw* 1273 RH (Y) where the reference must be to some other relative of the child. This type of name was not uncommon in Yorkshire and Lancashire and survives also in HICKMOTT, HITCHMOUGH and HUDSMITH. cf. Geoffrey *Cokesmahc* 1183 Boldon; Henry *Ravesmaughe* 1296 CalSc; William *Matheumogh* 1327 SRDb; Adam *Duncanmaugh*, Thomas *Nicholmagh*, William *Raynaldmagh* 1332 SRCu; Adam *Dyemogh*, William *Robertmogh*, John *Wilkemogh* 1332 SRLa; John *Godesmagh* 1351 FrY; John *Elysmagh*, Richard *Gepmouth*, John *Tailliourmoghe* 1379 PTY. *-moor*, *-more* are more or less phonetic spellings of *-maw*.

**Watrer:** *v.* WATERER

**Watsham:** Osbert *de Wachesham* c1190 BuryS (Sf). From Wattisham (Suffolk).

**Watson:** Richard *Watson* 1324 Wak (Y); Thomas *Watteson* 1327 SRWo; William *Wattesson*, *Wattysson* 1409, 1430 Crowland (C). 'Son of *Wat*.'

**Watt, Wattis, Watts:** *Watte* 1292 AssCh, 1306 Wak (Y); *Wat* of Carneyg 1446 Black; Paganus *Wat* 1176 P (D); William *Watte* 1317 AssK; William *Wattes* 1279 RH (O); Roger *Wattys* 1381 SRSf. A pet-form of *Walter* or *Water*. cf. Matilda *Walte* 1278 Oseney (O).

**Watten:** Robert *Watten* 1275 SRWo; David *Watten* 1332 SRWo; Nycholas *Watten* 1583 Musters (Sr). The West Midland genitive form of *Wat*, a pet-form of *Walter*. cf. GEFFEN.

**Watters:** *v.* WATER

**Watterson:** *v.* WATERSON

**Watthews, Watthey:** *v.* WALTHEW

**Watton:** John *de Wauton'* 1198 FFNf; Robert *de Watton'* 1219 AssY; Robert *de Wadeton* 1333 PN D 224. From Watton (Herts, Nf, ERY), or Watton' in Bere Ferrers (D).

**Watty, Watties:** William *Watti* 1297 SRY; John *Waties* 1327 SRWo. A diminutive of *Wat*, a pet-form of *Walter*.

**Waud:** *v.* WALD

**Waudby:** *v.* WALBY

**Waugh:** Robert, Thomas *Walgh(e)* 1296 Black (Roxburgh, Peebles); William *Wahh* 1379 PTY; William *Waugh* 1436 Black; Edward *Wauch* 1526 ib. Neither Black's derivation from *Wauchope* nor Harrison's from *waugh* 'wall' can be accepted. The only evidence for a place-name is David *de Waughe* 1296 Black (Lanarks) and this may be an error for *le Waughe*. The places along the Roman Wall are from OE *weall*, not from *wāg*: Walby, Walton, Wallhead, Wallsend. The surname is common on both sides of the Scottish border and is probably OE (Anglian) *walh* 'foreigner', a name given by the Anglians to Strathclyde Celts, parallel to *Gall* given by Highlanders to Lowlanders.

**Wauldby:** *v.* WALBY

**Waule:** May be for WALL or WALD.

**Waumsley:** *v.* WALMESLEY

**Wauton:** *v.* WALTON

**Wavell, Weavill, Wevell, Wevill:** William *de Waluile* 1086 DB (D); Robert *de Wavill'* 1220 Cur (L); Henry *de Wayvill*, *de Wavill* 1268 AssSo; Richard *de Weavile*, Henry *de Wevile* 1276 RH (Beds); Jannes *de Weuyle* 1296 SRSx; Thomas *de Weyuile* 1327 ib.; John *Wauel* 1376 ColchCt. From Vauville (La Manche).

**Waverley:** William *de Wauerlegh* 1220 Acc. From Waverley (Sr).

**Waw, Wawe:** *Wawe* 1285 Oseney, identical with *Walter* of Leicester (Ed); Walter *Wawe* 1249 AssW; Walter *Wawe* p1316 LuffCh; John *Wawe* 1340 NIWo. A pet-form of *Walter*.

**Wawne:** Peter *de Wagena* 1175 P (Y); Geoffrey *of Waghne* 1242 FFY; John *Waghen* 1403 IpmY. From Wawne (ERY).

**Wax:** Ralph *Wax* 1279 RH (O). Metonymic for *Waxman*.

**Waxman:** Adam *Wexman* 1185 P (Wo), OE *weax* 'wax' and *mann*. 'Seller of wax.' cf. Robert *le Waxmongere* 1310 MESO (Nf).

**Way, Waye, Waigh, Weigh, Wey, Whay:** Roger *de Waie* 1194 P (Do); William *Waye* 1236 Fees (Do); Richard *de la Weye* 1249 Fees (D); John *ate Wey* 1279 RH (C). From Waye (Devon, Dorset) or from residence near a road or path (OE *weg*).

**Waybrow:** *v.* WYBER

**Wayland, Waylen, Wayling, Weyland:** (i) *Welland* 1086 DB (D); *Weilandus* 1185 Templars (So); *Weyland* le Fevre 1318 Pat (Sf); John *Weland'* 1194 Cur (Sf); William *Welond* 1272 FFY; William *Weyland* 1327 SRSo. OG *Weland*. (ii) Haldan *de Wegland* 1196–1203 RegAntiquiss; Thomas *de Weylaund* 1280 CartAntiq; Robert *de Weylond* 1327 SRSo. From Wayland Hundred (Nf).

**Waylatt, Waylett:** Peter *Atteweilete* 1275 RH (Sf); Gundreda *atte Weylete* 1280 Bart (Ess); Robert *Weylate* 1327 *SR* (Ess). 'Dweller by the cross-roads' (OE *weg-gelǣte*).

**Wayler:** *v.* WHEELER

**Wayman:** *v.* WAITHMAN, WYMAN

**Waymark:** *v.* WYMARK

**Waymont:** *v.* WYMAN

**Wayne:** *v.* WAIN

**Wayre, Weyer:** Henry *ate Wayhore* 1296 SRSx. 'Dweller by the bathing-pond or stream', ME *weyour*, AFr *wayour*, OFr *gayoir*.

**Wayson:** *v.* WACE

**Wayt(e), Waytes:** *v.* WAIT

**Waythe:** *v.* WATH

**Weakley, Weekley:** Adam *de Wykele* 13th AssNth; Thomas *Weekelie* 1642 PrD. From Weekley (Nth).

**Weald:** *v.* WALD

**Wealder:** William *le Weildere* 1305 AssW. 'Dweller in the Weald', a derivative of OE *weald*.

**Wealthy:** *v.* WALTHEW

**Wear, Weare, Wears, Weir, Were, Where:** Peter *de la Were* 1242 Fees (He); John *atte Wer* 1332 SRSx. OE *wer* 'weir, dam, fish-trap'. 'Dweller by a dam' or 'keeper of the fishing-weir', fisherman.

**Wearham:** v. WAREHAM

**Wearing:** v. WAREING

**Wearn(e):** v WARNE

**Weary:** Hugh *Wery* 1221 AssWo; Ralph *Wery* 1275 SRWo; John *Wery* 1332 SRDo. OE *wearg* 'wicked, acursed'. cf. Jordan *le Cursede* 1284 Wak (Y).

**Weasel, Wessel, Wessell, Wessels:** William *Wesele* 1193 P (L); Matilda *Wesel'* 1206 AssL. A nickname from the weasel, OE *wesle*. cf. John *Wesilheued* 1384 AssL 'weasel-head'.

**Weate:** v. WHEAT

**Weather, Weathers:** Almer *Wether, Weder* 1180–2 P (K); William, John *le Wether* 1327 *SR* (Ess), 1332 SRSx. OE *weðer* 'wether, sheep'.

**Weatherall, Weatherell, Weatherhill, Weatherill, Weatherilt, Weatheritt:** v. WETHERALD

**Weatherby:** v. WETHERBEE

**Weathercock, Wethercock:** John *Wedercok* 1196–1237 Colch (Ess); John *Wethercok'* 1230 P (Lo); Peter *Wedircoke* 1327 SRC. OE *wedercoc* 'a weathercock', probably in the sense 'changeful, fickle'.

**Weatherhead:** v. WETHERHEAD

**Weathrall:** v. WETHERALD

**Weaver, Weavers, Weafer, Wheaver:** (i) John *le Weuere*, William *Weuere* 1296 SRSx. A derivative of OE *wefan* 'to weave', 'a weaver'. (ii) Simon *de Wevere* 1259 AssCh. From Weaver Hall (Ches).

**Weavill:** v. WAVELL

**Webb, Webbe:** Alger *se Webba* c1100–30 OEByn (D); Jernagan, Osbert, Wigor *Webbe* 1221 *ElyA* (Sf); Elyas *le Webbe* 1255 Oseney (O); Alice *la Webbe, Webbes* 1337 ColchCt. OE *webba* (m) or *webbe* (f), 'weaver'.

**Webber:** John *le Webber* 1255 FFEss; Hugo *le Webbere* 1327 SRSf. A derivative of OE *web* 'web', 'weaver' (c1440 NED).

**Webley:** Thomas *de Webbele* 1308 FFEss; Richard *de Webbeleye* 1363 IpmGl; William *Webley* 1663 HeMil. From Weobley (He), *Webbeleye* 1242.

**Webman:** Walter *Webbemon* 1382 AssWa. Probably for a weaver, OE *webb, mann*.

**Webster:** John *le Webestere* 1275 RH (Nf); William *le Webbester* 1284 AssLa. OE *webbestre* 'female weaver', though usually used of men. v. WEBB.

**Wederell:** v. WETHERALD

**Wedge:** v. WEGG

**Wedgewood, Wedgwood:** William *Weggewode* 1370–1 FFSr; William *Weggewode* 1424 Landwade; John *Wedgwood* 1642 PrD. From Wedgwood (St).

**Wedlake, Wedlock, Widlake:** . . . *Wedlocke* 1674 HTSf. From Wedlake in Petertavy (D).

**Wedmore:** William *Wedmor* 1276 AssW; Richard *Wedmore* 1641 PrSo. From Wedmore (So).

**Weech, Week(e)s:** v. WICH

**Weed, Weeds:** William *Wede* 1275 RH (Do); William *le Weed* 1296 SRSx. OE *wēd* 'fury, rage, madness'.

**Weeden, Weedon:** Robert *de Wedonia* c1160 EngFeud (Nth); Ralph *de Wedon'* 1207 Pleas (Bk); William *Wedon* 1396–7 FFWa. From Weedon (Bk), or Weedon Beck, Lois (Nth).

**Weeds:** v. WEED

**Weekley:** v. WEAKLEY

**Weems:** v. WEMYSS

**Weeper:** Henry *le Weper* 1237 Oseney; John *le Wepere* 1284 Balliol; William *Weper* 1428 *ERO*. Presumably a derivative of OE *wēpan* 'to weep', the weeper.

**Weet:** v. WHEAT

**Weetch:** v. WICH

**Weetman:** v. WHATMAN

**Wegg, Wegge, Wedge:** William *Wegge* 13th Cust; John *Wegge* 1327 SRSo; Robert *Wegge* 1391–2 FFSr. OE *Wegga*. v. PNDB 460.

**Weigall, Weighell, Weighill, Wiggall:** Katerina *de Wyghehale* 1379 PTY. From Wighill (WRYorks).

**Weigh:** v. WAY

**Weight:** v. WIGHT

**Weightman:** v. WIGHTMAN

**Weighton:** Godfrey *de Wycton'* 1219 AssY. From Little, Market Weighton (ERY), *Wihtun* 1156.

**Weild:** v. WALD

**Weir:** v. WEAR

**Welbeck:** Oliuer *de Wellebec* 1202 P (Nt). From Welbeck (Nt).

**Welbelove, Wellbeloved:** v. WELLBELOVE

**Welbourn, Welbourne, Welbon, Welburn, Wellborne, Wellburn:** Ailmer *de Welleburnia* 1185 Templars (L). From Welborne (Norfolk), Welbourn (Lincs) or Welburn (NRYorks).

**Welby, Wellby:** Thomas *de Wellebi* 1202 P (L); Henry *de Welleby* 1332 SRWa; Geoffrey *Welby* 1395 AssL. From Welby (Leics, Lincs).

**Welch:** v. WELSH

**Welchman:** v. WELSHMAN

**Welcome, Welcomme, Wellcome, Wellicome, Willicombe:** (i) Hynulfus *de Welecumb'* 1221 AssWa; William *de Wellecumb'* 1275 AssSo. From Welcombe (Devon) or Welcombe in Stratford (Warwicks). (ii) Picotus *Wilicom* 1279 RH (C); Roger *Welcum* Ed 3 Rydware (St). OE *wilcume* 'welcome'. (iii) Ralph *Welikempt* 1275 SRWo; Lovekyn *Welikembd* 1285 *Ass* (Ess); Edmund *Welkemt* 1332 SRWa. 'Well-kempt, well-combed', past part. of OE *cemban* 'to comb'. The original reference would be to well-kept hair or a well-trimmed beard, later to one of elegant appearance. The nickname was not uncommon and *Welkemt* would easily become *Welcome*.

**Weld:** v. WALD

**Welding:** v. WALDING

**Weldon, Welden, Wellden, Welldon:** William *de Welledon'* 1197 P (Nth); Henry *de Weldon* 1327 SRWo; William *Weldon* 1468 IpmNt. From Weldon (Northants), or 'dweller by a hill near a stream'.

**Weldrick, Wildrake:** Adam *de Queldrik* 1272 FrY. From Wheldrake (ERYorks).

**Weldsmith:** v. WILDSMITH

**Welfare, Welfear:** Simon *Welfare* 1275 RH (Nf). OE *\*wel-faru* 'well faring'. cf. FAREWELL.

**Welfitt:** William *Welifed* 1195 P (D); John *le Welyfed* 14th AD iii (Sx); William *Wellefed* 1379 PTY; William *Welfed* 1411 Finchale; Jeremiah *Welfitt* 1664 FrY. 'The well-fed', a nickname not uncommon and widely distributed. *Welfitt* might also be for WELFORD.

**Welford:** Ralph *de Welleford* 1190 P (Gl); Geoffrey *de Welleforde* 1313 Balliol (O). From Welford (Berks, Glos, Northants).

**Welham, Wellam, Wellum:** John *de Weleham* 1276 RH (Lei); Adam *de Wellome* 1363 FrY; Peter *Wellam* 1483 FrY. From Welham (Leics, Notts, ERYorks). *Welham, Wellam* and *Wellom* are common variants (1560–1617) in the Horringer Parish Register (Suffolk).

**Welland:** Thomas *de Welaund* 1284–5 FFSr; James *Welland* 1642 PrD. From Welland (Wo).

**Wellbelove, Wellbeloved, Wellbelow, Wellbeluff, Welbelove, Welbeloved:** Richard *Wellbyloved* 1327 SRWo. Self-explanatory.

**Wellborne, Wellbourn:** v. WELBOURN

**Wellby:** v. WELBY

**Wellcome:** v. WELCOME

**Wellden, Welldon:** v. WELDON

**Wellemin:** v. GILLMAN

**Weller, Wheller:** Richard *le Weller* 1272 FFSx, Gregory *Wellor* 1332 SRCu. A derivative of OE *wiellan, wellan* 'to boil', a salt-boiler. May also mean 'dweller by the spring or stream'. v. WELLS.

**Wellerman:** v. GILLMAN

**Wellesley, Wellersley, Wellsley:** Christiana *de Welleslegh* 1220 Cur (So); William *de Wellesleye* 1252 IpmSo; Robert *Wellesleghe* 1327 SRSo. From Wellesley in Wells (So).

**Welliam:** v. WILLIAMS

**Wellicome:** v. WELCOME

**Wellington:** Roger *de Welington* 1209 ForSa; John *de Welington* 1332 Glast (So); Thomas *Wellyngton* 1524 SRD. From Wellington (Ha, Sa, So).

**Wellinow:** (i) Robert *Welynogh* 1260 AssC; William *Welynough* 1327 SREss; John *Welenowe* 1414 FFEss. 'Well enough', OE *wel, genōh*. cf. John *Welgoud* 1332 SRDo 'very good'; John *Weliwar* 1317 AssK 'well aware'; Juliana *Welykenid* 1327 IpmGl 'well begotten'. (ii) Geoffrey *Weliknowe* 1302 AssW; Alice *Weliknowe* 1327 SRSo. 'Well known', OE *wel*, and *cnāwen*.

**Wellisch, Wellish:** v. WELSH

**Wellman, Welman:** Simon *le Welleman* 1308 MESO (Nf); John *Welman* 1327 SR (Ess). 'Dweller by a well or stream' (OE *wella*). v. also GILLMAN.

**Wellock:** Emma *Walhoc* 1313 Wak (Y); William *Walok* 1379 PTY; John *Wellock* 1707 WRS. A diminutive of a personal name in *Wealh-, Walh-*. v. also WHEELOCK.

**Wells, Wels:** Toke *de Welles* 1177 P (Nf); Roger *Attewell* 1200 Cur (Sx); Isabella *Welles* 1312 ColchCt. From Wells next the Sea (Norfolk) or Wells (Som), or from residence near a group of springs (OE *wella*). The singular *atte, del, de la Welle* is very common in medieval sources and survives in ATTWELL 'dweller by a spring or stream' but the modern *Well* seems to have been completely absorbed by *Wells*.

**Wellsley:** v. WELLESLEY

**Wellsman:** v. WELSHMAN

**Wellstead, Wellsteed, Wellsted, Wellstood, Welstead, Welsted, Willstead, Willsteed:** (i) Robert *de Wellested* 1305 LLB C. If the form is correct, this is 'dweller by the site of the well' and is unidentified. (ii)

Walter *atte Wellesheuede, atte Willeshefde* 1327, 1333 MELS (So). From Wellshead near Exford (Som). The variety of modern forms suggests that the name was not uncommon, 'dweller at the upper end of the stream', from OE *hēafod* 'head' and *wella*, preserving also the Somerset dialect *will* 'well'.

**Wellum:** v. WELHAM

**Wellwick, Welwick:** William *de Wellewyk* 1337 FFY. From Welwick (ERY).

**Welm:** Henry *atte Welme*, Stephen *de la Welme* 1275 MELS (Wo). 'Dweller by the spring or stream', OE *wielm, welm*.

**Welman:** v. WELLMAN

**Welsford:** Roger *de Wellesford* 1327 SRSo. From Welsford in Hartland (D), or Wellisford in Langford Budville (So).

**Welsh, Welch, Wellisch, Wellish:** Simon *Welsche* 1279 RH (Beds); Margery *Wellis* 1327 SR (Ess); Roger *Welch* 1334 ColchCt. ME *welshe*, OE *wēalisc, wēlisc*, 'foreign, Welsh', cf. WALSH.

**Welshman, Welchman, Wellsman, Welsman:** Davye *Welchman* 1544 Bardsley; John *Welsheman* 1564 ib. 'Welshman'. cf. WALSHMAN.

**Welstead, Welsted:** v. WELLSTEAD

**Welton:** William *de Welletun* c1155 Gilb; Isabella *de Welton* 1296 SRSx; Thomas *Welton* 1411–12 FFSr. From Welton (Cumb, Lincs, Northants, Northumb, Som, ERYorks).

**Welwick:** v. WELLWICK

**Wemyss, Weems, Wemes:** Michael *de Wemys* 1261 Black; David *Wemyss* 1286 ib.; David *de Wemes* 1314 ib. From the lands of Wemyss (Fife).

**Wenborn, Wenbourne, Wenban, Wenbon:** Gilbert *de Waneburne* 1296 SRSx; John *Wenbourne* 1410 CtH; Katherine *Wenborn* 1510 Ct. Thomas *Wenborne* 1772 SalehurstPR (Sx) signed the marriage register as *Wenban*. From Wenbon's Farm in Wadhurst (Sussex).

**Wenche:** Godwin *Wenche* 1204 Cur (D); Robert *le Wenche* 1278 AssSt; William *Wenche* 1332 SRWa. OE *wencel* 'child'.

**Wende:** William *Wende* 1275 RH (Nf); Roger *atte Wend* 1297 SRY. 'Dweller by the bend', OE *\*wende*.

**Wenden, Wendon:** John *de Wendon* 1179 P (Ess); William *de Wendenne* 1207 FFEss; Richard *ate Wendene* 1341 MELS (Sx). From Wendens Ambo, or Wendon Lofts (Ess).

**Wender:** Robert *Wender* 1332 SRCu. 'Dweller by a bend.' v. WENDE.

**Wendout, Wendut:** Richard *Wendhut* 1275 SRWo; John *Wendout* 1332 SRSx; Richard *Wendout* 1379 PTY. 'Go out', OE *wendan, ūt*, probably a nickname for a messenger. cf. Hugh *Wendowai* 1332 SRCu 'go away'.

**Wendover:** Richard *de Wendovr'* 1214 Cur (Bk); John *de Wendovre* 1341 FFEss; Nicholas *Wendover* 1470 FFEss. From Wendover (Bucks).

**Wenham:** William, Hawise *de Wenham* 1194 P (Sr), 1327 SRSo; Robert *Wenam* 1525 SRSx. From Wenham (Suffolk, Sussex).

**Wenlock:** John *de Wenlac* 1203 Pl (Sa); John *de Wenlak* 1323 IpmGl; William *Wenlocke* 1421 IpmNt. From Wenlock (Salop).

**Wenman:** v. WAINMAN

**Wenn:** John *atte Wenne* 1316 SR£o; Walter *atte Wen* 1327 SRSf. OE *wenn* 'wen, wart', used topographically of a hill or barrow and, as a surname, of a dweller by such.

**Wennell, Whennell:** Gilbert *Weynild* 1327 SRSf. OE *\*Wynhild* (f) 'joy-war'.

**Wennock:** Martin *de lâ Wennok* 1255 MELS (Sx). 'Dweller by the small hill', OE *\*wennoc*.

**Wensley, Wensly:** Jordan *de Wandesleye* 1204 AssY; Walter *de Wendeslay* 1339 FFY; John *Wensley* or *Wenslowe* 1509 LP (Y). From Wensley (Derby, NRYorks).

**Went, Whent:** Henry *de la, ate Wente* 1275 RH (Sf); Stephen *ad le Wente* 1279 RH (C); John *Wente* 1327 SRC. 'Dweller by the cross-road', ME *went(e)*.

**Wentbridge:** Robert *Wentbrigg* 1375 FFY. From Went Bridge in Snaith, in Kirk Seaton (WRY).

**Wentworth:** Robert *de Wintewrth'* 1219 AssY; William *de Wynteworth* 1303 IpmY; Richard *Wenteworth* 1422 IpmNt. From Wentworth (Cambs, WRYorks), both *Winteworde* DB.

**Were:** v. WEAR

**Werlock:** v. WARLOCK

**Werman:** For WARMAN or, possibly, WORKMAN.

**Werren, Werring:** Richard *Weryn* 1332 SRSx. OG *Werin*, *Guerin*, more commonly found as *Warin*. v. WAREING. cf. *Warinus*, *Werinus* (identical) LVD.

**Werry, Warry, Warriss:** *Guericus*, *Gueri* 1086 DB (Nf, Lo); *Werri* de Marinis 1166 RBE (Y); *Werreis* de Pilledona 1179 P (Do); *Werricus*, *Warricus*, *Werrei*, *Werreys* de Cadamo 1219–20 Cur (Sf); Nicholas, Geppe *Werri* 1206 P (C), 1228 FeuDu; John *Warry* 1260 AssC. OG *\*Werric*, OFr *Guerri*.

**Werth:** v. WORTH

**Werwell, Wherwell:** Richard *de Werewelle* 1285 WiSur. From Wherwell (Ha).

**Wesbroom:** v. WESTBROOM

**Wesby:** v. WESTBY

**Wescomb(e):** v. WESTCOMBE

**Wescot(t):** v. WESTCOTT

**Weske:** v. VECK

**Wesker:** Robert *Westker* 1275 RH (Nf). 'Dweller by the west fen', from ME *kerr* 'bog, fen', ON *kjarr* 'brushwood'. v. also WISHART.

**Weskett:** v. WESTCOTT

**Wesley, Westley, Westly:** Wlmar *de Westle* c1095 Bury; Adam *de Westeleg'* 1242 Fees (La); Richard *de Westleye* 1332 SRWa; William *Westly*, Robert *Wesly* 1674 HTSf. From Westleigh (Devon, Lancs), Westley Waterless (Cambs), or Westley (Suffolk).

**Wessel, Wessell, Wessels:** v. WEASEL

**Wesson:** v. WESTON

**West:** Maurice *de West* 1152–70 Colch (Ess); Goche *West* 1197 P (Nf); William *del West* 1262 *For* (Ess); Robert *bi Westen* de Copford 13th Colch (Ess); John *Bywest* de Welde LLB D; John *in le West* 1379 PTY. 'Man from the west' or 'dweller to the west' of a certain place.

**Westaby:** v. WESTOBY

**Westacott:** v. WESTCOTT

**Westbourn, Westbourne:** William *Westbourne* 1443 CtH; Agnes *Wesbourne* 1576 SRW. From Westbourne (Mx, Sx).

**Westbrook, Westbrooke, Westbrock:** Reginald *de Westbroc* 1255 FFK; John *de Westbroke* 1327 SRC; Edmund *Westbrook* 1440 FFEss. From Westbrook (Berks, Devon, Kent, Wight).

**Westbroom, Wesbroom:** Æilmer *de Westbrom* c1095 Bury (Sf). From a lost place in Woolpit (Sf).

**Westbury:** William *de Westbir'* 1208 Fees (Wo); John *de Westbyr'* 1242 Fees (Ha). From Westbury (Bk, Ha, Sa, So, W), or Westbury on Severn, on Trym (Gl).

**Westby, Westbey, Wesby:** William *de Westebi* 1219 AssL. From Westby (Lancs, Lincs, WRYorks). v. also WESTOBY.

**Westcastle:** Alan dictus *biwestcastel* 1263–4, Alan *Biwestcastel* 1274–5 RegAntiquiss. 'Dweller to the west of the castle', OE *bī*, *westan*, *castel*.

**Westcliff, Westcliffe:** John *de Westclive* 1219 P (K). 'Dweller by the western slope', OE *west*, *clif*.

**Westcoate:** v. WESTCOTT

**Westcombe, Wescomb, Wescombe:** (i) Henry *de Westecomb* 1333 PN D 436. From one of the three Westcombes in Devon. (ii) Thomas *atte Westcompe* 1327 SRSx. 'Dweller by the west field', OE *camp*.

**Westcott, Westcoate, Westacott, Westicott, Waistcoat, Wescot, Wescott, Weskett:** Turbertus *de Westcota* 1170 P (Berks); Robert *de Westecote* 1221 AssSt; Robert *atte Westcote* 1327 SRSx. From Westcot (Berks), Westcote (Glos, Hants, Warwicks), Westcott (Bucks, Surrey, Devon), Westacott (Devon), or from residence at the west cottage.

**Westdale:** Denis *Westdále* 1419 IpmY. From West Dale in Hunmanby (ERY).

**Wester:** Laurence *le Westre* 1327 SRSx. 'The westerner.' OE *\*wester*.

**Westerby:** From Westerby (Leics) or for WESTOBY.

**Westerdale:** John *de Westerdale* 1327 SRY; James *Westerdale* 1406, Nicholas *Westerdall* 1419 IpmY. From Westerdale (NRY).

**Westerman:** Richard *Westerman* 1383 AssL; Robert *Westerman* 1621 SRY; Arthur *Westerman* 1672 HTY. 'The man from the west', OE *westerne*, *mann*.

**Western, Westren, Westron:** Geoffrey *le Westerne* c1172–80 DC (L); Richard *le Westerne*, *le Westerneys* 1286 ForSt; Adam, Richard *le Westren* 1296, 1307 Wak (Y); Alan *Westerne* 1327 SRSf. OE *westerne* 'western', the man from the west.

**Westfield:** John *de Westfeld* 1276 RH (Beds); John *de Westfeld* 1376 IpmLa; William *Weistfeld* 1642 PrD. From Westfield (Nf, Sx), or 'dweller by the west field', OE *west*, *feld*.

**Westgate:** William *de Westgat'* 1199 Cur (Sr). From Westgate (Durham, Kent, Northumb, Wilts). Peter *de Westgate* of Canterbury (1198 P) no doubt lived near the west gate of the city.

**Westhead:** Robert *del Westheved* 1313 FFLa. From Westhead (Lancs) or one who lived near the west headland (OE *hēafod*).

**Westhope, Westop, Westopp:** John *de Westhop'* 1255 RH (Sa). From Westhope (Sa).

**Westhorp, Westhorpe:** Hugh *de Westorp* 1194 P (Db). From Westhorpe (Suffolk). cf. WESTROP.

**Westicott:** *v.* WESTCOTT

**Westie:** Andrew *de Westheie* 1185 Templars (Ess); Coleman *de Westhaie* c1188 BuryS (Sf); Peter *de Weysthagh* 1296 SRSx. From Westhay (Northants) or from residence near some western enclosure (OE *(ge)hæg*).

**Westington:** William *a Westethun*, *de Westetoune*, 1296, 1332 SRSx; Ralph *Biwestetoun* 1327 SRSx; William *Westetoun* 1332 SRSx. This was a common type of surname in Sussex, found in some 22 parishes between 1296 and 1332. Occasionally *west in tūne*, the man who lived 'west in the *tūn*', but usually from residence at a place so called, *Westi(n)ton, Westeton*, which occasionally became *Weston*. *v.* SINTON.

**Westlake, Weslake:** Martin *de Westlak'* 1211 Cur (St). From residence 'west of the streamlet'. Richard *Bywestelake* lived at Westlake (Devon) in 1333 (PN D 274).

**Westland:** Robert *de Westlonde* 1296 SRSx; John *Westland* 1379 PTY; Sibota *Westland* 1408 IpmY. 'Dweller by the west field', OE *west, land*. The Sussex example is from Westland Fm in Petworth (Sx).

**Westley, Westly:** *v.* WESLEY

**Westmacott, Westmancoat:** Elena *de Westmecote* 1327 SRWo. From Westmancote (Worcs).

**Westman:** Herebertus *filius Westman*, Herebert *Westman* 1148 Winton (Ha); Elyas *Westman* 1200 P (Ha). ON *Vestmaðr* 'man from the west', especially from Ireland, as a personal name.

**Westmead:** Simon *Westmed* 1327 SRY. 'Dweller by the west meadow', OE *west, mǣd*.

**Westmore:** Isabell' *de Westemere* 1332 SRSx. 'Dweller by the west mere.'

**Westmoreland, Westmorland, Westmarland:** Richard *de Westmerland* 1242 AssDu; William *de Westmarlonde* 1375 SRWo; John *Westmerland'* 1379 PTY. 'The man from Westmorland.'

**Westnage, Westnedge, Westnidge, Westney:** *v.* WASTNAGE

**Westoby, Westaby:** Tomas *Westibi* 1203 P (Y); Ernebertus *Westiby* 1208 FFL; Peter *Westinby* 1297 SRY; John *Westerbe* 1500 GildY; John *Westobie* 1607 FrY; William *Westaby* 1630 YWills. Scand *vestr i bý* (the man who lived) 'west in the village'. *Westinby* is partly anglicized. *v.* also SOTHEBY, EASTERBY, DUNNABY, and NORDABY. *Northinby* 1297 SRY, Robert *Northibi* 1301 SRY, John *Northobe* 1535 GildY, Nicholas *Northaby* 1617 FrY, which does not seem to have survived. For a similar English formation, *v.* SINTON.

**Weston:** (i) Godwinus *de Westuna* 1086 DB (Hu); Adestan *de Westuna* 1086 InqEl (C). From one of the numerous Westons. (ii) Adam *de la Weston* 1275 SRWo; William *Weston* 1296 SRSx; Alan *ate Weston* 1327 SRSx. 'Dweller at the west farm.' (iii) Robert *Biweston* 1200 P (Ess). OE *(be) westan tūne* (one who lived) 'to the west of the village'. *v.* WESTINGTON.

**Westop, Westopp:** *v.* WESTHOPE

**Westover:** John *de Westouer* 1296 SRSx; John *Westhouer*, Richard *Westouer* 1327, 1332 ib. From Westover (Som, Wight), Westovers in Twineham (Sussex).

**Westrap:** *v.* WESTROP

**Westray:** *v.* WESTRICH. Also from Westray (Cumb): Roger *de Westwra* 1292 PN Cu 384.

**Westren:** *v.* WESTERN

**Westrich, Westray:** Richard *le Westrais* 1206 AssL; William *le Westreys* 1219 Cur (Nth); Philip *Westrays* (*le Westreis*) 1230 P (Wa); John *Westrey* 1372 ColchCt. OFr *westreis* 'the westerner'.

**Westron:** *v.* WESTERN

**Westrop, Westrope, Westropp, Westrap, Westrup:** Gilbert *de Westrop* 1297 MinAcctCo (W); Martha *Westhrope* 1599 SfPR; Frances *Westroppe* 1624 ib. A metathesized form of WESTHORP, surviving in Westrop (Wilts).

**Westward:** Robert *a Westeward* 1279 RH (O). Dweller 'to the westward'.

**Westway:** George *Westway* 1642 PrD. 'Dweller by the west road', OE *west, weg*.

**Westwell:** Ranulf *de Westwelle* 1192 P (Ha); Thomas *de Westwell'* 1212 Cur (O); Henry *de Westwelle* 1297 MinAcctCo. From Westwell (K, O).

**Westwick:** Serlo *de Westwic'* 1219 AssY; Gamell *de Westwice* 1242 AssDu; Adam *Westwyke* 1304–5 IpmY. From Westwick (D, Du, Nf, WRY).

**Westwood:** Robert *de Westwod'* 1207 Pl (K); Alan *de Westwude* 1221 AssGl; John *Westwod'* 1371 CorEss. From Westwood (Kent, Wilts, Warwicks, Worcs), or 'dweller by the west wood'.

**Wetherald, Wetherall, Wetherell, Wetheril, Wetherill, Wetherilt, Weatherall, Weatherell, Weatherill, Weatherilt, Weatheritt, Weatherhill, Weathrall, Wederell:** Richard *de Wederhal* 1332 SRCu; William *Wethereld, Wetherald, Wetherold* 1429–33 Bacon (Sf); Elizabeth *Wetherill* 1648 Bardsley. From Wetheral (Cumb).

**Wetherbee, Wetherby, Weatherby, Witherby:** Ivo *de Werreby* 1219 AssY; Richard *de Wetherby* 1302 LLB B. From Wetherby (WRYorks).

**Wethercock:** *v.* WEATHERCOCK

**Wetherhead, Wethered, Weatherhead:** Thomas dictus *Wethyrhyrde* c1200 Black; Augustin *Wetherherde* 1214 FFK; John *le Wetherhirde* 1297 SRY; Henry *Weydurherd, Wedirhed* 1476 DbCh; George *Weddirheid* 1532 Black; Thomas *Weatherheade* 1634 LaWills. OE *weðer* 'wether, sheep, ram' and *hierde* 'shepherd'.

**Wetheril(l), Wetherilt:** *v.* WETHERALD

**Wetheringsett:** Robert *Wetheryngsete* 1408 FFEss. From Wetheringsett (Sf).

**Wetter, Wetters:** John *atte Wetere, Atwater* 1328 ArchC 33; Thomas *Wettas* 1524 SRSf. 'Dweller by the water.'

**Wetwan:** Richard *de Wetewang'* 1219 AssY. From Wetwang (ERYorks).

**Wetweather:** Ranulf *Wetweder* 1201 Pleas; Richard *Wetweder* 1392 IpmGl. A nickname, 'wet weather', OE *wǣt, weder*.

**Wevell, Wevill:** *v.* WAVELL

**Wewell:** *v.* WHEWELL

**Wey:** *v.* WAY

**Weyer:** *v.* WAYRE

**Weyland:** *v.* WAYLAND

**Weyman, Weymont:** *v.* WYMAN

**Whackum:** v. WAKEHAM

**Whaddon:** William *de Waddon'* 1182 NLCh (Bk). From Whaddon (Nk, C, Gl, W).

**Whaite(s):** v. WAIT

**Whale, Whales, Whall:** Hugh *le Whal'* 1249 FFEss; John *Whale* 1568 SRSf. OE *hwæl*, ME *whal* 'whale', used of any large fish including the walrus, grampus and the porpoise. The original sense was 'roller' and the name may refer to gait or to size and weight.

**Whaler:** v. WHEELER

**Whaley:** v. WHALLEY, WALEY

**Whalley, de Whalley, Whaly, Walley:** Adam *de Walleg'* 1185 P (La); Robert *de Walley* 1230 P (Nt). From Whalley (La) or Whaley (Db). v. WALEY.

**Wham:** Nicol *atte Wamme* 1296 SRSx; John *atte Whamme* 1327 ib. 'Dweller in the corner, angle or small valley' (OE *hwamm*).

**Whamond:** v. WYMAN

**Whant:** v. WANT

**Wharf, Wharfe:** Alan *ate Warf* 1320 LLB E. 'Worker at a wharf' (OE *hwerf*). cf. John *le Wharfager* 1322 LLB E.

**Wharin:** v. WAREING

**Wharram, Wharam:** Hog *de Warrum* 1219 AssY. From Wharram Percy, le Street (ERY).

**Wharrick:** v. WARWICK

**Wharrier:** v. WARRIOR

**Wharton:** Richard *de Wharton* 1324 IpmNt; Alexander, Thomas *Wharton* 1481 Past, 1672 HTY. From Wharton (Ches, Lincs, Westmorland). v. also WARTON.

**Whate:** v. WAIT, WHEAT

**Whateley, Whately, Whatley, Wheatleigh, Wheatley, Wheatly, Watley:** Lambert *de Watileia* 1084 GeldR (So); Martin *de Watelega* 1130 P (Nt); Roger *de Wetelea* 1182 P (St); Peter *de Watteleg'* 1196 P (Y); Elyas *de Hwatele* 1219 AssY; Peter *de Watteleg'* 1221 AssWa; Henry *de Whateleia* Hy 3 Colch (Ess); Walter *de Whetele* 1273 RH (Nt); Reginald *de Watele*, Robert *Watteleghe* 1296 SRSx. From Whateley (Warwicks), Whatley (Som), Wheatley (Essex, Lancs, Notts, Oxon, WRYorks), or Wattlehill Fm in Ewhurst (Sussex), all 'wheat-lea'. cf. WHITELEY.

**Whatham:** v. WALTHAM

**Whatkins:** v. WATKIN

**Whatlin, Whatling, Watling:** *Whatlingus, Watlin'* portarius, *Wathling* janitor 12th FeuDu; Edmund *Watlyng* 1524 SRSf. OE *\*Hwætling* 'son of *Hwætel'*, a diminutive of names in *Hwæt*- or a derivative of OE *hwæt* 'active, bold, brave'. The unrecorded *\*Hwætel* is the first element of Whatlington (Sussex).

**Whatman, Wheatman, Watman:** *Wateman, Wetman* 1066 DB (Do, He); Algar *filius Watman* 1168 P (D); *Wheteman* atte Brok 1359 LLB G; John *Watemon* 1264 Eynsham (O); Hugo *Weteman* 1279 RH (O); Richard *Whateman* (*Wateman*) 1327 *SR* (Ess); OE *Hwætmann* 'bold, brave man'. cf. WHATLIN.

**Whatmaugh, Whatmough, Whatmuff:** v. WATMOUGH

**Whatmoor, Whatmore, Watmore:** William *de Wetemor'* 1274 RH (Sa). From Whatmoor (Salop). Or for WATMOUGH.

**Whatnall, Whetnall, Whetnall, Whatner:** John *Watnowe* 1477, *Whatno* 1480, *Watno* 1512 FFEss. From Watnall (Nt).

**Whatrup:** v. WARDROBE

**Whattler:** v. WATLER

**Whatton:** John *Whatton* 1454 IpmNt. From Whatton (Nt), or Long Whatton (Lei).

**Whay:** v. WAY

**Whayman, Whaymand, Whaymond:** v. WYMAN

**Whealler:** v. WHEELER

**Whealock:** v. WHEELOCK

**Wheal(s):** v. WHEEL

**Wheat, Wheate, Whate, Weate, Weet:** Bernard *Hwette, Wete* c1157, 1180 Holme (Nf); Hugh, John *le Wete* 1249, 1327 PN Ess 483; Henry *Whete* 1440 ShefA; William *Whett, Wheate* 1484, 1563 ib. OE *hwæt* 'active, bold, brave'.

**Wheatcroft, Whitcroft:** Adam *de Wetecroft* 1191 P (Y); Richard *de Whatecrofth* 1327 SRSf; Thomas *de Whatecroft* son of Adam *Whetecroft* 1339 AssSt. 'Dweller by a croft where wheat was grown.' cf. Wheatcroft Fm in Cullompton (Devon), *Whitecroft* 1809 PN D 562.

**Wheater:** v. WHITER

**Wheatfill:** Ailmer *de Hwatefelde* 1121–38 Bury (Sf). From Whatfield (Suffolk), 'wheat-field'. Possibly also from Wheatfield (Oxon), 'white field'.

**Wheatland:** Richard *Whetlond* 1327 SRSx. 'Dweller by the wheat-land.' In Devon, Wheatland is from a family (*atte*) *Whytelond* 1281, 1333 PN D 375, 331, 'white land'.

**Wheatleigh, Wheatley, Wheatly:** v. WHATELEY

**Wheatlove:** Robert *Wetelof* 1221 ElyA (Sf); Richard *Wetelof* 1277 Ely (Sf); John *Whetelof* 1327 SRSf. 'Wheat loaf', OE *hwǣte, hlāf*, probably a nickname for a baker.

**Wheatman:** v. WHATMAN

**Weaver:** v. WEAVER

**Wheeker:** v. WICKER

**Wheel, Wheele, Wheels, Wheal, Wheale, Wheals:** Isabella *del Wele* 1297 MinAcctCo (Y); Hugh *atte Wheole* 1327 SRSo. One who lived near or was in charge of a water-wheel (OE *hwēol* 'wheel'). cf. WHEELHOUSE.

**Wheeldon, Wheelden, Wheildon, Whieldon, Wheldon, Wildon:** Hugo *de Hweldon'* 1279 RH (O); Peter *de Whilden* 1281 Pat (Bk). From Whielden Lane in Amersham (Bucks) 'curving valley' (PN Bk 212). Or from Wheeldon in North Huish (Devon), *Whyledon* 1520, *Wheldon* 1614 (PN D 304).

**Wheeler, Wheeller, Wheler, Whealler, Wailer, Wayler, Whaler:** Roger *le Weweler* 1249 MESO (Sx); John *le Whelare* 1275 SRWo; Hugh *le Welere* 1279 RH (C); Thomas *le Wegheler* 1284 FFSx; Stephen *le Whelere* 1317 FFEss; John *le Wheghler* 1327 SRSx; Gilbert *le Whygler* 1351 AD i (Sr). A derivative of OE *hweogol, hweowol, hwēol* 'wheel', 'wheel-maker, wheelwright' (1497 NED). *Wailer, Wayler*, both rare, are from *Wegheler; Whaler*, also rare, is from *Wheghler*, with spelling assimilated to that of the more common word.

**Wheeley:** William *de Wheleye* 1327 SRWo. From Weoley Castle (Wo).

**Wheelhouse:** William *de Whelehous* 1379 PTY; Martin *Whelas* 1617 FrY; Joseph *Wheelhouse* 1702 FrY. From residence near or employment at a wheelhouse, often, no doubt, near a dammed-up stream where the cutler ground his knives on a wheel driven by water. The name is especially a West Riding one. Sometimes equated with *Wheeler*: James *Wheelar*, inholder, son of John *Wheelas* alias *Wheelar*, flax-dresser 1740 FrY.

**Wheelock, Whellock, Whillock, Wellock, Whealock:** Richard *Whelelocke* 1582 Musters (Sr). From Wheelock (Ch). *v.* also WELLOCK.

**Wheelwright:** Brihtric *Hueluurihte* c1095 Bury (Sf); Hugo *le Wellewriche* 1256 AssNb; Walter *Welwryhte* 1274 RH (Ess); Henry *le welwrite* 1297 MinAcctCo; Richard *Whelwright* 1305 MESO (Y); Robert *le Whilwright* 1349 DbCh. OE *hwēol* 'wheel' and *wyrhta* 'wright', maker of wheels and wheeled vehicles (1281 NED).

**Wheildon, Wheldon:** *v.* WHEELDON

**Wheldale:** Thomas *de Queldale* 1349, John *de Wheldall* 1390 FrY. From Wheldale (WRY).

**Wheler:** *v.* WHEELER

**Wheller:** *v.* WELLER

**Whellock:** *v.* WHEELOCK

**Whelp:** Peter *Welp* 1253 ForNth; Nicholas *le Whelp* 1327 SRSf; Henry le Whelp 1334 SRK. OG *Hwelp*, ON *Hvelpr*, or a nickname from OE *hwelp* 'cub'.

**Whelpley:** Roger *de Whelpele* 1214 P (Nth); William *Whelpele* 1372 AssLo; Wylliam *Whelple* 1576 SRW. From Whelpley Hill in Ashley Green (Bk), Whelpley in Hailsham (Sx), or Whelpley in Whiteparish (W).

**Whenman:** *v.* WAINMAN

**Whennell:** *v.* WENNELL

**Whent:** *v.* WENT

**Where:** *v.* WEAR

**Wherwell:** *v.* WERWELL

**Whesker:** For *Wesker* or *Whisker*. *v.* WISHART.

**Whetman:** *v.* WHATMAN

**Whetnall:** *v.* WHATNALL

**Whetston, Whetstone:** William *de Wetstan* 12th DC (Lei); James *de Wetstan'* 1221 AssWa; John *Whetston* 1642 PrD. From Whetstone (Lei, Mx), or Wheston (Db), *Whetston* 1271.

**Whetter:** Richard *le Wetthere* 1332 SRSr. A derivative of OE *hwettan* 'to whet, sharpen', a sharpener of tools.

**Whewell, Wewell:** William *Wewel* 1296, John *le Wewel* 1332 SRSx. OE *hweogol* 'wheel'. Metonymic for a wheelwright. Probably also local: Adam *of Whewall* 1401 AssLa.

**Wheybrew:** *v.* WYBER

**Whicher:** *v.* WHITCHER

**Whick:** *v.* WICH

**Whicker:** *v.* WICKER

**Whickman:** *v.* WICKMAN

**Whidden, Whiddon:** Atwill *Whidden* 1642 PrD. From one or other of the seven places called Whiddon in Devon.

**Whiddup:** *v.* WIDDOP

**Whieldon:** *v.* WHEELDON

**Whight:** *v.* WIGHT

**Whild(e):** *v.* WILD

**Whilesmith:** *v.* WILDSMITH

**Whillis:** *v.* WILLIS

**Whillock:** *v.* WHEELOCK

**Whinfrey:** *v.* WINFREY

**Whinnerah, Whineray, Whinrae, Whinray, Winnery, Winrow:** Elizabeth *Whinnery* 1568 SRSf; John *Whinray* 1584 Bardsley (La); Hugo *Whynrowe* 1622 PrGR; John *Whinerall* 1682 ib. Probably from Whinneray in Gosforth (Cumb).

**Whinney:** John *de Whinhow* 1332 SRCu. 'Dweller on the whin-covered hill', perhaps Winnow in Thursby (Cumb), *Whinney, Whynna* 1589, or from Whinnah in Lamplugh or a Cumberland place named Whinny.

**Whinyates:** *v.* WINGATE

**Whipp, Whipps:** Alan *Wyppe* 1279 RH (C); Robert *Wyps* 1279 RH (O); William *Whyppe* 1329 IpmW. OE *Wippa*.

**Whippet, Whippett, Wippett:** Peter *Wypet* 1295 ParlR (Ess); Richard *Whippat* 1349 FFY; John *Whipet* 1478 ER 54. *Wipp-et*, a diminutive of OE *Wippa*.

**Whipple:** Thomas *Wypel* Ed I Battle; John *Whepyll* 1504 *PetreA*; George *Wipple* 1642 PrD. *Wipp-el*, a diminutive of OE *Wippa*.

**Whiscard, Whisker:** *v.* WISHART

**Whisken, Whiskin:** *v.* WISKEN

**Whissell:** (i) William *Wystle* 1247 AssBeds; Richard *Whistel* 1297 MinAcctCo (Y). OE *hwistle* 'pipe, flute', metonymic for WHISTLER. (ii) Simon *Whytsall* 1275 RH (Sx); John, William, Nicholas *Whitsaul(e)* 1279 RH (C). OE *hwīt* 'white' and *sāwol* 'soul'. The reason for the nickname is not clear. cf. GODSAL.

**Whisson, Wisson:** Gilbert *de Wiscand, de Wissand, de Witsand* 1086 DB (Sf); William *de Wisant* c1180 Bury (Sf); William *de Whitsand* 1197 P (Y); William *Whitsond* 1242 Fees (W); William *de Wytsonde* 1296 SRSx. From Wissant (Pas-de-Calais). Some of the forms may have a different origin. cf. Thomas *de la Witesand* 1236 MELS (Sr). 'Dweller by the white sand.'

**Whistance, Wistance:** Cissota *de Wistones* 1309 Wak (Y). 'Dweller by the white stones.' From Whitestones (NRYorks).

**Whistlecraft, Whistlecroft:** Thomas *Wyslylcroft* (sic) 1524 SRSf. Probably 'dweller at the croft in the river-fork' (OE *twisla*), ME *at twiselcroft* becoming *at wiselcroft*.

**Whistler, Wissler:** Osbert *le Wistler* 1243 AssSo; John *Whiseller*, Nat. *Whisler* 1674 HTSf. OE *hwistlere* 'piper, fluter'. In 1279 AssSo the scribe has altered William *le Wyzelere* to *le Vylur* 'fiddler'. *v.* VIELER.

**Whiston, Wiston:** Robert *de Whiston* 1297 AssSt; Robert *de Wyston* 1379 PTY; William *Whiston* 1449 FFEss. From Whiston (La, Nth, St, WRY), or Wiston (Sx).

**Whistondale:** *Wytstayndale* occurs in 1290 (PN NRY 204) for Whitestones in Hawnby (NRYorks). cf. WHISTANCE.

**Whitaker, Whiteaker, Whittaker, Whitticase, Widdaker:** (i) Richard *de Wetacra* 1177 P (Nf); Furmentinus *de Whetacre* 1254 ArchC xii; Walter *Weat Aker* 1296 SRSx; John *de Wheteacre* 1327 SRSf.

From Wheatacre (Norfolk) or Whiteacre in Waltham (Kent), both 'wheat-field'. (ii) Simon *de Wit Acra* 1180 P (Wa); Robert *de Witacra* 1189 P (Nth); Richard *de Whitacre* 1336 WhC (La); Henry *Wyteacre* 1379 PTY. From Whitacre (Warwicks) or High Whitaker (Lancs), 'white field'.

**Whitamore:** *v.* WHITEMORE

**Whitbourn, Whitbourne, Whiteborn:** (i) Thomas *de Wytebourne* 1327 SRSo. From Whitbourne (Hereford). (ii) Robert *Wytebarne* 1301 SRY; William *Whitbarn* 1438 LLB K. OE *hwīt* and *bearn* 'fair child'. cf. Richard *Whiteboye* 1277 AssSo, Roger *Wyitfelawe* 1264 AD iii (Beds).

**Whitbread, Whitebread:** (i) Roger *Wythbred* 1254–67 Rams (Hu), father of Richard *Witbred* 1267–85 ib.; William *Wytbred* 1275 SRWo; Robert *Witbred* 1279 RH (O); Robert *Whytbred* 1327 SRSf; Robert *Whytebrede* 1436 FFHu. OE *hwīt* and *brēad* 'white bread', used of a seller of white, i.e. the best bread, made of wheat. (ii) William *Watebred* 1221 ElyA (Ess); John *Wetebred* 1239 FrLeic; Robert *Whetbred* 1327 SRSx; John *Whatebred* 1327 *SR* (Ess). OE *hwǣte* and *brēad* 'wheat-bread'. For the development to *Whitbread*, cf. WHITAKER. (iii) William *Witberd* 1221 ElyA (Sf); Walter *Wyteberd* 1297 MinAcctCo; Adam *Whitberd* 1327 SRSx; John *Whitebeard* 1503 RochW. OE *hwīt* and *beard* 'white beard'.

**Whitby, Witby:** Tiece *de Witebi* 1181 P (Y); Thomas *de Whiteby* 1295 IpmY; Robert *Whytby* 1408–9 FFWa. Usually from Whitby (NRY), but sometimes from Whitby (Ch).

**Whitcher, Whicher, Witcher:** William, Richard *Wicher* 1176 P (Bk), 1279 RH (Beds); Robert *le Wiccher'* 1288 MESO (Sx); Richard *Whychere* 1327 SRSx; Robert *le Whicchere* 1333 MESO (Ha). A derivative of OE *hwicce* 'chest', a maker of chests. The early loss of the *h* is proved by the synonymous William *le Wyccewrichte* (1256 FFSo) who, no doubt, followed the same occupation as Richard *Le Wycher* 1305 MESO (So). As OE *wīc* became both *wike*, and *wiche* in ME, the surname may also mean 'dweller at a place called *Wich* (or *Wick*)' or 'dairy-farmer', as is proved by Peter *le Wycher* and John *Wych* (1327 SRWo) who were both assessed in the parish of Hambury juxta Wych. In view of the common interchange of *Wh* and *W*, this may also have become *Whi(t)cher*. Or, we may have 'dweller by the wych-elm enclosure', as at Witcha Fm in Ramsbury (Wilts), the home of Richard *atte Wycheheye* in 1332 (PN W 290).

**Whitchurch:** Leofric *æt Hwitecyrcan* a995 OEByn (D); William *de Witchurch* 1205 FFO; William *Whitchirche* 1413–14 FFWa. From Whitchurch (Bk, D, Ha, He, O, Sa, So, Wa), or Whitchurch Canonicorum (Do).

**Whitcomb, Whitcombe, Witcomb, Witcombe:** William *de Whitecumbe* 1201 P (So); Isoda *Wytecoumbe*, Robert *de Wythecoumb* 1332 SRSx. The first example is from Whitcombe (Som). As a place-name, Whitcombe 'wide or withy valley' is found in Dorset, Wilts and the Isle of Wight, and in Devon as 'wheat valley'. Witcombe 'wide or withy valley' occurs in Devon, Gloucester and Somerset. The surname may derive from any of these or from a lost place in Sussex and is sometimes identical in meaning with WIDDECOMBE.

**Whitcroft:** *v.* WHEATCROFT

**White, Whitt, Whyte, Witt, Witte, Witts:** (i) *Whita* 1066 DB (Sf); *Wit* filius Willelmi 1198 FFNf; Purcil *Hwita, Wite, Wita* 1038, 1066 OEByn (He); Ordgar *se Wite* c1070 ib. (So); Alestanus *hwit* 1066 Winton (Ha); Alwin *Wit* 1066, 1086 DB (Ha); Uuiaett *Hwite* c1097–1107 Black; Lewinus *Wite* c1114 Burton (St); Berwaldus *le White* Hy 2 DC (L); Hugo *Wit* 1190 BuryS (Sf); Walter *le Whyte* 1284 LLB A; William *le Wytt* 1327 SRY; John *le Whytt* 1327 SRSo. Occasionally this is from OE *Hwīta*, a short form of names in *Hwīt-*, but is more often a nickname from OE *hwīt* 'white', with reference to fair hair or complexion, a well-established name before the Conquest and very common thereafter. (ii) Ralf *de Wyte* 1279 AssSo; John *atte Wyte* 1296 SRSx. 'Dweller by a bend or curve' in a river or road (OE *\*wiht*), the source of Great Whyte (Hunts). *v.* WIGHT. *White* also appears to be a development of *wait.* cf. White (Devon), the home of Adam *atte Wayte* in 1330. This is explained in PN D 108 as 'ambush', 'place where one lies in wait'. It is more likely to denote 'a place where on watches', 'a look-out post'. cf. WAIT.

**Whiteaker:** *v.* WHITAKER

**Whitear:** *v.* WHITTIER

**Whiteaway:** *v.* WHITEWAY

**Whitebelt:** William *Wytbelt* 1277 Wak (Y); John *Whitebeld* 1379 PTY. 'The wearer of a white belt', OE *hwīt, belt.* cf. Robert *Witkirtel* 1316 Wak (Y) 'white kirtel'; William *Wythemantell'* 1279 AssNb 'white mantle'.

**Whitebread:** *v.* WHITBREAD

**Whitebrow:** John *Witbrowe* 1332 SRCu; William *Qwhitbrow* 1376 Black; Richard *Whitebrow* 1401 AssLa. 'With white eyebrows', OE *hwīt, brū.* cf. William *Whytebon* 1327 SRLei 'white bone'; Alice *Witfax* 1289 Misc (Y) 'white hair'; Robert *Wytfleis* 1223 Pat (Y) 'white flesh'; John *Whitshonk* 1401 AssLa 'white shanks'.

**Whitebuck:** John *Whitbok* 1313 AssNf. 'White buck', OE *hwīt, bucca.* cf. John *Whytebull'* 1379 PTY 'white bull'; Ralph *Witebullock* 1196 P (Co) 'white bullock'; Ralph *Whittecalf* 1340–1450 GildC 'white calf'; Adam *Witecolt* 1225 Cur (Berks) 'white colt'; Gilbert *Whitecou* 1327 SRY 'white cow'; John *Whytegos* 1334 SRK 'white goose'; John *Whitehorse* 1525 SRSx 'white horse'.

**Whitecock:** Adam *Whitcok* 1226 Cur (Berks); Robert *Wytcok* 1275 RH (W); Adam *Whitecok* 1327 SRSf. A diminutive of OE *Hwīta.*

**Whitefield, Whitfield:** Leonard *de Witefelde* 1154 Eynsham; William, John *de Whitefeld* 1230 P (So), 1338 CorLo; Richard *Whytefeld* 1396 IpmGl. From Whitefield (Glos, Lancs, Wight), Whitfield (Derby, Kent, Northants, Northumb), or 'dweller by the white field'.

**Whitefoot:** Ulfuine *Huitfot* c1095 Bury (Sf); John *Weytefot* 1327 SRC; Robert *Whitfott* 1311 ColchCt. A nickname. OE *hwītfōt* 'white foot'.

**Whiteford:** *v.* WHITFORD

**Whitehair:** *v.* WHITTIER

**Whitehall:** William *de Withalgh* 1332 SRLa; Gilbert *del Whithalgh* 1397 PrGR; James *Quhithall* 1585 Black. 'Dweller by the white nook or hall', OE *hwīt, halh/heall*.

**Whitehand:** Robert *Withand'* 1181 P (Nt); Richard *Whithand* 1204 P (Nf); Robert *Wytehand* SRY. A nickname. 'White hand.'

**Whitehart:** *v.* WHITTARD

**Whitehead, Whytehead:** John *Whithaued* 1219 Cur (Lei); Roger *Witheved* 1279 RH (Hu); William *Hwitheued* 1317 AssK; William *Whithefd* 1332 SRSx; Robert *Whithede* 1338 ShefA (Y). OE *hwīt, hēafod* 'white head', i.e. whitehaired or fair-haired. Occasionally from *hwīt, hōd* 'white hood': Agnes *Wythod* 1279 RH (O). Also occasionally local, from residence at the 'white head' of a field or hill; Henry *de Whiteheved* 1297 MinAcctCo.

**Whiteheard, Whiteheart:** *v.* WHITTARD

**Whitehorn, Whitehorne, Whithorn:** Martin *Withorn* 1275 RH (Sf); Thomas *Whithorn* 1327 SRSx. Clearly a nickname, perhaps one with a splendid trumpet or drinking-horn (OE *hwīt* 'white, fair, splendid' and *horn*).

**Whitehouse, Whithouse, Whitters:** (i) Stephen *atte Whitehous* 1327 SRSo. 'Dweller at the white (stone) house.' Whittas Park in Torpenhow (Cumb) is *Whitt house* in 1777 (PN Cu 326). (ii) William *Whitehals* 1369 FrY; Robert *Whithawse* 1551 ib.; Marmaducus *Whithaus* 1598 ib.; Anthonius *Whitus* 1615 ib. 'White neck.' cf. WHITTLES, SHORTHOSE.

**Whitehurst:** Richard *Whithurst* 1327 SRSx; Hawise *Whytehurst* 1387–8 Hylle. 'Dweller by the white wood', OE *hwīt, hyrst*.

**Whiteing:** *v.* WHITING

**Whitelam, Whitelum, Whitlam, Whitlum:** Alice *Whitlambe* 1379 PTY; Richard *Whitelam* 1488 GildY. 'White lamb', a nickname.

**Whiteland:** John *atte Wytlonde* 1296 SRSx; John *Whyteland* 1374 FFEss. From Whiteland in West Dean (Sx).

**Whitelark:** *v.* WHITELOCK

**Whitelaw, Whitlaw, Whytelaw, Whytlaw:** John *de Wytelowe* 1296 Black (Edinburgh); John *Whitelaw* 1430 ib. From Whitelaw (Morebattle, Bowden).

**Whitelegg, Whitelegge:** Agnes *Whytlegge* 1584 Bardsley. This cannot be identical with *Whiteley* as suggested by Bardsley, Harrison and Weekley. *-leg(ge)* is a common medieval form of *-ley* but never gives a modern *-leg* in place-names. The surname must be a nickname 'white leg'. cf. the ON nickname *Hvítbeinn* 'white leg', WHITEFOOT, WHITEHAND, WHITEHEAD, WHITESIDE and Robert *Wytfleis* 1223 Pat (Y); William *Witteflese* 1326 FeuDu 'white flesh'.

**Whiteley, Whitely, Whitla, Whitley, Whittla, Whittley, Witley:** William *de Witteleia* 1125 StCh; R. *de Witelay* 1190 P (Y); Hilda *de Whitelai* 1200 P (Nt); Henry *de Hwittele* 1221 AssWa; Richard *de Whiteleg* 1246 AssLa; John *del Wyteley* 13th Shef (Y). From Whitleigh (Berks) or Whitley (Ches, Northumb, Wilts, Warwicks, WRYorks). The place-name is common and all these are 'white *lēah*'; it is found in minor names as Whiteley (Devon) and Whiteley

Woods (Sheffield) and in many that have not survived. Occasionally it is 'wheat *lēah*'. *Whetelegh* (1333 PN D 344) is a surname derived from Whitley in Molland (Devon). Edith *de Wheteley* was probably of the same family as Thomas *de Whitteley* 1316 Wak (Y). cf. WHITAKER and WHEATELEY.

**Whitell:** *v.* WHITTLE

**Whitelock, Whitelocke, Whitlock, Whitlark:** (i) *Witlac* 1066 DB (L), *Wislak* ib. (Ha); *Willak* 1086 DB (Ha); *Withlac* de molend' 1202 P (Ha); *Wihtlac de Longo Vado* 1207 FineR (Ha); Emma *filia Witlok* 1279 RH (Hu); Toke *Wictlok* 1208 Cur (Nf); John *Witloc* 1243 AssSo; William *Whytlok* 1285 FFHu; John *Wyhtlok (Wytlok)* 1327 *SR* (Ess); John *Whytlak* 1524 SRSf; James *Whitlake* 1568 SRSf. Sometimes from OE *Wihtlāc* 'elf-play', but also a nickname from OE *hwīt* and *locc* 'white lock, white hair'. The Suffolk river Lark is either a back-formation from Lackford or was originally called *lacu* 'stream'. *Whitelark* may be due to a similar dialectal pronunciation of the 1524 *Whytlak*. cf. WOODLAKE. (ii) William *atte Whytelak* 1327 SRSo. 'Dweller by the white stream' (OE *lacu*).

**Whitelum:** *v.* WHITELAM

**Whiteman, Whitman, Wittman:** *Quithmanus* 1121–38 Bury (Sf); *Witeman fugitiuus* 1170 P (Gl); William *Witman* 1230 P (K); Richard *Wyteman* 1243 AssSt; Stephen *Whytman* 1243 AssSo; Michael *Whiteman* 1310 ColchCt; OE *Hwītmann*.

**Whitemore, Whitamore, Whitmer, Whitmore, Whittamore, Whittemore, Whittimore, Wittamore:** John *de Witemore* 1199 AssSt; Adam *de Whitemor* 1249 PN D 353; Gilbert *de Whitemere* 1275 SRWo. From Whitmore (Staffs), Whitmoor (Devon) or residence near a white moor or mere. *Whittamore*, etc. are Devonshire forms which do not seem to survive as place-names.

**Whitenow:** (i) Thomas *Wytynow* 1327 SRSo. 'White enough.' (ii) John *Whiteknafe* 1379 PTY. 'White, fair boy', or 'servant of White'. cf. GOODENOUGH. (iii) 'Dweller by the white knoll', as at Whiteknowe in Nicholforest (Cumb).

**Whiteoak:** *v.* WHITTOCK

**Whitepayn, Whitpayn:** John *Whytepayn* 1336 ChertseyCt (Sr); Thomas *Whytpayn* 1525 SRSx; Thomas *Whitepayne* 1642 PrD. 'White bread', OE *hwīt*, OFr *pain*, a nickname for a baker. cf. Roger *Witepese* 1183 P (K) 'white peas'.

**Whiter, Whitter, Whitta, Wheater:** John *Witer* 1181–6 Clerkenwell (Ess); Henry *le Witere, le Wytter'* 1221 AssWa; Andrew *le Whytere* 1310 ColchCt; Thomas *le Whittere* 1320 MESO (So). A derivative of OE *hwītian* 'to make white', 'whitewasher'. 'The Keep of the Tower of London was known as the White Tower from its being resplendent with whitewash.' Corfe Castle was all whitewashed outside. (Building 157). For *Wheater*, cf. WHITAKER.

**Whiteside, Whitesides:** Robert *Wytside* 1230 P (Wa); John *Witside* 1250 Fees (Ha); Richard *Whitside* 1279 RH (C). A nickname 'white side' (OE *hwīt* and *sīde* 'side, flank').

**Whitesmith:** Richard, William *le Wytesmith* 1260 Cl (O), 1279 RH (C); John *le Wytesmyt, le Whitesmyth* 1313, 1332 SRWa. OE *hwīt* and *smið* 'white-smith', a tin-smith (1302 NED).

**Whiteson:** v. WHITSON

**Whitestart:** John *Whytstert* 1327 SRSf. 'White tail', OE *hwīt, steort.* cf. Aldred *Quit Swire* 1207 Cur (Gl) 'white neck'; William *Whytetop* 1381 PTY 'white top', probably 'white-haired'.

**Whitestone, Whitston:** Leofwine *on Hwitastane* c1100–30 OEByn (D); Walter *atte Wyteston* 1279 PN Sx 8; John *atte Whitestone* 1332 SRSx. From Whitstone (Co, D), or Whitestone Fm in Bridham (Sx). In Scotland from Whitestone (Perth), or Whiteston (Aberdeen).

**Whiteway, Whiteaway:** William *Wyttewey, Whitewe* 1296, 1327 SRSx. From Whiteway House, Whiteway Barton (Devon), where the surname was *Whythaweye* in 1335 (PN D 491, 479), or from residence near a white road (OE *hwīt, weg*).

**Whitewhite:** Thomas *de Whitthuait* 1332 SRCu. 'Dweller by the white clearing' (OE *hwīt*, ON *þveit*).

**Whitewick, Whittick, Wittick:** Euarard *de Whitwic, de Witic,* 1208 AssY, Cur (Y). From Whitwick (Lei), pronounced *Wittick.*

**Whitewood, Whitwood:** William *de Whitewude* 1197 P (Y); Malger *of Wytewode* 1276 IpmY; Richard *Whytwoode* 1532 FFEss. From Whitwood (WRY), or 'dweller by the white wood', OE *hwīt, wudu.*

**Whitewright:** v. WHITTERIDGE

**Whitey, Whitie, Whittey, Whitty:** (i) John *Witege* 1186 Eynsham (O); Richard *Whiteye* 1276 RH (L). Either a nickname 'white eye' (OE *hwīt, ēage*) or OE *wītega* 'wise man, prophet, soothsayer'. (ii) Nicholas *de la Wytheg'* 1279 RH (O); Bartholomew *atte Withegh* 1332 SRSx; Edmund *de Whitehey* 1351 AssEss. 'Dweller by the white enclosure' (OE *hwīt, (ge)hæg*). v. also WITTY, with which there has, no doubt, been confusion.

**Whitfield:** v. WHITEFIELD

**Whitford, Whiteford:** Roger *de Witeford* 1221 AssWo; John *de Whyteford* 1327 SRWo; William *Whitford'* 1332 SRDo. From Whitford (D), or Whitford in Bromsgrove (Wo). In Scotland from Whiteford near Paisley (Renfrew).

**Whitgift:** William *de Wytegift* 1297 SRY; John *Whitgyft* 1415–16 IpmW; John *Whitgifte* 1441 TestEbor. From Whitgift (WRY).

**Whitgreave:** Robert *de Witegreve* 1155 StCh; William *de Whitegreve* 1293 AssSt; Robert *Whitgreve* 1460 IpmNt. From Whitgreave (St).

**Whithair, Whithear:** v. WHITTIER

**Whitham:** v. WITHAM

**Whithorn:** v. WHITEHORN

**Whithouse:** v. WHITEHOUSE

**Whitie:** v. WHITEY

**Whiting, Whiteing, Whitting, Witting:** Simundus *filius Witing* a1150 DC (L); Roger *Witenc* 1084 GeldR (So); John *Hwiteng, Witinge* 1128–34 Colch (Ess); William *Witting* 1194 Cur (Bk); William *Whiting* 1197 P (Bk); Giffard *Whytteng* 1230 P (So); Walter *le Witing* 1268 AssSo; Thomas, Matilda *Hwytyng* 1327 SRSx. OE *Hwīting*, a patronymic from OE *Hwīta* 'Hwita's son', a personal name still in use in the 12th century. Also, and probably more commonly, an original nickname from OE *hwīta* 'the white one'.

**Whitkirk:** William *Whytekyrke* 1408 IpmY. From Whitkirk (WRY).

**Whitla:** v. WHITELEY

**Whitlam:** v. WHITELAM

**Whitlaw:** v. WHITELAW

**Whitley:** v. WHITELEY

**Whitling:** (i) *Witling de Inskip* 1261 AssLa; William *Wytling* 1327 SRSf; John *Whitling* 1332 SRLa; Cecilia *Whytling* 1345 ColchCt. OE *\*Wihtling*, known only from the above example, but probably also the personal-name which lies behind Whitlingham (Norfolk). This Ekwall derives from OE *Wihthelm*, a compound not evidenced in the early forms: *Wislingeham* DB, *Wicthlingham* 1206 DEPN. OE *Wihtlāc* may well have had a short form *\*Wihtla*, from which was formed *\*Wihtling* 'son of Wihtla'. Or *Wihtling* may have been regarded as a pet-form of *Wihtlāc.* cf. the interchange of *Brihtling* and *Brihtric* in Brightlingsea (PN Ess 331). But, as forms are few and late, and show no sign of OE *-ht-*, we may have OE *\*Hwītling*, similarly formed from *Hwīta.* (ii) Walter *de Whittling'* 1221 AssWo. From Whitlinge (Worcs). v. WICKLING.

**Whitlock:** v. WHITELOCK

**Whitlow:** Robert *Whitloue* 1332 SRWa; John *de Whitlowe* 1415 PrGR. 'Dweller by the white hill', OE *hwīt, hlāw.* v. also WHITELAW.

**Whitlum:** v. WHITELAM

**Whitman:** v. WHITEMAN

**Whitmarsh:** Henry *de Wytemers* 1245 FFEss. 'Dweller by the white marsh', as at Whitemarsh Fm in Sedgehill or Witmarsh Bottom in West Dean (Wilts). v. PN W 192, 378.

**Whitmee, Whitmey:** Henry *Whitemay* 1255 RH (Sa); William *Wytemey* 1279 RH (O). ME *may* 'young man or maid'. The surname may mean 'fair youth' or may be a derogatory nickname 'fair maid'.

**Whitmer, Whitmore:** v. WHITEMORE

**Whitnall, Whitnell:** John *de Wytenhull* 1279 RH (O); William *Whitenel* 1327 SREss. From Whitehill Wood in North Leigh (O), or Witnells End in Upper Arley (Wo).

**Whitney:** William *de Wittenheia* 1210–11 PWi; John *de Whyteneye* 1327 SRSf; Emanuel *Whitney* 1642 PrD. From Whitney (Hereford), Whitney Wood in Stevenage (Herts), or 'dweller by or on the white island'.

**Whitpayn:** v. WHITEPAYN

**Whitrick, Whitridge:** v. WHITTERIDGE

**Whitson, Whitsun, Whiteson:** Nicholas *Witesone* Ed 1 Battle (Sx); Walter *Hwyttesone* 1318 AD iv (Sa); John *Whitsone* 1327 SRSx; William *Quhitsoun* 1369 Black (Perth). 'Son of White or Whitt.' Occasionally this may be a simplified pronunciation of *Whitston* 'white stone'. cf. Whitson Fm (PN D 243).

**Whitston:** v. WHITESTON

**Whitsunday:** William *Wytesoneday* 1273 RH (So). Probably a nickname for one born on that day.

**Whitt:** v. WHITE

**Whitta:** v. WHITER

**Whittaker:** v. WHITAKER

**Whittam:** v. WITHAM

**Whittamore:** v. WHITEMORE

**Whittard, Whitehart, Whiteheart, Whiteheard, Witard:** *Wiþardus de Foro* 1185 Templars (Ess);

Thomas *Wytard* 1275 SRWo; William *Whitherd*, Gilbert *Whitard* 1327 SRSx. Perhaps OE *Wihtheard* 'elf-brave', of which only one pre-Conquest example is known (in 825). cf. also *terra Wichardi* 1190 BuryS (Sf). Or OE* *Hwītheard*.

**Whittear:** *v.* WHITTIER
**Whittek:** *v.* WHITTOCK
**Whittell:** *v.* WHITETELL
**Whittemore:** *v.* WHITEMORE
**Whitter:** *v.* WHITER

**Whitteridge, Whittrick, Whitrick, Whitridge, Whitewright, Whitherick, Witrick, Witterick, Wittrick, Wittridge, Widrich:** (i) *Wihtric* (Sf), *Wictric, Witric* (Sa), *Wistricus* (St), *Wistrinc* (Ha) 1066 DB; John, William *Wickrik* 1276 RH (Berks); William *Wyterik* 1279 RH (C); Roger *Wythrich* 1327 SRSf; Simon *Wheterich* 1327 SRC; Henry *Wyhtrich* (*Whytrich*) 1327 *SR* (Ess). OE *Wihtrīc* 'sprite-ruler'. (ii) Robert *de Wyterik'* 1249 Cl (Cu); Robert *de Whiterig* 1332 SRCu. From Wheatridge (Northumb), and perhaps occasionally from one of the three places named Whitrigg in Cumberland; though the latter are from ON *hryggr*, forms in -*rik* are found.

**Whitters:** *v.* WHITEHOUSE
**Whittey:** *v.* WHITEY
**Whitticase:** For *Whittakers*: Thomas *Whitacres* 1526 FrY.
**Whittick:** *v.* WHITEWICK
**Whitticom:** *v.* WIDDECOMBE

**Whittier, Whittear, Whitear, Whitehair, Whithair, Whithear:** Ralf *Wittauuere* 12th AD ii (Nth); Norman *Wittowiere* 1224–46 Bart (Lo); Thomas *le Wytewere* 1279 RH (C); Eustace *le Wittowere* 1279 RH (Hu); William *le Wyttawyere* 1285 *Ass* (Ess); William *le Whythawere* 1309 SRBeds; John *Whitouer* 1364 LLB G; Grace *Whityer* 1634 Bardsley. 'White-leather dresser', one who taws skins into whitleather (1284 NED), from OE *hwīt* 'white' and ME *tawyere, towyere, tewere*, from WS *tāwian*, Anglian *tēwian* 'to taw'. *v.* MESO.

**Whittimore:** *v.* WHITEMORE
**Whitting:** *v.* WHITING
**Whittingham, Whittenham, Wittenham:** Vhtred *de Witingeham* 1163 P; Gilbert, Oliver *de Whitingham* c1214 Black, 1279 AssNb; Thomas *de Whytenham* 1339 CorLo; Adam *Whityngham* 1401 AssLa. From Whittingham (Lancs, Northumb), Whittinghame (East Lothian), or Wittenham (Berks).

**Whittington, Whitington:** Peter *de Witinton'* 1201 Pl (Co); John *de Whityngton'* 1327 SRLei; Richard *Whityngton* 1420–1 FFWa. From Whittington (Derby, Glos, Lancs, Northumb, Salop, Staffs, Warwicks, Worcs).

**Whittla:** *v.* WHITELEY
**Whittle, Whittell, Whitell:** Robert *de Withull* 1242 Fees (La); John *Whittle* 1581 Bardsley (Ch); Anthonius *Whitell* 1608 FrY; Christopherus *Whithill* son of Anthonius *Whithill* 1637 ib. From Whittle (Lancs, Northumb), Whittle-le-Woods (Lancs), or from residence near a white hill. Whittles Fm in Mapledurham (Oxon) is *Whitelewe* 1493 (PN O), 'white hill', from OE *hlāw*. Also 'dweller by a clear spring or stream': Whittles Hall in Springfield (Essex)

was the home of Hasculph *'de Whitewell* 1346 Cl. Whitwell (Cambs) is pronounced *Wittle*.

**Whittles:** Richard *Wythals* son of Ralph *Wythals* 1316 AD ii (Nf). A nickname, 'white neck', OE *hwīt, hals. v.* WHITEHOUSE.

**Whittlesea, Whittlesee, Whittlesey:** Roger *de Wytlesheye* 1279 RH (Hu). From Whittlesey (C).

**Whittley:** *v.* WHITELEY
**Whittock, Whittuck, Whytock, Whittek:** John *Wyttok* 1327 SRSo; William *Whittoc* 1334 AD ii (Wt); Joseph *Whiteoak* 1704 FrY. OE *Hwit(t)uc*, a diminutive of *Hwīt*.

**Whittome:** *v.* WITHAM
**Whitton:** (i) William *de Whyttun'* e 13th WoCh; Richard *Whitton* 1378 IpmGl; Thomas *Whitton* 1421 FFEss. From Whitton (Du, L, Mx, Nb, Sa, Sf). (ii) Michael *de Witton* 1296 (Selkirk), *de Whitton* 1303, David *Qwitton* 1361 (Roxburgh) Black. From Whitton in Morebattle (Selkirk).

**Whittred:** William *Witered* 1380 IpmNt; Bartholomew *Whitwright* (*Whitred* in margin) 1643 FrYar; Anthony *Whythred*, Richard *Witered*, Robert *Whitreed* 1674 HTSf. OE *\*Hwītrǣd*.

**Whittrick:** *v.* WHITTERIDGE
**Whittuck:** *v.* WHITTOCK
**Whitty:** *v.* WHITEY
**Whitwell, Whitwill, Witwell:** Henry *de Whitewell* 1197 P (R); Thomas *de Witewell'* 1219 AssY; William *de Wytewell* 1305–6 FFSr; John *de Whitewell* 1418 IpmLa. From Whitwell (Db, Do, Herts, Nf, R, We, Wt).

**Whitwood:** *v.* WHITEWOOD
**Whitworth:** Elyas *de Witewurde* 1194 P (Sr); John *de Whiteworth* 1336 FFY; Leonard *Whytworth* 1539 CorNt. From Whitworth (Durham, Lancs).

**Wholehouse:** *v.* WOOLHOUSE
**Whorall:** *v.* WORRAL
**Why(e):** *v.* GUY
**Whyard:** *v.* WYARD
**Whyatt:** *v.* WYATT
**Whyberd, Whybird:** *v.* WYBERD
**Whyborn:** *v.* WYBORN
**Whybra, Whybray, Whybrew, Whybro, Whybrow:** *v.* WYBER
**Whyman, Whymant:** *v.* WYMAN
**Whymark:** *v.* WYMARK
**Whysall:** Roger *de Wisho* 1199 P (Nt). From Wysall (Notts).
**Whyte:** *v.* WHITE
**Whytehead:** *v.* WHITEHEAD
**Whytelaw, Whytlaw:** *v.* WHITELAW
**Whytok:** *v.* WHITTOCK
**Whyton:** *v.* WYTON
**Wiatt:** *v.* WYATT
**Wibbe:** William *filius Wibbe* 1224 Pat (Nb); Geoffrey *Wibbe* 1210–11 PWi; Robert *Wybbe* 1363 FFY; Richard *Wybbe* 1381 AssWa. OE *Wibba*, OSw *Vibbe*.
**Wiberg:** *v.* WYBER
**Wiblin:** Jordan *Wibelin* 1203 P (L); Godfrey *Wybelyn* 1327 SRSx. A Norman double diminutive of *Wibbe.* cf. Willelmus *filius Wibbe* 1231 Pat (Nb), Richard *Wybbe* 1381 AssWa. OE *Wibba* is probably

the first element of Wibtoft (Warwicks). cf. (*to*) *wibban beorge* 959 BCS 1045.

**Wibrew, Wibroe, Wibrow:** *v.* WYBER

**Wich, Wych, Weech, Weetch, Wick, Wicke, Wickes, Wicks, Wix, Wike, Wyke, Wykes, Weekes, Weeks, Whick:** Alueredus *de Uuica* 1084 GeldR (So); Goscelin *del Wich* 1184 P (Wo); Jordan *de la Wike* 1194 Cur (Gl); Robert *de la Wyk* 1248 FFEss; Nicholas *Attewyche* 1270 AssSo; William *Wiche* c1280 *ERO* (Ess); Thomas *Wiches* 1302 ib.; Roger *atte Wykes* 1327 SRSo; Thomas *atte Whic* 1327 SRSx; William *Wyxe* ib.; John *Weekes* 1571 Bardsley. OE *wīc*, primarily 'dwelling-place, abode', then 'village, hamlet, town', was later used of a dairy-farm, as in Cowick, Gatwick (goat-farm), Oxwick, Shapwick (sheep-farm), Butterwick, Chiswick (cheese-farm), and in this sense the simple *wick* was very common in the 13th and 14th centuries and survived in common use in Essex as late as 1729 when we read of 'a wick or dairy of 20 cows' at St Osyth. The surname may derive from Wix (Essex) or any of the many places named Wick, Wyke or Week (a south-western, particularly Devon, form), or it may denote a dweller near or a worker at a dairy-farm. The final -*s* is a plural form. cf. WICKEN. There seem to be no certain examples of the palatalized form (*Wich*) surviving in the uncompounded place-name but the form certainly existed and may survive as a surname. Camden refers to 'making cheese of ewes' milk in their little dairy houses or huts [in Canvey Island] built for that purpose, which they call Wiches'. *Wich. Wych, Weech* and *Weetch*, may also, and, probably usually, mean 'dweller by the wych-elm' (OE *wice*), as at Weach Barton (Devon) and Wychstreet in Woking (Surrey).

**Wicherley, Wycherley:** John *Whycherley* 1465 Paston; Daniel *Wicherley* 1663 HeMil. From Wycherley (Sa).

**Wichford:** Robert *de Wycheford* 1256–7 FFL; Henry *de Wycheford* 1298 AssL; William *Wycheforde* 1369 FFW. From Witchford (C).

**Wicken, Wickens, Wickins:** Thomas *de le Wikin* 1275 RH (Nf); Henry *de Wikin* 1279 RH (Bk); Geoffrey *Wykin* 1327 *SR* (Ess). From Wicken (Cambs, Northants), Wicken Bonhunt (Essex), or 'dweller or worker at the dairy-farms', ME *atte wiken*. cf. WICH. Both the Essex and Cambridgeshire places are found as *Wykyng* in the 16th century which may be one source of *Wicking*.

**Wickenden:** Martin *de Wiggindenn* c1200 ArchC vi 210; John *de Wykendene* 1337 LLB F. From a lost place in Cowden (Kent), now represented by Polefields, first recorded in a charter of 1044 as *Wingindene* and in one of 1081 as *Wigendene* (KPN 324) and last mentioned as *Wykenden* in a list of church marks of 1542, but in that of 1663 as *Polefields*, long occupied by the Wickenden family; 'ould mother wickenden of powlfields' was buried in 1626. Prosperous landholders in Cowden, in the 16th century they were so numerous that their surname required a further attribute: Wickenden de Ludwells, Thomas Wykinden de Cowden Streate (1558), Thomas Wickenden de Bechinwoode (1571), Thomas Wickenden de la hole (1589). *v.* Ewing. The surname is still found in Tonbridge, on the Kent-Sussex border, and in Essex.

**Wicker, Wickers, Wheeker, Whicker:** Walter *le Wykere* 1225 AssSo; Henry *Wyker* Hy 3 Gilb (L); Thomas *Whicker* 1581 Oxon (D). 'Dweller or worker at the dairy-farm' (OE *wīc*). cf. WICH.

**Wickerley, Wickersley:** Roger *de Wikerlay* 1173 P (Y); Robert *de Wykerley* 1251 AssY; Richard *de Wykerslay* 1332 ShefA. From Wickersley (WRY).

**Wicket, Wickett, Wicketts:** *v.* WIGGETT

**Wickfield:** Nigel *de Wikefeld'* 1211 Cur (Berks). 'Dweller by the field near the dairy-farm.' OE *wīc, feld.* cf. *Wykfeld* 1253 Ch (St).

**Wickford:** William *de Wikeford'* c1200 Reg-Antiquiss; Stephen *de Wikford'* 1240–1 ForEss; Isaberd *of Wykeford* 1249 AssW. From Wickford (Ess).

**Wickham, Wykeham:** Wulfric *æt Wicham* 955 OEByn; Robert *de Wikam* 1218 FFO; William *de Wykeham* 1305 IpmY; Walter *Wykham* 1400 IpmGl. From one or other of the many places of this name, or from Wickhambreux (Kent), Wickhambrook (Suffolk), Wycomb (Leics), Wycombe (Bucks), or Wykeham (Lincs, NRYorks).

**Wicking, Wickings:** (i) *Wiching* (Wa), *Wikingus* (Sf), *Wichin* (D) 1066 DB; *Wiching* c1095 Bury (Sf, C). ON *Víkingr*, ODa, OSw *Viking*. Or OE *wīcing* 'pirate'. (ii) Richard *Wykyng* 1456 Ewing; Thomas *Wekyn, Wikyng* 1505–10 ib.; Thomas *Wycken* 1588 ib. From Wickens in Cowden (Kent) where the surname appears as *de Wyking'* in 1248 (KPN 324, n. 2).

**Wickins:** *v.* WICKEN

**Wickling:** William *de Wyklinge* 1275 SRWo. From Whitlinge in Hartlebury (Worcs) where William was assessed. *v.* WHITLING.

**Wicklow:** Richard *de Wikelaw'* 1212 Cur (Sf). From a lost *Wicklow*, the meeting-place of the Suffolk franchise of the Bishop of Ely.

**Wickman, Whickman:** Herbertus *filius Wycmanni* 1140–53 Holme (Nf); Vxor *Wichmanni* 1170 P (Nf); William *Wikeman* 1209 P (Nf); Alan *Wichman* 1275 RH (Sf); Richard *Wycman* 1275 RH (Nf). OE *wīc* and *mann* 'dairy-farm man'. cf. WICKER. For its use as a personal-name, cf. *Ewicman* 1066 DB (Nf), probably *\*eowu-wīcman* 'worker at the ewe-farm'.

**Wickstead, Wicksted, Wicksteed:** Nicholas *de Wykestede* 1279 PN W 27; John *Whicksteed* or *Weecksteede* 1602 Oxon; Edward *Wickstead* 1760 FrYar. From Wickstead (Ch), or Wicksted Fm in Highworth (W).

**Wiclif, Wicliffe, Wyclif, Wycliffe:** Robert *of Wyclyve* 1252 FFY; Robert *de Wyclyf* 1354 FFY; Robert *Wyclif* 1388 LLB H. From Wycliffe (NRY), or 'dweller by the white cliff', OE *hwīt, clif.*

**Widdaker:** *v.* WHITAKER

**Widdas:** *v.* WIDDOWES

**Widdecombe, Widdicombe, Withacombe, Withecombe, Withycombe, Whitticom:** Geoffrey *de Widecumbe* 1196 P (So); Hugh *Widecombe* alias *Withecomb* 1700 DKR 41 (D). From Widdacombe, Widdecombe, Widdicombe (Devon), or Withycombe (Devon, Somerset), all 'withy valley'.

**Widden:** Richard *de Wydden* 1269–70 FFSr; Richard *Widden* 1379 AssNu; Richard *Wyddene* 1386–8 FFSr. 'Dweller in the wide valley', OE *wīd, denu*. Sometimes, perhaps, for WHIDDON.

**Widders, Widdess:** *v.* WIDDOWES

**Widderson, Widdeson, Widdison:** *v.* WIDDOWSON

**Widdick, Widdicks:** Agnes *atte Whitedik* 1327 SRSx. From White Dyke in Hailsham (Sx).

**Widdington:** Thomas *de Widdintona* 1210–11 PWi; William *de Wydinton* 1284 IpmY; Robert *Wydyngton* 1426–7 FFSr. From Widdington (Ess, WRY).

**Widdiwiss:** *v.* WOODIWISS

**Widdop, Widdup, Widup, Whiddup:** Walter *Widuppe* 1652 RothwellPR (Y); Arthur *Widdop* 1672 HTY. From Widdop in Wadsworth (WRY).

**Widdowes, Widdows, Widdas, Widders, Widdess:** (i) Alice *Wedue* 1279 RH (C); Agnes *le Wydu*, Peter *le Wydoue* 1297 SRY; John *le Wydewe* 1327 SRSf. OE *widuwe, widewe* 'widow'. In ME, and perhaps also in OE, the word could have the sense 'widower'. (ii) William *Widders* 1576 FrLeic; Edward *Widdhouse* 1586 FrY; Thomas *Widdowes* 1619 ib.; John *Widhouse* 1681 ib.; George *Woodhouse* son of John *Widhouse* 1732 ib. A variant of WOODHOUSE, from OE *widu* 'wood', though the forms are late. As *Widdowson* exists, *Widdowes*, no doubt, sometimes means 'the widow's son', but the final *-s* has probably often been added to *Widdow* through association with *Widhouse*.

**Widdowson, Widderson, Widdeson, Widdison, Widowson:** Richard *Wyduesone* 1309 SRBeds; Peter, John *la Wydewesone* 1326 FFEss, 1327 SRDb; William *le Wydusone* 1332 SRSt. 'The widow's son.'

**Widdrington:** Girard *de Widrington* 13th FeuDu; Thomas *Widdrington* 1636 FrY. From Widdrington (Nb).

**Widdup:** *v.* WIDDOP

**Widger:** *Wihtgarus, Witgarus, Withgarus, Wisgar* 1066 DB (Sf); *Wyhtgarus* presbiter 1198 Colch (Ess); Ordric *Wihgar* c1095 Bury (Sf); Adam *Wydger* 1327 SRWo. OE *Wihtgār* 'elf-spear'.

**Widegrave, Widgrave, Widgrove:** Alan *Widegrave* 1202 FFY. 'Dweller by the wide grove', OE *wīd, grāf*.

**Widemouth:** Osbert *Widmuth* 1194 Cur (W); John *Widmouth* 1665 HTO. A nickname, OE *wīd, mūð*, 'wide mouth'.

**Widgrave, Widgrove:** *v.* WIDEGRAVE

**Widlake:** *v.* WEDLAKE

**Widmer, Widmore:** Geoffrey *Widimer* 1222 FFBk; John *Wydmere* 1380 AssLo. From Widmere in Ibstone (Bk). There was also a personal name: *Widmer* 1205 Cur.

**Widmerpole, Widmerpool:** Nicholas *Wydmerpole* 1454 IpmNt. From Widmerpool (Nt).

**Widmore:** *v.* WIDMER

**Widrich:** *v.* WHITTERIDGE

**Widup:** *v.* WIDDOP

**Wifeless:** John *le Wyfles* 1327, John *le Wyflese*, Stephen *Wyflese* 1332 SRSx. 'Without a wife', OE *wīf, lēas*. cf. FATHERLESS.

**Wigan:** Adam *de Wigain* 1209 P (La). From Wigan (Lancs).

**Wigan, Wigand, Wigens, Wigin, Wiggans, Wiggin, Wiggins:** *Wighen* 1086 ICC, DB (C); Radulfus *filius*

**Wigein** 1163 P (Lei); *Wygan* le Breton 1252 FFEss; William *Wygeyn* 1275 RH (Nf); John *Wygen* 1297 MinAcctCo (Co); James *Wiggans, Wiggins* 1752, 1756 FrY. An OFr personal name of Breton origin, OBret *Uuicon, Guegon* (Loth), introduced into England at the Conquest.

**Wigfall, Wigfull:** Henry *de Wigfall*, John *Wigfall* 1379 PTY. From Wigfall (WRYorks).

**Wigg, Wiggs:** Ailmar *Wigga* c1130 ELPN; Walter *Wigge* 1204 P (Ess). Probably a nickname from OE *wicga* 'beetle'. In later examples, perhaps also metonymic for WIGGER.

**Wigall:** *v.* WEIGALL

**Wiggans:** *v.* WIGAN

**Wigger:** (i) *Wiger* filius Roberti 1210 Cur (L); *Wigor' Buuebroc, Wigorus* Webbe 1221 ElyA (Sf); Roger *Wygor* ib. ON *Vigarr* or OE *\*Wīggār* (cf. PNDB 413). In 13th-century Suffolk, *Wigor* was a not uncommon peasant's name. The surname may also derive from OG *Wigger*. (ii) William *le Wygger* 1332 MESO (Ess). A derivative of ME *wygge*, MLG, MDu *wigge* 'wedge, wedge-shaped cake', a maker or seller of buns or cakes. cf. Matilda *la Wyggestr'* 1296 SRSx.

**Wiggett, Wicket, Wickett, Wicketts:** *Wigot, Wigod* 1066 DB; Willelmus *filius Wigot* 1212 Cur (L); John *Wiket, Wiget* 1180, 1182 P (Wo); William *Wigot* 1185 Templars (Wa); John *Wigod* 1255 RH (Sa); Walter *Wyket* 1284 FFSf. ODa, OSw *Vigot* (PNDB). Early examples of *Wiket* are not common.

**Wiggin(s):** *v.* WIGAN

**Wigginton, Wiginton:** Akeman *de Wigentona* 1170 P (Y); Ralph *de Wigenton* 1185 Templars (O); John *de Wigetone* 1319 SRLo. From Wigginton (Herts, O, St, NRY), Wigton (Cu, WRY), or Wiggaton in Ottery St Mary (D).

**Wigglesworth, Wiglesworth, Wigelsworth:** Geoffrey *de Wicleswrthe* 1202 FFY; William *de Wyglesworth* 1379 PTY; Henrie *Wigglesworth* 1629 FrY. From Wigglesworth (WRYorks).

**Wight, Weight, Whight:** Roger *Wicht* 1176 P (Gl); Richard *Wiht* 1200 P (Beds); Hugh *Wyght* 1222 Cur (Beds); William *le Wyhte* 1275 RH (Sx); Thomas *Whight* 1291 Black. ME *wiht, wight* 'agile, strong', from ON *vig-t*. Nicholas *de Wight* (1332 SRLo) probably came from the Isle of Wight. *v.* also WHITE.

**Wightman, Weightman:** *Wyctman* de Freton 1248 FFEss; William *Wihtman* 1227 AssBeds; William *Wightman* 1332 SRCu; Roger *Wightman* son of John *Whightman*, milloner 1639 FrY; Henry *Weightman* son of John *Weightman*, milner 1654 FrY. OE *\*Wihtmann* 'elf-man', or a nickname, 'brave, strong man', from ME *wiht* (*v.* WIGHT) and *man*.

**Wighton:** Alan *de Wihton* 1195 P (Sr); Peter *de Wihtton* 1248 FFK; Hugh *de Wyghton* 1251 AssY. From Wighton (Nf), or Market Weighton (ERY), *v.* WEIGHTON.

**Wigin:** *v.* WIGAN

**Wiginton:** *v.* WIGGINTON

**Wigley:** Hugh *de Wygeley* 1292 Sheff: William *Wygley* 1576 SRW. From Wigley (Db, Ha).

**Wigman:** Alexander *Wigman* 1275 RH (Nth); Geoffrey *Wyg(e)man* RH (Y). Either 'dairy-farm man' (*v.* WICKMAN) or 'cake-man' (*v.* WIGG). cf. CHEESEMAN.

**Wigmore:** Turstin *de Wigemore* 1066 DB (He); Roger *de Wigemore* 1199 P (Gl); John *Wygemore* 1407 KB (Mx). From Wigmore (He).

**Wignall, Wignell:** William *de Wigenhal'* 1197 P (Nf); John *de Wikenholt* 1219 Eynsham: Nicholas *de Wegenhale* 1327 SRC. From Wiggenhall (Nf), or Wiggenholt (Sx), *Wikeholt* 1212.

**Wigram:** Adam *Wytegrom* 1327 SRC; John *Whytegrom* 1341 FFW; William *Whytegrome* 1525 SRSx. 'Fair servant', OE *hwīt*, ME *grom*. cf. Ralph *le Whyteclerk* 1248 AssBerks 'fair cleric'; Henry *Wythkenep* 1254 *WAM* 'fair servant', OE *cnapa*.

**Wigsell:** *v.* WIGZELL

**Wigsley:** Adam *de Wyggesle* 1298 AssL. From Wigsley (Nt).

**Wigzell, Wigsell:** John *Wygeselle* 1392 CtH; Thomas *Wygsel* 1525 SRSx; Rychard *Wigsell* 1569 StaplehurstPR (K). From Wigsell in Salehurst (K).

**Wike:** *v.* WICH

**Wilber:** *v.* WILDBORE

**Wilberforce, Wilberfosse:** William *de Wilburfoss* temp. Stephen Whitby (Y). From Wilberfoss (ERYorks).

**Wilbert:** Henry *Wylbert* 1279 RH (Hu). OE *Wilbeorht* 'will-bright', an early but uncommon OE name, which is not recorded in DB or later. It must have remained in use, probably among the peasants.

**Wilbraham, Wilbram:** Aluric *de Wilburgeham* 1066 ICC (C); Richard *de Wilbirham* 1260 AssC; Ralph *Wilbram* 1435–6 IpmNt. From Great, Little Wilbraham (C).

**Wilby, Wilbye, Wilbe, Wilbee:** Robert *de Willeby* a1161 YCh; Adam *de Wilebi* 1208 P (R); Robert *Wilby* 1372 FFEss. From Wilby (Norfolk, Northants, Suffolk).

**Wilcock, Wilcocke, Wilcocks, Wilcox, Willcock, Willcocks, Willcox:** *Wilcok* 1246 AssLa; *Wilcoc* 1286 AssCh; *Wylecok* Hervy 1275 RH (Lo); William *Wylecok* 1254 AssSo; John *Wilcokes* 1316 Wak (Y); Godfrey *Willecok* 1327 SRSf. *Wilcoc*, a pet-name for *William*.

**Wilcockson, Wilcoxson:** William *Wilcokson* 1332 SRCu. 'Son of *Wilcock*.'

**Wild, Wilde, Whild, Whilde, Wyld, Wylde, Wyldes, Wylds:** (i) Uluricus *Wilde* 1066 DB (L); William *le Wilde* 1177 P (Lei); Henry *le Wylde* 1236 FFEss. OE *wilde* 'wild, violent'. (ii) William *de Wilde* 1200 P (Sx); Walter *de la Wylde* 1256 RamsCt (Hu); William *atte Wylde* 1347 LLB F. 'Dweller by the waste, uncultivated land' from OE *wilde* 'wild, waste', used as a noun. The above forms are much too early to be regarded as due to the development of *weald* to *wild*.

**Wildash:** *v.* WILDISH

**Wildblood, Wildeblood:** Roger, William *Wyldeblod* 1366 SRLa, 1381 SaltAS (OS) xiv. A nickname for an untamed spirit or a rake. Nicknames formed with a compound of *wild-* were common. cf. Richard *Wildecat* 1176 P (Wo), William *Wildebef* 1327 SRSx, William *Wildecnave* 1327 SRWo, Roger *Wildehog* 1246 FFK, Robert *Wildprest* 13th Guisb, Stephen *Wyldraven* 1300 ib.

**Wildbore, Wildeboer, Wyldbore, Wilber:** William *Wyldebar* 1242 AssLa; William *Wyldebor* 1307 Wak (Y); John *Wylbor* 1379 PTY. A nickname, 'wild-boar' from OE *bār*.

**Wildbuck:** Peter *Wildbuk* 1525 SRSX. 'Wild buck', OE *wilde, bucca.* cf. William *Wildefuhel* 1185 Templars (Y) 'wild bird'; Aedwin *Wildegrom* 1184 P (St) 'wild servant', or 'servant of *Wild*'.

**Wildee:** *v.* WALTHEW

**Wilden:** William *de Wilden'* 1221 Cur (Beds); John *Willeden* 1327 SRSx; Richard *Wylden* 1370–1 FFSr. From Wilden (Beds).

**Wilder, Wilders:** William *Wilder, Wyldere* 1327 SRDb. Probably OE *wildēor* 'wild animal'.

**Wildey:** *v.* WALTHEW

**Wildgoose, Wildgose, Wildgust, Willgoss:** Henry *Wildegos* 1201 Cur (Sa); Osbert *Wildgos* 1206 Cur (Sf); John *Wylegous* 1379 PTY. A nickname 'wild goose'. cf. William *Wildefuhel* 1185 Templars (Y), Nicholas *Wyldefoul* 1288 RamsCt (Hu). 'Wild bird.'

**Wilding, Willding:** *Wildingus* Prepositus 1224 Cur (He); Adam *Wilding* 1207 Cur (Nf), 1286 AssCh; John *Wylding* 1332 SRCu. OE *\*Wilding*, a derivative of *wilde*, 'the wild one', both as a personal name and a nickname. As Adekyn *Wylding* (1285 AssCh) was the son of Alice *Wildegos*, the nickname may have been regarded as synonymous with *wild-goose*. The modern surname may also be a late development of *Wheeldon* or *Wheldon*. Wilding's Copse in Aldbourne (Wilts) was *Wheleding, Whelden* in 1561 (PN W 293).

**Wildish, Wildash:** Simon *le Wealdessh* 1316 FFK; Robert *de* (sic) *Weldysh* 1317 FFK. OE *\*wealdisc* 'belonging to the Weald'; the inhabitants of the Weald were called Wealdish men. The name survives in Kent farms: Wildage Fm in Elham (from John *Weldisse* 1292, John *Wealdissh'* 1327); Welldishes Fm in Linton (William *Woldishe* 1542), Upper and Lower Woolwich in Rolvenden (Robert *de Waldeys, de Weldysh'* 1278, 1317, William *Weldisshe* 1327). *v.* PN K 434, 139, 355.

**Wildman:** William *Wyldman* 1379 PTY. 'The wild man.'

**Wildmer, Wildmore:** Robert *Wildemare* 1328 IpmNt. From Wildmore (L).

**Wildon:** *v.* WHEELDON

**Wildrake:** *v.* WELDRICK

**Wildridge, Willdridge, Willrich:** Hugo *filius Wilrici* 1203 Cur (Nth); *Wildricus* Mæl 1205 Cur (He); Henry *Wildriche* 1568 SRSf. OE *Wilrīc* 'will-powerful', a rare name, only two OE examples being known.

**Wildsmith, Wilesmith, Weldsmith, Whilesmith:** Euota *Welsmyth'* 1319 *SR* (Ess); Ivo *le Welsmyth* 1327 *SR* (Ess); Anne *Wilesmith* 1787 Bardsley. This surname has been variously explained as a corruption of *weld-smith*, a forger in iron, weald-smith, and the smith in the wild, none of which is satisfactory, nor is Bardsley's association of the name with wool. *Welsmyth* is from OE *\*hwēol-smiδ* 'wheel-smith', a maker of wheels, especially the iron parts. cf. WHEELWRIGHT. The vowel was raised to [ī] (*Wheelsmith*), then shortened to *Willsmith*, which, with an intrusive *d*, became *Wildsmith*. ME *W(h)īlsmith* developed normally to *W(h)ilsmith*. cf. *Whygler* for *Wheeler*, *Whilwright* for *Wheelwright*, and the development of *weald* to *wild*.

**Wildy:** v. WALTHEW

**Wiles, Wyles:** (i) William *de la Wile* 1185 P (Do); Osbert *de Wila* 1204 P (Sa); Adam *de la Wile* 1221 AssWo; Walter, Thomas *atte Wyle* 1296 SRSx, 1317 AD iii (Wa). From Wild (Berks), Monkton Wyld (Dorset), or Wylam (Northumb), all of which Ekwall (DEPN) derives from late OE *wīl*, used of some such mechanical device as a windmill or a trap. Löfvenberg (MELS) notes ME *wyle* 'a wicker trap for catching fish, especially eels' (1256 NED), and the frequency of the name suggests that this may have been a common meaning. (ii) Robert *le Wile* 1195 FFNf; John *Wiles* 1202 AssL; John, Robert *Wyles* 1251 AssY, 1327 SRLei. This, too, is probably from late OE *wīl* 'trick', and the almost invariable plural suggests that it is metonymic for a trapper, one in charge of the fishing traps, though it could be metaphorical, 'a man of many wiles'. For the type of name, cf. TRAPP, WRENCH.

**Wiley, Wylie, Wyley, Wyly:** John *de Wili*, *de Wylegh* 1201 P (W), 1230 Cur (Wa); William *de Wyly* 1299 IpmW; Simon *Wylegh* 1327 SRSx; Richard *Wyleye* 1390 FFEss. From Willey (Ches, Devon, Hereford, Salop, Warwicks), Wylye (Wilts), or Wyly in East Hoathly, Whiligh in Ticehurst (Sussex).

**Wilford, Willford:** Roger *de Willeford* 1199–1200 FFSf; Gervase *de Wylford* 1360 IpmNt; Thomas *Wilforde* 1576 SRW. From Wilford (Notts, Suffolk).

**Wilfred:** *Wilfrei* miles 1055 FeuDu; Robert *Wilfrith* c1280 SRWo. OE *Wilfrīð* 'will-peace'.

**Wilgress:** v. WILLGRASS

**Wilk, Wilke, Wilkes, Wilks, Wilck:** *Wylke de Chyrchele* 1246 AssLa; *Wilke* 1286 AssCh; Roger *Wylk* 1279 RH (Beds); John *Wilkys* 1327 SRWo. *Wilke* may be a shortening of *Willock*. cf. DILKE.

**Wilkerson:** v. WILKINSON

**Wilkie:** David *Wilke* 1495 Black (Fife); William *Wilkie* 1529 ib.; John *Wilky* 1580 ib. A distinctively Scottish double diminutive of *William*. v. WILLIAMS.

**Wilkin, Wilkins, Wilkens, Wilkings:** *Wilechin* 1166 P (Nb); *Wilekinus* 1191 P (Ha), 1207 Cur (Nth); Richard, William *Wilekin* 1180 P (Ha), 1220 Cur (Lo); Roger *Wylkyns* 1327 SRSt. *Wilkin*, a diminutive of *Will*. William de Ros 1201 Cur (K) and *William* Chubbe 1230 P (Lo) are both also called *Wilekin*.

**Wilkinson, Wilkerson:** Roger *Wyleconsesone* 1332 SRSx; Thomas *Wilkynson* 1332 SRCu. 'Son of *Wilkin*.'

**Wilkshire:** v. WILTSHIRE

**Will, Wille, Willes, Wills, Wyllys:** (i) *Wille* Walt' 1207 Cur (Nf); *Wille* Bret 1212 Cur (L); Robert, William *Wille* 1279 RH (C), 1323 AssSt; Cecily *Willes* 1279 AD v (W). *Will*, a short form of *William*. (ii) John *atte Wylle* 1296 SRSx; Walter *de la Wille* 1297 MinAcctCo. 'Dweller by the spring or stream.' v. WELLS.

**Willacy, Willersey:** Thomas *de Wyllereseye* 1275 SRWo. From Willersey (Gl).

**Willament, Willment:** v. GILLMAN

**Willan, Willans, Willen:** *Wilanus* de la Hele 1208 Cur (D); John *Willon* 1275 SRWo; Thomas *Willan* 1581, Frauncis *Willans*, son of Thomas *Willance* 1608 FrY. OG *Wiland*. Sometimes, perhaps, a diminutive of *Will*, a short form of *William*.

**Willard, Wyllarde:** *Wielardus* 1086 DB (Ess); *Willardus* de Wridlint' 1121–48 Bury (Sf); *Hwylardus* de Wytherendenn c1180 ArchC vi; *Wilardus* uinitor 1192 P (Y); William, Geoffrey *Wilard* 1166, 1190 P (Nf, Wo); Nicholas *Wypelard* 1279 RH (O); Nicholas *Wylard* 1296 SRSx; Richard *Withelard* (*Wythelard*) 1327 SR (Ess). OG *Widelard*. Occasionally, perhaps, OE *Wilheard*. v. WILLET.

**Willcock, Willcox:** v. WILCOCK

**Willday, Willdey:** v. WALTHEW

**Willding:** v. WILDING

**Willdridge:** v. WILDRIDGE

**Willer, Willers:** William *le Willer* 1327 SRSx; John *le Wylyare* 1327 MEOT (Ha); William *le Wyliere* 1332 ib.; Robert *le Wyliar* 1332 SRSx. A derivative of OE *wilige* 'basket', a basket-maker.

**Willerby:** Yuo *de Wilardebi* 1185 P (Y); Henry *de Wilardebi* 1208 AssY. From Willerby (ERY).

**Willersey:** v. WILLACY

**Willeson:** v. WILSON

**Willet, Willets, Willats, Willett, Willetts, Willitt, Willott:** *Wil(l)ot de Foxwist* 1286 AssCh; Matill', Symon *Wylot* c1248 Bec, 1269 FFSf; Thomas *Wilet* 1277 AssSo; William *Wyllet*, *Wyllot* (identical) 1327, 1332 SRSx; Stephen *Wilotes* 1327 SRWo; Thomas *Willard* alias *Willett* 1582 FFHu. *Wilot*, *Wilet*, diminutives of *Will* (William). Also a late development of WILLARD, and, occasionally, a variant of WAYLATT: Aspelon *atte Wylet* 1240 FFEss.

**Willey, Willy, Wyllie, Wylly:** *Willy* de Dalton, John *son of Willy* 1300 Misc (La); Henry *Willy* 1274 RH (Sa); William *Wylli* 1297 MinAcctCo; Roger *Willy* 1360 IpmW. A pet-form of *Will*, a short form of *William*. Sometimes for WILEY.

**Willford:** v. WILFORD

**Willgoss:** v. WILDGOOSE

**Willgrass, Willgress, Willgross, Wilgress:** Gilbert *Wildegris* Hy 2 DC (Nt); William *Wyldgryce* 1511 NorwW (Nf). 'Wild pig' (ON *gríss*).

**Williams, Willems, Willam, Fitzwilliam, Fitzwilliams:** Robertus *filius Willelmi* 1086 DB; Richard *Willam* 1279 RH (O); John *Wylyam* 1296 SRSx; Rauf *le fuiz William* 1299 Whitby (Y); Henry *Fitz William* 1300 LoCt; Ralph *Willem*, *Willeam* 1304, 1315 AD vi (K); Thomas *Willames* 1307 AssSt; Robert *Williames* 1309 ib.; Thomas *William* 1327 SRSf. OG *Willihelm*, *Willelm*, the Norman form of Fr *Guillaume* (v. GILLAM), after the Conquest the most popular christian name until superseded by *John*.

**Williamson:** Arnald *Williamssone* 1360 ColchCt; Roger *Williamson* 1386 NottBR. 'Son of *William*.'

**Williatt:** Roger *Williat* 1327 SRSf; Henry *Wylyot* 1342 LoPleas; John *Willyot* c1436 Paston. *Willy-ot*, a diminutive of *Willy*. v. WILLEY.

**Willicombe:** v. WELCOME

**Williman, Williment, Willimont:** v. GILLMAN

**Willing, Willings:** William *Willing* 1221 AssWo; William *Willing* 1279 RH (Bk); John *Willyng* 1327 SRSo. 'Son of *Willa*.'

**Willingale, Willingdale:** Richard *de Wylynghale* 1356 HPD. From Willingale Doe, Spain (Ess).

**Willingham:** Odo *de Willingeham* 1190 P (Sf); John *Wylyngham* 1394 TestEbor; William *Wyllyngham*

1445–6 IpmNt. From Willingham (C, Sf), or North, South, Cherry Willingham, Willingham by Stow (L).

**Willington:** John *de Wilentona* 13th, Nicholas *de Wilentona* 1251 Burton; Walter *Willington* 1662–4 HTDo. From Willington (Beds, Ch, Db, Du, Nb, Wa).

**Willis, Wyllys, Whillis:** Walter *Willys* 1327 SRSt; Roger *Wyllys* 1438 LLB K; Richard *Willys* 1517 FFEss. Forms of WILL retaining the vowel of the genitive ending.

**Willisher:** *v.* WILTSHIRE

**Willison:** *v.* WILSON

**Willman, Wilman:** Symon, Adam *Willeman* 1279 RH (C), 1379 PTY. 'Servant of *Will*.' *v.* also GILLMAN.

**Willmin:** *v.* GILLMAN

**Willmore:** *v.* WILMORE

**Willmot, Willmott, Willmett, Willmetts, Willimott, Wilmot, Wilmott, Wilmut:** *Willimot* Ric 1 Cur (L); *Wylimot* 1306 Wak (Y); Walter *Wilimot* 1252 Rams (Hu); Henry *Wilmot* 1279 RH (C); Thomas *Wilmet* 1317 AssK. A hypocoristic of *Willelm* (William). cf. OFr *Guillemot*.

**Willock, Willocks, Willox:** Rannulfus *filius Willoc* 1219 AssL; William *Willoc* 1221 AssWo. OE *Willoc* (rare).

**Willott:** *v.* WILLET

**Willoughby:** Robert *de Wilgebi* 1175–84 YCh; William *de Wylugby* 1301 FFY; Thomas *Willoughby* 1449 FFEss. From Willoughby (Leics, Lincs, Notts, Warwicks).

**Willows:** Robert *in le Willewys* 1290 Crowland (C); John *in le Welwes* 1327 SRC. 'Dweller among the willows', OE *wylig, welig*.

**Willrich:** *v.* WILDRIDGE

**Willshaw, Willshear, Willshire:** *v.* WILTSHIRE

**Willstead, Willsteed:** *v.* WELLSTEAD

**Willy:** *v.* WILLEY

**Wilmer, Wilmers:** William, Simon *Wilmer* 1296 Wak (Y), 1327 SRC; John *Wilmer* 1374–5 AssL. OE *Wilmær* 'will-famous'. *v.* also WILMORE.

**Wilmington:** John *de Wilminton'* 1200 P (Do); Bartholomew *Wylmynton* 1392 CtH; John *de Wilmynton* 1510 SaG. From Wilmington (D, K, Sa, So, Sx).

**Wilmore, Willmore:** William *de Wiltemore* 1221 AssWo; Thomas *de la Wildemore* 1275 MELS (Wo); William *de Wyldemor* 1327 SRSt; John *atte Wildemor'* 1327 MELS (Wo). From Wildmore (Lincs), Wildmore in Belbroughton (Worcs), or 'dweller by the waste fen'.

**Wilsden, Wilsdon:** Elyas *de Wulsingden'* 1219 AssY. From Wilsden (WRY).

**Wilsford:** Heldewin *de Wiuelesford'* 1177 P (L); Juetta *de Wiuelesford* 1201 P (W). From Wilsford (L, W).

**Wilsham:** Robert *de Wylesham, de Wylsham* 1296, 1332 SRSx lived at Wilson's Cross in Ashburnham (Sussex). There is also a Wilsham in Devon, *Willmersham* 1426.

**Wilshaw, Wilsher, Wilshere, Wilshire:** *v.* WILTSHIRE

**Wilson, Willeson, Willison, Willson:** Robert *Willeson* 1324 Wak (Y); Robert *Wilson* 1341 Kirkstall (Y); John *Willison* 1366 SRLa. 'Son of *Will*', i.e

William. Surnames derived from Wilson (Devon, Leics) may have contributed to the frequency of this name.

**Wilstrop, Wilsthorp, Wilsthorpe:** William *de Wilsthorp* 1327, Bartholomew *Wilstrupp* 1652 FrY. cf. John *Wyllesthorpman* 1366 FrY. From Wilsthorpe (Db, L, ERY), or Wilstrop (WRY).

**Wiltcher:** *v.* WILTSHIRE

**Wilthew:** *v.* WALTHEW

**Wilton:** Gerald *de Wiltune* 1066 DB (W); Hugh, Walter *de Wilton* 1162 P, 1273 IpmW; John *Wilton* 1390 FFEss. From Wilton (Cumb, Hereford, Norfolk, Som, Wilts, ER, NRYorks).

**Wiltshire, Wiltshear, Wiltsher, Wiltshier, Wiltshaw, Wilcher, Wilshaw, Wilsher, Wilshere, Wilshire, Wilkshire, Willisher, Willshaw, Willshear, Willsher, Willshere, Willshire:** Hunfr' *de Wilechier, de Wilecher* 1157, 1162 P (Sx); Nicholas *de Wiltesir'* 1207 Cur (W); Robert *Wylchar* 1275 SRWo; John *Wilteshire* 1298 *Ct* (Ha); William *Wylkeshire* 1440 LLB K; Richard *Wylshyre* 1456 Oriel (O); Thomas *Wylshere* 1483 FFC; Nathaniel *Wilksheire* 1674 HTSf; William *Willshaw* 1758 FrY. From Wiltshire.

**Wimbolt:** *v.* WINBOLT

**Wimborne, Wimbourne:** *v.* WINBORNE

**Wimbush:** Richard *de Wymbys* 1198 FFEss; Walter *de Wymbysse* 1326 CorLo; Robert *Wymbyssh* 1403, *Wymbussh* 1442 IpmNt. From Wimbish (Ess).

**Wimmer, Wimmers:** John *filius Winemeri* 1187 P (K); Hamo *filius Wymer'* 1219 AssY; William *Wymer* 1230 Cur (St); Richard, Roger *Winemer* 1279 RH (C); Henry *Wymer* 1301 CorNth. OE *\*Winemær* 'friend-famous'.

**Wimpenny:** *v.* WINPENNY

**Wimple:** cf. John *le Wymppelwebbe* 1325 AD ii (Herts); Simon *(le) Wimpler* 1183–4 P (Nth). *Wimple* is OE *wimpel* 'veil' and is metonymic for a weaver or maker of wimples.

**Winand, Winans, Winant, Wynands:** *Wynan* Tirel 1297 MinAcctCo; Thomas *Wynan* 1332 SRSx. OG *Wignand, Winand*, OFr *Guinant*. *v.* Forssner 257.

**Winbolt, Wimbolt, Winbow:** *Winebaldus* 1195 P (Nth); *Winebold'* 1197 P (Y); John *Winebald* 1210 P (Nf); Geoffrey *Wynebaud* 1229 Cl (W). OE *Winebeald* 'friend-bold'.

**Winborne, Winborn, Winbourne, Winburn, Wynburne, Wimborne, Wimbourne:** Philip *de Winburne* 1166 PN Do ii 267; Walter *of Winburn'* 1249 AssW; Walter *de Wymburn* 1276 FFEss. From Wimborne Minster, St Giles, Monkton, Up Wimborne (Do).

**Wincer:** *v.* WINDSOR

**Winch, Wynch, Wink, Winks:** (i) William *de la Winche* 1275 SRWo; Stephen *atte Wynke* 1327 SRSo; Thomas *atte Wynche* 1332 SRWo. OE *wince* 'winch, pulley' later developed the meanings 'well-wheel' and 'well'. In place-names it has been taken to mean 'a sharp bend in a river or valley'. Hence, 'dweller by the well from which the water is drawn by a winch' or 'dweller in a bend'. (ii) Walter *Winch* 1184 P (C); William *Wynk* 1312 FFSf; William *le Wynch* 1327 SRSx. A nickname from the lapwing, recorded as OE *hlēapewince*, from its leaping, twisted flight.

**Winchcombe, Winchecombe:** Vincent *de Winchecumbe* 1207 Pl (O); Richard *de Wynchecoumbe*, John *Wynchecombe* 1351, 1382 AssLo. From Winchcombe (Glos).

**Winchell:** (i) Hawis *Wenchel* 1203 AssNth; Puncok *le Wenchil* 1260 AssCh. OE *wencel* 'child'. *v.* also WENCHE. (ii) John *de Wyncheles* 1280, John *de Wyndesceles* 1389 Black. From the lands of Winscheill, now Windshiel (Berwick).

**Winchester:** Odo *de Wincestre* 1086 DB (Ha); William *de Wyncestre* 1286 IpmY; John *Wynchestre* 1360 IpmGl. From Winchester (Hants).

**Wincott, Wincote, Winnicott:** Hugh *de Withmecote* 1221 AssWa; William *de Wynnecote* 1253–4 FFWa; John *Wynecotte* 1332 SRWo. From Wincott in Whichford (Wa).

**Wind, Wynde:** (i) Walter *Winde* 1197 P (Ess); Geoffrey *Wynd* 1230 P (Hu); Clement *le Wynd* 1327 *SR* (Ess). OE *wind*, a nickname for one speedy as the wind. cf. Richard *Wyndswyft* 1301 SRY. (ii) John *de Wynd* 1268 Whitby (Y); William *de la Wynde* 1275 RH (Sx); Roger *atte Wynde* 1285 Ass (Ess); John *in ye Wyndde* 1297 SRY. 'Dweller by the winding path or ascent', OE *(ge)wind*.

**Winday, Windey:** *Wundai* 1066 Winton (Ha); Abraham *filius Wyneday* 1229 Pat (K); Richard *Windai* 1176–7 P (Nf/Sf); Hugh *Wynneday* 1317 AssK. OE *Wyndæg*.

**Windeat, Windeatt, Windeate, Windiate, Windyate:** Elizabeth *Windeyate* 1440 FrY; Richard *Windeatt* 1662–4 HTDo. From Windgate (Wt), Wingate (D, Du, Nb, Sr, WRY), or Winnats Pass (Db).

**Windebank, Windibank:** Henry *de Wyndibonk* c1300 WhC (La); Robert *del Wyndybankes* 1315 Wak (Y). 'Dweller on the windy hill.'

**Windell, Windle:** Alan *de Windhull'*, *de Windul* 1201 P, 1242 Fees (La); Walter *del Wyndhel* 1327 SRSf; Richard *atte Wyndhull* 1327 *SR* (Ess). 'Dweller on the windy hill', as at Windhill (WRYorks) or Windle (Lancs).

**Winder:** Richard, John *le Winder(e)* 1275, 1276 RH (L, D). A derivative of OE *windan* 'to wind', probably a winder of wool.

**Winders, Windes:** *v.* WINDUS

**Windey:** *v.* WINDAY

**Windham, Wyndham:** Thomas *de Wymundeham*, *de Wymondham* 1261–2 FFEss, 1305 LoCt; Robert *de Wyndeham* 1327 SRSx; Ralph *Wyndeham* 1332 SRSx. From Wymondham (Leics, Norfolk).

**Windiate:** *v.* WINDEAT

**Windley, Winley:** Henry *de Wynleye* 1320, *Wyndele* 1323 AssSt. From Windley (Db).

**Windmill, Winmill:** Isabella *atte Wyndmylle* 1366 Oseney (O). 'Dweller or worker at a windmill.'

**Windover:** Richard *de Wyndehover* 1345 FrY; William *de Wyndesouer* 1379 PTY; Robert *Wyndover* 1545 SRW. From Winds Over in Ilkley (WRY), or a late form of WENDOVER.

**Windows:** *v.* WINDUS

**Windridge:** *Wenric* 1066 DB (Gl); Goduine *Uuenric* c1095 Bury (Sf); Margrett *Wyndrych* 1524 SRSf. OE *Wynrīc* 'joyful-ruler'.

**Windriss, Windross:** Thomas *Wynderiche* 1419, John *Wyndrysce* 1467, *Wyndres* 1507 FrY. From Windros Laithe in Flasby (WRY).

**Windsor, Wincer, Winser, Winsor, Winzer:** Godfrey *de Windelesor* 1066 Winton (Ha); Reginald *de Windesor'* 1130 P (Bk). From Windsor (Berks) or Little Windsor and Broadwindsor (Dorset).

**Windswift:** Richard *Windswift* 1301 SRY. 'As swift as the wind', OE *wind, swifte*. cf. John *le Wyndwode* 1297 MinAcctCo 'as mad as the wind'.

**Windus, Windows, Winders, Windes:** William *de Wyndhows* 1379 PTY; Thomas *Wyndhouse* 1431 FrY; William *Wyndowes* 1458 ib.; William *Wyndes* 1530 ib.; the last three, all weavers. 'Worker at the winding-house' (for threads, yarn, etc.). cf. WINDER.

**Windyate:** *v.* WINDEAT

**Wine, Wines, Wyne, Wynes, Wyness:** Osketel *Wine* 1199 P (Sf); Thomas *le Wyne* 1296 SRSx. OE *wine* 'friend'.

**Winfarthing:** Thomas *de Wynneferthyn* 1279 RH (C). From Winfarthing (Norfolk).

**Winfield, Wingfield:** Nicholas *de Wynefeld'* 1228 Cur (Beds); Richard *de Winfeld* 1274 RH (Nb); John *de Wingefeld* 1343 FFY; Robert *Wyngefeld* 1465 Past. From Wingfield (Beds, Derby, Suffolk), the first two *Winfeld* c1200, *Winefeld* DB.

**Winford:** Felicia *attewenforde* 1303 PN Wt 173; John *Wynneford* 1312–13 FFEss. From Winford (So), or Winford in Newchurch (Wt).

**Winfrey, Winfrith, Winnefrith, Whinfrey, Wynfrey:** *Winfridus* de Ethona c1190 DC (Lei); *Winfred* serviens 1275 RH (Nf); *Winfrid'* 1279 RH (Bk). OE *Winfriδ*.

**Wing:** (i) Walter *Wenge* 1202 AssNth; John *Wynge* 1329 FFW; Geoffrey *Weng* 1401 AssLa. Old West Scandinavian *Wenge*. (ii) Alured *de Withunge* 1066 DB (Bk). From Wing (Bk, R).

**Wingar, Winger:** *Winegar* 1182–1211 BuryS (Sf); Henry *Wyngar* 1251 Rams (Hu); William *Wyneger* 1327 SRSf. OE *Winegār* 'friend-spear', found as the name of an East Anglian moneyer and in an 11th-century place-name.

**Wingard:** *v.* WINYARD

**Wingate, Wingett, Whinyates, Wynniatt, Wynyates:** Aldret *de Windegate* c1145–65 Seals (Nb); Henry *de Wingate* 1204 P (K); David *de Windyates* 1332 SRCu. 'Dweller in a windy pass' (OE *\*wind-geat* 'gate for the wind'), as at Wingate (Devon, Durham), Wingates (Northumb), and Winnats (Derby).

**Winger:** *v.* WINGAR

**Wingfield:** *v.* WINFIELD

**Wingham:** Ralph *de Wingeham* 1148 Winton (Ha); Henry *de Wingham* 1236 FFK; William *de Wyngeham* 1334 SRK. From Wingham (K).

**Wingold:** John *Wynnegold* 1327 SRSf; Simon *Wyngold'* 1331 FFK; John *Wynnegold* 1365 LoPleas. 'Win gold', OE *winnan, gold*, a nickname for a prosperous man. cf. John *Wynnefoddere* 1305 LLB B 'win food'.

**Wingood:** *Winegod* 1066 DB (W), 1208 ChR; Geoffrey *Winegod* 1221 *ElyA* (Hu); William *Wynegod* 1276 RH (Berks). OG *Winegot*.

**Wink:** *v.* WINCH

**Winkfield:** Henry *de Winkefeld'* 1230 P (Berks); John *Wynkefeld* 1380 AssLo. From Winkfield (Berks, W).

**Winley:** *v.* WINDLEY

**Winman:** *Wineman* de Wanetun' 1166 P (Nt); Walter *Wineman, Wyneman* 1225 Pat (W), 1250 Fees (Nth); William *Wynemon* 1274 RH (Sa). OE *Winemann* 'friend-man'.

**Winmill:** *v.* WINDMILL

**Winn, Wynn, Wynne, Wynnes:** Walter *Win* 1198 P (Ess); William *Wine* 1202 AssL; Robert *le Win* 1293 Fees (Db). OE *wine* 'friend'. Also a personal-name. OE *\*Wine*, a short form of names like *Winebeald*, etc. *v.* WINE.

**Winnall, Winnell:** (i) Richard *Winel* 1221 AssWa; Nicholas *Wynel* c1260 *ERO* (Ess); Thomas *Wynel* 1312 LLB D. *Win-el*, a diminutive of OE *Wine*. (ii) William *de Winhal'* 1214 Cur (Wa); Bartholomew *Wynhal* 1402 FA (Sf). From Winnall (He).

**Winnefrith:** *v.* WINFREY

**Winnell:** *v.* WINNALL

**Winnery, Winrow:** *v.* WHINNERAH

**Winney, Winny:** *Wengeue* uidua 1205 FFSf; Robert *Wyneue*, Benedict *Wynyeue* 1327 SRSf; John *Wyneue*, Thomas *Wynyff* 1478, 1479 SIA xii; John *Wenef, Wyny*, Robert *Wyndyff* 1524 SRSf; Nicholas *Wynnye* 1577 Musters (Nf). OE *\*Wyngeofu* (f) 'joy-battle'.

**Winnick, Winwick, Wynick:** Geruase *de Winewich* 1176 P (Ess/Herts); Fulco *de Wynewik'* 1241 FFHu; Augustin *de Wynewyk* 1276 AssLa. From Winwick (Hu, La).

**Winnicott:** *v.* WINCOTT

**Winnington:** Richard *Wynyngton* 1432–3 FFWa; Oliver *Wynnyngton* 1568 SRSf. From Winnington (Ch, St).

**Winpenny, Wimpenny:** Richard *Winepeni* 1219 SaG; William *Wynpeny* 1379 PTY; John *Winpenny* 1642 PrD. 'Win penny', a nickname for one of acquisitive habits. cf. John *Wynnegold* 1327 SRSf, 1365 LoPleas.

**Winser, Winsor:** *v.* WINDSOR

**Winship:** John *de Wenchep* 1275 RH (K); William *Wynchepe* 1457 ArchC 32; Robert *Winship* 1611 FrY. From Wincheap Street in Canterbury (K).

**Winskill, Winskell:** Thomas *de Wynscales*, John *de Wynschale* 1332 SRCu; Thomas *Wynteskelf* 1542 FrY. From Winskill, or from Winscales (Cu).

**Winsley:** Matthew *Wynselay* 1437 FrY. From Winsley (Db, He, W).

**Winslow, Winsloe:** William *de Wynselowe* 1332 SRWa; William *Wynselawe, Wynselowe* 1370, 1372 FFEss; Robert *Wynslowe* 1525 SRSx. From Winslow (Bucks).

**Winson:** Dola *Wines suna* c970 OEByn; Robert *Wynneson* 1330 NottBR; William *Wynson* 1525 SRSx. 'Son of *Wine*', *v.* WINN. Used as a christian name in the 17th century: *Winson* Risdon 1642 PrD.

**Winstanley, Winnstanley:** Roger *de Winstanesleg'* 1212 Fees (La); Hugh *de Wynstanlegh, de Wynstanley* 1387 IpmLa, 1401 AssLa. From Winstanley (Lancs).

**Winston, Winstone:** (i) Burewold *filius Wnstani* 1066 DB; Ælfwine *filius UUenstani* 1087–98 Bury; Amice *Wynston* 1303 IpmW. OE *Wynstān* 'joystone'. (ii) Emma, Robert *de Wineston* 1205 Pl (Sf); Robert *de Wynnestone* 1298 IpmGl; Adam *de Winstone* 1317 AssK. From Winston (Durham, Suffolk, Wight), Winston in Brixton (Devon), or Winstone (Glos). (iii) In Wales used as a translation of Welsh *Trewin* 'house of *Wyn*', cf. Drogo *de Wynston*, lord of the manor of *Trewyn* 1623 Morris.

**Winteney, Wintney:** Herteley *Wynteney* Hy 7 AD iii (Ha). From Wintney (Ha).

**Winter, Winters, Wintour, Wynter, Wynters:** *Winter* c1113 Burton (St); William, Roger *Winter* 1185 Templars (Wa), 1195 P (Berks). OE *Winter, Wintra*, or OG *Wintar*. Or a nickname. Medieval houses often had their walls painted with scenes from biblical history, romances, etc., or allegorical subjects such as the Wheel of Fortune or the representation of Winter 'with a sad and miserable face', which Henry III had painted over the fireplace in one of his rooms (EngLife 94).

**Winterborn, Winterborne, Winterbourn, Winterbourne, Winterburn:** Henry *de Winterburna* 1175 P (Do, So); James *de Winterburn'* 1230 Cur (O); John *de Wynterbourne* 1341 IpmW; William *Wynterbourne* 1372 IpmGl. From Winterborne (Dorset), Winterbourne (Berks, Wilts), or Winterburn (WRYorks).

**Winterflood:** Robert *de Winterflod* 1196–1237 Colch (Ess); Walter *Winterflood* 1274 RH (Ess); John *Wynterflode* 1535 FFEss. From Winterflood's Fm in Birch (Ess).

**Winterfold:** Thomas *de Winterfold* 1275 SRWo. From Winterfold in Chaddesley Corbett (Wo).

**Winteringham:** *v.* WINTRINGHAM

**Winterman:** John *Wynterman* 1312 LLB D. 'Servant of *Winter*', or a nickname for one cold and cheerless as winter.

**Winterscale, Wintersgill:** William *Wynterskale* 1584 FrLeic; Robert *Winterskill* 1721 FrY; George *Wintersgill* 1739 ib. From Winterscale 'winter hut' (ON *skáli*) in Rathmell, or Winterscales in Garsdale or Ingleton (WRYorks).

**Wintersett:** Richard *Wintresete* 1166 P (Y); Adam *of Wintersete* 1226 FFY; Hugh *de Wynterset'* 1379 PTY. From Wintersett (WRY).

**Winterslow:** John *de Wyntereslowe* 1375 IpmW. From Winterslow (W).

**Winterton:** Tochus *de Wintretune* 1066 DB (Nf); Osmund *de Winterton'* 1204 Pl (L); William *Wynterton* 1379 PTY. From Winterton (Lincs, Norfolk).

**Winthorpe, Winthrop, Winthrope, Winthrup, Wintrop, Wintrup:** John *de Wynthorp'* 13th RegAntiquiss; Simon *de Whynthorp'* 1298 AssL; John *Winthrop* 1568 SRSf. From Winthorpe (L, Nt).

**Wintney:** *v.* WINTENEY

**Winton, Wynton:** William *de Winton'* 1202 FFY; Nicholas *de Wynton'* 1277 FFEss; Richard *Wynton* 1525 SRSx. From Winton (Hants, Lancs, Westmorland, NRYorks).

**Wintringham, Winteringham:** Alan *de Wintringham* 1204 P (L); Reynold *de Wyntringham* 1300 FFY; William *Wyntryngham* 1392 IpmGl. From Wintringham (Hu, ERY), or Winteringham (L).

**Wintrop, Wintrup:** v. WINTHORPE

**Winwick:** v. WINNICK

**Winyard, Wingard, Wynyard:** Sarra *de Winiard'* 1212 Cur (Berks); Juliana *de la Wingarde* 1275 SRWo; William *ate Wyneard* 1327 SRSo. 'Worker in the vineyard' (OE *wīngeard*).

**Wiper, Wypers:** William *Wiper* 1202 Cur (Sx). An occupational name. The meaning is obscure but there is obviously no support for Weekley's derivation from Ypres.

**Wippe:** John *Wipe* 1203 AssNth; Peter *Wyppe* 1300 IpmW; John *Wippe* 1379 PTY. OE *Wippa*. v. also WHIPP.

**Wippett:** v. WHIPPETT

**Wire, Wyer, Wyers:** (i) Alicia *Wire* 1221 AssWo; Robert *le Wir* 1279 RH (Bk); William *Wier* 1576 SRW. OE *wīr* 'wire'. Metonymic for a wire-drawer. cf. William *le Wirdrawiere* 1320 LLB E. (ii) John *atte Wyre* 1367 LoPleas. 'Dweller at the place where bog-myrtle grows', OE *wīr*.

**Wirfauk:** John *Wirfax* 12th RegAntiquiss; Osbert *Wyrfauk* 1234, Thomas *Wirfauk* 1341 FFY. The first example suggests a nickname, 'wiry hair', OE *wīr, feax*.

**Wirth:** v. WORTH

**Wisbey:** v. WISBY

**Wisbidge:** John *Wysebech* 1379 LoCh; William *Wisbidg* 1662 HTEss. From Wisbech (C).

**Wisby, Wisbey:** Ralph *de Wiceby* 1200 Cur (L); Robert *de Wiscebi* 1202 AssL. From Whisby (L).

**Wiscard:** v. WISHART

**Wiscombe:** Nicholas *Wiscombe* 1662–4 HTDo. From Wiscombe Park in Southleigh (D).

**Wisdom:** Durand *Wisdom* 1198 P (Ess); Gilbert *Wysdom* 1243 AssSo. OE *wīsdōm* 'wisdom, learning'. Lower derives this from an estate in Cornwood (Devon). This is Wisdome, so named in 1618, and, possibly, earlier *Wymesdone* (PN D 271). Wisedom Fm in Drewsteignton is *Wisdoms* in 1464 (ib. 434), and is probably named from an owner called *Wisdom*. All the forms of the surname point clearly to the abstract noun.

**Wise, Wyse:** Johann *se Wisa* 11th OEByn; Ædwin *Wise* 1176 P (D); Roger *le Wis* 1203 P (Sx). OE *wīs* 'wise'.

**Wisebairn:** William *Wysebarne* 1304 IpmY, *Wysbarn* 1327 SRY. 'Son of *Wise*', or a nickname, 'wise child', OE *wīs, bearn*. cf. John *Wysheued* 1327 SRY 'wise head'; William *le Wyseprest* 1298 AssSt 'wise priest'.

**Wiseman:** *Wisman'* 1166 P (C); Lewin, Alexander *Wisman* 1154 Bury (Sf), 1202 AssL. OE *wīs* and *mann* 'the wise, discreet man', clearly used as a personal name. As an attribute, the meaning need not always be literal. Meanings recorded only later may have existed much earlier. The word is used ironically, applied to a fool, as in the 'wise men of Gotham'. cf. also 'Yonge *Wyseman* othyrwyse callyd *Foole*' (1471 NED). It was also used of a man skilled in the hidden arts, as magic, witchcraft and the like, 'magician, wizard' (1382 Wycliff).

**Wish:** Godric *Wisce* 1087 OEByn; Hugo *Wisc* 1199 P (Nf); John *de la Wisse* 1261 PN Sx 418; William *atte Wyshe* de la Rye 1305 LLB B. From residence near a damp meadow or marsh (OE *wisc*).

**Wishart, Wiskar, Wisker, Wesker, Whiscard, Whisker, Wysard, Vizard:** *Wischardus* Leidet 1176 P (Beds); *Guichard* de Charrun 1270 Ipm (Nb); Walter *Wiscard, Wishard* 1162, 1169 P (Sr); Rannulf *Wichard* 1212 Cur (Ess). ONFr *Wisc(h)ard*, OFr *Guisc(h)ard, Guiscart*, from ON *vizkr* 'wise' and the French suffix -(*h*)*ard*.

**Wisk, Wiske:** v. WISSOCK

**Wisken, Wiskin, Wisking, Whisken, Whiskin:** John *Wisekyn* 1327, Henry *Wiskin* 1568 SRSf; William *Whiskin*, Robert *Whisking* 1609 RamptonPR (C). *Wise-kin*, a diminutive of *Wise*.

**Wissler:** v. WHISTLER

**Wissock, Whisseck, Wisk, Wiske:** Roger *Wissok* 1275 SRWo; John *atte Wyseke* 1327 SRSx; Robert *atte Wiske* 1365 MELS (Sx). 'Dweller by the brook or water meadow', OE *\*wisoc*.

**Wisson:** v. WHISSON

**Wistance:** v. WHISTANCE

**Wiston:** v. WHISTON

**Wiswall:** Henry *de Wysewell* 1246 AssLa. From Wiswell (Lancs).

**Witard:** v. WHITTARD

**Witby:** v. WHITBY

**Witchell:** Perhaps OE *hwīt* and *cild* 'fair child': Alan *Wytechild, Whitechill* 1221 AssWo; Richard *Whitchild* 1322 FrLeic. Or from a place, as Witchells in Shirwell (Devon), *Wichehole* 1423 PN D 68.

**Witcher:** v. WHITCHER

**Witcomb(e):** v. WHITCOMB

**Withacombe:** v. WIDDECOMBE

**Witham, Withams, Wittams, Whitham, Whittam, Whittome:** Warin *de Whyteham* Ed 1 FFEss; Peter *de Wytham* 1295 FFEss; John *Witham* 1327 *SR* (Ess). From Witham (Essex, Devon, Lincs, Som).

**Withecombe:** v. WIDDECOMBE

**Wither, Withers:** (i) *Wither, Wider* 1066 DB (Nf, L); *Wyther* cognomento Turnel 1134–40 Holme (Nf); Richerus *filius Wither* 1153–68 ib.; William *Wither* c1160 Gilb (L); Robert *Wiðer* 1176 P (St); Geoffrey *Wider* 1192 P (Gl). ON *Viðarr*, ODa *Withar*. (ii) John *Wythiar'* 1327 SRSx; Thomas *le Wythier* 1332 ib. 'Dweller by the willow' (OE *wīþig*). cf. WITHEY.

**Witherby:** For WETHERBEE.

**Witherick:** v. WHITTERIDGE

**Witheridge:** Ralph *de Wyperug'* 1279 RH (O). From Witheridge Hill in Rotherfield Greys (Oxon). Also from Witheridge (Devon).

**Witherington, Withrington:** Hwylardus *de Wytherendenn* c1180 ArchC vi; Crystina *de Wytheryndenn* 1296 SRSx. From Witherenden in Ticehurst (Sussex).

**Witherley, Witherly:** Iuetta *de Widerleg'* 1201 P (Sa). From Witherley (Lei).

**Withey, Withy:** Adam *de la Wythye* 1241 Abbr (So); Richard *atte Widege*, Robert *Widie* 1279 RH (O). 'Dweller by the willow' (OE *wīþig*). cf. WITHER. (ii) John *de Wythye* and Alan *atte Wythye* (1296, 1332 SRSx) lived at Withy Fm in Danehill (Sussex).

**Withington:** Roger *de Withinton* 1231–2 FFWa; John *de Withyngton* 1290 AssCh; Henry *de Wythyngton'* 1332 SRDo. From Withington (Ch, Gl, He, La, Sa), or Witherington Fm in Standlynch (W), *Widintona* 1209.

Withnall, Withnell: Robert *de Withinhulle* 1332 SRLa. From Withnell (La).

Withycombe: *v.* WIDDECOMBE

Witless: Robert *Wytelas* 1275, John *Wytles* 1327 SRWo. 'Witless, foolish', OE *witt, lēas*.

Witley: *v.* WHITELEY

Witney: William *de Wittenheia* 1210–11 PWi; Simon *de Witteney* 1278–9 CtH; Baldwin *de Wytteneye* 1346–7 FFWa. From Witney (O).

Witrick: *v.* WHITTERIDGE

Witt, Witte, Witts: *v.* WHITE

Wittaker: *v.* WHITAKER

Wittamore: *v.* WHITEMORE

Wittams: *v.* WITHAM

Wittenham: *v.* WHITTINGHAM

Witterick: *v.* WHITTERIDGE

Wittering: William *de Wyghtryng* 1327 SRSx. From Wittering (Nth, Sx).

Wittey: *v.* WITTY

Wittick: *v.* WHITEWICK

Witting: *v.* WHITING

Wittman: *v.* WHITEMAN

Witton: Robert *de Wittun* 12th Gilb; Robert *of Wyton* 1245–6 FFY; John *Wytton* 1440 FFEss. From Witton (Ch, La, He, Nf, Wa, Wo), Witton Gilbert, le Wear (Du), Long, Nether Witton (Nb), or East, West Witton (NRY).

Wittrick, Wittridge: *v.* WHITTERIDGE

Witty, Wittey: Robert *Witie, Wity* 1221 AssWo; Symon *Wittie* 13th Riev (Y). OE *wit(t)ig* 'sagacious, wise', also in ME 'witty'. cf. Willelmus *Fascetus* (*Facetus*) 1200 Cur (Nf). *v.* also WHITEY.

Witwell: *v.* WHITWELL

Wix: *v.* WICH

Woburn: John *de Woburn* 1328 FFW. From Woburn (Beds, Sr).

Woby: Probably for *Waudby. v.* WALBY.

Wod, Wode: *v.* WOOD

Wodhams: *v.* WOODHAM

Wodehouse: *v.* WOODHOUSE

Wodeman: *v.* WADEMAN

Woffenden, Woffinden, Woffindin: *v.* WOLFENDEN

Woffitt: *v.* WOOLVETT

Wogan: John *Wogan* 1292 QW (Cu); John *Wougan* 1297 MinAcctCo (Y). An anglicizing of Welsh *Gwgan*, a name found in the *Mabinogion*, a diminutive of *gwg* 'a scowl, frown'.

Wolbold: *Wlbald, Wlbolt* 1066 DB (Sf); *Ulbold, Hulboldus* c1130 ELPN; Robert *filius Wulfbalt* 1177 P (Sf); *Wulbald* 1275 RH (K); Walter *Woulbould*, John *Wolbol* 1327 SRSo. OE *Wulfbeald* 'wolf-bold', found only in the 10th and 11th centuries in OE. The surname is rare. Catherine *Woolball* was a benefactor of Walthamstow (Essex) in 1786.

Wolde: *v.* WALD

Woledge: *v.* WOOLWICH, WORLEDGE

Wolf, Wolfe, Wolfes, Woolf, Woolfe, Wulff, Woof, Wooff: Robert *Wulf* 1166 P (L); John *le Wolf* 1279 RH (Beds). A nickname from the wolf (OE *wulf*), often latinized as *Lupus* and found in the French forms *Lou* and *Love. v.* LOW, LOVE. The derivation from OE *Wulf* is very doubtful. This personal name is late and not common in OE and Redin doubts if any

of the examples are really native. ON *Úlfr* was not uncommon and has survived. *v.* ULPH. But *Wolf* as a surname is very seldom without the article *le* in the 13th and 14th centuries. The frequency of the surname, and some of its forms, are due to recent immigrants from Germany.

Wolfarth: *v.* WOLLARD

Wolfendale: Richard *Woofendall, Woffendale* 1675 RothwellPR (Y). Perhaps a dialectal pronunciation of *Wolfenden*, a name not uncommon in Leeds and Bradford.

Wolfenden, Woolfenden, Woffenden, Woffendon, Woffinden, Woffindin, Wooffinden: James *Wolfenden* 1614 Bardsley. From Wolfenden in Newchurch-in-Rossendale (Lancs), earlier *Wolfhamdene*.

Wolfhunt: William *le Wlfhunte* 1249 AssW; Richard *le Wulfhunte* 1271 FFK; Walter *le Wolfhunte* 1339 IpmNt. 'Wolf hunter', OE *wulf, hunta*. cf. Geoffrey *Wolvesheved* 1279 AssNb 'wolf's head, i.e. outlaw'.

Wolfin: *v.* WOOLVEN

Wolfit: *v.* WOOLVETT

Wolfner: *v.* WOOLNER

Wolford, Wolforth, Wolfoot: *v.* WOOLLARD

Wolfram: *Vluerona* uxor Vlf 1130 P (L); Ralph *filius Wlfrun* 1209 Pl (Nf); John *Wulfrun* 1230 Cur (Sr); Thomas *Woulfran* 1297 MinAcctCo (O); Alice *Wolfroun* 1332 SRSr. OE *Wulfrūn* (f).

Wolfryd: *v.* WOOLFREY

Woll: *v.* WOOLL

Wollan: *v.* WOOLLAND

Wollard: *v.* WOOLLARD

Wollas: *v.* WOOLHOUSE

Wollaston, Woolaston, Wolston: Rolland *de Wlauestona* 1199 FFNth; Henry *de Wolaston* 1327 SRSt; William *Wolston* of Wollaston 1426 FFEss. From Wollaston (Nth, Sa, Wo).

Wollaton: Thomas *de Wolaton* 1406 IpmNt. From Wollaton (Nt).

Wolledge: *v.* WOOLWICH, WORLEDGE

Wollen: *v.* WOOLLAND

Woller: *v.* WOOLLER

Wollerton: *v.* WOOLERTON

Wollett: *v.* WOOLVETT

Wolley: *v.* WOOLLEY

Wollman, Wolman: *v.* WOOLMAN

Wollmer: *v.* WOOLMER

Wollop: *v.* WALLOP

Wolpole: *v.* WALPOLE

Wolrich, Wolrige: *v.* WOOLRICH

Wolseley, Wolsley, Woosley: William *de Wulfsieslega* 1177 P (St); Robert *de Wolsleg* 1285 FA (St). From Wolseley (Staffs).

Wolsey: *v.* WOOLSEY

Wolstencroft, Wolstoncroft, Woolstencroft, Worsencroft, Worstencroft, Wosencroft, Wozencroft: Thomas *Wustyncrofte*, John *Wustencroft* 1524 SRSf; James *Wolsoncroft* 1584 Bardsley; Thomas *Worsencroft* 1610 ib.; William *Woolstencroft* 1613 ib.; John *Wossencroft* 1635 ib. From Wolstancroft (Lancs).

Wolstenholme, Wolstanholme, Wolstenhulme, Woolstenhulme, Worstenholme, Wostenholm, Wusteman, Woosnam, Woosman, Worsman: Andrew *de Wolstanesholm* 13th WhC (La). From Wolstenholme in Rochdale (Lancs).

**Wolston:** *v.* WOLLASTON, WOOLSTON

**Wolters:** *v.* WALTER

**Wolton:** *v.* WOOLTON

**Wolveridge:** *v.* WOOLRICH

**Wolverton:** *v.* WOOLVERTON

**Wolvey:** (i) Robert *de Wlueia* 1200 P (Wa); Maurice *ate Wolfaghe* 1327 SRSx; Roger Wolfey 1327 *SR* (Ess). From Wolvey (Warwicks), or 'man in charge of the enclosure to protect flocks from wolves' (OE *wulfhaga*), or 'wolf-trapper'. (ii) Robertus *filius Wulfiue* 1193 P (Nf); Willelmus *filius Wulveve* 1279 RH (C); Hardwin *Wuluiue* 1205 P (Nf); Isabella *Wolfueue* 1327 SRSf. OE *Wulfgifu* (f) 'wolf-gift'. *v.* WOOLWAY.

**Wombe:** *v.* WAMBE

**Wombwell, Wombell, Woombell, Woombill, Woomble, Womwell:** Reiner *de Wambewell'* 1219 AssY; William *Wombell* 1666 FrY. From Wombwell (WRYorks).

**Womersley:** Thomas *Wymbersley* 1463 GildY; John *Wymbersley, Womersley* or *Wymbersla* 1509 LP (Y); Richard *Womerslay* 1567 GildY. From Womersley (WRY).

**Wontner:** Richard *Wantenor* 1450 SaG. A derivative of OE *wante* 'mole', a mole-catcher.

**Wood, Woodd, Woode, Woods, Wod, Wode:** (i) Walter *de la Wode* 1242 Fees (He); John *del Wode* 1274 Wak (Y); John *Atewode* 1274 RH (Ess); John *I the Wode,* Gilbert *a la Wode* 1275 SRWo; Elias *in le Wode* 1279 RH (C); Alexander *of the Wode* 1285 AssLa; John *at the Wode* 1293 Fees (D); William *Bythewode* 1296 SRSx; Richard *dilwod* 1327 Pinchbeck (Sf). From residence in or near a wood. (ii) Adam *le Wode* 1221 AssWo; Richard *Wod* 1230 P (So); William *le Wod* 1275 RH (Do). OE *wōd,* ME *wod(e),* wood 'frenzied, wild'. Much less common than the local origin. cf. Shakespeare's 'And heere am I, and *wood* within this wood'.

**Woodall:** *v.* WOODHALL

**Woodard, Huddart:** (i) *Wudardus* de Rosshebrok 1148–56 Bury (Sf); Johannes *filius Wdardi* 1166 RBE (Nb), *filius Wudardi* 1168 P (W); Willelmus *filius Vdard'* 1181 P (Cu); *Udardus* de Karelton 1214 Cur (Nt); Walterus *filius Wodard* 1227 Pat (Sf); Walter *Wudard* 1221 AssWa; William *Hudard* Hy 3 Colch (Sf); John *Wodard* 1279 RH (C). OE *\*Wuduheard,* unrecorded in OE, but regularly formed of the themes *wudu* 'wood' and *heard* 'hard'. (ii) Richard *le Wodehirde* 1275 RH (Nf); Robert *le Wodehirde* 1325 FFSf. OE *wudu* 'wood' and *hierde* 'herd', one who tends animals feeding in a wood; probably a swineherd.

**Woodberry, Woodbury:** Ralph *de Wudeburc* 1194 P (Nt); John *de Wodeburgh* 1324 FFEss; John *Woodberie* 1642 PrD. From Woodborough (Nt, W), or Woodbury (D).

**Woodbourn, Woodbourne:** *v.* WOODBURN

**Woodbridge:** Alan *de Wodebrig'* 1243 MELS (So); David *de Wdebregg'* 1275 RH (Nf). From Woodbridge (Sf), or 'dweller by the wooden bridge', OE *wudu, brycg.*

**Woodburn, Woodburne, Woodbourn, Woodbourne:** Jordan *de Wodeburn'* 1230 P (D); Richard *Wodebourn*

1433 AssLo; William *Wodburn* 1470 FrY. From Woodburn (Nb), or Woodburn in Oakford (D).

**Woodbury:** *v.* WOODBERRY

**Woodcock, Woodcocks:** William *Wdecoch'* 1175 P (Nf); Roger *Wudecoc* 1176 P (Herts); William *Widecoc* 1200 FFEss; Nicholas *Wodecok* 1288 AssCh. A nickname from the bird, OE *wuducocc* 'woodcock', later used to mean 'a fool, simpleton, dupe'. The frequency of this surname is probably partly due to the absorption of a local surname *Woodcot:* Richard *de Wdecote* 1198 FF (O); Adam *de Wudecota* 1193 P (Sa); Henry, Thomas *at(t)e Wodecote* 1327 SRSx, 1332 SRWo, from Woodcote (Hants, Oxon, Salop, Surrey, Warwicks), Woodcott (Ches, Hants), or from residence at a cottage in or near a wood.

**Woodcraft, Woodcroft:** Ascelinus *de Wudecroft* 1162 P (Nth). 'Dweller at a croft by a wood' (OE *wudu, croft*).

**Wooddin:** *v.* WOODING

**Wooddisse:** *v.* WOODHOUSE, WOODIWISS

**Woodell:** *v.* WOODHILL

**Woodend, Woodsend:** Clarcia *de la Wodeande* 1260 FFEss; Adam *de Wodeshende* 1273 RH (Do); Robert *de la Wode ende* 1275 SRWo. 'Dweller at the end of the wood'. cf. TOWNEND.

**Wooder:** Andrew, Geoffrey *le Wodere* 1275 RH (Sx), 1280 AssSo. OE *wudere* 'wood-man, woodcarrier'. *v.* also WADER.

**Wooders:** *v.* WOODHOUSE

**Woodey:** (i) Thomas *de la Wdhaye* 1275 RH (L); Robert *atte Wodeheye* 1333 MELS (So). 'Dweller by the enclosure in the wood', OE *wudu, (ge)hæg.* (ii) Roger *Wody* 1255 RH (Sa); Geoffrey *Wody* 1275 RH (Nth); Richard *le Wodye* 1296 SRSx; John *Woodye* 1568 SRSf. This seems to have been more common in ME than (i) and is apparently ME *\*wody,* a derivative of ME *wod,* wood, from OE *wōd,* 'frenzied, wild', formed on the analogy of *moody, mighty.* cf. WOOD.

**Woodfall:** Gilbert *de Wudefalde* 1193 P (Ess); Richard *de Wodefalle* 1285 AssLa 'Dweller by a fold in the wood' as at Woodfalls (Wilts), *Wudefolde* 1258 PN W 398, or at a place where trees have fallen, as at Woodfall Hall (Lancs), *Wudefal* a1230 PN La 109.

**Woodfield:** Margaret *de Wodefeld* 1316 FA (W). From Woodfolds Fm in Oaksey (W), *Woodfeldes* 1568. In Scotland from Woodfield (Dumfries).

**Woodfin, Woodfine, Woodvine:** George *Wodvine* 1642 PrD; Thomas *Woodfine* 1662 HTEss. 'Dweller by the pile of wood', OE *wudu, fin.*

**Woodford, Woodforde, Woodfords:** (i) Daniel *de Wudeford'* 1196 P (O); Geoffrey *de Wodeford* 1327 SRSo; William *Wodeford* 1423 FFEss. From Woodford (Ches, Cornwall, Essex, Northants, Wilts). (ii) Jordan *de Wodford* c1170 Black; Robert *de Wodforde, de Vodford* 1296, c1330 ib. From Woodford in St Boswells (Roxburgh).

**Woodfull:** *Wudefugel* 1195 P (Y), 1196 P (Nth); Avice, Robert *Wodefoul* 1279 RH (Bk), 1311 RamsCt (Hu). OE *wudu-fugol* 'a bird of the woods', perhaps a nickname, but certainly used as a personal-name like *Sæfugul. v.* SAFFELL.

**Woodgate, Woodgates, Woodget, Woodgett, Woodjetts, Woodyatt:** William *de Wudegat'* 1199 Cur (Do); Walter *de Wodiate* 1208 Cur (Do); Aluric *de la Wdegate* 1222 DBStP (Ess); Martin *atte Wodegate* 1296 SRSx; William *Wodiet* 1328 AD i (Bk); John *atte Wodeyate* 1332 SRWa. From Woodyates (Dorset) or residence by the gate to the wood (OE *wudu, geat*). *Woodgett* shows the same development of *di* as in *soldier*.

**Woodger:** *v.* WOODIER

**Woodgett:** *v.* WOODGATE

**Woodhall, Woodall:** Peter *de Wudehale* 1193 P (L); Jordan *de la Wodehalle* c1265 Calv (Y); John *del Wodhall* 1332 SRCu; Alice *atte Wodehalle* 1332 MELS (Sr). 'Dweller (or servant) at the hall in the wood.'

**Woodham, Woodhams, Wodhams:** Alexander *de Wudeham* 1170 P (Ess); Thomas *ate Wodehamme* 1332 SRSr; Thomas *Wodeham* 1351 AssEss. From Woodham (Essex, Surrey), *\*Wuduhām* 'village in the wood', or 'dweller at the enclosure in the wood', *\*Wuduhamm.*

**Woodhatch:** Robert *atte Wodehacche* 1369 MELS (Sr). 'Dweller by, or keeper of the gate to the wood', OE *wudu, hæcc,* as at Woodhatch (Essex, Surrey).

**Woodhead:** Edward *de Wodheved* 1243 AssSo; Christiana *del Wodeheved* 1316 Wak (Y); Isabel *Wodhed* 1440 ShefA (Y). 'Dweller at the head or top of the wood' (OE *hēafod*), as at Woodhead (Devon, WRYorks).

**Woodhill, Woodell:** Albin *de Wuduhull'* c1227 Fees (W); Symon *de Wodhill* 1275 RH (R); John *de Wodhull* 1325 FFEss; Michael *Woodhill,* cordwainer 1693, William *Woodwill,* cordwainer, son of Michael *Woodell,* cordwainer 1755 FrY. From Woodhill Park in Clyffe Pypard (W), Woodhill in Sandon (Ess), or 'dweller by the hill where woad grows', or 'by the wooded hill', OE *wād/wudu, hyll.*

**Woodhirst, Woodhurst:** Matilda *de Wodehurst* 1332 SRSx. From Woodhurst (Hu), or Woodhurst in Slaugham (Sx).

**Woodhorn:** Liulf *de Wudehorn* 1200 P (Nb), William *Wodhorn* 1280 AssSo; John *Wodehorne* 1327 SRSx. From Woodhorn (Nb), and perhaps also from a personal name, OE *\*Wuduhorn,* or OE *\*Wōdhorn. v.* OEByn 357.

**Woodhouse, Wodehouse, Woodus, Wooders, Wooddisse:** Helias *de Wudehus* 1170 P (Nt); Richard *del Wodehus* 1275 RH (Sf); Geoffrey *Wodehus* 1276 RH (Y); Roger *ate Wodehose,* Thomas *Wodehose* 1332 SRSr; Walter *atte Wodehous* 1332 SRSt; Thomas *Wodehouse* or *Wodesse* 1509 Ewen. 'Dweller at the house in the wood.' Woodhouse Fm in Upper Arley (Worcs) is *Woddus* 1387 PN Wo 33. *v.* WIDDOWES.

**Woodhurst:** *v.* WOODHIRST

**Woodier, Woodyear, Woodyer, Woodger:** Robert *le Wodehyewere* 1301 ParlR (Ess); Walter *le Wodehewer* 1309 SRBeds; Richard *Woodyere* 1605 Oxon; John *Woodger* 1663 Bardsley. OE *wuduhēawere* 'hewer of wood, wood-cutter'. For *Woodger,* cf. WOODGATE, GOODGER.

**Wooding, Woodings, Wooddin, Woodin, Wooden:** William *Woding* 1247 AssBeds; Richard *atte Woding* 1294 MELS (Wo); Abram *Woodinge,* Daniel *Wooden* 1674 HTSf. 'Dweller at a place where wood has been cut', OE *\*wudung* 'cutting of wood'. The first example, without a preposition, though not without parallels at this date, may be a nickname from OE *\*wōding,* from *wōd* 'mad', 'the mad one'.

**Woodiwiss, Wooddisse, Widdiwiss:** Robert *de* (sic) *Wudewuse* 1251 AssY. OE *wuduwāsa* 'faun, satyr'. cf. 'Wodewese (woodwose). *Silvanus, satirus*' PromptParv.

**Woodjetts:** *v.* WOODGATE

**Woodkirk:** Thomas *de Wodkyrc* 1379 PTY. From Woodkirk (WRY).

**Woodlake:** Turstin *filius Wudelach'* 1180 P (W); Goduine *Udelac* c1095 Bury (Sf); John *Wdeloch* 1225–50 Dublin. OE *\*Wudulāc,* an addition to the rare compounds of *Wudu-.* Roger *Wudeloc* 1245 Ct (Ha) gave name to Woodlark Fm in Bighton (Hants).

**Woodland, Woodlands:** Henry *de Wudeland'* 1195 P (L); William *Wudeland'* 1214 Cur (Sx); Thomas *atte Wodelond* 1327 SRSo. 'Dweller by the wood-land.'

**Woodleaf:** William *Wodelof* 1210–11 PWi; William *Wodelef* 1378–9 FFSr. OE *\*Wuduleof.*

**Woodley, Woodleigh:** Ælfnoð *at Wudeleage* 1008–12 OEByn (D); Osbert *de Wudeleg'* 1199 P (D); Walter *de Wodeleye* 1332 SRWo; Richard *Wodle* 1384 IpmGl. From Woodleigh (Devon), or 'dweller at a glade in the wood'.

**Woodlow:** Thomas *Wodelowe* 1432–3 FFWa. From Woodloes in Warwick St Nicholas (Wa).

**Woodman:** *Vdeman, Odeman* 1066 DB; *Wudemann* 1066–75 ASWrits; *Wdemannus* Hy 2 AD i (Nth); Hugo *Wudeman* 1166 P (Y); John *Wodeman* 1213 Cur (Do); Nicholas *le Wodeman* 1294 AssSt. OE *Wudumann,* from *\*wudumann* 'woodman', also used as an occupational name.

**Woodmouse:** Ailwin *Wudemus* 1188 P (Db); Thomas *Wdemus* 1233–45 RegAntiquiss; William *Wodemous* 1286 Wak (Y). 'Wood mouse', OE *wudu, mūs.*

**Woodnorth, Woodnoth, Woodnut, Woodnutt:** Henry *Wodenot* Hy 3 IpmY; William *Wodenot* 1289 AssCh; Robert *Wodenot* 1331 Pat. OE *\*Wudunoð.* cf. Shavington Woodnoth in Rope (Ch).

**Woodriff, Woodriffe:** This surname is probably from OE *wudurife* 'woodruff'. *v.* WOODROFF. It is just possible it may be for *wood-reeve:* Philip *Wodereve* 1253 AssSo (1579 NED). This is rare and no modern example has been noted, although Harrison gives *Woodreefe, Woodreeve* and *Woodreve.*

**Woodroff, Woodroffe, Woodroof, Woodroofe, Woodrooffe, Woodrough, Woodruff, Woodruffe, Woodrup:** *Wuderoua* 1170 P (Wa); Hugo *Wderoue* 1185 Templars (L); Robert *Woderove* 1225 AssSo; John *Woderofe* 1419 DbAS 36. OE *wudurofe* 'woodruff'. cf. 'Woderove, herbe. *Hasta regia, hastula, ligiscus'.* It was formerly a custom for women to carry woodruff, because of its sweet smell, with their prayer-books to church. 'The woodruff is also called *sweet* *woodruff* because of its strongly sweet-scented leaves. This may account for the nickname. It may have been given to a person who used perfumes, or the nicknames may be ironical' (ELPN 172). The usual etymology from 'wood-reeve' finds no support in the above forms.

**Woodrose:** Henry *Woderose* 1332 SRSx. 'Wild rose', OE *wudu*, Lat *rosa*.

**Woodrow:** William *Woderowe* 1260 AssC; Juliana *de Woderoue*, William, Matilda *Woderoue* 1296 SRSx. The last three were assessed in the same parish. OE *wudu* 'wood' and *rāw* 'row', probably from residence in a row of cottages in a wood. Also probably from Woodrow (Wilts, Worcs).

**Woodsend:** v. WOODEND

**Woodsford:** Henry *de Wodesford* 1327 SRSo. From Woodsford (Do).

**Woodshaw:** Roger *de Wodeshawe* 1299 AssW; Roger *Woodshawe* 1545 SRW. From Woodshaw in Wootton Bassett (W).

**Woodside:** Robert *del Wodsid* 1332 SRCu. 'Dweller by the wood-side.'

**Woodsom, Woodsome:** Adam *de Wodesom* 1335, Thomas *de Wodesum* 1368 FFY. From Woodsome (WRY).

**Woodstock:** Alisius *de Wodestoke* 1235 FFO; John *de Wodestok'* 1340 Glast (So); Robert *Wodestoke* 1359 AssD. From Woodstock (O).

**Woodthorpe, Woodthrop:** William *de Wudetorp* 1190 P (L); Thomas *de Wuttorp* 1220 Cur (L); William *of Wodethorp* 1251–2 FFY. From Woodthorpe (Db, L), Woodthorpe Hall in Handsworth, or Woodthorpe in Sandal (WRY).

**Woodus:** v. WOODHOUSE

**Woodvine:** v. WOODFIN

**Woodwall, Woodwell:** Robert *de Wudewelle* c1240 Fees (Beds); Geoffrey *de Wodewell* 1327 SRSf; John *Wodewale* 1409 CtH. 'Dweller by the stream or spring in the wood', OE *wudu, wiella*.

**Woodward:** Sewhal *le wudeward* 1208 P (Ha); Ralph *de* (sic) *Wodeward* 1230 P (Herts); Robert *Wodeward* 1296 SRSx. OE *wuduweard* 'forester'. Henry the reeve (*prepositus*) is also called *le wudeward'* 1221 AssWa.

**Woodyatt:** v. WOODGATE

**Woodyear, Woodyer:** v. WOODIER

**Woof(f):** v. WOLF

**Wooffinden:** v. WOLFENDEN

**Wooland:** v. WOOLLAND

**Woolard:** v. WOOLLARD

**Woolas(s):** v. WOOLHOUSE

**Woolaston:** v. WOLLASTON

**Woolaway:** v. WOOLWAY

**Woolbar:** Probably a corruption of WOLBOLD.

**Woolbridge:** v. WOOLL

**Wooldridge:** v. WOOLRICH

**Wooler:** v. WOOLLER

**Woolerton, Wollerton, Woolloton:** Robert *de Wuluriton'* 1221 AssSa. From Wollerton (Salop).

**Wooles:** v. WOOLL

**Wooley:** v. WOOLLEY

**Woolf(e):** v. WOLF

**Woolfall:** John *de la Wulfhal* 1242 FFLa; Robert *de Wolfalle* 1332 SRLa. From Wolfall Hall in Huyton (Lancs) or Wolf Hall in Grafton (Wilts).

**Woolfenden:** v. WOLFENDEN

**Woolfield:** Richard *Wolfild* 1334 SRK. OE *Wulfhild* (f) 'wolf-war', recorded four times in the 10th and 11th centuries but, though rare, must have survived the Conquest.

**Woolfit:** v. WOOLVETT

**Woolford(e), Woolfoot:** v. WOOLLARD

**Woolfrey, Woolfries, Wolfryd:** *Vlfert, Vluert* 1066 DB (D, So); Thomas *Wulfrith* 1260 AssC; William *Wlfrið* 1279 RH (C); John *Wolfrith*, Matilda *Woulfryth* 1297 MinAcctCo. OE *Wulffrið* 'wolf-peace'.

**Woolgar, Woolger:** *Wlgar, Vlgar* 1066 DB; *Wlfgarus* de Cokesale 1252 Colch (Ess); Brixi *Wulgar* 1188 P (Nf); Teobald *Wolgar* c1250 Rams (C). OE *Wulfgār* 'wolf-spear'.

**Woolham:** v. WOOLLAND

**Woolhouse, Woolas, Woolass, Wollas, Wholehouse:** Malyn *del Wllehus* 1277 Wak (Y); Ralph *del Wolehouse* 1301 SRY; John *Wllehous, Wollehous* 1377 ColchCt. 'Worker at the wool-(ware) house' (OE *wull, hūs*).

**Woolich, Woolidge:** v. WOOLWICH, WORLEDGE

**Wooll, Woolls, Wooles, Woll:** John *de Wolle* 1296 SRSx; Henry *atte Wolle* 1327 SRSx; Thomas, Alan *atte Wulle* 1327 SRSo, 1332 SRSr. 'Dweller by the spring or stream.' In the south-west, particularly in Dorset, OE *wielle, wyll* became ME *wull, woll*, instead of *well*, and this survives in Wool and in Wool Bridge in East Stoke which gave rise to a surname *Wullebrigg* in 1244 (PN Do 145). The same development has been noted in Somerset, West Sussex and West Surrey.

**Woollage:** v. WOOLWICH, WORLEDGE

**Woollam(s):** v. WOOLLAND

**Woolland, Woollan, Woollon, Woollons, Woollen, Wooland, Woolen, Wollan, Wollen, Woollam, Woollams, Woolham:** Samuel *Woolons* 1637 FrY; John *Woolland* 1674 HTSf. 'Dweller by the curved or crooked land(s)' (OE *wōh, land*). cf. Woollensbrook (PN Herts 212): *la Wowelond* 1235, *Wowlondes* 1466, *Woolans brook* 1689, *Woollens* c1840; Woolens Copse in Damerham, *Wolonde* (1518 PN W 402); *Woulond* 1230 PN C 335; and *v*. also Woolland (PN Do 192). This is a type of place-name likely to be common and accounts for all the forms. The final *n*, as often, became *m*, giving *Woollam*, and this was re-spelled *Woolham* on the analogy of names in -*ham*. No such place as *Woolham* is known. The occasional -*s* is from the plural, -*lands*.

**Woollard, Woolard, Woolatt, Wollard, Woolward, Wolfarth, Wolford, Wolforth, Woolford, Woolforde, Woolfoot:** (i) *Wluuard, Vluuard* 1066 DB; *Wluerdus* Legredere 1188 BuryS (Sf); Robertus *filius Wolfward* 1212 Fees (Sr); *Wulford* atte Heghelonde 1275 RH (K); Martin *Wlward* 1199 FFC; Richard *Wulward* 1206 P (Hu); Matilda *Woleward* 1275 SRWo; Robert *Wlfard, Wulvard* 1279 RH (O, C); Ralph *Wolford* 1317 AssK; John *Woolat* 1674 HTSf. OE *Wulfweard* 'wolf-ward'. (ii) Robert *de Wlfward* 1210 P (Wa);

Engerram *de Wolward'* 1212 Cur (Wa). From Wolford (Warwicks).

**Woollcott, Woolcot, Woolcott, Woolcock:** Philip *Woolcott* 1642 PrD; Anne *Woollcott* 1662–4 HTDo. From Wollacott in Thrushelton (D).

**Wooler, Wooler, Woller:** William *le Woller'* 1319 *SR* (Ess); John *Woller* 1327 SRSo. A derivative of OE *wull* 'wool', a dresser, weaver or seller of wool. cf. Thomas *le Wollestere* 1297 Wak (Y), where we have the feminine form. In John *Wollore* 1430 FrY, the name derives from Wooler (Northumb), *Wulloure* 1187 DEPN.

**Woollet, Woollett:** *v.* WOOLVETT

**Woolley, Wooley, Wolly:** *Hugo de Wuluele* 1219 AssY; Ralph *de Wullueleye* 1230 P (Berks); Roger *de Wolvele* 1279 RH (O); Nicholas *de Wolveleye* 1280 AssSt; Bate *de Wolflay* 1308 Wak (Y); Robert *de Woluelie* 1327 SRSx; John *de Wolley* 1332 SRSt. From Woolley (Berks, Devon, Hunts, Wilts, WRYorks) or Woolfly Fm in Henfield (Sussex), all 'wolves' wood', or lost places of the same name, or from Wooley in Slaley (Northumb), *Ulflawe* 1296 'wolf-hill', or from Woolley (Som), *Wilege* DB 'clearing by a stream'. cf. WOOLL.

**Woollon(s):** *v.* WOOLLAND

**Woolloton:** *v.* WOOLVERTON

**Woollston:** *v.* WOOLSTON

**Woollven, Woollvern, Woollvin:** *v.* WOOLVEN

**Woolman, Wolman, Wollman:** William *Wolleman* 1351 AssEss; Richard *Woolman* 1674 HTSf. OE *wull* 'wool' and *mann*, a dealer in wool, a wool-merchant (1390 NED).

**Woolmer, Woolmore, Wollmer, Ullmer:** (i) *Wlmer, Vlfmer, Wlmar* 1066 DB; *Ulmer* Ælltredes c1095 Bury (Sf); *Wlfmerus* filius Sirici 1101–7 Holme (Nf); *Wulmarus* Lamb 1230 P (Nf); William *Wulmare* 1260 AssC; Reginald *Wlmer* 1263 Rams (Hu). OE *Wulfmær* 'wolf-famous'. (ii) Richard *de Wulvemor*, Robert *de Wullemore* 1246 AssLa. This is from a lost *Wolmoor* in Ormskirk (Lancs) 'wolves' moor'. The surname may also derive from Woolmer Forest (Hants), Wolmer Fm in Ogbourne St George (Wilts), or Woolmore Fm in Melksham (Wilts), all 'wolves' pool'.

**Woolner, Woolnoth, Woolnough, Wolfner:** *Wlnod, Vlnoth, Vlnoht* 1066 DB; Godwine *Hulnodi filius* c1095 Bury (Sf); *Wulnoth* Koc 1296 SRSx; William *Wulnoð* 1221 ElyA (Sf); Ranolph *Wollenoth* 1284 RamsCt (Hu); Robert *Wolthnoth* 1327 SRSf; Thomas *Wolnoth* 1348 FFSf; William *Vlnawgh* son of Mawt *Vlnaugh*, William *Wulnow*, Richard *Wulnaw* 1524 SRSf; Edmund *Wolner* 1524 SRSf; Margaret *Woollnoth* 1525 SIA xii; John *Woolnough*, Francis *Woolnoe*, Henry *Woolner* 1674 HTSf. OE *Wulfnōð* 'wolf-boldness', a personal name which, though not common, remained in use until the beginning of the 14th century. As with *Elnough*, from *Æðelnōð*, the form *Woolnough* probably reflects the dialectal sound change of *ð* to *f*. *v.* OEByn 170. The *-er* of *Woolner* and *Wolfner* represents the pronunciation of unaccented *-nough*. cf. ALLNATT.

**Woolrich, Woolrych, Wooldridge, Woolridge, Wolrich, Wolrige, Wolryche, Wolveridge, Woolveridge,**

**Woolwright, Ullrich, Ulrich, Ulrik:** *Wlfric, Vlfric* 1066 DB; *Uluric* Stanilde filius c1095 Bury (Sf); *Wlricus* filius Actman 1198 Cur (K); William *Wulfric* 1212 Cur (Berks); William *Wlwrith* 1242 Rams (C); Ralph *Wolfrich* 1250 Fees (Sr); Robert *Wolfrick, Wolurich* 1279 RH (O); Robert *Wulrich* 1286 AssSt; Robert *Wolvrych* 1296 SRSx; Richard *Wolrich* 1307 AssSt; Ralph *Wolverych* 1318 AD vi (Berks); Robert *Wolryge* 1332 SRSt; John *Woldrych* 1524 SRSf; John *Wooldridge* 1663 SfPR. OE *Wulfrīc* 'wolf-powerful', a personal name fairly common after the Conquest.

**Woolsey, Wolsey, Woosey:** (i) Leofwine *Wulfsiges sunu* a1038 OEByn (Ha); *Wlsi, Vlsi* 1066 DB; Willelmus *filius Wulsi* (*Wlfsi*) 1166 P (Nf); William *Wulsy* 1219 AssY; Richard *Wolsy* 1313 FFEss; Robert *Wulcy, Wolcye* 1524 SRSf. OE *Wulfsige* 'wolf-victory'. The personal name was common in Suffolk where it gave rise to the surname of Cardinal Wolsey. (ii) Roger *Wulfesege, Wulfes æge* 1168, 1171 P (Ha); Waldevus *Wulfesega* 1176 P (Nb). A nickname, 'wolf's eye', which may have contributed to the frequency of the surname.

**Woolstencroft:** *v.* WOLSTENCROFT

**Woolstenhulme:** *v.* WOLSTENHOLME

**Woolston, Woolstone, Woollston, Wolston:** (i) *Vlfstan, Wlstan, Vlstan* 1066 DB; Ordric *Uulfstani filius* c1095 Bury (Sf); *Wlston* 1188 BuryS (Sf); Gilbert *Wlstan* 1199 P (L); Ralph *Wolstan* 1212 Cur (Nf). OE *Wulfstān* 'wolf-stone'. (ii) Hugelot *de Woluest011* 1131 FeuDu; William *de Wolstone* 1279 RH (Bk). The surname may also be local in origin, from several places of varied origin: Wolston (Warwicks), *Ulvricetone* DB; Woolston (Devon), *Ulsiston* DB; (Lancs), *Wolueston* 1242; (Hants), *Oluestune* DB; Woolston in North Cadbury (Som), *Wolston* 1316; Woolston in Bicknoller (Som), *Wolwardeston* 1225; Woolston Hall in Chigwell (Essex), *Ulfelmestuna* DB; Woolstone (Bucks), *Wlsiestone* DB; or Wolviston (Durham), locally pronounced *Wooston*.

**Woolterton, Woolterton:** Walter *de Wulterton'* 1220 Cur (Nf). From Wolterton (Norfolk).

**Woolton, Wolton:** John *de Wulton'* 1180 P (Lei); John *de Woleton* 1241 FFK; Thomas *de Wolton* 1382 IpmGl. From Woolton (Lancs).

**Woolven, Woollven, Woollvern, Woollvin, Wolfin, Woollen, Woollin, Woollings, Woollon, Woollons:** *Wluuine, Vlfuuinus, Vluuin* 1066 DB; Lefstan *Ulfuini filius* c1095 Bury (Sf); *Wlfwinus* Holepot 1182–1211 BuryS (Sf); *Wolvin* Cote 1340 Husting; Nicholas *Wolwin* c1236 Seals (W); William *Wulwyne* 1296 SRSx; John *Wolvyne* 1328 LLB E; Richard *Woluen* 1381 PTY. OE *Wulfwine* 'wolf-friend'.

**Woolveridge:** *v.* WOOLRICH

**Woolverton, Wolverton:** Of varied local origin. Nicholas *de Wulferton'* 1166 P (Nf), from Wolferton (Norfolk); Tomas *de Wulfrinton'* 1190 P (Sa), from Wollerton (Salop); Vivian *de Wluerton'* 1200 P (Bk), from Wolverton (Bucks); Alina *de Wulwardinton'* 1221 AssWa, from Wolverton (Warwicks); Richard *de Wulfrinton'* 1221 AssWo, from Wolverton (Worcs); or from Wolverton (Hants) or Woolverton (Som).

**Woolvett, Woolfit, Woollatt, Woollet, Woollett, Wolfit, Wollett, Ullett, Ulliott, Ullyatt, Ullyett,**

**Ullyott, Ulyate, Ulyatt, Ulyett, Woffitt:** *Wluiet, Wluiat, Vlfiet, Vluiet* 1066 DB; *Wlfgiet* 1066 InqEl; Goduuin *Ulfgeti filius* c1095 Bury (Sf); *Vlfgiet* Moder 1162 P (Nth); *Wulfiet* Mus 1176 P (Nt); *Wulfet de Branwic'*, Thoroldus *filius Wulfat* 1219 AssY; Johannes *filius Wulfoti* 1223 Pat (Y); Leuricus *Wlfuiet*, Eudo *Wluiet* 1199 P (L, Nf); Robert *Woluyet* 1306 FFEss; Robert *Woluet* 1315 AD iv (Ess); Margaria *Woliet* 1351 AssEss; John *Wlyet* 1367 AD iv (Ess); Widow *Ullet* 1674 HTSf. OE *Wulfgēat* 'wolf-Geat'.

**Woolving:** *Wlueing* 1227 AssBk. OE *\*Wulfing* 'son of *Wulf(a)*'.

**Woolward:** *v.* WOOLLARD

**Woolway, Woolaway:** *Wluui, Wlui* 1066 DB; Ricardus *filius Ulfui* c1095 Bury (Sf); *Vlfgiet filius Wlwi* 1202 P (Bk); Robert *Wulfwi* 1190 P (Lei); Lucas *Wulwy* 1230 Pat; William *Wolwy* 1275 SRWo; Symon *Wolvy, Wolwy* 1279 RH (C). OE *Wulfwīg* 'wolf-war'. This might also have contributed to WOLVEY.

**Woolwich, Woolich, Woolidge, Woollage, Woledge, Wolledge:** From Woolwich (Kent) 'wool-farm'. Also from Wollage Green and Woolwich Wood in Womenswold (Kent) where the surname was *de Wolfech'*, *de Wulfhecch* 1313 PN K 540, OE *wulf-hæcce*, probably denoting a wolf-pit, wolf-trap. *v.* also WORLEDGE.

**Woolwright:** *v.* WOOLRICH

**Woombell, Woombill, Woomble:** *v.* WOMBWELL

**Woor, Woore:** Nicholas *le Wowere* 1275 RH (Sf); Emma *Woweres* 1279 RH (O); Adam *le Wogher'* 1327 SRSx. OE *wōgere* 'wooer, suitor'.

**Woosey:** *v.* WOOLSEY

**Woosley:** *v.* WOLSELEY

**Woosman, Woosnam:** *v.* WOLSTENHOLME

**Wooster:** *v.* WORCESTER

**Wooton, Woottton, Wootten, Wotton:** Wagen *de Wotton* c1048 OEByn (Wa); Robert *de Wodeton'* 1242 Fees (Sa); John *de Wutton* 1256 AssNb; John *ate Wodeton* 1275 RH (Mx); Gilbert *de la Wotton* 1316 MELS (Wo). From one of the many places named Wootton or Wotton, or from residence at a farm by a wood (OE *\*wudu-tūn*).

**Wora(c)ker:** *v.* WARWICKER

**Worboys:** *v.* WARBOY

**Worcester, Worster, Wooster, Wostear:** Ralph *de Wircestr'* 1180 P (C); William *de Worcester* 1290 AssSt; Reynold *Woster* 1567 Bardsley; Alice *Wooster* 1658 ib. From Worcester.

**Wordman:** *v.* WARDMAN

**Wordsworth:** *v.* WADSWORTH

**Woricker:** *v.* WARWICKER

**Work, Worke:** Richard *del Worke* c1260 *ERO* (Ess); Simon *Werk* 1279 RH (Bk); Nicholas *del Werk* 1357 Whitby. OE *(ge)weorc* 'work, fortification'. A name, probably, for one working on fortifications or for a man employed in some capacity at a fort.

**Worker:** Richard *le Wercare* 1332 MEOT (Sr). A derivative of OE *wyrcan, weorcan* 'to work'. One who works or does work of any kind (13.. NED). A rare surname.

**Workman:** Adwordus *Wercman* 1214 StGreg (K); William *le Werkman* 1236 FFSx; Gilbert *le Worcman*

1279 RH (O). OE *weorcmann* 'workman'. In addition to this meaning, PromptParv has: 'Werkemanne, þat cann werke wythe bothe handys a-lyke. *Ambidexter*', which might well have given rise to a nickname.

**Worksop, Worsop:** Badelota *de Wirkeshop'* 1205 P (L); John *Wursop* 1429 AssLo. From Worksop (Nt).

**Worledge, Worlidge, Worlledge, Worllege, Woledge, Wolledge, Woolidge, Woollage:** John *Woorliche* 1468 SIA xii; John *Worlych, Wurlych* 1524 SRSf; Richard *Worleche* 1536 SIA xii; John *Worlich* 1551 SfPR; William *Worlege* 1568 SRSf; Maria *Worledge* 1671 SfPR; Philip *Worledge*, Robert *Wolladge*, William *Woollagge*, — *Woolage*, — *Worlike* 1674 HTSf; William *Wooledge* 1827 SfPR. Forms are late, but this is probably OE *weorþlīc* 'worthy, noble, distinguished'. cf. LOVELY, and *v.* WOOLWICH.

**Worlock:** *v.* WARLOCK

**Worm, Worms:** Water *Wormes* 1275 RH (D); John *le Werm*, William *le Wurm* 1296 SRSx. OE *wyrm* 'snake, dragon'.

**Wormald, Wormhall, Wormal, Wormell, Warmoll, Wormull:** Alexander *de Wormwall* 1379 PTY; Robert *Wormall, Wormell* 1592, 1598 RothwellPR (Y); Samuel *Wormald* 1748 FrY. From Wormald (WRYorks), or Wormhill (Derby).

**Wormington:** Roger *de Wrmitune* 1185 Templars (Gl); William *de Wurmynton* 1287–8 FFSx; William *de Wormyngton* 1296 SRSx. From Wormington (Gl).

**Wormwell:** Robert *de Wermewell'* 1214 Cur (Do); Widow *Wormwell* 1674 HTSf. From Warmwell (Do).

**Worral, Worrall, Worrell, Worrill, Whorall:** Roger *de Wyrhal'* 1219 AssY; Richard *de Wyrall* 1351 ShefA; Alan *de Worrell* 1388 IpmNt; William *Worrall* 1517 ShefA; Joseph *Worrell* 1705 FrY. From Wirral (Ches), or Worrall (WRYorks).

**Worsdale, Worsdall, Worsdell:** John *de Wiresdale* 1332 SRLa; Anne *Worsdall* 1667 Bardsley. From Wyresdale (Lancs).

**Worsell, Worssell, Woursell, Wurzal:** Peter *de Wirkeshal'* 1219 AssY; John *Worsell* 1439 FrY. From Worsall (NRYorks), *Wercesel* DB.

**Worsencroft:** *v.* WOLSTENCROFT

**Worship:** John *Worchipe* 1327 SRSf; Robert *Worshipe* 1327 SRSx; John *Worschep* 1332 SRCu. OE *weorþscipe* 'worship, honour, dignity'.

**Worsick:** *v.* WORSWICK

**Worsley, Worslay:** Geoffrey *de Wyrkesle* 1246 AssLa; Robert *de Worvesle* 1275 SRWo; Richard *Worseley* 1396 AssL; James *Wursley* 1548 CorNt. From Worsley (Lancs), *Werkesleia* 1196, or Worsley (Worcs), *Worvesleg* c1180.

**Worsman:** *v.* WOLSTENHOLME

**Worsop:** *v.* WORKSOP

**Worssell:** *v.* WORSELL

**Worstead, Worsted:** John *Worstede* 1367 AssLo; Simon *Wurstede* 1392 LoCh; William *Worstede* 1426 Paston. From Worstead (Nf).

**Worstencroft:** *v.* WOLSTENCROFT

**Worstenholme:** *v.* WOLSTENHOLME

**Worster:** *v.* WORCESTER

**Worswick, Worsick:** Anne *Worswick* 1670 LaWills; Alexander *Worsick* 1731 ib. From Urswick (Lancs).

**Wort, Worts, Wortt:** Æluuric *Uuort* c1095 Bury (Sf); Henry *Wortes* 1524 SRSf. OE *wyrt*, ME *wort* 'plant, vegetable'. Metonymic for WORTMAN.

**Worters:** *v.* WARTERS

**Worth, Werth:** John *de Wurde* 1195 P (K); Hawisia *de la Wurth* 1204 FFSo; Elyas *de la Worth'* 1243 AssSo; William *de Werthe* 1275 RH (Sx); Richard *atte Werthe* 1296 SRSx; William *atte Worthe* 1327 SRSx. OE *worð* (*wyrð*) 'enclosure, homestead'. From Worth (Ches, Devon, Kent, Sussex), Worth Matravers (Dorset), Highworth (Wilts) or Littleworth (Berks), both *Wurda* 1196 P, or from residence or employment at an outlying farm.

**Wortham:** Adam *de Wrtham* 1188 BuryS (Sf). From Wortham (Suffolk).

**Worthen:** Richard *de Worthin* 1274 RH (Sa); Richard *atte Worthyne* 1316 MELS (Wo). From Worthen (Salop), Worthing (Norfolk), or from residence at an enclosure or homstead, OE *worþign*, a term particularly common in the West Midlands. It sometimes becomes *Wortham*: William *atte Wurthen* lived at Wortham (Devon) in 1330 (PN D 212).

**Worthing:** (i) Robert *de Worthinge* 1276 RH (Beds). From Worthing (Sussex), probably the usual source; or a variant of WORTHEN. (ii) *Worthing'* Medicus, *Honorius* Medicus 1203 Cur (Nf); *Worthingus* 1206 Cur (Beds); *Urthing, Wrthing* de Marisco 1240 ER 61; John *Worthang'* 1219 Cur (Sx). An OE personal-name *Wurð* has been postulated to explain the place-names Worsthorne (Lancs), Worston (Lancs), Worthing (Sussex) and Worthington (Lancs, Leics). In the form *Weorð*, it has now been noted twice in 12th-century London: *Werth* Hy 1 ELPN, *Werd* 1111–38 ib.; in addition, three examples of an unrecorded compound *Weorðgifu* (f) are now known: Æilric *Uuordgiue filius* c1095 Bury (Sf); Ricardus *filius Wrthiue* 1212 Cur (Beds); Ricardus *nepos Wrtheve* 1222 DBStP (Ess). *Weorð* is from OE *weorð*, late OE *wurð* 'worthy'. From this a derivative *Weorðing* was formed, of which the above are the only known examples.

**Worthington:** Snarri *de Wurðintona* 1169 P (L); William *de Wurthington* 1246 AssLa; John *Worthyngton* 1439 IpmNt. From Worthington (Lancs, Leics).

**Worthy:** Godwine *æt Worðige* 1001 ASC (Ha); Chiping' *de Ordia* 1066 Winton (Ha); Geoffrey *de la Worthy* 1293 Fees (D). From Worthy (Hants, Devon).

**Wortley:** Henry *de Wortelay* 1204 Cur (Y); Thomas *de Wortley* 1299 IpmNt; Richard *Wortlay* 1428 Shef. From Wortley (WRY).

**Wortman:** Simon *Wurtman* 1297 MinAcctCo. 'Seller of vegetables', OE *wyrt* and *mann*.

**Worton:** Walter *de Worton* 1202–12 StP; Gilbert *de Worton* 1256 AssNb; Adam *de Worton* 1279 RH (O). From Worton (O, W, NRY).

**Wosencroft:** *v.* WOLSTENCROFT

**Wostear:** *v.* WORCESTER

**Wostenholm:** *v.* WOLSTENHOLME

**Wotton:** *v.* WOOTON

**Would(s):** *v.* WALD

**Wouldhave:** John *Waldhaue* 1396 FrY. Possibly a corruption of *Waldeve*. cf. WADDILOVE, WALTHEW.

**Woursell:** *v.* WORSELL

**Wozencroft:** *v.* WOLSTENCROFT

**Wragby:** Henry *de Wraggebi* 1202 AssL; Thomas *de Wraggeby* 1263 FFL; Robert *de Wragby* 1396 AssL. From Wragby (L, WRY).

**Wragg, Wragge:** *Wrag* 1193 P (Nth); William *Wraggi* 1192 P (Wo); William, John *Wrag* 1221 AssWa, 1323 AD i (L). ODa *Wraghi*.

**Wraight, Wrate:** *v.* WRIGHT

**Wraith:** John *Wrayth* son of Thomas *Wrath* 1587 FrY. OE *wrāð* 'angry, fierce'. A northern form of WROTH.

**Wrangell, Wrangle, Wrangles:** Alice *de Wrangel* 1275 RH (L); William *Wrangel* 1312–13 NorwLt; Robert *Wrangle* 1392 AssL. From Wrangle (L).

**Wrangham:** Thomas *de Wrangham* 1332 SRCu; James *Wrangham* 1479 Black. From Wrangham (Berwick).

**Wrangle, Wrangles:** *v.* WRANGELL

**Wrangways:** William *Wrangwis* 1219 AssY; Lecia *Wrangwyse* 1301 SRY; Thomas *Wrangwyssh* 1457 FrY. OE *wrangwīs* 'wicked'.

**Wraste:** Adam *Wraste* 1256 AssNth; Thomas *Wrast'* 1355–9 AssBeds; John *Wrast* 1379 PTY. OE *wrǣst* 'firm, strong, excellent'.

**Wray:** *v.* WREY, WROE

**Wrayford, Wraysford:** William *Wreyford* 1340–1 FFSr; Thomas *Wrayford, Sibbanes Wraysford* 1642 PrD. 'Dweller by the ford on the River Wray'.

**Wreak, Wreaks:** Robert *le Wrek* 1254 FFEss; Robert *Wrekkes* 1400–1 IpmY; Ralph *Wreekes* 1577 FrY. Perhaps OE *wrecca* 'fugitive, exile'. Sometimes, perhaps, from Wreak (Lei), or Wreaks in Hampsthwaite (WRY).

**Wreath, Wrede:** *v.* WRIDE

**Wreghitt:** Richard *Wreggett* 1653 FrY. From Wreggets in Wheldrake (ERYorks), *Wragate* 13th PN ERY 271.

**Wreight:** *v.* WRIGHT

**Wren, Wrenn:** Wlsy, William *Wrenne* 1275 RH (Nf), 1317 AssK; Ralph *Wranne* 1296 SRSx. A nickname from the wren, OE *wrenna, wrænna*.

**Wrench, Wrinch:** Alnodus *Wrench* 1176 P (D); Edwin *Wrench* 1199 P (Nf). ME *wrench*, OE *wrenc* 'wile, trick, artifice'.

**Wressell, Wressle:** Matilda *de Wresele* 1200 P (L); Stephen *Wresel'* 1327 SRLei; William *Wresyll* 1404 FrY. From Wressell (ERY).

**Wrestler:** William *le Wrastler* 1332 SRWa; Alice *Wresteler* 1379 PTY. OE *wrǣstlere* 'wrestler'.

**Wrey:** (i) Robert *de Wrey* 1275 RH (D). From Wray Barton (Devon). (ii) Stephen *le Wrey* 1313 PN D 212. 'The twisted, crooked', from ME *wry*(*e*) 'awry', from OE *wrīgian* 'to turn'.

**Wride, Wreath, Wrede:** Roger *Wrid* 1199 P (Sf); John *Wrid'* 1219 AssL; Elyas *le Wrede* 1274 RH (Ess); Hugh *Wride* 1279 RH (O); Roberd *Wrethe* 1522 BuryW (Sf). OE *wride* 'twist, turn', from OE *wrīþan* 'to twist', was used topographically of a winding stream. cf. Wryde (PN C 282). The term may also have been applied to a 'twisting' man, the colloquial 'slippery customer'. In ME the *i* was sometimes lowered to *e*, as in *Wrethefeld* 1318. Hence *Wrethe* and *Wreath*.

**Wridgway:** For RIDGEWAY.

**Wrigglesworth, Wriglesworth, Rigglesford, Riggulsford:** John *de Writhelfford* 1379 FrY; Robert *Wriglesforth* 1591 RothwellPR (Y); William *Wriglesworth* 1605 ib. From Woodlesford (WRYorks), *Wridelesford* 12th, *Wudelesford* 1425, *Wriglesforth* 1596, *Wriglesworth* 1620 (PN WRYorks ii, 141).

**Wright, Wrighte, Wraight, Wraighte, Wrate, Wreight:** Patere *le Writh* 1214 FFSx; Robert *le Wrichte* 1255 FFEss; Robert *Wricht* 1274 RH (Sa); Richard *le Wreȝte* 1317 AssK; Hugh *le Wreghte* 1327 SRSx; Thomas *le Wrighte* 1327 SRDb; Walter *le Wrytte* 1327 SRSf; Amiscia *la Wrihtes* 1333 ColchCt; Thomas *Wrayte* 1547 RochW. OE *wyrhta, wryhta* 'carpenter, joiner'.

**Wrightson:** Robert *Wryghtson* 1379 PTY, 'Son of the wright'.

**Wrigley, Rigley:** William *de Wriggeley* 1327 SRDb. From Wrigley Head (Lancs).

**Wrinch:** *v.* WRENCH

**Wrington, Rington:** Turbert *de Wrinctona* 1189 Sol; William *de Wryngton'* 1340 Glast (So). From Wrington (So).

**Writer:** Adam *le Wrytar* 1275 SRWo. OE *wrītere* 'writer, copier of manuscripts'. The writers of London formed two separate crafts in 1422: the writers of the ordinary book-hand, 'textscriveyns', and the writers of the court-hands (LoPleas 199).

**Wrixon:** Adam *Wrygson* 1379 PTY. ME *wright* had a variety of forms, and names like *Aldrich* often vary between *-rich* and *-rick*. Hence, probably for *Wrightson*, through such forms as *Wrichtson, Wrichson, Wrickson, Wrigson*.

**Wroath:** *v.* WROTH

**Wroe, Wray:** Tomas *de Wra* 1196 P (Y); Richard *del Wro* 1210 Cur (Nf); Ralph *in le Wra* 1260 AssY; John *in ye Wro* 1366 ShefA. ON *vrá* 'nook, corner, remote or isolated place'. In the north it became Wray (Lancs) and Wrea, Wreay (Cumb). At Wrayside (Cumbs) in 1285 lived Henry *del Wra*. Farther south this became ME *wro* which does not seem to have survived as a place-name. Both *wra* and *wro* were common in Yorks where both *Wray* and *Wroe* are still found. Here and in Norfolk they must derive from unidentified remote spots.

**Wrong:** (i) Osbert *Wrange* 1182 P (D); Ailnoth *Wrang* 1202 P (Sr); John *Wrong* 1275 RH (Sf); William *le Wronge* 1317 AssK. Late OE *wrang* of Scand origin, 'crooked'. (ii) Robert *ate Wrong* 1327 SRC; Richard *atte Wronge* 1348 FFC. OE *wrang* is found as a compound in place-names meaning 'crooked'. Here it is used as a substantive for a crooked, irregularly shaped piece of land.

**Wronger, Wronker:** Equivalent to *atte Wronge*.

**Wrongfoot:** Richard *Wrongfot* 1260 AssC. 'Crooked foot', OE *wrang, fōt.* cf. William *Wranghorn* 1276 AssLo 'crooked horn'.

**Wroot:** (i) John, William *Wrote* 1200 P (Ha), 1296 SRSx. A nickname from OE *wrōt* 'snout'. (ii) Algar *Wroth*, Walter *Wrot* 1202 AssL. These Lincs names must be early examples of the loss of the preposition and derive from Wroot (Lincs), formerly an island or a spur of land thought to resemble a pig's snout.

**Wrosen:** John *de la Wrosne* 1275, William *ate Wrosne* 1293 MELS (Wo). From Wrens Nest Hill in Dudley (Wo), *Wrosne* 1278.

**Wroth, Wroath:** W. *Wrothe* c1210 ELPN; Walter *le Wrothe* 1221 AssSa. OE *wrāð* 'angry, fierce'.

**Wrotham:** William *de Wroteham* 1212 P (K); Richard *de Wrotham* 1219 P (Do/So); William *de Wrotham* 1323 CorLo. From Wrotham (K).

**Wrottesley:** Adam *de Wrotteslega* 1170 P (St); William *de Wrottesleye* 1299 AssSt; Walter *Wrotchley* 1663 HeMil. From Wrottesley (St).

**Wrye:** Roger *Wrye* 1281 CtW; William *Wry* 1360 IpmGl. 'Twisted, bent', cf. OE *Wrigian* 'to twist, bend'.

**Wulff:** *v.* WOLF

**Wurzal:** *v.* WORSELL

**Wusteman:** *v.* WOLSTENHOLME

**Wyan, Wyant:** *Wianus* filius Jone 1198 P (Sa); *Wyon* son of Wervill 1246 AssLa; Hugh *Wyan* 1395 SaltAS (OS) ix. *Wy-un*, a diminutive of OG *Wido*, OFr *Guy*.

**Wyard, Wyartt, Whyard:** *Wiardus* de Nichole Hy 2 Gilb (L); Henry *Wiard, Wihard* 1188, 1193 P (Ha); Edmund *Wyard* 1230 P (Ess). OE *Wīgheard* 'war-brave'. This surname has, no doubt, frequently become WYATT. Wyatt's Lane in Walthamstow (Ess) owes its name to the family of Walter and Richard *Wiard* (1223–52 PN Ess 106), the latter of whom is also called Richard son of *Wyard* 1223–52 AD i.

**Wyatt, Wiatt, Whyatt, Guyat, Guyatt, Gyatt, Guyot, Giot:** *Wiot* de Acham 1192 P (L); *Wioth* de Cratella Hy 2 DC (Nt); *Gwiot* 1203 Cur (Gl); *Wyot* 1219 AssY; William *Wyot* 1274 RH (Sa); Robert *Wiot* 1279 RH (Beds); Thomas *Guyot* 1295 FFEss; Henry *Guyot* 1327 SRSo. A diminutive in *-ot* of OFr *Guy*, OG *Wido*. Occasionally we have a diminutive in *-et*: Richard *Guiet* 1141–51 Colch (Ess), Elias *Wyete* 1296 SRSx. *Giot* de Kemesinge 1317 AssK is also called *Guydo* Formage. Occasionally, *Wiot* is a diminutive of *William*: *Wiotus* or *Guillotus* 1175 P (Berks). The form *Wyatt* really represents *Wyard. Wyott* does not survive but seems to have become *Wyatt* commonly in the 16th century: John *Wyot* 1575 Oxon (D), William *Wyatt* 1576 ib. (D). The Norman *Wyatt* is much more common both in ME and today than the French *Guyatt.* On the other hand, *Guy* is the regular modern form of *Wido. Giot* survives in Jersey.

**Wyber, Wyberg, Wybergh, Wybrew, Wybrow, Waybrow, Wheybrew, Whybra, Whybray, Whybrew, Whybro, Whybrow, Wiber, Wiberg, Wibrew, Wibroe, Wibrow:** *Wyburgh* (f) 1182–1211 BuryS (Sf); Osbertus *filius Wiburge* 1219 Cur (Lei); *Wybur'* 1279 RH (O); Roger *Wybir'* 1279 RH (O); Margeria *Wyburgh'* 1327 SRC; Thomas *Wybourgwe* 1327 SRSf. OE *Wīgburh* 'war-fortress', a woman's name recorded once in 901 and then not until the 12th century. In view of the variety and frequency of the surnames it must have been more common than the records suggest.

**Wyberd, Wyburd, Whyberd, Whybird:** *Wibert* 1066 DB (Y); *Guibertus* 1086 DB (Ess); Rodbertus *filius Wiberti* 1121–48 Bury (Sf); *Wibertus* 1214 Cur (K); Robert, Amicia *Wyberd* 1327, 1332 SRSx;

Bartholomew *Wyberd* 1327 SRSf. OE *Wīgbeorht* 'war-bright' or OFr *Wibert, Guibert* from OG *Wigbert.* Both are probably represented.

**Wyborn, Wybourn, Wyburn, Wybron, Whyborn:** *Wibern* de Keistret' 1212 Cur (K); Ralph *Wybern* 1241 FFEss; Richard *Wyborn'* 1275 SRWo. OE *Wīgbeorn* 'war-hero', late and rare in OE and not common later.

**Wybrew, Wybrow:** *v.* WYBER

**Wyburn:** *v.* WYBORN

**Wych:** *v.* WICH

**Wycherley:** *v.* WICHERLEY

**Wyclif, Wycliffe:** *v.* WICLIF

**Wycombe:** John de *Wycoumbe* 1326 FFEss; John *Wycombe* 1391 IpmGl. From Wycomb (Lei), or Chipping, High, West Wycombe (Bk).

**Wye:** Hugh *de Wy* 1275 RH (K). From Wye (Kent). *v.* also GUY.

**Wyer, Wyers:** *v.* WIRE

**Wyke(s):** *v.* WICH

**Wykeham:** *v.* WICKHAM

**Wyld(e), Wyld(e)s:** *v.* WILD

**Wyldbore:** *v.* WILDBORE

**Wyles, Wyley:** *v.* WILES, WILEY

**Wyllie, Wylly:** *v.* WILLEY

**Wyllys:** *v.* WILLIS

**Wyly:** *v.* WILEY

**Wyman, Wymann, Wymans, Weyman, Weymont, Wayman, Waymont, Whayman, Whaymand, Whaymond, Whaymont, Whyman, Whymant, Whamond:** *Wimundus* 1066 DB (Nf); *Wymund* ater Walle 1296 SRSx; Robert *Wymund, Wimund* 1275 SRWo, 1279 RH (C); Alan *Wymand* 1275 SRWo; Robert *Waymon* 1275 SRWo; Peter, Robert *Weymund* 1279 RH (C); William *Wayman* 1357 AD ii (D); Robert *Whayemonde* 1568 SRSf. OE *Wīgmund* 'war-protection'. In Lincs, Leics and Norfolk the personal name may be ON *Vigmundr,* ODa, OSw *Vimund.*

**Wymark, Waymark:** Robertus *filius Wimarc* 1066 DB (Ess); *Wimarca* (f) Hy 2 DC (L); *Wimarc* (f) 1219 AssY; Robert *Wimarc* 1066 DB (Ess), 1199 P (Nf). OBret *Wiuhomarch,* both masculine and feminine. *v.* also WYMER.

**Wymer:** *Wimerus* 1066, 1086 DB (Sf); *Wymerus* de Westwyc 1153–66 Holme (Nf); Eche *filius Wymeri,* Eche *Wimer* c1160, 1186–1210 ib.; Willelmus *filius Wimari,* William *Wymer'* 1212, 1236 Fees (St); Adam *Wymer* 1327 SRSf. OE *Wīgmǣr* 'battle-famous' or OG *Wigmar.* Forms like *Wiomar, Wiomer* 1208 Cur (C), *Wyomer'* 1219 AssY are not uncommon in districts where Bretons are known to have been numerous and there has been confusion with OBret *Wiuhomarch* 'worthy to have a horse' which was originally masculine. This confusion is found already in 1086. *Wihumar'* dapifer comitis Alani, called also *Wiemar'; Wihemar'* 1086 ICC (C) appears in DB as *Wihomarc.* A similar assimilation to a more common ending *-ard* took place in France where OFr *Guiomarch* became *Guyomard.*

**Wyn, Wynne:** *v.* GWINN

**Wynands:** *v.* WINAND

**Wynburne:** *v.* WINBORNE

**Wynch:** *v.* WINCH

**Wynde:** *v.* WIND

**Wyndham:** *v.* WINDHAM

**Wyne(s), Wyness:** *v.* WINE

**Wynfrey:** *v.* WINFREY

**Wynick:** *v.* WINNICK

**Wynn(e), Wynnes:** *v.* WINN

**Wynniatt:** *v.* WINGATE

**Wynter(s):** *v.* WINTER

**Wynton:** *v.* WINTON

**Wynyard:** *v.* WINYARD

**Wynyates:** *v.* WINGATE

**Wyon, Wyons, Guyon:** Robertus *filius Guiun, Gwiun* 1203, 1209 P (Y); *Wyon, Wyun* 1246 AssLa; William *Wyonn,* Nicholas *Gyan* 1327 AssSo. Either the cas-régime of GUY, *Guyon,* usually in the Norman form *Wyon,* or a diminutive in *-un* of GUY. Rare surnames.

**Wyton, Whyton:** William *de Wyton'* 1219 AssY. From Wyton (Hu), or Wyton Hall (ERY).

**Wyvill, Wyville:** Hugo *de Wituile, de Widuile* 1086 DB; William *de Wituilla, de Wiuilla* Hy 2 DC (L). From Gouville (Eure), earlier *Wivilla. v.* OEByn.

# Y

---

**Yacksley, Yaxley:** John *Jakesley* 1327 SRSo; John *Yaxlee* 1503 Paston. From Yaxley (Hu, Sf).

**Yale:** Madog *Yale* 1391 Chirk. 'Dweller at the fertile upland', Welsh *iâl*. Elihu *Yale*'s family came from Plas-Gronow near Wrexham (Denbigh).

**Yalland, Yelland, Yolland:** John *de la Yaldelonde* 1275 RH (D). 'Dweller by the old cultivated land'. Common place-names in Devonshire, with surnames *de la Yalelande* 1238, *Attenoldelonde* 1244, *de la Yoldelonde* 1270, *Yollelonde* 1281, *Yeallelond* 1333 (PN D).

**Yapp:** Richard *Yap*, *Yape* 1200 P (Nt), 1297 SRY. A nickname from OE *gēap* 'bent'.

**Yarborough, Yarbrough, Yarbord, Yerburgh, Yerbury:** Gilbert *de Yerdeburc'* 1275 RH (L); John *de Yerdeburgh'* 1323 KB (L); John *Yearberye*, William *Yearburye* 1576 SRW. From Yarburgh, Yarborough Camp (L), or Yardbury Fm (So).

**Yard, Yarde:** Thomas *de la Yurda* 1225 AssSo; Robert *atte Yerde* 1309 LLB D. OE *gerd*, *gyrd* 'virgate, thirty acres', probably denoting a holder of a yardland.

**Yardley, Yeardley:** Richard *de Gerdelai* 1199 Pl (Nth); William *de Erdeleg'* 1229 Cur (Herts); Adam *de Jerdeleye* c1280 SRWo; Robert *Yerdeleye* 1372 IpmGl; John *Yardeley* 1499 FFEss. From Yardley (Essex, Northants, Worcs), or Yarley (Som), *Gyrdleg* 1065.

**Yarham, Yarm:** Stephen *de Yarum* 1275 RH (L). From Yarm (NRY), *Yarum* 1182.

**Yarker, Yorker:** John *le Yarker*, *Yurker* 1327 SRC; John *Yarkar* 1379 PTY. A derivative of ME *yerk* 'to draw stitches tight, to twitch, as a shoemaker in sewing' (c1430 NED), a shoemaker.

**Yarmouth, Yarmuth:** Lewine *de Gernemut'* 1168–75 Holme; Nicholas *de Yernemuthe* 1319 SRLo; Hugh *of Yarmouth* 1306 AssW. From Yarmouth (Wt), Great Yarmouth (Nf), or Little Yarmouth (Sf).

**Yarwood, Yarworth:** *v.* YORATH

**Yate, Yates, Yeates, Yeats, Yetts:** Hereward *de Jette* 1198 P (Gl); Philip *del Yate* 1260 AssCh; Roger *atte Yat* 1268 AssSo; John *atte Yete* 1327 SRSo; Robert *atte Yates* 1344 FrNorw. From Yate (Glos), or 'dweller by a gate', or 'gate-keeper' (OE *geat*). *v.* GATE.

**Yatman, Yeatman, Yetman:** John *Yateman* 1603 Bardsley; Ann *Yeatman* 1758 ib.; David *Yetman* 1775 ib. *v.* YATE.

**Yaxham:** Ralph *de Yaxham* 1308 AssNf. From Yaxham (Nf).

**Yaxley:** *v.* YACKSLEY

**Yea, Yeo:** Geoffrey *de la Ya* 1242 Fees (D); Alexander *de Ya* 1257 AssSo; Viel *de la Yo* 1260 MELS (So); Richard *atte ȝeo*, Adam *atte ȝaa* 1333 ib.; Robert *atte Yea* 1340 ib. (So); Hugo *de la Yeo* 1333 PN D 287; Nicholas *Yeo*, *Yoo* 1439 LLB K. From residence near a stream; OE *ēa* in Somerset and Devon became ME *eā*, *ya*, *yo*, which survives in various places named Yeo. *v.* ATYEO, REA, NYE.

**Yeadon:** Walter *de Yeadon* Ric I P (Y); John *of Yedon* 1226 FFY; Richard *Yeadon* 1459 Kirk. From Yeadon (WRY).

**Yealland:** Adam *de Yeland* 1230 P (La). From Yealand (Lancs).

**Yeaman:** *v.* YEOMAN

**Yeardley:** *v.* YARDLEY

**Yeates:** *v.* YATE

**Yeatman:** *v.* YATMAN

**Yeldham:** Gilbert *de Geldham* 1178 AC; Walter *de Gellam* 1197, William *de Geldham* 1203 FFEss. From Great, Little Yeldham (Ess).

**Yelland:** *v.* YALLAND

**Yelling:** John *Yellyng* 1447 CtH; Ephraim *Yelling* 1642 PrD. From Yelling (Hu).

**Yellow:** Jordan *le Yelewe* 1234 Oseney; Peter *le Yelewe* 1307 AssSt; John *Yelowes*, William *Yeallowe* 1576 SRW. OE *geolu* 'yellow', of hair or complexion.

**Yelverton:** John *de Yelvertune* 1275 RH (Nf); William *Yeluerton* 1443 Paston. From Yelverton (Nf).

**Yeman:** *v.* YEOMAN

**Yendall:** James *Yendall* 1641 PrSo; Thomas *Yendall*, Christopher *Yeandall* 1642 PrD. Perhaps from Yen Hall in West Wickham (C).

**Yendamore:** William *Beyundemyre* 1333 PN D 100. From Yendamore in Meeth (D).

**Yeo:** *v.* YEA

**Yeoman, Yeomans, Yeaman, Yeman:** William *Zeman* 1296 SRSx; John *Yemon* 1332 SRLa; Thomas *le Yomon* 1381 SRSt. ME *ȝoman*, *ȝeman* 'servant or attendant in a noble house, ranking between a sergeant and a groom or between a squire and a page'.

**Yeoward, Yeowart:** *v.* EWART

**Yeowell, Yeuell:** *v.* YOULE

**Yepp:** Richard *le Yepe* 1240–1 ForEss; William *Yep* 1355–9 AssBeds. OE *gēap* 'crooked, deceitful; intelligent, shrewd'. *v.* also YAPP.

**Yerburgh, Yerbury:** *v.* YARBOROUGH

**Yetman:** *v.* YATMAN

**Yetts:** *v.* YATE

**Yewdall, Yeudall, Yewdale, Youdale, Youdle, Udal, Udale, Udall, Udell:** Alice *de Yowdall* 1379 PTY; John *Yowdall* 1546 LP (Cu); Thomas *Vdall* 1662–4 HTDo. From Yewdale (La).

**Yewen:** *v.* EWAN

**Yoell:** *v.* YOULE

**Yoke:** Roger *ate Yoke* 1317 AssK; Roger *atte Yok* 1324–5 FFEss; William *Yoke* 1390 IpmGl. 'The holder of a certain measure of land', OE *geoc.*

**Yolland:** *v.* YALLAND

**Yondover:** Walter *Yondovere* 1327 SRSo. From Yondover in Loders (Do).

**Yong(e):** *v.* YOUNG

**Yonwin:** *Iunguine* c1095 Bury (Sf); Robertus *filius Iunguini* 1182 Oseney (O); *Yungwin'* de Tedford 1177 P (Sf); Simon *Yongwyne* 1327 SRSf; Roger *Youngwyne* 1327 *SR* (Ess). OE *\*Geongwine* 'young friend'.

**Yorath, Yorwarth, Yarwood, Yarworth:** Griffin *filius Yeruerth'* 1196 P (Sa); *Jarforth'* filius Ithell' 1221 AssSa; Tudur *ap Jorwerth* 1391 Chirk; Samuel *Yerworth* 1641 PrSo. OW *Iorwerth.*

**York, Yorke:** Ernisius *de Eboraco* a1160 YCh; John *de York* 1324 CorLo; Thomas *York* 1522 CorNt. From York.

**Yorkshire, Yorksher, Yorksheer:** John *de Euirwikescire* 1260 AssCh; William *Yorkescher* 1379 PTY. 'The man from Yorkshire.'

**Youatt:** *v.* EWART

**Youdale, Youdle:** *v.* YEWDALL

**Youens, Youings:** *v.* EWAN

**Youlden, Youldon:** 'Dweller on the old hill', as at Youlden and Youldon (Devon), with the same development as in Yalland. *v.* PN D.

**Youle, Youles, Youll, Youel, Youels, Yoell, Yeowell, Yeuell, Youhill, Yuell, Yule, Yuile, Yuill, Yuille:** William *Yol* 1199 P (L); William *Yoel* 1297 Wak (Y); Robert *Youle* 1379 PTY. OE *geōl*, ON *jól* 'Yule, Christmas', a name for one born at that time.

**Young, Younge, Youngs, Yong, Yonge:** Wilferð *seo Iunga* 744 ASC E; Richard *le Yunge* 12th Lichfield (St); Walter *Yonge* 1296 SRSx. OE *geong* 'young', a name often used, no doubt, to distinguish a younger from an older man.

**Youngbond:** John *Yongebonde* 1257 Ch (Lei); Nicholas *le Youngebonde* 1298, *le Yongebonde* 1316 IpmGl. 'The young householder', OE *geong, bōnda.*

**Younger:** Edmund *Yonger* 1379 PTY. 'The younger.' cf. Alanus *Junior* 1201 Cur. John *Yongehere* 1364 LoPleas was a Fleming and his surname is MDu *jonghheer* 'young nobleman', an earlier example of which is probably: William *Yunghare* 1297 Wak (Y).

**Younghusband:** Richard *le Yongehosebonde* 1275 SRWo; Robert *þe yengehusbonde* 1277 *Ely* (Sf); Robert *le Yungehusebonde* 1298 AssSt. 'Young farmer' (late OE *hūsbonda*).

**Youngman, Younkman:** Roger *Yungeman* 1235 FFEss; William *le Yongeman* 1302 Clerkenwell (Lo). 'Young servant.'

**Youngmay:** Martin *le Yungemey* 1275 RH (Sx). 'Young lad, servant.' OE *māg.*

**Youngson:** Peter *Yungeson* 1268 AssSo. 'Young son.'

**Younson:** John *Ewynsone* 1296 Black; William *Yewnsoun* 1603 ib. 'Son of *Ewan*.'

**Yoxhall, Yoxall:** Walter *de Yoxhale* 1272 FFSt; William *Yokisall* 1545 SRW. From Yoxall (St).

**Yuell, Yuill, Yule:** *v.* YOULE

# Z

Zeal, Zeale: v. SEAL
Zealey, Zealley: v. SEALEY
Zell, Zelle: v. SELL

Zeller: v. SELLAR
Zelley: v. SEALEY
Zouch: v. SUCH

OXFORD

## MORE OXFORD PAPERBACKS

This book is just one of nearly 1000 Oxford Paperbacks currently in print. If you would like details of other Oxford Paperbacks, including titles in the World's Classics, Oxford Reference, Oxford Books, OPUS, Past Masters, Oxford Authors, and Oxford Shakespeare series, please write to:

**UK and Europe:** Oxford Paperbacks Publicity Manager, Arts and Reference Publicity Department, Oxford University Press, Walton Street, Oxford OX2 6DP.

Customers in UK and Europe will find Oxford Paperbacks available in all good bookshops. But in case of difficulty please send orders to the Cash-with-Order Department, Oxford University Press Distribution Services, Saxon Way West, Corby, Northants NN18 9ES. Tel: 0536 741519; Fax: 0536 746337. Please send a cheque for the total cost of the books, plus £1.75 postage and packing for orders under £20; £2.75 for orders over £20. Customers outside the UK should add 10% of the cost of the books for postage and packing.

**USA:** Oxford Paperbacks Marketing Manager, Oxford University Press, Inc., 200 Madison Avenue, New York, N.Y. 10016.

**Canada:** Trade Department, Oxford University Press, 70 Wynford Drive, Don Mills, Ontario M3C 1J9.

**Australia:** Trade Marketing Manager, Oxford University Press, G.P.O. Box 2784Y, Melbourne 3001, Victoria.

**South Africa:** Oxford University Press, P.O. Box 1141, Cape Town 8000.

# OXFORD LIVES

## STANLEY

Volume I: The Making of an African Explorer
Volume II: Sorceror's Apprentice

*Frank McLynn*

Sir Henry Morton Stanley was one of the most fascinating late-Victorian adventurers. His historic meeting with Livingstone at Ujiji in 1871 was the journalistic scoop of the century. Yet behind the public man lay the complex and deeply disturbed personality who is the subject of Frank McLynn's masterly study.

In his later years, Stanley's achievements exacted a high human cost, both for the man himself and for those who came into contact with him. His foundation of the Congo Free State on behalf of Leopold II of Belgium, and the Emin Pasha Relief Expedition were both dubious enterprises which tarnished his reputation. They also revealed the complex—and often troubling—relationship that Stanley has with Africa.

'excellent . . . entertaining, well researched and scrupulously annotated' *Spectator*

'another biography of Stanley will not only be unnecessary, but almost impossible, for years to come' *Sunday Telegraph*

## POLITICS IN OXFORD PAPERBACKS
## GOD SAVE ULSTER!
### The Religion and Politics of Paisleyism
*Steve Bruce*

Ian Paisley is the only modern Western leader to have founded his own Church and political party, and his enduring popularity and success mirror the complicated issues which continue to plague Northern Ireland. This book is the first serious analysis of his religious and political careers and a unique insight into Unionist politics and religion in Northern Ireland today.

Since it was founded in 1951, the Free Presbyterian Church of Ulster has grown steadily; it now comprises some 14,000 members in fifty congregations in Ulster and ten branches overseas. The Democratic Unionist Party, formed in 1971, now speaks for about half of the Unionist voters in Northern Ireland, and the personal standing of the man who leads both these movements was confirmed in 1979 when Ian R. K. Paisley received more votes than any other member of the European Parliament. While not neglecting Paisley's 'charismatic' qualities, Steve Bruce argues that the key to his success has been his ability to embody and represent traditional evangelical Protestantism and traditional Ulster Unionism.

'original and profound . . . I cannot praise this book too highly.' Bernard Crick, *New Society*

# OXFORD LETTERS AND MEMOIRS
## RICHARD HOGGART

### A Local Habitation
### Life and Times: 1918–1940

With characteristic candour and compassion, Richard Hoggart evokes the Leeds of his boyhood, where as an orphan, he grew up with his grandmother, two aunts, an uncle, and a cousin in a small terraced back-to-back.

'brilliant . . . a joy as well as an education' Roy Hattersley

'a model of scrupulous autobiography' Edward Blishen, *Listener*

### A Sort of Clowning
### Life and Times: 1940–1950

Opening with his wartime exploits in North Africa and Italy, this sequel to *A Local Habitation* recalls his teaching career in North-East England, and charts his rise in the literary world following the publication of *The Uses of Literacy*.

'one of the classic autobiographies of our time' Anthony Howard, *Independent on Sunday*

'Hoggart [is] the ideal autobiographer' Beryl Bainbridge, *New Statesman and Society*

# POPULAR SCIENCE FROM OXFORD PAPERBACKS

## THE SELFISH GENE

### Second Edition

*Richard Dawkins*

Our genes made us. We animals exist for their preservation and are nothing more than their throwaway survival machines. The world of the selfish gene is one of savage competition, ruthless exploitation, and deceit. But what of the acts of apparent altruism found in nature—the bees who commit suicide when they sting to protect the hive, or the birds who risk their lives to warn the flock of an approaching hawk? Do they contravene the fundamental law of gene selfishness? By no means: Dawkins shows that the selfish gene is also the subtle gene. And he holds out the hope that our species—alone on earth—has the power to rebel against the designs of the selfish gene. This book is a call to arms. It is both manual and manifesto, and it grips like a thriller.

*The Selfish Gene*, Richard Dawkins's brilliant first book and still his most famous, is an international bestseller in thirteen languages. For this greatly expanded edition, endnotes have been added, giving fascinating reflections on the original text, and there are two major new chapters.

'learned, witty, and very well written . . . exhilaratingly good.' Sir Peter Medawar, *Spectator*

'Who should read this book? Everyone interested in the universe and their place in it.' Jeffrey R. Baylis, *Animal Behaviour*

'the sort of popular science writing that makes the reader feel like a genius' *New York Times*

# PAST MASTERS

*General Editor: Keith Thomas*

## SHAKESPEARE

*Germaine Greer*

'At the core of a coherent social structure as he viewed it lay marriage, which for Shakespeare is no mere comic convention but a crucial and complex ideal. He rejected the stereotype of the passive, sexless, unresponsive female and its inevitable concommitant, the misogynist conviction that all women were whores at heart. Instead he created a series of female characters who were both passionate and pure, who gave their hearts spontaneously into the keeping of the men they loved and remained true to the bargain in the face of tremendous odds.'

Germaine Greer's short book on Shakespeare brings a completely new eye to a subject about whom more has been written than on any other English figure. She is especially concerned with discovering why Shakespeare 'was and is a popular artist', who remains a central figure in English cultural life four centuries after his death.

'eminently trenchant and sensible . . . a genuine exploration in its own right' John Bayley, *Listener*

'the clearest and simplest explanation of Shakespeare's thought I have yet read' Auberon Waugh, *Daily Mail*

# THE WORLD'S CLASSICS
# THE WIND IN THE WILLOWS

*Kenneth Grahame*

*The Wind in the Willows* (1908) is a book for those 'who keep the spirit of youth alive in them; of life, sunshine, running water, woodlands, dusty roads, winter firesides'. So wrote Kenneth Grahame of his timeless tale of Toad, Mole, Badger, and Rat in their beautiful and benevolently ordered world. But it is also a world under siege, threatened by dark and unnamed forces—'the Terror of the Wild Wood' with its 'wicked little faces' and 'glances of malice and hatred'—and defended by the mysterious Piper at the Gates of Dawn. *The Wind in the Willows* has achieved an enduring place in our literature: it succeeds at once in arousing our anxieties and in calming them by giving perfect shape to our desire for peace and escape.

The World's Classics edition has been prepared by Peter Green, author of the standard biography of Kenneth Grahame.

'It is a Household Book; a book which everybody in the household loves, and quotes continually; a book which is read aloud to every new guest and is regarded as the touchstone of his worth.' A. A. Milne

# OXFORD BOOKS

## THE OXFORD BOOK OF ENGLISH GHOST STORIES

*Chosen by Michael Cox and R. A. Gilbert*

This anthology includes some of the best and most frightening ghost stories ever written, including M. R. James's 'Oh Whistle, and I'll Come to You, My Lad', 'The Monkey's Paw' by W. W. Jacobs, and H. G. Wells's 'The Red Room'. The important contribution of women writers to the genre is represented by stories such as Amelia Edwards's 'The Phantom Coach', Edith Wharton's 'Mr Jones', and Elizabeth Bowen's 'Hand in Glove'.

As the editors stress in their informative introduction, a good ghost story, though it may raise many profound questions about life and death, entertains as much as it unsettles us, and the best writers are careful to satisfy what Virginia Woolf called 'the strange human craving for the pleasure of feeling afraid'. This anthology, the first to present the full range of classic English ghost fiction, similarly combines a serious literary purpose with the plain intention of arousing pleasing fear at the doings of the dead.

'an excellent cross-section of familiar and unfamiliar stories and guaranteed to delight' *New Statesman*

# ILLUSTRATED HISTORIES IN OXFORD PAPERBACKS

## THE OXFORD ILLUSTRATED HISTORY OF ENGLISH LITERATURE

### *Edited by Pat Rogers*

Britain possesses a literary heritage which is almost unrivalled in the Western world. In this volume, the richness, diversity, and continuity of that tradition are explored by a group of Britain's foremost literary scholars.

Chapter by chapter the authors trace the history of English literature, from its first stirrings in Anglo-Saxon poetry to the present day. At its heart towers the figure of Shakespeare, who is accorded a special chapter to himself. Other major figures such as Chaucer, Milton, Donne, Wordsworth, Dickens, Eliot, and Auden are treated in depth, and the story is brought up to date with discussion of living authors such as Seamus Heaney and Edward Bond.

'[a] lovely volume . . . put in your thumb and pull out plums' Michael Foot

'scholarly and enthusiastic people have written inspiring essays that induce an eagerness in their readers to return to the writers they admire' *Economist*

# OXFORD REFERENCE

## THE CONCISE OXFORD COMPANION TO ENGLISH LITERATURE

### *Edited by Margaret Drabble and Jenny Stringer*

Based on the immensely popular fifth edition of the *Oxford Companion to English Literature* this is an indispensable, compact guide to the central matter of English literature.

There are more than 5,000 entries on the lives and works of authors, poets, playwrights, essayists, philosophers, and historians; plot summaries of novels and plays; literary movements; fictional characters; legends; theatres; periodicals; and much more.

The book's sharpened focus on the English literature of the British Isles makes it especially convenient to use, but there is still generous coverage of the literature of other countries and of other disciplines which have influenced or been influenced by English literature.

From reviews of *The Oxford Companion to English Literature*:

'a book which one turns to with constant pleasure . . . a book with much style and little prejudice' Iain Gilchrist, *TLS*

'it is quite difficult to imagine, in this genre, a more useful publication' Frank Kermode, *London Review of Books*

'incarnates a living sense of tradition . . . sensitive not to fashion merely but to the spirit of the age' Christopher Ricks, *Sunday Times*

# Oxford Reference

The Oxford Reference series offers authoritative and up-to-date reference books in paperback across a wide range of topics.

# OXFORD POPULAR FICTION
## THE ORIGINAL MILLION SELLERS!

This series boasts some of the most talked-about works of British and US fiction of the last 150 years—books that helped define the literary styles and genres of crime, historical fiction, romance, adventure, and social comedy, which modern readers enjoy.

| | |
|---|---|
| *Riders of the Purple Sage* | Zane Grey |
| *The Four Just Men* | Edgar Wallace |
| *Trilby* | George Du Maurier |
| *Trent's Last Case* | E C Bentley |
| *The Riddle of the Sands* | Erskine Childers |
| *Under Two Flags* | Ouida |
| *The Lost World* | Arthur Conan Doyle |
| *The Woman Who Did* | Grant Allen |

**Forthcoming in October:**

| | |
|---|---|
| *Olive* | Dinah Craik |
| *The Diary of a Nobody* | George and Weedon Grossmith |
| *The Lodger* | Belloc Lowndes |
| *The Wrong Box* | Robert Louis Stevenson |

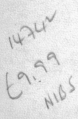